MOLECULAR BIOLOGY

Jones & Bartlett Learning Titles in Biological Science

Plant Structure: A Colour Guide
Brian G. Bowes, James D. Mauseth

Lewin's CELLS, Second Edition
Lynne Cassimeris
George Plopper
Vishwanath R. Lingappa

Environmental Science, Eighth Edition
Daniel D. Chiras

Human Biology, Seventh Edition
Daniel D. Chiras

Plants, Genes, and Crop Biotechnology
Maarten J. Chrispeels
David E. Sadava

Restoration Ecology
Sigurdur Greipsson

Evolution: Principles and Processes
Brian K. Hall

Strickberger's Evolution, Fourth Edition
Brian K. Hall
Benedikt Hallgrímsson

Essential Genetics: A Genomics Perspective, Fifth Edition
Daniel L. Hartl

Genetics: Analysis of Genes and Genomes, Seventh Edition
Daniel L. Hartl
Elizabeth W. Jones

Genetics of Populations, Fourth Edition
Philip W. Hedrick

Lewin's Essential GENES, Second Edition
Jocelyn E. Krebs
Elliott S. Goldstein
Stephen T. Kilpatrick

Lewin's GENES X
Jocelyn E. Krebs
Elliott S. Goldstein
Stephen T. Kilpatrick

Tropical Forests
Bernard A. Marcus

Botany: An Introduction to Plant Biology, Fourth Edition
James D. Mauseth

Environmental Science, Fourth Edition
Michael L. McKinney
Robert M. Schoch
Logan Yonavjak

RNA Interference and Model Organisms
Joanna A. Miller

Introduction to the Biology of Marine Life, Tenth Edition
John F. Morrissey
James L. Sumich

Exploring Bioinformatics: A Project-Based Approach
Caroline St. Clair
Jonathan E. Visick

Molecular Biology: Genes to Proteins, Fourth Edition
Burton E. Tropp

The Ecology of Agroecosystems
John H. Vandermeer

Mammalogy, Fifth Edition
Terry A. Vaughan
James M. Ryan
Nicholas J. Czaplewski

MOLECULAR BIOLOGY

FOURTH EDITION

Genes to Proteins

BURTON E. TROPP

Professor Emeritus
Queens College
City University of New York
Flushing, New York

World Headquarters
Jones & Bartlett Learning
40 Tall Pine Drive
Sudbury, MA 01776
978-443-5000
info@jblearning.com
www.jblearning.com

Jones & Bartlett Learning Canada
6339 Ormindale Way
Mississauga, Ontario L5V 1J2
Canada

Jones & Bartlett Learning International
Barb House, Barb Mews
London W6 7PA
United Kingdom

Substantial discounts on bulk quantities of Jones & Bartlett Learning publications are available to corporations, professional associations, and other qualified organizations. For details and specific discount information, contact the special sales department at Jones & Bartlett Learning via the above contact information or send an email to specialsales@jblearning.com.

Jones & Bartlett Learning books and products are available through most bookstores and online booksellers. To contact Jones & Bartlett Learning directly, call 800-832-0034, fax 978-443-8000, or visit our website, www.jblearning.com.

Production Credits
Chief Executive Officer: Ty Field
President: James Homer
SVP, Chief Operating Officer: Don Jones, Jr.
SVP, Chief Technology Officer: Dean Fossella
SVP, Chief Marketing Officer: Alison M. Pendergast
SVP, Chief Financial Officer: Ruth Siporin
Publisher, Higher Education: Cathleen Sether
Acquisitions Editor/Science: Molly Steinbach
Senior Associate Editor: Megan R. Turner
Editorial Assistant: Rachel Isaacs
Production Manager: Louis C. Bruno, Jr.
Senior Marketing Manager: Andrea DeFronzo
V.P., Manufacturing and Inventory Control: Therese Connell
Illustrations: Elizabeth Morales
Composition: Shepherd, Inc.
Cover Design: Kristin E. Parker
PhoResearch and Permission Supervisor: Christine Myaskovsky
Assistaearch and Permission Associate: Emily O'Neill
Associate oto Researcher: Elise Gilbert
Text Protein and Photo Researcher: Carolyn Arcabascio
Cover Image: ©ures: Generated using software programs from Accelrys Software, Inc.
Printing and Bindingopal Murti/Photo Researchers, Inc.
Cover Printing: Couriourier Kendallville
ndallville

About the Cover
A colored transmission electron micrograph of human DNA strands superimposed on a confocal light micrograph of a mammalian nucleus (bright blue and pink). The DNA is in its uncondensed form and measures about 2 meters in length. A mammalian cell nucleus, however, which holds a cell's DNA, only measures between 0.02 and 0.01 millimeters in diameter. To fit into the nucleus the DNA is first coiled around proteins (called histones) to form a complex known as a nucleosome. The nucleosomes then further coil and condense to form chromatin, the DNA-protein complex that makes up chromosomes. Magnification is approximately 30,000×.

To order this book, use IS 978-1-4496-0091-4

Library of Congress Catalogin Publication Data
Tropp, Burton E.
Molecular biology : genes to protei Burton E. Tropp. — 4th ed.
p. ; cm.
Includes bibliographical references and in
ISBN 978-0-7637-8663-2 (alk. paper)
1. Molecular biology. I. Title.
[DNLM: 1. Molecular Biology. QU 450]
QH506.F73 2012
572.8—dc22 2010 3850

6048

Printed in the United States of America
15 14 13 12 11 10 9 8 7 6 5 4 3 2

To my wife Roslyn, and to our family—Jonathan, Lauren, Matthew, Julie, Paul, Erica, Sarah, Rachel, Gabrielle, Katie, Gracie, Alice, Jane, and Charles—and to the memory of my parents, Sol and Renee Tropp.

Tribute to David Friefelder

David Friefelder taught biochemistry and molecular biology first at Brandeis Unviersity and later at the University of California, San Diego. His research interests and expertise were in a broad range of subjects, therefore, he was qualified to write both general and specialized textbooks. He was a gifted writer, and he devoted extensive time and energy to preparing a collection of textbooks and monographs. From his teaching and writing experience he developed a remarkable understanding of the ways in which students learn. He organized his textbooks using a layering approach for coping with biological complexity. This acknowledgment is, therefore, offered as a tribute to David's memory.

Brief Contents

Dedication v
Preface xxxiii

CHAPTER 1 **Introduction to Molecular Biology** 1

SECTION I **Protein Structure and Function** 27
CHAPTER 2 **Protein Structure** 28
CHAPTER 3 **Protein Function** 75

SECTION II **Nucleic Acids and Nucleoproteins** 107
CHAPTER 4 **Nucleic Acid Structure** 108
CHAPTER 5 **Techniques in Molecular Biology** 150
CHAPTER 6 **Chromosome Structure** 210

SECTION III **Genetics and Virology** 253
CHAPTER 7 **Genetic Analysis in Molecular Biology** 254
CHAPTER 8 **Viruses in Molecular Biology** 305

SECTION IV **DNA Metabolism** 364
CHAPTER 9 **DNA Replication in Bacteria** 365
CHAPTER 10 **DNA Replication in Eukaryotes and the Archaea** 415
CHAPTER 11 **DNA Damage** 448
CHAPTER 12 **DNA Repair** 468
CHAPTER 13 **Recombination** 511
CHAPTER 14 **Transposons and Other Mobile Elements** 574

SECTION V **RNA Metabolism** 621
CHAPTER 15 **Bacterial RNA Polymerase** 622
CHAPTER 16 **Regulation of Bacterial Gene Transcription** 661

CHAPTER 17 **RNA Polymerase II: Basal Transcription 723**

CHAPTER 18 **RNA Polymerase II: Regulation 763**

CHAPTER 19 **RNA Polymerase II: Cotranscriptional and Posttranscriptional Processes 829**

CHAPTER 20 **RNA Polymerases I and III and Organellar RNA Polymerases 895**

CHAPTER 21 **Small Silencing RNAs 936**

SECTION VI **Protein Synthesis 962**

CHAPTER 22 **Protein Synthesis: The Genetic Code 963**

CHAPTER 23 **Protein Synthesis: The Ribosome 1006**

Index 1065

Photo Credits 1097

Contents

Dedication v
Preface xxxiii

CHAPTER 1 **Introduction to Molecular Biology 1**

1.1 Intellectual Foundation 2
Two studies performed in the 1860s provided the intellectual underpinning for molecular biology. 2

1.2 Genotypes and Phenotypes 4
Each gene is responsible for the synthesis of a single polypeptide. 4

1.3 Nucleic Acids 5
Nucleic acids are linear chains of nucleotides. 5

1.4 DNA Structure and Function 6
Transformation experiments led to the discovery that DNA is the hereditary material. 12
Chemical experiments also supported the hypothesis that DNA is the hereditary material. 15
The blender experiment demonstrated that DNA is the genetic material in bacterial viruses. 16
RNA serves as the hereditary material in some viruses. 17
Rosalind Franklin and Maurice Wilkins obtained x-ray diffraction patterns of extended DNA fibers. 19
James Watson and Francis Crick proposed that DNA is a double-stranded helix. 19
The central dogma provides the theoretical framework for molecular biology. 22
Recombinant DNA technology allows us to study complex biological systems. 24
A great deal of molecular biology information is available on the Internet. 25

Suggested Reading 26

SECTION I Protein Structure and Function 27

CHAPTER 2 **Protein Structure 28**

2.1 The α-Amino Acids 29
α-Amino acids have an amino group and a carboxyl group attached to a central carbon atom. 29
Amino acids are represented by three-letter and one-letter abbreviations. 32

2.2 The Peptide Bond 33

α-Amino acids are linked by peptide bonds. 33

2.3 Protein Purification 36

Protein mixtures can be fractionated by chromatography. 36

Proteins and other charged biological polymers migrate in an electric field. 40

2.4 Primary Structure of Proteins 41

The amino acid sequence or primary structure of a purified protein can be determined. 41

Polypeptide sequences can be obtained from nucleic acid sequences. 43

The BLAST program compares a new polypeptide sequence with all sequences stored in a data bank. 45

Proteins with just one polypeptide chain have primary, secondary, and tertiary structures while those with two or more chains also have quaternary structures. 46

2.5 Weak Noncovalent Bonds 46

The polypeptide folding pattern is determined by weak noncovalent interactions. 46

2.6 Secondary Structures 51

The α-helix is a compact structure that is stabilized by hydrogen bonds. 51

The β-conformation is also stabilized by hydrogen bonds. 54

Loops and turns connect different peptide segments, allowing polypeptide chains to fold back on themselves. 54

Certain combinations of secondary structures, called supersecondary structures or folding motifs, appear in many different proteins. 55

We cannot yet predict secondary structures with absolute certainty. 55

2.7 Tertiary Structure 56

X-ray crystallography and nuclear magnetic resonance studies have revealed the three-dimensional structures of many different proteins. 56

Intrinsically disordered proteins lack an ordered structure under physiological conditions. 59

Structural genomics is a field devoted to solving x-ray and NMR structures in a high throughput manner. 60

The primary structure of a polypeptide determines its tertiary structure. 62

Molecular chaperones help proteins to fold inside the cell. 64

2.8 Proteins and Biological Membranes 66

Proteins interact with lipids in biological membranes. 66

The fluid mosaic model has been proposed to explain the structure of biological membranes. 68

Suggested Reading 71

CHAPTER 3 **Protein Function 75**

3.1 Myoglobin, Hemoglobin, and the Quaternary Structure Concept 76

Differences in myoglobin and hemoglobin function are explained by differences in myoglobin and hemoglobin structure. 76

Normal adult hemoglobin (HbA) differs from sickle cell hemoglobin (HbS) by only one amino acid. 80

3.2 Immunoglobulin G and the Domain Concept 83

Large polypeptides fold into globular units called domains. 83

3.3 Enzymes 85

Enzymes are proteins that catalyze chemical reactions. 85

Different methods can be used to detect enzyme activity. 85

Enzymes lower the energy of activation but do not affect the equilibrium position. 87

All enzyme reactions proceed through an enzyme-substrate complex. 88

Molecular details for enzyme-substrate complexes have been worked out for many enzymes. 90

Regulatory enzymes control committed steps in biochemical pathways. 91

Regulatory enzymes exhibit sigmoidal kinetics and are stimulated or inhibited by allosteric effectors. 92

Enzyme activity can be altered by covalent modification. 94

3.4 G-Protein-Linked Signal Transduction Systems 94

G-protein-linked signal transduction systems convert extracellular chemical or physical signals into intracellular signals. 94

3.5 The Ubiquitin Proteasome Proteolytic Pathway 99

The ubiquitin proteasome system is responsible for the specific degradation of intracellular proteins. 99

Suggested Reading 103

SECTION II Nucleic Acids and Nucleoproteins 107

CHAPTER 4 **Nucleic Acid Structure 108**

4.1 DNA Size and Fragility 109

DNA molecules vary in size and base composition. 109

DNA molecules are fragile. 110

4.2 Recognition Patterns in the Major and Minor Grooves 110

Enzymes can recognize specific patterns at the edges of the major and minor grooves. 110

4.3 DNA Bending 112
Some base sequences cause DNA to bend. 112
4.4 DNA Denaturation and Renaturation 112
DNA can be denatured. 112
Hydrogen bonds stabilize double-stranded DNA. 114
Base stacking also stabilizes double-stranded DNA. 115
Base stacking is a cooperative interaction. 115
Ionic strength influences DNA structure. 116
The DNA molecule is in a dynamic state. 116
Distant short patches of complementary sequences can base pair in single-stranded DNA. 117
Alkali denatures DNA without breaking phosphodiester bonds. 117
Complementary single strands can anneal to form double-stranded DNA. 118
4.5 Helicases 119
Helicases are motor proteins that use the energy of nucleoside triphosphates to unwind DNA. 119
4.6 Single-Stranded DNA Binding Proteins 121
Single-stranded DNA binding proteins (SSB) stabilize single-stranded DNA. 121
4.7 Topoisomers and Topoisomerases 123
Covalently closed circular DNA molecules can form supercoils. 123
Bacterial DNA usually exists as a covalently closed circle. 123
Plasmid DNA molecules are used to study the properties of circular DNA *in vitro*. 124
Circular DNA molecules often have superhelical structures. 124
Supercoiled DNA results from under- or overwinding circular DNA. 125
Superhelices can have single-stranded regions. 127
Topoisomerases catalyze the conversion of one topoisomer into another. 128
Enzymes belonging to the topoisomerase I family can be divided into three subfamilies. 128
Type II topoisomerases require ATP to convert one topoisomer into another. 132
4.8 Non-B DNA Conformations 133
A-DNA is a right-handed double helix with a deep major groove and very shallow minor groove. 133
Z-DNA has a left-handed conformation. 135
DNA conformational changes result from rotation about single bonds. 137
Several other kinds of non-B DNA structures appear to exist in nature. 137
4.9 RNA Structure 141
RNA performs a wide variety of functions in the cell. 141
RNA secondary structure is dominated by Watson-Crick base pairs. 141
RNA tertiary structures are stabilized by interactions between two or more secondary structure elements. 142

4.10 The RNA World Hypothesis 144
The earliest forms of life on earth may have used RNA as both the genetic material and the biological catalysts needed to maintain life. 144
Suggested Reading 145

CHAPTER 5 **Techniques in Molecular Biology 150**

5.1 Nucleic Acid Isolation 151
The method used for DNA isolation must be tailored to the organism from which the DNA is to be isolated. 151
Great care must be taken to protect RNA from degradation during its isolation. 153
Different physical techniques are used to study macromolecules. 153

5.2 Electron Microscopy 153
Electron microscopy allows us to see macromolecules. 153

5.3 Centrifugal Techniques 156
Velocity sedimentation can separate macromolecules and provide information about their size and shape. 156
Equilibrium density gradient centrifugation separates particles according to their density. 158

5.4 Gel Electrophoresis 160
Gel electrophoresis separates charged macromolecules by their rate of migration in an electric field. 160
Gel electrophoresis can be used to separate proteins and determine a polypetide's molecular mass. 162
Capillary gel electrophoresis is a rapid and automated process that provides quantitative data. 163
Pulsed-field gel electrophoresis (PFGE) can separate very large DNA molecules. 163

5.5 Nucleases and Restriction Maps 165
Nucleases are useful tools in DNA investigations. 165
Restriction endonucleases that cleave within specific nucleotide sequences are very useful tools for characterizing DNA. 167
Restriction endonucleases can be used to construct a restriction map of a DNA molecule. 170

5.6 Recombinant DNA Technology 173
DNA fragments can be inserted into plasmid DNA vectors. 173
Southern blotting is used to detect specific DNA fragments. 174
Northern and Western blotting are used to detect specific RNA and polypeptide molecules, respectively. 177
DNA polymerase I, a multifunctional enzyme with polymerase, 3′→5′ exonuclease, and 5′→3′ exonuclease activities, can be used to synthesize labeled DNA. 177
DNA polymerase I can synthesize DNA at a nick. 180
The polymerase chain reaction is used to amplify DNA. 182
Site-directed mutagenesis can be used to introduce a specific base change within a gene. 184

5.7 DNA Sequence Determination 185
The Maxam-Gilbert method uses controlled chemical degradation to sequence DNA. 185

The chain termination method for sequencing DNA uses dideoxynucleotides to interrupt DNA synthesis. 186
DNA molecules that are 1 to 8 kbp long can be sequenced by primer walking. 188
DNA sequences can be stitched together by using information obtained from a restriction map. 188
Shotgun sequencing is used to sequence long DNA molecules. 189
The Human Genome Project used hierarchical shotgun assembly to sequence the human genome. 189
Whole genome shotgun sequencing has also been used to sequence the human genome. 192
The human genome sequence provides considerable new information. 194
A new generation of DNA sequencers provide rapid and accurate information without the need for electrophoresis or *in vivo* cloning. 195
Reverse transcriptase can use an RNA molecule as a template to synthesize DNA. 201
DNA chips are used to follow mRNA synthesis, search for a specific DNA sequence, or to find a single nucleotide change in a DNA sequence. 203
Suggested Reading 206

CHAPTER 6 **Chromosome Structure 210**

6.1 Bacterial Chromatin 212
Bacterial DNA is located in the nucleoid. 212
MukB makes an important contribution to the compaction of the bacterial chromosome. 214
Additional nucleoid-associated proteins contribute to bacterial DNA compaction. 217

6.2 Mitosis and Meiosis 217
In higher animals, germ cells have a haploid number of chromosomes and somatic cells have a diploid number. 217
The animal cell life cycle alternates between interphase and mitosis. 218
Mitosis allows cells to maintain the chromosome number. 219
Cohesin is a four subunit complex that keeps sister chromatids together. 221
Meiosis reduces the chromosome number in half. 226

6.3 Karyotype 227
Chromosome sites are specified according to nomenclature conventions. 227
A karyotype shows an individual cell's metaphase chromosomes arranged in pairs and sorted by size. 228
A great deal of information can be obtained by examining karyotype preparations. 228
Fluorescent *in situ* hybridization (FISH) provides a great deal of information about chromosomes. 230

6.4 The Nucleosome 231
Five major histone classes interact with DNA in eukaryotic chromatin. 231

The first level of chromatin organization is the nucleosome. 232
X-ray crystallography provides high-resolution images of nucleosome core particles. 233
The precise nature of the interaction between H1 and the core particle is not known. 236

6.5 The 30-nm Fiber 237
A chain of nucleosomes appears to fold into a 30-nm fiber. 237

6.6 The Scaffold Model 238
The scaffold model was proposed to explain higher order chromatin structure. 238
Condensins and topoisomerase II help to stabilize condensed chromosomes. 239

6.7 The Centromere 241
The centromere is the site of microtubule attachment. 241

6.8 The Telomere 243
The telomere, which is present at either end of a chromosome, is needed for stability. 243

Suggested Reading 246

SECTION III **Genetics and Virology 253**

CHAPTER 7 **Genetic Analysis in Molecular Biology 254**

7.1 Introduction to Genetic Recombination 255
Genetic recombination involves an exchange of DNA segments between DNA molecules or chromosomes. 255
Recombination frequencies are used to obtain a genetic map. 256

7.2 Bacterial Genetics 257
Bacteria, which are often selected as model systems for genetic analyses, have complex structures. 257
Bacteria can be cultured in liquid or solid media. 259
Specific notations, conventions, and terminology are used in bacterial genetics. 261
Cells with altered genes are called mutants. 262
Some mutants display the mutant phenotype under all conditions, while others display it only under certain conditions. 262
Certain physical and chemical agents are mutagens. 263
Mutants can be classified on the basis of the changes in the DNA. 263
A mutant organism may regain its original phenotype. 265
Mutants have many uses in molecular biology. 265
A genetic test known as complementation can be used to determine the number of genes responsible for a phenotype. 268
E. coli cells can exchange genetic information by conjugation. 270
Approximately 40 F factor genes are needed for successful mating and DNA transfer to occur. 274
The F plasmid can integrate into a bacterial chromosome and carry it into a recipient cell. 276

Bacterial mating experiments can be used to produce an *E. coli* genetic map. 278
F′ plasmids contain part of the bacterial chromosome. 280
Plasmid replication control functions are usually clustered in a region called the basic replicon. 282
Plasmids often confer advantageous properties to their hosts. 283
7.3 Budding Yeast (*Saccharomyces cerevisiae*) 284
Yeasts are unicellular eukaryotes. 284
Specific notations, conventions, and terminology are used in yeast genetics. 286
Yeast cells exist in haploid and diploid stages. 287
The yeast mating type is determined by an allele present in the mating type (*Mat*) locus. 288
Yeast mating factors act as signals to initiate the mating process. 290
7.4 Restriction and Amplified Fragment Length Polymorphisms 290
Recombinant DNA techniques have facilitated genetic analysis in humans and other organisms. 290
7.5 Somatic Cell Genetics 295
Somatic cell genetics can be used to map genes in higher organisms. 295
Animal cells can be studied in culture. 295
Two different animal cells can fuse to form a heterokaryon. 297
Hybrid cells can be used to make monoclonal antibodies. 299
Suggested Reading 301
CHAPTER 8 **Viruses in Molecular Biology 305**
8.1 Introduction to Viruses 306
Viruses are obligate parasites that can only replicate in a host cell. 306
8.2 Introduction to the Bacteriophages 308
Bacteriophages were of interest because they seemed to have the potential to serve as therapeutic agents to treat bacterial diseases. 308
Investigators belonging to the "Phage Group" were the first to use viruses as model systems to study fundamental questions about gene structure and function. 308
Bacteriophages come in different sizes and shapes. 309
Bacteriophages have lytic, lysogenic, and chronic life cycles. 310
Bacteriophages form plaques on a bacterial lawn. 311
Bacteria and the phages that infect them are in a constant struggle for survival. 312
8.3 Virulent Bacteriophages 313
E. coli phage T4 DNA is terminally redundant and circularly permuted. 313
E. coli phage T7 DNA is terminally redundant but not circularly permuted. 325
E. coli phage ϕX174 contains a single-stranded circular DNA molecule. 328

Some phages have-single-stranded RNA as their genetic material. 332

8.4 Temperate Phages 334

E. coli phage λ DNA can replicate through a lytic or lysogenic life cycle. 334

E. coli phage P1 can act as generalized transducing particles. 341

8.5 Chronic Phages 342

After infection, a chronic phage programs the host cell for continued virion particle release without killing the cell. 342

8.6 Animal Viruses 348

Polyomaviruses contain circular double-stranded DNA. 349

Adenonviruses have linear blunt-ended, double-stranded DNA with an inverted repeat at each end. 352

Retroviruses use reverse transcriptase to make a DNA copy of their RNA genome. 355

Suggested Reading 359

SECTION IV **DNA Metabolism 364**

CHAPTER 9 **DNA Replication in Bacteria 365**

9.1 General Features of DNA Replication 366

DNA replication is semiconservative. 366

Bacterial and eukaryotic DNA replication is bidirectional. 369

DNA replication is semidiscontinuous. 370

DNA ligase connects adjacent Okazaki fragments. 375

RNA serves as a primer for Okazaki fragment synthesis. 376

The bacterial replication machinery has been isolated and examined *in vitro*. 376

Mutant studies provide important information about the enzymes involved in DNA replication. 379

9.2 The Initiation Stage 381

The replicon model proposes that an initiator protein must bind to a DNA sequence called a replicator at the start of replication. 381

E. coli chromosomal replication begins at *oriC*. 381

DnaA, the bacterial initiator protein, has four functional domains. 383

DnaA•ATP assembles to form a filament at *oriC*, causing the DNA unwinding element (DUE) to melt. 384

DnaB helicases have double-ring structures. 385

DnaC loads DnaB helicase onto the single-stranded DNA generated at the DUE. 386

DnaG (primase) catalyzes RNA primer synthesis. 388

9.3 The Elongation Stage 388

Several enzymes act together at the replication fork. 388

DNA polymerase III is required for bacterial DNA replication. 389

A polymerase's processivity can be determined by using a polymerase "trap" to bind the polymerase after it dissociates from its DNA substrate. 390

DNA polymerase holoenzyme has ten distinct subunits that form three subassemblies. 391
The core polymerase has one subunit with 5′→3′ polymerase activity and another with 3′→5′ exonuclease activity. 392
The β clamp forms a ring around DNA, tethering the remainder of the polymerase holoenzyme to the DNA. 394
The clamp loader places the sliding clamp around DNA. 396
The DNA polymerase III holoenzyme clamp loader has three τ subunits. 399
The replisome catalyzes coordinated leading and lagging DNA synthesis at the replication fork. 400
Core polymerase is released from the β clamp by a premature release (also called signaling release) or collision release mechanism. 403
Three models have been proposed to explain how helicase moves 5′ → 3′ on the lagging strand while primase moves in the opposite direction as it synthesizes primer. 404

9.4 The Termination Stage 405
Replication terminates when the two growing forks meet in the terminus region, which is located 180° around the circular chromosome from the origin. 405
The terminus utilization substance (TUS) binds to *Ter* sites. 406

9.5 Regulation of Bacterial DNA Replication 408
Three mechanisms regulate bacterial DNA replication at the initiation stage. 408
Suggested Reading 409

CHAPTER 10 **DNA Replication in Eukaryotes and the Archaea 415**

10.1 The SV40 DNA Replication System 416
The SV40 T antigen binds to the origin of replication and unwinds DNA. 416
SV40 T antigen helps to recruit DNA polymerase/α-primase (Pol α) to the proto-replication bubble. 418

10.2 Introduction to Eukaryotic DNA Replication 421
Eukaryotic replication machinery must replicate long linear duplexes with multiple origins of replication. 421

10.3 Eukaryotic Replication Initiation 422
Eukaryotic chromosomes have many replicator sites. 422
Autonomously replicating sequences (ARS) determine the site of DNA chain initiation in yeast. 422
Two-dimensional gel electrophoresis can locate origins of replication. 425
The origin of recognition complex (ORC) serves as the eukaryotic initiator. 426
CDC6 and Cdt1 help load MCM2-7 helicase onto the origin to form a pre-replication complex (pre-RC). 427
The licensed origin must be activated before replication can take place. 427

10.4 Eukaryotic Replication Elongation 429
Pol δ and Pol ε are primarily responsible for copying the lagging- and leading-strand templates, respectively. 429

10.5 The End-Replication Problem 432

Studies of the *Tetrahymena* and yeast telomeres suggested that a terminal transferase-like enzyme is required for telomere formation. 432

Telomerase uses an RNA template to add nucleotide repeats to chromosome ends. 434

Telomerase plays an important role in solving the end-replication problem. 437

Telomerase plays a role in aging and cancer. 438

10.6 Replication Coupled Chromatin Synthesis 439

Chromatin disassembly and reassembly are tightly coupled to DNA replication. 439

10.7 DNA Replication in the Archaea 441

The archaeal replication machinery is similar to that in eukaryotes. 441

Orc1/Cdc6 recruits MCM to the archaeal origin of replication. 442

The basic steps in archaeal elongation are very similar to those in bacteria and eukaryotes. 442

Suggested Reading 443

CHAPTER 11 **DNA Damage 448**

11.1 Radiation Damage 449

Ultraviolet light causes cyclobutane pyrimidine dimer (CPD) formation and (6-4) photoproduct formation. 449

X-rays and gamma rays cause many different types of DNA damage. 452

11.2 DNA Instability in Water 453

DNA is damaged by hydrolytic cleavage reactions. 453

11.3 Oxidative Damage 455

Reactive oxygen species damage DNA. 455

11.4 Alkylation Damage by Monoadduct Formation 457

Alkylating agents damage DNA by transferring alkyl groups to centers of negative charge. 457

Many environmental agents must be modified by cell metabolism before they can alkylate DNA. 458

11.5 Chemical Cross-Linking Agents 461

Chemical cross-linking agents block DNA strand separation. 461

Psoralen and related compounds can form monoadducts or cross-links. 462

Cisplatin combines with DNA to form intra- and interstrand cross-links. 464

11.6 Mutagen and Carcinogen Detection 464

Mutagens can be detected based on their ability to restore mutant gene activity. 464

Suggested Reading 464

CHAPTER 12 **DNA Repair 468**

12.1 Direct Reversal of Damage 469

Photolyase reverses damage caused by cyclobutane pyrimidine dimer formation. 469

O^6-Alkylguanine, O^4-alkylthymine, and phosphotriesters can be repaired by direct alkyl group removal by a suicide enzyme. 474

AlkB catalyzes the oxidative removal of methyl groups in 1-methyladenine and 3-methylcytosine. 476

12.2 Base Excision Repair 477

The base excision repair (BER) pathway removes and replaces damaged or inappropriate bases. 477

12.3 Nucleotide Excision Repair 482

Nucleotide excision repair removes bulky adducts from DNA by excising an oligonucleotide bearing the lesion and replacing it with new DNA. 482

UvrA, UvrB, and UvrC proteins are required for bacterial nucleotide excision repair. 482

Individuals with the autosomal recessive disease xeroderma pigmentosum have defects in enzymes that participate in the nucleotide excision repair pathway. 485

12.4 Mismatch Repair 493

The DNA mismatch repair system removes mismatches and short insertions or deletions that are present in DNA. 493

12.5 The SOS Response and Translesion DNA Synthesis 498

Error-prone DNA polymerases catalyze translesion DNA synthesis. 498

RecA and LexA regulate the *E. coli* SOS response. 499

The SOS signal induces the synthesis of DNA polymerases II, IV, and V. 502

Human cells have at least 14 different template-dependent DNA polymerases. 504

Suggested Reading 505

CHAPTER 13 **Recombination 511**

Hannah Klein

13.1 Introduction to Homologous Recombination 512

Homologous recombination is an essential process for repairing DNA breaks and for ensuring correct chromosome segregation in meiosis. 512

13.2 Early Clues from Bacteriophage 517

Crossing over involves an exchange of DNA between the two interacting DNA molecules. 517

13.3 Early Models of Homologous Recombination 518

The Holliday model of homologous recombination proposes a crossed strand intermediate called a Holliday junction. 518

The Meselson–Radding model of recombination—a second homologous recombination model—is based on one single-strand nick for initiation. 522

13.4 A Homologous Recombination Model Initiated by a Double-Strand Break 524

Yeast repair gapped plasmids by homologous recombination. 524

The double-strand break repair (DSBR) model is based on a double-strand break for initiation. 525

13.5 Bacterial Homologous Recombination Proteins 526
E. coli recombination mutants have reduced conjugation rates and are sensitive to DNA damage. 526
RecA is a strand exchange protein. 528
The RecBCD complex prepares double-strand breaks for homologous recombination and alters its activity at *chi* sites. 529
The RecFOR pathway repairs single-strand gaps. 532

13.6 Eukaryotic Homologous Recombination Proteins 533
Several key homologous recombination proteins are conserved between bacteria and eukaryotes, but there are additional novel proteins found only in eukaryotes. 533

13.7 A Variation of the Double-Strand Break Repair Model 540
The synthesis-dependent strand-annealing (SDSA) model is a gene conversion-only model. 540

13.8 Meiotic Recombination 540
Some aspects of meiotic recombination are novel. 540
Some recombination proteins are made only in meiotic cells. 541
Meiosis recombination models propose two different types of homologous recombination events. 543

13.9 Using Mitotic Recombination to Make Gene Knockouts 544
Mitotic recombination can be used in genetic engineering to make targeted gene disruptions. 544
Gene knockouts in yeast occur by homologous recombination with high efficiency. 544
Gene knockouts in mice also can be made by gene targeting methods. 545

13.10 Mitotic Recombination and DNA Replication 549
Mitotic homologous recombination is essential during DNA replication when replication forks collapse. 549
Recombination must be regulated to prevent chromosome rearrangements and genomic instability. 550
The single-strand annealing (SSA) mechanism results in deletions. 553

13.11 Repairing a Double-Strand Break without Homology 553
Nonhomologous end-joining is a model for rejoining ends with no homology. 553

13.12 Site-Specific Recombination 555
Site-specific recombination occurs at defined DNA sequences and is used for immunoglobulin diversity and by transposable elements. 555
Mating type switching in yeast occurs by synthesis-dependent strand-annealing initiated at a defined site. 556
V(D)J recombination produces the immune system diversity. 558
FLP/*FRT* and Cre/*lox* systems can be used to make targeted recombination events. 564

Suggested Reading 566

CHAPTER 14 **Transposons and Other Mobile Elements 574**
Joseph E. Peters

14.1 Transposition 577
The simplest mobile elements in bacteria are called insertion sequences. 577
The transposase forms a specific complex with the ends of the mobile element. 577
Coordinated breakage and joining events occur during transposition. 579
Some elements do cut-and-paste transposition, where the element is directly moved to a new location. 580
Transposition during DNA replication and host DNA repair allow cut-and-paste elements to increase in copy number. 582
Transposons are found at various levels of complexity in bacteria. 583
Replicative transposons leave one copy of the element at the donor site. 584
Transposons in eukaryotes are mechanistically similar to bacterial transposons. 587
Diverse systems allow transposition to be regulated. 589
Most transposons prefer DNA targets that are bent. 590
Transposons can target certain sequences. 591
Some transposons target specific molecular processes. 591
Some elements have evolved the ability to choose between certain target sites. 592
Transposons are important tools for molecular genetics. 594

14.2 Conservative Site-Specific Recombination 607
Two families of proteins do conservative site-specific recombination with different pathways. 607
Bacteriophage λ uses a conservative site-specific recombinase to integrate into the host genome. 609
Multiple other systems use the conservative site-specific recombinase reaction. 611

14.3 Target-Primed Reverse Transcription 612
Target-primed reverse transcription can mobilize information through an RNA intermediate. 612
Target-primed reverse transcription is used for LINE movement. 613
LINE movement affects genome stability and evolution. 614
Mobile group II introns move by target-primed reverse transcription. 616
Two transesterification reactions allow group II intron movement. 617
Homology to the target site determines if mobile group II introns move by retrotransposition or retro-homing. 617

14.4 Other Mechanisms of DNA Mobilization 618
Suggested Reading 618

SECTION V RNA Metabolism 621

CHAPTER 15 Bacterial RNA Polymerase 622

15.1 Introduction to the Bacterial RNA Polymerase Catalyzed Reaction 624

RNA polymerase requires a DNA template and four nucleoside triphosphates to synthesize RNA. 624

Bacterial RNA polymerases are large multisubunit proteins. 626

15.2 Initiation Stage 628

Bacterial RNA polymerase holoenzyme consists of a coreenzyme and sigma factor. 628

A transcription unit must have an initiation signal called a promoter for accurate and efficient transcription to take place. 628

The DNase protection method provides information about promoter DNA. 629

DNA footprinting shows that σ^{70}-RNA polymerase combines with promoter DNA to form a closed and an open complex. 630

Bacterial RNA polymerase crystal structures show how the enzyme is organized and provide insights into how it works. 633

Genetic and biochemical studies provide additional information about bacterial promoters. 636

Members of the σ^{70} family have four conserved domains. 638

RNA polymerase scrunches DNA during transcription initiation. 639

Transcription initiation is a stepwise process. 640

Alternative σ factors direct RNA polymerase to genes that code for proteins that bacteria require to survive under specific types of environmental stress. 643

The σ^{54}-RNA polymerase requires an activator protein. 645

15.3 Transcription Elongation Complex 645

The transcription elongation complex is a highly processive molecular motor. 645

The trigger loop and the β' bridge helix help to move RNA polymerase forward by one nucleotide during each nucleotide addition cycle. 648

Pauses influence the overall transcription elongation rate. 649

RNA polymerase can detect and remove incorrectly incorporated nucleotides. 651

15.4 Transcription Termination 652

Bacterial transcription machinery releases RNA strands at intrinsic and Rho-dependent terminators. 652

15.5 Antibiotics that Target Bacterial RNA Polymerase 654

RNA polymerase is a target for broad spectrum antibacterial therapy. 654

Suggested Reading 656

CHAPTER 16 **Regulation of Bacterial Gene Transcription 661**

16.1 Messenger RNA 662

Bacterial mRNA may be monocistronic or polycistronic. 662

Bacterial mRNA usually has a short lifetime compared to other kinds of bacterial RNA. 663

Controlling the rate of mRNA synthesis can regulate the flow of genetic information. 664

Messenger RNA synthesis can be controlled by negative and positive regulation. 665

16.2 Lactose Operon 665

The *E. coli* genes *lac*Z, *lac*Y, and *lac*A code for β-galactosidase, lactose permease, and β-galactoside transacetylase, respectively. 665

The *lac* structural genes are regulated. 667

Genetic studies provide information about the regulation of *lac* mRNA. 667

The operon model explains the regulation of the lactose system. 669

Allolactose is the true inducer of the lactose operon. 670

The Lac repressor binds to the *lac* operator *in vitro*. 672

The *lac* operon has three *lac* operators. 673

The Lac repressor is a dimer of dimers, where each dimer binds to one *lac* operator sequence. 674

16.3 Catabolite Repression 677

E. coli uses glucose in preference to lactose. 677

The inhibitory effect of glucose on expression of the *lac* operon is a complicated process. 678

The cAMP•CRP complex binds to an activator site (AS) upstream from the *lac* promoter and activates *lac* operon transcription. 680

cAMP•CRP activates more than 100 operons. 682

16.4 Galactose Operon 684

The galactose operon is also regulated by a repressor and cAMP•CRP. 684

16.5 The *araBAD* Operon 686

The AraC activator protein regulates the *araBAD* operon. 686

16.6 Tryptophan Operon 691

The tryptophan (*trp*) operon is regulated at the levels of transcription initiation, elongation, and termination. 691

16.7 Bacteriophage Lambda: A Transcription Regulation Network 696

Phage λ development is regulated by a complex genetic network. 696

The lytic pathway is controlled by a transcription cascade. 697

The lysogenic pathway is also controlled by a transcription cascade. 699

The CI regulator maintains the lysogenic state. 699

Ultraviolet light induces the λ prophage to enter the lytic pathway. 702

16.8 Messenger RNA Degradation 703

Bacterial mRNA molecules are rapidly degraded. 703

16.9 Ribosomal RNA and Transfer RNA Synthesis 706
Bacterial ribosomes are made of a large subunit with a 23S and 5S RNA and a small subunit with 16S RNA. 706
E. coli has seven rRNA operons, each coding for a 16S, 23S, and 5S RNA. 707
A promoter upstream element (UP element) increases *rrn* transcription. 708
Three Fis protein binding sites increase *rrn* transcription. 709

16.10 Regulation of Ribosome Synthesis 710
Amino acid starvation leads to the production of guanine nucleotides that inhibit rRNA synthesis. 710
E. coli rRNA and tRNA syntheses increase with growth rate. 711
E. coli regulates r-protein synthesis. 712

16.11 Processing rRNA and tRNA 713
Bacteria process the primary transcripts for rRNA and tRNA to form the physiologically active RNA molecules. 713

Suggested Reading 718

CHAPTER 17 **RNA Polymerase II: Basal Transcription 723**

17.1 Introduction to RNA Polymerase II 725
The eukaryotic cell nucleus has three different kinds of RNA polymerase. 725
RNA polymerases I, II, and III can be distinguished by their sensitivities to inhibitors. 727
Each nuclear RNA polymerase has some subunits that are unique to it and some that it shares with the two other nuclear RNA polymerases. 728

17.2 RNA Polymerase II Structure 731
High-resolution yeast RNA polymerase II structures help explain how the enzyme works. 731
The crystal structure has been determined for the complete 12-subunit yeast RNA polymerase II bound to a transcription bubble and product RNA. 734

17.3 Transcription Start Site Identification 734
Nuclear RNA polymerases have limited synthetic capacities. 734
Various techniques have been devised to locate RNA polymerase II transcription start sites. 736
Cap analysis of gene expression (CAGE) is a high throughput technique used to identify transcription start sites and their flanking promoters. 739

17.4 The Core Promoter 742
The core promoter extends from 40 bp upstream of the transcription start site to 40 bp downstream from this site. 742

17.5 General Transcription Factors: Basal Transcription 743
RNA polymerase II requires the assistance of general transcription factors to transcribe naked DNA from specific transcription start sites. 743
The core promoter allows a cell-free system to catalyze a low-level of RNA synthesis at the correct transcription start site. 744

When the core promoter has a TATA box, preinitiation complex assembly begins with either TFIID or TATA binding protein (TBP) binding to the core promoter. 745
TFIID can bind to core promoters of protein-coding genes that lack a TATA box. 746
TFIIA is not required to reconstitute the minimum transcription system. 748
TFIIB helps to convert the closed promoter to an open promoter. 749
Sequential binding of RNA polymerase II•TFIIF complex, TFIIE, and TFIIH completes preinitiation complex formation. 751

17.6 Transcription Elongation 752
The C-terminal domain of the largest RNA polymerase subunit must be phosphorylated for chain elongation to proceed. 752
A variety of transcription elongation factors help to suppress transient pausing during elongation. 753
Elongation factor SII reactivates arrested RNA polymerase II. 754
The transcription elongation complex is regulated. 754

17.7 Archaeal RNA Polymerase 756
The archaea have a single RNA polymerase that is similar to RNA polymerase II. 756

Suggested Reading 758

CHAPTER 18 **RNA Polymerase II: Regulation 763**

18.1 Regulatory Promoters, Enhancers, and Silencers 764
Linker-scanning mutagenesis reveals the regulatory promoter's presence just upstream from the core promoter. 764
Enhancers stimulate transcription and silencers block transcription. 767
The upstream activating sequence (UAS) regulates genes in yeast. 768

18.2 Transcription Activator Proteins 769
Transcription activator proteins help to recruit the transcription machinery. 769
A combinatorial process determines gene activity. 770
DNA affinity chromatography can be used to purify transcription activator proteins. 770
A transcription activator protein's ability to stimulate gene transcription can be determined by a transfection assay. 772

18.3 DNA-Binding Domains with Helix-Turn-Helix Structures 773
Homeotic genes assign positional identities to cells during embryonic development. 773
Homeotic genes specify transcription activator proteins. 774
The homeodomain contains a helix-turn-helix motif. 776
POU proteins have a homeobox and a POU domain. 776

18.4 DNA Binding Domains with Zinc Fingers 777

Many transcription activator proteins have Cys_2His_2 zinc fingers that bind to DNA in a sequence-specific fashion. 777
Nuclear receptors have Cys_4 zinc finger motifs. 780
Ligand-binding domain structure provides considerable information about nuclear receptor function. 784
Gal4, a yeast transcription activator protein belonging to the Cys_6 zinc cluster family, regulates the transcription of genes involved in galactose metabolism. 787

18.5 Loop-Sheet-Helix DNA Binding Domain 789
p53 has a loop-sheet-helix DNA binding domain. 789

18.6 DNA Binding Domains with Basic Region Leucine Zippers 791
Basic region leucine zipper (bzip) transcription activator proteins bind to DNA as dimers that are held together through coiled coil interactions. 791

18.7 DNA-Binding Domains with Helix-Loop-Helix Structures 794
Helix-loop-helix transcription regulatory proteins are dimers. 794
The bHLH zip family of transcription regulators have both HLH and leucine zipper dimerization motifs. 795

18.8 Activation Domain 797
The activation domain must associate with a DNA-binding domain to stimulate transcription. 797
Gal4 has DNA binding and activation domains. 798
The yeast two-hybrid assay permits us to detect polypeptides that interact through noncovalent interactions. 804

18.9 Mediator 805
Squelching occurs when transcription activator proteins compete for a limiting transcription machinery component. 805
Mediator is required for activated transcription. 806
The yeast Mediator complex associates with the UAS in active yeast genes. 808

18.10 Epigenetic Modifications 810
Cells remodel or modify chromatin to make the DNA in chromatin accessible to the transcription machinery. 810
DNA methylation plays an important role in determining whether chromatin will be silenced or actively expressed in vertebrates. 816
Epigenetics is the study of inherited changes in phenotype caused by changes in chromatin other than changes in DNA sequence. 816
Genomic imprinting in mammals determines whether a maternal or paternal gene will be expressed. 817
Pluripotent cells usually become more specialized during development. 818
A cell nucleus from a terminally differentiated cell can be reprogrammed by an enucleated egg cell to produce a live animal. 818
Lineage restricted cells can be programmed to produce induced pluripotent stem (iPS) cells. 820

Suggested Reading 821

CHAPTER 19 **RNA Polymerase II: Cotranscriptional and Posttranscriptional Processes 829**

19.1 Pre-mRNA 830

Eukaryotic cells synthesize large heterogeneous RNA (hnRNA) molecules. 830

Messenger RNA and hnRNA both have poly(A) tails at their 3′-ends. 830

19.2 Cap Formation 831

Messenger RNA molecules have 7-methylguanosine caps at their 5′-ends. 831

5′-m^7G caps are attached to nascent pre-mRNA chains when the chains are 20 to 30 nucleotides long. 835

All eukaryotes use the same basic pathway to form 5′-m^7G caps. 836

CTD must be phosphorylated on Ser-5 to target a transcript for capping. 837

19.3 Split Genes 840

Viral studies revealed that some mRNA molecules are formed by splicing pre-mRNA. 840

Amino acid–coding regions within eukaryotic genes may be interrupted by noncoding regions. 841

Exons tend to be conserved during evolution, whereas introns usually are not conserved. 845

A single pre-mRNA can be processed to produce two or more different mRNA molecules. 847

Combinations of the various splicing patterns within individual genes lead to the formation of multiple mRNAs. 848

Drosophila form an mRNA that codes for essential protein isoforms by alternative *trans*-splicing. 850

Pre-mRNA requires specific sequences for precise splicing to occur. 851

Two splicing intermediates resemble lariats. 852

Splicing consists of two coordinated transesterification reactions. 854

19.4 Spliceosomes 855

Aberrant antibodies, which are produced by individuals with certain autoimmune diseases, bind to small nuclear ribonucleoprotein particles (snRNPs). 855

snRNPs assemble to form a spliceosome, the splicing machine that excises introns. 856

U1, U2, U4, and U5 snRNPs each contains Sm polypeptides, whereas U6 snRNP contains Sm-like polypeptides. 857

Each U snRNP is formed in a multistep process. 859

U1, U2, U4, U5, and U6 snRNPs have been isolated as a penta-snRNP in yeast. 861

In vitro studies show that spliceosomes assemble on introns via an ordered pathway. 863

RNA and protein may both contribute to the spliceosome's catalytic site. 866

Cells use a variety of mechanisms to regulate splice site selection. 868

Splicing begins as a cotranscriptional process and continues as a posttranscriptional process. 870

19.5 Cleavage/Polyadenylation and Transcription Termination 871

Poly(A) tail synthesis and transcription termination are coupled, cotranscriptional processes. 871

Transcription units often have two or more alternate polyadenylation sites. 875

Alternative processing forces us to reconsider our concept of the gene. 876

Transcription termination takes place downstream from the poly(A) site. 877

RNA polymerase II transcription termination appears to involve allosteric changes and a 5′→3′ exonuclease. 878

In higher animals, most histone pre-mRNAs require a special processing mechanism. 878

19.6 RNA Editing 880

RNA editing permits a cell to recode genetic information in a systematic and regulated fashion. 880

The human proteome contains a much greater variety of proteins than would be predicted from the human genome. 883

19.7 Messenger RNA Export 884

Messenger RNA splicing and export are coupled processes. 884

Suggested Reading 886

CHAPTER 20 **RNA Polymerases I and III and Organellar RNA Polymerases 895**

20.1 Eukaryotic Ribosome 896

The eukaryotic ribosome is made of a small and a large ribonucleoprotein subunit. 896

20.2 RNA Polymerase I 897

The 5.8S, 18S, and 28S rRNA coding sequences are part of a single transcript. 897

Eukaryotes have multiple copies of rRNA transcription units arranged in clusters on just a few chromosomes. 899

The rRNA transcription unit promoter consists of a core promoter and an upstream promoter element (UPE). 900

Ribosomal RNA transcription and processing takes place in the nucleolus. 902

RNA polymerase I is a multisubunit enzyme with a structure similar to that of RNA polymerase II. 903

RNA polymerase I–associated factors are required for transcription. 904

RNA polymerase I requires two auxiliary transcription factors, upstream binding factor (UBF) and selectivity factor (SL1/TIF-1B). 905

The transcription initiation complex can be assembled *in vitro* by the stepwise addition of individual components. 906

RNA polymerase I acts through a transcription cycle that begins with the formation of a pre-initiation complex. 907

Pre-rRNA undergoes a complex series of cleavages and

modifications as it is converted to mature ribosomal rRNAs. 909

20.3 Self-Splicing Ribosomal RNA 911

Tetrahymena thermophila pre-rRNA contains an intron that catalyzes its own excision. 911

20.4 Ribosome Assembly 916

Eukaryotic ribosome assembly is a complex multistep process. 916

20.5 RNA Polymerase III 919

RNA polymerase III transcripts are short RNA molecules with a variety of biological functions. 919

RNA polymerase III transcription units have three different types of promoters. 920

The transcription factors required to recruit RNA polymerase III depend on the nature of the promoter. 922

RNA polymerase III does not appear to require additional factors for transcription elongation or termination. 923

Pre-tRNAs require extensive processing to become mature tRNAs. 924

20.6 Transcription in Mitochondria 927

Mitochondrial DNA is transcribed to form mRNA, rRNA, and tRNA. 927

20.7 Transcription in Chloroplasts 930

Chloroplast DNA is also transcribed to form mRNA, rRNA, and tRNA. 930

Suggested Reading 932

CHAPTER 21 **Small Silencing RNAs 936**

21.1 RNA Interference (RNAi) Triggered by Exogenous Double-Stranded RNA 937

The nematode worm *Caenorhabditis elegans* is an attractive organism for molecular biology studies. 9397

RNA interference was discovered in the nematode worm *Caenorhabditis elegans*. 938

In vitro studies helped to elucidate the RNA interference pathway. 940

Dicer acts as a molecular ruler that generates double-stranded RNA fragments of discrete size. 942

The guide strand's 5′-phosphate and 3′-ends bind to sites in the Argonaute protein's Mid and PAZ domains, respectively. 944

RISC loading complex is required for siRISC formation. 946

RNAi blocks virus replication and prevents transposon activation. 946

21.2 Transitive RNAi 947

In some organisms, RNA interference that starts at one site spreads throughout the entire organism. 947

SID-1, an integral membrane protein in *C. elegans*, assists in the systemic spreading of the silencing signal. 948

ERI-1, a 3′→5′ exonuclease in *C. elegans*, appears to be a negative regulator of RNA interference. 949

21.3 RNAi as an Investigational Tool 949

RNAi is a powerful tool for investigating functional genomics. 949

21.4 The MicroRNA Pathway 950

The miRNA pathway blocks mRNA translation or causes mRNA degradation. 950

21.5 Piwi Interacting RNAs (piRNAs) 954

piRNAs help to maintain germline stability in animals. 954

21.6 Endogenous siRNA 957

Animals have a functional endogenous siRNA pathway. 955

21.7 Heterochromatin Assembly 955

The RNAi machinery can establish and maintain heterochromatin. 955

Suggested Reading 957

SECTION VI Protein Synthesis 962

CHAPTER 22 **Protein Synthesis: The Genetic Code 963**

22.1 Discovery of Ribosomes and Transfer RNA 964

Protein synthesis takes place on ribosomes. 964

22.2 Transfer RNA 966

An amino acid must be attached to a transfer RNA before it can be incorporated into a protein. 966

All tRNA molecules have CCA_{OH} at their 3′-ends. 969

An amino acid attaches to tRNA through an ester bond between the amino acid's carboxyl group and the 2′- or 3′-hydroxyl group on adenosine. 971

The tRNA for alanine was the first naturally occurring nucleic acid to be sequenced. 972

Transfer RNAs have cloverleaf secondary structures. 973

Transfer RNA molecules fold into L-shaped three-dimensional structures. 975

22.3 Aminoacyl-tRNA Synthetases 977

Aminoacyl-tRNA synthetases can be divided into two classes, I and II. 977

Some aminoacyl-tRNA synthetases have editing functions. 979

Ile-tRNA synthetase can hydrolyze valyl-$tRNA^{Ile}$ and valyl-AMP. 980

A polypeptide insert in the Rossman-fold domain forms the editing site for Ile-tRNA synthetase. 982

Editing-defective alanyl-tRNA synthetase causes a neurodegenerative disease in mice. 982

Each aminoacyl-tRNA synthetase can distinguish its cognate tRNAs from all other tRNAs. 983

Many gram-positive bacteria and archaea use an indirect pathway for Gln-tRNA synthesis. 985

Selenocysteine and pyrrolysine are building blocks for polypeptides. 986

22.4 Messenger RNA and the Genetic Code 987

Messenger RNA programs ribosomes to synthesize proteins. 987

Three adjacent bases in mRNA that specify an amino acid are called a codon. 987
The discovery that poly(U) directs the synthesis of poly(Phe) was the first step in solving the genetic code. 989
Protein synthesis begins at the amino terminus and ends at the carboxyl terminus. 991
Messenger RNA is read in a 5′ to 3′ direction. 993
Trinucleotides promote the binding of specific aminoacyl-tRNA molecules to ribosomes. 994
Synthetic messengers with strictly defined base sequences confirmed the genetic code. 995
Three codons, UAA, UAG, and UGA, are polypeptide chain termination signals. 996
The genetic code is nonoverlapping, commaless, almost universal, highly degenerate, and unambiguous. 996
The coding specificity of an aminoacyl-tRNA is determined by the tRNA and not the amino acid. 999
Some aminoacyl-tRNA molecules bind to more than one codon because there is some play or wobble in the third base of a codon. 1000
The origin of the genetic code remains a puzzle. 1002
Suggested Reading 1002

CHAPTER 23 **Protein Synthesis: The Ribosome 1006**

23.1 Ribosome Structure 1007
Bacterial ribosome structure has been determined at atomic resolution. 1007
Arachael and eukaryotic ribosome structures appear to be similar to the bacterial ribosome structure. 1011

23.2 Initiation Stage in Bacteria 1012
Protein synthesis can be divided into four stages. 1012
Bacteria, eukaryotes, and archaea each have their own translation initiation pathway. 1012
Each bacterial mRNA open reading frame has its own start codon. 1012
Bacteria have an initiator methionine tRNA and an elongator methionine tRNA. 1013
The 30S subunit is an obligatory intermediate in polypeptide chain initiation. 1014
Initiation factors participate in the formation of 30S and 70S initiation complexes. 1016
The mRNA Shine-Dalgarno (SD) sequence interacts with the 16S rRNA anti-Shine-Dalgarno (anti-SD) sequence. 1018
Riboswitches regulate translation initiation of some bacterial mRNA molecules. 1020

23.3 Initiation Stage in Eukaryotes 1020
Eukaryotic initiator tRNA is charged with a methionine that is not formylated. 1020
Eukaryotic translation initiation proceeds through a scanning mechanism. 1021
Eukaryotes have at least twelve different initiation factors. 1022
Translation initiation factor phosphorylation regulates protein synthesis in eukaryotes. 1028

The translation initiation pathway in archaea appears to be a mixture of the eukaryotic and bacterial pathways. 1029

23.4 Elongation Stage 1029

Polypeptide chain elongation requires elongation factors. 1029

The elongation factors act through a repeating cycle. 1030

An EF-Tu•GTP•aminoacyl-tRNA ternary complex carries the aminoacyl-tRNA to the ribosome. 1030

An additional elongation factor, EF-P, is required to synthesize the first peptide bond. 1032

Specific nucleotides in 16S rRNA are essential for sensing the codon-anticodon helix. 1032

EF-Ts is a GDP-GTP exchange protein. 1034

The ribosome is a ribozyme. 1035

The hybrid-states translocation model offers a mechanism for moving tRNA molecules through the ribosome. 1040

EF-G•GTP stimulates the translocation process. 1041

23.5 Termination Stage in Bacteria 1043

Bacteria have three protein release factors. 1043

The class 1 release factors, RF1 and RF2, have one tripeptide that acts as an anticodon and another that binds at the peptidyl transferase center. 1044

RF3 is a nonessential G protein that stimulates RF1 or RF2 dissociation from the ribosome. 1045

A stalled ribosome translating a truncated mRNA that lacks a termination codon can be rescued by tmRNA. 1046

Mutant tRNA molecules can suppress mutations that create termination codons within a reading frame. 1050

23.6 Termination Stage in Eukaryotes 1050

Eukaryotic cells have bacteria-like release factors in their mitochondria and a different kind in their cytoplasm. 1050

23.7 Recycling Stage 1051

The ribosome recycling factor (RRF) is required for the bacterial ribosomal complex to disassemble. 1051

23.8 Nascent Polypeptide Processing and Folding 1052

Ribosomes have associated enzymes that process nascent polypeptides and chaperones that help to fold the nascent polypeptides. 1052

23.9 Signal Sequence 1053

The signal sequence plays an important role in directing newly synthesized proteins to specific cellular destinations. 1053

Suggested Reading 1056

INDEX 1065

PHOTO CREDITS 1098

Preface

The fourth edition of *Molecular Biology: Genes to Proteins* is a multipurpose textbook that follows the example set forth by David Freifelder in the first two editions, that is, "emphasizing basic molecular processes (such as the synthesis of DNA, RNA, and protein) and genetic phenomena in both prokaryotic and eukaryotic cells." The book is intended for undergraduate and first-year graduate students who have completed full-year courses in college biology, general chemistry, and organic chemistry. Whenever possible the book uses a discovery approach so students learn about the experimental evidence relevant to the concepts discussed. This pedagogical approach provides historical and experimental background information that permits the reader to see how molecular biologists examine clues and develop the hypotheses that ultimately lead to new advances in the field. In my experience, this approach helps the reader become part of the discovery process. In this way, the reader begins to understand the pleasure and sense of satisfaction investigators derive from solving a molecular biology problem. Of course, space does not allow every concept to be presented in this way. Therefore, when faced with a choice of simply presenting a concept or omitting it, I opted to present the concept. This book follows the practice of the third edition in introducing scientists into the discovery process. I chose to do so because it is important for students to appreciate the human effort involved in the discovery process and to recognize that molecular biology is an evolving field. Many students who enroll in a molecular biology course are interested in the medical sciences. In recognizing this interest, a major effort has been made to include interesting and important medical applications when they are relevant.

Chapter Organization

After an introduction to the field of molecular biology, the book is divided into six sections: (I) protein structure and function, (II) nucleic acids and nucleoproteins, (III) genetics and virology, (IV) DNA metabolism, (V) RNA metabolism, and (VI) protein synthesis. The material presented in the first two sections provides basic information required to understand the chemical basis of molecular biology, while that provided in the third section provides information required to understand the biological basis of molecular biology. The order of the remaining chapters reflects the fact that genetic information in cells normally flows from DNA to RNA to protein. A major challenge in presenting new information in molecular biology, and indeed in most science courses, is that information presented in a later chapter is sometimes necessary to fully understand information in earlier chapters. This edition, like the last one, has tried to minimize this problem by introducing basic concepts and techniques that are required to understand molecular biology in the first three sections.

The book follows Freifelder's philosophy that "molecular biology must emphasize both molecules and biology, and that to be molecular, it must also be chemical and physical." With this philosophy in mind, the chapters' contents are as follows. **Chaper 1** provides a brief historical introduction to the field of molecular biology, including the discovery of the double helix model, which provides a very clear example of structure-function relationships, a major theme in molecular biology and one that is emphasized throughout this book. **Chapter 2** introduces important information about the primary, secondary, tertiary, and quaternary structures of proteins and describes the structure of proteins and lipids in biological membranes. A section has been added describing structural genomics. **Chapter 3** describes structure-function relationships in proteins. The first part of Chapter 3 examines hemoglobin and myoglobin, proteins that provide important insights into structure-function relationships encountered throughout the text. The second part uses immunoglobulin G to explore the domain concept. The third part describes enzymes and enzyme kinetics, topics that are central to understanding the processes of DNA replication, repair, and recombination, RNA synthesis and processing, and protein synthesis. The fourth part describes the G-protein signal system. The fifth part, which is new to the fourth edition, examines the ubiquitin proteasome proteolytic pathway. The first part of **Chapter 4** builds on the information provided in Chapter 1 about DNA structure. It emphasizes conformational variations, helical stability, strand separation, helical reformation, circular DNA, and topoisomers. The last part of the chapter, which is new to the fourth edition, examines RNA structure. **Chapter5** includes an examination of the physical techniques that are used to isolate and characterize nucleic acids as well as a discussion of techniques that use enzymes to manipulate DNA in the laboratory. Information that is new to the Fourth Edition includes the Maxam-Gilbert method of controlled chemical degradation to sequence DNA, primer walking, hierarchical shotgun assembly, whole genome shotgun sequencing, and pyrophosphate sequencing. **Chapter 6** describes the interactions between specific proteins and DNA to form nucleoprotein complexes. A part has been added that examines how cohesin keeps sister chromatids together. **Chapter 7** introduces concepts in genetic analysis that are essential for working in molecular biology, including a brief introduction to genetic recombination and descriptions of basic techniques used by molecular biologists to locate genes in bacteria, yeast, and higher organisms. The part describing F factor transmission has been extensively revised and updated. **Chapter 8** provides some of the fundamental information that molecular biologists need to know about viruses. The information presented will help to place in perspective viral systems that are discussed in subsequent chapters. Some students and instructors may decide to skip this chapter but then refer back to appropriate sections when specific viruses are mentioned in subsequent chapters. **Chapter 9** presents both the general features of DNA replication and a detailed examination of the initiation, elongation, and termination stages of bacterial DNA replication. This chapter has been comprehensively revised and updated to reflect rapid advances in our understanding of bacterial DNA replication. For instance, the chapter examines the function of the Dia (DnaA-initiator association) protein and considers the significance of the recent discovery that the bacterial clamp loader has a $\gamma 3\delta\delta'\chi\psi$ structure. **Chapter 10** examines DNA replication in eukaryotes and the archaea. The telomere and telomerase part has been expanded and a new part describing replication coupled chromatin formation has been added. **Chapter 11** focuses on the different types of DNA damage that take place in cells. **Chapter 12** explores the mechanisms that cells use to reverse DNA damage, excise and replace damaged elements, or tolerate the damage.

A description of the mutasome has been added. **Chapter 13** explores homologous recombination models, based on genetic and biochemical data. Individual steps of homologous recombination and the enzymes that carry out these reactions are described. **Chapter 14** examines transposons and other mobile elements with particular emphasis on the three types of reactions that can explain the movement of all mobile genetic elements. Some instructors may wish to assign the material covered in Chapter 13, Chapter 14, or both after completing Chapter 23. Chapters 13 and 14 are written in a way that makes this possible. **Chapter 15** examines bacterial RNA polymerase and its function in RNA synthesis. **Chapter 16** examines the mechanisms that bacteria use to regulate the synthesis of the different classes of RNA, starting with the regulation of bacterial mRNA synthesis. The section describing mRNA degradation has been updated and extensively revised. **Chapter 17** examines the eukaryotic RNA polymerase II (the enzyme responsible for mRNA formation) and the general transcription factors that it requires for basal transcription. The chapter also describes how RNA polymerase II and the general transcription factors combine to form the basal transcription machinery. New material in this chapter includes (1) a description of Cap analysis of gene expression (CAGE), a high throughput technique used to identify transcription start sites and their flanking promoters, (2) an update of the description of the core promoter, (3) a part describing how TFIIB's role in converting the closed promoter to an open promoter, and (4) an examination of the archaeal RNA polymerase. **Chapter 18** describes two types of transcriptional factors, transcription activators and mediators, that interact with the basal transcription machinery to form the much more efficient transcription machine that is in fact responsible for eukaryotic mRNA synthesis. New material has been added describing p53 and its loop-sheet-helix DNA binding domain. A new part has been added that examines epigenetic modifications. **Chapter 19** explores the various stages in cotranscriptional processing. It also describes RNA editing and messenger RNA export. **Chapter 20** examines the role of RNA polymerase I and its accessory factors in synthesizing eukaryotic ribosomal RNA. It also describes the role of RNA polymerase III and its accessory factors in synthesizing transfer RNA and other small RNA molecules. The last part of the chapter examines the RNA polymerases present in mitochondria and chloroplasts. **Chapter 21**, new to the Fourth Edition, examines roles that siRNA, miRNA, and piwiRNA play in regulating gene expression. **Chapter 22** describes the structure and function of transfer RNA molecules and then examines how these molecules act, along with mRNA and ribosomes, to specify the amino acid sequences in proteins. New material has been added that describes how editing-defective alanyl-tRNA synthetase causes a neurodegenerative disease in mice. **Chapter 23** examines ribosome structure and function to determine how ribosomes are able act as universal translators.

Supplements to the Text

Jones & Bartlett Learning offers an array of ancillaries to assist instructors and students in teaching and mastering the concepts in this text. Additional information and review copies of any of the following items are available through your Jones & Bartlett Learning sales representative, or by going to www.jblearning.com.

For the Student

Developed exclusively for the fourth edition of *Molecular Biology: Genes to Proteins,* authored by Brent Nielsen of Brigham Young University, the student companion Web site offers a variety of resources to enhance understanding of molecular biology. Free

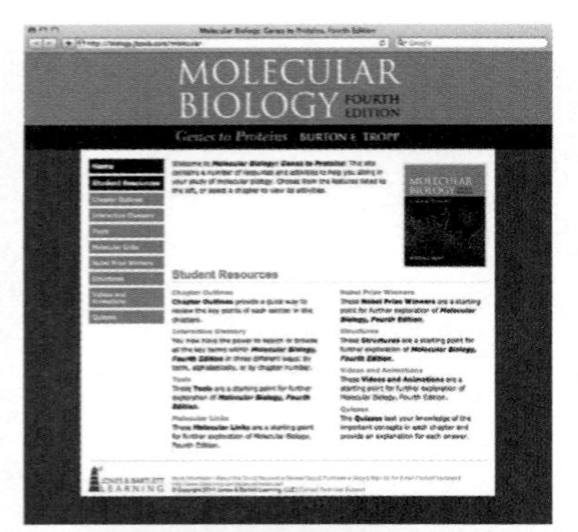

access is provided with each new copy of the text. This Web site contains chapter outlines, quizzes to test comprehension and retention, and an interactive glossary. The site also provides links to relevant materials such as animations, structural programs, and tutorials that are available on the Web. The URL for the Web site is http://biology.jbpub.com/molecular.

Laboratory Investigations in Molecular Biology presents well-tested protocols in molecular biology. The experiments are designed to guide students through realistic research projects and to provide students with instruction in methods and approaches that can be immediately translated into research projects conducted in modern research laboratories.

For the Instructor

Compatible with Windows and Macintosh platforms, the Instructor's Media CD-ROM provides instructors with the following traditional ancillaries:

The PowerPoint® Image Bank provides the illustrations, photographs, and tables (to which Jones & Bartlett Learning holds the copyright or has permission to reproduce digitally) inserted into the PowerPoint slides. You can quickly and easily copy individual images or tables into your existing lecture outlines.

The PowerPoint Lecture Outline Slides presentation package, authored by Cheryl Ingram-Smith of Clemson University, provides lecture notes and images for each chapter of *Molecular Biology: Genes to Proteins*. Instructors with the Microsoft PowerPoint software can customize the outlines, art, and order of presentation.

A Test Bank and Instructor's Manual, provided as a text file and including chapter overviews, complete lecture outlines, and key terms from each chapter, will be available to instructors through the Jones & Bartlett Learning Web site. Visit www.jblearning.com for more details.

A Note from the Author

I would like to tell a brief story that indicates just how far molecular biology has advanced in the past five decades. When I started out as a graduate student in the early 1960s we were all very excited about the rapid advances that were being made in molecular biology, especially the discovery of the genetic code. We viewed molecular biology as a discipline that soared above the other scientific disciplines that our classmates were studying. Several of us, however, were brought back down to earth one day when a young medical doctor, who was working as a postdoctoral fellow, challenged us to provide a single example of how molecular biology had actually benefited anybody in a clinical setting. After much consideration, we were only able to come up with one example. Molecular biology explained the cause of sickle cell anemia. The young doctor then correctly pointed out that this knowledge did not really provide any benefit to individuals suffering from sickle cell anemia. As I wrote this book I thought about that challenge so many years ago. Molecular biology remains an exciting and rapidly developing discipline, but it also is a discipline that has a direct impact on our lives in so many different ways. There are so many examples of how molecular biology has contributed to our medical well-being that it would take several books to describe them all. Therefore, a few examples will have to suffice. Techniques developed by molecular biologists help to detect bacterial and viral infections, produce new drugs and hormones, study the effectiveness of a chemotherapeutic agent used to treat a malignant disease, determine whether an individual has an inborn error of metabolism, and design drugs to treat diseases such as AIDS. Although initial attempts to cure inborn errors of metabolism by genetic engineering have been for the most part unsuccessful, and indeed some have proved dangerous to the subject, the next generation of molecular biologists likely will solve this and a host of other health-related problems.

Acknowledgments

I would like to acknowledge the assistance of the many people who helped to prepare the fourth edition of this molecular biology text. First and foremost, I wish to acknowledge the contributions made by two gifted molecular biologists—Hannah Klein (Departments of Biochemistry and Medicine, NYU Medical Center) and Joseph E. Peters (Department of Microbiology, Cornell University)—the authors of Chapter 13 (Recombination) and Chapter 14 (Transposons and Other Mobile Elements), respectively. I also thank the following talented scholars for reviewing one or more chapters and, in the process of doing so, correcting errors and making valuable suggestions to improve the book:

Steven Ackerman, University of Massachusetts, Boston
Charles Austerberry, Creighton University
James Bardwell, University of Michigan, Ann Arbor
David Bentley, University of Colorado Health Sciences Center
Steven Berberich, Wright State University
James Botsford, New Mexico State University
John Boyle, Mississippi State University
Phyllis Braun, Fairfield University
John Brunstein, University of British Columbia
Martin Buck, Imperial College of Science, Technology, and Medicine
Michael Campbell, Pennsylvania State University, Erie
George Chaconas, University of Western Ontario
Karl Chai, University of Central Florida
Mary Connell, Appalachian State University
John Cronan, Jr., University of Illinois, Urbana-Champaign
Helena Danielson, Uppsala University
Robert Dotson, Tulane University
John Elder, Jr., Valdosta State University
Mark Erhart, Chicago State University
Charles Gasser, University of California, Davis
Barbara Hamkalo, University of California, Irvine
Lynn Harrison, University of Louisiana, Health Sciences Center
Daniel Herman, University of Wisconsin, Eau Claire
Martinez Hewlett, University of Arizona
Deborah Hoshizaki, University of Nevada, Las Vegas
Deborah Hursh, American University
Stan Ivey, Delaware State University
Russell Johnson, Colby College
Susan Karcher, Purdue University
Jennifer Kugel, University of Colorado, Boulder
Paul Laybourn, Colorado State University
Mark Lee, Spelman College
Steven Matson, University of North Carolina, Chapel Hill
Philip Meaden, Heriot-Watt University
Isabel Mellon, University of Kentucky, College of Medicine

William Merrick, Case Western Reserve University
John Mullican, Washburn University
David Mullin, Tulane University
Melody Neumann, University of Toronto
Hao Nguyen, California State University, Sacramento
Desmond Nicholl, University of Paisley
Amy Pasquinelli, University of California, San Diego
Bruce Patterson, University of Arizona
Frank Pugh, Pennsylvania State University
William Reznikoff, University of Wisconsin, Madison
Randall Shortridge, State University of New York, Buffalo
Grace Ann Spatafora, Middlebury College
James Thompson, University of Oklahoma
Michael Tully, University of Bath
Barbara van Leeuwen, The Australian National University
Charles Vigue, University of New Haven
Karl Wilson, State University of New York, Binghamton
Mark Winey, University of Colorado, Boulder
Stephan Zweifel, Carleton College

It is a pleasure to thank the many outstanding people at Jones & Bartlett Learning who contributed so much to this edition. I have been very fortunate to work with Cathleen Sether, the editor in charge of this project. Her advice, assistance, and patience were most helpful. She recruited the outstanding production team that worked on this edition. I am pleased to acknowledge the contributions of individual members of this team. I am very grateful to Lou Bruno, who coordinated the copyediting, artwork, and other production activities and made the entire process run smoothly. I greatly appreciate his patience, dedication, and incredible organizational skills. Elizabeth Morales created instructive and beautiful figures in this edition and the previous one, often working from crude sketches. Shellie T. Newell copyedited the manuscript, making many improvements in the process. Jan Cocker served as proofreader, helping to assure that errors did not slip through from the copyedited version to the final version that appears in this book. It is a pleasure to acknowledge Alexandra Nickerson for her work in preparing the index. Special thanks go to Megan Turner for the support and assistance that she provided during manuscript preparation. I am very grateful to Emily O'Neill and Kimberly Potvin for their help in obtaining necessary permissions for the figures that appear in this book and to the authors, journal editors, and publishers for granting permissions to use copyrighted material.

Finally, I thank my wife Roslyn for her support, concern, and understanding while this book was in preparation.

Burton E. Tropp

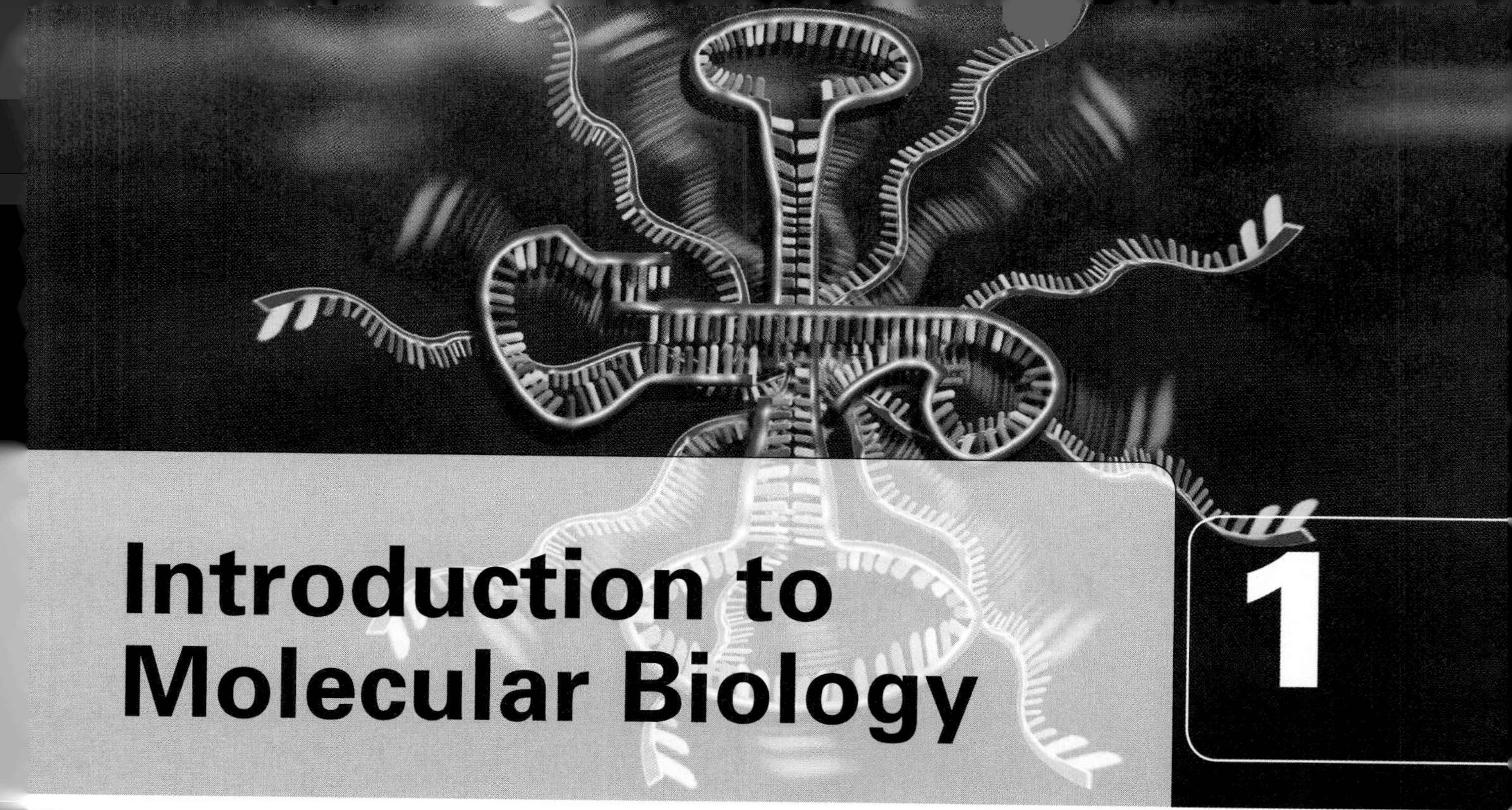

1 Introduction to Molecular Biology

OUTLINE OF TOPICS

1.1 Intellectual Foundation
Two studies performed in the 1860s provided the intellectual underpinning for molecular biology.

1.2 Genotypes and Phenotypes
Each gene is responsible for the synthesis of a single polypeptide.

1.3 Nucleic Acids
Nucleic acids are linear chains of nucleotides.

1.4 DNA Structure and Function
Transformation experiments led to the discovery that DNA is the hereditary material.
Chemical experiments also supported the hypothesis that DNA is the hereditary material.
The blender experiment demonstrated that DNA is the genetic material in bacterial viruses.
RNA serves as the hereditary material in some viruses.
Rosalind Franklin and Maurice Wilkins obtained x-ray diffraction patterns of extended DNA fibers.
James Watson and Francis Crick proposed that DNA is a double-stranded helix.
The central dogma provides the theoretical framework for molecular biology.
Recombinant DNA technology allows us to study complex biological systems.
A great deal of molecular biology information is available on the Internet.

Suggested Reading

Classic Papers

The term **molecular biology** first appeared in a report prepared for the Rockefeller Foundation in 1938 by Warren Weaver, then director of the Foundation's Natural Sciences Division. Weaver coined the term to describe a research approach in which physics and chemistry would be used to address fundamental biological problems. Weaver proposed that the Rockefeller Foundation fund research efforts to seek molecular explanations for biological processes. His proposal was remarkably farsighted, especially when considering that many of his contemporaries believed that living cells possessed a vital force that could not be explained by chemical or physical laws that govern the inanimate world. In fact, some physicists entered the field of biology in the hope that they might discover new physical laws.

At the time of Weaver's report, biology was on the threshold of major changes. Two new disciplines—biochemistry and genetics—had altered the way that biologists think about living systems. Biochemists had delivered a major blow to the vital force theory by demonstrating that cell-free extracts can perform many of the same functions as intact cells. Geneticists established that the functional and physical unit of heredity is the gene. However, they did not know the chemical nature of genes, the way that hereditary information is stored in genes, how genes are replicated so that they can be transmitted to the next generation, or how information stored in genes determines a specific physical trait such as eye color.

Neither biochemistry nor genetics had the power to solve these problems on its own. In fact, it took an interdisciplinary effort involving specialists in many fields of the life sciences, including biochemistry, biophysics, chemistry, x-ray crystallography, developmental biology, genetics, immunology, microbiology, and virology, to solve the heredity problem. This interdisciplinary effort resulted in the creation of a new discipline, molecular biology, which seeks to explain genetic phenomena in chemical and physical terms.

1.1 Intellectual Foundation

Two studies performed in the 1860s provided the intellectual underpinning for molecular biology.

The earliest intellectual roots of molecular biology can be traced back to the work of two investigators in the 1860s. No connection was apparent between the experiments performed by the two investigators for more than 75 years, but when the connection was finally made, the result was the birth of molecular biology and the beginning of a scientific revolution that continues today.

Mendel's Three Laws of Inheritance

The work of the first investigator, Gregor Mendel, an Austrian monk and botanist, is familiar to all biology students and so will only be summarized briefly here. Mendel discovered three basic laws of inheritance by studying the way in which simple physical traits are passed on from one generation of pea plants to the next. For convenience, Mendel's laws of inheritance will be described using two modern biological terms, **gene** for a unit of heredity and **chromosome** for a structure bearing several linked genes.

1. **The law of segregation**—A specific gene may exist in alternate forms called alleles. An organism inherits one allele for each trait from each parent. The two alleles, which may be the same or different, segregate (or separate) in germ cells (sperm or egg) and combine again during reproduction so that each parent transmits one allele to each offspring.
2. **The law of independent assortment**—Specific physical traits such as plant size and color are inherited independently of one another. Mendel was fortunate to have selected physical traits that were determined by genes that were on different chromosomes.
3. **The law of dominance**—For each physical trait, one allele is dominant so that the physical trait that it specifies appears in a definite 3:1 ratio. The alternative form is recessive. In Mendel's peas, tallness was dominant and shortness recessive. Therefore, three times as many pea plants were tall as were short. Today we know that there are exceptions to the law of dominance. Sometimes neither allele is dominant. For instance, a plant that inherits a gene for a red flower and a gene for a white flower may produce a pink flower.

Unfortunately, scientists failed to recognize the significance of Mendel's work during his lifetime. His paper remained obscure until about 1900 when scientists rediscovered Mendel's laws of inheritance, giving birth to the science of genetics.

Miescher and DNA

The second investigator, the Swiss physician Friedrich Miescher, performed experiments that led to the discovery of deoxyribonucleic acid (DNA), which we, of course, now know is the hereditary material. Miescher did not set out to discover the hereditary material but instead was interested in studying cell nuclei from white blood cells, which he collected from pus discharges on discarded bandages that had been used to cover infected wounds. Miescher used a combination of protease (enzymes that hydrolyze proteins) digestion and solvent extraction to disrupt and fractionate the white blood cells. One fraction, which he called **nuclein**, contained an acidic material with unusually high phosphorus content. Miescher later found that salmon sperm cells, which have remarkably large cell nuclei, are also an excellent source of nuclein. In 1889, Miescher's student, Richard Altmann, separated nuclein into protein and a substance with a very high phosphorous content that he named **nucleic acid**. Because of its high phosphorus content, investigators initially thought that nucleic acids might serve as storehouses for cellular phosphorus.

1.2 Genotypes and Phenotypes

Each gene is responsible for the synthesis of a single polypeptide.

Mendel's experiments showed that the genetic makeup of an organism, its **genotype**, determines the organism's physical traits, its **phenotype**. However, his experiments did not show how genes are able to determine complex physical traits such as plant color or size. Archibald Garrod, an English physician, was the first to provide an explanation for the relationship between genotype and phenotype. Garrod uncovered this relationship in the early 1920s while studying alkaptonuria, a rare but harmless inherited human disorder in which the urine of affected individuals becomes very dark upon standing due to the accumulation of homogentisic acid, a breakdown product of the amino acid tyrosine. Garrod correctly proposed that alkaptonuria results from a recessive gene, which causes a deficiency in the enzyme that normally converts homogentisic acid into colorless products.

Garrod's work was generally ignored until the early 1940s, when the American geneticists, George Beadle and Edward Tatum rediscovered it while seeking experimental proof for the connection between genes and enzymes. They believed that, if a gene really does specify an enzyme, it should be possible to create genetic mutants that cannot carry out specific enzymatic reactions. They, therefore, exposed spores of the bread mold *Neurospora crassa* to x-rays or UV radiation and demonstrated that the mutant molds had a variety of special nutritional needs. Unlike their wild-type parents, the mutants could not reproduce without having specific amino acids or vitamins added to their growth medium. A mutant that requires a specific supplement that is not required by the wild-type parent is called an **auxotroph**. Genetic analysis revealed that each auxotroph appeared to be blocked at a specific step in the metabolic pathway for the required amino acid or vitamin. Furthermore, the auxotrophs accumulated large quantities of the substance formed just prior to the blocked step. Beadle and Tatum thus had replicated in the bread mold the same type of situation that Garrod had observed in alkaptonuria. A defective gene caused a defect in a specific enzyme that resulted in the abnormal accumulation of an intermediate in a metabolic pathway. As a result of their work with the *N. crassa* mutants, Beadle and Tatum proposed the one gene–one enzyme hypothesis, which states that each gene is responsible for synthesizing a single enzyme. We now know that many enzymes are made of more than one type of polypeptide chain and that a single mutation may affect just one of the polypeptide chains. The original one gene–one enzyme hypothesis hence was modified to become a one gene–one polypeptide hypothesis. As we will see later, however, even the one gene–one polypeptide hypothesis is an oversimplification.

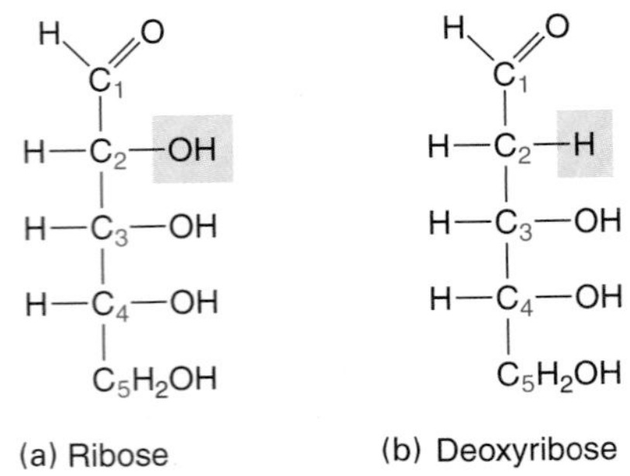

(a) Ribose (b) Deoxyribose

FIGURE 1.1 The two sugars present in nucleic acids. (a) Ribose and (b) deoxyribose each contain five carbon atoms and an aldehyde group and therefore belong to aldopentose family of simple sugars. The only difference between the two sugars is that ribose has a hydroxyl group at carbon-2 and deoxyribose as a hydrogen atom at carbon-2.

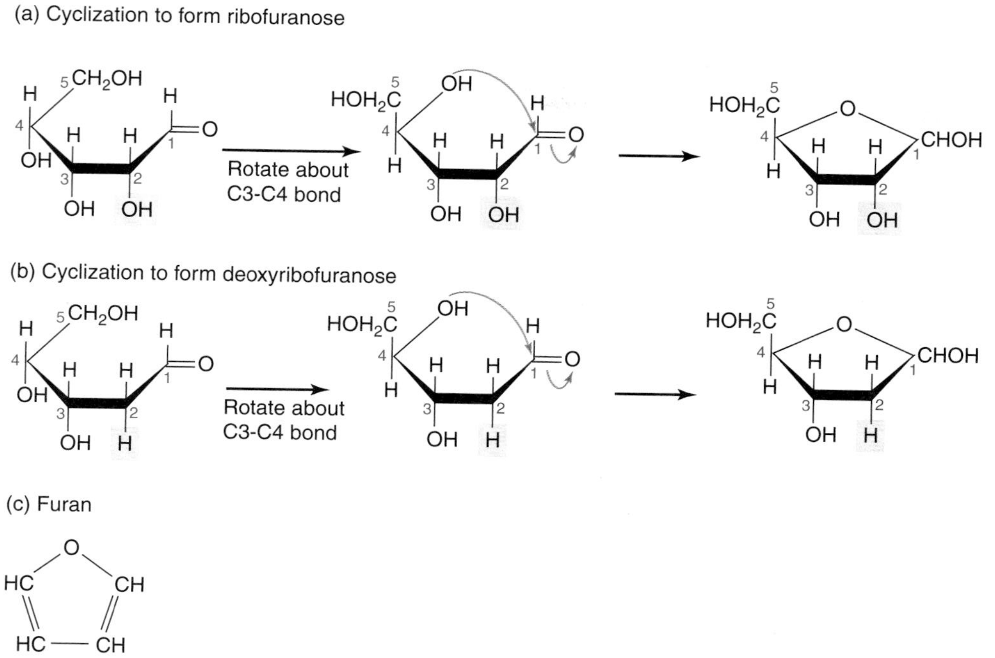

(a) Cyclization to form ribofuranose

Rotate about C3-C4 bond

(b) Cyclization to form deoxyribofuranose

Rotate about C3-C4 bond

(c) Furan

FIGURE 1.2 Conversion of straight chain ribose and deoxyribose to cyclic forms. (a) Conversion of ribose to ribofuranose and (b) conversion of deoxyribose to deoxyribofuranose. (c) Structure of furan, a five-atom ring system.

1.3 Nucleic Acids

Nucleic acids are linear chains of nucleotides.

Investigators slowly came to realize that nucleic acids could be divided into two major groups: deoxyribonucleic acid (DNA) and ribonucleic acid (RNA). Although DNA was initially thought to be present in animals and RNA in plants, investigators eventually detected both kinds of nucleic acids in all living systems. The principal difference between DNA and RNA is that the former contains deoxyribose, and the latter, ribose (FIGURE 1.1).

The two five-carbon sugars (pentoses) differ in only one substituent: deoxyribose has a hydrogen atom at carbon-2, whereas ribose has a hydroxyl group at this position. By convention, the aldehyde group in the pentose is C-1, the next carbon is C-2, and so forth. Each pentose chain can close to form a five-carbon ring in which an oxygen bridge joins C-1 to C-4 (FIGURE 1.2A and B).

Because the sugar rings are derivatives of furan (FIGURE 1.2C), they are called **furanoses**. Ribofuranose and deoxyribofuranose are often depicted as Haworth structures (named after Walter N. Haworth, the investigator who devised the representations). As illustrated in FIGURE 1.3, a Haworth structure represents a cyclic sugar as a flat ring

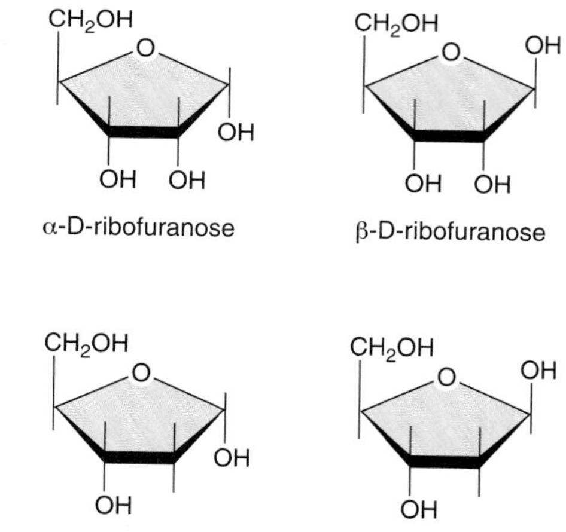

FIGURE 1.3 Haworth structures for ribofuranose and deoxyribofuranose. Haworth structure represents a cyclic sugar as a flat ring perpendicular to the plane of the page with the ring-oxygen in the back and C-1 to the right. A hydrogen atom attached to the sugar ring is represented by a line. Ring formation can lead to two different stereochemical arrangements at C-1; one in which the hydroxyl at C-1 points down (a) and another in which it points up (b).

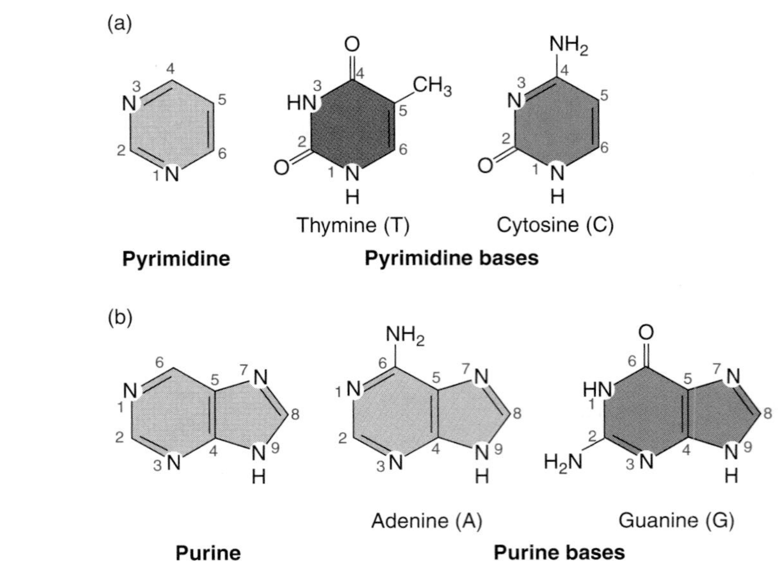

FIGURE 1.4 Pyrimidine and purine bases in DNA.

Uracil (U)

FIGURE 1.5 Uracil (U).

perpendicular to the plane of the page with the ring-oxygen in the back and C-1 to the right. The ring's thick lower edge projects toward the viewer, its upper edge projects back behind the page, and its substituents are visualized as being either above or below the plane of the ring. A line is used to represent a hydrogen atom attached to the sugar ring. Ring formation can lead to two different stereochemical arrangements at C-1. One arrangement, the α-anomer, is represented by drawing the hydroxyl group attached to C-1 below the plane of the ring and the other, the β-anomer, by drawing it above the plane of the ring. We will see in Chapter 4 that ribofuranose and deoxyribofuranose actually have puckered rather than planar conformations. Nevertheless, Haworth structures are convenient representations for sugar rings when precise three-dimensional information is not required.

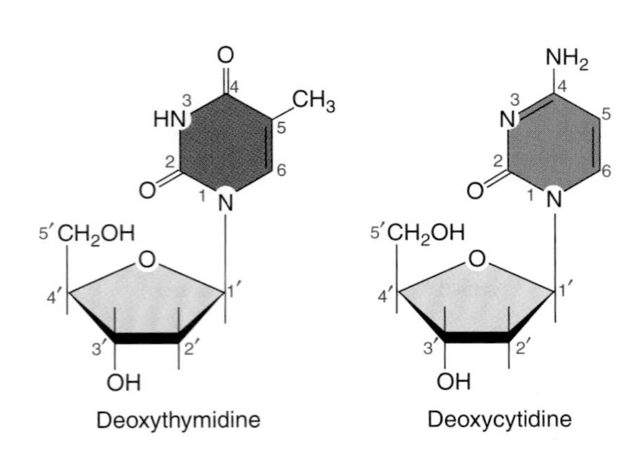

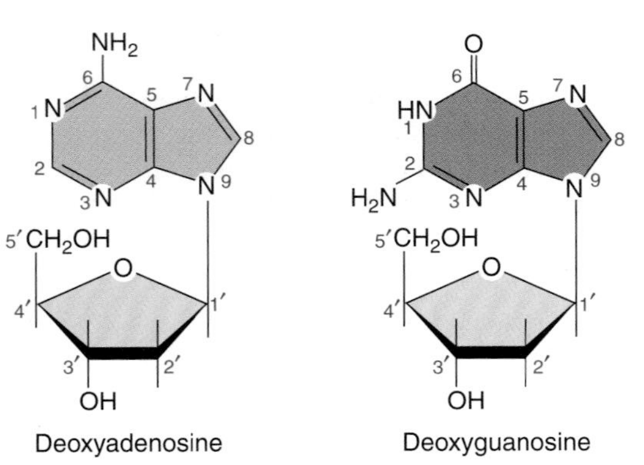

FIGURE 1.6 Deoxyribonucleosides.

1.4 DNA Structure and Function

An early pioneer in the study of nucleic acid chemistry, Phoebus A. Levene found that DNA contains four different kinds of heterocyclic ring structures, which are now known simply as bases because they can act as proton acceptors. Two of the bases, **thymine (T)** and **cytosine (C)**, are derivatives of **pyrimidine** (FIGURE 1.4A) and the other two **adenine (A)** and **guanine (G)**, are derivatives of **purine** (FIGURE 1.4B). RNA also contains cytosine, adenine, and guanine, but the pyrimidine **uracil (U)** replaces thymine (FIGURE 1.5). The only difference between uracil and thymine is that the latter contains a methyl group attached to carbon-5. Levene showed that T, C, A, and G combine with deoxyribose to form a class of compounds called deoxyribonucleosides (FIGURE 1.6) and that U, C, A, and G combine with ribose to form a related class of compounds called ribonucleosides (FIGURE 1.7). Each base is linked to the pentose ring by a bond that joins a specific nitrogen atom on the base (N-1 in pyrimidines and N-9 in purines) to C-1 on the furanose ring. This bond is termed an N-glycosylic bond. Because each nucleoside has two ring systems (the sugar and the base that is attached to it), a method is required to distinguish between atoms in each ring system. This problem is solved by adding a prime (′) after the sugar atoms. The first carbon atom in the sugar, thus, becomes 1′, the second 2′, and so forth.

Nucleosides that have a phosphate group attached to the sugar group are called nucleotides. Ribonucleoside and deoxyribonucleoside derivatives are called ribonucleotides and deoxyribonucleotides, respectively (**FIGURE 1.8**). The pentose carbon atom to which the phosphate group is attached is given as part of the nucleotide's name. Thus, the phosphate group is attached to C-3′ in uridine-3′-monophosphate (3′-UMP) and thymidine-3′-monophosphate (3′-dTMP) and to C-5′ in uridine-5′-monophosphate (5′-UMP) and thymidine-5′-monophosphate (5′-dTMP). As indicated in **Table 1.1**,

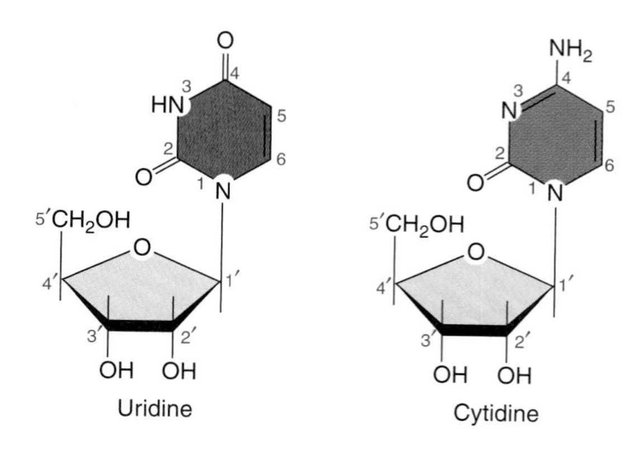

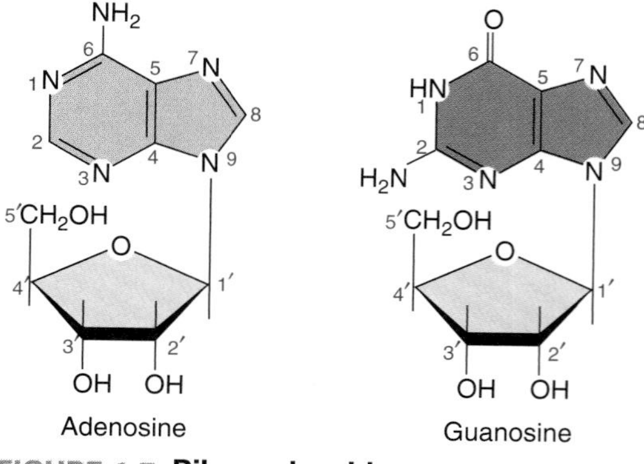

FIGURE 1.7 Ribonucleosides.

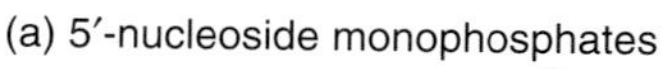

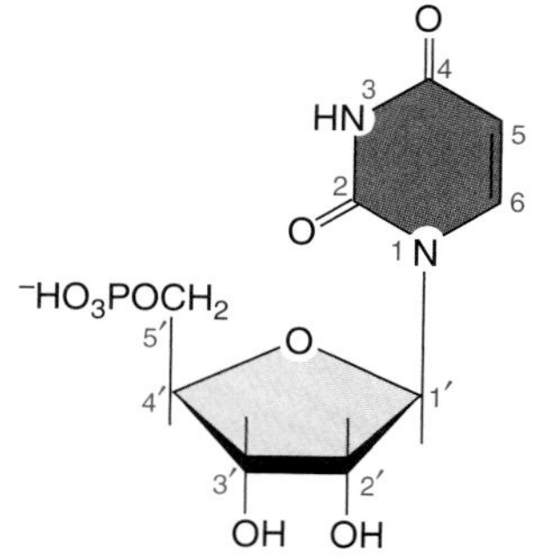

Uridine-5′-monophosphate (5′-UMP) or 5′-uridylate

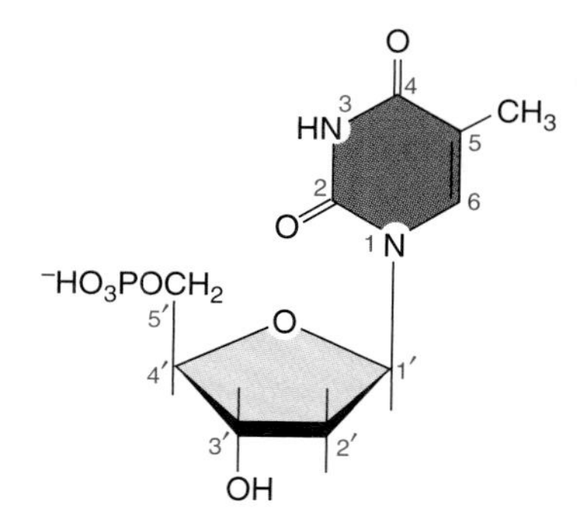

Thymidine-5′-monophosphate (5′-dTMP) or 5′-thymidylate

(b) 3′-nucleoside monophosphates

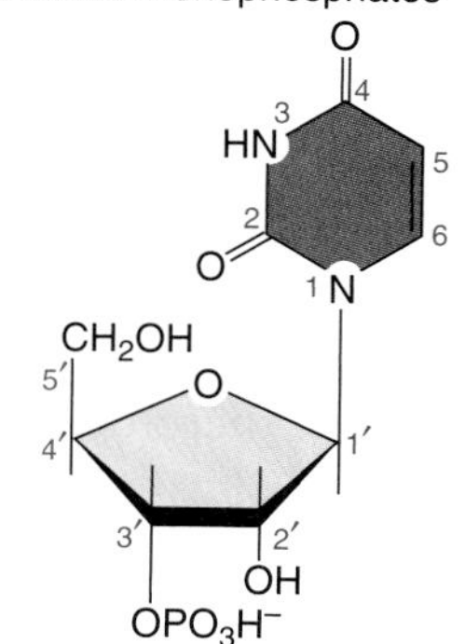

Uridine-3′-monophosphate (3′-UMP) or 3′-uridylate

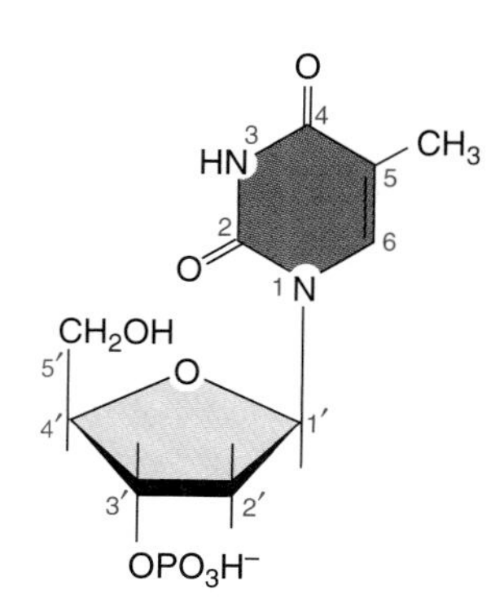

Thymidine-3′-monophosphate (3′-dTMP) or 3′-thymidylate

FIGURE 1.8 Nucleotides. Nucleotides are formed by adding a phosphate group to the pentose ring in a nucleoside. (a) Nucleotides formed by adding a phosphate group to the 5′-hydroxyl group in uridine or thymidine. (b) Nucleotides formed by adding a phosphate group to the 3′-hydroxyl group in uridine or thymidine.

TABLE 1.1 **Comparison of Major Features in A-, B-, and Z-Forms of DNA**

Base	Sugar	Nucleoside	5′-Mononucleotide
Uracil (U)	ribose	uridine	Uridine-5′-monophosphate or 5′-uridylate (5′-UMP)
Cytosine (C)	ribose	cytidine	Cytidine-5′-monophosphate or 5′-cytidylate (5′-CMP)
Adenine (A)	ribose	adenosine	Adenosine-5′-monophosphate or 5′-adenylate (5′-AMP)
Guanine (G)	ribose	guanosine	Guanosine-5′-monophosphate or 5′-guanylate (5′-GMP)
Thymine (T)	deoxyribose	deoxythymidine[1]	Deoxythymidine-5′-monophos-phate or 5′-deoxythymidylate (5′-dTMP)[1]
Cytosine (C)	deoxyribose	deoxycytidine	Deoxycytidine-5′-monophos-phate or 5′-deoxycytidylate (5′-dCMP)
Adenine (A)	deoxyribose	deoxyadenosine	Deoxyadenosine-5′-monophos-phate or 5′-deoxyadenylate (5′-dAMP)
Guanine (G)	deoxyribose	deoxyguanosine	Deoxyguanosine-5′-monophos-phate or 5′-deoxyguanylate (5′-dGMP)

[1]Deoxythymidine and deoxythymidine-5′-monophosphate are also called thymidine and thymidine-5′-monophosphate, respectively. When thymine is attached to ribose, the nucleoside is called ribothymidine and the nucleotide is called ribothymidylate. This nomenclature convention follows from the fact that thymine is most frequently attached to deoxyribose.

nucleoside monophosphates have two alternative names. For example, cytidine-5′-monophosphate (5′-CMP) is also known as 5′-cytidylate.

Levene also suggested that neighboring deoxyribonucleosides in DNA are joined to one another by 5′ to 3′ (5′→3′) phosphodiester bridges. Unfortunately, Levene is remembered more for a hypothesis that he proposed that turned out to be wrong than for all his important contributions to the field of nucleic acid chemistry. To place Levene's hypothesis in context, it is important to note that the analytical tools available to him were quite poor. Not surprisingly, therefore, Levene based his idea about DNA structure on the incorrect assumption that T, C, A, and G are present in DNA in equimolar concentrations. He proposed that DNA has a tetranucleotide structure (**FIGURE 1.9**). This hypothetical structure became untenable in the 1930s when the Swedish researchers Torbjörn Caspersson and Einar Hammersten showed that DNA has a very high molecular mass and therefore must be a very large molecule or macromolecule.

Nevertheless, scientists continued to accept the idea that DNA contains a repeating sequence of the four nucleotides. They were therefore forced to look elsewhere for a molecule that could carry genetic information because DNA, which they mistakenly thought had a repetitive and monotonous nucleotide sequence, seemed an unlikely candidate to carry much information. Proteins seemed to be much better candidates for the genetic material. Although little was known about protein structure in the early part of the twentieth century, there did seem to be many kinds of different proteins. Furthermore, proteins,

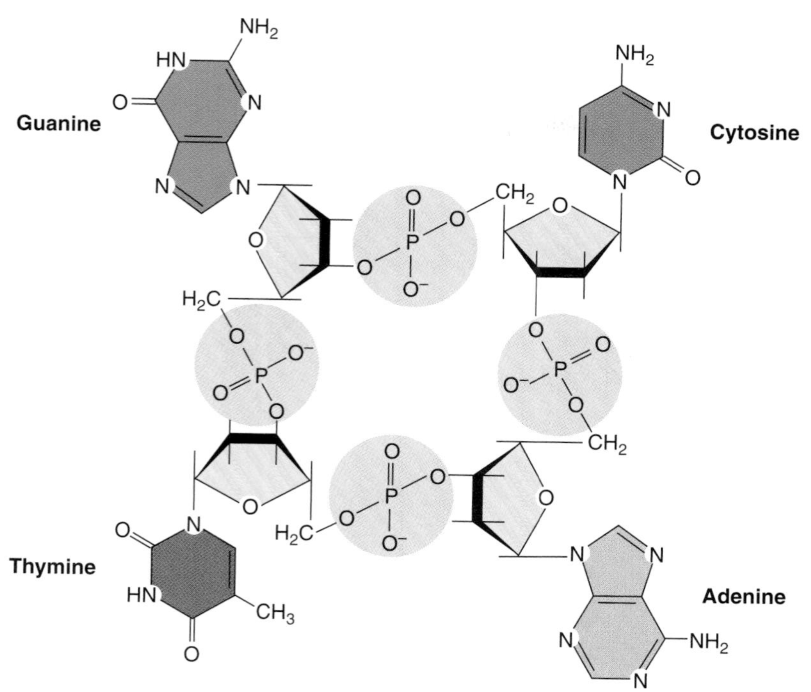

FIGURE 1.9 **The tetronucleotide structure proposed by Phoebus A. Levene.** This structure shows the correct connectivity between nucleotides but assumes that the structure closes on itself (is circular rather than linear). Although wrong, Levene's proposal had a marked influence on the field of nucleic acid chemistry for many years. (Adapted from W. Klug and M. R. Cummings. *Concepts of Genetics, Second edition*. Merrill Publishing Company, 1986.)

along with DNA, were known to be part of the chromosomes, distinct structures in the nucleus that had been demonstrated to carry genes. Once DNA had been eliminated as the genetic material because it seemed to be a "dumb" molecule, investigators focused on proteins.

It was not until 1952 that the distinguished British organic chemist Lord Alexander Robertus Todd and his associates showed the way that the various groups in DNA are joined together. These DNA groups form a long unbranched polymer in which phosphate groups join 5′- and 3′-carbons of neighboring deoxyribonucleosides (FIGURE 1.10a). Thus, *each linear DNA chain has a 3′- and a 5′-terminus*. This directionality will be very important in future discussions of DNA structure and function. The structure of RNA is very similar to that for DNA (FIGURE 1.11a).

Just three years after the discovery that nucleic acids are linear chains of nucleotides, Frederick Sanger working in England showed that each kind of polypeptide molecule is a linear chain of amino acids arranged in a specific order. The amino acid order is responsible for the unique physical and chemical properties of each type of protein, including its ability to catalyze specific reactions, protect cells from foreign organisms and substances, transport materials, or support the cellular infrastructure. The awareness that DNA molecules are made of linear chains of nucleotides and polypeptides are made of linear chains of amino acids led to the sequence hypothesis that proposes nucleotide sequences specify amino acid sequences.

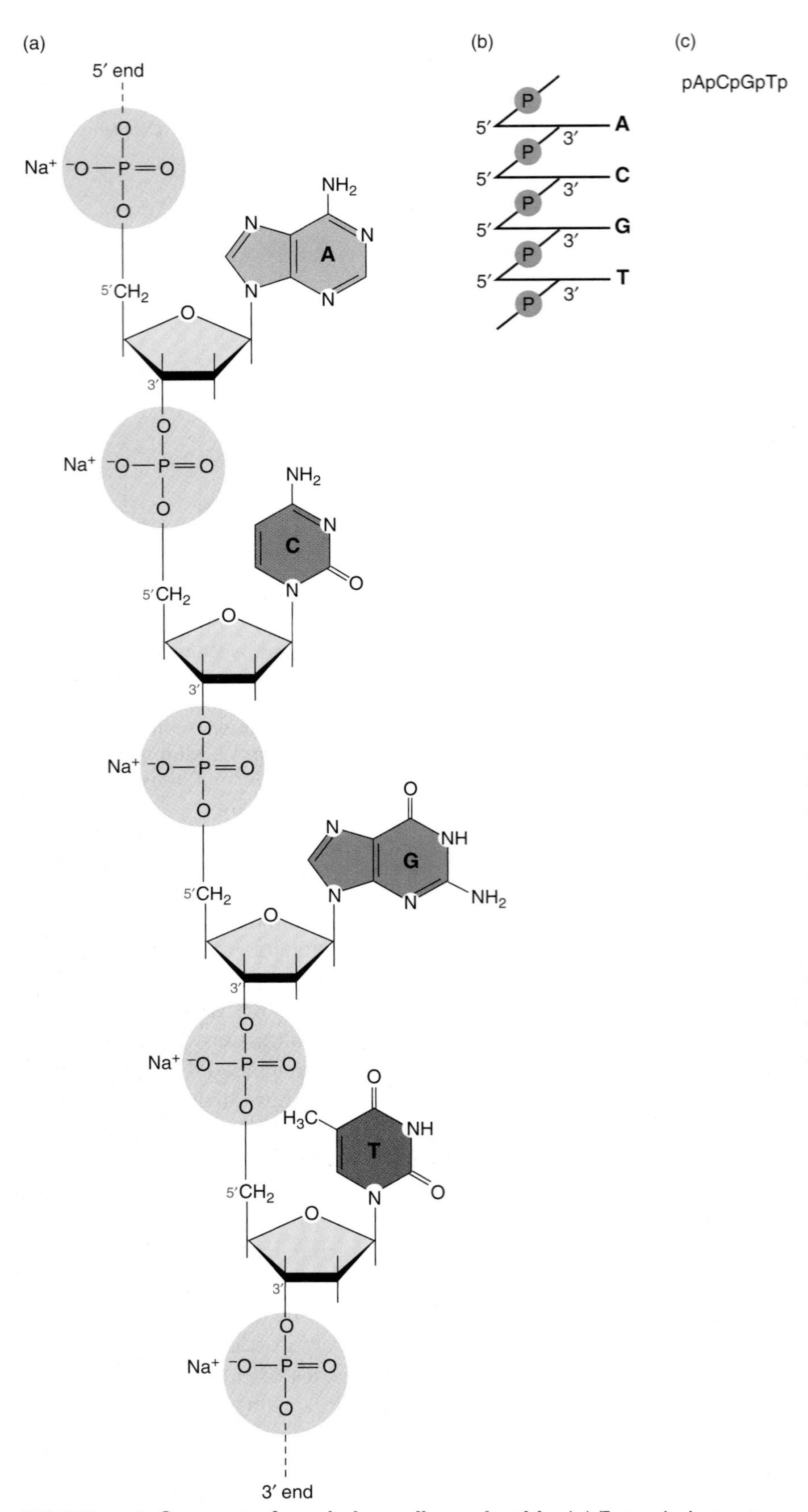

FIGURE 1.10 Segment of a polydeoxyribonucleotide. (a) Extended structure as a sodium salt, (b) stick figure structure, and (c) an abbreviated structure.

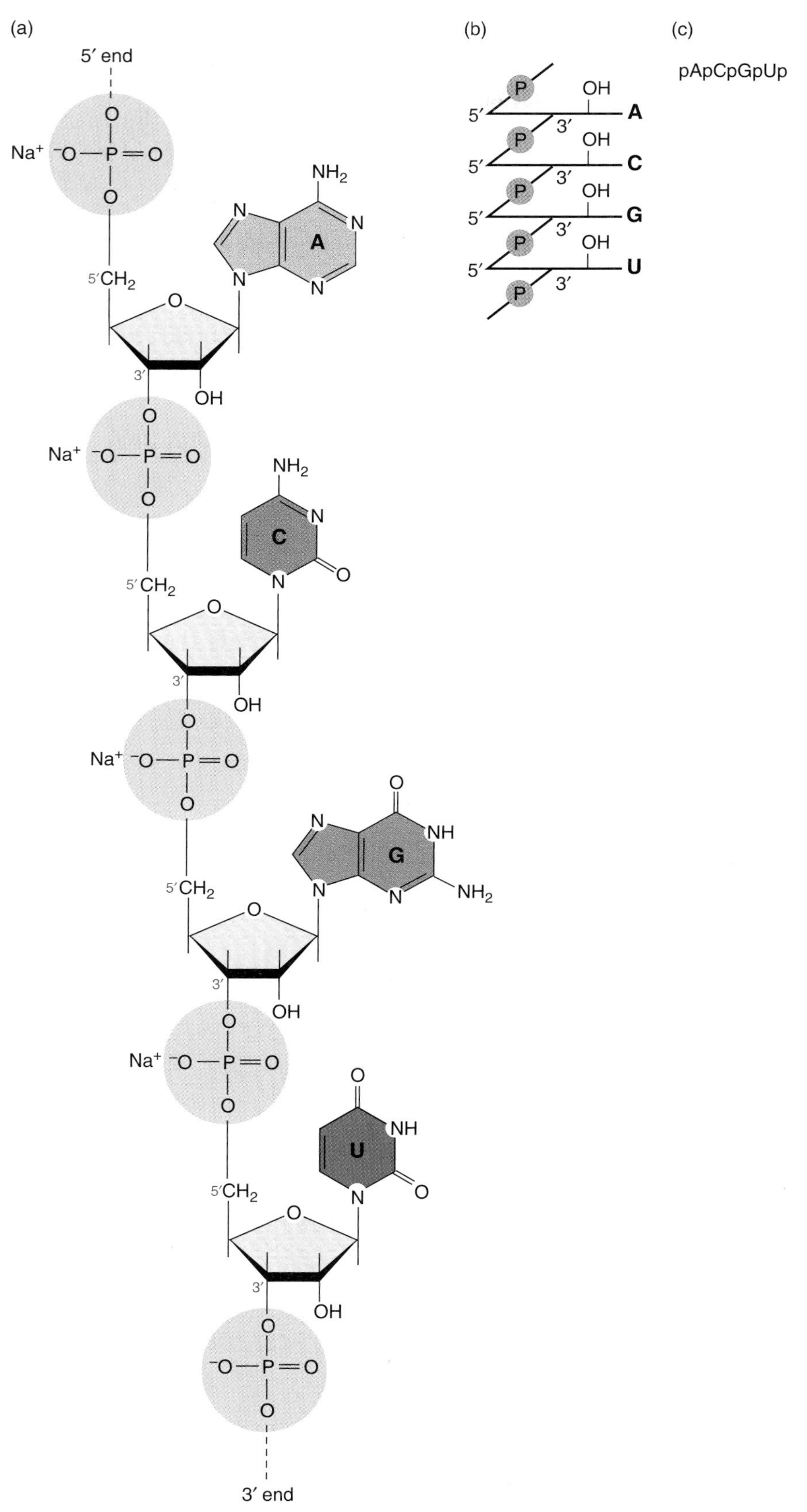

FIGURE 1.11 **Segment of a polyribonucleotide.** (a) Extended structure as a sodium salt, (b) stick figure structure, and (c) an abbreviated structure.

Drawing extended structures for DNA and RNA chains requires considerable time and space. It is more convenient to draw stick figure structures, which are adequate representations for many purposes (FIGURES 1.10b and 1.11b). The few simple conventions for drawing stick figure representations for DNA and RNA are as follows: (1) A single line (shown as horizontal in Figures 1.10b and 1.11b) represents the pentose ring. (2) The letter A, G, C, U or T, at one end of the line, represents the purine or pyrimidine attached to C-1′ of the pentose ring. (3) The letter P, connected by short diagonal lines to adjacent lines, represents the 5′→3′ phosphodiester bond. (4) The symbol OH represents a hydroxyl group. An even simpler method for indicating nucleotide sequence is to just write the letters corresponding to the bases (FIGURES 1.10c and 1.11c).

Transformation experiments led to the discovery that DNA is the hereditary material.

The first step on the path leading to the discovery that genes are made of DNA was an observation made in 1928 by Fred Griffith, who was studying *Streptococcus pneumoniae*, the bacterium responsible for human pneumonia. The virulence of this bacterium was known to depend on a surrounding polysaccharide capsule that protects the bacterium from the body's defense systems. This capsule also causes the bacterium to produce smooth-edged (S) colonies on an agar surface. It was known that S bacteria normally killed mice. Griffith isolated a rough-edged (R) colony mutant, which proved to be both

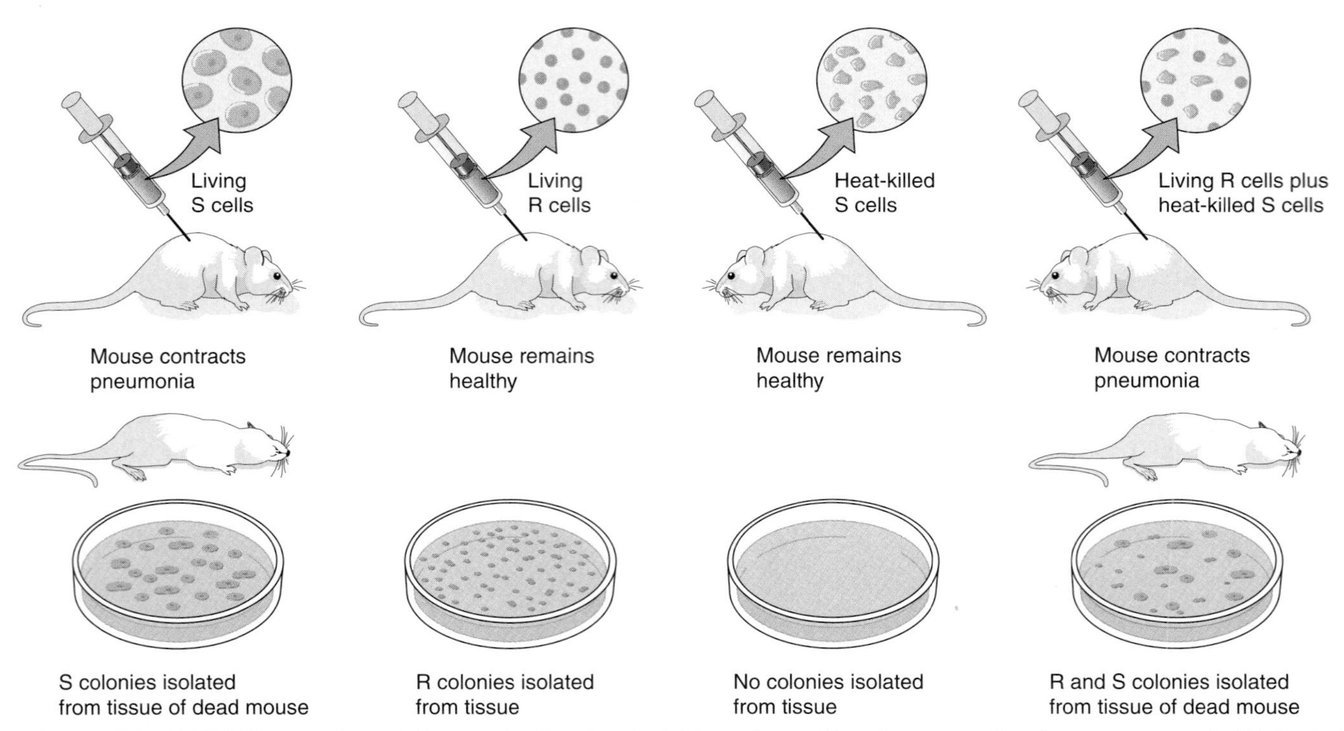

FIGURE 1.12 **The Griffith's experiment demonstrating bacterial transformation.** A mouse dies from pneumonia if injected with the virulent S (smooth) strain of Streptococcus pneumoniae. However, the mouse remains healthy if injected with either the nonvirulent R (rough) strain or the heat-killed S strain. R cells in the presence of heat-killed S cells are transformed into the virulent S strain, killing the mouse.

non-encapsulated and non-lethal. He then observed that while either live R or heat-killed S bacteria are non-lethal, a mixture of the two is lethal (FIGURE 1.12). Furthermore, when bacteria were isolated from a mouse that had died from such a mixed infection, the bacteria were live S and R. Live-R bacteria, therefore, had somehow either been replaced by or transformed to S bacteria.

Several years later, investigators showed that the mouse itself was not needed to mediate this transformation because when a mixture containing live R bacteria and heat-killed S bacteria was incubated in culture medium, living S cells were produced. One possible explanation for this surprising phenomenon was that the R cells restored the viability of the dead S cells. This hypothesis was eliminated, however, by the observation that living S cells appeared even when the heat-killed S culture in the mixture was replaced by a cell extract prepared from broken S cells, which had been freed from both intact cells and the capsular polysaccharide by centrifugation. It was concluded that the cell extract contained a **transforming principle**, the nature of which was unknown.

The next development occurred in 1944 when Oswald Avery, Colin MacLeod, and Maclyn McCarty determined the chemical nature of the transforming principle. They did so by isolating DNA from S cells and adding the DNA to live R bacterial cultures (FIGURE 1.13). After allowing the mixture to incubate for a period of time, they placed samples on an agar surface and incubated them until colonies appeared. Some of the colonies (about 1 in 10^4) that grew were S type. To show that this was a permanent genetic change, Avery and

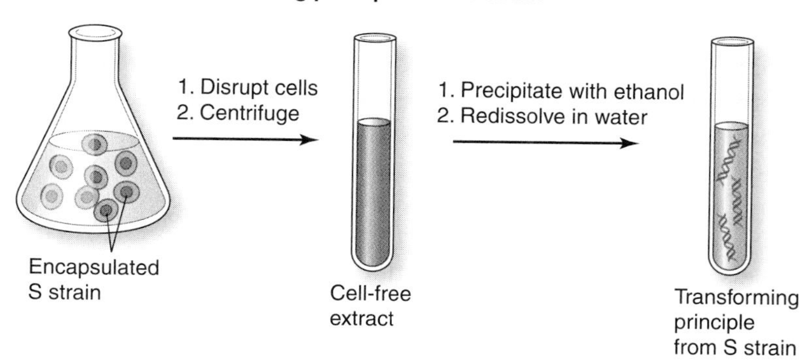

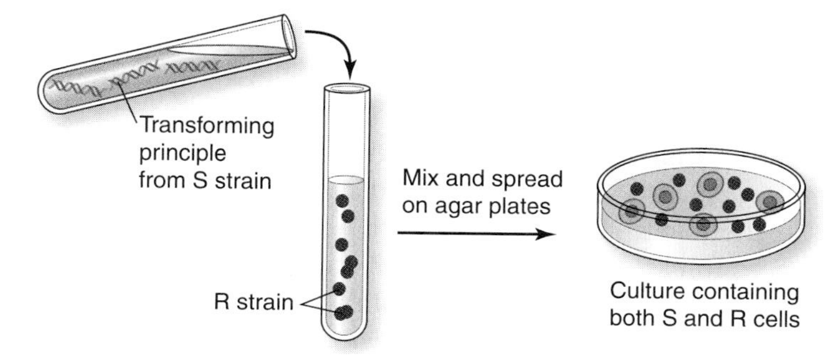

FIGURE 1.13 **The Avery, MacLeod, and McCarty transformation experiment.**

coworkers dispersed many of the newly formed S colonies and placed them on a second agar surface. The resulting colonies were again S type. If an R colony arising from the original mixture was dispersed, only R bacteria grew in subsequent generations. R colonies, hence, retained the R character, whereas the transformed S colonies bred true as S. Because S and R colonies differed by a polysaccharide coat around each S bacterium, Avery and coworkers tested the ability of purified polysaccharide to transform, but observed no transformation. Because the procedures for isolating DNA then in use produced DNA containing many impurities, it was necessary to provide evidence that the transformation was actually caused by DNA and not an impurity. This evidence was provided by the following four procedures.

1. Chemical analysis of samples containing the transforming principle showed that the major component was a deoxyribose-containing nucleic acid.
2. Physical measurements showed that the sample contained a highly viscous substance having the properties of DNA.
3. Incubation with trypsin or chymotrypsin, enzymes that catalyze protein hydrolysis, or with ribonuclease (RNase), an enzyme that catalyzes RNA hydrolysis, did not affect transforming activity. The transforming principle thus is neither protein nor RNA.
4. Incubation with deoxyribonuclease (DNase), an enzyme that catalyzes DNA hydrolysis, inactivated the transforming principle.

In drawing conclusions from their experiments, Avery, MacLeod, and McCarty avoided stating explicitly that DNA was the hereditary material and stated at the end of their work only that nucleic acids have "biological specificity, the chemical basis of which is as yet undetermined." The problem they faced in persuading the scientific community to accept their conclusion was that the hereditary material had to be a substance capable of enormous variation in order to contain the information carried by the huge number of genes. The tetranucleotide hypothesis, however, seemed incompatible with the idea that DNA could be the sole component of the hereditary material. Furthermore, the consensus was that genes were made of chromosomal protein, an idea that, in the course of a 40-year period, had logically evolved from the recognition that protein composition and structure varied greatly among organisms. For these reasons, the transformation experiments had little initial impact. Investigators who supported the genes-as-protein theory posed two alternative explanations for the transformation results. (1) The transforming principle might not be DNA but rather one of the proteins invariably contaminating the DNA sample. (2) DNA somehow affected capsule formation directly by acting in the metabolic pathway for biosynthesis of the polysaccharide and permanently altering this pathway. The first point should have already been discounted by the original work because the experiments showed insensitivity to proteolytic enzymes and sensitivity to DNase. Because the DNase was not a pure enzyme, however, the possibility could not be eliminated conclusively.

The transformation experiment was repeated five years later by Rollin Hotchkiss with a DNA sample with a protein content that was only 0.02%, and it was found that this extensive purification did not reduce the transforming activity. This result supported the view of Avery, MacLeod, and McCarty but still did not prove it. The second alternative, however, was clearly eliminated—also by Hotchkiss—with an experiment in which he transformed a penicillin-sensitive bacterial strain to penicillin-resistance. Because penicillin resistance is totally distinct from the rough-smooth character of the bacterial capsule, this experiment showed that the transforming ability of DNA was not limited to capsule synthesis. Interestingly enough, most biologists still remained unconvinced that DNA was the genetic material. It was not until Erwin Chargaff showed in 1950 that a wide variety of chemical structures in DNA were possible—thus allowing biological specificity—that this idea was accepted.

Chemical experiments also supported the hypothesis that DNA is the hereditary material.

The hypothesis of a tetranucleotide structure for DNA was based on the belief that DNA contained equimolar quantities of adenine, thymine, guanine, and cytosine. This incorrect conclusion arose for two reasons. First, in the chemical analysis of DNA, the technique used to separate the bases before identification did not resolve them very well, so the quantitative analysis was poor. Second, the DNA analyzed was usually isolated from animals, plants, and yeast in which the four bases are present in nearly equimolar concentration, or from bacterial species such as *Escherichia coli* that also happened to have nearly equimolar base concentrations. Using the DNA from a wide variety of organisms, Chargaff applied new separation and analytical techniques and showed that the molar content of bases (generally called the base composition) could vary widely. The base composition of DNA from a particular organism is usually expressed as a fraction of all bases that are G•C pairs. This fraction called the G + C content can be expressed as follows:

$$G + C \text{ content} = ([G] + [C])/[\text{all bases}]$$

where the square brackets ([ ]) denote molar concentrations.

Chargaff's studies also revealed one other remarkable fact about DNA base composition. In each of the DNA samples that Chargaff studied, he found that [A] = [T] and [G] = [C]. Although the significance of these equalities, known as Chargaff's rules, was not immediately apparent, they would later help to confirm the structure of the DNA molecule.

Base compositions of hundreds of organisms have been determined. Generally speaking, the value of the G + C content is near 0.50 for the higher organisms and has a very small range from one species to the next (0.49–0.51 for the primates). For the lower organisms, the value of the G + C content varies widely from one genus to another. For example, for bacteria the extremes are 0.27 for the genus *Clostridium* and 0.76 for the genus *Sarcina*; *E. coli* DNA has the value 0.50. Thus, it was demonstrated that DNA could have variable composition, a primary requirement for the hereditary material. Upon publication of Chargaff's results, the tetranucleotide hypothesis quietly died and the DNA-gene idea began to catch on. Shortly afterward, workers in several laboratories found that, for a wide variety of organisms, somatic cells have twice the DNA content of germ cells, a characteristic to be expected of the genetic material, given the tenets of classical chromosome genetics. Although it could apply just as well to any component of chromosomes, once this result was revealed, objections to the work of Avery, MacLeod, and McCarty were no longer heard, and the hereditary nature of DNA rapidly became the fashionable idea.

The blender experiment demonstrated that DNA is the genetic material in bacterial viruses.

An elegant confirmation that DNA is the hereditary material came from an experiment performed by Alfred Hershey and Martha Chase in 1952 in which viral replication was studied in the bacterium *E. coli*. Bacterial viruses are called **bacteriophages** or **phages** for short (see Chapter 8 for more information about bacteriophages and other viruses). The experiment, known as the blender experiment because a kitchen blender was used as a major piece of apparatus, demonstrated that the DNA injected by a bacteriophage T2 particle into a bacterium contains all of the information required to synthesize progeny phage particles. A single phage T2 particle consists of DNA (now known to be a single molecule) encased in a protein shell and a long protein tail by which it attaches to sensitive bacteria (FIGURE 1.14). Hershey and Chase showed that an attached phage can be torn from a bacterial cell wall by violent agitation of the infected cells in a kitchen blender. It thus was possible to separate an absorbed phage from a bacterium and determine the component(s) of the phage that could not be shaken free by agitation; presumably, those components had been injected into the bacterium.

DNA is the only phosphorus-containing substance in the phage particle. The proteins of the shell, which contain the amino acids methionine and cysteine, have the only sulfur atoms. Phage T2 containing radioactive DNA can be prepared by infecting bacteria with phage T2 in a growth medium that contains [^{32}P]phosphate as the sole source of phosphorus. If instead the growth medium contains radioactive sulfur as [^{35}S]sulfate, phage containing radioactive proteins are obtained. When these two kinds of labeled phages are used to infect a bacterial host, the phage DNA and the protein molecules can always be located by their radioactivity. Hershey and Chase used these phages to show that ^{32}P but not ^{35}S remains associated with the bacterium.

^{35}S-Labeled phage particles were attached to bacteria for a few minutes. The bacteria were separated from unattached phage and phage fragments by centrifuging the mixture and collecting the sediment (the pellet), which consisted of the phage-bacterium complexes.

(a)

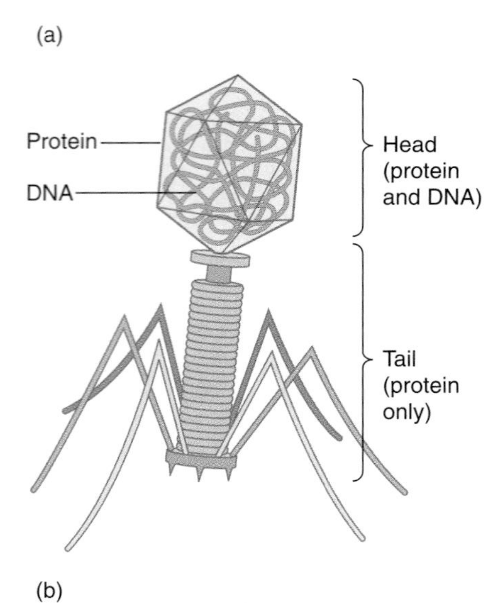

(b)

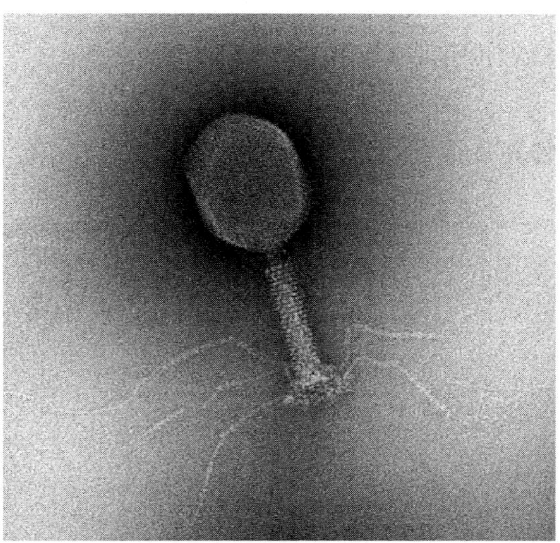

FIGURE 1.14 **Bacteriophage T2.** (a) Drawing of *E. coli* phage T2, showing various components. The DNA, which is confined to the interior of the head, is the only component labeled by ^{32}P. Methionine and cysteine in the proteins can be labeled with ^{35}S. (b) An electron micrograph of phage T4, a closely related phage. (Part b courtesy of Robert Duda, University of Pittsburgh.)

These complexes were resuspended in aqueous solutions and blended. The suspension was again centrifuged, and the pellet (consisting almost entirely of bacteria) and the supernatant were collected. It was found that 80% of the radioactive sulfur was in the supernatant and 20% was in the pellet. The 20% of the ^{35}S that remained associated with the bacteria was shown many years later to consist mostly of phage tail fragments that adhered too tightly to the bacterial surface to be removed by the blending. A very different result was observed when the phage population was labeled with radioactive phosphate. In this case, 70% of the ^{32}P remained associated with the bacteria in the pellet after blending and only 30% was in the supernatant. Of the radioactivity in the supernatant, roughly one third could be accounted for by breakage of the bacteria during the blending. (The remainder was shown some years later to be a result of defective phage particles that could not inject their DNA.) When the pellet material was resuspended in growth medium and reincubated, it was found to be capable of phage production. Thus, the ability of a bacterium to synthesize progeny phage is associated with transfer of ^{32}P, and hence of DNA, from parental phage to the bacteria.

Another series of experiments (FIGURE 1.15), known as **transfer experiments**, supported the interpretation that genetic material contains ^{32}P but not ^{35}S. In these experiments, progeny phage were isolated, after blending, from cells that had been infected with either ^{35}S- or ^{32}P-containing phage and the progeny were then assayed for radioactivity. The idea was that some parental genetic material should be found in the progeny. No ^{35}S but about half of the injected ^{32}P was transferred to the progeny. This result indicated that although ^{35}S might be residually associated with the phage-infected bacteria, it was not part of the phage genetic material. The interpretation (now known to be correct) of the transfer of only half of the ^{32}P was that progeny DNA is selected at random for packaging into protein coats and that all progeny DNA is not successfully packaged.

RNA serves as the hereditary material in some viruses.

The transformation and blender experiments settled once and for all the question of the chemical identity of the genetic material. The absolute generality of the conclusion remained in question, though, because several plant and animal viruses were known to contain single-stranded RNA and no DNA. These particles became understandable shortly afterward as the role of RNA in the flow of information from gene to protein became clear. That is, DNA stores genetic information for protein synthesis and the pathway from DNA to protein always requires the synthesis of an RNA intermediate, which is copied from a DNA template. A virus that lacks DNA, therefore, uses the base sequence of RNA both for storage of information and as a template from which the amino acid sequence of proteins can be obtained. RNA does not serve these two functions as efficiently as the DNA→RNA system, but it does work.

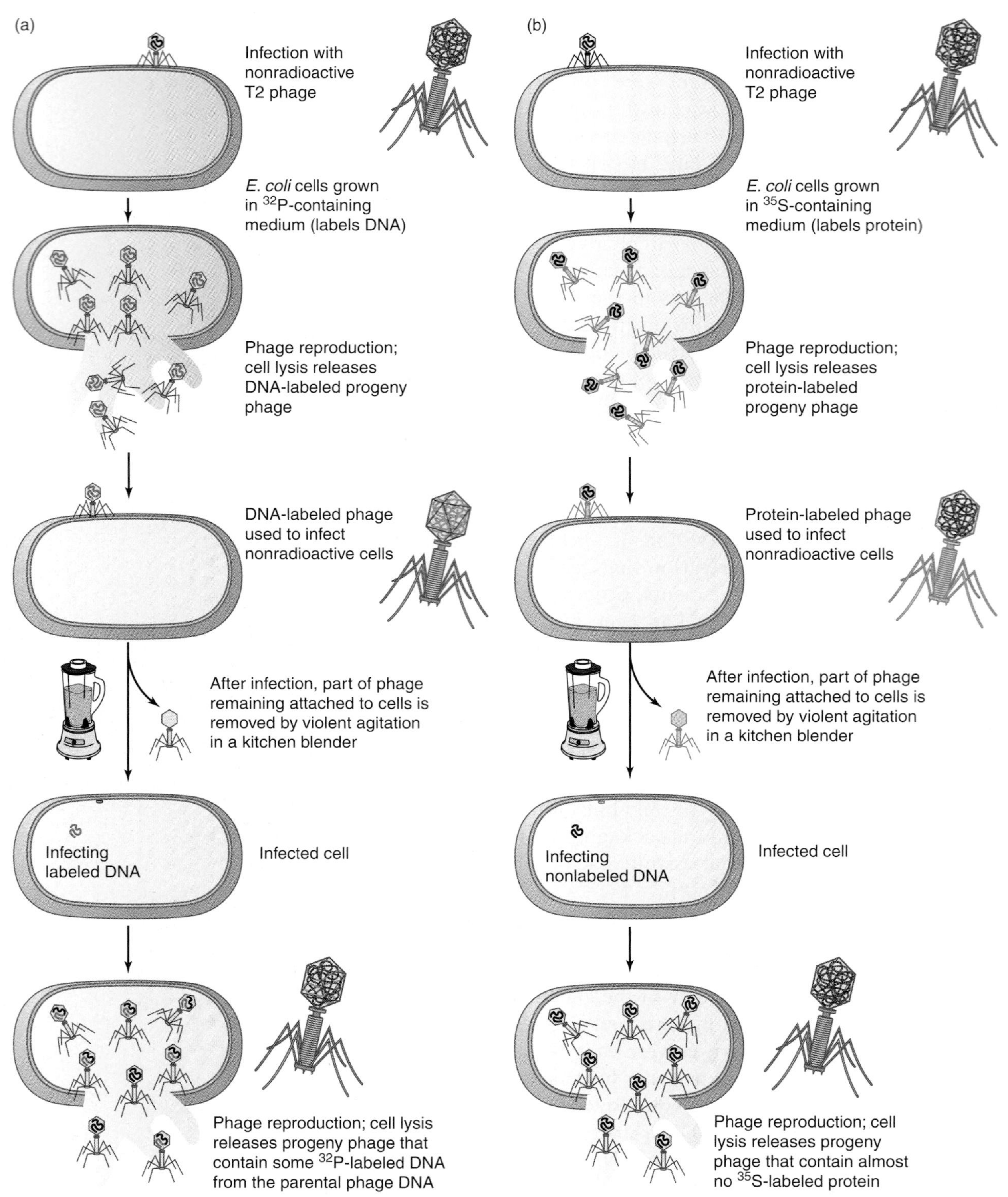

FIGURE 1.15 **The Hershey-Chase ("transfer") experiment demonstrating that DNA, not protein, is directing reproduction of T2 phage in infected *E. coli* cells.**

Rosalind Franklin and Maurice Wilkins obtained x-ray diffraction patterns of extended DNA fibers.

At the same time that chemists were attempting to learn something about the composition of DNA, crystallographers were trying to obtain a three-dimensional image of the molecule. Rosalind Franklin and Maurice Wilkins obtained some excellent x-ray diffraction patterns of extended DNA fibers in the early 1950s (FIGURE 1.16). One might predict that all the x-ray diffraction patterns would look alike, but this was not the case. DNA structure and, therefore, x-ray diffraction patterns depend on several variables. One of the most important of these is the relative humidity of the chamber in which DNA fibers are placed.

Two types of DNA structure are of particular interest. **B-DNA** is stable at a relative humidity of about 92%. **A-DNA** appears as the relative humidity falls to about 75%. Crystallographers did not know whether A-DNA or B-DNA is present in the living cell. Partly for this reason, Wilkins turned his attention toward taking x-ray diffraction pictures of DNA in sperm cells. Franklin focused her attention on x-ray diffraction patterns of A-DNA because they appeared to provide more detail. She believed that careful analyses of the detailed patterns would eventually lead to the solution of DNA's structure.

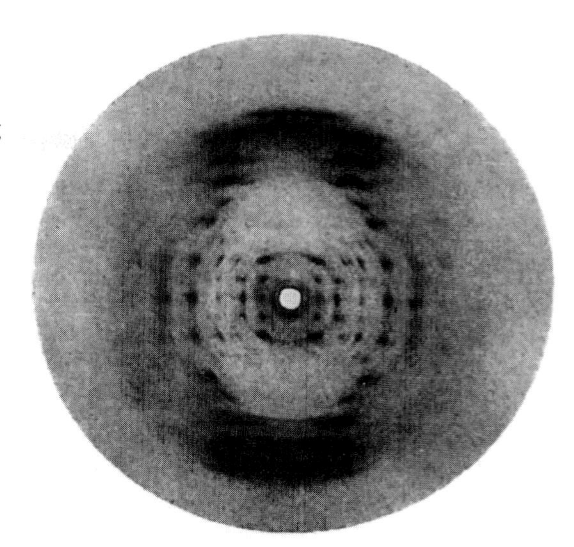

A-form DNA

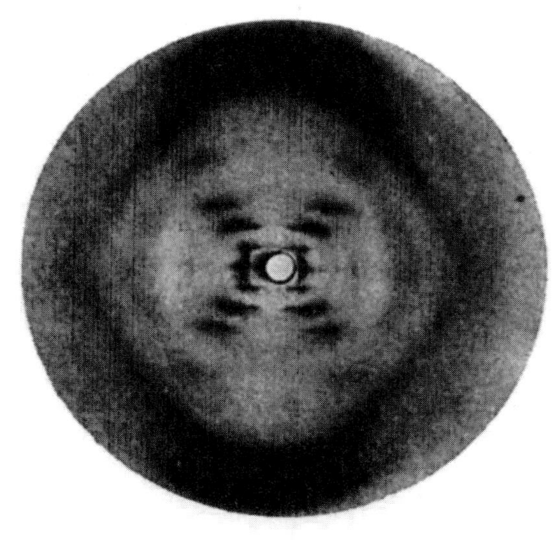

B-form DNA

FIGURE 1.16 **X-ray diffraction patterns of the A and B forms of the sodium salt of DNA.** (Reproduced from R. E. Franklin and R. G. Gosling, *Acta Crystallographica* 6 [1953]: 673–677. Photos courtesy of the International Union of Crystallography.)

James Watson and Francis Crick proposed that DNA is a double-stranded helix.

The American biologist, James D. Watson, and the English crystallographer, Francis Crick, working together in England, took a different approach to determining DNA's structure. They tried to obtain as much information as they could from the x-ray diffraction patterns and then to build a model consistent with this information.

The term *model* has a special meaning to scientists. A **model** is a hypothesis or tentative explanation of the way a system works, usually including the components, interactions, and sequences of events. A successful model suggests additional experiments and allows investigators to make predictions that can be tested in the laboratory. If predictions do not agree with experimental results, the model must be considered incorrect in its current form and modified. A model cannot be proved to be correct merely by showing that it makes a correct prediction. If it makes many correct predictions, however, it is probably nearly, if not completely, correct.

Watson and Crick focused their attention on Franklin's x-ray diffraction patterns of B-DNA. This pattern indicated that B-DNA has a helical structure, a diameter of approximately 2.0 nm, and a repeat distance of 0.34 nm. Their model would have to account for these structural features. Watson and Crick still had to work out the number of DNA chains in a DNA molecule, the location of the bases, and the position of the phosphate and deoxyribose groups. The density of DNA seemed to be consistent with one, two, or three chains per molecule. Watson and Crick tried to build a two-chain model with hydrogen bonds (weak electrostatic attractions; see Chapter 2) holding the bases together. They were unsuccessful until Jerry Donohue suggested that they use the *keto tautomeric* forms of T and G in their models. At first, Watson tried to link two purines together and two pyrimidines together in what he called a "like-with-like" model. A dramatic turning point occurred in 1953 when Watson realized that adenine forms hydrogen bonds with thymine and guanine forms hydrogen bonds with cytosine. Watson describes this turning point in his book, *The Double Helix*:

> *When I got to our still empty office the following morning, I quickly cleared away the paper from my desk top so that I would have a large, flat surface on which to form pairs of bases held together by hydrogen bonds. Though I initially went back to my like-with-like prejudices, I saw all too well that they led nowhere. When Jerry [Donohue] came in I looked up, saw that it was not Francis [Crick], and began shifting the bases in and out of various other pairing possibilities. Suddenly I became aware that an adenine-thymine pair held together by two hydrogen bonds was identical in shape to a guanine-cytosine base pair held together by at least two hydrogen bonds. All the hydrogen bonds seemed to form naturally; no fudging was required to make the two types of base pairs identical in shape.*

With the realization that adenine-thymine and cytosine-guanine base pairs have the same width, Watson and Crick were quickly able to construct a double helix model of DNA that fit Franklin's x-ray diffraction data (FIGURE 1.17).

The key features of the Watson-Crick Model for B-DNA are as follows:

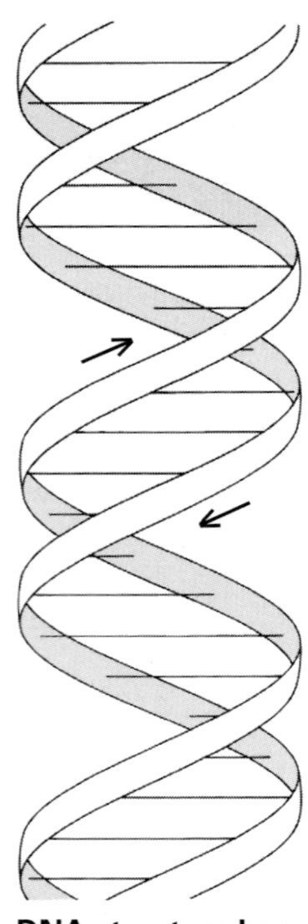

FIGURE 1.17 **DNA structure based on the 1953 paper by Watson and Crick in *Nature* showing their double helix model for DNA for the first time.**

1. Two polydeoxyribonucleotide strands twist about each other to form a double helix.
2. Phosphate and deoxyribose groups form a backbone on the outside of the helix.
3. Purine and pyrimidine base pairs stack inside the helix and form planes perpendicular to the helix axis and the deoxyribose groups.
4. The helix diameter is 2.0 nm (or 20 Å).
5. Adjacent base pairs are separated by an average distance of 0.34 nm (or 3.4 Å) along the helix axis. The structure repeats itself after about ten base pairs, or about once every 3.4 nm (or 34 Å) along the helix axis.
6. Adenine always pairs with thymine and guanine with cytosine. The original model showed two hydrogen bonds stabilizing each kind of base pair. Although this was an accurate description for A-T base pairs, later work showed that G-C base pairs are stabilized by three hydrogen bonds (FIGURE 1.18). (These base pairing relationships explained Chargaff's observation that the molar ratios of adenine to thymine and guanine to cytosine are one.)
7. The two strands are antiparallel, which means that the strands run in opposite directions. That is, one strand runs $3' \rightarrow 5'$ in one direction while the other strand runs $5' \rightarrow 3'$ in the same direction. Because the strands are antiparallel, a convention is

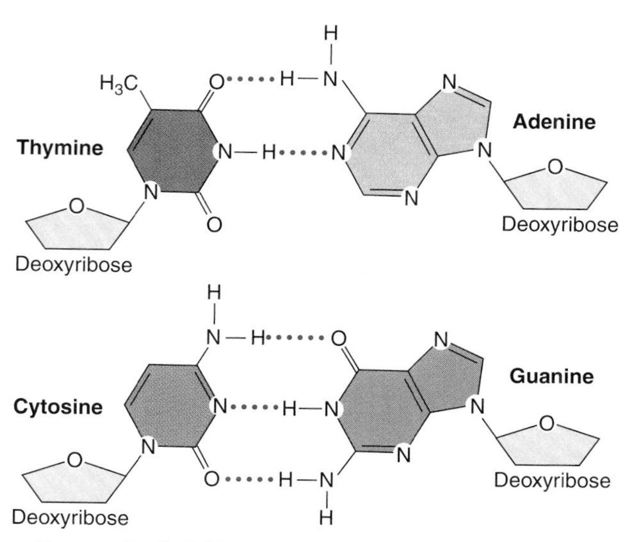

FIGURE 1.18 **Base pairs in DNA.**

needed for stating the sequence of bases of a single chain. The convention is to write a sequence with the 5′-P terminus at the left; for example, ATC denotes the trinucleotide 5′-ATC-3′. This is also often written as pApTpC, again using the conventions that the left side of each base is the 5′-terminus of the nucleotide and that a phosphodiester group is represented by a p between two capital letters.

8. A major and a minor groove wind about the cylindrical outer helical face. The two grooves are of about equal depth but the major groove is much wider than the minor groove.

The Watson-Crick Model indicates that when the nucleotide sequence of one strand is known, the sequence of the complementary strand can be predicted, providing the theoretical framework needed to understand the fidelity of gene replication. Each strand serves as a mold or **template** for the synthesis of the complementary strand (FIGURE 1.19). Watson and Crick ended their short paper announcing the double helix model with the following sentence that must be one of the greatest understatements in the scientific literature: "It has not escaped our notice that the specific pairing we have postulated immediately suggests a possible copying mechanism for the genetic material." The double helix showed how establishing a chemical structure can be used to understand biological function and to make predictions that guide new research.

Structure-function relationships remain a central theme of molecular biology. We will see, time and again, throughout this book how knowledge of structure helps us to understand function and leads us to new insights. Biological structures are sometimes quite complex and difficult to study. However, it is worth the effort to study the structures because the reward for doing so is so great. The Watson-Crick Model also serves as a powerful example of the fundamental principles that help to define molecular biology as a discipline. (1) The same physical

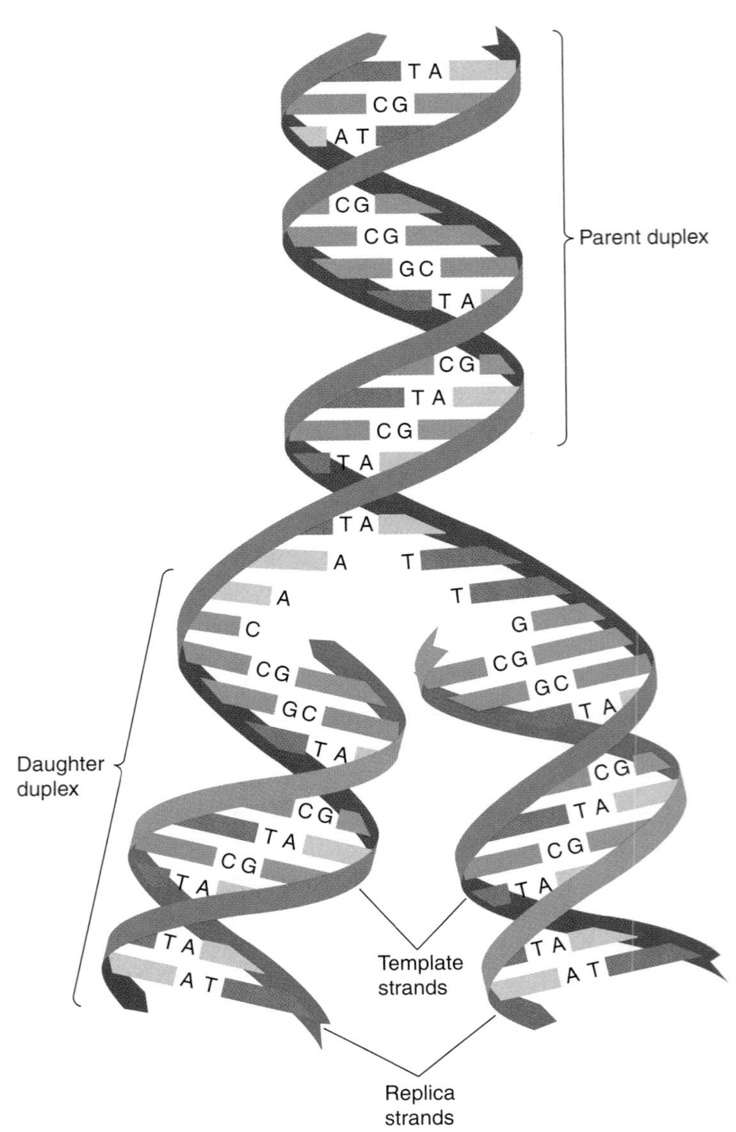

FIGURE 1.19 **Replication of DNA.** Replication of DNA duplex as originally envisioned by Watson and Crick. As the parental strands separate, each parental strand serves as a template for the formation of a new daughter strand by means of A-T and G-C base pairing.

and chemical laws apply to living systems and inanimate objects. (2) The same biological principles tend to apply to all organisms. (3) Biological structure and function are intimately related. The Watson-Crick Model provided an excellent starting point for the challenging job of elucidating the chemical basis of heredity.

The central dogma provides the theoretical framework for molecular biology.

Much of the research in molecular biology in the late 1950s was directed toward discovering the mechanisms by which nucleotide sequences specify amino acid sequences. In a talk given at a 1957 symposium, Crick suggested a model for information flow that provides a theoretical framework for molecular biology. The following excerpt from Crick's presentation states the theory known as the **central dogma** in a clear and concise fashion.

> *In more detail, the transfer of information from nucleic acid to nucleic acid, or from nucleic acid to protein may be possible, but transfer from protein to protein, or from protein to nucleic acid is impossible. Information means here the precise determination of sequence, either of bases in nucleic acid or of amino acid residues in protein.*

Crick thus was proposing that genetic information flows from DNA to DNA (**DNA replication**), from DNA to RNA (**transcription**), and from RNA to polypeptide (**translation**) (FIGURE 1.20).

By the mid-1960s, molecular biologists had obtained considerable experimental support for the central dogma. In particular, they had discovered enzymes that catalyze replication and transcription and elucidated the pathway for translating nucleotide sequences to amino acid sequences. With some variations in detail, all organisms use the same basic mechanism to translate information from nucleotide sequences to amino acid sequences. Three major components of the translation machinery—**messenger RNA** (mRNA), **ribosomes,** and **transfer RNA** (tRNA)—play essential roles in this information transfer (FIGURE 1.21). Each gene can serve as a template for the synthesis

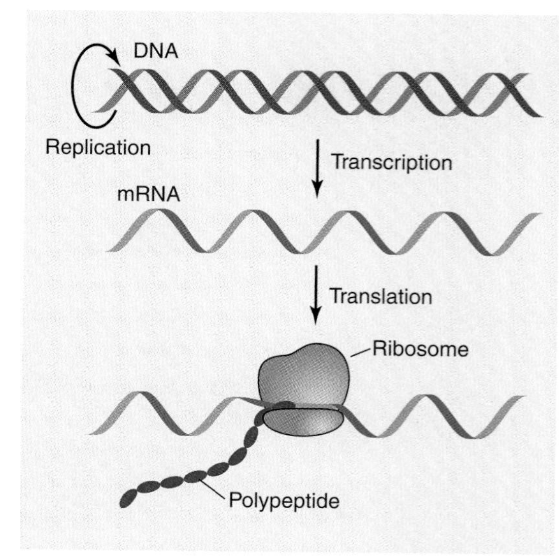

FIGURE 1.20 **The "central dogma."** The central dogma as originally proposed by Francis Crick postulated information flow from DNA to RNA to protein. The ribosome is an essential part of the translation machinery. Later studies demonstrated that information can also flow from RNA to RNA and from RNA to DNA (reverse transcription).

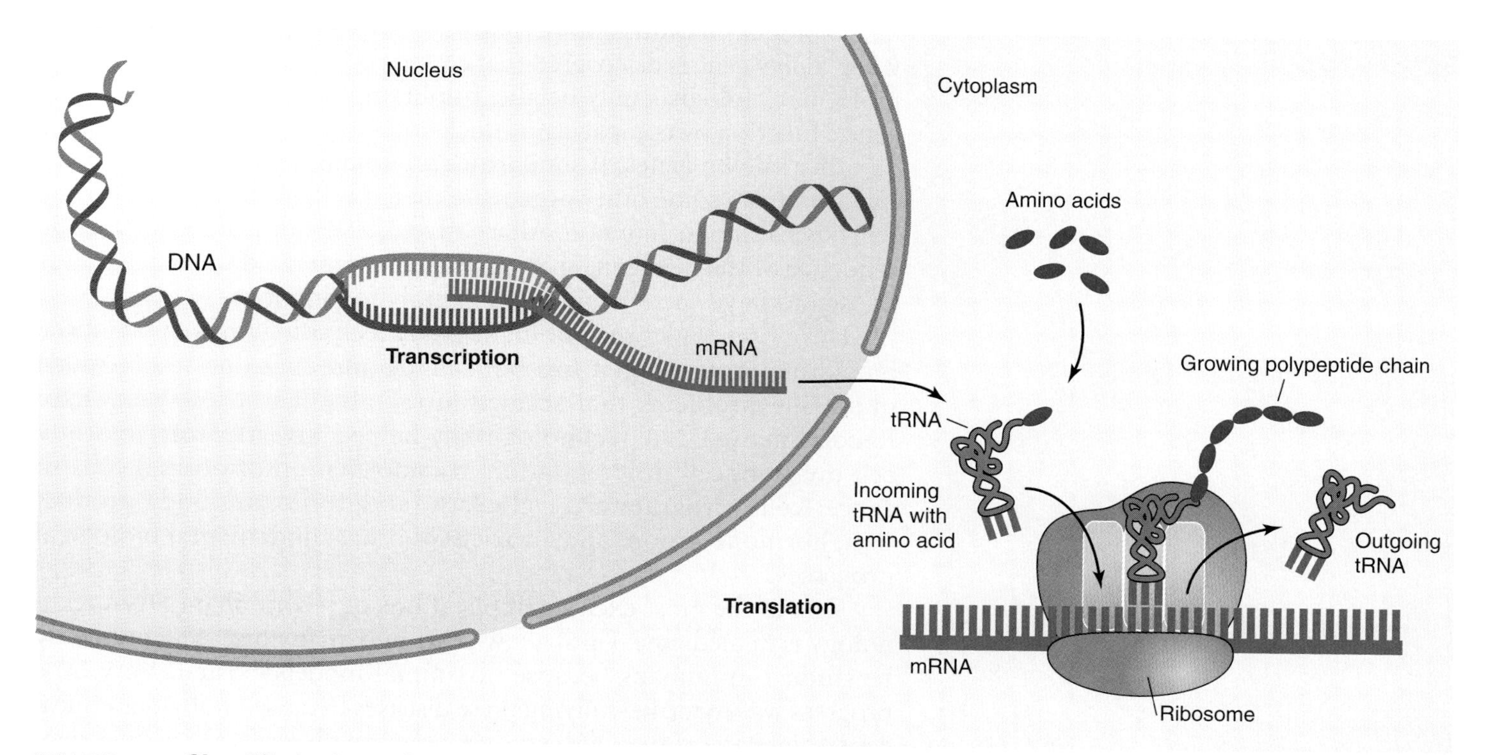

FIGURE 1.21 **Simplified schematic diagram of the protein synthetic machinery in an animal, plant, or yeast cell.** Transfer RNA (tRNA) carries the next amino acid to be attached to the growing polypeptide chain to the ribosome, which recognizes a match between a three nucleotide sequence in the mRNA (the codon) and a complementary sequence in the tRNA (the anticodon) and transfers the growing polypeptide chain to the incoming amino acid while it is still attached to the tRNA. (Adapted from D. Secko, *The Science Creative Quarterly*, 2007 [http://www.scq.ubc.ca/a-monks-flourishing-garden-the-basics-of-molecular-biology-explained/]. Accessed May 5, 2010.)

of specific mRNA molecules. Messenger RNA molecules program ribosomes (protein synthetic factories) to form specific polypeptides. Transfer RNA molecules carry activated amino acids to programmed ribosomes, where under the direction of the mRNA the amino acids join to form a polypeptide chain. The twelve years of extraordinary scientific progress that followed the discovery of DNA's structure were capped by a series of brilliant investigations by Marshal W. Nirenberg and others that culminated in deciphering the genetic code. By 1965, investigators could predict a polypeptide chain's amino acid sequence from a DNA or mRNA molecule's nucleotide sequence. Remarkably, the genetic code was found to be nearly universal. Each particular sequence of three adjacent nucleotides or **codon** specifies the same amino acid in bacteria, plants, and animals. Could one ask for more convincing support for the uniformity of life processes?

Recombinant DNA technology allows us to study complex biological systems.

The second major wave in molecular biology started in the late 1970s with the development of **recombinant DNA technology** (also known as *genetic engineering*). Thanks to recombinant DNA technology, genes can be manipulated in the laboratory just like any other organic molecule. They can be synthesized or modified as desired and their nucleotide sequence determined. Sequence information can save molecular biologists months and perhaps even years of work. For example, when a new gene is discovered that causes a specific disease in humans, molecular biologists may be able to obtain valuable clues to the new gene's function by comparing its nucleotide sequence with sequences of all other known genes. If similarities are found with a gene of known function from some other organism, then it is likely that the new gene will have a similar function. Alternatively, a segment of the nucleotide sequence of the new gene may predict an amino acid sequence that is known to have specific binding or catalytic functions.

Development of recombinant DNA technology has allowed the incredible progress that has been made and continues to be made in solving problems that seemed intractable just a few years ago. Recombinant DNA technology has helped investigators to study cell division, cell differentiation, transformation of normal cells to cancer cells, programmed cell death (apoptosis), antibody production, hormone action, and a variety of other fundamental biological processes.

One of the most exciting applications of recombinant DNA technology is in medicine. Until a few years ago, many diseases could only be studied in humans because no animal models existed. Very little progress was made in studying these diseases because of obvious ethical constraints associated with studying human diseases. Once investigators learned how to transfer genes from one species to another, they were able to create animal models for human diseases, thus facilitating study of the diseases. Recombinant DNA technology also has led to the production of new drugs to treat diabetes, anemia, cardiovascular disease, and cancer as well as to the development of diagnostic tools to detect a wide variety of diseases. The list of practical medical applications grows longer with each passing day.

Although recombinant DNA technology promises to change our lives for the better, it also forces us to consider important social, political, ethical, and legal issues. For example, how do we protect the interests of an individual when DNA analysis reveals that the individual has alleles that are likely to cause a serious physical or mental disease in the future, especially if there is no cure or treatment? What impact will the knowledge have on affected individuals and their families? Should insurance companies or potential employers have access to the genetic information? If not, how do we limit access to information and how do we enforce the limitation? Rapid progress in recombinant DNA technology also raises troubling ethical issues. Germ line therapy allows new genes to be introduced into fertilized eggs and thereby alter the genetic characteristics of future generations. This technique has been used to introduce desired traits into plants and animals, but it has not as yet been reported in humans.

An argument can be made for using germ line therapy to correct human genetic diseases such as Tay Sachs disease, cystic fibrosis, or Huntington disease so that affected individuals and their families can be spared the devastating consequences of these diseases. We must be very careful about application of germ line therapy to humans, however, because the technique has the power to do great harm. Who will decide which genetic characteristics are desirable and which are undesirable? Should the technique be used to change physical appearance, intelligence, or personality traits? Should anyone be permitted to make such decisions?

A great deal of molecular biology information is available on the Internet.

Recombinant DNA technology has generated so much information that it would be nearly impossible to share all of it in a timely fashion with the entire molecular biology community by conventional means such as publishing in professional journals or writing books. Fortunately, there is an alternate method for sharing large quantities of rapidly accumulating information that is both quick and efficient. A worldwide network of communication networks, the **Internet**, allows us to gain almost instant access to the information. The Internet also provides many helpful tutorials and instructive animations. Because addresses for Web sites tend to change over time, the publisher will maintain a site (http://biology.jbpub.com/book/molecular) that contains a variety of resources, including links to in-depth information on topics covered in each chapter, and review material designed to assist in your study of molecular biology.

Suggested Reading

General

Choudri, S. 2003. The path from nuclein to human genome: a brief history of DNA with a note on human genome sequencing and its impact on future research in biology. *Bull Sci Technol Soc* 23:360–367.

Crick, F. 1974. The double helix: a personal view. *Nature* 248:766–769.

Crick, F. 1988. *What Mad Pursuit: A Personal View of Scientific Discovery*. New York: Basic Books.

Judson, H. F. 1996. *The Eighth Day of Creation: Makers of the Revolution in Biology*. Cold Spring Harbor, NY: Cold Spring Harbor Laboratory.

Maddox, B. 2002. *Rosalind Franklin: The Dark Lady of DNA*. New York: Harper Collins.

McCarty, M. 1985. *The Transforming Principle: Discovering that Genes Are Made of DNA*. New York: W.W. Norton.

Olby, R. C. 1994. *The Path to the Double Helix: The Discovery of DNA*. New York: Dover.

Pukkila, P. J. 2001. Molecular biology: the central dogma. *Encyclopedia of Life Sciences*. pp. 1–5. London, UK: Nature.

Sayre, A. 1978. *Rosalind Franklin and DNA*. New York: W. W. Norton.

Stent, G. 1972. Prematurity and uniqueness in scientific discovery. *Sci Am* 227:84–93.

Summers, W. C. 2002. History of molecular biology. *Encyclopedia of Life Sciences*. pp. 1–8. London, UK: Nature.

Watson, J. D. 1968. *The Double Helix: A Personal Account of the Discovery of the Structure of DNA*. New York: Antheneum Books.

Wilkins, M. 2003. *The Third Man of the Double Helix: The Autobiography of Maurice Wilkins*. Oxford, UK: Oxford University Press.

Classic Papers

Avery, O. T., Macleod, C. M., and McCarty, M. 1944. Studies on the chemical nature of the substance inducing transformation of pneumococcal types. Induction of transformation by a deoxyribonucleic acid fraction isolated from *Pneumococcus* type III. *J Exp Med* 79:137–156.

Beadle, G. W., and Tatum, E., 1941. Genetic control of biochemical reactions in Neurospora. *Proc Nat Acad Sci USA* 27:499–506.

Hershey A. D., and Chase, M. 1952. Independent functions of viral protein and nucleic acid in growth of bacteriophage. *J Gen Physiol* 36:39–56.

Watson, J. D., and Crick F. H. 1953. Molecular structure of nucleic acids; a structure for deoxyribose nucleic acid. *Nature* 171:737–738.

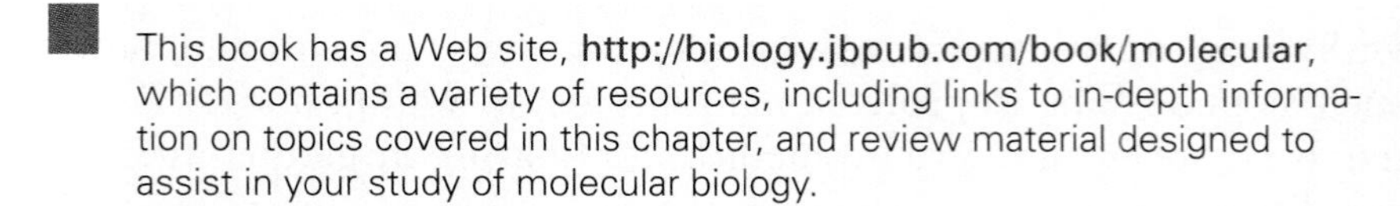

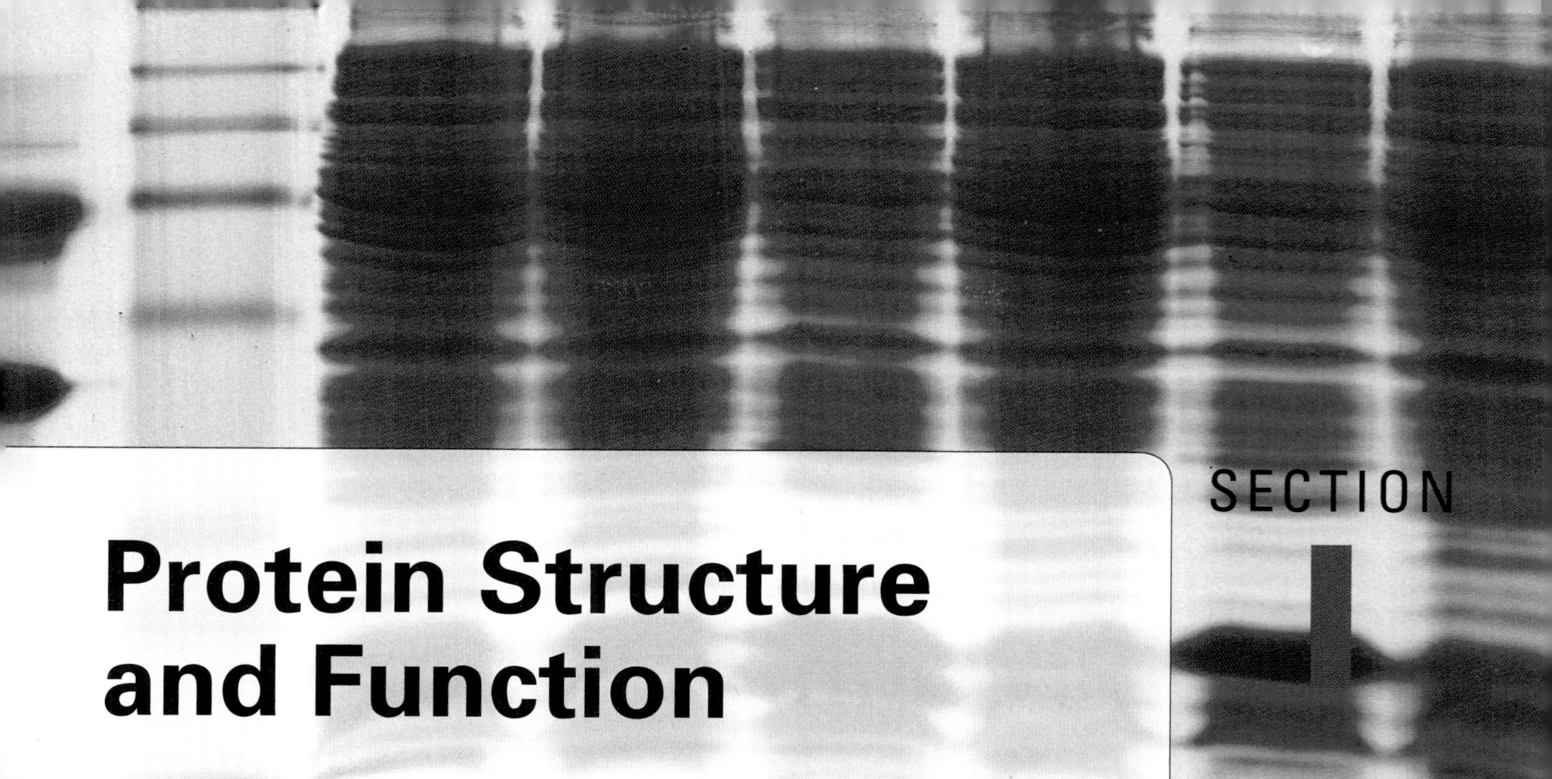

Protein Structure and Function

CHAPTER 2 **Protein Structure**

CHAPTER 3 **Protein Function**

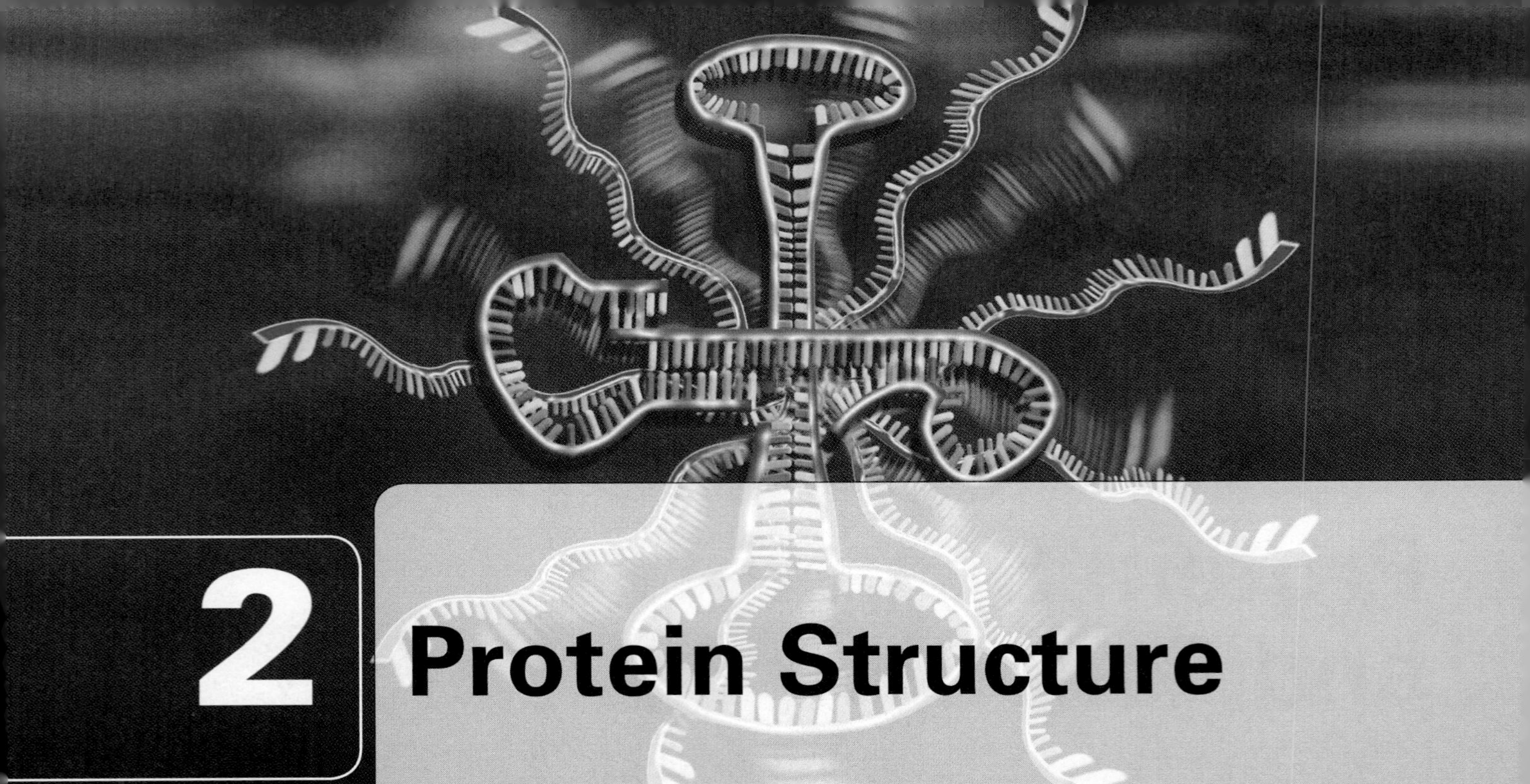

2 Protein Structure

OUTLINE OF TOPICS

2.1 **The α-Amino Acids**
α-Amino acids have an amino group and a carboxyl group attached to a central carbon atom.
Amino acids are represented by three-letter and one-letter abbreviations.

2.2 **The Peptide Bond**
α-Amino acids are linked by peptide bonds.

2.3 **Protein Purification**
Protein mixtures can be fractionated by chromatography.
Proteins and other charged biological polymers migrate in an electric field.

2.4 **Primary Structure of Proteins**
The amino acid sequence or primary structure of a purified protein can be determined.
Polypeptide sequences can be obtained from nucleic acid sequences.
The BLAST program compares a new polypeptide sequence with all sequences stored in a data bank.
Proteins with just one polypeptide chain have primary, secondary, and tertiary structures while those with two or more chains also have quaternary structures.

2.5 **Weak Noncovalent Bonds**
The polypeptide folding pattern is determined by weak non-covalent interactions.

2.6 **Secondary Structures**
The α-helix is a compact structure that is stabilized by hydrogen bonds.
The β-conformation is also stabilized by hydrogen bonds.
Loops and turns connect different peptide segments, allowing polypeptide chains to fold back on themselves.
Certain combinations of secondary structures, called supersecondary structures or folding motifs, appear in many different proteins.
We cannot yet predict secondary structures with absolute certainty.

2.7 **Tertiary Structure**
X-ray crystallography and nuclear magnetic resonance studies have revealed the three-dimensional structures of many different proteins.
Intrinsically disordered proteins lack an ordered structure under physiological conditions.
Structural genomics is a field devoted to solving x-ray and NMR structures in a high throughput manner.
The primary structure of a polypeptide determines its tertiary structure.
Molecular chaperones help proteins to fold inside the cell.

2.8 **Proteins and Biological Membranes**
Proteins interact with lipids in biological membranes.
The fluid mosaic model has been proposed to explain the structure of biological membranes.

Suggested Reading

As described in Chapter 1, the Watson-Crick Model helped to bridge a major gap between genetics and biochemistry, and in so doing helped to create the discipline of molecular biology. The double helix structure showed the importance of elucidating a biological molecule's structure when attempting to understand its function. This chapter and Chapter 3 extend the study of structure-function relationships to polypeptides, which catalyze specific reactions, transport materials within a cell or across a membrane, protect cells from foreign invaders, regulate specific biological processes, and support various structures.

The basic building blocks for polypeptides are small organic molecules called **amino acids**. Amino acids can combine to form long linear chains known as **polypeptides**. Each type of polypeptide chain has a unique amino acid sequence. Although a polypeptide must have the correct amino acid sequence to perform its specific biological function, the amino acid sequence alone does not guarantee that the polypeptide will be biologically active. The polypeptide must fold into a specific three-dimensional structure before it can perform its biological function(s). Once folded into its biologically active form, the polypeptide is termed a **protein**. Proteins come in various sizes and shapes. Those with thread-like shapes, the **fibrous proteins**, tend to have structural or mechanical roles. Those with spherical shapes, the **globular proteins**, function as enzymes, transport proteins, or antibodies. Fibrous proteins tend to be water-insoluble, while globular proteins tend to be water-soluble.

Polypeptides are unique among biological molecules in their flexibility, which allows them to fold into characteristic three-dimensional structures with specific binding properties. Mutations that alter a protein's ability to interact with its normal molecular partners often result in a loss of protein activity. One of the most common partners of a folded polypeptide is another folded polypeptide, which may be identical to it or different. A complex that contains two, three, or more identical polypeptides is called a **homodimer, homotrimer**, and so forth, whereas one that contains different polypeptides is called a **heterodimer, heterotrimer**, and so forth.

2.1 The α-Amino Acids

α-Amino acids have an amino group and a carboxyl group attached to a central carbon atom.

The typical amino acid building block for polypeptide synthesis has a central carbon atom that is attached to an amino ($-NH_2$) group, a carboxyl ($-COOH$) group, a hydrogen atom, and a side chain (–R). At pH 7, the amino group is protonated (i.e., the addition of a proton) to form $-NH_3^+$ and the carboxyl group is deprotonated to form $-COO^-$ so that the amino acid has the structure shown in FIGURE 2.1. These amino acids are termed **α-amino acids** in accordance with a

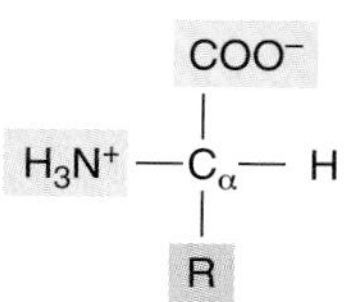

FIGURE 2.1 **Structure of an α-amino acid.** A typical α-amino acid in which the central carbon atom is attached to an amino ($-NH_3^+$) group, a carboxylate ($-COO^-$) group, a hydrogen atom, and a side chain (–R).

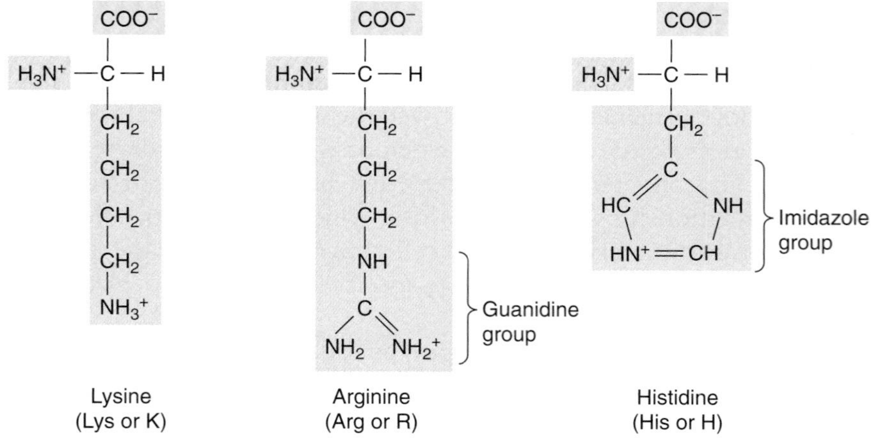

Lysine (Lys or K) — Arginine (Arg or R) — Guanidine group — Histidine (His or H) — Imidazole group

FIGURE 2.2 **Amino acids with basic side chains.**

pre-IUPAC nomenclature system, in which the atoms in a hydrocarbon chain attached to a carboxyl (–COOH) group are designated by Greek letters. The carbon atom closest to the carboxyl group is designated α, the next β, and so forth.

Each amino acid has characteristic physical and chemical properties that derive from its unique side chain. Amino acids with similar side chains usually have similar properties. This relationship is an important consideration when comparing amino acid sequences of two different polypeptides or when considering the effect that an amino acid substitution will have on protein function. Based on side chain structure, amino acids can be divided into four groups.

Side Chains with Basic Groups

Arginine, **lysine**, and **histidine** are called basic amino acids because their side chains are proton acceptors (FIGURE 2.2). The guanidino group in arginine's side chain is a relatively strong base. The amine group in lysine's side chain is a somewhat weaker base, and the imidazole group in histidine's side chain is the weakest of the three bases. Hence, at pH 7, arginine and lysine side chains are very likely to have positive charges, whereas histidine side chains have only about a 10% probability of having a positive charge.

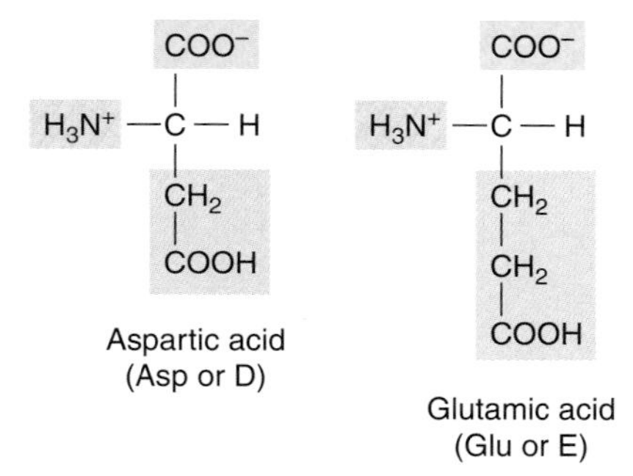

FIGURE 2.3 **Amino acids with acidic side chains.**

Side Chains with Acidic Groups

Aspartic acid and **glutamic acid** each has a carboxyl group as part of its side chain (FIGURE 2.3). Both the α-carboxyl and the side chain carboxyl groups are deprotonated and have negative charges at pH 7. The α-carboxyl group is a slightly stronger acid, however, because the α-carbon is also attached to a positively charged amino group. When the side chain is deprotonated, aspartic and glutamic acids are more appropriately called aspartate and glutamate, respectively. Because aspartic acid and aspartate refer to the same amino acid at different pH values, the names are used interchangeably. The same is true for glutamic acid and glutamate.

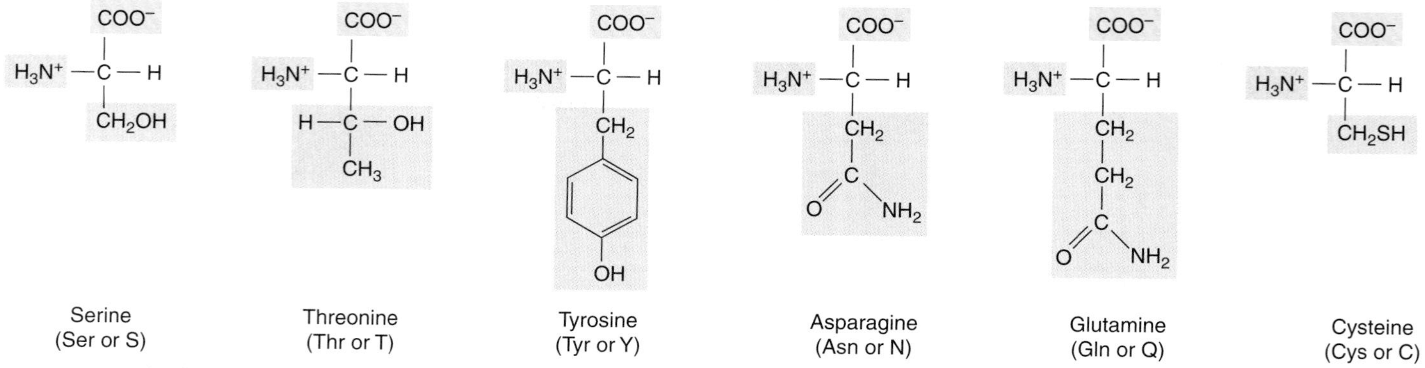

Serine (Ser or S) | Threonine (Thr or T) | Tyrosine (Tyr or Y) | Asparagine (Asn or N) | Glutamine (Gln or Q) | Cysteine (Cys or C)

FIGURE 2.4 Amino acids with polar but uncharged side chains at pH 7.0.

Side Chains with Polar but Uncharged Groups

Six amino acids have side chains with polar groups (FIGURE 2.4). **Asparagine** and **glutamine** are amide derivatives of aspartate and glutamate, respectively. **Serine, threonine,** and **tyrosine** have side chains with hydroxyl (–OH) groups. The tyrosine side chain also has another interesting feature; it is aromatic. Cysteine is similar to serine but a sulfhydryl (–SH) group replaces the hydroxyl group. When exposed to oxygen or other oxidizing agents, sulfhydryl groups on two cysteine molecules react to form a disulfide (–S–S–) bond, resulting in the formation of **cystine** (FIGURE 2.5). Cystine, which is not a building block for polypeptide synthesis, is formed by the oxidation of cysteine side chains after the polypeptide has been formed.

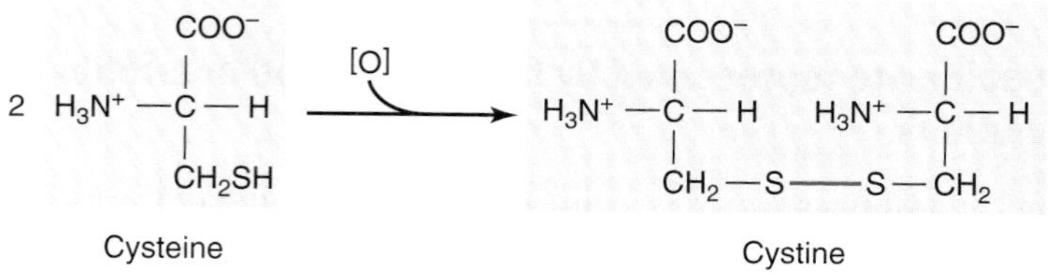

Cysteine → Cystine

FIGURE 2.5 Oxidation of cysteine to form cystine.

Side Chains with Nonpolar Groups

Nine amino acids have side chains with nonpolar groups (FIGURE 2.6). **Glycine,** with a side chain consisting of a single hydrogen atom, is the smallest amino acid and the only one that lacks a stereogenic carbon atom. Because its side chain is so small, glycine can fit into tight places and tends to behave like amino acids with polar but uncharged side chains when present in a polypeptide. **Alanine, isoleucine, leucine,** and **valine** have hydrocarbon side chains. **Phenylalanine** and **tryptophan** have aromatic side chains. **Methionine** and **proline** have side chains with unique features. The methionine side chain contains a thioether ($-CH_2-S-CH_3$) group. Proline's side chain is part of a five-member ring that includes the α-amino group, making the α-amino group a secondary rather than a primary amine group. The rigid ring structure can influence the way a polypeptide chain folds by introducing a kink into the structure.

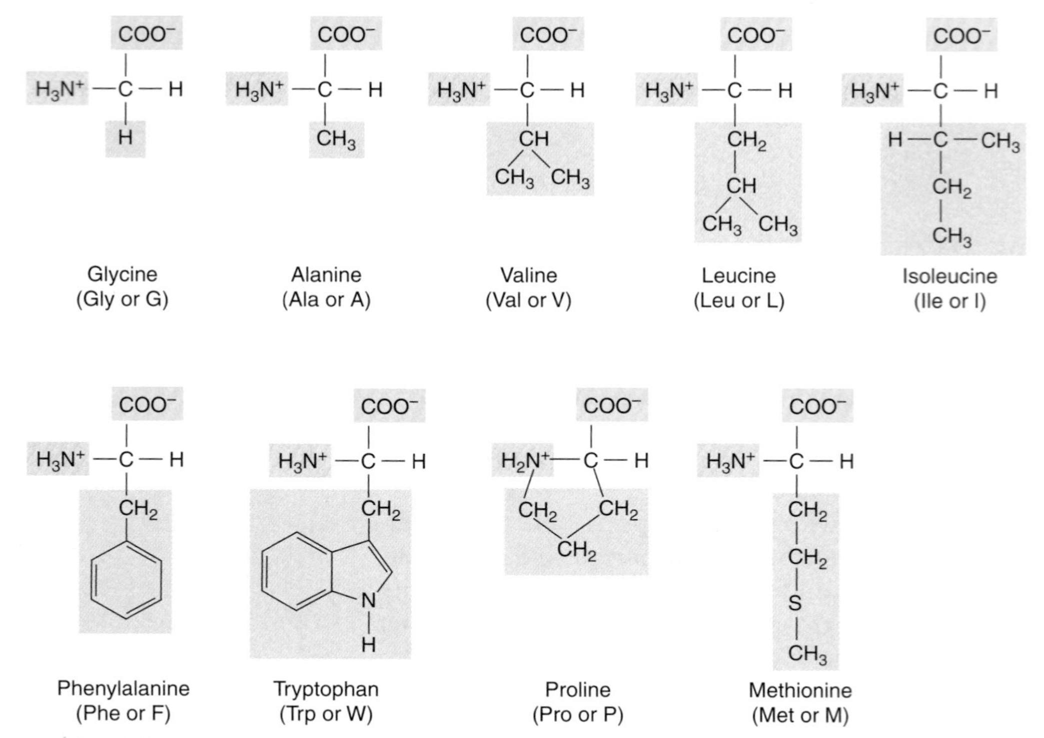

FIGURE 2.6 Amino acids with nonpolar side chains.

Amino acids are represented by three-letter and one-letter abbreviations.

Writing the full names of the amino acids is inconvenient, especially for polypeptide chains with many amino acids. Two systems of abbreviations listed in Table 2.1 offer more convenient methods for representing amino acids.

In the first system, each amino acid is represented by a three-letter abbreviation. For most amino acids, the first three-letters of the amino acid's name are used. For example, Arg is used for arginine, Phe for phenylalanine, and Lys for lysine. But four amino acids have unusual three-letter abbreviations; aspartate and asparagine have identical first three letters and the same is true for glutamate and glutamine. So the abbreviations for aspartate and glutamate are the expected Asp and Glu, respectively and those for asparagine and glutamine are Asn and Gln, respectively.

The other two amino acids with unusual three letter abbreviations are Trp for tryptophan and Ile for isoleucine. Even the three-letter abbreviation system requires too much space for many applications. So investigators devised the one-letter abbreviation system shown in Table 2.1.

TABLE 2.1 **Amino Acid Abbreviations**

Amino Acid	Three Letter Abbreviation	One Letter Abbreviation
Alanine	Ala	A
Arginine	Arg	R
Asparagine	Asn	N
Aspartic acid (Aspartate)	Asp	D
Cysteine	Cys	C
Glutamine	Gln	Q
Glutamic acid (Glutamate)	Glu	E
Glycine	Gly	G
Histidine	His	H
Isoleucine	Ile	I
Leucine	Leu	L
Lysine	Lys	K
Methionine	Met	M
Phenylalanine	Phe	F
Proline	Pro	P
Serine	Ser	S
Threonine	Thr	T
Tryptophan	Trp	W
Tyrosine	Tyr	Y
Valine	Val	V

2.2 The Peptide Bond

α-Amino acids are linked by peptide bonds.

The α-carboxyl group of one amino acid can react with the α-amino group of a second amino acid to form an amide bond and release water (FIGURE 2.7). Amide bonds that link amino acids are designated **peptide bonds** and the resulting molecules are called **peptides.** Peptides with two amino acids are dipeptides, those with three are tripeptides, and so forth.

Systematic physical studies by Linus Pauling and Robert R. Corey in the late 1930s provide important information about bond distances and angles in dipeptides. The results of their studies, summarized in FIGURE 2.8, show that the carbon–nitrogen peptide bond is 0.133 nm long, placing it between the length of a carbon–nitrogen single bond (0.149 nm) and a carbon–nitrogen double bond (0.127 nm). The peptide bond therefore has some double bond character, which produces an energy barrier to free rotation and imposes planarity on the peptide bond. Two configurations are possible, one in which adjacent C_α atoms are on opposite sides (the *trans* isomer; Figure 2.8) and another in which they are on the same side of the peptide bond (the *cis* isomer; FIGURE 2.9). The *trans* isomer is the more stable of the two

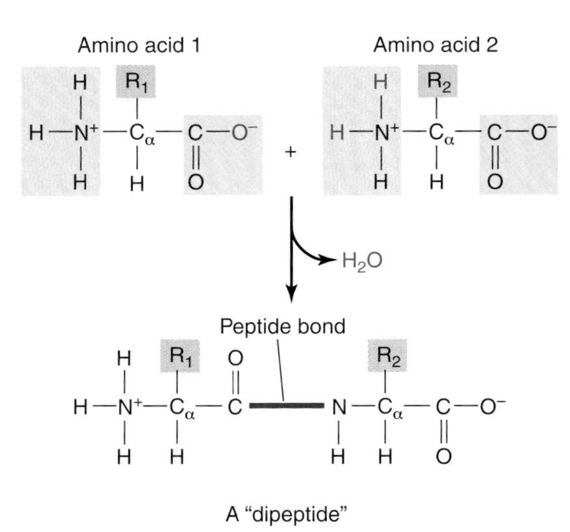

FIGURE 2.7 **Peptide bond formation in the laboratory.** Two amino acids combine with the loss of water to form a dipeptide. The peptide bond is shown in purple.

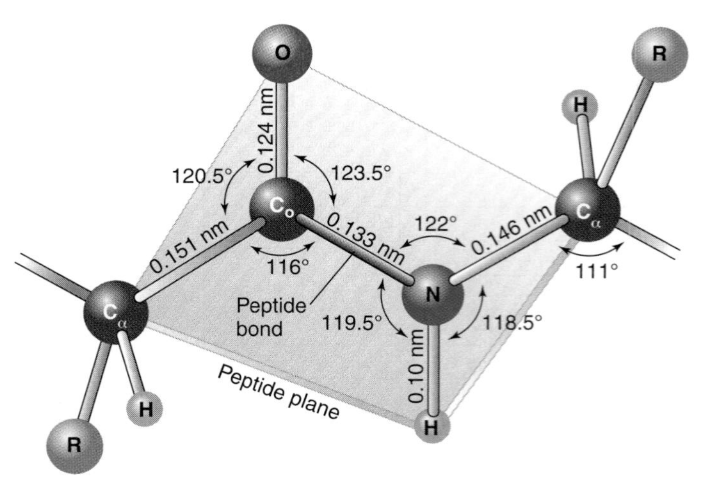

FIGURE 2.8 **The *trans* peptide bond.** The standard dimensions of bond lengths in nm and bond angles in degrees (°) of the planar dipeptide were determined by averaging the corresponding quantities in x-ray crystal structures of peptides. (Adapted from D. Voet and J. G. Voet. *Biochemistry, Third edition.* John Wiley & Sons, Ltd., 2005. Original figure adapted from R. E. Marsh and J. Donohue, *Adv. Protein Chem.* 22 [1967]: 235–256.)

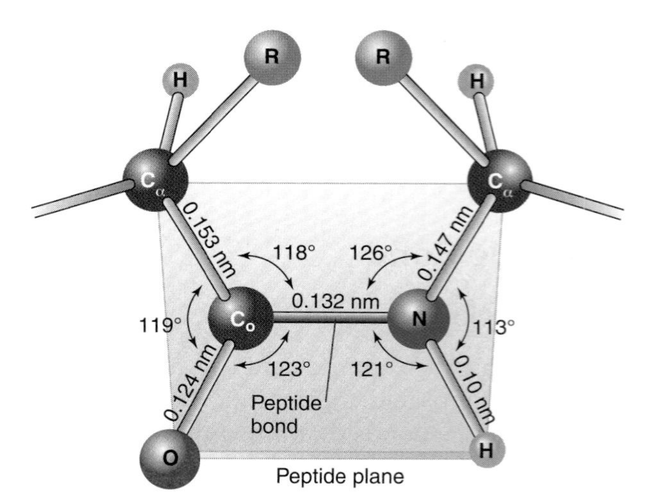

FIGURE 2.9 **The *cis* peptide bond.** (Adapted from D. Voet and J. G. Voet. *Biochemistry, Third edition.* John Wiley & Sons, Ltd., 2005.)

because there is less physical contact between the side chains of the two amino acids forming the peptide bond. Although rare, *cis* isomers do occur in polypeptides, especially when a proline is on the carboxyl side of a peptide bond.

Like polynucleotide chains, peptide chains have directionality. A free amino group is present at one end of the peptide chain and a free carboxyl group at the other end. By convention, the free amino group is drawn on the left. Once linked in a peptide chain, amino acids are called **amino acid residues** (or **residues** for short). Some important peptide nomenclature conventions are summarized in FIGURE 2.10.

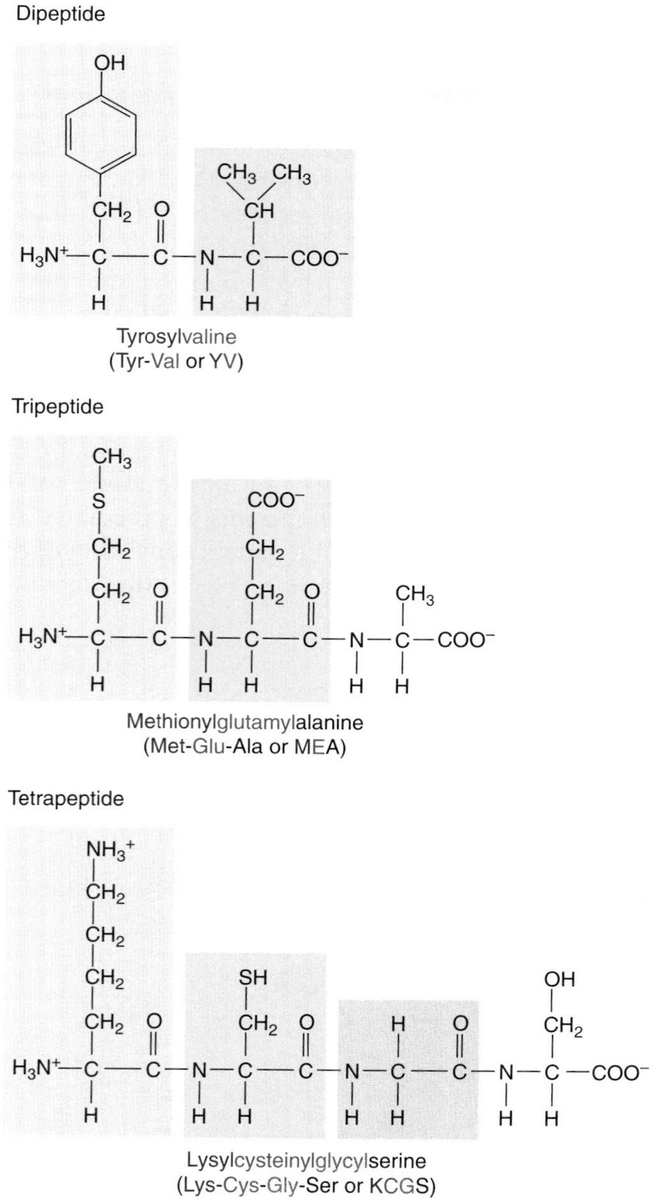

Dipeptide

Tyrosylvaline
(Tyr-Val or YV)

Tripeptide

Methionylglutamylalanine
(Met-Glu-Ala or MEA)

Tetrapeptide

Lysylcysteinylglycylserine
(Lys-Cys-Gly-Ser or KCGS)

FIGURE 2.10 **Conventions for drawing peptides.** By convention, the amino acid terminus (N-terminus) is on the left and the carboxyl terminus (C-terminus) is on the right. Peptides are named as derivatives of the carboxyl terminal amino acid.

Rather arbitrarily, peptides are divided by size into two major groups. Those with less than fifty amino acids are **oligopeptides**, while those with fifty or more residues are **polypeptides**. As indicated above, the term protein is usually reserved for a polypeptide chain (or set of associated polypeptide chains) in the biologically active conformation. Because peptides can vary in chain length, amino acid sequence, or both, one can imagine an almost limitless variety of peptides. For example, there are 20^{50} or slightly more than 1.12×10^{65} possible sequences for polypeptides with just 50 amino acid residues.

2.3 Protein Purification

Protein mixtures can be fractionated by chromatography.

A complex process such as DNA replication or RNA synthesis requires many different proteins that must work together. Each protein makes a specific contribution to the overall process. It's difficult to examine the structure and function(s) of an individual protein when it is present in a mixture of other proteins, however. Fortunately, most proteins are reasonably hardy and so retain their biological activity during purification. Nevertheless, it is usually desirable to fractionate proteins at 4°C and at or about pH 7 to prevent the loss of biological activity.

One general method for protein purification, called **column chromatography**, separates proteins in a mixture by repeated partitioning between a mobile aqueous solution and an immobile solid matrix. The solution containing the protein mixture is percolated through a column containing the immobile solid matrix consisting of thousands of tiny beads (FIGURE 2.11). As the solution passes through the column,

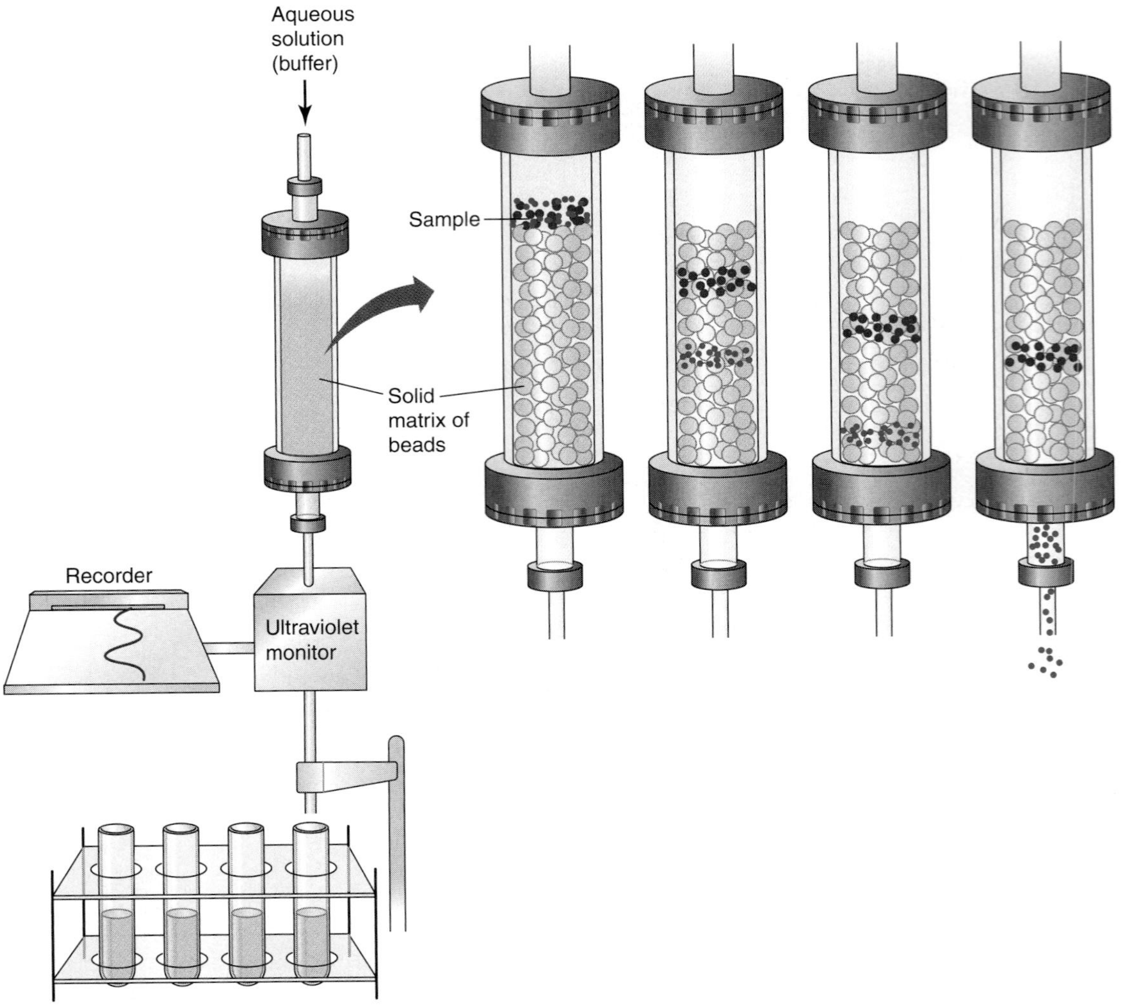

FIGURE 2.11 **Schematic for column chromatography.** (Adapted from an illustration by Wilbur H. Campbell, Michigan Technological University [http://www.bio.mtu.edu/campbell/bl4820/lectures/lec6/482w62.htm]. Accessed September 1, 2007.)

proteins interact with the immobile matrix as described below and are retarded. If the column is long enough, it can separate proteins that have different migration rates. Proteins released from the column can be detected by an ultraviolet monitor and then collected in tubes by a fraction collector.

Protein separation can be improved by increasing the solid matrix's surface area through the use of a longer column or finer beads. Both methods for increasing the surface area reduce solvent flow through the column, however. Initial attempts to increase the flow rate by forcing the aqueous solution through the column under pressure were unsuccessful because the pressure compressed the soft beads, impeding solvent flow.

The flow rate problem was solved by developing hard chromatographic beads that permit the aqueous solution to be forced through the column under high pressure, shortening the time required to achieve a separation and increasing resolution. This modified procedure known as **high-performance liquid chromatography** (HPLC) is now widely used in protein purification. Several different kinds of attractive interactions can be used to retard a protein migration through the solid matrix.

Ion-exchange chromatography uses electrostatic interactions between the protein and the solid matrix to fractionate proteins (FIGURE 2.12). A sample containing a mixture of proteins is allowed to percolate through a column packed with an immobile matrix, such as polysaccharide beads that are coated with positively (or negatively) charged groups. The beads' charged groups interact with the charged amino acid side-chains on the protein.

At pH 7, aspartate, glutamate, and carboxyl terminal residues will have negative charges and interact with positively charged resins (anion-exchange chromatography). Lysine, arginine, and amino terminal residues will have positive charges and interact with negatively charged resins (cation exchange resins).

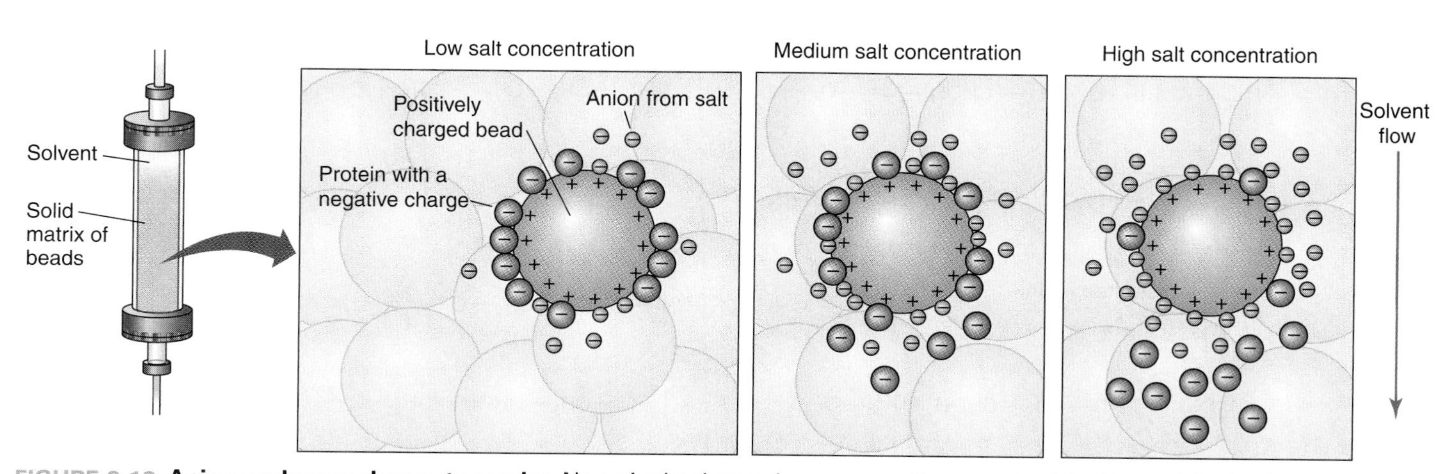

FIGURE 2.12 **Anion exchange chromatography.** Negatively charged groups on the proteins bind to positively charged groups on the anion exchange resin. Increasing salt concentrations produce anions that displace the proteins. Cation exchange resins work in a similar way except in this case, positively charged groups on the proteins bind to the negatively charged groups on the resin and cations displace the proteins.

Proteins are released by passing aqueous solutions with progressively higher salt concentrations through the column. The salt ions displace the charged side chains from the ion exchange beads. Proteins that interact with the column most weakly will migrate through the column fastest. Because proteins have both positively charged and negatively charged side chains on their surface, a specific protein may be fractionated by both anion and cation exchange chromatography.

Reverse phase chromatography (also called hydrophobic chromatography) uses the weak attractive interactions between nonpolar amino acid side chains and nonpolar groups such as phenyl or octyl groups attached to polysaccharide beads to retard protein migration (FIGURE 2.13). Proteins, dissolved in an aqueous solution with a high salt concentration, are applied to the column filled with the reverse-phase chromatography beads. Proteins stick to the beads more tightly when the salt concentration is high. The interaction of a given protein with the reverse phase resin depends on the number and placement of its nonpolar amino acid residues. Proteins are released by passing aqueous solutions with progressively lower salt concentrations through the column. Proteins with the lowest affinity for the beads will be released first. The conditions used for binding and eluting in reverse phase chromatography, therefore, are the opposite of those used in ion exchange chromatography.

Gel filtration or **molecular exclusion chromatography** separates protein molecules by size. This method depends upon special beads that permit small proteins to penetrate into their interior while excluding large proteins from this region (FIGURE 2.14a). A gel filtration column has two different water compartments: the internal compartment consists of the aqueous solution inside the beads and the external compartment consists of the aqueous solution outside the beads. Small protein molecules have access to both compartments, whereas large protein molecules only have access to the external compartment. The

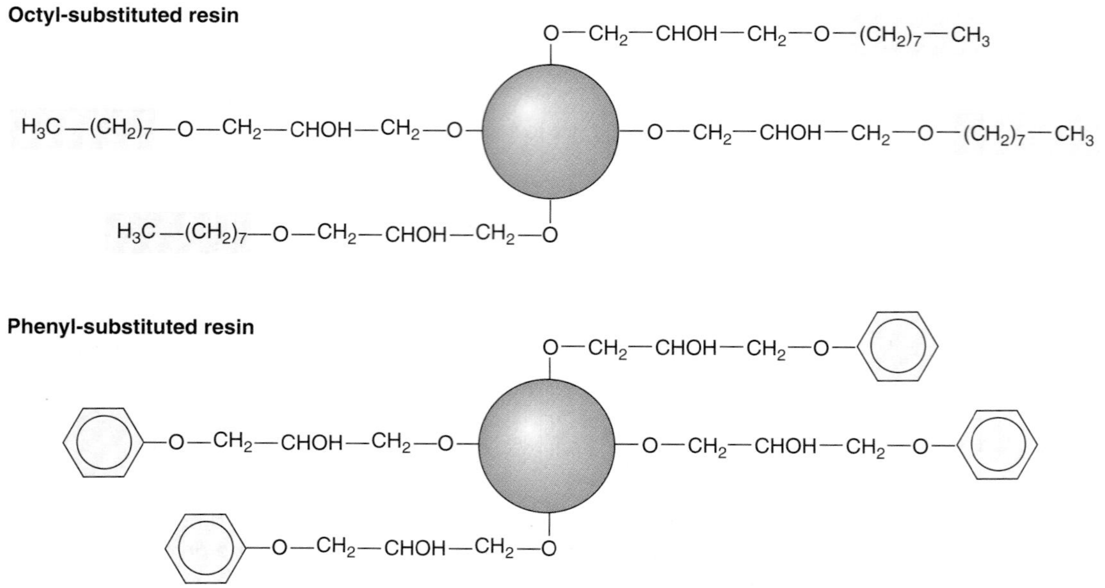

Octyl-substituted resin

$O—CH_2—CHOH—CH_2—O—(CH_2)_7—CH_3$

$H_3C—(CH_2)_7—O—CH_2—CHOH—CH_2—O—$ (bead) $—O—CH_2—CHOH—CH_2—O—(CH_2)_7—CH_3$

$H_3C—(CH_2)_7—O—CH_2—CHOH—CH_2—O$

Phenyl-substituted resin

$O—CH_2—CHOH—CH_2—O—$ (phenyl)

(phenyl) $—O—CH_2—CHOH—CH_2—O—$ (bead) $—O—CH_2—CHOH—CH_2—O—$ (phenyl)

(phenyl) $—O—CH_2—CHOH—CH_2—O$

FIGURE 2.13 **Resins used for reverse phase chromatography.**

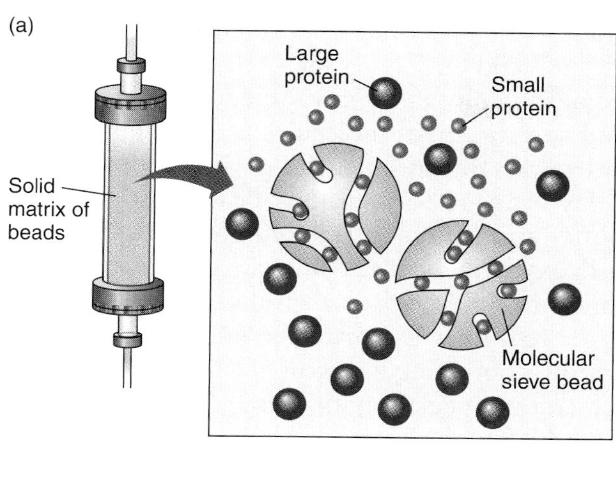

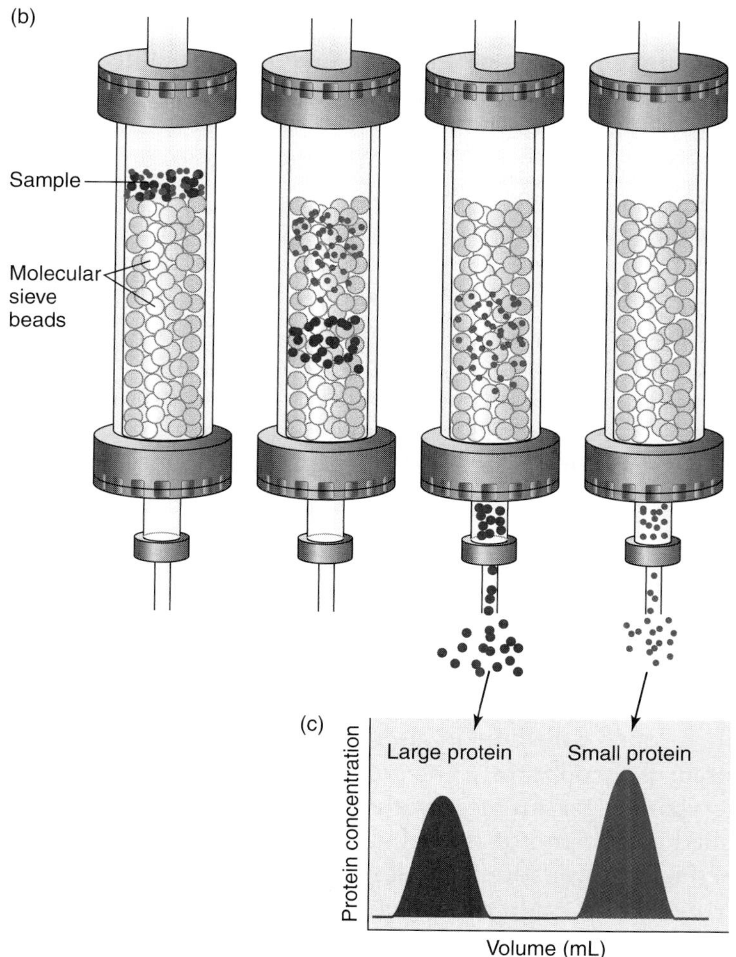

FIGURE 2.14 **Gel filtration chromatography.** (a) The gel filtration column has an internal compartment consisting of the aqueous solution inside specially designed beads and an external compartment consisting of the aqueous solution outside the beads. The beads permit small proteins to penetrate into their matrix while excluding large proteins from the region; the large proteins can only get into the external compartment. (b) Proteins are separated by size with the larger proteins appearing in earlier fractions and smaller proteins in later fractions. (c) Protein elution profile. (Adapted from B. E. Tropp. *Biochemistry: Concepts and Applications, First edition.* Brooks/Cole Publishing Company, 1997.)

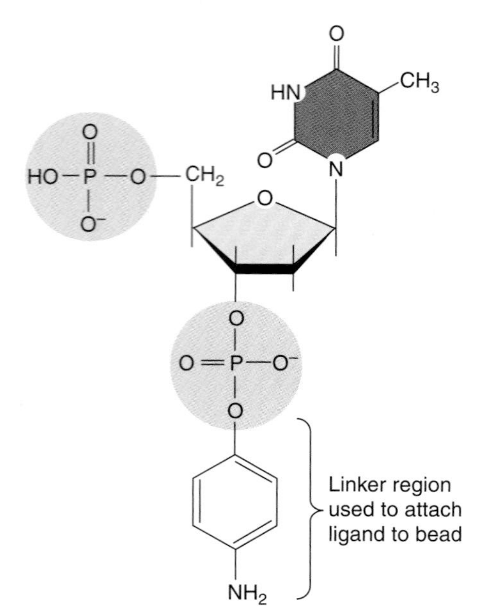

FIGURE 2.15 **Affinity chromatography.** Affinity chromatography exploits a protein's ability to bind to ligands. Nucleases bind to deoxyribonucleotide derivatives such as the one shown. Affinity resins are prepared by attaching the ligand to tiny water-insoluble beads. (Adapted from B. E. Tropp. *Biochemistry: Concepts and Applications, First edition.* Brooks/Cole Publishing Company, 1997.)

large proteins, therefore, appear in earlier fractions than do the small proteins (FIGURE 2.14b and c).

Affinity chromatography takes advantage of the fact that many proteins can bind specific small molecules termed **ligands**. Affinity chromatography exploits this specificity to purify the protein. Ligands are attached to tiny beads to form affinity beads that are suspended in an aqueous buffer and poured into a column. In one of the earliest experiments, the deoxythymidylic acid derivative shown in FIGURE 2.15 was attached to an insoluble polysaccharide to form affinity beads that bind nucleases (enzymes that catalyze nucleic acid hydrolysis). An aqueous solution, containing a mixture of bacterial proteins, was passed through a column packed with these affinity beads. Most of the proteins did not bind to the ligand and passed through the column. The nuclease, however, did bind to the ligand and was retained by the column. Active nuclease was recovered by washing the beads with a buffer solution at a low pH. Enzymes can also be eluted by washing the column with a solution that contains free ligand.

Affinity chromatography can provide a high degree of protein purification due to the specificity of the binding step. Thanks to recombinant DNA technology, it is possible to use a variant of affinity chromatography to purify almost any protein if the DNA that codes for it has been isolated. The basic approach is to modify the DNA so that the gene of interest now codes for the protein with an additional six histidine residues at its amino or carboxyl end. Specific resins have been devised that have a high affinity for proteins with a $(His)_6$ sequence. Because only the protein encoded by the modified DNA will have a $(His)_6$ sequence, the recombinant protein can be separated from all the other proteins in the mixture. Although addition of a $(His)_6$ tag usually does not alter a protein's biological properties, one must test to be certain that it does not.

Proteins and other charged biological polymers migrate in an electric field.

Multiply charged macromolecules such as proteins and nucleic acids migrate through a medium in response to an electric field (FIGURE 2.16). In **protein electrophoresis**, the medium is a porous matrix such as a polyacrylamide gel saturated with buffer solution. The protein sample is applied to one end of the gel and the electric field is generated by connecting a power source to electrodes attached at either end of the gel. Proteins that migrate through the gel at the fastest rate tend to have the greatest net charge, the most compact shape, and the smallest size. The net charge can be altered by changing the pH of the medium. The pH at which a protein has no net charge, and therefore will not migrate in an electric field, is called its **isoelectric pH**. Protein bands are visualized by staining with dyes. A protein free of all contaminants will appear as a single band. Electrophoresis, therefore, is a very useful method for monitoring protein purity.

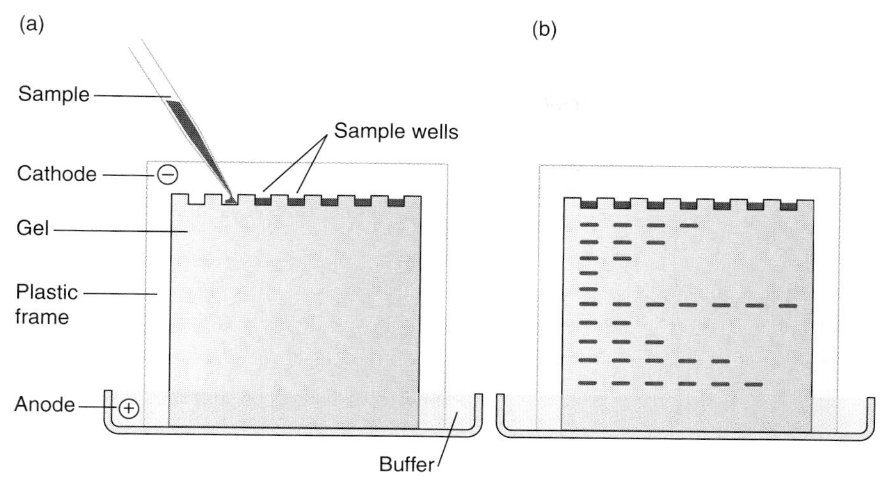

FIGURE 2.16 **Gel electrophoresis.** Samples are applied to slots in a porous matrix such as polyacrylamide gel, and an electric field is generated. The buffer pH is adjusted to ensure that the proteins have a net negative charge. Proteins that migrate through the gel at the fastest rate tend to have the greatest net negative charge, the most compact shape, and the smallest size. (Adapted from B. E. Tropp. *Biochemistry: Concepts and Applications, First edition.* Brooks/Cole Publishing Company, 1997.)

2.4 Primary Structure of Proteins

The amino acid sequence or primary structure of a purified protein can be determined.

Once a protein has been purified, it must be characterized to learn more about its chemical and biological properties. Until recently, a degradation technique devised by Pehr Edman in 1950, has been the most widely used method for amino acid sequence determination. Edman degradation involves a series of chemical steps that remove the amino acid from the amino terminal end of a polypeptide (FIGURE 2.17). The released amino acid derivative is then identified and the process repeated through several rounds of amino acid removal and identification.

Because cleavage efficiency is less than 100%, each successive cleavage cycle produces an increasingly heterogeneous peptide population. After about 50 cycles, the peptide population is so heterogeneous that it becomes nearly impossible to interpret the data. Cutting long polypeptides into well-defined fragments with a digestive enzyme such as trypsin or a chemical reagent such as cyanogen bromide (CNBr) solves this problem (FIGURE 2.18). Fragments produced by cleavage with a specific protease or CNBr are separated by chromatography and then sequenced by Edman degradation. Although the fragments are usually short enough to be sequenced completely, one must still determine the fragment order in the original polypeptide.

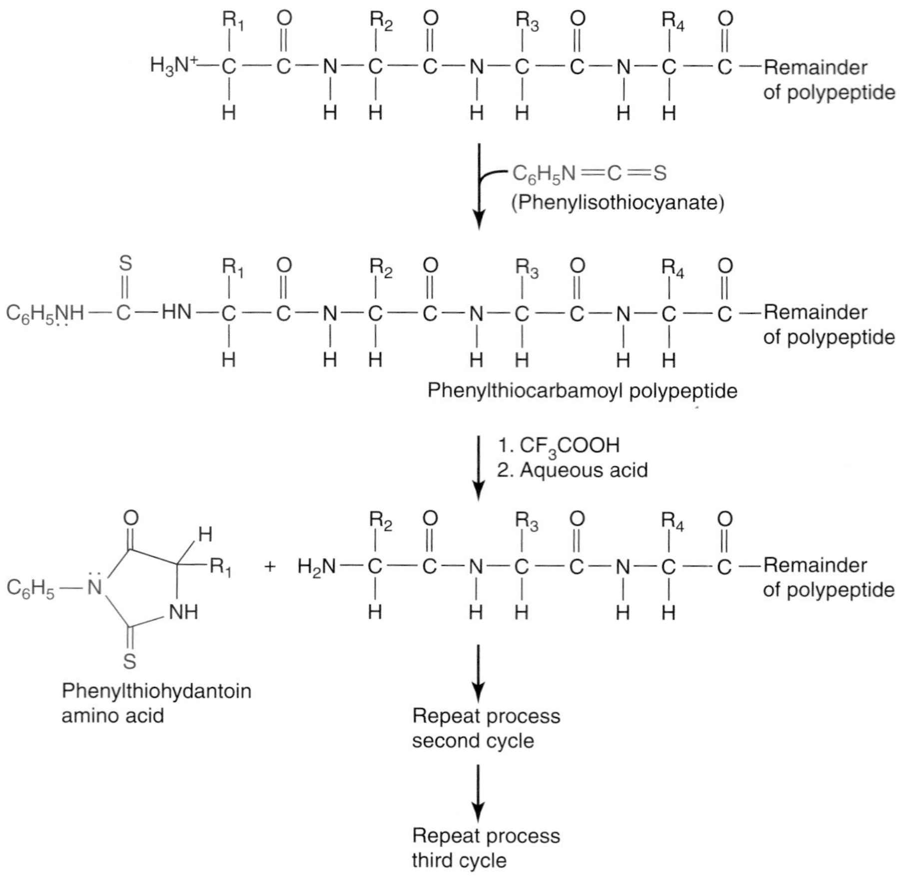

$C_6H_5N{=}C{=}S$
(Phenylisothiocyanate)

Phenylthiocarbamoyl polypeptide

1. CF_3COOH
2. Aqueous acid

Phenylthiohydantoin amino acid

Remainder of polypeptide

Repeat process second cycle

Repeat process third cycle

FIGURE 2.17 **The Edman degradation.** The Edman degradation, which selectively removes the N-terminal amino acid from a polypeptide chain, can be used to determine a polypeptide's amino acid sequence. (Adapted from B. E. Tropp. *Biochemistry: Concepts and Applications, First edition.* Brooks/Cole Publishing Company, 1997.)

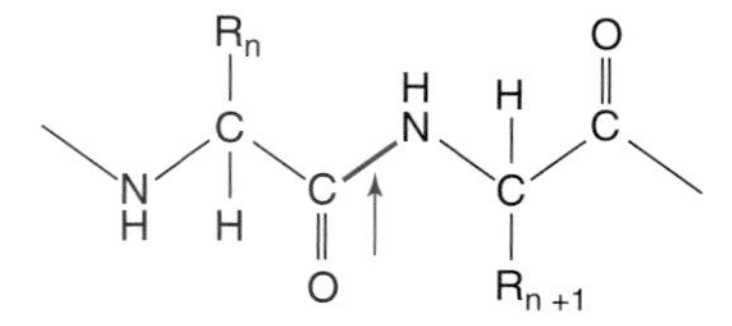

Enzyme or chemical agent	R_n
Trypsin	Lysine or arginine
Chymotrypsin	Phenylalanine, tyrosine, or tryptophan
Cyanogen bromide (CNBr)	Methionine

FIGURE 2.18 **Specific cleavage.** Trypsin, chymotrypsin, and cyanogen bromide (CNBr) cleave after specific amino acid residues. The arrow indicates the bond cleavage.

Fragment order is determined by the overlap method (FIGURE 2.19). This method depends upon having the sequences of two fragment collections, each generated by using an enzyme or chemical reagent that cuts after specific residues. For example, trypsin generates a set of fragments by cutting after arginine and lysine residues and CNBr generates a different set of fragments by cutting after methionine residues. The order of the fragments within the original polypeptide chain is determined by searching the two sets of fragments for overlapping sequences.

Despite its great value, Edman degradation also has limitations. It is time consuming, does not work when a peptide is blocked at its amino terminus, and does not provide information about amino acid residues that have been modified. A new approach that overcomes these difficulties takes advantage of mass spectrometry, a technique in which molecules are ionized and their masses are determined by following the specific trajectories of the ionized fragments in a vacuum system. Because mass spectrometry is very sensitive, it requires very little protein. Moreover, peptide fragmentation takes place in seconds

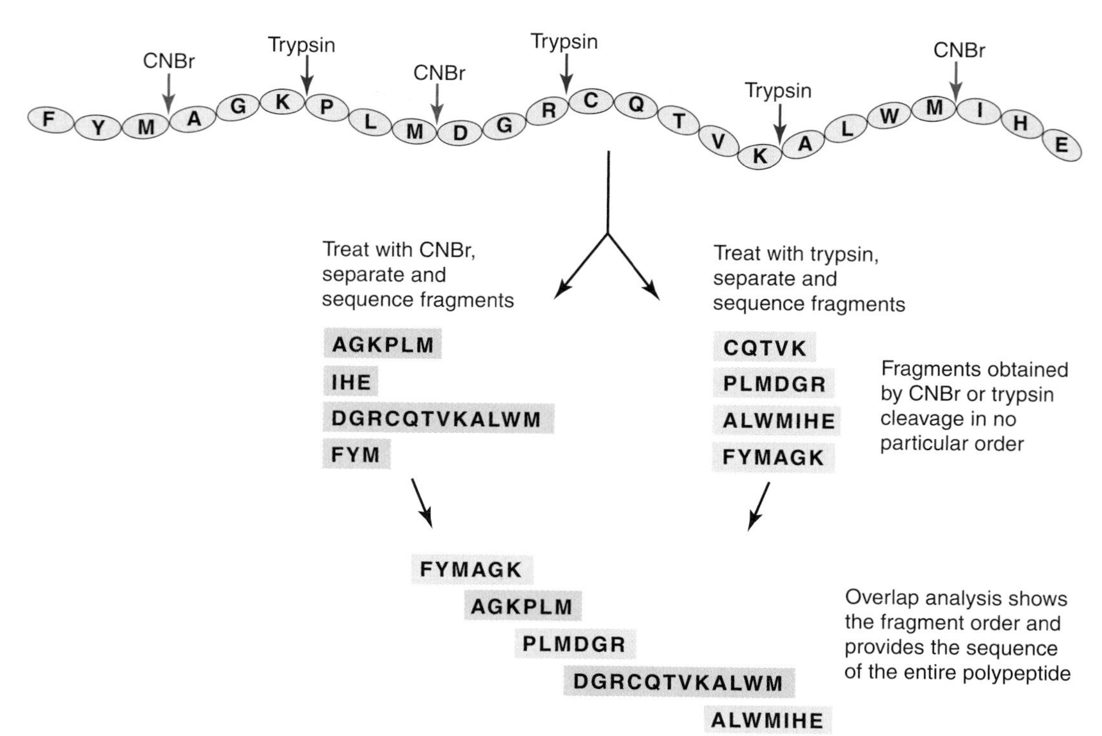

FIGURE 2.19 **The overlap method for amino acid sequence determination.** A polypeptide that has phenylalanine (F) as its N-terminal residue is divided into two samples. One sample is treated with CNBr, which cleaves after methionine (M) and the other is treated with trypsin, which cleaves after arginine (R) or lysine (K). The CNBr cleavage sites are indicated by the red arrows and the trypsin cleavage sites by the purple arrows. Then the fragments produced from each sample are resolved by chromatography and each purified fragment is sequenced by the Edman degradation procedure. Finally, the fragments are sequenced by searching for overlaps.

rather than hours, and sequencing is possible even if the protein is not completely pure.

FIGURE 2.20 summarizes the main steps in the mass spectrometer sequence technique. A protein population is prepared from a biological source and the individual polypeptides separated by electrophoresis. After separation, the gel lane is cut into several slices. A specific protease or chemical agent is added to the gel slice of interest to digest the trapped protein, converting the protein into peptides. The peptides in the mixture that is generated by this digestion are separated by HPLC and then analyzed by the mass spectrometer. The process is repeated by cutting the same protein with another specific protease or cleavage agent and the primary structure of the protein determined by the overlap method.

Polypeptide sequences can be obtained from nucleic acid sequences.

The development of rapid DNA sequencing methods, which will be described in Chapter 5, greatly accelerated the pace at which polypeptide sequences were determined. Instead of sequencing purified polypeptides, investigators determine DNA sequences of specific genes or all of the genetic material in the chromosomes of a particular organism (the **genome**) and then translate this information to obtain polypeptide sequences.

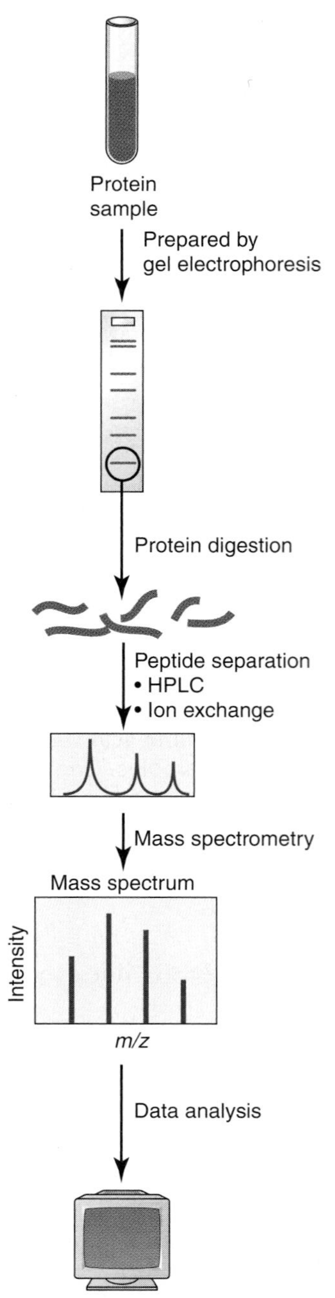

FIGURE 2.20 **Mass spectrometric determination of peptide sequence.** A protein population is prepared from a biological source such as a bacterial or cell culture. The protein of interest is purified; the last purification step is usually electrophoresis. The gel lane is cut to obtain the band that contains the desired protein. The protein is digested while still in the gel with specific enzymes, chemicals, or both. Then the peptide mixture that is generated is separated by chromatography and individual peptides are analyzed by mass spectrometry. The results are analyzed to determine the amino acid sequence within each peptide. The polypeptide sequence can be determined by digesting different samples of the same protein with different cleavage agents so that the overlap method can be used to order the proteins. (Adapted from H. Steen and M. Mann, *Nature Rev. Mol. Cell Biol.* 5 [2004]: 699–711.)

Although the DNA sequencing approach is very fast and quite accurate, extrapolating to polypeptide sequences does have serious limitations. A DNA sequence does not necessarily predict the chemical nature of the biologically active protein for the following reasons: (1) In eukaryotes large precursors of messenger RNA (mRNA) are converted to mRNA by a precise splicing mechanism in which intervening RNA sequences, called **introns**, are removed with concomitant joining of flanking sequences, called **exons** (FIGURE 2.21; see Chapter 19). The resulting mRNA molecules that are missing sequences present in DNA will program ribosomes to form polypeptides that are shorter than those predicted from the DNA sequences. In many cases, we cannot predict the sequences that will be lost during splicing and therefore cannot predict the sequence of the biologically active polypeptide. (2) Many polypeptides are converted into their biologically active form by cleavage at specific sites. For instance, the polypeptide precursor to insulin, preproinsulin, is converted into the active hormone by specific peptide bond cleavage. (3) Many proteins are subject to covalent modifications such as disulfide bond formation or the addition of phosphate, sugar, acetyl, methyl, lipid, or other groups. These modifications, which often influence the protein's biological activity and stability, can only be revealed by studying the purified protein. (4) Many polypeptides do not act alone but instead function as a part of a complex that contains other polypeptides of the same or different types. The true nature of these protein complexes can only be revealed by studying the purified complex.

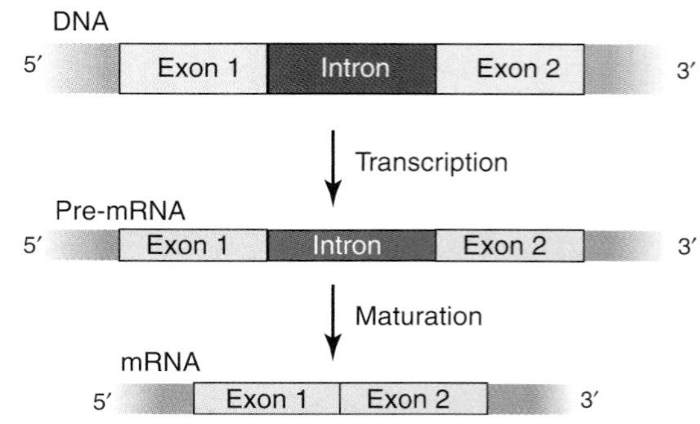

FIGURE 2.21 **Split genes.** Many eukaryotic genes have intervening sequences (introns) that are included in the precursor to messenger RNA (mRNA) molecules that are formed when the genes are transcribed but are removed during a maturation process in which precursor mRNA molecules are converted to mature mRNA molecules. The coding (or expressed) sequences, which are included in both the precursor mRNA and mature mRNA molecules, are called exons.

The BLAST program compares a new polypeptide sequence with all sequences stored in a data bank.

The genomes of approximately one thousand different organisms have now been sequenced. The world's most comprehensive catalog of protein sequence information is available from Uniprot. Comparison of a new sequence with sequences stored in the data bank is possible by using the BLAST (Basic Local Alignment Search Tool) program available at the Uniprot site.

Five different BLAST programs offer fast, sensitive, and relatively easy ways to compare specific nucleic acid or polypeptide sequences (the query sequences) with all sequences (the subject sequences) in the data bank. Each program permits a different type of search. Here we consider just one of the five BLAST programs, the *blastp program*, because it is the one that compares the amino acid query sequence with all polypeptide sequences in the data bank. The blastp program searches the data bank by first looking for every tripeptide in the data bank that is similar to tripeptides in the query polypeptide, and then extends initial regions of similarity into larger alignments without gaps. Once alignments have been created, the blastp program determines and reports the probability of their arising by chance, lists the data bank sequences that are most similar to the query sequence, and shows a local alignment of the query sequence with matched data bank sequences. A newly sequenced polypeptide may be so similar to a polypeptide with a known function in another organism that we can safely conclude that the two polypeptides have the same function.

However, a search of the data banks may also reveal no similar polypeptides or it may reveal polypeptides with similar sequences but with no known function. Now we are in the rather unsettling position of knowing that a polypeptide exists and knowing its sequence but not having any idea about what the polypeptide actually does. Solving this "function problem" is one of the major challenges for molecular biologists in the coming decades.

Proteins with just one polypeptide chain have primary, secondary, and tertiary structures while those with two or more chains also have quaternary structures.

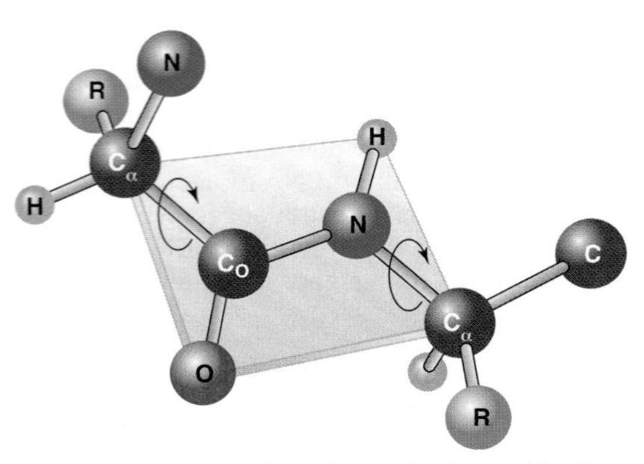

FIGURE 2.22 **Rotation about C_α–C_o and N–C_α.** Rotation does not take place about N–C_o but is free to take place about C_α–C_o and N–C_α. (Adapted from C. K. Mathews, et al. *Biochemistry: Third edition.* Prentice Hall, 2000.)

The primary structure is just the first of four possible levels of polypeptide structures. The other three levels are determined by the way that the polypeptide is arranged in space. The polypeptide backbone has three types of bonds, C_α–C_o, C_o–N (the peptide bond), and N–C_α (FIGURE 2.22). Although rotation about peptide bonds is severely limited, rotation does occur about N–C_α and C_α–C_o, the two single bonds in the polypeptide backbone. Rotation about the single bonds permits the polypeptide to fold into biologically active proteins. Because of this folding, proteins have three levels of structure in addition to their primary structure (FIGURE 2.23a). The **secondary structure** describes the folding pattern within a segment of a polypeptide chain containing neighboring residues. Among the many different possible secondary structures, the three most common are the α-helix, the β-conformation, and the loop or turn (FIGURE 2.23b). The **tertiary structure** provides a view of a protein's entire three-dimensional structure (FIGURE 2.23c), including spatial arrangements among different segments and among residues within the different segments. The **quaternary structure**, which will be examined in the next chapter, applies only to proteins with two or more polypeptide chains and indicates the way that the chains are arranged in space with respect to one another (FIGURE 2.23d).

2.5 Weak Noncovalent Bonds

The polypeptide folding pattern is determined by weak noncovalent interactions.

Secondary, tertiary, and quaternary structures are stabilized by four kinds of weak non-covalent interactions, the **hydrogen bond, ionic bond, hydrophobic interaction,** and **van der Waals interaction.** Because each kind of weak interaction has a range of binding energies there is considerable energy overlap among them. Although the noncovalent interactions are weak (1-40 kJ · mol^{-1}) compared to covalent bonds (200-1000 kJ · mol^{-1}), the combined effect of many weak interactions is sufficient to determine a protein's folding pattern. Moreover, the fact that the noncovalent interactions are weak means that they are

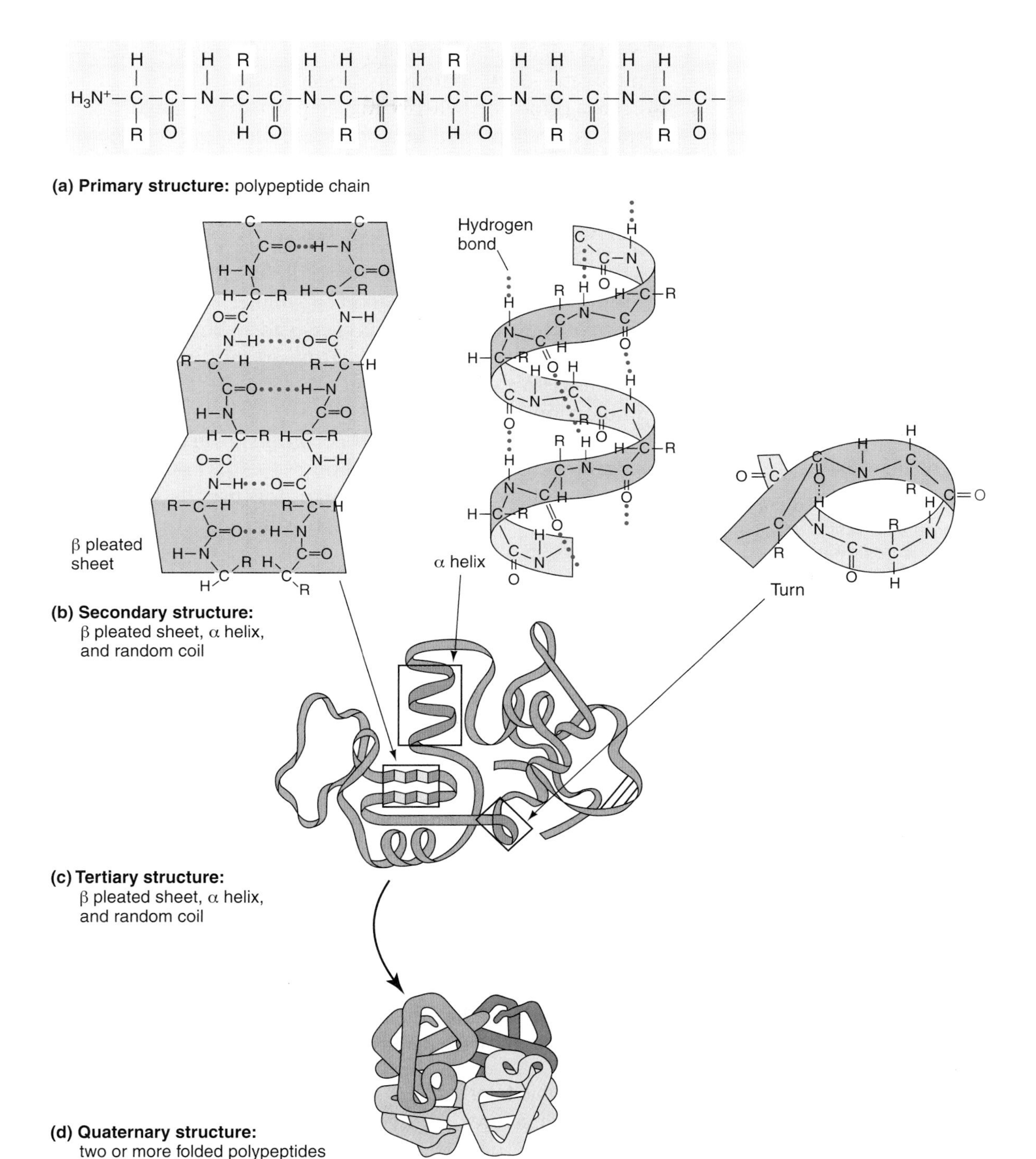

FIGURE 2.23 **Protein structure.** Polypeptides can be viewed at four different structural angles. The ribbon diagram (b and c) is a simple and effective way to represent secondary and tertiary structures. (Adapted from B. E. Tropp. *Biochemistry: Concepts and Applications, First edition.* Brooks/Cole Publishing Company, 1997.)

easily formed and easily broken, permitting proteins to assume different conformations as they perform their functions.

Hydrogen Bond

The hydrogen bond results from an attractive interaction between an electronegative atom and a hydrogen atom attached to a second

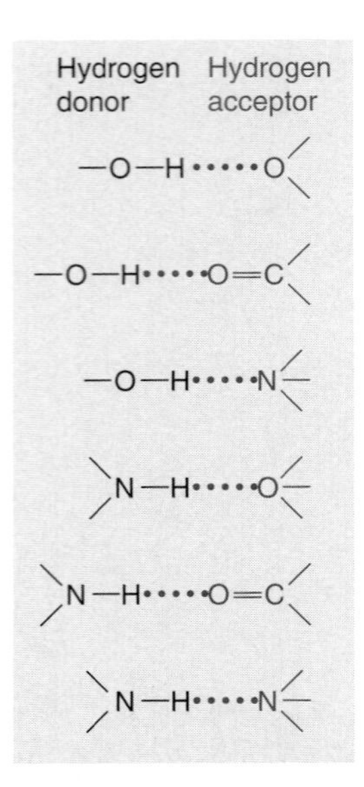

FIGURE 2.24 **Some typical hydrogen bonds in proteins.** The amino acid residue that supplies the hydrogen atom is designated the donor and the residue that binds the hydrogen atom is designated the acceptor.

electronegative atom. Only two electronegative atoms—oxygen and nitrogen—participate in hydrogen bond formation in biological molecules. (Fluorine, which can also participate in hydrogen bond formation, is rarely present in biological molecules.) A glance at the periodic table reveals that nitrogen, oxygen, and fluorine are in period two and therefore the smallest electronegative atoms. Hence, the partially negative charge that results when these atoms pull electrons from a covalently attached carbon or hydrogen atom is concentrated in a small region of space. The partially positive charge in the hydrogen atom is concentrated in an even smaller region of space. The strong electron pull exerted by oxygen and nitrogen together with charge concentration over the small atoms results in an attractive force that is considerably stronger than that resulting from other dipole-dipole interactions (below). In the strongest hydrogen bonds, the two electronegative atoms and the hydrogen atom lie on a straight line. Hydrogen bond strengths range from about 8 to 40 $kJ \cdot mol^{-1}$. The bond energy depends on physical environment. It is usually weaker on the protein surface than in the protein interior because it is much more likely to be subject to competing interactions with water on the protein surface than it is in the interior. Some typical examples of hydrogen bonds that occur in biological molecules are shown in FIGURE 2.24. The Watson-Crick Model (see Chapter 1) for DNA recognizes that hydrogen bonds between adenine-thymine base pairs and guanine-cytosine base pairs contribute to the stability of the double helix.

Ionic Bond

The **ionic bond** results from attraction between positively and negatively charged ionic groups. For example, the negatively charged side chain in aspartate or glutamate can form an ionic bond with the positively charged side chain in arginine or lysine. Ionic bonds can form between residue pairs that are near to one another in the primary structure or pairs that are far apart. The strength of the ionic bond varies depending on its surroundings. In an aqueous solution, ionic bond strength is about 2 $kJ \cdot mol^{-1}$. The ionic bond strength is about ten times greater in a protein's hydrophobic interior. Ionic bonds are easily disrupted by pH changes, which alter the charges on interacting side chains or by high concentrations of small ions that compete with the interacting side chains.

Hydrophobic Interaction

The driving force for hydrophobic interactions can be explained by considering the structural arrangement of surrounding water molecules. When a nonpolar molecule or group is placed in water, water molecules interact through hydrogen bonds to form highly ordered cages around the nonpolar molecule or group (FIGURE 2.25). The second law of thermodynamics tells us that disorder is favored over order. Stated another way, the second law indicates that **entropy**, the measure of disorder, increases for spontaneous processes. Thus, the ordering of water molecules to form a cage around a nonpolar molecule or group is an unfavorable process. Placing two nonpolar molecules or groups into water would require water molecules to form two highly ordered cages. If the nonpolar molecules or groups were close together, a single water cage would suffice. For strictly geometric reasons, the number of water molecules required to form a single cage around the pair of nonpolar molecules or groups is less than half the total number needed to form a cage around each molecule or group separately. In general, the number of ordered water molecules per nonpolar molecule or group is always smaller if the nonpolar molecules or groups are clustered or stacked. Hence, clustering or stacking of nonpolar molecules or groups is thermodynamically favored because a cluster or stack requires fewer water molecules to be arranged in highly ordered cages. Note that no bonds are formed; it is only that a cluster is the more probable arrangement. One can obtain an indication of the tendency of molecules or groups to aggregate in water, known as **hydrophobicity**, from their tendency to transfer from water to a nonpolar solvent. The energy released by transferring a methylene group from water to a nonpolar organic solvent is about 3 kJ • mol^{-1}. Many amino acid side chains are hydrophobic. For example, the hydrocarbon chains of alanine, leucine, isoleucine, and valine tend to form clusters and the aromatic side chains in tyrosine and tryptophan tend to form stacks in water. When interacting hydrophobic side chains are present on residues that are far apart, they bring distant parts of the polypeptide chain together. Thus, folding patterns of polypeptides are strongly influenced by hydrophobic interactions among the side chains of amino acid residues. Dozens of hydrophobicity scales have been proposed for amino acid side chains, each based on slightly different experimental data and theoretical principles. Unfortunately, values provided by each of the scales are only approximations. One widely used scale devised by J. Kyte and R. F. Doolittle is shown in FIGURE 2.26.

(a) Water molecules in bulk phase of water

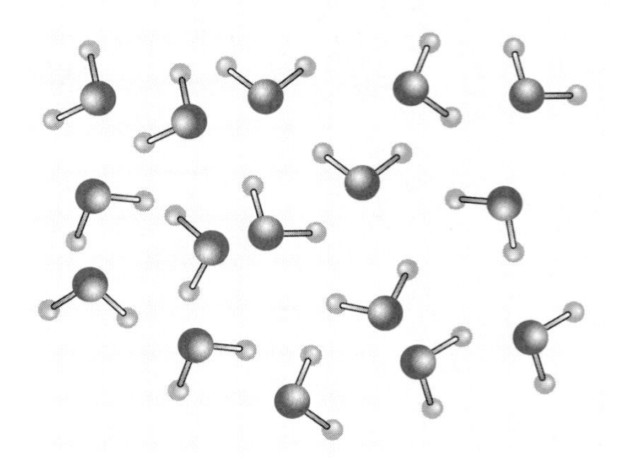

(b) Water molecules in cage around hydrocarbon

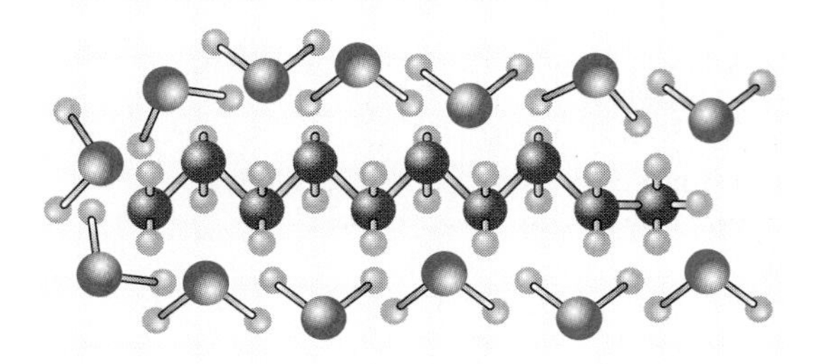

FIGURE 2.25 **The hydrophobic effect.** (a) Water molecules move randomly in the absence of nonpolar (hydrophobic) molecules. (b) Water molecules form ordered cages around hydrophobic molecules or groups. Because disorder is favored over order, the most favorable situation is one that uses the fewest possible water molecules to form cages. For strictly geometric reasons, the number of water molecules required to form a cage around a group of hydrophobic molecules is less than the number required to form separate cages around each molecule. Therefore, hydrophobic molecules (or groups) tend to aggregate in water. (Adapted from B. E. Tropp. *Biochemistry: Concepts and Applications, First edition.* Brooks/Cole Publishing Company, 1997.)

van der Waals Interactions

Van der Waals interactions are weak electrostatic interactions between two polar groups, a polar group and a nonpolar group, or two nonpolar groups. Electrostatic attraction between two polar groups results from the attraction between a partially positive atom on one polar group and a partially negative atom on another polar group. Electrostatic attractions between a polar and a nonpolar group result from the polar group's ability to induce a short-lived polarity in the nonpolar group, which in turn leads to a weak attractive interaction between the oppositely charged regions on the two groups. Very weak electrostatic attractions between two nonpolar groups arise from fluctuating charge densities in the nonpolar groups. At any given time, there is a small probability that a nonpolar group will have an asymmetric electron distribution. A nonpolar group that experiences such a transitory perturbation of charge distribution can induce polarity in neighboring nonpolar groups. The combination of fluctuating and induced polarity accounts for the very weak forces of attraction that hold nonpolar molecules together. The strength of a van der Waals interaction ranges from about 1 to 10 kJ • mol^{-1}. Because the attractive force between two

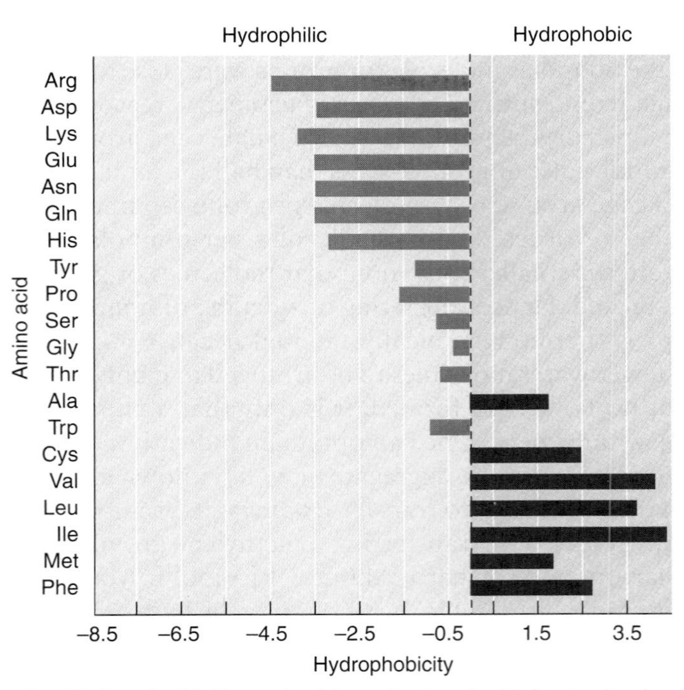

FIGURE 2.26 **Hydrophobicity scale.** Many hydrophobicity scales have been devised. The one shown here, which was devised by J. Kyte and R.F. Doolittle, is based on experimental data.

TABLE 2.2 **van der Waals Radii**

Atom	van der Waals radii (nm)
Hydrogen	0.120
Oxygen	0.152
Nitrogen	0.155
Carbon	0.170
Sulfur	0.180
Phosphorus	0.180

atoms is proportional to $l/r6$ (r is the distance between their nuclei), van der Waals interactions become significant only when two atoms are very near one another (0.1–0.2 nm apart). A powerful repulsive force also comes into play when the outer electron shells of the two atoms overlap. The **van der Waals radius** is defined as the distance at which the attractive and repulsive forces between the atoms balance precisely. Van der Waals radii differ from one kind of atom pair to another; some representative values are shown in Table 2.2. The shape of a molecule is in essence the surface formed by the van der Waals spheres of each atom. FIGURE 2.27 shows the shapes of alanine and proline when defined in this way. The average energy of thermal motion at room temperature is about 2.5 kJ • mol^{-1}. Therefore, van der Waals interaction between two atoms is usually not sufficient to maintain these atoms in proximity. However, if the interactions of *several* pairs of atoms are combined, the cumulative attractive force can be great enough to withstand being disrupted by thermal motion. Thus, two nonpolar molecules can attract one another if several of their component atoms can mutually interact. However, because of the $l/r6$-dependence, the intermolecular fit must be nearly perfect. Therefore, two nonpolar molecules will hold together if their shapes are complementary. Likewise, two separate regions of a polymer will hold together if their shapes match. Sometimes, the van der Waals attraction between two regions is not large enough to cause binding; however, it can significantly strengthen other weak interactions such as the hydrophobic interaction, if the fit is good.

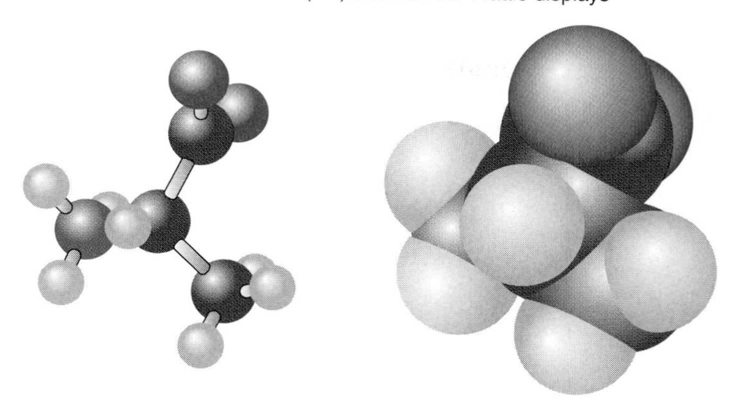

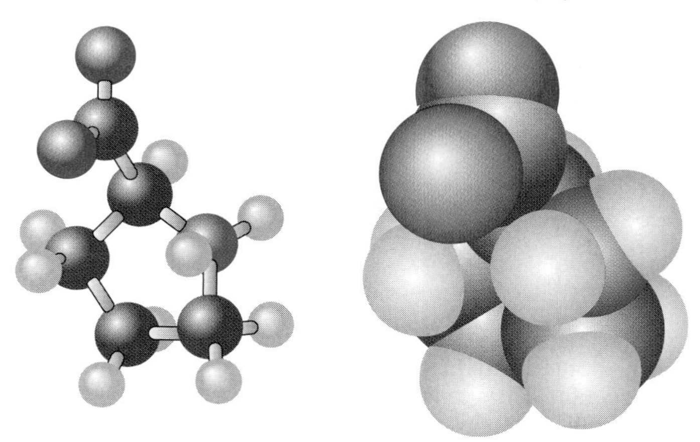

FIGURE 2.27 **Alanine and proline as ball and stick and van der Waals displays.** (a) Alanine shown in ball and stick (left) and van der Waals (right) displays. (b) Proline shown in ball and stick (left) and van der Waals (right) displays.

2.6 Secondary Structures

The α-helix is a compact structure that is stabilized by hydrogen bonds.

Adjacent residues on a polypeptide chain can fold into regular secondary structures. Linus Pauling predicted the existence of the first secondary structure, the **α-helix**, in 1951 while recovering from an illness at home. To occupy his time, Pauling drew a short peptide with correct bond angles and bond lengths on a piece of paper. Upon creasing the paper, he noticed that the peptide backbone folded into a helix. When he returned to the laboratory, Pauling and his colleague Robert Corey constructed a more accurate three-dimensional model that they named the α-helix. The model (FIGURE 2.28) has the following features: Each complete helical turn extends 0.54 nm along the

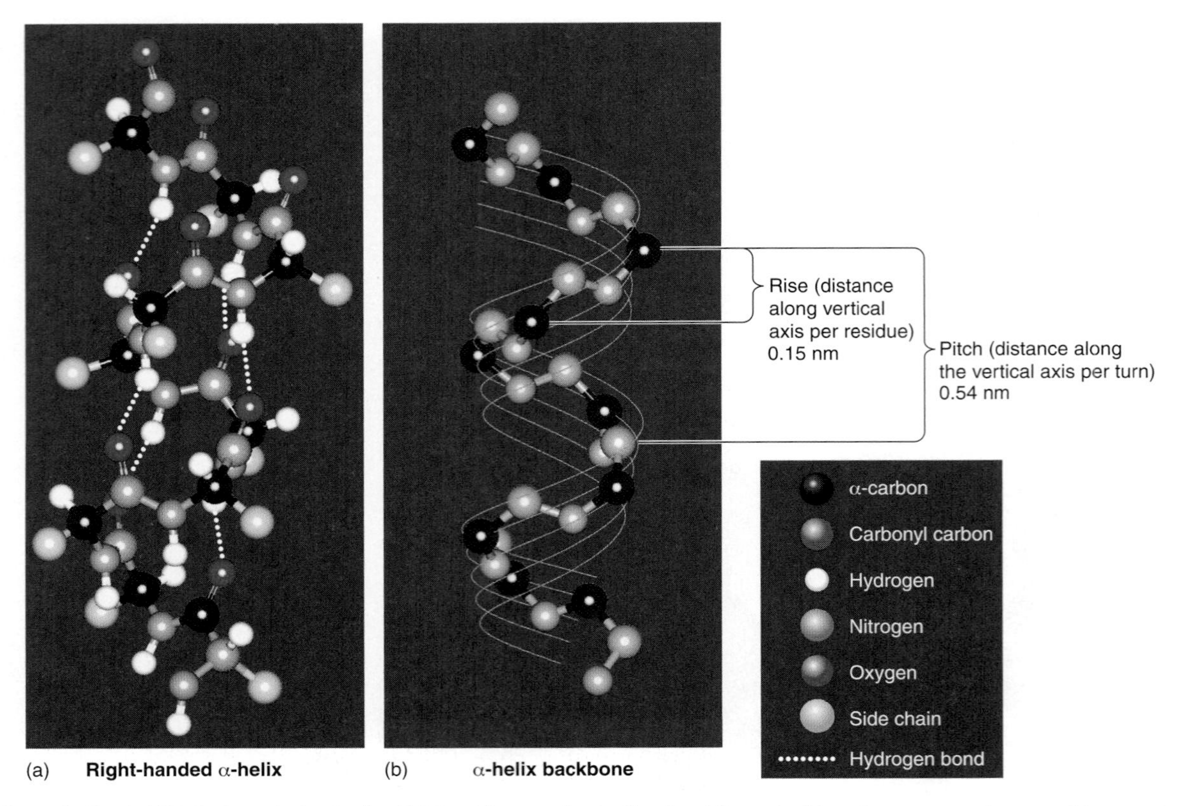

FIGURE 2.28 **The α-helix.** (a) Right-handed α-helix. (b) Backbone of α-helix. A white spiral has been superimposed to emphasize the backbone's helical structure. The N-terminal residue is at the bottom and the C-terminal residue at the top of the figure. The oxygen in each carbonyl group forms a hydrogen bond with the amide hydrogen that is four residues more toward the C-terminus, with the hydrogen bond approximately parallel to the long axis of the helix. All carbonyl groups point toward the C-terminus. In an ideal α-helix, equivalent positions reappear every 0.54 nm (the pitch of the helix) with each amino acid residue advancing the helix by 0.15 nm along the axis of the helix (the rise), and there are 3.6 amino acid residues per turn. In a right-handed helix the backbone turns in a clockwise direction when viewed along the axis from its N-terminus. (Structures from Protein Data Bank 1L64. D. W. Heinz, W. A. Baase, and B. W. Matthews, *Proc. Natl. Acad. Sci. USA* 89 [1992]: 3751–3755. Prepared by B. E. Tropp.)

vertical axis and requires 3.6 amino acid residues. The vertical rise per residue is 0.15 nm/residue (0.54 nm/3.6 residues).

Hydrogen bonds are located inside the helix, forming a regular repeating pattern. The oxygen atom on the carbonyl group of residue n forms a hydrogen bond with the hydrogen atom of the N–H group of residue n + 4. The first and last four residues in the helix cannot form a full set of hydrogen bonds because hydrogen bond partners are not available. All C=O bonds point in one direction and all N–H bonds point in the opposite direction, producing a significant net dipole for the α-helix with a partial positive charge at the amino end and a partial negative charge at the carboxyl end (FIGURE 2.29). These electrostatic charges are often offset by acidic residues at the amino end of the helical segment and basic residues at the carboxyl end of the helical segment.

The α-helix is a right-handed helix. The right-handed nature of the helix is most easily visualized by picturing the threads of a screw. In a right-handed screw threads are shaped so that clockwise rotation produces tightening. Threads could also be in a left-handed form, however, so that counterclockwise rotation produces tightening. Naturally

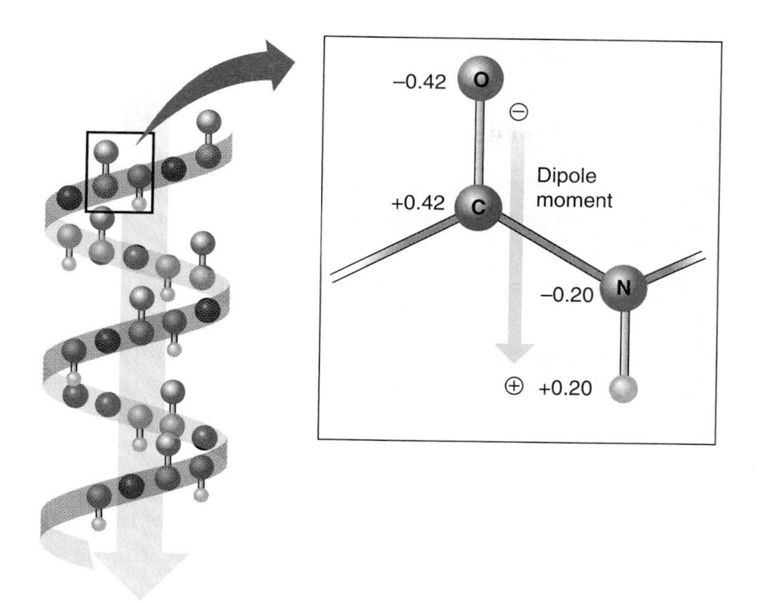

FIGURE 2.29 **Macroscopic dipole in the a-helix.** The individual dipole moments of the N-H and C=O groups along the helical axis generate a large net dipole for the helix. The numbers shown indicate fractional electric charges on the respective atoms. (Adapted from R. H. Garrett and C. M. Grisham. *Biochemistry, Third edition.* Brooks/Cole Publishing Company, 2005.)

occurring polypeptide chains, which are made of L-amino acids, fold into right-handed α-helices. When synthetic polypeptides are made of D-amino acids, the polypeptides fold into left-handed helices.

All amino acid residues can fit in the α-helix but they differ in their propensity to do so. Proline and glycine have the least tendency to fit into a helix. The nitrogen atom in a proline that is part of a polypeptide lacks the substituent hydrogen atom needed to form a hydrogen bond. Moreover, the proline side chain is rigid and so does not easily fit into the α-helix. Proline is occasionally present in the first helical turn, where its side chain geometry and inability to hydrogen bond do not create a problem. Glycine's tendency not to be part of an α-helix results from an entirely different problem. Glycine has great conformational freedom because its side chain, a single hydrogen atom, is so small. Furthermore, a single hydrogen atom is insufficient to protect the hydrogen bonds inside the helix from disruption by water. Certain combinations of adjacent residues also tend to disrupt or break the helical structure. When neighboring residues have bulky side chains, the side chains make physical contact that prevents them from fitting properly. For instance, neighboring isoleucine, tryptophan, and tyrosine residues would disrupt the helical structure. A run of positively charged or negatively charged side chains repel one another, destabilizing the helix. This phenomenon is best illustrated by comparing polyglutamate and polylysine structures at different pH values. Polyglutamate exists as an α-helix at pH 2 but not at pH 7 or pH 10. The explanation for this behavior is that the carboxylic acid side chains are uncharged at pH 2 but have negative charges at

(a) Parallel β pleated sheet

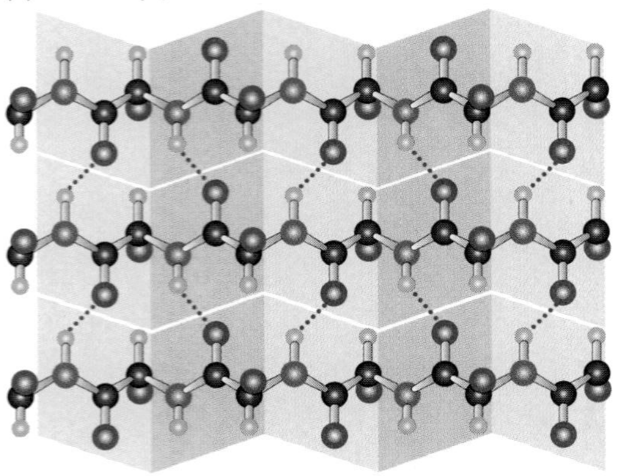

(b) Antiparallel β pleated sheet

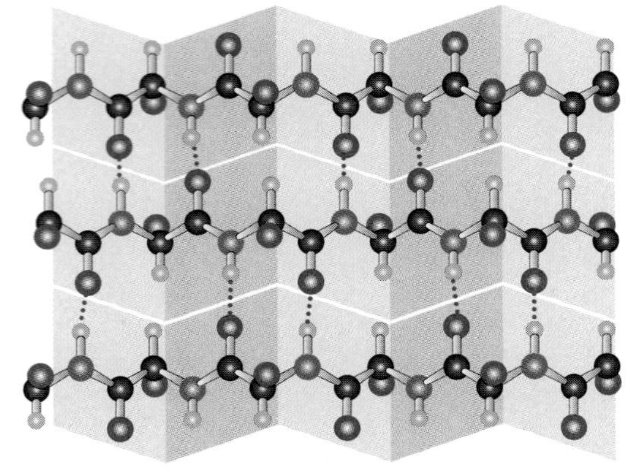

FIGURE 2.30 **The β-conformation of polypeptide chains.** Polypeptide chains with β-conformations can line up side by side so that C=O and N-H groups on adjacent chains interact through hydrogen bonds to produce an almost flat structure called a β-sheet. The polypeptide chains' zigzag structure gives the β-sheet the appearance of being pleated with consecutive C_α atoms slightly above or below the plane of the sheet. The sheets can be organized so that (a) all peptide chains have the same amino to carboxyl direction—parallel pleated sheets or (b) successive polypeptide chains are oriented in opposite directions—antiparallel pleated sheets. The color coding is the same as in Figure 2.28. (Adapted from D. C. Nelson and M. M. Cox. *Lehninger Principles of Biochemistry, Fourth edition.* W. H. Freeman & Company, 2004.)

pH 7 and pH 10. Side chain repulsion at pH 7 and pH 10 prevents helix formation.

In contrast, polylysine exists as an α-helix at pH 10 but not at the two lower pH values. Once again the explanation is charge repulsion. The positively charged lysine side chains repel one another at pH 2 and pH 7. Although the α-helix is the most common helix present in proteins, it is not the only one. For instance, the so-called 3_{10} helix, which is much less common, has hydrogen bonds between residue n and n + 3.

A few fibrous proteins, most notably α-keratin the major protein in hair, have no residues or combination of residues that disrupt the α-helix and so have a single regular structure throughout. Most proteins do have helix breakers, however, and so have helical segments surrounded by nonhelical segments. When present in a water-soluble protein, α-helical segments tend to be on the outside of the protein with one side facing the aqueous medium and the other the hydrophobic interior. The side facing the aqueous medium tends to have hydrophilic (acidic, basic, or polar) residues while the side facing the interior has mostly hydrophobic residues. Because the α-helix has 3.6 residues per turn, such an arrangement can be achieved by switching from hydrophobic to hydrophilic side chains with a 3 or 4 residue periodicity.

The β-conformation is also stabilized by hydrogen bonds.

Pauling and Corey also predicted the existence of a second type of secondary structure, the **β-conformation**, in which the polypeptide chain is almost fully extended. Polypeptide chains with β-conformation can line up side by side so that C=O and N–H groups on adjacent chains interact through hydrogen bonds to produce an almost flat structure called a β-sheet (FIGURE 2.30). The polypeptide chain's zigzag structure gives the β-sheet the appearance of being **pleated** with consecutive C_α atoms slightly above or below the plane of the sheet. A sheet can be organized so that all polypeptide chains have the same amino to carboxyl direction, **parallel pleated sheet** (Figure 2.30a), or so that successive polypeptide chains are oriented in opposite directions, **antiparallel pleated sheet** (Figure 2.30b). In a parallel pleated sheet, hydrogen bonds are evenly spaced and at angles to the long axes of the polypeptide chain. Hydrophobic side chains are present on both sides of the parallel pleated sheet. In an antiparallel pleated sheet, pairs of narrowly spaced hydrogen bonds alternate with pairs of more widely spaced hydrogen bonds, with all hydrogen bonds perpendicular to the long axes of the polypeptide strands. The polypeptide chains that comprise an antiparallel pleated sheet tend to have alternating hydrophilic and hydrophobic residues, so that hydrophobic side chains tend to be present on one side of the sheet and hydrophilic residues on the other.

Loops and turns connect different peptide segments, allowing polypeptide chains to fold back on themselves.

The average globular protein has a diameter of about 2.5 nm, corresponding to about eleven residues in an α-helix and only seven residues in the extended β-conformation. Secondary structures known as turns and loops connect α-helical and β-strand segments within a protein, allowing the polypeptide backbone to fold back upon itself and reverse direction. Turns and loops usually are on the protein surface, extending into the surrounding aqueous environment. They, therefore, tend to be made of hydrophilic residues but also exploit the special conformational properties of glycine and proline residues to reverse direction. The most common kind of turn, the β-turn, consists of four residues and allows the polypeptide chain to reverse direction. The carbonyl oxygen of the first residue in a β-turn forms a hydrogen bond with the amino group of the fourth residue. The peptide bonds of the middle two residues do not interact through hydrogen bonds. Glycine and proline are commonly present in β-turns. Glycine's conformational flexibility allows it to fit into the tight turn. Proline's steric constraints are also well suited to the β-turn. Two of the most common types of β-turns are shown in FIGURE 2.31. Loops contain five or more residues, tend to be quite flexible, and lack a defined structure.

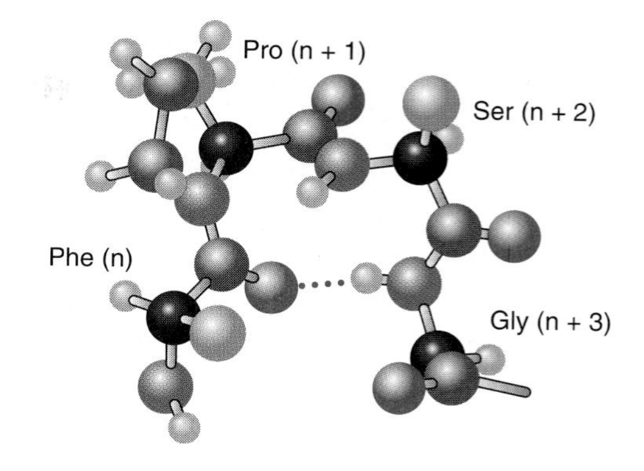

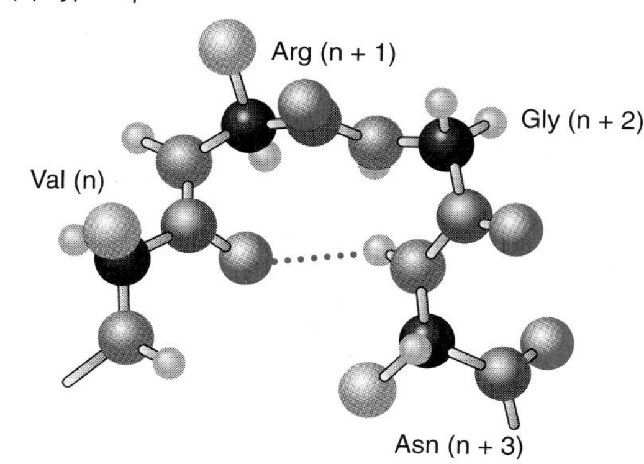

FIGURE 2.31 **Two types of β-turns.** The two most commonly occurring β-turns are called (a) type I and (b) type II. Type I β-turns occur with about twice the frequency of type II β-turns. In each case there is a hydrogen bond between residue n and residue n + 3. Type II β-turns always have a glycine at position n + 2. The color coding is the same as in Figure 2.28.

Certain combinations of secondary structures, called supersecondary structures or folding motifs, appear in many different proteins.

Folding patterns for a few thousand different proteins are now available. Although at first glance these patterns look unique, more careful analysis reveals that some unifying principles do exist. One of the most important of these is that certain combinations of secondary structures, called **supersecondary structures** or **folding motifs**, are present in many different proteins. Supersecondary structures are represented by a schematic topology diagram in which β-strands are represented by arrows pointing from the amino to the carboxyl terminus, α-helices by cylinders or helical structures, and loops and turns by ribbons. Some common supersecondary structures are shown in FIGURE 2.32.

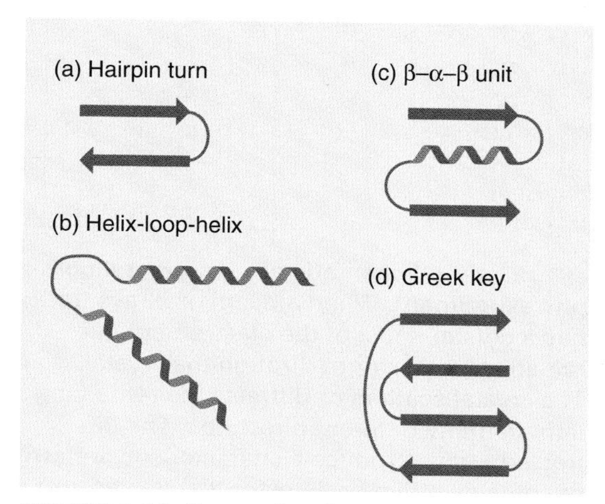

FIGURE 2.32 **A sample of supersecondary protein structures.**

We cannot yet predict secondary structures with absolute certainty.

Examination of amino acid side chain structures provides helpful insights into the way that a residue may fit into a secondary structure. For example, the methyl group in alanine fits well into an α-helix, whereas the larger isopropyl group in valine experiences some steric hindrance. Additional insights come from comparing folding patterns of many different polypeptides. A statistical analysis of data from such studies allows us to predict the likelihood that a particular residue will be in an α-helix or β-strand (FIGURE 2.33). However, residues do not always behave as expected. For example, studies by Daniel L. Minor, Jr. and Peter S. Kim in 1996 showed that an 11-amino acid sequence (Ala-Trp-Thr-Val-Glu-Lys-Ala-Phe-Lys-Thr-Phe) folds into an α-helix when in one position in the primary sequence of a polypeptide but as a β-sheet when in another position. Peptide sequences, thus, can form different secondary structures when placed in different protein contexts.

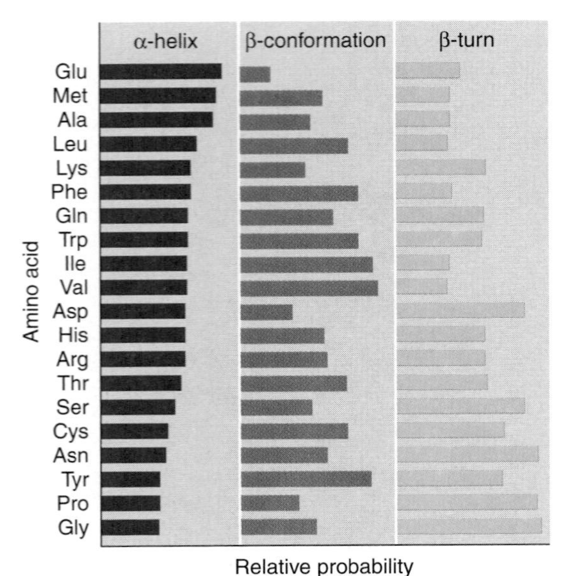

FIGURE 2.33 **Relative probabilities that a given amino acid will occur in the three common types of secondary structure.** (Adapted from D. C. Nelson and M. M. Cox. *Lehninger Principles of Biochemistry, Fourth edition.* W. H. Freeman & Company, 2004.)

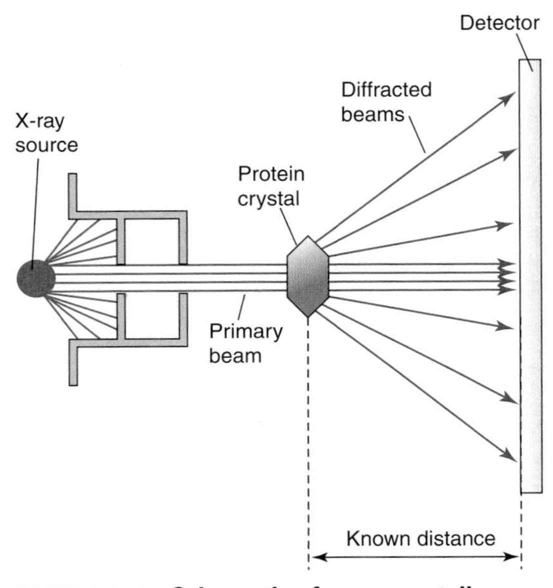

FIGURE 2.34 **Schematic of x-ray crystallography experiment.** When a beam of x-rays (red) hits a crystal, most of the electromagnetic radiation passes right through the crystal. The crystal scatters or diffracts the remaining light in many different directions. The diffracted x-rays produce a characteristic pattern of spots on a film or detector that is placed behind the crystal. Distances between spots and spot intensities provide the necessary information for determining protein structure. (Adapted from C. I. Branden and J. Tooze. *Introduction to Protein Structure, Second edition.* Garland Science, 1999.)

2.7 Tertiary Structure

X-ray crystallography and nuclear magnetic resonance studies have revealed the three-dimensional structures of many different proteins.

It is much more difficult to determine the three-dimensional structure of a globular protein that contains a combination of secondary structures than it is to determine the three-dimensional structure of a regular protein that is all α-helix or all β-conformation. Two physical methods, **x-ray crystallography** and **nuclear magnetic resonance (NMR) spectroscopy**, are used to elucidate the three-dimensional structure of proteins and nucleoproteins (protein-nucleic acid complexes) at atomic detail. A thorough examination of these two techniques is beyond the scope of this book, but a brief overview of how x-ray crystallography and NMR are used to study protein structure is provided.

X-Ray Crystallography

An ordinary optical instrument such as a light microscope does not permit us to see proteins in atomic detail because the distances between atoms are too small. In general, the wavelength required to resolve two objects (recognize the two objects as distinct entities) must be less than half the distance between the objects. Because distances between atoms linked by covalent bonds are about 0.15 nm, molecular resolution requires the very short wavelengths in x-rays. X-rays do not provide a direct image of protein molecules because currently available lenses and mirrors cannot focus such short wavelengths. X-ray crystallography allows interatomic distances in proteins to be measured by exploiting the fact that the electrons surrounding atoms in a crystalline protein scatter or diffract x-rays. The diffracted x-rays produce a characteristic pattern of spots on a film or detector that is placed behind the protein crystals (FIGURE 2.34). A heavy metal such as uranium is attached to a specific residue without altering protein structure to provide a reference point for data interpretation. Distances between spots and spot intensities provide the necessary information for determining protein structure.

The power of x-ray diffraction crystallography became apparent for all to see in 1957, when John Kendrew used the technique to determine the structure of myoglobin, an oxygen storage protein that is present in high concentrations in the muscles of diving mammals. A short time later, Max Perutz reported the three-dimensional structure of hemoglobin, the oxygen transport protein in red blood cells. Many improvements have been made in x-ray crystallography since the structures of myoglobin and hemoglobin were first reported. Three advances in particular are especially noteworthy: (1) faster computers facilitate data analysis; (2) high-intensity x-ray beams emanating from synchrotrons allow investigators to study protein crystals that are much smaller than those used in earlier studies of protein structure; and (3) synchrotron radiation at multiple wavelengths eliminates the need to attach heavy metals to specific sites in the protein. These advances have greatly accelerated the rate at which protein crystal structures can

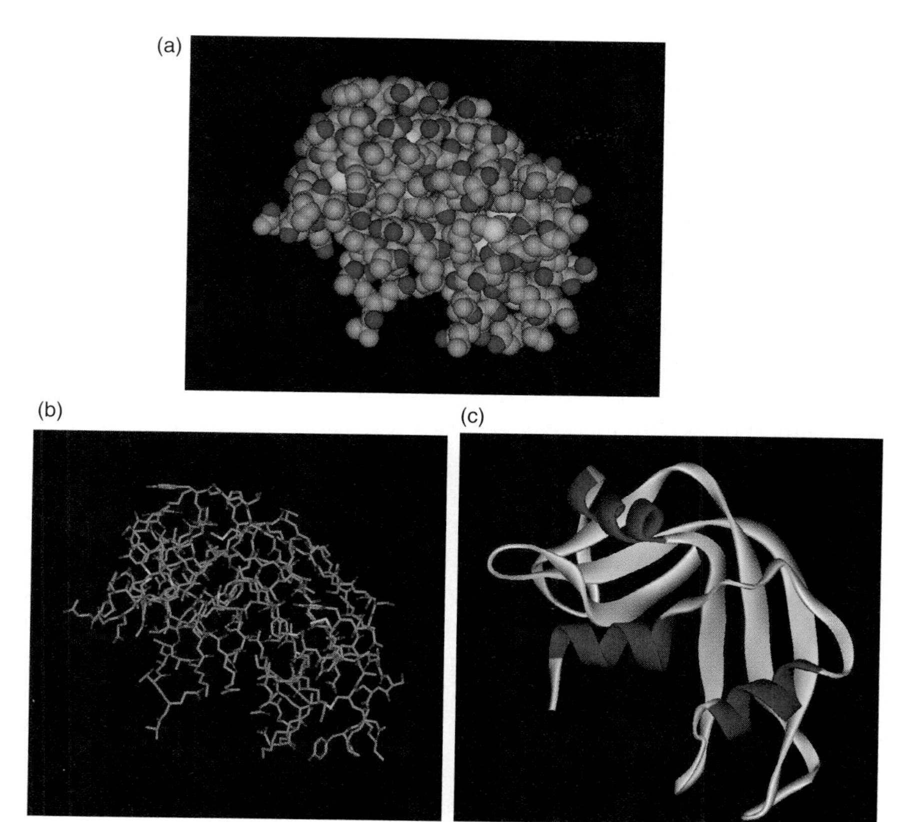

FIGURE 2.35 **Crystal structure of ribonuclease A.** (a) Crystal structure displayed in space filled form. The colors are in standard CPK (Corey, Pauling, Kultin) color scheme. Carbon is gray, hydrogen white, nitrogen blue, and sulfur yellow. (b) Crystal structure displayed in stick form. The standard CPK color scheme is used. The orientation is as in (a). (c) Crystal structure shown in ribbon form. α-helices are shown in red and β-conformations in yellow. (Structures from Protein Data Bank 1JVT. L. Vitagliano, et al., *Proteins* 46 [2002]: 97–104. Prepared by B. E. Tropp.)

be solved. Thanks to these technological advances, much less time is required to solve protein structure problems today. Recombinant DNA techniques have also played a critical role by allowing investigators to construct cells that produce large quantities of a desired protein.

Today, approximately 60,000 protein crystal structures are available at the Research Collaboratory for Structural Bioinformatics Protein Data Bank (RCSB PDB) and new structures are being added each day. The data are entered as a Protein Data Bank ("pdb") file. A program that serves as a viewer allows us to convert the pdb file into a three-dimensional image that can be manipulated on the computer. One excellent viewer, the Discovery Studio Visualizer 2.0® (from Accelrys), permits the protein structure to be displayed in different forms and colors. FIGURE 2.35 shows the crystal structure of pancreatic ribonuclease A (RNase A), an enzyme that digests ribonucleic acids, in a spacefill display, a stick display, and a ribbon display.

Despite its enormous power, x-ray crystallography does have some shortcomings. First and foremost, proteins that do not form crystals cannot be examined by this technique. It is usually difficult to obtain membrane proteins in a crystalline form. Most of the protein structures determined to date, therefore, are for water-soluble proteins. Second, protein molecules pack close together when they form a crystal. Sometimes this packing causes residues on the protein surface to assume positions that are slightly different from their position in solution. Third, many proteins have regions that are highly disordered, so the structures of these regions cannot be determined by x-ray crystallography.

Initially there was concern that protein crystal structures might differ from protein structures in aqueous solution. It is now clear, however, that protein crystals contain considerable water, which allows proteins to retain their biological activity. Furthermore, structures determined by NMR spectroscopy of proteins in solution are the same as the crystal structures.

Nuclear Magnetic Resonance Spectroscopy

Nuclear magnetic resonance (NMR) spectroscopy has been used to elucidate the structure of relatively small organic molecules since the 1950s. Thanks to the pioneering efforts of Kurt Wüthrich beginning in the 1980s, this physical technique can now also be used to determine the three-dimensional structure of a purified protein, provided that the protein has a molecular mass of less than 40 kDa and is sufficiently soluble in water so that a concentration of at least 1 mM can be achieved. As the technology improves, it is likely that NMR spectroscopy will be able to determine structures of larger proteins. NMR spectroscopy offers two major advantages: (1) proteins can be studied in solution in an environment similar to that in the living cell, and (2) the proteins do not need to form crystals.

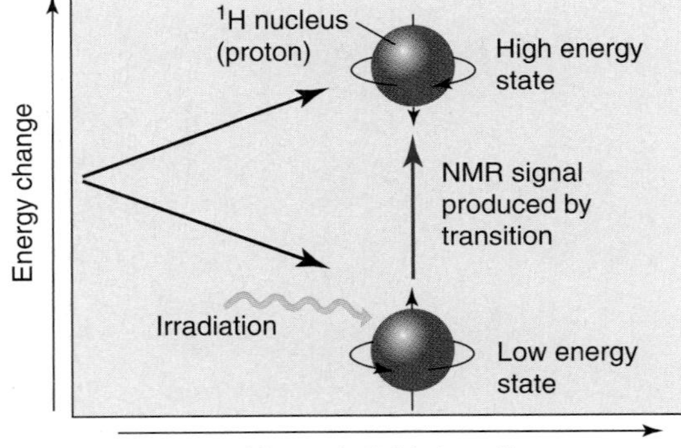

FIGURE 2.36 **Basis of NMR spectroscopy.** When a strong magnetic field is applied to proteins in solution, ^{1}H, ^{13}C, and ^{15}N nuclei line up parallel to the field (low energy) or antiparallel to the field (high energy). The energy difference between the two orientations is comparatively small so that the low energy state is populated by only slightly more protons than the high energy state. A pulse of electromagnetic radiation with an energy that exactly matches the energy difference between the two spin states can raise the lower energy state to the higher energy state. (Adapted from J. M. Berg, et al. *Biochemistry, Fifth edition.* W. H. Freeman and Company, 2002.)

NMR takes advantage of the fact that certain atomic nuclei such as ^{1}H, ^{13}C, and ^{15}N are intrinsically magnetic and display a property called magnetic spin that results in the generation of a magnetic dipole. In essence, ^{1}H, ^{13}C, and ^{15}N nuclei act as tiny magnets. The hydrogen nucleus (^{1}H) is especially important in NMR spectroscopy structural studies of proteins because ^{1}H atoms are distributed throughout naturally occurring proteins. C-13 and N-15 isotopes have to be introduced by culturing cells in medium containing nutrients enriched for these isotopes. When a strong magnetic field is applied to proteins in solution, the ^{1}H nuclei line up parallel to the field (low energy) or antiparallel to the field (high energy) (FIGURE 2.36). The energy difference between these two orientations is comparatively small, so that the low energy state is populated by only slightly more protons than the high energy state. Because the solution contains a large population of identical protein molecules the effect is amplified. A spinning ^{1}H nucleus in the lower energy state can be raised to the higher energy state by applying a pulse of electromagnetic radiation in the radio-frequency range with an energy that exactly matches the energy difference between the two spin states. This radio-frequency is said to be in **resonance** with the proton when its energy exactly matches that required to convert the lower energy spin state to the higher energy spin state. The transition energy required to induce a spin transition for a specific ^{1}H nucleus is influenced by the spin of protons that are covalently connected by only one or two other atoms. This type of interaction is called "interaction through bonds" (FIGURE 2.37a). The transition energy of a proton in a peptide is also influenced by protons that are located several hundred bonds away, provided that the peptide folds so that the interacting protons are within 0.5 nm of one another. This type of interaction is called interaction through distance (FIGURE 2.37b). Data obtained from NMR spectroscopy allow investigators to estimate the distances between specific pairs of distant atoms. The resulting set of distances, together with the amino acid sequence and known geometric constraints such as bond angles and distances, group planarity, stereoconfiguration, and van der Waals radii, are used to compute the protein's three-dimensional structure. Because distances between proton pairs are not precise, one obtains an ensemble of very similar structures rather than a single structure. The structure obtained for ribonuclease A by NMR spectroscopy, shown in FIGURE 2.38, is remarkably similar to the crystalline structure for this protein (Figure 2.35a). At present, the Protein Data Bank contains approximately 8500 structures that were determined by NMR.

Until recently virtually all protein structures were determined by studying highly purified proteins outside the cell. This limitation raised a question about whether a structure determined using isolated pure proteins would be the same as that inside the cell where there are many other kinds of proteins and thousands of different biomolecules. Investigators have now devised techniques that permit NMR spectroscopy to be used to determine protein structures inside the cell. The studies by Yutaka Ito and coworkers are particularly noteworthy. They examined the structure of a heavy metal binding protein normally produced by the bacterium *Thermus thermophilus*. They began by introducing the gene that codes for this protein into *Escherichia coli*. Then they placed the *E. coli* into a growth medium that contained [^{13}C]glucose, [^{15}N]NH_4Cl, and [^{2}H]H_2O under conditions in which the bacteria were forced to make large amounts of heavy isotope labeled metal binding protein. Finally they subjected a concentrated sample of the intact bacteria to NMR spectroscopy to determine the labeled protein's structure. Ito and coworkers did not need to be concerned about normal *E. coli* proteins because these proteins were not labeled with heavy isotopes. Comparison of *in vitro* and *in vivo* structures revealed marked similarities. Some structural differences were observed in the heavy metal binding site and in loops that change as the protein performs its functions, however. The former difference may have been due to metal ions inside the bacteria, but the latter difference probably reflects the crowding conditions of proteins inside the cell. These studies show that while a protein structure obtained *in vitro* provides valuable information, still more may be learned by examining a protein in its natural environment.

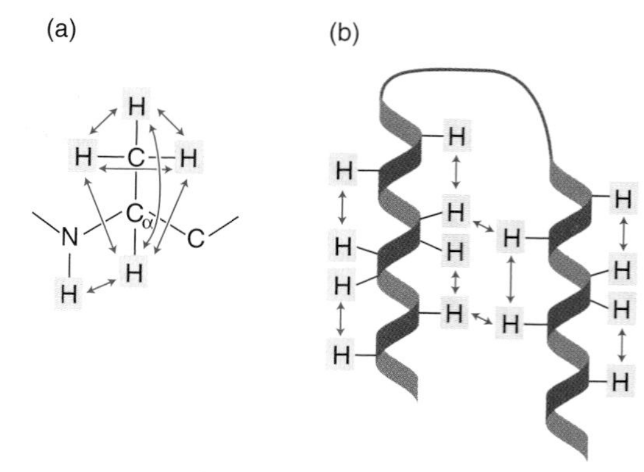

FIGURE 2.37 **Two types of NMR information.** The transition energy required to induce a spin transition for a specific ^{1}H nucleus is influenced by (a) the spin of protons that are covalently connected by only one or two other atoms (interaction through bonds) and (b) protons that are located several hundred bonds away provided that the polypeptide folds so that the interacting protons are within 0.5 nm of one another (interaction through distance). (Adapted from C. I. Branden and J. Tooze. *Introduction to Protein Structure, Second edition.* Garland Science, 1999.)

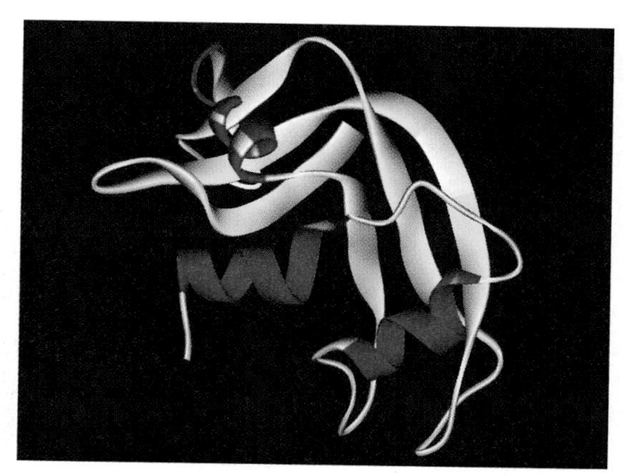

FIGURE 2.38 **High-resolution three-dimensional structure of ribonuclease A in solution by nuclear magnetic resonance spectroscopy.** (Structure from Protein Data Bank 2AAS. J. Santoro, et al., *J. Mol. Biol.* 229 [1993]: 722–734. Prepared by B. E. Tropp.)

Intrinsically disordered proteins lack an ordered structure under physiological conditions.

Not all proteins fold into ordered structures under physiological conditions. Those that do not, the **intrinsically disordered proteins**, are fairly common in eukaryotes, accounting for about 30% of the protein population. There are notable differences between the amino compositions of intrinsically disordered proteins and globular proteins. Intrinsically disordered proteins have fewer hydrophobic residues than globular residues. The hydrophobic residues, which help to stabilize globular protein structure by forming a hydrophobic core, are not available to provide the same stability to intrinsically disordered proteins. Furthermore, intrinsically disordered proteins are richer in polar residues that interact with water and in proline and glycine residues that promote disorder. Intrinsically disordered proteins contribute to cell maintenance and viability by participating in molecular recognition, assisting in molecular assembly, playing a role in protein modification, and helping to fold RNA. Some intrinsically disordered proteins assume stable three-dimensional structures upon interacting with other proteins, nucleic acids, or specific biological molecules. When an ordered complex forms, its structure can be determined by x-ray crystallography.

Structural genomics is a field devoted to solving x-ray and NMR structures in a high throughput manner.

Greatly encouraged by the success of genome sequencing projects and advances in protein structure determination methods, structural biologists met in the late 1990s to discuss ways that they might cooperate to begin a systematic worldwide effort to determine the three dimensional structure of proteins in a high throughput manner. A formal agreement in 2001 helped to launch an international effort in **structural genomics** (also called **structural proteomics**) dedicated to the large scale determination of protein structure. Government and private funding agencies in many countries helped to create major structural genomics facilities. Data and information are shared over the Internet. For example, the TargetDB, a target registration database for structural genomics projects, provides a list of proteins under study (target proteins) and the progress that has been made. The immediate goals of the structural genomics effort are as follows:

1. Organize known protein sequences into groups of evolutionarily related proteins—Members of each grouping, termed a **protein family**, share a common folding pattern and tend to have related functions. For instance, the animal proteins hemoglobin and myoglobin and the plant protein leghemoglobin are oxygen-binding proteins that have the same globin fold (see Chapter 3). Investigators plan to determine three-dimensional structures of representative proteins in each family.
2. Study proteins associated with disease states—Research centers are targeting various human pathogens such as *Mycobacterium tuberculosis*, the bacterium that causes tuberculosis. The functions of many of the proteins produced by pathogenic microorganisms and viruses are not known. The expectation is that structural information will provide clues to function, especially if the structure resembles that of a protein from another organism with a known function. In addition, establishing the

structures of proteins produced by pathogens will result in a better understanding of the disease process and lead to new methods of disease treatment and prevention. An example will help to illustrate the point. A knowledge of the structure of the protease produced by cells infected with the human immunodeficiency virus (HIV; see Chapter 8) led to the design of protease inhibitors. These inhibitors, in combination with other anti-retroviral drugs, have saved the lives of millions of individuals who might otherwise have died from acquired immune deficiency syndrome (AIDS).

3. Establish the structures of proteins that participate in a specific biochemical pathway—Considerable progress has already been made in determining the structures of enzymes that participate in DNA metabolism (replication, repair, and recombination), RNA synthesis, and protein synthesis. Nevertheless, structures are still not available for all of the proteins that are involved in these and other biochemical pathways. Establishing the structures of all the proteins in a biochemical pathway will add to our understanding of the pathway and the mechanisms that control it and may also lead to new drugs for treating various diseases such as cancer and diabetes.
4. Determine the structures of new folding patterns—The long range goal of structural proteomics is to determine three-dimensional structures of all proteins. Genome sequencing allows us to predict amino sequences for several hundred thousand proteins, many of which have not been isolated or studied. Because it costs about $100,000 to determine a protein structure, determining the structures of all these proteins would be prohibitively expensive, to say nothing of the time required to accomplish the task. In principle, it may be possible to predict the three-dimensional structure of a protein of known sequence. One approach is homology-based or comparative protein modeling. One begins by comparing the protein of interest (the query protein) to a protein already in the Protein Data Bank (the template protein). Various scoring systems are available to score for similarities. High scoring proteins can be used as templates for structural modeling. The chance of a successful structural prediction is greatest when query and template proteins are very similar. There is only a small probability that the structural prediction will be correct if there is less than a 30% amino acid identity between the query and template proteins. Information provided by structural proteomics should greatly assist in protein structure predictions by expanding the Protein Data Bank so that there are more potential template proteins and by adding proteins with new folding patterns. Information about structural genomics projects is available from the Structural Genomics Knowledge Base.

Although the efforts of the structural genomics research facilities will greatly increase our knowledge of protein structure, there will still be a need for investigators to examine the structure of proteins as they actually perform their biological function. Thomas A. Steitz, one of the leading authorities in the field of structural biology has explained why additional information is needed as follows: "In order for structural studies to provide understanding of a biological process, one must know the structures of the entire assembly that executes that process, captured at each step in the process. The structures of the pieces of a clock do not inform on how a clock works and even the structure of the whole clock—in a single state—does not show how it functions." We will examine protein structure–function relationships in more detail in Chapter 3.

The primary structure of a polypeptide determines its tertiary structure.

Heat and certain chemical agents such as acids, detergents, and urea (NH_2CONH_2) cause proteins to lose their biological activity by disrupting the weak noncovalent bonds that stabilize secondary and tertiary structures. Before the weak noncovalent bonds are disrupted, the protein is said to be in its **native state**. After disruption, the protein is said to be in a **denatured state**, existing as a mixture of **random conformations**. Hydrophobic segments that are normally buried in the core of water-soluble proteins become exposed after denaturation and bind to hydrophobic segments of other denatured proteins to form water-insoluble aggregates. Protein denaturation is a rather common phenomenon. For example, milk spoils as a result of bacterial growth that produces acidic waste products, which cause the milk to have a sour taste and curdle. The curd is composed of denatured milk proteins that have become water-insoluble.

Some polypeptides can refold to their native state after denaturation. Studies of this process known as **renaturation** have provided a great deal of information about the folding process. In the mid-1950s Christian Anfinsen and colleagues selected bovine pancreatic ribonuclease A (RNase A) as a model protein for studying *in vitro* renaturation (FIGURE 2.39). Eight of the 124 residues in the polypeptide chain are cysteines that pair to form four disulfide bonds. These disulfide bonds help to lock the tertiary structure into place, making it very difficult to denature the enzyme. RNase A can be denatured by simultaneous treatment with urea and β-mercaptoethanol (CH_2-$OHCH_2SH$). Neither chemical agent by itself is sufficient to denature RNase A. Urea interferes with hydrophobic interactions and disrupts hydrogen bonds, allowing the β-mercaptoethanol to gain access to the disulfide bonds. β-Mercaptoethanol disrupts the disulfide bonds, permitting further urea denaturation. If urea and the mercaptoethanol are slowly removed, perfect renaturation occurs, including formation of the four correct disulfide bonds (disulfide bonds can form spontaneously by oxidation in air). This latter finding is remarkable because there are 105 possible ways that eight cysteine residues can combine to form four disulfide bonds. Anfinsen's interpretation of this experiment was that the folding of RNase A is determined exclusively by its amino acid sequence and that the proper disulfide bonds are formed because, during folding, the cysteine residues are correctly placed for

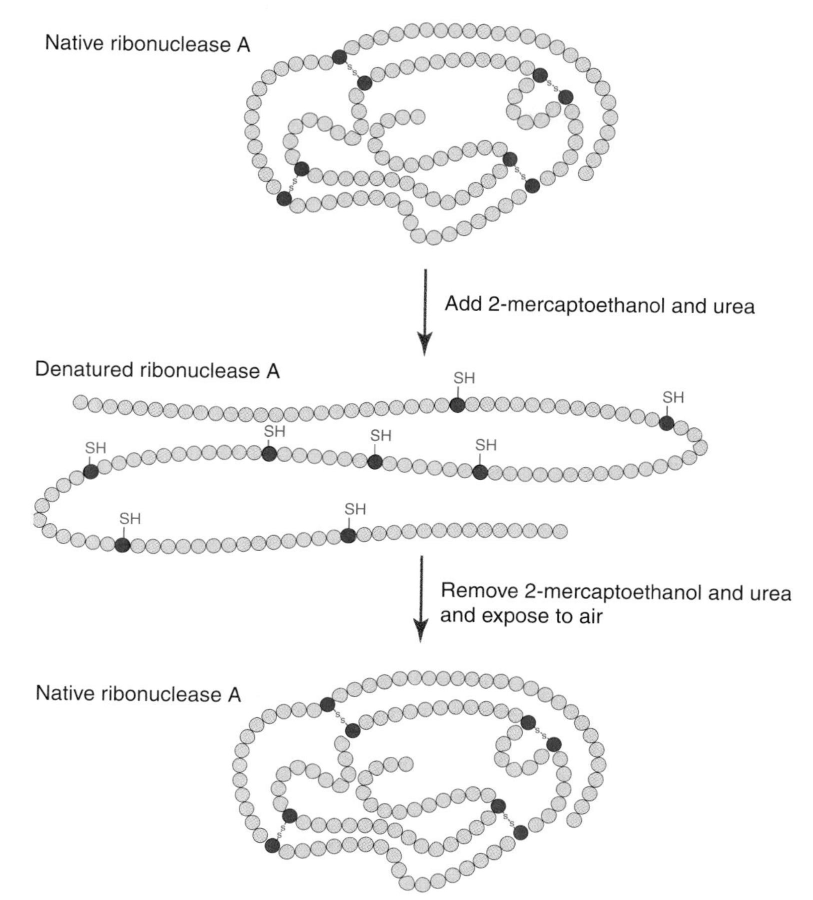

FIGURE 2.39 **Denaturation and renaturation of RNase A.** Denature pancreatic RNase A with urea in the presence of 2-mercaptoethanol. Renature enzyme by dialyzing denatured RNase to remove the urea and 2-mercaptoethanol, and then exposing the polypeptide chain to air to re-form disulfide bonds. (Adapted from L. A. Moran and K. G. Scrimgeour. *Biochemistry, Second edition.* Prentice Hall, 1994.)

joining. Evidence that disulfide bond formation does not direct the folding comes from an experiment in which the β-mercaptoethanol was removed first and oxidation was allowed to occur prior to removal of the urea, that is, while the RNase A was a random coil. With this protocol, the native molecule was not formed.

The Anfinsen experiment shows that, at least for ribonuclease A, the tertiary structure is determined by the amino acid sequence and is the one with the lowest energy. Similar observations have been made for many other proteins, but not for all. The fact that denatured RNase can fold to form its native structure is remarkable in view of the number of possible conformations. If we assume that each residue has ten possible conformations available to it, then the 124 residues would have to sample 10^{124} different conformations before achieving the correct one. If the polypeptide chain had to sample each possible conformation, the folding process could not possibly take place during the organism's lifetime. This problem, first recognized by Cyrus Levinthal, can only be solved if the polypeptide does not need to sample all possible conformations.

Many models have been proposed in an attempt to explain how the folding problem is solved. One recent proposal, the **zipping and assembly model**, incorporates features from a number of earlier models. According to this model, folding begins when small segments along a polypeptide chain fold into secondary structures that are sufficiently stable to survive for a short time. This segmental folding takes place in a pico- or nanosecond time frame with each peptide segment searching for its own conformation independently of other segments. The order of segment folding, therefore, differs from one polypeptide chain to another. Although not stable enough to retain their conformations on their own, some of the segment structures survive long enough to grow (zip) into more stable structures or coalesce (assemble) with other structures.

In addition to its great theoretical significance, the Anfinsen experiment also had practical applications. At the time of the Anfinsen experiment, no one had yet managed to synthesize a polypeptide chain as long as ribonuclease. Some investigators questioned whether it would be worth the effort, because a synthetic polypeptide might not be able to fold into a biologically active form outside of the cell. The Anfinsen experiment showed that polypeptides prepared in the laboratory would have a reasonable chance of folding into biologically active proteins, stimulating efforts to synthesize long polypeptides. Although it is now possible to synthesize polypeptides in the laboratory by standard organic chemical techniques, the more common practice is to isolate or synthesize DNA that codes for the desired protein and then introduce that DNA into a living cell so that the cell synthesizes the protein.

Molecular chaperones help proteins to fold inside the cell.

Protein folding is a much more complex process inside the living cell, which has high concentrations of proteins and other macromolecules. Instead of interacting with one another to form the correct tertiary structure, hydrophobic segments in an unfolded polypeptide may interact with hydrophobic segments in other unfolded polypeptide chains to form biologically inactive aggregates. The tendency to form aggregates is increased by the fact that stable tertiary structure formation usually requires a complete folding unit (100–300 amino acid residues). The complete folding unit, however, does not emerge from the protein synthetic factory, the ribosome, in a single discrete step. Instead, the folding unit slowly emerges from the ribosome as each succeeding amino acid is added to the growing end of the polypeptide chain. Correct folding takes place only after the entire folding unit has emerged from the ribosome. Exposure of hydrophobic segments on the growing polypeptide would make them susceptible to intermolecular aggregation were it not for the presence of specific proteins called **molecular chaperones** that help to stabilize the emerging polypeptide until the entire folding unit has been extruded by the ribosome. The chaperones assist the polypeptide to fold as it emerges

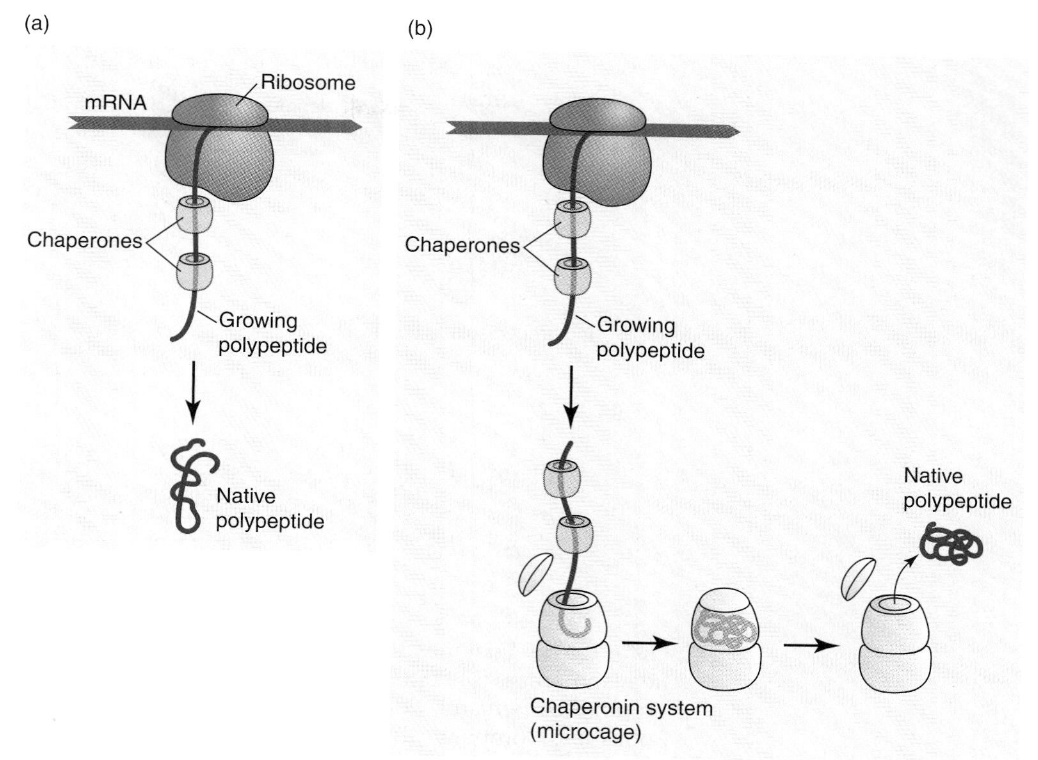

FIGURE 2.40 **Protein folding in the cell.** Chaperones assist polypeptides to fold as the polypeptides emerge from the ribosome. Chaperones bind to the hydrophobic segments as they emerge from the ribosome and prevent the hydrophobic segment from interacting with hydrophobic segments from other unfolded polypeptides. The chaperones are released after the entire folding unit has been extruded, freeing the polypeptide chain to (a) fold into its native state or (b) enter a microcage called a chaperonin system in which the polypeptide can fold without interference from other polypeptides. (Adapted from J. C. Young, et al., *Nat. Rev. Mol. Cell Biol.* 5 [2004]: 781–791.)

from the ribosome (FIGURE 2.40). Some chaperones work by binding to hydrophobic segments as they emerge from the ribosome, which prevents the hydrophobic segment from interacting with hydrophobic segments from other unfolded polypeptides. Molecular chaperones are released after the entire folding unit has been extruded freeing the polypeptide chain to fold into its native state or to enter a microcage called a **chaperonin system** in which the polypeptide is free to fold without making contact with other unfolded polypeptides, thereby preventing aggregation.

Two enzymes, **protein disulfide isomerase** and **peptidylprolyl *cis-trans* isomerase**, speed the polypeptide folding process in cells (FIGURE 2.41). Protein disulfide isomerases catalyze thiol/disulfide interchange, facilitating the formation of the correct set of disulfide bonds by reshuffling disulfide bonds when incorrect pairings are formed. Peptidylprolyl cis-trans isomerase catalyzes the isomerization of amino acid-proline peptide bonds, accelerating the refolding of polypeptide chains that contain proline.

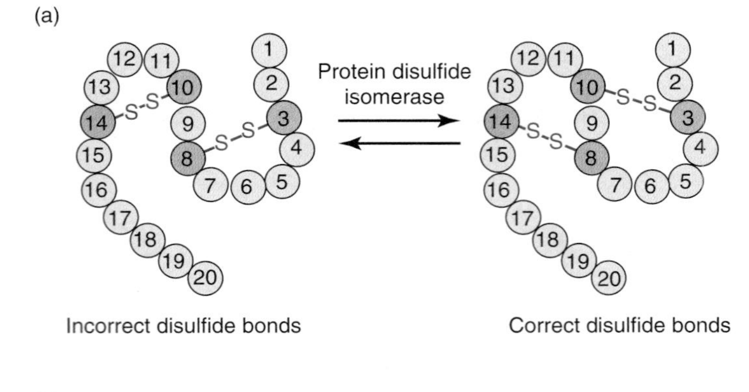

(b)

Peptidylprolyl *cis-trans* isomerase

trans

cis

FIGURE 2.41 Enzymes that assist polypeptide folding. (a) Protein disulfide isomerase. (b) Peptidylprolyl *cis-trans* isomerase. (Adapted from B. E. Tropp. *Biochemistry: Concepts and Applications, First edition.* Brooks/Cole Publishing Company, 1997.)

2.8 Proteins and Biological Membranes

Proteins interact with lipids in biological membranes.

Until this point, we have been considering only those proteins that are water-soluble. Many important proteins are present in biological membranes, however. The remainder of this chapter briefly examines the way that these proteins are organized in membranes, which contain lipids as their second major component. A **lipid** is defined as a biological molecule that is soluble in an organic solvent such as chloroform but not in water. The most notable feature of nearly all biological membranes is that they consist of two layers, called a **bilayer**. This structure is a consequence of the chemical nature of the lipids that form the bilayer. Three major lipid families, the glycerophospholipids, the sphingolipids, and the sterols are commonly found in biological membranes. We focus on the glycerophospholipids (FIGURE 2.42), which are glycerol-3-phosphate derivatives, because they are the only lipids present in all biological membranes. Members of this family are named as derivatives of phosphatidate, a lipid that is usually present in membranes in only trace amounts.

When placed in aqueous solution, glycerophospholipids aggregate to form bilayers. An examination of phosphatidylcholine structure shows why this is so (FIGURE 2.43). Each phosphatidylcholine molecule has a polar head group, phosphocholine, and two nonpolar hydrocarbon tail groups (Figure 2.43a). The two tails usually contain different

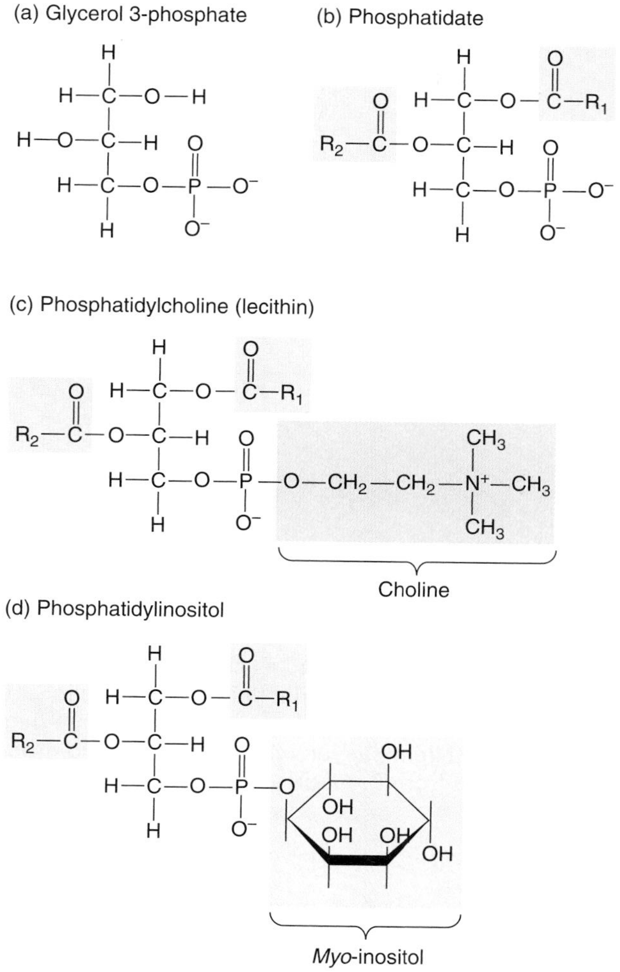

(a) Glycerol 3-phosphate

(b) Phosphatidate

(c) Phosphatidylcholine (lecithin)

Choline

(d) Phosphatidylinositol

Myo-inositol

FIGURE 2.42 **Some glycerophospholipids.** (a) Glycerol-3-phosphate is the building block for glycerophospholipids. (b) Phosphatidate is usually present in membranes in only trace amounts but other glycerophospholipids are named as its derivatives. (c) Phosphatidylcholine (lecithin) is present in eukaryotic membranes but not always in bacterial membranes. (d) Phosphatidylinositol is a membrane lipid that is converted into bioactive molecules.

numbers of carbon atoms (range, 14–24) and hence have different lengths. One tail typically has one or more *cis* double bonds, each of which causes a kink in the chain. At low concentrations, cylindrically shaped phosphatidylcholine molecules form a sheet that is one molecule thick called a **lipid monolayer** on the surface of an aqueous solution (Figure 2.43b). For convenience glycerophospholipids are represented as stick figures with hydrocarbon chains drawn as zigzag lines and polar head groups as circles. Hydrocarbon groups stick up in the air and polar head groups interact with water. This orientation avoids an unfavorable entropy situation in which water molecules would be required to form highly ordered cages around the hydrocarbon chains. The lipid monolayer gains additional stability from van der Waals

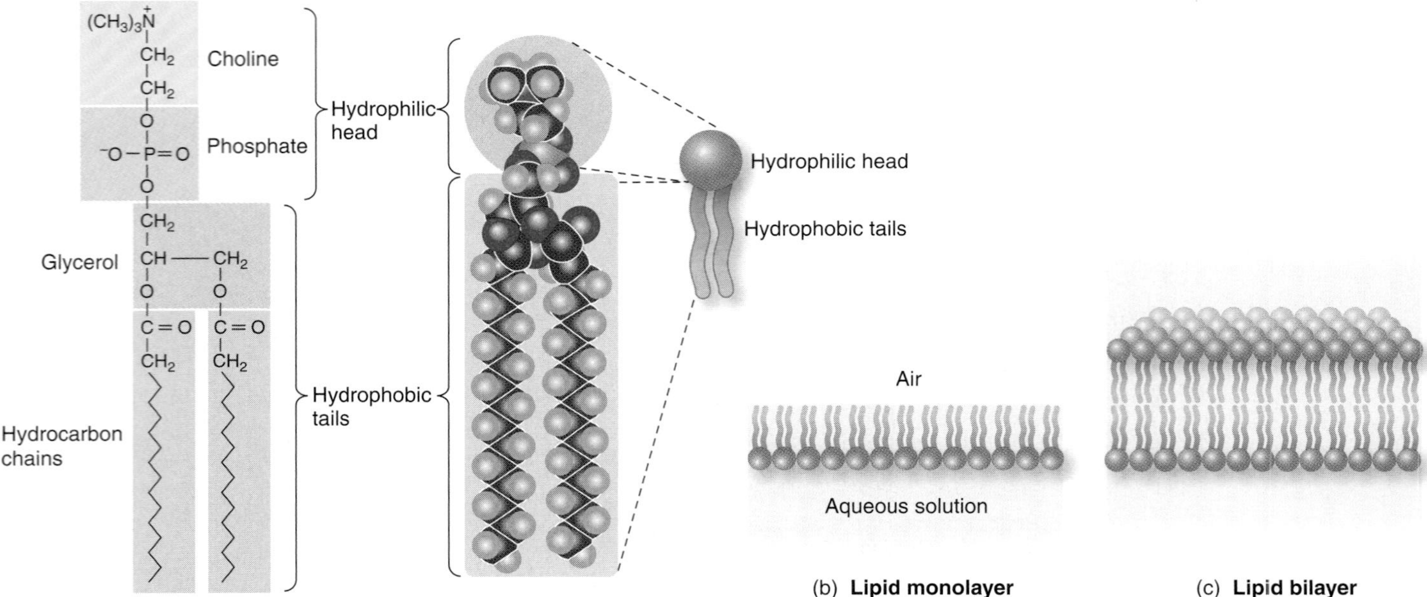

FIGURE 2.43 **Lipid monolayer and lipid bilayer.** (a) Glycerophospholipids are represented as stick figures with polar head groups shown as circles and hydrocarbon chains as zigzag lines. (b) When glycerophospholipids are placed in water, the polar head groups interact with water and the hydrocarbon tails stick up into the air to form a monolayer. (c) When there is no more room on the water's surface, glycerophospholipids tend to form a lipid bilayer with their polar head groups directed toward the water and their hydrophobic tails sequestered inside. (Adapted from B. Alberts, et al. *Molecular Biology of the Cell, Fourth edition.* Garland Science, 2002.)

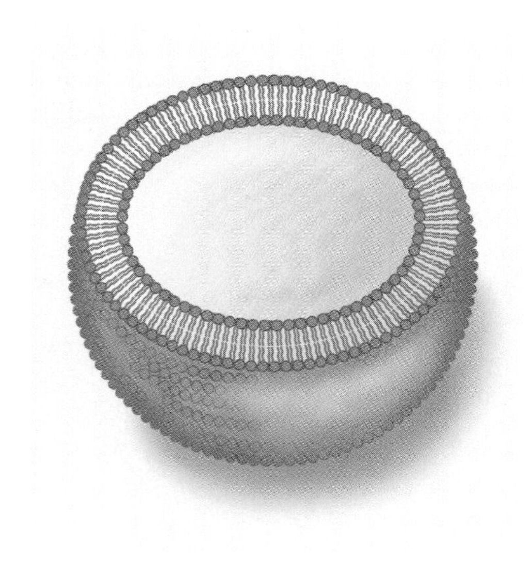

FIGURE 2.44 **Lipid bilayer vesicle.** Glycerophospholipids can form spherical lipid bilayer vesicles called liposomes. The hydrocarbon chains are sequestered within the bilayer and the polar head groups face the aqueous solution on either side of the membrane. The structure, which is actually spherical, is shown here in cross section.

interactions among hydrocarbon chains. When there is no more room on the surface of the aqueous solution, glycerophospholipids form a **lipid bilayer** with the hydrocarbon chains sequestered inside and the polar head groups directed toward the surrounding aqueous solution (Figure 2.43c). The entropy of water also provides the driving force for bilayer formation.

A variety of physicochemical studies of artificial membranes show that the glycerophospholipids in a membrane drift laterally, indicating that the membrane is fluid. However, glycerophospholipids seldom, if ever, move from one leaflet of the bilayer to the other. Such transverse or flip-flop movement is energetically unfavorable because it requires the glycerophospholipid's polar head group to pass through the bilayer's hydrophobic core. Biological membranes have proteins called flippases that catalyze the movement of glycerophospholipid molecules from one leaflet to the other. The ends of a lipid bilayer are unstable because the hydrocarbon chains are exposed to water. Lipid bilayers, thus, tend to close upon themselves to form hollow bilayer spheres known as **vesicles** or **liposomes** (FIGURE 2.44).

The fluid mosaic model has been proposed to explain the structure of biological membranes.

Naturally occurring biological membranes contain many protein molecules (varying from about 30%–70% by weight), and membranes having different functions contain different proteins. There are three kinds of membrane proteins.

1. **Integral membrane proteins** can only be removed from the membrane by harsh treatments such as extraction with detergents, organic solvents, or protein denaturing agents that destroy the membrane. Even when separated from the membrane, integral membrane proteins often retain bound lipids, suggesting that a similar association exists in the native membrane. When the lipids are removed, integral membrane proteins often form insoluble aggregates and lose their activity. Integral membrane proteins usually account for over 70% of the total membrane proteins. Some integral membrane proteins have a single region that is embedded in the membrane whereas others have two or more embedded regions. The most common type of secondary structure of embedded segments is the α-helix. However, some bacterial integral membrane proteins in the bacterial outer membrane have embedded segments with a β-sheet secondary structure.
2. **Peripheral membrane proteins** can be extracted from membranes by relatively mild treatments such as increasing the salt concentration or adding chelating agents. The former disrupts ionic bonds that link the proteins to the membrane, and the latter removes calcium and magnesium ions that serve as salt bridges between the proteins and the membrane. Once dissociated, peripheral membrane proteins are free of lipids and relatively soluble in aqueous solutions. These properties suggest that peripheral membrane proteins are held to the membrane by weak noncovalent bonds and are not strongly associated with membrane lipids.
3. **Lipoproteins** have a lipid group that is covalently attached to one of the amino acid residues. The hydrophobic lipid group is inserted into one of the monolayers in the membrane bilayer and anchors the lipoprotein to the membrane. The lipoproteins, like the integral membrane proteins, can only be extracted by detergents, organic solvents, or protein denaturing agents that destroy the membrane.

An experiment performed by L. D. Frye and Michael Edidin in 1970 demonstrated that many integral membrane proteins move randomly and freely. As shown in FIGURE 2.45, this experiment made use of a cell fusion technique in which human and mouse cells were stimulated to fuse by infection with Sendai virus inactivated by ultraviolet light. Antibodies were prepared against the integral membrane proteins of human cells and mouse cells. The antibody to mouse protein was covalently linked to rhodamine, a molecule that fluoresces red when activated with the appropriate wavelength of light. The human antibody was linked to fluorescein, which fluoresces green. Immediately after fusion, the two fluorescent antibodies were added and the joined cells (**heterokaryons**) were observed with a fluorescence microscope. Initially each heterokaryon had a red half and a green half. Observation at various times showed a gradual mixing of the colors. After

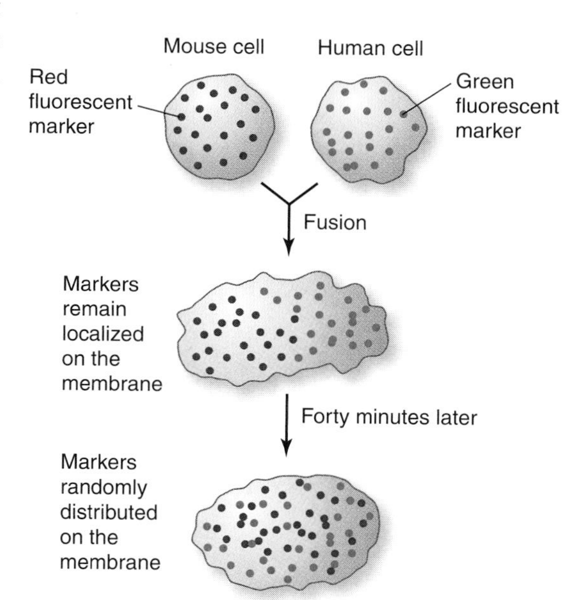

FIGURE 2.45 **Diffusion of integral membrane proteins in fused cells.** Mouse and human cells can be induced to fuse to form a hybrid cell. The experiment clearly indicates that many membrane proteins are able to move in the plane of the membrane. (Adapted from L. A. Moran and K. G. Scrimgeour. *Biochemistry, Second edition.* Prentice Hall, 1994.)

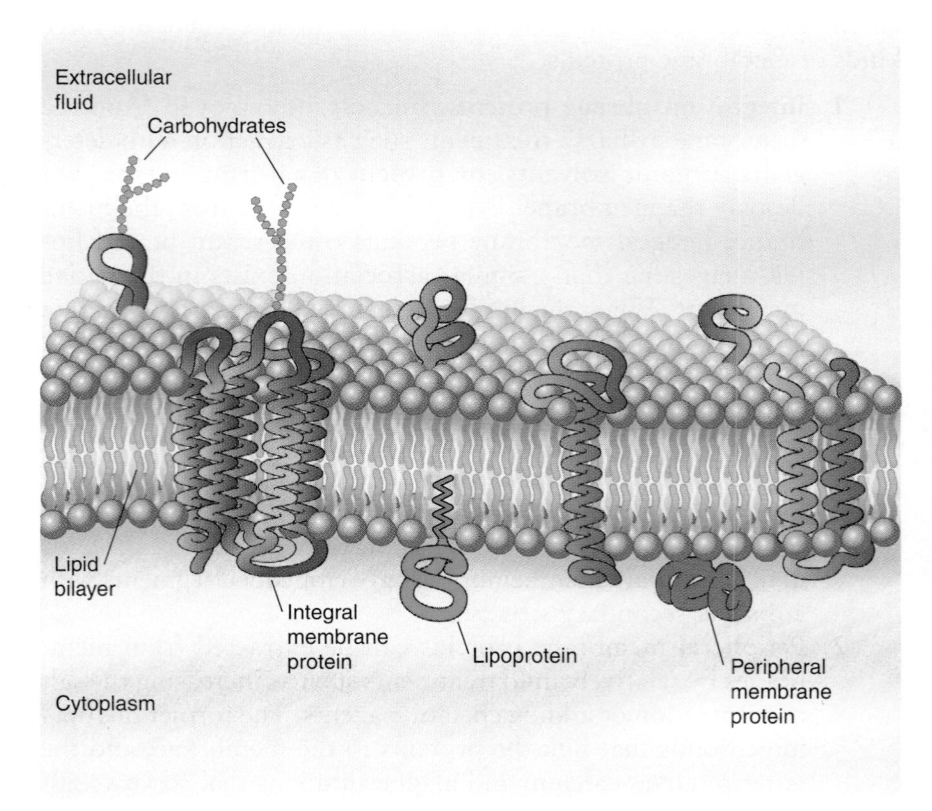

FIGURE 2.46 **Fluid mosaic model for the cell membrane.** Integral membrane proteins are embedded in the lipid bilayer. The embedded regions have amino acids with hydrophobic side chains, whereas regions that are exposed on either side of the bilayer tend to have amino acids with hydrophilic side chains. Lipoproteins have attached lipid groups such as fatty acids or glycolipids that are embedded in the bilayer. Peripheral membrane proteins associate with integral membrane proteins, lipoproteins, or lipids through weak, non-covalent interactions. (Adapted from G. M. Cooper and R. E. Hausman. *The Cell: A Molecular Approach, Second edition.* Sinauer Associates, Inc., 2000.)

40 minutes at 37°C, both red and green fluorescence covered the cell surface, indicating that the human and mouse integral membrane proteins were completely and randomly mixed.

In 1972, Jonathan S. Singer and Garth Nicholson proposed the fluid mosaic model to explain how proteins are organized in biological membranes (FIGURE 2.46). According to this model, lipids and integral membrane proteins are free to move laterally within a monolayer but not to move across the membrane in a transverse fashion (from one monolayer to another) known as flip-flop. Flip-flop motion does not occur because the polar regions of proteins and lipids cannot move through the hydrophobic core of the lipid bilayer without the assistance of special proteins. All copies of an integral membrane protein have the same amino-to-carboxyl orientation within the lipid bilayer. The two sides of a biological membrane, therefore, must be different. That is, a biological membrane is an asymmetric structure having (taking the cell as reference) an inside and an outside. Membrane asymmetry also extends to the lipid bilayer so that each monolayer has a unique lipid composition.

Although quite brief, this description of biological membranes allows us to examine molecular biology problems as they relate to biological membranes.

Suggested Reading

General Overview

Branden, C-I., and Tooze, J. 1999. *Introduction to Protein Structure* (2nd ed). New York: Garland.

Creighton, T. E. 1993. *Proteins: Structures and Molecular Properties.* New York: W. H. Freeman.

Pauling, L. 1993. How my interest in proteins developed. *Protein Sci* 2:1060–1063.

Petsko, G. A., and Ringe, D. 2003. *Protein Stucture and Function.* Sunderland, CT: Sinauer Associates.

Amino Acids and Peptide Bonds

Cozzone, A. J. 2010. Proteins: Fundamental chemical properties. *Encyclopedia of Life Sciences.* pp. 1–10. Hoboken, NJ: John Wiley & Sons.

Deber, C. M., and Brodsky, B. 2001. Proline residues in proteins. *Encyclopedia of Life Sciences.* pp. 1–6. London: Nature.

Dwyer, D. S. 2008. Chemical properties of amino acids. *Wiley Encyclopedia of Chemical Biology.* pp. 1–11. Hoboken, NJ: John Wiley & Sons.

Gilbert, H. F. 2001. Peptide bonds, disulfide bonds and properties of small peptides. *Encyclopedia of Life Sciences.* pp. 1–7. London: Nature.

Protein Purification

Deutscher, M. (ed). 1997. *Guide to Protein Purification.* San Diego: Academic Press.

Fassina, G. 2001. Affinity chromatography. *Encyclopedia of Life Sciences.* pp. 1–3. London: Nature.

Jennissen, H. P. 2001. Hydrophobic interaction chromatography. *Encyclopedia of Life Sciences.* pp. 1–8. London: Nature.

Mansoor, M. A. 2002. Liquid chromatography. *Encyclopedia of Life Sciences.* pp. 1–3. London: Nature.

Roe, S. (ed). 2001. *Protein Purification Techniques. A Practical Approach* (2nd ed). Oxford, UK: Oxford University Press.

Primary Structure

Aebersold, R., and Mann, M. 2003. Mass spectrometry-based proteomics. *Nature* 422:198–207.

Barker, W. C. 2001. Protein sequence databases. *Encyclopedia of Life Sciences.* pp. 1–3. London: Nature.

Smith, J. B. 2001. Peptide sequencing by Edman degradation. *Encyclopedia of Life Sciences* pp. 1–3. Hoboken, NJ: John Wiley & Sons.

Steen, H., and Mann, M. 2004. The abc's (and xyz's) of peptide sequencing. *Nat Rev Mol Cell Biol* 5:699–711.

Yates, J. R. III. 1998. Mass spectrometry and the age of the proteome. *J Mass Spect* 33:1–19.

Weak Noncovalent Bonds

Chan, H. S. 2002. Amino acid side-chain hydrophobicity. *Encyclopedia of Life Sciences.* pp. 1–7. London: Nature.

Fersht, A. R. 1987. The hydrogen bond in molecular recognition. *Trend Biochem Sci* 12:301–304.

Herzfeld, J., and Olbris, D. J. 2002. Hydrophobic effect. *Encyclopedia of Life Sciences*. pp. 1–9. London: Nature.

Hubbard, R. E. 2001. Hydrogen bonds in proteins: role and strength. *Encyclopedia of Life Sciences*. pp. 1–6. London: Nature.

Sharp, K. A. 2001. Water: structure and properties. *Encyclopedia of Life Sciences*. pp. 1–7. London: Nature.

Secondary Structure

Barton, G. J. 1995. Protein secondary structure prediction. *Curr Opin Struct Biol* 5:372–376.

Chandonia, J-M. 2001. Protein secondary structures: prediction. *Encyclopedia of Life Sciences*. pp. 1–6. London: Nature.

Eisenberg, D. 2003. The discovery of the α-helix and β-sheet, the principal structural features of proteins. *Proc Natl Acad Sci USA* 100: 11207–11210.

Eisenhaber, F., Persson, B., and Argos, P. 1995. Protein structure prediction: recognition of primary, secondary, and tertiary structural features from amino acid sequence. *Crit Rev Biochem Mol Biol* 30:1–94.

Minor, D. L. Jr, and Kim, P. S. 1996. Context-dependent secondary structure formation of a designed protein sequence. *Nature* 380:730–734.

Rost, B. 2001. Protein secondary structure prediction continues to rise. *J Struct Biol* 134:204–218.

Sansom, M. S. P. 2001. Hydrophobicity plots. *Encyclopedia of Life Sciences*. pp. 1–4. London: Nature.

Tertiary Structure

Anfinsen, C. B. 1973. The principles that govern the folding of protein chains. *Science* 181:223–230.

Burz D. S., and Shekhtman, A. 2009. Inside the living cell. *Nature* 485:37–38.

Carter, W. C. Jr, and Sweet, R. M. (eds). 1997. Macromolecular crystallography. *Methods in Enzymology*, vol 276 and 277. San Diego: Academic Press.

Clare, D. K., Bakkes, P. J., van Heerikhuizen, H., van der Vies, S. M., and Saibil, H. R. 2008. Chaperonin complex with a newly folded protein encapsulated in the folding chamber. *Nature* 457:107–110.

Consalvi, V., and Chiaraluce, R. 2005. Chaperones, chaperonin and heat shock proteins. *Encyclopedia of Life Sciences* pp. 1–7. Hoboken, NJ: John Wiley & Son.

Consalvi, V., and Chiaraluce, R. 2007. Chaperonins. *Encyclopedia of Life Sciences* pp. 1–6. Hoboken, NJ: John Wiley & Sons.

Daggett, V., and Fersht, A. R. 2003. Is there a unifying mechanism for protein folding? *Trend Biochem Sci* 28:18–25.

Dill, K. A., Ozkan, S. B. Shell, M. S., and Weikl, T. R. 2008. The protein folding problem. *Ann Rev Biophys* 37:289–316.

Ellis, R. J. 1993. The general concept of molecular chaperones. *Phil Trans Royal Soc Lond Series B* 339:257–261.

Feldman, D. E., and Frydman, J. 2000. Protein folding in vitro: the importance of molecular chaperones. *Curr Opin Struct Biol* 10:26–33.

Fersht, A. R. 2008. From the first protein structure to our current knowledge of protein folding: delights and scepticisms. *Nat Rev Mol Cell Biol* 9:650–654.

Fischer, G., and Schmid, F. X. 2001. Peptidylproline *cis*-trans-isomerases. *Encyclopedia of Life Sciences*. pp. 1–6. London: Nature.

Fuxreiter, M., Tompa, P., Simon, I., Uversky, V. N., Hansen, J. C., and Asturias, F. J. 2008. Malleable machines take shape in eukaryotic transcriptional regulation. *Nature Chem Biol* 4:728–737.

Goloubinoff, P., and De Los Rios, P. 2007. The mechanism of Hsp70 chapterones: (entropic) pulling the models together. *Trends Biochem Sci* 32:372–379.

Gronenborn, A. M. 2008. NMR for proteins. *Encyclopedia of Chemical Biology* pp. 1–11. Hoboken, NJ: John Wiley & Sons.

Guss, J. M., and King, G. F. 2002. Macromolecular structure determination: comparison of crystallography and NMR. *Encyclopedia of Life Sciences*. pp. 1–5. London: Nature.

Han, X., Aslanian, A. and Yates III, J. R. 2008. Mass spectrometry for proteomics. *Curr Opin Chem Biol* 12:483–490.

Hartl, F. U. 1996. Molecular chaperones in cellular protein folding. *Nature* 381:571–580.

Helliwell, J. R. 2001. X-ray diffraction at synchrotron light sources. *Encyclopedia of Life Sciences*. pp. 1–6. London: Nature.

Hock, R. J., Yeo, H. C., Kolatkar, P. R., and Clarke, N. D. Assessment of CASP7 structure predictions for template free targets. *Proteins* 69(Supp 8):57–67.

Inomata, K., Ohno, A., Tochio, H., Tenno, T., Nakase, I., Takeuchi, T., et al. 2009. High resolution multi-dimensional NMR spectroscopy of proteins in human cells. *Nature* 458:106–110.

Kang, T. S., and Kini, R. M. 2009. Structural determinants of protein folding. *Cell Mol Life Sci* 66:2341–2361.

Kay, L. E., and Gardner, K. H. 1997. Solution NMR spectroscopy beyond 25 kDa. *Curr Opin Struct Biol* 7:722–731.

Kyte, J., and Doolittle, R. F. 1982. A simple method for displaying the hydropathic character of a protein. *J Mol Biol* 157:105–132.

Nielsen, G., Stadler, M., Jonker, H., Betz, M., and Schwalbe, H. 2008. Nuclear magnetic resonance (NMR) spectroscopy: overview of applications in chemical biology. *Wiley Encyclopedia of Chemical Biology*. pp. 1–24. Hoboken, NJ: John Wiley & Sons.

Norin, M., and Sundstrom, M. 2006. Structural proteomics: large scale studies. Structural Genomics. *Encyclopedia of Life Sciences* pp. 1–6. Hoboken, NJ: John Wiley & Sons.

Norvell, J., Li, J., Saltsman, K., and Berg, J. 2006. Structural Genomics. *Encyclopedia of Life Sciences* pp. 1–4. Hoboken, NJ: John Wiley & Sons

Norvell, J. C., and Berg, J. M. 2007. Update on the protein structure initiative. *Structure* 15:1519–1522.

Pearl, F. M. G., Orengo, C. A., and Thornton, J. M. 2001. Protein structure classification. *Encyclopedia of Life Sciences*. pp. 1–7. London: Nature.

Rhodes, R. 1993. Crystallography made crystal clear. San Diego: Academic Press.

Sakakibara, D., Sasaki, A, Ikeya, T., Hamatsu, J., Hanashima, T., Mishima, M., et al. 2009. Protein structure determination in living cells by in-cell NMR spectroscopy. *Nature* 458:102–106.

Schumann, W. 2001. Heat shock response. *Encyclopedia of Life Sciences*. pp. 1–7. London: Nature.

Schwaller, M. D., and Gilbert, H. F. 2001. Protein disulfide isomerases. *Encyclopedia of Life Sciences*. pp. 1–7. London: Nature.

Skolnik, J. 2007. Protein structure prediction. *Encyclopedia of Life Sciences* pp. 1–7. Hoboken, NJ: John Wiley & Sons.

Steitz, T. A. 2007. Collecting butterflies and the protein structure initiative: the right questions? *Structure* 15:1523–1524.

Stevens, R. C. 2007. Generation of protein structures for the 21st century. *Structure* 15:1517–1519.

Tavaglini-Allocatelli, C., Ivarsson, Y. Jemth, P., and Gianni, S. 2008. Folding and stability of globular proteins and implications for function. *Curr Opin Struct Biol* 19:1–5.

Terwillliger, T. C., Stuart, D., and Yokoyama, S. 2009. Lessons from structural genomics. *Ann Rev Biophys* 38:371–383.

Tugarinov, V., Hwang, P. M., and Kay, L. E. 2004. Nuclear magnetic resonance spectroscopy of high-molecular weight proteins. *Ann Rev Biochem* 73:107–146.

Uversky, V. N., Oldfield, C. J., and Dunker, A. K. 2008. Intrinsically disordered proteins in human diseases: introducing the D^2 concept. *Ann Rev Biophys* 37:215–246.

Young, J. C., Agashe, V. R., Siegers, K., and Hartl, F. U. 2004. Pathways of chaperone-mediated protein folding in the cytosol. *Nat Rev Mol Cell Biol* 5:781–791.

Wedemeyer, W. J., and Scheraga, H. A. 2001. *Encyclopedia of Life Sciences* pp. 1–9. Hoboken, NJ: John Wiley & Sons.
Wery, J. P., and Scehvitz, R. W. 1997. New trends in macromolecular x-ray crystallography. *Curr Opin Chem Biol* 1:365–369.
Wickner, S., Maurizi, M. R., and Gottesman, S. 1999. Posttranslational quality control: folding, refolding, and degrading proteins. *Science* 286:1888–1893.
Wüthrich, K. 2001. Nuclear magnetic resonance spectroscopy of proteins. *Encyclopedia of Life Sciences*. pp. 1–5. London: Nature.
Wüthrich, K. 2003. NMR studies of structure and function of biological macromolecules. *J Biomol NMR* 27:13–39.
Yates, J. R. 2001. Mass spectrometry in biology. *Encyclopedia of Life Sciences*. pp. 1–5. London: Nature.
Zhang, H. 2002. Protein tertiary structures: prediction from amino acid sequences. *Encyclopedia of Life Sciences*. pp. 1–7. London: Nature.
Zhang, Y. 2008. Progress and challenges in protein structure prediction. *Curr Opin Struct Biol* 18:342–348.

Biological Membranes

Dzikovoski, B., and Freed, J. H. 2008. Membranes, fluidity of. *Wiley Encyclopedia of Chemical Biology*. pp. 1–14. Hoboken, NJ: John Wiley & Sons.
Edidin, M. 2003. Lipids on the frontier: a century of cell-membrane bilayers. *Nat Rev Mol Cell Biol* 4:414–418.
Pietzsch, J. 2004. Mind the membrane. *Horizon Symposia*. pp.1–4. London: Nature.
Raudino, A., and Sarpietro, M. G. 2001. Lipid bilayers. *Encyclopedia of Life Sciences*. pp. 1–6. London: Nature.
Singer, S. J., and Nicolson, G. L. 1972. The fluid mosaic model of the structure of cell membranes. *Science* 175:720–731.
Tanford, C. 1980. *The Hydrophobic Effect: Formation of Micelles and Biological Membranes*. Hoboken, NJ: John Wiley & Sons.
Yeagle, P. (ed). 2004. *The Structure of Biological Membranes* (2nd ed). Boca Raton, FL: CRC Press.
Yeagle, P. L. 2001. Cell membrane features. *Encyclopedia of Life Sciences*. pp. 1–7. London: Nature.

This book has a Web site, **http://biology.jbpub.com/book/molecular**, which contains a variety of resources, including links to in-depth information on topics covered in this chapter, and review material designed to assist in your study of molecular biology.

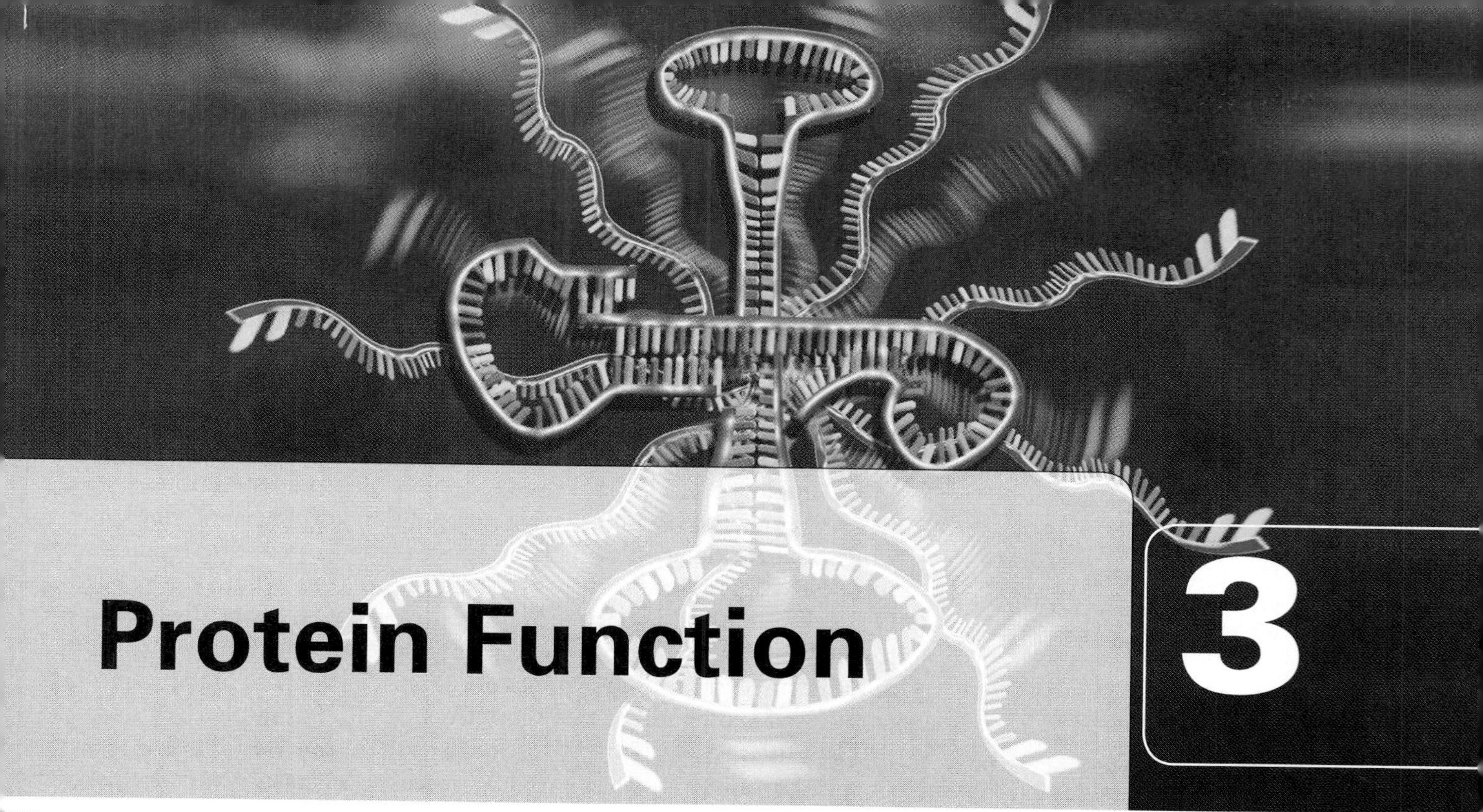

3 Protein Function

OUTLINE OF TOPICS

3.1 Myoglobin, Hemoglobin, and the Quaternary Structure Concept

Differences in myoglobin and hemoglobin function are explained by differences in myoglobin and hemoglobin structure.

Normal adult hemoglobin (HbA) differs from sickle cell hemoglobin (HbS) by only one amino acid.

3.2 Immunoglobulin G and the Domain Concept

Large polypeptides fold into globular units called domains.

3.3 Enzymes

Enzymes are proteins that catalyze chemical reactions.

Different methods can be used to detect enzyme activity.

Enzymes lower the energy of activation but do not affect the equilibrium position.

All enzyme reactions proceed through an enzyme-substrate complex.

Molecular details for enzyme-substrate complexes have been worked out for many enzymes.

Regulatory enzymes control committed steps in biochemical pathways.

Regulatory enzymes exhibit sigmoidal kinetics and are stimulated or inhibited by allosteric effectors.

Enzyme activity can be altered by covalent modification.

3.4 G-Protein-Linked Signal Transduction Systems

G-protein-linked signal transduction systems convert extracellular chemical or physical signals into intracellular signals.

3.5 The Ubiquitin Proteasome Proteolytic Pathway

The ubiquitin proteasome system is responsible for the specific degradation of intracellular proteins.

Suggested Reading

The functions of all proteins depend on their ability to specifically interact with other molecules. Such specificity is possible because polypeptides with different amino acid sequences fold into different tertiary structures. Each kind of protein evolved to interact with a specific molecule or ligand. Catalytic proteins—the enzymes—convert the ligands into other molecules. Structural proteins interact with specific molecules, often endowing the bound molecules with special biological properties. For instance, one class of proteins, the **histones**, binds to DNA to form compact nucleoprotein structures called **nucleosomes**, while a second class of proteins combines with RNA to form the ribonucleoprotein complex known as the **ribosome**. Transport proteins (such as hemoglobin) bind to specific ligands (in this case oxygen) and transport the ligand to a site where it is needed. Other kinds of transport proteins allow specific molecules to pass through biological membranes. Storage proteins such as myoglobin, another oxygen-binding protein, allow the cell to store higher concentrations of the ligand than otherwise would be possible. Regulatory proteins interact with other proteins or with nucleic acids to slow down or speed up some crucial biological process. Receptor proteins change conformation after binding a specific ligand and then trigger a series of metabolic changes. There are even toxic proteins such as cholera toxin that modify other proteins so that they no longer work properly.

Although it might seem that transport proteins and storage proteins would be of somewhat less interest to molecular biologists than other kinds of proteins, in fact this is not true. The oxygen storage protein myoglobin and the oxygen transport protein hemoglobin have a special place in the early history of molecular biology. They were the first two proteins to have their structures determined. Moreover, the lessons learned about structure-function relationships from studying these two proteins remain relevant today. We, therefore, begin our examination of protein function by considering the structure-function relationships that exist for myoglobin and hemoglobin.

3.1 Myoglobin, Hemoglobin, and the Quaternary Structure Concept

Differences in myoglobin and hemoglobin function are explained by differences in myoglobin and hemoglobin structure.

Myoglobin is a single polypeptide chain with 153 residues and it stores oxygen in muscles. Hemoglobin, which has two α- and two β-globin chains, transports oxygen in blood. Each α-globin chain has 141 residues and each β-chain has 146 residues. The folding patterns of the myoglobin and hemoglobin chains are nearly identical (FIGURE 3.1). Helical segments within each polypeptide form a pocket containing a large planar ring system with an Fe^{2+} ion at its center known as **heme** (Figure 3.1c), which binds oxygen.

An examination of myoglobin and hemoglobin structure reveals certain general rules that apply to other globular proteins. These rules

(a)

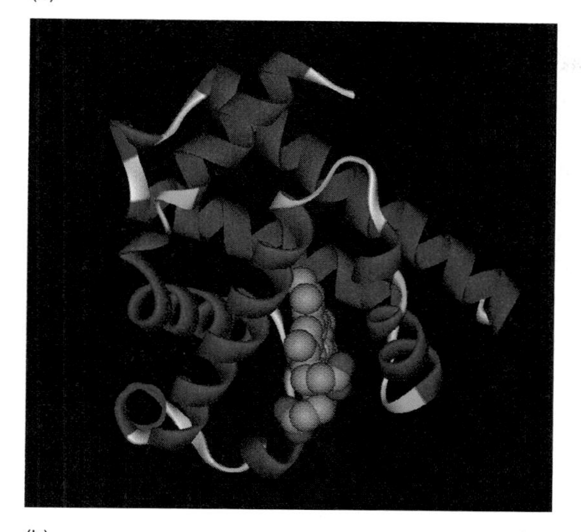

(b)

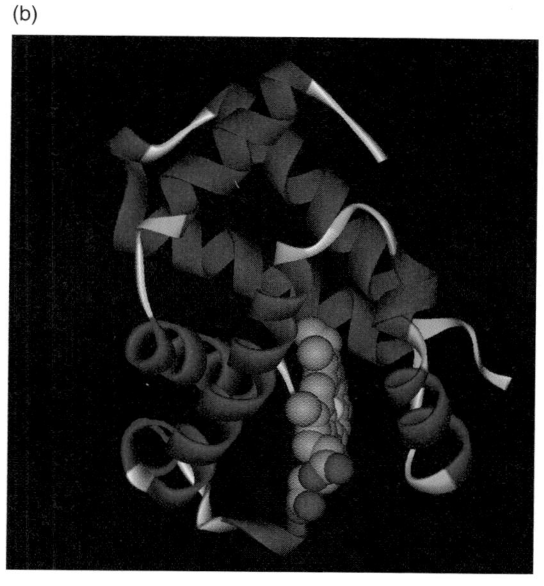

(c)

CH_3 $CH{=}CH_2$
$CH_2{=}CH$ CH_3
N N
Fe
N N
H_3C CH_3
CH_2 CH_2
CH_2 CH_2
COOH COOH

Heme

FIGURE 3.1 **Folding pattern of myoglobin and hemoglobin.** (a) Sperm whale myoglobin. (b) Human hemoglobin β-chain. (c) Heme. Polypeptide chains are shown in ribbon display and their heme groups in spacefill display. The accession number for human hemoglobin is 2HHB and that for sperm whale myoglobin is 1VXA. (Part A structure from Protein Data Bank 1VXA. F. Yang and G. N. Phillipsac, Jr., *J. Mol. Biol.* 256 (1996): 762-774. Prepared by B. E. Tropp. Part B structure from Protein Data Bank 2HHB. G. Fermi, et al., *J. Mol. Biol.* 175 [1984]: 159–174. Prepared by B. E. Tropp.)

include the following: (1) Globular proteins tend to be compact, with hydrophobic amino acid residues accounting for about 65% of the protein interior. (2) Hydrophilic residues are almost always on the protein surface. (3) Water is generally excluded from the protein core but fills cavities and crevices when they are present. (4) Hydrophobic interactions, hydrogen bonds, ionic bonds, and van der Waals interactions stabilize tertiary structures (and quaternary structures when they are present as in hemoglobin). (5) Disruption of just a few weak noncovalent bonds can initiate changes that eventually cause the conformation of the entire protein to change. This conformational flexibility is essential for protein function.

Comparisons of myoglobin and hemoglobin structure and function also help to spotlight the contribution that quaternary structure makes to protein function. Based on the similarities in their tertiary structures, one might expect the polypeptides in myoglobin and hemoglobin to behave in the same way. At the simplest level this expectation

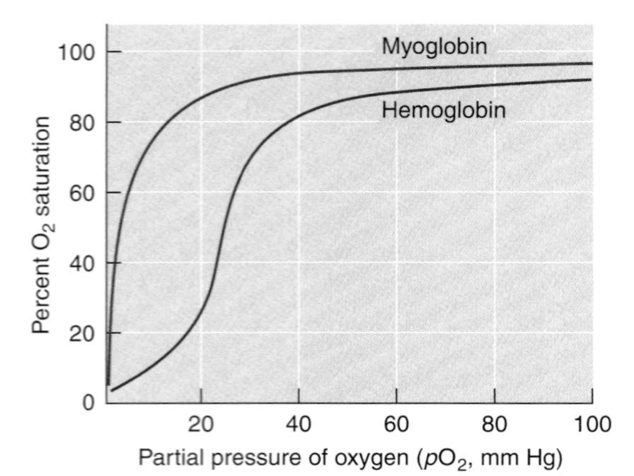

FIGURE 3.2 **Oxygen-binding curves for myoglobin and hemoglobin.**

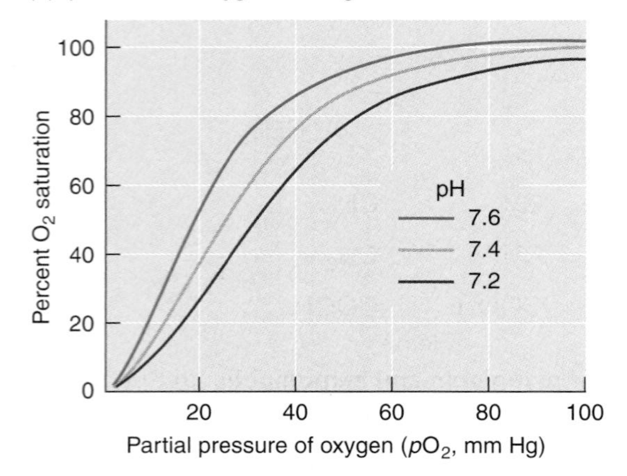

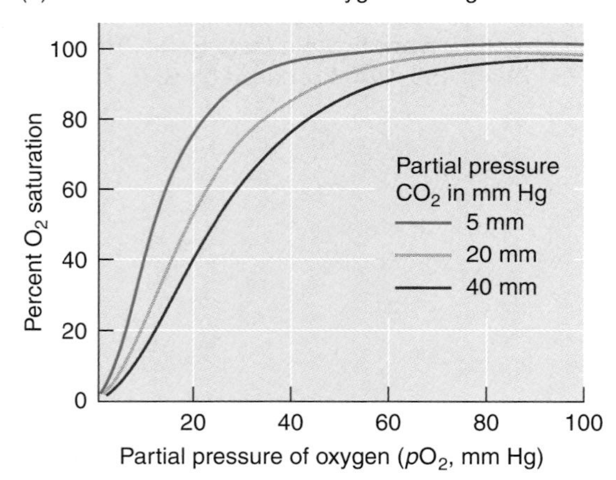

FIGURE 3.3 **The Bohr effect** (a) pH effect on oxygen-binding curve. (b) Carbon dioxide effect on oxygen binding curve.

is justified. Both proteins bind oxygen, changing from a red purple color to a scarlet color in the process.

A more detailed analysis of the oxygen-binding properties reveals profound differences, however. The oxygen-binding curve for myoglobin is hyperbolic, whereas that for hemoglobin is S-shaped or **sigmoidal** (FIGURE 3.2). The sigmoidal curve indicates a **cooperative** interaction among the oxygen-binding sites in the hemoglobin molecule. That is, the binding of oxygen to one of the hemes in a hemoglobin molecule increases the remaining hemes' affinity for oxygen. On the other hand, the loss of an oxygen molecule from one of the hemes in a hemoglobin molecule decreases the other hemes' affinity for oxygen. **Cooperativity**, which requires some kind of communication among the hemes, permits hemoglobin to bind oxygen in the lungs where oxygen is plentiful, and release oxygen near actively respiring cells where oxygen is scarce. Cooperativity is impossible in myoglobin because each myoglobin molecule has only one heme group.

Hemoglobin also has other remarkable properties that distinguish it from myoglobin. A drop in pH (FIGURE 3.3a) or an increase in carbon dioxide concentration (FIGURE 3.3b) causes hemoglobin to release oxygen but has no effect on myoglobin. The pH and carbon dioxide effects are known collectively as the **Bohr effect** after Christian Bohr, the investigator who first discovered them. The Bohr effect has important physiological consequences. As red blood cells pass through the capillary vessels of rapidly respiring tissue, they are exposed to the carbon dioxide and low pH produced by those tissues. The hemoglobin molecules inside the red blood cells release oxygen and pick up carbon dioxide. When the red blood cells reach the lungs hemoglobin releases carbon dioxide and picks up oxygen.

Hemoglobin also has one other remarkable property that is not shared by myoglobin. The small organic phosphate 2,3-bisphosphoglycerate (BPG) causes hemoglobin to release oxygen (FIGURE 3.4) but has no effect on myoglobin. The effect of BPG is a very important concern for stored red blood cells at blood banks. Over time, the BPG is broken down. When this takes place, the hemoglobin in red blood cells binds oxygen in much the same way as myoglobin. So, instead of releasing oxygen to rapidly respiring cells, the red blood cells retain the oxygen. The solution to this serious problem is to add nutrients that are converted to BPG to the stored blood.

The explanation for the cooperative effect and for each of hemoglobin's other special biological properties became clear when investigators compared the structures of the oxygenated and deoxygenated forms of hemoglobin and myoglobin. The conformation of deoxymyoglobin and deoxyhemoglobin polypeptide chains change when the heme group binds oxygen (FIGURE 3.5). The driving force for this change is Fe^{2+} ion movement.

In deoxymyoglobin (and each deoxyhemoglobin subunit), the Fe^{2+} ion is slightly out of the plane of heme's ring system for two reasons. First, Fe^{2+} is too large to fit into the center of the ring system and, second, electrons in the ring system repel electrons in the histidine residue attached to Fe^{2+}. The Fe^{2+} moves into the plane of the ring after it binds oxygen. This movement is possible because the effective

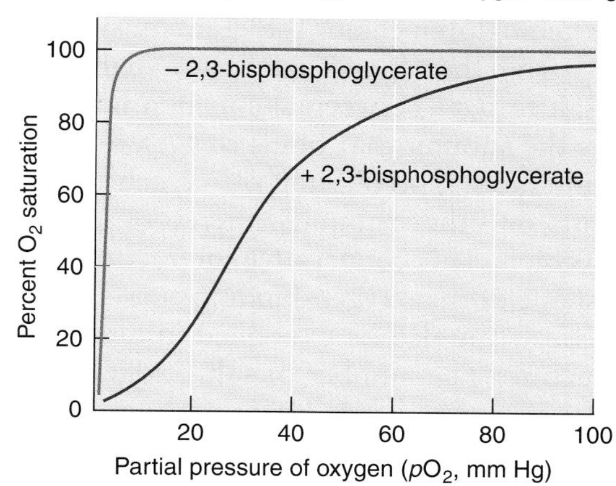

FIGURE 3.4 **Effect of 2,3-bisphosphoglycerate on oxygen binding in hemoglobin.** (a) Structure of 2,3-bisphosphoglycerate. (b) Effect of 2,3-bisphosphate on hemoglobin oxygen-binding curve.

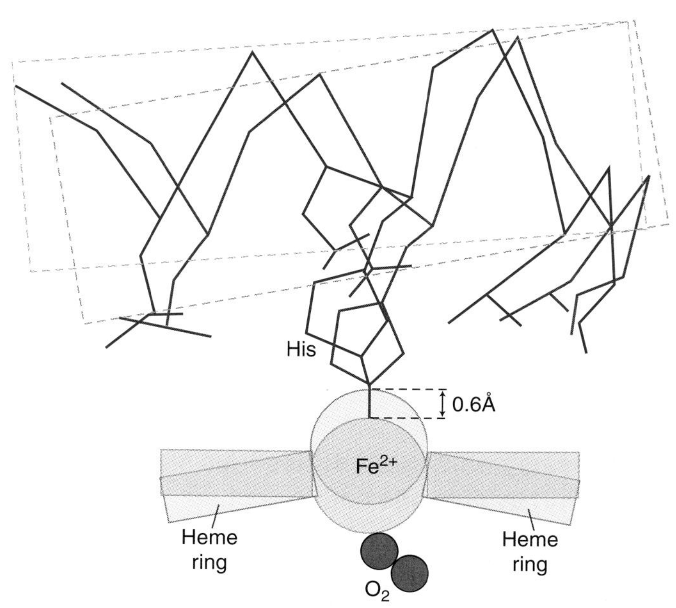

FIGURE 3.5 **Movement of Fe^{2+} and its associated helix during the transition from deoxymyoglobin (or deoxyhemoglobin) to oxymyoglobin (or oxyhemoglobin).** In deoxymyoglobin (and each deoxyhemoglobin subunit), the Fe^{2+} ion is slightly out of the plane of heme's ring system (blue). Fe^{2+} moves into the plane of the ring after it binds oxygen. As Fe^{2+} moves into the plane of the heme system, it pulls its attached histidine with its associated helix along with it (red). (Reproduced from *Biochemistry, Third edition*, D. Voet and J. G. Voet, Copyright © 2005 and reproduced with permission of John Wiley & Sons, Inc.)

radius of the Fe^{2+} ion decreases upon binding oxygen, and the binding energy of the Fe^{2+} ion for oxygen exceeds the repulsive electrostatic forces between the heme ring system and the histidine residue. As Fe^{2+} moves into the plane of the heme system, it pulls the attached histidine and its associated helix along with it. In myoglobin, the helix is not locked into place and so is free to move.

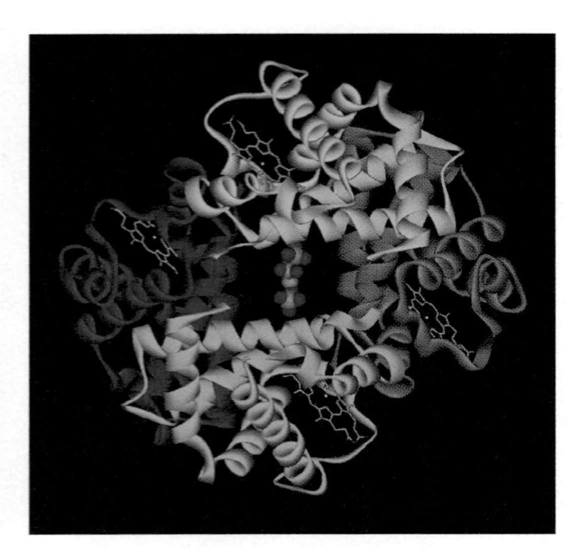

FIGURE 3.6 **Human deoxyhemoglobin with 2,3-bisphosphoglycerate bound to the two β-globin chains.** The globin chains are in ribbon display with each chain a different color. The heme groups are in stick structure display and the 2,3-bisphosphoglycerate bound to the two β-globin chains in the cavity formed by the four globin chains is in van der Waals display. (Structure from Protein Data Bank 1B86. V. Richard, G. G. Dodson, and Y. Mauguen, *J. Mol. Biol.* 233 [1993]: 270–274. Prepared by B. E. Tropp.)

The situation is different in hemoglobin because the conformation of each chain is stabilized by ionic bonds to other chains. The ionic bonds between hemoglobin chains therefore must be broken before a chain can change from the deoxygenated form to the oxygenated form. Disruption of salt bridges between chains in a hemoglobin molecule not only affects the subunit that binds oxygen but also the subunits with which it interacts. The binding of oxygen to one subunit also loosens restraints on other subunits, therefore making it easier for them to bind oxygen. This cooperative interaction is responsible for the sigmoidal shape of the oxygen-binding curve for hemoglobin.

The Bohr effect can also be explained in terms of hemoglobin's quaternary structure. The number of salt bridges between the polypeptide subunits increases when H^+ or carbon dioxide bind to the polypeptide subunits, stabilizing the structure of the deoxygenated form of hemoglobin. 2,3-Bisphosphoglycerate also interacts by stabilizing the structure of deoxygenated hemoglobin; it fits into a cavity between the β-subunits of deoxyhemoglobin forming an ionic bridge between them (FIGURE 3.6). Oxyhemoglobin, which lacks the cavity, cannot bind BPG. Deoxyhemoglobin must release BPG before it can bind oxygen and oxyhemoglobin must release oxygen before it can bind BPG. Thus, even though the oxygen and BPG binding sites are distinct and distant, binding at one site prevents binding at the other.

Carbon dioxide, H^+, and BPG all lower the ability of hemoglobin to bind oxygen by increasing the number of salt links between the subunits of hemoglobin. Each thus acts at a site distinct from the oxygen-binding site. Molecules that influence protein activity by binding to sites that are distinct from the functional or active sites are called **allosteric effectors**. Binding an allosteric effector to a site on the protein that is some distance from the active site therefore can modify a protein's activity, permitting the protein to "sense" its chemical environment and to act accordingly.

Normal adult hemoglobin (HbA) differs from sickle cell hemoglobin (HbS) by only one amino acid.

In 1904, a West Indian student of African origin who was suffering from anemia, recurrent pains, leg ulcers, jaundice, a low red blood cell count, and an enlarged heart sought the assistance of James Herrick, a Chicago physician. Upon examining the student's blood under the microscope, Herrick observed that the red blood cells had a sickle or crescent shape rather than the biconcave appearance normally observed. Herrick hypothesized that the sickle-shaped red blood cells might be the key to understanding the patient's anemia and other symptoms. The disease suffered by the student, which is now known to be an inborn error of metabolism, was given the name **sickle cell anemia.**

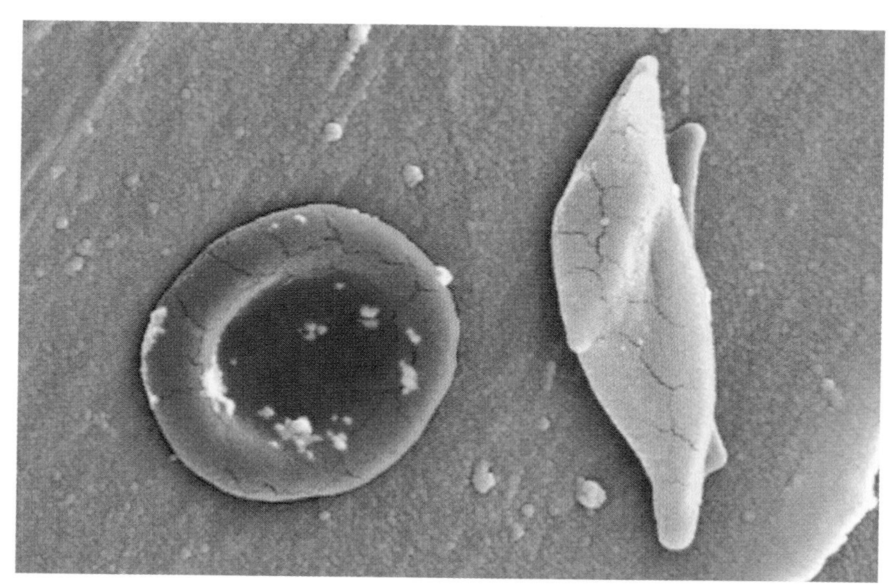

FIGURE 3.7 **Biconcave- and sickle-shaped red blood cells.** (Photo courtesy of the Sickle Cell Foundation of Georgia/Janice Haney Carr/CDC.)

Herrick's hypothesis received additional support in 1927, when investigators observed that red blood cells isolated from individuals with sickle cell anemia change from a biconcave to a sickle shape when deprived of oxygen, whereas red blood cells from normal individuals remain biconcave (FIGURE 3.7).

Linus Pauling and coworkers tried a new approach to the study of sickle cell anemia in 1949. They knew that hemoglobin, the major protein in red blood cells, was responsible for binding oxygen and suspected that individuals with sickle cell anemia have a variant form of hemoglobin. Their hypothesis was tested by comparing the electrophoretic mobilities of normal adult hemoglobin (**HbA**) and sickle cell hemoglobin (**HbS**). HbA migrated as though it were slightly more negative than HbS, suggesting that sickle cell anemia is caused by one or more amino acid substitutions in hemoglobin. However, the studies by Pauling and coworkers did not reveal whether the amino acid substitution(s) altered the α-globin chain, β-globin chain, or both, nor did the studies reveal which amino acid(s) was (were) altered.

The problem was solved in an ingenious fashion by Vernon Ingram in 1954. Modern techniques for sequencing polypeptide chains were not yet available, so Ingram could not compare the two forms of hemoglobin amino acid by amino acid. Instead, he used trypsin to cut HbA and HbS into well-defined fragments, which he spotted onto filter paper and partially separated by electrophoresis. Then he dried the paper, turned it on a right angle, and developed it in a second direction by chromatography. Because each type of protein treated in this fashion yields a unique two-dimensional pattern, the method is called **peptide fingerprinting** or **peptide mapping**. With one important exception, the "fingerprint" of HbS is identical to that of HbA (FIGURE 3.8a and b).

Amino acid and sequence analyses showed that the fragment obtained from HbA that does not match up with one from HbS is an

(a) Hemoglobin A (HbA)

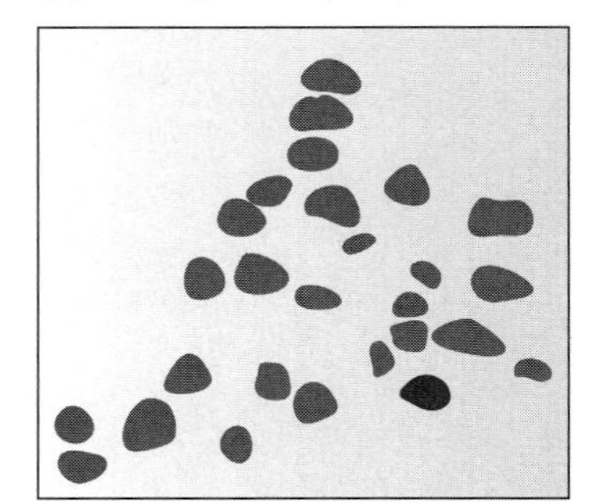

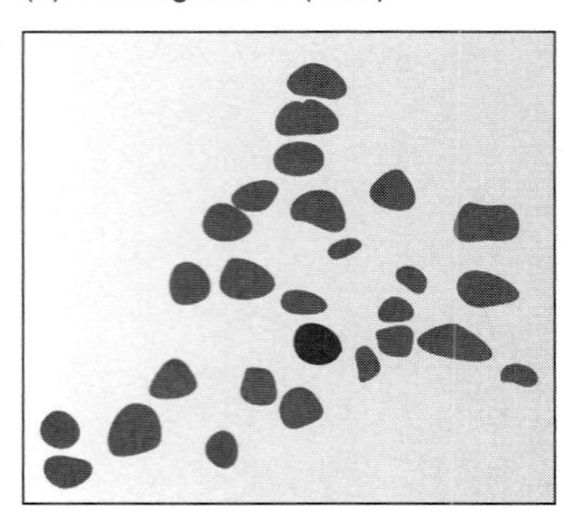

(c) Amino terminal fragment produced from the β-globin chains of HbA and HbS.

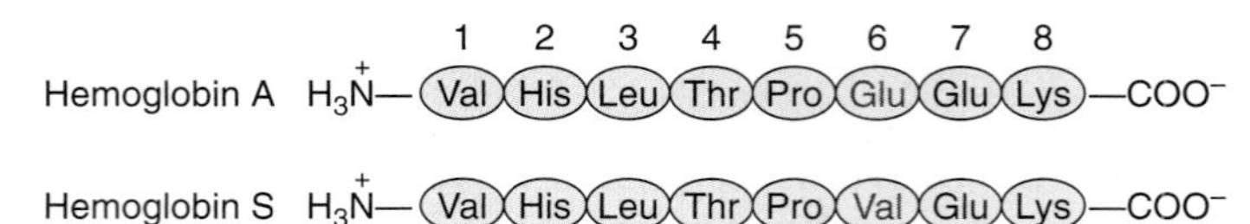

FIGURE 3.8 **Peptide map of hemoglobin A (HbA) and hemoglobin S (HbS).** Schematic diagram shows peptide maps created by trypsin digestion of (a) HbA and (b) HbS. All of the peptides but one are identical in HbA and HbS. The one peptide that is different, which is shown in red, has a single amino acid replacement. (c) Glutamate in position-6 of the β-globin chain is replaced by valine. (Adapted from C. Baglioni, *Biochim. Biophys. Acta* 48 [1961]: 392–396.)

octapeptide derived from the N-terminal end of the β-chain. The only difference between the fragments that do not match up is that the one derived from HbS contains a valine residue in place of the glutamate residue found in HbA. This substitution was shown to be in the sixth residue from the N-terminal residue of the β-chain (FIGURE 3.8c). The observed substitution is consistent with the observation by Pauling and coworkers that HbA migrates as a slightly more negatively charged protein than does HbS. Ingram's studies showed for the first time that a single amino acid substitution can cause a profound change in protein function. This finding stimulated great interest in the nascent field of molecular biology and reinforced the importance of studying structure-function relationships. Moreover, Ingram's studies introduced peptide mapping, a very powerful technique that is still used today.

Investigators originally thought that capillaries are blocked because the poorly deformable sickle cells cannot fit through them. More recent studies show that sickle red blood cells, but not normal red blood cells, adhere abnormally to the endothelial cells that line blood vessels. Research activity in the field of sickle cell anemia is now directed toward trying to correct the genetic error and developing an effective therapy to treat the symptoms.

Why have selective pressures not worked to eliminate sickle cell anemia? The rather surprising answer is that selective pressure for another factor actually favors the perpetuation of the sickle cell gene. Individuals who have inherited a nonsickling allele from one parent and a sickling allele from the other parent have the sickle-cell trait associated with a mild form of anemia. When exposed to *Plasmodium falciparum* or closely related protozoan parasites that cause malaria, they are more resistant to malaria than are individuals with two non-sickling alleles. As might be expected, the greatest incidence of sickle cell anemia and trait occurs in those regions of Africa where malaria is most prevalent.

3.2 Immunoglobulin G and the Domain Concept

Large polypeptides fold into globular units called domains.

Myoglobin and hemoglobin chains are relatively short and therefore fold into a single compact globular structure. Larger polypeptides tend to fold into two or more compact globular units, called **domains**, which are joined by short lengths of the peptide chain. Domains may either behave like completely independent structural units or exhibit varying degrees of structural interaction. Well-defined domains often act as independent folding units (see Chapter 2) and retain at least part of their normal biological activity even if they are split off from the rest of the protein.

The antibody **immunoglobulin G** or **IgG** provides an excellent example of a multidomain protein. IgG's function is to interact with specific foreign substances (**antigens**) and render them harmless. Vertebrates produce millions of different kinds of IgG molecules, each specific for a different antigen. IgG is a tetramer made of two identical **light chains**, each with approximately 214 amino acid residues, and two identical **heavy chains**, each with approximately 446 amino acid residues (FIGURE 3.9). Each L-chain is attached to an H-chain by a disulfide bond as well as by weak noncovalent interactions. Disulfide bonds and weak noncovalent interactions also join the two H-chains.

Each L-chain can be divided into two regions of about equal size. The first region, extending from residues 1 to 108, has different sequences in different IgG molecules and is therefore called the variable region of the L-chain (abbreviated V_L). The second region, extending from residue 109 to the end of the chain, has the same sequence in all IgG molecules and is therefore called the constant region of the L-chain (abbreviated C_L). The H-chain also has a variable region (V_H) extending from residue 1 to about residue 110. The remainder of the H-chain has three regions of about equal size, each with a constant amino acid sequence. These constant regions, designated C_H1, C_H2, and C_H3, have amino acid sequences that are quite similar, but not identical, to one another.

Each L- and H-region folds into a distinct compact globular structure so that the IgG molecule has a total of 12 domains (Figure 3.9). C_L and C_H domains have very similar structures, resembling collapsed barrels with four β-strands on one side and three on the other. A short

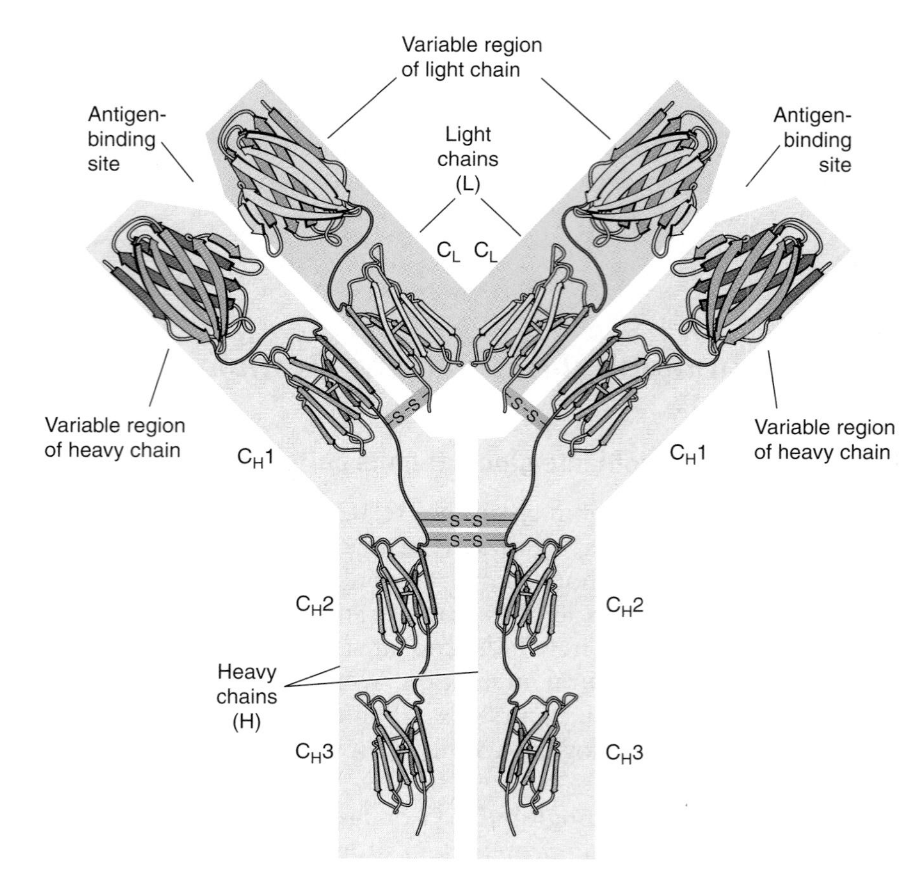

FIGURE 3.9 **Schematic of immunoglobulin G (IgG) structure.**

loop joins each β-strand to the one next to it. (FIGURE 3.10a). This folding pattern, known as the immunoglobin fold, is present in many other proteins. V_L and V_H domains also resemble collapsed barrels. However, one side of the barrel has five rather than three β-strands (FIGURE 3.10b). Although the major secondary structure in IgG domains is the β-strand, this is certainly not true for domains in all proteins. Domains are also formed from a mixture of α-helix and β-strands or mainly from α-helix.

Constant and variable region domains have different functions. Constant region domains are responsible for the overall structure of the molecule and for its recognition by other components of the immune system. Variable region domains act as antigen binding sites. Each antigen binding site contains one V_L domain and one V_H domain. Each IgG molecule thus contains two antigen binding sites (Figure 3.9). Three small regions within V_L and another three within V_H display much more variation than the rest of the variable regions. These so-called **hypervariable regions**, form loops that extend into the surrounding aqueous environment (Figure 3.10) to make contact with the antigen. The amino acid residues in the hypervariable region determine antibody specificity. The mechanism that allows vertebrates to produce so many different kinds of hypervariable regions is described in Chapter 13.

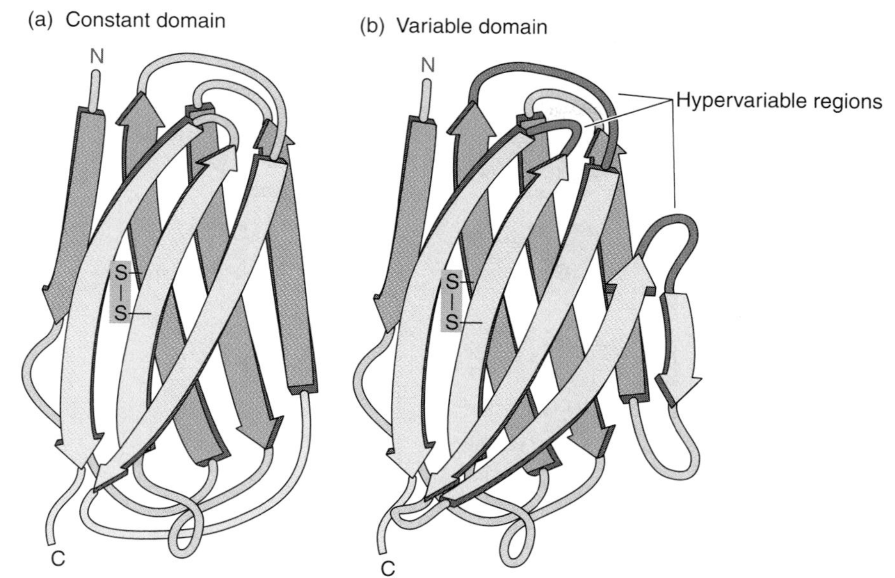

FIGURE 3.10 **Folding patterns of the IgG constant and variable domains.** (a) Constant region domains from either light or heavy chains fold in the same way, resembling a collapsed barrel with four β-strands on one side and three on the other. (b) Variable region domains of the L- and H-chains also resemble collapsed barrels. However, one side of the barrel has five β-strands rather than three strands. The hypervariable regions, which are shown in purple, are loops joining the β-strands in the V_L and V_H domains. (Modified from C. Branden and J. Tooze. *Introduction to Protein Structure, First edition.* Garland Science, 1999. Used with permission of John Tooze, The Rockefeller University.)

3.3 Enzymes

Enzymes are proteins that catalyze chemical reactions.

We now apply the lessons learned from studying protein structure to protein catalysts known as **enzymes**, which are from our perspective the most important class of proteins. An enzyme often requires one or more **cofactors** to catalyze a reaction. The cofactor may be a simple metal ion such as Mg^{2+} or a small organic molecule. Organic cofactors, known as **coenzymes**, are vitamin derivatives. For instance, the cofactor nicotinamide adenine dinucleotide (NAD^+) is a derivative of vitamin B_3 (niacin). Virtually all chemical reactions that take place in the cell are catalyzed by enzymes or RNA catalysts called **ribozymes** (see Chapter 4). Enzymes catalyze the vast majority of the reactions involved in DNA, RNA, and protein metabolism. A large number of different enzymes are required to catalyze a complex process such as replication, transcription, or translation. To understand these processes, we must learn how the enzymes involved in them work.

Different methods can be used to detect enzyme activity.

The molecule on which an enzyme acts is its **substrate**. Only a small number of substrate molecules, sometimes only one, participate in a single catalyzed reaction. Enzymatic activity is measured by following

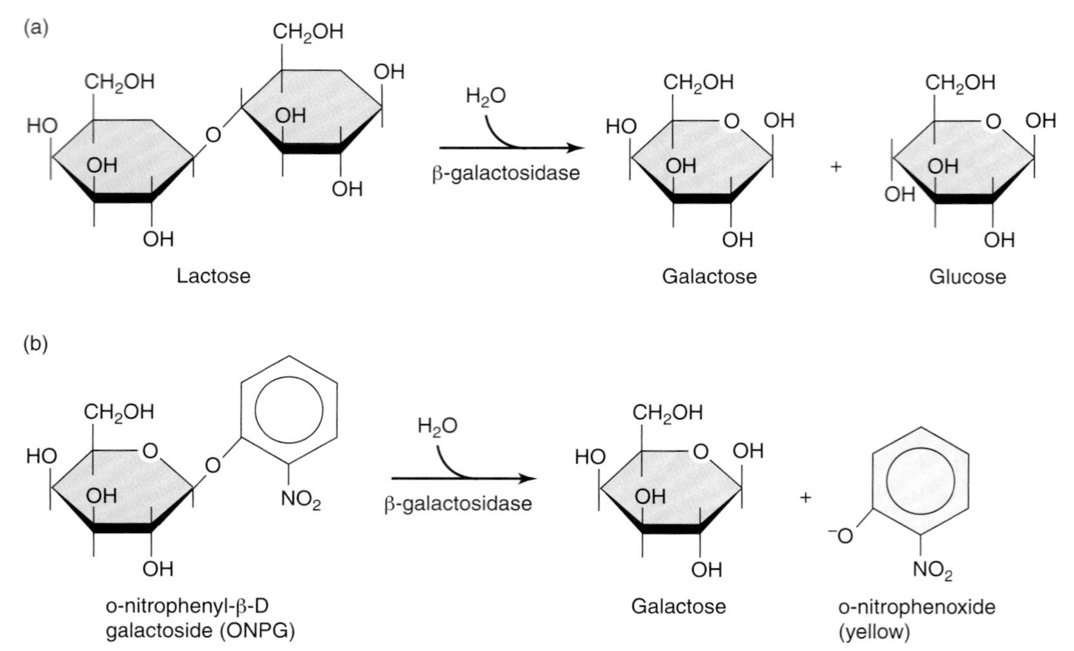

FIGURE 3.11 β-galactosidase assay. β-galactosidase converts (a) its natural substrate lactose to galactose and glucose and (b) a synthetic substrate o-nitrophenyl-β-D-galactoside to galactose and o-nitrophenoxide, which has an intense yellow color.

the rate of substrate breakdown or product formation. Several different physical techniques can be used to monitor an enzyme catalyzed reaction. The two most common techniques used by molecular biologists are colorimetric and radiotracer techniques.

Colorimetric assays are based on the fact that a substrate (or product) absorbs light of a particular wavelength. When a substance that absorbs visible or ultraviolet light is either a substrate or product of an enzyme catalyzed reaction, then substrate breakdown or product formation can be followed using a spectrophotometer. For this reason, molecular biologists often work with specially designed substrates that generate products with unique light absorption properties. One such specially designed substrate is o-nitrophenyl-β-D-galactoside (ONPG), which is used to assay β-galactosidase, an enzyme that cleaves lactose to form galactose and glucose (**FIGURE 3.11a**). The bond broken, a β-galactoside linkage, is also present in ONPG. β-Galactosidase hydrolyzes the colorless ONPG to form galactose and o-nitrophenoxide, which is intensely yellow (**FIGURE 3.11b**). β-Galactosidase activity, therefore, is readily followed by measuring the concentration of o-nitrophenoxide at a wavelength of 420 nm (blue light).

In a radioactivity assay, radioactive substrate is added to a reaction mixture and either appearance of radioactive product or the loss of radioactive substrate is measured. This type of assay is used to measure the conversion of a radioactive amino acid into a radioactive protein or a radioactive nucleotide into a radioactive nucleic acid. The assay used is based upon the following fact: proteins and nucleic acids are insoluble in 0.5 M trichloracetic acid (TCA), whereas amino acids and nucleotides are TCA-soluble. This property allows protein synthesis to be measured by adding a radioactive [^{14}C] or [^{3}H]amino acid to a reaction mixture that contains the other 19 nonradioactive amino acids, ribosomes, and the appropriate enzymes and factors. After a period of time, TCA is added and the mixture is filtered. The [^{14}C] amino acid is soluble and passes through the filter; any [^{14}C]protein that is made will precipitate and be retained on the filter. Counting the filter-bound radioactivity is then a measure of the extent of reaction.

Synthesis of DNA can also be measured in this way by using a mixture containing the four deoxyribonucleotide precursors of DNA, one of which is radioactive, the enzyme, and other appropriate components of the mixture. For example, using [^{14}C]thymidine triphosphate (TTP), after the reaction, TCA is added and the mixture is filtered. The [^{14}C]TTP passes through the filter but any DNA that has been synthesized is retained on the filter.

Enzymes lower the energy of activation but do not affect the equilibrium position.

Because enzymes play such a critical role in reactions of interest to molecular biologists, it is important to learn how they work. We begin by drawing a reaction coordinate diagram for an uncatalyzed reaction in which reactants A and B are converted to products C and D (FIGURE 3.12; plot in red). The extent of reaction, called the reaction coordinate, is plotted on the x-axis and the energies of reactants, intermediates, and products are plotted on the y-axis. Reactants require enough energy to reach the top of the energy barrier to form a molecular complex, called an **activated complex** or **transition state**, before they can be converted to products. The rate of a reaction is directly related to the fraction of reactant molecules that reach the transition state in a given period of time and depends on the **energy of activation** (**E_a**), that is, the energy needed to reach the transition state. A catalyst increases the reaction rate by providing an alternative reaction path with a lower energy of activation (Figure 3.12, plot shown in blue).

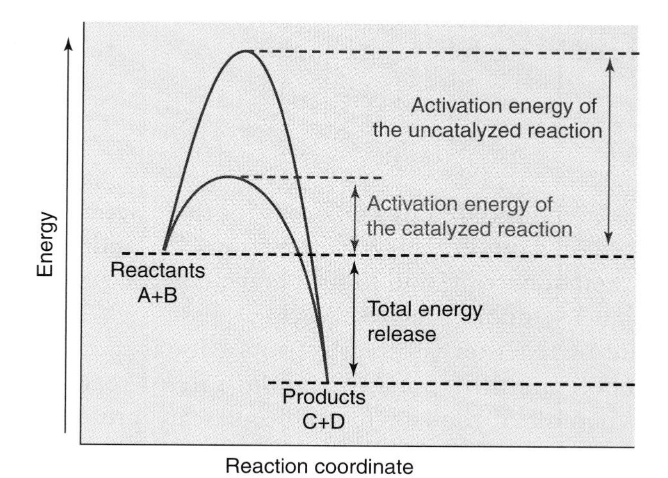

FIGURE 3.12 **Reaction coordinate diagram for an uncatalyzed reaction (red) and an enzyme catalyzed reaction (blue).** The enzyme lowers the activation energy but does not affect the energy released by the reaction. The enzyme therefore increases the reaction rate but does not affect the equilibrium.

Catalysts do not change the equilibrium position for a reaction only the rate at which the reaction takes place.

The catalytic power of enzymes exceeds all man-made catalysts. A typical enzyme accelerates a reaction 10^8- to 10^{10}-fold, though there are enzymes that increase the reaction rate by a factor of 10^{15}. Enzymes are also highly specific in that each catalyzes only a single reaction or set of closely related reactions. Several different hypotheses have been proposed to explain the catalytic properties of enzymes. These hypotheses, which are not mutually exclusive, include the following.

1. Enzymes lower the energy of activation by stabilizing the transition state. This hypothesis is supported by studies that show IgG molecules, which are formed when animals are injected with a stable analog of the transition state, act as catalysts. Catalytically active antibodies are called **abzymes.**
2. Enzymes lower the energy of activation by putting a strain on a susceptible bond. Distortion of the susceptible bond makes it easier to break the bond.
3. Enzymes lower the energy of activation for reactions involving two or more substrates by holding the substrates near to each other and in the proper orientation.
4. Enzymes lower the energy of activation by forming a covalent bond with a reactant molecule that destabilizes some other bond.
5. Enzymes lower the energy of activation by acting as proton donors and acceptors.

All enzyme reactions proceed through an enzyme-substrate complex.

In any enzyme-catalyzed reaction, the enzyme, E, always combines with its substrate, S, to form an **enzyme-substrate** (ES) **complex,** which can then either dissociate to reform substrate or go forward to product, P, and enzyme, E. After the ES complex forms, the substrate is usually altered in some way that facilitates further reaction. The process can be summarized by the following equation:

$$E + S \underset{k_{-1}}{\overset{k_{+1}}{\rightleftharpoons}} ES \overset{k_{+2}}{\longrightarrow} E + P$$

where k_{+1}, k_{-1}, and k_{+2} are rate constants for the reaction. (By convention, kinetic constants for forward reactions k_{+1} and k_{+2} have a (+) symbol in their subscript, and kinetic constants for reverse reactions (k_{-1}) have a (–) symbol in their subscript.)

The rate of reaction is directly proportional to the ES concentration. Hence, the **theoretical maximum rate of reaction** (V_{max}), is observed when all of the enzyme molecules are present in enzyme-substrate complexes. The ratio $(k_{-1} + k_{+2})/k_{+1}$ is called the **Michaelis constant** (K_M). For most enzymatic reactions, formation of ES is reversible in the sense that ES can dissociate, yielding free E and free S.

Usually, dissociation of the ES complex is more rapid than conversion of the complex to enzyme and product. When this is the case, the value of K_M is a measure of the strength of the ES binding. That is, when $k_{-1} >>> k_{+2}$, K_M approaches k_{-1}/k_{+1}, the dissociation constant for enzyme-substrate complex, and K_M is a measure of an enzyme's affinity for its substrate. A high value of K_M indicates low affinity (and weak binding), and a low value of K_M means high affinity (and strong binding). Strength of binding depends on several conditions, such as temperature, pH, the presence of particular ions, and the overall ion concentration. For most enzymes, K_M ranges from 10^{-6} to 10^{-1} M, which shows that the affinity of E and S varies widely for different enzymes. V_{max} and K_M values can be determined from a hyperbolic curve such as that shown in FIGURE 3.13, which is generated by plotting the initial rate of reaction (v_o) on the y-axis and substrate concentration ([S]) on the x-axis. The theoretical maximum velocity (V_{max}) is the limiting velocity obtained as the substrate concentration approaches infinity. The K_M value corresponds to the substrate concentration at which the reaction rate is $V_{max}/2$. It is the same as the concentration at which half of the enzyme molecules in the solution have their active sites occupied by a substrate molecule. Intracellular substrate concentrations tend to range between 0.1 K_M and K_M.

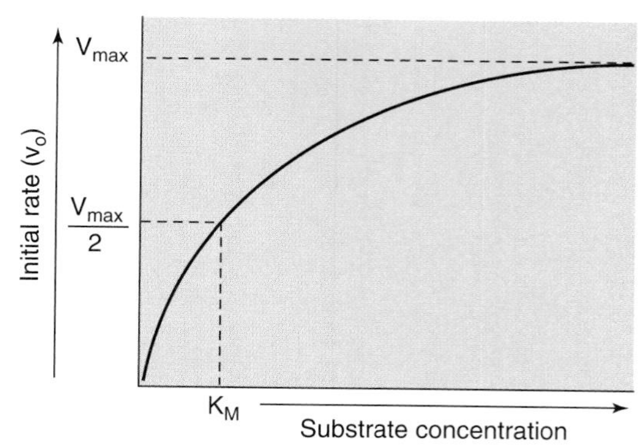

FIGURE 3.13 Calculation of V_{max} and K_M. V_{max} and K_M values can be determined from a hyperbolic curve generated by plotting the initial rates of reaction, v_o, on the y-axis, and substrate concentration on the x-axis. The theoretical maximum velocity, V_{max}, is the limiting velocity obtained as the substrate concentration approaches infinity. The K_M value corresponds to the substrate concentration at which the reaction rate is $V_{max}/2$.

The number of substrate molecules converted to product molecules by an enzyme molecule in a specified time is called the **turnover number**. The turnover number is determined by dividing V_{max} by the molar concentration of the total enzyme present in the reaction mixture.

The specific region on an enzyme responsible for binding the substrate(s) and catalyzing the reaction is called the **active site**. Despite the great variations that exist among enzymes, certain generalizations about the active site are possible. The active site is a cleft or crevice that is usually located on the surface of the protein and represents about 10–20% of the enzyme's total volume. The amino acid residues that form the active site come from different regions of the polypeptide chain, sometimes very far apart in the linear sequence. Much of the cleft is lined with hydrophobic side chains, creating an environment that excludes water molecules unless they are reactants. Polar and charged side chains are located at specific sites within the active site, where they contribute to substrate specificity and participate in the catalytic process. Amino acid residues within the active site that directly participate in making and breaking chemical bonds are called catalytic residues.

Two models for enzyme binding have been proposed to explain the extraordinary specificity of enzyme-substrate binding. The first of these, the **lock and key model** (FIGURE 3.14a), was proposed by Emil Fischer in 1894 to explain how an enzyme recognizes its specific substrate. According to this model, each substrate has a characteristic geometric shape that fits into a complementary geometric shape in the enzyme (Figure 3.14a). That is, the enzyme behaves like a rigid lock that will only accept a specific substrate key. As more information became available about ES interactions, the lock and key model's shortcomings became evident. For example, the model predicts that an enzyme should be able to act on a molecule with the same shape as its substrate but with less bulky substituent groups. Enzymes are usually quite specific for their substrates, however, and rarely act on smaller molecules.

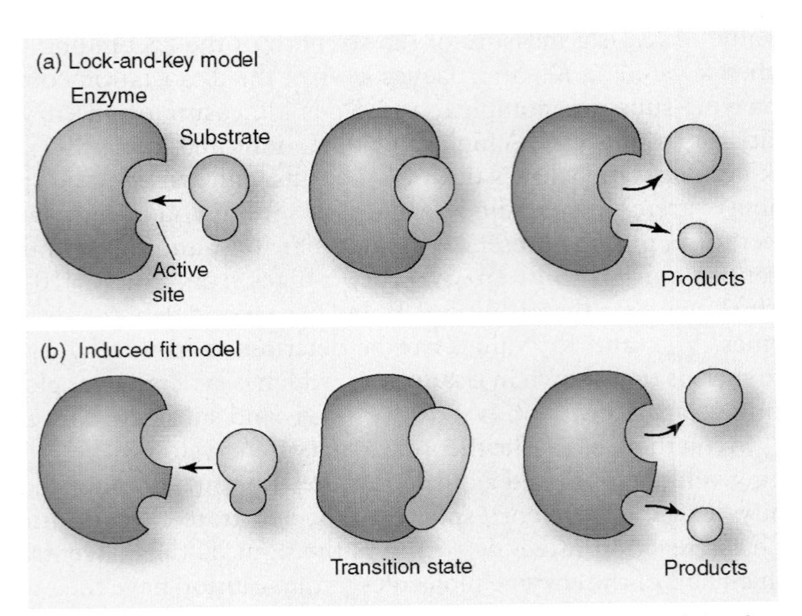

FIGURE 3.14 **Enzyme-substrate interaction.** There are two models of enzyme binding, (a) lock-and-key and (b) induced fit. In the lock-and-key model, the shape of the active site of the enzyme is complementary to the shape of the substrate. In the induced fit model, the enzyme changes shape upon binding the substrate and the active site has a shape that is complementary to that of the substrate only after the substrate is bound. (Adapted from C. K. Mathews, et al. *Biochemistry: Third edition.* Prentice Hall, 2000.)

The **induced fit model** (FIGURE 3.14b), proposed by Daniel Koshland in 1958, addressed the known shortcomings of the lock and key model. According to this model, the enzyme is a flexible structure that changes shape on binding substrate. The induced fit model explains an enzyme's inability to act on small substrate analogs as follows. A molecule with the same shape as the substrate but less bulky substituent groups would not be able to make all of the contacts needed to induce the conformational changes. Although the induced fit model is primarily concerned with the change in enzyme shape, it is important to note that the substrate's conformation also usually changes as it binds to the enzyme (Figure 3.14b). In fact, the strain to which the substrate is subjected is often the principal mechanism of catalysis, that is, the substrate is held in an enormously reactive conformation. The induced fit model was proposed before the three-dimensional structures of enzymes were known. X-ray crystallography and NMR spectroscopy made it possible to test the model. Studies of enzyme and enzyme-substrate complex structures show that most enzymes do change conformation on binding their substrate.

Molecular details for enzyme-substrate complexes have been worked out for many enzymes.

The first three-dimensional image of an enzyme was that for hen egg-white lysozyme. This enzyme cleaves certain bonds between sugar residues in some of the polysaccharide components of bacterial cell walls and is responsible for maintaining sterility within eggs.

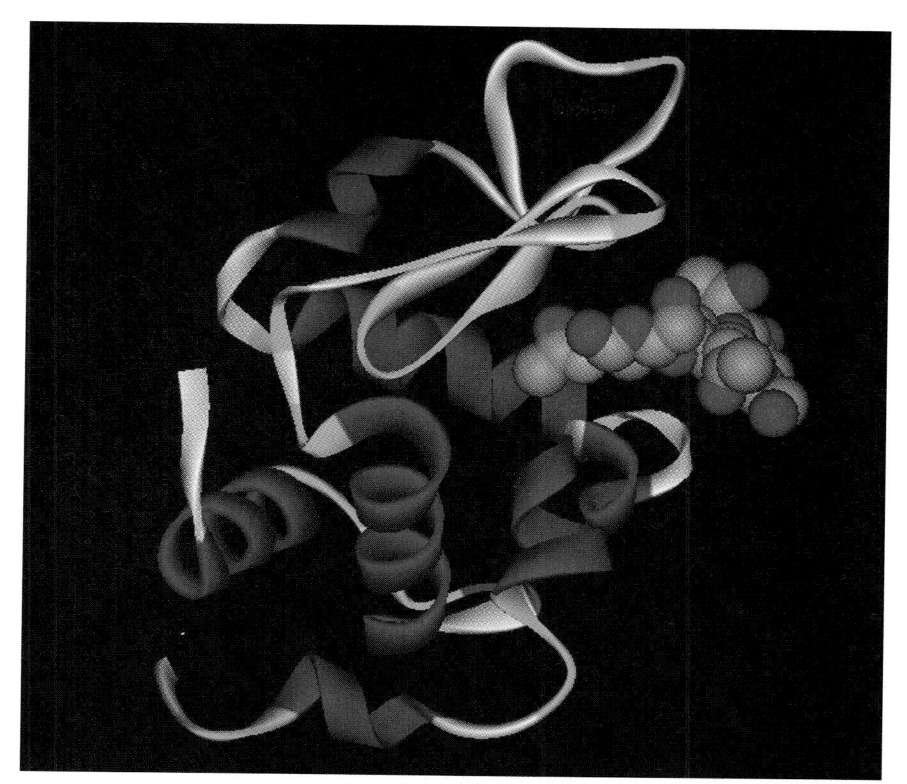

FIGURE 3.15 **Hen egg-white lysozyme in ribbon display.** Helical regions of the protein are shown in red and β-strands in cyan. The bound substrate analog is in van der Waals display. (Structure from Protein Data Bank 1LZB. K. Maenaka, et al., *J. Mol. Biol.* 247 [1995]: 281–293. Prepared by B. E. Tropp.)

FIGURE 3.15 shows egg-white lysozyme bound to a substrate analog. The amino acids that are part of the active site form widely separated clusters along the chain. Only when the chain folds do they come into proximity and form the active site. The true substrate is a hexasaccharide segment that fits into the cleft and is distorted upon binding. The enzyme itself changes shape when the substrate is bound. A variety of interactions (van der Waals, hydrogen, and ionic bonds) stabilize the binding. Another enzyme, yeast hexokinase A, which catalyzes the reaction of glucose and adenosine 5′-triphosphate (ATP) to form glucose-6-phosphate and adenosine 5′-diphosphate (ADP), has been studied in order to examine conformational changes that take place on substrate binding. The enzyme contains two domains, which move together when glucose binds. This movement creates a catalytic site that is inaccessible to the surrounding water molecules. These changes are shown in the pair of spacefilling models in FIGURE 3.16.

Regulatory enzymes control committed steps in biochemical pathways.

Enzyme regulation offers a quick and efficient way to adjust the flow through biochemical pathways, allowing a cell to synthesize products that are in short supply and to stop the synthesis of products that are in abundance. Regulation is usually achieved by control of flow through

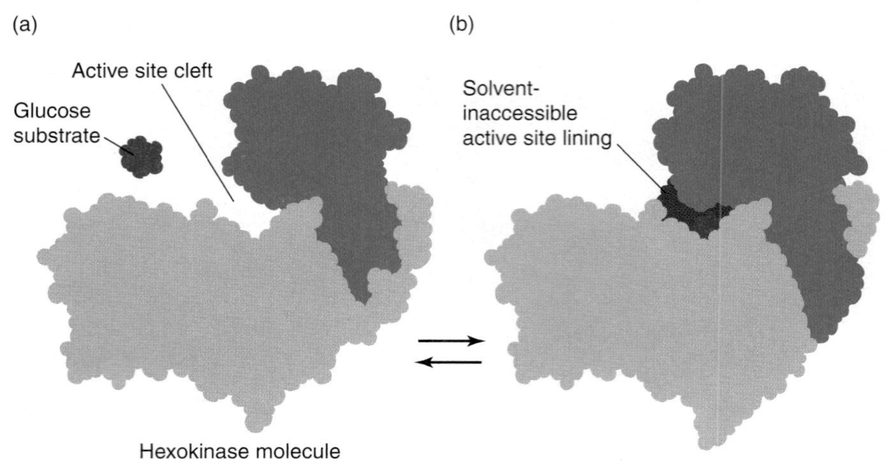

FIGURE 3.16 **Hexokinase, an example of induced fit.** A drawing, roughly to scale, of an idealized hexokinase molecule (a) without and (b) with bound glucose. The two hexokinase domains move together when glucose is bound, creating the catalytic site. The dark blue area in (b) represents solvent inaccessible surface area in the active site cleft that results when the enzyme binds glucose. (Photos courtesy of Thomas A. Steitz, Yale University.)

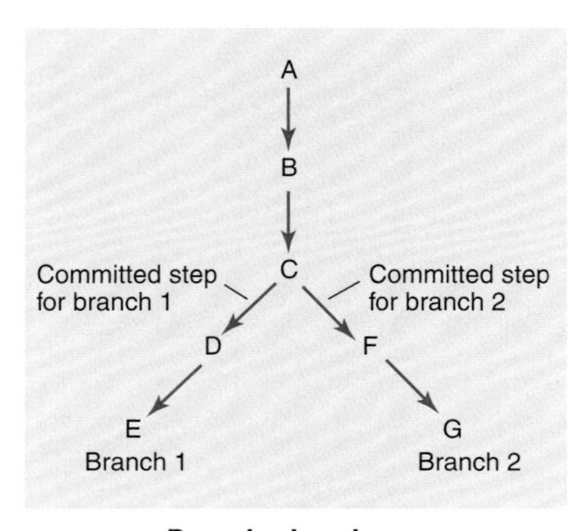

FIGURE 3.17 **Branched pathway.**

a few key steps in a biochemical pathway, such as the hypothetical example shown in FIGURE 3.17. In many cases, the cell requires only the end products of the pathway (in this example E and G). Pathway intermediates are essential to synthesize E and G but are not otherwise needed by the cell. The pathway has a branchpoint at C. The first reaction after a branchpoint is almost always irreversible, committing the flow of material to that branch. Thus, the first step after a branch is called the **committed step** for that branch. Enzymes that catalyze the committed steps in a reaction pathway are usually **regulatory enzymes** that are inhibited by the end product of the branch. Thus, E, the end product of the C → D → E branch, inhibits the enzyme that converts C to D. Likewise, G, the end product of the C → F → G branch, inhibits the enzyme that converts C to F. An abundance of both E and G can block the third key regulatory enzyme in the path, the enzyme that converts A to B. This can occur in one of three ways (FIGURE 3.18). C will accumulate when both branches are blocked and inhibit the enzyme that converts A to B (Figure 3.18a). Alternatively, E and G will act together to block the enzyme that converts A to B (Figure 3.18b). Sometimes both types of inhibition contribute (Figure 3.18c). The method of control, in which an end product inhibits specific steps in a biochemical pathway, is called **feedback inhibition.**

Regulatory enzymes exhibit sigmoidal kinetics and are stimulated or inhibited by allosteric effectors.

Regulatory enzymes often produce S-shaped or **sigmoidal kinetic curves** when the initial velocity (v_o) is plotted on the y-axis and substrate concentration ([S]) is plotted on the x-axis (FIGURE 3.19). The reason for the S-shaped curve is exactly the same as that given for the

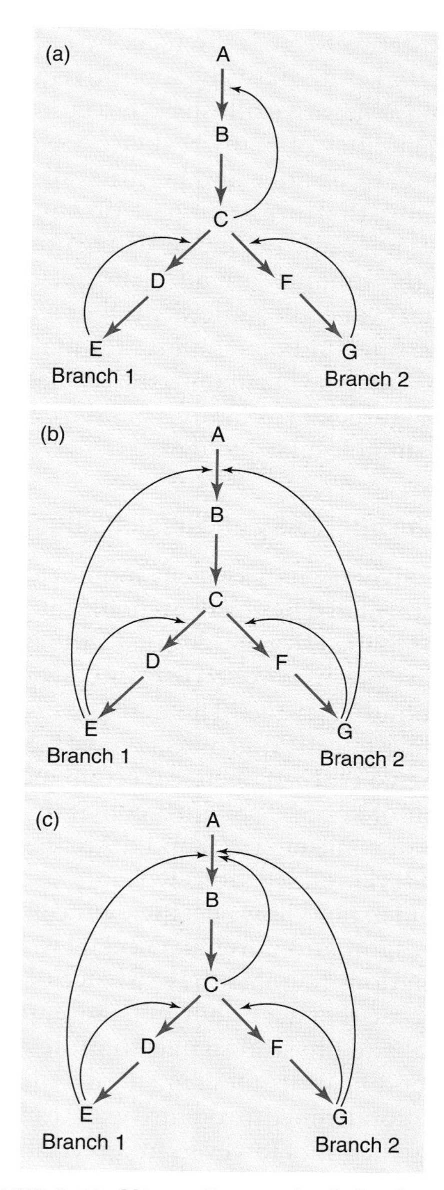

FIGURE 3.18 **Alternative methods for feedback inhibition to take place in a pathway with two branch points.**

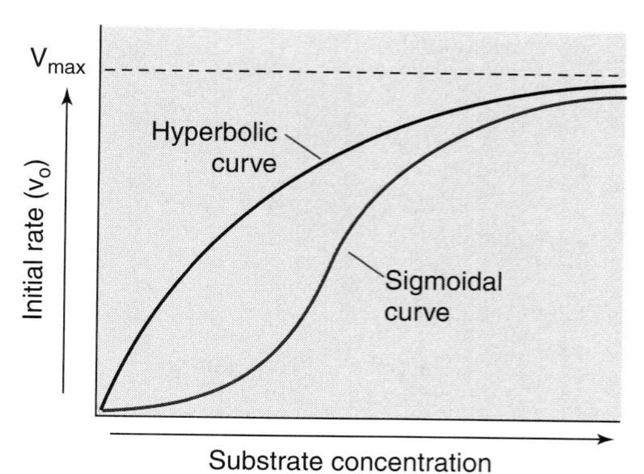

FIGURE 3.19 **Sigmoidal kinetics for regulatory enzymes.** Plots of substrate concentration versus initial velocity (v_o) generate hyperbolic curves for non-regulatory enzymes but usually generate sigmoidal curves for regulatory enzymes. The sigmoidal curve indicates that substrate binding is cooperative, that is, the binding of a substrate molecule to one enzyme subunit increases the chances that other enzyme subunits will also bind substrate molecules.

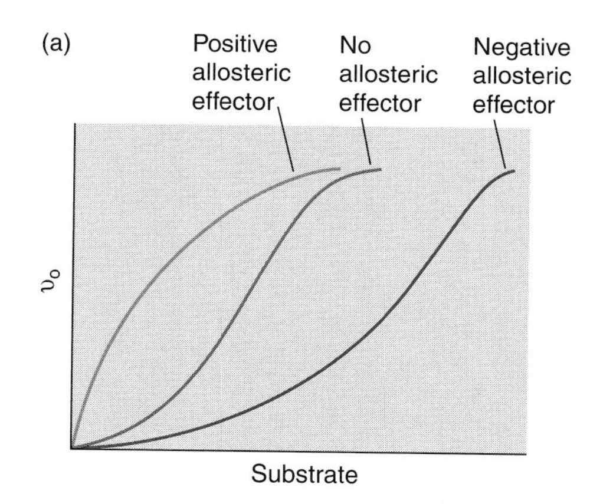

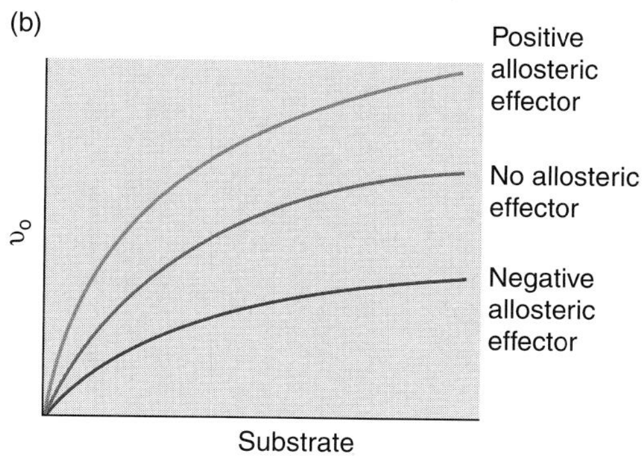

FIGURE 3.20 **Effects of positive and negative allosteric effectors.** (a) Allosteric effectors can influence concentrations required to give $V_{max}/2$ or (b) allosteric effectors can influence V_{max}.(Reproduced from B. E. Tropp. *Biochemistry: Concepts and Applications, First edition.* Brooks/Cole Publishing Company, 1997. Used with permission of B. E. Tropp.)

oxygen-binding curve for hemoglobin. The regulatory enzyme is made of subunits that bind substrate in a cooperative fashion.

Regulatory enzymes are also inhibited and stimulated by allosteric effectors. Once again, hemoglobin is a useful model. Subunits of an allosteric enzyme can exist in two different conformations, a more active conformation and a less active one. Allosteric effectors act by binding to sites on the enzyme that are distinct from the catalytic site. Positive allosteric effectors stabilize the more active form of the enzyme, whereas negative allosteric effectors stabilize the less active form. Allosteric effectors can be recognized by their effects on enzyme kinetics (FIGURE 3.20). Positive allosteric effectors often act by lowering the substrate concentration required to give half the maximal velocity,

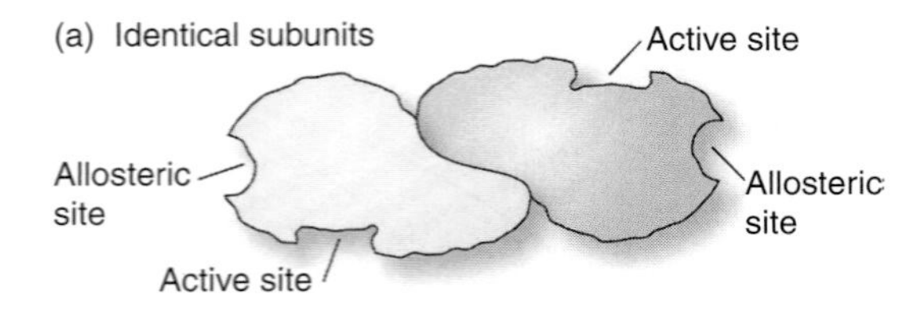

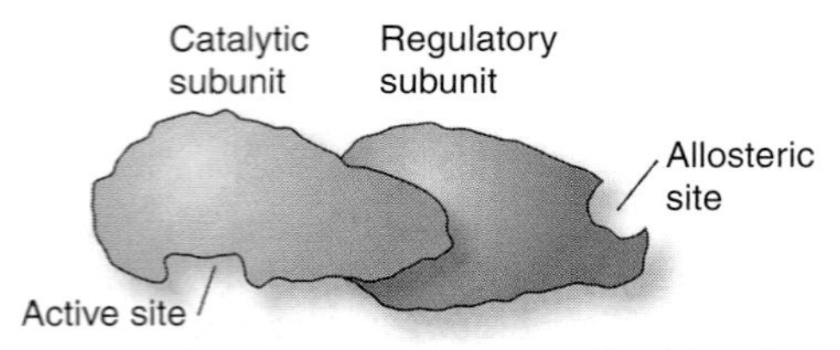

FIGURE 3.21 **Regulatory enzymes.** Nearly all regulatory enzymes are made up of two or more subunits. The subunits of a regulatory enzyme may be (a) identical to one another or (b) different from one another.

whereas negative allosteric effectors often act by increasing this concentration (Figure 3.20a). Alternatively, a positive allosteric effector may increase V_{max} and a negative allosteric effector may lower V_{max} (Figure 3.20b). The activities of many regulatory enzymes are influenced by a combination of positive and negative allosteric effectors.

Regulatory proteins that are inhibited or stimulated by allosteric effectors are almost always made of two or more subunits (FIGURE 3.21). When the subunits are all identical, each subunit has an active site and at least one allosteric site (Figure 3.21a). When the subunits are different, one kind of subunit, the **catalytic subunit**, has the active site and another type of subunit, the **regulatory subunit**, has the allosteric site(s) (Figure 3.21b). Both types of regulatory enzymes play important roles in molecular biology.

Enzyme activity can be altered by covalent modification.

Allosteric regulation involves noncovalent modification of enzymes. Enzymes also may be regulated by covalent modification. The activities of some enzymes are modified by peptide bond cleavages. For instance, trypsinogen, an inactive precursor of trypsin, is converted to the active enzyme by peptide bond cleavage. The activities of other enzymes are modified by adding chemical groups. We conclude this chapter by briefly examining two types of chemical group modifications. The first, phosphorylation of specific seryl, threonyl, or tyrosyl residues, increases the activity of some enzymes and decreases the activities of others. These phosphorylation reactions were discovered by biochemists who wished to understand how specific extracellular signals such as the hormones glucagon and adrenaline regulate cell metabolism. The resulting investigations revealed the existence of proteins embedded in the cell membrane that act as hormone receptors that transmit extracellular chemical and physical signals to proteins inside the cell. The second protein modification, attachment of a small protein called **ubiquitin** to specific condemned proteins, was originally shown to mark proteins for destruction. However, **ubiquitylation** (the addition of ubiquitin to a protein) now also appears to be a signal involved in the regulation of a variety of biological processes.

3.4 G-Protein-Linked Signal Transduction Systems

G-protein-linked signal transduction systems convert extracellular chemical or physical signals into intracellular signals.

For cells to interact with their environment and to communicate with other cells, they must be able to transfer information across their cell membrane. One method for doing so involves transport of materials across the cell membrane. Eukaryotic cells also use another method, in which receptor proteins embedded in the cell membrane transfer information (but not chemical substances) across the cell membrane.

These receptor proteins can be grouped into families based on structure and function. The largest family in vertebrates, the **G protein coupled-receptor** (**GPCR**) family, includes more than one thousand different proteins. The G protein-coupled receptor is part of the **G protein-linked signal transduction system.** In its most basic form this system also includes a **G protein complex**, and an **effector**.

FIGURE 3.22 is a computer-generated, composite image that shows how GPCR might interact with the G protein complex. The G protein complex contains three subunits: G_α, G_β, and G_γ. The G_α and G_γ subunits have attached lipids that insert into the lipid monolayer on the cytoplasmic side of the cell membrane. Further work is required to establish the precise three-dimensional relationship that exists between the G protein complex and GPCR. Progress will depend on preparing crystal preparations of the desired complex.

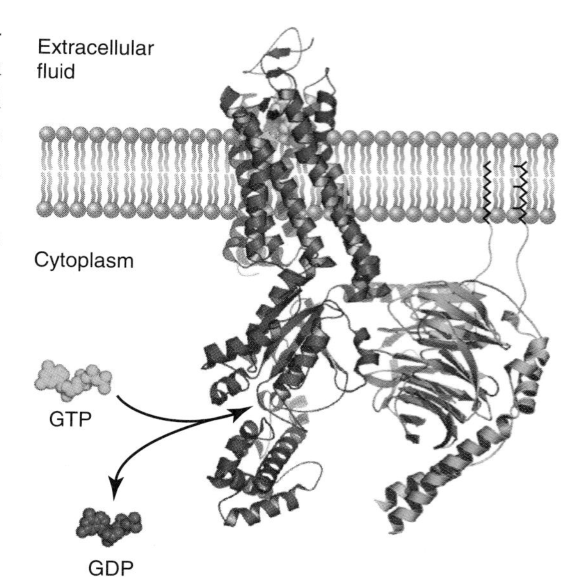

FIGURE 3.22 A model for G protein heterotrimer interaction with a G protein-coupled receptor (GPCR) and the lipid bilayer. The G-protein signal receptor is shown in red. The ligand is shown as a yellow spacefilling structure. The G_α subunit is blue, the G_β subunit is green, and G_γ is gold. GPCR and the $G_\alpha G_{\beta\gamma}$ complex structures are based on crystal structure data but their arrangement with respect to one another is speculative. (Photo courtesy of Heidi E. Hamm and Will Oldham, Vanderbilt University Medical Center.)

1. **G protein-coupled receptor (GPCR).** Each kind of GPCR responds to a unique extracellular physical or chemical signal. A receptor is said to be activated after interacting with the signal that is specific for it. These signals range in size from a single photon to a polypeptide such as the hormone glucagon. Although members of the GPCR family vary in amino acid sequences and specificities, they nevertheless share several structural features. These features include an extracellular N-terminal segment, an intracellular C-terminal segment, and seven α-helical segments embedded in the cell membrane to form a transmembrane core (Figure 3.22). Three extracellular loops and three intracellular loops connect the helical segments. Chemical signals with low molecular masses, such as nucleotides, amines, and lipids, tend to bind to sites within the hydrophobic core formed by the transmembrane α-helices. Those with high molecular masses such as proteins bind to extracellular N-terminal segments, loops, or both. Recent studies suggest that GPCRs are present in the cell membrane as homodimers or heterodimers. The functional implications of this dimeric structure are under investigation. For simplicity, GPCR is shown as a monomer in Figure 3.22.
2. **G protein complex.** The G protein complex transfers signals from an activated GPCR to an effector (see below). The "G" in the protein complex's name derives from the G_α subunit's ability to bind and hydrolyze guanosine 5′-triphosphate (GTP). Humans produce twenty different variants of G_α, six of G_β, and twelve of G_γ for a total of 1440 possible combinations. Many, but not all, of these combinations actually exist. The combination present in a heterotrimer is probably dictated by specific interactions among subunits. The G_α subunit exists in three conformational states, depending on whether it has GDP, GTP, or no nucleotide bound to it. G_α•GDP binds to the activated receptor whereas G_α•GTP regulates the activity of an effector protein.

Each G-protein-linked signal transduction system converts a specific extracellular signal into an intracellular response

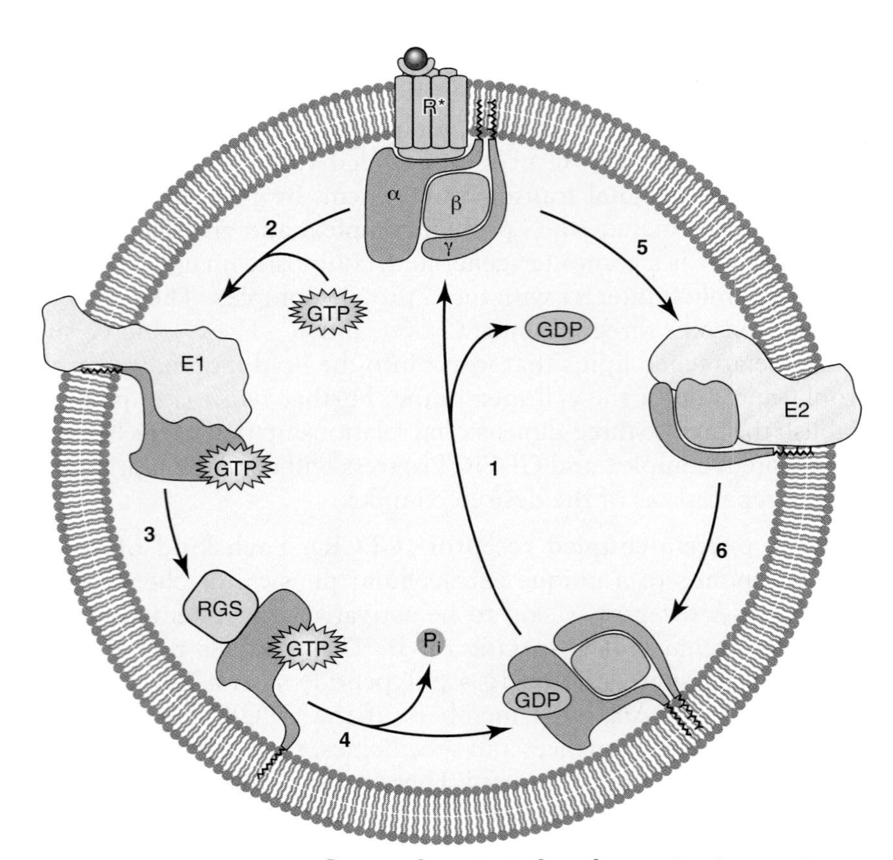

FIGURE 3.23 **Steps in the G-protein-linked signal transduction system.** (1) Activated GPCR (R*) interacts with the G_α(GDP)$G_{\beta\gamma}$ complex, causing G_α to release its bound GDP. (2) The G_α subunit binds GTP and the resulting G_α • GTP dissociates from the heterotrimer and binds to effector protein E1. (3) RGS (regulator of G protein signaling) protein interacts with the G_α • GTP complex. (4) This interaction accelerates the rate at which G_α hydrolyzes GTP. (5) The released $G_{\beta\gamma}$ complex binds to effector protein E2. (6) The G_α • GDP complex generated in step 4 combines with $G_{\beta\gamma}$ complex to re-form the G_α(GDP)$G_{\beta\gamma}$ complex and the cycle is ready to begin again. (Modified from W. H. Oldham and H. Hamm, Structural basis of function in heterotrimeric G proteins, *Q. Rev. Biophys.*, volume 39, issue 2, pp. 117–166, 2006 © Cambridge Journals, reproduced with permission.)

using the cyclic pathway shown in FIGURE 3.23. The steps in this pathway are as follows: (1) Activated GPCR interacts with the G_α(GDP)$G_{\beta\gamma}$ complex, causing G_α to release its bound GDP. (2) The G_α subunit binds GTP, which is present at a much higher intracellular concentration than GDP. The resulting G_α•GTP dissociates from the $G_{\beta\gamma}$ heterodimer and binds to effector protein E1, regulating its activity. (3) This regulation is short-lived because **RGS (regulator of G protein signaling)** protein interacts with the G_α•GTP complex. (4) This interaction accelerates the rate at which G_α hydrolyzes GTP. (5) The released $G_{\beta\gamma}$ complex binds to effector protein E2, regulating its activity. (6) The G_α•GDP complex generated in step 4 combines with $G_{\beta\gamma}$ complex to re-form the G_α(GDP)$G_{\beta\gamma}$ complex and the cycle is ready to begin again.

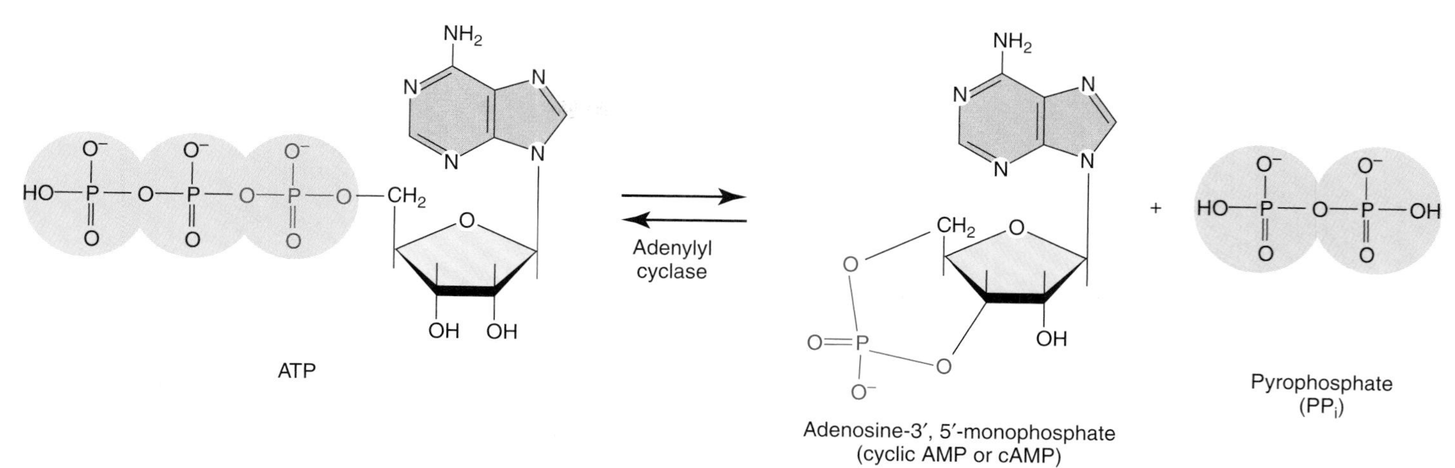

FIGURE 3.24 **Adenylyl cyclase.** Adenylyl cyclase catalyzes the reversible conversion of ATP to 3′,5′-cyclic AMP and pyrophosphate. The reaction is driven to completion when pyrophosphate is irreversibly hydrolyzed to form inorganic phosphate.

3. **Effectors.** Cells require a wide variety of effectors to allow them to respond effectively to the myriad chemical and physical stimuli they encounter. Table 3.1 lists some of the effector proteins that are regulated by the G_α•GTP and $G_{\beta\gamma}$ complexes. The present discussion is limited to one type of **adenylyl cyclase** and one type of **phospholipase C.**

   Adenylyl cyclase is an integral membrane protein that converts ATP to cyclic AMP (cAMP) and pyrophosphate (FIGURE 3.24). Nine isozymes of mammalian adenylyl cyclase have been cloned. Each appears to have two transmembrane regions with functions that are not yet known, and two cytoplasmic regions that associate to form the catalytic site. Once formed, the cyclic nucleotide activates **cAMP-dependent protein kinase A** (FIGURE 3.25). Each protein kinase A molecule has two catalytic subunits (C) and two regulatory subunits (R) and so has the general formula R_2C_2. Each R-subunit has two cAMP binding sites. When cAMP binds to these sites,

TABLE 3.1 **Effectors Regulated by G_α • GTP and $G_{\beta\gamma}$**

Effectors Regulated by G_α • GTP	Effectors Regulated by $G_{\beta\gamma}$
Adenylyl cyclase	Adenylyl cyclase
Ca^{2+} channels	Ca^{2+} channels
Phospholipase C	Phospholipase C
Phospholipase A_2	Phospholipase A_2
cGMP-phosphodiesterase	β-Adrenergic receptor kinase Mitogen-activated protein kinase

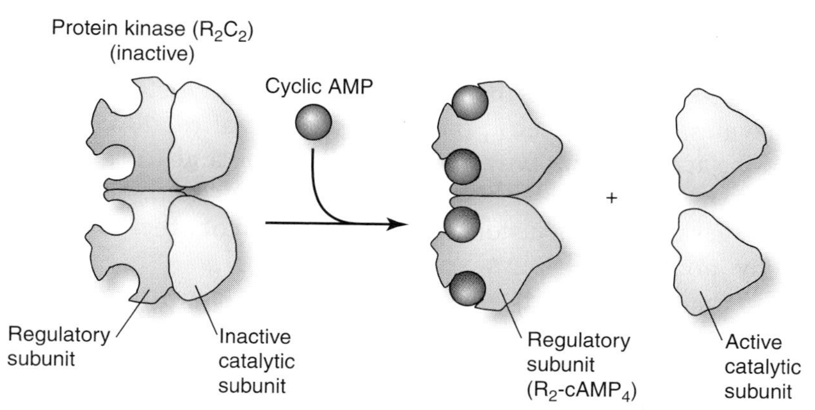

FIGURE 3.25 **Protein kinase activation by cAMP.** cAMP binds to the regulatory (R) subunits of protein kinase A, releasing the active catalytic (C) subunits. Then the C subunits catalyze the ATP-dependent phosphorylation of other proteins.

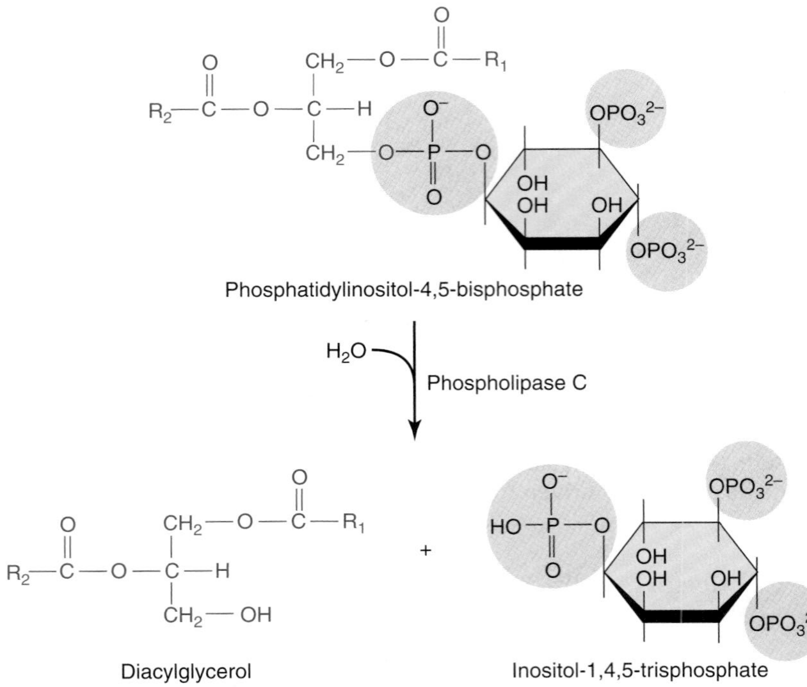

Phosphatidylinositol-4,5-bisphosphate

H_2O — Phospholipase C

Diacylglycerol + Inositol-1,4,5-trisphosphate

FIGURE 3.26 Phospholipase C activity.

the R_2 dimer is released and the C-subunits, now free of the regulatory proteins, catalyze phosphoryl group transfer from ATP to specific serine or threonine residues on target enzymes. Some enzymes become more active when phosphorylated while others become less active. A phosphodiesterase can hydrolyze cAMP, converting it to 5′-AMP and blocking further stimulation. Protein phosphatases can remove phosphate groups from phosphorylated proteins.

Phosphoinositide-specific phospholipase C (PLC) is also associated with the cell membrane. It catalyzes the hydrolytic cleavage of phosphatidylinositol 4,5-bisphosphate to form diacylglycerol and inositol 1,4,5-trisphosphate (FIGURE 3.26). Both products act as intracellular messengers to regulate biological processes (FIGURE 3.27). Inositol 1,4,5-trisphosphate diffuses away from the cell membrane and stimulates the endoplasmic reticulum to release its stored calcium ions to the cytoplasm, where the ions alter the activity of a wide variety of enzymes. Diacylglycerol, the other product of the PLC catalyzed reaction, acts together with phosphatidylserine and calcium ions to activate protein kinase C or C-kinase. Upon activation, protein kinase C catalyzes the ATP-dependent phosphorylation of other proteins and thereby modifies their physiological activity.

External signals such as photons, odorants, growth factors, hormones, and neurotransmitters act as "first messengers" to stimulate the G protein–linked signal transduction complex, while cAMP, inositol 1,4,5-trisphosphate, and diacylglycerol act as "**second messengers**" to alter the activity of intracellular enzymes as well as membrane proteins. This remarkable mechanism allows a cell to amplify an external signal consisting of a single molecule and to then respond to it in an appropriate fashion.

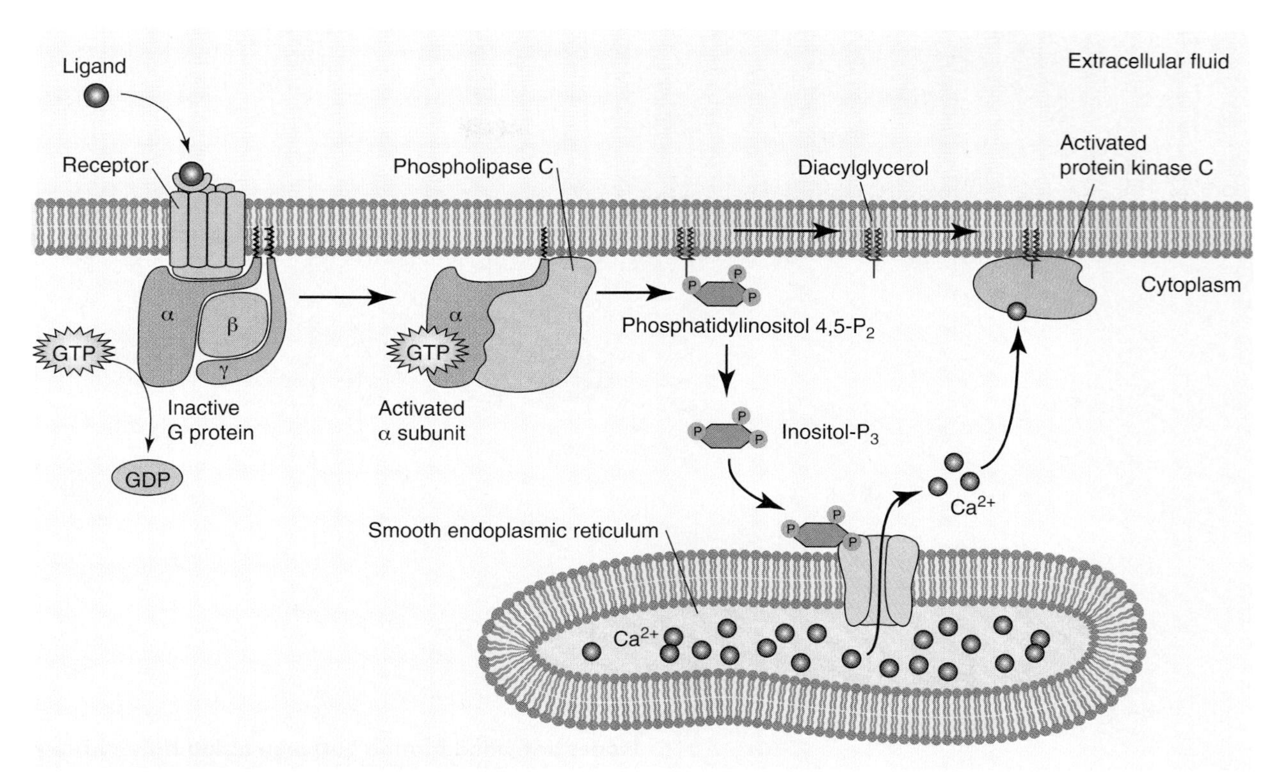

FIGURE 3.27 **Phosphoinositide cascade.** (Adapted from an illustration by George M. Helmkamp, Jr., School of Medicine, University of Kansas.)

3.5 The Ubiquitin Proteasome Proteolytic Pathway

The ubiquitin proteasome system is responsible for the specific degradation of intracellular proteins.

A recurring theme in molecular biology, almost from its inception as a scientific discipline, is that the intracellular level of each protein is determined by transcriptional and translational regulation. Advances over the last two decades show that specific protein degradation also makes an important contribution to regulating intracellular protein levels. One reason for the delay in recognizing that eukaryotes can degrade proteins in a specific manner is that for many years investigators thought that lysosomes, digestive organelles discovered by Christian de Duve in the mid-1950s, were responsible for all, or nearly all, protein degradation. It is now clear that lysosomal protein degradation is limited to extracellular proteins, membrane proteins, some cytoplasmic proteins, and autophagocytosed proteins. An entirely new system known as the **ubiquitin proteasome system** is required for the specific degradation of cytoplasmic and nuclear proteins.

An important clue to the discovery of the ubiquitin proteasome system came from studies performed by Avram Hershko and Aaron Ciechanover in the late 1970s. These investigators were aware that

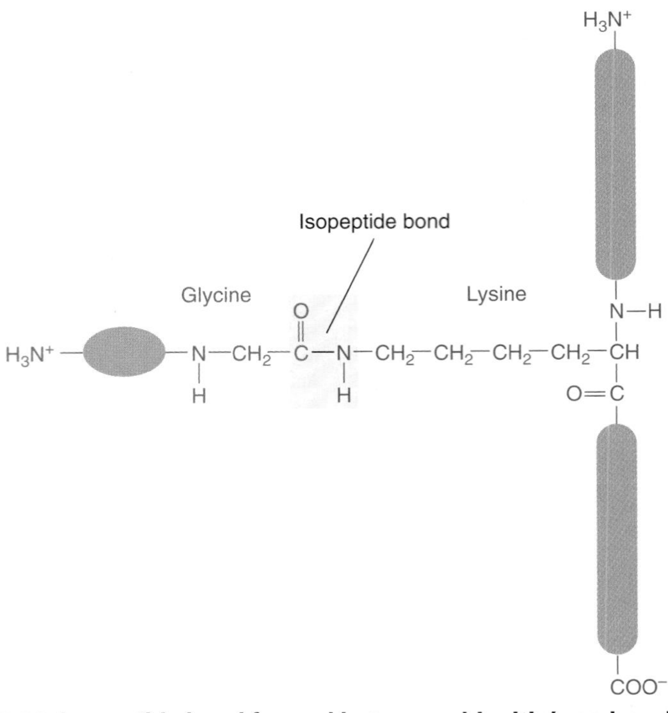

FIGURE 3.28 Isopeptide bond formed between ubiquitin's carboxyl terminal glycine residue and an ε-amine group of a lysine residue on the condemned protein. Ubiquitin is shown in blue and the condemned protein in red.

red blood cells lose the ability to degrade proteins when deprived of an energy source. Realizing that this energy-dependent protein degradation could not involve lysosomes because red blood cells lack this organelle, they decided to see if they could reproduce the effect *in vitro*. They observed that red blood cell extracts were able to degrade proteins if supplied with ATP. This was a surprising observation because protein hydrolysis is known to be a spontaneous process and therefore should not require energy to take place. They therefore decided to use classical enzyme fractionation techniques to discover why ATP is required for protein degradation. They were joined in this effort by Irwin Rose, a classically trained enzymologist. During the course of their investigations they discovered that **ubiquitin**, a short (76 residue), heat-stable protein must be present for protein degradation to take place. The term ubiquitin, which suggests the protein is found everywhere, is somewhat misleading. Although a highly conserved protein present in all eukaryotes, ubiquitin is not present in the other two domains of life, the bacteria or archaea. Further study by Hershko and coworkers revealed that ATP supplies the energy needed to attach a ubiquitin molecule to a condemned protein, marking the condemned protein for degradation. Attachment is through an **isopeptide bond** formed between ubiquitin's carboxyl terminal glycine residue and an ε-amine group of a lysine residue on the condemned protein

(FIGURE 3.28). In rare cases where the condemned protein lacks a lysine residue, ubiquitin's glycine is joined to the condemned protein's N-terminal residue. Ubiquitin attachment to a condemned protein is a three step process (FIGURE 3.29).

1. Ubiquitin's carboxyl terminal glycine is activated in an ATP-dependent reaction to form a thioester to a cysteine side chain on ubiquitin activating enzyme (E1). Most organisms have only one kind of E1.
2. Activated ubiquitin is transferred to a specific cysteine sulfhydryl group on ubiquitin carrier protein (E2). Mammals have more than 20 different kinds of E2.
3. Ubiquitin-E2 and the condemned protein bind to ubiquitin-protein ligase (E3), which attaches the ubiquitin to a lysine residue on the condemned enzyme to form the isopeptide bond. Mammals contain about a thousand different kinds of ubiquitin-protein ligases. These ligases may be divided into families based on structure. Each kind of ligase recognizes a specific set of proteins and marks them for degradation. The addition of a ubiquitin group to a condemned protein is called ubiquitylation (or ubiquitination).

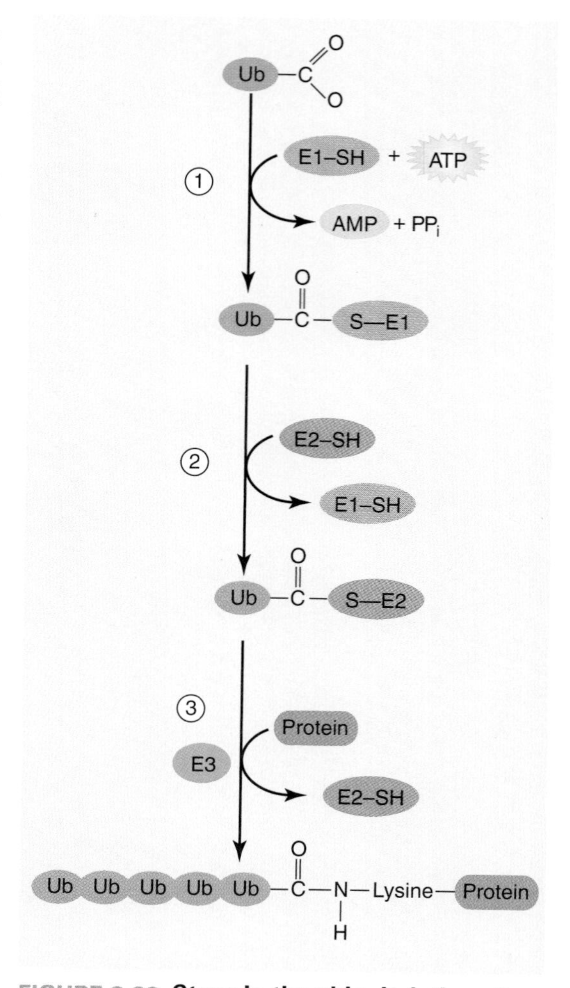

FIGURE 3.29 **Steps in the ubiquitylation of a condemned protein.** (1) Ubiquitin's carboxyl terminal glycine is activated in an ATP-dependent reaction to form a thioester to a cysteine side chain on ubiquitin activating enzyme (E1). (2) Activated ubiquitin is transferred to ubiquitin carrier protein (E2). (3) Ubiquitin-E2 and the condemned protein bind to ubiquitin-protein ligase (E3), which attaches the ubiquitin to a lysine residue on the condemned enzyme to form the isopeptide bond.(Adapted from T. Saric and A. L. Goldberg, *Encyclopedia of Life Sciences* [DOI: 10.1038/npg.els.ppp5722]. Posted January 27, 2006.)

For efficient protein degradation, the condemned protein must be attached to a chain of at least four ubiquitin molecules. The ubiquitin chain is formed by attaching the carboxyl terminal glycine of the incoming ubiquitin molecule to Lys 48 of the last ubiquitin molecule to be attached to the condemned protein. Each attachment is through an isopeptide bond just like that used to attach the first ubiquitin molecule to the condemned protein.

The discovery of ATP-dependent ubiquitin attachment to condemned proteins seemed to explain why ATP is needed to degrade proteins. It soon became clear, however, that ATP was also needed in the next stage of protein degradation. In 1987, groups led by Alfred L. Goldberg and Martin Rechsteiner reported the discovery of a high molecular weight protease that can degrade ubiquitylated proteins. Further study revealed that this large protease, now known as the **proteasome,** consists of a **core particle** and a **regulatory particle** (FIGURE 3.30). The core particle is shaped like a barrel with four rings stacked on top of each other. The inner two rings, the β rings, are made of seven distinct but homologous β proteins. Protease activity is associated with three of the β proteins and faces the cavity so that peptide bond cleavage takes place inside the core particle. This internal location is a very important structural feature because it means that cytoplasmic proteins outside the proteasome will not be degraded. The two outer rings, the α rings, also consist of seven distinct but homologous proteins. One or both α rings is capped by a regulatory particle made of 17 distinct subunits. The regulatory particle requires ATP to open a gate in the α ring so that the condemned protein can enter the core particle. ATP also provides energy needed to unfold the condemned protein so that it can pass through the narrow open

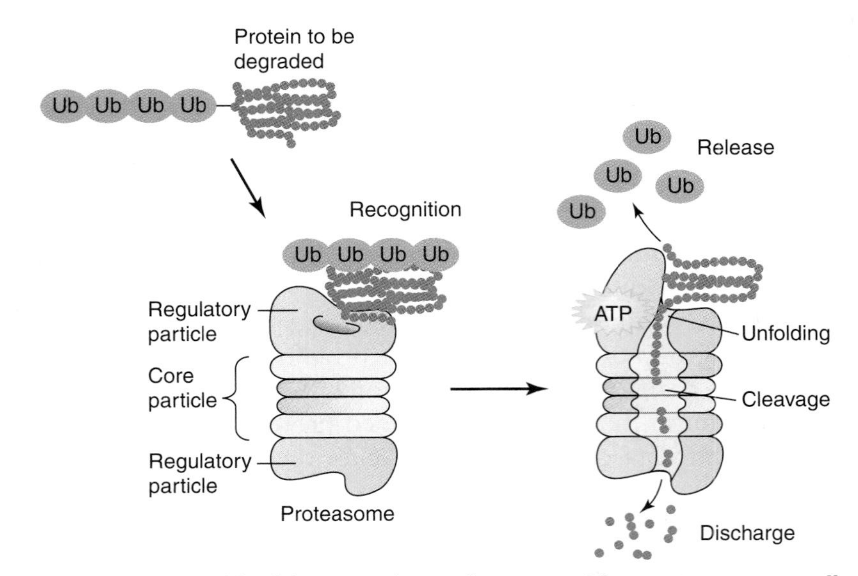

FIGURE 3.30 **Polyubiquitin-tagged protein targeted for proteasome mediated degradation.** A condemned protein (orange) marked with a polyubiquitin chain is targeted to the proteasome. Ubiquitin receptors in (or associated with) the regulatory particle associate with the polyubiquitylated condemned protein. As illustrated in the cut-away on the right, ATPases within the regulatory particle unfold the substrate and move it into the core particle. β-Subunit proteases cleave the condemned protein. Deubiquitylating enzymes that are part of the regulatory particle (or that bind to it) release ubiquitin so that it can be used again to mark other condemned proteins. The α and β rings in the core particle are blue and yellow, respectively. The regulatory particle is light purple. (Adapted from M. Hochstrasser, *Nature* 358 [2009]: 422–429.)

channel and reach the protease in β ring. Either the amino or carboxyl end of a condemned protein may enter the channel first. Short peptides are released during degradation. Deubiquitylating enzymes that are part of the regulatory particle (or that bind to the regulatory particle) release ubiquitin so that it can be used again to mark other condemned proteins.

Although ubiquitin was originally discovered as part of a protein degradation system, it has also been shown to modify proteins for other purposes. For instance, the attachment of a single ubiquitin molecule to a component of the DNA replication machinery triggers a specific kind of DNA damage repair. Cells also have ubiquitin like proteins that modify proteins and alter their activities. One such protein, SUMO (small ubiquitin-related modifier), alters the activity of an enzyme that removes thymine from a G-T mismatch in DNA.

Bacteria and archaea also have proteasome-like complexes. These large proteases also require ATP to unfold and degrade proteins in a central core. However, the condemned proteins are not marked with ubiquitin. For that reason, we will not examine the bacterial and archaeal proteasome-like structures.

Suggested Reading

General

Branden, C-I., and Tooze, J. 1999. *Introduction to Protein Structure* (2nd ed). New York: Garland.

Creighton, T. E. 1993. *Proteins: Structures and Molecular Properties.* New York: W. H. Freeman.

Petsko, G. A., and Ringe, D. 2003. *Protein Structure and Function.* Sunderland, MA: Sinauer Associates.

Myoglobin, Hemoglobin, and the Quaternary Structure Concept

Allison, A. C. 2002. The discovery of resistance to malaria of sickle cell heterozygotes. *Mol Biol Educ* 30:279–287.

Dickerson, R. E., and Geis, I. 1983. *Hemoglobin: Structure, Function, and Pathology.* Upper Saddle River, NJ: Benjamin Cummings.

Edsall, J. T. 1980. Hemoglobin and origins of the concept of allosterism. *Fed Proc* 39:226–235.

Garry, D. J., Ordway, G. A., Lorenz, J. N., et al. 1998. Mice without myoglobin. *Nature* 395:905–908.

Gelin B. R., Lee, A. W. M., and Karplus, M. 1983. Hemoglobin tertiary structural change on ligand binding: its role in the co-operative mechanism. *J Mol Biol* 171:489–559.

Ho, C., and Lukin, J. A. 2001. Haemoglobin: cooperativity in protein-ligand interactions. *Encyclopedia of Life Sciences.* pp. 1–11. London: Nature.

Hsia, C. C. W. 1998. Respiratory function of hemoglobin. *N Engl J Med* 338:239–248.

Ishimori, K. 2002. Myoglobin. *Encyclopedia of Life Sciences.* pp. 1–5. London: Nature.

Jones, S., and Thornton, J. M. 2001. Protein quaternary structure: subunit-subunit interactions. *Encyclopedia of Life Sciences.* pp. 1–7. London: Nature.

Manning, J. M., Dumoulin, A., Li, X., and Manning, L. R. 1998. Normal and abnormal protein subunit interactions in hemoglobins. *J Biol Chem* 273:19359–19362.

Perutz, M. F., 1978. Hemoglobin structure and respiratory transport. *Sci Am* 239:92–125.

Perutz, M. F., Wilkinson, A. J., Paoli, M., and Dodson, G. G. 1998. The sterochemical mechanism of the cooperative effects in hemoglobin revisited. *Ann Rev Biophys Biomol Struct* 27:1–34.

Springer, B.A., Sliger, S. G., Olson, J. S., and Phillips, G. N. Jr. 1994. Mechanism of ligand recognition in myoglobin. *Chem Rev* 94:699–714.

Strasser, B. J. 1999. Sickle cell anemia, a molecular disease. *Science* 286:1488–1490.

Stuart, M. J., and Nagel, R. L. 2004. Sickle cell disease. *Lancet* 364:1343–1360.

Wittenberg, J. B., and Wittenberg, B. A. 1990. Mechanisms of cytoplasmic hemoglobin and myoglobin function. *Ann Rev Biophys* 19:217–241.

Immunoglobulin Structure and the Domain Concept

Alzari, P. M., Lascombe, M. B., and Poljak, R. J. 1988. Three-dimensional structure of antibodies. *Ann Rev Immunol* 6:555–580.

Capra, J. D., and Edmundson, A. B. 1977. The antibody combining site. *Sci Am* 236:50–59.

Edelman, G. M. 1970. The structure and function of antibodies. *Sci Am* 223:34–42.

Kumagai, I., and Tsumoto, K. 2001. Antigen-antibody binding. *Encyclopedia of Life Sciences.* pp. 1–7. London: Nature.

Lucas, A. H. 2001. Antibody function. *Encyclopedia of Life Sciences.* pp. 1–8. London: Nature.

Wingren, C., Alkner, U., and Hansson, U. B. 2001. Antibody classes. *Encyclopedia of Life Sciences*. pp. 1–8. London: Nature.
Zanetti, M., and Capra, J (eds).1999. *The Antibodies*, vol. 6. New York: Harwood Academic.
Zouali, M. 2001. Antibodies. *Encyclopedia of Life Sciences*. pp. 1–8. London: Nature.

Enzymes

Bugg, T. D. H. 2001. Enzymes: general properties. *Encyclopedia of Life Sciences*. pp. 1–8. London: Nature.
Cleland, W. W. 2001. Enzyme kinetics: steady state. *Encyclopedia of Life Sciences*. pp. 1–5. London: Nature.
Copeland, R. A. 2001. Enzymology methods. *Encyclopedia of Life Sciences*. pp. 1–5. London: Nature.
Cornish-Bowden, A. 1995. *Fundamentals of Enzyme Kinetics*. London: Portland.
Fersht, A. (ed). 1998. *Structure and Mechanism in Protein Science: A Guide to Enzyme Catalysis and Protein Folding*. New York: W. H. Freeman.
Fisher, H. F. 2001. Protein-ligand interactions: molecular basis. *Encyclopedia of Life Sciences*. pp. 1–9. London: Nature.
Ghanem, E., and Raushel, F. M. 2007. Enzymes: The active site. *Encyclopedia of Life Sciences*. pp. 1–7. Hoboken, NJ: John Wiley and Sons.
Gigant, B., and Knossow, M. 2001. Catalytic antibodies. *Encyclopedia of Life Sciences*. pp. 1–7. London: Nature.
Hedstrom, L. 2001. Enzyme specificity and selectivity. *Encyclopedia of Life Sciences*. pp. 1–7. London: Nature.
Jencks, W. P. 1969. *Catalysis in Chemistry and Enzymology*. New York: McGraw-Hill.
Knowles, J. R. 1991. To build an enzyme. *Phil Trans R Soc Lond B Biol Sci* 332:115–121.
Koshland, D. K. Jr. 1994. The key-lock theory and the induced fit theory. *Angew Chem Intl Ed Engl* 33:2375–2378.
Martin, B. L. 2007. Regulation by covalent modification. *Encyclopedia of Life Sciences*. pp. 1–7. Hoboken, NJ: John Wiley and Sons.
Mobashery, S., and Kotra, L. P. 2002. Transition state stabilization. *Encyclopedia of Life Sciences*. pp. 1–6. London: Nature.
Nadaraia, S., Yohrling IV, J. V., Jiang, G. C-T., Flanagan. J. M., and Vrana, K. E. 2007. Enzyme activity: control. *Encyclopedia of Life Sciences*. pp. 1–8. Hoboken, NJ: John Wiley and Sons.
Pliška, V. K. 2001. Substrate binding to enzymes. *Encyclopedia of Life Sciences*. pp. 1–10. London: Nature.
Post, C. B. 2002. Transition states: substrate induced conformational transitions. *Encyclopedia of Life Sciences*. pp. 1–7. London: Nature.
Rossomando, E. F. 1990. Measurement of enzyme activity. *Meth Enzymol* 182:38–49.
Schultz, P. G., and Lerner R. A. 1995. From molecular diversity to catalysis: lessons from the immune system. *Science* 269:1835–1842.
Scopes, R. K. 2001. Enzyme activity and assays. *Encyclopedia of Life Sciences*. pp. 1–6. London: Nature.
Toney, M. D. 2001. Binding and catalysis. *Encyclopedia of Life Sciences*. pp. 1–6. London: Nature.
Traut, T. 2001. Enzyme activity: allosteric regulation. *Encyclopedia of Life Sciences*. pp. 1–11. London: Nature.
Yohrling, G. J. IV, Jiang, G. C-T., and Flanagan, J. M, and Vrana, K. E. 2007. Enzymatic activity: control. *Encyclopedia of Life Sciences*. pp. 1–8. Hoboken, NJ: John Wiley and Sons.

G-Protein-Linked Signal Transduction Systems

Bockaert, J. 2009. G-protein-coupled receptors. *Encyclopedia of Life Sciences*. pp. 1–10. Hoboken, NJ: John Wiley and Sons.

Bourne H. 1997. How receptors talk to trimeric G proteins. *Curr Opin Cell Biol* 9:134–142.

Cherezov, V., Rosenbaum, D. M., Hanson, M.A., et al. 2007. High-resolution crystal structure of an engineered human β2-adrenergic G-protein-coupled receptor. *Science* 318:1258–1265.

Chuang, T. T., Iacovelli, L., Sallese, M., and De Blasi, A. 1996. G protein coupled receptors: heterologous regulation of homologous desensitization and its implications. *Trends Pharmacol Sci* 17:416–421.

Gether, U., and Kobilka, B. K. 1998. G protein-coupled receptors. II. Mechanism of agonist activation. *J Biol Chem* 273:17979–17982.

Gilman, A. G. 1987. G proteins: transducers of receptor-generated signals. *Ann Rev Biochem* 56:615–649.

Hamm H. E., and Gilchrist, A. 1996. Heterotrimeric G proteins. *Curr Opin Cell Biol* 8:189–196.

Hanson, M. A., and Stevens, R. C. 2009. Discovery of new GPCR biology: one receptor structure at a time. *Structure* 17:8–14.

Karoor, V., and Malbon, C. C. 1998. G-protein-linked receptors as substrates for tyrosine kinases: cross-talk in signaling. *Adv Pharmacol* 42:425–428.

Ji, T. H., Grossmann, M., and Ji, I. 1998. G protein-coupled receptors. I. Diversity of receptor-ligand interactions. *J Biol Chem* 273:17299–17302.

Malbon, C. C., and Wang, H. Y. 2001. Adrenergic receptors. *Encyclopedia of Life Sciences*. pp. 1–8. London: Nature.

Milligan, G., and Kostenis, E. 2006. Heterotrimeric G-proteins: a short history. *Br J Pharmacol* 147:S46–S55.

Schwartz, W., and Hubbell, W. 2008. A moving story of receptors. *Nature* 455:473–474.

Sutherland, E. 1972. Studies on the mechanism of hormone action. *Science* 177:401–408.

Oldham, W. M., and Hamm, H. E. 2006. Structural basis of function in heterotrimeric G proteins. *Q Rev Biophys* 39:117–166.

Oldham, W. M., and Hamm, H. E. 2008. Heterotrimer G protein activation by G-protein–coupled receptors. *Nature Rev Mol Cell Biol* 9:60–71.

Strader, C, Fong, T, Tota, M., et al. 1994. Structure and function of G-protein–coupled receptors. *Ann Rev Biochem* 63:101–132.

Weis, W. I., and Kobilka, B. K. 2008. Structural insights into G-protein–coupled receptor activation. *Curr Opin Struct Biol* 18:734–740.

Wess, J. 1997. G protein-coupled receptors: molecular mechanisms involved in receptor activation and selectivity of G-protein recognition. *FASEB J* 11:346–354.

The Ubiquitin Proteasome Proteolytic Pathway

Ciechanover, A. 2005. Proteolysis: from the lysosome to ubiquitin and the proteasome. *Nat Rev Mol Cell Biol* 6:79–86.

Goldberg, A. L. 2005. Nobel committee tags ubiquitin for distinction. *Neuron* 45:339–344.

Haglund, K., and Dikic, I. 2005. Ubiquitylation and cell signaling. *EMBO J* 24:3353–3359.

Hershko, A. 2005. The ubiquitin system for protein degradation and some of its roles in the control of the cell division cycle. *Cell Death Differ* 12:1191–1197.

Hochstrasser, M. 2009. Origin and function of ubiquitin-like proteins. *Nature* 458:422–429.
Rose, A. I. 2005. Ubiquitin at Fox Chase. *Proc Natl Acad Sci USA* 102:11575–11577.
Saric, T., and Goldberg, A. L. 2005. Protein degradation and turnover. *Encyclopedia of Life Sciences*. pp. 1–8. Hoboken, NJ: John Wiley and Sons.
Schmidt, M., and Finley, D. 2007. Protease complexes. *Encyclopedia of Life Sciences*. pp. 1–11. Hoboken, NJ: John Wiley and Sons.
Schwartz, A. L., and Ciechanover, A. 2009. Targeting proteins for destruction by the ubiquitin system: implications for human pathobiology. *Ann Rev Pharmacol Toxicol* 49:73–96.
Smith, D. M., Benaroudj, N., and Goldberg, A. 2006. Proteasomes and their associated ATPases: a destructive combination. *J. Struct Biol* 156:72–83.
Varshavsky, A. 2006. The early history of the ubiquitin field. *Protein Sci* 15:647–654.

This book has a Web site, **http://biology.jbpub.com/book/molecular**, which contains a variety of resources, including links to in-depth information on topics covered in this chapter, and review material designed to assist in your study of molecular biology.

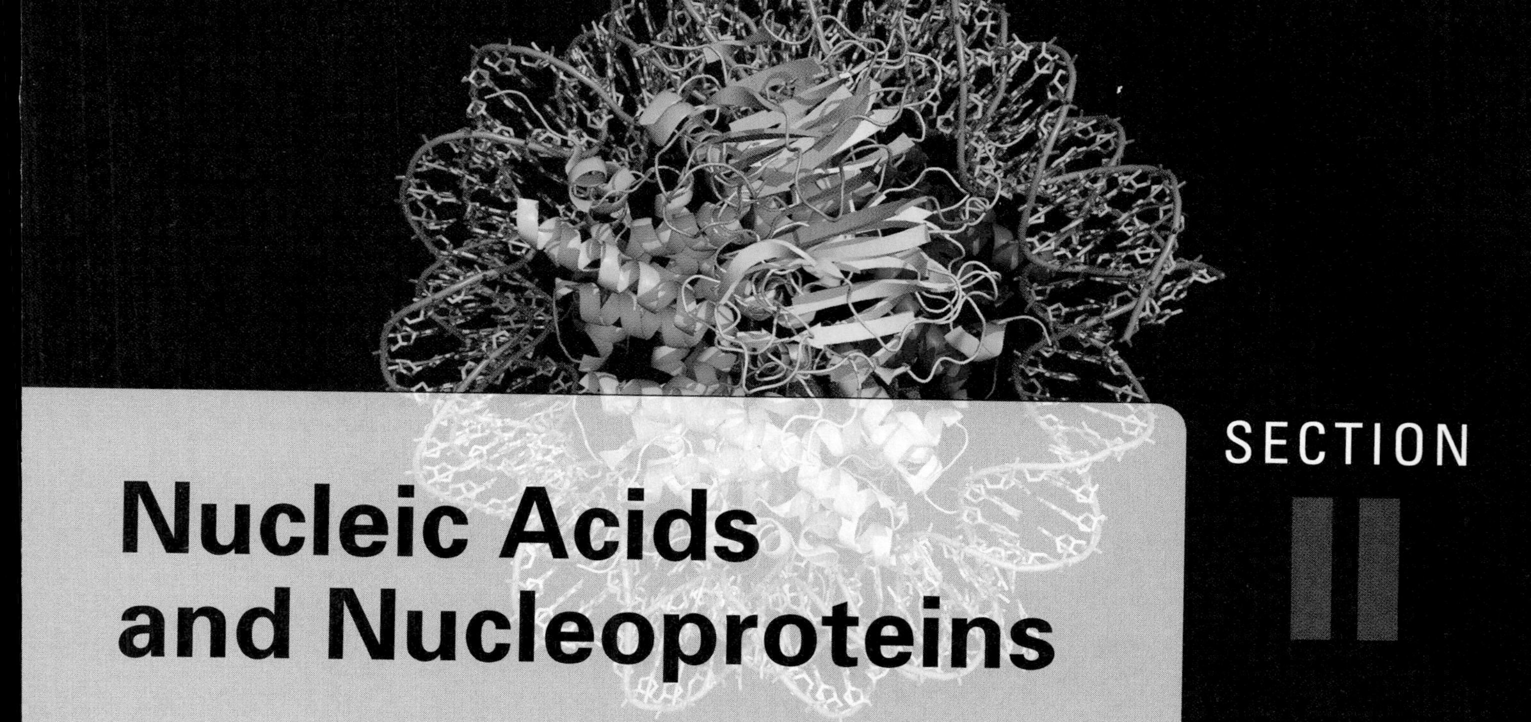

SECTION II

Nucleic Acids and Nucleoproteins

CHAPTER 4 **Nucleic Acid Structure**

CHAPTER 5 **Techniques in Molecular Biology**

CHAPTER 6 **Chromosome Structure**

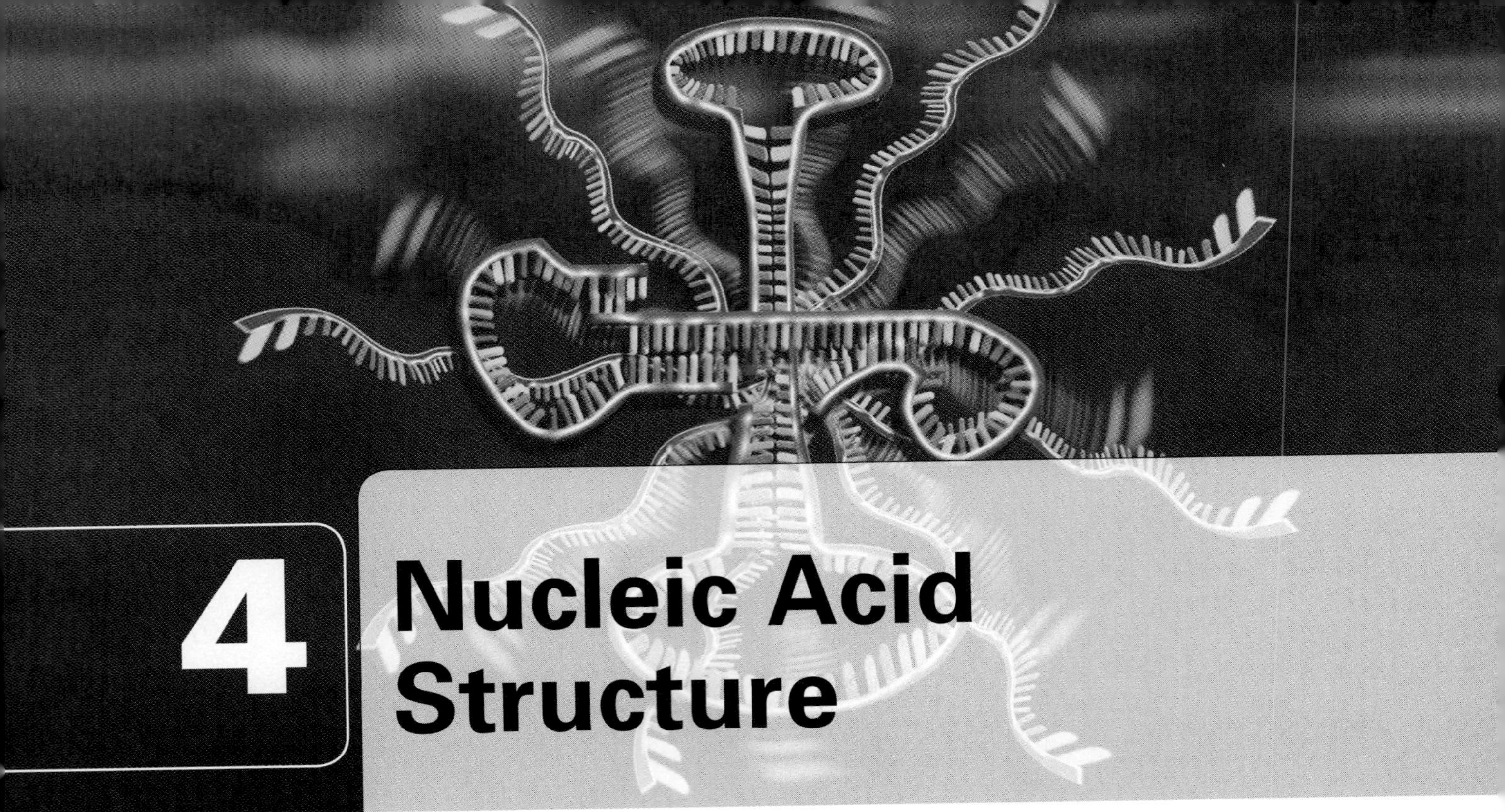

4 Nucleic Acid Structure

OUTLINE OF TOPICS

4.1 DNA Size and Fragility
DNA molecules vary in size and base composition.
DNA molecules are fragile.

4.2 Recognition Patterns in the Major and Minor Grooves
Enzymes can recognize specific patterns at the edges of the major and minor grooves.

4.3 DNA Bending
Some base sequences cause DNA to bend.

4.4 DNA Denaturation and Renaturation
DNA can be denatured.
Hydrogen bonds stabilize double-stranded DNA.
Base stacking also stabilizes double-stranded DNA.
Base stacking is a cooperative interaction.
Ionic strength influences DNA structure.
The DNA molecule is in a dynamic state.
Distant short patches of complementary sequences can base pair in single-stranded DNA.
Alkali denatures DNA without breaking phosphodiester bonds.
Complementary single strands can anneal to form double-stranded DNA.

4.5 Helicases
Helicases are motor proteins that use the energy of nucleoside triphosphates to unwind DNA.

4.6 Single-Stranded DNA Binding Proteins
Single-stranded DNA binding proteins (SSB) stabilize single-stranded DNA.

4.7 Topoisomers and Topoisomerases
Covalently closed circular DNA molecules can form supercoils.
Bacterial DNA usually exists as a covalently closed circle.
Plasmid DNA molecules are used to study the properties of circular DNA *in vitro*.
Circular DNA molecules often have superhelical structures.
Supercoiled DNA results from under- or overwinding circular DNA.
Superhelices can have single-stranded regions.
Topoisomerases catalyze the conversion of one topoisomer into another.
Enzymes belonging to the topoisomerase I family can be divided into three subfamilies.
Type II topoisomerases require ATP to convert one topoisomer into another.

4.8 Non-B DNA Conformations
A-DNA is a right-handed double helix with a deep major groove and very shallow minor groove.
Z-DNA has a left-handed conformation.
DNA conformational changes result from rotation about single bonds.
Several other kinds of non-B DNA structures appear to exist in nature.

4.9 RNA Structure
RNA performs a wide variety of functions in the cell.
RNA secondary structure is dominated by Watson-Crick base pairs.
RNA tertiary structures are stabilized by interactions between two or more secondary structure elements.

4.10 The RNA World Hypothesis
The earliest forms of life on earth may have used RNA as both the genetic material and the biological catalysts needed to maintain life.

Suggested Reading

The first part of this chapter builds on information provided in Chapter 1 about B-DNA structure. More specifically, it explores DNA size, fragility, grooves, bending, denaturation, renaturation, and superhelicity. This new information, together with information provided in Chapters 2 and 3 about proteins and enzymes, is applied to study enzymes that unwind double-stranded DNA, proteins that stabilize single-stranded DNA, and enzymes that catalyze changes in superhelical structures. Although B-DNA is the predominant DNA conformation inside the cell, other conformations also exist. Some of these conformations and their possible physiological significance are discussed.

The second part of the chapter examines RNA structure. As described in Chapter 1, ribonucleotide building blocks are linked by 5′→3′ phosphodiester bonds to form linear polyribonucleotide chains. Cells make many different kinds of RNA chains; each kind has a unique nucleotide sequence (primary structure) and size. Some RNA molecules can perform their functions as unstructured single strands. Many others, however, have distinct secondary and tertiary structures that must be formed for the molecule to perform its function. Although RNA chains lack the flexibility of polypeptide chains and their component nucleotides lack the variety of functional groups present in amino acid side chains, some RNA molecules fold into structures that bind specific substrates and catalyze chemical reactions. RNA molecules also interact with proteins to form stable ribonucleoprotein complexes. This chapter introduces some important aspects of RNA structure. More detailed information about particular RNA molecules and ribonucleoprotein complexes is presented in later chapters in which RNA functions are examined.

4.1 DNA Size and Fragility

DNA molecules vary in size and base composition.

DNA molecules exist in a wide range of sizes and base compositions in viruses, bacteria, archaea, and eukaryotes. In most prokaryotes, the total DNA content is usually included in a single DNA molecule. In eukaryotes, including the unicellular organisms such as algae, yeast, and protozoa, the DNA is partitioned into a number of chromosomes. The long DNA molecule in each chromosome winds around protein complexes made of basic proteins called **histones.** Exact sizes and sequences are known for many viral, bacterial, and eukaryotic DNA molecules. Table 4.1 lists the lengths of individual DNA molecules from various sources. Sizes vary greatly among viral DNA molecules but much less so for bacterial DNA molecules. The length of the duplex DNA molecules can be calculated from the 0.34 nm distance between base pairs (bp). Therefore, the DNA molecules listed in Table 4.1 range from approximately 1.7 μm to 83,500 mm (8.3 cm!). The width of a DNA molecule is 2.0 nm. In general, more complex organisms require

TABLE 4.1 Sizes of Various DNA Molecules

Source of DNA	Size in Base Pairs (bp)
Plasmid pBR322*	4,361
Simian virus 40 (SV40)	5,200
Phage T7*	39,937
Phage λ*	48,502
F plasmid*	99,159
Vaccinia virus strain WR	194,711
Fowlpox virus	266,145
Mycoplasma genitalium	580,073
Yeast chromosome IV	1,531,929
Escherichia coli	4,639,221
Human chromosome 1	245,522,847

Note: Phages (viruses that infect bacteria) and plasmids marked with an asterisk have *E. coli* as a host. *Mycoplasma genitalium* is the smallest known free-living bacterium. For yeast and humans the molecular mass of the largest DNA molecule in the organism is given.

much more DNA than simpler organisms (though the cells of both the toad and the South American lungfish have considerably more DNA than human cells).

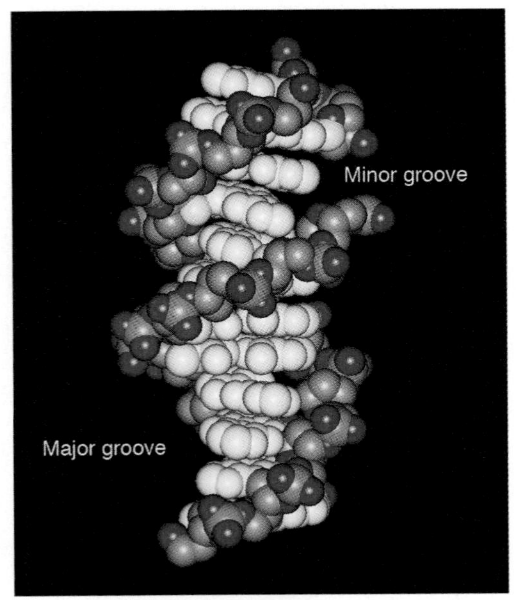

FIGURE 4.1 Major and minor grooves in B-DNA. B-DNA is shown as a spacefilling structure with the bases in light blue and the rest of the molecule in standard CPK element coloring. (Structure from Protein Data Bank 1BNA. H. R. Drew, et al., *Proc. Natl. Acad. Sci. USA* 78 [1981]: 2179–2183. Prepared by B. E. Tropp.)

DNA molecules are fragile.

The great lengths of DNA molecules make them extremely susceptible to breakage by the hydrodynamic shear forces resulting from such ordinary operations as pipetting, pouring, and mixing. Unbroken DNA molecules shorter than about 300,000 bp usually can be isolated from viruses. Unless great care is taken, larger DNA molecules are almost always broken during isolation so that the average length of isolated DNA is usually about 40,000 bp. Bacterial DNA, for instance, is fragmented into about 50 to 100 pieces. The fact that the DNA of bacteria and of higher organisms is invariably fragmented by manipulation has important experimental consequences.

4.2 Recognition Patterns in the Major and Minor Grooves

Enzymes can recognize specific patterns at the edges of the major and minor grooves.

DNA's length and fragility presents a challenge when investigators wish to study DNA *in vitro*. This challenge is technical, however, and does not influence our basic concept of how DNA works. A

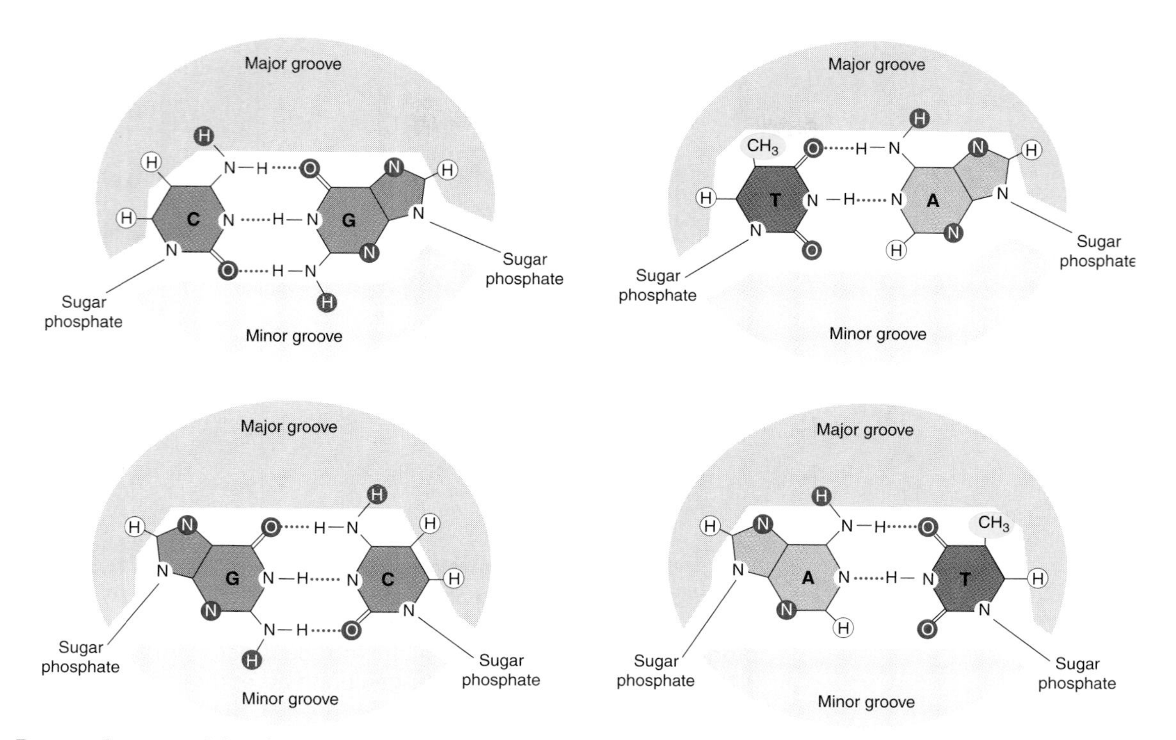

FIGURE 4.2 Base pair recognition from the edges in the major and minor grooves. (a–d) The four types of base pairs are shown. Hydrogen bonds between base pairs are shown as a series of short red lines. Potential hydrogen bond donors are shown in blue and potential hydrogen bond acceptors in orange. Nonpolar methyl groups in thymine are yellow and hydrogen atoms that are attached to carbon atoms and, therefore, unable to form hydrogen bonds are white. (Modified from C. Branden and J. Tooze. *Introduction to Protein Structure, First edition*. Garland Science, 1999. Used with permission of John Tooze, The Rockefeller University.)

more fundamental problem results from the fact that base pairs are located within the helix. It, therefore, was initially difficult to see how enzymes recognize and interact with specific base sequences. One possible solution to the problem is for DNA to unwind. Although DNA does unwind (see below), many enzymes appear to be able to recognize base sequences in the helical structure. Examination of the space filling structure shown in FIGURE 4.1 reveals that the edges of the major and minor grooves are accessible to enzymes. These grooves arise because deoxyribofuranose groups are attached to base pairs in an asymmetric fashion. That is, the sugar rings lie closer to one side of the base pair than to the other. The grooves' edges are lined with hydrogen bond donors, hydrogen bond acceptors, nonpolar methyl groups, and hydrogen atoms (FIGURE 4.2). Each of the four base pairs projects a unique pattern at the edge of the major groove, but T•A and A•T base pairs project the same pattern at the edge of the minor groove as do C•G and G•C base pairs (FIGURE 4.3). These patterns permit the enzymes to read the sequence from outside the helix.

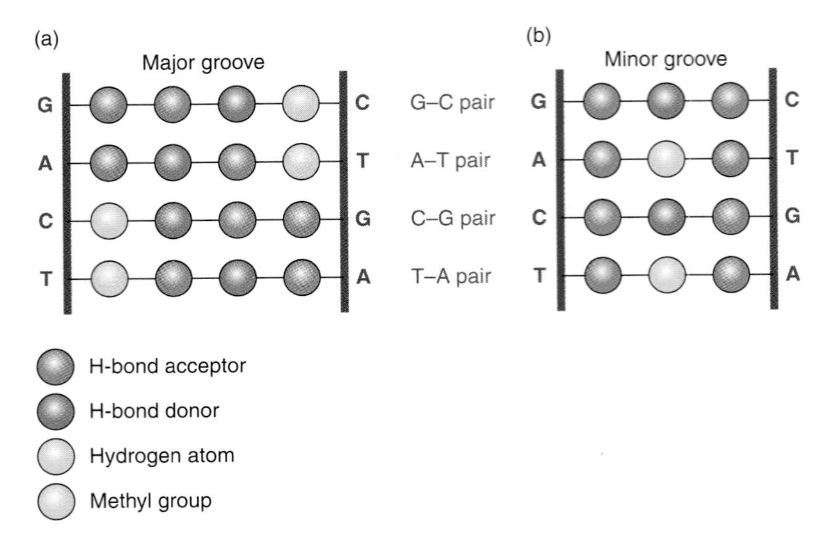

FIGURE 4.3 **DNA recognition code.** Distinct patterns of hydrogen bond donors, hydrogen bond acceptors, methyl groups, and hydrogen atoms are observed when looking directly at the edges of the base pairs in the major (a) or minor (b) grooves. Each of the four base pairs projects a unique pattern of hydrogen bond donors, hydrogen bond acceptors, methyl groups, and hydrogen atoms at the edge of the major groove. However, the patterns are similar at the edge of the minor groove for T•A and A•T as well as for C•G and G•C. (Modified from C. Branden and J. Tooze. *Introduction to Protein Structure, First edition*. Garland Science, 1999. Used with permission of John Tooze, The Rockefeller University.)

4.3 DNA Bending

Some base sequences cause DNA to bend.

An immense variety of base sequences have been observed in DNA. Although most sequences do not have any special features that cause them to influence DNA structure, some do. For instance, tracts consisting of 4 to 6 adjacent adenine residues, called **A-tracts**, cause DNA to bend. Each A-tract contributes 17° to 22.5° of curvature. When A-tracts are in phase within a DNA molecule so that they are repeated at 10 or 11 bp intervals, their contributions are additive and the DNA molecule bends back on itself. Other sequences such as 5′-RGCY-3′, where R is a purine and Y is a pyrimidine, can also cause bending. The local structure of B-DNA, thus, may differ slightly from the classic linear helix.

4.4 DNA Denaturation and Renaturation

DNA can be denatured.

When the Watson-Crick Model was first proposed, many investigators thought that the long DNA strands would not be able to unwind and therefore complete strand separation would be impossible. In an attempt to dispel this concern, biophysical chemists tried to show that the unwinding process can also take place in the test tube. The approach was to treat DNA with a physical or chemical agent that would disrupt the weak non-covalent interactions (see below) that hold base pairs together without disrupting covalent bonds. Early efforts by Paul Doty and coworkers in the late 1950s showed that DNA solutions undergo a striking drop in viscosity when heated. This observation was interpreted to mean that the double helical structure collapses when heated. This collapse was also accompanied by a change in the DNA's ability to rotate plane-polarized light, resulting from the loss of the right-handed helical structure. It seemed probable that the change in the secondary structure observed when the DNA solution was heated represented a conversion of the linear double helical structure into separate single strands. Several different kinds of experiments helped to establish that the two strands do in fact unwind to form separate strands when a DNA solution is heated. For instance, the mass/length ratio of DNA before heating is twice that of DNA after heating and a deoxyribonuclease specific for single-stranded DNA was shown to digest DNA after heating but not before. The transition from the double helical structure (the native state) to randomly coiled single strands (the denatured state) is called **denaturation.**

The simplest way to detect DNA denaturation is to monitor the ability of DNA in a solution to absorb ultraviolet light at a wavelength of 260 nm, λ_{260}. The nucleic acid purine and pyrimidine bases absorb 260 nm light strongly. The absorbance at 260 nm, A_{260}, is proportional to concentration. The A_{260} value for double-stranded DNA at a concentration of 50 $\mu g \cdot mL^{-1}$ is 1.00 unit. Furthermore, the amount of light absorbed by nucleic acids depends on the structure of the molecule. The more ordered the structure, the less light that is absorbed. Therefore, double-stranded DNA absorbs less light than the single-stranded chains that form it, and these chains in turn absorb less light than nucleotides released by hydrolysis. For example, three solutions of double-stranded DNA, single-stranded DNA, and free nucleotides, each at 50 $\mu g \cdot mL^{-1}$, have the following A_{260} values:

Double-stranded DNA:	$A_{260} = 1.00$
Single-stranded DNA:	$A_{260} = 1.37$
Free nucleotides:	$A_{260} = 1.60$

This relationship is often described by stating double-stranded DNA is **hypochromic** or free nucleotides are **hyperchromic.**

If a DNA in a solution that is about 0.15 M sodium chloride is slowly heated and the A_{260} is measured at various temperatures, a melting curve such as that shown in FIGURE 4.4 is obtained. The following features of this curve should be noted:

1. The A_{260} remains constant up to temperatures well above those encountered by most living cells in nature.
2. The rise in A_{260} occurs over a range of 6° to 8°C.
3. The maximum A_{260} is about 37% higher than the starting value.

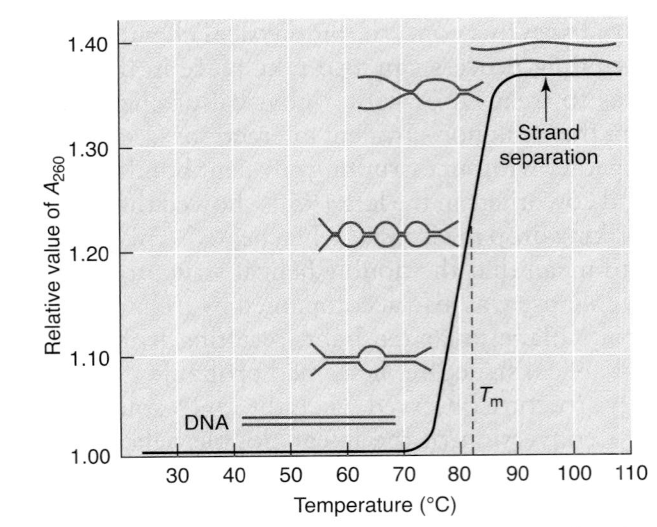

FIGURE 4.4 **DNA melting curve.** A melting curve of DNA showing T_m (the melting temperature) and possible molecular conformations for various degrees of melting.

The state of a DNA molecule in different regions of the melting curve is also shown in Figure 4.4. Before the A_{260} rise begins, the molecule is fully double-stranded. In the rise region, non-covalent interactions between base pairs in various segments of the molecule are disrupted; the extent of the disruption increases with temperature. In the initial part of the upper plateau a few non-covalent interactions remain to hold the two strands together until a critical temperature is reached at which the last remaining non-covalent interactions are disrupted and the strands separate completely.

A convenient parameter to characterize a melting transition is the temperature at which the rise in A_{260} is half-complete. This temperature is called the **melting temperature** and it is designated T_m.

In the course of studying strand separation, another important fact emerged. If a DNA solution is heated to a temperature at which most but not all non-covalent interactions are disrupted and then cooled to room temperature, A_{260} drops immediately to the initial undenatured value. Additional experiments show that the native structure is restored. Therefore, if strand separation is not complete and denaturing conditions are removed, the helix rewinds. Thus, if two separated strands were to come in contact and form even a single base pair at the correct position in the molecule, the native DNA molecule should re-form. We will encounter this phenomenon again when renaturation is described.

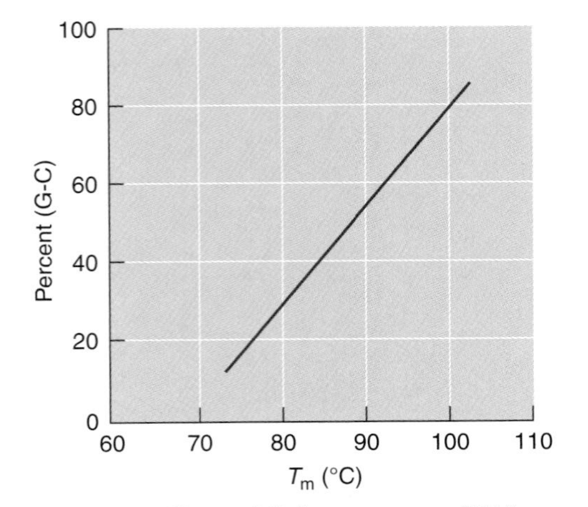

FIGURE 4.5 **Effect of G-C content on DNA melting temperature.** T_m increases with increasing percent of G + C. The DNA solution contained 0.15 M sodium chloride and 0.015 M sodium citrate.

Hydrogen bonds stabilize double-stranded DNA.

In 1962, Julius Marmur and Paul Doty isolated DNA from various bacterial species in which the base compositions vary from 20% G + C to 80% G + C. T_m values from many such DNA molecules are plotted versus percent G + C in FIGURE 4.5. Note that T_m increases with increasing percent G + C. This relationship is explained by proposing that hydrogen bonds are at least partially responsible for stabilizing the double-stranded structure. It requires more energy to disrupt the three hydrogen bonds in a G•C base pair than to disrupt the two hydrogen bonds in an A•T base pair.

A decrease of T_m values in the presence of a denaturing agent such as urea (NH_2CONH_2) or formamide ($HCONH_2$), which can form hydrogen bonds with DNA bases, supports the role of hydrogen bonds in stabilizing the double-stranded structure. Hydrogen bonds between base pairs have very low energies and so are easily broken. However, hydrogen bonds are also able to rapidly re-form (see below). Denaturing agents shift this equilibrium by forming hydrogen bonds with an unpaired base on one strand and thereby prevent the base from re-forming hydrogen bonds with the complementary unpaired base on the other strand. Denaturing agents, therefore, can maintain the unpaired state at a temperature at which complementary unpaired bases would normally be expected to pair again. Melting of a section of paired bases, therefore, requires less input of thermal energy and T_m is reduced.

Base stacking also stabilizes double-stranded DNA.

The planar bases in the double helix are stacked so that the pi electron ring systems in neighboring base pairs are in direct contact. The forces that stabilize stacking in the double helix include electrostatic interactions of interacting dipoles, van der Waals forces, and hydrophobic effects. We do not know the precise contribution that each of these weak interactions makes to the helical stability because it is difficult to modify the structure of DNA so that just one kind of interaction is altered.

Both base stacking and hydrogen bonds are weak non-covalent interactions and as such are easily disrupted by thermal motion. Stacking is enhanced if the bases are unable to tilt or swing out from a stacked array. Similarly, maximum hydrogen bonding occurs when all bases are pointing in the right direction. Clearly, the two weak interactions reinforce each other. Stacked bases are more easily hydrogen bonded and correspondingly, hydrogen-bonded bases, which are oriented by the bonding, stack more easily. If one of the interactions is eliminated, the other is weakened, explaining why T_m drops so markedly after the addition of an agent that destroys either type of interaction.

Base stacking is a cooperative interaction.

In a sequence of stacked bases, for example, ABCDEFGHIJ, it would be very unlikely for base E to swing out of the stacked array because the plane of the base tends to be parallel to the planes of both D and F. The tendency to conform to an orderly stacked array is not so great at the ends of the molecule, however. Only a single base, B, stabilizes the orientation of base A. Therefore, a rapidly moving solvent molecule might crash into A and cause it to rotate out of the stack more easily than a collision with E would cause disorientation of E. Because A has a lower probability of being stacked than B, then of course B must also be more easily disoriented than is C.

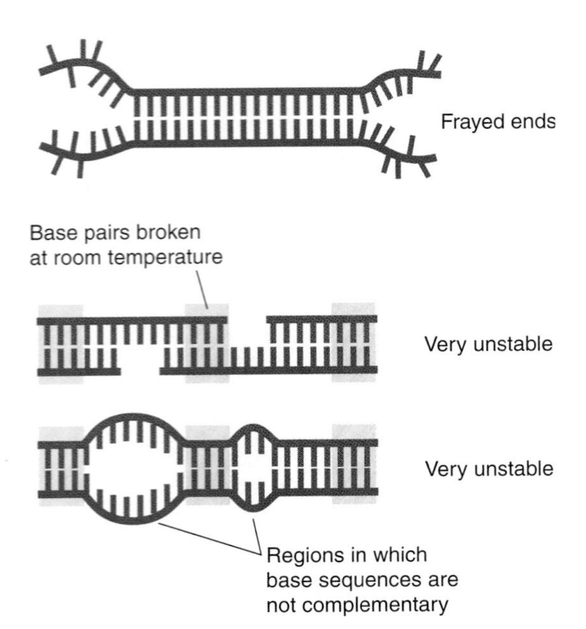

FIGURE 4.6 **Several effects of cooperativity of base-stacking.** Each shaded area indicates base pairs that would be broken at room temperature. However, if these tracts contained more than fifteen base pairs they would be stable.

This slight tendency toward instability, which is also present in a double-stranded molecule, is most noticeable in the value of T_m for double-stranded polynucleotides containing fewer than 20 base pairs (oligonucleotides). For example, if a molecule having 10^5 base pairs is broken down to fragments having 10^3 base pairs, there is no detectable change in T_m. However, under conditions in which T_m for a large DNA molecule is 90°C, T_m for a double-stranded hexanucleotide (six nucleotides per strand) can be as low as 30°C. The exact value depends on the base composition and sequence. This effect has the following consequences:

1. The ends of a linear double-stranded DNA molecule are usually not hydrogen-bonded, but are frayed (FIGURE 4.6), with about seven base pairs broken. However, some base sequences stack better than others and are even stacked at an end of a molecule.
2. Short double-stranded oligonucleotides, having fewer than 15 base pairs per molecule, have particularly low T_m values. A double-stranded trinucleotide (3 bases per strand) is not stable at room temperature.
3. Molecules in which the paired regions are very short and are flanked by unpaired regions (such as the two lower ones in Figure 4.6) cannot maintain the conformation shown at physiological temperatures.

Ionic strength influences DNA structure.

In addition to the cooperative attractive interactions between adjacent DNA bases and between the two strands, there is an interstrand electrostatic repulsion between the negatively charged phosphates. (There is also an intrastrand repulsion, which is probably not important for duplex structure.) This strong force would drive the two strands apart if the charges were not neutralized. Examining the variation of T_m as a function of the ionic concentration of the buffer solution, reveals that T_m decreases sharply as salt concentration decreases. Indeed, in distilled water, DNA denatures at room temperature.

The explanation for the effect of ionic strength on DNA structure is as follows. In the absence of salt, the strands repel one another. As salt is added, positively charged ions such as Na^+ form "clouds" of charge around the negatively charged phosphates and effectively shield the phosphates from one another. Ultimately, all of the phosphates are shielded and repulsion ceases; this shielding occurs near the physiological salt concentration of about 0.2 M. However, T_m continues to rise as the sodium chloride concentration increases because purine and pyrimidine solubility decreases, increasing hydrophobic interactions.

The DNA molecule is in a dynamic state.

An important structural feature of the DNA molecule becomes apparent when DNA is examined in the presence of formaldehyde (HCHO). Formaldehyde can react with the NH_2 groups of the bases and thus eliminate their ability to hydrogen bond. Adding formaldehyde, therefore, causes slow and irreversible DNA denaturation. Because the amino groups must be available to formaldehyde for the reaction to take place, bases must continually unpair and pair (i.e., hydrogen bonds must break and re-form).

A related phenomenon is observed when DNA is dissolved in tritiated water ($[^3H]H_2O$). There is a rapid exchange between the hydrogen-bonded protons of the bases and the $^3H^+$ ions in the water. These two observations indicate that DNA is a dynamic structure in which double-stranded regions frequently open to become single-stranded bubbles and then close again. This transient localized melting is called DNA "breathing." Because a G•C base pair has three hydrogen bonds and an A•T base pair has only two, transient melting occurs more often in regions rich in A•T pairs than in regions rich in G•C pairs.

Distant short patches of complementary sequences can base pair in single-stranded DNA.

To obtain the data for the melting curves of the sort that have been shown, A_{260} is measured at various temperatures that are plotted on the x-axis. Denaturation is usually complete at a temperature above 90°C. In most experiments, there is a total increase in A_{260} of about 37% and the solution consists entirely of single strands with unstacked bases.

If the solution is rapidly cooled to room temperature and the salt concentration is above 0.05 M, however, the value of A_{260} reached at the maximum temperature drops significantly but not totally (FIGURE 4.7). The reason is that in the absence of disrupting thermal motion, random intrastrand hydrogen bonds form between distant short tracts of bases with sufficiently complementary sequences. Typically, the value of A_{260} drops to 1.12 times the initial value for the native DNA, suggesting that, after cooling, about two thirds of the bases are either hydrogen-bonded or in such close proximity that stacking is restored. The molecule will be very compact (Figure 4.7).

The situation is quite different if the salt concentration is 0.01 M or less. In this case, the electrostatic repulsion due to negative phosphate groups keeps the single strands sufficiently extended that the bases cannot approach one another. Thus, after cooling no hydrogen bonds are formed and base-stacking remains at a minimum.

Alkali denatures DNA without breaking phosphodiester bonds.

Heat can be used to prepare denatured DNA, which is often an essential step in many experimental protocols. High temperature may break phosphodiester bonds, however, so the product of heat denaturation often is a collection of broken single strands. The degradation problem is avoided by using another method to denature DNA. Addition of a base such as sodium hydroxide to the DNA solution removes protons from the ring nitrogen atoms of guanine and thymine. This deprotonation, which occurs above pH 11.3, disrupts the hydrogen-bonded

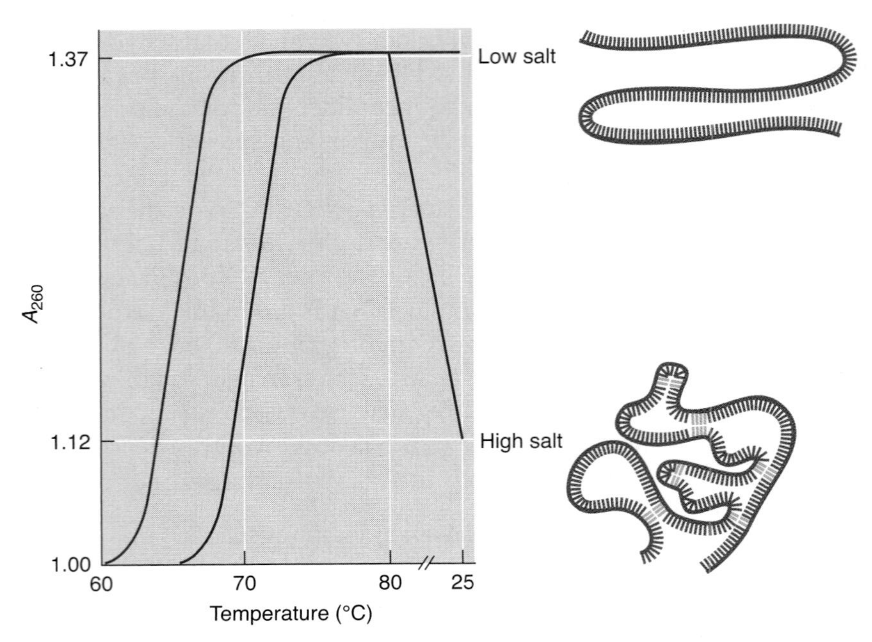

FIGURE 4.7 **The effect of lowering the temperature to 25°C after strand separation has taken place.** DNA molecules in a solution containing a high salt concentration (blue curve) or low salt concentration (red curve) are heated to 90°C so that the two strands completely unwind and separate. Then the solutions are rapidly cooled to 25°C. After cooling, the A_{260} for the DNA in the solution at high salt concentration is much lower than that for the DNA in the solution at low salt concentration because the DNA molecules in the high salt solution form intrastrand base pairs (shown in green) but those in the low salt solution do not.

double-helical structure and causes DNA to denature. Because DNA is quite resistant to alkaline hydrolysis, this procedure is the method of choice for denaturing DNA. Acid also causes denaturation but is seldom used for that purpose because acid also causes purine groups to be cleaved from the polynucleotide chain, a process known as depurination.

Complementary single strands can anneal to form double-stranded DNA.

A solution of denatured DNA can be treated in such a way that native DNA re-forms. The process is called **renaturation** or **reannealing** and the re-formed DNA is called **renatured DNA**. Renaturation has proved to be a valuable tool in molecular biology. It can be used to demonstrate genetic relatedness between different organisms, detect particular species of RNA, determine whether certain sequences occur more than once in the DNA of a particular organism, and locate specific base sequences in a DNA molecule. Two requirements must be met for renaturation to occur:

1. The salt concentration must be high enough so that electrostatic repulsion between the phosphates in the two strands is eliminated; usually 0.15 to 0.50 M NaCl is used.
2. The temperature must be high enough to disrupt random, intrastrand or interstrand hydrogen bonds. The temperature

cannot be too high, however, or stable interstrand base-pairing will not occur. The optimal temperature for renaturation is 20° to 25°C below the T_m value.

Renaturation is a slow process compared to denaturation. The rate-limiting step is not the actual rewinding of the helix (which occurs in roughly the same time as unwinding) but the precise collision between complementary strands such that base pairs are formed at the correct positions. Because renaturation is a result only of random motion, it is a concentration-dependent process. At concentrations normally encountered in the laboratory, renaturation takes several hours.

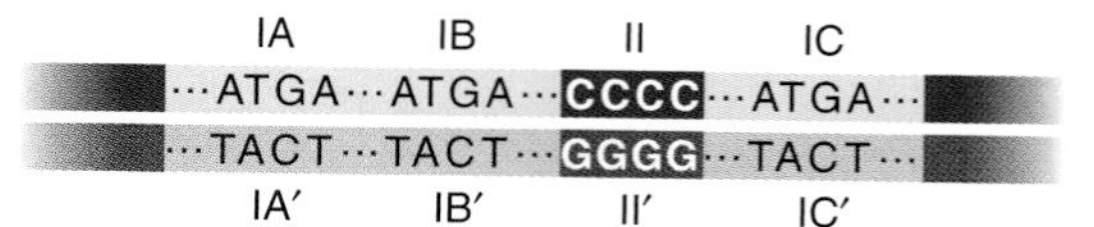

FIGURE 4.8 **Molecular details of renaturation using a hypothetical DNA molecule.** A hypothetical DNA molecule containing a sequence that is repeated several times. The roman numerals on either side of the DNA molecule refer to segments of the DNA molecule that are discussed in the text.

The molecular details of renaturation can be understood by referring to the hypothetical molecule shown in FIGURE 4.8, which contains a sequence that is repeated several times. Assume that each single strand contains 50,000 bases and that the base sequences are complementary. Any short sequence of bases (say, 4–6 bases long) will certainly appear many times in such a molecule and can provide sites for base-pairing. Random collision between non-complementary sequences such as IA and II′ will be ineffective but a collision between IA and IC′ will result in base-pairing. This pairing will be short-lived, however, because the bases surrounding these short complementary tracts are not able to pair and stacking stabilization will not occur. At the temperatures used for renaturation, these paired regions rapidly become disrupted. As soon as two sequences such as IB and IB′ pair, the adjacent bases will also rapidly pair and the entire double-stranded DNA molecule will "zip up" in a few seconds.

It is important to realize that each renatured native DNA molecule is not formed from its own original single strands. In a solution of denatured DNA, the single strands freely mix so that during renaturation original partner strands seldom find each other. This mixing was shown in an experiment using two DNA samples isolated from *E. coli* cultured, in one case, in a medium containing $^{14}NH_4Cl$, and in the other, $^{15}NH_4Cl$. The two DNA samples were mixed, denatured, and then renatured. The resulting mixture contained three types of renatured DNA molecules: 25% contained ^{14}N in both strands, 50% contained ^{14}N in one strand and ^{15}N in the other, and 25% contained ^{15}N in both strands. This result indicates random mixing of the strands during renaturation. Methods for distinguishing the three types of duplexes are described in Chapter 5.

4.5 Helicases

Helicases are motor proteins that use the energy of nucleoside triphosphates to unwind DNA.

Duplex DNA must unwind under physiological conditions during DNA replication. "Molecular motor" enzymes called **helicases** catalyze nucleoside triphosphate-dependent unwinding of double-stranded DNA in cells. Helicases are often part of larger protein complexes and their activities are influenced by other proteins in the complex.

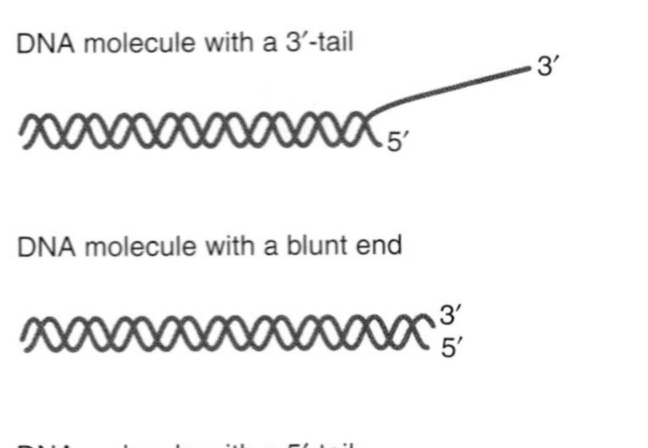

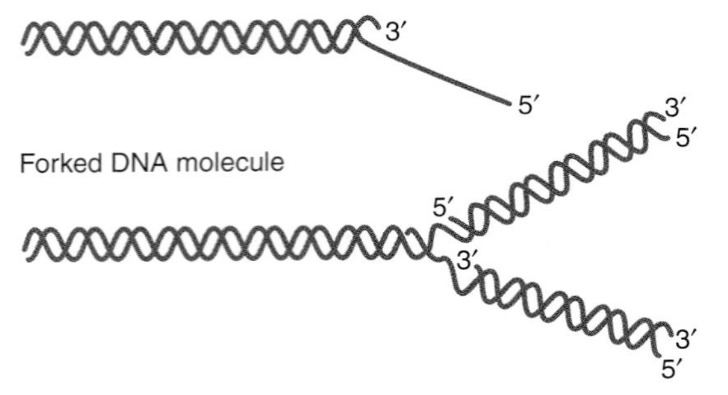

FIGURE 4.9 **DNA structural preferences of different types of helicases.**

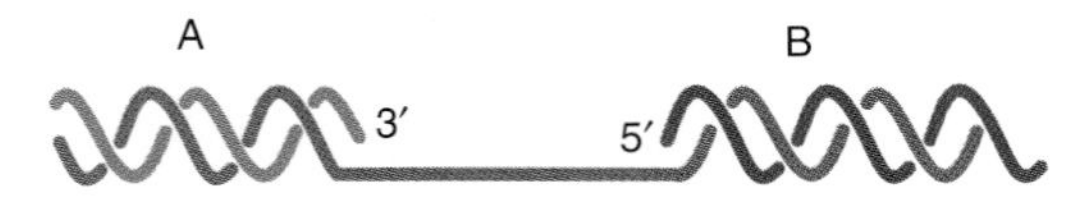

FIGURE 4.10 **DNA substrates used to assay helicase activity.**

The observation that a cell extract has DNA-dependent nucleoside triphosphatase activity usually, but not always, indicates a helicase is present. A more reliable indication is nucleoside triphosphate-dependent unwinding of double-stranded DNA to single strands, which can be detected by their susceptibility to single strand specific nucleases. DNA helicases tend to exhibit structural preferences for their DNA substrates (FIGURE 4.9); some require a forked DNA molecule, others act on DNA with 3′ or 5′ tails, and still others work on blunt end DNA.

Some DNA helicases move along single-stranded DNA in a 3′→5′ direction while others move in a 5′→3′ direction. The direction of movement can be determined by using a substrate in which a single-stranded DNA molecule (shown in red) has a complementary fragment at each end (FIGURE 4.10). Assuming that the helicase binds to the long single-stranded region, release of a fragment at the 3′-end of the single strand (Fragment A in Figure 4.10) indicates 5′→3′ movement along the single strand, while release of a fragment at the 5′-end (Fragment B in Figure 4.10) indicates 3′→5′ movement. Released short fragments can be distinguished from starting material and the long single-strand on the basis of size. Techniques for separating nucleic acids according to size are described in Chapter 5. Detection is simplified if the released fragment is made radioactive or tagged with a fluorescent label.

At least 14 different DNA helicases have been isolated from *E. coli*, six from bacterial viruses, 15 from yeast, eight from plants, and 24 from human cells. Some helicases are specific for double stranded regions in RNA. A classification system based on conserved sequence motifs has been devised that divides known DNA and RNA helicases into six superfamilies. Some helicases belonging to superfamily 1 and superfamily 2 function as monomers and others function as dimers. Superfamily 2 includes the largest number of known helicases. Most enzymes in this superfamily move in 3′→5′ direction but some move in the opposite direction.

Although it is not possible to study all the members of superfamilies 1 and 2, it is instructive to examine PcrA helicase, one of the best studied members of these two superfamilies, to see how a helicase with just one polypeptide subunit works. PcrA helicase, a member of superfamily 1, is an essential enzyme in gram-positive bacteria. It participates in DNA repair and a type of DNA replication known as rolling circle replication. PcrA helicase moves 3′→5′ along single-stranded DNA at a rate of about 50 nucleotides$\cdot$s^{-1}. The enzyme appears to use one ATP molecule for each nucleotide traversed. Dale B. Wigley and coworkers have obtained a crystal structure for PcrA helicase bound to a 3′-tailed double-stranded DNA (FIGURE 4.11). The enzyme contacts both single- and double-stranded regions of its DNA substrate, distorting the double-stranded region at the junction of the single and double strands and causing the two strands to begin separation. Based on the crystal structure and biochemical data, Wigley and coworkers have proposed an inchworm model like that shown in FIGURE 4.12 to explain how the PcrA helicase moves along the single strand and unwinds the double strand.

Members of superfamilies 3–6 function as hexameric rings. Two hexameric helicases belonging to superfamily 4 are of special interest

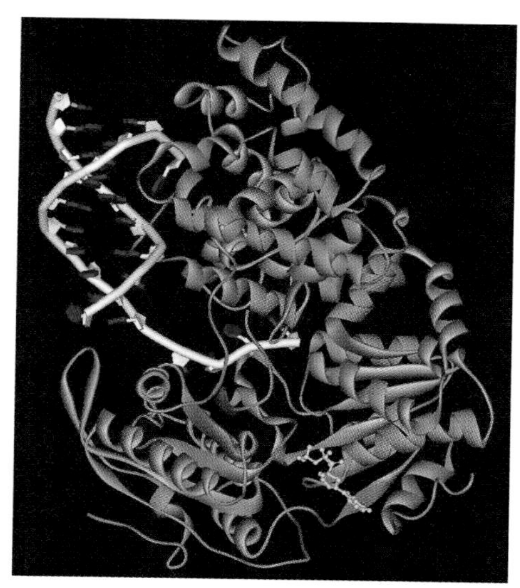

FIGURE 4.11 **Crystal structure of PcrA helicase from the gram-positive bacteria *Bacillus stearothermophilus*.** The helicase (pink) is bound to a 3-tailed double-stranded DNA (white tube form with colored bases) and an ATP analog (yellow stick form). The helicase contacts the double stranded DNA and destabilizes it. The double-stranded region nearest to the single strand/double strand junction is distorted and the two strands have started to separate. (Structure from Protein Data Bank 3PJR. S. S. Velankar, et al., *Cell* 97 [1999]: 75–84. Prepared by B. E. Tropp.)

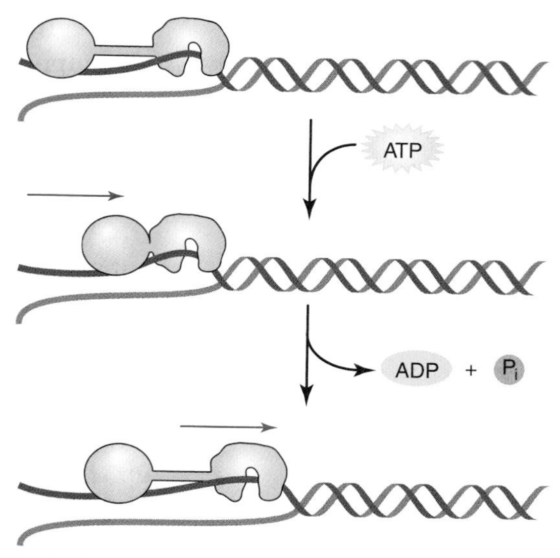

FIGURE 4.12 **The inchworm model proposed for PcrA helicase activity.** (1) At the start, the helicase does not have a bound nucleotide. (2) As a result of binding ATP, the helicase changes conformation, closing a cleft between two domains. (3) ATP hydrolysis reverses the conformational change. The resulting domain movements cause the helicase to move in a 3′→5′ direction along one strand and to displace the other strand. (Adapted from R. L. Eoff and K. D. Raney, *Biochem. Soc. Trans.* 33 [2005]: 1474–1478.)

because they have been extensively studied and play essential roles in DNA replication. These helicases, the phage T7 (bacterial virus) helicase and the *E. coli* DnaB helicase (named for the *dna*B gene that codes for it), bind to forked DNA molecules, encircling one DNA strand while excluding the other (FIGURE 4.13). They require nucleoside triphosphates to move in a 5′→3′ direction along the bound single-strand. The phage T7 helicase prefers dTTP and DnaB helicase prefers ATP, but each will also use other nucleoside triphosphates. DnaB helicase's participation in DNA replication is described in Chapter 9.

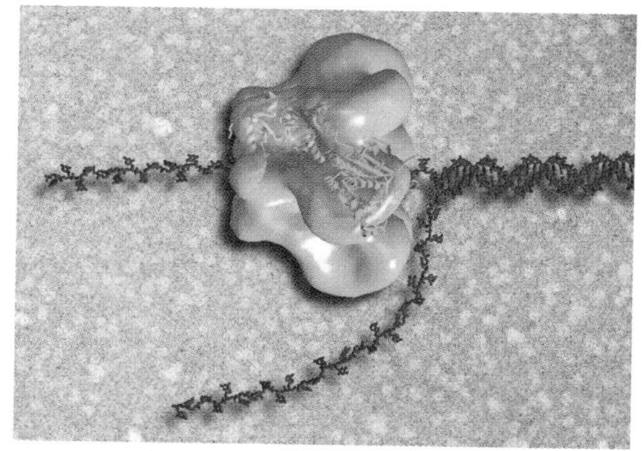

FIGURE 4.13 **Reconstruction of the three-dimensional structure of the bacteriophage T7 helicase in action.** The hexameric T7 gp4 helicase/primase is shown encircling one strand of DNA, while the second strand is displaced outside the ring. The helicase domain is represented by the green ribbon, while the primase domain is represented by the cyan ribbon. The helicase activity is generated by the ring walking along the single strand in the central channel in a 5′ to 3′ direction. (Reprinted from *J. Mol. Biol.*, vol. 311, M. S. VanLoock, et al., The primase active site is on the outside . . ., pp. 951–956, copyright 2001, with permission from Elsevier [http://www.sciencedirect.com/science/journal/00222836]. Photo courtesy of Edward H. Egelman, University of Virginia.)

4.6 Single-Stranded DNA Binding Proteins

Single-stranded DNA binding proteins (SSB) stabilize single-stranded DNA.

Proteins that bind to single-stranded DNA, **single-stranded DNA binding proteins (SSBs)**, stabilize the transient single-stranded regions that are formed by the action of helicases on double-stranded DNA. As essential participants in DNA metabolism, SSBs are present in all cells. Some SSBs consist of a single polypeptide, others contain two or more identical polypeptide subunits, and still others are made of different

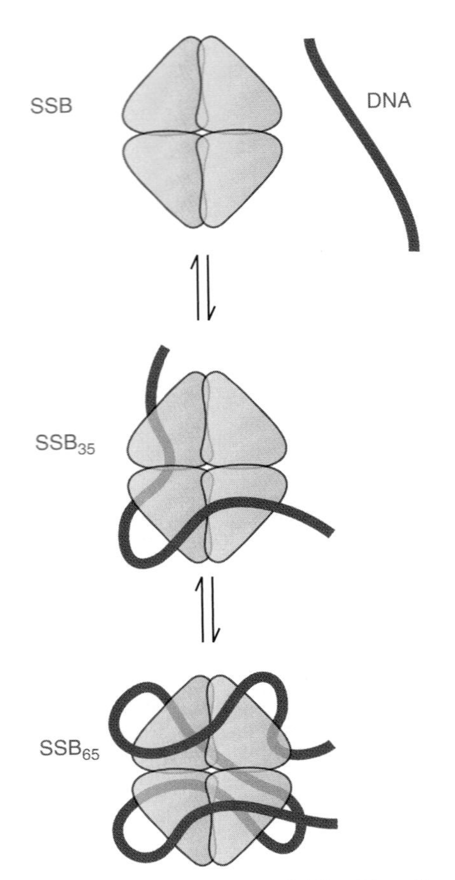

FIGURE 4.14 Proposed model for *E. coli* SSB•single-stranded DNA complexes. Two types of *E. coli* SSB•single-stranded DNA complexes have been observed. In one, SSB_{35}, 33–35 nucleotides bind to SSB and in the other SSB_{65}, 65 nucleotides bind to SSB. (Adapted from P. E. Pestrayakov and O. I. Lavrik, *Biochemistry* 73 [2008]: 1388–1404.)

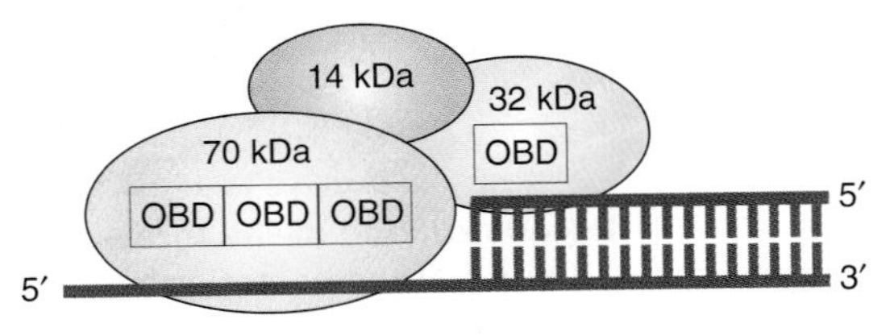

FIGURE 4.15 Proposed model for the RPA·DNA complex based on biochemical evidence. The 70 kDa, 32 kDa, and 14 kDa subunits are yellow, green, and pink, respectively. The oligonucleotide binding domains (OBDs) that have been predicted to interact with DNA are shown as three boxes in the 70 kDa subunit and one box in 32 kDa subunit. (Reproduced with kind permission from Springer Science+Business Media: *Biochemistry Mosc.*, Mechanisms of single-stranded DNA-binding protein. . ., vol. 73, 2008, pp. 1388–1404, P. E. Pestryakov and O. I. Lavrik, figure 8.)

polypeptide subunits. Despite their structural diversity, all SSBs share one important property—they bind more tightly to single-stranded DNA than to double-stranded DNA or RNA.

Bruce Alberts isolated the first SSB, gp32 (a product of the phage T4 gene 32), from an extract of phage T4 infected *E. coli* in 1970. The extract was passed through a column containing denatured DNA fixed to cellulose. Most proteins passed through the column but gp32 was retained because of its affinity for the denatured DNA. The gp32 was released by washing the column with a concentrated salt solution.

Purified gp32, a stable monomer (molecular mass = 33.5 kDa), destabilizes double-stranded DNA and lowers its T_m by 40°C or more. Destabilization involves cooperative binding of gp32 to DNA. The first gp32 molecule binds to a region of single-stranded DNA produced by transient melting. A segment of the single-stranded DNA consisting of about ten nucleotides fits into a large cleft in gp32 that is lined with arginine and lysine residues. Binding is very tight so the gp32 stays in place, freeing bases on the opposite strand to bind additional gp32 molecules and destabilizing adjacent base pairs on the same strand.

Destabilization results from the fact that paired bases adjacent to unpaired bases cease to be optimally stacked and their hydrogen bonds become less stable. Each succeeding gp32 tends to bind next to one that is already bound, breaking still other base pairs and enabling still other gp32 molecules to bind. The highly cooperative process continues, with individual gp32 molecules lining up next to each other along a single strand, until the duplex is totally denatured. Alberts' purification method has also been used to purify SSBs from other organisms. Other types of SSB also destabilize double-stranded DNA by shifting the melting equilibrium toward the single-stranded state.

Alberts isolated a second type of SSB, a homotetramer (molecular mass = 75 kDa), from uninfected *E. coli*. Each subunit has one **oligonucleotide binding domain (OBD)**. A bacterial cell has about 800 copies of the SSB tetramers. Although bacterial SSB also binds to single-stranded DNA in a cooperative fashion, the method of interaction is somewhat different from that of gp32. Two kinds of *E. coli* SSB•single-stranded DNA complexes, SSB_{35} and SSB_{65}, have been observed (FIGURE 4.14). In SSB_{35}, 33–35 nucleotides bind to the SSB tetramer and the single-stranded DNA only makes contact with OBDs in two subunits. In SSB_{65}, 65 nucleotides bind to the SSB tetramer and the DNA makes contact with OBDs in all four subunits.

Still another type of SSB, **replication protein A** or **RPA,** has been isolated from eukaryotes as different as yeast and humans. RPA is a heterotrimer, consisting of subunits of about 70, 30, and 14 kDa. Based on sequence analysis, RPA has six potential OBDs. Four of these are present in the 70 kDa subunit and one each in the 30 and 14 kDa subunits. A model for the RPA•DNA complex has been proposed based on biochemical data (FIGURE 4.15). According to this model, three OBDs in the 70 kDa subunit interact with single-stranded DNA on the 5′-side of the junction of the single and double strands. The OBD in the 32 kDa subunit interacts with DNA at the junction of the single- and double-strands. Structural confirmation of this model and further details await the determination of the crystal structure for the RPA•DNA complex.

4.7 Topoisomers and Topoisomerases

Covalently closed circular DNA molecules can form supercoils.

Linear double-stranded DNA molecules are free to unwind and completely separate. Complete separation cannot take place, however, if the ends of the chain are joined by a phosphodiester bond to form a closed covalent circular structure. As used here, *circular* means a continuous or unbroken DNA chain, rather than a geometric circle. The mathematical discipline of topology helps to understand the properties of closed covalent circular DNA. A topological property is one that remains unchanged, when the object of interest (covalently closed, circular double-stranded DNA) is distorted but not torn or broken. The two DNA strands in a covalently closed circular double stranded DNA circle are said to be topologically linked. Complete strand separation cannot take place unless one strand is nicked. Introduction of a nick would alter the DNA molecule's topological properties, however.

In a linear duplex, a base pair can be distorted without influencing the structure further along the molecule. In contrast, distorting a base pair when the strands are topologically linked influences the rest of the molecule's structure. Furthermore, topological constraints may force the DNA double helix to form a **supercoil**. There are two general forms of supercoil (FIGURE 4.16). In the first, the DNA axis repeatedly crosses over and under itself to form an **interwound supercoil** (Figure 4.16a). In the second, the DNA coils in a series of spirals around a ring to form a **toroidal supercoil** (Figure 4.16b). The circular DNA molecules described in this chapter all form interwound superhelicies. Chapter 6 describes how eukaryotic DNA wraps around a protein complex to form a toroidal structure. A double-stranded DNA does not have to form a closed covalent circle for the two strands to be topologically linked. Topological linkage also takes place when a double-stranded DNA loop is held together at its two ends by a protein complex.

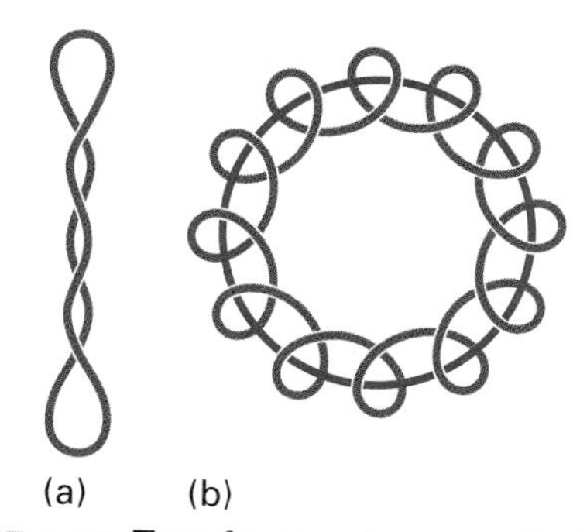

FIGURE 4.16 **Two forms of supercoiled DNA.** A double-stranded DNA molecule, shown for simplicity as a single line, is arranged in two different supercoiled forms. (a) The DNA axis repeatedly crosses over and under itself to form an interwound supercoil. (b) The DNA is arranged in a series of spirals around an imaginary ring (shown in magenta) to form a toroidal supercoil. (Modified from C. R. Calladine, et al. *Understanding DNA: The Molecule and How it Works, Third edition.* Elsevier Academic Press, 2004. Copyright Elsevier 2004.)

Bacterial DNA usually exists as a covalently closed circle.

The existence of circular DNA molecules was not noticed for many years because, as mentioned earlier, large DNA molecules usually break during isolation. John Cairns was the first to detect circular DNA molecules in bacteria. The experiment that he performed in 1963 was designed to obtain an image of an intact bacterial DNA molecule.

He hoped that the experiment would reveal whether bacteria have a single large chromosome or many smaller ones. Previous attempts by other investigators to obtain this information had failed because DNA is such a fragile molecule. Cairns knew that he required a very gentle method to avoid breaking the DNA. He decided to take advantage of the fact that tritium labeled DNA emits β-particles, which upon striking a photographic emulsion produce an image of the DNA. This technique of using a radioactively labeled substance to produce an image on a photographic emulsion is called **autoradiography**. Cairns cultured *E. coli* in a medium containing [^{3}H]thymidine, a specific precursor for DNA, and then gently released labeled DNA from the bacteria by treating the cells with a combination of lysozyme (an egg white enzyme) to digest the bacterial cell wall and detergent to disrupt the cell membrane. After collecting the released DNA on a dialysis membrane, he coated the dried membrane with a photographic emulsion and stored the preparation in the dark for two months to allow sufficient time for the β-particles to produce an image. Analysis of the array of dark spots, which appeared after developing the emulsion, revealed that *E. coli* DNA is a double-stranded circular molecule with a contour length of approximately 1.5 mm, or about 1000 times longer than the bacteria itself.

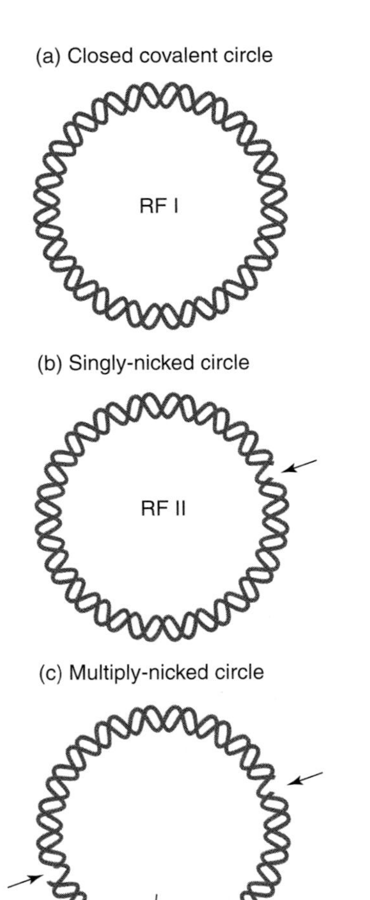

FIGURE 4.17 Closed covalent and nicked circles. Arrows point to the nicked site(s). (a) Closed covalent circle. (b) Singly-nicked circle. (c) Multiply-nicked circle.

Plasmid DNA molecules are used to study the properties of circular DNA *in vitro*.

Intact bacterial DNA is too long to be studied conveniently *in vitro*. Fortunately, there are readily available substitutes that allow us to study the properties of closed covalent circular DNA molecules in the laboratory. Many bacteria carry copies of an autonomously replicating small circular DNA molecule, called a **plasmid**, in addition to their large chromosomal DNA molecule. Plasmids, which range in size from 0.1 to 5% of the chromosome, replicate more or less independently of chromosomal DNA replication, and hence are transmitted from one generation to the next. We examine plasmids in more detail in later chapters. For now, we need only be concerned with the fact that plasmids are autonomously replicating double-stranded circular DNA molecules that are small enough to remain intact when pipetted or otherwise manipulated in the laboratory. They are therefore ideal subjects for studying circular DNA molecules.

Circular DNA molecules often have superhelical structures.

A circular DNA molecule may be a **covalently closed circle** with two unbroken complementary single strands, or it may be a **nicked circle** with one or more interruptions (**nicks**) in one or both strands (FIGURE 4.17). By convention, a closed covalent circular DNA molecule is called RFI (**RF = replicating form**) and a circular DNA molecule with a single nick is called **RFII**.

With few exceptions, the axis of a closed covalent circle crosses itself or writhes, as shown in Figure 4.16a. Such a circle is said to be a **superhelix** or a **supercoil**. Two double-stranded circular DNA molecules with identical base pair sequences but different degrees of supercoiling are said to be **topological isomers** or **topoisomers**. We will first examine the properties of topoisomers and then enzymes, called **topoisomerases**, which convert one topoisomer into another. Physical techniques used to recognize and separate topoisomers are described in Chapter 5.

Supercoiled DNA results from under- or overwinding circular DNA.

The two ends of a linear DNA helix can be brought together and joined in such a way that each strand is continuous. If, in so doing, one of the ends is untwisted 360° with respect to the other, some unwinding of the double helix will occur. When the ends are joined, the resulting covalent circle will, if the hydrogen bonds re-form, resemble a figure 8 with one crossover point or **node**. If the linear duplex is instead untwisted 720° prior to joining, the resulting superhelical molecule will have two nodes (FIGURE 4.18).

The reason for the superhelicity is as follows. In the case of a 720° unwinding of the helix, about 20 bp must be broken (because the linear molecule has 10.5 bp per turn of the helix). A DNA molecule has such a propensity for maintaining a right-handed (positive) helical structure, however, with about 10.5 bp per turn that it will deform itself to form a negative superhelix (Figure 4.18). If instead the initial rotation produces overwinding, the joined circle will writhe in the opposite sense to form a positive superhelix. The structures of a relaxed circle and two topoisomers (one that is a negative supercoil and another that is a positive supercoil) are shown in FIGURE 4.19.

Most bacterial DNA molecules are underwound and hence form negative superhelices. Overwinding exists in hyperthermophilic archaea and plasmid DNA molecules isolated from such organisms are positively supercoiled. Positive supercoiling is probably required to ensure that double-stranded DNA does not unwind at the high temperature required for optimum archaeal growth.

In bacteria, underwinding of superhelical DNA is not a result of unwinding prior to end-joining but is instead introduced into preexisting circles by an enzyme called DNA gyrase, which is described below. In eukaryotes, underwinding is due to the formation of a structure called a nucleosome in which about two turns of DNA are wound around a protein complex (see Chapter 6).

Topology provides a quantitative method for characterizing supercoils. One topological property, the linking number (*Lk*), is particularly useful for describing the topological forms that covalently closed, circular double-stranded DNA molecules assume. The *Lk* is closely related to the number of times that the two sugar phosphate backbones wrap around, or are "linked with" each other. It indicates how often two DNA strands twist about each other to form the helix or how often the helix axis writhes about itself to form the superhelix.

As a topological property, the *Lk* for a covalently closed DNA does not change unless a DNA strand breaks. Furthermore, the *Lk* must

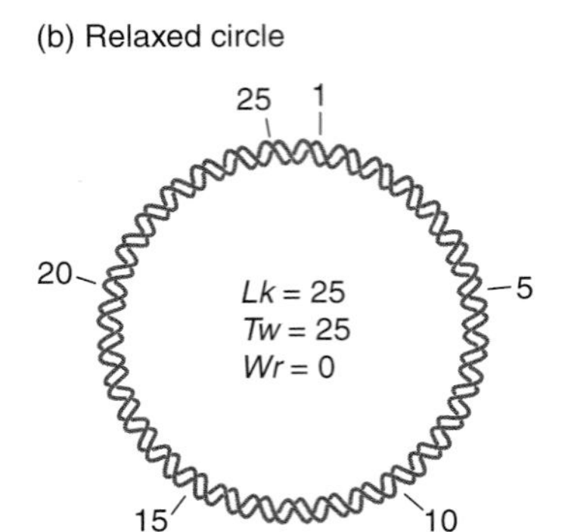

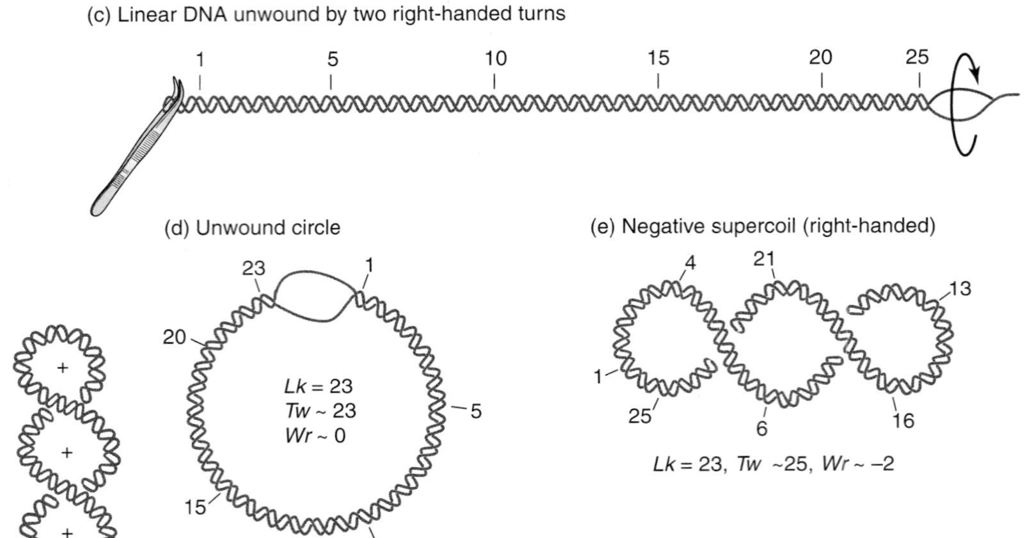

FIGURE 4.18 **Relations among the linking number (*Lk*), twisting number (*Tw*), and writhing number (*Wr*) of circular DNA revealed schematically.** (a) Linear double-stranded DNA is joined end to end to produce (b), a relaxed circle. (c) The same linear double-stranded DNA is unwound by two right-handed turns to produce (d), an unwound circle that then changes shape to form (e), a negative superhelix. Linking number (*Lk*), twisting number (*Tw*), and writhing number (*Wr*) are defined in the text. (Adapted from J. M. Berg, et al. *Biochemistry, Fifth edition.* W. H. Freeman and Company, 2002.)

(a) Relaxed circle (b) Negative superhelix (c) Positive superhelix

FIGURE 4.19 **The structure of supercoils.** (a) Relaxed circle—the helix axis does not cross itself and the DNA circle lies flat on the plane. (b) Negative superhelix—the front segment of a DNA molecule crosses over the back segment from right to left. (c) Positive superhelix—the front segment of a DNA molecule crosses over the back segment from left to right. (Adapted from J. B. Schvartzman and A. Stasiak, *EMBO Rep.* 5 [2004]: 256–261.)

be an integer. The linking number is the sum of the **twisting number, *Tw***, and the **writhing number, *Wr***, where:

$$Lk = Tw + Wr$$

The twisting number is determined from the total number of turns of the double-stranded molecule. For a nicked circle of known size, the value of Tw is calculated as the total number of base pairs divided by the number of base pairs per turn. A nicked double-stranded circle with 105 bp would have a linking number of 10 (105 bp/10.5 bp). The writhing number (the number of times the helix axis crosses itself) is zero for a nicked circle. Unlike the linking number, neither the twisting number nor the writhing number has to be an integer.

For a closed covalent, circular double-stranded DNA constrained to lie flat on a surface, *Lk* is the number of times one strand revolves around the other. The number is positive for revolutions in right-handed helical regions and negative for a left-handed helix or a left-handed segment such as that in Z-DNA (see below). The linking number enables one to distinguish positive from negative supercoiling.

When supercoiled DNA molecules are treated with small quantities of DNase to introduce a single nick into one of their strands, the molecules uncoil to form relaxed circles (rings that have no superhelical turns and are unconstrained when they lie flat on a surface).

The linking number cannot be changed without (1) breaking a strand, (2) rotating one strand about the other, and (3) rejoining. A change in the linking number (ΔLk) for a process, therefore, provides information about the mechanism of change. For example, ΔLk tells us something about how enzymes that affect supercoiling do their job. Changes in the linking number are related to changes in the twisting and writhing numbers as follows:

$$\Delta Lk = \Delta Tw + \Delta Wr$$

A decrease in *Lk* corresponds to some combination of underwinding and negative supercoiling, and an increase in *Lk* reflects some combination of overwinding and positive supercoiling.

The extent to which a DNA molecule is supercoiled is usually expressed in terms of supercoiling density (σ), which is defined by the following relationship:

$$\sigma = (Lk - Lk_0)/Lk_0$$

where Lk_0 is the linking number of the relaxed circular DNA molecule and *Lk* is the linking number of the supercoiled DNA. The supercoiling density of most bacterial DNA molecules is about –0.05. The negative sign indicates that the DNA molecule is underwound and therefore a negative superhelix.

Superhelices can have single-stranded regions.

Two arrangements can be envisioned to explain how DNA counteracts the strain of unwinding (FIGURE 4.20). (1) All of the strain of underwinding might be taken up by having one or more large single-stranded regions (Figure 4.20b). (2) All of the strain of underwinding might be taken up by writhing (Figure 4.20c). The actual situation is somewhere between these two extremes with approximately 75% of the underwinding strain being taken up by writhe. The reason that the strain is not more evenly distributed is that the B-DNA conformation is very stable.

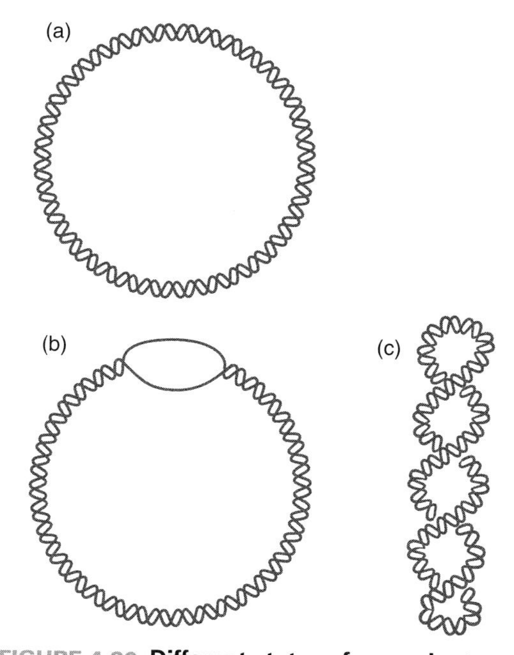

FIGURE 4.20 **Different states of a covalent circle.** (a) A nonsupercoiled or relaxed covalent circle having 36 turns of the helix. (b) An underwound covalent circle having only 32 turns of the helix. (c) The molecule in part (b) but with a writhing number of 4 to eliminate the underwinding. Solution (b) and (c) would be in equilibrium. The equilibrium would shift toward (b) with increasing temperature.

It is important to recall that a DNA molecule is a dynamic structure in which hydrogen bonds break and re-form so that bubbles continuously appear and disappear throughout the supercoiled DNA molecule. At any given instant the fraction of a supercoiled molecule that is single-stranded is greater than in a nicked circle. The sequences in a supercoil that are most likely to be unpaired and form bubbles are those that are more than 90% A•T pairs. As we will see in later chapters, A•T rich sequences play important roles in processes such as the initiations of genetic recombination, DNA replication, and messenger RNA synthesis.

Topoisomerases catalyze the conversion of one topoisomer into another.

DNA molecules encounter various topological challenges during virtually all stages of their metabolism. Two examples involving bacterial DNA replication will help to illustrate the point. A bacterial DNA molecule must unwind during replication, but because it is a closed covalent circle, unwinding in one region causes overwinding in another region. The resulting torsional strain must be relieved or replication will not be possible. Another topological challenge arises after bacterial DNA synthesis is complete because the two daughter DNA molecules form interlocking rings known as **catenanes**. A mechanism is needed that allows the interlocking rings to separate so that the DNA molecules can segregate to daughter cells.

Both of these replication problems as well as a variety of other twisting, writhing, and tangling problems, are solved by transient cleavage of the DNA backbone. The enzymes that catalyze this transient cleavage, called **topoisomerases**, convert one topoisomer into another. Topoisomerases are essential for cell viability because they manage DNA topology so that replication, transcription, and other processes involving DNA can take place. Furthermore, topoisomerases are of great practical interest because they are targets for a wide variety of drugs used as antimicrobial and anticancer agents.

Topoisomerases act by cutting DNA molecules and forming transient adducts in which a tyrosine at the active site attaches to the nicked DNA by a phosphodiester bond (FIGURE 4.21). The enzymes are divided into two broad types based on whether they form transient attachments to one or two strands of DNA. Type I topoisomerases form transient attachments to one strand, while type II topoisomerases form transient attachments to both strands. Consequently, type I topoisomerases change linking numbers by one unit at a time, whereas type II topoisomerases change them by two units. FIGURE 4.22 summarizes the activities of the various topoisomerases. You may wish to refer to it as we continue to examine these enzymes.

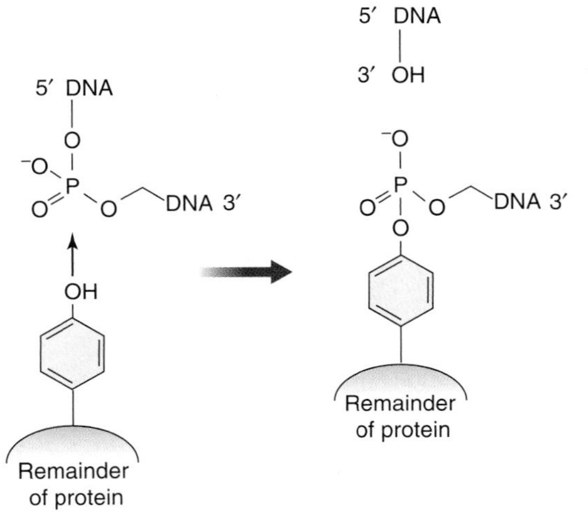

FIGURE 4.21 **Catalysis of transient breakage of DNA by DNA topoisomerases.** Transesterification takes place between a tyrosol residue on the enzyme and a DNA phosphate group, leading to cleavage of a DNA phosphodiester bond and formation of a covalent enzyme-DNA intermediate. The phosphodiester bond can be re-formed by a reversal of the reaction that is shown. A Type IA or Type II topoisomerase catalyzes a reaction in which a 3′-OH is the leaving group and the tyrosine at the active site is covalently linked to a 5′-phosphoryl group, as shown. A Type IB topoisomerase catalyzes a reaction in which a 5′-OH is the leaving group and the tyrosine at the active site is covalently linked to a 3′-phosphoryl group (not shown). (Adapted from J. C. Wang, *Nature Rev. Mol. Cell Biol.* 3 [2002]: 430–440.)

Enzymes belonging to the topoisomerase I family can be divided into three subfamilies.

Type I topoisomerases can be further divided into three subfamilies, type IA, type IB, and type IC. The first member of the type IA subfamily, *E. coli* topoisomerase I (molecular mass = 97 kDa) was

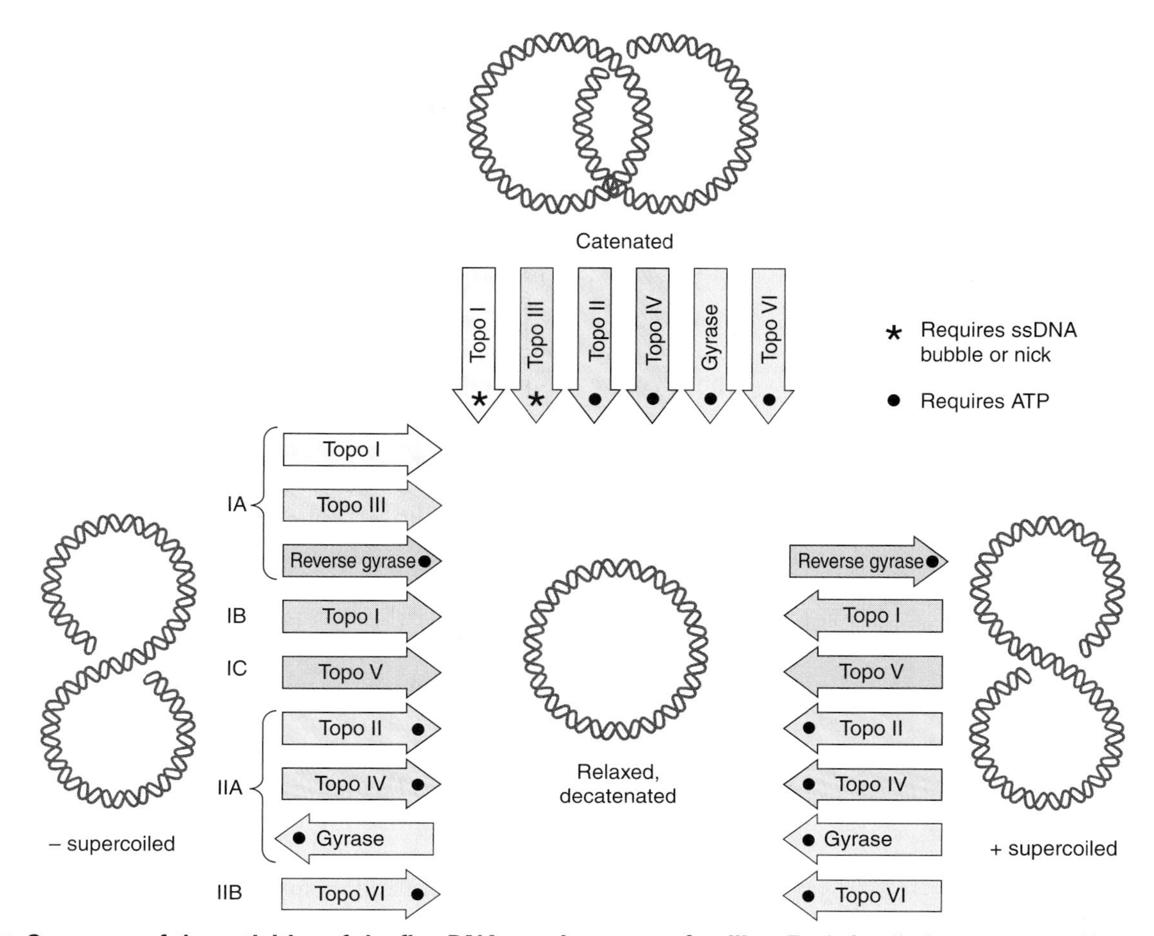

FIGURE 4.22 **Summary of the activities of the five DNA topoisomerase families.** Each family is represented by a labeled colored arrow. Arrows pointing from left to center indicate that the topoisomerase relaxes negative supercoils. The arrow pointing from center to left indicates that DNA gyrase introduces negative supercoils. The arrows that point from right to center indicate that the topoisomerase relaxes positive supercoils. The arrow that points from center to right indicates that reverse gyrase introduces positive supercoils. Several of these enzymes also catalyze the decatenation of linked rings. These enzymes are indicated by arrows that point from catenated ring structures at the top of the figure to the relaxed structure in the middle of the figure. Specific information about these enzymes is provided in the text. (Reproduced from A. J. Schoeffler and J. M. Berger, DNA topoisomerases: harnessing and constraining energy . . ., *Q. Rev. Biophys.,* volume 41, issue 1, pp. 41–101, 2008 © Cambridge Journals, reproduced with permission.)

discovered by James Wang in 1971. This enzyme also has the distinction of being the first topoisomerase to be discovered. Since Wang's discovery, a second type IA topoisomerase, topoisomerase III, has been isolated from *E. coli*, and similar enzymes have been isolated and characterized from other bacteria and eukaryotes. A defining characteristic of type IA topoisomerases is that the active site tyrosine forms a transient attachment to the 5′-phosphate end of the cleaved DNA strand (Figure 4.21). In general, type IA topoisomerases relax underwound DNA (negatively supercoiled) by first melting a short stretch of double-stranded DNA and then introducing a transient break in one of the strands in the melted region. The unbroken strand is then free to move through the transient break before the nick is resealed. The type IA enzyme's ability to melt supercoiled DNA decreases as the DNA becomes more relaxed, so the enzyme becomes less and less proficient as the reaction continues. This kinetic property helps type

IA topoisomerases maintain the circular DNA in a bacterial cell at its optimal supercoiling density.

Type IA topoisomerases do not alter overwound (positively supercoiled) DNA unless the DNA has a pre-existing single-stranded region. Type IA topoisomerases catalyze four types of topological conversions. (FIGURE 4.23): (1) they partially relax negatively supercoiled DNA; (2) they knot and unknot single-stranded DNA rings; (3) they link two complementary single-stranded DNA rings into a double-stranded DNA ring; and (4) they convert double-stranded circles with at least one nick into catenanes (interlocking rings). None of these conversions requires the addition of an outside energy source such as ATP.

The crystal structure for the N-terminal fragment of *E. coli* topoiomerase I (residues 2–590) has been determined (FIGURE 4.24). The structure resembles a padlock. Four domains surround a central hole with a diameter of about 2.7 nm that is large enough to encircle either a single- or double-stranded piece of DNA. Domains I, III, and IV surround the bottom half of the hole while domain II forms an arch at the top. The hole is lined with basic amino acids that have favorable electrostatic interactions with the negatively charged DNA backbone. Domains II and III behave as though they are connected to the rest of the protein by a hinge, allowing the enzyme to have an open and closed conformation. Tyrosine-319 in domain III, which is part of the active site, forms a transient phosphodiester bond with the 5′-end of the broken strand. Based on structural and kinetic information, Wang and coworkers proposed the mechanism shown in FIGURE 4.25 to explain how *E. coli* topoisomerase I and other type IA topoisomerases work. According to this mechanism, the enzyme acts by forming a transient

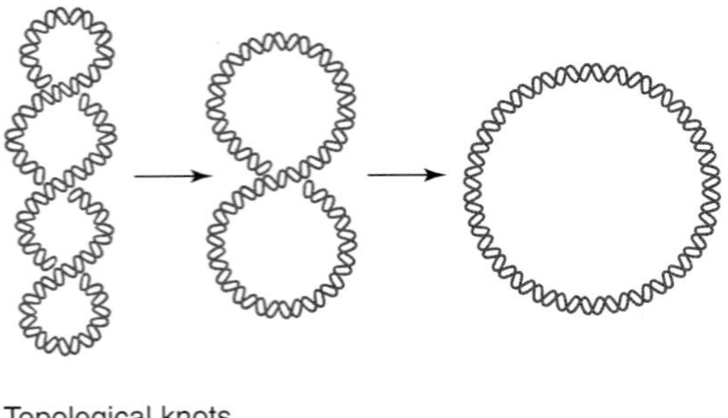

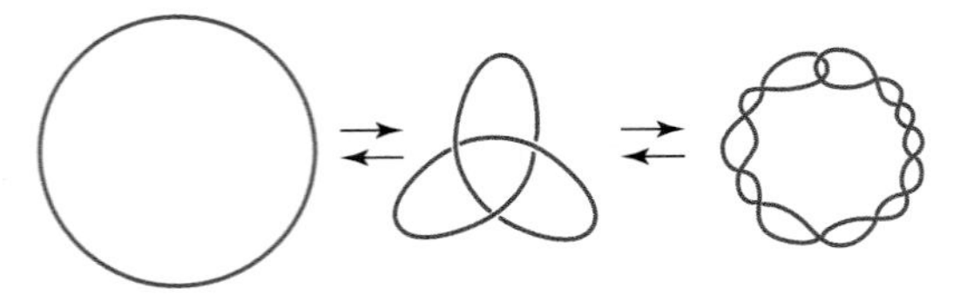

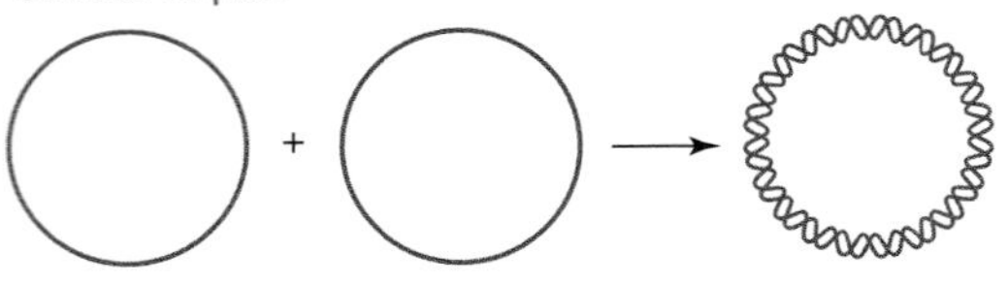

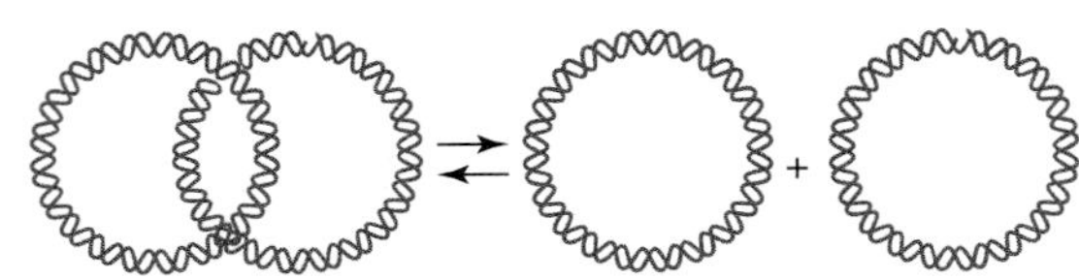

FIGURE 4.23 Four types of topological conversions catalyzed by topoisomerase I. (Adapted from A. Kornberg and T. A. Baker. *DNA Replication, Fourth edition.* W. H. Freeman & Company, 1991.)

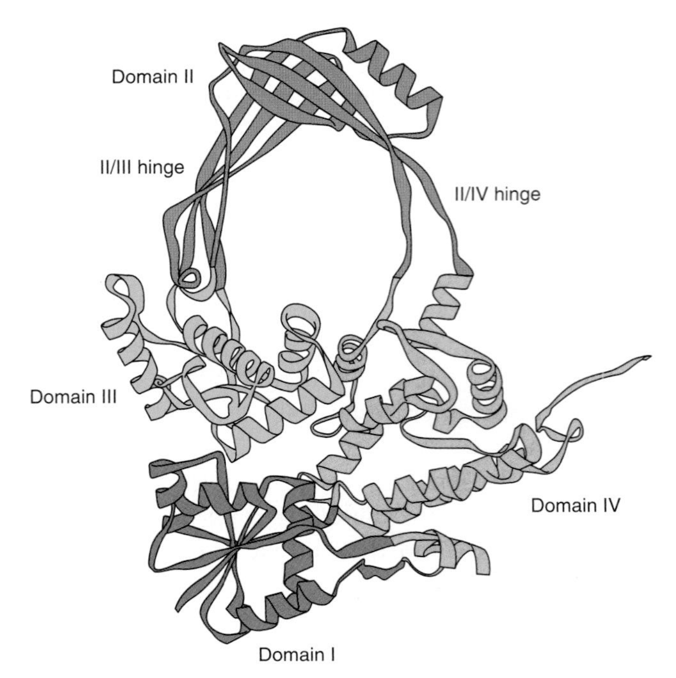

FIGURE 4.24 *Escherichia coli* topoisomerase I, a type IA topoisomerase. (Adapted from J. J. Champoux, *Ann. Rev. Biochem.* 70 [2001]: 369–413.)

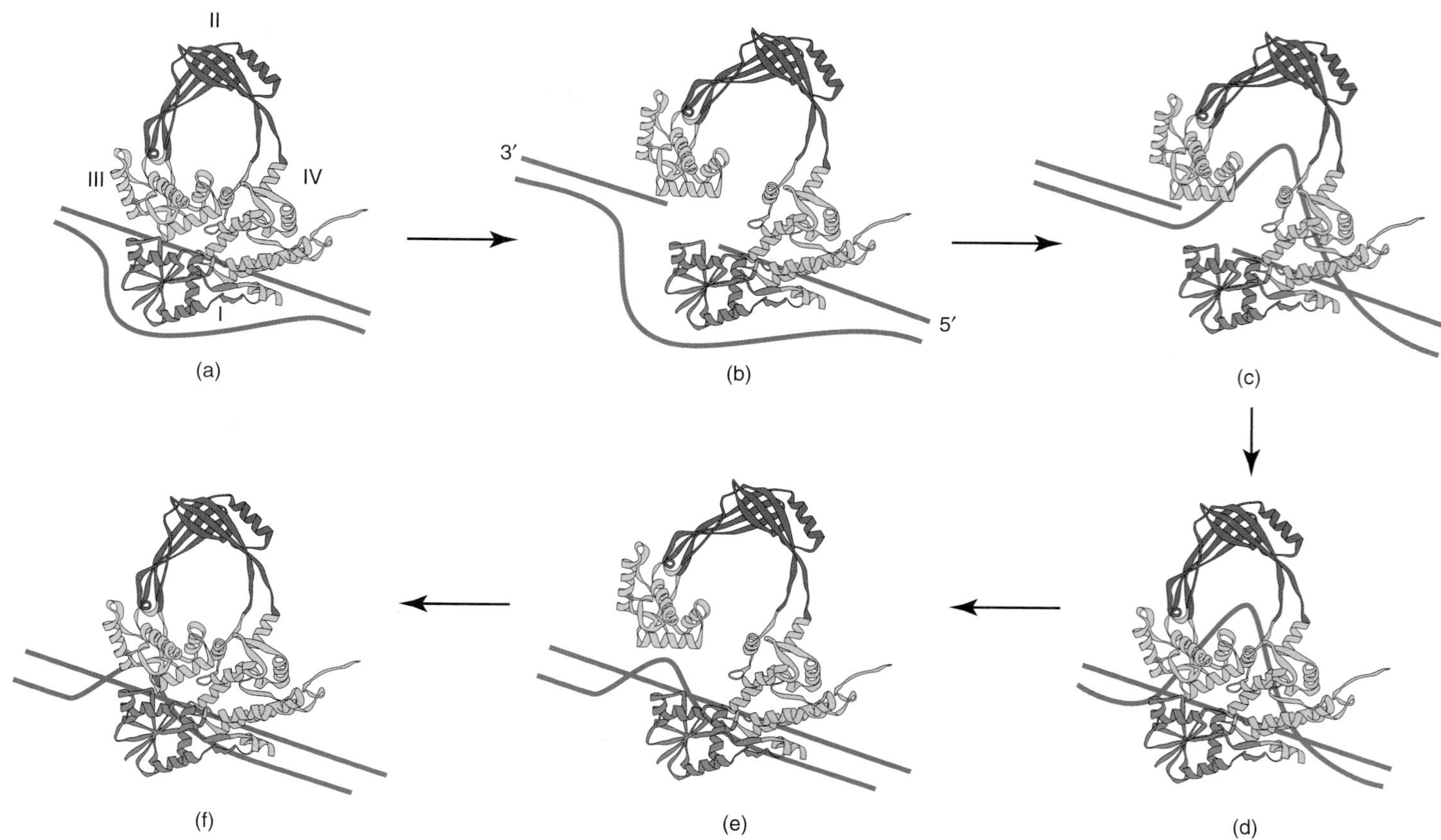

FIGURE 4.25 **Proposed mechanism of relaxation by *E. coli* topoisomerase I.** This schematic shows a series of steps that are proposed to take place during the relaxation of one turn of a negatively supercoiled plasmid DNA. The two strands of the DNA are shown as blue lines (not to scale). The color code used for the four domains, which are labeled in panel a, is the same as that used in Figure 4.24. The strand to be cleaved binds to the surface of the topoisomerase near the large cleft and its polarity is indicated in panel b. For simplicity, the length of the intact strand that passes through the open gate is exaggerated. The interaction with Tyr319 at the active site is not shown. The topoisomerase conformation is proposed to oscillate between a closed form (panels a, d, and f), and an open conformation (panels b, c, and e) that allows the DNA to enter the central hole. The conformation of the open form was modeled by permitting movement at both the II/III and II/IV hinges shown in Figure 4.24. The same mechanism could be applied for decatenation (separating linked rings) or knotting by replacing the intact strand in this figure with a DNA segment from another molecule or from another region of the same molecule, respectively. (Adapted from J. J. Champoux, *Ann. Rev. Biochem.* 70 [2001]: 369–413.)

gate in a single strand that allows another single strand or a double strand to pass through. Because the topoisomers that topoisomerase I acts on are supercoiled, they have more energy than the relaxed form. This energy difference drives the reaction toward the relaxed form.

Members of the topoisomerase type IB subfamily relax both positive and negative supercoiled DNA and relaxation goes to completion. A defining characteristic of type IB topoisomerases is that they form transient covalent intermediates with DNA in which the active site tyrosine attaches to the 3′-phosphate end of the cleaved strand. Human DNA topoisomerase I is the best-studied member of the type IB subfamily. The DNA segments that flank the transient nick are free to rotate relative to each other by turning around single bonds in the intact strand.

Only one member of the topoisomerase type IC family is known at present. This enzyme called topoisomerase V was isolated from

Methanopyrus, a member of the archaea domain. It appears to work in the same way as type IB topoisomerases but is listed as separate type because its structure differs from that of members of the type IB family.

Type II topoisomerases require ATP to convert one topoisomer into another.

Type II topoisomerases catalyze ATP-dependent transport of one intact double-stranded DNA molecule through another. The type IIA and IIB subfamilies appear to catalyze similar types of reactions but to have different structures. Both types act by creating phosphotyrosyl linkages to the 5′ end of DNA. Until 1998, all known type II topoisomerases belonged to the IIA subfamily, which includes mammalian topoisomerase II and the bacterial enzymes DNA gyrase and topoisomerase IV. So far type IIB enzymes have only been found in plants, certain algae, and the archaea. The first type IIB enzyme, topoisomerase VI, was isolated from an archaeon.

Eukaryotic topoisomerase II enzymes act as molecular clamps within which active-site tyrosyl residues bind to nicked DNA. ATP binding and hydrolysis cause the clamp to close and open, respectively. The two-gate mechanism for type II topoisomerase is shown in FIGURE 4.26. The enzyme makes staggered transient cuts in both strands of a double-stranded DNA molecule with the concomitant formation of a phosphomonoester bond between the active-site tyrosine and 5′-ends of the cut DNA. Then the enzyme undergoes a conformational change that allows the topoisomerase to pull the two ends of the cut duplex DNA apart to create an opening in the DNA.

The DNA region that contains the opening is called the gated (or **G-segment**) DNA. A second region of duplex DNA from either the same molecule or a different molecule passes through the open DNA gate. This second region of DNA is designated the transported or **T-segment**.

E. coli has two type IIA topoisomerases, **DNA gyrase** and **topoiomerase IV**. DNA gyrase has the unique catalytic ability to introduce negative supercoils into covalently closed, circular DNA in the presence of ATP. DNA gyrase activity is essential for *E. coli* viability. Other bacteria also require DNA gyrase for survival. Bacterial DNA gyrase works in a similar fashion to human topoisomerase II except that the DNA strands wrap around the gyrase. Topoisomerase IV catalyzes the ATP-dependent relaxation of negative and positive supercoils. Topoisomerase IV is required to decatenate interlocked DNA rings that are formed during bacterial DNA replication. Several antibiotics used to treat bacterial infections in humans target DNA gyrase, topoisomerase IV, or both.

Hyperthermic archaea have a remarkable **reverse gyrase** that introduces positive supercoils into DNA. The enzyme may be needed to somehow stabilize DNA when replication occurs at high temperatures.

In later chapters we see how the topoisomerases, single-stranded DNA binding proteins, and helicases participate in DNA replication, repair, and recombination as well as in transcription.

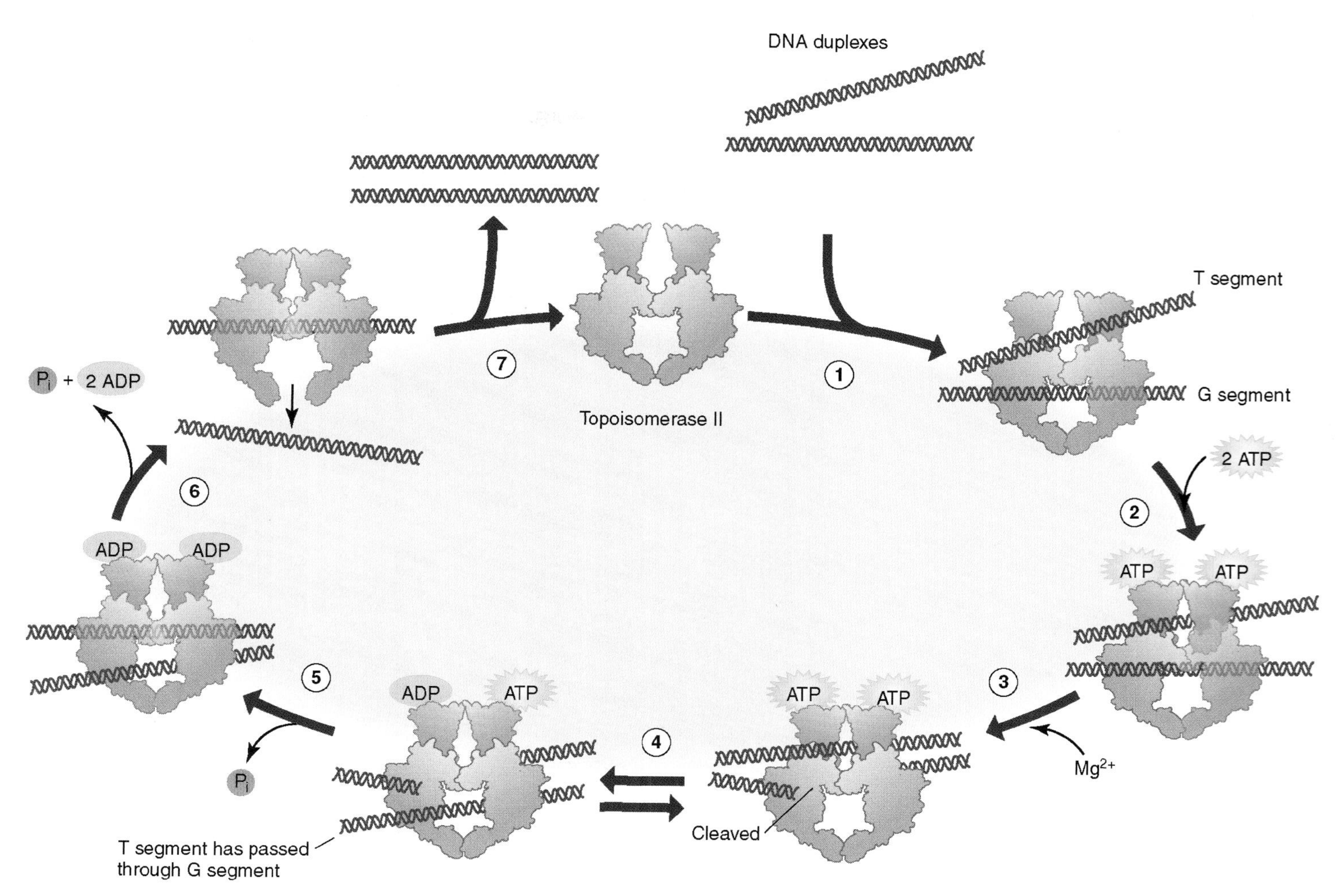

FIGURE 4.26 **Proposed mechanism for the catalytic cycle of DNA topoisomerase II.** The topoisomerase II ATPase domains and core domains are shown in green and light blue, respectively. (Step 1) The catalytic cycle begins when the topoisomerase binds to two double-stranded DNA segments, which are designated the gate segment or G segment (red) and the transported segment or T segment (dark blue). (Step 2) An ATP molecule binds to each ATPase domain, causing the domains to associate. (Step 3) The G segment is cleaved. (Step 4) The T segment is passed through the break in the G segment with concomitant hydrolysis of one ATP molecule. (Step 5) The G segment is rejoined and the remaining ATP molecule is hydrolyzed. (Step 6) After the two ADP molecules dissociate, the T segment is transported through the opening of the core domain. (Step 7) The opening in the core domain is closed and the ATPase domains separate, allowing the enzyme to dissociate from the DNA. (Adapted from A. K. Larsen, et al., *Pharmacol. Ther.* 2 [2003]: 167–181.)

4.8 Non-B DNA Conformations

A-DNA is a right-handed double helix with a deep major groove and very shallow minor groove.

B-DNA is the predominant DNA form in living systems but is not the only form. Several non-B DNA conformations also exist in nature, although often just fleetingly. The first non-B conformation to be identified was A-DNA. Recall from Chapter 1 that Rosalind Franklin obtained the x-ray diffraction pattern for A-DNA when she examined

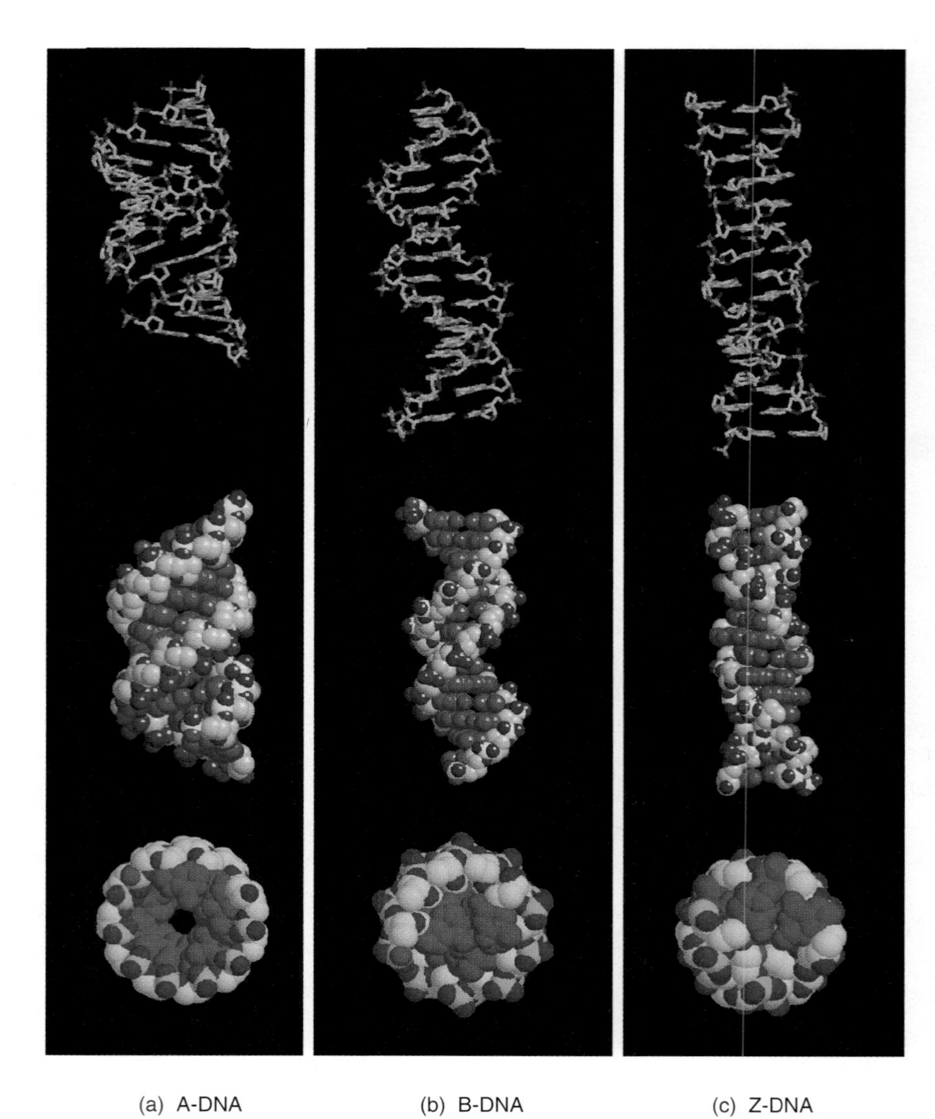

(a) A-DNA (b) B-DNA (c) Z-DNA

FIGURE 4.27 **DNA conformations.** Three conformations of DNA are shown (a) A-DNA, (b) B-DNA, and (c) Z-DNA. (Top) Structures are in stick display so that the orientation of the base pairs with respect to the helix axis is visible. (Middle) Structures are in a spacefilling display with the backbone in standard CPK colors and the base pairs in magenta to emphasize the grooves. (Bottom) Structures are once again in a spacefilling display but viewed from the top of the DNA molecules. Each DNA molecule contains twelve base pairs. (Top structures from Protein Data Bank 213D. B. Ramakrishnan and M. Sundaralingam, *Biophys. J.* 69 [1995]: 553–558. Prepared by B. E. Tropp; Middle structures from Protein Data Bank 1BNA. H. R. Drew, et al., *Proc. Natl. Acad. Sci. USA* 78 [1981]: 2179–2183. Prepared by B. E. Tropp; Bottom structures from Protein Data Bank 2ZNA. A. H.-J. Wang, et al., *Left-handed double helical DNA . . .* Prepared by B. E. Tropp.)

DNA fibers at a relative humidity of about 75%. Both A-DNA and B-DNA are right-handed double helices. That is, each spirals in a clockwise direction as an observer looks down its helical axis of symmetry. Despite this similarity, A- and B-DNA differ in many important respects (FIGURE 4.27 and Table 4.2), including the following: (1) A-DNA has 11 bp per helical turn, whereas B-DNA has 10.5 bp per

TABLE 4.2 Comparison of Major Features in A-, B-, and Z-Forms of DNA				
Parameter		A-DNA	B-DNA	Z-DNA
Helix sense		Right	Right	Left
Base pairs per turn		11	10.5	12
Axial rise per base pair (nm)		0.26	0.34	0.45
Base pair tilt (°)		20°	–6°	7°
Rotation per residue (°)		33°	36°	–30°
Diameter of helix (nm)		2.3	2.0	1.8
Configuration of glycosidic bond	dT, dC	anti	anti	anti
	dG, dA	anti	anti	syn
Sugar Pucker	dT, dC	C_3'-endo	C_2'-endo	C_2'-endo
	dG, dA	C_3'-endo	C_2'-endo	C_3'-endo

Adapted from D. W. Ussery, *Encyclopedia of Life Sciences* [DOI: 10.1038/npg.els.0003122]. Posted May 16, 2002.

helical turn. (2) The plane of a base pair in A-DNA is tilted 20° away from the perpendicular to the helix axis, whereas the corresponding value for B-DNA is –6°; (3) A-DNA has an axial hole when viewed down its long axis, whereas B-DNA does not; and (4) A-DNA has a deep major groove and very shallow minor groove, whereas both the major and minor grooves are about the same depth in B-DNA. Double-stranded DNA seldom, if ever, assumes the A-form in the aqueous environment present in living systems. However, double-stranded RNA (see below) and DNA-RNA hybrids (nucleic acids with one DNA strand and one RNA strand) form right-handed helices that closely approximate A-DNA.

Z-DNA has a left-handed conformation.

Data obtained by using x-ray diffraction analysis of DNA fibers do not provide information about the distances between specific base pairs. Such information only can be obtained by studying DNA crystals but DNA isolated from natural sources is too heterogeneous to form crystals. By the late 1970s, organic chemists had devised methods for DNA synthesis that permitted the synthesis of a homogenous DNA sample that formed a crystal. Alexander Rich and his colleagues took advantage of this advance to prepare crystals of $d(CG)_3$, a self-complementary hexadeoxyribonucleotide,

5′-CGCGCG-3′
3′-GCGCGC-5′

Rich and coworkers examined the x-ray diffraction pattern of the crystalline double-stranded DNA fragment, expecting to see the diffraction pattern for A- or B-DNA. To their great surprise, however, the diffraction pattern was one that had never been seen before, indicating an entirely new DNA conformation. Because the diffraction data showed that the sugar phosphate backbone has a zigzag appearance (FIGURE 4.28), Rich and coworkers called the new form of DNA **Z-DNA**. Subsequent studies showed that the alternating pyrimidine-purine repeat $d(CA)_3$ and its complement $d(TG)_3$ base pair to

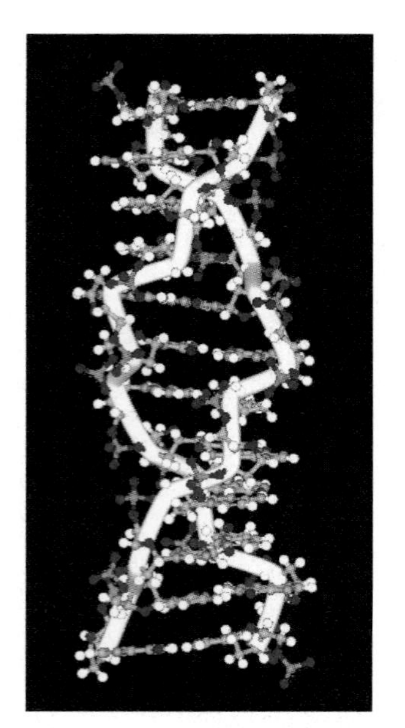

(a) Z-DNA with zig-zag sugar phosphate backbone shown in white

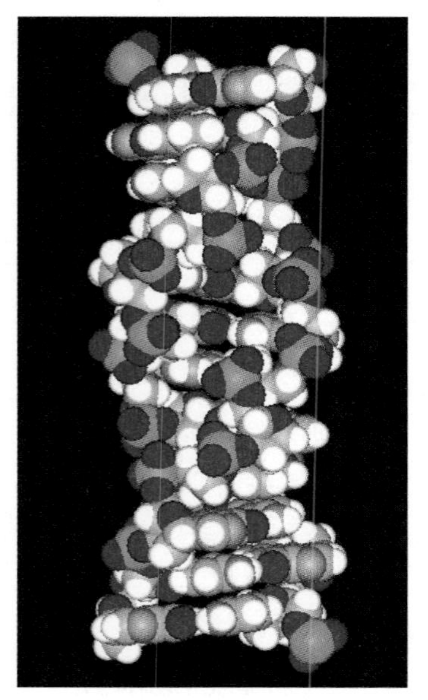

(b) The same Z-DNA with the zigzag sugar phosphate backbone shown in space filling display

FIGURE 4.28 **Z-DNA.** (a) Z-DNA with its zigzag backbones shown as white tubes. The bases, sugars, and phosphates are shown as ball and stick structures. (b) The same Z-DNA shown in a space filling display. (Structures from Protein Data Bank 2ZNA. A. H.-J. Wang, et al., *Left-handed double helical DNA* . . . Prepared by B. E. Tropp.)

(a) C_2'-endo

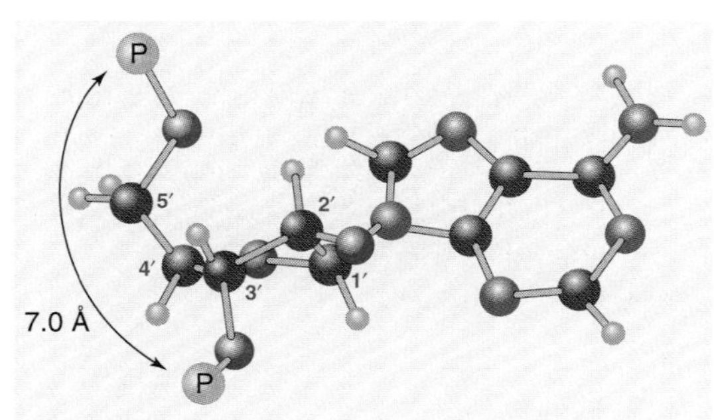

(b) C_3'-endo

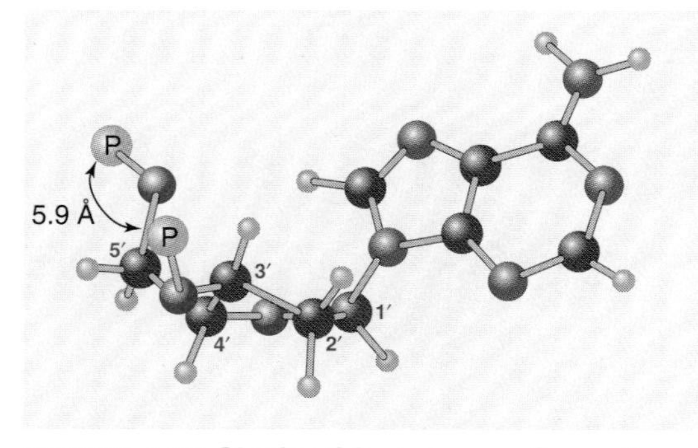

FIGURE 4.29 **Nucleotide sugar conformation.** (a) The C2′-endo conformation, which occurs in B-DNA. Carbon-2 in deoxyribose lies above the plane of the ring as oriented here. (b) The C3′-endo conformation, which occurs in A-DNA and double-stranded RNA. Carbon-3 in deoxyribose lies above the plane of the ring as oriented here. (Adapted from D. Voet and J. G. Voet. *Biochemistry, Third edition.* John Wiley & Sons, Ltd., 2005.)

form a duplex that also can adopt the Z-conformation. The self-complementary hexadeoxyribonucleotide $d(TA)_3$ does not assume a Z-conformation, however. Even when a duplex can assume the Z-conformation, it will not do so unless present in a solution with a high salt concentration. For instance, the 5′-CGCGCG-3′ hexanucleotide requires a sodium chloride concentration greater than 2 M or a magnesium chloride concentration greater than 0.7 M to assume the Z-conformation. The B-conformation is favored at lower salt concentrations. If an alternating d(CG) sequence is contained within a longer DNA tract, such as

5′-TGATCCGCGCGCGAGTCTT-3′
3′-ACTAGGCGCGCGCTCAGAA-5′

the alternating d(CG) sequence can assume the Z-conformation in a 2 M sodium chloride solution, but the rest of the DNA will be in the B-conformation. At least one base pair appears to be disrupted at the Z-DNA/B-DNA junction, forcing at least two bases out of the helix.

Z-DNA is a left-handed helix. The left-handed Z-DNA and right-handed B-DNA are not mirror images but entirely different structures. This difference is immediately obvious when examining the grooves. Z-DNA has a single groove, whereas B-DNA has a major and a minor groove (compare Figures 4.27c and b). The immediate question that arises is whether the left-handed structure has biological significance.

Certainly, the intracellular salt concentration does not ever approach 2 M, so *in vivo* salt could not cause the transition. Polyamines such as spermine and spermidine may help to stabilize the Z-conformation in the cell. There is considerable evidence that localized regions of Z-DNA do exist in cells at least for short periods of time. For instance, antibodies to Z-DNA strongly bind to *Drosophila* salivary gland chromosomes. Additional support comes from the isolation of Z-DNA binding proteins from bacteria, yeast, and animals. Although the functions of these Z-DNA binding proteins have not been completely established, existing evidence suggests that they play a role in regulating the expression of a few eukaryotic genes. Z-DNA may play a harmful role in human health. There is mounting evidence that regions with the potential to form Z-DNA are hot spots for DNA double-strand breaks, which can cause chromosomal rearrangements that result in malignant diseases.

(a) Deoxyguanylate in B-DNA in *anti* conformation

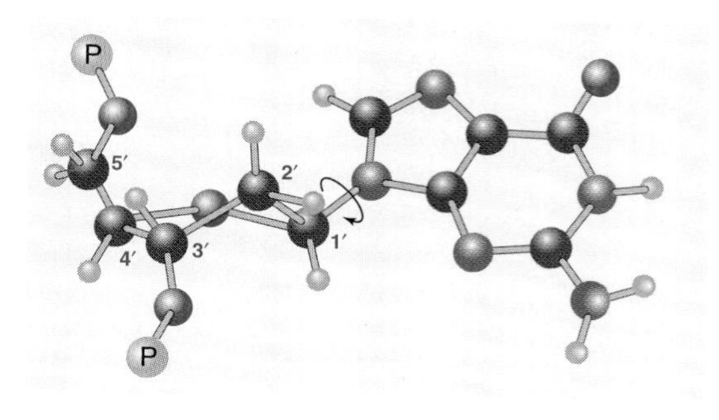

(b) Deoxyguanylate in Z-DNA in *syn* conformation

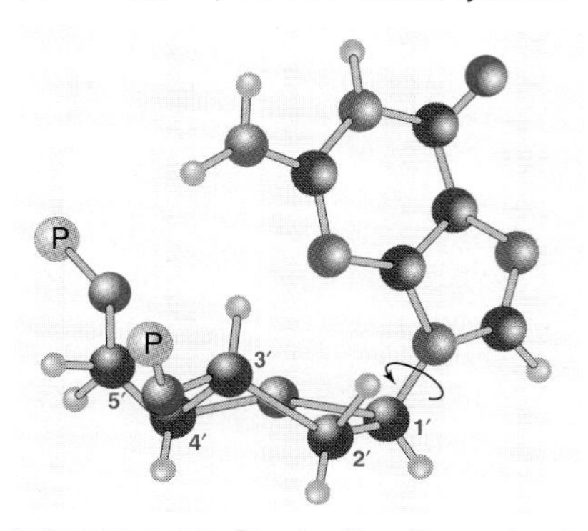

FIGURE 4.30 Sterically allowed orientations of purine bases with respect to the attached deoxyribose unit. Purine bases are (a) in the anti conformation in both A- and B-DNA but (b) in the syn conformation in Z-DNA.

DNA conformational changes result from rotation about single bonds.

Different DNA conformations are possible because polydeoxyribonucleotide chains have single bonds that permit rotations. Two types of single bonds are of special interest, those in the five-member sugar ring and the N-glycosylic bond that joins carbon-1 in the sugar to purine bases. Single bonds in a furanose permit the ring to assume a puckered conformation in which four atoms are nearly coplanar while the fifth atom is about 0.05 nm out of the plane of the ring. X-ray crystallography studies of deoxyribonucleotides indicate that either C-2′ or C-3′ is out of the plane of the deoxyribose ring on the same side as C-5′ (FIGURE 4.29). These conformations are called **C_2′-endo** and **C_3′-endo**, respectively. The C_2′-endo conformation is present in B-DNA, whereas the C_3′-endo conformation is present in A-DNA and in double-stranded RNA. The situation is a bit more complicated for Z-DNA where pyrimidine nucleotides have C_2′-endo conformations and purine nucleotides have C_3′-endo conformations. Rotation about the N-glycosylic bond results in two purine nucleoside conformations (FIGURE 4.30). In the **anti** conformation (Greek: against) the purine base is positioned away from the sugar, whereas in the **syn** (Greek: with) the purine base is positioned over the sugar. Pyrimidine deoxyribonucleosides are nearly always present in the anti conformation because of steric interference between the sugar and pyrimidine in the syn conformation. Purine deoxyribonucleosides are in the anti conformation in both A- and B-DNA but in the syn conformation in Z-DNA. These conformational relationships are summarized in Table 4.2.

Several other kinds of non-B DNA structures appear to exist in nature.

Some DNA regions can exist transiently in a cruciform, triplex, slipped (hairpin) structure, or quadruplex structure.

1. The cruciform structure—A cruciform structure can only be formed in a region that contains an **inverted repeat sequence** such as

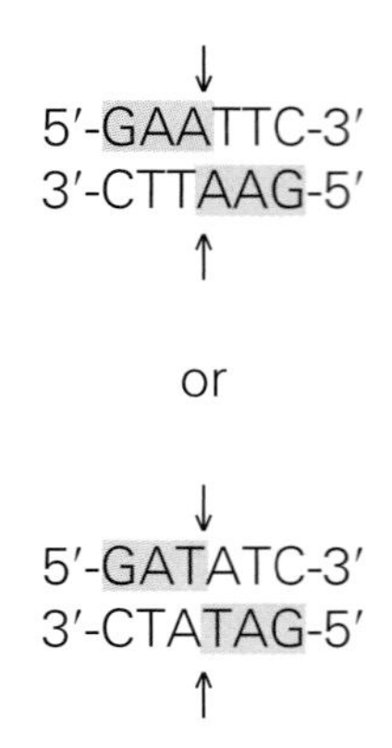

The arrows point to the vertical axis of symmetry: the double-stranded segment to the right of the axis can be superimposed on the one to the left by a 180° rotation in the plane of the page. The left to right sequence on the top strand hence is repeated right to left on the bottom strand. Crystallographers refer to this type of arrangement as dyad symmetry. Inverted repeats range in length up to about 50 base pairs. Molecular biologists use the term **palindrome** when referring to an inverted repeat sequence. A lexicologist might take issue with this use of the term, however, because a palindrome was originally defined as a word (such as "madam") or a phrase (such as "Able was I ere I saw Elba") that reads the same forward or backward on a single line.

In theory, DNA molecules that have inverted repeats can exist in two alternate forms, a normal duplex in which base pairs form between the two complementary strands, or a **cruciform structure** (FIGURE 4.31) in which base pairs form between complementary regions on the same strand to produce double-stranded branches. Model building and energy calculations show that cruciform structures are somewhat strained compared to normal duplexes. Cruciform structures were originally produced in the laboratory under special conditions, but they also exist in cells.

2. Triplex structures—Under certain conditions the DNA double helix can accommodate a third strand in its major groove to form a **triplex structure** (FIGURE 4.32). When one helical strand has a run of purines, the major groove can accommodate a pyrimidine- or purine-rich oligonucleotide (FIGURE 4.33). A pyrimidine-rich oligonucleotide orients parallel to the purine strand (Figure 4.33a). Each purine simultaneously engages in base pairing interactions with two pyrimidines, one with standard

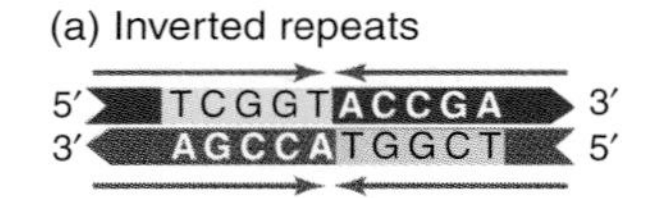

(b) Cruciform structure

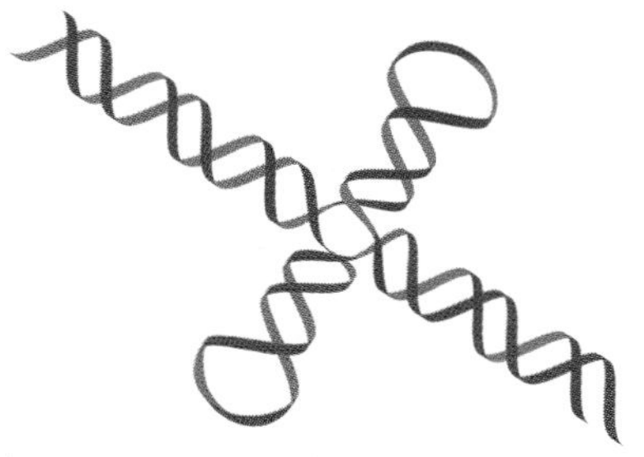

FIGURE 4.31 **Cruciform structure.** (a) A segment of DNA with inverted repeats. The arrows above and below the structure indicate the inverted repeats. (b) A DNA molecule with an inverted repeat can form a cruciform structure. (Adapted from A. Bacolla and R. D. J. Wells, *J. Biol. Chem.* 279 [2004]: 47411–47414.)

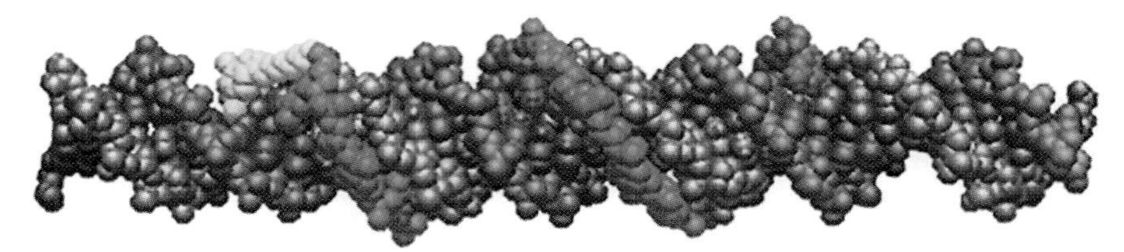

FIGURE 4.32 **A space-filling model of triplex DNA.** Strands in the Watson-Crick double helix are shown in dark green and purple. The triplet forming oligonucleotide (orange) in the major groove is tagged with a psoralen molecule (light yellow) at its 5′-end. (Reproduced from L. A. Christensen, et al., *Nucleic Acids Res.* 36 [2008]: 7136–7145, by permission of Oxford University Press. Photo courtesy of Karen Vasquez, University of Texas M.D. Anderson Cancer Center.)

Watson-Crick geometry and the other with a non-standard form of base pairing. The non-standard form is called Hoogsteen base pairing after Karst Hoogsteen, the investigator who first recognized the alternate geometry in 1963. When a purine-rich oligonucleotide fits into the major groove, the oligonucleotide orients antiparallel to the purine strand (Figure 4.33b). The two original strands continue to interact through standard Watson-Crick base pairing. However, the purine-rich oligonucleotide and the purine rich strand interact through a form of non-standard base pairing called reverse-Hoogsteen pairing. Investigators have used triplex-forming oligonucleotides as sequence-specific strips to bind to specific genes. When triplex formation takes place *in vivo*, there is increased likelihood of double-strand breaks and mutations appearing.

3. Quadruplex structure—One, two, or four G-rich DNA strands can form a **quadruplex** (or **tetraplex**) **structure** (FIGURE 4.34). The fundamental unit of a quadruplex structure is the quartet, a planar structure containing four guanine groups held together by eight hydrogen bonds. Two or more quartets stack on one another to form a quadruplex. The structure is stabilized by the hydrogen bonds and hydrophobic base stacking. Additional stability is provided by cation-dipole interactions between the eight guanine groups and a metal cation, usually Na^+ or K^+, which sits between two quartets. Intramolecular quadruplexes formed by a single strand are of special interest because the ends of eukaryotic chromosomes have G-rich 3′-overhangs that are rich in guanine. According to one hypothesis, quadruplexes formed from these overhangs influence chromosome replication and stability. Quadruplexes may also form in DNA regions that regulate the expression of a few eukaryotic genes.

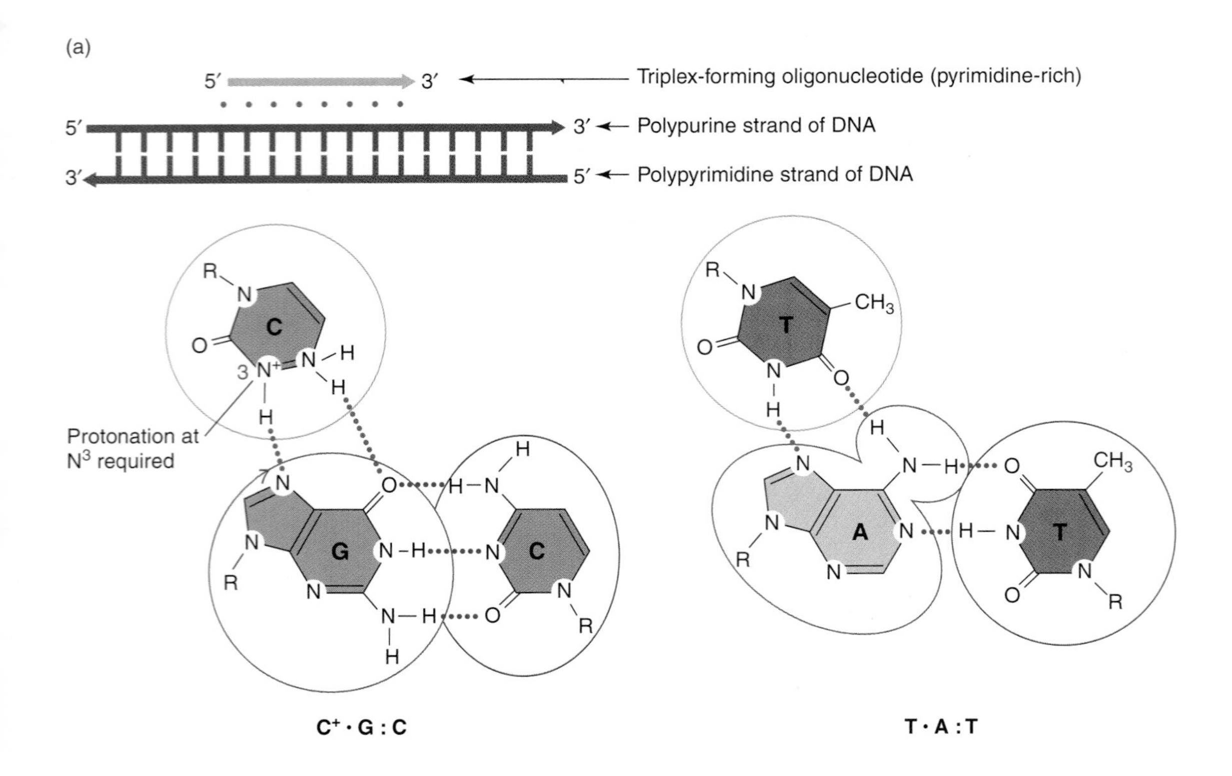

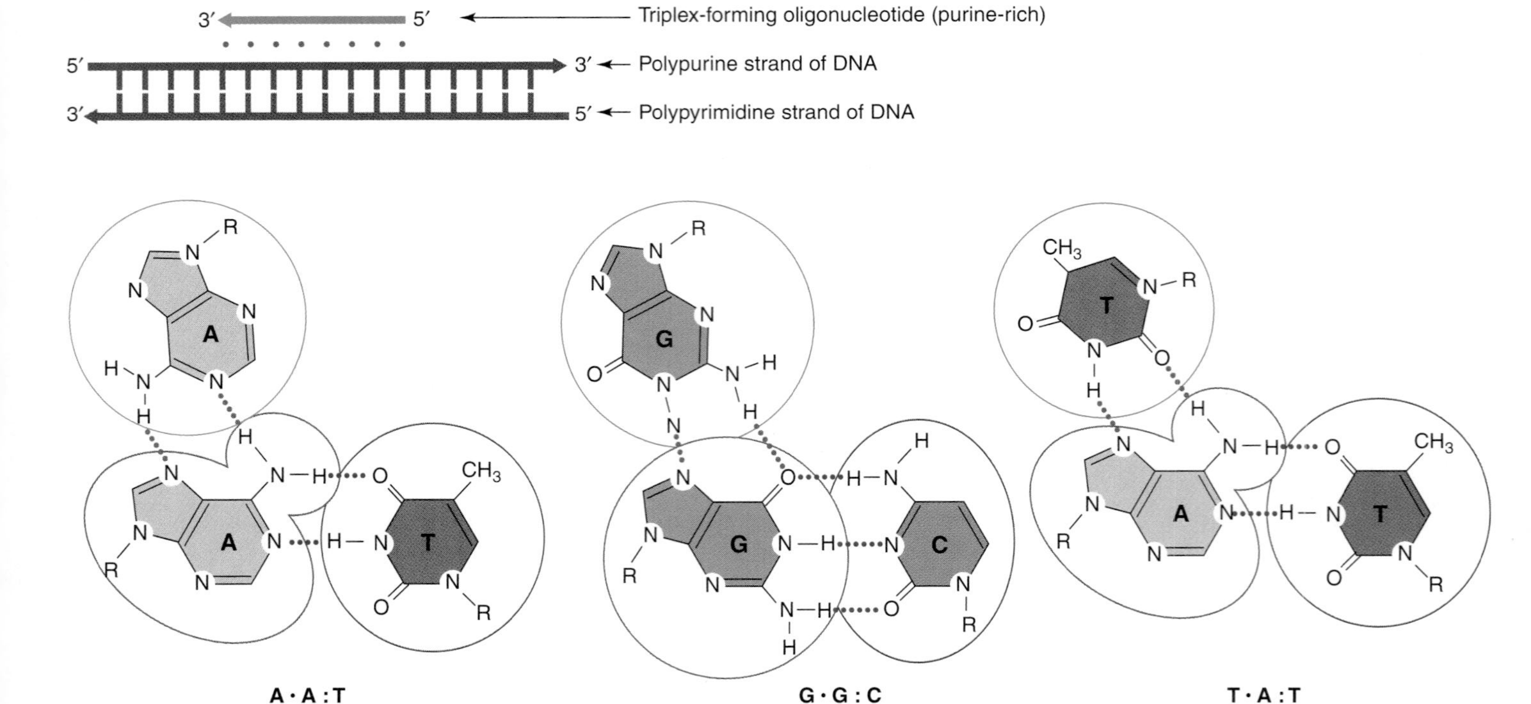

FIGURE 4.33 **Two arrangements for a DNA triplex.** (a) When one strand in a helical duplex has a run of purines, a pyrimidine rich oligonucleotide can fit in the major groove. When the triplex forming strand is pyrimidine-rich it orients antiparallel to the purine strand. Each purine base pairs with two pyrimidines, one with standard Watson-Crick geometry and the other with a non-standard form of base pairing. (b) A purine rich oligonucleotide can also fit into the major groove. In this case, the oligonucleotide orients antiparallel to the purine strand. The two original strands continue to interact through Watson-Crick base pairing. The purine interact through reverse Hoogsteen base pairing. Hydrogen bonds resulting from non-standard base pairing are shown in green. (Reproduced from K. M. Vasquez and P. M. Glazer, Triplex-forming oligonucleotides . . ., *Q. Rev. Biophys.*, volume 35, issue 1, pp. 89–107, 2002 © Cambridge Journals, reproduced with permission.)

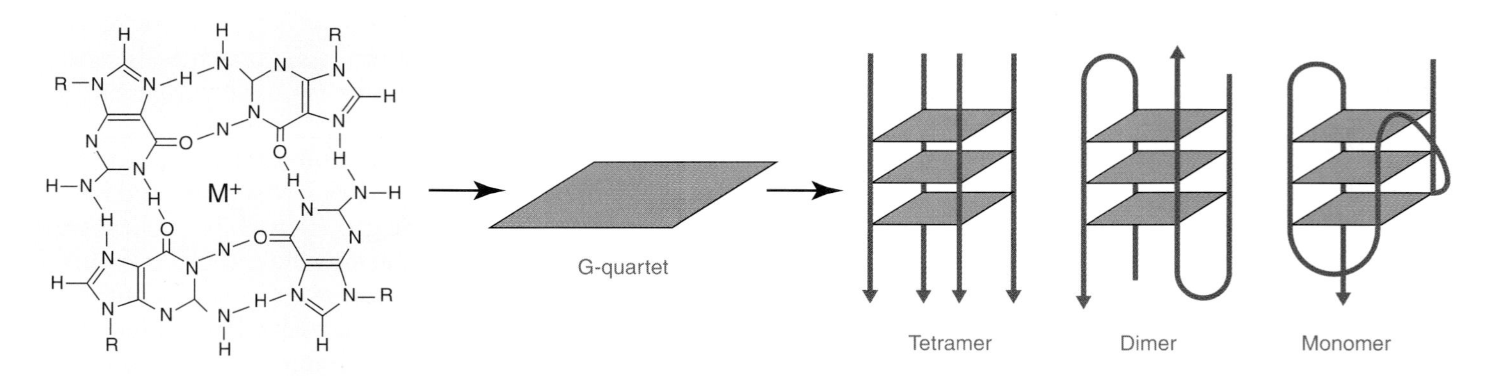

FIGURE 4.34 **Quadruplex structure.** (a) G-quartet. (b) Tetrameric, dimeric, and monomeric G-quadruplexes composed of three G-quartets. (Adapted from P. Bates, J.-L. Megny, and D. Yang, *EMBO Rep.* 11 [2007]: 1003–1010.)

4.9 RNA Structure

RNA performs a wide variety of functions in the cell.

When the Watson-Crick Model for B-DNA was proposed in 1953, very little was known about RNA function. Within the next few years, investigators showed that some viruses use RNA as the genetic material. This important discovery did not explain RNA's function(s) in cells, however, because cells use DNA as their hereditary material. The RNA function problem appeared to have been solved when three kinds of RNA molecules were shown to make essential contributions to polypeptide synthesis. Ribosomal RNA (rRNA) molecules combine with proteins to form ribosomes, the ribonucleoprotein complex that serves as the protein synthetic factory. Transfer RNA (tRNA) molecules carry activated amino acids to the ribosomes where messenger RNA (mRNA) specifies the order in which the ribosome adds amino acids to the growing polypeptide chain. Then in the early 1980s Sidney Altman and Thomas Cech, working independently, demonstrated that some RNA molecules act as catalysts, a role that molecular biologists had previously assumed to be limited to proteins. The list of RNA functions grows with each passing year. The focus for now is on RNA structure.

RNA secondary structure is dominated by Watson-Crick base pairs.

RNA structure, like protein structure, is divided into primary, secondary, tertiary, and quaternary structures. The primary sequence of an RNA molecule is its base sequence. RNA secondary structure consists of helical regions and various kinds of loops, bulges, and junctions within the helical regions, which are stabilized by Watson-Crick base pairing. The tertiary structure consists of the arrangements of these secondary structures into a three-dimensional structure, which is often compact and stabilized by metal cations. The quaternary structure describes the arrangement of an RNA molecule with respect to other

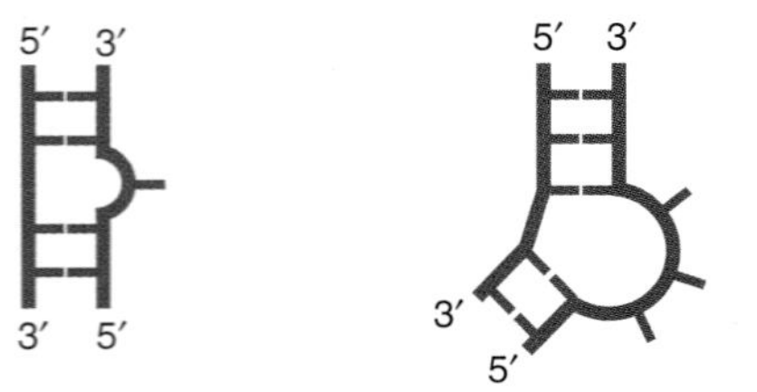

(a) Single nucleotide bulge (b) Three nucleotide bulge

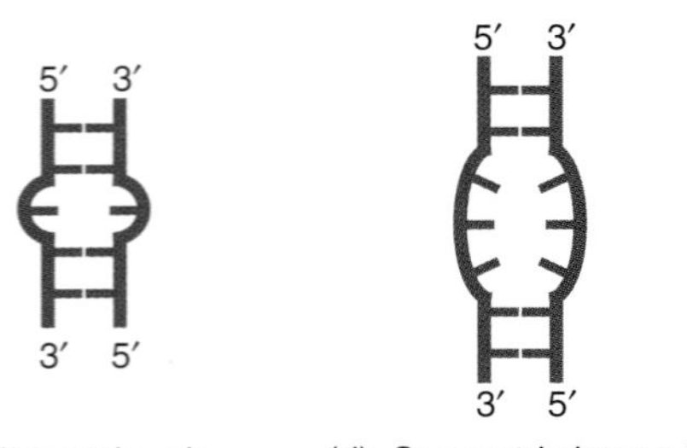

(c) Mismatch pair (d) Symmetric internal loop

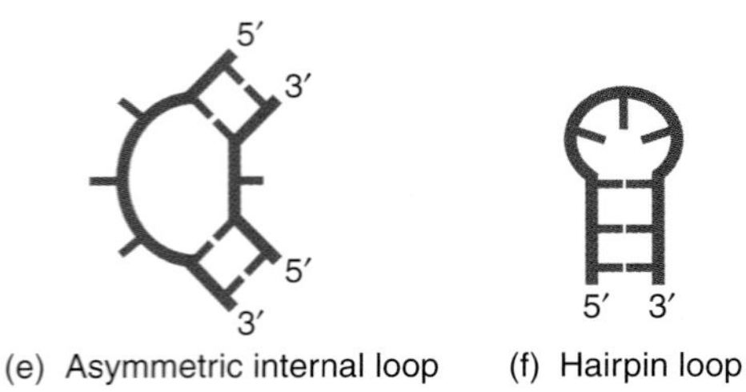

(e) Asymmetric internal loop (f) Hairpin loop

FIGURE 4.35 **RNA loop and bulge secondary elements.** (Reprinted from *Semin. Virol.*, vol. 8, J. Nowakowski and I. Tinoco, Jr., RNA structure and stability, pp. 153–165, copyright 1997, with permission from Elsevier [http://www.sciencedirect.com/science/journal/10445773].)

RNA molecules or with protein molecules. It is important to note that some kinds of RNA molecules do not assume a specific three-dimensional structure but instead function as unstructured single strands.

RNA's predominant secondary structure building blocks are helical tracts, which are stabilized by Watson-Crick base pairs and have a similar conformation to A-DNA. Ribose groups within a helical tract have C_3′-endo conformations, whereas those at the end of a tract (or in single-stranded RNA) are a mix of C_2′-endo and C_3′-endo conformations. A helical tract seldom exceeds ten successive base pairs before being interrupted by one or more **loops** or **bulges** (FIGURE 4.35). The smallest bulge results from the presence of a single unpaired base (Figure 4.35a). If the unpaired base is stacked within the helix, the helix bends. If the base is outside the helix, the helix does not bend but the base can interact with other parts of the same RNA molecule, other RNA molecules, or proteins. Larger bulges result from the presence of additional unpaired bases (Figure 4.35b). An internal loop forms when one or more nucleotides on each RNA strand are unpaired. The smallest loop forms when a single pair of non-complementary bases, a mismatch pair, interrupts the helical tract (Figure 4.35c). Larger loops form when additional unpaired bases interrupt the helical tract. A **symmetric internal loop** forms when each strand has the same number of opposing unpaired bases (Figure 4.35d) and an **asymmetric internal loop** forms when one strand has more unpaired bases than the other (Figure 4.35e). A **hairpin** (or **stem-loop**) **structure** forms when one strand folds back on itself to form a stem that contains Watson-Crick base pairs (Figure 4.35f). A hairpin loop may be as small as two nucleotides or it may be several nucleotides long.

Double helical stems may come together to form a **junction** (FIGURE 4.36). A junction is an important structural element because it helps to establish the overall structure of the RNA molecule. When two helical tracts meet end-to-end at a junction, they form a structure resembling a long helix. This end-to-end interaction, called **coaxial stacking**, helps to stabilize the junction.

RNA tertiary structures are stabilized by interactions between two or more secondary structure elements.

The RNA tertiary structure, which describes the three-dimensional structure of the RNA molecule, results from interactions between two or more secondary structures. Tertiary structures are stabilized by metal cations such as Na^+ and Mg^{2+} that offset charge repulsions that would otherwise prevent the negatively charged sugar phosphate backbone from folding into a condensed structure. Specific structural elements contribute to the tertiary structure. Some of these structural elements are described here and others will be introduced when the structures of specific RNA molecules are examined in later chapters.

A **pseudoknot** forms when a base sequence in a hairpin loop pairs with a complementary single stranded region that is adjacent to the hairpin stem (FIGURE 4.37). The two helical tracts stack end-to-end, forming a coaxial stack.

Several kinds of structural interactions can bring distant regions of a large RNA molecule together. Two of these interactions are shown in FIGURE 4.38. **Kissing hairpins** form when unpaired nucleotides in one hairpin base pair with complementary nucleotides in another hairpin (Figure 4.38a). **Hairpin loop-bulge** contacts form when unpaired nucleotides in a bulge base pair with complementary nucleotides in a hairpin (Figure 4.38b).

Crystal structures have been determined for many RNA molecules. Those for RNA catalysts are especially interesting. The smallest known RNA catalyst, the **hammerhead ribozyme** (FIGURE 4.39), is specified by virus-like RNA molecules that infect plants. The ribozyme's name derives from the shape of its secondary structure (Figure 4.39a). The minimal functional RNA consists of three short helices and a conserved junction sequence (Figure 4.39b). A functional hammerhead ribozyme can be constructed from two separate RNAs. One strand serves as the

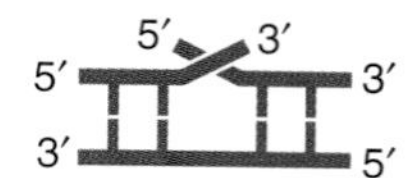

(a) Two-stem junction

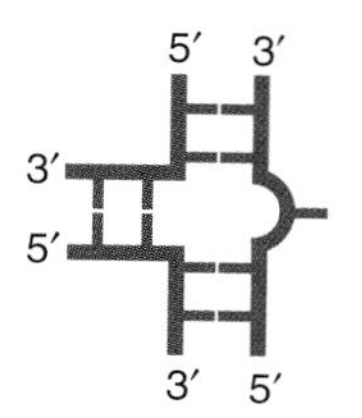

(b) Three-stem junction

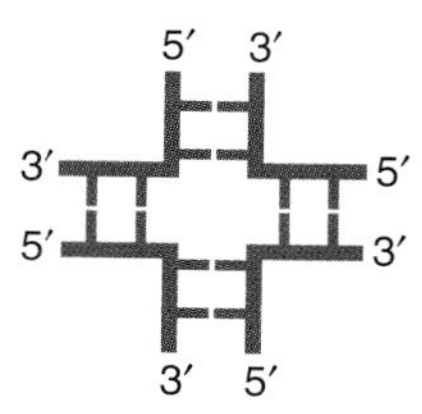

(c) Four-stem junction

FIGURE 4.36 **RNA junction secondary elements.** (Reprinted from *Semin. Virol.*, vol. 8, J. Nowakowski and I. Tinoco, Jr., RNA structure and stability, pp. 153–165, copyright 1997, with permission from Elsevier [http://www.sciencedirect.com/science/journal/10445773].)

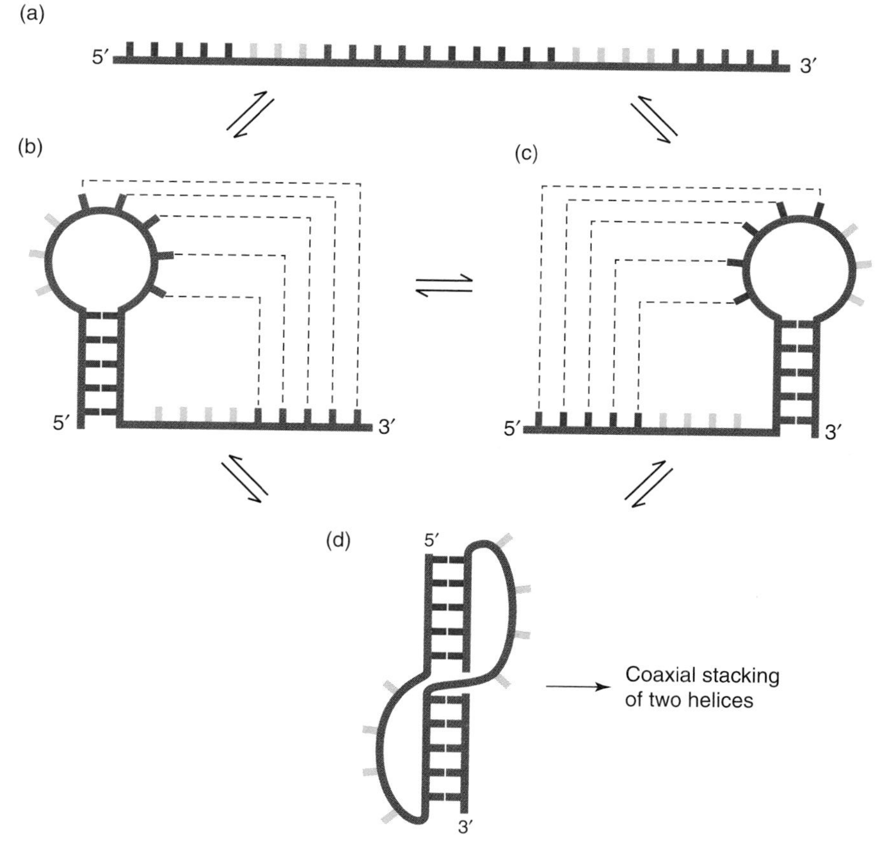

FIGURE 4.37 **RNA pseudoknot structure.** (a) Unstructured RNA with two pairs of complementary base sequences. One pair is shown in red and the other in blue. (b) The hairpin (stem-loop) structure formed when the complementary sequences shown in red base pair. (c) The hairpin (stem-loop) structure formed when the complementary sequences shown in blue base pair. (d) The pseudoknot that forms when a base sequence in a hairpin loop pairs with a complementary single stranded region that is adjacent to the hairpin stem. The two helical tracts stack end-to-end, forming a coaxial stack. (Modified from *Encyclopedia of Life Sciences*, F. Varani, Copyright © 2001 and reproduced with permission of John Wiley & Sons, Inc.)

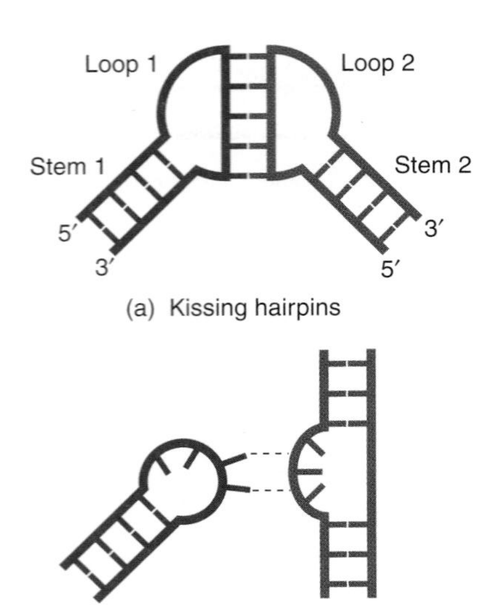

FIGURE 4.38 **Interactions that bring distant RNA segments together.** (a) Kissing hairpin interaction and (b) hairpin loop-bulge interaction. (Reprinted from *Semin. Virol.*, vol. 8, J. Nowakowski and I. Tinoco, Jr., RNA structure and stability, pp. 153–165, copyright 1997, with permission from Elsevier [http://www.sciencedirect.com/science/journal/10445773].)

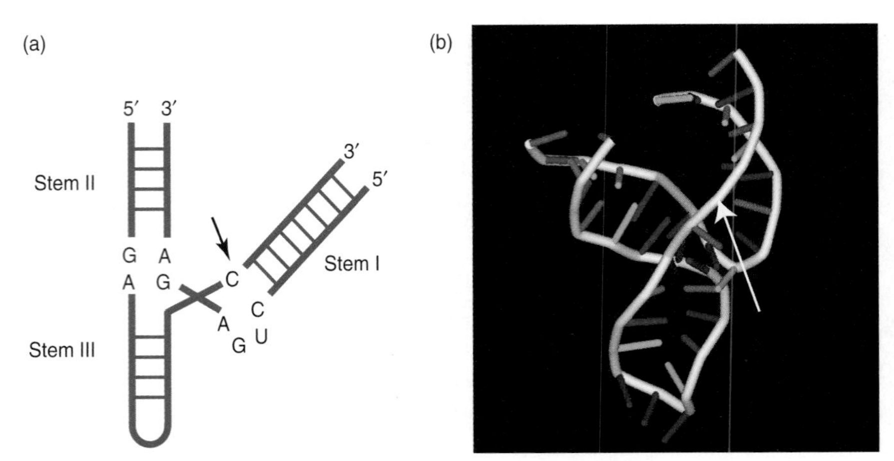

FIGURE 4.39 **Hammerhead ribozyme.** (a) Secondary structure of the hammerhead ribozyme. Nucleotides important for catalytic activity are indicated; the cleavage site is indicated by an arrow. (b) Crystal structure of the hammerhead ribozyme. The arrow points to the easily cut (scissile) phosphodiester bond. (Part a adapted from J. A. Doudna and T. R. Cech, *Nature* 418 [2002]: 222–228. Part b structure from Protein Data Bank 379D. J. B. Murray, et al., *Cell* 92 [1998]: 665–673. Prepared by B. E. Tropp.)

catalyst and the other as the substrate, permitting multiple turnover reactions. Divalent metal cations were originally thought to be essential for catalytic activity, but recent experiments suggest that they can be replaced by a high concentration of monovalent cations. Hence, it now appears that the metal ions are needed to achieve the proper conformation rather than to participate in the catalytic reaction.

4.10 The RNA World Hypothesis

The earliest forms of life on earth may have used RNA as both the genetic material and the biological catalysts needed to maintain life.

The discovery that RNA can act as a catalyst has profound implications for the way that we view biochemical evolution. Biologists have long speculated about whether the earliest progenitors of modern cells contained DNA or protein. It was hard to see how either macromolecule could work in the absence of the other. Protein molecules fold into stable tertiary structures to form catalytic sites but do not transmit genetic information. DNA stores and transmits genetic information but no naturally occurring DNA has yet been found that can act as a catalyst. The term *naturally occurring* is an important qualifier because DNA molecules have been synthesized in the laboratory that can act as catalysts. Even so, RNA molecules are more versatile because their additional 2′-hydroxyl group permits them to form stable tertiary structures that cannot be achieved by DNA.

The discovery that naturally occurring RNA molecules have catalytic properties suggested a solution to the "which came first, the chicken or egg" problem. The earliest progenitors to modern cells probably contained RNA, a molecule that can store genetic information and catalyze biochemical reactions. The earliest life forms thus may have lived in an "RNA world." This hypothesis, first proposed by Walter Gilbert in 1986, suggests that proteins eventually replaced RNA molecules as catalysts for most (but not all) biological reactions because proteins offered more possible sequence and structural alternatives, while DNA would eventually replace RNA molecules for genetic storage in most (but not all) biological systems because it is more stable and easy to repair.

Suggested Reading

General

Bloomfield, V. A., Crothers, D. M., and Tinoco, I. Jr. 2000. *Nucleic Acids: Structures, Properties and Functions*. Herndon, VA: University Science Books.

Bowater, R. P. 2005. DNA structure. In: *Encyclopedia of Life Sciences*. pp. 1–8. Hoboken, NJ: John Wiley and Sons.

Calladine, C. R., and Drew, H. R. 1997. *Understanding DNA*, 2nd ed. San Diego: Academic Press.

Sinden, R. R. 1994. *DNA Structure and Function*. San Diego: Academic Press.

Soukup, G. A. 2001. Nucleic acids: general properties. In: *Encyclopedia of Life Sciences*. pp. 1–9. Hoboken, NJ: John Wiley and Sons.

DNA Bending

Harvey, S. C., Diakic, M., Griffith, J., et al. 1995. What is the basis of sequence-directed curvature in DNAs containing A tracts? *J Biomol Struct Dyn* 13:301–307.

DNA Denaturation and Renaturation

Doty, P. 2003. DNA and RNA forty years ago. *J Biomol Struct Dyn* 21:311–316.

Kool, E. T. 2001. Hydrogen bonding, base stacking, and steric effects in DNA replication. *Ann Rev Biophys Biomol Struct* 30:1–22.

Marmur, J., and Doty, P. 1961. Thermal renaturation of deoxyribonucleic acids. *J Mol Biol* 3:585–594.

Marmur J., and Doty, P. 1962. Determination of the base composition of deoxyribonucleic acid from its thermal denaturation temperature. *J Mol Biol* 5:109–118.

Thomas, R. 1993. The denaturation of DNA. *Gene* 135:77–79.

Wetmur, J. G. 1976. Hybridization and renaturation kinetics of nucleic acids. *Ann Rev Biophys Bioeng* 5:337–361.

Helicases

Arnold, D. A., and Kowalczykowski, S. C. 2001. RecBCD Helicase/Nuclease. In: *Encyclopedia of Life Sciences*. pp. 1–6. London: Nature.

Bird, L. E., Subramanya, H. S., and Wigley, D. B. 1998. Helicases: a unifying structural theme? *Curr Opin Struct Biol* 8:14–18.

Caruthers, J. M., and McKay, D. B. 2002. Helicase structure and mechanism. *Curr Opin Struct Biol* 12:123–133.

Dillingham, M. S. 2006. Replicative helicases: a staircase with a twist. *Curr Biol* 16:R844–R847.

Donmez, I., and Patel, S. S. 2006. Mechanisms of a ring shaped helicase. *Nucleic Acids Res* 34:4216–4224.

Eoff, R. L., and. Raney, K. D. 2005. Helicase-catalysed translocation and strand separation. *Biochem Society Trans* 33:1474–1478.

Egelman, E. 2001. DNA helicases. In: *Encyclopedia of Life Sciences*. pp. 1–5. London: Nature.

Enemark. E. J., and Joshua-Tor. L. 2008. On helicases and other motor proteins. *Curr Opin Struct Biol* 18:243–257.

Frick, D. N. 2003. Helicases as antiviral drug targets. *Drug News Perspect* 16:355–362.

Lohman, T. M., and Bjornson K. P. 1996. Mechanisms of helicase-catalyzed DNA unwinding. *Ann Rev Biochem* 65:169–214.

Lohman, T. M., Tomko, E. J., and Wu, C. G. 2008. Non-hexameric DNA helicases and translocases: mechanisms and regulation. *Nat Rev Mol Cell Biol* 9:391–401.

Mackintosh, S. G., and Raney, K. D. 2006. DNA unwinding and protein displacement by superfamily 1 and superfamily 2 helicases. *Nucleic Acids Res* 34:4154–4159.

Matson, S. W., Bean, D. W., and George, J. W. 1994. DNA helicases: enzymes with essential roles in all aspects of DNA metabolism. *BioEssays* 16:13–22.

Nakura, J., Ye, L., Morishima, A., et al. 2000. Helicases and aging. *Cell Mol Life Sci* 57:716–730.

Niedziela-Majka, A., Chesnik, M. A., Tomko, E. J., and Lohman, T. M. 2007. *Bacillus stearothermophilus* PcrA monomer is a single-stranded DNA translocase but not a processive helicase *in vitro*. *J Biol Chem* 282:27076–27085.

Patel, S. S., and Picha, K. M. 2000. Structure and function of hexameric helicases. *Ann Rev Biochem* 69:651–697.

Pugh, R. A., and Spies, M. 2008. DNA helicases, chemistry and mechanisms of. In: *Wiley Encyclopedia of Chemical Biology*. pp. 1–11. Hoboken, NJ: John Wiley and Sons.

Pyle, A. M. 2008. Translocation and unwinding mechanisms of RNA and DNA helicases. *Ann Rev Biophys* 37:317–336.

Singleton, M. R., Dillingham, M. S., and Wigley, D. B. 2007. Structure and mechanism of helicases and nucleic acid translocases. *Ann Rev Biochem* 76:23–50.

Soultanas, P., and Wigley, D. B. 2001. Unwinding the 'Gordian knot' of helicase action. *Trends Biochem Sci* 26:47–54.

Tuteja, N., and Tuteja, R. 2004. Prokaryotic and eukaryotic DNA helicases essential molecular motor proteins for cellular machinery. *Eur J Biochem* 271:1835–1848.

von Hippel, P. H., and Delagoutte, E. 2001. A general model for nucleic acid helicases and their "coupling" within macromolecular machines. *Cell* 104:177–190.

Single-Stranded DNA Binding Proteins

Alberts, B. M., and Frey, L. 1970. T4 bacteriophage gene 32: a structural protein in the replication and recombination of DNA. *Nature* 227:1313–1318.

Bochkarev A., and Bochkareva, E. 2004. From RPA to BRCA2: lessons from single stranded DNA binding by the OB-fold. *Curr Opin Struct Biol* 14:36–42.

Haring, S. J., Mason, A. C., Binz, S. K., and Wold, M. S. 2008. Cellular functions of human RPA1. Multiple roles of domains in replication, repair, and checkpoints. *J Biol Chem* 283:19095–19111.

Iftode, C., Daniely, Y, and Borowiec, J. A. 1999. Replication protein A (RPA): the eukaryotic SSB. *Crit Rev Biochem Molec Biol* 34:141–180.

Lohman, T. M., and Ferrari, M. E. 1994. *Escherichia coli* single-stranded DNA binding protein: multiple DNA-binding modes and cooperativities. *Ann Rev Biochem* 63:527–570.

Pestryakov, P. E., and Lavrik, O. I. 2008. Mechanisms of single-stranded DNA-binding protein functioning in cellular DNA metabolism. *Biochemistry* (Moscow) 73:1388–1404.

Raghunathan, S., Kozlov, A., Lohman, T., and Waksman, G. 2000. Structure of the DNA binding domain of *E. coli* SSB bound to ssDNA. *Nat Struct Biol* 7:648–652.

Shamoo, Y. 2002. Single-stranded DNA-binding proteins. In: *Encyclopedia of Life Sciences*. pp. 1–7. London: Nature.

Wold, M. S. 2001. Eukaryotic replication protein A. In: *Encyclopedia of Life Sciences*. pp. 1–7. London: Nature.

Topoisomers and Topoisomerases

Bates, A. D. 2001. Topoisomerases. In: *Encyclopedia of Life Sciences*. pp. 1–9. London: Nature.

Bates, A. D., and Maxwell, A. 1993. DNA Topology. Washington, DC: IRL Press.

Bauer, W. R., Crick, F. H. C., and White, J. H. 1980. Supercoiled DNA. *Sci Am* 243:100–113.

Bowater, R. P. 2002. Supercoiled DNA structure. In: *Encyclopedia of Life Sciences*. pp. 1–9. London: Nature.

Changela, A., Perry, K., Taneja, B., and Mondragón, A. 2002. DNA manipulators: caught in the act. *Curr Opin Struct Biol* 13:15–22.

Corbett, K. D., and Berger, J. M. 2004. Structure, molecular mechanisms, and evolutionary relationships in DNA topoisomerases. *Ann Rev Biophys Biomol Struct* 33:95–118.

Hardy, C. D., Crisona, N. J., Stone, M. D., and Cozzarelli, N. R. 2004. Disentangling DNA during replication: a tale of two strands. *Phil Trans Royal Soc London B* 359:39–47.

Larsen, A. K., Escargueil, A. E., and Skaldanowski, A. 2003. Catalytic topoisomerase II inhibitors in cancer therapy. *Pharmacol Ther* 99:167–181.

Lebowitz, J. 1990. Through the looking glass: the discovery of supercoiled DNA. *Trends Biochem Sci* 15:202–207.

Lindsley, J. E. 2005. DNA topology: supercoiling and linking. In: *Encyclopedia of Life Sciences*. pp. 1–7. Hoboken, NJ: John Wiley and Sons.

Mirkin, S. M. 2002. DNA topology: fundamentals. In: *Encyclopedia of Life Sciences*. pp. 1–11. London: Nature.

Nadal, M. 2007. Reverse gyrase: an insight into the role of DNA topoisomerases. *Biochimie* 89:447–455.

Nöllmann, M., Crisona, N. J., and Arimondo, P. B. 2007. Thirty years of *Escherichia coli* DNA gyrase: from in vivo function to single-molecule mechanism. *Biochimie* 89:490–499.

Schoeffler, A.J., and Berger, J. M. 2008. DNA topoisomerases: harnessing and constraining energy to govern chromosome topology. *Quart Rev Biophys* 41:41–101.

Travers, A. A., and Thompson, J. M. T. 2004. An introduction to the mechanics of DNA. *Phil Trans R Soc Lond A* 362:1265–1279.

Wang, J. C. 1996. DNA topoisomerases. *Ann Rev Biochem* 65:635–692.

Wang, J. C. 2002. Cellular roles of DNA topoisomerases: a molecular perspective. *Nat Rev Mol Cell Biol* 3:430–440.

Wang, J. C. 2009. A journey in the world of DNA rings and beyond. *Ann Rev Biochem* 78:31–54.

Non-B DNA Conformations

Armitage, B. A. 2007. The rule of four. *Nature Chem Biol* 3:203–204.

Arnott, S. 2006. Historical article: DNA polymorphism and the early history of the double helix. *Trends Biochem Sci* 31:349–354.

Bacolla, A., and Wells, R. D. 2004. Non-B DNA conformations, genomic rearrangements, and human disease. *J Biol Chem* 279: 47411–47414.

Bacolla, A., and Wells, R. D. 2009. Non-B DNA conformations as determinants of mutagenesis and human disease. *Mol Carcinog* 48:273–285.

Bates, P., Megny, J-L., and Yang, D. 2007. Quartets in G-major. *EMBO Rep* 11:1003–1010.

Baumann, P. 2005. Taking control of G quadruplexes. *Nat Struct Mol Biol* 12: 832–833.

Burge, S., Parkinson, G. N., Hazel, P., Todd, A. K., and Neidle, S. 2006. Quadruplex DNA: sequence, topology and structure. *Nucleic Acids Res* 34:5402–5415.

Dickerson, R. E. 1992. DNA structures from A to Z. *Methods Enzymol* 211:67–111.

Dickerson, R. E., and Ng, H. L. 2001. DNA structure from A to B. *Proc Natl Acad Sci USA* 98:6986–6988.

Gagna, C. E., Kuo, H., and Lambert, W. C. 1999. Terminal differentiation and left-handed Z-DNA: a review. *Cell Biol Int* 23:1–5.

Huppert, J. L. 2008. Hunting G-quadruplexes. *Biochimie* 90:1140–1148.

Li, H., Xiao, J. Li, J., Lu, L., Feng, S., and Dröge, P. 2009. Human genomic Z-DNA segments probed by the Zα domain of ADAR1 *Nucleic Acids Res* 37:2737–2746.

Herbert, A., and Rich, A. 1999. Left-handed Z-DNA: structure and function. *Genetica* 106:37–47.

Maizels, N. 2006. Dynamic roles for G4 DNA in the biology of eukaryotic cells. *Nat Struct Mol Biol* 13:1055–1059.

Mirkin, S. M. 2008. Discovery of alternative DNA structures: a heroic decade (1979-1989). *Frontiers Biosci* 13:1064–1071.

Lane, A. N., Chaires, J. B., Gray, R. D., Trent, J. O. 2008. Stability and kinetics of G-quadruplex structures. *Nucleic Acids Res* 36:5482–515.

Potaman, V. N., and Sinden, R. R. 2005. DNA: alternative conformations and biology. In: *DNA Conformation and Transcription*, T. Ohyama, ed., pp. 1–16. New York: Springer-Verlag.

Rich, A. 2003. The helix: a tale of two puckers. *Nat Struct Biol* 10:247–249.

Rich, A., and Zhang, S. 2003. Timeline: Z-DNA: the long road to biological function. *Nat Rev Genet* 4:566–572.

Rogers, F. A., Lloyd, J. A., and Glazer, P. M. 2005. Triplex-forming oligonucleotides as potential tools for modulation of gene expression. *Curr Med Chem Anti-Cancer Agents* 5:319–326.

Ussery, D. W. 2002. DNA Structure: A-, B- and Z-DNA helix families. In: *Encyclopedia of Life Sciences*. pp. 1–6. London: Nature.

Vasquez, K. M., and Glazer, P. M. 2002. Triplex-forming oligonucleotides: principles and applications. *Quart Rev Biophys* 35:89–107.

Wang, G., and Vasquez, K. M. 2006. Non-B DNA structure-induced genetic instability. *Mutat Res* 598:103–119.

Wang, G., and Vasquez, K. M. 2007. Z-DNA, an active element in the genome. *Frontiers Biosci* 12:4424–4438.

Wells, R. D. 2007. Non-B DNA conformations, mutagenesis and disease. *Trends Biochem Sci* 32:271–278.

Wells, R. D., Dere, R., Hebert, M. L., Napierala, M., and Son, L. S. 2005 Advances in mechanisms of genetic instability related to hereditary neurological diseases *Nucleic Acids Res* 33:3785–3798.

RNA Structure

Batey, R. T., Rambo, R. P., and Doudna, J. A. 1999. Tertiary motifs in RNA structure and folding. *Angew Chem Int Ed* 38:2326–2343.

Bevilacqua, P. C., and Blose, J. M. 2008. Structures, kinetics, thermodynamics, and biological functions of RNA hairpins. *Ann Rev Biophys* 59:79–103.

Brierley, I., Gilbert, R. J. C., and Pennell, S. 2008. RNA pseudoknots and the regulation of protein synthesis. *Biochem Society Trans* 36:684–689.

Carter, R. J., and Holbrook, S. R. 2002. RNA structure: Roles of Me^{2+}. In: *Encyclopedia of Life Sciences*. pp. 1–7. Hoboken, NJ: John Wiley and Sons.

Cech, T. R. 2002. Ribozymes, the first 20 years. *Biochem Soc Trans* 30:1162–1166.

Cheong, C., and Cheong, H.-K. 2001. RNA structure: Tetraloops. In: *Encyclopedia of Life Sciences*. pp. 1–7. Hoboken, NJ: John Wiley and Sons.

Cruz, J. A., and Westhof, E. 2009. The dynamic landscapes of RNA architecture. *Cell* 136:604–609.

Doudna, J. A. and Cech, T. R. 2002. The chemical repertoire of natural ribozymes. *Nature* 418:222-228.

Draper, D. E. 2004. A guide to ions and RNA structure. *RNA* 10:335–343.

Draper, D. E., Grilley, D., and Soto, A. M. 2005. Ions and RNA folding. *Ann Rev Biophys Biomol Struct* 34:221–243.

Gultyaev, A. P., Pleij, C. W. A., and Westhof, E. 2005. RNA structure: pseudoknots. In: *Encyclopedia of Life Sciences*. pp. 1–7. Hoboken, NJ: John Wiley and Sons.

Hendrix, D. K., Brenner, S. E., and Holbrook, S. R. 2005. RNA structural motifs: building blocks of a modular biomolecule. *Quart Rev Biophys* 38:221–243.

Hermann, T., and Patel, D. J. 2000. RNA bulges as architectural and recognition motifs. *Structure* 8:R47–R54.

Holbrook, A. R. 2008. Structural principles from large RNAs. *Curr Opin Struct Biol* 15:302–308.

Hou, Y.-M. 2002. Base pairing in RNA: unusual patterns. In: *Encyclopedia of Life Sciences*. pp. 1–7. Hoboken, NJ: John Wiley and Sons.

Leontis, N. B., and Westhof, E. 2001. Geometric nomenclature and classification of RNA base pairs. *RNA* 7:499–512.

Leontis, N. B., and Westhof, E. 2003. Analysis of RNA motifs. *Curr Opin Struct Biol* 13:300–308.

Leontis, N. B., Lescoute, A., and Westhof, E. 2006. The building blocks and motifs of RNA architecture. *Curr Opin Struct Biol* 16:279–287.

Nowakowski, J., and Tinoco , I. Jr. 1997. RNA structure and stability. *Semin Virol* 8:153–165.

Schroeder, E., Barta, A., and Semrad, K. 2004. Strategies for RNA folding and assembly. *Nat Rev Mol Cell Biol* 5:908–919.

Staple, D. W., and. Butcher, S. E. 2005. Pseudoknots: RNA structures with diverse functions. *PLoS Biol* 3:956–959.

Tung. C.-S. 2002. RNA structural motifs. In: *Encyclopedia of Life Sciences*. pp. 1–4. Hoboken, NJ: John Wiley and Sons.

Varani, G. 2001. RNA structure. In: *Encyclopedia of Life Sciences*. pp. 1–8. Hoboken, NJ: John Wiley and Sons.

Walter, F., and Westhof, E. 2002. Catalytic RNA. In: *Encyclopedia of Life Sciences*. pp. 1–12. Hoboken, NJ: John Wiley and Sons.

The RNA World Hypothesis

Cech, T. R. 2009. Crawling out of the RNA world. *Cell* 136:599–602.

Gesteland, R. F. (ed). (2005) *The RNA World*, 3rd ed. New York: Cold Spring Harbor Press.

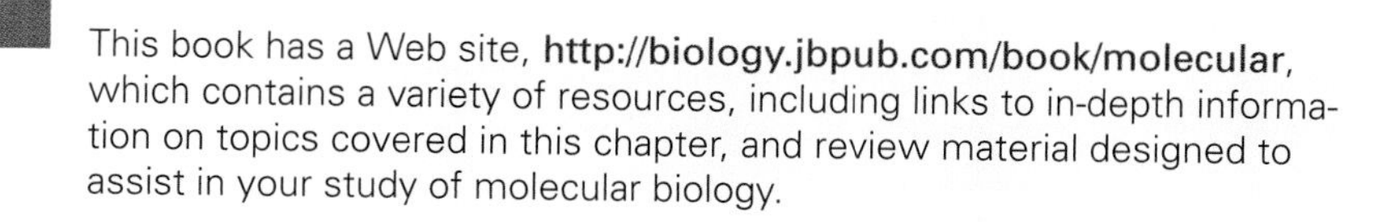
This book has a Web site, **http://biology.jbpub.com/book/molecular**, which contains a variety of resources, including links to in-depth information on topics covered in this chapter, and review material designed to assist in your study of molecular biology.

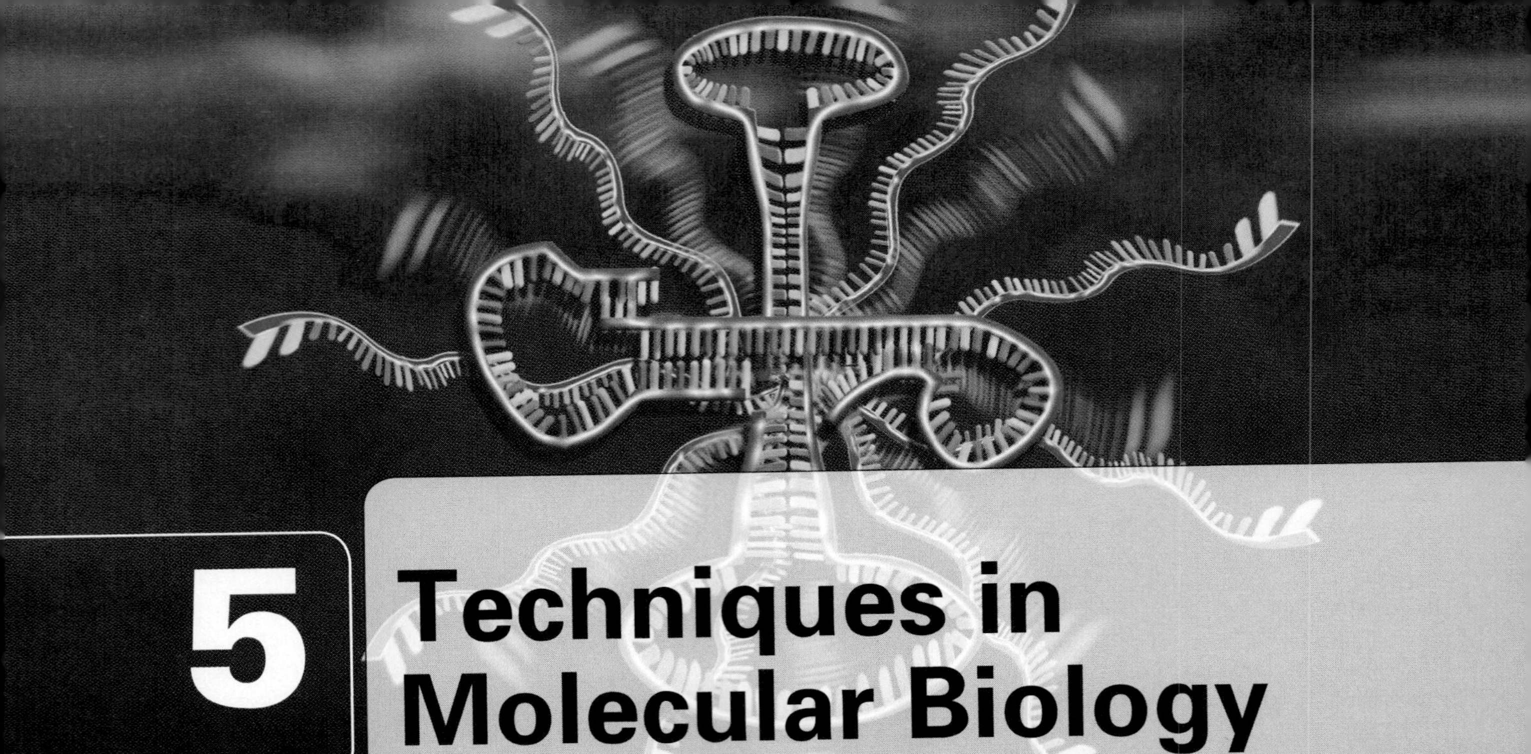

5 Techniques in Molecular Biology

OUTLINE OF TOPICS

5.1 Nucleic Acid Isolation

The method used for DNA isolation must be tailored to the organism from which the DNA is to be isolated.

Great care must be taken to protect RNA from degradation during its isolation.

Different physical techniques are used to study macromolecules.

5.2 Electron Microscopy

Electron microscopy allows us to see macromolecules.

5.3 Centrifugal Techniques

Velocity sedimentation can separate macromolecules and provide information about their size and shape.

Equilibrium density gradient centrifugation separates particles according to their density.

5.4 Gel Electrophoresis

Gel electrophoresis separates charged macromolecules by their rate of migration in an electric field.

Gel electrophoresis can be used to separate proteins and determine a polypetide's molecular mass.

Capillary gel electrophoresis is a rapid and automated process that provides quantitative data.

Pulsed-field gel electrophoresis (PFGE) can separate very large DNA molecules.

5.5 Nucleases and Restriction Maps

Nucleases are useful tools in DNA investigations.

Restriction endonucleases that cleave within specific nucleotide sequences are very useful tools for characterizing DNA.

Restriction endonucleases can be used to construct a restriction map of a DNA molecule.

5.6 Recombinant DNA Technology

DNA fragments can be inserted into plasmid DNA vectors.

Southern blotting is used to detect specific DNA fragments.

Northern and Western blotting are used to detect specific RNA and polypeptide molecules, respectively.

DNA polymerase I, a multifunctional enzyme with polymerase, 3′→5′ exonuclease, and 5′→3′ exonuclease activities, can be used to synthesize labeled DNA.

DNA polymerase I can synthesize DNA at a nick.

The polymerase chain reaction is used to amplify DNA.

Site-directed mutagenesis can be used to introduce a specific base change within a gene.

5.7 DNA Sequence Determination

The Maxam-Gilbert method uses controlled chemical degradation to sequence DNA

The chain termination method for sequencing DNA uses dideoxynucleotides to interrupt DNA synthesis.

DNA molecules that are 1 to 8 kbp long can be sequenced by primer walking.

DNA sequences can be stitched together by using information obtained from a restriction map.

Shotgun sequencing is used to sequence long DNA molecules.

The Human Genome Project used hierarchical shotgun assembly to sequence the human genome.

Whole genome shotgun sequencing has also been used to sequence the human genome.

The human genome sequence provides considerable new information.

A new generation of DNA sequencers provide rapid and accurate information without the need for electrophoresis or *in vivo* cloning.

Reverse transcriptase can use an RNA molecule as a template to synthesize DNA.

DNA chips are used to follow mRNA synthesis, search for a specific DNA sequence, or to find a single nucleotide change in a DNA sequence.

Suggested Reading

Molecular biologists use physical techniques and enzyme-catalyzed reactions to characterize and modify nucleic acids. These techniques have become so widely used that you have probably encountered many of them in other life science courses. The first part of this chapter examines physical techniques used to isolate nucleic acids and characterize nucleic acids, proteins, and nucleoprotein complexes. The second part reviews techniques that use enzymes to manipulate and sequence DNA in the laboratory. Of course, these enzymes have important biological functions in the cell. But for now, our attention is directed toward using these enzymes as catalytic agents to replicate, sequence, cut, or otherwise modify DNA in the laboratory. The biological functions of the enzymes are considered in later chapters.

5.1 Nucleic Acid Isolation

The method used for DNA isolation must be tailored to the organism from which the DNA is to be isolated.

DNA isolation is an essential step in many experiments. The common feature in all procedures is that a cell or virus is first broken, and then the released DNA is separated from other components such as protein, RNA, lipid, and carbohydrate molecules. Once DNA has been released from the cell or virus, care must be taken to avoid vigorous stirring or other procedures that would produce hydrodynamic shear forces and break the DNA.

Because the structure and composition of organisms vary, the particular procedure used to isolate DNA must be tailored to the organism from which the DNA is to be obtained. The technique of choice depends on how the DNA is enclosed and the percent of total dry weight that is DNA (which varies from about 1% in complex mammalian cells to about 50% in bacterial viruses). The five basic procedures are the following:

1. *Bacterial viruses.* The simplest procedure is employed with bacterial viruses. An aqueous suspension of virus particles is gently mixed with phenol, a reagent that is slightly miscible with water. A small amount of phenol enters the aqueous layer, breaking open the protein coat and denaturing individual protein molecules. (Sometimes chloroform and isoamyl alcohol are added to assist the phenol.) Most of the denatured protein either enters the phenol layer or precipitates at the phenol-water interface. The aqueous layer, containing the DNA, is carefully separated from the phenol layer and ethanol is added to the aqueous fraction to precipitate DNA. After collection by centrifugation, DNA is dissolved in an aqueous solution having the desired composition.
2. *Bacteria.* The contents of bacterial cells are enclosed in a multilayered cell envelope consisting of a cell membrane and cell

wall in gram-positive bacteria such as *Bacillus subtilis* and also an outer membrane in gram negative bacteria such as *Escherichia coli* (see Chapter 7). The cell envelope cannot simply be removed by exposure to phenol but can be degraded by successive treatment of a cell suspension with lysozyme to digest the cell wall, and one of several different detergents, the most common being sodium dodecyl sulfate [SDS; $CH_3(CH_2)_{11}OSO_3^-$ Na^+], to disrupt the cell membrane. Then phenol is added with gentle mixing and the phases allowed to separate. The aqueous fraction is collected and treated with ethanol to precipitate DNA and RNA. The precipitate containing DNA and some contaminating RNA is spooled out with a glass rod and excess alcohol is removed. Then the spooled DNA is dissolved in a buffered solution containing RNase to digest contaminating RNA. Chloroform is added to the mixture with gentle mixing and the phases allowed to separate. The aqueous fraction is collected and treated with ethanol to precipitate DNA, which is spooled out with a glass rod and dissolved in an aqueous solution with the desired composition.

3. *Plasmids.* Plasmids, which are small autonomously replicating DNA molecules, are released from bacteria by first incubating the bacteria with lysozyme and then suspending them in an SDS-sodium hydroxide solution. Lysozyme digests the cell wall and the detergent disrupts the cell membrane. Sodium hydroxide hydrolyzes RNA to form a mixture of 2′- and 3′-nucleoside monophosphates. DNA, which lacks 2′-hydroxyl groups, is denatured but not hydrolyzed. Bacterial DNA fragments separate into two strands, but the covalently closed circular plasmid DNA molecules remain intertwined. When acid is added to neutralize the extract, the intertwined plasmid DNA rewinds while the bacterial DNA strands form an insoluble aggregate that is removed, along with denatured protein, by centrifugation. The soluble plasmid DNA is precipitated by adding ethanol, collected by centrifugation, and redissolved in an aqueous buffer solution.
4. *Yeast and fungi.* The polysaccharides that make up yeast and fungi cell walls are resistant to lysozyme. However, other enzymes (such as cellulase isolated from snails) will break down these cell walls. Once cell walls have been disrupted, the DNA isolation procedure is similar to that used for bacteria.
5. *Higher eukaryotes.* Animal and plant cells have a low ratio of DNA to protein, and for technical reasons substantial loss of DNA occurs if the DNA is directly purified. To minimize this problem, nuclei usually are isolated first. This step increases the DNA to protein ratio and also avoids contamination of nuclear DNA by mitochondrial or chloroplast DNA. Nuclei are broken open, enzymes are added to digest RNA and protein molecules, and then ethanol is added to precipitate the DNA.

Great care must be taken to protect RNA from degradation during its isolation.

One of the greatest concerns when isolating RNA is that the RNA will be degraded by RNases released during the isolation procedure. One solution to this problem is to freeze the cells in liquid nitrogen, transfer the frozen cells to a mortar containing liquid nitrogen, and grind the cells with a pestle. Then ground cells are suspended in a lysis reagent that contains guanidinium thiocyanate and phenol, which denatures proteins including the RNases, dissociates nucleoprotein complexes, and causes any cells that are still intact to lyse. After a few minutes of incubation at room temperature, chloroform is added, and the phases are separated by centrifugation. RNA stays in the aqueous phase, while DNA and proteins move to the interphase and the organic phase. The upper aqueous phase containing the RNA is transferred to a clean tube and ethanol or isopropanol is added. The RNA, which precipitates out of solution, is collected by centrifugation, washed with alcohol, and dissolved in an aqueous buffer.

Different physical techniques are used to study macromolecules.

Several physical techniques for studying macromolecules have broad applications. Four of these techniques—electron microscopy, velocity sedimentation, equilibrium density gradient centrifugation, and electrophoresis—will be described briefly, so that the experiments presented in this book can be understood. The reader should consult the references at the end of this chapter for additional information.

We often encounter situations in which more than one approach can be used to solve a problem. For example, each of the methods that are described below can be used to characterize DNA molecules of different sizes and shapes. Some of these methods allow us to distinguish one topoisomer from another and therefore can be used to monitor topoisomerase activity. As you read the descriptions of these methods, decide on one that you would select to assay topoisomerase activity. In arriving at your decision, consider both the convenience of the method and the time required to obtain a result.

5.2 Electron Microscopy

Electron microscopy allows us to see macromolecules.

Macromolecules can be viewed directly by electron microscopy. Three of the many techniques of sample preparation will be described here.

The first technique, **metal shadowing**, provides a three-dimensional image of the surface of the biological sample (FIGURE 5.1). The biological sample is dried on a translucent film and then placed in a vacuum chamber, where a filament of a heavy metal such as tungsten is heated. The evaporated metal forms a thin metal coat on the surface of the

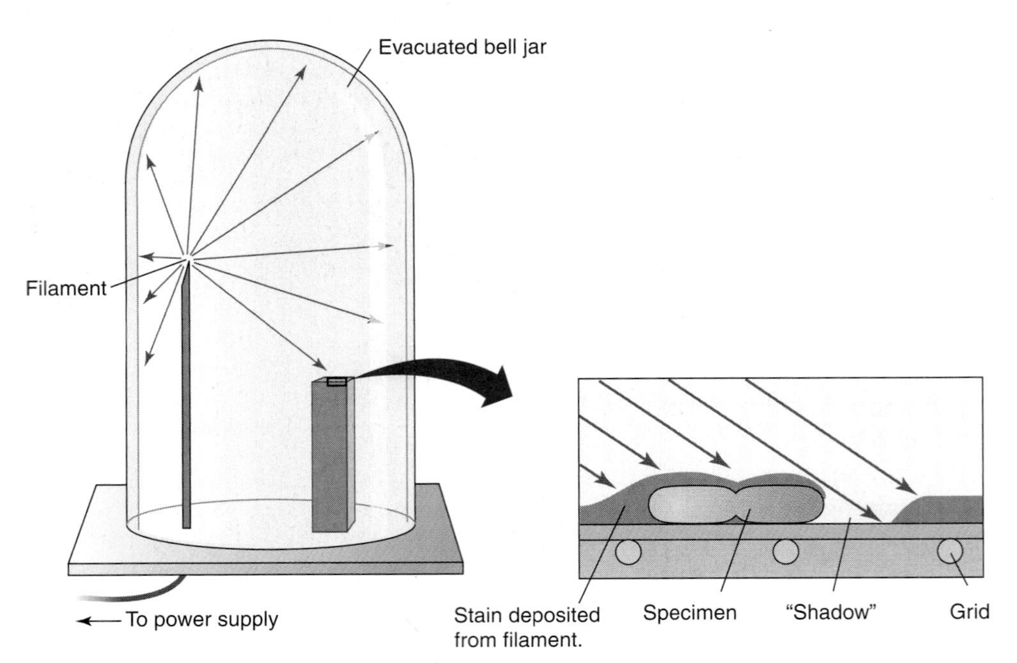

FIGURE 5.1 **Metal shadowing.** This technique allows the observer to view details on the surface of small particles. The sample is dried on a thin film, which is then placed in a vacuum chamber. A filament of a heavy metal such as tungsten is heated and the metal evaporates, forming a thin metal coat on the sample. The observer uses an electron microscope to view the metal coating, which is a replica of the biological sample (Adapted from J. L. Ingraham and C. A. Ingraham. *Introduction to Microbiology, Third edition.* Brooks/Cole Publishing Company, 2003.)

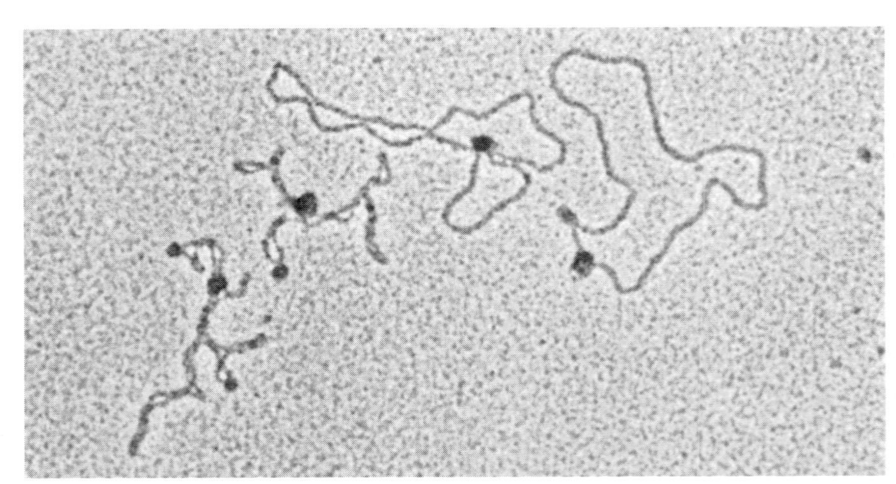

FIGURE 5.2 **Plasmid pBR322 DNA supercoils.** Relaxed covalently closed plasmid pBR322 DNA was incubated with DNA gyrase in the presence of ATP at 30°C. The reaction was stopped by adding formaldehyde and glutaraldehyde and the sample shadowed with tungsten. By chance, the plasmids viewed in this electron micrograph show increasing supercoiling from right to left. The globular structures associated with the DNA are probably DNA gyrase molecules. (Reproduced from C. L. Moore, et al., *J. Biol. Chem.* 258 [1983]: 4612–4617. © 1983, The American Society for Biochemistry and Molecular Biology. Photo courtesy of Jack D. Griffith, University of North Carolina at Chapel Hill.)

biological sample. The observer uses an electron microscope to view the metal coating, which is a replica of the biological sample. Metal shadowing is a very useful technique for observing very long molecules when one is primarily interested in their length and linear topology.

An example of an electron micrograph of DNA topoisomers is shown in FIGURE 5.2.

The second technique, **negative staining**, provides an alternative way to visualize molecules and more complicated structures. With this technique, the sample is mixed with a solution of phosphotungstic acid or uranium salts (usually uranyl acetate or uranyl formate), both of which contain atoms that absorb electrons strongly. Then the solution is deposited as microdroplets on a supporting film and allowed to dry. The thickness of metal ions in the droplet is less where the macromolecule is present than elsewhere in the droplet. As shown in FIGURE 5.3, when the sample is placed in a beam of electrons, more electrons pass through the droplet in the region of a macromolecule than elsewhere. Thus, a negative image of a protein or a small virus particle can be obtained. Electron micrographs of a protein molecule and a virus particle viewed in this way are shown in FIGURE 5.4.

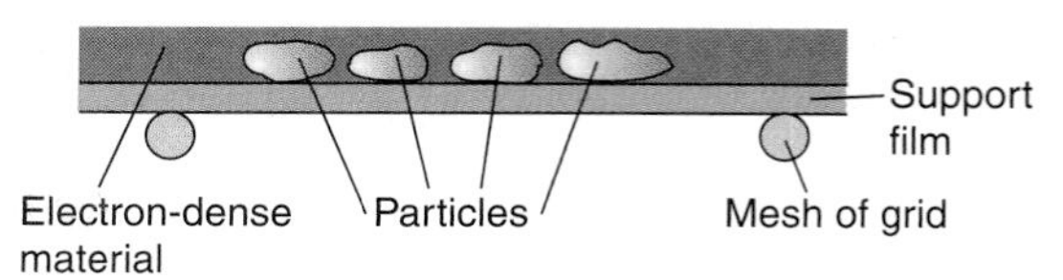

FIGURE 5.3 **Negative contrast method of visualizing particles by electron microscopy.** Four particles are embedded in a substance that absorbs electrons strongly. As the beam passes through the sample, the fraction of the electrons in the beam that is absorbed depends on the total thickness of the substance; therefore, more electrons will pass through the regions containing each particle and the particle will appear bright against a dark background.

Electron microscopy provides a method for determining the molecular mass of a DNA molecule that is between 5×10^6 Daltons (Da) and 10×10^7 Da. A measurement of the length of the molecule yields the value of molecular mass, because the mass per unit length of double-stranded DNA is 2×10^6 Da per micrometer (μm). Often errors arise in determining the value of the magnification of the molecules. A common technique that avoids the necessity of knowing this value is to add a molecule of precisely known molecular mass to the sample being studied, and to determine the ratio of the lengths of the standard molecule and the molecule of interest.

The third technique, **cryo-electron microscopy**, is very effective for obtaining high resolution images of a large protein complex such as RNA polymerase or a large nucleoprotein complex such as a ribosome. This technique involves rapidly freezing a droplet of a buffer solution containing the specific protein or nucleoprotein complexes so that a

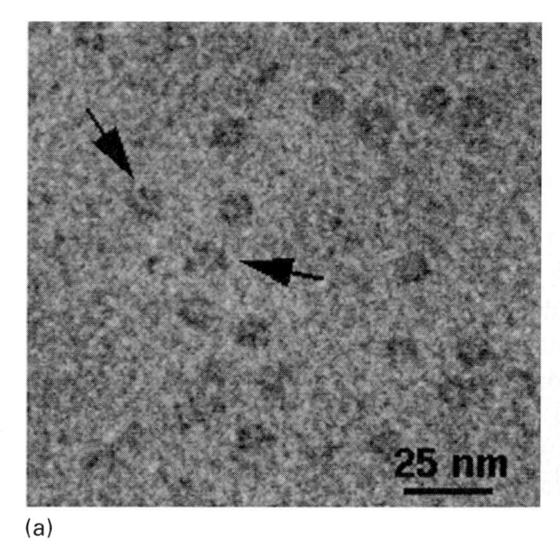

(a)

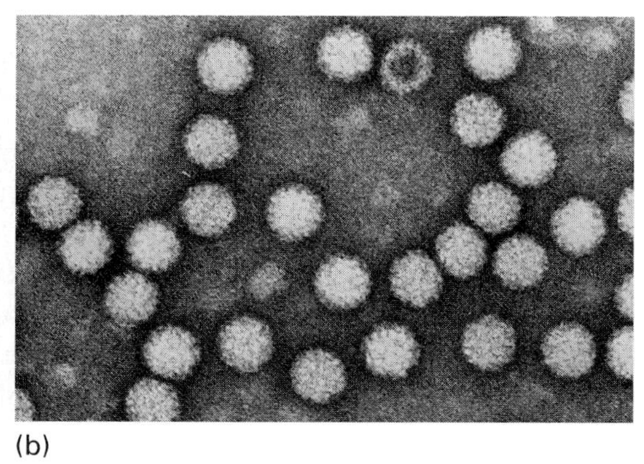

(b)

FIGURE 5.4 **Electron micrographs obtained by the negative contrast procedure.** (a) Image of negatively stained bacteriophage T7 helicase/primase (arrows) in the presence of dTDP. (b) Tomato bushy stunt virus particles—note the surface details, which shows the individual protein molecules that form the protein coat of which each particle is composed. (Part a reproduced from M. T. Norcum, et al., *Proc. Natl. Acad. Sci. USA* 102 [2005]: 3623–3626. Photo courtesy of J. Anthony Warrington, Florida State University. Part b courtesy of Robert G. Milne, Plant Virus Institute, National Research Council, Turin, Italy.)

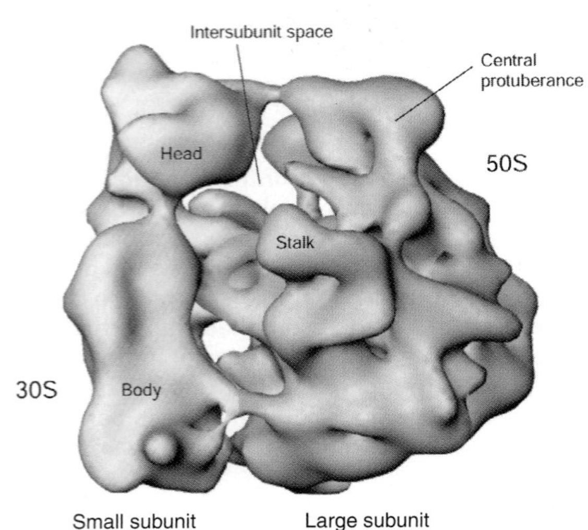

FIGURE 5.5 Cryo-electron microscopy reconstruction of the *Escherichia coli* ribosome. (Reproduced from *Encyclopedia of Life Sciences*, J. Frank and R. K. Agrawal, Copyright © 2001 and reproduced with permission of John Wiley & Sons, Inc. Photo courtesy of Rajendra K. Agrawal, Wadsworth Center.)

thin amorphous layer of ice forms that holds the complexes in random orientations. A picture is formed in the electron microscope by using a radiation dose that is so low that the complexes are not damaged. The resulting micrograph shows hundreds of the protein or nucleoprotein complexes lying in different orientations. Computer-assisted image processing converts the projections into three-dimensional constructs such as those shown for the bacterial ribosome in FIGURE 5.5.

5.3 Centrifugal Techniques

Velocity sedimentation can separate macromolecules and provide information about their size and shape.

Several important properties of macromolecules can be determined from the rates at which they sediment in a centrifugal field. Sedimentation studies cannot be done with an ordinary laboratory centrifuge, because this instrument cannot produce a great enough centrifugal force to sediment molecules sufficiently rapidly that their positions are not randomized by diffusion. Modern ultracentrifuges, however, can generate forces as great as 700,000× gravity, which is more than adequate to cause macromolecules to sediment through a solution.

The velocity with which a molecule moves is mainly a function of two properties of the molecule:

1. *Its molecular mass.* As molecular mass increases, the sedimentation velocity increases.
2. *Its shape.* The motion of any particle through a fluid is impeded by friction. If a ball and a stick with the same mass were moving through a liquid, the more compact ball would encounter less frictional resistance and, hence, move faster. This is true of macromolecules, too. For a given value of molecular mass, the less extended the shape of a polymer chain, the more rapidly it will move.

The ratio of sedimentation velocity to centrifugal force is called the sedimentation coefficient (s). That is,

$$s = \text{velocity/centrifugal force}$$

The value of s for a particular molecule is often the same in many different solutions, so an s value is frequently considered to be a constant that characterizes a molecule. Furthermore, because the value of s depends on molecular mass and shape, changes in the s value, as experimental conditions are varied, can be used to monitor changes in molecular aggregation or conformation.

For most macromolecules the value of s is between 1×10^{-13} and 100×10^{-13} sec. In honor of Theodor Svedberg, the ultracentrifuge's inventor, 10^{-13} seconds is called one **svedberg** or one S. The s value of a molecule formed by the association of two smaller molecules cannot be determined by simply adding the s values of the two smaller particles. For example, a 30S bacterial ribosomal subunit combines with

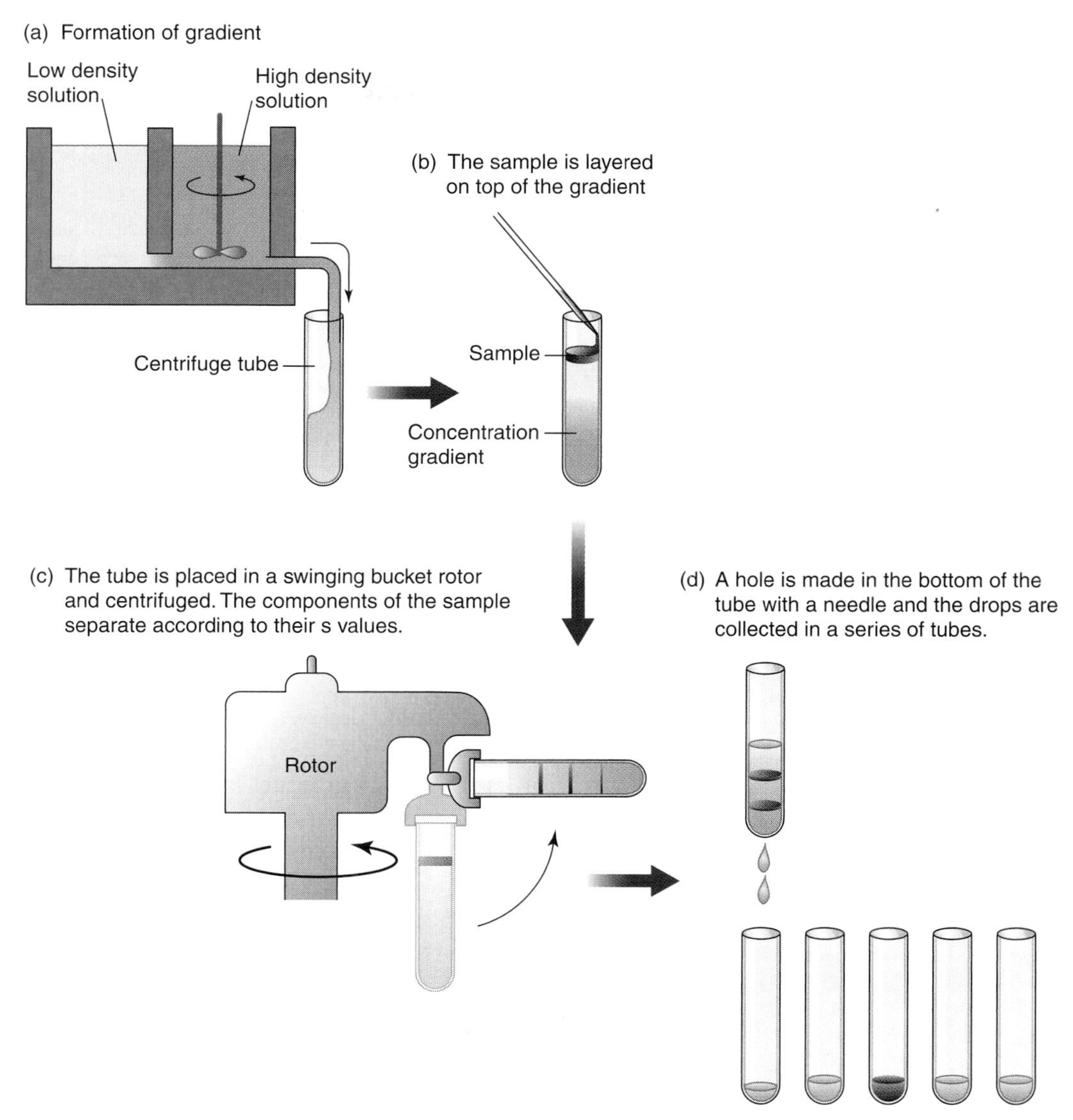

FIGURE 5.6 **Zonal centrifugation.**

a 50S bacterial ribosomal subunit to form a 70S bacterial ribosome (and not an 80S bacterial ribosome).

The most common type of sedimentation in use today is **zonal centrifugation**. In this procedure (FIGURE 5.6), a centrifuge tube is filled with a sucrose solution (occasionally, glycerol or other solutes are used) so that the concentration increases continuously from the top to the bottom of the centrifuge tube. Because solution density increases with solute concentration, it is greatest at the bottom of the tube. The density of the solution of molecules to be sedimented is adjusted to be lower than the density of the sucrose solution at the top of the tube, so that the sample can be layered on the surface of the sucrose solution to form a band (or zone). Because of the sucrose concentration gradient, this procedure is often called **sucrose gradient centrifugation**. After layering the sample, the tube is centrifuged in a swinging bucket rotor for a particular time. After centrifugation is completed, a tiny

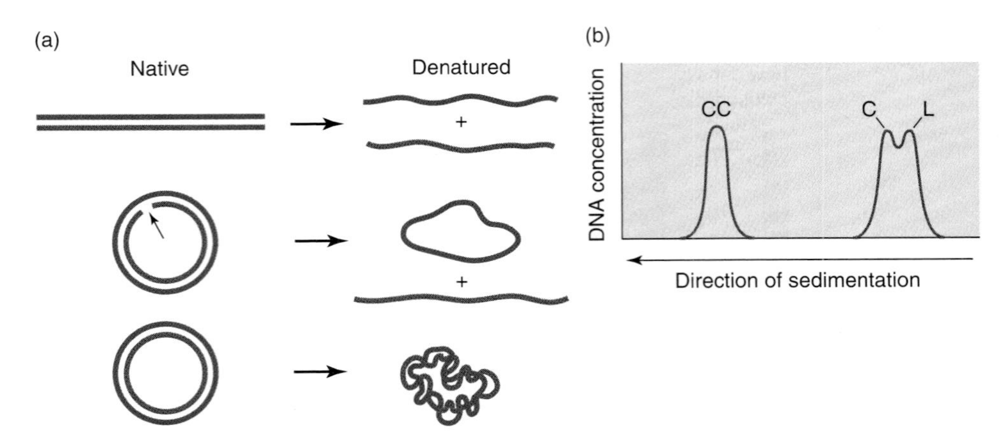

FIGURE 5.7 **Sedimentation at alkaline pH.** (a) Products of the denaturation of different forms of DNA. (b) Separation of covalently closed double stranded circles (CC), single stranded circles (C), and linear single strands (L) by sedimentation in alkali. The horizontal axis represents the length of a centrifuge tube. Sedimentation is from right to left.

hole is punched in the bottom of the tube and drops of the solution are collected. These drops, representing successive layers of solution in the tube, are analyzed to determine the macromolecule's concentration along the tube.

Zonal centrifugation in an alkaline sucrose solution separates covalently closed circles from linear duplexes or nicked circles. Above pH 11.3, all hydrogen bonds are broken and DNA molecules unwind. For a linear DNA molecule, two single strands result with s values that are about 30% greater than that of the native DNA if the salt concentration is 0.3 M or greater. In contrast, the two strands of a covalently closed circle cannot separate, so the molecules collapse in a tight tangle with an s value that is about three times as great as that of native DNA. If the circle has a single nick, one linear molecule and one single-stranded circle result (FIGURE 5.7a), the s value of the latter is 14% greater than that of the former. FIGURE 5.7b shows a sedimentation pattern for a mixture composed of linear molecules, single circles, and covalently closed circles in an alkaline sucrose gradient.

Equilibrium density gradient centrifugation separates particles according to their density.

Another centrifugation technique, **equilibrium density gradient centrifugation,** is also widely used. In this procedure, the macromolecules (usually nucleic acids or virus particles) are suspended in a CsCl (or Cs_2SO_4) solution at a concentration that is chosen so that the solution density is approximately equal to that of the macromolecules.

Under the influence of a powerful centrifugal force, Cs^+ and Cl^- (or SO_4^{2-}) ions move toward the bottom of the centrifuge to some extent. They do not accumulate on the bottom of the centrifuge tube, however, because the centrifugal force is not great enough to counteract the tendency for diffusion to maintain a uniform distribution of the ions. The result is that, after a period of several hours, the ions achieve an equilibrium concentration distribution in which there is a nearly linear concentration gradient and, hence, a nearly linear density gradient in the centrifuge tube.

The density is maximal at the bottom of the tube. As the density gradient forms, macromolecules begin to migrate. Those in the upper

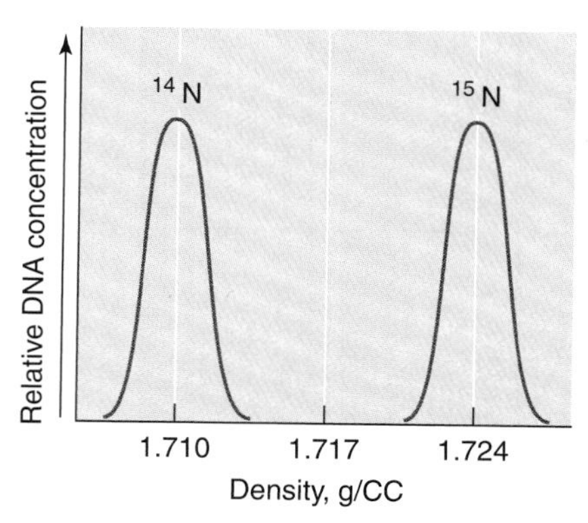

FIGURE 5.8 Demonstration of DNA strand separation by equilibrium density centrifugation in CsCl. ^{14}N indicates DNA isolated from bacteria cultured in a medium that contained [^{14}N]NH_4Cl. ^{15}N indicates DNA isolated from bacteria cultured in a medium that contained [^{15}N]NH_4Cl.

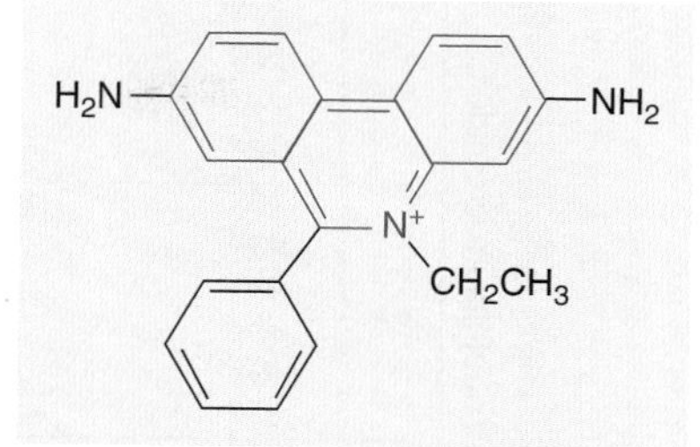

FIGURE 5.9 Ethidium bromide. Ethidium bromide contains a planar phenanthridium ring (shown in red) that inserts between the stacked bases in DNA. DNA absorbs ultraviolet light radiation at 260 nm and transmits it to the bound dye, which re-emits the light at 590 nm in the red-orange region of the spectrum.

reaches of the tube move toward the bottom, stopping at the position at which their density equals the solution density. Similarly, macromolecules in the lower part of the tube move upward, stopping at the same position. In this way, the macromolecules form a narrow band in the tube. If the solution contains macromolecules having different densities, each macromolecule forms a band at the position in the gradient that matches its own density, and thus the macromolecules can be separated.

The technique's resolution is extraordinary. As shown in FIGURE 5.8, DNA molecules with a density of 1.710 $g \cdot cm^{-3}$ can be separated from other DNA molecules in which the naturally occurring ^{14}N atoms have been replaced by ^{15}N atoms and which therefore have a density of 1.724 $g \cdot cm^{-3}$.

Equilibrium density gradient centrifugation can also be used to separate topoisomers by taking advantage of the fact that **ethidium bromide** (FIGURE 5.9) binds very tightly to DNA (in both dilute and concentrated salt solutions) and, in so doing, decreases the density of the DNA. Ethidium bromide must be handled with great care because it is a carcinogen (cancer causing agent).

Binding occurs with little or no sequence specificity as a result of the insertion of the planar phenanthridinium ring (Figure 5.9) between adjacent base pairs. This type of insertion, known as **intercalation**, causes the DNA molecule to unwind. No covalent bonds are altered and hydrogen bonds between base pairs remain intact. Because a covalently closed, circular DNA molecule has no free ends and ethidium bromide does not introduce breaks into the DNA backbone, the linking number (*Lk*) does not change as ethidium bromide unwinds the molecule. However, recall from Chapter 4 that the linking number is the sum of the twisting number (*Tw*) and the writhing number (*Wr*). When a relaxed circular DNA molecule binds enough ethidium bromide to

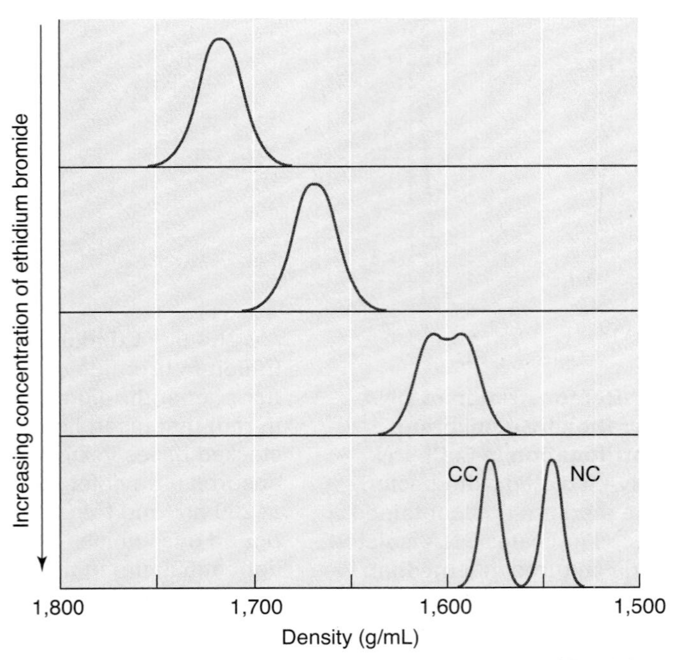

FIGURE 5.10 **Effect of ethidium bromide on the density of DNA in a CsCl solution.** A mixture of equal parts of nicked circles (NC) and covalently closed circles (CC) is centrifuged in CsCl solutions containing different concentrations of ethidium bromide. The densities of the DNA•ethidium bromide complexes decrease as more ethidium bromide is bound. A closed covalent complex cannot bind as much ethidium bromide as a nicked circle and therefore will have a higher density.

unwind by 360°, it therefore will assume a figure 8 shape. As more ethidium bromide molecules intercalate, supercoiling will increase.

At some point, topological constraints will prevent further supercoiling and no more ethidium bromide molecules will be able to bind. Same sized linear duplexes and nicked circles will bind more ethidium bromide because they lack these topological constraints. The densities of DNA•ethidium bromide complexes decrease as more ethidium bromide is bound. At a saturating ethidium bromide concentration, therefore, a closed covalent circle will have a higher density than a linear duplex or a nicked circle. Thus, covalent circles can be separated from the other forms in a density gradient, as shown in FIGURE 5.10.

5.4 Gel Electrophoresis

Gel electrophoresis separates charged macromolecules by their rate of migration in an electric field.

The migration of charged molecules (such as nucleic acids or proteins) through an agarose or polyacrylamide gel under the influence of an electric field is called **slab gel electrophoresis**. One widely used type of slab gel electrophoresis apparatus used to separate DNA molecules according to size is shown in FIGURE 5.11. A thin slab of an agarose

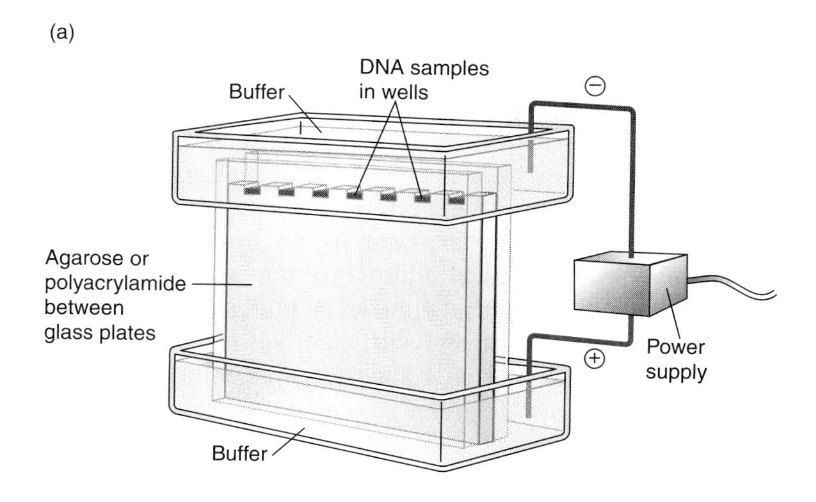

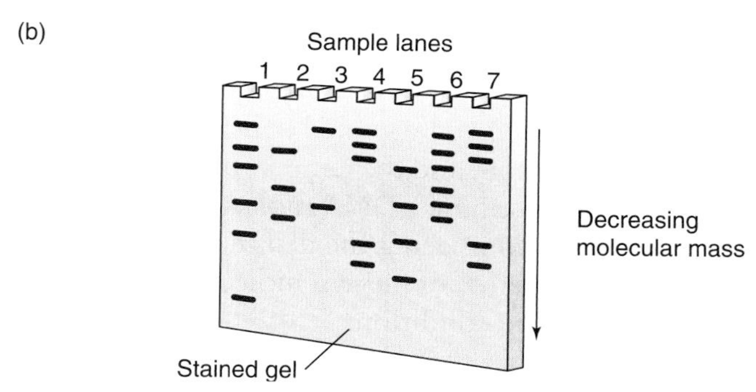

FIGURE 5.11 Gel electrophoresis. (a) Apparatus for gel electrophoresis capable of handling several samples simultaneously. An agarose or polyacrylamide suspension is placed in a mold fitted with glass or plastic plates on each side, and an appropriately shaped mold is placed on top of the gel during hardening in order to make "wells" for the samples. After the gel has hardened, the mold on top is removed and samples are placed in the wells. The power supply is connected to the electrophoresis apparatus and the nucleic acids migrate toward the positive electrode (anode). When electrophoresis is complete, the power is turned off and the nucleic acids in the gel are made visible by removing the glass or plastic plates and immersing the gel in a solution containing a reagent that binds to or reacts with the separated molecules. (b) The separated components of a sample appear as bands, which may either be colored or fluoresce. The region of a gel in which the components of one sample can move is called a lane. This gel, thus, has seven lanes.

or polyacrylamide gel is prepared containing small loading wells into which samples are placed. An electric field is applied and the negatively charged DNA molecules penetrate and move through the complex network of molecules that comprise the gel and toward the positive electrode or anode.

DNA's migration rate depends on its total charge and, as in sedimentation, on its shape (that is, its frictional resistance). Its migration rate is also influenced by its molecular mass because surface area, which affects frictional drag, increases with molecular mass. Smaller molecules squeeze through the narrow, tortuous passages in the gel network more easily than larger molecules. Migration rate, therefore, increases as the molecular mass decreases.

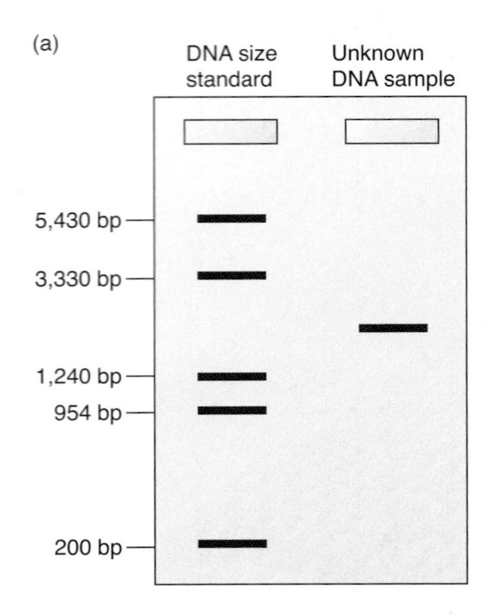

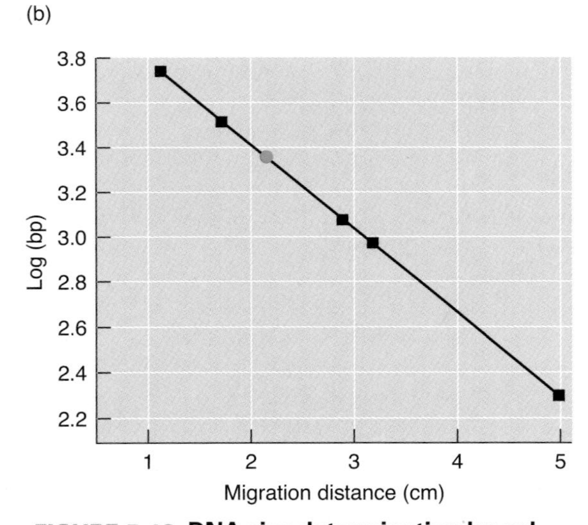

FIGURE 5.12 **DNA size determination by gel electrophoresis.** (a) A DNA size standard and a DNA of unknown size were placed in the left and right lanes, respectively. (b) A graph showing the migration distance in cm vs. log bp. The points shown in black squares are for the standard DNA and the point shown in a green circle is for the DNA of unknown size. (Modified from illustrations by Michael Blaber, Department of Biomedical Sciences, Florida State University [http://www.mikeblaber.org/oldwine/bch5425/lect20/lect20.htm]. Accessed June 18, 2010.)

Conditions can be adjusted so that the rate of migration depends on length in nucleotides for single-stranded DNA or base pairs (bp) for double-stranded DNA. Polyacrylamide gels can separate single-stranded DNA molecules that differ in size by just one nucleotide provided that the strands are between 5 and 750 nucleotides long. Agarose gels have lower resolving power but separate DNA molecules that range in size from 200 bp to about 50,000 bp. DNA bands can be viewed by staining the gel with ethidium bromide, which forms a DNA• ethidium bromide complex that fluoresces under 254 nm ultraviolet light. The sensitivity of detection is sufficiently great that samples with as little as 0.1 μg DNA are easy to see. Because ethidium bromide is carcinogenic, many workers prefer to work with commercially available safer substitutes.

Empirical studies show that the distance moved by DNA (D) during slab gel electrophoresis depends logarithmically on its length in bp, obeying the equation,

$$D = a - b \log bp$$

in which a and b are empirically determined constants that depend on the buffer, the gel concentration, and the temperature. This logarithmic relationship allows us to determine a DNA molecule's molecular mass (or length in base pairs) by comparing the distance it moves with distances moved by DNA standards of known molecular mass (or length in base pairs) under the same conditions (FIGURE 5.12).

Gel electrophoresis can also be used to separate topoisomers (FIGURE 5.13). Because superhelical molecules are compact, they move through the gel more rapidly than relaxed circles with the same mass. The difference is sufficiently great that it is possible to separate two DNA molecules with linking numbers that differ by only 1. Each intermediate present in a reaction mixture containing a type I or type II topoisomerase thus can be separated. Gel electrophoresis, therefore, serves as a convenient method for assaying topoisomerase activity. The advantages of this method are that it requires very little DNA, is very sensitive, can be used to analyze several samples at one time, requires relatively inexpensive equipment, and is simple to perform.

Gel electrophoresis can be used to separate proteins and determine a polypeptide's molecular mass.

As described in Chapter 2, proteins can also be separated by gel electrophoresis. Proteins can be either positively or negatively charged, however, and a sample containing several different proteins must be placed in a centrally located well so that migration can occur in both directions. The charge per unit mass, which is very small because most amino acids are uncharged, varies from one protein to the next (in contrast with DNA or RNA, which have one negative charge per nucleotide). Moreover, proteins come in a variety of shapes, so there is no simple way to predict the migration rate.

Proteins could be separated solely on the basis of size if they had a uniform charge and the shape factor could be eliminated. A technique known as **SDS-PAGE** (**sodium dodecyl sulfate-polyacrylamide** gel electrophoresis) does just that. The detergent SDS and the disulfide bond-breaking agent β-mercaptoethanol ($HSCH_2CH_2OH$) are added to the protein, causing all the polypeptide chains to be denatured, forming rod-shaped structures that are coated with negatively charged SDS. The net effect is that, as in the case of DNA, all the proteins migrate toward the anode, the migration rate increases as chain length decreases, and the dependence is logarithmic. SDS-PAGE, therefore, can be used to determine a polypeptide's molecular mass by comparing its migration rate with the migration rates of polypeptide standards.

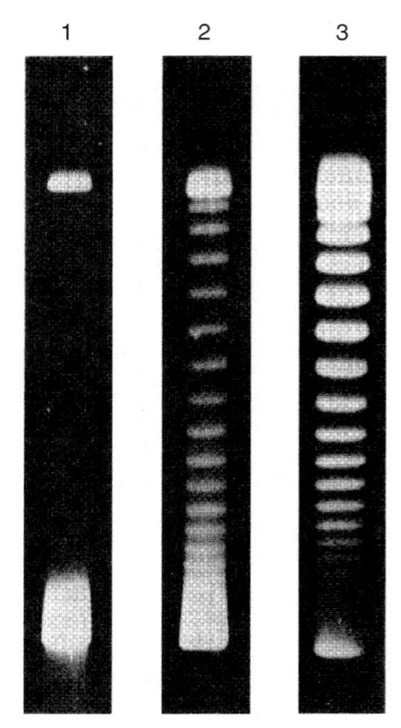

FIGURE 5.13 **Agarose gel electrophoresis pattern of covalently closed circular SV40 (simian virus 40) DNA.** The DNA was applied to the top of the gel. Lane 1 contains untreated negatively supercoiled native DNA (lower band). Lanes 2 and 3 contain the same DNA that was treated with a type I topoisomerase, which makes a single strand break in only one chain that relaxes negative supercoils by causing successive one unit increases in the linkage number ($\Delta Lk = +1$). The DNA in lanes 2 and 3 were treated for 5 and 30 minutes respectively. (Reproduced from W. Keller, *Proc. Natl. Acad. Sci. USA* 72 [1975]: 2550–2554. Photo courtesy of Walter Keller, University of Basel.)

Capillary gel electrophoresis is a rapid and automated process that provides quantitative data.

Although quite effective in separating DNA molecules by size, slab gel electrophoresis is (1) slow (usually requiring hours to complete a run), (2) difficult to automate, and (3) does not easily provide quantitative data. An alternate means for separating DNA in an applied electric field, called **capillary gel electrophoresis**, does not suffer from these disadvantages (FIGURE 5.14a). Separations are performed in very thin (20–100 μm) capillary tubes made of quartz, glass, or plastic filled with a replaceable high molecular mass linear polymers. Because the capillary tubes are so thin, heat is rapidly dissipated. This heat dissipation allows separations in electrical fields (typically 100–500 $V \cdot cm^{-1}$), which are about 10 times higher than those used in slab gel electrophoresis. Capillary gel electrophoresis, thus, allows high resolution separations comparable to those obtained by slab gels in minutes rather than hours (FIGURE 5.14b). Furthermore, capillary gel electrophoresis can be automated and provides quantitative data. Commercially available capillary gel electrophoresis systems can run as many as 96 capillaries at the same time, allowing for high throughput DNA sequence analysis (see below).

Pulsed-field gel electrophoresis (PFGE) can separate very large DNA molecules.

DNA molecules, which are 5×10^4 bp or longer, migrate at the same rate and appear as a single large diffuse band when analyzed by slab gel electrophoresis. This method, therefore, cannot be used to resolve very large DNA molecules. In 1982, David Schwartz and Charles Cantor devised a new method known as **pulsed-field gel electrophoresis** (PFGE) that can separate very large DNA molecules from one another.

In this technique, the DNA is forced to change direction by reorienting the electric field relative to the gel at regular time intervals. A change in field orientation can be achieved by using two electric fields that are arranged at a transverse angle (usually between 90° and 120°) and turned on and off in a pulsed manner so that they alternate in directing DNA migration (FIGURE 5.15a). With each reorientation, smaller DNA molecules begin moving in the new direction more rapidly than larger DNA molecules.

(a)

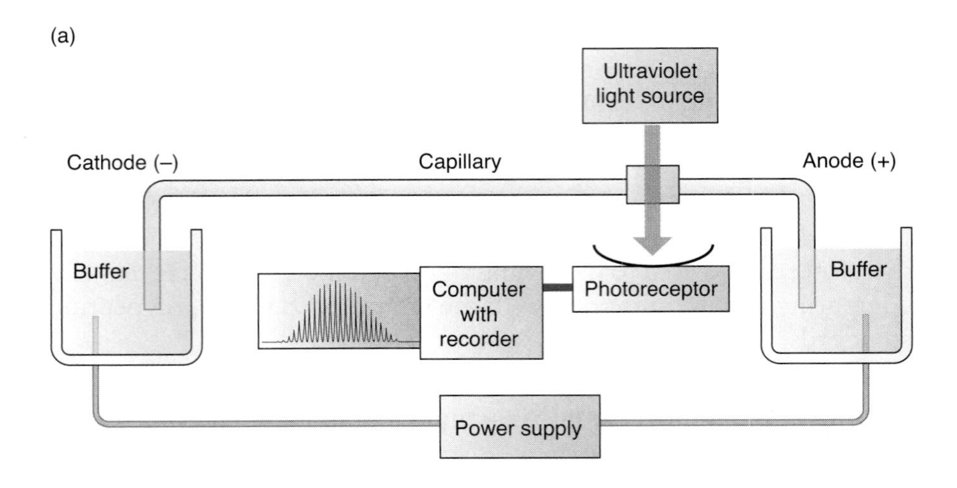

(b)

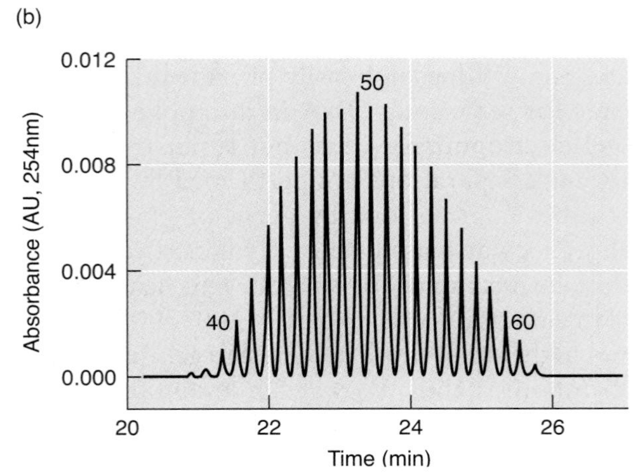

FIGURE 5.14 **Capillary gel electrophoresis.** (a) Schematic of the apparatus—DNA moves toward the positive electrode (anode) and is detected by its ability to absorb ultraviolet light. (b) Capillary gel electrophoresis separation of a single-stranded polyadenylic acid 40-60 on a polyacrylamide gel. (Part a adapted from an illustration from the Natural Toxins Research Center, Texas A&M University. Part b modified from A. Guttman, *LCGC N. America* 22 [2004]: 896–904. Used with permission of András Guttman, Horváth Laboratory of Bioseparation Sciences–Hungary.)

The explanation appears to be related to the fact that DNA molecules stretch out lengthwise in the direction of the electric field. Longer DNA molecules appear to require more time to reorient themselves and start moving in the new direction. Pulsed-field gel electrophoresis can be used to purify DNA molecules up to about 10^7 bp long. Even though large DNA molecules such as yeast chromosomes are easily broken by hydrodynamic shear forces, they can, nevertheless, be separated by embedding intact yeast cells in agarose plugs and then treating the cells with enzymes to digest cell walls and proteins while leaving the undamaged DNA behind. Once digestion is complete, the plugs are cut to size and sealed in the well of the electrophoresis gel with agarose. FIGURE 5.15b shows the application of pulse field gel electrophoresis to chromosomal DNA from *Saccharomyces cerevisiae*.

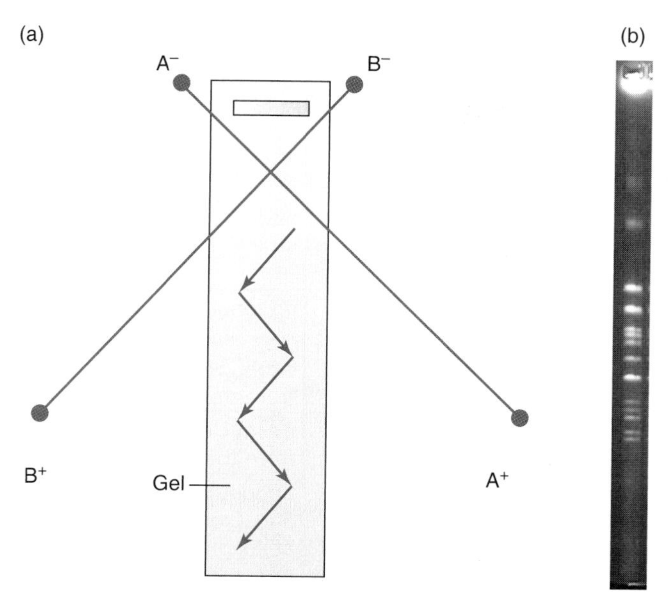

FIGURE 5.15 **Pulsed field gel electrophoresis.** (a) In this schematic, the electric fields (shown in blue lines) are at a transverse angle (usually between 90° and 120°). Each pulse forces the DNA molecules to reorient themselves so that they move down the gel following the zigzag path indicated by the red arrows. Larger DNA molecules reorient themselves more slowly than smaller molecules and, therefore, move forward more slowly. A^+ and B^+ represent positive electrodes and A^- and B^- represent the corresponding negative electrodes. (b) Pulsed-field gel electrophoresis separation of chromosomes from a diploid yeast cell. The chromosomes range in length from about 250 to 1500 kbp. DNA was stained with ethidium bromide. (Part a modified from B. Joppa, et al., *Probe: Newsletter for the USDA Plant Genome Research Program* 2 [1992]: 23–28 [http://www.nal.usda.gov/pgdic/Probe/v2n3/puls.html]. Accessed June 22, 2010. Used with permission of Hoefer, Inc. Part b reproduced from *Genetics: A Periodical Record of Investigations Bearing on Heredity and Variation* by R. B. Tennyson, et al., 160 [2002]: 1363–1373. Copyright 2002 by the Genetics Society of America. Reproduced with permission of the Genetics Society of America in the format Textbook via Copyright Clearance Center. Photo courtesy of Janet E. Lindsley, University of Utah.)

5.5 Nucleases and Restriction Maps

Nucleases are useful tools in DNA investigations.

Nucleases, enzymes that digest polynucleotides by cleaving the 5′→3′ phosphodiester bonds that link nucleotides, are valuable tools for studying nucleic acids. Those that digest DNA are designated deoxyribonucleases (DNases), while those that digest RNA are designated ribonucleases (RNases). Some nucleases digest both kinds of nucleic acids. Some DNases are specific for double-stranded (ds) DNA, others for single-stranded (ss) DNA, and still others act on both kinds of DNA.

Nucleases also differ in their specificity for where they cut the nucleic acid. Nucleases that act within a strand are called **endonucleases.** Some endonucleases are specific and cleave only between particular bases. Nucleases that act only at the end of a nucleic acid, removing a single nucleotide at a time, are called **exonucleases.** Exonucleases

TABLE 5.1 **Properties of Selected Nucleases**

Nuclease	Substrate	Site of Cleavage	Product
Pancreatic RNase	ssRNA in high salt; all RNA in low salt	endonuclease; after C or U	mono- or oligo-nucleotides with a 3′-P
T1 RNase	ssRNA in high salt; all RNA in low salt	endonuclease; after G	mono- or oligo-nucleotides with a 3′-P
Pancreatic DNase I	ss or dsDNA	endonuclease	oligonucleo-tides
Phosphodiesterase venom	RNA or DNA	exonuclease from the 3′ end	5′-P mono-nucleotides
Phosphodiesterase spleen	RNA or DNA	exonuclease from the 5′ end	3′-P mono-nucleotides
Micrococcal nuclease	DNA (or RNA) ss or ds (prefers ssDNA to dsDNA; ssRNA to ds RNA)	endonuclease and exonuclease; at AT or AU rich regions, requires Ca^{2+}	mononucleo-tides and oligo-nucleotides with 3′-P ends
S1 nuclease	ssDNA (or ssRNA. S1 nuclease is 5× more active on ssDNA than on ssRNA)	endonuclease	5′-P mono-nucleotides
Exonuclease I *E. coli*	ssDNA	3′→5′ exo-nuclease	5′-P mono-nucleotides plus a terminal dinucleotide
Exonuclease III *E. coli*	dsDNA	3′→5′ exo-nuclease	5′-P mono-nucleotides
Exonuclease VII *E. coli*	ssDNA	3′→5′ exo-nuclease and 5′→3′ exo-nuclease	5′-P mono-nucleotides

that begin cutting at the 5′- or 3′-ends are designated 5′→3′ or 3′→5′ exonucleases, respectively. Although nucleases have many important biological functions, the present focus is on using them as laboratory tools. Some specific nucleases are described in Table 5.1 and others are described when relevant in the text.

A specific example, the **S1 endonuclease** isolated from *Aspergillus oryzae*, illustrates how we use nucleases to investigate nucleic acid structure. The S1 endonuclease acts exclusively on single-stranded polynucleotides or on single-stranded regions of double-stranded nucleic acids. It differs from other single-strand-specific enzymes in that the single-stranded region can be as small as one or two bases. We noted in the previous chapter that negatively supercoiled DNA contains single-stranded bubbles resulting from localized transient melting. Supercoiled DNA can be cleaved by S1 nuclease because of these regions. In fact, this nuclease can be used to distinguish supercoiled from both non-supercoiled covalent circles and nicked circular DNA, both of which are resistant to the enzyme. S1 nuclease makes a double-strand break because it acts on both single-stranded branches of the bubble.

TABLE 5.2 **Types of Restriction Endonuclease**

Type	Structure	Cofactor(s)	Recognition Sequence	Cleavage
I	R_2M_2S	ATP (hydrolysis) AdoMet, Mg^{2+}	Asymmetric interrupted	Cut DNA at sites distant from the recognition sequence
Example	*Eco*B		-TGA(N_8)TGCT-	
II	R_2	Mg^{2+}	Palindrome 4–8 bp	Within the recognition sequence to produce blunt or staggered ends
Example	*Eco*RV		-GATATC-	
III	RM	ATP (no hydrolysis) Mg^{2+}, AdoMet	Asymmetric	Cut DNA close to the recognition sequence
Example	*Eco*PI		-AGACC-	

Abbreviations: R, restriction endonuclease subunit; M, modification subunit; S, specificity subunit; AdoMet, S-adenosylmethionine.
Source: Data from A. Pingoud, and A. Jeltsch, *Eur. J. Biochem.* 246 (1997): 1–22.

Restriction endonucleases that cleave within specific nucleotide sequences are very useful tools for characterizing DNA.

Molecular biologists recognized that they would be able to use nucleases that cut DNA at specific sites to study nucleic acids in the same way that protein chemists use trypsin and other specific proteases to study proteins. The problem of finding nucleases with sufficient sequence specificity to be useful was solved when investigators discovered a class of nucleases that cleave viral DNA molecules that enter a cell, impeding viral replication. These nucleases, which are called **restriction endonucleases** because they block or restrict viral replication, act only on DNA with specific recognition sequences and only when the recognition sequences are not modified. Host DNA is protected because it has methyl groups attached to specific bases within the recognition sequence.

Three major types of restriction-modification systems have been studied (Table 5.2). Type I restriction-modification systems consist of five polypeptide subunits: two identical restriction endonuclease subunits (R), two identical modification subunits (M), and a specificity subunit (S). If the sequence that is recognized by the specificity subunit does not have a methyl group, then one of two things will happen. The modification subunits will methylate the sequence and the DNA will be protected, or the restriction subunits will cleave the DNA at a nonspecific site, often 1 kb or more from the recognition sequence, and the DNA will be degraded. Type II restriction-modification systems are made of two independent enzymes, a homodimeric restriction endonuclease and a monomeric methyl transferase (methylase). Type II restriction-modification enzymes recognize sequences that are 4 to 8 bp long. Type II methylases transfer methyl groups to bases within the recognition sequence and type II endonucleases cleave DNA within the recognition sequence. Type III restriction-modification systems consist of two subunits, a modification subunit and a restriction subunit. Modification occurs within the recognition sequence but cleavage takes place about 25 bp away from this site. The discussion that follows is limited to the type II endonucleases because they are the only one of the three types that has been widely used to manipulate DNA.

Hamilton O. Smith was the first to isolate and characterize a Type II restriction endonuclease. The enzyme, isolated from *Haemophilus influenzae* and called HindII, recognizes the set of sequences,

5′····GTPy↓PuAC-····3′
3′····CAPu↑PyTG-····5′

where Py and Pu represent a pyrimidine and purine, respectively. The arrows indicate sites at which each strand is cut.

The discovery of HindII motivated investigators to seek other endonucleases that cut within specific nucleotide sequences. The search has been rewarded. More than 2500 different type II enzymes have been identified, and more than 50 of these have been sequenced and biochemically characterized. In many cases, two or more different restriction endonucleases recognize the same sequence. Different restriction endonucleases that recognize the same nucleotide sequence and that cleave it in the same position are called **isoschizomers**. Longer recognition sequences are statistically less likely to appear within a DNA molecule than are shorter ones. Restriction endonucleases that recognize 8 bp sequences, therefore, will make many fewer cuts in a DNA molecule than those that recognize 4 bp sequences.

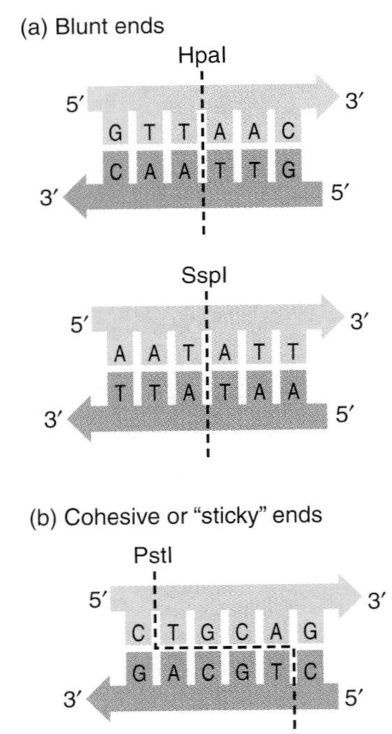

FIGURE 5.16 **Restriction endonuclease generated ends.** (a) Some restriction endonucleases such as HpaI and SspI cut on the line of symmetry to produce blunt ends. (b) Other restriction endonucleases such as PstI cut on either side of the line of symmetry to produce cohesive ends (also called "sticky" or staggered ends). (Adapted from an illustration by Maria Price Raposa, Carolina Biological Supply Company [http://www.carolina.com/].)

Type II restriction endonuclease recognition sites are inverted repeat sequences (FIGURE 5.16). When rotated 180° about the central point in the plane of the page, the recognition site reads exactly as it did before the rotation. Recall from Chapter 4 that molecular biologists call a sequence with such a dyad axis of symmetry a **palindrome**. Type II restriction endonucleases cut each strand within the palindrome sequence one time. Some type II enzymes cut each strand at the axis of symmetry to generate **flush** or **blunt ends** (Figure 5.16a). Others make staggered cuts (cuts that are symmetrically placed around the axis of symmetry) to generate **cohesive** or **sticky ends** (Figure 5.16b). In either case, phosphodiester bond cleavage generates 3′-hydroxyl and 5′-phosphate ends. Table 5.3 lists some Type II restriction endonucleases, along with their recognition sites.

Important insights concerning enzyme-DNA interactions have been gained by examining the way that restriction endonucleases act on DNA. To be effective in protecting the host cell from viral attack, the host's restriction endonuclease has to cut the viral DNA before its modification enzyme can methylate the DNA. A restriction endonuclease must therefore be able to reach its target site quickly. It does so by first binding to the foreign DNA in a nonspecific fashion and then scanning the DNA for recognition sequences. This linear diffusion along the DNA molecule allows the restriction endonuclease to find and cleave all recognition sites rapidly and efficiently. The alternative possibility, a three-dimensional diffusion process in which the endonuclease alternately associates with and dissociates from DNA

TABLE 5.3	Sequence Specificity of Some Restriction Endonucleases

Organism of Origin	Restriction Endonuclease	Recognition Sequence
Arthrobacter luteus	AluI	5′...A G↓CT ... 3′
Anebaena variabilis	Ava I	5′...C↓PyCGPuG...3′[a]
Bacillus amyloliquefaciens H	Bam HI	5′...G↓GATCC...3′
Bacillus globigii	Bgl HI	5′...A↓GATCT...3′
Escherichia coli RY13	Eco RI	5′...G↓AATTC...3′
Escherichia coli J62 pLG74	Eco RV	5′...GAT↓ATC...3′
Haemophilus aegyptius	Hae II	5′... PuGCGC↓Py...3′
Haemophilus aegyptius	Hae III	5′...GG↓CC...3′
Haemophilus haemolyticus	Hha I	5′...GCG↓C...3′
Haemophilus influenzae Rd	Hind II	5′...GTPy↓PuAC...3′
Haemophilus influenzae Rd	Hind III	5′...A↓AGCTT...3′
Haemophilus parainfluenzae	Hpa I	5′...GTT↓AAC...3′
Haemophilus parainfluenzae	Hpa II	5′...C↓CGG...3′
Kliebsiella pneumoniae	Kpn I	5′...GGTAC↓C...3′
Moraxella bovis	Mbo I	5′... ↓GATC...3′
Nocardia otitidis-caviarum	Not I	5′...GC↓GGCCGC...3′
Providencia stuartii	Pst I	5′...CTGCA↓G...3′
Serratia marcescens	Sma I	5′...CCC↓GGG...3′
Streptomyces stanford	Sst I	5′...GAGCT↓C...3′
Xanthomonas malvacearum	XmaI	5′...C↓CCGGG...3′

[a]Py, pyrimidine; Pu, purine.

until it finally binds to the recognition sequence, would be too slow to protect the host from viral replication.

Crystal structures have been obtained for several restriction endonucleases. Based on these structures, the enzyme's approach to its DNA substrate appears related to its cutting pattern. Restriction endonucleases that make blunt end cuts approach DNA from the minor groove side and loops from each monomer wrap around the DNA to contact the major groove. This type of interaction can be seen in the crystal structure of the blunt end cutter EcoRV (from *E. coli*) and a DNA molecule with the cognate (correct) recognition sequence (FIGURE 5.17a). A tight network of hydrogen bonds, ionic bonds, and van der Waals interactions is formed between the enzyme and its recognition sequence. The DNA is distorted from a regular B-DNA structure to a highly strained conformation in which the DNA unwinds, the central two base pairs in the recognition sequence (GATATC) unstack, and the DNA bends so that the major groove becomes narrower and deeper. The catalytic site on each subunit then cleaves the exposed phosphodiester bond at the center of the recognition site.

Restriction endonucleases that produce 5′-overhanging ends approach DNA from the major groove and make most of their base-specific contacts there. In most cases, the conformation of both the enzyme and its DNA substrate change when the enzyme•DNA complex is formed. The crystal structure of the stagger end cutter BamHI (from *Bacillus amyloli*), which binds at the recognition sequence

(a)

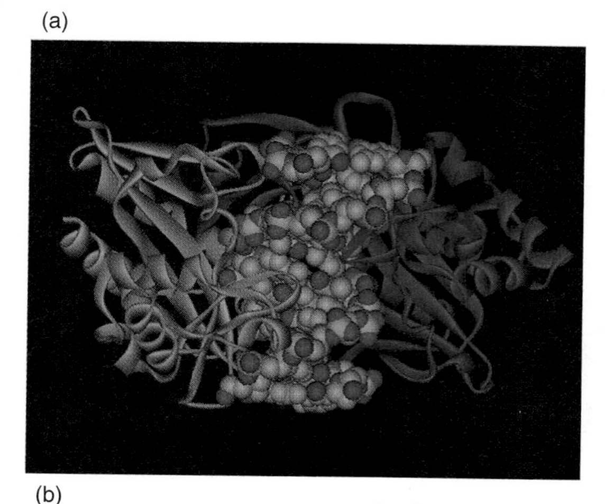

(b)

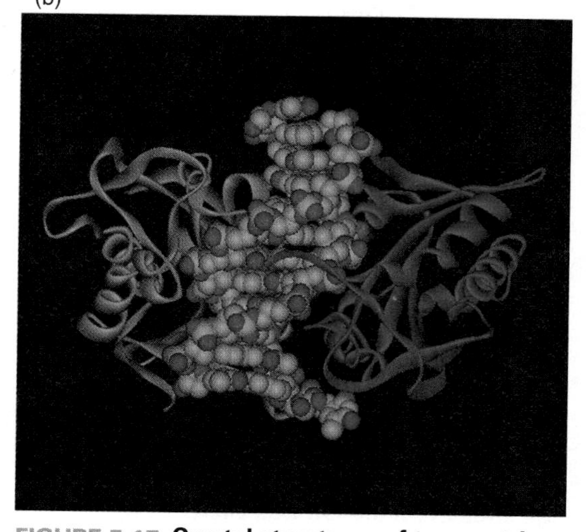

FIGURE 5.17 **Crystal structures of two restriction endonucleases.** The protein subunits are shown as ribbon structures and DNA molecules as a space filling structure. (a) The blunt end cutter EcoRV approaches the DNA from the minor groove side and loops from each polypeptide subunit wrap around the DNA to contact the major groove. Although not clearly visible in the perspective shown here, the enzyme has caused the DNA to bend. (b) The stagger end cutter BamHI approaches the DNA from the major groove side and makes most of its specific contacts there. (Part a structure from Protein Data Bank 1EOO. N. C. Horton and J. J. Perona, *Proc. Natl. Acad. Sci. USA* 97 [2000]: 5729–5734. Prepared by B. E. Tropp. Part b structure from Protein Data Bank 1BHM. M. Newman, et al., *Science* 269 [1995]: 656–663. Prepared by B. E. Tropp.)

5′-GGATCC-3′ and cleaves this sequence just after the 5′-guanylate on each strand, is shown in FIGURE 5.17b. BamHI undergoes a conformational change on binding to DNA but is somewhat unusual in not inducing a conformational change in the bound DNA.

Restriction endonucleases can be used to construct a restriction map of a DNA molecule.

Because a particular type II restriction endonuclease recognizes a unique sequence, it can only make a limited number of cuts in an organism's DNA. For example, an endonuclease that recognizes a 6 bp sequence will cut a typical bacterial DNA molecule, which contains roughly 3×10^6 bp into a few hundred to a few thousand fragments. Smaller DNA molecules such as phage or plasmid DNA molecules may have fewer than ten sites of cutting (frequently, one or two, and often none). Because of the specificity just mentioned, a particular restriction enzyme generates a unique family of fragments from a particular DNA molecule. Another enzyme will generate a different family of fragments from the same DNA molecule.

FIGURE 5.18a shows the sites of cutting of *E. coli* phage lambda DNA by the enzymes EcoRI and HindIII. The family of fragments generated by a single enzyme can usually be resolved by agarose gel electrophoresis (FIGURE 5.18b). A fragment's length is determined by comparing its mobility to the mobility of fragments of known length run concurrently.

Maps of restriction endonuclease cutting sites such as those shown in Figure 5.18a are called **restriction maps**. Distances between the cut sites correspond to the sizes of the DNA fragments produced by restriction endonuclease digestion. One approach to constructing a restriction map is to cut the DNA sample in three ways: with one restriction endonuclease, with a second restriction endonuclease, and with both

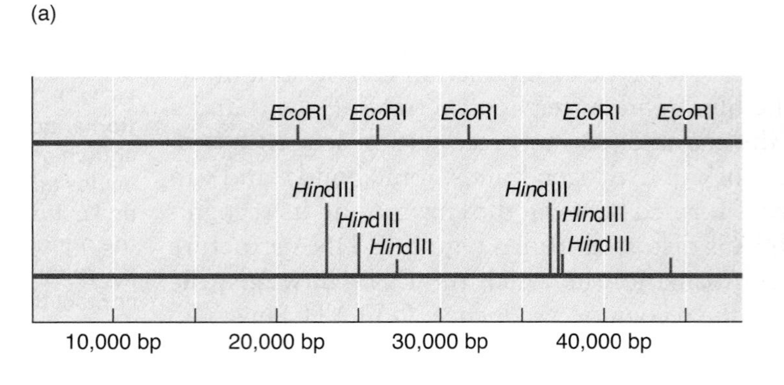

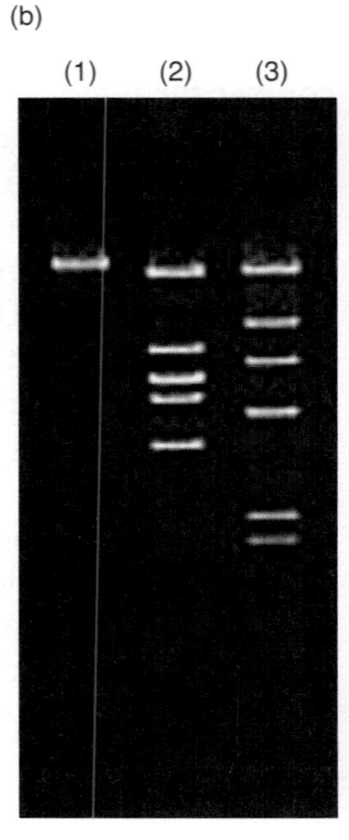

FIGURE 5.18 EcoRI and HindIII restriction endonuclease cleavage sites in phage lambda DNA. (a) (top line) EcoRI cleavage sites, (middle line) HindIII cleavage sites, and (bottom line) DNA base pair number starting from the left side. (b) A gel electrophoresis showing the restriction fragments that are produced: (lane 1) uncut lambda DNA, (lane 2) EcoRI digest, and (lane 3) HindIII digest. Two small DNA fragments produced by HindIII digestion are not visible in this gel. The largest fragments are at the top of the gel and the smallest at the bottom. (Part b courtesy of FOTODYNE Incorporated.)

enzymes (a double digest). The fragments generated are resolved by gel electrophoresis, which also reveals the number of fragments and their lengths in kilobase pairs (kbp). The fragmentation pattern and fragment lengths are used to deduce the order of restriction endonuclease cut sites and to assign intervals between them. Because actual lengths are determined by comparing fragment mobilities with the mobilities of standards of known length, all intervals in the map are additive.

A simple example will help to illustrate this approach (FIGURE 5.19). Samples of a 6-kbp linear duplex are cut by BamHI, EcoRV, or both endonucleases into fragments, which are separated by gel electrophoresis as shown in Figure 5.19a. EcoRV cleaves at one site to produce two fragments (2.3 and 3.7 kbp). Inspection of restriction fragments produced by the BamHI and EcoRV mixture reveals that BamHI cuts the 3.7 kbp fragment into two pieces (1.2 and 2.5 kbp) and the 2.3 kbp fragment into two pieces (0.5 and 1.8 kbp). BamHI cleaves at two sites to produce three fragments (4.3 kbp, 1.2 kbp, and 0.5 kbp). A comparison of the fragments produced by BamHI with those produced by the double digest indicates that EcoRV cuts the 4.3 kbp fragment into a 2.5 kbp and a 1.8 kbp fragment. This information allows us to determine how the 3.7 and 2.3 kbp fragments produced by EcoRV are joined, yielding the restriction map shown in Figure 5.19b.

Although the principles are simple, the approach does have practical limitations. Some fragments are so small that they are difficult to see or move off the gel. Under these conditions, the sum of the fragment sizes will not add up to the size of the original DNA. Two or more fragments may be the same size or have such similar sizes that the gel does not resolve them. When this occurs, one of the bands in the agarose gel will appear to have twice the fluorescence intensity as neighboring bands of similar size in the same gel. Finally, interpretation of results becomes progressively more difficult as the number of bands

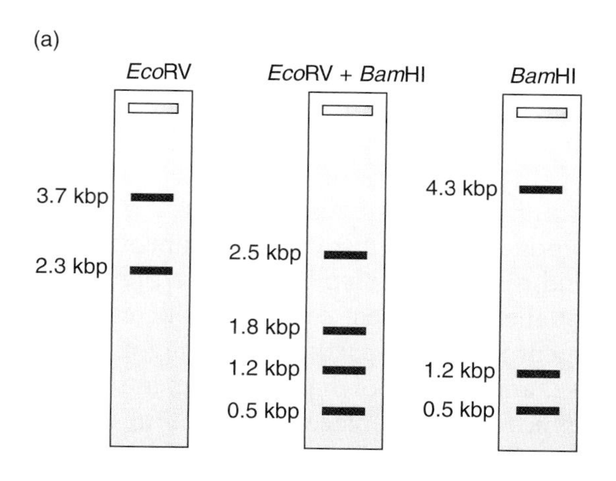

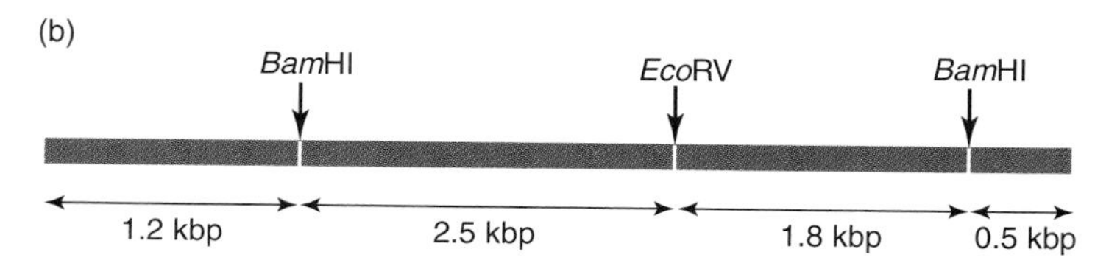

FIGURE 5.19 **Restriction map construction.** (a) Gel electrophoresis pattern of fragments produced by digesting the same DNA molecule with the restriction endonuclease(s) shown at the top of each lane. Fragment sizes are given in kilobase pairs (kbp). (b) The restriction map deduced from the fragment sizes indicated by gel electrophoresis. Specific restriction site(s) for each enzyme are shown by the vertical arrows.

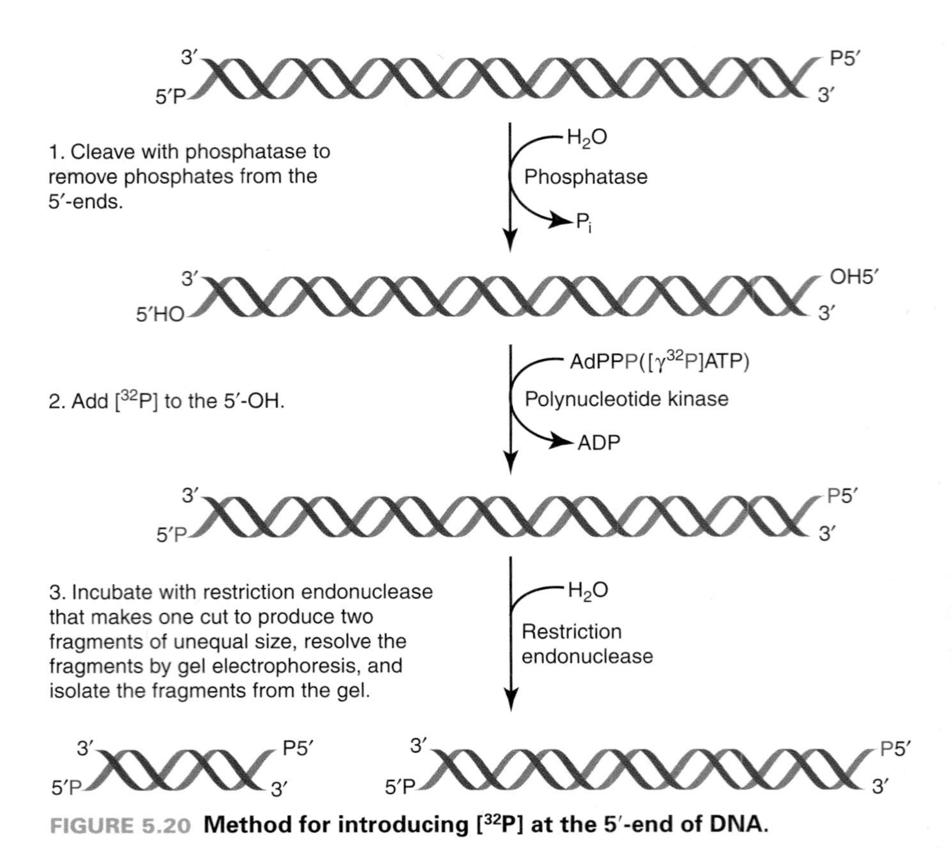

FIGURE 5.20 **Method for introducing [^{32}P] at the 5′-end of DNA.**

increase. Results become difficult, if not impossible to interpret, when each restriction endonuclease produces more than seven fragments.

In 1976, Hamilton Smith and Max L. Birnstiel devised a simple method for DNA restriction site mapping. This method reduces the computational complexity by starting with a DNA molecule with a radioactive label at just one end. It exploits the ability polynucleotide kinase, an enzyme that is encoded by bacteriophage T4, to transfer the γ–phosphoryl group from ATP to the free hydroxyl group at the 5′ end of a DNA or RNA molecule. Labeling DNA at one end requires time and effort but can be achieved by the approach shown in FIGURE 5.20.

The DNA is treated with phosphatase to remove phosphate groups from the 5′-ends, incubated with polynucleotide kinase and [γ-^{32}P]ATP to add labeled phosphate groups to the 5′-ends, and then cut with a restriction endonuclease to produce two DNA fragments of unequal size, each with a single labeled end that are purified by gel electrophoresis. The restriction map of each labeled fragment can then be determined by placing the labeled fragment in a tube, partially digesting it with a new restriction endonuclease, and then resolving the fragments by gel electrophoresis. A photographic film is then placed on top of the gel and stored in the dark. The β particles released by the ^{32}P reduce the silver ions in the photographic film, producing dark bands over each of the fragments. This autoradiographic detection technique reveals patterns that are easy to interpret, with each lane containing bands of fragments that start at the labeled end and extend to a site to be mapped.

The order of the sites is obtained by reading the bands on the gel from the bottom to the top, from the smallest fragment (site closest to the radioactively labeled end) to the largest fragment (site furthest from the labeled end). Intervals between cutting sites correspond to differences between fragment sizes. Although simple in principle, this method has several shortcomings. As fragments get longer, measurements of chain length become less precise. It is also difficult to find conditions that will give the desired level of cleavage. Furthermore, the rate of cutting is affected by the nucleotides around the cleavage site so that some cutting sites will be cleaved to a greater extent than others.

5.6 Recombinant DNA Technology

DNA fragments can be inserted into plasmid DNA vectors.

It is often desirable to generate large amounts of a particular DNA fragment, which we will designate DNA fragment I. Although one might guess that replication of DNA fragment I could be accomplished by first introducing the fragment into a cell and then using the cell's replication machinery to do the work of replicating the fragment, this approach rarely works. The reason for the high rate of failure is that DNA molecules must have specific replication sequences before a cell's replication machinery can act on them (see Chapter 9). Most DNA fragments lack these sequences. There is a simple way to circumvent the problem, however. Insert DNA fragment I into a DNA molecule that has the required replication sequence and then introduce the recombinant DNA molecule into the cell.

Plasmids, which are double-stranded circular DNA molecules capable of autonomous replication in bacteria, are excellent carriers or **vectors** for DNA fragments. As illustrated in FIGURE 5.21, the insertion of DNA fragment I into a plasmid is in principle a simple process. A plasmid bearing an antibiotic resistance gene is cut with a restriction endonuclease to produce cohesive ends that are complementary to the cohesive ends of DNA fragment I. The cut must be at a site that is outside both the antibiotic resistance gene and the sequence required for plasmid DNA replication. DNA fragment I is inserted into the plasmid DNA by joining the ends of DNA fragment I with the ends of the plasmid DNA. An enzyme called DNA ligase, which will be described in Chapter 9, catalyzes the joining or ligation process. The DNA ligase encoded by bacteriophage T4 is one of the most widely used enzymes for joining DNA molecules in the laboratory. A major reason that this enzyme is so frequently used to catalyze ligation reactions is that it can join a pair of cohesive ends or a pair of blunt ends. The **recombinant plasmid**, which contains DNA fragment I inserted within the plasmid vector, then is introduced into living bacterial cells that have been made competent to accept plasmid DNA.

One method for making *E. coli* **competent** to take up DNA from the surrounding medium is to incubate the cells with 100 mM calcium chloride solution at 4°C for 30 min, centrifuge, re-suspend the bacterial pellet in a 75 mM calcium chloride solution containing 15% glycerol, and then freeze 0.2 mL cell samples. The cells are transformed by thawing the samples on ice, adding recombinant DNA, incubating

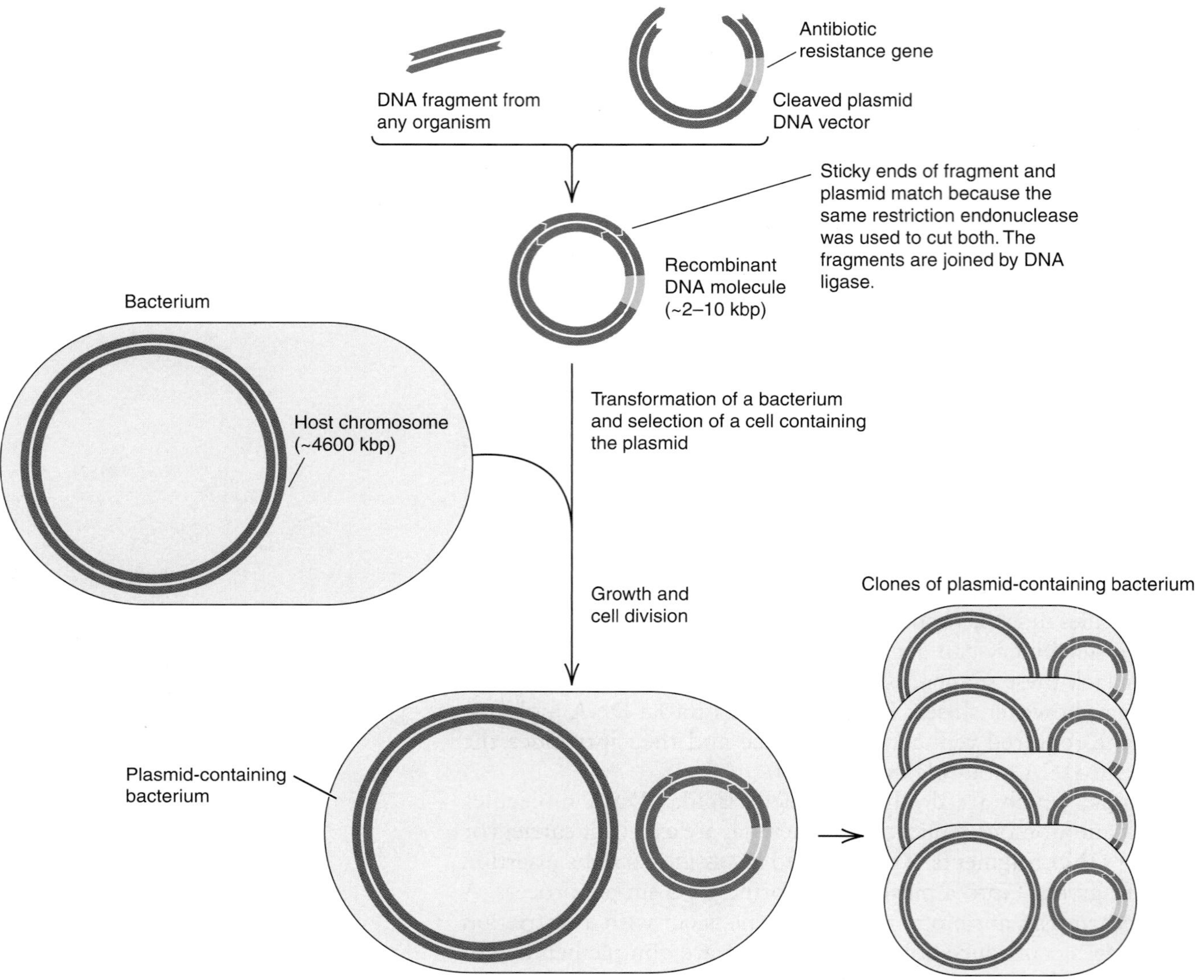

FIGURE 5.21 **An example of cloning.** A fragment of DNA from any organism is joined to a cleaved plasmid. The recombinant plasmid is then used to transform a bacterial cell, where the recombinant plasmid is replicated and transmitted to the progeny bacteria. The bacterial host chromosome is not drawn to scale. It is typically about 1000 times larger than the plasmid.

the suspension at 42°C for 90 s, and then spreading the suspension on an agar plate containing the selective antibiotic. Each cell containing a recombinant plasmid produces a colony or clone consisting of millions of progeny bacteria with the same recombinant plasmid. For this reason, the recombinant plasmid is said to be cloned and the process is called **cloning.** Other types of autonomously replicating DNA molecules such as viral DNA are also used as cloning vectors.

Southern blotting is used to detect specific DNA fragments.

The availability of restriction endonucleases that cleave DNA into unique and fairly small fragments makes it possible to detect small segments within large DNA molecules by taking advantage of the fact that complementary polynucleotide strands (DNA or RNA) can anneal to form double stranded molecules.

In the earliest experiments, individual DNA fragments were painstakingly isolated from electrophoretic gels and annealing, also called **hybridization**, was carried out with these purified fractions. This tedious procedure was replaced in 1975 by the **Southern transfer** or **Southern blot procedure**, named for Edwin M. Southern, the investigator who invented it. This procedure, which allows hybridization to a large number of particular DNA segments without the necessity of purifying individual DNA fragments, exploits the fact that very thin nitrocellulose or nylon membrane filters tightly bind single-stranded DNA fragments. The bases of the bound fragments remain free to form hydrogen bonds with complementary single strands of DNA or RNA.

In the Southern blot procedure (FIGURE 5.22), one or more restriction endonucleases completely digest DNA and the resulting fragments

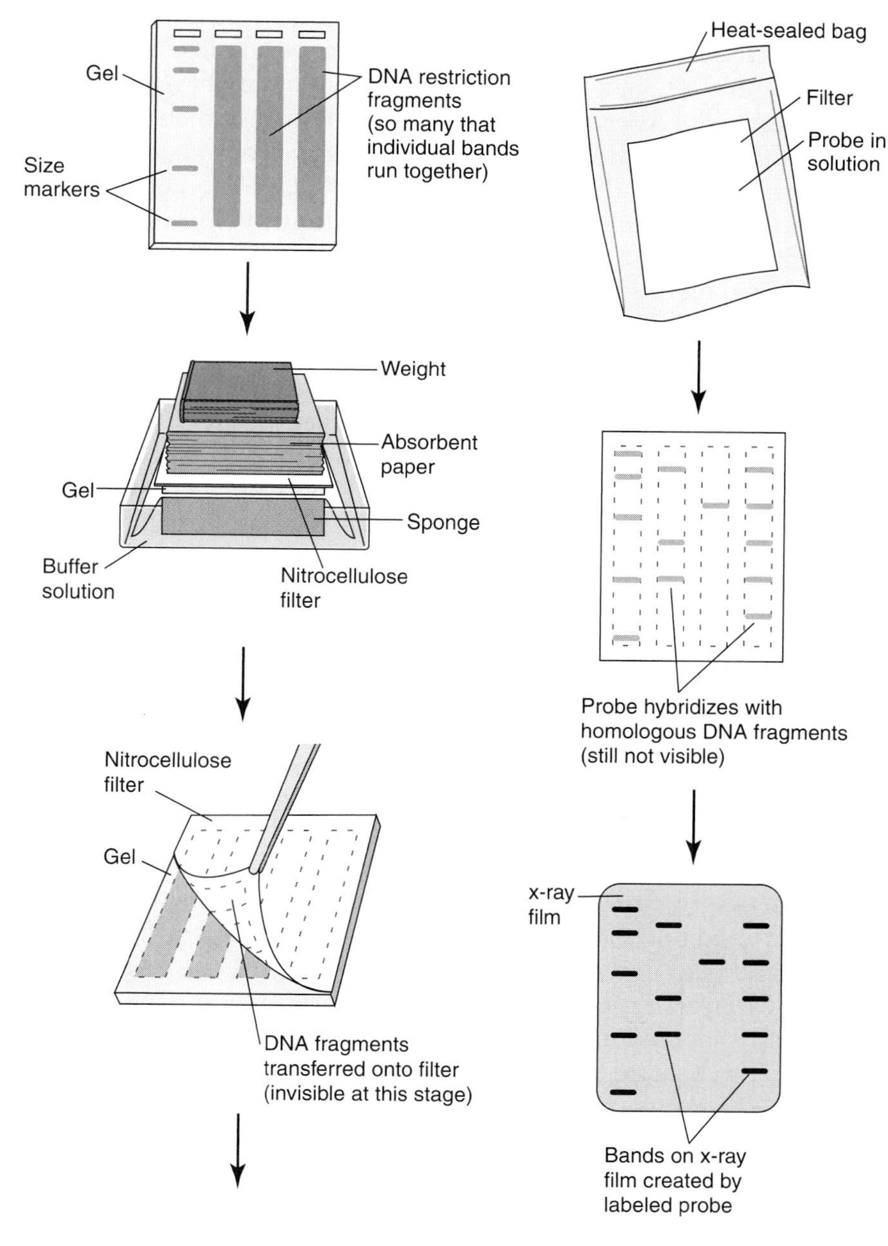

FIGURE 5.22 **Southern blot analysis: an experimental method for identifying a specific DNA fragment in a gel.** The method is divided into the following steps: (1) DNA is cleaved by one or more restriction endonucleases and the resulting fragments are separated by gel electrophoresis, (2) The resolved DNA fragments are denatured by soaking the gel in a basic solution and denatured fragments are blotted onto a nitrocellulose filter, (3) the nitrocellulose filter is exposed to a probe with a radioactive or other detectable label, and (4) when a radioactive probe is used, the filter is exposed to photographic film and the film is developed.

are separated according to size by gel electrophoresis. After separation is complete, the gel is soaked in a sodium hydroxide solution to denature the DNA. Then the gel is rinsed with distilled water, soaked in buffer solution to adjust the pH, and placed on a sponge that is itself in a reservoir of buffer solution. The top surface of the gel is covered with a nitrocellulose or nylon membrane, which in turn is covered with several layers of dry paper towels, and a heavy weight is placed on top of the paper towels. The paper towels act as a blotter, drawing buffer solution from the reservoir through the various layers by capillary action. As the buffer solution moves up, it carries single-stranded DNA from the gel to the nitrocellulose or nylon membrane. In a variation of this procedure, an electrophoretic process called **electroblotting** transfers denatured DNA. In either case, the single-stranded DNA molecules bind to positions on the membrane identical to their positions on the agarose gel, preserving the band pattern.

After transfer is complete, the setup is disassembled and the single-stranded DNA is fixed permanently to the nitrocellulose membrane by baking at 80°C in a vacuum or to the nylon membrane by ultraviolet light induced cross-linking. The membrane is then incubated in a solution containing bovine serum albumin, polysucrose, polyvinylpyrolidine, and denatured salmon sperm DNA to eliminate the membrane's inherent ability to bind single-stranded DNA. Denatured DNA that was transferred to the membrane from the electrophoresis gel, however, retains its ability to bind complementary single-stranded DNA or RNA molecules.

The DNA sequence(s) of interest bound to the membrane are identified by using a complementary single-stranded DNA or RNA probe with a detectable label. In the early studies the probe was labeled with a radioactive isotope, but today investigators often prefer to use a detectable chemical tag such as a **biotin** group attached to the base of a nucleotide. The nitrocellulose or nylon membrane is placed in a buffer solution containing the labeled probe and incubated for several hours at a suitable renaturation temperature to permit the probe to hybridize to its complementary sequence in the DNA bound to the nitrocellulose or nylon membrane. Then the membrane is washed with buffer to remove unbound probe and dried. Autoradiography is used to detect a radioactive probe. DNA fragments complementary to the probe appear as black or stained bands. Biotin containing probes are detected by using a two-component protein molecule. One component is **streptavidin**, which binds tightly to biotin. The other component is alkaline phosphatase. Once the step to bind the streptavidin-alkaline phosphatase protein to the biotin-labeled probe is complete, the membrane is rinsed to remove unbound streptavidin-alkaline phosphatase protein and placed in a color development system. The alkaline phosphatase catalyzes the conversion of soluble precursor into a dark blue precipitate that remains bound to the DNA probe, causing the band(s) of interest to become visible.

The Southern blot procedure is a very versatile and powerful tool that detects changes in DNA that alter restriction cutting sites or the lengths of segments between restriction cutting sites. It has become an important clinical laboratory tool for identifying genetic problems. For example, the Southern blot in FIGURE 5.23 shows that a band present in normal breast tissue is missing from tumor cells.

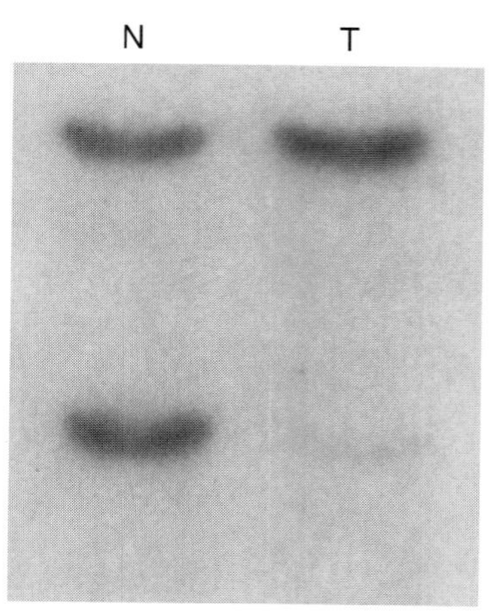

FIGURE 5.23 **Southern blot of genomic DNA from normal tissue (N) and from a tumor (T) from a patient with breast cancer.** Two bands are visible in the normal breast tissue but only the upper band is present in the tumor cells. The faint lower band represents contribution from nontumor cells in the tumor sample. (Reproduced from L. Tougas, et al., *Clin. Invest. Med.* 19 [1996]: 222–230. © 1996, Canadian Society for Clinical Investigation. Photo courtesy of Serge Jothy, University of Toronto.)

Northern and Western blotting are used to detect specific RNA and polypeptide molecules, respectively.

In a variant of Southern blotting, which is called the **Northern blotting** (in a play on words), RNA molecules are separated by gel electrophoresis and then transferred to nylon or nitrocellulose membranes. Labeled single-stranded RNA or DNA probes can hybridize with the bound RNA. Still another variant is **Western blotting**, in which proteins are separated by gel electrophoresis and then transferred to a nitrocellulose or polyvinyl membrane. Bound protein is detected with labeled or tagged antibody probes.

DNA polymerase I, a multifunctional enzyme with polymerase, 3′→5′ exonuclease, and 5′→3′ exonuclease activities, can be used to synthesize labeled DNA.

One of the most important enzymes in the molecular biologist's toolkit is an enzyme called **DNA polymerase I**, which was first detected in *E. coli* extracts by Arthur Kornberg and coworkers in 1956, just three years after the Watson-Crick Model was proposed. The Roman numeral I is used because *E. coli* has a few other DNA polymerases and Kornberg's enzyme was the first to be discovered. DNA polymerase I converts deoxynucleoside triphosphates into DNA in the presence of pre-formed DNA according to the following reaction.

$$\left\{\begin{matrix} \text{dATP} \\ + \\ \text{dGTP} \\ + \\ \text{dCTP} \\ + \\ \text{dTTP} \end{matrix}\right\} \underset{\text{DNA polymerase, } Mg^{2+}}{\overset{\text{DNA}}{\rightleftharpoons}} \text{DNA + pyrophosphate}$$

The enzyme is assayed by adding it to a mixture that also contains the four deoxyribonucleoside triphosphates, one of which is radioactive, a DNA template-primer (see below), and magnesium ions. The reaction is stopped by adding acid, which also precipitates newly formed radioactive DNA. Then the acid-insoluble DNA is separated from the acid-soluble deoxyribonucleoside triphosphates by centrifugation or filtration and its radioactivity is determined in a liquid scintillation counter, which is a device for detecting and counting light flashes produced by ionizing radiation. All four deoxynucleotides must be present for the reaction to occur. The source of the pre-formed double-stranded DNA does not matter; prokaryotic and eukaryotic DNA work equally well. The two strands of the pre-formed DNA have different functions at the site of DNA synthesis (FIGURE 5.24). One strand, the **primer strand**, is the site of attachment for incoming deoxynucleotides. The other strand, the **template strand**, determines the order of attachment to the primer strand according to Watson-Crick

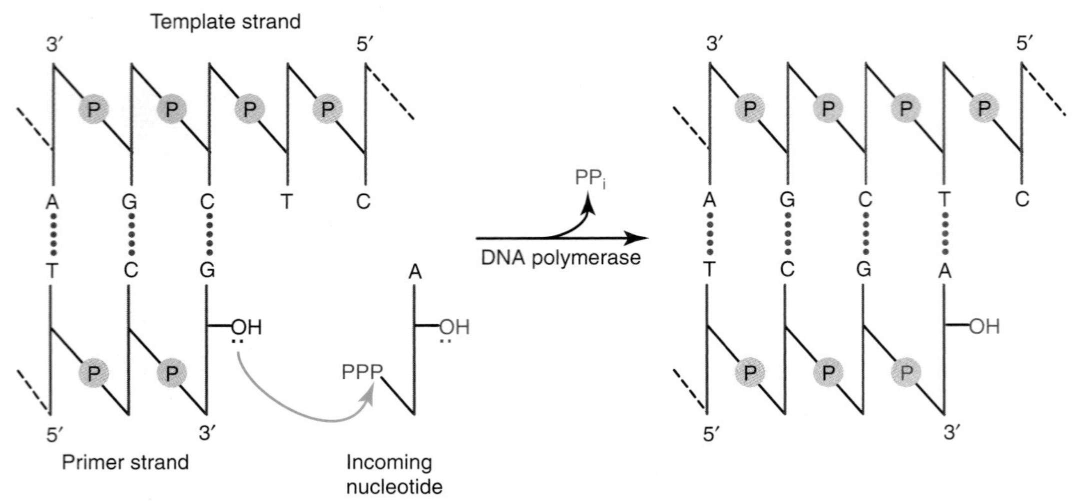

FIGURE 5.24 DNA polymerase I, polymerase function. DNA polymerase I uses the template strand to determine the next nucleotide that it adds to the primer strand.

base pairing rules. DNA polymerase I has binding sites for the primer strand, the template strand, and the incoming deoxynucleotide. The enzyme catalyzes the nucleophilic displacement of a pyrophosphate group, creating a phosphodiester bond between the 3′-end of the growing primer strand and the 5′- end of the incoming deoxynucleotide. Hence, the primer strand grows in a 5′→3′ direction, a characteristic of all known DNA and RNA polymerases. Inorganic pyrophosphate hydrolysis drives the reaction to completion inside the cell.

Early preparations of *E. coli* DNA polymerase I contained exonuclease activity, which Kornberg and coworkers expected to remove by further enzyme purification. Contrary to their expectation, highly purified DNA polymerase I, consisting of a single polypeptide with 928 amino acid residues (molecular mass = 103 kDa), has both 3′→5′ and 5′→3′ exonuclease activities. The discovery that a single enzyme has the ability to both synthesize and degrade DNA was mystifying. Further studies by Kornberg's group, however showed that the two-exonuclease activities, in fact, do make important contributions to DNA synthesis.

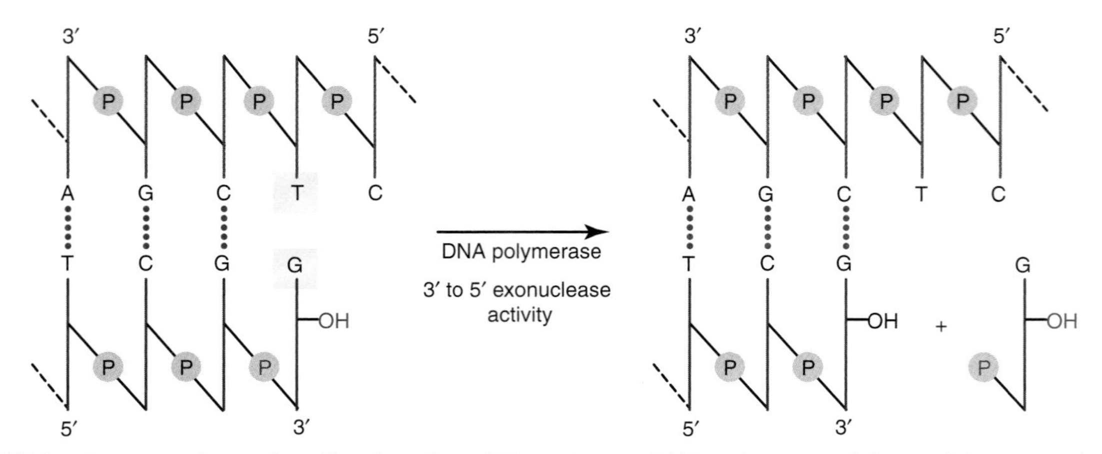

FIGURE 5.25 DNA polymerase I, proofreading function. Although rare, DNA polymerase I does misinsert nucleotides. The 3′→5′ exonuclease activity allows DNA polymerase to remove a misinserted deoxyribonucleotide before the next deoxyribonucleotide is added to the growing primer chain.

The 3′→5′ exonuclease, which catalyzes the sequential hydrolysis of one 5′-deoxynucleotide at a time from the 3′-end, has a proofreading function (FIGURE 5.25). Although DNA polymerase I has great specificity, it does make occasional errors either by misinserting a nucleotide or by misaligning the template-primer. Either type of error will eventually result in base pair substitution, addition, or deletion. Although errors are rare, occurring at a frequency of about one nucleotide for every 10^4 nucleotides added, failure to correct an error would be harmful to the cell. The 3′→5′ exonuclease activity removes the misinsertions before the next deoxynucleotide is added to the growing chain. The 5′→3′ exonuclease, which can remove either a mononucleotide or a short oligonucleotide from the 5′-end, plays an editing role in DNA synthesis that will be described in Chapter 9 and, as will be described below, serves as a useful tool for preparing labeled DNA.

DNA polymerase I is a multifunctional polypeptide (a single polypeptide with more than one catalytic activity) with three domains. Domain 1, which includes the first 325 or so residues at the amino terminus, has 5′→3′ exonuclease activity. Domain 2, consisting of the next 200 or so residues, has 3′→5′ exonuclease activity. Domain 3, which includes the remaining residues at the carboxy-terminus, has polymerase activity. In 1970, Hans Klenow demonstrated that subtilisin (a proteolytic enzyme produced by *B. subtilis*) cleaves DNA polymerase I into two fragments of unequal length (FIGURE 5.26). The smaller fragment (molecular mass = 35 kDa) containing domain 1 has 5′→3′ exonuclease activity. The larger one (molecular mass = 68 kDa), known as the **Klenow fragment**, contains domain 2 (3′→5′ exonuclease activity) and domain 3 (polymerase activity).

Thomas Steitz and coworkers have determined the three-dimensional structure of the Klenow fragment by x-ray crystallography. The polymerase domain, which is made of three subdomains, resembles a partially opened right hand (FIGURE 5.27). The polymerase

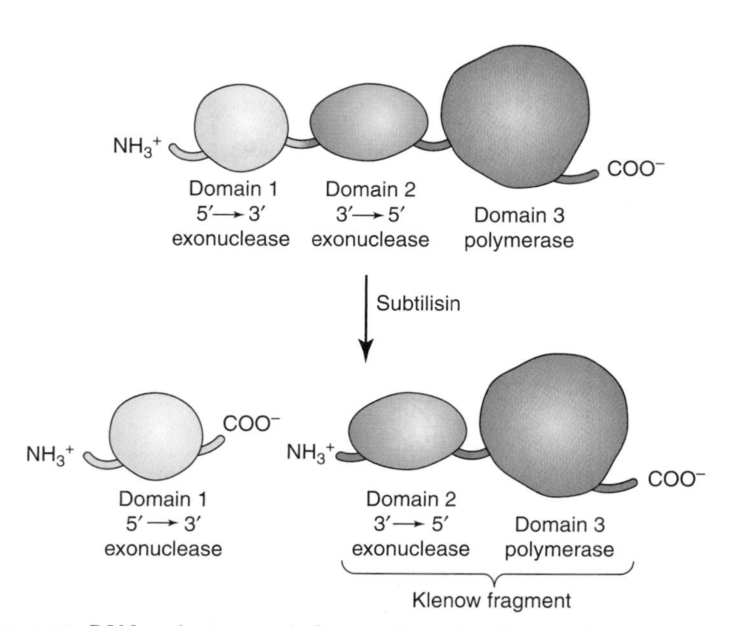

FIGURE 5.26 **DNA polymerase I.** Generation of a Klenow fragment.

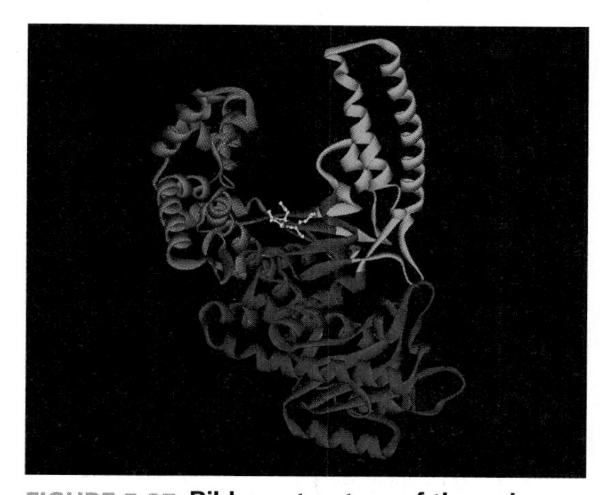

FIGURE 5.27 **Ribbon structure of the polymerase domain of the Klenow fragment.** The polymerase domain resembles a partially opened right hand. The subdomains corresponding to the thumb, fingers, and palm are colored green, purple, and red, respectively. The carboxylate triad, Asp-705, Asp-882, and Glu-883, at the polymerase active site is shown in yellow stick figure form. (Structure from Protein Data Bank 1KFD. L. S. Beese, J. M. Friedman, and T. A. Steitz, *Biochemistry* 32 [1993]: 14095–14101. Prepared by B. E. Tropp.)

catalytic site is located in the palm subdomain. As shown in the schematic diagram in FIGURE 5.28a, the thumb subdomain appears to contact the minor groove of the primer-template duplex while the finger subdomain binds the uncopied template strand. Three-acidic residues (Asp-705, Asp-882, and Glu-883) form a carboxylate triad that is essential for polymerase activity (shown in FIGURE 5.29 as carboxyl groups). Two magnesium ions appear to be integral parts of the catalytic site.

Kinetic studies suggest that the polymerase stalls when an incorrect nucleotide is attached to the growing end of the primer chain, allowing time for the 3′→5′ exonuclease to remove the incorrect nucleotide (FIGURE 5.28b). X-ray diffraction studies indicate that the polymerase and 3′→5′ catalytic sites are separated by about 3.5 nm. Proofreading can occur by an intra- or intermolecular mechanism. In the former case, the 3′-end moves from the polymerase catalytic site to the 3′→5′ exonuclease catalytic site whereas in the latter, enzyme-DNA complex dissociates and then a different DNA polymerase I removes the mispaired base. A zinc ion and a magnesium ion at the exonuclease catalytic site appear to position a water molecule for a nucleophilic attack on the 5′-phosphate of the mismatched nucleotide.

DNA polymerase I can synthesize DNA at a nick.

As shown in FIGURE 5.30, DNA polymerase I has the remarkable ability to synthesize DNA at a single-strand break (nick). This synthesis requires melting the DNA beyond the nick and progressive strand displacement of the 5′-end. When the 5′→3′ exonuclease activity is

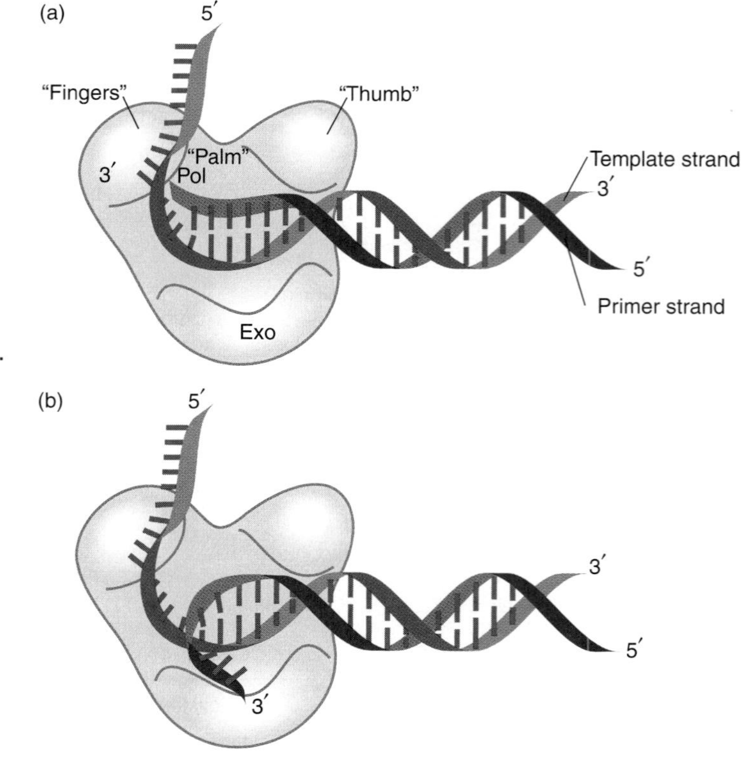

FIGURE 5.28 **Schematic of DNA polymerase I in the (a) polymerase mode and (b) proofreading mode.** (a) DNA polymerase I in the polymerase mode. The thumb subdomain contacts the minor groove of the template-primer duplex, the finger subdomain binds the uncopied template strand, and the 3′-OH end of the primer strand binds to the polymerase active site. (b) DNA polymerase I in the 3′→5′-exonuclease mode. The thumb subdomain still contacts the minor groove of the primer-template duplex and the finger subdomain still binds the uncopied template strand. However, the 3′-OH end of the primer strand with the misinserted nucleotide binds to the 3′→5′ exonuclease catalytic site active site. (Adapted from C. M. Joyce and T. A. Steitz, *J. Bacteriol.* 177 [1995]: 6321–6329.)

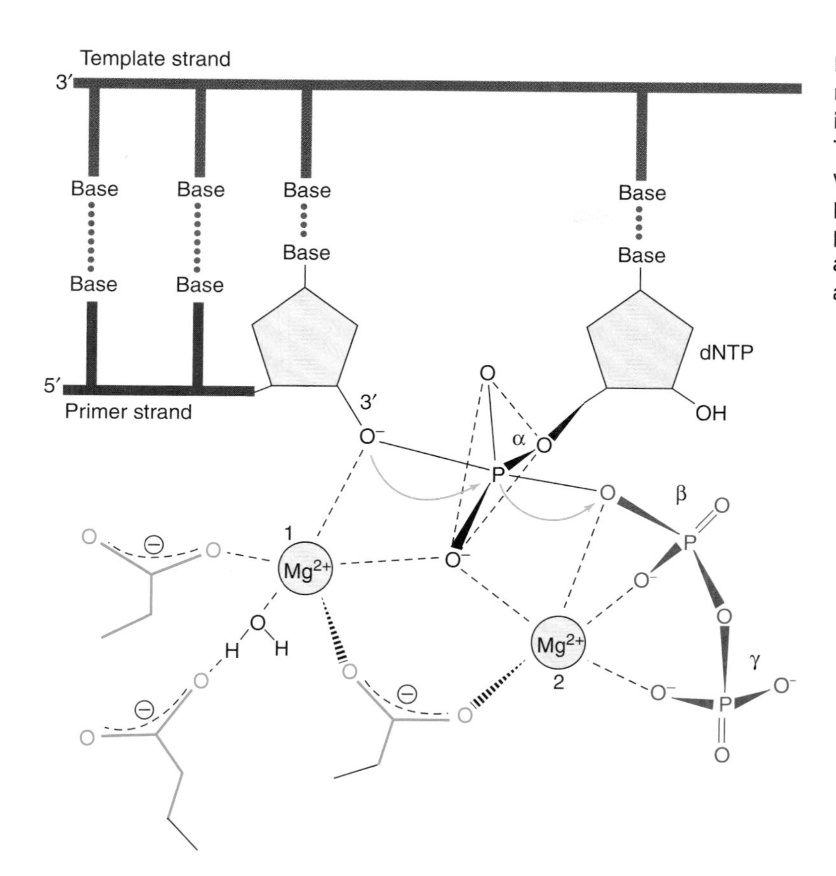

FIGURE 5.29 Intermediate (or transition state) in the mechanism proposed for the polymerase reaction, involving catalysis mediated by two magnesium ions. Three acidic residues (Asp-705, Asp-882, and Glu-883), which form a carboxylate triad that is essential for polymerase activity, are shown in green. The pyrophosphate leaving group is shown in red and mechanistic arrows are shown in blue. (Adapted from C. M. Joyce and T. A. Steitz, *J. Bacteriol.* 177 [1995]: 6321–6329.)

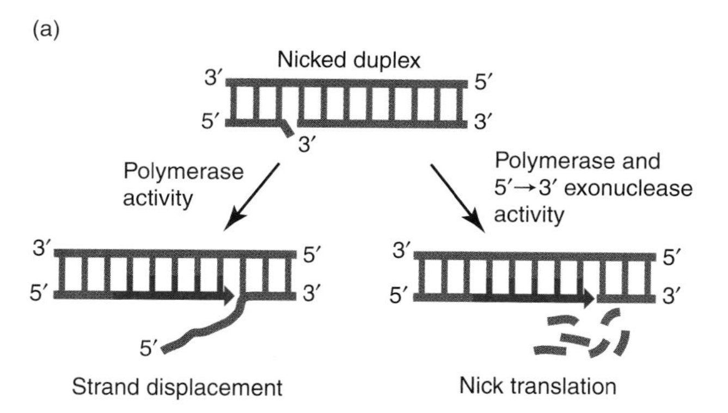

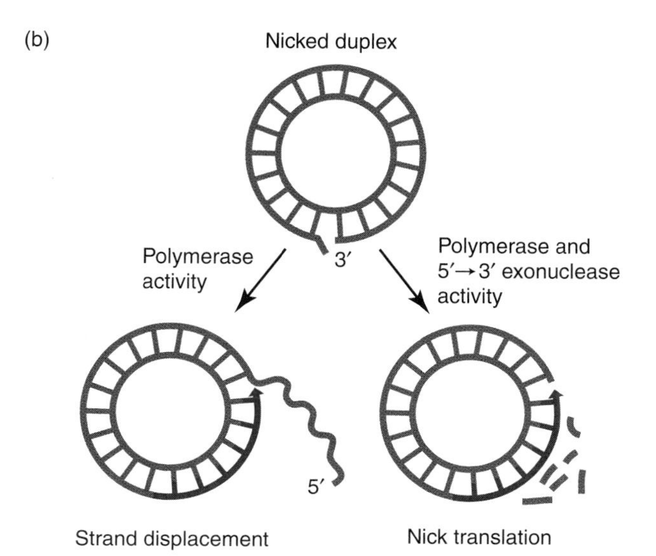

FIGURE 5.30 Strand displacement and nick translation on linear and circular molecules. In nick translation one nucleotide is removed by DNA polymerase I's 5′-exonuclease activity for each nucleotide added by the enzyme's polymerase activity. The growing strand is shown in red. (a) Strand displacement and nick translation on nicked linear DNA duplex. (b) Strand displacement and nick translation on nicked circular DNA duplex.

low or missing (as in the Klenow fragment), DNA synthesis proceeds with strand displacement. When 5′→3′ exonuclease activity is present, however, the 5′-end of the displaced strand is digested and the nick moves along the molecule in the direction of synthesis in a process known as **nick translation.** Using radioactive deoxyribonucleoside triphosphates during nick translation converts unlabeled DNA into radioactive DNA with the same nucleotide sequence. Hence, nick translation provides a convenient means for preparing radioactive DNA probes for Southern blotting. The same approach can be used to prepare DNA with fluorescent markers and other kinds of tags. For example, a modified dUTP in which a biotin group is attached to C-5 in the pyrimidine ring can be used in place of dTTP. Even though the biotin group replaces the methyl group normally present in thymine, DNA polymerase can still use the nucleotide analog as a substrate. Although nick translation allows us to synthesize DNA with a radioactive, fluorescent, or chemical tag, it does not lead to the net increase in total DNA that is present.

The polymerase chain reaction is used to amplify DNA.

It seems reasonable to suppose that DNA polymerase I should be able to catalyze the synthesis of many copies of a specific double-stranded DNA fragment. It was not until 1983, almost three decades after the discovery of DNA polymerase I, however, that a satisfactory method was finally devised to do so. The method, which was invented by Kary B. Mullis, called the **polymerase chain reaction** (**PCR**), is shown in FIGURE 5.31.

The linear duplex to be amplified is heat denatured. Then an oligonucleotide primer (about 20 to 30 nucleotides in length) is annealed to each of the denatured single strands. Each primer oligonucleotide sequence is selected so that it will anneal to a sequence in the outer region of the DNA segment to be amplified (the target sequence). The primers are oriented with their 3′-ends directed toward each other so that chain extension will copy the region between them. DNA polymerase I is then added to extend each primer until it reaches the end of the template. The new strands have defined 5′-ends (the 5′-ends of the oligonucleotide primers). This completes cycle 1 of PCR amplification. The same three-step process of (1) denaturing the linear duplex, (2) annealing the primer, and (3) extending DNA primer with DNA polymerase I is repeated in cycle 2 and each of the subsequent cycles. Repeated DNA polymerase I additions at each cycle can be avoided by using DNA polymerase isolated from *Thermus aquaticus*, a thermophilic bacterial strain. This polymerase, called Taq polymerase, retains its activity after the heat cycle used to denature DNA. Today, other thermostable polymerases such as Pfu polymerase from *Pyrococcus furiosus* and Vent polymerase from *Thermococcus litoralis* are often preferred because they have lower error rates than Taq polymerase. PCR machines known as **thermal cyclers**, which are programmed to shift their temperature up and down during different stages of the cycle, take advantage of thermostable DNA polymerase to allow PCR to be automated. In theory, amplification is exponential so that after n cycles the amplification yield would be 2^n. A 30-cycle amplification

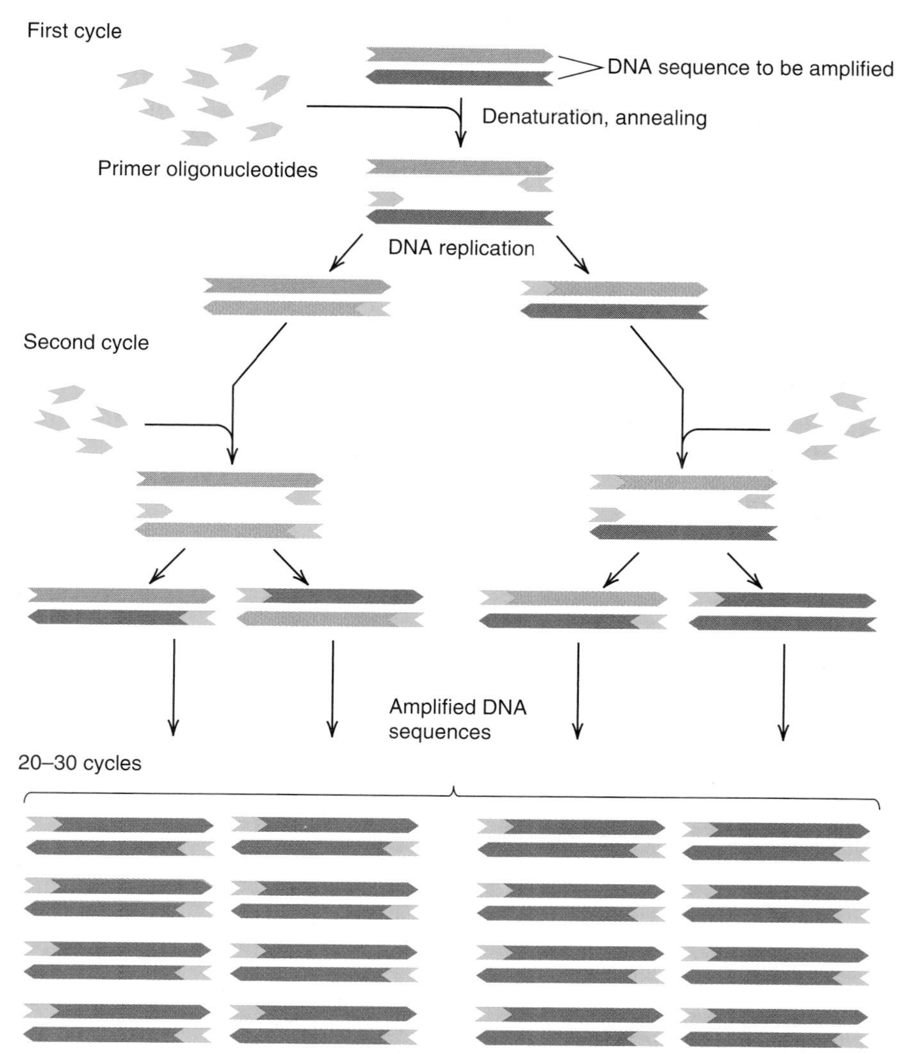

FIGURE 5.31 **Schematic for the polymerase chain reaction (PCR), a method for amplifying a specific target DNA sequence.** The linear duplex to be amplified (blue) is heat denatured. Then an oligonucleotide primer (green) is annealed to each of the denatured single strands. DNA polymerase is added to extend each primer. The new strands (red) have defined 5′-ends (the 5′-ends of the oligonucleotide primers). This completes cycle 1 of PCR amplification. The same three-step process of (1) denaturing the linear duplex, (2) annealing the primer, and (3) extending DNA primer with DNA polymerase I is repeated in cycle 2 and each of the subsequent cycles.

therefore would be expected to produce about 10^9 copies of the target segment. This calculation assumes that each cycle proceeds with a 100% efficiency of amplification. Because the efficiency of amplification is estimated to be 60% to 85%, actual amplification yields are lower than the calculated value. Nevertheless, extraordinary amplifications are possible in just a couple of hours because each cycle requires only 4 to 6 minutes. Among the many important practical applications of PCR are the following: (1) amplification of a segment of a large DNA molecule for subsequent use in genetic engineering; (2) rapid detection of pathogenic bacteria and viruses; (3) detection of inborn errors of metabolism; and the (4) detection of tumors. The popular press has described many situations in which PCR has been used in criminal investigations to amplify DNA from saliva, blood, sperm cells, or even, a single hair. The amplified DNA is then characterized by restriction endonuclease digestion followed by gel electrophoresis. Amplified DNA can also be sequenced (see below).

Site-directed mutagenesis can be used to introduce a specific base change within a gene.

Once a gene has been cloned and sequenced (see below) it is possible to introduce specific base changes anywhere within the gene. The technique for introducing the specific base changes, called **site-directed mutagenesis**, was first conceived by Michael Smith in the

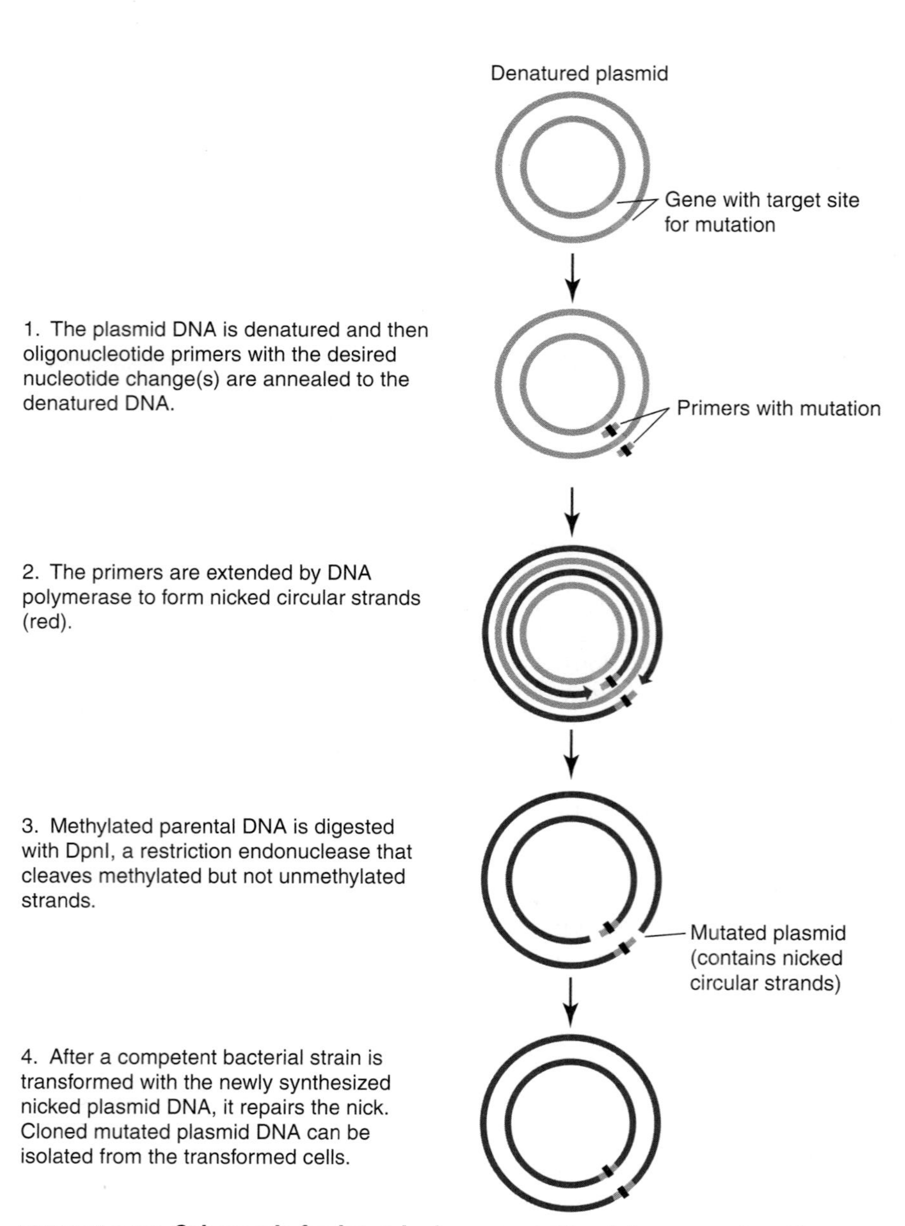

FIGURE 5.32 **Schematic for introducing a mutation into a gene by site-directed mutagenesis.**

mid-1970s, and is a very powerful tool for studying how specific amino acid changes influence protein function. A convenient method for performing site-directed mutagenesis for a gene that is present in a recombinant plasmid is shown in FIGURE 5.32. The recombinant plasmid is denatured by heating and a pair of oligonucleotide primers with the desired nucleotide change(s) are annealed to the complementary plasmid strands. Then the primers are extended by DNA polymerase to produce nicked circular strands. The denaturation, annealing, and primer extension steps are repeated several times in a thermal cycler to produce large quantities of nicked circles. Unlike the original plasmid DNA, which was methylated in the cell from which it had been isolated, the DNA strands in the nicked circles are unmethylated. This difference in methylation provides a simple means for removing all of the original plasmid DNA. The restriction endonuclease DpnI is specific for the following palindrome (arrows indicate cleavage sites).

```
    H3C
     |
5′—G A ↓ T C—3′
3′—C T ↑ A G—5′
         |
        CH3
```

DpnI has the unusual property of only cutting when the adenine groups are methylated. All of the original plasmid therefore can be removed by DpnI digestion. The undigested nicked circles bearing the mutation are introduced into competent bacteria, which repair the nick in the mutated plasmid and the cloned mutated plasmid can be isolated from the transformed bacterial strain.

5.7 DNA Sequence Determination

The Maxam-Gilbert method uses controlled chemical degradation to sequence DNA

Two approaches for sequencing DNA were introduced in 1977. One devised by Allan Maxam and Walter Gilbert relies on chemicals to degrade specific bases and the other devised by Frederick Sanger and coworkers relies on dideoxynucleotides to interrupt DNA synthesis. The Maxam-Gilbert controlled chemical degradation approach begins with a DNA fragment with a tag at its 5′-end. The tagged DNA is divided into four different test tubes. Each tube contains chemical agents that selectively attack one or two of the four bases in DNA. Strand cleavage takes place on the 5′-side of the attacked base. The chemical agents used in the Maxam-Gilbert procedure cleave DNA molecules at (G), (G+A), (C+T), and (C). A set of fragments of different sizes is generated by controlling the reaction so that only some of the strands are cleaved at any given base.

A simple example will help to illustrate how the Maxam-Gilbert method works. Suppose a DNA fragment has the sequence

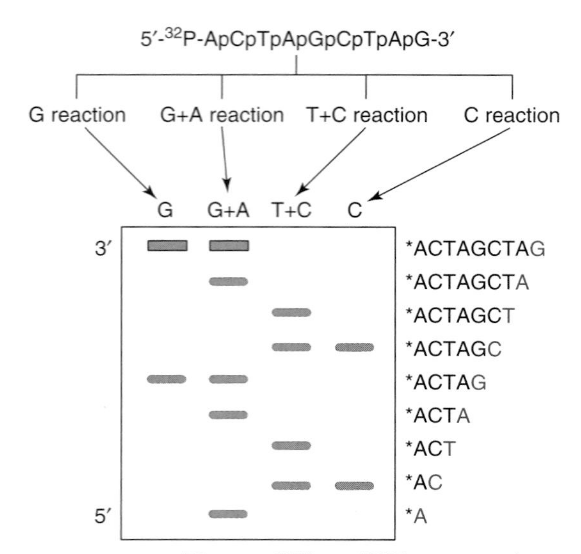

FIGURE 5.33 Maxam-Gilbert DNA sequencing method. Sequencing begins with a DNA fragment with a tag at its 5′-end. Tagged DNA is divided into four different test tubes. Chemical agents cleave DNA molecules at (G), (G+A), (C+T), and (C), generating a set of fragments of different sizes. Tagged fragments produced in each tube are separated according to size by polyacrylamide gel electrophoresis. Sequence information is read directly from the tagged bands (shown in red) in the gel.

5′-^{32}P-ACTAGCTAG-3′. The radioactive fragments produced in each of the tubes would be as follows:

Cleavage at G alone:	Cleavage at A + G:
^{32}P-ACTA	^{32}P-ACT
^{32}P-ACTAGCTA	^{32}P-ACTA
	^{32}P-ACTAGCT
	^{32}P-ACTAGCTA
Cleavage at C + T:	**Cleavage at C alone:**
^{32}P-A	^{32}P-A
^{32}P-AC	^{32}P-ACTAG
^{32}P-ACTAG	
^{32}P-ACTAGC	

The labeled fragments produced in each tube are then separated according to size by polyacrylamide gel electrophoresis. The radioactive label produces an image of the fragments on x-ray film. Sequence information is read directly from the gel (FIGURE 5.33).

The chain termination method for sequencing DNA uses dideoxynucleotides to interrupt DNA synthesis.

Frederick Sanger recognized that DNA polymerase's ability to add nucleotides to a primer under the direction of a template could be used to sequence DNA. The Sanger method uses small amounts of dideoxynucleoside triphosphates (ddNTP) to cause random termination of primer chain extension. Dideoxynucleoside triphosphates are deoxynucleoside triphosphate analogs that have a hydrogen atom attached to C-3′ in place of a hydroxyl group (FIGURE 5.34). Once DNA polymerase has transferred a dideoxynucleotide to the 3′-end of a growing primer chain, further chain extension is impossible because the primer chain must have a 3′-hydroxyl group for the next deoxynucleotide to be attached.

The major features of the chain termination method are summarized in FIGURE 5.35. Four sequencing reaction mixtures are prepared. The mixtures differ only in the dideoxynucleoside triphosphate that is added. Each kind of dideoxynucleoside triphosphate is labeled with a different color fluorescent dye. All other components including the single-stranded DNA molecule to be sequenced, DNA polymerase, the oligonucleotide primer, and the four deoxynucleoside triphosphates are the same in all four reaction mixtures. The primer oligonucleotide, which determines the particular region within the DNA molecule that will be sequenced, is usually synthesized by chemical methods. Primer chain extension continues in a 5′→3′ direction until the process is terminated by random dideoxynucleotide attachment. As a result of this random chain termination, each reaction mixture produces a nested population ladder of extended primer sequences, in which all extended chains end with the specific dideoxynucleotide that was added to that reaction mixture. Once synthesis is complete the newly synthesized DNA molecules are denatured and the contents of the four different tubes are combined so that the fragments can be separated by polyacrylamide gel electrophoresis or by capillary gel electrophoresis. After fragment separation is complete, sequence information can be read directly from the gel.

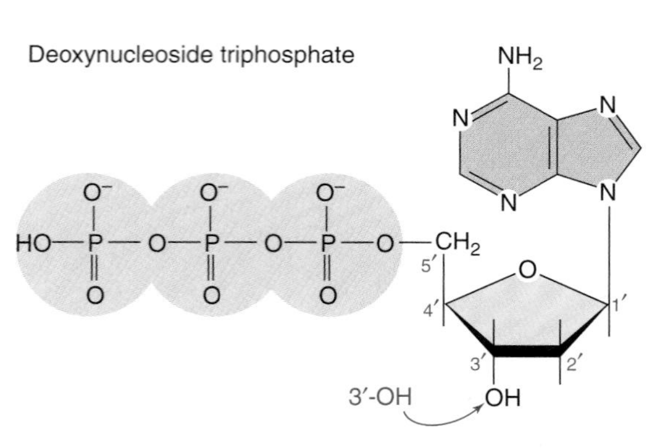

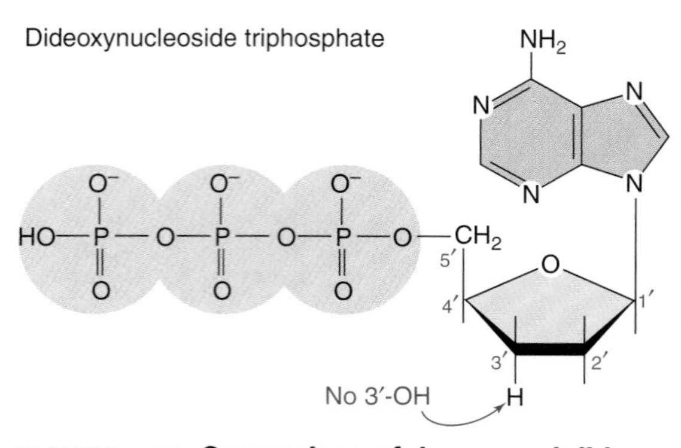

FIGURE 5.34 Comparison of deoxy- and dideoxynucleoside triphosphate structures.

A major advance in the dideoxynucleotide method, known as **dideoxy-terminator cycle sequencing,** uses a thermal cycler to amplify reaction products, allowing double-stranded DNA molecules to be sequenced. The reaction mixture, containing an oligonucleotide primer complementary to a segment of only one of the two strands, goes through the rounds of denaturation, annealing, and primer extension that were described for the polymerase chain reaction. Because the reaction mixture is heated during the denaturation part of the cycle, the DNA polymerase used must be thermostable. Cycle sequencing amplifies the production of chain extension products that

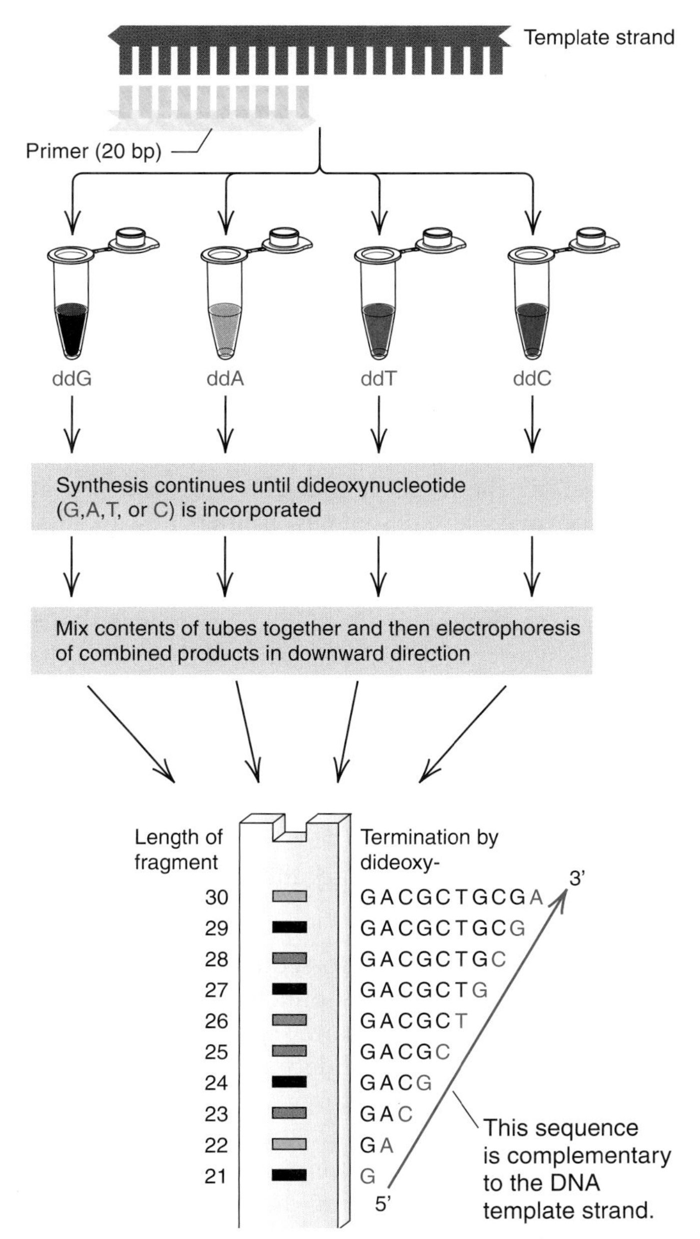

FIGURE 5.35 **Chain termination method for sequencing DNA.** Lengths of the terminated DNA fragments are shown at the left of the gel. The sequence of the daughter strand is read from the bottom of the gel according to the color of each band as 5′-GACGCTGCGA-3′.

are terminated by dideoxynucleotides labeled with fluorescent dyes. After the fragments have been separated by size using capillary electrophoresis, a scanner is used to detect the fragments and transfers the information to a computer that stores the nucleotide sequence data. Because the product is amplified, dideoxy-terminator cycle sequencing requires less DNA than the standard technique. Furthermore, both strands of a double-stranded DNA can be sequenced, diminishing the chance of an error.

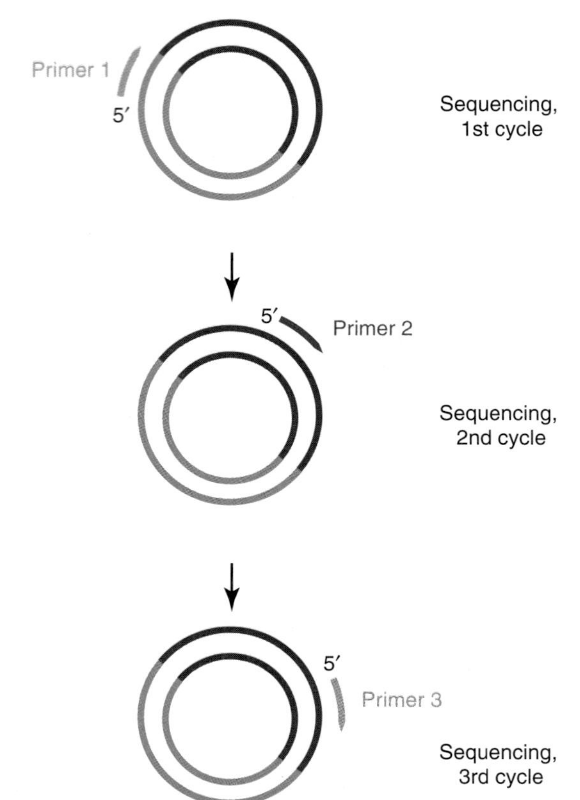

FIGURE 5.36 **Primer walking.** The DNA molecule to be sequenced (the target DNA shown in red) is inserted into a cloning vector (blue) with a known sequence at the insertion site. A primer specific to the known sequence (Primer 1) is annealed to the recombinant plasmid DNA and the chain termination method used to sequence the first 500–750 bp of target DNA. This sequence information is used to design Primer 2 that complements a region toward the end of the sequenced segment. Primer 2 is annealed to a new sample of recombinant plasmid DNA and the chain termination method is used to obtain sequence information further along the target DNA. The process is repeated, using new sequence information to design new primers to walk along the target DNA until overlapping sequences have been obtained for the entire DNA molecule. The complementary DNA strand can be sequenced in the same way to ensure accuracy.

DNA molecules that are 1 to 8 kbp long can be sequenced by primer walking.

The chain termination method can provide sequence reads that are about 500–750 bases long. Other approaches are required to sequence longer DNA molecules. One such approach, **primer walking** (FIGURE 5.36), can be used for DNA molecules that are about 1–8 kbp long. The DNA molecule to be sequenced, the target DNA, is inserted into a cloning vector with a known sequence at the insertion site. A primer specific to the known sequence is annealed to the recombinant plasmid DNA and the dideoxy-terminator cycle sequencing method used to sequence the first 500–750 bp of target DNA. This sequence information is used to design a new primer that complements a region toward the end of the sequenced segment. The new primer is annealed to a new sample of recombinant plasmid DNA and the dideoxy-terminator cycle sequencing method is used to obtain sequence information further along the target DNA. The process is repeated, using new sequence information to design new primers to walk along the target DNA until overlapping sequences have been obtained for the entire DNA molecule. Although a fairly straightforward procedure, primer walking does have the following disadvantages. Newly designed primers do not always work reliably. As the target DNA's length increases, so too does the possibility that a primer will bind to a sequence that is similar to the desired priming site, leading to mis-priming. Finally, the procedure is slow and expensive because a primer has to be synthesized after each sequencing step.

DNA sequences can be stitched together by using information obtained from a restriction map.

A DNA restriction map can help to sequence long DNA molecules. DNA is cut with the restriction endonuclease(s) used to produce the map. Then the resulting fragments are separated by gel electrophoresis and each is sequenced by the dideoxy-terminator cycle sequencing method. The restriction map provides the fragment order, which allows the sequences of the small fragments to be stitched together to provide the overall sequence. Although this approach seems fairly straightforward, it has the following drawbacks. The time required (and degree of difficulty) for restriction map construction increases with DNA length. Some restriction fragments are too long to permit sequencing without further cutting and others are so short that they are lost during purification by gel electrophoresis.

Shotgun sequencing is used to sequence long DNA molecules.

The preferred approach to sequence long DNA molecules is **shotgun sequencing**, a technique devised by Sanger and coworkers in the early 1980s. The approach begins with random DNA cleavage to produce small fragments. Cleavage can be conveniently achieved by subjecting DNA in solution to the shear forces generated by passing the solution through a narrow gauge syringe. Breakage sites are different for each DNA molecule, producing a heterogeneous population of small fragments. The term *shotgun* refers to the random shredding process. As a result of this random shredding, any given nucleotide in the original large DNA ends up in different small overlapping fragments. Many of the small fragments have 3′- or 5′-overhangs that must be converted to blunt ends before the next sequencing step is possible. This conversion is accomplished by incubating the DNA fragments with the four deoxyribonucleoside triphosphates and bacteriophage T4 DNA polymerase, which fills in 3′-recessed ends and removes protruding 3′-ends to produce blunt ended fragments (FIGURE 5.37). The reason for using T4 DNA polymerase rather than the Klenow fragment is the phage enzyme has a more active 3′→5′ exonuclease. Blunt end fragments are incubated with polynucleotide kinase and ATP to ensure that 5′-ends are phosphorylated (not shown) and inserted into a cloning vector with a universal primer sequence at its insertion site. Resulting recombinant plasmids are introduced into competent bacteria by transformation. Cloned DNA is isolated and the short fragments are sequenced using the universal primer. Sequencing reads (or reads) are stored in a computer that has a sequence assembly software program that searches for overlaps and uses the overlaps to assemble the reads into a long contiguous sequence or **contig.**

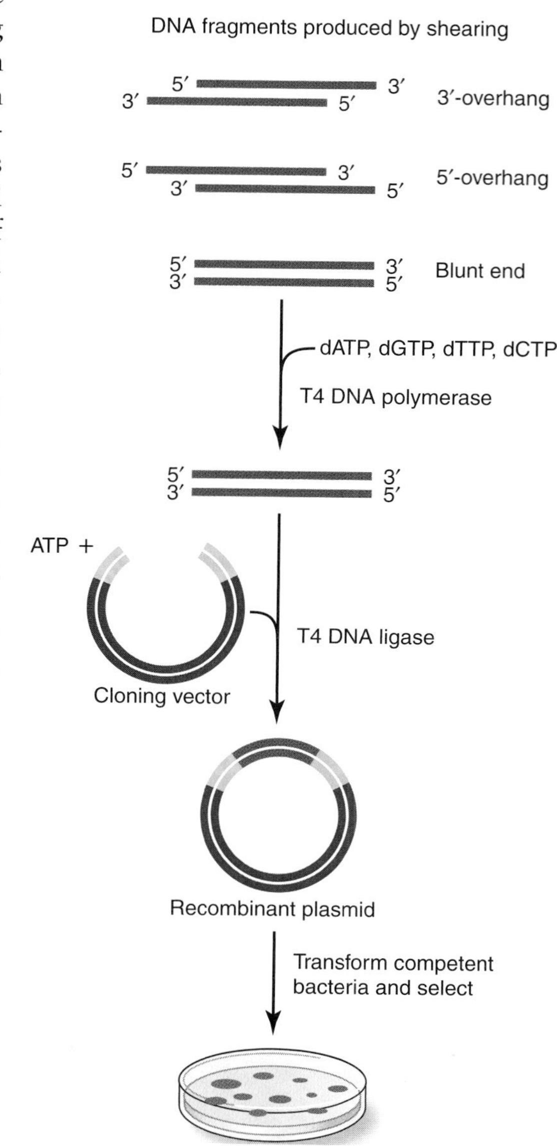

FIGURE 5.37 **Recombinant plasmid construction for shotgun sequencing.** Fragments formed by mechanical shearing (blue) are incubated with the four deoxyribonucleoside triphosphates and bacteriophage T4 DNA polymerase to fill 3-recessed ends and remove protruding 3-ends. Blunt end fragments are incubated with polynucleotide kinase and ATP to ensure that 5-ends are phosphorylated (not shown) and inserted into a cloning vector (green) with a universal primer sequence (red) at its insertion site. Resulting recombinant plasmids are introduced into competent bacteria by transformation.

In 1995, J. Craig Venter, Hamilton Smith, and coworkers used a variant of shotgun sequencing called **whole genome shotgun sequencing** to determine the complete 1.83×10^6 bp genome sequence of *Haemophilus influenzae Rd*, a nonpathogenic variant of a bacterial strain that can cause ear infections and meningitis. This was the first time that the genome of a free-living organism had been sequenced. As the term *whole genome shotgun sequencing* suggests, the whole bacterial genome is broken randomly into fragments that are cloned and sequenced. Approximately 20,000 clones were sequenced to ensure that every nucleotide in the genome was included in the final sequence. Based on the sequence information, the *H. influenzae Rd* genome is predicted to contain 1743 genes, Five years later, Venter and a very large group of coworkers used whole genome shotgun sequencing to determine the sequence of nearly 67% of the approximately 1.8×10^8 bp genome of the fruit fly *Drosophila melanogaster*, including all of the genetically active sequences.

The Human Genome Project used hierarchical shotgun assembly to sequence the human genome.

The possibility of sequencing the approximately 3 billion base pairs in a single set of human chromosomes (the haploid human genome) was first formally proposed at a meeting sponsored by the U. S. Department of Energy in 1985. The scientific community had a mixed reaction to the proposal. Some investigators argued that sequence information, which could only be obtained through a large scale effort, was essential for medical diagnosis and treatment. Others argued that the $3 billion needed for the project would be better spent if used to support investigator initiated projects designed to address specific medical problems. Investigators were also concerned that DNA techniques, sequencing methods, data handling, and data storage then available were not up to the task of sequencing the human genome. Those who supported the project thought that technological advances would produce the necessary tools and that the project should support efforts to make such advances. The Human Genome Project was officially launched in the United States in 1990 as a $3 billion, 15 year effort sponsored by the National Institutes of Health and Department of Energy. Participating laboratories in the United States, designated genome sequencing centers, were soon joined by sequencing centers in China, France, Germany, Japan, and the United Kingdom.

After careful consideration of various possible approaches, investigators participating in the Human Genome Project decided to use the **hierarchical shotgun sequencing** approach that is shown in FIGURE 5.38. Although this figure depicts an idealized scheme, it nevertheless provides a clear idea of the systematic clone-by-clone approach that was used. DNA, isolated from blood samples from a few anonymous donors, was partially digested with restriction endonucleases. Pieces between 100 kbp and 200 kbp in length were inserted into **bacterial artificial chromosomes** (**BAC**s), which are derived from the *E. coli* fertility (F factor) plasmid. We will examine the fertility plasmid in greater detail in Chapter 7. The important points for now are that BAC plasmids are stably maintained in *E. coli* at 1 or 2 copies per cell, accurately partition into daughter cells at each cell division, and are suitable for large scale DNA isolation. A recombinant BAC collection that includes the complete human genome is called a **BAC library** because recombinant BACs are like library books containing information that can be retrieved or duplicated. Bacterial clones bearing recombinant BACs were stored in freezers, making it possible to recover a particular recombinant BAC whenever it was needed.

Each BAC insert had to be assigned to a specific location in human genome. One way to accomplish this task is to match a unique landmark in the human genome with the same landmark in an insert. This approach required a high-density of evenly distributed landmarks. Although protein coding genes are useful landmarks, there are two few of them to order the BAC inserts. Another type of landmark was required. Recombinant DNA technology provided the solution to the problem. Any short piece of DNA with a unique sequence and a known chromosomal location can serve as a landmark. Thousands of such short (200–500 bp) unique sequences called **sequence tagged sites** (**STS**s) have been found and mapped. BAC clones with an STS of interest are detected by labor intensive PCR or hybridization techniques.

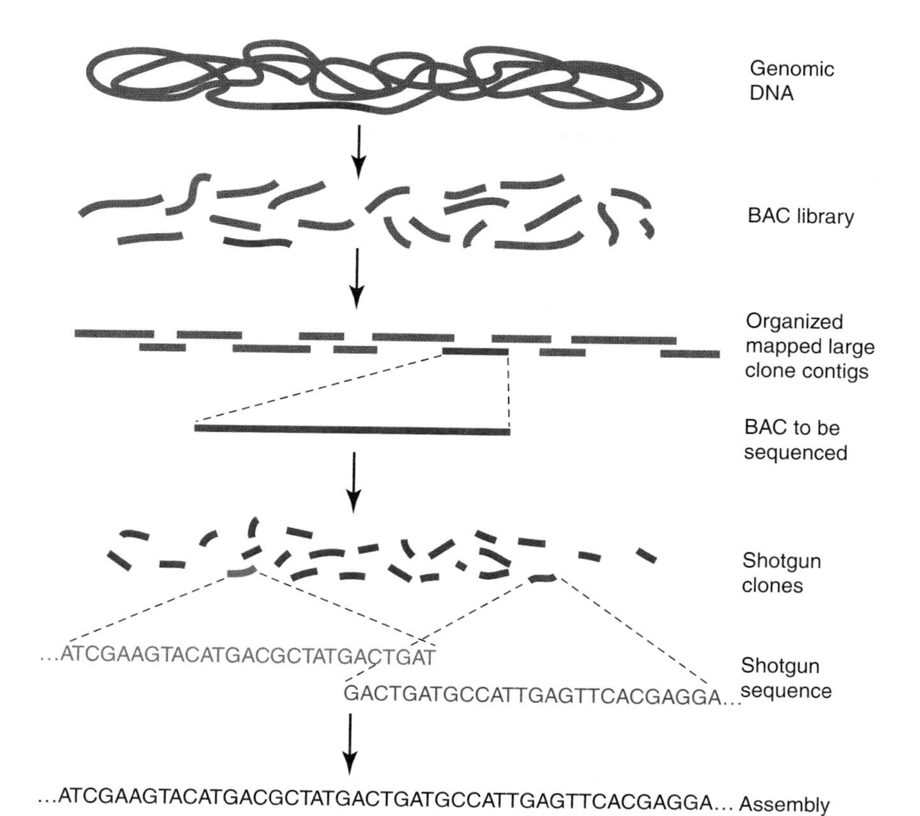

FIGURE 5.38 **Idealized representation of hierarchical shotgun sequencing.** DNA is partially digested with restriction endonucleases. Pieces between 100 kbp and 200 kbp are inserted into bacterial artificial chromosomes (BACs). Each BAC insert is assigned to a specific location in human genome. BAC clones that cover the entire genome with minimum overlap are selected for further analysis. DNA from each selected BAC clone is purified, mechanically sheared, and subcloned. The chain termination method is used to sequence DNA in each subclone. The overlap information from these sequences is used to assemble long contiguous sequences called contigs. (Adapted from International Human Genome Sequencing Consortium, *Nature* 409 [2001]: 860–921.)

BAC inserts were also ordered by **restriction fingerprint analysis** as follows. BAC inserts were digested with a restriction endonuclease and the resulting fragments separated by gel electrophoresis. If electrophoretic patterns ("fingerprints") produced by different BAC clones had bands in common, then the inserts were likely to contain overlapping fragments. Computer programs were designed to interpret the fingerprint data and order the BAC inserts.

BAC clones that covered the entire genome with minimum overlap were selected for further analysis. The set of clones that represents a whole chromosome with minimum possible overlap is called a **tiling path.** DNA from each selected BAC clone was purified, subjected to shotgun sequencing, and contigs assembled. As the project progressed, computing power improved, new sequence assembly software was written, and the entire process was automated. A single capillary DNA sequencer handled about a 4×10^5 bp in a day. By late 1999, the sequencing center at MIT, which was led by Eric Landers, sequenced and processed about 120,000 DNA fragments each day, producing sequence information for about 65 million nucleotides. Most routine work was done robotically. Sequences that were 2000 bases or longer were entered into a public data bank, accessible to all, within 24 hours of assembly.

Using the clone-by-clone approach the Human Genome Project completed initial sequencing in 2001. Further efforts closed gaps and improved accuracy. Approximately 2.85 billion of the 3 billion nucleotides in the human genome have been sequenced with an error rate that is estimated to be less than 1 bp in every 10,000 bp. The sequenced DNA includes virtually all active or potentially active genes. Work continues to fill the 300 or so remaining gaps in the human genome.

Whole genome shotgun sequencing has also been used to sequence the human genome.

At the start of the human genome project, investigators considered applying shotgun sequencing to the whole human genome without first constructing a BAC library. They decided that this approach known as **whole genome shotgun sequencing** would not work, however, because then-available computing power and software were not up to the task. They also thought that the human genome contains so much repetitive DNA that it would be nearly impossible to assemble contigs by searching for overlapping sequences.

In 1998, J. Craig Venter thought that it might be time to revisit the idea of applying the whole genome sequencing approach to the human genome. Several technological advances made whole shotgun sequencing appear to be a more realistic approach in 1998 than it had been in 1985. Venter's success in sequencing the *H. influenzae Rd* and *Drosophila* genomes showed that the approach worked.

Furthermore, several technological advances made the approach more attractive. Capillary DNA sequencers were much faster, sequencing operations were automated, computing power was much greater, and sequence assembly software was much better. There was also a strong financial incentive. Venter worked at a private company, Celera Genomics, which hoped to sequence important genes before the Human Genome Project was able to do so and then sell the proprietary information to drug and biotechnology companies.

The basic approach that Venter and coworkers applied to the human genome are shown in FIGURE 5.39. Purified genomic DNA was randomly sheared to produce fragments, which were subsequently separated into three size classes (2 kbp, 10 kbp, and 50 kbp) and inserted into plasmid vectors adjacent to a universal primer sequence. Universal primers were used to sequence the small inserts by chain termination sequencing. A new sequence assembly software program was created to analyze sequence reads, search for overlaps, and assemble contigs. Many gaps remained because the contigs could not be pieced together. Both ends of longer inserts were sequenced to provide information needed to bridge gaps and connect contigs. If one end of a long insert was part of one contig and the other end was part of another contig then the long insert must bridge the gap between the two contigs to

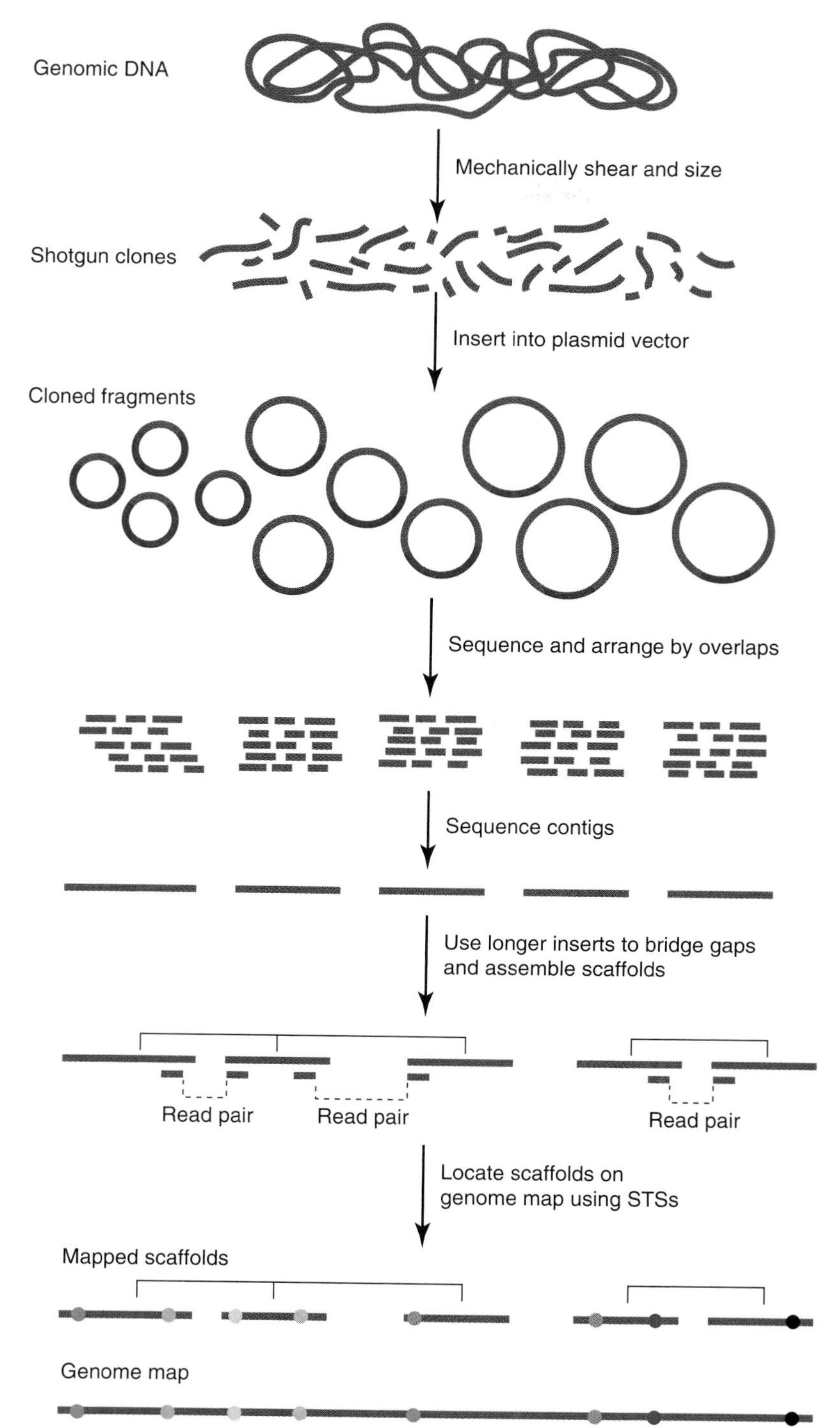

FIGURE 5.39 **Whole genome shotgun sequencing.** Purified genomic DNA is randomly sheared to produce fragments, which are subsequently separated into three size classes (2 kbp, 10 kbp, and 50 kbp) and inserted into plasmid vectors adjacent to a universal primer sequence. Universal primers are used to sequence the small inserts by chain termination sequencing. Overlapping segments are assembled into contigs. However, gaps remain because many contigs cannot be pieced together. Universal primers are used to sequence both ends of the long inserts. If one end of a long insert is part of a contig and the other end is part of another contig then the long insert must bridge the gap between the two contigs to form a supercontig or scaffold. The regions between the ends are shown as dashed lines. Sequence tagged sites (STSs), which are shown as different color circles, help to place the scaffolds on the genome map. (Adapted from E. D. Green, *Nat. Rev. Genet.* 2 [2001]: 573–583.)

form a **scaffold**. Thus, a scaffold is generated when a connection is made between two contigs that have no sequence overlap. Investigators experienced great difficulty dealing with the many long regions in the human genome that have repetitive DNA sequences. It is very hard to assemble such sequences using overlaps. The problem was particularly acute for the whole genome shotgun sequencing approach because it does not construct BAC clones that can be assigned to specific chromosomal regions. Therefore, Venter and coworkers were forced to use genetic and physical mapping information available from the Human Genome Project. The race to determine the human genome sequence ended in a tie with both investigative teams reporting their results in February 2001.

A controversy arose concerning whether whole genome shotgun sequencing could have determined the human genome sequence without the publically available information. It now seems clear that the most effective approach for sequencing large genomes such as those found in vertebrates is to use both hierarchical shotgun assembly and whole genome shotgun sequencing. The former provides a sequencing roadmap and the latter permits large amounts of sequence data to be rapidly collected and assembled. Genomes from many other organisms have been completed. At the time of this writing, approximately 100 eukaryotic, 900 bacterial, and 67 archaeal genomes have been sequenced and the number of sequenced genomes continues to grow at a rapid rate.

The human genome sequence provides considerable new information.

The human genome sequence has provided a great deal of important information including the following. (1) The human genome contains about 20,000 protein coding genes, a number that is much smaller than anticipated. (2) DNA that codes for the protein-coding genes accounts for only about 2% of the genome. (3) Gene density varies among chromosomes, with some chromosomes being more gene-rich than others. (4) Several thousand non-protein coding genes code for RNA molecules. (5) Gene-dense DNA regions within a chromosome usually have a high G-C content. (6) About half the protein-coding genes specify proteins of unknown function. (7) The human genome sequence is almost identical (99.9%) in all people. The finished human genome sequence serves as an invaluable reference that is now being used to learn how our genes work, influence metabolic processes, and are linked to diseases. Although the human genome sequence provides considerable valuable information, it is in some ways like a road map that does not show the names of the roads or the towns and cities that they pass through. The human genome requires **annotation**. That is, protein-coding genes and their regulatory signals need to be identified, introns and exons indicated, and regions that code for regulatory RNA molecules specified. The annotated human genome has the potential to provide molecular biologists with a factual base that is every bit as important as the periodic table is for chemists. We will return to the human genome sequence as we explore related topics throughout the remainder of the book.

A new generation of DNA sequencers provide rapid and accurate information without the need for electrophoresis or *in vivo* cloning.

The chain termination method has been the predominant DNA sequencing technique for the past 30 years. It is now being challenged by a new generation of sequencing techniques, which are faster and cost less to run because they don't require *in vivo* cloning or gel electrophoresis. The first new generation sequencer was introduced by 454 Life Science, a member of the Roche Group, in 2005 and two years later an improved version, the Genome Sequencer FLX™ (GS FLX), went on the market. This instrument performs about 400,000 parallel sequence reads with an average read length of 250 bp. It supplies about 100×10^6 bp of sequence information in 8 hours, enabling a single investigator to sequence an entire bacterial genome in a few days. Not surprisingly, many investigators have started to use this instrument. For that reason, and because the scientific principles behind the GS FLX are informative and interesting, we will examine its operation. GS FLX sequencing can be divided into the following four stages:

Stage 1—Create single-stranded DNA library (FIGURE 5.40). Purified genomic DNA is randomly broken by mechanical shearing. Fragments with chain lengths between 300 and 800 bp are collected and incubated with the four deoxyribonucleoside triphosphates in the presence of bacteriophage T4 DNA polymerase to convert overhangs to blunt ends. Short adaptors are attached to the blunt ends. Each adaptor has only one blunt end and so it can add in only one orientation. The adaptor's other end, a 5′-overhang, cannot attach to the blunt ended DNA fragment. Two kinds of adaptors are used (labeled A and B in Figure 5.40). B adaptors have biotin groups attached to their 5′-overhangs. Half of the recombinant DNA molecules produced have an A adaptor at one end and a B adaptor at the other. Twenty-five percent have A adaptors at both ends, and another 25% have B adaptors at both ends. The remaining steps in the first stage are designed to produce a DNA library of single-stranded template DNA molecules with each strand having an A adaptor at one end and a B adaptor at the other. This goal is achieved by mixing the recombinant DNA molecules with streptavidin-coated beads. Recombinant DNA molecules with a B adaptor at one or both ends bind to the beads very tightly. Those with A adaptors at both ends do not bind to the beads and are washed away. Washed beads are treated with a sodium hydroxide solution to denature bound double-stranded DNA. Single strands generated by denaturing recombinant DNA molecules that have B adaptors at both ends remain tightly bound to the streptavidin coated beads because each strand has a biotinylated 5′-end. The situation is different for recombinant DNA molecules with A and B adaptors at opposite ends. Biotin-free strands are released, forming the single-stranded DNA library used in the next sequencing stage.

Stage 2—Use emulsion PCR to produce millions of Sepharose® beads, each with its own unique clonally amplified DNA fragment

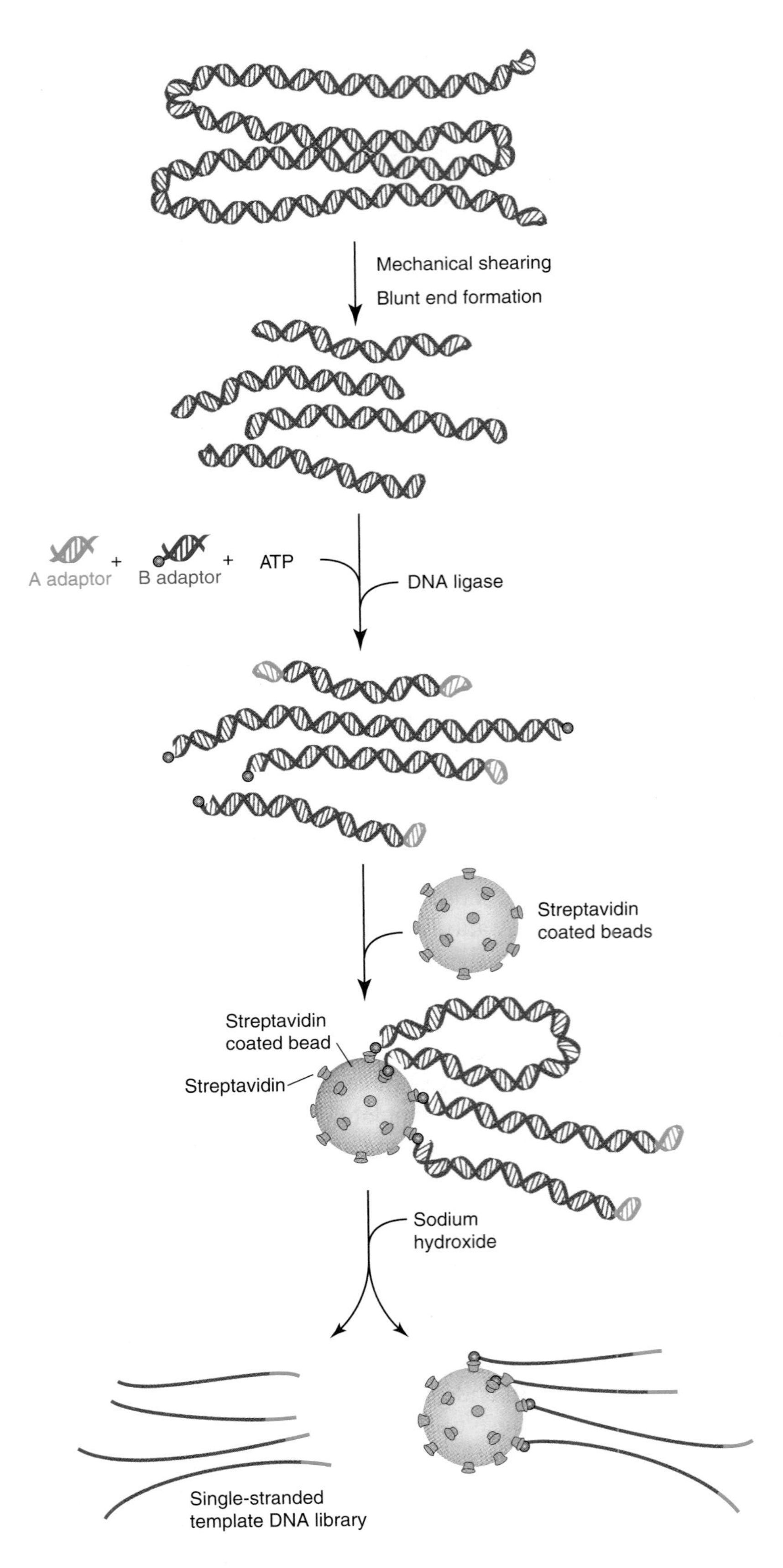

FIGURE 5.40 **Single-stranded DNA library construction.** Purified genomic DNA is randomly broken by mechanical shearing and fragments with chain lengths between 300–800 bp are collected, their ends made blunt, and short adaptors added to each blunt end. The recombinant DNA molecules are mixed with streptavidin-coated Sepharose® beads. Washed beads are treated with a sodium hydroxide solution to denature bound double-stranded DNA. Single strands generated by denaturing recombinant DNA molecules that have B adaptors at both ends remain tightly bound to the streptavidin coated beads because each strand has a biotinylated 5′-end. Biotin-free strands are released when recombinant DNA molecules with an A adaptor at one end and a B adaptor at the other end are denatured. The released strands form the single-stranded DNA library used in the next sequencing stage. (Adapted from Roche Diagnostics Corporation, *454 Sequencing: How Genome Sequencing is Done* [http://www.454.com/downloads/news-events/how-genome-sequencing-is-done_FINAL.pdf]. Accessed June 22, 2010.)

(FIGURE 5.41). This step requires specially designed Sepharose® beads called capture beads. Each capture bead has millions of copies of an oligonucleotide complementary to the B adaptor attached to it. The single-stranded DNA library is mixed with capture beads under conditions that are adjusted so that each capture bead anneals to only one DNA strand. Then a PCR reaction mixture and synthetic oil are added and the mixture is vigorously shaken. Shaking produces an emulsion in which a water droplet forms around each capture bead. The picoliter-size droplets function as reaction vessels. Each reaction vessel contains a single capture bead with an annealed DNA template strand, DNA polymerase, deoxynucleoside triphosphates, an oligonucleotide complementary to the A adaptor and another complementary to the B adaptor. DNA amplification in each droplet produces a capture bead with millions of identical double-stranded DNA molecules attached to it. That is, the DNA library strand annealed to the capture bead is clonally amplified. Because **emulsion PCR** takes place in parallel in millions of droplets, the emulsion mixture contains millions of capture beads, each with its own unique clonally amplified DNA. These beads are released by adding alcohol to disrupt the droplets and the beads are collected. After an enrichment procedure that is not shown in Figure 5.41, capture beads with clonally amplified DNA are incubated with NaOH to produce capture beads with clonally amplified single-stranded DNA attached to them. These beads are ready for the sequencing step.

Stage 3—Obtain DNA sequence reads. The single strands attached to the capture beads are annealed to an oligonucleotide with a sequence complementary to the A adaptor. Then a slurry containing the beads, DNA polymerase, and enzyme beads with bound ATP sulfurylase and luciferase (see below) is loaded onto a 7 cm × 7.5 cm fiber-optic slide that has 1.6 million wells (FIGURE 5.42a). Each well is so small (44 μm diameter) that only one Sepharose® bead (33 μm diameter) can fit inside. Enzyme beads are quite small and so many such beads can fit around the capture bead. After loading, the fiber-optic slide is placed in the GS FLX sequencer. Nucleotide solutions flow across the fiber-optic slide sequentially in a fixed order (FIGURE 5.42b).

The sequencing process is best understood by focusing on chemical events that take place in a single well. When the nucleotide that flows into a well is complementary to the next base in the template strand, the nucleotide will be added to the primer strand and pyrophosphate will be released. If the nucleotide is not complementary then no reaction will take place. Because the order in which nucleotides flow into a cell is known, sequencing can be performed by correlating pyrophosphate release with this known order of nucleotides that flow into the well. The problem of finding a quick and accurate way to monitor pyrophosphate release was solved by using three coupled enzyme reactions to

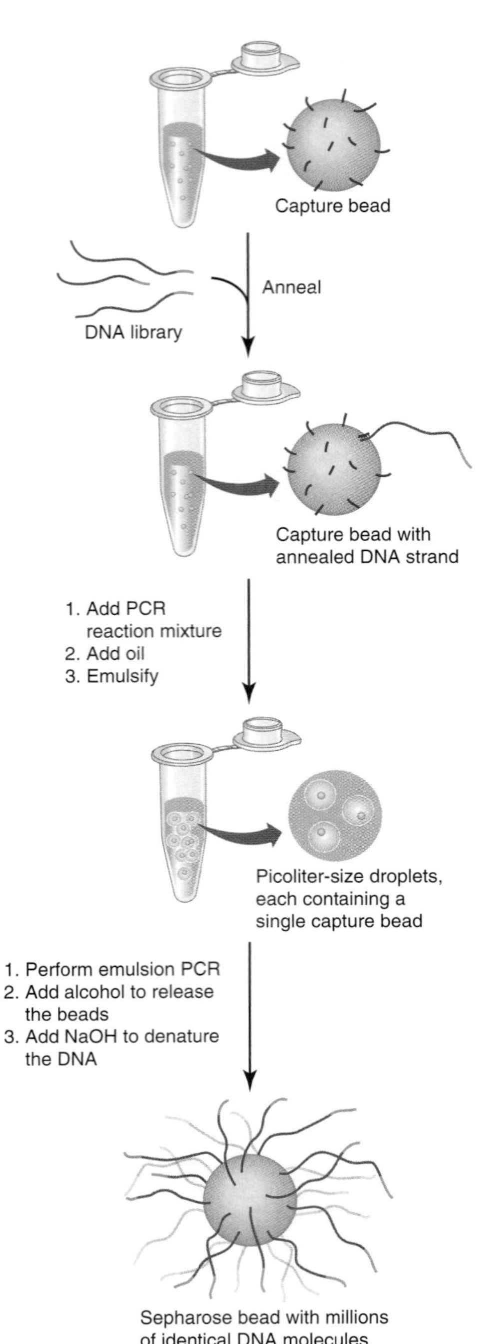

FIGURE 5.41 **Emulsion PCR.** The single-stranded DNA library is mixed with capture beads under conditions that are adjusted so that each capture bead anneals to only one DNA strand. After adding a PCR reaction mixture and synthetic oil, the mixture is vigorously shaken to produce an emulsion. Each picoliter-size droplet surrounding a bead functions as a PCR reaction vessel. DNA amplification produces a capture bead with millions of identical DNA molecules attached to it. Then the droplets are broken by adding alcohol to the mixture. Released capture beads are collected, enriched (not shown), and treated with a sodium hydroxide solution to denature the DNA. Each capture bead is now attached to millions of clonally amplified single-stranded DNA molecules. (Adapted from Roche Diagnostics Corporation, *454 Sequencing: How Genome Sequencing is Done* [http://www.454.com/downloads/news-events/how-genome-sequencing-is-done_FINAL.pdf]. Accessed June 22, 2010.)

monitor pyrophosphate (FIGURE 5.43). The three reactions are as follows: (1) DNA polymerase releases inorganic pyrophosphate as it adds each deoxynucleotide to the primer. (2) ATP sulfurylase (ATP:sulfate adenylyltransferase) catalyzes adenylyl group transfer from adenylyl sulfate to the released pyrophosphate to form ATP. (3) The firefly enzyme luciferase catalyzes an ATP-dependent chemiluminescent reaction in which luciferin conversion to oxyluciferin is accompanied by light emission. Although the enzyme assay is very fast, reproducible, and quite sensitive, it does have one drawback. dATP is a weak substrate for luciferase, causing background light emission. This difficulty is circumvented by replacing dATP with an analog that is a substrate for DNA polymerase but not for luciferase. The analog, dATPαS [deoxyadenosine 5′-(O-1-thiotriphosphate)], differs from dATP in only one atom; a nonbonding oxygen atom in the α-phosphate group is replaced by a sulfur atom (FIGURE 5.44). The light signal generated in a well each time a nucleotide adds to the primer chain is recorded by a CCD (charge-coupled device) camera. This sequencing technique, called **pyrosequencing**, has a single read accuracy that is greater than 99.5%. When errors do occur, they usually result from the presence of a homopolymer region in the template DNA. In a limited range (up to about 10 repeated nucleotides), the light signal strength is proportional to the number of nucleotides. The CCD camera can record light emission events as they occur in each of the 1.6 million wells of the picotiter plate and transmit the information to a computer for analysis.

Stage 4—Analyze data to assemble the DNA sequence. During the final stage, data generated by GS FLX sequencer is analyzed and used to assemble the DNA sequence. Data processing occurs

(a) Loading capture beads onto the fiber-optic slide

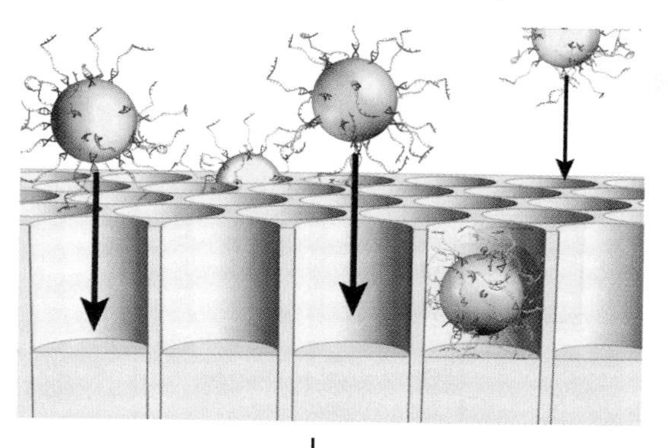

Load emzyme beads (ATP sulfurylase and luciferase)

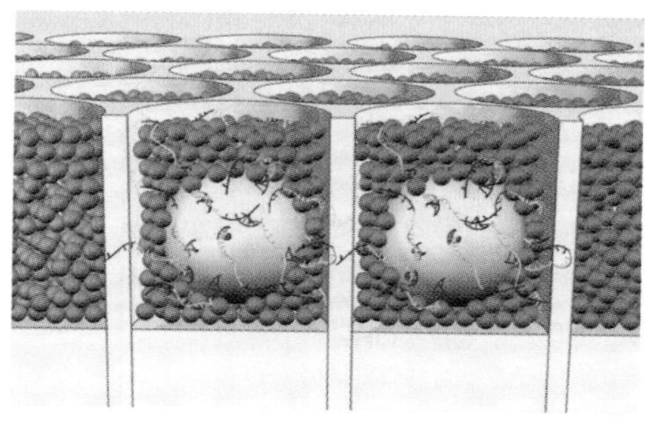

(b) Schematic showing the location of the fiber-optic slide in the GS FLX instrument

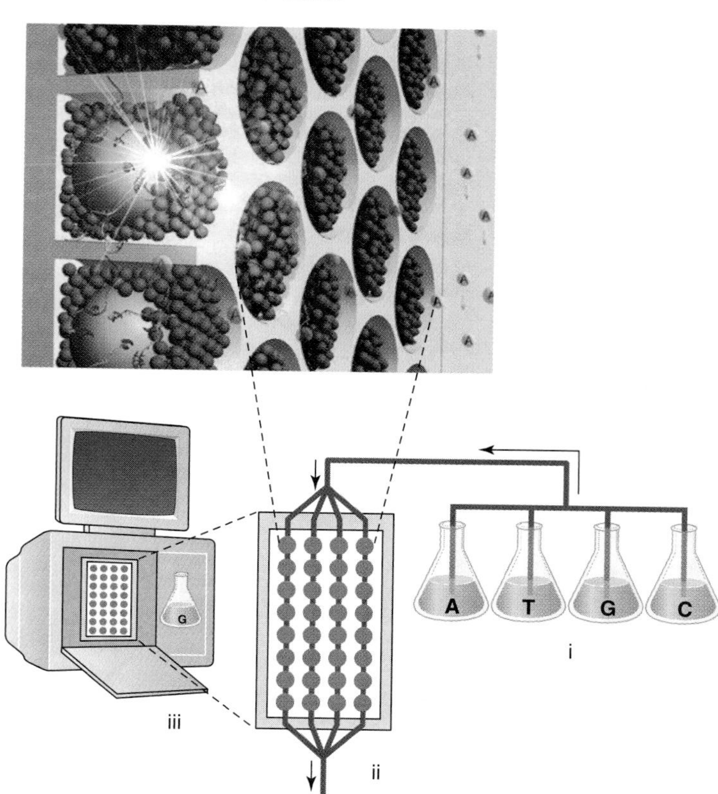

FIGURE 5.42 **The fiber-optic slide.** (a) Loading beads onto the fiber-optic slide. (b) Schematic showing the location of the fiber-optic slide in the GS FLX instrument. The GS FLX instrument has three major subsystems: (i) a flow cell that includes the fiber-optic slide with about one million wells, (ii) a CCD (charge-coupled device) camera to image the fiber-optic slide, and (iii) a computer containing proprietary software needed to collect and analyze the data and assemble the sequence. (Part a and part b [top panel] courtesy of 454 Sequencing. © 2010 Roche Diagnostics. Part b [bottom panel] adapted from 454 Life Sciences, *How Is Genome Sequencing Done?* Roche Diagnostics, 2010.)

during two phases, a runtime phase and a post runtime phase. During the runtime phase, the sequencer collects digital images, identifies the locations of wells in which sequencing reactions occur, and processes the signal information to generate a sequence read for each well. During the post runtime phase, the sequencer uses overlaps to assemble the sequence reads into contigs. When a run is performed on DNA from a species with a known DNA sequence, the sequence reads are mapped against the reference sequence. This process, called re-sequencing, permits investigators to determine the sequence of a single individual's DNA in a relatively short time. The first application of massively parallel DNA sequencing to human genome sequencing was completed in

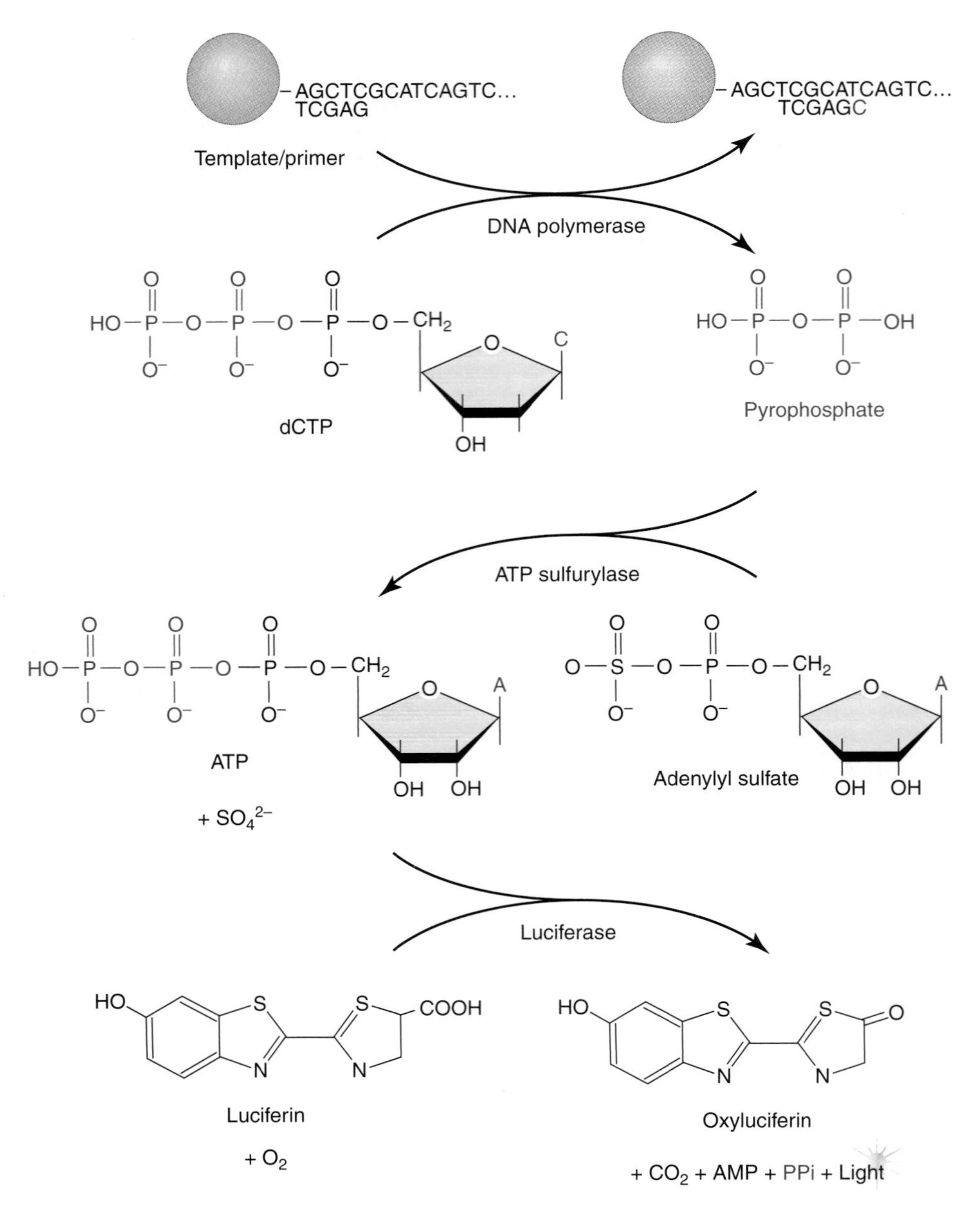

FIGURE 5.43 Pyrosequencing reactions.

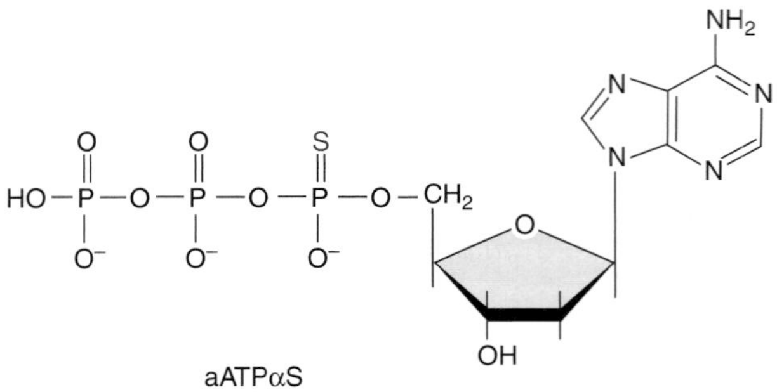

aATPαS

FIGURE 5.44 dATPαS deoxyadenosine 5′-(*O*-1-thiotriphosphate).

2008. Dr. James D. Watson, the co-discoverer of the double helical structure of DNA, provided a blood sample to a small team of investigators, who then used re-sequencing to determine his DNA sequence in less than four months at a cost of $1.5 million. New generation sequencing still has a way to go before it can be used in clinical laboratories. The race is now on to find a technique that will determine an individual's DNA sequence for under $1000 because DNA sequencing then might become a practical method for diagnosing disease and providing a rational basis for treatment. So-called "new-new generation" techniques offer some promise of reaching the $1000 goal.

In 2008, Stephan C. Schuster and coworkers sequenced nuclear DNA extracted from hair shafts collected from permafrost remains of the extinct woolly mammoth (*Mammuthus primigenius*). They used hair because it allowed them to remove bacteria and other contaminants without damaging the keratin-encased endogenous DNA. Based on C-14 dating the hair sample was calculated to be about 18,500 years old. Although fragmented and damaged, the ancient DNA could be sequenced using the GS FLX sequencer. Schuster and coworkers determined the sequence for 3.3 billion bases out of an estimated total of about 4 billion bases. Based on their data, they estimate that the extinct wooly mammoth and the African elephant are 99.78% identical at the amino-acid level. The estimated divergence rate between mammoth and African elephant is half of that between human and chimpanzee. The DNA of a second extinct species was sequenced at about the same time. Svante Pääbo and coworkers used the GS FLX sequencer to determine the complete sequence of mitochondrial DNA from a bone sample of a Neanderthal individual who lived about 38,000 years ago. Based on the sequence information, it appears that the divergence between modern humans and Neanderthals took place about 660,000 ± 140,000 years ago. Clearly, DNA sequencing's impact extends well beyond molecular biology. It is providing important new knowledge in all fields of biology and promises to continue doing so at an accelerating rate in the future.

Reverse transcriptase can use an RNA molecule as a template to synthesize DNA.

Many nucleic acid manipulations start with RNA rather than DNA. When this is the case, it is often convenient to make a DNA copy of the RNA molecule. Special enzymes, called **reverse transcriptases**, are required for RNA-directed DNA synthesis. Howard Temin and David Baltimore, working independently in 1970, discovered that these enzymes are present in a class of RNA viruses known as **retroviruses** (see Chapter 8). This class of viruses includes the HIV virus that causes AIDS in humans. Reverse transcriptases are not unique to retroviruses but in fact are present in other biological systems. The present discussion will be limited to examining reverse transcriptases as laboratory tools for studying nucleic acids.

Two kinds of reverse transcriptase, an enzyme prepared from avian myeloblastosis virus (AMV) and an enzyme from Moloney murine leukemia virus (Mo-MLV), are widely used as laboratory tools. Both of these enzymes, like DNA polymerase I, require a template-primer and catalyze deoxynucleotidyl group transfer from deoxynucleoside triphosphates to the 3′-end of the growing primer chain. The reverse transcriptases thus catalyze 5′→3′ chain extension to produce a new DNA strand complementary to the template strand. Although the similarities with DNA polymerase I are striking, important differences also exist. For instance, reverse transcriptases can use either an RNA or a DNA chain as a template, lack a 5′→3′ proofreading function, and have RNase H activities that digest RNA when it is present in either a DNA-RNA hybrid or a double-stranded RNA molecule. These properties make reverse transcriptase an ideal laboratory tool to copy an RNA template to produce either single- or double-stranded DNA. One of the most common applications is the synthesis of a DNA copy of mRNA, called **complementary DNA** or **cDNA** (FIGURE 5.45). Reverse transcriptase can be used to obtain cDNAs for the complete set of RNA molecules produced by the genome at any one time. This

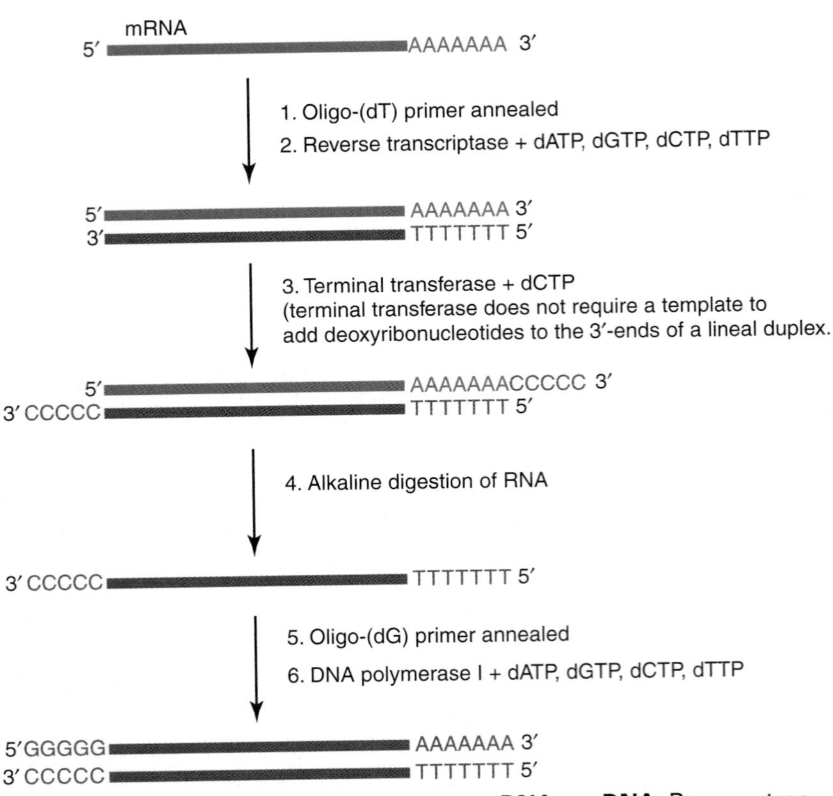

FIGURE 5.45 **Preparation of complementary DNA or cDNA.** Reverse transcriptase produces a complementary DNA strand (blue) to an mRNA molecule (green). The RNA can then be digested by alkaline hydrolysis as shown here or by RNase digestion and the resulting single-stranded DNA used as a template to make a complementary DNA (cDNA; also shown in blue). A primer is required for cDNA synthesis. In this example, terminal transferase is used to add oligo-(dC) to the 3′-end of the single stranded DNA so that it can anneal to an oligo-(dG) primer.

collection of transcripts is called the **transcriptome.** cDNA molecules prepared by using reverse transcriptase are readily sequenced by the methods described above.

DNA chips are used to follow mRNA synthesis, search for a specific DNA sequence, or to find a single nucleotide change in a DNA sequence.

A powerful technique known as **DNA chip** or **microarray technology** is currently being used to follow mRNA synthesis, search for DNA molecules with specific sequences, and detect DNA molecules with just one nucleotide change in their sequence. A DNA chip consists of a solid surface, usually glass, to which DNA fragments are attached. Copies of each kind of fragment are attached to the glass surface at a specific site to produce a regular pattern or array. DNA fragments attached to the chip act as probes that can hybridize with complementary DNA or RNA molecules (targets) in solution. Two different approaches are commonly used to construct a DNA chip (FIGURE 5.46). The first technique is a combination of oligonucleotide synthesis and

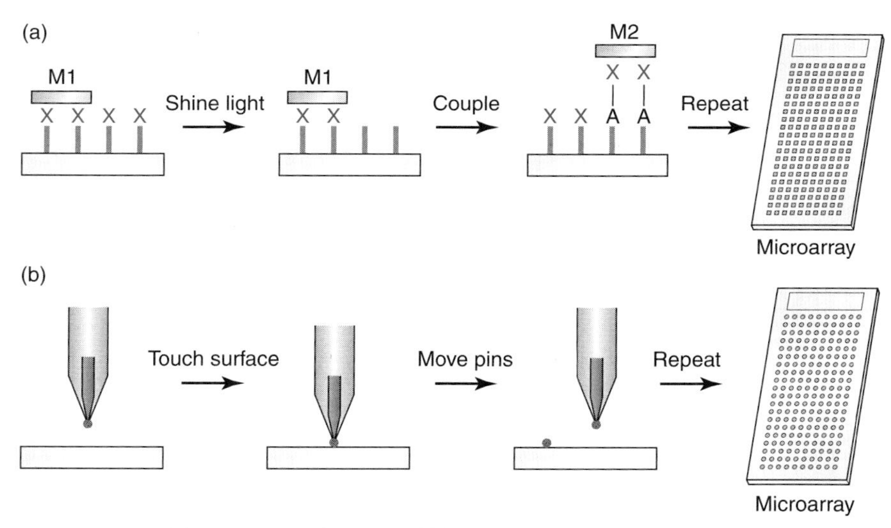

FIGURE 5.46 **Two approaches to DNA chip manufacturing.** (a) Photolithography. A glass chip, which is modified with photolabile protecting groups (X), is covered with a photomask (M1). Light that passes through specific openings in the photomask selectively activates exposed sites for DNA synthesis. Then the chip is flooded with a photoprotected DNA base (A–X), resulting in a spatially defined coupling on the surface of the glass chip. A second photomask (M2) replaces M1 and exposed sites are once again deprotected by shining light through the photomask. Repeated deprotection and coupling cycles allow high-density oligonucleotide chips to be prepared. (b) Mechanical microspotting. A biochemical sample is loaded onto a spotting pin by capillary action and the sample is transferred to the surface of the solid chip by physical contact with the pin. The pin is washed after the spotting cycle is complete. A second sample is transferred to the pin and transferred to an adjacent site on the chip. The process is automated by using robotic control systems and multiplexed print heads. (Reprinted from *Trends Biotech.*, vol. 16, M. Schena, et al., Microarrays: biotechnology's discovery platform . . ., pp. 301–306, copyright 1998, with permission from Elsevier [http://www.sciencedirect.com/science/journal/01677799].)

photolithographic technology (Figure 5.46a). The oligonucleotides are formed by using chemical reagents to attach one nucleotide at a time to a nucleotide that is attached to the glass surface. A key step in the synthetic process requires exposure to ultraviolet light before the next nucleotide can be attached. A mask that blocks exposure to ultraviolet light at a specific site will prevent the addition of a nucleotide during the next synthetic cycle. By using specific masks and adding nucleotide building blocks in a known order, the manufacturer can synthesize up to 400,000 different oligonucleotide sequences on a single glass chip with an area of 1.6 cm^2. The second method for constructing DNA chips deposits pre-formed DNA fragments as spots that are about 50 to 150 μm in diameter onto a coated glass surface (Figure 5.46b). The pre-formed DNA fragments are usually produced by the PCR technique or by using reverse transcriptase to make cDNA. Although it is much easier to construct DNA chips by this method, the chips produced have only about 10,000 different attached DNA fragments.

Of course, some method is required to detect DNA or RNA molecules that bind to the chips. The most popular method is to attach fluorescent probes to the target DNA or RNA molecules (FIGURE 5.47). Then fluorescence detection devices scan the chips to identify the sites where hybridization has occurred. Under optimal conditions, a single DNA chip can be used to detect up to 400,000 specific different DNA or RNA sequences in a mixture. This massive parallel processing, when coupled with improved methods of information handling and storage, offers great promise for discovering harmful changes in DNA leading to cancer and other genetic diseases.

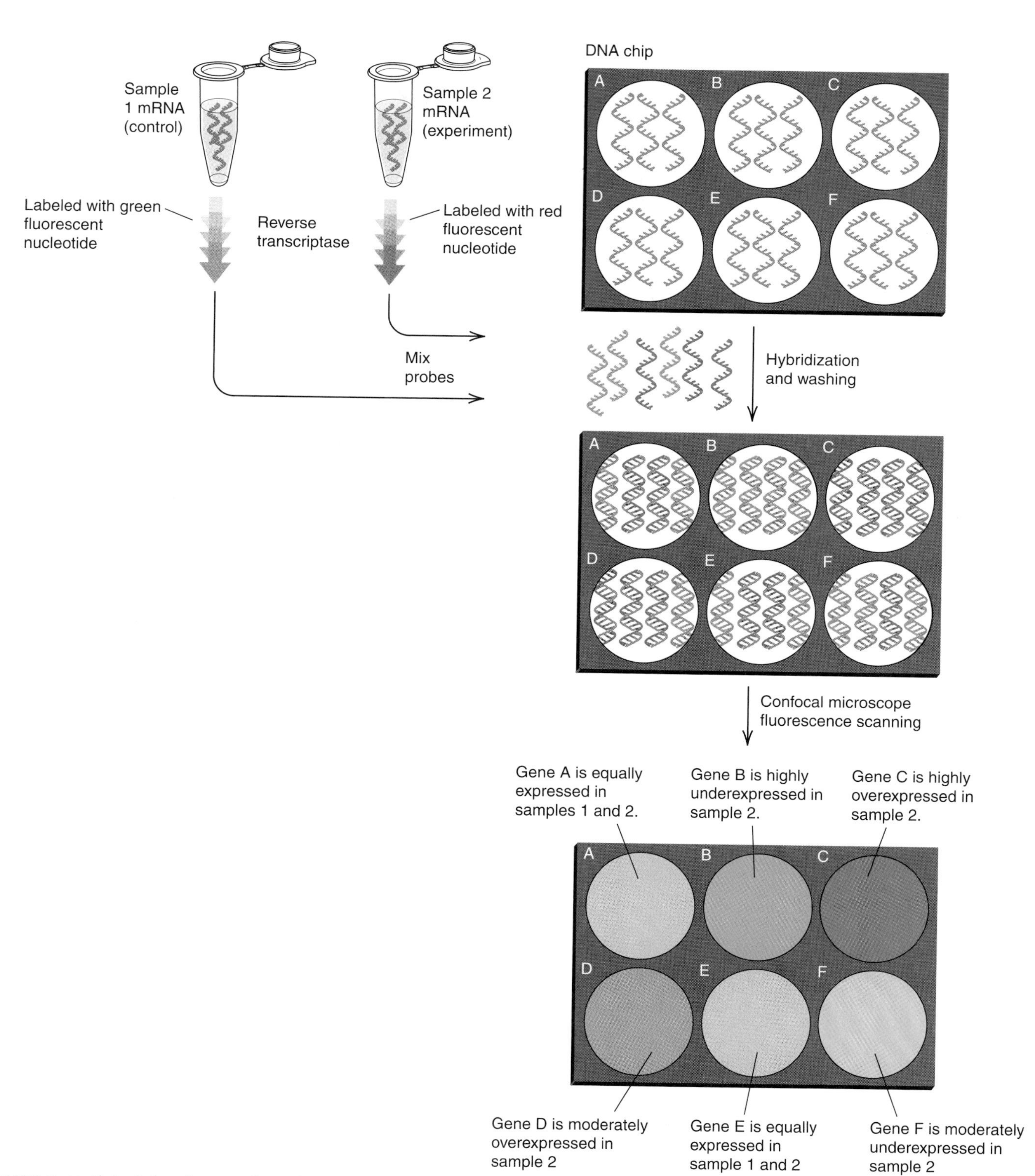

FIGURE 5.47 **Principle of operation of a mechanical microspotting DNA microarray.** At the upper right are dried microdrops, each of which contains immobilized DNA strands from a different gene (A–F). These are hybridized with a mixture of fluorescence-labeled DNA samples obtained by reverse transcription of cellular mRNA. Competitive hybridization of red (Sample 2—experimental) and green (Sample 1—control) label is proportional to the relative abundance of each mRNA species in the samples. The relative levels of red and green fluorescence of each spot are assayed by microscopic scanning and displayed as a single color. Red or orange indicates overexpression in the Sample 2 (experimental), green or yellow-green underexpression in the experimental sample, and yellow equal expression.

Suggested Reading

General

Brown, T. 1998. *Molecular Biology LabFax* (2nd ed). Frankfurt, Germany: Elsevier.

Karam, J. D., Chao, L., and Warr, G. W. 1990. *Methods in Nucleic Acid Research.* Boca Raton, FL: CRC Press.

Ream, W., and Field, K. G. 1999. *Molecular Biology Techniques: An Intensive Laboratory Course.* San Diego: Academic Press.

Surzycki, S. 2000. *Basic Techniques in Molecular Biology.* New York: Springer.

Thiel, T., Bissen, S. T., and Lyons, E. M. 2002. *Biotechnology DNA to Protein—A Laboratory Project in Molecular Biology.* New York: McGraw Hill.

Wilson, K., and Walker, J. (eds) 2000. *Principles and Techniques of Practical Biochemistry* (5th ed). Cambridge, UK: Cambridge University Press.

Nucleic Acid Isolation

Bowien, B., and Dürre, P. (eds). 2004. *Nucleic Acids Isolation Methods.* Stevenson Ranch, CA: American Scientific.

Chomczynski, P. 2004. Single-step method of total RNA isolation by guanidine-phenol extraction. In: *Encyclopedia of Life Sciences.* pp. 1–6. Hoboken, NJ: John Wiley and Sons.

Chomczynski, P. and Mackey, K. 2001. RNA: Methods for preparation. In: *Encyclopedia of Life Sciences.* pp. 1–2. Hoboken, NJ: John Wiley and Sons.

Farrell, R. E. Jr. 2009. *RNA Methodologies: A Laboratory Guide for Isolation and Characterization* (4th ed). San Diego: Academic Press.

Rapley, R. (ed). 2000. *The Nucleic Acids Protocol Book.* Totowa, NJ: Humana Press.

Roe, B. A., Crabtree, J. S., and Khan A., 1996. *DNA Isolation and Sequencing.* Hoboken, NJ: John Wiley & Sons.

Electron Microscopy

Bozzola, J. J., and Russell, L. D. 1998. *Electron Microscopy: Principles and Techniques for Biologists* (2nd ed). Sudbury, MA: Jones & Bartlett.

Braun, T., and Engel, A. 2005. Two-dimensional electron crystallography. In: *Encyclopedia of Life Sciences.* pp. 1–7. Hoboken, NJ: John Wiley and Sons.

Dubochet, J., and Stahlberg, H. 2001. Electron cryomicroscopy. In: *Encyclopedia of Life Sciences.* pp. 1–5. Hoboken, NJ: John Wiley and Sons.

Groscurth, P., and Ziegler, U. 2001. In: *Encyclopedia of Life Sciences.* pp. 1–5. London: Nature.

Hayat, M. A. 2002. *Principles and Techniques of Electron Microscopy: Biological Applications* (4th ed). Cambridge, UK: Cambridge University Press.

Ludtke, S. J., and Chiu, W. 2001. Electron cryomicroscopy and three-dimensional computer reconstruction of biological molecules. In: *Encyclopedia of Life Sciences.* pp. 1–5. Hoboken, NJ: John Wiley and Sons.

Maunsbach, A. B., and Afzelis, B. A. 1999. *Biomedical Electron Microscopy: Illustrated Methods and Interpretations.* San Diego: Academic Press.

McIntosh, J. R. 2001. Electron microscopy of cells: a new beginning for a new century. *J Cell Biol* 153:F25–F32.

Tao, Y., and Zhang, W. 2000. Recent developments in cryo-electron microscopy reconstruction of single particles. *Curr Opin Struct Biol* 10:616–622.

Unger, V. M. 2001. Electron cryomicroscopy methods. *Curr Opin Struct Biol* 11:548–554.

Centrifugal Techniques

Birnie, G. D., and Rickwood, D. (eds). 1978. *Centrifugal Separations in Molecular and Cell Biology.* Tolly, UK: Butterworths.

Graham, J. M., and Rickwood, D. 2001. *Biological Centrifugation.* London: BIOS Scientific Publishers.

Völkl, A. 2002. Ultracentrifugation. In: *Encyclopedia of Life Sciences*. pp. 1–7. London: Nature.
Price C. A. (ed.) 1982. *Centrifugation in Density Gradients*. San Diego: Academic Press.

Gel Electrophoresis

Birren, B., and Lai, E. 1993. *Pulsed Field Gel Electrophoresis: A Practical Guide*. San Diego: Academic Press.
Guttman, A. 2004. The evolution of capillary gel electrophoresis: from proteins to DNA sequencing. *LCGC N Am* 22:896–904.
Jones, P. 1996. *Gel Electrophoresis: Nucleic Acids: Essential Techniques*. New York: John Wiley & Sons.
Mitchelson, K. R., and Cheng, J. (eds). 2001. *Capillary Electrophoresis of Nucleic Acids* (vol 1): *Introduction to the Capillary Electrophoresis*. Totowa, NJ: Humana Press.
Perrett, D. 1999. Capillary electrophoresis in clinical chemistry. *Ann Clin Biochem* 36:133–150.
Righetti, P. G. 2005. Capillary electrophoresis. In: *Encyclopedia of Life Sciences*. pp. 1–4. Hoboken, NJ: John Wiley and Sons.
Southern, E. M. 2002. Pulsed-field gel electrophoresis of DNA. In: *Encyclopedia of Life Sciences*. pp. 1–7. Hoboken, NJ: John Wiley and Sons.
Westermeier, R. 2005. *Electrophoresis in Practice : A Guide to Methods and Applications of DNA and Protein Separations* (4th ed). New York: John Wiley & Sons.

Nucleases and Restriction Maps

Aggarwal, A. K. 1995. Structure and function of restriction endonucleases. *Curr Opin Struct Biol* 5:11–19.
Bickle, T. A. 1993. The ATP-dependent restriction enzymes. In: Linn S. M., Lloyd, R. S., and Roberts, R. J. (eds). *Nucleases*. Woodbury, NY: Cold Spring Harbor Laboratory Press, pp. 89–109.
Bickle, T. A., and Krüger, D. H. 1993. Biology of DNA restriction. *Microbiol Rev* 57:434–450.
Bjerrum, O. J., and Heegaard, N. H. H. 2009. Western blotting: immunoblotting. In: *Encyclopedia of Life Sciences*. pp. 1–8. Hoboken, NJ: John Wiley and Sons.
Chandrasegaran, S. 2001. Restriction enzymes. In: *Encyclopedia of Life Sciences*. pp. 1–7. London: Nature.
Gormley, N. A., Watson, M. A., and Halford, S. E. 2001. Bacterial modification-restriction systems. In: *Encyclopedia of Life Sciences*. pp. 1–11. London: Nature.
Mani, M., Kandavelou, K, and Chandrasegaran, S. 2007. Restriction enzymes. In: *Encyclopedia of Life Sciences*. pp. 1–8. Hoboken, NJ: John Wiley and Sons.
Nathans, D. 1992. Restriction endonucleases, simian virus 40, and the new genetics. In: Linsted, J. (ed). *Nobel lectures, Chemistry* 1971–1980. pp. 498–517. Hackensack, NJ: World Scientific Publishing.
Perona, J. J. 2002. Type II restriction endonucleases. *Methods* 28:353–364.
Pingoud, A., and Jeltsch, A. 1997. Recognition and cleavage of DNA by type-II restriction endonucleases. *Eur J Biochem* 246:1–22.
Roberts R. J., and Halford, S. E. 1993. Type II restriction enzymes. In: Linn S. M., Lloyd, R. S., and Roberts, R. J. (eds). *Nucleases*. pp. 35–88. Woodbury, NY: Cold Spring Harbor Laboratory Press.
Roberts, R. J., et al. 2003. A nomenclature for restriction enzymes, DNA methyltransferases, homing endonucleases and their genes. *Nucleic Acids Res.* 31:1805–1812.
Smith, H. O. 1992. Nucleotide sequence specificity of restriction endonucleases. In: Linsted, J. (ed). *Nobel lectures, Chemistry* 1971–1980. pp. 523–541. Hackensack, NJ: World Scientific Publishing.
Szybalski W, Kim S. C., Hasan, N., and Podhajska, A. J. 1991. Class-IIs restriction enzymes. *Gene* 100:13–26.
Tűmmler, B., and Mekus, F. 2005. Restriction mapping. In: *Encyclopedia of Life Sciences*. pp. 1–3. Hoboken, NJ: John Wiley and Sons.
Williams, R. J. 2003. Restriction endonucleases: classification, properties, and applications. *Mol Biotechnol* 23:225–243.

Recombinant DNA Technology

Bjerrum, O. J., and Heegaard, N. H. H. 2009. Western blotting: immunoblotting. In: *Encyclopedia of Life Sciences*. pp. 1–8. Hoboken, NJ: John Wiley and Sons.

Bowen, D. J. 2005. In Vitro Mutagenesis. In: *Encyclopedia of Life Sciences*. pp. 1–6. Hoboken, NJ: John Wiley and Sons.

Edelheit, O., Hanukoglu, A., and Hanukoglu, A. 2009. Simple and efficient site-directed mutagenesis using two single-primer reactions in parallel to generate mutants for protein structure-function studies. *BMC Biotechnol* 9:61.

Friedberg, E. C. 2006. The eureka enzyme: the discovery of DNA polymerase. *Nature Rev Mol Cell Biol.* 7:143–147.

Joyce, C. M. 2004. DNA polymerase I, bacterial. In: *Encyclopedia of Biological Chemistry* 1:720-725. New York: Academic Press.

Mullis, K. B. 1990. The unusual origin of the polymerase chain reaction. *Sci Am* 262:56–65.

Smith, M. 1997. Synthetic DNA and biology. In Malstrom, B. G. (ed). *Nobel lectures, Physiology or medicine* 1991–1995. Hackensack, NJ: World Scientific Publishing.

DNA Sequence Determination

Adams, M. D., et al. 2000. The genome sequence of *Drosophila melanogaster*. *Science* 287:2185–2195.

Ansorge, W. J. 2009. Next generation sequencing techniques. *New Biotechnol* 25:195–203.

Bentley, D. R., et al. 2008. Accurate whole human genome sequencing using reversible terminator chemistry. *Nature* 456:53–59.

Droege, M., and Hill, B. 2008. The Genome Sequencer FLX™ System—Longer reads, more applications, straight forward bioinformatics and more complete data sets. *J. Biotechnol.* 136:3–10.

Dunham, I. 2005. Genome sequencing. In: *Encyclopedia of Life Sciences*. pp. 1–6. Hoboken, NJ: John Wiley and Sons.

Fleischmann, R. D., et al. 1995. Whole-Genome random sequencing and assembly of *Haemophilus influenzae* Rd. *Science* 269:496–512.

França, L. T. C., Carrilho, E., and Kist, T. B. L. 2002. A review of DNA sequencing techniques. *Quart Rev Biophys* 35, 2:169–200.

Gilbert, W. 1992. DNA sequencing and gene structure. In: Linsted, J. (ed). *Nobel lectures, Chemistry* 1971–1980. pp. 408–426. Hackensack, NJ: World Scientific Publishing.

Graham, C. A., and Hill, A. J. M. (eds). 2001. *DNA Sequencing Protocols* (2nd ed). Totowa, NJ: Humana Press.

Green, E. D. 2001. Strategies for the systematic sequencing of complex genomes. *Nature Rev. Genetics* 2:573–583.

Green, R. E., et al. 2008. A complete Neandertal mitochondrial genome sequence determined by high-throughput sequencing. *Cell* 134:416–426.

Griffiths, A. D., and Tawfik, D. S. 2006. Miniaturising the laboratory in emulsion droplets. *Trends Biotechnol* 24:395–402.

Hall, N. 2007. Advanced sequencing technologies and their wider impact in microbiology. *J Exper Biol* 209:1518–1525.

Hardin, S. H. 2001. DNA sequencing. In: *Encyclopedia of Life Sciences*. pp. 1–6. London: Nature.

Hubbard, T. J. P. 2005. Human genome: draft sequence. In: *Encyclopedia of Life Sciences*. pp. 1–5. Hoboken, NJ: John Wiley and Sons.

Hutchison III, C. A. 2007. DNA sequencing: bench to bedside and beyond. *Nucleic Acids Res.* 35:6227–6237.

Kaiser, O., et al. 2003. Whole genome shotgun sequencing guided by bioinformatics pipelines—an optimized approach for an established technique. *J Biotechnol* 106:121–133.

Kehrer-Sawatzki, H. 2008. Sequencing the human genome: Novel insights into its structure and function. In: *Encyclopedia of Life Sciences*. pp. 1–9. Hoboken, NJ: John Wiley and Sons.

Lander, E. S., et al. 2001. Initial sequencing and analysis of the human genome. *Nature* 409:860–921.

Lockhart, D. J. 2005. DNA chip revolution. In: *Encyclopedia of Life Sciences*. pp. 1–6. Hoboken, NJ: John Wiley and Sons.

Mardis, E. R. 2008. Next-generation DNA sequencing methods. *Ann Rev Genomics Hum Genet* 9:387–402.

Margulies, M., et al. 2005. Genome sequencing in microfabricated high-density picolitre reactors. *Nature* 437:376–380.

Maxam, A. M., and Gilbert W. 1977. A new method for sequencing DNA. *Proc Natl Acad Sci USA* 74:560–564.

Meidanis, J. 2005. Sequence assembly. In: *Encyclopedia of Life Sciences*. pp. 1–4. Hoboken, NJ: John Wiley and Sons.

Meyers, B. C., Scalabrin, S., and Morgante, M. 2004. Mapping and sequencing complex genomes: Let's get physical! *Nature Rev Genet* 5:578–588.

Miller, W., et al. 2008. Sequencing the nuclear genome of the extinct woolly mammoth. *Nature* 456:387–390.

Mukhopadhyay, R. 2009. DNA sequencers: the next generation. *Anal Chem* 81:1736–1740.

Nyrén, P. 2007. The history of pyrosequencing®. *Methods Mol Biol* 373:1–13.

Osoegawa, K., et al. 1998. An improved approach for construction of bacterial artificial chromosome libraries. *Genomics* 52:1–8.

Osoegawa, K., et al. 2001. A bacterial artificial chromosome library for sequencing the complete human genome. *Genome Res* 11:483–496.

Pettersson, E., Lundeberg, J., and Ahmadian, A. 2009. Generations of sequencing technologies. *Genomics* 93:105–111.

Quail, M. A. 2005. DNA: mechanical breakage. In: *Encyclopedia of Life Sciences*. pp. 1–4. Hoboken, NJ: John Wiley and Sons.

Rothberg, J. M., and Leamon, J. H. 2008. The development and impact of 454 sequencing. *Nature Biotechnol* 26:1117–1124.

Sanger, F. 1992. Determination of nucleotide sequences in DNA. In: Linsted, J. (ed). *Nobel lectures, Chemistry* 1971–1980. pp. 431–447. Hackensack, NJ: World Scientific Publishing.

Sanger, F., Nicklen, S., and Coulson, A. R. 1977. DNA sequencing with chain-terminating inhibitors. *Proc Natl Acad Sci USA* 74:5463–5467.

Shendure, J., and Ji, H. 2008. Next-generation DNA sequencing. *Nature Biotechnol* 26:1135–1145.

Smith L. M., et al. 1986. Fluorescence detection in automated DNA sequence analysis. *Nature* 321:674–679.

Stratowa, C. 2007. Microarrays in disease diagnosis and prognosis. In: *Encyclopedia of Life Sciences*. pp. 1–6. Hoboken, NJ: John Wiley and Sons.

Sverdlov, E., and Azhikina, A. 2005. Primer walking. In: *Encyclopedia of Life Sciences*. pp. 1–3. Hoboken, NJ: John Wiley and Sons.

Teng, X., and Xiao, H. 2009. Perspectives of DNA microarray and next-generation DNA sequencing technologies. *Sci China C Life Sci* 52:7–16.

Venter, J. C., et al. 2001. The sequence of the human genome. *Science* 291:1204–1351.

Venter, J. C., et al. 2003. Massive parallelism, randomness, and genomic advances. *Nature Genet* 33:219–227.

Waterston, R. H., Lander, E. S., and Sulston, J. E. 2002. On the sequencing of the human genome. *Proc Natl Acad Sci USA* 99:3712–3716.

Weber, J. L., and Myers, E. W. 1997. Human whole-genome shotgun sequencing. *Genome Res* 7:401–409.

Wheeler, D. A., et al. 2008. The complete genome of an individual by massively parallel DNA sequencing. *Nature* 452:872–876.

Wilton, S. 2002. Dideoxy sequencing of DNA. In: *Encyclopedia of Life Sciences*. pp. 1–16. London: Nature.

This book has a Web site, **http://biology.jbpub.com/book/molecular**, which contains a variety of resources, including links to in-depth information on topics covered in this chapter, and review material designed to assist in your study of molecular biology.

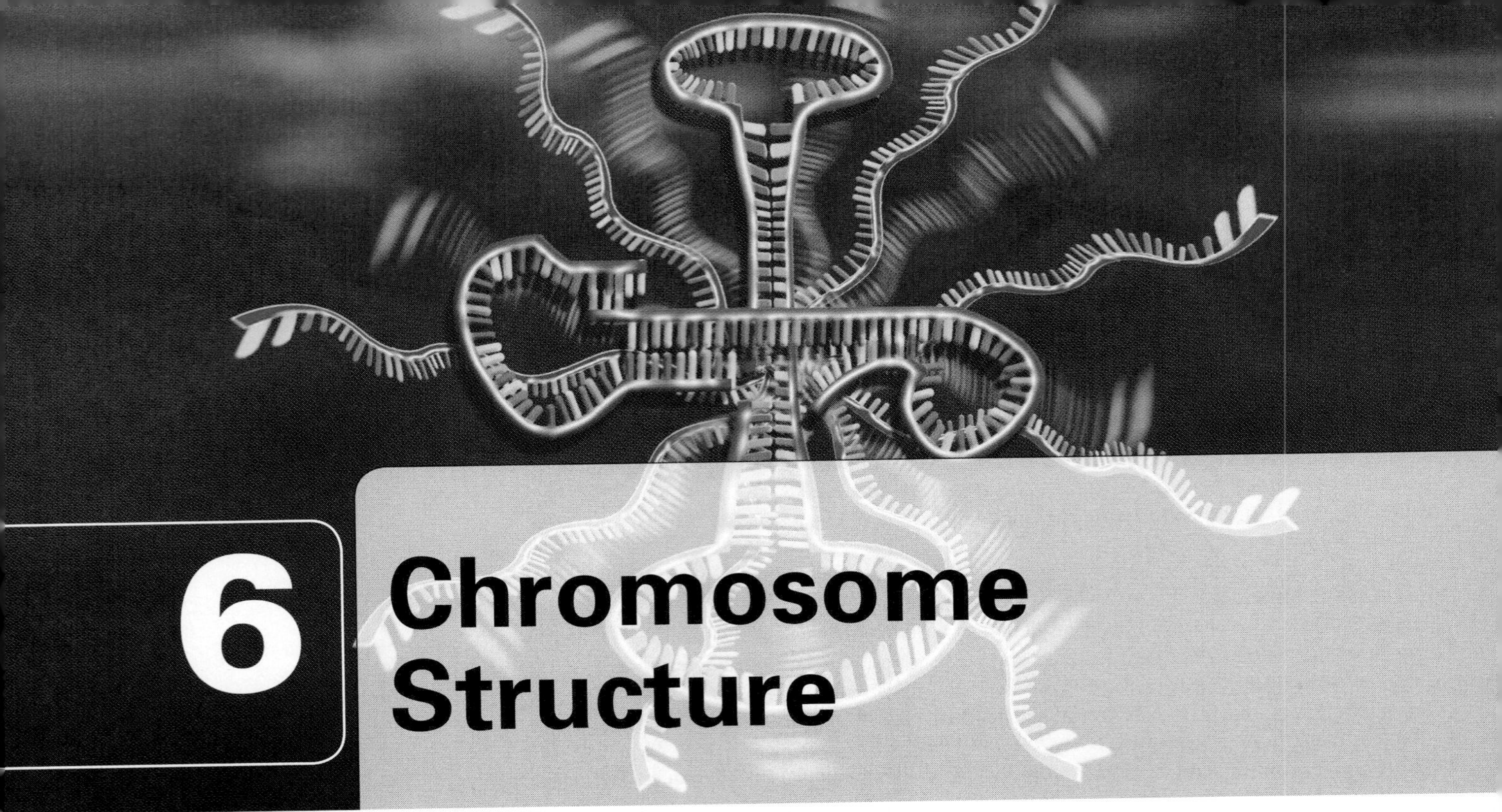

6 Chromosome Structure

OUTLINE OF TOPICS

6.1 Bacterial Chromatin

Bacterial DNA is located in the nucleoid.

MukB makes an important contribution to the compaction of the bacterial chromosome.

Additional nucleoid-associated proteins contribute to bacterial DNA compaction.

6.2 Mitosis and Meiosis

In higher animals, germ cells have a haploid number of chromosomes and somatic cells have a diploid number.

The animal cell life cycle alternates between interphase and mitosis.

Mitosis allows cells to maintain the chromosome number.

Cohesin is a four subunit complex that keeps sister chromatids together.

Meiosis reduces the chromosome number in half.

6.3 Karyotype

Chromosome sites are specified according to nomenclature conventions.

A karyotype shows an individual cell's metaphase chromosomes arranged in pairs and sorted by size.

A great deal of information can be obtained by examining karyotype preparations.

Fluorescent *in situ* hybridization (FISH) provides a great deal of information about chromosomes.

6.4 The Nucleosome

Five major histone classes interact with DNA in eukaryotic chromatin.

The first level of chromatin organization is the nucleosome.

X-ray crystallography provides high-resolution images of nucleosome core particles.

The precise nature of the interaction between H1 and the core particle is not known.

6.5 The 30-nm Fiber

A chain of nucleosomes appears to fold into a 30-nm fiber.

6.6 The Scaffold Model

The scaffold model was proposed to explain higher order chromatin structure.

6.7 The Centromere

The centromere is the site of microtubule attachment.

Condensins and topoisomerase II help to stabilize condensed chromosomes.

6.8 The Telomere

The telomere, which is present at either end of a chromosome, is needed for stability.

Suggested Reading

DNA molecules are extremely long in comparison with the size of living cells and so must be compacted to fit the available space inside a cell. Specific proteins interact with DNA, leading to the formation of a condensed nucleoprotein complex called **chromatin.** Until the mid-1970s, biologists' view of chromatin was colored by the belief that all life on Earth belonged to one of two primary lineages, the **eukaryotes** (animals, plants, fungi that have a defined nucleus) and the **prokaryotes** (all remaining microscopic organisms that lack a defined cell nucleus). Based on this classification scheme and what was then known about chromatin structure, it seemed likely that prokaryotes would have one type of chromatin structure and eukaryotes would have another.

Then in 1977, Carl Woese proposed that the prokaryotes actually contain two types of organisms, the **bacteria** and the **archaea.** Despite the fact that bacteria and archaea both lack defined nuclei, archaeal and bacterial metabolic machinery differ from one another as much as either differs from eukaryotic metabolic machinery. With the identification of the archaea, the living world was divided into three domains, the bacteria, the archaea, and the eukaryota.

Once the existence of the three domains was established, it seemed possible that each domain would have a characteristic chromatin structure. Experimental studies do not support this possibility, however. As the following examples illustrate, differences in the chromatin structures of organisms within a single domain are nearly as great as those among organisms belonging to different domains.

Most, but not all, eukaryotes contain chromatin formed by interactions between DNA and a family of basic proteins called **histones.** Dinoflagellates, a very large and diverse group of eukaryotic algae, lack histones completely.

Variations in chromatin structure also exist in bacteria and archaea. DNA molecules in most bacterial species that have been studied to date have circular structures like those of the DNA molecules in *Escherichia coli.* DNA molecules in some bacteria such as *Agrobacterium tumefaciens* (a species that infects plants) and *Streptomyces* species, however, are linear duplexes and in this respect resemble eukaryotic DNA. The archaea exhibit, if anything, even greater variations in chromatin structure than do organisms belonging to the other two domains. Some archaea appear to form nucleoprotein complexes that resemble those present in eukaryotes, whereas others seem to form nucleoprotein complexes that resemble those in bacteria.

As this brief overview shows, it is not possible to examine the chromatin structure from a single kind of organism and expect the information will apply to all organisms within the same domain. We can certainly obtain valuable information about chromatin structure in related organisms, however, and hope that this information will eventually permit us to obtain a coherent picture. With this thought in mind, we begin by examining bacterial chromatin structure and then examine chromatin structure in higher animals and plants.

6.1 Bacterial Chromatin

Bacterial DNA is located in the nucleoid.

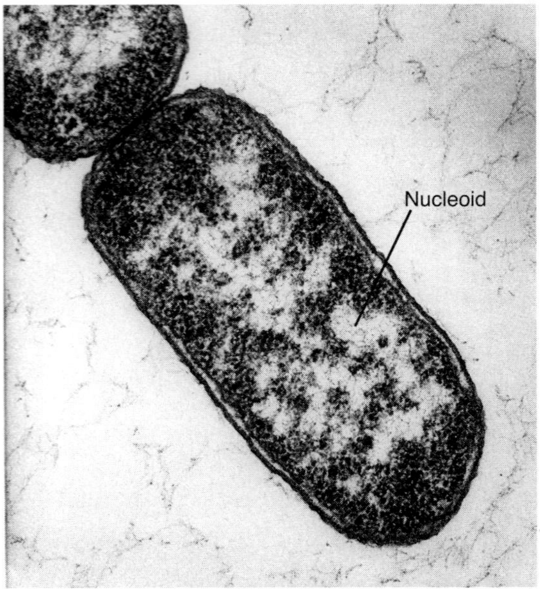

E. coli DNA, which appears as a closed covalent circle with a total length of about 1500 μm, must fit into a cylindrical cell with a diameter of about 0.5 μm and a length of about 1 μm. The intracellular DNA is therefore about 1000-fold more compact than the free DNA. Specific proteins interact with the bacterial DNA to form a highly condensed nucleoprotein complex called the **nucleoid** that occupies about a quarter of the cell's volume (FIGURE 6.1).

FIGURE 6.1 **An electron micrograph of a thin-section of *Escherichia coli*.** The nucleoid is the light region. (Photo courtesy of the Molecular and Cell Biology Instructional Laboratory Program, University of California, Berkeley.)

Bacterial chromatin can be released from the cell by a gentle cell **lysis** (disruption) technique that avoids DNA breakage or protein denaturation. The released DNA contains a fixed amount of protein and a variable amount of RNA. Most of the RNA is probably nascent RNA (RNA caught in the process of being synthesized) rather than an integral part of the *E. coli* chromatin.

An electron micrograph of released *E. coli* chromatin reveals multiple loops emerging from a central region with some loops supercoiled and others relaxed (FIGURE 6.2). Relaxed loops were probably formed as a result of a nick introduced into supercoiled loops by a cellular DNase during the isolation procedure. The fact that supercoiled and

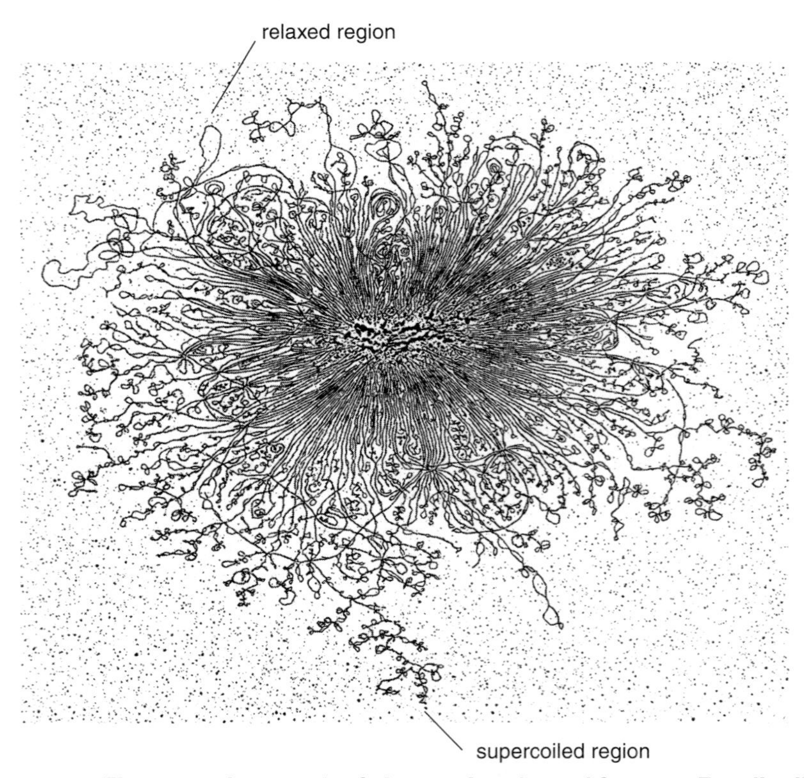

FIGURE 6.2 **Electron micrograph of chromatin released from an *E. coli* cell after gentle cell disruption.** Multiple loops can be seen emerging from a central region with some loops supercoiled and others relaxed. (Photo courtesy of Bruno Zimm and Ruth Kavenoff. Used with permission of Georgianna Zimm, University of California, San Diego.)

relaxed loops are both present indicates that each loop is somehow insulated from the others.

Further support for loop insulation comes from studies in which the released *E. coli* chromatin was observed at different times after adding trace quantities of DNase. If a supercoiled DNA molecule receives one nick, the strain of underwinding is immediately removed by free rotation about the opposing sugar-phosphate bond, and all supercoiling is lost. Because a nicked circle is much less compact than a supercoiled molecule of the same molecular mass, the nicked circle sediments much more slowly. One nick thus causes an abrupt decrease (by about 30%) in the sedimentation value. The sedimentation value decreases continuously, however, when DNase introduces one nick at a time into released *E. coli* chromatin; that is, the structure does not change in an all-or-none fashion but proceeds through a large number of intermediate states. This finding indicates that free rotation of the entire DNA molecule does not occur when a single nick is introduced. Electron microscopy studies confirm that as nuclease treatment continues, the number of nonsupercoiled loops increases.

A model of *E. coli* chromatin deduced from sedimentation and electron microscopy studies is shown in FIGURE 6.3. According to this model, the bacterial DNA is arranged in supercoiled loops that are fastened to a central protein matrix so that each supercoiled loop is topologically independent of all the others. A nick that causes one supercoiled loop to relax, therefore, would have no effect on other supercoiled loops. The *E. coli* chromosome is estimated to have about 400 such loops, each with an average length of about 10 to 20 kbp (kilobase pairs). Biochemical and genetic studies show that supercoiled loops are dynamic structures, which change during cell growth and division. This change allows the entire chromosome to be accessible to the transcription machinery and other enzymes throughout the cell cycle.

Although supercoiling makes an important contribution to bacterial DNA compaction, it is not the only factor. Macromolecular crowding and DNA-binding proteins also contribute to DNA compaction.

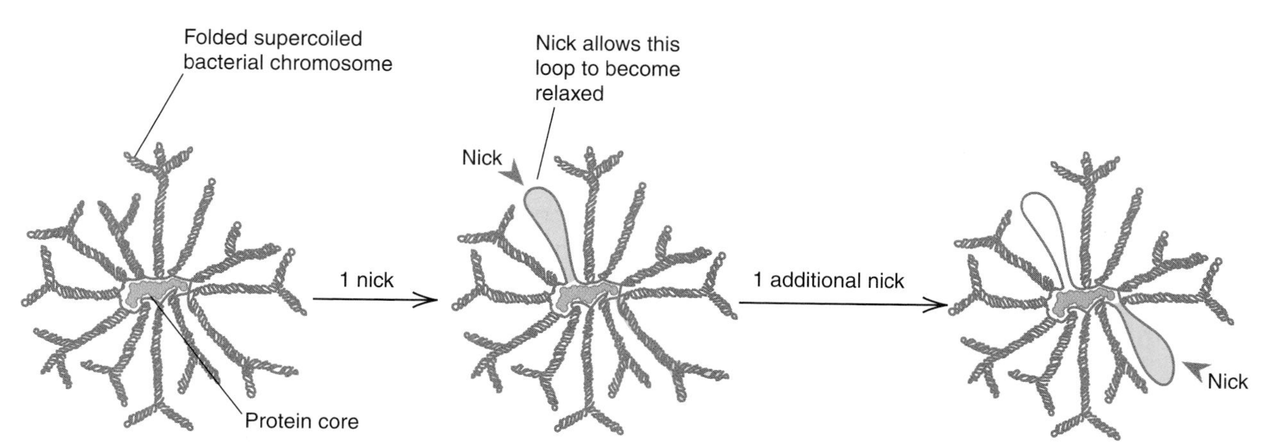

FIGURE 6.3 **Mechanism of folding of a bacterial chromosome.** The *E. coli* chromosome has about 400 negatively supercoiled loops attached to a central protein matrix. Each loop (10 to 20 kbp long) is topologically independent of the others.

High intracellular soluble macromolecule concentration limits the available aqueous volume, forcing the DNA to become more compact. Crowding also favors interactions between DNA and the proteins that bind to it. Some of these proteins help to determine the chromatin architecture.

MukB makes an important contribution to the compaction of the bacterial chromosome.

A major advance in our understanding of bacterial DNA compaction started with experiments performed by Sota Higara and coworkers in 1991 that were designed to find mutants that produce anucleate cells (cells without a nucleoid) at high frequency. Their work led to the discovery of a mutation in a previously unknown gene, which they named *mukB* (from the Japanese word *mukaku*, meaning anucleate) because mutant cells produced anucleate cells with a high frequency (5%–15% in *mukB* mutants vs. 0.03% in wild type cells) when cultured below 30°C. No growth was observed when cells were cultured above 30°C. The *mukB* gene codes for a 177-kDa protein called MukB. A *mukB* mutant grows more normally when a second mutation is introduced that lowers topoisomerase I activity. The explanation for this effect is that topoisomerase I removes negative superhelicity. The DNA in a cell with a defect in topoisomerase I, therefore, will be more negatively supercoiled and not be as dependent on MukB for compaction.

Mutations in two genes adjacent to *mukB* also result in mutants that produce anucleate cells at high frequency. These genes *mukE* and *mukF* code for the MukE and MukF proteins, respectively, which interact to form a $(MukE_2MukF)_2$ complex.

MukB belongs to a family of proteins called the **SMC (structural maintenance of chromosomes) proteins**, which organize and compact DNA. The MukB subunit, like other SMC family members, has a conserved N-terminal region followed by a variable region, a moderately conserved linker region, a second variable region, and a conserved C-terminal region (**FIGURE 6.4a**). The polypeptide chain folds back on itself. The N- and C-terminal interact to form an ATP-binding head group and the two variable regions interact to form a coiled-coil structure (**FIGURE 6.4b**). Two folded subunits combine to form a V-shaped dimer. The two subunits' linker regions interact at the apex of the V to form a hinge with an opening angle that can vary from 0° to 180° (**FIGURE 6.4c**). In bacteria, the two polypeptide subunits are identical, whereas in eukaryotes, they are similar but not identical.

A crystal structure is not yet available for an intact SMC protein. However, Byung-Ha Oh and coworkers have obtained a crystal structure for the SMC head group from the gram-negative bacteria *Pyrococcus furiosus*. The flat surface attached to the roots of the coiled-coil has many positively charged residues (**FIGURE 6.5**). Oh and coworkers used site-directed mutagenesis to replace these residues with glutamates. Several of these substitutions caused the head group to lose some or all of its ability to bind DNA. These results strongly suggest that DNA binds to the flat positively charged head group

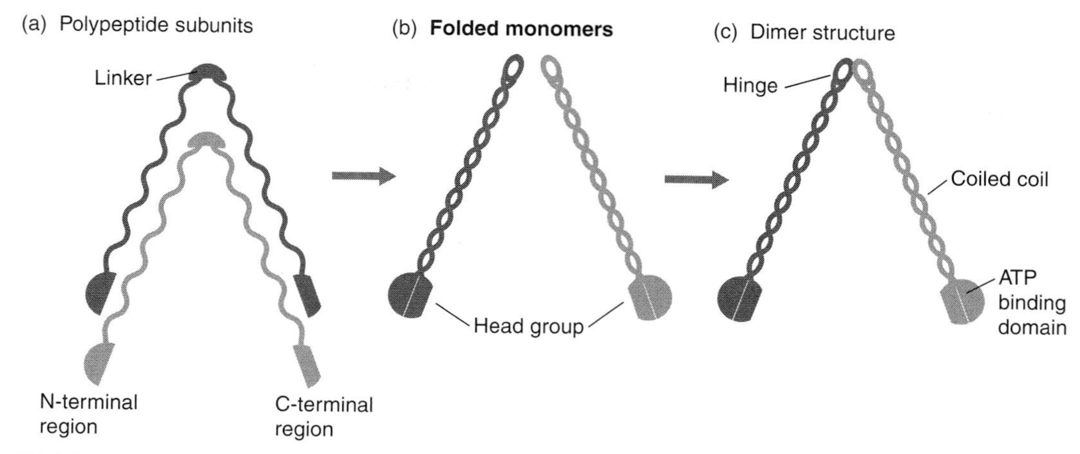

FIGURE 6.4 **The MukB protein.** (a) The MukB subunit has a conserved N-terminal region followed by a variable region, a moderately conserved linker region, a second variable region, and a conserved C-terminal region. (b) The polypeptide chain folds so that its N- and C-terminal regions come together to form a head group and its two variable regions form a coiled-coil. (c) Linker regions of two subunits combine to form a hinge. The resulting V-shaped dimer has an opening angle that can vary from 0° to 180°. (Modified from *J. Struct. Biol.*, vol. 156, M. Thanbichler and L. Shapiro, Chromosome organization and segregation . . ., pp. 292–303, copyright 2006, with permission from Elsevier [http://www.sciencedirect.com/science/journal/10478477].)

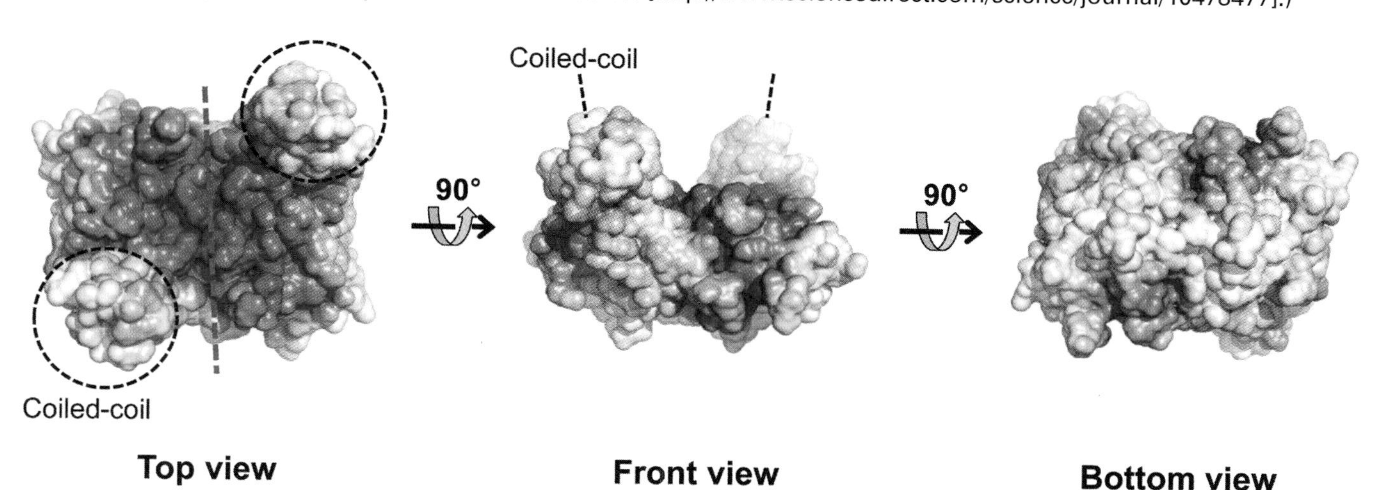

FIGURE 6.5 **Space-filling model of the bacterial SMC head group from the gram-negative bacteria *Haemophilus ducreyi*.** DNA binds to the predominantly positively charged flat surface (indicated by blue color) attached to the coiled-coil roots. The dashed red line indicates the dimer interface and the dashed black circle shows the roots of the coiled-coil. (Reproduced from *Biochem. Biophys. Res. Commun.*, vol. 386, J.-H. Lim and B.-H. Oh, Structural and functional similarities between . . ., pp. 415–419, copyright 2009, with permission from Elsevier [http://www.sciencedirect.com/science/journal/0006291X]. Photos courtesy of Jae-Hong Lim, Pohang Accelerator Laboratory, POSTECH.)

surface. DNA also binds to the hinge, which appears to stimulate ATPase activity in the head group.

Valentin V. Rybenkov and coworkers have used a single DNA molecule approach to examine how MukB works. They began by attaching one end of a double-stranded DNA molecule to a magnetic bead and the other to the surface of a glass capillary (FIGURE 6.6). MukB was added to the capillary in some experiments but not in others. A strong stretching force was applied to straighten out the DNA. Then the stretching force was reduced by about 30-fold. When MukB was present, the DNA contracted steadily over a period of a few minutes. When MukB was absent, no contraction was observed.

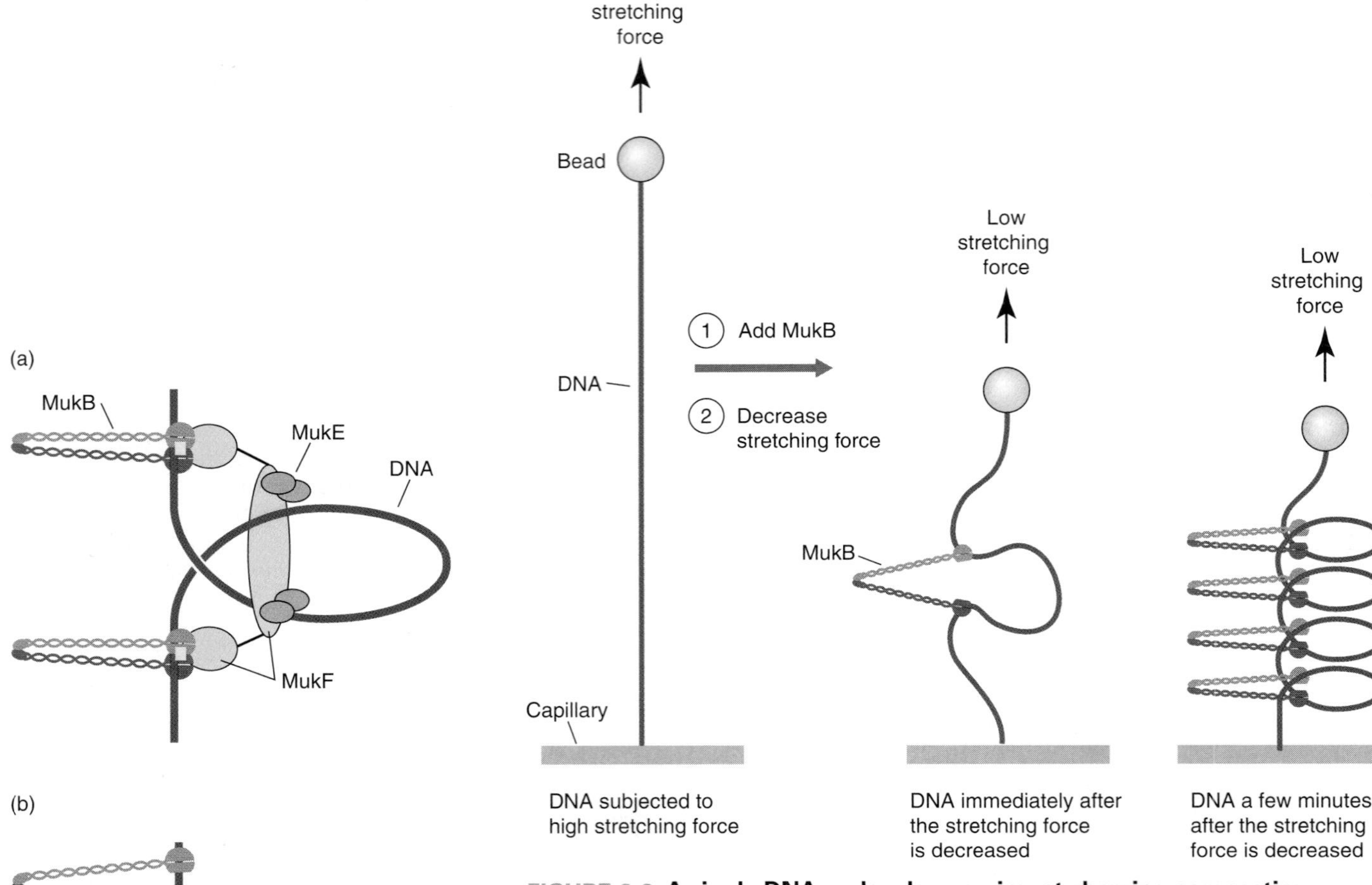

FIGURE 6.6 A single DNA molecule experiment showing compaction by MukB. A double-stranded DNA molecule is stretched after one end is attached to a bead and the other to the surface of a glass capillary. When the stretching force is decreased about 30-fold in the presence of MukB, DNA contraction is observed. (Adapted from Y. Cui, Z. M. Petrushenko, and V. V. Rybenkov, *Nat. Struct. Mol. Biol.* 15 [2008]: 411–418.)

FIGURE 6.7 A working model for the organization of MukBEF. Two possibilities are shown. (a) An ATP (yellow box) binds to each MukB head group. This interaction converts the MukB dimer to a ring structure. The MukEF complex bridges two ring-like structures and DNA is trapped as a loop. (b) An ATP (yellow box) binds to each MukB head group, causing head groups from different MukB dimers to come together to generate a MukB network. Once again, the MukEF complex interacts with non-flat head surface and DNA is trapped as a loop. (Adapted from V. V. Rybenkov, *Nat. Struct. Mol. Biol.* 16 [2009]: 104–105).

Reapplication of the stretching force caused the DNA to straighten out once again and MukBs to fall off the DNA (not shown). The length gained per released MukB was variable, suggesting that MukBs cause DNA looping but cannot maintain the loops when a strong stretching force is applied. ATP stimulates initiation but not propagation of DNA condensation. Additional experiments showed that MukB binds to DNA in a highly cooperative manner and causes compaction by stabilizing DNA loop formation.

Based upon his studies and the structural information available for MukB and MukEF, Rybenkov has proposed two possible working models for the organization of MukBEF (FIGURE 6.7). Both models begin with an ATP molecule binding to each MukB head group. In the first model, the two head groups in a single MukB dimer interact to form a ring-like structure and the $(MukE_2MukF)_2$ complex forms a bridge between two ring-like structures to trap a DNA loop. In the second model, a head group from one MukB dimer interacts with that from another dimer to form a MukB network and the $(MukE_2MukF)_2$ complex forms a bridge between head groups to trap a DNA loop.

Additional nucleoid-associated proteins contribute to bacterial DNA compaction.

The MukBEF complex is just one of many nucleoid-associated proteins that appear to participate in bacterial DNA compaction. These proteins can be classified into two groups, DNA bridging proteins and DNA bending proteins. The MukBEF complex belongs to the DNA bridging group. Another member of this group is the 15.4 kDa H-NS (histone-like nucleoid structuring) protein. The active form of H-NS appears to be a homodimer (or perhaps an oligomer). The polypeptide subunit has two functionally distinct domains separated by a flexible linker. Its C-terminal domain binds DNA and its N-terminal domain is required for dimer and oligomer formation. There is no crystal structure for the intact H-NS molecule and little is known about the way that the protein bridges DNA.

Three nucleoid associated proteins are known to bend DNA. Two of these, IHF (integration host factor) and HU (heat unstable) protein are closely related heterodimers. The crystal structure for the IHF•DNA complex is shown in FIGURE 6.8. The compact IHF body is made of intertwined α-helices from the two subunits. A flexible β-structure arm extending from each polypeptide subunit inserts into the minor groove, causing the DNA to bend. HU protein interacts with DNA in a similar fashion. The third nucleoid associated protein that bends DNA, FIS (factor for inversion stimulation), is a homodimer. A crystal structure is not available for the FIS•DNA complex. Additional nucleoid associated bridging and bending proteins have been identified but little is known about their structure or the specific contributions that they make to DNA compaction.

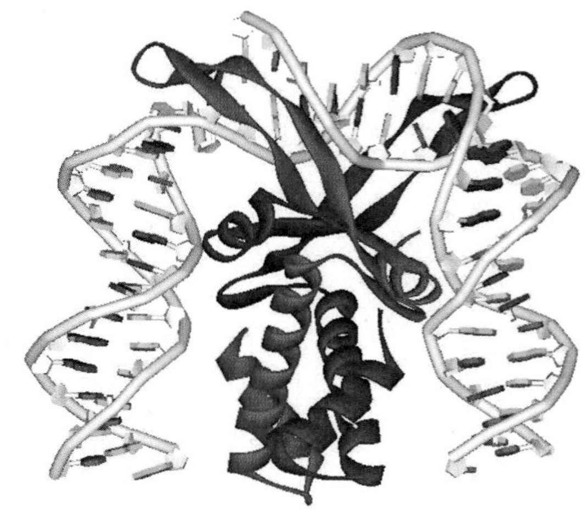

FIGURE 6.8 **Crystal structure for the IHF•DNA complex.** The compact IHF body is made of intertwined α-helices from the two polypeptide subunits (blue and magenta solid ribbons). Each polypeptide subunit has a flexible β-structure arm that inserts into the minor groove of the DNA, causing it to bend. (Structure from Protein Data Bank 1IHF. P. A. Rice, et al., *Cell* 87 [1996]: 1295–1306. Prepared by B. E. Tropp.)

6.2 Mitosis and Meiosis

In higher animals, germ cells have a haploid number of chromosomes and somatic cells have a diploid number.

When cells in higher animals start to divide, the chromatin in their cell nuclei can be seen under the light microscope as highly condensed structures called **chromosomes.** As biologists carefully examined chromosomes in many different kinds of cells from various higher animals, two important conclusions emerged. (1) Reproductive or **germ cells** have a characteristic number of chromosomes (n). This number, called the **haploid** number, is 3, 4, 23, and 30 for germ cells from the mosquito, fruit fly, human, and cattle, respectively. (2) Non-reproductive or **somatic cells,** such as those from lung, kidney, and brain, contain two versions of each chromosome (one from each parent) called **homologs.**

Because somatic cells have twice the number of chromosomes as germ cells they are said to be **diploid** (2n). Higher animals use one type of nuclear division, **mitosis**, to maintain the number of chromosomes during cell division and another, **meiosis**, to reduce the chromosome number in half during sexual reproduction. The diploid number is reestablished by zygote formation resulting from the union of a sperm and egg cell during sexual reproduction. Both mitosis and meiosis involve complex changes throughout the entire cell. Because we are primarily concerned with chromosomes, the discussion that follows will focus on chromosomal behavior during mitosis and meiosis in the cells of higher animals. Although details differ, similar types of nuclear division also take place in plants and fungi.

The animal cell life cycle alternates between interphase and mitosis.

A eukaryotic cell spends only a part of its **life cycle** (the time between its formation by parent cell division and its own division to form two daughter cells) in mitosis (M). The remainder of the time, often approaching 90% of its life cycle, is spent in **interphase**, a stage during which DNA, RNA, protein, and other biological molecules are synthesized.

Two forms of chromatin, a less condensed form called **euchromatin** and a more condensed form called **heterochromatin** are present in the eukaryotic nucleus during interphase (FIGURE 6.9). Euchromatin, the predominant form, is actively transcribed during interphase. In contrast, heterochromatin, which tends to be located near the nuclear membrane, is not actively transcribed. Nearly all regions of the human genome that remain to be sequenced are in the heterochromatin.

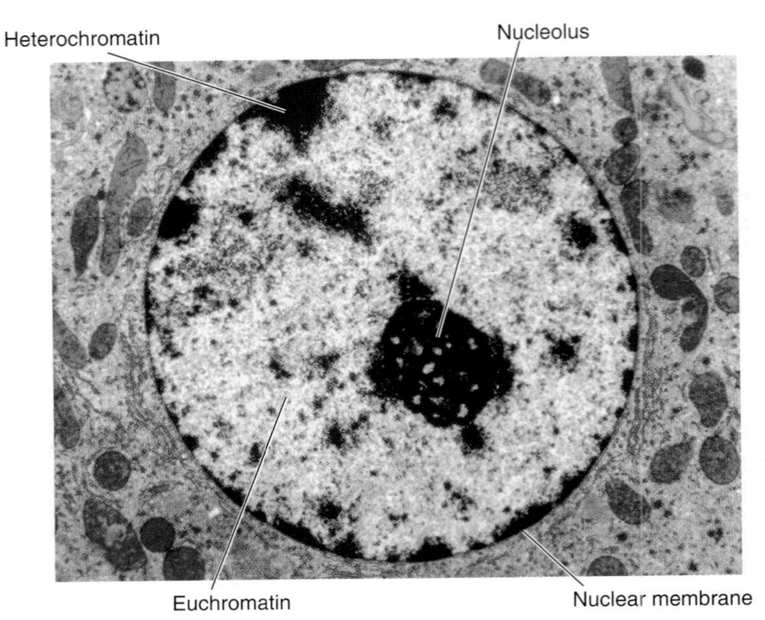

FIGURE 6.9 **Electron micrograph of a liver cell nucleus.** (© Phototake, Inc./Alamy Images.)

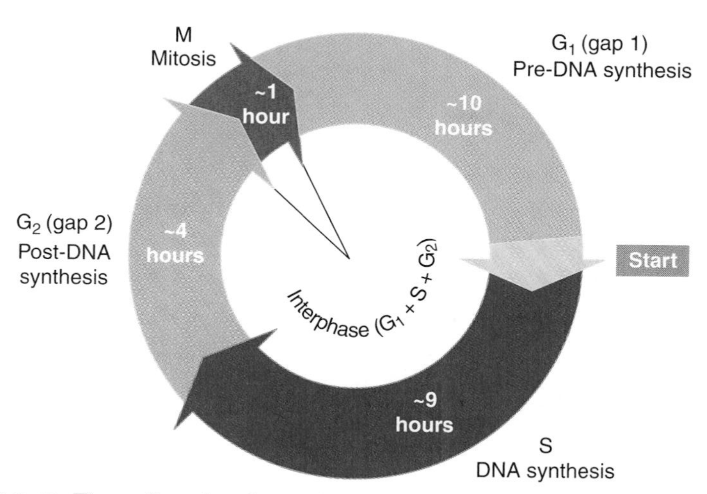

FIGURE 6.10 The cell cycle of a typical mammalian cell growing in tissue culture with a generation time of 24 hours.

DNA replication and histone synthesis occur during only a part of interphase, the DNA synthetic phase or **S phase**. The S phase is bracketed by two gap phases, **G1** and **G2**, so that the stages in the life cycle are in the order $G_1 \rightarrow S \rightarrow G_2 \rightarrow M$ (FIGURE 6.10). The timing of S, G_2, and M tend to be relatively uniform for a given type of somatic cell. However, time spent in G_1, a period of active protein, lipid, and carbohydrate synthesis, is quite variable. Some eukaryotic cells spend almost their entire life cycle in G_1.

Mitosis allows cells to maintain the chromosome number.

Even though mitosis is a continuous process, it is usually divided into four stages for convenience. Chromosomal changes during these four stages, which occur in the order prophase → metaphase →anaphase → telophase, are depicted in FIGURE 6.11 and summarized below.

Prophase

Chromatin, which was replicated during the S phase of interphase, condenses to form visibly distinct chromosomes. Each chromosome is divided along its long axis into two identical subunits called **sister chromatids** that are held together by a member of the SMC protein family called **cohesin** (see below). Cohesin is enriched about threefold in a 20–50 kb domain flanking a specific chromosomal region known as the **centromere**, relative to its concentration on chromosome arms. As prophase ends, the nuclear region involved in ribosomal RNA synthesis, called the **nucleolus**, disappears and the nuclear membrane disassembles to form membrane vesicles. A **mitotic spindle**, consisting of fiber-like bundles of protein molecules called **microtubules**, starts to form. Microtubule assembly involves aggregation of a protein called **tubulin**, which is itself made of two globular polypeptides, α-tubulin and β-tubulin, each with a molecular mass of about 55,000 Da (FIGURE 6.12).

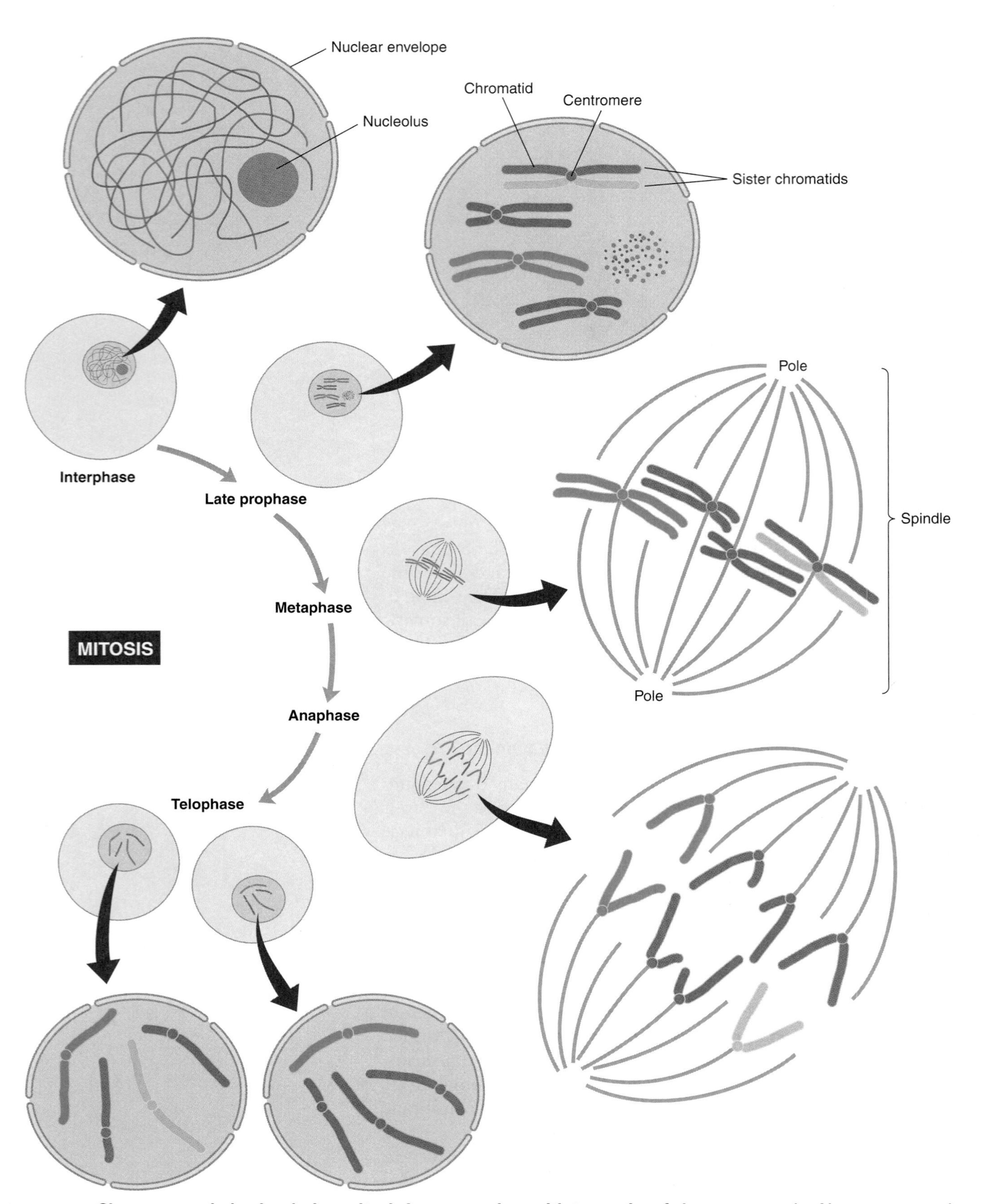

FIGURE 6.11 **Chromosome behavior during mitosis in an organism with two pairs of chromosomes (red/rose vs. green/blue).** At each stage, the smaller inner diagram represents the entire cell, and the larger diagram is an exploded view showing the chromosomes at that stage.

Metaphase

The assembly of the mitotic spindle is completed and the spindle moves to the region previously occupied by the nucleus. Several spindle fibers attach to each chromosome in a region of the centromere called the **kinetochore**. Once attachment is complete, the chromosomes move toward the center of the cell until the kinetochores lie in an imaginary plane equidistant from the two spindle poles. Each chromosome must be attached to both poles of the spindle, and the chromosomes must be properly aligned along the imaginary plane equidistant from the spindle poles for mitosis to proceed to the next stage.

We can learn a great deal by examining chromosomes during the metaphase (see below).

Anaphase

Cohesin is cleaved and cohesion between sister chromatids is dissolved. The two sister chromatids (now considered to be separate chromosomes) move toward opposite spindle poles so that an equal number of identical chromosomes are located at either end of the spindle as anaphase comes to a close. The number of chromosomes in each group is the same as that present in the cell nucleus at the start of interphase.

Telophase

The spindle disappears, nuclear membranes form around the two groups of chromosomes, and nucleoli re-form. Chromosomes become less and less condensed until they can no longer be seen with a light microscope. The cells divide to produce two identical daughter cells.

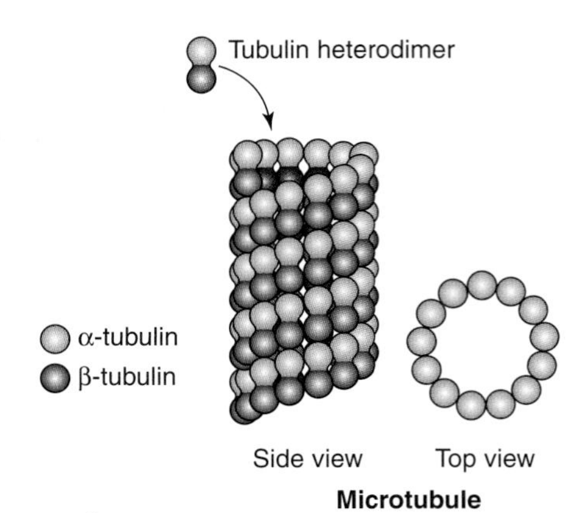

FIGURE 6.12 **Structure of microtubules.** Microtubules are long tubular structures built from a helical arrangement of the protein tubulin. Two kinds of tubulin are shown here, α tubulin and β tubulin, each having a molecular mass of 55,000 Da. The microtubule has a definite sense of direction, which arises from the asymmetry of the tubulin subunits and their defined and reproducible orientation within the fiber. (Adapted from J. A. Hadfield, et al. *Prog. Cell Cycle Res.* 5 [2003]: 309–325.)

Cohesin is a four subunit complex that keeps sister chromatids together.

During mitosis, sister chromatids must be held together from the time that they are synthesized in S phase until they segregate in anaphase. If they were not tethered, sister chromatids could separate from one another before attachment to both poles of the spindle, and would not be distributed equally to the daughter cells. Specific proteins called **cohesins** are required to hold the sister chromatids together. Although cohesin structure is the same in all eukaryotes, investigators use different nomenclature systems for the subunits from different organisms. To avoid confusion, we will use the nomenclature system for yeast cohesin.

Cohesin contains four subunits, which are arranged as shown in FIGURE 6.13. The core of the molecule is the Smc1•Smc3 heterodimer. The Smc subunits are similar to one another and to other members of the Smc family such as MukB in amino acid sequence and structure. Each head group in the heterodimer binds two ATP molecules. The third subunit, Scc1, bridges the two head groups and the fourth subunit, Scc3, binds to Scc1.

Three models have been proposed to explain how cohesin tethers sister chromatids (FIGURE 6.14). According to the ring model (Figure 6.14a), cohesin topologically traps sister chromatids. The two other models propose that cohesins bind to each sister chromatid and

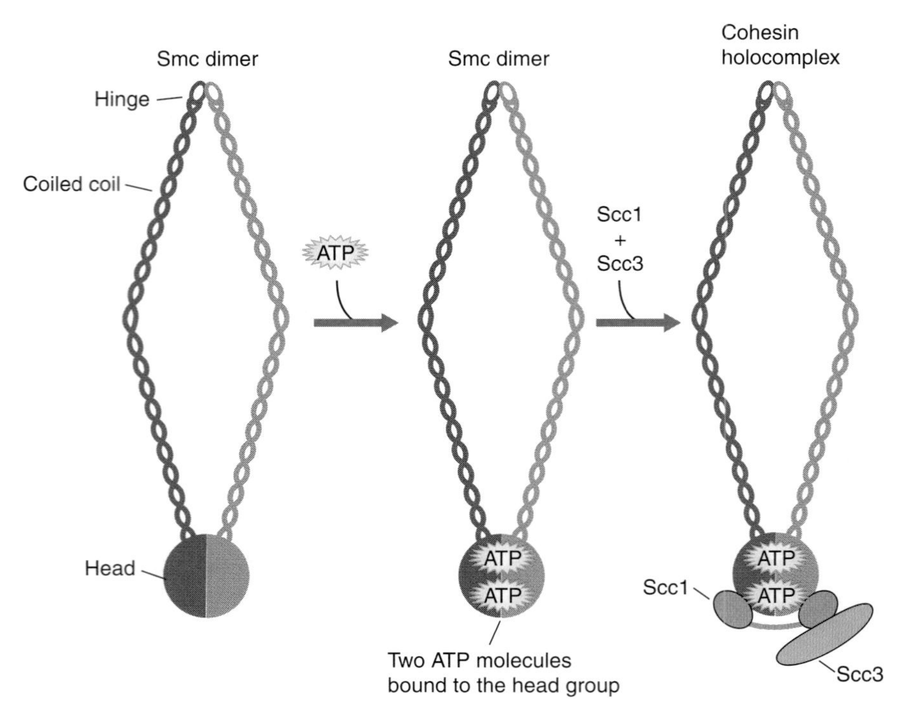

FIGURE 6.13 **Cohesin architecture.** Cohesin contains four subunits. The core of the molecule is an Smc1·Smc3 heterodimer. Each head group in the heterodimer binds two ATP molecules. The third subunit, Scc1, bridges the two head groups and the fourth subunit, Scc3, binds to Scc1. The names used in this figure are those for yeast cohesin. The architectures of cohesins from other organisms are the same but subunit nomenclature differs. (Reproduced from *Annual Review of Cell and Developmental Biology* 1997 (Hard) by I. Onn, et al. Copyright 2008 by ANNUAL REVIEWS, INC. Reproduced with permission of Annual Reviews, Inc. in the format Textbook via Copyright Clearance Center.)

then interact to form oligomers. In the snap model, cohesin interaction is through the coiled-coils (Figure 6.14b), whereas in the bracelet model it is between the hinges (Figure 6.14c). Further work is required to determine which, if any, of the three models is correct.

Loading cohesin onto chromatin is a complex process that is subject to both temporal and spatial regulation (FIGURE 6.15). In budding yeast, a cohesin loader made of Scc2 and Scc4 proteins starts to load cohesins onto chromatin during the G1/S phase transition and continues to do so until anaphase. Very little is known about how the cohesin loader accomplishes this task. Loaded cohesins are highly concentrated around the centromere but also present in the chromosome arms. Immediately after loading cohesins are in a noncohesive state and must be converted to a cohesive state to tether the sister chromatids. Conversion begins in S phase as new DNA is being synthesized. Establishment of cohesion requires the Eco1 protein. Further studies are needed to determine how Eco1 works and to identify the factors that regulate its activity.

Cohesins must be completely removed at the start of anaphase. In budding yeast this task is accomplished by the protease Esp1, which

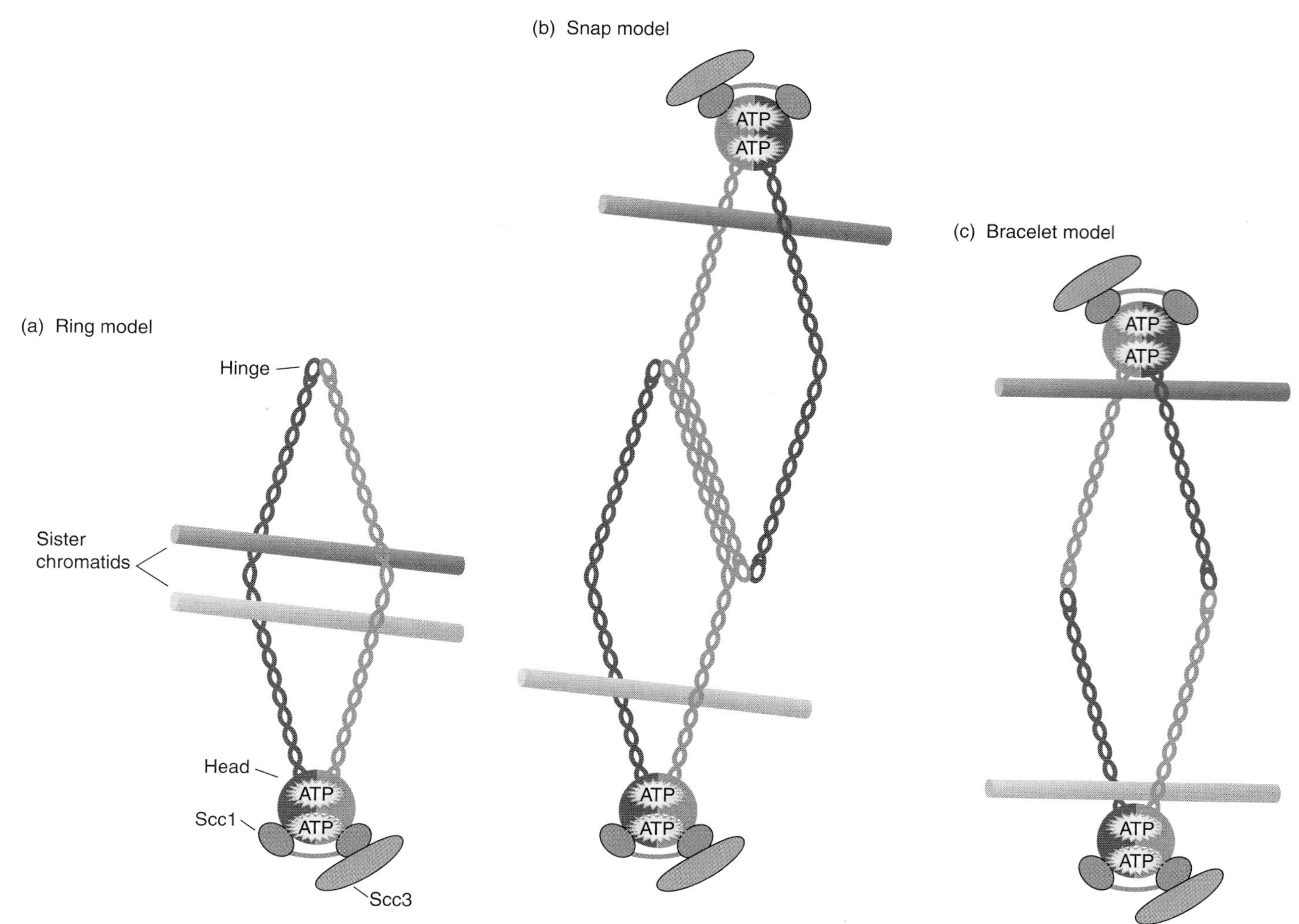

FIGURE 6.14 Three models to explain how cohesins tether sister chromatids. (a) The ring model proposes that cohesin topologically traps the sister chromatids. (b) The snap model proposes that cohesins bind each sister chromatid and then interact through coiled-coil domains. (c) The bracelet model proposes that cohesins bind each sister chromatid and then interact through hinge domains. (Reproduced from *Annual Review of Cell and Developmental Biology* 1997 (Hard) by I. Onn, et al. Copyright 2008 by ANNUAL REVIEWS, INC. Reproduced with permission of Annual Reviews, Inc. in the format Textbook via Copyright Clearance Center.)

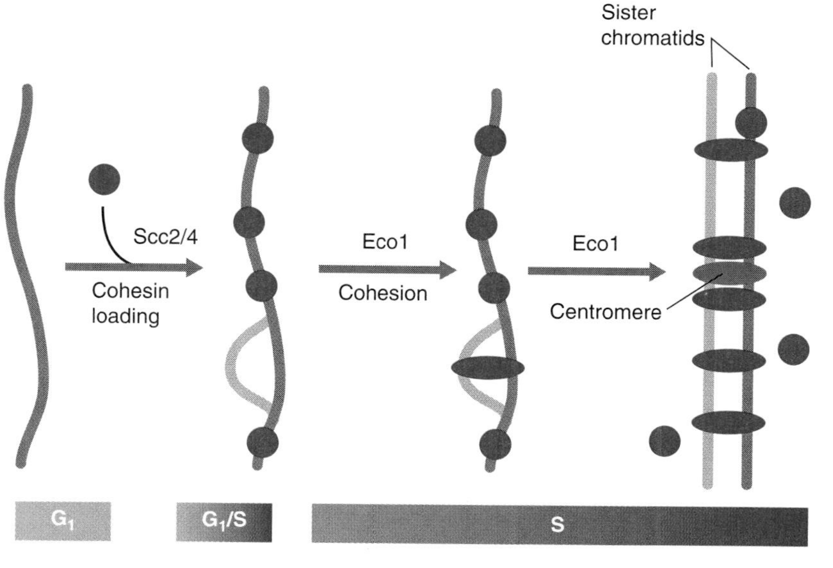

FIGURE 6.15 Loading cohesin onto chromatin in budding yeast. A cohesin loader (Scc2/Scc4) starts to load cohesins onto chromatin during the G_1/S phase transition and continues to do so until anaphase. Immediately after loading cohesins are in a noncohesive state (round) and must be converted to a cohesive state (oval) to tether the sister chromatids. Conversion, which begins in S phase as new DNA is being synthesized, requires the Eco1 protein. (Reproduced from *Annual Review of Cell and Developmental Biology* 1997 (Hard) by I. Onn, et al. Copyright 2008 by ANNUAL REVIEWS, INC. Reproduced with permission of Annual Reviews, Inc. in the format Textbook via Copyright Clearance Center.)

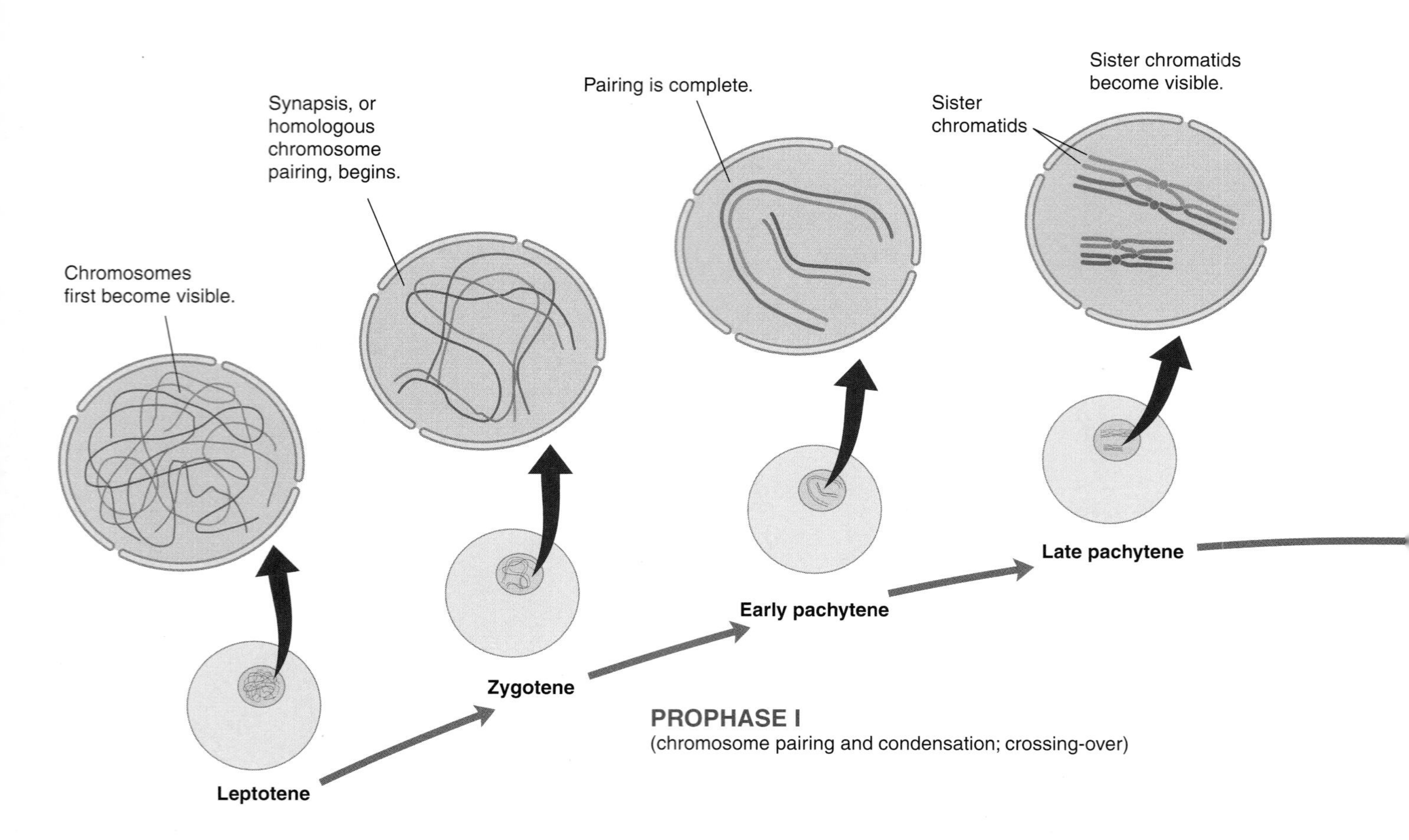

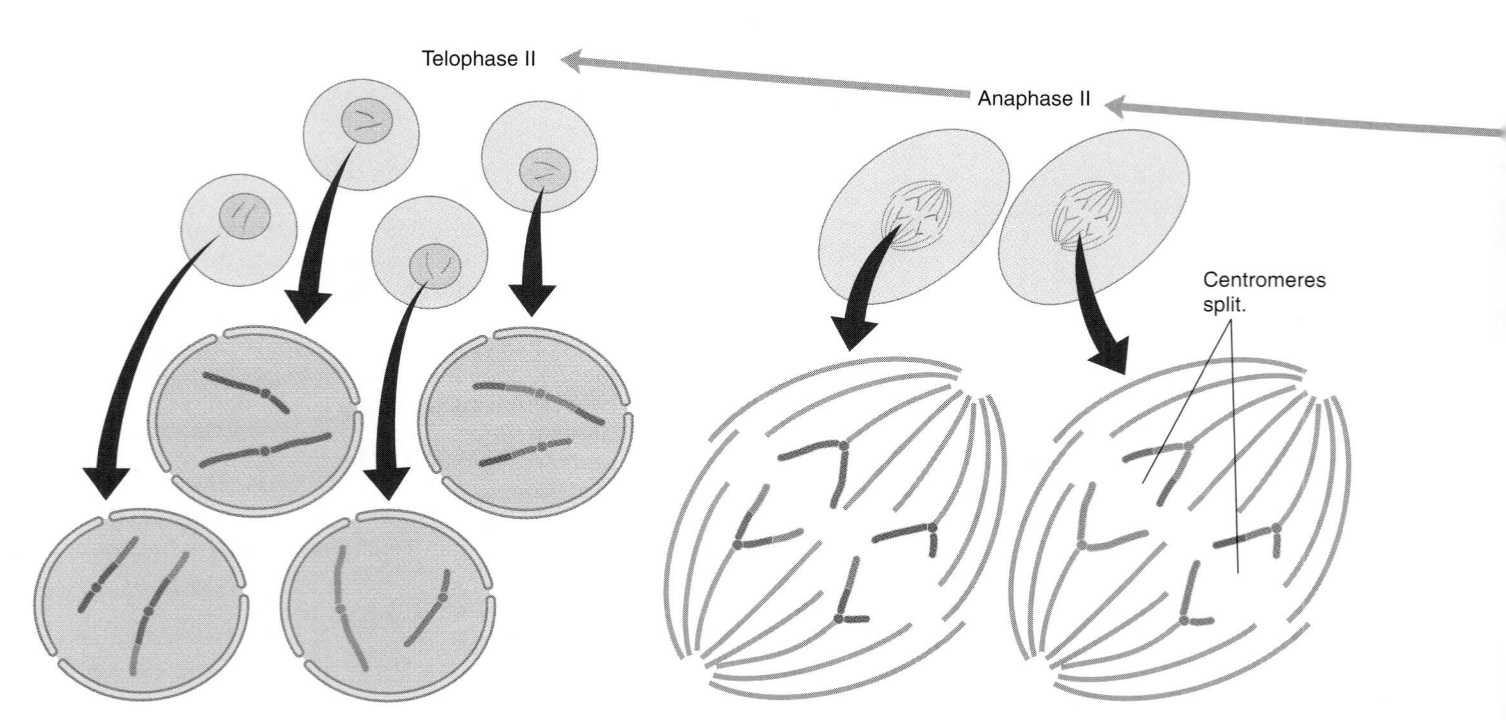

FIGURE 6.16 Chromosome behavior during meiosis in an organism with two pairs of homologous chromosomes (red/rose and green/blue). At each stage, the small diagram represents the entire cell and the larger diagram is an expanded view of the chromosomes at that stage.

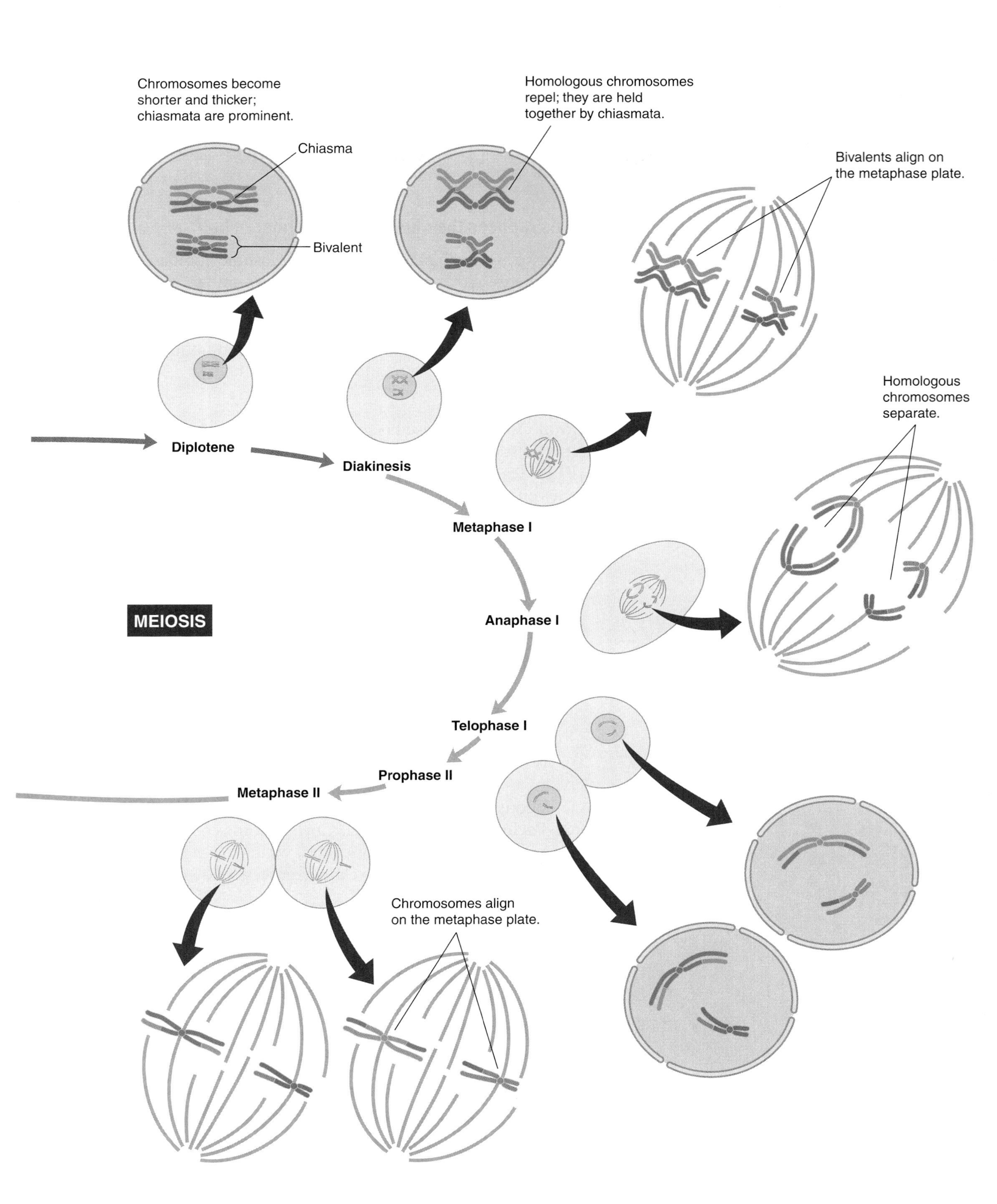
Chromosomes become shorter and thicker; chiasmata are prominent.
Chiasma
Bivalent
Homologous chromosomes repel; they are held together by chiasmata.
Bivalents align on the metaphase plate.
Homologous chromosomes separate.
Diplotene
Diakinesis
Metaphase I
Anaphase I
Telophase I
Prophase II
Metaphase II
MEIOSIS
Chromosomes align on the metaphase plate.

cleaves the Scc1 cohesin subunit, freeing the sister chromatids so that they can segregate to the opposite poles of the mitotic spindle. The cohesin pathway in other organisms is similar to that described for yeast but some details differ.

Meiosis reduces the chromosome number in half.

Unlike mitosis, which maintains the chromosome number, meiosis produces gametes with a haploid chromosome number. Meiosis requires two successive nuclear divisions to accomplish this task. The first and second meiotic divisions go through prophase, metaphase, anaphase, and telophase stages. Chromosomal changes that occur during meiosis are depicted in FIGURE 6.16 and summarized below.

First Meiotic Division

Prophase I: Prophase I begins after DNA replication is completed. It is of special interest because chromosome homologs exchange DNA leading to **genetic recombination** during this stage. For convenience, prophase I is divided into five substages. Each substage's name derives from chromosomal appearance during that substage. The substages are as follows:

1. **Leptotene** (*thin or fine thread*). As the chromatin starts to condense, beads, called **chromomeres**, appear at irregular intervals along the length of the chromosome. The number, size, and arrangement of chromomeres is unique for each kind of chromosome.
2. **Zygotene** (*paired threads*). Homologous chromosomes pair. These pairings (**synapsis**) begin at the chromosome tips and continue along the chromosome, establishing specific chromomere–chromomere pairing. Fully paired chromosomes are called **bivalents** (indicating that each pair consists of two types of chromosomes) or **tetrads** (indicating that each homologous pair contains four closely associated chromatids).
3. **Pachytene** (*thick thread*). Chromosome condensation continues. The paired chromosomes, which are very close together, begin a chromosome exchange process known as **crossing over** (see Chapter 13).
4. **Diplotene** (*double thread*). Condensation reaches a state in which the sister chromatids in each chromosome become visible. Homologous chromosomes start to separate. Complete separation cannot occur at this stage, however, because the homologous chromosomes are joined by cross-connections called **chiasmata** (singular, chiasma = crosspiece), which are produced when nonsister chromatids break and reunite during the crossing-over process. This is the stage in animal cell egg production in which ribosomes, proteins, fats, carbohydrates, and other cytoplasmic components are synthesized.
5. **Diakenesis** (*moving apart*). Chromosomes reach the stage of maximum condensation. Homologous chromosomes in a

bivalent start to move apart but remain connected through at least one chiasma. Spindle formation is initiated and the nuclear membrane disassembles.

Metaphase I: Spindle fibers from one pole make contact with one chromosome in a homologous pair while spindle fibers from the other pole make contact with the other chromosome in the pair. Each chromosome moves into the metaphase plate (the imaginary plane that is equidistant from each spindle pole).

Anaphase I: Homologous chromosomes are pulled apart and begin to move to opposite poles. The two members of each homologous pair are separated so that each pole has the haploid number of chromosomes.

Telophase I: The single spindle disassembles and two new spindles form in the region that had been occupied by the spindle poles. In some species, the nuclear membrane re-forms, whereas in others the chromosomes enter directly into the second meiotic division. There is a seamless transition from telophase I to prophase II. In fact, the transition is so seamless that prophase II is almost nonexistent in many organisms. No chromosomal replication occurs between the first and second meiotic divisions.

Second Meiotic Division

Prophase II: The chromosomes remaining at the two poles of the first meiotic division begin to move to the midpoints of the two newly formed spindles.

Metaphase II: Chromosomes align on the metaphase plate.

Anaphase II: Cohesins are cleaved, centromeres split, and the chromosomes move to opposite poles of the spindle. Each chromatid is now considered to be a separate chromosome.

Telophase II: The chromosomes decondense and nuclear membranes form around the four division products to produce four haploid nuclei.

6.3 Karyotype

Chromosome sites are specified according to nomenclature conventions.

Cytogenetics, the scientific discipline concerned with the study of chromosomes, allows us to correlate chromosomal changes with specific clinical problems. Chromosomes are examined during the mitotic metaphase, the stage in which each chromosome consists of two highly-condensed sister chromatids that are held together at their centromere by cohesins. The centromere's position, which may be near the center of the chromosome, off-center, or close to one end, determines the chromatid arm lengths. The shorter of the two arms is called the **p arm** (petite arm) and the longer is called the **q arm**.

(a)

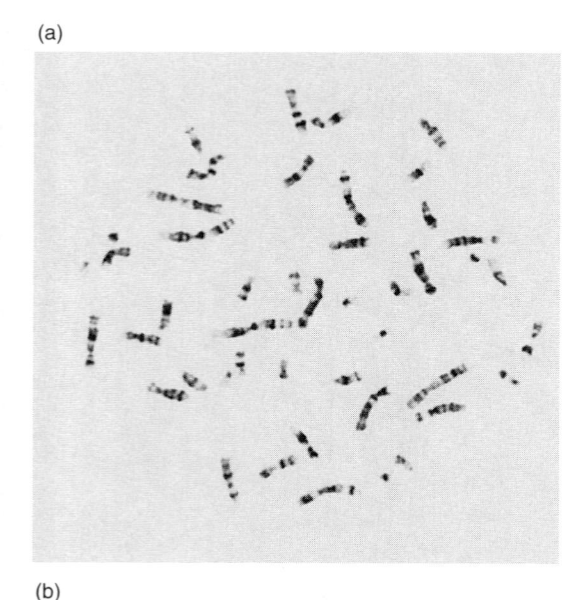

(b)

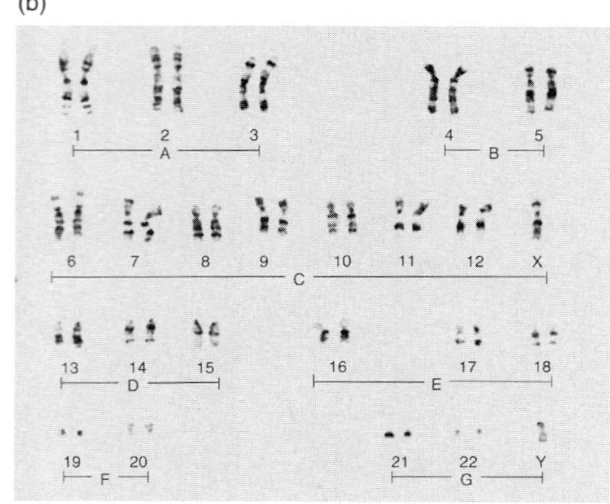

FIGURE 6.17 **A karyotype of a normal human male.** Blood cells arrested in metaphase were stained with Giesma and photographed under a microscope. (a) The chromosomes as seen in the cell by microscopy. (b) The chromosomes have been cut out of the photograph and paired with their homologs. Paired homologs are arranged according to size in groups A, B, and so forth. The largest homologous pair is in the upper left. X and Y sex chromosomes have not been paired. (Photos courtesy of Patricia A. Jacobs, Wessex Regional Genetics Laboratory, Salisbury District Hospital.)

Investigators attempted to use size and centromere position to classify human chromosome pairs, but these criteria do not allow unambiguous chromosome identification. With the discovery in the late 1960s and early 1970s that certain dyes stain chromosomes to produce unique and reproducible band patterns, investigators were able to identify specific chromosome pairs. For example, staining with Giemsa dye produces light and dark transverse bands along the length of the chromosome (FIGURE 6.17a).

Despite the fact that a typical band looks quite narrow when viewed with a light microscope, it actually extends over more than a million base pairs and dozens of genes. Even when different chromosomes are the same size and have their centromeres located in the same place, they can still be distinguished by their unique "bar code" patterns. The bands along a chromosome are part of clearly delimited regions. Nomenclature conventions for specifying sites in a chromosome, which are illustrated for human chromosome 1 in FIGURE 6.18, are as follows:

1. Each chromosome is assigned a number, with the largest being assigned number 1.
2. The short arm is designated p and the long arm is q.
3. Regions and bands are numbered consecutively from the centromere outward along each chromosome arm.
4. Chromosome number, arm, region number, and band number are written in order and without punctuation or spacing.

A karyotype shows an individual cell's metaphase chromosomes arranged in pairs and sorted by size.

A great deal of information can be obtained from digital images of stained chromosomes taken at metaphase. The digital images are cut and pasted with a computer, arranging chromosome pairs by size, shape, and banding pattern to facilitate interpretation. By convention, chromosome pairs are arranged in decreasing order of size to produce an arrangement called a **karyotype**. FIGURE 6.17b, a human karyotype preparation, reveals the presence of 23 pairs of chromosomes. Twenty-two chromosome pairs, called **autosomes**, are the same in males and females. The remaining pair, the **sex chromosomes**, determines the individual's sex. Male sex chromosomes consist of one X and one Y chromosome, whereas female sex chromosomes consist of the two X chromosomes. A single Y chromosome is sufficient to produce maleness while its absence is required for femaleness. The Y chromosome is smaller than the X chromosome and the two have different banding patterns.

A great deal of information can be obtained by examining karyotype preparations.

Human karyotype preparations provide important clinical information. Although almost any population of dividing cells can be used to obtain the metaphase cells required for such a preparation, blood, bone marrow, fibroblasts, and amniotic fluid are the most common sources of cells to be analyzed. Isolated cells are incubated with a plant or bacterial protein called a **mitogen** that binds to receptors on the outer surface of the cell membrane and induces mitosis. After three or four days of incubation, a drug such as colchicine, which disrupts mitotic spindles, is added to arrest dividing cells in metaphase. The population, now enriched in metaphase cells, is stained and analyzed.

Many congenital problems can be identified by chromosomal anomalies. **Down syndrome**, which occurs in about one of every 800 births, will serve as an illustrative example. Approximately 95% of individuals with Down syndrome have three copies of chromosome 21. Down syndrome is associated with mild to severe forms of mental retardation, which is often accompanied by various medical problems including epilepsy, heart defects, and a marked susceptibility to respiratory infections. Individuals with Down syndrome also appear to age at an accelerated rate and have a high probability of showing the clinical signs and symptoms of Alzheimer's-type dementia after age 35.

A condition in which cells have either more or less than the normal diploid number of chromosomes is called **aneuploidy**. Most cases of simple aneuploidy are caused by errors in chromosomal segregation during meiosis. Many of these errors result from defects in the cohesin pathway. If pairs of homologous chromosomes do not separate during the first meiotic division or if the centromere joining sister chromatids does not separate during the second meiotic division, the gametes formed will have too many or too few chromosomes.

It seems reasonable to suppose that cells with three copies of chromosome 21 overproduce specific proteins that somehow modify fetal development. The challenge is to discover which genes are involved and how their gene products work. The problem is complicated because several genes may be involved. In this regard, it is interesting to note that some forms of Down syndrome are caused by chromosomal rearrangements in which a segment of chromosome 21 is attached to another chromosome. In this case, the individual with Down syndrome has two copies of chromosome 21 and a part of chromosome 21 attached to another chromosome. Unfortunately, the chromosome 21 segment is still so large that it is not yet possible to identify the specific genes responsible for Down syndrome.

A rearrangement of chromosomal material in which part of one chromosome is joined to some other chromosome is called **translocation**. One of the best-characterized examples of translocation occurs in chronic myelogenous leukemia. Studies of chromosomes in tumor cells reveal reciprocal translocation of material from chromosome 9 to 22 to produce what is known as the Philadelphia chromosome (FIGURE 6.19). This translocation moves a gene (*abl*) from its normal location on chromosome 9 to a new location on chromosome 22, leading to an altered *abl* gene. This altered *abl* gene produces an abnormal protein that activates **constitutively** (all the time) a number of cell processes that normally are turned on only under special conditions.

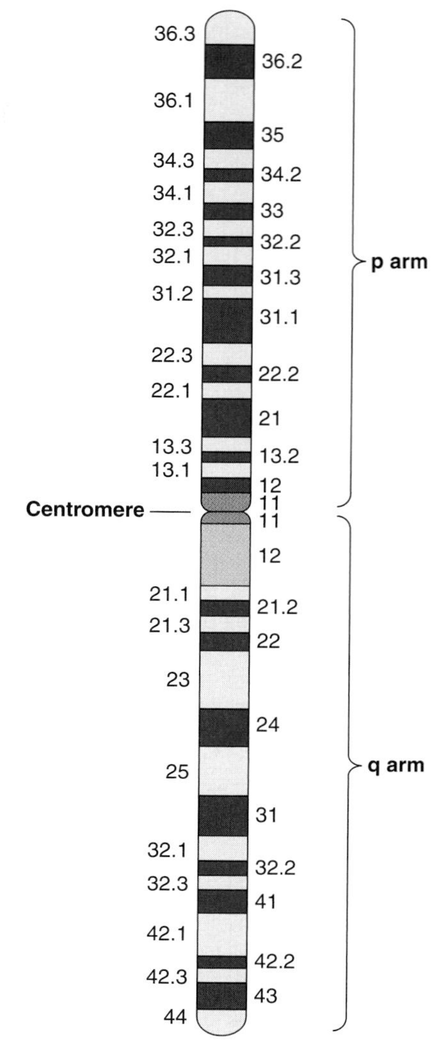

FIGURE 6.18 **Human chromosome 1.** The short arm is designated p and the long arm q. Regions and bands are numbered consecutively from the centromere outward along each chromosome arm. (Modified from HYPERLINK [http://www.genome.jp.kegg]. Used with permission of Kanehisa Laboratories Bioinformatics Center, Institute for Chemical Research, Kyoto University and the Human Genome Center, Institute for Medical Science, University of Tokyo.)

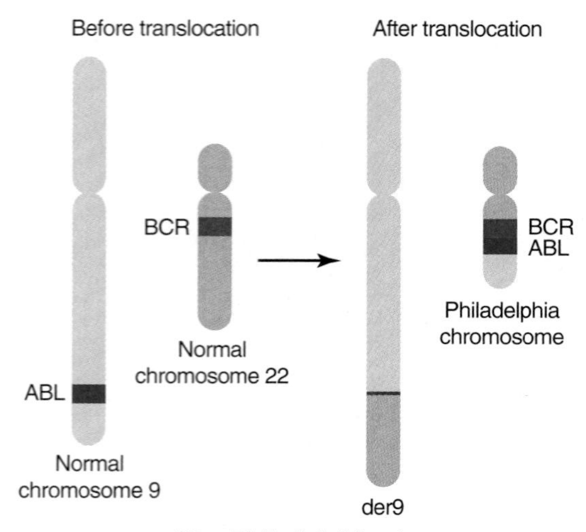

FIGURE 6.19 **The Philadelphia chromosome.** The Philadelphia chromosome, which is present in individuals suffering from chronic myelogenous leukemia, forms when a piece of chromosome 9 changes place with a piece of chromosome 22. It is the extra-short chromosome formed by the exchange that contains the abnormal *bcr-abl* gene. The other product of the exchange, the extra-long chromosome, is called der9.

Fluorescent in situ hybridization (FISH) provides a great deal of information about chromosomes.

Classical staining techniques do not provide sufficient sensitivity to detect translocations, deletions, or insertions that involve small segments within a chromosome. Investigators have taken advantage of lessons learned from molecular biology, however, to devise a very sensitive technique for detecting even very small chromosomal changes in a sample fixed to a microscope slide.

This technique, called **fluorescent in situ hybridization** (**FISH**), takes advantage of the fact that a DNA probe with an attached fluorescent dye will bind to a specific DNA sequence within a denatured chromosome (FIGURE 6.20). The fluorescent probes can be prepared by nick translation or by the polymerase chain reaction (see Chapter 5). FISH has a wide variety of applications, which include detecting aneuploidy, identifying chromosomal aberrations, and locating genes and other DNA segments on a chromosome. It can also be used to locate DNA segments during interphase when the chromosome is not visible.

A variation of **FISH**, called **chromosome painting**, uses a fluorescent dye bound to DNA pieces that bind all along a particular chromosome. The major shortcoming of FISH and chromosome painting is that they cannot be used to study all chromosomes at the same time, because there are not enough fluorescent dyes with sufficient color differences to mark all 23 chromosomes in a unique color.

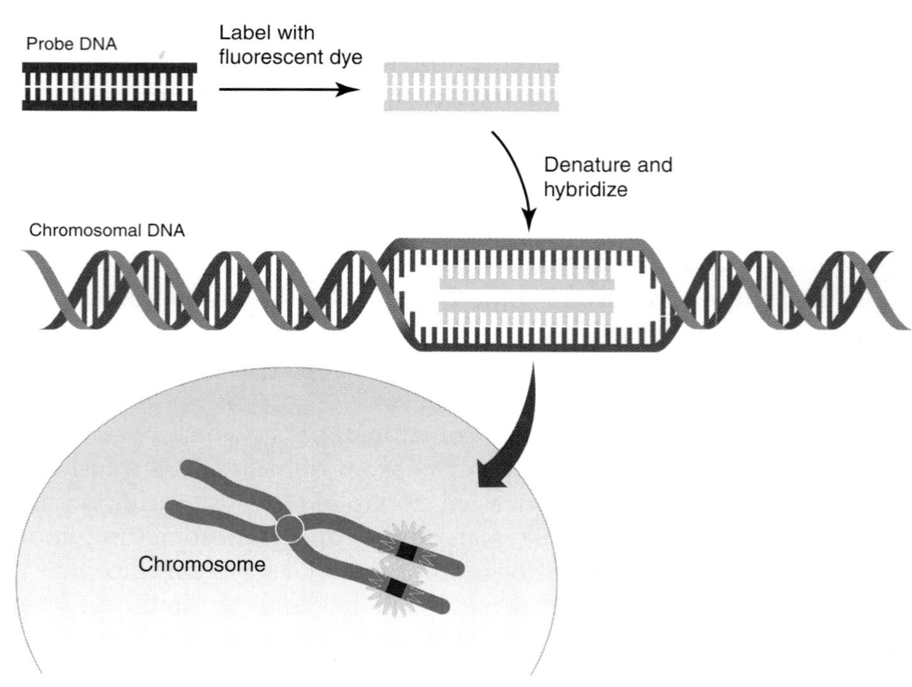

FIGURE 6.20 **The fluorescent in situ hybridization (FISH) technique.** A DNA probe with an attached fluorescent DNA (yellow) binds to a specific DNA sequence within a denatured chromosome. This technique is used to detect aneuploidy, identify chromosomal aberrations, and locate DNA segments or genes. (Adapted from an illustration by Darryl Leja, National Human Genome Research Institute [www.genome.gov]).

This problem was solved by labeling the painting probes for each chromosome with a different assortment of fluorescent dyes, a technique called **spectral karyotyping**. When the fluorescent probes hybridize to a chromosome, each kind of chromosome is labeled with a different assortment of fluorescent dye combinations. Stained chromosomes are then viewed through a series of filters, each of which transmits only light emitted by a single fluorescent dye. Alternatively, an interferometer determines the full spectrum of light emitted by the stained chromosome. In either case, a computer provides a composite picture that shows different chromosome pairs as if they were stained in different colors (FIGURE 6.21).

(a)

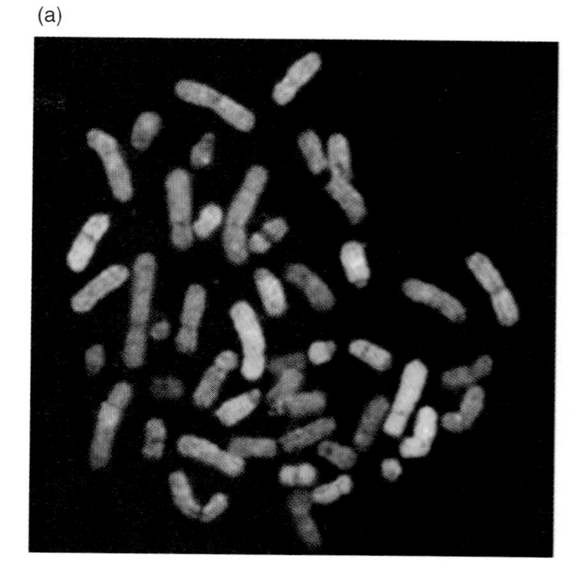

(b)

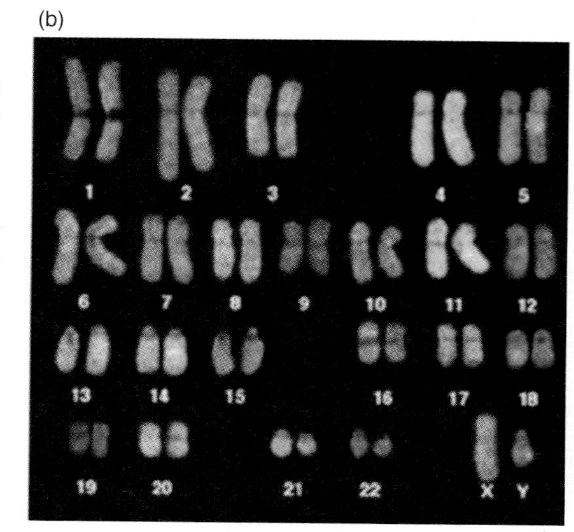

FIGURE 6.21 **Spectral karyotyping of human chromosomes.** (a) A normal metaphase plate after the simultaneous hybridization of 24 differentially labeled chromosome painting probes. Stained chromosomes are viewed through a series of filters, each of which transmits only light emitted by a single fluorescent dye. A computer provides a composite picture that shows each kind of chromosome in a distinct color. (b) The same chromosomes rearranged so that homologous chromosomes are shown as pairs. Somatic chromosomes are shown in decreasing size order with the largest chromosome pair in the upper left. The sex chromosome pair is shown to the right of the smallest chromosome pair in the bottom row. (Photos courtesy of Johannes Wienberg and Thomas Ried, National Institute of Health.)

6.4 The Nucleosome

Five major histone classes interact with DNA in eukaryotic chromatin.

Clearly, a chromosome has an elaborate structure and undergoes complex changes during the cell cycle. The task now before molecular and cell biologists is to describe chromosomal structure and explain chromosomal behavior at the molecular level. We are still a long way from understanding the folding process that condenses the DNA in a human cell so that DNA with a total length of over 2 m fits into the cell nucleus with a diameter of about 5 μm. Considerable progress has been made in understanding the first level of organization, however, the interactions between DNA and histones, which compacts DNA by about sevenfold.

Chromatin contains five major classes of histones: H1, H2A, H2B, H3, and H4 (Table 6.1). A typical human cell contains about 60 million copies of each kind of histone so that their combined mass in chromatin is about equal to that of the DNA. Each histone has a high percentage of lysine and arginine but the lysine-to-arginine ratio differs in each type of histone. With the possible exception of histone H4, higher organisms have variants of each histone subtype.

The positively charged side chains of lysine and arginine residues enable histones to bind to the negatively charged phosphates of the DNA. This electrostatic attraction is an important stabilizing force in chromatin because if chromatin is placed in solutions of high salt concentration (e.g., 0.5 M NaCl), which breaks down electrostatic

TABLE 6.1 **Histone Composition**

Histone	Molecular Mass (kDa)	% Lysine	% Arginine	% Lysine + Arginine
H1	~21.0	29	1.5	30.5
H2A	14.5	11	9.5	20.5
H2B	13.7	16	6.5	22.5
H3	15.3	10	13.5	23.5
H4	11.3	11	14.0	25.0

interactions, the chromatin dissociates to yield free histones and free DNA. Moreover, chromatin can be reconstituted by mixing purified histones and DNA in a concentrated salt solution and then gradually lowering the salt concentration by dialysis. Although this result shows that no other components are needed to form chromatin, the rate of chromatin assembly observed under these conditions is much slower than it is in the cell. Protein factors are needed for chromatin formation in the cell (see Chapter 10). Reconstitution experiments also have been carried out in which histones from different organisms are mixed. Usually, almost any combination of histones works, because, except for H1, the histones from different organisms are very much alike. In fact, H3 amino acid sequences are nearly identical from one organism to the next (sometimes one or two of the amino acids differ). The same is true for H4. For instance, H4 from a cow differs by only two amino acids from H4 from peas—arginine for lysine and isoleucine for valine—which shows that the structure of histones has not changed in the billion years since plants and animals diverged.

The first level of chromatin organization is the nucleosome.

When viewed under the electron microscope, uncondensed chromatin from interphase cells resembles beads on a string (FIGURE 6.22). Each bead is a nucleoprotein complex called a **nucleosome** formed by winding DNA around a protein assembly consisting of eight histone molecules. The DNA connecting two nucleosomes is called **linker DNA**. Micrococcal nuclease (see **Table 5.1**) can cleave the linker DNA but not the DNA that is wrapped around the protein assembly. Brief digestion with micrococcal nuclease therefore cleaves the chromatin to produce free nucleosomes (FIGURE 6.23). The length of DNA extracted from these nucleosomes varies from one organism to the next, ranging from about 170 to 240 bp. This variation results from different

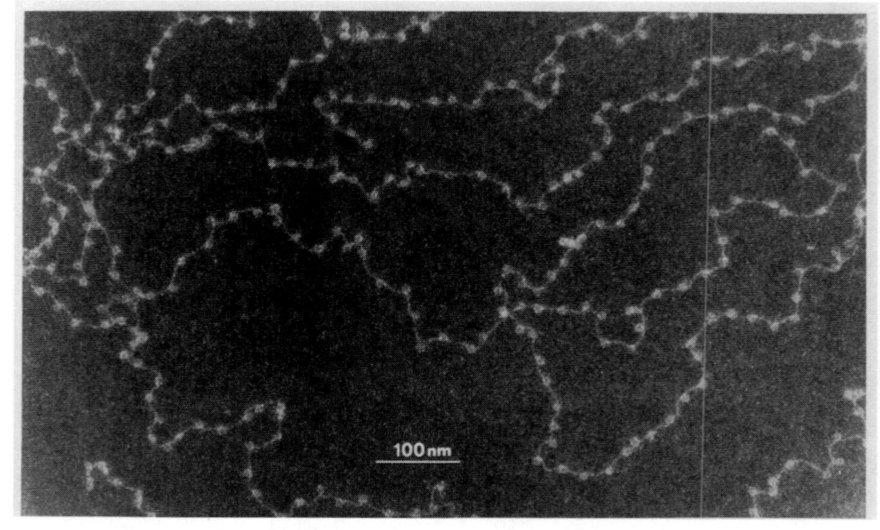

FIGURE 6.22 **Electron micrograph of chromatin.** The beadlike nuclosome particles have diameters of approximately 11nm. (Photo courtesy of Ada L. Olins and Donald E. Olins, Bowdoin College.)

sizes of the linker DNA between the nucleosomes. Surprisingly, linker length in the same organism may also vary from one tissue to another (for instance, brain versus liver). The significance of this variation is unknown.

Prolonged nuclease digestion gradually cleaves additional nucleotides until the DNA is about 146 bp long. The structure that remains, the **nucleosome core particle**, consists of an **octameric protein complex** (two copies each of H2A, H2B, H3, and H4) with a 146 bp DNA fragment wound around it.

X-ray crystallography provides high-resolution images of nucleosome core particles.

In 1997, Timothy J. Richmond and colleagues determined the structure of the nucleosome core particle at a resolution of 0.28 nm (FIGURE 6.24). This remarkable accomplishment was achieved by constructing nucleosome cores from a 146 bp palindromic DNA molecule with a defined sequence and H2A, H2B, H3, and H4 histones synthesized by *E. coli*. Although bacteria do not normally synthesize histones, recombinant DNA technology produced bacterial cells that could do so. The reason for going to the trouble of creating bacteria that synthesize histones was to obtain histone molecules that had not been modified by eukaryotic enzymes that acetylate, methylate, and phosphorylate the polypeptides at various sites, thereby producing a heterogeneous polypeptide population.

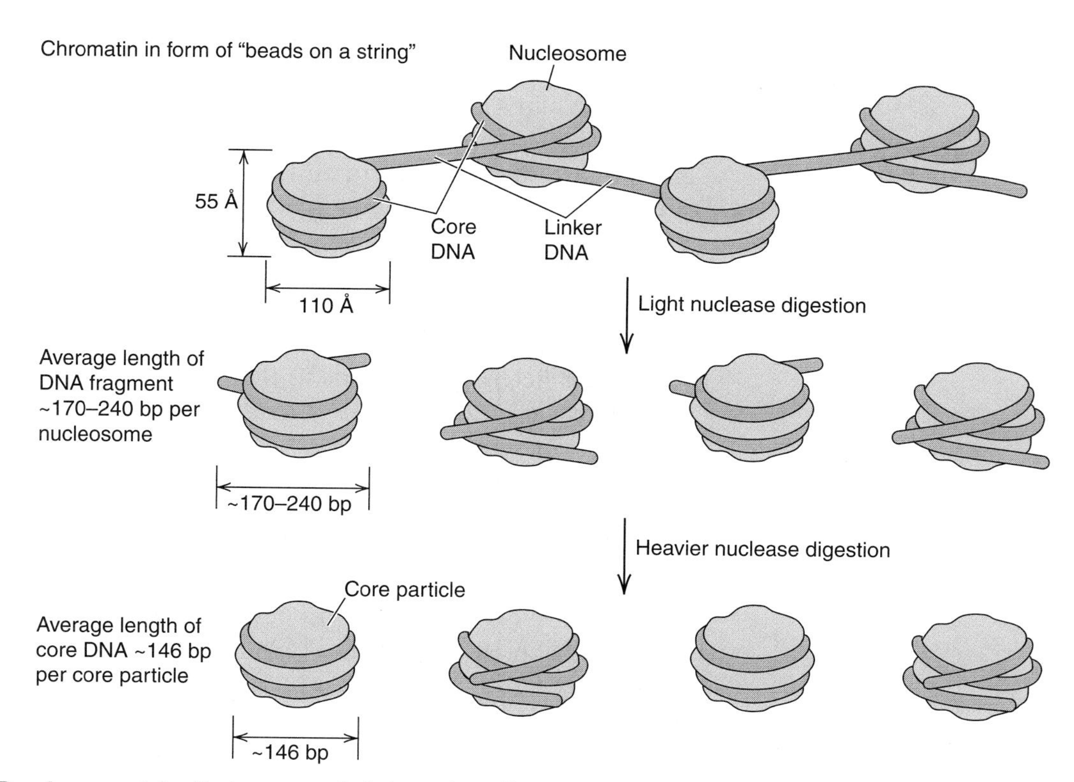

FIGURE 6.23 **"Beads on a string" structure of chromatin.** Effect of treatment with micrococcal nuclease. Brief treatment cleaves DNA in the linker region releasing free nucleosomes that contain 170 to 240 bp of DNA. More extensive treatment results in digestion of all but the 146 base pairs of DNA in intimate contact with the octameric protein complex.

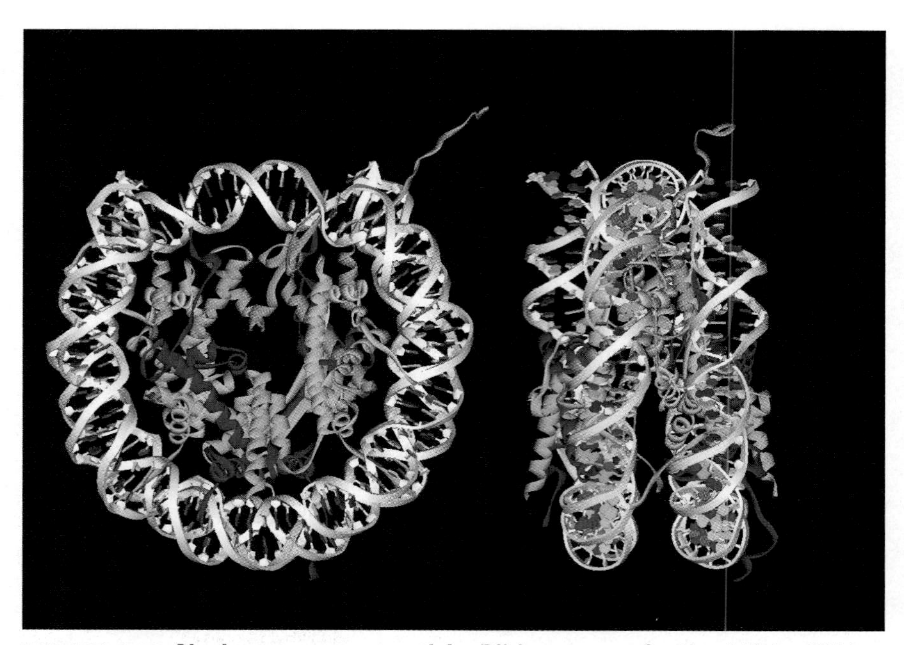

FIGURE 6.24 **Nucleosome core particle.** Ribbon traces for the 146-bp DNA phosphodiester backbone (white) and eight histone protein main chains (purple–H3; yellow–H4; red–H2A; and green–H2B) are shown from two different perspectives. (Structures from Protein Data Bank 1AOI. K. Luger, et al., *Nature* 389 [1997]: 251–260. Prepared by B. E. Tropp.)

Two copies of each histone protein, H2A, H2B, H3, and H4, assemble to form an octamer. All four histone proteins have a similar folding pattern known as the histone fold, which consists of three α-helices connected by two short loops. Structures for histones 3 and 4 are shown in FIGURES 6.25a and b, respectively. The central α-helix is about twice as long as the two flanking α-helices. An H3 subunit and H4 subunit interact to form an H3•H4 heterodimer in which loop 1 of one subunit is adjacent to loop 2 of the other subunit (FIGURE 6.25c). The dimer is stabilized by interactions between their antiparallel long α-helices. H2A and H2B subunits interact in a similar way to form H2A•H2B heterodimers. Two H3•H4 heterodimers pair to form a tetramer through contacts between their H3 subunits. Each H2A•H2B heterodimer binds to the $(H3H4)_2$ tetramer through contacts between H2B and H4. The resulting histone octamer has exact twofold symmetry.

Core DNA makes 1.65 turns as it wraps around a helical ramp in the octamer, to generate a left-handed toroidal supercoil with a radius of about 4.25 nm. The DNA averages about 10.2 bp per turn, but this value varies over the length of the DNA. This average value is less than the 10.5 to 10.6 bp per turn observed for naked DNA in solution. Each H2A•H2B and H3•H4 heterodimer binds 27 to 28 bp of DNA, generating bends that cause the DNA to follow a nonuniform path as it wraps around the histone octamer. DNA bending compresses the minor groove. A-T rich sequences accommodate this compression more readily than G-C rich sequences. Therefore, nucleosomes tend to position themselves on DNA so that A-T rich minor grooves contact the histone octamer. Extensive DNA-protein interactions take place

(a) Histone H3

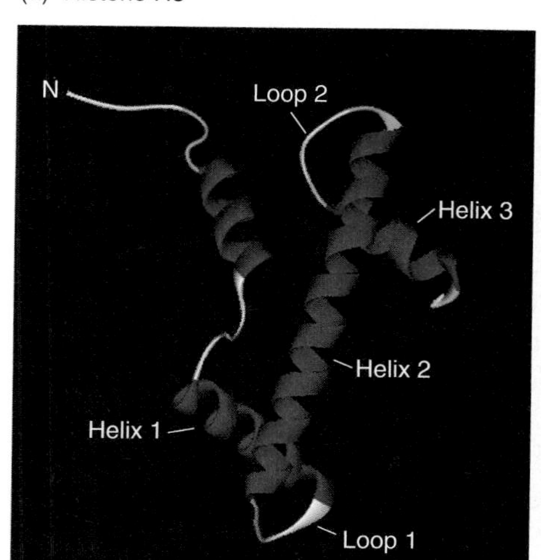

(b) Histone H4

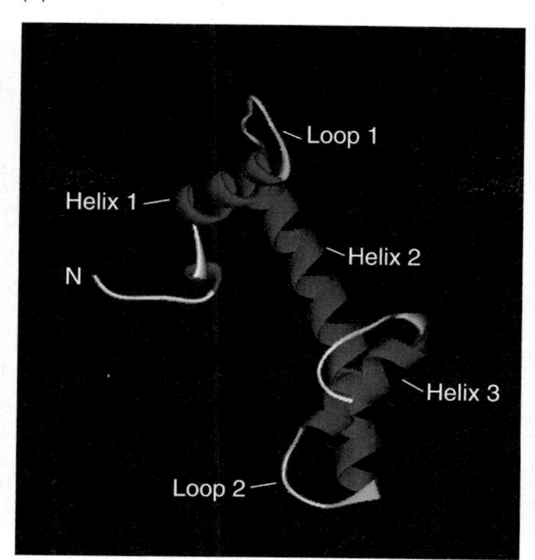

(c) Histone H3•H4 heterodimer

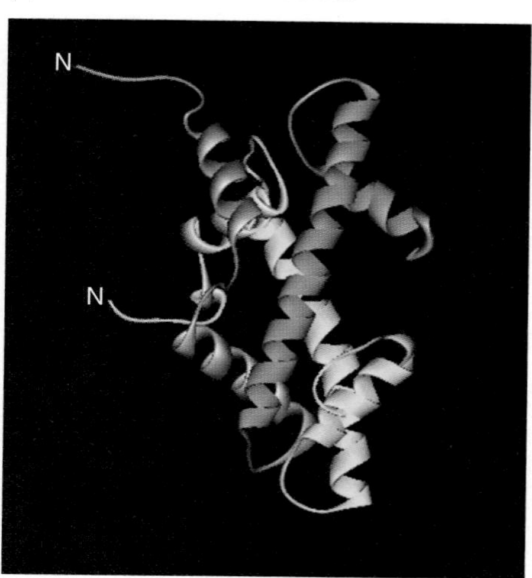

FIGURE 6.25 **The histone fold and histone heterodimer structure.** The histone fold in (a) H3 and (b) H4. The three α-helices (red) that are part of the histone fold are labeled helix 1, helix, 2, and helix 3. Other helices (red) are not labeled. Loop 1 connects helix 1 to helix 2 and loop 2 connects helix 2 to helix 3. (c) The structure of the H3•H4 heterodimer. Histone H3 is shown in blue and histone H4 in yellow. The orientation of H3 and H4 are identical to their orientations in (a) and (b), respectively. (Structures from Protein Data Bank 1AOI. K. Luger, et al., *Nature* 389 [1997]: 251–260. Prepared by B. E. Tropp.)

through hydrogen bonding between hydrogen atoms in the peptide bonds and oxygen atoms in phosphate groups that form the DNA backbone.

Because the histone octamer and the DNA each have a twofold axis of symmetry, one might expect the nucleosome core to also have this property. This is not so, however. The dyad of the histone octamer sits over a base pair rather than between two base pairs so that one half of the nucleosome core is 1 bp longer than the other half. This means that corresponding base pairs in each half of the DNA are in different physical environments.

As a general rule, histone octamers do not remain at a specific site on the DNA but are highly dynamic, moving from one position on DNA to another as the cell's requirements change. In a few rare cases, however, nucleosomes are located at specific positions on the DNA. Such specific positioning is often due to the presence of a non-histone protein that binds to a specific DNA sequence and then either stimulates or blocks nucleosome formation at adjacent sites.

N-terminal histone tails help to regulate chromatin structure and chromatin function and are therefore of considerable interest. Unfortunately, large segments of these tails are disordered and therefore cannot be seen by x-ray crystallography. The visible tail regions appear to be extended structures that interact with the DNA through the minor groove. For instance, the N-terminal tails of H3 and H2B pass through a channel created by the minor grooves in adjacent superhelical turns. Moreover, the N-terminal H4 tail makes many contacts with the face of an H2A•H2B dimer of a neighboring nucleosome core. This interparticle protein-protein contact probably helps to stabilize a higher order nucleosome packing arrangement, called the 30-nm fiber, (see below).

The precise nature of the interaction between H1 and the core particle is not known.

Thus far we have examined the contribution that four of the five histones make to chromatin structure. The fifth histone, histone H1, is present as part of a nucleoprotein particle called the **chromatosome** that is produced by briefly digesting chromatin with micrococcal nuclease. The chromatosome consists of a DNA that is 166 bp long that is wrapped around an octameric histone core and held in place by histone H1. A chromatosome can be converted to a nucleosome core by subjecting it to further micrococcal nuclease digestion to remove 10 bp of linker DNA from either end. Chromatosomes have not been prepared in crystalline form. The precise nature of the interaction between H1 and the rest of the chromatosome particle, therefore, is not known.

Eleven different human genes have been identified that code for histone H1 subtypes. All H1 isoforms have a common three-domain structure. A globular central domain is attached to a short N-terminal domain and a long lysine and arginine-rich C-terminal domain. Both the globular and C-terminal domains contribute to the protein's ability to bind to chromatosomal DNA. Based on biochemical, genetic, and physical evidence Tom Misteli and coworkers have proposed that H1 interacts with chromatosomal DNA as shown in FIGURE 6.26. According to this model, H1 seals about 1.6 turns of DNA by binding to an exit/entrance site on the surface of the chromatosome. Considerable evidence indicates that histone H1 contributes to the higher-order folding states of chromatin.

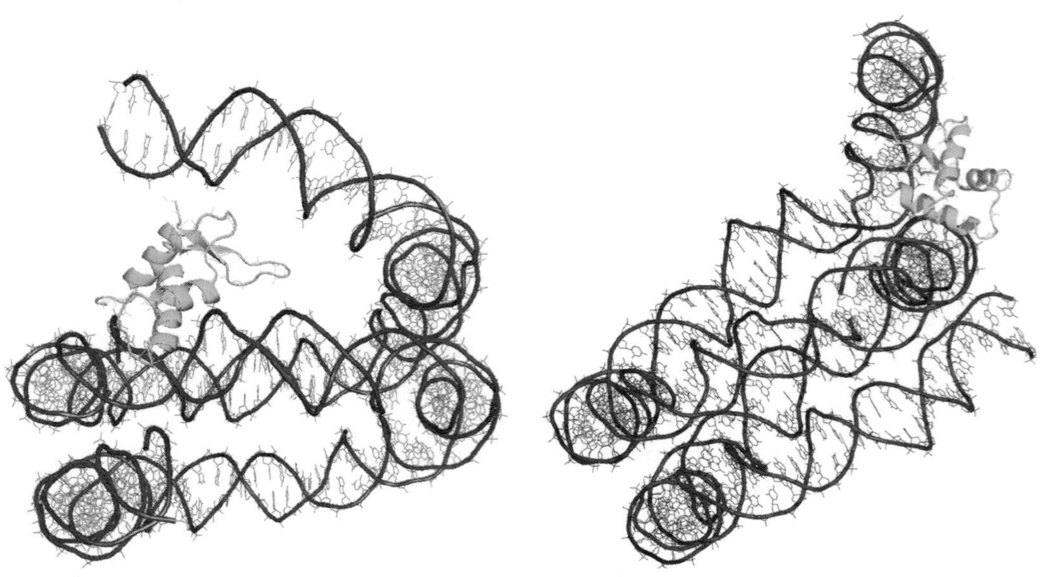

FIGURE 6.26 **Two views of a molecular model of H1 interaction within the chromosome.** The H1 globular domain is shown in green, nucleosomal DNA is red, and the dyad is blue. The other histones are not shown. (Photo courtesy of David T. Brown, University of Mississippi Medical Center.)

6.5 The 30-nm Fiber

A chain of nucleosomes appears to fold into a 30-nm fiber.

Despite extensive efforts over more than 30 years, we still do not know how a chain of nucleosomes folds into higher order structures. For some time, investigators have thought that the next level of organization is a 30-nm fiber first seen in electron micrographs of chromosomal material. FIGURE 6.27 shows the 30-nm chromatin fiber from a mouse chromosome.

Various models have been proposed to explain how nucleosomes fold to form the 30-nm fiber. The two that have gained the most support, the **zigzag model** and the **solenoid model,** are shown in FIGURE 6.28. The zigzag model predicts that the linker DNA forms a straight path between successive nucleosomes that lie on opposite

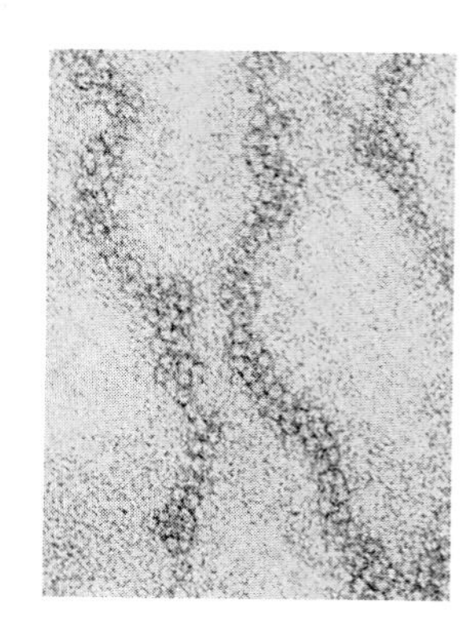

FIGURE 6.27 **The 30-nm chromatin fiber.** Electron micrograph of the 30-nm chromatin fiber in mouse chromosomes. (Courtesy of Barbara A. Hamkalo, University of California, Irvine.)

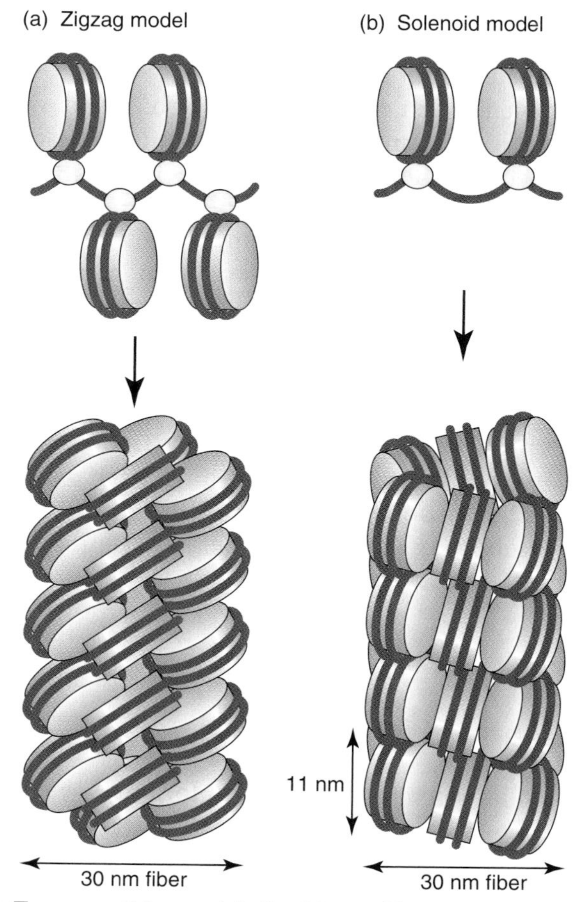

FIGURE 6.28 **Two possible models for 30-nm fiber.** (a) The zigzag model predicts a straight DNA linker will connect successive nucleosomes, which will be on opposite sides of the 30-nm fiber. (b) The solenoid model predicts the nucleosome chain forms a helical structure with a bent DNA linker connecting successive nucleosomes. (Reprinted from *Cell*, vol. 116, S. Khorasanizadeh, The nucleosome, pp. 259–272, copyright 2004, with permission from Elsevier [http://www.sciencedirect.com/science/journal/00928674]).

sides of the fiber, resulting in a three-dimensional zigzag linker pattern. The solenoid model predicts that the nucleosome chain forms a helical structure with about 6 nucleosomes per turn. The solenoid axis is perpendicular to the core particle axis (corresponding to the axis of the superhelical DNA winding around the histone octamer). Linker DNA must be bent to connect neighboring nucleosomes in the solenoid. Reconstitution experiments have been performed to discover which model is correct. Some of these experiments appear to support the zigzag model and others the solenoid model. Unfortunately, there are so many variables involved in reconstituting the chromatin fiber, that it is not possible to tell whether the zigzag or solenoid model more closely resembles the 30-nm fiber in the living cell. Two examples will help to illustrate the problem. Timothy J. Richmond and coworkers built a tetranucleosome and showed that its crystal structure fits the zigzag model. On the other hand, Daniela Rhodes and coworkers reconstituted fibers with up to 72 nucleosomes. Cryo-electron microscopy studies of the fibers indicate that the fiber has a solenoid type structure. It is possible that both structures exist in *vivo* or that neither model describes the actual structure in *vivo*.

The situation is even more unsettled than this brief account would suggest. Electron microscopy studies that first revealed the 30-nm fibers used samples that were fixed chemically, dehydrated, and embedded in plastic. Jacques Dubochet and coworkers have recently used cryo-electron microscopy to examine mitotic chromosomes in frozen hydrated sections of human cells that had not been fixed or stained. They did not see any evidence of the 30-nm fiber even though they could see microtubules of comparable size. Their experiments suggest that the mitotic chromosomes viewed in earlier electron microscopy experiments might have been altered during sample preparation. If so, the 30-nm fibers viewed in previous electron microscopy studies might have been artifacts. Even if the condensed chromosome does not have a discrete 30-nm fiber, it still seems likely that the 30-nm fiber is an intermediate in the compaction process.

6.6 The Scaffold Model

The scaffold model was proposed to explain higher order chromatin structure.

In the late 1970s, Ulrich K. Laemmli and coworkers proposed the **scaffold model** for chromosome structure. According to this model, non-histone proteins form a central scaffold along the long axis of a chromatid and somehow hold chromatin fibers in loops that extend out from the axis. This model predicts that loop structures will be maintained after histones are selectively removed. Laemmli and James R. Paulson tested the model by first removing histones from isolated mitotic chromosomes and then examining the histone depleted chromosomes under the electron microscope. As predicted by the model,

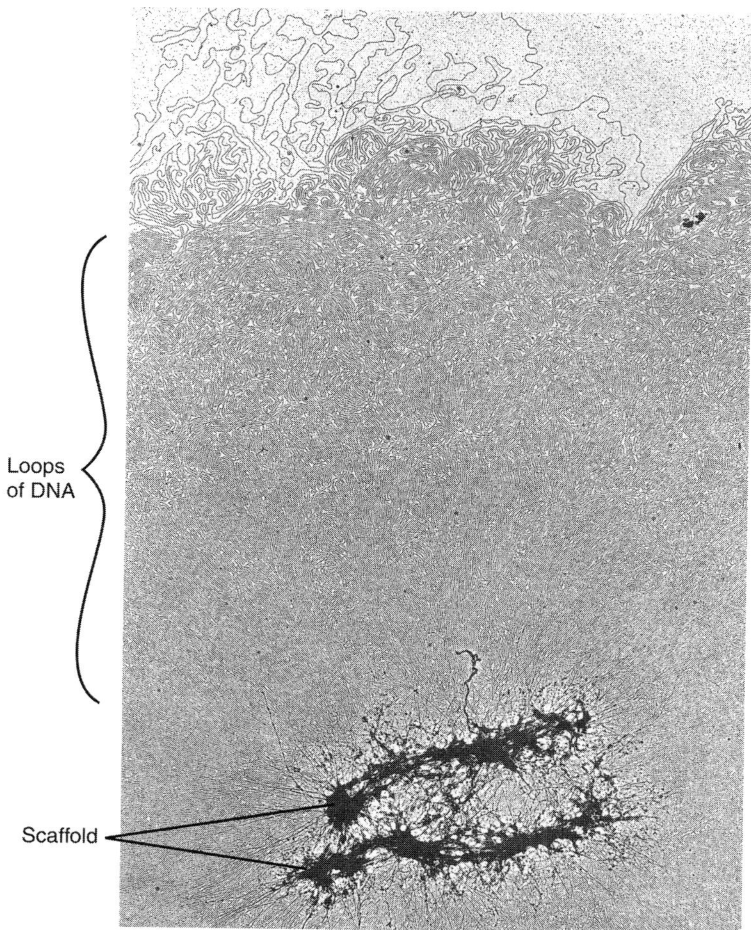

FIGURE 6.29 **Scaffold structure and part of surrounding DNA.** A structure called a scaffold becomes visible after the histones are depleted by treating the chromosome with 2 M NaCl. DNA loops extend from the scaffold. (Reprinted from *Cell*, vol. 12, J. R. Paulson and U. K. Laemmli, The structure of histone-depleted . . ., pp. 817–828, copyright 1977, with permission from Elsevier [http://www.sciencedirect.com/science/journal/00928674]. Photo courtesy of Ulrich K. Laemmli, University of Geneva, Switzerland.)

the histone-depleted chromosomes have DNA loops attached to a protein scaffold (FIGURE 6.29).

FIGURE 6.30 shows an updated version of the scaffold model. A-T rich DNA sequences called **scaffold attachment regions** (**SARs**) interact with specific scaffold proteins to form loops of the 30-nm fibers that may contain as much as 100 kbp of DNA.

Condensins and topoisomerase II help to stabilize condensed chromosomes.

In vitro studies show that topoisomerase II and SMC-related proteins are required for higher level chromatin condensation in amphibian egg cell extracts. Topoisomerase II disentangles DNA from different chromosomes, permitting each chromosome to fold into a discrete compact structure. Vertebrates have two SMC-related proteins, **condensin I** and

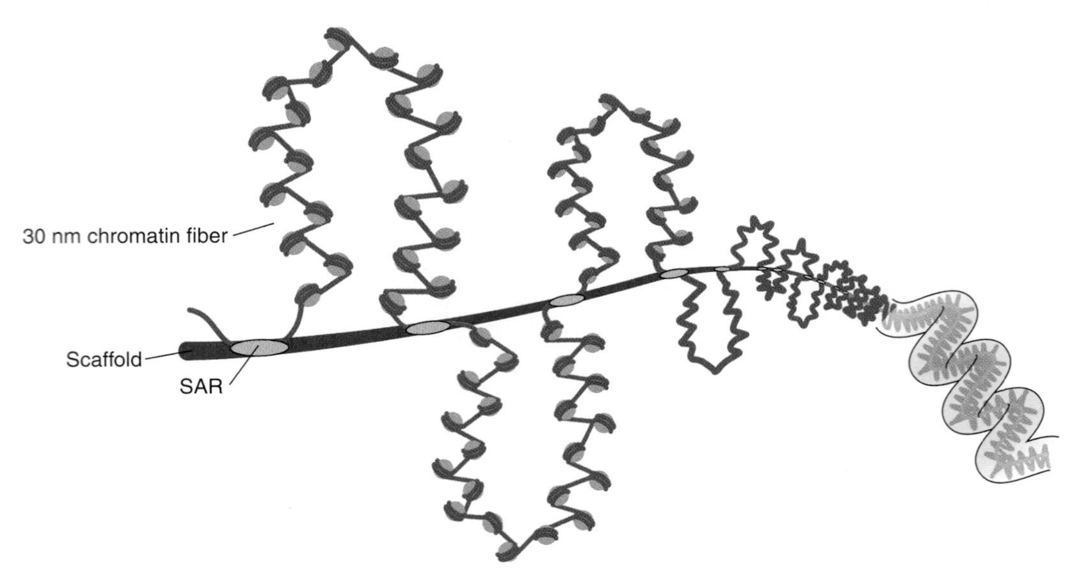

FIGURE 6.30 **Scaffold model.** Folding the 30-nm chromatin fiber into loops, mediated by proteins in the chromosome scaffold (red) and DNA segments (pink ovals) called scaffold attachment regions (SAR). (Adapted from J. R. Swedlow and T. Hirano, *Mol. Cell* 11 [2003]: 557–569.)

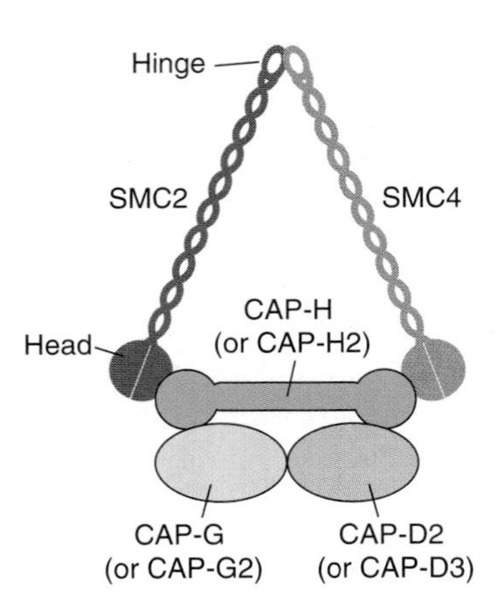

FIGURE 6.31 **Condensin structure.** Condensin I and II are heteropentamers. Both complexes have an SMC2•SMC4 heterodimer. The other three subunits are CAP-D2, CAP-G, and CAP-H in condensin I and CAP-D3, CAP-G2, and CAP-H2 in condensin II. (Adapted from K. Maeshima and M. Eltsov, *J. Biochem.* 143 [2008]: 145–153.)

condensin II. Both condensins are protein complexes containing five subunits (FIGURE 6.31). Two of the subunits, SMC2 and SMC4, form a heterodimer with a structure that is very similar to that described for the bacterial MukB protein. The other three subunits are CAP-D2, CAP-G, and CAP-H in condensin I and CAP-D3, CAP-G2, and CAP-H2 in condensin II. Condensin I has a DNA-dependent ATPase activity, which permits it to introduce positive supercoils into closed circular DNA. The contribution that this catalytic activity makes to chromosome structure remains to be determined. Further studies are required to determine whether condensin 2 also has this activity.

Histochemical studies of chromosomes from vertebrate cells reveal that condensin I, condensin II, and topoisomerase II are located along the long chromatid axis (FIGURE 6.32). Condensin I remains in the cytoplasm until the nuclear membrane is disassembled during prophase. In contrast, condensin II is present in the nucleus during interphase. Experiments have been performed using cells that cannot make condensin I or condensin II. As might be expected from their intracellular location, condensin I depletion does not influence prophase chromosome condensation, whereas condensin II depletion causes a significant delay in the initiation of prophase chromosome condensation. William C. Earnshaw and coworkers constructed chicken cells that do not make SMC2 (and therefore lack condensins I and II) when cultured in the presence of a tetracycline-like antibiotic. Chromosome condensation was delayed in drug-treated cells but eventually reached an almost normal condensation level. The chromosome structure was not stable, however. Earnshaw and coworkers also observed that nonhistone proteins such as topoisomerase II were not in their normal chromosomal locations. Their studies suggest that the condensins are not required for chromosome condensation but are instead needed to promote the correct association of nonhistone proteins with mitotic chromosomes and to form a stable chromosome structure.

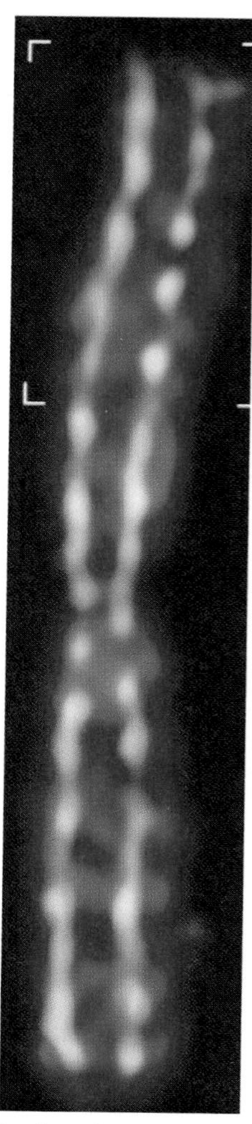

FIGURE 6.32 Isolated mitotic chromosome from a HeLa cell doubly stained for topoisomerase II and CAP-H. The immunostain for topoisomerase II is green and that for CAP-H is red. DNA is stained blue. (Reprinted from *Dev. Cell*, vol. 4, K. Maeshima and U. K. Laemmli, A two-step scaffolding model for mitotic . . ., pp. 467–480, copyright 2003, from Elsevier [http://www.sciencedirect.com/science/journal/15345807]. Photo courtesy of Ulrich K. Laemmli, University of Geneva, Switzerland.)

6.7 The Centromere

The centromere is the site of microtubule attachment.

Each chromosome has a distinct morphological region, the **centromere**, which is visible as a thin region in the metaphase chromosome (FIGURE 6.33). The centromere is the chromosomal substructure that is responsible for a eukaryotic cell's ability to accurately partition sister chromatids between two daughter cells during mitosis and meiosis. It is usually the last attachment site between sister chromatids during mitosis. The connection is usually broken at anaphase by cleaving the cohesins that hold the sister chromatids together.

In budding yeast cells, centromere DNA consists of a specific nucleotide sequence that is about 125 bp long. In contrast, human centromere DNA has a hierarchic organization. A 171 bp sequence is repeated with slight nucleotide sequence variations to form a higher repeat, which is in turn repeated with high sequence conservation to form a still higher order repeat called an α **satellite DNA array**, which ranges in size from 200 to 9000 kb (FIGURE 6.34).

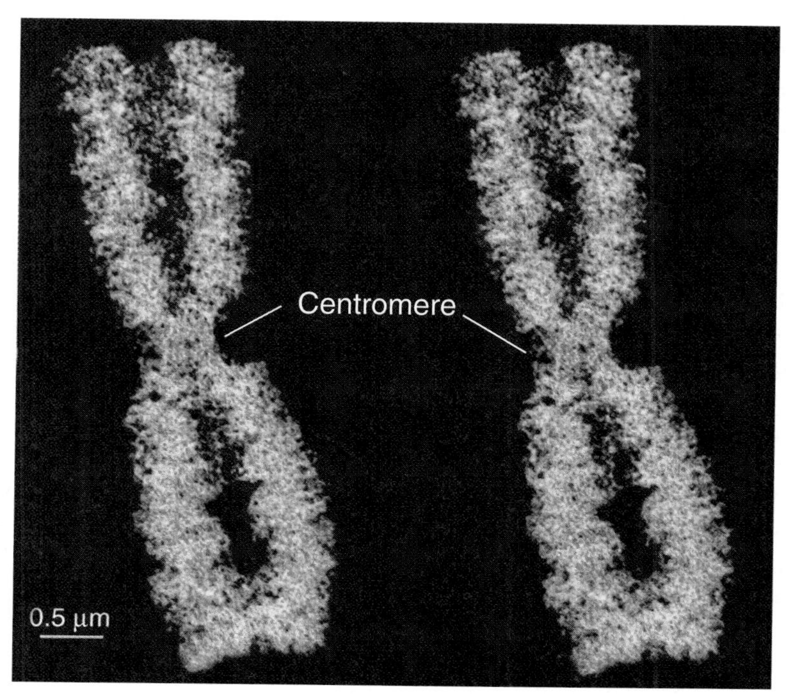

FIGURE 6.33 Electron tomography 3D reconstruction of Chinese hamster ovary (CHO) cell metaphase. (Photo courtesy of Peter Engelhardt, University of Helsinki, Finland.)

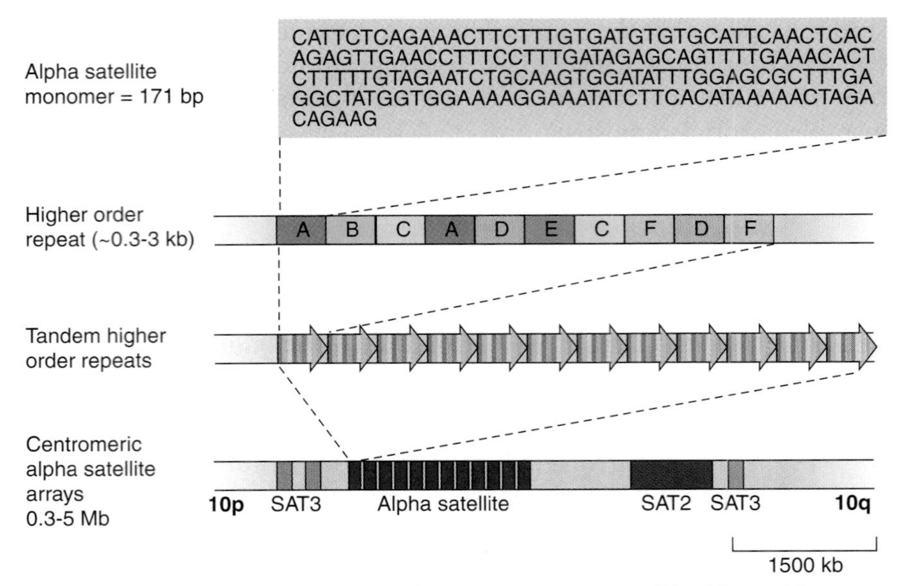

FIGURE 6.34 DNA organization in human centromeres. The hierarchic organization of a satellite is illustrated. A 171 bp sequence is repeated with slight nucleotide sequence variations to form a higher repeat, which in turn is repeated with high sequence conservation to form a still higher order repeat called an alpha satellite DNA array, which ranges in size from 200 to 9000 kb. At bottom is a diagram of the centromere region of chromosome 10, illustrating the ~2Mb alpha satellite array with surrounding pericentric satellite arrays (SAT2 and SAT3). (Adapted from D. W. Cleveland, Y. Mao, and K. F. Sullivan, *Cell* 112 [2003]: 407–421.)

The centromere contains a specialized form of chromatin that is assembled by packing the DNA with histones and other proteins. The nonhistone proteins can be divided into two groups, constitutive and passenger proteins. Constitutive proteins remain associated with the centromere throughout the entire cell life cycle, whereas passenger proteins associate with the centromere during one stage of the life cycle but not others. One of the constitutive proteins, **CENP-A** (**centromere protein A**) is a histone H3 variant that combines with the other histones to form a new type of nucleosome that is dispersed among normal nucleosomes in the centromere.

Other constitutive and passenger proteins assemble to form the microtubule attachment site called the **kinetochore**. Electron microscopic analysis has shown that a single spindle-protein fiber is attached to the kinetochore in chromosomes present in budding yeast (*Saccharomyces cerevisiae*). Human cell chromosomes, which are much larger than yeast chromosomes, have several spindle fibers attached to their kinetochores.

The kinetochore consists of two regions, an inner kinetochore region that interacts with the CENP-A containing nucleosomes and an outer kinetochore region that is the site of attachment for microtubules that form the mitotic spindle fiber. FIGURE 6.35 shows a model proposed by K. H. Choo and coworkers for human centromere and the kinetochore structures.

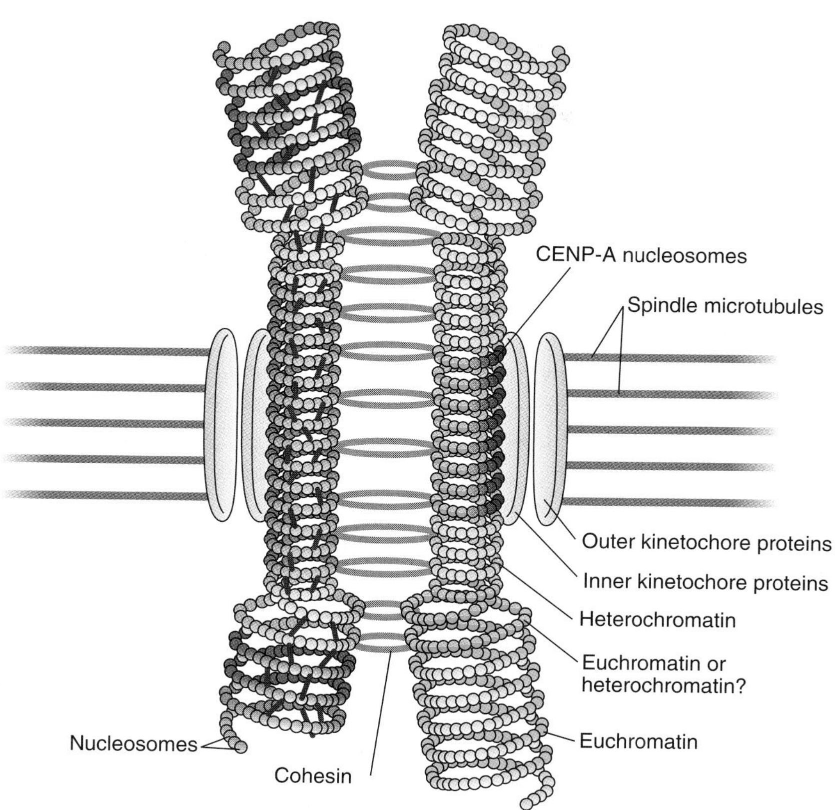

FIGURE 6.35 **Model for three-dimensional organization of a human centromere.** A chromatin fiber in the centromere may fold into a cylindrical coil as a zigzag (or solenoid) structure. This arrangement would allow the individual blocks of nucleosomes that contain centromere protein A (CENP-A) molecules to form a single domain consisting of a series of repetitive functional units on the metaphase chromosome face on the poleward side, where they can interact with the remaining foundation of kinetochore proteins. On the opposite cylindrical face, intervening histone H3 blocks would also form a single domain. The extent to which this inner domain is heterochromatic or euchromatic is uncertain. This uncertainty has implications for the localization of cohesin (shown in gray), a multi-subunit protein complex that is necessary to maintain chromatid cohesion. CENP-A is shown in red; related centromere proteins are shown in purple or green. Heterochromatin is shown in yellow; euchromatin is shown in blue; and regions of uncertain heterochromatin or euchromatin status are shown in half blue and half yellow. Outer kinetochore proteins are shown in cyan. (Adapted from D. J. Amor, et al. *Trends Cell Biol.* 14 [2004]: 359–368.)

6.8 The Telomere

The telomere, which is present at either end of a chromosome, is needed for stability.

Independent studies by two distinguished geneticists, Hermann J. Muller and Barbara McClintock, performed in the late 1930s revealed that the ends of a chromosome are essential for chromosome stability. Muller, who coined the term **telomere** (Gr. *telos* = end, *meros* = part) to describe chromosome ends, became aware of their importance as a result of experiments in which x-rays were used to produce deletions in *Drosophila* chromosomes. Because he was able to recover stable chromosomes with internal deletions but could not find chromosomes with end deletions, Muller concluded that the ends were essential for chromosome stability. McClintock reached a similar conclusion when she observed that broken chromosomes in maize fused together, underwent structural changes, or were degraded.

Very little progress was made in understanding telomere structure until 1978 when Elizabeth H. Blackburn and Joseph G. Gall succeeded in isolating telomeres from *Tetrahymena thermophilia*, a ciliated protozoan with two kinds of nuclei, a micronucleus and a macronucleus. The micronucleus, which has five chromosome pairs, serves as the germ-line nucleus, transferring genetic information from one generation to the next but is not involved in transcription. The macronucleus, which is transcriptionally active, is derived from the micronucleus. During a stage of the *T. thermophilia* life cycle, chromosomes in the micronucleus are split into segments that then are amplified to produce thousands of "minichromosomes." Blackburn and Gall showed that the ends of these minichromosomes have the short simple sequence, TTGGGG, repeated over and over.

The studies with *T. thermophilia* minichromosomes motivated investigators to determine whether chromosomes from other organisms also have simple sequence repeats at their ends. Telomeric DNAs from a variety of eukaryotes have now been characterized. In most cases, the DNA consists of a simple repeat array in which the repeat unit is usually 5 to 8 bp in length.

Most telomeric DNA sequences have a high G content in the strand that runs 5′→3′ toward the end of the chromosome. For example, the repeat array is TTAGGG in humans and other vertebrates and TTTAGGG in *Chorella*. The repeat pattern for these simple sequences varies from a few hundred base pairs in yeast and protozoans to a few thousand base pairs in vertebrates. The G-rich strand ends in a 3′ single stranded overhang (FIGURE 6.36). That is, the G-rich strand extends beyond the 5′-end of the complementary C-rich strand. The overhang contains from two to three repeat units in simple eukaryotes but approaches 30 repeat units in humans.

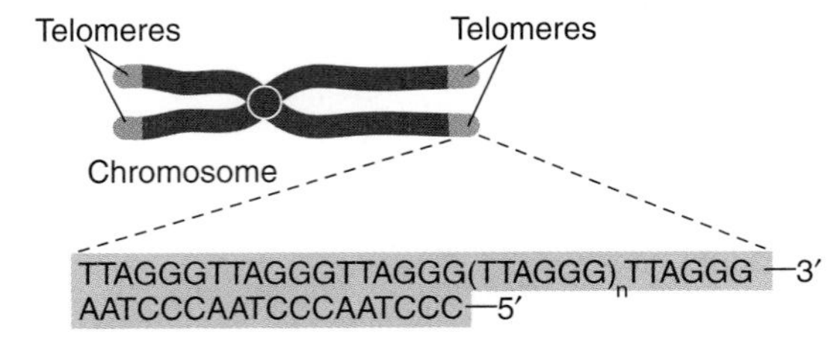

FIGURE 6.36 3′-Telomere overhang. Telomeres, shown in pink, are located at the ends of the sister chromatids. Each telomere has a G-rich 3′-overhang. In simple eukaryotes, n usually has a value of 1 or 2 but in humans, it may reach about 30.

Electron microscopy studies performed by Titia de Lange and coworkers in 1999 revealed that the G-rich strand's 3′-overhang folds back to invade the double-stranded region in the mammalian telomere to form a t-loop (telomere loop) and a D-loop (displacement loop; FIGURE 6.37). Specific proteins are required to establish and maintain telomere structure.

Although here we focus on proteins that protect the mammalian telomere, similar proteins are also present in other kinds of eukaryotes. Two of the telomere proteins, **TRF1** and **TRF2** (TTAGGG repeat binding factors 1 and 2), bind to the double-stranded region of the telomere as pre-formed homodimers. Although both proteins have similar structures, they have different functions. TRF1 is a negative

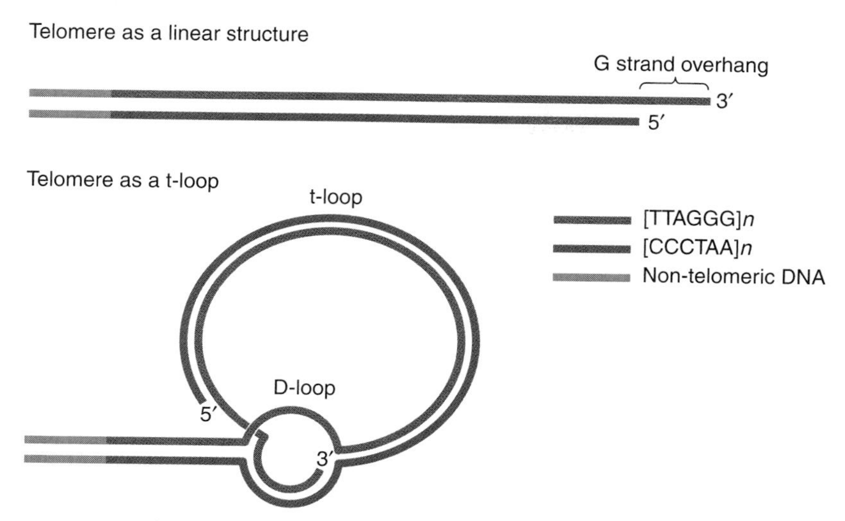

FIGURE 6.37 **Structure of telomere t- and D-loops.** A t-loop structure forms when the 3′ G-strand extension at the end of a chromosome (telomere) invades duplex telomeric repeats, thereby forming a displacement loop (D-loop). (Adapted from V. Lundblad, *Science* 288 [2000]: 2141–2142.)

regulator of telomere length. Cells that overproduce TRF1 appear to have shorter telomeres than normal. TRF2 participates in t-loop formation and also appears to be required to cap and protect the chromosome ends. TRF1 and TRF2 each recruit other proteins to the telomere.

The recruited proteins also influence telomere formation and structure. Defects in some of these proteins have been associated with inborn errors of metabolism. For example, one of the proteins recruited by TRF2, the Werner syndrome helicase (WRN), is defective in individuals with Werner syndrome, a genetic disease associated with premature aging. A second type of DNA binding protein, **POT1** (protection of telomeres) binds to single-stranded DNA containing a TTAGGGTTAGGG decamer and is probably associated with the single strand in the D-loop. FIGURE 6.38 shows a model of the mammalian telomere with associated proteins. TRF1, TRF2, POT1, and three other proteins form a complex called **shelterin** that enables cells to distinguish their natural chromosome ends from DNA breaks. The three additional proteins, TINT1 (TRF1 interacting protein), TPP1, and Rap1, are not shown in Figure 6.38.

The cell must maintain its telomeres. Failure to do so leads to a wide assortment of serious problems, ranging from chromosome fusion to programmed cell death. As described in detail in Chapter 10, eukaryotic chromosomes lose DNA from their ends as a result of replication and require a special enzyme called telomerase to replace the lost DNA. Most human somatic cells lack telomerase and so cannot replace the telomere DNA lost during replication.

Failure to replace this DNA is probably responsible for the fact that somatic cells go through a finite number of divisions when studied in tissue culture, a phenomenon that is directly related to the aging

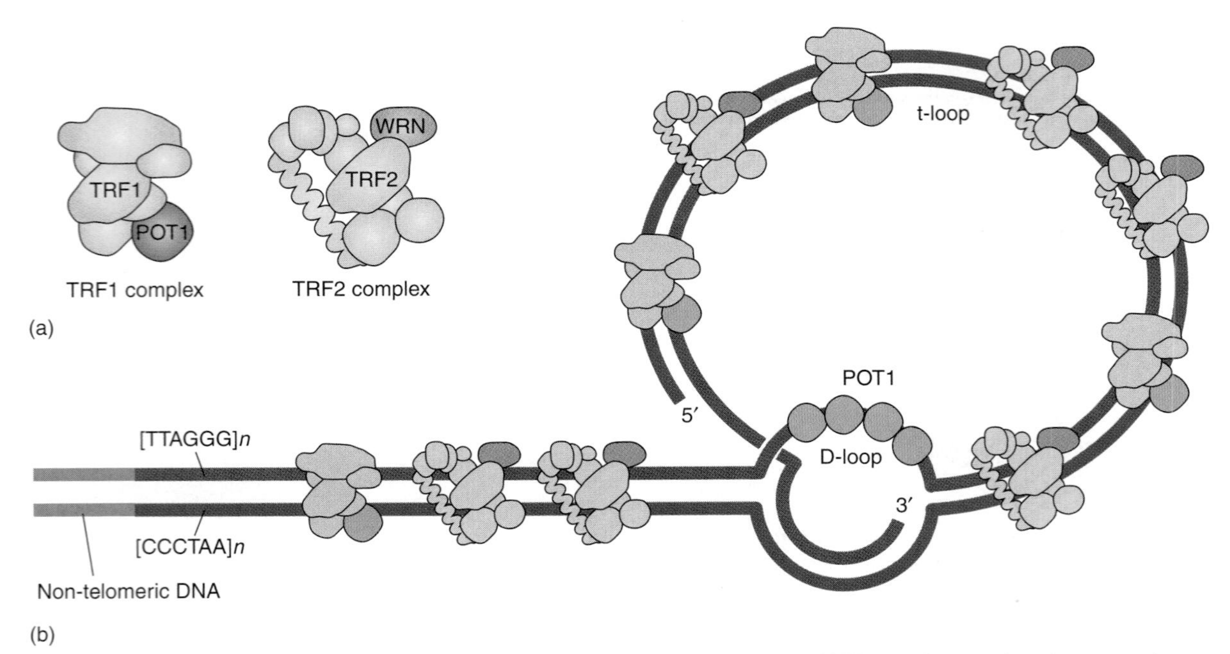

FIGURE 6.38 **Proposed structure of the human telomeric complex.** (a) TRF1 and TRF2 each recruit other proteins to the telomere. The major role of the TRF2 complex appears to be to protect the chromosome ends. The TRF1 complex appears to regulate telomerase-mediated telomere maintenance. (b) The 3′-G strand overhang in human telomeres, an array of TTAGGG repeats that is 100 to 200 nucleotides long, is connected to an array of TTAGGG repeats in the linear duplex region that is 2 to 30 kb long. The telomere DNA can exist as a t-loop in which the 3′-G strand overhang invades the duplex-repeat array to form a displacement or D-loop of TTAGGG repeats. Other configurations are also possible. POT1 (protection of telomeres 1) protein binds to the single-stranded TTAGGG-repeat DNA that results from t-loop formation. Two double-stranded TTAGGG repeat binding factors (TRFs), TRF1, and TRF2, interact with the duplex repeats. (Adapted from T. de Lange, *Nat. Rev. Mol. Cell Biol.* 5 [2004]: 323–329.)

process. Germ cells have telomerase and so can divide indefinitely. Most cancer cells also have telomerase. Telomerase is not an essential feature of tumor cells, however, because some tumor cells have an alternate means for maintaining their telomeres. Nevertheless, telomerase is an attractive target for treating many different types of cancers.

Suggested Reading

Bacterial Chromatin

Bendich, A. J., and Drlica, K. 2000. Prokaryotic and eukaryotic chromosomes: what's the difference. *Bioessays* 22:481–486.

Breier, A. M., and Cozzarelli, N. R. 2004. Linear ordering and dynamic segregation of the bacterial chromosome. *Proc Natl Acad Sci USA* 101:9175–9176.

Case, R. B., Chang, Y.-P., Smith, S. B., et al. 2004. The bacterial condensin MukBEF compacts DNA into a repetitive, stable structure. *Science* 305:222–227.

Cui, Y., Petrushenko, Z. M., and Rybenkov, V. V. 2008. MukB acts as a macromolecular clamp in DNA condensation. *Nat Struc Mol Biol* 15:411–418.

Cunha, S., Odjik, T., Süleymanoglu, E., and Woldringh, C. L. 2001. Isolation of the *Escherichia coli* nucleoid. *Biochimie* 83:149–154.

Dagupta, S., Masinier-Patin, S., and Nordström, K. 2000. New genes with old modus operandi. *EMBO Rep* 1:323–327.

Dame, R. T. 2005. The role of nucleoid-associated proteins in the organization and compaction of bacterial chromatin. *Mol Microbiol* 56:858–870.

Dame, R. T., and Goosen, N. 2002. HU: promoting or counteracting DNA compaction? *Fed Eur Biochem Sci* 529:151–156.

Dorman, C. J., and Deighan, P. 2003. Regulation of gene expression by histone-like proteins in bacteria. *Curr Opin Genet Dev* 13:179–184.

Dorman, C. J., Hinton, J. C. D., and Free, A. 1999. Domain organization and oligomerization among H-NS-like nucleoid-associated proteins in bacteria. *Trend Microbiol* 7:124–128.

Fenkiel-Krispin, D., Ben-Avraham, I., Englander, J., et al. 2004. Nucleoid restructuring in stationary-state bacteria. *Mol Microbiol* 51:395–405.

Gloyd, M., Ghirlando, R., Mattews, L. A., and Guarne, A. 2007. MukE and MukF form two distinct high affinity complexes. *J. Biol Chem* 282:14373–14378.

Hirano, T. 2005. SMC proteins and chromosome mechanics: from bacteria to humans. *Philo Trans R Soc Lond B Biol Sci* 360:507–514.

Hirano, T. 2006. At the heart of the chromosome: SMC proteins in action. *Nat Rev Mol Cell Biol* 7:311–322.

Holmes, V. F., and Cozzarelli, N. R. 2000. Closing the ring: links between SMC proteins and chromosome partitioning, condensation, and supercoiling. *Proc Natl Acad Sci USA* 97:1322–1324.

Joongbaek, K., Yoshimura, S. H., Hizume, K., et al. 2004. Fundamental structural units of the *Escherichia coli* nucleoid revealed by atomic force microscopy. *Nucl Acids Res* 32:1982–1992.

Kellenberger, E. 2006. Bacterial chromosome. In: *Encyclopedia of Life Sciences.* pp. 1–8. Hoboken, NJ: John Wiley and Sons.

Lim, J.-H., and Oh, B.-H. 2009. Structural and functional similarities between two bacterial chromosome compacting machineries. *Biochem Biophys Res Commun* 386:415–419.

Luijsterburg, M. S., Noom, M. C., Wuite, G. J. L., and Dame, R. T. 2006. The architectural role of nucleoid-associated proteins in the organization of bacterial chromatin: a molecular perspective. *J Struc Biol* 156:262–272.

Niki, H. Jaffe, A., Imamura, R., Ogura, T., and Hiraga, S. 1991, The new gene *mukB* codes for a 177 kd protein with coiled-coil domains involved in chromosome partitioning of *E. coli. EMBO J* 10:183–193.

Lovett, S. T., and Segall, A. M. 2004. New views of the bacterial chromosome. *EMBO Rep* 5:860–864.

Postow, L., Hardy, C. D., Arsuaga, J., and Cozzarelli, N. R. 2004. Topological domain structure of the *Escherichia coli* chromosome. *Genes Dev* 18:1766–1779.

Reyes-Lamothe, R., Wang, X., and Sheratt, D. 2008. *Escherichia coli* and its chromosome. *Trends Microbiol* 16:238–245.

Rimsky, S. 2004. Structure of the histone-like protein H-NS and its role in regulation and genome superstructure. *Curr Opin Microbiol* 7:109–114.

Rybenkov, V. V. 2009. Towards the architecture of the chromosomal architects. *Nat Struc Mol Biol* 16:104–105.

Saier, M. H. Jr. 2008. The bacterial chromosome. *Crit Rev Biochem Mol Biol* 43:89–134.

Sharpe, M. E., and Errington, J. 1999. Upheaval in the bacterial nucleoid: an active chromosome segregation mechanism. *Trends Genet* 15:70–74.

She, W., Wang, Q, Mordukhova, E. A. and Rybenkov, V. V. 2007. MukEF is required for stable association with the chromosome. *J Bacteriol* 189:7062–7068.

Sheratt, D. J. 2003. Bacterial chromosome dynamics. *Science* 301:780–785.

Tendeng, C., and Berlin, P. N. 2003. H-NS in gram-negative bacteria: a family of multifaceted proteins. *Trends Microbiol* 11:511–518.

Thanbichler, M., and Shapiro, L. 2006. Chromosome organization and segregation in bacteria. *J. Struct Biol* 156:292–303.

Thanbichler, M., Viollier, P. H., and Shapiro, L. 2005. The structure and function of the bacterial chromosome. *Curr Opin Genet Dev* 15:1–10.

Thanbichler, M., Wang, S. C., and Shapiro, L. 2005. The bacterial nucleoid: a highly organized and dynamic structure. *J Cell Biochem* 96:506–521.

Valens, M., Penaud, S., Rossignol, M., et al. 2004. Macrodomain organization of the *Escherichia coli* chromosome. *EMBO J* 23:4330–4341.
Woo, J.-S., Lim, J.-H., Shin, H.-C., et al. 2009. Structural studies of a bacterial condensin complex reveal ATP-dependent disruption of intersubunit interactions. *Cell* 136:85–96.
Wu, L. J. 2004. Structure and segregation of the bacterial nucleoid. *Curr Opin Genet Dev* 14:126–132.

Mitosis and Meiosis

Allison, D. C., Nestor, A. L., and Isaka, T. 2005. Chromosomes during cell division. In: *Encyclopedia of Life Sciences*. pp. 1–11. Hoboken, NJ: John Wiley and Sons.
Appels, R., Morris, R., Gill, B. S., and May C. E. 1998. *Chromosome Biology*. Norwell, MA: Kluwer Academic Publishers.
Barbero, J. L. 2009. Cohesins: chromatin architects in chromosome segregation, control of gene expression, and much more. *Cell Mol Life Sci* 66:2025–2035.
Barra, Y. 2004. FEAR pulls them apart. *Dev Cell* 6:608–610.
Criqui, M. C., and Genschik, P. 2002. Mitosis in plants: how far we have come at the molecular level? *Curr Opin Plant Biol* 5:487–493.
Eissenberg, J. C., and Elgin, S. 2005. Heterochromatin and euchromatin. In: *Encyclopedia of Life Sciences*. pp. 1–7. Hoboken, NJ: John Wiley and Sons.
Gartenberg, M. 2009. Heterochromatin and the cohesion of sister chromatids. *Chromosome Res* 17:229–238.
Gonczy, P. 2002. Nuclear envelope: torn apart at mitosis. *Curr Biol* 12:R242–R244.
Hassold T., and Hunt P. 2001. To err (meiotically) is human: the genesis of human aneuploidy. *Nat Rev Genet* 2:280–291.
John, B. 1990. *Meiosis*. Cambridge, UK: Cambridge University Press.
Jordan, M. A., and Wilson, L. 1999. The use and action of drugs in analyzing mitosis. *Methods Cell Biol* 61:267–295.
Kleckner, N. 1996. Meiosis: how could it work? *Proc Natl Acad Sci USA* 93:8167–8174.
Kohli, J., and Hartsuiker, E. 2008. Meiosis. In: *Encyclopedia of Life Sciences*. pp. 1–8. Hoboken, NJ: John Wiley and Sons.
Lichten M. 2001. Meiotic recombination: breaking the genome to save it. *Curr Biol* 11:R253–R256.
McKee, B. D. 2004. Homologous pairing and chromosome dynamics in meiosis and mitosis. *Biochim Biophys Acta* 1677:165–180.
McNairn, A. J., and Gerton, J. L. 2008. The chromosome glue gets a little stickier. *Trends Genet* 24:382–389.
Mitchison, T. J., and Salmon, E. D. 2001. Mitosis: a history of division. *Nat Cell Biol* 3:E17–E21.
Moore, C. M., and Best, R. G. 2007. Chromosome mechanics. In: *Encyclopedia of Life Sciences*. pp. 1–10. Hoboken, NJ: John Wiley and Sons.
Nasmyth, K., and Haering, C. H. 2009. Cohesin: its roles and mechanisms. *Ann Rev Genetics* 43:525–558.
Onn, I., Heidinger-Pauli, J. M. Guacci, V, Ünal, E., and Koshland, D. E. 2008. Sister chromatid cohesion: a simple concept with a complex reality. *Ann Rev Cell Dev Biol* 24:105–129.
Page, A. W., and Orr-Weaver, T. L. 1997. Stopping and starting the meiotic cell cycle. *Cur Opin Genet Dev* 7:23–31.
Page, S. L., and Hawley, R. S. 2003. Chromosome choreography: the meiotic ballet. *Science* 301:785–789.
Pines, J., and Rieder, C. L. Re-staging mitosis: a contemporary view of mitotic progression. *Nat Cell Biol* 3:E3–E6.
Peters, J.-M., and Hauf, S. 2005. Meiosis and mitosis: molecular control of chromosome separation. 2005. In: *Encyclopedia of Life Sciences*. pp. 1–7. Hoboken, NJ: John Wiley and Sons.
Peters, J.-M., Tedeschi, A, and Schmitz, J. 2008. The cohesin complex and its role in chromosome biology. *Genes Dev* 22:3089–3114.

Rieder, C. L., and Khodjakov, A. 2003. Mitosis through the microscope: advances in seeing inside live dividing cells. *Science* 300:91–96.

Roeder, G. S. 1997. Meiotic chromosomes: it takes two to tango. *Genes Dev* 11: 2600–2621.

Russell, P. 1998. Checkpoints on the road to mitosis. *Trends Biochem Sci* 23:399–402.

Stukenberg, P. T. 2003. Mitosis: long-range signals guide microtubules. *Curr Biol* 13:R848–R850.

Watrin, E., and Legagneux, V. 2003. Introduction to chromosome dynamics in mitosis. *Biol Cell* 95:507–513.

Uhlmann, F. 2001. Chromosome cohesion and segregation in mitosis and meiosis. *Curr Opin Cell Biol* 13:754–761.

Karyotype

Barch, M.J., Knutsen, T., and Spurbeck, J. L. (eds). 1997. *The AGT Cytogenetics Laboratory Manual*, 3rd ed. Baltimore, MD: Lippincott-Raven.

Bickmore, W. A. 2001. Karyotype analysis and chromosome banding. In: *Encyclopedia of Life Sciences*. pp. 1–7. London: Nature.

Blennow, E. 2005. Banding techniques. In: *Encyclopedia of Life Sciences*. pp. 1–5. Hoboken, NJ: John Wiley and Sons.

Craig, J. M., and Bickmore, W. A. 1993. Chromosome bands–flavours to savour. *Bioessays* 15:349–354.

Craig, J. M., and Bickmore, W. A. 1997. *Chromosome Bands: Patterns in the Genome*. New York: Springer-Verlag.

Gosden, J. R. 1994. Chromosome analysis protocols. In: Walker J. M. (ed). *Methods in Molecular Biology*, vol. 29. Totawa, NJ: Humana Press.

Mitelman, F. (ed). 1995. *An International System for Human Cytogenetic Nomenclature*. Basel, Switzerland: Karger.

Moore, C. M., and Best, R. G. 2001. Chromosome preparation and banding. In: *Encyclopedia of Life Sciences*. pp. 1–7. London: Nature.

Rooney, D. E., and Czepulkowski, B. H. (eds). 1992. *Human Cytogenetics: A Practical Approach, vol. I, Constitutional Analysis*, 2nd ed. Washington, DC: IRL Press.

Rooney, D. E., and Czepulkowski, B. H. (eds). 1994. *Human Cytogenetics: Essential Data*. Hoboken, NJ: John Wiley and Sons.

Sandberg, A. 1990. *The Chromosomes in Human Cancer and Leukemia*, 2nd ed. Frankfurt, Germany: Elsevier.

Schröck, E., du Manoir, S., Veldman, T. et al. 1996. Multicolor spectral karyotyping of human chromosomes. *Science* 273:494–497.

Shaffer, L. G. 2005. Karyotype interpretation. In: *Encyclopedia of Life Sciences*. pp. 1–7. Hoboken, NJ: John Wiley and Sons.

Speicher, M. R. 2005. Chromosome. In: *Encyclopedia of Life Sciences*. pp. 1–7. Hoboken, NJ: John Wiley and Sons.

Sumner, A. T. 1990. *Chromosome Banding*. London: Unwin Hyman.

Therman, E., and Susman, M. 1993. *Human Chromosomes: Structure, Behavior, and Effects*, 3rd ed. New York: Springer-Verlag.

Verma, R.S., and Babu, A. 1995. *Human Chromosomes: Principles and Techniques*, 2nd ed. New York: McGraw-Hill.

The Nucleosome

Brown, T. D., Izard, T., and Misteli, T. 2006. Mapping the interaction surface of linker histone $H1^0$ with the nucleosome of native chromatin in vivo. *Nature Struc Mol Biol* 13:250–255.

Doenecke, D. 2005. Histones: From gene organization to biological roles. In: *Encyclopedia of Life Sciences*. pp. 1–7. Hoboken, NJ: John Wiley and Sons.

Dutnall, R., and Ramakrishnan, V. 1997. Twists and turns of the nucleosome: tails without ends. *Structure* 5:1255–1259.

Happel, N., and Doenecke, D. 2009. Histone H1 and its isoforms: contribution to chromatin structure and function. *Gene* 431:1–12.

Kornberg, R. D., and Lorch, Y. 1999. Twenty-five years of the nucleosome, fundamental particle of the eukaryote chromosome. *Cell* 98:285–294.

Khorasanizadeh, S. 2004. The nucleosome: from genomic organization to genomic regulation. *Cell* 116:259–272.

Luger, K. 2001. Nucleosomes: structure and function. In: *Encyclopedia of Life Sciences*. pp. 1–8. London: Nature.

Luger, K., Mäder, A. W., Richmond, R. K., et al. 1997. Crystal structure of the nucleosome core particle at 2.8 A resolution. *Nature* 389:251–260.

Luger, K., and Richmond, T. J. 1998. DNA binding within the nucleosome core. *Curr Opin Struct Biol* 8:33–40.

Luger, K., and Richmond, T. J. 1998. The histone tails of the nucleosome. *Curr Opin Genet Dev* 8:140–146.

Ouzounis, C., and Kyprides, N. C. 1996. Parallel origins of the nucleosome core and eukaryotic transcription from archaea. *J Mol Evol* 42:234–239.

Sandman, K., Soares, D., and Reeve, J. N. 2001. Molecular components of the archaeal nucleosome. *Biochimie* 83:277–281.

van Holde, K.E., Zlatanova, J., Arents, G., and Moudrianakis, E. 1995. Elements of chromatin structure: histones, nucleosomes, and fibres. In: Elgin S.C.R. (ed). *Chromatin Structure and Gene Expression*, pp. 1–26. Washington, DC: IRL Press.

Widom, J. 1998. Structure, dynamics, and function of chromatin in vitro. *Ann Rev Biophys Biomol Struct* 27:285–327.

Wolfe, A. P. 2001. Nucleosomes: detailed structure and mutation. In: *Encyclopedia of Life Sciences*. pp. 1–8. London: Nature.

Woodcock, C. L., Skoultchi, A. I, and Fan, Y. 2006. Role of linker histone in chromatin structure and function: stoichiometry and nucleosome repeat length. *Chromosomal Res* 14:17–25.

Wu, J., and Grunstein, M. 2000. 25 years after the nucleosome model: chromatin modifications. *Trends Biochem Sci* 25:619–623.

The 30-nm Fiber

Bassett, A., Cooper, S., Wu, C. and Travers, A. 2009. The folding and unfolding of eukaryotic chromatin. *Curr Opin Genet Dev* 19:159–165.

Eltsov, M., MacLellan, K. M., Frangakis, A. S., and Dubochet, J. 2008. Analysis of cryo-electron microscopy images does not support the existence of 30-nm chromatin fibers in mitotic chromosomes in situ. *Proc Natl Acad Sci USA* 105:19732–19737.

Huynh, V. A. T., Robinson, P. J. J., and Rhodes, D. 2004. A method for in vitro reconstitution of a defined "30 nm" chromatin fibre containing stoichiometric amounts of the linker histone. *J Mol Biol* 345:957–968.

Kornberg, R., and Lorch, Y. 2007. Chromatin rules. *Nat Struct Mol Biol* 14:986–988.

Luger, K., and Hansen, J. C. 2005. Nucleosome and chromatin fiber dynamics. *Curr Opin Struct Biol* 15:1–9.

Robinson, P. J. J., Fairall, L., Huynh, V. A. T., and Rhodes, D. 2006. EM measurements define the dimensions of the "30-nm" chromatin fibre: evidence for a compact interdigitated structure. *Proc Natl Acad Sci USA* 103:6506–6511.

Robinson, P. J. J., and Rhodes, D. 2006. Structure of the "30 nm" fibre: a key role for the linker histone. *Curr Opin Struct Biol* 16:336–343.

Schalch, T., Duda, S., Sargent, D. F., and Richmond, T. J. 2005. X-ray structure of a tetranucleosome and its implications for the chromatin fibre. *Nature* 436:138–141.

Segal, E., and Widom, J. 2009. Poly(dA:dT) tracts: major determinants of nucleosome organization. *Curr Opin Struct Biol* 19:1–7.

Staynov, D. Z. 2008. The controversial 30 nm chromatin fibre. *BioEssays* 30:1003–1009.

Swedlow, J. R., and Hirano, T. 2003. The making of the mitotic chromosome: modern insights into classical questions. *Mol Cell* 11:557–569.

Tremethick, D. J. 2007. Higher-order structure of chromatin: the elusive 30 nm fiber. *Cell* 128:651–654.

Vaquero, A., Loyola, A., and Reinberg, D. 2003. The constantly changing face of chromatin. *Sci Aging Knowledge Environ* RE4:1–16.
Wu, C., Bassett, A., and Travers, A. 2007. A variable topology for the 30-nm chromatin fibre. *EMBO Rep* 8:1129–1134.

The Scaffold Model

Almagro, S., Riveline, D., Hirano, T., et al. 2004. The mitotic chromosome is an assembly of rigid elastic axes organized by structural maintenance of chromatin (SMC) proteins surrounded by a soft chromatin envelope. *J Biol Chem* 279:5118–5126.
Belmont, A. S. 2002. Mitotic chromosome scaffold structure: new approaches to an old controversy. *Proc Natl Acad Sci USA* 99:15855–15857.
Belmont, A. S. 2006. Mitotic chromosome structure and condensation. *Curr Opin Cell Biol* 18:632–638.
Hirano, T. 2005. Condensins: Organizing and segregating the genome. *Curr Biol* 15:R265–R275.
Losada, A., and Hirano, T. 2005. Dynamic molecular linkers of the genome: the first decade of SMC proteins. *Genes Devel* 19:1269–1287.
Maeshima, K., and Eltsov, M. 2008. Packaging the genome: the structure of mitotic chromosomes. *J Biochem* 143:145–153
Maeshima, K., and Laemmli, U. K. 2003. A two-step scaffolding model for mitotic chromosome assembly. *Dev Cell* 4:467–480.

The Centromere

Amor, D. J., Kalitsis, P., Sumer, H., and Choo, K. H. 2004. Building the centromere: from foundation proteins to 3D organization. *Trends Cell Biol* 14:359–368.
Black, B. E.. and Basset, E. A. 2008. The histone variant CENP-A and centromere specification. *Curr Opin Cell Biol* 20:91–100.
Black, B. E., Foltz, D. R., Chakravarthy, S., et al. 2004. Structural determinants for generating centromeric chromatin. *Nature* 430:578–582.
Cleveland, D. W., Mao, Y., and Sullivan, K. F. 2003. Centromeres and kinetochores: from epigenetics to mitotic checkpoint signaling. *Cell* 112:407–421.
Engelhardt, P. 2000. Electron tomography of chromosome structure. In: *Encyclopedia of Analytical Chemistry*. pp. 4948–4984. Hoboken, NJ: John Wiley and Sons.
Henikoff, S., and Dalal, Y. 2005. Centromeric chromatin: what makes it unique? *Curr Opin Genet Dev* 15:1–8.
Hill, E., and Willims, R. 2009. Super-coil me: Sizing up centromeric nucleosomes. *J. Cell Biol* 186:453–456.
Murphy, T. D., and Karpen, G. H. 1998. Centromeres take flight: alpha satellite and quest for the human centromere. *Cell* 93:317–320.
Przewloka, M. R., and Glover, D. M. 2009. The kinetochore and the centromere: a working long distance relationship. *Ann Rev Genet* 43:439–465.
Sullivan, B. A., and Karpen, G. H. 2004. Centromeric chromatin exhibits a histone modification pattern that is distinct from euchromatin and heterochromatin. *Nat Struct Mol Biol* 11:1076–1083.

The Telomere

Baumann, P., and Cech, T. R. 2001. POT1, the putative telomere end-binding protein in fission yeast and humans. *Science* 292:1171–1174.
Biasco, M. A. 2003. Mammalian telomeres and telomerase: why they matter for cancer and aging. *Eur J Cell Biol* 82:441–446.
Blackburn, E. H. 2001. Telomeres. In: *Encyclopedia of Life Sciences*. pp. 1–7. London: Nature.
Cech, T. R. 2004. Beginning to understand the end of the chromosome. *Cell* 116:273–279.
Court, R., Chapman, L., Fairall, L., and Rhodes, D. 2004. How the human telomeric proteins TRF1 and TRF2 recognize telomeric DNA: a view from high-resolution crystal structures. *EMBO Rep* 6:39–45.

de Lange, T. 2002. Protection of mammalian telomeres. *Oncogene* 21:532–540.
de Lange, T. 2004. T-loops and the origin of telomeres. *Nat Rev Mol Cell Biol* 5:323–329.
de Lange, T., and DePinho R. A. 1999. Unlimited mileage from telomerase? *Science* 283:947–949.
de Lange, T., and Jacks, T. 1999. For better or worse? Telomerase inhibition and cancer. *Cell* 98:273–275.
Fletcher, T. M. 2003. Telomere higher-order structure and genomic instability. *IUBMB Life* 55:443–449.
Greider, C. 1998. Telomerase activity, cell proliferation, and cancer. *Proc Natl Acad Sci USA* 95:90–92.
Greider, C. 1999. Telomeres do D-loop-T-loop. *Cell* 97:419–422.
Lei, M., Podell, E. R., Baumann, P., and Cech, T. R. 2003. DNA self-recognition in the structure of POT1 bound to telomeric single-stranded DNA. *Nature* 426:198–203.
Lei, M., Podell, E. R., and Cech, T. R. 2004. Structure of human POT1 bound to telomeric single-stranded DNA provides a model for chromosome end-protection. *Nat Struct Mol Biol* 11:1223–1229.
Palm, W., and de Lange, T. 2008. How shelterin protects mammalian telomeres. *Ann Rev Genet* 42:301–304.
Pardue, M.-L., and DeBaryshe, P. G. 1999. Telomeres and telomerase: more than the end of the line. *Chromosoma* 108:73–82.
Pardue, M. L., and DeBaryshe, P. G. 2001. Telomeres in cell function: cancer and ageing. In: *Encyclopedia of Life Sciences*. pp. 1–6. London: Nature.
Rhodes, D., Fairall, L., Simonsson, T., et al. 2002. Telomere architecture. *EMBO Rep* 3:1139–1145.
Shay, J. W. 1999. At the end of the millennium, a view of the end. *Nat Genet* 23:382–383.
Wei, C., and Price, C. M. 2003. Protecting the terminus: t-loops and telomere end-binding proteins. *Cell Mol Life Sci* 60:2282–2294.

This book has a Web site, **http://biology.jbpub.com/book/molecular**, which contains a variety of resources, including links to in-depth information on topics covered in this chapter, and review material designed to assist in your study of molecular biology.

SECTION III

Genetics and Virology

CHAPTER 7 **Genetic Analysis in Molecular Biology**

CHAPTER 8 **Viruses in Molecular Biology**

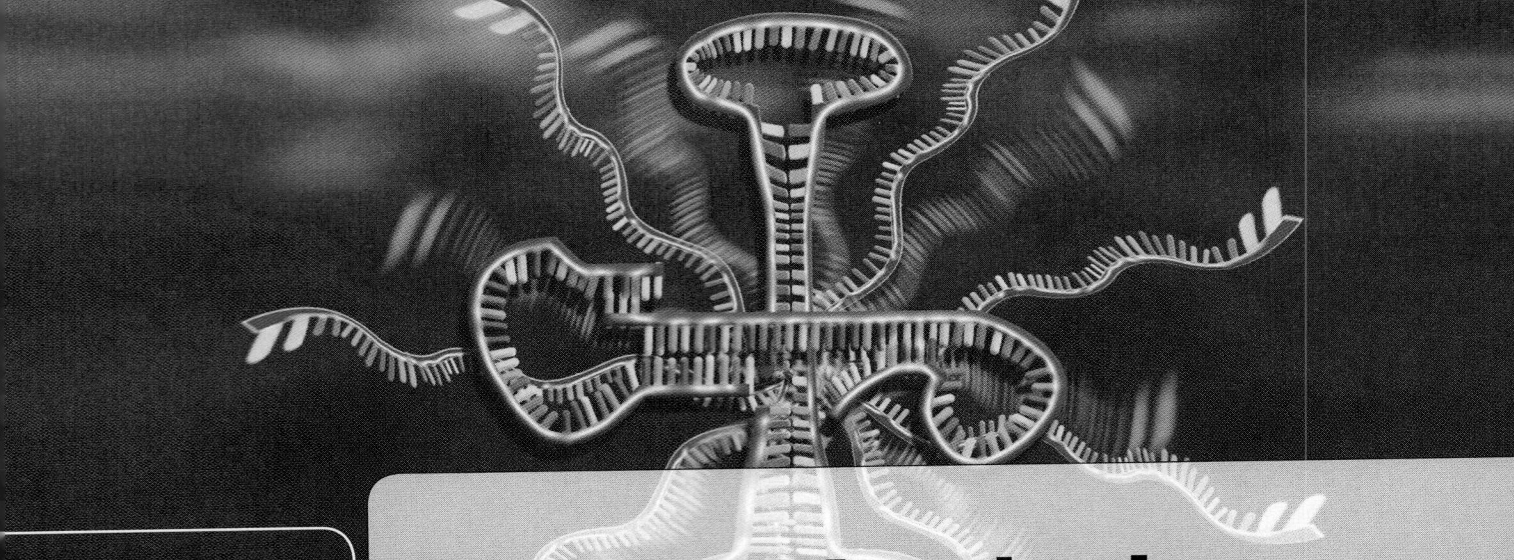

7 Genetic Analysis in Molecular Biology

OUTLINE OF TOPICS

7.1 Introduction to Genetic Recombination

Genetic recombination involves an exchange of DNA segments between DNA molecules or chromosomes.

Recombination frequencies are used to obtain a genetic map.

7.2 Bacterial Genetics

Bacteria, which are often selected as model systems for genetic analyses, have complex structures.

Bacteria can be cultured in liquid or solid media.

Specific notations, conventions, and terminology are used in bacterial genetics.

Cells with altered genes are called mutants.

Some mutants display the mutant phenotype under all conditions, while others display it only under certain conditions.

Certain physical and chemical agents are mutagens.

Mutants can be classified on the basis of the changes in the DNA.

A mutant organism may regain its original phenotype.

Mutants have many uses in molecular biology.

A genetic test known as complementation can be used to determine the number of genes responsible for a phenotype.

E. coli cells can exchange genetic information by conjugation.

Approximately 40 F factor genes are needed for successful mating and DNA transfer to occur.

The F plasmid can integrate into a bacterial chromosome and carry it into a recipient cell.

Bacterial mating experiments can be used to produce an *E. coli* genetic map.

F′ plasmids contain part of the bacterial chromosome.

Plasmid replication control functions are usually clustered in a region called the basic replicon.

Plasmids often confer advantageous properties to their hosts.

7.3 Budding Yeast (*Saccharomyces cerevisiae*)

Yeasts are unicellular eukaryotes.

Specific notations, conventions, and terminology are used in yeast genetics.

Yeast cells exist in haploid and diploid stages.

The yeast mating type is determined by an allele present in the mating type (*Mat*) locus.

Yeast mating factors act as signals to initiate the mating process.

7.4 Restriction and Amplified Fragment Length Polymorphisms

Recombinant DNA techniques have facilitated genetic analysis in humans and other organisms.

7.5 Somatic Cell Genetics

Somatic cell genetics can be used to map genes in higher organisms.

Animal cells can be studied in culture.

Two different animal cells can fuse to form a heterokaryon.

Hybrid cells can be used to make monoclonal antibodies.

Suggested Reading

The term **gene** (Greek *genos* = birth) was coined by the Danish botanist William Louis Johannsen in 1909 to describe the basic unit of heredity. The gene concept has changed considerably over the years as information has been obtained about the physical and chemical properties of genes. We will, for the purposes of the present discussion, define a gene as a DNA sequence that contains information needed to synthesize an RNA molecule, which in many (but not all) cases codes for a polypeptide chain. This definition will require further refinement after we have had an opportunity to look more closely at gene structure and function.

Genes are located at specific sites on chromosomes. Gene locations are usually summarized in the form of a **genetic map**, which can be quite helpful when one wishes to clone, move, disrupt, or otherwise manipulate a gene. This chapter examines some of the strategies used to construct a genetic map in bacteria and eukaryotic cells. Historically, the bacteria *Escherichia coli* (abbreviated as *E. coli*) and the budding yeast *Saccharomyces cerevisiae* (*S. cerevisiae*) have been very important model systems for genetic studies. We, therefore, describe techniques that are used to map genes in these two biological systems. Some techniques used to map genes in multicellular organisms also are introduced. Even though the discussion that follows will be limited to a few organisms, the concepts developed are applicable to all organisms.

7.1 Introduction to Genetic Recombination

Genetic recombination involves an exchange of DNA segments between DNA molecules or chromosomes.

Genetic recombination is the process of combining two genetic loci, initially on two different chromosomes, onto a single chromosome. The molecular mechanism, which is complex, is discussed in Chapter 13. For present purposes, it is sufficient to assume that two chromosomes align with one another, a cut is made in both chromosomes at random but matching points, and the four fragments are then joined together to form two new combinations of genes.

This crude model accounts for only some of the features of genetic exchange, but these features are in fact the only ones of concern at this time. According to this model, two parental chromosomes with the genotypes *Ab* and *aB* can align and recombine to form two recombinant chromosomes with the genotypes *AB* and *ab* (FIGURE 7.1). The

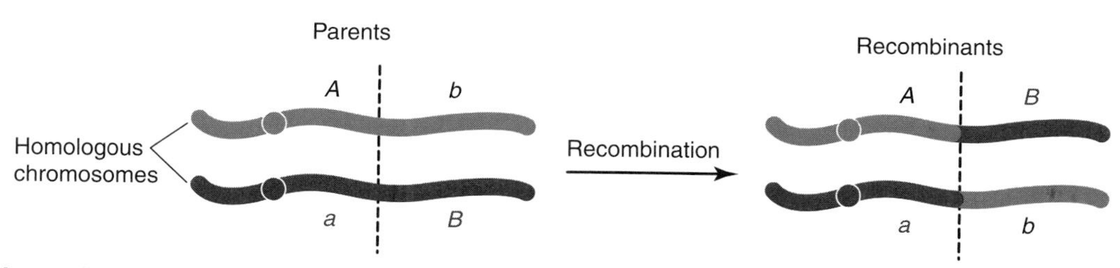

FIGURE 7.1 **A schematic diagram showing genetic exchange.**

process in which homologous chromosomes exchange parts normally reciprocally but sometimes unequally is called crossing over.

Recombination frequencies are used to obtain a genetic map.

Most of the genetic mapping studies that were performed in the first part of the twentieth century were based on the principle that the distance along the chromosome between two recombining genes determines the recombination frequency. As long as the two genes are not too close to one another, the recombination frequency is proportional to distance, because chromosomal crossovers take place at random. In the following crosses between chromosomes, the genotypes of which are *Abc* and *aBC*, and the genes of which are in alphabetical order and equally spaced,

$$\frac{A \quad\quad b \quad\quad c}{} \times \frac{a \quad\quad B \quad\quad C}{}$$

thus there will be twice as many *AC* recombinants as *AB* recombinants, because loci *A* and *C* are twice as far apart as loci *A* and *B*.

Because the recombination frequency is proportional to distance, recombination frequency can be used to determine the arrangement of genes on the chromosome. This can be seen in a simple example in which three genes—*A*, *B*, and *C*—are arranged along a single chromosome in an unknown order. If we assume that the recombination frequency between genes A and B is 1% ($A \times B = 1\%$) and the recombination frequency between genes *B* and *C* is 2% ($B \times C = 2\%$), then the two arrangements shown in FIGURE 7.2 are consistent with the data. The correct order can be determined from the recombination frequency between genes *A* and *C*. Let us assume that this is 1%. If that is the case, only arrangement 2 is possible. The order *C A B* for these genes and the relative separation constitute a genetic map.

Any number of genes can be mapped in this way. For instance, consider a fourth gene, *D*, in the preceding example. If $D \times B = 0.5\%$, *D* must be located 0.5 unit either to the left or to the right of *B*. If $A \times D = 1.5\%$, *D* is clearly to the right of *B* and the gene order is *C A B D*. If $A \times D = 0.5\%$, the gene order would be *C A D B*.

The analysis just given has been oversimplified because the occurrence of multiple exchanges has been ignored. If two crossover events occur between two genes, then recombination between the two genes

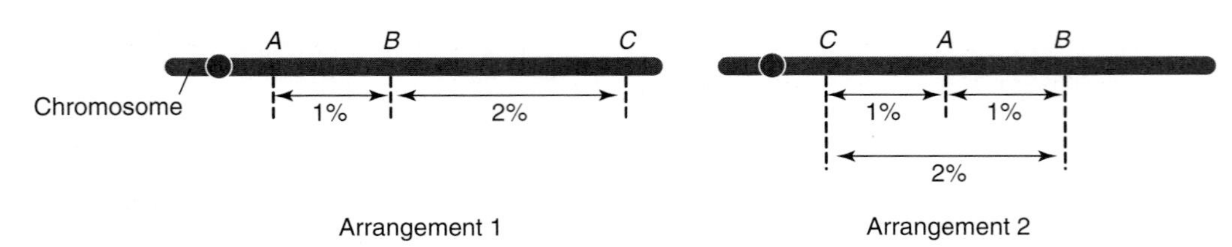

FIGURE 7.2 **The arrangements for which the recombination frequency between genes A and B is 1% and the recombination frequency between genes B and C is 2%.**

will not be observed, because the second event will cancel the effect of the first. Discussion of this important point can be found in any genetics textbook; however, the effect of multiple exchanges is unimportant for the simple considerations described here. The general approach outlined above is used to map the genes in plants and animals by performing mating experiments.

Many of the early recombination studies were performed with the fruit fly, *Drosophila melanogaster*. This organism offers considerable advantages for genetic studies because it is easy to handle, has a short life cycle of just two weeks, and does not cost much to keep in large numbers. Not surprisingly, as more genetic information was gained about the fruit fly there was more incentive to select this organism for further study. By the 1950s, interest started to shift toward other systems, particularly the bacteria *E. coli* and some of the viruses that infect it. The reason for this shift was that it is even easier to culture and maintain the bacteria and its viruses. Moreover, genetic mapping experiments can be performed in a day rather than two weeks. Bacterial genetic systems are discussed in the next section and bacterial viruses, also known as bacteriophages, are examined in Chapter 8.

7.2 Bacterial Genetics

Bacteria, which are often selected as model systems for genetic analyses, have complex structures.

Many basic concepts in molecular biology have their origins in studies of the genetics of bacteria and the viruses that infect them. Because a basic knowledge of bacterial structure, physiology, and growth is required to fully appreciate these genetic studies, we will start with a brief introduction to bacteria. Like all other cells, a bacterial cell is enclosed by a membrane, which separates the cytoplasm from the surroundings. A rigid multilayered cell wall made of **peptidoglycan** (a combination of polysaccharides and peptides) surrounds the membrane, protecting the cell from osmotic and mechanical damage and giving the cell its characteristic shape, which can be spherical, rod-like, and so forth (FIGURE 7.3).

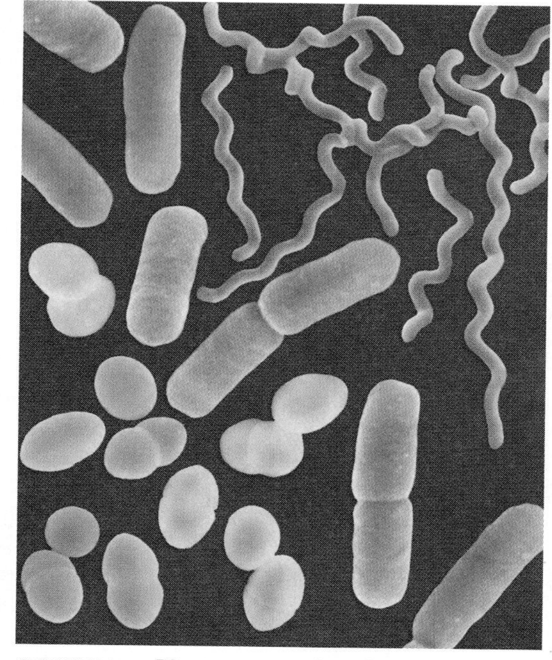

FIGURE 7.3 **Photocomposite of three common types of bacterial morphology coccus, bacillus, and spirillum.** These spherical, rod, and spiral shaped morphologies are typical of such genera as: *Streptococcus* or *Staphylococcus* (green), *Escherichia* or *Bacillus* (orange), and *Leptospira* or *Spirillum* (yellow), respectively. The colors shown here do not represent actual colors of the organisms. (© Phototake, Inc./Alamy Images.)

Bacteria can be divided into two major groups based on the thickness of their cell walls by using a staining procedure that was devised by the Danish bacteriologist Christian Gram in 1884. In brief, a bacterial suspension is placed on a microscope slide and air dried. Then the slide is covered with a crystal violet solution for a short time, washed with water, and exposed to an iodine solution. Both gram-positive and gram-negative bacteria now have a purple color due to the presence of a crystal violet-iodine complex in their cytoplasm. Washing the slide with an acetone-ethanol solution decolorizes the gram-negative bacteria but not the gram-positive bacteria. (This difference occurs because the thick cell walls of gram-positive bacteria prevent dye extraction, whereas the thin cell walls of gram-negative bacteria do not.) Finally, cells are treated with a second dye, safranin, which does not alter the purple color of gram-positive bacteria but causes gram-negative bacteria to stain pink.

The distinct staining properties of the gram-positive (purple) and gram-negative (pink) bacteria, thus, is determined by differences in their cell walls. One other important structural feature distinguishes the two groups of bacteria. Gram-negative bacteria have an additional membrane called the **outer membrane** that surrounds their cell wall but gram-positive bacteria do not (FIGURE 7.4). Small water-soluble molecules pass through protein channels called **porins** in the outer membrane. Secreted proteins are too large to pass through the

(a) Gram-positive bacterial envelope

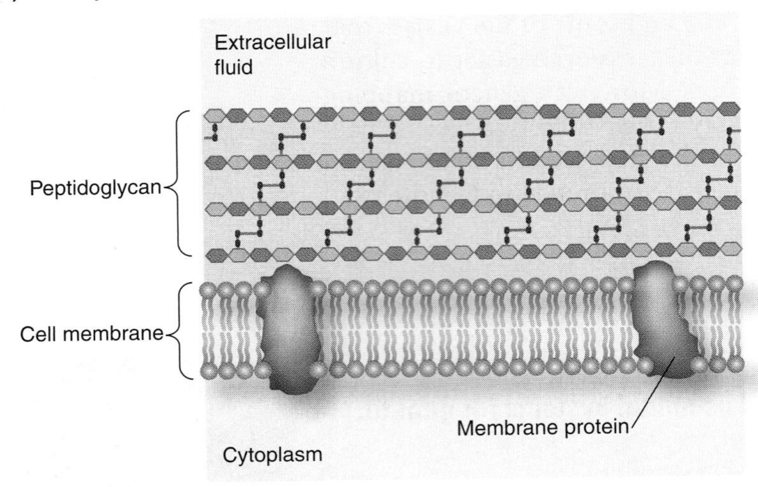

(b) Gram-negative bacterial envelope

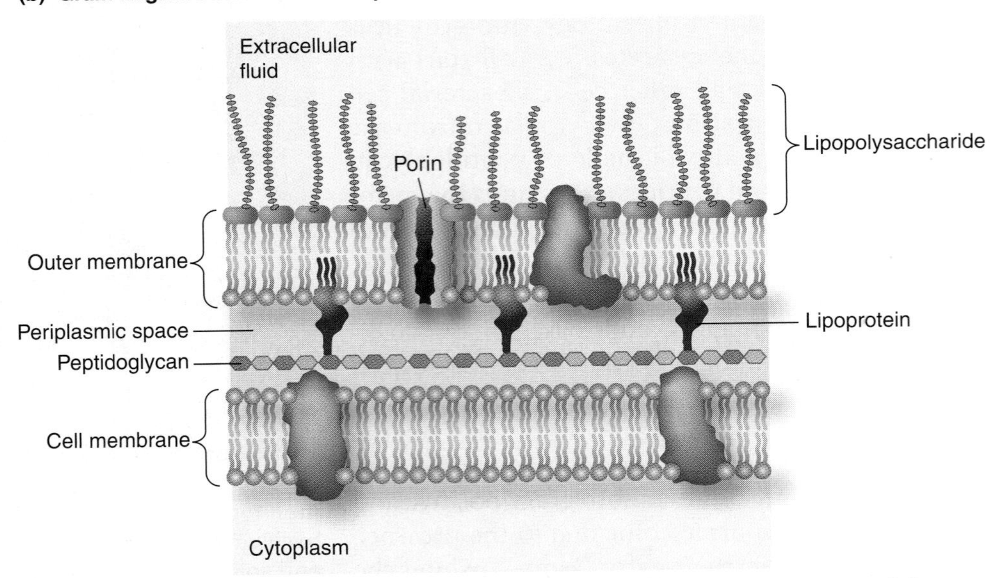

FIGURE 7.4 **Bacterial cell envelopes.** Bacteria are classified in two groups, the gram-negative and the gram-positive bacteria. (a) Gram-positive bacteria have a very thick cell wall made of peptidoglycan surrounding the cell or plasma membrane. (b) Gram-negative bacteria have a thinner cell wall made of peptidoglycan surrounding the cell (also called plasma or inner) membrane. A second membrane called the outer membrane surrounds the cell wall. The outer leaflet of the outer membrane contains lipopolysaccharides. Lipoproteins extend from the peptidoglycan to the outer membrane. The space between the inner and outer membranes is called the periplasmic space.

porins, however, and so most are retained in the space between the cell membrane and outer membrane called the **periplasmic space**. Some periplasmic proteins bind nutrients and assist in their transport across the cell membrane while others are digestive enzymes. The outer membrane serves as a barrier to detergents and hydrophobic antibiotics, protecting the cell from them. Gram-positive bacteria, which lack an outer membrane, are usually much more sensitive to detergents and hydrophobic antibiotics than are gram-negative bacteria.

When bacteria are treated with the enzyme **lysozyme**, which is isolated from chicken egg white, some of the cell wall components are removed and the rigidity of the wall is lost. All bacteria treated in this way become spherical and ultimately burst. The spherical form, which can be stabilized in suspending media with a high osmotic strength such as 20% sucrose, is called a **spheroplast** if some cell wall remains and a **protoplast** if the cell wall is completely removed. The cell membrane, which lies just below the cell wall, provides a permeability barrier that determines which substances can pass in and out of the cell.

Bacteria can be cultured in liquid or solid media.

Bacteria are easy to culture, divide rapidly, and have relatively simple nutritional requirements compared to cells in multicellular organisms. The bacterium that has served as perhaps the most important model system in molecular biology is *E. coli*, which divides about every 30 to 60 minutes at 37°C under laboratory conditions. A single bacterium, thus, multiplies exponentially to become 10^9 bacteria in less than a day.

Bacteria can grow in a **liquid growth medium** or on a solid surface. A population growing in a liquid medium is called a bacterial **culture**. If the liquid is a complex extract of biological material, it is called a **broth**. An example is tryptone broth, which contains the milk protein casein hydrolyzed by the digestive enzyme trypsin to yield a mixture of amino acids and small peptides. If the growth medium is a simple chemically defined mixture containing no organic compounds other than a carbon source such as a sugar, it is called a **minimal medium**. A typical minimal medium contains sodium, potassium, magnesium, calcium, ammonium, chloride, phosphate, sulfate, and a few trace metal ions as well as a carbon source such as glucose or glycerol. Bacteria grow more rapidly in a minimal medium containing glucose than in a minimal medium with any other single carbon source. If a bacterium can grow in a minimal medium—that is, if it can synthesize all necessary organic substances, such as amino acids, vitamins, and lipids—the bacterium is said to be a **prototroph**. If a nutrient such as the amino acid leucine or the vitamin thiamine must be added for growth to occur in the presence of a carbon source, the bacterium is termed an **auxotroph**.

Bacteria are frequently grown on solid surfaces. The earliest surface used for growing bacteria was a slice of raw potato. Media solidified by gelatin later replaced the potato. Because many bacteria excrete enzymes that digest gelatin, an inert gelling agent was sought. **Agar**, a gelling agent obtained from seaweed, which is resistant to enzymes from most microorganisms, is now universally used. A solid growth

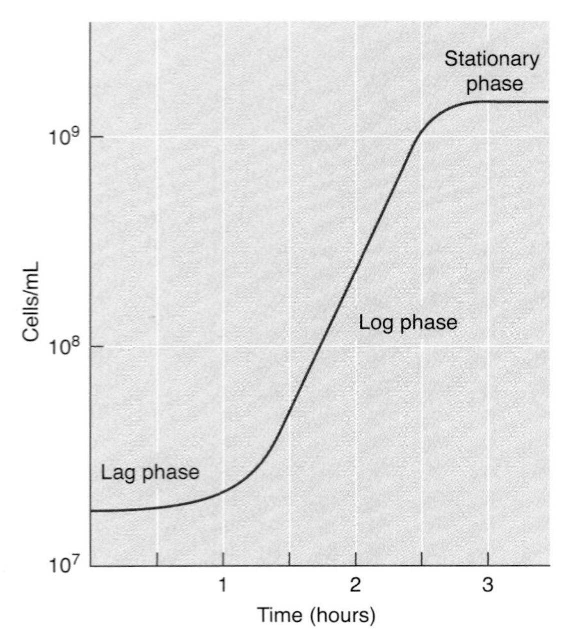

FIGURE 7.5 **Typical bacterial growth curve.**

medium is called a nutrient agar, if a broth medium is gelled, or a minimal agar, if a minimal medium is gelled. Solid media are typically placed in a shallow glass or plastic flat-bottomed dish with a lid called a **petri dish**. In laboratory jargon, a petri dish containing a solid medium is called a **plate** and the act of depositing bacteria on the agar surface is called **plating**.

When *E. coli* cells are placed in a liquid medium, they grow and divide. A typical growth curve for an *E. coli* culture growing in an Erlenmeyer flask that is placed on a shaker and maintained at 37°C is shown in FIGURE 7.5. After an initial period of slow growth called the **lag phase**, the bacteria begin a period of rapid growth in which they divide at a fixed time interval called the **doubling time**. The number of cells per milliliter, the **cell density**, doubles repeatedly, giving rise to a logarithmic increase in cell number. This stage of growth of the bacterial culture, called the log phase, continues until the *E. coli* achieve a cell density of about 10^9 cells·mL^{-1}. At this cell density, the O_2 supply and pH limit the rate of cell growth. *E. coli* growth usually ceases at a cell density of 2 to 3 × 10^9 cells·mL^{-1} and the bacteria enter the **stationary phase**. Considerably higher cell densities can be achieved by increasing the rate of aeration, maintaining the medium at optimal pH, or both. The terms used in this section are also used in discussing the growth of all microorganisms and frequently of animal cells.

A bacterium growing on an agar surface also divides. Because most bacteria are not very motile on a solid surface, the progeny bacteria remain very near the location of the original bacterium. The number of progeny increases so much that a visible cluster of bacteria appears. Such a cluster, which arises from a single cell, is called a **colony** or **clone** (FIGURE 7.6). Colony formation allows one

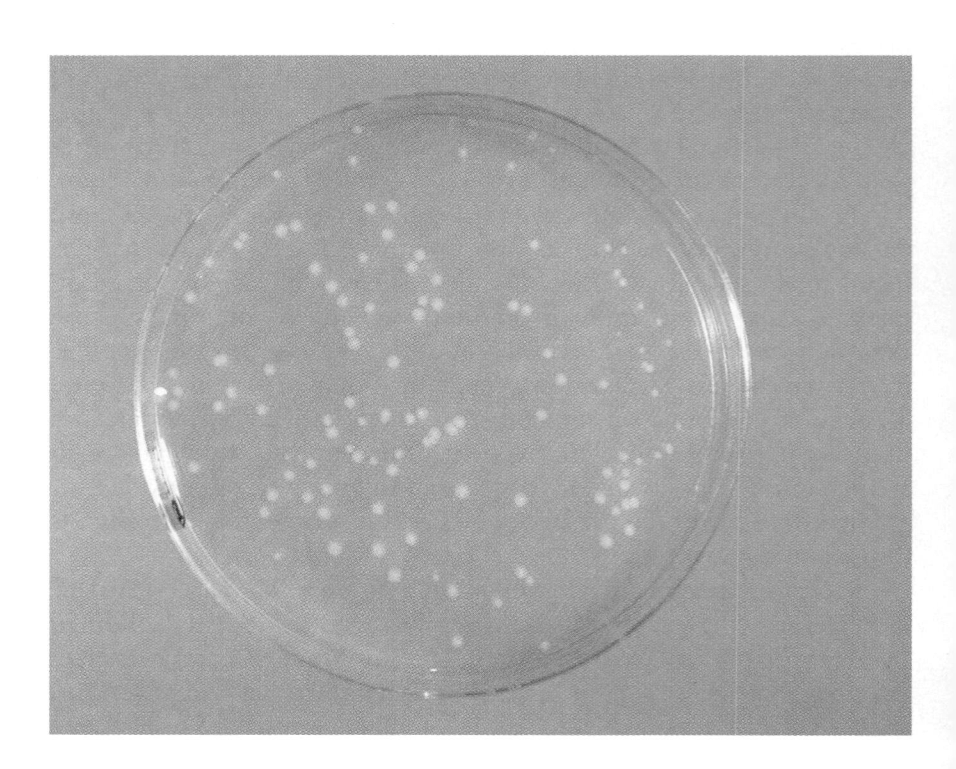

FIGURE 7.6 **A petri dish with bacterial colonies that have formed on agar.** (Photo courtesy of Dr. Jim Feeley/CDC.)

to determine the number of bacteria in a culture. For instance, if 100 cells are plated, 100 colonies will be visible the next day. If 0.1 mL of a 10^6-fold dilution of a bacterial culture is plated and 200 colonies appear, the cell density in the original culture is $(200/0.1)(10^6) = 2 \times 10^9$ cells/mL.

Plating is a convenient way to determine if a bacterium is an auxotroph. Several hundred bacteria are spread on minimal agar and nutrient agar plates, which are stored overnight in a constant-temperature incubator. Several hundred colonies are subsequently observed on the nutrient agar plate because it contains so many substances that can satisfy the requirements of nearly any bacterium. If colonies are also observed on the minimal agar, the bacterium is a prototroph; if no colonies are observed, it is an auxotroph and some required substance is missing from the minimal agar plates. Minimal agar plates are then prepared with various supplements. If the bacterium is a leucine auxotroph, the addition of leucine alone will enable a colony to form. If both leucine and histidine must be added, the bacterium is auxotrophic for both of these substances.

Specific notations, conventions, and terminology are used in bacterial genetics.

Bacterial geneticists use specific notations to indicate genotypic and phenotypic traits. As we will see below, another system of notations and conventions applies to yeast. Unfortunately, a variety of notations are used for eukaryotic organisms that have been studied for a long time, such as maize and *Drosophila*. The phenotype of a bacterial cell that can synthesize the amino acid leucine is denoted Leu^+, whereas the phenotype of a bacterial cell that cannot do so is denoted Leu^-. Note that the symbol for the phenotype has three letters, begins with an uppercase letter, and is not italicized. A Leu^- cell lacks a gene needed to synthesize leucine; this gene would be denoted leu^- (or in some books, *leu*, without the minus sign). Note that the genotype is written in lowercase letters and is italicized. A gene that specifies a particular protein is said to be the **structural gene** for that protein.

A bacterial cell may require several genes to synthesize leucine. These genes would usually be denoted *leuA*, *leuB*, . . . with all letters italicized. A Leu^+ (leucine-synthesizing) cell must have one functional copy of every requisite gene; thus, its genotype must be $leuA^+$, $leuB^+$, . . . and would normally be summarized by writing leu^+ unless it is important for some reason to state the genotype of each gene. A Leu^+ cell might also be diploid for *leu* genes and have a defective gene (e.g., *leuA*) in one chromosomal set. The two haploid sets are separated in their notation by a diagonal line; the genotype would therefore be $leu^+/leuA^-$. Some genes are responsible for resistance and sensitivity to certain extracellular substances such as antibiotics. The genotypes of an ampicillin-resistant cell and an ampicillin-sensitive cell are written amp^R and amp^S, respectively; the phenotypes are correspondingly Amp^R and Amp^S.

In summary, for bacteria the following conventions are used:

1. Abbreviations of phenotypes contain three regular typeface letters (the first uppercase) with a superscript (+) or (–) to denote presence or absence of the designated character and superscript "S" and "R" for antibiotic sensitivity and resistance, respectively.
2. The genotype designation is always lowercase and all components of the symbol are italicized.

We also have occasion to designate particular mutants of a gene. Mutants are usually numbered in the order in which they were isolated. The 58th and 79th leucine mutants discovered, thus, would be written *leu58* and *leu79*. To denote mutants of a particular gene, one would write *leuA58* and *leuB79* if the mutations were in the *leuA* and *leuB* genes, respectively.

The term **genome** is useful, though it has evolved to have various meanings. It is correctly defined as the genetic complement (the set of all genes and genetic signals) of a cell or virus. With eukaryotes, the term is often used to refer to one complete (haploid) set of chromosomes. In laboratory jargon, when discussing bacteria, phages, or most animal and plant viruses, the term often refers to their single DNA or RNA molecule(s), a classically incorrect but now accepted usage.

Cells with altered genes are called mutants.

We have seen that a gene can be either functional or nonfunctional and that these states are denoted by a superscript (+) or (–) after the gene abbreviation. The functional form of a gene is sometimes called **wild type** because presumably this is the form found in nature. This term is ambiguous though, because often the (–) form is the one that is prevalent. For example, many bacterial species isolated from nature carry the genes for lactose metabolism, yet in many strains the gene is either nonfunctional or missing, that is, *lac*$^-$. In this book we will try to use the term wild type as little as possible. The precise genetic term **allele** is used to indicate that there are alternative forms of a gene, and sometimes the (+) and (–) forms are called the (+) allele and the (–) allele, respectively. It is very common to call the nonfunctional (–) form of an allele a mutant form. Strictly speaking, a mutant is an organism with a genotype (or, more precisely, a DNA base sequence) that differs from that found in nature. It is more convenient (and definitely it is common jargon), however, to equate the terms mutant gene and nonfunctional gene, and we will use the word mutant in that sense in this book. A word of caution is needed, however, because in some instances a mutant has a more active enzyme than its parent.

Some mutants display the mutant phenotype under all conditions, while others display it only under certain conditions.

Mutants can be classified in several ways. One classification is based on the conditions in which the mutant character is expressed. An **absolute defective mutant** displays the mutant phenotype under all conditions;

that is, if a bacterium requires leucine for growth in all culture media and at all temperatures, it is an absolute defective. A **conditional mutant** does not always show the mutant phenotype; its behavior depends on physical conditions and sometimes on the presence of other mutations. An important example of a conditional mutant is a **temperature-sensitive (Ts) mutant**, which behaves normally below 30°C (the **permissive temperature**) and as a mutant above 42°C (the **nonpermissive temperature**); intermediate states are usually observed between these temperatures. Note that the gene does not mutate above 42°C; rather, the product of the gene is inactive above 42°C.

Temperature-sensitive mutants have been of great use in the laboratory because they enable one to turn off the activity of a gene product simply by raising the temperature. In many cases, the temperature-sensitive defect is reversible, so the activity of the gene product can be regained by lowering the temperature. Temperature-sensitive mutants are especially useful for studying the function(s) of an essential gene. The temperature-sensitive mutant can be maintained at a permissive temperature and studied at the nonpermissive temperature.

Another widely encountered type of conditional mutant is the **suppressor-sensitive mutant**, which exhibits the mutant phenotype in some strains but not in others. The difference is a consequence of the presence of particular gene products, called **suppressors**. These suppressor gene products either compensate for the defect in the mutant or, in a variety of ways, enable an altered gene to produce a functional gene product. In the jargon of molecular biology, one says that the phenotype of a suppressor-sensitive mutant "depends on the genetic background." A mutation is sometimes designated as suppressor-sensitive by adding the regular typeface letters Am (refers to amber), Oc (for ochre), or Op (for opal) in parentheses. (Amber, ochre, and opal are terms that arose from a private laboratory joke.) The symbols are regular typeface and capitalized because they represent a phenotype. If the mutant has a number, it is also added. Thus, if mutation 35 in the *leu* gene is a suppressor-sensitive Am type, it would be written *leu35*(Am). We will examine the molecular basis of suppressor mutations in Chapter 23.

Certain physical and chemical agents are mutagens.

The process of formation of a mutant organism is called mutagenesis. In nature and in the laboratory, mutants sometimes arise spontaneously, without any help from the experimenter. This process is called **spontaneous mutagenesis**. Mutagenesis can also be induced by adding chemicals called mutagens or by exposure to radiation, both of which result in chemical alterations of the genetic material. Mutagenesis and mutations will be discussed in detail in Chapter 11.

Mutants can be classified on the basis of the changes in the DNA.

Another method for classifying mutants is based on the number of changes that have occurred in the genetic material (that is, the number of DNA base pairs that have changed; FIGURE 7.7). If only one base pair

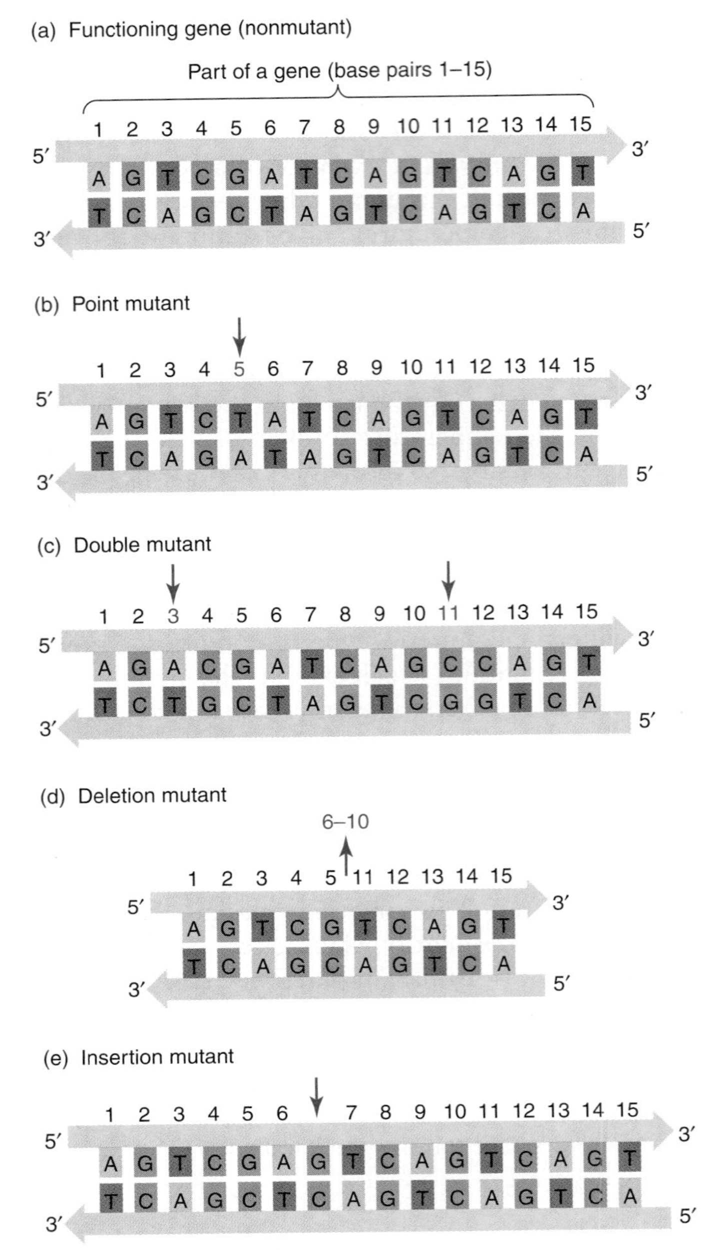

FIGURE 7.7 **Schematic of normal DNA molecule and various kinds of mutants.** (a) Part of a functioning (nonmutant) gene. (b) Point mutant—base pair 5 is changed from G-C to T-A. (c) Double mutant—base pair 3 is changed from T-A to A-T and base pair 11 is changed from T-A to C-G. (d) Deletion mutant—base pairs 6–10 are deleted. (e) Insertion mutant—a G-C base pair is inserted between base pairs 6 and 7.

change has occurred, the mutant is called a **point mutant.** Sometimes a mutation occurs by means of removal of all or part of a gene; in that case, the mutation is a **deletion**. Still another type of mutation, an **insertion**, involves placing additional DNA, ranging in size from a single base pair to a DNA segment that is thousands of base pairs long, within a gene.

A mutant organism may regain its original phenotype.

A mutant organism sometimes regains its original character. This occurs by means of chemical changes in the mutant genetic material, which restore the genetic material to a functional state. The process of regaining the original phenotype is called reversion, and an organism that has reverted is called a **revertant**. A point mutation will revert at measurable, albeit low, frequency. In contrast, a deletion mutation has virtually no chance to replace missing DNA with the needed sequence and reversions are not observed.

Mutants have many uses in molecular biology.

Some of the most significant advances in molecular biology have come about by the use of mutants. The experimental use of mutants falls into several categories.

A Mutant Defines a Function

The intake of Fe^{3+} ions by bacteria might be by passive diffusion through micropores in the cell membrane, or a particular system might be responsible for the process. Wild type *E. coli* can take in the Fe^{3+} ion from a 10^{-5} M solution but mutants have been found that cannot do so unless the ion concentration is very high. This finding indicates that a genetically determined system for Fe^{3+} intake exists, though the observation does not tell what this system is.

Temperature-sensitive mutants are especially useful in defining functions. For example, temperature-sensitive mutants of *E. coli* have been isolated that fail to synthesize DNA. These mutants fall into over a dozen distinct classes, suggesting that there may be at least a dozen different proteins required for DNA synthesis.

Mutants Can Introduce Biochemical Blocks that Aid in the Elucidation of Metabolic Pathways

The metabolism of the sugar galactose requires the activity of three distinct genes called *galK*, *galT*, and *galE*. If radioactive galactose ([^{14}C]Gal) is added to a culture of Gal$^+$ cells, many different radioactive compounds can be found as the galactose is metabolized. At very early times after addition of [^{14}C]Gal, three related compounds are detectable: [^{14}C]galactose-1-phosphate (Gal-1-P), [^{14}C]uridine diphosphogalactose (UDP-Gal), and [^{14}C]uridine diphosphoglucose (UDP-Glu). Different mutant genes will block different steps of the metabolic pathway. If the cell is a *galK*$^-$ mutant, the [^{14}C]Gal label is found only in galactose. The *galK* gene, therefore, is known to be responsible for the first metabolic step. If the mutant *galT*$^-$ is used, ([^{14}C]Gal-1-P accumulates. The first step in the reaction sequence, hence, is found to be the conversion of galactose to Gal-1-P by the *galK* gene product (namely, the enzyme galactokinase). If a *galE*$^-$ mutant is used, some ([^{14}C] Gal-1-P is found, but the principal radiochemical is ([^{14}C]UDP-Gal. Thus, the biochemical pathway must be,

$$\text{Gal} \xrightarrow[\text{product}]{galK} \text{Gal-1-P} \xrightarrow[\text{product}]{galT} \text{UDP-Gal} \xrightarrow[\text{product}]{galE} \text{X}$$

The identity of X cannot be determined from these genetic experiments, but one might guess that it is UDP-Glu (which it is).

Mutants Enable One to Learn about Genetic Regulation

Many mutants have been isolated that alter the amount of a particular protein that is synthesized or the way the amount synthesized responds to external signals. These mutants define regulatory systems. For example, the enzymes encoded by *galK*, *galT*, and *galE* are normally not detectable in bacteria but appear only after galactose is added to the growth medium. Mutants have been isolated in which these enzymes are always present, however, whether or not galactose is also present. This finding indicates that some gene is responsible for turning the system of enzyme production on and off, and the regulatory gene product must be responsive to the presence and absence of galactose (see Chapter 16).

Mutants Enable a Biochemical Entity to be Matched with a Biological Function or an Intracellular Protein

The *E. coli* enzyme DNA polymerase I was studied in great detail for many years. As described in Chapter 5, purified DNA polymerase I catalyzes DNA synthesis *in vitro*, so it was believed that this polymerase activity was solely responsible for *in vivo* bacterial DNA synthesis. However, an *E. coli* mutant (*polA*$^-$) was isolated in which the activity of DNA polymerase I was reduced 50-fold, yet the mutant bacterium grew and synthesized DNA normally. This observation strongly suggested that DNA polymerase I could not be the only enzyme that catalyzes DNA synthesis in the cell. Indeed, biochemical analysis of cell extracts of the *polA*$^-$ mutant showed the existence of two other enzymes—DNA polymerase II and DNA polymerase III—which, when purified, also could synthesize DNA. In further study, a temperature-sensitive mutation in a gene called *dnaE* was found to be unable to replicate DNA at 42°C, though replication was normal at 30°C. The three enzymes DNA polymerases I, II, and III were isolated from cultures of the *dnaE*$^-$ (Ts) mutant and each enzyme was assayed. Although DNA polymerases I and II were active at both 30° and 42°C, DNA polymerase III was active at 30°C but not at 42°C. DNA polymerase III, therefore, was determined to be the product of the *dnaE* gene and an enzyme that is essential for DNA replication (see Chapter 9).

Mutants Locate the Site of Action of External Agents

The antibiotic rifampicin, which is used to treat various bacterial diseases including tuberculosis, prevents synthesis of RNA. When first discovered, it was not known whether rifampicin might act by preventing synthesis of precursor molecules, by binding to DNA and thereby preventing the DNA from being transcribed into RNA, or by binding to RNA polymerase, the enzyme responsible for synthesizing RNA. Mutants were isolated that were resistant to rifampicin. These mutants were of two types: those in which the bacterial cell envelope was altered such that rifampicin could not enter the cell (an uninformative type of mutant in terms of RNA synthesis), and those in which the RNA polymerase was slightly altered. The finding of the latter mutants proved that the antibiotic acts by binding to RNA polymerase (see Chapter 15).

Mutants Can Indicate Relations Between Apparently Unrelated Systems

Bacteriophage λ, which normally attaches to and grows in *E. coli*, fails to attach to a bacterial mutant that cannot metabolize the sugar maltose. Such failure is not associated with mutants incapable of metabolizing other sugars or with any other phages, and this knowledge implicated some product or agent of maltose metabolism in the attachment of bacteriophage λ (see Chapter 8).

Mutants Can Indicate That Two Proteins Interact

A hypothetical example will help to illustrate how mutants can provide information about protein interactions. Suppose that mutants in two genes *a* and *b*, which are responsible for synthesizing the proteins A and B, fail to carry out some process. Both gene products, thus, are necessary for this process to occur. These gene products may act consecutively or interact to form a single functional unit consisting of both products. Interaction as a single unit is often indicated by reversion studies. When a^- mutant revertants are sought, one sometimes finds (by additional genetic analysis) that the revertants have a mutation in gene *b*. When this type of reversion occurs, one often finds that other b^- mutants (those not formed by reversion of an a^- mutant) revert as a result of (different) mutations in gene *a*. The interpretation of these results is as follows. Proteins A and B are subunits in an active AB protein complex (FIGURE 7.8). An alteration in either the A or B subunit that impedes subunit interaction will prevent active AB

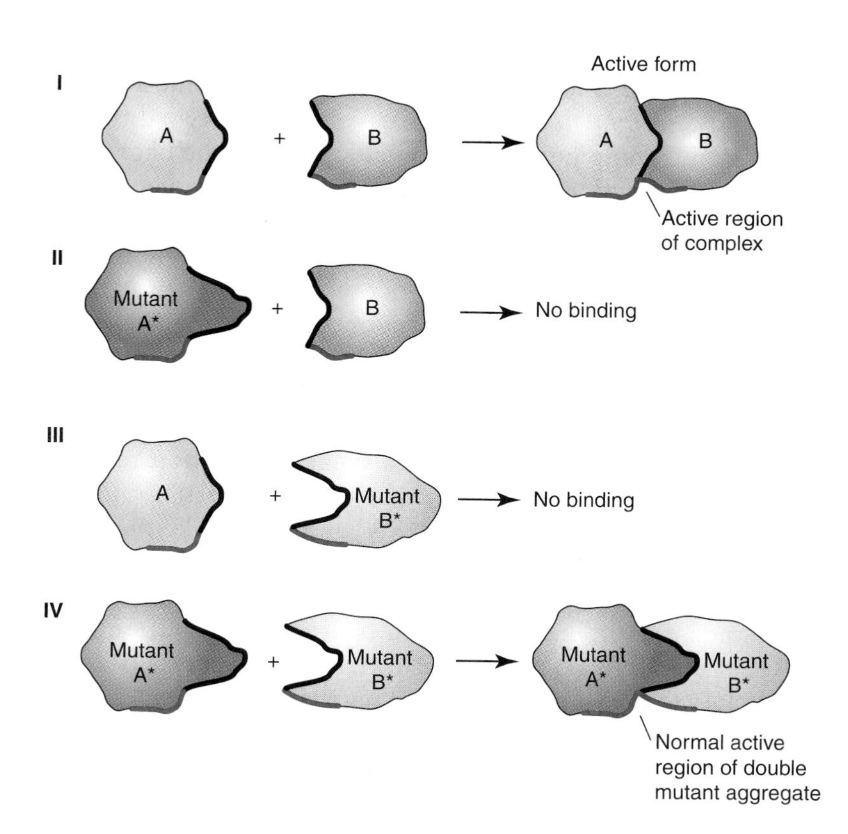

FIGURE 7.8 **Schematic diagram showing how two separately inactive mutant proteins can combine to form a functioning protein complex.** Sites of interaction of proteins A and B are denoted by heavy black lines. Asterisks indicate mutant polypeptides. Components of the AB complex active site are shown in red. Only subunits in I and IV can form a complex with a normal active region.

complex formation. A compensating alteration in the other subunit can then enable the interaction to occur again. Such an interpretation has frequently been found to be correct.

A genetic test known as complementation can be used to determine the number of genes responsible for a phenotype.

A particular phenotype is frequently the result of the activity of many genes. In the study of any genetic system, it is always important to know the number of genes and regulatory elements that constitute the system. The genetic test used to evaluate this number, called **complementation**, requires that two copies of the genetic unit to be tested be present in the same cell. This requirement can be met in bacteria by constructing a **partial diploid**, that is, a cell containing one complete set of genes and duplicates of some of these genes, (We see how cells of this type are constructed later in this chapter.) A partial diploid is described by writing the genotype of each set of alleles on either side of a diagonal line. For example, $b^+c^+d^+/a^+b^-c^-d^-e^+ \ldots z^+$, indicates that a chromosomal segment containing alleles *b*, *c*, and *d* is present in a cell with a single chromosome containing all of the alleles *a*, *b*, *c*, . . ., *z*. Usually, only the duplicated genes are indicated, so this partial diploid would be designated $b^+c^+d^+/b^-c^-d^-$.

Let's consider a hypothetical bacterium that synthesizes a green pigment from the combined action of genes *a*, *b*, and *c*. The genes code for the enzymes A, B, C, which we assume to act sequentially to form the pigment. If there were a mutation in any of these genes, no pigment would be made. Pigment is made by the partial diploid $a^-b^+c^+/a^+b^-c^+$, however, because the cell contains a set of genes that produce functional proteins A, B, and C. B and C will be made from the $a^-b^+c^+$ chromosome, and A and C from the $a^+b^-c^+$ chromosome. In a partial diploid $a_1^-b^+c^+/a_2^-b^+c^+$, in which a_1^- and a_2^- are two different mutations in gene *a*, no pigment can be made because the bacterium will not contain a functional A protein. The two mutations a^- and b^- of the diploid are said to complement one another because the phenotype of the partial diploid containing them is A^+B^+; the mutations a_1^- and a_2^- do not complement one another because the phenotype of the partial diploid containing these mutations is A^-.

Suppose that now a mutation x^-, which also blocks green pigment formation, has been isolated but the gene in which the mutation has occurred has not been characterized. By constructing a set of partial diploids, this gene can be identified by a complementation test. As a start, we might test the genes *a*, *b*, and *c* with the partial diploids a^-/x^- (I), b^-/x^- (II), and c^-/x^- (III). If diploids I and II make pigment, the mutation cannot be in genes *a* or *b*. If no pigment is made by diploid III, the mutation must be in gene *c*. If pigment were made in all three diploids, then the important conclusion that mutation x^- is in none of the genes *a*, *b*, or *c* could be drawn. Furthermore, because we have assumed that *a*, *b*, and *c* are each pigment genes, the fact that *x* is not in any of these genes but prevents pigment formation would be evidence that pigment formation requires at least four genes ("at least" because more genes might still be discovered).

A common approach to the initial characterization of a genetic system is to isolate about 50 mutants and perform complementation tests among them. In practice, complementation analysis is usually performed after experiments have been performed to determine the mutations' positions on a genetic map. This mapping information reduces the number of partial diploids that must be constructed because two mutations that map far from each other cannot be in the same gene.

Two rules are very helpful for interpreting data from complementation experiments.

Rule I. If two mutations complement, they are almost always in different genes. There is a rare exception to this rule. This exception must satisfy two conditions. The protein product must be made of two or more identical subunits and each mutant subunit must be able to provide a function that is missing in the other. This very rare phenomenon is called **intrallelic complementation.**

Rule II. If two mutations fail to complement, then one of the following is true:

a. They are in the same gene.

b. At least one of the mutations is in a regulatory site for the other gene.

c. At least one of the mutations yields an inhibitory gene product.

Several examples will help to show how complementation analysis can be used to identify mutations in different genes. Each entry in Table 7.1 designates a partial diploid constructed with the alleles shown at the top and side of the table. A (+) entry indicates that the partial diploid is able to grow and the mutations complement. In contrast, a (–) entry indicates that the partial diploid does not grow and the mutations do not complement. The (–) entries along the diagonal indicate that a mutation cannot complement itself. Rules I and IIa permit mutations

TABLE 7.1 Example of Complementation Data

Mutation Number	1	2	3	4	5	6	7	8	9	10
1	–									
2	+	–								
3	+	+	–							
4	+	–	+	–						
5	–	+	+	+	–					
6	+	+	–	+	+	–				
7	+	+	–	+	+	–	–			
8	+	–	+	–	+	+	+	–		
9	–	–	–	–	–	–	–	–	–	
10	±	–	±	–	±	±	±	–	–	–

Symbols are (+) complementation; (–) no complementation; (±) weak complementation. An entry at the intersection of the horizontal row and a vertical column represents the result of one complementation test between two mutations. For example, the + entry at the intersection of row 3 and column 2 indicates that mutations 2 and 3 complement; the corresponding entry for row 2 and column 3 is not given because it is the same as the one just stated—that is, the table is symmetric about the diagonal. The analysis first discussed in the text uses only mutations 1 to 8, which is the reason for the dashed line between rows 8 and 9.

1 to 8 to be placed in mutually exclusive complementation groups. For example, in column 1, mutations 1 and 5 fail to complement. Other noncomplementing mutations can be found in columns 2, 3, 4, and 6. Three complementation groups can be identified in this way. The first contains mutations 1 and 5; the second contains mutations 2, 4, and 8; and the third contains mutations 3, 6, and 7. Each of these complementation groups corresponds to a gene. For convenience, each gene will be given an identifying letter.

Gene	Mutation
A	1, 5
B	2, 4, 8
C	3, 6, 7

The simplest explanation for the data is that the phenotype being studied requires at least three genes. If complementation analysis only allowed the identification of individual genes it would not be very useful today because genes can be identified from DNA sequences. Complementation analysis, however, provides additional information that cannot be provided by DNA sequence analysis alone. Let's consider mutations 9 and 10 to see why this is so.

Suppose a ninth mutant had been isolated having the complementing property shown below:

1	2	3	4	5	6	7	8	9
–	–	–	–	–	–	–	–	–

Alone, these data might suggest that all of the mutations are in the same gene. The data indicates the presence of three genes, however: *A*, *B*, and *C*. When mutation 9 is present on a chromosome the products of genes *A*, *B*, and *C* cannot be made. Mutation 9 thus prevents related genes on the same chromosome from being expressed. Mutations of this type are said to be ***cis*-dominant**. Many examples of *cis*-dominant mutations will be seen in later chapters.

Let's now turn our attention to mutation 10. Because it fails to complement mutations 2, 4, and 8, mutation 10 might appear to be a new mutation in gene *B*. However, the (+) response is weak with mutants 1, 3, 5, 6, and 7. Mutant 10, therefore, cannot be considered complementary to gene *A* or *C*. The correct interpretation of these results is that mutation 10 is in gene *B* and that the mutant gene product inhibits the activity of the good copy of the B polypeptide. For example, the B protein may be a homotetramer that requires that all subunits are fully functional. The (+) response is weak in diploids that contain normal *A* and *C* genes because the total activity of the inhibited B is not adequate for a normal (+) response.

E. coli cells can exchange genetic information by conjugation.

Genetic studies of *E. coli* and its viruses provided an essential part of the foundation on which molecular biology is built. Genetics was already a mature discipline in 1946 when Joshua Lederberg, then a 21-year-old graduate student, and his advisor Edward L. Tatum, first showed that one might be able to perform genetic studies in *E. coli*.

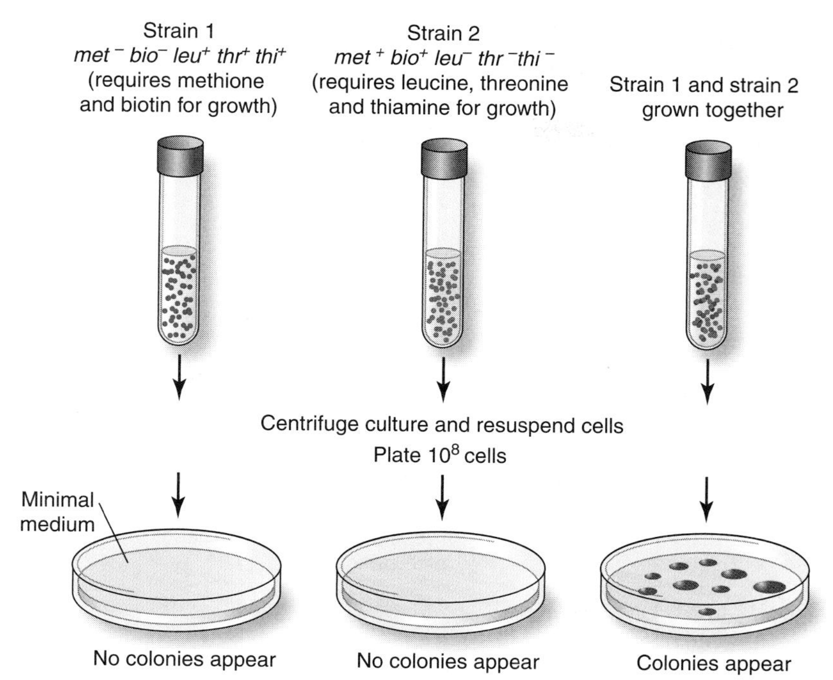

FIGURE 7.9 The Lederberg-Tatum experiment providing the first evidence for bacterial conjugation. The genotype for strain 1 is met^-, bio^-, leu^+, thr^+, thi^+, and that for strain 2 is met^+, bio^+, leu^-, thr^-, thi^-. The symbols bio^- and thi^- indicate requirements for the vitamins biotin and thiamine, respectively. Other abbreviations are standard for amino acids. Colonies are observed on minimal medium plates only after strains 1 and 2 have been permitted to grow together.

Their approach, which is shown in FIGURE 7.9, was simple and direct. They selected two strains of *E. coli* with different nutritional requirements. Strain 1 required methionine and biotin for growth, while Strain 2 required threonine, leucine, and thiamine. No colonies were formed when each strain was plated separately on minimal medium. This result was expected because the probability of two or three mutations all reverting in a single cell is extremely low. An entirely different result was observed when the two strains were mixed and incubated together before plating. In this case, a few colonies appeared on the minimal medium, suggesting that genetic material had somehow been exchanged between the two strains.

Subsequent studies by Lederberg and others demonstrated that *E. coli* possesses two mating types: donors (males), and recipients (females). Lederberg and Tatum were fortunate in their choice of strains because not all *E. coli* strains mate. Maleness is determined by a plasmid that is about 100-kb long known as the **fertility factor** (**F factor**) or **F plasmid.** A male cell, which has the F plasmid, is designated F^+, whereas a female cell, which lacks the F plasmid, is designated F^-. Male cells can transfer the F plasmid to female cells. This process, known as mating or **conjugation**, can be divided into four stages (FIGURE 7.10):

1. *Formation of specific donor-recipient pairs through effective contact.* Effective contact, begins when the sex pilus, a hair-like

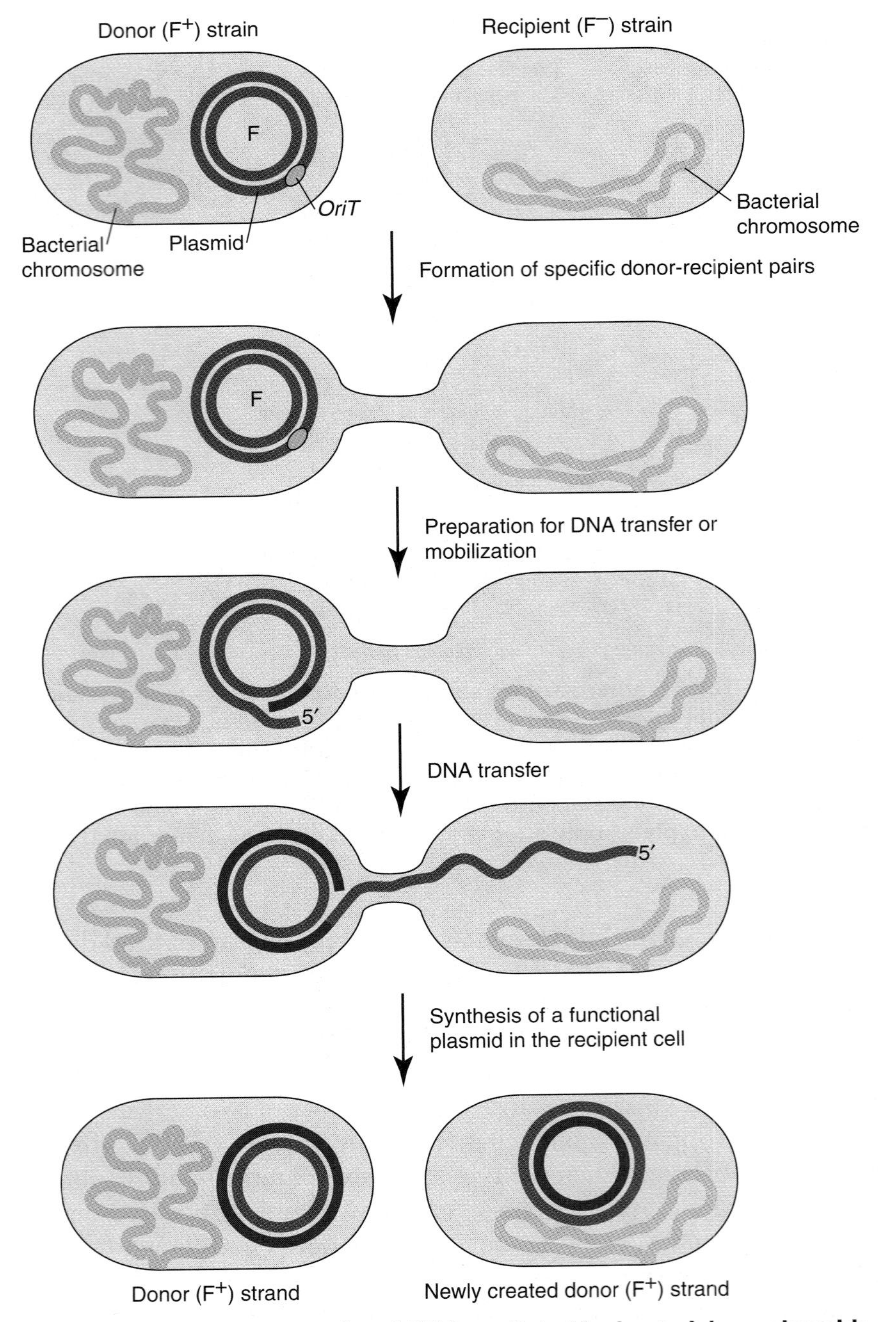

FIGURE 7.10 Conjugal transfer of DNA mediated by bacterial sex plasmids. Transfer of F plasmid during bacterial conjugation. The F plasmid, *oriT*, and bacterial chromosome are not drawn to scale. *E. coli* chromosomal DNA is about 50-times longer than F factor DNA.

appendage that projects from the outer surface of the male cell, binds to a receptor on the surface of the female cell (FIGURE 7.11). Pili are about 2–20 μm long and about 8–9 nm in diameter.

2. *Preparation for DNA transfer or mobilization*. A single-strand break is introduced in a unique base sequence within the F plasmid called the **transfer origin** or ***oriT***. The nicking

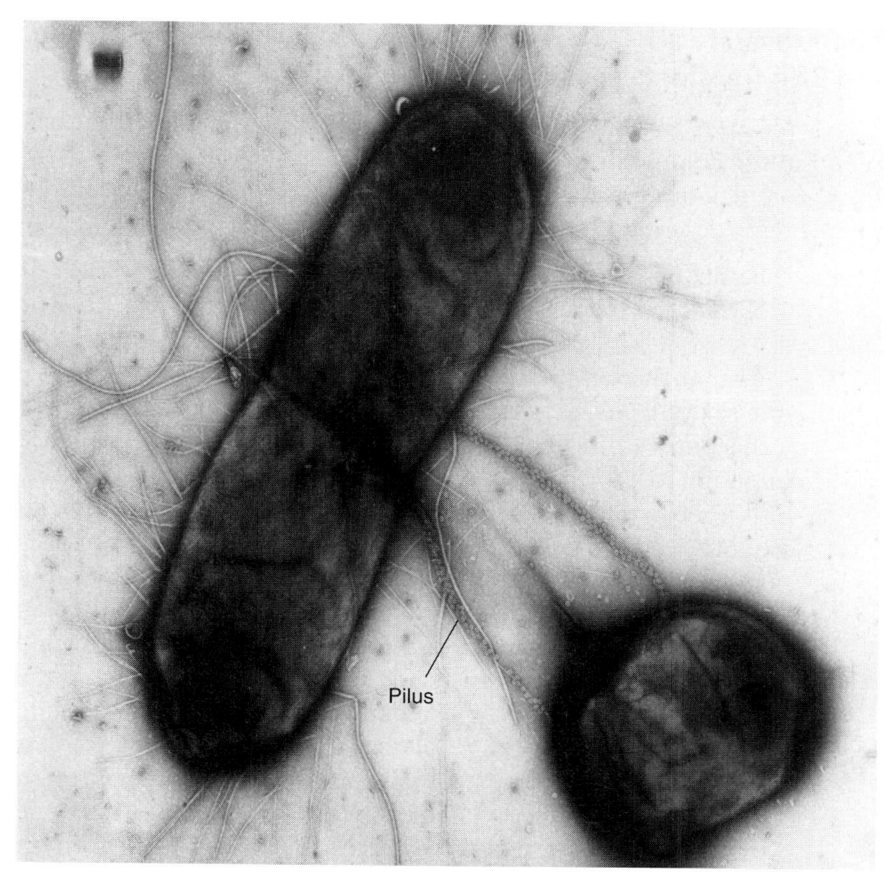

FIGURE 7.11 **F pili connecting mating *E. coli* cell.** Small bacteriophages that bind to F pili were added to mating bacteria to make the F pili more visible when viewed by electron microscopy. The phage coated F pili are the top and bottom appendages between the mating bacteria. Other appendages are not coated by the virus particles and are presumed to be common pili but not F pili. (Photo courtesy of Professor Ron Skurray, School of Biological Sciences, University of Sydney.)

enzyme (see below) is covalently attached to the 5′-end of the nicked chain.

3. *DNA transfer.* The nicked DNA strand is transferred in a 5′→3′ direction from the donor to the recipient cell. DNA replication in the donor cell takes place through a mechanism known as **rolling circle replication** in which the DNA replication machinery in the donor cell extends the 3′-end of the nicked strand by using the intact complementary strand as a template.
4. *Synthesis of a functional plasmid in the recipient cell.* The single strand that enters the recipient strain serves as a template for the synthesis of a complementary strand. Then the newly formed double strand cyclizes to form a new F plasmid, converting the recipient into a donor strain. Plasmid DNA replication also occurs in the donor cell so that the donor retains a copy of the F plasmid and its mating type remains unchanged.

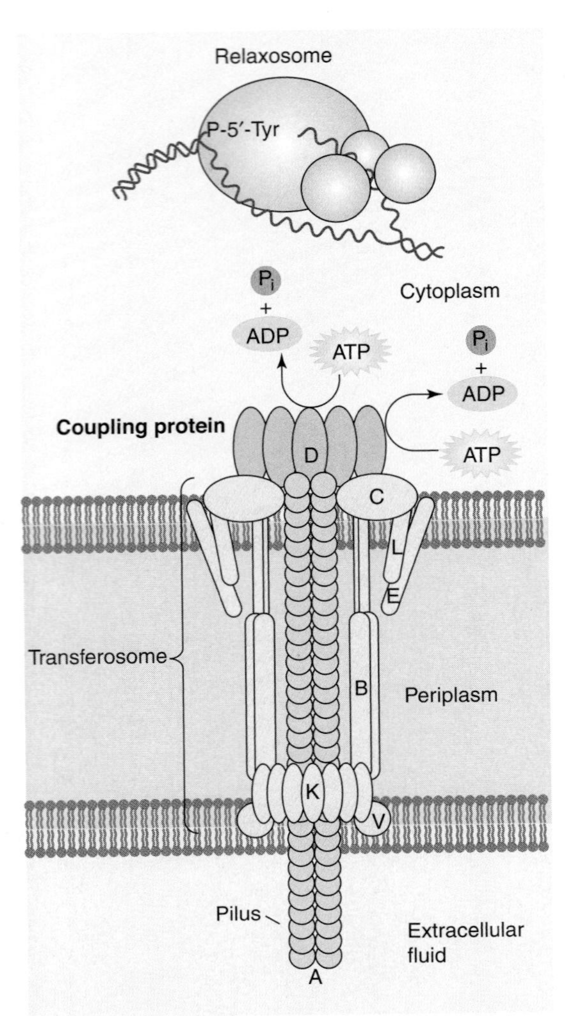

FIGURE 7.12 **Protein complexes required for F plasmid transfer.** The F-plasmid conjugation system is shown located in the inner and outer membranes and extending through the periplasm. Tra proteins are labeled with uppercase letters. The pilus is shown in tan, the relaxosome in gray, the transferosome in light tan, and the coupling protein in light purple. The relaxosome nicks the T-strand (transferred donor strand) at *oriT*. A tyrosine (Y) at the active site of the TraI subunit attaches to the phosphate at the 5′-end of the T-strand. The T-strand•TraI complex moves from the relaxosome, through a channel in the coupling protein, and then through the transferosome as it transfers to the female cell. Recent experiments (see text) suggest that the T-strand•TraI complex also passes through the sex pilus. (Adapted from L. S. Frost, et al., *Nat. Rev. Microbiol.* 3 [2005]: 722–732.)

Approximately 40 F factor genes are needed for successful mating and DNA transfer to occur.

The F plasmid has about 40 genes that code for proteins that are required for conjugation. The proteins specified by these genes form four protein complexes (FIGURE 7.12). Unless otherwise stated, each of the proteins described below is specified by an F plasmid gene.

1. The **sex pilus**. The sex pilus (shown in tan in Figure 7.12) is a hollow tube made of a single protein, pilin, which is encoded by *traA*. Two hypotheses have been proposed to explain the role that the sex pilus plays in mating. According to the first, the sex pilus retracts after contacting a receptor on the female cell's outer surface, bringing the mating cells together so that a junction can form between them. Although sex pili do retract, it is not clear if the cell envelopes of mating bacteria must make contact for DNA transfer to take place. The second hypothesis proposes that single-stranded DNA moves through the hollow sex pilus. Two recent studies support this hypothesis. Miroslav Radman and coworkers used fluorescent microscopy to distinguish donor DNA transferred into a recipient cell from the recipient cell's own DNA. Their approach was as follows. Male cells that methylate DNA were mated with female cells that do not. The female cells also produced a fluorescent protein that binds to hemimethylated double-stranded DNA (only one strand in the duplex is methylated) but not to fully methylated or fully unmethylated DNA. Upon transfer to the female cell, methylated single-stranded donor DNA was converted to hemimethylated double-stranded DNA, which remained that way because the female cell could not methylate it. Fluorescent foci appeared in the female cells as the diffuse fluorescent protein bound to the hemimethylated DNA. Approximately 6% of the female cells that received the male DNA were some distance away from the male donor. Distant DNA transfer was not observed when 0.01% SDS (which depolymerizes F pili) was present or when the F plasmid had a lesion in the *traA* gene. Tri-Rung Yew and coworkers, using an entirely different approach, also provided more direct evidence that the sex pilus is a conduit for DNA during mating. They used an atomic force microscopy (AFM) probe to image and dissect the sex pilus connecting a mating pair. Then they used an AFM probe functionalized with an anti-single-stranded DNA antibody to show that the dissected pilus had a transferring single-stranded DNA within it. Further work is needed to confirm and extend these recent experiments.
2. The **relaxosome**. The relaxosome (shown in gray in Figure 7.12) is a nucleoprotein complex that forms at the F plasmid's 360-bp long origin of transfer (*oriT*) and prepares the donor strand for transfer. The relaxosome consists of a relaxase/helicase (TraI) and three auxiliary DNA-binding proteins, IHF (see Chapter 6), TraY, and TraM. The gene that codes for IHF is on the bacterial chromosome. FIGURE 7.13 shows how

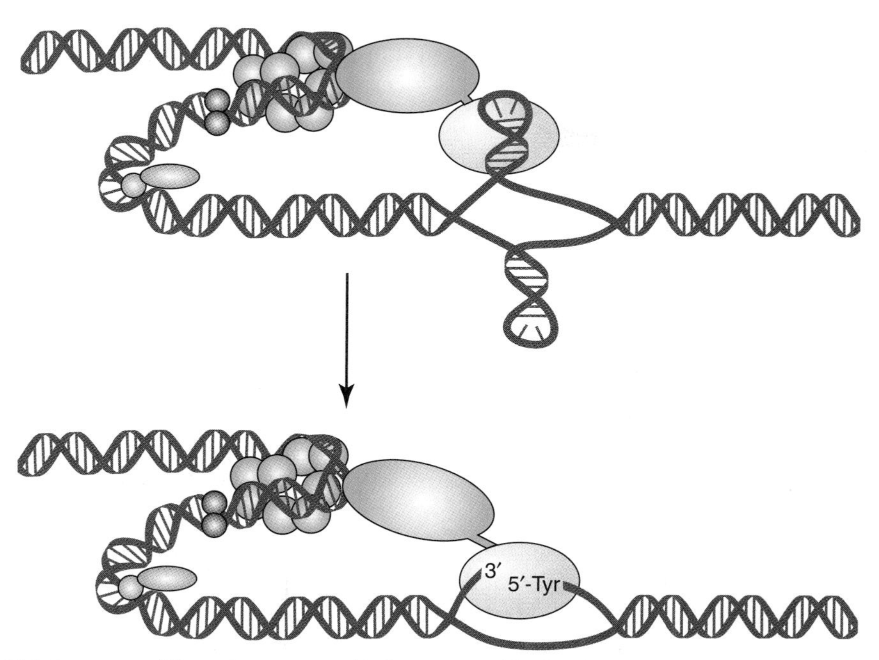

FIGURE 7.13 **The relaxosome.** Relaxase and helicase domains of TraI are shown in gray and light blue, respectively. The active site tyrosine (Tyr) in the relaxase domain nicks the T-strand (strand to be transferred) and attaches to the phosphate at the 5′-end of the nicked strand. TraM is shown in green, TraY in purple, and IHF in orange. (Reproduced from F. de la Cruz, et al., *FEMS Microbiol. Rev.* 34 [2009]: 18–40. Copyright © 2009 and reproduced with permission of John Wiley & Sons, Inc.)

the relaxosome is organized. TraI and the auxiliary proteins cause an AT-rich region within the DNA to melt. TraI has an N-terminal domain with relaxase activity and a C-terminal domain with helicase activity. The relaxase activity behaves like a type IA topoisomerase (see Chapter 4). In fact, the term relaxase reflects the enzyme's ability to relax supercoiled double-stranded plasmid DNA. An active site tyrosine introduces a nick in the T-strand (strand to be transferred) and attaches to the newly generated 5′-phosphate. TraI remains attached to the T-strand during the entire transfer process. A second TraI must also participate in the transfer process because the first TraI remains attached to the T-strand's 5′-end and moves along with it into the female cell. The second TraI is required to unwind plasmid DNA in the donor strain so that the T-strand can be transferred. This unwinding is catalyzed by the helicase domain, which moves in a 5′→3′ direction along the nicked DNA as it catalyzes the ATP-dependent unwinding process. The intact circular strand serves as a template for complementary strand biosynthesis by the host's DNA replication machinery using rolling circle replication.

3. The **transferosome.** The transferosome (shown in light tan in Figure 7.12) spans the cell envelope and actively transfers donor DNA into the recipient cell. The transferosome is a type IV secretion system, a versatile system that can secrete

DNA-protein complexes, single protein molecules, or protein complexes.

4. The **coupling protein.** The coupling protein (shown in light purple in Figure 7.12) is made of six identical TraD subunits. The coupling protein is required for conjugation because it links the relaxosome to the transferosome, forming a channel through which the T-strand passes as it moves to the female cell. The coupling protein has an ATPase activity that appears to be required for conjugation.

Plasmids such as the F plasmid that have all the genes needed for successful mating and DNA transfer are called **self-transmissible plasmids.** When a plasmid is missing one or more of these genes, the cell that bears it will not be able to perform some or all of the functions required for effective contact and DNA transfer. Plasmids that cannot make a functioning relaxosome but can make all of the other protein complexes needed for conjugation are called **conjugative plasmids** because cells that have them can make effective contact with recipient cells but cannot transfer plasmid DNA to the recipients. Plasmids with genes for DNA mobilization but that lack those needed to make the pilus or transferosome are called **mobilizable plasmids.** A cell with a mobilizable plasmid will be able to transfer the plasmid to a recipient if the cell has some other means for making effective contact. For instance, another plasmid within the cell may supply genes needed to make the pilus and transferosome. Plasmids lacking genes required for making effective contact and for mobilization are called **nontransmissible plasmids.**

The F plasmid can integrate into a bacterial chromosome and carry it into a recipient cell.

The F plasmid also has the ability to become integrated into the bacterial chromosome. (The mechanism for this integration will be discussed in Chapter 14.) When this integration occurs, the F plasmid and the bacterial chromosome become a single, circular DNA molecule, and F plasmid DNA behaves as if it were part of this chromosome, thereby increasing the size of the chromosome. The integration process happens rarely, but it is possible to purify cells that are the progeny of the cell in which integration occurred. Such cells are called **Hfr (high frequency of recombination) males.**

When an Hfr culture is mixed with an F^- culture, conjugation also occurs as described for the F plasmid, though the material transferred is different from that in an $F^+ \times F^-$mating. This time, under the influence of the F insert, DNA replication of the whole chromosome starts within the F insert and a replica of the chromosome is transferred to the F^- cell (FIGURE 7.14). Replication starts after the chromosome is nicked within *oriT*. Part of the F sequence is the first DNA to enter the female cell. The remaining part enters only after the entire bacterial chromosome has been transferred. Moreover, because bacteria are very small and in constant motion caused by constant water molecule bombardment (Brownian motion), and because it takes 100 minutes to

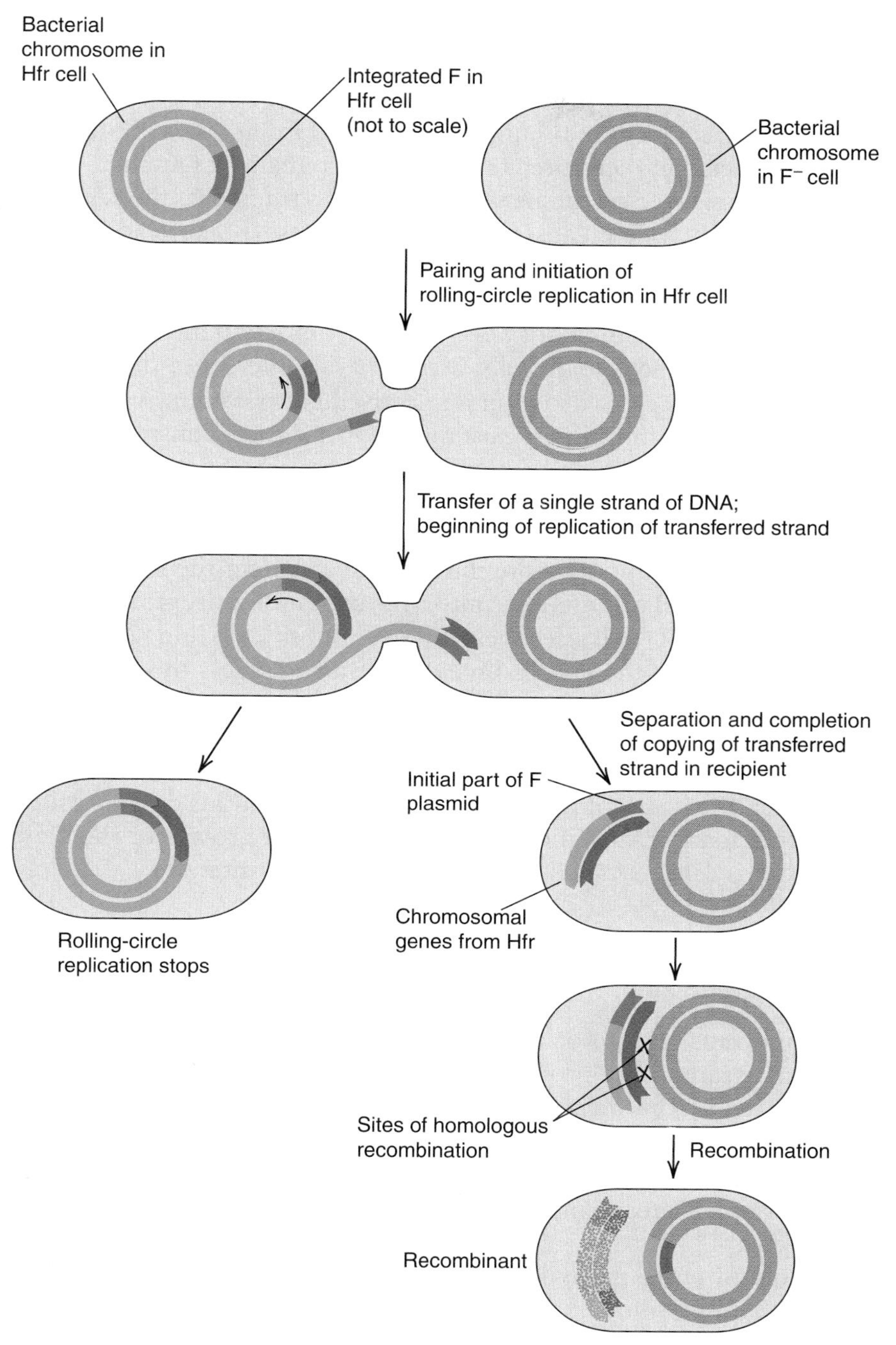

FIGURE 7.14 **Stages in transfer and production of recombinants in an Hfr X F⁻ mating.** Pairing (conjugation) initiated rolling-circle replication within the F sequence in the Hfr cell results in the transfer of a single strand of DNA. The single strand is converted into double-stranded DNA in the recipient. The mating cells usually break apart before the entire chromosome is transferred. Recombination takes place between the Hfr fragment and the F⁻ chromosome and leads to recombinants containing genes from the Hfr chromosome. Note that only a part of the F factor is transferred. This part of the F factor is not incorporated into the recipient chromosome. The recipient remains F⁻.

transfer an entire chromosome, the mating pair usually breaks apart before transfer is completed. The female, thus, receives both a small functionless fragment of F and a large fragment of the Hfr chromosome, which may contain hundreds of genes. In an Hfr × F^- mating, the mated female, therefore, almost always remains a female.

In the presence of the new chromosomal fragment, the female's recombination system causes genetic exchanges to take place, and a recombinant F^- cell often results. Thus, in a mating between an Hfr leu^+ culture and an F^-leu^- culture, recombinant F^-leu^+ cells arise. One can distinguish recombinant leu^+ females from leu^+ males by starting out with male and female cells with genetic differences that permit only the female cell to grow on a selective agar medium. A common method is to use antibiotic resistance. For instance, consider an Hfr that is not only Leu^+ but also streptomycin-sensitive (Str^S) and a female that is both Leu^- and streptomycin-resistant (Str^R). If the *leu* gene is near the origin of transfer but the *str* gene is not, the mating process is likely to be disrupted before the *str* gene can enter the F^- cell. Then plating the mated cell mixture onto agar containing streptomycin but lacking leucine (1) will selectively kill the Hfr Str^S cells and (2) will not lead to growth of the F^- cells unless these also possess the leu^+ allele. Thus, any cell that survives will have the genotype $leu^+\ str^R$ and will be a recombinant female. Only these $Leu^+\ Str^R$ recombinants can form a colony. When a mating is done in this way, the transferred allele that is selected by means of the agar conditions (leu^+ in this case) is called a **selected marker**, and the allele used to prevent growth of the male (str^R in this case) is called the **counterselective marker**.

Bacterial mating experiments have been used to produce an *E. coli* genetic map.

An important feature of an Hfr × F^- mating is that the transfer of the Hfr chromosome proceeds at a constant rate from a fixed point determined by the site at which F has been inserted in the Hfr chromosome. This means that the times at which particular genetic loci enter a female are directly related to the positions of these loci on the chromosome. Thus, a map can be obtained from the time of entry of each gene.

Time of entry mapping is performed in the following way. An Hfr with an $a^+b^+c^+d^+e^+\ str^S$ genotype is mated with an $a^-b^-c^-d^-e^-\ str^R$ female. At various times after mixing the cells, samples are removed and agitated violently in order to break apart all of the mating pairs simultaneously. Every sample is then plated on five different agars containing streptomycin, each of which also contains a different combination of four of the five substances A through E. Thus, colonies that grow on agar lacking A are $a^+\ str^R$; those growing without B are $b^+\ str^R$, and so forth. All of these data can be plotted on a single graph to give a set of time-of-entry curves, as shown in **FIGURE 7.15a**. Extrapolation of each curve to the time axis yields the time of entry of each gene a^+, b^+, . . ., e^+. These times can be placed on a map, as shown in **FIGURE 7.15b**. Use of a second female that is $b^-e^-f^-g^-h^-\ str^R$ can then provide the relative positions of three additional genes. These

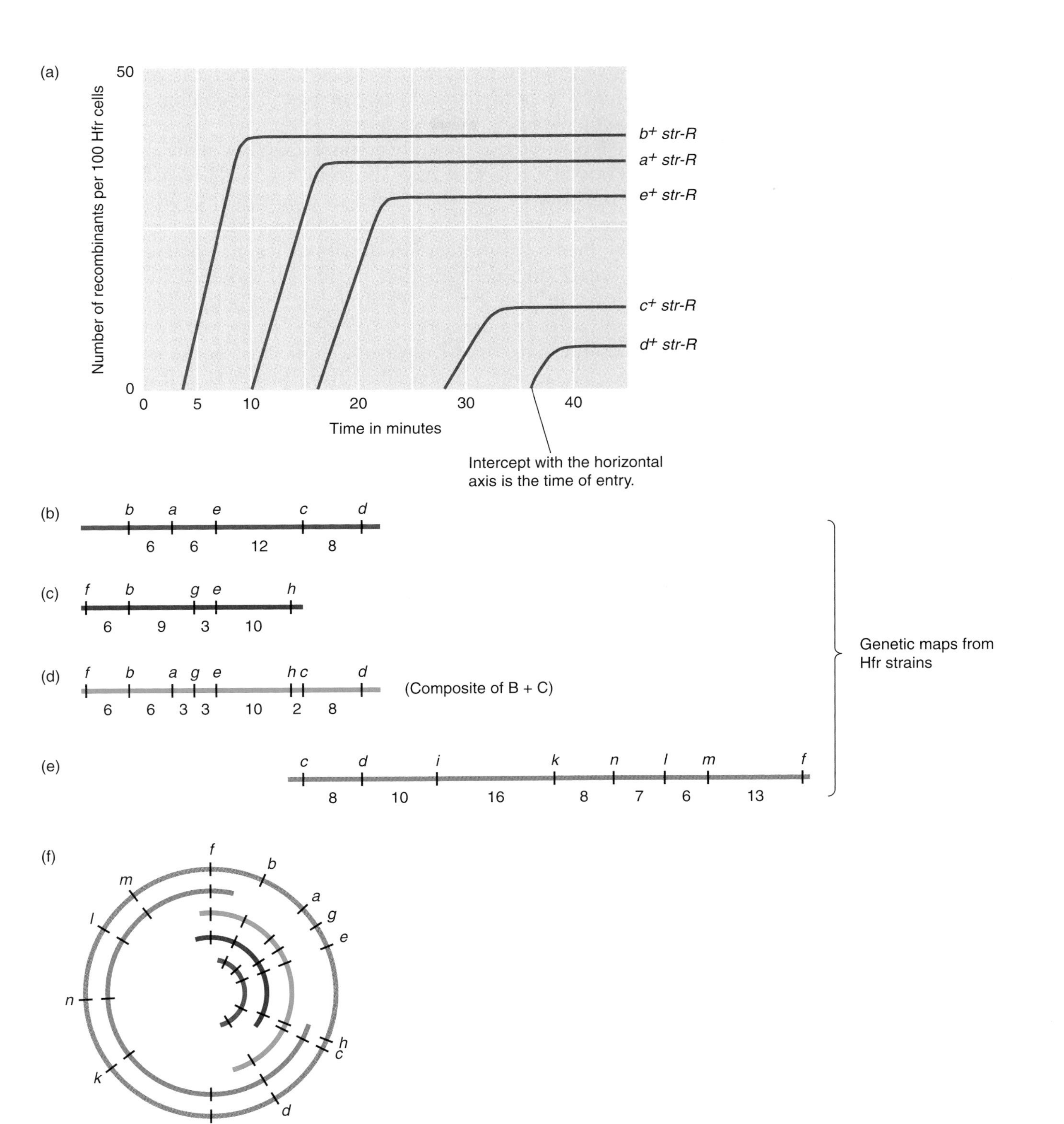

FIGURE 7.15 **Time-of-entry mapping.** (a) Time-of-entry curves for one Hfr strain. (b) The linear map derived from the data in part a. (c) A linear map obtained with the same Hfr but with a different F⁻ strain containing the alleles $b^- e^- f^- g^- h^-$. (d) A composite map formed from the maps in parts b and c. (e) A linear map from another Hfr strain. (f) The circular map (gold) obtained by combining the two maps (green and blue) of parts d and e.

might give a map such as that in FIGURE 7.15c. Because genes *b* and *e* are common to both maps, the two maps can be combined to form a more complete map, as shown in FIGURE 7.15d.

The F plasmid can integrate at numerous sites in the chromosome to generate Hfr cells that have different origins of transfer. Each of these Hfr strains can also be used to obtain maps. When separate maps are combined, a circle is eventually obtained. For instance, the map obtained from another Hfr might be that shown in FIGURE 7.15e, which, when combined with that in panel (d), would yield the circular map shown in FIGURE 7.15f. This mapping technique has been used repeatedly with hundreds of *E. coli* genes to generate an extraordinarily useful map, one with great significance in the development of molecular genetics. A simplified form of the *E. coli* genetic map is presented in FIGURE 7.16. The discovery in the mid-1950s that the *E. coli* genetic map is circular provided the first suggestion that the *E. coli* chromosome might be a circular DNA molecule, as was later found to be the case.

The *E. coli* chromosome has now been sequenced and the sequence data fit perfectly with the genetic map. Nucleotide sequence information has had a profound effect on molecular biology. Before the sequence was known, investigators isolated *E. coli* mutants on the basis of their phenotype and then used genetic techniques to determine the gene's position on the chromosome. Today we know the positions of all the *E. coli* genes but still do not know the functions of many of them. When mutants with interesting new phenotypes are discovered, the genes can be cloned (or amplified by the polymerase chain reaction [PCR]) and then sequenced. The gene's position on the genetic map can be determined by comparing its sequence with that of the entire *E. coli* chromosome.

F′ plasmids contain part of the bacterial chromosome.

As already stated, an Hfr cell is produced when the F plasmid stably integrates into the chromosome. The F plasmid can also be excised at a low frequency. When this happens, the excised circular DNA is sometimes found to contain genes that were adjacent to F in the chromosome (FIGURE 7.17). A plasmid containing both F genes and chromosomal genes is called an **F′ plasmid.** It is usual to describe an F′ plasmid by stating the genes it is known to possess—for example, F′ *lac pro* contains the genes for lactose utilization and proline synthesis. F′ plasmids can also be transferred from an F′ male to a female. This occurs sufficiently rapidly that the entire F′ is usually transferred before the mating pair breaks apart. Thus, the female recipient is converted to an F′ male in an F′ × F^- mating.

F′ plasmids have been used to construct partial diploid bacteria such as those that were described above for use in the complementation test. A lac^+/lac^- partial diploid can be constructed by mating an F′lac^+ str^S male with a lac^- str^R female and selecting for a Lac^+ Str^R colony. Cells in the colony will carry two copies of the *lac* gene: the lac^+ allele brought to the female in the F′ and the lac^- allele already present in the female chromosome. To denote this, the genotype of

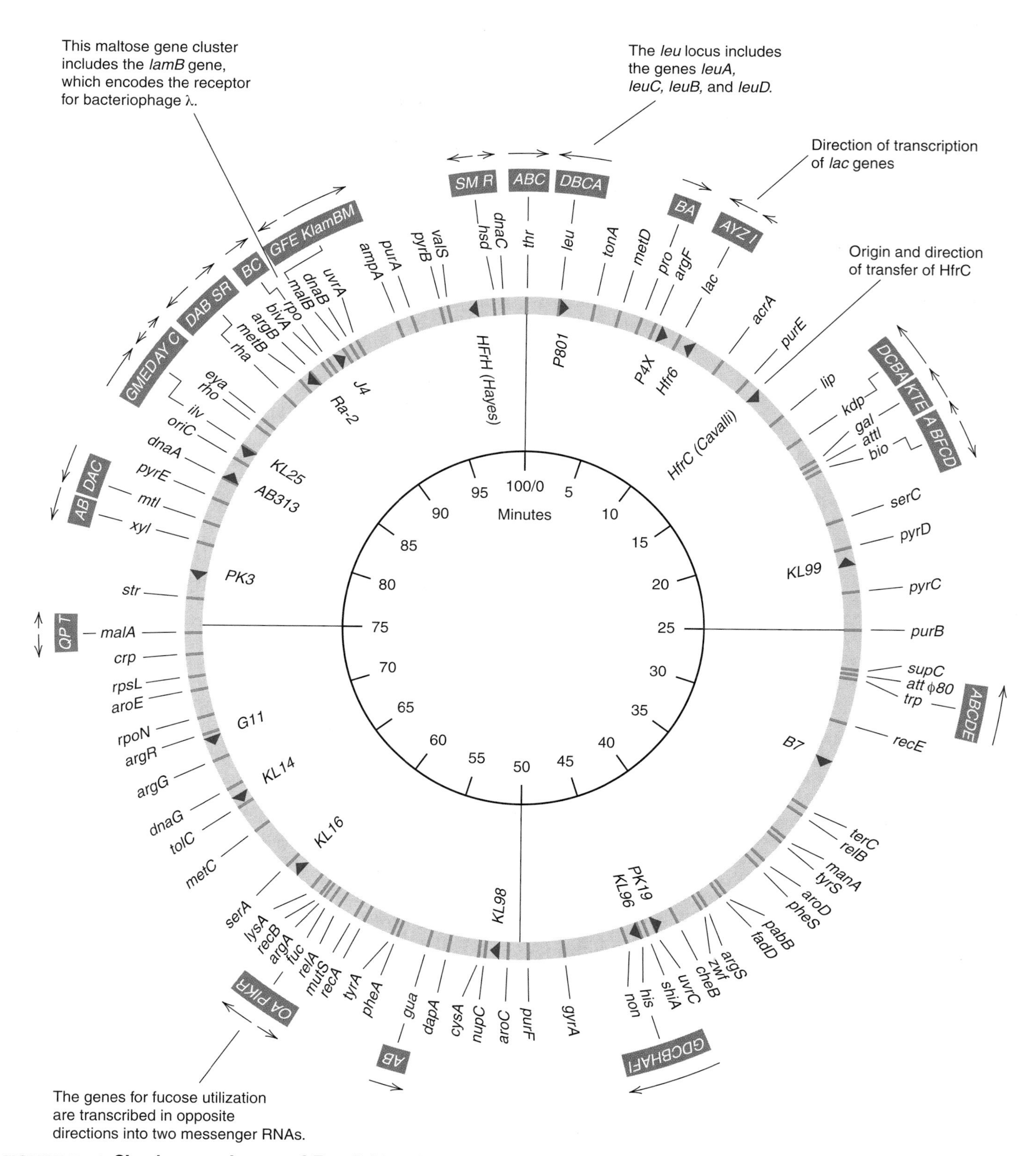

FIGURE 7.16 **Circular genetic map of *E. coli*.** Map distances are given in minutes; the total map length is 100 minutes. For some of the loci that encode functionally related gene products, the map order of the clustered genes is shown, along with the direction of transcription and length of transcript (black arrows). The purple arrowheads show the origin and direction of transfer of a number of Hfr strains. For example, HfrH transfers *thr* very early, followed by *leu* and other genes in a clockwise direction.

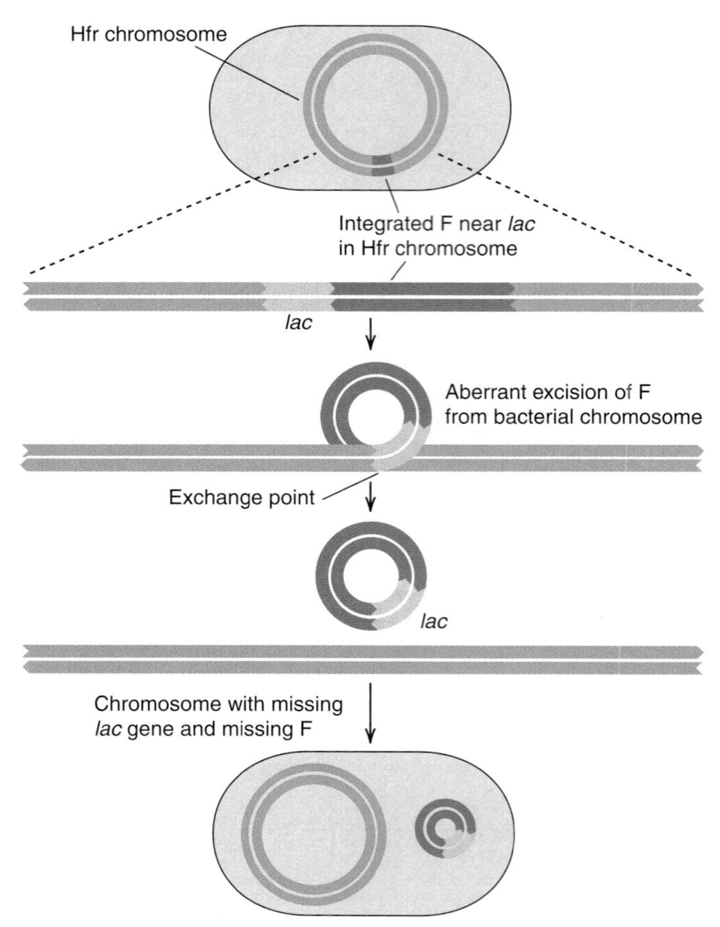

FIGURE 7.17 **Formation of an F′ lac plasmid by aberrant excision of F from an Hfr chromosome.** Breakage and reunion are between nonhomologous regions. The bacterial *lac* gene is shown in light blue.

the cell is written F′*lac*$^{+}$/*lac*$^{-}$ *str*^{R}, as is usual with partial diploid cells. By convention, genes carried on F′ plasmids are written at the left of the diagonal line.

Plasmid replication control functions are usually clustered in a region called the basic replicon.

When first discovered, the F factor seemed to be a unique type of hereditary unit. It soon became evident that the F factor is just one of many extrachromosomal DNA molecules that can replicate in *E. coli* and other organisms. The bacterial host normally supplies most of the enzymes needed for plasmid DNA replication. The plasmid controls its own replication. Plasmid replication control functions are usually clustered in a 1–3 kb region called the **basic replicon.** This region is defined experimentally as the smallest piece of plasmid DNA that can replicate with the normal copy number. Each kind of plasmid has its own unique basic replicon. A basic replicon that functions in one bacterial species usually will not function in other species. For example, most *E. coli* plasmids do not replicate in *Bacillus subtilis*. The basic replicon includes four elements. (1) The origin of replication region (about 50–200 bp long) is the initiation site for DNA synthesis. (2) The initiator region codes for protein and RNA molecules that are required to initiate DNA synthesis. (3) The replication copy control region codes for protein and RNA molecules that regulate the rate of plasmid replication. Some basic replicions are subject to much tighter regulation than others. These regulatory differences explain why some plasmids, the **stringent plasmids**, are maintained at only one or two copies per cell while others, the **relaxed plasmids**, are maintained at multiple copies per cell. Although most relaxed plasmids are maintained at between ten and twenty plasmids per cell, some are maintained at a few hundred copies per cell. The F factor, which is usually present at one or two copies per cell, is a stringent plasmid. (4) The partitioning system region codes for proteins that determines plasmid distribution at cell division. The partitioning system may be unnecessary when there is a high copy number (many plasmids per cell) because the probability that each daughter cell will receive some plasmids is very high even if the distribution is unequal.

Plasmids often confer advantageous properties to their hosts.

Plasmids may offer advantages to their host cell. **Drug resistance** or **R plasmids** have genes that make host cells resistant to one or more antibiotics or antibacterial drugs. The first R plasmid was discovered in Japan in the late 1950s when a *Shigella* strain, which was isolated from a patient suffering from dysentery, was found to be resistant to sulfanilamide, chloramphenicol, streptomycin, and tetracycline. Resistance to just one antibacterial agent might have been due to a rare mutation but the odds against four independent mutations arising in the same bacterial cell were astronomical. Investigators therefore suspected that the *Shigella* strain harbored a plasmid that made it resistant to all four drugs. Subsequent studies confirmed this suspicion. Some R plasmids are self-transmissible like the F factor, whereas others lack one or more of the genes required for self-transmission.

Transmissible R plasmids pose a serious public health threat even when they are present in a nonpathogenic host. Under normal circumstances, drug-sensitive nonpathogenic bacteria colonize the skin, mouth, and intestine. When exposed to antibacterial agents, these bacteria are eliminated and bacteria that bear R plasmids proliferate because they no longer have competition for the ecological niche. Serious problems arise when nonpathogenic drug-resistant bacteria transmit their R plasmids to invading pathogens, allowing the pathogens to become drug resistant and making clinical treatment difficult, if not impossible. The transfer of antibiotic resistance genes can take place between bacteria of the same species and also between bacteria with limited phylogenetic relatedness, including transfer between gram-negative and gram-positive species. Unfortunately, the problem extends beyond the unnecessary use of antibiotics in humans. Low doses of certain antibiotics are sometimes added to animal feed to prevent infection and promote growth. Animals that have been given these antibiotics often have bacteria with R plasmids, which can be transferred to bacteria that cause human diseases.

Virulence plasmids, which convert certain bacteria into pathogens, are largely responsible for the pathogenic effects of bacteria that cause dysentery, anthrax, plague, and tetanus. Virulence plasmids are of great concern because they, along with R plasmids, can be exploited to create pathogenic bacteria that can be used in germ warfare. Virulence plasmids are also present in bacteria that infect plants. *Agrobacterium tumafaciens*, a gram-negative bacteria that infects fruit trees and other broad leaf plants contains a tumor inducing plasmid called the Ti plasmid. Only a portion of the Ti plasmid, called T DNA, is transferred to the plant cell through a transferosome. These T DNA molecules, which are incorporated into plant cell chromosomes, are responsible for inducing the formation of tumors (also called crown galls). T DNA has been modified by recombinant DNA techniques so that it does not cause a tumor. This process, known as disarming, has created a plasmid that can be used for introducing foreign DNA into many plants.

Col plasmids program *E. coli* and other bacterial species to produce antibacterial proteins called colicins as well as proteins that protect the host from its own colicins. Many different kinds of colicins are known to exist. Examples in *E. coli* are as follows. Colicin E1 forms channels in the cell membrane that allow potassium ions to escape from the cell; colicin E2 degrades DNA; and colicin E3 blocks protein synthesis by making a specific cut in 16 S ribosomal RNA. The Col plasmids are advantageous to the host cell because the colicins they produce destroy bacteria that would otherwise compete for the same nutrients. Colicin producing bacteria are not killed because Col plasmids also produce immunity proteins that protect the cells. For instance, the immunity protein produced by Col E1, which is present in the cell membrane, prevents colicin E1 from forming channels in the cell membrane.

Degradative plasmids program host cells to make enzymes that degrade various organic molecules such as hydrocarbons and thereby allow the host cells to use biological fuels that they would not otherwise be able to use. Degradative plasmids may also protect the host from toxic organic molecules. Host cells harboring degradative plasmids are currently being used to clean up oil spills and to degrade toxic pesticides.

7.3 Budding Yeast (*Saccharomyces cerevisiae*)

Yeasts are unicellular eukaryotes.

Studies using *E. coli* as a model system over the past half-century have provided fundamental insights into the way that information flows from DNA to RNA to protein, thereby placing the discipline of molecular biology on a solid foundation. As a prokaryote, however, *E. coli* is not an effective model system for answering questions that are unique to eukaryotic cells. For instance, *E. coli* studies cannot provide information about nucleosome formation nor can they reveal how RNA is processed in the cell nucleus.

Investigators, searching for a model system that could be used to study fundamental problems such as these in eukaryotes, found that the unicellular yeasts offer many advantages. Like bacteria, yeast cells grow in liquid suspensions in either chemically defined media, in complex broths, or on a solid surface to form colonies. Moreover, yeasts grow very rapidly; the doubling time for most yeast strains is only about twice that of *E. coli*. The intracellular organization of a yeast cell is typical of a eukaryotic cell.

Despite these and other advantages, investigators encountered one problem with using yeast as a model system that hindered their studies. The yeast cell membrane is surrounded by a tough thick cell wall made of polysaccharides and polypeptides, making it very difficult to disrupt the cell without causing major damage to the cell organelles. This problem was solved when a snail gut enzyme was shown to degrade the yeast cell wall, allowing yeast cells to be gently disrupted.

S. cerevisiae (commonly known as bakers' yeast), the most thoroughly studied yeast from both a biochemical and genetic point of view, has an ovoid shape and a diameter of about 3-μm. *S. cerevisiae* (and other yeasts) exist in both a haploid and a diploid stage. The haploid cell contains 16 chromosomes, all of which have been sequenced. The yeast nucleus has about 3.5 times more DNA than is present in an *E. coli* cell but about 250-fold less DNA than is present in a human cell nucleus. Yeast, therefore, probably represents one of the simplest organisms with all of the genetic information required of a eukaryotic cell.

S. cerevisiae divide by budding during both their haploid and diploid stages (FIGURE 7.18). The daughter cell emerges as a small bud attached to the mother cell and enlarges until it is almost as big as the mother cell. Chromosomal replication takes place in the mother cell as

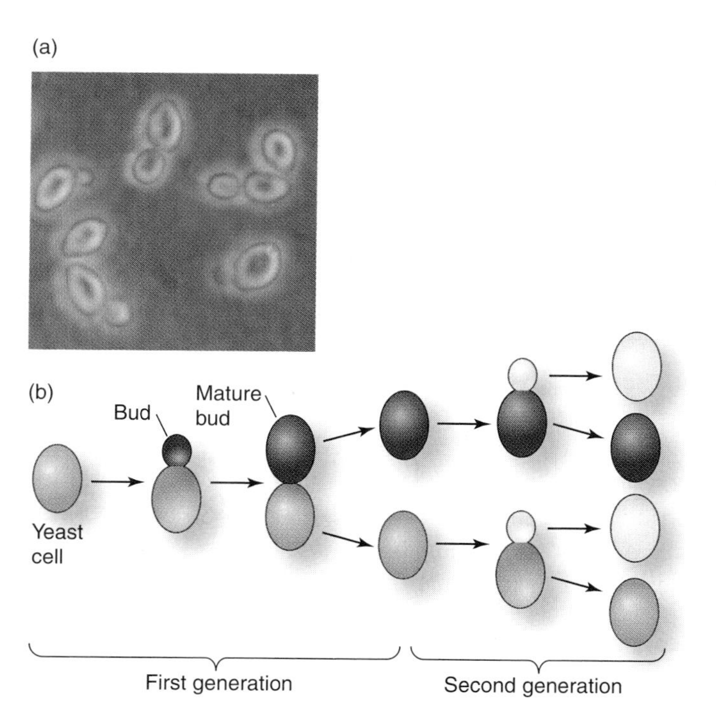

FIGURE 7.18 **Budding yeast.** (a) This light micrograph shows how yeast cells form small protuberances, called "buds," that gradually enlarge until they reach the size of the mother cell and become separated from it, ready to divide again in the same manner. (b) Budding, showing the mother-daughter relation. The first generation daughter, which is also the second generation mother, is shown in the darker color. (Part a courtesy of Breck Byers, University of Washington.)

the bud enlarges and is followed by mitosis. One of the two identical nuclei is transferred to the bud before it pinches off from the mother cell. In contrast to bacteria, which divide to yield twin progeny cells, with yeast there is a clear mother-daughter relation in the sense that the mother retains a scar on the cell wall at the site of budding. Not all yeasts divide by budding. For instance, *Schizosaccharomyces pombe*, another yeast species that has been extensively studied by molecular biologists, divides by binary fission. In most other ways, however, *S. pombe* is very similar to bakers' yeast.

An initial step in the study of *S. cerevisiae* has been to determine if biochemical mechanisms worked out for *E. coli* and that apply to many other bacteria are also valid for the simple eukaryote. The results are quite informative. The basic metabolism of yeast and bacteria are quite similar. For instance, both use the same pathways to synthesize purine and pyrimidine nucleotides. However, the overall process for synthesizing macromolecules (DNA, RNA, and protein) are not those of bacteria but instead those of higher eukaryotes. Yeast, thus, has become an outstanding model system for studying eukaryotic macromolecular synthesis, gene organization, and regulation of gene expression. So much is now known about yeast genetics and biochemistry that yeast is an attractive choice for fundamental studies in molecular biology. Other unicellular eukaryotes such as the alga *Chlamydomonas* and the protozoan *Tetrahymena* also are used in eukaryotic molecular biology.

Specific notations, conventions, and terminology are used in yeast genetics.

A nutritional requirement in yeast is usually specified in regular typeface letters followed by a superscript plus or minus sign. Thus, Leu^+ indicates that the yeast can make leucine and Leu^- indicates that it cannot. Three italicized letters are usually (but not always) used to specify a yeast gene. Many gene names are based on a mutant phenotype. For instance, the first three letters in an amino acid's name indicate a gene that is essential for the synthesis of that amino acid. Upper case letters such as *LEU* indicate that the gene is dominant and lowercase letters such as *leu* indicate that it is recessive.

Genes are also named for the proteins or RNA molecules they encode. For example, the gene that encodes calmodulin (a regulatory protein that binds calcium ions) is called *CMD1*. The genetic locus (position) is given by a number (rather than a letter) following directly after the gene symbol (for example, *leu2*). A wild type gene is designated with a superscript plus sign placed just after locus number ($LEU2^+$). The allele number is placed after the locus number and separated from it by a hyphen (for example, *leu2-5*). The name of a protein specified by a gene is indicated by writing the gene name in regular typeface as a proper noun (first letter in uppercase) often followed by the letter p. For instance the protein product of *STE3* is Ste3p or Ste3. The rules described for genetic nomenclature in yeast are summarized in Table 7.2.

TABLE 7.2 **Genetic Nomenclature, Using *LEU2* as an Example**

Gene Symbol	Definition
LEU$^+$	All wild type alleles controlling leucine requirement
leu2$^-$	Any *leu2* allele conferring a leucine requirement
LEU2$^+$	The wild type allele
leu2-9A	Specific allele or mutation
Leu$^+$	A strain not requiring leucine
Leu$^-$	A strain requiring leucine
Leu2p or Leu2	The protein encoded by *LEU2*
LEU2 mRNA	The mRNA transcribed from *LEU2*
Leu2-Δ1	A specific complete or partial deletion of *LEU2*
LEU2::ARG2	Insertion of the functional *ARG2* gene at the *LEU2* locus, and *LEU2* remains functional and dominant
leu2::ARG2	Insertion of the functional *ARG2* gene at the *LEU2* locus, and *leu2* is or becomes nonfunctional
leu2-5::ARG2	Insertion of the functional *ARG2* gene at the *LEU2* locus, and the specified *leu2-5* allele that is nonfunctional

Yeast cells exist in haploid and diploid stages.

One major advantage of yeast as an experimental system is that it has both haploid and diploid phases. There are two haploid mating types: Mata (**a**) and Matα (α), which can mate to form a stable **a**/α diploid. The existence of the mating types means that elegant genetic experiments can be done that complement physicochemical studies, and that both the mechanisms of meiosis and the chromosomal interactions responsible for genetic recombination can be studied. When starved for nutrients, an **a**/α diploid cell undergoes meiosis to form four haploid cells called **spores**, which are encased in a sac called an **ascus**. If removed from the ascus and separated from one another, each of the four spores (two **a** cells and two α cells) can grow and divide. If **a** cells and α cells are mixed together, however, they will mate to re-form the diploid state.

The yeast mating process occurs in discrete stages, which are summarized in the schematic representation in FIGURE 7.19. Cells of an opposite mating type agglutinate shortly after they are mixed. Glycoproteins on the cell surface act as gluelike substances, causing the cells to stick together. Many of the cells have buds at the beginning of the mating process. Buds are no longer evident 90 to 120 minutes later, however, because the cell division cycle is arrested at the G_1 stage. Arrested cells start to grow in an anisotropic fashion, assuming raindrop-like shapes. When the tips of the cells within a mating pair touch, enzymes remodel the cell walls and membranes so that a barrier no longer divides the two cells. The two nuclei, which are now in the same compartment, fuse to form a diploid cell.

Remarkably, most haploid *S. cerevisiae* strains found in nature can switch mating types. That is, a cell with an **a** mating type can switch to an α mating type and vice versa. A strain that can switch mating types is called a **homothallic strain** and one that cannot is called a

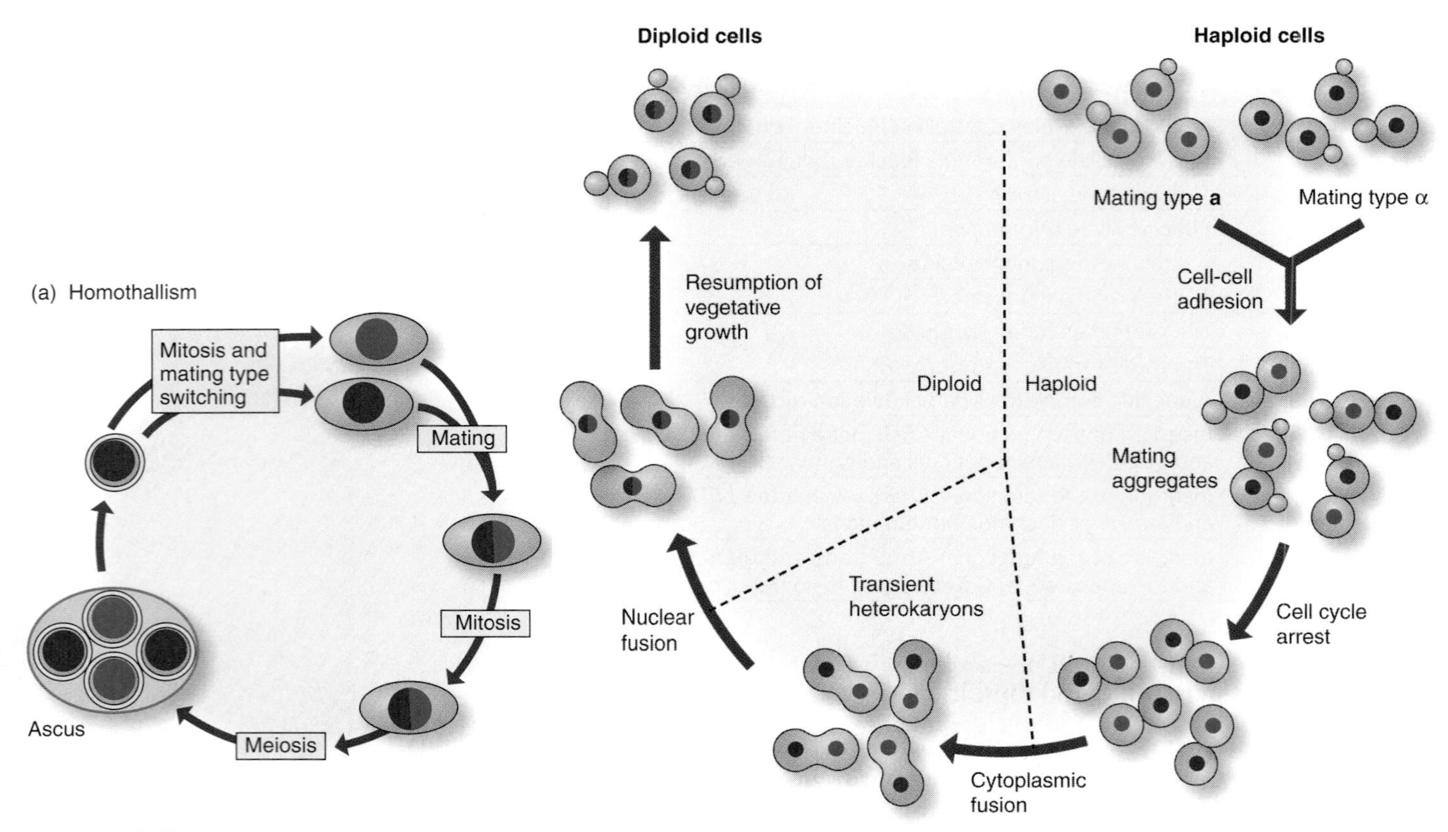

FIGURE 7.19 **Aspects of the *S. cerevisiae* cell cycle.** The blue and red colors indicate two different mating types (**a** and α). The colors are brought together in diploid cells to indicate that the duplicate nucleus has both **a** and α determinants. (Adapted from J. N. Strathern, J. R. Jones, and E. W. Broach. *The Molecular Biology of the Yeast Saccharomyces: Life Cycle and Inheritance (Monograph Series 11A).* Cold Spring Harbor Laboratory Press, 1981.)

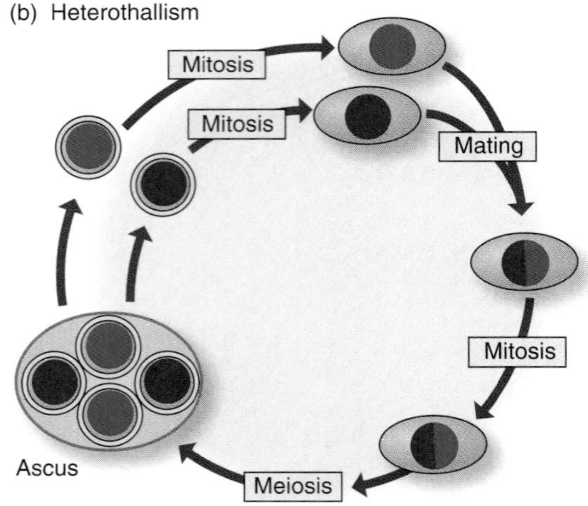

FIGURE 7.20 **Homothallic and heterothallic cells during mating.** A strain that can switch mating types is called a homothallic strain and one that cannot is called a heterothallic strain. (a) Progeny of a single homothallic spore colony can mate with each other to form diploids. For simplicity, the fate of only the spore shown in red is followed. (b) In contrast, progeny of a single heterothallic spore colony cannot mate with one another. (Adapted from J. N. Strathern, J. R. Jones, and E. W. Broach. *The Molecular Biology of the Yeast Saccharomyces: Life Cycle and Inheritance (Monograph Series 11A).* Cold Spring Harbor Laboratory Press, 1981.)

heterothallic strain (FIGURE 7.20). Heterothallic yeast strains are usually isolated in the laboratory by selecting for haploid cells that cannot mate. A single gene determines the homothallic trait. Homothallic strains possess a functional *HO* gene and heterothallic strains do not. The *HO* gene encodes a site-specific endonuclease that makes a double-strand break in DNA to initiate a DNA recombination process, which will be examined in Chapter 13.

The yeast mating type is determined by an allele present in the mating type (*Mat*) locus.

Haploid cells with **a** or α mating types differ from one another at only one genetic site. Cells with an **a** mating type have a *MATa* allele on chromosome III, while those with an α mating type have a *MAT*α allele in its place. The polypeptide product of *MATa*, Mata1, and the polypeptide products of *MAT*α, Matα1 and Matα2, regulate the transcription of three sets of genes. These three gene sets, the **a**-specific genes, the α-specific genes, and the haploid-specific genes, are represented by the symbols *asg*, α*sg*, and *hsg*, respectively. Both *asg* and *hsg* are transcribed unless a regulatory protein turns transcription off. In contrast, αsg is not transcribed unless a regulatory protein turns transcription on. A regulatory protein that turns transcription off is called a **negative regulator** while one that turns it on is called a **positive regulator.**

FIGURE 7.21 summarizes the effects that the regulatory polypeptides have on haploid cells with the **a**- or α-mating type and on diploid cells. A single segment will be used to represent each set of genes and a "sunburst" to indicate that transcription takes place. In a cell of mating type **a** (Figure 7.21a), *asg* and *hsg* are transcribed without the assistance of a regulatory protein and *αsg* is not transcribed. Thus, Mata1 has no role in regulating the transcription of the three gene sets in a haploid cell with an **a** mating type. In a cell of mating type α

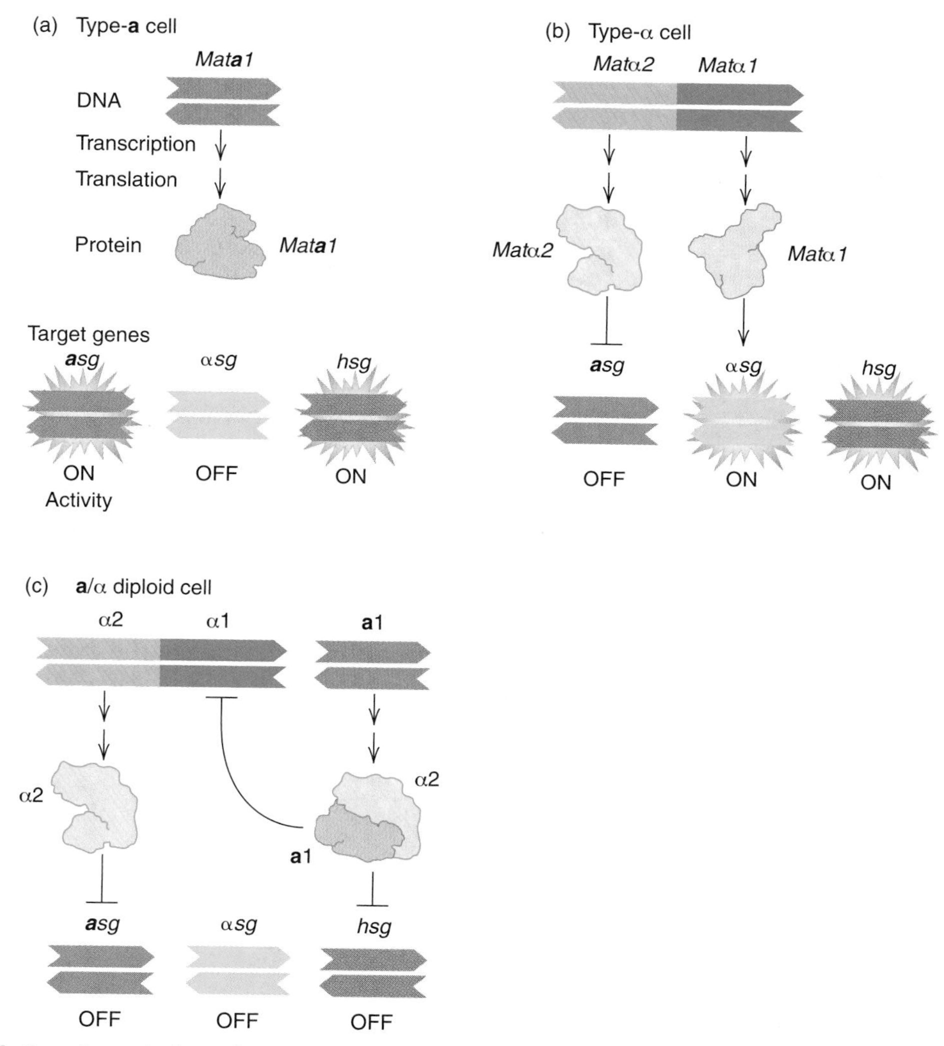

FIGURE 7.21 **Transcriptional regulation of mating type in yeast.** The symbols *asg*, *αsg*, and *hsg* denote sets of **a**-specific genes, α-specific genes, and haploid-specific genes, respectively. Sets of genes represented with a "sunburst" are "on," and those unmarked are "off." (a) In an **a** cell, Mat**a**1 has no role in regulation, and the sets of genes manifest their default states of activity (*asg* and *hsg* are on and *αsg* is off), so the cell is an **a** haploid. (b) In an α cell, Matα2 turns the *asg* off and Matα1 turns the *αsg* on, so the cell is an α haploid. (c) In an **a**/α diploid, Matα2 and Mat**a**1 form a complex that turns *hsg* off, the Matα2 turns the *asg* off, and the **a**sg manifest their default acitivity of off, so physiologically the cell is non-**a**, non-α, and non-haploid (that is, a normal diploid).

(Figure 7.21b), *hsg* is once again transcribed without the assistance of a regulatory protein. Matα1, however, stimulates the transcription of α*sg* while Matα2 blocks transcription of ***a**sg*. Still another method of control exists in an **a**/α diploid cell (Figure 7.21c). Transcription of ***a**sg* is again blocked by Matα2. Therefore, the diploid cell cannot have the **a**-mating type. Moreover, Mat**a**1 and Matα2 form a protein complex, which blocks the transcription of *hsg* and the gene within the *Mat*α locus that codes for Matα1. Thus, diploid cells cannot have an α mating type because they cannot transcribe α*sg* without Matα1. Furthermore, the Mat**a**1•Matα2 complex permits **a**/α diploid cells to initiate meiosis by blocking the expression of *RME1*, a gene that encodes a polypeptide that inhibits the transcription of genes needed for meiosis.

Yeast mating factors act as signals to initiate the mating process.

Haploid yeast cells with **a**- or α-mating types secrete **mating factors** that trigger complex responses in haploid cells of the opposite mating type. Each kind of mating factor is an oligopeptide with a unique amino acid sequence. The carboxyl terminal cysteine of the twelve residue **a**-mating factor is modified; the carboxyl group is methyl-esterified and the sulfhydryl group has a lipid attached to it. The α-mating factor is made of thirteen amino acid residues and is not modified. Each type of mating factor is originally part of a larger precursor protein. Two **a**-specific genes, *MFA1* and *MFA2*, encode protein precursors of **a**-factor and two α-specific genes, *MF*α*1* and *MF*α*2*, encode protein precursors of α-factor. Each of the protein precursors contains **a**- or α-factor repeat sequences separated from one another by spacers. Specific processing enzymes convert the precursors to biologically active mating factors.

Once released from the cell, both types of mating factors act on cells of the opposite mating type through a G protein–signal transduction complex (see Chapter 3). First the mating factor binds to a specific receptor embedded in the cell membrane, causing the receptor to change conformation and catalyze the exchange of GTP for GDP on the G protein complex. Once this nucleotide exchange is complete, G_{α}-GTP dissociates from the G complex, freeing $G_{\beta\gamma}$ to trigger a series of enzymatic reactions that prepare the cell for the mating process.

7.4 Restriction and Amplified Fragment Length Polymorphisms

Recombinant DNA techniques have facilitated genetic analysis in humans and other organisms.

Until the early 1980s, planned mating experiments such as those performed by Gregor Mendel in the mid-1860s were the primary (and usually the only) method available for the genetic analysis of higher organisms. Certain higher organisms such as the fruit fly and maize served as preferred model systems because they are easy to maintain and produce large numbers of progeny within a relatively short time. Geneticists also performed mating experiments with mice, rats, and other animals even though these experiments were more time-consuming and expensive. Because ethical considerations prohibit human experiments of this type, other methods had to be devised to study human genetics. Although considerable information was obtained by correlating family trees with specific traits such as color blindness or inborn errors of metabolism such as Tay Sachs disease, genetic studies in humans lagged behind those in many other biological systems.

New tools were needed to investigate human genetics. Recombinant DNA technology came to the rescue. Recombinant DNA tools work because homologous human chromosomes, although very similar, usually differ in approximately 0.2 to 0.5% of their base pairs. Two types of DNA sequence differences or **polymorphisms** are of special interest for genetic studies.

1. **Single nucleotide polymorphism (SNP).** A single nucleotide polymorphism or SNP (pronounced "snip") occurs when a single nucleotide at a specific DNA site is replaced by some other nucleotide in at least 1% of the total population. Approximately 1 in every 100 to 300 nucleotides along the 3-billion-base human genome is a SNP. SNPs occur in coding regions (regions that code for proteins and functional RNA molecules) as well as in noncoding regions. Even when present in a protein coding region, SNPs often do not affect cell function because the base change either does not produce an amino acid change or the amino acid change that it does produce does not affect protein function. Approximately 3 million unique human SNPs are now known and listed in various public and private data banks.
2. **Tandem repeat polymorphisms.** Tandem repeat polymorphisms occur when a DNA sequence, ranging from one to up to 100 bp, is repeated head to tail a different number of times at a distinct site in homologous chromosomes. These repeats occur in eukaryotic, bacterial, and archaeal chromosomes. FIGURE 7.22 shows DNA molecules that differ in the number of tandem repeats. In this example, the number of repeats varies from 1 to 10.

There are two types of tandem repeat polymorphisms. The first type, **microsatellites** have 1 to 6 bp long repeat sequences that are reiterated 5 to 100 or more times. They are distributed randomly throughout human and other vertebrate genomes and are present in both protein-coding noncoding regions. The second type, **minisatellites** have 7 up to about 100 nucleotide repeat sequences that are imperfectly reiterated 5 to 100 times. Human and other vertebrate genomes have many hundreds to thousands of minisatellites, each with its own distinct repeat unit.

SNPs sometimes create or eliminate restriction endonuclease cleavage sites in one of the two homologous chromosomes. In consequence, restriction endonucleases sometimes cleave corresponding regions of homologous chromosomes into fragments of different sizes. These

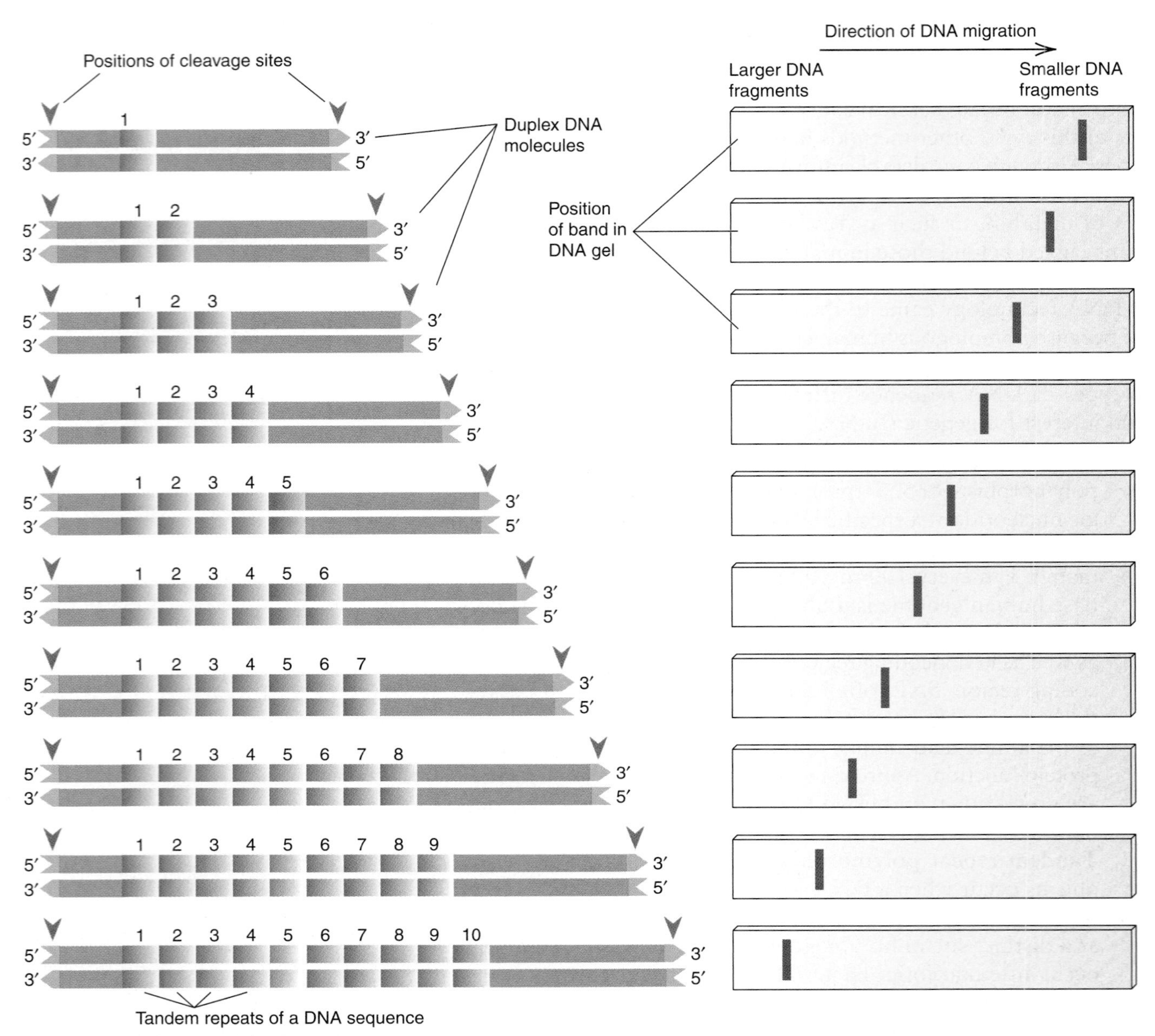

FIGURE 7.22 **Tandem repeats of a DNA sequence.** A genetic polymorphism in which the alleles in a population differ in the number of copies of a DNA sequence that is repeated in tandem along the chromosome. (a) This example shows alleles in which the repeat number varies from 1 to 10. (b) Cleavage at restriction sites flanking the repeat yields a unique fragment length for each allele that can be seen when the fragments are analyzed by gel electrophoresis. The alleles also can be distinguished by the size of the fragment amplified by PCR using primers that flank the repeat.

fragments, called **restriction fragment length polymorphisms** (**RFLPs**), are physical traits that can be followed as they are passed from one generation to the next. The gene that codes for the β-globin chain in human hemoglobin provides an important clinical application (FIGURE 7.23). The mutation responsible for sickle cell anemia, which converts Glu-6 to Val-6, also removes a cleavage site for the restriction endonuclease DdeI (Figure 7.23a and b). This change permits the gene

(a) Nucleotide sequences that code for residues 4–7 in the β globin chains of HbA (left) and HbS (right).

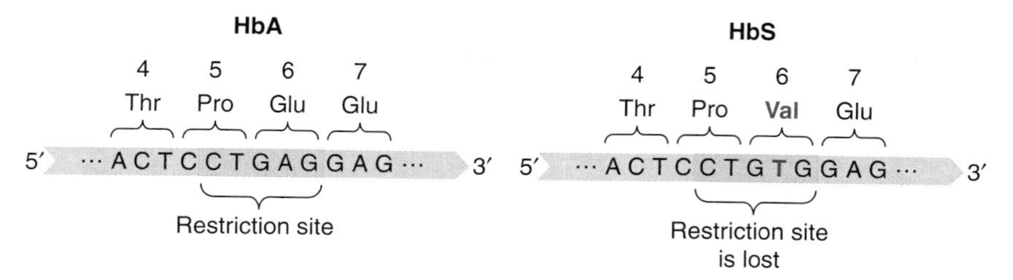

(b) *Dde*I restriction endonuclease site

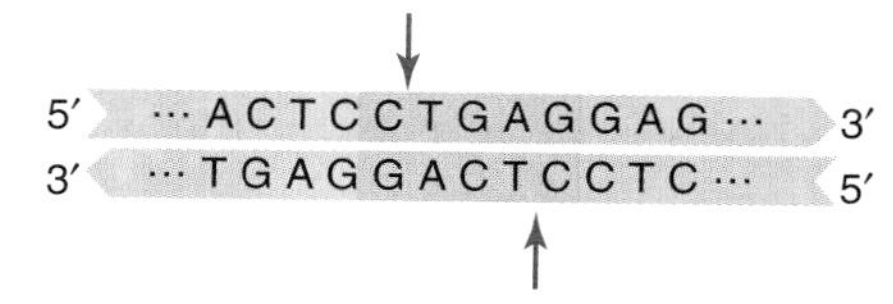

(c) Southern blot analysis of DNA from parents and children after digestion with *Dde*I restriction endonuclease

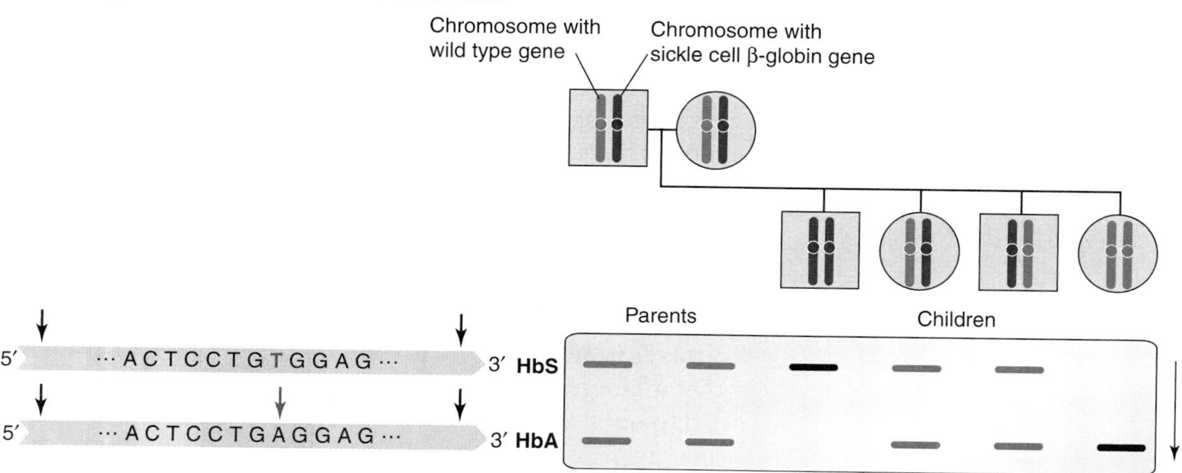

FIGURE 7.23 Use of restriction fragment length polymorphisms to distinguish chromosomes bearing a normal β-globin allele from one with a sickle cell β-globin allele. (a) The nucleotide sequences that code for residues 4 to 7 in the normal β-globin chain (HbA) and the sickle cell β-globin chain (HbS). Note that the restriction endonuclease site (CTGAG), which is present in the normal β-globin allele, is missing in the HbS β-globin allele. (b) Restriction site for the restriction endonuclease DdeI. The red arrows indicate cleavage sites. (c) Southern blot analysis of DNA from parents and children after digestion with the DdeI restriction endonuclease. Chromosomes bearing the normal and sickle cell β-globin alleles are shown in green and red, respectively. A radioactive probe is used to detect the DNA that codes for the β-globin protein. A DdeI restriction site is missing from the sickle cell β-globin gene. The restriction fragment from DNA with this allele is therefore larger than that for DNA with the normal allele. The electrophoretic bands are darker when an individual has two alleles of one type.

for sickle cell anemia to be tracked as it passes from one generation to the next (Figure 7.23c).

In 1980, David Botstein, Ronald W. Davis, and Mark Skolnick proposed that RFLPs could be used as markers to identify genes to which they are tightly linked; that is, they proposed that an RFLP can be used to find a gene that remains together with the RFLP after homologous chromosomes segregate into sperm or egg cells during meiosis. When such tight linkage exists, one can find the gene of interest by searching for the RFLP. Because SNPs normally produce only two RFLP variants, RFLPs that result from point mutations are of rather limited use for tracing alleles. Fortunately, tandem repeat polymorphisms do not suffer from this shortcoming. Hence, microsatellites or minisatellites located between two restriction endonuclease sites can help to follow the fate of tightly linked genes in homologous chromosomes. The approach is similar to that described for the β-globin gene. Instead of using a probe that binds to a specific gene sequence, however, the probe is selected for its ability to bind to the microsatellite or minisatellite during Southern blot analysis.

Tandem repeat polymorphisms can also be used to obtain a **DNA fingerprint** that is unique for each individual (with the possible exception of identical twins). Alec Jeffreys and coworkers discovered this application in 1985 while studying the myoglobin gene. As part of their studies, they observed that one of the introns in the myoglobin gene has a minisatellite, which has a different number of repeats in DNA from different people. Jeffreys and his coworkers hoped to use the minisatellite as a marker to locate the myoglobin gene on the chromosome. They therefore prepared a probe for the minisatellite, which they planned to use to identify the myoglobin gene. During one experiment, Jeffreys and coworkers digested DNA samples obtained from different individuals with a restriction endonuclease, resolved the fragments by gel electrophoresis, and then used the probe to detect the myoglobin gene. Jeffreys and coworkers expected to see two labeled bands, each corresponding to a different myoglobin allele. To their great surprise, they observed that every analyzed sample produced many labeled bands. They quickly realized that the probe was binding to a large number of different chromosomal regions from each person. Moreover, each individual's band pattern was unique. The probe, therefore, could be used to fingerprint the DNA. It soon became clear that probes that detect other minisatellites as well as those that detect microsatellites also can be used to obtain unique DNA fingerprints. While there is some ambiguity when a single kind of probe is used to obtain a DNA fingerprint, this ambiguity becomes negligible when several probes are used.

DNA fingerprinting based on restriction endonuclease sites can provide an unambiguous identification of an individual when one has an ample supply of DNA. For instance, one can determine paternity by taking DNA samples from the father and child. However, there are many situations in which the DNA that is to be tested is available in only trace quantities. For instance, there may be very little DNA left behind at a crime scene, perhaps just a hair follicle or some saliva. Trace amounts of DNA may also be available to identify a person who dies in a major catastrophe, perhaps just what is present on a hairbrush or toothbrush. Fortunately, PCR amplification (see Chapter 5) allows even trace quantities of DNA to be fingerprinted.

Tandem repeat polymorphisms are also used to diagnose genetic diseases. This diagnostic approach is especially important when the molecular basis of the genetic disease is not known. If a particular length (or version) of a microsatellite or minisatellite is very near to the defective allele, it can also be used to isolate and clone the gene. Once cloned, the gene can be sequenced and the information used to find the protein product. The protein products of the genes that cause muscular dystrophy and cystic fibrosis were found in this way.

7.5 Somatic Cell Genetics

Somatic cell genetics can be used to map genes in higher organisms.

A second approach for studying human genetics became possible with the discovery that cultured human and rodent cells fuse to form hybrid cells, which upon continued cultivation progressively lose human chromosomes. The reason human chromosomes are lost is still a mystery. As described in greater detail below, the availability of hybrid cells that contain mostly rodent chromosomes allows investigators to map genes in the few remaining human chromosomes. Because an understanding of this approach to genetic analysis, called **somatic cell genetics**, requires some knowledge of cell culture techniques, a brief description of these techniques will be presented before proceeding further.

Animal cells can be studied in culture.

Culture media have been developed that allow cells obtained from animal or plant cells to grow *in vitro*. The media have been formulated to keep the cells alive as long as possible, but do not always maintain a normal state for the cell. The behavior of most animal cells in culture is quite different from that in the animal. In culture, animal cells grow as individuals and the cells grow continually, but in nature, most animal cells grow in organized tissue units. That is, the cells are in contact with one another. Except for special circumstances, cultured animal cells require a substitute surface on which to grow (FIGURE 7.24). The usual surface that is provided in the laboratory is the bottom of a plastic culture flask or petri dish containing liquid growth medium. Cells slowly adhere to the plastic surface at the bottom of the flask or dish. Cell growth and division continues until the cells come in contact with one another to form a confluent monolayer on the bottom of the flask or dish.

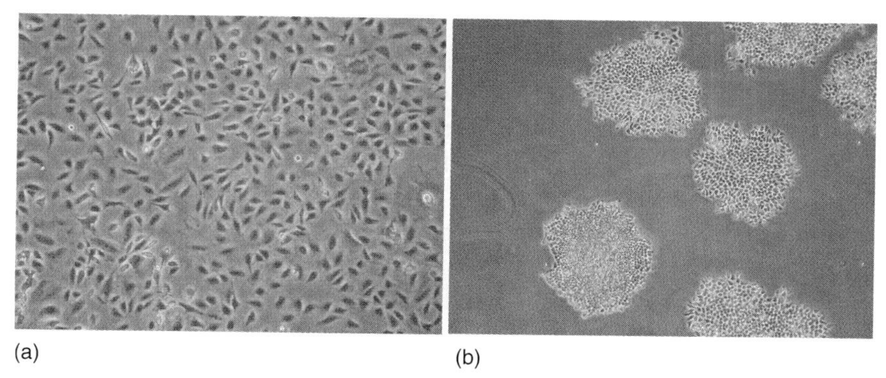

FIGURE 7.24 **Animal cells in tissue cultures.** (a) Chinese hamster fibroblasts, which have been growing for a few days in a petri dish. (b) HeLa cells, which have been growing for about one week on a petri dish. Cells were cultured in Dulbecco's modified Eagle medium containing 10% fetal bovine serum, viewed using an inverted microscope and pictures were taken using a digital camera. (Photos courtesy of Raghu G. Mirmira, Indiana University.)

At this stage growth and cell division stops, as in organized tissue. The cessation of cellular growth and division due to physical contact with other cells is called **contact inhibition**. Continued growth and division are obtained only by removing the confluent layer, separating the cells, and placing a small number of dispersed cells in another plastic culture flask. This handling is certainly an unnatural treatment and, in the course of many generations of growth, results in the accumulation of abnormal (mutant) cells that can tolerate these manipulations and the physiologically alien environment.

A related problem is that indefinite cell division of animal cells is itself exceptional. In an animal, most cells grow and divide only until adulthood is reached, a state that is attained in no more than 50 or 60 cell generations in a human. In adulthood, cell division is not a general phenomenon, except in specialized cells such as those in the skin, intestine, mucous membranes, and bone marrow, and those that divide during the healing of wounds. Inasmuch as cells obtained from many tissues other than an embryo have previously received signals that limit growth or terminate growth, their propagation in a culture medium is a novel cellular circumstance.

The result of forcing cells—which in their natural state grow only in contact with other cells and then only for a limited number of generations—to divide for an extended time in cell cultures, is that they become quite different from the cells originally taken from the animal tissue. Many of the cultured cells become tumor-like. The number of chromosomes in progeny cells is almost always increased beyond the normal diploid number and is usually not even the same in all cells of one culture. The longer these cells are grown, the more heterogeneous a culture becomes.

Two kinds of animal cell cultures are in common use: the **primary culture** and the **established cell line**. The distinction between these two types can be understood by examining how a cell culture is established and the characteristics of the culture in further growth. A culture is established by treating fresh tissue with a **proteolytic** (protein-digesting) enzyme, which disperses the cells, and placing a number of individual cells (and sometimes some of the undispersed tissue) in a plastic culture flask containing growth medium. The cells adhere to the plastic surface, begin to grow and divide after a few days, continue growth for one week to three months (the time depends on the cell type), and then gradually begin to disintegrate and die. During the final few weeks the culture consists mostly of a sick and dying population. A culture showing these characteristics is called a **primary culture** because each cell presumably has the same characteristics of the cells in the original tissue and is following the program of limited cell division set up when the tissue formed. If a primary culture is maintained for many months by removing some cells and transferring them to a fresh culture flask each time confluent growth is achieved, cells frequently arise that have gained the ability to grow and divide indefinitely (like a bacterium). A population derived from such a cell is called an **established cell line**.

Studying primary cultures has a particular advantage even though most of the cells eventually die, namely, that the cells obtained from a particular tissue retain, for at least a few generations, the properties of "normal" cells. Of course, there is a disadvantage that the primary population is short-lived. In any experiment that takes more than a few weeks, one does not know whether the results obtained apply to living or dying cells. An established cell line has the advantage of continued growth. However, the cells also are quite variable and often have the capacity to form tumors. Nevertheless, many experiments are done with established cell lines because of their immortality. Those cell lines that are used frequently in molecular biology experiments are HeLa (a human tumor cell), 3T3 (a mouse embryo cell), CHO (an apparently normal cell from Chinese hamster ovary), L (a mouse tumor), BHK (Syrian hamster kidney), Balb-c (a mouse tumor), and Vero (African green monkey kidney) cells.

In certain conditions, animal cells form colonies on the surface of a plastic culture flask or petri dish and this method is used occasionally to count the cells. It is more common, however, to remove the cells from the surface by treatment with trypsin and then either perform a count of a known microvolume with a microscope or use an electronic cell counter. The latter two procedures are always used when cells are grown in liquid suspension.

Two different animal cells can fuse to form a heterokaryon.

With the completion of this brief introduction to cell culture techniques, we are ready to resume our examination of hybrid cell formation and its application to somatic cell genetics. Although spontaneous cell fusion occurs too rarely to be of any practical use in the laboratory, certain agents induce the fusion process so that it becomes a common occurrence. The first fusion agent to be widely used was Sendai virus that had been inactivated by irradiation with ultraviolet light so that it could no longer replicate (FIGURE 7.25). Proteins in the lipoprotein envelope of the inactivated virus particle induce the fusion process. Today, polyethylene glycol, which destabilizes the cell membrane and thereby induces cell fusion, is often used as the fusing agent. A cell containing two separate nuclei, which is formed by the experimental fusion of two genetically different cells is called a **heterokaryon.** Cells from virtually any tissue and any species can be induced to fuse to form hybrid cells when incubated with the viral or chemical agent. The fusion process, which begins with cell membrane fusion, usually also involves the fusion of the cell nuclei from the two "parent" cells.

Some method is required to select newly formed hybrid cells from "parent" cells. One commonly used approach takes advantage of the fact that tetrahydrofolate (a derivative of the vitamin folic acid), is essential for de novo synthesis (synthesis from simple metabolic intermediates) of thymidylate and purine nucleotides. Aminopterin, a folic acid analog, blocks essential steps in the de novo pathways for thymidylate and purine nucleotide syntheses (FIGURE 7.26). Cells have alternate pathways, however, that allow them to synthesize thymidylate and purine nucleotides. These alternate pathways are called salvage pathways because their role is to convert pyrimidine and purine bases produced by nucleic acid degradation back to nucleotides.

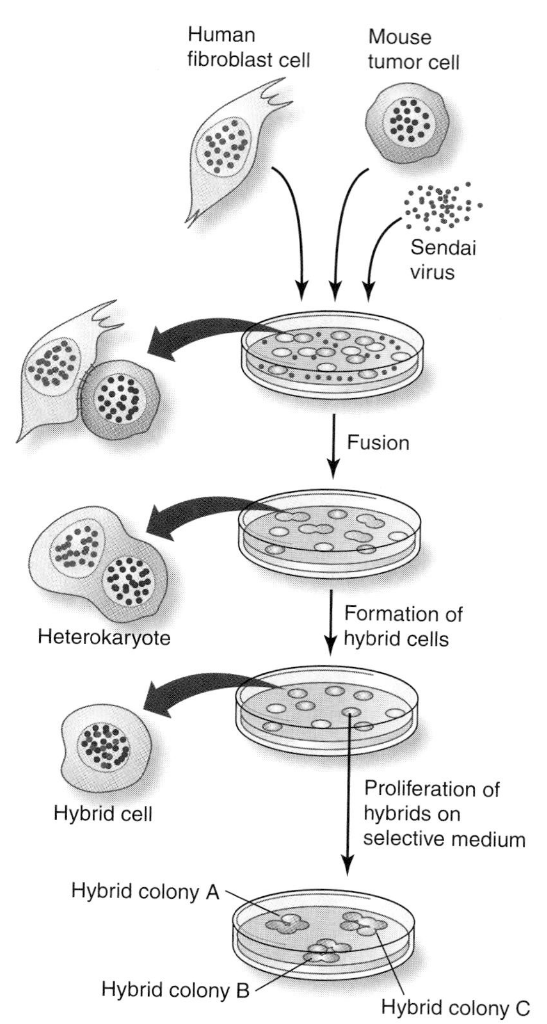

FIGURE 7.25 **Hybrid cell formation.** Human fibroblasts and mouse tumor cells are fused to produce hybrid cells, which contain a full complement of the mouse chromosomes (red) and a few human chromosomes (blue). Fibroblasts are derived from fibrous connective tissue. (Adapted from J. F. Griffiths, et al. *Modern Genetic Analysis: Integrating Genes and Genomes, Second edition.* W. H. Freeman & Company, 1999.)

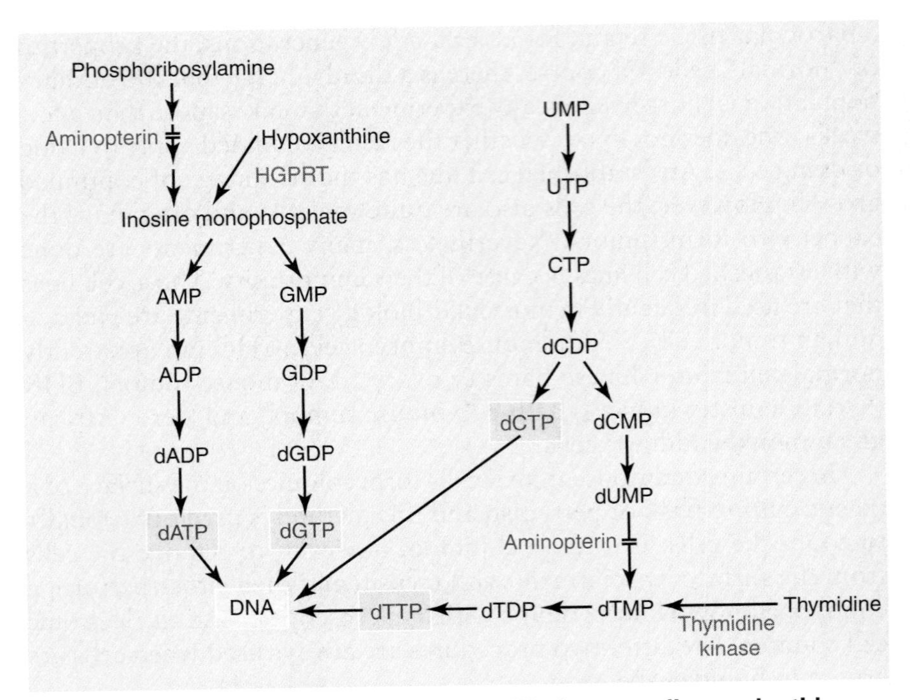

FIGURE 7.26 A schematic showing a simplified mammalian nucleotide biosynthesis pathway. Aminopterin, a folic acid analog, inhibits the conversion phosphoribosylamine to inosine monophosphate and the conversion of dUMP to dTMP. The salvage enzyme HGPRT (hypoxanthine guanine phosphoribosyltransferase) catalyzes the conversion of hypoxanthine and guanine (not shown here) to purine nucleotides. A second salvage enzyme, thymidine kinase, converts thymidine to dTMP. The arrows between reactants and products do not always represent the actual number of steps that take place.

Two reactions in the salvage pathway are of special interest (see Figure 7.26). The first, which converts thymidine to thymidylate, is catalyzed by thymidine kinase (TK). The second, which converts hypoxanthine (a purine) and 5-phosphoribosyl-1-pyrophosphate to inosine-5′-monophosphate, is catalyzed by hypoxanthine guanine phosphoribosyl transferase (HGPRT). Inosine-5′-monophosphate is then converted to adenosine-5′-monophosphate and guanosine-5′-monophosphate.

TK^- cells can be selected by growth in the presence of 5′-bromodeoxyuridine, a thymidine analog. The analog kills TK^+ cells but has no effect on TK^- cells because only TK^+ cells can convert bromodeoxyuridine to bromodeoxyuridine-5′-monophosphate, a toxic precursor for DNA synthesis. In a similar way, $HGPRT^-$ cells can be selected for their ability to grow in a medium that contains 6-thioguanine, a guanine analog. The analog kills $HGPRT^+$ cells but not $HGPRT^-$ cells because only the former can convert the purine analog into a toxic substance.

With this metabolic information as background, let's examine the procedure that is used to select hybrid cells (**FIGURE 7.27**). A TK^- "parent" cell will not grow in a medium that contains hypoxanthine, aminopterin, and thymidine (HAT medium) because the thymidine-5′-monophosphate salvage pathway will not work without thymidine

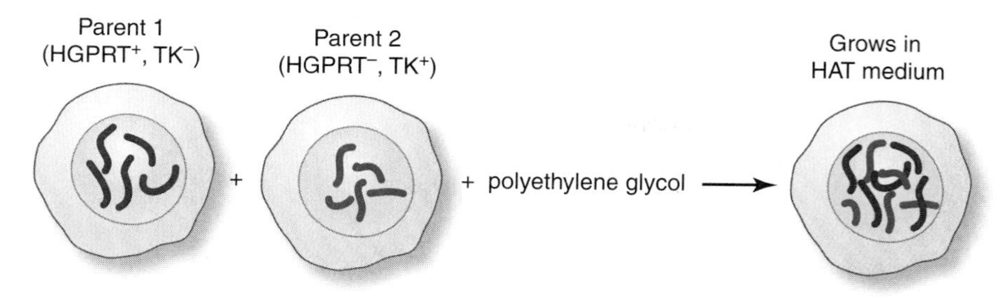

FIGURE 7.27 **Hybrid cell selection.** Neither parent strain will grow in HAT medium that contains hypoxanthine, aminopterin, and thymidine. The hybrid cell will grow in such a medium, however.

kinase and the de novo pathway will not work in the presence of aminopterin. A second "parent" cell that lacks HGPRT also will not grow in HAT medium. The salvage pathway will not work without HGPRT and the de novo pathway will not work in the presence of aminopterin. In contrast, hybrid cells that are both TK^+ and $HGPRT^+$ will grow in HAT medium.

Hybrids formed by fusing human and mouse cell are very useful for genetic analysis because they gradually lose their human chromosomes in random order as they grow and divide. Hybrid cells eventually lose all human chromosomes, while retaining all mouse chromosomes, when cultured in a medium that can support growth of either human or mouse cell growth. However, in a medium in which human cells can grow but mouse cells lack an enzyme required for growth, the one human chromosome that has the gene encoding the required enzyme will be retained while other human chromosomes are lost. Investigators take advantage of this approach to collect panels of cells that contain specific human chromosomes. One can then use these cells to locate a gene on a specific chromosome or to perform other kinds of genetic analyses.

Hybrid cells can be used to make monoclonal antibodies.

The benefits of hybrid cell formation are not limited to genetic analyses. As first shown by Georges J. F. Köhler and César Milstein in 1975, the technique can also be used to construct clones of immortal hybrid cells that synthesize a single type of antibody of predetermined specificity. These antibodies have proved so important to molecular biology that we will pause to examine how the cells that make them are created. Once again, a brief introduction to another area of biology—in this case the immune system—is required before we can continue.

The immune system protects vertebrates from foreign invaders through cellular and humoral responses. Cellular immunity involves the recognition and destruction of cancer cells or cells that have been infected by viruses by a type of white blood cell known as a T cell (or T lymphocyte). The humoral response is mediated by another type of white blood cell, the **B cell** (or **B lymphocyte**) that produces immunoglobulins. Foreign substances bind to receptors on the outer surface of the B cell, inducing the cell to differentiate into a **plasma**

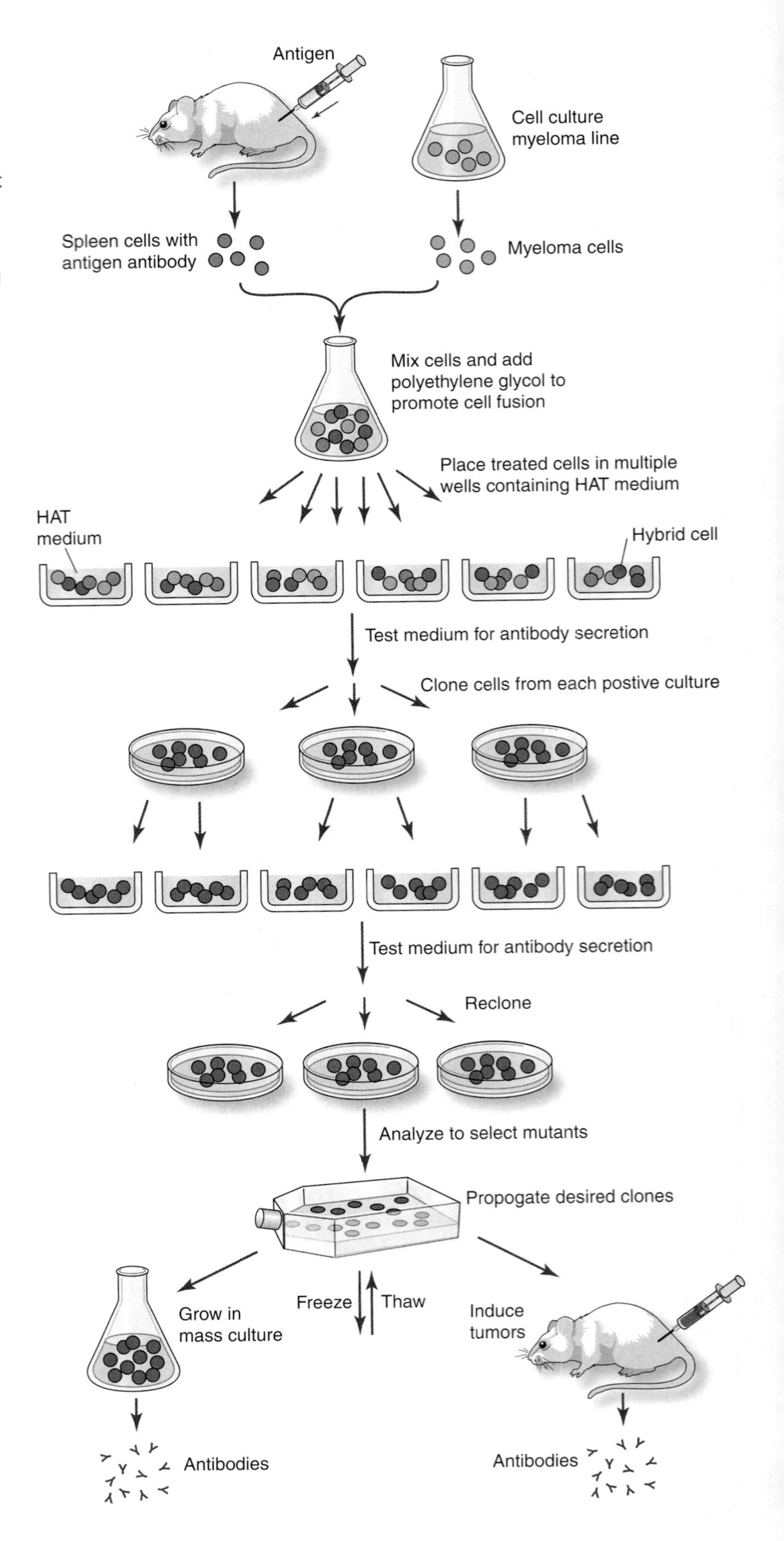

FIGURE 7.28 **Method for constructing hybridoma cells.** The first step in the procedure to obtain hybridoma cells, which produce monoclonal antibodies, is to fuse spleen cells from an immunized animal with myeloma cells in the presence of polyethylene glycol. HAT medium is used to select hybrids and the medium is tested for antibody secretion. Hybrids that test positive are cloned again and tested for the presence of immunoglobulin variants. Those finally selected are stored frozen. When thawed, the hybridomas can be grown in culture to produce antibody or injected in animals to secrete antibody. (Adapted from C. Milstein, *Sci. Am.* 241 [1980]: 66–74.)

cell, which divides to form a clone of plasma cells. Plasma cells do not divide very well when cultured in the laboratory and therefore cannot be maintained as cell lines.

Foreign substances used to immunize animals have many different chemical groups or determinants, each of which stimulates one or more different B cells to form plasma cell clones. The net effect is that immunization with foreign substances produces a polyclonal response (see Chapter 13). Cells belonging to each clone synthesize a unique type of immunoglobulin molecule. Immunoglobulin molecules synthesized by different clones may either bind to different determinants within the immunizing agent or bind to the same determinant in different ways. Under normal physiological conditions, therefore, vertebrates make a wide variety of antibodies in response to immunization by a single foreign agent.

There is, however, one condition in which humans have a very large population of a single kind of plasma cell and therefore produce extremely high concentrations of a single antibody. This condition, known as **multiple myeloma**, is a form of cancer in which a single plasma cell divides in an uncontrolled manner. In most cases, the determinant recognized by the overproduced antibody is not known. Myelomas also occur in other species, including mice and rats.

We are now ready to return to the fusion experiments performed by Köhler and Milstein (FIGURE 7.28). These workers mixed spleen cells from a mouse that had been immunized with sheep red blood cells, mouse myeloma cells, and an agent that promotes cell fusion. Then they obtained hybrid cells that grew in a selective medium. From these hybrid cells, they isolated clones that secreted a single type of antibody. The immortal cells derived from the myelomas, which can replicate in animals or in culture medium, are called **hybridomas**. Cells derived from a specific hybridoma cell all produce the same immunoglobulin (or antibody), which is known as a **monoclonal antibody**. Monoclonal antibodies have a large number of applications in molecular biology and in clinical practice. For instance, they are used to detect trace quantities of biological molecules, to bind specific membrane receptors, and as drugs to treat certain forms of cancer and some autoimmune diseases. These and other applications will be described, where appropriate, in the remaining chapters.

Suggested Reading

Introduction to Genetic Recombination

Brooker, R. 2004. *Genetics: Analysis and Principles*. New York: McGraw-Hill.

Hartl, D. L., and Jones, E. W. 2005. *Genetics: Analysis of Genes and Genomes*, 6th ed. Sudbury, MA: Jones and Bartlett.

Hartwell, L., Hood, L., Goldberg, M. L., et al. 2003. *Genetics: From Genes to Genomes*, 2nd ed. New York: McGraw-Hill.

Judd, B. H. 2001. Experimental organisms used in genetics. In: *Encyclopedia of Life Sciences*. pp. 1–7. Hoboken, NJ: John Wiley and Sons.

Keeney, J. B. 2007. Microorganisms: applications in molecular biology. In: *Encyclopedia of Life Sciences*. pp. 1–8. Hoboken, NJ: John Wiley and Sons.
Lewontin, R. C., Griffiths, A. J. F., Miller, J. H., and Gelbart, W. 2002. *Modern Genetic Analysis: Integrating Genes and Genomes*, 2nd ed. New York: W. H. Freeman.

Bacterial Genetics

Babi , A., Lindner, A. B., Vuli , M., Stewart, E. J., and Radman, M. 2008. Direct visualization of horizontal gene transfer. *Science* 319:1533–1536.
Beckwith, J., and Silhavy, T. J. 1992. *Power of Bacterial Genetics: A Literature-Based Course*. Woodbury, NY: Cold Spring Harbor Laboratory Press.
Black, J. Q. 2001. *Microbiology*, 5th ed. Hoboken, NJ: John Wiley and Sons.
Chilton, M.-D. 2001. Agrobacterium. A memoir. *Plant Physiol* 125:9–14.
Cascales, E., Buchanan, S. K., Duche, D. et al. 2007. Colin biology. *Microbiol Mol Biol Rev* 71:158–229
Clarke, M., Maddera, L., Harris, R. L., and Silverman, P. M. 2008. F-pili dynamics by live-cell imaging. *Proc Natl Acad Sci USA* 105:17978–17981.
Clewell, D. B. 2005. Antibiotic resistance plasmids in bacteria. In: *Encyclopedia of Life Sciences*. pp. 1–6. Hoboken, NJ: John Wiley and Sons.
Cupples, C. G. 2001. *Escherichia coli* and development of bacterial genetics. In: *Encyclopedia of Life Sciences*. pp. 1–6. Hoboken, NJ: John Wiley and Sons.
Dale, J. W., and Park, S. F. 2004. *Molecular Genetics of Bacteria*. Hoboken, NJ: John Wiley and Sons.
de la Cruz, F., Frost, L. S., Meyer, R. J., and Zechner, E. L. 2010. Conjugative DNA metabolism in gram-negative bacteria. *FEMS Microbiol Rev* 34:18–40.
Filutowicz, M., Burgess, R., Gamelli, R. L., et al. 2008. Bacterial conjugation-based antimicrobial agents. *Plasmid* 60:38–44.
Fronzes, R., Christie, P. J., and Waksman, G. 2009. The structural biology of type IV secretion systems. *Nat Rev Microbiol* 7:703–714.
Frost, L. S., Leplae, R., Summers, A. O., and Toussaint, A. 2005. Mobile genetic elements: the agents of open source evolution. *Nat Rev Microbiol* 3:722–732.
Funnell, B. E., and Phillips, G. J. (eds). 2004. *Plasmid Biology*. Washington, DC: American Society of Microbiology Press.
Garcillán-Barcia, M. P., Francia, M. V., and de la Cruz, F. 2009. The diversity of conjugative relaxases and its application in plasmid classification. *FEMS Microbiol Rev* 33:657–687.
Gomis-Rüth, F. X., and Coll, M. 2006. Cut and move: protein machinery for DNA processing in bacterial conjugation. *Curr Opin Struct Biol* 16:744–752.
Klotz, M. G. 2005. Bacterial genetics. In: *Encyclopedia of Life Sciences*. pp. 1–6. Hoboken, NJ: John Wiley and Sons.
Lawley, T. D., Klimke, W. A., Gubbins, M. J., and Frost, L. S. 2003. F factor conjugation is a true type IV secretion system. *FEMS Microbiol Lett* 224:1–15.
Liosa, M., Gomis-Rüth, F. X., Coll, M., and de la Cruz F. 2002. Bacterial conjugation: a two-step mechanism for DNA transport. *Mol Microbiol* 45:1–8.
Miller, J. H. 1999. *A Short Course in Bacterial Genetics: A Laboratory Manual and Handbook for* Escherichia coli *and Related Bacteria*. Woodbury, NY: Cold Spring Harbor Laboratory Press
Mobashery, S., and Azucena, E. F. Jr. 2002. Bacterial antibiotic resistance. In: *Encyclopedia of Life Sciences*. pp. 1–6. Hoboken, NJ: John Wiley and Sons.
Nikaido, H. 2009. Multidrug resistance in bacteria. *Ann Rev Biochem* 78:119–146.
Novick, R. P. 2001. Plasmids. In: *Encyclopedia of Life Sciences*. pp. 1–8. Hoboken, NJ: John Wiley and Sons.
Otten, L. Ti plasmids. 2001. In: *Encyclopedia of Life Sciences*. pp. 1–5. Hoboken, NJ: John Wiley and Sons.
Shu, A.-C., Wu, C.-C., Chen, Y.-Y., Peng, H.-L. et al. 2008. Evidence of DNA transfer through F-pilus channels during *Escherichia coli* conjugation. *Langmuir* 24:6796–6802.

Snyder, L., and Champness, W. 2002. *Molecular Genetics of Bacteria*. Washington, DC: American Society of Microbiology Press.

Streips, U. N., and Yashin, R. E. (eds). 2002. *Modern Microbial Genetics*, 2nd ed. Hoboken, NJ: John Wiley and Sons.

Summers, D. K. 1996. *The Biology of Plasmids*. Malden, MA: Blackwell Science, Inc.

Thomas, C. M., and Summers, D. 2008. Bacterial plasmids. In: *Encyclopedia of Life Sciences*. pp. 1–9. Hoboken, NJ: John Wiley and Sons.

Tortora, G. J., Funke, B. R., and Case, C. L. 2003. *Microbiology: An Introduction*, 8th ed. Upper Saddle River, NJ: Benjamin Cummings.

Trempy, J. 2003. *Fundamental Bacterial Genetics*. Malden, MA: Blackwell Science.

Budding Yeast (*Saccharomyces cerevisiae*)

Abelson, J. N., Simon, M. I., Guthrie, C., and Fink, G. R. (eds). 2004. *Methods in Enzymology*, Volume 194: *Guide to Yeast Genetics and Molecular Biology, Part A*. San Diego: Academic Press.

Adams, A., Gottschling, D. E., Kaiser, C. A., and Stearns, T. (eds). 1997. *Methods in Yeast Genetics*. Woodbury, NY: Cold Spring Harbor Laboratory Press.

Amberg, D. C., Burke, D. J., and Strathern, J. N. 2005. *Methods in Yeast Genetics*. Woodbury, NY: Cold Spring Harbor Laboratory Press.

Burke, D., Dawson, D., and Stearns, T. 2004. *Methods in Yeast Genetics*. Woodbury, NY: Cold Spring Harbor Laboratory Press.

Guthrie, C., and Fink, G. R. 2002. *Methods in Enzymology*, Volume 350: *Guide to Yeast Genetics and Molecular Cell Biology, Part B*. San Diego: Academic Press.

Mell, J. C., and Burgess, S. M. 2002. Yeast as a model genetic organism. In: *Encyclopedia of Life Sciences*. pp. 1–8. Hoboken, NJ: John Wiley and Sons.

Pausch, M. H., Kirsch, D. R., and Silverman, S. J. 2005. *Saccharomyces cerevisiae*: applications. In: *Encyclopedia of Life Sciences*. pp. 1–7. Hoboken, NJ: John Wiley and Sons.

Phaff, H.J. 2001. Yeasts. In: *Encyclopedia of Life Sciences*. pp. 1–11. Hoboken, NJ: John Wiley and Sons.

Scheiner-Bobis, G. 2009. Gene expression in yeast. In: *Encyclopedia of Life Sciences*. pp. 1–5. Hoboken, NJ: John Wiley and Sons.

Restriction and Amplified Fragment Length Polymorphisms

Butler, J. M. 2005. *Forensic DNA Typing: Biology, Technology, and Genetics Behind STR Markers*, 2nd ed. San Diego: Academic Press.

Debrauwere, H., Gendrel, C. G., Lechat, S., and Dutreix. M. 1997. Differences and similarities between various tandem repeat sequences: minisatellites and microsatellites. *Biochimie* 79:577–586.

Epplen, J. T., and Böhringer, S. 2005. Microsatellites. In: *Encyclopedia of Life Sciences*. pp. 1–4. Hoboken, NJ: John Wiley and Sons.

Epplen, J. T., and Kunstmann, E. M. 2005. Minisatellites. In: *Encyclopedia of Life Sciences*. pp. 1–4. Hoboken, NJ: John Wiley and Sons.

Housman, D. E. 1995. DNA on trial—the molecular basis of DNA fingerprinting. *N Engl J Med* 332:534–535.

Jeffreys, A. J., Wilson, V., and Thein, S. L. 1985. Hypervariable minisatellite regions in human DNA. *Nature* 314:67–73.

Lee, H., and Tirnady, F. 2003. *Blood Evidence: How DNA Is Revolutionizing the Way We Solve Crimes*. Jackson, TN: Perseus Publishing.

Nakamura, Y., Leppert, M., O'Connell, P., et al. 1987. Variable number of tandem repeat (VNTR) markers for human gene mapping. *Science* 235:1616–1622.

Rudin, N., and Inman, K. 2001. *An Introduction to Forensic DNA Analysis*, 2nd ed. Boca Raton, FL: CRC Press.

Somatic Cell Genetics

Davis, J. M. 2002. *Basic Cell Culture: A Practical Approach*. Oxford, UK: Oxford University Press.

Freshney, R. I. 2000. *Culture of Animal Cells: A Manual of Basic Technique*, 4th ed. Wilmington, DE: Wiley-Liss.

Girffiths, A., Doyle, J. B., and Newell, D. G. 1998. *Cell and Tissue Culture: Laboratory Procedures*. Hoboken, NJ: John Wiley and Sons.

Harris, H. 1995. *The Cells of the Body: A History of Somatic Cell Genetics*. Woodbury, NY: Cold Spring Harbor Laboratory Press.

Köhler, G. 1985. Derivation and diversification of monoclonal antibodies. *EMBO J* 4:1359–1365.

Milstein, C. 1985. From the structure of antibodies to the diversification of the immune response. *EMBO J* 4:1083–1092.

Morrison, C., and Takeda, S. 2000. Genetic analysis of homologous DNA recombination in vertebrate somatic cells. *Int J Biochem Cell Biol* 32:817–831.

Sedivy, J. M., and Dutriaux, A. 1999. Gene targeting and somatic cell genetics—a rebirth or a coming of age? *Trends Genet* 15:88–90.

Smith, R. H. 2000. *Plant Tissue Culture: Techniques and Experiments*, 2nd ed. San Diego: Academic Press.

This book has a Web site, **http://biology.jbpub.com/book/molecular**, which contains a variety of resources, including links to in-depth information on topics covered in this chapter, and review material designed to assist in your study of molecular biology.

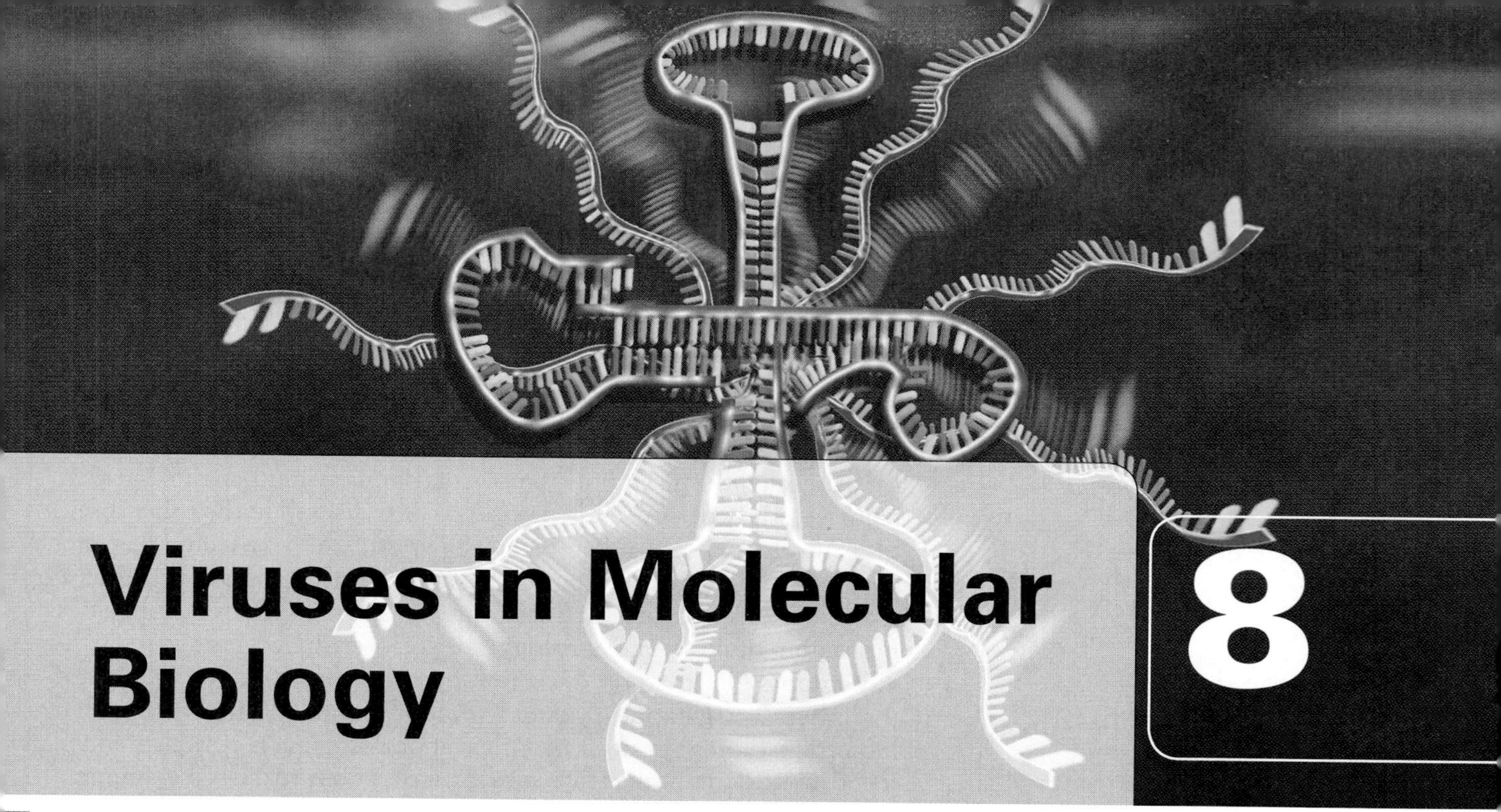

Viruses in Molecular Biology

8

OUTLINE OF TOPICS

8.1 Introduction to Viruses

Viruses are obligate parasites that can only replicate in a host cell.

8.2 Introduction to the Bacteriophages

Bacteriophages were of interest because they seemed to have the potential to serve as therapeutic agents to treat bacterial diseases.

Investigators belonging to the "Phage Group" were the first to use viruses as model systems to study fundamental questions about gene structure and function.

Bacteriophages come in different sizes and shapes.

Bacteriophages have lytic, lysogenic, and chronic life cycles.

Bacteriophages form plaques on a bacterial lawn.

Bacteria and the phage that infect them are in a constant struggle for survival.

8.3 Virulent Bacteriophages

E. coli phage T4 DNA is terminally redundant and circularly permuted.

E. coli phage T7 DNA is terminally redundant but not circularly permuted.

E. coli phage ϕX174 contains a single-stranded circular DNA molecule.

Some phages have single-stranded RNA as their genetic material.

8.4 Temperate Phages

E. coli phage λ DNA can replicate through a lytic or lysogenic life cycle.

E. coli phage P1 can act as generalized transducing particles.

8.5 Chronic Phages

After infection, a chronic phage programs the host cell for continued virion particle release without killing the cell.

8.6 Animal Viruses

Polyomaviruses contain circular double-stranded DNA.

Adenonviruses have linear blunt-ended, double-stranded DNA with an inverted repeat at each end.

Retroviruses use reverse transcriptase to make a DNA copy of their RNA genome.

Suggested Reading

Many of the fundamental principles that helped to establish molecular biology as a distinct scientific discipline were discovered by studying viruses. For example, we learned in Chapter 1 that Alfred Hershey and Martha Chase confirmed that DNA is the genetic material by following the fate of radioactive viral DNA and protein in infected bacteria. Viruses also provide us with important enzymes such as DNA ligase and reverse transcriptase (see Chapter 5), which are used in recombinant DNA technology. Viruses are also used as vectors to introduce foreign DNA into organisms.

Although the discipline of molecular biology has benefited greatly from the study of viruses, benefits have also flowed in the other direction. Virologists take advantage of the principles and techniques discovered by molecular biologists to examine how viruses cause disease and devise treatments for these diseases. In fact, the disciplines of molecular biology and virology are so intertwined that it is virtually impossible to study one discipline without some knowledge of the other. This chapter provides some of the basic information that molecular biologists need to know about viruses. It is not intended to provide a comprehensive view of viruses but rather to examine a few of the virus model systems that have had and continue to have a profound influence on the study of molecular biology.

8.1 Introduction to Viruses

Viruses are obligate parasites that can only replicate in a host cell.

We begin our examination of viruses by describing some of the properties that viruses have in common. Viruses are obligate parasites that invade cells and then use the cells' metabolic machinery to replicate. By itself, a virus particle or **virion** can persist, but it can neither grow nor replicate except within a living host cell. Virions come in a wide variety of sizes and shapes (FIGURE 8.1). Some kinds of virions have membrane envelopes surrounding them, whereas others do not. Virions, which are nearly always too small (10–200 nm) to be seen with the light microscope, can be viewed with an electron microscope.

A virion consists of a nucleic acid, either DNA or RNA, surrounded by a protective protein coat or **capsid**. Depending on the kind of virion, the nucleic acid may be either single- or double-stranded DNA or RNA and either linear or circular. The lengths of viral nucleic acids vary greatly from one type of virus to another but even the longest viral nucleic acid is much shorter than its host's DNA and, hence, codes for many fewer proteins than the host genome. Viral genes code for proteins that form the capsid as well as for enzymes that are essential for viral reproduction inside the host. Viruses that have very few genes rely on the host to provide nearly all the machinery that they require for reproduction whereas viruses with many genes

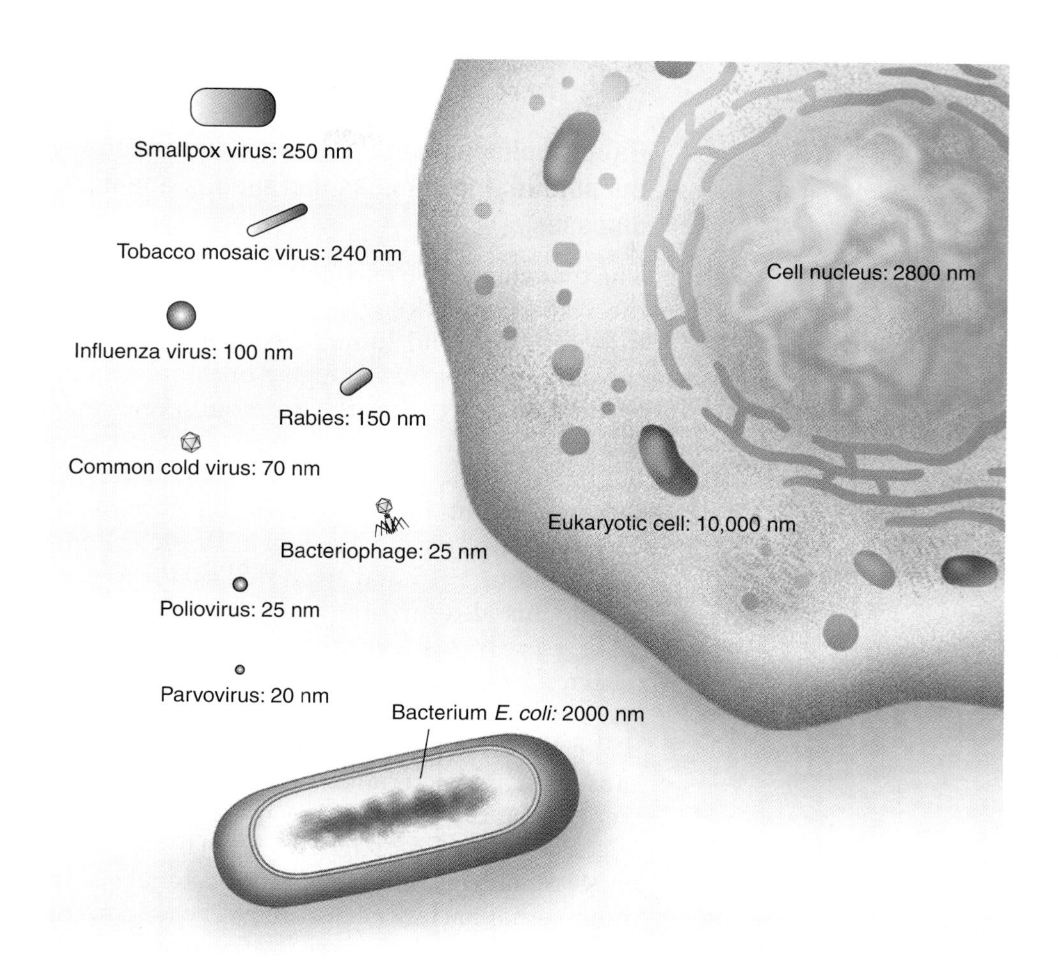

FIGURE 8.1 **Size relationships among viruses.** The sizes of various viruses relative to a eukaryotic cell, a cell nucleus, and the bacterium *E. coli*. The smallpox virus approximates the smallest prokaryote, the mycoplasmas, in size.

are less reliant on the host. At a very minimum, a virus requires the host cell's ribosomes, protein-synthesizing factors, amino acids, and energy-generating systems to reproduce. A virus, therefore, is not a living organism but rather a piece of genetic material that can only reproduce in a metabolizing cell.

Viruses are classified according to their host organisms, particle morphology, and nucleic acid composition. Despite their enormous diversity, all viruses must perform certain essential functions. They must be able to deliver their nucleic acid to the inside of a cell, protect their nucleic acid from harmful physical and chemical agents, convert the host cell to a virus factory, and create some means to allow progeny viruses to escape from infected cells. These functions are carried out in a variety of ways by different virus species. Each kind of virus infects a limited number of different kinds of organisms. Some kinds of viruses appear to infect only a single kind of organism, others infect closely related species, and still others infect widely different organisms, even jumping between plant and animal kingdoms.

8.2 Introduction to the Bacteriophages

Bacteriophages were of interest because they seemed to have the potential to serve as therapeutic agents to treat bacterial diseases.

The first clue to the existence of bacterial viruses came in 1896 when the British bacteriologist Ernest Hankin observed that the sewage-ridden Ganges and Jumna Rivers in India contained an unidentified substance that passed though a fine porcelain filter and had the ability to kill *Vibrio cholerae*, the bacterial species that causes the often fatal disease cholera. Hankin recognized that the filterable agent helped to control the level of the pathogenic bacteria in the river by killing the bacteria but did not attempt to determine the nature of the agent. The British physician Frederick Twort encountered a similar filterable agent in 1915 and suggested that it might be a virus but did not pursue this idea further.

Two years later, the French physician Felix d'Herelle noticed that a bacteria-free filtrate prepared from the stools of French troops, who were recovering from dysentery, had the ability to kill the bacteria that cause dysentery. Further studies showed that the filterable agent caused turbid bacterial cultures to become clear almost as if the agent were eating the bacteria. For this reason, d'Herelle called the agent a **bacteriophage** (Gr. *phago*, to eat). This term and its shortened version **phage** remain in common use for bacterial viruses to this day. d'Herelle thought that bacteriophages might be used to treat bacterial infections. Although initial clinical trials appeared to offer some support for this idea, more extensive clinical studies did not produce reproducible cures. The failure of bacteriophages to cure bacterial diseases was due in part to the appearance of phage-resistant bacterial mutants and in part to the destruction of the bacteriophages by the patients' immune or digestive systems. With the discovery of antibiotics, interest in using bacteriophages to treat bacterial infections waned and very little was done to pursue this clinical application for many years. Now that so many bacterial pathogens have become antibiotic-resistant, however, there is renewed interest in the potential of bacteriophages to prevent or treat bacterial diseases. For example, the food industry is currently exploring the use of bacteriophages to disinfect foods and the United States Food and Drug Administration has approved a physician initiated trial to determine whether a bacteriophage cocktail can effectively destroy *Staphylococcus*, *Pseudomonas*, and *E. coli* in skin wounds and burns.

Investigators belonging to the "Phage Group" were the first to use viruses as model systems to study fundamental questions about gene structure and function.

Even though bacteriophages did not seem to hold great promise as therapeutic agents, they remained fascinating subjects for scientific investigations. By the 1940s, a loosely knit, informal network of investigators decided to use bacteriophages to address fundamental questions related to gene structure and function. This network, which came to be known as the *Phage Group*, owed much to the intellectual leadership of two outstanding members, Max Delbrück and Salvador Luria. They convinced other talented scientists to join them in using a specific bacteriophage that infects *Escherichia coli* as a model system so that results from different laboratories could be compared and investigators could build on one another's work. The bacteriophage that they selected, **bacteriophage T4**, was one of seven different types of phages that had been isolated from raw sewage in Brooklyn, New York several years earlier. The bacteriophages were named bacteriophage type 1, type 2, and so forth. The names were later shortened to bacteriophage T1, T2, T3, T4, T5, T6, and T7, where "T" is short for type. By chance, bacteriophages T2, T4, and T6 turned out to be very similar to one another and so are called the **T-even phages**. Two of the T-odd bacteriophages, T3 and T7, are also similar to one another. T1 and T5 differ from the other five bacteriophages and also from each other. Many additional phages have now been isolated that are closely related to one or another of the seven original phage types.

Bacteriophages come in different sizes and shapes.

The bacteriophages that will be described in this chapter fall into three structural categories (FIGURE 8.2): icosahedral tailless, icosahedral head

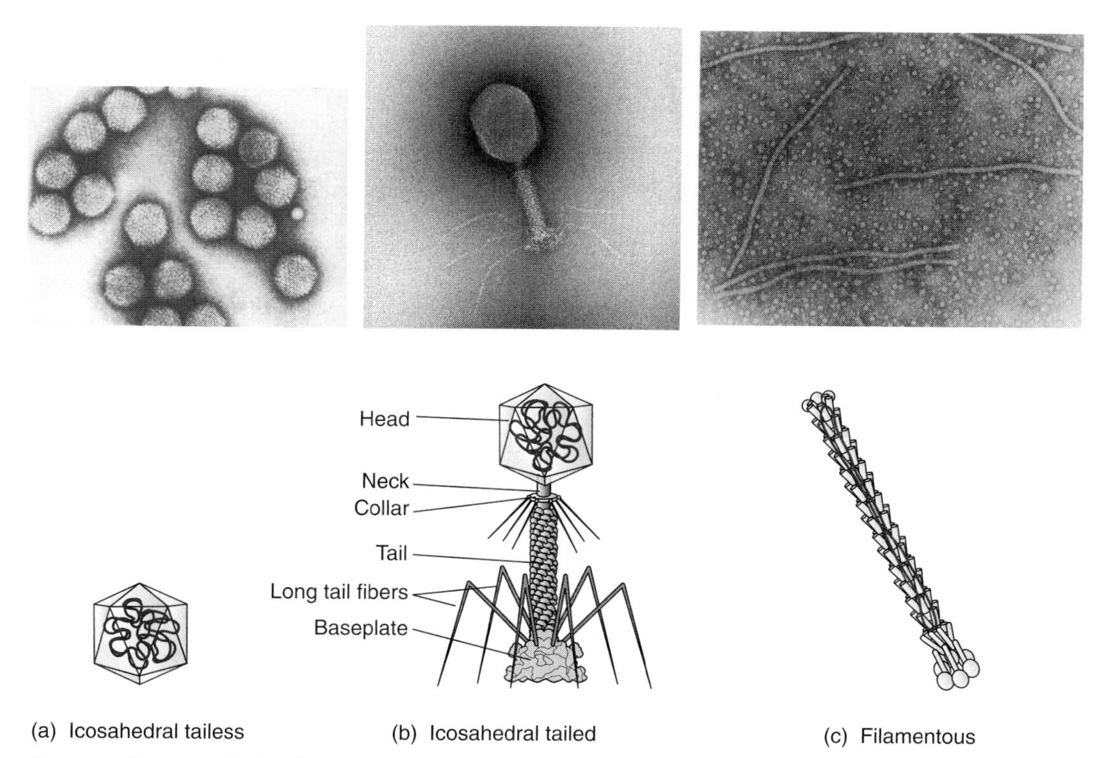

FIGURE 8.2 **The three major morphological classes of phages.** (a) Icosahedral, tailless: bacteriophage ϕX174. (b) Icosahedral, tailed: bacteriophage T4. (c) Filamentous: bacteriophage M13. (Top) Electron micrographs. (Bottom) Schematic diagrams of the phages. The tailed phages do not always have a collar and can have from 0 to 6 tail fibers, the number depending on the phage type. (Part a photo courtesy of Cornelia Büchen-Osmond, ICTVdB, Columbia University. Part b photo courtesy of Robert Duda, University of Pittsburgh. Part c photo copyright Rothamsted Research Ltd, Centre for Bioimaging.)

with tail, and filamentous. (An icosahedron is a quasispherical polyhedron having 20 triangular faces, 30 edges where two faces meet, and 12 corners where five edges meet.) Approximately 5600 phages have been examined in the electron microscope over the last 50 years. About 96% of these phages have tails.

The following are general points about phage architecture:

1. In both tailed and icosahedral tailless phages, the nucleic acid is contained in a hollow region formed by the capsid and is highly compact. In a filamentous phage, the nucleic acid is embedded in the capsid and is present in an extended helical form.
2. When present, the tail is a complex multicomponent structure often terminated by tail fibers.
3. In icosahedral phages, the length of the DNA molecule is very much greater than any dimension of the head.

There are many variations on the basic structure of the tailed phages. For example, the length and width of the head may either be the same or the length may be greater than the width; however, short fat heads are not seen. The tail may be very short (barely visible in electron micrographs) or up to four times the length of the head and it may be flexible or rigid. A complex baseplate may also be present on the tail; when present, it typically has from one to six tail fibers.

Bacteriophages have lytic, lysogenic, and chronic life cycles.

Bacteriophages, like all other viruses, are not considered to be living organisms because they depend on the host cells metabolic system and ribosomes for their reproduction. Nevertheless, the term phage life cycle is commonly used to describe the various stages of viral infection from the initial contact of the virus with its host cell to the ultimate release of progeny virus particles. Three types of phage life cycles have been observed.

Lytic Life Cycle

The bacteriophage's nucleic acid enters the host cell and in a complicated but understandable way converts the bacterium to a phage-synthesizing factory. Within about an hour (the time varying with the phage species), the infected bacterium's cell envelope (cell wall and membrane) is disrupted and a hundred or more progeny phages are released. This disruptive release is termed **lysis** and the suspension of newly synthesized phages is called a **phage lysate**. A phage that is capable only of lytic growth, designated a **virulent phage**, always kills its host.

Lysogenic Life Cycle

Bacteriophages that go through the lysogenic life cycle contain double-stranded DNA, which has two alternate fates upon entering the host cell. It can convert the bacterium to a phage-synthesizing factory, which produces large numbers of progeny phages that are released after cell lysis. Alternatively, the phage DNA can be integrated into the bacterial chromosome and replicate as part of the bacterial chromosome or in some cases form an autonomously replicating circular

DNA molecule. Integrated DNA, which is noninfectious, is called a **prophage.** Under normal growth conditions, a prophage remains integrated in the bacterial chromosome. On occasion, however, a prophage is excised from the bacterial chromosome and converts the bacterium into a phage-synthesizing factory, which eventually ruptures to release the progeny phage. Because a bacterial cell carrying a prophage has the potential for cell lysis, it is termed a **lysogenic cell.** A phage that sometimes exists as a prophage is called a **temperate phage.**

Chronic Infection Life Cycle

Bacteriophages that go through the chronic infection life cycle are continuously released without disrupting the cell envelope (wall and membrane) or killing the host. Such phages are called **chronic phages.**

Bacteriophages form plaques on a bacterial lawn.

Phages multiply much faster than bacteria. A typical *E. coli* cell in rich medium doubles in number in about half an hour, while a single virulent phage particle such as T4 gives rise to more than 100 progeny in the same time period. Released phage particles can then infect more bacteria. Phages released in the second cycle of infection can infect even more bacteria. Therefore, in two hours, there are four cycles of infection for both a bacterium and a phage. A single bacterium, however, becomes $2^4 = 16$ bacteria and a single phage becomes $100^4 = 10^8$ phage particles.

Virulent phage particles are easily counted by a technique known as the **plaque assay** in the following way. When 10^8 bacteria are plated on nutrient agar, the resulting colonies appear as a confluent, turbid layer of bacteria called a lawn. To achieve maximum uniformity of the turbidity of the **lawn,** the bacteria are suspended in a small volume of warm liquid agar that then is poured onto the surface of the solid medium. The liquid agar, which contains 0.7% agar compared to 1.5% in agar plates, forms a soft agar layer on cooling. Known as **top agar,** this soft layer provides a very smooth surface so the bacteria can grow with much uniformity (FIGURE 8.3a).

If phage particles are added along with the bacteria, they will attach to the bacteria in the agar. Shortly afterward, each infected bacterial cell lyses and releases about 100 phage particles, each of which will attach to nearby bacteria. These infected bacteria in turn will release a burst of phage particles that then can infect other bacteria in the vicinity. These multiple cycles of infection continue and after several hours, the phage will have destroyed all of the bacteria at a single localized area in the agar, giving rise to a clear, transparent circular region in the turbid, confluent layer (FIGURE 8.3b). This clear region is called a **plaque.** Because one phage forms one plaque, the individual phage particles put on the plate can be determined by counting plaques. One, thus, can determine the number of infective phage particles present in the original phage suspension. A similar approach can be used when studying temperate or chronic phage.

Temperate phages produce cloudy or turbid plaques instead of clear plaques. The turbidity arises because temperate phages have a

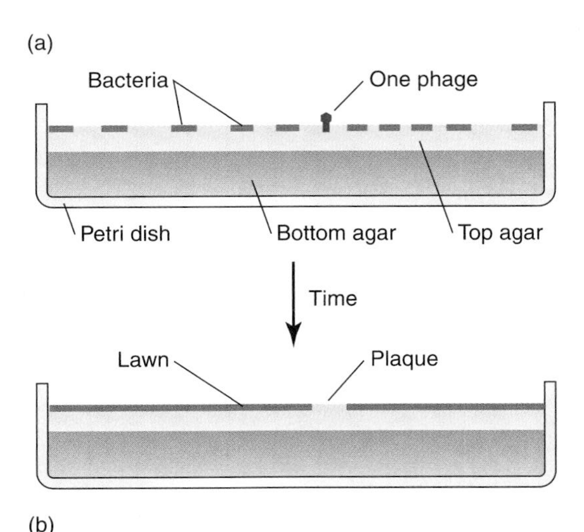

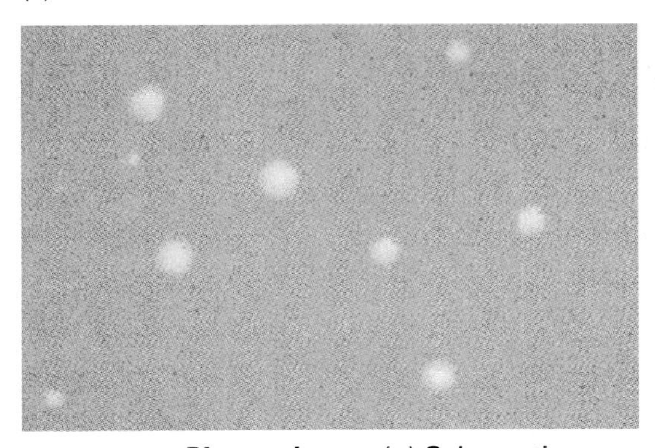

FIGURE 8.3 **Phage plaque.** (a) Schematic drawing of plaque formation. Bacteria grow and form a translucent lawn. There are no bacteria in the vicinity of the plaque, which remains transparent. Neither phage nor bacteria are drawn to scale; both are much smaller. (b) Phage T4 plaques on a bacterial lawn. (Part b © Ken Wagner/Visuals Unlimited.)

lysogenic as well as lytic life cycle. Most infected bacteria are lysed and killed by the temperate phages but some cells become lysogens and survive. The growth of these lysogenic cells gives the plaques their turbid appearance. Chronic phages also produce plaques even though there is no cell death. Phage infection slows the rate of bacterial growth, so the plaques are regions of low cell density.

Plaque assays are useful not only because they permit us to determine the number of phages in a suspension, but also because they allow us to isolate pure phage strains (given that each plaque arises from a single phage). The isolation of pure phage strains is very important for genetic studies and preparing phages to be used in recombinant DNA experiments.

Bacteria and the phage that infect them are in a constant struggle for survival.

It is estimated that about 10^{30} phage infections take place throughout the biosphere each day. Phages appear to have the upper hand in this struggle because they replicate at a much faster rate than their bacterial hosts. Nevertheless, bacteria have several methods to protect themselves from phage infection.

A phage must attach to a specific bacterial receptor such as a membrane protein or lipopolysaccharide before it can infect the cell. Mutations in these receptors protect against phage infection. The phages also mutate, however, and some of the mutants can attach to the altered receptors. If the attached phage can transfer its DNA into the cell and the DNA can replicate, then the cell will either be killed or stop growing. In the course of evolution, cell envelope components have evolved to be unrecognizable by most phage species. Phages have also evolved in order to be able to bind some bacterium. Failure of a particular phage to evolve in this way would result in its extinction.

Once the phage DNA has entered most bacteria, it must also be able to survive a **host restriction–modification system** (see Chapter 5). This system allows a bacterium of type X to distinguish a phage that has replicated in a type X bacterium (P•X) from a phage that has replicated in a type Y bacterium (P•Y). The data presented in Table 8.1 shows the restriction-modification pattern for phage λ infecting *E. coli* strains B, C, and K. Phage λ•K, obtained from an *E. coli* strain K lysate, forms plaques at a low efficiency in strain B. That is, phage λ•K is **restricted** by strain B. The rare plaques that do appear contain phage λ•B, which have been modified so that they now replicate efficiently in strain B but are restricted in strain K.

The molecular explanation for the restriction phenomenon is as follows. *E. coli* strain B contains EcoBI, a type I restriction endonuclease, that recognizes a specific DNA sequence and then cleaves the DNA (see Chapter 5). Phage λ•K contains this sequence, and when its DNA is injected into *E. coli* B, the phage DNA is cleaved. A few phage λ•K DNA molecules in the large population of infected cells escape this fate because EcoB methylase methylates an adenine in the recognition sequence before EcoBI restriction endonuclease has a chance to cleave the phage DNA. EcoBI restriction endonuclease cannot cleave the

TABLE 8.1 The Restriction and Modification Pattern of *E. coli* Phage λ

Bacterial Strain	Phage λ·K	λ·B	λ·C
K	1	10^{-4}	10^{-4}
B	10^{-4}	1	10^{-4}
C	1	1	1

Numbers indicate relative plating efficiencies.

methylated DNA molecule. Subsequent replication of the protected DNA molecules leads to phage λ•B production. EcoB methylase also protects the bacterial chromosome from the EcoBI restriction endonuclease by methylating all recognition sequences.

E. coli K contains EcoKI restriction endonuclease, which recognizes a base sequence that differs from that recognized by EcoBI restriction endonuclease. A λ phage that has replicated in strain K—namely, phage λ•K—is methylated in the EcoKI–specific target sequence and is resistant to EcoKI restriction endonuclease. Phage λ•B, however, has unmethylated EcoKI recognition sequences. Phage λ•B DNA, therefore, is usually cleaved when it enters a strain K cell. Occasionally, a phage λ•B DNA molecule escapes restriction, is methylated by EcoK methylase, and replicates. The rare progeny phages that result when phage λ•B successfully infects *E. coli* K are converted to phage λ•K. Because these λ•K phages now lack the B modification, they are restricted when infecting strain B.

Both phages λ•B and λ•K replicate efficiently in strain C. However, phage λ•C replicates very poorly in both strains B and K. The explanation for these results is that strain C lacks a restriction-modification system that recognizes any phage λ DNA sequence. Strain C, therefore, cannot restrict the replication of any phage λ DNA and phage λ•C DNA is not protected by methylation.

With this general introduction to bacteriophages as a background, we are ready to examine a few specific phages that are instructive model systems in molecular biology. Our focus is on phages that infect *E. coli*, because, as a general rule, these phages have been the most thoroughly studied and made the greatest contributions to the development of the field of molecular biology. Although we do not examine the many fascinating phages that infect other bacteria, most of the information presented on coliphages is directly applicable to phages that infect other bacteria.

8.3 Virulent Bacteriophages

E. coli phage T4 DNA is terminally redundant and circularly permuted.

We begin our examination of the coliphages by considering bacteriophage T4. Unless otherwise indicated, the information presented also applies to bacteriophages T2 and T6. Bacteriophage T4 is a tailed

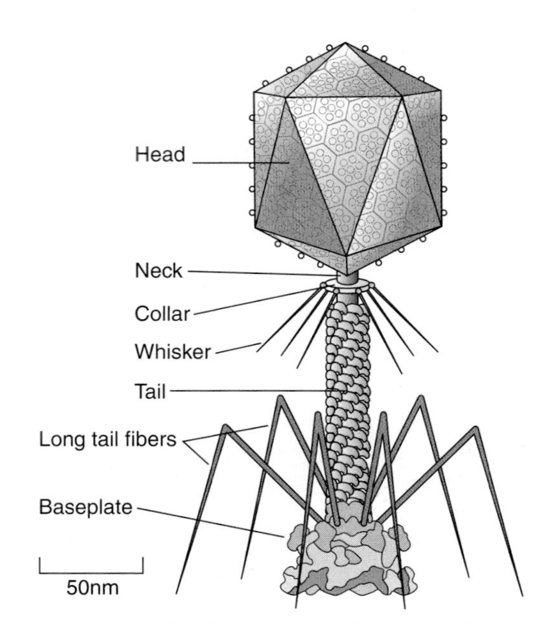

FIGURE 8.4 Basic structural features of bacteriophage T4. (Adapted from P. G. Leiman, et al., *Cell. Mol. Life Sci.* 60 [2003]: 2356–2370.)

virus (FIGURE 8.4). Its head is an elongated icosahedron that is connected to a neck surrounded by a collar with attached protein whiskers. The neck is also attached to the tail, which contains a rigid inner core or tube that is surrounded by a contractile sheath made of 144 identical polypeptide subunits, each with a bound ATP molecule. The tail ends in a multiprotein baseplate with tail fibers attached to it.

DNA molecules purified from T4 virions are linear duplexes with about 172 kilobase pairs (kbp) and a length of about 55 μm. The DNA in the virion, therefore, is quite long compared to phage head dimensions (approximately 0.06 by 0.09 μm), indicating that the DNA is tightly packed in the phage head. T4 DNA contains adenine, thymine, and guanine but not cytosine. Instead, there is a modified form of cytosine called **5-hydroxymethylcytosine** (abbreviated **HMC**; FIGURE 8.5a), which base-pairs with guanine. DNA molecules that contain HMC residues are subject to further modification. Glucoses are attached to the hydroxymethyl groups (see below). Glucosylation patterns differ in the T-even phages. In T4 DNA, every HMC residue has an attached glucose. Approximately 70% of the glycosidic bonds are in the α-configuration (FIGURE 8.5b) and 30% in the β-configuration (not shown). In T2 and T6 DNAs, some HMCs are not glucosylated and some have attached gentiobiosyl groups. Gentiobiose (6-O-β-D-glucopyranosyl-D-glucose) is a disaccharide containing two glucose units.

Phage T4 DNA has two other remarkable properties. First, it is **terminally redundant**; approximately 2% of the viral DNA is repeated at each end of the molecule. As shown in FIGURE 8.6, a terminally-redundant linear duplex can form a circle under laboratory conditions in which a 3′-exonuclease is used to trim back a single strand at each end.

Phage T4 DNA's second remarkable feature is that the terminally repeated genes differ from one phage particle to another (FIGURE 8.7). This difference is observed even within a phage population from a plaque derived from a single phage particle. The gene order is the same in the DNA from all phage particles but the DNA molecules are **circularly permuted.** Circular permutation is a property of the phage T4 population, whereas terminal redundancy is a property of an individual

(a) 5-Hydroxymethylcytosine (HMC)

(b) Glucosylated HMC

FIGURE 8.5 5-Hydroxymethylcytosine (HMC) and glucosylated 5-hydroxymethylcytosine. (a) If the CH_2OH group (red) in HMC were replaced by hydrogen, the molecule would be cytosine. (b) The glucose group in the glucosylated HMC is shown in blue.

phage T4 DNA molecule. Many phage species in addition to those from the T-even family produce terminally redundant and circularly permuted populations of DNA molecules. Terminal redundancy can also be present without circular permutation but the reverse has not been observed. Terminal redundancy and circular permutation are a natural byproduct of the mechanism used by the T-even phage DNA

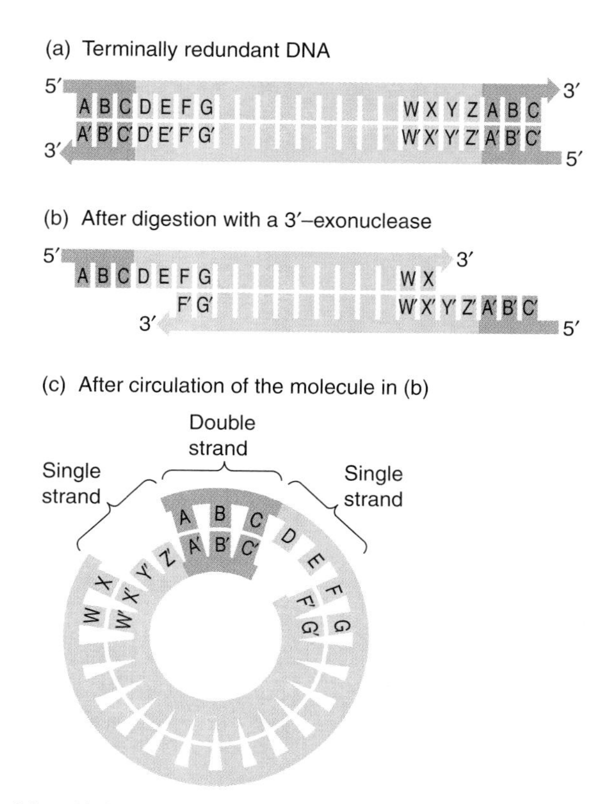

FIGURE 8.6 Identifying a terminally redundant linear duplex. (a) A terminally redundant DNA molecule can be identified (b) by means of exonucleolytic digestion and (c) circularization. A nonredundant DNA molecule cannot be circularized this way. The terminally redundant region is shown in red.

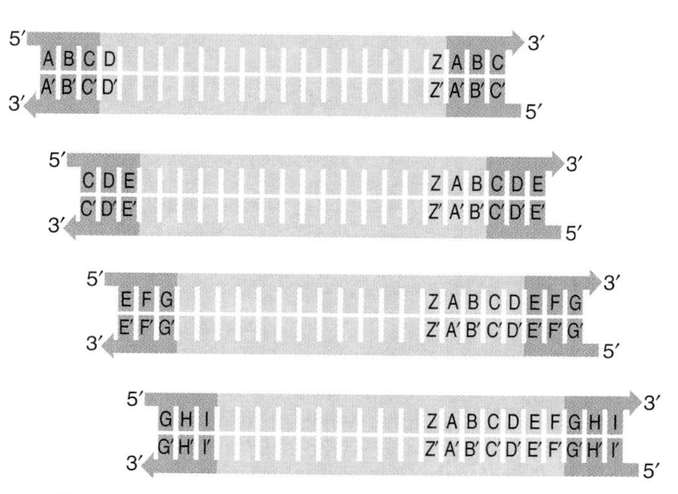

FIGURE 8.7 Circularly permuted collection of terminally redundant DNA molecules. Terminally redundant regions are shown in red.

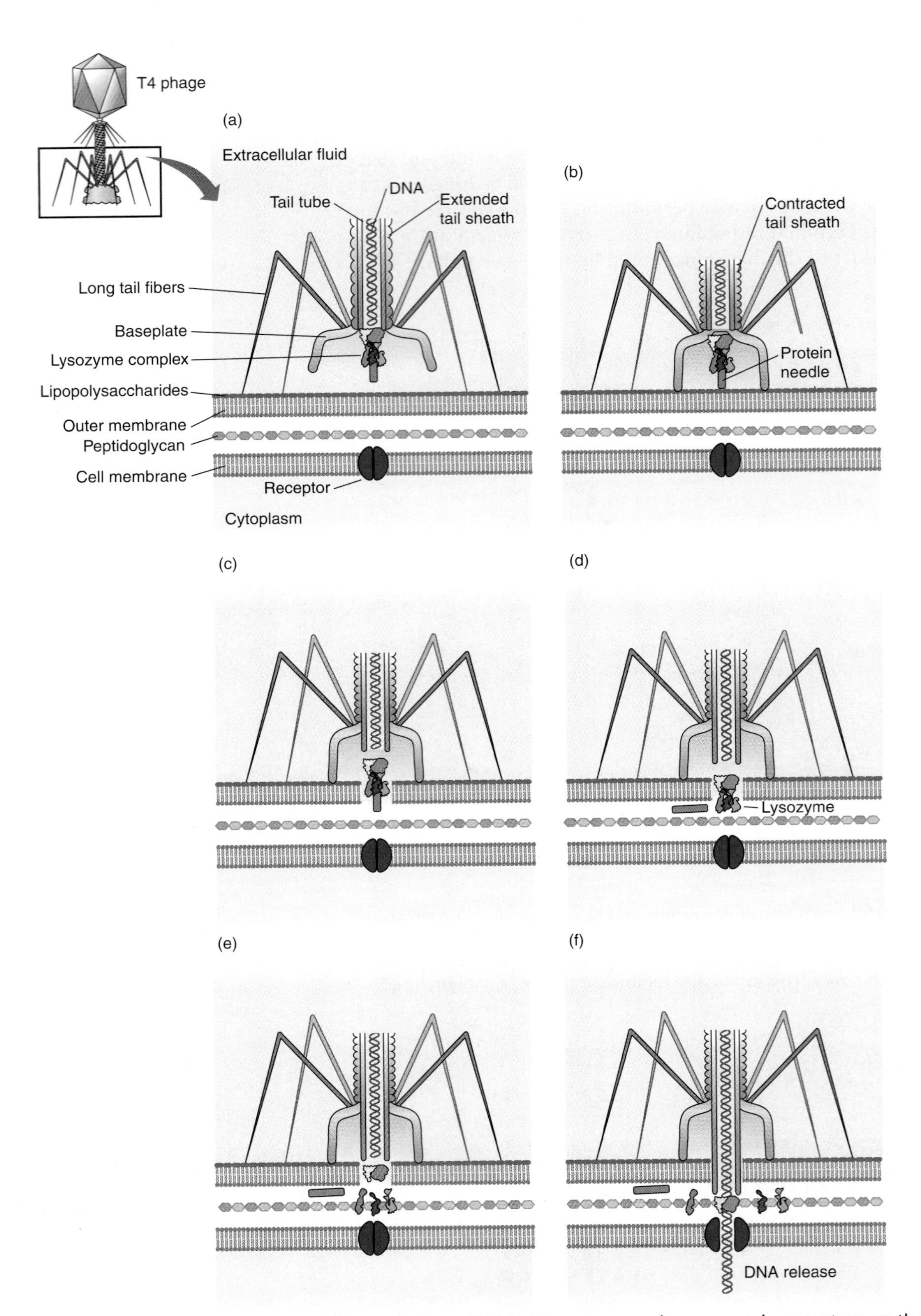

FIGURE 8.8 **Injection process.** (a) The long tail fibers of phage T4 bind to outer membrane protein receptors on the *E. coli* cell surface. (b) This binding triggers a large conformational change in the baseplate, which in turn causes the sheath that surrounds the tail to contract. (c) This contraction pushes a protein needle at the end of the rigid inner core through the outer cell membrane. (d) The protein needle dissociates from the tail tube, activating lysozyme. (e) Lysozyme catalyzes cell wall degradation. (f) A phage protein associates with a receptor on the inner membrane, initiating DNA release into the cytoplasm. (Adapted from P. G. Leiman, et al., *Cell. Mol. Life Sci.* 60 [2003]: 2356–2370.)

packaging machinery to move DNA into the phage head during virion particle formation (see below).

The phage T4 infection process begins with phage tail fibers contacting specific outer membrane receptor proteins on the cell surface. When the multiplicity of infection (ratio of infecting phage particles to bacterial cells) is low, most phages are observed to be attached at the bacterial poles or at the mid-cell, where the next generation of poles will form. Attachment somehow triggers a large conformational change in the baseplate, which in turn causes the sheath that surrounds the tail to contract. This contraction pushes a protein needle at the end of the rigid inner core through the outer membrane (**FIGURE 8.8**). Then the protein needle dissociates from the tail tube to expose lysozyme, which catalyzes cell wall degradation. A phage protein associates with a receptor on the inner membrane, initiating DNA release into the cytoplasm. The T4 DNA transfer rate has been estimated to be between 3 and 10 kbp per second.

Phage T4 DNA codes for approximately 290 different polypeptides, 8 tRNA molecules, and at least two other small stable RNA molecules of unknown function. Most of the known polypeptide coding genes are either metabolic genes or particle-assembly genes. Of the 82 metabolic genes, only 22 genes are essential, namely, those coding for proteins that participate in replication, transcription, and cell lysis. The remaining 60 metabolic genes duplicate bacterial genes. Particles in which these genes are mutated replicate, though occasionally they will have a smaller burst size.

The phage T4 life cycle is shown in the schematic diagram in **FIGURE 8.9**. Specific events take place at distinct times in the life cycle. Table 8.2 indicates some of the key events that take place during the T4 life cycle and the times at which they occur. The main feature to be noticed is the orderly sequence of events.

Shortly after infection, the phage DNA causes changes that turn off bacterial functions necessary for continued bacterial growth. For example, the host's RNA polymerase is modified so that host genes are recognized poorly. Furthermore, the first phage mRNA formed directs the synthesis of DNases that rapidly degrade host DNA. (Similar DNA degradation is also observed after infection by other T-even phages but is not observed after infection by many other lytic phages.) In time, the bacterial DNA is almost totally degraded to nucleotides and is not available as a template for replication or transcription.

After infection, T4 DNA is replicated but bacterial DNA is not. Five aspects of T4 DNA replication are especially interesting: (1) the source of nucleotides; (2) the synthesis of 5-hydroxymethylcytosine (HMC), which substitutes for cytosine (C); (3) the prevention of incorporation of cytosine; (4) glucosylation of T4 DNA; and (5) the enzymology of replication. The first four are discussed below. The enzymology of DNA replication is described in the next chapter.

1. Source of T4 DNA Nucleotides and Degradation of Host DNA

An early event in the phage T4 life cycle is the degradation of host DNA to deoxynucleoside monophosphates (dNMP). Degradation is initiated by two viral DNA endonucleases, which are the products of

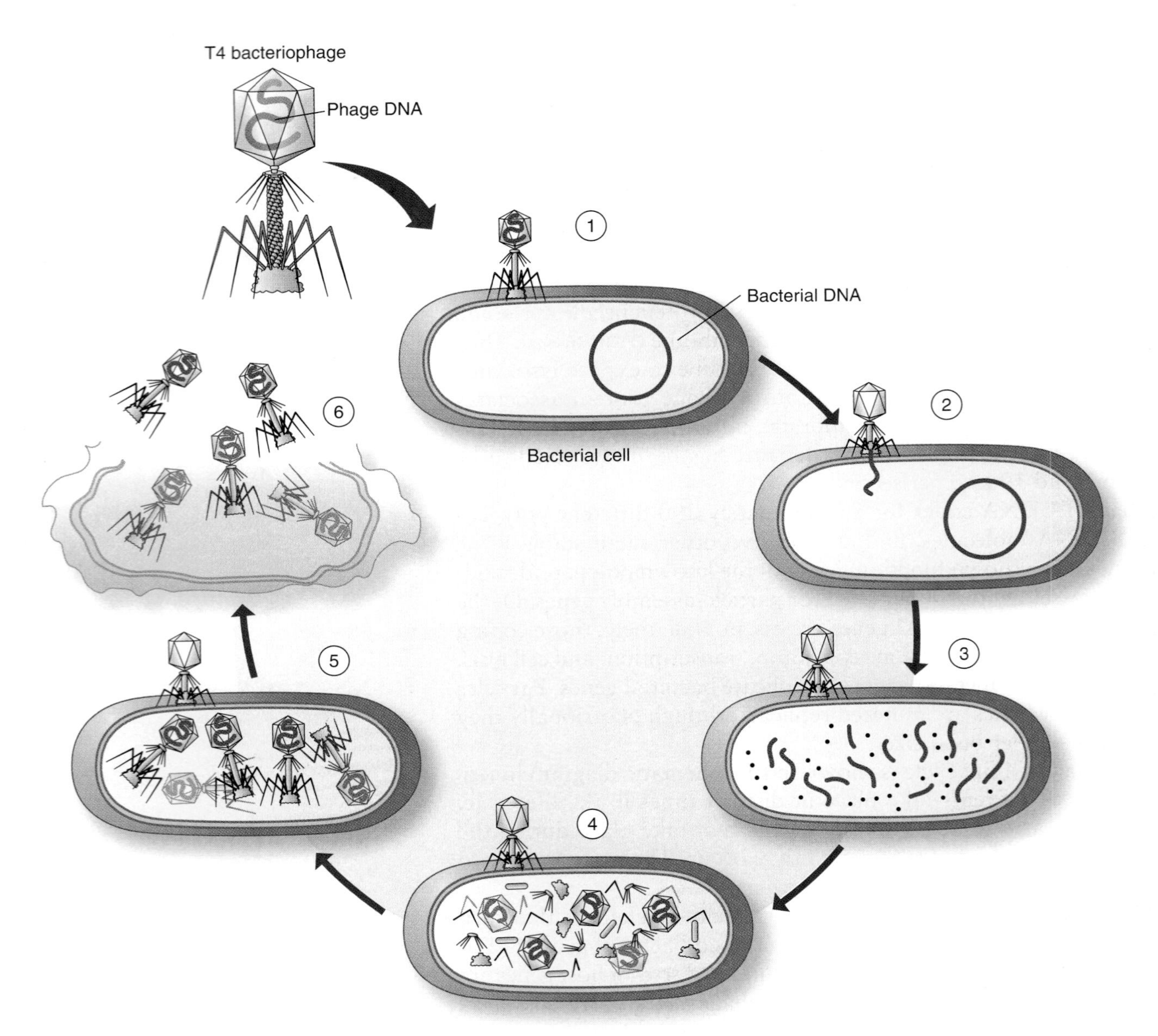

FIGURE 8.9 **Phage T4 life cycle.** (1) Phage attaches to a specific host bacterial cell. (2) Phage injects its DNA into the host cell. (3) Phage DNA codes for proteins that digest the bacterial DNA, killing the bacterial cell. (4) Phage takes over the bacterial machinery to make new phage parts. (5) The process ends with the assembly of new phages and (6) the disruption of the bacterial cell wall to release about one hundred new phages into the culture medium. (Adapted from K. Thiel, *Nat. Biotechnol.* 22 [2004]: 31–36.)

the genes *denA* and *denB*. These two enzymes, which are active only on cytosine-containing DNA, cleave the host DNA to double-stranded fragments, which are then degraded to dNMP by a phage-encoded exonuclease. The mononucleotides are built up to dATP, dTTP, dGTP, and dCTP by the usual *E. coli* enzymes, providing sufficient dNTP to synthesize about 30 T4 DNA molecules. DNA precursors are also formed by de novo synthesis but, to ensure an abundant supply of

TABLE 8.2 T4 Bacteriophage Life Cycle at 37°C

Time (min)	Change That Takes Place
0	Tail fibers attach to a specific outer membrane protein. Injection of phage DNA probably occurs within seconds of attachment.
1	Host DNA, RNA, and protein syntheses are totally turned off.
2	Synthesis of first mRNA begins.
3	Degradation of bacterial DNA begins.
5	Phage DNA synthesis is initiated.
9	Synthesis of "late" mRNA begins.
12	Completed heads and tails appear.
15	First complete phage particle appears.
22	Lysis of bacteria; release of progeny phage.

dNTP, five phage-encoded enzymes, which are virtually identical in activity to the *E. coli* enzymes, are also synthesized.

2. Synthesis of 5-hydroxymethylcytosine (HMC)

T4 DNA contains HMC instead of cytosine. *E. coli* does not possess enzymes for forming HMC, however. Two phage encoded enzymes convert dCMP to dHDP (deoxyhydroxymethylcytidine diphosphate). The *E. coli* enzyme, nucleoside diphosphate kinase, which forms all triphosphates in *E. coli*, then converts dHDP to dHTP, the immediate precursor of hydroxymethylcytosine in the DNA.

3. Prevention of Incorporation of Cytosine into T4 DNA

T4 DNA polymerase cannot distinguish between dCTP and dHTP, both of which hydrogen-bond to guanine. It is essential that no cytosine be incorporated into daughter T4 DNA strands, because such cytosine-containing DNA would be a substrate for the T4 endonucleases that degrade host DNA. Also, cytosine-containing DNA is not a template for the transcription that occurs late in the phage T4 life cycle. *E. coli* does not have enzymes that prevent incorporation of cytosine into DNA. The phage, therefore, encodes these enzymes.

For cytosine to become part of daughter DNA molecules, dCMP must be converted to dCDP and then to dCTP. A phage enzyme, called dCTPase, degrades both dCDP and dCTP to dCMP. This process seems somewhat wasteful because ATP is consumed in forming dCTP. The more economical process of preventing formation of dCTP would be difficult to carry out in *E. coli*, however, because nucleoside diphosphate kinase is responsible for the production of all triphosphates. Inhibiting the dCMP→dCDP reaction also would be ineffective because in *E. coli* most of the deoxynucleoside diphosphates are formed by enzymatic reduction of ribonucleoside diphosphates by a single enzyme.

Another phage enzyme, dCMP deaminase, converts dCMP to dUMP (in preparation for synthesis of dTMP). This enzyme duplicates the activity of a similar *E. coli* enzyme (and hence is a product of a nonessential T4 gene) but has an interesting economic function. The base compositions of *E. coli* DNA and T4 DNA are 50% (A + T) and 66% (A + T), respectively. In *E. coli*, the ratio of dTTP and dCTP is about 1:1, which is in keeping with the T:C ratio in DNA. The bacterial and phage dCMP deaminases, acting together, increase the amount of dTMP with respect to dCMP, so the ratio of dTTP to dHTP is 2:1, as is the T:HMC ratio in T4 DNA.

4. Glucosylation of T4 DNA

The presence of HMC in T4 DNA creates another problem for the phage because *E. coli* possesses an endonuclease that attacks certain sequences of nucleotides containing HMC. To avoid this damage, the HMC residues in T4 DNA are glucosylated. This is accomplished by two phage enzymes—α-glucosyl transferase and β-glucosyl transferase—each of which catalyzes glucose transfer from uridine diphosphoglucose (UDPG) to HMC that is already in DNA. Glucosylation, a post-replicative modification, has a single essential function. It protects phage T4 DNA from the *E. coli* nuclease that cleaves DNA molecules with unmodified HMC residues.

Production of complete phage particles can be separated into two parts: assembly of heads, tails, and other structures, and packaging of DNA of a sufficient length to provide a little more than one set of genes in the phage head. Assembly of T4 (and many other phages) has been studied by two techniques, both of which require a large collection of phage mutants unable to produce finished particles. In one method, different cultures of cells, each infected with a particular mutant, are lysed and examined by electron microscopy. This procedure shows that heads are made in the absence of tail synthesis and that tails are made by a mutant unable to synthesize heads; head and tail assembly are independent processes.

The second technique is a complementation assay of a type widely used to purify proteins. If two extracts of infected cells, one lacking heads and the other lacking tails, are mixed, phage particles form *in vitro* (FIGURE 8.10). The "headless" extract also can be fractionated and a component can be isolated that allows a "tailless" extract to make tails. In this way, a protein in the tail assembly pathway can be isolated and identified.

Assembly studies have shown that there are two types of components: **structural proteins** and **morphogenetic enzymes**. Some of the structural components assemble spontaneously to form phage structures, whereas others need the help of enzymes. A few host-encoded factors are also needed for head assembly. A schematic for the assembly pathway for T4 phage is shown in FIGURE 8.11.

Packaging DNA into the phage head requires the prior formation of a long DNA molecule called a **concatemer**, which is a multimer of the T4 chromosome, linked head to tail through shared terminally redundant sequences. Concatemer formation requires both DNA replication and recombination. After the phage DNA is injected into the bacterial cell, the linear duplex is replicated in both directions starting

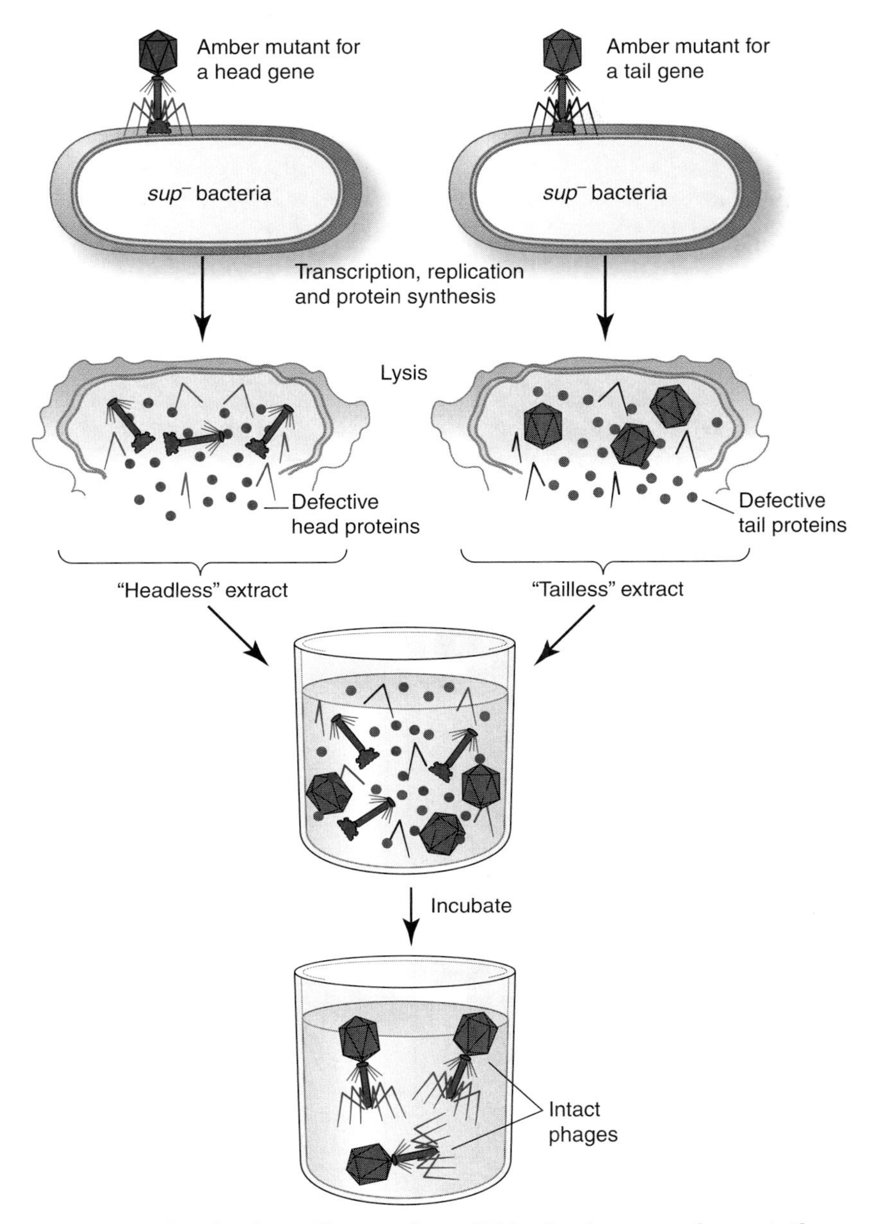

FIGURE 8.10 Production of intact phage T4 by in vitro complementation.

at a site within the linear duplex (FIGURE 8.12a). For reasons that will become clear in the next chapter, this type of bidirectional replication produces linear duplexes that have 3′-single strand overhangs at each end. The next stage in concatemer formation involves recombination among the newly formed linear duplexes. This process begins when the 3′-end of one linear duplex invades another linear duplex and ultimately leads to concatemer formation (FIGURE 8.12b–d).

The phage protein **terminase** recognizes the concatemeric DNA and makes cuts that initiate and then terminate DNA packaging into the preformed protein shell or **procapsid.** Terminase consists of small

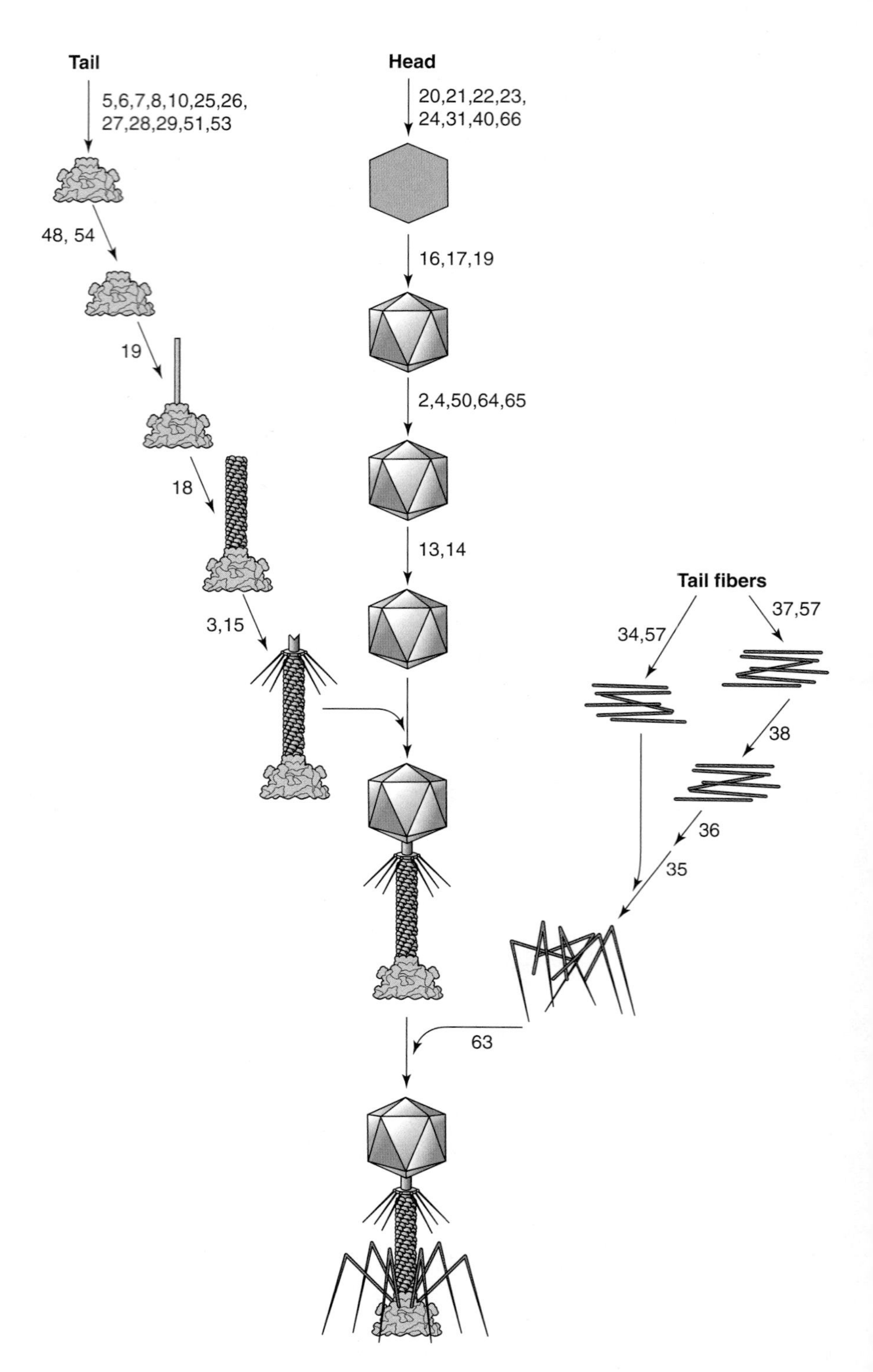

FIGURE 8.11 **Morphogenetic pathway of phage T4**. The numbers designate *T4* genes.

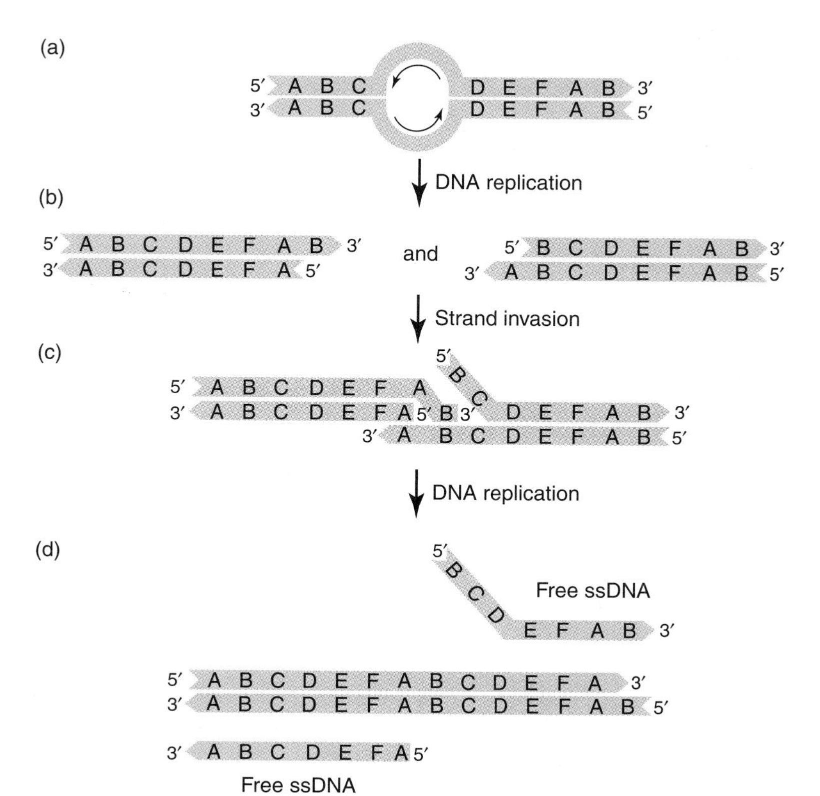

FIGURE 8.12 Concatemer formation. (a) T4 DNA is injected into the host cell. (b) DNA replication begins at a specific site on the DNA and proceeds bidirectionally from that site to the ends. This replication results in the production of 3′-overhangs at each end. (c) The 3′-overhang on one newly replicated DNA invades another synthesized DNA duplex at a homologous region. (d) DNA replication of this molecule leads to concatemers of the phage genome. Depending on the location of the strand invasion, branched molecules may also be formed. The DNA is packaged into new phage particles from the concatemers. (Adapted from N. J. Trun and J. E. Trempy. *Fundamental Bacterial Genetics, First edition.* Blackwell Publishing, 2003.)

(18 kDa) and large (70 kDa) subunits, which are gene products of phage T4 genes 16 and 17, respectively. Large terminase (gp17) subunits also are an essential part of the **phage T4 DNA packaging motor**, which uses chemical energy stored in ATP to package phage DNA into the procapsid. Small terminase subunits (gp16) stimulate the motor but are not required for DNA packaging. Five gp17 subunits dock at a special portal vertex on the procapsid that is formed by twelve gp20 subunits (FIGURE 8.13).

Large terminase has an N-terminal domain with ATPase activity, a C-terminal domain with nuclease activity, and a flexible linker that connects the two domains. The N- and C-terminal domains must be

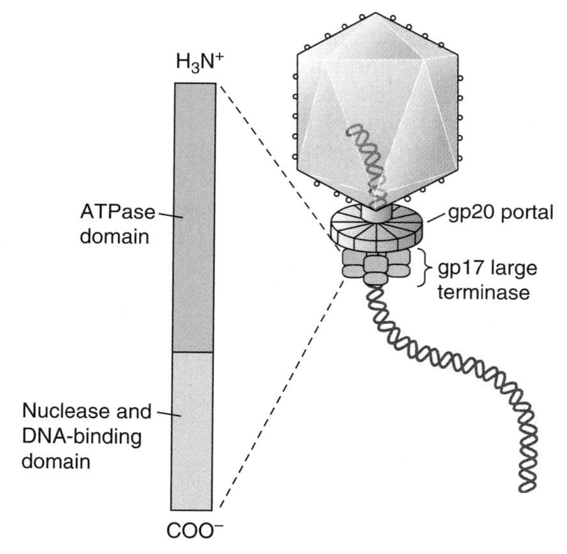

FIGURE 8.13 DNA—The bacteriophage T4 packaging motor. The packaging motor is made of twelve gp20 portal subunits (purple) and five gp17 large terminase subunits. Each large terminase subunit has an N-terminal domain (green) with ATPase activity and a C-terminal nuclease domain (blue) with nuclease activity and a DNA binding site. (Adapted from T. I. Alam, et al. *Mol. Microbiol.* 69 [2008]: 1180–1190.)

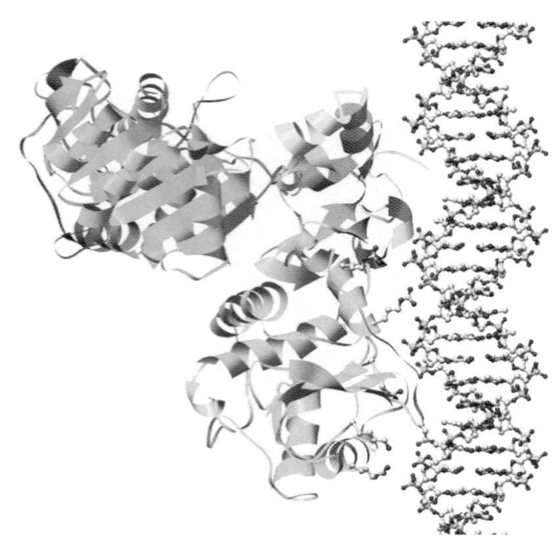

FIGURE 8.14 **A model showing interactions between the gp17 large terminase subunit and double-stranded DNA.** The colors used for the gp17 large terminase subunit are as follows: the C-terminal domain is cyan, the N-terminal subdomain I is green, and the N-terminal subdomain II is yellow. DNA is shown in ball-and-stick form using standard CPK colors. The arrow points to the ATP binding site. (Reprinted from *Cell*, vol. 135, S. Sun, et al., The structure of the phage T4 DNA . . ., pp. 1251–1262, copyright 2008, with permission from Elsevier [http://www.sciencedirect.com/science/journal/00928674]. Photo courtesy of Venigalla B. Rao, The Catholic University of America.)

connected for the phage T4 DNA packaging motor to work. Venigalla B. Rao, Michael G. Rossman, and coworkers recently determined the gp17 subunit's crystal structure. Based on this structure, they propose the model shown in FIGURE 8.14 for the interaction between the gp17 subunit and a double-stranded DNA molecule. This model, like the crystal structure, shows the N- and C-domains in close contact in a conformation that is called the "tensed" state.

Rao, Rossman, and coworkers also obtained a cryo-electron microscopy construct of gp17 complexed with gp20 subunits at the portal apex of the phage procapsid. It seems reasonable to suppose that gp17's size and shape would be the same in the crystal structure and the cryo-electron microscopy construct. However, when Rao, Rossman, and coworkers attempted to fit the gp17 subunit's crystal structure into the cryo-electron microscopy construct, they could not do so. They explained this inconsistency by proposing that the N- and C-terminal domains are in close contact in a "tensed" post-translocation conformation in the crystal structure but separated in a "relaxed" pretranslocation conformation in the cryo-electron microscopy construct.

The fact that gp17 exists in two distinct conformations suggested how the phage T4 packaging motor might work (FIGURE 8.15). The gp17 C-terminal domain binds DNA. Then the N-terminal domain catalyzes ATP hydrolysis, triggering conformational changes that cause the C-terminal domain to move toward the N-terminal domain. This movement, which is driven by electrostatic interactions between the C- and N-terminal domains, moves the bound DNA toward the procapsid by 0.68 nm or two base pairs per cycle of ATP hydrolysis. ADP and Pi are released and the C-terminal domain returns to its original position. Released DNA is aligned to bind the C-terminal domain of a neighboring gp17 subunit. The five gp17 subunits work in a coordinated fashion somewhat like the cylinders in a car engine. When scaled up to the mass of a car engine, the T4 DNA packaging motor has even more power than a typical car engine. The T4 DNA packaging motor can package as many as 2000 bp in a second. The packaging rate, however, decreases as the internal pressure within the procapsid increases due to packed DNA. When no further DNA can be packaged into the T4 capsid, the terminase makes a second cut in the DNA. This is known as the **headful mechanism** and it explains how both terminal redundancy and circular permutation arise. Further studies are required to determine how the nuclease is activated and how it makes a cut in each strand.

The gp17 subunits and the unpackaged DNA then dissociate from the filled head and interact with another empty procapsid. Terminase cuts are not made in unique base sequences in the DNA because if they were, T4 DNA could not be circularly permuted. Instead, the cuts are made at positions that are determined by the amount of DNA that can fit in a head. The essential point is that the DNA content of a T4 particle is greater than the length of DNA required to encode the T4 proteins. When cutting a headful from a concatemeric molecule, however, the final segment of DNA that is packaged is a duplicate of the DNA that is packaged first, which means that the packaged DNA

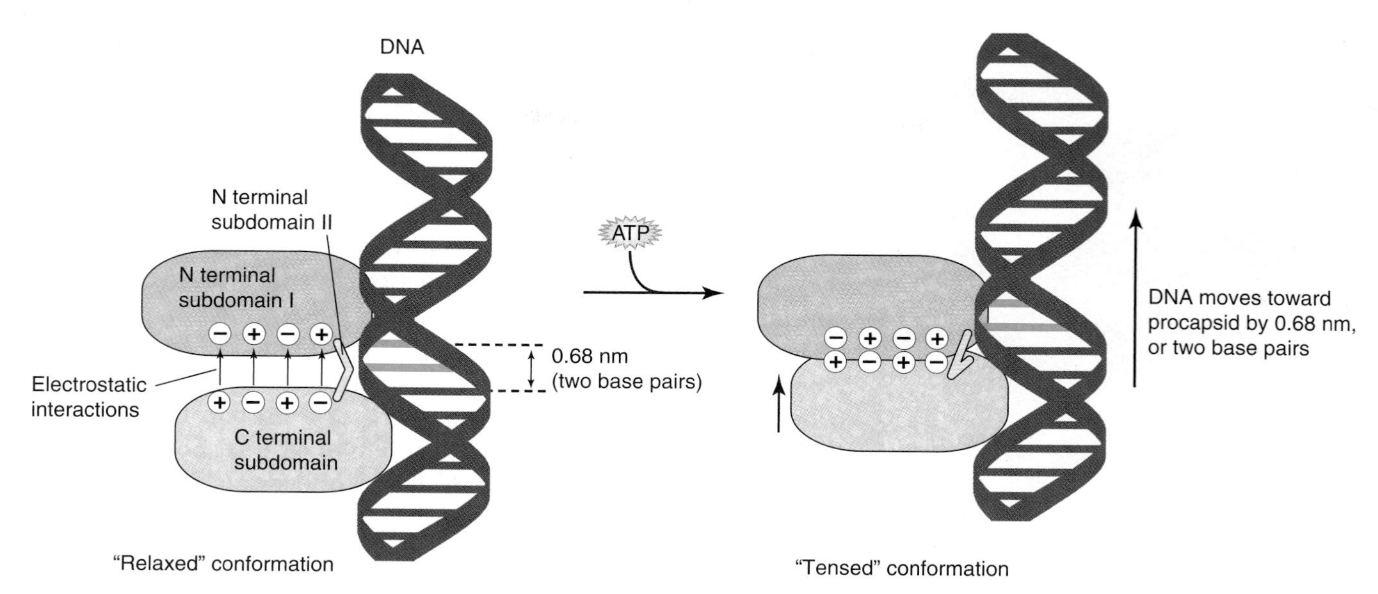

FIGURE 8.15 **A possible mechanism of action for the phage T4 packaging motor.** The N- and C-terminal domains in gp17 are separated in a "relaxed" pretranslocation conformation (left) and in close contact in a "tensed" posttranslocation conformation (right). The power cycle begins when the gp17 C-terminal domain binds DNA. Then the N-terminal domain catalyzes ATP hydrolysis, triggering conformational changes that cause the C-terminal domain to move toward the N-terminal domain. This movement, which is driven by electrostatic interactions between the C- and N- terminal domains, moves the bound DNA toward the procapsid by 0.68 nm or two base pairs per cycle of ATP hydrolysis. ADP and Pi are released and the C-terminal domain returns to its original position. (Reprinted from *Cell*, vol. 135, R. S. Williams, G. J. Williams, and J. A. Tainer, A charged performance by gp17, pp. 1169–1171, copyright 2008, with permission from Elsevier [http://www.sciencedirect.com/science/journal/00928674].)

is terminally redundant. The first segment of the second DNA molecule that is packaged is not the same as the first segment of the first phage. Furthermore, because the second phage must also be terminally redundant, a third phage-DNA molecule must begin with still another segment. The collection of DNA molecules in the phages produced by a single infected bacterium, therefore, is a circularly permuted set.

The final step in the T4 life cycle, lysis, is an abrupt event. The timing must be tightly regulated because early lysis would yield too few phages and late lysis would diminish the opportunity for an explosive reproduction cycle in a new host cell. Phage T4, like other tailed phages, requires two proteins encoded by the phage DNA for lysis. The first, an **endolysin,** is an enzyme that degrades the cell wall. The second, **holin,** assembles to form a pore in the cell membrane through which the endolysin moves to reach the cell wall. After endolysin degrades the cell wall, internal osmotic pressure causes cell lysis.

E. coli phage T7 DNA is terminally redundant but not circularly permuted.

E. coli phage T7 has an icosahedral head and a short tail (FIGURE 8.16). Its linear double-stranded DNA contains 37,937 bp, is terminally redundant but not circularly permuted, and codes for 56 known or potential gene products. T7 DNA contains the normal bases, adenine, thymine, guanine, and cytosine. Examination of the base sequence

(a)

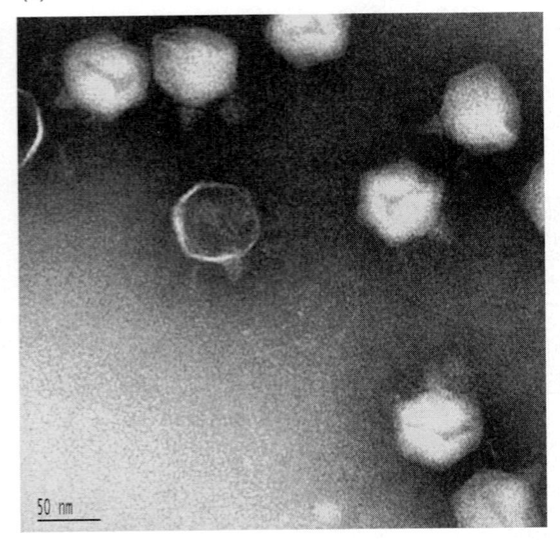

(b)

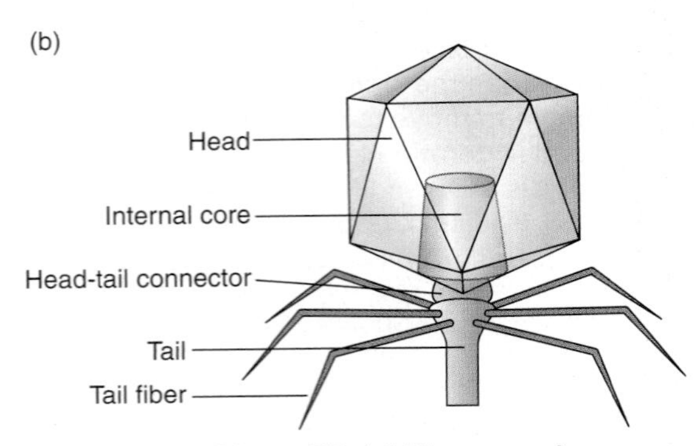

FIGURE 8.16 Phage T7. (a) Electron micrograph of phage T7. The tail fibers are not visible. The bar is 50 nm. (b) Schematic of the T7 virion indicating its protein components. (Part a courtesy of Dean Scholl, National Institute of Health. Part b adapted from S. T. Abedon and R. L. Calendar. *The Bacteriophages, Second edition.* Oxford University Press, 2005.)

indicates economical use of DNA. For example, spacers between genes are generally only a few nucleotides long, when present at all, whereas in bacterial genetic systems hundreds of nucleotides are usually used for this purpose. Furthermore, the catalytic proteins made early in the life cycle tend to be much smaller than the average bacterial protein, ranging from 29 to 883 amino acids.

A notable feature of the organization of the phage T7 genes is that they are clustered according to function and are arranged to provide a continuous order according to the time of their function (FIGURE 8.17). Phage T7 genes are numbered in order of their position on the DNA from left to right and the gene products are synthesized in the same order. The first gene, gene 0.3, codes for a protein that protects T7 DNA from the bacterial restriction endonuclease EcoKI by binding to the nuclease and inhibiting its ability to degrade the T7 DNA. The next few genes control transcription. These genes are followed by genes needed for DNA replication and then by genes encoding the structural proteins of the phage particle. The final genes are used for packaging newly synthesized DNA in the phage head. Table 8.3 indicates some of the key events that take place during the T7 life cycle and the times at which they occur. Most of these stages are discussed below.

Phage T7 first makes contact with the bacterial cell through the interaction of its tail fibers with lipopolysaccharides on the bacterial cell surface. T7 DNA enters the bacterial cell in three stages. At the start of the first stage, phage T7, which is tightly bound to the bacterial cell, releases two proteins that form a channel across the bacterial envelope. Then this channel appears to act as a motor to drive approximately 880 bp of T7 DNA into the bacterial cell at a rate of approximately 75 bp·s^{-1} at 30°C. During the second stage, *E. coli* RNA polymerase, which is assumed to be stationary, reels an additional 7 kbp of T7 DNA into the bacterial cell at a rate of approximately 40 bp·s^{-1}. During the third stage, newly synthesized T7 RNA polymerase reels in the remaining DNA at about 200 to 300 bp·s^{-1}. This tripartite mechanism for T7 DNA transport appears to allow the phage DNA sufficient time to direct the synthesis of the protein that blocks the EcoKI restriction endonuclease. Slow injection is a significant feature of the T7 life cycle because a gene cannot be transcribed until it has been injected. In fact, the timing of transcription, which is discussed below, depends in part on the kinetics of injection.

The life cycle of T7 is normally divided into two transcriptional stages called early and late. The early stage uses *E. coli* RNA polymerase to transcribe genes 0.3 through 1.3; the late stage uses a newly synthesized phage T7 RNA polymerase. Transcription of phage T7 DNA is temporally regulated. There are three major classes of transcripts, labeled I, II, and III in Figure 8.17. Each of these is transcribed from the same DNA strand, which is termed the *r* strand (because transcription occurs in the rightward direction when the DNA molecule is drawn in a standard orientation). The class I transcripts are synthesized by *E. coli* RNA polymerase and comprise the early mRNA.

Synthesis of class I transcripts begins two minutes after infection. The delay probably results from slow phage DNA entry into the bacterial cell. Transcription is mediated by the bacterial RNA polymerase

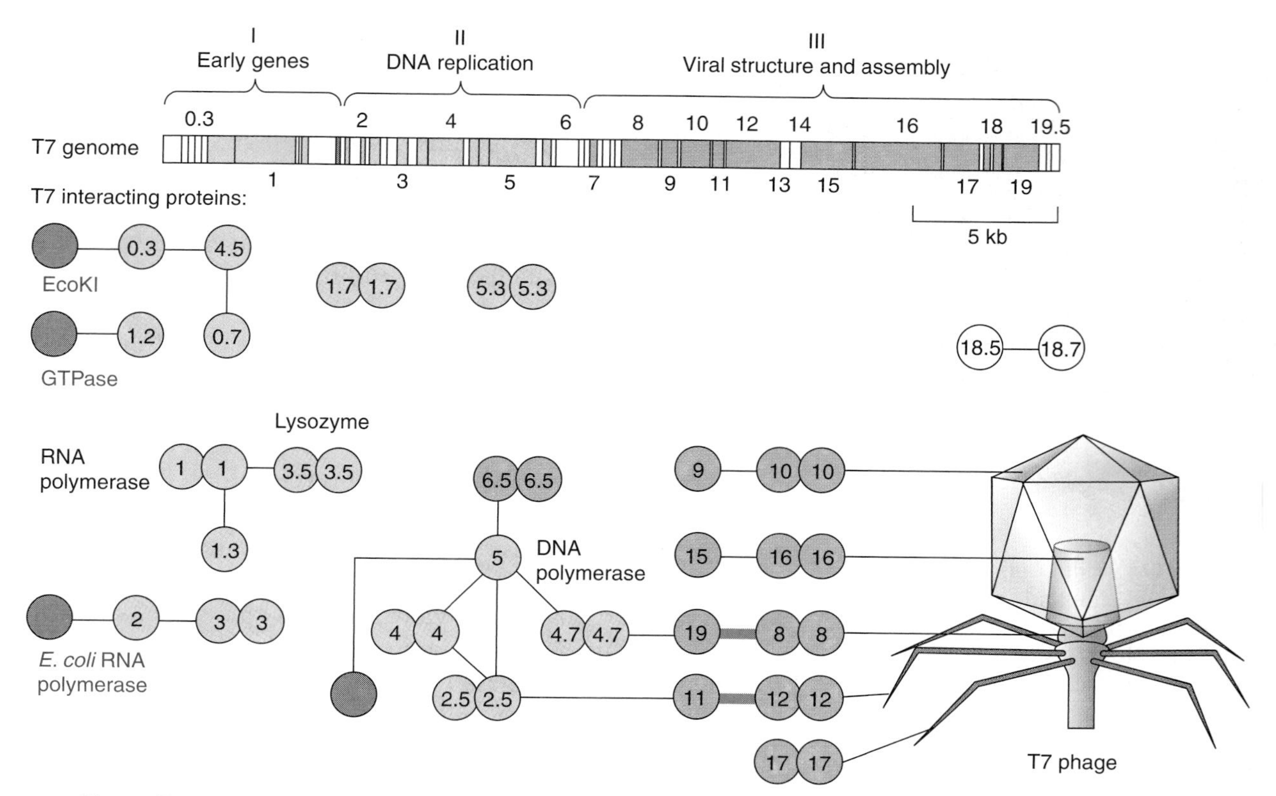

FIGURE 8.17 **Phage T7 genetic map.** The top panel shows a simplified version of the T7 genetic map with gene positions indicated by numbers. Box colors indicate a simplified assignment to functional classes as shown by the heading. Interactions between proteins are shown as lines, with self-interactions shown as dimers. Thick gray lines indicate expected interactions which have not been found yet. Gray proteins correspond to *E. coli* proteins. Blue proteins are involved in virus assembly and structure, and their rough location in the virus particle is shown on the right, if known. (Modified from P. Uetz, et al., *Genome Res.* 14 [2004]: 2029–2033. Copyright 2004, Cold Spring Harbor Laboratory Press. Used with permission of Peter Uetz, Forschungszentrum Karlruhe, Germany.)

TABLE 8.3 **The T7 Life Cycle at 30°**

Time (min)	Change That Takes Place
1	Phage attachment.
0–1	Start of phage DNA entry into the cell.
2	Initiation of synthesis of phage mRNA.
4	Turning off of host transcription begins.
8–9	Initiation of phage DNA synthesis.
8–10	Initiation of synthesis of structural proteins; completion of injection of phage DNA.
15	First phage appears.
25	Lysis and release of progeny phage.

and stops at a termination site just after gene 1.3. Class I transcripts are cleaved by *E. coli* RNase III into five mRNA molecules. However, this processing is not essential under laboratory conditions. If an *E. coli* mutant lacking RNase III is used as the host, the transcript remains intact and phage production seems to be normal. Possibly in nature or in other bacterial hosts processing is required.

Several proteins are translated from class I transcripts. One of these, protein kinase (gene 0.7) inactivates *E. coli* RNA polymerase

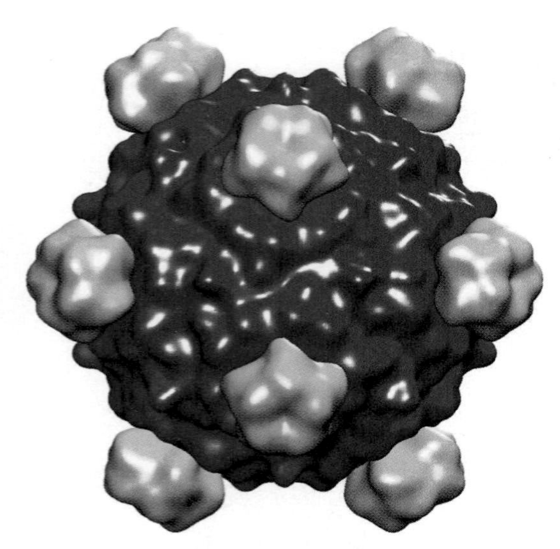

FIGURE 8.18 **ϕX174 virion.** Capsid and spike proteins are colored blue and yellow, respectively. (Reprinted from *J. Mol. Biol.*, vol. 337, R. Bernal, et al., The Phi-X174 protin J mediates DNA . . ., pp. 1109–1122, copyright 2004, with permission from Elsevier [http://www.sciencedirect.com/science/journal/00222836]. Photo courtesy of Michael G. Rossmann, Purdue University, and Ricardo A. Bernal, University of Texas, El Paso.)

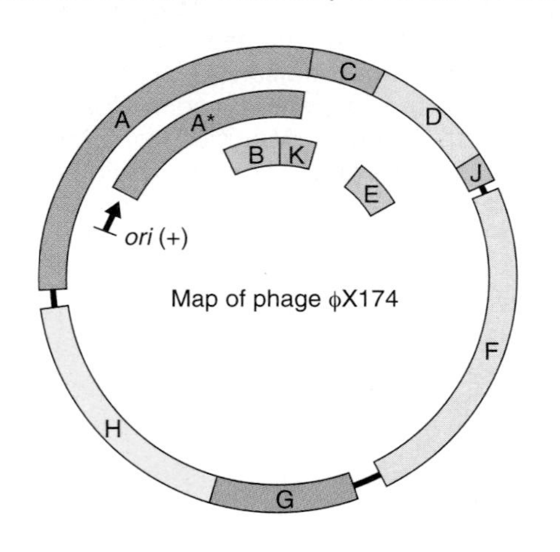

Gene	Function
ori	Origin of plus strand replication
Gene A	RF replication
Gene A*	Shut off host DNA synthesis
Gene B	Capsid morphogenesis
Gene C	DNA maturation
Gene D	Capsid morphogenesis
Gene E	Cell lysis
Gene F	Major coat protein
Gene G	Major spike protein
Gene H	Minor spike protein, adsorption
Gene J	Core protein, DNA condensation
Gene K	Function unknown

FIGURE 8.19 **Map of phage ϕX174.** (Adapted from an illustration by New England BioLabs, Inc.)

by phosphorylating it. This inactivation reduces transcription of *E. coli* DNA and is the first step in the takeover of the bacterium by the phage. The protein kinase is not necessary for phage T7 infection of most *E. coli* laboratory strains. As we will see shortly, another *E. coli* RNA polymerase inhibitor is made later. A second protein translated from a class I transcript is phage T7 RNA polymerase (gene 1).

Phage T7 RNA polymerase is an essential enzyme because it catalyzes the formation of all class II and III transcripts. Class II transcripts are synthesized first and then class III transcripts. The difference in time of transcription for the two classes is due to the fact that class II genes enter the cell several minutes before class III genes. Class II transcripts direct the synthesis for phage DNA polymerase and a second *E. coli* RNA polymerase inhibitor. The action of this inhibitor, combined with inhibition caused by the gene 0.7-protein kinase, completely blocks *E. coli* RNA polymerase. Thus, by eight minutes after infection there is no longer any synthesis of *E. coli* RNA and takeover of the bacterium is complete. Class I transcripts, which are made by the *E. coli* RNA polymerase, are also no longer made. At this stage, sufficient phage T7 RNA polymerase has been produced to complete the life cycle and no other products of class I transcripts are needed.

Class III genes that code for structural proteins are transcribed shortly after class II genes that code for proteins needed for DNA replication. The timing of transcriptional events during phage T7 infection contrasts with that in phage T4, in which structural protein synthesis is delayed with respect to replication enzyme synthesis. At first glance, the timing in phage T7 may seem to be inefficient because premature assembly of phage particles and packaging of phage DNA might result in termination of replication or at least a vast reduction of the number of templates for replication. However, the phage T7 DNA molecule is small and can be replicated rapidly (in nearly 20 seconds), while the maturation proteins (genes 18 and 19) are translated late and head assembly is slow. Replication of the original phage T7 DNA generates a long, linear concatemer, which is cut and packaged into the T7 procapsid.

Synthesis of class II transcripts stops 15 minutes after infection. The product of one of these genes, gene 3.5, is a bifunctional protein that binds to and inhibits phage T7 RNA polymerase and also acts as an endolysin to cleave the cell wall. Although endolysin activity is detectable in the cytoplasm shortly after infection, it does not cleave the cell wall because it cannot pass through the cell membrane to reach its intended target. This situation changes late in infection when holin is synthesized. The pores formed when holin assembles in the cell membrane allow endolysin to reach the cell wall, triggering cell lysis.

E. coli phage ϕX174 contains a single-stranded circular DNA molecule.

E. coli phage ϕX174 has a circular, single-stranded DNA molecule enclosed in an icosahedral head that contains spikes and has a diameter of about 32 nm (FIGURE 8.18). The viral DNA, which is 5386 nucleotides long, codes for only eleven polypeptides (FIGURE 8.19), each of which has been isolated. The number of amino acids contained in these polypeptides exceeds this very small DNA molecule's predicted coding capacity. This finding is explained by the fact that ϕX174 has efficiently evolved to have overlapping genes that are translated in different reading frames. For instance, gene A shares nucleotides with genes B and K.

Phage ϕX174 binds to a lipopolysaccharide receptor on the *E. coli* outer cell membrane. Once attached to the bacterium, the single-stranded DNA and at least one spike protein are ejected through a channel in the spike into the bacterial cell's periplasmic space (the space between the inner and outer membranes). After ϕX174 DNA penetrates the inner membrane, DNA and protein synthesis take place and new virions are assembled. A great deal is known about phage ϕX174 infection, but we only examine a few aspects of DNA replication in this section.

The mode of ϕX174 DNA replication is especially interesting because the phage has the problem of making an identical copy of a single strand. Clearly, the template cannot yield progeny molecules in a single step. The following terminology is used when discussing the replication of single-stranded DNA molecules. The strand contained in the virus particle has the same sequence as mRNA and it or any strand with the same base sequence is called a (+) strand; a strand having the complementary base sequence is called a (–) strand.

Replication of phage ϕX174 DNA takes place in several steps:

1. *Synthesis of a covalently closed double-stranded DNA molecule.* The parental single-stranded DNA molecule, the viral or (+) strand, is converted to a covalently closed, double-stranded molecule called **replicative form I** (**RFI**). This conversion occurs before transcription begins and hence depends on host enzymes exclusively. The newly synthesized strand, the (–) strand, is the only strand that is transcribed. The enzymatic mechanism for converting the (+) strand to RFI is described in the next chapter.
2. *Synthesis of many copies of RFI.* The (–) strand is transcribed and a multifunctional protein, the gene-A protein, is made. As shown in FIGURE 8.20, the A protein makes a single-strand break in the (+) strand between bases 4305 and 4306 and remains covalently linked to the 5′-P terminus. The nicked molecule is called **replicative form II** (**RFII**). *E. coli* proteins (synthesized prior to infection) then cause the parental (+) strand to be displaced from RFII by looped rolling-circle replication. When one round of replication is completed, the displaced (+) strand is cleaved from the looped rolling circle, recircularized, and used as a template for the synthesis of another (–) strand; the result is a new **RFI**.
3. *Synthesis of (+) strands for encapsidation.* The packaging system captures progeny (+) strands before they can serve as templates for further synthesis of (–) strands. The capture is delayed until nearly completed phage heads have been synthesized and begins when there are about 30 to 40 copies of RFI. At this point, most of the DNA is engaged in looped

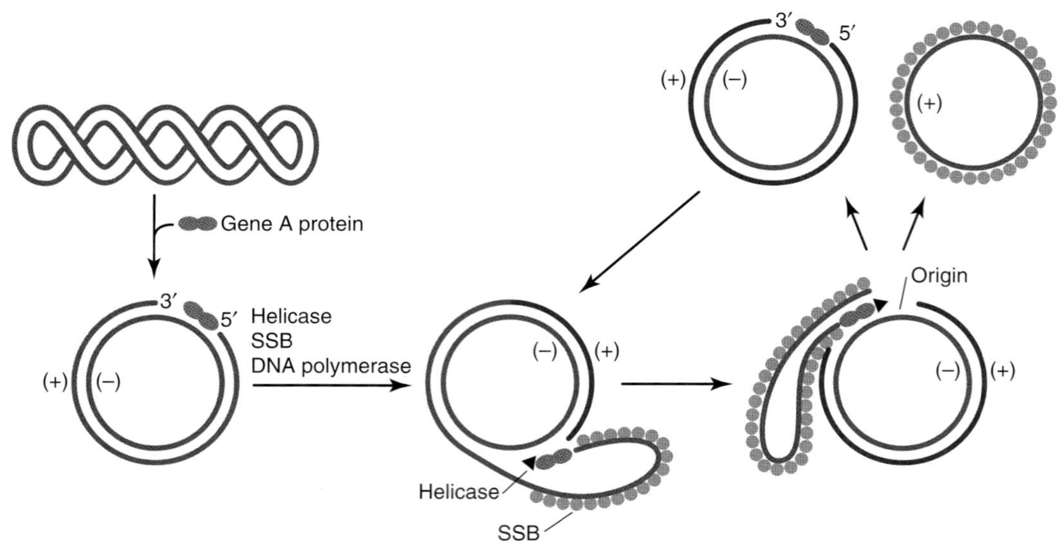

FIGURE 8.20 **Looped rolling-circle replication of ϕX174 DNA.** The parental single-stranded DNA molecule, the viral or (+) strand, is converted to a covalently closed, supercoiled double-stranded molecule. The gene A protein nicks the supercoiled DNA and binds to the 5′-terminus of the (+) strand. Rolling circle replication takes place to generate a daughter strand (red) and a displaced (+) single strand that is coated with single-stranded DNA binding protein (SSB) and still covalently linked to the A protein. When the entire (+) strand is displaced, it is cleaved by the joining activity of the A protein. The cycle is ready to begin anew. Note that the (–) strand is never cleaved.

rolling-circle replication, producing new RFI and some (+) strands for packaging. Host DNA synthesis finally stops at this time, partly as a result of an unknown function of the A protein and partly because all of the host DNA polymerase is engaged in ϕX174 DNA synthesis. As more heads are made, all the displaced (+) strands are packaged and synthesis of RFI stops for lack of a template. Packaging is described below.

Phage ϕX174 transcription starts at three sites, all of which are activated simultaneously. Except for a delay in the appearance of lysozyme activity, no temporal regulation is required. All transcription is from the (–) strand, so transcription cannot take place until the first (–) strand is made. From then on, RFI synthesis continues unabated until a sufficient number of heads are formed to start packaging. Because the ϕX174 DNA molecule is very small, many rounds of replication are completed before head proteins are synthesized and assembled. (The first completed particle appears about 10 minutes after infection.) Lysozyme activity is delayed because translation is slow and the enzyme is not particularly active. Lysis does not occur until 30 minutes after infection, by which time about 500 phage particles have been synthesized.

Packaging of ϕX174 DNA requires seven phage proteins, four of which are present in the finished particle. The steps of assembly, shown in FIGURE 8.21, are as follows:

1. The coat proteins F and G form pentamers that, in the presence of the B protein, form an aggregate called a 12S particle.
2. The H protein adds to the 12S particle.

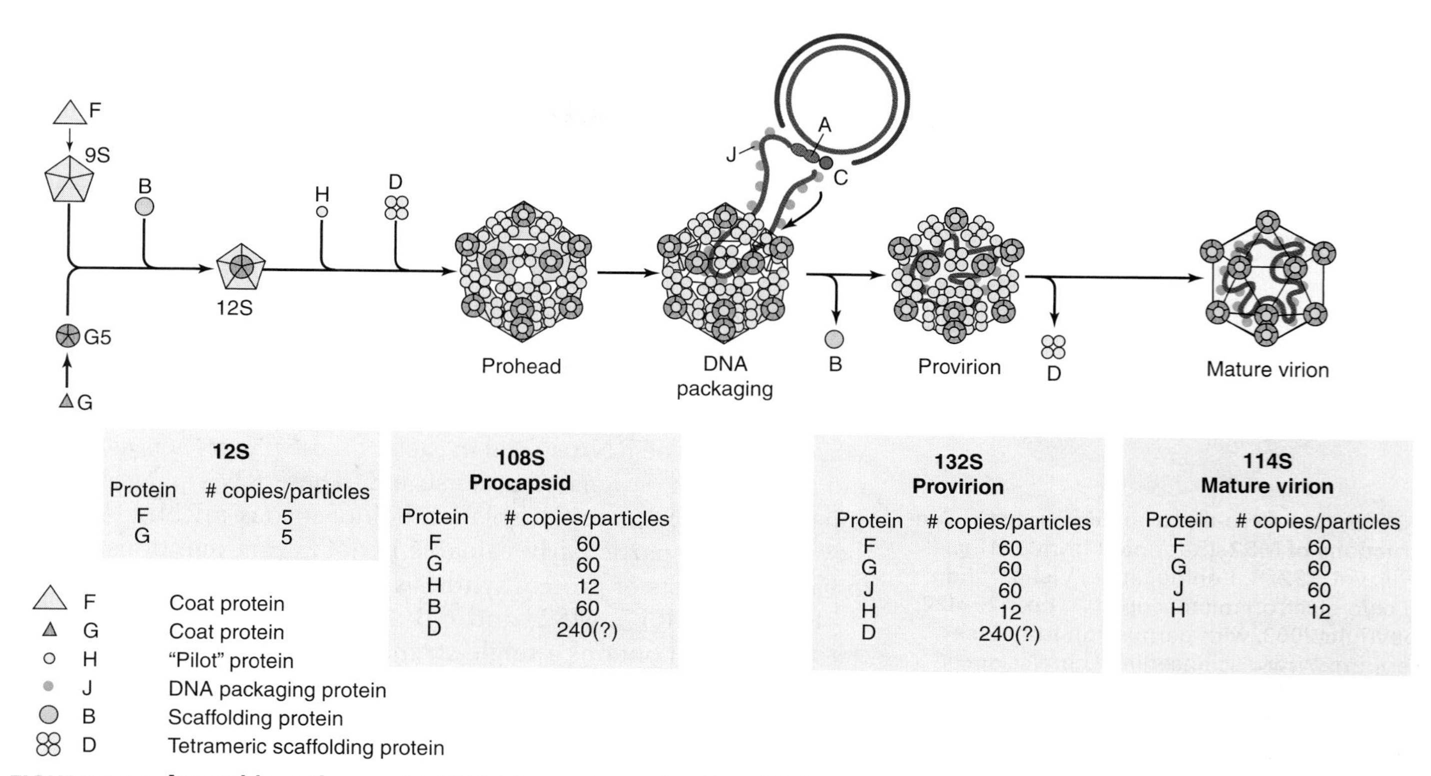

12S

Protein	# copies/particles
F	5
G	5

108S Procapsid

Protein	# copies/particles
F	60
G	60
H	12
B	60
D	240(?)

132S Provirion

Protein	# copies/particles
F	60
G	60
J	60
H	12
D	240(?)

114S Mature virion

Protein	# copies/particles
F	60
G	60
J	60
H	12

FIGURE 8.21 **Assembly pathway of ϕX174.** The coat proteins F and G form pentamers that, in the presence of the B scaffolding protein, form an aggregate called a 12S particle. H and D proteins combine with the 12S particle to form the procapsid. The C protein somehow directs the 5′ end of the displaced (+) strand into the procapsid. The J protein organizes the DNA inside the procapsid and consolidates the structure. The A protein cuts the displaced (+) strand from the looped rolling circle and circularizes the displaced strand. D protein is removed, leading to the formation of the completed phage particle. (Adapted from L. L. Ilag, et al., *Structure* 3 [1995]: 353–363.)

3. The gene D protein forms a frame on which a procapsid is built from the 12S particles. The procapsid, thus, contains the proteins B, D, F, G, and H.
4. The C protein shown schematically in the figure as binding to the 5′ end of the displaced (+) strand somehow directs the 5′ end into the procapsid. The J protein, which probably initially binds to the DNA outside the procapsid, organizes the DNA inside the procapsid and consolidates the structure.
5. B protein is released from the procapsid.
6. The A protein cuts the displaced (+) strand from the looped rolling circle and circularizes the displaced strand.
7. The D protein is then removed (probably spontaneously following completion of and closing of the capsid) leading to formation of the finished phage particle.

The final stage of infection, cell lysis, requires a 91-residue membrane protein that is encoded by the phage gene E. The E protein does not appear to be able to degrade the cell wall and therefore must be acting in some other way. An important clue to its still unknown mode of action comes from the observation that the host cell must be growing for lysis to take place. This growth requirement suggests

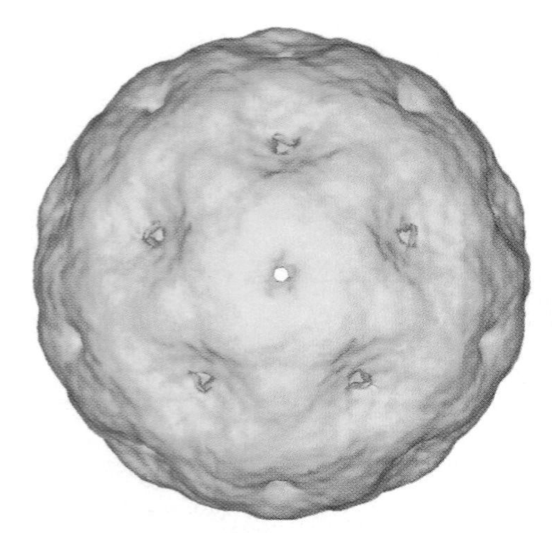

FIGURE 8.22 **Cryo-electron microscopy reconstructions of MS2.** (Reprinted from *J. Mol. Biol.*, vol. 332, R. Koning, et al., Visualization by cryo-electron microscopy . . ., pp. 415–422, copyright 2003, with permission from Elsevier [http://www.sciencedirect.com/science/journal/00222836]. Photo courtesy of Roman Koning, Leiden University Medical Center, The Netherlands.)

that the E protein causes cell wall degradation by somehow modifying a bacterial enzyme that is normally involved in cell wall synthesis so that it can no longer make an intact cell wall.

Some phages have single-stranded RNA as their genetic material.

In 1960, Norton Zinder and Timothy Loeb set out to discover whether phages exist that infect only F^+ and Hfr but not F^- *E. coli*. They found several such phages in the sewers of New York City. One of these, the f2 phage, was later shown to have single-stranded RNA rather than DNA as its genetic material. Subsequent studies led to the discovery of additional examples of single-stranded RNA phages such as R17, MS2, and Qβ. The RNA strand in each of these phages acts as mRNA and therefore is designated a (+) strand. These RNA phages are an important source of easily isolated, homogeneous mRNA. As such, they have been particularly valuable in answering questions related to various aspects of protein synthesis.

Phages f2, R17, MS2, and Qβ are tailless icosahedrons (FIGURE 8.22). Each contains a single-stranded linear RNA molecule with about 3600 to 4300 nucleotides and a great deal of intramolecular hydrogen-bonding. As shown in FIGURE 8.23, MS2 RNA contains four genes encoding a maturation protein (A), a coat protein (C), an RNA-dependent RNA polymerase (P), and lysin (L). The MS2 virion contains 180 identical coat protein subunits and a single copy of the A protein. The phage RNA serves as both a replication template and an mRNA. A typical burst size is 5000 to 10,000, which is very large compared to the burst of one hundred to a few hundred for DNA phages. The particles form huge crystalline arrays within each bacterium. Lysin disrupts the cell wall and lysis of the bacterial population is gradual, taking place from 30 to 60 minutes after infection.

The individual stages in the life cycle of an RNA phage are as follows:

1. *Attachment to the F pilus*. The RNA phages bind to the bacterial pili, the long filamentous extensions that are made of multiple copies of the protein pilin and extend from the surface of F^+ and Hfr cells. Large numbers of RNA phages attach along the sides of the pili. The A protein appears to facilitate binding to the pilus and to assist in the movement of the RNA into the bacterial cell at the base of the pilus. The details of this process remain to be determined.
2. *Phage RNA translation*. Immediately after phage RNA enters the cell, a ribosome attaches to the beginning of the C gene. Ribosome binding sites for the A and P genes are blocked by the presence of stem-and-loop structures in the RNA. However, translation of the C gene opens up the binding site for the P gene. Thus, both proteins are made initially but increasing amounts of the C protein efficiently block P gene translation. Approximately 2×10^6 coat protein copies are needed as structural components for 10,000 phages, whereas RNA-dependent

FIGURE 8.23 **Genetic map of the RNA phage MS2.** A is the maturation gene, C is the coat protein, L is lysin, and P is RNA-dependent RNA polymerase. The L gene overlaps the 3′-end of C and the 5′-end of P.

RNA polymerase is needed only in catalytic amounts. The synthesis of the A protein will be described shortly.

3. *Phage RNA replication* (FIGURE 8.24). Qβ RNA-dependent RNA polymerase, which is the best studied of the RNA phage polymerases, is a heterotetramer consisting of the phage P gene product and three host proteins. The three bacterial protein subunits are the translation elongation factors Tu and Ts (EF-Tu and EF-Ts), which place aminoacyl-tRNA molecules on the ribosome during protein synthesis, and the ribosomal protein S1 (see Chapter 23). All four subunits are required to copy the (+) strand but the S1 factor is not required to copy the (–) strand. An additional factor, which is encoded by the bacterial *hfq* gene and called the host factor (HF), is also required to copy the (+) strand. The RNA-dependent RNA polymerase copies the viral (+) strand to generate a (–) strand. While synthesis is proceeding, the (–) strand is in contact with the

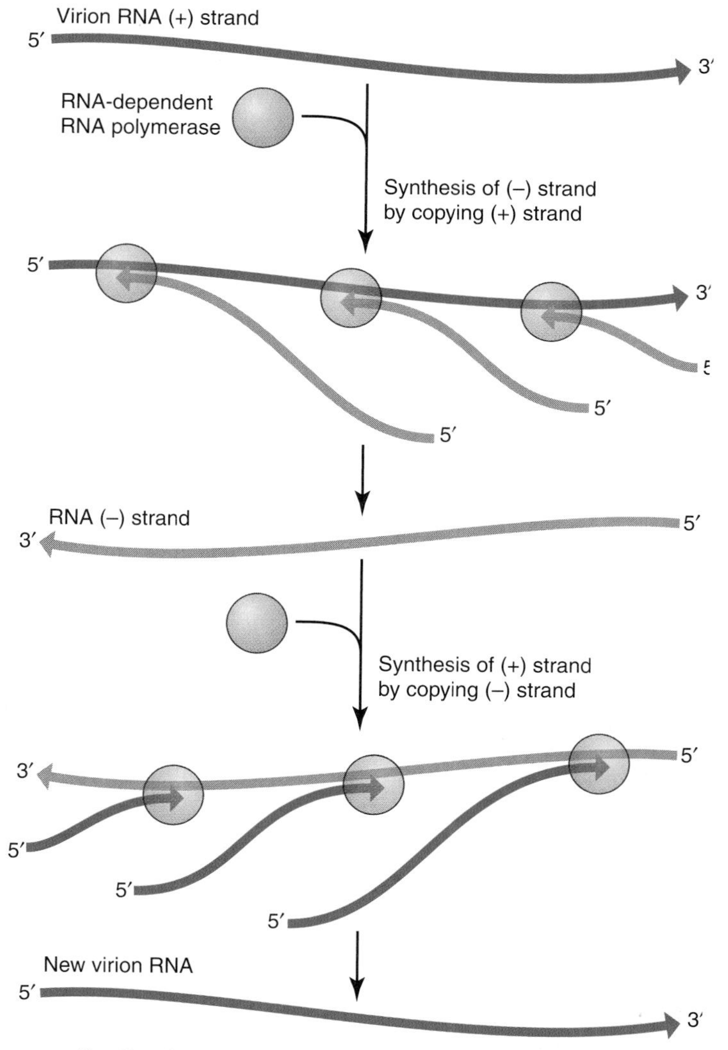

FIGURE 8.24 **Replication of phage RNA by RNA-dependent RNA polymerase.**

(+) strand only at the polymerization site. For the most part, the replicative form is, therefore, single-stranded. Initiation of several (–) strands occurs before the first (–) strand is complete, and the replicative form is branched. The (–) strands are released and immediately used by the RNA-dependent RNA polymerase to form (+) strands. Some of the (+) strands return to the ribosomes for synthesis of more coat proteins and others are packaged. All progeny contain (+) strands exclusively.

4. *Synthesis of the A protein.* The ribosomal binding site for the A protein is never available on a free (+)-strand because of intrastrand base pairing (not shown in Figure 8.24). However, just after (+) strand synthesis begins at the 3′-terminus adjacent to the A gene, there is a brief period during which the ribosomal binding site for the A protein is free because the complementary segment of the (+) strand has not yet been replicated. The A gene is translated during this time. Only one or two passages are possible before the binding site becomes closed. The number of A proteins, therefore, is maintained roughly equal to the number of (+) strands. This arrangement is economical because each virus particle contains one A protein molecule. Each A protein later becomes bound to one RNA molecule. The A protein remains bound to the RNA and enters the cell in a subsequent infection.
5. *Particle assembly.* Coat protein molecules spontaneously aggregate around the newly synthesized (+) strand and form an icosahedral shell.
6. *Cell lysis.* Lysis occurs after about 10,000 phage particles have formed. Some RNA phages code for a lysin that degrades the cell wall, others code for a protein that blocks cell wall synthesis.

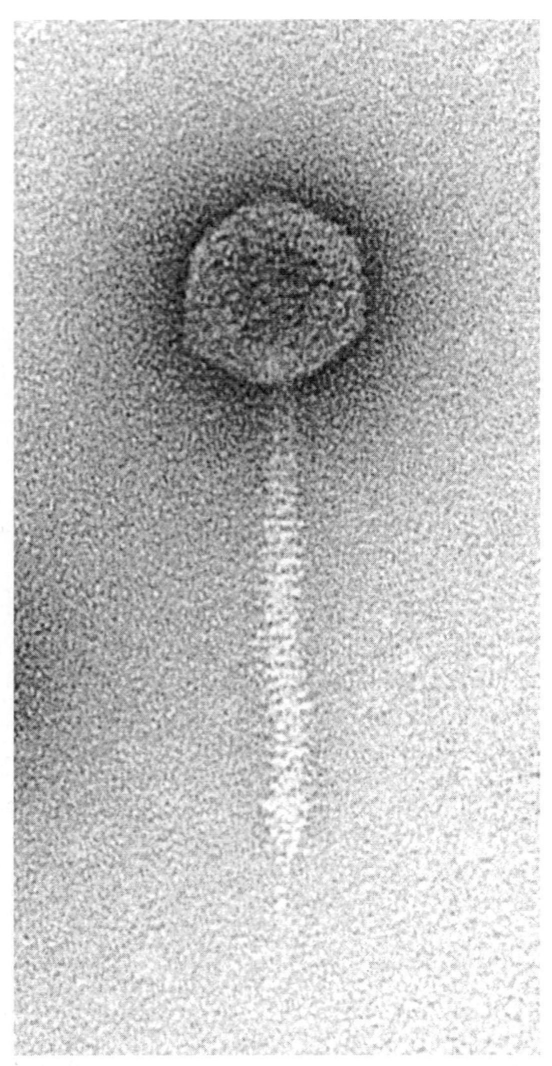

FIGURE 8.25 **Electron micrograph of phage λ.** (Photo courtesy of Robert Duda, University of Pittsburgh.)

8.4 Temperate Phages

E. coli phage λ DNA can replicate through a lytic or lysogenic life cycle.

E. coli phage λ has an icosahedral head and a long, noncontractile, flexible tail that ends in a tip structure (FIGURE 8.25). Four nonessential tail fibers extend from the junction of the tail and the tip structure. The virion's DNA, which is located in its head, is a 48,502 bp linear duplex with a 12 nucleotide 5′-overhang at each end. Phage λ genes are clustered according to function (FIGURE 8.26). For example, the head, tail, replication, and recombination genes form four distinct clusters. Although the present discussion is limited to phage λ, this phage is just one member of a very large family of similar phages that infect a wide variety of bacteria and are collectively known as the **lambdoid phages.**

The tip structure at the end of the phage λ tail plays a critical role in bacterial recognition. It binds to a bacterial protein called LamB on the outer surface of the cell envelope. *E. coli* normally uses its LamB

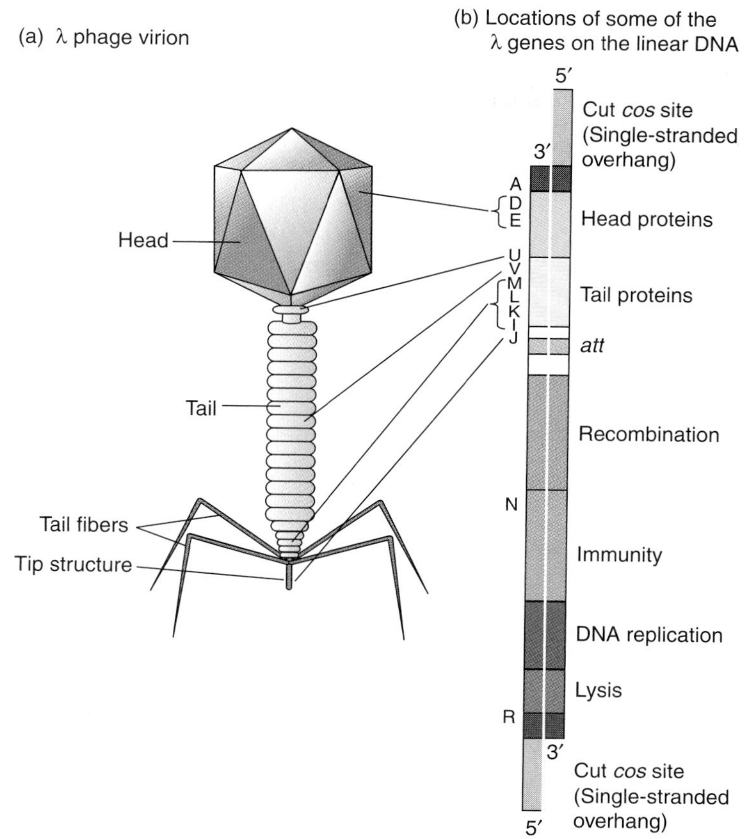

FIGURE 8.26 **Bacteriophage λ and the location of some genes for structural proteins.** (a) Schematic diagram of bacteriophage λ. (b) Locations for some genes that code for proteins in the virion.

protein to form nonspecific channels that allow small hydrophilic molecules such as maltose and maltodextrins to diffuse through the outer membrane. The phage, thus, appropriates for its own use a bacterial protein that has an entirely different function in the uninfected cell. The entire λ DNA enters the bacterium in less than five minutes at 37°C, but the mechanism of entry remains to be determined.

After λ DNA enters the cell, it takes advantage of an unusual structural feature to form a circular structure. The base sequences of the 5′-overhangs, which are known as **cohesive ends** or ***cos*** **elements,** are complementary to one another (FIGURE 8.27a). By forming base pairs between the single-stranded ends, the linear DNA molecule can circularize, yielding a circle containing two single-strand breaks (FIGURE 8.27b). Circularization is easily performed in the laboratory and also occurs in an infected cell within a few minutes following infection. No bases are missing in the newly formed double-stranded region of this circle, so DNA ligase can convert the molecule to a closed covalent circle (FIGURE 8.27c). Then DNA gyrase converts the relaxed closed covalent circular DNA into supercoiled DNA.

The supercoiled λ DNA can either enter the lytic or lysogenic cycle (FIGURE 8.28). The lytic cycle, which begins with λ DNA replication, produces about 150 new phage λ particles. The lysogenic cycle involves

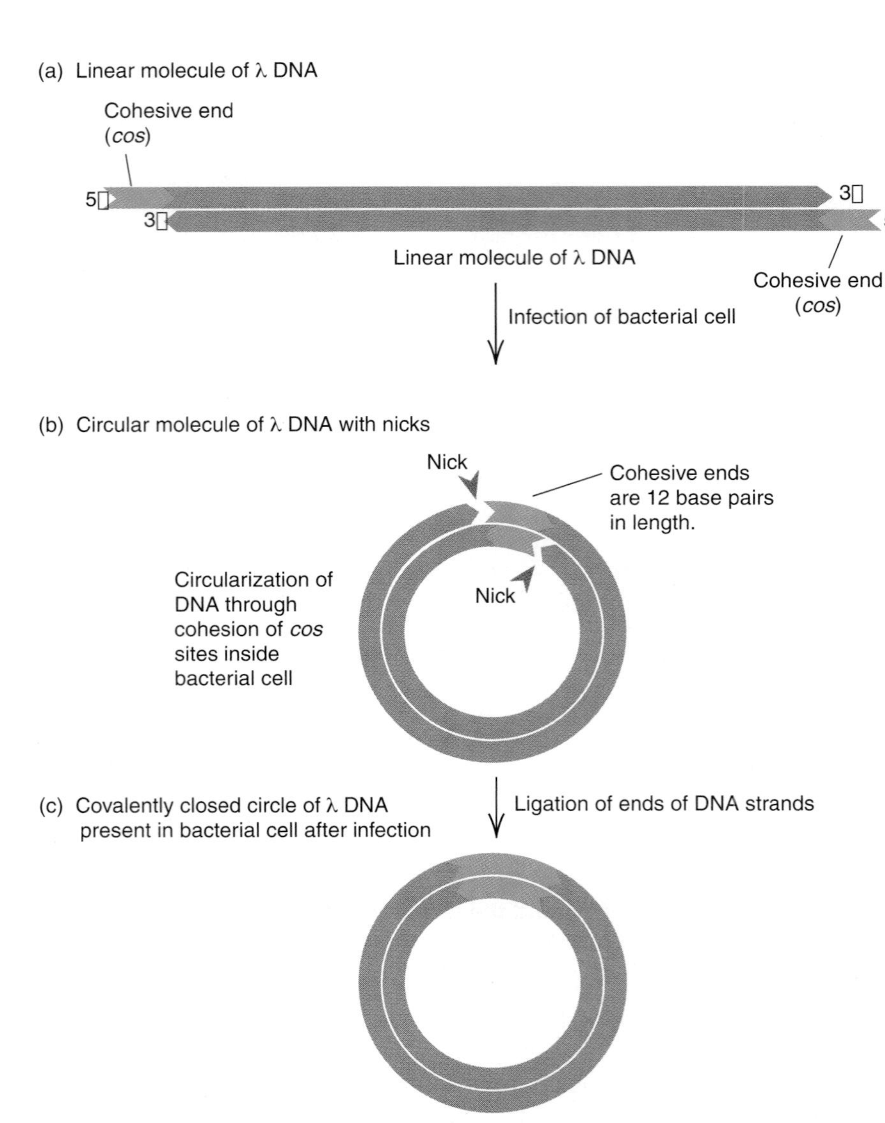

FIGURE 8.27 **A diagram of the λ DNA molecule showing the cohesive ends (complementary single-stranded ends).** (a) The DNA circularizes by means of base pairing between the cohesive ends. (b) and (c) The nicked circle that forms is converted into a covalently closed circle by sealing (ligation) of the single-stranded breaks. The length of the cohesive ends is 12 bp in a total molecule of 48,502 bp.

λ DNA insertion into a specific site in the bacterial chromosome. Commitment to the lytic or lysogenic cycle is determined by genetic switches that act through the transcription machinery. We examine these switches in Chapter 16 after we have had a chance to learn about the transcription machinery and its regulation.

The Lytic Cycle

Twenty-eight of the λ genes are essential for lytic reproduction. Most of these genes code for structural proteins or proteins needed to assemble the phage. Two types of phage λ DNA replication, θ **replication** and

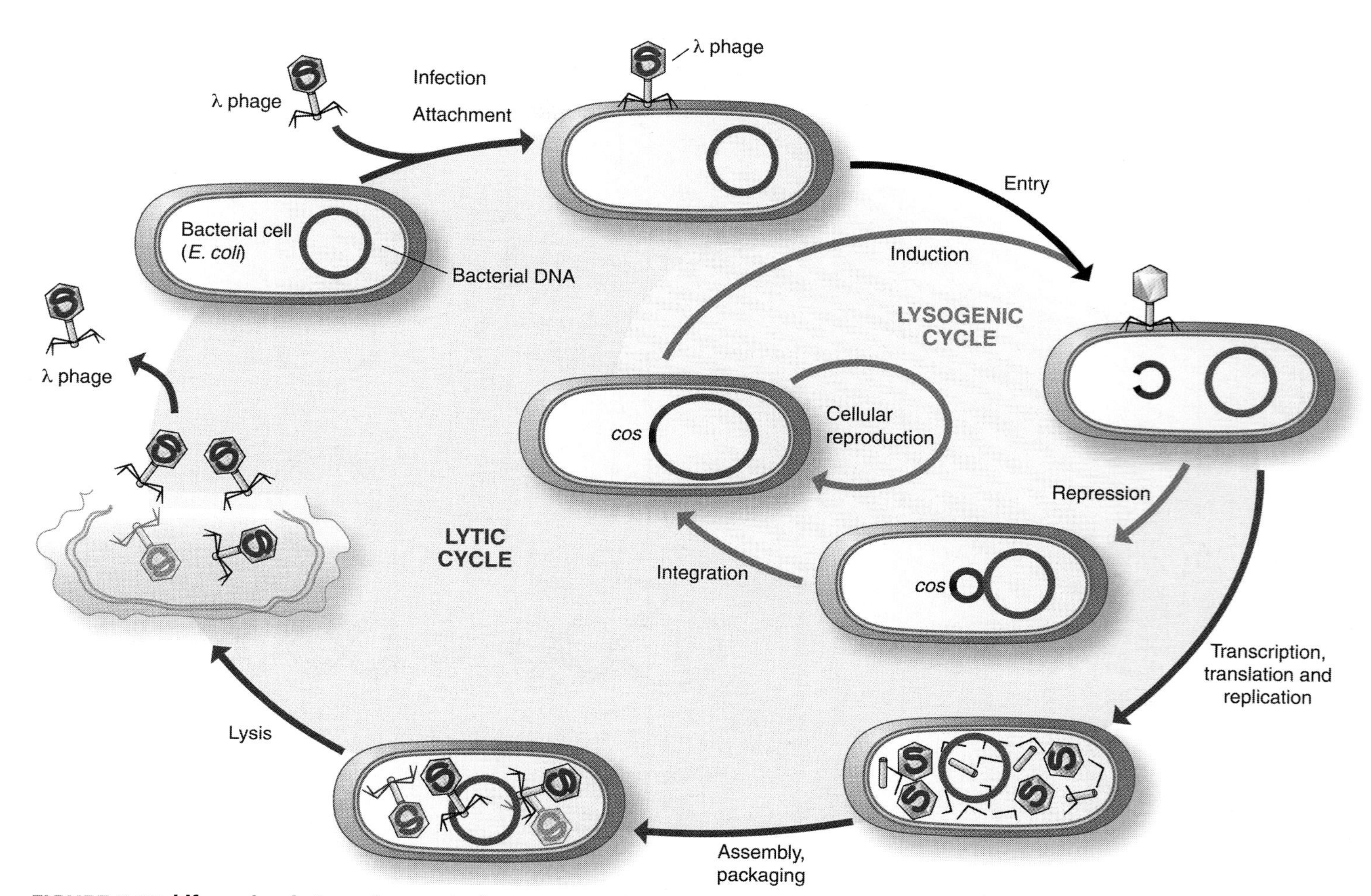

FIGURE 8.28 **Life cycle of phage λ, a typical temperate phage.** The tip of the phage tail attaches to a receptor on the bacterial cell surface and the phage DNA enters the cell. The empty phage protein shell remains outside the bacterial cell. After λ DNA enters the cell it takes advantage of its complementary 5′-overhangs, which are known as cohesive ends or *cos* elements, to form a circle. In the lytic cycle, the DNA is transcribed, translated, and replicated. As the lytic cycle continues, DNA synthesis switches from theta (θ) replication to rolling circle replication. Rolling circle replication generates long concatemers that are required to package DNA into the phage head. After DNA packaging, phage tails add to the protein shell and new phages are released after cell lysis. In the lysogenic cycle, phage development is repressed and the circular phage DNA integrates into the bacterial chromosome. The resulting lysogenic bacteria can replicate indefinitely. However, they also can be induced to return to the lytic cycle by ultraviolet light (or chemical mutagens) with the excision of phage DNA from the chromosome. (Adapted from A. Campbell, *Nat. Rev. Genet.* 4 [2003]: 471–477.)

rolling circle replication, take place during the lytic cycle (FIGURE 8.29). The term θ replication was coined because the replication intermediate resembles the Greek letter θ (theta). We examine the enzymatic aspects of θ replication in Chapter 9. For now, it is important to note that θ replication of phage λ DNA begins at a specific site on the circular DNA and produces two replication forks, which move in opposite directions around the λ DNA circle. θ replication's primary function is to increase the number of templates available for transcription and to provide circular DNA molecules for the next stage of replication, rolling circle replication. As the lytic cycle continues, DNA synthesis

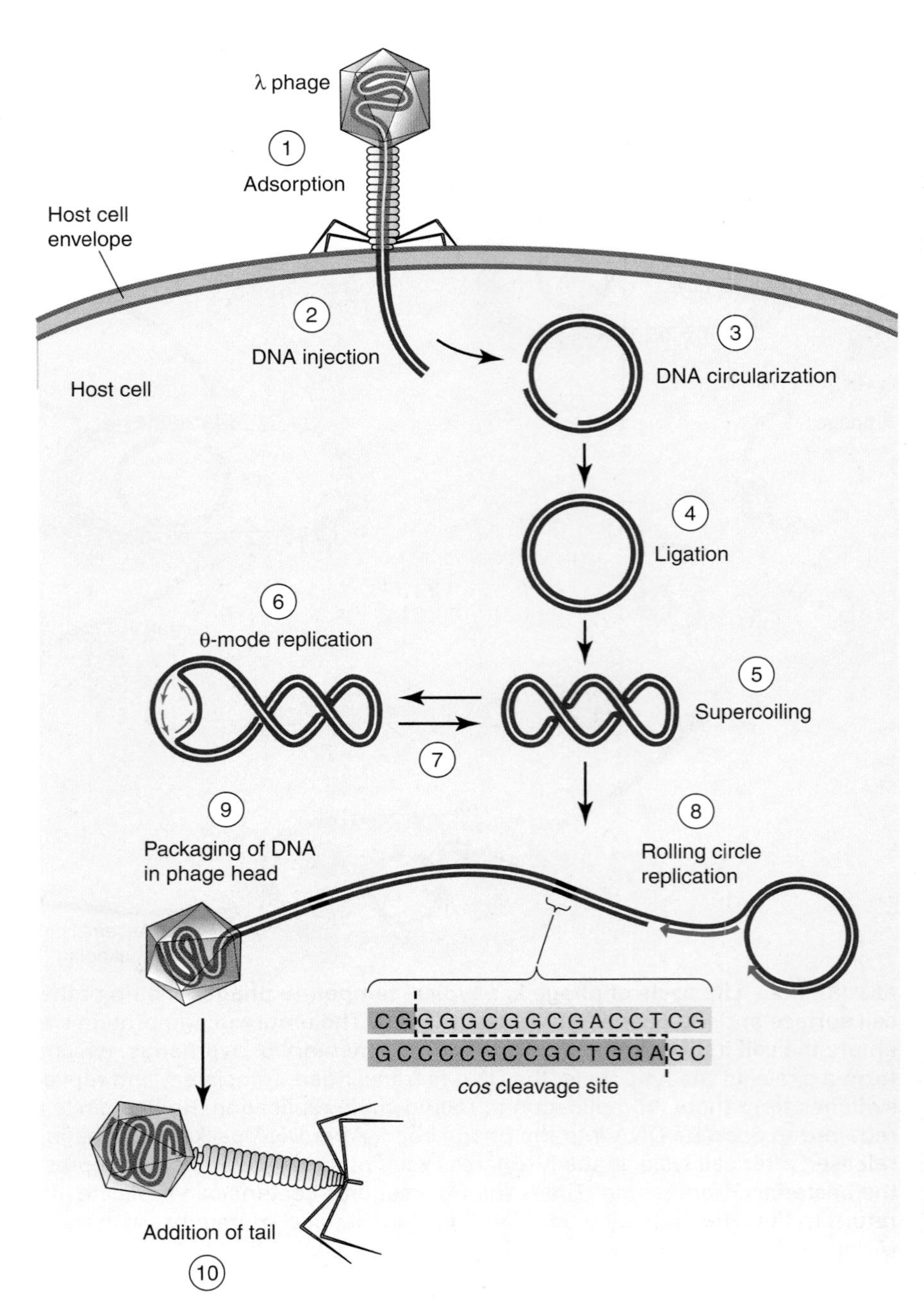

FIGURE 8.29 **DNA replication in the lytic mode of phage λ.** (1) The virion attaches to the host cell. (2) The linear DNA enters the cell. (3) The complementary *cos* elements pair to form a nicked circle. (4) DNA ligase converts the nicked circle to a relaxed covalent circle. (5) Topoisomerases convert the relaxed circular DNA to a supercoil. (6 and 7) Bidirectional or θ (theta) replication produces new DNA. (8) θ replication continues and rolling circle replication starts. In the later stages of infection, replication takes place exclusively through the rolling circle mode. The curved blue arrows indicate the most recently synthesized DNA at the replication forks and the arrowheads represent the 3′-ends of the growing DNA chains. (9) The concatemeric DNA produced by the rolling circle mode is specifically cleaved at its *cos* site and is packaged into phage heads. (10) Tails are added to complete the assembly of the mature phage particles, which are each capable of initiating a new round of infection. (Adapted from D. Voet and J. G. Voet. *Biochemistry, Third Edition*. John Wiley & Sons, Ltd., 2005.)

switches from the θ replication to rolling circle replication, which generates long concatemers that are required to package DNA in the phage head.

Because the ends of the DNA molecule in the phage λ particle always have single-stranded 5′-ends, long concatemers must be cut at their *cos* sites to generate these termini. This cutting is accomplished by a sequence-specific terminase, which is associated with newly formed empty heads. Note that because a rolling circle has two classes of *cos* sites—the one in the circle and those in the linear branch—some mechanism must exist for preventing cleavage of the one in the circle. If it were cleaved, replication would cease. This difficulty is avoided by a site requirement in the λ terminase system. Terminase-cutting requires two *cos* sites or one *cos* site and a free cohesive end on a single DNA molecule. Efficient cleavage of a single *cos* site does not occur.

Cutting at the *cos* sites and packaging of λ DNA are somehow coupled. In fact, the terminase system is virtually inactive unless the terminase proteins are components of an empty λ head. When a bacterium, therefore, is infected with a λ mutant that cannot make an intact phage head (for example, a λ mutant that fails to make the major head protein), the linear branch of a rolling circle of DNA is not cleaved.

Phage assembly also takes place *in vitro*, allowing us to examine individual steps in the process. Of even more importance from our point of view, the *in vitro* packaging system has helped to make phage λ a convenient cloning vector for recombinant DNA experiments. *In vitro* assembly requires an extract that contains empty heads and free tails as well as the various protein factors required for assembling the intact virion. One way to obtain a suitable extract is to expose a λ lysogen with a defective terminase gene to ultraviolet light, inducing the prophage to enter the lytic cycle. The phage heads will remain empty because λ DNA cannot be packaged without terminase and the tails will remain free because they cannot connect to empty heads. When the induced cells are disrupted, the resulting extracts, therefore, contain all of the components required to assemble an intact virion except terminase. With the addition of terminase, the extract gains the ability to package the λ DNA into the empty heads and carry out all of the subsequent steps required to produce biologically active virion particles. Bacterial DNA that is present in the extract is not packaged because it lacks *cos* sites. One can also prevent phage λ DNA from being packaged by using a lysogen with a prophage λ that lacks *cos* sites.

The Lysogenic Cycle

Infection with λ phage can have a second outcome. Instead of replicating by the lytic mode, the λ DNA can insert at a specific site of the bacterial chromosome, becoming a prophage. When λ DNA integrates, it is inserted at a preferred position in the *E. coli* chromosome. This site, between the *gal* and *bio* (biotin) genes, is called the λ attachment site and designated *att*. The resulting λ lysogen is generally quite stable and can replicate nearly indefinitely without release of phage. If a lysogenic bacterium were to become damaged, however, it would be to the advantage of the prophage to initiate the lytic cycle. This initiation does occur and the signal to enter the lytic mode is DNA damage caused by ultraviolet light, x-rays, or a chemical agent that disrupts or modifies the DNA (see Chapter 12). Phage λ DNA excision is usually a very precise process but on rare occasions (one cell in 10^6 or 10^7 cells) an excision error is made and a small piece of bacterial DNA is excised along with the λ DNA. The recombinant λ DNA is replicated and packaged into a λ particle, which can deliver the bacterial DNA to a new bacterial host in a process known as **specialized transduction.** Only genes adjacent to the prophage are transferred from the lysogen to another bacterial cell. We examine the mechanism of λ DNA integration into the bacterial chromosome and the mechanism of λ DNA excision from the bacterial chromosome as well as some important consequences of these two processes in Chapter 14.

Cloning Vector

Phage λ has been genetically modified to make it a versatile cloning vector (FIGURE 8.30). The key modification is the removal of restriction endonuclease sites throughout the phage DNA and the introduction of unique restriction sites on either side of nonessential genes located in the central one third of the genome between the *J* and *N* genes. Isolated λ DNA is cut with restriction endonuclease and left and right arms purified by sucrose density gradient centrifugation or gel

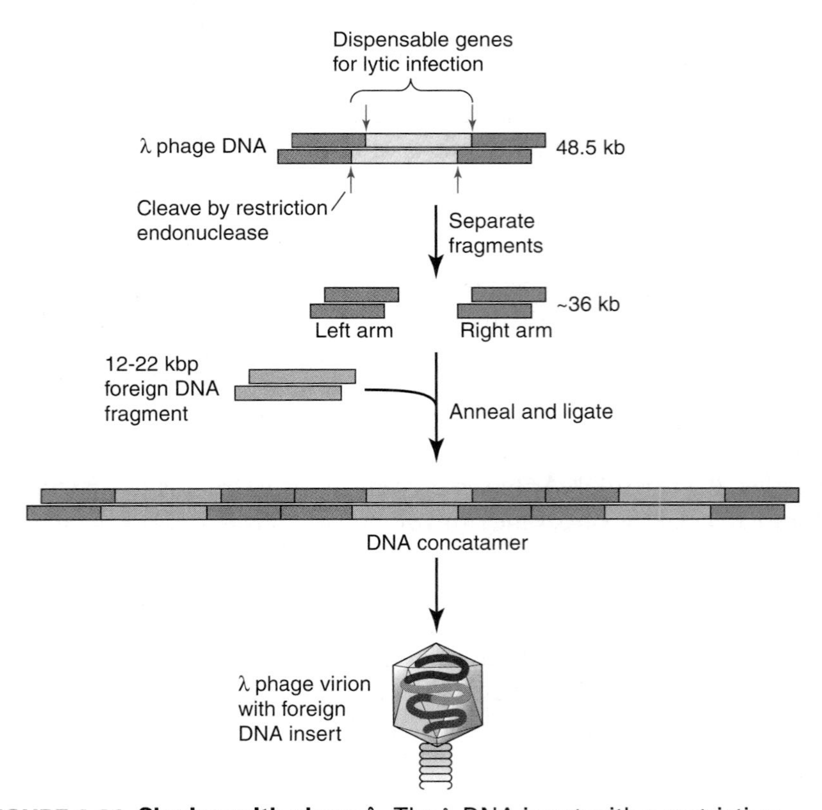

FIGURE 8.30 **Cloning with phage λ.** The λ DNA is cut with a restriction endonuclease at sites on either side of the nonessential region. The left and right arms are purified and the foreign DNA segment is inserted between them. The resulting concatemer is placed in a packaging extract to form intact virions with foreign DNA inserts.

electrophoresis. Then the foreign DNA fragment is inserted between the two arms and joined to the arms by DNA ligase. Inserts usually range from 12- to 22-kbp long so that newly created λ-like DNA can be packaged by terminase. The ligase also joins the *cos* site on the left arm of one recombinant λ DNA molecule to the *cos* site on the right arm of another recombinant λ DNA molecule to produce a long, linear DNA chain. If the insert is too long or too short, then terminase will not be able to package the DNA. Once the DNA has been constructed, it is added to a packaging extract and used to prepare virion particles that can infect *E. coli* to produce millions of copies of the virion bearing recombinant DNA.

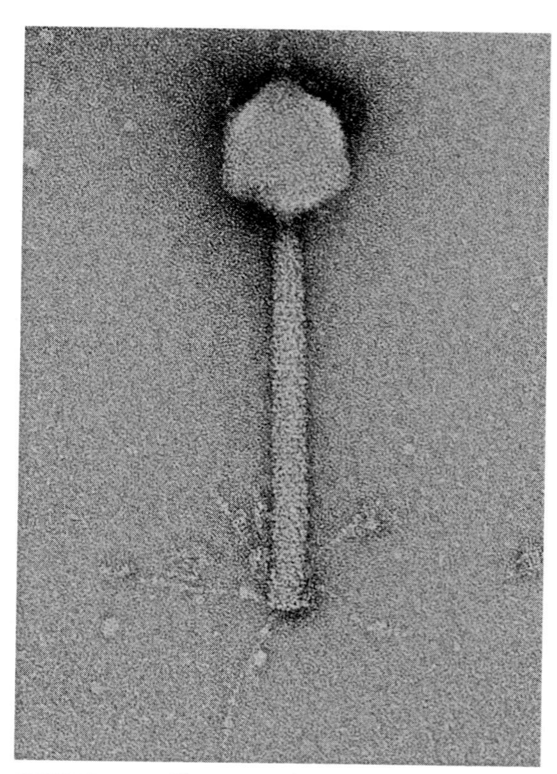

FIGURE 8.31 Electron micrograph of phage P1. (Photo courtesy of Michael Wurtz and the Biocenter at the University of Basel.)

E. coli phage P1 can act as generalized transducing particles.

The phage P1 particle has a tail that consists of a tube surrounded by a contractile sheath and is connected to an icosahedral head at one end and a baseplate with six kinked tail fibers at the other (FIGURE 8.31). The head contains a linear duplex DNA molecule, which is 93,601 bp, has a terminal redundancy of 10 to 15 kbp, and is circularly permuted. The virion binds to a lipopolysaccharide on the outer surface of the bacterial envelope. Phage P1 has a fairly broad host range, infecting *E. coli* as well as several other gram-negative bacterial species.

Based on the presence of the contractile sheath, P1 DNA is thought to be injected into the bacterial cell. Once inside the cell, the linear duplex is converted into a closed covalent circle by a host recombination system that acts on the terminally redundant ends. Bacterial DNA gyrase acts on this relaxed circular DNA, converting it into a supercoil that can enter the lytic or lysogenic cycle. Commitment to the lytic or lysogenic cycle is determined by genetic switches that act through the transcription machinery.

The Lytic Cycle

In the lytic cycle, DNA synthesis starts with θ replication and then switches to rolling circle replication. At approximately 45 minutes after infection, the cells are filled with DNA concatemers, empty phage heads, and phage tails. DNA packaging starts when the phage P1 terminase cleaves a specific 162 bp, termed the *pac* sequence, in the DNA concatemer. Then cut DNA is packaged unidirectionally from the cleaved *pac* end into an empty head until that head is full. When packaging is complete, the DNA is cut at a nonsequence dependent site. DNA end remaining outside the head is used in the next P1 sequential packaging event. Each headful of DNA is approximately 110 to 115 kbp or about 10 to 15% more DNA than is present in the viral genome. Because the virus packages this headful from a concatemer, the DNA present in each virus particle is terminally redundant and circularly permuted.

The Lysogenic Cycle

The circular P1 DNA produced after infection can also enter a lysogenic cycle in which it behaves like a plasmid and is maintained at one or two copies per cell. The phage DNA replicates once per bacterial life cycle so that when the bacterium divides, each daughter cell receives a P1 DNA molecule.

Generalized Transduction

Phage P1 has been a very important tool in molecular biology because it allows us to transfer genes from one bacterium to another (FIGURE 8.32). This process in which a phage particle carries random DNA fragments from one bacterium to another, called **generalized transduction**, was discovered by Joshua Lederberg and Norton D. Zinder in 1952. The phage particles that carry bacterial genes are designated **generalized transducing particles.**

Phage P1 has two characteristics that make it well suited to function as a transducing particle. First, it codes for a nuclease that slowly degrades bacterial DNA, so that when packaging begins, the host DNA is present as large fragments. Second, the P1 packaging system is not very fastidious, so the size of the DNA molecule does not have to be exactly the same in all particles. The result is that in the population of particles produced when infected bacteria lyse, rare particles contain a bacterial DNA fragment. Because fragmentation of the host DNA is a random process, these rare particles contain fragments derived from all regions of the host DNA. A sufficiently large population of phage P1 progeny, therefore, will contain at least one particle possessing each host gene. On the average, for any particular gene, roughly one virion particle out of 10^6 viable phages contains a bacterial fragment. These rare particles do not produce P1 phage progeny when they infect bacteria because they contain no P1 DNA. Instead, the bacterial DNA is injected into the host cell.

Let us now examine the events that ensue when one of these rare particles carrying a bacterial DNA fragment, attaches to a bacterial cell. Consider a particle that has emerged from an infected wild type *E. coli* and that contains the gene for leucine (*leu*) synthesis. Such a particle is denoted P1 *leu*$^+$. Consider further that the P1 *leu*$^+$ particle attaches to a bacterium with a *leu*$^-$ genotype and injects its DNA. The bacterium will survive because no phage genes are injected and its nucleases indeed may degrade the injected linear DNA fragment. Another possibility is that the *leu*$^+$ segment will be incorporated into the host DNA by genetic recombination, resulting in replacement of the *leu*$^-$ allele by a *leu*$^+$ allele to produce a *leu*$^+$ **transductant**. In this way, the genotype of the recipient cell would be converted from *leu*$^-$ to *leu*$^+$. Transducing phage particles are detected by their ability to transfer genes.

8.5 Chronic Phages

After infection, a chronic phage programs the host cell for continued virion particle release without killing the cell.

The three best studied chronic coliphages, M13, fd, and f1, are very similar filamentous phages with 98% identical DNA sequences. Each is a male-specific phage with a single-stranded closed covalent circular

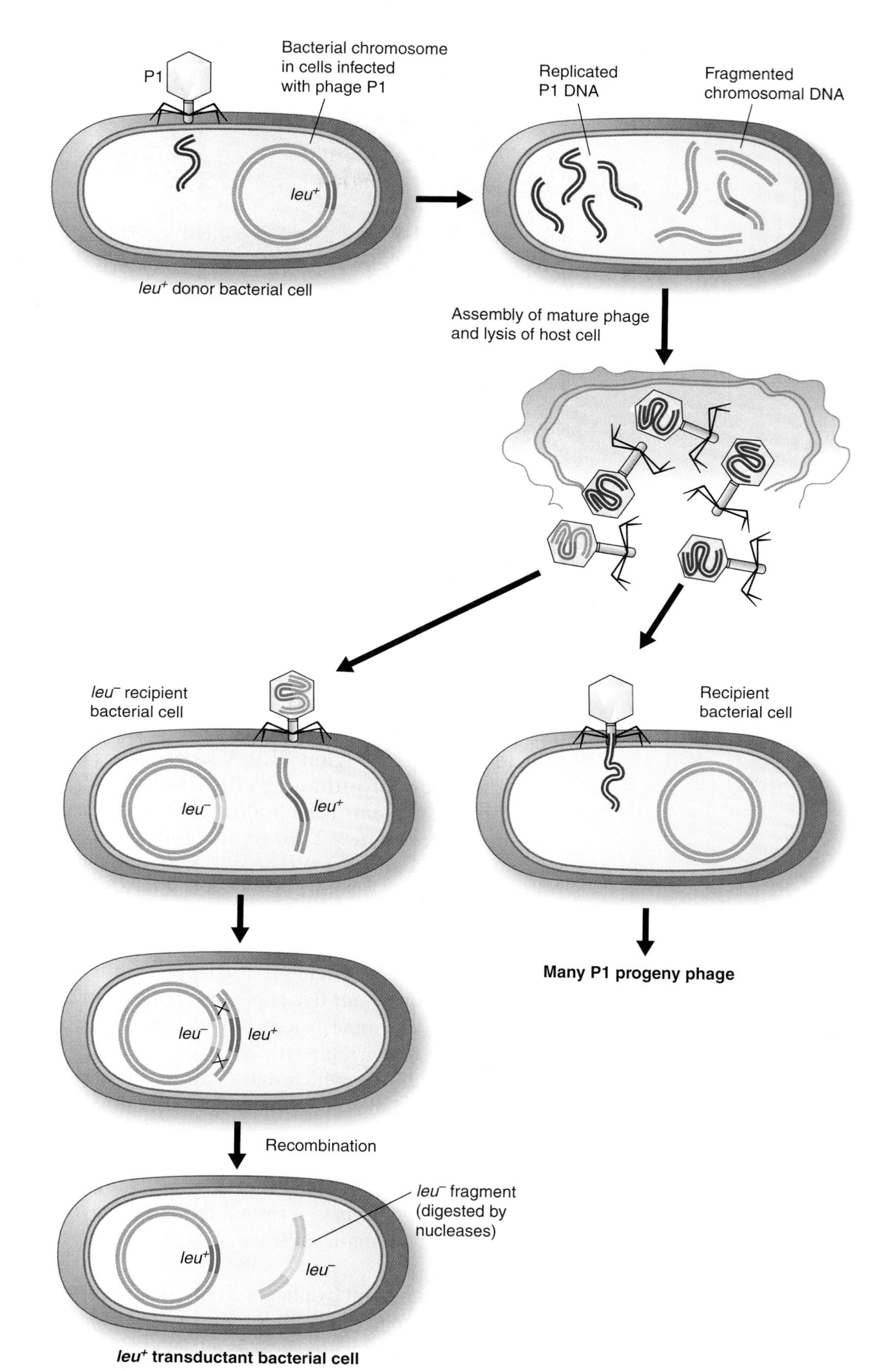

FIGURE 8.32 **Generalized transduction.** Phage P1 infects a *leu*+ donor, yielding predominantly normal P1 progeny but an occasional particle carries *leu*+ (or any other small segment of bacterial DNA) instead of phage DNA. When the phage population infects a *leu*– bacterial culture, the transducing particle can yield a *leu*+ transductant. Note that the recombination step requires two exchanges.

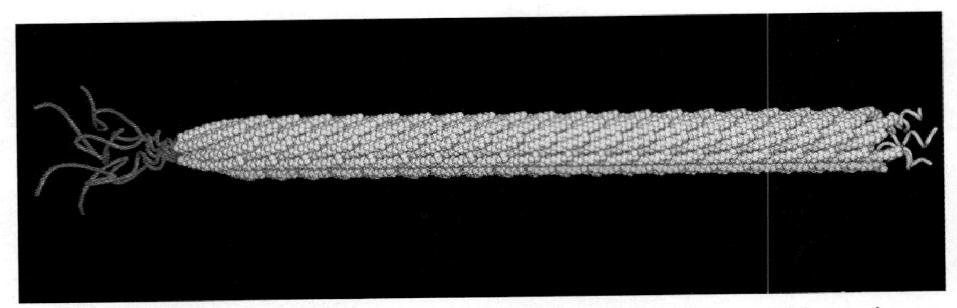

FIGURE 8.33 Model of phage M13. Phage proteins III, VIII, and IX are shown in orange, gray, and red, respectively. The proteins are not drawn to scale. (Reproduced from C. Mao, et al., *Science* 303 [2004]: 213–217 [http://www.sciencemag.org/cgi/content/abstract/303/5655/213]. Reprinted with permission from AAAS. Photo courtesy of Angela Belcher, Massachusetts Institute of Technology.)

DNA that is approximately 6400 nucleotides long. The DNA molecule is encased in a 930 nm fibrous particle (FIGURE 8.33). Unlike the other phages described, the filamentous phages neither kill their host cell nor cause it to lyse. Instead, the newly formed virions are continuously released as the cells continue to grow, albeit at a slower rate than uninfected cells.

The virion codes for eleven protein products, pI to pXI (FIGURE 8.34). The space between the genes that code for pII and pIV is called an intergenic region (IR). A second IR is located between the genes that code for pIII and pVIII. Genes III, VI, VII, VIII, and IX code for coat proteins. The hollow tube that encapsulates the DNA contains approximately 2700 copies of pVIII subunits, which overlap one another like shingles on the side of a house. The end of the phage that emerges from the host cell first contains five copies each of pVII and pIX. The circular single-stranded viral DNA has an imperfect but stable hairpin that is positioned at the pVII/pIX end. The other end of the virion, which attaches to the F pilus at the start of the infection process, contains about five copies each of pIII and pVI. Further work is required to determine exactly how the protein subunits are arranged at the ends of the virion particle.

The pIII subunit, which makes an important contribution to the infection process, has two N-terminal subdomains, N1 and N2, and a C-terminal domain. The C-terminal domain, which has a short hydrophobic region, is connected to the N2 subdomain by a flexible, glycine-rich linker (FIGURE 8.35a). The pIII subunit is anchored to the phage coat through its C-terminal domain. A hydrophobic region in the C-terminal domain appears to be covered by other phage coat proteins.

Bacterial infection by phages M13, fd, and f1 is a multistep process that begins when the N2 subdomain attaches to the tip of the F pilus. This attachment somehow acts as a trigger for the pilus to retract and allows the N1 subdomain to bind to the bacterial receptor TolA, which extends from the cell membrane into the periplasmic space. Nicholas J.

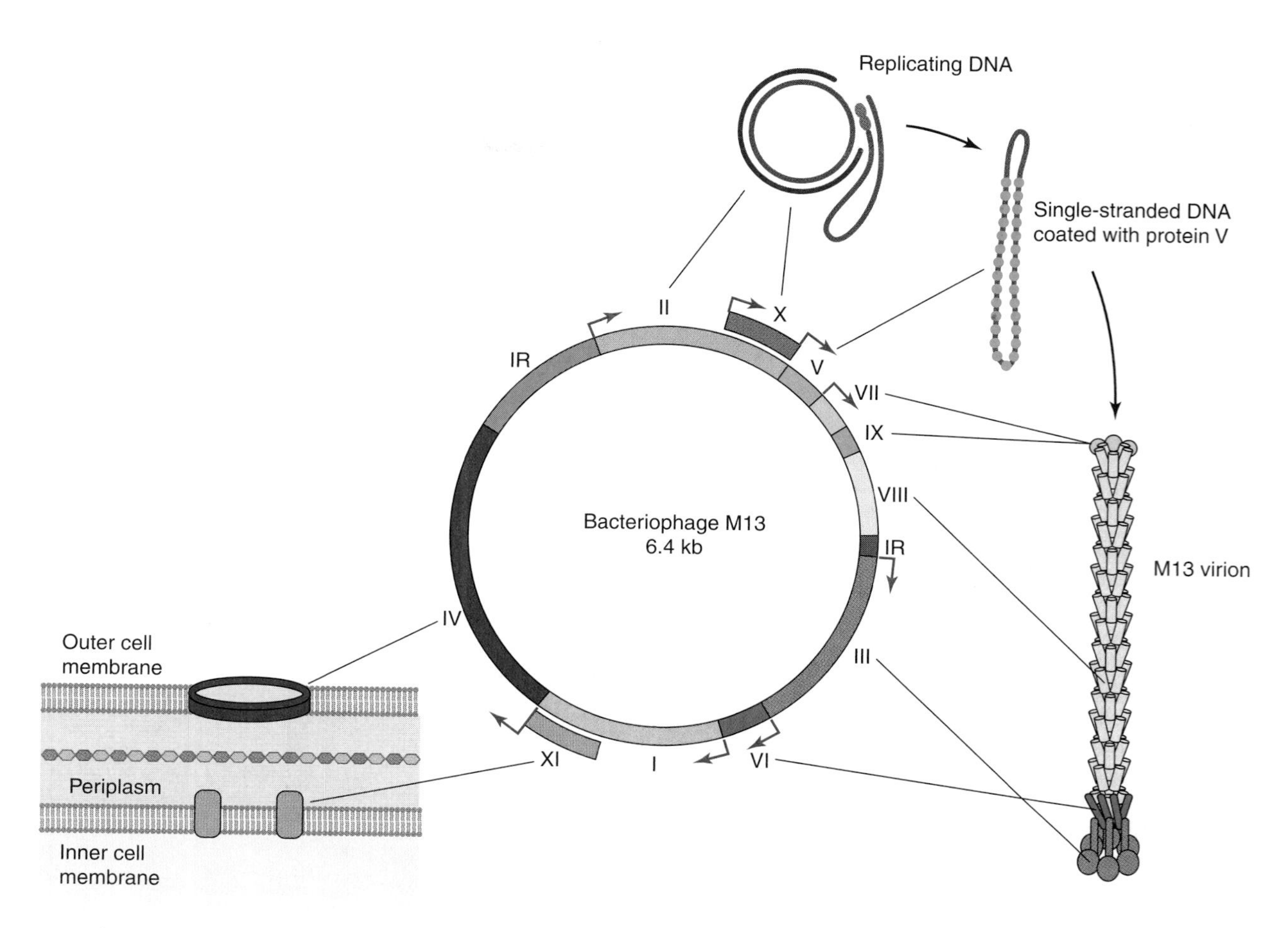

FIGURE 8.34 **Genetic map and gene products of phage M13.** Phage M13 is a single-stranded circular DNA, 6407 nucleotides long. The genes are numbered *I–XI*. The start sites for gene transcription are indicated by arrows, which point in the directions of transcription. The intergenic region (IR) between genes *IV* and *II* is essential for phage replication. The product of gene *II*, pII, binds to this IR of the double-stranded DNA (replicating form) and makes a nick in the (+) strand, initiating rolling circle replication by host proteins. A second IR is present between genes *VIII* and *III*. The product of gene *V*, pV, is required later in infection for the switch to single-stranded DNA accumulation. Genes *III* and *VI* encode pIII and pVI, respectively. pIII and pVI are located at the end of the virion, which is the first to make contact with the bacterial cell during the early stages of infection. Genes *I* and *XI* encode pI and pXI, respectively. pI and pXI end up in the bacterial cell membrane. Gene *IV* encodes pIV, a multimeric protein that forms an outer membrane channel that allows the maturing phages particle to exit the bacterial cell.

Bennett and Jasna Rakonjac have proposed that the next steps in the infection process, shown in FIGURE 8.35b, are as follows. (1) The interaction between the N1 subdomain and TolA produces a conformational change in the C-terminal domain. (2) This conformational change uncovers the C-terminal domain's hydrophobic region. (3) The exposed hydrophobic region inserts into the bacterial cell membrane, anchoring the virion. (4) This insertion triggers the integration of major coat protein (pVIII) into the cell membrane, which ultimately leads to the release of the single-stranded DNA into the host cytoplasm. The bacterial membrane protein TolQRA assists the major coat protein to integrate in the cell membrane. M13, fd, and f1 phages, thus, exploit bacterial proteins that normally help to maintain the cell envelope, to suit their own needs.

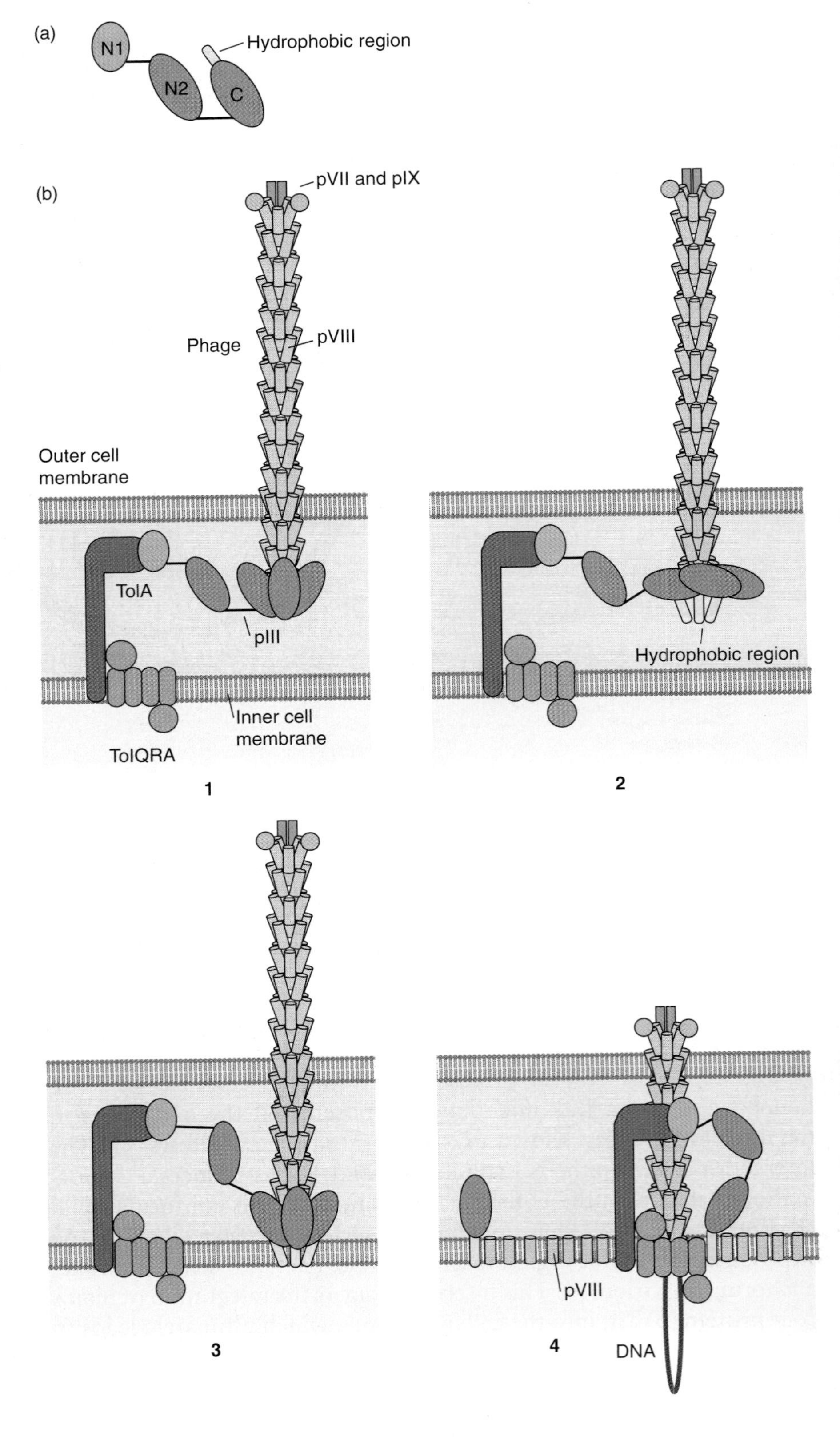

FIGURE 8.35 Phage infection by M13, fd, and f1. (a) The pIII subunit—the two N-terminal subdomains, N1 and N2, are shown in light green and blue, respectively. The C-terminal domain is shown in red and its short hydrophobic region in yellow. (b) The infection process. (1) Interaction between the N1 subdomain and the bacterial protein TolA produces a conformational change in the C-terminal domain. (2) The conformational change uncovers the C-terminal's hydrophobic region. (3) The exposed hydrophobic region inserts into the bacterial cell membrane, anchoring the virion. (4) This insertion triggers the integration of major coat protein into the cell membrane, which ultimately leads to the release of the single-stranded DNA into the host cytoplasm. pIII and its subunits are colored as in part a. For simplicity, only three pIII subunits are shown and only one of these is shown with the two N-terminal subdomains. Also, for simplicity pVI is not shown, pVIII is shown in tan and DNA is in blue. The bacterial proteins TolA and TolQRA are purple and bronze, respectively. (Part a adapted from B. J. Bennett and J. J. Rakonjac, *J. Mol. Biol.* 356 [2006]: 266–273. Part b reprinted from *J. Mol. Biol.*, vol. 356, N. J. Bennett and J. Rakonjac, Unlocking of the filamentous bacteriophage virion . . ., pp. 266–273, copyright 2006, with permission from Elsevier [http://www.sciencedirect.com/science/journal/00222836].)

After the DNA has entered the bacterial cell, it is coated with single-stranded DNA binding protein (SSB). The SSB does not bind to the hairpin, however, which then serves as the start site for (–) strand synthesis. Although some details are different, the filamentous phages use the same fundamental strategy for DNA replication as that described above for ϕX174.

Once sufficient pV has been formed, it binds cooperatively to newly synthesized (+) strands, preventing the (+) strand from serving as a template for further DNA synthesis. A large pool of (+) strand•pV complexes, thus, is present in the cell. The major coat protein (pVIII) is synthesized in large quantities and immediately deposited in the inner cell membrane; pVIII is not found free in the cytoplasm. This protein is an α helix with a hydrophobic center, and an acidic end. It is hypothesized that the hydrophobic center enables the protein to be situated in the hydrophobic membrane, the N-terminus is in periplasm, and the C-terminus is in the cell interior (**FIGURE 8.36**), perhaps displacing pV and binding to the (+) strand. The phage particle is assembled in the inner membrane. For each completed particle, a bud forms on the cell

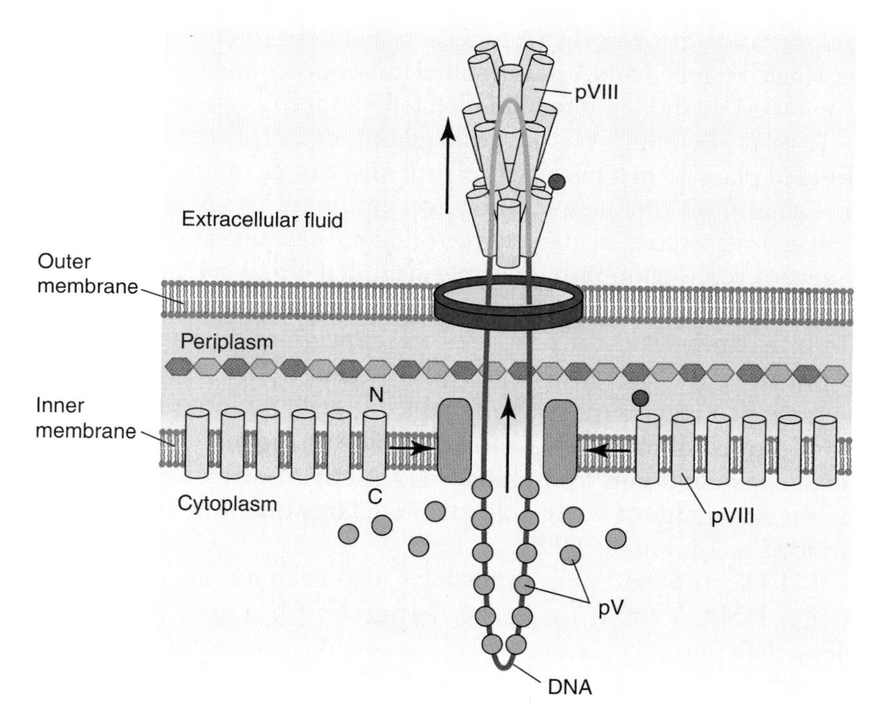

FIGURE 8.36 A simplified view of filamentous bacteriophage assembly. Newly synthesized coat proteins, pVIII (tan cylinders), are imbedded in the inner membrane (IM) with their N termini in the periplasm and their C termini in the cytoplasm. Single-stranded viral DNA is extruded through a pore complex. Coat proteins also interact with the pore complex, where they envelop the DNA as they move from the bacterial membrane into the assembling phage coat. By this means, newly assembled phage particles are extruded from the cytoplasm to the surrounding medium without *E. coli* lysis. When fused to a coat protein that can efficiently insert into the phage coat, a foreign particle (red circles) will be displayed on the virion surface. (Adapted from S. S. Sidhu, *Biomol. Eng.* 18 [2001]: 57–63.)

surface and a phage is extruded. This process can continue indefinitely without damage to the cell. The first progeny virions appear in the culture medium about 10 minutes after infection at 37°C. The number of virion particles increases exponentially for the next 40 minutes and then continues at a linear rate. After about one hour, the medium contains about 1000 virions per cell.

Joachim Messing recognized that phage M13 might be a very useful cloning vector if it were possible to introduce the foreign DNA into the phage DNA without disrupting an essential phage gene. This was a difficult task in the early 1970s because not much was known about phage M13 genes. Messing succeeded in obtaining the desired vector by inserting a cloning sequence with unique restriction endonuclease sites into the IR between genes II and IV. The basic approach to constructing an M13 vector with an inserted foreign DNA segment is as follows: (1) Double-stranded M13 vector DNA is isolated from infected cells and cut with a restriction endonuclease. (2) The foreign DNA is inserted between the restriction sites and the DNA molecules joined by DNA ligase. (3) Competent bacteria are transformed with the recombinant DNA. (4) Phage progeny are collected and the single-stranded DNA containing the inserted foreign DNA is isolated. There is no strict limitation to the size of the foreign DNA fragment that can be inserted into the phage DNA because the phage DNA is not inserted into a preformed structure. The single-stranded DNA is well suited for sequence analysis by the chain termination method and for site-directed mutagenesis (see Chapter 5).

A related cloning vector, a **phagemid**, can replicate as a double-stranded plasmid in a bacterial cell but also can be induced to produce large quantities of single-stranded recombinant DNA when desired. A phagemid is constructed by introducing plasmid and phage M13 origins of replication into a double-stranded circular DNA molecule. Under normal growth conditions, the phagemid replicates as a plasmid in bacteria. It can be induced to replicate as a phage, however, by infecting the bacterial cell with a helper phage, which supplies the gene products required to produce a virion containing a single-stranded recombinant DNA molecule. The phagemid, thus, offers the convenience of doing recombinant DNA experiments with a plasmid and the advantage of being able to obtain large quantities of the single-stranded recombinant DNA.

M13 vectors and phagemids have also been modified so that the foreign DNA is fused to a gene that codes for a coat protein. The released virions display the peptide encoded by the foreign DNA (see Figure 8.36), making it easy to identify the recombinant virion and to purify it. This technique, known as **phage display**, also provides a convenient means for identifying and modifying specific cloned peptides.

8.6 Animal Viruses

Molecular biologists are interested in eukaryotic viruses for much the same reasons that we are interested in bacteriophages. Fundamental questions concerning the molecular biology of eukaryotic cells can be answered by studying virus infected eukaryotic cells. For instance, we gain insights into how eukaryotic cells replicate their own DNA by examining how they replicate viral DNA and how eukaryotic cells synthesize their own mRNA and proteins by studying how they synthesize viral mRNA and proteins. The choice of the specific virus to be used is usually based on the ease of maintaining and propagating the virus in the laboratory, and its suitability for answering a specific set of questions. Investigators, however, are also mindful of the fact that the new information they hope to obtain might lead to better methods to treat specific viral diseases.

The three kinds of animal viruses that we examine in this section have each contributed to our basic understanding of the molecular biology of eukaryotic cells and each can cause serious illnesses. We focus on these three kinds of animal viruses because you will need to be familiar with them when examining eukaryotic replication, transcription, translation, and related processes in subsequent chapters. You may wish to consult a textbook on virology to obtain further information about these viruses as well as others that infect animals, plants, and fungi.

Polyomaviruses contain circular double-stranded DNA.

The **polyomavirus** was detected in mice in 1953. Its name derives from its ability to cause solid tumors at multiple sites in rodents under laboratory conditions. A closely related virus, **SV40** (**simian virus 40**), was discovered in monkey kidney cells in 1960 and later shown to contaminate early preparations of the Salk and Sabin polio vaccines, which were prepared by propagating the poliovirus on monkey kidney cells. Fortunately, the contaminating SV40 does not appear to have caused tumors in vaccinated individuals. The mouse polyomavirus and SV40 belong to a family of viruses called *Polyomaviridae*, which have now been recovered from a wide variety of additional organisms including cows, birds, and humans, and appear to be species-specific.

The *Polyomaviridae* contain a circular double-stranded DNA molecule (about 5.2 kbp), which is located in an icosahedral capsid with a diameter of about 60 nm (FIGURE 8.37). The capsid consists of three proteins: VP1, VP2, and VP3. VP1 is the major capsid protein, accounting for about 80% of the total capsid protein, while VP2 and VP3 together account for the remaining 20%. The DNA in the virus particle is wound around nucleosome cores made of histones H2A, H2B, H3, and H4 so that the DNA in the resulting minichromosome is highly condensed.

The SV40 genome is divided into a regulatory region, an early region, and a late region (FIGURE 8.38). The regulatory region (about 400 bp) includes the origin of replication and signals that regulate the transcription of the other two regions. The early and late regions are transcribed from different strands and in opposite directions. The early region codes for three proteins, large T antigen, small t antigen, and 17kT protein, that help to regulate early and late gene transcription. The late region codes for the coat proteins VP1, VP2, and VP3.

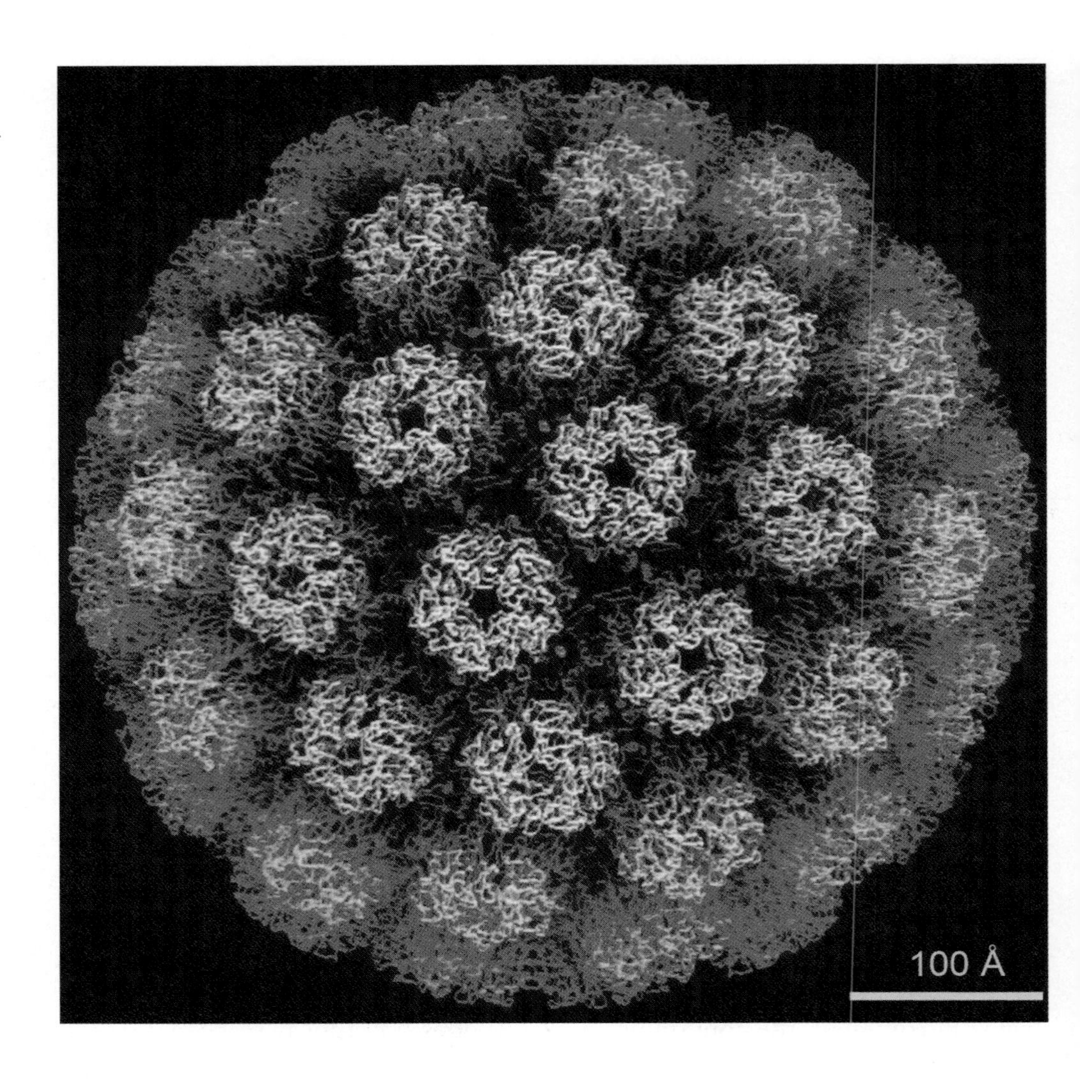

FIGURE 8.37 **SV40 virion.** (Photo courtesy of Stephen C. Harrison, Harvard Medical School and Howard Hughes Medical Institute.)

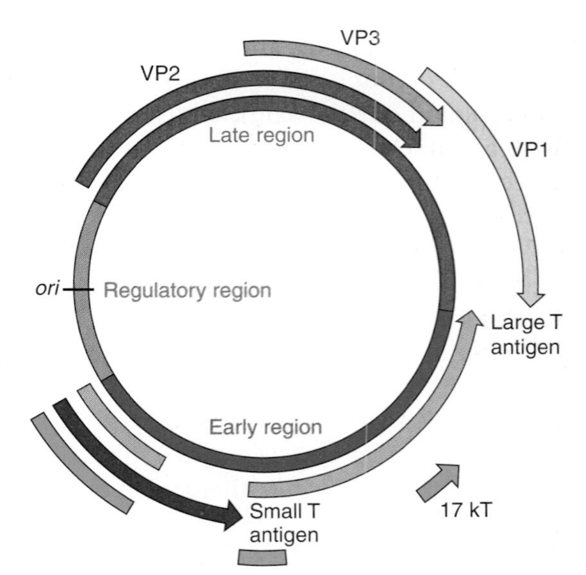

FIGURE 8.38 **Simplified map of SV40 genome.** The SV40 genome is divided into an early region (genes transcribed early in infection), a late region (genes transcribed late in infection), and a regulatory region (genes that control transcription of the early and late regions). The regulatory region also contains the origin of replication (*ori*). The early region codes for the large T antigen, small t antigen, and 17 kT antigen. The large T antigen is the only SV40 product required for viral DNA replication. The late region codes for the coat proteins VP1, VP2, and VP3. The early and late regions are transcribed from different strands and in opposite directions. The SV40 genome, which contains only 5243 bp, is able to code for so many proteins because of a process called alternative splicing that allows a nucleotide sequence to be incorporated into two or more different mRNA molecules. (Adapted from C. S. Sillivan and J. M. Pipas, *Microbiol. Mol. Biol. Rev.* 66 [2002]: 179–202.)

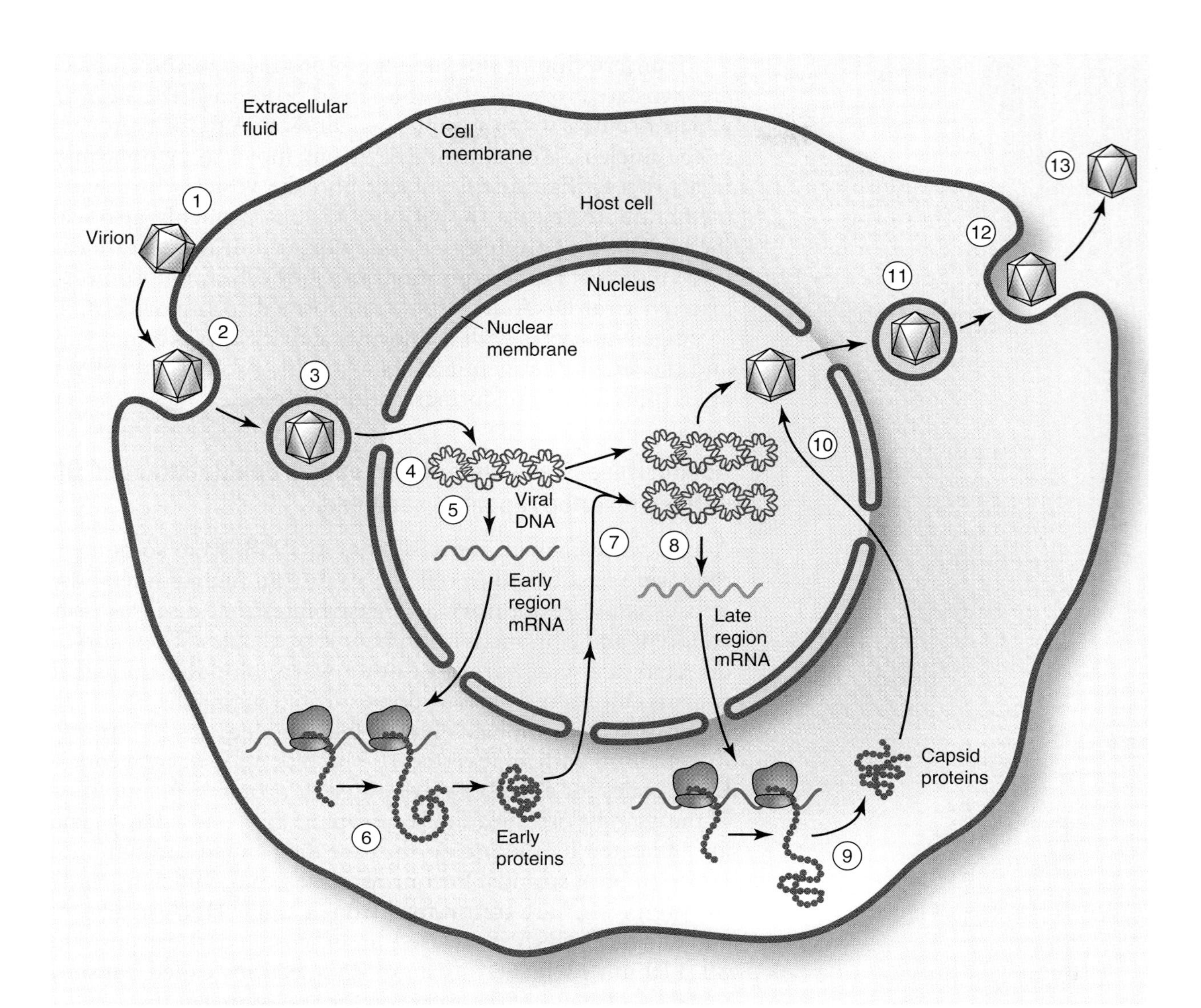

FIGURE 8.39 **Simplified polyomavirus replication cycle.** (1) The virion binds to the cell surface, (2) enters the cell by endocytosis, and (3) is transported to the cell nucleus by a route that remains to be determined, and (4) is uncoated. (5) The released viral DNA is transcribed to form early region mRNA molecules, which are (6) translated to form early proteins (T antigens). (7) The viral DNA is also replicated and (8) transcribed to produce late region mRNA molecules, which are (9) translated to form capsid proteins (VP1, VP2, and VP3). (10) Progeny virions are assembled in the nucleus and (11) the assembled virions move to cytoplasmic vesicles by a process that is still unclear. (12) The vesicles fuse with the cell membrane, (13) releasing the virions. Virions probably also leak out of the nucleus and are released following cell death. (Adapted from D. M. Knipe, et al. *Fundamental Virology, Third Edition.* Lippincott Williams & Wilkins, 1996.)

A simplified version of the SV40 virus replication cycle is shown in FIGURE 8.39. The virion attaches to a glycolipid receptor on the cell surface, is transported into the host cell by endocytosis and then moves to the cell nucleus through a pathway that will not be described here. Transcription of the early region produces mRNA molecules that are translated to form T antigens. The large T antigen is required to initiate viral DNA replication at the origin of replication. All other proteins needed for DNA replication are supplied by the host cell. Like host cell DNA replication, viral DNA replication takes place during the S phase of the cell cycle. Studies of SV40 DNA replication provide considerable information about the host cell's replication apparatus (see Chapter 10).

Transcription of the late region produces mRNA molecules that are translated to synthesize the capsid proteins (VP1, VP2, and VP3), which assemble with the replicated viral DNA to form progeny virions in the nucleus. The assembled virions move to cytoplasmic vesicles in a process that is still unclear and the vesicles fuse with the cell membrane to release the virions. Virions probably also leak out of the nucleus and are released following cell death. In rare cases, viral DNA inserts at random positions in a host cell chromosome. Cells with inserted viral DNA are often transformed to tumor cells. The large T antigen interferes with the normal activity of two tumor suppressors and the small t antigen inhibits a specific protein phosphatase. Both are required for transformation to take place.

Adenoviruses have linear blunt-ended, double-stranded DNA with an inverted repeat at each end.

Adenoviruses, which were isolated in 1953, were so-named because they were first found in cells derived from human adenoids. Adenoviruses cause respiratory and gastrointestinal diseases primarily in children, and conjunctivitis in people of all ages. They also have been detected in a wide variety of other warm-blooded animals including rodents, birds, and various domesticated mammals.

A typical virion has a linear, blunt-ended, double-stranded DNA (35–45 kbp) with an inverted 103 bp repetition at each end. The virus DNA codes for at least twelve different proteins. A simplified version of the genome organization is shown in FIGURE 8.40. Early genes, which are indicated by the prefix "E," are distributed throughout the viral DNA on both strands. In contrast, late genes, which are indicated by the prefix "L," are transcribed from a single DNA strand.

The viral DNA is encased in an icosahedral capsid (diameter of 80–110 nm), which consists of 252 sections called **capsomers** (FIGURE 8.41). Twelve of these capsomers are located at the vertices and are called **pentons** because they face five neighbors. Each penton contains five copies of the viral encoded polypeptide III and is attached to a

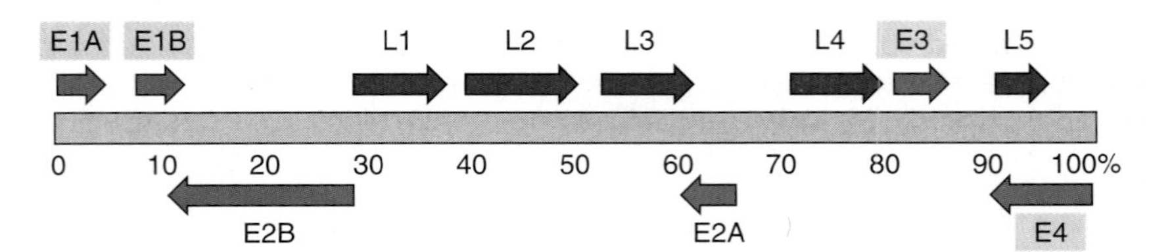

FIGURE 8.40 **Representative map of the genome of adenovirus.** Early genes, designated by the "E" prefix, are shown in green and late genes, designated by the "L" prefix, are shown in purple. Arrows show the direction of transcription. The genes that are in boxes are those that can be removed during the production of a replication-defective virus for gene therapy protocols. The *E1A* gene (which encodes the initial viral transcription unit) must be removed to prevent the recombinant virus from replicating. Other genes can be deleted to make more space for the insertion of larger foreign DNA fragments. (Reproduced from K. J. Wood and J. Fry, Gene therapy: potential applications in clinical transplantation, *Expert Rev. Mol. Med.*, volume 1, issue 11, pp. 1–20, 1999 © Cambridge Journals, reproduced with permission.)

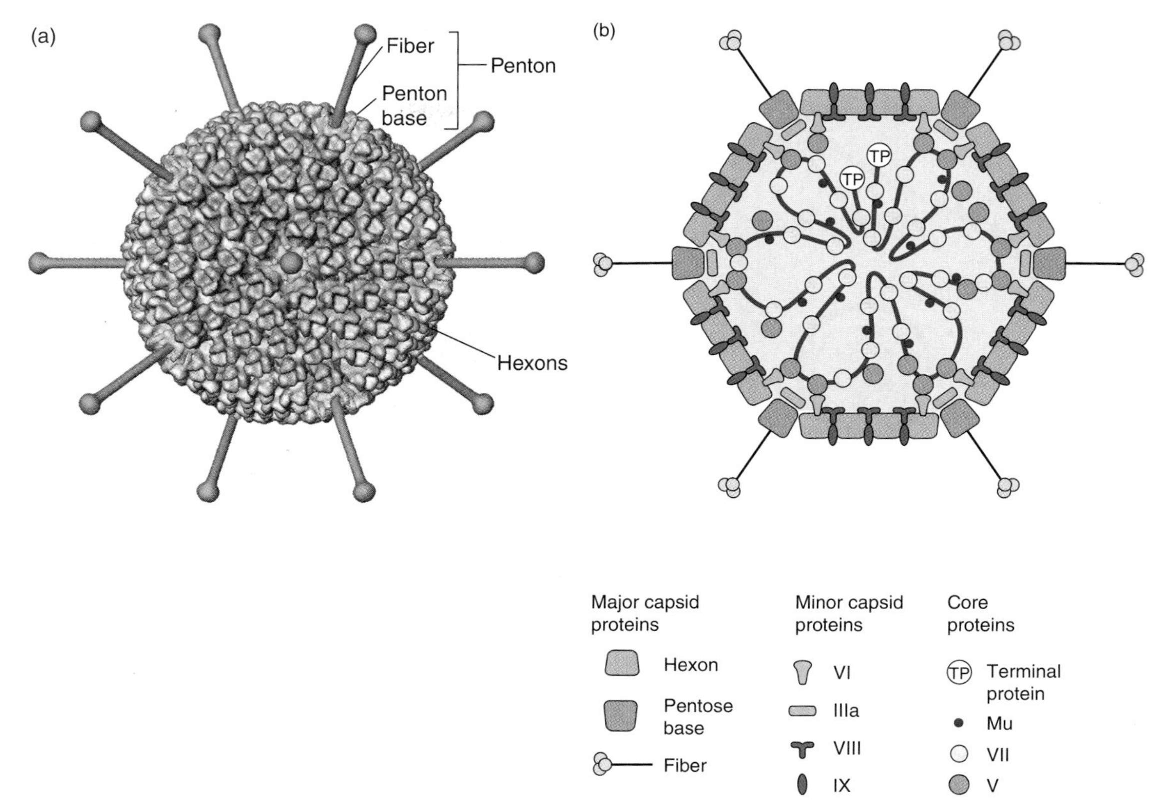

FIGURE 8.41 **Structure of adenovirus.** (a) Surface view of the adenovirus obtained by cryo-electron microscopy and image reconstruction. (b) Schematic showing the locations of the eleven structural virion proteins. The major capsid proteins are in the hexon, penton, and fiber. The minor protein pIX, which is located on the outside of the capsid, stabilizes the capsid. Although only a subset of hexons in the diagram are shown associating with protein VI, the cryo-electron microscope construct indicates that protein VI is associated with every hexon. The remaining capsid proteins pIIIa, pVI, and pVIII are on the inner surface of the capsid. The core proteins, terminal protein, Mu, V and VII, interact with the double-stranded DNA in the interior of the virion. The structural arrangement of the core proteins is not known. (Part a photo courtesy of Carmen San Martín, Centro Nacional de Biotecnología [CNB-CSIC]. Part b reprinted from *Virology*, vol. 384, G. R. Nemerow, et al., Insights into adenovirus host cell . . ., pp. 380–388, copyright 2009, with permission from Elsevier [http://www.sciencedirect.com/science/journal/00426822].)

fibrous protein that projects outward. The remaining 240 capsomers face six neighbors and are called **hexons.** Each hexon is made of three copies of the viral encoded polypeptide II. Additional virally encoded cement proteins help to stabilize the pentons and hexons as well as the entire capsid. The inner cavity of the capsid contains three virally encoded, arginine-rich proteins: V, VII, and μ (mu). All three proteins contact the virion DNA. Protein VII, the major core protein, appears to serve a histone-like function, forming a hub around which viral DNA is wound. Polypeptide V helps to position the nucleoprotein complex within the capsid. The function of the μ protein is not known. A fourth virally encoded protein is also present in the core. This protein, designated the **terminal protein,** is covalently attached to the DNA by a phosphodiester bond that links a tyrosyl side chain to the 5′-end of the DNA. Hence, a terminal protein is attached to each 5′-end of the DNA. Terminal protein serves as a primer for the synthesis of a new DNA strand. Adenovirus DNA replication has some unique features that are beyond the scope of this book.

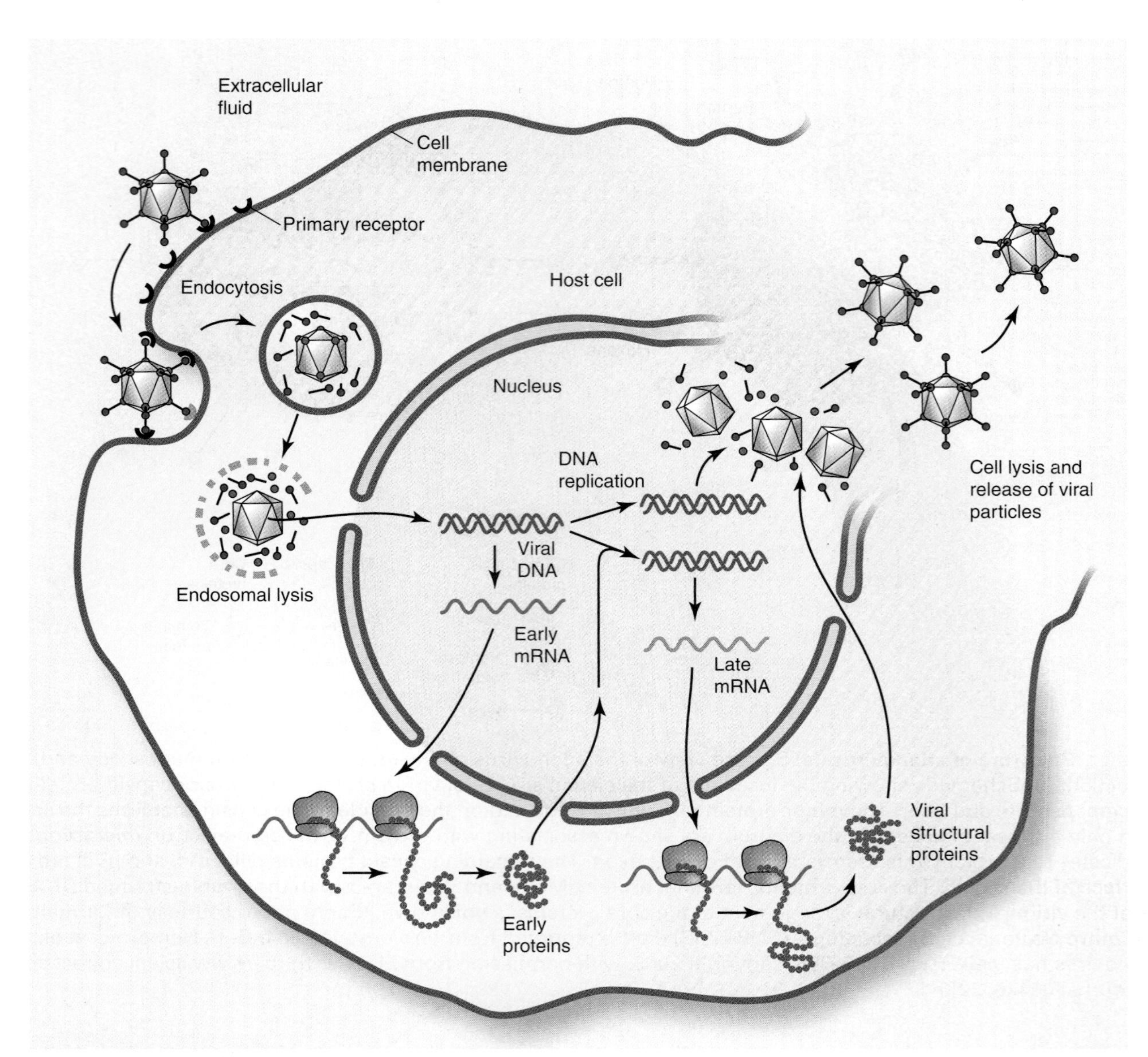

FIGURE 8.42 **Simplified adenovirus replication cycle.** The fiber attached to the penton has a knob at its end that contacts a receptor on the cell surface. This interaction facilitates the binding of the penton to a secondary receptor on the cell surface. The bound virion enters the cell by receptor-mediated endocytosis and is later released into the cytoplasm by acid-enhanced lysis of the endosomal vesicle. The virion particle dissociates and the released DNA moves to the nucleus, where it is transcribed by the host cell's transcription machinery. The first virus encoded protein that is formed, E1A, activates transcription of other early viral transcriptional units. Two of the early proteins formed, DNA polymerase and single-stranded DNA binding protein, work with host cell enzymes to replicate the viral DNA. Then late genes are transcribed and translated to form viral coat proteins, which assemble with newly formed viral DNA to form progeny virions. The virus blocks protein synthesis during the late stage of infection, resulting in cell lysis and the release of progeny virion particles. (Adapted from M. E. Mysiak, *Molecular architecture of the preinitiation complex in adenovirus DNA replication* [master's thesis, Universiteit Utrecht], 2004.)

A simplified version of the adenovirus replication cycle is shown in schematic form in FIGURE 8.42. The fiber attached to the penton has a knob at its end that contacts a receptor on the cell surface. This interaction facilitates the binding of the penton to a secondary receptor on the surface of the host cell. The bound virion is taken into the cell by receptor-mediated endocytosis and later released from the cytoplasmic vesicle by acid-enhanced lysis of the endosomal vesicle. The virion particle dissociates in a series of steps, starting with the initial binding to the cell surface receptors. The virion DNA moves to the nucleus, where it is transcribed by the host cell's transcription machinery. The first virus encoded protein that is formed, E1A, activates transcription of other early viral transcription units. Mutations that prevent E1A formation block all further viral gene transcription. Two of the early proteins formed, DNA polymerase and single-stranded DNA binding protein, work together with host cell enzymes to replicate the viral DNA. Then late genes are transcribed and translated to form viral coat proteins, which assemble with the newly formed viral DNA to form progeny virions. The virus blocks protein synthesis during this late stage of infection, resulting in cell lysis and the release of progeny virion particles.

Adenoviruses have not only been important model systems for studying mRNA synthesis, but have also served as vectors for carrying foreign DNA into host cells. The *E1A* gene must be removed so that the virus cannot replicate (Figure 8.40) but other genes can also be removed to make more room for the foreign DNA. Tragically, an early attempt to use a modified adenovirus as a vector in a gene therapy experiment resulted in the death of a young man, who was being treated for an inborn error in metabolism. Considerable work is now being done to make the adenovirus vector and other vectors used for human gene therapy experiments safe.

Retroviruses use reverse transcriptase to make a DNA copy of their RNA genome.

The retroviruses are probably the most extensively studied of all animal viruses at the present time. The primary reason for the intense activity in retrovirology is that one member of the retrovirus family, the **human immunodeficiency virus** (**HIV**), causes **acquired immune deficiency syndrome** (**AIDS**). Much of the research on retroviruses is driven by the desire to find effective methods to prevent and treat AIDS.

There are many other compelling reasons to study the retroviruses, however. At the most fundamental level, the retrovirus replication cycle requires information flow from RNA to DNA, termed **reverse transcription**, an entirely unexpected pathway when it was first proposed by Howard Temin in the mid-1960s (see below). The prefix "retro" in the family name derives from this "backward" flow of genetic information. As pointed out in Chapter 5, the retroviral enzyme that catalyzes RNA-dependent DNA synthesis, better known as reverse transcriptase, is an important tool in molecular biology and recombinant DNA technology. Furthermore, genetically altered retroviruses, which are constructed so that they are missing genetic information needed for the virus to propagate in host cells, are used as vectors to introduce foreign genes into animal cells and may, with continued improvements, prove to be safe vectors for use in human gene therapy.

Members of the retrovirus family all share certain common features. First, all family members contain two identical RNA molecules (7,000–10,000 kb), which are called (+) strands because they have the same sequence information as mRNA. These RNA molecules serve as templates for the reverse transcriptase to make DNA but are not required to direct ribosomes to make viral proteins (see below). All retrovirus RNA molecules contain three protein coding regions, which are designated *gag*, *pol*, and *env* (FIGURE 8.43a). The *gag* (group specific antigen) region codes for a polypeptide, which is called a **polyprotein,** because it is eventually cleaved to produce the following three distinct proteins:

1. Matrix protein. The matrix protein is closely associated with the membrane surrounding the virus.
2. Capsid protein. The capsid protein forms the core shell, the virion's major internal structure.
3. Nucleoprotein. The nucleoprotein is a basic protein that binds to the RNA.

The locations of these three proteins as well as other proteins in the retrovirus particle are shown in FIGURE 8.43b. One other protein should be mentioned in connection with the *gag* region. In some retroviruses the *gag* region also codes for protease, in others the *pol* region codes for this enzyme, and in still others the protease coding region is between the *pol* and *gag* regions. Protease is essential for viral reproduction because it is required for polyprotein cleavage. In fact, protease inhibitors have proven to be effective drugs for treating individuals with AIDS.

The *pol* (polymerase) region codes for a second polyprotein, which is cleaved to produce reverse transcriptase and integrase (see below).

(a) Virion RNA

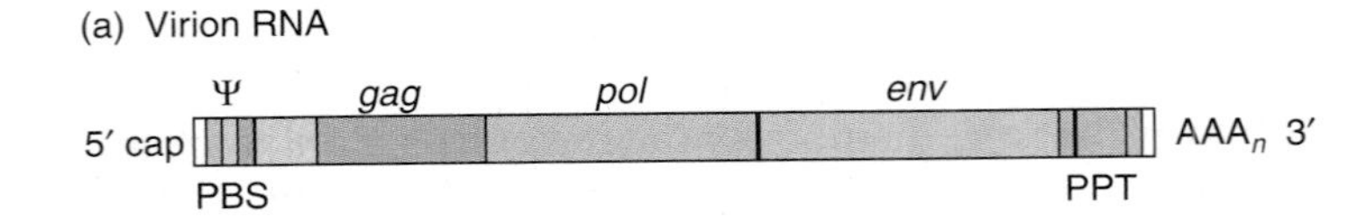

(b) Schematic of retrovirus particle

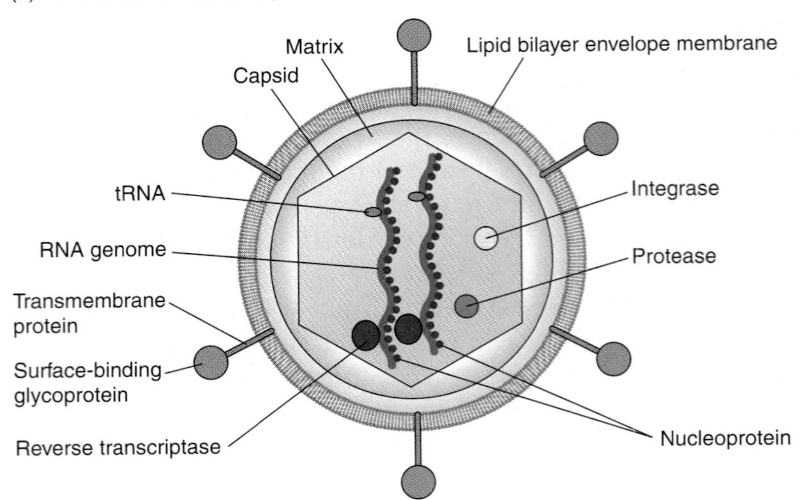

FIGURE 8.43 **Retroviral particle and genome structure.** (a) Genome organization of a simple retrovirus. (b) Retrovirus particle showing the approximate location of its components. (Adapted from F. S. Pedersen and M. Duch, *Encyclopedia of Life Sciences* [DOI: 10.1038/npg.els.0003843]. Posted May 3, 2005.)

The *env* (envelope) region codes for still another polyprotein, which is cleaved to produce transmembrane protein, an integral membrane protein located in the membrane envelope surrounding the virion, and surface glycoprotein, a peripheral membrane protein that binds to the transmembrane protein on the external membrane surface.

Retroviral RNA molecules also share other common features. They all have a modified guanosine at their 5′-end and a poly(A) sequence at their 3′-end that are called a cap and a tail, respectively. Cap and tail structures are typically found in eukaryotic mRNA and are described in greater detail in Chapter 19. The 5′-end of the virion RNA has two other sites that are essential for viral propagation. The first of these, the primer binding site (PBS) is required for DNA synthesis. The second, the ψ site, is a packaging signal that must be present for newly synthesized viral RNA to be packaged into virus particles. The 3′-end also has a unique feature called the polypurine tract (PPT). Retroviruses with RNA coding for just the proteins that have been described so far are termed **simple retroviruses.** Those with RNA that codes for additional proteins, most of which play important regulatory roles, are termed **complex retroviruses.** HIV is a complex retrovirus.

Retroviruses enter the target cell in much the same way that other enveloped viruses enter their host cells. The virion attaches to a specific receptor on the cell surface and then the virion membrane fuses with the cell membrane so that the core of the particle can enter the cell cytoplasm. Each member of the retrovirus family binds to cell receptors that are specific for it. For example, the HIV virus binds to a receptor called CD4, which is present on the surface of human T helper cells, immature thymocytes, monocytes, and macrophages. HIV must also bind to a co-receptor on the surface of the target cell before it can enter that cell. Individuals lacking the co-receptor are more resistant to HIV infection than are those who have the co-receptor, suggesting that the co-receptor might be a target for new drugs to treat AIDS.

The next steps in retrovirus infection are quite remarkable because they require an RNA template to synthesize DNA. Until the mid-1960s, investigators believed that all RNA viruses use an RNA-dependent RNA polymerase to replicate their RNA. This method of RNA replication is certainly used by a wide variety of RNA viruses including RNA phages, poliovirus, and tobacco mosaic virus (TMV). An entirely different situation was observed for the **Rous sarcoma virus**, an RNA virus discovered by Peyton Rous in 1911 while investigating the origins of a tumor in chickens. Rous observed that healthy chickens developed tumors when injected with an extract prepared from an excised tumor. Further experiments showed that the causative agent must be very small because the extract retained its ability to cause tumors even after being passed through a fine filter that did not let bacteria pass through. Based on this and related experiments, Rous concluded that the chicken tumor must be caused by a virus.

Contemporary investigators questioned the relevance of Rous observations to human cancer because most forms of human cancer are not caused by infectious agents. However, later studies performed by J. Michael Bishop and Harold E. Varmus showed that sequences in Rous sarcoma virus RNA are occasionally replaced by host cell genes that

FIGURE 8.44 A schematic view of the retroviral life cycle. (Adapted from S. Nisole and A. Saïb, *Retrovirology* 1 [2004]:1–9.)

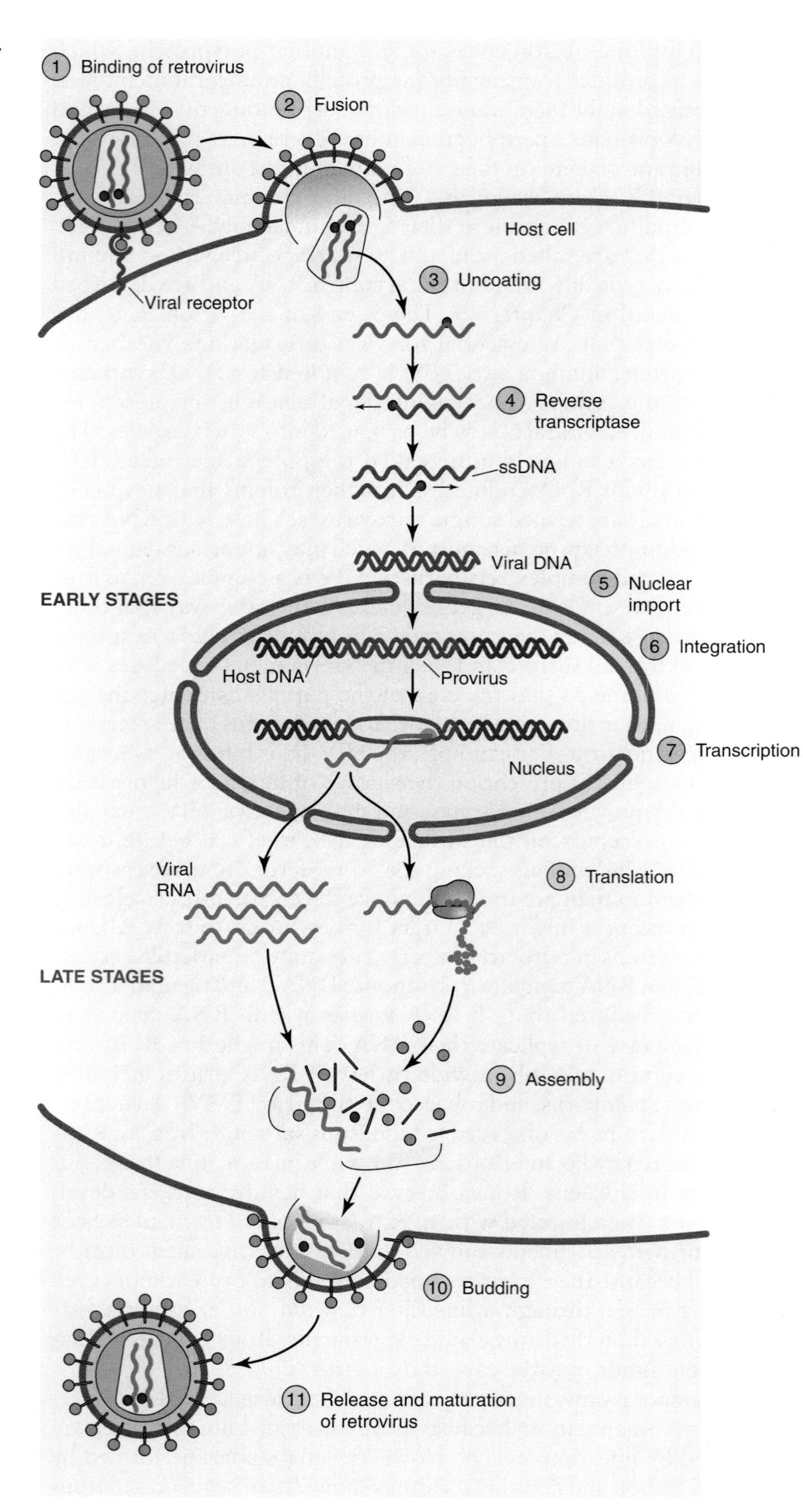

code for protein products that stimulate cell growth and division. These rare defective viruses cannot replicate without the assistance of other viruses. They can, however, transform normal cells into tumor cells.

Howard Temin's hypothesis that DNA synthesis is a required step in Rous sarcoma virus replication was subject to considerable criticism when first proposed in the mid-1960s. The hypothesis gained quick acceptance in 1970, however, when Temin and David Baltimore independently demonstrated that the virion particle contains an RNA-dependent DNA polymerase. Temin proposed the "Provirus Model" to explain the requirement for DNA synthesis. According to this model, retrovirus replication can be divided into an early and a late stage (FIGURE 8.44).

The virus particle supplies reverse transcriptase and integrase. Reverse transcriptase acts on viral RNA in the cytoplasm to synthesize a complementary DNA strand. An RNase H activity associated with reverse transcriptase degrades the viral RNA in the RNA-DNA hybrid. Then the reverse transcriptase uses the newly formed DNA strand as a template to form double-stranded DNA. The double-stranded DNA moves into the nucleus, where it encounters integrase, which integrates the DNA into a random site in one of the host chromosomes to produce the provirus. The provirus continues to replicate as part of the host chromosome and, if insertion takes place in a germ cell, the provirus will be passed from one generation to the next. Human DNA contains many sequences that appear to have been derived from retroviral DNA insertions (see Chapter 14).

The late stage of the replication cycle requires host cell enzymes. Cellular transcription machinery transcribes the inserted viral DNA to form mRNA that directs viral protein synthesis and viral RNA that assembles with the viral proteins to form progeny virus particles. Assembly takes place at the cell membrane and the mature virions are released from the cell by budding. This release process does not result in cell death.

You may wish to refer back to sections in this chapter when you encounter a virus that is described in later chapters. Information about other viruses and a more detailed examination of those viruses described here can be found in standard virology textbooks.

Suggested Reading

General Overview

Ackerman, H.-W. 2009. Phage classification and characterization. *Methods Mol Biol* 501:127–140.

Ackermann, H.-W. 2007. Bacteriophages: Tailed. In: *Encyclopedia of Life Sciences.* pp. 1–7. Hoboken, NJ: John Wiley and Sons.

Breitbart, M., and Rohwer, F. 2005. Here a virus, there a virus, everywhere the same virus? *Trends Micriobiol* 13:278–284.

Brüssow, H., Canchaya, C., and Wolf-Dietrich, H. 2004. Phages and the evolution of bacterial pathogens: from genomic rearrangements to lysogenic conversion. *Microbiol Mol Biol Rev* 68:560–602.

Cairns J, Stent G. S., and Watson, J. (eds) 1966. *Phage and the Origins of Molecular Biology*. Woodbury, NY: Cold Spring Harbor Laboratory Press.

Calendar R. (ed). 2005. *The Bacteriophages*. Oxford, UK: Oxford University Press.

Callanan, M. J., and Klaenhammer, T. R. 2008. Bacteriophages in industry. In: *Encyclopedia of Life Sciences*. pp. 1–8. Hoboken, NJ: John Wiley and Sons.

Campbell, A. 2003. The future of bacteriophage biology. *Nat Rev Genet* 4:471–477.

Campbell, A. M. 1996. Bacteriophages. In: F.C. Neidhardt, R. Curtiss III, J.L. Ingraham, et al. (eds). Escherichia coli *and* Salmonella typhimurium*: Cellular and Molecular Biology*, 2nd ed. pp. 2325–2338. Washington, DC: ASM Press.

Fujisawa, H., and Morita, M. 1997. Phage DNA packaging. *Genes Cells* 2:537–545.

Garcıa, P., Martınez, P. B., Obeso, J. M., and Rodrıguez, A. 2008. Bacteriophages and their application in food safety. *Lett App Microbiol* 47:479–485.

Guttman, B. S., and Kutter, E. 2002. Bacteriophage genetics. In: U. N. Streips, and R. E. Yasbin (eds). *Modern Microbial Genetics*, 2nd ed. pp. 85–126. Hoboken, NJ: John Wiley and Sons.

Guttman, B., Raya, R., and Kutter, E. 2005. *Bacteriophages: Biology and Applications*. Boca Raton, FL: CRC Press.

Harper, D. R., and Kutter, E. 2008. Bacteriophage: Therapeutic uses. In: *Encyclopedia of Life Sciences*. pp. 1–7. Hoboken, NJ: John Wiley and Sons.

Hendrix, R. W. 2003. Bacteriophage genomics. *Curr Opin Microbiol* 6:506–511.

Mattley, M., and Spencer, J. 2008. Bacteriophage therapy-cooked goose or Phoenix rising? *Curr Opin Biotechnol* 19:608–612.

McGrath, S., Fitzgerald, G. F., and van Sinderen, D. 2004. The impact of bacteriophage genomics. *Curr Opin Biotechnol* 15:95–99.

Minor, P. D. 2007. Viruses. In*: Encyclopedia of Life Sciences*. pp. 1–10. Hoboken, NJ: John Wiley and Sons.

Poranen, M. M., Daugelavičius R., and Bamford, D. H. 2002. Common principles in viral entry. *Ann Rev Microbiol* 56:521–538.

Sharp, R. 2001. Bacteriophages: biology and history. *J Chem Technol Biotechnol* 76:667–672.

Summers, W. C. 2001. Bacteriophage therapy. *Ann Rev Microbiol* 55:437–451.

Thiel, K. 2004. Old dogma, new tricks—21st century phage therapy. *Nat Biotechnol* 22:31–36.

Uetz, P., Rajagopala, S. V., Dong, Y.-A., and Haas, J. 2004. From ORFeomes to protein interaction maps in viruses. *Genome Res* 14:2029–2033.

Wagner, E., and Hewlett, M. 2003. *Basic Virology*, 2nd ed. Malden, MA: Blackwell Publishing.

Virulent Bacteriophages

Abedon, S. T. 2000. The murky origin of Snow White and her T-even dwarfs. *Genetics* 155:481–486.

Ackerman, H. W., and Krisch, H. M. 1997. A catalogue of T4-type bacteriophages. *Arch Virol* 142:2329–2342.

Alam, T. I., Draper, B., Kondabagil, K., et al. 2008. The headful packaging nuclease of bacteriophage T4. *Molec Microbiol* 69:1180–1190.

Bernal, R. A., Hafenstein, S., Esmeralda, R., et al. 2004. The ϕX174 protein J mediates DNA packaging and viral attachment to host cells. *J Mol Biol* 337:1109–1122.

Bernhardt, T. G., Roof, W. D., and Young, R. 2000. Genetic evidence that the bacteriophage ϕX174 lysis protein inhibits cell wall synthesis. *Proc Natl Acad Sci USA* 97:4297–4302.

Edgar, R., Rokney, A., Feeney, M., et al. 2008. Bacteriophage infection is targeted to cellular poles. *Molec Microbiol* 68:1107–1116.

Epstein R. H., Bolle, A., Steinberg, C., et al. 1963. Physiological studies of conditional lethal mutants of bacteriophage T4. *Cold Spring Harb Symp Quant Biol* 28:375–392.

Hayashi, M., Aoyama, A., Richardson, D. L., and Hayashi, M. N. 1988. Biology of the bacteriophage ϕX174. In: Calendar, R. (ed). *The Bacteriophages*, pp. 1–71. New York: Plenum Press.

Hendrix, R. 2008. Cell architecture comes to phage biology. *Molec Microbiol* 68:1077–1078.

Ilagit, L. L., Olson, N. H., Dokland, T., et al. 1995. DNA packaging intermediates of bacteriophage ϕX174. *Structure* 3:353–363.

Johnson, J. E., and Chiu, W. 2007. DNA packaging and delivery machines in tailed bacteriophages. *Curr Opin Struct Biol* 17: 237–243.

Karam J, D. (ed.) 1994. *Molecular Biology of Bacteriophage T4*. Washington, DC: American Society for Microbiology.

Kreuzer, K. N. 2000. Recombination-dependent DNA replication in phage T4. *Trends Biochem Sci* 25:165–173.

Leiman, P. G., Kanamarua, S., Mesyanzhinov, V. V., et al. 2003. Structure and morphogenesis of bacteriophage T4. *Cell Mol Life Sci* 60:2356–2370.

Loeb T., and Zinder, N.D. 1961. A bacteriophage containing RNA. *Proc Natl Acad Sci USA* 47:282–289.

Mathews, C. K. 2005. Bacteriophage T4. In: *Encyclopedia of Life Sciences*. pp. 1–9. Hoboken, NJ: John Wiley and Sons.

Mathews, C. K., Kutter, E. M., Mosig, G., and Berget, P. (eds). 1983. *Bacteriophage T4*. Washington, DC: American Society for Microbiology.

Miller, E. S., Kutter, E., Mosig, G., et al. 2003. Bacteriophage T4 genome. *Microbiol Mol Biol Rev* 67:86–156.

Molineux, I. J. 2001. No syringes please, ejection of phage T7 DNA from the virion is enzyme driven. *Mol Microbiol* 40:1–8.

Nechaev, S., and Severinov, K. 2008. The elusive object of desire interactions of bacteriophages and their hosts. *Curr Opin Microbiol* 11:186–193.

Ortin, J., and Parra, F. 2006. The structure and function of RNA replication. *Ann Rev Microbiol* 60:305–326.

Rao, V. B., and Feiss, M. 2008. The bacteriophage DNA packaging motor. *Ann Rev Genet* 42:647–681.

Ricardo A., Bernal, R. A., Hafenstein, S., et al. 2004. The fX174 protein J mediates DNA packaging and viral attachment to host cells. *J Mol Biol* 337:1109–1122.

Rossman, M. G., Mesyanzhinov, V. V., Arisaka, F., and Leiman, P. G. 2004. The bacteriophage T4 DNA injection machine. *Curr Opin Struct Biol* 14:171–180.

Sinsheimer, R. L. 1959. A single-stranded deoxyribonucleic acid from bacteriophage ϕX174. *J Mol Biol* 1:43–53.

Sun, S., Kondabagil, I., Draper, B., et al. 2008. The structure of the phage T4 DNA packaging motor suggests a mechanism dependent on electrostatic forces. *Cell* 135:1251–1262.

van Duin, J. 2005. Bacteriophages with ssRNA. In: *Encyclopedia of Life Sciences*. pp. 1–6. Hoboken, NJ: John Wiley and Sons.

van Duin, J., and Tsareva, N.A. 2005. Single-Stranded RNA phages. In: R. Calendar (ed). *The Bacteriophages*. Oxford, UK: Oxford University Press.

Wang, I.-N., Smith, D. L., and Young, R. 2000. Holins: the protein clocks of bacteriophage infections. *Ann Rev Microbiol* 54:799–825.

Williams, R. S., Williams, G, J., and Tainer, J. A. 2008. A charged performance by gp17 in viral packaging. *Cell* 135:1169–1171.

Temperate Bacteriophage

Bertani, G. 2004. Lysogeny at mid-twentieth century: P1, P2, and other experimental systems. *J Bacteriol* 186:595–600.

Bläsi, U., and Young, R. 1996. Two beginnings for a single purpose: the dual start holins in the regulation of phage lysis. *Mol Microbiol* 21:675–682.

Campbell, A. 1994. Comparative molecular biology of lambdoid phages. *Ann Rev Microbiol* 48:193–222.

Campbell, A. 1996. Cryptic prophages. In: F.C. Neidhardt, R. Curtiss III, J.L. Ingraham, et al. (eds). *Escherichia coli and Salmonella typhimurium: Cellular and Molecular Biology*, 2nd ed. pp. 2041–2046. Washington, DC: American Society of Microbiology Press.

Casjens, S. R., and Hendrix, R. W. 2001. Bacteriophage lambda and its relatives. In: *Encyclopedia of Life Sciences*, pp. 1–8. London, UK: Nature Publishing Co.

Catalano, C., Cue, D., and Feiss, M. 1995. Virus DNA packaging: the strategy used by phage lambda. *Mol Microbiol* 16:1075–1086.

Chauthaiwale, V., Therwath, A., and Deshpande, V. 1993. Bacteriophage lambda as a cloning vector. *Microbiol Rev* 57:290–301.

Dhar, A., and Feiss, M. 2005. Bacteriophage λ terminase: alterations of the high-affinity ATPase affect viral DNA packaging. *J Mol Biol* 347:71–80.

Friedman, D. I., and Court, D. L. 2001. Bacteriophage lambda: alive and still doing its thing. *Curr Opin Micriobiol* 4:201–207.
Gottesman, M. 1999. Bacteriophage λ: the untold story. *J Mol Biol* 293:177–180.
Gottesman, M. E., and Weisberg, R. A. 2004. Little lambda, who made thee? *Microbiol Molec Biol Rev* 68:796–813.
Hendrix R., Roberts J., Stahl F., and Weisberg R. (eds). 1983. *Lambda II.* Woodbury, NY: Cold Spring Harbor Laboratory Press.
Hoess, R. H. 2002. Bacteriophage lambda as a vehicle for peptide and protein display. *Curr Pharm Biotechnol* 3:23–28.
Lederberg, E. M., and Lederberg, J. 1953. Genetic studies of lysogenicity in *E. coli. Genetics* 38:51–64.
Szmelcman S., Hofnung, M. 1975. Maltose transport in *Escherichia coli* K12. Involvement of the bacteriophage lambda receptor. *J Bacteriol* 124:112–118.
Yarmolinsky, M. B. 2004. Bacteriophage P1 in retrospect and in prospect. *J Bacteriol* 186:7025–7028.
Zinder, N.D., and Lederberg, J. 1952. Genetic exchange in *Salmonella. J Bacteriol* 64:679–699.

Chronic Phages

Benhar, I. 2001. Biotechnological applications of phage and cell display. *Biotechnol Adv* 19:1–33.
Bennett, N. J., and Rakonjac, J. 2006. Unlocking of the filamentous bacteriophage virion during infection is mediated by the C domain of pIII. *J Mol Biol* 356:266–273.
Karlsson, F., Borrebaeck, C. A. K, Nilsson, N., Malmorg-Hager, A. C. 2003. The mechanism of bacterial infection by filamentous phage involves molecular interactions between TolA and phage protein 3 domains. *J Bacteriol* 185:2628–2634.
Marvin, D. A. 1998. Filamentous phage structure, infection, and assembly. *Curr Biol* 8:150–158.
Rodi, D. J., and Makowski, L. 1999. Phage-display—finding a needle in a vast molecular haystack. *Curr Opin Biotechnol* 10:87–93.
Russel, M., Lowman, H. B., and Clackson, T. Introduction to phage biology and phage display. In: Lowman, H. B., and Clackson, T. 2004. *Phage Display: A Practical Approach*. Oxford, UK: Oxford University Press.
Russell, M., and Model, P. 2005. Filamentous phage. In: R. Calendar (ed). *The Bacteriophages*. Oxford, UK: Oxford University Press.
Sidhu, S. S. 2001. Engineering M13 for phage display. *Biomol Eng* 18:57–63.

Animal Viruses

Atwood, W. J., and Shah, K. V. 2001. Polyomaviruses. In: *Encyclopedia of Life Sciences*, pp. 1–6. London: Nature.
Barré-Sinoussi, F. 1996. HIV as the cause of AIDS. *Lancet* 348:31–35.
Benjamin, T. L. 2001. Polyoma viruses: old findings and new challenges. *Virology* 289:167–173.
Blattner, W., Gallo, R. C., and Temin, H. M. 1988. HIV causes AIDS. *Science* 241:515–516.
Bleul, C. C., Wu, L., Hoxie, J. A., et al. 1997. The HIV coreceptors CXCR4 and CCR5 are differentially expressed and regulated on human T lymphocytes. *Proc Natl Acad Sci USA* 94:1925–1930.
Bukrinsky, M. I. 2001. HIV life cycle and inherited coreceptors. In: *Encyclopedia of Life Sciences*, pp. 1–6. London: Nature.
Coffin, J. M., Hughes, S. H., and Varmus, H. E. (eds). 1997. *Retroviruses*. Woodbury, NY: Cold Spring Harbor Laboratory Press.
Cole, C. N. 1996. Polyomavirinae: the viruses and their replication. In: B. N. Fields, P. M. Howley, D. M. Knipe et al. (eds). *Fundamental Virology*, 3rd ed. pp. 917–945. Baltimore: Lippincott, Williams and Wilkins.
Crandall, K. A. 2001. Human immunodeficiency viruses (HIV). In: *Encyclopedia of Life Sciences*, pp. 1–7. London: Nature.

D'Souzak, V., and Summers, M. F. 2004. Structural basis for packaging the dimeric genome of Moloney murine leukaemia virus. *Nature* 431:586–590.
Flint, S. J. 2001. Adenoviruses. In: *Encyclopedia of Life Sciences*, pp. 1–14. London: Nature.
Gallo, R. C. 2004. HIV-1: a look back from 20 years. *DNA Cell Biol* 23:191–192.
Gallo, R. C., and Montagnier, L. 2003. The discovery of HIV as the cause of AIDS. *N Engl J Med* 349:2283–2285.
Gazdar, A. F., Butel, J. S., and Carbone, M. 2002. SV40 and human tumors: myth and association or causality? *Nat Rev Cancer* 2:957–964.
Garcea, R. L., and Imperiale, M. J. 2003. Simian virus 40 infection of humans. *J Virol* 77:5039–5045.
Gottlieb, K. A., and Villarreal, L. P. 2001. Natural biology of polyomavirus middle T antigen. *Microbiol Mol Biol Rev* 65:288–318.
Graham, F. L., and Hitt, M. M. 2007. Adenovirus vectors in gene therapy. In*: Encyclopedia of Life Sciences*. pp. 1–6. Hoboken, NJ: John Wiley and Sons Ltd.
Hu, W.-S., and Pathak,V. K. 2000. Design of retroviral vectors and helper cells for gene therapy. *Pharm Rev* 52:493–511.
Hunter, E.1997. Viral entry and receptors. In: J. M. Coffin, S. H. Hughes, and H. E. Varmus (eds). *Retroviruses*, pp. 71–120. Woodbury, NY: Cold Spring Harbor Laboratory Press.
Lednicky, J. A., and Butel, J. S. 1999. Polyomaviruses and human tumors: a brief review of current concepts and interpretations. *Front Biosci* 4:D153–D164.
Löwer, R., Löwer, J., and Kurth, R. 1996. The viruses in all of us: characteristics and biological significance of human endogenous retrovirus sequences. *Proc Natl Acad Sci USA* 93:5177–5184.
Nelson, P. N., Carnegie, P. R., Martin, J., et al. 2003. Demystified . . . human endogenous retroviruses. *J Clin Pathol Mol Pathol* 56:11–18.
Nemerow, G. R., Pache, L., Reddy, C., and Stewart, P. L. 2009. Insights into adenovirus host cell interactions from structural studies. *Virology* 384:380–388.
Neu, U., Stehle, T, and Atwood, W. J. 2009. The *Polyomavirdae*: contributions of virus structure to our understanding of virus receptors and infectious entry. *Virology* 384:389–399.
Novembre, F. J. 2001. Simian retroviruses. In: *Encyclopedia of Life Sciences*, pp. 1–8. London: Nature.
Pedersen, F. S., and Duch, M. 2001. Retroviruses in human gene therapy. In: *Encyclopedia of Life Sciences*, pp. 1–10. London: Nature.
Pedersen, F. S., and Duch, M. 2006. Retroviral replication. In*: Encyclopedia of Life Sciences*. pp. 1–8. Hoboken, NJ: John Wiley and Sons.
Pipas, J. M. 2009. SV40: Cell transformation and tumorigenesis. *Virology* 384:294–303.
Rambaut, A., Posada, D., Crandall, K. A., and Holmes, E. C. 2004. The causes and consequences of HIV evolution. *Nat Rev Genet* 5:52–61.
Russell, W. C. 2009. Adenoviruses: update on structure and function. *J. Gen Virol* 90:1–20.
Shenk, T. E. 1996. Adenoviridae: the viruses and their replication. In: B. N. Fields, P. M. Howley, D. M. Knipe et al. (eds). 1996. *Fundamental Virology*, 3rd ed. pp. 979–1016. Baltimore: Lippincott, Williams and Wilkins.
Sullivan, C. S., and Pipas, J. M. 2002. T antigens of simian virus 40: molecular chaperones for viral replication and tumorigenesis. *Microbiol Mol Biol Rev* 66:179–202.
Turner, B. G., and Summers, M. F. 1999. Structural biology of HIV. *J Mol Biol* 285:1–32.
Whittaker, G. R., Kann, M., and Helenius, A. 2000. Viral entry into the nucleus. *Ann Rev Cell Dev Biol* 16:627–651.

This book has a Web site, **http://biology.jbpub.com/book/molecular**, which contains a variety of resources, including links to in-depth information on topics covered in this chapter, and review material designed to assist in your study of molecular biology.

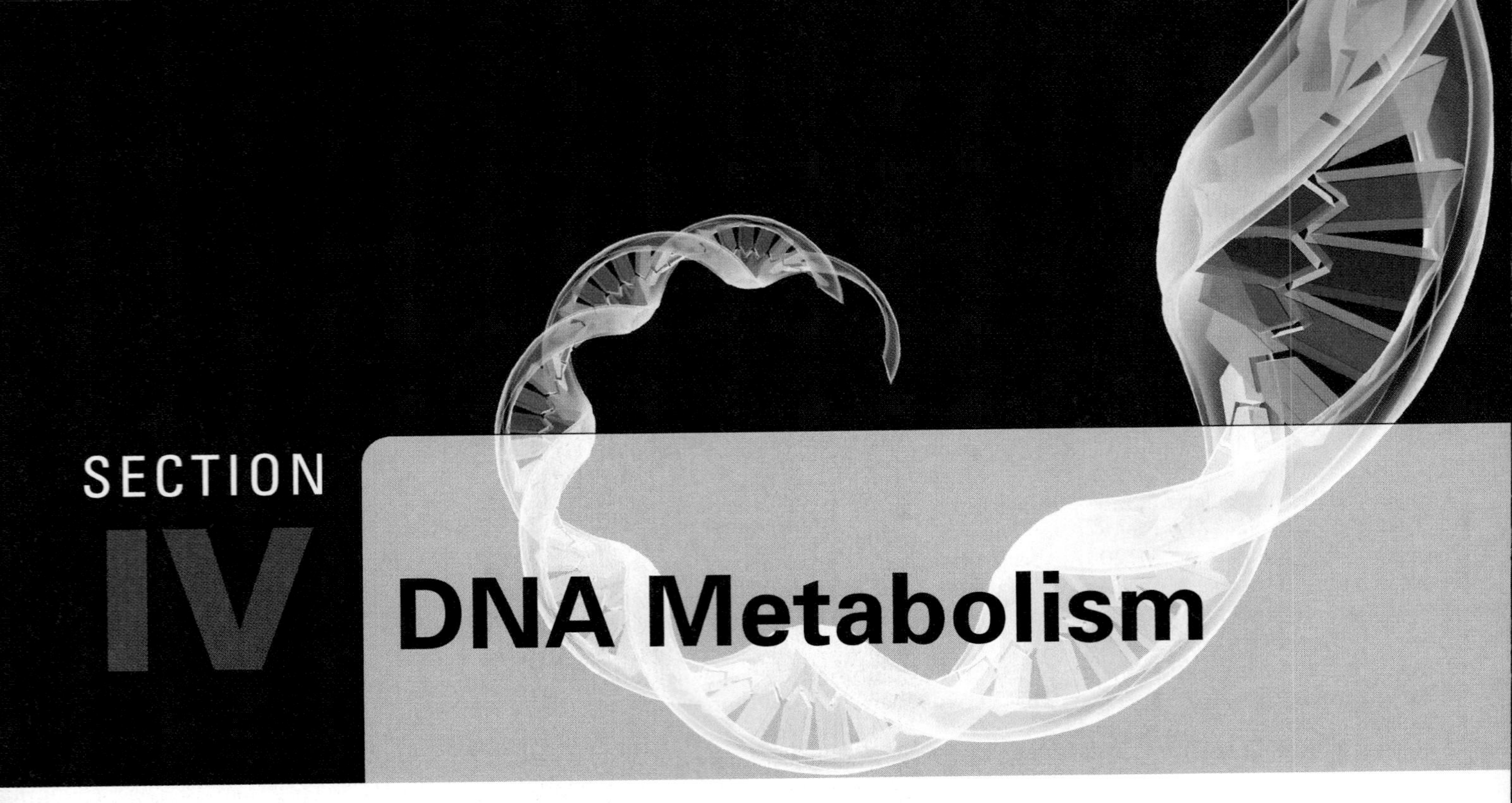

DNA Metabolism

CHAPTER 9 **DNA Replication in Bacteria**

CHAPTER 10 **DNA Replication in Eukaryotes and the Archaea**

CHAPTER 11 **DNA Damage**

CHAPTER 12 **DNA Repair**

CHAPTER 13 **Recombination**

CHAPTER 14 **Transposons and Other Mobile Elements**

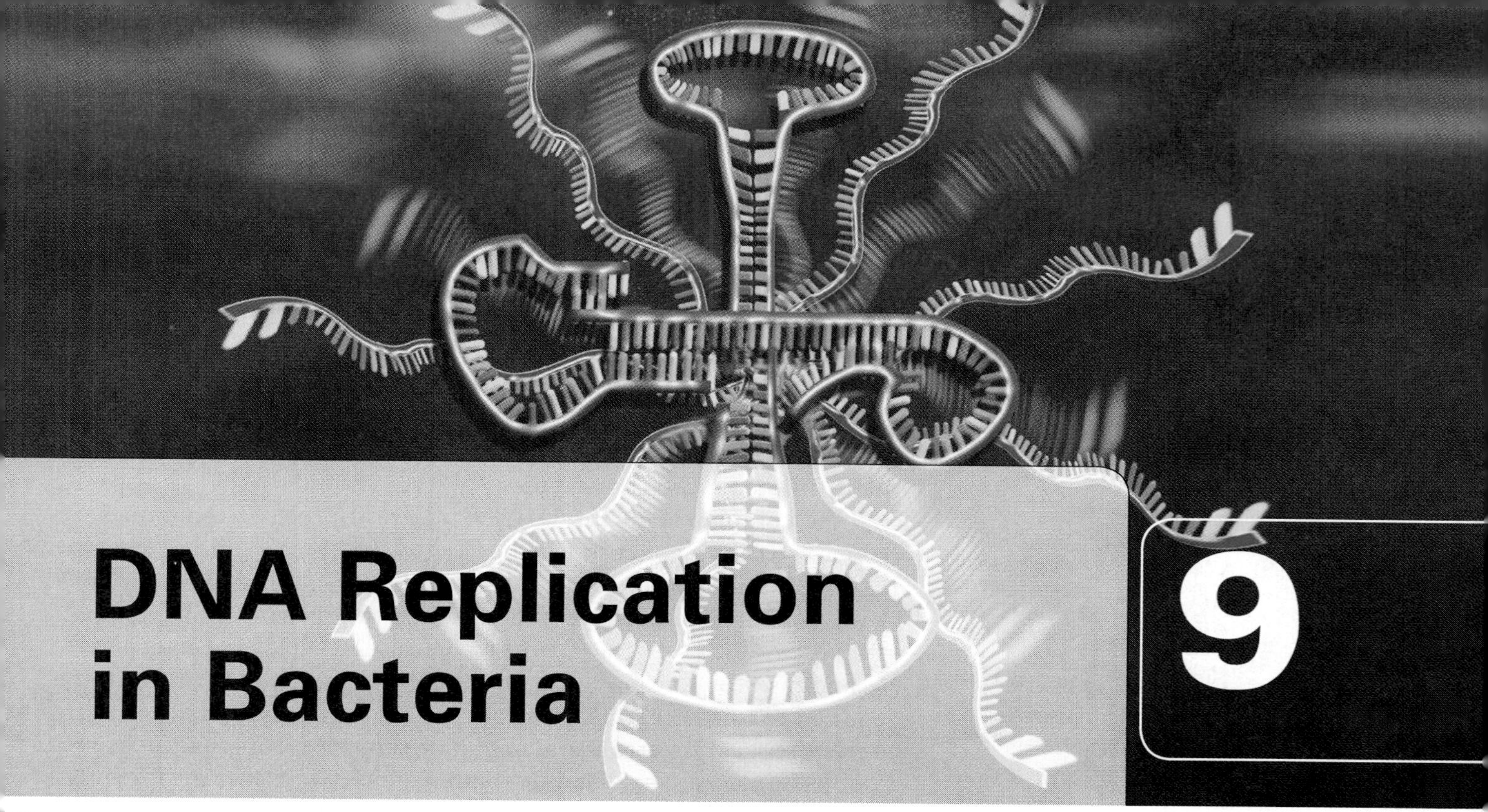

DNA Replication in Bacteria

9

OUTLINE OF TOPICS

9.1 General Features of DNA Replication

DNA replication is semiconservative.

Bacterial and eukaryotic DNA replication is bidirectional.

DNA replication is semidiscontinuous.

DNA ligase connects adjacent Okazaki fragments.

RNA serves as a primer for Okazaki fragment synthesis.

The bacterial replication machinery has been isolated and examined *in vitro.*

Mutant studies provide important information about the enzymes involved in DNA replication.

9.2 The Initiation Stage

The replicon model proposes that an initiator protein must bind to a DNA sequence called a replicator at the start of replication.

E. coli chromosomal replication begins at *oriC.*

DnaA, the bacterial initiator protein, has four functional domains.

DnaA•ATP assembles to form a filament at *oriC,* causing the DNA unwinding element (DUE) to melt.

DnaB helicases have double-ring structures.

DnaC loads DnaB helicase onto the single-stranded DNA generated at the DUE.

DnaG (primase) catalyzes RNA primer synthesis.

9.3 The Elongation Stage

Several enzymes act together at the replication fork.

DNA polymerase III is required for bacterial DNA replication.

A polymerase's processivity can be determined by using a polymerase "trap" to bind the polymerase after it dissociates from its DNA substrate.

DNA polymerase holoenzyme has ten distinct subunits that form three subassemblies.

The core polymerase has one subunit with 5′→3′ polymerase activity and another with 3′→5′ exonuclease activity.

The β clamp forms a ring around DNA, tethering the remainder of the polymerase holoenzyme to the DNA.

The clamp loader places the sliding clamp around DNA.

The DNA polymerase III holoenzyme clamp loader has three τ subunits.

The replisome catalyzes coordinated leading and lagging strand DNA synthesis at the replication fork.

Core polymerase is released from the β clamp by a premature release (also called signaling release) or collision release mechanism.

Three models have been proposed to explain how helicase moves 5′→3′ on the lagging strand while primase moves in the opposite direction as it synthesizes primer.

9.4 The Termination Stage

Replication terminates when the two growing forks meet in the terminus region, which is located 180° around the circular chromosome from the origin.

The terminus utilization substance (TUS) binds to *Ter* sites.

9.5 Regulation of Bacterial DNA Replication

Three mechanisms regulate bacterial DNA replication at the initiation stage.

Suggested Reading

The purpose of DNA replication is to copy a DNA molecule so that the two resulting daughter DNA molecules have the same genetic information as the original parent DNA. The process must proceed with great fidelity so that information is not lost or changed. When the Watson-Crick Model of double-stranded DNA was first proposed in 1953, many investigators assumed that just a few enzymes might suffice to catalyze DNA replication. As investigators studied DNA replication, however, they soon became aware that the process is considerably more complex than first assumed and requires many different kinds of enzymes.

The complexity of the DNA replication process results in part from the following facts: (1) helicases (see Chapter 4) and energy are required to unwind the double helix; (2) single-strand DNA binding (SSB) proteins (see Chapter 4) are needed to prevent the single DNA strands produced by the action of helicases from being hydrolyzed by nucleases or forming intrastrand base pairs; (3) DNA ligases are required to join nicked DNA strands; (4) safeguards are necessary to prevent replication errors and eliminate the rare errors that do occur; (5) topoisomerases (see Chapter 4) are required to release the torsional strain that builds up as a circular DNA or long-linear duplex replicates; and (6) special protein factors and enzymes are needed to initiate DNA replication and to terminate it. This chapter begins by examining a few general features of the replication process and then examines the sequence of events that take place during the initiation, elongation, and termination stages of bacterial DNA replication.

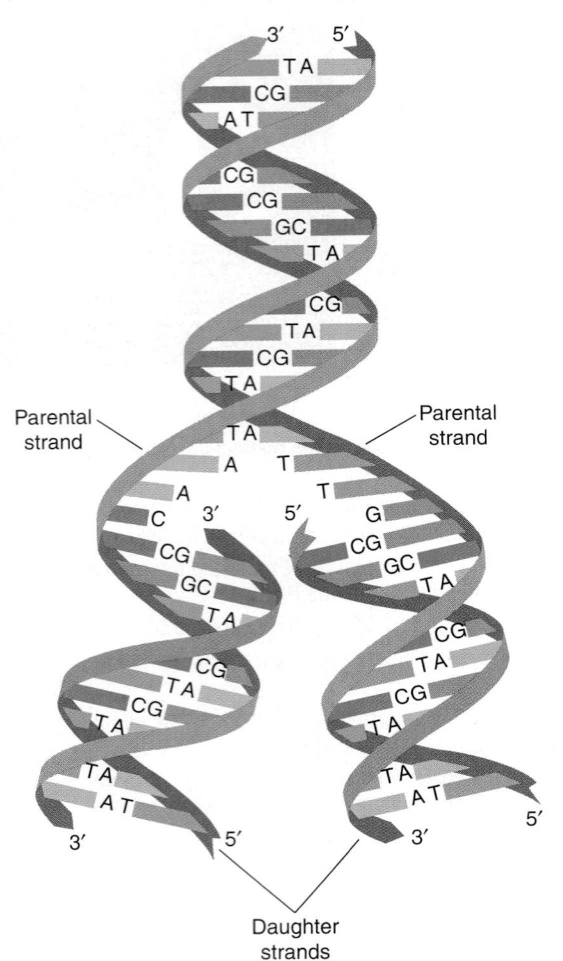

FIGURE 9.1 **Watson-Crick model of DNA replication.** Watson and Crick proposed that DNA replicates by separating the two parental strands and creating two new daughter strands by pairing new nucleotides with template nucleotides using the A:T and G:C pairing rule. Conserved parental strands are in blue and newly synthesized strands in red. The details of DNA replication are considerably more complex than shown in this figure because all new DNA strands are synthesized by adding new nucleotides to the 3′-end of existing primer chains. All new chains, therefore, are synthesized in a 5′→3′ direction.

9.1 General Features of DNA Replication

DNA replication is semiconservative.

Watson and Crick recognized that their DNA model suggests a replication mechanism in which the two parental strands separate, allowing each separated strand to serve as a template for the synthesis of a complementary strand (FIGURE 9.1). According to this replication mechanism, which is termed the **semiconservative model for DNA replication**, each double-stranded daughter DNA molecule will have a conserved DNA strand that is derived from the parental DNA and a newly synthesized strand (FIGURE 9.2a). At the time the semiconservative model was proposed, DNA denaturation was not understood and strand separation was, for a variety of reasons, considered to be impossible. Two alternative models, the conservative and dispersive models, thus also seemed possible.

The **conservative model of replication** makes two assumptions. First, the two strands of the double helix unwind at the replication site only to the extent needed for the base sequence there to be read by the polymerizing enzyme. Second, the two original strands remain entwined after replication so that one of the two DNA molecules present after replication contains both original strands (is conserved) and the other DNA molecule is made of two new strands (FIGURE 9.2b).

The **dispersive model of replication** shares some of the features of the conservative model but predicts that each strand of the daughter DNA molecules has interspersed sections of both old and new DNA (FIGURE 9.2c). Comparison of the three mechanisms of replication shown in Figure 9.2 reveals that they make different predictions about the composition of daughter DNA molecules after one or two rounds of replication. It, therefore, would be possible to establish the correct model, if some method could be devised to distinguish between new and old DNA strands.

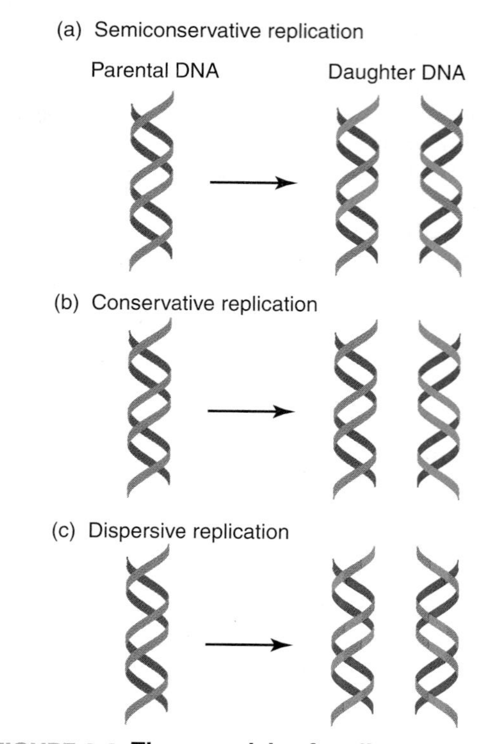

FIGURE 9.2 **Three models of replication.** (a) Semiconservative DNA replication predicts that each daughter helix will have one conserved (parental) DNA strand and one newly synthesized DNA strand. (b) Conservative DNA replication predicts that one daughter DNA molecule will have two conserved (parental) DNA strands while the other daughter DNA molecule will have two new DNA strands. (c) Dispersive DNA replication predicts that each daughter DNA molecule will have interspersed sections of both old and new DNA.

In 1958, just five years after the Watson-Crick Model was proposed, Matthew Meselson and Franklin Stahl realized that they might be able to use equilibrium density gradient centrifugation to demonstrate that DNA replication is semiconservative. They started by culturing *E. coli* for many generations in a growth medium containing [^{15}N]NH_4Cl as the sole source of nitrogen so that the purine and pyrimidine bases in DNA were uniformly labeled with the heavy isotope. Then they (1) transferred the bacteria to a new growth medium containing [^{14}N]NH_4Cl, (2) removed samples from the culture at various times, (3) extracted DNA from the samples, (4) added the extracted DNA to a concentrated cesium chloride solution, and (5) subjected the mixture to high-speed centrifugation for about one day to establish a cesium chloride equilibrium density gradient (see Chapter 5). Photographs of the gradient solutions were taken using ultraviolet light with a wavelength of 260 nm to reveal dark DNA bands. The results of the experiment are shown in FIGURE 9.3. The column labeled generations indicates the number of generations that the cells were incubated in the growth medium containing [^{14}N]NH_4Cl. The photographs are oriented so that cesium chloride density increases from left to right (top to bottom of the centrifuge tube). The tracing to the right of each photograph shows the 260 nm light absorption of the DNA in the photograph.

The data are fully consistent with the semiconservative model of replication and rule out the two other replication models. The semiconservative model predicts that one generation after transfer to the [^{14}N]NH_4Cl medium, each daughter DNA molecule should have one new [^{14}N]DNA strand and one parental [^{15}N]strand, producing a hybrid DNA molecule with a density between that of a DNA molecule with two [^{15}N]strands and one with two [^{14}N]strands. This is exactly what is observed at generation 1. The conservative model predicts the appearance of two DNA molecules, one with two [^{15}N] strands and one with two [^{14}N]strands. The data at generation 1, therefore, clearly rule out the conservative model but do not rule out the dispersive model. However, data from the next generation do so. The semiconservative model predicts that after two generations, two DNA molecules should have a hybrid density (one [^{15}N]DNA strand and one [^{14}N]DNA strand) and two DNA molecules should be made entirely of [^{14}N]DNA strands. The photograph of the DNA

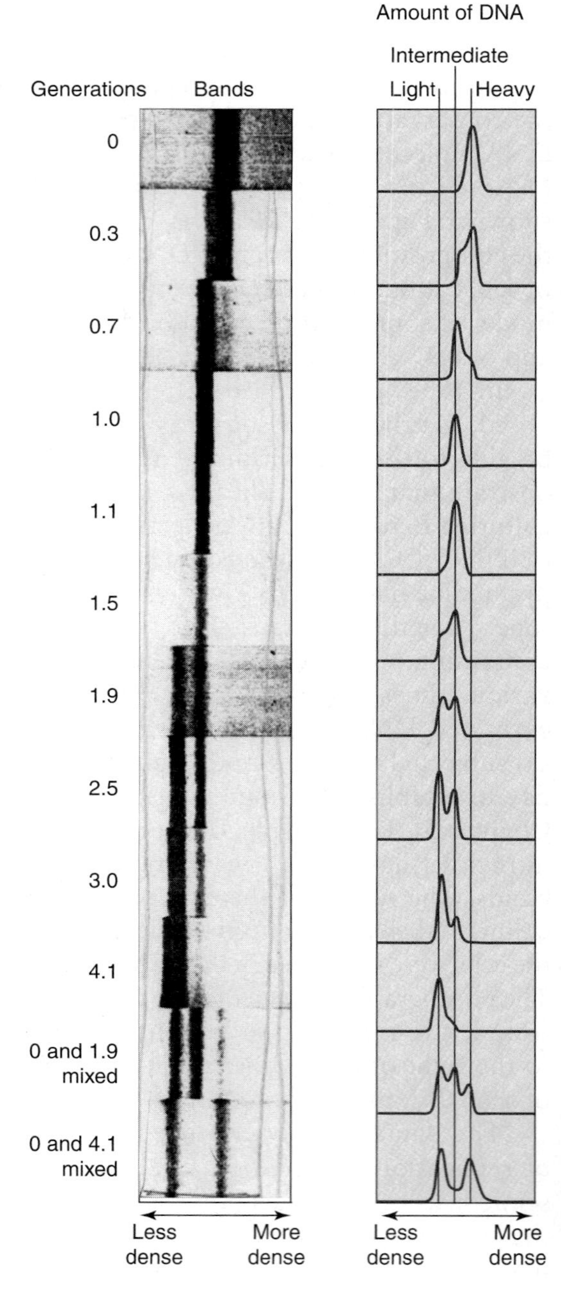

FIGURE 9.3 **Meselson-Stahl experiment showing semiconservative DNA replication.** *E. coli* were cultured in a growth medium containing $[^{15}N]NH_4Cl$ as the sole source of nitrogen for several generations so that purine and pyrimidine bases in DNA were uniformly labeled with the ^{15}N heavy isotope. Then the cells were transferred to a new growth medium containing $[^{14}N]NH_4Cl$. Samples were removed at intervals and their extracted DNA was subjected to equilibrium density centrifugation. During centrifugation, each type of DNA molecule moves until it comes to rest at a position in the centrifuge tube at which its density equals the density of the CsCl solution at that position. The column labeled generations indicates the number of generations that the cells were incubated in the growth medium containing $[^{14}N]NH_4Cl$. Photographs of the centrifuge tubes taken with ultraviolet light are shown at the left. The photographs are oriented so that the cesium chloride density increases from left to right (top to bottom of the centrifuge tube). The tracing to the right of each photograph shows the 260 nm light absorption of the DNA in the photograph. (Photo reproduced from M. Meselson and F. W. Stahl, *Proc. Natl. Acad. Sci. USA* 44 [1958]: 671–682. Photo courtesy of Matthew S. Meselson, Harvard University. Graph adapted from M. Meselson and F. W. Stahl, *Proc. Natl. Acad. Sci. USA* 44 [1958]: 671–682.)

sample after two generations of growth (shown as generation 1.9 in Figure 9.3) reveals two bands containing equal quantities of DNA. One band has the density predicted for hybrid DNA and the other band has the density predicted for DNA with two new strands. In contrast, the dispersive model predicts that all the DNA molecules should contain DNA strands with interspersed segments of $[^{15}N]$DNA and $[^{14}N]$DNA. Clearly, the photograph of DNA extracted from cells at generation 1.9 argues against the dispersive model. The semiconservative model is also the only one that is fully consistent with the results observed after four generations of growth (shown as generation 4.1 in Figure 9.3).

A second experiment confirmed the structure of the hybrid DNA found after one generation. In this experiment, Meselson and Stahl denatured the hybrid DNA by heating it to 100°C and then centrifuged the denatured DNA in cesium chloride. The heated DNA yielded two bands; one had the density expected for single-stranded [^{14}N]DNA and the other the density expected for single-stranded [^{15}N]DNA. This denaturation experiment proves that the hybrid duplex does in fact consist of one ^{14}N-strand and one ^{15}N-strand. Subsequent studies showed that DNA replication in other organisms also follows the semiconservative model.

Bacterial and eukaryotic DNA replication is bidirectional.

At the time that Meselson and Stahl performed their experiment, investigators mistakenly believed that naturally occurring *E. coli* DNA molecules were linear duplexes. Autoradiography studies performed by John Cairns in 1963 (described in Chapter 4), however, showed that *E. coli* DNA is circular. Cairns also obtained autoradiograms of bacterial DNA molecules in the process of replication. Because images of these molecules resembled the Greek letter θ (theta), Cairns called the replicating intermediates **θ-structures.** One of the most famous autoradiograms from the Cairns' collection is shown in FIGURE 9.4 along with Cairns interpretation.

The θ-structure can be explained by **unidirectional** (FIGURE 9.5a) or **bidirectional replication** (FIGURE 9.5b). In unidirectional replication, a single growing point moves around the circular DNA until replication is complete. In bidirectional replication, two growing points start at the same site and move in opposite directions until they meet at the opposite side of the circle. A region called the **replication bubble,** which contains newly synthesized DNA, grows as DNA synthesis continues. Cairns' autoradiography experiments do not distinguish between unidirectional and bidirectional replication and so a new approach was required.

In 1973, Elizabeth B. Gyurasits and R. Gerry Wake devised a labeling protocol to distinguish between unidirectional and bidirectional replication. They first added [^{3}H]thymidine with a relatively low specific activity to germinating spores of the gram-positive bacteria *Bacillus subtilis* to allow the replication bubble to become lightly labeled. Then the cells were diluted into a medium that contained [^{3}H]thymidine with a much higher specific activity and the mixture was incubated for a short time. Unidirectional replication predicts a lightly labeled bubble with a short heavily labeled segment at one end, whereas bidirectional replication predicts a lightly labeled bubble with heavily labeled segments at both ends (FIGURE 9.6). As shown in FIGURE 9.7, the data support bidirectional bacterial DNA replication. FIGURE 9.8 provides a simple schematic of bidirectional θ-replication. Subsequent experiments performed in many different laboratories have shown that DNA replication in other kinds of bacteria and in the nucleus of eukaryotes is also bidirectional. However, not all DNA replication is bidirectional. For example, *E. coli* plasmid ColE1 replication is unidirectional.

FIGURE 9.4 Demonstration of a θ structure intermediate in *E. coli* chromosome replication. Bottom figure: *E. coli* cells were cultured in the presence of [^{3}H]thymidine for one entire generation and part of a second. Then the DNA was gently extracted and covered with a photographic gel. Weak β-rays released by the tritium produce dark spots on the photographic gel. Top figure: Cairns diagrammatic interpretation shows that the structure can be divided into three sections (A, B, and C) that arise at two forks (X and Y). (Reproduced from J. Cairns, *Cold Spring Harb. Symp. Quant. Biol.* 28 [1964]: p. 44. Copyright 1963, Cold Spring Harbor Laboratory Press. Used with permission of John Cairns, University of Oxford.)

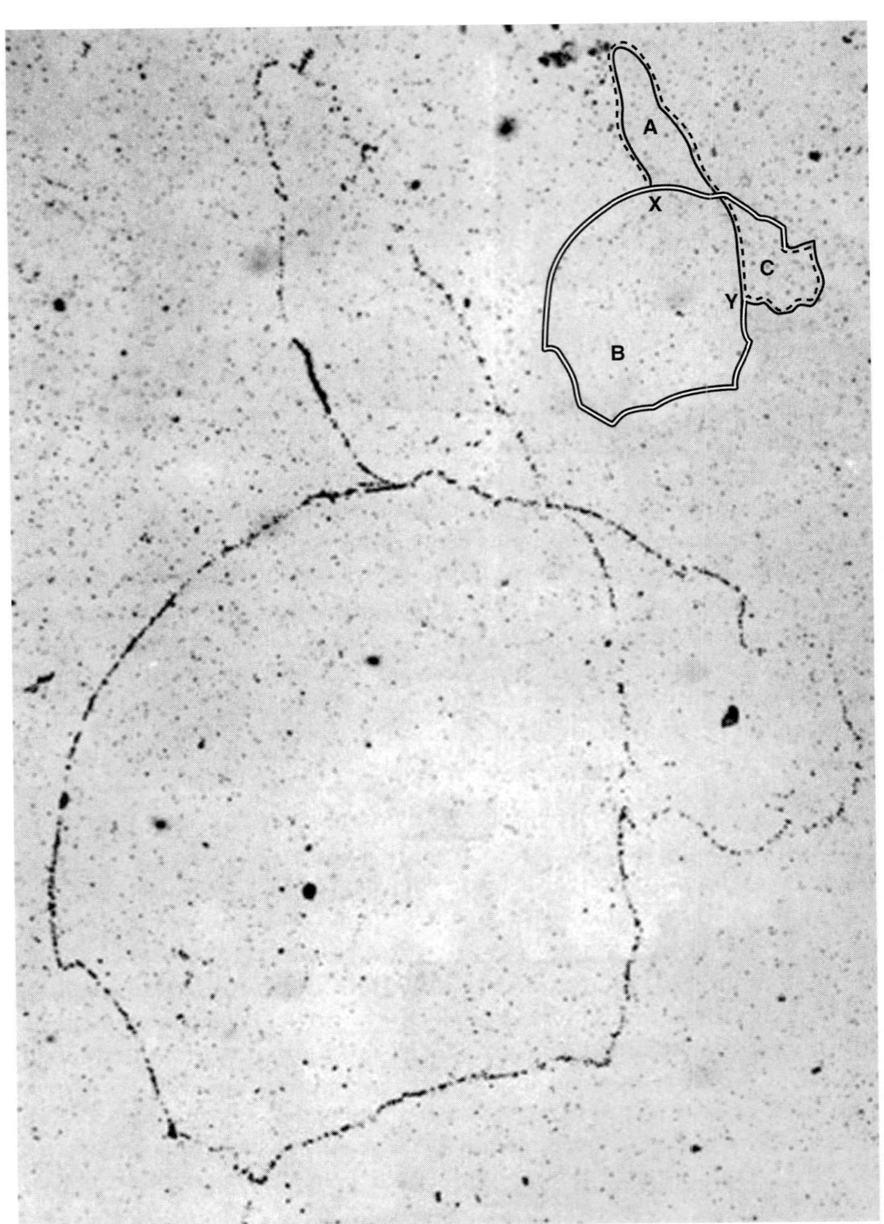

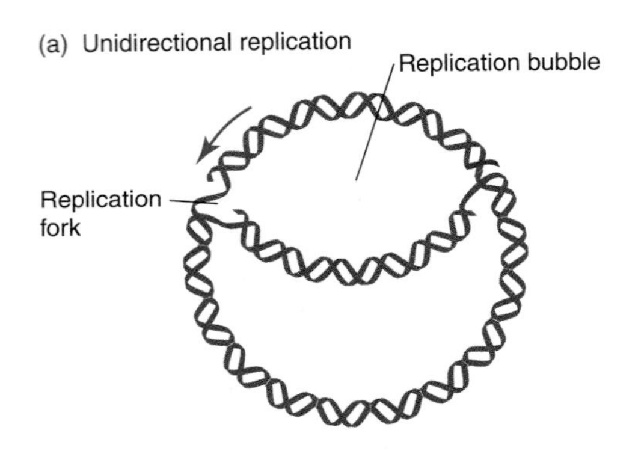

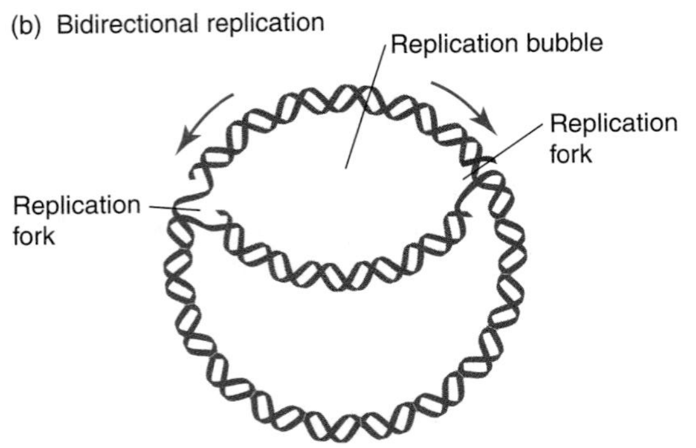

FIGURE 9.5 Two alternate methods of replication. (a) In unidirectional replication, a single replication fork (growing point) moves in the direction shown by the red arrow. (b) In bidirectional replication, two replication forks (growing points) move in opposite directions, as shown by the red arrows. Parental DNA is shown in blue and newly replicated DNA is shown in red.

DNA replication is semidiscontinuous.

In vivo studies of bacterial DNA replication led investigators to conclude that both new DNA strands grow in the same direction at the replication fork. This conclusion presents a problem because it requires one new chain to grow 3′→5′ and the other to grow 5′→3′. All known DNA polymerases add new nucleotides to the 3′-end of the growing chain and therefore catalyze 5′→3′ chain growth. Despite extensive efforts to find a DNA polymerase that adds nucleotides to the 5′-end of a growing chain, no such enzyme has ever been found. How then does the replication machinery synthesize the daughter strand that grows 3′→5′?

In 1968, Reiji Okazaki, working in Japan, proposed that 3′→5′ chain growth is **discontinuous**; that is, the replication machinery first

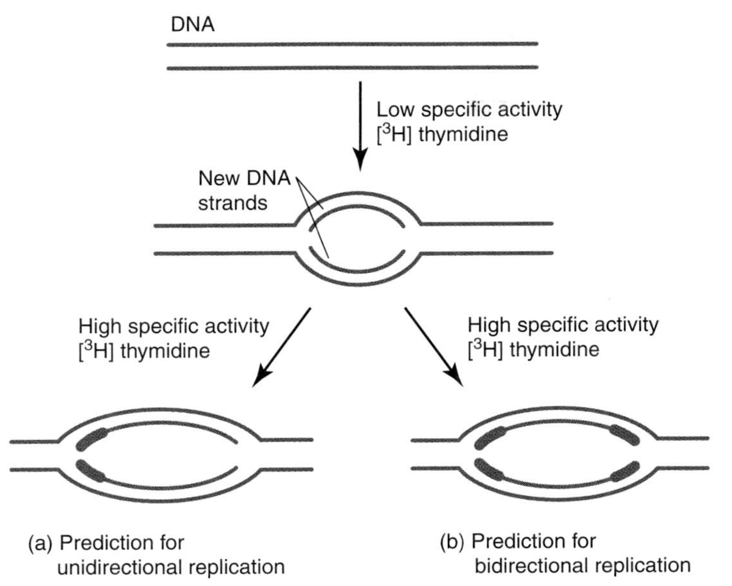

FIGURE 9.6 **Test for bidirectional bacterial DNA replication.** Germinating *B. subtilis* spores are incubated in a medium containing [^{3}H]thymidine with low specific activity to lightly label new strands in the replication bubble (thin red lines). The cells are then diluted into a medium containing [^{3}H]thymidine with a much higher specific activity and incubated for a short time. (a) Unidirectional replication predicts the two lightly labeled strands in the replication bubble will each have one short highly labeled strand on the same side of the replication bubble (thick red lines). (b) Bidirectional replication predicts the two lightly labeled strands (thin red lines) will have two highly labeled segments (thick red lines)–one on each side.

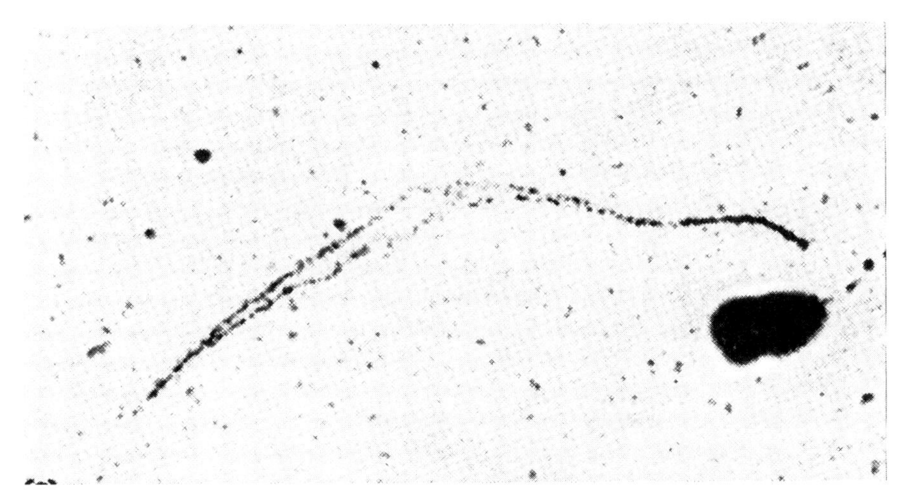

FIGURE 9.7 **Evidence to support bidirectional θ-replication.** [^{3}H]Thymidine with a low specific activity was added to germinating *B. subtilis* spores. A sufficient incubation period was allowed to permit the replication bubble to be lightly labeled. Then the cells were diluted into a medium containing [^{3}H] thymidine with a high specific activity. As predicted by bidirectional replication, the grain tracks revealed by autoradiography show the replication bubble has a lightly labeled segment, which is flanked by heavily labeled segments. (Reprinted from *J. Mol. Biol.*, vol. 73, E. B. Gyurasits and R. G. Wake, Bidirectional chromosome replication . . ., pp. 55–58, copyright 1973, with permission from Elsevier [http://www.sciencedirect.com/science/journal/00222836]. Photo courtesy of R. G. Wake, Professor Emeritus, The University of Sydney, Australia.)

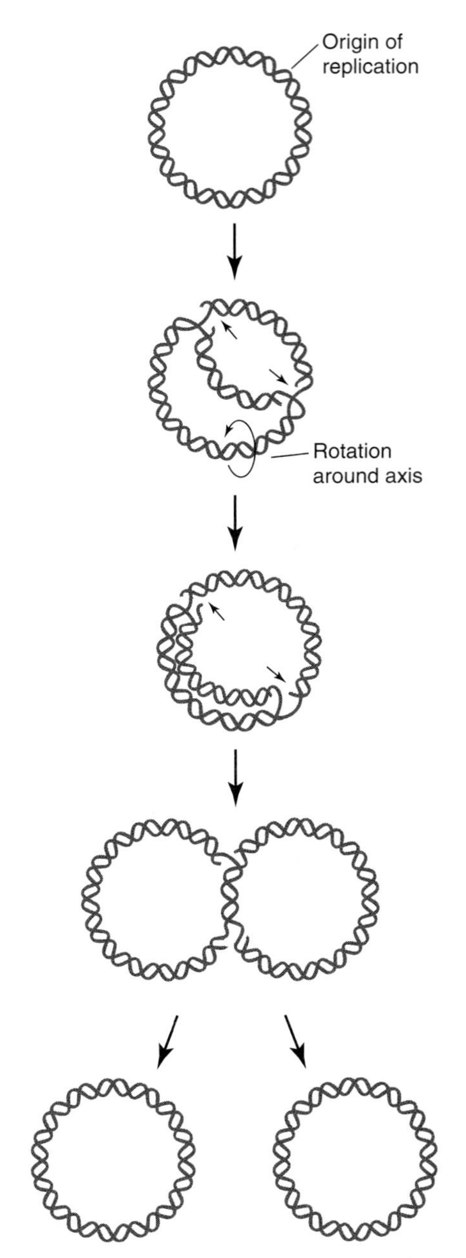

FIGURE 9.8 **Bidirectional replication of circular DNA molecules gives rise to θ structures.** (Adapted from an illustration by Kenneth G. Wilson, Miami University.)

makes small DNA fragments in a 5′→3′ direction and then joins the fragments. Because there was no logical requirement for discontinuous synthesis of the strand growing in an overall 5′→3′ direction, Okazaki could not be sure whether its synthesis was continuous or discontinuous. He therefore proposed the two alternative models shown in FIGURE 9.9. The first, called the **discontinuous model of replication,** postulates discontinuous synthesis of both new strands (FIGURE 9.9a). The second, the **semidiscontinuous model of replication,** also postulates discontinuous synthesis of the chain growing in an overall 3′→5′ direction but continuous synthesis of the chain growing in an overall 5′→3′ direction (FIGURE 9.9b).

The discontinuous model predicts that all newly formed DNA should exist as small fragments, whereas the semidiscontinuous model predicts that half should exist as fragments and half as long DNA strands. Okazaki followed fragment formation by subjecting DNA, which had been extracted from *E. coli* cultured in the presence of [^{3}H] thymidine for various times, to sucrose gradient centrifugation (see Chapter 5). Cells were incubated at 20°C to slow down the replication process, making it easier to follow changes in DNA as a function of time. The results of one particularly informative experiment are shown in FIGURE 9.10. In this experiment, the sucrose solution was made alkaline to cause the DNA strands to separate. Nearly all of the radioactive label incorporated into DNA after the cells were incubated with [^{3}H] thymidine for 5 or 10 seconds was present in a slowly sedimenting fraction near the top of the centrifuge tube. The radioactivity in this

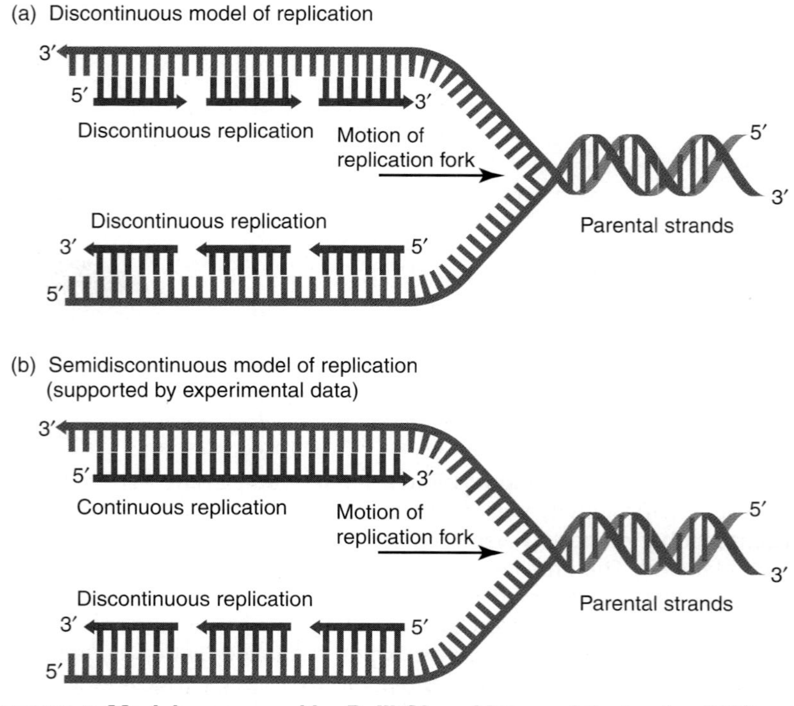

FIGURE 9.9 **Models proposed by Reiji Okazaki to explain *in vivo* DNA replication.** Parental strands are shown in blue and newly replicated strands are shown in red. The arrowheads indicate the 5′→3′ direction of DNA polymerase synthesis.

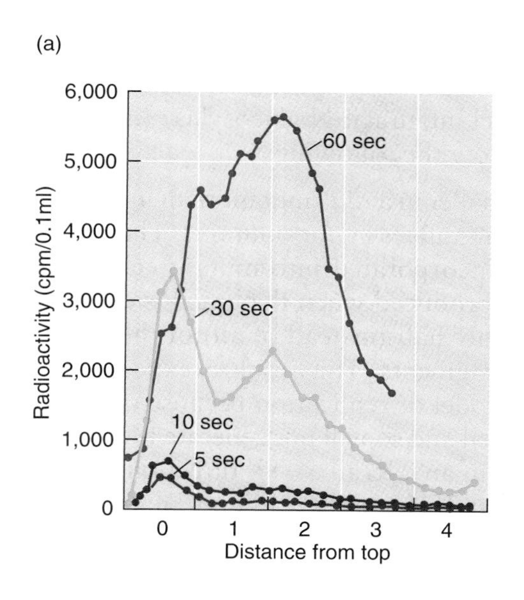

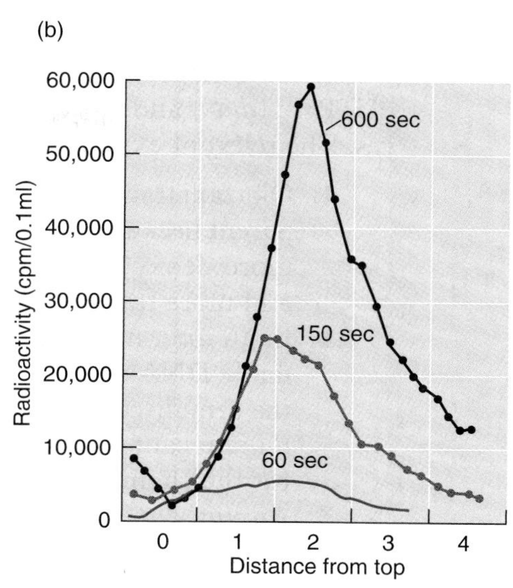

FIGURE 9.10 Okazaki's pulse labeling experiment. *E. coli* were cultured in the presence of [^{3}H]thymidine at 20°C for the indicated times and then the DNA was extracted and analyzed on alkaline sucrose gradients. The scales shown on the y-axis are different in (a) and (b) because more radioactive label was incorporated into DNA during the longer incubation periods shown in (b). The top of the gradient is on the left and the bottom is on the right. (Adapted from K. Sugimoto, T. Okazaki, and R. Okazaki, *Proc. Natl. Acad. Sci. USA* 60 [1968]: 1356–1362.)

fraction continued to increase until about 30 seconds and then leveled off. Based on sedimentation rates, Okazaki estimated the DNA in the slowly sedimenting fraction was between 1000 and 2000 nucleotides long. Under similar conditions, denatured *E. coli* DNA is usually 20 to 50 times longer (it would of course be longer if the strands were not broken in the course of isolation). A second labeled DNA fraction, which is barely noticeable in the 10 second sample, continues to increase in size in each subsequent sample and by 60 seconds it is the predominant form. These results appear to fit the discontinuous model of replication (Figure 9.9a), in which all newly synthesized DNA first appears as small fragments that are only later stitched together to form longer strands.

Okazaki's results were surprising because one would expect the strand growing in an overall 5′→3′ direction to do so in a continuous fashion. In fact, this strand is made continuously and fragmented after synthesis. The reason for the post-synthetic fragmentation is as follows. *E. coli* synthesizes small quantities of dUTP. Even though the bacteria have a hydrolytic enzyme, dUTPase, that converts dUTP to dUMP, trace quantities of dUTP are always present. Because the DNA replication machinery cannot distinguish between TTP, its normal substrate, and dUTP some uracil ends up in DNA. The cell's repair machinery acts to remove these uracil groups. The enzyme **uracil N-glycosylase** participates in the repair process by removing uracil from the deoxyribose group to which it is attached. Then the modified DNA strand is cut at the site lacking the pyrimidine base and the damaged region is removed and the excised nucleotides replaced by new ones. The details of the repair process will be described in Chapter 12. For now, the important point is that the repair system removes uracils from DNA and the repair steps following uracil removal are fairly slow. Furthermore, once the uracil is removed, the phosphodiester bond is readily hydrolyzed by alkali. Thus, newly synthesized DNA will be fragmented in the alkaline sucrose gradient. The question then is whether any of the fragments observed in Figure 9.10 are true precursor fragments arising as suggested in Figure 9.9b. The answer, which comes from studies with two bacterial mutants, one lacking dUTPase (*dut*$^-$) and the other lacking uracil N-glycosylase (*ung*$^-$), is yes. The relevant experiments show the following:

1. Okazaki fragments are smaller in a *dut*$^-$ mutant than in its *dut*$^+$ parent because dUTP accumulates in *dut*$^-$ mutants. Therefore, more deoxyuridylate is incorporated into mutant cell DNA and more fragments are produced when the uracil is excised.
2. In an *ung*$^-$ mutant, roughly half (instead of all) of the newly made DNA consists of fragments. This result is expected for the semidiscontinuous model of replication because the *ung*$^-$ mutant cannot excise uracil and so will not generate fragments for the chain that grows in an overall 5′→3′ direction
3. An *ung*$^-$*dut*$^-$ double mutant behaves like an *ung*$^-$ mutant.

These results argue that about half of the DNA fragments produced in wild type cells are true precursor fragments (now called **Okazaki fragments** in honor of their discoverer) and that the semidiscontinuous model of replication shown in Figure 9.9b is correct.

High-resolution electron micrographs of replicating phage λ DNA molecules, showing a short single-stranded region on one side of the replication fork (FIGURE 9.11), also support the semidiscontinuous model of replication. This short single-stranded region results from the fact that synthesis of the discontinuous strand is initiated only

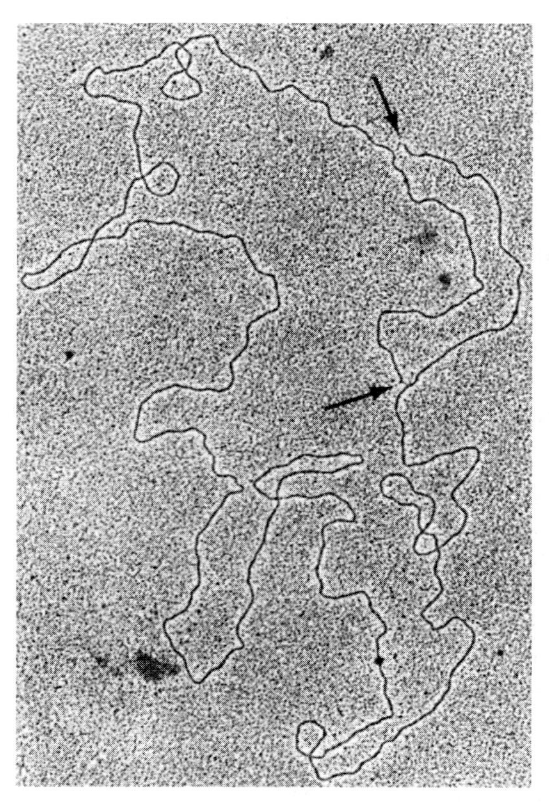

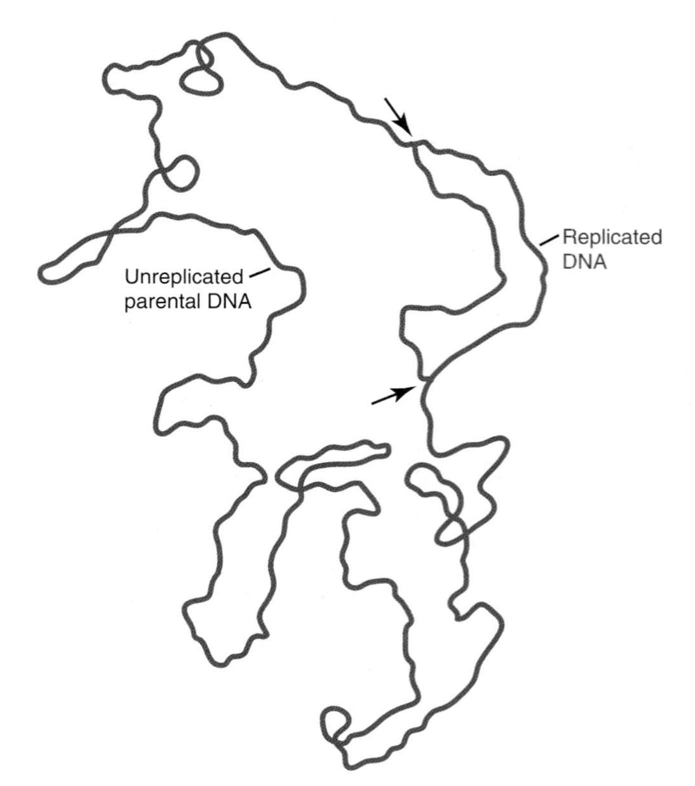

FIGURE 9.11 **θ-Replication of phage λ DNA.** (a) A replicating θ molecule of phage λ DNA. The arrows show the two replicating forks. The segment between each pair of thick lines at the arrows is single-stranded; note that it appears thinner and lighter. (b) An interpretive drawing. (Part a photo courtesy of Manuel S. Valenzuela, Division of Cancer Biology, Meharry Medical College.)

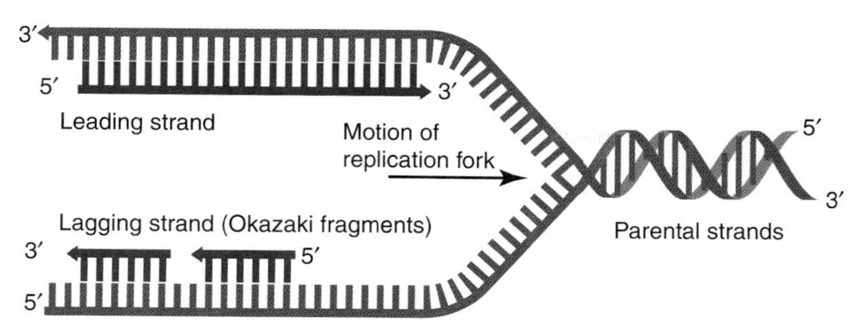

FIGURE 9.12 **Semidiscontinuous model of replication.** Both daughter strands (red) are synthesized in a 5′→3′ direction. The leading strand is synthesized continuously, however, and the lagging strand is synthesized discontinuously as Okazaki fragments.

periodically. In fact, the 3′-OH terminus of the continuously replicating strand is always ahead of the discontinuously replicating strand. For this reason, the convenient terms, **leading strand** and **lagging strand,** are used for the continuously and discontinuously replicating strands, respectively (FIGURE 9.12). Semidiscontinuous replication has been observed in all organisms and DNA viruses studied to date.

DNA ligase connects adjacent Okazaki fragments.

The semidiscontinuous model of replication requires an enzyme to join adjacent Okazaki fragments. DNA ligase, an enzyme discovered in 1967 by investigators interested in DNA recombination and repair, seemed to be an excellent candidate. Okazaki established that DNA ligase is the enzyme that is required to connect DNA fragments formed during semidiscontinuous replication by demonstrating that mutants with a defective DNA ligase accumulate large quantities of fragments.

DNA ligases fall into two groups, the ATP- and NAD^+-dependent ligases. All known eukaryotic DNA ligases require ATP. Phage T4 DNA ligase is also ATP-dependent. Bacterial DNA ligases are NAD^+-dependent. NAD^+ is a somewhat surprising energy source because it usually serves as a cofactor in redox reactions. Bacterial DNA ligases use the energy present in the phosphoanhydride bond that links the nicotinamide mononucleotide (NMN) and adenylate groups.

Several methods can be used to detect DNA ligase *in vitro*. One of the most convenient and informative uses a substrate formed by annealing oligo(dT) molecules with 5′-[^{32}P]phosphates to poly(dA) (FIGURE 9.13). The substrate is incubated with DNA ligase and ATP (or NAD^+) and the reaction stopped by placing the tube containing the reaction mixture in a boiling water bath. Then alkaline phosphatase is added to release any 5′-[^{32}P]phosphates that have not been converted into phosphodiester bonds. DNA ligase activity is monitored by measuring radioactivity that remains attached to the DNA after acid precipitation.

DNA ligation takes place in three nucleotidyl transfer steps (FIGURE 9.14). (1) An adenylyl group is transferred from ATP (or NAD^+) to an ε-amine group on an active site lysine in DNA ligase. (2) Then the adenylyl group is transferred to the 5′-phosphate at the DNA nick site. This transfer activates the 5′-phosphate. (3) The 3′-hydroxyl group at the nick

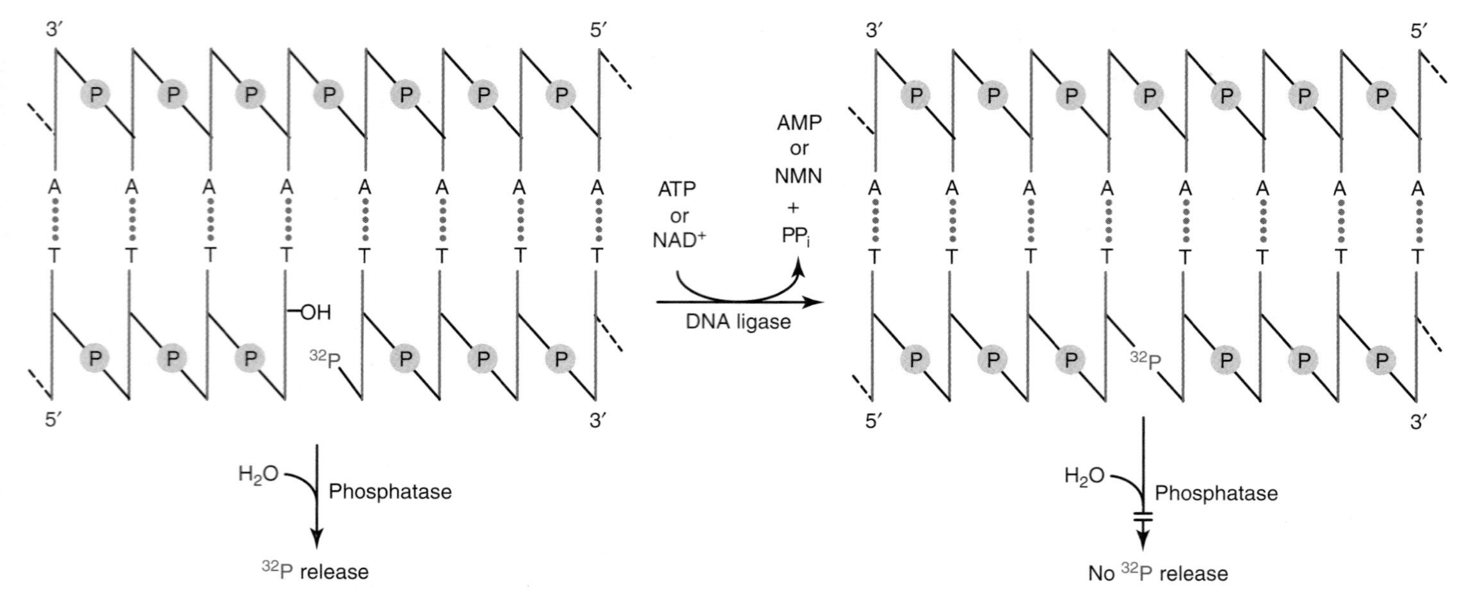

FIGURE 9.13 **A method for measuring DNA ligase activity.** The substrate, 5′-[^{32}P]-oligo(dT) molecules annealed to poly(dA), is incubated with DNA ligase and ATP (or NAD$^+$). The reaction is stopped by placing the tube containing the reaction mixture in a boiling water bath. Then alkaline phosphatase is added to release any 5′-[^{32}P]phosphates that have not been converted into phosphodiester bonds. DNA ligase activity is monitored by measuring radioactivity that remains attached to the DNA after acid precipitation.

site makes a nucleophilic attack on the activated 5′-phosphate, displacing the AMP and forming a phosphodiester bond and repairing the nick.

RNA serves as a primer for Okazaki fragment synthesis.

The semidiscontinuous model of replication requires the repeated initiation of DNA synthesis. However, bacterial DNA polymerases cannot initiate the synthesis of a new DNA strand; they can only add deoxyribonucleotides to the 3′-ends of preexisting primers. Then how does Okazaki fragment synthesis begin? Okazaki answered this question by showing that bacteria synthesize RNA oligonucleotides that serve as primers for DNA polymerase (FIGURE 9.15). The ribonucleotides have to be removed and replaced by deoxyribonucleotides before DNA ligase joins adjacent fragments. The pathway for this process is shown in FIGURE 9.16. **Ribonuclease H** (**RNase H**) removes all of the ribonucleotides at the 5′-terminus of the Okazaki fragment except for the ribonucleotide that is linked to the DNA end. **DNA polymerase I** (see Chapter 5) catalyzes nick translation. Its polymerase activity extends the 3′-end of the newly formed Okazaki fragment and its 5′→3′ exonuclease removes the ribonucleotide from the 5′-terminus of the neighboring Okazaki fragment. Finally, **DNA** ligase joins adjacent Okazaki fragments to form the lagging strand.

The bacterial replication machinery has been isolated and examined *in vitro*.

While the *in vivo* experiments described above provide considerable information about bacterial DNA replication, they do not shed much

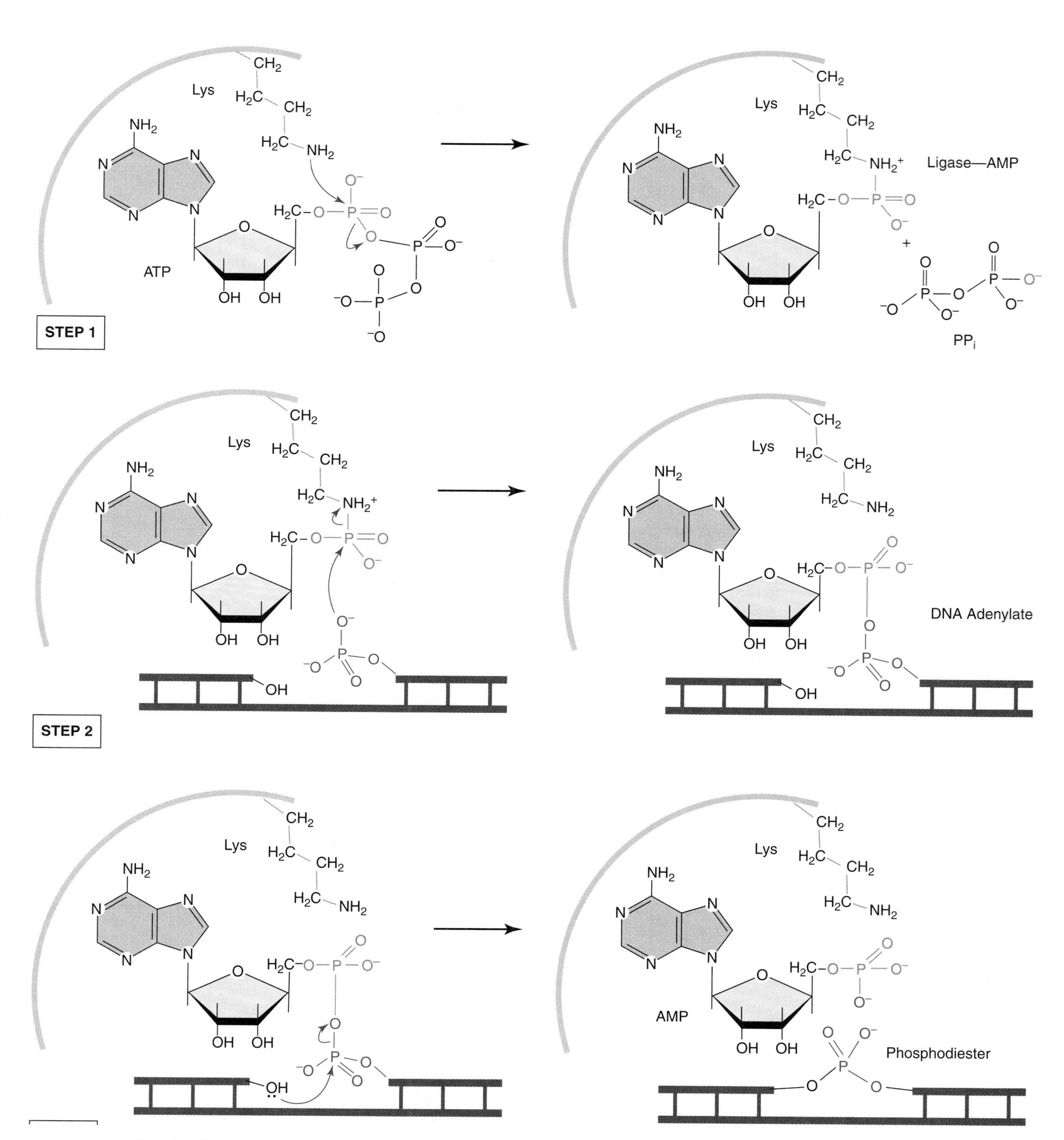

FIGURE 9.14 **Mechanism of nick sealing by DNA ligase.** DNA ligation takes place in three nucleotidyl transfer steps. (1) An adenylyl group is transferred from ATP (or NAD^+) to an ε-amine group on an active site lysine in DNA ligase. (2) Then the adenylyl group is transferred to the 5′-phosphate at the DNA nick site. This transfer activates the 5′-phosphate. (3) The 3′-hydroxyl group makes a nucleophilic attack on the activated 5′-phosphate, displacing the AMP and forming a phosphodiester bond and repairing the nick. (Adapted from S. Shuman, *J. Biol. Chem.* 284 [2009]: 17365–17369.)

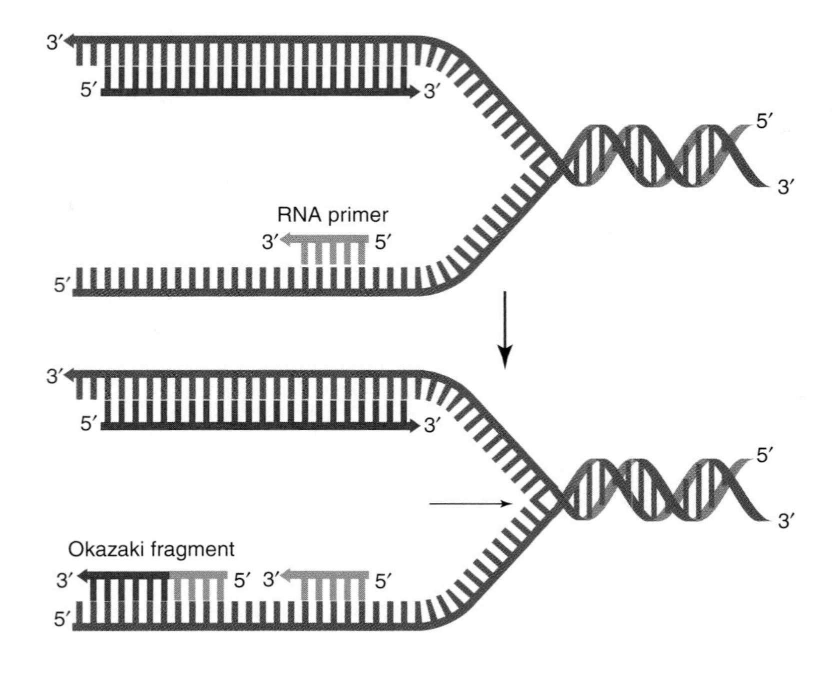

FIGURE 9.15 **RNA primers for Okazaki fragments.** The synthesis of Okazaki fragments is primed by short RNA segments. RNA primers are shown in green, parental DNA in blue, and new DNA in red.

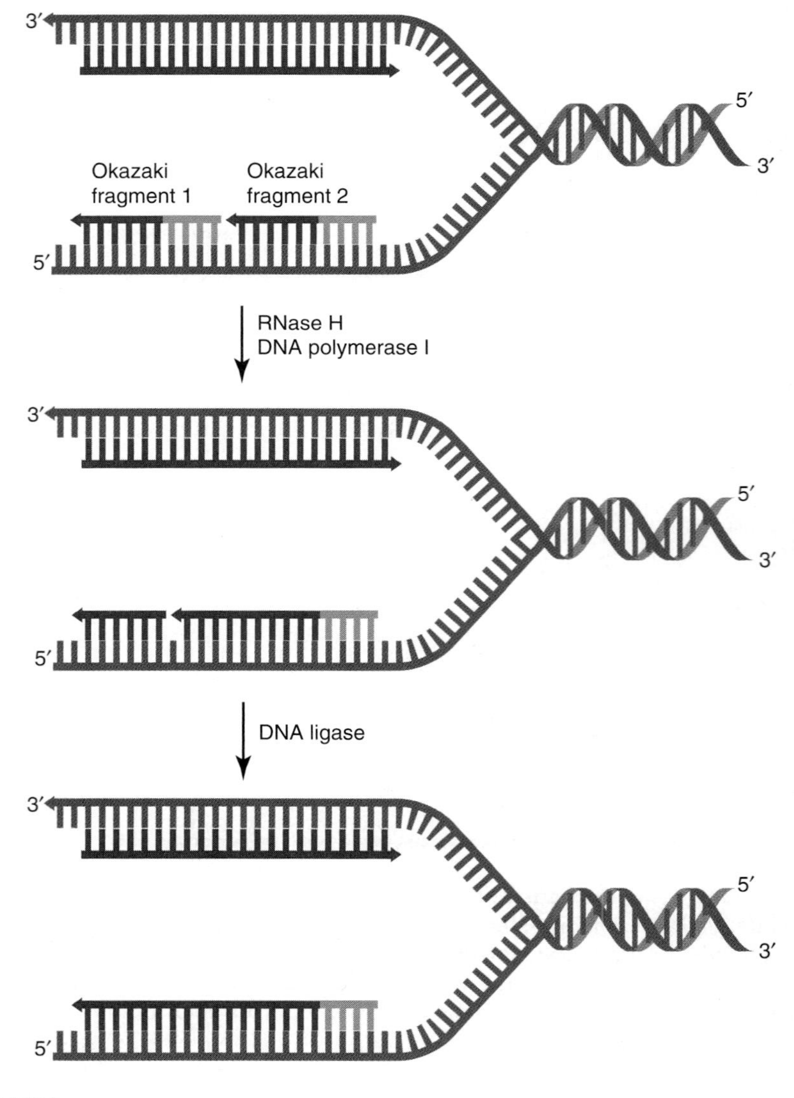

FIGURE 9.16 **Pathway for processing Okazaki fragments.** Ribonuclease H (RNase H) removes all of the ribonucleotides at the 5′-terminus of the Okazaki fragment except for the ribonucleotide that is linked to the DNA end. Then DNA polymerase I catalyzes nick translation. Its 5′→3′ exonuclease removes the ribonucleotide from the 5′-terminus of the neighboring Okazaki fragment. Finally, DNA ligase joins adjacent Okazaki fragments to form the lagging strand. RNA primers are shown in green, parental DNA in blue, and new DNA in red.

light on the cellular apparatus responsible for DNA replication. The best way to learn about the replication apparatus is to isolate it and then examine how each part works by itself and in concert with the other parts. This approach has been used to study the replication apparatus of viruses, bacteria, eukaryotes, and the archaea. Because the *E. coli* replication apparatus is probably the one that has been most thoroughly investigated, we begin by examining it. It is important to note that fundamental lessons learned from studying the *E. coli* replication apparatus apply to the replication apparatus of other organisms.

The replication apparatus is made of many different protein components that must be assembled on DNA before replication can begin. The first stage of DNA replication, **initiation**, involves the assembly of the replication apparatus at a unique site on the bacterial chromosome. Special proteins help to assemble the replication apparatus but do not participate further in the replication process. Initiation is followed by **elongation**, a process in which the leading strand is synthesized continuously and the lagging strand is synthesized discontinuously. A complex replication machine known as **DNA polymerase III holoenzyme** is responsible for DNA synthesis during this stage. The final stage of replication, **termination**, begins when the two-replication forks meet about half way around the DNA molecule; it also requires specific proteins.

Mutant studies provide important information about the enzymes involved in DNA replication.

Molecular biologists have learned a great deal about the replication process by studying mutants with blocks in specific steps in replication. Table 9.1 lists several essential genes for DNA replication in *E. coli* and the proteins that they encode along with some information about the proteins. Because a defect in DNA synthesis would be lethal, most of the replication mutants that have been studied are temperature-sensitive. Early replication mutants were isolated before the functions of the altered genes were known, and therefore were named *dnaA*, *dnaB*, and so forth. A major reason for the gaps in the alphabetical listing is that some genes were renamed when they were later shown to influence DNA synthesis indirectly rather than directly. For instance, the gene originally named *dnaF* was renamed *nrdA* when it was shown to be the structural gene for a subunit in ribonucleoside diphosphate reductase, the enzyme that converts ribonucleotides to deoxyribonucleotides.

Important clues to replication gene function were obtained by determining the effects that temperature shifts have on DNA synthesis in conditional mutants. A mutant with a defect in an enzyme required for chain extension will stop DNA synthesis immediately after being switched from a permissive to a restrictive temperature. For this reason they are called **quick-stop mutants**. In contrast, a mutant with a defect in an enzyme required for the initiation of DNA replication will complete the ongoing round of DNA synthesis but will not initiate a new round. These mutants are called **slow-stop mutants** because

TABLE 9.1	Essential Genes for DNA Replication		
Gene	**Mutant Phenotype**	**Protein**	**Primary Function**
dnaA	Slow stop	DnaA	*oriC* recognition
dnaB	Quick stop	DnaB	helicase
dnaC	Slow stop or quick stop	DnaC	Delivers DnaB helicase to *oriC*
dnaE	Quick stop	α subunit of DNA polymerase III holoenzyme	Polymerase activity
dnaG	Quick stop	primase	RNA primer synthesis
dnaN	Quick stop	Sliding clamp	Tethers core polymerase subassembly to DNA
dnaQ	Hypermutation	ε-subunit DNA polymerase III holoenzyme	3′→5′ exonuclease activity
dnaX	Quick stop	γ and τ subunits DNA polymerase III holoenzyme	γ subunit—ATP binding site in clamp loader subassembly τ subunit—central organizer for DNA polymerase III holoenzyme
holA	Quick stop	δ subunit DNA polymerase III holoenzyme	Subunit in clamp loader subassembly—opens sliding clamp
holB	Essential	δ′ subunit DNA polymerase III holoenzyme	Subunit in clamp loader subassembly—regulates activity
holC	Slow growth	χ subunit DNA polymerase III holoenzyme	Subunit in clamp loader subassembly—switching mechanism
hol*D*	Unknown	ψ subunit DNA polymerase III holoenzyme	Subunit in clamp loader subassembly—function is unknown
holE	None	θ subunit DNA polymerase III holoenzyme	Function is unknown
ssb	Quick stop	SSB	Binds to single-stranded DNA
gyrA	Quick stop	Subunit in DNA gyrase	ATP-driven negative supercoiling
gyrB	Quick stop	Subunit in DNA gyrase	ATP-driven negative supercoiling
polA	Defective DNA repair	DNA polymerase I	Gap filling and primer removal
ligA	Accumulates nascent fragments	DNA ligase	Seal DNA nicks

they continue to synthesize DNA for some time after the temperature switch. As more was learned about the replication process and additional mutants were discovered, investigators named the new genes according to their function. For instance, *gyrA* and *gyrB*, the structural genes for the two subunits in DNA gyrase, were named for their function. We now turn our attention to the initiation stage of the replication process.

9.2 The Initiation Stage

The replicon model proposes that an initiator protein must bind to a DNA sequence called a replicator at the start of replication.

Although genes replicate as part of larger chromosomes, most cannot replicate as independent units. An independently replicating DNA molecule such as a viral, bacterial, or eukaryotic chromosome or a plasmid that can maintain a stable presence in a cell is called a **replicon.**

In the early 1960s François Jacob and Sydney Brenner proposed the **replicon model** to explain how DNA molecules replicate autonomously. The replicon model requires two specific components, an **initiator protein** and a **replicator**. The initiator protein binds to the replicator, a specific set of sequences within the DNA molecule that is to be replicated. Because the replicator must be part of the DNA that is to be replicated, it is said to be ***cis*-acting**. Once bound to the replicator, the initiator helps to unwind the DNA and recruit components of the replication machinery. The specific site within the replicator at which replication is initiated is called the **origin of replication (*ori*)**. Bacterial cells usually require just one origin of replication. The *E. coli* origin of replication, *oriC*, maps at minute 84 of the *E. coli* genetic map (FIGURE 9.17). Because a bacterial replicator is usually quite short (200–300 bp), the terms replicator and origin of replication tend to be used interchangeably even though technically the origin of replication is just a part of the replicator. Eukaryotes, which have much longer chromosomes than bacteria, require many origins of replication.

E. coli chromosomal replication begins at *oriC*.

The *E. coli* chromosome's large size and fragility precludes studying *oriC* structure and function in the intact chromosome *in vitro*. Fortunately, many of the desired goals can be achieved by studying a plasmid that depends on *oriC* for its replication. The method used to construct this plasmid is shown in FIGURE 9.18. A restriction endonuclease is used to cut the bacterial chromosome into fragments. The same enzyme is also used to cut a plasmid DNA into a fragment containing a drug-resistance gene (ampicillin in this example) but lacking the plasmid origin of replication. After mixing, bacterial and plasmid DNA fragments are joined by DNA ligase, the resulting recombinant plasmid DNA is used to transform bacteria, and transformed bacteria are spread onto growth medium that contains ampicillin. Only cells bearing recombinant plasmid DNA with a drug-resistance gene and *oriC* will form colonies under these growth conditions. Intracellular nucleases eventually destroy DNA molecules that cannot replicate autonomously. The minimal *oriC* sequence needed for plasmid replication, which

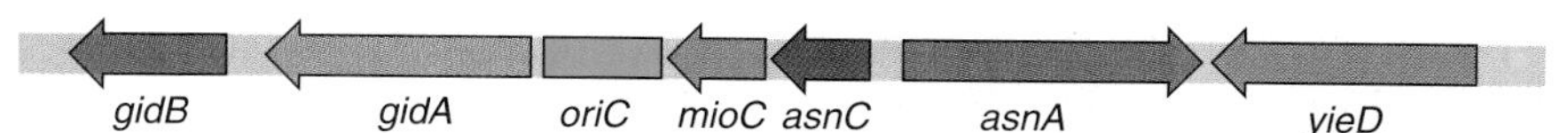

FIGURE 9.17 **The *oriC* region of the *E. coli* chromosome.** The arrows indicate the direction of gene transcription. (Adapted from S. Dasgupta and A. Løbner-Olesen, *Plasmid* 52 [2004]: 151–168.)

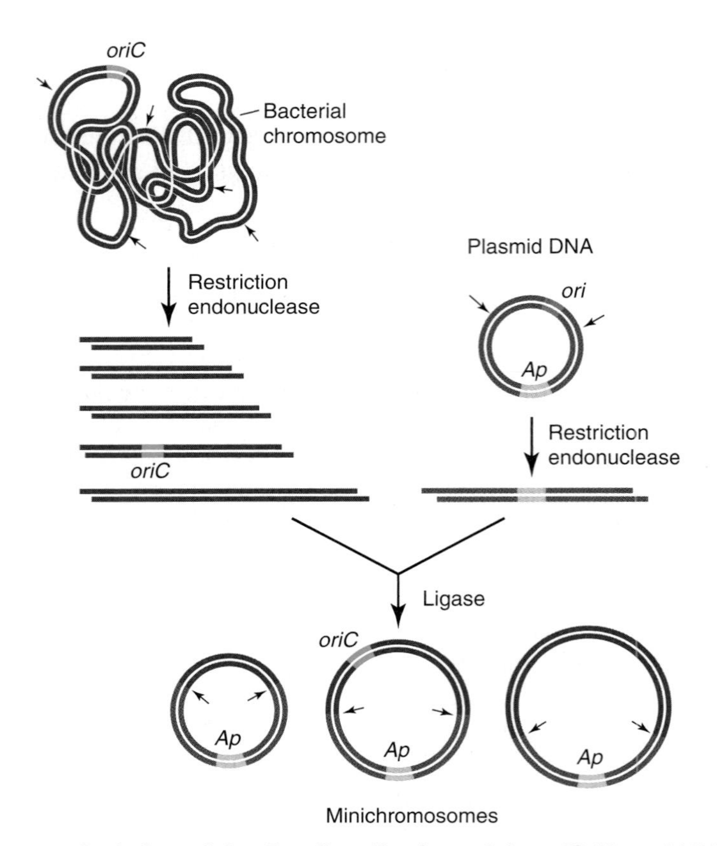

FIGURE 9.18 **Isolation of the *E. coli* replication origin *oriC*.** Plasmid DNA is digested with a restriction endonuclease to obtain a fragment that lacks the plasmid's origin of replication but retains its drug-resistance gene (ampicillin in this example). The bacterial chromosome is also digested with the restriction endonuclease to obtain a pool of different fragments. The plasmid and bacterial fragments are mixed and then joined by DNA ligase. The resulting recombinant plasmids are used to transform *E. coli*. Only transformants with a recombinant plasmid bearing both the drug-resistance gene and *oriC* will form colonies. (Adapted from A. Kornberg and T. A. Baker, *DNA Replication, Fourth edition*. W.H. Freeman & Company, 1991.)

was determined by trimming the ends of the cloned *oriC* region, is 245 bp long. Recent studies by David Bates and coworkers show that bacterial mutants remain viable with a truncated *oriC* that is missing most of the right side of *oriC* and is only 163 bp long. Further work is required to determine why plasmid DNA replication and bacterial chromosome replication have different *oriC* requirements. One possibility is that regions adjacent to *oriC* in the bacterial chromosome can somehow compensate for the deletion.

Once *oriC* had been cloned it became possible to study the initiation of DNA replication *in vitro*. Arthur Kornberg and coworkers performed the difficult but rewarding task of purifying the enzymes that participate in the initiation process and elucidating their functions. They discovered that four proteins, DnaA (initiator), DnaB (helicase), DnaC (loader), and DnaG (primase), participate in the DNA replication initiation stage. The functions of these proteins are as follows. Several DnaA•ATP molecules bind to specific 9 bp (9-mer) sites in *oriC*, causing an A-T rich region in *oriC* to melt. DnaC loader helps to load a DnaB helicase onto each of the single strands generated by the melting process. The DnaB helicases extend the single-stranded region and then bind DnaG primase, which catalyzes RNA primer synthesis. Finally, the DNA replication machinery assembles at the primer-template site to complete the initiation stage of replication. A fifth protein, Dia (DnaA-initiator association) protein, discovered by Tsutomu Katayama and coworkers in 2004, also participates in initiation (see below). We will now examine the contribution that each protein makes to the initiation stage in greater detail.

DnaA, the bacterial initiator protein, has four functional domains.

The *E. coli* initiator protein, DnaA, has four domains, numbered I to IV starting at the amino terminus (FIGURE 9.19a). Domain I interacts with other proteins, including DnaB helicase, Dia, and other DnaA molecules. Domain II is a flexible connection between domains I and III in *E. coli* DnaA. Domain II is probably not essential because the fully functional DnaA from the gram-negative thermophile *Aquifex aeolicus* lacks this domain (FIGURE 9.19b). Domain III has two subdomains, IIIa and IIIb. Both are necessary to bind ATP or ADP and for ATPase activity. Domain IV recognizes and binds to specific 9 bp sequences (9-mers) in *oriC* (see below).

James M. Berger and coworkers used recombinant DNA techniques to construct a bacterial strain that produces a truncated *A. aeolicus* DnaA containing just domains III and IV and then prepared crystals of this construct with a bound ADP or a bound ATP analog (β,γ–methylene adenosine triphosphate). The ATP analog was selected because it cannot be hydrolyzed by domain III's ATPase activity. The III/IV construct has a slightly different conformation when bound to the ATP analog than when bound to ADP. We will focus on the conformation with the bound ATP analog because DnaA•ATP is the active form.

There are two important reasons to think that the crystal structure of the domain III/IV construct bound to ATP analog has a direct bearing on the way that *E. coli* DnaA works. First, domains III and IV are highly conserved and therefore should have the same structure in the *A. aeolicus* and *E. coli* DnaA molecules. Second, the domain III/IV

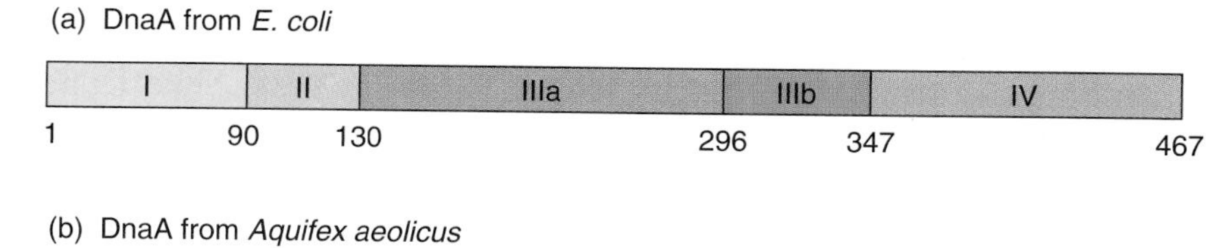

FIGURE 9.19 DnaA protein. (a) *E. coli* DnaA has four domains. (b) DnaA from *Aquifex aeolicus* is fully active even though it lacks domain II. Domain numbers are indicated by Roman numerals. Amino acid residue numbers are indicated below each figure. (Adapted from J. P. Erzberger, et al., *EMBO J.* 21 [2002]: 4763–4773.)

FIGURE 9.20 Crystal structure of a pair of *A. aeolicus* domain III/IV constructs. Each construct has an ATP analog (spacefill red) tucked into a space between subdomains IIIa and IIIb. Arg-230 (spacefill pink) in the domain III/IV construct on the right side of the figure contacts the ATP analog bound to the domain III/IV construct on the left side. This highly conserved residue is called the arginine finger. Domain IIIa is gray, domain IIIb is dark blue, and domain IV is tan. (Structure from Protein Data Bank 2HCB. J. P. Erzberger, et al., *Nat. Struct. Mol. Biol.* 13 [2006]: 676–683. Prepared by B. E. Tropp.)

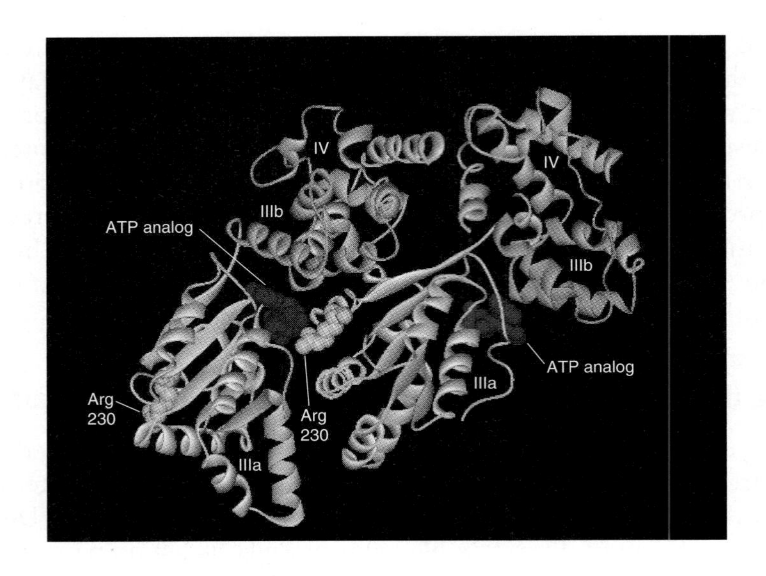

construct bound to the ATP analog forms a filament assembly on *oriC* that is very similar to that formed by DnaA•ATP. Furthermore, both filament assemblies cause nearby double-stranded DNA to melt (see below). It is important to note, however, that the III/IV construct differs from intact DnaA in one very important respect. DnaB helicase can be loaded onto the single-strands generated by DnaA•ATP filament assembly, but it cannot be loaded onto single strands generated by the domain III/IV construct assembly because the III/IV construct is missing domain I. The *A. aeolicus* domain III/IV crystal structure is an assembly of four constructs. FIGURE 9.20 shows the interactions between two of these constructs. Each construct has an ATP analog in a pocket between subdomains IIIa and IIIb. Arg-230 in the construct on the right contacts the ATP analog bound to the construct on the left. This highly conserved residue, called the **arginine finger**, makes an essential contribution to filament assembly (see below).

DnaA•ATP assembles to form a filament at *oriC*, causing the DNA unwinding element (DUE) to melt.

E. coli oriC has a DnaA assembly region on its right side and a DNA unwinding element (DUE) on its left side (FIGURE 9.21). The DnaA assembly region has ten 9-bp (9-mer) DnaA binding sites. The first five sites to be discovered, called **DnaA boxes**, have slight variations of the sequence 5′-TTATNCACA (where N is any nucleotide). Individual DnaA boxes, designated R1–R5, have different affinities for DnaA•ATP, which are as follows R1 = R4 > R2 > R3 ≈ R5. R1 and R4 also have a high affinity for DnaA•ADP. The other three DnaA boxes have a higher affinity for DnaA•ATP than for DnaA•ADP. The binding properties of the five remaining DnaA binding sites, three I sites (I1–I3) and two τ sites (τ_1 and τ_2), are similar to those of the weaker DnaA boxes.

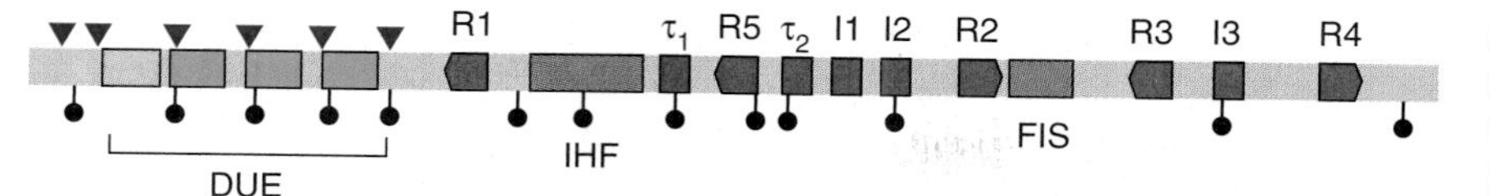

FIGURE 9.21 Minimal *oriC* region required for minichromosome replicaton. (Adapted from N. Stepankiw, et al. *Molec. Microbiol.* 74 [2009]: 467–479.)

Dia (DnaA-initiator association) protein, a homotetramer, helps DnaA•ATP bind to the eight weak binding sites. Each Dia subunit has a binding site for domain I in DnaA, allowing the tetramer to bridge DnaA molecules (FIGURE 9.22). This bridging effect leads to cooperative DnaA binding at the weak 9-mer sites. DnaA•ATP occupies all ten DnaA binding sites during the initiation stage of replication. Once the initiation stage is completed, however, DnaA•ATP is converted to DnaA•ADP, which only occupies DnaA boxes R1 and R4.

The DNA unwinding element (DUE) has three AT-rich elements, each containing 13 bp (13-mer) with slight variations of the sequence GATCTNTTNTTTT (where N is any nucleotide). DnaA•ATP complexes that occupy the 9-mer sites form a helical filament assembly that somehow causes the DUE to melt. This melting makes additional DnaA binding sites available on the upper single strand generated at the DUE and DnaA•ATP binds to these sites.

E. coli oriC also has binding sites for two other proteins that influence DnaA•ATP binding. **IHF (integration host factor)** binds to one of these sites and **FIS (factor for inversion stimulation)** binds to the other. These two proteins were initially identified because of their involvement in other processes, so their names do not provide any information about how the factors function at *oriC*. Although both proteins cause double-stranded DNA to bend (see Chapter 6), IHF and FIS have opposite effects on initiation. IHF stimulates initiation and FIS inhibits it.

Bacterial *oriC* has one other very notable feature. Based on completely random nucleotide sequences, one might expect *oriC* to have one or two GATC sites. The fact that it has so many of these sites strongly suggests that GATC sites have an important function that has been conserved during evolution. These GATC sites help to regulate DNA replication (see below).

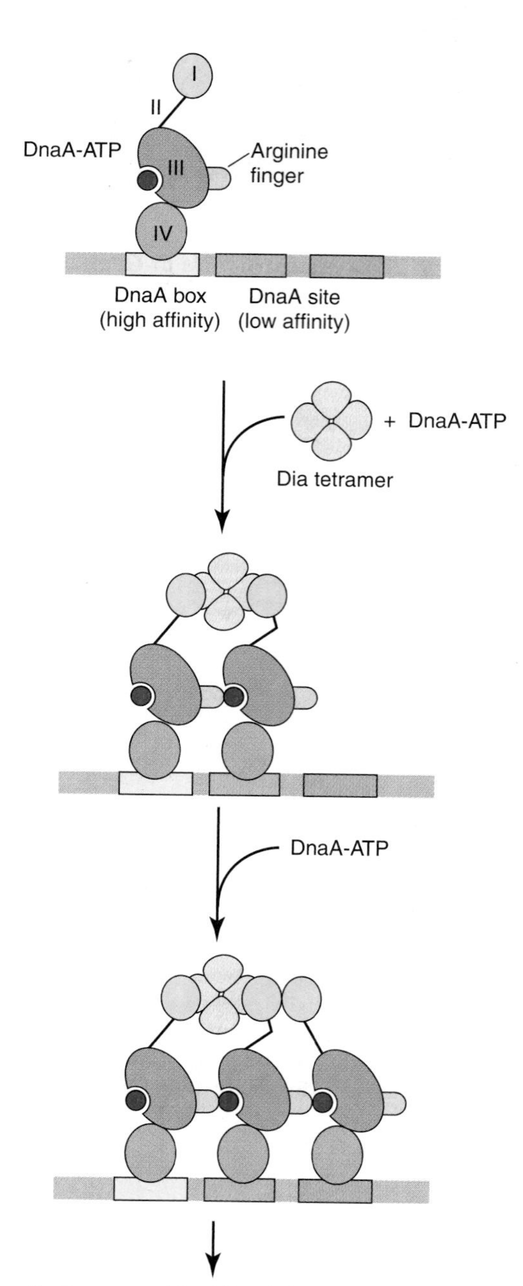

FIGURE 9.22 Model of DnaA-initiator association protein (DIA) assisting in DnaA•ATP aggregation at *oriC*. A Dia tetramer stimulates cooperative binding to *oriC* by bridging DnaA molecules. DnaA domains are indicated by Roman numerals. ATP is represented by a red circle. (Adapted from T. Katayama, *Biochem. Soc. Trans.* 36 [2008]: 78–82.)

DnaB helicases have double-ring structures.

DnaB helicase is a ring-shaped homohexamer. Each subunit contains an N-terminal domain that can interact with DnaG primase (the enzyme that catalyzes primer RNA synthesis—see below) and a C-terminal domain that binds ATP and has ATPase activity.

Both domains are required for helicase activity. Thomas Steitz and coworkers determined the crystal structure of the hexameric DnaB

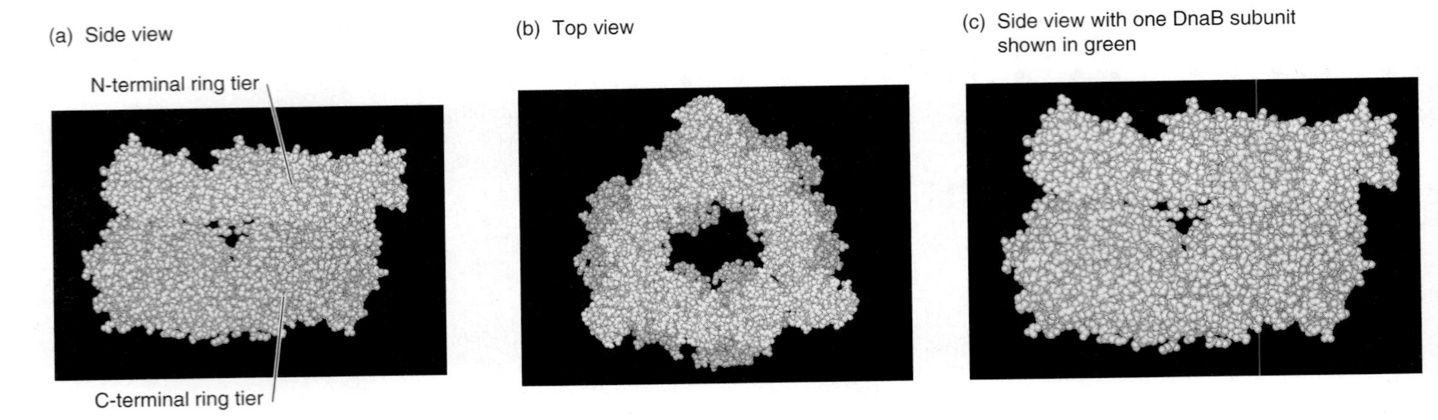

FIGURE 9.23 Crystal structure of *Geobacillus stearothermophilus* DnaB helicase. DnaB helicase has a two tier ring structure with one tier formed by N-terminal domains (yellow) and the other by C-terminal domains (blue). (a) Side view of DnaB hexamer. (b) Top view of DnaB hexamer. (c) Side view of DnaB hexamer with one DnaB subunit shown in green. (Structures from Protein Data Bank 2R6D. S. Bailey, et al., *Science* 318 [2007]: 459–463. Prepared by B. E. Tropp.)

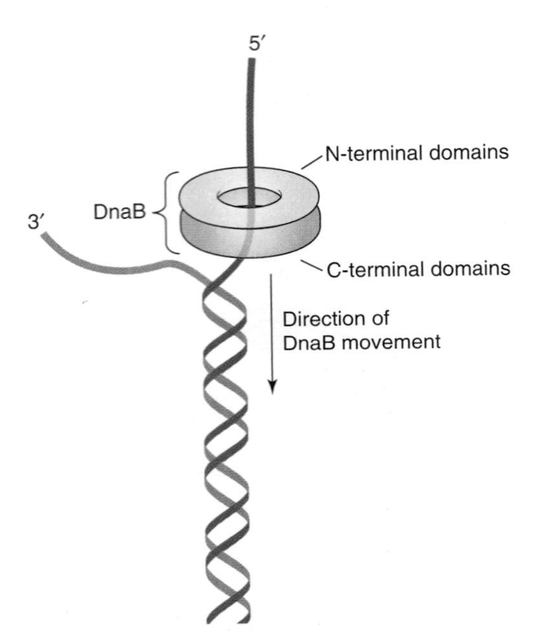

FIGURE 9.24 DnaB orientation at the replication fork. (Adapted from K. Marians, *Nat. Struct. Mol. Biol.* 15 [2008]: 125–127.)

helicase from *Geobacillus stearothermophilus*, a gram-positive thermophile. The enzyme has a two tier ring structure: the C-terminal tier is formed by C-terminal domains and the N-terminal tier by N-terminal domains (FIGURE 9.23). DnaB helicase binds to a forked DNA molecule, encircling one strand while excluding the other. The C-terminal ring tier is in front, as the DnaB helicase moves along the encircled strand in a 5′→3′ direction (FIGURE 9.24). DNA unwinding is achieved by excluding the other strand from the ring's interior and thereby forcing the duplex apart. The double-stranded circular DNA molecule now has two nascent replication forks.

DnaC loads DnaB helicase onto the single-stranded DNA generated at the DUE.

A bacterial DnaB helicase is loaded onto each of the single strands generated at the DUE. Before loading, the hexameric helicase exists as part of a $DnaB_6$•$DnaC_6$ complex with an ATP bound to each DnaC subunit. The two hexameric helicases are loaded onto the nascent replication forks in opposite orientations. DnaB hexamer orientation appears to depend on the fact that DnaA's N-terminal domain (domain I) binds to DnaB's N-terminal face, while DnaC's N-terminal domain binds to DnaB's C-terminal face. James M. Berger and coworkers used this and related information to construct a model that explains why DnaB is loaded in opposite orientations on the lower and upper single strands generated at the DUE (FIGURE 9.25). DnaB orientation on the lower single strand is determined by interaction between DnaA and DnaB N-terminal domains. The situation is somewhat different on the upper DNA strand. Additional DnaA•ATP molecules assemble to form a filament on this strand. DnaC•ATP appears to be able to add to the end of the resulting filament. DnaB hexamer orientation on the upper strand is determined by the fact that the DnaC N-terminal domain binds to the C-terminal DnaB face. Further experimental work is required to (1) verify the model, (2) determine how Dia fits into the model, and (3) elucidate the mechanism that is used to open the helicase hexamer.

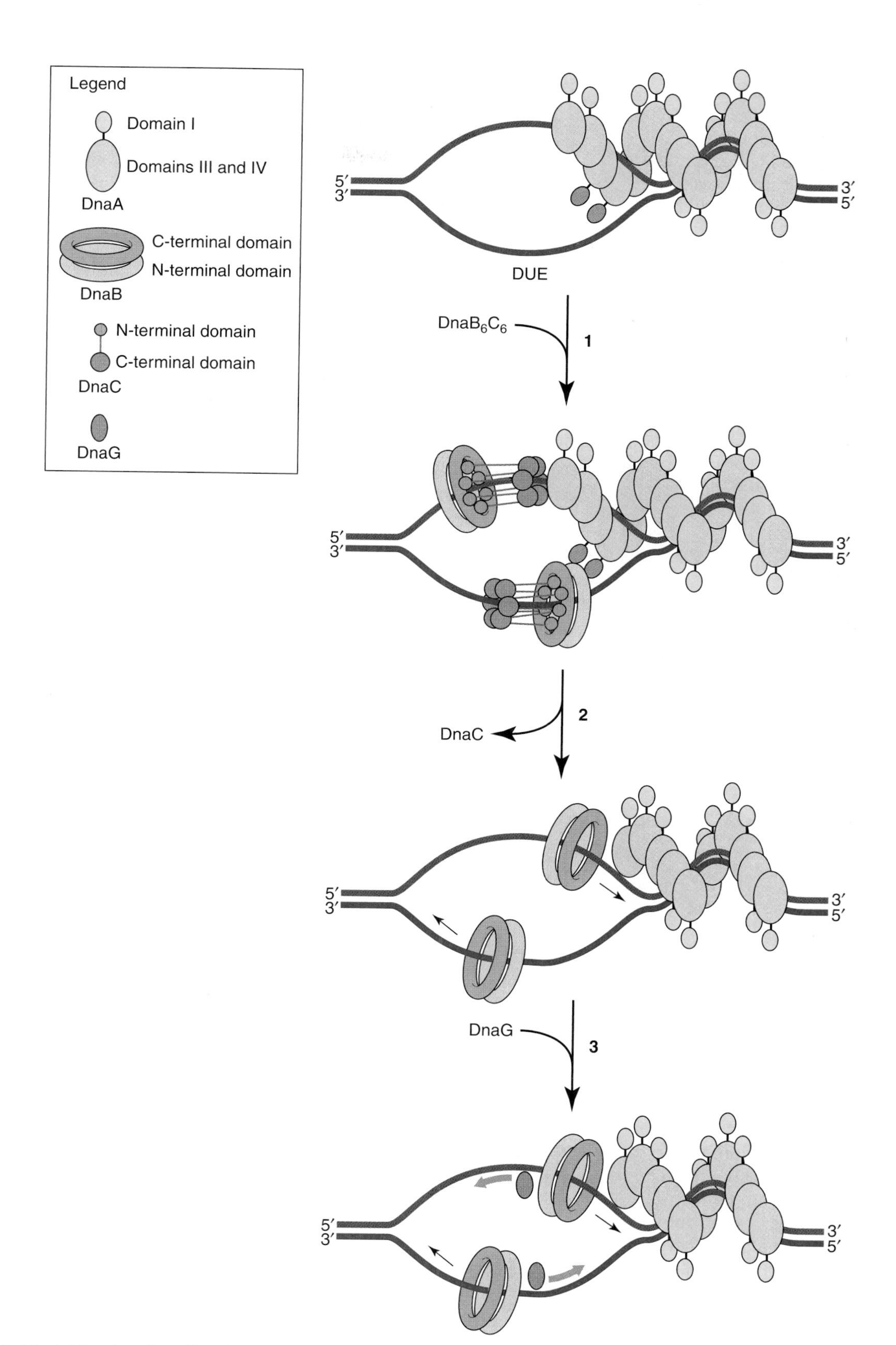

FIGURE 9.25 **Model for loading DnaB onto single strands at DUE.** (1) Two $DnaB_6{\bullet}DnaC_6$ complexes load onto the unwound DUE strands so that they end up in opposite orientations. The upper and lower DUE strands are blue and red, respectively. (2) ATP hydrolysis (not shown) leads to DnaC release, freeing each DnaB helicase to move in a 5′→3′ direction and thereby extend the unwound region of *oriC* to about 65 nucleotides. (3) DnaG primase binds to DnaB helicases and synthesizes primers (green). See text for additional details. (Adapted from M. L. Mott, et al., *Cell* 135 [2008]: 623–634.)

FIGURE 9.26 **Three domains in DnaG (primase).** (Reprinted from *J. Mol. Biol.*, vol. 300, M. Podobnik, et al., A TOPRIM domain in the crystal structure . . ., pp. 353–362, copyright 2000, with permission from Elsevier [http://www.sciencedirect.com/science/journal/00222836].)

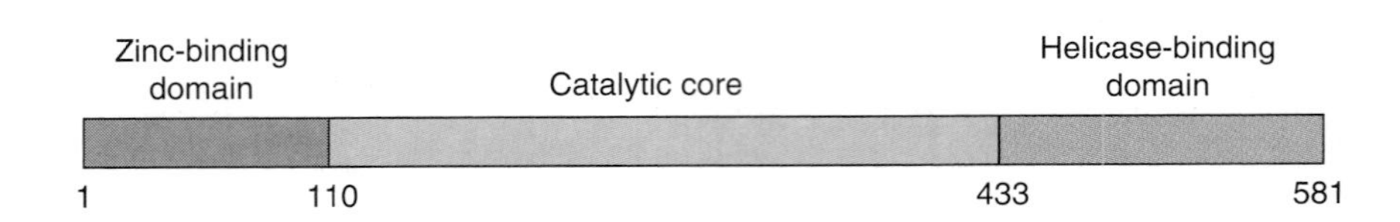

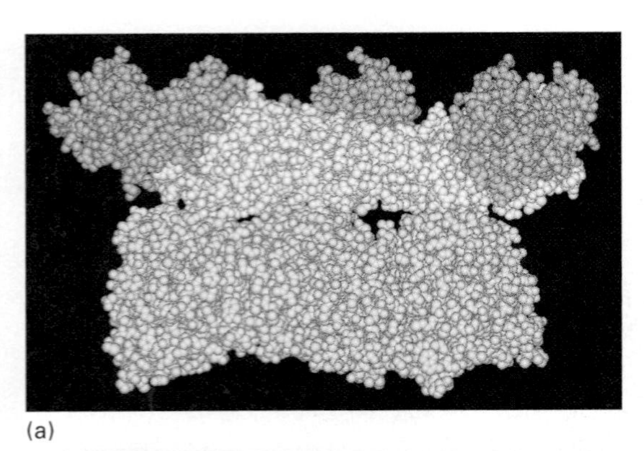

(a)

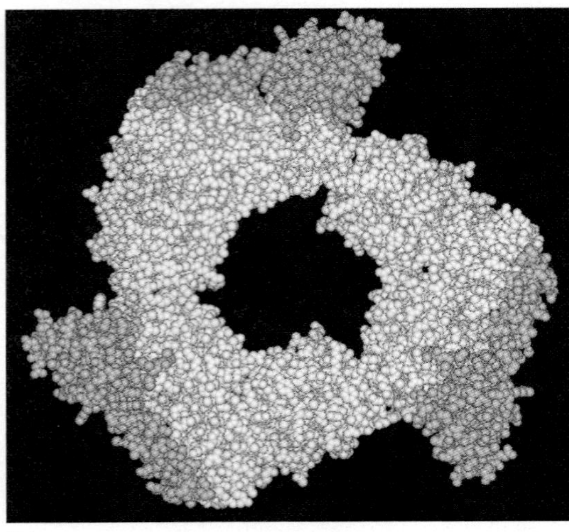

(b)

FIGURE 9.27 **Crystal structure of *Geobacillus stearothermophilus* DnaB helicase with bound helicase binding domain (HBD).** (a) Side view of DnaB hexamer with bound HBD. (b) Top view of DnaB hexamer with bound HBD. N- and C-terminal domains in the DnaB hexamer are yellow and blue, respectively. The helicase binding domain from primase is orange. (Structures from Protein Data Bank 2R6C. S. Bailey, et al., *Science* 318 [2007]: 459–463. Prepared by B. E. Tropp.)

DnaG (primase) catalyzes RNA primer synthesis.

DnaB is now ready to bind DnaG (primase), which catalyzes the formation of short RNA oligomers (about 10–12 nucleotides long) that serve as primers for DNA polymerase. Bacterial primase has three independent domains (FIGURE 9.26). The N-terminal or zinc binding domain has a tightly bound zinc ion. The middle or RNA polymerase domain can catalyze primer synthesis *in vitro* even when free of the other two domains. The C-terminal or helicase binding domain (HBD) binds DnaB helicase. The crystal structure has not yet been determined for the intact primase.

Biochemical and biophysical studies suggest that three DnaG molecules bind to each DnaB hexamer. This stoichiometric ratio has been confirmed by the crystal structure of a complex containing HBD bound to the N-terminal face of DnaB (FIGURE 9.27). Primase interaction with DnaB probably triggers a change in DnaB conformation that in turn induces DnaC to dissociate from DnaB.

Primase and DnaB helicase stimulate one another. Primase activity stimulation is due, at least in part, to the fact that DnaB helicase functions as a mobile docking station that increases the local concentration of the single-stranded DNA template. Primase activity is probably also stimulated by the interaction of the zinc binding domain of one primase molecule with the RNA polymerase domain of another primase molecule bound to the same DnaB helicase. DnaG may stimulate helicase activity by stabilizing the active helicase conformation.

Primase initiates leading strand synthesis only one time during each DNA replication cycle. Once leading strand synthesis starts, it continues in a 5′→3′ direction until replication is complete (or the replication machinery encounters a problem such as damaged DNA). Primase must act once in every 1000 to 2000 nucleotides during lagging strand synthesis to initiate Okazaki fragment synthesis. Primase's contribution to lagging strand synthesis is examined in the next section describing the elongation stage of DNA replication. The initiation stage ends with the addition of DNA polymerase III holoenzyme to the replication fork.

9.3 The Elongation Stage

Several enzymes act together at the replication fork.

DNA elongation is a complex process that involves cooperative interactions among many different proteins to synthesize two new DNA chains at an astonishing rate. Tania Baker and Stephen Bell provide the following insightful analogy to illustrate the amazing job that the

E. coli replication enzymes perform while synthesizing DNA. They begin by changing the scale so that the DNA duplex is 1 m in diameter. On this scale, the replication fork would move at about 600 km/h (375 mph) and the replication machinery would be about the size of a FedEx® delivery truck.

> "Replicating the *E. coli* genome would be a 40 min, 400 km (250 mile) trip for two such machines, which would, on average make an error only once every 170 km (106 miles). The mechanical prowess of this machine is even more impressive given that it synthesizes two chains simultaneously as it moves. Although one strand is synthesized in the same direction as the fork is moving, the other chain (the lagging strand) is synthesized in a piecemeal fashion (as Okazaki fragments) and in the opposite direction of the overall fork movement. As a result, about once a second one delivery person (i.e., polymerase active site) associated with the truck must take a detour, coming off and then rejoining the template DNA strand to synthesize the 0.2 km (0.13-mile) fragments." (Baker, T. A, and Bell, S. P. 1998. Polymerases and the replisome: machines within machines. *Cell* 92:295–305.)

This analogy underscores the enormous complexity and incredible speed of the DNA replication process. DNA polymerase, the bacterial counterpart of the FedEx® delivery truck, acts as the replication machine. It extends the leading strand by continuous nucleotide attachment to the 3′-hydroxy terminus, while helping to form the lagging strand by extending the primer until the resulting Okazaki fragment reaches the 5′-end of the adjacent fragment.

DNA polymerase III is required for bacterial DNA replication.

Investigators initially assumed that DNA polymerase I was the sole polymerase required for bacterial DNA synthesis. The first indication that this assumption was wrong came in 1969 when Paula Delucia and John Cairns isolated an *E. coli* mutant that lacked DNA polymerase I activity but continued to synthesize DNA and grow normally. The possibility that the mutant might have a low level of DNA polymerase I activity that allowed it to synthesize DNA was ruled out when an *E. coli* mutant with a deletion in ***polA*** (the structural gene for DNA polymerase I) was shown to also synthesize DNA.

The most likely explanation for the *polA* mutant's ability to synthesize DNA is that some other DNA polymerase is present and that enzyme is responsible for DNA synthesis. In support of this hypothesis, two new enzymes—DNA polymerase II and DNA polymerase III—were detected in *polA* mutant extracts when gapped DNA (created by partial hydrolysis of nicked DNA with an exonuclease) was used as a template. The two new polymerases add nucleotides to the 3′-end of the primer strand in the order specified by the template strand. Neither enzyme had been detected in bacterial extracts before because DNA polymerase I is so active that it masks their activity.

The next task was to determine what role, if any, the new DNA polymerases play in DNA replication. Once again, a genetic approach helped to provide the answer. Mutants lacking DNA polymerase II synthesize DNA normally, indicating that this enzyme is not essential for bacterial DNA replication. In contrast, temperature-sensitive DNA polymerase III mutants replicate DNA at 30°C but not at 42°C, indicating that DNA polymerase III is required for bacterial DNA synthesis.

Although genetic studies indicated that DNA polymerase III plays an essential role in bacterial DNA synthesis, the purified enzyme had few of the properties expected of the replication enzyme. For instance, DNA polymerase III could not extend a unique primer completely around a single-stranded circular DNA template even when large quantities of enzyme and substrate were added to reaction mixtures for long periods of time. Furthermore, DNA polymerase III synthesized DNA at about 20 nucleotides·s^{-1}, an exceptionally slow rate when compared to the 1000 nucleotides·s^{-1} observed in the living cell. The reason for this slow rate is that the enzyme dissociates from its DNA template rather frequently.

A polymerase's processivity can be determined by using a polymerase "trap" to bind the polymerase after it dissociates from its DNA substrate.

Enzymes that remain tightly associated with their template through many cycles of nucleotide addition are said to be highly processive. **Processivity** provides a quantitative measure of a polymerase's ability to remain tightly associated with its DNA template during DNA (or RNA) synthesis. A polymerase's processivity equals the average number of nucleotides the enzyme attaches to a growing chain before it dissociates from the DNA template. A polymerase with a low processivity value dissociates from its template after only a few nucleotides are added, catalyzing nonprocessive or **distributive replication.** Distributive replication is inefficient because the polymerase requires considerable time to associate with its template after each dissociation event.

One common procedure that is used to determine processivity is as follows. A DNA primer-template substrate is prepared in which the primer is relatively short, of a known and fixed length, and labeled with either a radioactive isotope or a fluorescent marker. DNA polymerase is incubated with the labeled primer-template in the absence of deoxynucleoside triphosphates, allowing the enzyme to bind to the primer-template but not to extend the primer chain. Then deoxynucleoside triphosphates and a large excess of unlabeled DNA primer-template (the "trap") are added. The pre-bound DNA polymerase extends the labeled primer until the polymerase dissociates from the primer-template. After dissociation, the polymerase is free to bind to either labeled substrate or the unlabeled "trap" DNA. It almost always binds to the unlabeled "trap" DNA, which is present in large excess. Because the "trap" prevents DNA polymerase from associating with the labeled substrate again, the number of nucleotides added to the labeled primer corresponds to the product of one round of processive synthesis catalyzed by the polymerase. After extension is complete, products are separated by gel electrophoresis, and the distribution of extended labeled primer is visualized by autoradiography or scanning for the fluorescent marker. Even though the "trap" DNA is extended, it cannot be seen because it is not labeled.

DNA polymerase holoenzyme has nine distinct subunits that form three subassemblies.

A possible explanation for DNA polymerase III's low activity is that additional proteins are needed to increase its processivity. Arthur Kornberg and coworkers thought they might be able to detect these additional proteins by using a circular single-stranded phage DNA as a template. Their choice of template was influenced by the knowledge that the circular single-stranded DNA molecules in bacteriophage such as M13 and ϕX174 are so small (5–6 kb long) that they lack sufficient genetic information to code for most of the enzymes needed for DNA replication. The bacteriophage, therefore, use bacterial enzymes to catalyze the first stage of DNA replication, the conversion of circular single-stranded DNA to the double-stranded replication form. It, thus, seemed reasonable to expect that bacterial extracts would be able to extend a primer annealed to circular single-stranded bacteriophage DNA completely around the template.

This primer extension is exactly what Kornberg and coworkers observed (FIGURE 9.28). Taking advantage of this assay, Kornberg's laboratory used standard protein fractionation procedures to purify the fully functional **replicase**, DNA polymerase III holoenzyme that contains three distinct subassemblies. One of these subassemblies, the core polymerase subassembly, is responsible for the DNA polymerase III activity that was originally detected in polA mutant extracts.

Because each cell contains only about 10 to 20 copies of DNA polymerase III holoenzyme, considerable effort was required to obtain sufficient quantities of the enzyme for study. This effort was rewarded when purified holoenzyme was shown to catalyze DNA synthesis at a rate of about 750 nucleotides·s^{-1}. The rapid synthetic rate, a result of the holoenzyme's high processivity, allows the enzyme to add several thousand nucleotides to a primer annealed to a circular

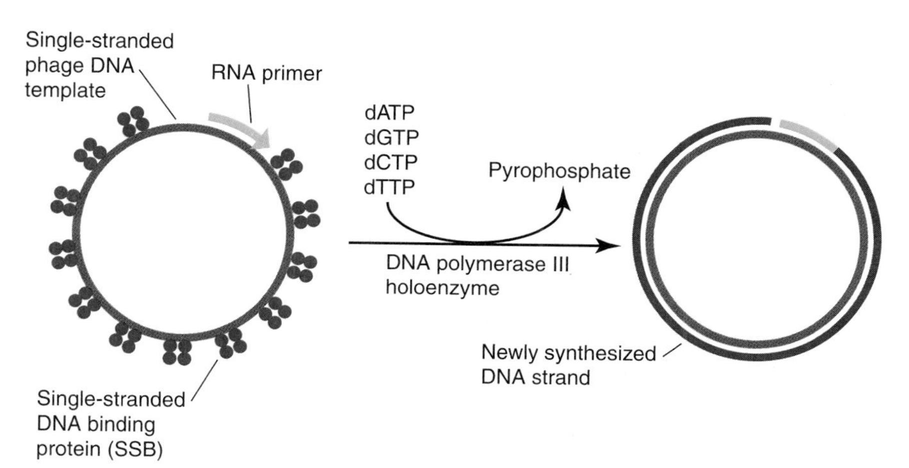

FIGURE 9.28 **Action of DNA polymerase III holoenzyme on a primed SSB-coated circular single-stranded phage DNA molecule.**

TABLE 9.2 **DNA Polymerase III Holoenzyme**

Component [Stoichiometry]	Gene	Molecular Mass in kDa	Comments
Core polymerase subassembly		166.0	Monomeric polymerase/ exonuclease
α [2]	*dnaE*	129.9	DNA polymerase
ε [2]	*dnaQ*	27.5	3′→5′ exonuclease
θ [2]	*holE*	8.6	Stimulates ε exonuclease
Clamp loader subassembly		297.1	ATP-dependent clamp loader
τ [3]	*dnaX*	71.1	τ binds and hydrolyzes ATP. It also interacts with core polymerase, the β clamp, and the template strand.
δ [1]	*holA*	38.7	Opens the β clamp
δ′[1]	*holB*	36.9	Regulates δ subunit
χ [1]	*holC*	16.6	Binds SSB
ψ [1]	*holD*	15.2	Connects χ to clamp loader
β clamp subassembly [2 dimers]	*dnaN*	40.6	Homodimeric ring required for processivity

Adapted from A. Johnson and M. O'Donnell, *Ann. Rev. Biochem.* 74 (2005):283–315.

single-stranded bacteriophage DNA molecule without dissociating from the DNA. DNA polymerase III holoenzyme thus synthesizes DNA at a much faster rate and with a much higher processivity than DNA polymerase I. One other important difference was noted. DNA polymerase III holoenzyme requires ATP for full activity, whereas DNA polymerase I does not. We will return to this ATP requirement when we examine the way that the holoenzyme works.

DNA polymerase III holoenzyme has nine distinct polypeptides (α, β, δ, δ′, ε, θ, τ, χ, and Ψ). Structural genes for these polypeptides have been cloned, allowing investigators to construct bacteria that overproduce the polypeptides. Isolated polypeptide subunits, which are available in ample supply, have been used to constitute DNA polymerase III holoenzyme as well as various subassemblies. Table 9.2 summarizes some important information about DNA polymerase III holoenzyme's polypeptide subunits, the **core polymerase**, **clamp loader**, and **β clamp** subassemblies they form, and their structural genes. Each subassembly makes a unique contribution to holoenzyme function. We can gain considerable insight into the way the replication machine works by examining subassembly function. Therefore, we begin by examining the core polymerase, the subassembly responsible for adding nucleotides to the 3′-hydroxyl end of growing polynucleotide chains.

The core polymerase has one subunit with 5′→3′ polymerase activity and another with 3′→5′ exonuclease activity.

The subassembly responsible for chain extension, the **core polymerase** (previously known as DNA polymerase III), consists of three polypeptide subunits (Table 9.2).

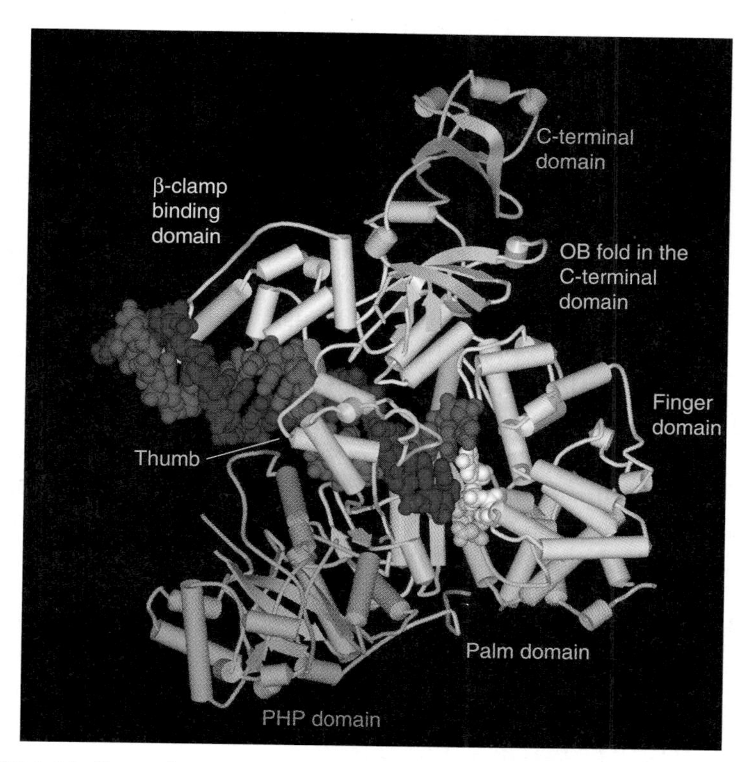

FIGURE 9.29 **Crystal structure of a PolIIIα•DNA•dATP ternary complex.** The protein structure is shown in a schematic display. The six domains are as indicated. The oligonucleotide binding (OB) fold (light tan) is part of the C-terminal domain (dark tan). Primer and template DNA strands are shown as a red and blue spacefill structures, respectively. The thumb domain is green. The incoming deoxyribonucleoside triphosphate is shown as a white spacefill structure. The three aspartate residues at the catalytic site in the palm domain are shown as gold spacefill structures. (Structure from Protein Data Bank 3E0D. R. A. Wing, et al., *J. Mol. Biol.* 382 [2008]: 859–869. Prepared by B. E. Tropp.)

1. The **α-subunit (polIIIα)**, encoded by *dnaE*, catalyzes 5′→3′ chain growth and is essential for DNA synthesis. Steitz and coworkers determined the crystal structure of a PolIIIα•DNA•dATP ternary complex (FIGURE 9.29). The enzyme was isolated from *Thermus aquaticus*, a gram-negative thermophilic bacteria. Three of the six PolIIIα domains, the palm, finger, and thumb domains are present in all DNA polymerases. The finger domain interacts with the incoming deoxynucleoside triphosphate, the palm domain contains the catalytic site, and the thumb domain grips the DNA substrate. The organization and structure of these domains differs from the comparable domains in DNA polymerase I (see Chapter 5). Thus, polIIIα and DNA polymerase I belong to different protein families. Two of the additional PolIIIα domains, the β clamp binding domain and the C-terminal domain, both contain binding sites for the β clamp, a ring shaped protein that increases core polymerase processivity by tethering the core polymerase to its DNA substrate (see below). The sixth PolIIIα domain, the histidinol phosphatase (PHP) domain, was so-named

because this domain was first observed in histidinol phosphatase. Initial attempts to determine the crystal structure of the PolIIIα•DNA•dATP ternary complex were unsuccessful because the PHP domain has a 3′→5′ exonuclease activity that digests DNA. The ternary complex shown in Figure 9.29 was obtained by using a PolIIIα preparation in which two aspartate residues required for nuclease activity were replaced with asparagine residues. The physiological function of PHP nuclease activity is unclear because the PHP domain does not contact the DNA substrate except possibly through a loop located below the thumb domain.

2. The **ε-subunit,** encoded by *dnaQ* (also called mutD), has 3′→5′ exonuclease activity. On those rare occasions when the polymerase makes an error, the exonuclease removes the mispaired nucleotide, thereby providing a proofreading function. Consistent with this function, cells with defective ε subunits have high mutation rates.
3. The **θ-subunit,** encoded by *holE,* seems to stimulate the ε subunit. This function is not essential, however, because DNA polymerase III holoenzyme isolated from *holE* deletion mutants is fully active. Thus, θ does not appear to have a unique function at this time.

Binding studies provide some information about the way the subunits of the core polymerase are organized. The α- and ε-subunits form a 1:1 complex in which each subunit's activity is greater than it is in the free subunit. The θ-subunit joins this complex by binding to ε to form a linear α-ε-θ arrangement. Associations among core polymerase subunits are so tight that denaturing agents are required for dissociation. Detailed information about core polymerase structure awaits crystallographic studies. The core polymerase does not bind tightly to DNA. We now examine the subassembly that tethers the core polymerase to DNA.

The β clamp forms a ring around DNA, tethering the remainder of the polymerase holoenzyme to the DNA.

The **β clamp** (also known as the **β-dimer** or the **sliding clamp**) is made of two identical polypeptide subunits that are encoded by ***dnaN***. Mike O'Donnell and coworkers have determined the crystal structure of the *E. coli* β clamp on DNA (FIGURE 9.30). Two semicircular shaped polypeptide subunits are arranged head-to-tail with a 3.5 nm diameter central opening. Each subunit has three domains (Figure 9.30a). Each domain within a subunit has a different amino acid sequence. Nevertheless, all three domains have the same folding pattern: two β-strands are on the outside and two α-helices on the inside. The net effect is that the β-dimer appears to have a sixfold rotational axis of

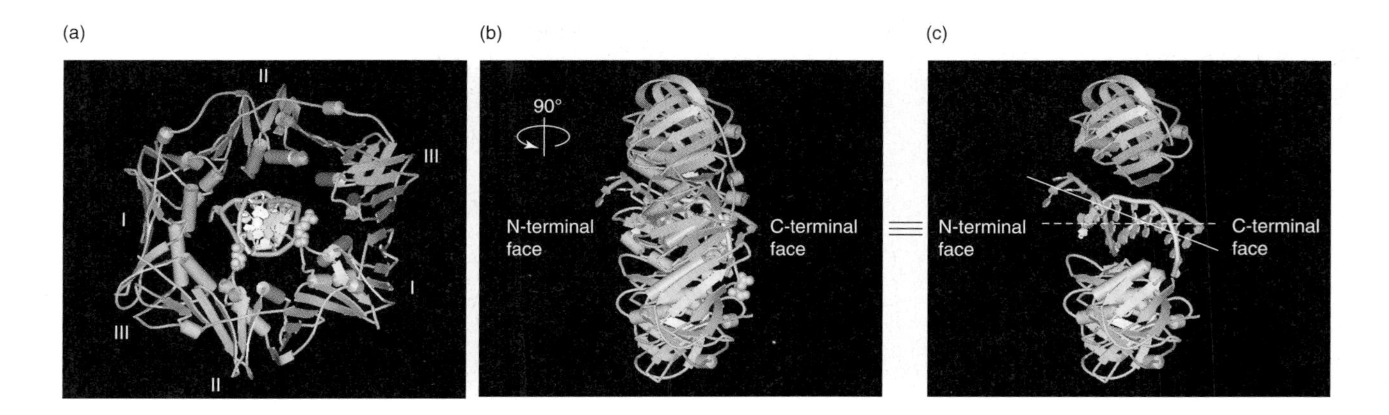

FIGURE 9.30 Structure of an *E. coli* sliding clamp on DNA. The β subunits are shown in schematic form with one subunit in orange and the other in green. The DNA strands are in tube form in light blue and cyan. (a) Crystal structure of the β dimer (sliding clamp) with bound DNA viewed looking down the DNA axis. Domains in each subunit are numbered I, II, and III. Arg-24 (pink) and Gln-149 (yellow) in the lower β subunit are shown in spacefill form. (b) The same sliding clamp with bound DNA viewed from the side with the N- and C-terminal faces labeled. The C-terminal face has been implicated in sliding clamp interactions with other proteins. (c) The same as (b) but amino acid residues that block the view of the DNA are not shown so that it is possible to see that the DNA axis (solid white line) is at a 22° angle with respect to the ring axis (dashed white line). (Structures from Protein Data Bank 3BEP. R. E. Georgescu, et al., *Cell* 132 [2008]: 43–54. Prepared by B. E. Tropp.)

symmetry even though its true rotational axis of symmetry is only twofold. Twelve α-helices (2 polypeptides × 3 domains/polypeptide × 2 α-helices/domain) line the inside of the β clamp. The arrangement of the α-helices in the β clamp allows it to move along the encircled DNA.

The β clamp has two distinct faces, a C-terminal face and an N-terminal face (Figure 9.30b). As the names suggest, the former is characterized by protruding C-termini and the latter by protruding N-termini. The core polymerase and the clamp loader bind to the C-terminal face. The β clamp, therefore, increases the holoenzyme's processivity by acting as a tether for the remainder of the holoenzyme. Other proteins also bind to the C-terminal face, including four additional DNA polymerases, various DNA repair enzymes, and DNA ligase. A single β clamp may bind at least one different protein to each subunit.

The crystal structure of the *E. coli* β clamp on DNA shows that the DNA axis is at a 22° angle to the ring axis (Figure 9.30c). This tilt may be caused by contact between the double-stranded DNA and two protruding loops on the C-terminal face. One loop's point of contact is Arg-24 and the other loop's point of contact is Gln-149 (Figure 9.30a). These contacts appear to be physiologically significant. When Arg-24 and Gln-149 are replaced by alanines, the altered β clamp has weak activity when tested in an *in vitro* replication assay using a primed M13 single-stranded DNA coated with SSB. O'Donnell and coworkers showed that the β clamp with the double mutation is deficient at the clamp loading stage rather than the elongation stage.

DNA can switch from being a substrate for an enzyme bound to one β clamp subunit to being a substrate for a different enzyme bound to the other β clamp subunit. The DNA tilt suggests that this switch takes place as follows. DNA initially tilts to contact protruding loops of the C-terminal face on one β subunit, allowing the enzyme bound to that subunit to act on the DNA. If some event causes the DNA to tilt the other way, the DNA will contact the protruding loops of the other β subunit, allowing the enzyme bound to that β subunit to act on the DNA.

The clamp loader places the sliding clamp around DNA.

The clamp loader uses energy provided by ATP to load the sliding clamp onto a DNA template-primer with a 5′-overhang. It exists as a subassembly of the DNA polymerase III holoenzyme and also in a free form. The free form does not appear to arise from the dissociation of the holoenzyme because (1) the holoenzyme is stable when studied *in vitro* and can only be dissociated under harsh nonphysiological conditions and (2) the two forms have different subunit compositions: $\gamma_3\delta\delta'\chi\psi$ in the free form and $\tau_3\delta\delta'\chi\psi$ in the holoenzyme. Differences between the two forms are not as great as they may at first appear because both γ and τ subunits are encoded by the same *dnaX* gene (**FIGURE 9.31a**). The τ subunit, the full-length *dnaX* protein product, has five domains (**FIGURE 9.31b**), which all are required for DNA replication. Domains I and II form an ATP binding site with ATPase activity. Domain III allows τ subunits to form oligomers (see below). Domains IV and V bind to DnaB helicase and PolIIIα, respectively. The γ subunit, a truncated form of τ, lacks domains IV and V (**FIGURE 9.31c**). Mutants that make τ but don't make γ have normal

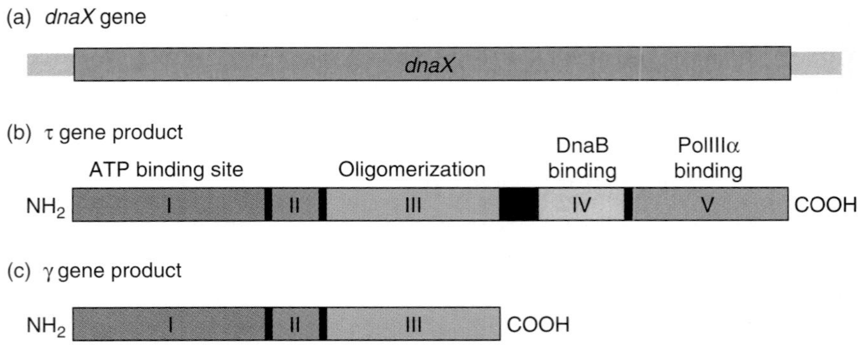

FIGURE 9.31 The *dnaX* gene and gene products. (a) *dnaX* gene. (b) The *dnaX* gene product τ has five domains. Domains IV and V bind to DnaB helicase and PolIII , respectively. (c) The *dnaX* gene product γ has three domains. Domains I and II in γ and τ bind ATP and have ATPase activity. Domain III in γ and τ are required for assembly into τ_3 and γ_3, respectively. (Adapted from P. McInerney, et al., *Mol. Cell* 27 [2007]: 527–538.)

DNA replication and no known phenotype. Therefore, the γ subunit appears to be a nonessential protein. The $\tau_3\delta\delta'\chi\psi$ clamp loader thus appears to be able to perform all essential functions. Perhaps the $\gamma_3\delta\delta'\chi\psi$ clamp loader participates in DNA repair or helps to remove β clamps from recently replicated DNA. Despite the uncertainty about its physiological role, *in vitro* studies using the $\gamma_3\delta\delta'\chi\psi$ clamp loader have helped us to understand how the clamp loader works.

Mike O'Donnell and coworkers obtained considerable information about clamp loader structure and function by studying the structure and functions of a $\gamma_3\delta\delta'$ complex they constructed by combining subunits *in vitro*. This $\gamma_3\delta\delta'$complex is called the **minimal clamp loader** because it is the smallest complex that can load sliding clamps onto DNA. Neither the χ nor ψ subunit is required to load the clamp. The ψ subunit stabilizes the clamp loader and the χ subunit displaces primase from SSB.

O'Donnell and coworkers obtained an important clue to the minimal clamp loader's mechanism of action by studying the effect that the δ subunit has on the β clamp. In one key experiment, they demonstrated that the δ subunit, working without the other subunits or ATP, causes the β dimer to be released from a topologically linked β clamp•circular DNA complex (FIGURE 9.32). This release is not a catalytic or reversible process because the δ subunit remains tightly

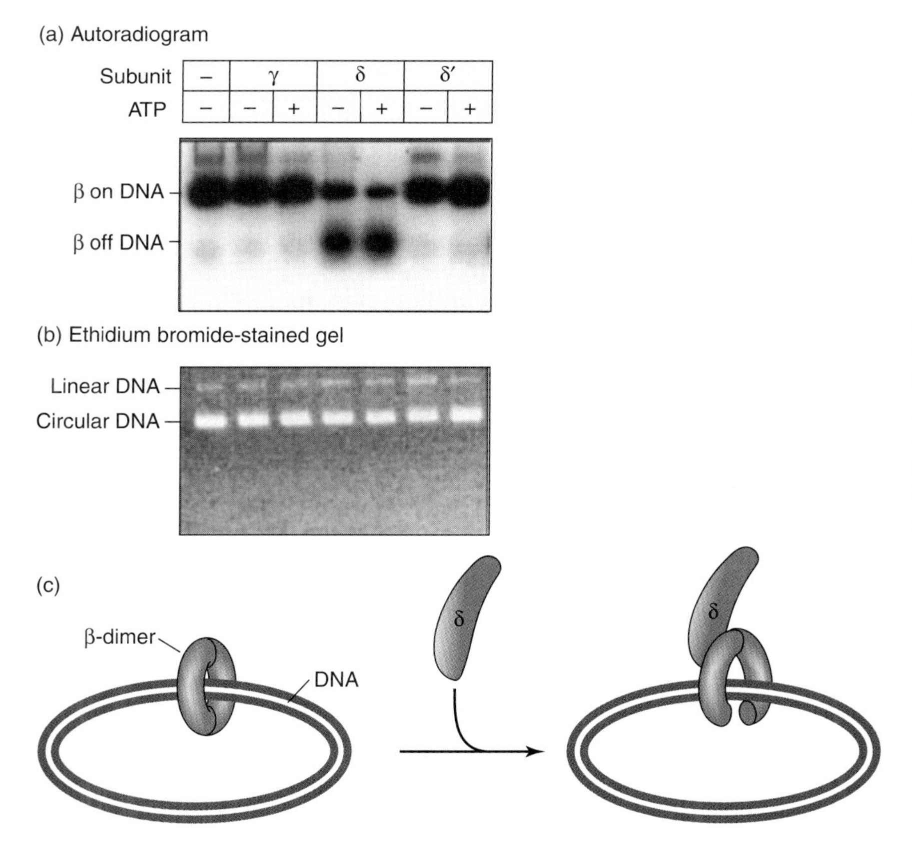

FIGURE 9.32 **The δ (delta) subunit opens the β clamp.** A radioactive β clamp was loaded onto circular double-stranded DNA and incubated with γ, δ, or δ' in the absence or presence of ATP and analyzed by agarose gel electrophoresis. (a) An autoradiogram of the agarose gel shows that only the δ subunit released the radioactive sliding clamp subassembly from the DNA and that it does so in the presence or absence of ATP. (b) The gel stained with ethidium bromide shows that the DNA is not degraded during this reaction. (c) A schematic interpretation of these results. (Parts a and b reprinted by permission from Macmillan Publishers Ltd: *Nature*, J. Turner, et al., vol. 18, copyright 1999. Photos courtesy of Michael O'Donnell, Rockefeller University. Part c adapted from J. Turner, et al., *EMBO J.* 18 [1999]: 771–783.)

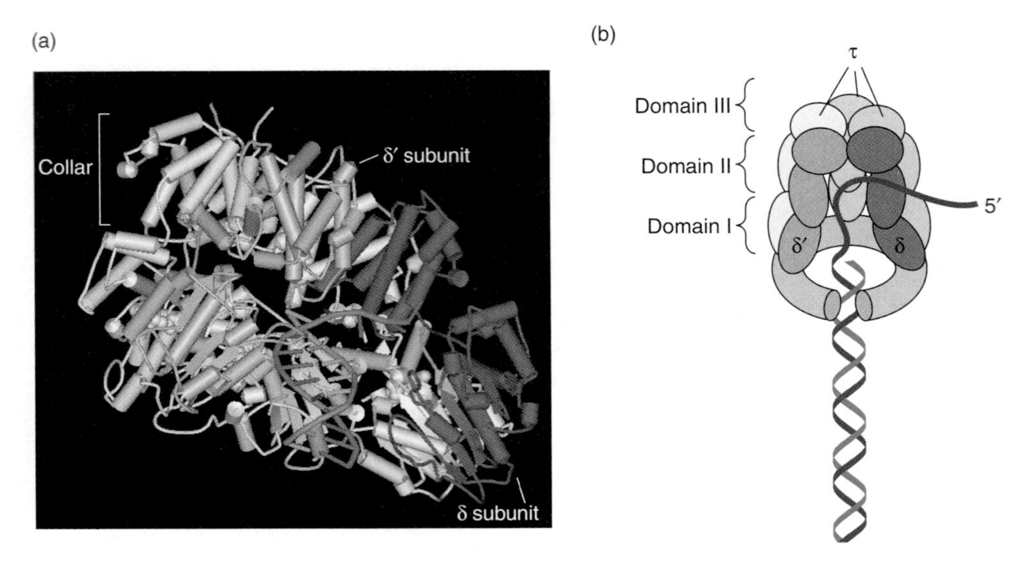

FIGURE 9.33 **The minimal clamp loader structure ($\gamma_3\delta\delta'$) with bound primer-template and adenine nucleotides.** (a) A side view of the minimal bacterial clamp loader ($\gamma_3\delta\delta'$) with a bound primer-template and bound adenine nucleotides. The clamp loader shown in schematic form is colored by subunit. The γ subunits are shown in different shades of blue, the δ subunit is rust color, and the δ′ subunit is yellow. The template strand, which has a 5′-overhang, is blue and the primer strand is green. Bound adenine nucleotides are shown as yellow spacefill structures. A continuous collar formed by the five subunits's C-terminal domains holds the pentamer together. The other two domains (domains I and II) in each subunit assemble to form a ring-like structure with a gap between δ and δ′. The collar presents a steric barrier to the 5′ template strand overhang, forcing it to make a sharp bend and exit the central chamber through the gap. (b) Schematic diagram of the minimal clamp loader structure with bound primer-template. Colors are the same as in (a). (Part a structure from Protein Data Bank 3GLF. K. R. Simonetta, et al., *Cell* 137 [2009]: 659–671. Prepared by B. E. Tropp. Part b adapted from C. Indiana and M. O'Donnell, *Nat. Rev. Mol. Cell Biol.* 7 [2006]: 751–761.)

associated with the released clamp and the free δ subunit cannot load the clamp onto DNA. The δ subunit appears to work as a "wrench," forcing one of the two interfaces in the β dimer to open.

O'Donnell and coworkers also have determined the high-resolution structure for the minimal clamp loader bound to a primer-template containing a double-stranded segment with a 5′ template strand overhang (FIGURE 9.33). The δ, δ′, and γ subunits have similar structures. The clamp loader is held together by a continuous collar that is formed by the five subunits' C-terminal domains. The other two domains (domains I and II) assemble to form a ring-like structure with a gap between δ and δ′. The double-stranded region of the primer-template fits inside the central chamber. Only the template strand makes contact with the interior walls of the clamp loader, however. The collar presents a steric barrier to the 5′ template strand overhang, forcing it to make a sharp bend and exit the central chamber through the gap.

The minimal clamp loader behaves like a machine with moving parts when it loads the β clamp onto a DNA primer-template complex (FIGURE 9.34). The γ (or τ) subunits, which are the only polypeptide subunits that bind and hydrolyze ATP, act as the machine's motor.

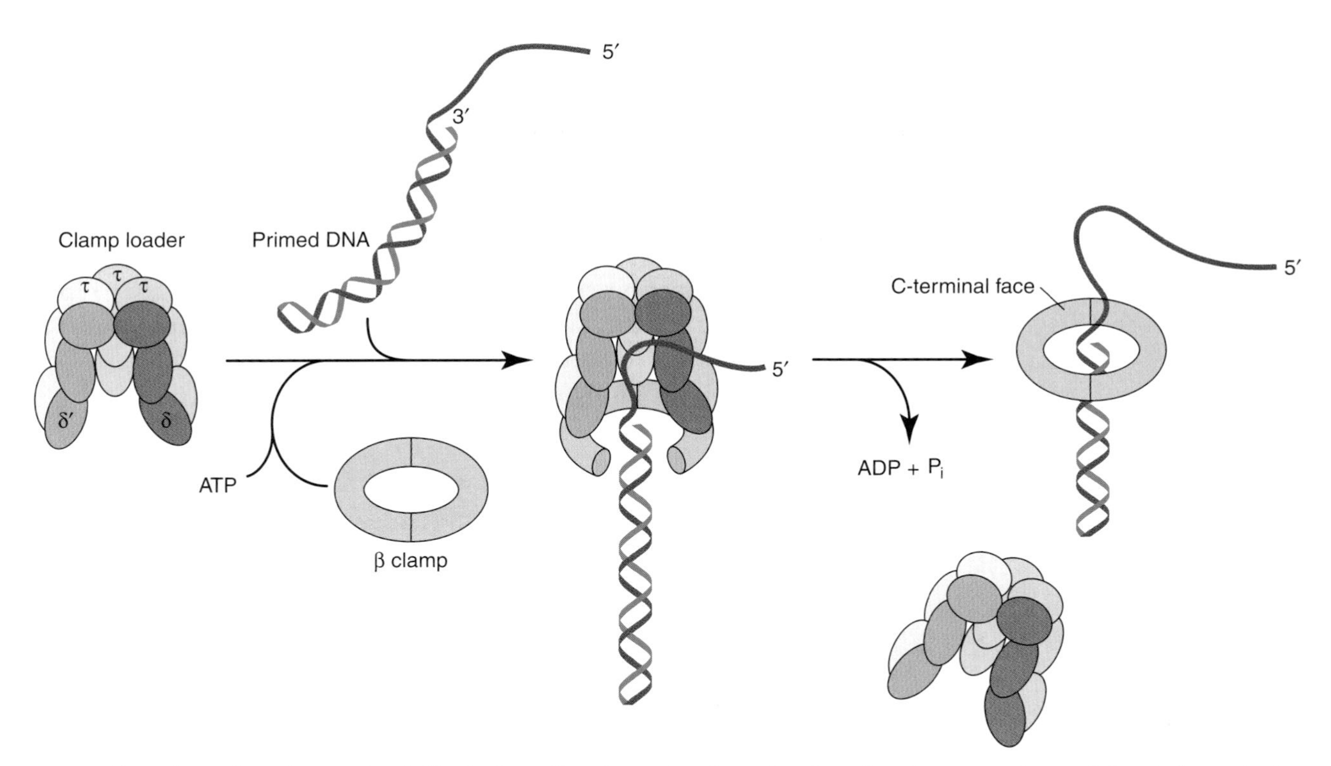

FIGURE 9.34 **The clamp loading cycle.** ATP binds to the clamp loader, allowing the clamp loader and the β clamp to form a tight complex. When bound to ATP, the clamp loader has a high affinity for primer-template junctions with 5′-overhangs. Binding to these junctions stimulates ATP hydrolysis and release of the DNA encircled by the β clamp. (Reprinted from *Trends Microbiol.*, vol. 15, R. T. Pomerantz and M. O'Donnell, Replisome mechanics: insights into a twin DNA . . ., pp. 156–164, copyright 2007, with permission from Elsevier [http://www.sciencedirect.com/science/journal/0966842X].)

The δ′ subunit, which appears to be stationary, regulates the δ subunit's ability to bind the β clamp. In the absence of ATP, δ′ prevents δ (the wrench) from binding the β clamp and opening it. When ATP binds to the γ subunits, the δ subunit pulls away from δ′ and thereby gains the ability to bind the β clamp. Clamp loader•β clamp complex has a high affinity for a DNA primer-template complex with a 5′-overhang. After the clamp loader•β clamp complex interacts with the primed DNA, bound ATP molecules are hydrolyzed, causing the clamp loader•β clamp complex to dissociate and allowing the β clamp to close around the DNA.

The DNA polymerase III holoenzyme clamp loader has three τ subunits.

Initial studies indicated that the clamp loader in the DNA polymerase III holoenzyme has two τ subunits in place of two γ subunits. Recent investigations, however, suggest that all three γ subunits are replaced by τ subunits. Evidence for complete replacement is as follows:

(1) Proteases hydrolyze the τ subunit under *in vitro* conditions to produce a protein that is about the same size as γ. Investigators, therefore, may have misidentified the degradation product as γ. (2) Under *in vitro* conditions, a mixture of τ, γ, δ, and δ' preferentially form $\tau_3\delta\delta'$ and $\gamma_3\ \delta\delta'$ rather than mixed complexes. If the same preference exists *in vivo*, then the bacterial clamp loader will be $\tau_3\delta\delta'\chi\psi$. (3) As mentioned, mutants that make τ but not γ subunits exhibit normal DNA replication. The τ subunits, like the γ subunits they replace, are the motor for the clamp loader. However, the τ subunits also interact with helicase to stimulate its catalytic efficiency and with PolIIIα to tether the core polymerase to DNA (see below).

The replisome catalyzes coordinated leading and lagging strand DNA synthesis at the replication fork.

By the late 1970s, data generated from autoradiography and electron microscopy experiments showed that leading and lagging strand synthesis are coordinated at the replication fork. However, investigators found it difficult to see how the replication machinery could coordinate this synthesis because DNA polymerases at the replication fork must move in opposite directions along the two antiparallel DNA template strands. In 1980, Bruce Alberts proposed the **trombone model** of replication to explain how coordination of replication of leading and lagging strands might take place. As shown in an updated version of this model in FIGURE 9.35, the lagging template strand loops out as each Okazaki fragment is formed. This looping allows the pair of core polymerase subassemblies to coordinate leading and lagging strand synthesis while each moves in a 5′→3′ direction along its template strand. Recent electron microscopy studies show that the trombone structure exists in cells.

When Alberts proposed the trombone model, relatively little was known about DNA polymerase III holoenzyme structure. The discovery that each holoenzyme has three core polymerases, however, fits the trombone model and explains how the DNA replication machinery coordinates leading and lagging strand synthesis at the replication fork. At a minimum, this replication machinery or **replisome** consists of DNA polymerase III holoenzyme, helicase, primase, and SSB. Other enzymes also contribute to DNA replication, however. The functions of DNA polymerase I, RNase H, and DNA ligase were described earlier. DNA gyrase and topoisomerase I (see Chapter 4) also make an essential contribution to bacterial DNA replication. Because the bacterial chromosome is a covalently closed circular DNA molecule, unwinding at the replication fork leads to tighter winding ahead of the replication fork. DNA gyrase and topoisomerase I act cooperatively to relieve the resulting torsional strain, allowing the replication fork to continue moving along the bacterial chromosome.

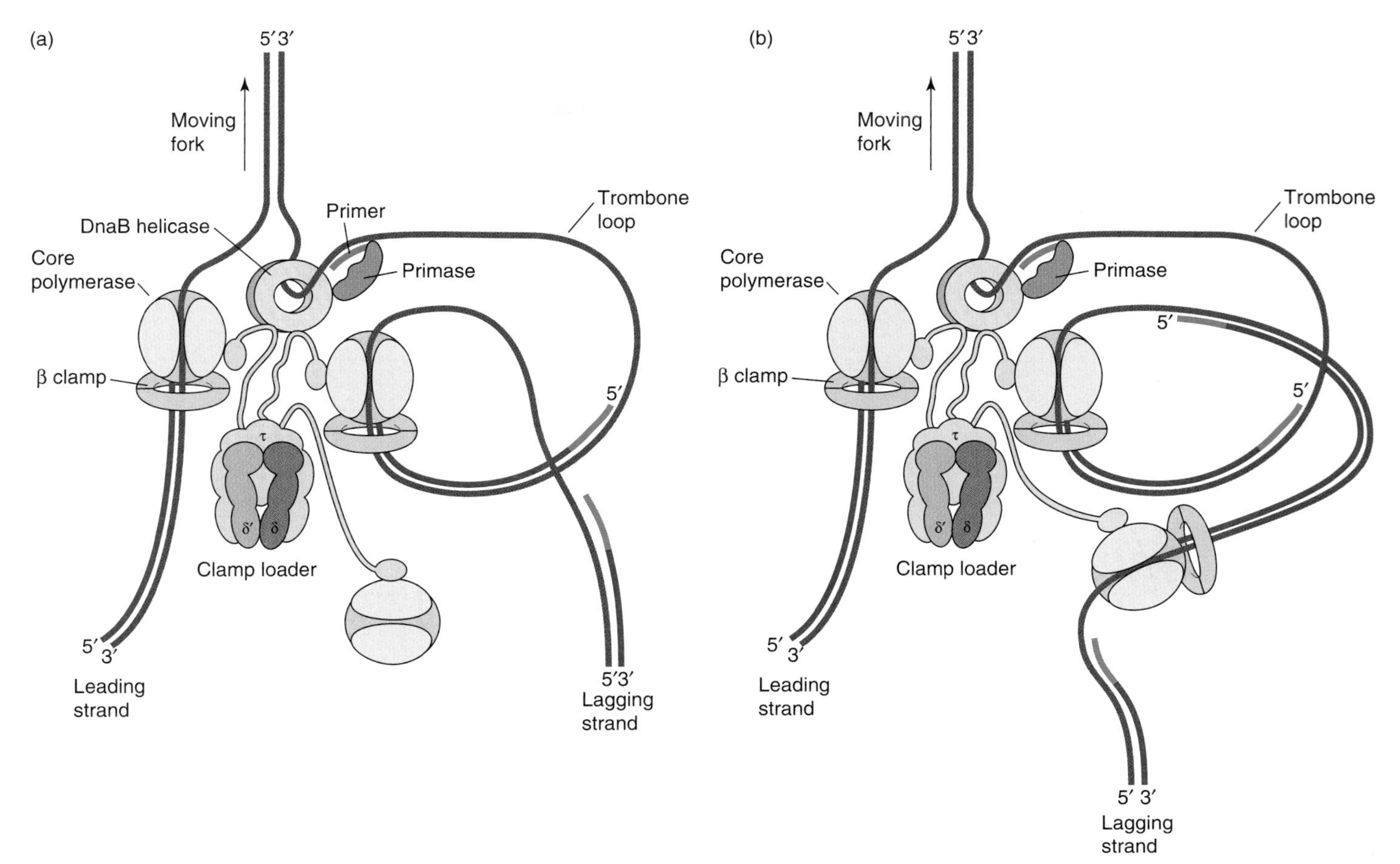

FIGURE 9.35 DNA replisome with three core polymerases. (a) The third core polymerase is off DNA. This situation may take place when only one core polymerase is required for lagging strand synthesis. (b) Two core polymerases are involved in lagging strand synthesis. The trombone loop and replisome components are labeled. For simplicity, only a single primase is shown associated with DnaB helicase and the ψ and χ clamp loader subunits are not shown. Parental DNA is dark blue, newly synthesized DNA is red and primer is green. (Reprinted from *Mol. Cell,* vol. 27, P. McInerney, et al., Characterization of a Triple DNA Polymerase Replisome, pp. 527–538, copyright 2007, with permission from Elsevier [http://www.sciencedirect.com/science/journal/10972765].)

Continuous replication of the leading strand is readily explained by the β clamp's ability to serve as a tether for the core polymerase. However, this ability presents a problem for lagging strand synthesis because the β clamp must release the core polymerase after each Okazaki fragment is completed. FIGURE 9.36 shows the core polymerase cycle on the lagging strand. The three steps in the lagging strand cycle are as follows: (1) Core polymerase synthesizes the Okazaki fragment and the clamp loader loads a new β clamp; (2) core polymerase dissociates from the β clamp; (3) and core polymerase binds to the new β clamp and synthesizes DNA. The trombone loop is reset after each Okazaki fragment has been completed.

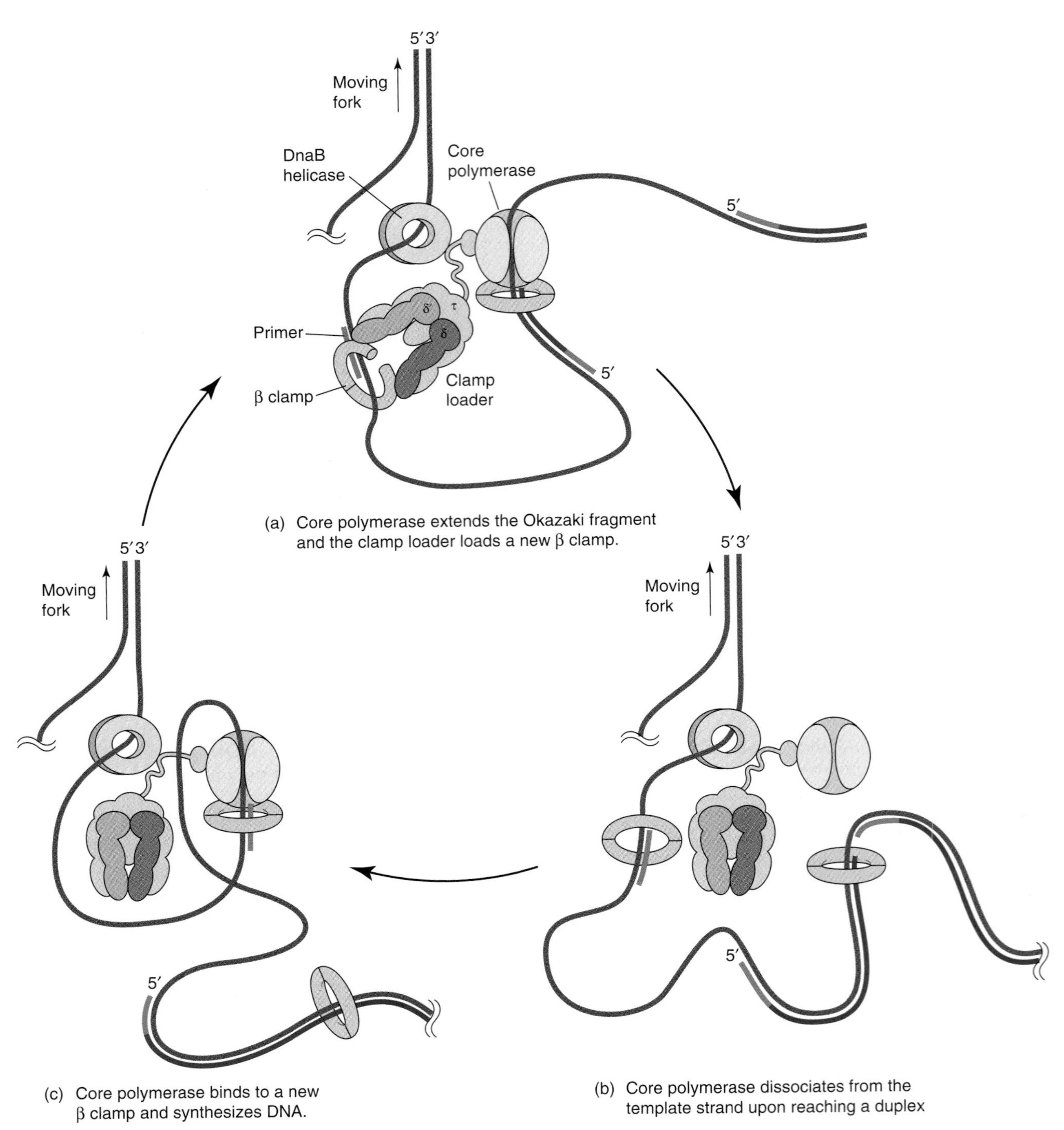

FIGURE 9.36 **Core polymerase cycle on the lagging strand.** Parental DNA is dark blue, newly synthesized DNA is red, and primers are green. For simplicity, leading strand replication is not shown and only one core polymerase is shown to be involved in lagging strand synthesis. (Reprinted from *Mol. Cell,* vol. 11, F. P. Leu, R. Georgescu, and M. O'Donnell, Mechanism of the *E. coli* τ processivity switch . . ., pp. 315–327, copyright 2003, with permission from Elsevier [http://www.sciencedirect.com/science/journal/10972765].)

Core polymerase is released from the β clamp by a premature release (also called signaling release) or collision release mechanism.

Some mechanism is required to permit the core polymerase to remain attached to the β clamp during Okazaki fragment synthesis and then dissociate when this synthesis is complete. The *E. coli* core polymerase uses two different release pathways. The first is called "premature release" or "signaling release." These names derive from observations that (1) core polymerase is released before Okazaki fragment formation is complete and (2) the presence of a new primer or newly assembled β clamp appears to signal release. The premature release pathway allows a replication fork to move beyond a damaged template.

The second release pathway is called "**collision release**" because core polymerase remains bound to the β clamp until it "collides" with a duplex. Recent studies by O'Donnell and coworkers provide important insights into how collision release is accomplished. They began by testing the possibility that core polymerase must recognize the 5′-end of the duplex in front of it for release to take place. Their studies, however, showed that such recognition is not essential because release occurs even when a protein blocks the 5′-end of the duplex. They then turned their attention to events that take place in a region of PolIIIα called the **oligonucleotide binding (OB) fold** (Figure 9.29).

The OB fold binds single-stranded DNA just ahead of the catalytic site in PolIIIα but does not bind double-stranded DNA. O'Donnell and coworkers constructed a mutant strain with three amino acid substitutions in the OB fold. The altered fold lost the ability to bind single-stranded DNA, but the PolIIIα catalytic site remained functional. O'Donnell and coworkers used the mutant PolIIIα to reconstitute an OB-mutant PolIII core polymerase. Then they compared the reconstituted mutant core polymerase with its wild type counterpart for their ability to catalyze processive DNA synthesis and bind the β clamp. OB-mutant Pol III core polymerase had a much lower processivity than OB wild type Pol III core polymerase. The OB-mutant PolIII also exhibited much weaker binding to the β clamp. These experiments suggest the following explanation for core polymerase's ability to bind the β clamp during Okazaki fragment formation and to release β clamp when this synthesis is complete. PolIIIα exists in two conformational states, one with a high affinity for β clamp and the other with a low affinity. The high affinity conformation is favored when the OB fold interacts with single-stranded DNA and the low affinity conformation is favored when the single-stranded DNA is converted to double-stranded DNA or a defective OB fold cannot interact with single-stranded DNA. The OB fold, therefore, acts as a sensor for the presence of single-stranded DNA. As the Okazaki fragment nears completion, the single-stranded DNA in the OB fold is converted to double-stranded DNA and interaction between single-stranded DNA and the OB fold becomes impossible, triggering a change from the high to low affinity β clamp conformation.

The τ subunit of the clamp loader also contributes to β clamp binding and release. The C-terminal domain of τ, τ_C, binds to single-stranded DNA near the OB fold. This binding somehow increases PolIIIα's affinity for a primed site. Once Okazaki fragment synthesis is complete, τ_C no longer has a single-stranded DNA to which it can bind. Therefore, it can no longer increase PolIIIα's affinity for a primed site. However, the core polymerase is not free to diffuse away because it remains bound to the τ subunit. The replication fork is now ready to begin a new cycle of Okazaki fragment formation.

The β clamps that are left behind after completion of Okazaki fragment synthesis bind DNA polymerase I and DNA ligase, which participate in the pathway that joins adjacent fragments (Figure 9.16). Abandoned β clamps must eventually be recycled, however, because each cell has about 300 β clamps to produce about ten times that number of Okazaki fragments. Free δ subunits, which are in excess over other clamp loading subunits in the cell, are the most likely clamp unloading agents. The free δ subunit is very active in unloading β clamps from DNA but cannot load clamps onto DNA. The $\gamma_3\delta\delta'\psi\chi$ clamp loader may also unload abandoned β clamps.

Three models have been proposed to explain how helicase moves 5′→3′ on the lagging strand while primase moves in the opposite direction as it synthesizes primer.

DnaB helicase moves in a 5′→3′ direction along the lagging strand as it unwinds double-stranded DNA. Primase, however, which needs to be associated with helicase to function, must move in the opposite direction to synthesize the primer. Three models have been proposed to solve this problem (**FIGURE 9.37**).

1. *The pausing model.* Helicase movement pauses to allow primase to synthesize the primer and then resumes. This model appears to be consistent with observations for the phage T7 replisome.
2. *The disassembly model.* One or more primase subunits dissociate from the helicase and then stay behind to synthesize a primer while the helicase and its remaining primase subunits continue to move along the lagging strand. The helicase needs to recruit replacement primase subunits for each cycle of Okazaki fragment formation. This model appears to be consistent with observations for the *E. coli* replisome.
3. *The priming loop model.* The helicase and primase remain associated. The helicase continues to move in a 5′→3′ direction along the lagging strand as it unwinds the double-stranded DNA while the associated primase moves in the opposite direction to synthesize primer. This arrangement results in the formation of a transient single-stranded DNA loop in the lagging-strand template. This loop, designated the **priming loop,** is subsequently released to become part of the trombone loop when the primer is passed to the core polymerase associated with the lagging strand. This model seems to fit data

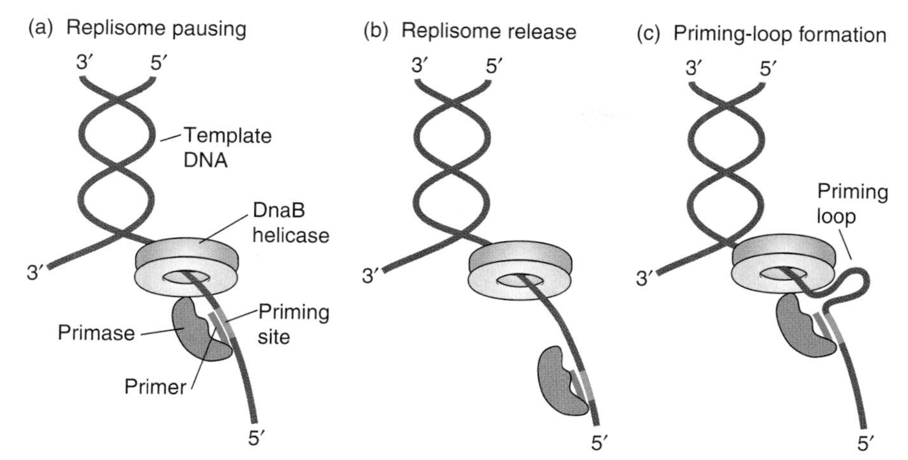

FIGURE 9.37 **Three priming mechanisms.** (a) In the pausing model, helicase movement pauses to allow primase to synthesize the primer and then resumes. (b) In the disassembly model, one or more primase subunits dissociate from the helicase and then stay behind to synthesize a primer while the helicase and its remaining primase subunits continue to move along the lagging strand. (c) In the priming loop model, the helicase and primase remain associated. The helicase continues to move in a 5′→3′ direction along the lagging strand as it unwinds the double-stranded DNA while the associated primase moves in the opposite direction to synthesize primer. This arrangement results in the formation of a transient single-stranded DNA loop, the priming loop, in the lagging-strand template. This loop is released later to become part of the trombone loop. (Adapted from N. E. Dixon, *Nature* 462 [2009]: 854–855.)

for the phage T4 replisome but may also apply to bacterial replisomes.

9.4 The Termination Stage

Replication terminates when the two growing forks meet in the terminus region, which is located 180° around the circular chromosome from the origin.

The two replication forks initiated from *oriC* (min 84) eventually meet at a termination region on the opposite side of the chromosome, triggering a series of events that lead to completion of chromosome synthesis and then chromosome separation. The movement of the replication fork within the termination region is arrested or at least forced to pause for a long time at specific **termination sites** (***Ter* sites**). *Ter* sites are conserved 11 bp sequences that may be present in two orientations. When present in one orientation, they allow the replication machinery to pass through but when present in the opposite orientation they stop replication. *E. coli* has a total of ten Ter sites that are present in two clusters, each containing five *Ter* sites. As shown in FIGURE 9.38, *TerC*, *TerB*, *TerF*, *TerG*, and *TerJ* block the progress of a replication fork that moves in a clockwise direction, whereas *TerA*, *TerD*, *TerE*, *TerI*, and *TerH* block the progress in the opposite direction.

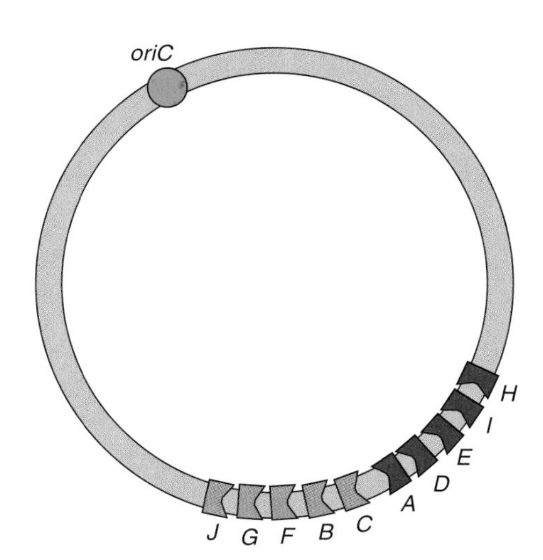

FIGURE 9.38 **Locations of the *E. coli* replication terminator (Ter) sites.** Ten terminators flank the terminus region. Clockwise replication is arrested by the terminators *TerC, TerB, TerF, TerG,* and *TerJ* and counterclockwise replication by terminators and *TerA, TerD, TerE, TerI,* and *TerH.* (Adapted from S. Mulugu, et al., *Proc. Natl. Acad. Sci. USA* 98 [2001]: 9569–9574.

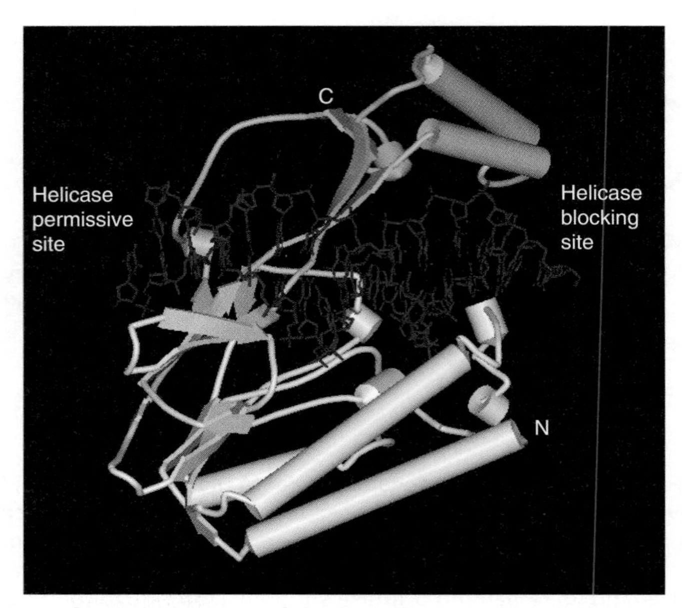

FIGURE 9.39 **Crystal structure of the Tus•*Ter* complex of *E. coli* showing the DNA-binding region of β-strands and the helicase blocking end.** Notice that the protein has a polarity and the side that blocks the DnaB helicase is on the right. The protein is shown in schematic form with the N-terminal domain in gray and the C-terminal domain in orange. The amino and carboxyl ends are indicated by N and C, respectively. The DNA is shown as a red stick figure. (Structure from Protein Data Bank 1ECR. K. Kamada, et al., *Nature* 383 [1996]: 598–603. Prepared by B. E. Tropp.)

The terminus utilization substance (TUS) binds to *Ter* sites.

A protein called the **terminus utilization substance** (**Tus**) binds to the *Ter* sites. The Tus protein binds to *Ter* as a monomer with a very high affinity, ensuring the polar function of the Tus•*Ter* complex (FIGURE 9.39). When present in the proper orientation, the Tus•*Ter* complex arrests the progress of the replication fork by interfering with DnaB helicase's ability to unwind DNA at the replication fork. The gene that encodes the Tus protein, *tus*, is located at minute 36.3 and so lies within the termination region. The physiological significance of the Tus•*Ter* system remains an open question because null mutants lacking the Tus protein grow normally under a variety of growth conditions. *Bacillus subtilis* uses an entirely different protein in place of Tus, one that acts as a dimer to bind to termination sites that are approximately 30-bp long. This difference is surprising because *E. coli* and *B. subtilis* replication systems are quite similar in other respects.

Two additional enzymes, **topoisomerase IV** (a type-2 topoisomerase; see Chapter 4) and **recombinase** (FIGURE 9.40), are needed to allow the newly formed sister chromosomes to separate. Topoisomerase IV allows the interlinked or catenated chromosomes to separate. It cannot help, however, when an odd number of recombination events occur during the replication process causing the two chromosomes to become joined by covalent bonds and form a dimer. Dimerization presents a problem because the two sister chromosomes will not be

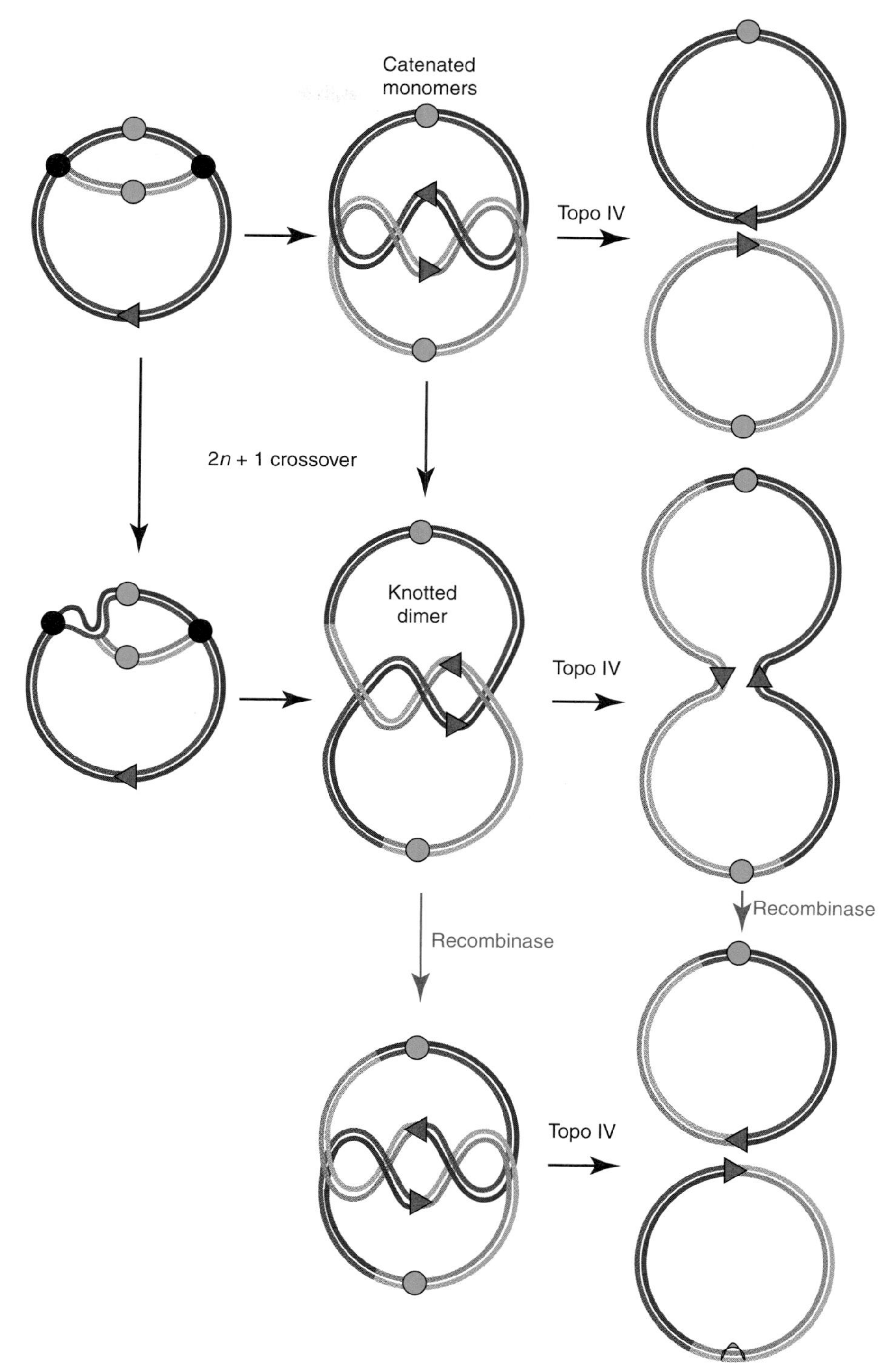

FIGURE 9.40 **Links between replication, recombination, and chromosome segregation.** Topoisomerase IV (Topo IV) is sufficient to produce separated monomers when the newly formed sister chromosomes form catenanes (interlocking rings) but are not covalently joined. When an odd number of crossovers takes place between homologous regions in newly formed sister chromosomes, it generates a knotted dimer in which the sister chromosomes are covalently joined. Then both recombinase and topoisomerase IV (Topo IV) are required to separate monomers. Origins are shown as green circles. Replication forks and associated replication machinery are shown as black circles. The recombination site *dif* that is recognized by recombinase is shown as a red triangle. Newly synthesized strands are shown in lighter shades. (Reprinted from *Curr. Opin. Microbiol.*, vol. 4, D. J. Sherratt, I. F. Lau, and F.-X. Barre, Chromosome segregation, pp. 653–659, copyright 2001, with permission from Elsevier [http://www.sciencedirect.com/science/journal/13695274].)

able to separate and move to daughter cells. Recombinase acts at a specific site within the termination region called *dif* (deletion-induced filamentation), converting the dimer into two separate daughter chromosomes. The site's name derives from the observation that *dif* deletion mutants continue to grow lengthwise, form long filaments, and eventually die because they cannot partition their joined chromosomes to complete cell division.

9.5 Regulation of Bacterial DNA Replication

Three mechanisms regulate bacterial DNA replication at the initiation stage.

Bacterial DNA replication is regulated at the initiation stage of replication, ensuring that the bacterial chromosome is replicated only one time during the cell cycle. DnaA plays an important role in this regulation. The approximately 1000 DnaA molecules present in each *E. coli* molecule are stable. However, intracellular DnaA•ATP levels rise and fall during the cell cycle. Approximately 80% of the DnaA is present in the active DnaA•ATP level just prior to the initiation of a new round of DNA replication but falls to about 20% after replication is initiated.

Three systems regulate the timing and synchrony of replication initiation: (1) regulatory inactivation of DnaA, (2) titration of free DnaA molecules, and (3) sequestration by the SeqA protein.

1. **Regulatory inactivation of DnaA (RIDA).** RIDA is a post-initiation regulatory mechanism that blocks further initiations by converting active DnaA•ATP into inactive DnaA•ADP. The two essential components of the RIDA system are the β clamp and a protein called Hda (homologous to Dna*A*). Hda binds to the β clamp and stimulates DnaA to hydrolyze ATP. This hydrolysis causes a rapid decrease in the intracellular DnaA•ATP level, preventing further DNA replication initiations.
2. **DnaA Titration.** The DnaA titration system also lowers the intracellular DnaA concentration. The *E. coli datA* (DnaA titration A) locus, which maps at 94.7 min of the chromosomal map, is essential for DnaA titration. The *datA* locus contains five 9-mers with very high affinities for DnaA. In fact, their affinity for DnaA is so high that the *datA* locus can titrate eight times more DnaA molecules than the region spanning *oriC* and the neighboring gene *mioC*. Because the *datA* locus is a relatively short distance from *oriC*, it replicates early in the replication cycle to double the number of high affinity DnaA binding sites. The *datA* locus appears to function as a

reservoir that collects excess DnaA molecules until the next initiation event. In support of this mechanism, genetic studies show that cells with a *datA* deletion have extra initiation events.

3. **Sequestration.** As mentioned, *oriC* has a much higher frequency of GATC sequences than would be predicted from completely random nucleotide sequences. DNA adenine methyltransferase (DAM) transfers methyl groups from S-adenosylmethionines to the adenines in these GATC sequences. Because methylation occurs after DNA replication, only the parental strand is methylated in newly synthesized DNA. The hemimethylated *oriC* is sequestered so that it cannot participate in a new round of replication. This sequestration requires **SeqA protein,** which binds to the hemimethylated GATC sites and thereby prevents DnaA•ATP from binding to DnaA binding sites in *oriC*. SeqA therefore acts as a regulatory protein that keeps all origins inactivated for about one-third of a generation. Further study is required to determine how sequestration ends. It is possible, however, that spontaneous dissociation of SeqA and subsequent methylation by DNA adenine methyltransferase may suffice.

Specific *E. coli* genomic sequences promote replication initiation by converting DnaA•ADP to DnaA•ATP. Two **DnaA-reactivating sequences, DARS1** and **DARS2,** appear to promote the release of ADP from DnaA•ADP. The DARS sites have a cluster of DnaA binding sites. Recent studies suggest that DnaA•ADP assembles on DARS1 (and perhaps also on DARS2) in a unique way that forces DnaA to assume a conformation with a low affinity for ADP. Once ADP dissociates from DnaA, the protein can bind ATP. Further studies are required to determine if the pathway requires an additional protein factor or enzyme. Cells that lack both DARS1 and DARS2 appear viable but have impaired replication initiation. These mutant cells may be able to use anionic glycerophospholipids in place of the DARS sequences to exchange ADP for ATP.

Suggested Reading

Overview

Baker, T. A., and Bell, S. P. 1998. Polymerases and the replisome: machines within machines. *Cell* 92:295–305.

Benkovic, S. J., Valentine, A. M., and Salinas, F. 2001. Replisome-mediated DNA replication. *Ann Rev Biochem* 70:181–208.

Johnson, A., and O'Donnell, M. 2005. Cellular DNA replicases: components and dynamics at the replication fork. *Ann Rev Biochem* 74:283–315.

Kornberg, A., and Baker, T. 1992. *DNA Replication*, 2nd ed. New York: W. H. Freeman.

General Features of DNA Replication

Cairns, J. 1963. The chromosome of *Escherichia coli. Cold Spring Harb Symp Quant Biol* 18:43–46.

Gyurasits, E. B., and Wake, R. G. 1973. Bidirectional chromosome replication in *Bacillus subtilis. J Mol Biol* 73:55–63.

Lehman, I. R. 1974. DNA ligase: Structure, mechanism, and function. *Science* 186:790–797.

Mesleson, M., and Stahl, F. 1958. The replication of DNA in *Escherichia coli. Proc Natl Acad Sci USA* 44:671–682.

Okazaki, R., Okazaki, T., Sakabe, K., et al. 1968. Mechanism of DNA chain growth. I. Possible discontinuity and unusual secondary structures of newly synthesized chains. *Proc Natl Acad Sci USA* 59:598–605.

Shuman, S. 2009. DNA ligases: Progress and prospects. *J Biol Chem* 284:17365–17369.

Initiation

Bailey, S., Eliason, W. K., and Steitz, T. A. 2007. Structure of the hexameric DnaB helicase and its complex with a domain of DnaG primase. *Science* 318:459–463.

Crooke, E. 2001. *Eschericia coli* DnaA protein—phospholipids interactions: in vitro and in vivo. *Biochimie* 83:19–23.

Cunningham, E. L., and Burger, J. M. 2005. Unraveling the early steps of prokaryotic replication. *Curr Opin Struct Biol* 15:68–76.

Dasgupta, S, and Løbner-Olesen, A. 2004. Host controlled plasmid replication: *Escherichia coli* minichromosomes. *Plasmid* 52:151–168.

Duderstadt, K. E., and Berger, J. M. 2008. AAA+ ATPases in the initiation of DNA replication. *Crit Rev Biochem Mol Biol* 43:163–187.

Erzerberger, I. P., Mott, M. L., and Berger, J. M. 2006. Structural basis for ATP-dependent DnaA assembly and replication-origin remodeling. *Nat Struct Mol Biol* 13:676–683.

Erzberger, J. P., Pirruccello, M. M., and Berger, J. M. 2002. The structure of bacterial DnaA: implications for general mechanisms underlying DNA replication initiation. *EMBO J* 21:4763–4773.

Fujimitsu, K., Senriuchi, T., and Katayama, T. 2009. Specific genomic sequences of *E. coli* promote replicational initiation by directly reactivating ADP-DnaA. *Genes Dev* 23:1221–1233.

Gilbert, D. M. 2004. In search of the holy replicator. *Nat Rev Mol Cell Biol* 3:1–8.

Kaguni, J. M. 2006. DnaA: controlling the initiation of bacterial DNA replication and more. *Ann Rev Microbiol* 60:351–371.

Katayama, T. 2008. Roles for the AAA+ motifs of DnaA in the initiation of DNA replication. *Biochem Soc Trans* 36:78–82.

Keyamura, K., Abe, Y., Higashi, M., Ueda, T. and Katayama, T. 2009. DiaA dynamics are coupled with changes in initial origin complexes leading to helicase loading. *J Biol Chem* 284:25038–25050.

Leonard, A. C., and Grimwade, J. E. 2005. Building a bacterial orisome: emergence of new regulatory features for replication origin unwinding. *Mol Microbiol* 55:978–985.

Leonard, A. C., and Grimwade, J. E. 2009. Initiating chromosome replication in *E. coli*: it makes sense to recycle. *Genes Dev* 23:1145–1150.

Makowska-Grzyska, M., and Kaguni, J. M. 2010. Primase directs the release of DnaC from DnaB. *Mol Cell* 37:90–101.

McGarry, K. C., Ryan, V. T., Grimwade, J. E., and Leonard, A. 2004. Two discriminatory binding sites in the *Escherichia coli* replication origin are required for DNA strand opening by initiator DnaA-ATP. *Proc Natl Acad Sci USA* 101:2811–2816.

Messer, W. 2002. The bacterial replication initiator DnaA. DnaA and *oriC*, the bacterial mode to initiate DNA replication. *FEMS Microbiol Rev* 26:355–374.

Messer, W. 2005. Prokaryotic replication origins: Structure and function in the initiation of DNA replication. In: *Encyclopedia of Life Sciences*. pp. 1–7. Hoboken, NJ: John Wiley and Sons.

Messer, W., Blaesing, F., Jakimowicz, D., et al. 2001. Bacterial replication initiator DnaA. Rules for DnaA binding and roles of DnaA in origin unwinding and helicase loading. *Biochimie* 83:5–12.

Miller, D. T., Grimwade, J. E., Betteridge, T., et al. 2009. Bacterial origin recognition complexes direct assembly of higher-order DnaA oligiomeric structures. *Proc. Natl Acad Sci USA* 106:18479–18484.

Molt, K. L., Sutera, V. A. Jr., Moore, K. K., and Lovett, S. T. 2009. A role for nonessential domain II of initiator protein, DnaA, in replication control. *Genetics* 183:39–49.

Mott, M. L., and Berger, J. M. 2007. DNA replication initiation: mechanisms and regulation in bacteria. *Nature Rev Microbiol* 5:343–354.

Ozaki, S., and Katayama, T. 2009. DnaA structure, function, and dynamics in the initiation at the chromosomal origin. *Plasmid* 62:71–78.

Ozaki, S. Kawakami, H., Nakamura, K., et al. 2008. A common mechanism for the ATP-DNA-dependent formation of open complexes at the replication origin. *J Biol Chem* 283:8351–8362.

Robinson, N. P., and Bell, S. D. 2005. Origins of DNA replication in the three domains of life. *FEBS J* 272:3757–3766.

Stepankiw, N., Kaido, A, Boyle, E., and Bates, D. 2009. The right half of the *Escherichia coli* replication origin is not essential for viability, but facilitates multi-forked replication. *Mol Microbiol*. 74:467–479.

Xu, Q., McMullan, D., Abdubek, P., et al. 2009. A structural basis for the regulatory inactivation of DnaA. *J Mol Biol* 385:368–380.

Elongation

Bailey, S., Wing, R. A., and Steitz, T. A. 2006. The structure of *T. aquaticus* DNA polymerase III is distinct from eukaryotic replicative DNA polymerase. *Cell* 126:893–904.

Bailey, S., Eliason, W. K., and Steitz, T. A. 2007. The crystal structure of the *Thermus aquaticus* DnaB helicase monomer. *Nucleic Acids Res* 35:4728–4736.

Bailey, S., Eliason, W. K., and Steitz, T. A. 2007. Structure of hexameric DnaB helicase and its complex with a domain of DnaG primase. *Science* 318:459–463.

Bamabara, R. A., Fay, P. J., and Mallaber, L. M. 1995. Methods of analyzing processivity. *Meth Enzymol* 262:270–280.

Bloom, L. F., and Goodman, M. F. 2001. Polymerase processivity: measurement and mechanisms. In: *Encyclopedia of Life Sciences*. pp. 1–6. Hoboken, NJ: John Wiley and Sons.

Bowman, G. D., Goedken, E. R., Kazmirski, S. L., et al. 2005. DNA polymerase clamp loaders and DNA recognition. *FEBS Lett* 579:863–867.

Bruck, I., and O'Donnell, M. 2001. The ring-type polymerase sliding clamp family. *Genome Biol* 2:1–3001.

Coman, M. M., Jin, M., Ceapa, R., et al. 2004. Dual functions, clamp opening and primer-template recognition, define a key clamp loader subunit. *J Mol Biol* 342:1457–1469.

Corn, J. E., and Berger, J.M. 2006. Regulation of bacterial priming and daughter strand synthesis through helicase-primase interactions. *Nucleic Acids Res* 34:4082–4088.

Corn, J. E., Pelton, J. G., and Berger, J. M. 2008. Identification of a DNA primase template tracking site redefines the geometry of primer synthesis. *Nature Struct Mol Biol* 15:163–169.

Davey, M. J., Jeruzalmi, D., Kuriyan, J., and O'Donnell, M. 2002. Motors and switches: AAA+ machines within the replisome. *Nat Rev Mol Cell Biol* 3:1–10.

Dixon, N. E. 2009. Prime-time looping. *Nature* 462:854–855.

Ellison, V., and Stillman, B. 2001. Opening of the clamp: an intimate view of an ATP-driven biological machine. *Cell* 106:655–660.

Evans, R. J., Davies, D. R., Bullard, J. M., et al 2008. Structure of PolC reveals unique DNA binding and fidelity determinants. *Proc Natl Acad Sci USA* 105:20695–20700.

Georgescu, R. E., Kim, S.-S., Yurieva, O., et al 2008. Structure of a sliding clamp on DNA. *Cell* 132:43–54.

Georgescu, R. E., Kurth, I., Yao, N Y., et al 2009. Mechanism of polymerase collision release from sliding clamps on the lagging strand. *EMBO J* 28:2981–2991.

Goedken, E. R., Kazmirski, S. L., Bowman, G. D., et al. 2005. Mapping the interaction of DNA with *Escherichia coli* DNA polymerase clamp loader complex. *Nat Struct Mol Biol* 12:183–189.

Hamdan, S. M., and Richardson, C. C. 2009. Motors, switches, and contacts in the replisome. *Ann Rev Biochem* 78:205–243.

Indiani, C., and O'Donnell, M. 2003. Mechanism of the δ wrench in opening the β sliding clamp. *J Biol Chem* 278:40272–40281.

Indiani, C., and O'Donnell, M. 2006. The replication clamp-loading machine at work in the three domains of life. *Nat Rev Mol Cell Biol* 7:751–761.

Jeruzalmi, D., O'Donnell, M., and Kuriyan, J. 2002. Clamp loaders and sliding clamps. *Curr Opin Struct Biol* 12:217–224.

Johnson, A., and O'Donnell, M. 2003. Ordered ATP hydrolysis in the γ complex clamp loader AAA+ machine. *J Biol Chem* 278:14406–14413.

Kuchta, R. D., and Stengel, G. 2010. Mechanisms and evolution of DNA primases. *Biochim Biophys Acta* 1804:1180–1189.

Langston, L. D., and O'Donnell, M. 2006. DNA replication: Keep moving and don't mind the gap. *Mol Cell* 23:155–160.

Leu, F. P., Georgescu, R., and O'Donnell, M. 2003. Mechanism of the *E. coli* τ processivity switch during lagging-strand synthesis. *Mol Cell* 11:315–327.

López deSaro F., Georgescu, R. E., Leu, F., and O'Donnell, M. 2004. Protein trafficking on sliding clamps. *Philos Trans R Soc Lond Biol Sci* 359:25–30.

Lovett, S. T. 2007. Polymerase switching in DNA replication. *Mol Cell* 27:523–526.

Marians, K. J. 2008. Understanding how the replisome works. *Nat Struct Mol Biol* 15:125–127.

McInerney, P., Johnson, A., Katz, F., and O'Donnell, M. 2007. Characterization of a triple DNA polymerase replisome. *Mol Cell* 27:527–538.

Mott, M. L., Erzberger, J. P., Coons, M. M., and Berger, J. M. 2008. Structural synergy and molecular crosstalk between bacterial helicase loaders and replication inhibitors. *Cell* 134:623–634.

Nossal, N. G., Makhov, A. M. Chastain, P. D., Jones, C. E., and Griffith, J. D. 2007. Architecture of the bacteriophage T4 replication complex revealed with nanoscale biopointers. *J Biol Chem* 282:1098–1108.

O'Donnell, M. 2006. Replisome architecture and dynamics in *Escherichia coli*. *J Biol Chem* 281:10653–10656.

O'Donnell, M., Jeruzalmi, D., and Kuriyan, J. 2001. Clamp loader structure predicts the architecture of DNA polymerase III holoenzyme and RFC. *Curr Biol* 11:R935–R946.

Pandey, M., Syed, S., Donmez, I., et al. 2009. Coordinating DNA replication by means of priming loop and differential synthesis rate. *Nature* 462:940–944.

Park, M. S., and O'Donnell, M. 2009. The clamp loader assembles the β clamp onto either a 3′ or 5′ primer terminus. *J. Biol Chem* 284:31473–31483.

Podobnik, M., McInerney, P., O'Donnell, M., and Kuriyan, J. 2000. A TOPRIM domain in the crystal structure of the catalytic core of *Escherichia coli* primase confirms a structural link to DNA topoisomerases. *J Mol Biol* 300:353–362.

Pomerantz, R. T., and O'Donnell, M. 2007. Replisome mechanics: insights into a twin DNA polymerase machine. *Trends Microbiol* 15:156–164.

Simonetta, K. R., Kazmirski, S. L., Geodken, E. R., et al. The mechanism of ATP-dependent primer-template recognition by a clamp loader complex. *Cell* 137:659–671.

Soultanas, P. 2005. The bacterial helicase-primase interaction: a common structural/functional module. *Structure* 13:839–844.

Syson, K, Thirlway, J., Houslow, A. M., et al. Solution structure of the helicase interaction domain of the primase DnaG: a model helicase activation. *Structure* 13:609–616.

Wing, R. A., Bailey, S., and Steitz, T. A. 2008. Insights into the replisome form the structure of a ternary complex of DNA polymerase III α-subunit. *J Mol Biol* 382:859–869.

Yao, N. Y., and O'Donnell, M. 2008. Replisome structure and conformational dynamics underlie fork progression past obstacles. *Curr Opin Cell Biol* 21:336–343.

Termination

Bastia, D., Zzaman, S, Krings. G., et al. 2008. Replication termination mechanism as revealed by *Tus*-mediated polar arrest of sliding helicase. *Proc Natl Acad Sci USA* 105:12831–12836.

Bussiere, D. E., and Bastia, D. 1999. Termination of DNA replication of bacterial and plasmid chromosomes. *Mol Microbiol* 31:1611–1618.

Duggin, I. G., and Bell, S. D. 2009. Termination structure in the *Escherichia coli* chromosome replication fork trap. *J Mol Biol* 387:532–539.

Duggin, I. G., Wake, R. G., Bell, S. D., and Hill, T. M. 2008. The replication fork trap and termination of chromosome replication. *Mol Microbiol* 70:1323–1333.

Duggin, I. G., and Wilce, J. A. 2005. Termination of replication in bacteria. In: *Encyclopedia of Life Sciences*. pp. 1–7. Hoboken, NJ: John Wiley and Sons.

Kaplan, D. L., and Bastia, D. 2009. Mechanism of polar arrest of a replication fork. *Mol Microbiol* 72:279–285.

Mulcair, M. D., Schaeffer, P. M., Oakley, A. J., et al. 2006. A molecular mousetrap determines polarity of termination of DNA replication in *E. coli*. *Cell* 125:1309–1319.

Mulugu, S., Potnis, A., Shamsuzzaman, et al. 2001. Mechanism of termination of DNA replication of *Escherichia coli* involves helicase-contrahelicase interaction. *Proc Natl Acad Sci USA* 98:9569–9574.

Sherratt, D. J., Lau, I. F., and Barre, F. X. 2001. Chromosome segregation. *Curr Opin Microbiol* 4:653–659.

Regulation of Bacterial Replication

Bach, T., Morigen, and Skarstad, K. 2008. The initiator protein DnaA contributes to keeping new origins inactivated by promoting the presence of hemimethylated DNA. *J Mol Biol* 384:1076–1085.

Chung, Y. S., Brendler, T., Austin, S., and Guarné, A. 2009. *Nucleic Acids Res* 37:3143–3152.

Fujimitsu, K, Senriuchi, T., and Katayama, T. 2009. Specific genomic sequences of *E. coli* promote replicational initiation by directly reactivating ADP-DNA. *Genes Dev* 23:1221–1223.

Kaguni, J. M. 2006. DnaA: controlling the initiation of bacterial DNA replication and more. *Ann Rev Microbiol* 60:351–371.

Leonard, A. C., and Grimwade, J. E. 2009. Initiating chromosome replication in *E. coli*: it makes sense to recycle. *Genes Dev* 23:1145–1150.

Nozaki, S., Yamada, Y., and Ogawa, T. 2009. Initiator titration complex formed at *data* with the aid of IHF regulates replication timing in *Escherichia coli.* 14:329–341.

Su'etsugu, M., Nakamura, K., Keyamura, K., Kudo, Y., and Katayama, T. 2008. Hda monomerization by ADP binding promotes replicase clamp-mediated DnaA-ATP hydrolysis. *J Biol Chem* 283:36118–36131.

Waldminghaus, T., and Skarstad, K. 2009. The *Escherichia coli* SeqA protein. *Plasmid* 61:141–150.

Xu, Q., McMullan, D., Abdubek, P., et al. 2009. A structural basis for the regulatory inactivation of DnaA. *J Mol Biol* 385:368–380.

This book has a Web site, **http://biology.jbpub.com/book/molecular**, which contains a variety of resources, including links to in-depth information on topics covered in this chapter, and review material designed to assist in your study of molecular biology.

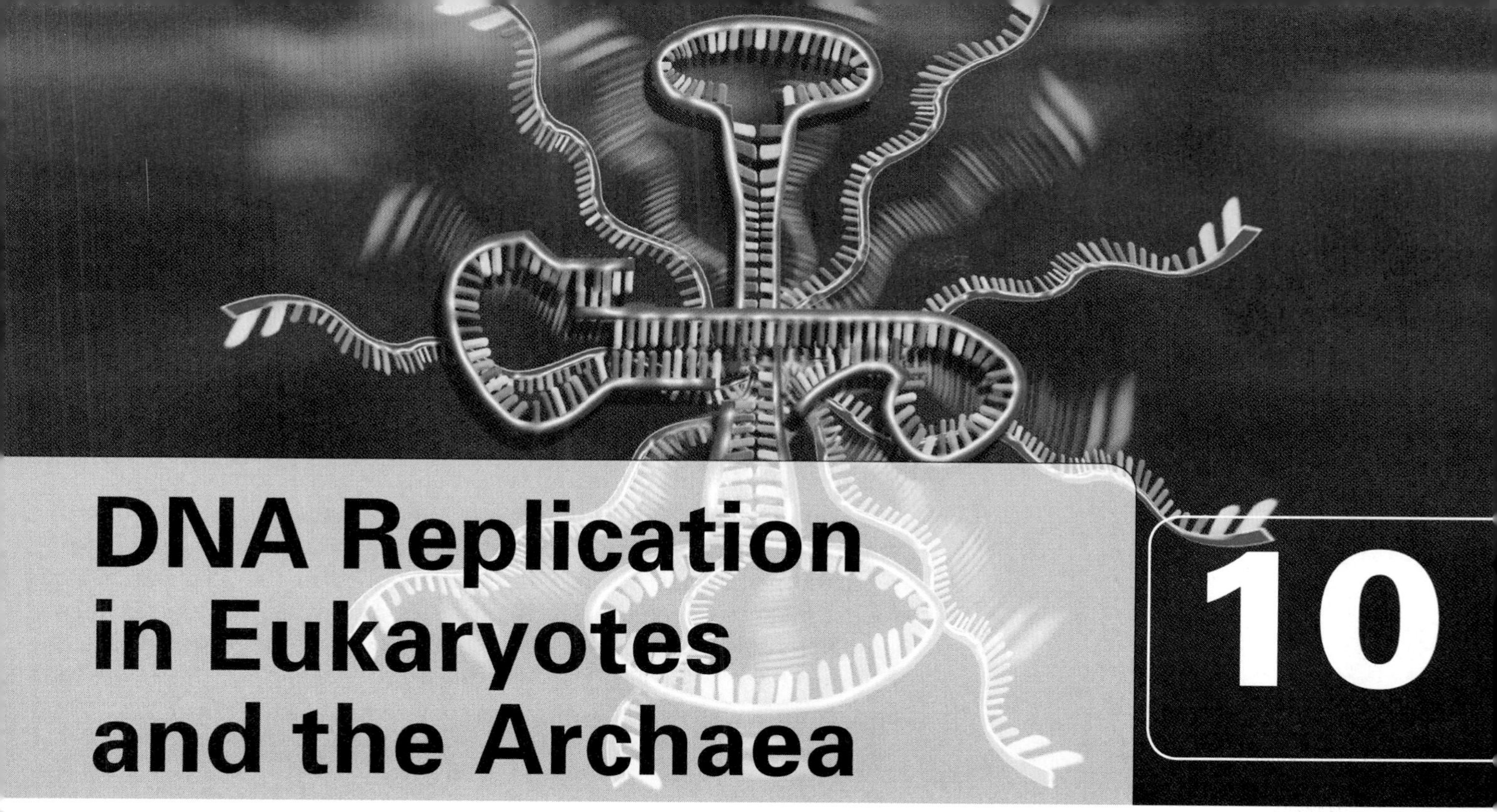

DNA Replication in Eukaryotes and the Archaea

10

OUTLINE OF TOPICS

10.1 The SV40 DNA Replication System.

The SV40 T antigen binds to the origin of replication and unwinds DNA.

SV40 T antigen helps to recruit DNA polymerase/α-primase (Pol α) to the proto-replication bubble.

10.2 Introduction to Eukaryotic DNA Replication

Eukaryotic replication machinery must replicate long linear duplexes with multiple origins of replication.

10.3 Eukaryotic Replication Initiation

Eukaryotic chromosomes have many replicator sites.

Autonomously replicating sequences (ARS) determine the site of DNA chain initiation in yeast.

Two-dimensional gel electrophoresis can locate origins of replication.

The origin of recognition complex (ORC) serves as the eukaryotic initiator.

CDC6 and Cdt1 help load MCM2-7 helicase onto the origin to form a pre-replication complex (pre-RC).

The licensed origin must be activated before replication can take place.

10.4 Eukaryotic Replication Elongation

Pol δ and Pol ε are primarily responsible for copying the lagging- and leading-strand templates, respectively.

10.5 The End-Replication Problem

Studies of the *Tetrahymena* and yeast telomeres suggested that a terminal transferase-like enzyme is required for telomere formation.

Telomerase uses an RNA template to add nucleotide repeats to chromosome ends.

Telomerase plays an important role in solving the end-replication problem.

Telomerase plays a role in aging and cancer.

10.6 Replication Coupled Chromatin Synthesis

Chromatin disassembly and reassembly are tightly coupled to DNA replication.

10.7 DNA Replication in the Archaea

The archaea replication machinery is similar to that in eukaryotes.

Orc1/Cdc6 recruits MCM to the archaeal origin of replication.

The basic steps in archaeal elongation are very similar to those in bacteria and eukaryotes.

Suggested Reading

There is no single eukaryotic or archaeal model system in which all aspects of DNA replication have been studied. Different systems therefore must be examined to obtain a comprehensive view of the replication process. Where possible, we will focus on two of the best-studied eukaryotic DNA replication systems. The first, the simian virus 40 (SV40) replication system, provides information about the enzymes that participate in DNA replication under *in vitro* conditions. The second, the yeast (*Saccharomyces cerevisiae*) replication system, takes full advantage of yeast genetics and biochemistry to examine DNA replication and is particularly useful for examining the initiation process. Additional eukaryotic systems will be introduced when necessary to consider a fundamental aspect of eukaryotic DNA replication that has not or cannot be studied in the SV40 or yeast model systems. Archaeal DNA replication is in many respects very similar to eukaryotic DNA replication. Many of the lessons learned from studying eukaryotic DNA replication systems can, therefore, be applied to archaeal DNA replication systems.

10.1 The SV40 DNA Replication System

The SV40 T antigen binds to the origin of replication and unwinds DNA.

We begin our study of eukaryotic DNA replication by examining SV40 DNA replication because *in vitro* studies of this system provide many important insights into eukaryotic DNA replication. Viral encoded **large tumor** (T) **antigen** (see Chapter 8) serves as both an initiator and helicase. The host cell provides all other enzymes required for SV40 DNA replication. Because SV40 DNA is a covalently closed circular duplex and has a single origin of replication (see Figure 8.38), its replication does not require components of the eukaryotic replication machinery that coordinate replication from multiple origins or extend telomeres that are shortened during replication. In 1984, Joachim J. Li and Thomas J. Kelly demonstrated that a soluble extract derived from SV40-infected monkey cells can replicate SV40 DNA. Extracts from uninfected monkey cells also support SV40 DNA replication if they are supplemented with T antigen.

The SV40 origin of replication contains about 300 bp, but only 64 of these, the so-called **core sequence**, are essential for DNA replication. Nonessential elements on either side of the core sequence facilitate the initiation process. The core sequence is divided into three functional regions (FIGURE 10.1a): a 17-bp AT-rich region (AT), a 23-bp pentanucleotide palindrome (PEN), and an early palindrome (EP). The PEN contains two pairs of GAGGC pentanucleotides (P1, P2 and P3, P4), which are inverted with respect to one another. The pentanucleotides are separated by a single cytidylate. Cytosine may be replaced by another base but the single nucleotide spacing between pentanucleotides pairs must be maintained for replication to occur. All four pentanucleotides must be present for the initiation of DNA replication to take place.

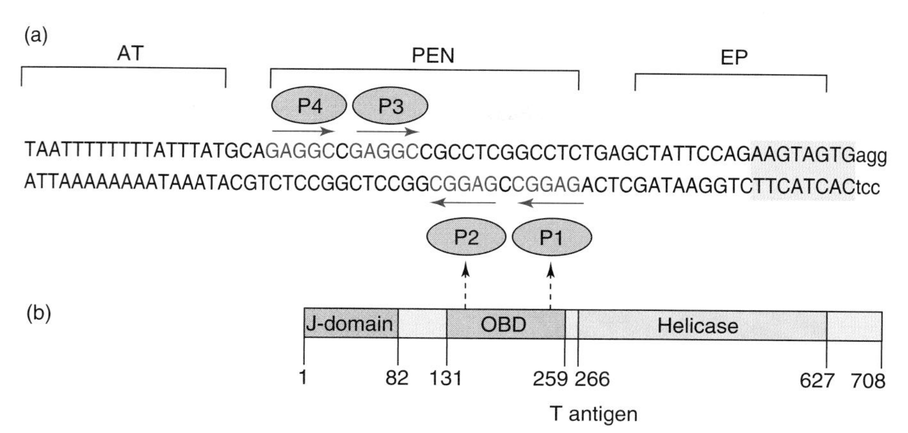

FIGURE 10.1 **Structural and functional elements of the SV40 DNA replication system.** (a) The 64-bp core sequence is divided into three functional regions: a 17-bp AT-rich region, a 23-bp pentanucleotide palindrome (PEN), and an early palindrome (EP). The PEN contains two pairs of GAGGC pentanucleotides (P1–P4) that are inverted with respect to one another (as indicated by the four red arrows). The T antigen OBDs (green ellipses) bind to the GAGGC pentanucleotides. The eight EP nucleotides that melt after the T antigen assembles are highlighted in pink. (b) The T antigen (708 amino acid residues) has three functional domains. The origin binding domain (OBD) binds to the GAGGC pentanucleotide in the PEN. The helicase domain uses energy provided by ATP to unwind DNA. The DnaJ domain is not required for *in vitro* DNA replication. (Adapted from E. Bochkareva, et al., *EMBO J.* 25 [2006]: 5961–5969.)

The T antigen (708 amino acid residues) has three functional domains (FIGURE 10.1b). Two of these, the origin-binding domain (OBD) and the helicase domain are essential for SV40 DNA replication under *in vitro* conditions. The OBD (residues 131–259) binds tightly to a GAGGC sequence in the PEN. This binding is required to recruit T antigen to the origin of replication. The helicase domain (residues 265–626), which is carried to the origin of replication by the OBD, uses energy provided by ATP to unwind DNA. The DnaJ domain (residues 1–82) is not required for *in vitro* DNA replication but, like a homologous bacterial chaperone protein for which it is named, participates in remodeling protein complexes.

Crystal structures are not available for the intact T antigen but have been determined for each of its domains. The oligonucleotide binding and helicase domains, the two domains required for DNA synthesis, are of special interest. Alexey Bochkarev and coworkers have determined the crystal structure for a nucleoprotein complex that has one OBD bound to the P1 pentanucleotide and another bound to the P3 pentanucleotide (FIGURE 10.2). Each OBD has two structural motifs, the so-called A1 loop and B2 elements, that contact the pentanucleotide. Although not shown, another OBD binds in the same way to the P2 pentanucleotide and still another to the P4 pentanucleotide. The OBDs do not make physical contact with one another. Xiaojian S. Chen and coworkers have obtained crystal structures for the T antigen helicase domain as a monomer and a homohexamer (FIGURE 10.3). Additional studies are required to determine how the T antigen

FIGURE 10.2 **The crystal structure of OBD bound to P1 and P3.** One OBD is bound to the P1 pentanucleotide and another OBD is bound to the P3 pentanucleotide. Each OBD has two structural motifs, the A1 loop (green) and the B2 element (yellow), that make contact with the pentanucleotide. DNA is shown as a spacefill model with one strand in dark blue and the other in red. The P1 and P3 pentanucleotides are shown in pink and white, respectively. Although not shown, two other OBDs bind to the P2 and P4 pentanucleotides at the bottom of the DNA in the same way. (Structure from Protein Data Bank 2ITL. E. Bochkareva, et al., *EMBO J* 25 [2006]: 5961–5969. Prepared by B. E. Tropp.)

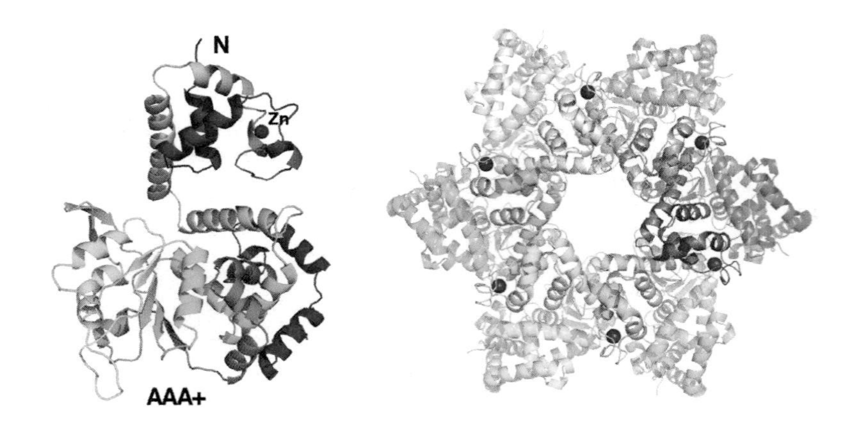

(a) T antigen helicase domain monomer (b) T antigen helicase domain hexamer

FIGURE 10.3 **Structure of the replicative helicase of the SV40 large tumor antigen.** (a) T antigen helicase domain (residues 251–627) monomer. Zn is a bound zinc ion. (b) T antigen helicase domain (residues 251–627) hexameric structure. (Adapted from D. Li, et al., *Nature* 423 [2003]: 512–518. Photos courtesy of Dr. Xiaojiang Chen's group, University of Southern California.)

unwinds DNA and elucidate the steps leading to the assembly of the replication machinery at the replication fork.

SV40 T antigen helps to recruit DNA polymerase/α-primase (Pol α) to the proto-replication bubble.

Replication protein A (RPA), the eukaryotic single-stranded DNA binding protein (see Chapter 4), and T antigen help to recruit **DNA polymerase/α-primase** (**Pol α**), a four-subunit protein complex, to the proto-replication bubble. As the only eukaryotic enzyme that can synthesize primer, Pol α is an essential participant in initiation. Its crystal structure has not yet been determined. Pol α has two distinct catalytic activities, primase and DNA polymerase, which reside in separate subunits. The other two subunits appear to play important structural roles, but no enzyme activities have as yet been assigned to them. Mammalian Pol α does not have proofreading activity. Primase forms RNA oligomers that are 8 to 12 nucleotides long, which serve as primers for DNA synthesis. Then the weakly processive Pol α DNA polymerase incorporates 15 to 25 deoxynucleotides before dissociating from the DNA template. The oligonucleotide synthesized by Pol α thus consists of an RNA primer joined to a short DNA segment (**initiator DNA**). The first pair of initiator DNA molecules, which are formed on either side of the core origin, serve to prime leading strand formation. Initiator DNA molecules formed by Pol α later in the replication process prime lagging strand synthesis.

Further work in Thomas J. Kelly's and other laboratories led to the identification and characterization of the following additional cellular proteins required for SV40 DNA replication.

1. **DNA polymerase δ** (**Pol δ**) catalyzes both leading and lagging strand DNA synthesis in the SV40 DNA replication system.

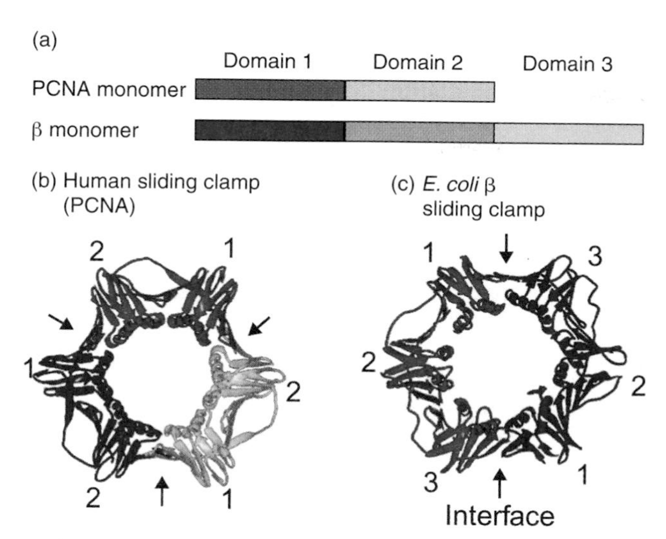

FIGURE 10.4 **Human and *E. coli* β clamps.** (a) The human sliding clamp (PCNA) subunit is about two thirds the size of the bacterial β subunit. The former contains two domains (shown in shades of blue) and the latter three domains (shown in shades of purple). (b) The crystal structure of the human sliding clamp (PCNA), which is similar to PCNAs from other eukaryotes. The three subunits are shown in red, yellow, and blue. (c) The crystal structure of the *E. coli* sliding clamp. The two subunits are shown in red and blue. Interfaces between subunits are indicated by arrows and domains within a subunit are numbered (1 or 2 for PCNA and 1, 2, or 3 for the bacterial clamp). (Part a adapted from I. Bruck and M. O'Donnell, *Genome Biol.* 2 [2001]: reviews 3001.1–3001.3. Parts b and c reproduced from I. Bruck and M. O'Donnell, *Genome Biol.* 2 [2001]: reviews 3001.1–3001.3. Photos courtesy of Michael O'Donnell, Rockefeller University.)

2. **PCNA**, the eukaryotic sliding clamp, was initially identified as a proliferating cell nuclear antigen. PCNA has a similar quaternary structure to that of the bacterial sliding clamp. This similarity is remarkable because the protein subunits in the eukaryotic and bacterial sliding clamps are not homologous and the eukaryotic clamp has three subunits, whereas the bacterial sliding clamp has two subunits (FIGURE 10.4). Pol δ, FEN1 (see below) and DNA ligase all bind to PCNA.
3. **Replication factor C (RFC)**, the eukaryotic clamp loader, couples ATP hydrolysis to the assembly of the PCNA sliding clamp onto a primed template. RFC has five subunits that are homologous to one another and to the subunits of the *Escherichia coli* clamp loader. RFC works in a similar (but not identical) way to the *E. coli* clamp loader.
4. **Flap endonuclease (FEN1)**, a 5′-endo/exonuclease, FEN1, recognizes a double-stranded DNA with a 5′-unannealed flap and makes an endonucleolytic cleavage at the base of the flap (FIGURE 10.5). FEN1 acts together with Pol δ and DNA ligase during Okazaki fragment maturation (see below).
5. **DNA ligase** seals the nick between adjacent Okazaki strands.

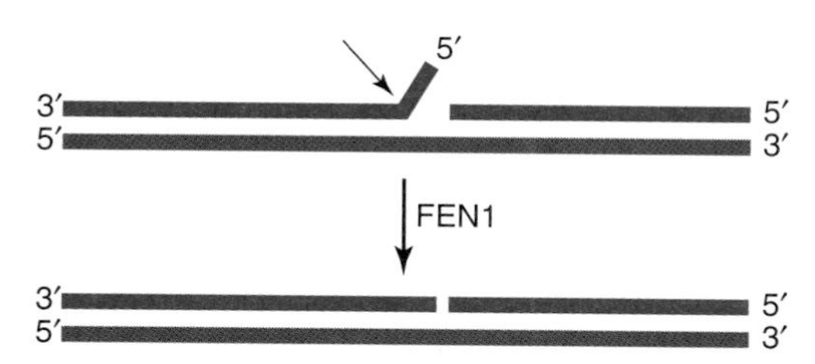

FIGURE 10.5 **Flap endonuclease 1 (FEN1) cleavage.** FEN1 recognizes a double-stranded DNA with a 5′-unannealed flap and makes an endonucleolytic cleavage at the base of the flap.

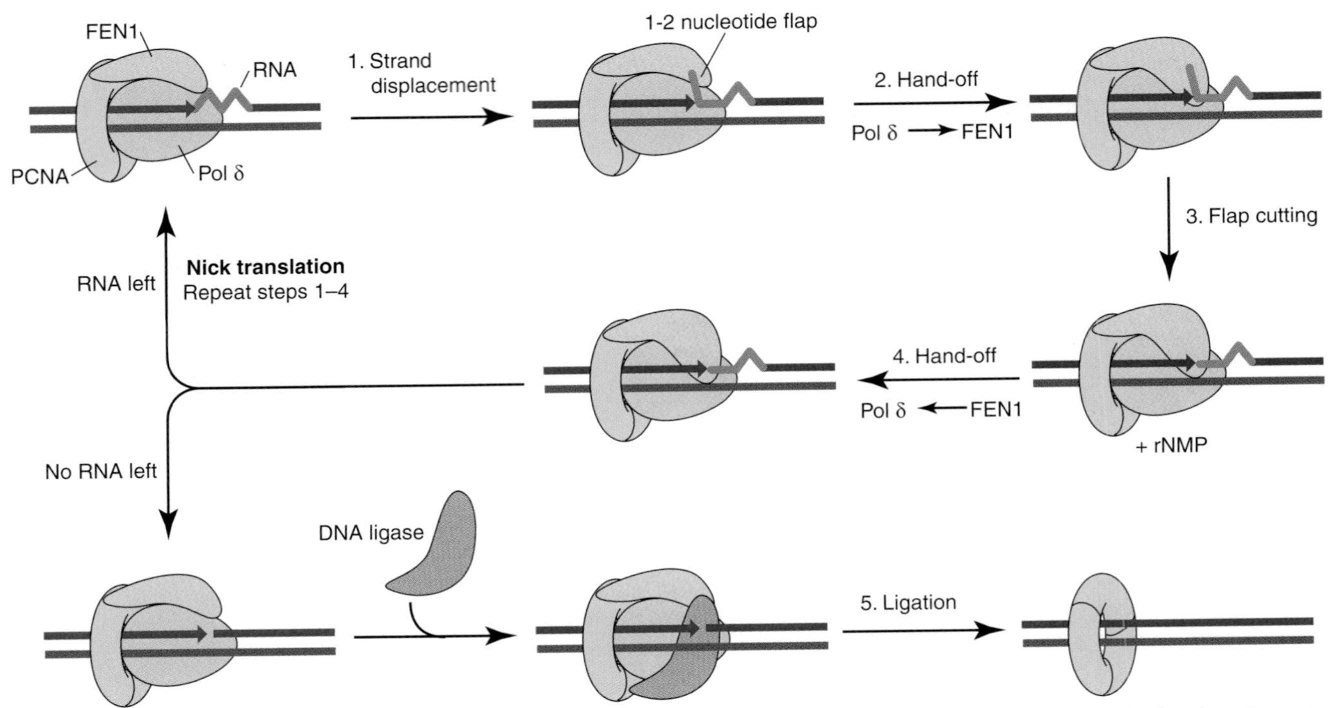

FIGURE 10.6 **Okazaki fragment maturation.** Pol δ, FEN1, and DNA ligase are all associated with PCNA during lagging strand synthesis. Pol δ continues DNA synthesis after running into the RNA primer of the previous Okazaki fragment and displaces one or two nucleotides at its 5′-end to produce a flap. FEN1 removes the 5′ flap. Pol δ synthesis continues to generate another flap, which is also cleaved by FEN1. Nick translation continues until all the RNA has been removed. Then DNA ligase joins the two fragments. The RNA primer is shown in green, the lagging strand template is in blue, and the lagging strand is in red. (Adapted from P. M. J. Burgers, *J. Biol. Chem.*, 284 [2009]: 4041–4045.)

After extending an initiator DNA on the lagging strand by about 100 to 200 nucleotides, Pol δ runs into the RNA primer of the previous Okazaki fragment. The RNA primer must be removed before DNA ligase joins the two fragments. Continued Pol δ synthesis displaces one or two nucleotides at the 5′-end of the previous Okazaki fragment. Endonucleolytic cleavage by FEN1 removes the flap and Pol δ synthesis continues to generate another flap, which is again cleaved by FEN1. This nick translation process continues until all the RNA primer has been removed. Then DNA ligase joins the two fragments (FIGURE 10.6). Other pathways probably also contribute to RNA primer removal. For example, RNase H may participate in primer removal.

FIGURE 10.7 shows how several of the important enzymes and proteins required for SV40 DNA elongation may be organized at the replication fork.

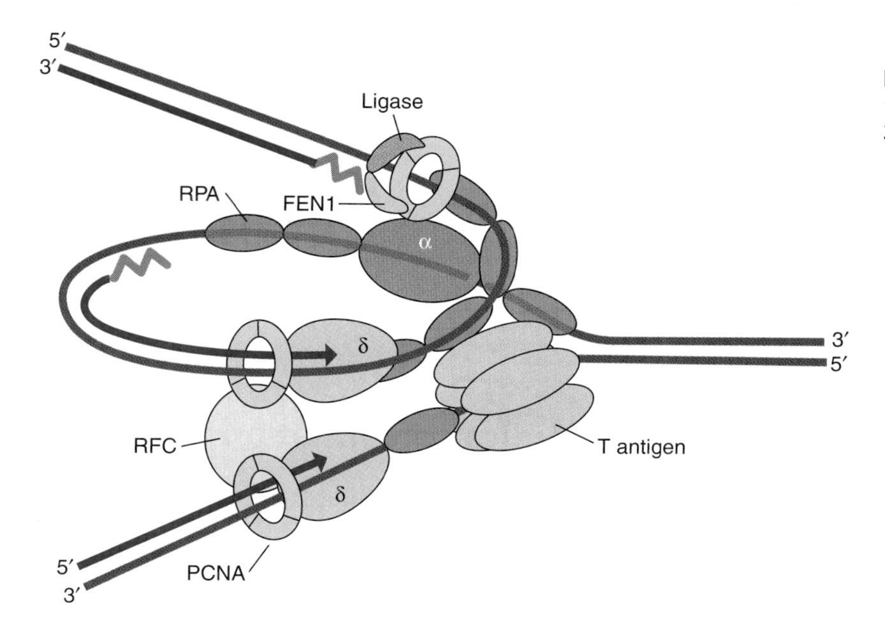

FIGURE 10.7 **Model for the organization of proteins and enzymes at SV40 replication fork.** (Adapted from B. Stillman, *Mol. Cell* 30 [2008]: 259–260).

10.2 Introduction to Eukaryotic DNA Replication

Eukaryotic replication machinery must replicate long linear duplexes with multiple origins of replication.

Major differences between bacterial and eukaryotic DNA replication arise from the fact that the bacterial chromosome is a closed covalent circle with a single origin of replication, whereas a eukaryotic chromosome is a long linear duplex. These differences lead to the following important consequences.

1. A bacterial chromosome has a single origin of replication, whereas each eukaryotic chromosome has several origins of replication. Eukaryotes have regulatory mechanisms that ensure each origin fires once, and only once, during the S phase of the life cycle.
2. Bacteria require only one DNA polymerase, DNA polymerase III holoenzyme, for normal replication fork propagation, whereas eukaryotes require three enzymes: Pol α, Pol δ, and Pol ε (see below).
3. Bacterial DNA replication terminates when the two replicating forks meet about half way around the bacterial chromosome and then are joined. Linear eukaryotic chromosomes require a special enzyme, telomerase, to form their ends.

We must keep these differences in mind as we examine the different stages of eukaryotic replication: initiation, elongation, and termination. Although there is no single eukaryotic model system in which all aspects of DNA replication have been studied, much of our current knowledge of eukaryotic DNA replication has come from studies with yeast. Where possible, we will focus on the yeast replication system, but we must be mindful that DNA replication in other eukaryotes may differ in some details.

10.3 Eukaryotic Replication Initiation

Eukaryotic chromosomes have many replicator sites.

The long linear eukaryotic DNA molecules initiate DNA replication at multiple sites. Evidence for this comes from electron micrographs of replicating eukaryotic DNA, which reveal multiple replication bubbles (FIGURE 10.8) and from autoradiograms of tritium labeled DNA fibers isolated from proliferating cells. Based on the size and location of the replication bubbles, initiation sites are estimated to be from 10 to 300 kb apart. Initiation at each site is restricted to once per S phase during normal cell growth. Bidirectional replication bubble growth continues until neighboring bubbles fuse (FIGURE 10.9). Rare multiple initiation events at the same site, which lead to gene amplification, are observed in tumor cells and for specific genes during normal cell development.

Autonomously replicating sequences (ARS) determine the site of DNA chain initiation in yeast.

In 1980, Ronald W. Davis and coworkers discovered that yeast DNA sequences, called **autonomously replicating sequence elements (ARS elements)**, allow plasmids containing selectable genes to replicate autonomously in yeast cells. The experiment shown in FIGURE 10.10 illustrates one method for demonstrating the existence of ARS elements. Two sets of circular plasmids, one containing an ARS element

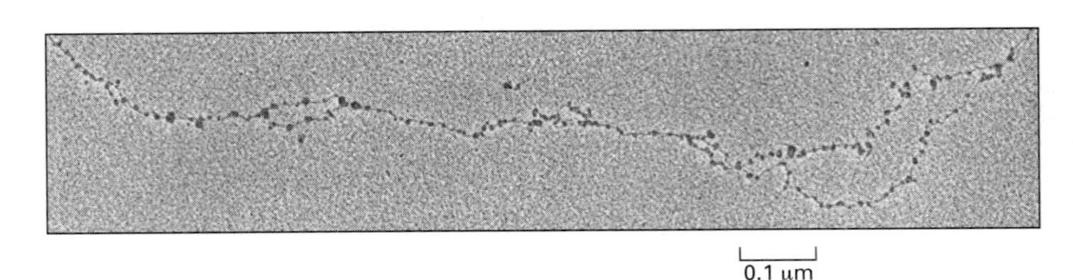

FIGURE 10.8 **An electron micrograph of a replicating chromosome in an early embryo of *Drosophila melanogaster*.** Each origin is apparent as a replication bubble along the DNA strand. (Courtesy of Victoria E. Foe, Center for Cell Dynamics, University of Washington.)

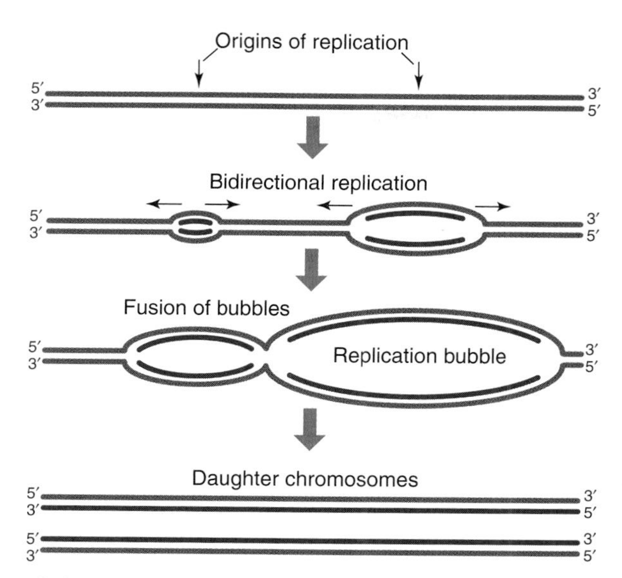

FIGURE 10.9 **Schematic showing the fusion of a pair of replication bubbles.**

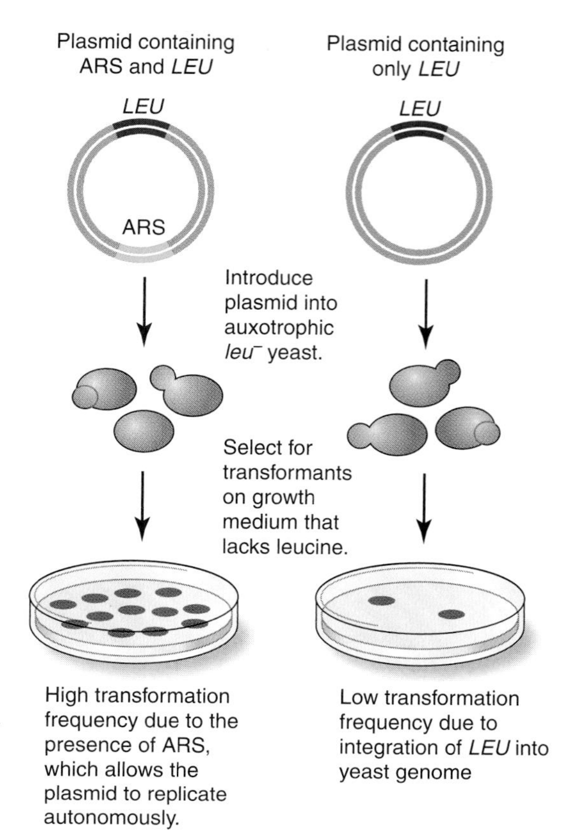

FIGURE 10.10 **Method for detecting yeast autonomously replicating sequences (ARS).** Two sets of plasmids, one with an ARS element and *LEU* and the other with just *LEU*, are introduced into leu^- yeast cells. Then the transformed yeast cells are spread on agar growth medium lacking leucine. Many colonies appear when cells transformed with ARS-plasmids are spread because these plasmids can replicate autonomously. In contrast, very few colonies appear when plasmids lacking ARS are spread and the colonies that do form contain cells in which *LEU* is integrated into the yeast genome.

and *LEU* and the other just containing *LEU*, are introduced into leu^- yeast cells, which are then spread on agar growth medium lacking leucine. A large number of colonies form when cells transformed with ARS-plasmids are spread on the agar medium because the plasmids can replicate autonomously. In contrast, very few colonies appear when cells transformed with plasmids lacking ARS are spread on the agar medium and the colonies that do form contain cells in which *LEU* is integrated into the yeast genome.

Yeast cells have approximately 300 ARS elements spread over their 16 chromosomes, while cells from higher animals probably have thousands. Most yeast ARS elements are named by using a number that designates their chromosome and position on the chromosome (FIGURE 10.11). Although the ARS elements were initially discovered because of their ability to support autonomous plasmid replication in yeast, they function as replicators in yeast and overlap sites of replication initiation. There are two main reasons why each yeast chromosome requires several ARS elements while the *E. coli* chromosome requires just one replicator. First, yeast forks migrate about

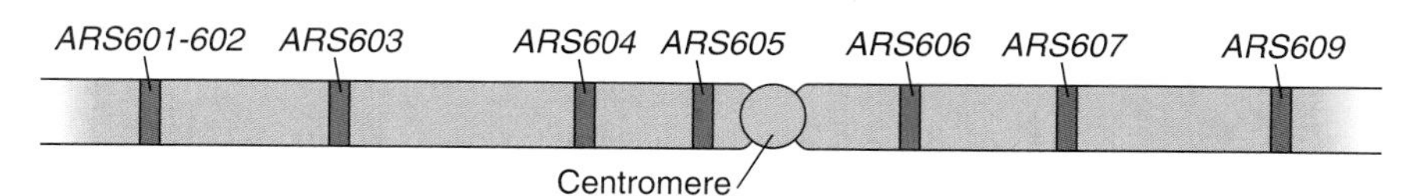

FIGURE 10.11 **The location of some, but not all, of the autonomously replicating sequence (ARS) elements in yeast chromosome VI.** ARS elements are shown in green. Origins near the centromere initiate replication earlier than those near the ends of the chromosome. (Adapted from M. Weinreich, M. A. P. DeBeer, and C. A. Fox, *Biochim. Biophys. Acta* 1677 [2004]: 142–157.

30 times more slowly than those in *E. coli*. Second, yeast chromosomes are much longer than the *E. coli* chromosome.

Several different observations suggest that ARS elements direct the initiation of yeast DNA replication. For instance, ARS plasmids and yeast chromosomes both are replicated only during the S phase of the cell growth cycle and both require the same gene products for their replication. One may therefore hope to obtain considerable information about the function of ARS elements by studying them in ARS plasmids, where they are easily manipulated. Two ARS elements that have been studied in great detail, ARS1 and ARS307, are shown in FIGURE 10.12. Sequence analysis reveals that ARS elements are more A/T rich than surrounding chromosomal DNA. Furthermore, all contain an 11-bp element [5′-(A/T)TTTA(T/C)(A/G)TTT(A/G)-3′] known as the **A-element** or **ARS consensus sequence** (**ACS**).

When ARS plasmid transforming ability and stability were used to identify sequences within ARS that are required for function, investigators observed that the A-element and short sequences on either side of it are essential for function. Nucleotide changes within the A-element usually abolish ARS function. A-element, however, is not sufficient for ARS function and 5′- and 3′-neighboring sequences also contribute to ARS function. Sequences 3′ to the T-rich strand of A-elements are called B elements, while those 5′ to the T-rich strand are called C elements. Systematic linker substitution analysis, using linkers in the range of 5 to 10 bp to replace sequences of the same size, revealed that ARS1 has three B elements (B1, B2, and B3) and ARS307 has two B elements (B1 and B2). The B1 elements of ARS1 and ARS307 have entirely different sequences but nevertheless can substitute for one another.

It has been difficult to extend lessons learned from studying replicators in *S. cerevisiae* to other eukaryotes. For instance, ARS elements in the fission yeast *Schizosaccharomyces pombe* are much longer than those in *S. cerevisiae* and do not appear to have conserved A-elements. Our current understanding of replicator structures in higher organisms

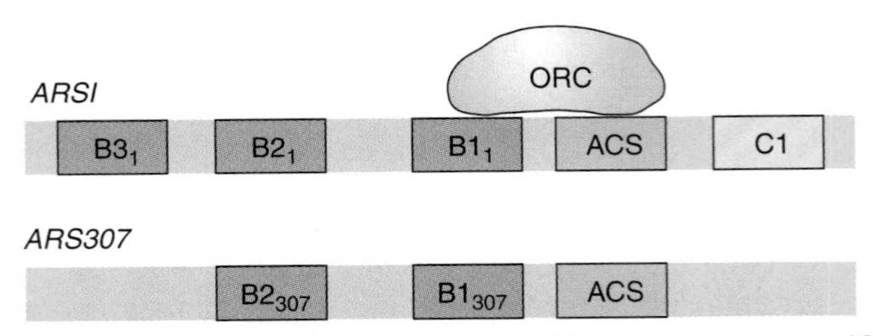

FIGURE 10.12 **Anatomy of autonomously replicating sequences (ARSs) from *S. cerevisiae*.** The ARSs have a module structure consisting of a highly conserved A-element also known as an ARS-consensus sequence (ACS). AT-rich regions in ARS1 and ARS307 are further divided into discrete functional elements called B1 and B2. The A-element and B1-element bind the origin recognition complex (ORC), the yeast initiator protein. Some ARSs including ARS1 have a third B element, B3, which binds transcription factors that appear to influence the initiation of replication (but these are not described further in this text). (Adapted from W. M. Toone, et al., *Ann. Rev. Microbiol.* 51 [1997]: 125–149.)

is at an even more primitive state. Relatively little is known about origin sequence specificity in animal cells. Mammalian replicators appear to be made of several nonredundant, sequence-specific elements that work cooperatively to direct initiation. The few mammalian replicators that have been examined to date appear to have different sequences.

Two-dimensional gel electrophoresis can locate origins of replication.

In 1987, Bonita J. Brewer and Walton L. Fangman devised a technique known as neutral/neutral two-dimensional gel electrophoresis that locates origins of replication in chromosomes (FIGURE 10.13). This technique was used to show that ARSs act as initiation sites for DNA replication in living cells. DNA isolated from unsynchronized rapidly proliferating cells is digested with a restriction endonuclease. Only a small fraction of any given fragment population will be derived from DNA caught in the act of replication and will contain replication forks (Figure 10.13a). These replicating fragments, which vary in size from n to 2n (where n is the length of a nonreplicating fragment), are first

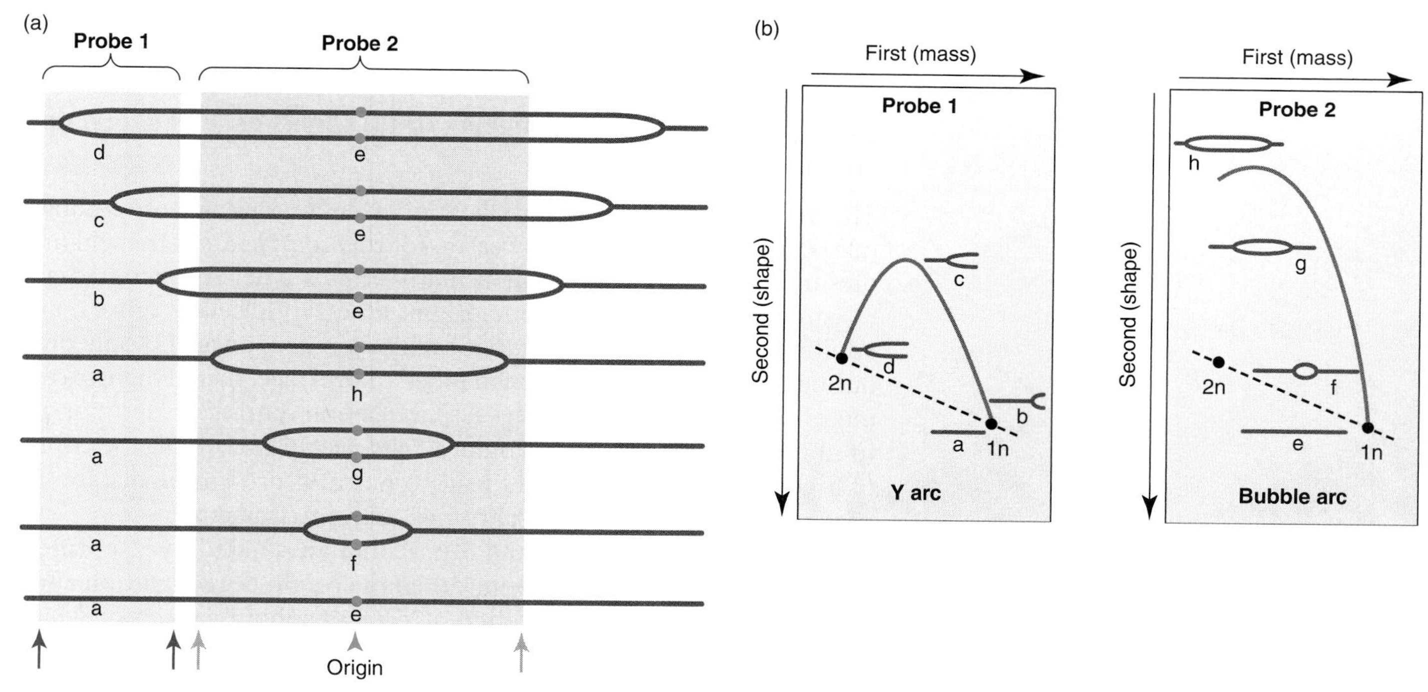

FIGURE 10.13 **Mapping origins using two-dimensional gels.** (a) Seven stages during the replication of one region of a chromosome are shown. Each line represents a duplex, and all stages are present in an unsynchronized cell population. Regions around the origin are duplicated before those at the ends. The DNA is isolated and cut with a restriction endonuclease (red and gray arrows). The fragments produced (labeled a–h) are resolved by neutral/neutral two-dimensional gel electrophoresis, transferred to a positively charged nylon membrane and detected by hybridization with the indicated DNA probe that is labeled with a radioactive or fluorescent tag. (b) A diagram illustrating the position of different fragments in the gel. Linear fragments of different sizes lie along the dotted black lines; most do not hybridize with either probe. Some replication intermediates that hybridize with probes 1 and 2 lie on one of the two blue curves, and the positions of the specific intermediates are shown. The curve produced by probe 1 (left) contains the Y-shaped replication forks and is therefore called a Y arc. The curve produced by probe 2 (right) contains bubble-shaped replication intermediates and is therefore called a bubble arc. Repeating the experiment with other probes hybridizing with different regions along the chromosome allows precise mapping. (Part a adapted from P. R. Cook, *Principles of Nuclear Structure and Function, First edition.* John Wiley & Sons, Ltd., 2001. Part b adapted from W. K. Fangman and B. J. Brewer, *Ann. Rev. Cell Biol.* 7 [1991]: 375–402.)

separated according to size by low-voltage agarose gel electrophoresis at neutral pH. Then the lane containing the separated fragments is cut from the gel, rotated 90°, and placed at the top of a new gel containing a higher agarose concentration for separation in a second dimension.

The new gel is run at high voltage at neutral pH, allowing it to separate fragments by shape. Fragments that migrate together in the first dimension, therefore, separate from one another in the second. Those fragments that contain bubbles migrate at the slowest rate, those with a single replication fork (Y-shaped) at an intermediate rate, and those that are linear duplexes at the fastest rate. Once migration is complete, the fragments are transferred to a positively charged nylon membrane and the migration of a specific fragment is detected by hybridization with a DNA probe that is labeled with a radioactive or fluorescent tag.

Two types of replication intermediates are readily distinguished from the observed patterns (Figure 10.13b). When a replication fork is passively generated from an origin of replication that is outside the fragment, the Y-shaped replication intermediates generate a Y-arc (Probe 1). When a replication origin is located within the boundaries of the restriction fragment, the bubble-shaped replication intermediates produce a bubble arc pattern (Probe 2).

The origin of recognition complex (ORC) serves as the eukaryotic initiator.

The initiation stage of DNA replication in eukaryotes is more complex than that in bacteria. A major reason for this complexity is that eukaryotes have to control initiation at many origins whereas bacteria have to control initiation at just one origin. Eukaryotic cells require many additional proteins to ensure that each origin fires once and only once during each cell cycle. If an origin failed to fire once, then daughter cells might have harmful or perhaps lethal deletion mutations. If an origin fired more than once, the daughter cells would have multiple copies of a genetic region that could lead to harmful protein or metabolite accumulation or to harmful chromosomal rearrangements.

In 1992, Stephen P. Bell and Bruce Stillman isolated a yeast protein that binds to ARS. This protein, called the **origin of recognition complex** (**ORC**) consists of six different protein subunits, each of which has been conserved during evolution. At least five of the ORC subunits bind ATP and have ATPase activity. The genes that code for each of the ORC subunits have been cloned from yeast and other eukaryotes, facilitating genetic and biochemical studies. When yeast ORC•ATP binds to ARS1, it protects the A-element and part of the B1 element from DNase digestion, indicating direct contact between ORC and these elements. ORCs from other eukaryotes don't bind to a specific DNA sequence but do show a preference for AT-rich sequences. Two lines of evidence suggest that ORC serves as the eukaryotic initiator. First, yeast mutants with altered ORC subunits have a problem initiating chromosomal DNA synthesis and maintaining ARS plasmids. Second, nucleotide substitutions in the A-element, which reduce binding to ORC *in vitro*, also lower plasmid stability *in vivo*.

CDC6 and Cdt1 help load MCM2-7 helicase onto the origin to form a pre-replication complex (pre-RC).

Yeast ORC, which is bound to DNA throughout the yeast life cycle, recruits a cell division cycle (CDC) protein called **Cdc6** during late M/early G_1 phase. Cdc6 is yet another protein that binds ATP and has ATPase activity. The gene that codes for this protein, *CDC6*, was first identified by using a genetic screen designed to identify mutants that interfere with normal progression through the cell cycle. Yeast mutants that lack a functional Cdc6 do not replicate DNA. Cdc6 is degraded by proteases once initiation is complete, helping to ensure that each origin fires just one time during the cell cycle. Two additional proteins are required to form the **pre-replication complex (pre-RC)**. These proteins, **MCM2-7** (helicase) and **Cdt1** combine to form a complex during M phase. MCM2-7 helicase, like bacterial DnaB helicase, has six subunits that each bind ATP and have ATPase activity. The similarity ends there, however. MCM2-7 is a heterohexamer rather than a homohexamer and must be assisted by other proteins before it can act as a helicase (see below).

Working independently, the laboratories of Christian Speck and John F. X. Diffley have used a combination of biochemical and electron microscopy techniques to study pre-RC formation. The results of their studies are as follows. MCM2-7's quaternary structure changes from a single hexamer before loading to a double hexamer after loading (FIGURE 10.14). The double-hexamer encircles double-stranded DNA and remains topologically linked to a covalently closed circular double-stranded DNA molecule or a linear duplex with both ends blocked by an attached protein but slides off a linear duplex when either end is unblocked. These results indicate that MCM2-7 is free to slide along a double-stranded DNA molecule in either direction. ATP hydrolysis is required during at least two steps in pre-RC formation: (1) Cdc6 hydrolyzes ATP while assisting Cdt1 to load MCM2-7 onto DNA. (2) ORC1 hydrolyzes ATP to release MCM2-7 from ORC so that the process of pre-RC formation can be completed. This process by which pre-RC is assembled is called **licensing**. Approximately 5 to 10 double MCM2-7 complexes are loaded onto each origin. It is likely that only one of these complexes is activated during the next stage of initiation, **pre-initiation complex** formation. MCM2-7 complexes that have not been activated may be needed to form a new replication fork if a replication fork collapses upon encountering damaged DNA.

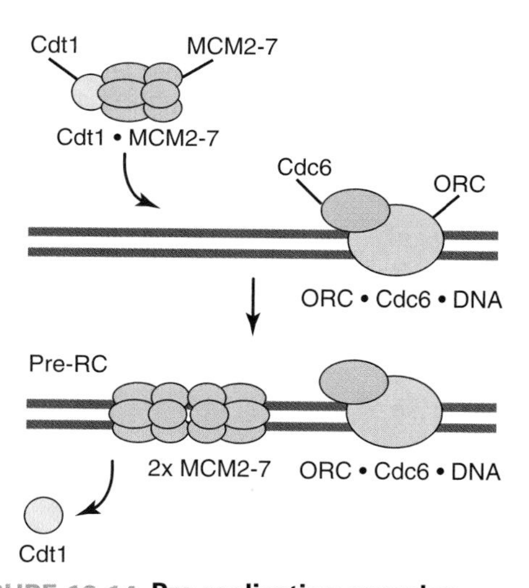

FIGURE 10.14 **Pre-replication complex (pre-RC) formation.** An ORC•Cdc6•DNA complex is joined by a Cdt1•MCM2•7 complex. MCM2-7's quaternary structure changes from a single hexamer before loading to a double hexamer after loading. Although not shown, ATP hydrolysis is required during at least two steps in pre-RC formation: (1) Cdc6 hydrolyzes ATP while assisting Cdt1 to load MCM2-7 onto DNA. (2) ORC1 hydrolyzes ATP to release MCM2-7 from ORC so that the process of pre-RC formation can be completed. ORC is shown in blue, Cdc6 in tan, Cdt1 in yellow, and MCM2-7 in purple. (Reproduced from C. Evrin, et al., *Proc. Natl. Acad. Sci. USA* 106 [2009]: 20240–20245.)

The licensed origin must be activated before replication can take place.

The pre-RC must be activated before the initiation of DNA replication can take place during the S phase. As shown in FIGURE 10.15a, the process begins with a licensed origin. Then a cyclin-dependent kinase, Cdc7, phosphorylates MCM2-7 and a second cyclin-dependent kinase, CDK, phosphorylates Sld2 and Sld3 proteins, which in turn form a complex with Dpb11 (FIGURE 10.15b). Sld2 interacts with the heterotetrameric GINS protein and DNA polymerase ε (see below), while Sld3 interacts with the Cdc45 protein and the resulting complex interacts with MCM2-7 (FIGURE 10.15c). Additional proteins including DNA polymerases α and δ (see below) interact with the pre-initiation complex, which somehow forms two replication forks and is now ready to begin DNA synthesis (FIGURE 10.15d).

Under *in vitro* conditions, the Cdc45•MCM2-7•GINS complex is an active helicase. Cdc45 and GINS appear to somehow trigger a conformational change in MCM2-7 that is essential for helicase activity. Although many of the proteins involved in the activation process have been identified, others probably still need to be identified. Many

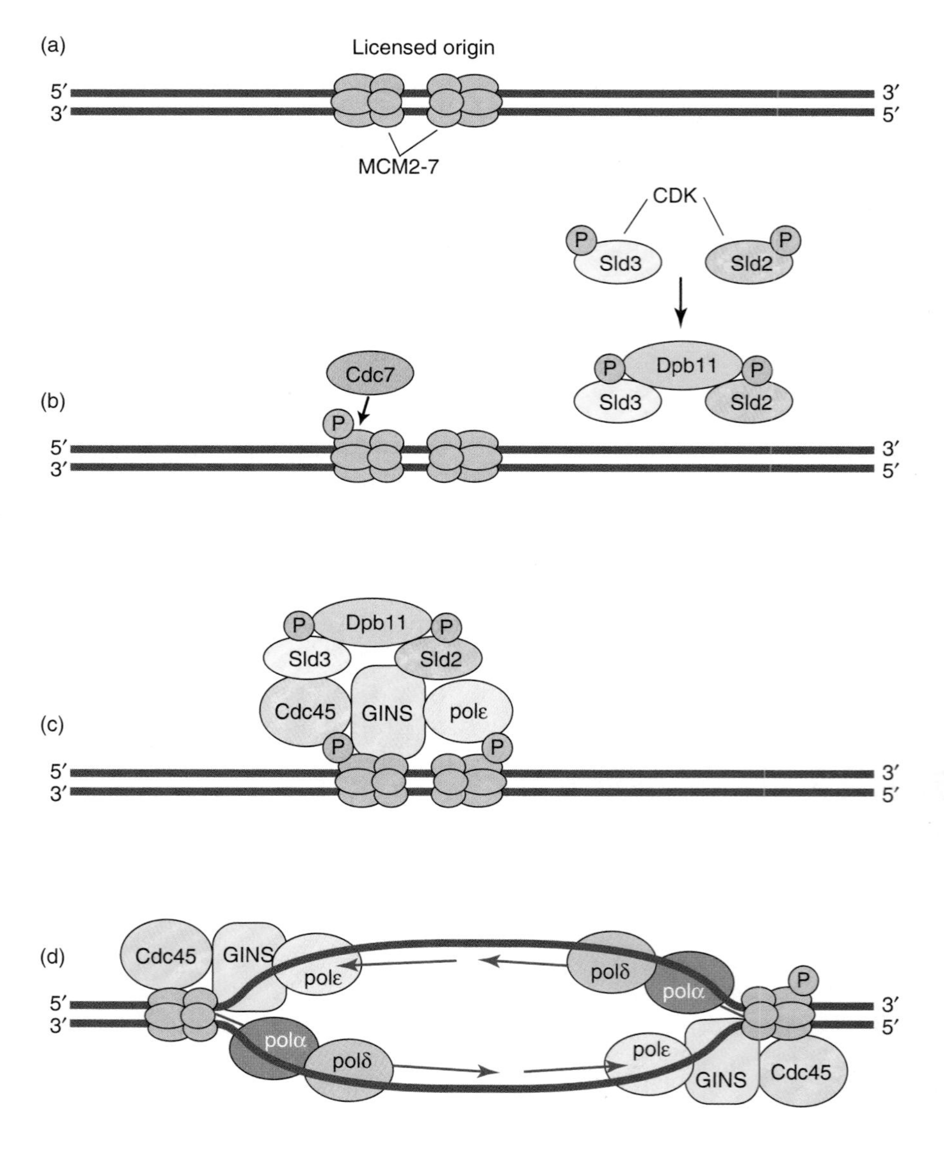

FIGURE 10.15 Activation of a licensed origin and formation of replication forks. (a) The process begins with a licensed origin. (b) A cyclin-dependent kinase, Cdc7, catalyzes the phosphorylation of MCM2-7. A second cyclin-dependent kinase, CDK, catalyzes the phosphorylation of Sld2 and Sld3 proteins, which then form a complex with Dpb11. A phosphate group is indicated by a P inside an orange circle. (c) Sld2 interacts with the heterotetrameric GINS protein and DNA polymerase ε, while Sld3 interacts with the Cdc45 protein. (d) Additional proteins including DNA polymerases α and δ interact with the pre-initiation complex, to complete replication fork formation. (Modified from L. S. Cox. *Molecular Themes in DNA Repair.* Royal Society of Chemistry, 2009. Adapted with permission of The Royal Society of Chemistry. Available at http://www.rsc.org/shop/books/2009/9780854041640.asp.)

questions remain to be answered. For example, we still do not know the mechanism that is used by MCM2-7 to unwind DNA or how replication forks are established. A detailed understanding of origin licensing and pre-RC activation will undoubtedly lead to a better understanding of various diseases that arise when the process goes awry and perhaps lead to new methods to treat these diseases.

10.4 Eukaryotic Replication Elongation.

Pol δ and Pol ε are primarily responsible for copying the lagging- and leading-strand templates, respectively.

Eukaryotes require three different DNA polymerases for normal replication fork propagation. Two of these, Pol α and Pol δ, are also essential for SV40 replication. The third, Pol ε, is not required for SV40 DNA replication and was therefore identified later than the other two. All three polymerases have four subunits in most of the eukaryotes that have been studied to date. One known exception, yeast Pol δ, has only three subunits. FIGURE 10.16 shows how the subunits are organized in yeast Pol α, Pol δ, and Pol ε. Crystal structures are not yet available for these enzymes or their subunits.

Pol α has limited processivity and lacks 3′→5′ exonuclease activity. Its Pri1 subunit has primase activity that forms short RNA primers (8–12 nucleotides). Its largest subunit, Pol1 adds 15 to 25 deoxynucleotides to the 3′-end of the primer to form initiator DNA. The two other Pol α subunits stabilize and regulate the catalytic subunits.

Pol δ is responsible for lagging strand synthesis (see below). Its large subunit has both 5′→3′ DNA polymerase activity and 3′→5′ exonuclease activity. Pol δ has a low processivity value, which becomes quite high when Pol δ is tethered to DNA by a sliding clamp.

Pol ε is primarily responsible for catalyzing leading strand synthesis (see below). Although Pol ε is not required for SV40 DNA replication, *S. cerevisiae* requires it to remain viable. Its largest subunit, Pol2, has both 5′→3′ DNA polymerase activity and 3′→5′ exonuclease activity. Pol ε has considerable binding affinity for single- and double-stranded DNA and may be able to function independently of PCNA. Cryo-electron microscopy studies by Erik Johansson and coworkers suggest a model to explain Pol ε processivity in the absence of PCNA. This model is based on the observation that the catalytic domain in the largest subunit is connected by a flexible linker to the C-terminal domain, which in turn interacts with the three other subunits. According to the model, Pol ε exists in an open and closed form (FIGURE 10.17). The flexible linker permits the closed form to open so that DNA can

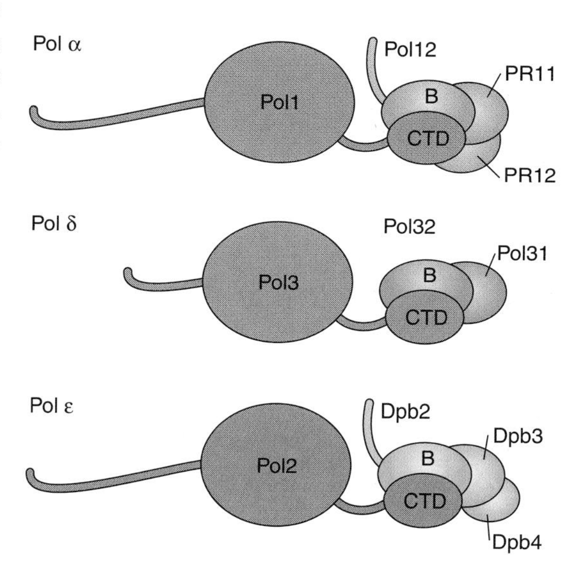

FIGURE 10.16 Subunit organization for yeast Pol α, Pol δ, and Pol ε. Each polymerase has a catalytic subunit, a conserved regulatory (or B) subunit, and other subunits with functions that remain to be determined. The catalytic subunits in Pol α, Pol δ, and Pol ε are called Pol1, Pol3, and Pol2, respectively. The regulatory subunits in Pol α, Pol δ, and Pol ε are called Pol12, Pol32, and Dbp2, respectively. Other subunits in each polymerase are as follows: PR11, and PR12 in Pol α; Pol31 in Pol δ; and Dpb2, Dpb3, and Dpb4 in Pol ε. (Adapted from S. Klinge, et al., *EMBO J.*, 28 [2009]: 1978–1987.)

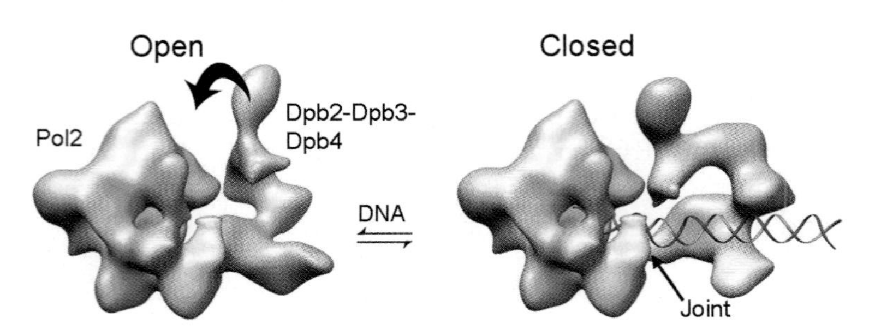

FIGURE 10.17 Model for Pol ε processivity in the absence of PCNA. Pol ε exists in an open form (left) and closed form (right). A flexible linker permits the closed form to open so that DNA can interact with the active site. Then the flexible linker permits Pol ε to close again, encircling the DNA to prevent its dissociation from Pol ε. The catalytic subunit, Pol2, is flexibly connected to an extended structure consisting of three other subunits Dpb2, Dpb3, and Dpb4. (Adapted from F. J. Asturias, *Nat. Struct. Mol. Biol.* 13 [2006]: 35–43. Photo courtesy of Francisco J. Asturias, The Scripps Research Institute.)

interact with the active site. Then the flexible linker permits Pol ε to close again, encircling the DNA to prevent its dissociation from Pol ε in the same way that closed sliding clamp prevents dissociation. Further studies are required to determine what, if any, role the PCNA clamp plays in DNA synthesis that is catalyzed by Pol ε.

Studies with the *in vitro* SV40 DNA replication system seemed to indicate that Pol δ is the only replicative enzyme required for eukaryotic DNA synthesis. The subsequent discovery of Pol ε raised troubling questions about the function of the two enzymes. The specific contribution that each enzyme makes to DNA synthesis might be found if it were possible to place each polymerase on a given strand but that is not yet possible. Peter M. J. Burgers, Thomas Kunkel and their coworkers realized that it would be possible to obtain useful information about the roles that Pol δ and Pol ε play at the replication fork by taking an indirect approach. Their approach relies on analyzing the mutation spectra produced by DNA polymerases with reduced replication fidelity (mutator DNA polymerases). The basic idea is to use a yeast cell with a mutator Pol δ (or mutator Pol ε) with an altered active site that causes the enzyme to occasionally pair a specific template base with a single incorrect incoming deoxynucleoside triphosphate. For example, a mutator Pol ε was used that permits a dT on the template strand to occasionally pair with an incoming dGTP. The mutator DNA polymerase retains its 3′→5′ proofreading function and robust catalytic activity that allows normal DNA replication and cell growth. Burgers and coworkers still needed to find a method to determine if the wrong incoming nucleotide was incorporated into the leading or lagging strand. They solved this problem by using yeast strains with a *URA3* gene inserted adjacent to ARS306 in either the forward or backward direction. They selected ARS306 because it fires in early S phase in >90% of yeast cells in a population. Burgers and coworkers identified the leading and lagging strands based on the known arrangement of *URA3* with respect to ARS306. Mutation spectra analysis revealed that the *altered Pol δ produced mutations along the lagging strand*, whereas the *altered Pol ε produced mutations along the leading strand*. Pol δ, therefore, is primarily responsible for copying the lagging-strand template and Pol ε is primarily responsible for copying the leading-strand template. Mutation spectra analysis does not rule out the possibility that Pol δ participates in leading strand synthesis if the replication machinery encounters a problem such as damaged DNA. The questions still remains: why are Pol δ and Pol ε both required for eukaryotic DNA replication but only Pol δ is required in SV40 DNA replication? The answer is probably related to the fact that SV40 replication uses the T antigen as both the initiator protein and helicase. Pol ε may need to interact with a eukaryotic protein that is not used for SV40 DNA replication.

FIGURE 10.18 shows how several of the important enzymes and proteins required for eukaryotic DNA elongation appear to be organized at the replication fork. Leading and lagging strand synthesis are coordinate processes in eukaryotic DNA replication just as they are in bacterial DNA replication. For convenience, we will consider the two processes separately.

Leading strand synthesis begins after Pol α completes initiator DNA synthesis. Then Pol ε somehow replaces Pol α at the replication fork and continues synthesis until two replication bubbles meet.

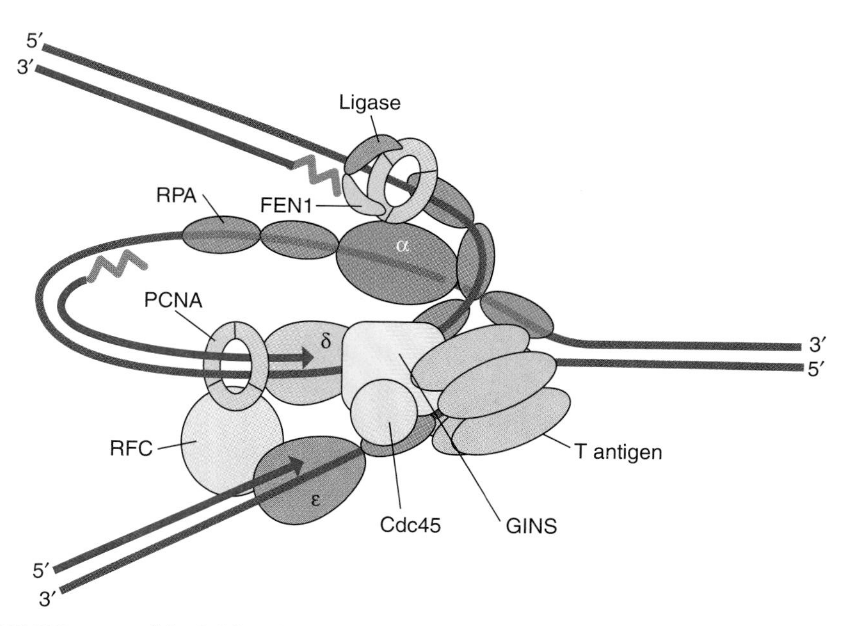

FIGURE 10.18 Model for the organization of proteins and enzymes at the eukaryotic replication fork. The functions of the various components are described in the text. (Adapted from B. Stillman, *Mol. Cell* 30 [2008]: 259–260.)

At present, it is not clear whether Pol ε interacts with PCNA during replication.

Lagging strand synthesis involves Okazaki fragment formation and maturation. As the replication machinery moves, the replication bubble continues to grow and Pol α binding sites become available on the lagging strand. Pol α synthesizes initiator DNA. Then a switch occurs in which Pol δ replaces Pol α. The steps in Okazaki fragment maturation are the same as those described for SV40 replication (see Figure 10.6).

10.5 The End-Replication Problem

Studies of the *Tetrahymena* and yeast telomeres suggested that a terminal transferase-like enzyme is required for telomere formation.

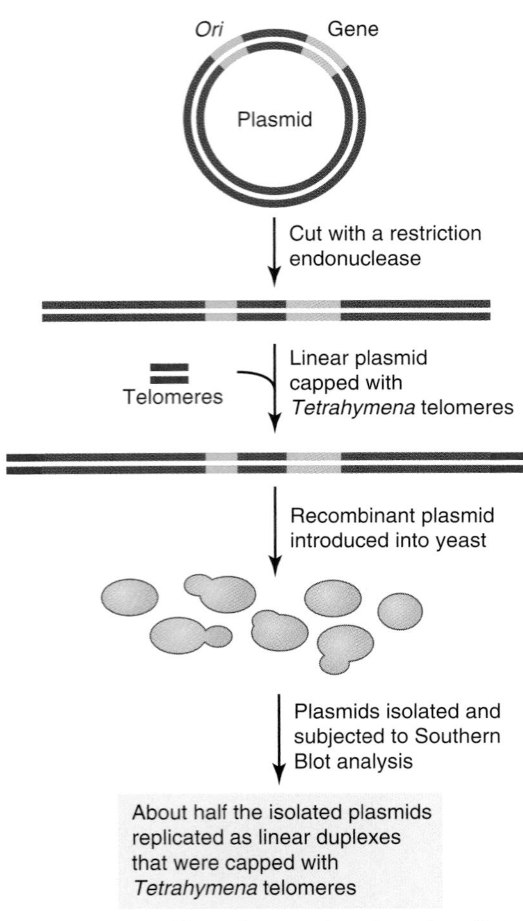

FIGURE 10.19 **Experiment showing that *Tetrahymena* telomeres function in yeast cells.** A circular plasmid with an origin of replication (*ori)* and a selectable marker (gene) was cut with a restriction endonuclease. A *Tetrahymena* telomere was added to each end of the linear plasmid and the recombinant plasmid introduced into yeast. Plasmids were isolated from the transformed cells and subjected to Southern blot analysis. Approximately half the isolated plasmids were linear duplexes with a *Tetrahymena* telomere at each end.

Eukaryotic DNA polymerases work by extending a pre-formed primer and therefore cannot copy the very end of a linear duplex. Molecular biologists needed to learn more about the DNA structure at the end of eukaryotic chromosomes before they could address this "end replication problem." As indicated in Chapter 6, Elizabeth H. Blackburn and Joseph G. Gall isolated telomeres from minichromosomes present in macronuclei of the ciliated protozoan *Tetrahymena thermophilia* in the late 1970s. They showed that these telomeres consist of an average of 50 tandem repeats of the simple hexanucleotide unit TTGGG and that this G-rich strand extends beyond the complementary C-rich strand, forming a 3′-overhang. Additional studies by Blackburn and others revealed that minichromosomes from related ciliated protozoa had similar telomeres.

Further studies were needed to learn whether telomeres are unique structures that are limited to minichromosomes of ciliated protozoa or are evolutionarily conserved structures that are present in chromosomes from other eukaryotes. Blackburn and Jack Szostak performed an experiment to address this issue in 1982. Szostak had been studying plasmid replication and recombination in yeast. Circular plasmids with an origin of replication replicate in yeast. However, yeast will not support the replication of the same plasmid if it is present as a linear duplex. If the ends of a linear duplex are homologous to yeast DNA, the plasmid is integrated into a yeast chromosome. Yeast cells usually degrade linear plasmids that do not share a region of homology with a yeast chromosome. On rare occasions yeast cells somehow join the ends of the linear duplex to form a circular DNA molecule that can replicate in yeast. Blackburn and Szostak thought that yeast cells might be tricked into replicating a linear plasmid with a *Tetrahymena* telomere at each end. Although they considered the possibility to be a long shot, the experiment was worth doing because it was easy to perform and a successful outcome would provide important insights into telomere function. They therefore constructed the desired recombinant plasmid and introduced it into yeast as shown in FIGURE 10.19. Then they isolated plasmids from the transformed yeast cells and subjected them to Southern blot analysis. About half the isolated plasmids migrated as linear duplexes, indicating that the *Tetrahymena* telomeres function in yeast. The fact that telomeres work in evolutionarily distant organisms suggests extraordinary functional conservation.

Szostak and Blackburn realized that they could construct an ideal cloning vector for a yeast telomere by removing a *Tetrahymena* telomere from one end of the recombinant linear plasmid. The steps involved in cloning the yeast telomere, which are shown in **FIGURE 10.20**, are as follows: (1) Remove a *Tetrahymena* telomere (t) from one end of the linear plasmid. (2) Digest yeast DNA with a restriction endonuclease to produce a DNA fragment pool with only rare fragments derived from chromosome ends. (3) Mix the DNA fragments with the modified linear plasmid and add DNA ligase to join the two types of fragments. (4) Transform yeast cells with the recombinant plasmids. Most of the recombinant DNA molecules produced could not replicate in yeast because they were attached to a *Tetrahymena* telomere at one end and a non-telomeric yeast DNA fragment at the other end. The rare recombinant DNA molecule with a *Tetrahymena* telomere (t) at one end and a yeast telomere (y) at the other end, however, was able to replicate in yeast. Sequencing studies showed that the yeast telomere has TG_{1-3} repeats. Szostak and Blackburn were able to construct a *yeast artificial chromosome* (*YAC*) by attaching yeast telomeres to both ends of a long linear duplex that contained an origin of replication and yeast centromere.

Szostak and Blackburn also made the remarkable discovery that yeast cells add about 200 bp of yeast telomere DNA to the ends of

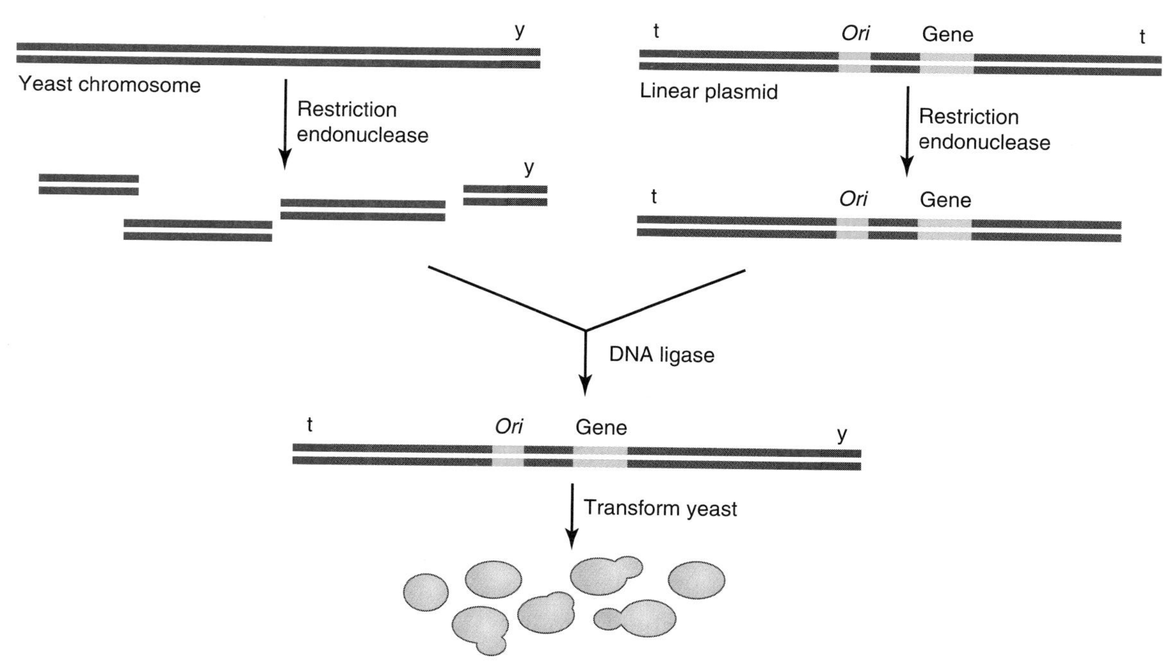

FIGURE 10.20 Method for cloning yeast telomere. A recombinant linear plasmid, constructed as shown in Figure 10.21, was digested with a restriction endonuclease to remove a *Tetrahymena* telomere (t) from one end. Yeast DNA (shown in red) was digested with a restriction endonuclease to produce a DNA fragment pool, which was mixed with the modified linear plasmid. DNA ligase was added to the mixture to join yeast fragments to the end of the linear plasmid that lacked a telomere and the recombinant DNA was introduced to yeast. Rare recombinant DNA molecules that had a *Tetrahymena* telomere (t) at one end and a yeast telomere (y) at the other end replicated in yeast.

linear plasmids capped with *Tetrahymena* telomeres. They thought that the most likely explanation for this addition was that yeast cells have a terminal transferase-like enzyme that adds nucleotides to the ends of the linear duplex. However, it was difficult to see how this enzyme could have the required specificity to add nucleotide repeats to the ends.

Telomerase uses an RNA template to add nucleotide repeats to chromosome ends.

Carol Greider joined Blackburn's laboratory as a graduate student in 1984 and set out to find the terminal transferase-like enzyme that had been predicted to add nucleotide repeats to chromosome ends. She succeeded by using an assay mixture that contained synthetic DNA oligonucleotide $(TTGGGG)_4$ as substrate, [^{32}dGTP], dTTP, and crude *Tetrahymena* cell extract. She followed the progress of the reaction by using gel electrophoresis to separate labeled DNA. The gel showed that hundreds of d(TTGGGG) repeats were added to the primer. Greider called the enzyme that catalyzes the addition of the nucleotide repeats **telomerase**.

Further studies by Greider and Blackburn revealed that telomerase contains an RNA molecule that is essential for its function. They isolated the 159-nucleotide RNA subunit from *Tetrahymena* telomerase and discovered that it contains an internal 5′-CAACCCCAA-3′ sequence, which is complementary to the $d(TTGGGG)_n$ telomeric repeat in *Tetrahymena*. Subsequent studies revealed that telomerase RNA molecules from other organisms differ in size, sequence, or both (**FIGURE 10.21**). Nevertheless, each telomerase RNA molecule has a sequence that is complementary to the telomeric DNA of the organism from which the telomerase is isolated. These results suggest that a short sequence within telomerase RNA acts as a template for telomeric DNA synthesis. Proof for this hypothesis comes from experiments that show telomerase activity is lost after RNase digestion but can be restored by removing RNase and adding back intact RNA. When the added RNA has a nucleotide substitution in its template sequence, the telomeric DNA produced has the predicted change. Genetic studies with *S. cerevisiae* also indicate that telomerase RNA is essential for telomerase function because the gene that specifies telomerase RNA (*TLC1*) is essential for telomere maintenance. Telomerase, therefore, acts as an RNA-dependent DNA polymerase or reverse transcriptase (see Chapter 5) that supplies its own template RNA molecule.

Joachim Lingner and Thomas Cech purified *Euplotes aediculatus* telomerase in 1996 by taking advantage of the fact that an oligonucleotide with a telomere sequence can bind to the RNA subunit in fully functional telomerase. Their approach was to pass a partially purified nuclear extract through a column containing beads linked to an oligonucleotide with the telomere sequence so that telomerase would bind to the beads. Then they displaced the bound telomerase by adding a solution that contained oligonucleotides with an even greater affinity for the oligonucleotides on the beads. Purified telomerase had a molecular mass of about 230-kDa and contained a 123-kDa

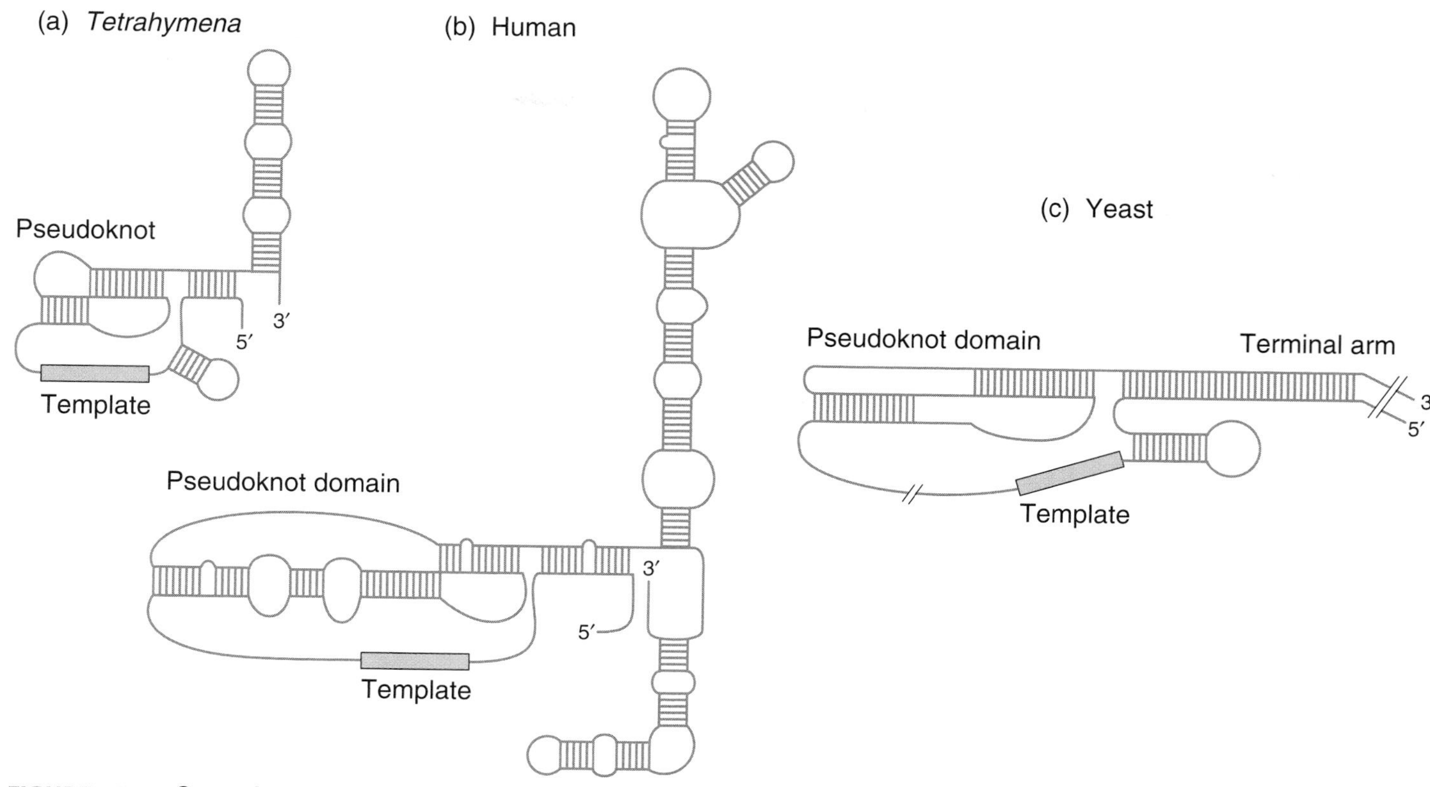

FIGURE 10.21 Secondary structures of telomerase RNA. (a) *Tetrahymena*, (b) human, and (c) yeast. Some parts of the RNA molecules that are not shown are indicated by slashes. (Adapted from C. Auxetier, and N. F. Lue. *Ann. Rev. Biochem.*, 75 [2006]: 493–517.)

polypeptide, a 43-kDa polypeptide, and a 66-kDa RNA subunit. When the amino acid sequence of the large polypeptide subunit was compared to sequences available in the genetic data bank, it was found to resemble a yeast polypeptide. Although the yeast polypeptide had not been isolated, genetic studies showed that the gene that specifies it, *EST2* (ever shorter telomeres), is essential for telomere maintenance. This sequence similarity suggested that the 123-kDa subunit is probably essential for telomerase activity and therefore likely to be the reverse transcriptase.

This conclusion was supported when sequence comparisons with known reverse transcriptases revealed that the 123 kDa polypeptide subunit has a reverse transcriptase domain at its carboxyl terminus. Furthermore, conserved amino acid residues in the catalytic site of reverse transcriptase are also present in both the 123-kDa polypeptide subunit and the *EST2* gene product. Yeast mutants with alterations in these conserved residues cannot maintain their telomeres. The catalytic subunit of telomerase has now been identified in many other organisms including humans. While the function of the 43 kDa polypeptide in *E. aediculatus* is not yet known, it seems likely to have an ancillary function in telomere maintenance. *S. cerevisiae* genetic studies also indicate the need for ancillary proteins *in vivo* because at least three additional yeast genes (*EST1*, *EST3*, and *EST4*) specify polypeptides that are required for telomere maintenance.

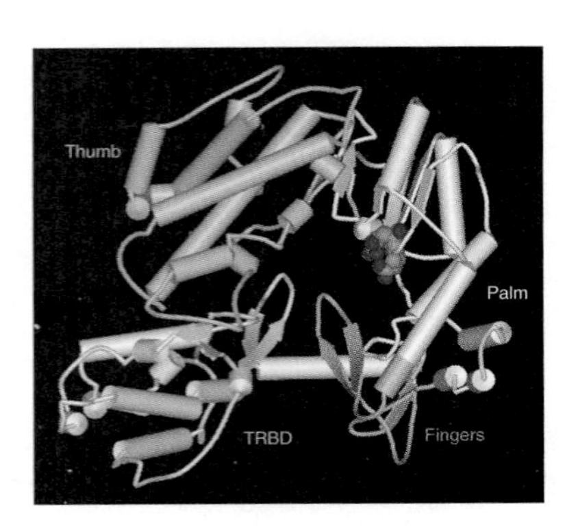

FIGURE 10.22 **Crystal structure for the catalytic telomerase subunit isolated from the red flour beetle, *Tibolium castaneum*.** The catalytic subunit contains three distinct domains, a reverse transcriptase domain (yellow and orange), an RNA-binding domain or TRBD (cyan), and a thumb domain (magenta). The reverse transcriptase domain consists of a palm subdomain (yellow) and a finger subdomain (orange). The three glutamates characteristic of DNA polymerases and reverse transcriptases, which mark the catalytic site in the palm subdomain, are shown in spacefilling form in standard colors for elements. The remainder of the structure is shown in schematic form. (Structure from Protein Data Bank 3DU5. A. J. Gillis, et al., *Nature* 455 [2008]: 633-638. Prepared by B. E. Tropp.)

Emmanuel Skordalakes and coworkers have determined the crystal structure for the catalytic telomerase subunit isolated from the red flour beetle, *Tibolium castaneum*. The catalytic subunit contains three distinct domains, a reverse transcriptase domain, an RNA-binding domain (TRBD), and a thumb domain (FIGURE 10.22). The reverse transcriptase domain consists of a palm subdomain and a finger subdomain. The active site is located in the palm subdomain and has three glutamates and so resembles the active sites of DNA polymerases and reverse transcriptases. The finger subdomain appears to be involved in RNA and nucleotide binding. The RNA binding domain binds to single- and double stranded regions of telomerase RNA. The thumb and reverse transcriptase domains make extensive contacts with the DNA substrate. More detailed information about the way that telomerase works will probably have to await a crystal structure for a complex that contains the catalytic subunit, the RNA subunit, and the DNA substrate.

FIGURE 10.23 is a schematic for a possible telomerase reaction cycle. According to this schematic, telomeric DNA binds to the template sequence in telomerase RNA. At the start of the reaction cycle,

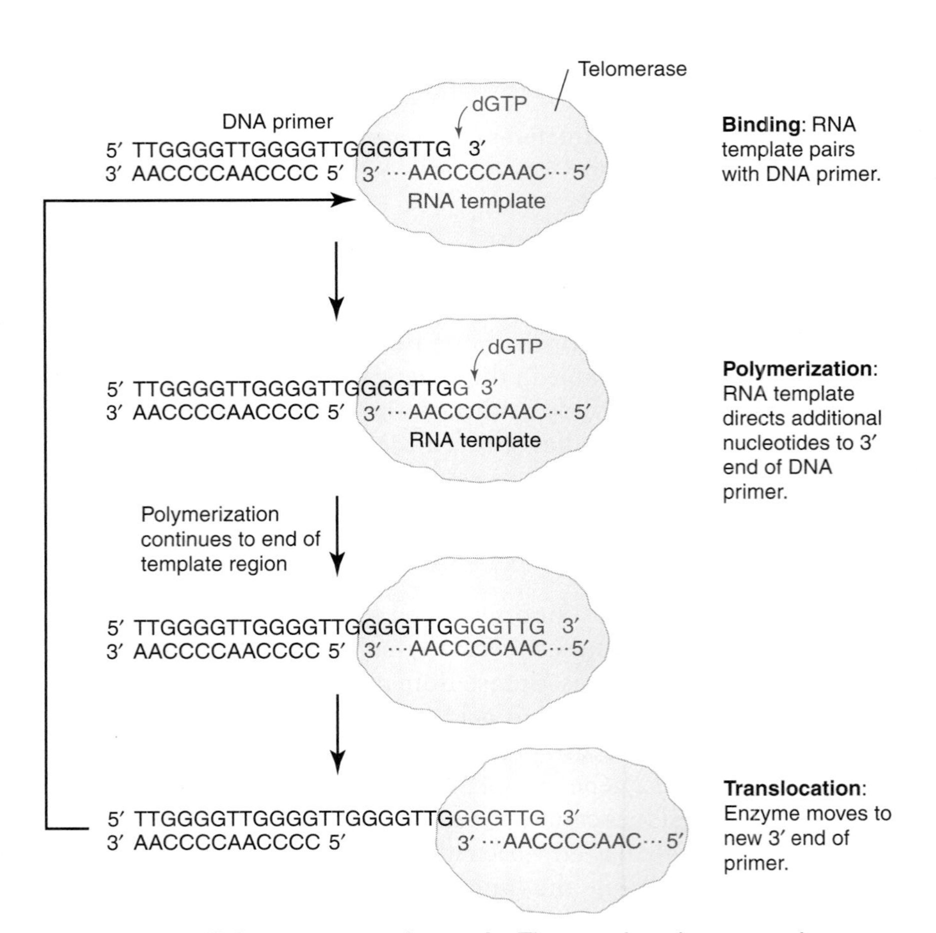

FIGURE 10.23 **Telomerase reaction cycle.** The reaction shown are those revealed by studying the *Tetrahymena* telomerase. The cycle consists of three steps, binding, polymerization, and translocation to reposition the RNA template. (Adapted from V. D. Chatziantoniou, *Pathol. Oncol. Res.* 7 [2001]: 161–170.)

telomerase binds to the DNA primer with the telomeric RNA template base pairing with the primer's 3′-end. Then the active site of the polymerase adds deoxynucleotides onto the primer 3′-end. Once this addition is complete, the enzyme moves to the new 3′-end of the primer strand. Telomerase's ability to carry out a single round of DNA synthesis is termed **nucleotide addition processivity.** Realignment of the same enzyme for a second round of addition is termed **repeat addition processivity.**

Telomerase plays an important role in solving the end-replication problem.

Telomere structure and telomerase's role in telomere synthesis provide important new insights into the end-replication problem. Semiconservative replication initiates at origins internal to the telomeric repeats and the replication forks move toward the chromosome ends (FIGURE 10.24a). Because of the structural arrangement in telomeres, the C-rich strand (denoted by C) is always assembled by lagging-strand synthesis and the G-rich strand (denoted by G) is always assembled by leading-strand synthesis (FIGURE 10.24b). Before the structure of the telomere was elucidated, investigators thought that lagging strand synthesis might present a problem because removing

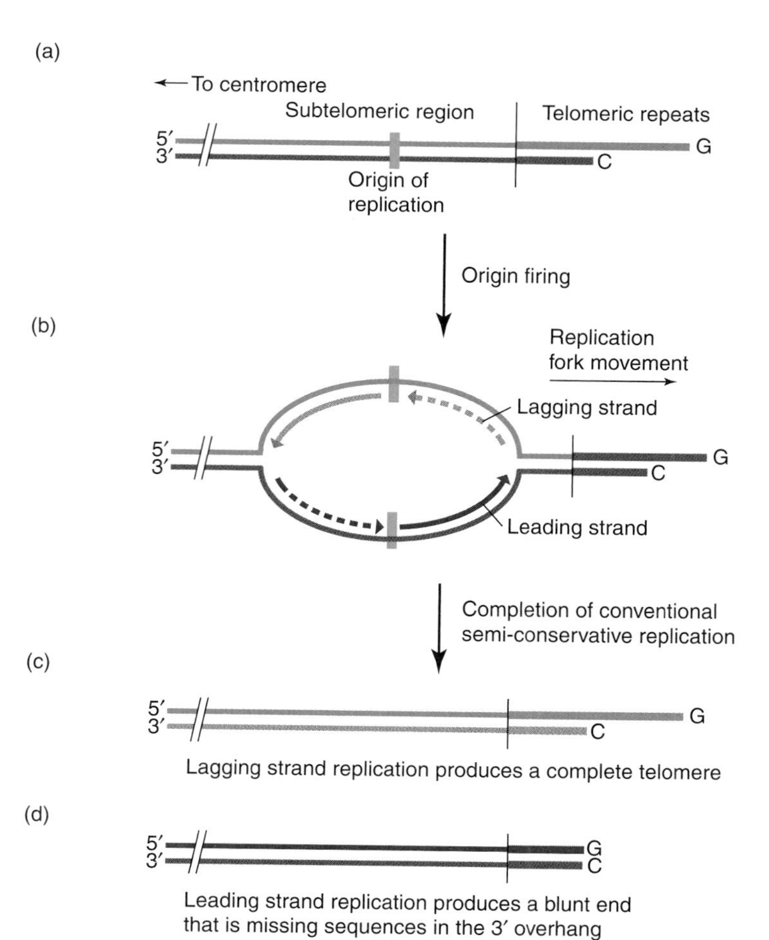

FIGURE 10.24 **Leading strand problem.** (a) The telomere G-rich strand (G, shown in light blue) ends in a 3′ single strand that overhangs the C-rich strand (C, shown in dark blue). (b) Semiconservative replication initiates at origins internal to the telomeric repeats and the replication forks move toward the chromosome end. (c) Synthesis of lagging strands does not necessarily present a problem for the replication machinery because the RNA primer at the end of the last Okazaki fragment can be removed without the loss of information. (d) Synthesis of leading strands introduces a problem because replication stops at the 5′-end of the C-rich template to produce a blunt end with a resulting loss of sequence information. (Adapted from M. Chakhparonian and R. J. Wellinger. *Trends Genet.* 19 [2003]: 439–446.)

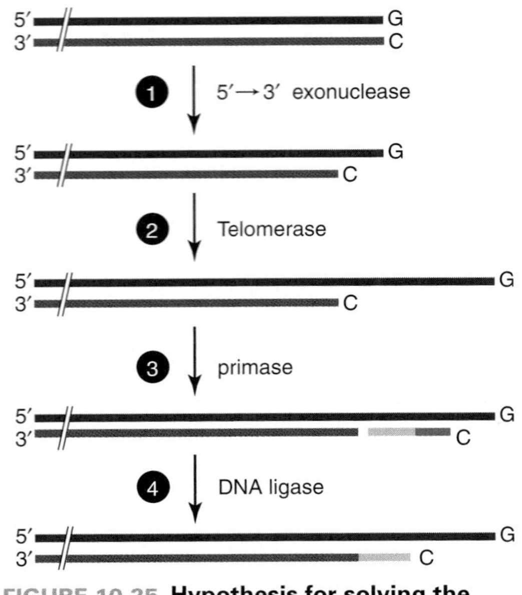

FIGURE 10.25 Hypothesis for solving the leading strand problem.

the RNA primer at the 5′-end of the last Okazaki fragment would produce a shortened new C-rich strand. New C-rich strand synthesis, however, does not necessarily present a problem for the replication machinery because the G-rich stand has a 3′-overhang. The RNA primer at the end of the last Okazaki fragment, therefore, can be removed without the loss of information (FIGURE 10.24c). In contrast, leading strand replication presents a serious problem because replication of the new G-rich strand will stop at the 5′-end of the C-rich template strand to produce a blunt end with a resulting loss of sequence information (FIGURE 10.24d). The leading strand problem is thought to be solved by using a 5′→3′ exonuclease to generate a 3′ overhang. Unless something is done to restore the lost sequence, this processing step will cause the telomere to become shorter. If this shortening process were to continue through several more replication cycles, then the telomere would be lost entirely and the chromosome would no longer be able to survive as an independent replicating unit.

Although telomerase cannot extend a DNA chain from a blunt end, the pathway shown in FIGURE 10.25 shows how the leading strand can be extended. The steps in this pathway are as follows: (1) a strand-specific 5′→3′ exonuclease removes nucleotides from the 5′-end of the C-rich strand to generate a G-rich 3′-overhang; (2) telomerase extends the G-rich 3′-overhang generated by the exonuclease; (3) the newly synthesized region serves as a template for standard semiconservative replication, restoring the 5′-end of C-rich strand.

Telomerase plays a role in aging and cancer.

Somatic cells from humans and other multicellular animals divide a variable but limited number of times when cultured and then enter senescence, a state in which they are alive but no longer dividing. Some animal cells, most notably germline cells and most cancer cells, can replicate indefinitely without entering senescence and so are said to be immortal. Experiments performed by Howard Cooke in 1986 indicated that germline cells have longer telomeres than somatic cells. Cooke speculated that telomerase might not be active in normal somatic human cells. Subsequent studies revealed that somatic cells have very low telomerase activity.

Studies by Andrea G. Bodnar and coworkers have shown that when vectors bearing genes for the telomerase reverse-transcriptase subunit are introduced into human fibroblasts or retinal pigment epithelial cells, which have little if any telomerase activity, the cells continue to divide well beyond their normal lifespan without entering senescence and appear normal when viewed under the microscope. Moreover, their telomeres are much longer than those of normal human fibroblasts or retinal pigment epithelial cells. These results establish a causal relationship between telomere shortening and cellular senescence, suggesting the existence of a "mitotic clock" that regulates telomere size. Similar conclusions have been reached by studying yeast cells, which normally have active telomerase and so do not enter a senescent state. However, yeast mutants that lack either telomerase RNA or telomerase reverse transcriptase become senescent.

As indicated above, most cancer cells have telomerase activity, which raises the possibility that telomerase may be a target for chemotherapeutic agents. There is also reason, however, to suspect that telomerase will not prove to be an effective target. Maria Blasco and coworkers have used recombinant DNA techniques to construct a strain of mice that lack the telomerase RNA gene. Remarkably, mice that are homologous for the missing gene appear to be normal and fertile. Continuous inbreeding has produced six generations of the mutant mice. Moreover, somatic cells from mutant mice are readily converted to tumor cells. One must be cautious in applying the information obtained with mice to humans because mouse telomeres are on average 5 to 10 times longer than human telomeres. The mutant mouse experiments raise the possibility of alternative mechanisms for maintaining telomeres such as genetic recombination. Most organisms use telomerase to maintain their telomeres; however, a few organisms do not require telomerase to solve the end-replication problem. For instance, *Drosophila* use arrays of retrotransposons (see Chapter 14) to maintain chromosome length.

10.6 Replication Coupled Chromatin Synthesis

Chromatin disassembly and reassembly are tightly coupled to DNA replication.

The DNA synthetic machinery must be able to make direct contact with its DNA substrate for replication to take place. However, the basic unit of structure in eukaryotic chromatin, the nucleosome (see Chapter 6), blocks such direct contact. It, therefore, seems reasonable to propose that cells can disassemble chromatin just before the replication fork and reassemble the chromatin just after the replication fork. Electron microscopy studies using the SV40 minichromosome as a model system are consistent with this hypothesis. These studies show that a stretch of about 300 bp of nucleosome-free (naked) DNA is present just before the replication fork and another stretch of at least 250 bp of naked DNA is present just after the replication fork.

Histone chaperones play a central role in replication coupled chromatin formation by preventing nonspecific interactions between positively charged histones and negatively charged DNA. They thereby permit ordered chromatin disassembly or assembly without themselves becoming a permanent part of nucleosomes. **Chromatin modifiers** also participate in chromatin formation by either adding acetyl, methyl, phosphate, or other groups to histones or removing these groups from modified histones. FIGURE 10.26 summarizes how some histone chaperones and chromatin modifiers interact during replication coupled chromatin formation. For convenience, we divide replication coupled chromatin formation into two stepwise processes, chromatin disassembly and chromatin assembly.

Chromatin disassembly begins when the histone chaperone **FACT** (facilitates chromatin transcription) removes H2A•H2B dimers from

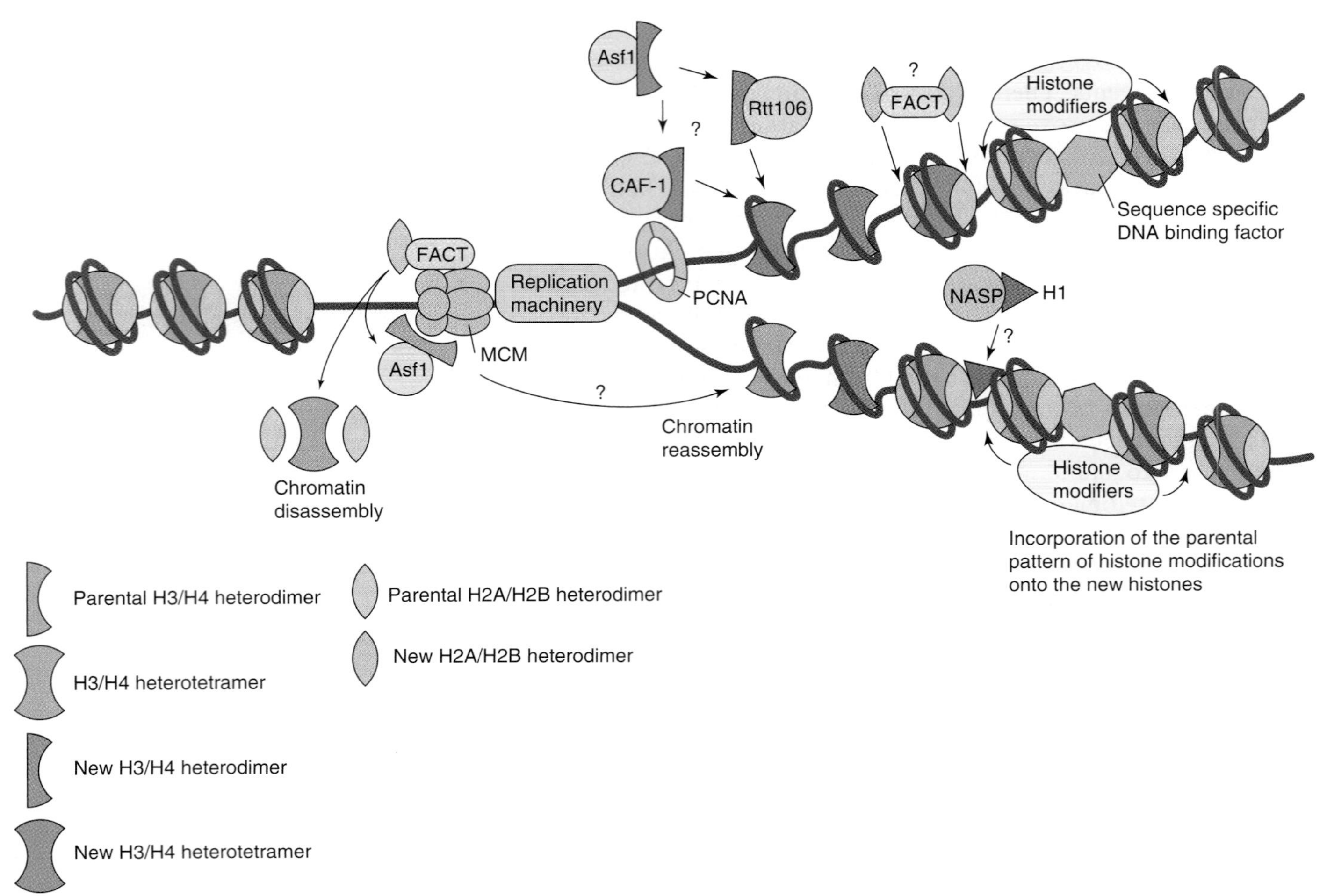

FIGURE 10.26 **Replication coupled chromatin disassembly and assembly. Chromatin disassembly**–Chromatin disassembly begins when the histone chaperone FACT (facilitates chromatin transcription) removes H2A•H2B dimers from nucleosomes. Then the MCM2-7 helicase releases the $(H3•H4)_2$ tetramer that is left behind as it unwinds the Asf1 (antisilencing factor) can split an $(H3•H4)_2$ tetramer into 2 dimers. The MCM2-7 helicase probably binds the released tetramer, ensuring that it remains close to the replicating fork. **Chromatin assembly**–The histone chaperones CAF-1 (chromatin assembly factor) and Asf1 appear to work together to deposit an $(H3•H4)_2$ tetramer (or two H3•H4 dimers) on newly replicated DNA during the first step in nucleosome assembly. Rtt106, another histone chaperone that binds H3•H4, also appears to participate in nucleosome assembly but its specific contribution remains to be determined. During the next step in nucleosome assembly, FACT appears to deposit H2A•H2B dimers onto $(H3•H4)_2$ tetramers. Finally, yet another histone chaperone, NASP (nuclear autoantigenic sperm protein), helps deposit histone H1 onto linker DNA between adjacent nucleosomes. Question marks indicate uncertainty about participation in a specific step. (Reprinted from *Cell*, vol. 140, M. Ransom, B. K. Dennehey and J. K. Tyler, Chaperoning Histones during DNA Replication and Repair, pp. 183–195. copyright 2010, with permission from Elsevier [http://www.sciencedirect.com/science/journal/00928674]).

nucleosomes. At least one other histone chaperone also appears able to perform this function. Then the MCM2-7 helicase releases the $(H3•H4)_2$ tetramer that is left behind as it unwinds the double-double stranded DNA. The MCM2-7 helicase probably binds the released tetramer, ensuring that it remains close to the replicating fork. The histone chaperone **Asf1** (anti-silencing factor) can split an $(H3•H4)_2$ tetramer into 2 dimers (see below).

Histones from disassembled nucleosomes ("old" histones) and newly synthesized histones ("new" histones) are used to assemble nucleosomes on newly formed daughter duplexes. The stable form for newly synthesized and disassembled H2A and H2B is the H2A•H2B dimer. The stable form of newly synthesized H3 and H4 is the H3•H4 dimer. There is uncertainty, however, about the stable form of H3 and H4 derived from nucleosome disassembly. The observation that Asf1 can split $(H3•H4)_2$ tetramers obtained from disassembled nucleosomes into H3•H4 dimers suggests that the dimer is the stable form. However, metabolic labeling studies suggest that the $(H3•H4)_2$ tetramer is the stable form because nearly all nucleosomes assembled on newly formed daughter duplexes contain either all newly synthesized H3 and H4 histones or all old H3 and H4 histones. It is difficult to explain this observation if nucleosome disassembly produces H3•H4 dimers because then a new H3•H4 dimer would be expected to combine with an old H3•H4 dimer to form a mixed tetramer. As shown in Figure 10.26, new and old $(H3•H4)_2$ tetramers appear to distribute randomly on the two new daughter DNA duplexes.

During the first stage of chromatin assembly, the histone chaperones **CAF-1** (chromatin assembly factor) and Asf1 appear to work together to deposit an $(H3•H4)_2$ tetramer (or two H3•H4 dimers) on newly replicated DNA. Almost all newly formed H3 is acetylated on lysine 56. This acetylation appears to help recruit H3•H4 dimers to the histone chaperones that are positioned at the replication fork to promote H3•H4 dimer assembly into DNA. **Rtt106**, another histone chaperone that binds H3•H4, also appears to participate in nucleosome assembly but its specific contribution remains to be determined. During the next step in chromatin assembly, FACT deposits H2A•H2B dimers onto $(H3•H4)_2$ tetramers. Finally, yet another histone chaperone, **NASP** (nuclear autoantigenic sperm protein), helps deposit histone H1 onto linker DNA between adjacent nucleosomes.

Although *in vitro* studies show that histone factors can work together to produce nucleosomes on duplex DNA, the nucleosomes are irregularly spaced. ATP-dependent **chromatin remodelers** such as **ACF** (ATP-utilizing chromatin assembly and remodeling factor) and **CHRAC** (chromatin accessibility complex) are required to generate regular physiological nucleosome spacing. Both chromatin remodelers contain a common subunit, **ISWI** (imitation switch), which can promote nucleosome spacing by itself, although not as efficiently as when it is part of the chromatin remodeler.

10.7 DNA Replication in the Archaea

The archaea replication machinery is similar to that in eukaryotes.

DNA replication in the archaea has not been as extensively investigated as it has in bacteria or eukaryotes. Furthermore, investigators working on archaeal DNA replication in different laboratories have not devoted their efforts to studying a single organism or even a few closely related organisms but have instead examined a variety of different organisms. Much less, therefore, is known about DNA replication in the archaea than in the other two domains. Nevertheless, a picture of archaeal DNA replication has started to emerge. Investigators initially expected that the archaeal replication machinery would be more similar to the bacterial replication machinery than the eukaryotic replication machinery because the archaea, like the bacteria, are prokaryotes that for the most part have circular chromosomes.

This initial expectation has proven to be wrong. DNA sequences have been determined for a few of the archaea. Based on these sequences, it is clear that the components of the archaeal replication machinery are homologous to the components of the eukaryotic and not the bacterial replication machinery.

DNA replication in archaea follows the same general pattern as it does in the other two domains. Archaeal DNA replication is semiconservative, bidirectional, and semidiscontinuous. It also can be divided into three stages: initiation, elongation, and termination. Very little is known about replication termination in the archaea and so the discussion that follows is limited to the initiation and elongation stages.

Orc1/Cdc6 recruits MCM to the archaeal origin of replication.

Some archaeal chromosomes appear to have a single origin of replication, whereas others have two or more origins. Homologs of some of the proteins involved in the initiation of eukaryotic DNA synthesis have been identified in archaeal genomes. Different archaeal species, however, appear to have different variants of the replication machinery components. These differences may reflect the great variation in environmental conditions under which the archaea live.

The archaea produce a single protein, designated Orc1/Cdc6, which binds to the archaeal origin of replication. This protein shares some sequences in common with the eukaryotic ORC. Once bound to the origin, Orc1/Cdc6 recruits MCM to the origin (FIGURE 10.27). The MCM helicase appears to work in the same way as the eukaryotic helicase. The helicase opens and unwinds the double-stranded DNA. RPA (single-stranded DNA binding protein) binds to the exposed single-stranded DNA. Primase associates with the RPA•DNA complex and synthesizes the short RNA primers required to initiate DNA synthesis. Then DNA polymerase associates with the replication bubble, initiating rapid and processive bidirectional DNA synthesis.

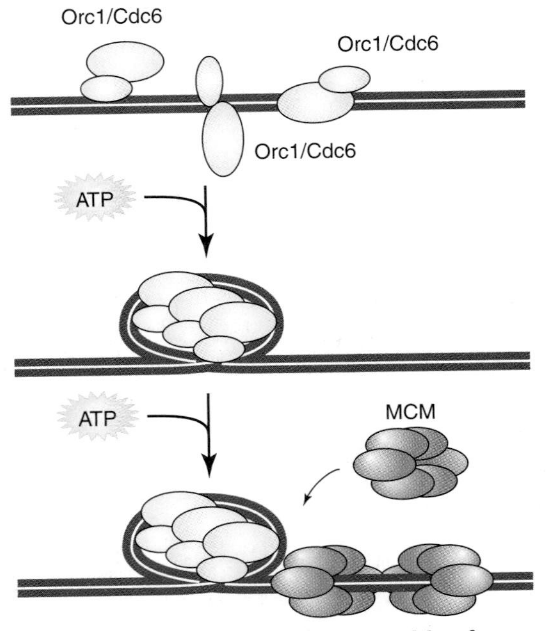

FIGURE 10.27 **Model for the assembly of MCM proteins at archaeal origins of DNA replication.** The archaea produce a protein complex designated Orc1/Cdc6, which binds to specific sequences in the archaeal origin of replication. The steps that follow remain to be established. Orc1/Cdc6 may oligomerize in an ATP-dependent manner to form a structure resembling the eukaryotic origin recognition complex (ORC) or one resembling the bacterial DnaA protein complex. Then the Orc1/Cdc6 complex probably recruits MCM helicase to the origin. (Adapted from B. Stillman, *FEBS Lett.*, 579 [2005]: 877–884.)

The basic steps in archaeal elongation are very similar to those in bacteria and eukaryotes.

The machinery required for the elongation stage of DNA synthesis in the archaea is very similar to that used by the bacteria and eukaryotes. Lagging and leading strand synthesis are coordinated. Processive DNA synthesis requires tethering DNA polymerase to a sliding clamp. The archaeal sliding clamp is similar in sequence and structure to PCNA in eukaryotes. The clamp loader appears to also be similar to its eukaryotic counterpart. Two different kinds of archaeal DNA polymerases, PolB and PolD, participate in the elongation process; however, not all archaeal species have PolB. Studies of DNA replication in the hyperthermophile *Pyrococcus abyssi* by Jean-Paul Raffin and coworkers indicate that Pol B and Pol D participate in leading and lagging strand synthesis, respectively. The archaeal counterparts of eukaryotic Fen-1 and RNase H remove RNA from the 5′-ends of the Okazaki fragments and DNA ligase joins the fragments. Archaeal ligase, like its eukaryotic counterpart, requires ATP for ligation. It therefore differs from the NAD^+-dependent bacterial ligase.

The archaeal replication system has one major surprise. Thermophiles require an ATP-dependent reverse gyrase to introduce positive supercoils. The reverse gyrase works by introducing transient nicks into a single strand and so functions by a different mechanism from the bacterial gyrase. Thermophiles may require reverse gyrase because they grow at high temperatures that tend to unwind DNA. Undoubtedly, the archaeal replication machinery will have other surprises in store as we learn more about its components and how they work together. Perhaps the most remarkable surprise of all is how similar the replication process is in all three domains of life.

Suggested Reading

The SV40 DNA Replication System

Bochkareva, E., Martynowski, D., Seitova, A., and Bochkarev, A. 2006. Structure of the origin-binding domain of simian virus 40 large T antigen bound to DNA. *EMBO J* 25:5961–5969.

Fanning, E., and Zhao, K. 2009. SV40 DNA replication: from the A gene to a nanomachine. *Virology* 384:352–359.

Gai, D., Li, D., Finkielstein, C. V., et al. 2004. Insights into the oligomeric states, conformational changes, and helicase activities of SV40 large tumor antigen. *J Biol Chem* 279:38952–38959.

Gai, D., Zhao, R., Li, D., Finkielstein, C. V., and Chen, X. S. 2004. Mechanisms of conformational change for a replicative hexameric helicase of SV40 large tumor antigen. *Cell* 119:47–60.

Herendeen, D., and Kelly, T. J. 1996. SV40 DNA replication In: J. Julian Blow (ed.) *Eukaryotic DNA Replication.* Oxford, UK: Oxford University Press.

Li, D., Zhao, R., Lileystrom, W. et al. 2003. Structure of the replicative helicase of the oncoprotein SV40 large tumor antigen. *Nature* 423:512–518.

Meinke, G., Phelan, P., Moine, S., et al. 2007. The crystal structure of the SV40 T-antigen origin binding domain in complex with DNA. *PLOS Biol* 5:144–156.

Valle. M., Chen, X. S., Donate, L. E., Fanning, E., and Carazo, J. M. 2006. Structural basis for the cooperative assembly of large T antigen on the origin of replication. *J Mol Biol* 357:1295–1305.

Eukaryotic Replication Initiation

Aladjem, M. I., and Fanning, E. 2004. The replicon revisited: an old model learns new tricks in metazoan chromosomes. *EMBO Rep* 5:666–691.

Bell, S. P. 2002. The origin recognition complex: from simple origins to complex functions. *Genes Dev* 16:659–672.

Bell, S. P., and Dutta, A. 2002. DNA replication in eukaryotic cells. *Ann Rev Biochem* 71:333–374.

Bielinsly, A.-K., and Gerbi, S. A. 2001. Where it starts: eukaryotic origins of DNA replication. *J Cell Sci* 114:643–651.

Bryant, J. A. 2010. Replication of nuclear DNA. *Progress in Botany*. 71:25–60. New York: Springer.

Costa, A., and Onesti, S. 2009. Structural biology of MCM helicases. *Crit Rev Biochem Mol Biol* 44:326–342.

Cvetic, C., and Walter, J. C. 2005. Eukaryotic origins of DNA replication: could you please be more specific? *Cell Dev Biol* 16:343–353.

Evrin, C., Clarke, P., Zech, J., et al. 2009. A double-hexameric MCM2-7 complex is loaded onto origin DNA during licensing of eukaryotic DNA replication. *Proc Natl Acad Sci USA* 106:20240–20245.

Françon, P., and Méchali, M. 2006. DNA replication origins. In: *Encyclopedia of Life Sciences*. pp. 1–5. Hoboken, NJ: John Wiley and Sons.

Ge, X. Q., and Blow, J. J. 2009. Conserved steps in eukaryotic DNA replication. In: Lynne S. Cox (ed.). *Molecular Themes in DNA Repair*. Cambridge, UK: Royal Society of Chemistry.

Iives, I, Petojevic, T., Pesavento, J. J., and Botchan, M. R. 2010. Activation of MCM2-7 helicase by association with Cdc45 and GINS proteins. *Mol Cell* 37:247–258.

Kawakami, H., and Katayama, T. 2010. DnaA, ORC, and CDC6: similarity beyond the domains of life and diversity. *Biochem Cell Biol* 88:49–62.

Kreigstein, H. J., and Hogness, D. S. 1974. Mechanism of DNA replication in *Drosophila* chromosomes: structure of replication forks and evidence for bidirectionality. *Proc Natl Acad Sci USA* 71:135–139.

MacNeill, S. A. 2010. Structure and function of GINS complex, a key component of the eukaryotic replisome. *Biochem J* 425:489–500.

Nawotka, K. A., and Huberman, J. 1988. Two-dimensional gel electrophoretic method for mapping DNA replicons. *Mol Cell Biol* 8:1408–1413.

Newlon, C. S., and Theis, J. F. 2002. DNA replication joins the revolution: whole-genome views of DNA replication in budding yeast. *Bioessays* 24:300–304.

Pospiech, H., Grosse, F., and Pisani, F. M. 2010. The initiation step of eukaryotic DNA replication. *Subcell Biochem* 50:79–104.

Remus, D., and Diffley, F. X. 2009. Eukaryotic DNA replication: lock and load, then fire. *Curr Opin Cell Biol* 21:771–777.

Remus, D., Beuron, F., Tolun, G., et al 2009. Concerted loading of Mcm2-7 double hexamers around DNA during DNA replication origin licensing. *Cell* 139:719–730.

Schwob, E. 2004. Flexibility and governance in eukaryotic DNA replication. *Curr Opin Microbiol* 7:680–690.

Sclafani, R. A., and Holzen, T. M. 2007. Cell cycle regulation of DNA replication. *Ann Rev Genet* 41:237–280.

Takara, T. J., and Bell, S. P. 2009. Putting two heads together to unwind DNA. *Cell* 139:652–654.

Toone, W, M., Aerne, B. L., and Morgan, B. A. 1997. Getting started: regulating the initiation of DNA replication in yeast. *Ann Rev Microbiol* 51:125–149.

Weinreich, M., DeBeer, M. A. P., and Fox, C. A. 2003. The activities of eukaryotic replication origins in chromatin. *Biochim Biophys Acta* 1677:142–157.

Wigley, D. B. 2009. ORC proteins: marking the start. *Curr Opin Struct Biol* 19:72–78.

Eukaryotic Replication Elongation

Asturias, F. J., Cheung, I. K., Sabouri, N. et al., 2006. Structure of *Saccharomyces cerevisiae* DNA polymerase epsilon by cryo–electron microscopy. *Nature Struct Biol* 13:35–43.

Bowman, G. D., O'Donnell, M., and Kuriyan, J. 2004. Structural analysis of a eukaryotic sliding DNA clamp-clamp loader complex. *Nature* 420:724–730.

Burgers, P. M. J. 2009. Polymerase dynamics at the eukaryotic DNA replication fork. *J Biol Chem* 284:4041–4045.

Cerritelli, S., and Crouch, R. J. 2009. Ribonuclease H: the enzymes in eukaryotes. *FEBS J* 276:1494–1505.

Ellenberger, T., and Tomkinson, A. E. 2008. Eukaryotic DNA ligases: structural and function insights. *Ann Rev Biochem* 77:313–338.

Garg, P., and Burgers, M. J. 2005. DNA polymerases that propagate the eukaryotic DNA replication fork. *Crit Rev Biochem Mol Biol* 40:115–128.

Hübscher, U., and Seo, Y.-S. 2001. Replication of the lagging strand: a concert of at least 23 polypeptides. *Mol Cell* 12:149–157.

Indiani, C., and O'Donnell, M. O. 2006. The replication clamp-loading machine at work in the three domains of life. *Nat Rev Mol Cell Biol* 7:751–761.

Klinge, S., Núñez-Ramírez, R., Llorca, O., and Pellegrini, L. 2009. 3D architecture of DNA Pol α reveals the functional core of multi-subunit replicative polymerases. *EMBO J* 28:1978–1987.

Kunkel, T. A., and Burgers, P. M. 2008. Dividing the workload at a eukaryotic replication fork. *Trends Cell Biol* 18:521–527.

Labib, K., and Gambus, A. 2007. A key role for the GINS complex at DNA replication forks. *Trends Cell Biol* 17:271–278.

Li, J. J., and Kelly, T. J. 1984. Simian virus 40 DNA replication in vitro. *Proc Natl Acad Sci* USA 81:6973–6977.

Liu, H., Kao, H.-I., Bambara, R. A. 2004. Flap endonuclease I: a central component of DNA metabolism. *Ann Rev Biochem* 73:589–615.

Maga, G., and Hübscher, U. 2003. Proliferating cell nuclear antigen (PCNA): a dancer with many partners. *J Cell Sci* 116:3051–3060.

McElhinny, S. A. N., Gordenin, E. A, Stith, C. M., Burgers, P. M. J., and Kunkel, T. A. 2008. Division of labor at the eukaryotic replication fork. *Mol Cell* 30: 137–144.

Moldovan, G.-L., Pfander, B., and Jentsch, S. 2007. PCNA, the maestro of the replication fork. *Cell* 129:665–679.

Pavlov, Y. I., and Shcherbakova, P. V. 2010. DNA polymerases at the eukaryotic replication fork—20 years later. *Mutation Res* 685:45–53.

Pursell, Z. F., and Kunkel, T. A. 2008. DNA polymerase ε: a polymerase of unusual size (and complexity). *Prog Nucleic Acid Res Mol Biol* 82:101–145.

Stith, C. M., Sterling, J., Resnick, M. A., Gordenin, D. A., and Burgers, P. M. 2008. Flexibility of eukaryotic Okazaki fragment maturation through regulated strand displacement synthesis. *J Biol Chem* 283:34129–34140.

Stillman, B. 2008. DNA polymerases at the replication fork in eukaryotes. *Mol Cell* 30:259-260.

Takahashi, T. S., Wigley, D. B., and Walter, J. C. 2005. Pumps, paradoxes, and ploughshares: mechanism of the MCM2-7 DNA helicase. *Trends Biochem Sci* 30:437–444.

Waga, S., and Stillman, B. 1998. The DNA replication fork in eukaryotic cells. *Ann Rev Biochem* 67:721–751.

Walther, A. P., and Wold, M. S. 2001. Eukaryotic replication fork. In: *Encyclopedia of Life Sciences*. pp 1–8. London: Nature.

Yao, N., Coryelli, L., Zhang, D., et al. 2003. Replication factor C clamp loader subunit arrangement within the circular pentamer and its attachment points to proliferating cell nuclear antigen. *J Biol Chem* 278:50744–50753.

Yuzhakov, A., Kelman, Z., Hurwitz, J., and O'Donnell, M. 1999. Multiple competition reactions for RPA order the assembly of the DNA polymerase δ holoenzyme. *EMBO J* 18:6189–6199.

Telomere and Telomerase

Autexier, C., and Lue, N. F. 2006. The structure and function of telomerase reverse transcriptase. *Ann Rev Biochem* 75:493–517.

Blackburn, E. H. 2001. Telomeres. In: *Encyclopedia of Life Sciences*. pp 1–7. London: Nature.

Blackburn, E. H. 2005. Telomeres and telomerase: their mechanisms of action and the effects of altering their functions. *FEBS Lett* 579:859–862.

Blackburn, E. H., Greider, C. W., and Szostak, J. W. 2006. Telomeres and telomerase: the path from maize, *Tetrahymena* and yeast to human cancer and aging. *Nat Med* 12:1133–1138.

Cech, T. R. 2004. Beginning to understand the end of the chromosome. *Cell* 116:273–279.

Chakhparonian, M., and Wellinger, R. J. 2003. Telomere maintenance and DNA replication: how closely are these two connected? *Trend Genet* 19:439–446.

Chan, S. R., and Blackburn, E. H. 2004. Telomeres and telomerase. *Philos Trans R Soc Lond B Biol Sci* 359:109–121.

Chatziantoniou, V. D. 2001. Telomerase: biological function and potential role in cancer management. *Pathol Oncol Res* 7:161–170.

de Lange, T. 2006. Lasker laurels for telomerase. *Cell* 126:1017–1020.

Gillis, A. J., Schuller, A. P., and Skordalakes, E. 2008. Structure of the *Tribolium castaneum* telomerase catalytic subunit TERT. *Nature* 455:633–638.

Gilson, E., and Géli, V. 2007. How telomeres are replicated. *Nat Rev Mol Cell Biol* 8:825–838.

Gilson, E., and Ségal-Benirdjian, E. 2010. The telomere story or the triumph of an open-minded research. *Biochimie* 92:321–326.

Harrington, L. 2003. Biochemical aspects of telomerase function. *Cancer Lett* 194:139–154.

Lansdorp, P. M. 2005. Major cutbacks at chromosome ends. *Trends Biochem Sci* 30:388–395.

Lingner, J., Cooper, J. P., and Cech, T. R. 1995. Telomerase and DNA end replication: no longer a lagging strand problem? *Science* 269:1533–1534.

Lue, N. F. 2004. Adding to the ends: what makes telomerase processive and how important is it? Bioessays 26:955–962.

Makarov, V. L., Hirose, Y, and Langmore, J. P. 1997. Long G tails at both ends of human chromosomes suggest a C strand degradation mechanism for telomere shortening. *Cell* 88:657–666.

Osterhage, J. L., and Friedman, K. L. 2009. Chromosome end maintenance by telomerase. *J Biol Chem* 284:16061–16065.

Oulette, M. M., and Choi, K. H. 2007. Telomeres and telomerase in ageing and cancer. In: *Encyclopedia of Life Sciences*. pp. 1–6. Hoboken, NJ: John Wiley and Sons.

Pardue, M.-L., and DeBaryshe, G. 2001. Telomeres in cell function: cancer and ageing. In: *Encyclopedia of Life Sciences*, pp. 1–6. London: Nature.

Sekaran, V. G., Soares, J., and Jarstfer, M. B. 2010. Structures of telomerase subunits provide functional insights. *Biochim Biophys Acta* 1804:1190–1201.

Sfeir, A. J., Chai, W., Shay, J. W., and Wright, W. E. 2005. Telomere-end processing: the terminal nucleotides of human chromosomes. *Mol Cell* 18:131–138.

Shore, D., and Bianchi, A. 2009. Telomere length regulation: coupling DNA processing to feedback regulation of telomerase. *EMBO J* 28:2309–2322.

Theimer, C. A., and Feigon, J. 2006. Structure and function of telomerase RNA. *Curr Opin Struct Biol* 16:307–318.

Vega, L. R., Mateyak, M. K, and Zakian, V. A. 2003. Getting to the end: telomerase access in yeast and humans. *Nat Rev Mol Cell Biol* 4:948–959.

Wellinger, R. J., Ethier, K. Labrecque, P., and Zakian, V. A. 1996. Evidence for a new step in telomere maintenance. *Cell* 85:423–433.

Replication Coupled Chromatin Synthesis

Corpet, A., and Almouzni, G. 2008. Making copies of chromatin: the challenge of nucleosomal organization and epigentic information. *Trends Cell Biol* 19:29–41.

Groth, A. 2009. Replicating chromatin: a tale of histones. *Biochem Cell Biol* 87:51–63.

Groth, A., Rocha, W., Verreault, A., and Almouzni, G. 2007. Chromatin challenges during DNA replication and repair. *Cell* 128:721–733.

Jasencakova, Z., and Groth, A. 2010. Restoring chromatin after replication: how new and old histone marks come together. *Semin Cell Dev Biol* 21:231–237.

Krebs, J. E., and Peterson, C. L. 2000. Understanding "active" chromatin: a historical perspective of chromatin remodelling. *Crit Rev Eukaryot Gene Expr* 10:1–12.

Ransom, M., Dennehey, B. K., and Tyler, J. K. 2010. Chaperoning histones during replication and repair. *Cell* 140:183–195.

Tyler, J. K. 2002. Chromatin assembly. *Eur J Biochem* 269:2268–2274.

The Archaea

Barry, E. R., and Bell, S. D. 2006. DNA replication in the archaea. *Microbiol Mol Biol Rev* 70:876–887.

Böhlke, K., Pisani, F. M., and Rossi, M. 2002. Archaeal DNA replication: spotlight on a rapidly moving field. *Extremophiles* 6:1–14.

Cann, I. K. O., and Ishino, Y. 1999. Archaeal DNA replication: identifying the pieces to solve the puzzle. *Genetics* 152:1249–1267.

Grabowski, B., and Kelman, Z. 2003. Archaeal DNA replication: eukaryl proteins in a bacterial context. *Ann Rev Microbiol* 57:487–516.

Henneke, G., Flament, D., Hübscher U., Querellou, J., and Raffin, J.-P., and Jean-Paul Raffin. 2005. The hyperthermophilic euryarchaeota *Pyrococcus abyssi* likely requires the two DNA polymerases D and B for DNA replication. *J Mol Biol* 350:53–64.

Jenkinson, E. R., and Chong, J. P. J. 2003. Initiation of archaeal DNA replication. *Biochem Soc Trans* 31:669–673.

Kelman, Z. 2000. DNA replication in the third domain (of life). *Curr Protein Pept Sci* 1:139–154.

Majernik, A. I., Jenkinson, E. R., and Chong, J. P. J. 2004. *Biochem Soc Trans* 32:236–239.

This book has a Web site, **http://biology.jbpub.com/book/molecular**, which contains a variety of resources, including links to in-depth information on topics covered in this chapter, and review material designed to assist in your study of molecular biology.

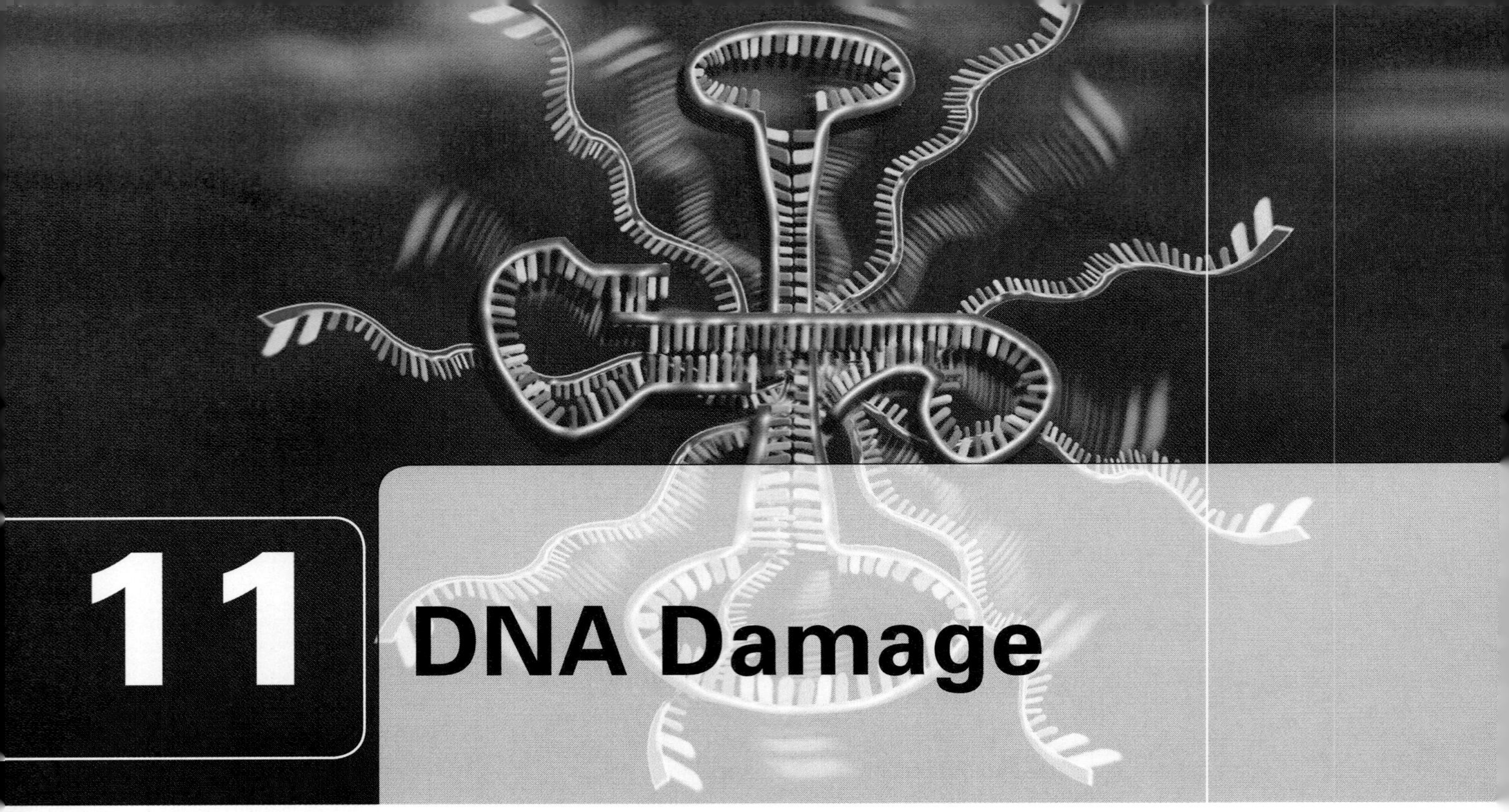

11 DNA Damage

OUTLINE OF TOPICS

11.1 Radiation Damage

Ultraviolet light causes cyclobutane pyrimidine dimer (CPD) formation and (6-4) photoproduct formation.

X-rays and gamma rays cause many different types of DNA damage.

11.2 DNA Instability in Water

DNA is damaged by hydrolytic cleavage reactions.

11.3 Oxidative Damage

Reactive oxygen species damage DNA.

11.4 Alkylation Damage by Monoadduct Formation

Alkylating agents damage DNA by transferring alkyl groups to centers of negative charge.

Many environmental agents must be modified by cell metabolism before they can alkylate DNA.

11.5 Chemical Cross-Linking Agents

Chemical cross-linking agents block DNA strand separation.

Psoralen and related compounds can form monoadducts or cross-links.

Cisplatin combines with DNA to form intra- and interstrand cross-links.

11.6 Mutagen and Carcinogen Detection

Mutagens can be detected based on their ability to restore mutant gene activity.

Suggested Reading

It may seem reasonable to suppose that the genetic material would be quite stable to ensure that information is passed accurately from one generation to the next, but quite the opposite is true. DNA is easily damaged under normal physiological conditions. Many different kinds of physical and chemical agents damage DNA. Some of these are **endogenous agents**; that is, they are formed inside the cell by normal metabolic pathways. Others are **exogenous agents** that come from the surrounding environment. Some exogenous agents act directly on DNA, whereas others must be modified by the cell's enzymes before they can damage DNA. This chapter examines the different types of DNA damage that can take place in cells and the next chapter examines how cells repair the damage.

11.1 Radiation Damage

Ultraviolet light causes cyclobutane pyrimidine dimer (CPD) formation and (6-4) photoproduct formation.

Cells exposed to high energy electromagnetic radiation, which includes ultraviolet light (wavelengths 100–400 nm) and two forms of ionizing radiation, x-rays (wavelengths 0.01–100 nm) and gamma rays (wavelengths < 0.01 nm), experience considerable damage to their DNA. We begin by examining the major types of damage caused by ultraviolet (UV) light, which is divided into three bands: UV-A (321–400 nm), UV-B (296–320 nm), and UV-C (100–295 nm). The majority of UV light reaching earth is UV-A, but this is the least energetic band and so does little damage to DNA. Although UV-B accounts for about 10% of the UV radiation reaching the earth's surface, it is responsible for most of the DNA damage in skin. UV-C includes the wavelength of maximum DNA absorbance (260 nm) and so would cause a great deal of DNA damage to exposed organisms if it were able to penetrate the earth's atmosphere. Fortunately, very little UV-C reaches the earth's surface because the ozone in the stratosphere prevents it from doing so. The ozone layer cannot be taken for granted, however, because it can be depleted by the release of chlorofluorohydrocarbons and other industrial chemicals into the atmosphere. Ozone depletion would lead to a much greater incidence of cancer of the skin, the only human tissue subject to direct UV damage. Laboratory studies that are designed to examine the way that UV light damages DNA are usually conducted with germicidal lamps that produce UV-C light, which affects DNA in the same way as UV-B, although much more efficiently.

Two major photoproducts account for nearly all of the UV-induced DNA damage. Their synthetic pathways involve dimer formation between adjacent pyrimidine bases on the same DNA strand. The first pathway, **cyclobutane pyrimidine dimer** (**CPD**) formation, accounts for about 75% of UV-induced damage. The cyclobutane ring is generated by forming one bond between C-5 atoms, and another between C-6 atoms, on adjacent pyrimidine rings (FIGURE 11.1a). The most common cyclobutane pyrimidine dimer is the thymine–thymine (T<>T)

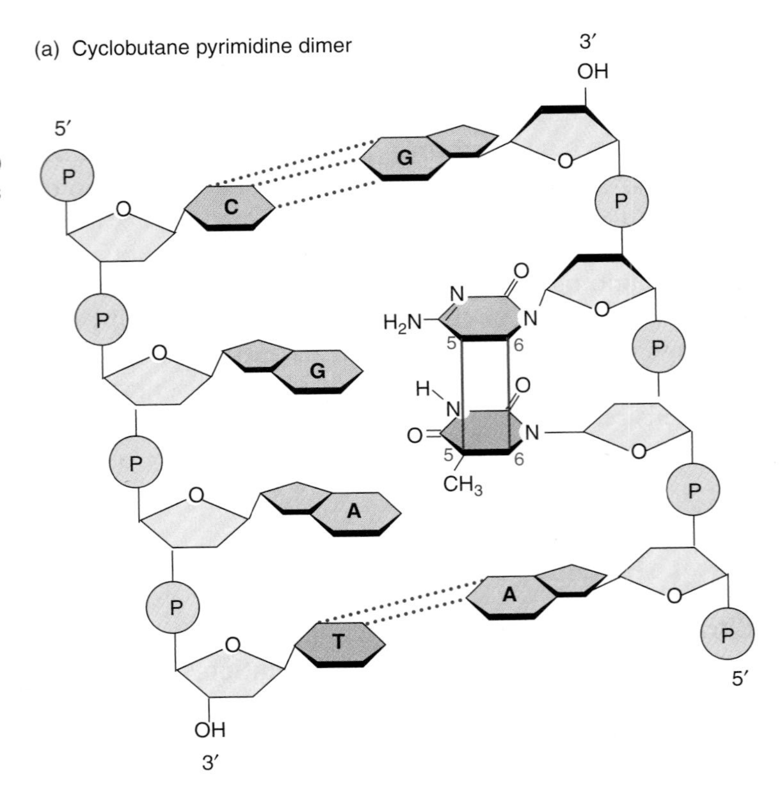

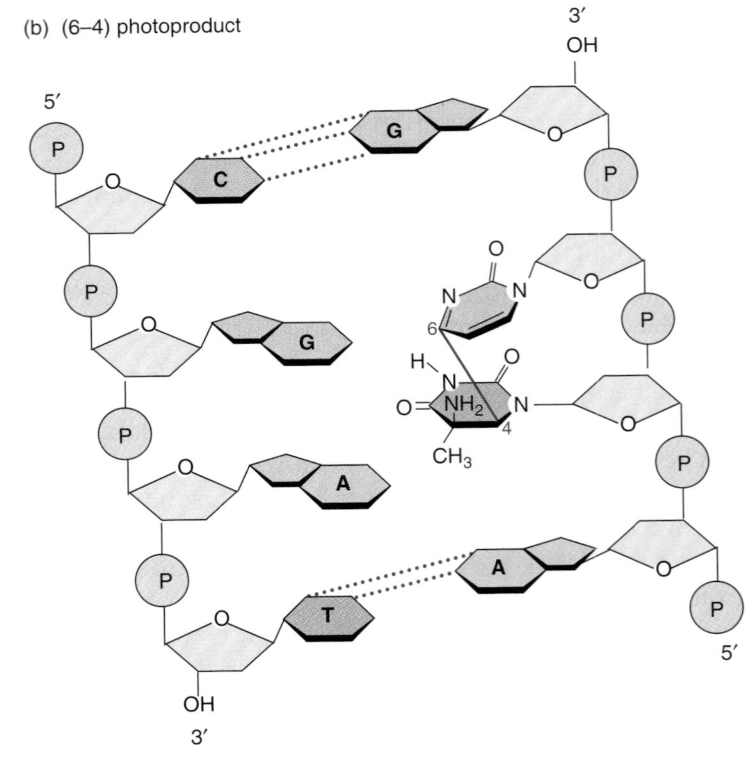

FIGURE 11.1 Ultraviolet light promoted cyclobutane pyrimidine dimer and (6-4) photoproduct formation. (a) Ultraviolet light promotes the formation of cyclobutane pyrimidine dimers by introducing two new bonds between adjacent pyrimidines (in this case a cytosine and a thymine) on the same DNA strand. One bond connects the C-5 atoms and the second connects the C-6 atoms. (b) Ultraviolet light also promotes the formation of (6-4) photoproducts by introducing a bond between the C-6 atom of one pyrimidine and the C-4 atom of an adjacent pyrimidine (in this case a cytosine and a thymine) on the same DNA strand. Although this figure shows the bonds between the pyrimidine rings, it does not show the way the pyrimidine rings are arranged in space. (Adapted from E. C. Friedberg, *DNA Repair and Mutagenesis, Second edition*. ASM Press, 2005.)

dimer. Cytosine–thymine (C<>T) and cytosine–cytosine (C<>C) dimers also form but at slower rates. Structural studies show: (1) B-DNA can accommodate a single T<>T dimer, forcing the helical axis to bend by about 30° toward the major groove (FIGURE 11.2); (2) the dimer's 3′-thymine can form a normal Watson-Crick base pair with its adenine partner on the complementary strand; and (3) the interaction between the 5′-thymine and its complementary adenine partner will be weaker than normal because only a single hydrogen bond can be formed. Thymine–thymine cyclobutane dimers are often used to study DNA repair systems because they are stable, easy to form, and easy to detect. The second pathway, which accounts for most of the remaining UV-induced DNA damage, produces **(6-4) photoproducts** (FIGURE 11.1b). A bond is formed between the C-6 atom of the 3′-pyrimidine (either thymine or cytosine) and the C-4 atom of the 5′-pyrimidine (usually cytosine). The (6-4) photoproduct causes a major distortion in B-DNA because the two pyrimidine rings are perpendicular to each other. If not removed, a pyrimidine dimer or (6-4) photoproduct can interfere with the normal operation of the replication and transcription machinery, resulting in mutations and cell death. Even if the lesion is removed, the result can be a mutation.

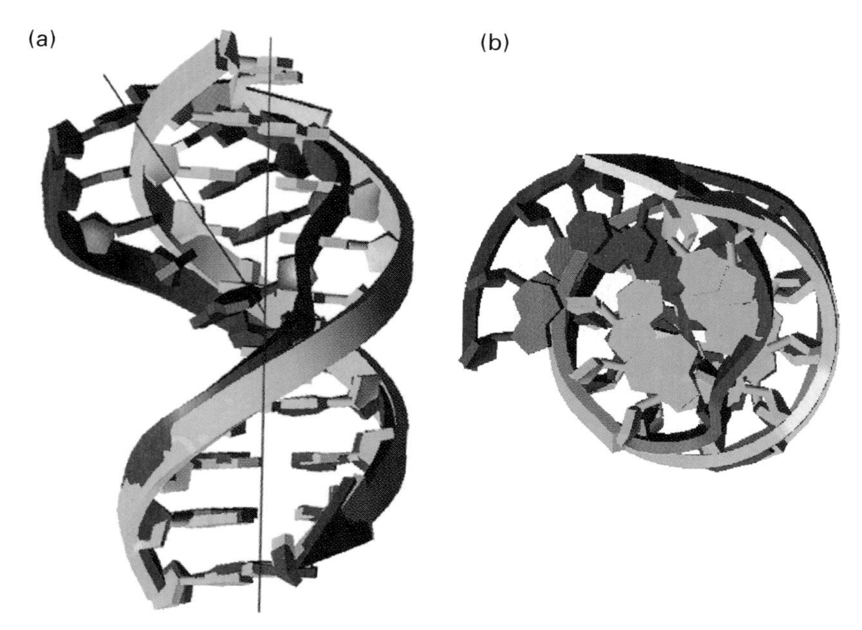

FIGURE 11.2 **Schematic diagram illustrating a kink in DNA resulting from a cyclobutane pyrimidine dimer.** A DNA molecule with a kink resulting from a cyclobutane pyrimidine dimer (red) is superimposed onto regular B-DNA (green). (a) Side view with a helical axis. (b) Top view. (Reproduced from H. Park, et al., *Proc. Natl. Acad. Sci. USA* 99 [2002]: 15965–15970. Copyright 2002, National Academy of Sciences, USA. Photo courtesy of ChulHee Kang, Washington State University.)

X-rays and gamma rays cause many different types of DNA damage.

Ionizing radiation directly or indirectly generates many different kinds of DNA lesions. Direct damage takes place when DNA or water tightly bound to it absorbs the radiation. Indirect damage takes place when water molecules or other molecules surrounding the DNA absorb the radiation and form reactive species that then damage the DNA. Lesions may be isolated or clustered (many lesions within a few helical turns). One type of clustered lesion, the double-strand break, is generally thought to be the primary reason that ionizing radiation is so lethal to cells. Double-strand breaks are also responsible for various chromosomal aberrations such as **deletions, duplications, inversions** (segments breaking away from the chromosome, inverting from end to end, and then re-inserting at the original breakage site), and **translocations** (segments breaking away from the chromosome and then re-inserting at new sites).

Approximately 65% of the damage to DNA that is caused by x-rays and γ-rays is due to indirect effects, primarily through the transfer of photons to water. The photon transfer activates the water, causing it to undergo two types of primary reactions. In the first of these, which accounts for about 80% of the energy transfer to water, the water molecule is ionized.

$$H_2O \xrightarrow[h\nu]{} H_2O^{\bullet +} + e^- \qquad \text{(Eq. 11.1)}$$

The $H_2O^{\bullet +}$ that is formed rapidly dissociates, releasing a proton and a hydroxyl radical (**•OH**).

$$H_2O^{\bullet +} \rightarrow H^+ + {}^\bullet OH \qquad \text{(Eq. 11.2)}$$

The electron generated by the reaction shown in Equation 11.1 can combine with any molecular oxygen that is present to from a superoxide radical ($^\bullet O_2^-$).There is, however, very little free oxygen outside the mitochondria.

$$e^- + O_2 \rightarrow {}^\bullet O_2^- \qquad \text{(Eq. 11.3)}$$

In the second type of primary reaction, which accounts for the remaining 20% of the energy transferred to water, excited water (H_2O^*) splits into a hydrogen atom (H·) and a hydroxyl radical (•OH).

$$H_2O \xrightarrow[h\nu]{} H_2O^* \rightarrow H^\bullet + {}^\bullet OH \qquad \text{(Eq. 11.4)}$$

The three highly reactive chemical species produced by the two primary pathways—•OH, $^\bullet O_2^-$, and •H—each attacks and damages whatever biomolecule they encounter. A wide variety of chemical changes take place when that molecule happens to be DNA. Specific changes caused by hydroxide and superoxide radicals will be described later in this chapter when we examine oxidative damage.

11.2 DNA Instability in Water

DNA is damaged by hydrolytic cleavage reactions.

DNA has three kinds of bonds with the potential for hydrolytic cleavage, namely: (1) phosphodiester bonds, (2) N-glycosyl bonds, and (3) bonds linking exocyclic amine groups to bases (FIGURE 11.3). Spontaneous phosphodiester bond cleavage, which introduces a nick into a DNA strand, is a very rare occurrence and probably does not make a significant contribution to DNA damage. N-glycosyl bond cleavage leads to the formation of an abasic site, which is also known as an **AP** (for apurinic and apyrimidinic) site (FIGURE 11.4). According to current estimates, about 10,000 purine and 500 pyrimidine bases are lost from DNA in a mammalian cell nucleus each day. These observations are consistent with *in vitro* experiments showing that purine N-glycosyl bonds are more easily hydrolyzed than pyrimidine N-glycosyl bonds. AP site formation sensitizes the neighboring 3′-phosphodiester bond to cleavage. In part, this increased sensitivity results from the ability of a deoxyribose group at an AP site to change from a ring structure to an open chain form with the production of a free aldehyde group (Figure 11.4), which makes the 3′-phosphodiester bond more sensitive to cleavage. A DNA strand with one or more AP sites makes a poor template because it lacks the information required to direct accurate replication and transcription.

Water-mediated deamination converts cytosine, guanine, and adenine to uracil, xanthine, and hypoxanthine, respectively (FIGURE 11.5). Hydrolytic cytosine deamination is estimated to take place about 100 to 500 times a day in a mammalian cell, whereas combined guanine and adenine deaminations are estimated to occur at about 1% or 2% of that value. The conversion of guanine to xanthine may result in mutations or arrested DNA synthesis because xanthine does not form stable base pairs with either cytosine or thymine. The conversion of adenine to hypoxanthine will, if not repaired, cause a T–A base pair to be replaced by a C–G base pair. Likewise, an uncorrected deamination

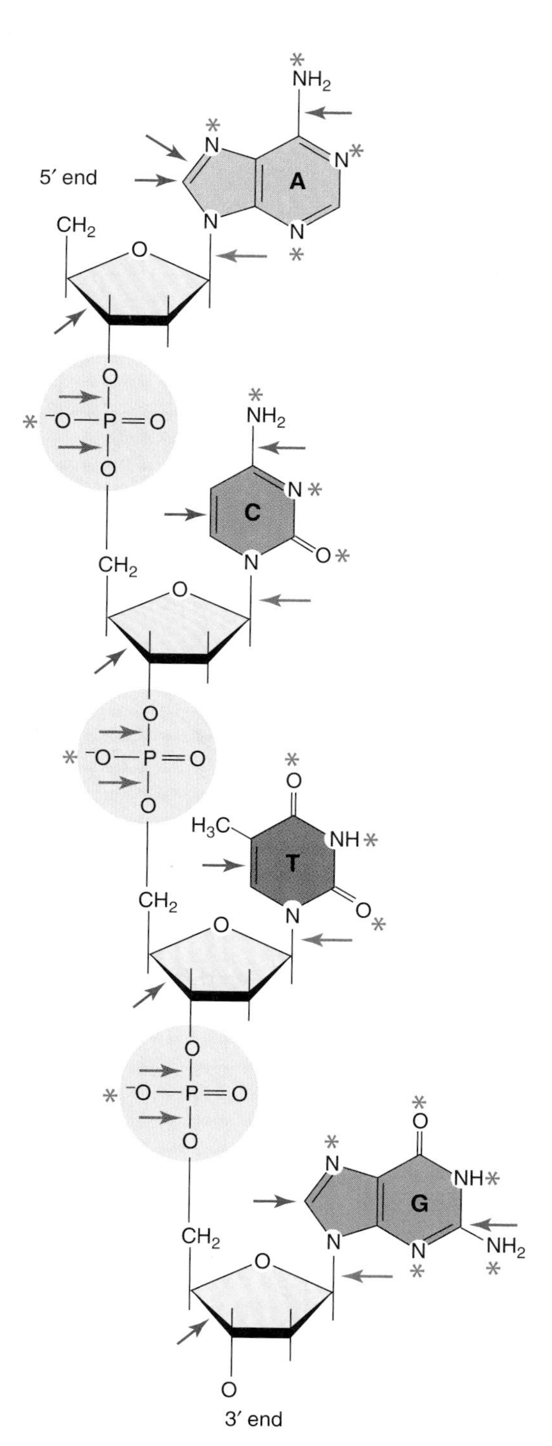

FIGURE 11.3 **Sites of chemical damage to DNA.** The bases (A, T, G, and C), deoxyribose groups, and phosphodiester bonds are vulnerable to attack by many exogenous and endogenous agents. Brown, green, and blue arrows point to phosphodiester bonds, N-glycosylic bonds, and bonds to exocyclic amine groups, respectively, that can be cleaved by water. Red arrows point to sites of attack by reactive oxygen species on the deoxyribose groups and bases. Green asterisks indicate electron-rich atoms that are attacked by alkylating agents. (Adapted from P. W. Doetsch, *Encyclopedia of Life Sciences*. [DOI: 11.1038/npg.els.0000557]. Posted April 2001.)

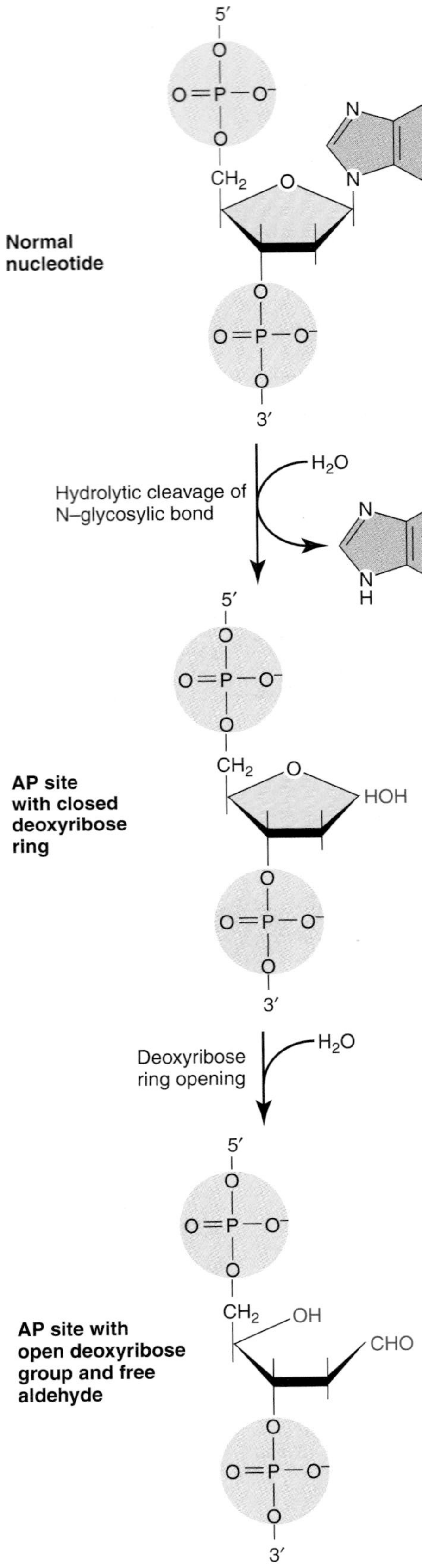

FIGURE 11.4 Hydrolytic cleavage of an N-glycosylic bond and subsequent generation of an open chain deoxyribose with a free aldehyde group.

Normal nucleotide

Hydrolytic cleavage of N–glycosylic bond

AP site with closed deoxyribose ring

Deoxyribose ring opening

AP site with open deoxyribose group and free aldehyde

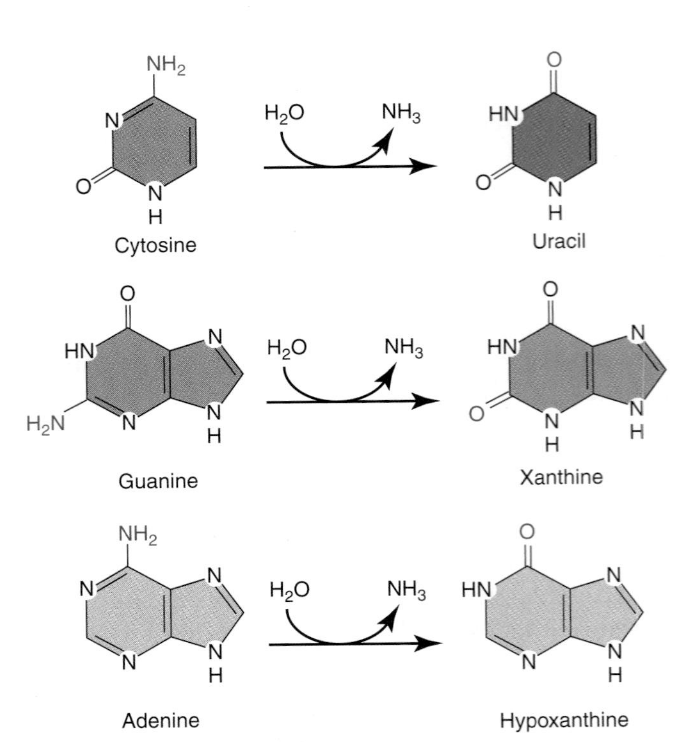

FIGURE 11.5 Water-mediated hydrolysis of the amino groups attached to cytosine, guanine, and adenine.

that converts C to U will cause a C–G base pair to be replaced by a T–A base pair. Mutations of this type in which a pyrimidine on one strand is replaced by a different pyrimidine and a purine on the other strand is replaced by a different purine are called **transition** mutations (FIGURE 11.6a). Another type of replacement mutation, termed a **transversion** mutation, which involves replacing a pyrimidine on one strand with a purine and a purine on the other strand with a pyrimidine (FIGURE 11.6b), will be encountered later in this chapter. A few cytosine bases in eukaryotic DNA are converted to the modified base 5-methylcytosine. This modified base is concentrated in so-called **CpG islands**, which are small segments of DNA often present in regulatory elements termed **promoters** that are located just before the transcription unit that they regulate. The term CpG island derives from the fact that the CpG segment is present in these DNA sections at a much higher frequency than in the rest of the DNA. The frequency of spontaneous deamination of 5-methylcytosine bases in CpG islands is even greater than that for cytosine. The product in this case, however, is thymine and not uracil, resulting in the conversion of a C–G base pair to a T–A base pair. Nitrous acid (HNO_2), which is formed from nitrites used as preservatives in processed meats such as bacon, sausage, and hot dogs, reacts with the amine groups attached to the ring structures in cytosine, adenine, and guanine, greatly increasing their rate of deamination. Bisulfite (HSO_3^-), an additive that is sometimes present in wine, beer, fruit juices, and dried fruits, also greatly increases the rate of cytosine deamination but does not affect purine or 5-methylcytosine deamination.

(a) Transition mutations

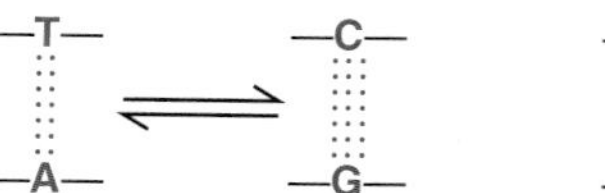

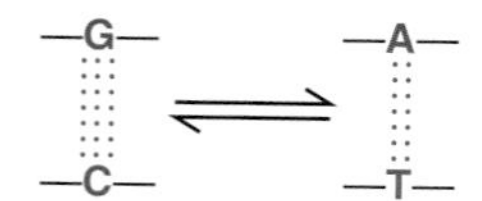

(b) Transversion mutations

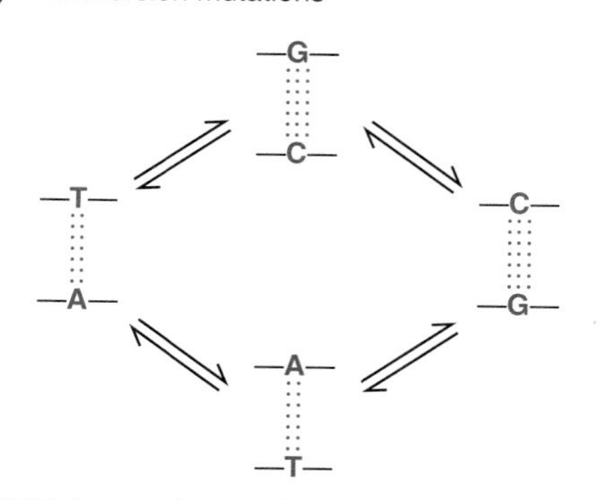

FIGURE 11.6 Transition and transversion mutations. (a) In a transition mutation, one pyrimidine-purine base pair replaces another base pair in the same pyrimidine-purine relationship. (b) In a transversion mutation, a pyrimidine-purine base pair is replaced by a purine-pyrimidine base pair (or vice versa). Purine bases are blue and pyrimidine bases are red.

11.3 Oxidative Damage

Reactive oxygen species damage DNA.

Cellular respiration involves electron transfer from various metabolic intermediates to a terminal electron acceptor, which is molecular oxygen (O_2) in aerobic organisms. The electron transport or respiratory chain, which is present in the inner mitochondrial membrane in eukaryotes and the cell membrane in bacteria, carries out the following sequential electron transfer to molecular oxygen.

$$O_2 \xrightarrow{e^-} \cdot O_2^- \xrightarrow{e^-,\ 2H^+} H_2O_2 \xrightarrow[\ \ H_2O]{e^-, H^+} \cdot OH \xrightarrow{e^-, H^+} H_2O \quad \text{(Eq. 11.5)}$$

Although the reactive oxygen species produced during cellular respiration have the potential to damage DNA, it is unlikely that they do so because: (1) the respiratory chain does not normally release these reactive oxygen species; (2) cells contain superoxide dismutase to convert superoxide radicals into molecular oxygen and hydrogen peroxide and catalase to convert hydrogen peroxide into oxygen and water; and (3) superoxide and hydroxyl radicals are so reactive that, if released by the respiratory chain, they would react with nearby biomolecules before they had a chance to reach nuclear DNA.

Reactive oxygen species produced by cellular respiration probably do not damage nuclear DNA under normal physiological conditions, but reactive oxygen species produced by other mechanisms can cause considerable nuclear DNA damage. The primary culprit appears to be the hydroxyl radical, which is produced by ionizing radiation as described above. Hydroxyl radicals, however, also can be produced chemically from hydrogen peroxide, a normal metabolic product of many biochemical pathways such as the one that converts guanine to uric acid. Hydrogen peroxide is not nearly as reactive as superoxide and hydroxyl radicals, so it has a much longer half-life in the cell provided that it escapes catalase and peroxidase, enzymes that can destroy it. If it does escape, hydrogen peroxide can be converted to a hydroxide radical by the following reaction:

$$Fe^{2+} + H_2O_2 \rightarrow Fe^{3+} + \cdot OH + OH^- \quad \text{(Eq. 11.6)}$$

This reaction, first reported by Henry J. H. Fenton in the 1890s, is called the Fenton reaction in his honor. Other transition metal ions such as copper, manganese, and cobalt can replace iron. Hydroxyl

(a) 8-oxoguanine

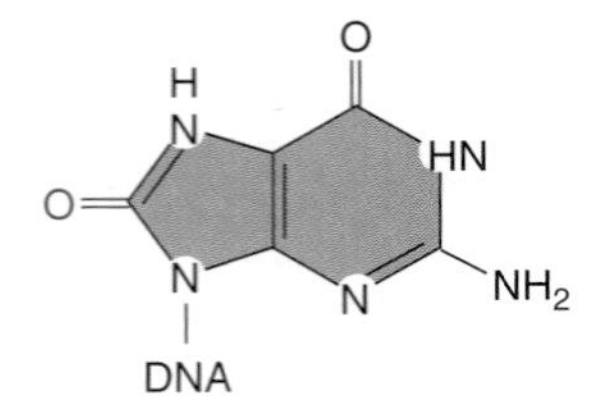

(b) Thymine glycol

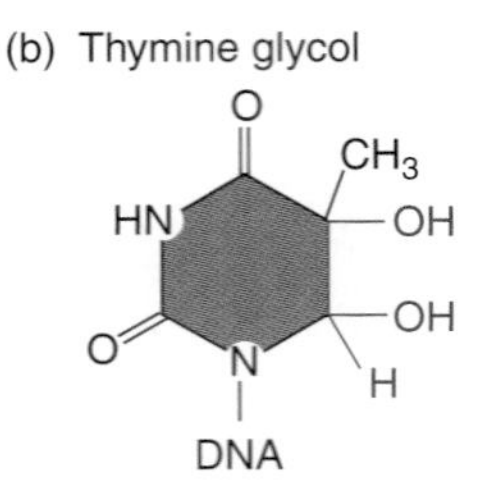

FIGURE 11.7 Two of the modified bases produced by oxidative damage to DNA caused by the hydroxyl radical.

radicals, whether generated by ionizing radiation or by the Fenton reaction, are known to cause more than 80 different kinds of base damage. Two of the oxidized base products, 8-oxoguanine (oxoG) and thymine glycol, are shown in FIGURE 11.7. 8-Oxoguanine can base pair with adenine or cytosine (FIGURE 11.8). If uncorrected, the resulting 8-oxoG–A base pair will be replicated to form a T–A base pair, causing a transversion mutation. Thymine glycol inhibits DNA replication and is therefore cytotoxic.

Hydroxyl radicals produced by the Fenton reaction tend to be more widely dispersed than those caused by ionizing radiation and, therefore, much less likely to produce double-strand breaks. Cells can repair single-strand breaks much more easily than they can repair double-strand breaks. Reactive oxygen species can also damage bases or sugars or convert various biomolecules into reactive species that can then damage DNA. For example, polyunsaturated fatty acid oxidation produces two aldehyde products, malondialdehyde and 4-hydroxynonenal, which contribute to base damage (FIGURE 11.9).

(a) 8-oxoguanine–cytosine base pair

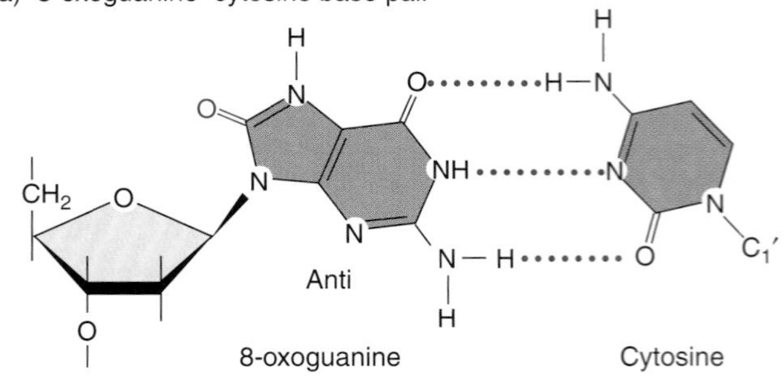

(b) 8-oxoguanine–adenine base pair

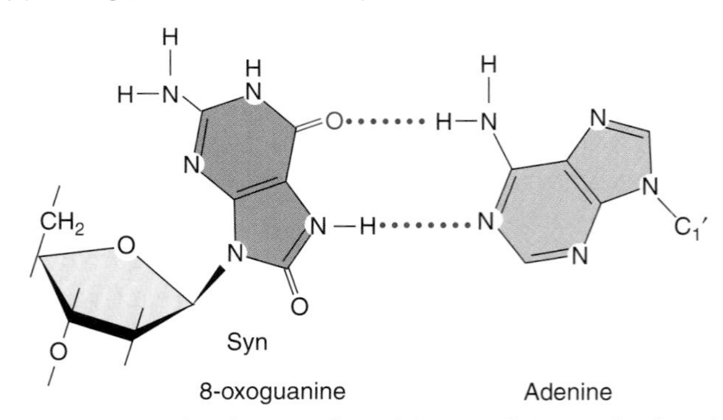

FIGURE 11.8 **8-oxoguanine base pairs with cytosine or adenine.** (a) In the anti conformation 8-oxoguanine base pairs with cytosine and (b) in the syn form it base pairs with adenine. If uncorrected, the 8-oxoguanine-adenine base pair leads to a transversion mutation in which a G–C base pair is replaced by a T–A base pair.

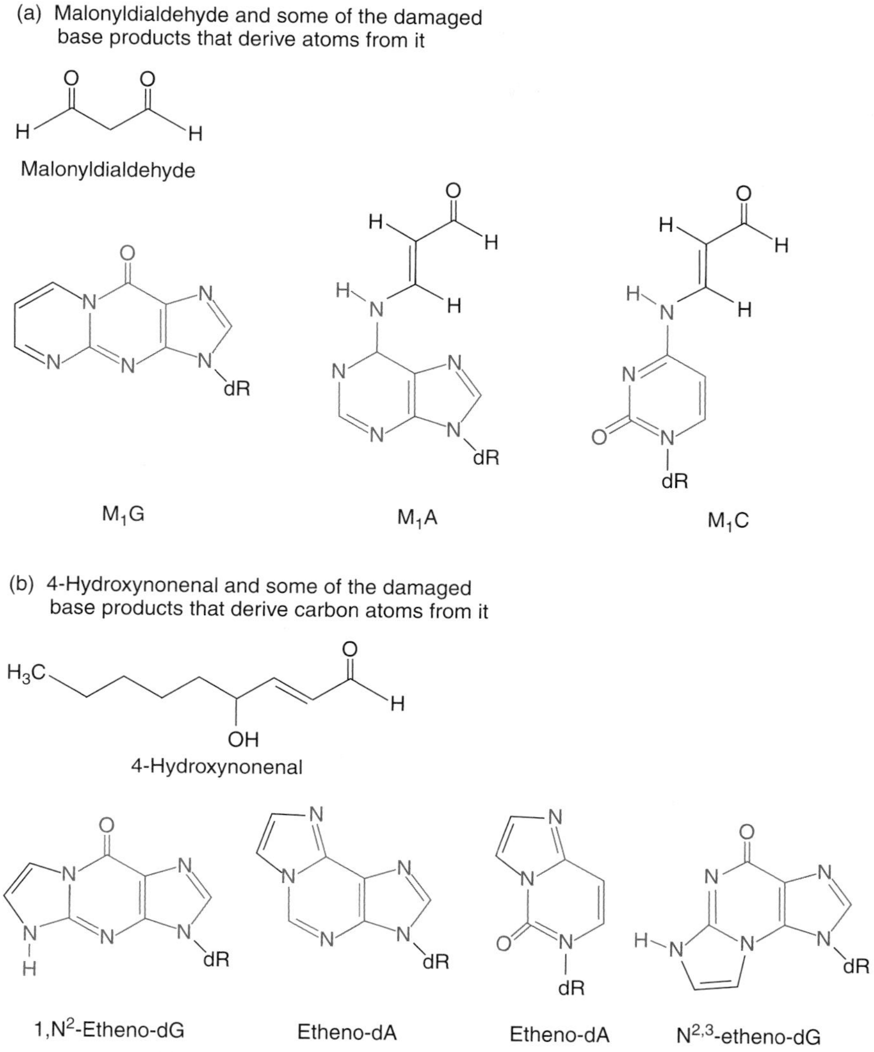

(a) Malonyldialdehyde and some of the damaged base products that derive atoms from it

Malonyldialdehyde

M_1G M_1A M_1C

(b) 4-Hydroxynonenal and some of the damaged base products that derive carbon atoms from it

4-Hydroxynonenal

1,N^2-Etheno-dG Etheno-dA Etheno-dA $N^{2,3}$-etheno-dG

FIGURE 11.9 **Malonyldialdehyde and 4-hydroxynonenol, two aldehyde products of polyunsaturated oxidation that contribute to DNA base damage.** (a) Malonyldialdehyde and some of the damaged base products that derive atoms from it. (b) 4-Hydroxynonenol and some of the damaged base products that derive atoms from it. Atoms and bonds shown in blue in the damaged bases are derived from the parent base. (Adapted from L. J. Marnett, *Carcinogenesis* 21 [2000]: 361–370.)

11.4 Alkylation Damage by Monoadduct Formation

Alkylating agents damage DNA by transferring alkyl groups to centers of negative charge.

DNA has electron-rich atoms that are readily attacked by electron-seeking chemicals called **electrophiles** (Gr. *electros* = electron and *philos* = loving) or electrophilic agents. This section examines DNA

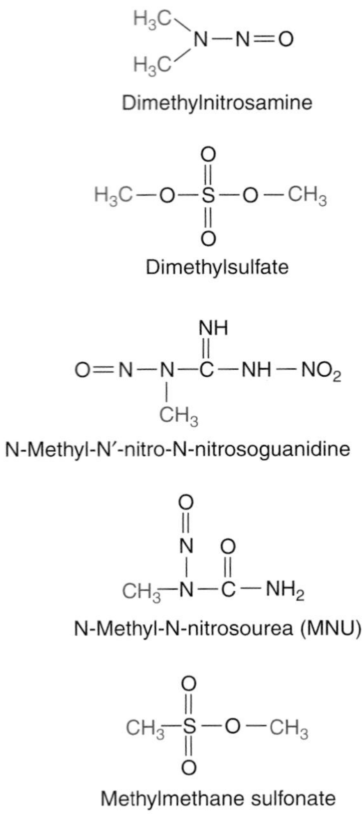

Dimethylnitrosamine

Dimethylsulfate

N-Methyl-N′-nitro-N-nitrosoguanidine

N-Methyl-N-nitrosourea (MNU)

Methylmethane sulfonate

FIGURE 11.10 **Chemical structures of a few simple DNA methylating agents.** Methyl groups that are transferred to DNA are shown in red.

damage caused by a highly reactive group of electrophiles called **alkylating agents** because they transfer methyl, ethyl, or larger alkyl groups to the electron-rich atoms in DNA. Alkylation takes place at: (1) nitrogen and oxygen atoms external to the base ring systems; (2) nitrogen atoms in the base ring systems except those linked to deoxyribose; and (3) non-bridging oxygen atoms in phosphate groups. Many different kinds of naturally occurring and synthetic chemical agents are known to transfer alkyl groups to DNA.

The product formed by attaching a chemical group to DNA is called an **adduct.** If the chemical group attaches to a single site on the DNA then the product is termed a **monoadduct.** For example, the exposure of DNA to dimethylnitrosamine leads to the production of a monoadduct in which a single methyl group attaches to DNA. The structures of dimethylnitrosamine and a few other simple agents that methylate DNA are shown in FIGURE 11.10. These methylating agents represent only a small fraction of the known alkylating agents, many of which transfer much larger alkyl groups to DNA.

FIGURE 11.11 shows the major methyl adducts formed when DNA is exposed to methylmethane sulfonate (MMS) and N-methyl-N-nitrosourea (MNU). In both cases, base methylation takes place most frequently at the N-7 position in guanine and next most frequently at N-3 in adenine. N7-Methylguanine forms a base pair with cytosine, but the modified guanine is readily removed from DNA with the resultant formation of an abasic site. Methylation at N-3 in adenine is of great practical significance because N3-methyladenine formation blocks DNA replication but does not appear to lead to mutations. Therefore, a methylating agent that could transfer a methyl group exclusively to the N-3 position in adenine would have the potential to kill cancer cells without causing mutations.

Methylations at O-6 in guanine and O-4 in thymine are much less frequent events than either of the N-methylations described above. Nevertheless, O^6-methylguanine and O^4-methylthymine formation are quite important because the methylated bases mispair during DNA replication (FIGURE 11.12), resulting in transition mutations. The phosphate groups in the DNA backbone can also be methylated. The resulting neutral phosphotriester is easily cleaved by water to produce single strand breaks.

Many environmental agents must be modified by cell metabolism before they can alkylate DNA.

Many environmental agents become active alkylating agents only after they are metabolized in the cell. The first clue to the existence of these compounds came from a discovery made in 1775 by the English surgeon Percival Potts. Potts noticed that several male patients with cancer of the scrotum had worked as chimney sweeps as young boys. This observation led him to propose a causal relationship between exposure to soot in the chimney and cancer. Potts, thus, became the first to point out the relationship between a hazardous workplace environment and cancer. We now know that the hazardous environmental

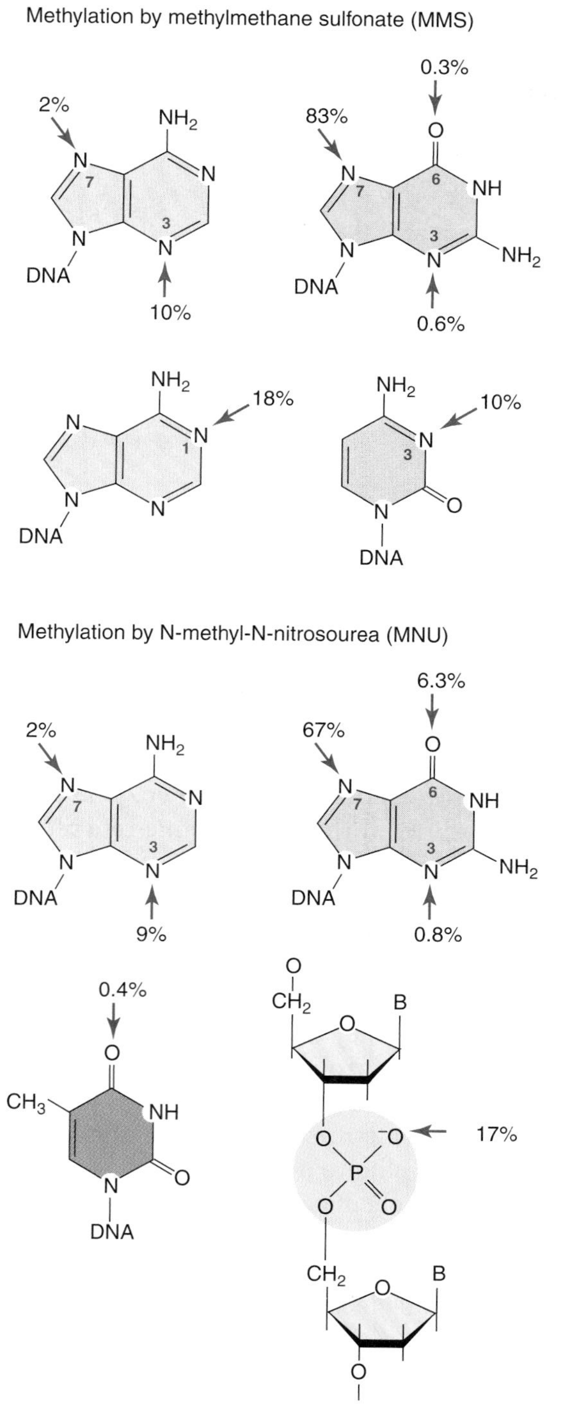

Methylation by methylmethane sulfonate (MMS)

2% NH2 N 7 N 3 N N DNA 10%

83% 0.3% O N 7 6 NH 3 N N NH2 DNA 0.6%

NH2 18% N 1 N N N DNA

NH2 10% N 3 N O DNA

Methylation by N-methyl-N-nitrosourea (MNU)

2% NH2 N 7 N 3 N N DNA 9%

67% 6.3% O N 7 6 NH 3 N N NH2 DNA 0.8%

0.4% O CH3 NH N O DNA

O CH2 B O O ⁻O 17% P O O CH2 B O O

FIGURE 11.11 **Methylation sites of the DNA bases and phosphate backbone with methylmethane sulfonate (MMS) and N-methyl-N-nitrosourea (MNU).** Red arrows indicate sites that are methylated in double strand DNA and blue arrows sites methylated in single strand DNA. The percent values adjacent to the arrows indicate the frequency of methylation at that site. (Adapted from Y. Mishina, E. M. Duguid, and C. He. *Chem. Rev.*, 106 [2006]: 215–232.)

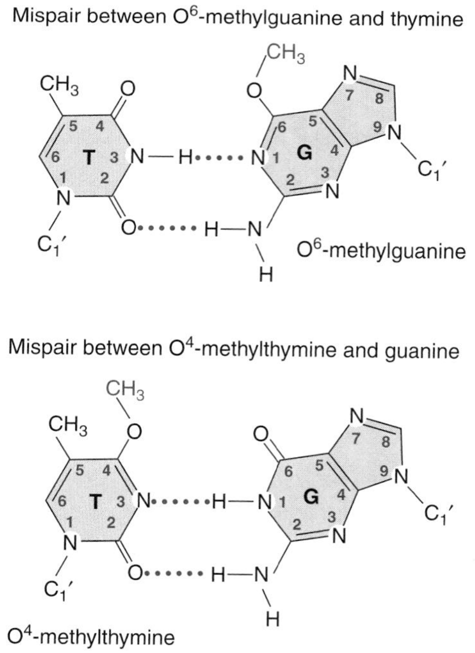

Mispair between O^6-methylguanine and thymine

CH_3 O CH_3 O N 7 8 5 4 6 5 9 N 6 T 3 N—H•••••N 1 G 4 C_1' 1 2 2 3 N N O••••••H—N C_1' H O^6-methylguanine

Mispair between O^4-methylthymine and guanine

CH_3 CH_3 O O N 7 8 5 4 6 5 9 N 6 T 3 N••••••H—N 1 G 4 C_1' 1 2 2 3 N N O••••••H—N C_1' H

O^4-methylthymine

FIGURE 11.12 **Mispairs that may form after methylation at position O-6 in guanine and O-4 in thymine.** If the methyl groups are not removed, replication of DNA containing the mispairs will result in transition mutations. (Adapted from A. Memisoglu and L. D. Samson, *Encyclopedia of Life Sciences* [DOI: 10.1038/npg.els.0000579]. Posted April 19, 2001.)

material in soot is a mixture of **polycyclic aromatic hydrocarbons (PAHs)** formed by the incomplete combustion of the burning wood or coal used as fuel. Similar types of polycyclic aromatic hydrocarbons are also present in tobacco smoke and charbroiled meats. The structures of a few of the more than 100 different members of the polycyclic aromatic hydrocarbon family are shown in FIGURE 11.13.

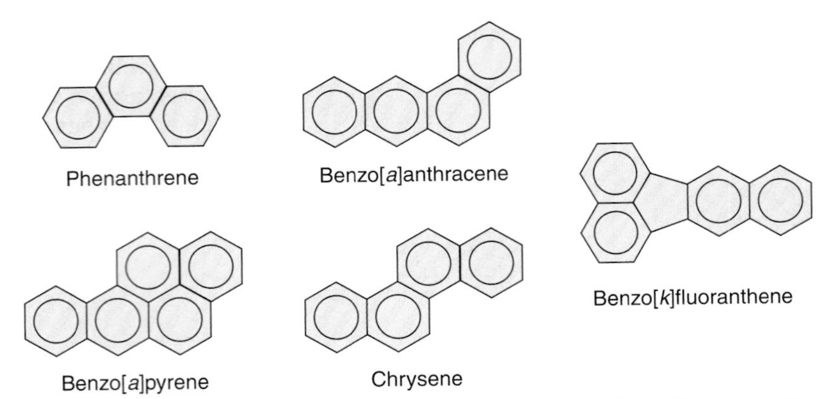

FIGURE 11.13 **A few representative members of the polycyclic aromatic hydrocarbon (PAH) family.**

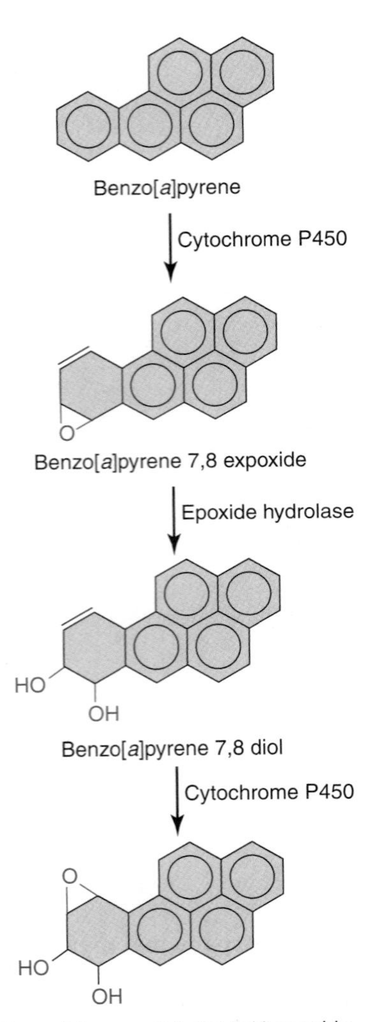

FIGURE 11.14 **Metabolic activation of benzo[a]pyrene, a polycyclic aromatic hydrocarbon.** Benzo[a]pyrene activation requires two kinds of enzymes, cytochrome P450 and epoxide hydrolase. The 9,10 epoxide product attacks DNA to form an adduct that is not shown in this figure.

The common structural feature in these and all other PAHs is two or more fused aromatic rings. These hydrophobic hydrocarbons are not able to damage DNA unless they are metabolically activated. The pathway for converting one of the polycyclic aromatic hydrocarbons, benzo[a]pyrene, into an active epoxide alkylating agent is shown in FIGURE 11.14. This pathway requires cytochrome P450 enzymes and epoxide hydrolase, which are located in the endoplasmic reticulum. Both enzymes normally make a valuable contribution to the cell's survival and well-being. Cytochrome P450 enzymes catalyze oxidation reactions that convert metabolic precursors to essential biomolecules and also help to detoxify harmful drugs by making them water soluble so that they can be excreted in the urine. Cytochrome P450 enzymes are not very specific, however, and so they are able to act on polycyclic aromatic hydrocarbons such as benzo[a]pyrene, adding oxygen atoms to form reactive three-membered epoxide rings. The epoxides then alkylate DNA, causing replication errors that result in mutations, which ultimately convert a normal cell into a cancer cell.

The **aflatoxins**, another class of chemical carcinogens that must be activated before damaging DNA, are produced by *Asperigillus flavus* and *Asperigillus parasiticus*, fungi that grow on peanuts and grains such as rice and corn. Animals feeding on contaminated peanuts or grains containing aflatoxins exhibit markedly increased rates of liver diseases including liver cancer. Aflatoxin B1, the most potent toxin produced by *A. flavus*, presents a particularly serious health threat in the United States. Cytochrome P450 converts aflatoxin B1 into an epoxide derivative that damages DNA (FIGURE 11.15). Under ideal conditions, a small tripeptide called **glutathione** will attack the epoxide ring, making the aflatoxin derivative soluble so that it can be excreted in the urine. Some of the reactive epoxide derivatives, however, escape attack by glutathione, and, therefore, are free to attack guanine rings in DNA. The flat aflatoxin ring system inserts between DNA bases (FIGURE 11.16), causing helical distortion that in turn leads to replication errors.

Aflatoxin B_1

Cytochrome P450

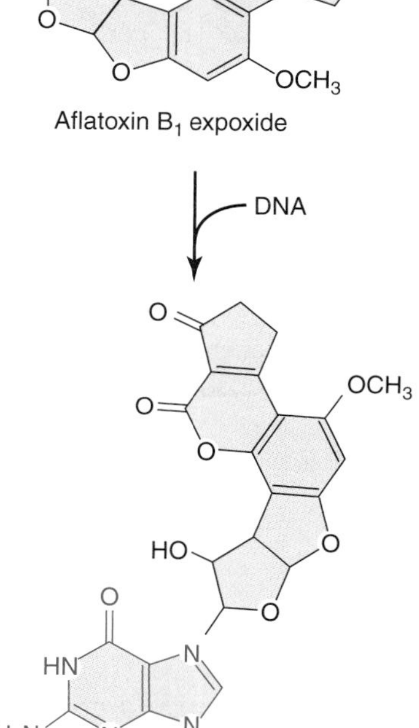

Aflatoxin attached to the N7 position of the guanine base, which is shown in red

FIGURE 11.15 **Metabolic activation of aflatoxin B1.** Aflatoxin B1, a toxic product of certain fungi that grow on grains, must be metabolically activated before it can add to DNA. Cytochrome P450 is required to introduce an epoxide group that activates aflatoxin B1, which then forms an adduct to the N-7 position in guanine.

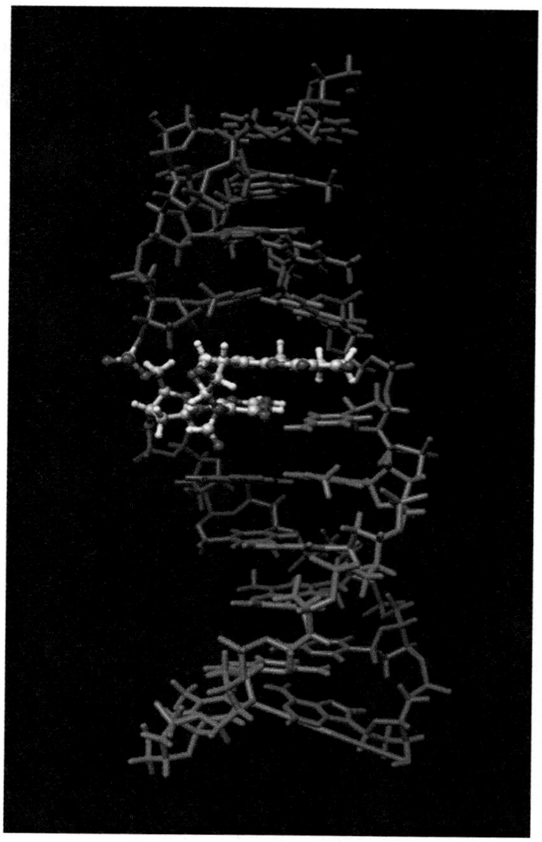

FIGURE 11.16 **Aflatoxin B1 derivative attached to DNA.** Activated aflatoxin attaches to a guanine in the DNA helix and inserts its bulky ring system between the DNA bases. This insertion distorts the helix and leads to replication errors when the DNA is replicated. The guanine imidazole ring has opened under the conditions used in this experiment. One DNA strand is colored blue and the other magenta. Atoms derived from aflatoxin and the guanine nucleotide to which it binds are in standard CPK colors. (Photo courtesy of Kyle L. Brown and Michael P. Stone, Vanderbilt University.)

11.5 Chemical Cross-Linking Agents

Chemical cross-linking agents block DNA strand separation.

Each of the alkylating agents described to this point forms a monoadduct. Many alkylating agents, however, have two reactive sites and therefore can form intrastrand (within the same strand) or interstrand (connecting opposite strands) cross-links. Interstrand cross-links are of special interest because they prevent strand separation and, if not corrected, are lethal. One of the simplest cross-linking agents, nitrogen mustard gas (bis[2-chloroethyl]methylamine; FIGURE 11.17a) was originally developed by the military as a chemical warfare agent that attacks the central nervous system. Later studies showed that nitrogen

FIGURE 11.17 **Nitrogen mustard gas, an agent that causes crosslink formation.** (a) Structure of bis(2-chloroethyl) methylamine (Nitrogen mustard gas). (b) Nitrogen mustard gas forms an interstrand crosslink through N7 positions of two guanine bases on opposite strands of a DNA double helix. (Part b adapted from E. C. Friedberg, et al. *DNA Repair and Mutagenesis, Second edition.* ASM Press, 2005.)

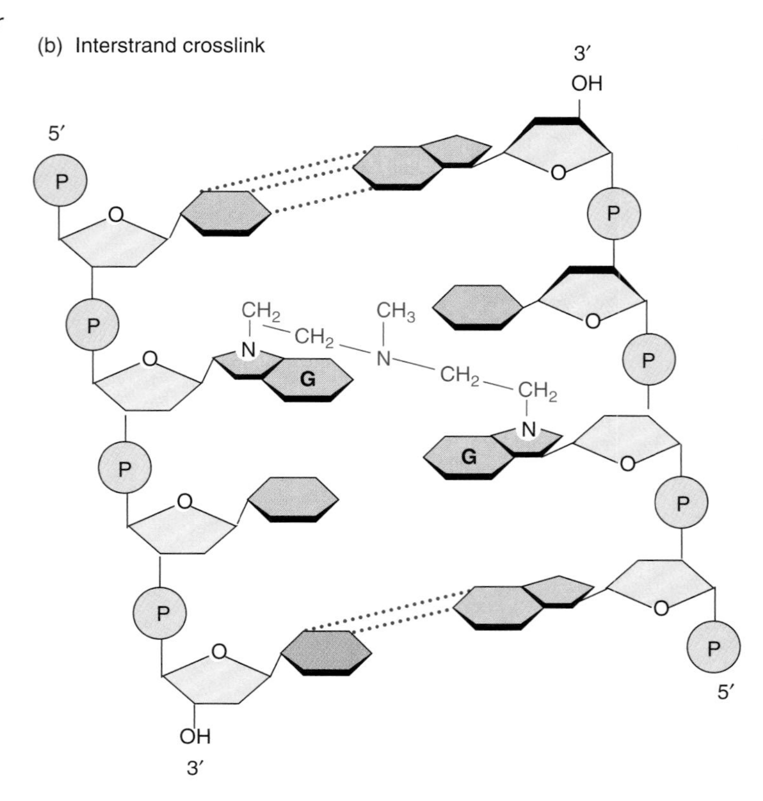

mustard gas also damages DNA by forming interstrand cross-links. It does so by attacking N-7 on two guanines, which are on opposite strands of DNA double helix (FIGURE 11.17b). Although a very toxic substance, nitrogen mustard gas has found clinical application as a chemotherapeutic agent for treating certain forms of lymphoma and leukemia.

Psoralen and related compounds can form monoadducts or cross-links.

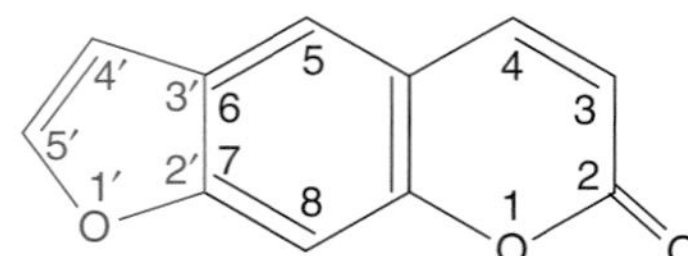

FIGURE 11.18 **Psoralen.** Psoralen is a planar tricyclic compound, which consists of a furan ring (red) fused to a coumarin ring system (black). Atoms in the ring system are numbered as indicated. Note that the positions of atoms in the furan ring are indicated by placing a prime (′) after the number.

Psoralen, a naturally occurring substance synthesized by some members of the carrot plant family, must be photoactivated before it can alkylate DNA. The planar psoralen molecule, which consists of a furan ring fused to a heterobicyclic ring system called **coumarin** (FIGURE 11.18), intercalates into the DNA molecule. Upon exposure to light with a wavelength of 400 to 450 nm, the furan ring in psoralen becomes activated and adds across the 5,6 double bond in a pyrimidine base (usually thymine) to form the 4′,5′, monoadduct (where

4′ and 5′, refer to positions on the psoralen furan ring) as shown in FIGURE 11.19a and b. The planar tricyclic psoralen derivative in the monoadduct is in position to combine with a second pyrimidine base on the opposite DNA strand, but it must be activated by light with a wavelength of 320 to 400 nm before it can do so. The resulting photoproduct (FIGURE 11.19c and d) contains a cross-link between pyrimidine bases on opposite strands of the DNA duplex. If not properly repaired, psoralen damage causes mutations and is lethal to cells. Psoralen plus UV-A is used to treat psoriasis, a chronic inflammatory and proliferative skin cell disease, as well as other skin disorders. This treatment must be administered with great care, however, because, although quite effective, it increases the risk of developing squamous

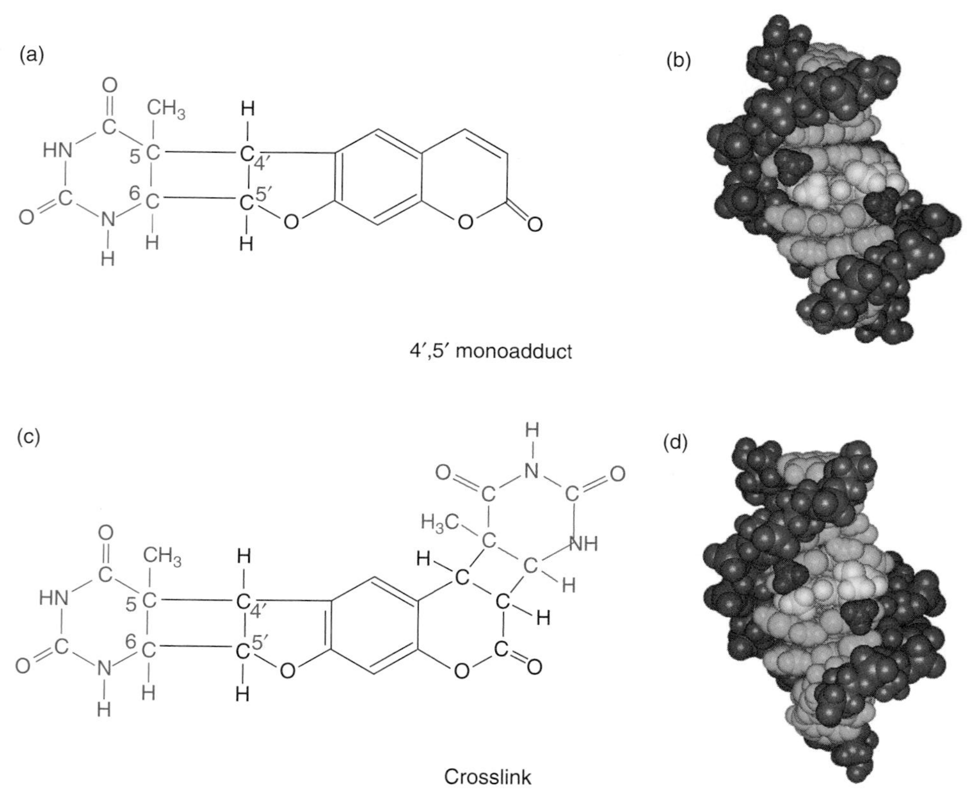

FIGURE 11.19 Psoralen photoproducts. (a) When activated by light with a wavelength of 400-450 nm, psoralen reacts with DNA to form a monoadduct. The main product is the 4′,5′ monoadduct photoproduct in which psoralen adds to a thymine base (shown in red). (b) Space-filling model of the structure of monoadduct viewed into the major groove showing the edge of the psoralen derivative between the T–A base pairs. The psoralen derivative is in yellow, and the thymine methyl groups are in red. (c) When activated by light with a wavelength of 320–400 nm, the psoralen group in the 4′,5′ monoadduct is activated and can add to a pyrimidine base (shown in blue) on the opposite DNA strand to form a crosslink. (d) Space-filling model of the structure of crosslink product viewed into the major groove showing the edge of the psoralen derivative between T–A base pairs. The psoralen derivative is in yellow and the thymine methyl groups are in red. (Parts a and c adapted from D. Bethea, *J. Dermatol. Sci.* 19 [1999]: 78–88. Parts b and d reproduced from H. P. Spielmann, et al., *Biochemistry* 34 [1995]: 12937–12953. Photo courtesy of H. Peter Spielmann, University of Kentucky.)

cell skin cancer. Related furocoumarins, which have been isolated from plants or synthesized in the laboratory, are currently under study in the hope of finding a drug with psoralen's beneficial effects but without its harmful side effects.

Cisplatin combines with DNA to form intra- and interstrand cross-links.

While examining the effects that electric currents have on *E. coli* growth in the mid-1960s, Barnett Rosenberg observed that treated bacteria grew to 300 times their normal length because the treatment blocked cell division but not other growth processes. Further studies showed that the growth effect was not caused by the electric field but by a compound that was formed in a reaction between the supposedly unreactive platinum electrodes and components in the bacterial suspension. The compound was later shown to be *cis*-diamminedichloroplatinum, which is better known as **cisplatin** (FIGURE 11.20a). Efforts that were made to see if cisplatin would also inhibit the division of other kinds of cells revealed that cisplatin blocks the division of tumor cells.

After cisplatin enters a cell by passive diffusion or active transport, it undergoes hydrolysis to produce $Pt(NH_3)_2ClH_2O^+$, a highly reactive, charged complex that coordinates to the N-7 atom of either a guanine or adenine base in DNA. Then the remaining chloride ligand is displaced by hydrolysis, allowing the platinum to coordinate to a second purine base on the same or opposite strand of the double stranded DNA (FIGURE 11.20b–d). Cisplatin's cytotoxic effects appear to be due to interstrand cross-linking, which blocks the replication and transcription machinery. Although a very effective chemotherapeutic agent for treating cancer of the bladder, ovaries, and testicles, cisplatin's clinical application is limited by significant side effects and drug resistance. A great deal of effort is currently being devoted to finding other metallo-organic cytotoxic agents that cause cross-linking in DNA but that do not have the same side effects as cisplatin.

(a) Cisplatin

(b) Covalent structure of a cisplatin cross-link

(c) Stereoview of a cisplatin intrastrand cross-link

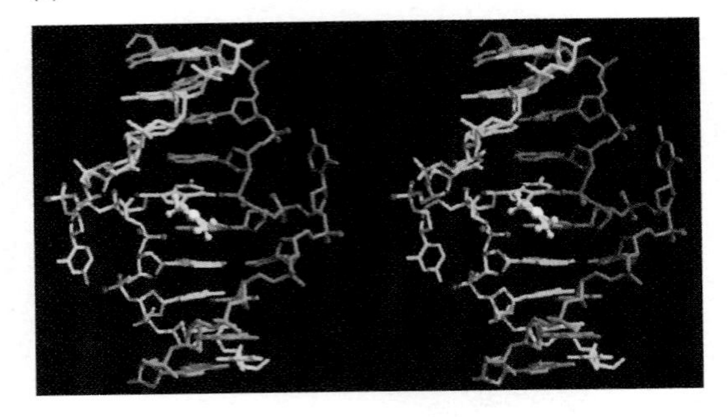

(d) Oligonucleotide used to obtain data for the model shown in (c)

```
    1  2  3  4  5  6  7  8  9  10
5′  C  A  T  A  G  C  T  A  T  G  3′

3′  G  T  A  T  C  G  A  T  A  C  5′
   10′ 9′ 8′ 7′ 6′ 5′ 4′ 3′ 2′ 1′
```

FIGURE 11.20 **Cisplatin.** (a) The chemical structure of cisplatin. (b) The chemical structure of the crosslink formed by interaction of cisplatin with two purine bases on the same or opposite strands of a DNA double helix. (c) A stereoview of cisplatin interstrand crosslink. The platinum ion is shown as a light blue sphere. (d) Oligonucleotide used to obtain data required to solve the structure shown in (c). (Part b adapted from H. Huang, et al., *Science* 270 [1995]: 1842–1845. Part c reproduced from H. Huang, et al., *Science* 270 [1995]: 1842–1845. Reprinted with permission from AAAS. Photo courtesy of Paul B. Hopkins, University of Washington.)

11.6 Mutagen and Carcinogen Detection

Mutagens can be detected based on their ability to restore mutant gene activity.

DNA damaging agents that cause mutations, called **mutagens**, also cause cancer in higher animals and so are said to be **carcinogens**. Carcinogenic agents work by generating mutations in two different groups of genes. The first group, the **proto-oncogenes**, code for products that promote normal cell growth and division. Mutations that alter proto-oncogenes convert these genes into oncogenes, which synthesize a larger quantity of the normal gene product or a more active mutant gene product. The second group, the **tumor suppressor genes**, codes for products that cause cell division and replication to slow down. Mutations that inactivate tumor suppressor genes remove a brake that helps to regulate cell division. A single mutation is usually not sufficient to convert a normal cell to a cancer cell. Many mutations, however, which can accumulate over a long time, may be sufficient to do so. Hence, it is advisable to keep exposure to exogenous DNA damaging agents to a minimum.

The pharmaceutical, food, textile, petroleum, and other industries synthesize tens of thousands of new chemicals each year in the hope of finding marketable products. Most of the chemicals do not pass the initial screening tests and so are not studied further. The few that do pass need to be subjected to a wide variety of additional tests. A major concern is that a chemical may serve the purpose for which it was intended but nevertheless not be marketable because it is a mutagen. Therefore, it is necessary to have a rapid and effective test available to detect chemical mutagens and thereby ensure that they do not come to market. Mutagen detection is important because DNA damage in a germ cell can result in birth defects and DNA damage in somatic cells can result in cancer.

Bruce Ames started to think about this safety issue in 1964 while reading the ingredients on a box of potato chips. He wondered how he could tell whether preservatives and other additives might be carcinogenic. His curiosity eventually led to his developing a test, which, after undergoing several modifications and improvements, is still used today to determine whether a chemical substance has a high probability of being a carcinogen.

In its most basic form, the **Ames test** involves determining whether the chemical to be tested causes a histidine-requiring mutant of the gram-negative bacteria *Salmonella typhimurium* that has a base substitution or frameshift mutation in a *his* gene to revert to the His^+ phenotype. The test is performed by adding approximately 10^8 bacteria, the chemical to be tested, and a trace quantity of histidine to molten top agar. Histidine is included to allow cells to divide a few times so that the mutagens that require DNA replication have a chance to work but is at too low a concentration to permit His^- bacteria to form visible colonies. After gentle agitation, top agar mixtures are poured onto glucose minimal agar plates. Colonies are counted after the plates have been incubated for 48 hours. A significant increase in the reversion frequency above that obtained in the absence of the tested chemical identifies the tested chemical as a mutagen. The reversion frequency depends on the concentration of the substance being tested and, for a known mutagen or carcinogen, correlates roughly with its known effectiveness.

In its original form, the Ames test failed to demonstrate the mutagenicity of several potent carcinogens. The reason for this failure is that some substances such as benzo[a]pyrene and aflatoxin B1 must be activated by cytochrome P450 enzymes before they can damage DNA. Because bacteria lack these enzymes, their DNA is not damaged by chemical agents that must be activated. Ames modified his test to correct for this problem by adding rat liver **microsomes** (a cell fraction that contains the endoplasmic reticulum) and required coenzymes to the top agar. The addition of the microsomal fraction, which is rich in cytochrome P450 enzymes, makes the Ames test sensitive to chemicals that must be activated by cytochrome P450 enzymes before they can damage DNA and cause mutagenesis.

Ames increased the sensitivity of the test by making three changes to the tester strains. First, he inactivated the *uvrB* gene, which codes for a protein required for nucleotide excision repair (see Chapter 12). Second, he introduced a mutation that makes the bacterial outer membrane more permeable to large molecules, increasing the likelihood that large chemical mutagens will be able to reach DNA inside the cell. Third, he introduced plasmids with genes that code for DNA polymerases that catalyze error prone DNA synthesis (see Chapter 12), which greatly increases the mutation rate. Although other tests also have been developed to detect mutagens and carcinogens, the Ames test remains the most widely used.

The Ames test has now been used with tens of thousands of substances and mixtures (such as industrial chemicals, food additives, pesticides, hair dyes, and cosmetics) and numerous unsuspected substances have been found to stimulate reversion in this test. A high frequency of reversion does not mean that the substance is definitely a carcinogen but only that it has a high probability of being so. As a result of these tests, many industries have reformulated their products. For example, the cosmetic industry has changed the formulation of many hair dyes and cosmetics to render them nonmutagenic. Ultimate proof of carcinogenicity is determined by testing for tumor formation in laboratory animals. The Ames test and several other microbiological tests are used to reduce the number of substances that have to be tested in animals because to date only a few percent of the substances known from animal experiments to be carcinogens failed to increase the reversion frequency in the Ames test. A basic assumption behind the Ames test and other tests that use bacteria as test organisms is that a chemical that causes mutations in bacteria will also do so in other organisms. This assumption appears to be well founded because the basic mechanisms of DNA damage are the same in all organisms.

Suggested Reading

General

Friedberg, E. C., Walker, G. C., Siede, W., et al. 2006. *DNA Repair and Mutagenesis*, 2nd ed. Washington, DC: ASM Press.

DNA Damage

Bethea, D., Fullmer, B., Syed, S., et al. 1999. Psoralen photobiology and photochemotherapy: 50 years of science and medicine. *J Dermatol Sci* 19:78–88.

Bordin, F. 1999. Photochemical and photobiological properties of furocoumarins and homologous drugs. *Int J Photoenergy* 1:1–6.

Cooke, M. S., Evans, M. D., Dizdaroglu, M., and Lunec, J. 2003. Oxidative DNA damage: mechanisms, mutation, and disease. *FASEB J* 17:1195–1214.

Doetsch, P. W. 2001. DNA damage. In: *The Encyclopedia of Life Sciences*. London: Nature; pp. 1–7.

Gasparro, F. P., Liao, B., Foley, P. J., et al. 1998. Psoralen photochemotherapy, clinical efficacy, and photomutagenicity: the role of molecular epidemiology in minimizing risks. *Environ Mol Mutagen* 31:105–112.

Goodsell, D. S. 2001. The molecular perspective: cytochrome P450. *Oncologist* 6:205–206.

Guengerich, F. P. 2003. Cytochrome P450 oxidations in the generation of reactive electrophiles: epoxidation and related reactions. *Arch Biochem Biophys* 409:59–71.

Jamieson, E. R., and Lippard, S. J. 1999. Structure, recognition, and processing of cisplatin DNA adducts. *Chem Rev* 99:2467–2498.

Kartalou, M., and Essigmann, J. M. 2001. Mechanisms of resistance to cisplatin. *Mutat Res* 478:23–43.

Marnett, L. J. 2000. Oxyradicals and DNA damage. *Carcinogenesis* 21:361–370.

McCann, J., Choi, E., Yamasaki, E, and Ames, B. N. 1975. Detection of carcinogens as mutagens in the *Salmonella*/microsome test: assay of 300 chemicals. *Proc Natl Acad Sci USA* 72:5135–5139.

Moore, B. S., Morris, L. P., and Doetsch, P. W. 2009. DNA damage. In: *Encyclopedia of Life Sciences*. pp. 1–10. Hoboken, NJ: John Wiley and Sons.

Park, H., Zhang, K., Ren, Y. C., et al. 2002. Crystal structure of a DNA decamer containing a cis–syn thymine dimer. *Proc Natl Acad Sci USA* 99:15965–15970.

Pierard, F., and Kirsch-De Mesmaeker, A. K. 2006. Bifunctional transition metal complexes as nucleic acid photoprobes and photoreagents. *Inorg Chem Commun* 9:111–126.

Rosenberg, B., Van Camp, L., and Krigas, T. 1965. Inhibition of cell division in *Escherichia coli* by electrolysis products from a platinum electrode. *Nature* 205:698–699.

Rosenberg, B., Van Camp, L., Grimley, E. B., and Thompson A. J., 1967. The inhibition of growth or cell division in *Escherichia coli* by different ionic species of platinum complexes. *J Biol Chem* 242:1347–1352.

Rosenberg, B., Van Camp, L., Trosko, J. E., and Mansour, V.H. 1969. Platinum compounds: a new class of potent antitumor agents. *Nature* 222:385–386.

Spielmann, H. P., Dwyer, T. J., Hearst, H. E., and Wemmer, D. E. 1995. Solution structures of psoralen monoadducted and cross-linked DNA oligomers by NMR spectroscopy and restrained molecular dynamics. *Biochemistry* 34:12937–12953.

Spratt, T. E., and Levy, D. E. 1997. Structure of the hydrogen bonding complex of O^6-methylguanine with cytosine and thymine during DNA replication. *Nucleic Acids Res* 25:3354–3361.

Tsutomu Shimada, T., and Fujii-Kuriyama, Y. 2004. Metabolic activation of polycyclic aromatic hydrocarbons to carcinogens by cytochromes P450 1A1 and 1B1. *Cancer Sci* 95:1–6.

Williams, J. H., Phillips, T. D., Jolly, P. E., et al. 2004. Human aflatoxicosis in developing countries: a review of toxicology, exposure, potential health consequences, and interventions. *Am J Clin Nutr* 80:1106–1122.

Zhang, C. X., and Lippard, S. J. 2003. New metal complexes as potential therapeutics. *Curr Opin Chem Biol* 7:481–489.

Detection of Mutagens

Ames, B. N., Durston, W. E., Yamasaki, E., and Lee, F. D. 1973. Carcinogens are mutagens: a simple test system combining liver homogenates for activation and bacteria for detection. *Proc Natl Acad Sci USA* 70:2281–2285.

Ames, B. N. 2003. An enthusiasm for metabolism. *J Biol Chem* 278:4369–4380.

Mortelmans, K., and Zeiger, E. 2000. The Ames *Salmonella*/microsome mutagenicity assay. *Mutat Res* 455:29–60.

Webrzyn, G., and Czyz, A. 2003. Detection of mutagenic pollution of natural environment using microbiological assays. *J Appl Microbiol* 95:1175–1181.

This book has a Web site, **http://biology.jbpub.com/book/molecular**, which contains a variety of resources, including links to in-depth information on topics covered in this chapter, and review material designed to assist in your study of molecular biology.

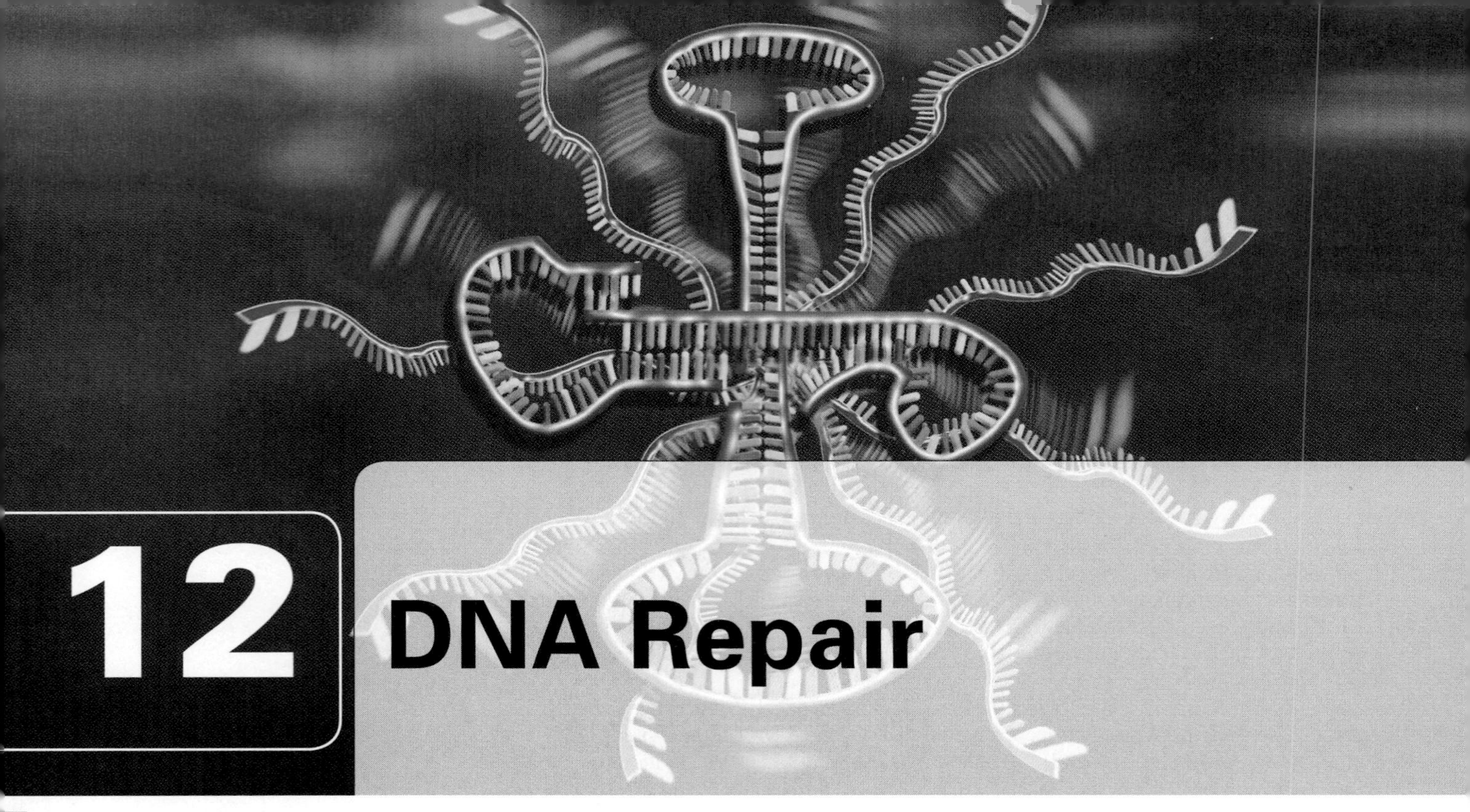

12 DNA Repair

OUTLINE OF TOPICS

12.1 Direct Reversal of Damage

Photolyase reverses damage caused by cyclobutane pyrimidine dimer formation.

O^6-Alkylguanine, O^4-alkylthymine, and phosphotriesters can be repaired by direct alkyl group removal by a suicide enzyme.

AlkB catalyzes the oxidative removal of methyl groups in 1-methyladenine and 3-methylcytosine.

12.2 Base Excision Repair

The base excision repair (BER) pathway removes and replaces damaged bases.

12.3 Nucleotide Excision Repair

Nucleotide excision repair removes bulky adducts from DNA by excising an oligonucleotide bearing the lesion and replacing it with new DNA.

UvrA, UvrB, and UvrC proteins are required for bacterial nucleotide excision repair.

Individuals with the autosomal recessive disease xeroderma pigmentosum have defects in enzymes that participate in the nucleotide excision repair pathway.

12.4 Mismatch Repair

The DNA mismatch repair system removes mismatches and short insertions or deletions that are present in DNA.

12.5 The SOS Response and Translesion DNA Synthesis

Error-prone DNA polymerases catalyze translesion DNA synthesis.

RecA and LexA regulate the *E. coli* SOS response.

The SOS signal induces the synthesis of DNA polymerases II, IV, and V.

Human cells have at least 14 different template-dependent DNA polymerases.

Suggested Reading

The previous chapter described many different types of DNA damage caused by physical and chemical agents. If a cell's damaged DNA is not repaired, then the cell's most likely fate is to produce daughter cells with harmful mutations or to die. Several different pathways have evolved that allow cells to repair damaged DNA. This chapter examines pathways that cells use to reverse DNA damage, excise and replace damaged elements, or tolerate the damage. The next chapter examines pathways that cells use to repair double-strand breaks. Although each repair pathway tends to correct a specific kind of DNA damage, there is considerable overlap in this specificity so that two or more repair pathways can correct the same type of damage.

12.1 Direct Reversal of Damage

Photolyase reverses damage caused by cyclobutane pyrimidine dimer formation.

Our examination of DNA repair begins with a look at enzymes that catalyze the direct reversal of DNA damage. The first clue to the existence of an enzyme that catalyzes the direct reversal of DNA damage was reported by Albert Kelner in 1949, four years before the Watson-Crick Model was proposed. Kelner's research was motivated by his desire to isolate antibiotic-resistant bacterial mutants. To do so, he first irradiated bacteria with UV light at doses that killed most of the bacteria and then tested the survivors to isolate the desired mutants. Even though Kelner was very careful to keep the experimental conditions the same, he noticed a great deal of variation in the number of survivors from one experiment to another. After considerable effort, Kelner finally discovered the explanation for this puzzling phenomenon. Cells that were placed in the dark after UV-irradiation had a very low survival rate, whereas those exposed to the light coming through the laboratory window had a high survival rate. Exposure to visible light thus reversed UV light's bactericidal effects.

Within several weeks of Kelner's discovery, Renato Dulbecco observed a similar phenomenon while studying UV-irradiated phage T2. Once again, the phenomenon manifested itself as an unexpected lack of reproducibility. Dulbecco prepared multiple medium agar plates, each containing the same numbers of UV-irradiated phages and sensitive bacteria, and placed the plates in a stack. Each of the stacked plates should have had about the same number of plaques. The plaque number decreased dramatically, however, going from top to bottom. Dulbecco explained this observation by proposing that the plates on the top of the pile were exposed to more light from the fluorescent bulb used to illuminate the laboratory than were the plates on the bottom. He tested this hypothesis by exposing some of the plates to fluorescent light while keeping others in the dark. As predicted, the numbers of plaques on plates exposed to the light were a great deal higher than those left in the dark. The bacteria were somehow using the visible light to repair the damaged phage DNA. The chemical basis for this light-dependent phenomenon, which Dulbecco called **photoreactivation**, remained to be elucidated.

Claud S. Rupert and coworkers devised an *in vitro* photoreactivation system in 1957, taking a major step toward determining the chemical mechanism of photoreactivation. Their straightforward approach was to isolate DNA from the gram-negative bacteria *Haemophilus influenzae*, irradiate the bacterial DNA with UV light to inactivate its transforming ability, and then demonstrate that a cell-free *Escherichia coli* extract, acting in the presence of visible light, restores transforming activity. Although this study had the potential to open the way for the purification and characterization of the photoreactivation enzyme, investigators still needed to establish the chemical nature of the DNA damage that was repaired.

This problem was solved over the next few years when investigators demonstrated that UV irradiation induces cyclobutane pyrimidine dimer formation. Further studies showed that the photoreactivation enzyme reverses UV-induced damage by using the energy provided by blue light (350–450 nm) to drive cyclobutane ring disruption in cyclobutane pyrimidine dimers (FIGURE 12.1). With the recognition that the photoreactivation enzyme catalyzes the disruption of carbon–carbon bonds, it was given the more descriptive name of **CPD photolyase.** Bacterial cells that lack CPD photolyase cannot repair cyclobutane pyrimidine dimer lesions by photoreactivation.

CPD photolyases, which are present in a wide variety of organisms including bacteria, the archaea, plants, and animals but not humans or other placental mammals, are monomeric proteins ranging in size from about 450 to 550 amino acid residues. All CPD photolyases have two domains, designated the N- and C-terminal domains. A light absorbing pigment, or **chromophore factor** as it is also known, binds

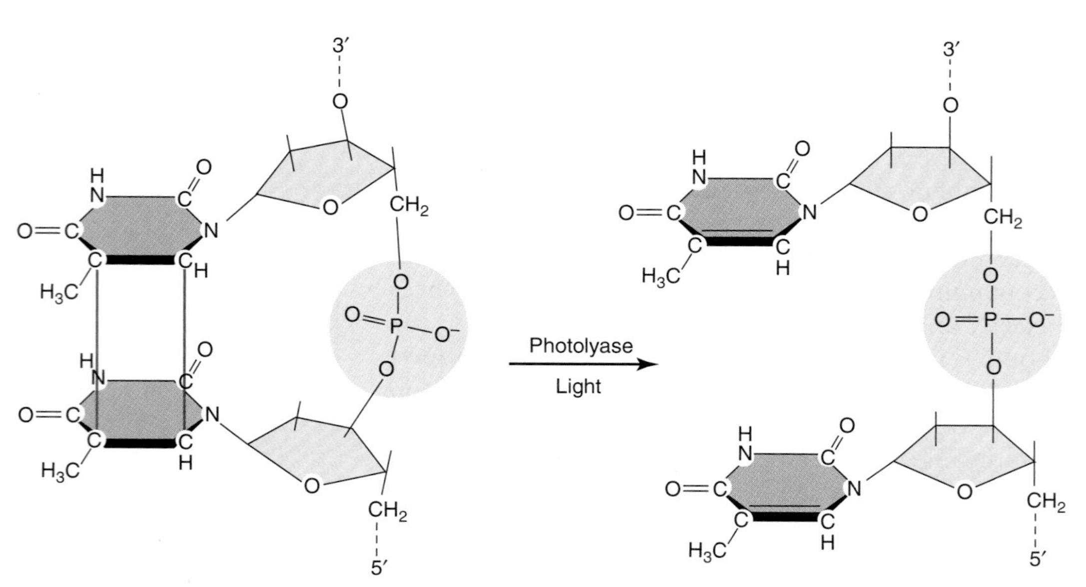

FIGURE 12.1 Photolyase catalyzes a light driven reaction that disrupts the cyclobutane ring in cyclobutane pyrimidine dimers, reversing the damaging effect of UV-irradiation.

to each domain through non-covalent bonds. The chromophore factor bound to the more highly conserved C-terminal domain, which is the deprotonated form of the reduced coenzyme flavin adenine dinucleotide ($FADH^-$; FIGURE 12.2a), must be present for the enzyme to have activity. The chromophore factor bound to the N-terminal domain, which is usually a tetrahydrofolate derivative (FIGURE 12.2b) in most bacteria and a riboflavin derivative (FIGURE 12.2c) in most animals and plants, acts as a photoantenna to capture light with wavelengths that would not otherwise be available to $FADH^-$. The photoantenna allows CPD photolyase to use available light energy much more efficiently than would otherwise be possible but is not required for enzyme activity. The crystal structure for the *E. coli* CPD photolyase (FIGURE 12.3) shows the relationship between the N- and C-terminal domains and the spatial arrangement of the two chromophores.

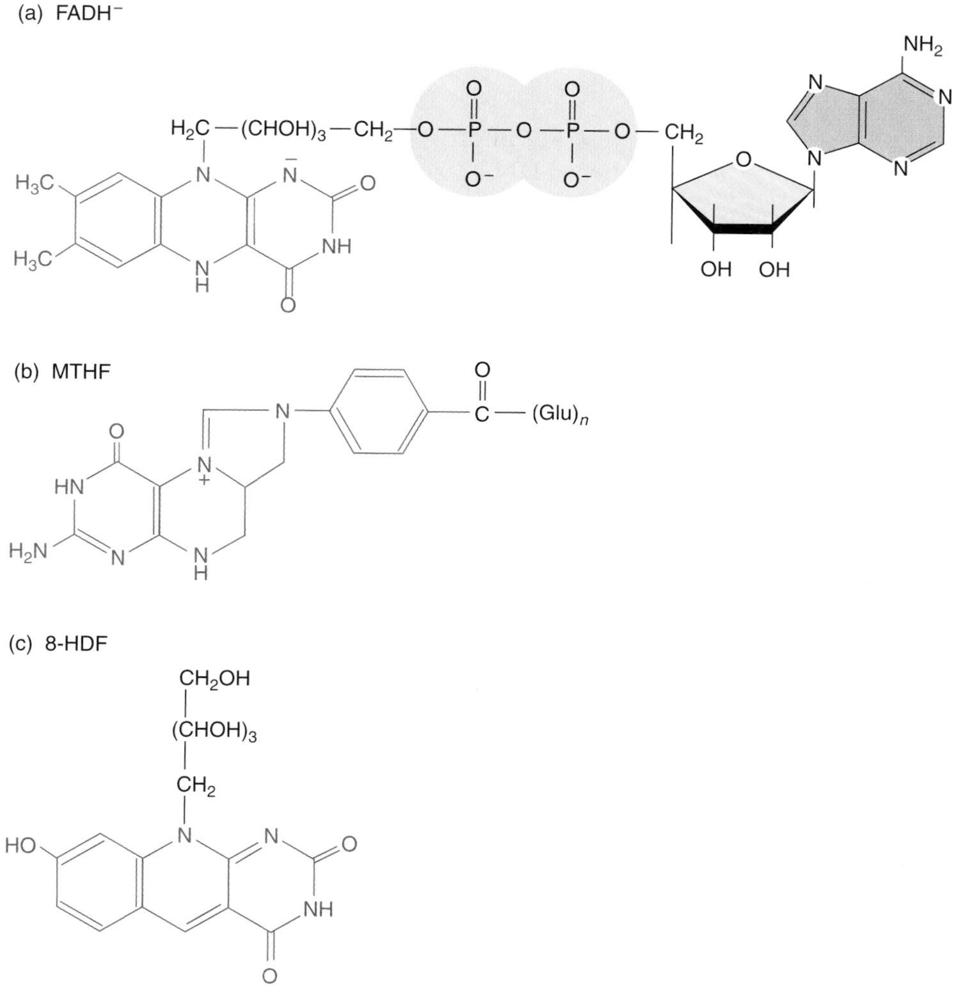

FIGURE 12.2 **Structures of the photolyase chromophores.** (a) $FADH^-$, the deprotonated form of reduced flavin adenine dinucleotide, must be bound to the C-terminal domain for catalytic activity. (b) MTHF, methenyltetrahydrofolate binds to the N-terminal domain in *E. coli* and yeast photolyases and acts as a photoantenna. Glu indicates glutamate residues. (c) 8-HDF, the riboflavin derivative 8-hydroxy-7,8-didemethyl-5-deazariboflavin, binds to the N-terminal domain in the cyanobacteria *Anacystis nidulans* and the bacteria *Streptomyces griseus* and acts as a photoantenna. Ring structures that absorb blue light are shown in blue. (Adapted from A. Sancar, *Chem. Rev.* 103 [2003]: 2203–2238.)

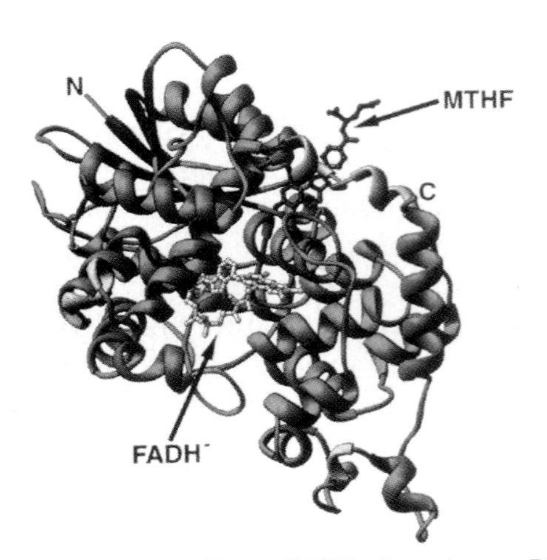

FIGURE 12.3 Crystal structure of *E. coli* CPD photolyase. Ribbon diagram representation showing MTHF (methenyltetrahydrofolate) bound to the N-terminal domain and $FADH^-$ bound to the C-terminal domain. (Reproduced from A. Sancar, *Chem. Rev.* 103 [2003]: 2203–2238. Photo courtesy of Aziz Sancar, University of North Carolina at Chapel Hill.)

CPD photolyase can bind to DNA in the dark, recognizing the altered DNA structure caused by a CPD rather than a specific nucleotide sequence. Binding is about 10^5 times tighter when a DNA segment contains a CPD than when it does not. Half of the binding energy appears to come from interactions between the enzyme and the DNA backbone and the other half from interactions between the $FADH^-$ at the active site and the CPD. The light-harvesting antenna pigment does not influence binding. Once the enzyme•DNA complex is formed, the CPD is flipped out of the DNA double helix and into the enzyme's active site. The flipped out pyrimidine bases are clearly visible in the crystal structure of the cyanobacteria *Anacystis nidulans* photolyase•DNA complex shown in FIGURE 12.4. After the CPD flips into the active site, the energy of an absorbed photon is transferred from the light-harvesting antenna pigment to the $FADH^-$, which then transfers an electron to the CPD to induce cyclobutane ring cleavage (FIGURE 12.5). The catalytic cycle is completed by back electron transfer from the repaired thymine to the FADH cofactor.

UV irradiation also induces the formation of a second type of pyrimidine dimer, the (6-4) photoproduct. Although the name of this lesion derives from the chemical bond linking carbon-6 of one pyrimidine ring to carbon-4 of an adjacent pyrimidine ring on the same strand, additional chemical changes also take place during dimer formation. Taking note of the additional chemical changes, investigators thought that it would be quite unlikely that a single enzyme would be able to catalyze direct reversal of a (6-4) photoproduct lesion.

It therefore came as quite a surprise when Takeshi Todo, Taisei Nomura, and coworkers reported in 1993 that *Drosophila melanogaster* has a photolyase that reverses (6-4) photoproduct lesions. This photolyase, which was designated the **(6-4) photolyase** to distinguish it

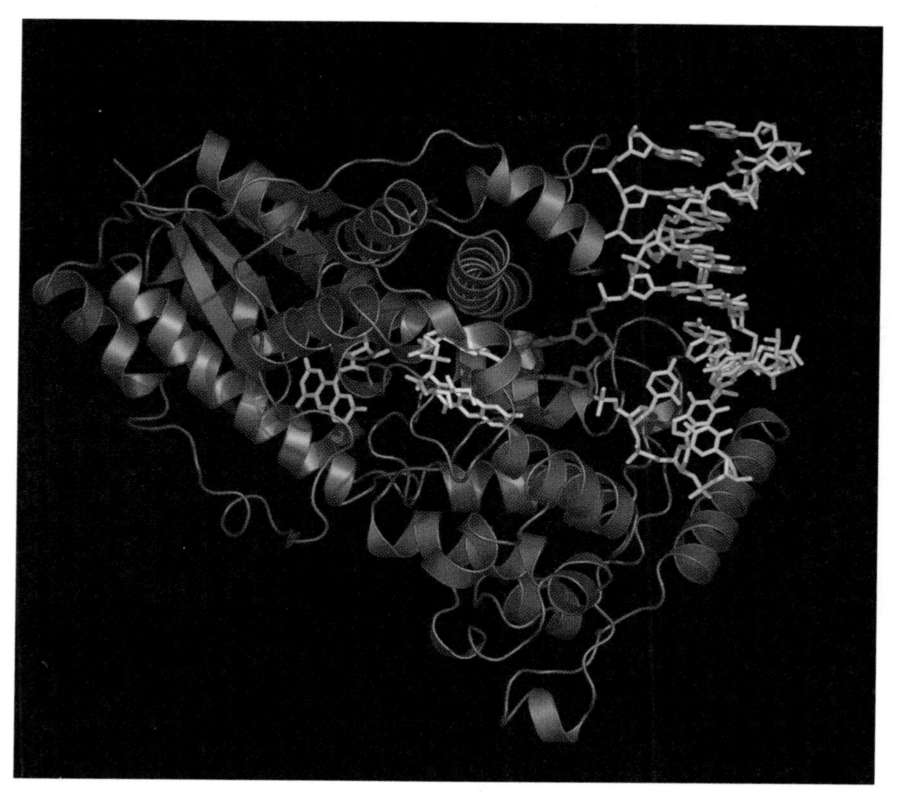

FIGURE 12.4 **Cyclobutane pyrimidine dimer flip out.** A crystal structure for the cyanobacteria *Anacystis nidulans* photolyase provides evidence that the T<>T dimer (red) is flipped out of the DNA helix (green) and into the active site of photolyase (gray ribbon), where it can react with FADH⁻, the catalytic flavin cofactor (yellow). The light-harvesting antenna pigment is shown on the left side in green. Remarkably, crystalline photolyase can use the x-rays generated by the synchrotron to drive the photolyase catalyzed reaction, converting the cyclobutane pyrimidine dimer into the two thymine residues shown here. The thymine residues do not flip back into DNA because they are frozen in position under the very low temperature conditions used in this experiment. (Reproduced from A. Yarnell, *Chem. Eng. News* 83 [2005]:85. Photo courtesy of Johannes Gierlich, Ludwig-Maximilians-University.)

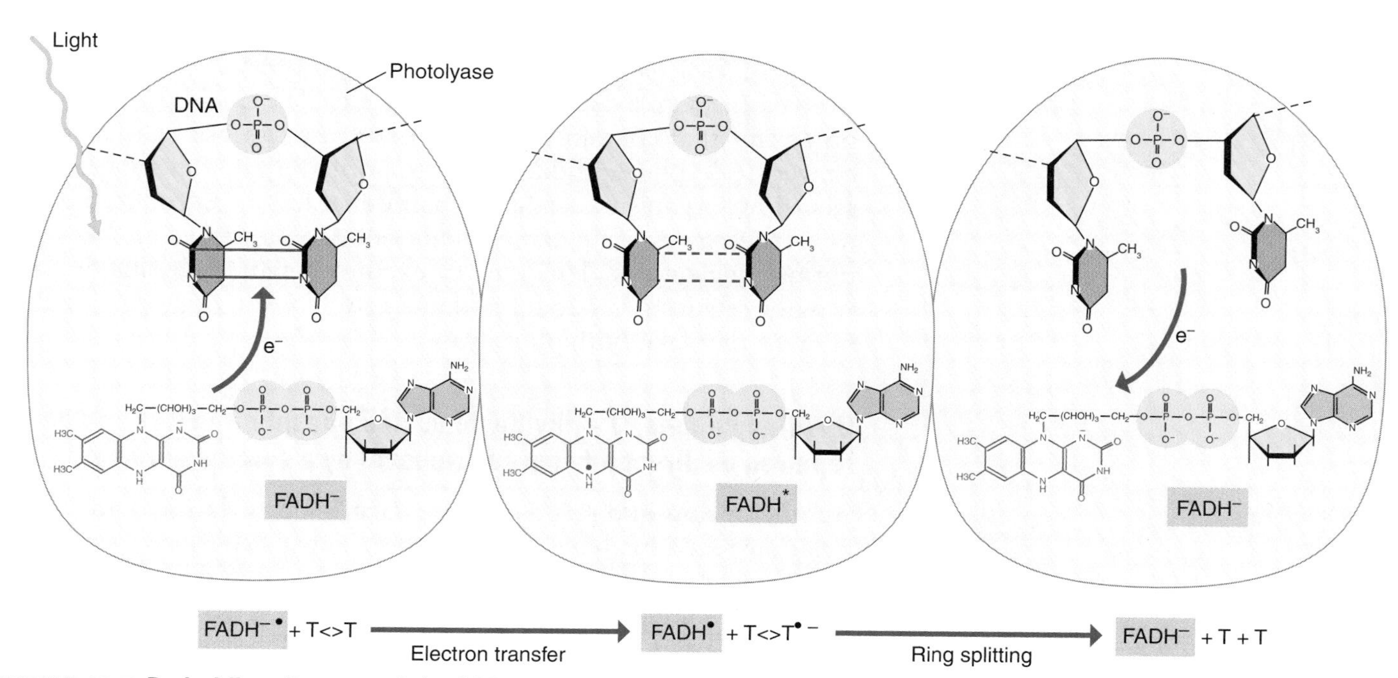

FIGURE 12.5 **Pyrimidine dimer repair by CPD photolyase.** A schematic representation showing the CPD photolyase catalyzed electron-transfer radical mechanism for pyrimidine dimer repair. The key catalytic steps, electron transfer and ring splitting, are shown at the bottom. (Adapted from Y. Kao, et al., *Proc. Natl. Acad. Sci. USA* 102 [2005]: 16128–16132.)

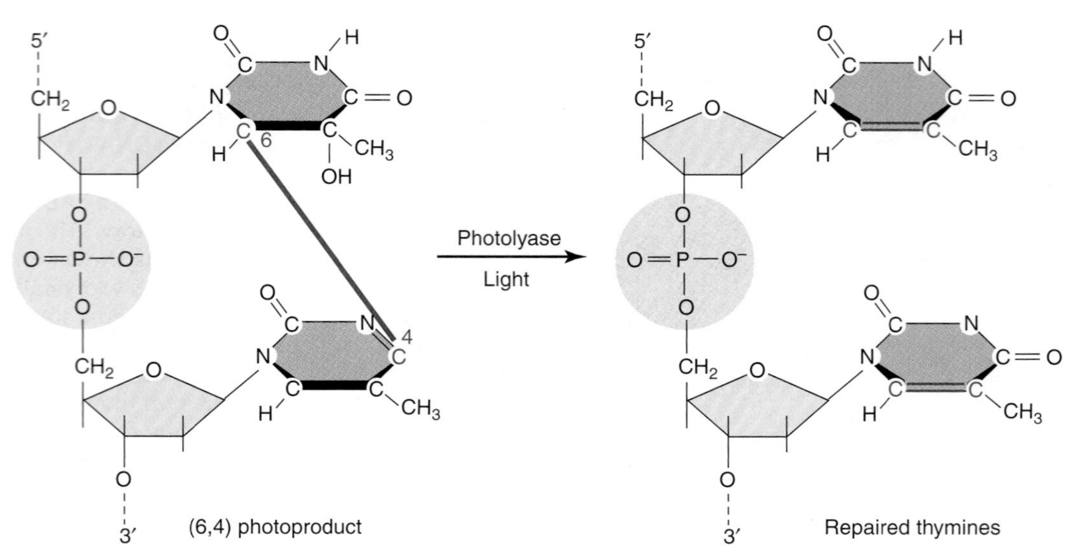

FIGURE 12.6 Reaction catalyzed by the (6-4) photolyase. In addition to cleaving the bond shown in red, the (6-4) photolyase also has to reverse other changes to the DNA that were caused by the UV damage.

from the CPD photolyase, catalyzes the reaction shown in FIGURE 12.6. Although widely distributed in plants and animals, the (6-4) photolyase has not been detected in bacteria or mammals that have been tested to date. Considerably less is known about the mechanism of action of the (6-4) photolyase than is known about that of the CPD photolyase; however, the two enzymes seem to work in a similar way. The (6-4) photolyase binds to damaged DNA, causing the (6-4) photoproduct to flip out of the DNA and into the active site of the enzyme, where it undergoes a rearrangement to produce a product that receives an electron from an excited $FADH^-$ molecule. The final outcome is that organisms that contain the (6-4) photolyase can use light energy to convert (6-4) photoproducts back to normal pyrimidine rings. As we see below, organisms can also repair dimer lesions introduced by UV irradiation by excising damaged nucleotides and replacing them with normal nucleotides. This type of excision repair is the major pathway for repairing UV-induced damage to DNA in organisms such as humans that lack both photolyases.

O^6-Alkylguanine, O^4-alkylthymine, and phosphotriesters can be repaired by direct alkyl group removal by a suicide enzyme.

Another means of direct damage reversal is dealkylation. Direct dealkylation reactions have probably been most extensively studied in *E. coli.* Three different proteins can catalyze the direct removal of alkyl groups attached to oxygen atoms in DNA. Two of these proteins, O^6-alkylguanine DNA alkyltransferases I and II, selectively remove the alkyl groups by transferring them to their own cysteine residues. As its name indicates, **O^6-alkylguanine DNA alkyltransferase I** can remove methyl and other alkyl groups attached to O-6 in guanine. The enzyme, however, also can remove alkyl groups attached to O-4 in thymine and to phosphotriesters. O^6-alkylguanine DNA alkyltransferase I is a

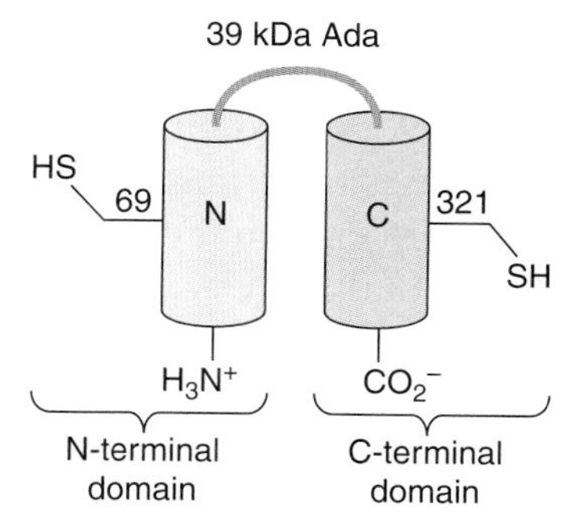

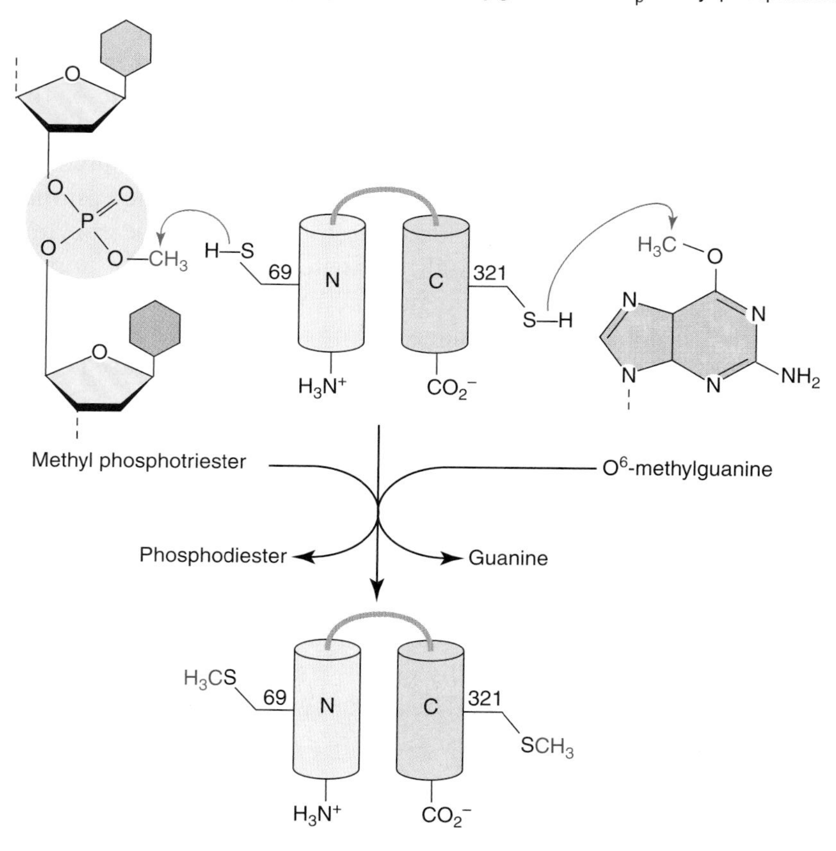

FIGURE 12.7 **O^6-alkylguanine and phosphotriester lesions repaired by O^6-alkylguanine DNA alkyltransferase.** (a) O^6-alkylguanine DNA alkyltransferase I contains N- and C-terminal domains that are connected by a flexible linker. (b) The N-terminal domain transfers an alkyl group from an S_P phosphotriester to one of its own cysteine residues, Cys-69. The C-terminal domain transfers an alkyl group from either O^6-alkylguanine or O^4-alkylthymine (not shown) to one of its own cysteine residues, Cys-321. (Adapted from L. C. Myers, et al., *Biochemistry* 31 [1992]: 4541–4547.)

monomer that has a flexible linker connecting its N- and C-terminal domains (FIGURE 12.7a). Each domain has an active site that performs a specific function. The N-terminal domain transfers an alkyl group from an S_P phosphotriester to one of its own cysteine residues, Cys-69 (FIGURE 12.7b). The S_P notation indicates that the enzyme is specific for an alkyl group attached to only one of the two possible nonbridging oxygen atoms. The C-terminal domain transfers an alkyl group from either O^6-alkylguanine (Figure 12.7b) or O^4-alkylthymine (not shown) to one of its own cysteine residues, Cys-321. Once alkylated, the protein cannot be regenerated and therefore behaves more like an alkyl transfer agent than a classical enzyme. Proteins such as

O^6-alkylguanine DNA alkyltransferase I, which lose activity after acting only one time, are called **suicide enzymes.**

O^6-Alkylguanine DNA alkyltransferase I has one additional remarkable function. After methylation at Cys-69, the protein is converted to a transcriptional activator that stimulates the transcription of the gene that codes for it as well as a few other genes that code for proteins that repair DNA damage. This additional activity permits the bacteria to adapt to environments in which they are exposed to alkylating agents by synthesizing more copies of O^6-alkylguanine DNA alkyltransferase I, which then repair the damage. O^6-alkylguanine DNA alkyltransferase I that is synthesized after all alkylation damage has been repaired will remain unmethylated and in this form blocks transcription of the same genes that were activated by the methylated protein.

O^6-Alkylguanine DNA alkyltransferase II, the second *E. coli* alkyl transfer protein, has properties that are very similar to those of the C-terminal domain of O^6-alkylguanine DNA alkyltransferase I. It also transfers a single alkyl group from the O-6 position in guanine or the O-4 position in thymine to one of its own cysteine residues and then loses activity and so is also classified as a suicide enzyme. However, O^6-alkylguanine DNA alkyltransferase II does not appear to be subject to genetic regulation. Its function appears to be to protect bacteria from alkylation damage during the time that it takes for the gene that codes for O^6-alkylguanine DNA alkyltransferase I to be fully expressed.

The eukaryotic protein that removes alkyl groups from O^6-methylguanine and O^4-methylthymine is similar to the bacterial O^6-alkylguanine DNA alkyltransferase II. Human **alkylguanine DNA alkyltransferase** is of considerable clinical interest because many chemotherapeutic agents used to destroy cancer cells are alkylating agents. When alkyltransferase activity in tumor cells is very high, these agents are not effective. On the other hand, when alkyltransferase activity in the surrounding healthy cells is too low, the chemotherapeutic agents will kill these cells. The human protein is also important because it helps to protect against carcinogenic alkylating agents.

AlkB catalyzes the oxidative removal of methyl groups in 1-methyladenine and 3-methylcytosine.

An entirely different type of activity that directly repairs alkylation damage to DNA was discovered in *E. coli* in 2002. In this case, the protein called AlkB catalyzes the direct conversion of 1-methyladenine, 1-methylguanine, 3-methylcytosine, and 3-methylthymine to adenine, guanine, cytosine, and thymine respectively. The AlkB catalyzed reaction requires Fe^{2+}, molecular oxygen, and α-ketoglutarate (FIGURE 12.8). A similar type of enzyme has been found in other organisms including humans and other mammals.

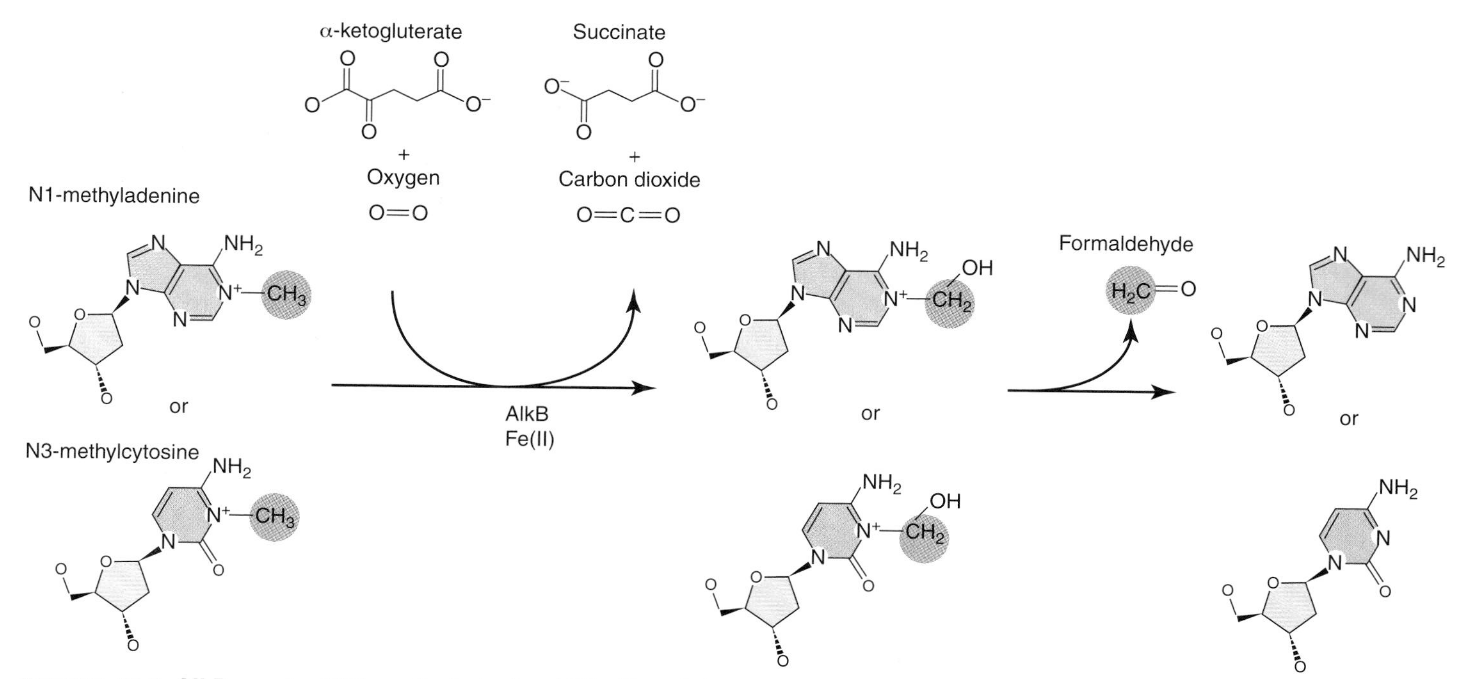

FIGURE 12.8 **AlkB repair of N1-methyladenine (1MeA) and N3-methylcytosine (3MeC) lesions.** AlkB repairs N1-methyladenine and N3-methylcytosine by catalyzing their oxidative demethylation. The cytotoxic methyl group in each lesion, shown in pink, is first oxidized to a hydroxymethyl group, which is removed in the form of formaldehyde to generate adenine (top) or cytosine (bottom). The reaction, which requires α-ketoglutarate, O_2, and $Fe2^+$, also produces succinate and carbon dioxide. (Modified from *Trends Biochem. Sci.*, vol. 28, T. J. Begley and L. D. Samson, AlkB mystery solved: oxidative demethylation . . ., pp. 2–5, copyright 2003, with permission from Elsevier [http://www.sciencedirect.com/science/journal/09680004].)

12.2 Base Excision Repair

The base excision repair (BER) pathway removes and replaces damaged or inappropriate bases.

Many types of DNA damage cannot be repaired by a single enzyme that catalyzes direct damage reversal. Instead, repair requires participation of several different enzymes, each performing a specific task in a multistep pathway. Damage to DNA bases caused by deamination, oxidation, and alkylation is mainly repaired by one such multistep pathway, **base excision repair (BER)**, which is essentially the same in all organisms. Enzymes in the base excision repair pathway also participate in single-strand break repair. Base excision repair derives its name from the first step in the pathway, N-glycosyl bond cleavage, which excises the damaged or inappropriate base from the DNA to form an abasic site. Because no single enzyme can distinguish the four bases normally present in DNA from the wide variety of altered bases generated by alkylation, deamination, and oxidation, cells must use many different enzymes to perform this function. Some N-glycosylases are monofunctional enzymes with only the ability to excise a damaged base (FIGURE 12.9a). Others also have an AP lyase activity that cleaves the bond between the sugar and the phosphate 3′ to the damaged site (FIGURE 12.9b). An enzyme with just N-glycosylase activity is designated a **DNA glycosylase**, whereas one that also has AP lyase activity is designated a **DNA glycosylase/lyase**. Both types of enzymes detect the damaged base, flip it out of the DNA helix into an active site pocket, and then cleave the N-glycosyl bond. Although some glycosylases excise a specific base, most have a somewhat broader specificity. The *E. coli* enzyme uracil N-glycosylase (Ung) (see Chapter 9) is specific for uracil, whereas another *E. coli* N-glycosylase, 3-methyladenine DNA glycosylase (also called AlkA), acts on 3- or 7-methylpurines, 3- or 7-ethylpurine, ethenoadenine, and O^2-methyl pyrimidines. A

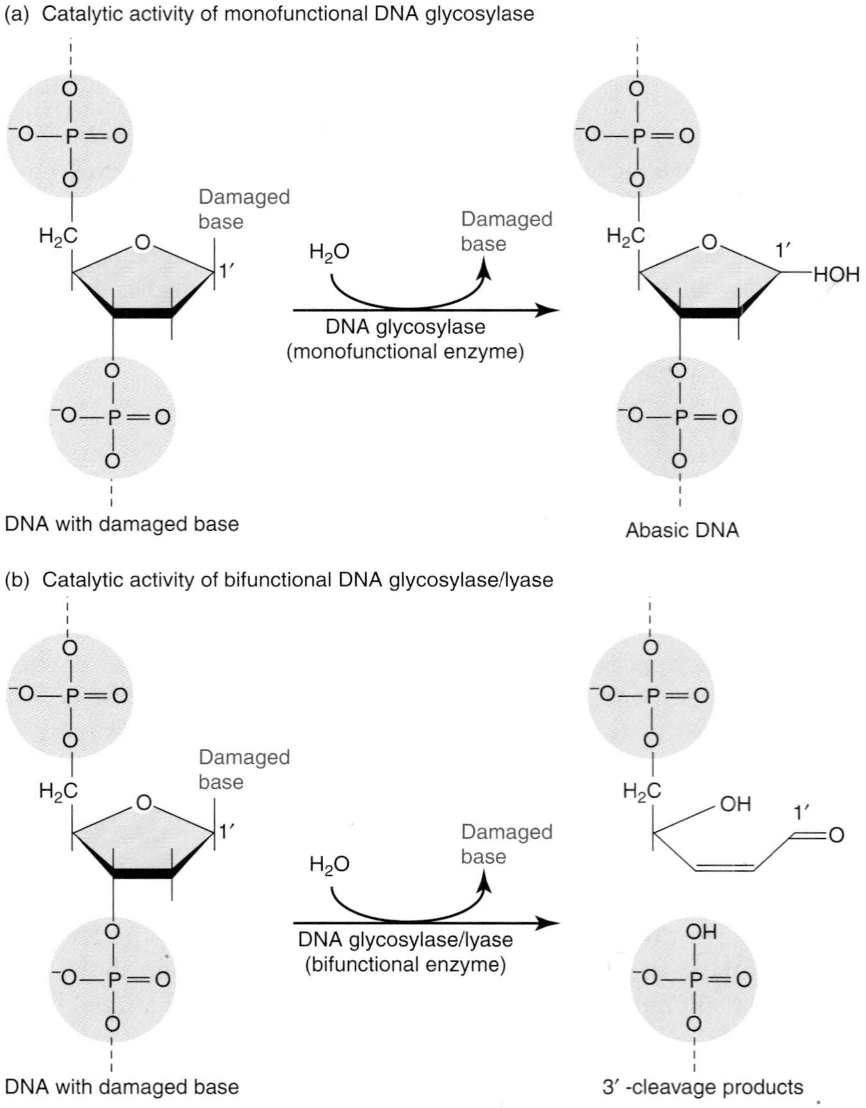

(a) Catalytic activity of monofunctional DNA glycosylase

(b) Catalytic activity of bifunctional DNA glycosylase/lyase

FIGURE 12.9 **Monofunctional and bifunctional DNA glycosylases.** (a) Monofunctional DNA glycosylases excise a damaged base. (b) Bifunctional DNA glycosylases also have an AP lyase activity that cleaves the bond between the sugar and the phosphate group 3′ to the AP site.

null mutation in a gene that codes for a DNA glycosylase or DNA glycosylase/lyase is not lethal, probably reflecting overlapping functional abilities among the glycosylases. Base excision repair follows a slightly different path when it starts with a DNA glycosylase than when it starts with a DNA glycosylase/lyase. Although the two pathways have common intermediates, it is simpler to treat them separately.

FIGURE 12.10 shows the base excision repair pathway that starts with a monofunctional DNA glycosylase. For simplicity, only eukaryotic

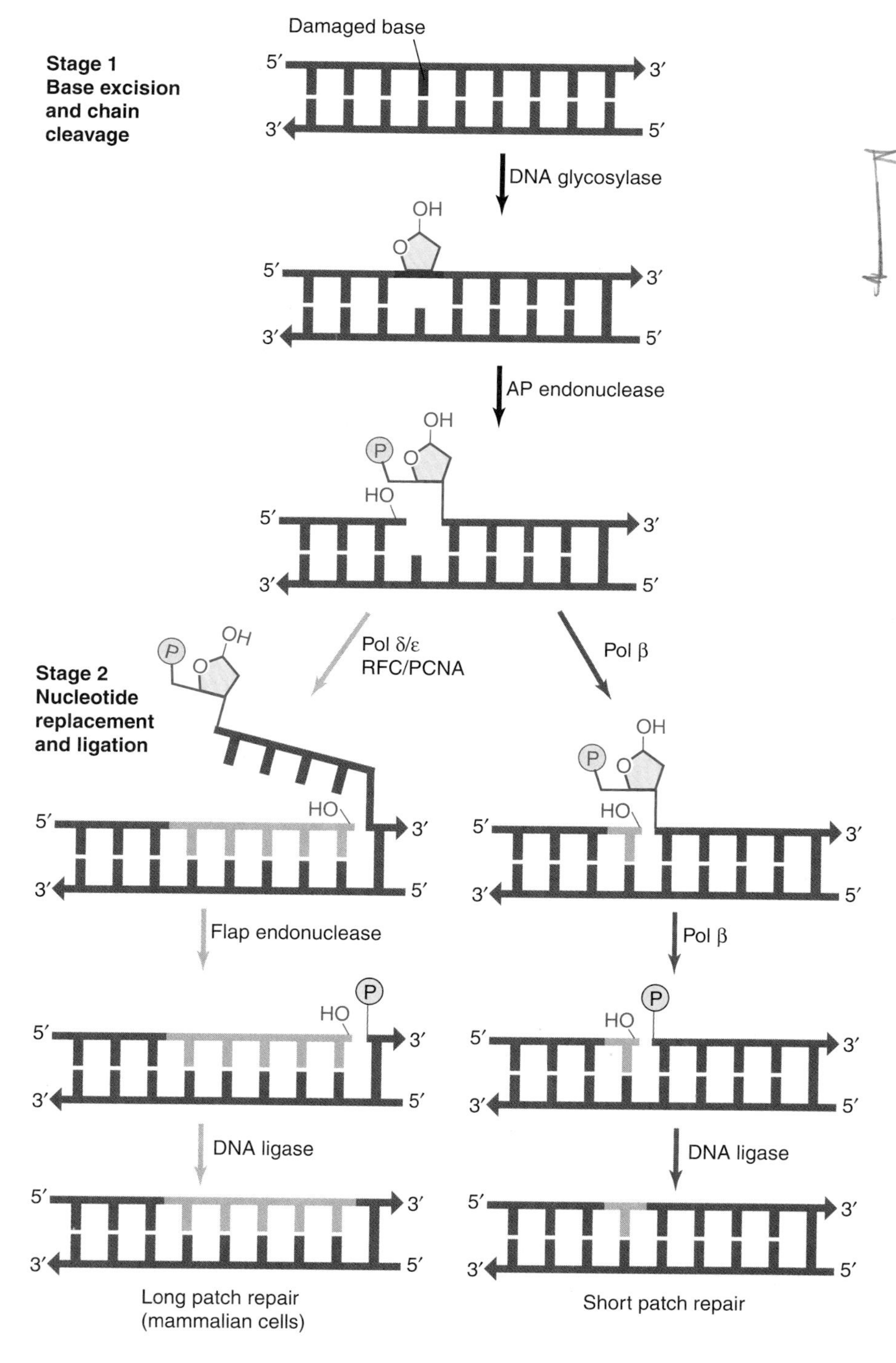

FIGURE 12.10 **Base excision repair in eukaryotes starting with a monofunctional DNA glycosylase.** Stage 1 (black arrows): A DNA glycosylase excises a damaged or inappropriate base to produce an abasic or AP site and then an AP endonuclease cleaves the DNA backbone 5′ of the AP site to generate a 5′-deoxyribose phosphate end and a free 3′-OH end. Stage 2 (short patch repair, dark blue arrows): DNA polymerase β (Pol β) adds a nucleotide to the 3′-OH end and removes the 5′-deoxyribose phosphate group. Then DNA ligase joins the ends to form an intact strand. Stage 2 (long patch repair, light blue arrows): DNA polymerase δ or ε acting with the clamp loader (RFC) and sliding clamp (PCNA) adds 2 to 8 nucleotides to the 3′-hydroxyl group while at the same time displacing the 5′-deoxyribosephosphate group to generate a flap. Flap endonuclease removes the flap and DNA ligase connects the ends. (Adapted from O. D. Schärer, *Angew. Chem. Int. Ed. Engl.* 34 [2003]: 2946–2974.)

enzymes are indicated. The DNA glycosylase catalyzes base excision to produces an **AP** (apurinic and apyrimidinic) site. The next enzyme in the pathway, **AP endonuclease,** hydrolyzes the phosphodiester bond 5′ to the AP site to generate a nick. FIGURE 12.11 shows this reaction in greater detail. *E. coli* has two well-characterized AP endonucleases, exonuclease III (Xth), which despite its name accounts for most of the bacterial AP endonuclease activity, and endonuclease IV (Nfo). Both are multifunctional enzymes that have 3′-phosphatase and 3′-repair phosphodiesterase activities. The former activity removes phosphate groups from the 3′-end of a DNA strand and the latter activity removes the 3′-unsaturated aldehydic group produced by DNA glycosylase/

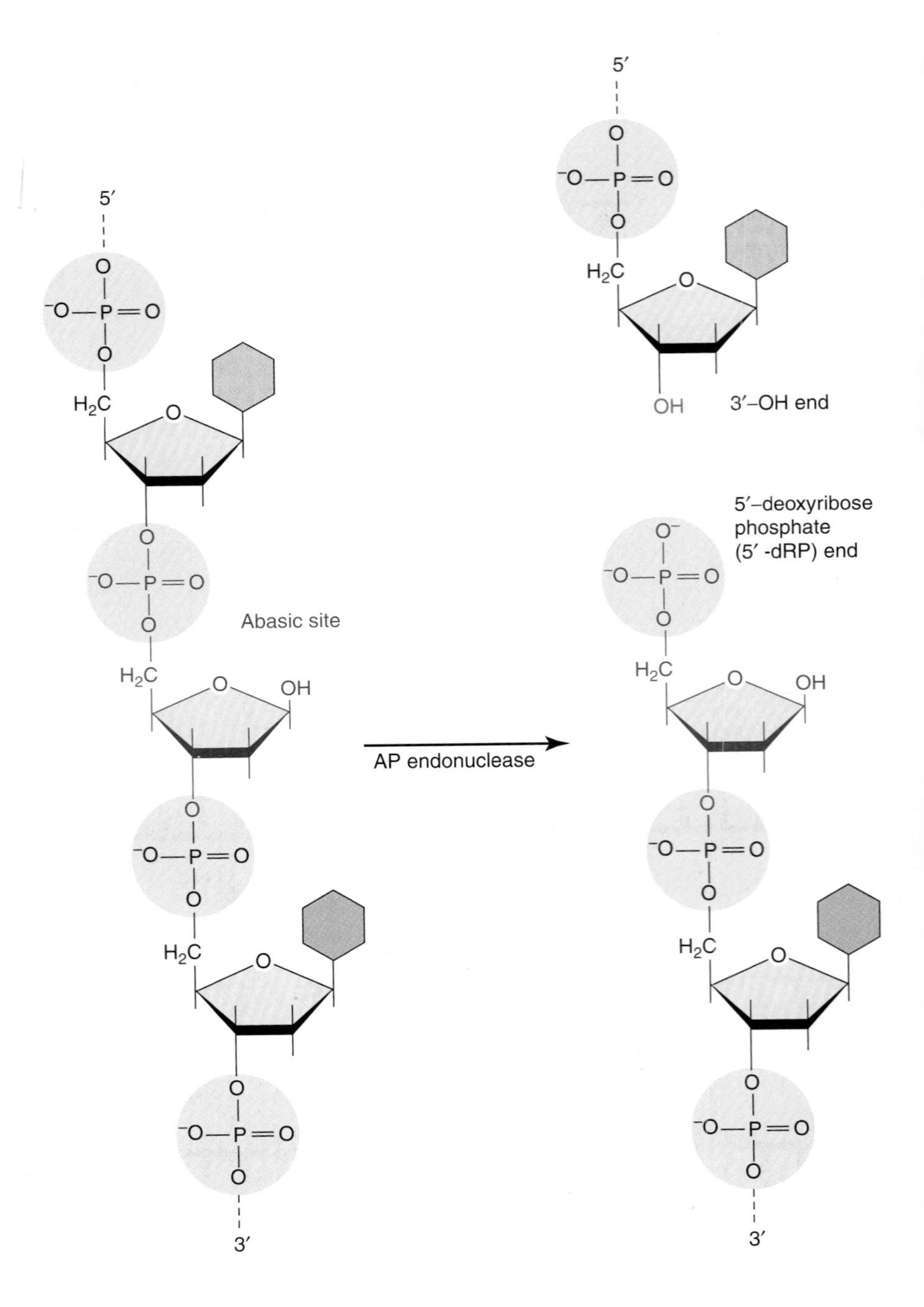

FIGURE 12.11 **AP endonuclease.** AP endonuclease hydrolyzes the phosphodiester bond 5′ to an AP site to generate a nick. In *E. coli* two well-characterized enzymes, exonuclease III and endonuclease IV, have this activity. The mammalian enzyme AP endonuclease I (APE1) is similar to exonuclease III. (Adapted from S. Boiteux, and M. Guillet, *DNA Repair* 3 [2004]: 1–12.)

lyase action (see below). These activities are important because DNA polymerase cannot attach nucleotides to a blocked 3′-end. In addition, exonuclease III has a 3′→5′ exonuclease activity that is relatively specific for double-stranded DNA from which its name is derived. The mammalian AP endonuclease, APE1 is homologous to *E. coli* exonuclease III.

Base excision by DNA glycosylase and strand cleavage by AP endonuclease introduces a gap with a 5′-deoxyribose phosphate (5′-dRP) on one side and a 3′-OH on the other. Additional enzymes are required to fill in the gap and remove the 5′-deoxyribose phosphate. Cells can repair the damage by two different pathways. The first of these, called **short patch repair** because only a single nucleotide is replaced (shown in Figure 12.10 by dark blue arrows), involves the following enzyme catalyzed reactions: (1) DNA polymerase adds a deoxyribonucleotide to the 3′-OH end; (2) deoxyribose phosphate lyase (dRPase) removes 5′-deoxyribose phosphate from the 5′-end; and (3) DNA ligase joins adjacent ends. In *E. coli*, DNA polymerase I fills in the gap. In eukaryotes, a single highly conserved enzyme, **DNA polymerase β (Pol β)** fills in the gap using the undamaged DNA strand as the template. The 39 kDa protein has a 31 kDa domain with polymerase activity and an 8 kDa domain with 5′-deoxyribose phosphate lyase activity (FIGURE 12.12). The eukaryotic enzyme differs from its bacterial counterpart in one very important respect; it lacks 3′→5′ proofreading activity. Two mammalian 3′ exonucleases, TREX1 and TREX2, may correct errors introduced by Pol β. TREX1 defects are associated with a severe brain disease.

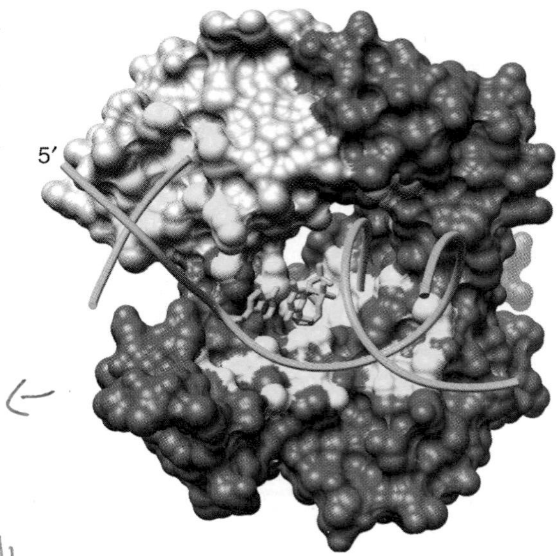

FIGURE 12.12 **DNA polymerase β and substrates.** The solvent-excluded molecular surfaces of DNA polymerase β and its substrates are shown. The protein is shown as a space-filling model with the polymerase domain in purple and the lyase domain in gray. The gapped DNA substrate is shown as a green backbone structure and an incoming ddCTP as a green stick figure structure. The 5′-end of the template strand is indicated. The surface of atoms that come within 3.5 Å of the DNA or ddCTP are highlighted in yellow. (Reproduced from W. A. Beard and S. H. Wilson, *Chem. Rev.* 106 [2006]: 361–382. Photo courtesy of Samuel H. Wilson, National Institute of Environmental Sciences.)

Repair may also proceed through an alternative pathway in mammalian cells termed **long patch repair** when two to eight nucleotides are replaced (shown in Figure 12.10 by light blue arrows). In this case, DNA polymerase δ or ε catalyzes chain extension with the assistance of the RFC clamp loader and the PCNA sliding clamp. As the polymerase adds nucleotides to the 3′-OH end on one side of the gap it displaces the 5′-deoxyribose phosphate end on the other side of the gap. The flap endonuclease cleaves the displaced strand and DNA ligase seals the remaining nick. The regulatory mechanism that selects short or long patch repair is not well understood. According to one hypothesis selection depends on the 5′ deoxyribose phosphate formed by the AP endonuclease. If this group can be removed Pol β, then the cell will use short patch repair and if it cannot, then the cell will use long patch repair.

When eukaryotic base excision repair begins with a DNA glycosylase/lyase, the pathway is as shown in FIGURE 12.13. Three distinct DNA glycosylase/lyase enzymes have been identified in *E. coli*, three in *S. cerevisiae*, and six in human cells. Virtually all oxidized bases are removed by bifunctional DNA glycosylases in mammals. DNA glycosylase/lyase excises the damaged base and cleaves the DNA strand 3′ of the AP site. The resulting sugar residue at the 3′-end is removed by AP endonuclease catalyzed cleavage. Then DNA polymerase β adds a nucleotide and DNA ligase seals the remaining nick to complete short patch repair.

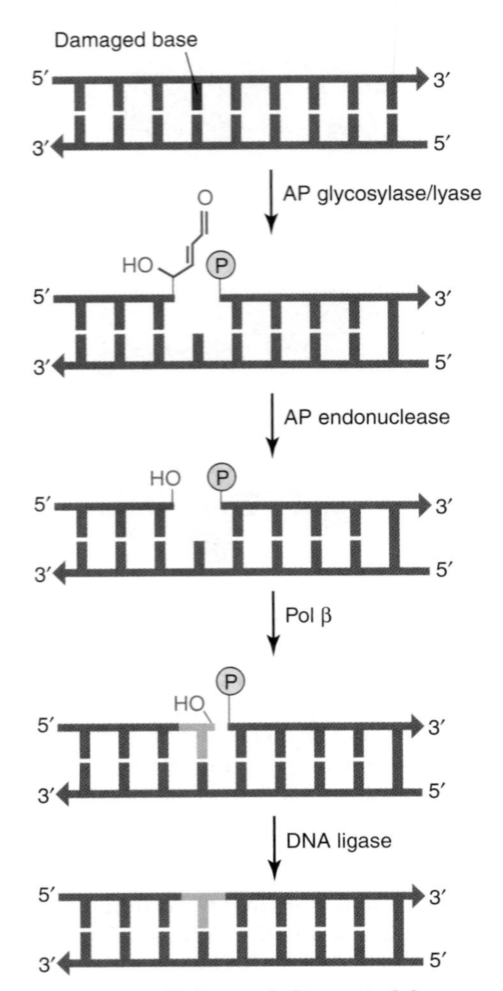

FIGURE 12.13 **Eukaryotic base excision repair starting with a bifunctional DNA glycosylase/lyase.** DNA glycosylase/lyase excises the damaged base and cleaves the DNA strand 3′ of the AP site and AP endonuclease removes the 3′ blocking group. Then DNA polymerase β adds a nucleotide and DNA ligase joins the two ends to complete short patch repair. (Adapted from O. D. Schärer, *Angew. Chem. Int. Ed. Engl.* 34 [2003]: 2946–2974.)

12.3 Nucleotide Excision Repair

Nucleotide excision repair removes bulky adducts from DNA by excising an oligonucleotide bearing the lesion and replacing it with new DNA.

The nucleotide excision repair pathway removes bulky adducts from DNA, correcting for many different types of structurally unrelated DNA damage. For instance, nucleotide excision repair excises UV-induced cyclobutane pyrimidine dimers, (6-4) photoproducts, damaged bases formed by alkylating agents such as psoralen, aflatoxin B, and certain types of cross-links. The efficiency of repair for different kinds of lesions can vary over several orders of magnitude. In general, there is a direct correlation between the amount of helical distortion produced by the lesion and the efficiency with which the lesion is removed. The basic nucleotide excision repair pathway, which is the same in all organisms, involves

1. damage recognition,
2. an incision (cut) in the damaged DNA strand on each side of the lesion,
3. excision (removal) of the oligonucleotide created by the incisions,
4. synthesis of new DNA to replace the excised segment using the undamaged DNA strand as a template, and
5. ligation of the remaining nick. Although the basic nucleotide excision pathways in all organisms are very similar, there are considerable differences in the proteins that carry out the various steps.

We will, therefore, treat the bacterial and eukaryotic nucleotide excision repair systems separately.

UvrA, UvrB, and UvrC proteins are required for bacterial nucleotide excision repair.

UV-irradiated *E. coli* can regain their ability to survive after incubation in the dark, although they recover more slowly than when incubated in the light. This observation suggests that the bacteria use some process other than photoreactivation to repair UV-induced light damage. Richard Setlow and William Carrier and, working independently, Richard Boyce and Paul Howard-Flanders, used a similar approach to investigate this alternative process in 1964. Both groups cultured *E. coli* in the presence of [^{3}H]thymine to label the DNA and then irradiated the cells with UV light to induce thymine cyclobutane dimer formation. Then they (1) incubated the UV-irradiated cells in the dark so that photoreactivation could not take place; (2) removed samples after various incubation times and added trichloracetic acid to them; (3) separated acid-insoluble DNA from acid-soluble oligonucleotides; (4) digested the DNA and oligonucleotides to release intact thymine cyclobutane dimers; and (5) detected the released dimers by chromatography. The experiments revealed that as the incubation time in the dark increases, cyclobutane thymine dimers disappear from the acid-insoluble DNA and appear in the acid-soluble oligonucleotide fraction. These results were correctly interpreted to mean that bacteria can excise an oligonucleotide containing a lesion and replace the excised oligonucleotide with newly synthesized DNA. Subsequent studies showed that eukaryotes and the archaea also have nucleotide excision repair pathways.

Genetic studies revealed that three *E. coli* genes—*uvrA*, *uvrB*, and *uvrC* (for UV radiation)—code for proteins that are essential for damage recognition, incision, and excision. All three genes have been cloned and the proteins that they encode (UvrA, UvrB, and UvrC) have been purified and characterized. Under normal physiological conditions, *E. coli* has about 25 molecules of UvrA, 250 molecules of UvrB, and 10 molecules of UvrC. Following DNA damage, UvrA and UvrB levels increase about ten- and fourfold, respectively, but the UvrC level remains the same. (The explanation for the increased levels of UvrA and UvrB following UV irradiation will be given later in this chapter in the section that describes the SOS response.) Although UvrA, UvrB, and UvrC do not combine to form a stable ternary complex, the polypeptides nevertheless are said to be part of a **UvrABC damage-specific endonuclease** or **UvrABC endonuclease** for short. Some investigators prefer the term **UvrABC excinuclease** to indicate that the proteins participate in incision and excision reactions. Fortuitously, the three polypeptides work in the order suggested by their names—that is, their order of action is UvrA, UvrB, and then UvrC.

Gregory L. Verdine and coworkers have determined the crystal structure of the UvrA homodimer isolated from the gram-positive bacteria *Bacillus stearothermophilus* (FIGURE 12.14). Each subunit has a UvrB binding site and three bound zinc ions. Each subunit also has two nucleotide binding sites with ATPase activity. These sites are occupied by ADP because the protein crystal was prepared in the presence of this nucleotide. Marcin Nowotony and coworkers

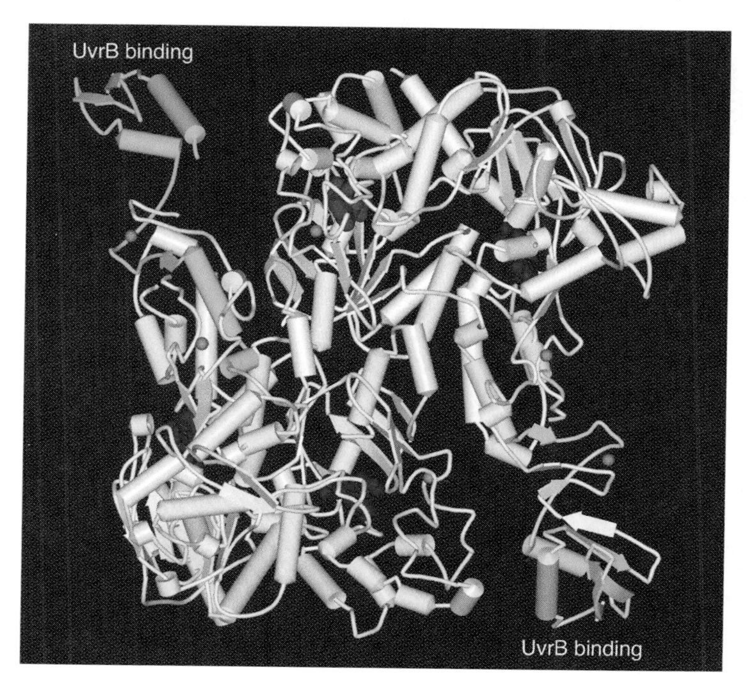

FIGURE 12.14 **Crystal structure of the UvrA homodimer isolated from the gram-positive bacteria *Bacillus stearothermophilus*.** One subunit is shown in light green and the other in light yellow. UvrB binding sites are in darker shades of green and yellow. The ADP molecules bound to the two ATPase active sites in each subunit are shown as red spacefill structures. The three zinc ions bound to each subunit are shown as purple spheres. (Structure from Protein Data Bank 2R6F. D. Pakotiprapha, et al., *Mol. Cell.* 29 [2008]: 122–133. Prepared by B. E. Tropp.)

recently determined the crystal structure of an UvrA•DNA complex. The UvrA was isolated from the rod-shaped hyperthermophilic bacterium *Thermotoga maritime*, and the DNA, which was 32 bp long, had a fluorescein group attached to thymine-14. The DNA region containing the bulky fluorescein adduct was deformed as a result of localized bending and unwinding. The crystal structure reveals that UvrA does not make direct contact with the modified thymine but does bind to DNA regions on either side of the lesion. Based on this information, it appears that UvrA makes an important contribution to DNA lesion recognition. However, it is important to keep in mind that the UvrA•DNA complex was formed *in vitro* in the absence of UvrB. The recognition process appears to be a bit more complicated *in vivo* where UvrA is part of a $(UvrA)_2$•UvrB complex (see below).

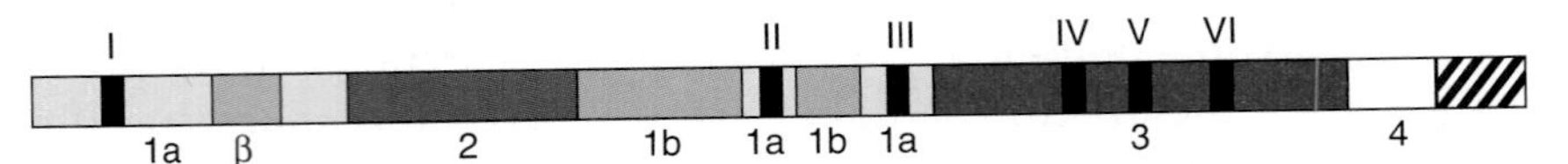

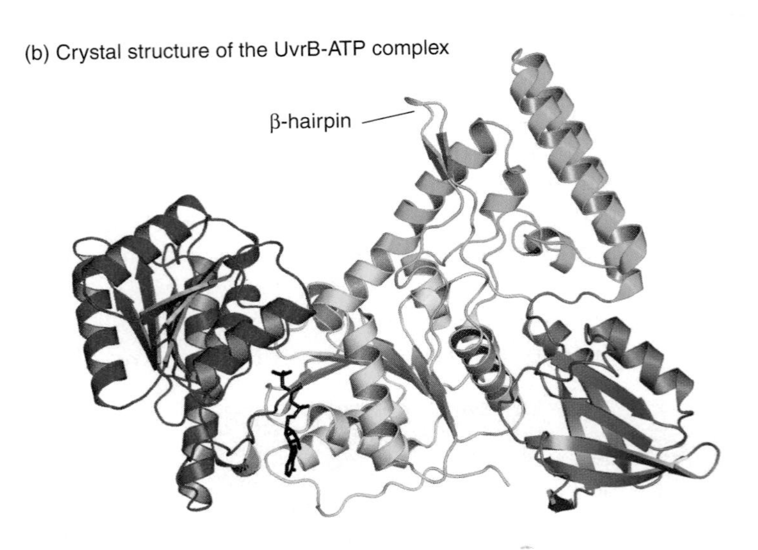

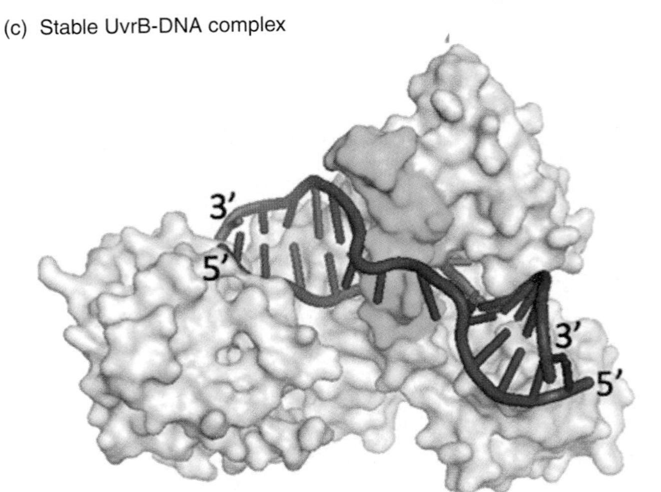

FIGURE 12.15 The structure of UvrB. (a) Functional domains in UvrB. UvrB is color coded according to its domain architecture. Domain 1a is yellow, 1b is green, 2 is blue, 3 is red, 4 is white (the crosshatch is the UvrC interacting region), and the β-hairpin is cyan. The six helicase motifs (I to VI) present in domains 1a and 3 are indicated and shown in black. (b) Crystal structure of the UvrB•ATP complex. The color coding is as in (a). The ATP molecule, located between domains 1a and 3, is shown in a stick representation and color-coded by element. (c) A model of the UvrB•DNA complex that shows the inner strand of the DNA (red) located between the β-hairpin (cyan) and domain 1b and the outer strand (blue) located around the outside of the β-hairpin. (Part a modified with permission from J. J. Truglio, et al., *Chem. Rev.* 106 [2006]: 233–252. Copyright 2006 American Chemical Society. Part b reprinted with permission from J. J. Truglio, et al., *Chem. Rev.* 106 [2006]: 233–252. Copyright 2006, American Chemical Society. Photo courtesy of Caroline Kisker, Rudolf Virchow Center for Experimental Biomedicine, University of Würzburg. Part c adapted from J. J. Truglio, et al., *Nat. Struct. Mol. Biol.* 13 [2006]: 360–364. Photo courtesy of Caroline Kisker, Rudolf Virchow Center for Experimental Biomedicine, University of Würzburg.)

UvrB plays a central role in nucleotide excision repair, interacting with both UvrA and UvrC, although not at the same time. The functional domains and crystal structure of UvrB are shown in FIGURE 12.15a and b, respectively. UvrB has at least two catalytic sites that are essential for its function: six helicase motifs (I to VI) are present in domains 1a and 3 and an ATPase active site is located at the interface of these two domains. Residues in domains 2 and 4 appear to participate in binding UvrA and UvrC.

UvrB combines with UvrA to form a $(UvrA)_2$•UvrB complex either in solution or on DNA. The protein complex initially binds to DNA at some distance from the damaged site. Once $(UvrA)_2$•UvrB•DNA forms, the UvrB helicase catalyzes ATP-dependent movement of $(UvrA)_2$•UvrB along the DNA until the protein complex encounters a bulky adduct or helix distortion. Then UvrA is released in an ATP-dependent reaction with a concomitant conformational change in UvrB that produces a stable UvrB•DNA complex (FIGURE 12.16). UvrA functions as a "molecular matchmaker" in this pathway, in much the same way that DnaC performs this function during the initiation of bacterial DNA replication (see Chapter 9). It uses energy provided by ATP to facilitate the formation of a stable UvrB•DNA complex that would not otherwise form and then dissociates from the complex. Bennet Van Houten and coworkers have proposed a padlock model to explain how UvrB binds so tightly to the damaged site. According to this model, UvrB has a β-hairpin structure that inserts between the two DNA strands so that one of the strands is clamped between the β-hairpin and domain 1b (FIGURE 12.15c).

UvrC, which has a flexible linker that connects its N- and C-terminal domains, binds to the UvrB•DNA complex and makes two incisions, one on each side of the lesion (FIGURE 12.17). The first incision of the damaged strand is four nucleotides toward the 3′-end from the lesion and the second is seven nucleotides toward the 5′-end from the lesion. A more detailed understanding of the way that UvrC interacts with UvrB and performs its functions awaits the solution of a crystal structure for the UvrB•UvrC•DNA complex. The following three additional steps are required to complete the repair process: (1) UvrD, a helicase, excises the damaged oligonucleotide; (2) DNA polymerase I uses the undamaged strand as a template to fill in the gap; and (3) DNA ligase seals the remaining nick to complete the repair process (Figure 12.17).

Individuals with the autosomal recessive disease xeroderma pigmentosum have defects in enzymes that participate in the nucleotide excision repair pathway.

The first hint that mammalian cells also carry out nucleotide excision repair came from experiments performed in the early 1960s by Robert Painter and coworkers, who had developed an autoradiographic technique to study $[^3H]$thymidine incorporation into mammalian cell DNA. Cells in the S-stage of the cell cycle incorporate large quantities of $[^3H]$thymidine into DNA, whereas cells in other stages of the cell cycle incorporate little, if any, radioactive label into DNA. Painter and coworkers observed that mammalian cells that were not in the S-stage of the cell cycle gained the ability to incorporate low levels of

FIGURE 12.16 **Schematic diagram of the formation of the stable UvrB•DNA complex.** UvrA and UvrB form an ATP-dependent heterotrimer that binds at some distance from the damaged DNA site (yellow star). Once the $(UvrA)_2$•UvrB•DNA complex forms, the UvrB helicase moves $(UvrA)_2$•UvrB along the DNA until the protein complex encounters the lesion. Then UvrA induces ATP-dependent conformational changes in UvrB that result in the release of UvrA and the formation of a stable UvrB•DNA complex in which the DNA is kinked and the strands are separated. The red region in UvrB represents the ATPase. (Adapted from J. J. Truglio, et al., *Chem. Rev.* 106 [2006]: 233–252.)

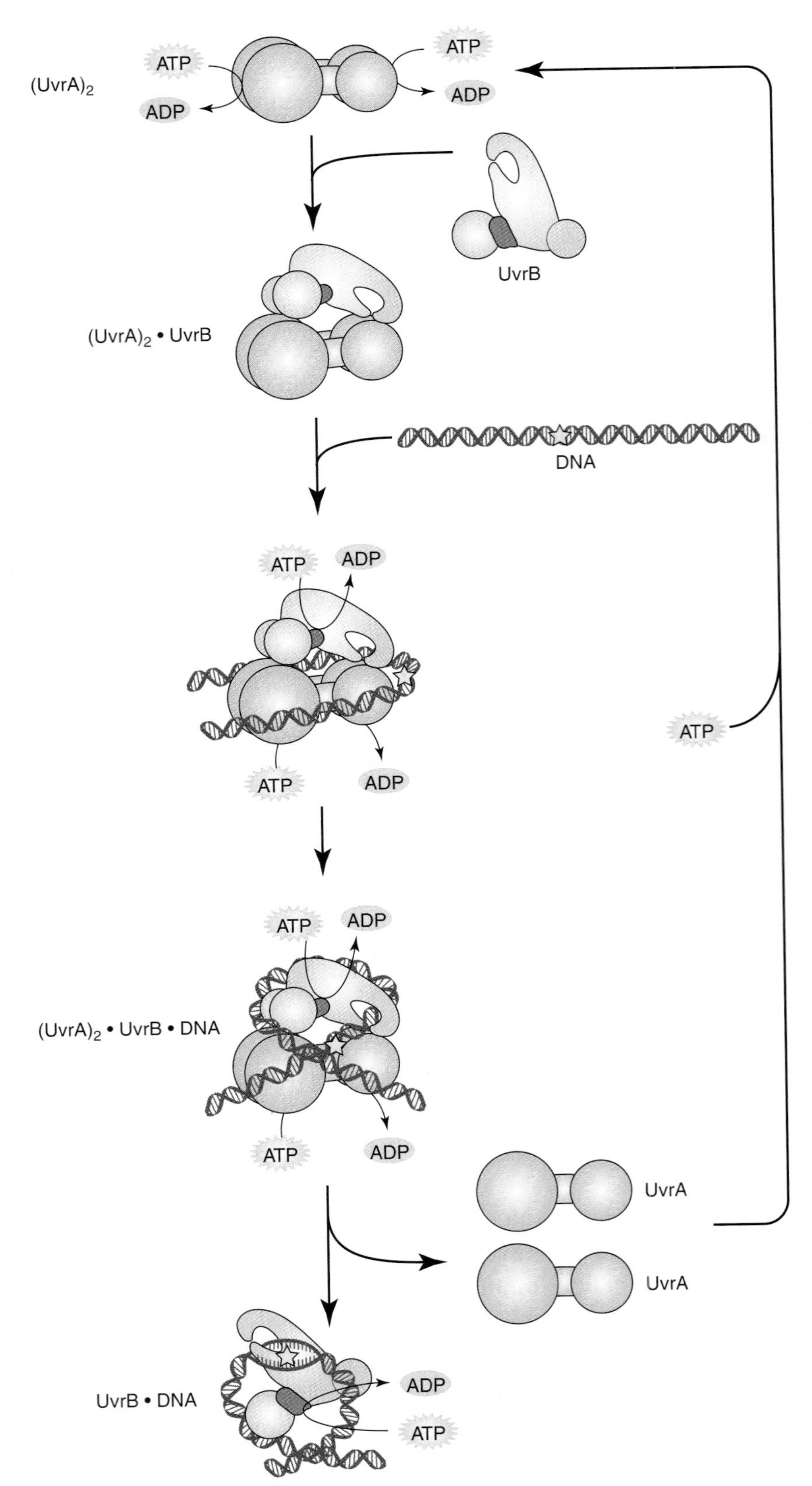

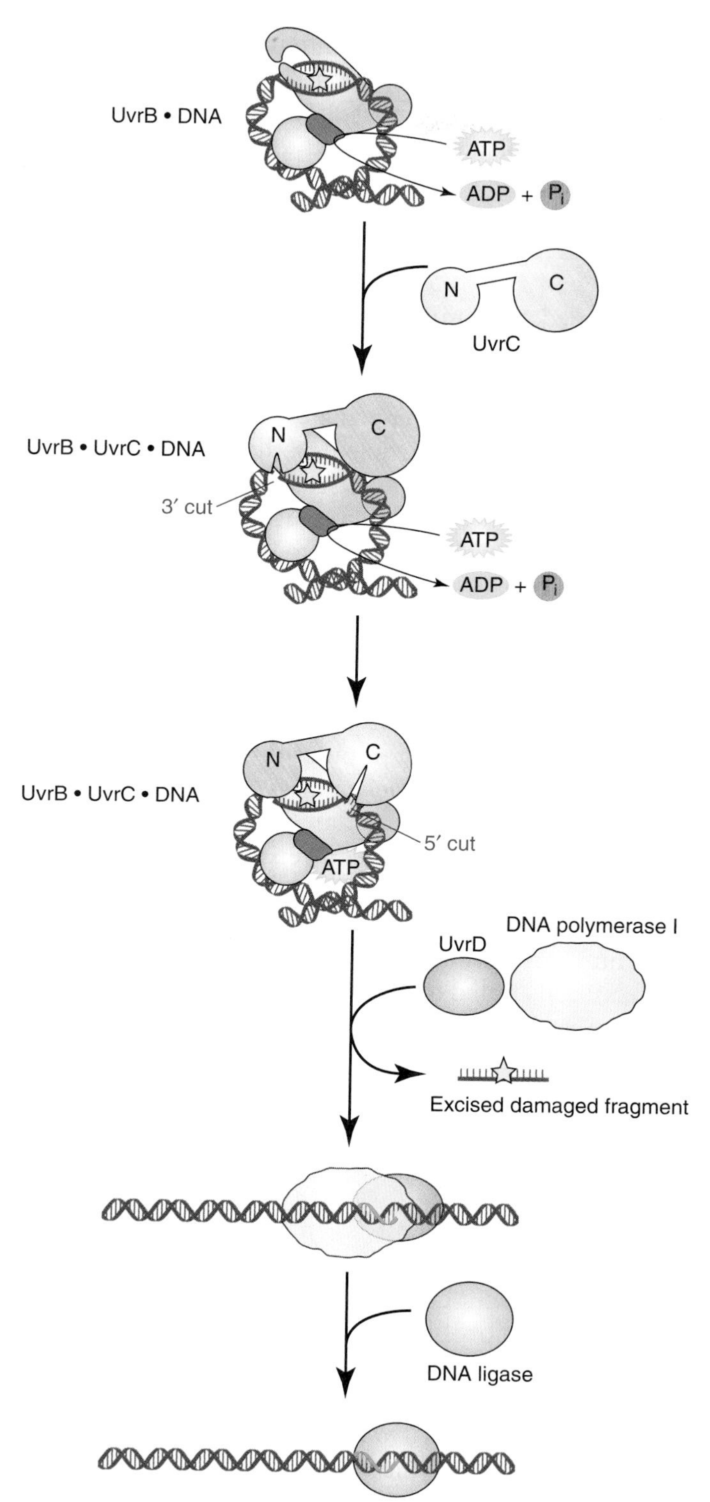

FIGURE 12.17 **Schematic diagram for incision, excision, and repair reactions.** UvrC binds to the UvrB•DNA complex and makes two incisions, one on each side of the lesion (yellow star). Then three additional enzymes complete the repair process. UvrD, a helicase, excises the damaged oligonucleotide, DNA polymerase I uses the undamaged strand as a template to fill in the resulting gap, and DNA ligase joins the strand ends to complete the repair process. The red region in UvrB represents the ATPase. (Adapted from J. J. Truglio, et al., *Chem. Rev.* 106 [2006]: 233–252.)

[^{3}H]thymidine into DNA after UV-irradiation, and proposed that this "unscheduled" DNA synthesis was part of a repair pathway.

James E. Cleaver, a young investigator working in Painter's laboratory, helped to establish the connection between unscheduled DNA synthesis and nucleotide excision repair in 1968. Cleaver was aware of the then recently published work showing that bacteria have a nucleotide excision repair pathway and suspected that the unscheduled DNA synthesis was due to a mammalian nucleotide excision repair pathway. He hoped to obtain evidence to support his hypothesis by first isolating UV-sensitive mammalian cell lines and then showing that they did not incorporate low levels of [^{3}H]thymidine into DNA after UV irradiation. When preparing to isolate the desired cell lines, he came across a newspaper article describing **xeroderma pigmentosum (XP)**, an autosomal recessive disease. Each parent of an XP child carries an XP mutation but displays no obvious symptoms of the disease. In contrast, the XP child has a 1000-fold greater likelihood of developing skin cancer, especially in parts of the body exposed to sunlight such as the hands, face, and neck. The incidence of skin cancer is directly related to the sunlight exposure. Without considerable precautions to avoid sunlight, the median age at which skin cancer first appears is about eight years old, which is roughly fifty years earlier than that for the general population. The relationship between sunlight exposure and the incidence of skin cancer may be explained by proposing that sunlight causes damage to DNA in skin cells in all individuals but that normal cells can repair the damage, whereas XP cells cannot. Cleaver correctly deduced that XP cells could be used in place of the UV-sensitive mutants that he had planned to isolate to test whether mammalian cells have a nucleotide excision repair system. As anticipated, XP cells failed to incorporate low levels of [^{3}H]thymidine after they were irradiated with UV light. Cleaver also demonstrated that wild-type human cells have a nucleotide excision repair system, which is missing in XP cells.

Dirk Bootsma and colleagues provided important new insights into the mammalian nucleotide excision repair system in the early 1970s after noticing significant variations in the levels of DNA repair in cells obtained from different XP patients. The existence of these variations suggested the possibility that XP may result from a defect in one of several different genes. Bootsma and coworkers verified this possibility by fusing fibroblasts isolated from one patient with those isolated from another patient. When a fibroblast, for example XP-A, from one patient was fused to a fibroblast, for example XP-B, from another patient, the fused cell was not hypersensitive to UV light and was able to perform unscheduled DNA synthesis. These fusion experiments indicated that each cell makes a protein required for nucleotide excision repair that is not made by the other. In other words, the XP-A cell makes a normal XPB protein and the XP-B cell makes a normal XPA protein. Systematic cell fusion experiments using all available XP cell lines eventually revealed that human cells have seven genes, *XPA* to *XPG*, which code for polypeptides that are essential for nucleotide excision repair.

Depending on the precise location of the DNA lesion, the nucleotide excision repair system may act on an untranscribed DNA strand or an actively transcribed strand. The pathway for repairing untranscribed DNA is designated **nucleotide excision repair.** Some investigators refer to it as the **global genome excision repair pathway** because most of the cell's DNA is not being transcribed at any given time. The pathway for repairing transcribed DNA, the **transcription-coupled nucleotide excision repair pathway,** was discovered when investigators noticed that lesions that block the progress of the transcription machinery are repaired more rapidly in transcriptionally active DNA regions than in transcriptionally silent DNA regions. The transcription-coupled excision repair pathway is specific to the transcribed strand.

Progress in understanding the mammalian nucleotide excision repair system required purification and characterization of the XP proteins. *In vitro* assays were devised to detect these proteins. In one commonly used assay, XP protein activity is followed by first preparing a DNA substrate with a radioactively labeled cyclobutane thymine dimer and then monitoring the excision of the oligonucleotide bearing the thymine dimer. Isolation of XP proteins was greatly facilitated by the availability of XP cell lines with specific XP protein defects. Cell-free extracts from normal cells were subjected to standard protein fractionation techniques and the fractions tested to determine if they could complement the activity of an extract prepared from an XP cell line with a known defect. For instance, the XPC protein was isolated by fractionating cell extracts that complemented the activity of XP-C cell extracts. Investigators were eventually able to purify all the mammalian proteins required to support *in vitro* nucleotide excision repair. The only XP protein that does not appear to be required for *in vitro* reconstitution of the nucleotide excision repair system is XPE. Because the *in vitro* assay uses free DNA rather than chromatin, it is possible that XPE is required under physiological conditions where chromatin is present. Alternatively, XPE may be required to repair some lesions but not others. There is some evidence to support each of these hypotheses. Proteins that were not originally identified by studies of XP cell lines were also shown to be required for nucleotide excision repair. One of these proteins, **hHR23B** (human homolog of the yeast repair protein Rad23), is always found tightly associated with XPC in an XPC•hHR23B complex. Another of these proteins, **ERCC1,** is tightly associated with XPF. The ERCC1 subunit does not appear to have any catalytic activity of its own but is required to stabilize XPF endonuclease activity. ERRC1 is named for the *ERRC1* (excision repair cross-complementing 1) gene, which was originally defined by its ability to complement a repair-defective Chinese hamster ovary (CHO) cell. Finally, several proteins that were found to be essential to *in vitro* nucleotide excision repair also make essential contributions to other pathways. Included among these are single-stranded RPA (DNA binding protein), RFC (clamp loader), PCNA (sliding clamp), DNA polymerase δ (or ε), and DNA ligase. The structures and functions of these proteins were described in Chapter 10.

One final protein that makes an essential contribution to nucleotide excision repair is general **transcription factor IIH** (**TFIIH**), a large multiprotein complex containing ten subunits, which assists RNA polymerase II (the enzyme that transcribes eukaryotic protein coding genes). TFIIH is described in greater detail in Chapter 17. For now, our focus will be on the contributions that just two of its subunits, XPB and XPD, make to nucleotide excision repair. XPB is a 5′→3′ ATP-dependent helicase and XPD is a 3′→5′ ATP-dependent helicase.

The various components of the nucleotide excision repair system appear to be recruited to the damaged site in the sequence shown in FIGURE 12.18. XPC•hHR23B recognizes the helical distortion caused by the bulky adduct and binds to the lesion. An additional protein factor, UV-damaged DNA binding protein (UV-DDB—not shown in figure), is required to recognize cyclobutane pyrimidine dimers, which cause only a subtle distortion in DNA structure. UV-DDB binds tightly to UV-damaged DNA and helps to recruit the XPC complex to the damaged site. After binding, XPC•hHR23B initiates localized unwinding, which helps to recruit TFIIH. The XPB and XPD subunits in TFIIH help extend the DNA opening so that other proteins can add to the site. XPD appears to unwind the DNA until it stalls upon reaching a chemically modified base. XPD has been proposed to have a proofreading function because if it does not encounter a chemically modified base, then the bound proteins can disassemble. This additional recognition step is consistent with a **bipartite model** of DNA damage recognition that was proposed by Hanspeter Naegeli and coworkers in 1997. According to that model, a lesion must both introduce a helical distortion into DNA and chemically modify a base before it can be removed by nucleotide excision repair. XPB continues to unwind the DNA until the lesion is surrounded by a 20-bp open "bubble." RPA, XPA, and XPG are recruited next. The binding of XPG appears to cause XPC•hHR23B release, resulting in the formation of the **preincision complex**. The XPF endonuclease nicks the damaged strand on the 5′-side of the lesion and the XPG endonuclease cleaves it on the 3′-side of the lesion. The resulting 24 to 32 nucleotide long oligonucleotide that contains the lesion is excised. Finally, DNA polymerase δ or ε, RFC, and PCNA work together to fill in the gap using the undamaged DNA strand as template and DNA ligase seals the remaining nick.

Although the *XP* genes were originally identified in patients with xeroderma pigmentosum, certain mutations in *XPB* and *XPD* are also associated with **trichothiodystrophy**, another autosomal recessive disease. The primary clinical characteristic of trichothiodystrophy is brittle hair and nails resulting from a deficiency in a class of sulfur-rich proteins. Many individuals who suffer from trichothiodystrophy also exhibit growth and mental retardation as well as scaly skin. Although nucleotide excision repair is defective in many but not all cases, a propensity toward skin cancer has not been observed. Trichothiodystrophy may result from the failure of a defective *XPB* or *XPD* to perform a function required for transcription rather than one required for nucleotide excision repair.

With the exception of XPC•hHR23B, transcription coupled nucleotide excision repair requires the same enzymes and protein factors as

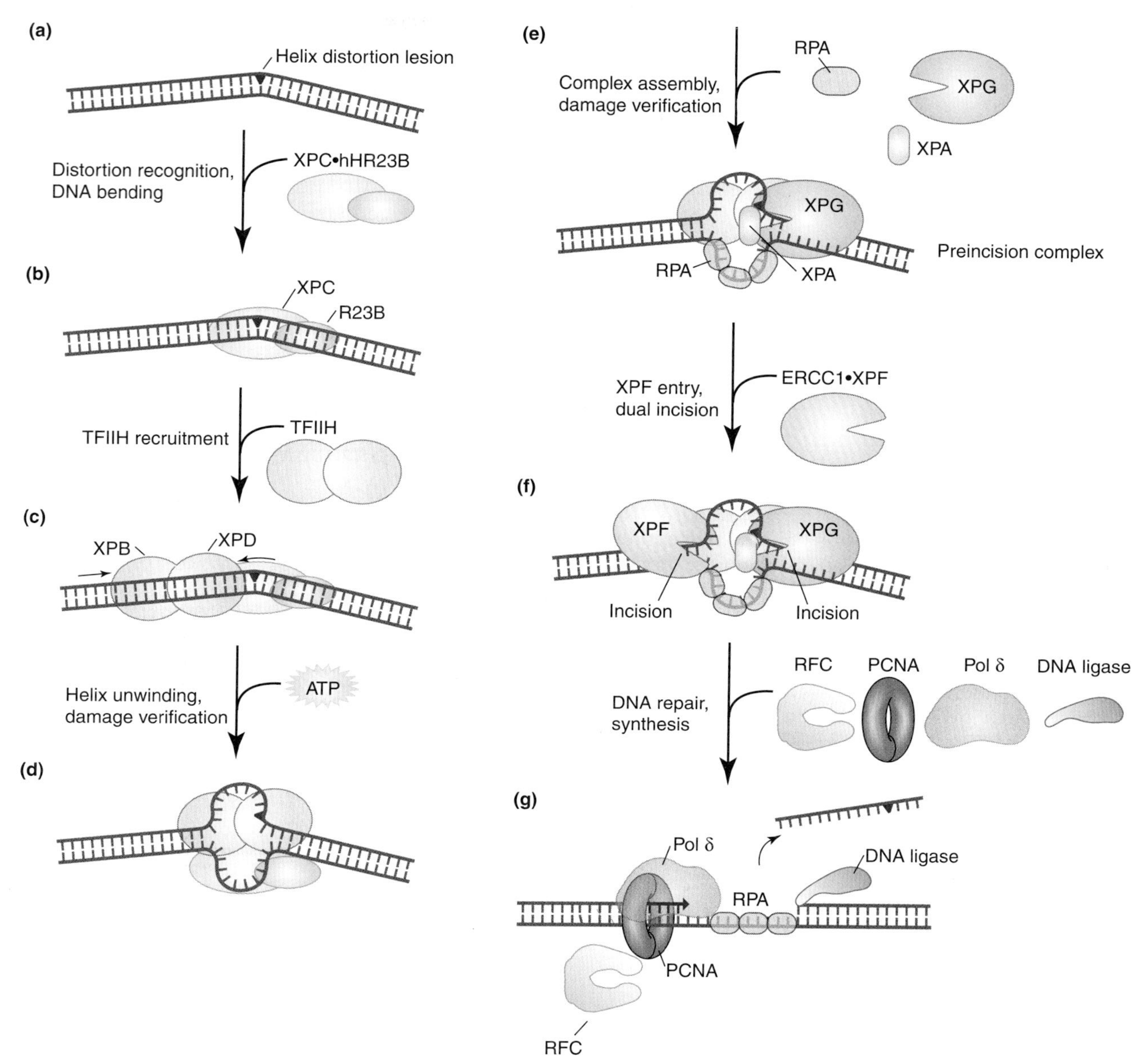

FIGURE 12.18 **Proposed pathway for eukaryotic nucleotide excision repair.** (a) A lesion (red triangle) induces a helical distortion in DNA. (b) XPC•hHR23B detects the helix distortion and stabilizes the DNA bend. The XPE complex (not shown) appears to participate in this recognition process when some lesions such as a cyclobutane pyrimidine dimer are present. (c) Transcription factor IIH (TFIIH) is recruited to the lesion site by XPC•hHR23B. (d) TFIIH catalyzes ATP-dependent unwinding of the DNA helix surrounding the lesion until its XPD subunit encounters a chemically modified base. The second helicase subunit, XPB, continues unwinding the DNA to generate a 20 bp open "bubble" structure. (e) RPA (single-stranded DNA binding protein), XPA, and XPG then add to the complex to form a "preincision" complex. (f) The ERCC1•XPF heterodimer adds to the complex and dual incision (5′ by ERCC1•XPF and 3′ by XPG) takes place. (g) RPA remains bound to the ssDNA, facilitating transition to repair synthesis by DNA polymerase δ (or ε), which is assisted by the clamp loader RFC and the sliding clamp PCNA. DNA ligase seals the remaining nick. (Adapted from L. C. J. Gillet, and O. D. Schärer, *Chem. Rev.* 106 [2006]: 253–276.)

nucleotide excision repair. In addition, transcription-coupled nucleotide excision repair requires CSA and CSB. The genes that code for these proteins, ***CSA*** and ***CSB***, were first detected in mutant form in individuals with **Cockayne syndrome**, a rare autosomal recessive disorder that decreases the mean life expectancy to about 12 years. Clinical symptoms include retarded growth, neurological dysfunction, and photosensitivity but not a tendency to skin cancer. The reason that XPC•hHR23B is not required for transcription-coupled nucleotide excision repair is that RNA polymerase II, the enzyme required to transcribe all protein coding genes in eukaryotes, helps to recognize the lesion. The pathway proposed for transcription-coupled nucleotide excision repair in mammals is shown in FIGURE 12.19. According to this model, RNA polymerase II moves along the template DNA strand until it encounters the lesion, which blocks its forward progress. The stalled transcription machinery triggers the recruitment of XPG and CSB

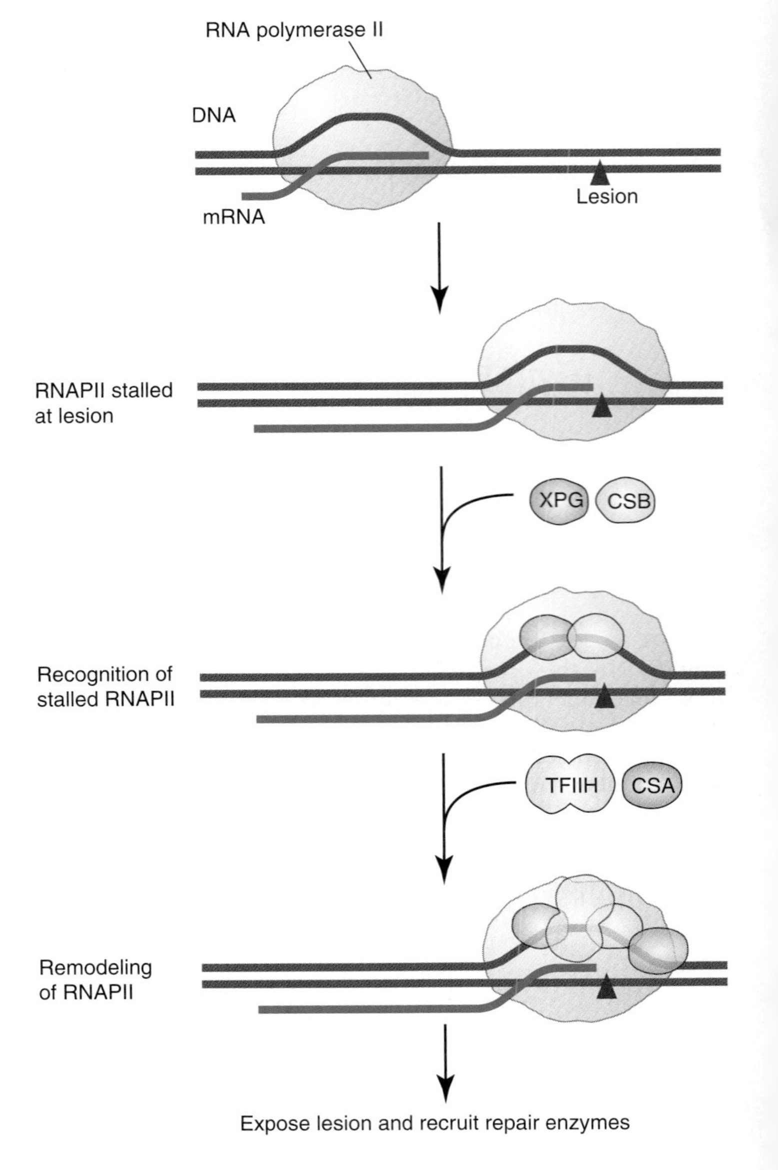

FIGURE 12.19 **Eukaryotic transcription coupled nucleotide excision repair.** RNA polymerase II moves along the template DNA strand until it encounters the lesion (red triangle), which blocks the progress forward. The stalled transcription machinery triggers the recruitment of XPG and CSB and then TFIIH and CSA. TFIIH modifies the interactions between RNA polymerase II and the DNA, exposing the lesion on the template strand and allowing XPG to gain access to it. Subsequent recruitment of XPA, RPA, and ERCC1•XPF allows excision of the lesion without removal of RNA polymerase II. (Adapted from A. H. Sarker, et al., *Mol. Cell* 20 [2005]: 187–198.)

and then TFIIH and CSA. TFIIH modifies the interactions between RNA polymerase II and the DNA, exposing the lesion on the template strand and allowing XPG to gain access to it. Subsequent recruitment of XPA, RPA, and ERCC1•XPF allows excision of the lesion without removal of RNA polymerase II. Two problems can arise if the lesion is not repaired. First, an essential gene may not be transcribed preventing cell survival and second, RNA polymerase II may be trapped at the damaged site, preventing it from transcribing other genes. The difference in clinical symptoms between XP and Cockayne syndrome are probably due to the fact that RNA polymerase II is tied up in the latter condition and therefore cannot perform its fundamental task of transcribing protein coding genes in an efficient manner.

12.4 Mismatch Repair

The DNA mismatch repair system removes mismatches and short insertions or deletions that are present in DNA.

Mismatch repair corrects rare base pair mismatches and short deletions or insertions that appear in DNA following replication. DNA polymerases introduce about one mispaired nucleotide per 10^5 nucleotides. The 3′→5′ proofreading exonuclease increases replication fidelity by 100-fold by removing mispaired nucleotides before the growing chain can be extended further. Although an error frequency of 1 nucleotide in 10^7 may seem extremely low, it would result in a high mutation rate. The second type of error that arises during replication, short insertions and deletions, result from the fact that repeated-sequence motifs such as $[CA]_n$ or $[A]_n$ in microsatellites (see Chapter 7) sometimes dissociate and then re-anneal incorrectly. As a result of this slippage, the newly synthesized strand will have a different number of repeats than the template strand. Introduction of an insertion or deletion into the newly synthesized DNA is likely to produce a mutation if it affects a region that codes for a protein or an essential RNA molecule. Cells with a nonfunctional mismatch repair system have a high rate of mutation due to their inability to efficiently repair base pair mismatches, short insertions, or short deletions that arise during replication.

We begin our examination of mismatch repair by considering the *E. coli* mismatch repair system because this system has been the most extensively studied. Although this system provides valuable lessons for studying mismatch repair in other organisms, it differs from the mismatch repair systems used by gram-positive bacteria and eukaryotes in one important respect. The *E. coli* mismatch repair system can distinguish a newly synthesized strand from a parental strand because only the latter has methyl groups attached to sites with the sequence GATC.

E. coli has a **deoxyadenosine methylase** that transfers methyl groups from S-adenosylmethionine molecules to deoxyadenosines in GATC sequences. The timing of methylation by deoxyadenosine methylase, however, lags behind that of nucleotide addition at the replication fork by about two minutes, so the newly synthesized strand is transiently unmethylated. The *E. coli* mismatch repair system exploits this period of transient unmethylation to identify and cut GATC sites in a newly synthesized strand with a mismatch.

Genetic and biochemical studies have demonstrated that three *E. coli* proteins, MutS, MutL, and MutH, are dedicated to mismatch repair. Although these proteins are essential for mismatch repair, they are not sufficient. Several additional enzymes and protein factors also make important contributions. Among these enzymes and protein factors are: DNA helicase II (UvrD), single-stranded DNA binding protein

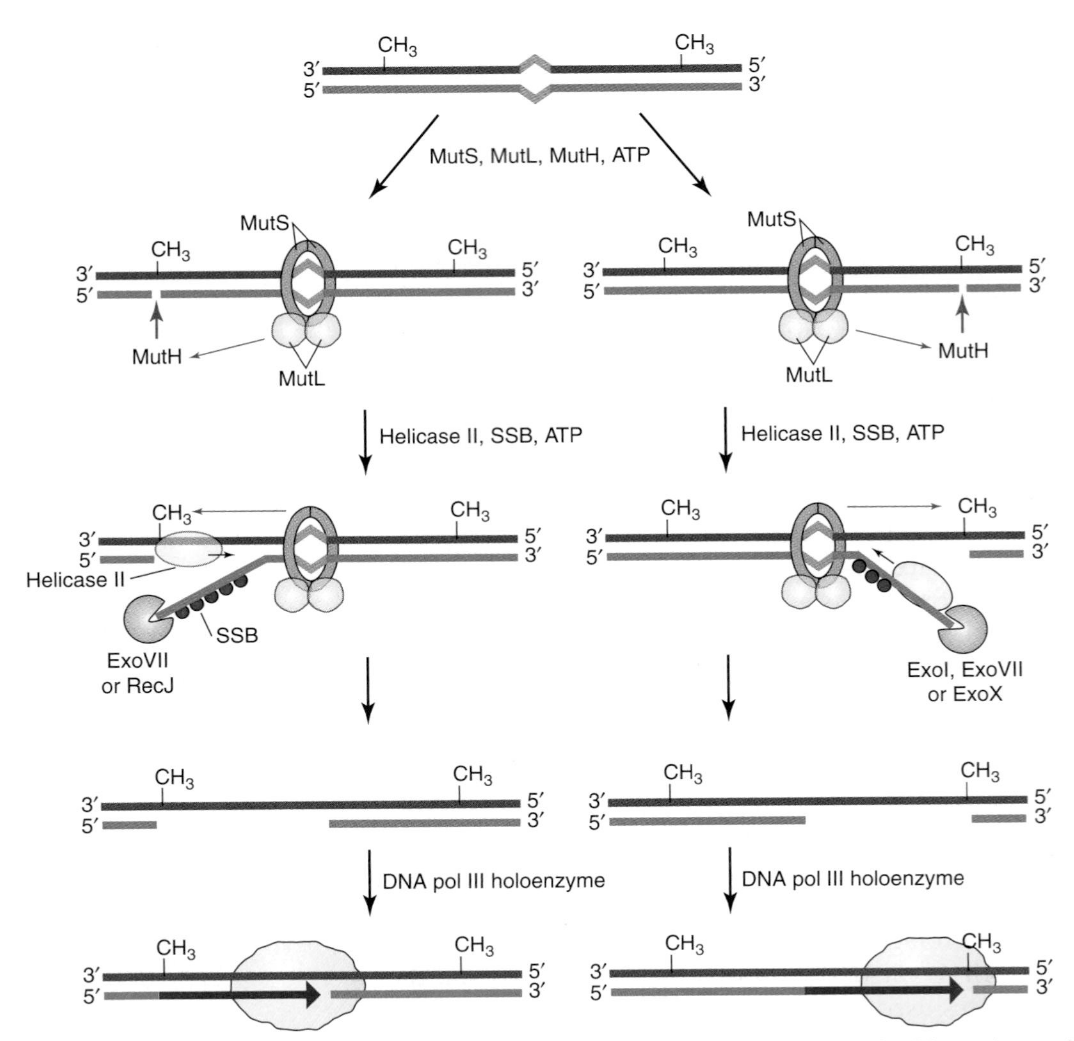

FIGURE 12.20 ***E. coli* mismatch repair system.** The newly synthesized DNA strand (light blue) with a mismatch (orange) is transiently unmethylated at GATC sites. The parent strand (dark blue) has methylated GATC sites. The mismatch-activated MutS•MutL•ATP complex stimulates the MutH endonuclease to incise the nearest unmethylated GATC sequence (either 5′ or 3′ from the mismatch). Helicase II (UvrD) unwinds the DNA and SSB binds to the single strands. If the incision is 5′ to the mismatch (left), then ExoVII or RecJ endonucleases hydrolyze the nicked strand in a 5′→3′ direction. If the incision is 3′ to the mismatch (right), then ExoI, ExoVII, or EcoX exonucleases hydrolyze the nicked strand in a 3′→5′ direction. DNA polymerase III holoenzyme fills the gap with new DNA (shown in red). DNA ligase seals the remaining nick and deoxyadenosine methylase adds a methyl group to the GATC site (not shown in figure). (Modified with permission from R. R. Lyer, et al., *Chem. Rev.* 106 [2006]: 302–323. Copyright 2006 American Chemical Society.)

(SSB), 5′→3′ exonucleases (ExoVII or RecJ), 3′→5′ exonucleases (ExoI, ExoVII, or ExoX), DNA polymerase III holoenzyme, DNA ligase, and deoxyadenosine methylase. The *E. coli* mismatch repair system has been reconstituted *in vitro* from purified proteins, permitting us to determine how each enzyme and protein factor contributes to the process. The assay system is based on the recovery of a restriction endonuclease site that is lost because of the mismatch but recovered by the action of the mismatch repair system. The results of the *in vitro* studies are summarized in FIGURE 12.20.

The process begins when MutS as either a homodimer or homotetramer binds to the mismatch. For simplicity, Figure 12.20 shows a homodimer but this is not firmly established. MutS recruits MutL (a homodimer) in an ATP-dependent fashion. Then the MutS•MutL complex activates MutH, which makes an incision at the nearest unmethylated GATC site, either 5′ or 3′ to the mismatch, in the newly synthesized strand. MutH shares sequence homology with the type II restriction endonuclease, Sau3AI. Although both enzymes recognize and cleave GATC sequences, MutH does not bind or cleave fully methylated GATC sites, whereas Sau3AI cleaves fully, hemi, and unmethylated GATC sites. Wei Yang and coworkers have isolated MutH from the gram-negative bacteria *H. influenzae* and used this enzyme to obtain the crystal structure of a MutH•DNA complex (FIGURE 12.21). Two MutH monomers bind to the 22 bp DNA, which has two hemimethylated GATC sites.

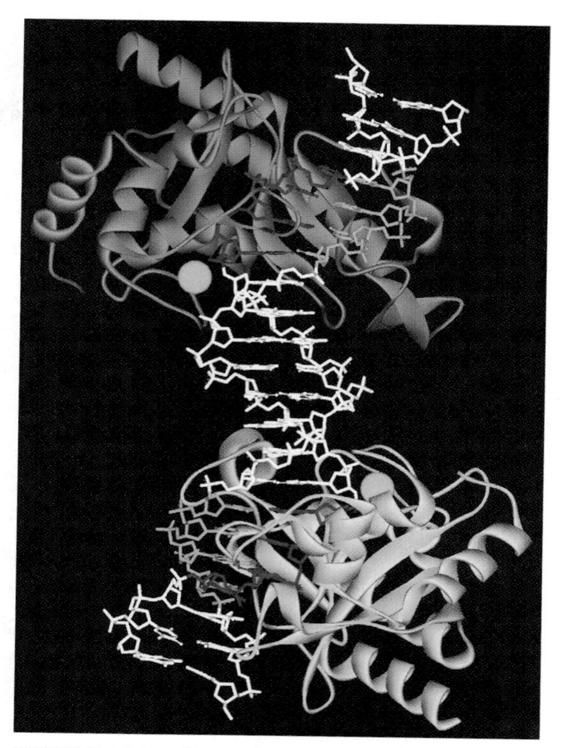

FIGURE 12.21 **Crystal structure of a MutH•hemimethylated DNA complex.** Two MutH monomers from *Haemophilus influenzae* form a complex with a 22 bp DNA, which has two hemimethylated GATC sites. The MutH monomers are shown as blue and orange ribbon structures. The DNA backbone is shown as a white stick structure with unmethylated GATC sequences shown as red stick structures and methylated GATC sequences shown as green stick structures. Calcium ions are shown as green spheres. (Structure from Protein Data Bank 2AOR. Y. H. Li, et al., *Mol. Cell.* 20 [2005]: 155–166. Prepared by B. E. Tropp.)

As shown in Figure 12.20, helicase II (UvrD) unwinds the DNA and SSB binds to the resulting single strands. When the incision is 5′ to the mismatch, Exo VII or RecJ exonucleases hydrolyze the nicked strand in a 5′→3′ direction. When the incision is 3′ to the mismatch, ExoI, ExoVII, or EcoX exonucleases hydrolyze the nicked strand in a 3′→5′ direction. DNA polymerase III holoenzyme fills the gap with new DNA (shown in red). DNA ligase seals the remaining nick and deoxyadenosine methylase adds a methyl group to the GATC site. All organisms that have a mismatch repair system have MutS and MutL homologs. MutH, however, is present only in *E. coli* and some other gram-negative bacteria. Organisms that lack MutH require some means other than the recognition of the unmethylated GATC site to distinguish between newly synthesized DNA strands and parental strands. The *E. coli* mismatch repair system suggests a possible solution to the cleavage recognition problem. MutH is not required if the DNA molecule already has a nick on either the 5′ or 3′ side of the mismatch.

Titia K. Sixma and coworkers have solved the crystal structure for a C-terminal truncated *E. coli* MutS homodimer in complex with a 30 bp DNA oligomer with a G/T mismatch (FIGURE 12.22a). The deletion of the last 53 amino acid residues in the truncated MutS changes the protein from a tetramer to a dimer in solution, perhaps explaining why the truncated MutS does not support mismatch repair. The DNA oligomer passes through the larger of two channels formed by the MutS monomers. The function of the other channel, which is large enough to accommodate another double-stranded DNA segment, is not known. The MutS dimer also has two ATPase domains, which are located at the far end of the structure with respect to the

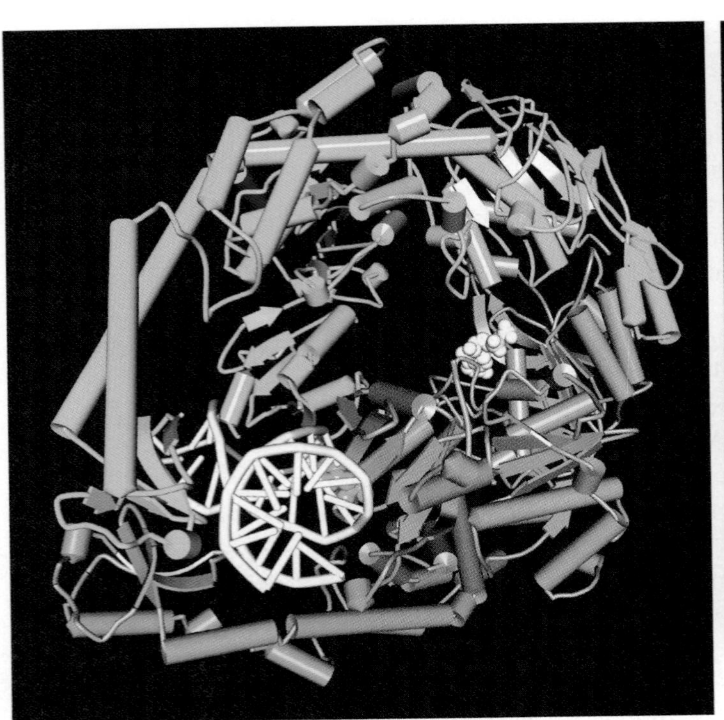

(a)

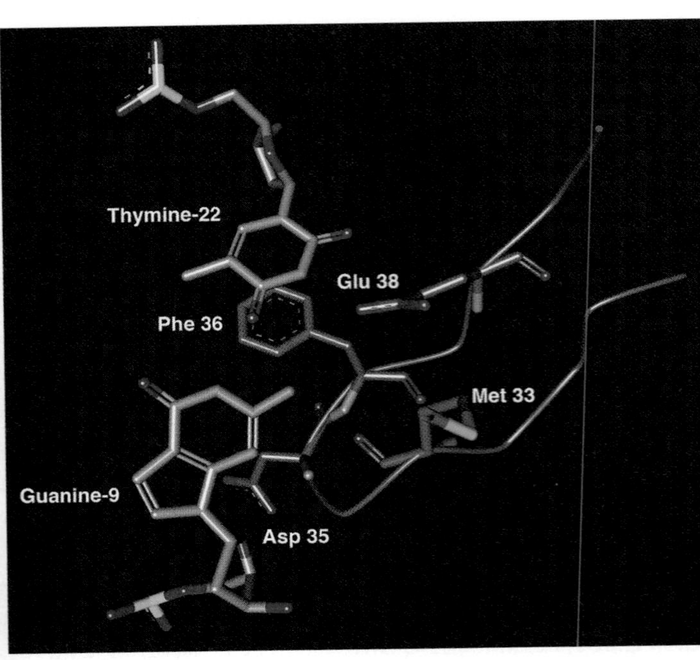

(b)

FIGURE 12.22 **Structure of a C-terminal truncated *E. coli* MutS homodimer in complex with a 30 bp DNA oligomer with a G/T mismatch at position 9.** (a) The MutS dimer forms a clamp containing two channels. The DNA oligomer (white) passes through the larger channel at the bottom of the dimer. Only one subunit (orange) makes contact with the G/T mismatch. Phe-36 in this subunit is shown as a green spacefill structure. The same subunit also binds ADP (white spacefill structure) at a site that is distal to the DNA channel. The other subunit (blue) does not have a nucleotide in its ATPase site. (b) The phenylalanine (Phe-36) in the highly conserved Phe-X-Glu motif stacks with a mispaired base after intercalating into the helix through the minor groove. Thy and Gua indicate thymine and guanine, respectively. (Structures from Protein Data Bank 1E3M. M. H. Lamers, et al., *Nature* 407 [2000]: 711–717. Prepared by B. E. Tropp.)

DNA channel. Although the MutS dimer is made of two identical polypeptides, the MutS•DNA oligomer complex is asymmetric. Only one MutS subunit makes contact with the mismatch base pair. The Phe residue in a highly conserved Phe-X-Glu (where X is any amino acid residue) motif of one of the subunits inserts into the minor groove of the DNA helix at the mismatch site, causing a 60° bend in the DNA (FIGURE 12.22b). The subunit that makes contact with the mismatch has an ADP bound to its ATPase site. The ATPase site in the other subunit appears to be empty. Although ATP is not required for the MutS to bind DNA, it is required for specificity.

MutL exists as a homodimer in solution. The crystal structure of the intact protein has not yet been solved. MutL interacts with MutS but it is not yet known whether this interaction takes place at the mismatch site or after MutS has moved from this site. In fact, there is conflicting evidence about whether MutS actually moves from the mismatch site or remains bound to this site. In the former case, MutS, with or without MutL, could use the energy provided by ATP to slide to the GATC site. In the latter case, some type of DNA bending would be required to ensure that the MutS•MutL complex bound to the mismatch site was also able to interact with the GATC site. In either case, the MutS•MutL complex would need to activate MutH so that it could nick the unmethylated GATC site.

Eukaryotes have proteins that are homologous to MutS and MutL but lack homologs to MutH. Three human MutS homologs, designated MSH2, MSH3, and MSH6, participate in mismatch repair. MSH2 and MSH6 combine to form a heterodimer called MutSα, and MSH2 and MSH3 combine to form a second heterodimer called MutSβ. MutSα initiates mismatch repair at single mismatches and small insertion/deletion loops, whereas MutSβ only initiates mismatch repair at insertion/deletion loops of various sizes. The structures of MutSα and MutSβ are thought to be similar to the MutS homodimer in bacteria. The MSH6 subunit, but not the MSH2 subunit, in Mutα is the source of the Phe-X-Glu motif that makes contact with the mismatch. Mammalian homologs of the bacterial MutL protein that participate in mismatch repair are designated MLH1 and PMS2 and the heterodimer containing these two subunits is called MutLα.

Paul Modrich and coworkers have reconstituted the human mismatch repair system *in vitro*. A strand break on either side of the mismatch is sufficient to direct repair. Purified human MutSα, MutLα, ExoI (a 5′→3′ exonuclease), RPA, PCNA, RFC, and DNA polymerase δ are required for bidirectional repair. MutSα, ExoI, and RPA suffice to excise a mismatch when the nick is on the 5′ side of the mismatch but MutLα, RFC, and PCNA also are required when the nick is on the 3′-side of the mismatch. The observation that the mismatch repair system can degrade newly synthesized strands with a nick on the 3′-side of the mismatch was very puzzling because ExoI, the only exonuclease added to the system, degrades DNA in a 5′→3′ direction. Modrich and coworkers solved the puzzle by demonstrating that MutLα is a latent endonuclease that is activated in a mismatch-, MutSα-, RFC-, PCNA-, and ATP-dependent fashion. Once activated, MutLα preferentially makes incisions in the strand that already has a nick, that is, the discontinuous strand during replication. The endonuclease activity appears to require an amino acid motif present in the PMS2 but not the MLH1 subunit. This motif is also present in MutL homologs from other organisms, including many bacteria. The motif, however, is not present in MutL proteins from bacteria such as *E. coli* with mismatch repair systems that depend on GATC methylation. FIGURE 12.23 presents a schematic diagram that shows 5′- and 3′-directed human mismatch repair pathways. MutSα, PCNA, and RFC cooperate to activate the latent MutLα endonuclease, which nicks the discontinuous strand of a DNA strand on both the 5′ and 3′ sides of the mismatch. When the original nick is on the 3′-side of the mismatch, MutLα incisions produces a new 5′ terminus on the far side of the mismatch that can serve as an entry site for MutSα-activated ExoI, allowing it to remove the mismatch using its 5′→3′ exonuclease activity. RPA stimulates ExoI activity in the presence of MutSα as long as the mismatch is present. Once the mismatch has been removed, RPA inhibits the exonuclease, probably by displacing ExoI from the DNA.

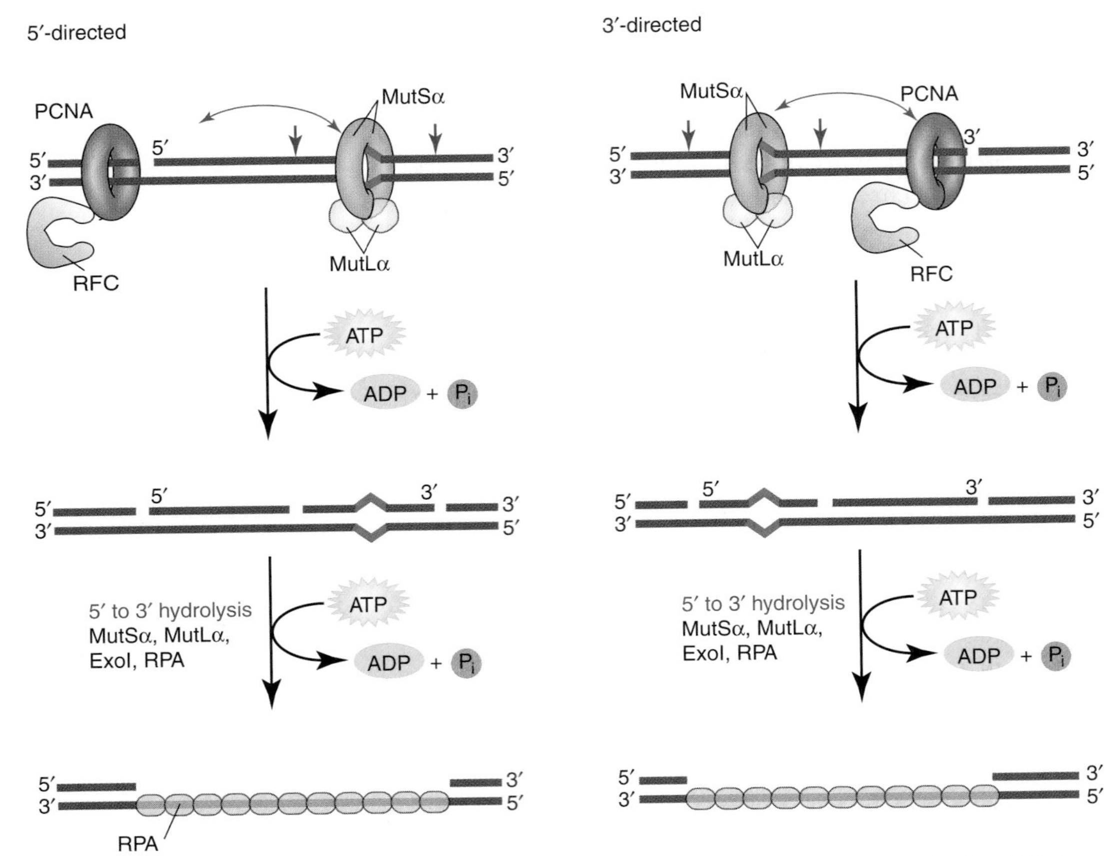

FIGURE 12.23 **Incision of the discontinuous heteroduplex strand in human mismatch repair.** MutSα, PCNA, and RFC activate a latent MutLα endonuclease, which nicks the discontinuous strand of a DNA with a nick on the 5′ or 3′ side of the mismatch in an ATP-dependent reaction (red arrows). For simplicity, protein components are shown only in the top structures. For a nick on the 3′ side, this produces a new 5′ terminus on the distal side of the mismatch that serves as an entry site for MutSα-activated ExoI, which removes the mismatch in a 5′ to 3′ hydrolytic reaction controlled by RPA. (Adapted from F. A. Kadyrov, et al. *Cell* 126 [2006]: 297–308.)

Individuals with a nonfunctional mismatch repair system due to a defective MutSα or MutLα suffer from **nonpolyposis colon cancer (HNPCC)**, an autosomal recessive syndrome that greatly increases their predisposition to develop intestinal cancer.

12.5 The SOS Response and Translesion DNA Synthesis

Error-prone DNA polymerases catalyze translesion DNA synthesis.

Replicative DNA polymerases such as the DNA polymerase III holoenzyme in *E. coli* and DNA polymerases δ and ε in eukaryotes occasionally encounter DNA lesions such as abasic (AP) sites or cyclobutane pyrimidine dimers that have escaped DNA repair systems. Such lesions can derail the replication complex, causing the replicative polymerase subassembly to dissociate from the DNA and possibly also from the sliding clamp. Cells survive this event by using a special class of DNA polymerases, the **error-prone DNA polymerases**, to catalyze DNA synthesis across the lesion. Once this **translesion DNA synthesis** is complete, replicative DNA polymerases can resume their normal task. Error-prone DNA polymerases catalyze DNA synthesis with a much lower fidelity than the replicative DNA polymerases. Because of their lower fidelity, the error-prone DNA polymerases are able to extend the primer strand even when the template strand is damaged. The advantage of using the error-prone DNA polymerases is that the cell is usually able to survive DNA damage that has not been repaired. This survival comes at a cost, however, because translesion DNA synthesis introduces mutations. Important insights into error-prone DNA synthesis and the enzymes that catalyze it come from studies of the SOS response in *E. coli*.

RecA and LexA regulate the *E. coli* SOS response.

The term **SOS response** may seem like a strange name for a complex biological response to DNA damage. Miroslav Radman suggested that the international distress signal, SOS, be applied to the signal that activates a multigene response to DNA damage in *E. coli* after UV irradiation or a chemically induced chromosomal change. Although the total number of genes induced by the SOS signal was not known in 1973 when Radman made his suggestion, subsequent studies revealed that the SOS signal induces more than 40 *E. coli* SOS genes. Some induced genes such as *uvrA*, *uvrB*, and *uvrD* code for proteins that are used to repair damaged DNA, others code for proteins that participate in recombinational DNA repair (see Chapter 13), and still others code for error-prone DNA polymerases.

The SOS response is regulated by two proteins, **LexA** and **RecA**. The LexA protein is a homodimer that binds to regulatory sites that lie just 5′ of the transcription initiation sites of genes that respond to the SOS signal. The regulatory sites, 20 bp sequences that vary slightly from one transcription unit to another, are called **operators**. Each operator has a twofold axis of symmetry, permitting one LexA subunit to bind to the left half-site and another to bind to the right half-site. The properties of operators will be described in much greater detail in Chapter 16. The important point for understanding how the SOS signal works is that the LexA dimer represses gene transcription when bound to the operator. Moreover, the degree of repression is directly related to the LexA dimer's affinity for the SOS gene operator. Each LexA subunit folds into two domains that are joined by a short connecting peptide. The N-terminal domain binds to an operator half-site and the C-terminal domain has a dimerization element that holds the two LexA subunits in the dimer together. The C-terminal domain also has a latent protease activity that is an important part of the signal response. Pheobe A. Rice and coworkers have determined the structure for a LexA dimer bound to DNA (FIGURE 12.24).

Two interesting classes of LexA mutants have been isolated. The first class, LexA[Ind$^-$] mutants, code for a LexA variant that binds

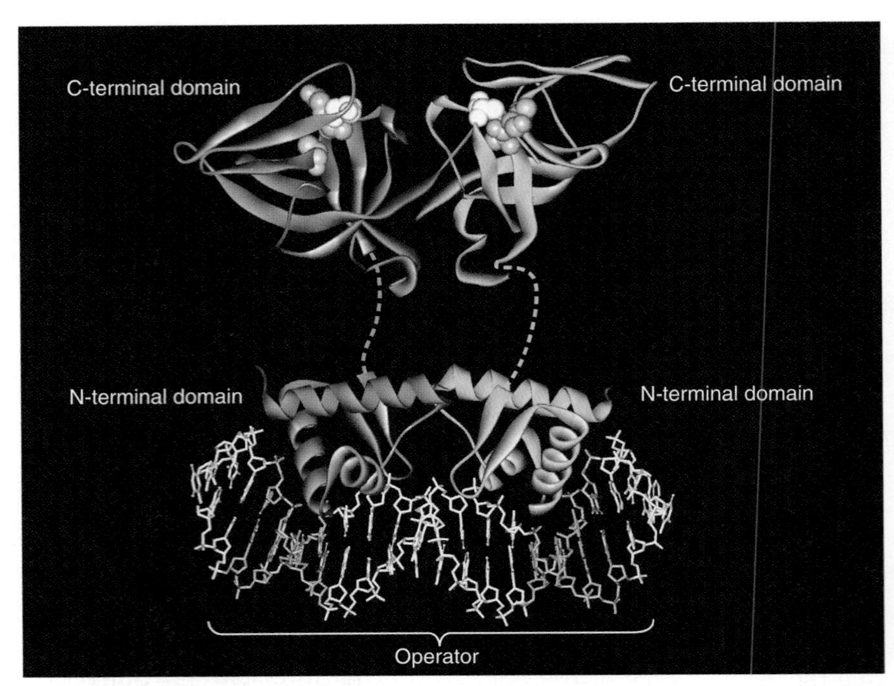

FIGURE 12.24 Model of LexA dimer docked on a recA operator site. LexA subunits are shown as blue and green ribbons. The N-terminal domain of each LexA subunit binds to a DNA sequence called the operator. The five-residue linker connecting the two domains, which is disordered and so not visible, is shown as a dashed line. The operator DNA has the sequence CTGTN$_8$ACAG (where N can be any nucleotide). CTGT and ACAG are shown as orange stick figures and the other nucleotides as white stick figures. The C-terminal domain has elements necessary for dimerization. Ser-119 and Lys-156 are part of the active site for a latent protease. Ser-119 is shown as a white spacefill structure and Lys-156 (here mutated to alanine) is shown as an orange spacefill structure. Cleavage takes place between Ala-84 and Gly-85, which are shown as orange spacefill structures. (Structure from Protein Data Bank 3JSO. A. P. P. Zhang, Y. Z. Pigli and P. A. Rice, *Nature* 466 [2010]: 883–886. Prepared by B. E. Tropp.)

to SOS gene operators with such high affinity that the SOS genes are turned off even in the presence of an SOS signal. The second class, LexA[Def] mutants, cannot make LexA proteins or make defective LexA proteins with little, if any, affinity for SOS gene operators. SOS gene operators in LexA[Def] mutants are not occupied and as a result the SOS response is permanently turned on.

RecA is a multifunctional protein that is essential for regulation of the SOS response, error prone DNA synthesis, and as indicated by its name, homologous recombination. Its role in homologous recombination is described in Chapter 13. The first step in the SOS response is the appearance of a single-stranded region of DNA, which is formed in response to a DNA lesion that blocks the progress of the replicating fork. RecA combines with the single-stranded DNA region to form a nucleoprotein filament that acts as the SOS response signal. Filament assembly does not take place during active replication because the SSB protein is usually present in high enough concentration to successfully compete with RecA for the relatively short single-stranded regions present at the normal replication fork. RecA nucleoprotein filament

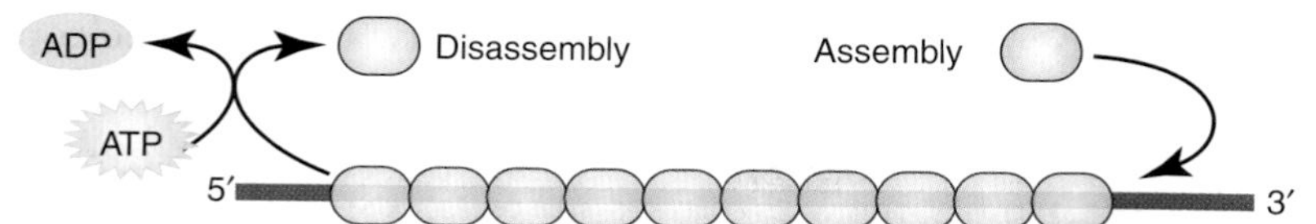

FIGURE 12.25 Assembly and disassembly of RecA nucleofilaments. The assembly and disassembly of RecA nucleofilaments are unidirectional processes, each moving in a 5′ to 3′ direction along a single stranded DNA segment. More specifically, RecA adds at the 3′-proximal end and leaves from the 5′-proximal end. Disassembly requires ATP hydrolysis. (Modified with permission from K. Schlacher, et al. *Chem. Rev.* 106 [2006]: 406–419. Copyright 2006 American Chemical Society.)

assembly, which requires ATP but not its hydrolysis, proceeds in a 5′ to 3′ direction with RecA molecules adding to the 3′-end of the single-stranded DNA segment (FIGURE 12.25). RecA has an ATPase activity that can convert bound ATP to ADP. This hydrolysis is accompanied by the conversion of RecA from a conformation with a high affinity for DNA to one with a low affinity for DNA. ATP hydrolysis is required for nucleoprotein filament disassembly, which also takes place in a 5′ to 3′-direction.

Key steps in the SOS response are shown in FIGURE 12.26. RecA in the nucleoprotein filament acts as a co-protease for LexA by activating the latent protease in the LexA C-terminal domain, without directly participating in proteolytic cleavage. Once activated, the LexA protease catalyzes its own cleavage, splitting into an N-terminal fragment and a C-terminal fragment. The N-terminal fragment, which contains the DNA binding domain, no longer binds to the operator with high affinity. Cleavage of LexA activates transcription of the SOS genes including the *recA* and *lexA* genes. When the *recA* gene is fully

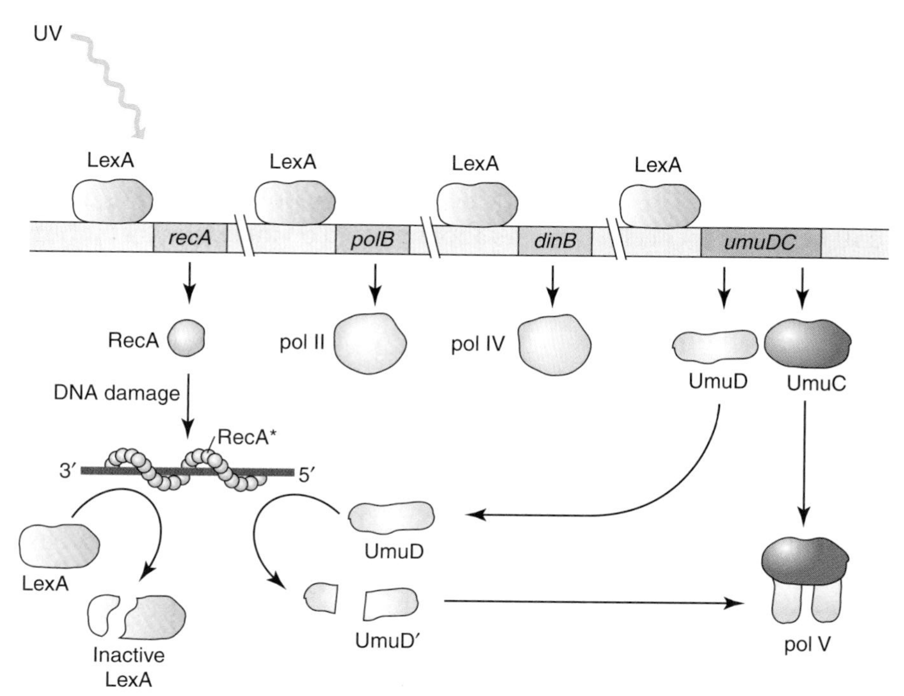

FIGURE 12.26 Regulation of the SOS response. LexA protein (green) binds to the operator, which lies just before the SOS response genes, repressing their transcription. After UV irradiation, RecA (orange ball) adds to single-stranded regions of DNA to form a nucleoprotein filament. RecA in the nucleoprotein filament acts as a co-protease to activate the latent protease activity in C-terminal domain of LexA. The activated protease splits LexA into two fragments. The N-terminal fragment containing the DNA binding domain loses its high affinity for the SOS response gene operators. As a result, the SOS genes are no longer repressed. The product of one SOS response gene, dinB (pink), forms DNA polymerase IV. The products of two other SOS response genes, umuC and umuD (blue), contribute to the formation of DNA polymerase V. Both DNA polymerases IV and V can catalyze translesion synthesis. (Adapted from K. Schlacher, et al., *Chem. Rev.* 106 [2006]: 406–419.)

induced, approximately 1000 RecA monomers will form a nucleoprotein filament on a segment of DNA that is about as long as an Okazaki fragment. Even a short segment of single-stranded DNA bound by 50 to 60 RecA monomers produces an SOS response signal that is strong enough to ensure the inactivation of LexA within a couple of minutes. The SOS response can be divided into three stages:

1. Lesions are removed by base excision repair and nucleotide excision repair.
2. RecA participates in recombinational repair, which corrects lesions that remain after the first stage (see Chapter 13).
3. Translesion DNA synthesis bypasses remaining lesions, permitting normal DNA synthesis to resume.

The LexA dimer cannot accumulate as long as RecA is present in the nucleoprotein filament. The nucleoprotein filament, however, disassembles after DNA repair or translesion DNA synthesis is complete, allowing the LexA dimer to once again accumulate and repress the SOS genes.

The SOS signal induces the synthesis of DNA polymerases II, IV, and V.

DNA damage induces the synthesis of three DNA polymerases that help the cell to deal with DNA damage that escapes the repair systems. The first of these enzymes, **DNA polymerase II**, the *dnaB* gene product, is a high fidelity polymerase that contains a 3′→5′ exonuclease activity. Normally present at about 30 to 50 molecules per cell, its level increases about sevenfold in response to the SOS signal. DNA polymerase II appears to play an important role in restarting DNA replication after lesion bypass DNA synthesis is complete. It does so by extending a mismatched primer terminus when DNA polymerase III has difficulty in doing so. DNA polymerase II can also synthesize across an abasic (AP) site and bypass some bulky adducts by primer realignment that takes advantage of repeated nucleotide sequences and introduces frameshift mutations.

The second DNA polymerase that is induced in response to the SOS signal is **DNA polymerase IV**, a product of the *dinB* (damaged induced) gene. The uninduced level of DNA polymerase IV, about 250 molecules per cell, increases about tenfold on induction by the SOS signal. DNA polymerase IV can perform either error-free or error-prone translesion DNA synthesis. Its fidelity is determined by the nature of the lesion and the nucleotides that surround the lesion. Remarkably, the DNA polymerase III core subassembly and DNA polymerase IV both bind to the same β_2-sliding clamp. A choice, therefore, must be made between the two polymerases at a template-primer junction. The DNA polymerase III core subassembly is selected during normal and uninterrupted replication. DNA polymerase IV gains control when some impediment stalls the progress of the DNA polymerase III core subassembly. Once the stall is relieved, the DNA polymerase III core subassembly, which remains bound to the β_2 sliding clamp, regains control and resumes normal replication. The precise mechanism that leads to switching between the two DNA polymerases remains to be elucidated.

The third DNA polymerase induced in response to the SOS signal is **DNA polymerase V**, a heterotrimer. The protein subunits in DNA polymerase V are products of the *umuC* and *umuD* genes (where *umu* indicates UV mutagenesis). DNA polymerase V is below detectable levels in uninduced bacteria. After induction, UmuC and UmuD protein levels increase to about 200 and 2400 copies per cell, respectively. The RecA nucleoprotein filament activates a latent protease activity in UmuD in much the same way that it activates the latent protease in LexA. Once activated, the protease cleaves UmuD, converting it to UmuD′. Then two molecules of UmuD′ combine with a molecule of UmuC to form DNA polymerase V (UmuD′$_2$C). DNA polymerase V requires the presence of the clamp loader, the β_2 sliding clamp, SSB, and RecA for efficient DNA translesion synthesis. DNA polymerase V is converted to an active UmuD′$_2$C•RecA•ATP complex called the **mutasome**. The UmuC subunit contains the DNA polymerase activity. Myron F. Goodman and coworkers have proposed the model shown in FIGURE 12.27 to explain how the mutasome is formed. According to this model, the UmuD′$_2$C complex has little activity in the absence of a RecA nucleoprotein filament (RecA*). The mutasome is formed by the transfer of RecA•ATP from the nucleoprotein filament's 3′ end to the UmuD′$_2$C complex. After translesion DNA synthesis is complete, the mutasome dissociates from primer/template DNA. Goodman and coworkers propose that RecA•ATP binds to UmuD′$_2$ in the active

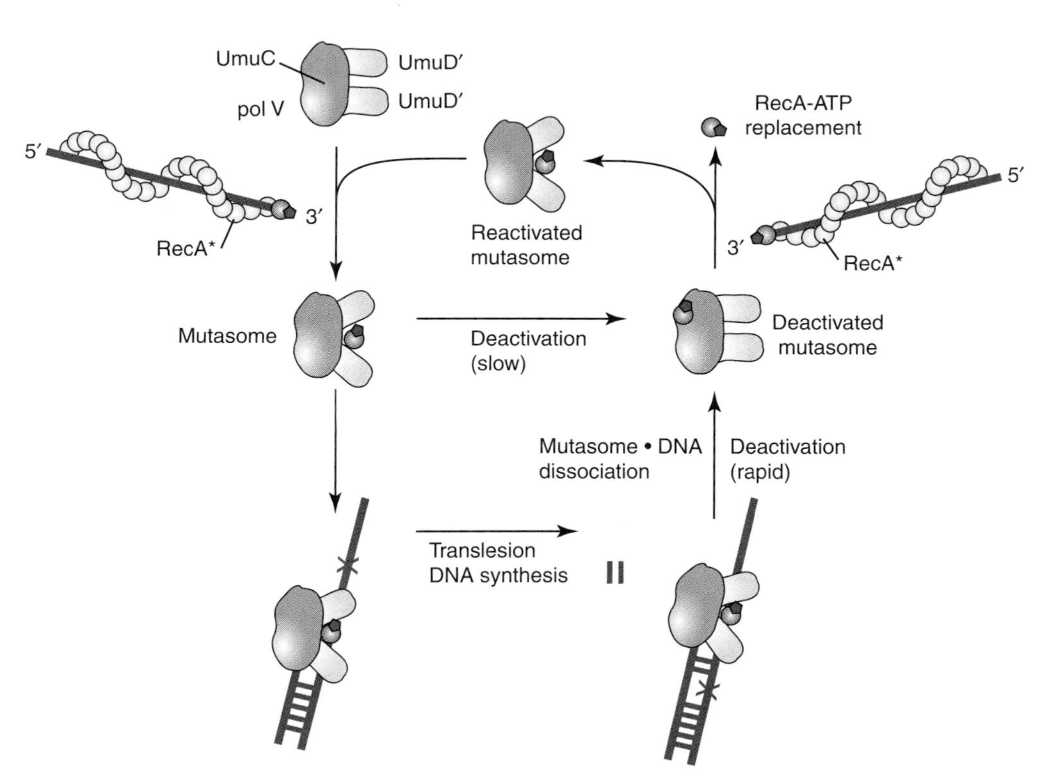

FIGURE 12.27 **The mutasome is formed by the transfer of RecA•ATP from the 3′ end of a nucleoprotein filament to the UmuD′$_2$C complex.** The active mutasome catalyzes translesion synthesis. After translesion synthesis is complete, the mutasome dissociates from primer/template DNA, which somehow triggers the movement of RecA•ATP from UmuD′$_2$ in the active mutasome to UmuC in the deactivated complex. RecA is indicated by a tan circle and ATP by a red pentagon. X indicates an uncorrected DNA lesion. (Adapted from Q. Jiang, et al., *Nature* 460 [2009]: 359–363.)

mutasome but that dissociation from primer/template DNA somehow causes RecA•ATP to move to UmuC and thereby deactivate the complex. The mutasome can catalyze translesion synthesis through several different types of DNA lesions, including cyclobutane thymine dimers, (6-4) photoproducts, abasic (AP) sites, and large covalent adducts. Its low fidelity permits it to insert a nucleotide opposite a lesion and then add nucleotides to the resulting 3′-end. Bacteria with a nonfunctional mutasome do not accumulate mutations in response to UV light or to various chemical mutagens. Hence, such bacteria cannot be used in the Ames test (see Chapter 11). The mutasome not only adds nucleotides across from a lesion but it also introduces untargeted mutations in sites that appear to be undamaged. Thus, cells require the mutasome to survive damage that is not repaired by DNA repair systems but this enzyme must be highly regulated so that it cannot introduce mutations into undamaged DNA when replicative DNA polymerases are not blocked by DNA damage.

Human cells have at least 14 different template-dependent DNA polymerases.

The discoveries of *E. coli* DNA polymerases IV and V motivated investigators to search the eukaryotic genome data banks to determine if eukaryotes also contained additional DNA polymerases. As a result of this type of search and related biochemical and genetic studies, it has now been established that human cells have at least 14 different template-dependent DNA polymerases. The names and some physiological properties of the eukaryotic DNA polymerases are given in Table 12.1. The original Greek names have been replaced by names suggested by the Human Genome Organization (HUGO). Unfortunately,

TABLE 12.1 Proposed Nomenclature for Eukaryotic DNA Polymerases

Greek Name	HUGO name[1]	Class	Proposed Main Function
α (alpha)	POLA	B	DNA replication
β (beta)	POLB	X	Base excision repair
γ (gamma)	POLG	A	Mitochondrial replication
δ (delta)	POLDI	B	DNA replication
ε (epsilon)	POLE	B	DNA replication
ζ (zeta)	POLZ	B	Bypass synthesis
η (eta)	POLH	Y	Bypass synthesis
υ (theta)	POLQ	A	DNA repair
ι (iota)	POLI	Y	Bypass synthesis
κ (kappa)	POLK	Y	Bypass synthesis
λ (lambda)	POLL	X	Base excision repair
μ (mu)	POLM	X	Non-homologous end joining
σ (sigma)	POLS	X	Sister chromatid cohesion
	REVIL	Y	Bypass synthesis
	TDT	X	Antigen receptor diversity

Note: [1]HUGO is an acronym for Human Genome Organization.
Source: Modified from P. M. J. Burgers, et al., *J Biol. Chem.* 276 (2001): 43487–43490.

the HUGO recommendations are not always followed in the literature and the older Greek names are often used. DNA polymerases are grouped into classes depending on sequence homologies and structure similarities. The *E. coli* DNA polymerases I, II, III, and IV/V are founding members of Classes A, B, C, and Y, respectively. Eukaryotic PolB (DNA polymerase β) is the founding member of Class X. Although each kind of eukaryotic DNA polymerase appears to have a specific function, some polymerases appear able to stand in for others.

We now describe the function of just one of the new eukaryotic DNA polymerases. POLH (DNA polymerase η) is of special interest because it has a direct bearing on DNA repair. Individuals who lack a functional form of POLH suffer from a variant form of xeroderma pigmentosum known as XP-V (XP-variant). The reason they have an increased likelihood of developing skin cancer is that human POLH decreases susceptibility to skin cancer by promoting DNA synthesis past UV-induced cyclobutane thymine dimers that escape nucleotide excision repair. The key to POLH's ability to reduce the level of mutations is that POLH can incorporate two adenylate molecules opposite the thymine bases in the thymine dimer. The functions of some of the other eukaryotic DNA polymerases will be discussed in the next chapter in connection with their participation in recombination.

Suggested Reading

Historical

Friedberg, E. C. *Correcting the Blueprint of Life*. 1997. Woodbury, NY: Cold Spring Harbor Laboratory Press.

General

Dalhus, B., Laerdahl, J. K., Backe, P. H., and Bjørås, M. 2009. DNA base repair–recognition and initiation of catalysis. *FEMS Microbiol Rev* 33:1044–1078.

Friedberg, E. C., Walker, G. C., Siede, W., et al. 2006. *DNA Repair and Mutagenesis*, 2nd ed. Washington, DC: ASM Press.

Sancar, A., Lidsey-Boltz, L. A., Ünsel-Kaçmaz, and Linn, S. 2004. Molecular mechanisms of mammalian DNA repair and the DNA damage checkpoints. *Ann Rev Biochem* 73:39–85.

Schärer, O. D. 2003. Chemistry and biology of DNA repair. *Angewandte Chemie Int Ed* 42:2946–2974.

Damage Reversal

Begley, T. J., and Samson, L. D. 2002. AlkB mystery solved: oxidative demethylation of N1-methyladenine and N3-methylcytosine adducts by a direct reversal mechanism. *Trends Biochem Sci* 28:2–5.

Begley, T. J., and Samson, L. D. 2004. Reversing DNA damage with a directional bias. *Nature Struct Mol Biol* 11:688–690.

Falnes, P.O., Johansen, R.F., and Seeberg, E. 2002 AlkB-mediated oxidative demethylation reverses DNA damage in *Escherichia coli*. *Nature* 419:178–182.

Goosen, N., and Moolenaar, G. F. 2008. Repair of UV damage in bacteria. *DNA Repair* 7: 353–379.

He, C., Hus, J. C., Sun, L. J., et al. 2005. A methylation-dependent electrostatic switch controls DNA repair and transcriptional activation by *E. coli* Ada. *Mol Cell* 20:117–129.

Kao, Y.-T., Saxena, C., Wang, L., et al. 2005. Direct observation of thymine dimer repair in DNA by photolyase. *Proc Natl Acad Sci USA* 102:16128–16132.

Kelner, A.1949. Effect of visible light on the recovery of *Streptomyces griseus conidia* from ultraviolet irradiation injury. *Proc Natl Acad Sci USA* 35:73–79.

Lee, D.-H., Jin, S.-J., Cai, S., et al. 2005. Repair of methylation damage in DNA and RNA by mammalian AlkB homologues. *J Biol Chem* 280:39448–39459.

Mees, A., Klar, T., Gnau, P., et al. 2004. Crystal structure of a photolyase bound to a CPD-like DNA lesion after in situ repair. *Science* 306:1789–1793.

Memisoglu, A., and Samson, L. D. 2001. DNA repair by reversal of damage. In: *Encyclopedia of Life Sciences*. London: Nature, pp. 1–8.

Mishina, Y., Duguid, E. M., and He, C. 2006. Direct reversal of DNA alkylation damage. *Chem Rev* 106:215–232.

Müller, M., and Carell, T. 2009. Structural biology of DNA photolyases and cryptochromes. *Curr Opin Struct Biol* 19:277–285.

Myers, L. C., Terranova, M. P., Nash, H. M., et al. 1992. Zinc binding by the methylation signaling domain of the *Escherichia coli* Ada protein. *Biochemistry* 31:4541–4547.

Rupert, C. S., Goodgal, S. H., and Herriott, R. M. 1957. Photoreactivation in vitro of ultraviolet inactivated *Hemophilus influenzae* transforming factor. *J Gen Physiol* 41:451–471.

Sancar, A. 2003. Structure and function of DNA photolyase and cryptochrome blue-light photoreceptors. *Chem Rev* 103:2203–2237.

Sancar, A. 2008. Structure and function of photolyase and *in vivo* enzymology: 50th anniversary. *J Biol Chem* 283:32153–32157.

Scährer, O. D. 2003. Chemistry and biology of DNA repair. *Angewandte Chemie Int Ed* 42:2946–2974.

Todo, T.,Takemori, H., Ryo, H., et al. 1993. A new photoreactivating enzyme that specifically repairs ultraviolet light-induced (6-4) photoproducts. *Nature* 361:371–374.

Trewick, S. C., Henshaw, T. F., Hausinger, R. P., et al. 2002. Oxidative demethylation by *Escherichia coli* AlkB directly reverses DNA base damage. *Nature* 419:174–178.

Yarnell, A. 2005. Light shed on photolyases. *Chem Engineer News* 83:35.

Base Excision Repair

Barnes, D. E., and Lindahl, T. 2004. Repair and genetic consequences of endogenous DNA base damage in mammalian cells. *Ann Rev Biochem* 38:445–476.

Baute, J., and Depicker, A. 2008. Base excision repair and its role in maintaining genome stability. *Crit Rev Biochem Mol Biol* 43:239–276.

Beard, W. A., and Wilson, S. H. 2006. Structure and mechanism of DNA polymerase. *Chem Rev* 106:361–382.

Boiteux, S., and Guillet, M. 2004. Abasic sites in DNA: repair and biological consequences in *Saccharomyces cerevisiae*. *DNA Repair* 3:1–12.

Hegde, M. L., Hazra, T., and Mitra, S. 2008. Early steps in the DNA base excision/single-strand interruption repair pathway in mammalian cells. *Cell Res* 18:27–47.

Krokan, H. E., Nilsen, H., Skorpen, F., et al. 2000. Base excision repair of DNA in mammalian cells. *FEBS Lett* 476:73–77.

Memisoglu, A., and Samson, L. 2000. Base excision repair in yeast and mammals. *Mutat Res* 451:39–51.

Nilsen, H., and Krokan, H. E. 2001. Base excision repair in a network of defense and tolerance. *Carcinogenesis* 22:987–998.

Robertson, A. B., Klungland, A., Rognes, T., and Leiros, I. 2009. Base excision repair: the long and short of it. *Cell Mol Life Sci* 66:981–993.

Sander, M., and Wilson, S. H. 2002. Base excision repair, AP endonucleases, and DNA glycosylases. In: *Encyclopedia of Life Sciences*. London: Nature, pp. 1–9.

Seeberg, E., Eide, L., and Bjørås, M. 1995. The base excision pathway. *Trends Biochem Sci* 20:391–397.

Nucleotide Excision Repair

Batty, D. P., and Wood, R. D. 2000. Damage recognition in nucleotide excision repair of DNA. *Gene* 241:193–204.

Cleaver, J. E. 2001. Xeroderma pigmentosum: the first of the cellular caretakers. *Trends Biochem Sci* 26:398–401.

Cleaver, J. E., Karplus, K., Kashani-Sabet, M., and Limoli, C. L. 2001. Nucleotide excision repair "a legacy of creativity." *Mutat Res* 485:23–36.

Dinant, C., Houtsmuller, A. B., and Vermeulen, W. 2008. Chromatin structure and DNA damage repair. *Epigenetics & Chromatin* 1.

Doetsch, P. W. 2001. DNA repair disorders. In: *Encyclopedia of Life Sciences*. London: Nature, pp. 1–6.

Friedberg, E. C. 2000. Nucleotide excision repair and cancer disposition. *Am J Pathol* 157:693–701.

Friedberg, E. C. 2001. How nucleotide excision repair protects against cancer. *Nat Rev Cancer* 1:22–33.

Gillet, L. C. J., and Schärer, O. D. 2006. Molecular mechanisms of mammalian global genome nucleotide excision repair. *Chem Rev* 106:253–276.

Hoeijmakers, J. H. 2001. From xeroderma pigmentosum to the biological clock contributions of Dirk Bootsma to human genetics. *Mutat Res* 485:43–59.

Hutsell, S. Q., and Sancar, A. 2005. Nucleotide excision repair, oxidative damage, DNA sequence polymorphisms, and cancer treatment. *Clin Cancer Res* 11:1355–1357.

Jaciuk, M., Nowak, E., Skowronek, K. et al. 2011. Structure of UvrA nucleotide excision repair protein in complex with modified DNA. *Nat Struct Mol Biol* (In press).

Lainé, J. P., and Egly, J.-M. 2006. When transcription and repair meet: a complex system. *Trends Genet* 22:430–436.

Lehmann, A. R. 1995. Nucleotide excision repair and the link with transcription. *Trends Biochem Sci* 20:402–405.

Mitchell, J. R., Hoeijmakers, J. H. J., and Niedernhofer, L. J. 2003. Divide and conquer: nucleotide excision repair battles cancer and ageing. *Curr Opin Cell Biol* 15:232–240.

Pakotiprapha, D., Inuzuka, Y., Bowman, B. R, et al. 2008. Crystal structure of *Bacillus stearothermophilus* UvrA provides insight into ATP-modulated dimerization, UvrB interaction, and DNA binding. *Mol Cell* 29:122–133.

Park, C.-J., and Choi, B. S. 2006. The protein shuffle: sequential interactions among components of the human nucleotide excision repair pathway. *FEBS J* 273:1600–1608.

Rasmussen, R. E., and Painter R. B. 1964. Evidence for repair of ultra-violet damaged deoxyribonucleic acid in cultured mammalian cells. *Nature* 203:1360–1362.

Reed, S. H. 2005. Nucleotide excision repair in chromatin: the shape of things to come. *DNA Repair (Amst)* 4:909–918.

Robertson, A. B., Klungland, A., Rognes, T., and Leiros, I. 2009. Base excision repair: the long and short of it. *Cell Mol Life Sci* 66:981–993.

Sarker, A. H., Tsutakawa, S. E., Kostek, S., et al. 2005. Recognition of RNA polymerase II and transcription bubbles by XPG, CSB, and TFIIH: insights for transcription-coupled repair and Cockayne syndrome. *Mol Cell* 20:187–198.

Schärer, O. D. 2008. XPG: its products and biological roles. *Adv Exp Med Biol.* 637:83–92.

Sugasawa, K. 2009. UV-DDB: A molecular machine linking DNA repair with ubiquitination. *DNA Repair (Amst)* 8:969–972.

Truglio, J. J., Croteau, D. L., Van Houten, B., and Kisker, C. 2006. Prokaryotic nucleotide excision repair: the UvrABC system. *Chem Rev* 106:233–252.

Truglio, J. J., Karakas, E., Rhau, B., et al. 2006. Structural basis for DNA recognition and processing by UvrB. *Nat Struct Mol Biol* 13:360–364.

Van Houten, B., Croteau, D. L., DellaVecchia, J., et al. 2005. "Close fiting sleeves": DNA damage recognition by the UvrABC nuclease system. *Mutat Res* 577:92–117.

Mismatch Repair

Ban, C., and Yang, W. 1998. Structural basis for MutH activation in *E. coli* mismatch repair and relationship of MutH to restriction endonucleases. *EMBO J* 17:1526–1534.

Constantin, N., Dzantiev, L., Kadyrov, F. A., and Modrich, P. 2005. Human mismatch repair reconstitution of a nick-directed bidirectional reaction. *J Biol Chem* 280:39752–39761.

de Wind, N., and Hays, J. B. 2001. Mismatch repair: praying for genome stability. *Curr Biol* 11:R545–R548.

Fishel, R. 1999. Signaling mismatch repair in cancer. *Nat Med* 5:1239–1241.

Fleck, T. M., Kunz, C., and Fleck, O. 2002. DNA mismatch repair and mutation avoidance pathways. *J Cell Physiol* 191:28–41.

Harfe, B. D., and Jinks-Robinson, S. 2000. DNA mismatch repair and genetic instability. *Ann Rev Genet* 34:359–399.

Iyer, R. R., Pluciennick, A., Burdett, V., and Modrich, P. L. 2006. DNA mismatch repair: functions and mechanisms. *Chem Rev* 106:302–323.

Jiricny, J. 2000. Mismatch repair: the praying hands of fidelity. *Curr Biol* 10:R788–R790.

Jiricny, J. 2006. The multifaceted mismatch-repair system. *Nat Rev Mol Cell Biol* 7:335–346.

Jiricny, J. 2006. MutLa: at the cutting edge of mismatch repair. *Cell* 126:239–241.

Kadyrov, F. A., Dzantiev, L., Constantin, N., and Modrich, P. 2006. Endonucleolytic function of MutLα in human mismatch repair. *Cell* 126:297–308.

Kolodner, R. D. 1995. Mismatch repair: mechanisms and relationship to cancer susceptibility. *Trends Biochem Sci* 20:397–401.

Kunkel, T. A., and Erie, D. A. 2005. DNA mismatch repair. *Ann Rev Biochem* 74:681–710.

Lamers, M. H., Perrakis, A., Enzlin, J. H., et al. 2000. The crystal structure of DNA mismatch repair protein MutS binding to a G•T mismatch. *Nature* 407:711–717.

Lebbink, J. H. G., Georgijevic, D., Natrajan, G., et al. 2006. Dual role of MutS glutamate 38 in DNA mismatch discrimination and in the authorization of repair. *EMBO J* 25:409–419.

Lee, J. Y., Chang, J., Joseph, N., et al. 2005. MutH complexed with hemi- and unmethylated DNAs: coupling base recognition and DNA cleavage. *Mol Cell* 20:155–166.

Li, G.-M. 2008. Mechanisms and functions of DNA mismatch repair. *Cell Res* 18:85–98.

Marti, T. M., Kunz, C., and Fleck, O. 2002. DNA mismatch repair and mutation avoidance pathways. *J Cell Physiol* 191:28–41.

Modrich, P. 2006. Mechanisms in eukaryotic mismatch repair. *J Biol Chem* 281:30305–30309.

Peltom ki, P. 2001. DNA mismatch repair and cancer. *Mutat Res* 488:77–85.

Polosina, Y. Y., and Cupples, C. G. 2009. MutL: conducting the cell's response to mismatched and misaligned DNA. *BioEssays* 32:51–59.

Schofield, M. J., and Hsieh, P. 2003. DNA mismatch repair: molecular mechanisms and biological function. *Ann Rev Microbiol* 57:579–608.

Sixma, T. K. 2001. DNA mismatch repair: MutS structures bound to mismatches. *Curr Opin Struct Biol* 11:47–52.

SOS Response and Translesion DNA Synthesis

Baynton, K., and Fuchs, R. P. P. 2000. Lesions in DNA: hurdles for polymerases. *Trends Biochem Sci* 25:74–79.

Burgers, P. M. J., Koonen, E. V., Bruford, E., et al. 2001. Eukaryotic DNA polymerases: proposal for a revised nomenclature. *J Biol Chem* 276:43487–43490.

Butala, M., Žgur-Bertok, D., and Busby, S. J. W. 2009. The bacterial LexA transcriptional repressor. *Cell Mol Life Sci* 66:82–93.

Chattopadhyaya R., Ghosh, K., and Namboodiri, V. M. 2000. Model of a LexA repressor dimer bound to recA operator. *J Biomol Struct Dyn* 18:181–197.

Chattopadhyaya, R., and Pal, A. 2004. Improved model of a LexA repressor dimer bound to recA operator. *J Biomol Struct Dyn* 5:681–689.

Clark, A. J., and Margulies, A. D. 1965. Isolation and characterization of recombination-deficient mutants of *Escherichia coli* K12. *Proc Natl Acad Sci USA* 53:451–459.

Defals, M., and Devoret, R. 2001. SOS response. In: *Encyclopedia of Life Sciences*. London: Nature, pp. 1–9.

Friedberg, E. C., Lehmann, A. R., and Fuchs, R. P. P. 2005. Trading places: how do DNA polymerases switch during translesion DNA synthesis? *Mol Cell* 18:499–505.

Friedberg, E. C., Wagner, W., and Radman, M. 2002. Specialized DNA polymerases, cellular survival, and genesis of mutations. *Science* 296:1627–1630.

Goodman, M. F. 2000. Coping with replication 'train wrecks' in *Escherichia coli* using Pol V, Pol II, and RecA proteins. *Trends Biochem Sci* 25:189–195.

Goodman, M. F. 2002. Error-prone repair DNA polymerases in prokaryotes and eukaryotes. *Ann Rev Biochem* 71:17–50.

Goodman, M. F., and Tippin, B. 2000. Sloppier copier DNA polymerases involved in genome repair. *Curr Opin Genet Dev* 10:162–168.

Guengerich, F. P. 2006. Interactions of carcinogen-bound DNA with individual RNA polymerases. *Chem Rev* 106:420–452.

Hübscher, U. 2005. DNA polymerases: Eukaryotic. In: *Encyclopedia of Life Sciences*. pp. 1–4. John Wiley and Sons.

Janion, C. 2001. Some aspects of the SOS response system—a critical survey. *Acta Biochim Polon* 48:559–610.

Jiang, Q., Karata, K., Woodgate, R., et al 2009. The active form of DNA polymerase V is UmuD'$_2$C –RecA–ATP. *Nature* 460:259–263.

Kunkel, T. A., Pavlov, Y., and Bebenek, K. 2003. Functions of human DNA polymerases and suggested by their properties, including fidelity with undamaged DNA templates. *DNA Repair (Amst)* 2:135–149.

Livneh, Z. 2001. DNA damage control by novel DNA polymerases: translesion replication and mutagenesis. *J Biol Chem* 276:25639–25642.

López de Saro, F. J., Georgescu, R. E., Goodman, M. F., and O'Donnell, M. 2003. Competitive processivity-clamp usage by DNA polymerases during DNA replication and repair. *EMBO J* 22:6408–6418.

McIherney, P., and O'Donnell, M. 2004. Functional uncoupling of twin polymerases. *J Biol Chem* 279:21543–21551.

Miller, J. H. 2005. Perspective on mutagenesis and repair: the standard model and alternate modes of mutagenesis. *Crit Rev Biochem Mol Biol* 40:155–179.

Plosky, B. S., and Woodgate, R. 2004. Switching from high-fidelity replicases to low-fidelity lesion-bypass polymerases. *Curr Opin Genet Dev* 14:113–119.

Prakash, S., Johnson, R. E., and Prakash, L. 2005. Eukaryotic translesion synthesis DNA polymerases: specificity of structure and function. *Ann Rev Biochem* 74:317–353.

Ramadan, K., Shevelev, I., and Hübscher, U. 2004. The DNA-polymerase-X family: controllers of DNA quality? *Nat Rev Mol Cell Biol* 5:1038–1043.

Rattray, A. J., and Strathern, J. N. 2003. Error-prone DNA polymerases: when making a mistake is the only way to go ahead. *Ann Rev Genet* 37:31–66.

Rosenberg, S. M. 2001. Evolving responsively: adaptive mutation. *Nat Rev Genet* 2:504–515.

Scherbakova, P. V., and Fijalkowska, I. J. 2006. Translesion synthesis DNA polymerases and control of genome stability. *Front Biosci* 11:2496–2517.

Schlacher, K., and Goodman, M. F. 2007. Lessons from 50 years of SOS DNA-damage-induced mutagenesis. *Mol Cell Biol* 8:587–594.

Schlacher, K., Pham, P., Cox, M. M., and Goodman, M. F. 2006. Roles of DNA polymerase V and RecA protein in SOS damage-induced mutation. *Chem Rev* 106:406–419.

Showalter, A. K., Lamarch, B. J., Bakhtina, M., et al. 2006. Mechanistic comparison of high-fidelity and error prone DNA polymerases and ligases involved in DNA repair. *Chem Rev* 106:340–360.

Steitz, T. A., and Yin, Y. W. 2003. Accuracy, lesion bypass, strand displacement and translocation by DNA polymerases. *Philos Trans R Soc (Lond B)* 359:17–23.

Tippin, B., Pham, P., and Goodman, M. F. 2004. Error-prone replication for better or worse. *Trends Microbiol* 12:288–295.

Yang, W. 2003. Damage repair DNA polymerases Y. *Curr Opin Struct Biol* 13:23–30.

Zhang, A. P. P., Pigli, Y. Z., and Rice, P. A. 2010. Structure of the LexA-DNA complex and implications for SOS box measurement. *Nature* 466:883–886.

This book has a Web site, **http://biology.jbpub.com/book/molecular**, which contains a variety of resources, including links to in-depth information on topics covered in this chapter, and review material designed to assist in your study of molecular biology.

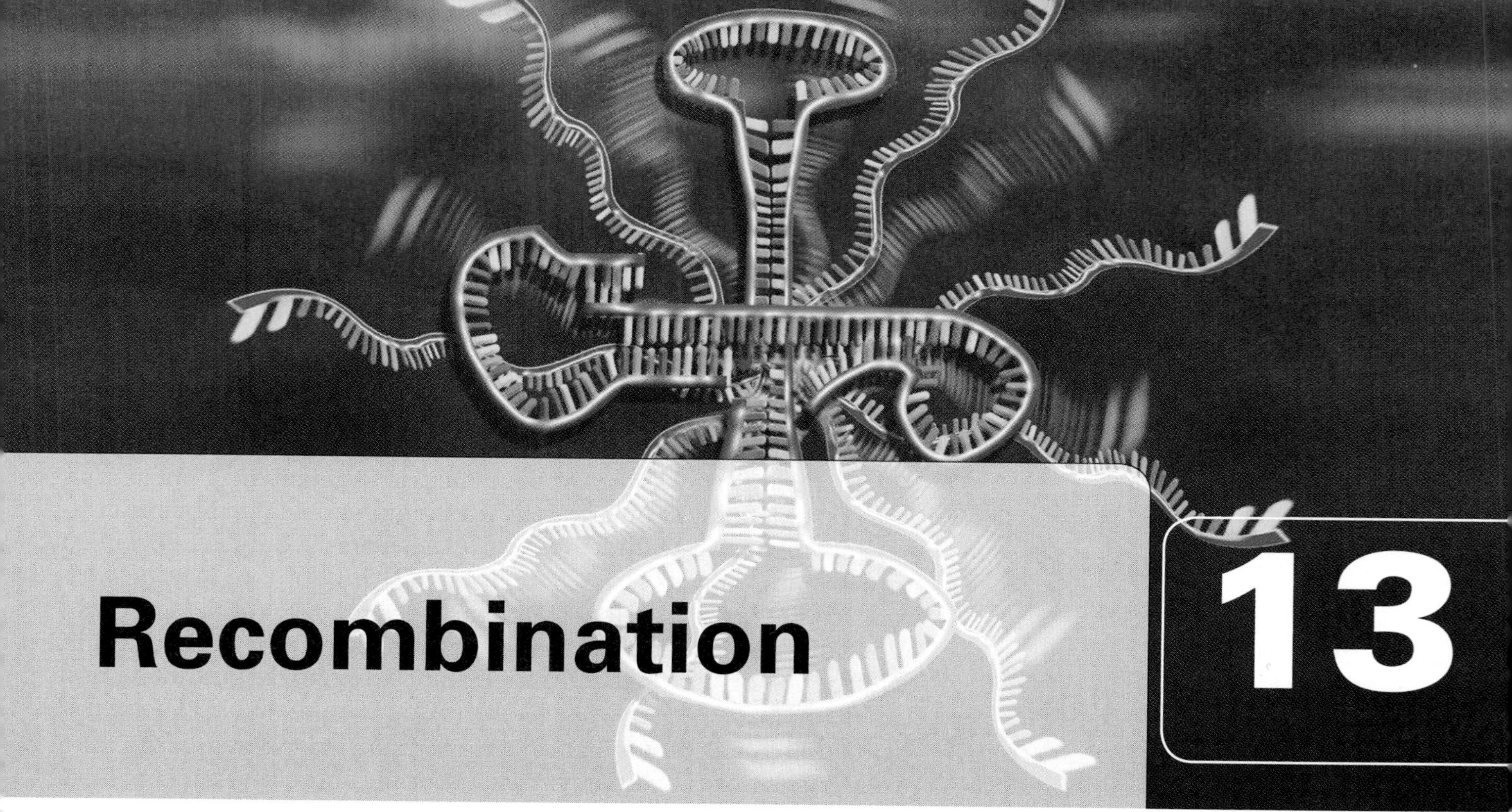

Recombination

13

OUTLINE OF TOPICS

13.1 Introduction to Homologous Recombination
Homologous recombination is an essential process for repairing DNA breaks and for ensuring correct chromosome segregation in meiosis.

13.2 Early Clues from Bacteriophage
Crossing over involves an exchange of DNA between the two interacting DNA molecules.

13.3 Early Models of Homologous Recombination
The Holliday model of homologous recombination proposes a crossed strand intermediate called a Holliday junction.
The Meselson-Radding model of recombination—a second homologous recombination model—is based on one single strand nick for initiation.

13.4 A Homologous Recombination Model Initiated by a Double-Strand Break
Yeast repair gapped plasmids by homologous recombination.
The double-strand break repair (DSBR) model is based on a double strand break for initiation.

13.5 Bacterial Homologous Recombination Proteins
E. coli recombination mutants have reduced conjugation rates and are sensitive to DNA damage.
RecA is a strand exchange protein.
The RecBCD complex prepares double-strand breaks for homologous recombination and alters its activity at *chi* sites.
The RecFOR pathway repairs single strand gaps.

13.6 Eukaryotic Homologous Recombination Proteins
Several key homologous recombination proteins are conserved between bacteria and eukaryotes, but there are additional novel proteins found only in eukaryotes.

13.7 A Variation of the Double-Strand Break Repair Model
The synthesis-dependent strand-annealing (SDSA) model is a gene conversion-only model.

13.8 Meiotic Recombination
Some aspects of meiotic recombination are novel.
Some recombination proteins are made only in meiotic cells.
Meiotic recombination models propose two different types of homologous recombination events.

13.9 Using Mitotic Recombination to Make Gene Knockouts
Mitotic recombination can be used in genetic engineering to make targeted gene disruptions.
Gene knockouts in yeast occur by homologous recombination with high efficiency.
Gene knockouts in mice also can be made by gene targeting methods.

13.10 Mitotic Recombination and DNA Replication
Mitotic homologous recombination is essential during DNA replication when replication forks collapse.
Recombination must be regulated to prevent chromosome rearrangements and genomic instability.
The single-strand annealing (SSA) mechanism results in deletions.

13.11 Repairing a Double-Strand Break without Homology
Nonhomologous end-joining is a model for rejoining ends with no homology.

13.12 Site-Specific Recombination
Site-specific recombination occurs at defined DNA sequences and is used for immunoglobulin diversity and by transposable elements.
Mating type switching in yeast occurs by synthesis-dependent strand-annealing initiated at a defined site.
V(D)J recombination produces the immune system diversity.
FLP/*FRT* and Cre/*lox* systems can be used to make targeted recombination events.

Suggested Reading

In Chapter 12 we discussed a number of different repair pathways used by cells to repair DNA damage such as UV damage, DNA adducts, and base pair mismatches. Here we examine repair of **double strand breaks (DSBs)** by a process known as **homologous recombination.** Homologous recombination is responsible for much of the genetic diversity among progeny of common parentage and is also essential for correct segregation of homologous chromosome pairs at the first meiotic division (see Chapter 6). When the process goes awry, homologous chromosomes are not held together and therefore segregate randomly at meiosis I, giving rise to aneuploid gametes that are not functional. Somatic cells use homologous recombination to (1) repair double-strand breaks arising from harmful endogenous or exogenous agents (see Chapter 11), (2) restart a replication fork that has stalled at a lesion on a template strand, and (3) reinitiate a replication fork that has collapsed at a nick or other single strand interruption on a template strand. Finally, homologous recombination can maintain telomeres when telomerase is missing. Recent awareness of homologous recombination's importance has increased because cancers such as hereditary breast, ovarian, and colon cancers have been associated with defects in homologous recombination proteins.

After a brief historic view, this chapter describes genetic and biochemical evidence for models of homologous recombination and examines the enzymes and proteins that participate in this process. It next considers various aspects of homologous recombination that are unique to meiosis and then returns to homologous recombination in somatic cells, general genome maintenance, and specialized site-directed reactions.

13.1 Introduction to Homologous Recombination

Homologous recombination is an essential process for repairing DNA breaks and for ensuring correct chromosome segregation in meiosis.

Gregor Mendel was very fortunate in the physical traits that he chose to study in pea plants because the genes responsible for these traits were on different chromosomes and therefore were inherited independently of one another. In 1905, shortly after Mendel's work had been rediscovered, William Bateson, Edith Rebecca Saunders, and Reginald C. Punnett performed genetic crosses with sweet pea plants that showed that different physical traits may also be inherited together. That is, they discovered exceptions to Mendel's law of independent assortment. Based on their findings, Bateson and coworkers proposed that certain alleles must somehow be linked but they were unable to provide a physical explanation for this linkage.

A few years later Thomas Hunt Morgan, the leader of the *Drosophila* laboratory at Columbia University, used fruit flies to show that linked genes must be real physical objects that are located near one another on the same chromosome. He proposed that chromosomes sometimes exchange segments in a process that he called **crossing over.** He further suggested that crossing over takes place during meiosis when homologous chromosomes pair, and involves a physical exchange between the chromosomes (FIGURE 13.1). This hypothesis was confirmed in 1931 by studies in maize and *Drosophila* that showed a correlation between crossing over and an exchange between chromosomes by using cytologically visible chromosomal variations.

Morgan also proposed that the frequency of recombinant meiotic products resulting from the genetic exchange would increase as the distance between the genes increases. In 1911, Alfred H. Sturtevant, then an undergraduate student working in the *Drosophila* laboratory, used recombination frequencies to construct a linkage map that showed the order and spacing of genes on the *Drosophila* X chromosome. The unit of distance along a genetic map that corresponds to a 1% frequency of recombination is called a centimorgan (cM) in honor of T. H. Morgan. In humans, a cM corresponds to an average physical distance of about 1×10^6 bp.

Although crossing over was first discovered to occur during meiosis, later studies by Curt Stern showed that crossing over also can occur during mitosis, FIGURE 13.2 shows how nonsister chromatids of homologous chromosomes in *Drosophila* can recombine to produce adjacent somatic cell clones of different genotypes.

According to Mendel's law of segregation, alleles should segregate equally into gametes. In most fungi, the four spore products of a single meiosis are recovered together in a sac called an **ascus.** If a diploid parent cell has two different alleles for a single gene

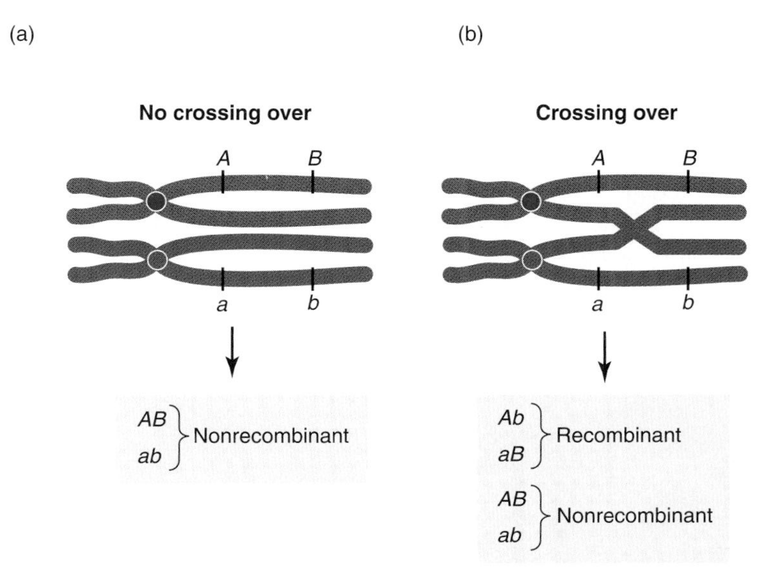

FIGURE 13.1 **Unlinked and linked genes.** (a) No crossing over between the *A* and *B* genes gives rise to only nonrecombinant gametes. (b) Crossing over between the *A* and *B* genes gives rise to the recombinant gametes *Ab* and *aB* and the nonrecombinant gametes *AB* and *ab*.

FIGURE 13.2 **Mitotic crossing over.** Homologous chromosomes in mitotic prophase are shown in yellow and dark red. Rare crossing over between nonsister chromatids can result in homozygosis of the recessive genes *y* (yellow) and *sn* (singed). Homozygosis occurs when one nonexchanged and one exchanged nonsister chromatids segregate together, as shown. This results in one daughter cell being homozygous recessive (*y*/*y*) for yellow, and the other daughter cell being homozygous recessive (*sn*/*sn*) for singed. The panel on the left shows the twin spot that has a patch of yellow body color (*y*) and normal bristles (sn^+) next to a patch of normal body color (y^+) and singed (*sn*) bristles. (Adapted from J. F. Griffiths, et al. *Modern Genetic Analysis, Second Edition.* W. H. Freeman and Company, 1999.)

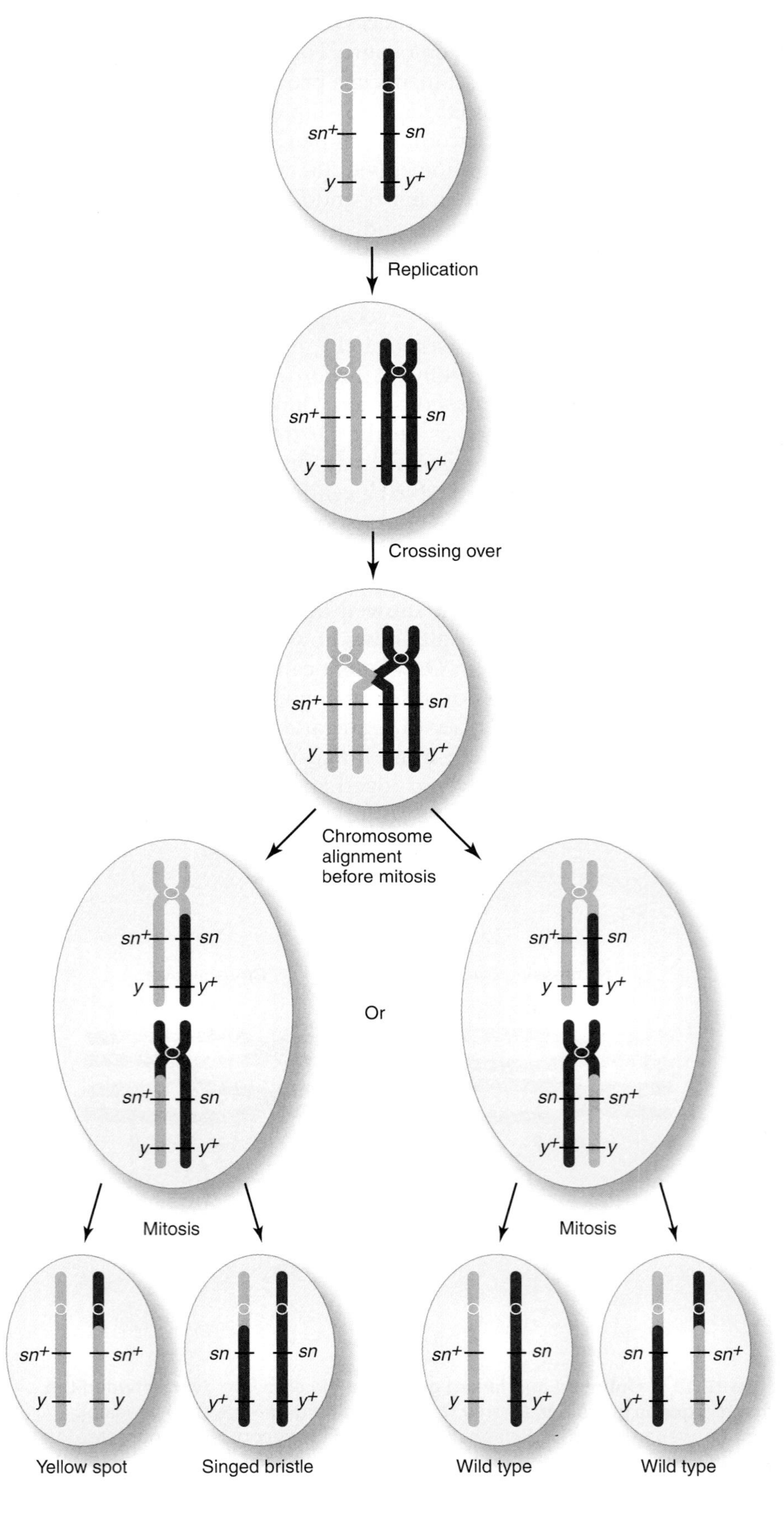

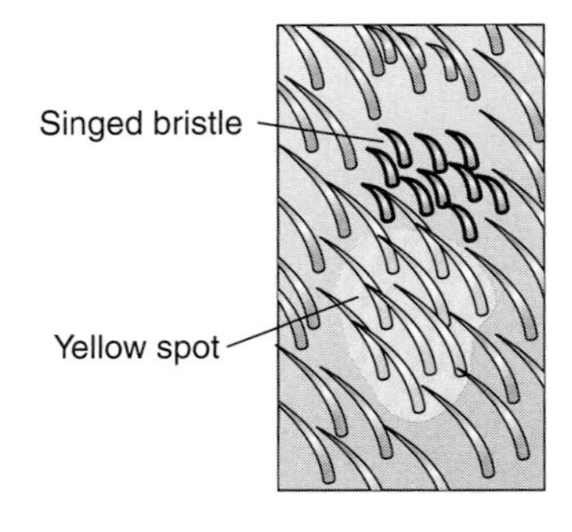

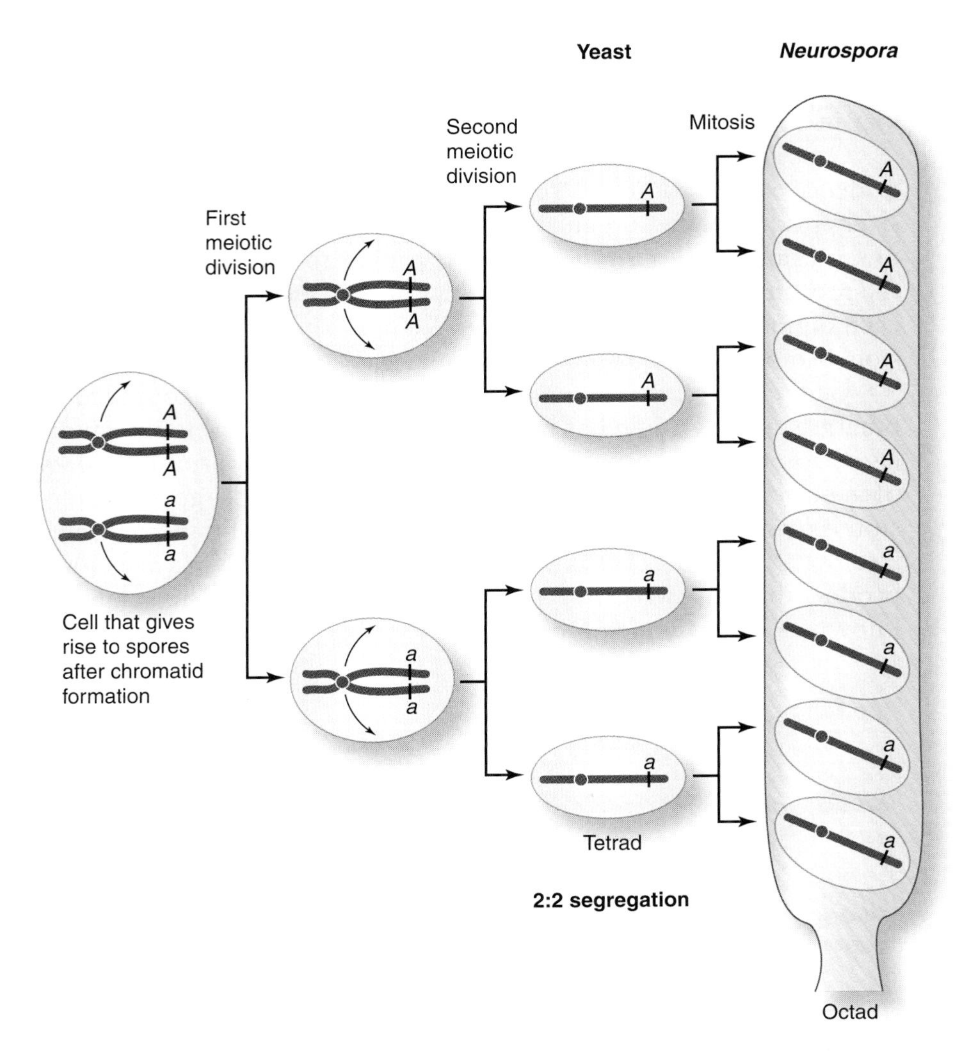

FIGURE 13.3 Meiosis in yeast and *Neurospora*. Meiosis in the yeast *Saccharomyces cerevisiae* gives 2*A*:2*a* spores in the tetrad. A mitotic division occurs in *Neurospora* before spores are formed in an octad, giving 4*A*:4*a* spores. (Adapted from J. F. Griffiths, et al. *Modern Genetic Analysis, Second Edition.* W. H. Freeman and Company, 1999.)

(*A* and *a*) then Mendel's first law predicts that two spores should have the *A* allele and two should have the *a* allele (FIGURE 13.3). This 2:2 segregation reflects the constancy of the gene, which changes only at a very low frequency due to mutation. In yeast and many other fungi, meiotic products, called tetrads, occasionally deviate from the 2:2 segregation and have a 3:1 or 1:3 segregation (FIGURE 13.4). This deviation occurs at about a frequency of 1% for any given gene, but can reach up to 50% in some cases. In other fungi such as *Neurospora*, where there is one mitotic division following meiosis prior to ascus formation, the segregation

Yeast			*Neurospora*		
Normal segregation	Gene conversion		Normal segregation	Gene conversion	
A	*A*	*A*	*A*	*A*	*A*
A	*A*	*a*	*A*	*A*	*A*
a	*A*	*a*	*A*	*A*	*a*
a	*a*	*a*	*A*	*A*	*a*
			a	*A*	*a*
			a	*A*	*a*
			a	*a*	*a*
			a	*a*	*a*
2:2	3:1	1:3	4:4	6:2	2:6

FIGURE 13.4 Gene conversions. Gene conversion in yeast gives tetrads with 3*A*:1*a* or 1*A*:3*a* spores. In *Neurospora,* gene conversion gives 6*A*:2*a* or 2*A*:6*a* octads.

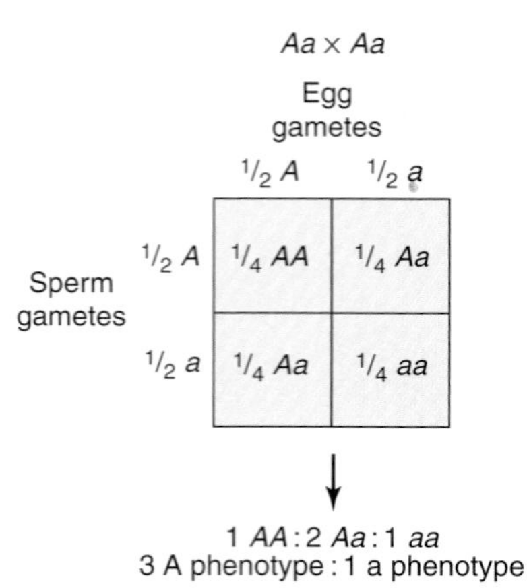

FIGURE 13.5 **Cross of heterozygous parents.** In a cross of *Aa* heterozygous parents, each parent will have 1/2 of the gametes being *A* and 1/2 being *a*. Random fertilization gives approximately 1*AA*:2*Aa*:1*aa* progeny, or 3:1 A phenotype:a phenotype in the progeny. Low gene conversion rates of a few percent cannot be detected, as these will not change the 3:1 ratio of the A to a phenotype in the progeny.

is 4:4. Then the occasional deviants have 6:2 and 2:6 segregation. This irregular segregation was called **gene conversion** by Hans Winkler in 1930.

Gene conversion remained controversial for many years because no evidence existed for it in *Drosophila* and maize, the two most extensively studied higher organisms at that time. Of course, Mendel's law of segregation in *Drosophila* and maize gives an approximate ratio in the progeny, because only one product of each meiosis is recovered in each diploid progeny (FIGURE 13.5). We now know that gene conversion also occurs in *Drosophila* and maize.

Gene conversion is an important outcome of homologous recombination, and its origin has been very important for developing models of homologous recombination. The two most important features of gene conversion for recombination models are: (1) gene conversion is often associated with a reciprocal crossover of flanking genes (FIGURE 13.6) and (2) it is sometimes incomplete, giving a meiotic product that has two genotypes, with one or a few DNA mismatches known as **heteroduplex DNA** (FIGURE 13.7).

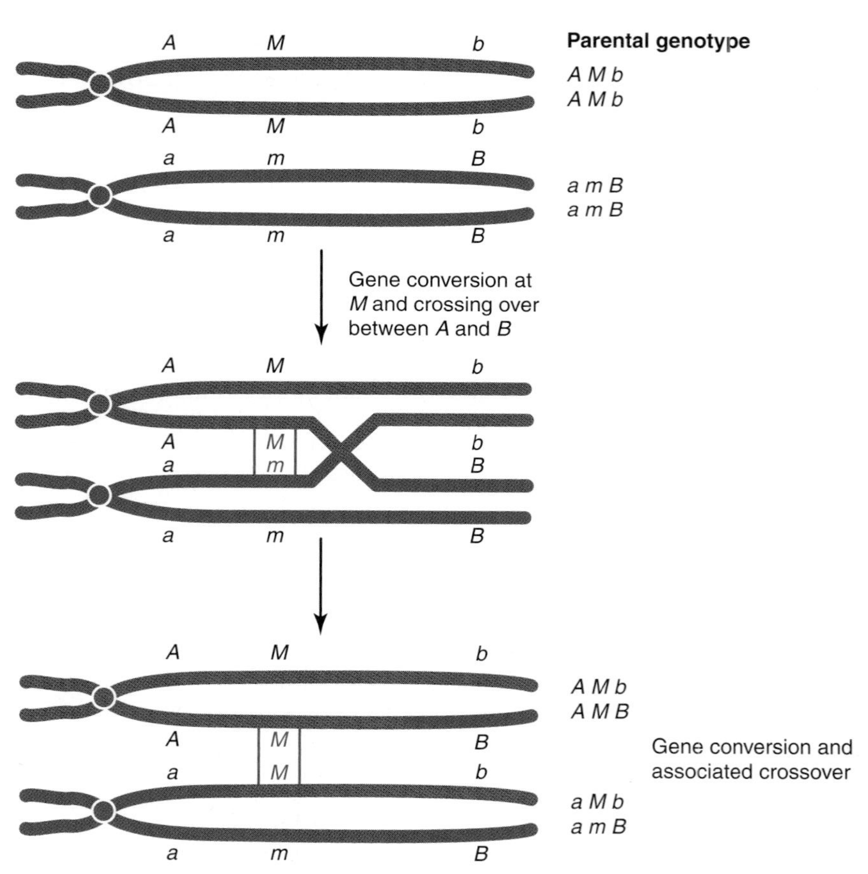

FIGURE 13.6 **Gene conversion associated with a crossover.** A gene conversion takes place at *M* to give 3*M*:1*m*. The gene conversion is associated with a crossover between *A* and *B*, seen in the *AMB* and *aMb* products. Note that genes *A* and *B* still segregate as 2*A*:2*a* and 2*B*:2*b*.

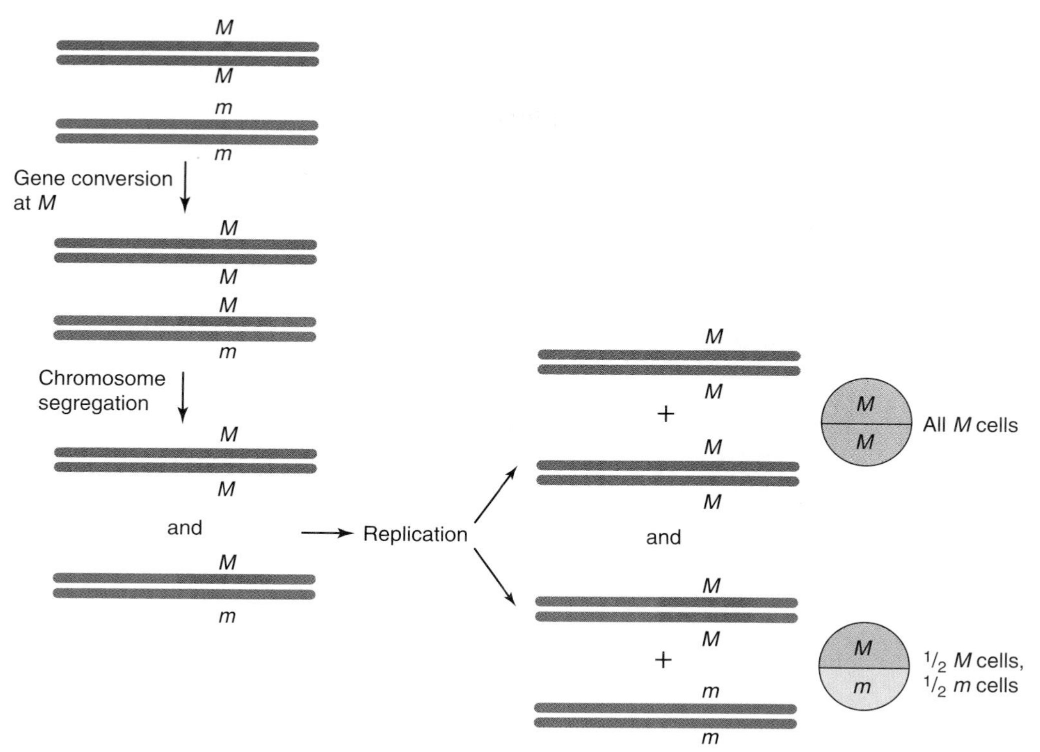

FIGURE 13.7 **Phenotypes resulting from gene conversion.** Failure to mismatch repair heteroduplex DNA results in progeny of different genotypes. The single strand of two interacting DNA duplexes are shown. Incomplete gene conversion at *M* leaves heteroduplex DNA, shown as a red segment in the blue DNA, and a *M*/*m* genotype after meiosis. If the mismatch is not corrected by the mismatch repair system, then after DNA replication and the first mitotic division the heteroduplex is resolved to give two daughter cells of differing genotypes at the *M* gene, one being *M* and the other being *m*. This gives a colony of cells with two phenotypes, *M* and *m*.

13.2 Early Clues from Bacteriophage

Crossing over involves an exchange of DNA between the two interacting DNA molecules.

Homologous recombination is not restricted to eukaryotic organisms. Bacterial and viral chromosomes also recombine. Studies of homologous recombination in these biological systems have provided important information about the physical nature and enzymology of homologous recombination. Recall from Chapter 8 that phage λ produces plaques on *Escherichia coli* lawns. Plaque morphology is determined by several phage λ genes, which can be used for recombination studies. Phage λ recombination can be followed by examining the plaques formed when *E. coli* is infected with a pair of λ phages that differ in plaque morphology. The morphologies of some plaques are those expected for the infecting phages. However, other plaques have new morphologies resulting from the formation of recombinant phages (FIGURE 13.8).

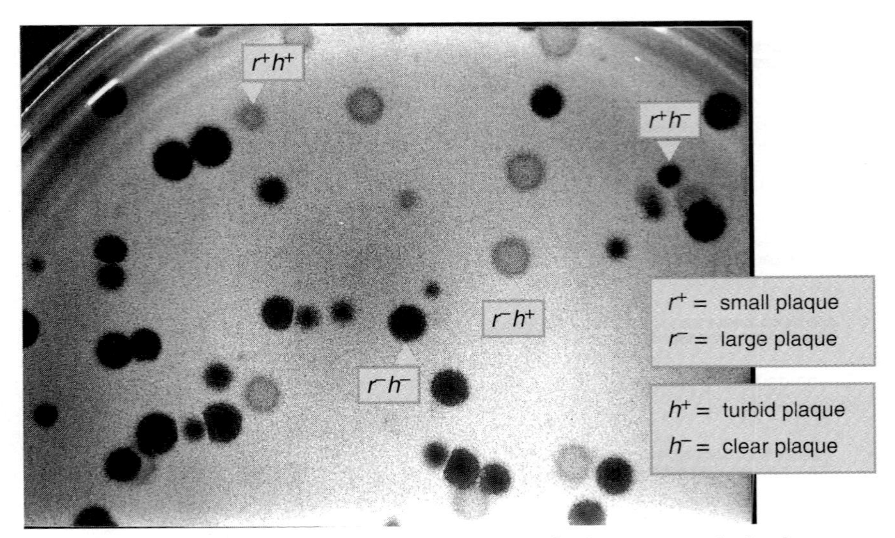

FIGURE 13.8 **Parental and recombinant phage λ plaque morphologies.** Parental and recombinant phage λ plaque morphologies can be seen on a plate from a single cross. The parental plaque types are h^-r^+ and h^+r^-. The recombinant types are h^-r^- and h^+r^+. (Photo courtesy of Leslie Smith and John W. Drake, National Institute of Environmental Health Sciences.)

In 1961, Matthew Meselson and Jean Weigle coinfected *E. coli* with a pair of λ phages that differed in genetic markers and in DNA density. One phage λ was used to infect bacterial cells cultured in a "light" medium that contained nutrients with normal isotopes ^{12}C and ^{14}N, while the other phage λ was used to infect bacterial cells cultured in a "heavy" medium that contained nutrients with the heavy isotopes ^{13}C and ^{15}N. The λ phages released when the cells in each culture lysed were used to coinfect *E. coli* cultured in "light" medium. The cell lysate containing the progeny λ phages was centrifuged in a cesium chloride gradient, which separated the λ phages based on density. The λ phages were collected and then tested for parental or recombinant genotype (FIGURE 13.9). The major conclusion of this experiment is that recombinant DNA can be formed by breakage and rejoining as indicated by the appearance of genetic recombinants that had both heavy and light DNA.

13.3 Early Models of Homologous Recombination

The Holliday model of homologous recombination proposes a crossed strand intermediate called a Holliday junction.

In 1964, Robin Holliday proposed a model for homologous recombination in meiosis (FIGURE 13.10). According to this model, recombination begins after DNA replication. An endonuclease is proposed to introduce a single nick into each duplex so that strands with the same

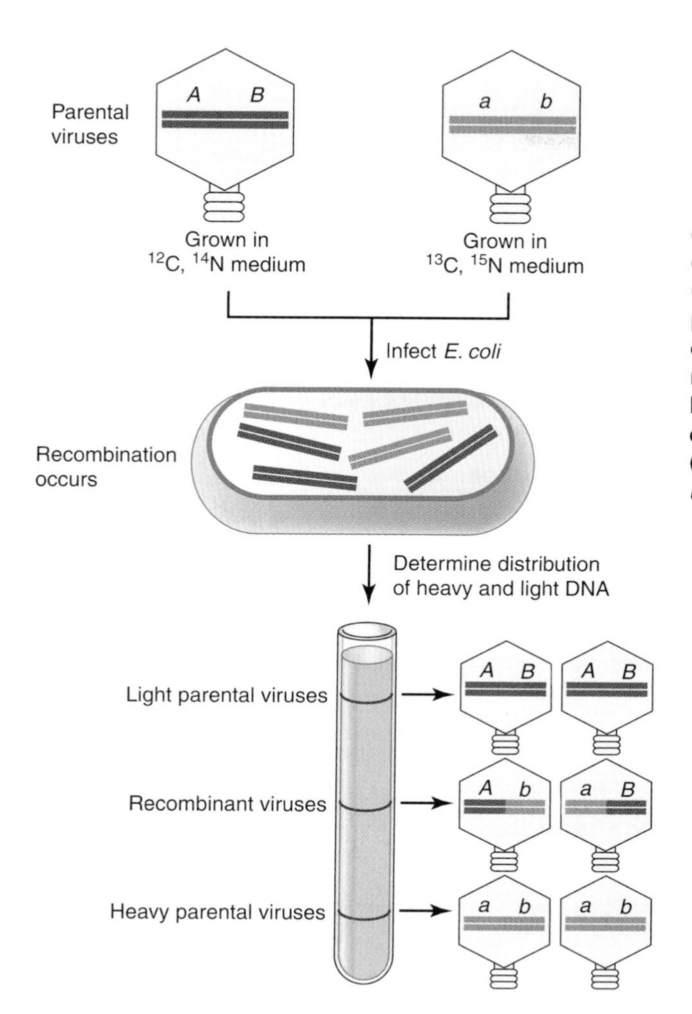

FIGURE 13.9 **The Meselson Weigle experiment.** Bacteria were infected with phage λ under conditions where DNA replication was inhibited. Parental λ phages (*AB*) had been obtained from infecting bacteria cultured in medium containing ^{14}N and ^{12}C. Mutant parental λ phages (*ab*) had been obtained from infecting bacteria cultured in medium containing ^{15}N and ^{13}C. Phages from the cross outlined at the top were centrifuged in a cesium chloride gradient to separate λ phages based on density. The parental *AB* (nonrecombinational) λ phages were primarily light in density and parental *ab* (nonrecombinational) λ phages were primarily heavy in density. The recombinant λ phages *Ab* and *aB* were primarily in the range of intermediate densities, showing that they are derived from DNA of both parents and that recombination involves a physical exchange between DNA duplexes through breaking and rejoining. (Adapted from G. M. Cooper and R. E. Hausman. *The Cell: A Molecular Approach, Second Edition.* Sinauer Associates, Inc., 2000.)

orientation in each duplex are nicked at the same site. Next, symmetric exchange of single DNA strands between duplexes occurs to form what is called **symmetric heteroduplex DNA.** The crossed strand structure is an important intermediate and is called the **Holliday junction** (HJ). This structure has four-way symmetry and can move down the paired chromatids by a process called **branch migration**, forming more heteroduplex DNA. At some point, the Holliday junction structure is cleaved to generate two separate nicked duplexes, a process known as **resolution.** Because the Holliday junction has four-way symmetry, it can be **resolved** in an east–west direction, giving noncrossover products, or in a north–south direction, giving crossover products. Mismatch repair of the heteroduplex DNA region can either restore the original genotype or change the genotype of one (e.g., *B*) to the other (e.g., *b*). This change gives a 1:3 (or 3:1) segregation and explains the association of gene conversion with crossovers in the flanking region.

While the Holliday model explained some of the features of meiotic recombination in the fungi, it did not explain all of them nor did it explain some observations in other recombination systems. Foremost was the observation that recombination does not always result in a

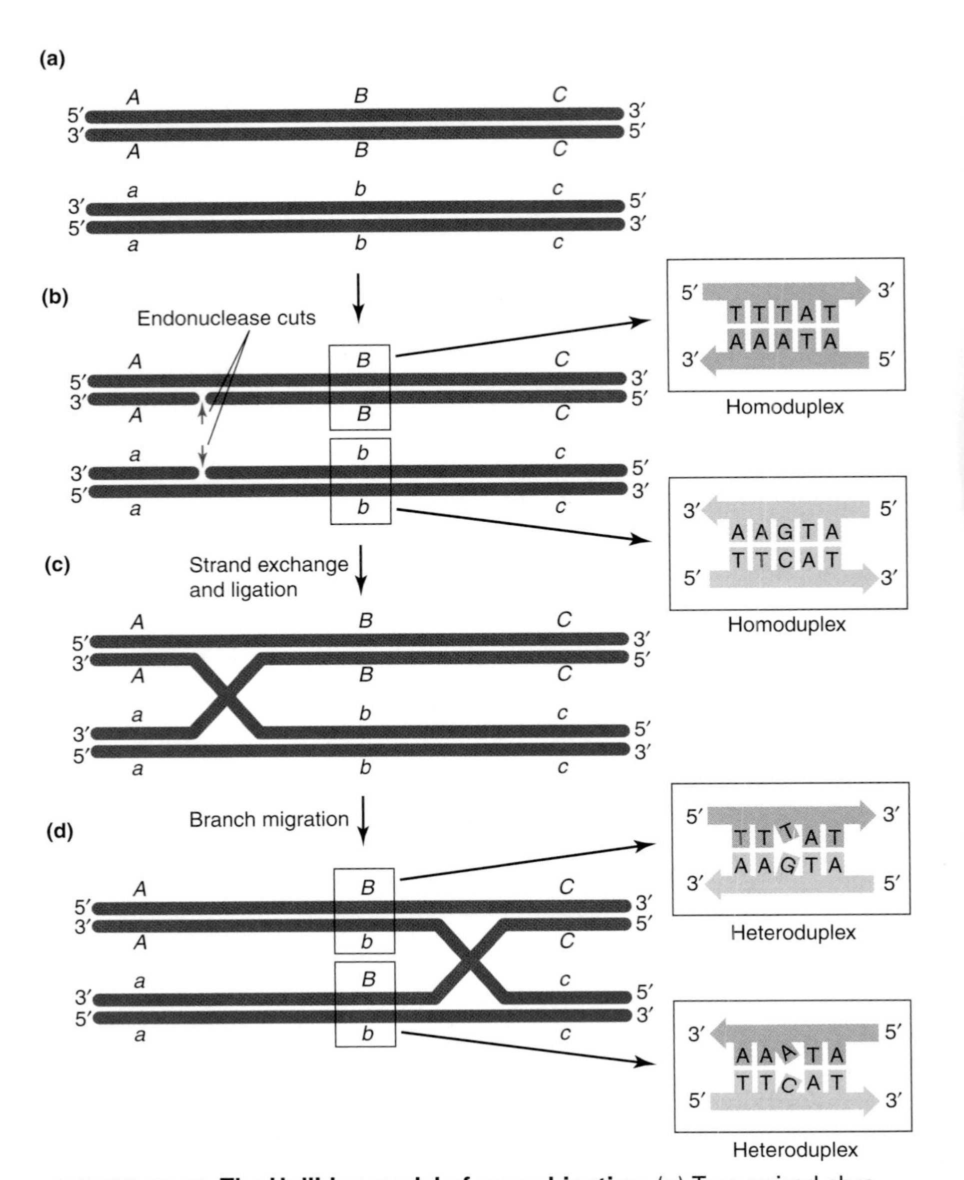

FIGURE 13.10 **The Holliday model of recombination.** (a) Two paired chromatids from a homolog pair are shown, with a DNA sequence difference. (b) Symmetric nicks are made in the two chromatids. (c) Strand exchange between the chromatids and ligation results in a crossed strand or Holliday junction structure. (d) Branch migration of Holliday junction extends the heteroduplex DNA region, which is symmetric and formed on both chromatids. The consequences of heteroduplex DNA at the *B* gene are shown in the insets. (e) The Holliday junction can undergo isomerization, a rotation of the DNA strands around the crossed strand junction as indicated, to produce the open structure shown on the left. Resolution of the Holliday junction in a north–south (vertical) direction yields the splice product chromatids on the left, which are crossover for the markers A and C. Resolution of the Holliday junction in an east–west (horizontal) direction yields the patch chromatids on the right, which are noncrossover for the markers A and C. (Adapted from J. Watson, et al. *Molecular Biology of the Gene, Fifth Edition.* Benjamin Cummings, 2004.)

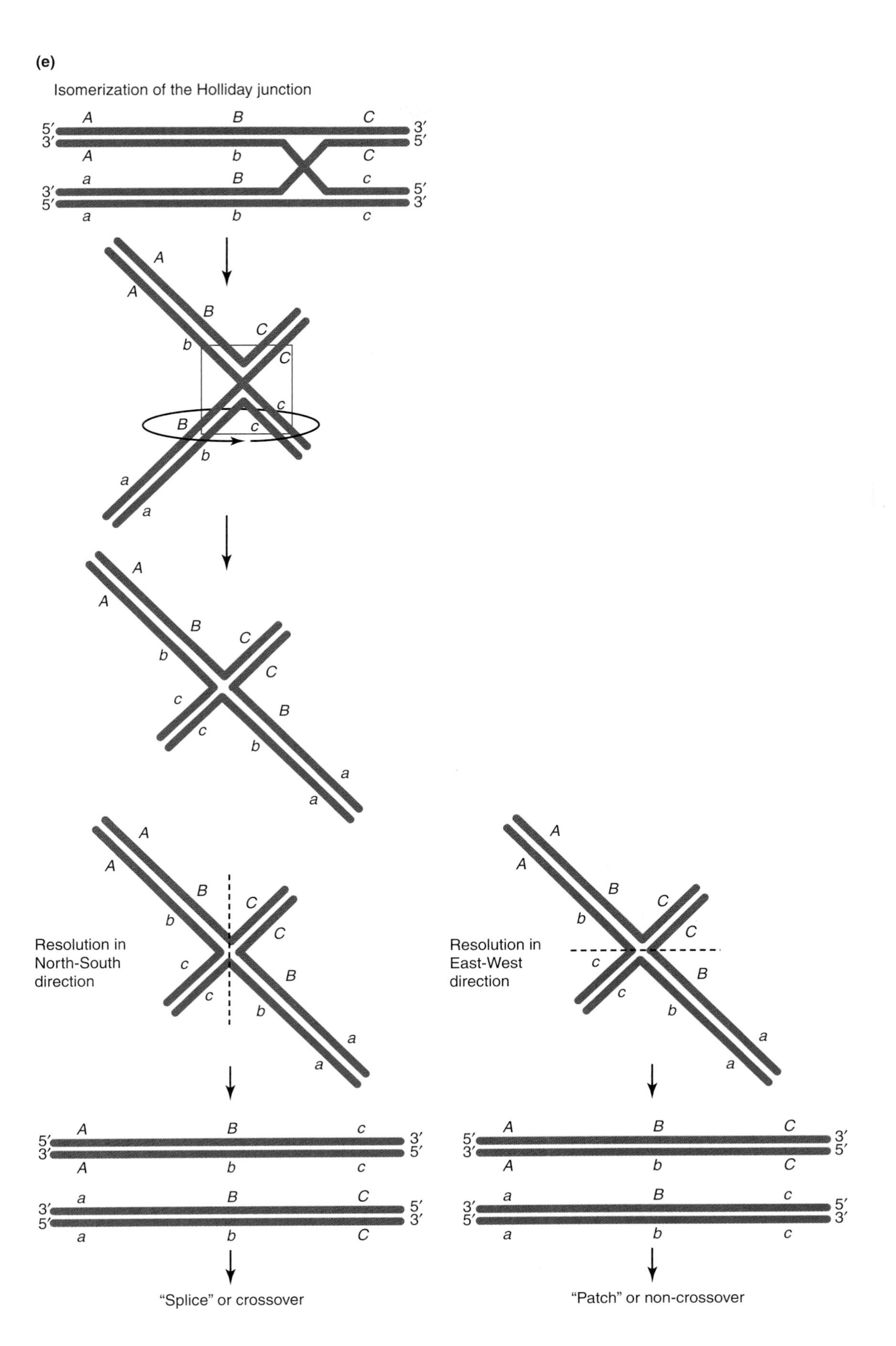
(e)
Isomerization of the Holliday junction
A
B
C
5′
3′
3′
5′
A
b
C
a
B
c
3′
5′
5′
3′
a
b
c
Resolution in North-South direction
Resolution in East-West direction
"Splice" or crossover
"Patch" or non-crossover

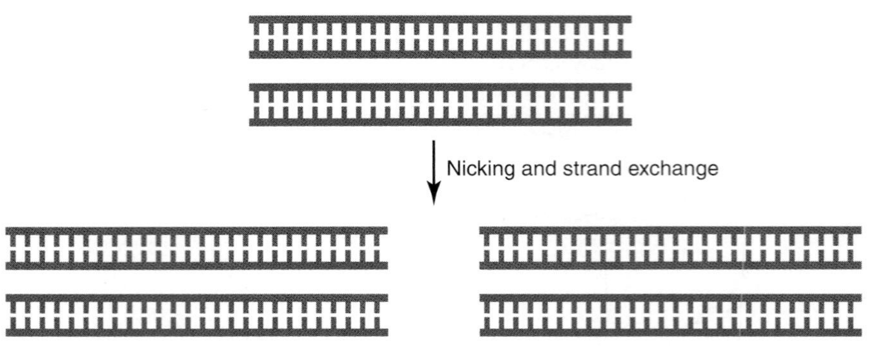

FIGURE 13.11 **Symmetric versus asymmetric heteroduplex DNA.** Heteroduplex DNA that is present on the same region of two chromatids is called symmetric heteroduplex DNA, whereas heteroduplex DNA that covers the same region of only one chromatid is called asymmetric heteroduplex DNA. Both symmetric and asymmetric heteroduplex DNA can be formed during homologous recombination. Symmetric heteroduplex DNA (shown on the left) is formed on both participating DNA duplexes at the same region. Branch migration also can lead to symmetric heteroduplex DNA (Figure 13.10). Asymmetric heteroduplex DNA is formed on only one of the two participating DNA duplexes. It is formed by strand invasion.

reciprocal DNA segment exchange to produce a symmetric pair of heteroduplexes (FIGURE 13.11). Instead, only one of the duplexes had a recombinant strand, resulting in the formation of an asymmetric heteroduplex. Another serious problem with the Holliday model was the need to introduce nicks in homologous duplexes at similar if not identical locations. Even though extensive efforts have been made to detect an endonuclease with the specificity required by the model, no such enzyme has been found in eukaryotes. Despite these shortcomings, the Holliday model has made an important contribution to our understanding of DNA recombination and several features of the model, including the heteroduplex DNA intermediate, branch migration, the Holliday junction, and resolution, remain an important part of the models that followed.

The Meselson-Radding model of recombination—a second homologous recombination model—is based on one single-strand nick for initiation.

Biochemical studies show that RecA (see Chapter 12) promotes a set of DNA strand exchange reactions. Based on this and related information, Matthew Meselson and Charles Radding proposed a new recombination model in 1975. The Meselson-Radding model (sometimes called the Aviemore model because it was first proposed at a meeting in Aviemore, Scotland) retains some key features of the Holliday model, namely the formation of heteroduplex DNA as a homologous recombination intermediate, mismatch repair of the heteroduplex DNA to give 3:1 and 1:3 gene conversion events, and the Holliday junction intermediate that could be resolved as noncrossover or crossover. The main features of the Meselson-Radding model are as follows (FIGURE 13.12):

1. A single-strand nick is introduced into a strand of one of the recombining duplexes.

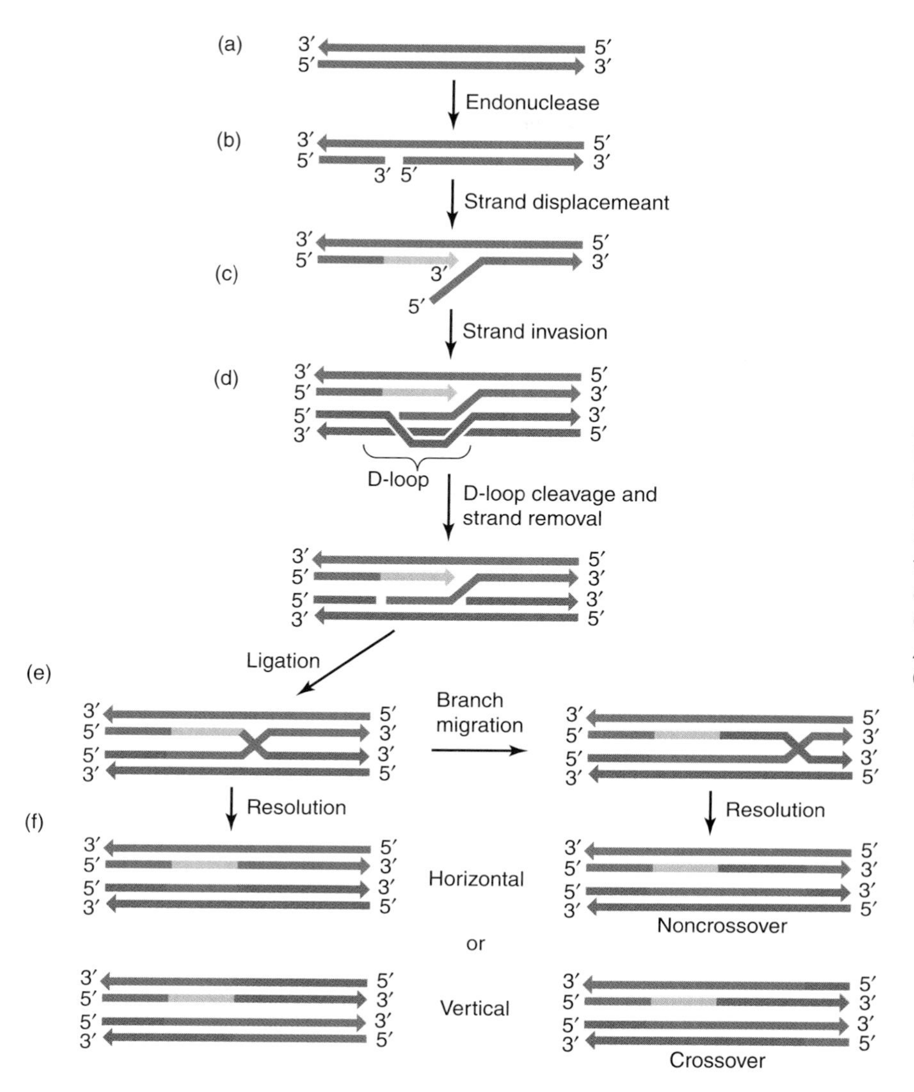

FIGURE 13.12 **The Meselson-Radding model of homologous recombination.** (a) Recombination is initiated by a single-strand nick on a duplex DNA molecule. (b) Synthesis from the nick displaces a single DNA strand with a 5′-P tail, which can (c) participate in strand invasion into a homologous DNA sequence. (d) Processing of the strand invasion intermediate by nicking removes the D-loop sequence. (e) The intermediate is then stabilized by ligation, forming a Holliday junction. (f) The intermediate on the left contains asymmetric heteroduplex DNA. Resolution of the Holliday junction results in noncrossover or crossover products. Alternatively, the Holliday junction can move by branch migration to form symmetric heteroduplex adjacent to the asymmetric heteroduplex region (right). Resolution of the Holliday junction results in noncrossover or crossover products. (Adapted from J. F. Griffiths, et al. *An Introduction to Genetic Analysis, Eighth Edition.* W. H. Freeman and Company, 2004.)

2. DNA polymerase extends the newly created 3′-end, causing the displacement of the strand on the other side of the nick. With the assistance of RecA, the displaced strand with a 5′-P tail invades a homologous region in the second duplex to form a **D-loop**.
3. The D-loop is cleaved, forming a single-strand link between the two DNA duplexes. It should be noted that this intermediate has heteroduplex DNA on only one of the two DNA duplexes and so is an asymmetric heteroduplex DNA.
4. The strands are isomerized or rotated around the crossed strand point to allow the free ends at the exchange site to be ligated, forming a Holliday junction.
5. The Holliday junction can branch migrate. Branch migration away from the initial site of strand invasion forms symmetric heteroduplex DNA. It was important to include the possibility to form symmetric heteroduplex DNA because, as discussed above, there are rare occurrences of postmeiotic segregation in two spores of a tetrad in yeast. In other fungal systems, such as

Ascobolus immersus, the occurrence of postmeiotic segregation in two spores of a tetrad is more frequent. Flexibility of the model to allow for the formation of symmetric heteroduplex made it applicable to the recombination mechanism in a wide range of organisms.

6. Last, the Holliday junction can be resolved to give noncrossover products or crossover products.

The Meselson-Radding model accounts for some features of homologous recombination that were not readily explained by the Holliday model. It also has, however, some major shortcomings. Double-strand breaks are known to promote homologous recombination but the Meselson-Radding model does not explain how homologous recombination repairs double-strand breaks. Additionally, the Meselson-Radding model uses the broken or damaged strand as the donor of information during homologous recombination, whereas genetic evidence suggests that the broken strand is the recipient of information. The model also suggests that crossovers can only take place 3′ to a gene conversion event, whereas crossovers are observed both 5′ and 3′ to a gene conversion event. Finally, the model postulates strand invasion by a 5′-P tailed single-strand DNA, whereas later experiments showed that the invading single strand in homologous recombination has a 3′-OH end. These and other shortcomings remained a problem until models using a double-strand break as the initiating event were developed.

13.4 A Homologous Recombination Model Initiated by a Double-Strand Break

Yeast repair gapped plasmids by homologous recombination.

Additional information about homologous recombination was gained by transforming yeast cells with recombinant plasmids that have yeast genes but lack a yeast replicator sequence. Because the plasmids lack a yeast replicator, their yeast genes can only be maintained if they integrate into the yeast genome. Early experiments showed that plasmids integrated into the yeast genome by homologous recombination between the yeast gene on the plasmid and the homologous yeast gene in the chromosomal DNA.

In 1981, Terry Orr-Weaver, Jack Szostak, and Rodney Rothstein performed experiments with a yeast transforming plasmid that lacks a yeast replicator and is therefore incapable of autonomous replication. The yeast plasmid carried two yeast markers derived from different chromosomal genes. In some experiments, they used a restriction endonuclease to make one double-strand break in one of the two genes and in other experiments they used two restriction endonucleases to make two double-strand breaks in one of the genes, removing an internal sequence and creating a gap (FIGURE 13.13). Surprisingly,

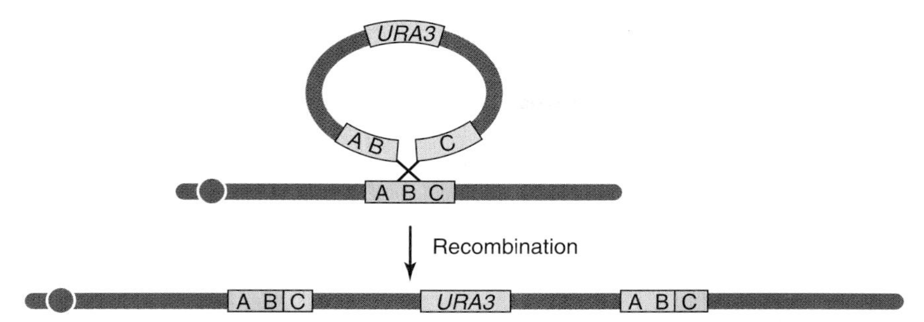

FIGURE 13.13 **Plasmid transformation in yeast by homologous recombination.** The plasmid is constructed to have two chromosomal DNA regions, *URA3* and *ABC*. When the *ABC* region has a double-strand break, introduced by restriction endonuclease digestion prior to transformation, the plasmid pairs with the *ABC* sequence and integrates there, forming a duplication of the *ABC* region.

the broken plasmids transformed at much higher frequencies than the uncut plasmid. All transformants resulted from plasmid integration into chromosomal DNA. The plasmid integrated at the homologous gene sequence in the chromosome. Moreover, the gapped plasmid was repaired during the transformation and integration process. These experiments, which show that double-strand breaks stimulate homologous recombination, formed the basis for a new homologous recombination model.

The double-strand break repair (DSBR) model is based on a double-strand break for initiation.

Recognizing that the plasmid integration process resembled meiotic homologous recombination of gene conversion and crossing over, Jack Szostak and coworkers proposed the **double-strand break repair (DSBR)** model for homologous recombination (FIGURE 13.14) in 1983. Key features of the double-strand break repair model are:

1. A double-strand break is introduced into the DNA, initiating homologous recombination.
2. The double-strand break is processed by resection to produce single strands with 3′-OH ends. Double-strand break ends were known to be processed during meiosis to give 3′-OH tails of approximately 200 nucleotides in length.
3. The 3′-end of single-strand DNA invades a homologous DNA sequence to form a D-loop, leading to the formation of asymmetric heteroduplex DNA.
4. DNA polymerase extends from the 3′-end of the invading strand.
5. The second 3′-end of the double-strand break is captured when the extended D-loop reaches the single stranded region on the other double-strand break end and can anneal to it. This

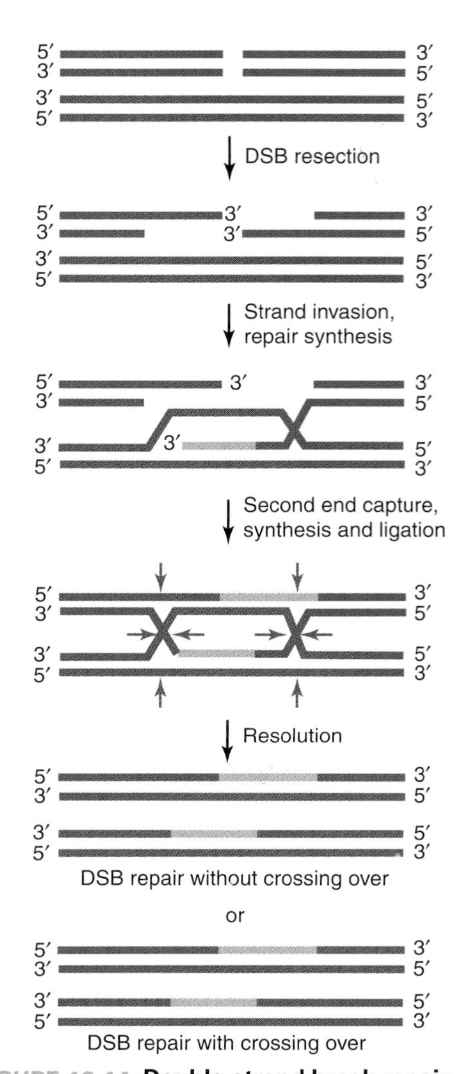

FIGURE 13.14 **Double-strand break repair model of homologous recombination.** Recombination is initiated by a double-strand break. Following nuclease degradation of the ends (called DNA resection), single strand tails with 3′-OH ends are formed. Strand invasion by one end into homologous sequences forms a D-loop. Extension of the 3′-OH end by DNA synthesis enlarges the D-loop. Once the displaced loop can pair with the other side of the break, the second double-strand break end is captured. DNA synthesis, which fills in the gaps, is followed by ligation to form two Holliday junctions. Resolution at the blue arrowheads produces a noncrossover product. Resolution of one Holliday junction at the blue arrowheads and the other Holliday junction at the red arrowheads results in a crossover product. The dotted lines represent DNA synthesis.

second 3′-OH is extended by DNA synthesis and the ends ligated. This pathway produces an intermediate in which gene conversion can occur by mismatch repair if the DNA sequence of the gene or marker under study is in the heteroduplex DNA formed during recombination, or by replication if the repaired sequence is in the region between the double-strand break ends. The double-strand break repair model also explains the gap repair phenomenon in plasmid transformation, as gaps can be repaired by replication in this model using the homologous chromosomal DNA sequence.

6. It should be noted that two Holliday junctions are formed in the double-strand break repair model. Evidence for a double Holliday junction recombination intermediate in meiotic and mitotic recombination has come from two-dimensional DNA gels that separate molecules on the basis of size and shape. These gels are similar to those used to detect replication intermediates in yeast (see Chapter 10).
7. Resolution of the double Holliday junction gives crossover and noncrossover outcomes. It was proposed that one Holliday junction is always resolved in the noncrossover east–west direction, while the other Holliday junction can be resolved in either the east–west noncrossover direction or the north–south crossover direction.

The attraction of the double-strand break repair model is that it explains many features of fungal meiotic recombination. Later studies showed that, indeed, double-strand breaks are formed in meiosis and that they are the initiators of meiotic homologous recombination. In mitosis, double-strand breaks also initiate homologous recombination, but single strand interruptions in duplex DNA can also initiate homologous recombination.

13.5 Bacterial Homologous Recombination Proteins

E. coli recombination mutants have reduced conjugation rates and are sensitive to DNA damage.

After conjugation and recombination were shown to take place in *E. coli* (see Chapter 7), researchers tried to isolate recombination-defective mutants. Simply collecting mutants defective in conjugation would not specifically target the recombination process, however, because mutants unable to make pili or form a conjugation junction also would be defective in conjugation. Alvin J. Clark and Ann Dee Margulies devised a double screen process to isolate recombination (*rec*) mutants. They reasoned that *E. coli rec* mutants would be (1) defective in conjugation (not fertile) and (2) very sensitive to UV light, because the UV DNA damage repair process seemed to involve similar steps to those involved in homologous recombination, namely, degradation of single-stranded DNA flanking the damaged region or the recombination junction, repair synthesis to fill in gaps, and ligation. Using these criteria, Clark and Margulies isolated the first recombination-defective mutant, *recA*, in 1965. Subsequently, many more *rec* mutants and other mutants defective in bacterial homologous recombination were isolated. The list now extends to more than 30 mutants (Table 13.1).

TABLE 13.1 ***E. coli*** **Recombination Proteins**

Step in Recombination	Protein	Activity
Initiation	RecBCD	DNA helicase, DNA exonuclease, DNA endonuclease
	SSB	Single-strand DNA binding protein
	RecE	Double-stranded DNA exonuclease
	RecQ	DNA helicase
	RecJ	Single-stranded DNA exonuclease
Presynapsis	RecA	Single-stranded DNA renaturation
ATPase	RecF	DNA binding
	RecO	Part of a complex with RecR
	RecR	Part of a complex with RecF or RecO
	SSB	Single-strand DNA binding protein
	RecT	DNA renaturation
Heteroduplex extension	RecA	Strand exchange, DNA renaturation ATPase
	PolA	DNA polymerase I
	TopA	DNA topoisomerase I
	GyrA	DNA gyrase
	GyrB	DNA gyrase
	RecJ	Single-stranded DNA exonuclease RuvA Holliday junction binding
	RuvB	DNA helicase, interaction with RuvA to branch migrate Holiday junctions
	RecG	DNA helicase, branch migrate Holiday junctions
Resolution	RuvC	Holiday junction cleavage
	Rus	Holiday junction cleavage
	Lig	DNA ligase
Other	RecN	
	SbcB	Single-stranded DNA exonuclease
	SbcBC	
	DinI	Acts with RecX to affect the stability of RecA-DNA filaments
	RecX	Acts with DinI to affect the stability of RecA-DNA filaments
	Pol II	DNA polymerase
	Pol III	DNA polymerase
	UvrD	DNA helicase, removes RecA filaments from DNA
	Hel IV	DNA helicase
	RdgC	Negative regulator of RecA
	MgsA	Unknown function, some cellular overlap with RecA

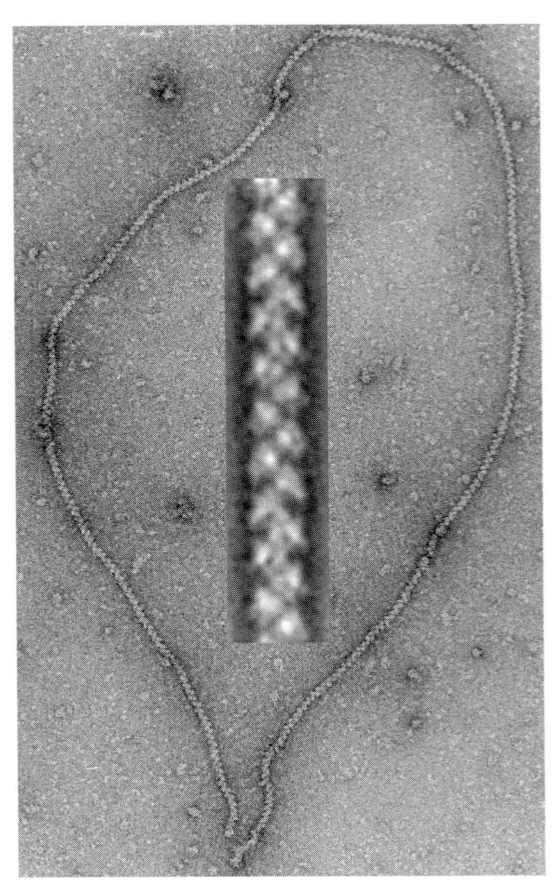

FIGURE 13.15 **RecA filament polymerized on double-stranded circular DNA.** The inset image shows the helical nature of the filament. (Photo courtesy of Edward H. Egelman, University of Virginia.)

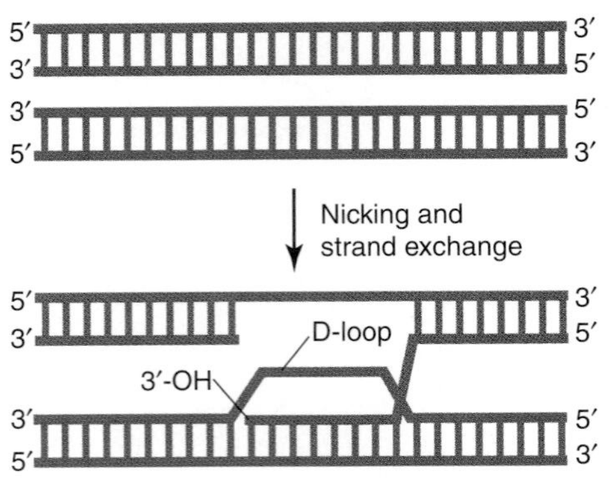

FIGURE 13.16 **A D-loop formed after strand invasion by a homologous single strand from another DNA duplex.** Strand invasion, promoted by RecA binding to the invading red single strand, displaces homologous sequence in the blue double-stranded DNA. The displaced single-strand structure is called a D-loop.

RecA is a strand exchange protein.

E. coli uses several recombination pathways and the RecA protein, which was introduced in Chapter 10, is required for all of them with the exception of a specialized RecE pathway. RecA binds to single-stranded DNA and double-stranded DNA as a helical filament (FIGURE 13.15). Binding of RecA to single-stranded DNA promotes the early steps of strand exchange between homologous DNA sequences. Approximately 6 to 8 RecA monomers are present in each helical turn, stretching the DNA by about 1.5-fold and covering about 20 nucleotides.

RecA promotes the pairing of single-stranded DNA with identical or very similar DNA sequences *in vitro* and *in vivo*. *In vitro* studies have shown that RecA binds to single-stranded DNA and promotes **strand invasion** to homologous sequences that can be in a single-stranded form or a double-stranded form. An important intermediate in strand invasion is the formation of the displacement loop (D-loop), which forms when one end of the invading strand displaces a strand with an identical or nearly identical sequence in double-stranded DNA (FIGURE 13.16). D-loop formation requires a free end for the invading strand. During strand exchange, ATP hydrolysis occurs through RecA activity. This hydrolysis promotes RecA dissociation from the single-strand DNA after strand exchange has occurred.

The three stages of RecA-mediated strand exchange are (FIGURE 13.17):

1. *Presynapsis*. RecA binds to single-stranded DNA in the presence of ATP or dATP to form a helical nucleoprotein filament. Presynaptic complex formation is enhanced by single-stranded DNA binding (SSB) protein, which helps by binding to single-stranded DNA and removing any secondary structure. This prepares the single-stranded DNA for association with RecA protein.
2. *Synapsis*. RecA-coated single-stranded DNA contacts double-stranded DNA, allowing the single-stranded DNA to search for sequence homology. Initial contacts are proposed to be random, but homologous sequences eventually become aligned. Strand exchange is initiated by local denaturation of the double-stranded DNA molecule in a region of homology. The invading single strand forms side by side base pairs with its complementary strand in the denatured region. This region of base pairing in which the two strands are not yet intertwined (topologically linked) is an unstable intermediate called a **paranemic joint**. A transition that begins at the free end of the invading strand converts the paranemic joint into a double helical region termed a **plectonemic joint**. A heteroduplex DNA forms when the complementary DNA strands in the plectonemic joint have one or more base pair mismatches or short insertion/deletion loops.
3. *DNA heteroduplex extension*. Once the plectonemic joint molecule is established, heteroduplex DNA formation can be extended by unidirectional branch migration in a $5' \rightarrow 3'$ direction. In the case shown in Figure 13.17, branch migration proceeds by local denaturation of the double-stranded DNA and annealing of the complementary single strand with the circular single-stranded DNA until a linear single-stranded DNA molecule is completely displaced.

In the example shown in Figure 13.17, the double-stranded DNA has free ends. In the recombination models to be discussed, the double-stranded DNA does not have two free ends. The branch migration reaction is limited, but can extend for several hundred nucleotides.

The RecBCD complex prepares double-strand breaks for homologous recombination and alters its activity at *chi* sites.

Double-strand breaks occur mainly during conjugation or from collapse of a replication fork (see below). Recombination promoted by these double-strand breaks proceeds through the RecBCD pathway. The double-strand break that precedes recombination must be processed before recombination can take place. This processing is performed by the RecBCD protein complex, which cuts back or resects the

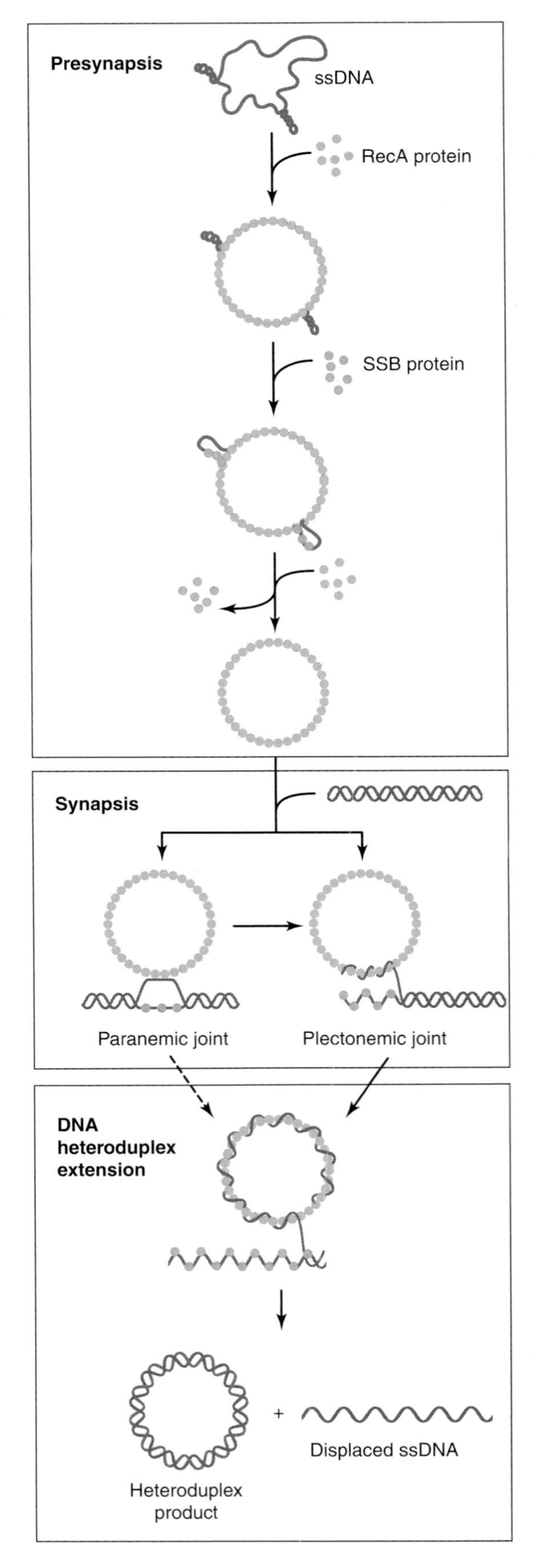

FIGURE 13.17 **RecA protein during homologous recombination functions at multiple steps.** In the presynapsis stage, RecA forms a filament on single-stranded DNA with the aid of single-strand binding protein (SSB). The RecA-coated ssDNA complex searches for homology during the synapsis stage. Once homology is found, side by side pairing is formed (called paranemic pairing), which then transitions to plectonemic pairing where the paired DNA strands are in a double helix configuration. Note that these pairing stages involve strand invasion and D-loop formation. During the heteroduplex extension stage the pairing between the RecA•single-strand DNA complex and the homologous sequence in the double-stranded linear molecule is increased, with displacement of the single strand from the original double-stranded linear DNA. (Adapted from P. R. Bianco and S. C. Kowalczykowski, *Encyclopedia of Life Sciences* [DOI:10.1038/npg.els.0003925]. Posted September 23, 2005.)

double-strand break DNA end. *recB* and *recC* mutants were isolated in initial screens for conjugation-defective and UV-sensitive *E. coli* mutants. Later studies showed that the RecB and RecC proteins act in a complex that also contains the RecD protein. The RecBCD complex has 3′→5′ exonuclease, a 5′→3′ exonuclease, and DNA helicase activities, which work together to prepare the double-strand break end for homologous recombination. The RecBCD complex binds to double-strand break ends and then unwinds the double strand through its helicase activity while degrading each single strand separately (FIGURE 13.18). Degradation of two single strands occurs at different rates, with the 3′→5′ exonuclease activity being predominant, so that the 3′ end is degraded more rapidly. This degradation pattern continues until the RecBCD complex reaches a sequence called a ***chi*** (crossover hotspot instigator) site. Investigators discovered *chi* sites while studying phage λ mutants that had defects in their recombination system. These recombination-deficient (*red*) mutants depend on the bacterial RecBCD complex to perform the essential recombination functions needed for viral DNA replication. Because the bacterial recombination system is a poor substitute for the phage system, bacterial infection produces small plaques. However, a few large plaques sometimes are formed. Lambda phages isolated from these large plaques were shown to have an additional mutation that created hot spots (now called *chi* sites), which allowed the bacterial RecBCD complex to function more efficiently. Subsequent studies showed that *chi* sites have the sequence 5′-GCTGGTGG-3′ and are quite common in the bacterial genome, occurring about once in every 5,000 bp in the *E. coli* chromosome (a much higher frequency than would be predicted for a random nucleotide sequence). When the RecBCD complex encounters a *chi* site, its 3′→5′ exonuclease activity decreases, while its 5′→3′ exonuclease

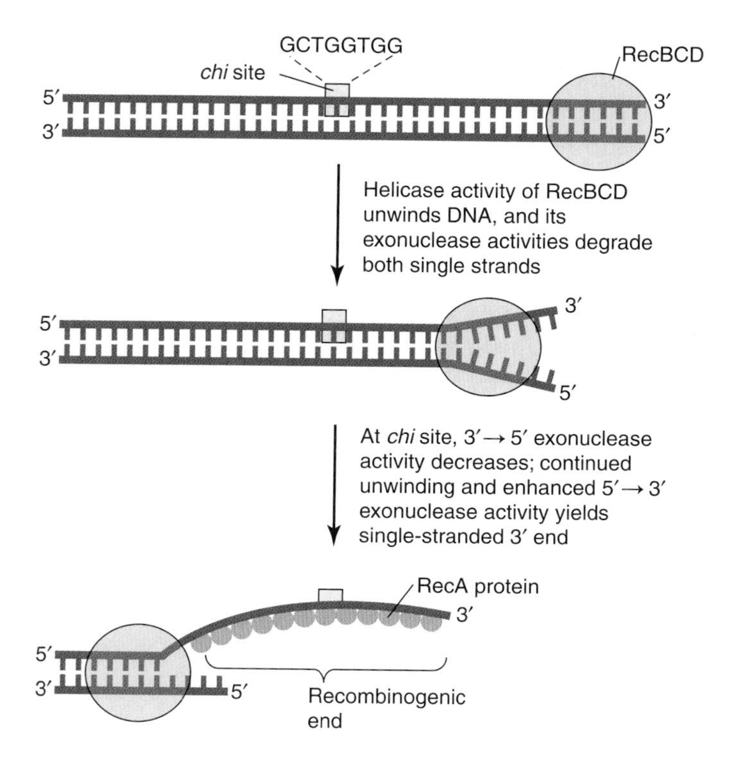

FIGURE 13.18 RecBCD processing double-strand break ends. RecBCD binds to a double-strand end and unwinds the end in concert with its 3′→5′ and 5′→3′ exonuclease activities. When RecBCD reaches a *chi* site (shown in yellow), the exonuclease activities change. The 3′→5′ exonuclease activity is decreased while the 5′→3′ exonuclease activity is increased. This generates a single-strand end with a 3′-OH end, which is a substrate for RecA binding. The RecA ssDNA filament promotes recombination as detailed in Figure 13.17. (Adapted from H. Lodish, et al. *Molecular Cell Biology, Fifth Edition.* W. H. Freeman and Company, 2003.)

activity increases, leading to the formation of a 3′ single-strand overhang. RecA is loaded by the RecBCD complex and binds to the overhang to form a nucleoprotein filament, facilitating its invasion into the homologous duplex and initiating homologous recombination. Because the *chi* site is not symmetric, it stimulates recombination in only one direction. The RecA coated nucleoprotein filament then engages in a search for homology and upon encountering a homologous DNA duplex, promotes strand invasion (FIGURE 13.19). The outcome of this is to form a Holliday junction. The bacterial resolution enzymes have been well identified and characterized. The Ruv proteins, RuvA, RuvB, and RuvC act on the Holliday junction. RuvA and RuvB form a complex that binds to Holliday junctions and moves it through branch migration. RuvC binds to the RuvAB complex and resolves the Holliday junction by making single-strand nicks in the opposite strands of the Holliday junction that are subsequently sealed through ligation. If RuvABC does not act on the Holliday junction, branch migration can take place through the action of the RecG DNA helicase. RecG has no endonucleolytic activity and so is thought to promote resolution by branch migration of the Holliday junction to a preexisiting nick.

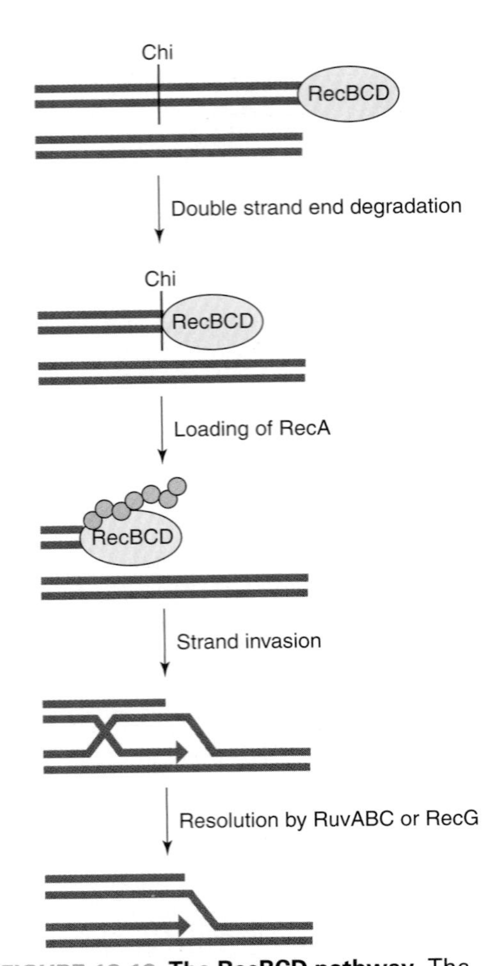

FIGURE 13.19 **The RecBCD pathway.** The RecBCD pathway acts on double-strand breaks. The double-strand break end is degraded by the combined action of DNA helicases and nucleases of the RecBCD complex. After reaching a *chi* site, the activity of the RecBCD complex changes, as shown in Figure 13.13, to generate a single-strand DNA end with a 3′-OH. RecA protein is loaded onto the single-strand DNA to form the RecA nucleoprotein filament. The RecA nucleoprotein filament engages in a search for homology and strand invasion to form the D-loop intermediate. The resulting Holliday junction is resolved by the RuvABC helicase/resolvase or the RecG helicase.

The RecFOR pathway repairs single-strand gaps.

E. coli with *recB* or *recC* null mutations have a low level of residual recombination, which can be restored to a near normal level by suppressor mutations that permit recombination to take place through an alternate pathway. Because the mutants were discovered as suppressors, their genes were called *sbc* (suppressor of *recBC*) genes. The *sbcB* gene codes for exonuclease 1 and the *sbcC* and *sbcD* code for the subunits of the SbcCD nuclease. The primary function of the alternate recombination pathway, the **RecFOR pathway** (named for its major components RecF, RecO and RecR), is to repair any single-strand gaps that remain in the *E. coli* chromosome after DNA replication. Under conditions where the RecBCD pathway is not functional, however, the RecFOR pathway can repair double-strand breaks if the SbcB and SbcCD nucleases do not prevent it from doing so.

The RecFOR pathway is shown in FIGURE 13.20. The RecJ exonuclease appears to act on nicks or small gaps to enlarge them in preparation for homologous recombination.

RecJ acts in concert with a DNA helicase, RecQ, so that unwinding and degradation occur in a coordinated fashion. RecQ homologs exist in eukaryotes, most notably the product of the human Bloom syndrome gene BLM. The RecO and RecR proteins form a RecOR complex, which binds to single-strand DNA that is coated with the SSB proteins and helps RecA displace the SSB from the single-strand DNA, thereby promoting RecA binding. RecF has DNA binding activity and *in vivo* binds to the 5′ end of a single-strand gap when complexed with RecO and RecR. This promotes loading of RecA to the single-strand gap. Once RecA has formed a nucleoprotein filament on single-stranded DNA, it engages in a search for homology similar to that in the RecBCD pathway and results in a Holliday junction. The Holliday junction is resolved by RuvABC or RecG similar to the RecBCD pathway intermediates.

13.6 Eukaryotic Homologous Recombination Proteins

Several key homologous recombination proteins are conserved between bacteria and eukaryotes, but there are additional novel proteins found only in eukaryotes.

Let's now consider the proteins involved in homologous recombination with particular emphasis on the DSBR pathway. It is important to understand the activities of these proteins, because mutations in several of them have been associated with human diseases. Other proteins that participate in homologous recombination have not been associated with human diseases, most likely because these proteins are so important for cell viability that cells cannot survive without them. A list of eukaryotic homologous recombination proteins is found in Table 13.2.

Many, but not all, of the eukaryotic homologous recombination genes are called ***RAD* genes**, because they were first isolated in screens for yeast mutants with increased sensitivity to x-rays. Because x-rays make double-strand breaks in DNA, it is not surprising that *rad* mutants sensitive to x-rays also are defective in mitotic and meiotic recombination. You may wish to refer to the double-strand break repair model shown in Figure 13.15 as we consider the enzymes that participate in different steps of the pathway.

1. End Processing/Presynapsis

In mitotic cells, double-strand breaks are produced by exogenous sources such as irradiation or chemical treatment, and from endogenous sources such as topoisomerases and nicks on the template strand. During replication these nicks are converted to double-strand breaks. The ends of these breaks are processed by exonucleolytic degradation to have 3′ overhangs. In yeast, this processing requires the Mre11•Rad50•Xrs2 (MRX) complex, which has helicase, endonuclease, and exonuclease activities. Mutations in *MRE11*, *RAD50*, or *XRS2* render cells sensitive to ionizing radiation and diploids have a poor meiotic outcome.

Mre11 has 3′→5′ exonuclease activity. Rare mutations that produce MRE11 with low activity are associated with **ataxia-telangiectasia-like disorder (ATLD)** in humans. Patients with this disorder have developmental problems and show defects in DNA damage checkpoint signaling but do not appear to be cancer prone.

Rad50 has a coiled-coil domain similar to structural maintenance of chromosome (SMC) proteins (see Chapter 6) and a globular domain (FIGURE 13.21). The globular domain, which has two ATP binding sites and ATPase activity, binds Mre11 and Xrs2. Rad50 is thought to help hold double-strand break ends together via dimers connected at the tips by a hook structure that becomes active in the presence of zinc ion.

Xrs2 is replaced by a structurally unrelated DNA-binding protein called Nbs1 in humans. Both proteins bind DNA. Nbs1 is so named because a mutant allele was first discovered in individuals with **Nijmegen breakage syndrome**, a rare DNA damage syndrome that is associated with defective DNA damage checkpoint signaling

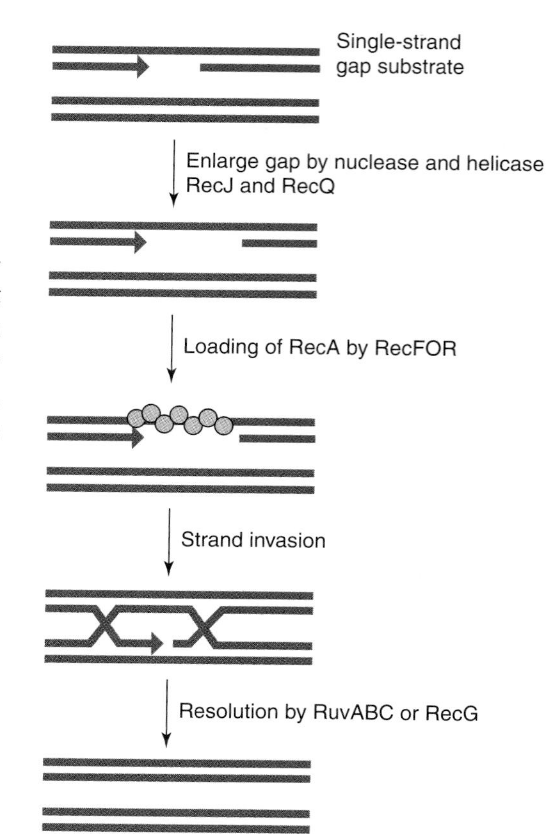

FIGURE 13.20 **The RecFOR pathway.** The RecFOR pathway is used to repair single-strand gaps. Single-strand gaps are enlarged through the combined action of the RecJ nuclease and the RecQ helicase to form a single-strand region large enough to engage in a search for homology. The single-strand region is then bound by RecA, using RecFOR to load RecA, to form a nucleoprotein filament. The RecA nucleoprotein filament is used in strand invasion to form a D-loop and the resulting Holliday junction(s) are resolved by RuvABC or RecG. The RecFOR pathway can also use double-strand break ends in recBC sbcB sbcCD mutants, processing the end with RecJ and RecQ or another helicase, and then proceeding with the loading of RecA and strand invasion as shown in Figure 13.19.

TABLE 13.2 Eukaryotic Recombination Proteins

Step in Recombination	*S. cerevisiae* Protein	Human Protein	Activity
Mitotic Recombination			
Initiation	Mre11	MRE11	DNA endonuclease and exonuclease, forms a complex with Rad50 and Xrs2/Nbs1
	Rad50	RAD50	DNA binding, ATPase
	Xrs2	NBS1	DNA binding
	RPA	RPA	Single-strand DNA binding protein
Presynapsis	RPA	RPA	Single-strand DNA binding protein
	Rad51	RAD51	Strand exchange, ATPase
	Rad52	RAD52	Single-strand DNA binding and annealing, interacts with Rad51 as a mediator, interacts with self and Rad59 (Rad52 paralog)
		BRCA1	
	Brh2 (*Ustilago maydis*)	BRCA2	Interacts with RAD51, mediator activity
	Rad55		Rad55 and Rad57 form a complex with mediator activity
	Rad57		
		RAD51B	RAD51B and RAD51C form a complex with
		RAD51C	mediator activity
		RAD51D	RAD51B and XRCC2 form a complex
		XRCC2	
		XRCC3	
	Rad54	RAD54	ATPase, double-strand DNA translocase branch migration
	Rdh54/Tid1		ATPase, double-strand DNA translocase
	Rad59		Single-strand DNA binding and annealing in SSA
Heteroduplex extension	Rad51	RAD51	Strand exchange, ATPase
	Pol3	POL δ	DNA polymerase
	Rad30	POL η	DNA polymerase
Resolution	Sgs1	BLM	DNA helicase, together with Top3 or hTOPOIIIα can dissolve Holliday junctions
	Top3	hTOPOIIIα	DNA topoisomerase
	Rmi1	BLAP75	Interacts with Top3/ hTOPOIIIα
	Mus81	MUS81	Structure-specific endonuclease, forms a complex with Mms4/Eme1
	Mms4	EME1	
		XRCC3•RAD51C	Holliday junction resolution
	Lig1	LIG1	DNA ligase
Meiotic-Specific Recombination Factors			
	Spo11	SPO11	Related to type II topoisomerases, DSB formation
	Dmc1	DMC1	Strand exchange, ATPase
	Mei5•Sae3		Interacts with Dmc1, mediator activity
	Hop2•Mnd1	HOP2•MND1	Interacts with Dmc1, DNA binding
	Rdh54•Tid1	RAD54•RAD54B	ATPase, double-strand DNA translocase interacts with Dmc1
	Mer3		DNA helicase
	Msh4•Msh5	MSH4•MSH5	Related to mismatch repair proteins of the MutS family, needed for crossover formation

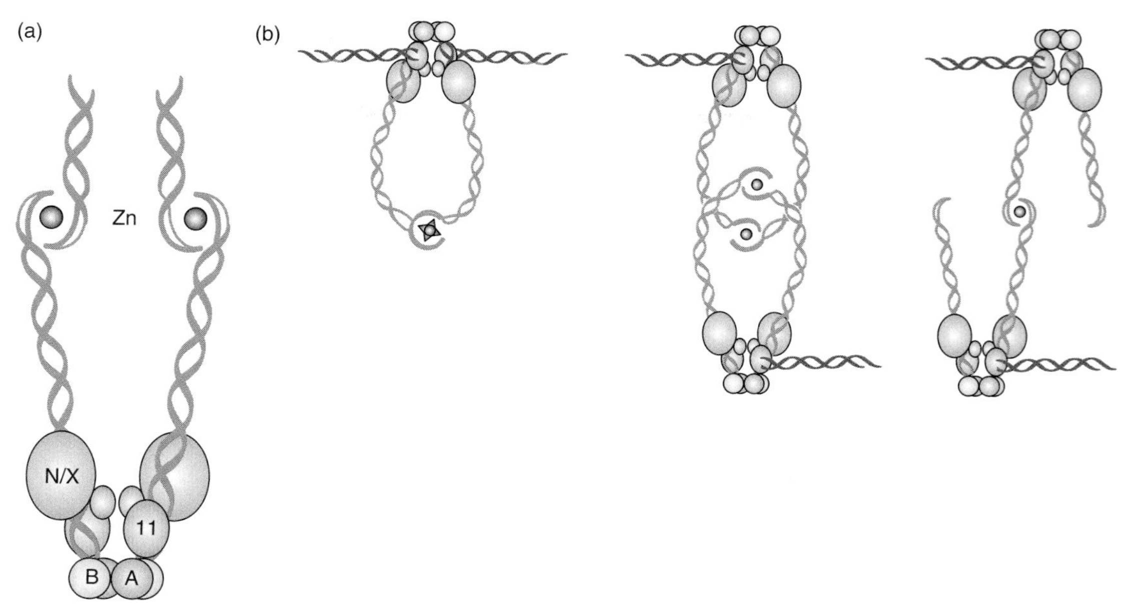

FIGURE 13.21 **Structure of Rad50 and model for the MRX (or MRN) complex binding to double-strand breaks.** (a) Rad50 has a coiled-coil domain similar to SMC proteins. The globular end contains two ATP binding and hydrolysis regions (A and B) and forms a complex with Mre11 (11) and Nbs1 (N) or Xrs2 (X). The other end of the coil binds zinc cation and forms a dimer with another MRX (or MRN) complex. The globular end binds to chromatin. (b) The MRX (or MRN) complex binds to double-strand breaks and can bring them together in a reaction involving two ends and one MRX (or MRN) complex (top right figure) or through an interaction between two MRX (or MRN) dimers as depicted in the bottom right figure. (Adapted from M. Lichten, *Nat. Struct. Mol. Biol.* 12 [2005]: 392–393.)

and lymphoid tumors. Null mutations of *NBS1* are lethal in mice as are null mutations of *MRE11* and *RAD50*, or *NBS1*.

The MRX complex in yeast cells (or MRN complex in human cells) makes an important contribution to double-strand break resection in meiosis and mitosis (FIGURE 13.22). The MRX complex works with an endonuclease called Sae2 in yeast. The MRX•Sae2 complex catalyzes an endonucleolytic cleavage that removes a 50 to 100 nucleotide block from the 5′ end of the double-stranded DNA break. The complex also processes blocked DNA ends such as those with an attached Spo11 protein (a type II topoisomerase-like protein; see below). The processed end now serves as substrates for further resection by the two following pathways: (1) Exo1, a 5′→3′ exonuclease and flap endonuclease, removes additional nucleotides from the recessed 5′ end (CtIP, a related endonuclease replaces Exo1 in human cells) or (2) a complex containing Dna2, another 5′→3′ exonuclease, and Sgs1, a helicase, removes the additional nucleotides from the recessed 5′ end.

Rad51, which has 30% identity with bacterial RecA, binds to the 3′-OH single-strand tails of the processed DNA to form a right-handed helical nucleoprotein filament in an ATP-dependent process. This binding stretches the DNA by approximately 1.5-fold compared to B-form DNA so that there are six Rad51 molecules and 18 nucleotides of single-strand DNA per helical turn. Rad51 is required for all homologous recombination processes, with the exception of single-strand annealing and nonhomologous end-joining mechanisms (see below). Yeast *rad51* null mutants are reduced in mitotic recombination, sensitive to ionizing radiation, and fail to undergo recombination

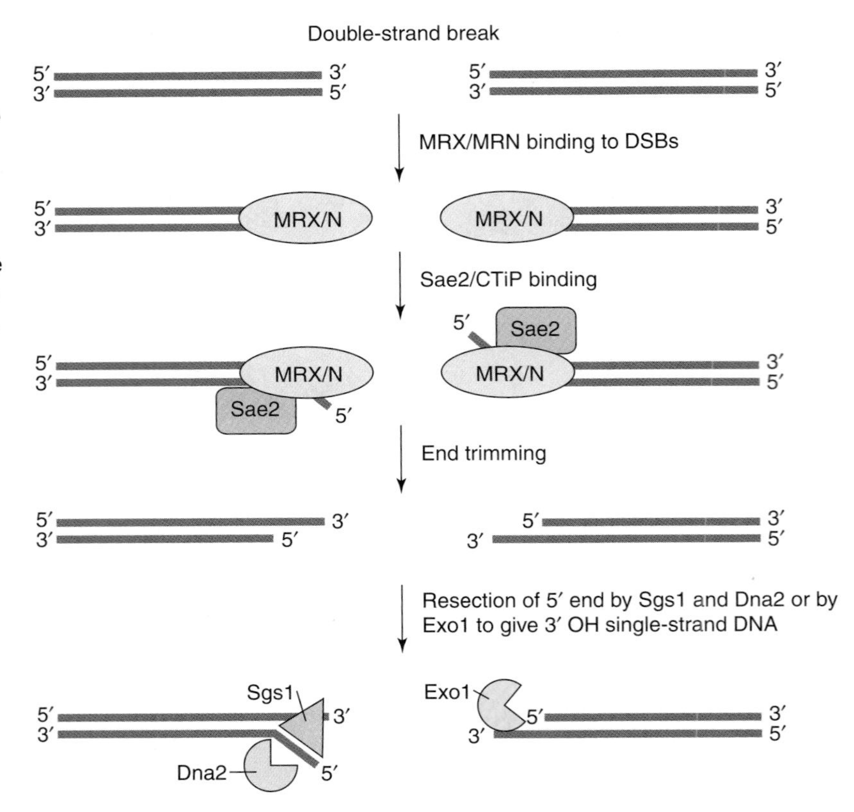

FIGURE 13.22 **Model for double-strand break end processing.** After a double-strand break is formed, the MRX/MRN complex binds to the ends. If the ends have a protein bound to it, or do not have 5′-P and 3′-OH ends, they must be first processed by the Sae2/CtIP nuclease to make ends available for further processing. In the second step, the 5′ end is resected by the Exo1 nuclease or the combination of the Sgs1 helicase and Dna2 nuclease to give single-strand tails with 3′-OH ends that can engage in homologous recombination. (Adapted from E. P. Mimitou, E. P. and L. S. Symington, *DNA Repair* 8 [2009]: 983–995.)

during meiosis. Double-strand breaks form but become degraded. In mice, Rad51 is essential. A fetus that is homozygous for mutant *rad51* does not survive past early stages of embryogenesis. This failure to survive is thought to reflect the fact that in vertebrates, at least one double-strand break occurs spontaneously during every replication cycle. As discussed later in Section 13.10, replication of an unrepaired nick on the template strand creates a double-strand break.

Rad51 nucleoprotein filament assembly is aided by RPA, the eukaryotic single-strand binding protein, which removes secondary structure in the single-stranded DNA, and recombination factors called **mediators**, which help to remove RPA, assemble Rad51 on single-stranded DNA, and promote strand exchange reactions. The yeast mediators are Rad52, Rad55, and Rad57. Rad52-deficient yeast mutants are extremely sensitive to ionizing radiation, defective in all types of homologous recombination, and never complete meiosis. In contrast, Rad52 does not appear to function as a mediator or to be essential for homologous recombination in mammals. Yeast Rad55 and Rad57 have some homology to Rad51 and combine to form a Rad55•Rad57 complex, which has no strand exchange activity when tested *in vitro*. Yeast *RAD55* or *RAD57* deletion mutants have a temperature-dependent sensitivity to ionizing radiation, reduced levels of homologous recombination, and do not undergo successful meiosis.

The human mediators RAD51B, RAD51C, RAD51D, XRCC2, and XRCC3 are also related to Rad51, with 20 to 30% sequence identity. Genes that code for these mediators appear to have arisen by duplication and therefore are related by sequence but have evolved to have different functions. Genes of this type are termed **paralogs**.

Human mediator proteins form three complexes: RAD51B•RAD51C, RAD51D•XRCC2, and RAD51C•XRCC3. RAD51 paralog genes have been deleted in chicken cell lines. The resulting cells are viable but are subject to numerous chromosome breaks rearrangements, and have reduced viability compared to normal cell lines. When the RAD51 paralog genes are deleted in mice, the animals undergo early embryonic death.

The **human breast cancer susceptibility gene 2 (*BRCA2*)** product, BRCA2, has mediator activity. Mutations in the *BRCA2* gene are associated with familial breast and ovarian cancers, and the DNA damage syndrome **Fanconi anemia**. BRCA2 interacts with RAD51 protein and can bind to single-strand DNA. The related Brh2 protein of the pathogenic fungus *Ustilago maydis* initiates Rad51 nucleoprotein filament formation by binding to Rad51 and recruiting it to single-strand DNA coated with RPA. Genetic studies in mouse cells have shown that Brca2 is required for homologous recombination.

2. Synapsis

Once the Rad51 filament has formed on single-strand DNA, in the double-strand break repair, a search for homology with another DNA molecule begins. When homology is found, strand invasion to form a D-loop occurs. Strand invasion requires the Rad54 protein and the related Rdh54 (also called Tid1) protein in yeast, and the Rad54B protein in mammalian cells. Rad54 and Rdh54, members of the SWI/SNF chromatin remodeling superfamily (see Chapter 18), possess a double-strand DNA-dependent ATPase activity, translocate on double-stranded DNA, and can induce superhelical stress in double-stranded DNA. Although Rad54 and Rdh54 from yeast, and Rad54 and Rad54B from mammalian cells are not DNA helicases, their translocase activity causes local opening of double strands that may serve to stimulate D-loop formation. Rad54-deficient yeast cells are sensitive to ionizing radiation and DNA damaging compounds. Rdh54-deficient yeast cells have a modest defect in recombination and are slightly DNA damage sensitive. Sensitivity is enhanced when both *RAD54* and *RDH54* are deleted. Yeast *rad54* mutants can complete meiosis, but have reduced spore viability. Yeast *rdh54* mutants are deficient in meiosis, and have a greatly reduced spore viability. The *rad54 rdh54* double mutant does not complete meiosis. Chicken and mouse cells with deletions in both *RAD54* and *RAD54B* are viable, but have increased sensitivity to ionizing radiation and reduced rates of recombination.

3. DNA Heteroduplex Extension

The proteins involved in heteroduplex extension are not as well defined as those required in the early steps of homologous recombination. D-loop formation results in a Rad51 filament being formed on double-strand DNA. Rad54 protein has the ability to remove Rad51 from double-strand DNA. This step might be important for DNA polymerase extension from the 3′-OH terminus. DNA polymerase δ is thought to be the polymerase for repair synthesis in double-strand

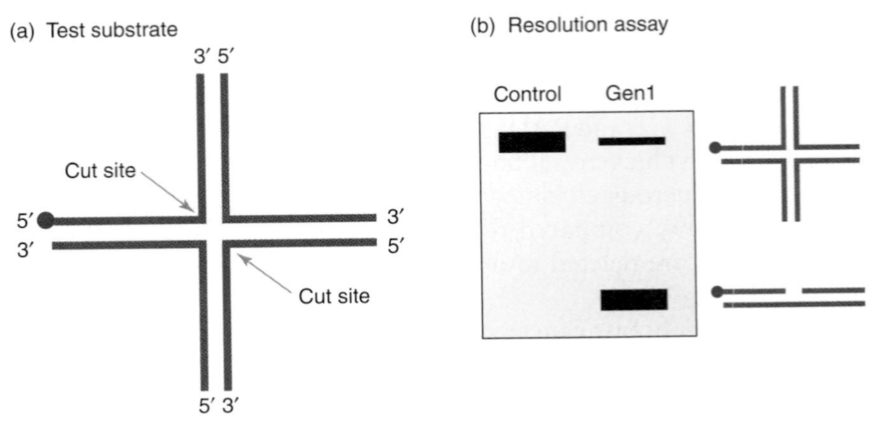

FIGURE 13.23 **Holliday junction resolution assay.** (a) A test four-way junction substrate for Holliday junction resolution is shown. The red circle indicates a 5′-P that is radioactively labeled to detect substrate and product. The green arrows indicate where a Holliday junction resolvase would cut symmetrically. (b). The product on an agarose gel is diagrammed after treatment with the human Gen1 Holliday junction resolvase. The control lane has no enzyme added and shows the migration position of the intact four-way junction substrate. After treatment with Gen1 most of the product is cut. Only one of the two products can be detected as the radioactive tag is associated with only one of the two products. This product differs from the substrate in size and so migrates at a different position on the agarose gel. (Adapted from C. Y. Ip, et al., *Nature* 456 [2008]: 357–361.)

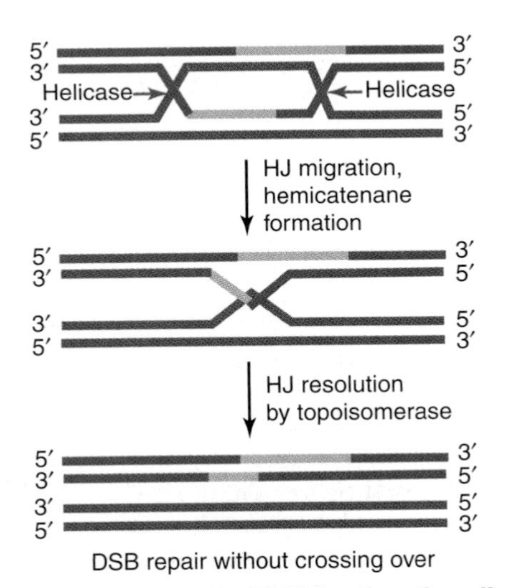

FIGURE 13.24 **Double Holliday junction dissolution by the action of a DNA helicase and topoisomerase.** The two Holliday junctions are pushed toward each other by branch migration using the DNA helicase activity. The resulting structure is a hemicatenane where single strands from two different DNA helices are wound around each other. This is cut by a DNA topoisomerase, unwinding and releasing the two DNA molecules and forming noncrossover products.

break-mediated recombination. Some recent studies, however, also suggest that DNA polymerase eta (POLH; see Chapter 12) is able to extend from the strand invasion intermediate terminus.

4. Resolution

The search for eukaryotic resolvase proteins equivalent to the bacterial RuvC protein (Table 13.1) has been long and, until recently, not definitive. Recent studies have shown that several proteins can resolve Holliday junctions (FIGURE 13.23). The first resolvase, the Mus81•Eme1 complex, cleaves Holliday junctions that contain an exposed 5′ DNA strand end at or close to the junction crossover point. A second resolvase is called the SLX1•SLX4 complex in yeast and the SLX1•BTBD12 complex in humans. SLX1 is highly conserved and contains endonuclease activity. SLX4, which is is not well conserved, associates with other nucleases in addition to SLX1. It is the SLX1•SLX4 and SLX1•BTBD12 heterodimers, however, that have the Holliday junction resolving activity. A third resolvase is called GEN1 in humans and Yen1 in yeast.

The double-strand break repair model proposes two Holliday junctions. Similar structures might be formed during restart of stalled replication forks. Because mutants with a defective DNA helicase Sgs1 in yeast or a defective BLM helicase in humans result in higher crossover rates, these helicases are proposed to prevent crossover formation by Holliday junction resolution. This prevention is proposed to occur by branch migration of the double Holliday junctions to convergence, through the DNA helicase action (FIGURES 13.24 and 13.25). The end structure is suggested to be a **hemicatenane**, where one DNA strand

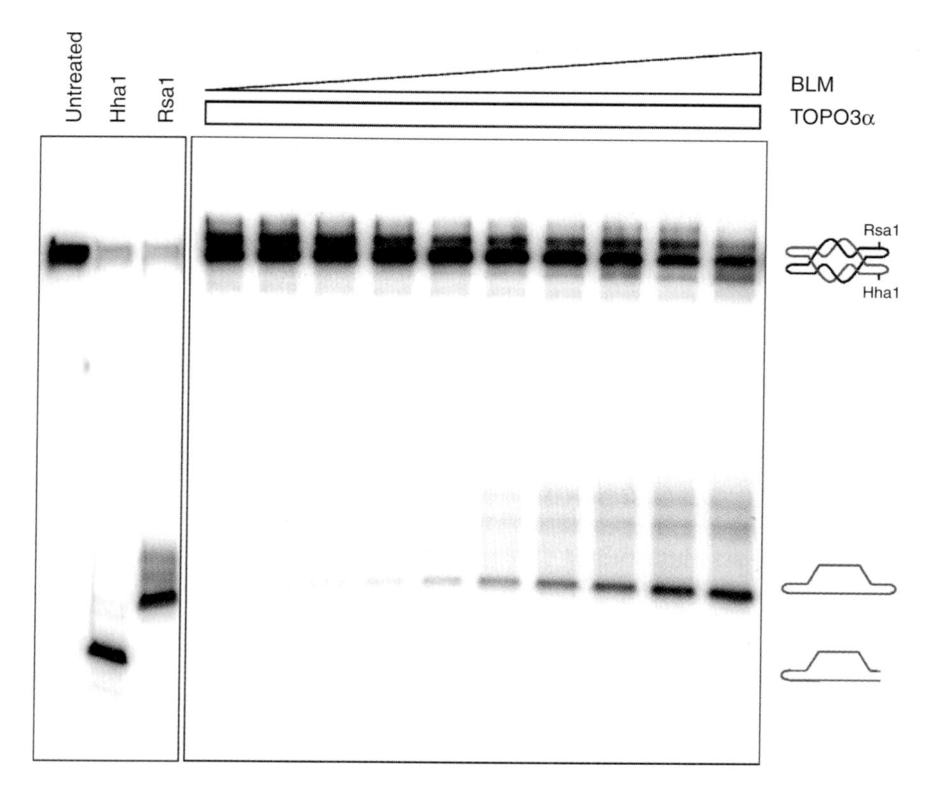

FIGURE 13.25 **Holliday junction dissolution by BLM helicase and hTOPO IIIα topoisomerase.** Artificial Holliday junctions are formed by annealing two sequences (shown in black and red) with hairpin ends. The black and red ends contain different restriction endonuclease sites for identification. *Left box*: digestion of the substrate with Hha1 or Rsa1 restriction endonucleases results in cleaved or closed red molecules, respectively, which are radioactively labeled for detection (see drawings of the structures on the right). *Right box*: adding hTOPOIIIα alone does not release the red molecule, but adding hTOPOIIIα with increasing amounts of the BLM helicase (indicated by the triangle above the lanes) causes release of the intact red molecule. (Photo courtesy of Ian D. Hickson, Weatherall Institute of Molecular Medicine, University of Oxford.)

of a duplex is wound around a single DNA strand of another duplex to form a looped link. This structure is then resolved by the action of an associated DNA topoisomerase, Top3 in the case of Sgs1 and hTOPOIIIa in the case of BLM. *In vitro*, BLM and hTOPOIIIa can resolve a double Holliday junction without any crossing over. This process, which is called **dissolution**, takes place in two steps. First, the topoisomerase nicks one DNA strand and then the DNA helicase disconnects the interconnected strands.

Enzyme(s) that can resolve Holliday junctions as crossovers during meiosis have not been fully identified. Additional endonuclease activities associated with the Mus81•Mms4 complex in yeast and Mus81•Eme1 complex in mammalian cells can cleave nicked Holliday junction-like structures and branched DNA structures *in vitro*. The relationship of this activity to meiotic crossover formation is not fully defined, however.

Chromatin structure has a profound effect on homologous recombination by providing a signal to repair a double-strand DNA break and influencing the recombination machinery's ability to gain access to the broken DNA strands. Two types of chromatin-associated factors influence recombination. The first is composed of factors that directly modify histones. The key histone in this process, H2AX, a variant of H2A, comprises 2% to 25% of the H2A pool and is unevenly distributed throughout chromatin. A DNA break in mammalian cells leads to the rapid phosphorylation of serine-139 on H2AX histones in nucleosomes that are estimated to be located 1 to 2 Mb around the break. The phosphorylated form of H2AX, γ-H2AX serves as a signal to accumulate factors needed for DNA repair and cell cycle arrest. Further modifications of histones, through ubiquitylation (see Chapter 3) and methylation, influence histone removal and double-strand break formation in meiosis. The second type of chromatin-associated factors are the chromatin remodeling complexes (see Chapter 18), which can remove nucleosomes from DNA to allow the DNA to be resected to produce a 3′ overhang that can pair with a homologous sequence. In particular, the chromatin remodeling complexes associated with DSBR remove γ-H2AX that is located right next to the double-strand break to allow MRX (or MRN) binding and end resection. The chromatin remodeling complexes that play a role in DSBR are called INO80, SWR1, TIP60 and RSC.

13.7 A Variation of the Double-Strand Break Repair Model

The synthesis-dependent strand-annealing (SDSA) model is a gene conversion-only model.

Even though the double-strand break repair model is an excellent model for meiotic homologous recombination, it falls short of serving as a universal model for homologous recombination because mitotic gene conversions are rarely associated with crossing over. The **synthesis-dependent strand-annealing (SDSA) model** was proposed to solve this problem. It is the accepted model for homologous recombination gene conversion events that are initiated by a double-strand break but are not associated with any crossovers. The SDSA model grew out of observations on the nature of mating-type switching events in yeast (see Chapter 7) and recombinational repair of gaps in DNA. Mating-type switching is initiated by a programmed double-strand break at the *MAT* locus, resulting in a gene conversion event that changes the broken *MAT* locus to a new sequence. It is not associated with crossing over.

The initial steps in the synthesis-dependent strand-annealing pathway are similar to those described for the double-strand break repair model (FIGURE 13.26). After strand invasion and DNA synthesis, however, the second end is not captured. Instead, the invading strand is displaced and anneals with the other broken end. The final steps are synthesis and ligation. This pathway results in repair of the double-strand break, using the homologous chromosome sequence as a template for repair synthesis and donation of genetic information, but without any associated crossover. The synthesis-dependent strand-annealing model accounts for mitotic gene conversions without crossovers. In addition, synthesis-dependent strand-annealing has an important role in meiotic recombination.

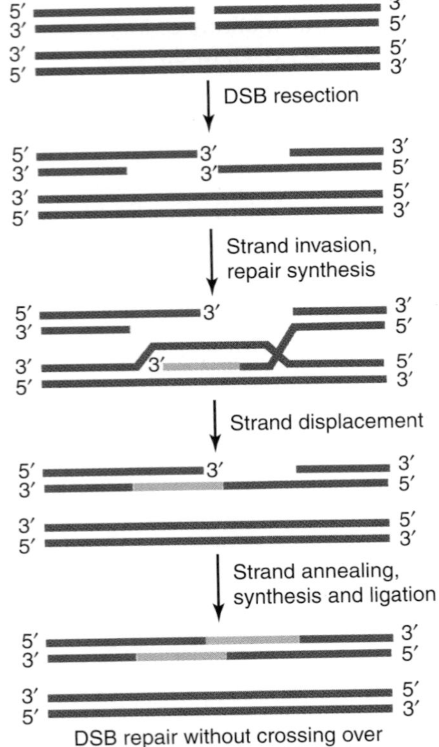

FIGURE 13.26 **Synthesis-dependent strand-annealing model of homologous recombination.** Recombination is initiated by a double-strand break and is followed by end resection to form single-strand tails with 3′-OH ends. Strand invasion and DNA synthesis repair one strand of the break. Instead of second strand capture as depicted in Figure 13.15, the strand in the D-loop is displaced. The single strand can anneal with the single strand of the other end. Repair synthesis then completes the double-strand break repair process. No Holliday junction is formed and the product is always noncrossover. The light red and blue lines represent DNA synthesis.

13.8 Meiotic Recombination

Some aspects of meiotic recombination are novel.

As discussed earlier in this chapter, meiotic recombination occurs at much higher frequencies than mitotic recombination. This difference is primarily due to the specific initiation of recombination at double-strand breaks catalyzed by the **Spo11** protein. In *Saccharomyces cerevisiae* Spo11 makes about 150 to 200 breaks per genome in meiosis, which is a sufficient number of breaks to ensure at least one crossover per chromosome arm in meiosis. In addition to Spo11, there are meiotic-specific homologous recombination proteins (discussed below) that promote homologous recombination from the Spo11-induced double-strand breaks in pathways that result in a significant number of crossovers. The high crossover frequency is due to both the meiotic-specific proteins such as Spo11, and the use of recombination pathways that result in a high frequency of crossover products. In contrast to mitosis, where the sister chromatid is the preferred donor of information for homologous recombination and repair of a double-strand break, in meiosis one of the nonsister chromatids on the homologous chromosome is the preferred donor of information for recombination.

The distribution of crossovers along a chromosome is not spaced randomly, but rather crossovers are separated by large distances. When a crossover occurs at one position on a chromosome, it is highly unlikely that a second crossover occurs nearby. This phenomenon, known as **interference**, limits the number of crossovers per chromosome arm, but also helps to ensure that there is at least one crossover per chromosome. Because crossovers are required for proper chromosome segregation at meiosis I, interference is one mechanism to regulate chromosome segregation.

Meiotic recombination occurs in the context of a proteinaceous structure called the **synaptonemal complex**. The synaptonemal complex forms along chromosome arms at the pachytene stage of meiosis and serves as a scaffold to connect the interacting homologous chromosomes (FIGURE 13.27). Meiotic recombination starts before the synaptonemal complex is formed along the chromosomes, but synaptonemal complex proteins are needed for crossover formation, and mutation of the synaptonemal complex components abolishes the formation of viable meiotic products. It is thought that synaptonemal complex formation starts at sites of meiotic recombination that will result in crossovers, and some of the synaptonemal complex proteins interact with the homologous recombination proteins, possibly linking synaptonemal complex formation to homologous recombination.

Some recombination proteins are made only in meiotic cells.

Spo11 is expressed only in meiosis and is absolutely required for formation of meiotic double-strand breaks. Its mode of action is similar to that of DNA topoisomerases and indeed Spo11 resembles type II topoisomerases. Spo11 breaks both strands of DNA through a covalent attachment to DNA through the tyrosine residue in the catalytic site of the protein. Then Spo11 is removed from the 5′ ends of the double-strand breaks by the action of MRX and Sae2 (sporulation in the absence of spo eleven 2) in yeast, which prepares the ends for the next step of recombination, end resection to give the single-strand 3′-OH tails (Figure 13.22). Single-strand tails in yeast meiosis are approximately 1 kb in length.

FIGURE 13.27 Synaptonemal complex. (a) A model of synaptonemal complex with the light blue lateral elements aligned by a network of black transverse filaments and dark blue longitudinal filaments, and the chromatin loops on the outside. Recombination nodules, depicted as a gray oval, promote crossing over. (b) Synaptonemal complexes along the chromosomes of tomato chromosomes in meiosis. (Reproduced from J. Hey, *PLoS Biol.* 2 [2004]: e190 [DOI:10.1371/journal.pbio.0020190]. Photo courtesy of Daniel G. Peterson, Mississippi State University.)

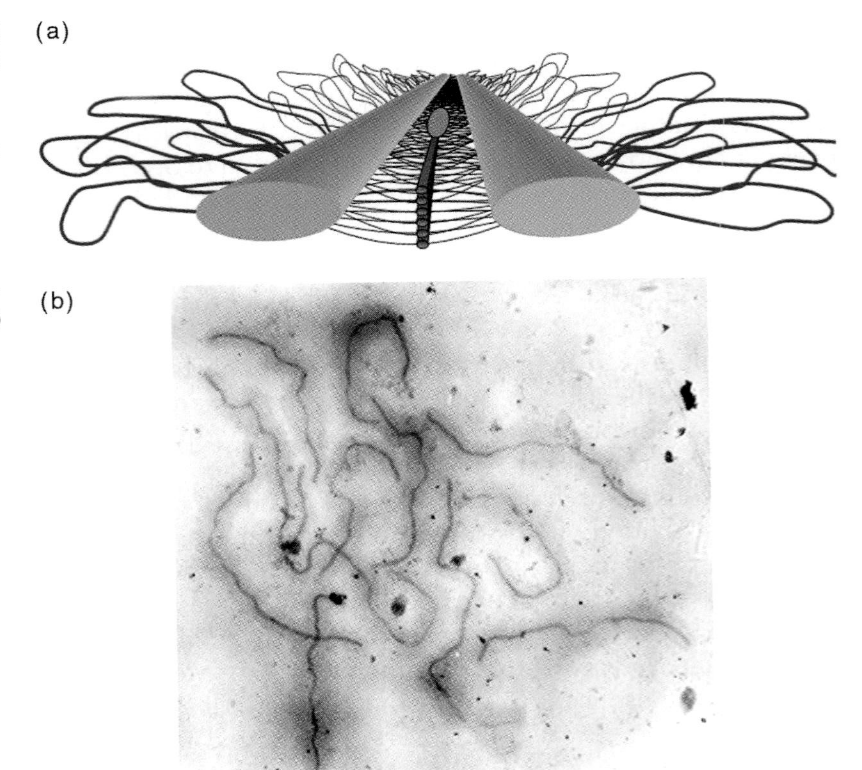

Two recombinases of the RecA family, Rad51 and Dmc1, are active in meiosis. Mutation of the gene that codes for either protein results in a failure to complete meiosis, and a reduction in crossover formation as assayed by physical methods. Dmc1 has 45% identity to RecA and, unlike Rad51, is expressed only in meiosis. *In vitro*, human DMC1 can promote strand exchange and forms a filament on single-strand DNA. It is thought that the heterodimers Mei5•Sae3 and Hop2•Mnd1 act as mediators in the Dmc1-promoted strand exchange reaction. The Hop2•Mnd1 complex stimulates D-loop formation by Dmc1 *in vitro*, and mutant studies suggests that the Hop2•Mnd1 complex functions with Dmc1 in meiotic homologous recombination. Overexpression of Rad51 can partially suppress *dmc1* mutants, showing that Rad51 and Dmc1 perform related roles *in vivo*.

Mutations in the genes that code for Mus81 and Eme1 in the fission yeast *S. pombe* have a strong effect on meiosis and result in spore inviability. These mutations can be complemented by overexpression of a bacterial resolvase called RusA, suggesting that the Mus81•Eme1 complex has a role in resolution in meiosis in *Schizosaccharomyces pombe*. Crossing over in the rescued spores is reduced up to 50-fold, suggesting that the Mus81•Eme1 complex is required for crossover formation. The *S. cerevisiae* Mus81•Eme1 complex does not seem to be required for crossover formation. In the next section, we discuss various models for the production of meiotic crossovers, which may explain the differing need for the Mus81•Eme1 complex.

Meiotic recombination models propose two different types of homologous recombination events.

The Holliday model, the Meselson-Radding model, and the double-strand break repair model of homologous recombination all link crossing over to gene conversion through the resolution of Holliday junction intermediates as either crossover or noncrossover events with equal probability. This linkage explained the association of meiotic gene conversion with crossing over, but did not explain the observation that most mitotic DSBR was through gene conversion not associated with crossing over. To explain this observation, the synthesis-dependent strand-annealing model (see Section 13.7) was developed that proposed repair of double-strand breaks through gene conversion without any associated crossover.

More recent studies on meiotic recombination in the yeast *S. cerevisiae* have found that there are two distinct phases of meiotic recombination (FIGURE 13.28). In the first phase, most double-strand breaks are repaired by a synthesis-dependent strand-annealing mechanism, giving rise to meiotic gene conversion but no crossovers. In the second phase, the remaining double-strand breaks are repaired by a double-strand break repair mechanism that is preferentially resolved in the crossover mode. The nonrandom spacing of these two types of

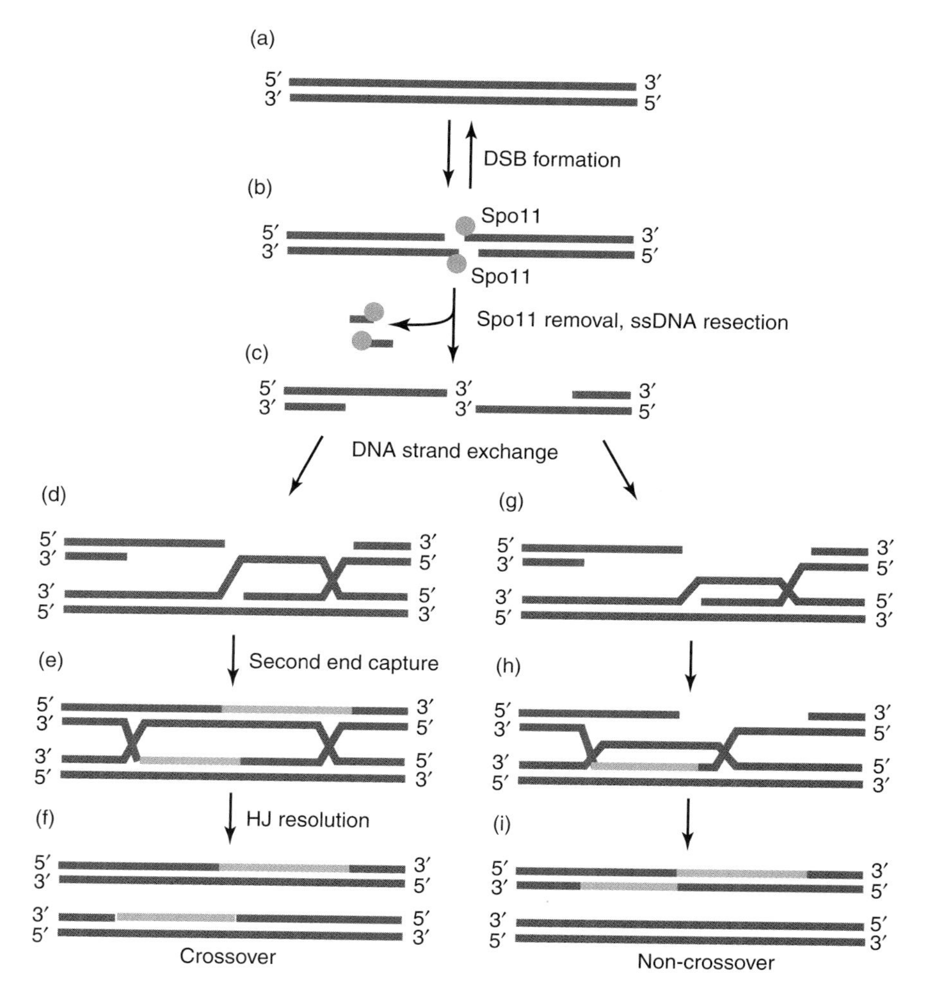

FIGURE 13.28 **Model of meiotic homologous recombination.** A DNA duplex (a) is cleaved by Spo11 to form a double-strand break with Spo11 covalently attached to the ends (b). After Spo11 is removed the ends are resected by the MRX (or MRN) complex to give single-strand tails with 3′-OH ends (c), which form a complex with Rad51 and Dmc1. Strand exchange occurs by strand invasion (d and g). Second end capture results in a double Holliday junction, which is resolved to form crossover products (e and f). Most of the double-strand breaks do not engage in a second end capture mechanism and instead engage in a synthesis-dependent strand-annealing mechanism (h and i), which results in non-crossover products. (Adapted from M. J. Neale and S. Keeney, *Nature* 442 [2006]: 153–158.)

meiotic recombination along the chromosomes results in the apparent association of gene conversion with crossing over in meiosis.

The second phase of meiotic recombination, the DSBR mode with preferential resolution of Holliday junctions as crossovers, requires a set of proteins called the ZMM proteins. The ZMM proteins include the synaptonemal complex proteins Zip1, Zip2, and Zip3, and Msh4•Msh5, which bind to Holliday junctions and the helicase Mer3, which can bind to and unwind a Holliday junction. The ZMM pathway is proposed to be more active in some species than others. *S. cerevisiae* uses the ZMM pathway for meiotic crossovers, while *S. pombe* is proposed to use a pathway dependent on Mus81•Eme1 to resolve branched structures that form during homologous recombination. This difference in activities accounts for the different effects of *mus81* mutants on meiosis in these two organisms. ZMM proteins are not expressed in mitosis, limiting the occurrence of crossovers. A further limitation comes from the action of Sgs1/BLM and Top3/hTOPO IIIa, which act to resolve Holliday junctions as noncrossovers.

13.9 Using Mitotic Recombination to Make Gene Knockouts

Mitotic recombination can be used in genetic engineering to make targeted gene disruptions.

With the realization that plasmids or DNA fragments can be integrated into genomes by homologous recombination, recombinant DNA techniques were devised to make a gene inoperative. A gene can be eliminated or "knocked out" by **gene replacement** or **gene insertion** (FIGURE 13.29). In gene replacement, all or part of the coding sequence is deleted and replaced with a selectable marker. In **gene insertion**, a selectable marker is inserted into the coding region of a gene, thereby disrupting its activity.

A gene knockout DNA construct can integrate randomly into the genome of the recipient cell, or it can integrate by homologous recombination, using homology between the gene sequences on the DNA fragment and the same gene sequences in the recipient cell DNA. When the latter occurs, the introduced DNA will integrate at the gene of interest, and will change the gene sequence. This type of integration is called **gene targeting**.

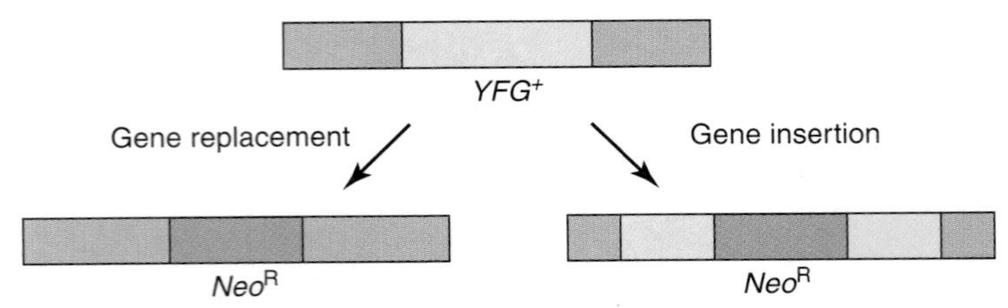

FIGURE 13.29 **Gene knockout by gene replacement or gene insertion.** In gene replacement, all or part of *YFG* (*y*our *f*avorite *g*ene) is deleted and replaced with a selectable marker (here Neo^R, which confers resistance to neomycin). In gene insertion, the selectable marker is inserted into *YFG*, thereby disrupting its activity.

Gene knockouts in yeast occur by homologous recombination with high efficiency.

Early studies on yeast transformation showed that double-strand break ends are recombinogenic and allow plasmids to integrate by homologous recombination into the yeast genome. This work was soon followed by several methods to make knockouts in yeast. One of the most effective ways was to make a construct that had 500 bp to 1 kbp of flanking homology to a gene of interest, with a nutritional selectable marker between these regions of homology (FIGURE 13.30). When the knockout construct was excised from a plasmid vector by restriction enzyme digestion and introduced into yeast by transformation, the colonies that had now acquired the selectable marker were disrupted for the gene of interest.

This approach was soon modified to use much shorter regions of flanking homology and drug resistance selectable markers. The method was further streamlined to eliminate the plasmid and cloning step and to make the entire construct using the **polymerase chain reaction (PCR)** (FIGURE 13.31). PCR primers that have approximately 50 bp of homology to the beginning or end of the gene of interest, plus 20 bp of homology to the beginning or end of the selectable marker DNA sequence are used to amplify the selectable marker sequence. The PCR product therefore will also have the 50 bp of sequence flanking the selectable marker product that is homologous to the yeast gene region of interest. Fifty base pairs turn out to be sufficient for homologous recombination in yeast. Because transformation occurs by homologous recombination and there is little random integration of sequences in yeast transformations, transformant colonies that express the selectable marker have gene replacements of the gene of interest with the selectable marker.

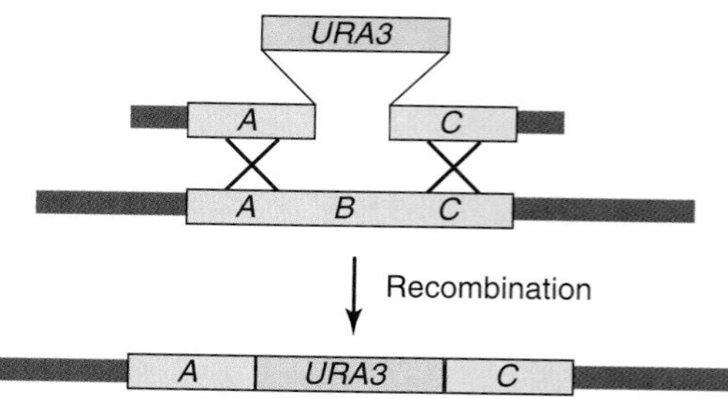

FIGURE 13.30 **One step gene knockout in yeast.** Transformation with a linear fragment containing a substitution of part or all of the *ABC* gene by a marker, here *URA3*, results in insertion of the *URA*3 marker at the *ABC* gene by homologous recombination. The intact *ABC* gene is replaced by the disrupted *ABC* gene with the *URA3* marker.

Gene knockouts in mice also can be made by gene targeting methods.

Homologous recombination to make gene knockouts is not as efficient in mouse cells as it is in yeast cells, but the methods have been modified to make it feasible to create accurate and efficient gene knockouts. These methods were pioneered by Mario Capecchi and Oliver Smithies, who developed gene targeting methods in the 1980s. The first important step in the ability to make gene knockouts in mice was the development of a selection system for the gene knockout while being

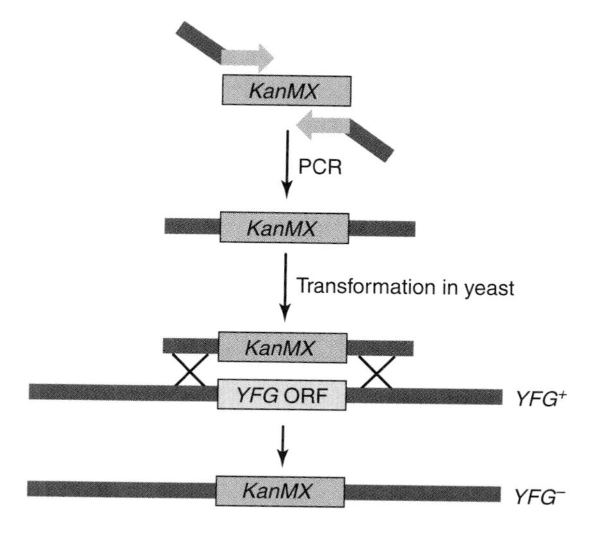

FIGURE 13.31 **Gene knockout by PCR.** Primers using homology flanking the *YFG* open reading frame (ORF) shown in blue, and having short regions of homology to the *KanMX* gene at the 3′ end, shown in red with arrows, are used to amplify the *KanMX* gene by PCR. The *KanMX* gene is the version of the *Neo*^{R} gene that expresses in yeast. The PCR product, the *KanMX* gene flanked by short regions of homology outside the *YFG* ORF, is used in transformation of yeast. Insertion of *KanMX* at the *YFG* ORF occurs by homologous recombination and results in replacement of *YFG* ORF with *KanMX* and a knockout of the *YFG* ORF.

able to eliminate the background of random insertions. The second important step was the use of **embryonic stem (ES) cells**, which are derived from the inner cell mass of the mouse blastocyst and can be grown in culture. When reintroduced into mouse blastocysts, the modified ES cells can develop into germline or other tissues of the mouse. Martin Evans and Elizabeth Robertson, pioneers in ES cell research, showed that ES cells can grow stably in culture, be modified by transformation, and develop into germline cells when injected into mouse blastocysts. The last feature is important because the gene knockout made in cells in culture has to be heritable in mice in order to study the consequences of the knockout.

Capecchi and coworkers showed that it was possible to make a knockout of the hypoxanthine guanine phosphoribosyl transferase (HGPRT) gene (see Chapter 7) in mouse cells, while Smithies and coworkers made disruptions of the human β-globin gene locus by homologous recombination. The first gene disruption DNAs were introduced into cells by injection but soon electroporation became standard. Electroporation involves the application of a brief electrical shock to cells. This shock transiently opens pores in the plasma membrane and allows cells to readily take up DNA.

In the first gene knockout experiments, Capecchi and coworkers replaced one exon of the *HGPRT* gene, exon 8, with the selectable marker *Neo*^{R}, which confers resistance to the aminoglycoside antibiotic G418 (FIGURE 13.32). Homologous recombination between the sequences of the *HGPRT* gene flanking the *Neo*^{R} insertion and the chromosomal *HGPRT* gene resulted in a replacement of exon 8 in the genome with *Neo*^{R}. Confirmation of the *HGPRT* knockout was made by showing that the knockout cells could grow in special medium containing 6-thioguanine (6-TG), which is a poison to *HGPRT*1 cells. The efficiency of targeted knockout of the *HGPRT* gene to random insertion of the *Neo*^{R} gene was about 1:1,000.

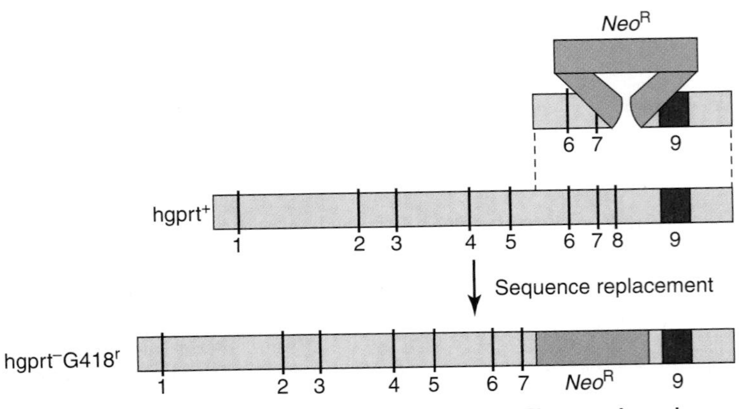

FIGURE 13.32 Disruption of the mouse *HGPRT* gene. A region of the mouse *HGPRT* gene containing exons 6 to 9 was cloned into a vector. Exon 8 was disrupted by replacement with the *Neo*^{R} gene. Transfection of cells with DNA containing a *Neo*^{R} insert results in homologous recombination to replace exon 8 with *Neo*^{R}. The hgprt^{-} cells are resistant to the aminoglycoside antibiotic G418, provided by the *Neo*^{R} gene. (Adapted from M. Capecchi, *Nat. Med.* 7 [2001]: 1086–1090.)

To increase this ratio, Capecchi developed a selection called a **positive-negative selection**, which enriched the targeted disruption (FIGURE 13.33). The positive-negative selection could be applied to any gene knockout and thus made gene targeting in mouse a realistic protocol. The positive-negative selection has two components, the positive selection for the *Neo*^{R} marker that disrupts the gene of interest and a negative selection for cells that carry the herpes simplex virus thymidine kinase (*HSV-tk*) gene. The vector construct has an insertion of the *Neo*^{R} gene to replace an exon of the gene, and also carries the *HSV-tk* gene at the end. Cells that have the *HSV-tk* gene are sensitive to [1-(2-deoxy-2-fluoro-b-D-arabino-furanosyl)-5-iodouracil] (FIAU). Because the *HSV-tk* gene is located outside the gene homology on the knockout construct, only random insertion

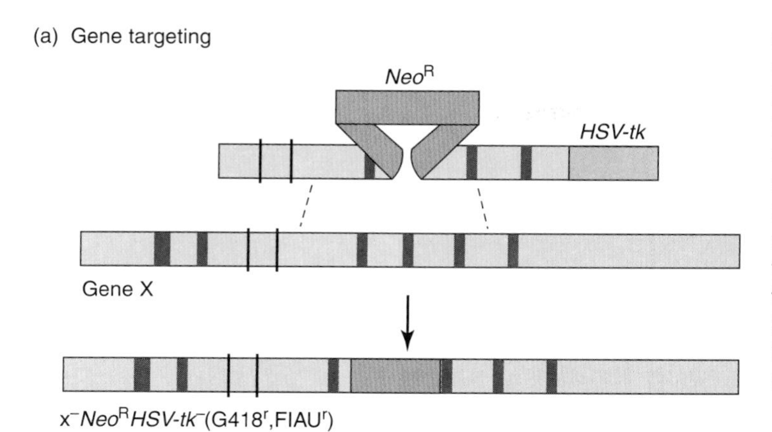

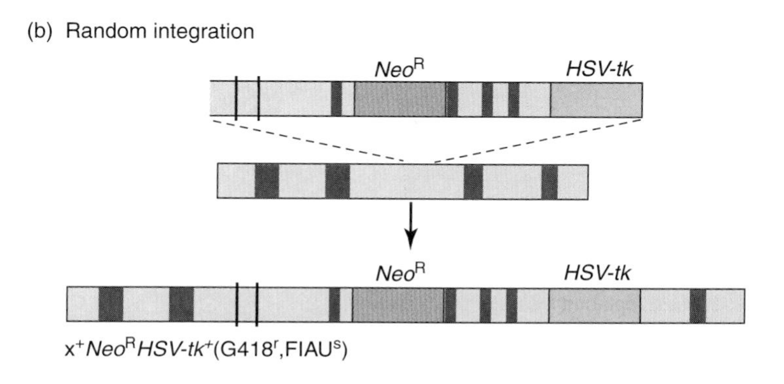

FIGURE 13.33 **Positive-negative selection to select precise targeting events.** The vector construct has an insertion of the *Neo*^{R} gene to replace an exon of the gene, and also carries the *HSV-tk* gene at the end. (a) Homologous recombination between the targeting vector and the chromosomal region results in replacement of one exon of gene X, G418 resistance and FIAU resistance. (b) If the vector inserts randomly, the cells will be G418-resistant, but will also carry the *HSV-tk* gene and will be sensitive to FIAU. Growth of the cells in G418 and FIAU after introduction of the vector DNA selects for targeted integration and against random insertion as those cells will be killed by the FIAU drug. (Adapted from M. Capecchi, *Nat. Med.* 7 [2001]: 1086–1090.)

events will have the *HSV-tk* gene inserted in the genome. Random integration events therefore will have both the *Neo*^{R} and *HSV-tk* genes inserted and will be resistant to the aminoglycoside antibiotic G418 but sensitive to FIAU. Gene targeting events will have only the *Neo*^{R} gene inserted by homologous recombination and will be resistant to both G418 and FIAU. This allows for the selection of the rare gene targeting events against the background of random insertion events.

After the appropriate gene targeted knockout has been made in ES cells, the cells are injected into a mouse blastocyst. The strain of mouse used for the blastocysts differs from that of the injected ES cells by genes that determine mouse coat color so that it will be possible to identify mice that have the disrupted ES cells (FIGURE 13.34).

Once the blastocysts are injected with the ES cells, they are implanted into pseudopregnant female mice. The mice that come from the implanted blastocysts are called *chimeric mice* because they are a mixture of cells derived from the injected ES cells and cells from the blastocyst. Chimeric mice show both coat colors, those from the ES cells and those from the blastocyst cells. Mating of these mice with test mice will give the next generation of mice that carry a germline knockout of the gene of interest. These are called **transgenic mice.** The transgenic mice are heterozygous for the knockout gene of interest and are used for further studies and matings to generate homozygous knockout mice.

FIGURE 13.34 Gene replacement in mice. (a) Embryonic stem (ES) cells are altered by gene targeting to express an altered version of a gene. The cells are expanded into colonies and cells from each colony are tested for the correct replacement. (b) The chosen ES cells are injected into an early mouse embryo, the blastocyst, which is then introduced into a pseudopregnant recipient mouse. Mice produced from the ES cells will have a coat color derived from the injected ES cells, here shown in pink. In a few cases, the ES cells will populate the germline. When these mice are bred with normal mice, all of the cells in the progeny mice will have the altered gene. (Adapted from B. Alberts, et al., *Molecular Biology of the Cell, Fourth Edition.* Garland Science, 2002.)

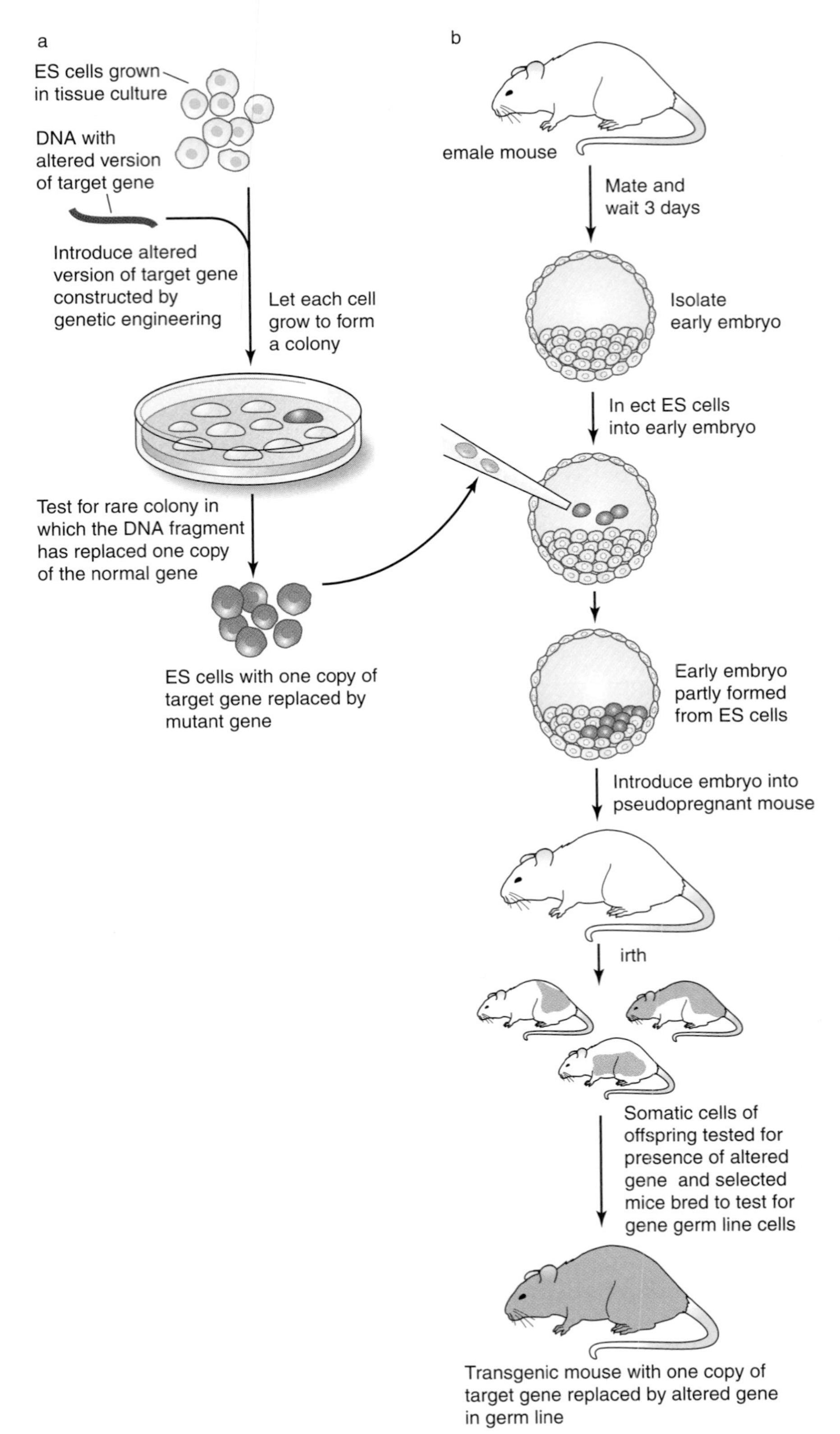

13.10 Mitotic Recombination and DNA Replication

Mitotic homologous recombination is essential during DNA replication when replication forks collapse.

The DNA template is subject to multiple types of damage, some of which have been discussed in Chapter 12. If certain forms of damage such as bulky adducts to nucleotides or single-strand nicks are not repaired prior to replication, they can put the replication fork at risk for stalling. When a replication fork encounters a nick, it collapses and a double-strand break is formed, which has the potential for generating DNA rearrangements. In systems such as *E. coli*, which have a circular chromosome with only one replication origin and two replication forks, stalling or collapse of the fork may compromise replication of the entire genome. Cells have developed several mechanisms to restart stalled or collapsed replication forks, some of which are dependent on homologous recombination (FIGURE 13.35).

In 1966, Philip Hanawalt proposed that replication forks could collapse at single-strand gaps in the template DNA and in 1976, Bernard Strauss and coworkers proposed that replication forks could collapse at nicks through breakage of stalled forks. He proposed that replication past damage on the template strand could be accomplished by reverse branch migration in which the original template strands

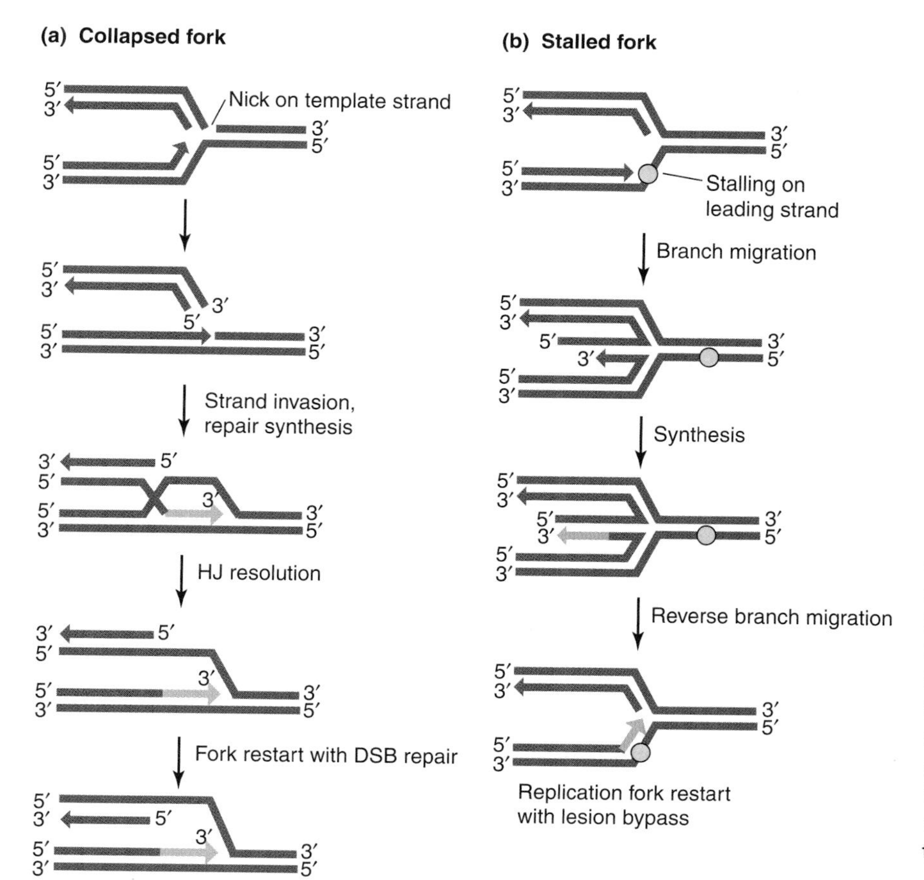

FIGURE 13.35 **Mechanisms of replication fork restart.** (a) Creation of a replication fork by strand invasion and homologous recombination. Replication past a nick on the template strand collapses the replication fork and creates a one-ended double-strand break. Strand invasion by homologous recombination into the sister chromatid forms a Holliday junction. Resolution of the Holliday junction re-establishes the replication fork. The bottom structure is the same as the top structure, except that the nick on the template strand has been repaired. (b) Stalled replication can lead to the formation of reversed replication forks. Replication past a lesion on the template strand can cause a dissociation between leading and lagging strand synthesis, with lagging strand synthesis continuing on in this example. Branch migration causes the fork to reverse and a branched structure is formed. Note that the two nascent sister strands are now paired. Repair synthesis of the leading strand, using the lagging strand as a template, permits the leading strand to be extended. Note that the lesion in the template strand is not involved in the repair synthesis. Reversal of the branched structure allows replication to resume, with the lesion on the template strand having been bypassed.

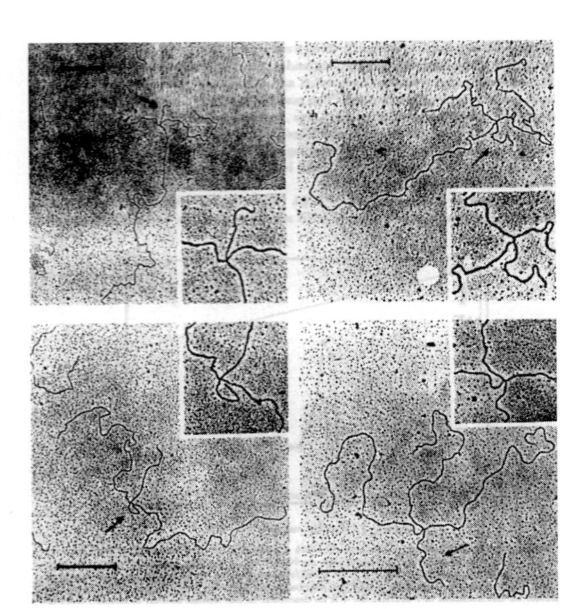

FIGURE 13.36 **Examples of reversed fork intermediates.** Four-pronged replication forks were isolated from mammalian cells after DNA replication was arrested by DNA damage treatment. Replication intermediates were enriched for on density cesium chloride gradients. The replication intermediates were visualized by electron microscopy. (Reproduced from *J. Mol. Biol.*, vol. 101, N. P. Higgins, K. Kato, and B. Strauss, A model of replication repair . . . , pp. 417–425, copyright 1976, with permission from Elsevier [http://www.sciencedirect.com/science/journal/00222836]. Photo courtesy of Bernard S. Strauss, University of Chicago.)

were reannealed while the newly synthesized DNA strands were displaced as short branches (Figure 13.35b). The displaced strand with the free 3′-OH was then used to prime repair synthesis using the displaced sister strand as a template. Once repair synthesis reached to the end of the displaced sister strand, the fork was reversed, and normal semiconservative replication resumed, having bypassed the lesion in the template strand. Evidence for this mode of replication came from electron microscopy observations of four-pronged structures or reversed forks, and the finding of sister chromatid pairing by density labeling (FIGURE 13.36). The use of the sister chromatid instead of the template strand for replication is called **template switching**. Although template switching or fork reversal is not a homologous recombination event, formation of the reversed fork may require branch migration enzymes and the reversed fork intermediate resembles a Holliday junction and can be acted on by Holliday junction resolving enzymes.

The underlying principle of recombination-dependent DNA replication is that replication can be initiated at sites other than replication origins, without the assistance of origin binding factors such as DnaA in *E. coli* and ORC binding complex in eukaryotes (see Chapter 10). In *E. coli*, DnaB is loaded at the oriC through its interaction with DnaA. Another protein called PriA can load DnaB at sites other than oriC. PriA is not required for initiation of DNA replication, but it is required for replication restart. Further studies have shown that PriA can load DnaB and hence other replication factors at non-oriC to assemble a replication fork at a recombination intermediate similar to the one shown in Figure 13.37a.

The rescue of stalled forks is a special problem in bacteria when cells do not activate the SOS pathway and hence cannot use translesion DNA polymerases to bypass damage in the template DNA strand. When a replication fork collapses, for example, when the fork encounters a nick, it can be reassembled by double-strand break repair mechanisms using homology. Both single-strand annealing and strand invasion mechanisms have been invoked for replication fork repair (FIGURE 13.37).

Recombination to repair the collapsed fork occurs between the collapsed fork and the sister chromatid. The proximity of these two molecules most likely promotes this interaction, but other factors such as protein complexes that hold sister chromatids together may also promote sister strand recombination.

The dependence of DNA replication on recombination factors to restart replication at stalled or collapsed forks might explain the observation that many homologous recombination genes of mammalian cells are essential. These genes are not essential in organisms with smaller genomes, perhaps due to the less frequent replication fork collapse and the more robust DNA repair pathways to restart replication.

Recombination must be regulated to prevent chromosome rearrangements and genomic instability.

Repair and recombination factors promote genome stability and prevent deleterious rearrangements. In cells defective in such factors, there

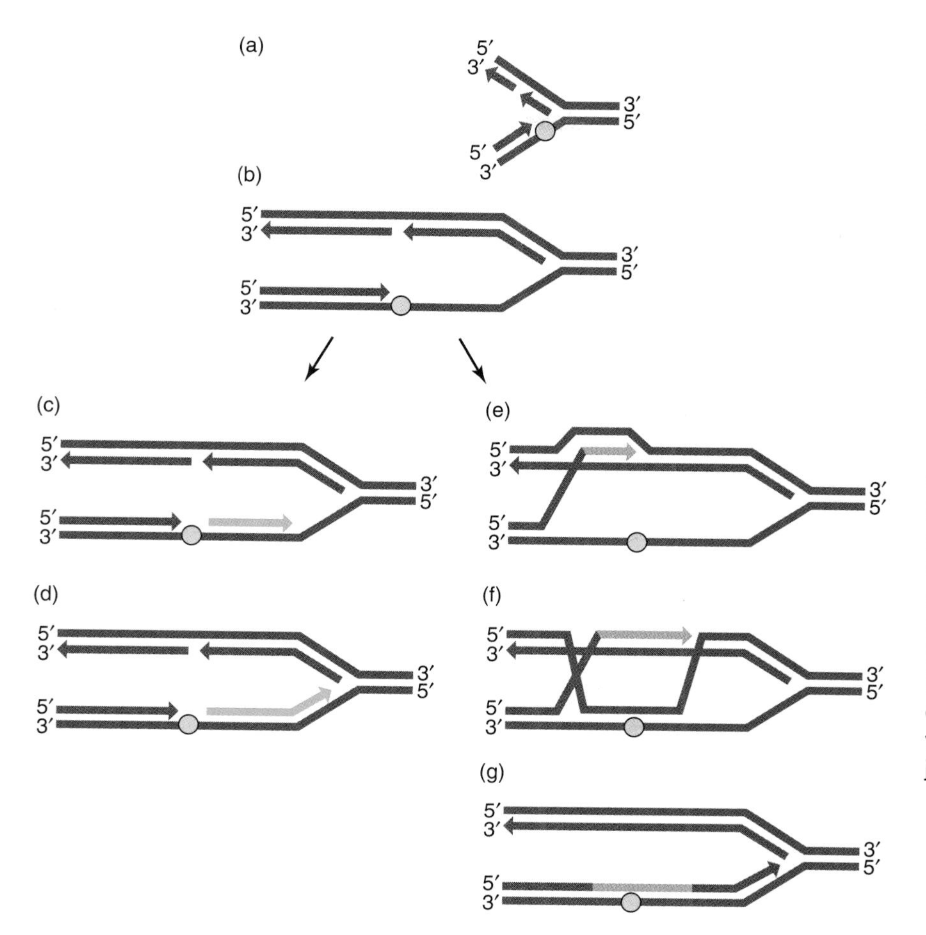

FIGURE 13.37 **Replication fork restart by homologous recombination.** (a–b) Damage on the leading strand template DNA results in stalling of the leading strand and uncoupling of the lagging strand synthesis. (c) Replication can be restarted by new priming and synthesis past the lesion, (d) leaving a gap in the DNA at the lesion site. (e) Alternatively, the stalled newly synthesized strand can invade the intact sister chromatid to form a D-loop. (f) Extension from the D-loop displaces the blue template strand, which can then pair with the other template strand, forming a Holliday junction. (g) Branch migration of the Holliday junction re-establishes the replication fork and bypasses the template strand lesion. Note that this proposed recombination is initiated by a single-strand gap, not a double-strand break, and the template strands are never broken during the recombination event. (Modified from *Exp. Cell Res.*, vol. 312, B. Eppink, C. Wyman and R. Kanaar, Multiple interlinked mechanisms . . . , pp. 2660–2665, copyright 2006, with permission from Elsevier [http://www.sciencedirect.com/science/journal/00144827].)

is an increase in sister chromatid exchanges, nonreciprocal translocations, chromatid breaks, and other abnormal chromosome structures. Many of these rearrangements occur at sites of repeated DNA sequences, suggesting that uncontrolled double-strand breaks can be very dangerous to the integrity of the genome. The BLM DNA helicase, which we have already discussed, prevents homologous recombination intermediates from being resolved as crossovers. In cells that are defective in the BLM DNA helicase, there is an increased frequency of sister chromatid exchange as well as other rearrangements. The sister chromatid exchanges are a measure of the amount of homologous recombination that is normally ongoing in cells. This shows that double-strand breaks occur during normal cell growth, but are repaired in ways that do not generate crossovers, such as the SDSA pathway, or repair pathways such as the translesion synthesis pathway discussed in Chapter 12. *BRCA2*, one of the genes defective in heritable breast cancers, also controls genome integrity. As we have discussed, BRCA2 interacts with Rad51 and is thought to bring Rad51 to double-strand breaks for homologous recombination. BRCA2-deficient cells are characterized by multiple chromosome rearrangements, showing that misregulated homologous recombination is deleterious (FIGURE 13.38).

Cells contain some sequences that are especially sensitive to forming gaps or breaks on metaphase chromosomes when DNA replication is disturbed. These sequences, termed **fragile sites**, represent

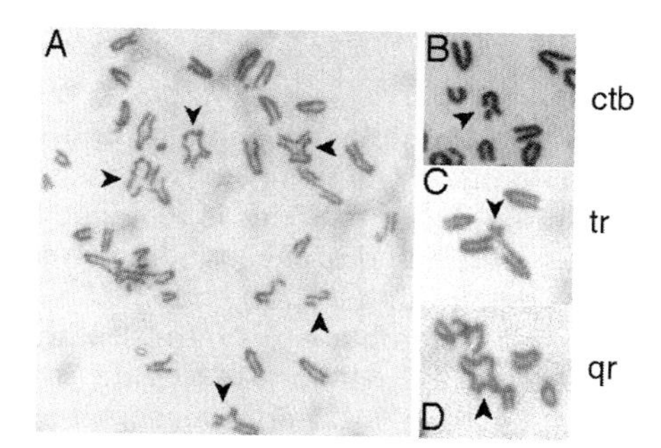

FIGURE 13.38 **Chromosome rearrangements in *BRCA2* mutant cells.** (a) The arrows point to chromatid breaks (ctb) and rearranged chromosomes in metaphase spreads. (b), (c), and (d) show higher magnifications of chromatid breaks (ctb) and chromosome structures with multiple arms, triradial (tr) and quadriradial (qr) structures. (Reproduced from *Mol. Cell*, vol. 1, K. J. Patel, et al., Involvement of Brca2 in DNA repair, pp. 347–357, copyright 1998, with permission from Elsevier [http://www.sciencedirect.com/science/journal/10972765]. Photo courtesy of Ashok R. Venkitaraman, University of Cambridge and The Medical Research Council, UK.)

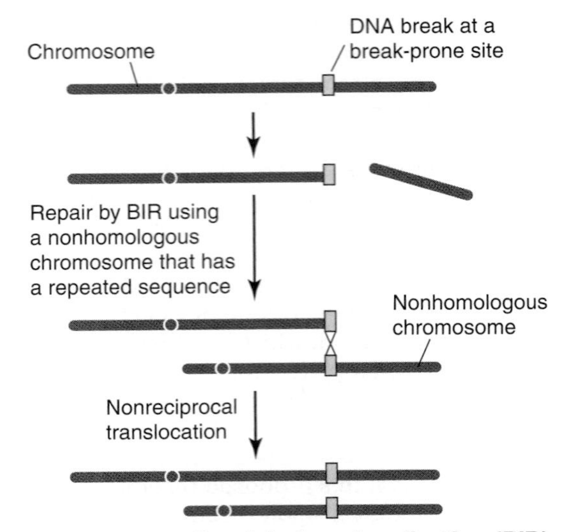

FIGURE 13.39 **Break-induced replication (BIR) at a double-strand break gives translocations.** Break-prone sites can break during DNA replication. They frequently contain repeated DNA sequences related to transposable elements. Homologous recombination using a homologous sequence of the transposable element family on a different chromosome leads to rearrangements and translocations. The lines represent double-stranded DNA.

chromosome regions where DNA replication is slowed down, which may allow the formation of single-strand DNA gaps. Fragile sites can be substrates for phosphodiester bond cleavage to make double-strand breaks. Many translocations are associated with fragile sites. A rare class of fragile sites is caused by simple tandem repeats of trinucleotides such as CCG and AGC. From studies in model systems, it has been proposed that double-strand breaks at repeated DNA sequences are repaired by a nonreciprocal, recombination-dependent replication process called **break-induced replication** (**BIR**; FIGURE 13.39).

Break-induced replication begins as a recombination event at one end of the DNA, either because there is only one free DNA end or because only one of the two ends of the double-strand break succeeds in engaging in the recombination intermediate. Substrates for break-induced replication occur at broken replication forks and telomeres. The BIR pathway begins with processing of the double-strand break end to give a 3′-OH single-strand tail. This step is followed by strand invasion into a homologous sequence. One possible mechanism is shown in FIGURE 13.40. The 3′ end of the invading strand serves as a primer for DNA replication, leading to a migrating D-loop bubble. Then the displaced newly synthesized DNA strand serves as a template to synthesize double-stranded DNA. The genetic outcome of break induced replication may be a gene conversion event that is several hundred kbp long when a homologous chromosome is used for break induced replication repair, or a nonreciprocal translocation when a nonhomologous chromosome with homology to the double-strand break end is used for break induced replication repair as shown in Figure 13.39.

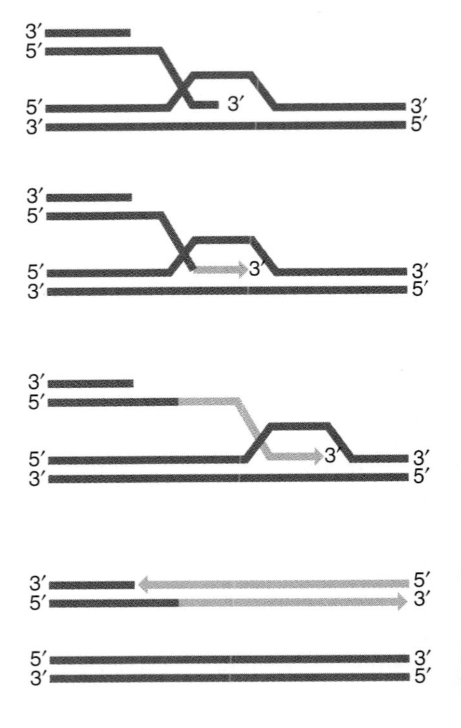

FIGURE 13.40 **Possible mechanisms of break-induced replication.** The break-induced replication pathway begins with processing of the double-strand break end to give a 3′-OH single strand tail. This is followed by strand invasion into homologous sequences. The 3′ end of the invading strand serves as a primer for DNA replication, leading to a migrating D-loop "bubble." The displaced newly synthesized DNA strand then serves as a template to synthesize double-stranded DNA. (Modified from M. J. McEachern and J. E. Haber, *Annu. Rev. Biochem.* 75 [2006]: 111–135. Copyright 2006 by Annual Reviews, Inc. Reproduced with permission of Annual Reviews, Inc., in the format Textbook via Copyright Clearance Center.)

The single-strand annealing (SSA) mechanism results in deletions.

A specialized mechanism is used to repair double-strand breaks that occur between two repeated sequences that have the same orientation. This mechanism, called **single-strand annealing (SSA)**, was identified in studies showing that some double-strand break repair events do not require the strand exchange protein Rad51 that was discussed above, yet still use DNA sequence homology for repair. Single-strand annealing occurs by end processing of double-strand break ends to give 3′-OH single strand tails that anneal at the repeated sequences (FIGURE 13.41). The overhanging 3′-OH single-strand tails are processed by endonucleases and the reaction is finished by ligation. Because there is no strand invasion involved, SSA is independent of strand exchange proteins. The end product of single-strand annealing is a deletion of the sequences between the direct repeats, and reduction of the duplication to a single copy. Single-strand annealing has been observed in eukaryotic cells using test substrates with direct repeat duplications, and may occur *in vivo* between direct repeats if essential genes are not positioned between the repeats. Some heritable cases of human diseases such as familial hypercholesterolemia, α-thalassemia, insulin-dependent diabetes, and Fabry disease result from the deletion of sequences between direct repeat sequences. The deletions could have occurred during mitotic growth of the germ cell precursor cells, during maturation of the germ cells, or during meiosis. Nonetheless, it seems that intervening sequences between direct repeats are at risk for deletion, presumably through a single-strand annealing process.

Clearly then, homologous recombination is essential to mitotic cells, but must be regulated to promote replication fork restart without generating the deleterious rearrangements that can occur through double-strand break repair.

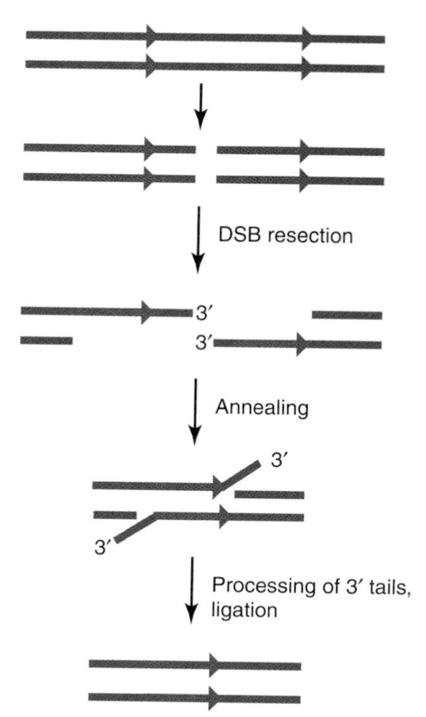

FIGURE 13.41 **Single-strand annealing model of homologous recombination.** A double-strand break occurs between direct repeats, depicted as red arrows. Following end processing to form single-strand tails with 3′-OH ends, the single strands anneal by homology at the red arrows. The single-strand tails are removed by endonucleases that recognize branch structures. The end product is double-strand break repair with a deletion of the sequences between the repeats and loss of one repeat sequence.

13.11 Repairing a Double-Strand Break without Homology

Nonhomologous end-joining is a model for rejoining ends with no homology.

Nonhomologous end-joining (NHEJ) is used to repair double-strand breaks where there is no homology or a limited "microhomology" of 1 to 10 nucleotides. It is used primarily in the G_1 phase of the cell cycle when there is no closely associated homologous sequence such as the sister chromatid, which is present after DNA replication. Nonhomologous end-joining is sometimes called **illegitimate recombination** because no homology is used in end joining. The Ku protein complex, a heterodimer consisting of a 70 and 80 kDa polypeptide, plays an essential role in nonhomologous end-joining. The protein was first discovered in the cell nucleus as a target of autoantibodies in patients with scleroderma, an autoimmune disease. (The name "Ku" was derived from the patient's name). The Ku70 and Ku80 subunits are defective in x-ray-sensitive mammalian cells with mutations in the *XRCC6* and *XRCC5* genes, respectively. Homologous genes are present in a wide

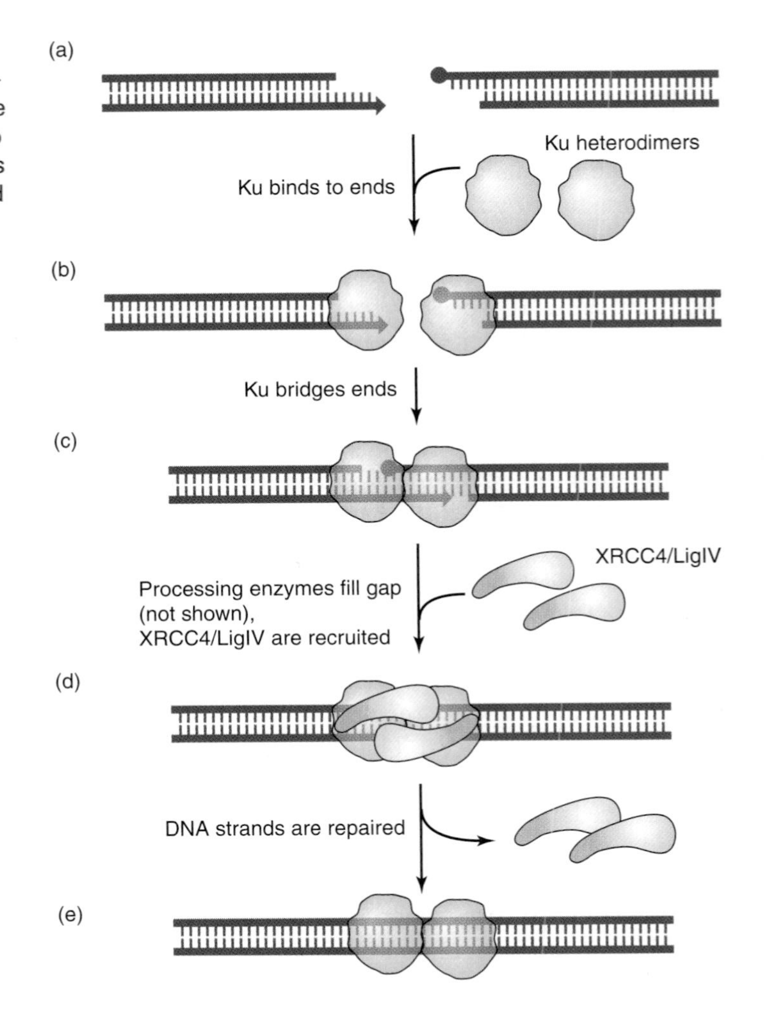

FIGURE 13.42 **Nonhomologous end-joining.** (a) The blue dot on one of the two double-strand break ends signifies a nonligatable end. (b) The double-strand break ends are bound by the Ku heterodimer. (c) The Ku•DNA complexes are juxtaposed to bridge the ends and the gap is filled in by processing enzymes and POLL or POLM. (d) The ends are ligated by the specialized DNA ligase LigIV with its partner XRCC4 to (e) repair the double-strand break. (Adapted from J. M. Jones, M. Gellert, and W. Yang, *Structure* 9 [2001]: 881–884.)

variety of organisms including yeast and bacteria. The nonhomologous end-joining pathway begins with the binding of the Ku70•Ku80 complex to the broken DNA ends (FIGURE 13.42). In humans, another protein called DNA-dependent protein kinase (DNA-PK) is essential in nonhomologous end-joining. The Ku complex binds to the ends as a barrel ring structure. The two ends bound by the Ku complex are brought together, assisted by the MRX (Mre11•Rad50•Xrs2) complex of yeast or the MRN (Mre11•Rad50•Nbs1) complex in humans, which forms a bridge between the two ends. The ends are then sealed by a specialized ligase called DNA ligase IV. Ligase IV is associated with another protein called Lif1 in yeast and Xrcc4 in humans. Other proteins, listed in Table 13.3, complete the reaction.

Occasionally the broken ends are "dirty" because they lack either a 5′-phosphate end or a 3′-hydroxy end and so cannot serve as a substrate for DNA ligase. "Dirty" ends are produced by radiation, oxidation, and radiomimetics (chemicals that make double-strand breaks) or by enzymes such as nucleases and defective topoisomerases. Additional processing is required to convert the "dirty" ends into ends that can be joined together by DNA ligase. This processing can lead to small mutations consisting of a few nucleotides of insertion or deletion at the end joints (FIGURE 13.43).

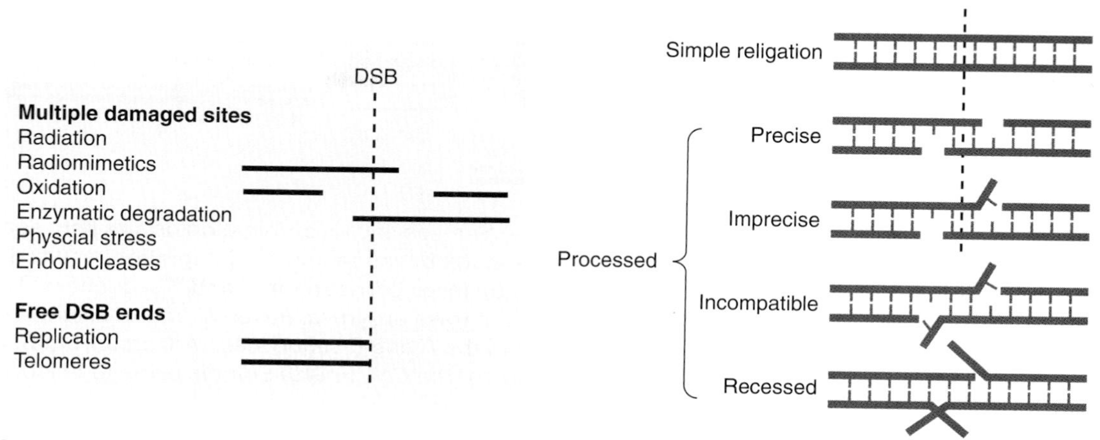

FIGURE 13.43 **Nonhomologous end-joining of dirty ends.** Many double-strand break ends are damaged (described in the left panel) and cannot engage in a simple ligation reaction. These must be processed first to remove the damaged ends. In other cases, the ends require a few nucleotides to be added or removed to allow annealing of two to three nucleotides prior to ligation. (Modified from J. M. Daley, et al., *Annu. Rev. Genet.* 39 [2005]: 431–451. Copyright 2005 by Annual Reviews, Inc. Reproduced with permission of Annual Reviews, Inc., in the format Textbook via Copyright Clearance Center.)

TABLE 13.3 **NHEJ proteins**

Step in NHEJ	*S. cerevisiae* protein	Mammalian protein
End binding and juxtaposition	Ku70/80 Mre11/Rad50/Xrs2	Ku70/80 DNA-PKCs
End joining without processing	Dnl4 Lif1 Nej1	DNA ligase IV XRCC4 XLF
End joining with processing	Dnl4 Lif1 Nej1 Pol4	DNA ligase IV XRCC4 Cernunnos/XLF Artemis Polynucleotide kinase (PNK) POLL POLM

The NHEJ process is related to the mechanism of DNA transposition, discussed in Chapter 14. It is also related to *V(D)J* recombination, which is the process of making a mature immunoglobulin gene from separated modular parts. *V(D)J* recombination, discussed later in this chapter, uses some of the core nonhomologous end-joining factors in addition to novel factors specific for *V(D)J* recombination.

13.12 Site-Specific Recombination

Site-specific recombination occurs at defined DNA sequences and is used for immunoglobulin diversity and by transposable elements.

The process of site-specific recombination is the exchange, insertion, or deletion of DNA segments at one or two defined DNA sequences. The reaction is promoted by various families of recombinases that

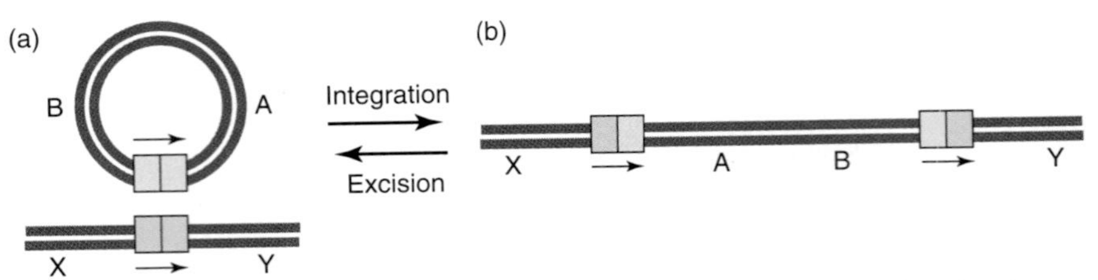

FIGURE 13.44 (a) Site-specific recombination occurs between the circular and linear DNAs at the boxed region. (b) Integration results in an insertion of the A and B sequences between the X and Y sequences. The reaction is promoted by integrase enzymes. Reversal of the reaction results in a precise excision of the A and B sequences. (Adapted from B. Alberts, et al., *Molecular Biology of the Cell, Fourth Edition.* Garland Science, 2002.)

recognize one or more sequences and use different types of chemistry to cleave DNA. The recognition sequences are often complex such that the nuclease makes only a few cuts per genome. Site-specific recombination mechanisms are known by both the specific site and the enzyme that recognizes the site.

Three broad families of site-specific recombination reactions called transposition, conservative site-specific recombination, and target-primed reverse transcription will be discussed in the following chapter. A simple example of conservative site-specific recombination is shown in FIGURE 13.44. Several site-specific recombination pathways have been modified for use in gene targeting and conditional gene expression or knockouts in different model systems. Examples of some of these are discussed below.

Mating type switching in yeast occurs by synthesis-dependent strand-annealing initiated at a defined site.

The mating type switching system of yeast is a mechanism to quickly change cell type (see Chapter 7). Mating type is determined by the information at the *MAT* locus, which encodes transcription activators that turn on cell-specific genes. Mating type switching is active in haploid cells that have a single mating type gene, and is shut off in diploid cells that are heterozygous at the *MAT* locus. Mating type switching is induced by expression of the *HO* gene, which encodes an endonuclease that recognizes a 24 bp sequence that is not palindromic. Induction of the double-strand break at the *MAT* locus triggers a gene conversion event between *MAT* and one of the silent cassettes of mating type information, called *HML* and *HMR*. Gene conversion without crossing over results in a change of mating type, using the information from *HML* or *HMR* to repair the double-strand break at MAT (FIGURE 13.45) in a synthesis-dependent strand-annealing reaction. Because mating type switching is a DSBR by a homologous recombination process, the genes that are needed for double-strand break repair are also needed for mating type switching.

The HO system has become a useful tool to make a specific double-strand break in mitotic cells to induce homologous recombination. Either the *MAT* sequence or the *HO* recognition site is inserted into

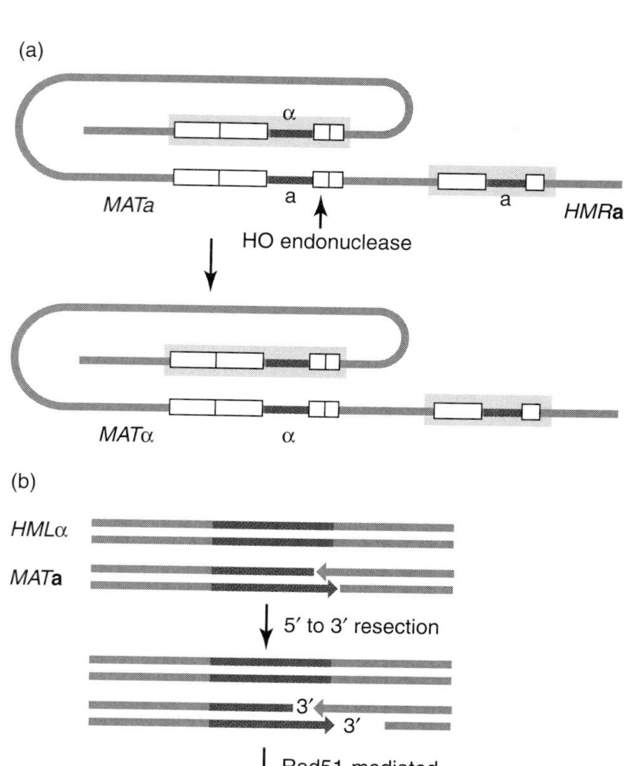

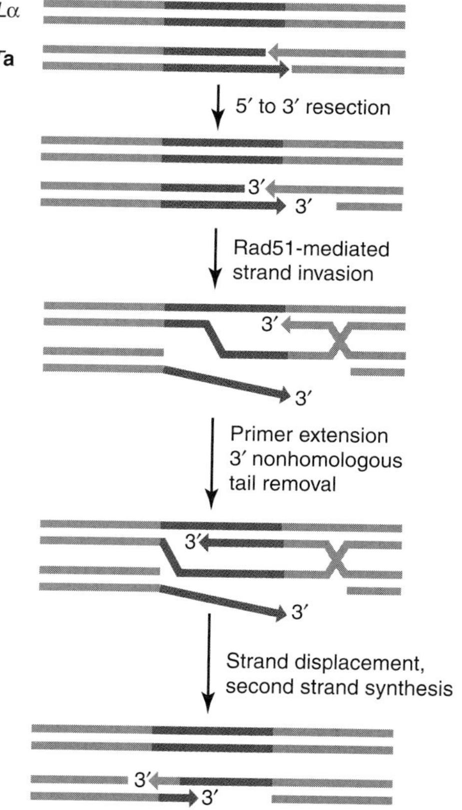

FIGURE 13.45 **Site-specific HO endonuclease and mating type switching in *Saccharomyces cerevisiae*.** (a) The mating type locus *MAT* encodes **a** (red) or α (blue) information. The two silent cassettes (light gray), *HML*α and *HMR***a** carry mating type information but do not express it. Switching is initiated by HO endonuclease cutting of the *MAT* locus at a specific recognition sequence. (b) Mating type switching occurs by homologous recombination, using a synthesis-dependent strand-annealing process to change the red *MAT***a** information to the blue *MAT*α information, without any associated crossing over. (Modified from *DNA Repair*, vol. 5, J. E. Haber, Transpositions and translocations induced . . . , pp. 998–1009, copyright 2006, with permission from Elsevier [http://www.sciencedirect.com/science/journal/15687864].)

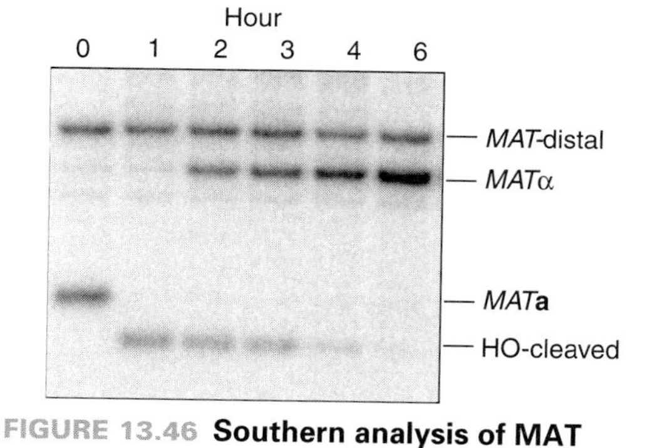

FIGURE 13.46 **Southern analysis of MAT cleavage by HO endonuclease.** At 0 hours, *MAT* cleavage is initiated by HO endonuclease in a synchronized fashion. Cleavage of *MAT***a** can be seen by 1 hour. The HO-cleaved product disappears as the double-strand break is repaired by synthesis-dependent strand-annealing and mating type is switched to *MAT*α. The *MAT***a** and *MAT*α genes migrate at different positions on the gel because they have different DNA sequence and hence different restriction enzymes recognition sites. (Reproduced from *Meth. Enzymol.*, vol. 408, N. Sugawara and J. E. Haber. Repair of DNA double strand . . . , pp. 416–429, copyright 2006, with permission from Elsevier. [http://www.sciencedirect.com/science/bookseries/00766879]. Photo courtesy of Neal Sugawara, Brandeis University.)

the chromosome region of interest. *HO* gene activity can be modified so that it is regulated by the carbon source, being shut off when cells are cultured in medium containing glucose and turned on when cells are cultured in medium containing galactose but no glucose. This regulatory response has allowed investigators to examine the fate of a double-strand break at the molecular level in normal and homologous recombination mutant cells (FIGURE 13.46).

More recently, the I-Sce1 system from yeast has been used in mouse cells to make a specific double-strand break in somatic cells. The I-Sce1 endonuclease, which is derived from the intron in the 21S rRNA gene

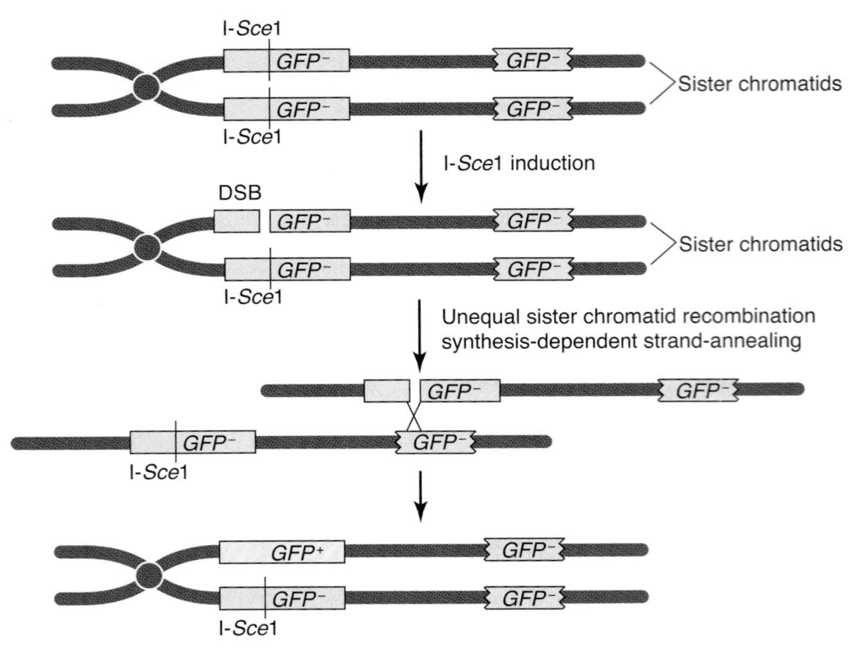

FIGURE 13.47 **Homologous recombination in mouse cells promoted by a I-Sce1-induced double-strand break (DSB).** The recombination reporter consists of two green fluorescent protein (GFP) genes, both of which are nonfunctional. The left copy is rendered nonfunctional by the insertion of the I-Sce1 cutting site, which contains two in-frame stop codons. The right copy is truncated at the 5′ and 3′ ends. Induction of I-Sce1 expression results in I-Sce1 protein expression. The protein recognizes the I-Sce1 site and cuts it, but the cutting is not 100% efficient; therefore, often only one of two sister chromatids is cut by the endonuclease. The double-strand break is most often repaired by mitotic gene conversion using the truncated right copy of GFP, here shown using the copy of the sister chromatid. Synthesis-dependent strand-annealing results in gene conversion repair of the double-strand break and removal of the I-Sce1 site, giving a functional copy of GFP.

of yeast mitochondria, has a recognition site that is 18 bp in length. Expression of I-Sce1 in mouse cells is more efficient than expression of the HO endonuclease. Maria Jasin and colleagues constructed a mouse cell with the I-Sce1 recognition site inserted into a recombination reporter (green fluorescent protein) and then introduced an I-Sce1 expression plasmid. They observed that the double-strand breaks generated at the I-Sce1 sites are preferentially repaired in somatic cells using the sister chromatid (FIGURE 13.47).

V(D)J recombination produces the immune system diversity.

The adaptive immune system in vertebrates can generate a seemingly limitless number of T-cell receptors and antibodies against diverse pathogens and antigens. It does this through programmed site-specific recombination of the immunoglobulin gene precursors, generating up to 10^{11} different antibodies. The immunoglobulin genes are organized as repeated segments of variable (*V*), diversity (*D*), and joining (*J*) segments, which are used to form the light and heavy chains of immunoglobulins (FIGURE 13.48). The idea that the immunoglobulin genes were formed from noncontiguous gene segments was first proposed in 1965 by William J. Dreyer and J. Claude Bennett. They suggested that: (1) the variable and constant regions of the immunoglobulin genes are encoded by different DNA sequences; (2) some process brings the variable and constant region sequences together to form a single transcriptional unit, one for the heavy chain and another for the light chain of an immunoglobulin; and (3) each heavy or light chain gene locus has multiple *V* region sequences but only one *C* region sequence prior to rearrangement in B cells. In 1976 Susumu Tonegawa showed that this suggested organization was correct by demonstrating that the

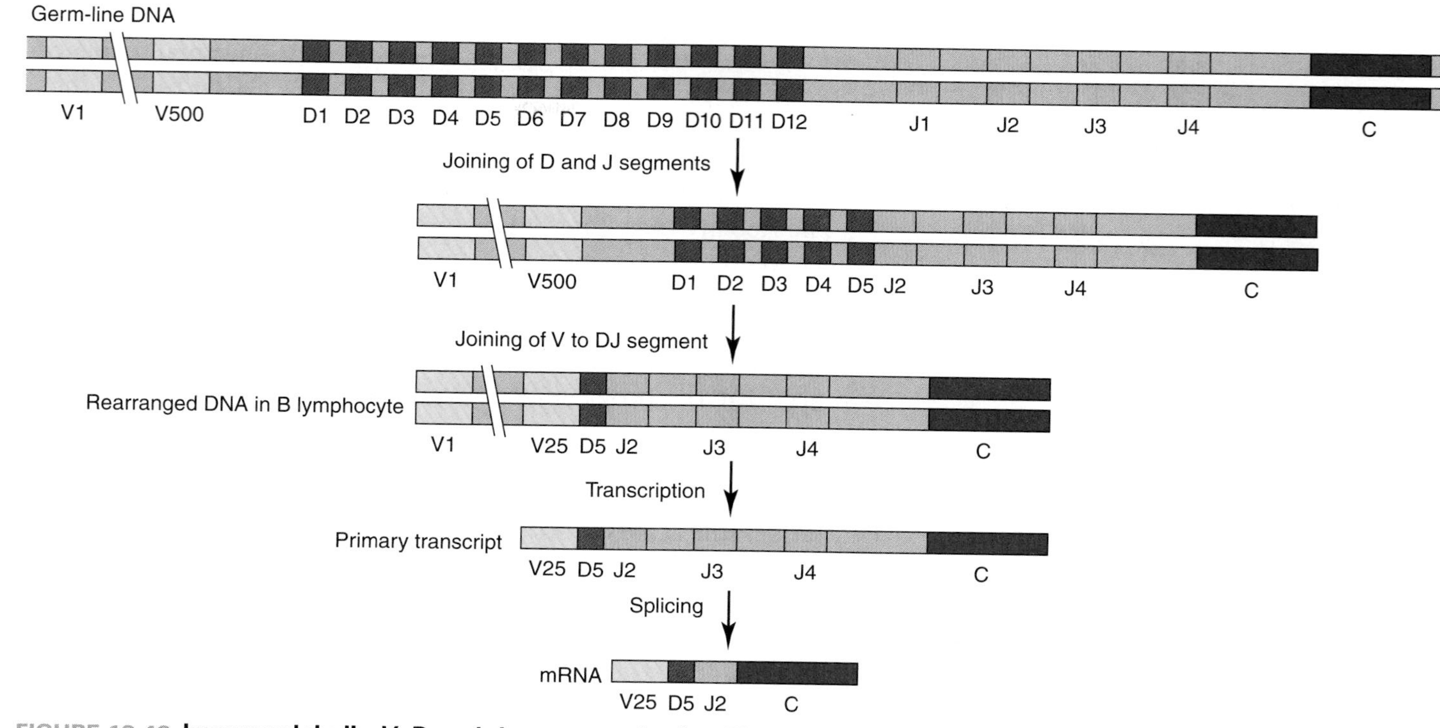

FIGURE 13.48 **Immunoglobulin *V, D,* and *J* gene organization.** The germline contains multiple *V, D,* and *J* regions. In immunoglobulin heavy chain genes, the *D* and *J* regions are joined first. Next, a *V* region is joined to the *DJ* region to form a *VDJ* gene. The *C* region is joined to the *VDJ* region at the RNA level by splicing to make mRNA encoding the *VDJC* regions. (Adapted from G. M. Cooper and R. E. Hausman. *The Cell: A Molecular Approach, Second Edition.* Sinauer Associates, Inc., 2000.)

variable and constant regions are separated in the DNA from mouse embryos, but are adjacent in mature B cells. Tonegawa's work set the stage for the demonstration of programmed somatic DNA rearrangements in B cells.

Subsequent studies revealed the existence of additional sequences, the joining (*J*) sequences in light and heavy immunoglobulin chains and the diversity (*D*) sequences in heavy immunoglobulin chains. The light chains of mature immunoglobulins are formed by joining *V* and *J* sequences to the constant region sequence, while the heavy chains are formed by joining *V*, *D*, and *J* segments to the constant region segment. Joining of these segments occurs during B-cell development by somatic recombination. Diversity in the immune system is achieved by different recombination reactions. There are three levels of diversity. The first level, **combinatorial diversity**, involves selecting specific *V*, *D*, and *J* segments. The second level, **junctional diversity**, results from the fact that the process that joins the *V*, *D*, and *J* coding segments is imprecise resulting in the deletion or addition of nucleotides at the joining junction. The third level, **pairing diversity**, results from the fact that heavy and light chains combine to make an antigen receptor protein. Additionally, gene conversion and somatic hypermutation (discussed below) add to the diversity of the mature immunoglobulin genes.

V(*D*)*J* recombination occurs by programmed double-strand breaks. Because double-strand breaks have the potential to be joined to incorrect double-strand break ends, the process of *V*(*D*)*J* recombination

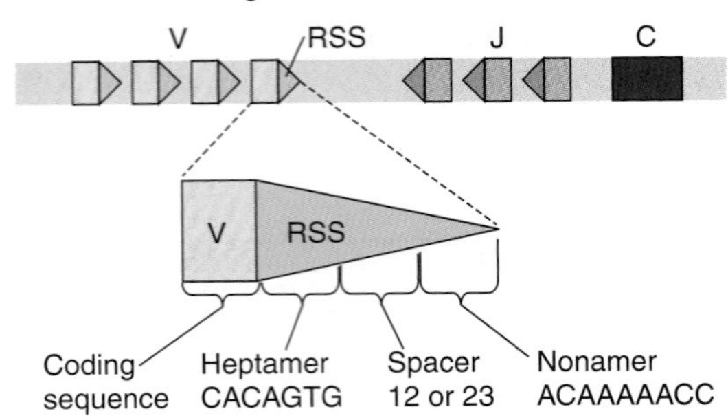

FIGURE 13.49 **The recombination signal sequences (RSS) in *V(D)J* recombination.** Each *V*, *D*, and *J* gene coding sequence is flanked by a RSS that contains conserved heptamer and nonamer sequences flanking a spacer of 12 or 23 base pairs. (Adapted from D. B. Roth, *Nat. Rev. Immunol.* 3 [2003]: 656–666.)

must be carefully regulated to guard against this problem. The *V*, *D*, and *J* segments are flanked by special **recombination signal sequences (RSS)**, which participate in the recombination reaction to join them together (FIGURES 13.49 and 13.50). RSS are formed from conserved heptamer and nonamer elements that are separated by a spacer region of 12 or 23 nucleotides.

The RSS are the target of two enzymes that cleave at the RSS in a defined manner. These enzymes are called **RAG-1** and **RAG-2**, for the recombination-activating genes that code for them. The *RAG* genes are expressed only during lymphocyte development, thus limiting the production of dangerous double-strand breaks to a stage where there is a need for the programmed rearrangements. The RAG proteins make a single strand nick at the junction between the RSS and the coding sequence, forming a free 3′-OH group. The RAG•RSS complex then pairs with another RAG•RSS complex, and the free 3′-OH ends attack the opposite strand through transesterification to form a

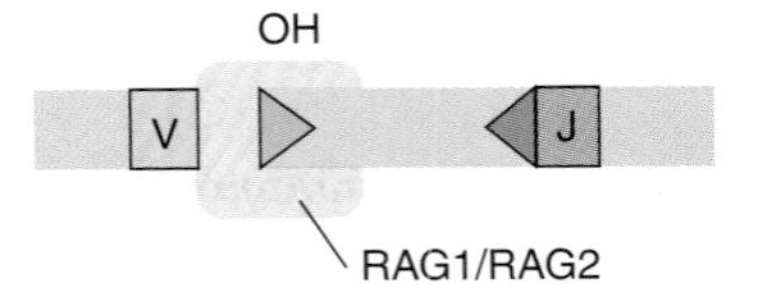

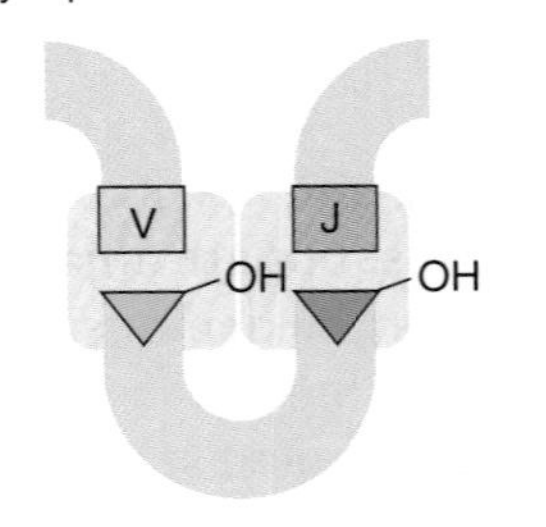

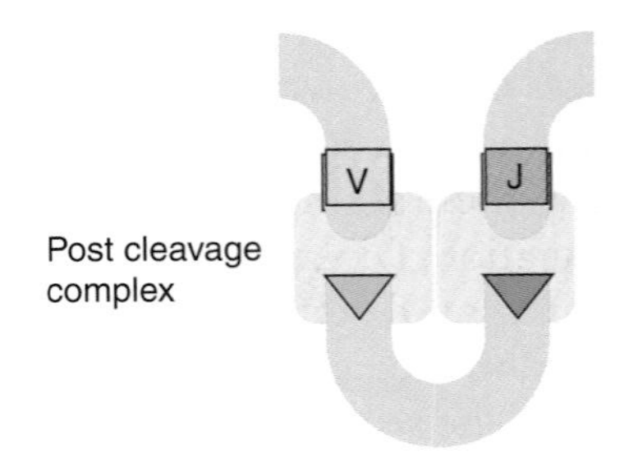

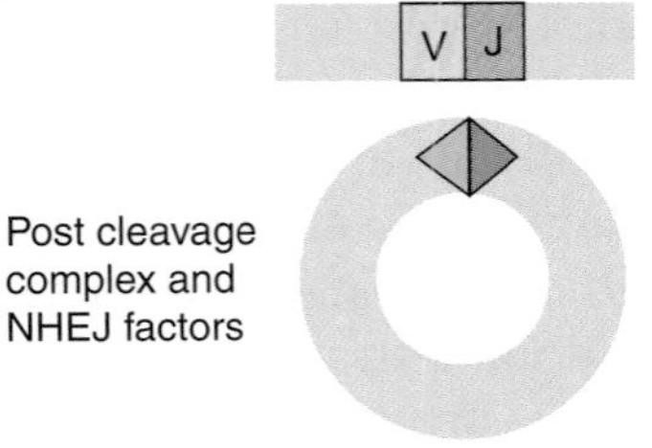

FIGURE 13.50 **The *V(D)J* reaction recombination reaction.** (a) RAG1 and RAG2 bind to a recombination signal sequence (RSS) and make a single-strand nick. (b) The RAG1•RAG2•RSS complex synapses with another RSS and makes another nick on the opposite strand to make a double-strand break. (c) The coding ends are formed into hairpins by the cleavage, and remain associated with RAG1•RAG2 complex and the RSS ends in a post-cleavage complex. (d) The hairpins are opened by the Artemis protein and nonhomologous end-joining factors to join the two coding ends together and the two signal ends together. (Adapted from D. B. Roth, *Nat. Rev. Immunol.* 3 [2003]: 656–666.)

double-strand break at the RSS/coding sequence junction. During the process of forming the double-strand break, the ends of the coding sequence region become sealed by a hairpin. To prevent joining of two *V* sequences, two *D* sequences, or two *J* sequences together, the recombination uses the **12/23 rule**, whereby a RSS with a 12 bp spacer can only be joined to a RSS with a 23 bp spacer. The *V* and *J* sequences have 23 bp spacers while the D sequences have a 12 bp spacer. This rule forces the joining to be V sequence to D sequence to J sequence.

Once the two coding ends with hairpin ends are in a complex with the RAG proteins and other factors, the hairpins are cleaved by a protein called Artemis. Cleavage of the hairpin sequences is not at a specific sequence, so the opened ends are of variable length. The nicked hairpin ends may be further processed by exonuclease digestion, terminal deoxynucleotidyl transferase catalyzed nucleotide addition (called **N nucleotides**), or by the addition of **palindromic (P) nucleotides** complementary to the last nucleotides of the coding end, thus creating the junctional diversity mentioned above. Junctional diversification can alter the reading frame, rendering the immunoglobulin transcription unit nonfunctional. However, selection for functional antigen receptors balances the formation of inactive transcription units. The enzymes for NHEJ described above join the coding sequences. These enzymes include Ku70•Ku80, DNA-PKcs, DNA ligase IV, Xrcc4, Cernunnos•XLF and also DNA polymerase mu (POLM; see Table 12.1).

Defects in nonhomologous end-joining genes are found in patients suffering from one form of the disease called **severe combined immunodeficiency (SCID)**. Patients with this form of SCID have few mature antigen receptors and cannot mount a proper immune response against infections and pathogens. SCID is sometimes called the bubble boy disease, recalling the young boy with SCID who during the 1970s lived in a sterile bubble environment to avoid exposure to infectious agents. SCID patients who have a defect in NHEJ and cannot carry out double-strand break repair by nonhomologous end-joining are sensitive to ionizing radiation. Patients with another form of SCID, a deficiency in adenosine deaminase, are not sensitive to ionizing radiation.

The RAG protein mechanism of *V(D)J* recombination uses the same chemistry as transposases and retroviral integrases (discussed in Chapter 14). The *V(D)J* recombination process brings the recombining sequences together before the double-strand breaks are formed, thus protecting the double-strand breaks from joining with other double-strand breaks and generating genomic instability.

In some lymphomas, an RSS becomes joined to an oncogene by mistake, with the result that an oncogene can be expressed in the wrong circumstances under the control of a strong transcription regulatory element (FIGURE 13.51). This type of mistake might account for the chromosome translocations that are characteristic of different lymphomas. **Burkitt lymphoma** is characterized by a translocation between chromosome 8 and chromosome 14, which fuses the *c-myc* oncogene (see Chapter 18) to the Ig heavy chain locus. This fusion deregulates *c-myc* expression, resulting in uncontrolled cell growth.

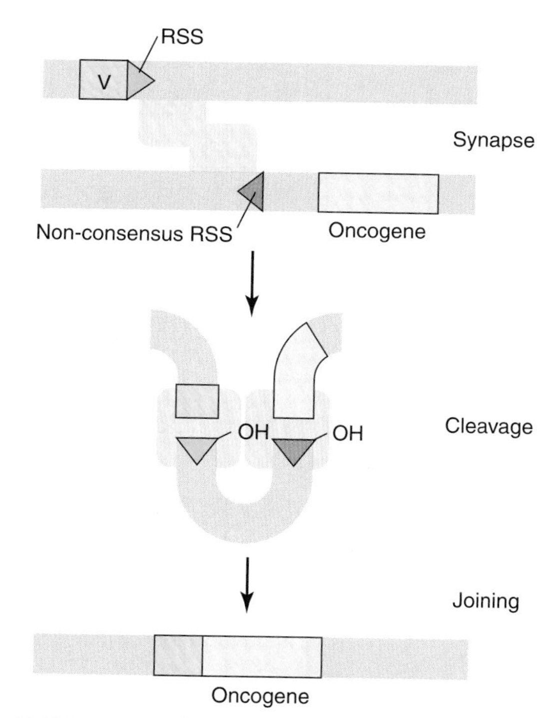

FIGURE 13.51 **Incorrect RSS recombination.** Occasionally, the RAG1•RAG2 complex recognizes a non-consensus recombination signal sequence (RSS) for a true RSS with the 12 base pair or 23 base pair spacer and cleaves this sequence. The non-consensus RSS then engages in a joining reaction with a true RSS flanking a V, D, or J coding sequence. The reaction proceeds as outlined in Figure 13.50 and can join a V, D, or J region to an oncogene. (Adapted from D. B. Roth, *Nat. Rev. Immunol.* 3 [2003]: 656–666.)

The process of *V(D)J* recombination and the added or deleted nucleotides at the coding joint contribute to immunoglobulin diversity. In some systems such as chicken and rabbit, diversity is achieved by gene conversion between incomplete *V* gene segments called **pseudogenes** and the rearranged *V* gene sequence. These systems use only one complete *V* gene sequence, so diversity is not provided by *V(D)J* recombination. We have reviewed the process of mitotic gene conversion by synthesis-dependent strand-annealing. Gene conversions between the *V* gene sequences occur by this mechanism as the events are not associated with crossing over (FIGURE 13.52).

The final mechanism of immunoglobulin diversity occurs after the mature immunoglobulin gene has been formed by *V(D)J* recombination. This occurs in only B cells in response to antigenic selection and affects the B cell receptors. The process, which is called **somatic hypermutation,** occurs by mutation of the *V* sequences at a high frequency. Somatic hypermutation makes point mutations in the immunoglobulin variable region and is thought to increase the specificity of

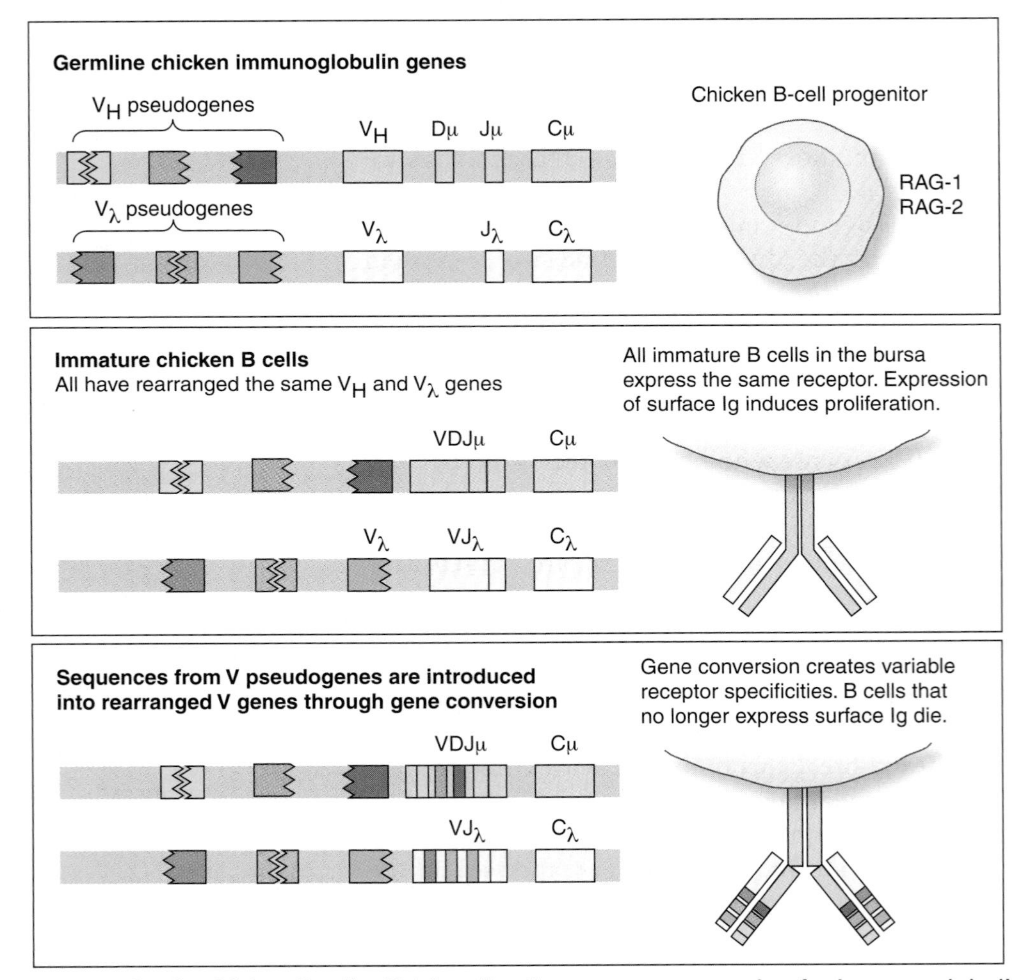

FIGURE 13.52 **Gene conversion in chicken B cells.** Chicken B cells use gene conversion for immunoglobulin diversity. Immature chicken B cells all express the same *VDJ* heavy-chain gene and *VJ* light-chain gene. The V pseudogenes are used in gene conversion events with the active *VDJ* or *VJ* genes to produce diversity. The gene conversion reaction occurs by a synthesis-dependent strand-annealing mechanism. If the gene conversion event inactivates the *VDJ* or *VJ* gene, that B cell is eliminated. (Adapted from C. Janeway, et al., *Immunobiology, Fifth Edition.* Garland Science, 2002.)

an antibody. It occurs through the introduction of uracil residues into the *V* sequences, by the action of an enzyme called **activation-induced cytidine deaminase** (**AID**; FIGURE 13.53). During transcription of the immunoglobulin genes, the DNA single strand region on the coding strand (the strand that is not transcribed) becomes modified by AID, deaminating C bases to U bases. If the uracil base is not removed, it can be replicated, inserting an A base opposite the U base, leading to a C:G to T:A transition. Uracil in DNA is recognized by uracil N-glycosylase (Ung; see Chapter 12). In a mechanism that is a base excision repair reaction, the uracil bases are excised by the Ung enzyme. Replication past the abasic site by translesion DNA polymerases also results in C:G to T:A transitions and C:G to A:T transversions. Last, the U:G mispair can be recognized by the MSH2•MSH6 mismatch repair

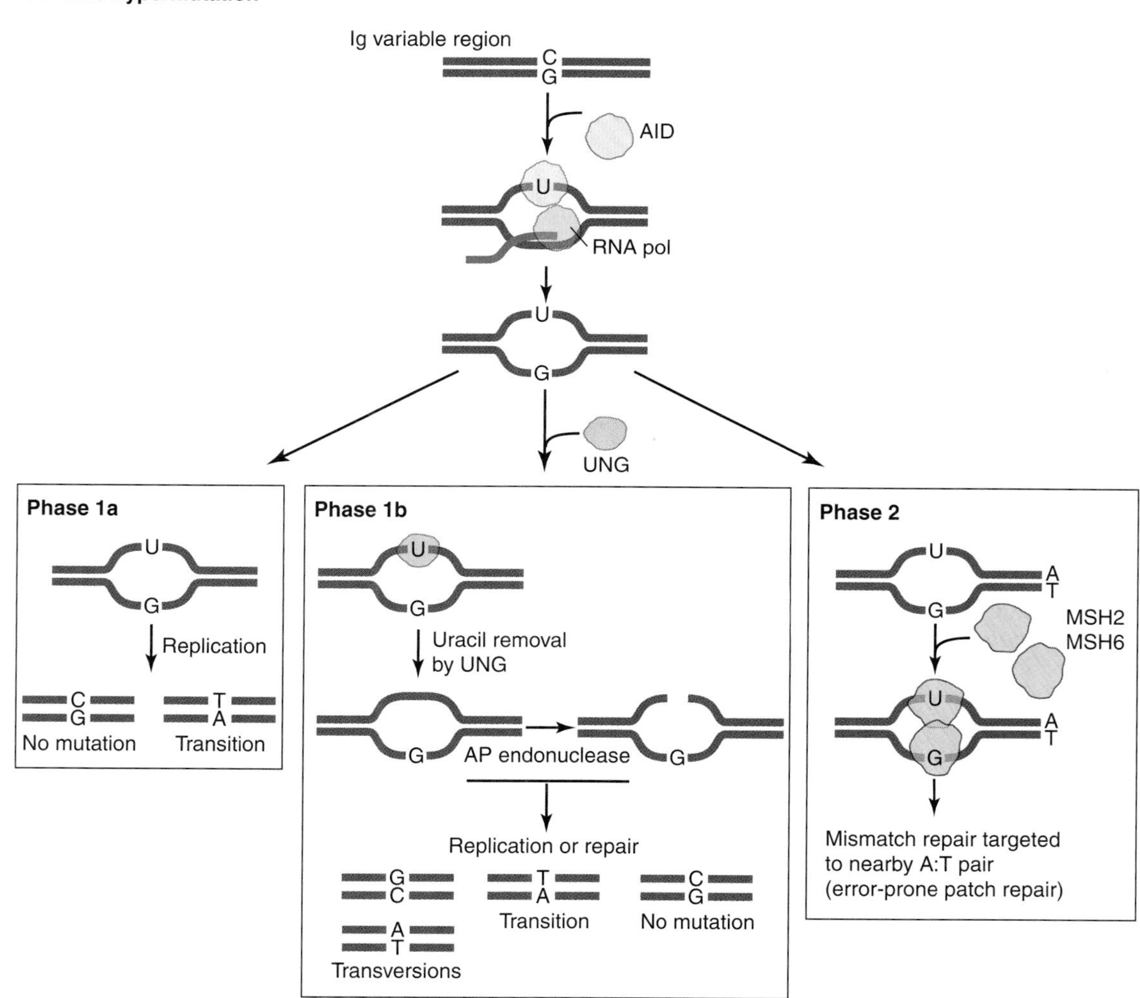

FIGURE 13.53 **Somatic hypermutation.** Deamination of cytosines (C) to uracils (U) occurs by the activation induced cytidine deaminase (AID) during transcription of the immunoglobulin gene variable region to form U:G mispairs. Replication forms a T:A transition. Alternatively, the U:G mispair is recognized by the uracil N-glycosylase (UNG) and the U residue is removed. Replication or repair results in transition and transversion mutations. The U:G mispair also may be recognized by the MSH2/MSH6 mismatch repair system and can stimulate repair of nearby A:T pairs in an error-prone manner. (Modified from *Mol. Cell*, vol. 16, G. S. Lee, V. L. Brandt, and D. B. Roth, B Cell Development Leads Off with a Base Hit, pp. 505–508, copyright 2004, with permission from Elsevier [http://www.sciencedirect.com/science/journal/10972765].)

system and the ensuing repair tract could introduce new mutations at A:T base pairs.

FLP/*FRT* and Cre/*lox* systems can be used to make targeted recombination events.

The yeast *S. cerevisiae* has an endogenous plasmid called the two-micron plasmid, which is present in about fifty copies per cell. The two-micron plasmid has inverted repeated sequences called *FRT* (FLP recombinase target) sites. As the name suggests, these sequences are recognized by a recombinase called FLP (pronounced "flip"). FLP promotes recombination between the inverted repeats during replication, promoting a rolling circle type of replication and amplification of the two-micron genome.

Investigators have learned to exploit the FLP/*FRT* system to induce site-specific mitotic recombination. An example of a site-specific induction in mitotic recombination in *Drosophila* is shown in FIGURE 13.54. To use the FLP/*FRT* system in *Drosophila*, the FLP gene expression

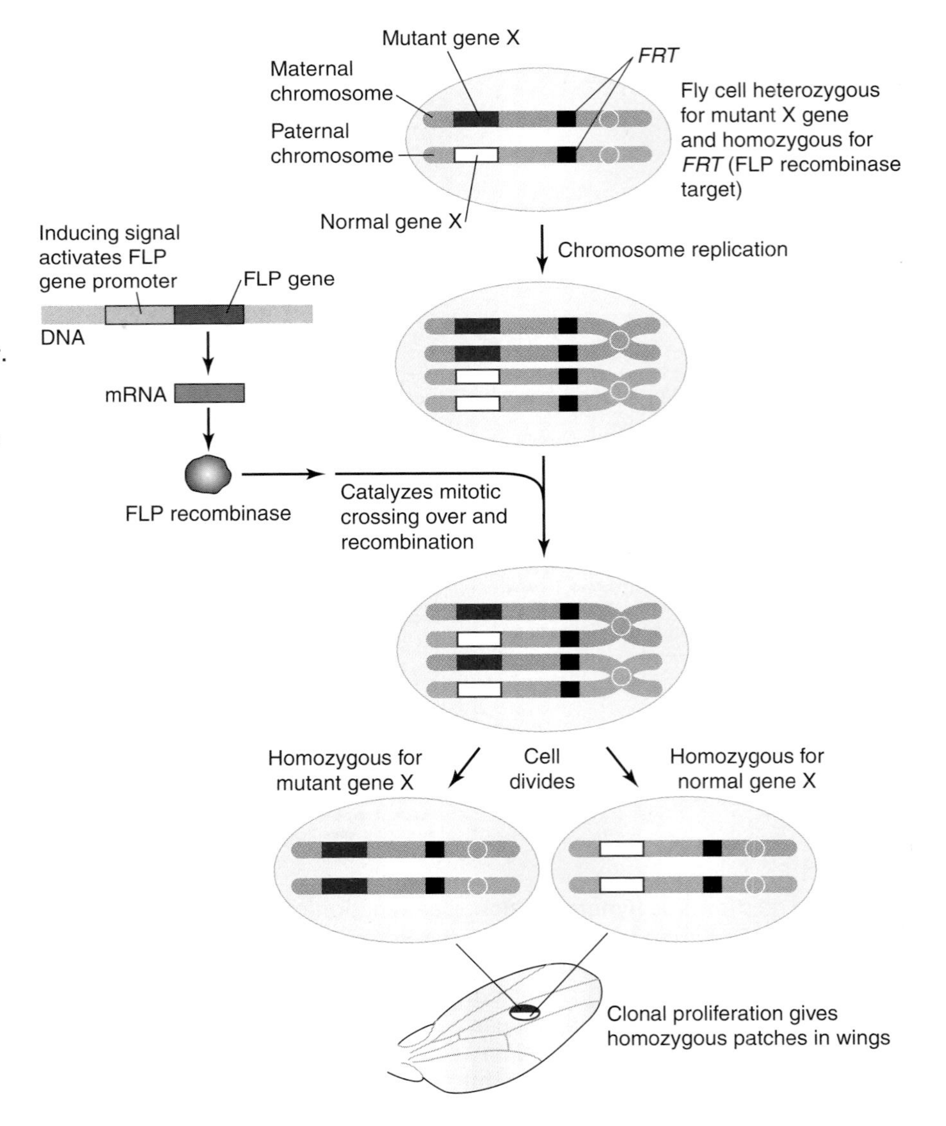

FIGURE 13.54 **Using *FLP/FRT* to make homozygous recessive cells by homologous recombination.** A fly is heterozygous for a mutant gene and homozygous for the inserted FRT site on the same chromosome. Induction of the *FLP* gene allows the FLP recombinase protein to be made. FLP recognizes the *FRT* site and makes a double-strand break, which promotes homologous recombination. Some of the recombination events occur by the double-strand break repair mechanism and result in crossing over. Following chromosome segregation, one daughter cell receives two mutant copies of the gene and the other daughter cell receives two normal copies of the gene. In the example shown, a patch of mutant cells is formed on the wing of a *Drosophila*. This technique allows assessment of a recessive mutant phenotype at a late stage in development. (Adapted from B. Alberts, et al., *Molecular Biology of the Cell, Fourth Edition.* Garland Science, 2002.)

is regulated. When FLP is expressed, it cuts the *FRT* sites which have been inserted on a chromosome where there is a gene of interest centromere distal to the *FRT* site. The cutting of the *FRT* site, which is not 100% efficient, induces a double-strand break at the *FRT* site. The double-strand breaks are repaired by homologous recombination, and some of them will result in crossing over. Depending on how the chromosomes then segregate, some cells will now be homozygous for the mutant gene. This is analogous to the "twin spots" mitotic crossing over products shown in Figure 13.2. In genetic studies, the chromosome is often marked by a gene that affects a pigment, to give a visual readout for the recombination. The mitotic recombination uncovers the recessive pigmentation mutation and the mutant gene of interest, making them homozygous recessive. One use of this system is to see the effects of a recessive mutation that is lethal. When the mutation is homozygous recessive in the zygote, it will be lethal. If the mutation is carried in the heterozygous state, however, the organism will be viable. Then the gene is rendered homozygous in clones of cells by induction of FLP, either by temperature or a tissue-specific transcription regulation, enabling the investigator to ask questions about effects of loss of the gene in specific cells at a specific time during development.

The Cre/*lox* system, which is derived from bacteriophage P1, functions in a similar manner. The Cre enzyme recognizes and cleaves *loxP* sites. One of the most common uses of the Cre/*lox* system is in gene targeting in mouse (FIGURE 13.55). Cre/*lox* can be used to conditionally

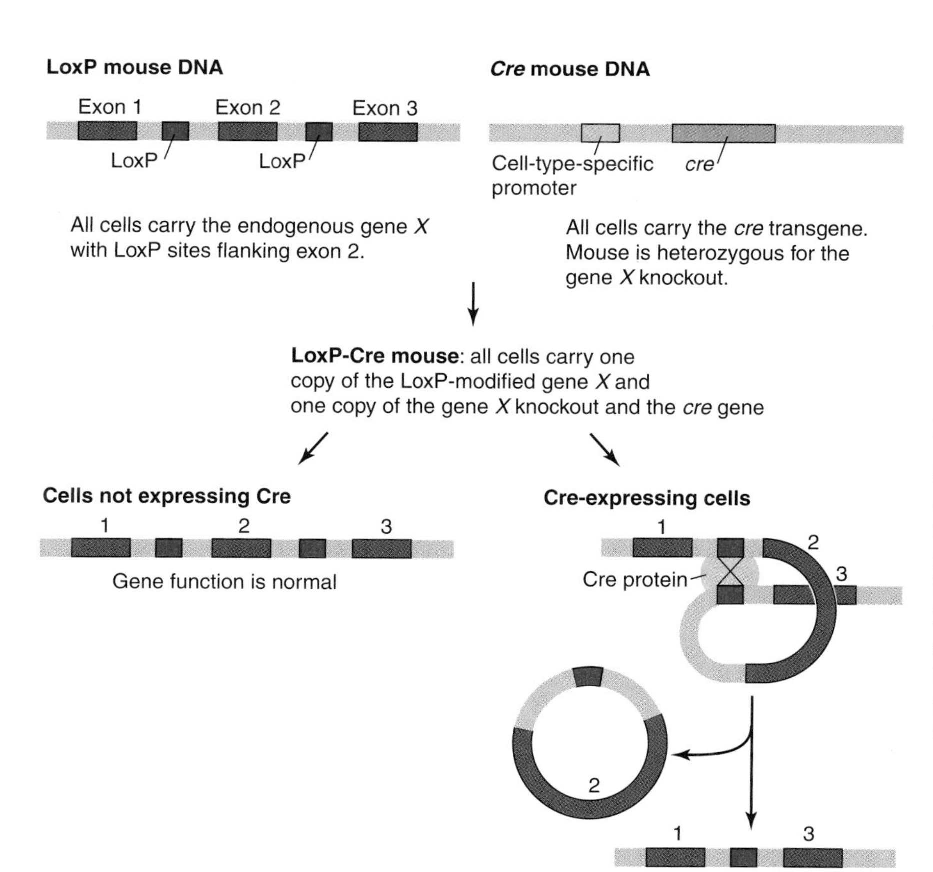

FIGURE 13.55 **Using Cre/*lox* to make cell-type specific gene knockouts in mouse.** *LoxP* sites are inserted into the chromosome to flank exon 2 of gene *X* of one chromosome. The second copy of the *X* gene has been knocked out. The mouse formed with this construct is called the LoxP mouse. Another mouse, called the Cre mouse, has the *cre* gene inserted into the genome. Adjacent to the *cre* gene is a promoter (transcription regulatory region) that directs expression of the *cre* gene only in certain cell types. This mouse also carries a knockout of one copy of gene *X*. When the two mice are crossed, progeny that carry the *LoxP* construct, the gene *X* knockout, and the *cre* gene are produced. When Cre protein is expressed in cells that activate the promoter, it catalyzes site-specific recombination between the *LoxP* sites, and exon 2 of gene *X* is deleted. This inactivates the one functional copy of gene *X* in those cells expressing Cre. (Adapted from H. Lodish, et al., *Molecular Cell Biology, Fifth Edition.* W.H. Freeman and Company, 2003.)

turn off or turn on a gene in mouse. The construct is designed to be flanked by *lox* sites, with the *cre* gene under control of a promoter that can be turned on by temperature or hormones. A **promoter** is a DNA sequence at the beginning of a gene that functions as a signal to initiate transcription (see Chapter 15). Expression of *cre* results in production of the Cre protein, which then recognizes the *loxP* sites, cuts the *loxP* sites, and promotes rejoining of the cut *loxP* sites to leave behind a single *loxP* site with the material between the *loxP* sites having been excised.

The Cre/*lox* system can be used to conditionally remove an exon from a mouse gene, making what is called a gene knockout or inactivation on that chromosome, or it can fuse the gene of interest to a promoter (a transcription regulatory sequence) and thereby control expression of the gene of interest. Expression of a gene in tissues where it is not normally expressed, or at a time when the gene is not normally expressed, is called **ectopic expression**. Ectopic expression studies can reveal information about gene redundancy, specificity, and cell autonomy. In the next chapter, other types of site-specific recombination reactions will be discussed.

Suggested Reading

General

Cao, L., Alani, E., and Kleckner, N. 1990. A pathway for generation and processing of double-strand breaks during meiotic recombination in *S. cerevisiae*. *Cell* 61:1089–1101.

Cox, M. M. 1991. The RecA protein as a recombinational repair system. *Mol Microbiol* 5:1295–1299.

Haber, J. E. 1992. Exploring the pathways of homologous recombination. *Curr Opin Cell Biol* 4:401–412.

Haber, J. E., Ira, G., Malkova, A., and Sugawara, N. 2004. Repairing a double-strand chromosome break by homologous recombination: revisiting Robin Holliday's model. *Philos Trans R Soc Lond B Biol Sci* 359:79–86.

Keeney, S. 2001. Mechanism and control of meiotic recombination initiation. *Curr Top Dev Biol* 52:1–53.

Keeney, S., and Neale, M. J. 2006. Initiation of meiotic recombination by formation of DNA double-strand breaks: mechanism and regulation. *Biochem Soc Trans* 34:523–525.

Krogh, B. O., and Symington, L. S. 2004. Recombination proteins in yeast. *Ann Rev Genet* 38:233–271.

Liu, Y., and West, S. C. 2004. Happy Hollidays: 40th anniversary of the Holliday junction. *Nat Rev Mol Cell Biol* 5:937–944.

Michel, B., Flores, M. J., Viguera, E., et al. 2001. Rescue of arrested replication forks by homologous recombination. *Proc Natl Acad Sci USA* 98:8181–8188.

Orr-Weaver, T. L., Szostak, J. W., and Rothstein, R. J. 1981. Yeast transformation: a model system for the study of recombination. *Proc Natl Acad Sci USA* 78:6354–6358.

Paques, F., and Haber, J. E. 1999. Multiple pathways of recombination induced by double-strand breaks in *Saccharomyces cerevisiae*. *Microbiol Mol Biol Rev* 63:349–404.

Radding, C. M. 1978. Genetic recombination: strand transfer and mismatch repair. *Ann Rev Biochem* 47:847–880.

Radding, C. M. 1981. Recombination activities of *E. coli* recA protein. *Cell* 25:3–4.

Radding, C. M. 1982. Homologous pairing and strand exchange in genetic recombination. *Ann Rev Genet* 16:405–437.

Schwacha, A., and Kleckner, N. 1995. Identification of double Holliday junctions as intermediates in meiotic recombination. *Cell* 83:783–791.

Sun, H., Treco, D., Schultes, N. P., and Szostak, J. W. 1989. Double-strand breaks at an initiation site for meiotic gene conversion. *Nature* 338:87–90.

Sung, P., Krejci, L., Van Komen, S., and Sehorn, M. G. 2003. Rad51 recombinase and recombination mediators. *J Biol Chem* 278:42729–42732.

Symington, L.S. 2002. Role of RAD52 epistasis group genes in homologous recombination and double-strand break repair. *Microbiol Mol Biol Rev* 66:630–670.

Recombination Models

Ferguson, D. O., and Holloman, W. K. 1996. Recombinational repair of gaps in DNA is asymmetric in *Ustilago maydis* and can be explained by a migrating D-loop model. *Proc Natl Acad Sci USA* 93:5419–5424.

Heyer, W. D. 2004. Recombination: Holliday junction resolution and crossover formation. *Curr Biol* 14:R56–R58.

Higgins, N. P., Kato, K., and Strauss, B. 1976. A model for replication repair in mammalian cells. *J Mol Biol* 101:417–425.

Holliday, R. 1964. A mechanism for gene conversion in fungi. *Genet Res Camb* 5:282–304.

Hollingsworth, N. M., and Brill, S. J. 2004. The Mus81 solution to resolution: generating meiotic crossovers without Holliday junctions. *Genes Dev* 18:117–125.

Ivanov, E. L., Sugawara, N., Fishman-Lobell, J., and Haber, J. E. 1996. Genetic requirements for the single-strand annealing pathway of double-strand break repair in *Saccharomyces cerevisiae*. *Genetics*142:693–704.

Kraus, E., Leung, W. Y., and Haber, J. E. 2001. Break-induced replication: a review and an example in budding yeast. *Proc Natl Acad Sci USA* 98:8255–8262.

McEachern, M. J., and Haber, J. E. 2006. Break-induced replication and recombinational telomere elongation in yeast. *Ann Rev Biochem* 75:111–135.

Meselson, M. S., and Radding, C. M. 1975. A general model for genetic recombination. *Proc Natl Acad Sci USA* 72:358–361.

Nassif, N., Penney, J., Pal, S., et al. 1994. Efficient copying of nonhomologous sequences from ectopic sites via P-element-induced gap repair. *Mol Cell Biol* 14:1613–1625.

Osman, F., Dixon, J., Doe, C. L., and Whitby, M. C. 2003. Generating crossovers by resolution of nicked Holliday junctions: a role for Mus81-Eme1 in meiosis. *Mol Cell* 12:761–774.

San Filippo, J, Sung, P., and Klein, H. 2008. Mechanism of eukaryotic homologous recombination. *Ann Rev Biochem* 77:229–257.

Strathern, J. N., Klar, A. J., Hicks, J. B., et al. 1982. Homothallic switching of yeast mating type cassettes is initiated by a double-stranded cut in the *MAT* locus. *Cell* 31:183–192.

Szostak, J. W., Orr-Weaver, T. L., Rothstein, R. J., and Stahl, F. W. 1983. The double-strand-break repair model for recombination. *Cell* 33:25–35.

Whitby, M. C. 2005. Making crossovers during meiosis. *Biochem Soc Trans* 33:1451–1455.

Recombination Proteins in Bacteria

Handa, N., Morimatsu, K., Lovett, S. T., and Kowalczykowski, S. C. 2009. Reconstitution of initial steps of dsDNA break repair by the RecF pathway of *E. coli*. *Genes Dev* 23:1234–1245.

Ivancic-Bace, I., Salaj-Smic, E., and Brcic-Kostic, K. 2005. Effects of *recJ*, *recQ*, and *recFOR* mutations on recombination in nuclease-deficient *recB recD* double mutants of *Escherichia coli*. *J Bacteriol* 187:1350–1356.

Kowalczykowski, S. C. 2000. Initiation of genetic recombination and recombination-dependent replication. *Trends Biochem Sci* 25:156–165.

Nowosielska, A. 2007. Bacterial DNA repair genes and their eukaryotic homologues: 5. The role of recombination in DNA repair and genome stability. *Acta Biochim Pol* 54:483–494.

Sakai, A., and Cox, M. M. 2009. RecFOR and RecOR as distinct RecA loading pathways. *J Biol Chem* 284:3264–3272.

Recombination Proteins in Eukaryotes

Alani, E., Subbiah, S., and Kleckner, N. 1989. The yeast *RAD50* gene encodes a predicted 153-kD protein containing a purine nucleotide-binding domain and two large heptad-repeat regions. *Genetics* 122:47–57.

Anderson, D. G., and Kowalczykowski, S. C. 1997. The translocating RecBCD enzyme stimulates recombination by directing RecA protein onto ssDNA in a *chi*-regulated manner. *Cell* 90:77–86.

Baumann, P., Benson, F. E., and West, S. C. 1996. Human Rad51 protein promotes ATP-dependent homologous pairing and strand transfer reactions in vitro. *Cell* 87:757–766.

Benson, F. E., Baumann, P., and West, S. C. 1998. Synergistic actions of Rad51 and Rad52 in recombination and DNA repair. *Nature* 391:401–404.

Bianchi, M., DasGupta, C., and Radding, C. M. 1983. Synapsis and the formation of paranemic joints by *E. coli* RecA protein. *Cell* 34:931–939.

Bishop, D. K., Park, D., Xu, L., and Kleckner, N. 1992. DMC1: a meiosis-specific yeast homolog of *E. coli recA* required for recombination, synaptonemal complex formation, and cell cycle progression. *Cell* 69:439–456.

Boddy, M. N., Gaillard, P. H., McDonald, W. H., et al. 2001. Mus81-Eme1 are essential components of a Holliday junction resolvase. *Cell* 107:537–548.

Carreira, A. and Kowalczykowski, S. C. 2009. BRCA2. *Cell Cycle* 136:1032–1043.

Cox, M. M. 2007. Motoring along with the bacterial RecA protein. *Nat Rev Mol Cell Biol* 8:127–138.

Cox, M. M., and Lehman, I. R. 1981. RecA protein of *Escherichia coli* promotes branch migration, a kinetically distinct phase of DNA strand exchange. *Proc Natl Acad Sci USA* 78:3433–3437.

Cox, M. M., Soltis, D. A., Livneh, Z., and Lehman, I. R. 1983. On the role of single-stranded DNA binding protein in recA protein-promoted DNA strand exchange. *J Biol Chem* 258:2577–2585.

Davies, A.A., Masson, J.Y., McIlwraith, M. J., et al. 2001. Role of BRCA2 in control of the RAD51 recombination and DNA repair protein. *Mol Cell* 7:273–282.

Davis, A. P., and Symington, L. S. 2001. The yeast recombinational repair protein Rad59 interacts with Rad52 and stimulates single-strand annealing. *Genetics* 159:515–525.

Davis, A. P., and Symington, L. S. 2003. The Rad52-Rad59 complex interacts with Rad51 and replication protein A. *DNA Repair (Amst)* 2:1127–1134.

Dunderdale, H. J., Benson, F. E., Parsons, C. A., et al. 1991. Formation and resolution of recombination intermediates by *E. coli* RecA and RuvC proteins. *Nature* 354:506–510.

Eggleston, A. K., and West, S. C. 2000. Cleavage of Holliday junctions by the *Escherichia coli* RuvABC complex. *J Biol Chem* 275:26467–26476.

Fekairi, S., Scaglione, S., Chahwan, C., et al. 2009. Human SLX4 is a Holliday junction resolvase subunit that binds multiple DNA repair/recombination endonucleases. *Cell* 138:78–89.

Flory, J., Tsang, S. S., and Muniyappa, K. 1984. Isolation and visualization of active presynaptic filaments of recA protein and single-stranded DNA. *Proc Natl Acad Sci USA* 81:7026–7030.

Gravel, S., Chapman, J. R., Magill, C., and Jackson, S. P. 2008. DNA helicases Sgs1 and BLM promote DNA double-strand break resection. *Genes Dev* 22:2767–2772.

Hopfner, K. P., Craig, L., Moncalian, G., et al. 2002. The Rad50 zinc-hook is a structure joining Mre11 complexes in DNA recombination and repair. *Nature* 418:562–566.

Ip, S. C., Rass, U., Blanco, M. G., Flynn, H. R., Skehel, J. M., and West, S. C. 2008. Identification of Holliday junction resolvases from humans and yeast. *Nature* 456:357–361.

Ira, G., Malkova, A., Liberi, G., et al. 2003. Srs2 and Sgs1-Top3 suppress crossovers during double-strand break repair in yeast. *Cell* 115:401–411.

Kahn, R., and Radding, C. M. 1984. Separation of the presynaptic and synaptic phases of homologous pairing promoted by RecA protein. *J Biol Chem* 259:7495–7503.

Kaliraman, V., Mullen, J. R., Fricke, W. M., et al. 2001. Functional overlap between Sgs1-Top3 and the Mms4-Mus81 endonuclease. *Genes Dev* 15:2730–2740.

Karow, J. K., Constantinou, A., Li, J. L., West, S. C., and Hickson, I. D. 2000. The Bloom's syndrome gene product promotes branch migration of Holliday junctions. *Proc Natl Acad Sci USA* 97:6504–6508.

Keeney, S., Giroux, C. N., and Kleckner, N. 1997. Meiosis-specific DNA double-strand breaks are catalyzed by Spo11, a member of a widely conserved protein family. *Cell* 88:375–384.

Keeney, S., and Kleckner, N. 1995. Covalent protein-DNA complexes at the 5′ strand termini of meiosis-specific double-strand breaks in yeast. *Proc Natl Acad Sci USA* 92:11274–11278.

Kodadek, T., and Alberts, B. M. 1987. Stimulation of protein-directed strand exchange by a DNA helicase. *Nature* 326:312–314.

Krejci, L., Van Komen, S., Li, Y., et al. 2003. DNA helicase Srs2 disrupts the Rad51 presynaptic filament. *Nature* 423:305–309.

Lengsfeld, B. M., Rattray, A. J., Bhaskara, V., Ghirlando, R., and Paull, T. T. 2007. Sae2 is an endonuclease that processes hairpin DNA cooperatively with the Mre11/Rad50/Xrs2 complex. *Mol Cell* 28:638–651.

McIlwraith, M. J., Vaisman, A., Liu, Y., et al. 2005. Human DNA polymerase eta promotes DNA synthesis from strand invasion intermediates of homologous recombination. *Mol Cell* 20:783–792.

Mimitou, E. P., and Symington. L. S. 2008. Sae2, Exo1 and Sgs1 collaborate in DNA double-strand break processing. *Nature* 455:770–774.

Morrison, A. J., and Shen, X. 2009. Chromatin remodelling beyond transcription: the INO80 and SWR1 complexes. *Nat Rev Mol Cell Biol* 10:373–384.

Munoz, I. M., Hain, K., Declais, A. C., et al. 2009. Coordination of structure-specific nucleases by human SLX4/BTBD12 is required for DNA repair. *Mol Cell* 35:116–127.

New, J. H., Sugiyama, T., Zaitseva, E., and Kowalczykowski, S. C. 1998. Rad52 protein stimulates DNA strand exchange by Rad51 and replication protein A. *Nature* 391:407–410.

Ogawa, T., Yu, X., Shinohara, A., and Egelman, E. H. 1993. Similarity of the yeast RAD51 filament to the bacterial RecA filament. *Science* 259:1896–1899.

Petukhova, G., Stratton, S., and Sung, P. 1998. Catalysis of homologous DNA pairing by yeast Rad51 and Rad54 proteins. *Nature* 393:91–94.

Petukhova, G., Sung, P., and Klein, H. 2000. Promotion of Rad51-dependent D-loop formation by yeast recombination factor Rdh54/Tid1. *Genes Dev* 14:2206–2215.

Ralf, C., Hickson, I. D., and Wu, L. 2006. The Bloom's syndrome helicase can promote the regression of a model replication fork. *J Biol Chem* 281:22839–22846.

Register, J. C. 3rd, Christiansen, G., and Griffith, J. 1987. Electron microscopic visualization of the RecA protein-mediated pairing and branch migration phases of DNA strand exchange. *J Biol Chem* 262:12812–12820.

Roman, L. J., and Kowalczykowski, S. C. 1986. Relationship of the physical and enzymatic properties of *Escherichia coli* recA protein to its strand exchange activity. *Biochem* 25:7375–7385.

Rudin, N., Sugarman, E., and Haber, J. E. 1989. Genetic and physical analysis of double-strand break repair and recombination in *Saccharomyces cerevisiae*. *Genetics* 122:519–534.

San Filippo, J., Chi, P., Sehorn, M. G., et al. 2006. Recombination mediator and Rad51 targeting activities of a human BRCA2 polypeptide. *J Biol Chem* 281:11649–11657.

Savic, V., Yin, B., Maas, N. L., et al. 2009. Formation of dynamic γ-H2AX domains along broken DNA strands is distinctly regulated by ATM and MDC1 and dependent upon H2AX densities in chromatin. *Mol Cell* 34: 298–310.

Sehorn, M. G., Sigurdsson, S., Bussen, W., et al. 2004. Human meiotic recombinase Dmc1 promotes ATP-dependent homologous DNA strand exchange. *Nature* 429:433–437.

Seigneur, M., Bidnenko, V., Ehrlich, S.D., and Michel, B. 1998. RuvAB acts at arrested replication forks. *Cell* 95:419–430.

Shinohara, A., Ogawa, H., and Ogawa, T. 1992. Rad51 protein involved in repair and recombination in *S. cerevisiae* is a RecA-like protein. *Cell* 69:457–470.

Shinohara, A., and Ogawa, T. 1998. Stimulation by Rad52 of yeast Rad51-mediated recombination. *Nature* 391:404–407.

Sigurdsson, S., Van Komen, S., Bussen, W., et al. 2001. Mediator function of the human Rad51B-Rad51C complex in Rad51/RPA-catalyzed DNA strand exchange. *Genes Dev* 15:3308–3318.

Soltis, D. A., and Lehman, I. R. 1983. An unpaired 3′ terminus stimulates recA protein-promoted DNA strand exchange. *J Biol Chem* 258:14073–14075.

Sonoda, E., Sasaki, M. S., Buerstedde, J. M., et al. 1998. Rad51-deficient vertebrate cells accumulate chromosomal breaks prior to cell death. *EMBO J* 17:598–608.

Stasiak, A., Egelman, E. H., and Howard-Flanders, P. 1988. Structure of helical RecA-DNA complexes. III. The structural polarity of RecA filaments and functional polarity in the RecA-mediated strand exchange reaction. *J Mol Biol* 202:659–662.

Stasiak, A., Stasiak, A. Z., and Koller, T. 1984. Visualization of RecA-DNA complexes involved in consecutive stages of an in vitro strand exchange reaction. *Cold Spring Harb Symp Quant Biol* 49:561–570.

Stewart, G. S., Maser, R. S., Stankovic, T., et al. 1999. The DNA double-strand break repair gene hMRE11 is mutated in individuals with an ataxia-telangiectasia-like disorder. *Cell* 99:577–587.

Sugawara, N., Wang, X., and Haber, J. E. 2003. In vivo roles of Rad52, Rad54, and Rad55 proteins in Rad51-mediated recombination. *Mol Cell* 12:209–219.

Sung, P. 1994. Catalysis of ATP-dependent homologous DNA pairing and strand exchange by yeast RAD51 protein. *Science* 265:1241–1243.

Sung, P. 1997. Function of yeast Rad52 protein as a mediator between replication protein A and the Rad51 recombinase. *J Biol Chem* 272:28194–28197.

Sung, P., and Robberson, D. L. 1995. DNA strand exchange mediated by a RAD51-ssDNA nucleoprotein filament with polarity opposite to that of RecA. *Cell* 82:453–461.

Svendsen, J. M., Smogorzewska, A., Sowa, M. E., et al. 2009. Mammalian BTBD12/SLX4 assembles a Holliday junction resolvase and is required for DNA repair. *Cell* 138:63–77.

Tsaneva, I. R., Muller, B., and West, S. C. 1992. ATP-dependent branch migration of Holliday junctions promoted by the RuvA and RuvB proteins of *E. coli*. *Cell* 69:1171–1180.

Tsuzuki, T., Fujii, Y., Sakumi, K., et al. 1996. Targeted disruption of the Rad51 gene leads to lethality in embryonic mice. *Proc Natl Acad Sci USA* 93:6236–6240.

van Attikum, H., Fritsch, O., and Gasser, S. M. 2007. Distinct roles for SWR1 and INO80 chromatin remodeling complexes at chromosomal double-strand breaks. *EMBO J* 26:4113–4125.

van Attikum, H., Fritsch, O., Hohn, B., and Gasser, S.M. 2004. Recruitment of the INO80 complex by H2A phosphorylation links ATP-dependent chromatin remodeling with DNA double-strand break repair. *Cell* 119:777–788.

van Attikum, H., and Gasser, S. M. 2009. Crosstalk between histone modifications during the DNA damage response. *Trends Cell Biol* 19:207–217.

Van Komen, S., Petukhova, G., Sigurdsson, S., et al. 2000. Superhelicity-driven homologous DNA pairing by yeast recombination factors Rad51 and Rad54. *Mol Cell* 6:563–572.

West, S. C., Cassuto, E., and Howard-Flanders, P. 1981. recA protein promotes homologous-pairing and strand-exchange reactions between duplex DNA molecules. *Proc Natl Acad Sci USA* 78:2100–2104.

West, S. C., and Connolly, B. 1992. Biological roles of the *Escherichia coli* RuvA, RuvB and RuvC proteins revealed. *Mol Microbiol* 6:2755–2759.

Whitby, M. C., and Lloyd, R. G. 1995. Branch migration of three-strand recombination intermediates by RecG, a possible pathway for securing exchanges initiated by 3′-tailed duplex DNA. *EMBO J* 14:3302–3310.

Whitby, M. C., Osman, F., and Dixon, J. 2003. Cleavage of model replication forks by fission yeast Mus81-Eme1 and budding yeast Mus81-Mms4. *J Biol Chem* 278:6928–6935.

Whitby, M. C., Ryder, L., and Lloyd, R. G. 1993. Reverse branch migration of Holliday junctions by RecG protein: a new mechanism for resolution of intermediates in recombination and DNA repair. *Cell* 75:341–350.

Wiltzius, J. J., Hohl, M., Fleming, J. C., and Petrini, J. H. 2005. The Rad50 hook domain is a critical determinant of Mre11 complex functions. *Nat Struct Mol Biol* 12:403–407.

Wu, A. M., Kahn, R., DasGupta, C., and Radding, C. M. 1982. Formation of nascent heteroduplex structures by RecA protein and DNA. *Cell* 30:37–44.
Wu, L., Davies, S. L., North, P. S., et al. 2000. The Bloom's syndrome gene product interacts with topoisomerase III. *J Biol Chem* 275:9636–9644.
Wu, L., and Hickson, I. D. 2003. The Bloom's syndrome helicase suppresses crossing over during homologous recombination. *Nature* 426:870–874.
Zhu, Z., Chung, W. H., Shim, E. Y., Lee, S. E., and Ira, G. 2008. Sgs1 helicase and two nucleases Dna2 and Exo1 resect DNA double-strand break ends. *Cell* 134:981–994.

Nonhomologous End-Joining

Daley, J. M., Palmbos, P. L., Wu, D., and Wilson, T. E. 2005. Nonhomologous end joining in yeast. *Ann Rev Genet* 39:431–451.
Lieber, M. R., Ma, Y., Pannicke, U., and Schwarz, K. 2003. Mechanism and regulation of human non-homologous DNA end-joining. *Nat Rev Mol Cell Biol* 4:712–720.
Lieber, M. R., Ma, Y., Pannicke, U., and Schwarz, K. 2004. The mechanism of vertebrate nonhomologous DNA end joining and its role in V(D)J recombination. *DNA Repair (Amst)* 3:817–826.

Meiotic Recombination

Allers, T., and Lichten, M. 2001. Differential timing and control of noncrossover and crossover recombination during meiosis. *Cell* 106:47–57.
Borner, G. V., Kleckner, N., and Hunter, N. 2004. Crossover/noncrossover differentiation, synaptonemal complex formation, and regulatory surveillance at the leptotene/zygotene transition of meiosis. *Cell* 117:29–45.
Borts, R. H., and Haber, J. E. 1989. Length and distribution of meiotic gene conversion tracts and crossovers in *Saccharomyces cerevisiae*. *Genetics* 123:69–80.
Hunter, N., and Kleckner, N. 2001. The single-end invasion: an asymmetric intermediate at the double-strand break to double-Holliday junction transition of meiotic recombination. *Cell* 106:59–70.
Lichten, M., Goyon, C., Schultes, N. P., et al. 1990. Detection of heteroduplex DNA molecules among the products of *Saccharomyces cerevisiae* meiosis. *Proc Natl Acad Sci USA* 87:7653–7657.
Liu, J., Wu, T. C., and Lichten, M. 1995. The location and structure of double-strand DNA breaks induced during yeast meiosis: evidence for a covalently linked DNA-protein intermediate. *EMBO J* 14:4599–4608.
Martini, E., Diaz, R. L., Hunter, N., and Keeney, S. 2006. Crossover homeostasis in yeast meiosis. *Cell* 126:285–295.
Neale, M. J., and Keeney, S. 2006. Clarifying the mechanics of DNA strand exchange in meiotic recombination. *Nature* 442:153–158.
Rockmill, B., Sym, M., Scherthan, H., and Roeder, G. S. 1995. Roles for two RecA homologs in promoting meiotic chromosome synapsis. *Genes Dev* 9:2684–2695.
Schwacha, A., and Kleckner, N. 1994. Identification of joint molecules that form frequently between homologs but rarely between sister chromatids during yeast meiosis. *Cell* 76:51–63.
Shinohara, A., Gasior, S., Ogawa, T., et al. 1997. *Saccharomyces cerevisiae recA* homologues *RAD51* and *DMC1* have both distinct and overlapping roles in meiotic recombination. *Genes Cells* 2:615–629.
Storlazzi, A., Xu, L., Cao, L., and Kleckner, N. 1995. Crossover and noncrossover recombination during meiosis: timing and pathway relationships. *Proc Natl Acad Sci USA* 92:8512–8516.

Targeted Gene Knockouts

Baudin, A., Ozier-Kalogeropoulos, O., Denouel, A., et al. 1993. A simple and efficient method for direct gene deletion in *Saccharomyces cerevisiae*. *Nucleic Acids Res* 21:3329–3330.
Capecchi, M. R. 2005. Gene targeting in mice: functional analysis of the mammalian genome for the twenty-first century. *Nat Rev Genet* 6:507–512.

Muller, U. 1999. Ten years of gene targeting: targeted mouse mutants, from vector design to phenotype analysis. *Mech Dev* 82:3–21.

Rothstein, R. J. 1983. One-step gene disruption in yeast. *Methods Enzymol* 101:202–211.

Smithies, O., Gregg, R. G., Boggs, S. S., et al. 1985. Insertion of DNA sequences into the human chromosomal beta-globin locus by homologous recombination. *Nature* 317:230–234.

Thomas, K. R., and Capecchi, M. R. 1986. Introduction of homologous DNA sequences into mammalian cells induces mutations in the cognate gene. *Nature* 324:34–38.

V(D)J Recombination

Chaudhuri, J., and Alt, F. W. 2004. Class-switch recombination: interplay of transcription, DNA deamination and DNA repair. *Nat Rev Immunol* 4:541–552.

Chaudhuri, J., Khuong, C., and Alt, F. W. 2004. Replication protein A interacts with AID to promote deamination of somatic hypermutation targets. *Nature* 430:992–998.

Gellert, M. 2002. *V(D)J* recombination: RAG proteins, repair factors, and regulation. *Ann Rev Biochem* 71:101–132.

Ma, Y., Pannicke, U., Schwarz, K., and Lieber, M. R. 2002. Hairpin opening and overhang processing by an Artemis/DNA-dependent protein kinase complex in nonhomologous end joining and *V(D)J* recombination. *Cell* 108:781–794.

Ma, Y., Schwarz, K., and Lieber, M. R. 2005. The Artemis:DNA-PKCs endonuclease cleaves DNA loops, flaps, and gaps. *DNA Repair (Amst)* 4:845–851.

Moshous, D., Callebaut, I., de Chasseval, R., et al. 2001. Artemis, a novel DNA double-strand break repair/V(D)J recombination protein, is mutated in human severe combined immune deficiency. *Cell* 105:177–186.

Neiditch, M. B., Lee, G. S., Huye, L. E., et al. 2002. The V(D)J recombinase efficiently cleaves and transposes signal joints. *Mol Cell* 9:871–878.

Roth, D. B. 2003. Restraining the V(D)J recombinase. *Nat Rev Immunol* 3:656–666.

Schwarz, K., Ma, Y., Pannicke, U., and Lieber, M. R. 2003. Human severe combined immune deficiency and DNA repair. *Bioessays* 25:1061–1070.

Somatic Hypermutation

Di Noia, J., and Neuberger, M. S. 2002. Altering the pathway of immunoglobulin hypermutation by inhibiting uracil-DNA glycosylase. *Nature* 419:43–48.

Harris, R. S., Sale, J. E., Petersen-Mahrt, S. K., and Neuberger, M. S. 2002. AID is essential for immunoglobulin V gene conversion in a cultured B cell line. *Curr Biol* 12:435–438.

Neuberger, M. S., Di Noia, J. M., Beale, R. C., et al. 2005. Somatic hypermutation at A-T pairs: polymerase error versus dUTP incorporation. *Nat Rev Immunol* 5:171–178.

Williams, G. T., Jolly, C. J., Kohler, J., and Neuberger, M. S. 2000. The contribution of somatic hypermutation to the diversity of serum immunoglobulin: dramatic increase with age. *Immunity* 13:409–417.

Mitotic Recombination and DNA Replication

Branzei, D., and Foiani, M. 2005. The DNA damage response during DNA replication. *Curr Opin Cell Biol* 17:568–575.

Carney, J. P., Maser, R. S., Olivares, H., et al. 1998. The hMre11/hRad50 protein complex and Nijmegen breakage syndrome: linkage of double-strand break repair to the cellular DNA damage response. *Cell* 93:477–486.

Cox, M. M., Goodman, M. F., Kreuzer, K. N., et al. 2000. The importance of repairing stalled replication forks. *Nature* 404:37–41.

Heller, R. C., and Marians, K. J. 2005. The disposition of nascent strands at stalled replication forks dictates the pathway of replisome loading during restart. *Mol Cell* 17:733–743.

Ira, G., Pellicioli, A., Balijja, A., et al. 2004. DNA end resection, homologous recombination and DNA damage checkpoint activation require CDK1. *Nature* 431:1011–1017.

Lopes, M., Cotta-Ramusino, C., Pellicioli, A., et al. 2001. The DNA replication checkpoint response stabilizes stalled replication forks. *Nature* 412:557–561.

Marians, K. J. 1999. PriA: at the crossroads of DNA replication and recombination. *Prog Nucleic Acid Res Mol Biol* 63:39–67.

Marians, K. J. 2004. Mechanisms of replication fork restart in *Escherichia coli. Philos Trans R Soc Lond B Biol Sci* 359:71–77.

Sandler, S. J., and Marians, K. J. 2000. Role of PriA in replication fork reactivation in *Escherichia coli. J Bacteriol* 182:9–13.

Wu, L., and Hickson, I. D. 2006. DNA helicases required for homologous recombination and repair of damaged replication forks. *Ann Rev Genet* 40:279–306.

DNA Rearrangements

Bosco, G., and Haber, J. E. 1998. Chromosome break-induced DNA replication leads to nonreciprocal translocations and telomere capture. *Genetics* 150:1037–1047.

Kuppers, R., and Dalla-Favera, R. 2001. Mechanisms of chromosomal translocations in B cell lymphomas. *Oncogene* 20:5580–5594.

Lieber, M. R., Yu, K., and Raghavan, S. C. 2006. Roles of nonhomologous DNA end joining, V(D)J recombination, and class switch recombination in chromosomal translocations. *DNA Repair (Amst)* 5:1234–1245.

Mieczkowski, P. A., Lemoine, F. J., and Petes, T. D. 2006. Recombination between retrotransposons as a source of chromosome rearrangements in the yeast *Saccharomyces cerevisiae. DNA Repair (Amst)* 5:1010–1020.

Site-Specific Recombination

Chou, T. B., and Perrimon, N. 1992. Use of a yeast site-specific recombinase to produce female germline chimeras in *Drosophila. Genetics* 131:643–653.

Connolly, B., White, C. I., and Haber, J. E. 1988. Physical monitoring of mating type switching in *Saccharomyces cerevisiae. Mol Cell Biol* 8:2342–2349.

Dang, D. T., and Perrimon, N. 1992. Use of a yeast site-specific recombinase to generate embryonic mosaics in *Drosophila. Dev Genet* 13:367–375.

Egli, D., Hafen, E., and Schaffner, W. 2004. An efficient method to generate chromosomal rearrangements by targeted DNA double-strand breaks in *Drosophila melanogaster. Genome Res* 14:1382–1393.

Golic, K. G., and Golic, M. M. 1996. Engineering the *Drosophila* genome: chromosome rearrangements by design. *Genetics* 144:1693–1711.

Lakso, M., Pichel, J. G., Gorman, J. R., et al. 1996. Efficient in vivo manipulation of mouse genomic sequences at the zygote stage. *Proc Natl Acad Sci USA* 93:5860–5865.

Le, Y., and Sauer, B. 2001. Conditional gene knockout using Cre recombinase. *Mol Biotechnol* 17:269–275.

Sauer, B., and Henderson, N. 1989. Cre-stimulated recombination at *loxP*-containing DNA sequences placed into the mammalian genome. *Nucleic Acids Res* 17:147–161.

White, C. I., and Haber, J. E. 1990. Intermediates of recombination during mating type switching in *Saccharomyces cerevisiae. EMBO J* 9:663–673.

Xu, T., and Rubin, G. M. 1993. Analysis of genetic mosaics in developing and adult *Drosophila* tissues. *Development* 117:1223–1237.

This book has a Web site, **http://biology.jbpub.com/book/molecular**, which contains a variety of resources, including links to in-depth information on topics covered in this chapter, and review material designed to assist in your study of molecular biology.

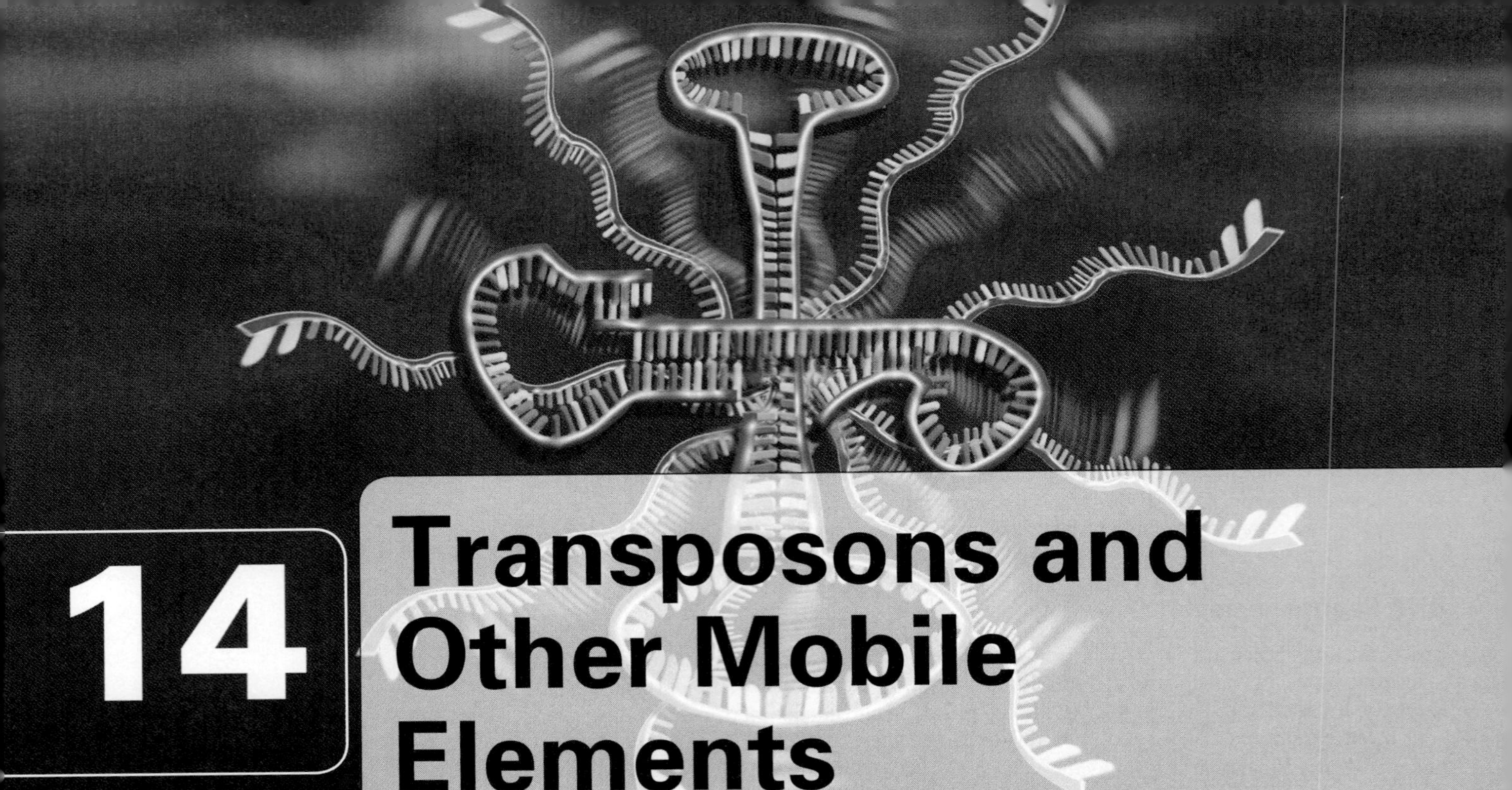

14 Transposons and Other Mobile Elements

OUTLINE OF TOPICS

14.1 Transposition

The simplest mobile elements in bacteria are called insertion sequences.

The transposase forms a specific complex with the ends of the mobile element.

Coordinated breakage and joining events occur during transposition.

Some elements do cut-and-paste transposition, where the element is directly moved to a new location.

Transposition during DNA replication and host DNA repair allow cut-and-paste elements to increase in copy number.

Transposons are found at various levels of complexity in bacteria.

Replicative transposons leave one copy of the element at the donor site.

Transposons in eukaryotes are mechanistically similar to bacterial transposons.

Diverse systems allow transposition to be regulated.

Most transposons prefer DNA targets that are bent.

Transposons can target certain sequences.

Some transposons target specific molecular processes.

Some elements have evolved the ability to choose between certain target sites.

Transposons are important tools for molecular genetics.

14.2 Conservative Site-Specific Recombination

Two families of proteins do conservative site-specific recombination with different pathways.

Bacteriophage λ uses a conservative site-specific recombinase to integrate into the host genome.

Multiple other systems use the conservative site-specific recombinase reaction.

14.3 Target-Primed Reverse Transcription

Target-primed reverse transcription can mobilize information through an RNA intermediate.

Target-primed reverse transcription is used for LINE movement.

LINE movement affects genome stability and evolution.

Mobile group II introns move by target-primed reverse transcription.

Two transesterification reactions allow group II intron movement.

Homology to the target site determines if mobile group II introns move by retrotransposition or retro-homing.

14.4 Other Mechanisms of DNA Mobilization

Suggested Reading

Previous chapters described processes that cells use to stably maintain genetic information over generations. It may therefore come as some surprise that most genomes contain a significant number of discrete DNA elements that can move between various locations. The mindset of a static genome was one of the barriers faced by Barbara McClintock when her studies with maize in the 1940s indicated that certain chromosomal elements can "jump" in specific genetic crosses. Her first clue to the existence of jumping genetic elements came from her observation that in some maize strains DNA breaks always occur at the same position in chromosome 9. Further studies revealed that two elements are required for the DNA breaks to take place. A break-inducing genetic entity, the dissociation element (*Ds*) has to be present at the breakage site and a second discrete entity, the activator element (*Ac*) has to be present elsewhere in the genome. McClintock's next major breakthrough came from the discovery that alteration in kernel pigment patterns correlated with *Ds* element movement (**FIGURE 14.1**). The gene responsible for pigment color is inactivated when the *Ds* element inserts into it and is reactivated when the *Ds* element is excised. The *Ac* element is required to supply the recombinase needed for *Ds* movement. Additional studies showed that the *Ds* element is an *Ac* element that has lost the ability to make the recombinase, more specifically called a **transposase**, needed to move the element. The *Ac* element can also move and does not require the

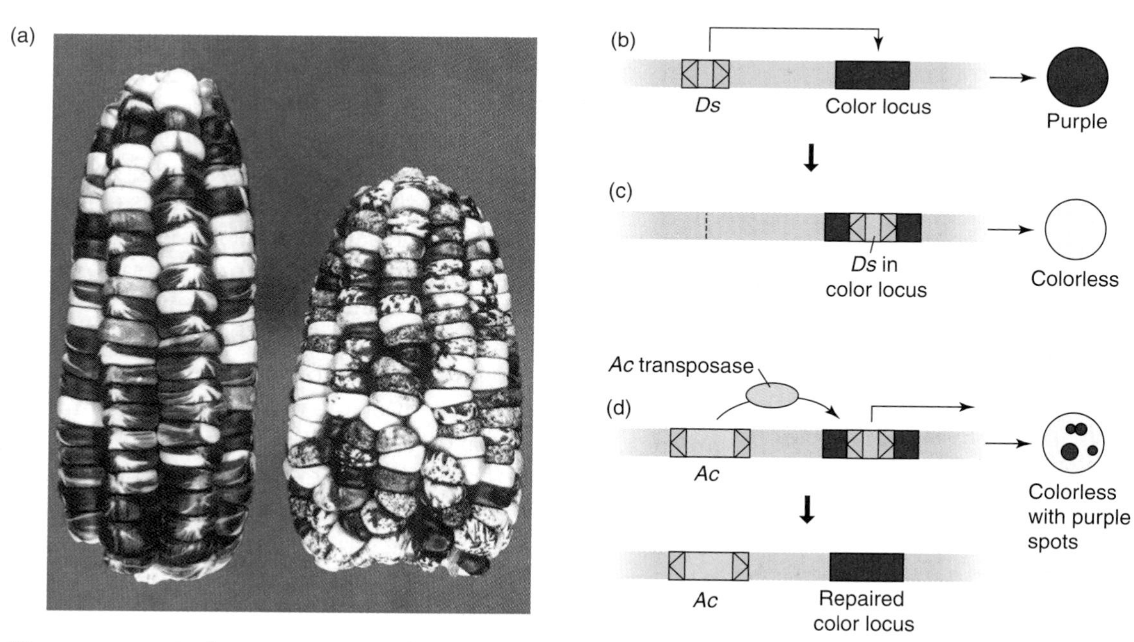

FIGURE 14.1 The movement of autonomous and nonautonomous transposable elements can alter the pigmentation of maize. (a) Changes in kernel coloration could be explained by the movement of DNA elements in the chromosome. (b) Maize kernels are purple when the responsible genes are not interrupted by endogenous DNA elements like the nonautonomous transposable element called *Ds*. (c) An insertion element like *Ds* can jump into a locus responsible for purple pigmentation prior to development of the coat. (d) If the *Ds* element were again mobilized by *Ac*, an autonomous element that produces transposase, the progeny could revert back to purple during development of the kernel if the site was correctly repaired. (Part a reproduced from R. N. Jones. *Cytogenet. Genome Res.* 109 [2005]: 90–103, with permission from S. Karger AG, Basel. Photo courtesy of Neil Jones, The University of Wales, Aberystwyth. Parts b–d adapted from R. N. Jones, *Cytogenet. Genome Res.* 109 [2005]: 90–103.)

assistance of any other genetic element to do so. McClintock coined the term **transposon** for mobile genetic elements such as the *Ac* element. We now know that virtually all living organisms contain transposons within their DNA.

Transposons can move between sites in the genome in many different ways. As expected, over time many of these elements have evolved within their host to do as little harm as possible and sometimes even provide some benefits so as to favor their maintenance. The potential negative effect that mobile elements have on the host is made most clear with the observation that most genomes contain many inactivated elements. The high incidence of inactive elements suggests that the most fit organisms are those where the elements no longer function. Our own genome is estimated to be comprised of 30% to 50% transposable element sequence from two types of elements, the **LINEs** (long interspersed nucleotide elements) and **SINEs** (short interspersed nucleotide elements). As the following examples show, mobile genetic elements can have important physiological functions. Telomeres in *Drosophila* are primarily maintained by transposons and not by telomerase as in most other eukaryotes. The variable (*V*), diversity (*D*), and joining (*J*) [*V*(*D*)*J*] recombination that is responsible for antibody diversity (see Chapter 13), is derived from an ancient transposon that was integrated into the inner workings of our immune system. Mobile genetic elements can impart a variety of beneficial functions to bacteria through the genes they transport. For instance, these elements account for the high incidence of antibiotic resistance in bacteria. Mobile elements also are often responsible for the bacteria's ability to develop the metabolic capacity to degrade essentially every organic substance produced on the planet. In addition to their impact on organisms in the environment, mobile genetic elements are of interest as laboratory tools to isolate mutants and construct new strains.

In very broad terms, three types of reactions can generally explain the movement of all mobile genetic elements (FIGURE 14.2): (1) transposition and the related process of retroviral integration; (2) conservative site-specific recombination; and (3) target-primed reverse transcription.

Transposition is a process in which a discrete DNA entity moves between DNA sites that lack homology using a self-encoded transposase. One large class of elements that is very common in bacteria, archaea, and many eukaryotes always move as a DNA element. Another class called **long terminal repeat (LTR) retrotransposons** move with an RNA intermediate using a mechanism that is identical in almost all respects to that used during retrovirus infection. LTR retrotransposons are common in eukaryotes, but are rarely found in bacteria.

Conservative site-specific recombination is a process in which a segment of DNA moves between specific recombination sites as DNA. Recombinases involved in conservative site-specific recombination are classified as tyrosine or serine recombinases based on whether a tyrosine or serine residue transiently links to DNA during the recombination process.

Target-primed reverse transcription is a process in which the free 3′ end of a broken target DNA is used to prime a replication process, allowing a DNA copy to be made of an element that moves as a

How do elements move to new positions?

1. Transposition and retroviral integration
   - IS elements, transposons, bacteriophage Mu
   - LTR elements (retrotransposons, retroviruses)
   - DNA transposons (*Tc/mariner*, *Ac*, *Tam*, P)
2. Conservative site-specific recombination
   - Tyrosine recombinases
     - Lambda phage, Cre, Flp, XerCD
     - Conjugative transposons
     - Integrons
   - Serine recombinases
     - Tn3-resolvase, γδ
3. Target-primed reverse transcription
   - R2
   - Non-LTR elements (LINES, SINES)
   - Group II mobile introns

FIGURE 14.2 There are three broad families of mechanisms that account for how DNA can move to a new location in genomes.

segment of RNA. This process involves the so-called non-LTR elements that predominate in our genome, the LINEs, and SINEs.

This chapter examines the three ways that mobile genetic elements can move in genomes. A few examples from each group are described in some detail. As part of this treatment, we also consider some of the ways that these elements control where and when they insert to maximize their dispersal while doing as little harm as possible to their host. Additionally, we discuss some of the ways that these mobile elements are used in the laboratory to study living organisms.

14.1 Transposition

The simplest mobile elements in bacteria are called insertion sequences.

Although the idea of transposition was first established in maize, work with bacterial elements has likely made the greatest contribution to our understanding of these elements. Well-studied elements are also derived from many types of organisms, such as the P element in flies, the Tc elements in worms, and *mariner* elements in insects. These types of elements are sometimes referred to as **DNA transposons** because they move without an RNA intermediate. A more inclusive term, **DDE transposons**, includes related elements called the retrotransposons that travel via an RNA intermediate. The DDE indicates the two aspartic acid (D) residues and one glutamic acid (E) residue in the active site of the transposase.

In the simplest form, transposable elements in bacteria are usually called **insertion sequences**. Insertion sequences are grossly similar to the *Ac* element originally discovered by McClintock. Insertion sequences are comprised of the transposase-encoding gene flanked by the DNA sequences on which the transposase protein acts to move the element (FIGURE 14.3). These DNA sequences include one or more binding sites for the transposase proteins. In many cases, the number and spacing of the transposase binding sites are important factors for regulating transposition. A second set of essential DNA sequences, those at the very ends of the element, are used to identify the place where the transposase breakage event will occur. The elaborate array of sequence information that denotes the ends of the element is often simply referred to as the **inverted repeats** because these regions must be in opposite orientation to allow transposition of the element. The inverted repeats can be identical, but if not they must always be similar. We can often find an insertion sequence in a genome by searching for its characteristic inverted sequences.

The transposase forms a specific complex with the ends of the mobile element.

A critical step in the transposition reaction involves the transposase binding to the inverted repeat ends and then forming a stable complex

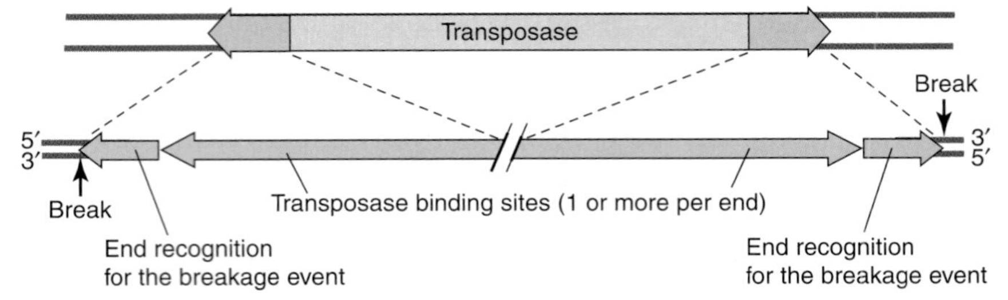

FIGURE 14.3 **Anatomy of a transposon.** The simplest transposons, also known as insertions sequences, have inverted repeats at their ends (shown as green arrows in this figure) and a single gene that codes for a recombinase known as transposase. The transposase binds specific sequences found in inverted orientation fashion outside the transposase. An additional sequence in these elements is a short region recognized as the cleavage site for the transposase. There may be many transposase binding sites, but only the most terminal transposase binding site will be adjacent to a cleavage site. Breaks are initially only made on the 3′ ends of the element and the fate of the 5′ ends of the element differs with the many transposition systems.

where the protein-bound ends interact with each other to form a synaptic complex (FIGURE 14.4). The highly coordinated reactions in this nucleoprotein complex have been shown in a number of systems to involve catalysis *in trans*, where a protein bound to one end of the mobile element actually catalyzes the chemistry on the other end of the element within the nucleoprotein complex; that is, a transposase molecule bound to the left end of the element will actually cut on the right end of the element and vice versa. When the element inserts into a target DNA, the joining event to the target DNA almost always occurs at staggered positions on the sugar–phosphate backbone, leading to another distinctive feature of transposition called **target site duplication**. The staggered nature of the joining event is easier to think about if we consider the three dimensional double-helical structure of B-form DNA (Figure 14.4). Specifics in the recognition of each element's transposase and the local changes they impart when they bind the target DNA leads to a characteristic stagger in the position

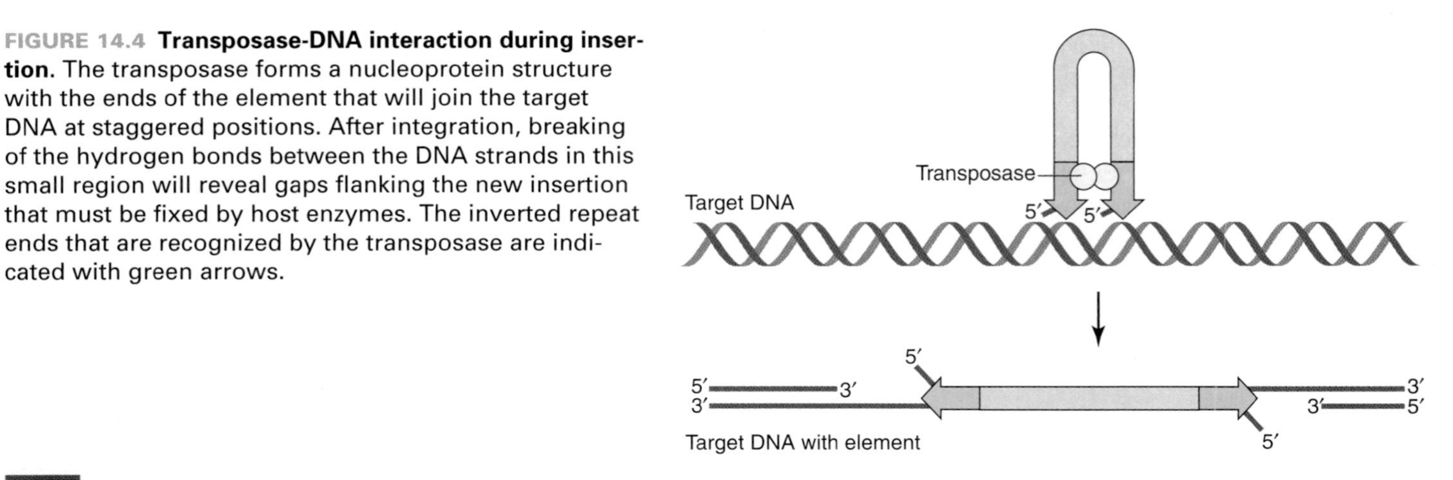

FIGURE 14.4 **Transposase-DNA interaction during insertion.** The transposase forms a nucleoprotein structure with the ends of the element that will join the target DNA at staggered positions. After integration, breaking of the hydrogen bonds between the DNA strands in this small region will reveal gaps flanking the new insertion that must be fixed by host enzymes. The inverted repeat ends that are recognized by the transposase are indicated with green arrows.

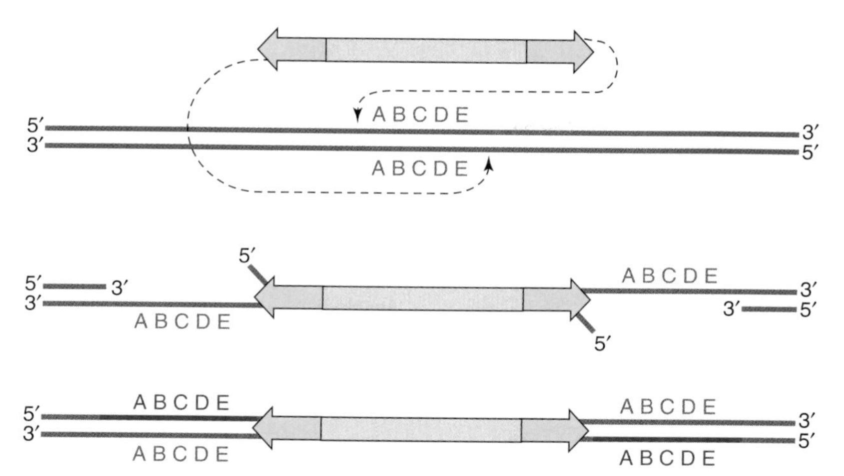

FIGURE 14.5 **Gap repair and target duplication.** Repair of the gaps flanking the new insertion event result in a short target site duplication. This is an effect of the staggered cuts made during insertion. The free 3′ end of the gap is used to prime repair with a host enzyme making the second strand of the small region indicated with "A-B-C-D-E." Host enzymes are also responsible for degrading any sequences brought from the donor site. Ligase is required to seal the remaining nick in the DNA.

of the joining events. Small gaps at the ends of the element will result after transposition because of the staggered joining events to the target DNA (Figure 14.4). Repair of the gaps by a host DNA polymerase will therefore result in a small target site duplication flanking the element (FIGURE 14.5). These directly repeated sequences that flank the ends of the element after transposition are often 2 to 10 bp in length and are characteristic for each element. In some cases, a few nucleotides at the 5′ ends of the element that are carried over from the position where the element previously resided also must be processed by host proteins (Figures 14.4 and 14.5). A host-encoded ligase is required to seal the remaining nicks in the DNA backbone.

Coordinated breakage and joining events occur during transposition.

The chemistry used by the DDE transposons is conserved and well understood. The transposases that catalyze this chemistry have conserved aspartic acid (D) and glutamic acid (E) residues, which are required for catalytic activity as evidenced by the loss of activity after site-directed mutagenesis. Crystal structures of bacterial transposases have further confirmed that the DDE residues coordinate divalent metals to catalyze the multi-step transposition reaction (FIGURE 14.6). At each end of the element within the synaptic complex, a multi-step process occurs to transfer DNA strands by a concerted breaking and joining reaction (FIGURE 14.7). The process begins when an activated water molecule cleaves a phosphodiester bond in one DNA strand. The newly generated 3′-OH group attacks the opposite DNA strand to form a **hairpin**. In the case of the bacterial transposon Tn*5*, the hairpin forms on the end of the element and results in the formation of a double-strand DNA break in the host chromosome DNA. Which DNA receives the hairpin and which DNA receives the double-strand break differs by element, however (see below). A second activated water molecule resolves the hairpin intermediate. In the case of the Tn*5* transposase, the resulting post-cleavage synaptic complex captures a target DNA molecule and strand transfer takes place by a

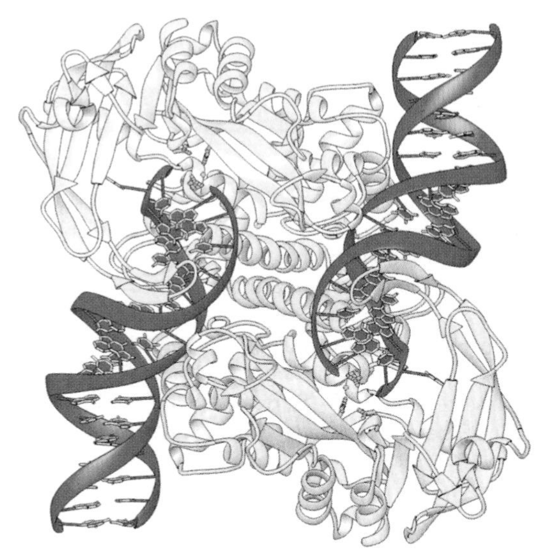

FIGURE 14.6 **Ribbon representation of the bacterial transposon Tn*5* transposase/DNA dimer as determined by X-ray crystallography.** The dimer subunits are colored in yellow and blue. The synapsed inverted repeats (20 bp) are shown in purple. The catalytic aspartic acid and glutamic acid residues of the DDE motif are shown in green in the ball-and-stick format. The associated Mn^{2+} ion is shown as a green dot. (Reproduced from D. R. Davies, et al., *Science* 289 [2000]: 77–85. Reprinted with permission from AAAS. Photo courtesy of Ivan Rayment, University of Wisconsin.)

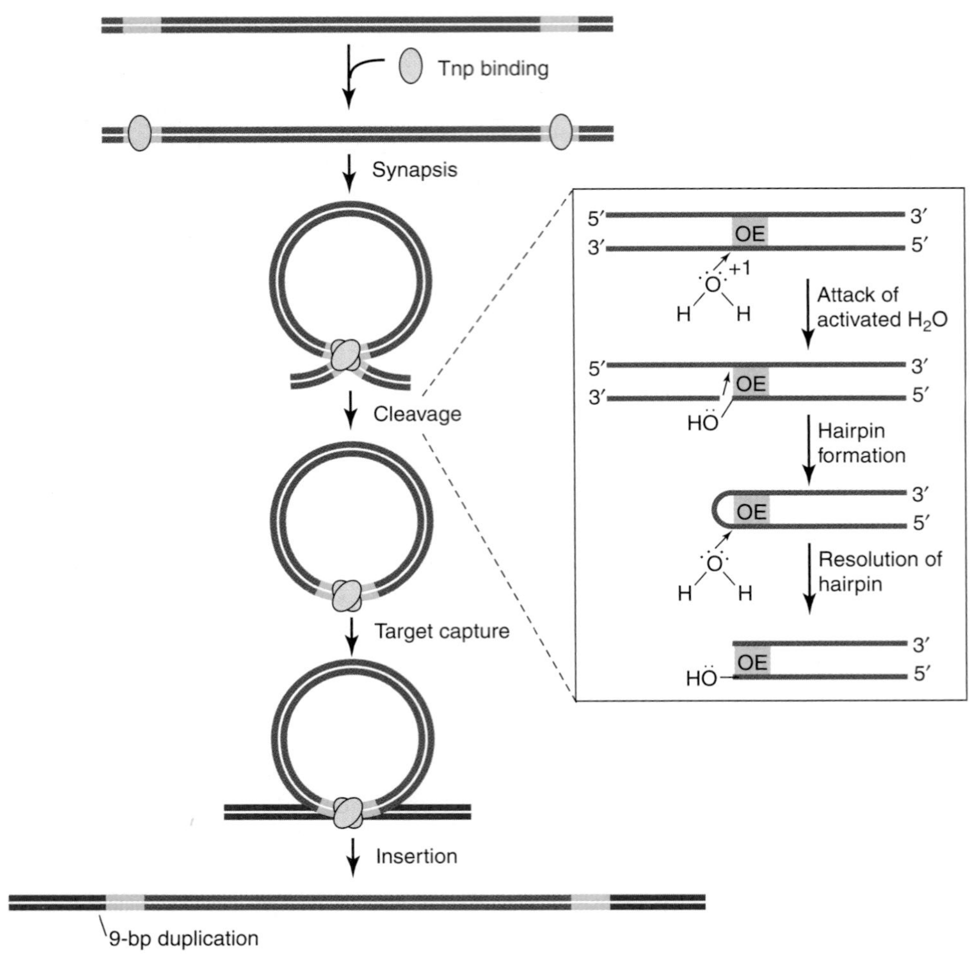

FIGURE 14.7 **Steps in the hairpin-mediated transposition process of bacterial transposon Tn*5*.** Transposase (light blue) binds inverted repeats (green) at the ends of the Tn*5* element to form a synapse where DNA is cleaved and joined to a target DNA. The underlying chemistry at the outer end (OE) of the element first involves cleavage at the 3′ ends of the element using an activated water molecule that liberates the 5′ ends of the element while forming a hairpin at the transposon ends (only one is shown). A second cleavage event resolves the hairpin and joins the element to a target DNA molecule. Staggered joining events made in the target DNA lead to gaps at the ends of the element. The repair of these gaps leaves the 9 base pair target site duplication that is characteristic of Tn*5* transposition. (Modified from *Curr. Opin. Struct. Biol.*, vol. 14, M. Steiniger-White, I. Rayment, and W. S. Reznikoff, Structure/function insights into Tn5 transposition, pp. 50–57, copyright 2004, with permission from Elsevier [http://www.sciencedirect.com/science/journal/0959440X].)

transesterification reaction in which the transposon's 3′-OH groups attack the target DNA's phosphodiester bonds in a staggered fashion. The transposon integrates by covalent bond formation between the 3′-OH groups of the transposon ends and the 5′-phosphate groups of the target DNA. When the synaptic complex captures the target DNA depends on the transposon. In some cases, elements capture the target DNA before cleavage while in other cases target DNA capture can happen either before or after cleavage.

Some elements do cut-and-paste transposition, where the element is directly moved to a new location.

The basic reaction carried out by the transposases of the DDE transposons has been harnessed in very different ways in nature to result

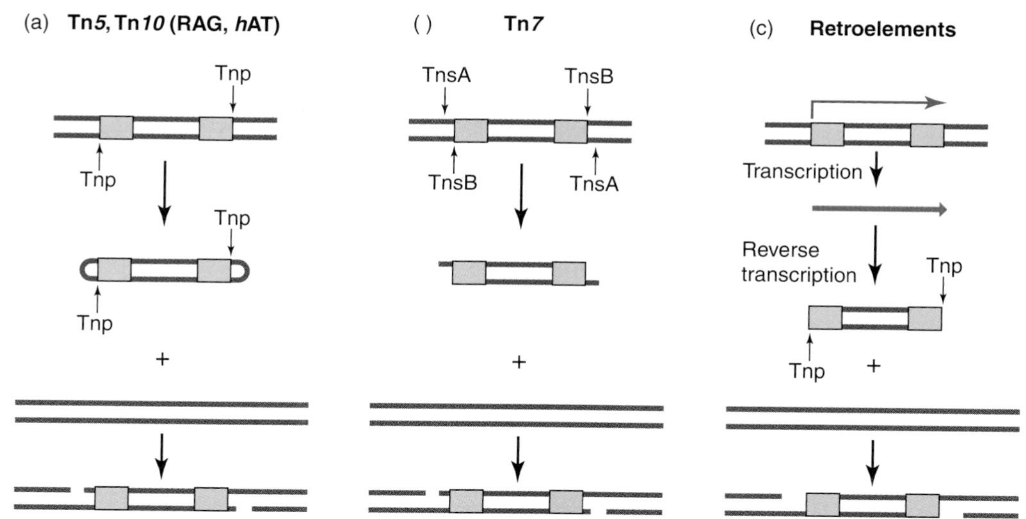

FIGURE 14.8 **Cut-and-paste transposition.** Cut-and-paste transposition occurs by multiple mechanisms and occurs by a related process with the movement of retroelements. (a) The bacterial transposons Tn*5* and Tn*10* (shown with green inverted repeats) break the DNA backbone using a transposase (Tnp) at the 3′ ends of the element. The 5′ end of the element is liberated when joined as a hairpin at the end of the element. An additional break resolves the hairpin structure when the element is joined to a target DNA (shown in red). A similar process happens with the RAG recombinases and the *hAT* family transposases, but the hairpin is formed on the host end of the DNA (or the equivalent end in V(D)J recombination). (b) The bacterial transposon Tn*7* uses its TnsB protein to join the 3′ ends of the element directly to the target DNA (shown in red). A second subunit of the transposase is responsible for nicking at the 5′ ends of the element to completely free the element from the donor DNA. (c) Retroelements use a similar mechanisms as the cut-and-paste transposons except a cDNA element is utilized that is made from an RNA copy of the element. (Adapted from N. C. Craig, et al., eds. *Mobile DNA II*. ASM Press, 2002.)

in DNA double-strand breaks that completely free the element from the DNA (FIGURE 14.8) or joining events that occur directly to the target DNA (see replicative transposition section below). It has been shown in a variety of systems that the transposase can form DNA double-strand breaks at the ends of the element by joining one strand of DNA to its complementary DNA strand forming a hairpin at the end of the element (Figures 14.7 and 14.8a). If these hairpins are at the ends of the transposon, they can be opened in a subsequent reaction where the activated 3′-OH allows the joining reaction with the target DNA (Figures 14.7 and 14.8a). This reaction pathway results in the element excising from one DNA site and inserting into a new DNA site. The reaction pathway is sometimes referred to as **cut-and-paste transposition** because the element physically moves to a new position. Cut-and-paste transposition has been shown to occur both with bacterial and eukaryotic transposons and in the recombination steps carried out by the RAG proteins responsible for *V*(*D*)*J* recombination. A byproduct of this hairpin-forming reaction is a DNA double-strand break that remains at the old position of the element and, presumably, must be repaired (see Chapter 13 and below).

While other elements use the reaction mechanism described above to do cut-and-paste transposition, they apparently can do so without forming a hairpin structure at the ends of the element. For example, the bacterial transposon Tn7 makes breaks at the 3′ ends of the element while directly joining to a target DNA with the TnsB protein, which is one subunit of the transposase protein (Figure 14.8b). While this reaction would still result in the 5′ ends of the element remaining joined to the donor DNA molecule, a second subunit of the transposase, called TnsA, is capable of nicking at the 5′ ends of the element to form the DNA double-stand breaks required for cut-and-paste transposition. Interestingly, the TnsA crystal structure indicates a folding pattern that differs from that of DDE family transposases, but similar to that of a restriction endonuclease. Recall that restriction endonucleases cut at specific DNA sequences only when they lack methylation on a certain residue and function as a rudimentary protection system from invading DNA. Other transposons have transposases that cause breaks at the 3′ ends of the element without cleaving the 5′ ends of the element (i.e., without using a hairpin structure or an accessory protein like TnsA). These transposases can use extensive DNA replication to leave one copy of the element behind in a process called **replicative transposition** that is described below. Eukaryotic retrotransposons use a similar chemistry as the cut-and-paste transposons from bacteria (Figure 14.8c). However, retrotransposons do not lose the original element. Instead, an RNA copy of the element is subjected to reverse transcription and this DNA entity is inserted into a new target site.

Transposition during DNA replication and host DNA repair allow cut-and-paste elements to increase in copy number.

Because all DNA elements need to multiply to be maintained over time, it stands to reason that transposons using cut-and-paste transposition need a mechanism to ensure more copies of it are made if it is to persist. This likely explains why some bacterial cut-and-paste transposable elements have evolved molecular mechanisms to upregulate transposition with DNA replication (FIGURE 14.9). One possibility is that the ability to transpose from behind a DNA replication fork (i.e., the region that has been replicated) to a position ahead of the replication fork would give the element the opportunity to go from two copies in the cell to three copies in the cell (Figure 14.9a). The double-strand break can be repaired by using any of a number of mechanisms. Another possibility

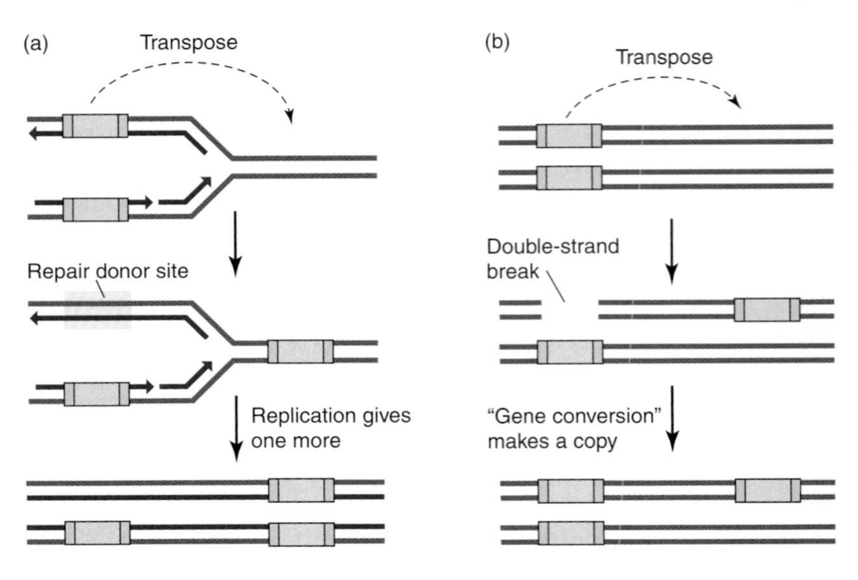

FIGURE 14.9 **Regulating transposition with DNA replication can allow transposons that utilize cut-and-paste transposition to increase in copy number.** (a) If a transposon moves to a new location soon after being replicated to a position ahead of DNA replication, the copy number can increase. (b) If two copies of the chromosome reside in the cell, repair of the DNA break left in the donor site can be templated by a sister chromosome. This will result in a "copy" of the element replacing the one lost by cut-and-paste transposition.

is for transposition to take place after DNA replication (Figure 14.9b). By regulating transposition to occur after DNA replication, the element may be able to ensure that repair involves a recently duplicated section of the chromosome and hence allows restitution of the element that moved to a new site. Gene conversion, therefore, will reproduce the original transposition event at the donor site, and transposing immediately after replication likely facilitates this process by ensuring an intact chromosome is available and nearby.

Transposons are found at various levels of complexity in bacteria.

Insertion sequences (IS) have adapted in nature so that they can carry additional genetic information by various means (FIGURE 14.10). Insertion sequences can form **composite transposons** where two insertion sequence elements flank genes that can benefit the bacterial host, such as genes encoding antibiotic resistance. The outermost ends of the two IS elements will be used for transposition and allow the movement of the structure (two IS elements and the intervening DNA) (FIGURE 14.11). Over time, one of the cognate elements can lose its autonomy and subsequently require the other IS element to mobilize the pair of elements. In addition, considerable adaptation occurs with these elements,

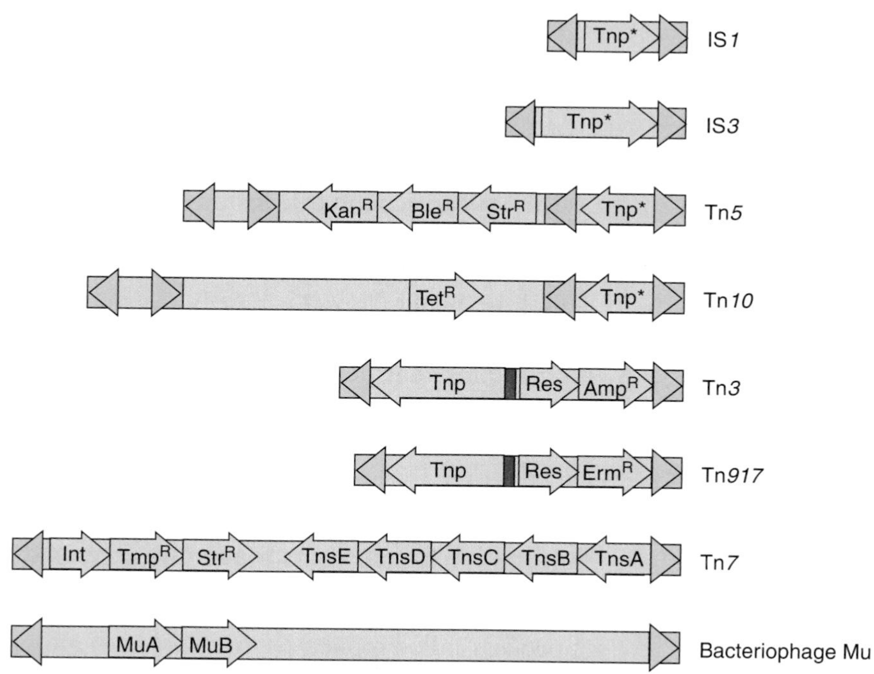

FIGURE 14.10 **Examples of transposable elements found in bacteria.** The simplest transposable elements are the IS elements, containing only a transposase (Tnp) flanked by inverted repeats (shown in green). In many cases truncated translation products are utilized for regulating transposition (indicated as a Tnp*). Tn*5* and Tn*10* are composite transposons where two IS elements surround and mobilize additional genes. Tn*3* and Tn*917* are replicative transposons that require extensive DNA replication and a resolution site (red boxes) and a resolvase (Res) to complete transposition. Tn*7* is an elaborate transposon that utilizes five proteins to control transposition (TnsA, TnsB, TnsC, TnsD, and TnsE) and also contains an integron system with its cognate integrase (Int). Bacteriophage Mu utilizes replicative transposition to replicate its genome. Transposition requires two proteins MuA and MuB. The indicated transposons encode resistance to many antibiotics; kanamycin (Kan^R), bleomycin (Ble^R), streptomycin (Str^R), tetracycline (Tet^R), ampicillin (Amp^R), erythromycin (Erm^R), and trimethoprim (Tmp^R) resistance. The inverted repeats that are *cis*-acting entities for transposition are shown in green (see text for details).

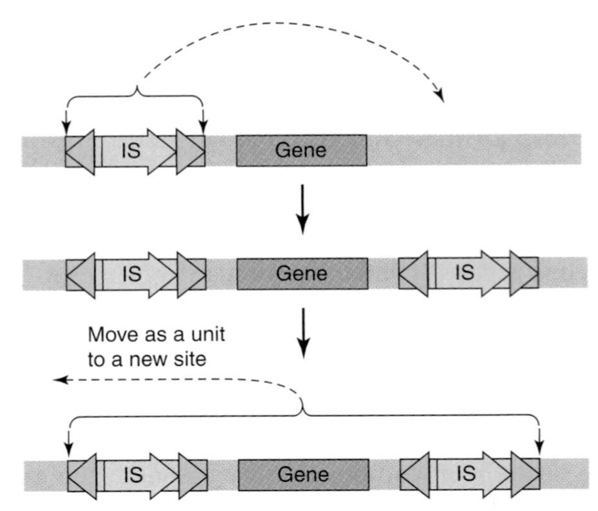

FIGURE 14.11 **Gene mobilization by composite elements.** Composite elements result when two IS elements transpose using the outside ends of the elements to mobilize both elements and intervening DNA. These elements co-adapt in the case of Tn*5* and Tn*10* to work as a functional unit with one of the elements losing its capacity to express transposase. The green triangles indicate the inverted repeats that are *cis*-acting entities for transposition. (Adapted from L. Snyder and W. Champness. *Molecular Genetics of Bacteria, Second edition*. ASM Press, 2003.)

producing regulatory systems that coordinate transposition with DNA replication (see below). Complex noncomposite transposons also have evolved where functions important for the host such as antibiotic resistance can accumulate inside these elements (Figure 14.10). Transposons that encode functions that benefit the host can help maintain a parasitic element. Many elaborate elements have evolved this strategy. This host-benefit process can be further enhanced by gene systems called **integrons** that allow the acquisition of gene cassettes that code for proteins that perform functions that benefit the host bacterium. The gene cassettes move by a conservative site-specific recombination mechanism, catalyzed by a tyrosine recombinase called integrase (int) (see below).

Replicative transposons leave one copy of the element at the donor site.

A process called replicative transposition allows a transposon to move to a new site without technically being excised from its original site. After replicative transposition, there is one mobile element at the old site and another at a new site. The replicative transposition process uses a combination of extensive DNA replication and conservative site-specific recombination to compensate for never cutting at the 5′ ends of the transposon. Similar to the process carried out by Tn7, the DDE transposase joins the 3′ ends of the element directly to a target DNA (FIGURE 14.12). In fact, if the 5′ end breaking activity is inactivated in the TnsA protein, Tn7 is capable of forming cointegrates where the

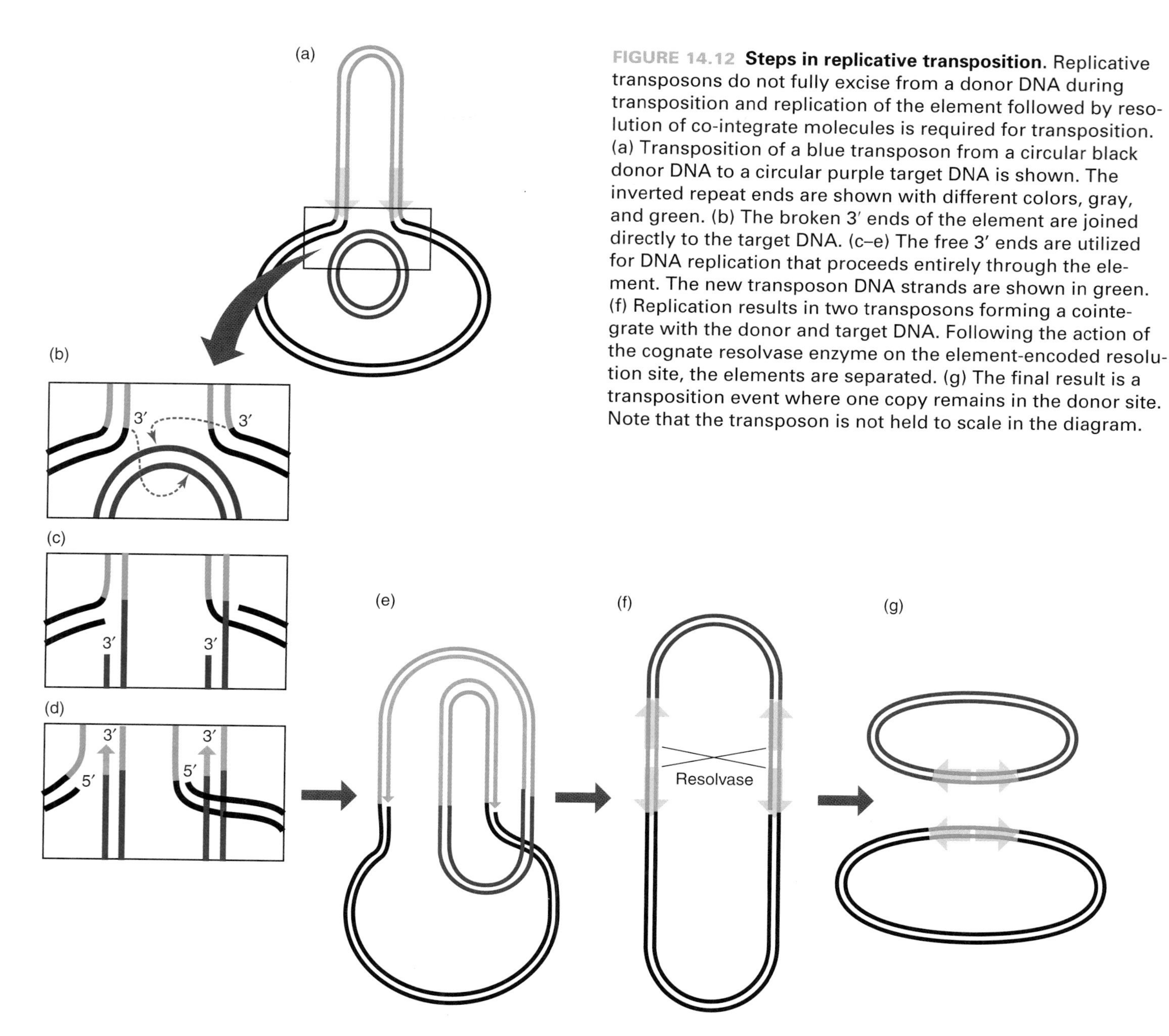

FIGURE 14.12 **Steps in replicative transposition.** Replicative transposons do not fully excise from a donor DNA during transposition and replication of the element followed by resolution of co-integrate molecules is required for transposition. (a) Transposition of a blue transposon from a circular black donor DNA to a circular purple target DNA is shown. The inverted repeat ends are shown with different colors, gray, and green. (b) The broken 3′ ends of the element are joined directly to the target DNA. (c–e) The free 3′ ends are utilized for DNA replication that proceeds entirely through the element. The new transposon DNA strands are shown in green. (f) Replication results in two transposons forming a cointegrate with the donor and target DNA. Following the action of the cognate resolvase enzyme on the element-encoded resolution site, the elements are separated. (g) The final result is a transposition event where one copy remains in the donor site. Note that the transposon is not held to scale in the diagram.

DNA elements are physically joined to one another, presumably using these same steps (Figure 14.8).

The mechanism of replicative transposition is diagrammed in Figure 14.12. Transposition from a donor DNA shown in black occurs into a target DNA shown in purple (Figure 14.12a). DNA from the original element is shown in blue. No hairpins need to be invoked in the process of replicative transposition; the 3′ ends of the element are joined directly to the target DNA (Figure 14.12b). Because there are no breaks at the 5′ ends of the element, the donor and target DNA are covalently linked (Figure 14.12c). The free 3′ ends are recognized and extended by the host DNA polymerase (Figure 14.12d and e). DNA polymerase extends both 3′ ends of the donor DNA to allow replication of the entire transposon (see the green lines in Figure 14.12). The result of this extensive replication is two transposons, with one old and one new strand of DNA, forming a cointegrate between the donor and target DNA (Figure 14.12e). Transposons that use replicative transposition contain a resolution site and a cognate resolvase that will allow the separation of the donor and target molecules (Figure 14.12f and g).

When multiple copies of a transposon with a resolvase system reside in the genome, there can be consequences for genome stability. If these resolvase containing elements are in the same chromosome, recombination between resolution sites can lead to loss of the intervening sequences or inversion of the intervening DNA region depending on the orientation of the resolution sites used by the systems. Alternatively, if the resolvase containing transposons reside on different DNA molecules, for example, a plasmid and the chromosome, a plasmid can now integrate into the chromosome at the resolution sites of these genetic elements. This integration explains how the F factor integrates into the bacterial chromosome (see Chapter 7). If a mobile plasmid integrates into the chromosome using transposon-encoded resolution sites on the plasmid and chromosome or simply the homology provided by two transposons, the subsequent mobilization of the plasmid will transport segments of the bacterial chromosome to a new bacterial host. This exact process led to the original identification of the Fertility (F) plasmid. Chromosome mobilization with an integrated F factor allowed Joshua Lederberg to show that bacteria had genetic information and evolved in the same way as plants and animals.

The barrier between moving as a replicative transposon or a cut-and-paste transposon may be very low. For example, the bacterial insertion sequence IS*903* has the unusual ability to transpose by the cut-and-paste mechanism and by replicative transposition. The majority of products are formed by the cut-and-paste mechanism. A single mutation in the nucleotide immediately flanking the transposon can tip the balance from moving as a cut-and-paste element to moving as a replicative element. This could allow a diversity of recombination reactions each with a different evolutionary outcome.

Another way that the DDE recombinase has adapted to move DNA elements using extensive DNA replication is found with IS*3* family elements. While the details of the reaction steps are yet to be elucidated, these elements use an asymmetric reaction where the break takes place only at one 3′ end of the element. The newly broken end is then joined to the same DNA strand outside the other end of the element liberating a single-stranded copy of the element. Replication must be used to make the complementary strand of the element at the old site and to establish the full double-stranded element for integration into a new site. Because no cointegrate is formed in this process a resolvase and resolution site are not needed after the element moves.

Transposons in eukaryotes are mechanistically similar to bacterial transposons.

Transposons found in eukaryotes can be very similar to those found in bacteria with some modest adaptations, such as the presence of nuclear localization sequences in the transposase (FIGURE 14.13). A large family of elements called the "*hAT*" transposons is found in a wide variety of eukaryotes including fungi, plants, and invertebrate and vertebrate animals. The term *hAT* derives from the first letters in the names of the three founding members of this family, the *hobo* element in *Drosophila*, the *Ac* element in maize, and the *Tam3* element in the snapdragon plant. The transposition mechanism used by these elements is of particular interest because it reveals striking mechanistic similarities with the RAG recombinases used in *V(D)J* recombination. The mechanism of *hAT* element transposition was elucidated with the *hAT* family member element *Hermes* found in the housefly, *Musca domestica*. Work with a reconstituted *Hermes* transposition system revealed that the cut-and-paste element excised from the donor backbone via the formation of a hairpin, something that was broadly similar to what had been shown for bacterial elements Tn*10* and Tn*5* (Figure 14.8). However, an interesting difference was found between the *Hermes* element and Tn*10*/Tn*5*. With the *Hermes* transposase, the hairpin was not formed at the ends of the element, but instead on the donor DNA leaving a clean break on the ends of the element (FIGURE 14.14). This observation is of interest because this is the mechanistic progression that is found during *V(D)J* recombination catalyzed by the RAG proteins (Figure 14.14). Of further interest, the specific nucleotide position that was used to join the ends of the *Hermes* element to the target DNA is highly conserved in *hAT* elements, and is also used by the RAG recombinase during *V(D)J* recombination. Furthermore, multiple similarities exist in the distribution of the essential DDE amino acids and other amino acids that are believed to be important for altering the structure of B-form DNA to allow formation of the hairpin structures. *In vitro* reactions with the *Hermes* transposase indicated that the 8 bp target site duplication occurs that is characteristic of the *hAT* family. After *Hermes* transposition *in vivo* the broken donor site needs to be repaired, likely by a mechanism similar to that for *V(D)J* recombination (see Chapter 13).

The crystal structure of the *Hermes* transposase reveals a catalytic domain similar to the one that is found in the other DDE type proteins and retroviral integrases (FIGURE 14.15). The crystal structure also reveals the presence of a conserved domain within the *hAT* transposases that is responsible for allowing the transposase to form a multimeric structure. Interestingly, the evidence for the multimeric nature of the protein came from a complexity in the purification of the *Hermes* transposase. Purification of the *Hermes* transposase was bedeviled by the presence of an ~100 amino acid peptide that contaminated the protein preparations. Further analysis indicated that the

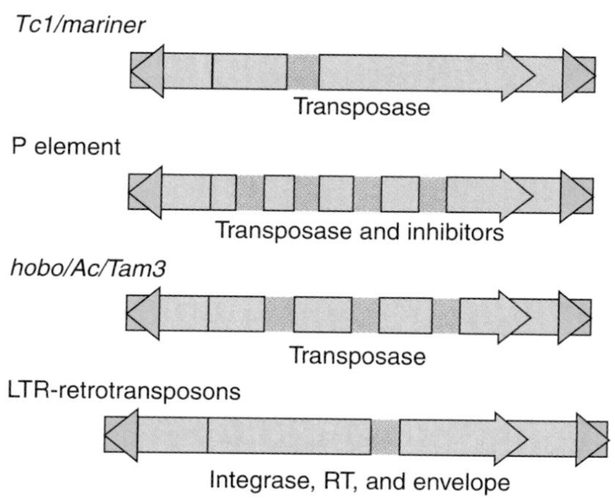

FIGURE 14.13 **Examples of transposable elements found in eukaryotes.** Pink sections indicate the translation product with its associated function listed below. The gray squares indicate the introns within the mRNA encoding sequences. The green triangles indicate the inverted repeats that are *cis*-acting entities for transposition. RT indicates reverse transcriptase. (Adapted from F. Bushman, *Lateral DNA Transfer: Mechanisms and Consequences*. Cold Spring Harbor Laboratory Press, 2002.)

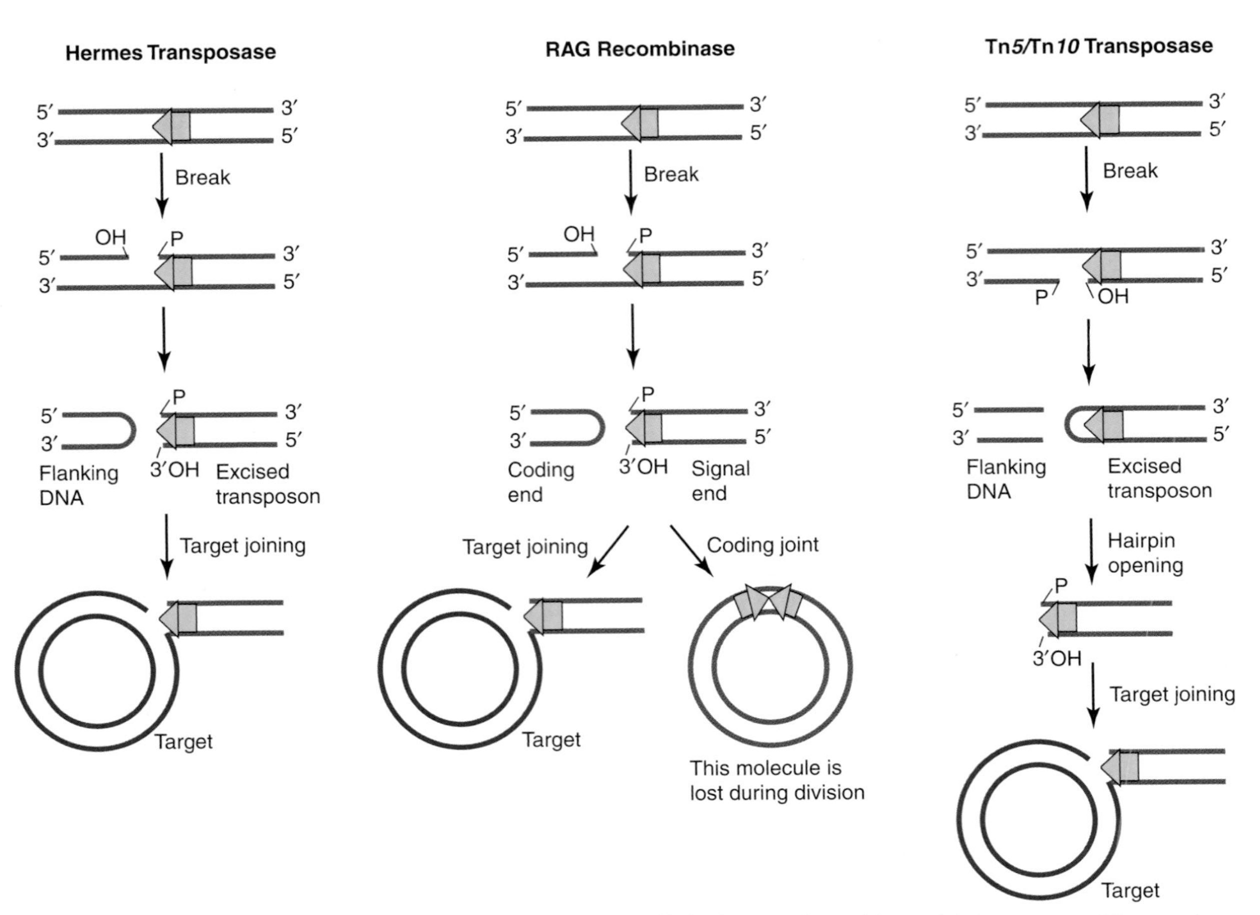

FIGURE 14.14 **Transposition systems that use DDE transposase.** Hairpins are found in multiple transposition systems using DDE transposase. This figure focuses on one end of the mobile element as it is broken and subsequently joined to a small circular target DNA (shown in red). With the *Hermes* element and in *V(D)J* recombination hairpins are formed in the flanking DNA, or the correlating region, the coding DNA, during *V(D)J* recombination. In the case of the *Hermes* transposon, the element ends can then carry out transposition. In the case of *V(D)J* recombination the DNA region containing the inverted repeats circularizes and is lost from the cell by division. The bacterial transposons Tn*5* and Tn*10* use a similar procedure, but the hairpin is formed on the transposon end of the DNA. (Adapted from L. Zhou, et al., *Nature* 432 [2004]: 995–1001.)

contaminating fragment was actually a fragment of the transposase, which was likely required to allow purification of the transposase. This fragment was presumably nonspecifically cleaved from other monomers during isolation and purification. Further studies showed that the protein fragment contains a multimerization domain. Computer reconstruction of the monomers supports the view that the *Hermes* transposase operates as a hexamer (Figure 14.15). This conclusion receives support from work showing that when *Hermes* protein preparations are separated by size, a large structure consistent with the hexamer is the active form for *in vitro* transposition. Furthermore, when the predicted multimerization domain was subjected to mutational analysis it was found to be required for multimer formation. The hexameric structure has channels leading to the active site, which are lined with basic residues that could in theory accommodate DNA. It is unclear how widespread hexameric transposases are and if this structural feature is a characteristic of all eukaryotic transposases.

Diverse systems allow transposition to be regulated.

While transposons need to move to reproduce and to distribute to new hosts, the process of transposition inevitably jeopardizes the host organism. Transposition at many places in the genome will inactivate important, if not essential, genes. Transposition could activate genes also when insertion events occur before host genes that are subsequently expressed by element-encoded promoters and enhancers. Most elements, therefore, regulate transposition in some way. As described above, the Tn*5* and Tn*10* elements probably upregulate transposition as a mechanism to make additional copies of the element. These elements upregulate transposition when they have been replicated by responding to the methylation state of the DNA. In bacteria like *Escherichia coli* and its relatives, the DNA adenine methylase protein will methylate adenine bases in the sequence GATC (see Chapters 9 and 12). Normally, all GATC sites in the chromosome on both DNA strands will be methylated on the adenine at these sites. After DNA replication, however, the chromosome will be transiently hemimethylated where the old strand of DNA contains a methyl group at the GATC site, but the newly replicated strand has not yet had a chance to be methylated. The hemimethylated state of the element increases the frequency of transposition in two distinct ways in Tn*5* and in other elements. For one, the end sequences are better bound by the transposase when they are hemimethylated. In addition, promoters within the elements are also sensitive to the hemimethylation state. The fully methylated state of the element favors transcription from promoters that will subsequently lead to translation of N-truncated forms of the element's transposase protein that act as inhibitors of transposition (FIGURE 14.16). After replication the element will be hemimethylated, favoring the production of a full-length transposase that catalyzes transposition to new positions in the cell. Tn*5* is also sensitive to the

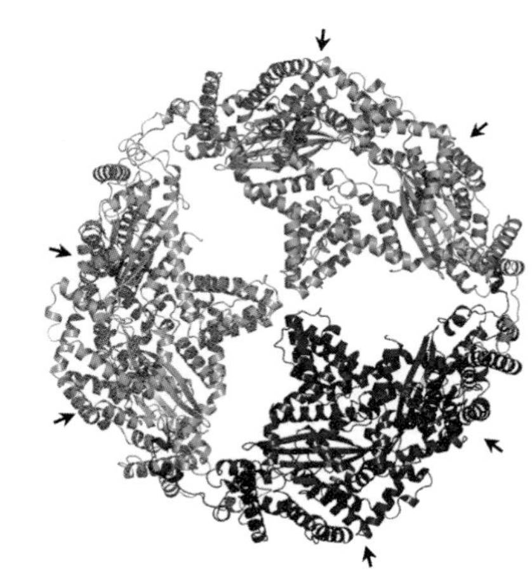

FIGURE 14.15 **The hexameric form of the *Hermes* transposase.** The *Hermes* transposase crystal structure was used to model the hexameric complex of the proteins. The DDE active sites are on the periphery of the structure and are indicated with black arrows. (Reprinted by permission from Macmillan Publishers Ltd: *Nature Structural and Molecular Biology*. A. B. Hickman, et al., 12: 715–721, copyright 2005. Photo courtesy of Fred Dyda, NIDDK, National Institutes of Health.)

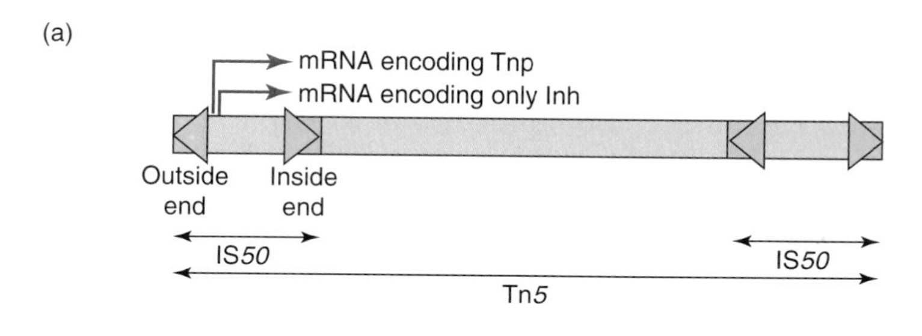

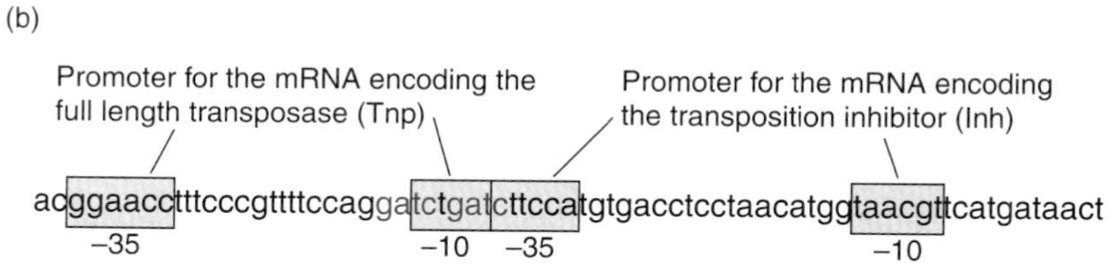

FIGURE 14.16 **Tn*5* regulates transposition with DNA replication.** (a) Tn*5* has promoters within one of its IS*50* elements that transcribes mRNAs encoding either a full-length transposase or an N-truncated inhibitor of transposition. (b) Transcription from the full-length transposase promoter is favored after DNA replication via the GATC sites (see text for details).

concentration of DnaA, the *E. coli* initiator protein (see Chapter 9). This sensitivity likely involves a DnaA box found in the outer ends of the element that is recognized by the DnaA protein, but the exact mechanism of this behavior remains unclear.

Other elements appear to respond to host factors thereby allowing these genetic elements to repress or stimulate transposition in response to the regulatory networks of the host. The ability to monitor host function is therefore similar to the behavior of bacteriophage λ in how it coordinates its replication with the incidence of DNA damage using the SOS response in bacteria (see Chapter 12). Finally, it has been suggested that transposons can also sense and avoid transposition into highly transcribed genes in many systems. In bacteria avoiding highly transcribed regions may stem in part from steric interactions in these regions. The histone-like protein, H-NS, is also known to be important in regulating the frequency and position of horizontal transfer. In the case of some bacterial transposons H-NS is known to interact directly with the transposase•DNA complex to encourage transposition. In eukaryotes, histones also appear to play a role in determining when and where transposition occurs as well as the host's ability to silence these elements.

Eukaryotic elements also have ways of regulating transposition that can involve a truncated form of the transposase. One example of this type of regulation is found with the P element in *Drosophila* (Figure 14.13). The P element transposase contains four exons, the region of the RNA message that encodes the protein. Exons are assembled together in one contiguous messenger RNA in a process called splicing (see Chapter 19). A host-encoded splicing repressor protein can inhibit removal of the noncoding intron region between exons 2 and 3. The action of the splicing repressor leads to the production of a truncated transposase that is a repressor of P element transposition. This mechanism is broadly similar to the way that Tn*5* expresses a truncated transposase product that inhibits transposition (Figure 14.16). Because the host splicing repressor is only expressed in somatic cells, transposition is restricted to the germ line where the full-length transposase is made. Host mechanisms also control the level of transposition in multicellular animals. For example, a high number of one type of mobile element can activate special systems found in many multicellular organisms that will not allow these elements to be expressed (see Chapter 21).

Most transposons prefer DNA targets that are bent.

One common characteristic of many transposases and retroviral integrases seems to be a preference for insertion into bent DNAs. This preference is unlikely to be an adaptation and probably instead arises from the inability of the recombinase to bring an unbent target DNA into a confirmation that is competent to engage in the reaction. With transposon Tn*10*, *in vivo* and *in vitro* studies show that preferred DNA targets are inherently bent. Furthermore, Tn*10* transposase mutants that show less of a preference for such sequences are inherently better at bending DNA substrates. Special host-encoded DNA bending proteins, like integration host factor (IHF), are essential for transposition with some bacterial transposons (including Tn*10*). IHF was originally identified by its requirement for the conservative site-specific recombination process of bacteriophage λ integration into the chromosome (see below). Proteins found in eukaryotes that bend DNA stimulate RAG-mediated *V*(*D*)*J* recombination *in vitro* and may play a similar bending role to that played by IHF with some bacterial transposases.

Transposons can target certain sequences.

The sites where a transposon can insert probably influence its frequency of transposition. Successful transposable elements probably must set a balance between the frequency of transposition and the level of control they have over where they insert so they do not compromise the host. Part of the explanation for why certain specific DNA sequences are targeted may stem from the bent nature of these DNAs. However, multiple examples exist where consensus target site sequences are favored, suggesting that the transposase itself has a preference for certain sequences. These sequences are usually of modest size and specificity. For example, preferred Tn*10* insertion sites have a 5′-GCTNAGC-3′ Tn*10* or a slight variation of this sequence. Tn*5*/IS*50* preferentially transposes into sites that have a 5′-AGNTYWRANCT-3′ sequence (where R = purine, Y = pyrimidine, W = adenine or thymine, and N = any nucleotide) or a slight variation of this sequence. Comparison of a number of insertion sites favored by the element IS*903* identified an insertion site with a larger palindromic sequence that was experimentally confirmed by demonstrating that IS*903* transposes into this insertion site in plasmids. The Tc/*mariner* elements that are present in a variety of multicellular organisms have an almost absolute requirement for insertion site with the 5′-TA-3′ dinucleotide sequence. One bacterial element was shown to preferentially insert into a very large consensus sequence that is found in repetitive DNA in *E. coli* and other bacteria. The preference of this element for repeat DNA might serve as a mechanism to limit insertions into essential genes (which do not contain these repeat sequences). The strict use of a rare target site, however, will be at the expense of lateral transfer because elements capable of transfer between bacteria will be unlikely to have the rare consensus sequence.

Some transposons target specific molecular processes.

Transposons have also adapted to target certain biological processes. The Tn*5053* family transposons are replicative transposons found in bacteria that have adapted to selectively target certain plasmids. The impetus to test this idea came from the observation that the Tn*5053* elements were frequently found inserted within certain regions on plasmids called **resolution sites**. Resolution sites are the site where conservative site-specific recombination can occur to convert dimer plasmids into monomers (see below). These resolution sites are important for the stable maintenance of circular plasmids because dimer plasmids arise frequently via recombination. In the laboratory it was revealed that a Tn*5053* family element would direct transposition into the resolution sites only when the cognate resolvase that acts at these sites was present. Interestingly, insertion of the transposon into these resolution sites will also inactivate the site on the plasmid. While loss of the endogenous plasmid resolution site would normally destabilize the plasmid, this is not the case because the Tn*5053* elements encode their own resolvase because they are replicative transposons. The transposon-encoded resolvase, therefore, can replace the resolvase that is normally encoded by the plasmid. This strategy appears to have two benefits: (1) It allows the transposon to target transposition into plasmids. Inserting into plasmids is beneficial because plasmids have a higher copy number than do chromosomes, can frequently move to other bacteria, and are rarely essential for the host. (2) The plasmid now is likely to be reliant on the transposon-encoded resolution system and therefore much less likely to lose the element by deletion.

Examples also exist of transposons in eukaryotes that target transposition into DNAs that undergo certain biological processes. Ty elements are LTR retrotransposon elements that are common in *Saccharomyces cerevisiae*. Two of these elements, Ty1 and Ty3, target transposition to sites that are just before genes transcribed by RNA polymerase III (see Chapter 20). The targeting mechanism used by these elements must differ given that the distribution of the insertion events is markedly different between Ty1 and Ty3. Both Ty1 and Ty3 probably benefit from directing transposition into a region of the chromosome that is largely devoid of genes and sequences that would otherwise act as promoters and therefore is unlikely to be essential. Similarly, the yeast LTR retrotransposon Ty5 can direct transposition into regions of the chromosome near the telomeres (see Chapter 10) and mating loci (see Chapter 13) in *S. cerevisiae* that use a special system that decreases expression of genes in these regions. Finally, the Tf element of *Schizosaccharomyces pombe* are targeted to the 5′-end of a gene that codes for a protein.

Some elements have evolved the ability to choose between certain target sites.

The bacterial transposon Tn*7* is particularly adapted to choose certain target sites through the use of multiple target site selecting proteins. Tn*7* uses five transposon-encoded proteins for transposition (Figure 14.10). These proteins allow Tn*7* to target two very different types of target sites. Three of the Tn*7* proteins, TnsA, TnsB and TnsC, are required for all transposition events, but will not function without one of two target site selecting proteins TnsD or TnsE (FIGURE 14.17). The TnsD protein is a sequence-specific DNA binding protein that is able to recognize a single insertion site found in the bacterial chromosome called its attachment site (*attTn7*). The region within *attTn7* that is bound by TnsD is within the C-terminal encoding region of the highly conserved *glmS* gene that codes for the enzyme L-glutamine: D-fructose 6 phosphate aminotransferase. Cleverly, even though sequences within the *glmS* gene are recognized by TnsD, actual transposition is directed

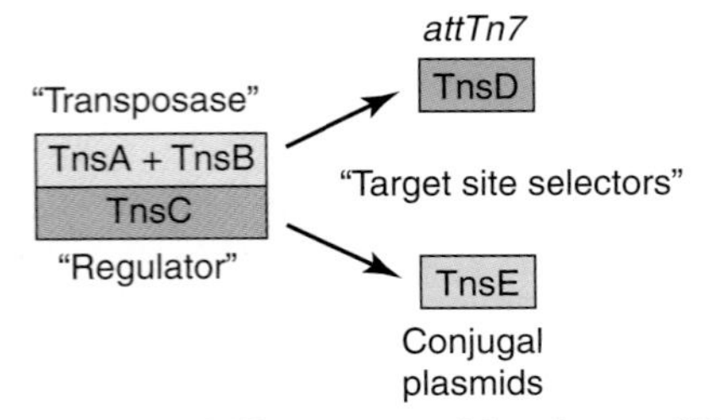

FIGURE 14.17 Different combinations of Tn*7*-encoded proteins allow the element to choose between certain target sites.

to a single location about 50 bp away within the transcriptional terminator of the gene preventing any detectable adverse effect on the bacterium. Tn*7* relatives are surprisingly widespread and almost found exclusively in the *attTn7* site in the host bacterium. Of further interest, the human genome contains *glmS* genes that will also be used as a TnsABC+D target when they are introduced into an *E. coli* strain! While it is highly unlikely that there is a biological reason why Tn7 would target human *attTn7*, it is probably true that evolution favored a target site that would be found in virtually any bacterium. The ability to direct transposition into the *attTn7* site is of great value because it virtually eliminates the possibility of transposition into essential host genes. Probably as an offshoot of this adaptation, transposition with the TnsD pathway occurs at a very high frequency, approximately 1,000-fold higher than transposition found with other elements.

Tn*7* has a second pathway of transposition that can direct transposition to specific plasmids. The TnsE protein preferentially directs transposition into mobile plasmids, called conjugal plasmids, when they move between bacteria. The TnsABC+E proteins can recognize a form of DNA replication that is associated with the process used by plasmids when they move between bacteria (FIGURE 14.18). TnsE recognizes multiple features of lagging-strand DNA replication, including an interaction with the β clamp (see Chapter 9). Plasmids and some bacteriophage use rolling circle DNA replication which appears to allow Tn*7* to recognize DNA molecules that can transport the transposon to new bacterial hosts. Once in a new host bacterium, Tn*7* can use the TnsD pathway to quickly and safely move into the new host's *attTn7* site. The continued use of the TnsABC+D and the TnsABC+E

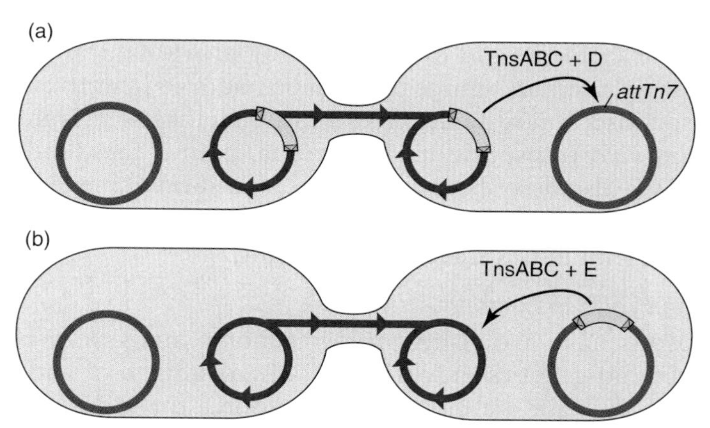

FIGURE 14.18 **Tn*7* transposition.** Tn*7* transposition is adapted to maximize dispersal while minimizing the chances of inactivating host genes. (a) When Tn*7* enters a new host on a conjugal plasmid or by some other mechanism (not shown) it can quickly insert into the *attTn7* site using the TnsABC+D proteins. (b) Once in the *attTn7* site, Tn*7* can reside without harm to the host until a conjugal plasmid enters the cell. Conjugal plasmids specifically stimulate TnsABC+E transposition events and the overwhelming majority of these insertions occur into the conjugal plasmid. The conjugal plasmid is shown as a red circle depicted with arrows to show unidirectional DNA replication. The movement of Tn*7* is shown with a black arrow. (Adapted from J. E. Peters, and N. L. Craig, *Nat. Rev. Mol. Cell Biol.* 2 [2001]: 806–814.)

would presumably allow Tn7 to quickly disseminate to new hosts without the risk of inserting into essential bacterial genes.

Transposons are important tools for molecular genetics.

General Attributes of Transposons as Tools

Transposons have long been important tools for classical genetics in a variety of organisms. It would also be hard to overemphasize, however, the importance of transposons in the post-genomic era. As more genome sequences become available, it remains clear that we do not know the function of many genes. Transposons offer an important tool for inactivating genes or otherwise manipulating chromosomes to understand gene function. As natural mutagens, transposons can be used to activate or inactivate genes. One of the most important advantages for using a transposon as a mutagenesis tool is the ability to locate its position by using genetic analysis or polymerase chain reaction (PCR) techniques. Transposon-based mutations are normally null mutations; however, transposons have been adapted in numerous ways to expand their utility.

Some basic themes apply to all transposon systems regardless of the organism in which they are used. Often experiments are set up so that transposition can be induced at a high level, making it easy to isolate cells with new transposition events. It is also important that transposition is inhibited following this induction to ensure that the element is stable. This goal can be achieved by using **nonautonomous mobile elements. Autonomous mobile elements** encode both the transposase and the end sequences required to mobilize the element (such as McClintock's *Ac* element in maize). Nonautonomous elements have the end sequence recognized by the transposase, but do not encode the transposase (such as McClintock's *Ds* element in maize). Nonautonomous elements can be mobilized by providing transposase *in trans* from an alternate source that can be removed later to ensure the transposon insertion is stable. Here we discuss the use of transposons in cells (*in vivo* transposition) in bacteria, plants, flies, and human cell lines. We also describe the use of transposition in the test tube (*in vitro* transposition) that offers numerous advantages and has become extremely useful in less tractable organisms.

Transposon Tools in Bacteria

Transposons have a rich history for genome analysis in bacteria. Early bacterial genetics unwittingly took advantage of endogenous IS elements that are the most common source for spontaneous null mutations in bacteria. Eventually, multiple transposons that occurred naturally with antibiotic resistance markers were used. Many of these systems still find use in the contemporary laboratory in a number of types of bacteria.

The construction of synthetic nonautonomous mobile elements marked an important turning point in the use of transposons (FIGURE 14.19). **Synthetic elements** are engineered to no longer encode the transposase, but instead simply carry a selectable genetic marker such as an antibiotic resistance gene that is flanked by the *cis*-acting transposon

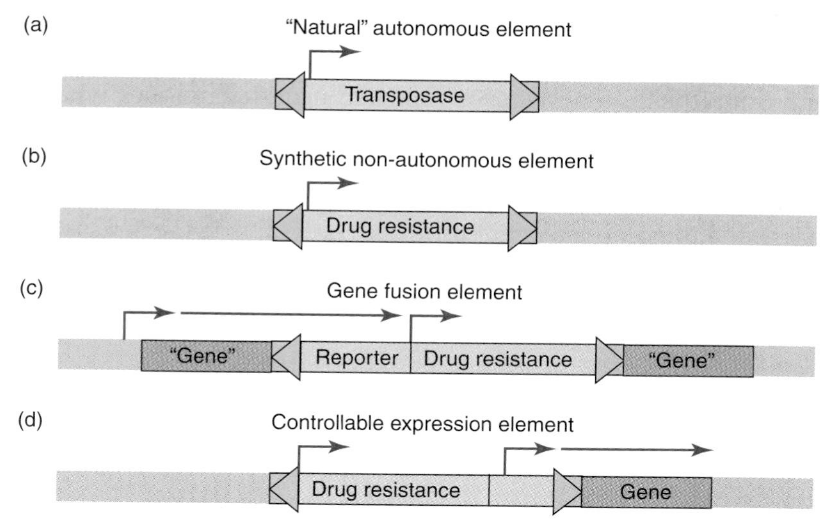

FIGURE 14.19 **Canonical transposon and a number of synthetic elements used as tools in genetics and genomics.** (a) A naturally occurring insertion sequence type transposon will encode a transposase and the cognate inverted repeats (green triangles) recognized by the enzyme. (b) Synthetic elements can be produced by replacing the transposase with a selectable marker such as a drug resistance determinant or a marker that can be screened that changes the phenotype of the transposon-containing organism, a pH change that can be detected in a colony of bacteria in a Petri dish, or the eye or body color of a fly. (c) Gene fusion elements can be designed where a promoter-less gene (the reporter) is included in the element that will only be expressed if the element inserts downstream of an active promoter. Similar strategies can be used to create protein fusions or to capture enhancers of transcription. (d) Controllable expression elements can be designed as tools to activate genes in a controllable fashion if a promoter or enhancer is included in the elements that can drive the expression of downstream genes.

ends (Figure 14.19b). These elements, sometimes termed defective transposons, are introduced using a vector, such as a bacteriophage or plasmid, which cannot replicate or integrate in the target host. In a typical experiment, a bacteriophage or plasmid vector containing a defective transposon that confers resistance to a specific antibiotic is introduced into bacteria. The gene specifying the transposase is present in a separate vector that is already present in the bacteria. Bacteria are spread on a solid growth medium containing the antibiotic. Bacteria that are resistant to the transposon-encoded marker can only produce colonies if the element moves from the vector to the host genome. Antibiotic resistant colonies are washed from the plate to create a transposon mutagenized pool. All of the bacteria in the population have a transposition insertion at a different random position in their genome. A sufficiently large pool size is collected to be reasonably sure that a transposon has inserted in every gene in the genome somewhere in the pool of cells. Then the inserted transposons are transferred to a second bacterial culture by generalized transduction (see Chapter 8). The transduced bacteria retain the inserted selectable marker but lack transposase activity. Because defective transposons do not encode their own transposase and the bacteria do not provide this activity *in trans*, the issue of transposon stability is solved.

Typically, the transduced bacterial pool will be screened for a particular phenotype or, where possible, a specific trait is selected. Conditional phenotypes can be screened using techniques where an archive copy of a transposon-containing clone is maintained when testing a variety of lethal conditions. One possible downside to these types of experiments is that transposon insertion mutations are almost exclusively complete null mutations. However, transposon insertion mutations in the 3′ end of an essential gene will sometimes occur that do not inactivate the gene, and occasionally these insertions will reveal interesting phenotypes.

Large-scale transposon mutagenesis experiments also have been useful for identifying essential genes. If the insertions site of a sufficient number of transposon insertions can be analyzed in an organism, and the genome is sufficiently small, one can begin to get a picture of the essential genes because they are never inactivated in the experiment. This type of experiment involves sequencing a large number of insertions and also requires that transposon insertion be very random. In these experiments a gene was assumed to be essential if a transposon insertion could never be identified in that particular gene.

Reporter fusions have made transposon screens immensely powerful in the study of bacteria. This type of synthetic transposon will usually contain a selectable genetic marker such as a gene encoding an antibiotic resistance marker (Figure 14.19c). The element also will encode a reporter gene that only will be expressed if it inserts into a gene that is already being expressed in the bacterial cell. The most common type of reporter is a promoter-less gene whose expression can easily be monitored (often by a color change in the colony growing in a Petri dish) when it inserts just after a host promoter (Figure 14.19c). One popular gene that has proven to be widely applicable as a reporter is the *lacZ* gene encoding β-galactosidase. The β-galactosides product allows the cleavage of the β–1-4 linkage between the galactose and glucose of lactose (see Chapter 3). However, β-galactosidase also can cleave the synthetic compound, 5-bromo-4-chloro-3-indolyl-β-D-galactopyranoside (X-gal). When X-gal is cleaved it changes from colorless to blue. In a common experiment, a pool of cells would be produced where the experimenter monitors bacterial colonies under different growth conditions to determine those genes that are induced under a specific set of conditions. Genes that are activated in response to various stresses such as DNA-damaging agents, temperature extremes, nutrient starvation, or toxic compounds can easily be isolated. The location of the individual transposon mutants can then be determined to discover the gene that is induced. These elements also can be used on a more quantitative basis to determine the regulation of the gene. A reporter element can also be designed such that the report protein is physically fused to the protein encoded in the target gene. In these constructs, the transposon must occur in the correct orientation and reading frame so that the reporter is fused to a gene to give a positive signal from the reporter construct. This type of protein fusion is useful in a variety of settings, but has been historically very important in early work on protein translocation in bacteria.

As described above, transposons predominantly cause null mutations when they inactivate genes. A number of synthetic elements, however, have been made that contain controllable promoters that can be used to activate genes that are located just after it (Figure 14.19d). These constructs can be used to activate cryptic genes but they have also been used in novel experiments to search for essential genes. Essential genes can be identified by techniques that identify transposon insertion events in the promoters of these genes. Insertion events into the promoters of an essential gene will render the isolate dependent on the inducer for the transposon-encoded promoter. Techniques are used where an archive copy of the isolate is maintained with the inducer while the isolate is tested for survival without the inducer.

In early experiments with bacterial transposons, mating experiments were performed to map the location of transposition events. In later experiments, elements were cloned using a selectable transposon-encoded marker. The element could be cloned from the host chromosome into a plasmid vector where the host DNA flanking the element was sequenced to determine the element's location. This process was further improved by using synthetic mobile elements that contain conditional DNA replicons. These transposons are almost self-cloning once the DNA is purified, digested with a restriction enzyme that does not cut within the element, ligated, and then introduced into a permissive host that replicates the plasmid. In a further improvement, various PCR strategies virtually eliminate the time consuming procedure of cloning the element from the chromosome.

The latest innovation in mapping transposons involves using microarray technology (see Chapter 5) to map transposons in a cell population. The specific microarray used is called a **whole genome array** because every sequence in the genome with the potential to specify a polypeptide is attached to a specific site on a solid surface. A sequence with the potential to specify a polypeptide is called an **open reading frame** (**ORF**). The whole genome array technique is useful because it can identify candidate genes that are conditionally essential. A typical experiment is shown in FIGURE 14.20. A large transposon mutagenized pool is prepared and split into test and reference (or control) cells. Test cells are cultured in minimal medium containing only a carbon source and salts, while reference cells are cultured in medium that also contains nutrients of interest such as amino acids. Cells with transposons in genes required for amino acid synthesis will not grow in minimal medium but will grow in the complete medium. Therefore, these cells will be diluted when cultured in minimal medium but not when cultured in complete medium. Then DNA from each population is isolated and sequences adjacent to the transposition events are amplified using modified PCR techniques under conditions in which the PCR products are labeled with fluorescent tags. The labeled PCR products derived from each pool are analyzed using a whole genome array. PCR preparations from the reference and test cells will differ in one important respect. The former will have transposon insertions in genes that are required for amino acid synthesis, while the latter will not. This difference is easily identified as an ORF that gives a

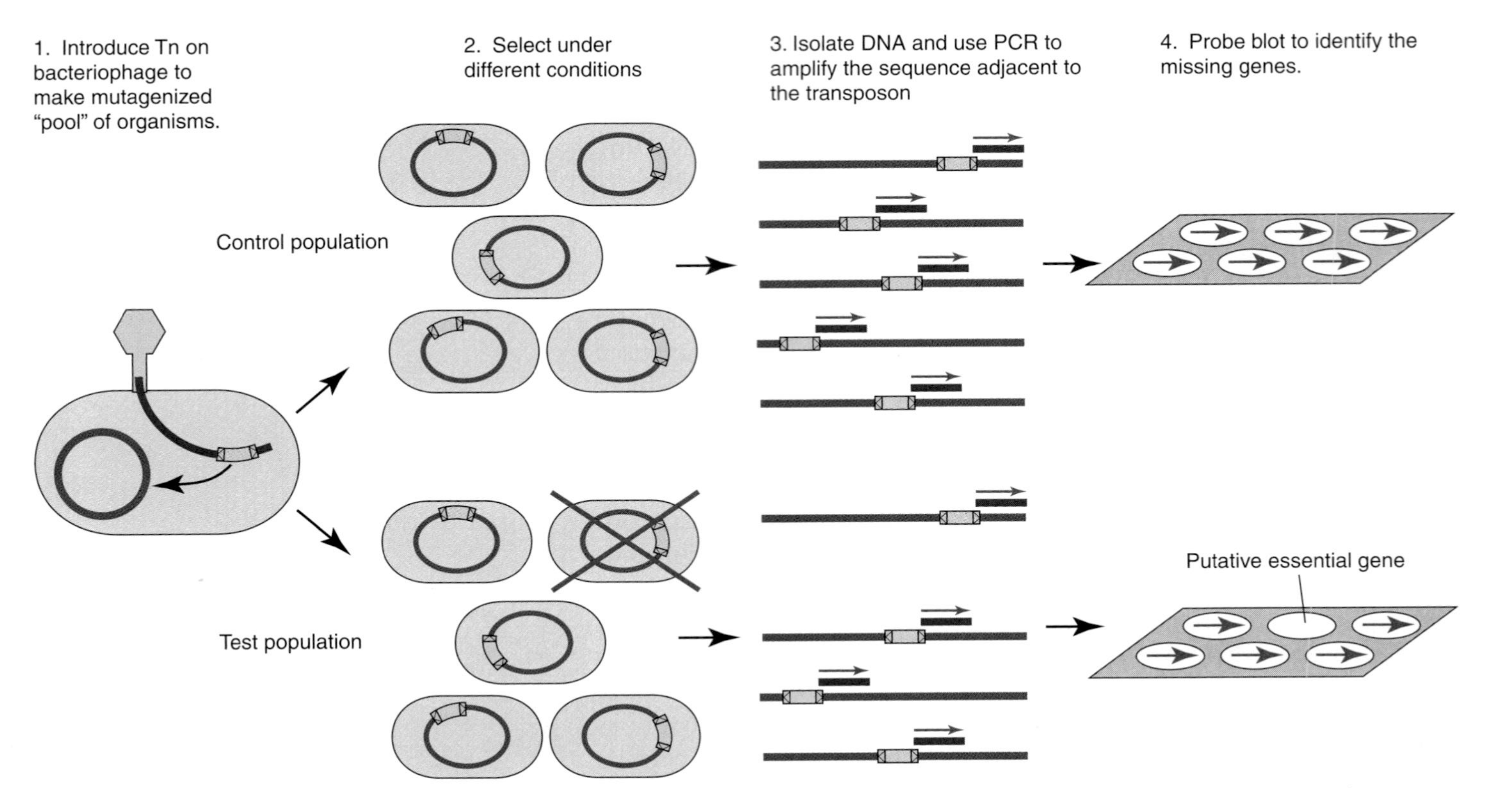

FIGURE 14.20 **Whole genome arrays can be used to map transposons in a population of bacteria.** (1) A collection of bacteria is made where individuals in the population have insertions each at a random location in the chromosome, called a pool or library. (2) The pool of cells is split where one part of the population is maintained under the same conditions that were used to make the library (the reference or control group) while the other population is subjected to a more stringent growth condition, like minimal media, at high temperature, or growth in a specific host or environment. (3) DNA is collected from the control and test populations and subjected to PCR, which allows the sequence adjacent to the transposon insertion to be amplified and labeled. (4) The transposon proximal sequences are used to probe an array to discover the genes that have a signal in the control population, but not in the test population. These are the putative essential genes.

fluorescent signal on the control array, but not the test array. Candidate genes identified with this procedure are confirmed using other strategies. In theory, this strategy can be used to identify essential genes by finding ORFs that never produce a signal in the experiment.

Transposon Tools in Plants

Transposons were first identified in plants and the genomes of plants appear to be greatly affected by these genetic elements. The genomes of some grasses can be 40% to 80% transposon DNA. As is found to be the case in many multicellular animals, however, almost all of the endogenous elements are inactive. An inactive element with *cis*-acting repeats at its ends that lacks a transposase encoding gene is called a **minimal inverted transposon repeat element** (**MITE**). Plants often contain many such elements. Plants contain DDE transposons as DNA transposons and as LTR retrotransposons. They also contain non-LTR retrotransposons, which are discussed below.

Transposons remain a very important tool in maize genetics and genomics. The transposon **Mutator** is a popular tool because it transposes at a very high frequency. The plant of interest is bred with one that contains a Mutator transposon (FIGURE 14.21). After a plant with

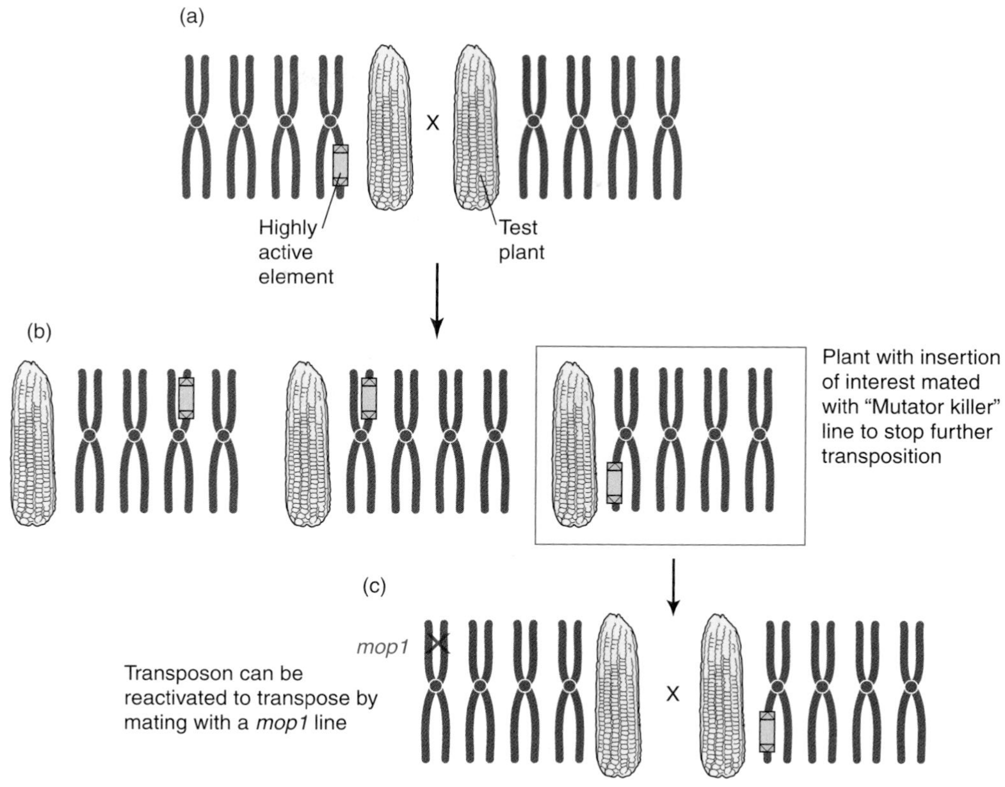

FIGURE 14.21 **Mutator element is an important transposon tool in maize.** (a) Corn is shown next to representative chromosomes from the organism. The pink and green box indicates a Mutator element. (b) Transposition in the test plant occurs when it is mated with a plant that has a highly active Mutator element. (c) Plants with an insertion of interest are identified by PCR techniques or by a phenotype. Mating this plant with a "Mutator killer" line or other techniques will inhibit any further transposition. The Mutator element can be reactivated for transposition by mating to a *mop1* strain that will release silencing. This is useful for gathering additional insertion events in the same region or for getting reversion alleles when the position where the element previously resided was repaired.

the desired insertion is identified, it can be mated with a plant that has a suitable background to prevent further transposition. This mating involves crossing into a Mutator killer line that contains an inverted copy of Mutator. Mutator elements in this configuration activate a host system that actively destroys RNAs expressed from Mutator. High Mutator copy number also can reduce Mutator transposition presumably by a similar mechanism. Both forms of Mutator inactivation are incomplete and residual Mutator movement can still occur. Mutator elements that were turned off using this global mechanism can be reactivated by crossing the strain with lines with an inactivated mediator of paramutation1 (*mop1*) gene. The *mop1* allele encodes for an RNA-dependent RNA polymerase that may be otherwise required to turn off the Mutator elements.

The original *Ac* element identified by McClintock still provides an important tool in maize. One property of the *Ac* element that can be viewed as a positive or a negative is its propensity to transpose to regions of the chromosome that are genetically close. Although local hopping can be a frustration, it also can be a tool for localized mutagenesis of a region. Genetic strategies have been used to isolate

Ac elements across all ten maize chromosomes. These elements should prove useful for more focused *Ac* mutagenesis for subregions of the maize genome. Both *Ac* and Mutator elements are useful also for generating mutations at a position after they transpose out of a gene. Mutations that occur in the donor site after transposition result from repair of the host chromosome through nonhomologous end-joining. This repair process produces footprints at the site of the original transposon insertion that can be used to obtain a variety of different mutations.

Transposon tools like the *Ac/Ds* system are used less frequently in two other popular plant model systems, the small flowering plant *Arabidopsis* and rice. These organisms are more easily modified using a bacterial DNA delivery mediated by *Agrobacterium*. However, the plant retroelement *Tos17*, which can be activated for transposition using cell culture techniques, has been a useful tool in rice. Plant stocks made from the modified cultured cells can be analyzed for useful insertion lines.

Transposon Tools in Insects

Drosophila melanogaster is an important experimental system for understanding multicellular animals. The **P element** has been an invaluable tool for genetic manipulations in *D. melanogaster*. P elements were originally discovered in crosses between males from the wild with laboratory maintained females. Such crosses led to high amounts of apparent recombination during spermatogenesis in progeny male flies, where normally recombination is essentially nonexistent. This phenomenon is known as **hybrid dysgenesis**. Subsequent work indicated that widespread movement of the P element had caused chromosome breakage in males, and subsequent repair events by nonhomologous end-joining caused genetic combinations that resembled recombination. Interestingly, P elements themselves are likely new to *D. melanogaster*, introduced less than 200 years ago from another *Drosophila* species, but are now almost ubiquitous in wild *D. melanogaster* populations. The newness of P elements to *D. melanogaster* probably explains its ability to transpose at such a high frequency. *D. melanogaster* strains that live in the wild presumably evolved a way to repress P element transposition. The high transposition frequency of the element in laboratory strains has allowed the P element to become a very valuable tool for genetics.

Nonautonomous elements with markers that change a fly's eye or body color have been constructed to allow easy identification of transposon-containing animals. These elements can be delivered to the organisms by co-injecting DNA containing the element with DNA encoding a transposase source; injection is into the posterior region of pre-blastoderm embryos, where the germ line will form (FIGURE 14.22). Alternatively, one can inject DNA containing the transposon into pre-blastoderm embryos from flies that have been engineered to express transposase in their embryos. Progeny from flies into whose germ line the transposon inserted are easily identified by the phenotype conveyed by the encoded marker gene. P elements have been adapted as tools for flies in essentially the same ways as other transposons have

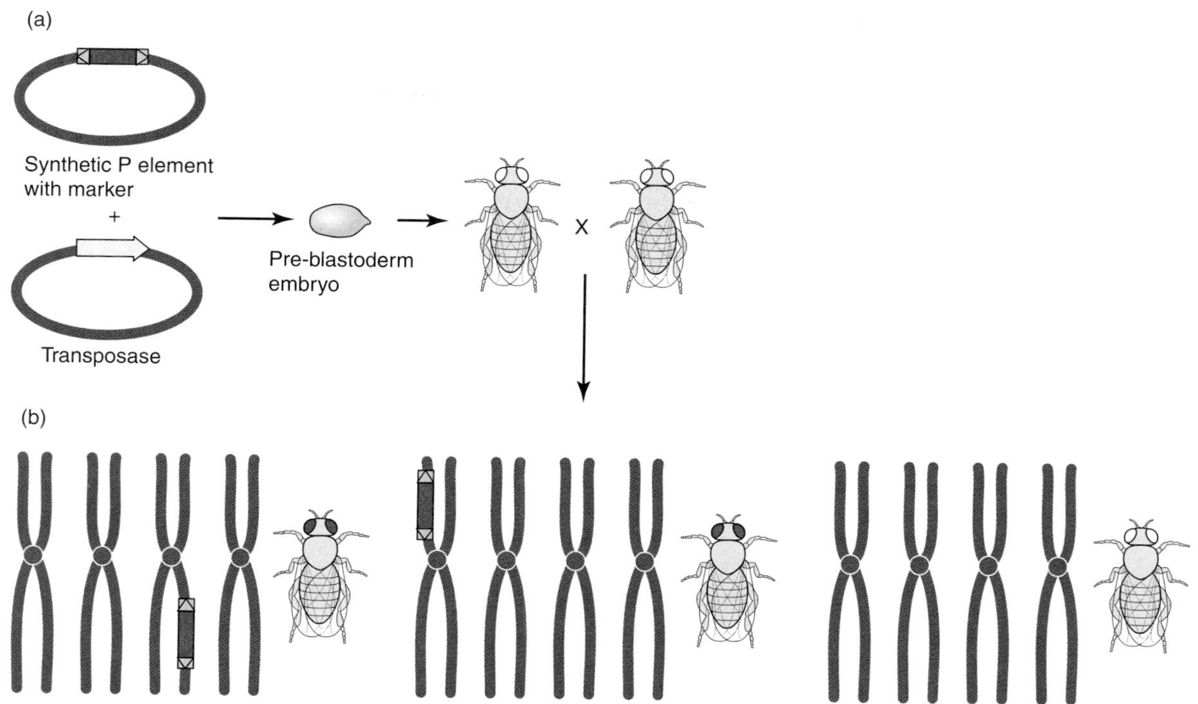

FIGURE 14.22 **Co-injection of a synthetic P element with a marker and a plasmid that encodes transposase.** Injecting DNAs encoding the P element transposase and a synthetic P element into embryos will allow a transposition event to be isolated in progeny. Flies are shown next to representative chromosomes from the organism. The green and red box indicates a synthetic P element. (a) A synthetic P element is introduced into a pre-blastoderm embryo with a source for expression of the P element transposase. (b) A portion of the progeny from these flies will have chromosomal P element insertions. Flies with the P element insertion are identified by a phenotypic marker such as eye color.

been adapted in bacteria. For example, a P element can be used as an insertion mutagen. P element insertions across the genome can be used for genomic screens. Projects also have been advanced to attempt to isolate mutant animals with P element insertions in each of the genes in *D. melanogaster*. Transposons have been constructed also as tools to identify genes with specific expression profiles, splice sites, and to make fusion proteins (Figure 14.19). Mobilization of P elements can be used to revert—or make deletions—in a P element-containing gene through perfect or imperfect excision, respectively. While the majority of donor sites are repaired by gene conversion from a sister chromosome (Figure 14.9), various forms of mutagenic nonhomologous end-joining also can lead to mutations at the old donor site.

One extremely powerful tool is the use of P elements to insert genes from one organism into another. A gene of interest can be cloned between the inverted repeat sequences of a P element and easily introduced into the fly genome as noted above. Transgenes introduced into the fly genome have a wide range of applications. These include simple transgenic-rescue (complementation) experiments to ascertain if a cloned gene is indeed the gene identified by a mutant phenotype, or to determine the full extent of the sequences needed to express that gene.

P element constructs also can be used to introduce specifically tagged versions of the protein of interest. For example, fusion derivatives with the green fluorescent protein allow the investigator to see where the protein localizes within the cell or organism under the microscope. Inserting a gene on a P element also can be used to allow expression in an inappropriate place or time under the control of the investigator (ectopic expression studies). Through the use of RNA-based techniques, genes expressed from P elements can be used to make functional knockouts of any gene of interest. Because the site of insertion of P elements cannot be controlled, multiple independent transgenic lines with different insertions of the same element are normally tested to account for position effects (e.g., that an element inserted into a region of the genome that prevents or changes its pattern of expression).

Another valuable application of P elements involves the use of a heterologous expression system from yeast. Transgenic lines are available that express the yeast transcription activator protein called Gal4 (see Chapter 18) under the control of tissue-, stage-specific or other useful *Drosophila* promoters. When these lines are crossed to transgenic lines carrying P elements containing the *GAL4* recognition sequences fused just before a gene of interest, the progeny of this cross will express the gene of interest in the pattern dictated by the promoter attached to *GAL4*. The great utility of this system stems from a large collection of available *Drosophila* strains that allow expression of the Gal4 activator in specific tissues or at specific times in development.

While the P element has shown the greatest utility in *D. melanogaster*, it appears to have a relatively restricted host range. Consequently, other transposons have been harnessed for other insects. Three elements used in insects outside *D. melanogaster* are *hobo*, *minos*, and *piggyBac*. The *piggyBac* element has been extremely useful over a wide host range and is described in more detail below.

Transposons for Mammalian Cells from Unexpected Sources

Transposons can be important tools to study various multicellular animals. For a long time, however, no useful transposable elements existed for work in mammalian systems. An interesting story for the production of a transposon that works in vertebrate cells starts with a family of ancient nonfunctional transposons identified in fish. This family of elements is estimated to have been nonfunctional for over 14 million years. Zsuzanna Izsvák and coworkers analyzing these defective transposons could identify numerous mutations that prematurely terminated translation of the transposase. By comparing the many nonfunctional copies of the element, they also could predict some of the missense mutations that occurred over time. Izsvák and coworkers used site-directed mutagenesis to restore the element to full function. Initial work was directed toward re-establishing transposase activity. Further efforts involved correcting missense mutations to gradually establish all the biochemical activities required for a functional eukaryotic transposon; namely nuclear localization signals for entry of the protein into the nucleus after translation, DNA binding ability to the transposon ends, and finally full integration capability. The restored functional transposon is called ***Sleeping Beauty*** because the element had been nonfunctional (or asleep) for millions of years (FIGURE 14.23).

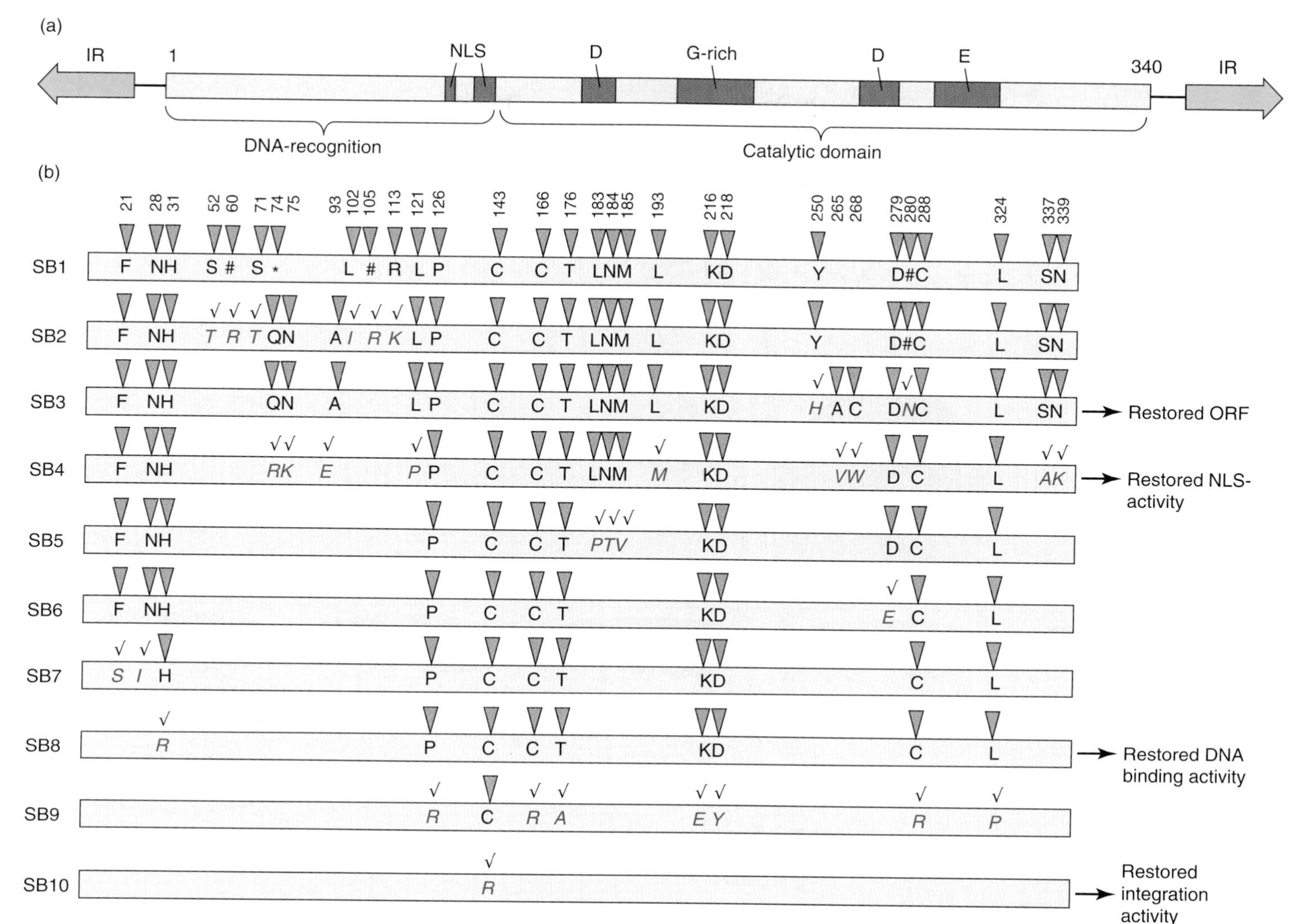

FIGURE 14.23 Reconstruction of salmon element *Sleeping Beauty* by site directed mutagenesis. (a) Diagram of the *Sleeping Beauty* element showing the inverted repeats (IR), the location of the nuclear localization signal (NLS), a region rich in glycine residues (G-rich), and the amino acids of the DDE motif (D, D, E). (b) Derivative SB1–SB10 indicate version of the protein at various steps during the site-directed mutagenesis process. Triangles indicate an amino acid residue that was altered. In each case the amino acid is indicated with its one letter code. Black letters indicate amino acids that differed from the consensus sequence; the asterisk indicates a translational stop codon and the number symbol (#) indicates a frameshift mutation. Red letters indicate the final amino acid. Landmark phenotypes are shown to the right. (Reproduced from *Cell*, vol. 91, Z. Ivics, et al., Molecular Reconstruction of *Sleeping Beauty*. . . , pp. 501–510, copyright 1997, with permission from Elsevier [http://www.sciencedirect.com/science/journal/00928674].)

Izsvák and coworkers constructed a synthetic element that has *cis*-acting *Sleeping Beauty* transposon ends flanking a neomycin-resistant gene and a simian virus 40 (SV40) promoter but that lacks a gene encoding transposase (**FIGURE 14.24**). They simultaneously introduced a DNA molecule bearing this synthetic element with another bearing the restored *Sleeping Beauty* transposase gene into cultured human cells to produce cell lines that contained the transposed synthetic element (Figure 14.24). They observed that transposition can only take place if (1) the synthetic element has inverted repeats flanking the neomycin-resistance gene, (2) the transposase gene is in the correct

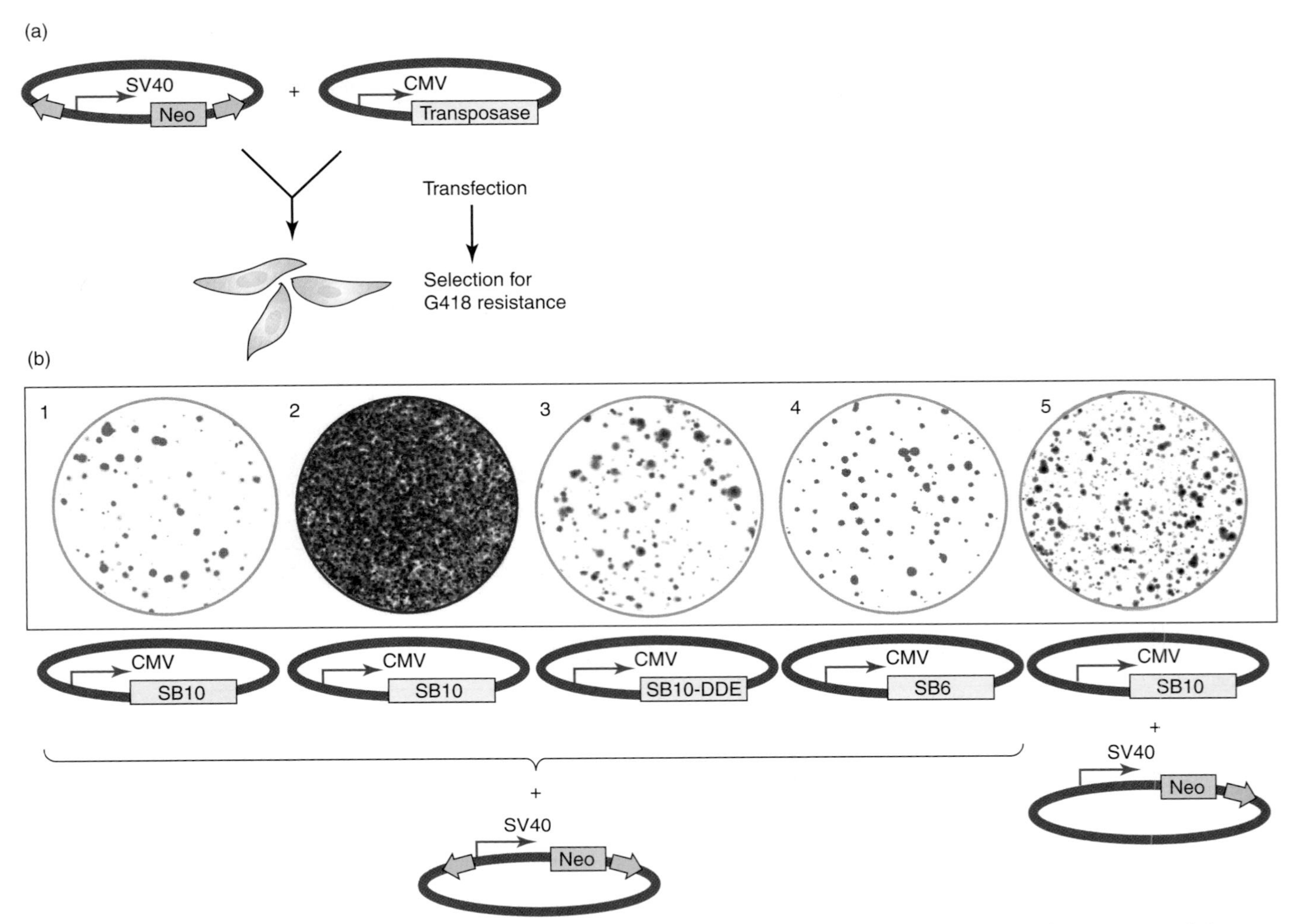

FIGURE 14.24 ***Sleeping Beauty*** **element transposition in human HeLa cells.** (a) A plasmid construct that expressed the *Sleeping Beauty* transposase from a promoter from the cytomegalovirus (CMV) and a plasmid construct with a synthetic element encoding neomycin resistance (Neo) and a promoter derived from the simian virus 40 (SV40) were cotransformed into human cells. (b) The neomycin resistance gene allows mammalian cells to be resistant to the drug G418. NeoR cells were only identified when the fully functional transposase was produced (SB10) in the correct orientation. Transposase derivatives with mutations in the DDE proteins (SB10-DDE) or other mutant derivatives of the transposase (SB6) did not allow transposition in the cells. In addition to requiring the SB10 version of the protein transposition also required that the synthetic element had both ends of the element flanking the neomycin resistance gene. (Reproduced from *Cell*, vol. 91, Z. Ivics, et al., Molecular Reconstruction of *Sleeping Beauty*. . . , pp. 501–510, copyright 1997, with permission from Elsevier [http://www.sciencedirect.com/science/journal/00928674]. Photo courtesy of Zoltan Ivics, Max Delbruck Center for Molecular Medicine.)

orientation with respect to the promoter, and (3) the transposase is fully functional. Further modification of *Sleeping Beauty* has allowed it to be used for transposition in mouse somatic cells, where it can serve as a tool for gene inactivation. As with *Sleeping Beauty*, similar strategies have subsequently been used to reanimate transposons from a wide variety of species.

Another element that has been adapted for use in mammalian cells is the *piggyBac* transposon. The *piggyBac* element is endogenous to the cabbage looper moth, *Trichoplusia ni*, and was identified in baculovirus vectors that were used to express genes *in trans* in these cells. Investigators identified small insertions in the baculovirus vector that later were shown to be an autonomous transposon. Coding information added to the *piggyBac* element can allow it to function to construct gene fusions. While *piggyBac* elements almost exclusively transpose into the sequence 5′-TTAA-3′ in largely AT-rich regions, they insert widely across the mouse chromosome.

Future work with *Sleeping Beauty*, *piggyBac*, and other elements holds great promise for picking up the pace of vertebrate genomics and possibly as tools for gene therapy.

In vitro Transposition Tools

In vitro transposition reactions using purified transposition proteins and synthetic transposons are useful in various types of experiments. Many *in vitro* systems are currently available. These systems can be used as sequencing aids or as mutagenesis tools when plasmid-containing constructs are analyzed in *E. coli*. Moreover, these systems also can be used as tools to extend transposon mutagenesis to an organism without having to rely on establishing transposase expression in the host (or to repress or remove transposase activity after mutagenesis to stabilize the transposition events).

One of the early uses of *in vitro* transposition systems was as an aid for sequencing large cloned DNA fragments. The original strategy for sequencing large cloned stretches of DNA involved using a primer within the plasmid backbone to read sequence into the cloned fragment (FIGURE 14.25). Because sequences longer than about 700 to 900 bp cannot be read from a primer, special techniques must be used to determine longer sequences. One way to solve this problem is to use primer walking (see Chapter 5). *In vitro* transposition offers another approach by permitting us to isolate individual plasmids containing transposition events at various

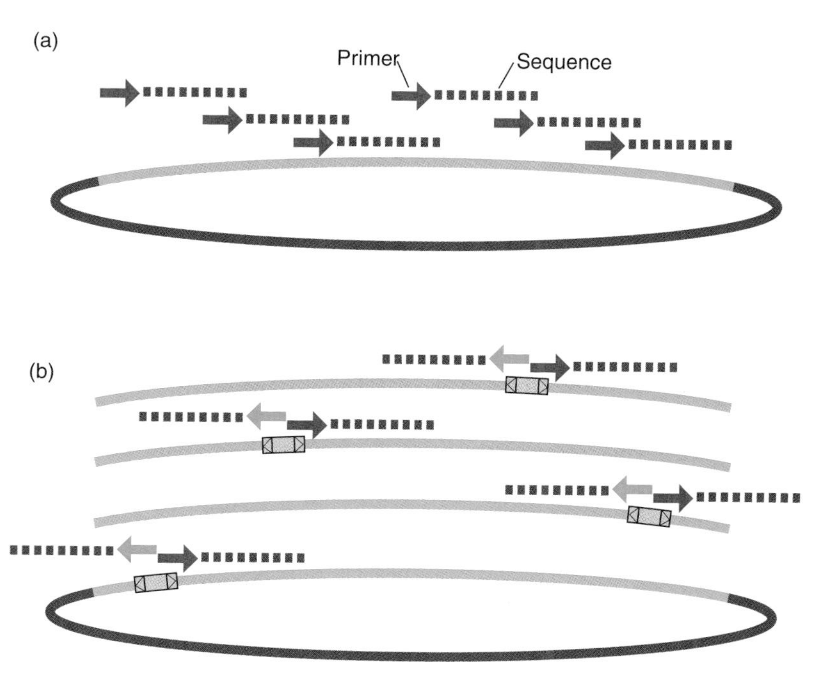

FIGURE 14.25 **Two methods of sequencing large DNA fragments cloned into a plasmid vector.** (a) In primer walking, primers specific to a plasmid are used to start to sequence the fragment. Subsequently an additional primer will need to be synthesized to obtain DNA sequence further into the cloned fragment. This process can be done multiple times to span the entire cloned sequence. (b) By isolating clones with transposon insertion at random locations throughout a cloned fragment, primers that are specific to the element can be used for DNA sequencing. This sequence information can be assembled to quickly derive sequence for the entire large cloned fragment.

positions. Sequence can be read from primers specific to either end of the transposon. Patches of sequence, therefore, can be read across the whole cloned fragment. Assembly of the information allows us to sequence large fragments, eliminating the need for multiple sequencing iterations as well as the need to design and prepare new primers. Cloned stretches of DNA also can be mutagenized and screened in *E. coli* using a library of insertions if a suitable phenotype can be monitored in this organism.

In vitro transposition systems can extend transposition to a variety of organisms. In one procedure, DNA from the target host is purified and then sheared or digested using restriction endonucleases. The resulting fragments are incubated with a DNA molecule containing a synthetic mobile element with a selectable marker inserted between the inverted repeats (Figure 14.19), and the corresponding transposase. Other genetic information can be included in the synthetic element such as reporters or origins of replication to facilitate downstream cloning events in *E. coli*. Many of these systems allow highly random transposition at very high frequencies. Often the best systems are adapted from elements that display target site immunity, a process in which the transposase has evolved the ability to not insert proximal to an element that has already inserted in a target DNA. This attribute greatly decreases the probability of multiple elements ending up in a single DNA. DNA fragments containing transposition events are then transformed into the target host. The target host must be capable of being transformed and have an efficient homologous recombination system. Homology to the host genome flanking the element allows the insertions to be genetically crossed into the genome. Because the linear fragments will not be maintained in the host, only organisms with the transposon crossed into the chromosome will be isolated. This strategy has been very successful in a range of bacteria as well as the eukaryotes *S. cerevisiae* and *Candida glabrata*. One downside to this procedure is that it is only useful in systems where a robust recombination system exists for crossing genetic information into the host genome.

An interesting adaptation of the *in vitro* procedure described above is sometimes possible where the transposase can bind to a synthetic element *in vitro* to produce a very stable nucleoprotein complex (FIGURE 14.26). Then the complex containing the transposase bound to the transposon ends is transformed into the host organism where it can undergo transposition into the host chromosome. This procedure has two distinct advantages. First, the investigator does not have to set up a transposase expression system in the target organism (and then devise some method to repress the transposase so that the insertion remains stable). Second, transposition takes place without the need for homologous recombination. Although this procedure may be subject to various forms of targeting bias, it has proven useful in a variety of organisms.

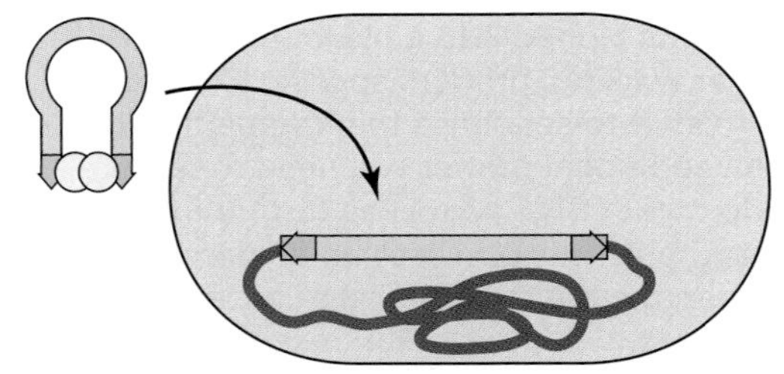

FIGURE 14.26 **Synaptic complex assembly *in vitro* for transportation *in vivo*.** In some systems, a "transpososome" can be assembled where a synthetic transposable element and the purified transposase enzyme are incubated for a stable synaptic complex that is competent for transposition when it is transformed into the cells.

14.2 Conservative Site-Specific Recombination

Two families of proteins do conservative site-specific recombination with different pathways.

Conservative site-specific recombination is a process in which a DNA segment moves between specific sequences recognized as DNA recombination sites using a cognate recombinase (FIGURE 14.27). One functional difference from transposition is that this process almost exclusively involves the same flanking site found in both the target and donor DNA. Conservative site-specific recombination also uses a chemistry that is very different from that used by transposons and involves a covalent protein-DNA linkage. These reactions play a very important role in stabilizing autonomously replicating circular DNAs that can recombine to form dimers in plasmids and chromosomes. As described above, this chemistry is also important after replicative transposition to resolve cointegrates that are made during the process. Bacteriophage λ and other bacteriophages that integrate into a host chromosome as a prophage typically use conservative site-specific recombination. These reactions can be important in gene shuffling and inversion reactions that influence the expression of genes.

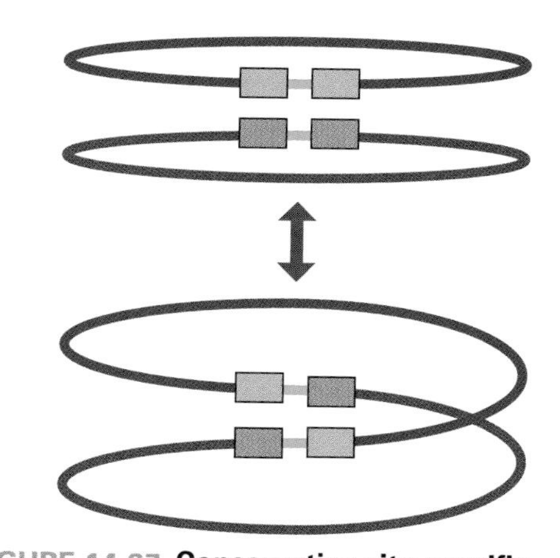

FIGURE 14.27 **Conservative site-specific reactions involving a site on the donor DNA with a homologous site on the target DNA.** This site contains sequences bound by the recombinases (boxes) that surround the actual region where the crossover takes place (green). This crossover fuses or separates the circular DNAs across the region recognized by the recombinase.

The conservative site-specific recombination reaction involves two inverted DNA sites in both the donor and target DNA molecules that are about 30 to 40 bp in size and are recognized by a cognate recombinase (FIGURE 14.28). These recombinase binding sites flank a crossover region. Four monomer enzyme subunits are involved in the reaction that make either sequential or concerted cuts in the DNAs, allowing a crossover event between the recombinase binding sites (Figure 14.28). Although the sequences at the crossover regions must be identical, there seems to be little constraint to the exact sequence found in this region. The enzymes involved in conservative site-specific recombination belong to two large and unrelated families, the tyrosine-recombinases and the serine-recombinases. The family is determined by the amino acid that transiently links to DNA during the reaction. The two families are unrelated and differ in whether the breaks underlying recombination occur sequentially or at the same time. Although the recombination pathways for the serine and tyrosine recombinases are different, the products at the end of the reaction are the same (FIGURE 14.29). In the case of both tyrosine and serine recombinases, the nucleophile is the active site amino acid residue for which they are named. This residue forms a covalent bond with the DNA in a strand breakage process and there is no requirement for a metal ion cofactor. Common to both the tyrosine and serine recombinases is the lack of a high-energy cofactor leading to the name **conservative site-specific recombination**.

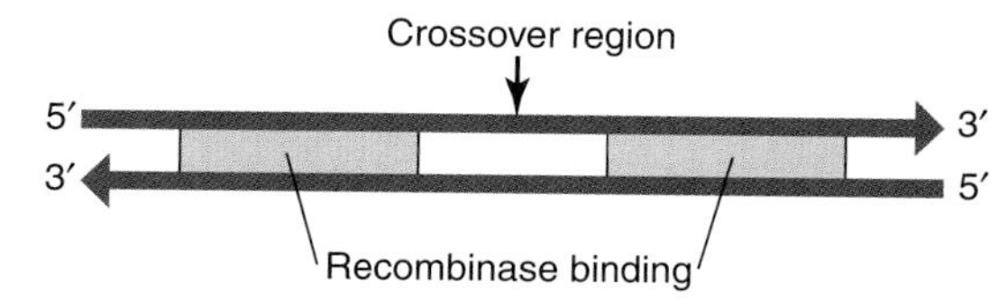

FIGURE 14.28 **Sites used for conservative site-specific recombination.** The sites utilized for conservative site-specific recombination involves two recombinases binding sites separated by a crossover region. The crossover region must be identical in the donor and target DNA sites.

Only two monomers of tyrosine recombinase are active at a time. The first reaction involves the formation of a Holliday junction between sites of recombination. In the reaction catalyzed by the

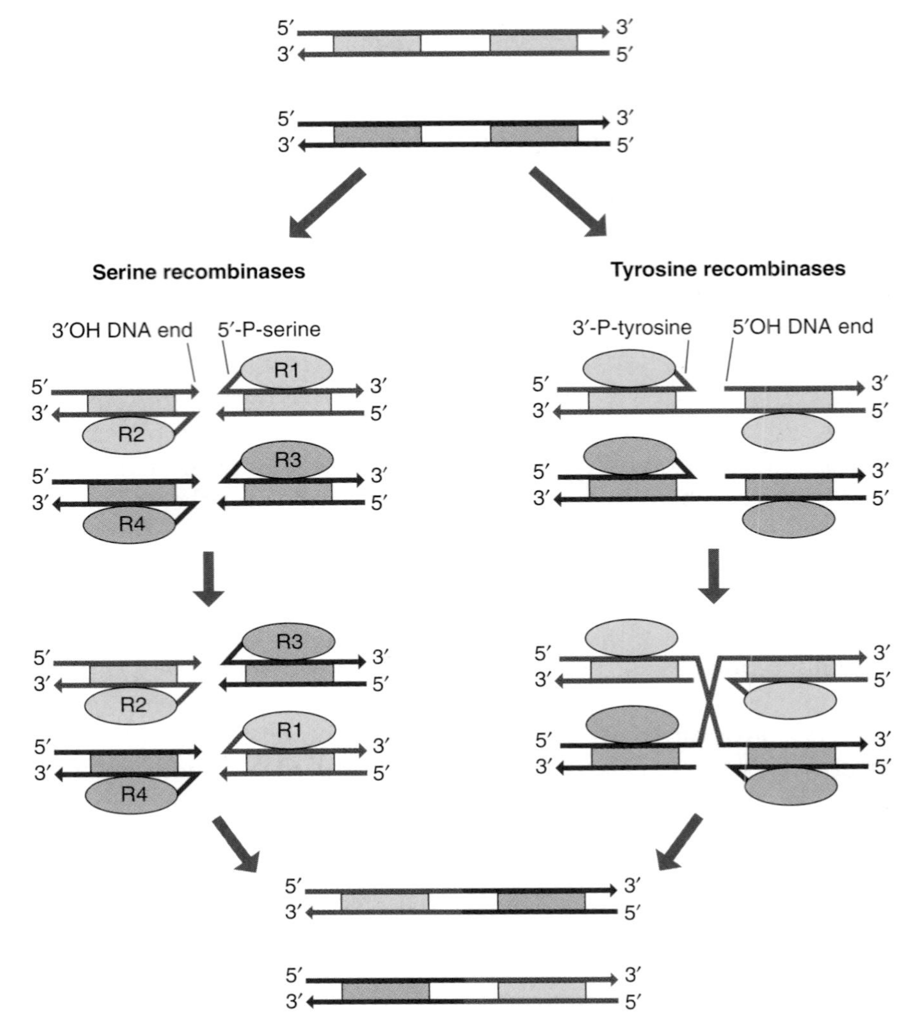

FIGURE 14.29 Comparison of the tyrosine conservative site-specific recombinases and the serine conservative site-specific recombinases. See the text for details.

tyrosine-recombinases, the active site tyrosine acts as nucleophile forming a 3′-phosphotyrosine linkage to the DNA thereby liberating a 5′-OH. A conserved arginine-histidine-arginine triad appears to activate the scissile phosphate. The histidine is the catalyst for general acid/base chemistry in the reaction while the arginine polarizes the phosphate-oxygen bond and stabilizes the transition state. The free 5′-OH attacks the opposing integrase-bound site to form the Holliday junction. A conformational change must then take place to activate the partner recombinases to allow a second set of reactions that resolves the Holliday structure (Figure 14.29). Many of the tyrosine recombinases require accessory proteins that serve an architectural function in the reaction. The accessory proteins may be host-encoded or encoded on a bacteriophage or plasmid element where the conservative site-specific recombination system resides. In the case of the lesser-understood serine recombinases, all enzymes are active at the same time and a DNA double-strand break is an intermediate in the reaction (Figure 14.29).

Bacteriophage λ uses a conservative site-specific recombinase to integrate into the host genome.

While conservative site-specific recombination occurs frequently in a variety of organisms, this reaction was first appreciated in the case of integration of the bacteriophage λ into the *E. coli* chromosome. Bacteriophage λ DNA can replicate through a lytic or lysogenic life cycle (see Chapter 8). In the lysogenic life cycle, bacteriophage λ DNA integrates into the host chromosome at a specific attachment site using a conservative site-specific recombinase called an **integrase**. Bacteriophages have different attachment sites, but the attachment sites tend to be at highly conserved sequences, such as transfer RNA (tRNA) genes. The bacteriophage itself contains a portion of the tRNA gene; therefore, no sequence information is lost in the process of integration and essential parts of the tRNA gene can be used as attachment sites. The attachment site for λ DNA is located between the galactose (*gal*) and biotin (*bio*) genes in the *E. coli* chromosome (FIGURE 14.30). The sequence recognized by the integrase in the bacterial λ DNA attachment site is called *attB*, for attachment site in the bacteria. The site recognized in λ DNA is called *attP* for attachment site phage.

In addition to the integrase (Int), the host-encoded protein aptly named integration host factor (IHF) also is required to make the necessary DNA bends (Figure 14.30). The bacteriophage Int protein recognizes recombination signals in *attP* and *attB*. The *attP* region contains extra DNA sequences that are bound by N-terminal domains in the integrase called arms (not shown). As with all conservative site-specific recombination reactions, following integration into the chromosome, the *attP* and *attB* sites are split to make new hybrid sites termed *attL* and *attR* (Figure 14.30). This difference in the sequences that flank the *att* sites in the excised and integrated states sets the stage for the different protein requirements needed for each set of recombination steps. An additional bacteriophage encoded protein **excisionase (Xis)** is required for the bacteriophage to excise from the chromosome. The integration process is very efficient and specific with insertions only occurring in a single orientation.

The λ DNA integration and excision system has been adapted into a commercially available, *in vitro* directional cloning system (FIGURE 14.31). This system allows the user to clone a gene of interest once into a so-called entry vector containing *att* sequences that will greatly facilitate subsequent subcloning steps into various expression and gene fusion constructs (Figure 14.31a). This initial cloning step into the entry clone can be done using restriction endonuclease and ligase (Figure 14.31a). Alternatively, the target gene can be cloned by lambda recombination (see below).

Arguably the greatest power of the lambda-based system is the ease with which one can subclone genes into new vectors, that is, from the entry clone into various destination vectors (Figure 14.31). When the gene of interest is present in the entry clone, it is flanked by two complete *attL* sites that have been specifically engineered by altering the crossover region. Recall from above that although the actual binding sites for the recombinase can differ, the crossover region must be

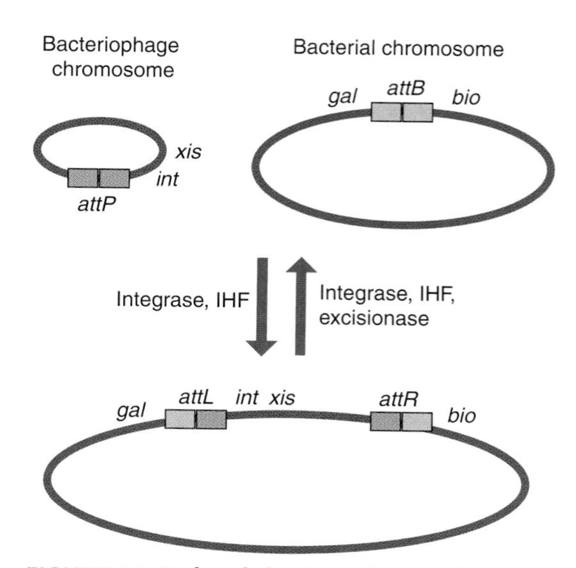

FIGURE 14.30 Lambda phage integration into an excision from the bacterial chromosome. The *attP* site found on the bacteriophage and the *attB* site found on the bacterial chromosome allow the orientation-specific introduction of the bacteriophage genome into the chromosome. This process requires the conservative site-specific recombinase and the integration host factor (IHF) encoded in the host. New sites are made in the process called *attL* and *attR*. Excision of the element from the chromosome requires an additional acidic bacteriophage encoded protein called excisionase. Galactose (*gal*), biotin (*bio*), integrase (*int*) and excisionase (*xis*) genes are shown.

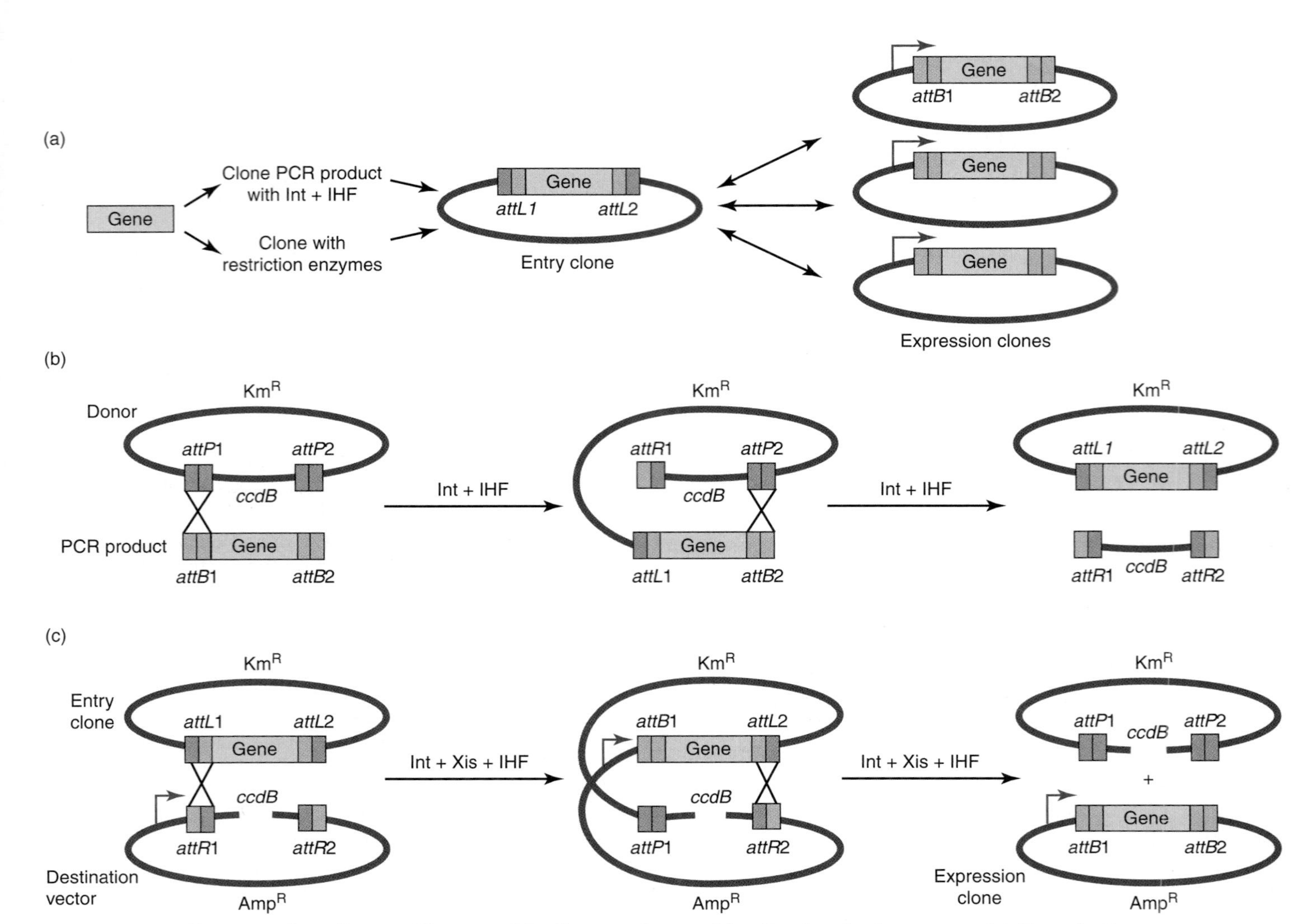

FIGURE 14.31 Conservative site-specific recombination can be used *in vitro* for cloning genes. (a) Steps in the conservative site-specific recombination system. A linear DNA fragment is cloned by some means into the "entry clone" and easily moved into any of a number of "expression cones" for analysis. (b) A gene that is amplified by PCR is cloned into a plasmid entry clone vector using *att* sites produced synthetically in the PCR primers. The plasmid is selected using kanamycin resistance (Kan^R). Unreacted donor plasmids are counter-selected because the protein encoded by the *ccdB* gene is toxic to *E. coli* strains that lack the appropriate mutation in *gyrA*. Modified *attP* and *attB* sites force recombination to progress though the correct sites, namely *attP1* and *attB1* or *attP2* and *attB2*. (c) Once the gene of interest is cloned into the entry clone it can be easily transferred to another plasmid that encodes ampicillin resistance Amp^R. Unreacted DNAs are again counter-selected using the CcdB toxin. (Adapted from J. L. Hartley, G. F. Temple, and M. A. Brasch, *Genome Res.* 10 [2000]: 1788–1795.)

identical for the reaction to work. Recombination, therefore, will only occur between the *attL1* and *attR1* sites or the *attL2* and *attR2* sites. These two reactions will generate *attP1* and *attP2* sites or *attB1* or *attB2* sites after the reaction (Figure 14.31b). To facilitate the process, the entry and destination vectors encode different antibiotic resistance markers to select the correct vector. Of additional utility, any unreacted vector that failed to receive the DNA fragment of interest can be excluded because it would contain the *ccbB* gene. The *ccbB* gene encodes a product that is toxic to the cell unless the host encodes a specific mutation in the *gyrA* gene that encodes gyrase.

Traditional cloning can be totally circumvented if the gene of interest is amplified with primers that have extra sequences added that can be recognized by the integrase proteins (Figure 14.31b). As with the system described above, recombinant molecules that have gained the fragment of interest will lose the *ccbB* gene that is toxic to cells that do not contain the specific host-encoded gyrase mutation.

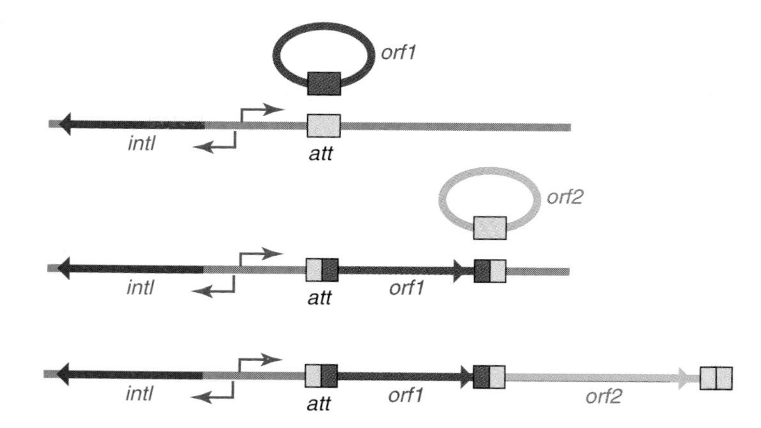

FIGURE 14.32 Natural integron systems utilizing gene cassettes and integrase protein. Integrons present as natural cloning systems where an integrase (IntI) can allow the integration of genes with circular cassettes containing compatible *att* sites. Multiple cassettes (i.e., *orf1, orf2, orf3* etc.) can accumulate in the integron system.

Multiple other systems use the conservative site-specific recombinase reaction.

In Chapter 13, multiple examples were given for how the conservative site-specific recombinases have been adapted to engineer genomes. The FLP recombinase, which is derived from a circular yeast plasmid, allows site-specific recombination at its cognate *frt* site. The Cre protein and its cognate *lox* site are found in nature in bacteriophage P1. Other interesting examples exist for how this reaction has been adapted by genetic elements that led to the spread of antibiotic resistance.

One interesting system is the **integron system** that presents as a natural gene cloning system in bacteria. Integrons are natural genetic systems that encode various antibiotic resistances and pathogenicity factors that can be found in plasmids, transposons, and the chromosome of bacteria. The integron consists of a tyrosine-recombinase integrase and DNA molecules called cassettes that encode individual ORFs usually lacking an endogenous promoter (FIGURE 14.32). The integrase can mobilize the cassettes to shuffle the position of the elements. The position of these cassettes is important because the cassettes often do not contain their own promoter and, therefore, only those cassettes close to the promoter near the integrase gene, but transcribed by a divergent promoter, will be actively expressed. Hence, the integron system expresses functions that can be beneficial to the host and also store a library of genetic information that can be easily accessed in the future. These cassettes also can be swapped between integrons in the same cell. A stunning example of how the integron system has been used by a pathogen is found in *Vibrio cholera* where a chromosomally encoded integron was found to contain over 100 cassettes.

There is not an absolute requirement that all conservative site-specific recombination enzymes use identical sites in the donor and target DNA. This adaptation to have less specificity for sites accounts for some of the important properties of some family members. For example, integrative conjugal elements, such as Tn*916*, can insert into many sites that only share weak homology in the donor and target DNAs. Because a conserved tyrosine is used to crosslink to the DNA in a process allowing mobilization to non-identical sites this type of recombinase is sometimes referred to as a Y-transposase (proteins that use a conserved serine residue have been called S-transposases for the same reason). An additional adaptation with Tn*916* and some other elements is the ability to transmit as a circular element to other cells through conjugation. As expected, this type of genetic transmission also is responsible for the spread of antibiotic resistance in bacteria. A eukaryotic version of this Y-transposase process may be found with a class of retrotransposon that appears to be able to form circular substrates that may integrate by using an element encoded tyrosine recombinase.

14.3 Target-Primed Reverse Transcription

Target-primed reverse transcription can mobilize information through an RNA intermediate.

The final broad category of mobile DNA to be discussed in this chapter are DNA elements that are copied to new locations via an RNA•protein complex that can break a DNA site to allow them to be copied into a new insertion site as DNA. While the LTR retrotransposons mentioned above also move via an RNA intermediate, these elements are reverse transcribed into DNA prior to integration (like retroviruses). Elements that move using target-primed reverse transcription do not require specific repeats at their ends because priming comes from the target DNA itself. Because elements that move by target-primed reverse transcription lack LTRs they are sometimes referred to as non-LTR elements. Non-LTR elements have poly(A) sequences at one end of the integrated element that result from the direct use of an RNA in the mobility of the element (FIGURE 14.33). Functional non-LTR elements typically encode a single protein with at least two activities: a reverse transcriptase activity and an endonuclease activity. However, they can encode various other functions like nucleic-acid binding proteins that also may act as chaperones. The coding region is flanked by 5′ and 3′ untranslated regions (5′ UTR and 3′ UTR).

Two types of elements that prime reverse transcription using breaks in the host chromosome, **long interspersed nucleotide elements (LINEs)** and **mobile group II introns**, are discussed. LINEs are found in mammalian genomes and comprise the bulk of mammalian DNA

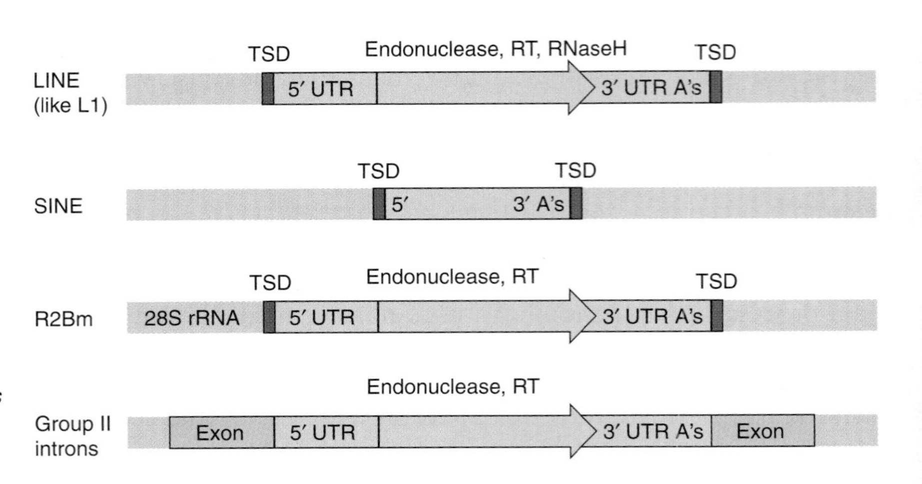

FIGURE 14.33 **Examples of genetic elements that move via a target primed reverse transcription mechanism.** The long interspersed nucleotide elements (LINE) like L1 and its non-autonomous derivative the short interspersed nucleotide element (SINE) are shown. The R2Bm element from the silkworm found in the rRNA locus and a representative group II intron are also shown. RT indicates the coding region for the reverse transcriptase. The 5′ and 3′ untranslated regions are shown (UTRs) as is the target site duplication (TSD). (Adapted from F. Bushman. *Lateral DNA Transfer: Mechanisms and Consequences*. Cold Spring Harbor Laboratory Press, 2002.)

(Figure 14.33). Mobile group II introns, which are genetic elements that can self-cleave out of an mRNA, are found predominantly in bacteria and bacterial derived organelles in eukaryotes.

Target-primed reverse transcription is used for LINE movement.

Mammalian genomes such as the human genome have hundreds of thousands of LINEs, although fewer than 100 of these elements are thought to function autonomously (Figure 14.33). In mice, the number of active elements of this class element is estimated to be around 3000. The mechanism of LINE movement was derived from earlier work with the R2Bm element from the silkworm *Bombyx mori* (Figure 14.33). This class of element uses an RNA intermediate, but does not require long terminal repeats because reverse transcription is primed by the 3′ ends of the target DNA itself generated from a break in the host chromosome (FIGURE 14.34). LINEs, typified by the element LINE-1 (L1), are about four to seven thousand base pairs in length. Functional L1 encodes an endonuclease for cutting the host chromosome and a reverse transcriptase activity that is responsible for making a DNA copy of the RNA element as well as an RNA

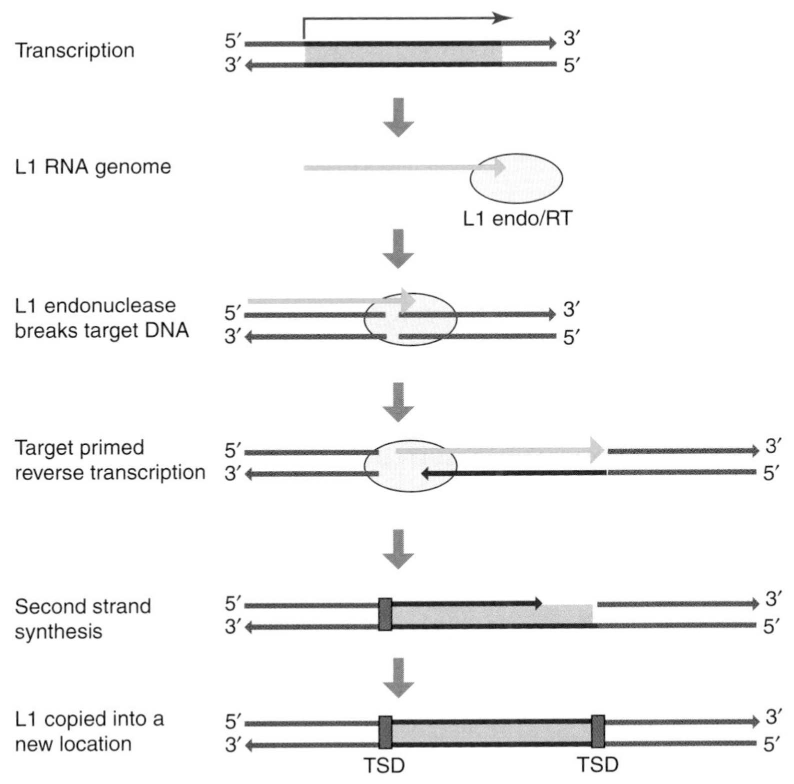

FIGURE 14.34 **Steps in the movement of L1 into a new location in a genome.** The RNA copy (green) of the element complexed with a protein with endonuclease activity to cut at the target site and reverse transcriptase to make a DNA copy of the RNA element (endo/RT). The broken chromosome is used to prime reverse transcription of the element (red). Second strand synthesis replaces the RNA portion of the element with DNA. A target site duplication (TSD) results because of the staggered nature of the cuts in the genome.

binding protein (Figure 14.34). After the L1 element is transcribed and its protein translated, the nucleoprotein complex (L1 RNA•L1 endonuclease/reverse transcriptase) can break a target DNA. This reaction usually does not occur at a specific sequence, but instead, appears to target DNA sequences that are inherently bent. The free 3′ end generated by the break is used to initiate reverse transcription of the element. Because the breaks made by the endonuclease are staggered, like those found with the DDE transposons, movement of LINEs also will result in a target site duplication.

The **short interspersed nucleotide elements** (**SINE**s) are less than a thousand base pairs in size and are nonfunctional LINEs that have been subjected to deletions over time. The extremely high copy number of defective elements helps to reduce the frequency of transposition of the active elements. This reduction is because the elements encoding a functional protein will be shared among all of the nonfunctional elements. The contribution of SINEs to mammalian genomes is significant; one common type of SINE called *Alu* is estimated to make up about 11% of the human genome. The 1.1 million *Alu* elements in the human genome are believed to trace back to a single insertion event that occurred into the *7SL* RNA gene in an ancient primate (**FIGURE 14.35**). This event occurred in the past 65 million years prior to the spread of primates. Comparing *Alu* elements can provide an important tool in human population genetics and primate comparative genomics. A controversial view holds that the diversity generated by LINEs and SINEs could have contributed in an essential way to the expansion of primates including humans by the mechanisms described below.

LINE movement affects genome stability and evolution.

Elements using target-primed reverse transcription are extremely common in multicellular organisms. LINEs can negatively and positively impact genomes. It is clear that LINEs will occasionally move; it is estimated that one in fifty humans will experience a new L1 integration event in their genome that occurred either in the parental germ cells or early in their own embryonic development. *Alu* retrotransposition is more common and insertions occur in an estimated one in thirty individuals. In all cases of retroelement movement, the new insertion events can act as an insertion mutagen or alter the expression of genes and cause disease. It is estimated that approximately 0.1% of human genetic disorders are caused by the movement of *Alu* elements.

Functional LINEs can move more elements than just nonfunctional LINEs and SINEs. It also appears that cellular mRNAs can be processed by the L1 machinery and inserted into new positions in the chromosome. This comes from the observation that nonfunctional genes called **processed pseudogenes**, which contain a poly(A) end and are flanked by target site duplications, appear to reside at preferred L1 insertion sites (Figure 14.33). These processed pseudogenes comprise an estimated 0.5% of the human genome.

Retroelements such as L1 also can provide regions of homology large enough for recombination enzymes to catalyze destructive

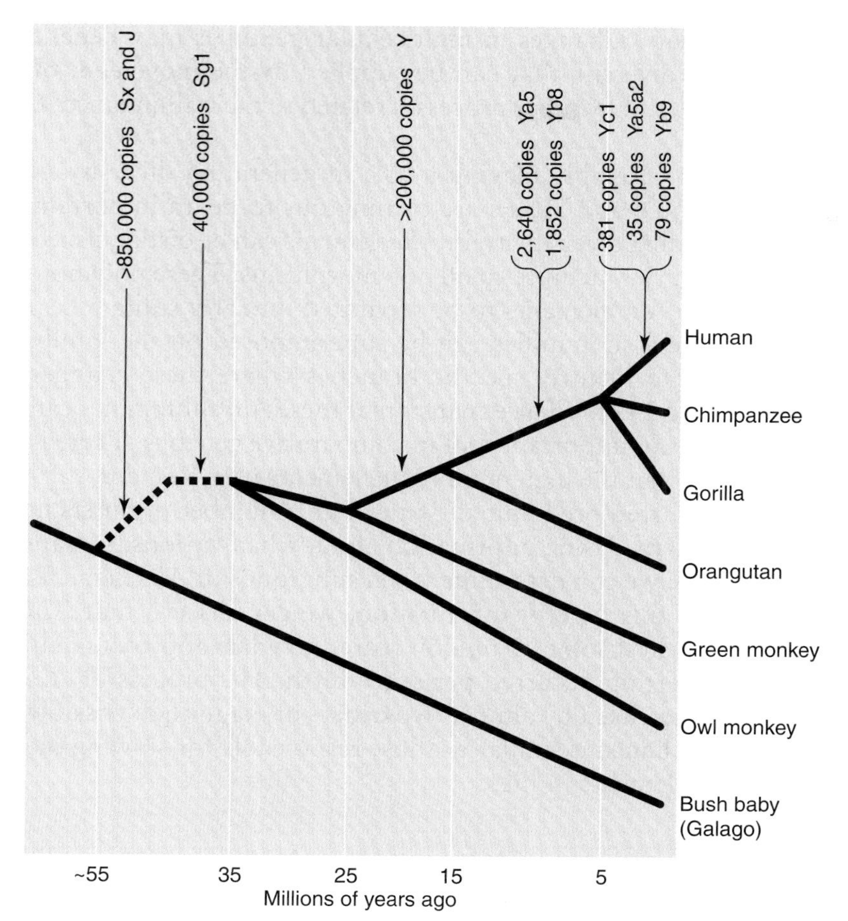

FIGURE 14.35 ***Alu* elements expanded in primates.** The increase in *Alu* elements in primates is shown with the subfamilies (Yc1, Ya5a2, Yb9, Ya5, Yb8, Y, Sg1, Sx, and J) are mapped on the evolutionary tree for primates. The approximate number of elements is shown. Mya means millions of years ago. The colored boxes indicate the time frame over which the first change occurred in an element to define a new subfamily until the new element reached its current copy number. (Adapted from M. A. Batzer and P. L. Deininger, *Nat. Rev. Genet.* 3 [2002]: 370–379.)

recombination events, such as inversions, deletions, and duplications. Such events have been detected in mammalian cells in nature and in culture. The relative paucity of L1 elements in gene-rich areas of the human genome may result from an evolutionary selection against these events because they can cause template destructive forms of recombination in these regions.

Retroelements could provide benefits to the host genome, too. It is clear that retroelements can take advantage of breaks in the chromosome that are made from other processes as a target for insertion. It remains unclear if they aid in the repair process or if they merely insert in a break that would normally be fixed through other means such as nonhomologous end-joining or replication-mediated repair. It is clear that L1-mediated events that do not require endonuclease activity are evident in cultured cells that are defective in nonhomologous end-joining. L1 elements also can participate in a gene fusion event with adjacent genes, a process that can contribute to the formation of new hybrid genes. Template switching could increase the number of hybrid genes. These types of processes have left a mark in mammalian genomes; hundreds of examples exist where human proteins contain portions of proteins derived from L1 and *Alu*. It remains to be determined whether these proteins are part of the normal functioning of our cells. LINEs and SINEs also can impact gene expression by altering the expression of genes through shuffling transcription regulatory elements. Changes in multicellular genomes may benefit greatly from the diversity that can be supplied by the movement of LINEs and SINEs and by other processes related to their abundance in genomes.

In addition to possibly playing a role in generating diversity in primates, the LINEs and SINEs are turning out to be an important tool in understanding diversity itself. *Alu* elements can be distinguished from each other if they acquire small polymorphisms. These polymorphisms allow the *Alu* elements to be separated into *Alu* subfamilies. The genesis of new subfamilies can be superimposed on the family tree of primates to identify specific branches where these changes took place (Figure 14.35). The expansion of these *Alu* subfamilies can therefore provide an important tool in comparative genetics. There is reason to believe that the high number of elements now found in various primates occurred predominantly through a number of bursts in retroelement activity. Using information about when various primate lineages diverged, we can determine when subgroups of *Alu* elements expanded. From this type of information, we can discern that *Alu* replication was fastest some 40 million years ago when approximately one new *Alu* insertion occurred per every birth. The process of *Alu* expansion has decreased by almost two orders of magnitude possibly as a result of mechanisms similar to those responsible for slowing the movement of P elements in flies.

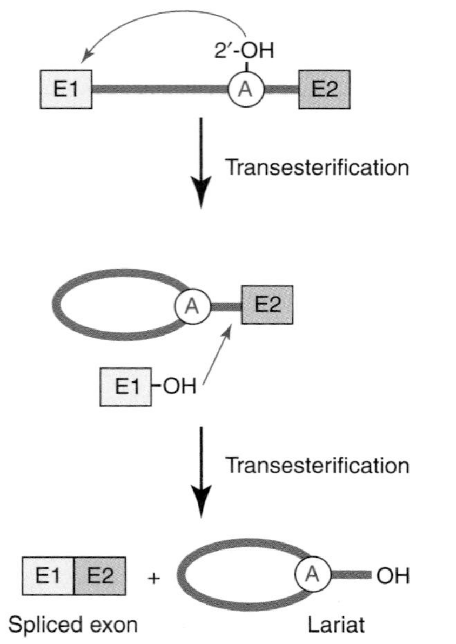

Mobile group II introns move by target-primed reverse transcription.

Mobile group II introns are elements that mobilize via an RNA intermediate by a mechanism that is similar to non-LTR elements like LINEs. It remains possible that LINEs and group II introns share a common ancestor. When a messenger RNA containing a mobile group II intron is transcribed, the element can use its splicing activity to escape from the RNA molecule as an intron. This intron codes for a protein that stabilizes the catalytic RNA and has reverse transcription activity to make a DNA copy of the RNA element in the target DNA. The intron-encoded protein sometimes has endonuclease activity that will allow a break in the target DNA that could facilitate reverse transcription of the RNA element.

Mobile group II introns are common in bacteria and by genome sequences, appear to be present in about a quarter of organisms. A number of examples exist where the elements have been found in the archaea. While formally no group II mobile introns have been found in nuclear eukaryotic genomes, these elements can be found in eukaryotic organelles believed to have evolved from bacteria. In fact, in some cases, in what are sometimes referred to as lower eukaryotes, group II mobile introns can make up a very significant part of the organellar DNA. The meganuclease I-Sce1 described in Chapter 13 is from a mobile intron found in *S. cerevisiae*. The primary target for integration is an empty site like the one from which they are transcribed in

FIGURE 14.36 Excision of group II mobile introns. Group II mobile introns excise from exons using two transesterification reactions. In the first reaction a 2′-OH is used to attack the 5′ exon junction. In a second reaction the free OH group attacks the 3′ exon junction releasing the element as a lariat on splicing the exons. E1 and E2 indicate exons, coding regions for the host proteins within the mRNA. (Modified from *Trends Genet.*, vol. 17, L. Bonen and J. Vogel, The ins and outs of group II introns, pp. 322–331, copyright 2001, with permission from Elsevier [http://www.sciencedirect.com/science/journal/01689525].)

a process called retro-homing. At a much lower frequency, however, these elements can recognize sites that are similar to the primary target site in a process called retrotransposition.

Two transesterification reactions allow group II intron movement.

Group II mobile introns can use their own ribozyme activity to self-excise from the messenger RNA where they reside. The first step in the process involves an adenine ribonucleotide that can use its 2′-OH in a transesterification reaction at the intron-exon junction at the 5′ end of the intron (FIGURE 14.36). The 3′-OH that is liberated on the exon can then be used in a second transesterification reaction at the intron-exon junction at the 3′ end of the intron, thus joining the exons and totally liberating the intron as a lariat. A nucleoprotein that is formed with this lariat structure and a protein encoded by the intron is competent for integration (FIGURE 14.37). Important features in the intron's RNA are regions that will recognize positions in the target DNA that encode the exon. Reverse splicing will join the RNA element to the top strand at the insertion site. The endonuclease encoded by the intron is then capable of cleaving the bottom strand to form a free 3′ end that can prime replication of a DNA copy of the RNA element. The fate of the remaining RNA strand at the insertion site is not as clear and differs depending on the element and host. In many cases, host repair activities may be responsible for processing the final RNA strand to DNA.

Homology to the target site determines if mobile group II introns move by retrotransposition or retro-homing.

Group II mobile introns are extremely efficient at inserting into an empty site like the one from which they were transcribed. This is because there are homologous sequences in the RNA and the insertion site. Interestingly, a new technique using group II mobile introns involves making a synthetic element using PCR where sequences of a particular region of the host genome can be inserted into a group II element. This synthetic element can then **retrohome** to that particular site. This process appears to have a broad host range and has been shown to work in human cells grown in the laboratory, suggesting this procedure might provide an important tool in the future. In nature, group II mobile elements will insert at a low frequency into chromosomal sites that have a low level of homology to the natural insertion site. This process, called **retrotransposition**, could be important in the impact and distribution of these elements.

It has been suggested that group II introns evolved early in life and that they could be the progenitors to introns in higher organisms. In addition, given the similarities in the chemistry of the reaction, it also remains possible that the retroelements that use target-primed reverse transcription could be descendants from these group II mobile elements.

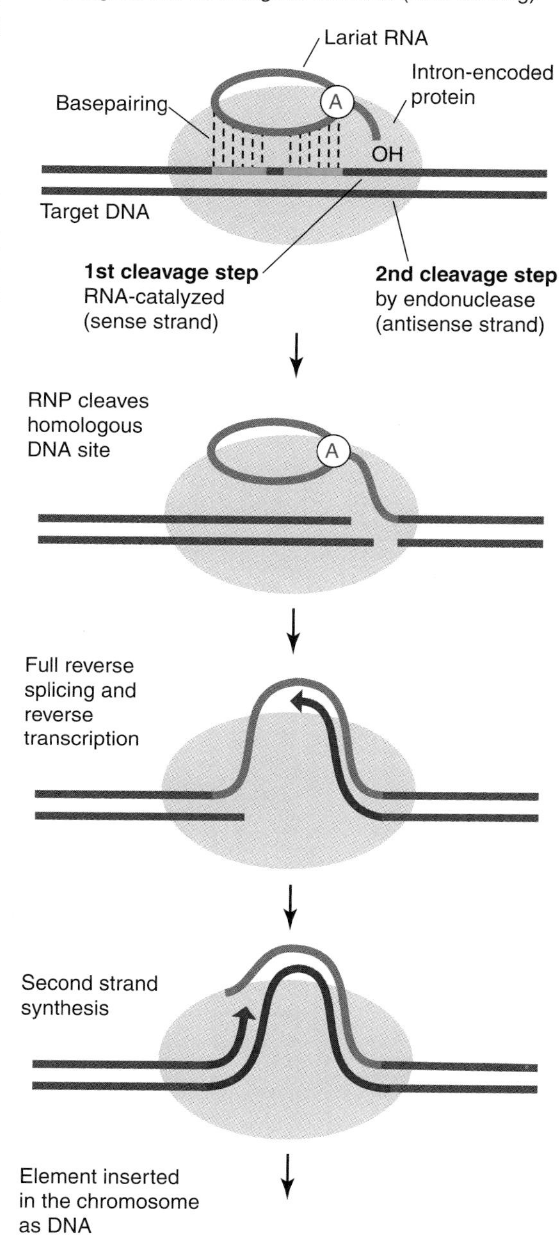

FIGURE 14.37 **Integration of group II introns.** Retro-homing involves extensive specific interaction between the target DNA and the lariat RNA within the ribonucleoprotein (RNP) complex. RNP-directed cleavage allows the RNA element to reverse splice into the genome. The free 3′ OH is used to prime reverse transcription of the RNA genome. Second strand synthesis replaces the RNA stand. (Adapted from L. Bonen and J. Vogel, *Trends Genet.* 17 [2001]: 322–331.)

14.4 Other Mechanisms of DNA mobilization

Other mechanisms of mobilization have been discovered that do not fit into the three broad classes of elements described. One relatively new class of mobile element uses a type of initiator protein involved in rolling-circle DNA replication that is normally found associated with plasmid and bacteriophage replication (see Chapter 8). This process is believed to use a conserved tyrosine to crosslink to one strand of DNA and therefore these elements are sometimes referred to as the Y2-transposons (to differentiate them from the Y-transposons described above). The eukaryotic versions of these elements are called *Helitrons* because of a helicase domain found in the element-encoded protein that presumably helps to unwind the strand to be mobilized. These elements do not form target-site duplications like many of the other elements. The ends of the element correspond to sequences in the element where DNA replication is initiated and terminated. Computational studies have identified *Helitron* DNA transposons in plants, fungi, insects, nematatodes, and vertebrates. Evolutionarily, *Helitrons* are proving interesting because of their ability to capture genes from the host. Presumably these elements can mobilize host DNA sequences when the normal replication termination signals are not recognized.

New twists on previously investigated families of recombination as well as completely new systems for mobilizing DNA will certainly continue to be discovered. Work with mobile elements promises to broaden our appreciation of the mechanisms shaping the genomes of all living organisms and provide important new tools for genome analysis and manipulation.

Suggested Reading

Historic Interest

Comfort, N. C. 2003. *The Tangled Field: Barbara McClintock's Search for the Patterns of Genetic Control*. Cambridge, MA: Harvard University Press.

Keller, E. F., and Mandelbrot, B. B. 1984. *A Feeling for the Organism: The Life and Work of Barbara McClintock*. New York: W. H. Freeman.

General Interest

Bushman, F. 2002. *Lateral DNA Transfer, Mechanisms and Consequences*. Woodbury, NY: Cold Spring Harbor Press.

Craig, N. L., Craigie R., Gellert M., and Lambowitz, A. M. (eds). 2002. *Mobile DNA*, 2nd ed. Washington, DC: ASM Press.

Curcio, M.J., and Derbyshire, K.M. 2003. The outs and ins of transposition: from mu to kangaroo. *Nat Rev Mol Cell Biol* 4:865–877.

Shuman, H. A., and Silhavy, T. J. 2003. The art and design of genetic screens: *Escherichia coli*. *Nat Rev Genet* 4:419–431.

Transposition

Davies, D. R., Goryhin, I. Y., Reznikoff, W. S., and Rayment, I. 2000. Three-dimensional structure of the Tn*5* synaptic complex transposition intermediate. *Science* 289:77–85.

Gellert, M. 2005. V(D)J recombination: RAG proteins, repair factors, and regulation. *Annu Rev Biochem* 71:101–132.

Hickman, A. B., Perez, Z. N., Zhou, P., et al. 2005. Molecular architecture of a eukaryotic DNA transposase. *Nat Struct Mol Biol* 12:715–721.

Ivics, Z., Hackett, P. B., Plasterk, R. H., Izsvák, Z. 1997. Molecular reconstruction of *Sleeping Beauty*, a Tc1-like transposon from fish, and its transposition in human cells. *Cell* 91:501–510.

Jones, R.N. 2005. McClintock's controlling elements: the full story. *Cytogenet Genome Res* 109:90–103.

Kumar, A., Seringhaus, M. Biery, M. C., et al. 2004. Large-scale mutagenesis of the yeast genome using a Tn7-derived multipurpose transposon. *Genome Res* 14:1975–1986.

Parks, A. R., Li, Z., Shi, Q., Owens, R. M., Jin, M. M., and Peters, J. E. 2009. Transposition into replicating DNA occurs through interaction with the processivity factor. *Cell* 138:685–695

Peters, J. E., and Craig, N. L. 2001. Tn7: smarter than we thought. *Nat Rev Mol Cell Biol* 2:806–814.

Reznikoff, W. S. 2008. Transposon Tn*5*. *Annu Rev Genet* 42:269–286.

Rousseau, P., and Chandler, M. 2008. DNA transposition: topics in chemical biology. In: *Encyclopedia of Chemical Biology*. pp. 1–8. Hoboken, NJ: John Wiley and Sons Ltd.

Singleton, T.L., and Levin, H.L. 2002. A long terminal repeat retrotransposon of fission yeast has strong preferences for specific sites of insertion. *Eukaryotic Cell* 1:44–55.

Zhou, L., Mitra, R., Atkinson, P. W., et al. 2004. Transposition of the *hAT* element links transposable elements and V(D)J recombination. *Nature* 432:995–1001.

Conservative Site-Specific Recombination

Boucher, Y., Labbate, M., Koenig, J. E., and Stokes, H. W. 2007. Integrons: mobilizable platforms that promote genetic diversity in bacteria. *Trends Microbiol* 15:301-309.

Hartley, J. L., Temple, G. F., and Brasch, M. A. 2000. DNA cloning using *in vitro* site-specific recombination. *Genome Res* 10:1788–1795.

Target-Primed Reverse Transcription

Batzer, M. A., and Deininger, P. L. 2002. Alu repeats and human genomic diversity. *Nat Rev Genet* 3:370–380.

Beauregard, A., Curcio, M. J., and Belfort, M. 2008. The take and give between retrotransposable elements and their hosts. *Annu Rev Genet* 42:587–617.

Furano, A. V., and Boissinot, S. 2008. Long interspersed nuclear elements (LINES): evolution. In: *Encyclopedia of Life Sciences*. pp. 1–6. Hoboken, NJ: John Wiley and Sons.

Goodier, J. L., and Kazazian, H. H. Jr. 2008. Retrotransposons revisited: the restraint and rehabilitation of parasites. *Cell* 135:23–35.

Kazazian, H. H. Jr. 2004. Mobile elements: drivers of genome evolution. *Science* 303:1626–1632.

Morrish, T. A., Gilbert, N. Myers, J. S., et al. 2002. DNA repair mediated by endonuclease-independent LINE-1 retrotransposition. *Nat Genet* 31:159–165.

Other Mechanisms of DNA Mobilization

Jurka, J., Kapitonov, V.V., Kohany, O., and Jurka M. V. 2007. Repetitive sequences in complex genomes: structure and evolution. *Annu Rev Genomics Hum Genet* 8:241–259.

Kapitonov, V. V., and Jurka, J. 2007. *Helitrons* on roll: eukaryotic rolling-circle transposons. *Trends Genet* 23:521–529.

This book has a Web site, **http://biology.jbpub.com/book/molecular**, which contains a variety of resources, including links to in-depth information on topics covered in this chapter, and review material designed to assist in your study of molecular biology.

SECTION V

RNA Metabolism

CHAPTER 15 **Bacterial RNA Polymerase**

CHAPTER 16 **Regulation of Bacterial Gene Transcription**

CHAPTER 17 **RNA Polymerase II: Basal Transcription**

CHAPTER 18 **RNA Polymerase II: Regulation**

CHAPTER 19 **RNA Polymerase II: Cotranscriptional and Posttranscriptional Processes**

CHAPTER 20 **RNA Polymerases I and III and Organellar RNA Polymerases**

CHAPTER 21 **Small Silencing RNAs**

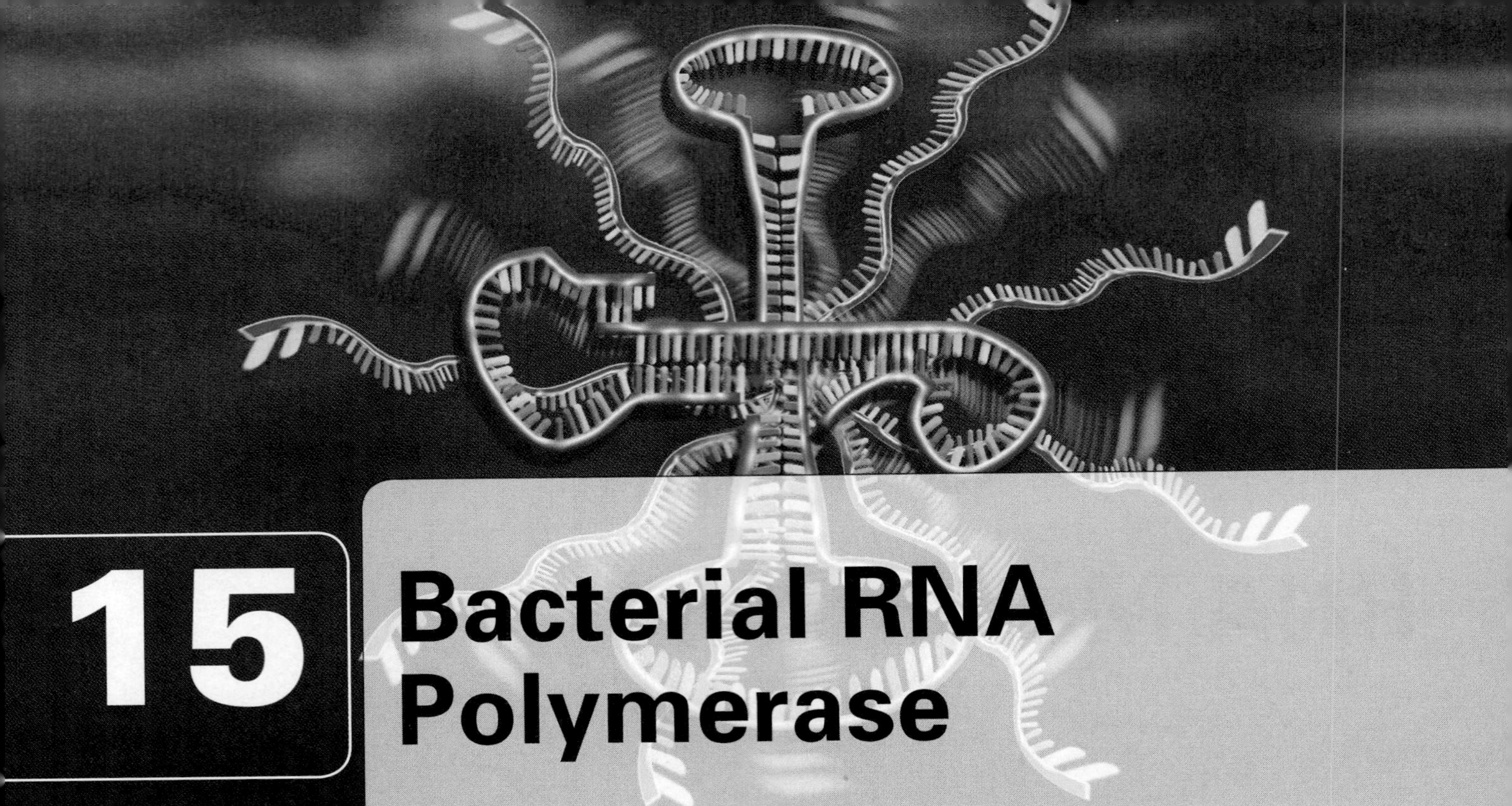

15 Bacterial RNA Polymerase

OUTLINE OF TOPICS

15.1 Introduction to the Bacterial RNA Polymerase Catalyzed Reaction

RNA polymerase requires a DNA template and four nucleoside triphosphates to synthesize RNA.

Bacterial RNA polymerases are large multisubunit proteins.

15.2 Initiation Stage

Bacterial RNA polymerase holoenzyme consists of a core enzyme and sigma factor.

A transcription unit must have an initiation signal called a promoter for accurate and efficient transcription to take place.

The DNase protection method provides information about promoter DNA.

DNA footprinting shows that σ^{70}-RNA polymerase binds to promoter DNA to form a closed and an open complex.

Bacterial RNA polymerase crystal structures show how the enzyme is organized and provide insights into how it works.

Genetic and biochemical studies provide additional information about bacterial promoters.

Members of the σ^{70} family have four conserved domains.

RNA polymerase scrunches DNA during transcription initiation.

Transcription initiation is a stepwise process.

Alternative σ factors direct RNA polymerase to genes that code for proteins that bacteria require to survive under specific types of environmental stress.

The σ^{54}-RNA polymerase requires an activator protein.

15.3 Transcription Elongation Complex

The transcription elongation complex is a highly processive molecular motor.

The trigger loop and the β' bridge helix help to move RNA polymerase forward by one nucleotide during each nucleotide addition cycle.

Pauses influence the overall transcription elongation rate.

RNA polymerase can detect and remove incorrectly incorporated nucleotides.

15.4 Transcription Termination

Bacterial transcription machinery releases RNA strands at intrinsic and Rho-dependent terminators.

15.5 Antibiotics that Target Bacterial RNA Polymerase

RNA polymerase is a target for broad spectrum antibacterial therapy.

Suggested Reading

Cells express their genes by transferring information from DNA to RNA and then from RNA to polypeptides. Enzymes that copy information present in DNA templates into RNA molecules are called **DNA-dependent RNA polymerases** or **RNA polymerases** for short. RNA polymerases were purified first from bacteria and later from the eukaryotes and archaea. RNA synthesis in all three domains of life takes place in three stages: initiation, elongation, and termination.

Bacterial, eukaryotic, and archaeal RNA polymerases are large multisubunit enzymes that require the assistance of additional factors to recognize specific genes. Initial studies suggested that RNA polymerases from organisms belonging to each domain might be unique to that domain. While the RNA polymerases do in fact differ in several important respects, more recent studies show that the catalytic core of the enzymes from all three domains are remarkably similar. Therefore, lessons learned from studying an RNA polymerase from an organism belonging to one domain can provide useful information for studying RNA polymerases from organisms belonging to the other two domains.

Despite the fact that bacterial, eukaryotic, and archaeal transcription share many important similarities, important differences also exist. For example, details of the bacterial and eukaryotic initiation and termination processes differ, involving different factors and different mechanisms. The mechanisms that regulate bacterial and eukaryotic RNA synthesis also differ, as do the pathways for converting the primary transcripts into mature RNA molecules. Because of these differences, bacterial and eukaryotic transcription will be examined in separate chapters. The archaeal transcription machinery will be examined in Chapter 17, which describes the eukaryotic RNA polymerase that synthesizes mRNA, because this enzyme is similar to the archaeal enzyme.

This chapter examines how bacterial RNA polymerases initiate, elongate, and terminate RNA synthesis. *In vitro* and *in vivo* studies, mostly involving the *Escherichia coli* RNA polymerase, have provided important insights into how bacterial RNA polymerases work. Unfortunately, nobody has yet succeeded in preparing *E. coli* RNA polymerase in a crystalline form and so no high resolution image is available for this enzyme. Crystal structures, however, have been determined for RNA polymerases isolated from two other gram-negative bacteria, the extreme thermophiles *Thermus aquaticus* and *Thermus thermophilus*. The high degree of sequence homology present in bacterial RNA polymerases permits us to apply information obtained from the RNA polymerase from one bacterial strain to RNA polymerases of other bacterial strains. This chapter begins with an introduction to the RNA polymerase catalyzed reaction and then examines the initiation, elongation, and termination stages of bacterial RNA synthesis. Additional transcription factors that regulate bacterial RNA synthesis are described in the next chapter.

15.1 Introduction to the Bacterial RNA Polymerase Catalyzed Reaction

RNA polymerase requires a DNA template and four nucleoside triphosphates to synthesize RNA.

Several different groups detected bacterial RNA polymerases at about the same time in the early 1960s. Under most growth conditions, *E. coli* has about 1000 to 2000 RNA polymerase molecules per cell. The enzyme catalyzes nucleoside monophosphate group transfer from a nucleoside triphosphate (NTP) to the 3′-end of the growing RNA chain (or the first nucleoside triphosphate) so that chain growth proceeds in a 5′→3′ direction (FIGURE 15.1). The essential chemical characteristics of RNA synthesis are as follows:

1. Phosphodiester bond formation takes place as the result of a nucleophilic attack of the 3′-hydroxyl group on the growing chain (or first nucleoside triphosphate) on the α phosphoryl group of the incoming NTP. This reaction is similar to the one that takes place during DNA synthesis. Also, as with DNA synthesis, pyrophosphate hydrolysis drives the reaction to completion.
2. The DNA template sequence determines the RNA sequence. Each base added to the growing 3′-end of the RNA chain is chosen by its ability to pair with a complementary base in the template strand. Thus, the bases C, T, G, and A in a DNA strand cause G, A, C, and U, respectively, to appear in the newly synthesized RNA molecule.
3. All four ribonucleoside triphosphates (adenosine 5′-triphosphate [ATP], guanosine 5′-triphosphate [GTP], cytidine 5′-triphosphate [CTP], and uridine 5′-triphosphate [UTP]) are required for RNA synthesis. When a single nucleotide is omitted RNA synthesis stops at the point where that nucleotide must be added.
4. The RNA chain grows in the 5′→3′ direction; that is, nucleotides are added only to the 3′-OH end of the growing chain. This direction of chain growth is the same as that in DNA synthesis.
5. RNA polymerases, in contrast with DNA polymerases, are able to initiate chain growth so that *no primer is needed*.
6. Only ribonucleoside 5′-triphosphates participate in RNA synthesis. The first base to be laid down in the initiation event is a triphosphate. Its 3′-OH group is the point of attachment for the second nucleotide and its 5′-triphosphate group remains at the 5′-end throughout chain elongation.

Only one of the two DNA strands in a given chromosome region acts as the **template strand**, dictating the sequence of the newly synthesized RNA molecule. The complementary DNA strand, the **nontemplate strand**, has the same base sequence as the RNA molecule (except that U replaces T) and is commonly referred to as the **coding**

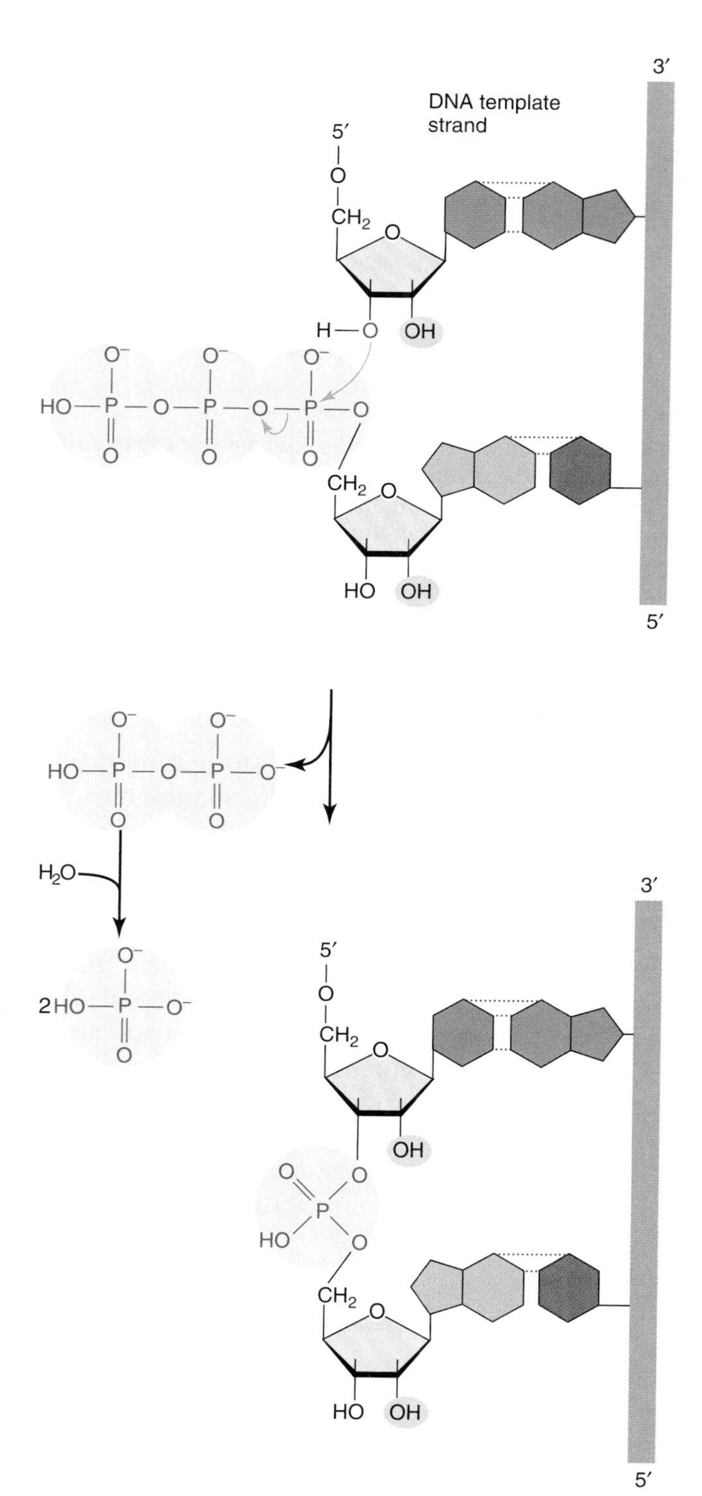

FIGURE 15.1 RNA polymerase catalyzed phosphodiester bond formation. RNA polymerase catalyzes nucleoside monophosphate group transfer from a nucleoside triphosphate (NTP) to the 3′-end of a growing RNA chain (or the first nucleotide triphosphate) so that chain growth proceeds in a 5′→3′ direction. (Adapted from J. M. Berg, et al. *Biochemistry, Fifth edition.* W. H. Freeman and Company, 2002.)

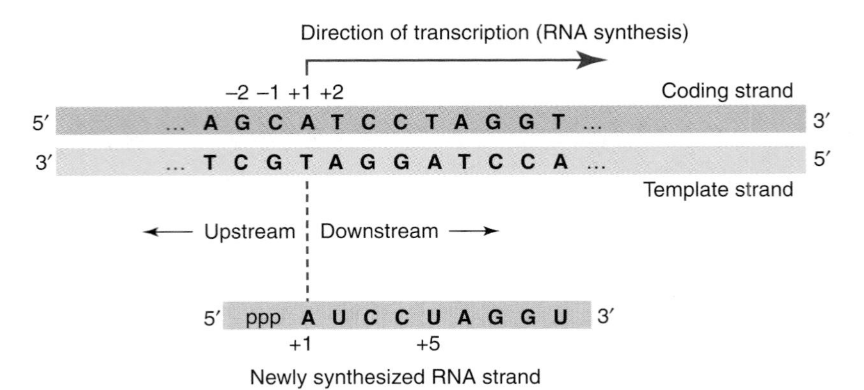

FIGURE 15.2 **Rules for numbering nucleotides on the sense strand.** The template strand (blue) dictates the nucleotide sequence in the newly formed RNA strand (green). The complementary DNA strand (red) is called the coding or sense strand. By convention the nucleotide at the transcription start site on the coding strand is designated as position +1; the next nucleotide as at position +2, and so forth. Sequences that come after the transcription start site (on the 3′-side) are downstream while those that come before it (on the 5′-side) are upstream.

or **sense strand** (FIGURE 15.2). Sequence information is usually given for just the coding strand because this information also provides the sequence of the **primary transcript** (the newly synthesized RNA molecule before it is processed.) Template strand sequence is easily obtained from Watson-Crick base pairing rules.

By convention, the nucleotide at the transcription start site on the coding strand is designated as position +1, the next nucleotide is at position +2, and so forth (Figure 15.2). The nucleotide that immediately precedes the transcription start site on the coding strand is at position −1, the nucleotide before that is at position −2, and so forth. Sequences that come *after* the transcription start site (on the 3′-side) are **downstream** while those that come *before* it (on the 5′-side) are **upstream.** RNA polymerase moves downstream as it transcribes a DNA template. With this background information in mind, let us now examine bacterial RNA polymerase.

Bacterial RNA polymerases are large multisubunit proteins.

Purification and characterization studies show that the fully active form of *E. coli* RNA polymerase, the **RNA polymerase holoenzyme,** is a large multisubunit protein (molecular mass = 459 kDa) that has six subunits ($\alpha_2\beta\beta'\omega s$) (Table 15.1). RNA polymerase holoenzymes from other bacteria have the same multisubunit structure and each subunit sequence appears to have been conserved during evolution.

RNA polymerase activity is usually assayed using a reaction mixture that contains RNA polymerase (which may be in a crude extract), DNA, Mg^{2+}, three nonradioactive NTPs, and one radioactive NTP labeled either in the base with ^3H or ^{14}C or in the phosphate attached to the ribose with ^{32}P. The reaction is stopped by adding trichloroacetic

TABLE 15.1 *E. coli* RNA Polymerase Subunits

Subunit	Number of Subunits in Holoenzyme	Gene	Map Position	Molecular Mass (Da)	Function
alpha (α)	2	*rpoA*	74.10	36,511	Required for enzyme assembly; interacts with some regulatory proteins
beta (β)	1	*rpoB*	90.08	150,616	Forms a pincer and is the site of rifampicin action
beta′ (β′)	1	*rpoC*	90.16	155,159	Forms a pincer and provides an absolutely conserved –NADFDGD– motif that is essential for catalysis
omega (ω)	1	*rpoZ*	82.34	10,105	Helps in enzyme assembly but is not required for enzyme activity
sigma (σ)	1	*rpoD*	69.21	70,263	Directs enzyme to promoters but is not required for phosphodiester bond formation.

acid, which causes the newly formed RNA but not the nucleoside triphosphate precursors to become insoluble, and the precipitate is collected by filtration or centrifugation. The amount of radioactivity in the precipitate is proportional to the amount of RNA synthesized.

The large size of the bacterial holoenzyme compared to typical enzymes raises the question of whether some complex feature of the polymerization reaction requires such a large enzyme. In fact, bacteriophage T7 RNA polymerase, which has only one polypeptide subunit with a molecular mass 99 kDa and a tertiary structure (FIGURE 15.3) reminiscent of the Klenow fragment (see Chapter 5), synthesizes RNA very efficiently. Clearly, the polymerization reaction itself does not require a huge multisubunit bacterial enzyme.

Clues to understanding reasons for the size differences between the *E. coli* and the T7 phage enzymes come from (1) attempting to use the T7 phage enzyme to transcribe *E. coli* DNA and (2) studying the gene organization of the T7 phage. First, the T7 phage RNA polymerase can transcribe at best a small fraction of the *E. coli* genes. Second, T7 phage genes are arranged in only a few transcription units, so there are only a few RNA polymerase binding sites. In *E. coli*, RNA polymerase must be able to recognize approximately 4,300 genes that are signaled by about a thousand different binding sites and respond to a large number of regulatory proteins that alter the polymerase's ability to recognize a binding site. It seems likely that these multiple requirements necessitate the large multisubunit *E. coli* enzyme.

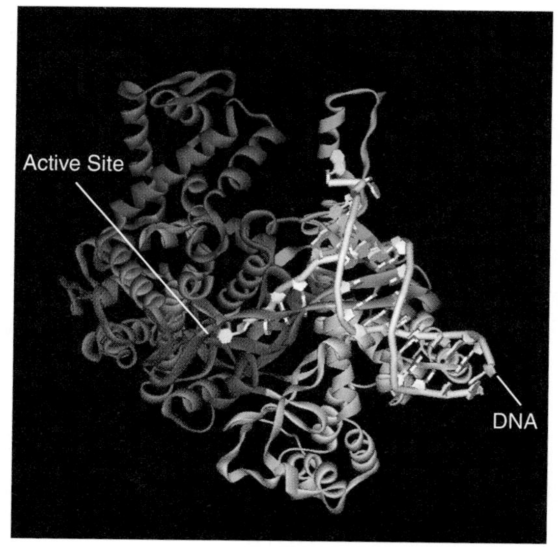

FIGURE 15.3 **Structure of the T7 RNA polymerase-DNA complex.** The polymerase subdomains are colored red "palm," green "thumb," and blue "fingers." The N-terminal domain (which is specific to all phage-like RNA polymerases) is yellow, and DNA is shown in white.(Structure from Protein Data Bank ID: 1CEZ. G. M. Cheetham, D. Jeruzalmi, and T. A. Steitz, *Nature* 399 [1999]: 80–83. Prepared by B. E. Tropp.)

15.2 Initiation Stage

Bacterial RNA polymerase holoenzyme consists of a core enzyme and sigma factor.

Early studies with an *E. coli* RNA polymerase holoenzyme that was missing the ω subunit showed that $\alpha_2\beta\beta'\sigma$ can dissociate under physiological conditions to form a core polymerase ($\alpha_2\beta\beta'$) and sigma (σ) factor as follows:

$$\underset{\text{RNA polymerase holoenzyme}}{\alpha_2\beta\beta'\sigma} \rightleftharpoons \underset{\text{core polymerase}}{\alpha_2\beta\beta'} + \underset{\text{sigma factor}}{\sigma}$$

The equilibrium lies far to the left with a dissociation constant of 10^{-9}. Richard Burgess and colleagues were able to separate the two proteins using phosphocellulose ion exchange chromatography, which shifts the equilibrium because core polymerase binds to the resin but σ factor does not (**FIGURE 15.4**).

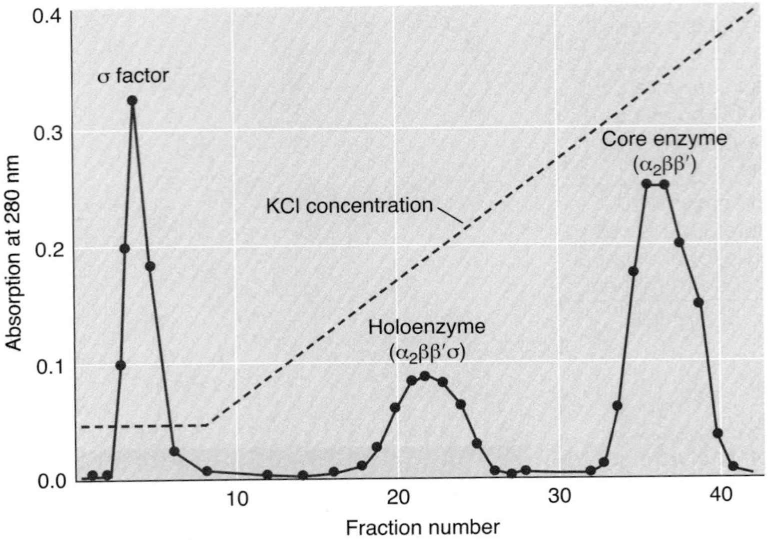

FIGURE 15.4 **Chromatographic separation of core polymerase from σ factor.** After being placed on a column containing phosphocellulose ion exchange resin, the protein was eluted with aqueous KCl solution at the indicated concentrations. Three protein peaks were observed by monitoring the light absorption at 280 nm. Analysis of the three peaks revealed that most of the holoenzyme dissociated into σ factor and core enzyme ($\alpha_2\beta\beta'$). Some of the original enzyme remained in its associated form ($\alpha_2\beta\beta'\sigma$), however, and is present in the central peak. The ω subunit was not detected in this study. (Adapted from R. R. Burgess, et al., *Nature* 221 [1969]: 43–46.)

The $\alpha_2\beta\beta'$ core polymerase can synthesize RNA molecules from single-stranded and nicked DNA templates but cannot use covalently closed circular double-stranded DNA molecules as templates. Hence, neither the σ factor nor the ω subunit is required for phosphodiester bond formation. The reason that the $\alpha_2\beta\beta'$ core polymerase can use nicked DNA templates is that DNA unwinds at nicks to generate nonphysiological initiation sites. The $\alpha_2\beta\beta'$ core polymerase remains intact under physiological conditions. When teased apart under nonphysiological conditions and purified, α, β, and β′ each lacks the ability to catalyze RNA synthesis by itself. The $\alpha_2\beta\beta'$ complex can be reassembled by mixing individual polypeptides in the order shown in **FIGURE 15.5**. The ω subunit helps fold the β′ subunit and then recruits it to the $\alpha_2\beta$ sub-assembly.

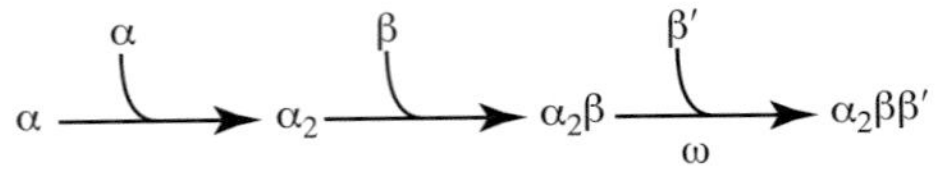

FIGURE 15.5 **Order of *in vitro* reassembly $\alpha_2\beta\beta'$ complex.** The β and β′ subunits are added after the α_2 dimer is formed. The ω subunit helps fold the β′ subunit and then recruits it to the $\alpha_2\beta$ sub-assembly.

A transcription unit must have an initiation signal called a promoter for accurate and efficient transcription to take place.

In contrast to the core polymerase ($\alpha_2\beta\beta'$), the RNA polymerase holoenzyme ($\alpha_2\beta\beta'\sigma$) can use intact DNA as a template for RNA synthesis. Transcription begins when the holoenzyme recognizes and binds to specific initiation signals called **promoters** at the beginning of a transcription unit. The existence of promoters was first demonstrated by isolation of a particular class of mutations in *E. coli* that prevent cells from synthesizing enzymes required for lactose metabolism. These mutations, termed **promoter mutations**, not only result in a lack of gene activity but also cannot be complemented because they are *cis*-dominant.

Sigma factor is essential for promoter DNA recognition but does not bind to promoter DNA on its own. *E. coli* has several different kinds of σ factors (see below). The primary *E. coli* σ factor, called σ^{70} because its molecular mass is about 70 kDa, helps core polymerase to bind to promoters in genes that code for housekeeping enzymes (enzymes required for essential metabolic steps in the cell) and destabilizes nonspecific interactions between RNA polymerase and DNA. Other bacterial species have σ^{70} homologs that perform the same function. The fact that core RNA polymerase can transcribe nicked DNA but not intact DNA suggested that σ^{70} might function by introducing transient nicks into DNA. Despite extensive efforts to observe such a nicking activity, none has ever been observed. For instance, σ^{70} does not alter the linking number of supercoiled DNA.

The DNase protection method provides information about promoter DNA.

RNA polymerase holoenzyme is sufficiently large to contact many deoxyribonucleotides within the promoter simultaneously. An estimate of the size of the region of the DNA where contact is made is obtained by selectively degrading adjacent DNA bases with DNase, a procedure known as the **DNase protection method** (FIGURE 15.6). RNA polymerase holoenzyme is bound to DNA and then a DNA endonuclease

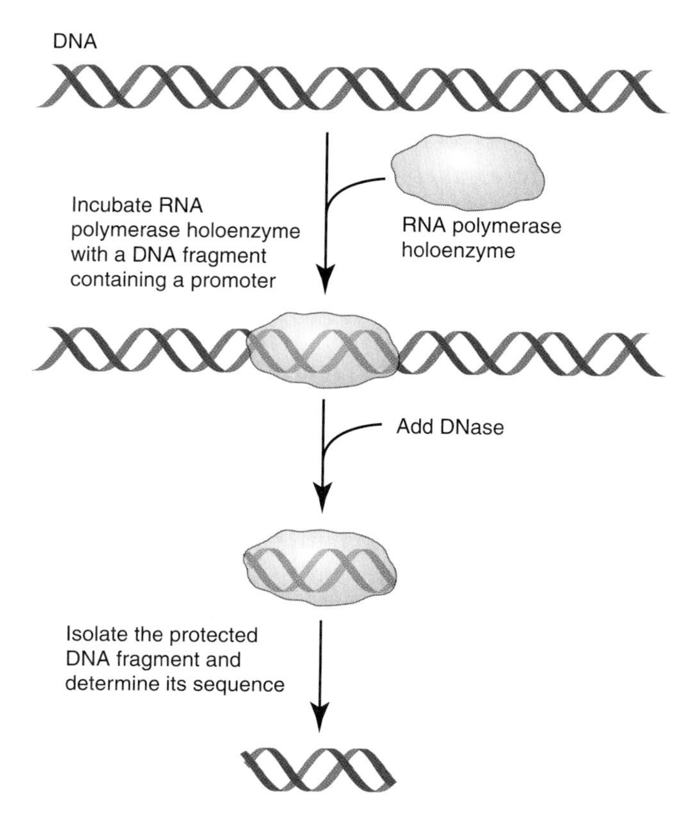

FIGURE 15.6 **DNase protection assay.** RNA polymerase holoenzyme binds to a DNA fragment containing a promoter. The enzyme protects the promoter from DNase hydrolysis. The protected region is isolated and sequenced. This technique can be used to characterize any sequence-specific protein–DNA interaction.

is added to the mixture. The endonuclease degrades most of the DNA to mono- and dinucleotides but leaves untouched DNA segments in close contact with RNA polymerase; protected segments vary in size from 41 to 44 base pairs (bp). If RNA polymerase holoenzyme were to be added to the total DNA complement of *E. coli* and then DNase were added, the protected promoter fragments would consist of about a thousand different DNA segments, each derived from a particular gene or set of adjacent genes. To study details of the DNA-enzyme binding process, it is obviously desirable to examine a single binding sequence. This can be accomplished by using a cloned gene in a protection experiment.

In 1975 David Pribnow used the DNase protection method to obtain DNA fragments containing promoters from genes that had been protected by RNA polymerase holoenzyme. These fragments were sequenced as were RNA molecules synthesized from each gene *in vitro*. The 5′-terminus of the RNA molecule revealed the initiation start site on the complementary DNA template strand. Pribnow recognized an important common feature in the protected DNA fragments. A six base pair sequence centered about 10 bp before (upstream from) the transcription start site is conserved. This hexamer is now known as the **–10 box** for its location or the **Pribnow box** for its discoverer. Examination of 300 *E. coli* promoters has shown that the frequency of occurrence of bases in –10 boxes is as follows (the subscript is the frequency):

$$T_{77}A_{76}T_{60}A_{61}A_{56}T_{82}$$

If the sequences were totally unrelated, one would expect each base to occur at each position 25% of the time, but instead the sequences are AT-rich. This AT-rich region melts during the early stage of transcription initiation.

Many, but not all, bacterial promoters have a second conserved six-base sequence, centered about 35 bp upstream from the transcription initiation site and therefore called the **–35 box**. Examination of 300 *E. coli* promoters revealed that the frequency of occurrence of bases in the –35 box is as follows (the subscript is the frequency):

$$T_{69}T_{79}G_{61}A_{56}C_{54}A_{54}$$

An idealized sequence such as that specified for the –10 box (or the –35 box), which indicates the most frequently found base in each position of many actual sequences, is termed a **consensus sequence**. Sequences of actual –10 and –35 boxes usually differ from the consensus sequence. These differences, which are evident in the *E. coli* promoter sequences recognized by σ^{70} holoenzyme (FIGURE 15.7), allow the cell to regulate genes based on the strength of RNA polymerase binding—an important regulatory mechanism.

DNA footprinting shows that σ^{70}-RNA polymerase binds to promoter DNA to form a closed and an open complex.

We can learn a great deal about the sequence specificity of a DNA binding protein by using the **DNA footprinting technique** (FIGURE 15.8). A particular piece of double-stranded DNA, labeled in one strand at its

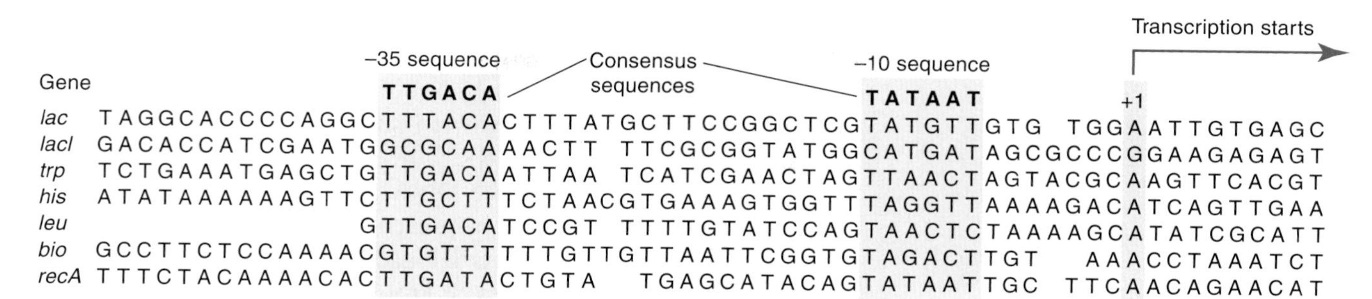

FIGURE 15.7 **Base sequences in promoter regions of several *E. coli* genes.** The consensus sequences located 10 and 35 nucleotides upstream from the transcription start site (+1) are indicated. Promoters vary tremendously in their ability to promote transcription. Much of the variation in promoter strength results from differences between the promoter elements and the consensus sequences in the −10 and −35 boxes.

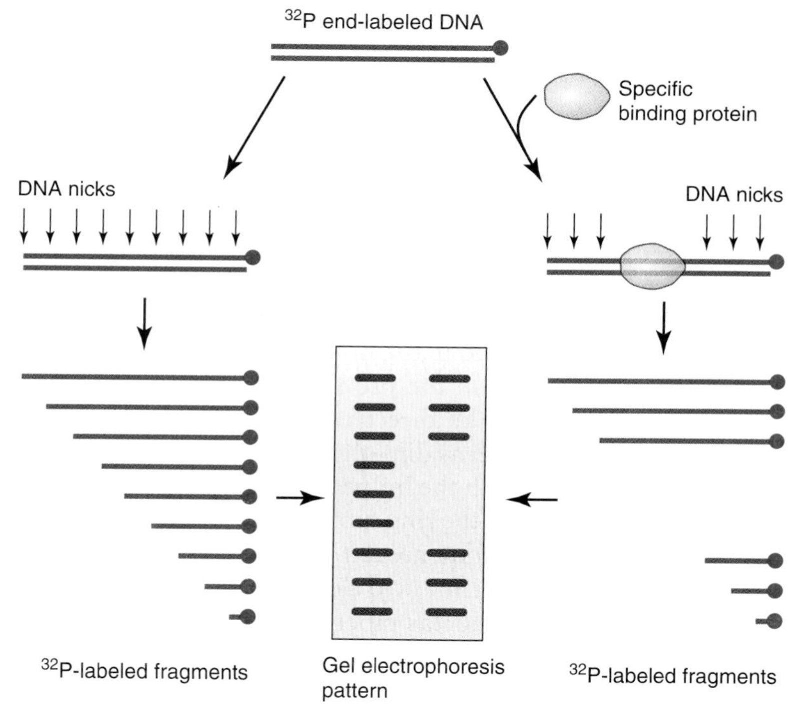

FIGURE 15.8 **DNA footprinting technique.** Double-stranded DNA, labeled in one strand at its 5′-terminus with ^{32}P (shown in red), is treated with DNase I for a very short time so that, on average, each DNA molecule receives no more than one single strand break. The same experiment is performed in the presence of a protein that binds to specific sites on the DNA. The DNA sites with the bound protein are protected from the action of DNase I. Certain fragments, therefore, will not be present. The missing bands in the gel pattern identify the binding site(s) on the DNA. (Adapted from L. Stryer. *Biochemistry, Fourth edition.* W.H. Freeman and Company, 1995.)

5′-terminus with ^{32}P (or in its 3′-terminus with a labeled nucleotide), is mixed with a DNA-binding protein of interest. Then DNase I is added, but so briefly that, on average, each DNA molecule receives no more than one single-strand break. (This brief exposure to DNase I is in marked contrast to the long exposure used in the DNase protection method.) Nicking occurs at all positions except those protected by

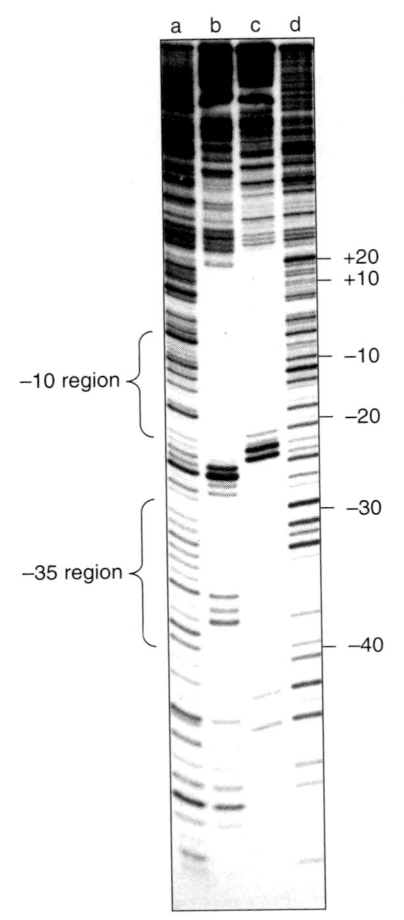

FIGURE 15.9 RNA polymerase holoenzyme footprints at 37°C on both strands of the lacUV5 promoter. The template strand was labeled in one DNA sample and the nontemplate strand in another DNA sample. Lanes a and d are DNase digests in the absence of RNA polymerase holoenzyme of the template and nontemplate strands, respectively. Lanes b and c are DNase digests in the presence of RNA polymerase holoenzyme of the template and nontemplate strands, respectively. The final RNA polymerase holoenzyme concentration was 1.4×10^{-7} M. The DNA or DNA and the RNA polymerase mixture were preincubated at 37°C for 10 min before a 1-min digestion with DNase I at a concentration of 5.4×10^{-10} M. (Reproduced from R. T. Kovacic, *J. Biol. Chem.* 262 [1987]: 13654–13661. © 1987 The American Society for Biochemistry and Molecular Biology. Photo courtesy of Roger Timothy Kovacic.)

the DNA-binding protein. Then the DNA is isolated, denatured, and analyzed by gel electrophoresis.

The radioactive bands observed after gel electrophoresis correspond to a set of molecules with sizes that are determined by the nick positions in relation to the radioactively labeled end. If the DNA contains *n* base pairs and the DNA-binding protein is *not* added, *n* sizes of DNA fragments will be present. However, if the DNA-binding protein is added and prevents DNase I from gaining access to *x* base pairs, only *n–x* different sizes of DNA fragments will be represented. Two DNA samples are compared in Figure 15.8, one without the DNA-binding protein (to obtain the positions of the *n* bands), and one with the DNA-binding protein (to determine the positions of the missing bands). The missing bands in the gel pattern identify the binding site(s) on the DNA.

FIGURE 15.9 shows the results of a DNA footprinting experiment in which a modified promoter for lactose genes designated *lac*UV5 was examined. The *lac*UV5 promoter has the same nucleotide sequence as the normal promoter except that two bases in the –10 box have been changed so that the *lac*UV5 –10 box now has a sequence that matches the consensus sequence, which makes it a stronger promoter. As expected, RNA polymerase holoenzyme protects nucleotides in the –10 and –35 boxes.

RNA polymerase holoenzyme need not make direct contact with a base to protect it from DNase because the large DNase molecule cannot pass through tiny gaps in the holoenzyme•promoter complex. Therefore, some bases within the promoter that are protected from DNase cleavage do not make direct contact with the holoenzyme. Small chemical reagents such as dimethyl sulfate can distinguish bases that make direct contact with the holoenzyme from those that do not. Dimethyl sulfate methylates the ring nitrogens N7 of G (major groove) and N3 of A (minor groove) that are not protected by RNA polymerase holoenzyme. Piperidine in formic acid cleaves both methylated guanine and adenine nucleotides, whereas piperidine alone selectively cleaves the methylated guanine nucleotides. DNA fragments produced by the chemical cleavage are separated by gel electrophoresis. Fragment sizes indicate methylation positions in relation to the labeled 5′-end. The bases protected by RNA polymerase holoenzyme tend to be clustered around the –10 and –35 boxes.

DNA footprinting experiments reveal that the RNA polymerase holoenzyme•promoter complex changes conformation during the transcription initiation process. These changes are evident when one compares the DNA footprint pattern obtained from an RNA polymerase holoenzyme•promoter DNA complex formed at 0°C with the same complex formed at 37°C. The low temperature prevents the conformational change from taking place. When the complex forms at 0°C, the protected DNA region extends from positions –55 to –10. Because the DNA remains completely double helical, this complex is designated a **closed promoter complex** (FIGURE 15.10a). When the complex forms at 37°C, the protected DNA region is longer, extending from positions –55 to +20 (FIGURE 15.10b). Furthermore, positions –12 to +2 become sensitive to $KMnO_4$, a chemical reagent that attacks single

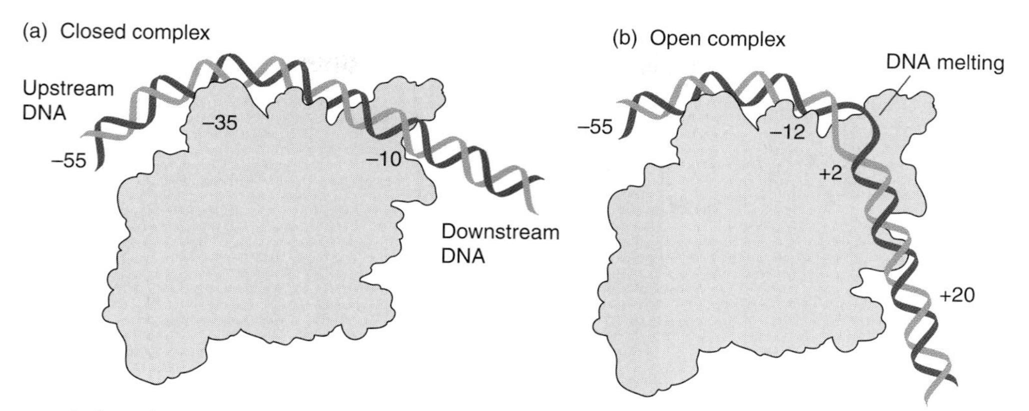

FIGURE 15.10 **Open and closed promoter complexes.** (a) Closed complex (RNA polymerase holoenzyme bound to promoter DNA at 0°C). The DNase footprint extends from positions −55 to −10. The entire promoter DNA remains as a double helix. (b) Open complex (RNA polymerase holoenzyme bound to promoter DNA at 37°C). The DNase footprint extends from positions −55 to +20 and a bubble opens from positions −12 to +2. The position numbers shown in the figure are at approximate locations. (Adapted from K. S. Murakami and S. A. Darst, *Curr. Opin. Struct. Biol.* 131 [2003]: 31–39.)

but not double-stranded DNA. This sensitivity indicates the strands have separated (opened) in positions −12 to +2 to form a bubble that exposes the transcription start site at position +1. Because the DNA has melted, this complex is designated an **open promoter complex.** Note that the AT-rich −10 box accounts for most of the melted region, explaining why a mutation in the −10 box that changes an AT base pair to a GC base pair causes a decrease in promoter strength. Because the protected DNA region in the open complex is about 26 nm and the longest holoenzyme dimension is just 15 nm, the protected DNA segment must wrap around the holoenzyme.

Bacterial RNA polymerase crystal structures show how the enzyme is organized and provide insights into how it works.

In 1999, Seth Darst and coworkers determined the crystal structure for the *T. aquaticus* (*Taq*) core RNA polymerase (FIGURE 15.11). The protein has a total of five subunits that are present in the stoichiometry $\alpha_2\beta\beta'\omega$. Each subunit is homologous with its *E. coli* counterpart. The ω subunit, which is not always present in isolated *E. coli* core RNA polymerase, helps to assemble the core RNA polymerase but is not required for RNA synthesis. As evident from the orientation presented in Figure 15.11, the core polymerase resembles a crab claw. One pincer is almost entirely β subunit and the other almost entirely β′ subunit. The core RNA polymerase is about 15 nm long (from the tips of the claws to the back) and 11 nm wide. Some other noteworthy features of the *Taq* core RNA polymerase are as follows:

1. The β and β′ subunits form a claw-like structure that functions as a DNA clamp.
2. The N-terminal domain (NTD) of one α subunit forms significant contacts with the corresponding domain in the other α subunit allowing the two α subunits to form a dimer. The two NTDs also bind the β and β′ subunits. The arrangement is

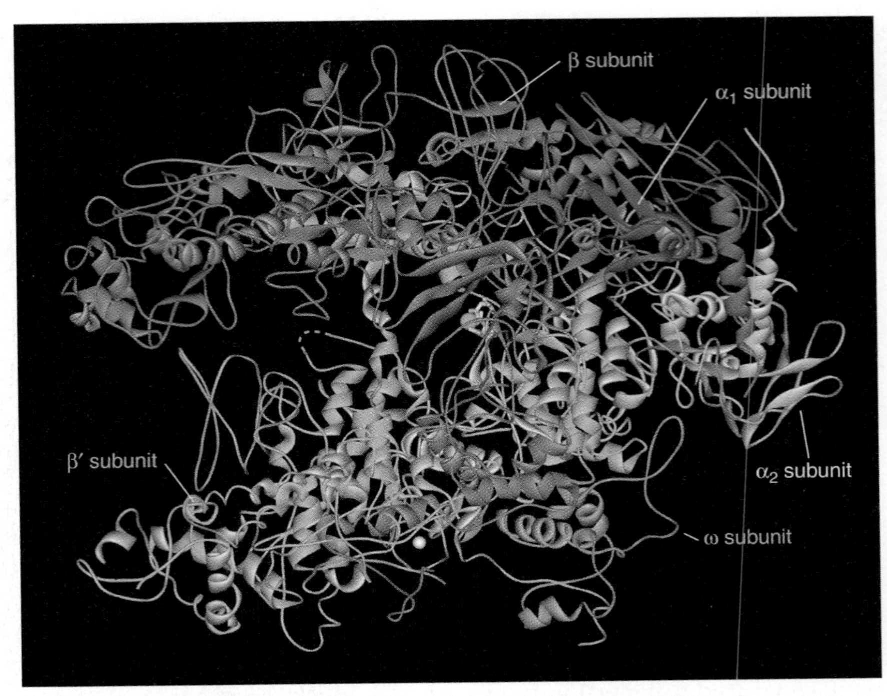

FIGURE 15.11 Structure of the *Thermus aquaticus* core RNA polymerase. The crystal structure of the *T. aquaticus* core RNA polymerase is shown as a ribbon structure. The three aspartates that are part of the absolutely conserved –NADFDGD–motif at the catalytic site are shown as pink stick structures. The magnesium ion bound to these aspartates is shown as a light green sphere. The bridge helix and trigger loop in the β′ subunit are shown in yellow and cyan, respectively. The part of the trigger loop shown as a dashed line is not resolved in the crystal structure. The zinc ion associated with the β′ subunit is shown as a white sphere.(Structure from Protein Data Bank 1HQM. L. Minakhin, et al., *Proc. Natl. Acad. Sci. USA* 98 [2001]: 892–897. Prepared by B. E. Tropp.)

not symmetrical, however, because the NTD of one α subunit, α_1, binds the β subunit while the NTD of the other α subunit, α_2, binds the β′ subunit. No residue in either α subunit has access to the internal channel of the core RNA polymerase where catalysis occurs.

3. The β and β′ subunits, which together account for about 60% of the core RNA polymerase mass, interact extensively with each other. The catalytic site is formed by one such interaction in the active site channel. The β′ subunit provides an absolutely conserved –NADFDGD–motif that is essential for catalysis. The three aspartates in this motif, along with the magnesium ion that is bound to them, are required for phosphodiester bond formation. The incoming NTP carries a second required magnesium ion to the active site.
4. The β′ subunit has a bridge helix and trigger loop adjacent to the catalytic site that play essential roles in the nucleotide addition cycle (see below).

The high-resolution image obtained for the *Taq* core polymerase is thought to closely approximate core RNA polymerase structures from

E. coli and other bacteria because: (1) The subunits in core RNA polymerases are conserved in bacteria. (2) The *Taq* core RNA polymerase crystal structure corresponds in both size and shape to low-resolution electron microscopy images of the *E. coli* protein, indicating that the two proteins have comparable structures. (3) The secondary structures predicted from *E. coli* β and β′ subunits sequences are consistent with the secondary structures actually observed in the high-resolution *Taq* polymerase crystal structure. (4) A high-resolution image of a fragment from the *E. coli* α subunit that contains the subunit's NTD has the same structure as its counterpart in the *Taq* protein. (5) Amino-acid substitutions that make *E. coli* RNA polymerase resistant to rifamycin, an antibiotic that inhibits the enzyme (see below), are scattered along the β chain, but these same amino acids cluster around a pocket in core polymerase. (6) Genetic and biochemical studies show that residues that bind the initiating NTP substrate are scattered throughout the *E. coli* β subunit sequence but are clustered together in the core RNA polymerase structure.

In 2002, Shigeyuki Yokoyama and coworkers determined the crystal structure for the *T. thermophilus* RNA polymerase holoenzyme (**FIGURE 15.12**). The most important structural difference between the core polymerase and the holoenzyme is that the holoenzyme contains the σ^{70} subunit, which is located almost entirely on the core's surface, with the exception of a segment that is buried within the core molecule.

Although core polymerase and holoenzyme crystal structures provide considerable information about RNA polymerase structure, they provide no direct information about interactions that exist among core polymerase, σ factor, and DNA. Darst and coworkers addressed these and related structural issues by preparing crystals that contain the *Taq* holoenzyme bound to a synthetic DNA with a fork-junction

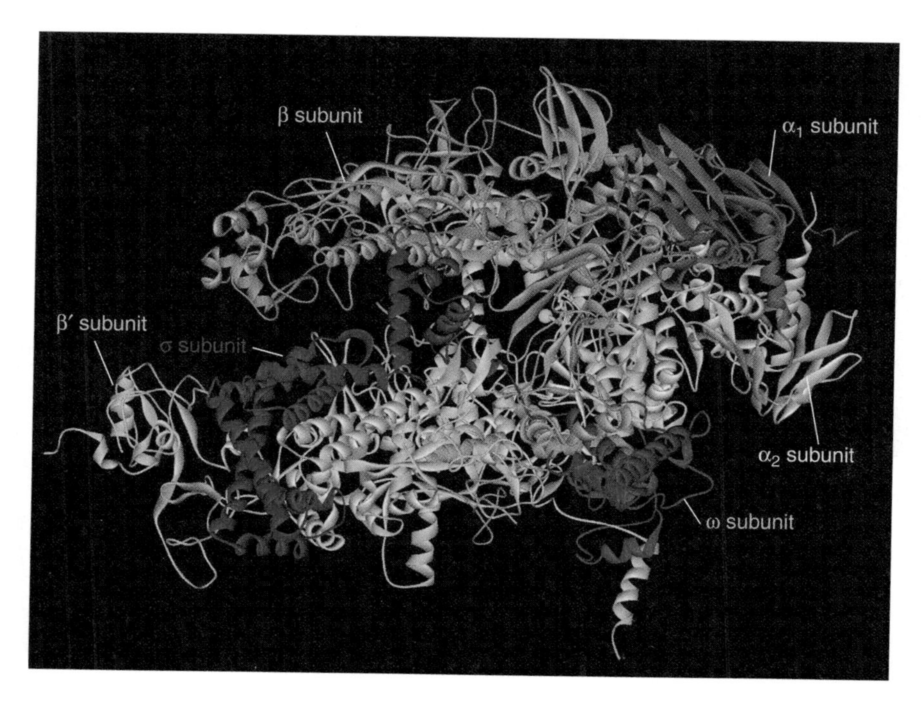

FIGURE 15.12 Structure of the *Thermus thermophilus* RNA polymerase holoenzyme. The crystal structure of the *T. thermophilus* RNA polymerase holoenzyme is shown as a ribbon structure. The three aspartates that are part of the absolutely conserved –NADFDGD– motif at the catalytic site are shown as pink stick structures. The magnesium ion bound to these aspartates is shown as a light green sphere. The bridge helix and trigger loop in the β′ subunit are shown in yellow and cyan, respectively. (Structure from Protein Data Bank 1IW7. D. G. Vassylyev, et al., *Nature* 417 [2002]: 712–719. Prepared by B. E. Tropp.)

FIGURE 15.13 Structure of *Thermus aquaticus* RNA polymerase holoenzyme bound to fork-junction DNA. (a) The synthetic fork-junction DNA used to prepare the crystal that contained the RNA polymerase holoenzyme•fork-junction DNA complex. (b) Structure of the complex between the RNA polymerase holoenzyme and the synthetic fork-junction DNA. The core RNA polymerase is shown as a spacefill structure. DNA is shown as a blue tube with the –35 and –10 boxes in yellow. The magnesium ion at the active site is shown as a light green sphere. (Part a modified from K. S. Murakami, et al., *Science* 296 [2002]: 1285–1290. Reprinted with permission from AAAS. Part b structure derived from Protein Data Bank 1L9Z. K. S. Murakami, et al., *Science* 296 [2002]:1285–1290. Prepared by B. E. Tropp.)

(a) Synthetic fork-junction DNA

Nontemplate strand

← Upstream   Downstream →

–40 –30 –20 –10

5′ **GGCCGCTTGACAAAAGTGTTAAATTGTGCTATACT** 3′
3′ CCGGCGAACTGTTTTCACAATTTAACACGA 5′

–35 element   Template strand   –10 element

(b) Structure of the complex between the RNA polymerase holoenzyme and fork-junction.

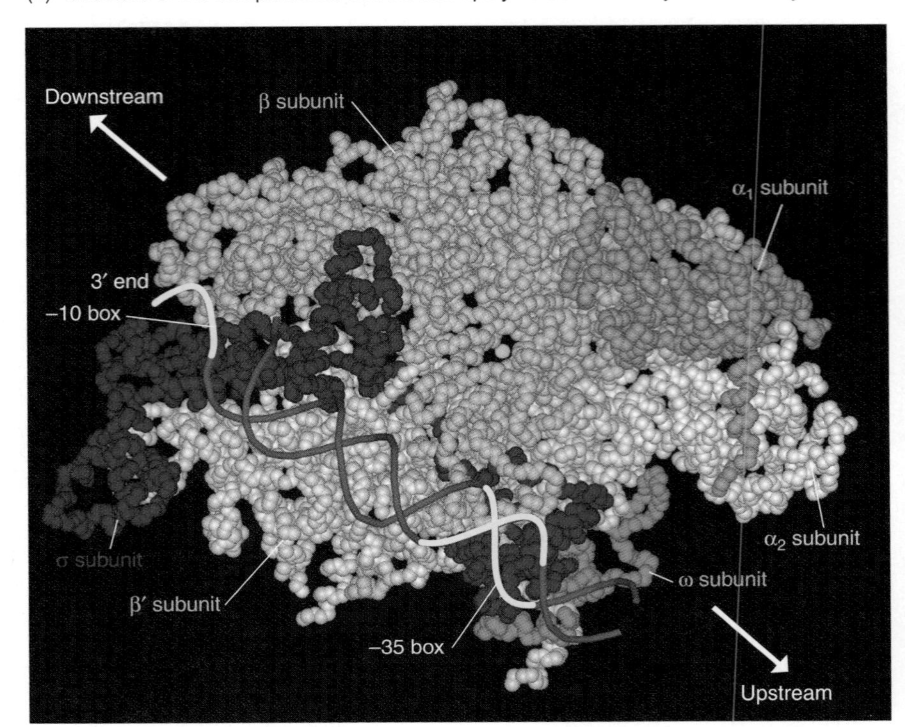

sequence (the junction between double-stranded DNA and single-stranded DNA in the transcription bubble). The DNA molecule they synthesized for this purpose (FIGURE 15.13a) has a double-stranded –35 box and a mostly open –10 box. The crystal structure of the RNA polymerase holoenzyme•fork-junction DNA complex, which appears to resemble the open complex, is shown in FIGURE 15.13b. The fork-junction DNA lies across one face of the holoenzyme, entirely outside the RNA polymerase active site. All sequence specific –10 and –35 box contacts with the RNA polymerase take place through interactions with the σ subunit.

Genetic and biochemical studies provide additional information about bacterial promoters.

Additional information about promoters has been obtained by introducing mutations into promoters that alter transcription initiation. The rationale is that if a base change affects promoter activity, that base must be contained in a critical promoter region. Four such mutations are shown for the *E. coli lac* promoter in FIGURE 15.14. Two of these mutations are in the –10 box and stimulate transcription initiation, whereas the other two are in the –35 box and inhibit transcription initiation. In general, promoters that support active transcription, termed *strong promoters*, have –10 and –35 boxes that are close to the consensus sequences and a spacer between the two boxes that is 17 ± 1 bp. Mutations in the –10 or –35 boxes that cause their sequences to differ from the consensus sequences usually lead to weaker promoter activity as evidenced by lower rates of initiation of RNA synthesis.

Further biochemical and genetic analysis revealed that bacterial promoters can have three additional control elements. These elements, which are shown in FIGURE 15.15, are as follows:

1. The **up element** is located just upstream of the –35 box (positions –57 to –38). UP elements are present in ribosomal RNA promoters and are responsible for a marked increase in promoter strength (see Chapter 16).
2. The **extended –10 element** is directly upstream from the –10 box and has the consensus sequence 5′-TGn-3′, where n is any base. The extended –10 element permits the σ^{70}-holoenzyme to bind to a small subset of *E. coli* promoters that would not otherwise be transcribed because they lack a –35 box.
3. The **discriminator region**, which is located between the transcription start site and the –10 box (positions –6 to –4) and has the optimal sequence 5′-GGG-3′, is present in ribosomal RNA and transfer RNA promoters.

The bacterial promoter therefore has a modular organization. Different promoters have different element combinations. No single promoter is likely to have all of the elements, however, because some combinations

FIGURE 15.14 Mutations that alter the *E. coli lac* promoter. Four nucleotide substitutions are shown in the coding strand for the *lac* promoter. The two substitutions shown in red weaken the promoter and are therefore called down mutations. The two shown in green strengthen the promoter and are therefore called up mutations. The –10 box is shaded in yellow and the –35 box in cyan. Many base changes that are known to alter promoter activity are located in or near the –10 box or are clustered around base –35 and thus define an important site. Deletions or additions in the spaces between the –10 and –35 boxes usually decrease promoter activity.

UP element | –35 | Ext | –10 | Dis | +1
TTGACA | TGn | TATAAT | GGG

FIGURE 15.15 DNA promoter elements. Moving downstream, promoter elements are as follows: the UP element, –35 box (–35), extended element (Ext), –10 box (–10), and discriminator (Dis). The consensus sequence for each promoter element is shown below that element (w indicates the presence of T or A and n indicates the presence of any base). (Adapted from S. P. Haugen, et al., *Nat. Rev. Microbiol.* 6 [2008]: 507–519.)

do not work well together. For example, when an extended –10 element is introduced into a promoter with a fully functional –35 box, the newly constructed promoter is weaker than the original promoter. The explanation may be that transcription initiation is hindered by too many contacts between the σ^{70}-RNA polymerase and the promoter.

Members of the σ^{70} family have four conserved domains.

Sequences are known for more than one hundred members of the σ^{70} family. Aligning sequences reveals four conserved domains, each of which can be subdivided into smaller highly conserved regions (FIGURE 15.16). Numbering for σ factor domains begins at the N-terminus. Small highly conserved regions are specified by two numbers that are separated by a decimal. The first number indicates the domain and the second indicates a small highly conserved region within the domain. For example, $\sigma_{2.4}$ indicates the fourth small highly conserved region in the second domain. With the exception of region 1.1, all highly conserved regions are present in all σ^{70} family members. Region 1.1 is present only in the primary σ factor of each bacterial species, which for *E. coli* is σ^{70} itself. The σ^{70} factor assumes a different conformation when bound to core polymerase. One notable change is the disruption of an interaction between region 1.1 and domain 4. This interaction prevents the free σ^{70} factor from binding DNA. In essence, negatively charged region 1.1 acts as a DNA mimic, which competes with promoter DNA for the binding site on domain 4. FIGURE 15.17 shows the specific interactions between *E. coli* σ^{70}-RNA polymerase and the various promoter elements.

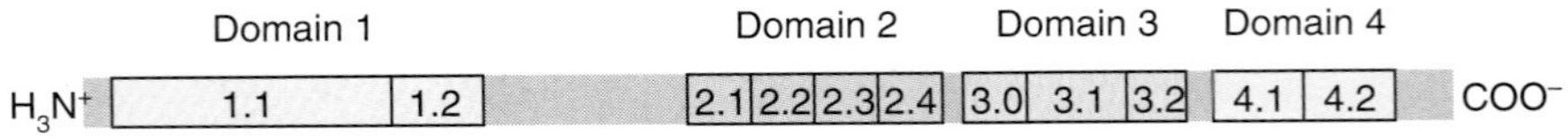

Domain 1
Domain 2
Domain 3
Domain 4
H_3N^+
1.1
1.2
2.1 2.2 2.3 2.4
3.0 3.1 3.2
4.1 4.2
COO^-

FIGURE 15.16 Representation of the intact σ^{70} factor showing the four conserved domains and small highly conserved regions within these domains. Numbering for σ factor domains begins at the N-terminus. Smaller regions within each domain are specified by two numbers that are separated by a decimal point: the first number specifies the domain and the second the small highly conserved region within the domain. (Adapted from B. A. Young, T. M. Gruber, and C. A. Gross, *Cell* 109 [2002]: 417–420.)

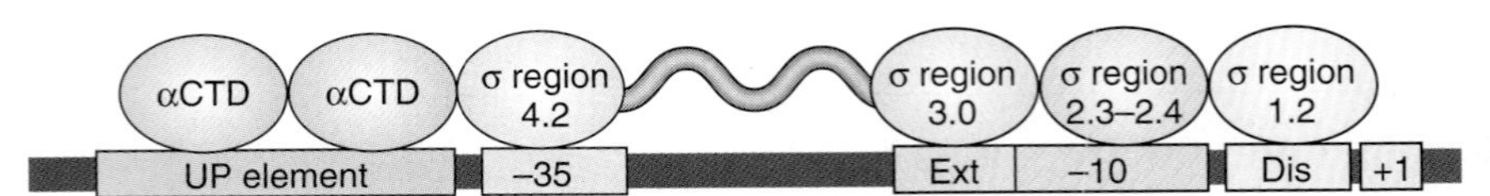

FIGURE 15.17 Interactions between DNA promoter elements and σ^{70}-RNA polymerase. Moving downstream, promoter elements are as follows: the UP element, –35 box (–35), extended element (Ext), –10 box (–10), and discriminator (Dis). The σ^{70} factor is drawn with the N-terminus on the right. CTD specifies the carboxyl terminal domain of the σ subunit. (Adapted from S. P. Haugen, et al., *Nat. Rev. Microbiol.* 6 [2008]: 507–519.)

RNA polymerase scrunches DNA during transcription initiation.

When an assay mixture contains σ^{70}-RNA polymerase, DNA, and only the first two NTP substrates, RNA synthesis comes to a halt and the nascent RNA is released. Although the cessation of RNA synthesis is expected because the full complement of nucleoside triphosphates is not present, the release of nascent RNA is surprising. Even more surprising, nascent RNA ranging from about 8 to 10 nucleotides is also released even when the reaction mixture contains all four NTP. These experiments indicate that a significant amount of **abortive initiation** takes place during the early stage of the σ^{70}-RNA polymerase catalyzed reaction.

The observation that abortive initiation leads to the formation of RNA products that are up to 8 to 10 nucleotides in length suggests that the RNA polymerase active center moves relative to the DNA that it acts on. Yet in apparent contrast, DNA-footprinting experiments indicate that the enzyme does not move during abortive transcription because it protects the same upstream DNA fragment before and after abortive RNA synthesis. Three models have been offered to reconcile these seemingly contradictory observations (FIGURE 15.18). The **scrunching model** proposes that RNA polymerase holoenzyme unwinds adjacent DNA segments and pulls the unwound DNA into itself during initial transcription ("scrunching") and then rewinds the unwound DNA when

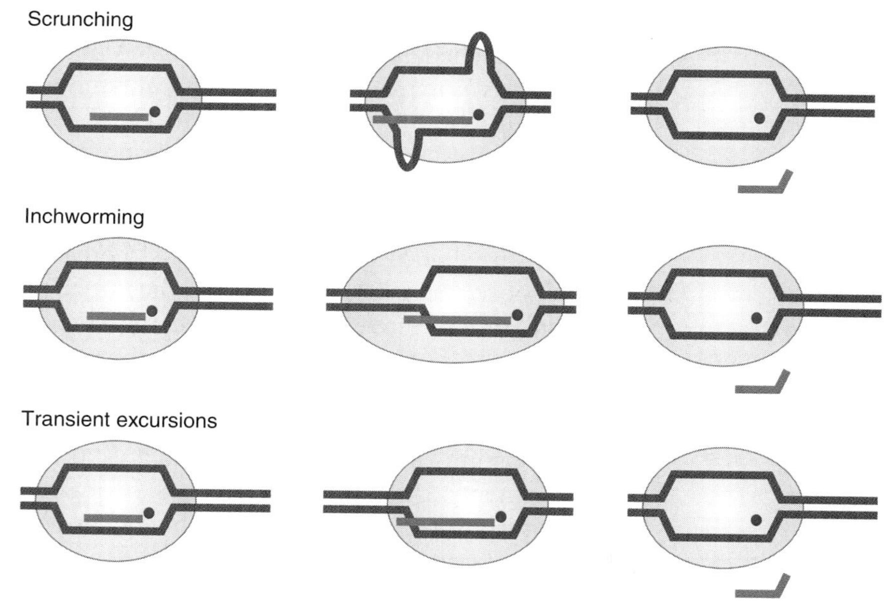

FIGURE 15.18 **Models to reconcile footprinting pattern during abortive initiation.** The transcription initiation complex is shown with RNA polymerase (red oval), DNA (blue lines), nascent RNA (green line), and a magnesium ion (purple sphere) at the active site. The scrunching model (top) proposes that RNA polymerase holoenzyme unwinds adjacent DNA segments and pulls the unwound DNA into itself during initial transcription. The energy stored in the scrunched DNA is then used for promoter escape. The inchworming model (middle) proposes that the leading edge of RNA polymerase advances during the early stage of transcription initiation to allow the active center to move forward, while the other end of the enzyme remains anchored to the upstream DNA region. The energy stored in the stretched protein is used for promoter escape. The transient excursion model (bottom) proposes transient cycles of forward enzyme motion during abortive RNA synthesis and backward enzyme movement after abortive transcript release with long intervals between cycles. The long interval times between cycles was postulated to explain the footprinting observations. (Modified from K. M. Herbert, et al., *Annu. Rev. Biochem.* 77 [2008]: 149–176. Copyright 2008 by Annual Reviews, Inc. Reproduced with permission of Annual Reviews, Inc., in the format Textbook via Copyright Clearance Center.)

the RNA polymerase leaves the initiation site and begins to move down the DNA ("unscrunching"). The energy stored in the scrunched DNA is then used for promoter escape. The **inchworming model** proposes that the leading edge of RNA polymerase advances during the early stage of transcription initiation to allow the active center to move forward, while the other end of the enzyme remains anchored to the upstream DNA region. The energy stored in the stretched protein is used for promoter escape. Finally, the **transient excursion** model proposes transient cycles of forward enzyme motion during abortive RNA synthesis and backward enzyme movement after abortive transcript release with long intervals between cycles. The long interval times between cycles was postulated to explain the footprinting observations.

Richard H. Ebright and coworkers performed two types of experiments to distinguish among the models. In the first, they attached pairs of fluorescent tags to specific sites on the RNA polymerase holoenzyme or DNA and then used fluorescence resonance energy transfer to monitor distances within single molecules of abortively initiating transcription initiation complexes. They observed that the RNA polymerase holoenzyme does not stretch to reach adjacent DNA segments (eliminating the inchworm model) and does not move to reach adjacent DNA segments (eliminating the transient excursions model). Instead, their observations indicated that the RNA polymerase holoenzyme remains stationary and pulls adjacent DNA into itself. In the second type of experiment, Ebright and coworkers monitored the end-to-end extension of a mechanically stretched, supercoiled, single DNA molecule. They demonstrated that RNA polymerase holoenzyme unwinds adjacent DNA segments and pulls the unwound DNA into itself during initial transcription ("scrunching") and then re-winds the unwound DNA when the RNA polymerase leaves the initiation site and begins to move down the DNA ("unscrunching"). Energy stored in the system during the scrunching stage is used during promoter escape to break interactions between the RNA polymerase holoenzyme and the initiation site and to allow RNA polymerase to move forward and catalyze transcription elongation.

Transcription initiation is a stepwise process.

Based on information obtained from crystal structures and biochemical experiments, Darst and Katsuhiko S. Murakami proposed a transcription initiation pathway. The pathway shown in FIGURE 15.19 differs slightly from the one they proposed to accommodate scrunching, which had not been discovered at the time of their proposal.

1. **Closed Promoter Complex**—The σ^{70} factor interacts with the core polymerase and DNA as follows: $\sigma_{1.1}$ is positioned in the active site channel, helping to open it; σ_2 interacts with the –10 box; the $\sigma_{3.2}$ loop projects into the RNA polymerase active site channel (active site is marked by the Mg^{2+}); and the σ_4 interacts with the –35 box.
2. **Intermediate Stage**—Promoter opening probably begins when the first adenine in the TATAAT consensus sequence flips out into a hydrophobic pocket in σ, triggering recognition of the

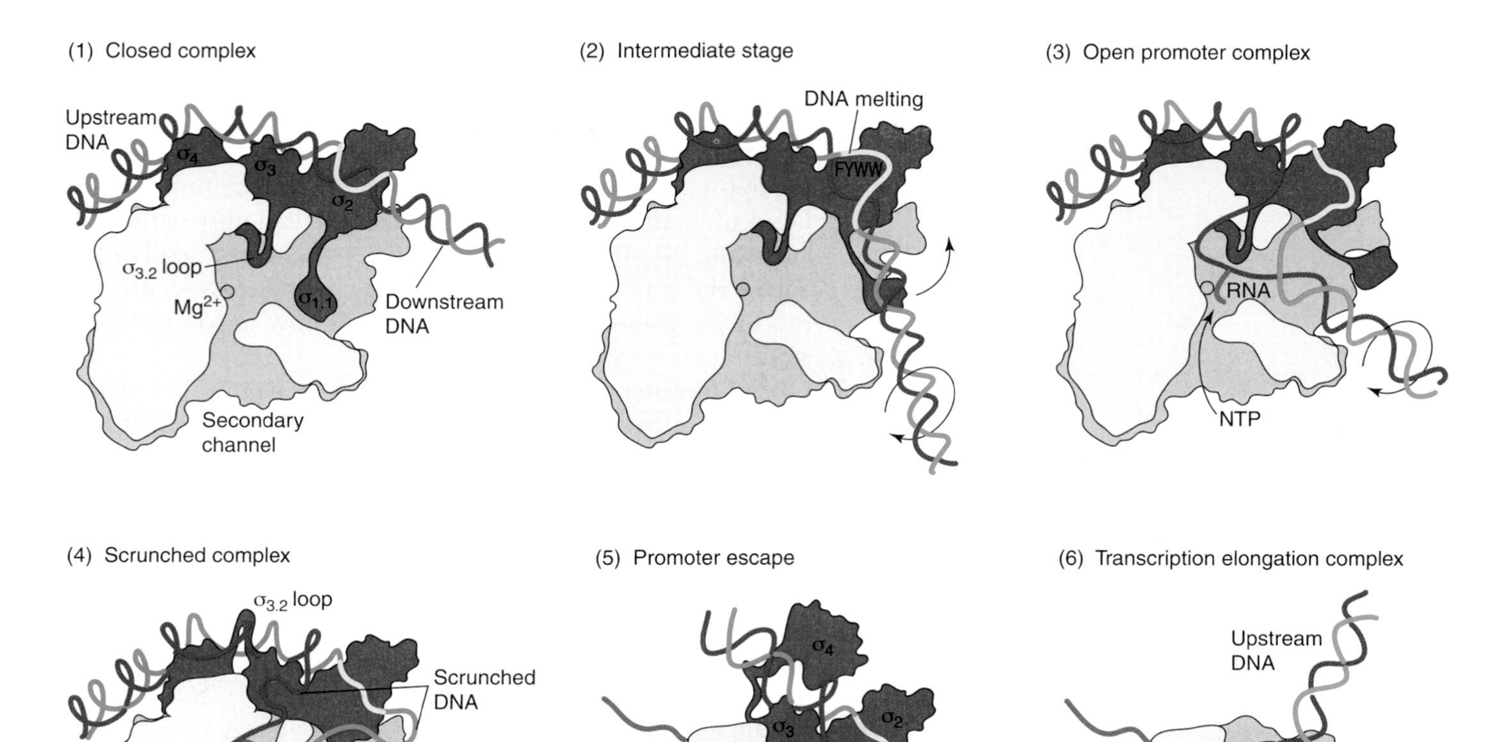

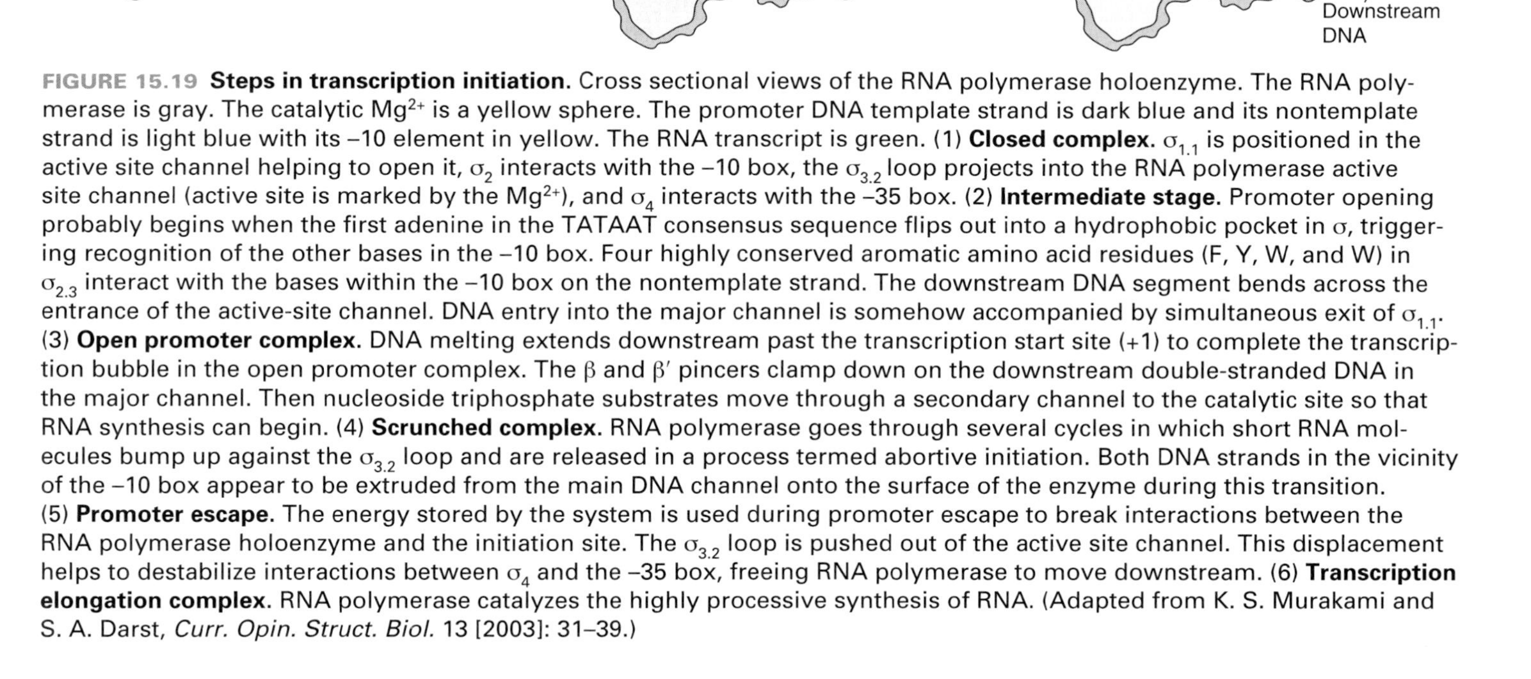

FIGURE 15.19 **Steps in transcription initiation.** Cross sectional views of the RNA polymerase holoenzyme. The RNA polymerase is gray. The catalytic Mg^{2+} is a yellow sphere. The promoter DNA template strand is dark blue and its nontemplate strand is light blue with its –10 element in yellow. The RNA transcript is green. (1) **Closed complex.** $\sigma_{1.1}$ is positioned in the active site channel helping to open it, σ_2 interacts with the –10 box, the $\sigma_{3.2}$ loop projects into the RNA polymerase active site channel (active site is marked by the Mg^{2+}), and σ_4 interacts with the –35 box. (2) **Intermediate stage.** Promoter opening probably begins when the first adenine in the TATAAT consensus sequence flips out into a hydrophobic pocket in σ, triggering recognition of the other bases in the –10 box. Four highly conserved aromatic amino acid residues (F, Y, W, and W) in $\sigma_{2.3}$ interact with the bases within the –10 box on the nontemplate strand. The downstream DNA segment bends across the entrance of the active-site channel. DNA entry into the major channel is somehow accompanied by simultaneous exit of $\sigma_{1.1}$. (3) **Open promoter complex.** DNA melting extends downstream past the transcription start site (+1) to complete the transcription bubble in the open promoter complex. The β and β′ pincers clamp down on the downstream double-stranded DNA in the major channel. Then nucleoside triphosphate substrates move through a secondary channel to the catalytic site so that RNA synthesis can begin. (4) **Scrunched complex.** RNA polymerase goes through several cycles in which short RNA molecules bump up against the $\sigma_{3.2}$ loop and are released in a process termed abortive initiation. Both DNA strands in the vicinity of the –10 box appear to be extruded from the main DNA channel onto the surface of the enzyme during this transition. (5) **Promoter escape.** The energy stored by the system is used during promoter escape to break interactions between the RNA polymerase holoenzyme and the initiation site. The $\sigma_{3.2}$ loop is pushed out of the active site channel. This displacement helps to destabilize interactions between σ_4 and the –35 box, freeing RNA polymerase to move downstream. (6) **Transcription elongation complex.** RNA polymerase catalyzes the highly processive synthesis of RNA. (Adapted from K. S. Murakami and S. A. Darst, *Curr. Opin. Struct. Biol.* 13 [2003]: 31–39.)

other bases in the –10 box. Four highly conserved aromatic amino acid residues (F, Y, W, and W) in $\sigma_{2.3}$ interact with the bases within the –10 box on the nontemplate strand, helping to stabilize a short single-stranded DNA segment corresponding to the upstream edge of the emerging transcription bubble. DNA unwinding (circular arrow) creates flexibility in the DNA at the bubble, allowing the downstream DNA to bend or kink across the entrance of the active-site channel. DNA entry into the RNA polymerase channel is somehow accompanied by simultaneous exit of the $\sigma_{1.1}$.

3. **Open Promoter Complex**—DNA melting extends downstream past the transcription start site (+1) to complete the transcription bubble in the open promoter complex. The template strand is directed to the active site through a positively charged tunnel, which is completely enclosed on all sides by protein. The β and β′ pincers clamp down on the downstream double-stranded DNA in the major channel. This clamping extends from position +5 to about +12. After the open complex forms, NTP substrates move through a secondary channel to the catalytic site so that RNA synthesis can begin.
4. **Scrunched Complex**—RNA polymerase often goes through several abortive initiation cycles. One reason for this abortive initiation is that the transcript, which is just a few nucleotides in length bumps into the $\sigma_{3.2}$ loop. The elongating RNA chain must either displace the $\sigma_{3.2}$ loop out of its path, or else dissociate from the complex and be released (probably through the secondary channel). Both DNA strands in the vicinity of the –10 box appear to be extruded from the main DNA channel onto the surface of the enzyme during this stage of transcription. That is, the DNA is scrunched.
5. **Promoter Escape**—Energy stored by the system is used during promoter escape to break interactions between the RNA polymerase holoenzyme and the initiation site. The $\sigma_{3.2}$ loop displacement, which occurs when the nascent RNA chain is 12 nucleotides long, may help to destabilize interactions between σ_4 and the β subunit. Release of σ_4 from the β subunit would then destabilize interactions between σ_4 and the –35 box, allowing the RNA polymerase to move downstream as it elongates the **RNA**. This movement of RNA polymerase away from the promoter is termed **promoter escape**.
6. **Transcription Elongation Complex**—Transition into transcription elongation complex does not require the complete release of σ because the paths of the nucleic acids in the transcription elongation complex do not preclude the binding σ_2 and σ_3 to the core polymerase. RNA polymerase catalyzes the highly processive synthesis of RNA (see below).

Alternative σ factors direct RNA polymerase to genes that code for proteins that bacteria require to survive under specific types of environmental stress.

E. coli has seven different σ factors, which were initially distinguished based on their molecular masses. At first, each σ factor was designated by using the Greek letter σ followed by a superscript specifying the σ factor's approximate molecular mass in kDa (Table 15.2). Because this nomenclature system becomes ambiguous when different σ factors have similar molecular masses, two alternative nomenclature systems were introduced. In the first, each σ factor is indicated by the Greek letter σ followed by a superscript letter. For example, *E. coli* σ^{70} is called σ^{D} in this system. In the second, each σ factor is named for the gene that encodes it. For example, σ^{70} becomes RpoD. Each kind of σ factor combines with core polymerase to form a holoenzyme that recognizes a unique set of promoter DNA sequences.

Based on sequence homologies, *E. coli* σ factors can be divided in two major families, the σ^{70} family and the σ^{54} (σ^{N} family). The only σ^{N} family member present in *E. coli* is the σ^{N} factor. All other *E. coli* σ factors belong to the σ^{70} family. *E. coli* σ factors compete with one another for the core RNA polymerase. In the absence of additional regulatory factors, the rate of transcription depends solely on promoter strength and the concentration of specific holoenzymes. The σ^{70}-RNA polymerase transcribes genes that code for proteins required for normal exponential growth, which account for most of the bacteria's genes. Therefore, σ^{70} is an essential protein. RNA polymerase holoenzymes containing one of the other σ factors recognize specific sets of genes distributed throughout the chromosome that code for proteins needed for cell survival under some special environmental condition such as heat shock (σ^{H}) or nitrogen deprivation (σ^{N}). Table 15.2 summarizes

TABLE 15.2 ***E. coli* Sigma Factors**

Sigma Factor	Molecular Mass (kDa)	Gene	Map Position (min)	Functions	Consensus Sequence[a]		
					–35 region	spacer bp	–10 region
σ^{70} or σ^{D}	70	*rpoD*	69.21	Major sigma factor during exponential growth	TTGACA	16–18 bp	TATAAT
σ^{32} or σ^{H}	32	*rpoH*	77.55	Transcription heat-shock genes	CTTGAA	13–15 bp	CCCCAT**n**T
σ^{24} or σ^{E}	24	*rpoE*	58.36	Response to periplasmic stress	GAACTT	16–17 bp	TCTGA
σ^{28} or σ^{F}	28	*rpoF*	43.09	Expression late flagellar genes	CTAAA	15 bp	GCCCATAA
σ^{18} or σ^{FecI}	18	*FecI*	92.35	Iron-isocitrate transport	GAAAAT	15 bp	TGTCCT
σ^{38} or σ^{S}	38	*rpoS*	61.75	Expression stationary phase genes	TTGACA	14–18 bp	CTAYACTT
σ^{54} or σ^{N}	54	*rpoN*	72.05	Nitrogen metabolism genes	–25 CTGGCAC	6 bp	–12 TTGCA

[a]**n** is any nucleotide and Y is a pyrimidine.

the properties and functions of *E. coli* σ factors. All σ factors bind to RNA polymerase core enzymes, allowing the holoenzyme to interact with specific promoter elements. Many, possibly all, σ factors also are targets of accessory ligands that regulate their activity.

The number of alternative σ factors present in a bacterial cell varies greatly from one bacterial species to another. For instance, *Mycoplasma sp.* has only one kind of σ factor, whereas *Streptomyces coelicolor* has at least 66 different kinds of σ factors. Sigma factors sometimes have novel functions. For instance, four sporulation-specific σ factors allow *Bacillus subtilis* to respond to certain types of nutrient deprivation by becoming metabolically dormant cells or spores that are surrounded by a protective multilayer envelope. Spore formation is a complex process, requiring many enzymes and proteins that are not synthesized by exponentially growing cells. During spore formation, preexisting sigma subunits are destroyed and new sigma subunits formed, which combine with the core enzyme to form a holoenzyme that recognizes promoters of genes needed for spore formation.

Proteins known as **anti-sigma factors** regulate transcription by preventing alternative σ factors from binding to the RNA polymerase core enzyme or promoter. A specific example, the FlgM anti-sigma factor, helps to illustrate how an anti-sigma factor works. The genes that code for the proteins that make up flagellum (the large extracellular appendage responsible for cell motility) can be divided into early, middle, and late genes based on their order of transcription. The σ^{70}-holoenzyme transcribes the early and middle genes. The middle genes encode the σ^F factor, the FlgM anti-sigma factor, and proteins required to form the so-called hook-basal body (HBB) flagellar substructure. FlgM wraps around σ^F, preventing σ^F from binding to the core polymerase until HBB formation is complete. Once complete, HBB acts as an export apparatus to secrete FlgM out of the cell, freeing σ^F to combine with the core polymerase. Then the σ^F holoenzyme transcribes the gene that encodes flagellin, the major structural protein in flagellum. In some cases, cells contain proteins that stimulate or inhibit the anti-sigma factor.

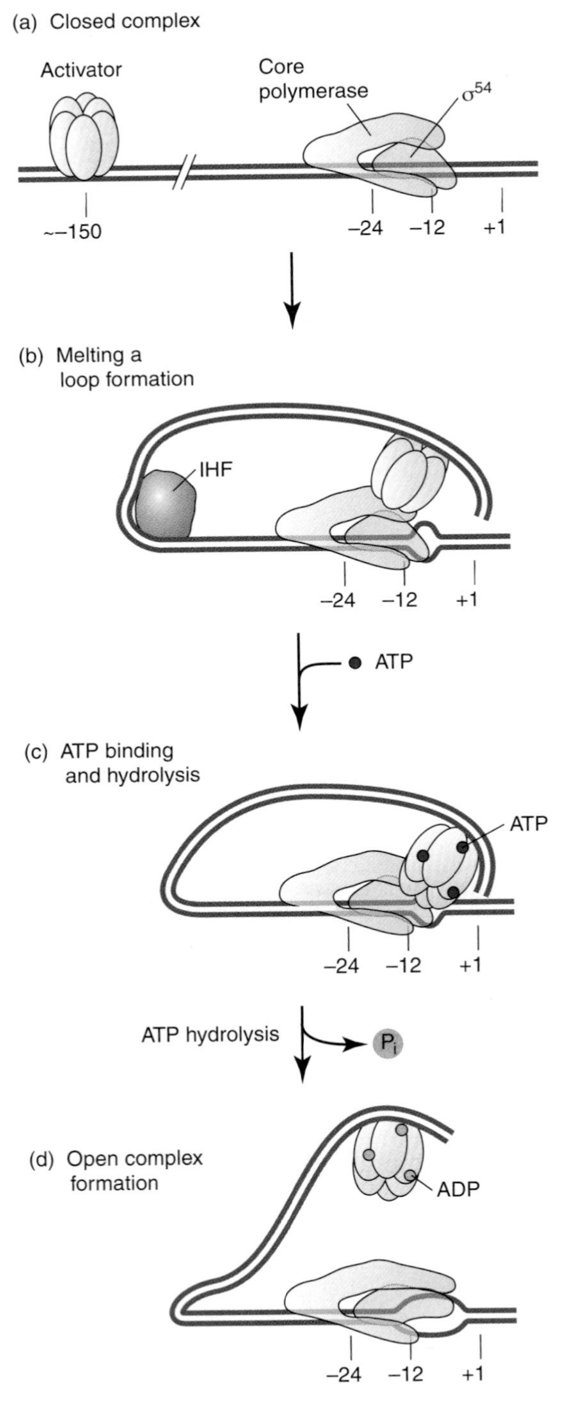

FIGURE 15.20 **Schematic illustrating the route to open complex formation by the σ^{54}-RNA polymerase.** (a) The activator (yellow), a ring-shaped hexamer, binds to enhancer sequences that are about 150 base pairs upstream of the transcription start site. The σ^{54}-RNA polymerase holoenzyme binds to its cognate promoter sequences to form the closed complex. (b) A site of localized melting is generated around position −12. The nucleotide sequence between the two protein-binding sites may produce DNA bending but an additional protein, such as the integration host factor (IHF) is sometimes required to assist the bending. (c) The addition of nucleoside triphosphates (NTPs, shown as small red circles) is required for a stable interaction between activator and σ^{54}-RNAP. (d) After nucleotide hydrolysis, the activator•polymerase complex becomes destabilized, resulting in open complex formation. (Adapted from P. C. Burrows, *Bioessays* 25 [2003]: 1150–1153.)

The σ^{54}-RNA polymerase requires an activator protein.

The σ^{54}-RNA polymerase is unique among bacterial RNA polymerase holoenzymes because it cannot initiate transcription without the assistance of an activator protein. The activation process, which is shown schematically in FIGURE 15.20, begins with the binding of activator protein to a site on the bacterial chromosome called an **enhancer**, which is usually 100 bp or more upstream from the σ^{54} promoter. Interaction between the activator proteins and σ^{54}-holoenzyme is possible because the DNA segment between the two protein-binding sites bends to form a loop. Although the nucleotide sequence between the two protein-binding sites may produce DNA bending, an additional protein such as the integration host factor (IHF) is sometimes required to assist the bending. Activator proteins are themselves subject to regulation by the binding of a small effector molecule or, more commonly, by the addition of a phosphate group to a specific site on the activator protein. Activation induces an ATPase activity within the activator protein that is essential for unwinding DNA in the promoter region so that transcription can begin.

15.3 Transcription Elongation Complex

The transcription elongation complex is a highly processive molecular motor.

As the initiation stage comes to an end, the RNA polymerase conformation changes to generate a transcription elongation complex (TEC), consisting of core RNA polymerase, template DNA, and a growing RNA chain. RNA chain elongation involves a catalytic cycle in which:

1. The NTP moves through the secondary channel to reach the binding site. The passage of nucleoside triphosphates through the secondary channel may limit the elongation rate because there is only a one in four chance that the correct nucleotide will move through the secondary channel and reach the binding site.
2. A pair of electrons on the 3′ hydroxyl at the growing end of the RNA strand displaces the pyrophosphate group from the NTP to form the 5′ to 3′ phosphodiester bond.
3. The core RNA polymerase moves one nucleotide downstream. RNA polymerase moves along the DNA template at about 30 nucleotides per second. The incoming nucleoside triphosphates provide sufficient energy to synthesize the phosphodiester bond and drive the RNA core polymerase one nucleotide downstream.

Biochemical studies provide important insights into the transcription elongation complex's structure. The DNase footprint of the transcription elongation complex is about 35 bp shorter than that of the initiating complex. Nevertheless, the TEC is the more stable of the

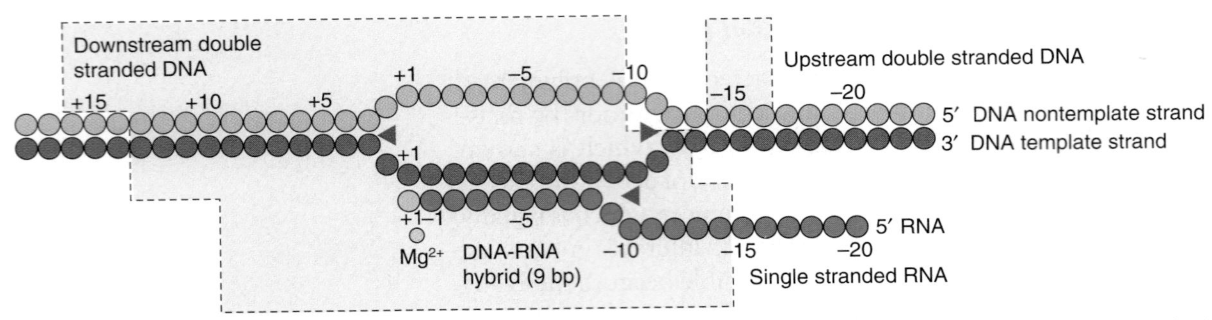

FIGURE 15.21 A schematic representation of nucleic acids within the transcription elongation complex. Transcription is taking place from right to left. The DNA is represented by dark blue circles for the template strand, and light blue circles for the nontemplate strand. The RNA chain is represented by dark green circles and the incoming nucleotide is shown in light green. Nucleotide residue positions are numbered relative to the position of the incoming nucleotide substrate, which is designated +1. DNA and RNA segments that are protected by RNA polymerase during chemical and enzymatic footprinting experiments are indicated by tan shading. Brown triangles indicate that RNA polymerase actively controls the length of the transcription bubble and the RNA-DNA hybrid. (Adapted from S. A. Darst, *Curr. Opin. Struct. Biol.* 11 [2001]: 155–162.)

two complexes (see below). Approximately 14 bp within the region protected from DNase are melted, forming a transcription bubble (FIGURE 15.21). The first eight nucleotides within this bubble are paired with the RNA chain. The transcription bubble size appears to remain the same during the elongation process as RNA polymerase moves downstream because double-stranded DNA opens in front of the bubble and re-forms behind it. Conventions for numbering nucleotides in the transcription elongation complex are as follows: The entry position for the incoming nucleotide is +1 and nucleotides downstream from this position are +2, +3, and so forth. The 3′-terminus of the RNA strand is at position −1 and nucleotides upstream from this position are −2, −3, and so forth. The transcription bubble extends from about −12 to +2 bp. This transcription elongation complex must be stable to synthesize RNA molecules that are 10^4 nucleotides or longer. High processivity is obligatory because a growing RNA chain cannot be extended after the transcription elongation complex dissociates. One indication of TEC stability is that it remains intact in 0.5 *M* potassium chloride. The initiating complex dissociates at this salt concentration.

An understanding of how the TEC works requires information about its architecture. In the absence of a crystal structure for the TEC, investigators performed chemical crosslinking experiments to learn how the transcription elongation complex is organized. Their basic approach involved first attaching a reactive chemical group or **affinity probe** to a specific site on the DNA template or RNA product in the TEC and then activating the probe so that it crosslinked with nearby segments of core RNA polymerase subunits. The presence of crosslink between two components indicated that the components were near one another in the TEC. Then, in 2007 Dimitry G. Vassylyev and coworkers determined the crystal structure of a *T. thermophilus* TEC with a core polymerase bound to a synthetic scaffold, which contained 14 bp of downstream DNA, 9 bp of an RNA-DNA

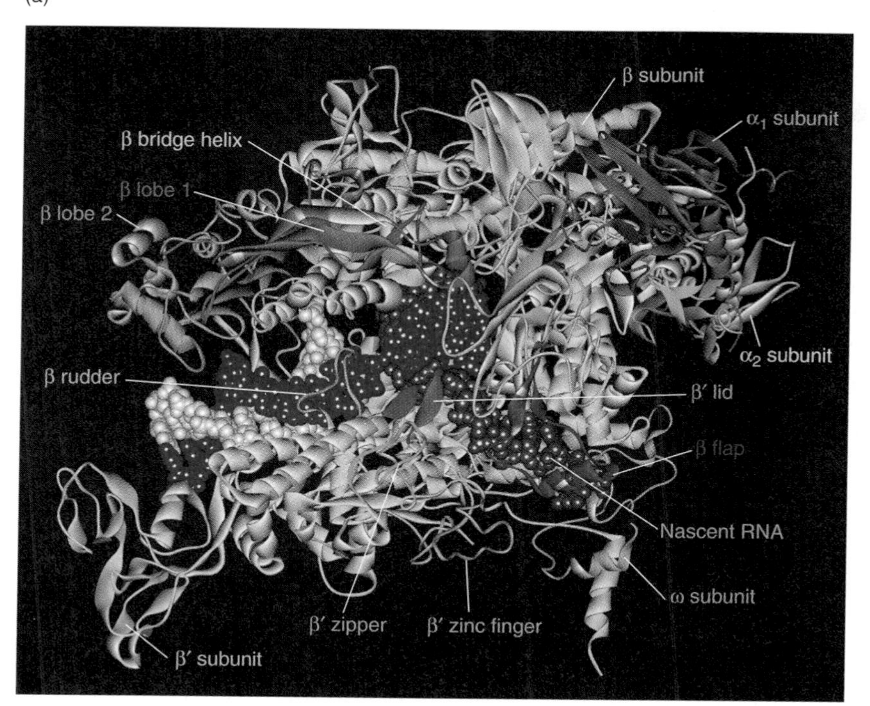

FIGURE 15.22 **The transcription elongation complex.** (a) The crystal structure of a *Thermus thermophilus* transcription elongation complex. The structure is a main channel view of RNA polymerase. Core polymerase is bound to a synthetic nucleic acid scaffold, which contains 14 bp of downstream DNA, 9 bp of an RNA-DNA hybrid and seven single-stranded nucleotides of displaced RNA transcript. Protein subunits are shown as ribbon structures and nucleic acids as space-fill structures. (b) A schematic model of the transcription elongation complex based on the crystal structure and biochemical studies. (Part a structure from Protein Data Bank 2O5I. D. G. Vassylyev, et al., *Nature* 448 [2007]: 157–162. Prepared by B. E. Tropp. Part b modified from E. Nudler, *Annu. Rev. Biochem.* 78 [2009]: 335–361. Copyright 2009 by Annual Reviews, Inc. Reproduced with permission of Annual Reviews, Inc., in the format Textbook via Copyright Clearance Center.)

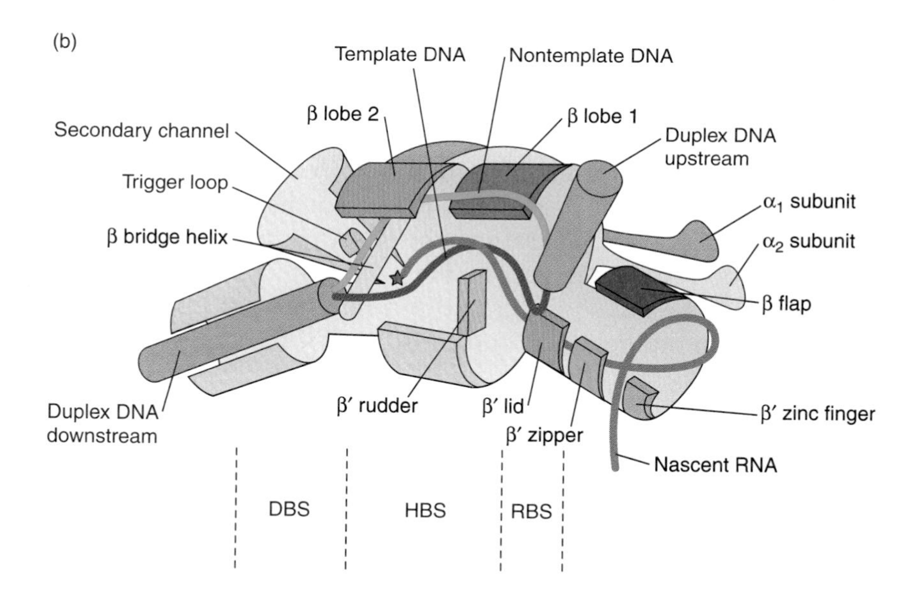

hybrid and seven single-stranded nucleotides of the displaced RNA transcript (FIGURE 15.22a). This crystal structure, when taken together with data from the chemical crosslinking studies, provides sufficient information to build the model for the transcription elongation complex shown in FIGURE 15.22b). The model shows that three adjacent nucleic acid-binding sites, the downstream **duplex-binding site (DBS)**, the **RNA-DNA hybrid binding site** (**HBS**), and the **RNA-binding site** (**RBS**) hold the TEC together.

1. **Duplex-binding site**—The duplex binding site is a deep cleft formed by the β′ subunit that encircles approximately 9 bp of downstream double-stranded DNA. The DNA makes direct contact with the protein through van der Waals and electrostatic interactions that are probably weak enough to allow the DNA to slide through the cleft. The two DNA strands begin to separate at about position +2 and the DNA makes a sharp 90° bend.
2. **RNA-DNA hybrid binding site**—The RNA-DNA hybrid binding site is part of the main channel formed by the β and β′ subunits. It begins slightly after the β′ bridge helix and extends to the β′ lid, surrounding 8 to 9 bp of RNA-DNA hybrid. Weak interactions between the protein and the RNA-DNA hybrid permit the hybrid to slide through the channel. The β′ rudder may help to stabilize the hybrid through contacts with the template DNA at positions –9, –10, and –11. The β′ lid acts as a wedge that helps the RNA to separate from the DNA and thereby prevent the hybrid from becoming overextended. It may also help stabilize the hybrid through contacts with the last bases in the hybrid. Furthermore, the lid's flexibility appears to allow the hybrid region to vary from 7 to 10 bp.
3. **RNA binding site**—The RNA binding site, which contains evolutionarily conserved amino acids, is located between the RNA-DNA hybrid-binding site and RNA exit channel. It interacts with the single-stranded RNA that has separated from the hybrid. After separation, the RNA passes through a narrow pore, formed in part by the β′ lid, enters a wider channel bounded by the β′ zipper and β′ zinc finger on one side and the β flap on the other, and exits the RNA polymerase.

The trigger loop and the β′ bridge helix help to move RNA polymerase forward by one nucleotide during each nucleotide addition cycle.

The fact that the duplex DNA, DNA-RNA hybrid, and RNA binding sites each surround the nucleic acids ensures that transcription elongation is a highly processive process. Moreover, the weak protein-nucleic acid interactions permit lateral enzyme movement along the nucleic acids. Molecular biologists initially thought nucleoside triphosphate hydrolysis served as the direct energy source for forward enzyme movement. Recent studies suggest, however, that RNA polymerase movement along the nucleic acids is driven by thermal energy so that the transcription elongation complex resembles a Brownian ratchet machine. Because the terms used to describe RNA polymerase movement derive from mechanical systems, it will be helpful to consider the **stationary pawl** and **reciprocating pawl** mechanical ratchet wheel machines shown in FIGURE 15.23. In the first machine (Figure 15.23a), a stationary pawl (wedge) blocks reverse wheel motion and thereby prevents random back and forth movement. In the second machine (Figure 15.23b), the pawl oscillates back and forth, pushing the ratchet wheel forward. With these two mechanical systems in mind, we are ready to examine how RNA polymerase moves forward during transcription elongation.

RNA polymerase's active site has two moving parts, the trigger loop and β′ bridge helix (Figure 15.22b), that contribute to nucleoside triphosphate binding and RNA polymerase movement. FIGURE 15.24 shows a model proposed by Evgeny Nudler and coworkers that reveals the contributions that trigger loop and β′ bridge helix make to the nucleotide addition cycle. The trigger loop is a flexible structure that can assume an unfolded conformation or a folded α-helical conformation. At the start of the cycle (the posttranslocational state), the trigger loop is in the unfolded or open conformation. When the correct nucleotide binds to the active site, the trigger loop assumes the folded α-helical or closed conformation and helps to stabilize nucleotide binding to the active site. The tightly bound NTP acts as a stationary pawl to block lateral RNA polymerase movement. After phosphodiester bond formation, pyrophosphate is released and the trigger loop returns to its unfolded state. The elongation complex is now in the pretranslocational state. The trigger loop oscillates between the unfolded conformation and an intermediate form, which presses against the bridge helix. This pressure causes the bridge helix to push against the RNA-DNA hybrid, forcing the elongation complex to enter an active posttranslocational state. Thus, the trigger loop and bridge helix function like a reciprocating pawl, facilitating RNA polymerase movement in the absence of any additional NTP.

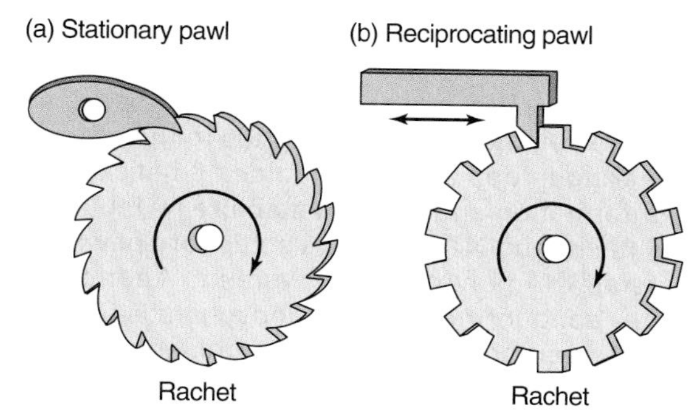

FIGURE 15.23 **Two mechanical ratchet wheel machines.** (a) A stationary pawl (wedge) blocks reverse wheel motion and thereby prevents random back and forth movement. (b) A reciprocating pawl oscillates back and forth, pushing the ratchet wheel forward. (Adapted from G. Bar-Nahum, et al., *Cell* 120 [2005]: 183–193.)

Pauses influence the overall transcription elongation rate.

New techniques have been devised to visualize the movement of individual RNA polymerase molecules along a DNA template. These techniques all involve immobilizing a single RNA polymerase molecule on a surface and then observing DNA movement by some form of microscopy. The technique that causes the least disturbance to the transcription elongation complex, the **tethered particle motion** approach, is shown in FIGURE 15.25. RNA polymerase is bound to a glass or mica surface covered by an aqueous solution of NTPs. The DNA template has a plastic bead or colloidal gold particle attached to its upstream end. The bead's Brownian motion is constrained by its attachment to DNA. The spatial extent of Brownian motion, as revealed by observation with a light microscope, is used to determine how the length of the DNA segment linking the bead to the polymerase changes during the transcription elongation process.

RNA polymerase does not move along the DNA template at a fixed rate. Instead, it spends proportionally more time at some template positions, termed **pause sites**, than at others. Pause sites were first detected in experiments in which RNA polymerase was stopped at specific sites along the DNA template, usually by deprivation of

FIGURE 15.24 Nucleotide addition cycle. At the start of the cycle (the posttranslocated state), the trigger loop (light blue) is in the unfolded or open conformation. When the correct NTP binds to the active site, the trigger loop assumes a folded β-helical or closed conformation and helps to stabilize NTP binding. The tightly bound NTP acts as a stationary pawl to block lateral RNA polymerase movement. After phosphodiester bond formation, pyrophosphate is released and the trigger loop returns to its unfolded state. The elongation complex is now in the posttranslocated state. The trigger loop oscillates between the unfolded conformation and an intermediate form, which presses against the bridge helix (yellow). This pressure causes the bridge helix to push against the RNA-DNA hybrid, forcing the elongation complex to enter an active posttranslocated state. Thus, the trigger loop and bridge helix function like a reciprocal pawl, facilitating RNA polymerase movement in the absence of NTP. Template DNA is in dark blue stick figures and RNA is in dark green stick figures. Magenta circles indicate magnesium ions I and II. (Modified from E. Nudler, *Annu. Rev. Biochem.* 78 [2009]: 335–361. Copyright 2009 by Annual Reviews, Inc. Reproduced with permission of Annual Reviews, Inc., in the format Textbook via Copyright Clearance Center.)

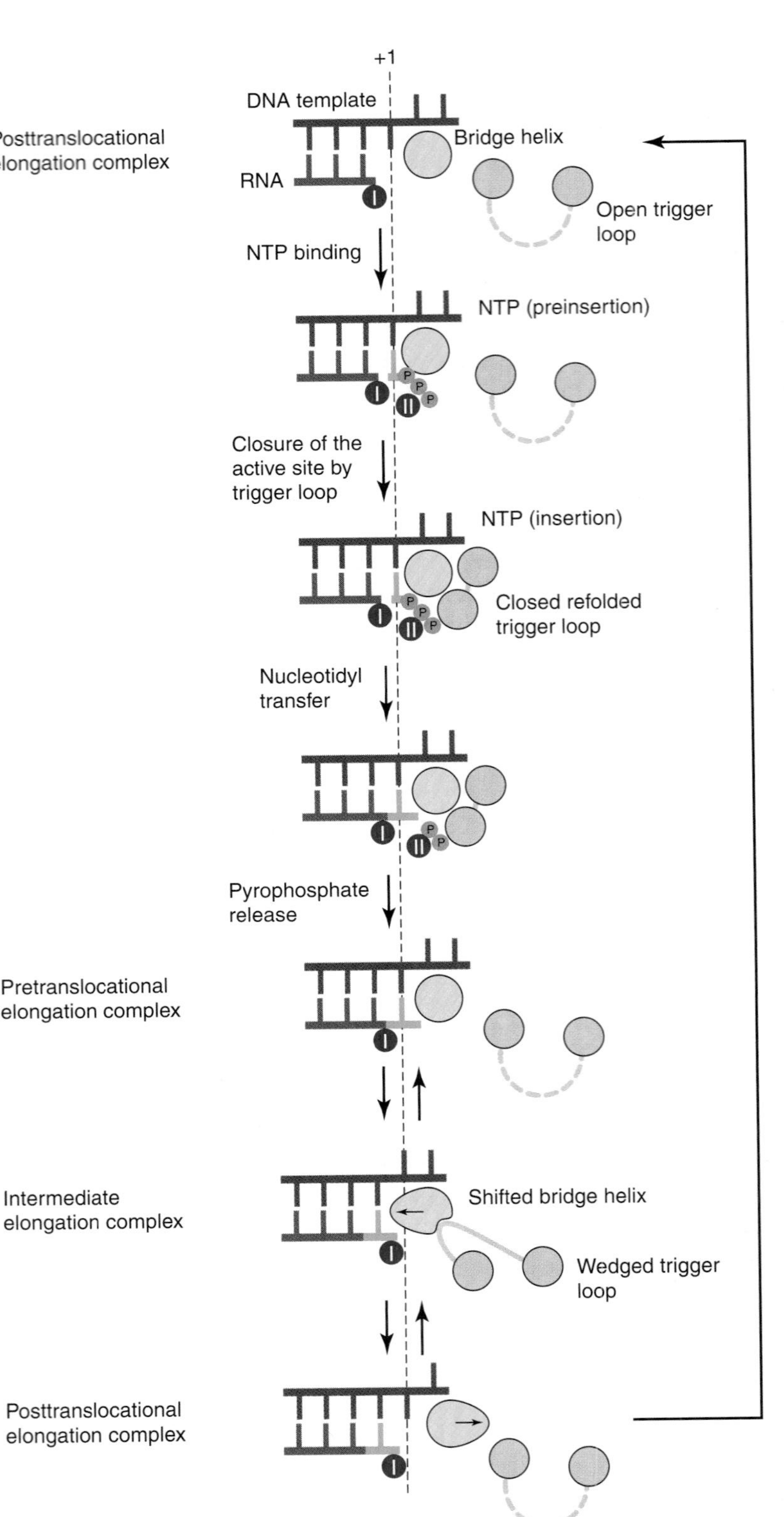

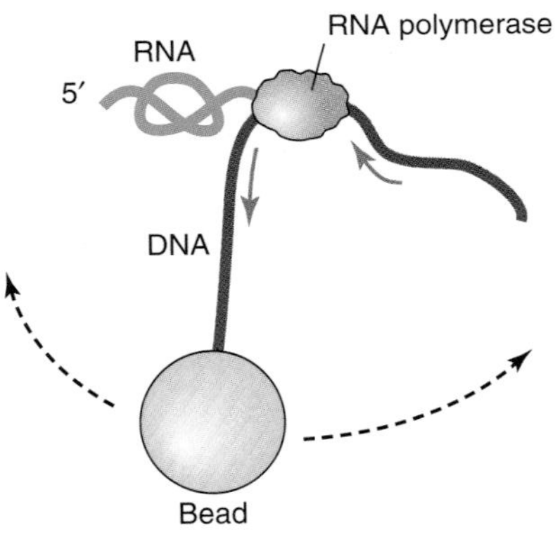

FIGURE 15.25 Tethered particle motion method. An RNA polymerase molecule is bound to a glass or mica surface that is coated with an aqueous solution of nucleoside triphosphates. A plastic bead or colloidal gold particle is attached to the DNA template's upstream end and undergoes rapid Brownian motion (dashed curved arrows), which is restricted by the bead's attachment to the DNA template. As the DNA template moves through the RNA polymerase in the direction that is indicated by the small green solid arrows, the length of the DNA segment linking the bead to the RNA polymerase changes. This change is revealed by observing the spatial extent of the bead's (or particle's) Brownian motion with a light microscope. (Adapted from J. Gelles and R. Landick, *Cell* 93 [1998]: 13–16.)

a single nucleotide, and then allowed to resume synthesis. Although transcription usually resumes right after the missing nucleotide is added to the mixture, pauses were observed at some sites. *Pausing* (the temporary delay in chain elongation) helps to synchronize transcription and translation, slows RNA polymerase movement to allow regulatory proteins to interact with the complex, and probably leads to both **arrest** (the complete halt of transcription without complex dissociation) and **termination** (complex dissociation).

RNA polymerase can detect and remove incorrectly incorporated nucleotides.

Transcriptional pausing is an important step in **proofreading.** When a nucleotide mismatch is present in the DNA-RNA hybrid region, RNA polymerase pauses and then backtracks so that the 3′-end of the RNA chain is displaced from the enzyme's active site (FIGURE 15.26). Backtracking increases as RNA-DNA hybrid stability decreases. Mismatches, therefore, promote backtracking. A backtracked complex can be rescued by internal hydrolytic cleavage and release of the transcript 3′-fragment to generate a new 3′-end that is in register with the RNA polymerase catalytic center. The RNA polymerase active site catalyzes this cleavage reaction but at a very slow rate.

Cleavage takes place much more rapidly when the transcription elongation factor **GreA** or **GreB** is present. Although the two protein factors have similar sequences, they have different functions. GreA induces removal of fragments that are two to three nucleotides long, whereas GreB does the same for fragments as large as 18 nucleotides long. Furthermore, GreA can only prevent the formation of an arrested complex, whereas GreB can rescue a preexisting arrested complex. Seth Darst and coworkers have used cryo-electron microscopy to reconstruct the structure of the *E. coli* RNA polymerase•GreB complex. The globular C-terminal domain of GreB binds to the rim of the secondary channel in RNA polymerase, while the NTD extends 45 Å into the channel directly to the RNA polymerase active site that contains a bound magnesium ion (FIGURE 15.27). Darst and coworkers propose that the GreB N-terminal domain helps to stabilize the binding of a second magnesium ion required for the cleavage reaction.

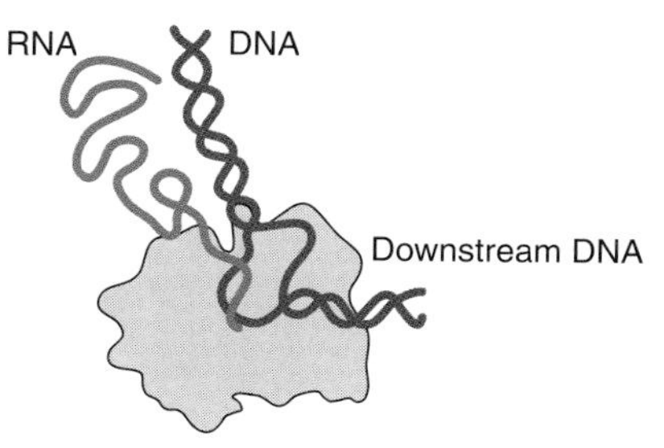

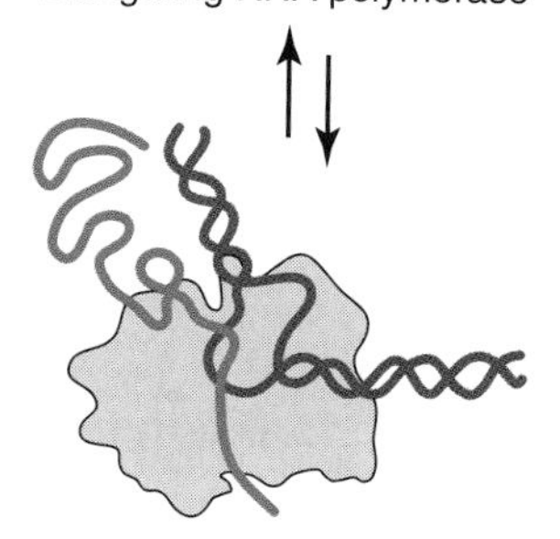

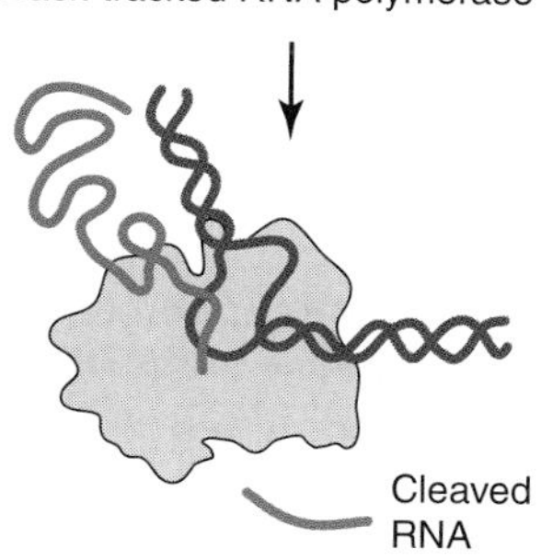

FIGURE 15.26 **RNA polymerase transcription and proofreading.** Under normal conditions, RNA polymerase (gray) elongates the nascent RNA chain (green) as the enzyme moves downstream on the DNA (blue). When a nucleotide mismatch is present in the DNA-RNA hybrid region, RNA polymerase pauses and then backtracks so that the 3′-end of the RNA chain is displaced from the enzyme's active site. The backtracked RNA polymerase can either slide forward again, returning to its previous elongating state (top) or cleave the nascent RNA (bottom) and then resume transcriptional elongation. (Adapted from J. W. Shaevitz, et al., *Nature* 426 [2003]: 684–687.)

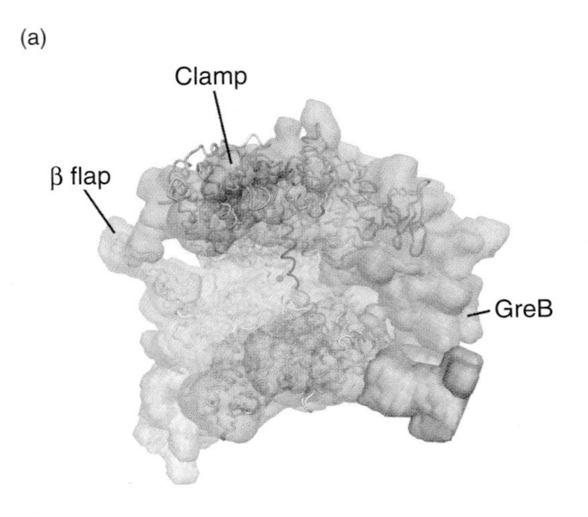

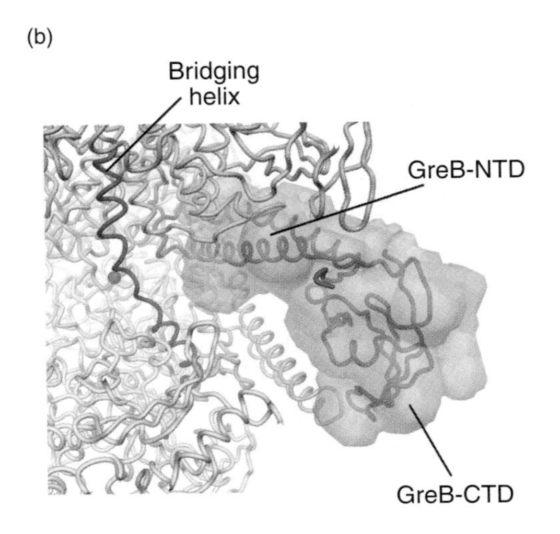

FIGURE 15.27 Cryo-electron microscopy reconstruction of the *E. coli* core RNA polymerase-GreB complex. (a) The reconstructed RNA polymerase•GreB complex. GreB is shown in pink. The α-carbon backbones for the β and β′ subunits are shown as cyan and pink worms, respectively. The β′ bridging helix is magenta and the active site magnesium ion is a magenta sphere. The region shown in blue is the *Thermus aquaticus* (*Taq*) core RNA polymerase, whereas the region shown in yellow corresponds to large insertions in the *E. coli* enzyme that are not present in the *Taq* enzyme. (b) A larger magnification of the RNA polymerase region that interacts with GreB. The colors are the same as in panel (a). However, the α-carbon backbone of GreB is shown as a red-orange worm. The N- and C- terminal domains of GreB are indicated by the notations NTD and CTD, respectively. (Reproduced from *Cell*, vol. 114, N. Opalka, et al., Structure and function of the transcription . . ., pp. 335–345, copyright 2003, with permission from Elsevier [http://www.sciencedirect.com/science/journal/00928674]. Photos courtesy of Seth Darst, Rockefeller University.)

15.4 Transcription Termination

Bacterial transcription machinery releases RNA strands at intrinsic and Rho-dependent terminators.

RNA polymerase is a highly processive macromolecular machine. Specific mechanisms are required to release RNA polymerase from the transcription elongation complex. Two transcription termination pathways, **intrinsic termination** and **Rho-dependent termination**, contribute about equally to this release in *E. coli*.

The intrinsic termination pathway is so named because it takes advantage of the core RNA polymerase's intrinsic catalytic activity to terminate transcription. Although nucleotide sequences specifying intrinsic terminators are present on DNA, it is actually the nascent RNA (and not DNA) that triggers the transcription termination response. Two sequence motifs on the nascent RNA strand are essential for intrinsic terminator function (FIGURE 15.28). The first motif, a G-C rich inverted repeat, allows the RNA to fold into a stem and loop structure that reaches to within seven to nine nucleotides of the 3′-end of the nascent RNA strand. Mutations that maintain the stable stem structure are usually tolerated. Those that decrease the stem structure's stability tend to reduce or eliminate termination. Multiple mutations within the stem region also may lead to loss of terminator activity even though the stem structure retains its stability. Secondary RNA structure, therefore, is important in determining intrinsic terminator activity, but nucleotide sequence within the stem also appears to make a contribution. The second motif, a run of 8 to 10 nucleotides that consists mostly of uridines, comes immediately after the stem and loop structure. Intrinsic terminators appear to act by first causing the TEC to pause and then to release the nascent RNA chain. Replacing

Terminate
G≡CUUUUUUUU G
C≡G
C≡G
C≡G
G≡C
C≡G
C A
U G
A A U

FIGURE 15.28 **Transcription termination signal for the intrinsic termination pathway.** (Adapted from R. A. Mooney, I. Artsimovitch, and R. Landick, *J. Bacteriol.* 180 [1998]: 3265–3275.)

the uridines in the second motif with other nucleotides converts an intrinsic terminator into a pause signal.

The Rho-dependent transcription termination pathway requires a hexameric helicase called the **Rho factor.** Jeffrey W. Roberts originally detected the Rho factor in purified bacterial extracts in 1969 because of the factor's ability to terminate transcription of bacteriophage lambda genes. Rho factor, which consists of six identical subunits (subunit molecular mass = 46 kDa), can be detected *in vitro* as an ATP-dependent helicase that releases RNA from an RNA-DNA hybrid. Each subunit has an NTD (residues 1–130) that contains the primary binding site for single-stranded RNA and a C-terminal domain (residues 131–419) that contains a secondary binding site for single-stranded RNA. The C-terminal domain also has an ATP binding site that is essential for helicase activity. Under physiological conditions Rho factor loads onto nascent mRNAs at a cytosine-rich region that contains 40 or more nucleotides known as the **Rho utilization (*rut*) site.**

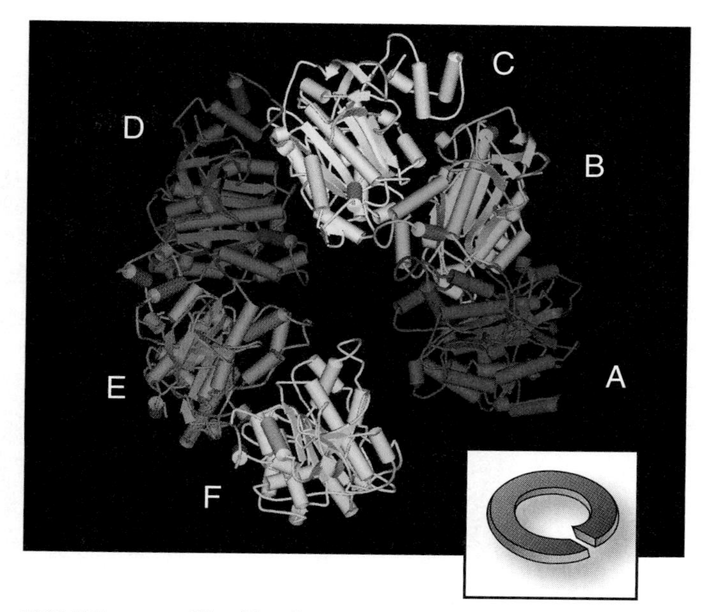

FIGURE 15.29 **The Rho factor.** The six subunits of the Rho factor pack into an open hexameric ring that resembles a lock washer (inset). Subunits are shown in different colors and labeled A–F. (Part a structure from Protein Data Bank ID: 1PVO. E. Skordalakes and J. M. Berger, *Cell* 114 [2003]: 135–146. Prepared by B. E. Tropp.)

Electron microscopy studies revealed that the Rho factor can exist as a closed or open ring structure. It was impossible, however, to tell whether the open ring contained all six subunits or was missing one. Crystal structures determined by Emmanual Skordalakes and James M. Berger in 2003 show that the open form of the Rho factor contains six subunits and resembles a "lock washer" (FIGURE 15.29). The ring opening is large enough to permit mRNA to enter the central cavity, which has a diameter that varies from 20 to 35 Å. The arrangement between the N- and C-terminal domains of each subunit is such that the primary RNA binding sites on the N-termini face inward, toward the center of the central cavity (FIGURE 15.30). The total length of RNA required to span the ring is 70 to 80 nucleotides. This arrangement forces the 3′-end of the mRNA into the central cavity, where the RNA binds to the secondary binding site in the C-terminal domain. This binding leads to a series of events that result in the closure of the hexameric ring and the hydrolysis of ATPs bound to the C-domains (FIGURE 15.31). The conformational change resulting from ring closure may activate the ATPase sites so that Rho moves in a 5′→3′ direction in pursuit of the RNA polymerase. Upon reaching RNA polymerase, Rho unwinds RNA from DNA and allows the RNA polymerase to disengage from the transcription elongation complex.

Transcription termination of some *E. coli* mRNAs requires a 21 kDa monomeric protein called NusG to assist Rho. NusG associates with both Rho factor and core RNA polymerase. It accelerates the rate of RNA polymerase elongation by 20 to 30% by stimulating escape of RNA polymerase at pause sites that contain weak RNA-DNA hybrids, which allow RNA polymerase to slide backwards on the template. Based on this and other information, Zvi Passman and Peter H. von Hippel have proposed that NusG may enhance both the rates of transcription elongation and Rho-mediated termination by

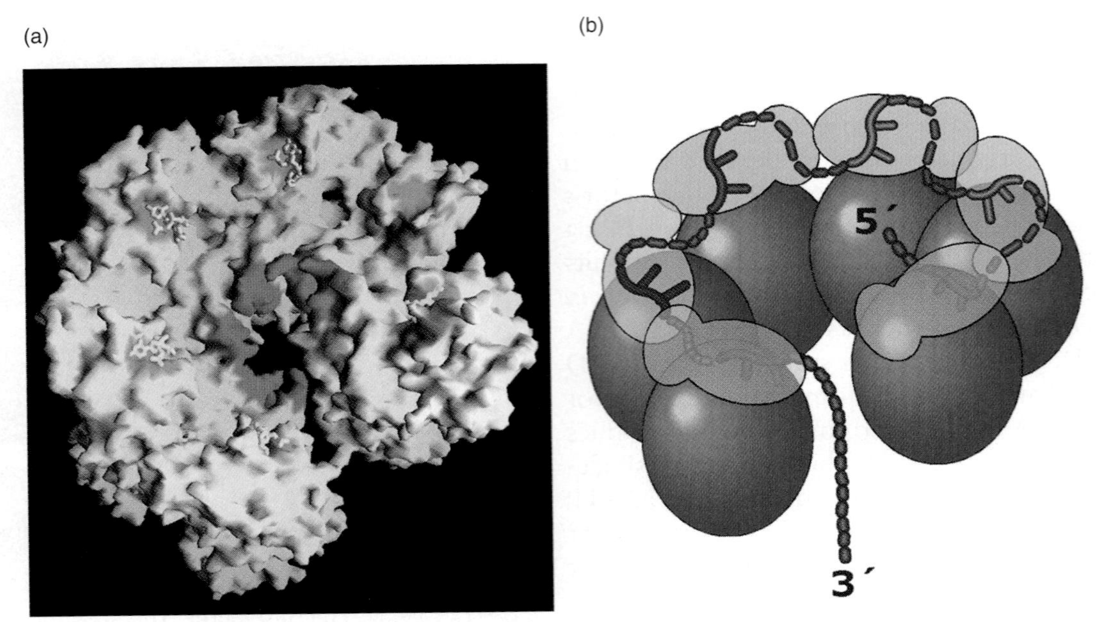

FIGURE 15.30 **Rho binding sites.** (a) Molecular structure of the rho hexamer. Primary RNA binding sites in the N-terminal domain are colored cyan. Secondary RNA binding sites in the C-terminal domain are colored magenta. Nucleic acid bound at the primary RNA binding sites is shown as yellow rods. (b) Schematic of the primary (N-terminal) RNA binding site configuration. The N- and C-terminal domains are colored green and red, respectively. Solid black lines represent the positions for the single-stranded nucleic acid, which binds across the primary RNA binding site so that the 3′-end is oriented toward the hole in the protein ring. The path that the nucleic acid needs to traverse between adjacent binding sites is shown by black broken lines. (Reproduced from *Cell*, vol. 114, E. Skordalakes and J. M. Berger, Structure of the rho transcription . . ., pp. 135–146, copyright 2003, with permission from Elsevier [http://www.sciencedirect.com/science/journal/00928674]. Photos courtesy of James M. Berger, University of California, Berkeley.)

preventing RNA polymerase from backtracking. Recent studies by Paul Rösch and coworkers indicate that NusG interacts with NusE, a protein identical to a ribosomal protein called S10. This finding, which provides a direct physical link between the transcription and translation processes, suggests that the ribosome plays an important role in determining the speed of transcription.

Rho factor is not present in all bacteria. Moreover, the Rho factor does not seem to be essential in some bacteria that contain it such as *Bacillus subtilis* or *Staphylococcus aureus*. Furthermore, Rho-like factors have not been observed in eukaryotes or archaea.

15.5 Antibiotics that Target Bacterial RNA Polymerase

RNA polymerase is a target for broad spectrum antibacterial therapy.

Even though bacterial, eukaryotic, and archaeal RNA polymerases use the same fundamental mechanism for RNA synthesis, there are sufficient structural differences among the three kinds of RNA polymerases to permit antibiotics to specifically target the bacterial enzyme.

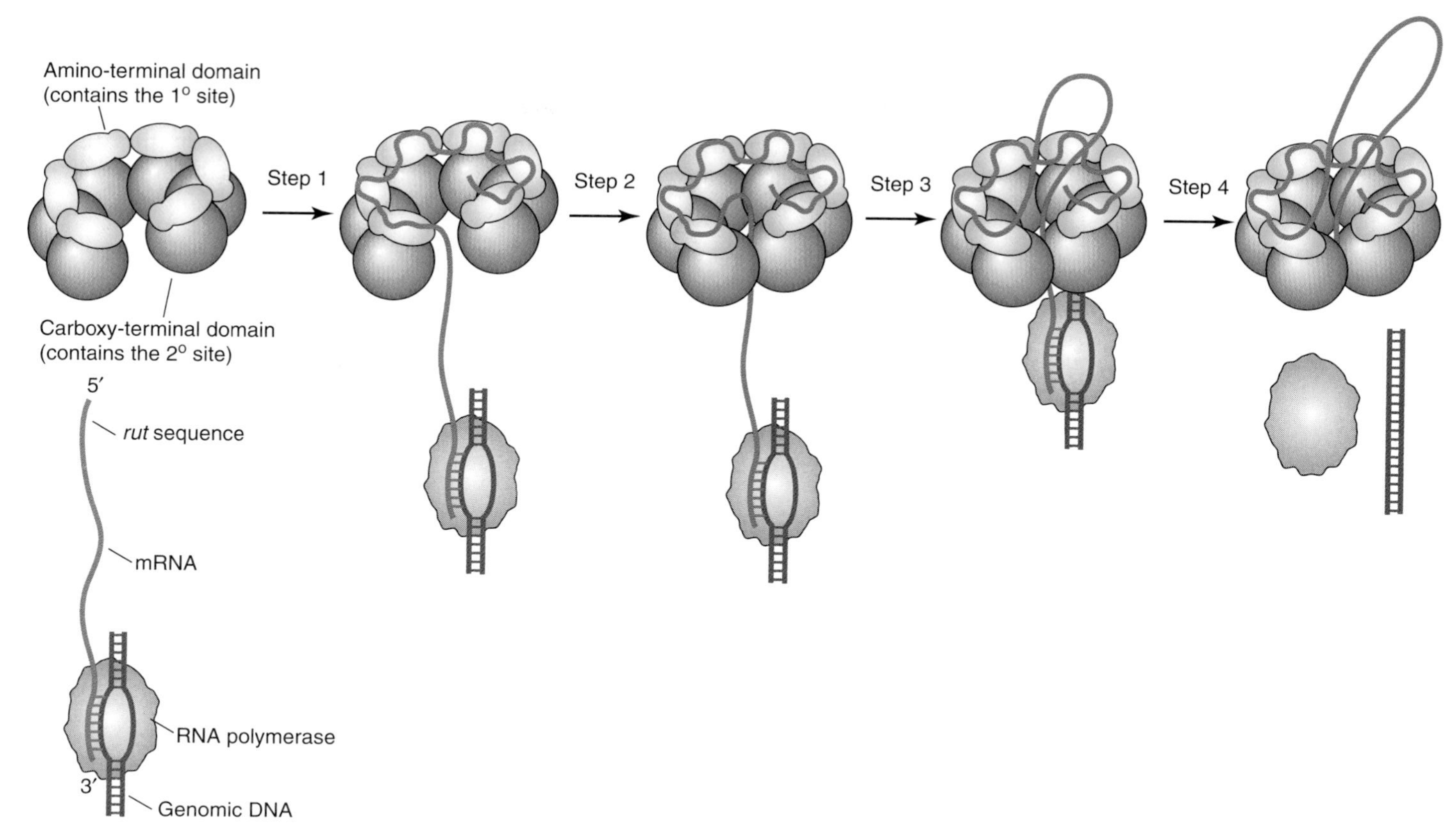

FIGURE 15.31 **Rho factor transcription termination in four steps.** Step 1: The Rho factor N-terminal domain (blue), which contains the primary RNA binding site, binds to the nascent mRNA *rut* sequence. Although Rho factor is shown as an open ring, its open and closed forms may be at equilibrium at this step. Step 2: The C-terminal Rho domain (red), which contains the secondary mRNA binding site, binds the mRNA downstream of *rut*, and the Rho hexameric ring closes. Step 3: The Rho carboxyl terminal domain cyclically binds and hydrolyzes ATP, driving itself along the mRNA in a 5′ to 3′ direction. The amino-terminal domain probably disengages from rut during this step. Step 4: The Rho factor functions as a helicase, coupling ATP binding and hydrolysis to the unwinding of nucleic acid. The net effect is that mRNA disengages from the genomic DNA and RNA polymerase (gray). (Modified from *Curr. Biol.*, vol. 13, D. L. Kaplan and M. O'Donnell, Rho Factor: Transcription Termination in Four Steps, pp. R714–R716, copyright 2003, with permission from Elsevier [http://www.sciencedirect.com/science/journal/09609822].)

Antibiotics that inhibit bacterial RNA polymerase have also been very helpful tools for studying how the enzyme works. Three antibiotics that have contributed to our understanding of bacterial RNA polymerase are as follows.

The **rifamycins**, a family of antibiotics produced by strains of *Streptomyces*, are used to treat a wide variety of bacterial infections. Unfortunately, many bacterial strains, including some that cause tuberculosis, have become resistant to rifamycins and its derivatives. A semisynthetic derivative of rifamycin, **rifampicin**, is widely used to study bacterial transcription. Rifampicin-resistant mutants map within *rpoB*, the structural gene for the β subunit. Rifampicin binds in a pocket within the β subunit close to the active site and obstructs the growing RNA chain's path, blocking synthesis after the first or second phosphodiester bond is formed. The inhibited enzyme remains bound to the promoter, preventing uninhibited enzyme from initiating transcription.

Sorangicin, which is produced by the mycobacterium *Sorangium cellulosum*, binds to the same site as the rifamycins and, like the rifamycins, inhibits transcription initiation but not transcription elongation. Although, sorangicin-resistant mutants also map within *rpoB*, many bacterial strains that are resistant to the rifamycins remain sensitive to sorangicin. Sorangicin, therefore, has the potential to work as an anti-tuberculosis drug for rifamycin-resistant strains.

Streptolydigin, which is produced by *Streptomyces lydicus*, binds to a site that includes the trigger loop and the β′ bridge helix. Resistant mutants map within *rpoB* and *rpoC*, the structural genes for the β and β′ subunits, respectively. Streptolydigin appears to inhibit transcription by blocking conformational changes during the nucleotide addition cycle by trapping the trigger loop in its unfolded conformation and the bridge helix in its straight conformation.

Now that we have become familiar with how bacterial RNA polymerase works, we are ready to examine the synthesis of the three major kinds of bacterial RNA—messenger RNA, ribosomal RNA, and transfer RNA—and the factors that influence the rates of synthesis of these RNA molecules. Chapter 16 examines these and related aspects of bacterial RNA synthesis.

Suggested Reading

Historical

Hurwitz, J. 2005. The discovery of RNA polymerase. *J Biol Chem* 280:42477–42488.

Initiation Stage

Artsimovitch, I. 2008. Post-initiation control by the initiation factor sigma. *Mol Microbiol* 68:1–3.

Borukhov, S., and Lee, J. 2005. RNA polymerase structure and function at lac operon. *C R Biol* 328:576–587.

Borukhov, S., and Nudler, E. 2003. RNA polymerase holoenzyme: structure, function, and biological implications. *Curr Opin Microbiol* 6:93–100.

Burgess, R. R., Travers, A. A., Dunn, J. J., and Bautz, E. K. 1969. Factor stimulating transcription by RNA polymerase. *Nature* 221:43–46.

Burrows, P. C. 2003. Investigating protein–protein interfaces in bacterial transcription complexes: a fragmentation approach. *Bioessays* 25:1150–1153.

Busby, S. J. W. 2009. More pieces in the promoter jigsaw: recognition of –10 regions by alternative sigma factors. *Mol Microbiol* 72:809–811.

Campbell, E. A., Korzheva, N., Mustaev, A., et al. 2001. Structural mechanism for rifampicin inhibition of bacterial RNA polymerase. *Cell* 104:901–912.

Campbell, E. A., Muzzin, O., Chlenov, M., et al. 2002. Structure of the bacterial RNA polymerase promoter specificity s subunit. *Mol Cell* 9:527–539.

Campbell, E. A., Westblade, L. F., and Darst, S. A. 2008. Regulation of bacterial RNA polymerase s factor activity: a structural perspective. *Curr Opin Microbiol* 11:121–127.

Cheetham, G. M. T., Jeruzalmi, D., and Steitz, T. A. 1999. Structural basis for initiation of transcription from an RNA polymerase-promoter complex. *Nature* 399:80–83.

Craig, M. L., Tsodikov, O. V., McQuade, K. L., et al. 1998. DNA footprints of the two kinetically significant intermediates in formation of an RNA polymerase-promoter open complex. *J Mol Biol* 283:741–756.

Cramer, P. 2002. Multisubunit RNA polymerases. *Curr Opin Struct Biol* 12:89–97.

Darst, S. A. 2001. Bacterial RNA polymerase. *Curr Opin Struct Biol* 11:155–162.

deHaseth, P. L., Zupancic, M. L., and Record, M. T. Jr. 1998. RNA polymerase-promoter interactions: the comings and goings of RNA polymerase. *J Bacteriol* 180:3019–3025.

Enz, S., Mahren, S., Menzel, C., and Braun, V. 2003. Analysis of the ferric citrate transport gene promoter of *Escherichia coli*. *J Bacteriol* 185:2387–2391.

Erie, D. A. 2002. The many conformational states of RNA polymerase elongation complexes and their roles in the regulation of transcription. *Biochim Biophys Acta* 1577:224–239.

Feklistov, A., and Darst, S. A. 2009. Promoter recognition by bacterial alternative s factors: the price of high selectivity. *Genes Dev* 23:2371–2375.

Finn, R. D., Orlova, E. V., Gowen, B., et al. 2000. *Escherichia coli* core and holoenzyme structures. *EMBO J* 19:6833–6844.

Gralla, J. D. 2000. Signaling through sigma. *Nat Struct Biol* 7:530–532.

Greive, S. J., and von Hippel, P. H. 2005. Thinking quantitatively about transcriptional regulation. *Nat Rev Mol Cell Biol* 6:221–232

Gruber, T. M., and Gross, C. A. 2003. Multiple sigma subunits and the partitioning of bacterial transcription space. *Ann Rev Microbiol* 57:441–466.

Haugen, S. P., Ross, W., and Gourse, R. L. 2008. Advances in bacterial promoter recognition and its control by factors that do not bind DNA. *Nat Rev Microbiol* 6:507–519.

Hellman, J. D. 2005. Sigma factors in gene expression. In: *Encyclopedia of Life Sciences*, pp. 1–7. Hoboken, NJ: John Wiley & Sons.

Hinton, D. M. 2005. Molecular gymnastics: distortion of an RNA polymerase s factor. *Trend Microbiol* 13:140–143.

Hsu, L. M. 2002. Open season on RNA polymerase. *Nat Struct Biol* 9:502–504.

Hsu, L. M. 2002. Promoter clearance and escape in prokaryotes. *Biochim Biophys Acta* 1577:191–207.

Hsu, L. M. 2009. Monitoring abortive initiation. *Methods* 47:25–36.

Kapnidis, A. N., Margeat, E., Ho, S. O., et al. 2006. Initial transcription by RNA polymerase proceeds through a DNA-scrunching mechanism. *Science* 314:1144–1147.

Koo, B-M., Rhodium, V. A., Nonaka, G., deHaseth, P. L., and Gross, C. A. 2009. Reduced capacity of alternative ss to melt promoters ensures stringent promoter recognition. *Genes Dev* 23:2426–2436.

Malhotra, A., Severinova, E., and Darst, S. A. 1996. Crystal structure of a s^{70} subunit fragment from *E. coli* RNA polymerase. *Cell* 87:127–136.

Mekler, V., Kortkhonjia, E., Mukhopadhyay, J., et al. 2002. Structural organization of bacterial RNA polymerase holoenzyme and the RNA polymerase-promoter open complex. *Cell* 108:599–614.

Miroslavona, N. S., and Busby, S. J. W. 2006. Investigations of the molecular structure of bacterial promoters. *Biochem Soc Symp* 73:1–10.

Mooney, R. A., and Landick, R. 1999. RNA polymerase unveiled. *Cell* 98:687–690.

Murakami, K, Masuda, S., Campbell, E. A., et al. 2002. Structural basis of transcription initiation: an RNA polymerase holoenzyme-DNA complex. *Science* 296:1285–1290.

Murakami, K. S., and Darst, S. A. 2003. Bacterial RNA polymerases: the whole story. *Curr Opin Struct Biol* 13:31–39.

Murakami, K. S., Masuda, S., and Darst, S. A. 2002. Structural basis of transcription initiation: RNA polymerase holoenzyme at 4 Å resolution. *Science* 296:1280–1284.

Nickels, B. E., and Hochschild, A. 2004. Regulation of RNA polymerase through the secondary channel. *Cell* 118:281–284.

Polyakov, A., Severinova, E., and Darst, S. A. 1995. Three-dimensional structure of *E. coli* core RNA polymerase: promoter binding and elongation conformations of the enzyme. *Cell* 83:365–373.

Record, T. M. Jr., Reznikoff, W. S., Craig, M. L., et al. 1996. *Escherichia coli* RNA polymerase (Eσ^{70}), promoters, and the kinetics of the steps of transcription initiation. In: Neidhardt, F. C. (ed). *Escherichia coli and Salmonella: Cellular and Molecular Biology*, 2nd ed., pp. 792–821. Washington, DC: ASM Press.

Revyakin, A., Liu, C., Ebright, R. H., and Strick, T. R. 2006. Abortive initiation and productive initiation by RNA polymerase involve scrunching. *Science* 314:1139–1143.

Roberts, J. W. 2006. RNA polymerase, a scrunching machine. *Science* 314:1097–1098.

Ross, W., and Gourse, R. L. 2009. Analysis of RNA polymerase-promoter complex formation. *Methods* 47:13–24.

Schmitz, A., and Galas, D. J. 1979. The interaction of RNA polymerase and lac repressor with the lac control region. *Nucl Acids Res* 6:111–137.

Typas, A., Becker, G., and Hengge, R. 2007. The molecular basis of selective promoter activation by the s^S subunit of RNA polymerase. *Molec Microbiol* 63:1296–1306.

Vassylyev, D. G., Sekine, S., Laptenko, O., et al. 2002. Crystal structure of a bacterial RNA polymerase holoenzyme at 2.6 Å resolution. *Nature* 417:712–719.

Wösten, M. M. S. M. 1997. Eubacterial sigma-factors. *FEMS Microbiol Rev* 22: 127–150.

Woychik, N. A. M., and Reinberg, D. 2001. RNA polymerases: subunits and functional domains. In: *Encyclopedia of Life Sciences*, pp. 1–8. London: Nature.

Young, B. A., Gruber, T. M., and Gross, C. A. 2002. Views of transcription initiation. *Cell* 109:417–420.

Zhang, G., Campbell, E. A., Minakhin, L., et al. 1999. Crystal structure of *Thermus aquaticus* core RNA polymerase at 3.3 Å resolution. *Cell* 98:811–824.

Zhang, X., Chaney, M., Wigneshweraraj, S. R., et al. 2002. Mechanochemical ATPases and transcriptional activation. *Mol Microbiol* 45:895–903.

Transcription Elongation Complex

Artsimovitch, I., and Landick, R. 2000 Pausing by bacterial RNA polymerase is mediated by mechanistically distinct classes of signals. *Proc Natl Acad Sci USA* 97:7090–7095.

Bar-Nahum, G. Epshtein, V., Ruckenstein, A. E., et al. 2005. A ratchet mechanism of transcription elongation and its control. *Cell* 120:183–193.

Borukhov, S., Lee, J., and Laptenko, O. 2005. Bacterial transcription elongation factors: new insights into molecular mechanism of action. *Mol Microbiol* 55:1315–1324.

Borukhov, S., and Nudler, E. 2007. RNA polymerase: the vehicle of transcription. *Trends Microbiol* 16:126–134.

Bureckner, F., Oritz, J., and Cramer, P. 2009. A movie of the RNA polymerase nucleotide addition cycle. *Curr Opin Struct Biol* 19:294–299.

Fish, R. N., and Kane, C. M. 2002. Promoting elongation with transcript cleavage stimulatory factors. *Biochim Biophys Acta* 1577:287–307.

Gelles, J., and Landick, R. 1998. RNA polymerase as a molecular motor. *Cell* 93:13–16.

Herbert, K. M., Greenleaf, W. J., and Block, S. M. 2008. Single-molecule studies of RNA polymerase: motoring along. *Annu Rev Biochem* 77:149–176.

Kireeva, M., Kashlev, M, and Burton, Z. F. 2010. Translocation by multi-subunit RNA polymerases. *Biochim Biophys Acta* 1799:389–401.

Korzheva, N., Mustaev, A., Kozlov, M., et al. 2000. A structural model of transcription elongation. *Science* 289:619–625.

Landick, R. 2001. RNA polymerase clamps down. *Cell* 105:567–570.

Marr, M. T., and Roberts, J. W. 2000. Function of transcription cleavage factors GreA and GreB at a regulatory pause site. *Mol Cell* 6:1275–1285.

Mooney, R. A., Artisimovitch, I., and Landick, R. 1998. Information processing by RNA polymerase: recognition of regulatory signals during RNA chain elongation. *J Bacteriol* 180:3265–3275.

Neuman, K. C., Abbondanzieri, E. A., Landick, R., et al. 2003. Ubiquitous transcriptional pausing is independent of RNA polymerase backtracking. *Cell* 115:437–447.

Nudler, E. 2009. RNA polymerase active center: the molecular engine of transcription. *Ann Rev Biochem* 78:335–361.

Opalka, N., Chlenov, M., Chacon, P., et al. 2003. Structure and function of the transcription elongation factor GreB bound to bacterial RNA polymerase. *Cell* 114:335–345.

Proshkin, S., Rahmouni, A. R., Mironov, A., and Nudler, E. 2010. Cooperation between translating ribosomes and RNA polymerase in transcription elongation. *Science* 328:504–508.

Roberts, J. W., Shankar, S., and Filter, J. J. 2008. RNA polymerase elongation factors. *Annu Rev Biochem* 62:211–233.

Shaevitz, J. W., Abbondanzieri, E. A., Landick, R., and Block, S. M. 2003. Backtracking by single RNA polymerase molecules observed at near-base-pair resolution. *Nature* 426:684–687.

Sosunova, E., Sosunova, V., Kozlov, M., et al. 2003. Donation of catalytic residues to RNA polymerase active center by transcription factor Gre. *Proc Natl Acad Sci USA* 100:15469–15474.

Svetlov, V., and Nudler, E. 2009. Macromolecular micromovements; how RNA polymerase translocates. *Curr Opin Struct Biol* 19:1–7.

Sydow, J. F., and Cramer, P. 2009. RNA polymerase fidelity and transcriptional proofreading. *Curr Opin Struct Biol* 19:732–739.

Vassylyev, D. G. 2009. Elongation by RNA polymerase: a race through roadblocks. *Curr Opin Struct Biol* 19:1–10.

Vassylyev, D. G., Vassylyeva, M. N., Perederina, A., Tahirov, T. H., and Artsimovitch, I. 2007. Structural basis for transcription elongation by bacterial RNA polymerase. *Nature* 448:157–162.

Vassylyev, D. G., Vassylyeva, M. N., Zhang, J., Palangat, M., Artsimovitch, I., and Landick, R. 2007. Structural basis for substrate loading in bacterial RNA polymerase. *Nature* 448:163–168.

Zhang, J., Palangar, M., and Landick, R. 2010. Role of the RNA polymerase trigger loop in catalysis and pausing. *Nat Struct Mol Biol* 17:99–104.

Transcription Termination

Burmann, B. M., Schweimer, K., Luo, X, et al. 2010. A NusE:NusG complex links transcription and translation. *Science* 238:501–504.

Kaplan, D. L., and O'Donnell, M. 2003. Rho factor: transcription termination in four steps. *Curr Biol* 13:R714–R716.

Passman, Z., and von Hippel, P. H. 2000. Regulation of Rho-dependent transcription termination by NusG is specific to the *Escherichia coli* elongation complex. *Biochemistry* 39:5573–5585.

Richardson, J. P. 2001. Transcript elongation and termination. pp. 1–7. In: *Encyclopedia of Life Sciences*. London: Nature.

Richardson, J. P. 2002. Rho-dependent termination and ATPases in transcript termination. *Biochim Biophys Acta* 1577:251–260.

Richardson, L. V., and Richardson, J. P. 2005. Identification of a structural element that is essential for two functions of transcription factor NusG. *Biochim Biophys Acta* 1729:135–140.

Skordalakes, E., and Berger, J. M. 2003. Structure of the Rho transcription terminator: mechanism of mRNA recognition and helicase loading. *Cell* 114:135–146.

Antibiotics that Target Bacterial RNA Polymerase

Darst, S. A. 2004. New inhibitors targeting bacterial RNA polymerase. *Trends Biochem Sci* 29:159–162.

Ho, M. X., Hudson, B. P., Das, K., Arnold, E., and Ebright, R. H. 2009. Structures of RNA polymerase-antibiotic complexes. *Curr Opin Struct Biol* 19:715–723.

Mariani, R., and Maffioli, S. I. 2009. Bacterial RNA polymerase inhibitors: an organized overview of their structure, derivatives, biological activity and current clinical development. *Curr Med Chem* 16:430–454.

Villain-Guillot, P., Bastide, L., Gualtieri, M., and Leonetti, J-P. 2007. Progress in targeting bacterial transcription. *Drug Discovery Today* 12:200–208.

This book has a Web site, **http://biology.jbpub.com/book/molecular**, which contains a variety of resources, including links to in-depth information on topics covered in this chapter, and review material designed to assist in your study of molecular biology.

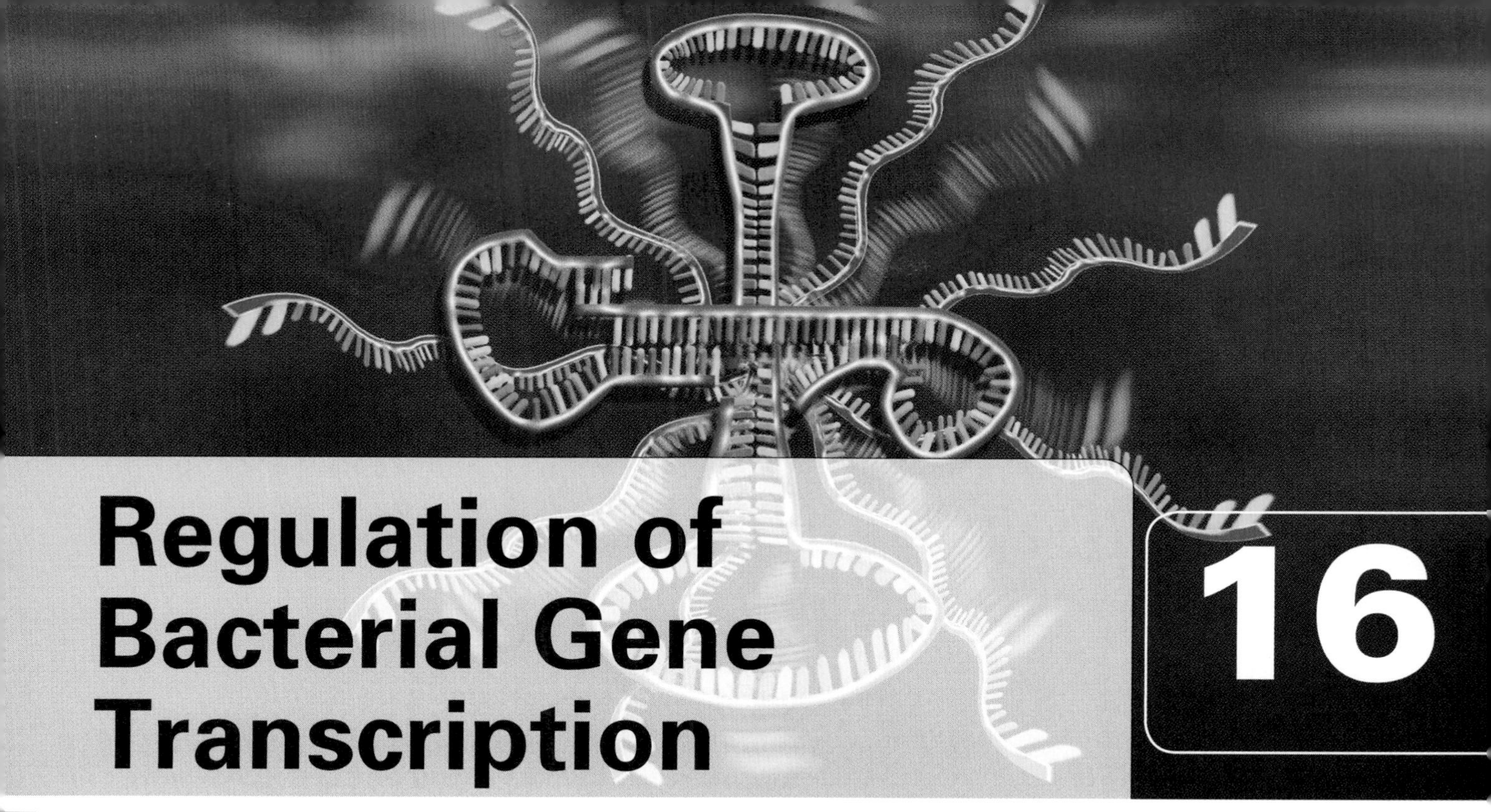

16 Regulation of Bacterial Gene Transcription

OUTLINE OF TOPICS

16.1 Messenger RNA

Bacterial mRNA may be monocistronic or polycistronic.

Bacterial mRNA usually has a short lifetime compared to other kinds of bacterial RNA.

Controlling the rate of mRNA synthesis can regulate the flow of genetic information.

Messenger RNA synthesis can be controlled by negative and positive regulation.

16.2 Lactose Operon

The *E. coli* genes *lacZ*, *lacY*, and *lacA* code for β-galactosidase, lactose permease, and β-galactoside transacetylase, respectively.

The *lac* structural genes are regulated.

Genetic studies provide information about the regulation of *lac* mRNA.

The operon model explains the regulation of the lactose system.

Allolactose is the true inducer of the lactose operon.

The Lac repressor binds to the *lac* operator *in vitro*.

The *lac* operon has three *lac* operators.

The Lac repressor is a dimer of dimers, where each dimer binds to one *lac* operator sequence.

16.3 Catabolite Repression

E. coli uses glucose in preference to lactose.

The inhibitory effect of glucose on expression of the *lac* operon is a complicated process.

The cAMP•CRP complex binds to an activator site (AS) upstream from the *lac* promoter and activates *lac* operon transcription.

cAMP•CRP activates more than 100 operons.

16.4 Galactose Operon

The galactose operon is also regulated by a repressor and cAMP•CRP.

16.5 The *araBAD* Operon

The AraC activator protein regulates the *araBAD* operon.

16.6 Tryptophan Operon

The tryptophan (*trp*) operon is regulated at the levels of transcription initiation, elongation, and termination.

16.7 Bacteriophage Lambda: A Transcription Regulation Network

Phage λ development is regulated by a complex genetic network.

The lytic pathway is controlled by a transcription cascade.

The lysogenic pathway is also controlled by a transcription cascade.

The CI regulator maintains the lysogenic state.

Ultraviolet light induces the λ prophage to enter the lytic pathway.

16.8 Messenger RNA Degradation

Bacterial mRNA molecules are rapidly degraded.

16.9 Ribosomal RNA and Transfer RNA Synthesis

Bacterial ribosomes are made of a large subunit with a 23S and 5S RNA and a small subunit with 16S RNA.

E. coli has seven rRNA operons, each coding for a 16S, 23S, and 5S RNA.

A promoter upstream element (UP element) increases *rrn* transcription.

Three Fis protein binding sites increase *rrn* transcription.

16.10 Regulation of Ribosome Synthesis

Amino acid starvation leads to the production of guanine nucleotides that inhibit rRNA synthesis.

E. coli rRNA and tRNA syntheses increase with growth rate.

E. coli regulates r-protein synthesis.

16.11 Processing rRNA and tRNA

Bacteria process the primary transcripts for rRNA and tRNA to form the physiologically active RNA molecules.

Suggested Reading

Bacterial RNA polymerase holoenzyme is required to synthesize the three major classes of bacterial RNA: messenger RNA (mRNA), ribosomal RNA (rRNA), and transfer RNA (tRNA). Additional enzymes and protein factors regulate the synthesis of these RNA molecules, however, and, in the case of the rRNA and tRNA molecules, are needed to convert the immediate product of transcription or **primary transcript** into the functional RNA product. We now examine the mechanisms that bacteria use to regulate the synthesis of the different classes of RNA starting with the regulation of bacterial mRNA synthesis.

16.1 Messenger RNA

Bacterial mRNA may be monocistronic or polycistronic.

Our examination of the regulation of mRNA synthesis begins by considering some characteristics of mRNA, the sequence of which is determined by a specific template DNA sequence within the bacterial chromosome. Because transcription and translation proceed in a 5′→3′ direction and both processes take place in the same cellular compartment, the bacterial protein synthetic machinery can start to read the 5′-end of mRNA before the 3′-end is formed. A bacterial cell, therefore, does not have a chance to alter a nascent mRNA molecule before the protein synthetic machinery begins to translate it. The situation is different in eukaryotes, which carry out transcription in the cell nucleus and translation in the cytoplasm. The eukaryotic cell thus can convert the primary transcript to mature mRNA in the cell nucleus before the mRNA is required to direct protein synthesis in the cytoplasm.

Protein synthetic machinery reads the mRNA nucleotide sequence in groups of three bases or **codons**. Each codon specifies an amino acid or a termination signal. The protein synthetic machinery begins polypeptide synthesis at a start codon located toward the 5′-end of the mRNA and continues synthesis in a 5′→3′ direction until it encounters a termination codon. The segment of mRNA that codes for a polypeptide chain is called an **open reading frame** (**ORF**) because the protein synthetic machinery begins reading the segment at a specific start codon and stops reading it at a specific termination codon. A DNA segment corresponding to an open reading frame plus the translational start and stop signals for protein synthesis is called a **cistron** and an mRNA encoding a single polypeptide is called *monocistronic mRNA*. Although the terms *cistron* and *gene* are sometimes used interchangeably to describe bacterial DNA segments that specify polypeptides, the term *gene* has a broader meaning because it also includes the promoter region and applies to DNA segments that code for RNA molecules such as tRNA and rRNA that are not translated.

Bacterial mRNA molecules often contain two or more cistrons. In fact, bacterial **polycistronic mRNA** molecules are actually more common than bacterial monocistronic mRNA molecules. Each cistron within a polycistronic mRNA specifies a specific polypeptide chain. Furthermore, cistrons contained in polycistronic mRNA often specify proteins for a single metabolic pathway. For example, one *Escherichia coli* mRNA has three cistrons, each coding for a different protein required for lactose metabolism and another contains eight cistrons, each coding for a different enzyme required for histidine synthesis.

Using polycistronic mRNA is a way for a cell to regulate synthesis of related proteins coordinately. With a polycistronic mRNA molecule, the synthesis of several related proteins—in similar quantities and at the same time—can be regulated by a single signal. The sizes of bacterial mRNA molecules vary within a broad range. A monocistronic mRNA with a 1,500-nucleotide ORF (500 codons) would be needed to specify a polypeptide that is 500 amino acid-residues long. Of course, a polycistronic mRNA would be much longer.

In addition to reading frames and start and stop sequences for translation, other regions in mRNA are significant. For example, mRNA translation rarely, if ever, starts exactly at the 5′ end and stops exactly at the 3′-end. Instead, initiation of synthesis of the first polypeptide chain of a polycistronic mRNA may begin hundreds of nucleotides from the 5′-RNA terminus. The untranslated RNA sequence before the coding region is called the **5′-untranslated region** (**5′-UTR**) or **5′-leader**. The untranslated sequence after the coding region is called the **3′-untranslated region** (**3′-UTR**). Polycistronic mRNA molecules also usually contain intercistronic sequences (**spacers**) that are tens of bases long.

Bacterial mRNA usually has a short lifetime compared to other kinds of bacterial RNA.

An important characteristic of bacterial mRNA is that its lifetime is short compared to other types of bacterial RNA molecules. The half-life of a typical bacterial mRNA molecule is a few minutes. Although mRNA's short lifetime may seem wasteful, it has an important regulatory function. A cell can turn off the synthesis of a protein that is no longer needed by turning off synthesis of the mRNA that encodes the protein. Soon after, none of that particular mRNA will remain and synthesis of the protein will cease. Of course, bacterial cells save energy by not being forced to synthesize proteins that they no longer need.

The short lifetime of bacterial mRNA is one criterion used to identify mRNA in bacteria. A common experimental technique to determine whether a particular RNA molecule or class of RNA molecules is mRNA is the **pulse-chase experiment**. Bacteria are briefly cultured in a medium that contains a radioactive precursor for RNA such as [^{3}H]uridine. Then the bacteria are switched to a medium that contains a high concentration of nonradioactive uridine (and no [^{3}H] uridine) and samples are removed at specific times for analyses. The RNA is isolated and different species are separated by gel electrophoresis and detected by their radioactivity. A typical radioactive mRNA molecule will decrease with a half-life of a few minutes, whereas radioactive rRNA and tRNA molecules will remain through many generations. One difficulty with this technique is that bacteria contain some long-lived mRNA molecules and these would be misclassified. A better criterion for identifying a molecule as mRNA would be to isolate it and determine in an *in vitro* protein-synthesizing system whether the particular RNA can direct protein synthesis.

Controlling the rate of mRNA synthesis can regulate the flow of genetic information.

Bacterial cells can control **gene expression** (the flow of genetic information) by regulating the rate at which specific genes are transcribed. We have already encountered one method that bacterial cells use to do so. They vary the promoter sequence so that RNA polymerase initiates transcription at some genes more efficiently than at others. However, this mechanism is not ideally suited for all situations. For example, the products of many genes are needed under certain physiological conditions but not under others. The products of such genes must be synthesized only when circumstances demand it.

Natural selection improves efficiency. Among the unicellular organisms, any mutation that increases the overall efficiency of cellular metabolism should enable a mutant cell to grow slightly faster than a wild-type organism; thus, if enough time is allowed, a mutant cell line will outgrow a wild-type one. For example, in a population of 10^9 bacteria with a 30.0-minute doubling time, if one bacterium is altered such that it has a 29.5-minute doubling time, in about 80 days of continued growth, 99.9% of the population will have a 29.5-minute doubling time. (In this calculation, it is assumed that the growing culture is repeatedly diluted. Otherwise, unless the volume was greater than that of the Earth, the culture would stop growth in a day or so.) This time frame may seem very long on the laboratory time scale but it is infinitesimal on the evolutionary scale. It, therefore, is reasonable on this basis alone that regulated systems, in which efficiency has been improved, should have evolved.

Bacterial cells increase their efficiency (and therefore their growth rates) by (1) selecting the catabolic pathway for energy production that yields the greatest amount of energy per unit time and (2) synthesizing molecules only as the need arises. Bacteria accomplish both of these objectives by turning on the transcription of specific genes when their products are needed and turning off their transcription when their products are not needed. Actually, there are no known examples of switching a system completely off. When transcription is in the "off" state, there always remains a basal level of gene expression. This basal level often amounts to only one or two transcription events per cell generation and thus very little mRNA synthesis. For convenience, when discussing transcription, we use the term *off*, but it should be kept in mind that what is meant is *very low*. We will also see examples in this chapter of systems in which activity is switched from fully on to partly on (or even slightly on) rather than to off.

In bacterial systems in which several enzymes act in sequence in a single metabolic pathway, it is often the case that either all of these enzymes are present or all are absent. This phenomenon, which is called **coordinate regulation**, results from control of the synthesis of

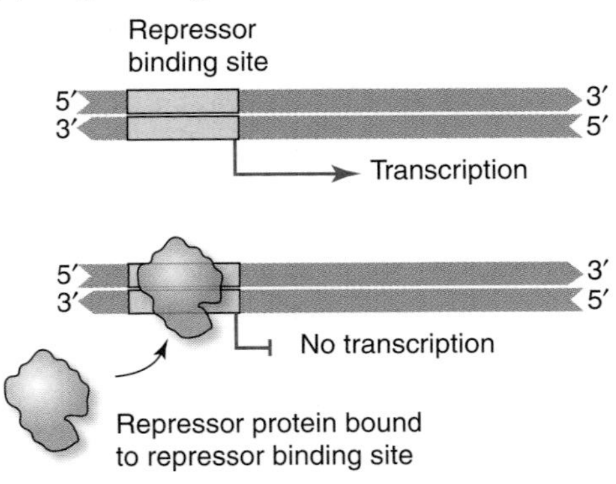

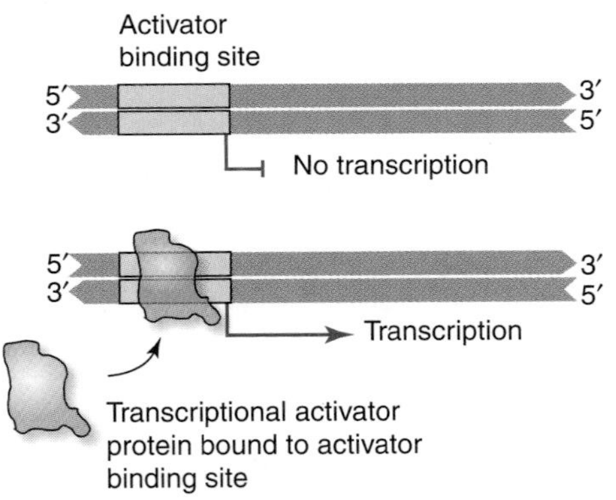

FIGURE 16.1 **The distinction between negative and positive regulation.** (a) In negative regulation, the "default" state of the gene is one in which transcription takes place. The binding of a repressor protein to the DNA molecule prevents transcription. (b) In positive regulation, the default state is one in which transcription does not take place. The binding of a transcriptional activator stimulates transcription. A single genetic element may be regulated both positively and negatively; in such a case, transcription requires the binding of the transcriptional activator and the absence of the repressor protein.

TABLE 16.1	mRNA Regulation	
Binding of Regulator to DNA	**Positive**	**Negative**
Yes	On	Off
No	Off	On

Note: If a system is both positively and negatively regulated, it is "on" when the positive regulator is bound to the DNA and the negative regulator is not bound to the DNA.

a single polycistronic mRNA that encodes all of the gene products. There are several mechanisms for this type of regulation, as will be seen below.

Messenger RNA synthesis can be controlled by negative and positive regulation.

The molecular mechanism of mRNA regulation can be divided into two major categories: **negative regulation** and **positive regulation** (FIGURE 16.1). In negative regulation, a **repressor** turns off the transcription of one or more genes. In positive regulation, an **activator** turns on the transcription of one or more genes. Negative regulation and positive regulation are not mutually exclusive. Many genes respond to both types of regulation. Table 16.1 summarizes the properties of negative and positive control. We now examine a variety of regulated bacterial systems beginning with the best-understood system, the genes responsible for lactose utilization. Studies of the lactose system have provided both the language and the principles required to understand genetic regulation.

16.2 Lactose Operon

The *E. coli* genes *lacZ*, *lacY*, and *lacA* code for β-galactosidase, lactose permease, and β-galactoside transacetylase, respectively.

In *E. coli*, two proteins are necessary for lactose metabolism. These proteins are the enzyme **β-galactosidase**, which cleaves lactose to yield galactose and glucose (FIGURE 16.2), and a carrier molecule, **lactose permease**, which transports lactose (and other galactosides) into the cell. The existence of the two proteins was first shown by a combination of genetic experiments and biochemical analysis.

First, hundreds of Lac$^-$ mutants (unable to use lactose as a carbon source) were isolated. By genetic manipulation, some of these mutations were moved from the *E. coli* chromosome to an F′*lac* plasmid (a plasmid carrying the genes for lactose utilization) and then partial diploids having the genotypes F′*lac*$^-$/*lac*$^+$ or F′*lac*$^+$/*lac*$^-$ were constructed. (The relevant genotype of the plasmid is given to the left of the diagonal line and that of the chromosome to the right.) It was observed that these diploids always have a Lac$^+$ phenotype (i.e., they

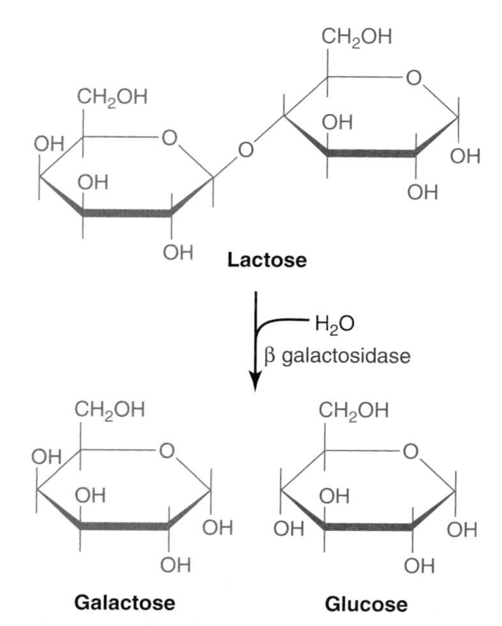

FIGURE 16.2 β-galactosidase catalyzed lactose hydrolysis.

make β-galactosidase), which shows that none of the *lac*⁻ mutants make an inhibitor that blocks *lac* gene function.

Partial diploids were also constructed in which both the chromosome and the F′*lac* plasmid were *lac*⁻. Using different pairs of *lac*⁻ mutants, some pairs were observed to have a Lac⁺ phenotype while others were observed to have a Lac⁻ phenotype. This complementation test (see Chapter 7) showed that all of the mutants initially isolated fell into one of two groups, which were called *lacZ* and *lacY*. Mutants in the two groups *Z* and *Y* have the property that the partial diploids F′*lacY*⁻*lacZ*⁺/*lacY*⁺*lacZ*⁻ and F′*lacY*⁺*lacZ*⁻/*lacY*⁻*lacZ*⁺ have a Lac⁺ phenotype and the genotypes F′*lacY*⁻*lacZ*⁺/*lacY*⁻*lacZ*⁺ and F′*lacY*⁺*lacZ*⁻/*lacY*⁺*lacZ*⁻ have the Lac⁻ phenotype. The existence of two complementation groups was good evidence that there are at least two genes in the *lac* system.

The *lacZ* gene is the structural gene for β-galactosidase. This enzyme is readily detected by a simple colorimetric assay that takes advantage of the fact that β-galactosidase catalyzes the hydrolysis of *o*-nitrophenyl-β-galactoside, a lactose analog (**FIGURE 16.3**). The product *o*-nitrophenoxide, which is yellow, can be detected by light absorption at a wavelength of 420 nm. The function of the *lacY* gene product as lactose permease was strongly suggested by experiments that showed that *lacZ*⁺ *lacY*⁻ cells cannot transport [^{14}C]lactose into the cell but *lacZ*⁻ *lacY*⁺ cells can do so.

Investigators discovered a third gene, *lacA*, while studying lactose transport. The *lacA* gene product is a β-galactoside transacetylase, which transfers an acetyl group from acetyl-CoA to lactose analogs (**FIGURE 16.4**). The reason that *lacA* was not detected at the same time as *lacZ* and *lacY* is that its gene product, transacetylase, is not required for lactose catabolism. The precise role of *lacA* is still a matter of

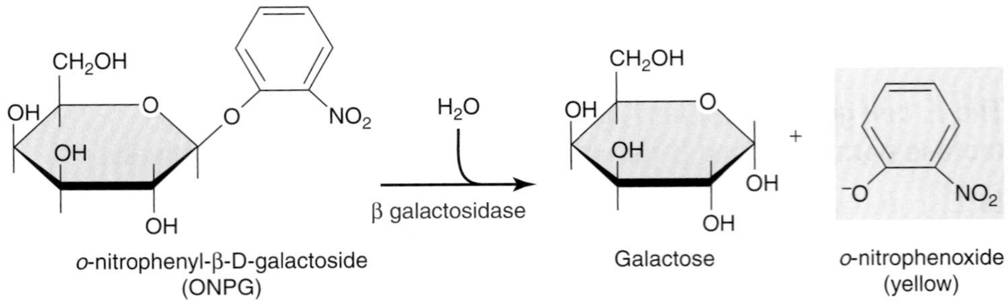

FIGURE 16.3 β-galactosidase assay using *o*-phenyl-β-D-nitrophenylgalactoside as a substrate.

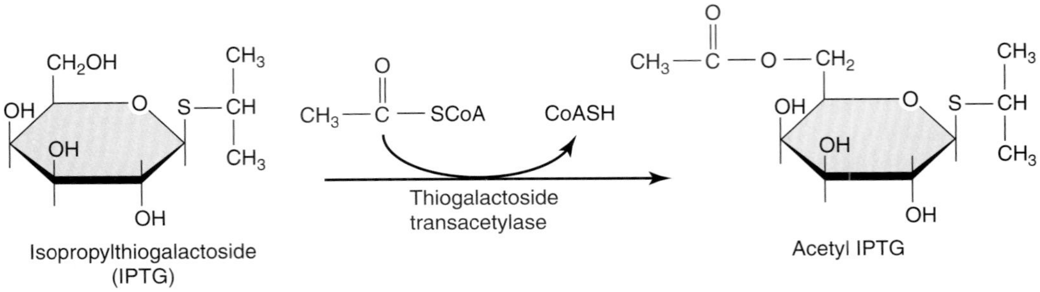

FIGURE 16.4 Reaction catalyzed by thiogalactoside transacetylase.

conjecture. One hypothesis is that transacetylase detoxifies lactose analogs that would harm cells.

The *lac* structural genes are regulated.

When *E. coli* with a *lac*$^+$ genotype is cultured in a lactose-free medium (also lacking glucose, a point that will be discussed shortly), the intracellular concentrations of β-galactosidase, permease, and transacetylase are exceedingly low—roughly one or two molecules of each protein per bacterial cell. When lactose is added to the growth medium, however, the concentration of these proteins increases simultaneously to about 10^5 molecules per cell (or about 1% of the total cellular protein). This phenomenon is shown for β-galactosidase and permease in FIGURE 16.5. Furthermore, lactose addition triggers the synthesis of *lac* mRNA as evidenced by studies in which mRNA, labeled with [^{32}P]phosphate at various times after lactose addition, is hybridized to DNA that carries *lac* genes (Figure 16.5).

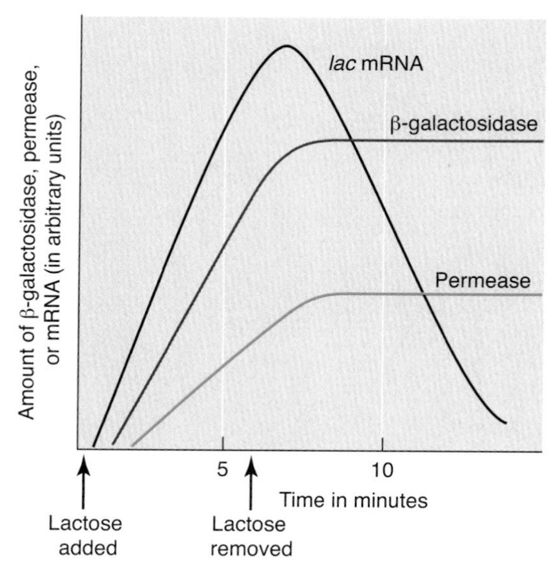

FIGURE 16.5 **The on-off nature of the *lac* system.** *Lac* mRNA appears soon after lactose is added; β-galactosidase and permease appear at about the same time but are delayed with respect to mRNA synthesis because of the time required for translation. When lactose is removed, no more *lac* mRNA is made, and the amount of *lac* mRNA decreases because of the usual degradation of mRNA. Both β-galactosidase and lactose permease are stable. Their concentrations remain constant even though no more enzymes can be synthesized.

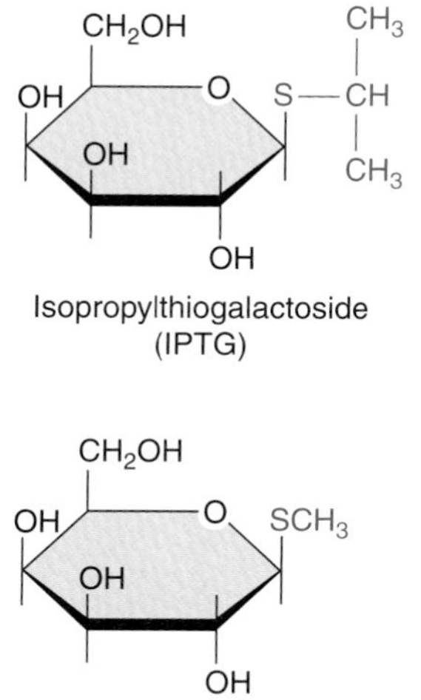

Enzymes such as β-galactosidase, lactose permease, and transacetylase are said to be **inducible enzymes** because their rate of synthesis increases in response to the addition of a small molecule (lactose) to the medium. Other enzymes, called **repressible enzymes,** exhibit a decreased rate of synthesis in response to the addition of a small molecule in the medium. For instance, the addition of tryptophan to the growth medium causes *E. coli* to greatly decrease the rate at which it produces enzymes needed for tryptophan synthesis. Still other enzymes, called **constitutive enzymes,** are synthesized at fixed rates under all growth conditions. Constitutive enzymes usually perform basic cellular "housekeeping" functions needed for normal cell maintenance.

Lactose is rarely used in experiments to study induction of the lactose enzymes because the β-galactosidase that is synthesized catalyzes lactose cleavage. As a result, the lactose concentration continually decreases, which complicates the analysis of kinetic experiments. Instead, two sulfur-containing lactose analogs are used, isopropylthiogalactoside (IPTG) and thiomethylgalactoside (TMG), which are effective inducers without being substrates of β-galactosidase (FIGURE 16.6). Inducers having this property are called **gratuitous inducers.**

Isopropylthiogalactoside (IPTG)

Thiomethylgalactoside (TMG)

FIGURE 16.6 Two lactose analogs that are gratuitous inducers.

Genetic studies provide information about the regulation of *lac* mRNA.

Two French investigators, Jacques Monod and François Jacob, sometimes working together and sometimes with other investigators, performed a series of genetic and biochemical experiments in the late 1950s that helped to elucidate the mechanism that regulates the *lac* system. They began by isolating constitutive *E. coli* mutants that make *lac* mRNA (and hence β-galactosidase, permease, and transacetylase) in the presence as well as absence of an inducer. Then they constructed a variety of partial diploid cells containing constitutive mutants and observed the cell's ability to synthesize β-galactosidase.

These mutations appeared to be of two types, termed ***lacI*** and ***lacO^C^*** (Table 16.2). The *lacI*$^-$ mutations behave like typical minus

TABLE 16.2 Characteristics of Partial Diploids Having Various Combinations of *lacI* and *lacO* Alleles

	Genotype	Constitutive or Inducible Synthesis of *lac* mRNA
1.	F′$lacO^{c}lacZ^{+}/lacO^{+}lacZ^{+}$	Constitutive
2.	F′$lacO^{+}lacZ^{+}/lacO^{c}lacZ^{+}$	Constitutive
3.	F′$lacI^{-}lacZ^{+}/lacI^{+}lacZ^{+}$	Inducible
4.	F′$lacI^{+}lacZ^{+}/lacI^{-}lacZ^{+}$	Inducible
5.	F′$lacO^{c}lacZ^{-}/lacO^{+}lacZ^{+}$	Inducible
6.	F′$lacO^{c}lacZ^{+}/lacO^{+}lacZ^{-}$	Constitutive

mutations in most genes and are recessive (entries 3, 4). Because *lac* mRNA synthesis is off in a $lacI^{+}$ cell and on in a $lacI^{-}$ mutant, the *lacI* gene is apparently a regulatory gene that codes for a product that acts as an inhibitor to keep the *lac* structural genes turned off. A $lacI^{-}$ mutant lacks the inhibitor and thus is constitutive. A $lacI^{+}/lacI^{-}$ partial diploid has one good copy of the *lacI*-gene product, so the system is inhibited. Monod and Jacob called the *lacI*-gene product the **Lac repressor.** Their original genetic experiments did not indicate whether the Lac repressor is a protein or an RNA molecule.

This question was answered when an *E. coli* mutant with a polypeptide chain termination codon inside *lacI* was isolated and found to synthesize β-galactosidase constitutively. The most likely explanation for the constitutive lactose system is that the mutant strain synthesizes a truncated repressor protein that cannot block transcription of the *lac* genes. This conclusion was confirmed when the Lac repressor was purified and characterized (see below). Genetic mapping experiments placed the *lacI* gene adjacent to the *lacZ* gene and established the gene order *lacI lacZ lacY lacA*.

A striking property of the $lacO^{c}$ mutations is that in certain cases, they are dominant (entries 1, 2, and 6 in Table 16.2). The significance of the dominance of the $lacO^{c}$ mutations becomes clear from the properties of the partial diploids shown in entries 5 and 6. Both combinations are Lac^{+}, because there is a functional *lacZ* gene. Entry 5, however, shows that β-galactosidase synthesis is inducible even though a $lacO^{c}$ mutation is present. The difference between the two combinations in entries 5 and 6 is that, in entry 5, the $lacO^{c}$ mutation is carried on a DNA molecule that also has a $lacZ^{-}$ mutation, whereas in entry 6, $lacO^{c}$ and $lacZ^{+}$ are carried on the same DNA molecule. Thus, $lacO^{c}$ causes constitutive synthesis of β-galactosidase only when $lacO^{c}$ and $lacZ^{+}$ are on the same DNA molecule, that is, are in *cis*.

Confirmation of this conclusion came from an important biochemical observation. An immunological test capable of detecting a mutant β-galactosidase showed that the mutant enzyme is synthesized constitutively in a $lacO^{c}lacZ^{-}/lacO^{+}lacZ^{+}$ partial diploid (entry 5), whereas the wild-type enzyme is synthesized only if an inducer is added. This experiment takes advantage of the fact that a purified antibody to β-galactosidase will also react with the mutant protein as long as the structural differences between wild-type and mutant protein are not too great. A reaction of this type in which an antibody that is raised in response to one protein is used to detect a closely related protein is called a **cross-reaction** and the closely related protein (mutant β-galactosidase in this experiment) is called **cross-reacting material (CRM)**. Thus the presence of CRM, which can be detected by a variety of standard immunological procedures, is indicative of the presence of mutant protein.

Genetic mapping experiments showed that all *lacO*^{c} mutations are located between genes *lacI* and *lacZ*, so the gene order of the five elements of the *lac* system is *lacI lacO lacZ lacY lacA*. Together these experiments lead to the conclusion that *lacO*^{c} mutations define a site or a noncoding region of the DNA rather than a gene (because mutations in coding genes should be complementable) and that the *lacO* region determines whether synthesis of the product of the adjacent *lacZ* gene is inducible or constitutive. The *lacO* site is called the **operator**.

The operon model explains the regulation of the lactose system.

Monod and Jacob proposed the **operon model** in 1961 to explain how the *lac* system is regulated. The term operon refers to two or more contiguous genes and the genetic elements that regulate their transcription in a coordinate fashion. Promoters had not yet been discovered when Monod and Jacob proposed the operon model but were readily incorporated into the operon model after their discovery. FIGURE 16.7 shows a revised version of the original *lac* operon model that includes the *lac* promoter. The five major features of the model are:

1. The products of the *lacZ*, *lacY*, and *lacA* genes are encoded in a single polycistronic *lac* mRNA molecule.
2. The promoter for this mRNA molecule is immediately adjacent to the *lacO* region. Promoter mutations (p^-) that are completely incapable of making β-galactosidase, permease, and transacetylase have been isolated. The promoter is located between *lacI* and *lacO*.
3. The operator is a sequence of bases (in the DNA) to which the repressor protein binds.
4. When the repressor protein is bound to the operator, *lac* mRNA transcription cannot take place.
5. Inducers stimulate *lac* mRNA synthesis by binding to the repressor. This binding alters the repressor's conformation so it cannot bind to the operator. In the presence of an inducer, therefore, the operator is unoccupied and the promoter is available for initiation of mRNA synthesis. This state is called **derepression**.

This simple model explains many of the features of the *lac* system and of other negatively regulated genetic systems. We will see in a later section, however, that this explanation is incomplete as the *lac* operon is also subject to positive regulation.

FIGURE 16.7 ***Lac* operon model.** (a) A map of the *lac* operon not drawn to scale. The *P* and the *O* sites are actually much smaller than the other regions and together comprise only 84 bp. (b) A diagram of the *lac* operon in the repressed state. (c) A diagram of the *lac* operon in the induced state. The inducer alters the repressor's conformation so that the repressor can no longer bind to the operator. The abbreviations *I, p, o, z, y,* and *a* are used instead of *lacI, lacO,* and so on.

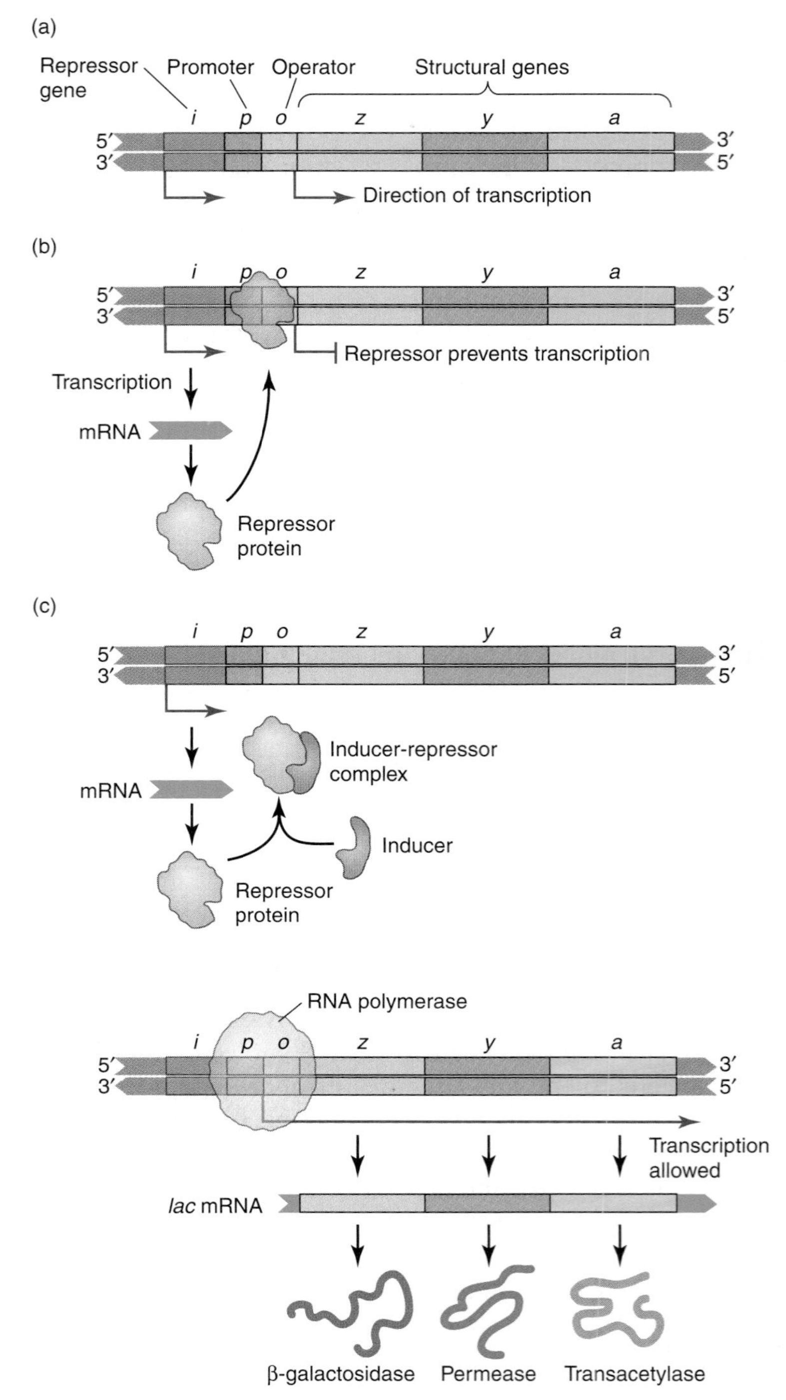

Allolactose is the true inducer of the lactose operon.

Two related problems became evident as the operon model was tested. First, inducers must enter a cell if they are to bind to repressor molecules, yet lactose transport requires permease, and permease synthesis requires induction. We, thus, must explain how the inducer gets into a cell in the first place. Second, the isolated Lac repressor does not bind lactose (4-O-β-D-galactopyranosyl-D-glucose) but does bind a lactose isomer called allolactose (6-O-β-D-galactopyranosyl-D-glucose). Remarkably, β-galactosidase, the enzyme that catalyzes lactose hydrolysis, also converts a small proportion of lactose to allolactose (FIGURE 16.8). Therefore, induction of the synthesis of β-galactosidase by lactose requires that β-galactosidase be present.

Both problems are solved in the same way; in the uninduced state, a small amount of *lac* mRNA is synthesized (roughly one mRNA molecule per cell per generation). This synthesis, called **basal synthesis**, occurs because the binding of repressor to the operator is never infinitely strong. Thus, even though the repressor binds tightly to the operator, it occasionally comes off and an RNA polymerase molecule can initiate transcription during the instant that the operator is free.

We can now describe in molecular terms the sequence of events following addition of a small amount of lactose to a growing Lac$^+$ culture. Consider bacteria growing in a medium in which the carbon source is glycerol. Each bacterial cell contains a few molecules of β-galactosidase and of lactose permease. When lactose is added, the few permease molecules transport a few lactose molecules into the cell and the few β-galactosidase molecules convert some of these lactose molecules into allolactose. An allolactose molecule then binds to a repressor molecule that is sitting on the operator, and the repressor is inactivated and falls off the operator. Synthesis of *lac* mRNA begins and these mRNA molecules are translated to produce hundreds of β-galactosidase and permease molecules. The new permease molecules allow lactose molecules to pour into the cell. Most of the lactose molecules are cleaved to yield glucose and galactose, but some are converted to allolactose molecules, which bind to and inactivate all of the intracellular repressor molecules. (Repressor is made continuously, though at a very low rate, so there is usually sufficient allolactose to maintain the cell in the derepressed state.) Thus, *lac* mRNA is synthesized at a high rate and the permease and β-galactosidase concentrations become quite high. The glucose produced by the cleavage reaction is used as a source of carbon and energy. (The galactose formed by the cleavage is converted to glucose-1-phosphate by a set of enzymes, the synthesis of which is also inducible. This inducible system in which galactose is the inducer, called the *gal* operon, is discussed in a later section of this chapter.)

Ultimately, all of the lactose in the growth medium and within the cells is consumed. Then the allolactose concentration within the cell drops so that there is not sufficient allolactose to bind to a repressor. The repressor binds to the operator, reestablishing repression and thereby blocking further synthesis of *lac* mRNA. In bacteria, most mRNA molecules have a half-life of only a few minutes. Hence, in less than one generation there is little remaining *lac* mRNA and synthesis of β-galactosidase and permease ceases. These proteins are quite stable but are gradually diluted out as the cells divide. Note that if lactose were added again to the growth medium one generation after the original lactose had been depleted, cleavage of lactose would begin immediately because the cells would already have adequate permease and β-galactosidase.

Lactose

β galactosidase

Allolactose

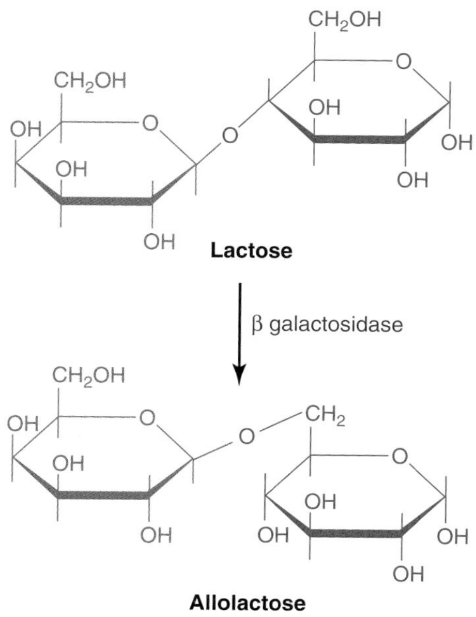

FIGURE 16.8 Synthesis of allolactose, the actual inducer of the *lac* operon.

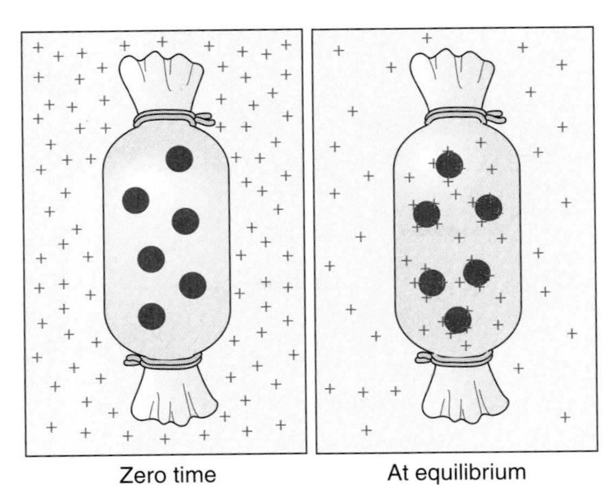

FIGURE 16.9 **Equilibrium dialysis.** A dialysis bag filled with a cell extract containing repressor (red circles) is placed in a solution containing radioactive IPTG (blue crosses). The radioactive IPTG can bind to the repressor. In the absence of a repressor, the concentrations of free IPTG would be the same inside and outside the bag. Because the repressor binds some of the radioactive IPTG, the concentration of radioactive IPTG is greater inside the bag than outside.

The Lac repressor binds to the *lac* operator *in vitro*.

An important step in proving the principal hypothesis of the operon model was the isolation of the Lac repressor and the demonstration of its expected properties. Walter Gilbert and Benno Müller-Hill succeeded in isolating the Lac repressor from *E. coli* extracts in 1966. Their approach was to fractionate proteins by standard techniques and then assay individual fractions for their ability to bind [^{14}C]IPTG (one of the gratuitous inducers). Binding was detected by equilibrium dialysis, as shown in FIGURE 16.9.

The Lac repressor is a homotetramer with a molecular mass of 154 kDa. Each subunit, which is made of 360 amino acids, can bind one molecule of IPTG. Crude cell extracts bind about 20 to 40 molecules of IPTG per cell, so there are roughly 5 to 10 repressor molecules per cell. Support for the idea that the IPTG-binding protein is the Lac repressor comes from the observation IPTG-binding protein is absent in extracts of *lacI*$^-$ mutants. Still stronger support comes from the observation that *lacI* mutants, which introduce amino acid substitutions in *lac* repressor, alter the repressor's affinity for IPTG.

Because the number of repressor molecules is extremely small, these molecules must be translated from no more than one or two repressor mRNA molecules transcribed per generation time. The number of mRNA molecules is so small that either repressor synthesis itself is regulated or the mRNA is transcribed from a weak promoter. Both mechanisms have been observed for regulation of repressor synthesis in other operons, but for the Lac repressor the second explanation is correct, that is, repressor mRNA is transcribed constitutively from a weak promoter. The reason for the small number of repressor molecules is made clear from the properties of several mutants in which the weak *lacI* promoter is converted to a strong promoter. These mutants are noninducible because it is not possible to fill a cell with enough inducer to overcome repression.

Repressor-overproducers have been extremely valuable experimentally because high concentrations of repressor (about 1% of the cellular protein) have in turn meant that very large amounts of repressor could be purified, providing sufficient amounts for physical study and characterization. With purified repressor, the specific binding of repressor to the operator sequence and the inhibition of this binding by an inducer have been demonstrated. An important procedure for studying repressor-operator binding is the **nitrocellulose filter assay**. Proteins stick to these filters but DNA does not. If a mixture

TABLE 16.3 **Demonstration of Repressor-Operator Binding by the Filter-Binding Assay**

Mixture Applied to Filter	^{14}C Bound to the Filter
[^{14}C]*lac* DNA	No
[^{14}C]*lac* DNA + repressor	Yes
[^{14}C]*lac* DNA + repressor + IPTG	No
[^{14}C]*lacO*^{c} DNA + repressor	No

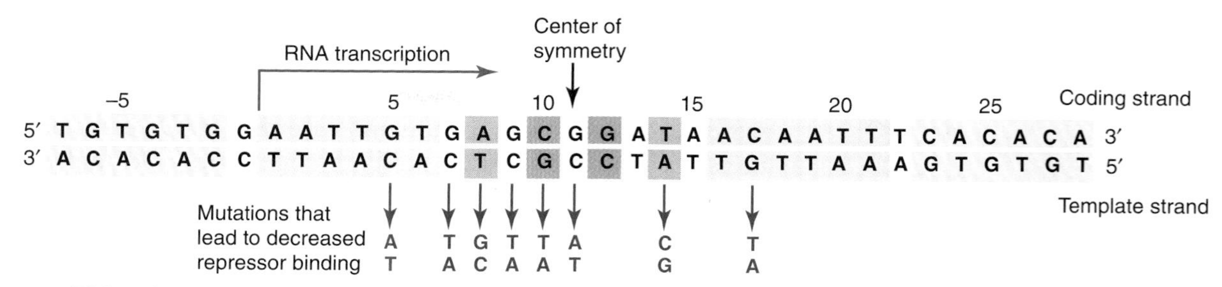

FIGURE 16.10 **Thirty-five base pairs protected by the *lac* operator.** Footprinting studies reveal that the *lac* repressor protects 35 base pairs from –7 to +28 so that the operator includes both the transcriptional start site at +1 and part of the *lac* promoter's –10 box. Regions of symmetry are shown in the same color. Some of the *lac* operon mutations that decrease the *lac* repressor's affinity for operon DNA are shown in magenta.

of repressor and radioactive *lac* DNA is passed through such a filter, radioactivity will be retained on the filter if the protein and the *lac* DNA form a complex. The data shown in Table 16.3 were obtained by means of this test. The results indicate that Lac repressor binds to DNA with a normal *lac* operator but fails to bind to DNA with a *lacO*^{c} mutant operator. Furthermore, IPTG prevents the Lac repressor from binding to DNA with a normal *lac* operator. These studies confirm the major predictions of the operon model.

Investigators obtained additional information about the binding of the Lac repressor to the operator by studying the process *in vitro*. Footprinting studies reveal that repressor protects 35 base pairs from –7 to +28 so that the operator includes both the transcriptional start site and part of the *lac* promoter's –10 box (FIGURE 16.10). Sequencing studies reveal that the operator has a twofold axis of symmetry that passes through the base pair at position +11 (Figure 16.10). The sequence is not a perfect palindrome, however, because only 28 of the 35 base pairs have the required symmetry. Mutations in the *lac* operator interfere with the Lac repressor's ability to bind to the operator and lead to constitutive enzyme production (Figure 16.10). Thus, Lac repressor makes important contacts with bases in the *lac* operator.

The *lac* operon has three *lac* operators.

After the *lac* operon was sequenced, it eventually became clear that it has two additional operators. The original operator with its center of symmetry at position +11 is now designated $lacO_1$. Auxiliary operators $lacO_2$ and $lacO_3$ have their centers of symmetry at positions +412 and –82, respectively. Thus, $lacO_3$ is upstream of the *lac* promoter while $lacO_2$ is located in *lacZ* (FIGURE 16.11).

The discoveries of $lacO_2$ and $lacO_3$ led investigators to ask whether these auxiliary operators participate in *lac* operon regulation. In 1990, Müller-Hill and coworkers performed a series of genetic experiments that provided an affirmative answer to this question. Their approach was to alter one or more *lac* operators and then determine the alteration's effect on repression. Point mutations in $lacO_1$ caused a 5- to 50-fold decrease in repression, but some repression was still observed. Destruction of either $lacO_2$ or $lacO_3$ caused a twofold decrease in

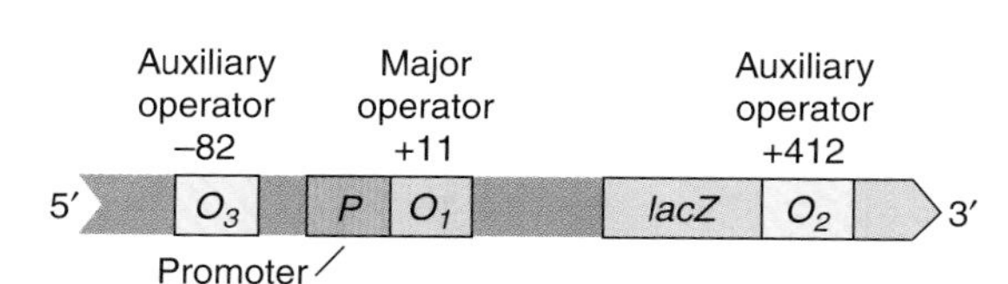

FIGURE 16.11 **The three *lac* operators.** The *lac* operon has three operators. The major operator, $lacO_1(O_1)$ is centered at position +11 and two auxiliary operators, $lacO_2(O_2)$ and $lacO_3(O_3)$, are centered at positions +412 and –82, respectively.

repression and destruction of both auxiliary operators caused a 70-fold decrease in repression. The ability to repress the *lac* system was completely lost in cells in which *lac*O_1 and either of the auxiliary operators were destroyed. These results indicate that *lac*O_1 plays a major role in repression but the two auxiliary operators also make important contributions. We will now examine the structure of the Lac repressor to see how it interacts with the *lac* operators.

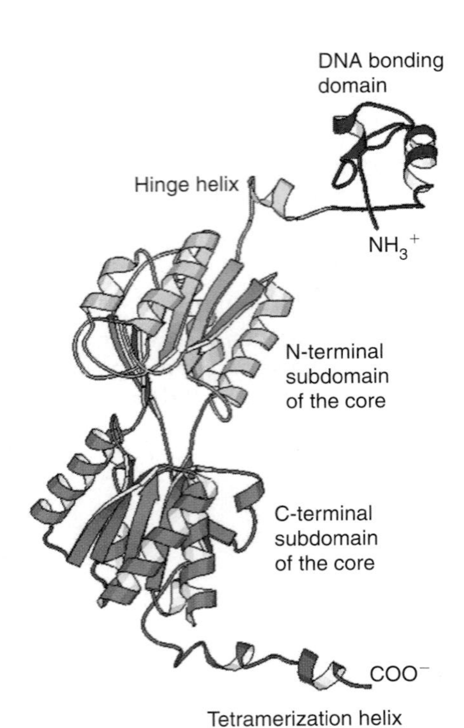

FIGURE 16.12 **A view of the *lac* repressor monomer from the complex with DNA.** The *lac* repressor monomer has four functional domains. The first, starting from the N-terminal domain or headpiece (residues 1–45) is colored red. The second, the hinge region (residues 46–62) is colored yellow. The third, the ligand binding domain or core (residues 63–329), which has distinct N- and C-terminal subdomains is colored two shades of blue. The fourth, the tetramerization helix (residues 341–357) is colored purple. (Protein Data Bank ID: 1LBG. M. Lewis, et al., *Science* 271 [1996]: 1247–1254. Structure rendered by Hongli Zhan and provided by Kathleen S. Matthews, Rice University.)

The Lac repressor is a dimer of dimers, where each dimer binds to one *lac* operator sequence.

Mitchell Lewis and coworkers obtained the crystal structure for the Lac repressor bound to IPTG or *lac* DNA in 1996. FIGURE 16.12 shows the structure of one of the four identical Lac repressor subunits. As indicated in the figure, the monomer has four distinct functional units.

1. The **N-terminal domain** or **headpiece** (residues 1–45) binds *lac* operator DNA. A helix-turn-helix motif within the headpiece recognizes and binds appropriate operator DNA (FIGURE 16.13). Helix-turn-helix motifs, which are often present in proteins that bind to DNA, consist of two short α-helical segments containing between 7 and 9 residues that are separated by a β-turn (a tight turn involving four amino acid residues). One of the two α-helices, designated the **recognition helix**, is positioned in the major groove of the DNA so that amino acid residues within this helix make specific contacts with bases in the DNA. The other helix, which is designated the **stabilization helix**, helps to stabilize the complex.

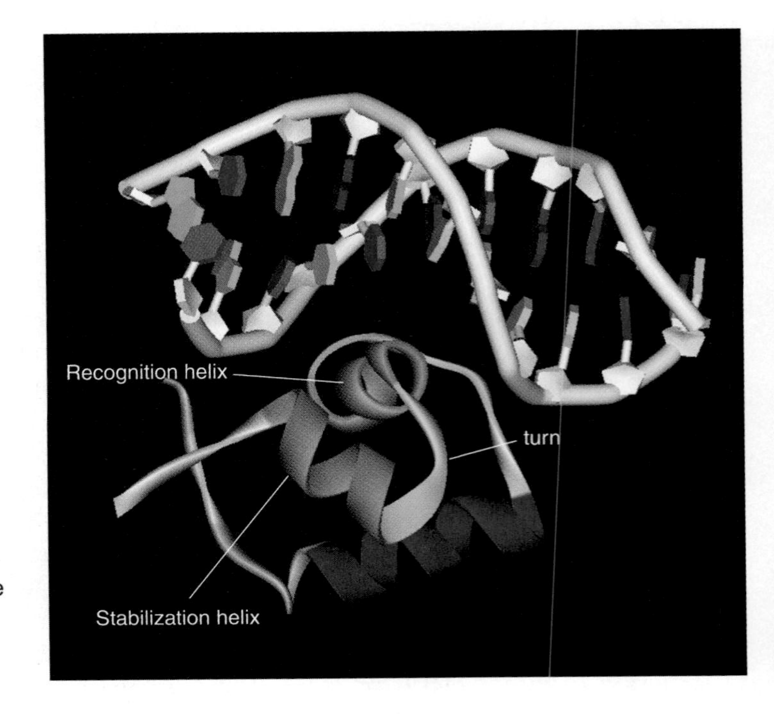

FIGURE 16.13 **A complex of Lac repressor N-terminal domain ("headpiece") with an 11 bp half-operator corresponding to the left half of wild-type *lac* operator.** Recognition and stabilization helices in the helix-turn-helix motif are shown in pink and the turn is shown in yellow. (Structure from Protein Data Bank 1LCC. V. P. Chuprina, et al., *J. Mol. Biol.* 234 [1993]: 446–462. Prepared by B. E. Tropp.)

2. The **hinge region** (residues 46–62), which joins the N-terminal headpiece to the core domain, is disordered in the absence of DNA but folds into an α-helix upon binding to operator DNA. Two hinge helices in a Lac repressor dimer associate through van der Waals interactions and the paired hinge helices bind to the center of the *lac* operator in the minor groove and bend the DNA (FIGURE 16.14).
3. The **ligand binding** or **core domain** (residues 62–340) is divided into an N-terminal subdomain and a C-terminal subdomain. The two kinds of subdomain have similar folding patterns. Weak non-covalent interactions between N-terminal subdomains as well as between C-terminal subdomains make important contributions to Lac repressor dimer formation. IPTG fits in a pocket between the two subdomains in each polypeptide (FIGURE 16.15a).
4. The **tetramerization helix** (residues 341–357) associates with corresponding regions of the other three subunits of the Lac repressor to form a four-helix bundle that produces a Lac repressor tetramer with a V-shape (Figure 16.15a). Mutations or deletions that alter the tetramerization helix result in the formation of dimers rather than tetramers. The dimers have partial repressor function. Although the tetramer retains its V-shape in the Lac repressor•DNA complex (FIGURE 16.15b), its conformation differs from that in the repressor•IPTG complex (Figure 16.15a).

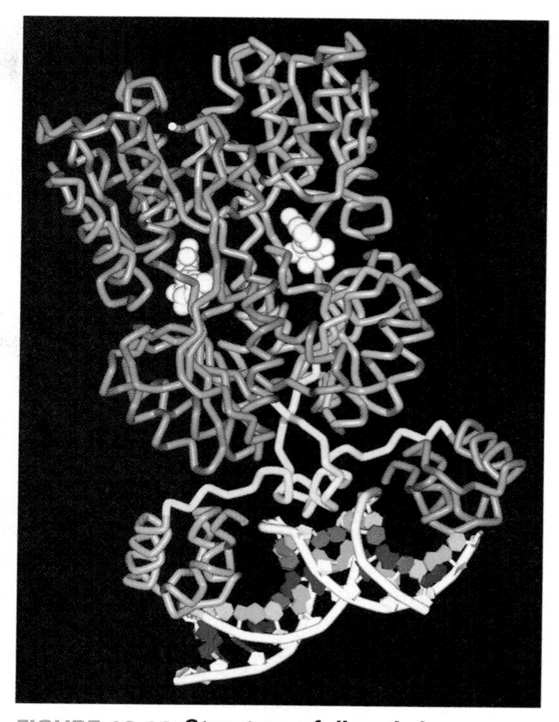

FIGURE 16.14 **Structure of dimeric Lac repressor created by recombinant DNA techniques so that the carboxyl terminal domains are missing.** The two hinge regions (one on each subunit), which are shown in yellow, fold into α-helices that associate with one another through van der Waals interactions and bind to DNA in the minor groove. Other parts of the repressor subunits are shown in green for one subunit and orange for the other subunit. Each subunit has a bound O-nitrophenyl-β-fucoside (ONPF), a galactoside analog, which stabilizes the Lac repressor in the conformation that binds to the *lac* operator. ONPF is shown as a white spacefill structure. (Structure from Protein Data Bank 1EFA. C. E. Bell and M. Lewis, *Nat. Struct. Biol.* 7 [2000]: 209–214. Prepared by B. E. Tropp.)

The crystal structures of Lac repressor bound to DNA or IPTG suggest a possible mechanism for the Lac repressor to act as a molecular switch that turns the Lac system off and on. Kathleen Matthews and colleagues have suggested the model shown in FIGURE 16.16 to explain how a Lac repressor dimer might work. According to this model, the helix-turn-helix motif in each N-terminal domain makes contact with bases in the major groove of a Lac operator half-site. Furthermore, the hinge helices insert into the minor groove of the central region of the structure, causing the DNA to bend. Binding IPTG to the Lac repressor causes the N-terminal core subdomain to move with respect to the C-terminal core subdomain. This movement is similar to screwing on a cap (with both translation and rotation), closing the IPTG-binding site and inducing a conformational change that disrupts the structures of the hinge helices. Disruption of the hinge helix structure frees the helix-turn-helix motif to leave the operator DNA.

The switching mechanism shown in Figure 16.16 requires a dimeric Lac repressor and therefore does not explain why the Lac repressor has evolved to be a tetramer. The explanation is that the Lac repressor is a more effective genetic switch when it binds to two operators at the same time. Recall that the *lac* operon has two auxiliary operators, $lacO_2$ and $lacO_3$, in addition to the major operator, $lacO_1$. DNA looping permits the Lac repressor to bind to $lacO_1$ and one of the auxiliary operators (FIGURE 16.17). The cell would need one super operator to achieve the same level of repression that it achieves with two weak cooperating operators.

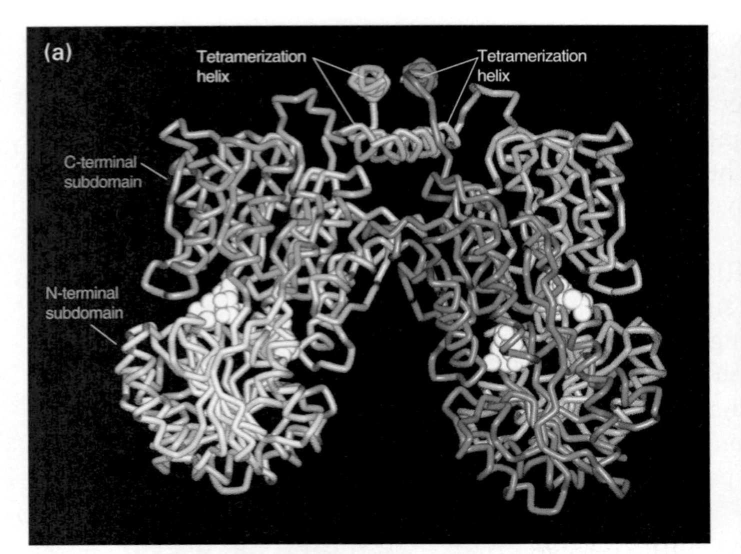

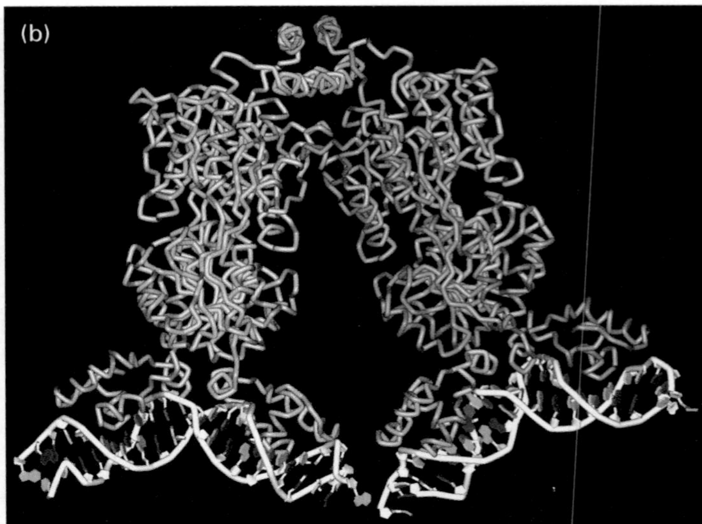

FIGURE 16.15 Tetrameric Lac repressor structure. Each monomer in the Lac repressor tetramer is shown as a different color tube structure (magenta, blue, green, or orange). (a) Repressor•IPTG complex. The N- and C-terminal core subdomains are shown for the magenta subunit. IPTG (white space-fill structure) fits in a pocket between the two subdomains in each polypeptide. DNA-binding domains are not shown because these domains are not well ordered and so not visible in the crystal structure. (b) Repressor•DNA complex. The 21 base-pair DNA duplexes that are bound by each dimer are shown as tubes and rings. The Lac tetramer appears to be a tethered dimer of dimers. (Part a structure from Protein Data Bank 1LBH. M. Lewis, et al., *Science* 271 [1996]: 1247–1254. Prepared by B. E. Tropp. Part b structure from Protein Data Bank 1LBG. M. Lewis, et al., Science 271 [1996]: 1247–1254. Prepared by B. E. Tropp.)

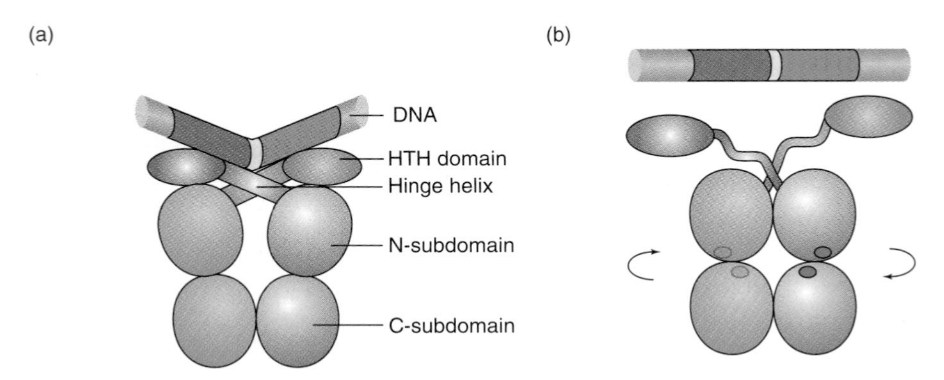

FIGURE 16.16 Model showing the reactions that are elicited when IPTG binds to the Lac repressor. (a) Lac repressor dimer bound to DNA. The Lac repressor makes contact with similarly colored DNA half-sites. The N-terminal helix-turn-helix (NTH) domains (ovals) contact B-form DNA in the major groove. The hinge helices insert into the minor groove at the central region of the structure, causing a significant bend in the DNA. This contact region is indicated by the yellow color in the central segment of operator DNA. (b) When IPTG (blue ovals) binds to the Lac repressor, the N-terminal core subdomain moves with respect to the C-terminal core subdomain in a manner that suggests screwing on a cap (with both rotation and translation), closing the sugar binding site. The conformational change disrupts the hinge-helix structures (indicated by thick wavy lines) and frees the helix-turn-helix domains (ovals), allowing them to dissociate from the DNA. (Adapted from K. S. Matthews, C. M. Falcon, and L. Swint-Kruse, *Nat. Struct. Biol.* 7 [2000]: 184–187.)

Experiments designed to explain how the Lac repressor prevents the transcription of the *lac* operon have led to conflicting hypotheses. Gilbert and coworkers originally proposed that the Lac repressor prevents RNA polymerase holoenzyme from binding to the promoter. Later studies by other investigators suggested that the Lac repressor does not prevent RNA polymerase holoenzyme from binding to the promoter but instead prevents the conversion of a closed *lac* promoter to an open *lac* promoter. Recent thermodynamic studies of repression suggest that Gilbert's original proposal may have been correct after all.

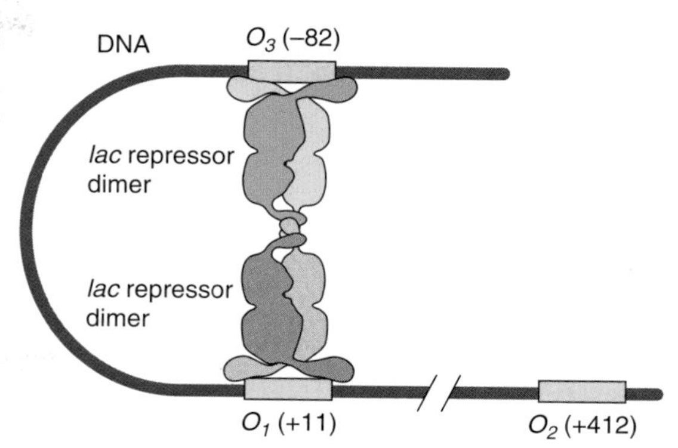

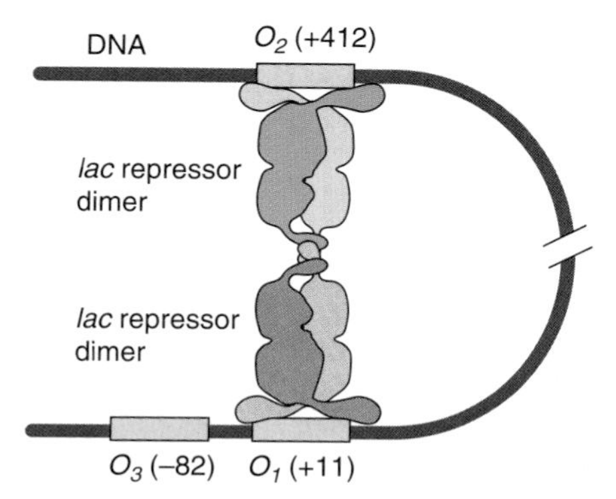

FIGURE 16.17 **DNA looping to permit Lac repressor to bind to two operators.** The *lac* operon DNA is shown in blue, *lac* operators in yellow, and the *lac* repressor dimers in shades of green or blue. The slanted double lines in the DNA indicate that not all of the DNA is shown. (Adapted from B. Müller-Hill, *Curr. Opin. Microbiol.* 1 [1998]: 145–151.)

16.3 Catabolite Repression

E. coli uses glucose in preference to lactose.

The function of β-galactosidase in lactose metabolism is to hydrolyze lactose to form glucose and galactose. If the growth medium contains both glucose and lactose, then in the interest of efficiency, there is no need for a cell to turn on the *lac* operon. Experiments performed by Monod in the mid-1940s demonstrated that cells behave according to this logic. *E. coli* cells incubated in the presence of glucose and lactose do not start to make β-galactosidase until all of the exogenous glucose is consumed (FIGURE 16.18). These findings were later extended to lactose permease and transacetylase, the two other proteins specified by the *lac* operon. The reason that lactose enzymes are not made when glucose is present is that no *lac* mRNA is made. Transcription-level inhibition of the lactose enzymes and a variety of other inducible enzymes by glucose (or other readily used carbon sources) is called **catabolite repression.**

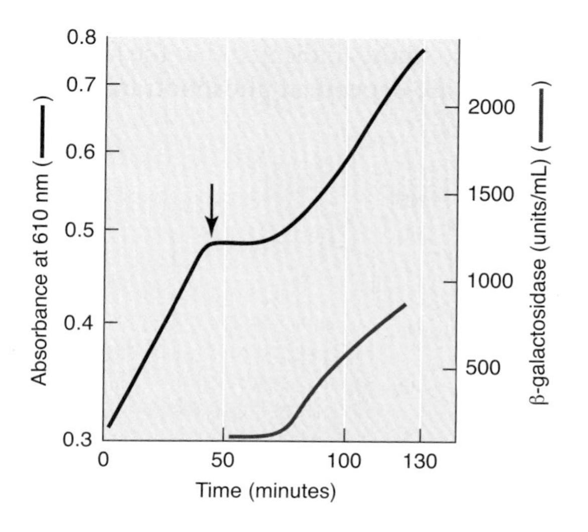

FIGURE 16.18 **Effect of changing the carbon source from glucose to lactose.** *E. coli* are cultured in medium that contains glucose and lactose. The cells initially use glucose as the carbon source. At the time indicated by the arrow, glucose is in such short supply that the *E. coli* cell mass stops increasing. The absorbance at 610 nm is a measure of the total cell mass (black curve). A short time after glucose depletion, the bacterial cells start to make β-galactosidase and use lactose as the carbon source. Enzyme activity is indicated in arbitrarily defined units (red curve).

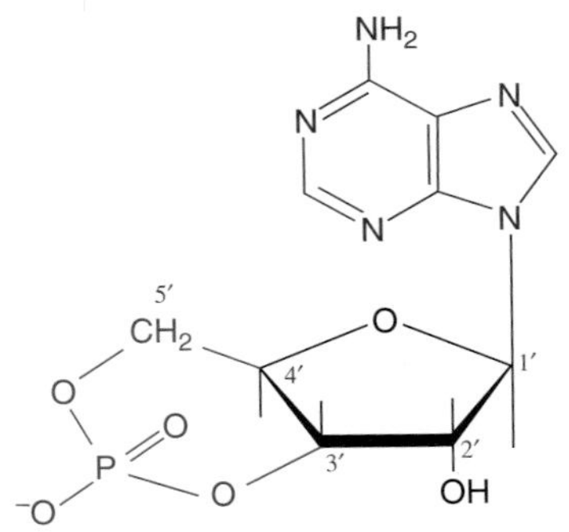

3′, 5′– cyclic adenylate (cAMP)

FIGURE 16.19 3′,5′-Cyclic adenylate (cAMP).

The inhibitory effect of glucose on expression of the *lac* operon is a complicated process.

The mechanism by which glucose inhibits β-galactosidase synthesis remained a complete mystery for about 20 years after Monod first observed the phenomenon. Richard S. Makman and Earl W. Sutherland found an important clue to the mystery in 1965 when they observed that the intracellular concentration of **3′, 5′-cyclic adenylate** or **cAMP** (FIGURE 16.19) drops from about 10^{-4} M to 10^{-7} M when glucose is added to a growing culture of *E. coli*.

Genetic studies confirmed cAMP's involvement in catabolite repression. Two mutant classes were isolated that could not synthesize *lac* enzymes when cultured in a medium containing lactose but no glucose. Class I mutants regained the ability to synthesize *lac* enzymes when cAMP was added to the growth medium but class II mutants did not.

Subsequent studies showed that class I mutants have defects in **adenylate cyclase**, the enzyme that converts ATP to cAMP (FIGURE 16.20). The structural gene for adenylate cyclase, *cya*, maps at minute 85.98. Adenylate cyclase exists in an active form that is *phosphorylated* (it contains an attached phosphate group) and an inactive form that is dephosphorylated. Class II mutants have defects in a protein that binds cAMP. This protein, called the **cAMP receptor protein** (**CRP**) or the **catabolite activator protein** (**CAP**), is encoded by the *crp* gene, which maps at minute 75.09. *In vitro* studies have shown that CRP and cAMP form a **cAMP•CRP complex**, which is needed to activate the *lac* system. The cAMP•CRP requirement is independent of the repression system because *cya* and *crp* mutants cannot make *lac* mRNA even if a *lacI*$^-$ or *lacO*^{c} mutation is present. Thus, cAMP•CRP is a **positive regulator** or **activator**, in contrast to the repressor, and the *lac* operon is independently regulated both positively and negatively.

Based on the information presented above, it seems reasonable to propose that glucose somehow inhibits phosphorylation of adenylate cyclase, thereby preventing cAMP formation. The next challenge was to find the link between glucose metabolism and adenylate kinase phosphorylation. Biochemical and genetic studies indicated that the link is a glucose-specific, **phosphoenolpyruvate-dependent phosphotransferase**

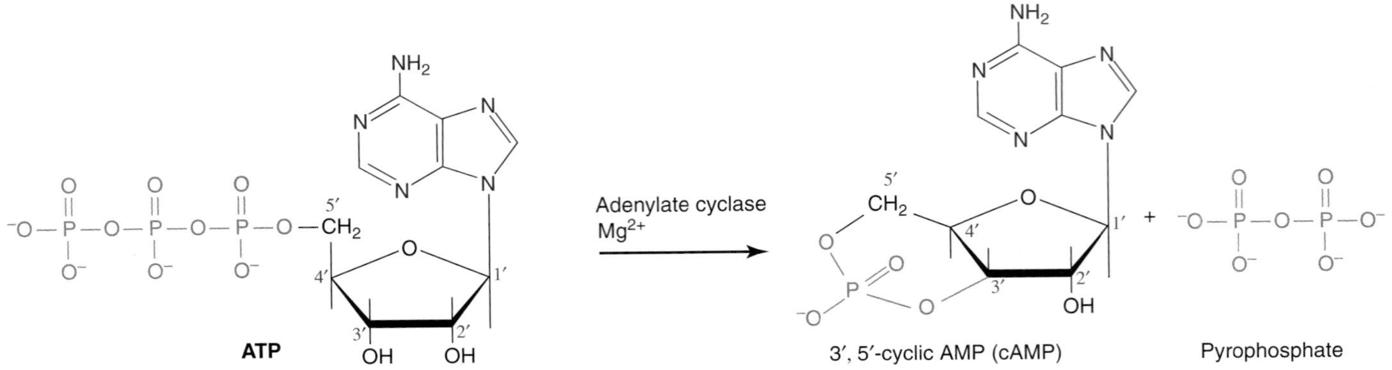

FIGURE 16.20 Cyclic AMP (cAMP) formation.

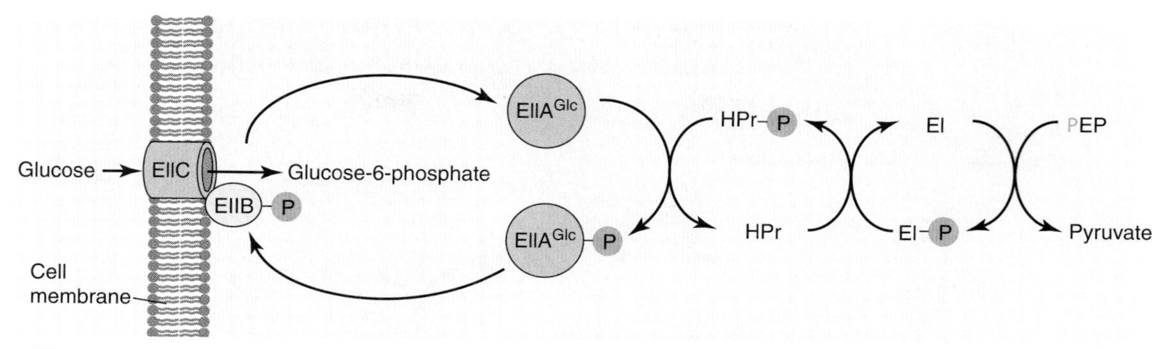

FIGURE 16.21 **Glucose-specific, phosphoenolpyruvate-dependent phosphotransferase system (PTS).** Phosphate is transferred from phosphoenol pyruvate (PEP) to the exogenous glucose simultaneously with glucose transport. This flow is mediated by EI, HPr, and carbohydrate-specific EIIA and EIIBC. (Adapted from J. Stülke and W. Hillen, *Curr. Opin. Microbiol.* 2 [1999]: 195–201.)

system (PTS). This system, which is depicted in FIGURE 16.21, uses energy supplied by phosphoenolpyruvate (PEP) to phosphorylate glucose as it transports the sugar across the inner cell membrane.

The system requires four proteins. Two of these, enzyme I (EI) and the histidine-containing protein (HPr), are also components of other sugar transport systems and therefore are unlikely to be direct participants in a glucose-specific phenomenon. The two other proteins, enzyme IIA (EIIA) and enzyme IIBC (EIIBC), are specific for the glucose transport system and are more likely participants in a glucose-specific phenomenon.

EIIA participates in catabolite repression by two different mechanisms (FIGURE 16.22). The first mechanism is based on the fact that EIIA can transfer a phosphoryl group from HPr-P to either EIIBC or adenylate cyclase. The preferred substrate is EIIBC, which then transfers the phosphoryl group to glucose to form glucose-6-phosphate. When glucose is unavailable EIIBC will be fully phosphorylated and EIIA-P has no other alternative but to transfer its phosphoryl group to adenylate cyclase. Phosphorylation changes the inactive dephosphorylated form of adenylate cyclase to the active phosphorylated form of adenylate cyclase, which then converts ATP to cAMP. Thus, glucose interferes with the conversion of the inactive form of adenylate cyclase to the active form.

Investigators initially believed that glucose's influence on adenylate cyclase activity was solely responsible for glucose's effect on *lac* operon transcription. Additional observations, however, suggested that cAMP could not be the only modulator of the glucose effect. For example, several investigators were unable to demonstrate a correlation between intracellular cAMP concentration and β-galactosidase synthesis. Furthermore, cells lacking adenylate cyclase activity but having a mutant CRP that does not require cAMP to activate the *lac* operon are subject to catabolite repression. Because the cells synthesized much less β-galactosidase in the presence of glucose, cAMP cannot be the sole modulator of *lac* mRNA synthesis.

Further study revealed a second mechanism by which glucose influences *lac* operon transcription. This mechanism, called **inducer**

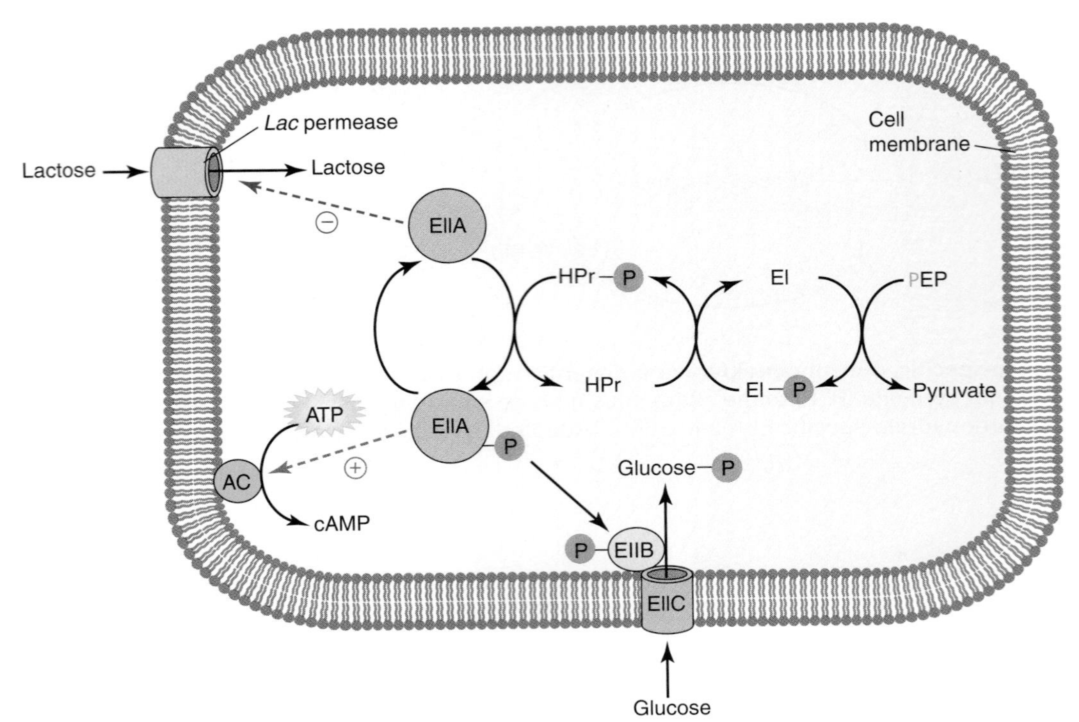

FIGURE 16.22 **cAMP-modulated and induced exclusion mechanisms for catabolite repression.** EIIA participates in catabolite repression by two different mechanisms. (1) cAMP modulated mechanism, where EIIA can transfer a phosphoryl group from HPr-P to either EIIBC or adenylate cyclase (AC). The preferred substrate is EIIBC, which then transfers the phosphoryl group to glucose to form glucose-6-phosphate. When glucose is unavailable, EIIBC will be fully phosphorylated, forcing EIIA-P to transfer its phosphoryl group to adenylate cyclase. Phosphorylation changes the inactive dephosphorylated form of adenylate cyclase to the active phosphorylated form of adenylate cyclase, which then converts ATP to cAMP. (2) Inducer exclusion mechanism where the dephosphorylated form of EIIA binds to lactose permease and inactivates it.

exclusion, also involves the glucose transport system. When glucose is present, EIIA-P transfers its phosphate group through EIIBC to the sugar and the dephosphorylated form of EIIA binds to lactose permease and inactivates it. Lactose permease inactivation prevents lactose from entering the cell and being converted to allolactose. In the absence of allolactose, the repressor remains bound to the operator and the *lac* operon is turned off. The cAMP modulation and inducer exclusion mechanisms may not be the only ones that contribute to glucose's ability to inhibit *lac* mRNA formation. For example, regulation of CRP synthesis may also be important. Thus, *lac* operon regulation, one of the best studied of all gene regulation systems, remains a very active area of inquiry. One productive approach for examining *lac* operon regulation has been to investigate the mechanism of cAMP•CRP action.

The cAMP•CRP complex binds to an activator site (AS) upstream from the *lac* promoter and activates *lac* operon transcription.

In the absence of cAMP•CRP, the *lac* promoter is quite weak because its –10 box differs significantly from the consensus sequence. A mutant *lac* promoter with a –10 box that has the consensus sequence does not require cAMP•CRP for transcription activation. It therefore seems reasonable to propose that interactions between cAMP•CRP complex and RNA polymerase holoenzyme increase the holoenzyme's affinity for the *lac* promoter. Biochemical and genetic studies support this proposal.

Thomas Steitz and coworkers determined the crystal structure for the cAMP•CRP complex bound to DNA (FIGURE 16.23). CRP consists of two chemically identical polypeptide chains of 209 amino acid residues. Each chain consists of an N-terminal domain and a C-terminal domain, which are connected by a hinge region containing four amino acids. The N-terminal domain consists of a series of antiparallel β-sheets that form a pocket for binding cAMP. The C-terminal domain contains a helix-turn-helix motif that binds to DNA. In the absence of cAMP, CRP•DNA interactions are nonspecific and weak. The cAMP•CRP complex, however, binds very tightly to a specific DNA sequence designated the **activator site (AS)**. In the *lac* operon, the center of AS is 61.5 bp upstream from the transcription start site (i.e., between nucleotides –61 and –62). Many other bacterial operons are activated by cAMP•CRP (see below). Promoters like the *lac* promoter that have AS at position –61.5 are designated **class I cAMP-dependent promoters.**

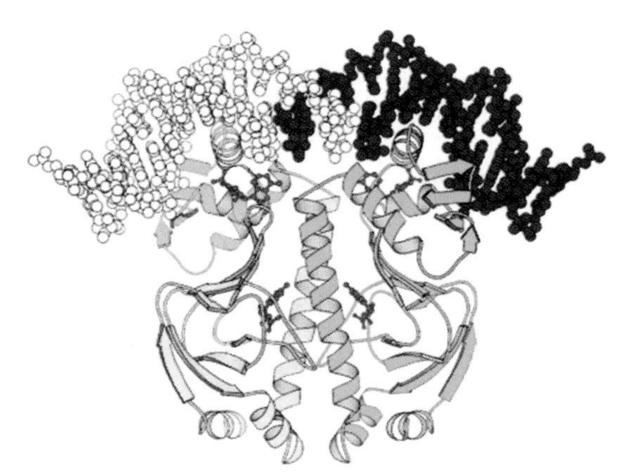

FIGURE 16.23 **Ribbon drawing of the CRP dimer bound to DNA and two cAMP molecules per monomer.** The N-terminal and C-terminal domains in one CRP monomer are colored yellow and blue, respectively. The DNA half-site bound by this CRP monomer is light gray. The other CRP monomer is shown in green and the DNA bound to it is black. Each monomer is bound to two cAMP molecules shown in magenta. In each CRP, one cAMP lies on the helix-turn-helix and close to the DNA and the other lies in the C-terminal domain. (Reproduced from J. M. Passner and T. A. Steitz, *Proc. Natl. Acad. Sci. USA* 72 [1997]: 2843–2847. Copyright 1997 National Academy of Sciences, U.S.A. Photo courtesy of Thomas A. Steitz, Yale University.)

Each of these operons also contains at least one AS. Comparing AS sequences reveals the following 22 bp consensus sequence with twofold symmetry.

5′-AAATGTGATCTAGATCACATTT-3′
3′-TTTACACTAGATCTAGTGTAAA-5′

The most highly conserved nucleotides in AS are the two TGTGA motifs. Mutations that alter nucleotides in the TGTGA motifs lead to decreased *lac* operon transcription. Each CRP subunit binds to half of the AS. The interactions between the C-terminal domain's helix-turn-helix motif and AS are similar to those described for Lac repressor and operator DNA. In fact, helix-turn-helix interactions with DNA are a recurring theme in DNA-protein recognition.

Binding of cAMP•CRP to the consensus sequence is so tight that an operon containing this sequence would be permanently switched on. It is therefore not surprising that actual cAMP•CRP activator sites differ from the consensus sequence. Activator sites in different operons compete for the cAMP•CRP with the activator preferentially binding to sequences that most closely resemble the consensus sequence. cAMP•CRP bound to DNA migrates in an anomalous fashion when subjected to nondenaturing gel electrophoresis. The most likely explanation for this behavior is that the protein causes the DNA to bend when it binds to it. X-ray crystallography studies confirmed this interpretation, showing that the cAMP•CRP complex sharply bends DNA by an angle of between 80° and 90°.

In addition to specific contacts with AS, the cAMP•CRP complex also makes specific contacts with RNA polymerase holoenzyme. Genetic studies show that the contact site in RNA polymerase holoenzyme is located on the α subunit. Recall from Chapter 15 that the α subunit consists of an N-terminal domain (αNTD), a C-terminal domain (αCTD), and a flexible linker region that joins the two domains. A mutant RNA polymerase with truncated αCTDs cannot transcribe the *lac* operon in the presence or absence of cAMP•CRP. However, the same enzyme can transcribe a *lac* operon with a consensus sequence in its promoter's –10 box. Furthermore, chemical crosslinking experiments show that CRP and αCTD are adjacent to one another. These observations indicate that cAMP•CRP interacts with αCTD. Experiments with RNA polymerases, which have one full-length α subunit and one truncated α subunit lacking αCTD, reveal that transcription activation requires RNA polymerase to have just one intact α subunit. It does not seem to matter if the intact subunit is the one associated with the β subunit or the one associated with the β′ subunit. The interactions between cAMP•CRP and RNA polymerase holoenzyme are summarized in the cartoon shown in FIGURE 16.24.

CRP mutants that have substitutions in residues 156 to 164 reduce or eliminate *lac* operon transcription activation but do not affect CRP's ability to bind or bend DNA. These residues that immediately precede the helix-turn-helix motif, therefore, are essential for *lac* operon activation. Examination of cAMP•CRP structure suggests an explanation. Residues 156 to 164 form a β turn on the CRP surface (designated activating region 1 or AR1) that interacts with RNA polymerase holoenzyme. Additional experiments were performed to distinguish between the possibility that AR1s of both CRP subunits are required for transcription activation and the possibility that the AR1 of just one subunit (promoter proximal or promoter distal) is all that is required (FIGURE 16.25). CRP heterodimers were constructed in which one subunit has a wild-type DNA binding motif but a mutated AR1 while the other subunit has a DNA binding motif with altered specificity but a wild-type AR1. Hybrid activator sites that had one wild-type half-site and one half-site that is only compatible with the CRP with altered DNA specificity forced the heterodimers to bind to *lac* promoters in two possible orientations. Transcription activation occurred only when wild type AR1 was present on the promoter proximal subunit.

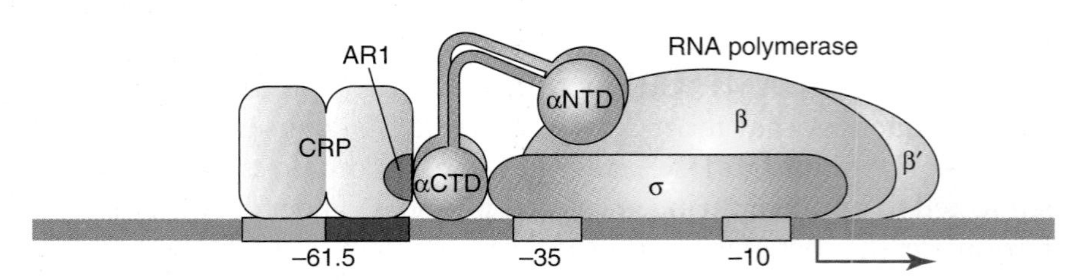

FIGURE 16.24 Transcriptional activation of the *E. coli lac* operon. The symbols are CRP, cAMP receptor protein; σ, σ factor; αCTD, carboxyl terminal domain α subunit; αNTD, amino terminal domain α subunit; β, β subunit; and β′, β′ subunit. The magenta oval represents activating region 1 (AR1) of CRP. The promoter in the *lac* operon is a prototype of a class of promoters known as class 1 cAMP-dependent promoters. Transcription activation requires direct protein–protein interaction between AR1 of the downstream subunit of CAP and one copy of αCTD. (Adapted from S. Busby and R. H. Ebright, *J. Mol. Biol.* 293 [1999]: 199–213.)

cAMP•CRP activates more than 100 operons.

Enzymes responsible for the catabolism of many other organic molecules, including galactose, arabinose, sorbitol, and glycerol, are synthesized by inducible operons. Each of these operons cannot be induced if glucose is present. These are called **catabolite-sensitive operons.** A network of operons that is under the control of a single regulatory protein such as cAMP•CRP is called a **modulon.**

A simple genetic experiment shows that cAMP•CRP participates in the regulation of many operons. A single spontaneous mutation in a gene for the catabolism of a sugar such as lactose, galactose, arabinose, or maltose arises with a frequency of roughly 10^{-6}. A double mutation, *lac⁻mal⁻*, would arise at a frequency of 10^{-12}, which for all practical purposes cannot be measured. However, double mutants that are phenotypically Lac⁻Mal⁻ or Gal⁻Ara⁻ do arise at a measurable frequency. These apparent double mutants are not the result of

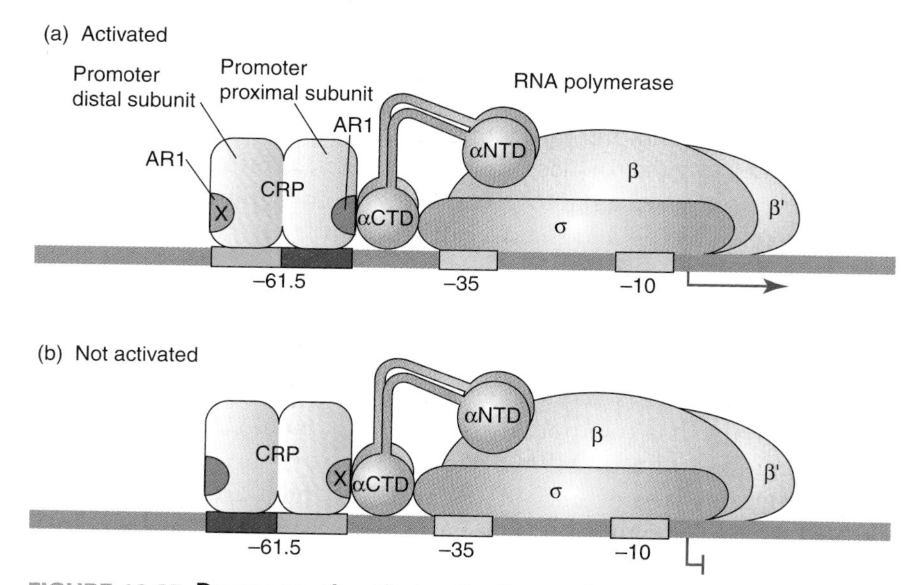

FIGURE 16.25 **Demonstration that activating region 1 (ARI) in the promoter proximal subunit of the cAMP receptor protein (CRP) is required for transcription activation.** CRP heterodimers are constructed so that one subunit (light green) has a wild-type DNA binding motif but a mutated AR1 (orange with an x) while the other subunit (light blue) has a DNA binding motif with altered specificity but a wild type AR1 (magenta). Two hybrid activator sites were also constructed. In the first (a), the promoter distal activator site is wild type (dark green) and the promoter proximal activator site is altered (dark blue) and in the second (b), the two activator sites are reversed. The orientation of the two activator sites forces an orientation on the cAMP•CRP complex. Activation occurs only when the orientation is as shown in (a), indicating the AR1 in the promoter proximal subunit is essential for activation while the AR1 in the promoter distal subunit is not needed. The symbols are CRP, cAMP receptor protein; σ, σ factor; αCTD, carboxyl terminal domain α subunit; αNTD, amino terminal domain α subunit; β, β subunit; and β′, β′ subunit. (Adapted from S. Busby and R. H. Ebright, *J. Mol. Biol.* 293 [1999]: 199–213.)

mutations in the two sugar operons but always turn out to be *crp*⁻ or *cya*⁻. Furthermore, if a Lac⁻Mal⁻ mutant appears as a result of a single mutation, the protein products of the other catabolite-sensitive operons are also not synthesized. Biochemical experiments with a few of these catabolite-sensitive operons indicate that binding of cAMP•CRP occurs in the promoter region in each of these systems.

16.4 Galactose Operon

The galactose operon is also regulated by a repressor and cAMP•CRP.

Although one might expect that bacteria would regulate other catabolite-sensitive operons in exactly the same way that they regulate the *lac* operon, in fact they use a wide variety of different mechanisms. Some of these mechanisms resemble the one that regulates the *lac* operon but others are markedly different. One of the best-studied catabolite-sensitive operons, the *gal* operon, contains four structural genes, *galK*, *galT*, *galE*, and *galM*, which specify the enzymes galactokinase, galactose transferase, galactose epimerase, and mutarotase, respectively (see Chapter 7). Mutarotase converts the β-D-galactose formed when β-galactosidase hydrolyzes lactose into α-D-galactose. Then galactokinase, galactose transferase, and epimerase act in a sequence of steps to yield the overall reaction

$$\text{Galactose} + \text{ATP} \rightarrow \text{Glucose-1-phosphate} + \text{ADP}$$

The galactose (*gal*) operon, like the *lac* operon, is regulated by a repressor and cAMP•CRP. The structural gene for the repressor, *galR*, is located far from the structural genes for the galactose enzymes. The *gal* operon also has two operators, $galO_E(O_E)$ and $galO_I(O_I)$, a promoter region, and a cAMP•CRP AS (FIGURE 16.26). The promoter region contains two promoters, P_1 and P_2, which are not shown as distinct regions in Figure 16.26 because there is extensive overlap. One operator, $galO_E$, is upstream from the promoter region and the other, $galO_I$, is in *galE*. A mutation in either operator causes at least partial constitutivity, indicating that both operators participate in repression.

The Gal repressor, which binds to both operators, is made of two identical polypeptide chains. A null mutation in *galR* makes both promoters constitutive. The amino acid sequence of the Gal repressor is similar to that of the Lac repressor. There is one important difference, however; the Gal repressor does not have the C-terminal residues that allow the Lac repressor to form a four-helical bundle and become a stable tetramer. The Gal repressor has a C-terminal domain that binds the inducer galactose and an NTD with a helix-turn-helix motif that binds to a $galO_E$ or $galO_I$ half-site.

Initial attempts to demonstrate DNA looping in response to Gal repressor under *in vitro* conditions were unsuccessful. This failure can be explained by the fact that the Gal repressor is a dimer and not a tetramer. Nonetheless, *in vivo* studies, which indicated that both

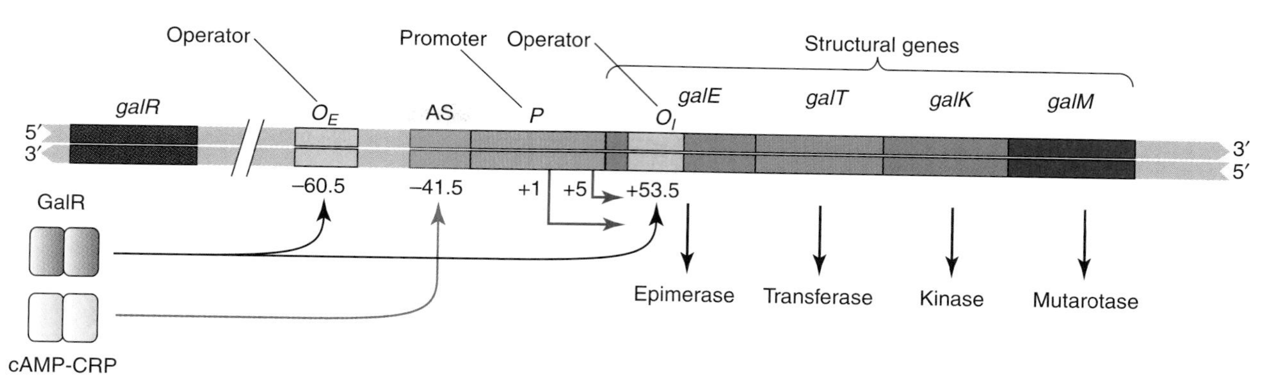

FIGURE 16.26 Structural genes, regulatory genes, and control elements in the gal operon. The four structural genes, *galE*, *galT*, *galK*, and *galM*, specify the enzymes galactose epimerase, galactose transferase, galactokinase, and mutarotase, respectively. The structural genes for the repressor, *galR*, is located far from the structural genes for the galactose enzymes. The *gal* operon also has two operators, $galO_E(O_E)$ and $galO_I(O_I)$, a promoter region, and a cAMP•CRP AS. The promoter region contains two promoters, P1 and P2, which are not shown as distinct regions because there is extensive overlap. The operon is not drawn to scale.

$galO_E$ and $galO_I$ are required for repression, suggested some type of cooperative interaction. Sankar Adhya and coworkers solved the problem by showing that a histone-like protein called HU binds to the *gal* promoter region only when the Gal repressors are bound at both $galO_E$ and $galO_I$, causing the DNA to loop (FIGURE 16.27). The specific nature of the protein-to-protein interactions still must be determined.

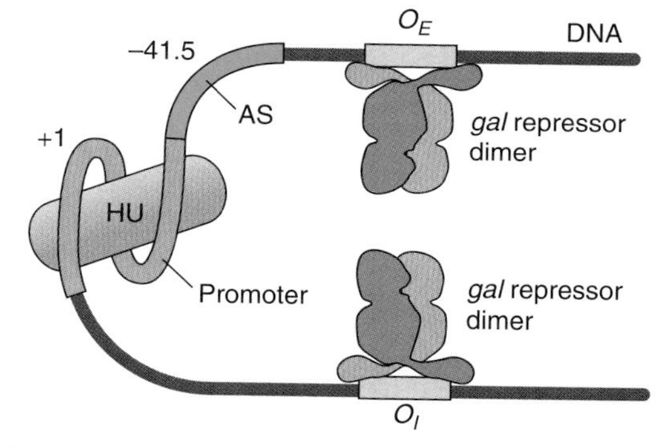

FIGURE 16.27 Gal repressor and HU action at the *gal* operon. The Gal repressor is a dimer. One dimer binds to $galO_E$ and another to $galO_I$. The histone-like protein, HU, binds to the gal promoter region (+1), causing the DNA to form a loop. This loop formation permits cooperative interactions between the Gal repressor dimer bound to $galO_E$ and that bound to $galO_I$, resulting in transcription repression. (Adapted from B. Müller-Hill, *Curr. Opin. Microbiol.* 1 [1998]: 145–151.)

The two *gal* promoters (P_1 and P_2) are separated by 5 bp (or one half-turn of B-DNA) and as a result, the transcription initiation site for P_1 is 5 bp upstream from that for P_2. Polycistronic mRNA molecules formed in response to either promoter contain all the information required to synthesize the four galactose enzymes. Binding cAMP•CRP to AS, which lies 41.5 bp upstream from the P_1 transcription start site, has opposite effects on the two promoters. Binding cAMP•CRP to AS represses transcription from P_2 but activates transcription from P_1. Moreover, P_2 can support *gal* mRNA transcription in the presence of glucose when intracellular cAMP•CRP levels are low but P_1 cannot do so. The fact that AS is 20 bp closer to P_1 in the *gal* operon than it is to P_{lac} in the *lac* operon leads one to predict that cAMP•CRP and RNA polymerase interactions should be different in the *gal* and *lac* operons. This prediction is correct. Promoters like P_1 in the *gal* operon that have AS at position –41.5 are called **class II cAMP-dependent promoters**. The organization of cAMP•CRP and RNA polymerase holoenzyme at class II cAMP-dependent operons is illustrated in the cartoon shown in FIGURE 16.28. Three activator regions on CRP—AR1, AR2, and AR3—interact with RNA polymerase holoenzyme. AR1 on the upstream subunit of the CRP dimer, and AR2 and AR3 on the downstream subunit of dimer interact with αCTD, αNTD, and the σ^{70} subunit, respectively.

Why does the *gal* operon have two promoters? The answer appears to be that galactose has two roles in cellular metabolism. It is both a carbon source and a precursor for lipopolysaccharide synthesis. When galactose is not available in the growth medium, cells require the

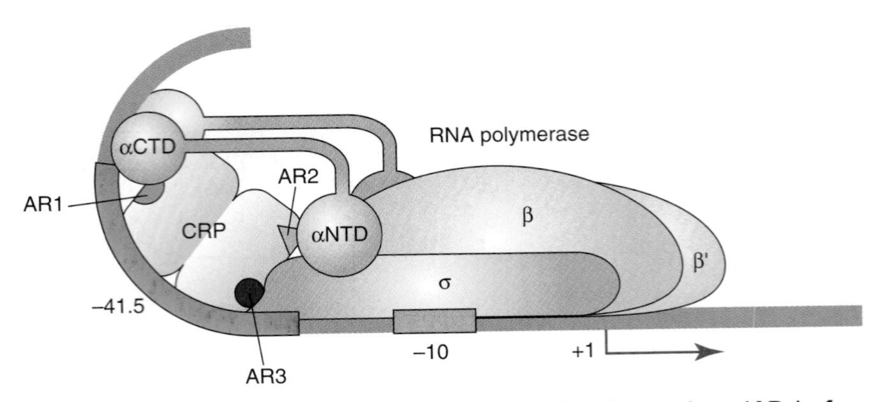

FIGURE 16.28 Organization of the functional activating regions (ARs) of cAMP receptor protein (CRP) and their target subunits in RNA polymerase at class II CRP-dependent promoters. The different activating regions of CRP, AR1, AR2, and AR3 are shown as an orange semicircle, a gray triangle, and a purple circle, respectively, displayed on the adjacent subunits of CRP dimer. AR1, on the upstream subunit of CRP dimer, and AR2 and AR3, on the downstream subunit of the dimer, interact with αCTD, αNTD, and the σ subunit, respectively, of the RNA polymerase holoenzyme. (Adapted from V. A. Rhodius and S. J. Busby, *Curr. Opin. Microbiol.* 299 [2000]: 295–310.)

epimerase specified by the *gal* operon to convert glucose to galactose, which is then used to make lipopolysaccharide. Synthesis from the P_2 promoter permits the low level of epimerase formation required to convert glucose to galactose-1-phosphate so that lipopolysaccharide can be formed.

If P_1 were the only promoter, then epimerase could not be made when glucose is present because P_1 requires cAMP•CRP activation. On the other hand, if P_2 were the only promoter, then galactose could not fully induce the operon when galactose was the sole carbon source because cAMP•CRP inhibits P_2. Thus, for the sake of both necessity and economy, a cAMP•CRP-independent promoter (P_2) is needed for background constitutive synthesis and a cAMP•CRP-dependent promoter (P_1) is needed to regulate high-level synthesis. Furthermore, the regulation is efficient only if P_2 is inhibited by cAMP•CRP.

16.5 The *araBAD* Operon

The AraC activator protein regulates the *araBAD* operon.

L-Arabinose, another sugar (a pentose) that can serve as a carbon source for metabolism, is present in the cell walls of many plants and is released in the human intestine after vegetables are eaten. It is not absorbed by the intestine and hence provides a source of carbon for bacterial inhabitants of the intestine such as *E. coli*. Because arabinose availability is sporadic, one would expect the ability to utilize this sugar to be inducible, as indeed it is.

The arabinose system allows *E. coli* and related bacteria to take up arabinose from the growth medium and convert it to

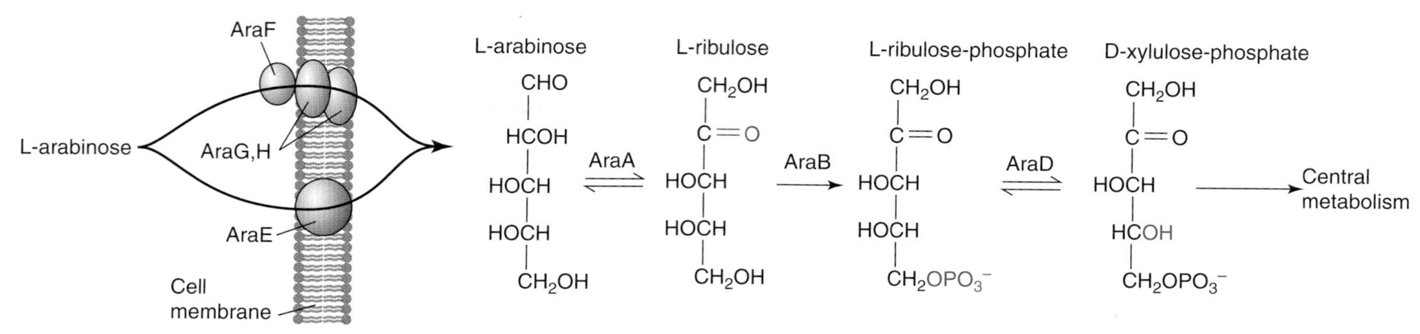

FIGURE 16.29 L-arabinose catabolism pathway in *E. coli.* (Adapted from R. Schleif, *Trends Genet.* 16 [2000]: 559–565.)

D-xylulose-5-phosphate, which the bacteria then convert to ribose-5-phosphate and other sugars (FIGURE 16.29). Arabinose transport proteins are encoded by *araE* at minute 64.2 and *araF*, *araG*, and *araH*, which are part of the *araFGH* operon at minute 42.7. The enzymes that convert arabinose to xylulose-5-phosphate are encoded by *araB*, *araA*, and *araD*, which are part of the *araBAD* operon at minute 1.4. Thus, the genes required for arabinose metabolism are located at three separate sites on the *E. coli* chromosome that are quite far apart from one another. We will focus our attention on the regulation of the *araBAD* operon.

The *araBAD* operon is inducible and the inducer is arabinose itself; that is, in a wild-type operon *araBAD* mRNA is made only when arabinose is present. Arabinose acts by binding to a regulatory protein to form a complex that stimulates the *araBAD* operon. This regulatory protein is called the AraC protein because it is encoded by *araC*, a gene that is near to *araBAD*. FIGURE 16.30 shows the relevant details of this region of the *E. coli* chromosome. The *araC* promoter (P_C) and the *araBAD* promoter (P_{BAD}) are transcribed in opposite directions. Wild type cells cannot synthesize *araBAD* mRNA when cultured in the presence of glucose, suggesting that transcription from P_{BAD} requires cAMP•CRP and sequence analysis shows a cAMP•CRP AS just upstream from P_{BAD}.

The AraC protein represses transcription from its own promoter, P_C, in either the presence or absence of arabinose. This control is an example of negative **autoregulation** because the gene product regulates its own synthesis. The AraC protein also represses transcription from

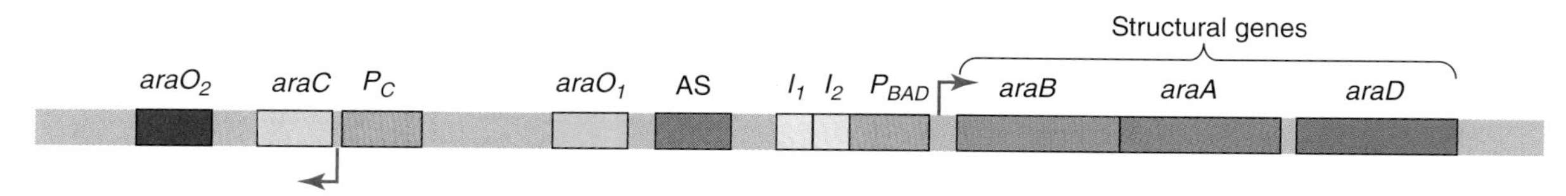

FIGURE 16.30 **The *araBAD* operon.** The *araC* gene codes for a regulatory protein. The *araB*, *araA*, and *araD* genes code for enzymes required for arabinose degradation. P_C and P_{BAD}, promoters for *araC* and *araBAD*, respectively, are transcribed in opposite directions. AS is a cAMP•CRP activator sequence. The regulatory elements O_2, I_1, and I_2 are 17 bp half sites of similar sequence that bind to one subunit of AraC, and O_1 is formed of two half sites, O_{1L} and O_{1R} (not shown), that bind two subunits.

the *araBAD* promoter, P_{BAD}, when arabinose is absent. As indicated above, however, the AraC protein stimulates transcription from P_{BAD} when arabinose is present. Thus, the AraC protein becomes a positive regulator of *araBAD* mRNA transcription when arabinose is present. This contrasts with the *lac* and *gal* repressors, which only function as negative regulators. The positive regulatory activity of the AraC protein is shown by the following results:

1. All point and deletion mutations in the *araC* gene, which are denoted *araC*$^-$, are unable to synthesize *araBAD* mRNA. Knowing nothing else, one might conclude either that an *araC*$^-$ cell contains a mutant repressor that cannot bind inducer or that the *araC* locus is the promoter. These possibilities are eliminated by the next two results.
2. Partial diploids having the genotype F′ *araC*$^+$/*araC*$^-$ are fully inducible. This means that the *araC*$^-$ allele is recessive. If *araC*$^-$ mutants were noninducible repressors, then *araC*$^-$ should be dominant to *araC*$^+$.
3. Partial diploids of the genotype F′*araC*$^+$*araA*$^-$/*araC*$^-$ *araA*$^+$ are fully inducible for synthesis of the *araA* product. If *araC* were a promoter, the *araA*$^+$ gene, being on the same DNA molecule as the *araC*$^-$ mutation, could not be transcribed. Because the *araC* product complements in a *trans* configuration, it must be a diffusible molecule.

Other genetic evidence—namely, that there are amber mutants in the *araC* gene—indicates that the *araC* product is a protein. These genetic results have been confirmed by purification and sequencing of the AraC protein. The important point of these experiments is that *araBAD* mRNA synthesis requires a functional AraC protein.

The conclusion that both cAMP•CRP and arabinose•AraC are required to initiate *araBAD* mRNA transcription has been confirmed by *in vitro* experiments of two sorts. The first experiments show that no *araBAD* mRNA is synthesized in a reaction mixture containing either (a) only *ara* DNA and RNA polymerase holoenzyme or (b) *ara* DNA, RNA polymerase holoenzyme, and only one of the regulatory complexes (arabinose•AraC or cAMP•CRP).Both regulatory complexes, therefore, must be present for RNA polymerase to synthesize *araBAD* mRNA. In the second series of experiments, *ara* DNA was examined by electron microscopy (which is capable of visualizing bound RNA polymerase molecules). One result of these experiments was the observation that no RNA polymerase is bound unless both regulatory complexes are present.

The structure of the AraC protein, a homodimer, provides important clues to the regulatory protein's mechanism of action. The subunit structure in the absence of arabinose is shown in cartoon form in FIGURE 16.31a. A flexible linker joins the N- and C-terminal domains. N-terminal domains of each subunit interact to stabilize the dimer structure. The NTD, hence, is commonly referred to as the dimerization domain. The N-terminal domain also contains an arabinose-binding pocket, which does not appear to undergo a major conformational change upon binding arabinose. However, an arm containing approximately 18 residues that is attached to the N-terminal domain does undergo a major change when arabinose binds to the NTD (**FIGURE 16.31b**). When arabinose is absent, the N-terminal arm binds to the side of the C-terminal (DNA-binding) domain away from the DNA. The combination of the arm plus the linker holds each DNA-binding domain relatively rigidly to its N-terminal (dimerization) domain. When arabinose is added, the N-terminal arm shifts position from the C-terminal (DNA-binding) domain to the NTD. This shift frees the N- and C-terminal domains to move with respect to one another. Thus, the N-terminal arm behaves like a "light switch," causing a major change in AraC structure that affects AraC's affinity for DNA binding sites.

In the absence of arabinose, the rigid AraC protein binds to *araI*$_1$ and *araO*$_2$ half-sites to form a DNA loop that is about 210 bp long (Figure 16.31a). The importance of this loop was indicated by a helical-twist experiment in which a half turn (5 bp) or a whole turn (10 bp)

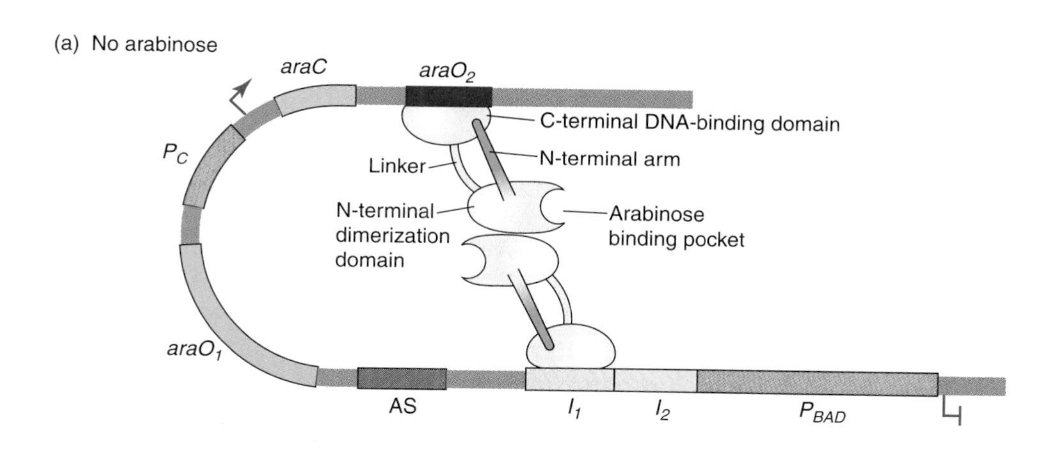

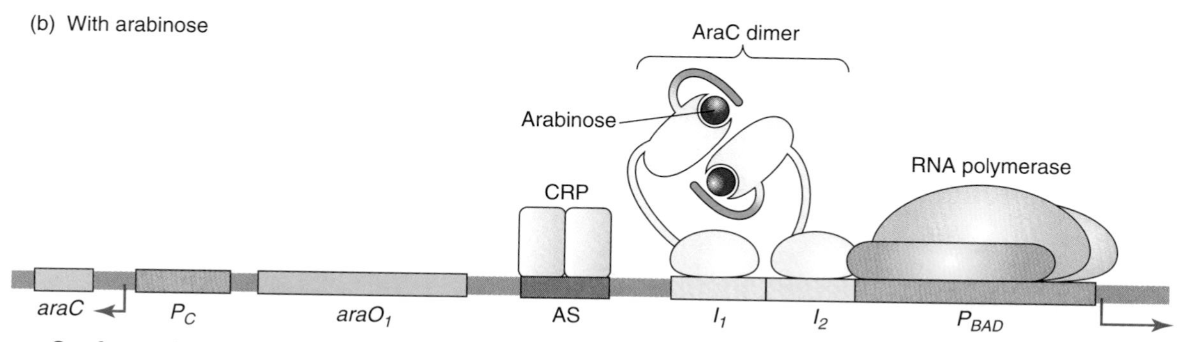

FIGURE 16.31 Conformational change in AraC after binding arabinose. (a) The *araP*$_C$ and *araP*$_{BAD}$ regulatory regions in the absence of arabinose. The regulatory elements *araO*$_2$, I_1, and I_2 are each 17-bp half-sites with similar sequences that can bind one AraC subunit. In the absence of arabinose, AraC binds to *araO*$_2$ and I_1, causing the DNA to form a loop. RNA polymerase is hindered from binding to *araP*$_{BAD}$ and *araP*$_C$ under these conditions. The cAMP receptor protein (CRP) is probably also hindered from binding to its DNA site. (b) The *araP*$_{BAD}$ and *araP*$_C$ regulatory regions in the presence of arabinose (red sphere). When arabinose is present, AraC binds primarily to the adjacent I_1 and I_2 half-sites and so the DNA does not form a loop. Hence, RNA polymerase is free to bind to P_{BAD} and cAMP•CRP is free to bind to its DNA site. *araO*$_1$ contains two half-sites, and so can bind an AraC dimer. AraC regulation of the *araC* structural gene is a complex process that involves *araO*$_1$ and *araP*$_C$ and is not shown here. DNA elements are not drawn to scale. (Adapted from R. Schleif, *Trends Genet.* 16 [2000]: 559–565.)

DNA was inserted between *$araI_1$* and *$araO_2$*. A half-turn insert would be expected to interfere with loop formation because it places *$araI_1$* or *$araO_2$* sites on opposite sides of the helix so that the AraC protein could not contact both sites at the same time unless the DNA experiences a half-twist, an energetically unfavorable event. A full-turn insert would increase the loop size but would allow the AraC protein to contact both sites at the same time because they remain on the same side of the helix. As predicted for a loop structure, a half-turn insert diminishes repression of P_{BAD} transcription, whereas a full-turn insert does not. The DNA loop has the following functions: (1) it sterically blocks the access of RNA polymerase holoenzyme to P_{BAD}, ensuring that the basal level of *araBAD* transcription will be low, (2) it blocks access of RNA polymerase to P_C, and (3) it may interfere with the ability of cAMP•CRP to bind to the activator site.

As indicated above, when arabinose binds to the pocket in the NTD, the N-terminal arm shifts position from the C-terminal domain to the N-terminal domain. As a result of this shift, AraC dissociates from the *$araO_2$* site and binds to the *$araI_2$* site, opening the DNA loop (Figure 16.31b). Because the AraC protein remains tethered to the *$araI_1$* site, this shift occurs very rapidly. With the opening of the DNA loop, RNA polymerase gains access to P_C and P_{BAD}. Transcription from P_C leads to *araC* mRNA formation, which then is translated to form AraC protein. After a short time, however, AraC binds to *$araO_1$*, repressing further *araC* mRNA formation.

Events at P_{BAD} follow a different course. Binding of the AraC protein to *$araI_2$*, which overlaps P_{BAD} by four bases, increases RNA polymerase's affinity for P_{BAD}. Interactions with cAMP•CRP increases the affinity still further. FIGURE 16.32 shows probable contacts among the AraC protein, DNA, RNA polymerase holoenzyme, and cAMP•CRP. Bacteria produce more than a hundred different homologs of the AraC protein. The details of the mechanism of most of these AraC homologs is not yet known, but it seems probable that the mechanism proposed for the *E. coli* AraC protein also will apply to many of its homologs. Thus, the *araBAD* operon is a useful prototype for the examination of bacterial systems that are regulated primarily by a positive control element. It may seem reasonable to ask why the *ara_{BAD}* operon is regulated positively rather than by a repressor. About 50 years ago, when the *lac* operon was considered to be the model for all operons, such a

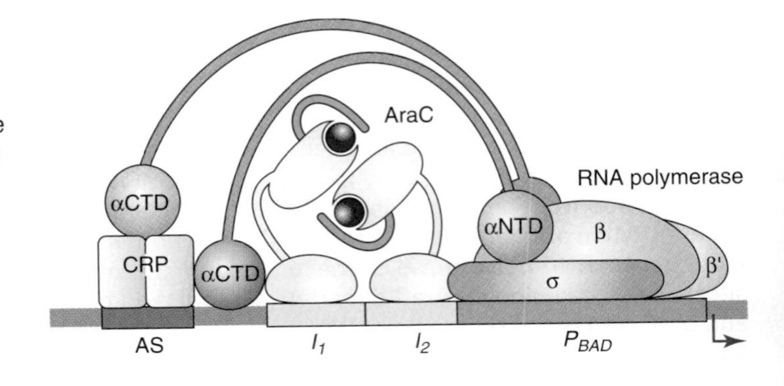

FIGURE 16.32 Probable contacts between the C-terminal domain of the α-subunit of RNA polymerase, the CRP protein, AraC, and DNA. The C-terminal domain of one α-subunit (purple) is shown as contacting the DNA between CRP and the AraC binding domain because this domain often does contact DNA. At present, however, there are no experimental data to support such contact. Additional contacts are made between RNA polymerase and the polymerase proximal AraC subunit. (Adapted from R. Schleif, *Trends Genet.* 16 [2000]: 559–565.)

question would have seemed very important. Over time, however, it has been realized that a variety of regulatory strategies have evolved and the criterion for survival of a system over millions of years is merely that the mechanism works.

16.6 Tryptophan Operon

The tryptophan (*trp*) operon is regulated at the levels of transcription initiation, elongation, and termination.

Some transcription regulatory systems work at the elongation and termination stages rather than at the initiation stage. The tryptophan (*trp*) operon, which we now examine, is regulated at all three transcriptional stages. The *trp* operon consists of a promoter, an operator, a leader (*trpL*), an attenuator, and five structural genes, designated *trpE*, *trpD*, *trpC*, *trpB*, and *trpA* (FIGURE 16.33). Consistent with the fact that the *trp* operon specifies biosynthetic rather than degradative enzymes, it does not have a cAMP•CRP activation site. The operator is part of a coarse on-off control, and the leader and attenuator allow

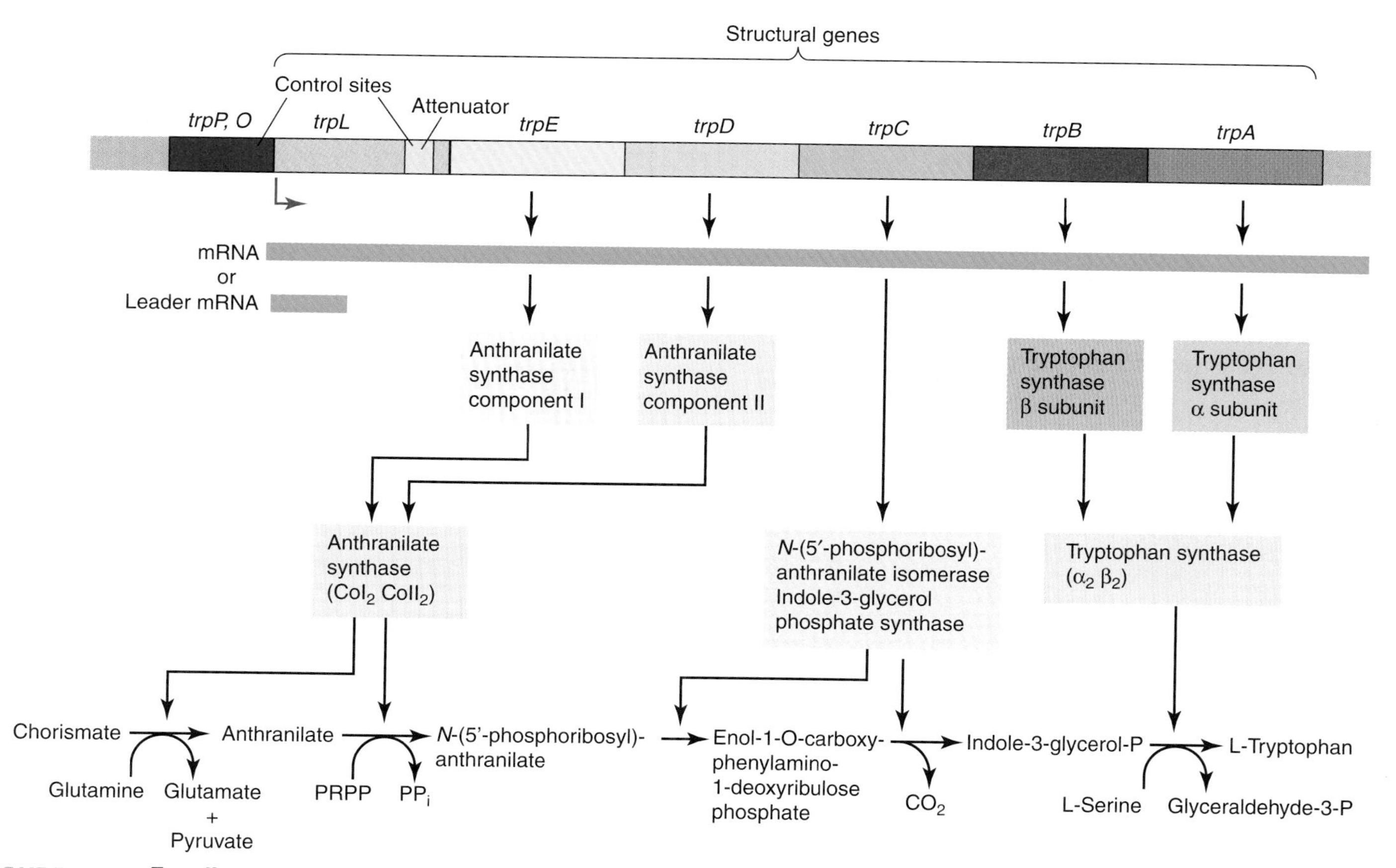

FIGURE 16.33 ***E. coli* tryptophan (*trp*) operon, the enzyme it specifies, and the reactions they catalyze.** PRPP is an abbreviation for phosphoribosylpyrophosphate. (Adapted from C. Yanofsky, *J. Am. Med. Assoc.* 218 [1971]: 1026–1035).

for finer control. The leader sequence, which codes for a short peptide (see below), is the first region of the *trp* operon to be transcribed. RNA polymerase then moves forward to transcribe the five structural genes in the order *trpE* → *trpA* to form the polycistronic *trp* mRNA. The five polypeptides specified by the structural genes form three enzymes that are essential for tryptophan biosynthesis.

Early studies revealed that the *trp* system is turned off when tryptophan is added to an *E. coli* culture, suggesting that the operon is repressed rather than induced. Repression, like induction, involves negative regulation of transcription initiation by a regulatory protein. However, the regulatory protein does not bind to the operator until after forming a complex with tryptophan (FIGURE 16.34). The structural gene for the Trp regulatory protein, *trpR*, is located a considerable distance away from the *trp* operon. This distance does not present a problem because the regulatory protein diffuses throughout the cell. The biologically active form of the regulatory protein is a homodimer. Each subunit contains a helix-turn-helix motif that can bind to a *trp* operator half-site. The *trp* operator and promoter regions have significant overlap. Therefore, binding of the TrpR•tryptophan complex and RNA polymerase are mutually exclusive.

The TrpR•tryptophan complex functions as a coarse on-off switch that turns the tryptophan operon off when tryptophan levels are high. A fine control mechanism also exists that allows cells to regulate their tryptophan enzyme concentration according to the tryptophan concentration. An important clue to the existence of this fine control mechanism came from an experiment performed

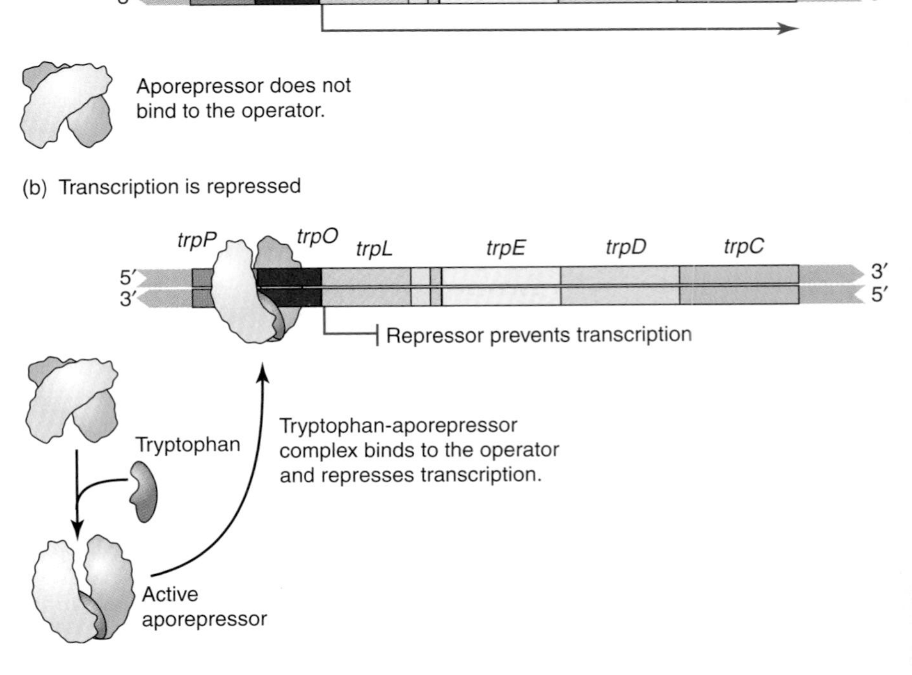

FIGURE 16.34 **Regulation of the *E. coli trp* operon.** (a) By itself, the Trp aporepressor does not bind to the operator, and transcription occurs. (b) In the presence of sufficient tryptophan, the combination of aporepressor and tryptophan forms the active repressor that binds to the operator and transcription is repressed.

by Charles Yanofsky in 1972, which showed that *E. coli* mutants that lack a functional TrpR protein increase *trp* operon transcription after being starved for tryptophan. If the TrpR•tryptophan complex were the only regulatory factor, then transcription of the *trp* operon should not have increased.

Further studies revealed that *trp* mRNA has a 162 nucleotide sequence before the first codon in *trpE*, designated the **leader** or *trpL*, which plays an essential role in the fine control mechanism. Constitutive mutants exhibit a sixfold increase in tryptophan enzyme synthesis when bases 123 to 150 within the leader are deleted. This 28 base sequence is called the **attenuator.** The attenuator can fold into a stem-and-loop structure with the potential to function as a rho-independent transcription terminator (FIGURE 16.35). Evidence that the attenuator actually terminates transcription comes from the observation that wild-type cells cultured in the presence of tryptophan terminate synthesis of most *trp* mRNA molecules when they are 140 nucleotides long, well short of full-length polycistronic *trp* mRNA. Deleting the attenuator removes the transcription termination site, allowing RNA polymerase to complete *trp* mRNA synthesis.

The *trp* leader has four complementary segments that can interact to form two sets of mutually exclusive hairpin structures (FIGURE 16.36). Segments 1 and 2 can base pair to form hairpin 1•2 while segments 3 and 4 base pair to form hairpin 3•4 (Figure 16.36a). Alternatively,

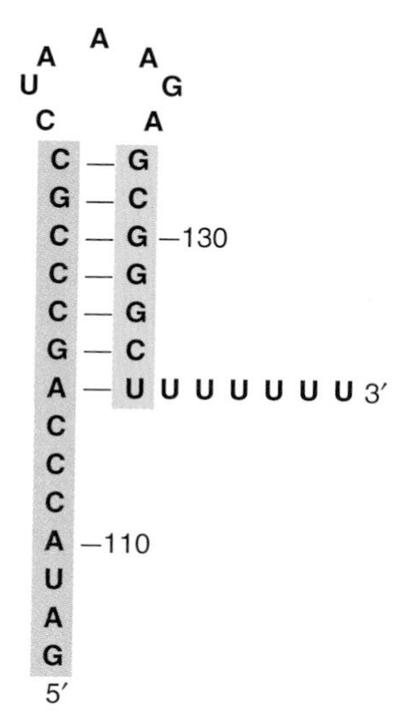

FIGURE 16.35 **The terminal region of the *trp* attenuator region.** The attenuator can fold into a stem-and-loop structure with the potential to function as a rho-independent transcription terminator. (Adapted from D. Voet, J. G. Voet, and C. W. Pratt. *Fundamentals of Biochemistry, First edition.* John Wiley & Sons, Ltd., 2000.)

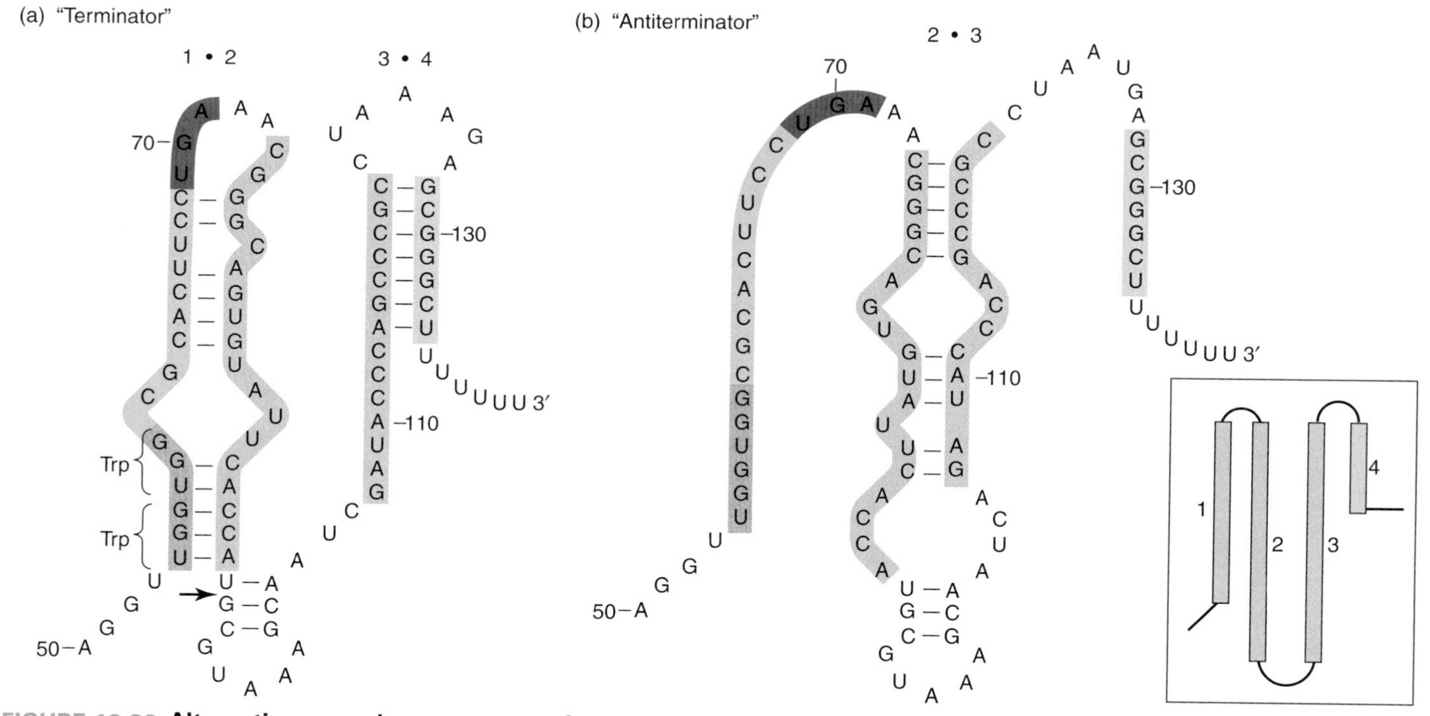

FIGURE 16.36 **Alternative secondary structures of trp_L mRNA.** Bases are numbered from the 5′-end of the transcript. Segments that base pair to form intrastrand secondary structure are numbered according to the inset. (a) 1•2 and 3•4 (terminator) hairpins signal transcriptional pausing and termination, respectively. The arrow indicates the site at which RNA polymerase pauses until it is approached by the moving ribosome. (b) Alternative 2•3 (antiterminator) hairpin that signals transcriptional read-through at the attenuator by preventing terminator 3•4 formation. (Adapted from D. Voet, J. G. Voet, and C. W. Pratt. *Fundamentals of Biochemistry, First edition.* John Wiley & Sons, Ltd., 2000.)

segments 2 and 3 can base pair to form hairpin 2•3 (Figure 16.36b). The purified *trp* leader mRNA forms the 1•2 and 3•4 hairpins because this secondary structure has the highest degree of hydrogen bonding and therefore the greatest stability. The 3•4 hairpin, which contains the attenuator sequence, is a transcription terminator. It follows that RNA polymerase would be able to synthesize full-length *trp* mRNA if conditions were somehow favorable for 2•3 hairpin formation rather than 3•4 hairpin formation.

The attenuation model, proposed by Yanofsky, explains how cells block 3•4 hairpin formation when tryptophan concentrations are low. This model builds on the observation that bacterial ribosomes start to translate mRNA molecules before RNA synthesis is complete. Coupled transcription-translation is possible in bacteria because (1) ribosomes and DNA are in the same cell compartment and (2) mRNA is transcribed in a 5′→3′ direction and translated in the same direction. The attenuation model is based on the fact that the leader sequence codes for a peptide that is 14 amino acids long with tryptophan residues at positions 10 and 11 (**FIGURE 16.37**). The presence of a pair of adjacent tryptophans is unusual because tryptophan normally accounts for about 1% of the amino acid residues in a protein.

RNA polymerase molecules that escape repression begin synthesizing *trp* mRNA. RNA polymerase continues transcribing the *trp* operon until it encounters a pause site located just after segment 2 of the leader sequence. Because the leader sequence is at the 5′-end of *trp* mRNA, it is the first sequence available for translation. Ribosomes begin translating the leader sequence at its AUG start codon. When the moving ribosome reaches the paused RNA polymerase, the paused RNA polymerase is released. Pausing serves the important role of synchronizing the transcription and translation processes. The subsequent fate of the RNA polymerase depends on the tryptophan concentration (**FIGURE 16.38**).

1. When the tryptophan concentration is high, the ribosome moves past the tryptophan codons in segment 1 and continues to translate the leader region until it encounters the stop codon (UGA) between segments 1 and 2 and falls off the nascent mRNA molecule. Once free of the ribosome, segment 1 pairs with segment 2 to form the 1•2 hairpin. RNA polymerase continues to transcribe the leader region, synthesizing segment 3 and then segment 4. These two segments pair to form the rho independent transcription terminator, which causes RNA

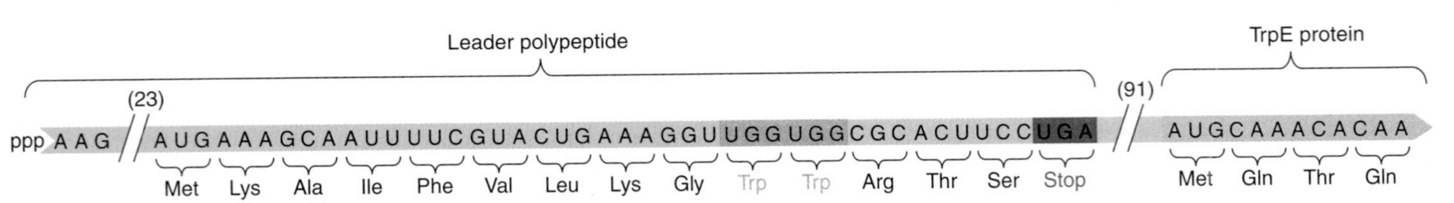

FIGURE 16.37 The sequence of nucleotides in *trp* leader mRNA, showing the leader polypeptide, the two tryptophan codons (orange), and the beginning of the TrpE protein. The numbers 23 and 91 are the numbers of nucleotides in the sequence that, for clarity, are not shown.

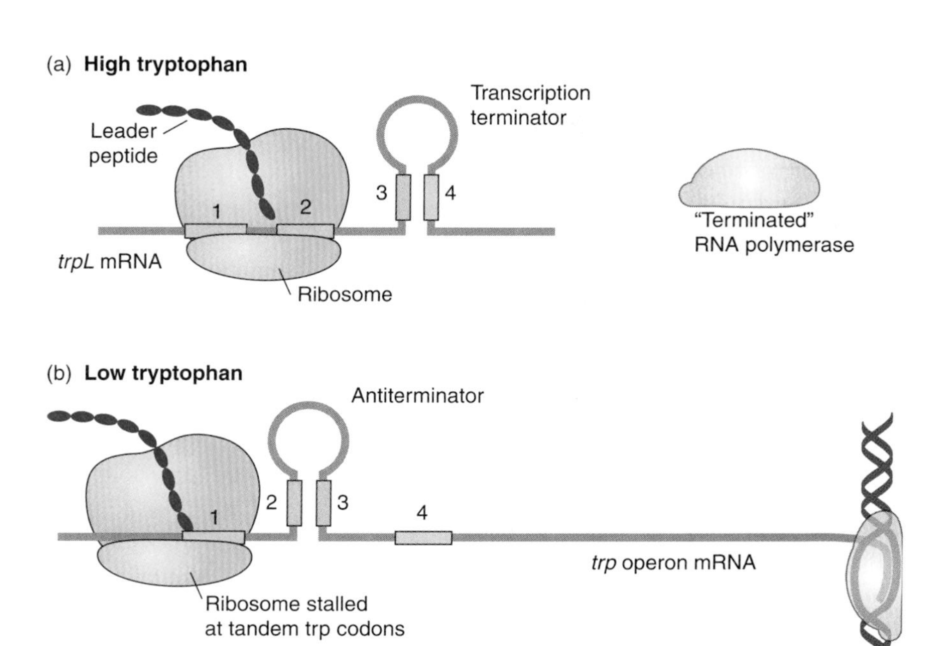

FIGURE 16.38 **Attenuation in the *trp* operon.** (a) When tryptophan is abundant, the ribosome reads through the tandem Trp codons in segment 1 of *trpL* mRNA, reaching segment 2 and preventing formation of the base-paired 2•3 hairpin. The 3•4 hairpin, an essential component of the transcriptional terminator, can then form, thus aborting transcription. (b) When tryptophan is scarce, the ribosome stalls on the tandem Trp codons of segment 1, permitting the formation of the 2•3 hairpin. This formation prevents the formation of the 3•4 hairpin, allowing RNA polymerase to transcribe through the unformed terminator and continue transcribing the *trp* operon. (Adapted from D. Voet, J. G. Voet, and C. W. Pratt. *Fundamentals of Biochemistry, First edition.* John Wiley & Sons, Ltd., 2000.)

polymerase to fall off the DNA template, preventing *trpE* transcription.

2. When the tryptophan concentration is very low, the bulky ribosome will pause at the tryptophan codons (UGG) on segment 1, preventing segment 1 from pairing with segment 2 to form the 1•2 hairpin. Segment 2 is therefore free to interact with segment 3 to form the 2•3 hairpin (the so-called **antiterminator**) as soon as RNA polymerase completes the synthesis of segment 3. Then RNA polymerase can continue to synthesize the complete *trp* mRNA because the 3•4 hairpin (the transcription terminator) is not formed.

Thus, if tryptophan is present in excess, transcription termination occurs at the attenuator and little enzyme is synthesized; if tryptophan is absent, transcription termination does not occur and tryptophan enzymes are made. Many operons responsible for amino acid biosynthesis are regulated by attenuators equipped with the base-pairing mechanism for competition as described for the *trp* operon. For example, this method of transcription regulation has been described for the histidine, threonine, leucine, isoleucine-valine, and phenylalanine operons of the bacteria *E. coli*, *Salmonella typhimurium*, and *Serratia marcescens*. Except for the phenylalanine operon, each lacks a repressor-operator system and is regulated solely by attenuation.

Because these operons are regulated adequately (although the range of expression is not as great as that of the *trp* operon), one might ask why the *trp* and *phe* operons have dual regulatory systems. This question has given rise to considerable speculation. The most obvious explanation is that the separate effects expand the range of tryptophan concentration in which regulation occurs. Whereas this is certainly true, there is more to it because the *trp* repressor also regulates the activity of four other operons. Such an operon network, which consists of two or more operons that are (1) regulated by a common regulatory protein and its effector(s) and (2) associated with a single pathway, function, or process, is called a **regulon**.

One hypothesis to explain the origin of the repression mechanism in the *trp* operon is that the *trp* operon was regulated only by attenuation in the distant past. The Trp aporepressor, however, actually evolved to serve a different function, to combine with tryptophan to regulate *aroH* (one of the five operons regulated by the Trp repressor complex). The existence of a tryptophan repressor complex allowed the possibility for a region adjacent to the *trp* promoter to evolve into an operator that can bind it.

16.7 Bacteriophage Lambda: A Transcription Regulation Network

Phage λ development is regulated by a complex genetic network.

Thus far, our discussion of bacterial transcription factors has been limited to the effect(s) that these factors have on a specific biochemical step or pathway. We now examine their contributions to a complex genetic network that determines which of two developmental pathways a phage λ will follow after it infects *E. coli*.

Recall from Chapter 8 that phage λ can operate in one of two ways. It can (1) enter the lytic pathway, directing the host cell to replicate viral DNA, make viral proteins, package the viral DNA and viral proteins to make new viral particles, and release the new virus particles after causing cell lysis. Alternatively, (2) it can enter a lysogenic pathway and direct the host cell to insert the viral DNA into the bacterial chromosome to form a prophage that remains part of the bacterial chromosome and makes the cell immune to infection by other phage particles.

Environmental factors influence the decision to enter the lytic or lysogenic state. The lytic pathway is favored when a single phage particle infects a bacterial cell in rich medium that supports rapid cell division. The lysogenic pathway is favored when two or more phage particles infect a bacterial cell in minimal medium that supports slow cell division. The environmental influences make sense if we consider the alternate pathways in terms of phage λ survival. A phage λ infects a rapidly growing cell to produce about 100 progeny phages, which in turn can infect other cells. However, a phage λ infecting a slowly growing cell might have a better chance of survival if it inserts its DNA into the bacterial chromosome and divides along with the cell. Ultraviolet light, which damages the bacterial chromosome and threatens host cell survival (see Chapter 12), induces the prophage to enter the lytic state so that newly made virus particles can escape the doomed host cell.

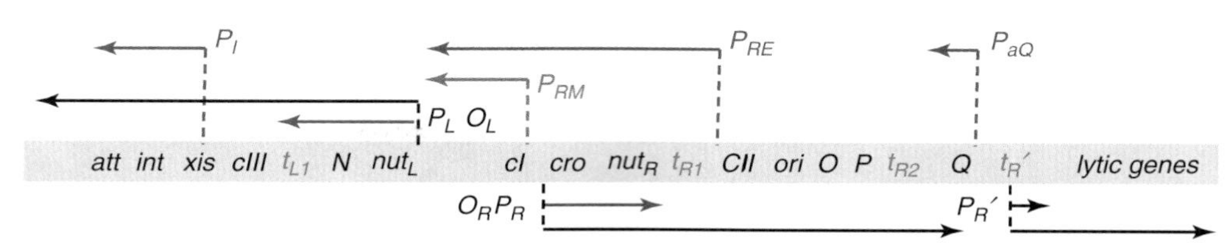

FIGURE 16.39 Key genes and signals in the regulatory region of the phage lambda chromosome. Early transcripts are shown by red arrows. Extended delayed early transcripts and the late transcripts are shown in black arrows. Transcripts initiated from the P_I, P_{RE}, and P_{aQ} (blue arrows) are required for lysogeny. The P_{RM} promoter (green) is activated by the CI repressor and required for maintenance of the lysogenic state. Critical transcription terminators, t_{L1}, t_{R1}, t_{R2}, and t_R', are marked in orange. Leftward promoters are indicated above and the rightward ones below the map. P_L and P_R are the early promoters and P_R' is the late lytic promoter. The operators O_L and O_R that regulate P_L and P_R, respectively, are also shown. The immunity region of the lambda chromosome includes P_LO_L, *cI*, O_RP_R, and *cro*. *ori* is the origin of replication. The protein products of genes *O* and *P* are required for DNA replication. The *int* gene codes for integrase, the enzyme required for the site-specific integration reaction. The *xis* gene codes for excisionase, which, along with integrase, is required for the excision reaction. (Reproduced from A. B. Oppenheim, *Annu. Rev. Genet.* 39 [2005]: 409–429. Copyright 2005 by Annual Reviews, Inc. Reproduced with permission of Annual Reviews, Inc., in the format Textbook via Copyright Clearance Center.)

FIGURE 16.39 shows the key genes within the regulatory region of the λ chromosome that determine whether the phage λ enters the lytic or lysogenic pathway. Some genes in this region were given letter names, for example *N* and *O*, which do not provide any information about their function. Other genes have names based on their function. For example, the names of some genes are related to plaque morphology. Mutants that produce nonfunctional *cI*, *cII*, or *cIII* gene products enter the lytic pathway and produce clear plaques. In contrast, wild-type phages produce plaques that appear turbid because they contain some surviving λ lysogens. Another regulatory gene in this region, *cro*, was also named on the basis of its function. The term *cro* is an acronym for control of repressor and other things.

The lytic pathway is controlled by a transcription cascade.

The first two genes transcribed after phage λ infection code for the N and Cro proteins. The promoters for these early transcripts are P_L (left promoter) and P_R (right promoter), respectively (Figure 16.39). Transcription from the P_L promoter terminates at transcription termination site t_{L1}, whereas that from the P_R promoter terminates at transcription termination site t_{R1}. P_R and P_L are regulated by the O_R and O_L operator regions, respectively. Each operator region contains three 17-bp operator sites, designated O_L1, O_L2, and O_L3 in O_L and O_R1, O_R2, and O_R3 in O_R. The sequences of the six operator sites are similar but not identical, allowing regulatory proteins to distinguish among them. Each site forms a palindrome, although none is perfect. The Cro protein is a homodimer. Each subunit contains a single domain with three α helices. Helices 2 and 3 form a helix-turn-helix structure that can bind to O_L and O_R sites (FIGURE 16.40). Helix 3 functions as the recognition helix, fitting into the major groove of the operator DNA site, while helix 2 helps to stabilize the protein-DNA complex. Cro must be functional for phage λ to enter the lytic cycle. Several proposals have been offered to explain why Cro is essential for the lytic cycle but none has yet proved to be consistent with all of

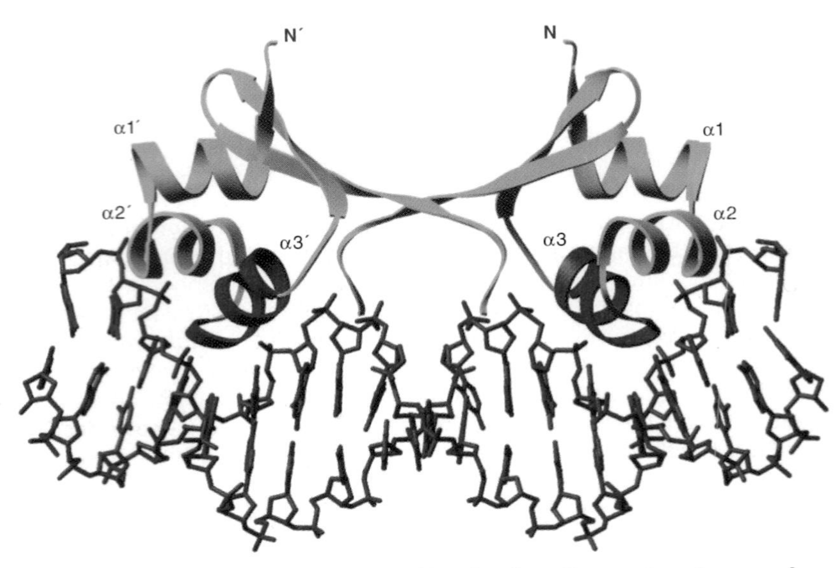

FIGURE 16.40 **Conformation adopted by the Cro dimer when in complex with operator DNA.** Recognition helices (α3 and α3′) are shown in red, while the remainder of the Cro dimer is shown in green. The operator DNA is shown in blue. (Reproduced from *J. Mol. Biol.*, vol. 280, R. A. Albright and B. W. Matthews. Crystal structure of λ-Cro . . ., pp. 137–151, copyright 1998, with permission from Elsevier [http://www.sciencedirect.com/science/journal/00222836]. Photo courtesy of Brian W. Matthews, Institute of Molecular Biology, University of Oregon.)

the experimental data. It appears at this time that the critical role of Cro in lytic development is to turn down the P_L and P_R promoters so that the infected cell's ability to make CII is limited.

The N-protein functions as an antitermination factor, promoting the assembly of a transcription complex that can continue transcription beyond the t_{L1} or t_{R1} transcription terminators. The assembly of this complex takes place at two sites on the nascent RNA called NUT_L and NUT_R. The upper case letters indicate that the NUT sites are on RNA rather than DNA. The DNA sequences that code for these RNA sites are nut_L (between the *N* gene and P_L) and nut_R (downstream from the *cro* gene), respectively. After N-protein binds to nascent RNA at NUT_L or NUT_R, it recruits four N-utilization substances, NusA, NusB, NusD, and NusE, to RNA polymerase. The resulting transcription complex reads through the t_{L1} and t_{R1} transcription termination sites to produce so-called delayed early transcripts. The proteins produced by translating the delayed early transcripts include the lysogenic regulators CII and CIII, the lytic replication functions O and P (required for DNA replication), and the late protein regulator Q. Once sufficient Q protein has accumulated in the infected cell, it modifies RNA polymerase that has just initiated transcription at promoter P_R'. Unlike N, which binds to a site on the nascent RNA, Q binds to a DNA sequence in P_R'. The modified RNA polymerase can read through transcription terminators downstream from the P_R' promoter to produce late mRNA. Translating the resulting late mRNA produces the head and tail proteins required to complete the lytic pathway.

The lysogenic pathway is also controlled by a transcription cascade.

The gene expression cascade leading to the lysogenic pathway also starts with divergent transcription initiated from the P_L and P_R promoters. Accumulation of the CII regulator, however, prevents the expression of the lytic regulators. The CII regulator is a tetramer. Each of the four monomers has a helix-turn-helix motif but only two of these motifs bind to DNA. The function of the other two motifs is not known.

The CII regulator activates transcription initiated from P_{RE}, P_I, and P_{aQ} (shown in blue in Figure 16.39). Activation of P_{RE} (promoter for repressor establishment) leads to rapid CI regulator synthesis. The CI regulator that accumulates in response to this activation binds to O_L and O_R and represses transcription from the early promoters P_L and P_R. The structure of the CI regulator protein and its interaction with operator DNA is described below. The intracellular concentration of the CI regulator that accumulates in response to the activation at P_{RE} is 10 to 20 times higher than that present in an established λ lysogen. The initial high concentration of the CI regulator probably ensures that all infecting phage DNA becomes repressed. Activation from P_I stimulates transcription of the *int* gene. The product of this gene, integrase (see Chapter 14), is needed to insert λ DNA into the host chromosome to form the prophage. The P_{aQ} promoter is located within the *Q* gene. The CII regulator, therefore, inhibits transcription of the *Q* gene when it binds to this promoter. Furthermore, the P_{aQ} transcript appears to function as an antisense RNA that inhibits the translation of the Q transcript. As should now be clear, the CII protein is very important for establishing lysogeny. An ATP-dependent host protein called FtsH can prevent the infected cell from entering the lysogenic pathway by cleaving CII. The CIII protein prevents this from happening by inhibiting the FtsH protease.

The CI regulator maintains the lysogenic state.

The CI regulator maintains λ DNA in the prophage state by regulating transcription initiation from P_L, P_R, and P_{RM} (promoter for repressor maintenance). P_R and P_{RM} are adjacent to each other but do not overlap (**FIGURE 16.41**). They are regulated by the O_R1, O_R2, and O_R3

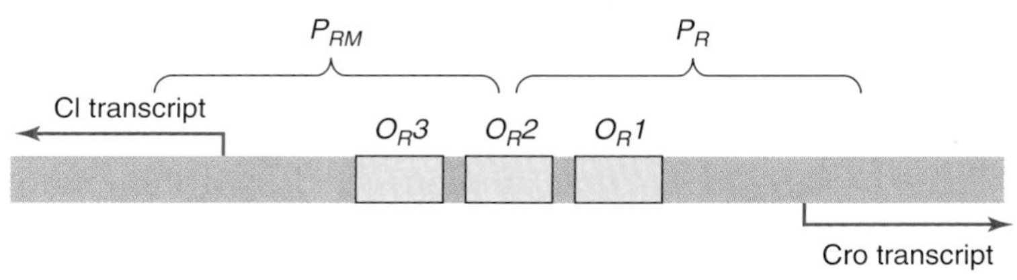

FIGURE 16.41 Operon region that regulates CI and Cro transcription. O_R (operator right) that contains three operator sites, O_R1, O_R2, and O_R3. O_R1 and O_R3 lie entirely within P_{RM} and P_R, respectively. O_R2 overlaps both P_{RM} and P_R. (Adapted from M. Ptashne and A. Gann, *Curr. Biol.* 8 [1998]: R812–R822.)

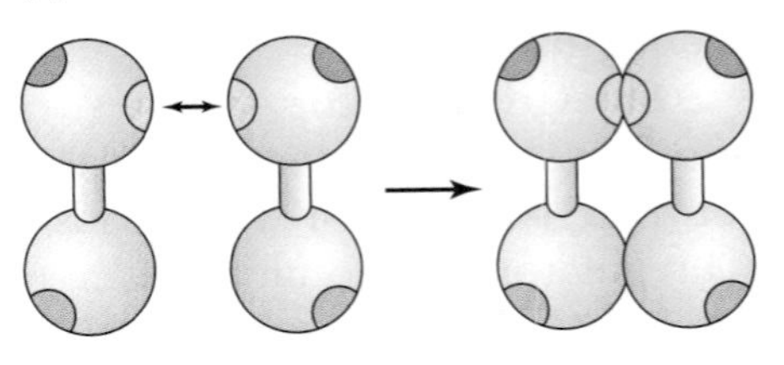

FIGURE 16.42 CI regulator monomer and dimer. (a) Cartoon of the CI regulator monomer, showing the N-terminal domain (residues 1–92) and C-terminal domain (132–236). These two domains are connected by a 40 amino acid long linker region. The N-terminal domain contains a helix-turn-helix motif that binds to the operator site and a site that interacts with RNA polymerase (green). The C terminal domain contains oligomerization sites (yellow). (b) Cartoon showing how two C terminal domains interact to form a CI dimer. (Adapted from M. Ptashne and A. Gann, *Curr. Biol.* 8 [1998]: R812–R822.)

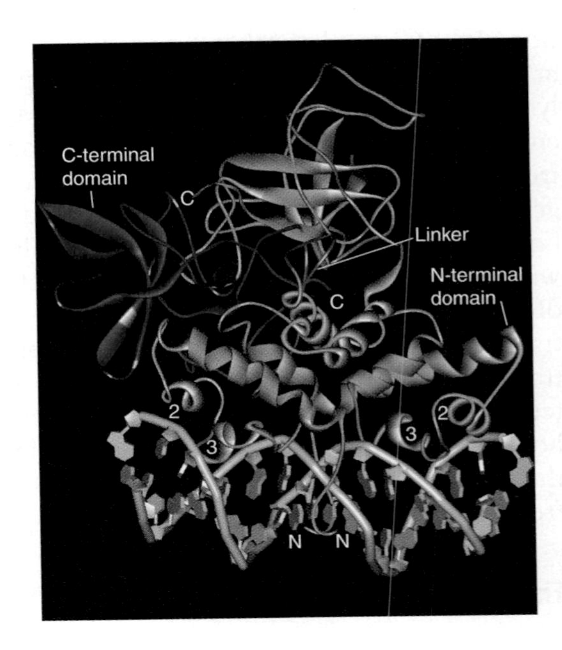

FIGURE 16.43 CI regulator•operator DNA complex. One protein homodimer subunit is shown as a solid bronze ribbon structure. The other subunit is also shown as a solid ribbon structure. Its N-terminal domain, which interacts with the operator DNA, is light blue. Its C-terminal domain, which contains oligomerization sites that are required for dimerization, is dark blue. The linker region between the N- and C-terminal domains is green. N- and C-terminal ends of each subunit are indicated by N and C, respectively. Recognition and stabilization helices in each N-terminal domain are labeled 3 and 2, respectively. DNA is shown as a gray tube and ring structure. The recognition helix inserts into the major groove of the operator DNA. (Structure from Protein Data Bank 3BDN. S. Stayrook, et al., *Nature* 452 [2008]: 1022–1025. Prepared by B. E. Tropp.)

operator sites. The O_R1 and O_R3 sites lie entirely within the P_R and P_{RM} promoters, respectively. The O_R2 site is located between O_R1 and O_R3 and overlaps both P_{RM} and P_R. The CI regulator binds to all three operator sites but its affinity for O_R1 is about ten times greater than its affinity for the other two operator sites. The nature of this binding will now be described.

The CI regulator contains 236 amino acids and folds into two nearly equal size domains that are connected by a 40 amino acid linker (FIGURE 16.42a). The CTD contains oligomerization sites that are required for dimerization (FIGURE 16.42b) and higher order oligomerization (see below). The N-terminal domain contains five successive α helices. Helices 2 and 3 form a helix-turn-helix structure that binds to the operator site (FIGURE 16.43). Helix 3, the recognition helix, is positioned in the major groove of the DNA, allowing amino acids within

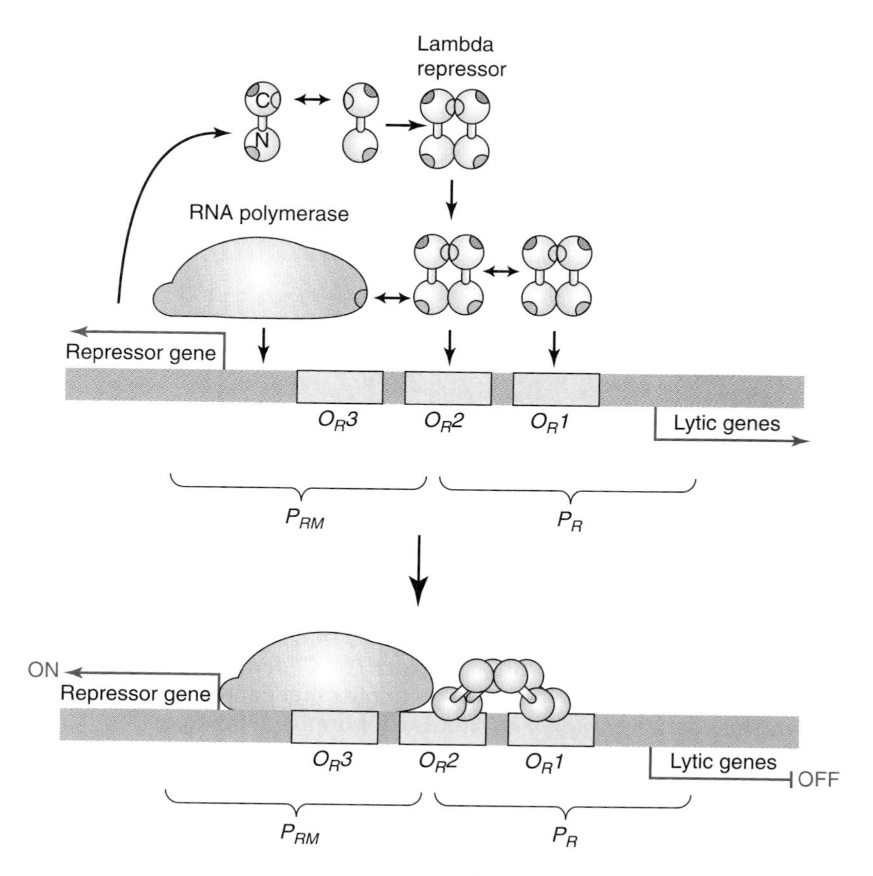

FIGURE 16.44 The phage lambda switch. Two CI regulator dimers bind cooperatively to the adjacent operator sites, O_R1 and O_R2. The dimer at O_R1 prevents RNA polymerase from binding to P_R and thereby blocks transcription initiation from this promoter, which would otherwise function as a strong promoter for a transcript required for the lytic pathway. Simultaneously, the dimer at O_R2 activates the weak promoter of the repressor gene itself, P_{RM}, by contacting RNA polymerase. At higher concentrations, a CI dimer also binds to O_R3 and turns off transcription from P_{RM}, preventing transcription of the *cI* gene. The three surfaces on the repressor involved in the three examples of cooperativity—repressor dimerization, interaction between dimers, and interaction with polymerase to activate P_{RM}—are shaded in yellow, magenta, and green, respectively. (Modified from *Curr. Biol.*, vol. 8, M. Ptashne and A. Gann, Imposing specificity by localization . . ., pp. R812–R822, copyright 1998, with permission from Elsevier [http://www.sciencedirect.com/science/journal/09609822].)

this helix to make specific contacts with bases in the DNA. Helix 2, the stabilization helix, helps to stabilize the protein-DNA complex.

The CI dimer binds first to O_R1, creating a favorable environment for a second CI dimer to bind to O_R2 (**FIGURE 16.44**). The C terminal domains of the dimer at O_R1 make contact with their counterparts in the dimer at O_R2 to form a tetramer. The dimer bound to the O_R1 site functions as a repressor, blocking RNA polymerase from binding to P_R so that RNA polymerase cannot initiate transcription from this otherwise strong promoter. This repression prevents the formation of early transcripts required for the lytic pathway. The dimer bound to O_R2 functions as an activator, stimulating the transcription from P_{RM} so that RNA polymerase makes CI transcripts. However, when the CI level becomes too high because of P_{RM} activation, a CI dimer binds to O_R3, turning off CI production (see below). Thus, the CI regulatory protein is an autoregulator, activating its own gene when present at low concentrations but repressing it when present at high concentrations.

The CI regulator also inhibits transcription initiation from P_L by binding to two operator sites, O_L1 and O_L2, in the O_L operator (**FIGURE 16.45a**). One CI dimer binds to O_L1 and another to O_L2. Once again, binding is cooperative and leads to the formation of a tetramer, which is stabilized by interactions between C terminal domains. The bound CI regulators repress synthesis of the early transcript that codes for the N protein. Because the O_L operator is separated from

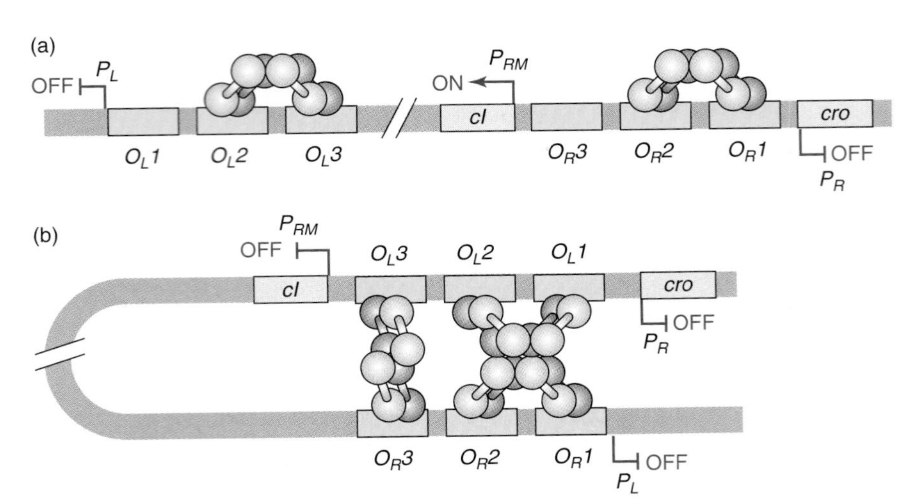

FIGURE 16.45 CI regulator interactions at O_L and O_R. (a) Each CI dimer is shown with one orange subunit and one blue subunit. Two CI dimers bind cooperatively to O_R and another two bind cooperatively to O_L. The dimer pair that binds cooperatively to O_R1 and O_R2 represses transcription from P_R, while the dimer pair that binds cooperatively to O_L1 and O_L2 represses transcription from P_L. (b) Interactions between the CI dimer pairs at O_R and O_L lead to the formation of an octameric complex that contains a 2.4 kb DNA loop. This looped complex facilitates cooperative binding of another pair of CI dimers to O_L3 and O_R3, resulting in the formation of the tetrameric complex that is shown. The CI dimer bound at O_R3 represses transcription from P_{RM} (negative autoregulation). (Adapted from A. Hochschild, *Curr. Biol.* 12 [2002]: R87–R89.)

the O_R operator by 2.3 kb, it was assumed that each operator functioned independently of the other. However, the DNA between the two operators forms a loop that allows the C-terminal domains of the tetramer bound to O_R to interact with the C-terminal domains of the tetramer bound O_L to form a CI octamer (**FIGURE 16.45b**).

The discovery that a cooperatively bound pair of CI dimers at O_R interacts with a cooperatively bound pair at O_L to form an octameric complex with a 2.3 kb DNA loop provides important new insights into how CI regulates its own synthesis. CI dimer affinity for O_R3 is too low to explain physiological transcription repression at P_{RM}. The looped complex, however, allows cooperative dimer binding by locking O_L3 and O_R3 into positions that are near enough so that a CI dimer binding to O_L3 can interact with another dimer binding to O_R3 to form a tetramer. This cooperative binding, which is mediated by oligomerization sites in the C-terminal domains, greatly increases the CI dimer's affinity for O_R3, allowing autorepression to take place at physiological CI concentrations. The looped complex also explains why O_L3 has been conserved during evolution: It is required to ensure that the CI dimer binds to O_R3.

Ultraviolet light induces the λ prophage to enter the lytic pathway.

The λ lysogen maintains about 30 CI dimers per prophage, a ratio that is high enough to ensure strong repression of the early promoters, P_L

and P_R, under normal growth conditions but not so high as to prevent CI inactivation when the lysogen's survival is threatened by exposure to a DNA-damaging agent such as ultraviolet light. DNA damaged by ultraviolet light activates RecA, a host cell protein that is part of the DNA repair and recombination system (see Chapters 12 and 13). Once activated, RecA stimulates a latent protease in CI. This protease activity cleaves CI in the linker that connects the N- and C-terminal domains. Hence, CI becomes a self-cleaving protease.

Once detached from their C-terminal domains, N-terminal domains have much lower affinities for O_L and O_R operator sites than they did before cleavage because they have lost their ability to bind in a cooperative manner. Hence, the N-terminal domains dissociate from the operator sites, freeing RNA polymerase to begin making transcripts for the N and Cro proteins. The former functions as an antiterminator, allowing transcription through the t_L and t_R transcription terminators, and the latter binds to the O_R3 site to turn off *cI* gene transcription. Further steps in the gene expression cascade lead to the transcription of the *xis* and *int* genes. The products of these two genes, excisionase and integrase, catalyze the excision of the prophage from the bacterial chromosome (see Chapter 14). The free circular λ DNA enters the lytic pathway to produce new phage particles, which are released by cell lysis.

16.8 Messenger RNA Degradation

Bacterial mRNA molecules are rapidly degraded.

Intracellular mRNA content depends on the rate at which mRNA is degraded as well as the rate at which it is synthesized. The typical *E. coli* mRNA molecule has a half-life of about two to three minutes. Some mRNA half-lives, however, are about 30 minutes while others are as short as a few seconds. Bacteria derive an important advantage from rapid mRNA degradation. If mRNA molecules were stable, newly synthesized mRNA molecules would have to compete with preexisting mRNA molecules for the protein synthetic machinery. Such competition would limit the cell's ability to synthesize proteins that are needed to respond to physiological changes. Rapid mRNA degradation frees the protein synthetic machinery to translate the newly formed mRNA molecules, which are formed in response to the cell's changing physiological requirements.

Until recently, investigators interested in bacterial mRNA degradation have tended to use *E. coli* as a model system, assuming that results obtained from *E. coli* would also apply to all other bacteria. Recent studies, however, show that the mRNA degradation pathway used by *Bacillus subtilis* differs from that used by *E. coli*. Thus, there are at least two different bacterial pathways for mRNA degradation and still others may exist. We will focus our attention on the better studied *E. coli* pathway.

E. coli use **ribonucleases (RNases)** and **polynucleotide phosphorylase** to degrade mRNA. RNases, which cleave phosphodiester bonds by adding water to them, can be divided into two classes. Endoribonucleases cut internal phosphodiester bonds to produce RNA fragments and exoribonucleases start cutting at one end of the RNA chain and continue down the chain, removing one nucleoside monophosphate at a time. To date, the only kind of exoribonucleases found in *E. coli* begin cutting at the 3′-end and move in a 3′→5′ direction. Polynucleotide phosphorylase degrades RNA by adding phosphate groups to phosphodiester bonds to form nucleoside diphosphates. Phosphorolytic cleavage begins at the 3′-end and continues sequentially in a 3′→5′ direction. We now examine how these and other enzymes participate in the *E. coli* mRNA degradation pathway.

E. coli mRNA has a triphosphate at its 5′ end that helps to protect the RNA from degradative enzymes (see below). Joel G. Belasco and coworkers have recently discovered a new enzyme, RNA pyrophosphohydrolase (RppH), which efficiently cleaves pyrophosphate from the 5′ end of triphosphorylated RNA. This cleavage increases the rate of mRNA degradation by converting the 5′-triphosphate end that protects mRNA from degradation into a 5′-monophosphate that does not. RNA pyrophosphohydrolase action is hindered when the 5′ end is part of a stem-loop structure.

After its 5′-triphosphate end has been converted to a 5′-monophosphate end, mRNA becomes a substrate for a large multisubunit protein complex called the **RNA degradosome**. The largest polypeptide in the complex, RNase E, is a multidomain protein that acts as both the most important endonuclease in mRNA degradation and an attachment site for other enzymes in the RNA degradosome. Temperature-sensitive *E. coli* mutants for RNase E are not viable at nonpermissive temperatures, indicating RNase E is an essential protein. Its essential function, however, may be to process tRNA, rRNA, or some other indispensable RNA. Bacterial cells may be able to use other ribonucleases to degrade mRNA when RNase E is missing. RNase E forms a homotetramer that is a dimer of dimers. Each polypeptide subunit can be divided into an N-terminal and C-terminal half, each containing about 500 amino acid resides. N-terminal halves of the four subunits interact to form the enzyme's catalytic core. The C-terminal halves, which are intrinsically disordered, have binding sites for at least three different proteins that are part of the RNA degradosome. Based on stoichiometric calculations, each C-terminal region appears to bind an RhlB helicase monomer, a polynucleotide phosphorylase homotrimer, and an enolase homodimer (FIGURE 16.46). The mRNA's 5′-monophosphate end fits in a pocket in the N-terminal half of one RNase E subunit, facilitating entry of the adjacent single-stranded RNA into a channel that directs the RNA to the cleavage site on another subunit. The enzyme cleaves mRNA on the 5′ side of an AU dinucleotide in

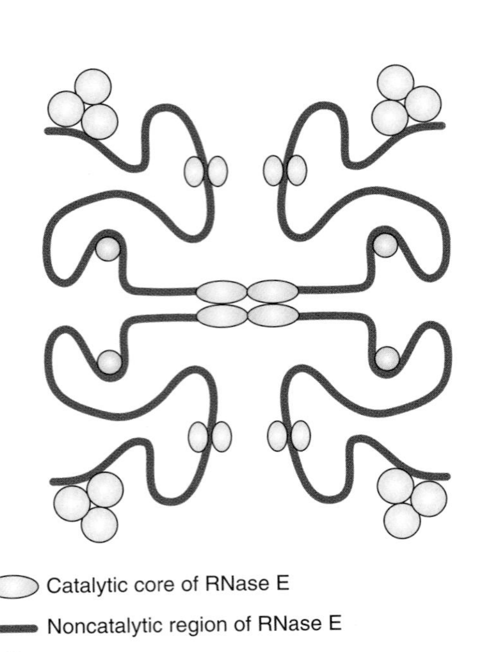

FIGURE 16.46 **A structural model for the RNA degradosome.** RNase E has an N-terminal domain of about 500 residues and a C-terminal domain of about equal size. Four RNase E molecules combine to form a homotetramer (a dimer of dimers). The N-terminal domains interact to form the ribonucleases catalytic core. Each C-terminal domain, which is intrinsically disordered, appears to bind an RhlB helicase monomer, a polynucleotide phosphorylase homotrimer, and an enolase homodimer. The entire protein complex is called an RNA degradosome. (Reproduced from A. J. Carpousis, *Annu. Rev. Microbiol.* 61 [2007]: 71–87. Copyright 2007 by Annual Reviews, Inc. Reproduced with permission of Annual Reviews, Inc., in the format Textbook via Copyright Clearance Center.)

single-stranded segments but does not appear to have other sequence constraints.

Polynucleotide phosphorylase degrades the RNA fragments produced by RNase E by phosphorolytic cleavage, which begins at the 3′-end of the polyribonucleotide chain and continues sequentially in a 3′→5′ direction. A phosphate group, rather than a water molecule, is added across the phosphodiester bond to form a nucleoside diphosphate. The RNA fragment's 3′-end must be single-stranded for polynucleotide phosphorylase to grab hold of an RNA chain. Therefore, a stem-loop structure such as a rho-independent terminator interferes with polynucleotide phosphorylase activity. *E. coli* solves this problem in a remarkable way. The bacteria create a binding site for polynucleotide phosphorylase by adding 14 to 60 adenylate groups to the 3′-end of the mRNA molecule. A single enzyme, poly(A) polymerase, is responsible for transferring these adenylate groups from ATP to form the poly(A) tail. Once bound to the poly(A) tail, polynucleotide phosphorylase can degrade the mRNA in a 3′→5′ direction. Remarkably, poly(A) tails seem to decrease mRNA stability in bacteria and increase mRNA stability in eukaryotes. Mutants that lack polynucleotide phosphorylase are viable if they have RNase II, a 3′→5′ exonuclease but mutants that lack both enzyme activities are not viable. This observation suggests that 3′→5′ RNA degradation is essential for cell survival and that either polynucleotide phosphorylase or RNase II can perform this essential function.

RhlB is an ATP-dependent helicase that unwinds double-stranded RNA that might otherwise interfere with the activity of RNase E or other degradative proteins that are specific for single-stranded RNA. Mutants that cannot make RhlB and other related RNA helicases are viable, making it difficult to determine the precise function of these enzymes in mRNA degradation. The enolase's role in mRNA degradation is even less clear. Its normal physiological role is to convert 2-phosphoglycerate to phosphoenol pyruvate during glycolysis. This function has led some investigators to speculate that enolase may somehow make the RNA degradosome sensitive to the cell's energy state.

Although the study of mRNA degradation in *E. coli* is still a work in progress, it is possible to propose a pathway for the degradation process based on what is known about the enzymes that form the RNA degradosome and other enzymes that appear to participate in mRNA degradation. For simplicity the pathway shown in **FIGURE 16.47** includes only a single stem-loop structure at the 3′ end of the mRNA molecule. *B. subtilis* lacks RNase E. Other endoribonucleases and a recently discovered 5′→3′ exonuclease appear to play important roles in *B. subtilis* mRNA degradation.

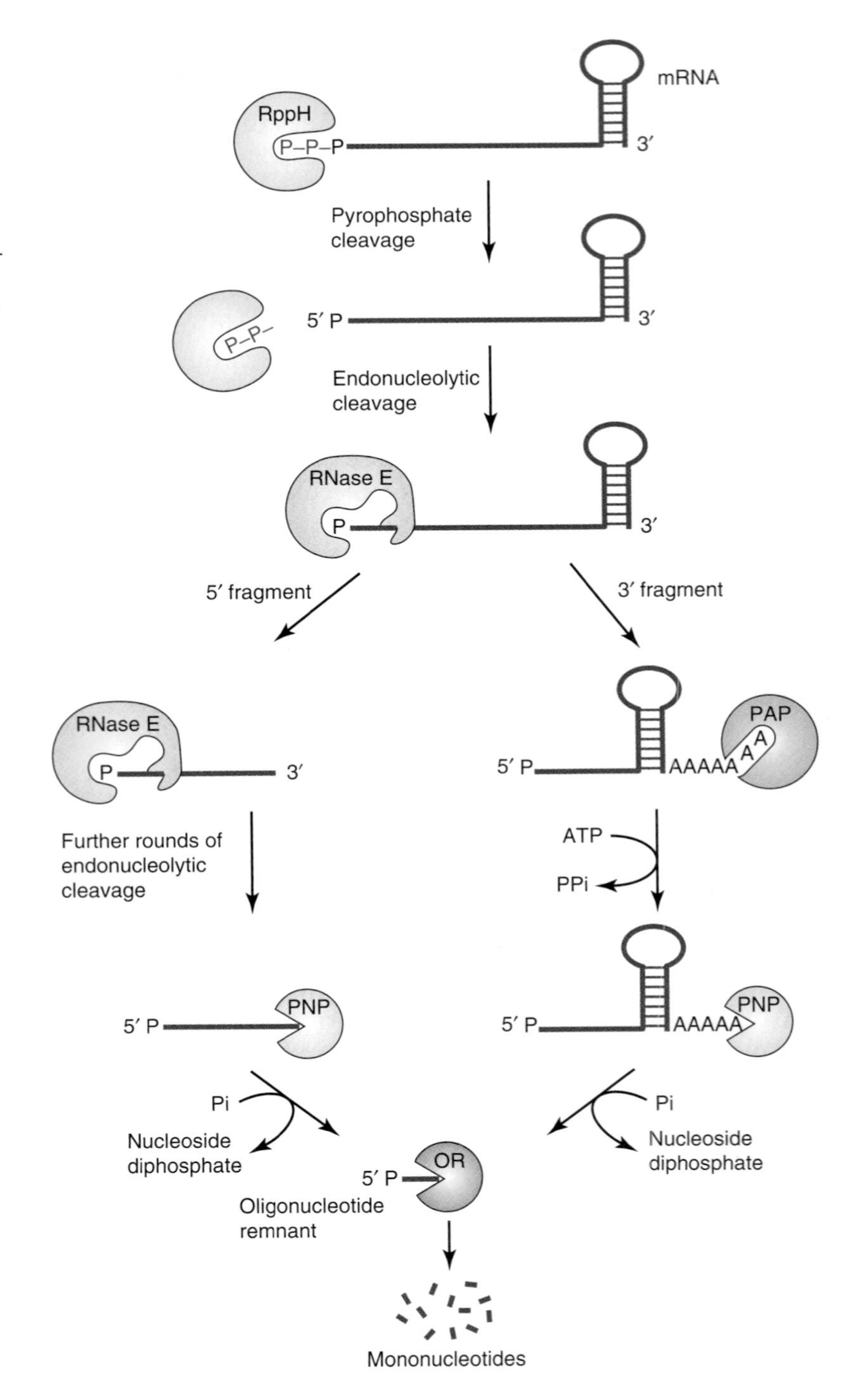

FIGURE 16.47 ***E. coli* mRNA degradation pathway.** The abbreviations used for the enzymes that participate in mRNA degradation are as follows: RppH, the RppH helicase; PNP, polynucleotide phosphorylase; PAP, poly(A) polymerase; and OR, oligoribonuclease. For simplicity, the full RNA degradosome is not shown but the established functions of two RNA degradosome components, RNase E and polynucleotide phosphorylase are shown. Also, only a single stem-loop structure is shown in the mRNA. (Adapted from J. Richards, et al., *Biochim. Biophys. Acta* 1779 [2008]: 574–582.)

16.9 Ribosomal RNA and Transfer RNA Synthesis

Bacterial ribosomes are made of a large subunit with a 23S and 5S RNA and a small subunit with 16S RNA.

Ribosomes serve as the cells protein synthetic factories. A typical bacterial ribosome has a diameter of about 25 nm, a molecular mass of 2.5×10^6 Da, and contains about twice the amount of RNA by mass

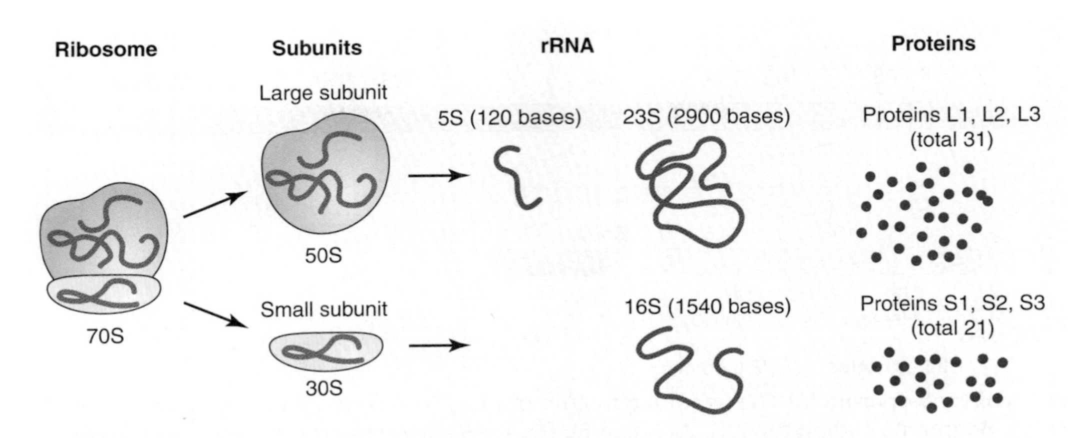

FIGURE 16.48 The ribosome and its components. (Adapted from an illustration by Kenneth G. Wilson, Department of Botany, Miami University.)

as protein. Each bacterial cell requires about 70,000 ribosomes to produce the proteins it needs. The production of such a large number of ribosomes makes enormous demands on a cell's energy resources and so cells have evolved regulatory mechanisms that allow them to regulate ribosome synthesis according to their needs. Ribosomal RNA (rRNA) synthesis is the rate limiting process in ribosome production.

Isolated bacterial ribosomes dissociate into large and small subunits when divalent cations are removed from the surrounding medium. Thus, a bacterial ribosome with a sedimentation coefficient of 70S dissociates into a 30S and a 50S subunit (FIGURE 16.48). Each *E. coli* 30S subunit contains one 16S RNA and 21 different polypeptide molecules while each 50S subunit contains a 23S RNA, a 5S RNA, and 31 different polypeptide molecules. The ribosomal subunits have distinct three-dimensional structures (FIGURE 16.49, which are examined in more detail in Chapter 23).

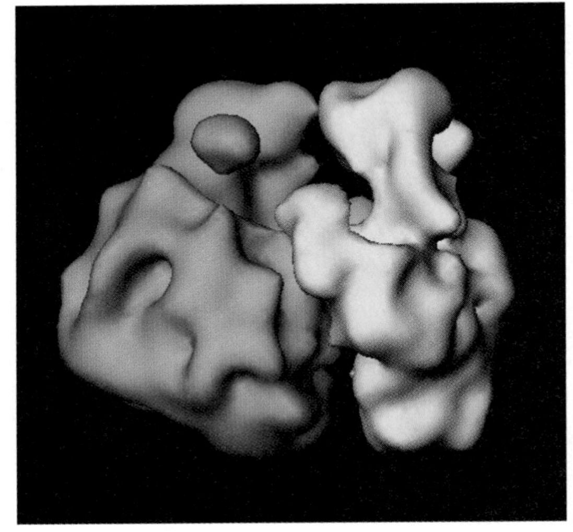

FIGURE 16.49 **Cryo-electron micrograph of the *E. coli* ribosome.** The 30S subunit is in yellow and the 50S subunit in blue. (Reproduced from J. Frank, *Am. Sci.* 86 [1998]: 428–439. Photo courtesy of Joachim Frank, Columbia University.)

E. coli has seven rRNA operons, each coding for a 16S, 23S, and 5S RNA.

E. coli uses seven nearly identical rRNA operons—*rrnA*, *rrnB*, *rrnC*, *rrnD*, *rrnE*, *rrnG*, and *rrnH*—that are scattered throughout the chromosome to make the large number of ribosomes it requires. FIGURE 16.50a provides key features of the rRNA operons. Each operon contains two promoters, P1 and P2, which are separated by about 120 bp. The 16S, 23S, and 5S RNA genes follow the promoters in that order. Transfer RNA genes are located between the 16S and 23S rRNA genes in all seven *E. coli* rRNA operons. Some rRNA operons also have tRNA genes after the 5S gene. Although many tRNA genes are part of an rRNA operon, most are not. Once transcription of an rRNA operon is complete, the primary transcript is processed to make mature rRNA and tRNA molecules.

These seven rRNA operons are responsible for about 50% of the cell's total RNA synthesis even though they account for less than 0.5% of the total operons in *E. coli*. It therefore seems reasonable

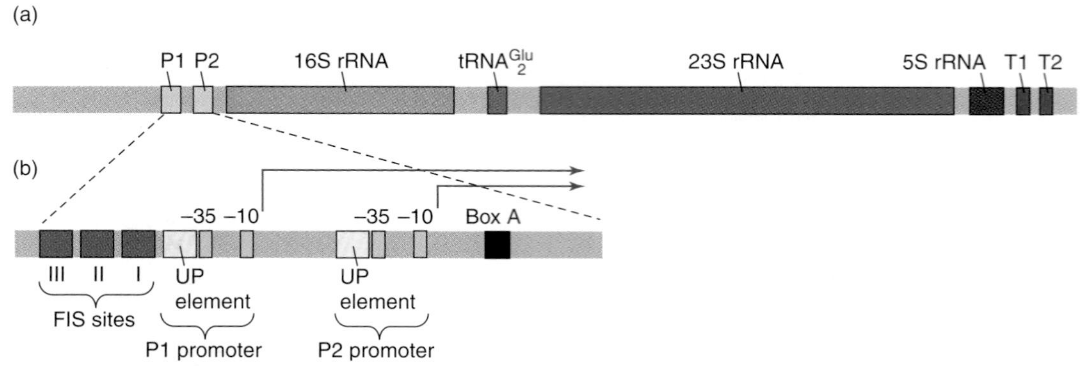

FIGURE 16.50 **The *E. coli rrnB* operon.** (a) The general structure of the *E. coli rrnB* operon, which codes for the mature ribosomal RNA molecules. Segments coding for 16S, 23S, and 5S RNA are shown in orange, blue, and brown, respectively. A segment that codes for $tRNA^{Glu}$ (magenta) is located between the segments that code for the 16S and 23S rRNAs. Spacer regions are shown in gray. The two promoters, P1 and P2, are shown in cyan, and the two terminators, T1 and T2, are shown in red. (b) The promoter region is expanded to show the FIS binding sites I, II, and III (purple) upstream of P1 and the UP elements (yellow). The −10 and −35 consensus hexamers (green) and the BoxA sequence (black) required for antitermination are also indicated. Transcription initiation sites are shown by lines with arrows. (Adapted from R. L. Gourse, et al., *Annu. Rev. Microbiol.* 50 [1996]: 645–677.)

to suppose that *rrn* promoters are much stronger than the typical *E. coli* promoter. Considerable evidence supports this supposition. Even though each *rrn* operon has two promoters, P1 and P2, only the P1 promoter appears to regulate the initiation of *rrn* transcription under most growth conditions; it will be the focus of our attention.

Part of the reason that the *rrn* P1 promoter is so strong is that its −10 and −35 box sequences closely match the consensus sequences. In fact, the −10 box in each of the seven *rrn* P1 promoters is the consensus sequence TATAAT. The −35 boxes vary slightly but all are good matches to consensus sequence. The spacing between the −10 and −35 hexamers is only 16 bp rather than the 17 bp found in most strong promoters. Increasing the spacing to 17 bp by recombinant DNA techniques creates a much stronger promoter. This leads one to ask why the P1 promoters did not evolve to have the 17 bp spacing. One possibility is that the P1 promoters evolved to allow for their own regulation and not just for maximal promoter activity.

A promoter upstream element (UP element) increases *rrn* transcription.

The region corresponding to the −10 and −35 boxes is designated the **core promoter**. A region that is just upstream of the core promoter, called the **promoter upstream element** or **UP element** (FIGURE 16.50b), makes P1 a much stronger promoter than it would otherwise be, increasing transcription by about 340-fold. Footprinting reveals that RNA polymerase holoenzyme protects two sub-sites within the UP element. The first of these, the proximal sub-site extends from position −38 to −46 and has the sequence 5′-AAAAAARNR-3′ and

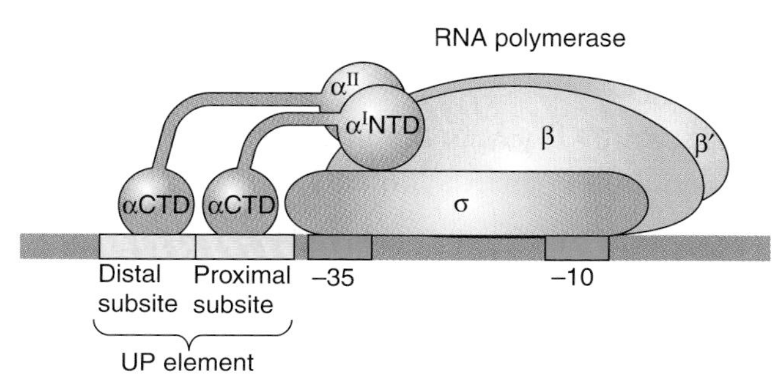

FIGURE 16.51 RNA polymerase holoenzyme interaction with UP element. (Adapted from R. L. Gourse, W. Ross, and T. Gaal, *Mol. Microbiol.* 37 [2000]: 687–695.)

the second, the distal sub-site extends from position −47 to −59 and has the sequence 5′-NNAWWWWWTTTTTN-3′ (N = any base; R = purine; and W = A or T). The proximal and distal sub-sites can each stimulate transcription in the absence of the other. Acting on its own, the proximal sub-site causes a 170-fold transcription increase. The distal sub-site, which is less effective, causes a 16-fold transcription increase. The UP element also stimulates transcription of other promoters when placed before them by recombinant DNA techniques. For instance, *lacZ* transcription is greatly increased by placing an UP element before the *lac* promoter.

RNA polymerase holoenzyme makes contact with the UP element through its α subunit and more specifically through the α C-terminal domain (αCTD). A holoenzyme with mutant α subunits that are missing their C-terminal domains will neither make contact with the UP element nor use the UP element to stimulate transcription. FIGURE 16.51 is a cartoon that shows the interaction between the RNA polymerase holoenzyme and the UP element.

Three Fis protein binding sites increase *rrn* transcription.

A small protein called the Fis protein (factor for inversion stimulation) stimulates *rrn* operon transcription by another five- to tenfold. This protein's unusual name derives from the fact that it was originally discovered as a host factor involved in several site-specific recombination events. The Fis protein binds to three sequences upstream from the UP element that are designated Site I, Site II, and Site III and centered at positions −71, −102, and −143, respectively (Figure 16.50b). A Fis protein molecule binds to each of these sites and somehow interacts with RNA polymerase holoenzyme. This interaction is mutually dependent because the binding of either protein to the *rrn* promoter stimulates the binding of the other. The Fis protein does not appear to depend on the presence of a functional UP element. Further studies are required to learn the exact nature of the interaction between RNA polymerase holoenzyme and the Fis protein.

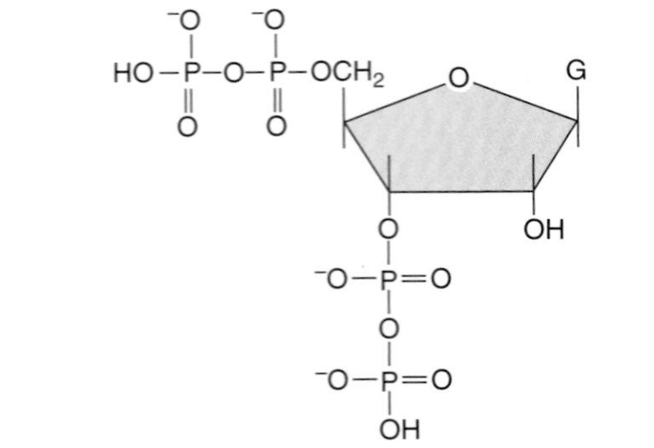

Guanosine-5′-diphosphate-3′-diphosphate (ppGpp)

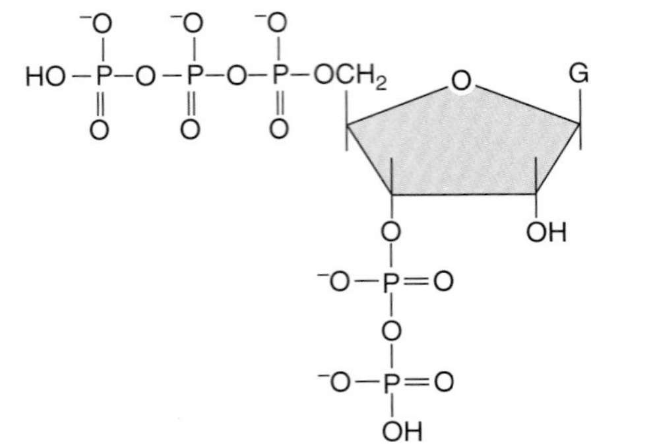

Guanosine-5′-triphosphate-3′-diphosphate (pppGpp)

FIGURE 16.52 Guanosine-5′-diphosphate-3′-diphosphate (ppGpp) and guanosine-5′-triphosphate-3′-diphosphate (pppGpp).

16.10 Regulation of Ribosome Synthesis

Amino acid starvation leads to the production of guanine nucleotides that inhibit rRNA synthesis.

If a growing *E. coli* culture is deprived of a required amino acid then protein, rRNA, and tRNA synthesis stop. The effect of amino acid deprivation on protein synthesis is explained by the translation machinery's inability to move beyond codons for the missing amino acid (or aminoacyl-tRNA). The relationship between amino acid deprivation and the cessation of rRNA and tRNA synthesis, known as the **stringent response**, is not obvious.

E. coli that exhibit stringent control of rRNA and tRNA synthesis synthesize two unusual guanine ribonucleotides: guanosine-5′-diphosphate-3′-diphosphate (ppGpp) and guanosine-5′-triphosphate-3′-diphosphate (pppGpp) (FIGURE 16.52). Factors responsible for the stringent response have been examined by studying mutants in which rRNA and tRNA syntheses continue unabated during amino acid starvation. The phenotype of these mutants is described as **relaxed.** The mutations map in a few genes designated *rel.* Here we examine only one of these genes, the *relA* gene.

The *relA* gene product, a protein known as the stringent factor, is an enzyme that catalyzes pyrophosphoryl group transfer from ATP to GTP (or GDP) to form pppGpp (or ppGpp). The stringent factor is located in about 1 in 200 ribosomes. During normal protein synthesis, the stringent factor is inactive and virtually no (p)ppGpp is synthesized. Activation of the stringent factor occurs only when free tRNA molecules replace aminoacyl-tRNA molecules in the ribosome. This rarely occurs when amino acids are present because most of the tRNA molecules would be charged with an amino acid. When cells are starved for amino acids, however, the tRNA species that normally carries that amino acid will be uncharged. Thus, amino acid deprivation leads to (p)ppGpp production.

A second *E. coli* gene, *spoT*, codes for a protein that removes the 3′-pyrophosphoryl group from pppGpp (or ppGpp) to form GTP (or GDP). The stringent factor is not the only bacterial enzyme that catalyzes (p)ppGpp formation because cells with a null mutation in *relA* still synthesize small quantities of (p)ppGpp. Remarkably, the *spoT* gene product also has the ability to transfer a pyrophosphoryl group from ATP to GTP to form (p)ppGpp. Double mutants with a *relA spoT* genotype do not synthesize (p)ppGpp. Both pppGpp and ppGpp inhibit rRNA and tRNA synthesis *in vivo*. Initial efforts to show that (p)ppGpp also inhibits rRNA and tRNA using an *in vitro* transcription system were unsuccessful. The problem was solved in 2004 when Richard L. Gourse and coworkers showed that adding the protein DksA to the *in vitro* transcription system yielded results that were consistent with the *in vivo* results. DksA's name derives from the fact that it was originally discovered as a *dnaK* suppressor. Like GreA and GreB (see Chapter 15), which it resembles in structure but not amino acid sequence, DksA appears to extend a globular domain into the secondary channel of RNA polymerase. (p)ppGpp appears to bind near the active site. Further work is required to establish the nature of the interactions that exist among DksA, (p)ppGpp, and RNA polymerase. (p)ppGpp's influence extends far beyond its ability to regulate rRNA and tRNA synthesis. The nucleotide increases the expression of genes specific for amino acid synthesis when cells that are deprived of the amino acids and influences replication, lipid metabolism, and protein synthesis.

E. coli rRNA and tRNA syntheses increase with growth rate.

The growth rates of all bacterial species vary with the composition of the growth medium. In a minimal medium with a good carbon source such as glucose, *E. coli* cells divide roughly every 45 minutes at 37°C. With a poorer carbon source such as proline, the doubling time is about 500 minutes. In rich media containing glucose, amino acids, purines, pyrimidines, vitamins, and fatty acids, a cell does not have to synthesize these substances and hence it can grow very rapidly. Typically, the generation time in a rich medium is about 25 minutes.

A ribosome has a limited capacity for protein synthesis (15 amino acids per second at 37°C, independent of carbon source). Thus, at different overall growth rates, at which different rates of protein synthesis occur, the number of ribosomes per cell varies (Table 16.4). If a bacterial culture is transferred from a growth medium in which growth is rapid to one in which growth is slow (this is called a "downshift"), the ribosome content of each cell decreases from the higher value for the rapid medium to the lower value characteristic of the slow medium. This decrease is reasonable because otherwise each cell would possess more ribosomes than it would need. The decrease is accomplished by allowing DNA synthesis to proceed without rRNA synthesis. Each slowly growing cell must sense some signal and this signal must inhibit transcription of the rRNA genes. The fact that in any growth medium the rate of synthesis of rRNA is proportional to the rate of production of ribosomes makes it clear that synthesis of rRNA is regulated.

One hypothesis for rRNA regulation postulates that (1) ribosomes are made in slight excess of that which is needed to attain the appropriate rate of protein synthesis, and (2) free nontranslating ribosomes inhibit (directly or indirectly) the synthesis of rRNA. Evidence supporting some type of ribosome feedback regulation comes from comparing rRNA synthesis in *E. coli* strains with different numbers of rRNA operons (Table 16.5).

TABLE 16.4 Some Characteristics of *E. coli* Growing at Different Growth Rates

Doubling Time (minutes)	Ribosomes per Cell	Ribosomes per DNA Molecule*
25	69,750	15,500 (4.5)
50	16,500	6,800 (2.4)
100	7,250	4,200 (1.7)
300	2,000	1,450 (1.7)

*The number in parentheses is the number of DNA molecules per cell.

TABLE 16.5 Amount of rRNA Made by Bacterial Cultures Containing Different Numbers of rRNA Operons

	Number of Operons per Cell				
Strain	Chromosomal	Plasmid (normal rRNA)	Plasmid (defective rRNA)	Amount of rRNA Made per Operon*	Total rRNA Made*
1	7	0	0	1	7
2	7	7	0	0.5	7
3	7	0	7	1	14

*Arbitrary units

Strain 1 contained the usual seven rRNA operons. Strain 2 contained the seven chromosomal rRNA operons plus seven copies of a plasmid with a single copy of one rRNA operon (marked to make it distinguishable from the chromosomal operons). Strain 3 was the same as Strain 2 except that the rRNA made by the seven plasmids had a deletion that prevented it from being assembled into ribosomes. The three strains were cultured in the same medium. The seven operons in Strain 1 each made roughly the same amount of rRNA, which was assembled into ribosomes. Even though Strain 2 had twice as many rRNA operons as Strain 1, it made the same amount of rRNA per cell (not twice as much). Furthermore, half of the rRNA was transcribed from the chromosomal operons and half from the plasmids. Thus, each of the 14 operons was being transcribed at half the normal rate, indicating that repression of some kind was occurring. Strain 3 made twice as much rRNA per cell as Strains 1 or 2. The plasmid rRNA operons were responsible for half of this rRNA. However, the plasmid-encoded (defective) rRNA did not contribute to regulating transcription of rRNA operons, which implies that only rRNA that can be assembled into ribosomes can lead to repression.

The interpretation of this and other experiments is that: (1) all functional rRNA is assembled into ribosomes; (2) ribosomes are made in slight excess; and (3) the excess (nontranslating) ribosomes repress transcription of the rRNA operons. How the repression occurs (e.g., direct interaction of a free ribosome with a promoter versus some indirect effect) is not yet known.

Because ribosomes function only in protein synthesis, production of tRNA also should be coupled to that of ribosome assembly. Many tRNA genes are part of the rRNA operons, so synthesis of these tRNA molecules is necessarily connected to that of rRNA. Experiments similar to those described above showing that nontranslating ribosomes control transcription of the rRNA operons have also been carried out for the tRNA species that are not made by the rRNA operons. Transcription of these tRNA genes also appears to be regulated by the number of nontranslating ribosomes.

E. coli regulates r-protein synthesis.

A gene dosage experiment also shows that the synthesis of ribosomal proteins (r-proteins) is not regulated at the transcriptional level. The genes encoding the 52 r-proteins are organized into 20 operons. Addition of an appropriate plasmid can increase the number of copies of one of these operons in a cell, say, tenfold. This would increase the rate of synthesis of the corresponding mRNA tenfold, but *the rate of production of r-proteins encoded in the mRNA remains that observed with one copy of the operon*. Thus, the amount of the r-proteins that is synthesized is not proportional to the amount of mRNA encoding them, so clearly translation and not transcription is being limited.

An understanding of the mode of translational regulation came from *in vitro* translation experiments. In these experiments, a single species of r-protein mRNA was translated and the inhibitory effect of each r-protein encoded in the mRNA was tested. It was observed that translation of each mRNA could be inhibited by the addition of *one* (a particular one) of the encoded r-proteins. Further analysis showed that the translational repression is a result of binding of that r-protein to a base sequence near the site at which ribosomes initially bind to mRNA. A variant of the *in vitro* experiment completes the story, namely, the addition of the particular rRNA (5S, 18S, or 23S) to which the r-protein binds in the ribosome prevents translational repression. A base sequence analysis of the mRNAs and the rRNA shows a similar sequence and a common stem-and-loop structure in the binding sites of each RNA molecule for a particular r-protein. Competition, therefore, exists between rRNA and r-protein mRNA for a particular r-protein. Binding studies show that each repressing r-protein species binds preferentially to the rRNA; hence, *as long as rRNA is available for ribosome production, r-proteins will bind to rRNA and synthesis of r-proteins will continue*. Studies with each r-protein mRNA show that when translational repression is not occurring, each mRNA is completely translated, yielding equal numbers of each r-protein encoded in a particular mRNA species. Furthermore, translational repression ensures that all r-proteins are synthesized at the same rate and coupled to the synthesis of ribosomes. This means that the synthesis of all r-proteins is ultimately regulated by the rate-limiting component in ribosome synthesis, that is, the rRNA.

16.11 Processing rRNA and tRNA

Bacteria process the primary transcripts for rRNA and tRNA to form the physiologically active RNA molecules.

Physiologically active rRNA and tRNA molecules differ from their primary transcripts (the newly synthesized RNA molecule) in three important respects.

1. Mature rRNA and tRNA molecules are terminated by a 5′-monophosphate rather than the expected triphosphate found at the ends of all primary transcripts.
2. Both rRNA and tRNA molecules are much smaller than the primary transcripts.
3. All tRNA molecules contain bases other than A, G, C, and U, and these "unusual" bases (as they are called) are not present in the original transcript.

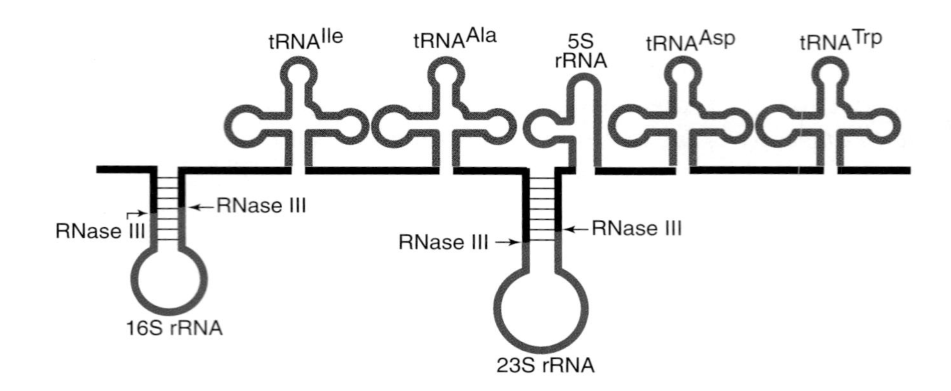

FIGURE 16.53 A pre-rRNA synthesized from information in one of the *E. coli* rRNA operons.

All of these molecular changes are made after transcription by processes collectively called **posttranscriptional modification** or, more commonly, **processing**. We begin by examining rRNA processing.

rRNA Processing

As shown in Figure 16.48, bacterial ribosomes contain 16S rRNA (1541 nucleotides), 23S rRNA (2904 nucleotides), and 5S rRNA (120 nucleotides). These molecules, plus several tRNA molecules, are cleaved from a continuous transcript having more than 5000 nucleotides of known sequence. The seven rRNA operons in *E. coli* differ by the identity of the tRNA molecules and the location of the tRNA sequences with respect to the rRNA sequences. A diagram of one of these transcripts is shown in FIGURE 16.53. This transcript contains four different tRNA molecules and the segments are, from the 5′ end to the 3′ end.

16S rRNA–$tRNA^{Ile}$–$tRNA^{Ala}$–23S rRNA–5S RNA–$tRNA^{Asp}$–$tRNA^{Trp}$

The general pattern, 16S–spacer–23S–5S–spacer, in which tRNA is in the spacer regions, is retained in primary transcripts of other rRNA operons, but the number of tRNAs varies.

Several enzymes that act in sequence cut the rRNA transcript as it is being synthesized. The first cuts are ordinarily made by RNase III, which cleaves double-stranded RNA in the double-stranded stem regions by making two single-strand breaks in complementary sequences but not opposite one another. A 5′-P and a 3′-OH group are generated but these are not the termini of the rRNA. Several enzymes are required to complete the processing. The processing sequence is not the same in all rRNA transcripts or in all bacteria, but the basic pattern of excision of all rRNA components from a single precursor seems to be a general phenomenon. Although RNA sequences are discarded, which appears somewhat wasteful, this mechanism provides a constant ratio of the 16S, 23S, and 5S RNA molecules. Because one molecule of each of these three is present in a ribosome and these molecules are used nowhere else in the cell, efficiency would demand that if each were transcribed separately, some means would be needed to maintain a 1:1:1 ratio. Once present in ribosomes, rRNA molecules are very stable. Mature tRNA molecules are also very stable.

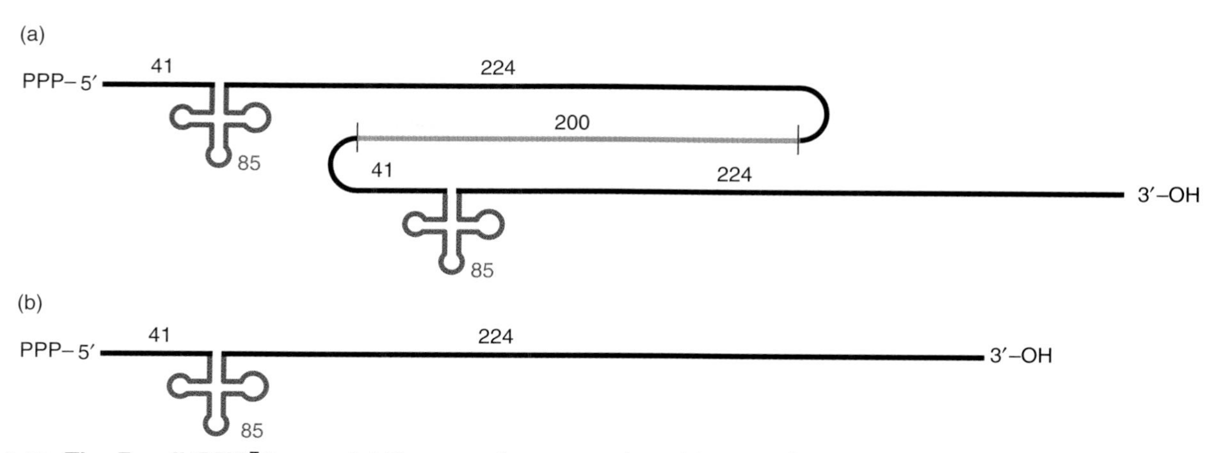

FIGURE 16.54 The *E. coli* tRNA$^{Tyr}_1$ gene. (a) The complete transcript with two adjacent identical tRNA segments and the spacer region (purple) and (b) the single genetic unit used to study processing. The numbers indicate the number of bases in each segment of the transcript. The tRNA sequences are shown in red.

Bacterial cells also process primary transcripts for tRNA. A well-understood tRNA molecule from the point of view of its synthesis is the *E. coli* tRNA$^{Tyr}_1$ molecule, a molecule containing 85 nucleotides of known sequence. In *E. coli*, there are two copies of the tRNA$^{Tyr}_1$ gene, that is, two identical adjacent copies of the DNA from which this tRNA is transcribed. Each gene consists of about 350 (not 85) bp separated by a "spacer" of 200 bp (FIGURE 16.54). The two genes are transcribed as a single RNA molecule that is cut up after transcription is complete. In order to simplify the study of the synthesis of tRNA$^{Tyr}_1$, genetic techniques have been used to create a transcription unit containing only a single 350-bp gene. The transcription start site is 41 bp before (upstream from) the 5′ end of the tRNA base sequence and a stop site is 224 bp downstream from the 3′ terminus of the tRNA.

The primary tRNA transcript is processed by a series of steps shown in FIGURE 16.55, which may be grouped into the following three stages:

1. **Formation of the 3′-OH Terminus** This process involves the action of an endonuclease that recognizes a hairpin loop (I in Figure 16.55) and an exonuclease that recognizes the three-base sequence CCA. After endonuclease digestion at site 1, the seven bases upstream are removed by an exonuclease called RNase D (step 2). This enzyme initially stops two bases short of the CCA terminus, though it later removes these two bases after the 5′ end is processed. This leaves a molecule called **pre-tRNA** that is easily isolated from *E. coli* and that has the structure

   5′-P—(41 bases)—tRNA—(2 bases)—3′-OH

2. **Formation of the 5′-P Terminus** The 5′-P terminus is formed by an enzyme called RNase P, which is probably responsible for generation of the 5′-P terminus of all *E. coli* tRNA molecules. Evidence for this comes from studies with an

FIGURE 16.55 **The stages in processing of *E. coli* tRNA$^{Tyr}_{1}$ gene transcript.** Stages 1 and 2 involve cleavage by endonuclease and nucleotide removal by RNase D. Stage 3 generates the 5′-P end. Stage 4 generates the 3′-OH end (the CCA end). In stage 5, six bases, all in or near the loops of the tRNA molecule, are modified to form pseudouridine (ψ), 2-isopentenyladenosine (2ipA), 2-O-methylguanosine (2mG), and 4-thiouridine (4tU). The continuous sequence that forms the final tRNA molecule is given in black.

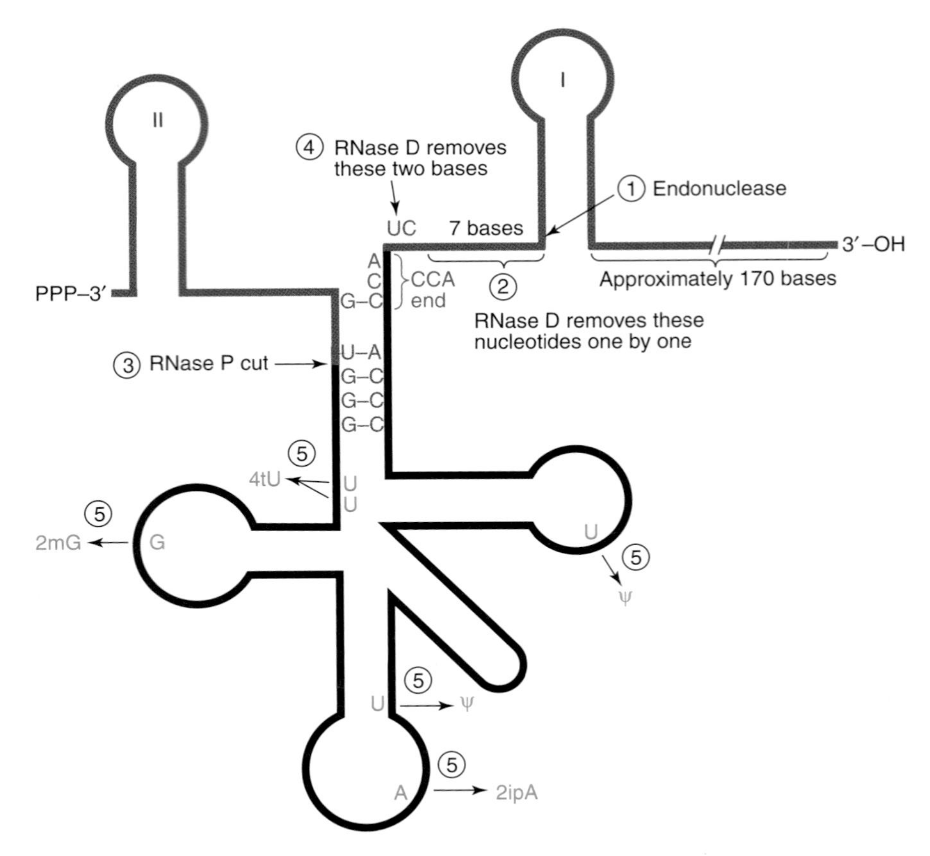

E. coli mutant in which RNase P is inactive at 42°C. When this mutant is grown at 42°C, large RNA molecules accumulate that contain tRNA sequences and the hairpin loop II (Figure 16.55) at the 5′ terminus. RNase P removes the excess RNA from the 5′ end of a precursor molecule by an endonucleolytic cleavage (step 3) that generates the correct 5′ end and a single fragment. RNase P does not recognize a specific base sequence at the cleavage site or anywhere else, but instead responds to the overall three-dimensional conformation of the tRNA molecule with its several hairpin loops and then makes a cut at just the right place. (Evidence for this statement is that base changes in the segment to be removed do not affect the cleavage unless the alteration causes extensive disruption of the stem-and-loop arrangement.) Once the 5′-P terminus has been formed, RNase D removes the two 3′-terminal nucleotides (step 4), leaving a tRNA molecule having the correct length. RNase P is an unusual enzyme in that it contains 86% RNA and 14% protein by weight. Furthermore, *the RNA possesses the catalytic activity* and the protein serves instead to ensure the correct folding of the RNA, in order to maximize the catalytic activity.

3. **Production of the Modified Bases** The final modification is to produce the altered nucleosides in the tRNA (step 5). Enzymes that act only on nucleosides at specific positions in tRNA

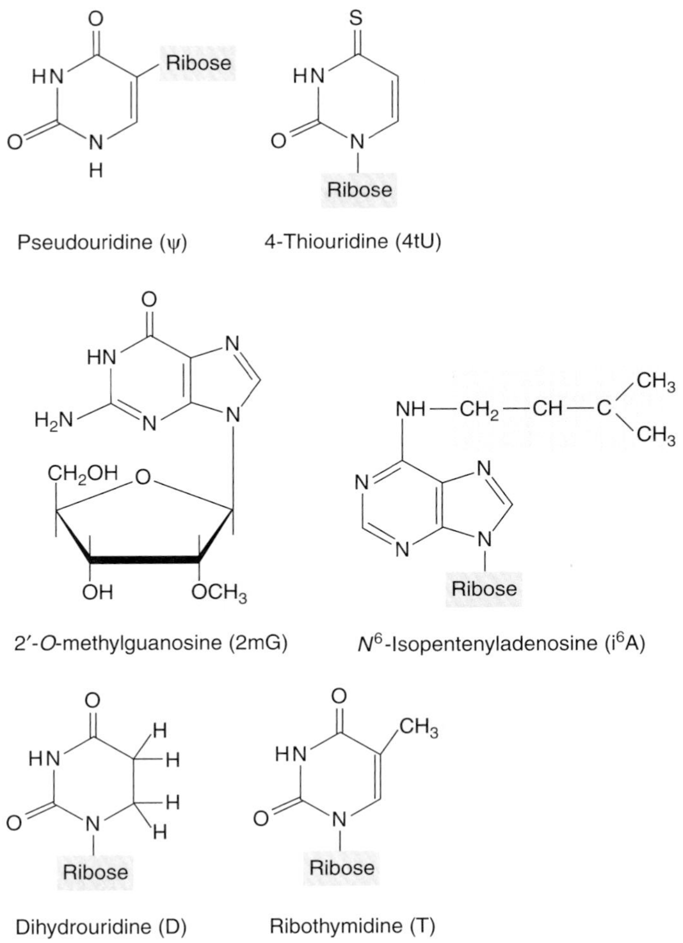

Pseudouridine (ψ) 4-Thiouridine (4tU)

2′-*O*-methylguanosine (2mG) N^6-Isopentenyladenosine (i^6A)

Dihydrouridine (D) Ribothymidine (T)

FIGURE 16.56 Some of the modified nucleosides that are present in tRNA molecules.

produce the necessary changes. Among the posttranscriptional changes that take place in tRNA are the following: uridines are converted to pseudouridine (Ψ), ribothymidine, dihydrouridine, and 4-thiouridines (4tU); guanosine is converted to 2′-O-methylguanosine (2mG); and adenosine is converted to isopentenyladenosine (2ipA) (**FIGURE 16.56**).

All tRNA molecules are terminated by CCA-3′-OH. The precursor shown in Figure 16.55 contains this sequence, so the terminus is generated by the appropriate cut. However, the precursors of some tRNA molecules lack a terminal CCA. With these molecules the CCA is added by the enzyme **tRNA nucleotidyl transferase.**

Multiple copies of a particular tRNA molecule are commonly found in a single transcription unit; for example, there are four copies of one of the $tRNA^{Leu}$ molecules in its precursor molecule. The occurrence of different tRNA molecules in a single transcript is also frequent. For example, one $tRNA^{Ser}$ and one $tRNA^{Thr}$ are present in a single unit in *E. coli.*

Although a great deal remains to be learned about the processing of bacterial RNA, the following facts seem to be well established:

1. All stable RNA molecules are processed.
2. About ten nucleases account for all of the cuts. The endonucleases always generate a 5′-P and a 3′-OH group.
3. The 3′-OH ends are generally formed by exonucleases; 5S rRNA may be an exception.
4. Many processing enzymes are not sequence-specific but recognize large structural features.
5. Bases other than A, U, G, and C are formed by enzymatic modification of bases already present in otherwise completed molecules.

Now that we have completed our examination of bacterial transcription, let's continue on to the more complex process of eukaryotic transcription in Chapter 17.

Suggested Reading

Messenger RNA

Grunberg-Manago, M. 1999. Messenger RNA stability and its role in control of gene expression in bacteria and phages. *Ann Rev Genet* 33:193–227.

The Lactose Operon

Adhya, S. 1996. The *Lac* and *Gal* operons today. In: Lin, E. C. C. and Lynch, S., eds. *Regulation of Gene Expression in Escherichia coli*. Georgetown, TX: R. G. Landes.

Beckwith, J. 1996. The operon: an historical account. In: Neidhardt, F. C., ed. *Escherichia coli and Salmonella typhimurium*. pp. 1227–1231. Washington, DC: ASM Press.

Bell, C. E., and Lewis, M. 2000. A closer view of the conformation of the Lac repressor bound to operator. *Nat Struct Biol* 7:209–214.

Bell, C. E., and Lewis, M. 2001. The Lac repressor: a second generation of structural and functional studies. *Curr Opin Struct Biol* 11:19–25.

Borukhov, S., and Lee, J. 2005. RNA polymerase structure and function at *lac* operon. *C R Biol* 328:576–587.

Chuprina, V. P., Rullmann, J. A., Lamerichs, R. M., et al. Structure of the complex of lac repressor headpiece and an 11 base-pair half-operator determined by nuclear magnetic resonance spectroscopy and restrained molecular dynamics. *J Mol Biol* 234:446–462.

Jacob, F., and Monod, J. 1961. Genetic regulatory mechanisms in the synthesis of proteins. *J Mol Biol* 3:318–356.

Kercher, M. A., Lu, P., and Lewis, M. 1997. *Lac* repressor–operator complex. *Curr Opin Struct Biol* 7:76–85.

Lewis, M. 2005. The *lac* repressor. *C R Biol* 328:521–548.

Lewis, M., Chang, G., Horton, N. C., et al. 1996. Crystal structure of the lactose operon repressor and its complexes with DNA and inducer. *Science* 271: 1247–1254.

Matthews, K. S., Falcon, C. M., and Swint-Kruse, L. 2000. Relieving repression. *Nat Struct Biol* 7:184–187.

Müller-Hill, B. 1998. Some repressors of bacterial transcription. *Curr Opin Microbiol* 1:145–151.

Ullmann, A. 2009. *Escherichia coli* lactose operon. In: *Encyclopedia of Life Sciences*. pp. 1–8. Hoboken, NJ: John Wiley and Sons.

Catabolite Repression

Busby, S., and Ebright, R. H. 1999. Transcription activation by catabolite activator protein (CAP). *J Mol Biol* 293:199–213.

Busby, S., and Kolb, A. 1996. The CAP modulon. In: Lin, E. C. C., and Lynch, S., eds. *Regulation of Gene Expression in E. coli*. Georgetown, TX: R. G. Landes Company.

Deutscher, J. 2008. The mechanisms of carbon catabolite repression in bacteria. *Curr Opin Microbiol* 11:87–93.

Deutscher, J., Francke, C., and Postma, P. W. 2006. How phosphotransferase system-related protein phosphorylation regulates carbohydrate metabolism in bacteria. *Microbiol Mol Biol Rev* 70: 939–1031.

Harman, J. G. 2001. Allosteric regulation of the cAMP receptor protein. *Biochim Biophys Acta* 1547:1–17.

Kimata, K., Takahashi, H., Inada, T., et al. 1997. cAMP receptor protein plays a crucial role in glucose-lactose diauxie by activating the major glucose transporter gene in *Escherichia coli*. *Proc Natl Acad Sci USA* 94:2914–2919.

Lawson, C. L., Swigon, D., Murakami, K. S., et al. 2004. Catabolite activator protein: DNA binding and transcription activation. *Curr Opin Struct Biol* 14:1–11.

Lengeler, J. W. 1996. The phosophoenolpyruvate-dependent carbohydrate: phosphotransferase system (PTS) and control of carbon source utilization. In: Lin, E. C. C., and Lynch, S., eds. *Regulation of Gene Expression in Escherichia coli*. Georgetown, TX: R. G. Landes.

Passner, J. M., and Steitz, T. A. 1997. The structure of a CAP-DNA complex having two cAMP molecules bound to each monomer. *Proc Natl Acad Sci USA* 94:2843–2847.

Rhodius, V. A., and Busby, S. J. W. 2000. Transcription activation by the *Escherichia coli* cyclic AMP receptor protein: determinants within activating region 3. *J Mol Biol* 299:295–310.

Stülke, J., and Hillen, W. 1999. Carbon catabolite repression in bacteria. *Curr Opin Microbiol* 2:195–201.

The Galactose Operon

Aiba, H., Adhya, S., and de Crombrugghe, B. 1981. Evidence for two functional gal promoters in intact *Escherichia coli* cells. *J Biol Chem* 256:11905–11910.

Choy, H. E., Hanger, R. R., Aki, T., et al. 1997. Repression and activation of promoter-bound RNA polymerase activity by Gal repressor. *J Mol Biol* 272:293–300.

Semsey S., Tolstorukov, M. Y., Virnik, K., et al. DNA trajectory in the Gal repressosome. *Genes Dev* 18:1898–1907.

Semsey, S., Virnik, K., and Adhya, S. 2005. A gamut of loops: meandering DNA. *Trends Biochem Sci* 30:334–341.

The Arabinose Operon

Dhiman, A., and Schleif, R. 2000. Recognition of overlapping nucleotides by AraC and the sigma subunit of RNA polymerase. *J Bacteriol* 182:5076–5081.

Gallegos, M. T., Schleif, R., Bairoch, A., et al. 1997. AraC/XylS family of transcriptional regulators. *Microbiol Mol Biol Rev* 61:393–410.

Schleif, R. 1996. Two positively regulated systems, ara and *mal*. In: Neidhardt, F. C., ed. *Escherichia coli and Salmonella typhimurium*. pp. 1300–1309. Washington, DC: ASM Press.

Schleif, R. 2000. Regulation of the L-arabinose operon of *Escherichia coli*. *Trends Genet* 16:559–565.

Schleif, R. 2003. AraC proteins: a love–hate relationship. *Bioessays* 25:274–282.

Seabold, R. R., and Schleif, R. F. 1998. Apo-AraC actively seeks to loop. *J Mol Biol* 278:529–538.

Soisson, S. M., MacDougall-Shackleton, B., Schleif, R., and Wolberger, C. 1997. Structural basis for ligand-regulated oligomerization of AraC. *Science* 276:421–425.

The Tryptophan Operon

Fisher, R. F., and Yanofsky, C. 1983. Mutations of the β subunit of RNA polymerase alter both transcription pausing and transcription termination in the *trp* operon leader region *in vitro*. *J Biol Chem* 258:8146–8150.

Henkin, T. M., and Yanofsky, C. 2002. Regulation by transcription attenuation in bacteria: how RNA provides instructions for transcription termination/antitermination decisions. *Bioessays* 24:700–707.

Landick R, Turnbough, C. L. Jr, and Yanofsky C. 1996. Transcription attenuation. In: Neidhardt F. C., et al. eds. *Escherichia coli and Salmonella: Cellular and Molecular Biology*. pp. 1263–1286. Washington, DC: ASM Press.

Merino, E., Jensen, R. A., and Yanofsky, C. 2008. Evolution of bacterial *trp* operons and their regulation. *Curr Opin Microbiol* 11:78–86.

Merino, E., and Yanofsky, C. 2005. Transcription attenuation: a highly conserved regulatory strategy used by bacteria. *Trends Genet* 21:260–264.

Oxender, D. L., Zurawski, G., and Yanofsky, C. 1979. Attenuation in the *Escherichia coli* tryptophan operon: role of RNA secondary structure involving the tryptophan codon region. *Proc Natl Acad Sci USA* 76:5524–5528.

Yanofsky, C. 2000. Transcription attenuation: once viewed as a novel regulatory strategy. *J Bacteriol* 182:1–8.

Yanofsky, C. 2003. Using studies on tryptophan metabolism to answer basic biological questions. *J Biol Chem* 278:10859–10878.

Lambda Phage: A Transcription Regulation Network

Albright, R. A., and Matthews, B. W. 1998. Crystal structure of λ-Cro bound to a consensus operator at 3.0 Å resolution. *J Mol Biol* 280:137–151.

Beamer, L. J., and Pabo, C. O. 1992. Refined 1.8 Å crystal structure of the lambda repressor-operator complex. *J Mol Biol* 227:177–196.

Casjens, S. R., and Hendrix, R. W. 2001. Bacteriophage lambda and its relatives. In: *Encyclopedia of Life Science*. London: Nature; pp. 1–8.

Court, D. L., Oppenheim, A. B., and Adhya, S. L. 2007. A new look at bacteriophage l genetic networks. *J Bacteriol* 189:298–304.

Gottesman, M. 1999. Bacteriophage λ: the untold story. *J Mol Biol* 293:177–180.

Gottesman, M., and Weisberg, R. A. 2004. Little lambda, who made thee? *Microbiol Mol Biol Rev* 68:796–813.

Hendrix, R., Roberts, J., Stahl, F., and Weisberg, R. (eds). 1983. *Lambda II*. Woodbury, NY: Cold Spring Harbor Laboratory Press.

Hochschild, A. 2002. The λ switch: cI closes the gap in autoregulation. *Curr Biol* 12:R87–R89.

Hochschild, A., and Lewis, M. 2009. The bacteriophage l CI protein finds an asymmetric solution. *Curr Opin Struct Biol* 19:1–8.

Jain, D., Kim, Y., Maxwell, K., et al. 2005. Crystal structure of bacteriophage λ cII and its DNA complex. *Mol Cell* 19:259–269.

Jain, D., Nickels, B, Sun, L., et al. 2004. Structure of a ternary transcription activation complex. *Mol Cell* 13:45–53.

Koudelka, G. B. 2000. Action at a distance in a classic system. *Curr Biol* 10:R704–R707.

Lederberg, E. 1951. Lysogenicity in *E. coli* K12. *Genetics* 36:560 (abstract).

Little, J. W. 2005. Threshold effects in gene regulation: when some is not enough. *Proc Natl Acad Sci USA* 102:5310–5311.

Oppenheim, A. B., Kobiler, O., Stavans, J., et al. 2005. Switches in bacteriophage lambda development. *Annu Rev Genet* 39:409–429.

Ptashne, M. 2004. *Genetic Switch: Phage Lambda Revisited*, 3rd ed. Woodbury, NY: Cold Spring Harbor Laboratory Press.

Ptashne, M. 2005. Regulation of transcription: from lambda to eukaryotes. *Trends Biochem Sci* 30:275–278.

Ptashne, M., and Gann, A. 1998. Imposing specificity by localization: mechanism and evolvability. *Curr Biol* 8:R812–R822.

Stayrook, S., Jaru-Ampornpan, P., Ni, J., Hochschild, A., and Lewis, M. 2008. Crystal structure of the l repressor and a model for pairwise cooperative operator binding. *Nature* 452:1022–1026.

Svenningsen, S. L., Costantino, N., Court, D. L., and Adhya, S. 2005. On the role of Cro in λ prophage induction. *Proc Natl Acad Sci USA* 102:4465–4469.

Messenger RNA Degradation

Anderson, J. S. J., and Parker, R. 1996. RNA turnover: the helicase story unwinds. *Curr Biol* 6:780–782.

Callaghan, A. J., Marcaida, M. J., Stead, J. A. et al. 2005. Structure of the *Escherichia coli* RNase E catalytic domain and implications for RNA turnover. *Nature* 437:1187–1191.

Carpousis, A. J. 2002. The *Escherichia coli* RNA degradosome: structure, function and relationship to other ribonucleolytic multienyzme complexes. *Biochem Soc Trans* 30(part 2):150–155.

Carpousis, A. J. 2002. The RNA degradosome of *Escherichia coli*: an mRNA-degrading machine assembled on RNase E. *Ann Rev Microbiol* 61:71–87.

Condon, C. 2007. Maturation and degradation of RNA in bacteria. *Curr Opin Microbiol* 10:271–278.

Deana, A., Celesnik, H., and Belasco, J. G. 2008. The bacterial enzyme RppH triggers messenger RNA degradation by 5-pyrophosphate removal. *Nature* 451:355–358.

Dickson, A. M., Wilusz, C. J., and Wilusz, J. 2009. mRNA Turnover. In: *Encyclopedia of Life Sciences*. pp. 1–9. Hoboken, NJ: John Wiley and Sons.

Nicholson, A. W. 1999. Function, mechanism and regulation of bacterial ribonucleases. *FEMS Microbiol Rev* 23:371–390.

Rauhut, R., and Klug, G. 1999. mRNA degradation in bacteria. *FEMS Microbiol Rev* 23:353–370.

Régnier, P., and Arralano, C. M. 2000. Degradation of mRNA in bacteria: emergence of ubiquitous features. *Bioessays* 22:235–244.

Richards, J., Sundermeier, T., Svetlanov, A., and Karzai, A. W. 2008. Quality control of bacterial mRNA decoding and decay. *Biochim Biophys Acta* 1779:574–582.

Sarkar, N. 1996. Polyadenylation of mRNA in bacteria. *Microbiology* 142:3125–3133.

Ribosomal RNA and Transfer RNA Synthesis

Estrem, S. T., Ross, W., Gaal, T., et al. 1999. Bacterial promoter architecture: subsite structure of UP elements and interactions with the carboxy-terminal domain of the RNA polymerase α subunit. *Genes Dev* 13:2134–2147.

Gourse, R. L., Ross, W., and Gaal, T. 2000. Ups and downs in bacterial transcription initiation: the role of the alpha subunit of RNA polymerase in promoter recognition. *Mol Microbiol* 37:687–695.

Regulation of Ribosome Synthesis

Barker, M. M., Gaal, T., Josaitis, C. A., and Gourse, R. L. 2001. Mechanism of regulation of transcription initiation by ppGpp. 1. Effects of ppGpp on transcription initiation *in vivo* and *in vitro*. *J Mol Biol* 305:673–688.

Gourse, R. L., Gaal, T., Bartlett, M. S., et al. 1996. rRNA transcription and growth rate-dependent regulation of ribosome synthesis in *Escherichia coli*. *Ann Rev Microbiol* 50:645–677.

Gourse, R. L., and Ross, W. 1996. Control of rRNA and ribosome synthesis. In: Lin, E. C. C. and Lynch, S., eds. *Regulation of Gene Expression in Escherichia coli*. Georgetown, TX: R. G. Landes.

Keener, J., and Nomura, M. 1996. The regulation of ribosome synthesis. In: Neidhardt, F.C., ed. *Escherichia coli and Salmonella typhimurium*. pp. 1417–1431. Washington, DC: ASM Press.

Potrykus, K., and Cashel, M. 2008. (p)ppGpp: still magical? *Ann Rev Microbiol* 62:35–51.

Srivatsan, A., and Wang, J. D. 2008. Control of bacterial transcription, translation and replication by (p)ppGpp. *Curr Opin Microbiol* 11:100–105.

Processing rRNA and tRNA

Kurz, J. C., and Fierke, C. A. 2000. Ribonuclease P: a ribonucleoprotein enzyme. *Curr Opin Chem Biol* 4:553–558.

Li, A., Pandit, S., and Deutscher, M. P. 1999. RNase G (CafA protein) and RNase E are both required for the 5′ maturation of 16S ribosomal RNA. *EMBO J* 18:2878–2885.

Schön, A. 1999. Ribonuclease P: the diversity of a ubiquitous RNA processing enzyme. *FEMS Microbiol Rev* 23:391–406.

Tollervey, D., and Lafontaine, D. L. J. 2001. Ribosomal RNA. In: *Encyclopedia of Life Sciences*. London: Nature. pp. 1–7.

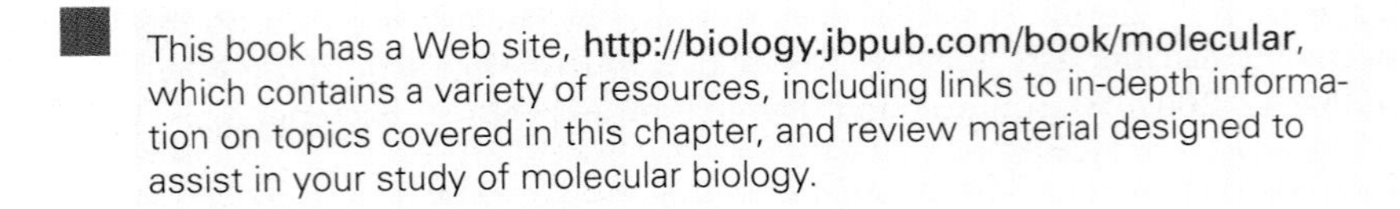

This book has a Web site, **http://biology.jbpub.com/book/molecular**, which contains a variety of resources, including links to in-depth information on topics covered in this chapter, and review material designed to assist in your study of molecular biology.

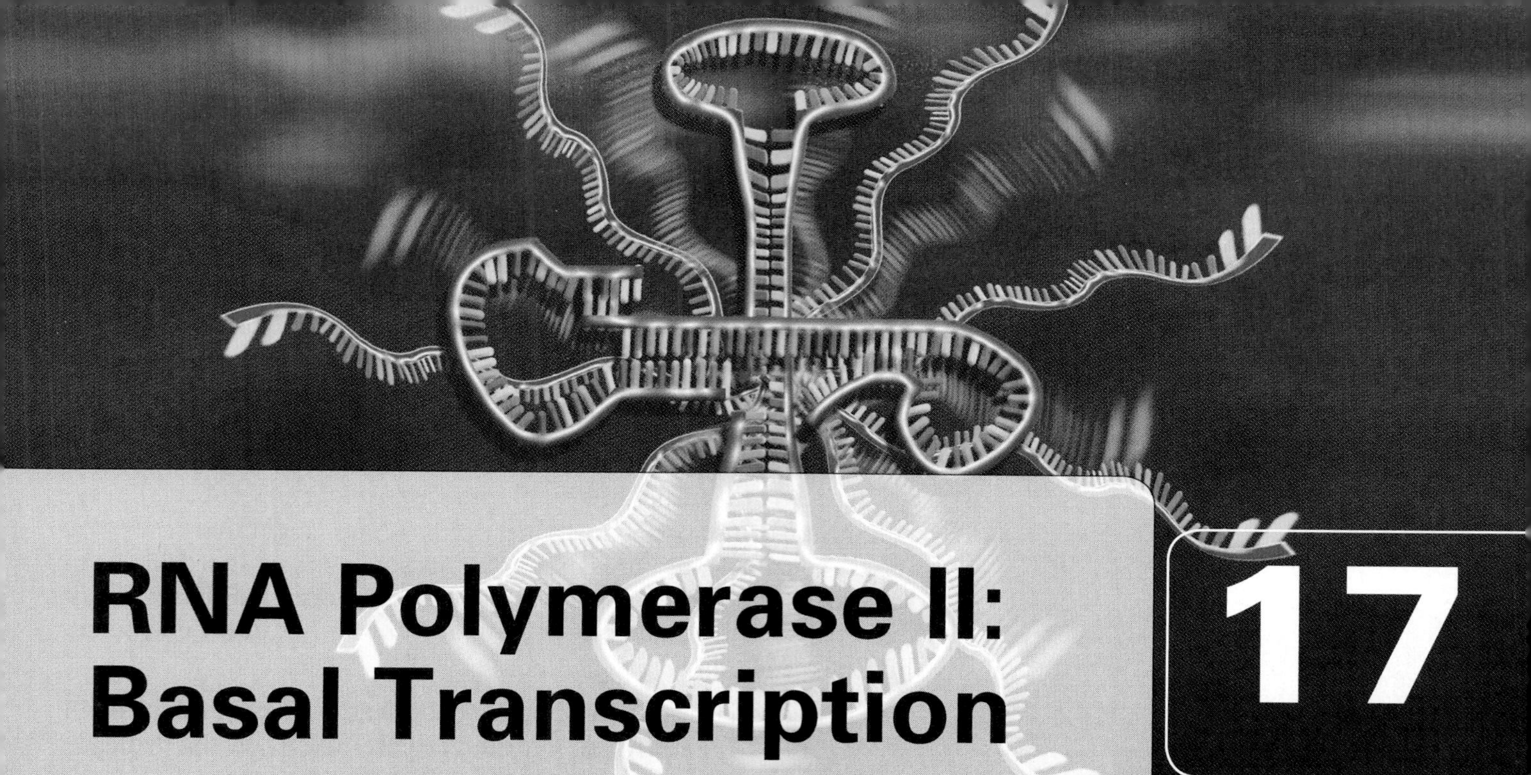

RNA Polymerase II: Basal Transcription

17

OUTLINE OF TOPICS

17.1 Introduction to RNA Polymerase II

The eukaryotic cell nucleus has three different kinds of RNA polymerase.

RNA polymerases I, II, and III can be distinguished by their sensitivities to inhibitors.

Each nuclear RNA polymerase has some subunits that are unique to it and some that it shares with one or both of the other nuclear RNA polymerases.

17.2 RNA Polymerase II Structure

High-resolution yeast RNA polymerase II structures help explain how the enzyme works.

The crystal structure has been determined for the complete 12-subunit yeast RNA polymerase II bound to a transcription bubble and product RNA.

17.3 Transcription Start Site Identification

Nuclear RNA polymerases have limited synthetic capacities.

Various techniques have been devised to locate RNA polymerase II transcription start sites.

Cap analysis of gene expression (CAGE) is a high throughput technique used to identify transcription start sites and their flanking promoters.

17.4 The Core Promoter

The core promoter extends from 40 bp upstream of the transcription start site to 40 bp downstream from this site.

17.5 General Transcription Factors: Basal Transcription

RNA polymerase II requires the assistance of general transcription factors to transcribe naked DNA from specific transcription start sites.

The core promoter allows a cell-free system to catalyze a low-level of RNA synthesis at the correct transcription start site.

When the core promoter has a TATA box, preinitiation complex assembly begins with either TFIID or TATA binding protein (TBP) binding to the core promoter.

TFIID can bind to core promoters of protein-coding genes that lack a TATA box.

TFIIA is not required to reconstitute the minimum transcription system.

TFIIB helps to convert the closed promoter to an open promoter.

Sequential binding of RNA polymerase II•TFIIF complex, TFIIE, and TFIIH completes preinitiation complex formation.

17.6 Transcription Elongation

The C-terminal domain of the largest RNA polymerase subunit must be phosphorylated for chain elongation to proceed.

A variety of transcription elongation factors help to suppress transient pausing during elongation.

Elongation factor SII reactivates arrested RNA polymerase II.

The transcription elongation complex is regulated.

17.7 Archaeal RNA Polymerase

The archaea have a single RNA polymerase that is similar to RNA polymerase II.

Suggested Reading

At the most fundamental chemical level, eukaryotic and bacterial RNA syntheses are similar. Both processes begin when two ribonucleoside triphosphates, which are lined up on a DNA template strand, join to form a dinucleoside tetraphosphate with a hydroxyl group at its 3′-end and a triphosphate group at its 5′-end. Both processes continue by extending the dinucleotide chain in a 5′→3′ direction when ribonucleotides are added to the 3′-end of the growing chain in the order specified by the DNA template until a termination signal is reached.

Based on these chemical similarities, one might expect bacterial and eukaryotic transcription machinery to have similar catalytic sites. Bacterial and eukaryotic core RNA polymerases do in fact share important structural and functional features. Many important differences also exist between the bacterial and eukaryotic transcription machinery, however, including:

1. Bacteria use a single RNA polymerase to synthesize rRNA, mRNA, and tRNA, while eukaryotes use a specific dedicated enzyme to synthesize each kind of RNA.
2. Bacterial RNA polymerase requires the assistance of at most one or two accessory factors to transcribe genes. Eukaryotic RNA polymerases require several such factors.
3. Bacterial RNA polymerase holoenzyme has direct access to its DNA template, whereas the eukaryotic transcription machinery has difficulty in reaching its DNA template because eukaryotic DNA interacts with histones to form nucleosomes, which in turn form more compact chromatin structures (see Chapter 6). An important consequence of chromatin structure is that eukaryotic genes tend to be turned off in the absence of regulatory proteins. In contrast, bacterial genes tend to be turned on or require only one or two regulatory proteins such as cAMP receptor protein (CRP) to become fully active (see Chapter 16). Additional factors are required to move and modify histone octamers that block access to eukaryotic genes before RNA synthesis can occur. Hence, chromatin structure introduces a level of transcriptional complexity in eukaryotes that does not exist in bacteria.
4. Bacteria make only one modification in primary transcripts for mRNA. They add poly (A) tails to the 3′-ends, marking the mRNA for degradation (see Chapter 16). Eukaryotes also add a poly (A) tail to the mRNA 3′-end; however, this tail helps to protect the mRNA from degradation rather than mark it for destruction. Eukaryotes also add a guanine nucleotide **cap** to the 5′-end of the growing mRNA chain and remove nucleotide sequences from within the **precursor mRNA** or **pre-mRNA** (FIGURE 17.1). Walter Gilbert coined the terms **intron** for an intervening sequence that is removed during the conversion of a primary transcript to a mature RNA molecule, and **exon** for a sequence that appears in the mature RNA molecule. Eukaryotes have special metabolic machinery that removes introns with great precision. Remarkably, eukaryotic cells can splice different

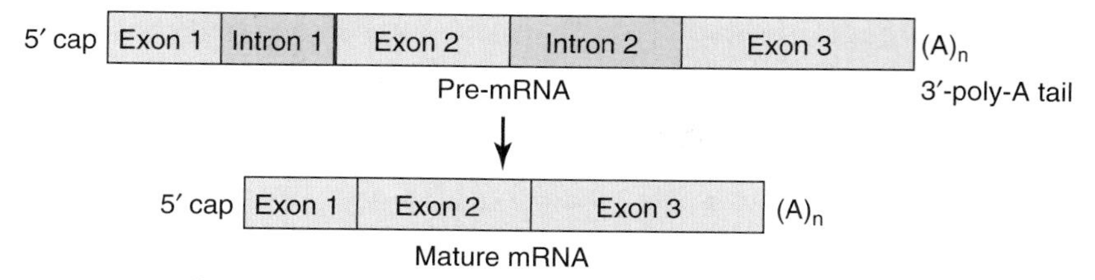

FIGURE 17.1 **Conversion of precursor mRNA to mRNA in eukaryotes.** Eukaryotes add a guanine nucleotide cap to the 5′-end of the growing mRNA chain, remove nucleotide sequences, from within the precursor mRNA (pre-mRNA), and add a poly (A) tail to the 3′-end of the mRNA. The excised nucleotide sequences are called introns, while those that are included within the mature mRNA are called exons.

combinations of exons within the same transcription unit to form alternate mRNA molecules that each codes for its own unique polypeptide. This **alternative splicing** mechanism, which accounts for the eukaryotic cell's ability to synthesize a wider variety of polypeptides than would be predicted from its DNA sequence, will be described in greater detail in Chapter 19.

We begin this chapter by describing early experiments that demonstrated eukaryotic cells have three different kinds of nuclear RNA polymerases and then focus on **RNA polymerase II**, the enzyme responsible for mRNA formation.

17.1 Introduction to RNA Polymerase II

The eukaryotic cell nucleus has three different kinds of RNA polymerase.

Samuel B. Weiss and Leonard Gladstone provided the first clear evidence for the existence of eukaryotic DNA-dependent RNA polymerase in 1959, about one year before the enzyme was detected in bacteria. Their studies showed that disrupted rat liver cell nuclei convert radioactive ribonucleoside triphosphates into an acid-insoluble, RNase-sensitive product with a base composition that is similar to rat liver DNA. Moreover, this conversion required the presence of all four ribonucleoside triphosphates and did not occur in the presence of DNase. Further progress was hindered by the inability to extract the RNA polymerase activity from disrupted nuclei. Several different factors contributed to this failure. One of the most important—the presence of inhibitory proteins—was eventually minimized by adjusting the extraction buffer's ionic strength so that RNA polymerase was extracted preferentially.

During the 1960s, the research groups of John J. Furth, Jamshed R. Tata, and Pierre M. Chambon each studied distinct RNA polymerase activities, never realizing the true tripartite nature of eukaryotic RNA synthesis. Although there were fleeting suggestions that eukaryotes

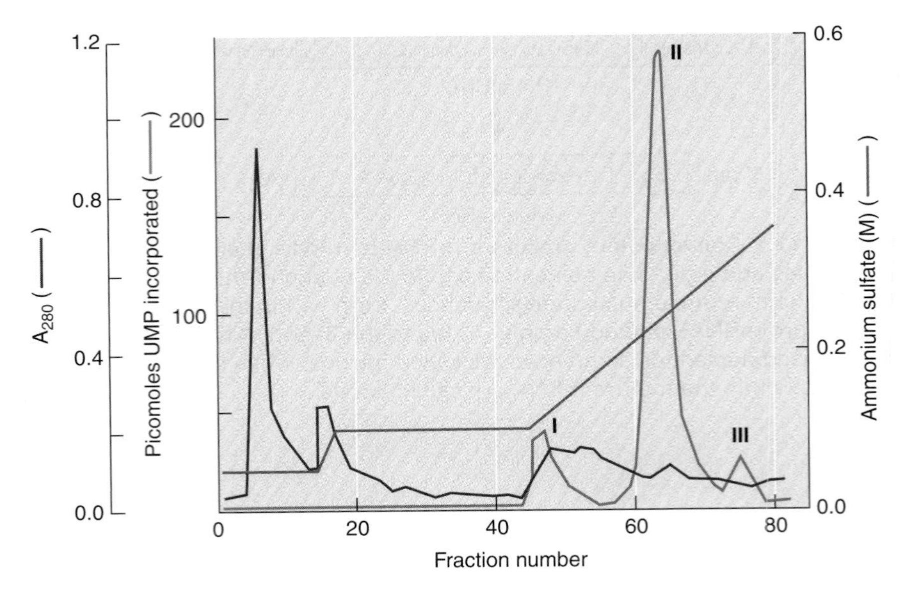

FIGURE 17.2 **Separation of RNA polymerase activities by DEAE-Sephadex chromatography.** Robert G. Roeder and William J. Rutter separated RNA polymerases from sea urchin embryos by DEAE-Sephadex chromatography. Red, protein measured by absorbance at 280 nm; green, RNA polymerase activity measured by incorporation of radioactively labeled UMP; blue, ammonium sulfate concentration used in elution buffer. (Adapted from R. G. Roeder and W. J. Rutter, *Proc. Natl. Acad. Sci. USA* 65 [1970]: 675–682.)

might have multiple RNA polymerases, this was considered to be quite unlikely because *Escherichia coli* has only a single kind of core RNA polymerase. It was not until the late 1960s that a graduate student, Robert G. Roeder, hypothesized that there were three distinct RNA polymerase activities. He proposed a series of experiments to prove his hypothesis both to himself and his doubting mentor, William J. Rutter.

A major breakthrough occurred in 1970 when Roeder and Rutter used DEAE-Sephadex chromatography to fractionate proteins in the nuclear extract (FIGURE 17.2). This fractionation technique takes advantage of the fact that negative charges on the protein surface interact with positive charges on the DEAE-Sephadex beads so that a protein's ability to bind to the beads tends to increase with its net negative charge (see Chapter 2). Roeder and Rutter detected three separate peaks of RNA polymerase activity when they applied sea urchin embryo or rat liver nuclear extracts to DEAE-Sephadex and then released the bound protein with a buffer solution containing increasing ammonium sulfate concentrations (Figure 17.2). The activities, which were named **RNA polymerase I, RNA polymerase II**, and **RNA polymerase III** in the order of their release from the column, subsequently were shown to be three distinct enzymes. Once these experiments were completed, it became clear that Tata had been studying RNA polymerase I, Chambon RNA polymerase II, and Furth RNA polymerase III. Recently Craig S. Pickard and coworkers discovered two new nuclear RNA polymerases, RNA polymerases IV and V, are present in land plants and some algae. Some of the subunits in the new RNA polymerases are identical to those in RNA polymerase II and others are homologous to those in RNA polymerase II. RNA polymerases IV and V appear to synthesize small interfering RNA (siRNA) molecules that are used for siRNA-mediated gene silencing (see Chapter 21). Because RNA polymerases IV and V have not yet been subject to extensive *in vitro* studies, they will not be discussed further.

TABLE 17.1 Comparing the Three Eukaryotic RNA Polymerases

Enzyme	Location	RNA Products	Sensitivity to α-amanitin	Sensitivity to actinomycin D
RNA polymerase I	Nucleolus	Pre-rRNA (leading to 5.8S, 18S, and 28S rRNA	Resistant	Very sensitive
RNA polymerase II	Nucleoplasm	Pre-mRNA and some snRNAs	50% inhibition at 0.02 μg/mL	Slightly sensitive
RNA polymerase III	Nucleoplasm	tRNA, 5S rRNA, U6 snRNA (spliceosome), and 7SL RNA (signal recognition particle)	50% inhibition at 20 μg/mL	Slightly sensitive

For convenience, we will use the term "nuclear RNA polymerases" to distinguish RNA polymerases I, II, and III, the three RNA polymerases present in all eukaryotic cell nuclei from RNA polymerases that are present in the mitochondria of all eukaryotes and in the chloroplasts of plants. The three nuclear RNA polymerases have different physiological roles (Table 17.1). RNA polymerase I, which is located in the nucleolus, transcribes genes specifying the precursors to 5.8S, 18S, and 28S rRNA. RNA polymerase II is located in the nucleoplasm (nuclear sap) and transcribes protein-coding genes and some genes that code for small nuclear RNA (snRNA). RNA polymerase III, also located in the nucleoplasm transcribes genes coding for small RNA molecules such as 5S rRNA and tRNA.

RNA polymerases I, II, and III can be distinguished by their sensitivities to inhibitors.

One way in which the three RNA polymerases can be distinguished from one another is their sensitivity to **α-amanitin** (Table 17.1), a toxic octapeptide produced by *Amanita phalloides*, a poison mushroom known as the *Death Cap* (FIGURE 17.3). When added at a concentration as low as 0.02 μg/mL, α-amanitin causes 50% inhibition of elongation by RNA polymerase II. A thousand times greater concentration is required to cause equivalent inhibition of RNA polymerase III. The toxic octapeptide has no effect on RNA polymerase I. The sensitivities of the three RNA polymerases can be exploited to block the synthesis of specific types of RNA in the living cell. Thus, very low α-amanitin concentrations block mRNA synthesis while permitting the continued synthesis of most other kinds of RNA. Higher concentrations also block the synthesis of 5S rRNA, tRNA, and most other small RNA molecules while permitting continued synthesis of other rRNA.

The transcription reactions catalyzed by the three RNA polymerases also exhibit differential sensitivity to **actinomycin D**. In this case, however, the reaction catalyzed by RNA polymerase I is the most sensitive. In contrast to α-amanitin, which binds to the enzyme, actinomycin D (FIGURE 17.4a) binds to the double-stranded DNA template. The

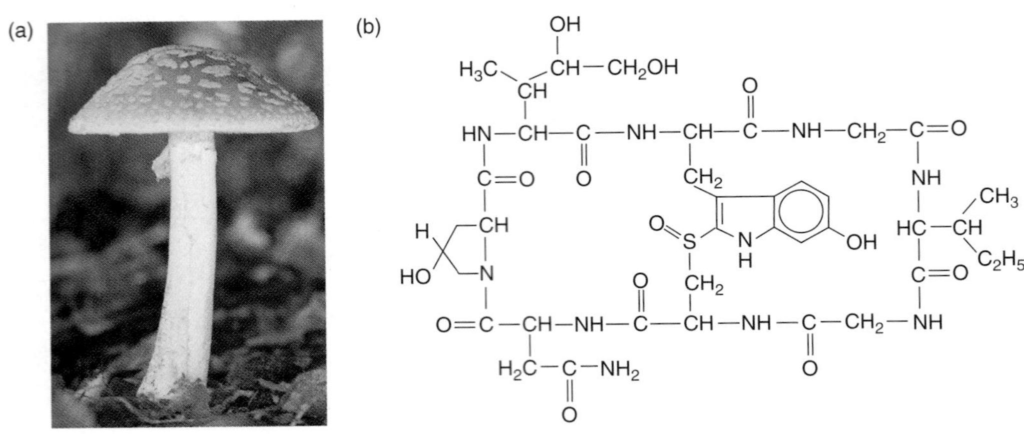

FIGURE 17.3 **α-Amanitin.** (a) *Amanita phalloides* (also known as the Death Cap). (b) Structural formula of α-amanitin. (Part a © Niels-DK/Alamy Images. Part b adapted from C. Defendenti, et al., *Forensic Sci. Int.* 92 [1998]: 59–68.)

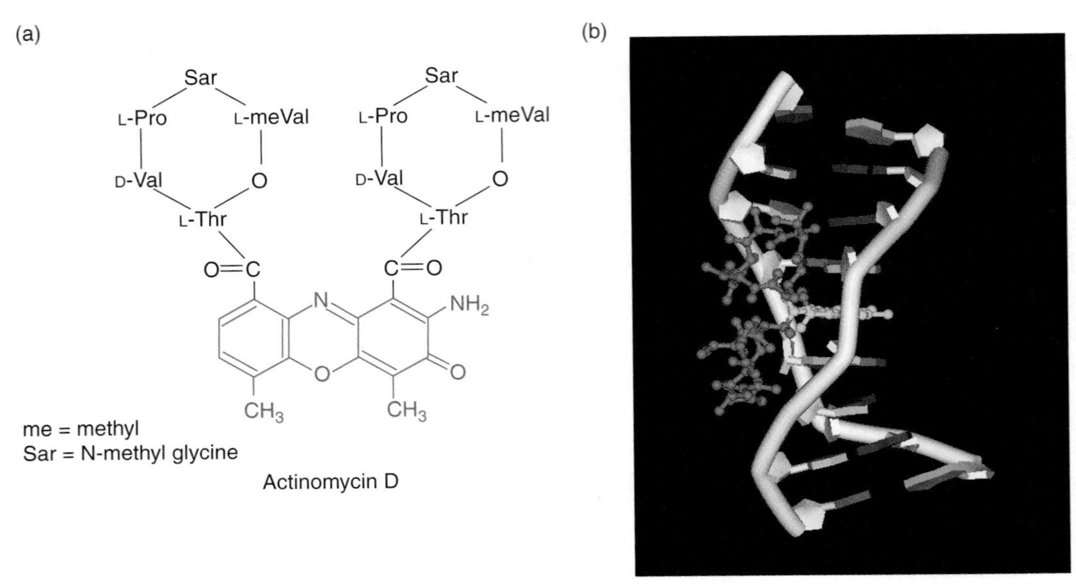

FIGURE 17.4 **Actinomycin D.** (a) The ring system (green) in actinomycin D binds between C-G and G-C base pairs in double-stranded DNA molecules. (b) DNA (5′-d[GpApApGpCpTpTpC]-3′) complexed with actinomycin D. The ring system of actinomycin D that intercalates between C-G and G-C base pairs in DNA is shown in yellow. The rest of the actinomycin D molecule is shown in orange. Actinomycin D inhibits transcription by preventing strand separation. (Part b structure from Protein Data Bank 2D55. M. Shinomiya, et al., *Biochemistry* 34 [1995]: 8481–8491. Prepared by B. E. Tropp.)

heterocycle ring system in actinomycin D inserts itself between C-G and G-C base pairs, preventing the DNA strand separation required for transcription and replication (FIGURE 17.4b). RNA polymerase I catalyzed reactions tend to be the most sensitive to actinomycin D because the rRNA genes that it transcribes are GC-rich.

Each nuclear RNA polymerase has some subunits that are unique to it and some that it shares with one or both of the two other nuclear RNA polymerases.

RNA polymerases I, II, and III have 14, 12, and 17 polypeptide subunits, respectively, making them considerably more complex and difficult to study than their bacterial counterpart. Characterization of the three enzymes has proven to be quite a challenge because of the difficulty in isolating such large complexes without subunit degradation or dissociation. In fact, it seems reasonable to ask if all the polypeptides in isolated RNA polymerases are genuine subunits (as opposed to contaminants that somehow bind to the enzyme during the purification process) because the fully active bacterial RNA polymerase holoenzyme has only five different kinds of subunit (α, β, β′, ω, and σ). The discussion that follows highlights the subunit composition of RNA polymerase II because it is the best studied of the three nuclear RNA polymerases and it transcribes protein-coding genes.

Sodium dodecyl sulfate-polyacrylamide gel electrophoresis (SDS-PAGE) analysis reveals that RNA polymerase II isolated from yeast, humans, or other eukaryotes contains 12 different polypeptides, of which 10 are present in stoichiometric amounts as would be expected for authentic RNA polymerase II subunits. The remaining two polypeptides appear to dissociate during enzyme purification. Support for the authenticity of the 12 polypeptides also comes from the observation that RNA polymerase II preparations from various eukaryotic species have homologous subunits. It would be quite unlikely for the same contaminating polypeptide to co-purify with RNA polymerase II molecules from different species.

Genes that code for each of the 12 RNA polymerase II subunits, designated *RPB1* to *RPB12*, have been identified in yeast and cloned. (The *RPB* gene designation is a holdover from an alternate system for RNA polymerase nomenclature in which the letters A, B, and C were used in place of Roman numerals I, II, and III. Genes that code for subunits in RNA polymerases I and III are designated *RPA* and *RPC*, respectively, followed by a number to indicate the specific subunit.) Yeast deletion mutants for *RPB4* and *RPB9* are temperature and cold sensitive, whereas those for the other ten subunits are nonviable (Table 17.2). These results are consistent with the fact RNA

TABLE 17.2 Genes for Subunits in *Saccharomyces cerevisiae* RNA Polymerase II

Gene	Amino Acid Residues in Polypeptide Subunit	Deletion Mutant Phenotype
RPB1	1733	Nonviable
RPB2	1224	Nonviable
RPB3	318	Nonviable
RPB4	221	Conditional
RPB5	215	Nonviable
RPB6	155	Nonviable
RPB7	171	Nonviable
RPB8	146	Nonviable
RPB9	122	Conditional
RPB10	70	Nonviable
RPB11	120	Nonviable
RPB12	70	Nonviable

polymerase II is an essential enzyme and support the view that each of the polypeptide subunits (Rpb1–Rpb12) is an authentic enzyme component. Remarkably, at least 10 of the human RNA polymerase II genes can be substituted for their counterparts in yeast. The significance of this result is that information obtained by studying RNA polymerase II in yeast applies to RNA polymerase II obtained from other eukaryotes including humans.

Although the large number of subunits present in each of the nuclear RNA polymerases presents serious challenges to our understanding of how each enzyme works, homology studies indicate that the situation is simpler than it at first seems. To begin, five subunits in *Saccharomyces cerevisiae* RNA polymerase II (Rpb5, Rpb6, Rpb8, Rpb10, and Rpb12) are also in RNA polymerases I and III. Information gained by studying the functions of each of these five subunits in one nuclear RNA polymerase will undoubtedly provide information about their functions in the other two. Furthermore, this information tells us that one or more of the other subunits must be responsible for the specificity differences that exist among the nuclear RNA polymerases. Homologies not only exist among nuclear RNA polymerase subunits but also extend to bacterial and archaeal RNA polymerases (Table 17.3). The archaeal RNA polymerase is especially noteworthy because its subunit number is similar to that of RNA polymerase II. The extra subunits present in the nuclear RNA polymerases thus did not evolve to solve some problem due to the presence of the cell nucleus.

TABLE 17.3 RNA Polymerase Subunits

Eukaryotic					
Pol I	Pol II	Pol III	Archaeal		*E. coli*
Rpa190	Rpb1	Rpc160	Rpo 1N Rpo 1C	(A′) (A″)	β′
Rpa135	Rpb2	Rpc128	Rpo2	(B)	β
Rpc40	Rpb3	Rpc40	Rpo3	(D)	α
Rpc19	Rpb11	Rpc19	Rpo11	(L)	α
Rpb6	Rpb6	Rpb6	Rpo6	(K)	ω
Rpb5	Rpb5	Rpb5	Rpo5	(H)	
Rpb8	Rpb8	Rpb8	Rpo8	(G)	
Rpb10	Rpb10	Rpb10	Rpo10	(N)	
Rpb12	Rpb12	Rpb12	Rpo12	(P)	
Rpa12.2	Rpb9	Rpc11			
Rpa14	Rpb4	Rpc17	Rpo4	(F)	
Rpa43	Rpb7	Rpc25	Rpo7	(E)	
Rpa49		Rpc37			
Rpa34.5		Rpc53			
		Rpc82			
		Rpc34			
		Rpc31			
			Rpo13		
14 subunits	12 subunits	17 subunits	13 subunits		5 subunits

Pol I, Pol II, and Pol III are RNA polymerases I, II, and III, respectively. The subunits in each row are either homologous or identical to one another. The five identical subunits in the three eukaryotic nuclear RNA polymerases are shown in red. The two identical subunits in Pol I and Pol III are shaded in blue. When subunits in the same row are not identical, they are homologous. For example, subunits Rpa135, Rpb2, Rpc128, B, and β are homologous. The archaeal RNA polymerase requires two subunits, A and A′, to replace Rpb1. It has been recommended that the standard archaeal nomenclature system (shown in parentheses) be replaced by that shown, which is consistent with the eukaryotic nomenclature system. The area shaded in yellow contains the ten subunits that make up the core of the eukaryotic nuclear RNA polymerases. The areas shaded in blue, pink, and green contain the Rpb4/7 subcomplex, the transcription factor IIF (TF IIF)-like complex, and the RNA polymerase III subcomplex, respectively. The structures and functions of RNA polymerases I and III are described in Chapter 20.

Although the archaeal RNA polymerase bears a closer structural relationship to nuclear RNA polymerases than to bacterial RNA polymerase, the two prokaryotic RNA polymerases share one important functional feature: both enzymes synthesize mRNA, rRNA, and tRNA. Homology studies also reveal that each bacterial core RNA polymerase subunit (β, β′, and α) has at least one homolog in each nuclear RNA polymerase. For instance, comparison of the bacterial subunits with those in RNA polymerase II reveals β′ is homologous to Rpb1, β to Rpb2, α to both Rpb3 and Rpb 11, and ω to Rpb6 (Table 17.3). It therefore seems reasonable to conclude that the conserved amino acid sequences have especially important roles in RNA synthesis and that a structural relationship exists between the bacterial and nuclear enzymes. Comparisons of crystal structures to be described in the next section show that this is indeed true.

17.2 RNA Polymerase II Structure

High-resolution yeast RNA polymerase II structures help explain how the enzyme works.

The large size and complex structure of RNA polymerase II presented a major challenge for investigators attempting to determine its structure. In the early 1980s, Roger Kornberg launched a program to study RNA polymerase II from *S. cerevisiae*, believing that the yeast enzyme would be similar to that from other eukaryotes and that yeast offered the enormous advantage of combining the power of yeast genetics with biochemistry.

Kornberg and coworkers were able to purify yeast RNA polymerase II, but finding the right conditions for protein crystallization presented a major stumbling block. Initial attempts to crystallize yeast RNA polymerase II were unsuccessful because the enzyme extracted from wild-type cells was not homogeneous. Some enzyme molecules contained Rpb4 and Rpb7 subunits, whereas others did not. Kornberg's group initially solved the problem by isolating RNA polymerase II from a yeast mutant with a deletion in the *RPB4* gene so that the purified enzyme lacked both Rpb4 and Rpb7 subunits. The resulting ten-subunit enzyme complex, designated Δ4/7 RNA polymerase II, formed crystals suitable for x-ray diffraction studies. This ten-subunit complex can catalyze RNA chain elongation and, using the proper synthetic DNA template, initiate RNA synthesis. The high-resolution image of Δ4/7 RNA polymerase II obtained by Kornberg and coworkers shows the structural features of the enzyme complex and the interactions that exist among the ten subunits (FIGURE 17.5). As predicted from amino acid sequence homologies, the Δ4/7 RNA polymerase II bears a striking resemblance to *Thermus aquaticus* RNA polymerase (FIGURE 17.6). The similarities are greatest in the region surrounding the cleft, where the active site is located, and in regions that interact with the nascent RNA. As expected for a region that interacts with nucleic acids, a relatively high proportion of the amino acids that form the

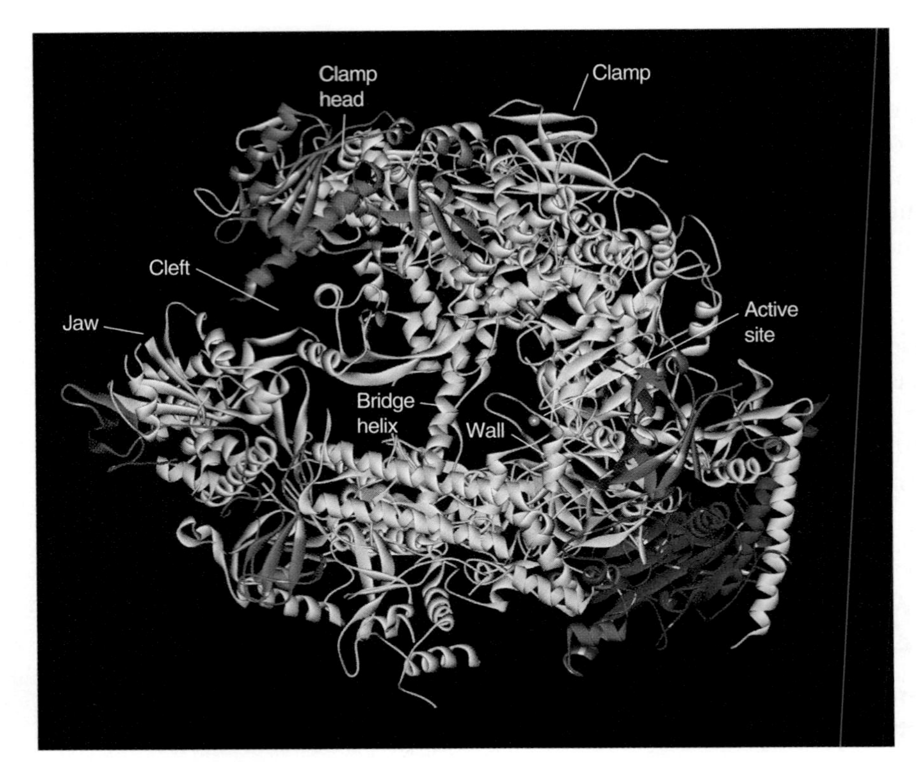

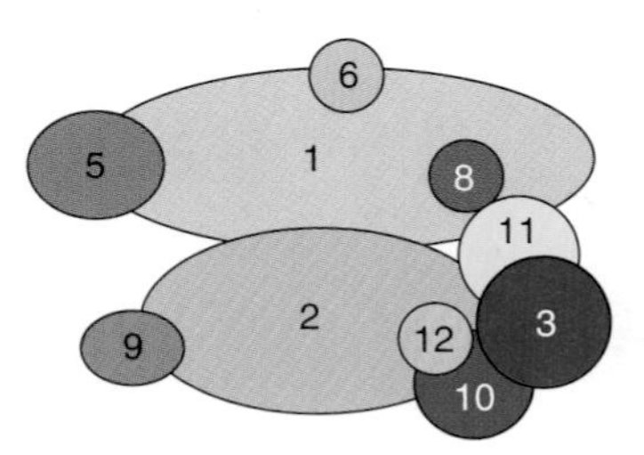

FIGURE 17.5 **Ribbon representation of yeast 4/7 RNA polymerase II structure.** Colors of specific subunits are indicated in the color code diagram. Numbers indicate the Rpb subunits. Some specific features of the 4/7 RNA polymerase II are indicated. The carboxyl terminal domain (CTD) of Rpb1 is largely unstructured and so is not visible. The active site magnesium is shown as a pink sphere. (Structure from Protein Data Bank 1I50. P. Cramer, et al., *Science* 292 [2001]: 1863–1876. Prepared by B. E. Tropp.)

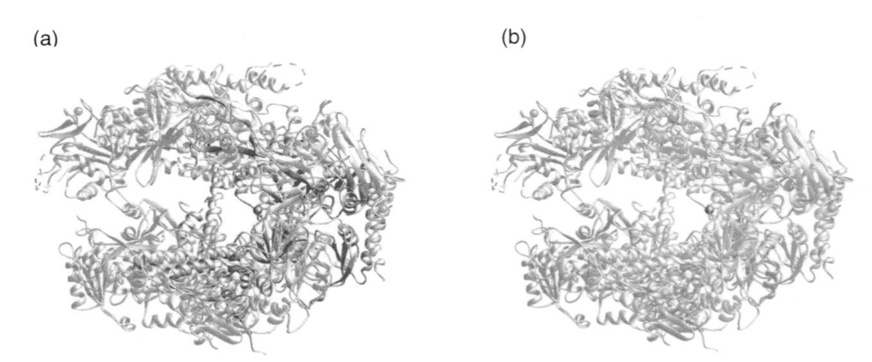

FIGURE 17.6 **A conserved RNA polymerase core structure.** (a) Blocks of sequence homology between the two largest subunits of bacterial and eukaryotic RNA polymerases are in red. (b) Regions of structural homology between RNA polymerase II and bacterial RNA polymerase, as judged by a corresponding course of the polypeptide backbone, are in green. (Reproduced from P. Cramer, D. A. Bushnell, and R. D. Kornberg, *Science* 292 [2001]: 1863–1876. Reprinted with permission from AAAS. Photos courtesy of Roger D. Kornberg, Stanford University School of Medicine.)

cleft in both the bacterial and yeast RNA polymerases have positively charged side chains. The inner core region of RNA polymerase II shares many structural features with bacterial RNA polymerases, including the following: the cleft, active site, bridge helix, trigger loop, rudder, lid, and secondary channel. The first three of these features are indicated in Figure 17.5 but the others are not.

The major problem in crystallizing the 12 subunit RNA polymerase II—the fact that subunits Rpb4 and Rpb7 are present in substoichiometric quantities—was solved using two different approaches. Kornberg and coworkers purified the complete RNA polymerase II from a yeast strain with a recombinant Rpb4 that carried a tag, which allowed the 12-subunit RNA polymerase II to be separated from Δ4/7 RNA polymerase II. Patrick Cramer and coworkers added excess Rpb4 and Rpb7 subunits to the ten-subunit RNA polymerase II. The crystal structure of the 12-subunit RNA polymerase II (FIGURE 17.7) reveals that the tip of the Rpb7 subunit inserts into a

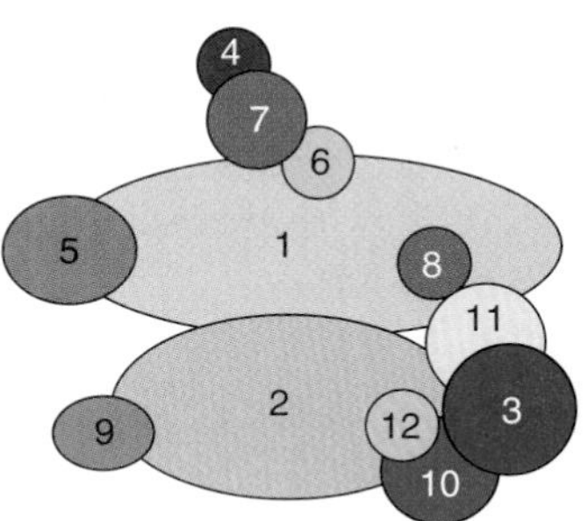

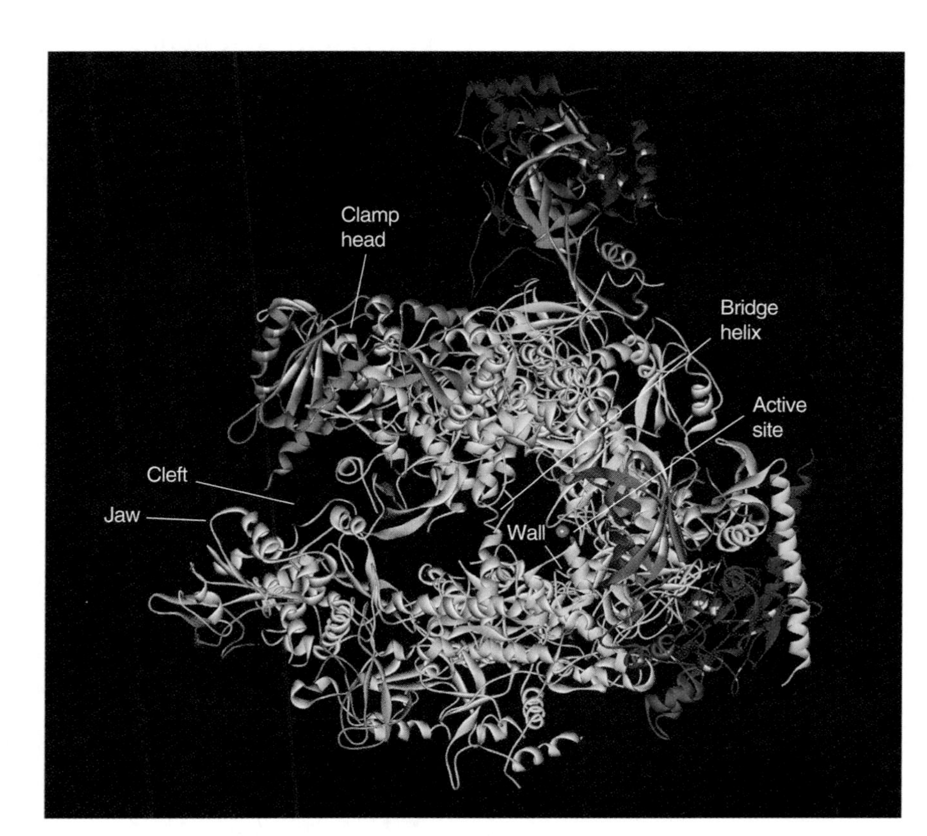

FIGURE 17.7 **Ribbon representation of the complete 12-subunit yeast RNA polymerase II structure.** Colors for specific subunits are indicated in the color code diagram. Numbers indicate the Rpb subunits. Some specific features of RNA polymerase II are indicated. The carboxyl terminal domain (CTD) of Rpb1 is largely unstructured and so is not visible. The active site magnesium is shown as a pink sphere. (Structure from Protein Data Bank 1WCM. K.-J. Armache, et al., *J. Biol. Chem.* 280 [2005]: 7131–7134. Prepared by B. E. Tropp.)

conserved pocket formed by subunits Rpb1, Rpb2, and Rpb6, acting as a wedge that locks the clamp (a part of the Rpb1) into a closed position. The cleft, thus, is actually much narrower than observed in the Δ4/7 RNA polymerase II, making it impossible for the double-stranded DNA to fit in the cleft during the initiation stage of transcription.

As we see below, several additional transcription factors are required for transcription initiation to take place inside the cell. RNA polymerases I and III have similar structures to the 12-subunit RNA polymerase II because corresponding subunits in all three nuclear RNA polymerases exhibit extensive homology.

The crystal structure has been determined for the complete 12-subunit yeast RNA polymerase II bound to a transcription bubble and product RNA.

Investigators realized that they might be able to obtain a "snapshot" of the transcription elongation complex if they could prepare the complex in crystal form. Cramer and coworkers approached the problem by assembling a complex that contained the 12-subunit RNA polymerase II bound to a transcription bubble and product RNA (FIGURE 17.8a). The crystal structure and a schematic cut-away view of the 12-subunit RNA polymerase II elongation complex are shown in FIGURE 17.8b and c, respectively. Protein "jaws" grip the downstream DNA that enters the enzyme. The DNA extends along the cleft toward the catalytic site and starts to unwind at position +3, where +1 is the entry position for the incoming nucleotide. Nucleoside triphosphates reach the active site by passing through a funnel-shaped secondary channel on the underside of the enzyme and then through a pore near the active site. The 3′-end of the growing RNA chain lies above the pore. The active site resembles that described for bacterial RNA polymerase (see Chapter 15). The DNA-RNA hybrid, consisting of 7 to 8 bp that emerges from the active site, is forced to change direction and turn by about 90° with respect to the downstream DNA because the "wall" at the end of the cleft blocks straight passage through the enzyme. The enzyme helps to separate the DNA-RNA strands. The nucleotide addition cycle for RNA polymerase II is the same as that described for the bacterial RNA polymerases (see Chapter 15).

17.3 Transcription Start Site Identification

Nuclear RNA polymerases have limited synthetic capacities.

Despite their large size and complexity, the nuclear RNA polymerases have very limited synthetic capacity. For instance, RNA polymerase II, acting on its own, can initiate a low level of transcription from random sites when the DNA template has a nick, a single-stranded gap, or a 3′-overhang. It cannot, however, initiate transcription from specific

(a)

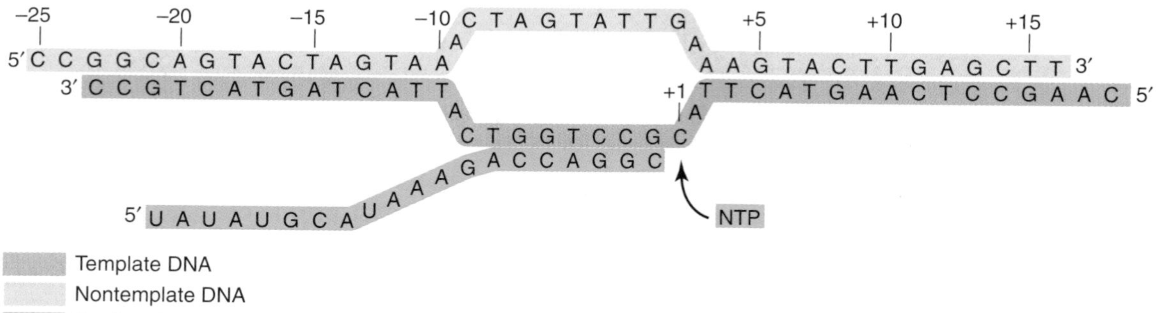

(b)

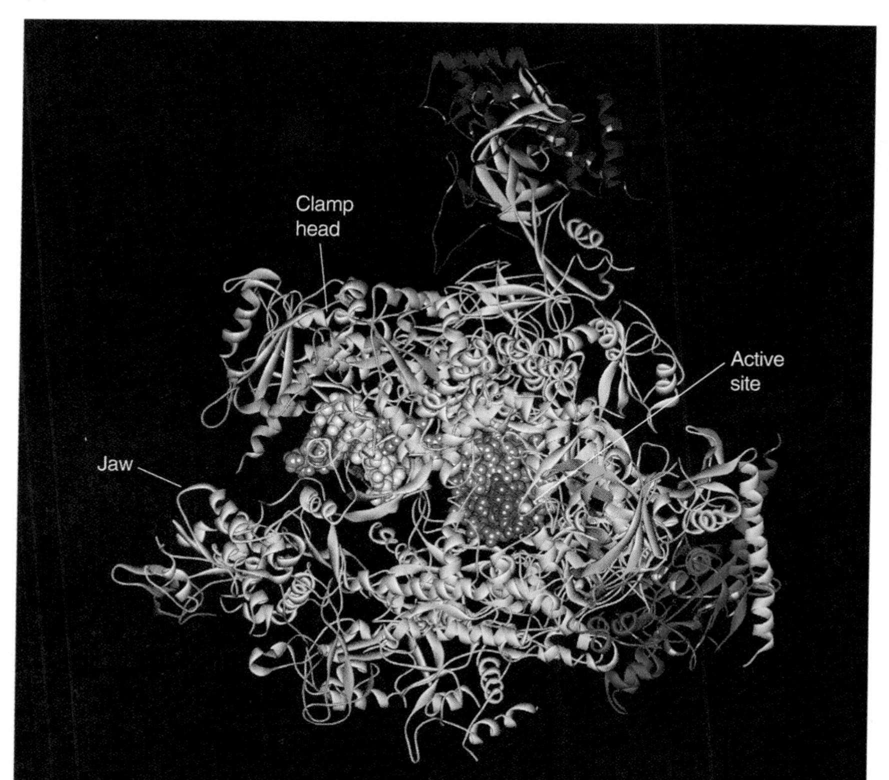

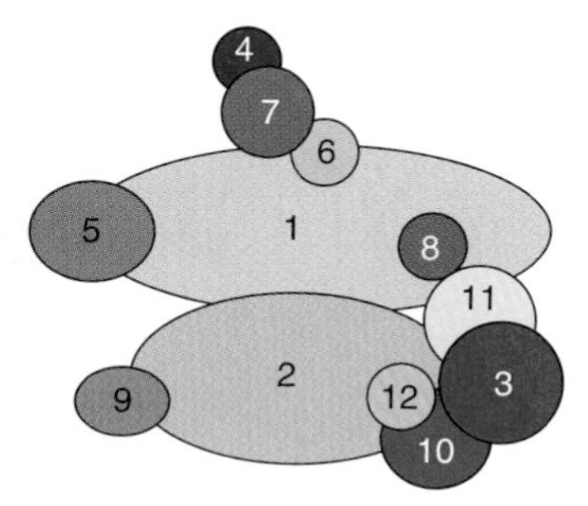

(c)

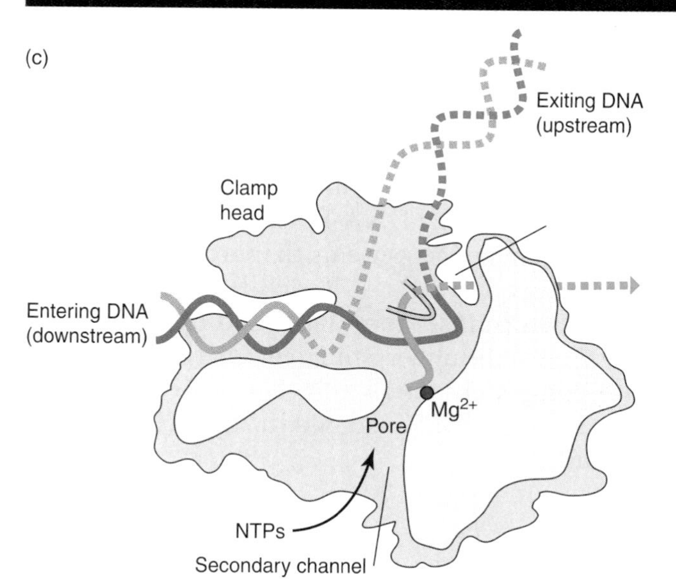

FIGURE 17.8 **The yeast RNA polymerase II transcription elongation complex.** (a) Schematic diagram of the transcription bubble and product mRNA. The bubble, formed by a 41-mer DNA duplex with an 11 nucleotide mismatch region, has an RNA 20-mer with eight 3′-terminal nucleotides complementary to the DNA template in the bubble. (b) A ribbon model of the RNA polymerase II subunits and nucleic acids in the transcription elongation complex. Colors of specific subunits are indicated in the color code diagram above. Numbers indicate the Rpb subunits. Nucleic acids are shown as spacefill structures. Template DNA, nontemplate DNA, and product RNA are shown as dark blue, light blue, and green space fill structures, respectively. The incoming nucleoside triphosphate is shown as an orange spacefill structure and the magnesium ion at the active site is shown as a pink sphere. (c) A schematic cut-away view of the 12 subunit RNA polymerase II transcription elongation complex. The view is the same as that shown in the ribbon model in (b). The DNA template and nontemplate strands are shown in dark blue and light blue, respectively. The nascent RNA chain is shown in green. Dashed lines indicate uncertainty about exact strand position. Nucleotide triphosphates, NTPs, enter the active site by first passing through the secondary channel and then through a pore. The magnesium ion shown as a pink sphere indicates the active site's location. (Part a adapted from K. J. Armache, H. Kettenberger, and P. Cramer, *Curr. Opin. Struct. Biol.* 15 [2005]: 197–203. Part b structure from Protein Data Bank 1Y77. H. Kettenberger, K.-J. Armache, and P. Cramer, *Mol. Cell* 16 [2004]: 955–965. Prepared by B. E. Tropp. Part c adapted from S. Hahn, *Nat. Struct. Mol. Biol.* 11 [2004]: 394–403.)

start sites within an intact double-stranded DNA template without the assistance of other proteins.

RNA polymerases I and III also require assistance from additional protein factors to catalyze specific transcription. Moreover, each of the nuclear RNA polymerases requires its own specific set of transcription factors to assist it in locating the transcription start site and making RNA. Because the details of the transcription process differ for each of the nuclear RNA polymerases, it is necessary to examine each one separately. Here, we examine RNA polymerase II and the transcription factors that assist it. RNA polymerases I and III and the transcription factors that assist them are explored in Chapter 20.

Various techniques have been devised to locate RNA polymerase II transcription start sites.

Several different techniques have been devised to locate transcription start sites. Three commonly used techniques assume that we already have a rough idea of the transcription start site's location. The first two, **S1 mapping** (also known as **S1 nuclease analysis**) and **primer extension**, identify transcription start sites in living cells as well as in cell-free systems. The third technique, **run-off analysis**, although experimentally less demanding, can only be used in cell-free systems.

S1 Mapping

S1 mapping requires a labeled, single-stranded DNA probe that spans the transcription start site. The labeled probe is added to a mixture of RNA molecules, which have been extracted from the cell sample, under conditions favorable to hybrid formation. The probe is designed to specifically anneal to complementary mRNA to produce a DNA-RNA hybrid with a 3′-polyribonucleotide overhang at one end and a 3′-polydeoxyribonucleotide overhang at the other (FIGURE 17.9). **S1 nuclease** (an endonuclease obtained from *Aspergillus oryzae* that hydrolyzes single-stranded DNA and RNA but that has no effect on a DNA-RNA hybrid) is then added to remove the single-strand polynucleotide overhangs and the resulting DNA hybrid is denatured. The length of the single-stranded DNA (275 nucleotides in the present example), which is determined by polyacrylamide gel electrophoresis (PAGE), indicates the distance from the 5′-end of the probe to the transcription start site, thereby identifying the transcription start site's location.

S1 mapping has two important limitations that both relate to the S1 nuclease reaction. The ends of the DNA-RNA hybrid tend to fray, allowing the S1 nuclease to make deeper cuts than expected. Fraying is an especially important concern for A-T rich transcription start sites. If fraying occurs, the distance from the 5′-end of the probe to the transcription start site will be underestimated. On the other hand, if S1 nuclease fails to completely remove the single-stranded DNA overhang, the distance between the 5′-end and the transcription start sites will be overestimated.

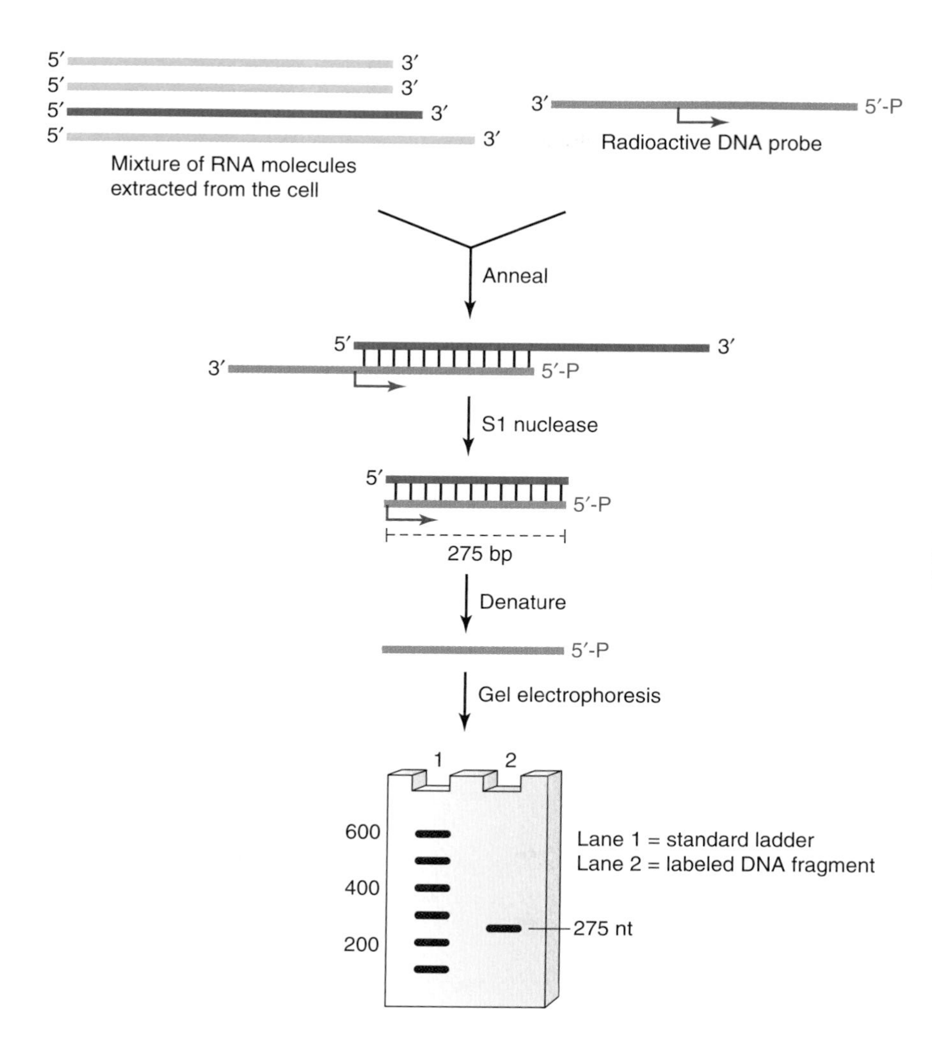

FIGURE 17.9 **S1 mapping.** A labeled DNA probe is added to a mixture of RNA molecules, which have been extracted from the cell sample, under conditions favorable to hybrid formation so that the probe specifically anneals to complementary mRNA (shown in dark green) to produce a DNA-RNA hybrid with a 3′-polyribonucleotide overhang at one end and a 3′-polydeoxynucleotide overhang at the other. Then S1 nuclease is added to remove the single-strand polynucleotide overhangs. The length of the single-stranded DNA (275 nucleotides in the present example), which is determined by polyacrylamide gel electrophoresis (using a standard ladder for reference) indicates the distance from the 5′-end of the probe to the transcription start site and therefore identifies the transcription initiation site's location.

Primer Extension Method

The primer extension method (FIGURE 17.10) involves three steps. (1) A labeled DNA primer of about 18 nucleotides, which has a sequence identical to a region that is 50 to 150 nucleotides downstream from the transcription start site, is annealed to a specific RNA within a mixture of RNA molecules extracted from the cell. (2) The primer is extended to the RNA template's 5′-end by reverse transcriptase (see Chapter 5). (3) The DNA-RNA hybrid is denatured and the DNA product's size is established by gel electrophoresis.

The exact location of the transcription start site can be determined by comparing the mobility of the extended primer with that of products formed by the dideoxynucleotide DNA sequencing method (see Chapter 5) in which the same primer is used. This result follows from the fact that the 3′-end of the extended primer should coincide with the 5′-end of the RNA template. The DNA product's length, therefore, indicates the distance from the labeled 5′-end of the primer

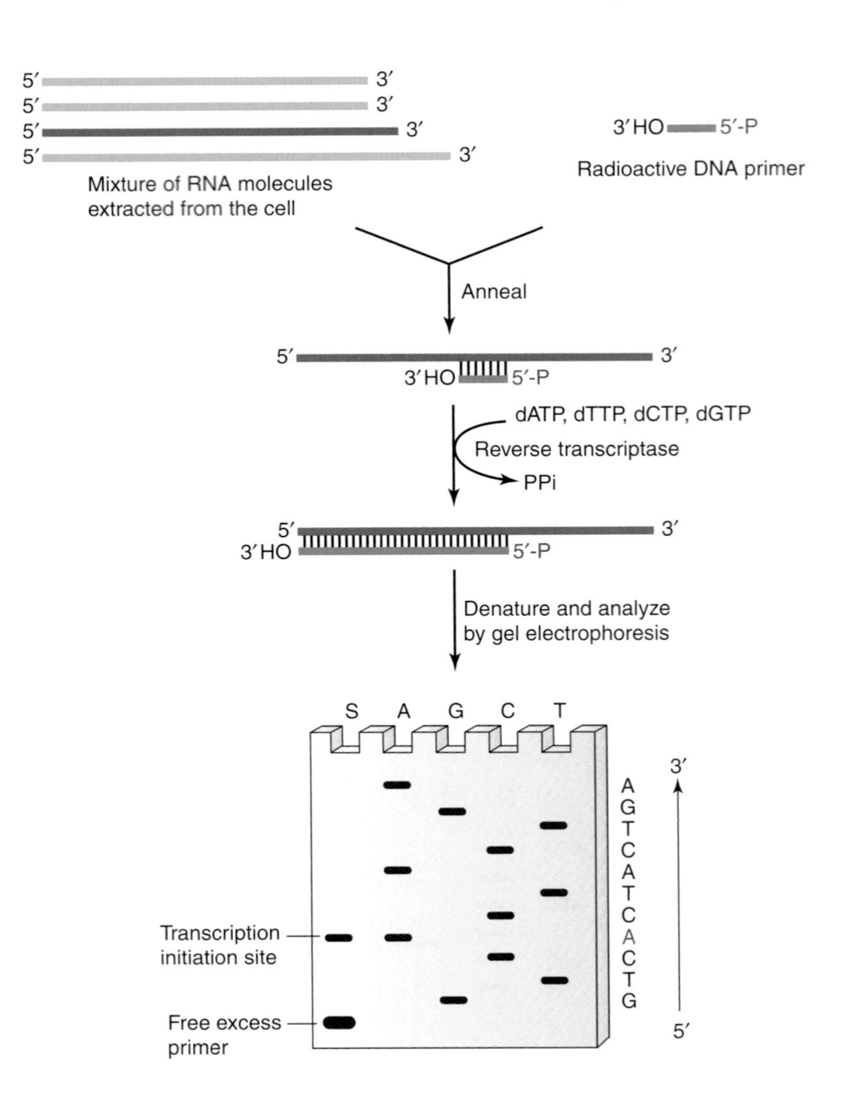

FIGURE 17.10 **Primer extension method.** A labeled DNA primer of about 18 nucleotides, which has a sequence identical to a region that is about 50 to 150 nucleotides downstream from the transcription initiation site, is synthesized by chemical means and annealed to a specific mRNA (shown in dark green) within a mixture of mRNA molecules (light green) extracted from the cell. The primer is extended to the RNA template's 5′-end by reverse transcriptase and the DNA-RNA hybrid is denatured and the size of the DNA product (blue) is established by gel electrophoresis. The exact location of the transcription initiation site can be determined by comparing the mobility of the extended primer with that of products produced by the dideoxynucleotide DNA sequencing method in which the same primer is used.

to the transcription start site. Although primer extension may seem to be simpler and more accurate than S1 mapping, this is not necessarily so because it is sometimes difficult to find a primer that works well with a new gene and reverse transcriptase does not work well with some primers.

Run-Off Transcription Method

Run-off transcription is usually the method of choice for identifying a transcription start site for *in vitro* studies of a cloned gene because it is experimentally less demanding than the other two methods. Run-off transcription takes advantage of the fact that RNA polymerase falls off a disrupted gene at the end of a linear template to produce a truncated transcript (FIGURE 17.11). The linear template, which can be constructed by cleaving a plasmid bearing the gene of interest at a restriction site about 200 bp downstream from the transcription start site, is added to a nuclear extract (or purified transcription system) along with radioactive NTPs. The length of the radioactive RNA product is determined by gel electrophoresis. Because the position of

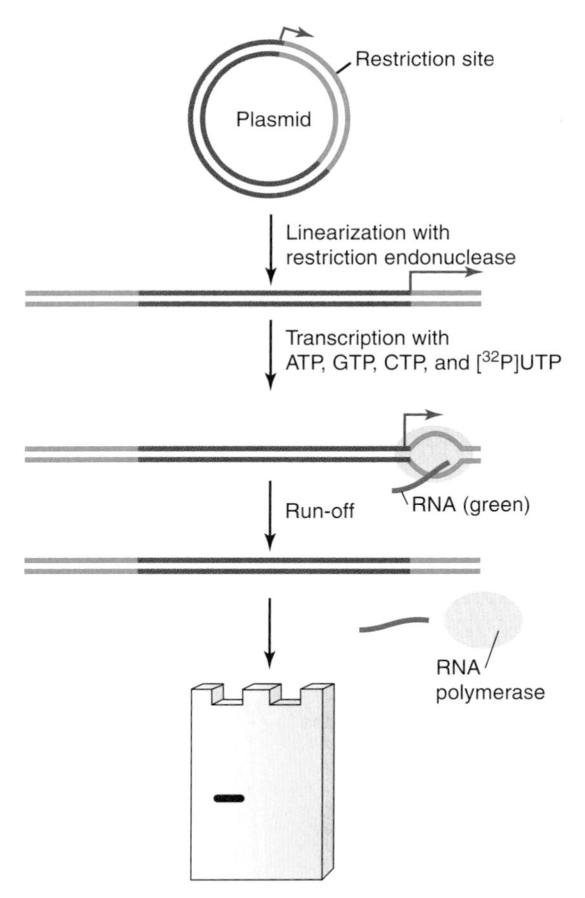

FIGURE 17.11 Run-off transcription method. A linear template, which can be constructed by cleaving a plasmid bearing the gene of interest (shown in light blue) at a restriction site about 200 bp downstream from the transcription initiation site, is added to a nuclear extract (or purified transcription system) along with radioactive nucleoside triphosphates. RNA polymerase is shown as a tan oval. The length of the radioactive RNA product (green) is determined by gel electrophoresis. Because the position of the restriction site at the 3′-end of the truncated gene is known, the length of the run-off transcript indicates the location of the transcription start site.

the restriction site at the 3′-end of the truncated gene is known, the length of the run-off transcript indicates the location of the transcription start site.

Cap analysis of gene expression (CAGE) is a high throughput technique used to identify transcription start sites and their flanking promoters.

The S1 mapping, primer extension, and run-off transcription methods permit us to investigate transcription start sites in a specific gene of interest. They do not, however, permit us to learn the sequences of large numbers of transcription start sites in a single experiment. Several new high throughput techniques have been developed to experimentally identify transcription start sites and their flanking core promoters. A widely used approach developed by Piero Carninci and coworkers, **cap analysis of gene expression** (CAGE), takes advantage of the fact that eukaryotic mRNA molecules have 7-methylguanosine caps at their 5′-ends (see Chapter 19). For convenience, we divide the CAGE technique into two parts. FIGURE 17.12 shows the first part, the preparation of the DNA copy of the mRNA. Oligo (dT) primers are annealed to the poly(A) tail of the mRNA (see Chapter 5). Reverse transcriptase

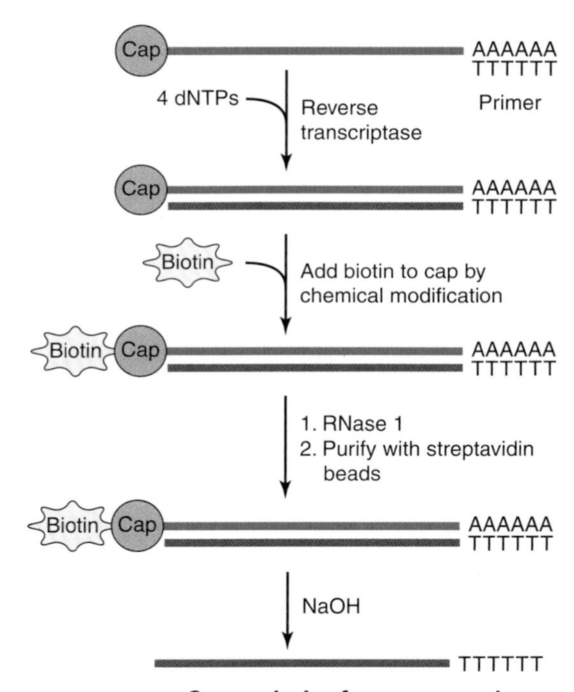

FIGURE 17.12 Cap analysis of gene expression—preparation of DNA. Oligo (dT) primers are annealed to the poly(A) tail of the mRNA. Reverse transcriptase extends the primer, using mRNA as a template. Biotin is added to the mRNA cap group and then the DNA-RNA hybrid is incubated with RNase I, which cleaves single-stranded RNA but not the DNA-RNA hybrid. Biotinylated DNA-RNA hybrids are separated from other nucleic acids by binding to streptavidin coated beads. DNA is released from the streptavidin coated beads by placing the beads in a sodium hydroxide solution, which degrades the RNA but not the DNA.

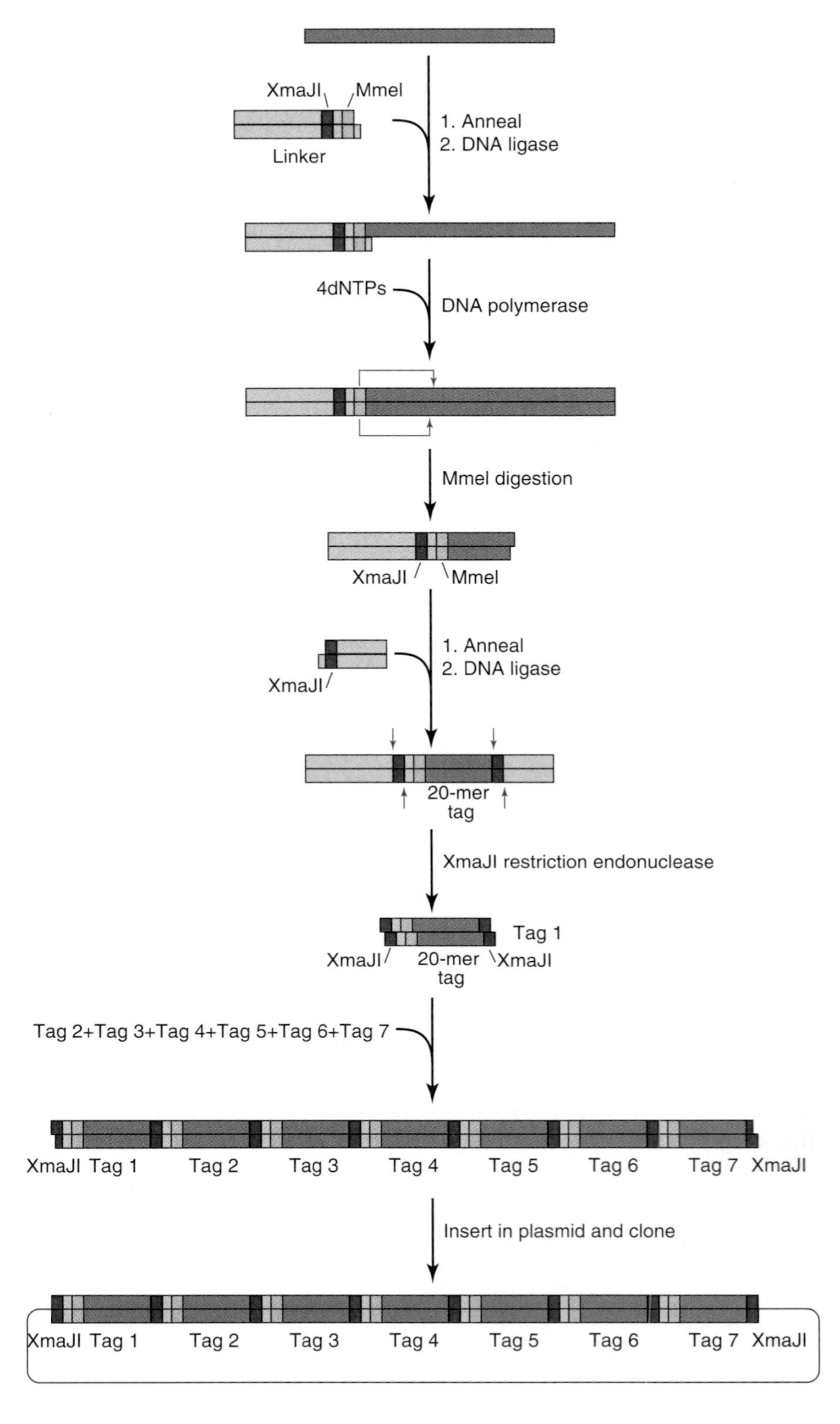

FIGURE 17.13 **Cap analysis of gene expression: Identification of flanking promoters.** DNA is joined to a linker that has an MmeI restriction signal and an XmaJI restriction site. DNA polymerase extends the 3′ end of the linker. The resulting double-stranded DNA is treated with MmeI, which makes a staggered cut about 20 nucleotides from the recognition site. The new end is ligated to a second linker that also has an XmaJI restriction site so that the resulting DNA molecule contains a 20-mer nucleotide tag bracketed by XmaJI restriction sites. The CAGE technique is performed in PCR microtiter plates that contain 96 wells so that 96 reactions are being run simultaneously. Furthermore, each well contains many different kinds of mRNA molecules. Ligating different Tags produces a large DNA molecule that contains sequence information for the 5′-ends of many different mRNA molecules. This information is obtained by inserting the large DNA molecule into a plasmid vector, cloning the vector, and determining the Tag sequences. Comparing this sequence information with the entire DNA genome sequence identifies the flanking promoter for each transcription start site.

extends the primer, using mRNA as a template. The next few steps ensure that the DNA copy extends to the 5′-end of the mRNA. Biotin is added to the mRNA cap group and then the DNA-RNA hybrid is treated with RNase I, which cleaves single-stranded RNA but not the DNA-RNA hybrid. If reverse transcriptase completes DNA synthesis to the 5′-end of the mRNA, then the mRNA retains its biotinylated cap, otherwise it does not. Biotinylated DNA-RNA hybrids are separated from other nucleic acids by binding to streptavidin coated beads. After separation, DNA is released from the streptavidin coated beads by placing the beads in a sodium hydroxide solution, which degrades the RNA but not the DNA.

The purified DNA is now ready for the second part of the CAGE technique, which is shown in FIGURE 17.13. The DNA is joined to a linker that has an MmeI restriction signal and an XmaJI restriction site. DNA polymerase extends the 3′ end of the linker. The resulting double-stranded DNA is treated with MmeI, an unusual Type II restriction endonuclease because it makes a staggered cut about 20 nucleotides from the TCCRAC (where R is a purine) recognition site.

$$5' \ldots \mathrm{TCCRAC(N)_{20}}\blacktriangledown \ldots 3'$$
$$3' \ldots \mathrm{AGGYTG(N)_{18}}\blacktriangle \ldots 5'$$

This property permits MmeI to generate a long sequence tag that includes the transcription start site. The new end is ligated to a second linker with an XmaJI restriction site. The resulting DNA molecule contains a 20-mer nucleotide tag bracketed by XmaJI restriction sites. Then the DNA is treated with XmaJI restriction endonuclease to generate staggered ends.

Although we have been focusing on events that take place with a single mRNA, the CAGE technique is performed in a PCR microtiter plate that contains 96 wells. Therefore, 96 reactions are run simultaneously. Furthermore, each well contains many different kinds of mRNA molecules. Ligating the so-called Tags thus produces a large DNA molecule that contains sequence information for the 5′-ends of many different mRNA molecules. This information is obtained by inserting the large DNA molecule in a plasmid vector, cloning the vector, and determining the Tag sequences. Comparing this sequence information with the entire DNA genome sequence identifies the region around each transcription start site, which is called the **core promoter**.

In the most extensive core promoter identification study to date, Carninci and coworkers used the CAGE technique to identify about 184,000 human and 177,000 mouse core promoters. The large number of core promoters can be explained, at least in part, by the fact that most mouse and human protein-coding genes have more than one core promoter and that many core promoters flank transcription start sites for non-coding RNA molecules. Vertebrate studies reveal the existence of two different types of transcription initiation: focused and dispersed (FIGURE 17.14). In **focused transcription initiation**, transcription starts at a single site or a cluster of sites that are located within a few nucleotides of one another. In **dispersed transcription initiation**, transcription starts at many weak sites distributed over 50 to 100 nucleotides. Most eukaryotic core promoters appear to be focused promoters. Vertebrate core promoters are an exception. About 70% of the known vertebrate core promoters are dispersed promoters, which usually are responsible for constitutive transcription and located in CpG islands (CG-rich genomic regions that contain a high frequency of CG dinucleotides).

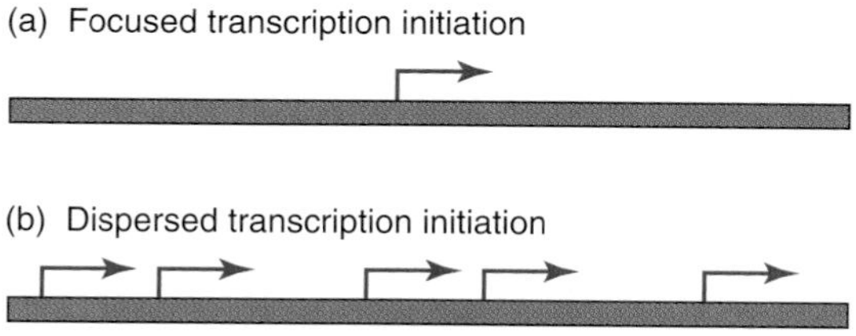

(a) Focused transcription initiation

(b) Dispersed transcription initiation

FIGURE 17.14 **Focused and dispersed transcription initiation.** (a) Focused transcription. Transcription starts at a single site or in a cluster of sites within a segment of a few nucleotides. (b) Dispersed transcription initiation. Transcription starts at several weak sites spread over about 50 to 100 nucleotides. (Modified from *Develop. Biol.*, vol. 339, T. Juven-Gershon and J. T. Kadonaga, Regulation of gene expression via the core promoter . . ., pp. 225–229, copyright 2010, with permission from Elsevier [http://www.sciencedirect.com/science/journal/00121606].)

17.4 The Core Promoter

The core promoter extends from 40 bp upstream of the transcription start site to 40 bp downstream from this site.

Investigators assumed that eukaryotic promoters, like bacterial promoters, would be just upstream from the transcription start site. Once methods became available for transcription start site identification, investigators started to search for promoters just upstream of these sites. Initial efforts to characterize the eukaryotic promoter concentrated on highly expressed protein-coding genes, such as those that code for hemoglobin, histone, and ovalbumin, because these genes were the easiest to study at the time. These efforts were rewarded in 1977 when David Hogness and coworkers discovered the consensus sequence TATAXAX (where X is an A or T) just before the transcription start site. This sequence motif, which is called the **TATA box**, is usually located 25 to 30 bp upstream from the transcription start site. The general transcription factor TFIID or one of its components, the **TATA binding protein** (**TBP**) binds to the TATA box (see below). As additional highly expressed protein coding-genes from animal cells and viruses became available for study, they too were shown to have TATA boxes 25 to 30 bp upstream from their transcription start sites. The TATA box also was observed just upstream from the transcription start site in protein-coding genes from plants and fungi. Its position in *S. cerevisiae*, however, was observed to vary from 30 to 120 bp upstream from the transcription start site.

The common occurrence of the TATA box suggested that it might be essential for the transcription of all protein coding genes. But as less highly expressed genes became available for study, it became apparent that most promoters lack a TATA box. The discovery of TATA-less promoters suggested that other short DNA sequences might be able to replace the TATA box function in transcription initiation. Several elements have been identified that either replace the TATA box in transcription initiation or act along with it during transcription initiation. The DNA region that includes these elements, the **core promoter**, extends from about 40 bp upstream of the transcription start site to about 40 bp downstream from this site. FIGURE 17.15 shows the TATA box and other elements in metazoan core promoters. The **initiator** (**Inr**), which is probably the most commonly occurring element in the core promoter, flanks the transcription start site and is recognized by the general transcription factor TFIID (see below). Three other core promoter elements, **DPE** (**downstream core promoter element**), **MTE** (**motif ten element**), and **DCE** (**downstream core element**) are also recognized by TFIID. The **BRE** (**TFIIB recognition element**) was initially identified as a site immediately upstream from the TATA box that binds the general transcription factor TFIIB (see below). Later, a second TFIIB binding site was identified immediately downstream of the TATA box. The upstream and downstream TFIIB binding sites are designated **BRE^u** and **BRE^d**, respectively. **XCPE1** (**X core promoter element 1**), which was identified from extensive analyses of the hepatitis B virus X gene promoter, is present in poorly characterized TATA-less

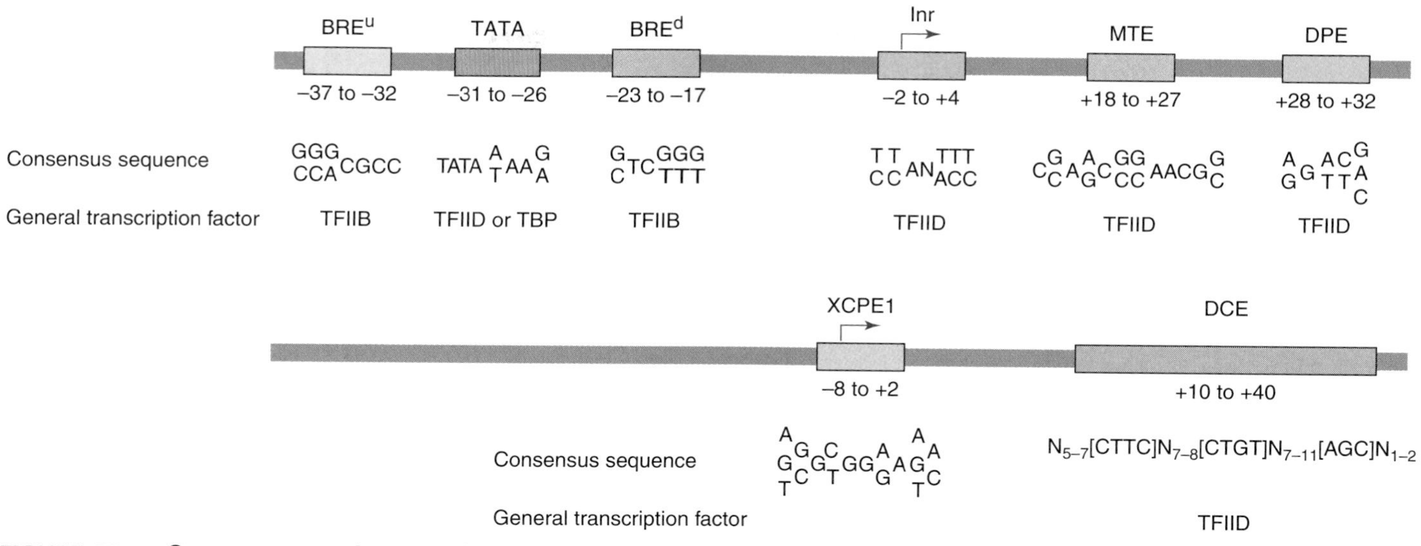

FIGURE 17.15 Core promoter elements that contribute to basal transcription in multicellular animals. (Adapted from G. A. Maston, et al., *Annu. Rev. Genomics Hum. Genet.* 7 [2006]: 29–59.)

human core promoters. It is estimated to be present in about 1% of human genes and, in contrast to the other core promoter elements, does not appear to act as a binding site for a general transcription factor.

Approximately 10,000 known human core promoters were surveyed for four core promoter elements (TATA, Inr, DPE, and BRE). Inr, the most common element, was present in nearly half of the core promoters, whereas DPE and BRE were each present in about a quarter of the core promoters, and TATA boxes were present in about one eighth of the core promoters. Remarkably, almost a quarter of the analyzed promoters had none of the four elements, raising the possibility that additional elements remain to be discovered.

17.5 General Transcription Factors: Basal Transcription

RNA polymerase II requires the assistance of general transcription factors to transcribe naked DNA from specific transcription start sites.

RNA polymerase II requires the assistance of protein factors to bind to the core promoter. This requirement was first demonstrated by studying specific initiation at the major late promoter of adenovirus DNA, which controls highly expressed genes for structural proteins in the virus particle (see Chapter 8). RNA polymerase II cannot catalyze specific initiation at this promoter but gains the ability to do so when a soluble cell-free extract from human KB cells (cells derived from an oral epidermoid carcinoma) is added to it. In 1979, Robert G. Roeder and coworkers used classical protein fractionation techniques

TABLE 17.4 **General Transcription Factors**

Factor	No. of Subunits	Functions
TFIIA	2	Stabilizes TBP and TFIID binding. Blocks the inhibitory effects of TAF1 and other proteins.
TFIIB	1	Stabilizes TFIID-promoter binding. Contributes to transcription start site selection. Helps recruit RNA polymerase II TFIIF to the core promoter.
TFIID (TBP and TAFs)	1 14	Binds TATA element and deforms promoter DNA. Platform for the assembly of TFIIB and TAFs. Binds Inr, MTE, DPE, and DCE promoter elements.
TFIIE	2	Helps to recruit TFIIH to the core promoter and is required for promoter melting.
TFIIF	3	Binds RNA polymerase II and is involved in recruiting the polymerase to the pre-initiation complex. Required to recruit EFIIE and EFIIH to the pre-initiation complex.
TFIIH	10	Functions in transcription and DNA repair. It has kinase and helicase activities and is essential for open complex formation.

to isolate protein factors from the KB cell extract that assist RNA polymerase II. These protein factors, or **general transcription factors (GTFs)** as they are now known, were named **TFIIA, TFIIB, TFIID, TFIIE, TFIIF,** and **TFIIH.** The first two letters, TF, indicate the protein is a general transcription factor; the Roman numeral II signifies the factor supports RNA polymerase II transcription; and the final letter was assigned based on the protein fractionation scheme rather than on protein function. (The letters C and G are missing because later studies showed that the proteins originally assigned these letters are not transcription factors for RNA polymerase II.) Subsequent studies by Roeder and other investigators demonstrated that these general transcription factors are present in all eukaryotes from yeast to humans. Table 17.4 summarizes the functions of yeast general transcription factors. Counterparts in other eukaryotes serve the same functions. The general transcription factors are described below in greater detail.

The core promoter allows a cell-free system to catalyze a low-level of RNA synthesis at the correct transcription start site.

RNA polymerase II, acting together with TFIIA, TFIIB, TFIID, TFIIE, TFIIF, and TFIIH, forms a transcription machine that can initiate transcription at the correct start site of a linear duplex that has a core promoter with a TATA box. This transcription machine also determines the correct direction of transcription. Because the level of transcription catalyzed by RNA polymerase II together with the general transcription factors is much lower than that observed in the cell, it is called **basal transcription**. For this reason, some investigators prefer to use the term basal transcription factor instead of general transcription factor. RNA polymerase II and general (basal) transcription factors assemble at the transcription start site to form a **preinitiation complex** or **PIC**. When the core promoter has a TATA box, assembly takes place through the multi-step pathway shown in FIGURE 17.16. The order of addition and general transcription factor requirements may vary at other kinds of core promoters.

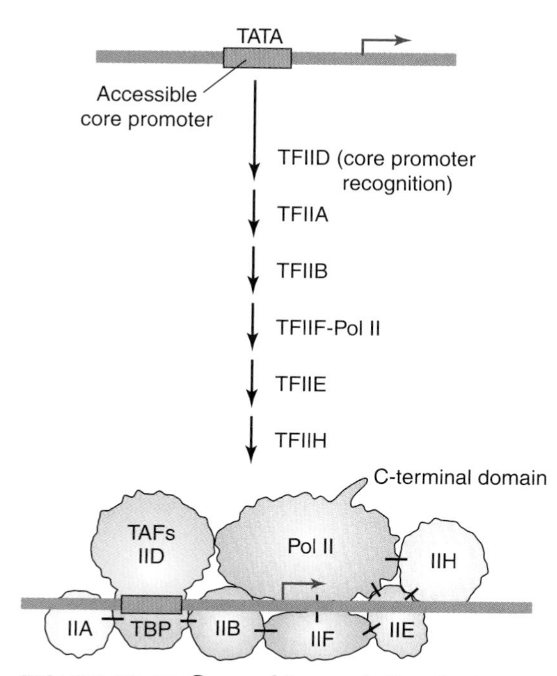

FIGURE 17.16 **General transcription factors and preinitiation complex (PIC) pathway for RNA polymerase II promoters with a TATA containing core promoter.** The preinitiation complex is assembled on DNA in a multistep process in which TFIID (D) binds first, followed by TFIIA (A), TFIIB (B), a pre-formed complex of RNA polymerase II (Pol II), and TFIIF (F), TFIIE (E), and TFIIH (H). Some of the known interactions among the general transcription factors and RNA polymerase II are indicated by lines. The red arrow indicates the transcription initiation site. (Adapted from R. G. Roeder, *Nat. Med.* 9 [2003]: 1239–1244.)

When the core promoter has a TATA box, preinitiation complex assembly begins with either TFIID or TATA binding protein (TBP) binding to the core promoter.

Michele Sawadogo and Robert Roeder elucidated the first step in preinitiation complex assembly in 1985 when they demonstrated that TFIID binds to the TATA box. TFIID is the only general transcription factor that can bind specifically to a core promoter that has a TATA box without the assistance of other factors. Initial attempts to study TFIID–TATA interactions were hampered by difficulties in purifying the scarce protein and in determining which of its many subunits actually binds to TATA.

A major breakthrough in our understanding of TATA box recognition occurred in 1988 when Stephen Buratowski and coworkers showed that a yeast protein can substitute for mammalian TFIID in a reconstituted mammalian transcription system. This demonstration led to the purification of a yeast protein called the **TATA-binding protein** (**TBP**) that specifically binds to the TATA box. The yeast gene encoding TBP was cloned and sequenced, providing the information needed to locate homologous genes in other organisms.

Yeast TBP has a 60-residue N-terminal region and a 180-residue C-terminal region. Mutants that lack the N-terminal region are viable, indicating that this region is not required for transcription. As might be expected for a nonessential region, the N-terminal region varies in size and sequence from one type of organism to another. In contrast, the C-terminal region is both essential for transcription and highly conserved. The overall folding pattern of the C-terminal region is the same for all TBP molecules that have been studied to date. Its structure from the flowering plant Arabidopsis thaliana, which is shown in FIGURE 17.17a, resembles a molecular saddle with two stirrups. The carboxyl terminal region has two identically folded domains, each containing a β-sheet and a long and a short α-helix. This high degree of internal symmetry suggests the C-terminal region of TBP probably evolved by the duplication of an ancestral gene. If so, this duplication must have occurred before the archaea and eukaryotes diverged because the C-terminus of archaeal TBP is approximately 40% identical to that of eukaryotic TBP and the two kinds of TBP have very similar structures.

The crystal structure of the C-terminal region of the TBP bound to the TATA box also has been determined (FIGURE 17.17b). Protein-DNA interactions occur through an induced fit mechanism involving conformational changes in both TBP and the TATA box. Amino acids on the underside of the saddle are very hydrophobic and interact with the hydrophobic

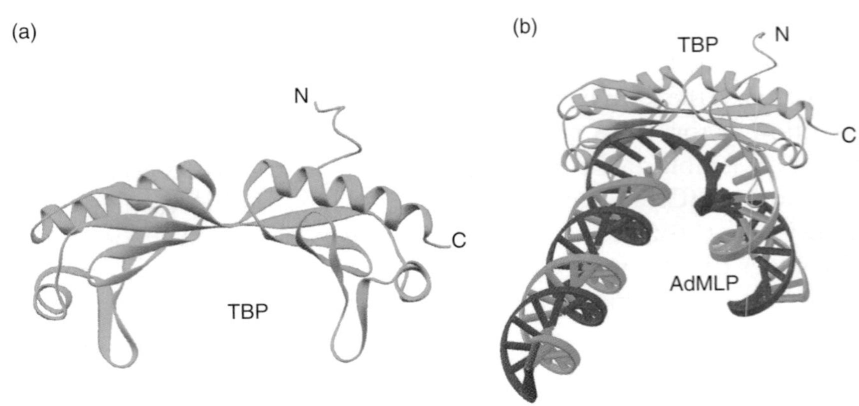

FIGURE 17.17 Crystal structure for the C-terminal region of TBP. (a) The three-dimensional structure of the C-terminal region of TBP from the flowering plant *Arabidopsis thaliana.* (b) The three-dimensional structure of the C-terminal DNA region of TBP from *Arabidopsis thaliana* bound to the TATA box of the adenovirus major late promoter (AdMLP). (Photos courtesy of Stephen Burley.)

groups in the DNA minor groove. This interaction is noteworthy because proteins nearly always interact with the major groove. A pair of phenylalanine side chains insert after the first T-A base pair in the TATA box, causing the minor groove to widen (FIGURE 17.18). A kink in the DNA forms as a result of the abrupt transition to a partially unwound right-handed double helix. A second pair of phenylalanine side chains insert before the last A-T base pair of the TATA box, producing a second kink and an equally abrupt return to B-form DNA. The net effect is that the DNA bends by about 80°. The two "stirrups" assist in this bending. The convex (upper) side of TBP binds additional protein factors that are required for transcription (see Chapter 18).

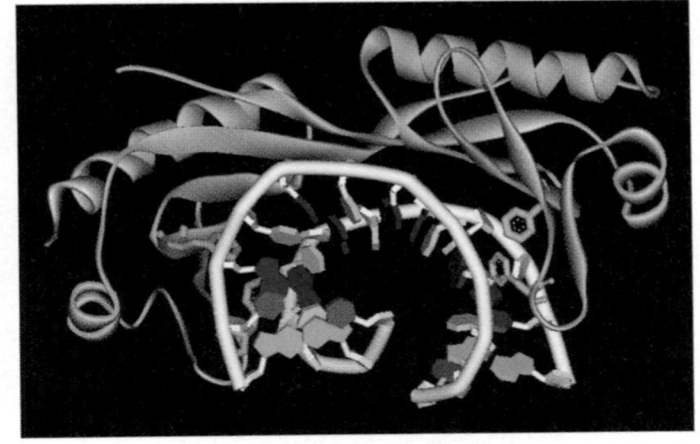

FIGURE 17.18 The C-terminal region of human TBP bound to a DNA duplex with the sequence CGTATATATACG. This binding causes a sharp bend in the DNA duplex. A pair of phenylalanine side chains (Phe 284 and Phe 301) shown as orange stick structures insert after the first T-A base pair in the TATA box, causing the minor groove to widen. A kink in the DNA forms as a result of the abrupt transition to a partially unwound right-handed double helix. A second pair of phenylalanine side chains (Phe 193 and Phe 210) shown as pink stick structures insert before the last A-T base pair of the TATA box, producing a second kink and an equally abrupt return to B-form DNA. (Structure from Protein Data Bank 1TGH. Z. S. Juo, et al., *J. Mol. Biol.* 261 [1996]: 239–254. Prepared by B. E. Tropp.)

TFIID can bind to core promoters of protein-coding genes that lack a TATA box.

Further study showed that TBP is the TATA binding subunit in TFIID. The additional proteins present in TFIID, called **TBP-associated factors (TAFs)** are required to transcribe genes that lack a TATA box as well as for the high levels of transcription that occur within the cell. TFIID in all organisms from yeast to humans contain a core set of 13 TAFs that is evolutionarily highly conserved. These TAFs are designated TAF1 to TAF13. TFIID is horseshoe shaped with its TBP subunit just above the central cavity of the horseshoe (FIGURE 17.19). Although crystal structures for the TFIID•core promoter complex are not yet available, the TBP subunit in TFIID probably binds to the TATA box pretty much as described for free TBP.

TFIID also participates in the transcription of protein-coding genes that lack a TATA box. In this case, however, it is the TBP associated factors or TAFs (and not TBP) that bind to regulatory elements in the core promoter. FIGURE 17.20 is a schematic showing the interactions of human TAFs with Inr and DPE in a promoter that lacks the TATA box. As depicted, TAF2 interacts with Inr while TAF6 and TAF9 interact with DPE. This binding assures that TBP will be positioned so that the upper (convex) part of its saddle will be available to interact with other proteins required for transcription. The TFIID conformation may vary depending on the nature of the contacts made with the core promoter, exposing different regions of TBP and the TAFs. Even more variation is possible because cells appear to have more than one kind of TFIID. This variation results in part from the fact that eukaryotes have TBP-like proteins that can replace TBP and in part from the fact that noncore TAFs and TAF-like factors may also be present.

Many factors influence TBP recruitment or activity. Two protein factors in yeast, **negative cofactor 2** (**NC2**) and **modifier of transcription 1** (**Mot1**), are of special interest. Similar protein factors are present in other eukaryotes. NC2, a heterodimer, binds to the TBP•TATA box complex, and prevents TFIIA and TFIIB from adding to the complex. NC2 also helps TBP to dissociate from the TATA box and slide along DNA. Mot1 uses ATP hydrolysis to release TBP from DNA. These actions of NC2 and Mot1 are consistent with the observation that the proteins inhibit TATA box-dependent transcription. Further studies showed that NC2 and Mot1 have the opposite effect on DPE-dependent transcription. One explanation for the proteins' ability to stimulate transcription from TATA-less promoters is that TBP molecules bind much more tightly to TATA boxes than they do to TATA-less promoters. NC2 and Mot1 therefore help to release tightly bound TBP from TATA boxes (and from non-productive binding sites) so that the TBP is available to bind where needed.

One final point needs to be made before we can examine the way that the other GTFs and RNA polymerase II become part of the preinitiation complex. TBP•TATA box (or TFIID•TATA box) complex formation and nucleosome formation are mutually exclusive processes; that is, TBP (or TFIID) cannot bind to DNA regions that are part of a nucleosome and histones cannot interact with DNA that is part of a TBP•TATA box (or TFIID•TATA box) complex. As we see in

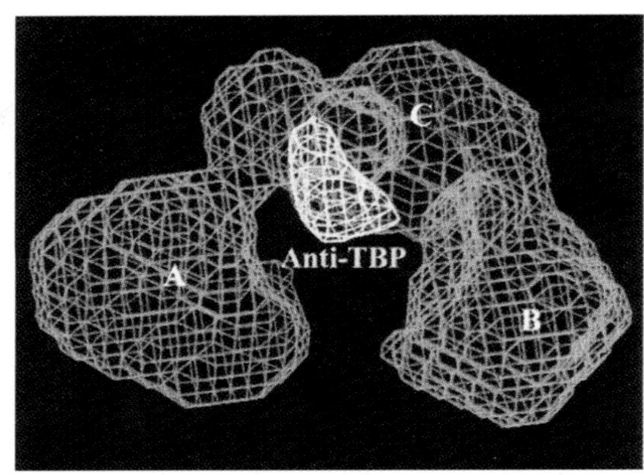

FIGURE 17.19 **Mapping of TBP on TFIID.** The blue mesh shows the horseshoe shape of the TFIID complex, which contains thirteen TBP associated factors (TAFs). The three lobes in the horseshoe are indicated by the letters A, B, and C. The yellow mesh shows the location of antibody that binds to TBP. (Reproduced from F. Andell III, et al., *Science* 286 [1999]: 2153–2156. Reprinted with permission from AAAS. Photo courtesy of Robert Tijan, University of California, Berkeley.)

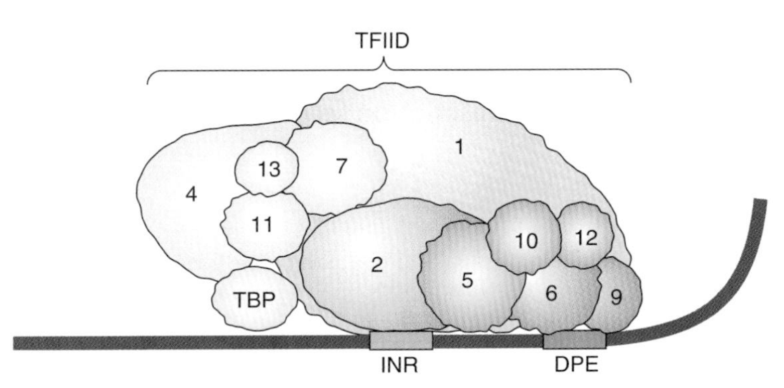

FIGURE 17.20 **Schematic composite showing the interactions of various TFIID subunits with the core promoter.** Several TFIID subunits have been implicated in binding to the core promoter, including TBP (yellow), which binds to the TATA box when it is present, TAF2 (blue), which interacts with the initiator (Inr), and TAF6 (orange) and TAF9 (orange), which interact with the downstream promoter element (DPE). Many, but not all, of the 13 highly conserved TAFs are shown in this figure. (Adapted from A. M. Näär, B. D. Lemon and R. Tijian, *Annu. Rev. Biochem.* 70 [2001]: 475–501.)

Chapter 18, special protein factors and enzymes are required to move or remove nucleosomes so that transcription can start. We will steer clear of this problem for now by considering naked double-stranded DNA templates.

TFIIA is not required to reconstitute a minimum transcription system.

Once investigators had established that TBP binds to the TATA box, it was only natural that they would try to determine the order in which the other general transcription factors bind to the core promoter. The **gel mobility shift assay** proved to be quite helpful in this endeavor. The underlying principle behind this assay is that binding a protein to a radioactive DNA fragment will produce a complex that will migrate more slowly than the free DNA fragment when both are subjected to gel electrophoresis (FIGURE 17.21). Using this assay, it is possible to show that TBP (or TFIID) bind to DNA fragments that contain a core promoter with a TATA box but that the other general transcription factors do not do so. DNase footprint studies provide further conformation for this binding.

The next step was to determine the order in which the other general transcription factors bind to the TBP•DNA complex. The TFIID•DNA complex was not used in early studies because investigators had not yet learned how to obtain sufficient quantities of pure TFIID. In any event, the TFIID•DNA complex would not have been suitable because the gel mobility shift assay is not sensitive enough to detect the small migration shift that would accompany the binding of a general transcription factor to the very large TFIID•DNA complex. The assay, however, was sufficiently sensitive to reveal that either TFIIA or TFIIB can bind to the TBP•DNA complex and, hence, the order of addition is TBP followed by either TFIIA or TFIIB.

Human TFIIA (which has three subunits, molecular masses 13, 19, and 37 kDa) is needed for basal transcription when TFIID is used to assemble the preinitiation complex but not when TBP is used. TFIIA's stimulation in the presence of TFIID appears to be related to its ability to block the inhibitory effects of TAF1 and other proteins. This finding suggests that TFIIA should be classified as a regulatory protein rather than as a general transcription factor. Additional studies show

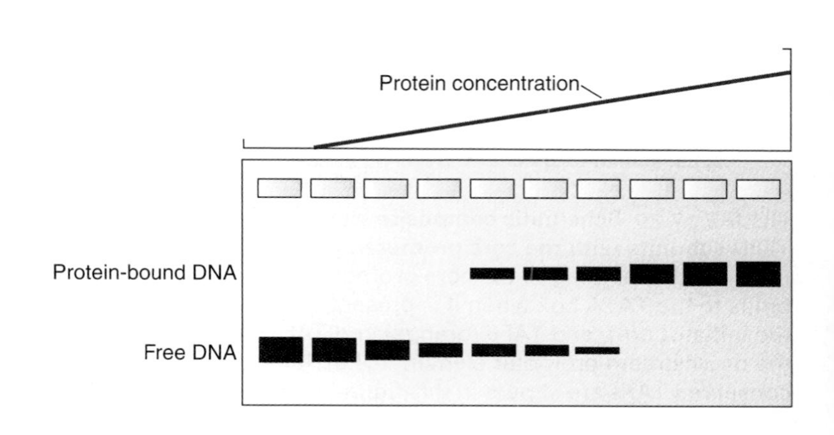

FIGURE 17.21 **Electrophoretic mobility shift assay.** The electrophoretic mobility shift assay is based on the observation that DNA fragments migrate more rapidly through a polyacrylamide gel than the same DNA fragment that binds to a protein. The dissociation constant for the DNA-protein complex can be calculated by measuring the amount of DNA bound to the protein as a function of the protein concentration.

that TFIIA's contribution to transcription initiation depends on the experimental conditions that are used to study it. Thus, TFIIA does not seem to fit into a simple classification scheme. Because TFIIA is not required to reconstitute a minimum transcription system, it will not be considered further.

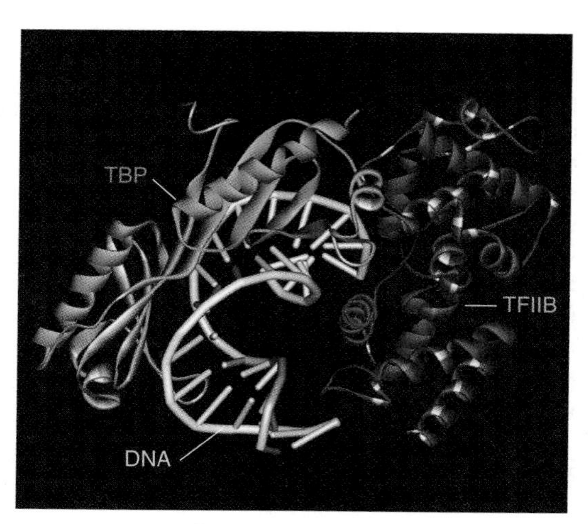

FIGURE 17.22 Crystal structure of a ternary complex of the human TFIIB bound the C-terminal region of TBP from *Arabidopsis thaliana*, which is in turn bound to TATA box of the adenovirus late major promoter. (Structure from Protein Data Bank 1VOL. D. B. Nikolov, et al., *Nature* 377 [1995]: 119–128. Prepared by B. E. Tropp.)

TFIIB helps to convert the closed promoter to an open promoter.

Human TFIIB is a single polypeptide. Homologs have been identified in a wide variety of other organisms including plants, yeast, and the archaea. The molecular mass of human TFIIB is about 35 kDa. In 1995, Stephen K. Burley and coworkers determined the crystal structure of a ternary complex of human TFIIB bound the C-terminal region of TBP from *Arabidopsis thaliana*, which was in turn bound to TATA box of the adenovirus late major promoter (FIGURE 17.22). Fourteen years later Patrick Cramer and coworkers attempted to determine the crystal structure for the yeast RNA polymerase II•TBP•TFIIB•DNA complex but were unable to do so because TBP and DNA dissociated from the complex during crystal formation. Despite this setback, their effort proved to be quite worthwhile because the crystals they did obtain contained the RNA polymerase II•TFIIB complex. FIGURE 17.23a shows the crystal structure of this complex with RNA polymerase II subunits in ribbon form and TFIIB in space-fill form. Note that the RNA polymerase II in this figure has been rotated around a vertical axis by 180° with respect to the RNA polymerase II shown in Figure 17.7 so that the enzyme is pointing in the opposite direction. This rotation permits us to see important interactions between TFIIB and RNA polymerase II. FIGURE 17.23b shows TFIIB in the same orientation as in Figure 17.23a but in ribbon form with the RNA polymerase subunits omitted. Moving from the N- to the C-terminus, TFIIB structural features are as follows: (1) B-ribbon with a bound zinc ion, (2) B-reader, (3) B linker, and (4) B-core N-terminal fold. A fifth structural feature, the B-core C-terminal fold, is not visible in the crystal structure. The RNA polymerase II•TFIIB crystal structure reveals (1) the B ribbon contacts a docking site on Rpb1 near the RNA exit site, (2) The B reader is located near the active site, where it is proposed to help read the DNA sequence during transcription start site selection, (3) The B-linker is located near the polymerase's rudder, and (4) the B-core is located above polymerase's wall. Chemical crosslinking and mutant studies are consistent with the crystal structure.

Based on the TFIIB•RNA polymerase II crystal structure, the TBP•TFIIB•TATA box crystal structure, and biochemical data, Cramer and coworkers propose the model shown in FIGURE 17.24 for the transcription initiation complex and its conversion to the transcription elongation complex. The basic steps in this model are as follows.

1. **Closed complex formation**—The TATA box is positioned above the cleft in RNA polymerase II near the B-core.
2. **Open complex formation**—The B-linker assists the bound DNA to melt about 20 bp downstream of the TATA box to initiate transcription bubble formation. The released template

FIGURE 17.23 Interaction between TFIIB and yeast RNA polymerase II. (a) RNA polymerase•TFIIB crystal structure. RNA polymerase is shown in ribbon form. The subunits are as indicated by the color code of the insert below. Numbers in the insert indicate Rpb subunits. TFIIB is shown in space-fill form with its specific regions labeled. (b) Ribbon form of TFIIB in the same orientation as in (a) but with the RNA polymerase subunits omitted. The colors of TFIIB structural features are the same in (a) and (b). Dashed lines indicate uncertainty about exact strand position. (Structures from Protein Data Bank 3K1F. D. Kostrewa, et al., *Nature* 462 [2009]: 323–330. Prepared by B. E. Tropp.)

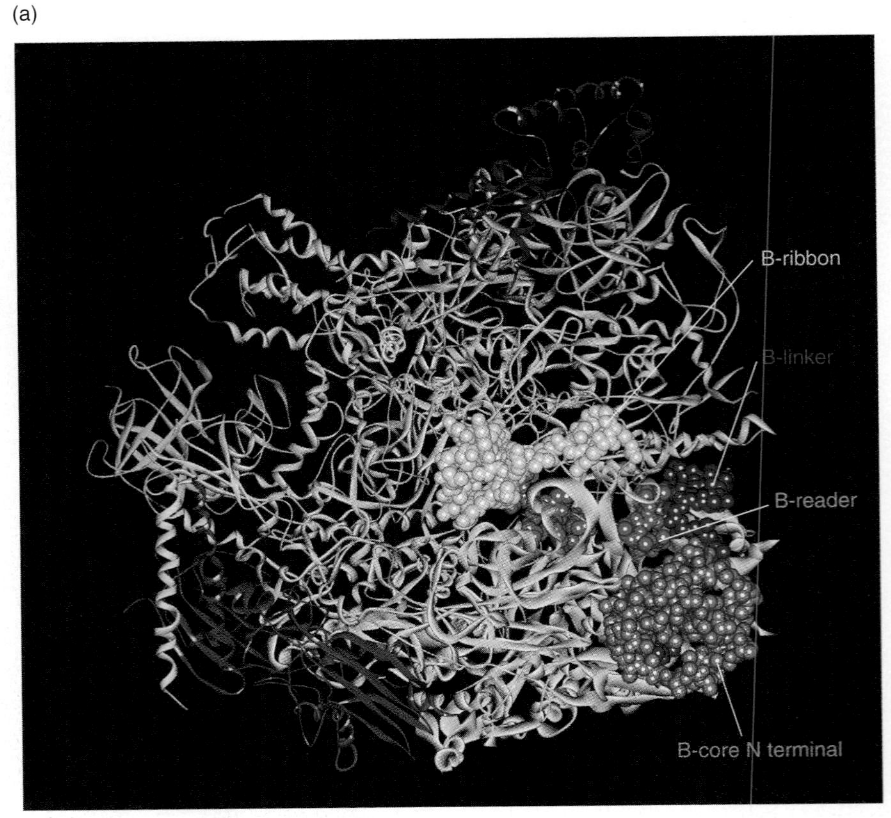

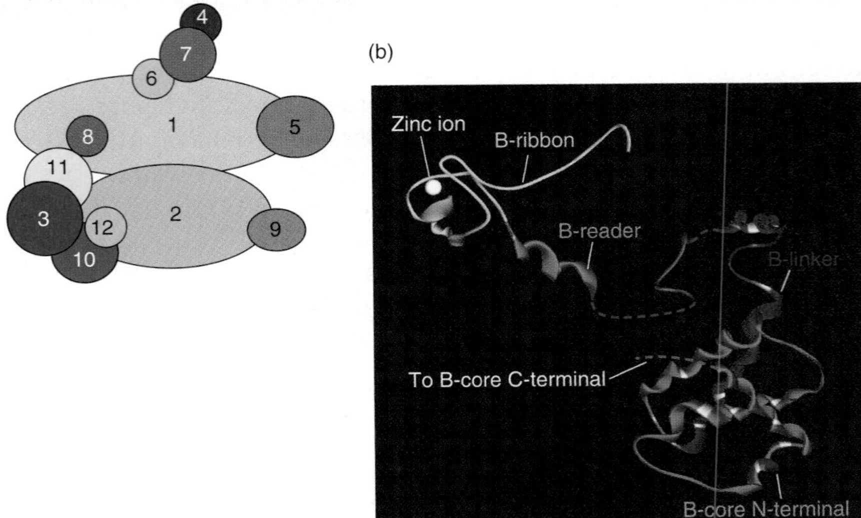

strand slides into the RNA polymerase II cleft, filling the template tunnel, and the downstream double-stranded DNA enters the downstream cleft. The B-reader helps stabilize the transcription bubble near the active site.

3. **DNA start site scanning**—The template strand makes its way through the template tunnel beside the active site and is scanned for an Inr element with the assistance of the B-reader.
4. **RNA chain initiation**—RNA synthesis begins at the open promoter when the first two nucleotides line up on the template

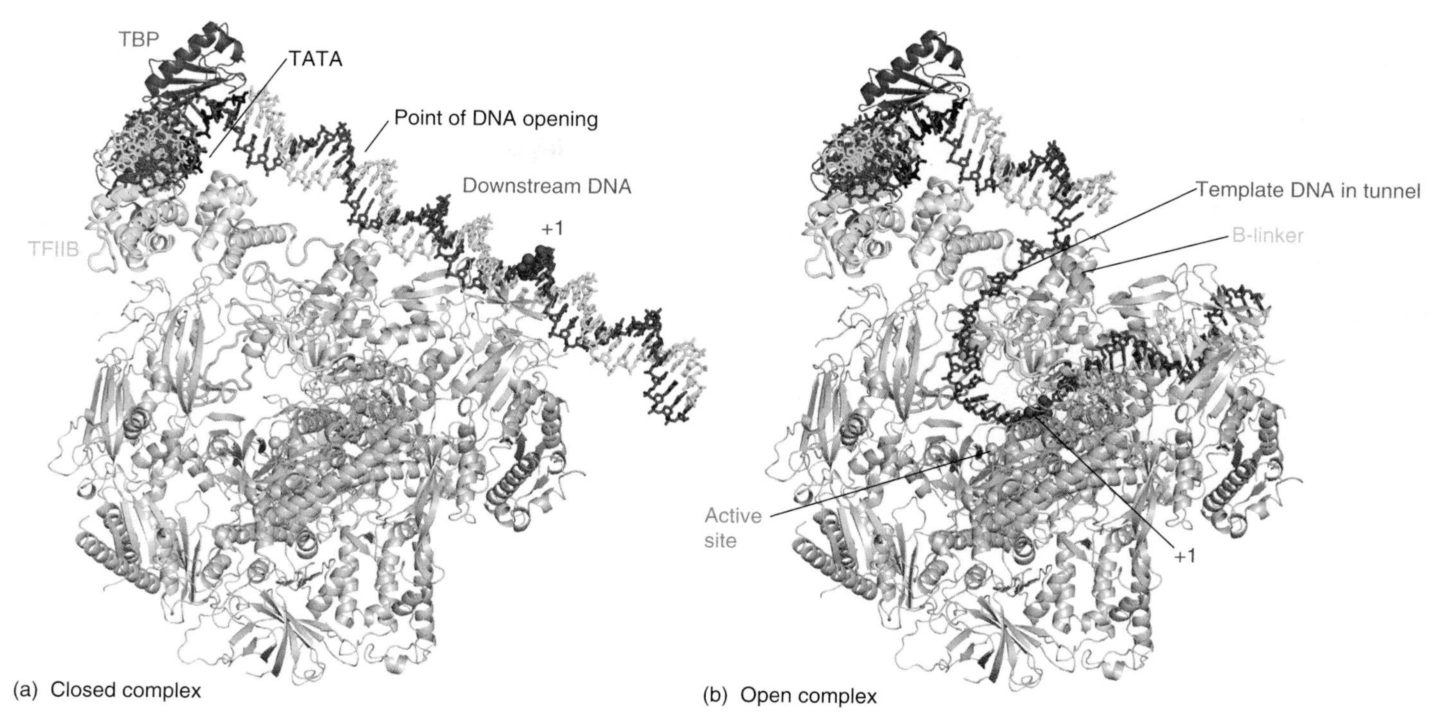

FIGURE 17.24 **Models of closed and open complexes.** (a) Closed complex. DNA template and non-template strands are in blue and cyan, respectively. TFIIB is shown as a green ribbon and TBP as a magenta ribbon. RNA polymerase subunits are shown as gray ribbon structures. The TATA box is in black and magnesium ion at the active site is shown as a magenta ball. (b) Open complex. The nucleotide in the template strand corresponding to position +1 is shown in space-fill form. (Adapted from D. Kostrewa, et al., *Nature* 462 [2009]: 323–330. Photos courtesy of Patrick Cramer, Ludwig-Maximilians-Universität München.)

strand in the newly formed transcription bubble and RNA polymerase II catalyzes phosphodiester bond formation.

5. **Abortive transcription**—Continued ribonucleotide addition leads to the formation of short transcripts. Many of these short transcripts are released, possibly due to the fact that the B-reader loop blocks chain growth.
6. **Promoter escape**—Chain extension beyond seven nucleotides triggers TFIIB release and the formation of the transcription elongation complex.

When the DNA being transcribed has a negatively supercoiled core promoter with a TATA box, TBP and TFIIB alone are sufficient to permit RNA polymerase II to catalyze basal transcription. When the core promoter is not negatively supercoiled, RNA polymerase II also requires TFIIF, TFIIE, and TFIIH for basal transcription.

Sequential binding of RNA polymerase II•TFIIF complex, TFIIE, and TFIIH completes preinitiation complex formation.

The binding of a pre-formed RNA polymerase II•TFIIF complex follows binding of TBP and TFIIB. TFIIF is a heterotetramer of the $\alpha_2\beta_2$ type. Because the subunits were originally identified as RNA polymerase-associated proteins, $TFIIF_\alpha$ and $TFIIF_\beta$ were originally

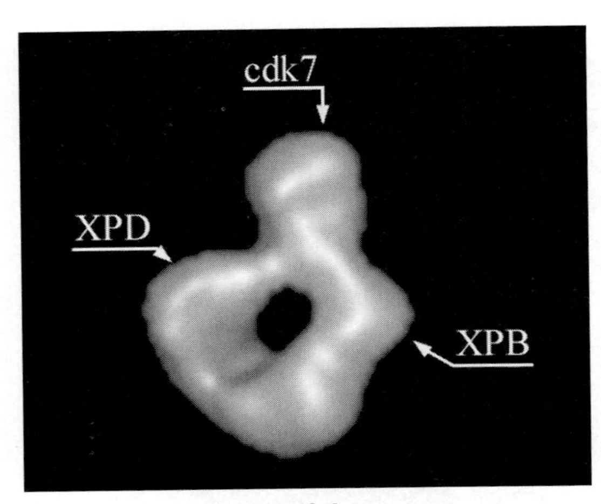

FIGURE 17.25 Model of the quaternary organization of TFIIH. The positions of the subunits cdk7, XPD, and XPB are inferred from immunolabeling experiments and indicated by arrows on the three-dimensional model of human TFIIH. (Reproduced from *Cell*, vol. 102, P. Schultz, et al., Molecular structure of human TFIIH, pp. 599–607. Copyright 2000, with permission from Elsevier [http://www.sciencedirect.com/science/journal/00928674]. Photo courtesy of Patrick Schultz, Institut de Génétique et de Biologie Moléculaire et Cellulaire.)

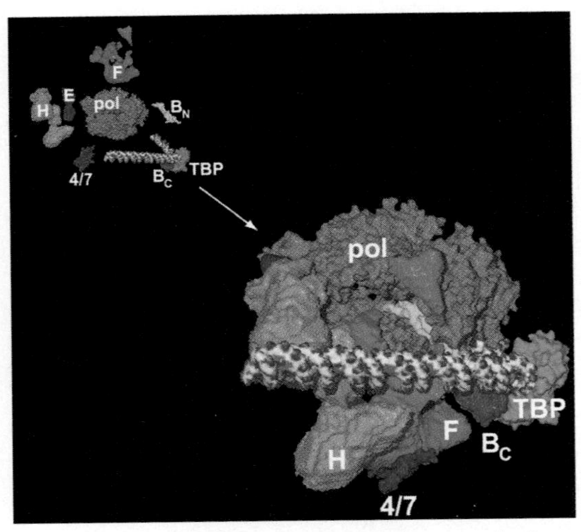

FIGURE 17.26 RNA polymerase II transcription initiation complex. This model was assembled by combining results from x-ray diffraction and electron microscopy. Bc and Bn are the C and N-termini of EFIIB. E, F, and H are EFIIE, EFIIF, and EFIIH, respectively. 4/7 is the Rpb4/7 heterodimer. (Reproduced from *FEBS Lett.*, vol. 579, H. Boeger, et al., Structural basis of eukaryotic gene transcription, pp. 899–903. Copyright 2005, with permission from Elsevier [http://www.sciencedirect.com/science/journal/00145793]. Photo courtesy of Roger D. Kornberg, Stanford University School of Medicine.)

named RAP30 and RAP74, respectively. The TFIIFα subunit possesses a serine/threonine kinase activity, allowing it to phosphorylate a serine and threonine residue on its own chain. This autophosphorylation may play some role in transcription regulation.

TFIIE enters the complex next, binding to TFIIB and then recruiting TFIIH to complete preinitiation complex formation. Human TFIIE is a heterotetramer consisting of two α and two β subunits, which have molecular masses of 57 and 34 kDa, respectively. TFIIE has a dumbbell shape and seems to interact with the RNA polymerase II "jaws" at the downstream end of the active center cleft, placing it in a position to interact with promoter DNA about 25 bp downstream of the transcription start site. Its major functions appear to be to recruit TFIIH to the RNA polymerase II complex and then regulate TFIIH's helicase and kinase activities (see below).

Human TFIIH has ten polypeptide subunits and a molecular mass of about 500 kDa. Two of the subunits, designated XPB and XPD, have $3' \rightarrow 5'$ and $5' \rightarrow 3'$ helicase activities, respectively. The XP designation derives from the fact that individuals who lack a functional XPB or XPD have an inborn error in metabolism known as xeroderma pigmentosum, which results from an inability to repair certain types of DNA damage (see Chapter 12). The helicases help unwind the DNA in the region over the active center cleft. TFIIH also has a third enzymatic activity, **cyclin-dependent protein kinase** (**CDK**). A **cyclin** is a protein that is expressed at different levels throughout the cell cycle. When a threshold level of the cyclin is reached, it interacts with a specific protein kinase such as CDK. This interaction stimulates the protein kinase to phosphorylate one or more specific proteins, enabling the phosphorylated proteins to perform functions required for cell division. Jean-Marc Egly and coworkers have used a combination of electron microscopy and antibody labeling techniques to investigate TFIIH and determine the locations of XPB, XPD, and CDK within the complex. The results of their studies, which are shown in FIGURE 17.25, reveal that the TFIIH is ring-shaped. Combining results from x-ray diffraction and electron microscopy, Kornberg and coworkers have proposed a model for the preinitiation complex that includes RNA polymerase II and the general transcription factors (FIGURE 17.26).

17.6 Transcription Elongation

The C-terminal domain of the largest RNA polymerase subunit must be phosphorylated for chain elongation to proceed.

The carboxyl terminal domain, CTD, of the largest subunit in RNA polymerase II, Rpb1, has an important role in the transition from the initiation complex to the elongation complex. This domain has a highly unusual amino acid sequence, consisting of tandem repeats of the heptapeptide Tyr-Ser-Pro-Thr-Ser-Pro-Ser. RNA polymerases I and III do not have comparable domains. The CTD is unstructured and so not visible in RNA polymerase II crystal structures. Although the basic repeating unit in the CTD is the same in all eukaryotes, the number of repeats varies from one species to another. For instance, human, *Drosophila*, and yeast RNA polymerase II molecules have 52, 43, and 26 repeats, respectively. RNA polymerase II requires the CTD to transcribe some genes in a cell-free system but not to transcribe others. For example, TATA-less promoters appear to require CTD, whereas promoters that have a TATA box usually do not. Yeast mutants that lack some heptapeptide repeats in Rpb1 can survive but exhibit cold- and temperature-sensitive phenotypes. Mutants, however, lose viability when more than half of the heptapeptide repeats are deleted.

Five of the seven residues within each heptapeptide have hydroxyl groups in their side chains that can be phosphorylated by protein kinases. The level of phosphorylation varies greatly during transcription. The residues must be dephosphorylated for RNA polymerase II to assemble into the preinitiation complex but must be phosphorylated for RNA polymerase II to work efficiently during transcription elongation. Different protein kinases are responsible for phosphorylation events that occur at different stages of the transcription process. The CDK subunit associated with TFIIH catalyzes phosphorylation of Ser-5 residues in CTD, permitting promoter clearance to occur so that transcription can enter the elongation phase. Ser-2 and Ser-7 residues are phosphorylated during the elongation stage. RNA polymerase II must be dephosphorylated after each round of transcription is completed before the enzyme can reassemble into an initiation complex to begin the next round of transcription. Specific protein phosphatases catalyze the dephosphorylation. A great deal remains to be learned about kinase and phosphatase specificity and regulation. More also needs to be learned about the influence that different combinations of phosphorylated residues in the CTD have on transcription.

A variety of transcription elongation factors help to suppress transient pausing during elongation.

RNA polymerase II does not move along the template DNA strand in a continuous manner. Instead, the enzyme oscillates between forward and backward movements. Reverse movement of RNA polymerase II along its DNA template called "backtracking" was described for bacterial RNA polymerase (see Chapter 15). Backtracking by RNA polymerase II leads to transcriptional pausing and arrest. Backtracking is usually less extensive in the paused state (2–4 nucleotides) than in the arrested state (7–14 nucleotides). Hence, paused RNA polymerase II can return to the transcription state without the assistance of other proteins, whereas arrested RNA polymerase II cannot.

The average rate of nucleotide addition, which is approximately 1,500 nucleotides per minute, appears to be limited by the fraction of time RNA polymerase II spends in the paused state rather than to the average rate of nucleotide addition. A variety of different transcription factors have been observed to increase the rate of transcription elongation by decreasing the time spent in the paused state. Ronald C. Conaway and coworkers propose that one of these, **elongin**, stimulates transcription by stabilizing the active conformation of RNA

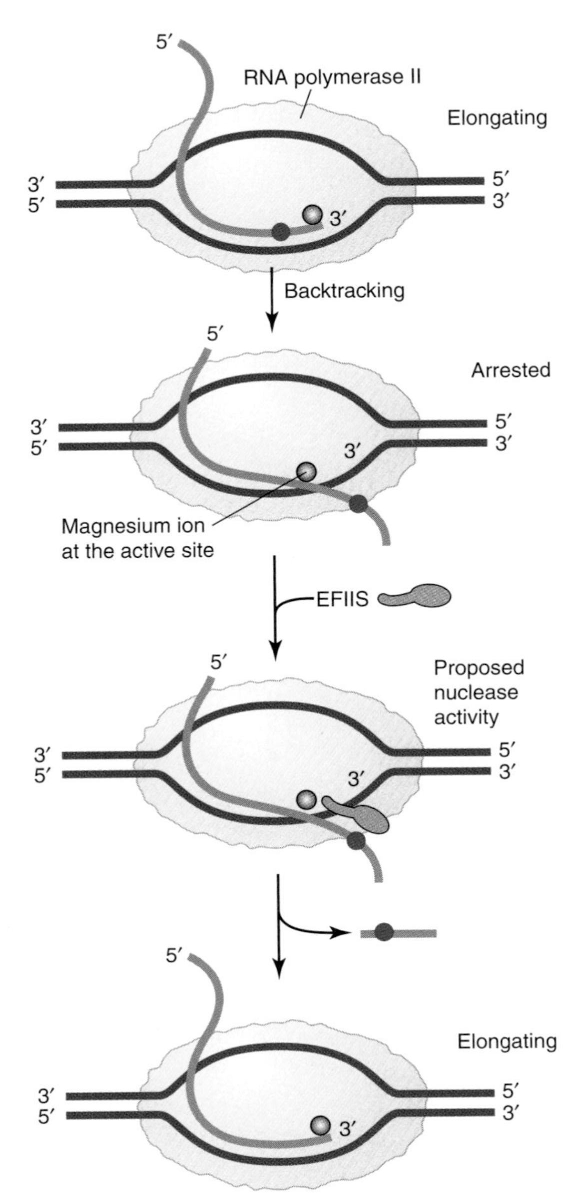

FIGURE 17.27 **Backtracking and TFIIS action.** RNA polymerase II attaches the wrong nucleotide (blue circle) to the growing RNA chain, producing a base pair mismatch. RNA polymerase somehow senses the mismatch and moves backward forcing the 3′-end of the nascent RNA chain to disengage from the active site so that a short segment containing the incorrectly incorporated nucleotide is forced out through the secondary channel and into the funnel. This extrusion causes RNA synthesis to be arrested. The transcription elongation factor (EFIIS) (orange amoeboid shape) binds to RNA polymerase II, inducing the enzyme to cleave the RNA stretch that includes the incorrect nucleotide. Cleavage of the RNA segment containing the incorrect nucleotide creates a new 3′-RNA end at the active site so that chain extension can continue. (Adapted from J. W. Conaway, et al., *Trends Biochem. Sci.* 25 [2000]: 375–380.)

polymerase II, by increasing the rate at which the inactive form is converted back to the active form, or by both methods. Other transcription elongation factors such as the members of the ELL family may act in a similar way.

Elongation factor SII reactivates arrested RNA polymerase II.

Arrested RNA polymerase II requires the assistance of an additional transcription factor, **TFIIS** (also called SII), before it can resume transcription. TFIIS is the eukaryotic counterpart to bacterial GreA and GreB proteins (see Chapter 15).

RNA polymerase II occasionally makes a mistake and attaches the wrong nucleotide to the growing RNA chain. Such an error produces a mismatched base pair within the transcription bubble that causes a distortion within the DNA-RNA hybrid and destabilizes the elongation complex. RNA polymerase II senses the distortion and backtracks (FIGURE 17.27). TFIIS binds to the backtracked complex. Cramer and coworkers determined the crystal structure for the RNA polymerase II•TFIIS complex (FIGURE 17.28). TFIIS extends along the surface of the polymerase and passes through the funnel into the pore to reach the active site at the end of the cleft. The active site of RNA polymerase II undergoes extensive structural changes that are consistent with a realignment of the RNA in the active site due to TFIIS binding. As a result of these changes, the active site is converted from a nucleotidyl transferase to a nuclease. Two essential and highly conserved acidic residues in TFIIS complement the RNA polymerase II active site, allowing a metal ion and water molecule to be positioned for hydrolytic RNA cleavage. RNA cleavage creates a new 3′-end at the active site, which allows transcription to resume.

The transcription elongation complex is regulated.

Transcription elongation by RNA polymerase II is subject to regulation. The proteins involved in this regulation were uncovered in the course of studies that were designed to determine why the nucleoside analog 5,6-dichloro-1-β-D-ribofuranosylbenzimidazole (DRB; FIGURE 17.29) blocks transcription elongation catalyzed by crude nuclear extracts but not by partially pure nuclear extracts. The likely explanation seemed to be that DRB interacts with some factor(s) present in the crude nuclear extracts but missing from the partially purified preparations. Investigators therefore used the partially purified transcription system to assay for transcription elongation factors required for DRB sensitivity. Three transcription elongation factors were eventually isolated. Two of these, the **DRB sensitivity-inducing factor** (**DSIF**) and the **negative elongation factor** (**NELF**), bind to RNA polymerase II and block transcription elongation (FIGURE 17.30). Because DSIF and NELF inhibit transcription elongation when the Rpb1 subunit in RNA polymerase II lacks CTD, the inhibitory proteins probably bind to some other site on the enzyme. Although the inhibition of transcription elongation by DSIF and NELF does not depend on a specific nucleotide sequence, it does appear to require the growing RNA chain to be

FIGURE 17.28 **RNA polymerase II-TFIIS•RNA complex.** (a) A side view of a modeled RNA polymerase II-TFIIS•RNA complex. Domains II and III in TFIIS are shown in green and yellow, respectively. The 12-subunit RNA polymerase II is shown in silver. The magnesium ion in the active site is shown as a purple sphere. Structural zinc ions in TFIIS and RNA polymerase are shown in cyan. The DNA template (blue) and RNA transcript (red) are positioned according to their locations in crystal structure of transcription elongation complex. (b) A cutaway model of (a) as viewed from the front. TFIIS, DNA, and RNA transcript are shown as ribbon models. The proposed path of the backtracked RNA, which is cleaved at the active site, is shown as a dashed red ribbon. (Reproduced from *Cell*, vol. 114, H. Kettenberger, K. -J. Armache, and P. Cramer, Architecture of the RNA Polymerase II-TFIIS . . ., pp. 347–357. Copyright 2003, with permission from Elsevier [http://www.sciencedirect.com/science/journal/00928674]. Photos courtesy of Robert Sims, New York University, and used with permission of Patrick Cramer, Ludwig-Maximilians-Universität München.)

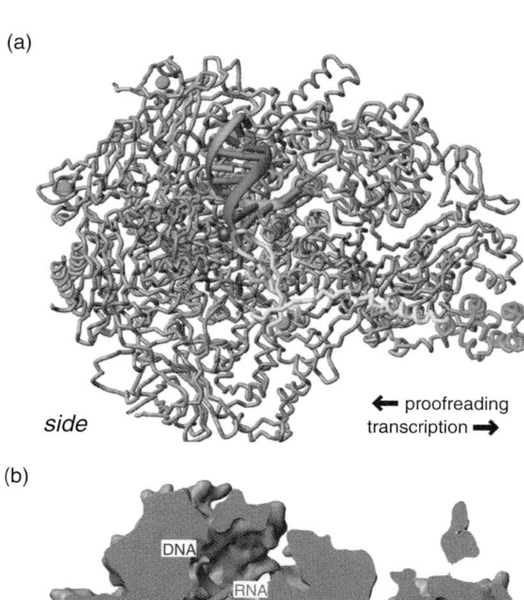

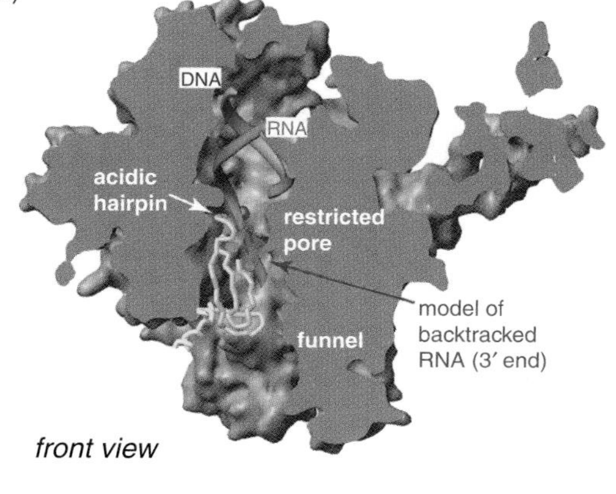

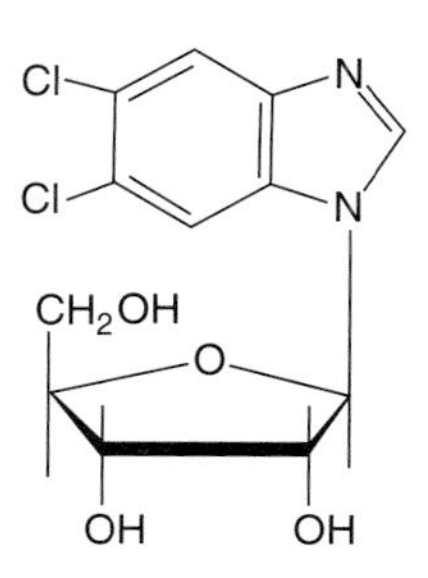

FIGURE 17.29 5,6-Dichlorobenzimidazole 1-β-D-ribofuranoside (DRB).

FIGURE 17.30 **Interplay between DSIF, NELF, and P-TEFb.** Negative elongation factor (NELF) and DRB-sensitivity-inducing factor (DSIF) bind to RNA polymerase II and inhibit transcription elongation. The positive elongating transcription factor (P-TEFb), a cyclin-dependent protein kinase, offsets this inhibition by phosphorylating CTD. The nucleoside analog DRB inhibits the kinase activity, preventing P-TEFb from phosphorylating CTD. (Adapted from Y. Yamaguchi, et al., *Cell* 97 [1999]: 41–51.)

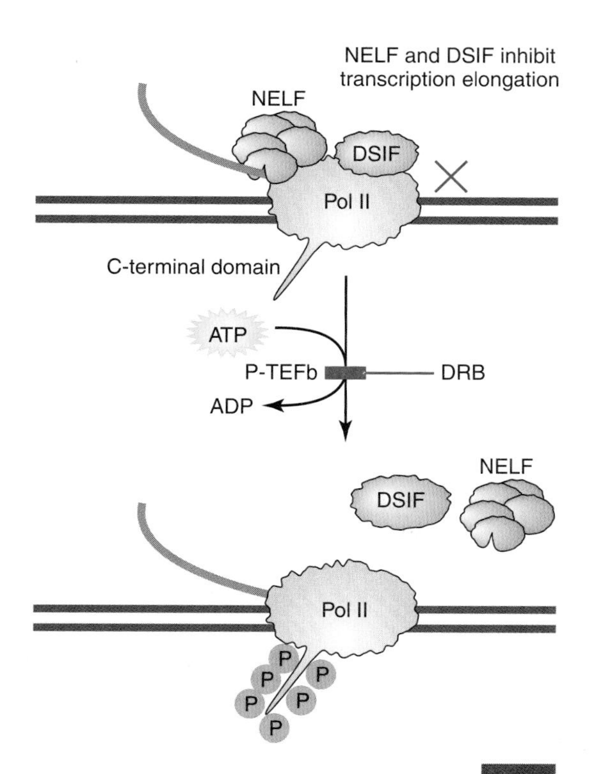

at least 50 to 60 nucleotides long. The third protein involved in the regulation of chain extension, the **positive transcription elongation factor b (P-TEFb)**, is a cyclin-dependent protein kinase that phosphorylates CTD at Ser-2 of the Tyr-Ser-Pro-Thr-Ser-Pro-Ser repeats. Phosphorylated CTD somehow reverses the inhibitory effects of DSIF and NELF. DRB blocks transcription elongation by inhibiting P-TEFb kinase activity.

The final step in basal transcription is the release of the newly synthesized RNA molecule. This release, however, is part of a complex process that involves modifications of the newly synthesized transcript. We therefore examine transcription termination in Chapter 19, which describes the processing of the primary transcript.

17.7 Archaeal RNA Polymerase

The archaea have a single RNA polymerase that is similar to RNA polymerase II.

Archaeal RNA polymerase has a subunit composition and molecular architecture that is very similar to RNA polymerase II. FIGURE 17.31 shows the crystal structure for the RNA polymerase from *Sulfolobus shibatae*, a hyperthermophilic archaeon that was first identified living in acidic geothermal hot springs. The archaeal RNA polymerase crystal structure bears a remarkable resemblance to the yeast RNA polymerase II crystal structure shown in Figure 17.7. With three exceptions, the subunits in the two structures are very similar. The three exceptions are as follows: (1) the largest RNA polymerase II subunit, Rpb1, is divided into two subunits, Rop1N and Rop1C, in the archaeal enzyme; (2) archaeal RNA polymerase does not have a subunit equivalent to Rpb9; and (3) archaeal RNA polymerase has an Rpo13 subunit that is not present in RNA polymerase II. Until recently, each archaeal RNA polymerase subunit was designated by an upper case letter (Table 17.3). Nicola G. A. Abrescia and coworkers propose replacing this nomenclature system with one that calls attention to the similarities between archaeal RNA polymerase and RNA polymerase II. According to their proposal, the name of each archaeal RNA polymerase subunit begins with the prefix Rpo followed by a number. The number is assigned so that each archaeal RNA polymerase subunit has the same number as its RNA polymerase II counterpart (Table 17.3).

In 2002, Finn Werner and Robert O. J. Weinzierl used recombinant DNA technology to insert genes for RNA polymerase subunits from *Methanococcus jannaschii*, a hyperthermophilic archaeon, into plasmids that replicate in *E. coli*. Then they purified the subunits from the different clones and mixed the subunits together under *in vitro* conditions to assemble a functional archaeal RNA polymerase. The success of this reconstitution experiment allowed Weinzierl and coworkers to

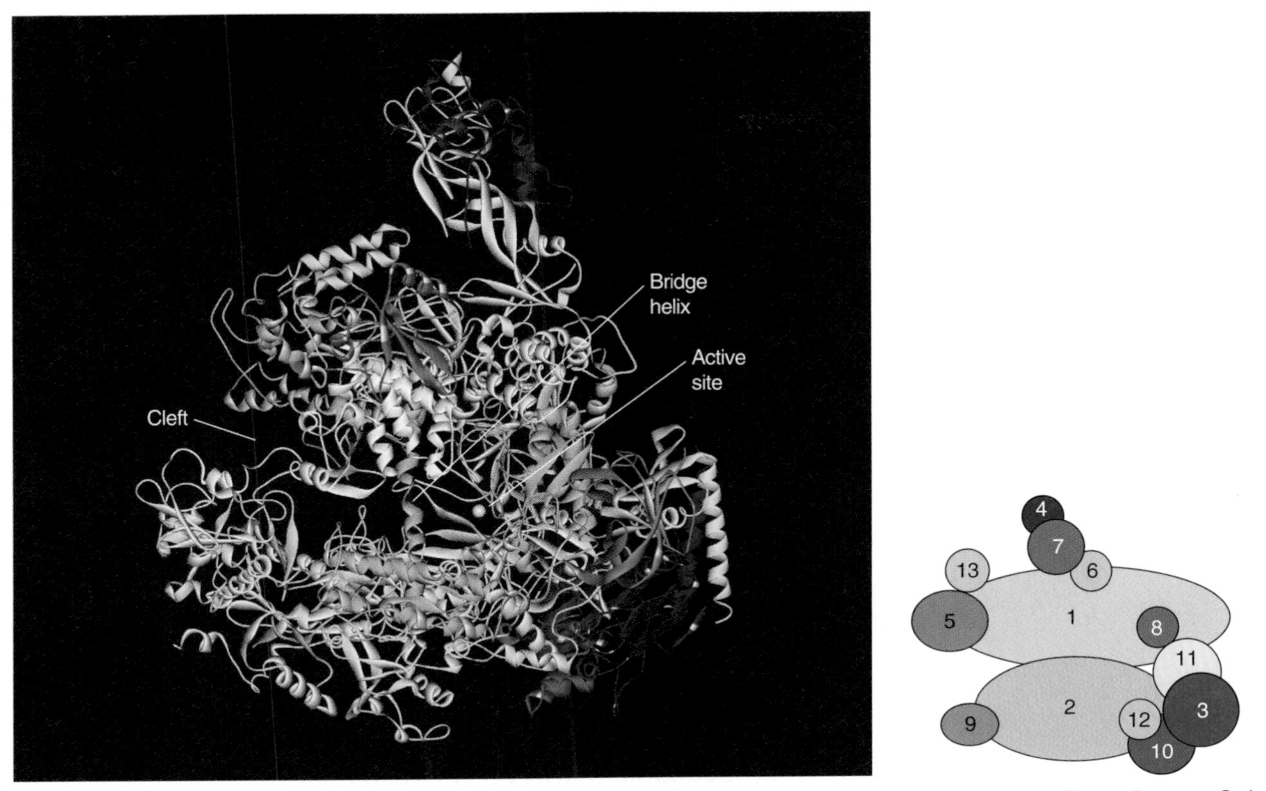

FIGURE 17.31 **Crystal structure for the RNA polymerase from *Sulfolobus shibatae*, a hyperthermophilic archaeon.** Colors of specific Rpo subunits are indicated in the color code diagram. Rpo1N and Rpo1C are shown in light and dark gray, respectively. (Structure from Protein Data Bank 1WAQ. Y. Korkhin, et al., *PLoS Biol.* 7 [2009]: 1–10. Prepared by B. E. Tropp.)

examine structure–function relationships for archaeal RNA polymerase in a way that has not been possible with RNA polymerase II because all attempts to assemble RNA polymerase II under *in vitro* condition have thus far been unsuccessful. Two factors may have contributed to the successful *in vitro* archaeal RNA polymerase assembly. First, the archaeal RNA polymerase is derived from a hyperthermophilic organism, which may lead to greater enzyme stability. Second, the two largest RNA polymerase II subunits are each split into two subunits in *M. jannaschii* RNA polymerase, which may make it easier for the enzyme to assemble. Note that the Rpo2 subunit from *M. jannaschii* RNA polymerase differs in this respect from the Rpo2 subunit in the *S. shibatae* RNA polymerase. For convenience, we will refer to Rpo1N and Rpo1C subunits and the two Rpo2 subunits in *M. jannaschii* RNA polymerase as Rpo1 and Rpo2, respectively. Rpo3, Rpo10, Rpo11, and Rpo 12 form an assembly platform for Rpo1 and Rpo2. Rpo3 and Rpo11 are homologous to the bacterial α subunits, which form the α_2 assembly platform for the β and β' subunits (see Chapter 15). The assembly platform (Rpo3, Rpo10, Rpo11, and Rpo 12) and catalytic subunits (Rpo1 and Rpo2) form the minimal subunit complex that is necessary and sufficient for promoter-directed transcription.

The remaining subunits may contribute to archaeal RNA polymerase stability, regulation, or both. Weinzierl and coworkers have adapted their reconstitution assembly technique to devise an automated, high throughput method for producing and characterizing RNA polymerase variants. This technique permits them to create 19 amino acid substitutions at any single site of interest and study the effect that each substitution has on enzyme activity. They can also assemble RNA polymerase molecules with desired mutations in two or more subunits. This important new technique has provided considerable information about the way that the bridge helix and trigger loop interact during RNA synthesis.

The archaeal RNA polymerase offers one additional research advantage when compared to RNA polymerase II. It requires just three general transcription factors, TBP, TFB (homologous to the eukaryotic TFIIB), and TFE (homologous to the N-terminal region of the largest eukaryotic TFIIE subunit). An understanding of the archaeal transcription system undoubtedly will lead to a better understanding of the eukaryotic transcription system.

Suggested Reading

Introduction to RNA Polymerase II

Feng, X., Jiang, Y., Meltzer, P., and Yen, P. M. 2000. Thyroid hormone regulation of hepatic genes *in vivo* detected by complementary DNA microarray. *Mol Endocronol* 14:947–955.

Ream, T. S., Haag, J. R., Wierzbicki, A. T., et al. 2009. Subunit composition of the RNA-silencing enzymes Pol IV and Pol V reveal their origins as specialized form of RNA polymerase II. *Mol Cell* 33:192–203.

Roeder, R. 2003. The eukaryotic transcriptional machinery: complexities and mechanisms unforeseen. *Nat Med* 9:1239–1244.

Roeder, R. G., and Rutter, W. J. 1970. Specific nucleolar and nucleoplasmic RNA polymerases. *Proc Natl Acad Sci USA* 65:675–682.

RNA Polymerase II Structure

Armache, K.-J., Kettenberger, H., and Cramer, P. 2003. Architecture of initation-competent 12-subunit RNA polymerase II. *Proc Natl Acad Sci USA* 100:6964–6968.

Armache, K.-J., Kettenberger, H., and Cramer, P. 2004. Complete RNA polymerase II elongation complex structure and its interactions with NTP and TFIIS. *Mol Cell* 16:955–965.

Armache, K.-J., Mitterweger, S., Meinhart, A., and Cramer, P. 2005. Structures of complete RNA polymerase II and its subcomplex, Rpb4/7. *J Biol Chem* 280:7131–7134.

Bell, S. D., and Jackson, S. P. 2000. Charting a course through RNA polymerase. *Nat Struct Biol* 7:703–705.

Boeger, H., Bushnell, D. A., Davis, R., et al. 2005. Structural basis of eukaryotic gene transcription. *FEBS Lett* 579:899–903.

Bushnell, D. A., and Kornberg, R. D. 2003. Complete, 12-subunit RNA polymerase II at 4.1-Å resolution: implications for the initiation of transcription. *Proc Natl Acad Sci USA* 100:6969–6973.

Chower, M. 2004. Rpb4 and Rpb7: subunits of RNA polymerase II and beyond. *Trends Biochem Sci* 29:674–681.

Chung, W.-H., Craighead, J. L., Chang, W.-H., Ezeokonkwo, C., Bareket-Samish, A., Kornberg, R. D., and Asturias, F. J. 2003. RNA polymerase II/TIIF structure and conserved organization of the initiation complex. *Mol Cell* 12:1003–1013.

Cramer, P. 2002. Multisubunit RNA polymerases. *Curr Opin Struct Biol* 12:89–97.
Cramer, P. 2004. Structure and function of RNA polymerase II. *Adv Protein Chem* 67:1–31.
Cramer, P. 2004. RNA polymerase II structure: from core to functional complexes. *Curr Opin Genet Dev* 14:218–226.
Cramer, P. 2010. Towards molecular biology systems of gene transcription and regulation. *Biol Chem* 391:731–735.
Cramer, P., Armache, K.-J., Baumli, S., et al. 2008. Structure of eukaryotic RNA polymerases. *Ann Rev Biochem* 37:337–352.
Cramer, P., Bushnell, D. A., Fu, J., et al. 2000. Architecture of RNA polymerase II and implications for the transcription mechanism. *Science* 288:640–649.
Cramer, P., Bushnell, D. A., and Kornberg, R. D. 2001. Structural basis of transcription: RNA polymerase II at 2.8 Ångstrom resolution. *Science* 292:1863–1876.
Gnatt, A. L., Cramer, P., Fu, J., et al. 2001. Structural basis of transcription: an RNA polymerase II elongation complex at 3.3 Å resolution. *Science* 292:1876–1882.
Hahn, S. 2004. Structure and mechanism of the RNA polymerase II transcription machinery. *Nat Struct Mol Biol* 11:394–403.
Kettenberger, H., Armache, K.-J., and Cramer, P. 2004. Complete RNA polymerase II elongation complex structure and its interactions with NTP and TFIIS. *Mol Cell* 16:955–965.
Kornberg, R. 2001. The eukaryotic gene transcription machinery. *Biol Chem* 382: 1103–1107.
Kornberg, R. D. 2007. The molecular basis of eukaryotic transcription. *Proc Natl Acad Sci USA* 104:12955–12961.
Kostrewa, D., Zeller, M. E., Armache, K.-J., et al. 2009. RNA polymerase II-TFIIB structure and mechanism of transcription initiation. *Nature* 462:323–330.
Näär, A. M., Lemon, B. D., and Tjian, R. 2001. Transcriptional coactivator complexes. *Ann Rev Biochem* 70:475–501.
Nikolov, B. D., and Burley, S. K. 1997. RNA polymerase II transcription initiation: a structural view. *Proc Natl Acad Sci USA* 94:15–22.
Wang, D., Bushnell, D. A., Westover, K. D., et al. 2006. Structural basis of transcription: role of the trigger loop in substrate specificity and catalysis. *Cell* 127:941–954.
Westover, K., Bushnell, D. A., and Kornberg, R. D. 2004. Structural basis of transcription: nucleotide rotation in the RNA polymerase II active center. *Cell* 119:481–489.
Westover, K. D., Bushnell, D. A., and Kornberg, R. D. 2004. Structural basis of transcription separation of RNA from DNA by RNA polymerase II. *Science* 303:1014–1016.
Woychik, N. A., and Hampsey, M. 2002. The RNA polymerase II machinery: structure illuminates function. *Cell* 108: 453–463.
Woychik, N. A., and Reinberg, D. 2001. RNA polymerases: subunits and functional domains. In: *Encyclopedia of Life Sciences Nature Publishing Group*. pp. 1–8. London: Nature.

Transcription Start Site Identification

Carninci, P., Kvam, C., Kitamura, A., et al. 1996. High-efficiency full-length cDNA cloning by biotinylated CAP trapper. *Genomics* 37:327–336.
Carninci, P., Sandelin, A., Lenhard, B., et al. 2006. Genome-wide analysis of mammalian promoter architecture and evolution. *Nat Genet* 38:626–635.
Harbers, M., and Carninci, P. 2005. Tag-based approaches for transcriptome research and genome annotation. *Nat Methods* 7:495–502.
Juven-Gershon, T., Hsu, J.-Y., Theisen, J. W. M., and Kadonaga, J. T. 2008. The RNA polymerase II core promoter—the gateway to transcription. *Curr Opin Cell Biol* 20:253–259.
Juven-Gershon, T., and Kadonaga, J. T. 2010. Regulation of gene expression via the core promoter and the basal transcriptional machinery. *Dev Biol* 339:225–229.
Kodzius, R., Kojima, M., Nishiyori, H., et al. 2006. CAGE: cap analysis of gene expression. *Nat Methods* 3:211–222.

Liu, X., Bushnell, D. A., and Wang, D., et al. 2010. Structure of an RNA polymerase II–TFIIB complex and the transcription initiation mechanism. *Science* 327:206–209.

Lockhart, D. J., and Winzeler, E. A. 2000. Genomics, gene expression and DNA arrays. *Nature* 405:827–836.

Maston, G. A., Evans, S. K., and Green, M. R. 2006. Transcriptional regulatory elements in the human genome. *Ann Rev Genomics Human Genet* 7:29–59.

Sambrook, J., and Russell, D. W. 2001. *Molecular Cloning—A Laboratory Manual*, 3rd ed. Woodbury, NY: Cold Spring Harbor Press.

Sandelin, A.l., Carninci, P., Lenhard, B., et al. 2007. Mammalian RNA polymerase II core promoters: Insights from genome-wide studies. *Nat Rev Genet* 8:424–436.

Schena, M., Heller, R. A., Theriault, T. P., et al. 1998. Microarrays: biotechnology's discovery platform for functional genomics. *TIBTECH* 16:301–306.

General Transcription Factors: Basal Transcription

Asturias, F. J. 2004. RNA polymerase II structure, and organization of the preinitiation complex. *Curr Opin Struct Biol* 14:121–129.

Asturias, F. J. 2009. TFIID: a closer look highlights its complexity. *Structure* 17:1423–1424.

Asturias, F. J., and Craighead, J. L. 2003. RNA polymerase II at initiation. *Proc Natl Acad Sci USA* 100:6893–6895.

Auble, D. T. The dynamic personality of TATA-binding protein. *Trends Biochem Sci* 34:49–52.

Buratowski, S., Hahn, S., Sharp P. A., and Guarente, L. 1988. Function of a yeast TATA element-binding protein in a mammalian transcription system. *Nature* 334:37–42.

Bushnell, D. A., Westover, K. D., Davis, R. E., and Kornberg, R. D. 2004. Structural basis of transcription: an RNA polymerase II–TFIIB cocrystal at 4.5 Angstroms. *Science* 303:983–988.

Butler, J. E. F., and Kadonaga, J. T. 2002. The RNA polymerase II core promoter: a key component in the regulation of gene expression. *Genes Dev* 16:2583–2592.

Cler, E., Papai, G., Schultz, P., and Davidson, I. 2009. Recent advances in understanding the structure and function of general transcription factor TFIID. *Cell Mol Life Sci* 66:2123–2134.

Coulombe, B. 1999. DNA wrapping in transcription initiation by RNA polymerase II. *Biochem Cell Biol* 77:257–264.

Green, M. R. 2000. TBP-associated factors (TAFIIs): multiple, selective transcriptional mediators in common complexes. *Trends Biochem Sci* 25:59–63.

Hahn, S. 2009. New beginnings for transcription. *Nature* 462:292–293.

Hampsey, M., and Reinberg, D. 2001. RNA polymerase II holoenzyme and transcription factors. In: *Encyclopedia of Life Sciences*. London: *Nature*. pp. 1–7.

Hayashi, K., Watanabe, T., Tanaka, A., et al. 2005. Studies of *Schizosaccharomyces pombe* TFIIE indicate conformational and functional changes in RNA polymerase II at transcription initiation. *Genes Cells* 10:207–224.

Hochheimer, A., and Tjian, R. 2003. Diversified transcription initiation complexes expand promoter selectivity and tissue-specific gene expression. *Genes Dev* 17:1309–1320.

Høiby, T., Zhou, H., Mitsiou, D. J., and Stunnenberg, H. G. 2007. A facelift for the general transcription factor TFIIA. *Biochim Biophys Acta* 1769:429–436.

Juo, Z. S., Chiu, T. K., Leiberman, P. M., et al. 1996. How proteins recognize the TATA box. *J Mol Biol* 261:239–254.

Kostrewa, D., Zeller, M. E., Armache, K.-J., et al. 2009. RNA polymerase II–TFIIB structure and mechanism of transcription initiation. *Nature* 462:323–330.

Krishnamurthy, S., and Hampsey, M. 2009. Eukaryotic transcription initiation. *Curr Biol* 19:R153–R156.

Liu, X., Bushnell, D. A., Wang, D., et al. 2010. Structure of an RNA polymerase II-TFIIB complex and the transcription initiation mechanism. *Science* 237:206–209.

Müller, F., and Tora, L. 2004. The multicoloured world of promoter recognition complexes. *EMBO J* 23:2–8.

Nikolov, D. B., Chen, H., Halay, E. D., et al. 1995. Crystal structure of a TFIIB-TBP-TATA-element ternary complex. *Nature* 377:119–128.
Nogales, E. 2000. Recent structural insights into transcriptional preinitiation complexes. *J Cell Sci* 113:4391–4397.
Reese, J. C. 2003. Basal transcription factors. *Curr Opin Genet Dev* 13:114–118.
Schultz, P., Fribourg, S., Poterszman, A., et al. 2000. Molecular structure of human TFIIH. *Cell* 102:599–607.
Sikorski, T. W., and Buratwoski, S. 2009. The basal initiation machinery: beyond the general transcription factors. *Curr Opin Cell Biol* 21:344–351.
Smale, S. T. 1997. Transcription initiation form TATA-less promoters within eukaryotic protein-coding genes. *Biochim Biophys Acta* 1351:73–88.
Smale, S. T., and Kadonaga, J. T. 2003. The RNA polymerase II core promoter. *Ann Rev Biochem* 72:449–479.
Tora, L. 2002. A unified nomenclature for TATA box binding protein (TBP)-associated factors (TAFs) involved in RNA polymerase II transcription. *Genes Dev* 16:673–675.
Tora, L., and Timmers, H. T. M. 2010. The TATA box regulates TATA-binding protein (TBP) dynamics *in vivo*. *Trends Biochem Sci* 35:309–314.
Warren, A. 2002. Eukaryotic transcription factors. *Curr Opin Struct Biol* 12:107–114.
Zurita, M., and Merino, C. 2003. The transcriptional complexity of the TFIIH complex. *Trends Genet* 19:578–584.

Transcription Elongation

Armache, K.-J., Kettenberger, H., and Cramer, P. 2005. The dynamic machinery of mRNA elongation. *Curr Opinion Struct Biol* 15:197–203.
Bar-Nahum, G., Epshtein, G., Ruckenstein, A. E., et al. 2005. A ratchet mechanism of transcription elongation and its control. *Cell* 120:183–193.
Brès, V., Yoh, S. M., and Jones, K. A. 2008. The multi-tasking P-TEFb complex. *Curr Opin Cell Biol* 20:334–340.
Conaway, J. W., Shilatifard, A., Dvir, A., and Conaway, R. C. 2000. Control of RNA polymerase II elongation. *Trends Biochem Sci* 25:375–380.
Fujita, T., Piuz, I., and Schlegel, W. The transcription elongation factors NELF, DSIF, and P-TEFb control constitutive transcription in a gene-specific manner. *FEBS Lett* 583:2893–2898.
Gnatt, A. 2002. Elongation by RNA polymerase II: structure-function relationship. *Biochim Biophys Acta* 1577:175–190.
Hartzog, G. A. 2003. Transcription elongation by RNA polymerase II. *Curr Opin Genet Dev* 13:119–126.
Kamenski, T., Heilmeir, S., Meihhart, A., and Cramer, P. 2004. Structure and mechanism of RNA polymerase II CTD phosphatases. *Mol Cell* 15:399–407.
Kim, D., Yamaguchi, Y., Wada, T., and Handa, H. 2001. The regulation of elongation by eukaryotic RNA polymerase II: a recent view. *Mol Cells* 11:267–274.
Kobor, M. S., and Greenblatt, J. 2002. Regulation of transcription elongation by phosphorylation. *Biochim Biophys Acta* 1577:261–275.
Kohoutek, J. 2009. P-TEFb-the final frontier. *Cell Div* 4:19.
Meinhart, A., Kamenski, T., Hoeppner, S., et al. 2005. A structural perspective of CTD function. *Genes Dev* 19:1401–1415.
Palancade, B., and Bensaude, O. 2003. Investigating RNA polymerase II carboxyl-terminal domain (CTD) phosphorylation. *Eur J Biochem* 270:3859–3870.
Price, D. H. 2002. P-TEFb, a cyclin-dependent kinase controlling elongation by RNA polymerase II. *Mol Cell Biol* 20:2629–2634.
Price, D. H. 2004. RNA polymerase II elongation control in eukaryotes. In: *Encyclopedia of Biological Chemistry*. Lennarz, W.J., and Lane, W. D. (eds.) pp. 766–769. New York: Academic Press.
Selth, L. A., Sigurdsson, S., and Svejstrup, J. Q. 2010. Transcript elongation by RNA polymerase II. *Ann Rev Biochem* 79:271–293.
Shilatifard, A. 2004. Transcriptional elongation control by RNA polymerase II: a new frontier. *Biochim Biophys Acta* 1677:79–86.

Shilatifard, A., Conaway, R. C., and Conaway, J. W. 2003. The RNA polymerase II elongation complex. *Ann Rev Biochem* 72:693–715.

Sims III, R. J., Belotserkovskaya, R., and Reinberg, D. 2004. Elongation by RNA polymerase II: the short and long of it. *Genes Dev* 18:2437–2468.

Sousa, R. 2005. Machinations of a Maxwellian demon. *Cell* 120:155–158.

Svejstrup, J. Q. 2007. Contending with transcriptional arrest during RNAPII transcript elongation. *Trends Biochem Sci* 32:165–171.

Wade, J. T., and Struhl, K. 2008. The transition from transcriptional initiation to elongation. *Curr Opin Genet Dev* 18:130–136.

Wang, D., Bushnell, D. A., Huang, X., et al. 2009. Structural basis of transcription: backtracked RNA polymerase II at 3.4 Angstrom resolution. *Science* 324:1203–1206.

Yamaguchi, Y., Takagi, T., Wada, T., et al. 1999. NELF, a multisubunit complex containing RD, cooperates with DSIF to repress RNA polymerase II elongation. *Cell* 97:41–51.

Archaeal RNA Polymerase

Grohmann, D., Hirtreiter, A, and Werner, F. 2009. Molecular mechanisms of archaeal RNA polymerase. *Biochem Soc Trans* 37:12–17.

Hirata, A., Klein, B., and Murakami, K. S. 2008. The x-ray crystal structure of RNA polymerase from archaea. *Nature* 451:851–854.

Hirata, A., and Murakami, K. S. 2009. Archaeal RNA polymerases. *Curr Opin Struct Biol* 19:1–8.

Kaplan, C. D., and Kornberg, R. D. 2008. A bridge to transcription by RNA polymerase. *J Biol* 7:39.1–39.4.

Korkhin, Y., Unligil, U. M., Littlefield, O., et al. 2009. Evolution of complex RNA polymerases: the complete archaeal RNA polymerase structure. *PLoS Biol* 7:1–10.

Kusser, A. G., Bertero, M. G., Naji, S., et al. 2008. Structure of an archaeal RNA polymerase. *J Mol Biol* 376:303–307.

Tan, L., Wiesler, S., Trzaska, D., Carney, H. C., and Weinzierl, R. O. J. 2008. Bridge helix and trigger loop perturbations generate superactive RNA polymerases. *J Biol* 7:40.1–40.15.

Thomm, M., Reich, C., Grünberg, S, and Naji, S. 2009. Mutational studies of archaeal RNA polymerase and analysis of hybrid RNA polymerases. *Biochem Soc Trans* 37:18–22.

Werner, F. 2008. Structural evolution of multisubunit RNA polymerases. *Trends Microbiol* 16:247–250.

Werner, F., and Weinzierl, R. O. J. 2002. A recombinant RNA polymerase II-like enzyme capable of promoter-specific transcription. *Mol Cell* 10:635–646.

This book has a Web site, **http://biology.jbpub.com/book/molecular**, which contains a variety of resources, including links to in-depth information on topics covered in this chapter, and review material designed to assist in your study of molecular biology.

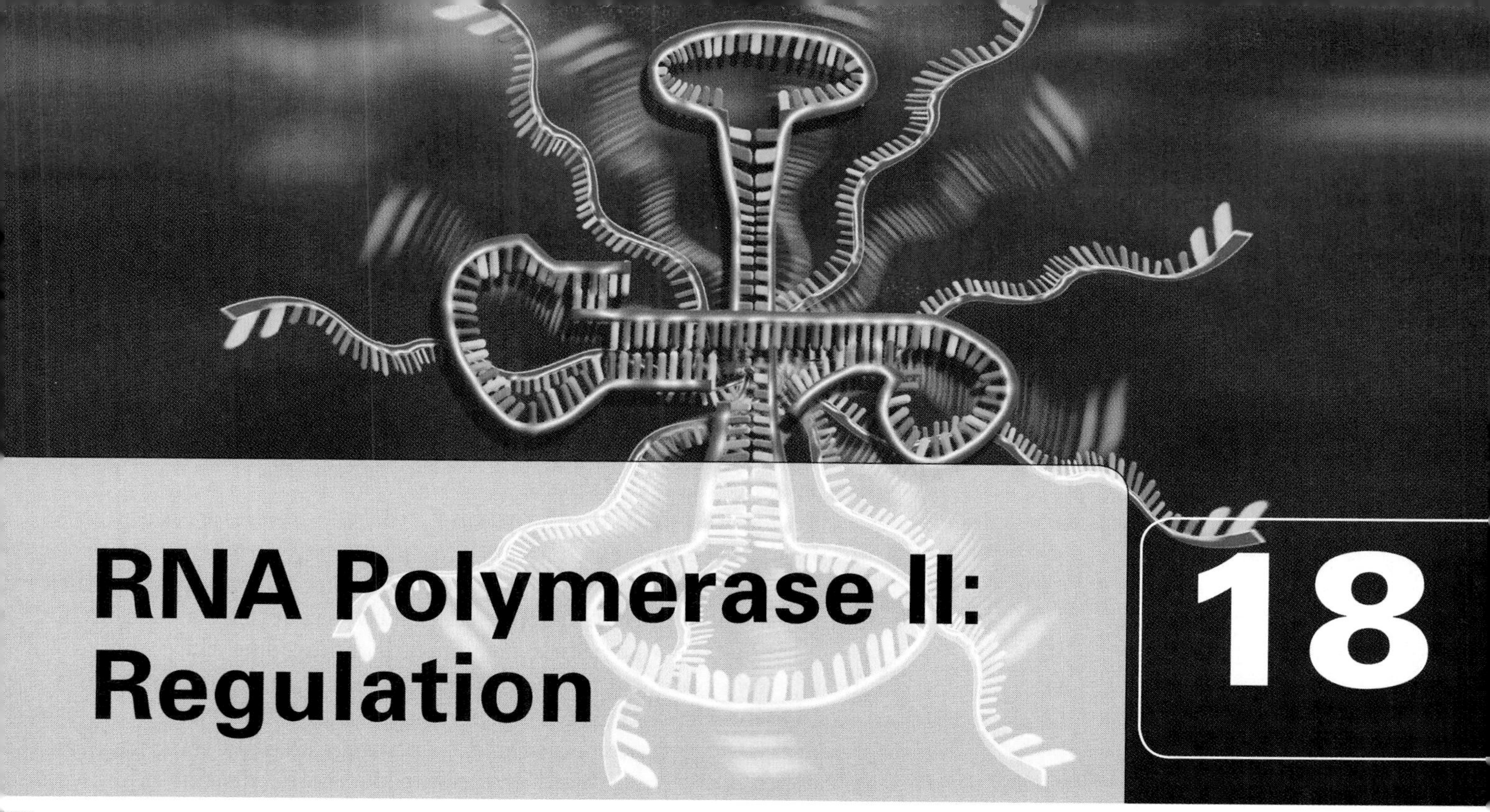

RNA Polymerase II: Regulation

18

OUTLINE OF TOPICS

18.1 Regulatory Promoters, Enhancers, and Silencers

Linker-scanning mutagenesis reveals the regulatory promoter's presence just upstream from the core promoter.

Enhancers stimulate transcription and silencers block transcription.

The upstream activating sequence (UAS) regulates genes in yeast.

18.2 Transcription Activator Proteins

Transcription activator proteins help to recruit the transcription machinery.

A combinatorial process determines gene activity.

DNA affinity chromatography can be used to purify transcription activator proteins.

A transcription activator protein's ability to stimulate gene transcription can be determined by a transfection assay.

18.3 DNA-Binding Domains with Helix-Turn-Helix Structures

Homeotic genes assign positional identities to cells during embryonic development.

Homeotic genes specify transcription activator proteins.

The homeodomain contains a helix-turn-helix motif.

POU proteins have a homeobox and a POU domain.

18.4 DNA-Binding Domains with Zinc Fingers

Many transcription activator proteins have Cys_2His_2 zinc fingers that bind to DNA in a sequence-specific fashion.

Nuclear receptors have Cys_4 zinc finger motifs.

Ligand-binding domain structure provides considerable information about nuclear receptor function.

Gal4, a yeast transcription activator protein belonging to the Cys_6 zinc cluster family, regulates the transcription of genes involved in galactose metabolism.

18.5 Loop-Sheet-Helix DNA-Binding Domain

p53 has a loop-sheet-helix DNA-binding domain.

18.6 DNA-Binding Domains with Basic Region Leucine Zippers

Basic region leucine zipper (bzip) transcription activator proteins bind to DNA as dimers that are held together through coiled coil interactions.

18.7 DNA-Binding Domains with Helix-Loop-Helix Structures

Helix-loop-helix transcription regulatory proteins are dimers.

The bHLH zip family of transcription regulators have both HLH and leucine zipper dimerization motifs.

18.8 Activation Domain

The activation domain must associate with a DNA-binding domain to stimulate transcription.

Gal4 has DNA-binding and activation domains.

The yeast two-hybrid assay permits us to detect polypeptides that interact through non-covalent interactions.

18.9 Mediator

Squelching occurs when transcription activator proteins compete for a limiting transcription machinery component.

Mediator is required for activated transcription.

The yeast Mediator complex associates with activators at the UAS in active yeast genes.

18.10 Epigenetic Modifications

Cells remodel or modify chromatin to make the DNA in chromatin accessible to the transcription machinery.

DNA methylation plays an important role in determining whether chromatin will be silenced or actively expressed in vertebrates.

Epigenetics is the study of inherited changes in phenotype caused by changes in chromatin other than changes in DNA sequence.

Genomic imprinting in mammals determines whether a maternal or paternal gene will be expressed.

Pluripotent cells usually become more specialized during development.

A cell nucleus from a terminally differentiated cell can be reprogrammed by an enucleated egg cell to produce a live animal.

Lineage restricted cells can be programmed to produce induced pluripotent stem (iPS) cells.

Suggested Reading

The basal RNA polymerase II transcription machinery is very inefficient and by itself accounts for little, if any, of the mRNA synthesized in the cell. Two additional types of transcriptional factors are needed to stimulate RNA synthesis. The first of these, the **transcription activator proteins**, have DNA-binding domains that recognize specific DNA sequence motifs or modules and activation domains that bind to one or more components of the transcription machinery. Unlike components of the basal RNA polymerase II transcription machinery, many transcription activator proteins are tissue specific, gene specific, or both.

The second type of transcription factor, the **Mediator**, is a multisubunit protein complex that acts as a bridge between activator proteins, repressor proteins (proteins that block transcription), and the basal RNA polymerase transcription machinery, serving as a control panel to regulate transcription. Mediator is required for the transcription of protein coding genes and is therefore of comparable importance to RNA polymerase II and the general transcription factors.

Eukaryotic gene expression is also influenced by the fact that eukaryotic DNA is wrapped around an octameric histone complex and the resulting beadlike structures—the nucleosomes—in turn are organized into higher-level structures that allow for the considerable compaction of DNA in chromatin (see Chapter 6). Nucleosomes tend to make DNA unavailable to the transcription machinery and thereby turn off gene expression. Two kinds of enzymes help to make the DNA available for transcription. The first kind modify nucleosomes by adding acetyl or other groups to specific sites on histones and, thereby, destabilize the nucleosome or mark it for further modification, movement, or removal. The second kind remodel chromatin by moving or removing nucleosomes to allow the transcription machinery to act on the DNA.

This chapter examines transcription activators (and repressors), Mediator structure and function, histone modifying enzymes, chromatin remodeling complexes, and the effects that histone modifications and chromatin remodeling have on phenotype.

18.1 Regulatory Promoters, Enhancers, and Silencers

Linker-scanning mutagenesis reveals the regulatory promoter's presence just upstream from the core promoter.

Based on their knowledge of bacterial gene regulation, investigators anticipated that eukaryotic regulatory elements for protein-coding genes would be located just upstream from the transcription initiation site. They, therefore, attempted to detect these regulatory elements by first introducing mutations in this region and then determining the mutations' effects on gene expression. Deleting several base pairs (bp) at the same time might seem to offer a rapid method for locating regulatory elements within a gene, but it has a serious shortcoming.

Deletion mutations not only remove the nucleotides of interest, they also alter the spacing between flanking DNA sequences. The loss of gene activity that results from deleting a DNA segment thus might be due to the fact that an essential segment was removed, the spacing between flanking sequences was changed, or both.

In 1982, Stephen McKnight and Robert Kingsbury devised a new technique that eliminated the spacing problem, facilitating the search for regulatory elements. This technique, called **linker-scanning mutagenesis**, involves systematic replacement of short DNA segments (usually 3–10 bp) in a region of interest with a DNA linker containing a random sequence of exactly the same size. Although a linker mutation, like a deletion mutation, changes a short DNA segment, it has the advantage of preserving the spacing between nucleotide sequences on either side of the altered segment. Retention of spacing is very important because it allows us to distinguish between effects due to sequence alterations and those due to space changes between flanking sequences.

Linker scanning mutagenesis was first used to search for the promoter of the thymidine kinase gene in the herpes simplex virus (an icosahedral, enveloped DNA virus responsible for cold sores and genital herpes). Linker mutations were introduced just upstream from the transcription initiation site of a cloned thymidine kinase gene that had been inserted into a plasmid. After mutagenesis, the plasmid was microinjected into oocytes of the frog *Xenopus laevis* to permit the *Xenopus* transcription machinery to transcribe the thymidine kinase gene (FIGURE 18.1). Gene activity was monitored by the primer extension method (see Chapter 17).

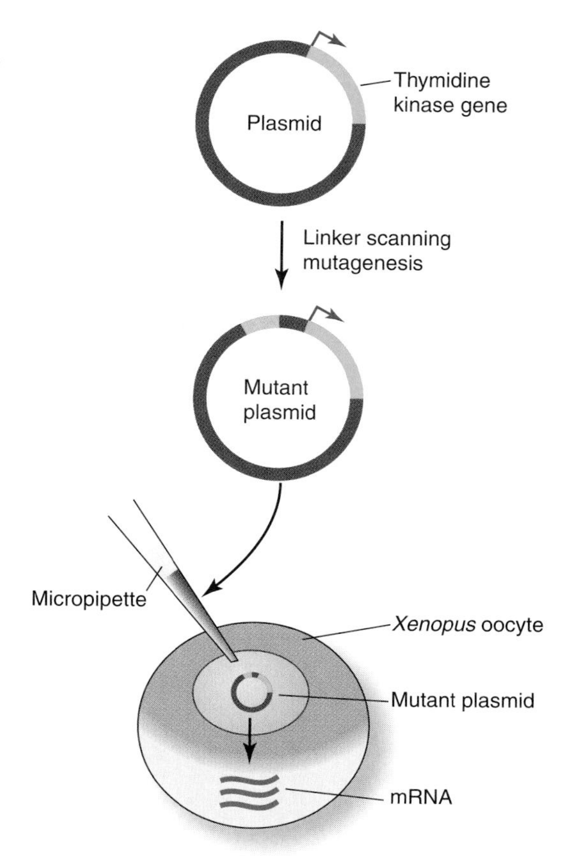

FIGURE 18.1 **Linker scanning mutagenesis technique for studying the SV40 thymidine kinase gene.** Replace a DNA segment upstream from the transcription initiation site (*arrow*) of a thymidine kinase gene (green), which has been inserted in a plasmid, with a linker DNA (yellow) of equal size. Then microinject mutagenized plasmid DNA into *Xenopus* oocyte to determine whether or not the oocyte transcription machinery can transcribe the thymidine kinase gene to produce mRNA. Structures are not drawn to scale.

The experimental results, which are summarized in the schematic diagram in FIGURE 18.2, indicate that thymidine kinase gene transcription is blocked by mutations in three distinct sequence motifs that are just upstream from the transcription initiation site. Mutations at other sites in this region have no effect on transcription. The first of these sequence motifs, the TATA box, is part of the core promoter, which was discussed in detail in Chapter 17. The other two sequence motifs, which were later found to be present in many other viral and eukaryotic protein coding genes, were named the **CCAAT box** and the **GC box** because their consensus sequences are GGC*CAAT*CT and GG*GCG*G, respectively. CCAAT and GC boxes are usually present somewhere between 50 and 200 bp upstream from the transcription initiation site. This region, which is called either the **regulatory promoter** because of its function or the **proximal promoter** because of its location, is just upstream from the core promoter (FIGURE 18.3).

Investigators considered the possibility that the CCAAT box is the eukaryotic counterpart of the bacterial –35 box but were forced to reject the idea for the following reasons: (1) Many protein-coding genes lack the CCAAT box. (2) When present, the precise position of the CCAAT box within the regulatory promoter varies from one kind of protein-coding gene to another. (3) Regulatory promoters often have more than one CCAAT box. The GC box was also ruled out as a possible eukaryotic counterpart for the –35 box for similar reasons.

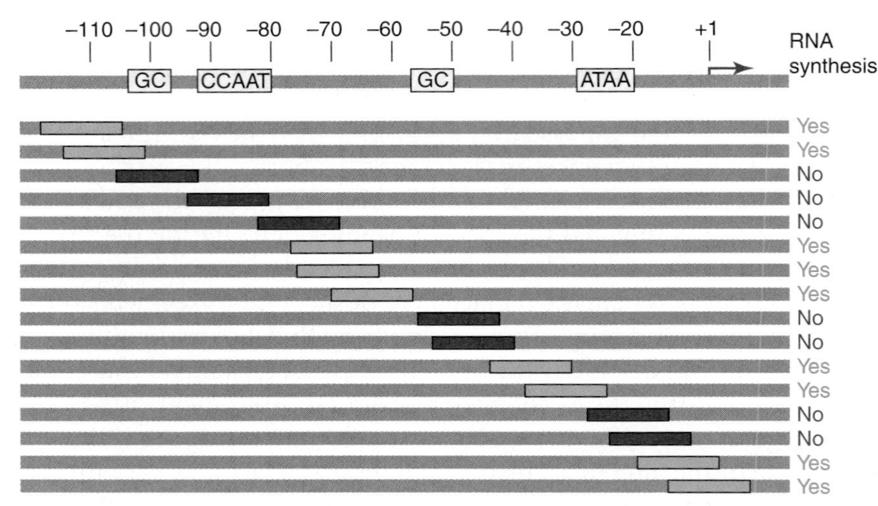

FIGURE 18.2 **Linker scanning mutagenesis studies of the herpes simplex virus (HSV) thymidine kinase gene.** The region of the HSV thymidine kinase gene, which is just upstream from the transcription initiation site, is organized as shown in the top line. The numbers above the line specify the distance in base pairs from the transcription initiation site (+1). The GC, CCAAT, and TATA regulatory elements are shown in yellow boxes. The mutants created by linker-scanning are shown in the series of lines below. Each rectangle represents a position in which a linker replaced a 6–10 nucleotide segment. Mutant DNAs were injected into *Xenopus* oocytes, where the cell's enzymes transcribe the thymidine kinase gene. Green rectangles indicate mutations that produce normal amounts of RNA. Red rectangles indicate mutations that cause reduced levels of RNA synthesis because of a substantial decrease in promoter activity. All mutations that changed sequences in the GC, CCAAT, and TATA regulatory elements decreased promoter activity. (Adapted from J. D. Watson, et al. *Recombinant DNA, Second edition*. W. H. Freeman and Company, 1992.)

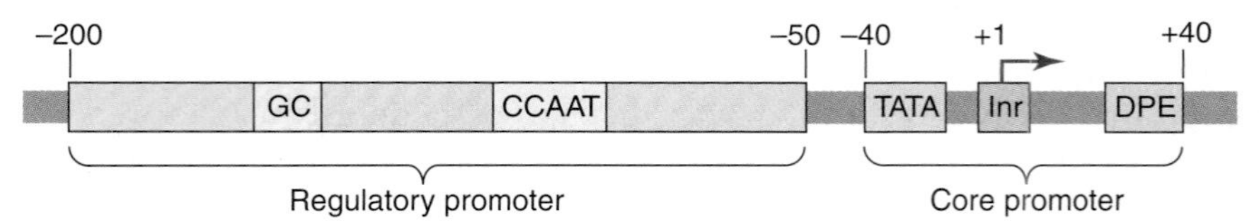

FIGURE 18.3 **The regulatory promoter.** The regulatory promoter (also called the proximal promoter) contains sequence motifs known as promoter proximal elements such as the CCAAT and GC boxes that bind specific transcription activator proteins. The regulatory promoter usually extends from about −200 to −50 with respect to the transcription start site (+1). The core promoter, which consists of a TATA box (TATA), initiator (Inr), and downstream promoter element (DPE), is also shown. The figure is not drawn to scale and does not represent a specific gene.

The question thus remained: What functions do the CCAAT and GC perform? The discovery that many protein-coding genes require additional modules within their regulatory promoters for full expression provided an important clue. The fact that some regulatory promoters have only these additional modules and lack both GC and CCAAT boxes was still another clue. What emerges is a picture where (1) each protein-coding gene has its own characteristic regulatory promoter that is made of some unique combination of modules, and (2) full gene expression occurs when transcription activator proteins bind to each module within the regulatory promoter. Regulatory sites are also present for repressor proteins. The focus for now, however, will be on activation.

Enhancers stimulate transcription and silencers block transcription.

George Khoury and coworkers discovered a remarkable new regulatory element in 1981 while studying transcription in simian virus 40 (SV40). Recall from Chapter 8 that this virus has a circular double-stranded DNA containing about 5.2 kbp. SV40 is particularly well suited for transcription studies because it has only two transcription units, the early and late transcription units, which are transcribed in opposite directions from a common control region (FIGURE 18.4). The early transcription unit is activated soon after the virus coat is removed in the cell nucleus and codes for the large tumor (large T) antigen required for DNA replication (see Chapter 10). The early transcription unit's core promoter contains a TATA box and its regulatory promoter contains six GC boxes (Figure 18.4). Khoury and coworkers detected two identical 72 bp sequences just upstream from the regulatory promoter (Figure 18.4). Removing either 72 bp sequence causes a slight inhibition of the early transcription unit but removing both sequences causes a 100-fold inhibition.

Experiments were performed to determine how the 72-bp sequence influences the transcription of a mammalian β-globin gene when the 72-bp sequence is inserted at different positions with respect to the gene's transcription initiation site. This objective was accomplished by inserting the 72-bp sequence into various sites within a plasmid bearing the β-globin gene and then using the recombinant plasmids to transfect cells that do not normally synthesize β-globin. Remarkably, the 72-bp sequence stimulated β-globin mRNA synthesis from the correct initiation site by more than 100-fold when it was inserted

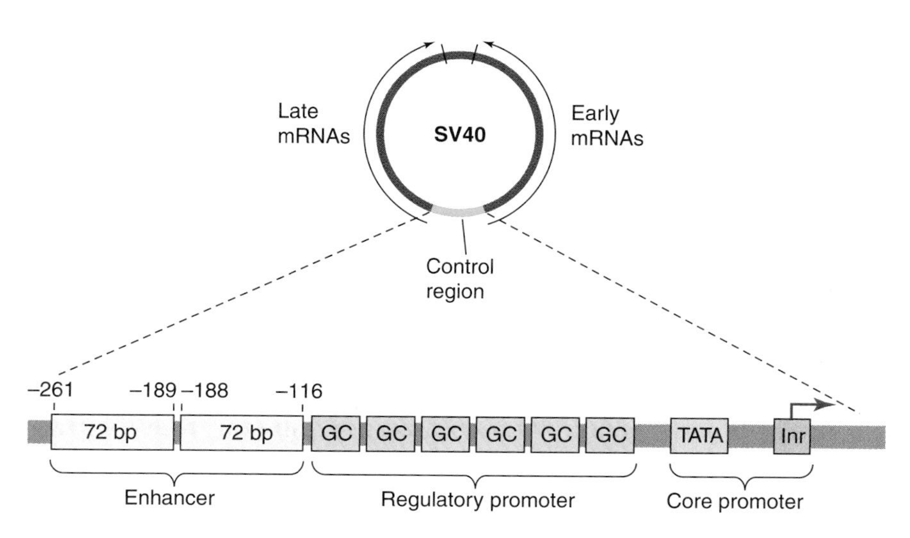

FIGURE 18.4 **SV40 early transcription unit enhancer and promoter regions.**

a great distance (1 kbp or more) upstream or downstream from the transcription initiation site, and even when it was inserted within the transcription unit. Furthermore, stimulation was the same when the 72-bp sequence was inserted in the forward or backward orientation. The 72-bp sequence's properties were so remarkable that it at first seemed possible that it might be a unique genetic element present only in SV40 and perhaps a few related viruses. Additional investigations, however, revealed that animal and plant cells also have regulatory sequences that stimulate transcription in the same way as the 72-bp sequence. These regulatory sequences, which were called **enhancers** because of their ability to stimulate transcription, have two unique features that distinguish them from other regulatory sequences: (1) Enhancers stimulate transcription from the correct transcription initiation site even when they are located several thousand bp upstream or downstream from that site; and (2) enhancers stimulate transcription when inserted in either orientation.

Many different enhancers have now been characterized. They range in size from about 50 bp to 1.5 kbp and, like regulatory promoters, consist of a cluster of modules. In fact, many of these modules, including the CCAAT and GC boxes, are the very same ones that are present in regulatory promoters. Each type of enhancer has a characteristic combination of modules. A major structural difference between enhancers and regulatory promoters is the modules appear to be closer together in enhancers. A cell must have transcription activator proteins capable of binding to the modules within an enhancer or a regulatory promoter for full gene expression to occur. This requirement provides a means for gene regulation by controlling the level of functional transcription activator proteins within the cell nucleus.

Eukaryotes also have a negative regulatory element called the **silencer**. Silencers are sequence-specific elements that repress transcription of a target gene. Most silencers function independently of distance and orientation but some silencers are position dependent. For example, one class of *Drosophila* silencers must be within about 100 bp of the target gene to repress transcription. Silencers are binding sites for negative transcription factors, repressors, which act by establishing a repressive chromatin structure (see below), preventing a nearby activator from binding to its DNA binding site, or in a few cases blocking pre-initiation complex formation.

The upstream activating sequence (UAS) regulates genes in yeast.

Most protein-coding genes in yeast have a single regulatory region, called an **upstream activating sequence** (UAS), which is located within a few hundred base pairs of the transcription initiation site. Like enhancers in higher organisms, the UAS works at variable distances from the transcription initiation site and in either the forward or reverse orientation. However, all the UAS elements that have been studied to date differ from enhancers in one important respect; UAS elements do not function when located downstream from the transcription initiation site.

18.2 Transcription Activator Proteins

Transcription activator proteins help to recruit the transcription machinery.

Each of the thousands of protein-coding genes within a eukaryotic cell competes for the limited transcription machinery that is available. The basal transcription machinery requires the assistance of a special class of transcription factors called **transcription activator proteins** to locate protein-coding genes that will be transcribed. Each transcription activator protein has at least two independently folding domains, a **DNA-binding domain** and an **activation domain** (FIGURE 18.5a). The DNA-binding domain makes sequence-specific contacts with control elements in a gene's regulatory promoter or enhancer (FIGURE 18.5b). For instance, DNA-binding domains of Sp1 (selective promoter factor 1) and the C/EBP (CAAT box and enhancer binding protein) transcription activator proteins bind the GC box and the CAAT box, respectively. X-ray crystallography and nuclear magnetic resonance (NMR) spectroscopy have provided considerable information about the DNA-binding domains.

Several different kinds of folding patterns have been observed in the DNA-binding domains, allowing us to assign most transcription factors to a structurally defined family. The activation domain recruits components of the transcription machinery to the gene and then interacts with various components of the transcription machinery to stimulate transcription (see below). We know very little about activation domain structure at this time.

Many transcription activator proteins also have additional structural features. Some of the most important of these are as follows. A short basic sequence containing arginine and lysine residues known as the **nuclear localization signal** allows transcription activator proteins to move through the nuclear pore complex from the cytoplasm to the nucleoplasm. Another signal, the **nuclear export signal**, permits the transcription activator proteins to move in the opposite direction. The direction of movement can be controlled by masking an import or export signal by covalent modification. The so-called **dimerization**

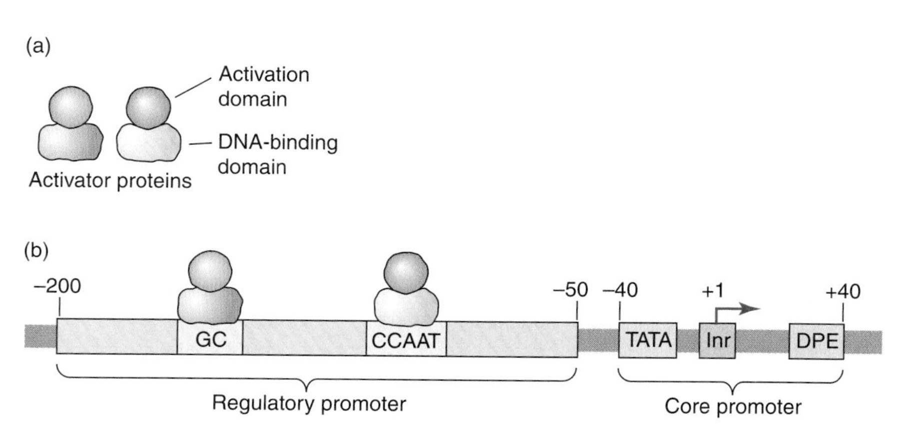

FIGURE 18.5 **A cartoon showing how promoter-selective DNA transcription factors called activator proteins are organized and interact with sequence motifs in the regulatory promoter.** (a) The promoter selective DNA-binding transcription factor called an activator has a DNA-binding domain and an activation domain. (b) The activators' DNA-binding domains bind to specific sequence motifs in the regulatory promoter, leaving the activation domains free to act together to bind transcription factors and thereby recruit the transcription machinery to the gene.

domain allows one transcription activator protein to pair with an identical transcription activator protein to form a homodimer or with a different transcription activator protein to form a heterodimer. Some transcription activator proteins have specific binding sites for small molecules or **ligands**. The **ligand-binding domains** allow the transcription activator proteins to respond to small bioactive molecules such as steroid hormones (see below).

A combinatorial process determines gene activity.

Initially, investigators thought that regulatory promoters and enhancers were completely different types of regulatory elements. The distinction between the two has become blurred, however, because the same sequence motifs are often present in both regulatory promoters and the enhancers. This means that the same activator protein may bind to both regulatory regions. Furthermore, DNA can form loops that bring activator proteins bound to regulatory promoters near activator proteins bound to enhancers (FIGURE 18.6), so that the activation domains can work cooperatively to recruit the transcription machinery to the gene.

Because each gene has a characteristic set of sequence motifs in its regulatory promoter and enhancer(s), it will bind a unique combination of activator proteins. Gene activity is determined in large part by how well the activation domains work together to recruit the transcription machinery to the gene. Hence, gene activation results from **combinatorial control**, which allows a cell to use relatively few polypeptides to regulate many protein coding genes in response to diverse signals.

DNA affinity chromatography can be used to purify transcription activator proteins.

Transcription activator proteins must be purified before they can be characterized. However, it was difficult to purify the transcription activator proteins because they are present at very low intracellular concentrations. In 1986, James T. Kadonaga and Robert Tjian solved the problem by using **DNA affinity chromatography** to purify

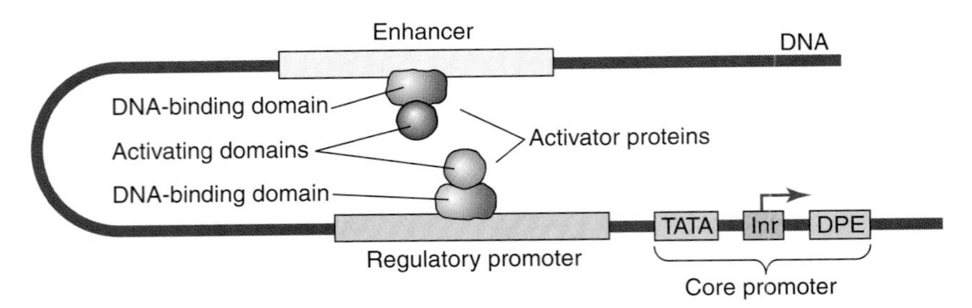

FIGURE 18.6 **DNA looping.** The DNA between the enhancer and the regulatory promoter loops, allowing the activation domains of activator proteins bound to regulatory promoters to be near one another.

transcription activator proteins from crude nuclear extracts or partially purified preparations (FIGURE 18.7). They prepared DNA affinity beads by synthesizing DNA fragments with tandem repeats of the sequence motif that binds the transcription activator protein and then linked these DNA fragments to inert polysaccharide beads. The crude nuclear extract or partially purified protein solution containing the transcription activator protein was mixed with sonicated calf thymus competitor DNA, and then the protein-DNA mixture was passed through a column filled with the DNA affinity beads. Proteins that bound nonspecifically to DNA passed through the column after binding to the sonicated thymus DNA. Activator proteins that bound specifically to the DNA affinity beads were eluted from the column by washing the column with a high salt buffer solution. Fractions were assayed by testing eluted proteins for their ability to bind to the sequence motif by DNase footprinting (see Chapter 15) or by a gel

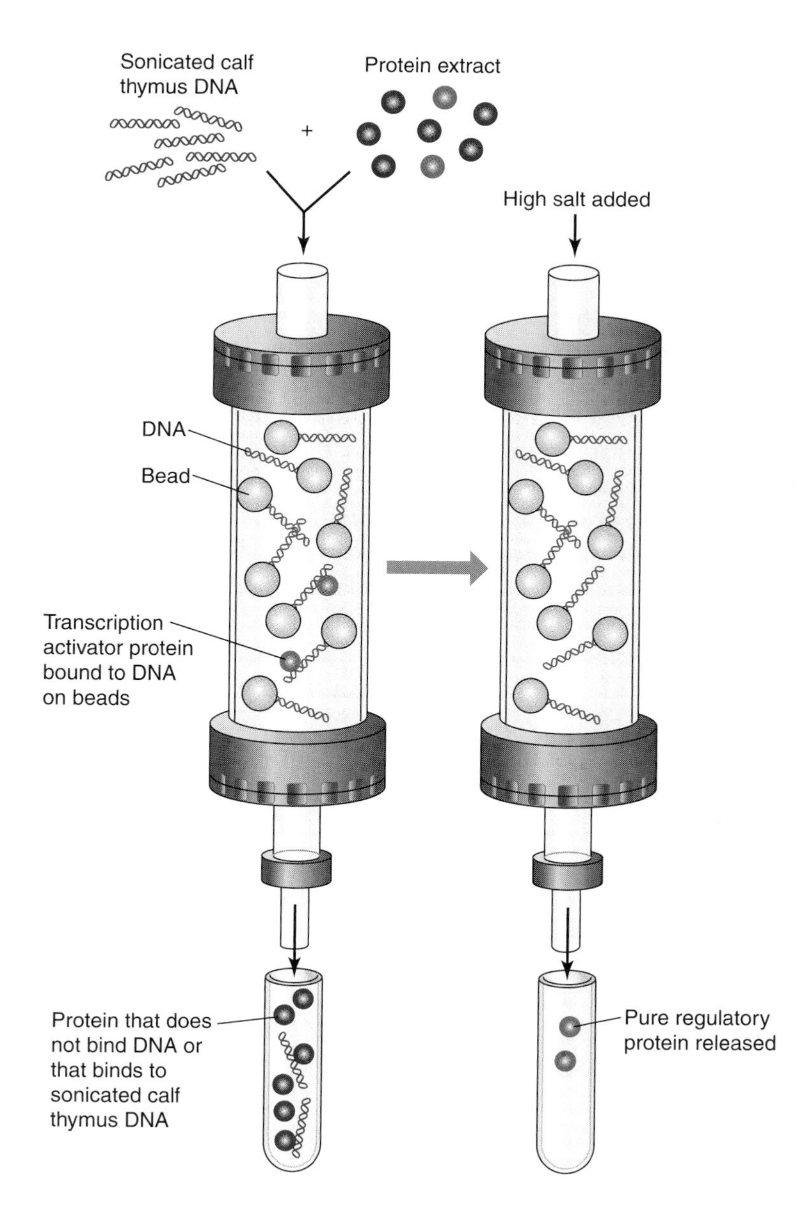

FIGURE 18.7 **Schematic of DNA affinity chromatography.** Sonicated calf thymus DNA is mixed with a protein extract that contains a few transcription activator proteins (green) and many other proteins (red). The activator proteins do not bind the sonicated calf thymus DNA, but some of the other proteins do bind to it. The protein-DNA mixture is passed through a column containing DNA affinity beads. The beads (yellow) have attached DNA fragments (blue) that have tandem repeats of the sequence motif that binds the transcription activator protein. The transcription activator protein binds specifically to the DNA attached to the beads. The other proteins pass through the affinity column. Transcription activator protein is eluted from the column with buffer that contains a high salt concentration. (Adapted from an illustration from the Natural Toxins Research Center, Texas A&M University.)

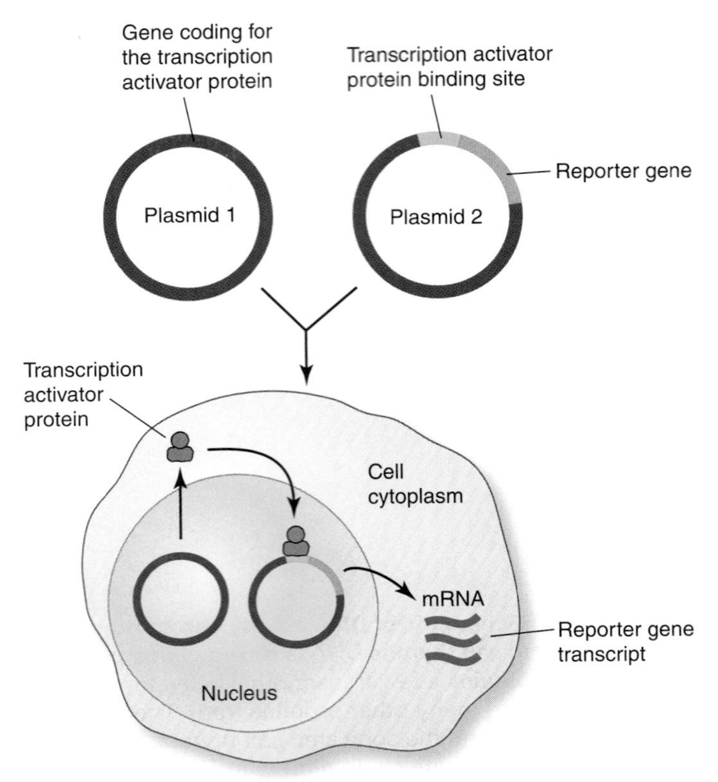

FIGURE 18.8 **Cell-based assay for the transcription activator protein.** Two recombinant plasmids are required. The first contains the gene for the transcription activator protein and the second contains a reporter gene and a transcription activator binding site. If the transcription activator protein is synthesized and able to bind to the transcription activator binding site then the reporter gene will be transcribed to produce mRNA that can be translated to produce a protein such as green fluorescent protein or β-galactosidase that is easy to detect. (Adapted from H. Lodish, et al. *Molecular Cell Biology, Fourth edition.* W. H. Freeman and Company, 2000.)

mobility shift assay (see Chapter 17). Although the amount of a transcription activator protein obtained by DNA affinity chromatography was quite low, it was sufficient for partial or complete amino acid sequence determination. Sequence information then was used to predict possible nucleotide sequences for a segment of the activator protein gene so that a probe could be synthesized to search for the activator protein gene.

Several transcription activator protein genes have been cloned in this fashion. Once a transcription activator protein gene has been cloned from one organism, it is usually fairly straightforward to find its counterpart in other organisms by launching a search for homologous sequences. The rapid pace of gene identification and sequence determination in many different eukaryotes including humans has facilitated this process.

A transcription activator protein's ability to stimulate gene transcription can be determined by a transfection assay.

A cloned transcription activator protein gene can be used to test the transcription activator protein's ability to stimulate transcription in a cell-based assay (FIGURE 18.8). Two recombinant plasmids are required to perform this assay. The first recombinant plasmid bears the gene that specifies the transcription factor and the second bears a reporter gene and the control element recognized by the transcription factor. Several different coding sequences have been used to construct reporter genes. Included among these are β-galactosidase from *Escherichia coli*, **luciferase** from firefly, and **green fluorescent protein** from jellyfish. Assays for β-galactosidase and luciferase are sensitive over a wide range of enzyme concentrations and easy to perform. β-Galactosidase activity is detected in a spectrophotometer by taking advantage of the fact that the enzyme converts the artificial substrate *o*-nitrophenylgalactoside to *o*-nitrophenol, a yellow substance that absorbs light at 450 nm. Luciferase activity is monitored in a luminometer or liquid-scintillation counter by exploiting the enzyme's ability to catalyze the ATP-dependent oxidative carboxylation of luciferin with the release of light. The green fluorescent protein has the advantage of not requiring any additional substrates. It emits a green light when irradiated with ultraviolet light or blue light, allowing direct measurement of its concentration in living cells.

A host cell that lacks both the reporter gene and the transcription activator protein gene is transfected with one or both recombinant plasmids. If the cloned gene in fact does code for the transcription activator protein, expression in the host cell will produce the transcription activator protein, which in turn will stimulate the synthesis of the reporter protein. No stimulation will be observed if only one of the two recombinant plasmids is used. Activator protein can also be tested *in vitro* with the same recombinant reporter gene and required general transcription factors, RNA polymerase II, and other required components of the transcription machinery (see below).

18.3 DNA-Binding Domains with Helix-Turn-Helix Structures

Homeotic genes assign positional identities to cells during embryonic development.

Transcription activator proteins are commonly grouped according to the structures of their DNA-binding domains. Although this organizational approach is based on structural similarities in just one domain, it often groups transcription activator proteins with similar biological functions together. This structure-function relationship is clearly evident in the first group of transcription activator proteins that we examine, which all have a helix-turn-helix motif in their DNA-binding domain. As described in Chapter 16, this motif is also present in the *E. coli lac* repressor and cAMP receptor protein (CRP), and is commonly used for DNA binding.

The first clue to the existence of the helix-turn-helix transcription activator proteins came from studies of eight developmental genes in the fruit fly *Drosophila*, the so-called **homeotic** (***HOM***) genes, which "inform" embryonic cells that they will become part of the fly's head, thorax, or abdomen; that is, *HOM* genes assign distinct positional identities to cells in different regions along the fly's anterior-posterior axis. As shown in FIGURE 18.9, the *HOM* genes are located in two

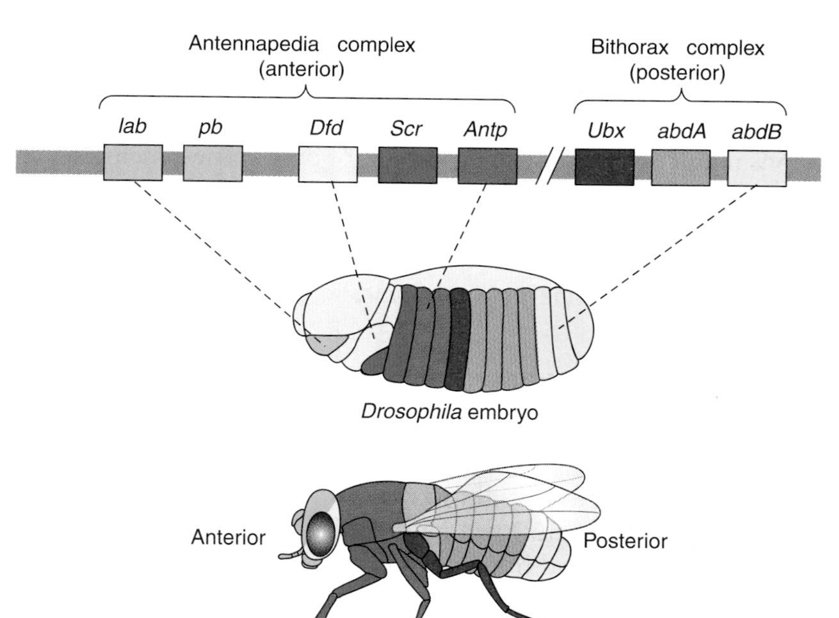

FIGURE 18.9 **Arrangement of *Drosophila HOM* genes.** The *HOM* genes are located in two small clusters on chromosome 3, designated the *Antennapedia* complex (five genes) and the *bithorax* complex (three genes). *Antennapedia* genes regulate the head and anterior thoracic segments, while bithorax genes regulate the posterior thoracic and abdominal segments. Gene order on the chromosome parallels body region order along the anterior to posterior axis of the embryo. (Adapted from P. H. Raven and G. B. Johnson. *Biology, Sixth edition.* McGraw-Hill Higher Education, 2001.)

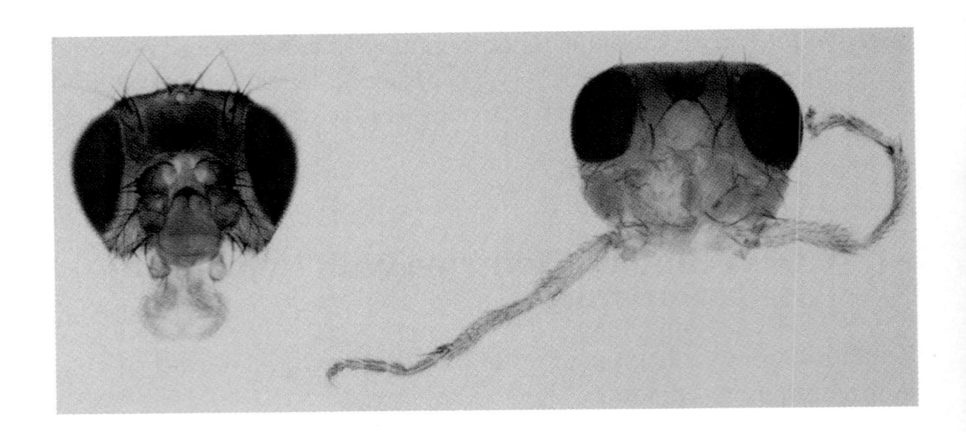

FIGURE 18.10 Head of a wild-type fruit fly (*Drosophila melanogaster*) compared to an *Antennapedia* mutant in which the antennae are replaced by a pair of middle legs. (Photo courtesy of Walter J. Gehring, Biozentrum, University of Basel.)

small clusters on chromosome 3, designated the *Antennapedia* complex (5 genes) and the *bithorax* complex (3 genes). *Antennapedia* genes regulate the head and anterior thoracic segments, while *bithorax* genes regulate the posterior thoracic and abdominal segments. Remarkably, gene order on the chromosome parallels body region order along the anterior to posterior axis of the embryo. Although usually lethal, *HOM* gene mutations can produce incredible phenotypic changes in surviving offspring. For instance, *Antennapedia* gene mutations cause legs to grow on the head where antennae should be located (FIGURE 18.10).

Vertebrates also have homeotic genes known as **Hox genes** (a contraction of the term homeobox, see below). Mammalian cells have 38 *Hox* genes, which are located in four gene clusters, designated *HoxA*, *HoxB*, *HoxC*, and *HoxD*. Each gene cluster is about 100 kbp long and is located on a separate chromosome. Genes within each cluster are homologous to those within the two *Drosophila* gene clusters (FIGURE 18.11). Moreover, gene arrangements within each cluster parallel the order of the body regions that they specify along the anterior-posterior embryo axis. *Hox* gene mutations cause structural deletions and transformations that are analogous to those caused by *HOM* gene mutations in fruit flies. For instance, *Hox A-3* deletion mutations in mice cause a complicated set of deformities, including an incomplete heart and the absence of thymus and parathyroid glands. As might be expected, most mutants with such major anatomical and physiological abnormalities die at birth. In fact, most homeotic gene mutations result in nonviable organisms.

Homeotic genes specify transcription activator proteins.

In 1984, independent studies by Walter Gehring and Matthew Scott showed that homeotic genes in *Drosophila* have a common DNA sequence within them that is about 180 bases long and specifies the 60-amino acid residue helix-turn-helix DNA binding domain. Subsequent studies by these and other workers showed that this conserved sequence, which is designated the **homeobox**, is also present in vertebrate, plant, and fungi homeotic genes as well as in non-homeotic genes such as the yeast mating type genes *MATa* and *MAT*α.

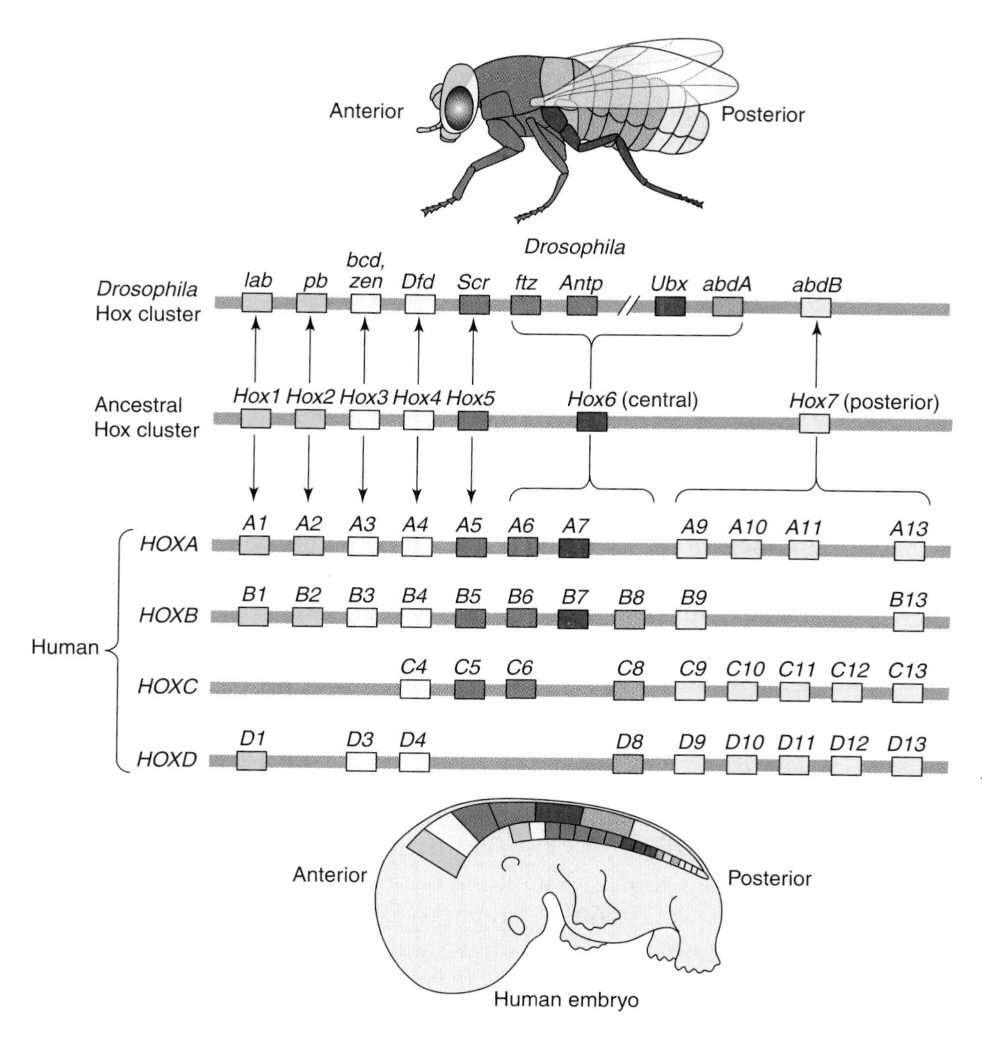

FIGURE 18.11 **Comparison of homeotic gene arrangements in *Drosophila* and humans.** Similar genes, which are arranged in the same order, control the development of the anterior and posterior parts of the bodies of flies and humans. These homeotic genes are located on a single chromosome in the fly (top row of colored squares) and on four separate chromosomes in mammals (lower rows of squares). Studies show that an ancestor of all bilateral animals may have had seven *Hox* genes. A hypothetical ancestral Hox cluster is shown in the middle, with arrows indicating the predicted origins of fly and mammalian *Hox* genes. The genes are color coded to match the parts of the body in which they are located. (Modified from *Mol. Genet. Metab.*, vol. 69, A. Veraksa, M. Del Campo, and W. McGinnis, Developmental Patterning Genes and Their Conserved . . ., pp. 85–100, copyright 2000, with permission from Elsevier [http://www.sciencedirect.com/science/journal/10967192].)

Because *MATa* and *MATα* previously had been demonstrated to be transcription activator proteins, it seemed reasonable to propose that homeobox proteins might also have the same function. Many different lines of evidence support this proposal. Among the most important of these are the following:

1. Electrophoretic mobility shift assays (see Chapter 17, Figure 17.21) show that homeobox proteins bind to regulatory promoters and enhancers in a sequence-specific manner.
2. Addition of a homeobox protein to a cell-free extract stimulates or inhibits transcription of genes bearing the cognate regulatory promoter, enhancer, or silencer element, suggesting that homeobox proteins act as transcription activator proteins or repressor proteins.
3. A positive correlation exists between a homeobox protein's ability to bind to a regulatory promoter, enhancer, or silencer *in vitro* and its ability to regulate the transcription of a gene with that same regulatory promoter, enhancer, or silencer *in vivo*.

The homeodomain contains a helix-turn-helix motif.

The homeobox protein folds into two domains (FIGURE 18.12). The N-terminus forms the activation domain and the C-terminus, consisting of 60-amino acid residues, forms the DNA-binding domain or **homeodomain**. The homeodomain can be synthesized by standard organic chemistry techniques or isolated from bacteria that have been transformed with a recombinant plasmid bearing the homeobox region.

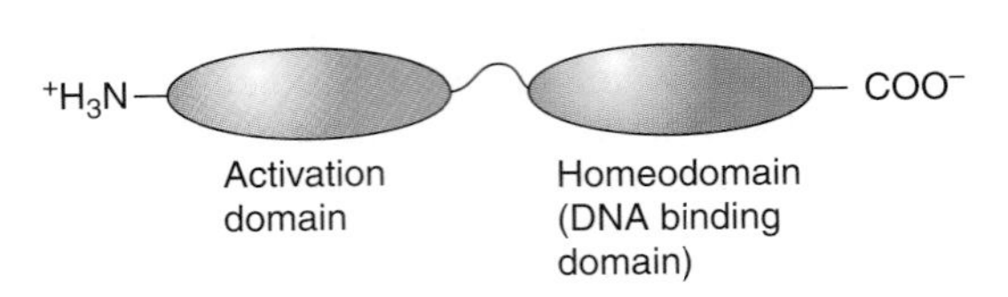

FIGURE 18.12 **Homeodomain protein.**

High-resolution crystal structures are not yet available for intact homeobox proteins (or for any other intact eukaryotic transcription activator protein) because of difficulties in preparing suitable crystals. Peptide fragments containing homeodomains do fold into stable structures, however, which form crystals suitable for x-ray crystallography.

Carl O. Pabo and coworkers obtained the first-high resolution crystal structure for a homeodomain in 1994 by subjecting a crystalline peptide fragment from a homeobox protein to x-ray analysis. The peptide fragment they studied, which was derived from the *Drosophila* homeobox protein engrailed, has a compact structure containing three α-helices (FIGURE 18.13). Helices 1 and 2 are joined by a short loop. Helices 2 and 3 are connected by a turn, forming a helix-turn-helix structural motif that makes sequence specific contacts with DNA. Helix 3 is the recognition helix; its hydrophilic face fits into the major groove of a DNA segment containing a 5′-ATTA-3′ (5′-TAAT-3′) sequence motif (Figure 18.13). The hydrophobic face of helix 3 packs against helices 1 and 2. The homeodomain also has an N-terminal arm that helps to fasten the protein to DNA by fitting into the minor groove. Other homeodomains have similar structures.

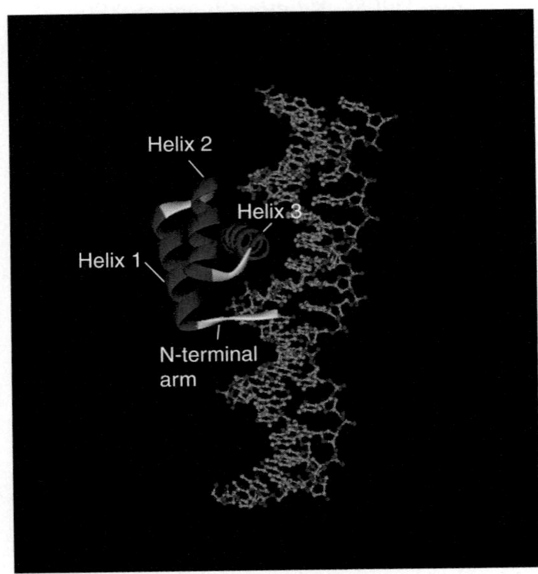

FIGURE 18.13 **Structure of engrailed homeodomain DNA complex.** The protein is presented in ribbon form (colored according to structure) and DNA (TTTTGCCATGTAATTACCTA) in ball and stick form. Helices 1 and 2 are joined by a short loop. Helices 2 and 3 are connected by a turn, forming a helix-turn-helix structural motif that makes sequence-specific contacts with DNA. Helix 3 is the recognition helix; its hydrophilic face fits into the major groove of a DNA segment containing a 5′-ATTA-3′ (5′-TAAT-3′) sequence motif. The hydrophobic face of helix 3 packs against helices 1 and 2, but this is not completely visible in this perspective. The homeodomain's N-terminal arm helps to fasten the protein to DNA by fitting into the minor groove. (Structure from Protein Data Bank 3HDD. E. Fraenkel, et al., *J. Mol. Biol.* 284 [1998]: 351–361. Prepared by B. E. Tropp.)

Under *in vitro* conditions, different homeobox proteins appear to have similar affinities for DNA fragments with a 5′-ATTA-3′ sequence motif. The same proteins, however, show definite preferences for specific promoters *in vivo*. The greater degree of specificity that is observed *in vivo* appears to be due to the ability of additional proteins to influence binding specificity. A monomeric homeodomain protein can bind to DNA in a sequence-specific manner, but it often attains even greater specificity as a result of interacting with other proteins.

POU proteins have a homeobox and a POU domain.

POU transcription activator proteins, which regulate the expression of genes for histones, immunoglobulins, and growth factors, have two helix-turn-helix motifs. Because the first members of this superfamily were discovered based on their ability to bind the octanucleotide sequence 5′-ATGCAAAT-3′, they were named the Oct-1 and Oct-2 transcription activator proteins. Oct-1 is present in all kinds of mammalian cells, whereas Oct-2 is specific for blood cells. Subsequent studies revealed related transcription activator proteins that regulate pituitary-specific genes and coordinated movement in the nematode *Caenorhabditis elegans*. The former was named Pit-1 because it activates growth hormone and prolactin genes in the pituitary gland and the latter was named Unc-86 because mutant *C. elegans* exhibited

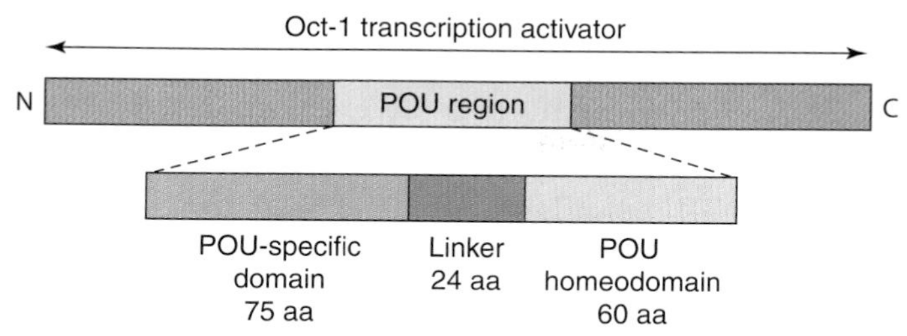

FIGURE 18.14 **Oct-1 transcription factor.** The Oct-1 transcription factor contains a POU region that binds DNA (blue) and activation domains (orange). The POU region contains two DNA-binding sites, the 75 amino acid POU-specific domain and the 60 amino acid POU homeodomain. A 24 amino acid linker connects the two DNA-binding domains. (Modified from C. Branden and J. Tooze. *Introduction to Protein Structure, First edition.* Garland Science, 1999. Used with permission of John Tooze, The Rockefeller University.)

uncoordinated movement caused by abnormal nervous system development. Transcription activator proteins that belong to the Pit, Oct, and Unc superfamily are called **POU proteins.** The distinguishing feature of a POU protein is a tandem pair of helix-turn-helix motifs, the **POU-specific sub-domain** (POU_S) and the **POU homeodomain** (POU_H), which can each bind DNA (FIGURE 18.14). A flexible linker joins POU_S to POU_H, forming the so-called **POU region**, which has an activation domain on either side.

Crystal structures have been obtained for a few POU domain-DNA complexes. We examine just one of these, the Oct-1 POU region bound to an octanucleotide (FIGURE 18.15). POU_H and POU_S bind on opposite sides of the DNA. POU_H has a typical homeodomain structure, permitting helix 3 (the recognition helix) to make sequence-specific contacts with ATGC in the major groove. POU_S is a bundle of four helices and two of these, helices 2 and 3, are part of the helix-turn-helix motif that binds DNA. POU_S helix 2 (the recognition helix) makes sequence-specific contacts with ATGC in the major groove. The presence of two DNA-binding sites on a single polypeptide results in much tighter binding than would be possible if only one binding site were present. Although POU_H and POU_S are similar in all POU proteins, the flexible linker that connects them differs, permitting various arrangements of POU_H and POU_S with respect to one another. This variation accounts in great part for the ability of each kind of POU protein to recognize its cognate DNA control element.

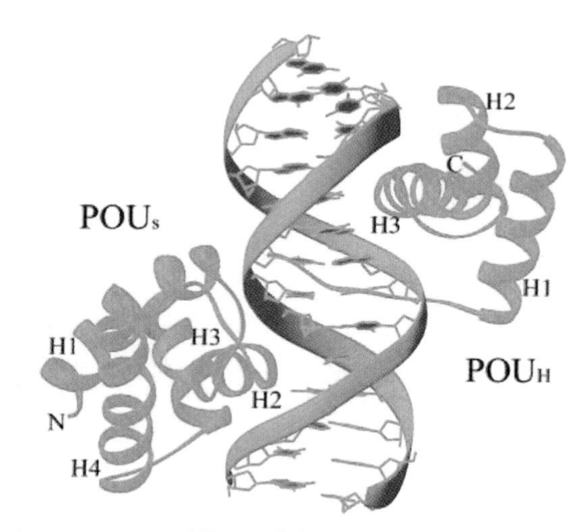

FIGURE 18.15 **View of the Oct-1 crystal structure looking into the major groove of the promoter DNA.** The POU_H (POU homeodomain) and POU_S (POU-specific) sub-domains are bound to the two DNA half-sites. The helices are labeled for the individual sub-domains and the N and C termini are indicated. (Reproduced from *J. Mol. Biol.*, vol. 302, K. Phillips and B. Luisi, The virtuoso of versatility . . ., pp. 1023–1039, copyright 2000, with permission from Elsevier [http://www.sciencedirect.com/science/journal/00222836]. Photo courtesy of Ben Luisi, University of Cambridge.)

18.4 DNA-Binding Domains with Zinc Fingers

Many transcription activator proteins have Cys_2His_2 zinc fingers that bind to DNA in a sequence-specific fashion.

Many transcription activator proteins have DNA binding structural motifs that are stabilized by zinc ions. Aaron Klug predicted the existence of these structural motifs in 1985 while studying TFIIIA, a transcription factor that regulates 5S RNA synthesis by RNA polymerase III. Klug selected TFIIIA for study because each *Xenopus laevis* oocyte has about 20,000 TFIIIA molecules together with an equal number of 5S RNA molecules in a 7S ribonucleoprotein complex, making it readily available. Klug and coworkers expected to be able to extract active TFIIIA from the 7S ribonucleoprotein using a buffer that contained a chelating agent but were unable to do so. The reason for their failure became clear when they discovered that the chelating agent removed about nine zinc ions from each TFIIIA molecule, thereby inactivating it.

A very different type of experiment helped to understand how the zinc ions interact with the polypeptide chain. Klug and coworkers observed that digesting TFIIIA with a protease produced progressively shorter fragments, ending in the buildup of tightly folded protease-resistant fragments of about 30 amino acid residues. Klug interpreted the digestion results to mean that TFIIIA consists almost entirely of tandem segments, each of which contains 30 amino acid residues and folds around a zinc ion into a small compact DNA-binding domain.

This hypothesis received strong support when Robert Roeder and coworkers sequenced TFIIIA and showed that (1) the first three quarters of the polypeptide chain contained nine similar sequence motifs of about 30 residues and (2) each of these motifs had a pair of cysteine residues at its N-terminus and a pair of histidine residues at its C-terminus (FIGURE 18.16a). Klug, who was aware that polypeptide chains normally use four amino acids, frequently some combination of cysteine and histidine, to bind zinc ions, proposed that a pair of cysteines and a pair of histidines act together to bind a zinc ion in each repeating sequence, causing each unit to fold independently (FIGURE 18.16b). The sequence motif was designated a **zinc finger** because it helps to grip DNA.

Subsequent investigations revealed the existence of other types of zinc fingers. Therefore, structural motifs that are similar to those in TFIIIA are now called **Cys_2His_2 zinc fingers** or **classical zinc fingers.** The sequences of several hundred classical zinc fingers have been determined. All share the consensus sequence (F/Y)-X-C-X_{2-5}-C-X_3-(F/Y)-X_5-Ψ-X_2-H-X_{3-5}-H, where X represents any amino acid and Ψ is a hydrophobic amino acid. SP1, the transcription factor that binds to the GC box, belongs to the classical zinc finger family.

Despite considerable efforts to crystallize TFIIIA, suitable crystals are still not available for x-ray diffraction studies and the intact protein is too large for structural analysis by NMR spectroscopy. Nevertheless, investigators have been able to determine the three-dimensional structure of

(a)

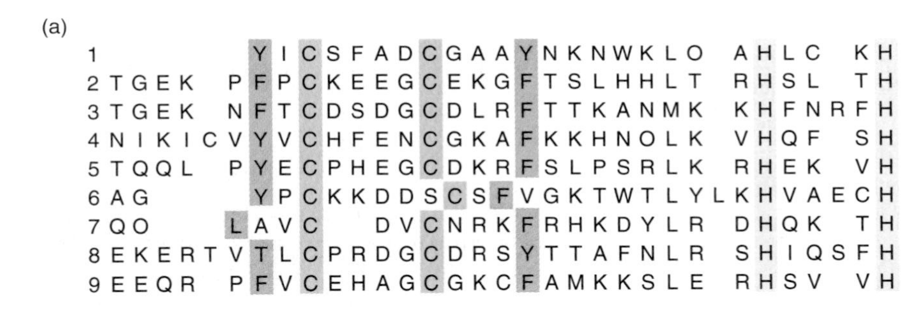

(b)

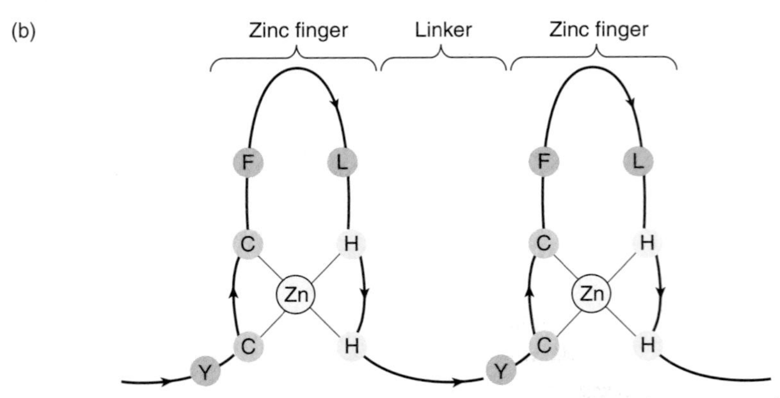

FIGURE 18.16 **The zinc-finger sequence motif.** (a) The nine successive zinc finger sequence motifs in TFIIIA. Histidine and cysteine residues are colored blue and green, respectively. Hydrophobic residues are colored orange. (b) A schematic diagram depicting a pair of zinc fingers that are joined by a linker. (Adapted from L. Stryer. *Biochemistry, Fourth edition.* W. H. Freeman and Company, 1995.)

Cys_2His_2 zinc fingers by examining protein fragments that contain them. The essential features of the Cys_2His_2 zinc finger, which are depicted in schematic form (FIGURE 18.17), are:

1. An antiparallel β-motif is followed by an α-helix with the zinc ion buried in the interior. The β strands are packed against the α-helix.
2. The zinc ion binds with tetrahedral geometry to Cys-3 and Cys-6 in the β-strand and to His-19, and His-23 in the α-helix.
3. The 12-residue sequence X_3-(F/Y)-X_5-Ψ-X_2 connects Cys-6 and His-19 to form a loop.
4. Two conserved hydrophobic residues (which are not visible in Figure 18.17), are buried in the interior. The zinc ion is required to stabilize the structural motif because two hydrophobic residues are insufficient to form a stable hydrophobic core.

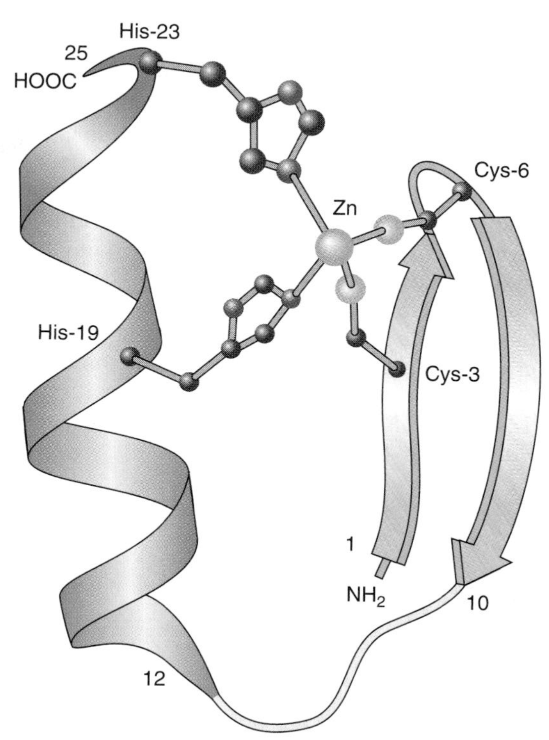

FIGURE 18.17 **Schematic diagram of the three-dimensional structure of a zinc finger.** The zinc finger contains an antiparallel β hairpin (residues 1–10 shown in green), which is followed by an α-helix (residues 12–24 shown in blue). The α-helix is fixed to the β-sheet by the interaction of the four zinc ligands (Cys-3, Cys-6, His-19, and His-23) with the zinc ion. (Adapted from C. K. Mathews, et al. *Biochemistry, Third edition*. Prentice Hall, 2000.)

The next step was to determine how zinc finger proteins interact with DNA. DNase footprinting experiments showed that zinc finger proteins make repeated contact with DNA. X-ray crystallography and NMR spectroscopy provided more detailed information about the nature of these interactions. Studies using a peptide fragment from the mouse transcription activator protein Zif268 are particularly instructive. This peptide fragment consists of three Cys_2His_2 zinc fingers, each connected to its neighbor by a short linker. The three zinc fingers interact with adjacent 3-bp sub-sites on DNA as the zinc-finger region winds around almost one turn of the DNA helix (FIGURE 18.18). The zinc ion is prevented from making contact with the DNA by the polypeptide chain that surrounds it.

Although each zinc finger binds to its DNA sub-site as an independent unit, overall binding specificity and affinity depends on contributions made by all three zinc fingers. Removing one zinc finger causes the binding affinity to decrease by about 100-fold and removing a second zinc finger causes a further decrease of about 100-fold. Further increases in binding affinity would result in such tight binding that the transcription activator protein could not dissociate from the DNA, an unsatisfactory state of affairs if transcription is to respond to changing developmental and physiological conditions. Many transcription activator proteins have several Cys_2His_2 zinc fingers. If only three of these zinc fingers participate in DNA binding then what is the function of the other zinc fingers? One possible answer to this question is that the remaining zinc fingers interact with other proteins.

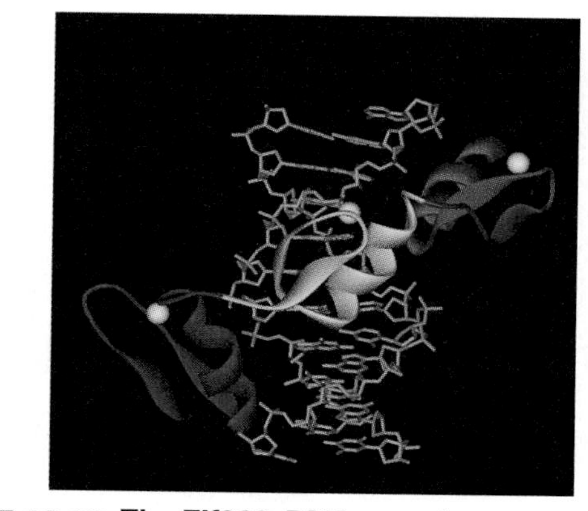

FIGURE 18.18 **The Zif268•DNA complex showing Zif268's three zinc fingers bound in the DNA major groove.** Fingers 1, 2, and 3 are colored red, yellow, and purple, respectively. Coordinated zinc ions are shown as light blue spheres. DNA is shown in blue. Base contacts made by each zinc finger are indicated by the color coded DNA sequence of the Zif268 site, which is shown on the left. (Structure from Protein Data Bank 1ZAA. N. P. Pavletich and C. O. Pabo, *Science* 252 [1991]: 809–817. Prepared by B. E. Tropp.)

The zinc-finger region that binds to DNA includes the second β-strand, the N-terminal half of the α-helix, and the two-residue turn between the β-strand and the α-helix. The α-helix of each zinc finger fits into the major groove, with its N-terminus closest to the base pairs. Residues at positions –1, 2, 3, and 6 with respect to the helix start make sequence-specific contacts with bases in the major groove of the DNA molecule (FIGURE 18.19). A short linker region connecting adjacent zinc fingers, which has the consensus sequence TGEKP, goes from a disordered to an ordered state when the zinc finger region binds to DNA, suggesting that it also participates in binding. Further support for this participation is provided by the observation of

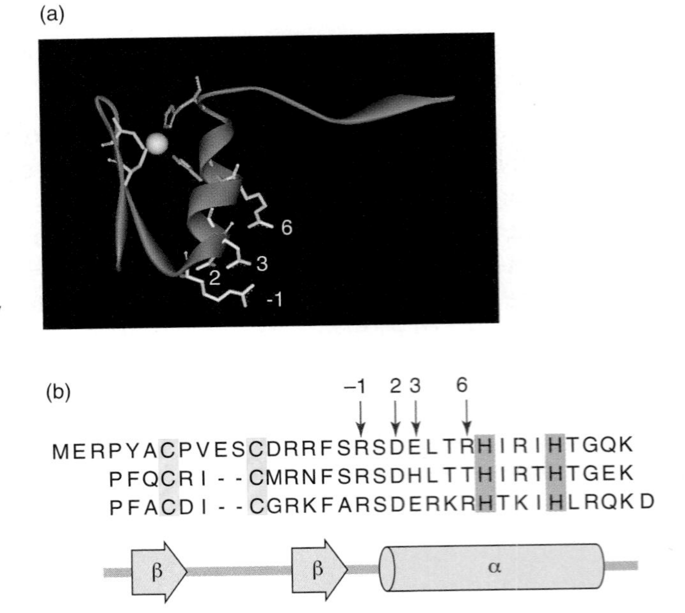

FIGURE 18.19 The Cys_2His_2 zinc finger motif and the Zif268 finger sequences. (a) The first zinc finger in Zif268 is shown in a ribbon diagram. The cysteine and histidine side chains that coordinate the zinc ion are colored yellow and pink, respectively. The zinc ion is shown as a light cyan sphere. Serine, the first residue in the recognition helix is assigned position 1. The numbers to the right of the structure identify key residues (shown in white) that contact bases in the major groove of the DNA. (b) The amino acid sequence alignment for the three Zif268 zinc fingers. Conserved cysteine and histidine ligands are in yellow and pink boxes, respectively. Secondary structure elements are shown below the sequences and residues −1, 2, 3, and 6 are also shown. (Part a structure from Protein Data Bank 1ZAA. N. P. Pavletich and C. O. Pabo, *Science* 252 [1991]: 809–817. Prepared by B. E. Tropp.)

altered binding by mutant proteins with amino acid substitutions in the TGEKP linker region.

Nuclear receptors have Cys_4 zinc finger motifs.

An entirely different kind of zinc finger motif is present in the **nuclear receptor superfamily**. Nuclear receptors allow cells in higher organisms to respond to a variety of external and internal chemical signals by increasing or decreasing the transcription of specific genes. Nuclear receptors were initially detected as nuclear proteins that bind steroid hormones, thyroid hormone, vitamin D, or retinoic acid. Investigators use this binding detection method to identify nuclear receptors as they are being purified from cellular extracts by protein fractionation techniques.

Purified nuclear receptor-ligand complexes bind to regulatory promoters and enhancers, indicating that they are probably transcription activator proteins. Some nuclear receptors such as the **thyroid receptor (TR)** appear to be located in the cell nucleus in both the free receptor form and in the receptor-ligand complex form. Although both forms of the TR bind to DNA, the free receptor inhibits gene expression, while receptor-ligand complex stimulates gene expression. Other nuclear receptors such as the **glucocorticoid receptor** (**GR**) bind in the free form to a cytoplasmic protein complex. Ligand binding induces the cytoplasmic protein complex to release the glucocorticoid receptor, which then moves to the cell nucleus and stimulates transcription of specific genes.

Once nuclear receptor genes were cloned and sequenced in the mid-1980s, it became apparent that they have a conserved 70-residue sequence that contains eight cysteine residues, suggesting the possibility of two zinc-binding sites that each contain four cysteine

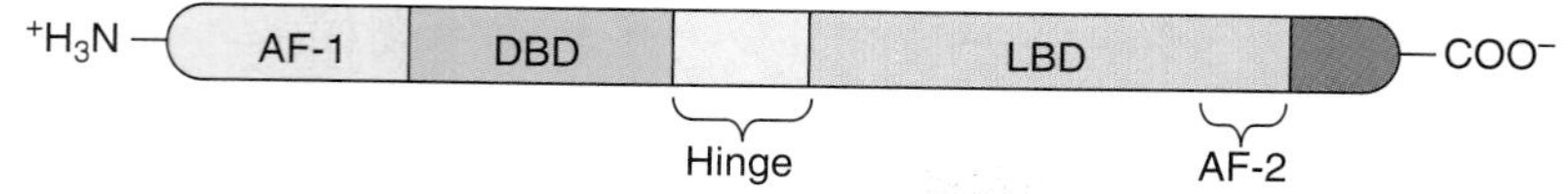

FIGURE 18.20 Schematic representation of a nuclear receptor. A typical nuclear receptor contains several functional domains. The variable amino-terminal region contains an activation function domain (AF-1). The conserved DNA-binding domain (DBD) recognizes specific DNA sequences. A variable hinge region connects the DNA-binding domain to the ligand-binding domain (LBD). A second activation function domain (AF-2) is located at the end of the ligand-binding domain.

residues. A search of the gene data bank for the conserved 70-amino acid sequence revealed that previously unidentified polypeptides have a similar sequence, signifying that they too might be nuclear receptors. These polypeptides were designated **orphan nuclear receptors** because the ligands that bind to them were unknown. Based on gene sequences from the human genome project, the best current estimate of the total number of human nuclear receptor genes is 49. However, many more nuclear receptors exist than this number suggests for two reasons. First, the active form for many nuclear receptors is a heterodimer, allowing for combinatorial variations. Second, alternate splicing of the primary transcript for a nuclear receptor often produces two or more different mRNA molecules (see Chapter 19) and, therefore, two or more nuclear receptor isoforms.

Nuclear receptors have a common structural design, which is shown in schematic form in FIGURE 18.20. Starting at the amino terminus and moving toward the carboxyl end, the major features are: (1) a poorly conserved activation function domain (AF-1); (2) a highly conserved DNA-binding domain; (3) a variable hinge; (4) a conserved ligand binding domain (the hormone binding site); and (5) a second activation domain (AF-2).

Investigators have determined the three-dimensional structure of DNA-binding domains for a few different nuclear receptors by using recombinant DNA technology to prepare peptide fragments containing the domains and NMR spectroscopy to study their structures. FIGURE 18.21 shows the sequence of a fragment containing the DNA binding domain of the glucocorticoid receptor and the DNA sequence to which the domain binds. Notice that the DNA-binding region has two Cys_4 zinc finger motifs.

The crystal structure of the glucocorticoid receptor DNA-binding domain shows the arrangement of the two zinc fingers, each consisting of an irregularly looped string of amino acids followed by an α-helix (FIGURE 18.22). The two α-helices cross one another near their midpoints so that the zinc fingers are intertwined. Hydrophobic interactions between conserved residues on the interacting helical faces stabilize the compact globular core. The intertwined structure clearly distinguishes the Cys_4 zinc finger from the classical zinc finger.

Steroid receptors such as the glucocorticoid receptor and the estrogen receptor (ER) bind to DNA as homodimers. The DNA-binding unit, therefore, consists of two DNA-binding domains, one from each

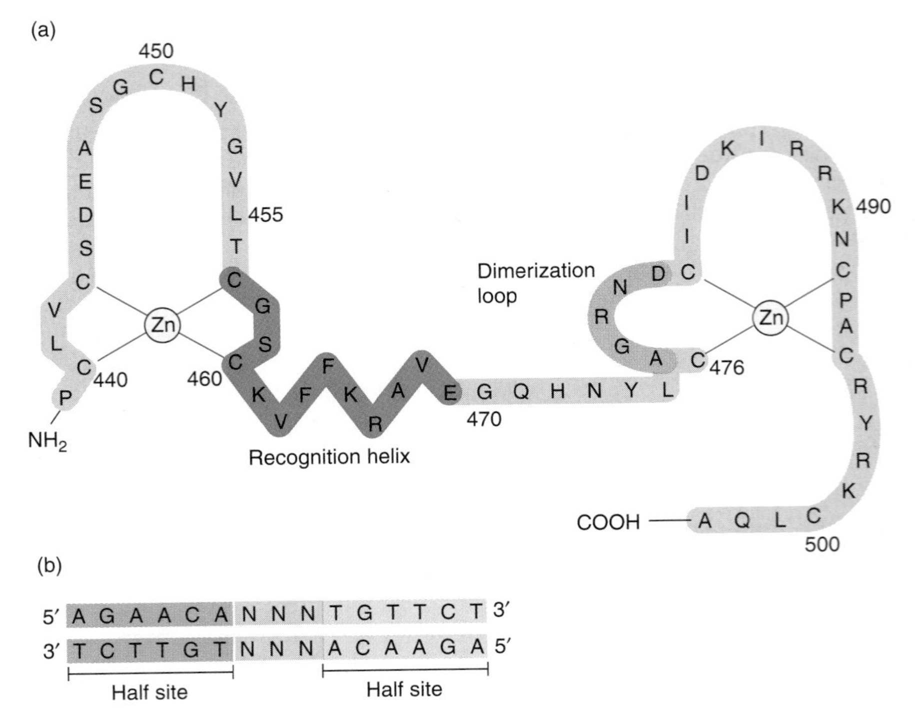

FIGURE 18.21 The glucocorticoid receptor. (a) The amino acid sequence of the zinc-containing DNA-binding domain of the glucocorticoid receptor. The two zinc atoms in this domain are shown in light blue. Each zinc atom binds to four cysteine residues. One of the zinc atoms stabilizes the recognition helix (red), which makes sequence-specific contact with the DNA. The other zinc atom stabilizes a loop involved in formation of the dimeric receptor molecule. (b) The glucocorticoid receptor binds to a DNA region known as the glucocorticoid response element, GRE. The GRE has two palindromic half sites (light and dark blue) that are separated by a three base pair spacer (NNN, where N is any nucleotide). (Part a modified from C. Branden and J. Tooze. *Introduction to Protein Structure, First edition*. Garland Science, 1999. Used with permission of John Tooze, The Rockefeller University. Part b adapted from C. I. Branden and J. Tooze. *Introduction to Protein Structure, Second edition*. Garland Science, 1999.)

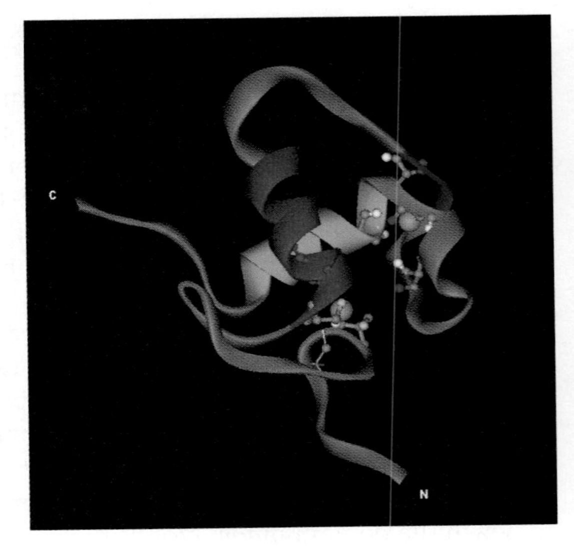

FIGURE 18.22 Crystal structure of the glucocorticoid receptor DNA-binding domain. The glucocorticoid receptor DNA-binding domain has two Cys_4 zinc fingers, each consisting of an irregularly looped string of amino acids followed by an α-helix. The two α-helices cross one another near their midpoints so that the zinc fingers are intertwined. Hydrophobic interactions between conserved residues on the interacting helical faces stabilize the compact globular core. The first α-helix (red) binds to bases in DNA and is therefore called the recognition helix. The second α-helix (green) functions as a support to hold the recognition helix in place. The loop colored purple contacts its counterpart on a second glucocorticoid receptor DNA-binding domain, stabilizing the homodimer structure. Zinc atoms are shown as magenta spheres and the cysteine residues as ball and stick structures with the sulfur atom in yellow. Amino- and carboxy-termini are indicated by the letters N and C, respectively. (Structure from Protein Data Bank 1R4R. B. F. Luisi, *Nature* 352 [1991]: 497–505. Prepared by B. E. Tropp.)

steroid receptor. FIGURE 18.23 shows the crystal structure for the glucocorticoid receptor DNA-binding domain interacting with DNA. The two DNA-binding domains are arranged head-to-head so that the overall structure is symmetric. A **dimerization loop** formed by the residues between the first two cysteine residues in the second zinc finger (Cys-476 and Cys-482) in one DNA-binding domain interacts with the dimerization loop in the other DNA-binding domain to stabilize the homodimer (Figure 18.23).

The two DNA-binding domains in the homodimer contact bases in the major groove on the same side of the DNA molecule. The

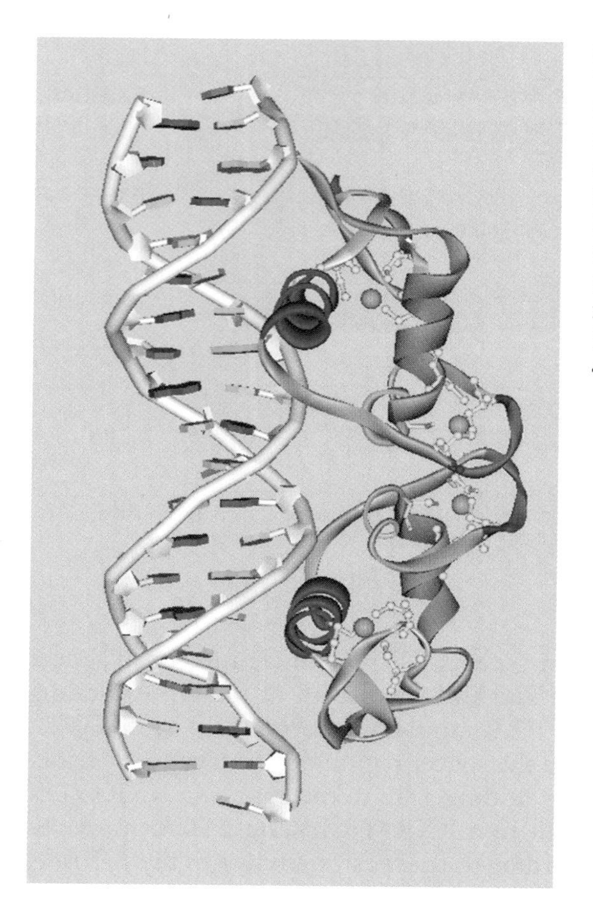

FIGURE 18.23 Glucocorticoid receptor DNA-binding domain bound to DNA. The glucocorticoid receptor binds to DNA as a homodimer that is stabilized by interactions between the dimerization loops (purple) in each monomer. Each subunit also contributes a DNA-binding domain that contains a pair of intertwined Cys_4 zinc fingers. The cysteine residues are shown as yellow ball and stick structures and the zinc atoms as magenta spheres. The two helices in a DNA-binding domain perform different structures. The α-helix (red) in the first zinc finger functions as a recognition helix to make specific contacts with bases in the major groove. The α-helix (green) in the second zinc finger functions as a support to hold the recognition helix in place. (Structure from Protein Data Bank 1R4R. B. F. Luisi, *Nature* 352 [1991]: 497–505. Prepared by B. E. Tropp.)

intertwined Cys_4 zinc fingers in each DNA-binding domain act as a unit but have different functions. The α-helix in the first zinc finger is a **recognition helix** that makes specific contacts with bases in the major groove; the α-helix in the second acts as a support to hold the recognition helix in place. The homodimer binds to a DNA sequence called the **hormone response element**, which is an imperfect palindrome consisting of inverted 6 bp repeats that are separated by a 3 bp spacer (FIGURE 18.24). This spacer must be 3 bp but the nucleotides within it do not appear to matter. Each DNA binding domain interacts with one 6 bp repeat or half-site. Consensus sequences for glucocorticoid and estrogen receptor half-sites are 5′-AGAACA-3′ and 5′-AGGTCA-3′, respectively.

5′ A G A A C A N N N T G T T C T 3′
3′ T C T T G T N N N A C A A G A 5′
Glucocorticoid response element

5′ A G G T C A N N N T G A C C T 3′
3′ T C C A G T N N N A C T G G A 5′
Estrogen response element

FIGURE 18.24 Glucocorticoid and estrogen response elements. Both the glucocorticoid and estrogen response elements are palindromic repeats, consisting of two 6-bp half sites separated by three nucleotides, N (where N can be any nucleotide).

Active forms of many other nuclear receptors are heterodimers. One important family consists of receptors that form heterodimers with the retinoid X receptor (RXR), which is the receptor for 9-*cis*-retinoic acid. Dimerization partners include the thyroid hormone receptor (TR), the vitamin D receptor (VDR), the retinoic acid receptor (RAR), and the peroxisome proliferator-activated receptor (PPAR). The DNA-binding domain of RXR only interacts with its heterodimeric partner in the presence of the hormone response element. When hormone response elements are present, DNA binding domains produce the same dimerization and selectivity patterns as full-length receptors. Heterodimer formation is induced by specific hormone response

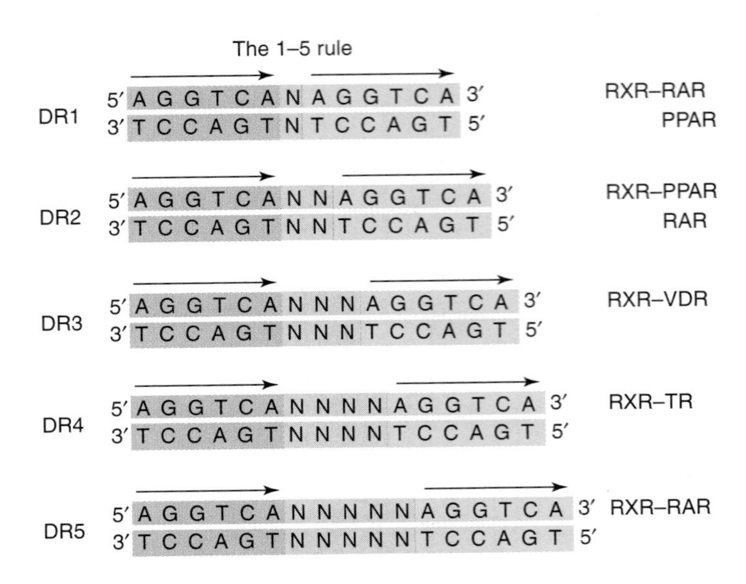

FIGURE 18.25 **The 1–5 rule of DNA direct repeat binding by RXR and its nuclear receptor partners.** The base pair size of the spacing between the AGGTCA sequences can vary from one to five. (Adapted from F. Rastinejad, *Curr. Opin. Struct. Biol.* 11 [2001]: 33–38.)

elements that contain direct repeats (DRs) with characteristic spacer regions between half-sites. The half-site for all of the DNA-binding domains that belong to the RXR family of heterodimers is ATTTCA. The number of base pairs in the spacer region determines the specificity of binding. For instance, adding 1 bp to the spacer in RXR•VDR response element converts it to a RXR•TR response element. Rules that govern specificity according to spacer size, known as the *1–5 rule*, are illustrated in FIGURE 18.25.

The number of combinatorial possibilities is increased because the heterodimer can switch polarities. For instance, RXR•RAR has one polarity when it binds to a response element with a 1 bp spacer and the opposite polarity when it binds to a response element with a 5 bp spacer. Hormone response changes as a result of the polarity switch. Protein-protein interactions between RXR and each of its partners help to ensure correct spacing recognition. Further stabilization is achieved through sequence-specific contacts with bases in the major groove at each of the half-sites. The RXR DNA-binding domain's ability to accommodate so many different dimerization partners is probably due to the fact that it has many combinatorial binding sites on its surface. FIGURE 18.26 shows the structures for RXR•TR and RXR•RAR DNA-binding domain heterodimers bound to DNA.

Ligand-binding domain structure provides considerable information about nuclear receptor function.

Structural studies of **ligand-binding domains** (LBDs) from nuclear receptors also have provided considerable information about the way that nuclear receptors work. As the name suggests, the LBD contains the hormone-binding site. Crystal structures have been solved for LBDs from many nuclear receptors. The different LBDs have very similar structures. Structures also have been solved for LBDs with a bound ligand or antagonist. Structural comparisons reveal that the LBD undergoes characteristic conformational changes when it binds

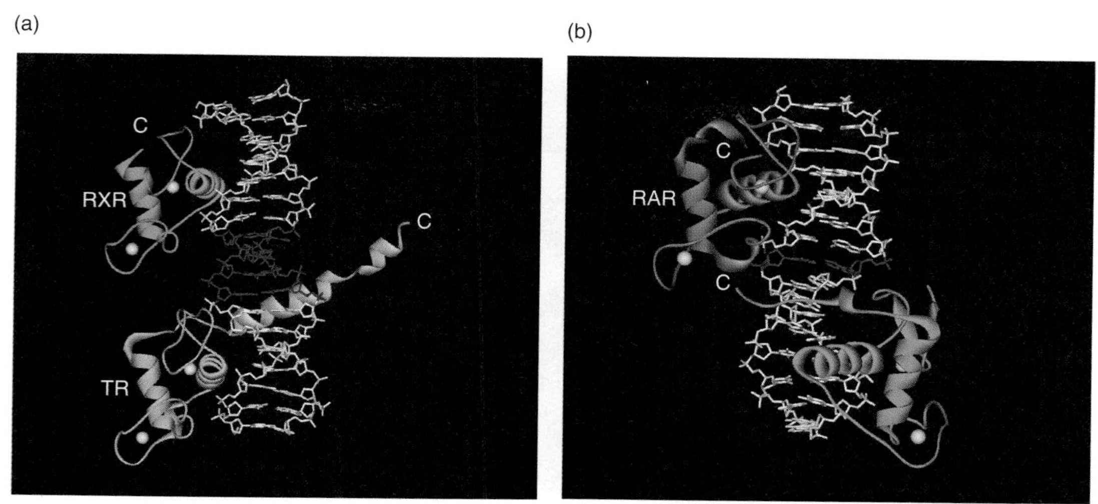

FIGURE 18.26 **Structures of DNA-binding complexes involving RXR and their dimer interfaces.** DNA is shown as a stick figure. Nucleotides belonging to the spacing element between direct repeats (DR) are shown in red. All other nucleotides are shown in white. Zinc ions are shown as light blue spheres. RXR (retinoid X receptor) is shown in green, TR (thyroid hormone receptor) in yellow, and RAR (retinoic acid receptor) in blue. In each case, the protein to protein contacts are formed directly over the minor groove of the spacing, with several protein-DNA phosphate contacts stabilizing the assembly. (a) The RXR•TR DNA-binding domain heterodimeric complex with four nucleotide spacing and (b) the RXR•RAR DNA-binding heterodimeric complex with one nucleotide spacing. (Part a structure from Protein Data Bank 2NLL. F. Rastinejad, et al., *Nature* 375 [1995]: 203–211. Prepared by B. E. Tropp. Part b structure from Protein Data Bank 1DSZ. F. Rastinejad, et al., *EMBO J.* 19 [2000]: 1045–1054. Prepared by B. E. Tropp.)

a ligand or an antagonist. FIGURE 18.27 is a schematic drawing that shows the major structural features for a ligand-binding domain by itself, with a bound ligand, and with a bound antagonist. The ligand-binding domain contains 12 conserved α-helical regions numbered H1 to H12. In the absence of ligand, the helices are arranged in three layers to form an "anti-parallel α-helical sandwich" (Figure 18.27a). Helices H1 to H3 form one side of the ligand-binding domain; helices H4, H5, H8, and H9 form the middle layer; and helices H6, H7, and H10 form the other side. Helix H11 is almost perpendicular to helix H10. Helix H12 extends away from the ligand-binding domain and contains the AF-2 region. The lower part of the ligand-binding domain, which varies from one kind of ligand-binding domain to another, forms the ligand-binding pocket. The ligand induces major conformational changes when it binds to the ligand-binding domain (Figure 18.27b). Most notable among these changes are that helices H10 and H11 line up to form a continuous helix and helix H12 swings underneath helix H4, enclosing the ligand in the **ligand-binding pocket** (**LBP**). Antagonists also induce major conformational changes when they bind to LBD (Figure 18.27c). This conformation, however, is different from that assumed when the normal ligand binds. The reason for this difference appears to be that antagonist molecules have bulky-side chains that prevent helix H12 from swinging underneath helix H4, causing AF-2 to be misaligned.

Hormone antagonists are used as drugs to treat a variety of serious health problems. For instance, two drugs that are used to counteract the effects of estrogen, **tamoxifen**, and **raloxifene** (FIGURE 18.28) have important clinical applications. Tamoxifen is used to treat hormone-dependent forms of breast cancer after surgery. The rationale for this

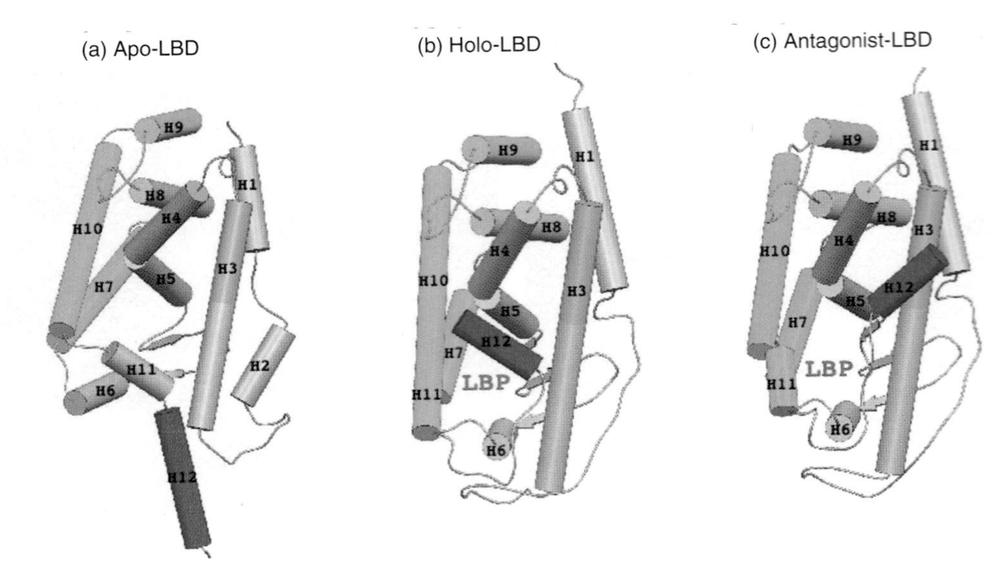

FIGURE 18.27 Schematic drawing of nuclear receptor ligand binding domains (LBDs) in three different conformations. (a) The retinoid X receptor (RXR) ligand binding domain in the absence of a ligand (apo-LBD). (b) The retinoic acid receptor (RAR) ligand binding domain with a bound retinoic acid (holo-LBD). The ligand induces major conformational changes when it binds to the ligand binding domain. Most notably, helices H10 and H11 line up to form a continuous helix and helix H12 swings underneath helix H4, enclosing the ligand in the ligand binding pocket (LBP). (c) The retinoic acid receptor protein ligand binding domain with a bound antagonist. Antagonists also induce major conformational changes when they bind to LBDs. α-helices (H1–H12) are shown as rods and β-strands as broad arrows. The various regions of the ligand binding domain are colored according to function. The dimerization surface is green. The co-activator and co-repressor binding site is shown in orange and the activation helix H12, which includes the residues of the core activation function (AF-2) activation domain (AD) is colored red. The other structural elements are shown in mauve. (Reproduced from *Trends Pharmacol. Sci.*, vol. 21, W. Bourguet, P. Germain, and H. Gronemeyer, Nuclear receptor ligand-binding . . ., pp. 381–388, copyright 2000, with permission from Elsevier [http://www.sciencedirect.com/science/journal/01656147]. Photo courtesy of Hinrich Gronemeyer, Institut de Génétique et de Biologie Moléculaire et Cellulaire.)

use is that estrogen stimulates the proliferation of normal and neoplastic breast cells. Tamoxifen binds to the estrogen receptor's LBD and blocks estrogen's ability to stimulate cell proliferation. Tamoxifen is sometimes also given to women at high risk for breast cancer to decrease the risk. The decision to use tamoxifen must be considered carefully, however, because tamoxifen may cause serious side effects. Although tamoxifen behaves as an antagonist that blocks estrogen's effects on breast cells, it behaves like estrogen in other tissues such as the uterus. Hence, tamoxifen may increase the risk of uterine cancer by stimulating uterine cell proliferation. Tamoxifen is a "**selective estrogen-receptor modulator**" (**SERM**) because it affects some parts of the body in the same way as estrogen but acts differently from estrogen in other parts of the body.

Raloxifene, another SERM, sometimes is given to postmenopausal women because it appears to act like estrogen in bone, helping to maintain bone strength and density. It also appears to mimic estrogen's ability to lower low density lipoprotein (LDL) cholesterol levels and thereby lower the risk for arteriosclerosis. However, raloxifene appears to increase the risk of blood clots. Clearly, a physician must consider many factors when prescribing a selective estrogen-receptor modulator.

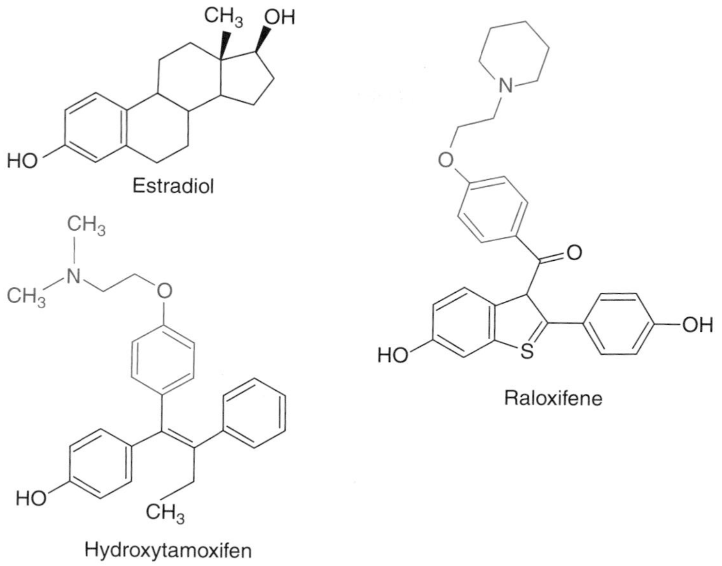

Estradiol

Hydroxytamoxifen

Raloxifene

FIGURE 18.28 The structure of estradiol, a normal ligand for estrogen receptors, and hydroxytamoxifen and raloxifene, two selective estrogen-receptor modulators. "Antagonistic" substitutions are represented in red. (Adapted from W. Bourguet, P. Germain, and H. Gronemeyer, *Trends Pharmacol. Sci.* 21 [2000]: 381–388.)

Gal4, a yeast transcription activator protein belonging to the Cys_6 zinc cluster family, regulates the transcription of genes involved in galactose metabolism.

An important family of transcription activator proteins in yeast and other fungi uses still another arrangement of zinc ions and cysteine residues to stabilize the DNA-binding domain. This arrangement is called the **Zn_2C_6 binuclear cluster** because two zinc ions are coordinated to six cysteine residues. The best-studied member of the Zn_2C_6 binuclear cluster family is **Gal4**, a transcription activator protein that regulates genes coding for galactose metabolism in *Saccharomyces cerevisiae.*

The active form of Gal4, which consists of a pair of identical 881-residue polypeptide subunits, binds to *Gal4* sites on DNA. Gal4's DNA-binding domain is at its amino terminus, extending from residues 1 to 64. A peptide fragment bearing this amino terminal sequence is a monomer in solution but forms a homodimer upon binding to DNA. The crystal structure for the complex formed when the amino terminal peptide fragment binds to DNA reveals three structural modules (FIGURE 18.29). Residues 1 to 40 form the Zn_2C_6 binuclear cluster, residues 51 to 64 form a weak dimerization region, and residues 41 to 50 form a linker that connects the C_6 zinc cluster to the weak dimerization region. The amphipathic α-helix formed by residues 51 to 64 of one subunit interacts with the corresponding residues in the other subunit to form a coiled coil that helps to stabilize the dimer (see the next section on leucine zippers for a more detailed description of coiled coils).

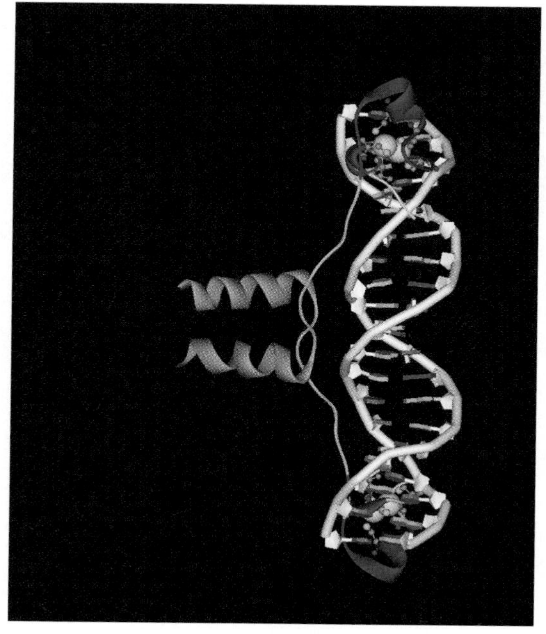

FIGURE 18.29 The Gal4-DNA complex. Residues 1–40 fold to form the Zn_2C_6 binuclear cluster (red) that interacts with DNA. Residues 51-64 form a weak dimerization region (blue). The binuclear cluster and weak dimerization region are connected by a linker (yellow) consisting of residues 41–50. Zinc atoms are shown in yellow. (Structure from Protein Data Bank 1D66. R. Marmorstein, et al., *Nature* 356 [1992]: 408–414. Prepared by B. E. Tropp.)

The Gal4 Zn_2C_6 binuclear cluster is shown in schematic form in FIGURE 18.30. Two of the cysteines bind to both zinc ions, forming a pair of bridges between the zinc ions (Figure 18.30a). Folding within the Zn_2C_6 binuclear cluster creates two short α helices (Figure 18.30b). The C-terminus of the first helix (shown in red in Figure 18.30b) makes specific contacts with bases in the major groove of the *Gal4* site, which is a 17-bp semi-palindrome with highly conserved CCG triplets at either end, reading in a 5′ to 3′ direction from the center toward the end. Each Zn_2C_6 binuclear cluster in the homodimer makes specific contacts with one of the CCG triplets at the end of the *Gal4* site. The linker and dimerization region also contribute to binding by contacting the phosphate backbone in the 11-bp spacer region. Although the Gal4 DNA-binding domain tolerates base pair substitutions in the 11-bp spacer, it has very little tolerance for deletions or insertions in the spacer.

Other transcription activator proteins belonging to the Zn_2C_6 binuclear cluster family also bind to semi-palindromes with CCG triplets at either end but exhibit varying specificities for the spacer lengths between the triplets. For instance, the yeast transcription activator protein Ppr1 (pyrimidine pathway regulator 1), which regulates genes involved in pyrimidine nucleotide metabolism, binds to a 12-bp semi-palindrome with a CCG triplet at either end. However, binding will only take place if the spacer is 6-bp long.

Richard J. Reece and Mark Ptashne took advantage of the differences in binding specificity between Gal4 and Ppr1 to demonstrate that the linker region (and not the Zn_2C_6 binuclear cluster) determines the spacer length that will be recognized. They did so by studying the specificity of chimeric DNA-binding domains, which they constructed

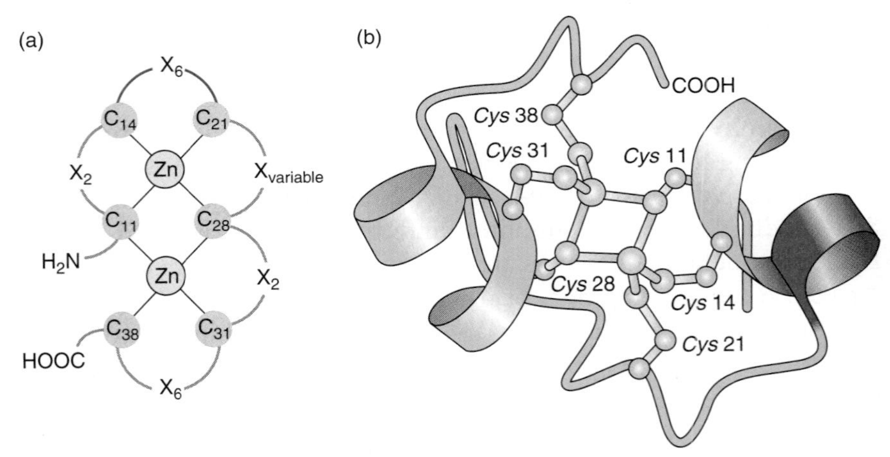

FIGURE 18.30 DNA-binding zinc cluster region of the Gal4 subunit. (a) Schematic diagram showing that the zinc cluster contains two zinc atoms each bound to four cysteine residues, two of which bridge the zinc atoms. The diagram also shows the number of amino acids in the loop regions between the cysteine ligands. (b) Three-dimensional representation of the Zn_2C_6 binuclear cluster. Folding with the Zn_2C_6 binuclear cluster creates two short α-helices. The region shown in red is involved in the sequence-specific DNA interactions. The zinc cluster stabilizes the structure to give the proper fold for DNA binding. (Modified from C. Branden and J. Tooze. *Introduction to Protein Structure, First edition.* Garland Science, 1999. Used with permission of John Tooze, The Rockefeller University.)

by using recombinant DNA techniques (FIGURE 18.31). They observed that a chimeric peptide that contains the Ppr1 zinc cluster, the Gal4 linker, and the Gal4 dimerization domain, binds to the Gal4 DNA site, whereas a chimeric peptide that contains the Ppr1 zinc cluster, the Ppr1 linker, and the Gal4 dimerization domain, binds to the Ppr1 DNA site. Thus, the highly conserved Zn_2C_6 binuclear cluster recognizes the CCG triplet at either end of the semi-palindrome and the linker region selects among different semi-palindrome sites based on their spacer size.

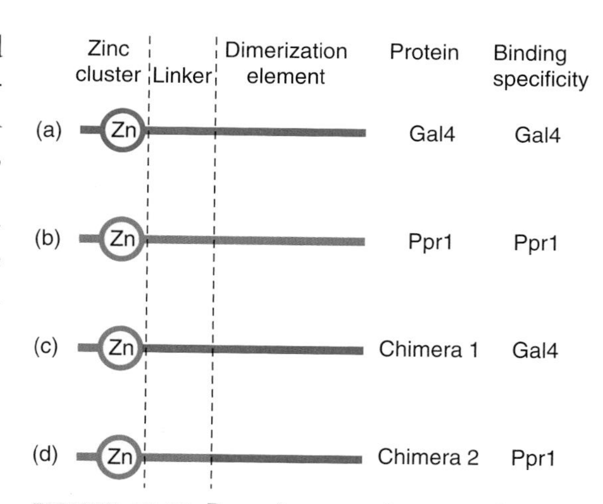

FIGURE 18.31 **Domain swapping experiment with Gal4 and the related transcription activator protein Ppr1.** (a) The Gal4 transcription activator protein binds to the Gal DNA site. (b) The Ppr1 transcription activator protein binds to the Ppr1 DNA site. (c) Chimera 1, which contains the Ppr1 zinc cluster, the Gal4 linker, and the Gal4 dimerization domain, binds to the Gal4 DNA site. (d) Chimera 2, which contains the Ppr1 zinc cluster, the Ppr1 linker, and the Gal4 dimerization domain, binds to the Ppr1 DNA site. Thus, the linker region determines binding specificity. (Modified from C. Branden and J. Tooze. *Introduction to Protein Structure, First edition*. Garland Science, 1999. Used with permission of John Tooze, The Rockefeller University.)

18.5 Loop-Sheet-Helix DNA-Binding Domain

p53 has a loop-sheet-helix DNA-binding domain.

Zinc also helps to stabilize transcription activator proteins with DNA binding domains with **loop-sheet-helix** structures. The best studied protein of this type, **p53**, is activated in response to DNA damage and other cellular stresses. Activation often involves some form of covalent modification such as phosphorylation or acetylation. Once modified, p53 binds to a decameric DNA element and activates or represses various genes involved in DNA repair, cell cycle arrest, apoptosis (programmed cell death), and senescence. p53 has three major domains (FIGURE 18.32). The N-terminal domain includes an activation region and a proline-rich region, which probably mediates protein-protein interactions. The DNA-binding domain (or core domain) is located between the N- and C-terminal domains. The C-terminal domain contains a nuclear localization sequence (NLS), which is required for p53 to gain entry into the cell nucleus and a tetramerization domain, which helps intact p53 subunits to assemble into a dimer of dimers. The extreme C-terminal region is intrinsically disordered but may become ordered upon interacting with other proteins.

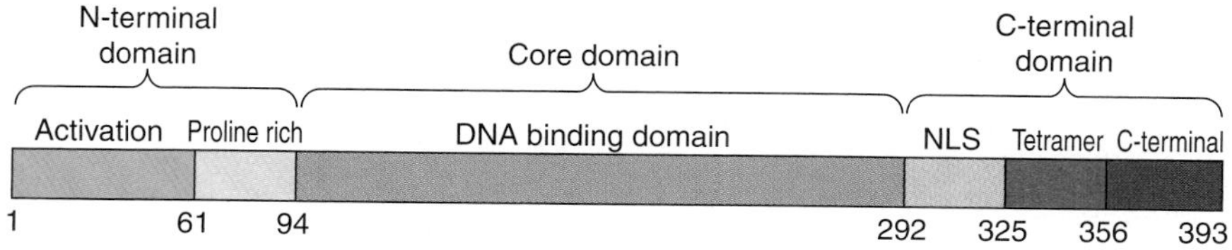

FIGURE 18.32 **The domain structure of p53.** The N-terminal domain contains an activation region and a proline-rich region, which probably mediates protein-protein interactions. The core domain folds into a sheet-loop-helix structure that binds DNA. The C-terminal domain contains a nuclear localization sequence (NLS), which is required for p53 to gain entry into the cell nucleus and a tetramerization domain (Tetramer), which helps p53 to assemble into a dimer of dimers. The extreme C-terminal region is intrinsically disordered but may become ordered upon interacting with other proteins. (Adapted from A. C. Joerger and A. R. Fersht, *Ann. Rev. Biochem.* 77 [2008]: 557–582.)

The DNA binding site consists of two decameric motifs (or half-sites) that are separated by 0 to 13 bp. The consensus sequence for the decameric motifs is RRRCWWGYYY (where R = A or G, W = A or T, and Y = C or T). Zippora Shakked and coworkers determined crystal structures for p53 core domains bound to a pair of decameric motifs. The DNA-binding domain folds into a β sandwich-like structure with a DNA-binding surface consisting of a loop-sheet-helix motif. Loops 2 and 3 are stabilized by a zinc ion that interacts with a histidine and three cysteines (**FIGURE 18.33a**). Four DNA-binding domains self-assemble into a tetramer on two DNA half sites (**FIGURE 18.33b**). The gene that codes for the p53 protein is altered in about 50% to 55% of human cancers. Approximately 80% to 90% of the missense mutations present in these human tumors are located in the DNA-binding domain. Not surprisingly, many laboratories are actively investigating p53 and trying to find ways to correct problems that arise when the protein is altered.

(a)

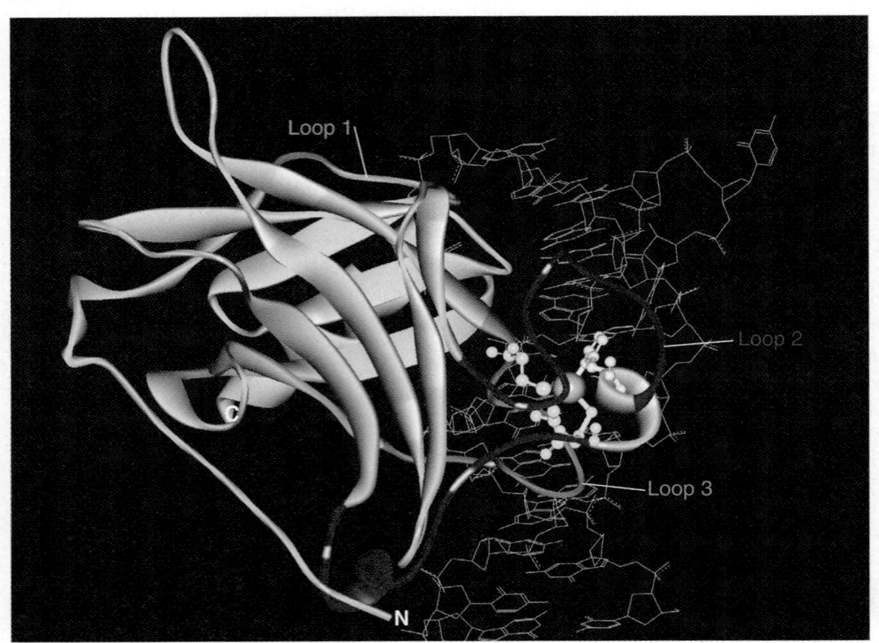

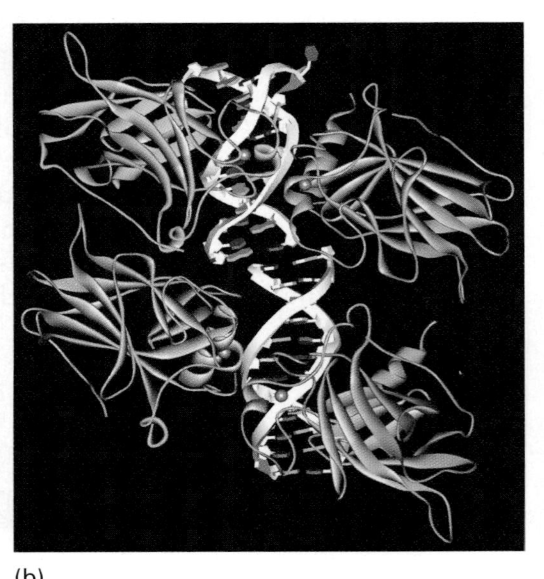

(a) (b)

FIGURE 18.33 Crystal structure of the p53 core domain bound to DNA. (a) Crystal structure of a single p53 DNA-binding domain bound to DNA. The domain folds into a β sandwich-like structure with a DNA binding surface consisting of a loop-sheet-helix motif. Two of the loops are stabilized by a zinc ion. The zinc ion (blue sphere) interacts with a histidine (ball and stick structure in light green) and three cysteines (ball and stick structures in yellow). DNA with the sequence cGGGCAT-GCCCg is shown as a line structure in light blue. The first c and last g are in lower case to indicate that they are not part of the decameric half site sequence motif. Amino- and carboxy-termini are indicated by the letters N and C, respectively. (b) Crystal structure of the tetrameric p53 core structure bound to two DNA half sites. The four DNA-binding domains are shown as ribbon structures. The DNA-binding domain shown in cyan is oriented as in (a). DNA has the same sequence as in (a) but is shown as a tube and ring structure. (Structures from Protein Data Bank 2AC0. M. Kitayner, et al., *Mol. Cell* 22 [2006]: 741–753. Prepared by B. E. Tropp.)

18.6 DNA-Binding Domains with Basic Region Leucine Zippers

Basic region leucine zipper (bzip) transcription activator proteins bind to DNA as dimers that are held together through coiled coil interactions.

Each of the transcription activator proteins described so far has a DNA-binding domain that can best be described as a globular structure. We now examine a family of transcription activator proteins called **basic region leucine zipper** (**bzip**) **proteins** that have fibrous DNA-binding domains. Members of this family, which bind to DNA as homo- or heterodimers, were initially identified because their polypeptide chains have α-helical segments that are about 30 residues long with a leucine in every seventh position. Steven L. McKnight and coworkers recognized that the repeating leucine pattern could explain how the polypeptide pair forms dimers. Because a typical α-helix has 3.6 residues per turn (see Chapter 2), the presence of a leucine at every seventh position means that the leucine side chains are on one face of the helix. McKnight and coworkers therefore proposed that two helical polypeptide chains line up side by side, allowing the leucine side chains to interdigitate like the teeth on a zipper (FIGURE 18.34).

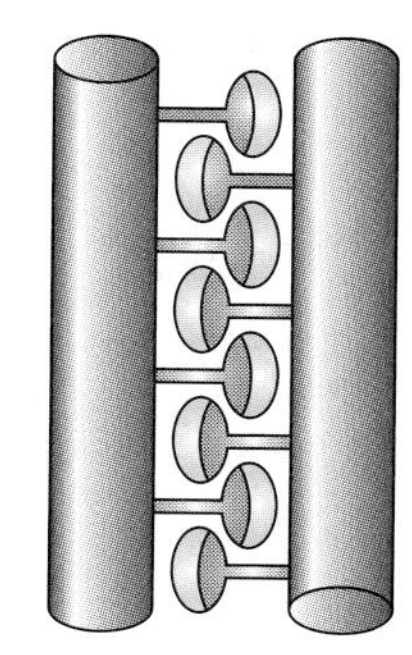

FIGURE 18.34 **Original leucine zipper model.** Schematic diagram showing the hypothetical interdigitation of leucine side chains between two α-helices. The two parallel tubes rep resent the approximate dimensions of the α carbon backbone of idealized α-helices. Interdigitating protrusions symbolize leucine side chains. The spheres located at the tip of each residue represents the two methyl groups attached to the γ carbon atom of the leucine side chain. (Modified from W. H. Landschulz, P. F. Johnson, and S. L. McKnight, *Science* 240 [1988]: 1759–1764. Reprinted with permission from AAAS and Steven McKnight, UT Southwestern Medical Center.)

The **leucine zipper model** was tested by synthesizing a peptide fragment corresponding to the leucine repeat region in a yeast transcription activator protein that coordinately increases the transcription of several genes involved in amino acid biosynthesis. Because the regulatory mechanism is known as general control of nitrogen metabolism, the transcription activator protein was named **Gcn4**. Although the synthetic Gcn4 fragments paired in a parallel orientation to form dimers in solution, the crystalline dimer structure differed from the predicted zipper model. Instead of lining up next to each other, the two polypeptides coiled about each other with a slight left-handed supertwist, to produce a **coiled coil** in which the smoothly bent α-helices made tight contacts over the length of the dimer (FIGURE 18.35). The observed helical repeat in the coiled coil is 3.5 residues per turn (0.1 residues per turn less than that in a free α-helix), so that leucine side chains extend out from the same face of the helix after every two turns. Thus, the leucine side chains do *not* interdigitate like the teeth of a zipper, but instead make side-to-side contacts in every other turn. Nevertheless, the term **leucine zipper** is still used for the dimerization region even though it does not precisely describe the true structure.

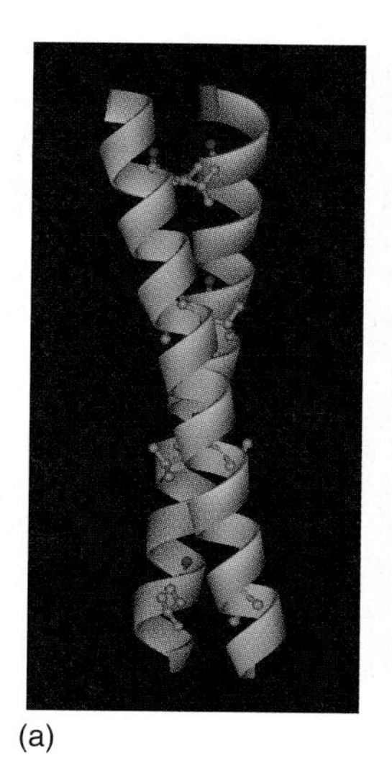

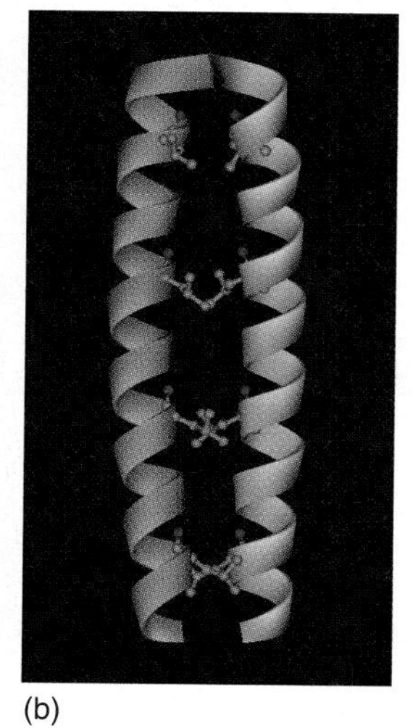

(a) (b)

FIGURE 18.35 **Crystal structure of Gcn4 leucine zipper, a two-stranded parallel coiled coil.** Backbones of the two peptide chains that form the Gcn leucine zipper are shown as red and blue helices. The side chains of the leucine residues are shown as ball and stick structures. The perspective in (a) was chosen to stress the coiled coil, whereas that shown in (b) was chosen to stress the interactions among the leucine side chains. (Structures from Protein Data Bank 2ZTA. E. K. O'Shea, et al., *Science* 254 [1991]: 539–544. Prepared by B. E. Tropp.)

We can gain some additional insight into how side chains in a coiled coil structure interact by looking down the helix axis of each polypeptide. FIGURE 18.36 shows how this approach applies to the 30 amino acid residues in the Gcn4 leucine zipper. We begin by drawing two **helical wheels**, one for each peptide in the dimer. Each helical wheel contains seven positions, a–g (or a′–g′), corresponding to the seven amino acids required to complete two helical turns. The first seven residues (MKQLEDK) in the 30-residue Gcn4 leucine zipper are placed in order at positions a–g (or a′–g′). Then the next seven residues

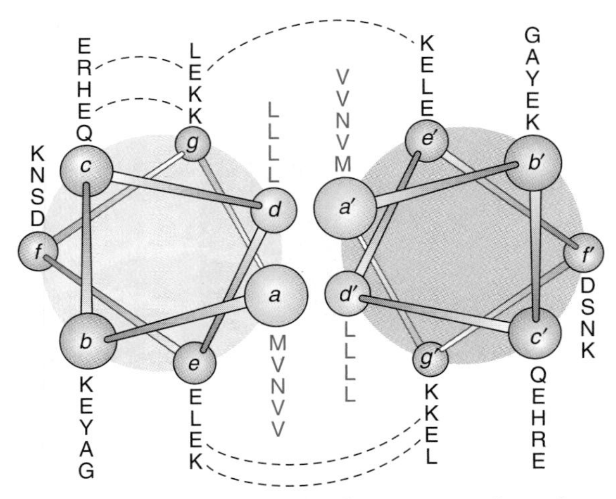

FIGURE 18.36 Helical wheel representation of Gcn4 domain that forms the coiled coil. Each strand of a coiled-coil protein may be viewed as repeated seven residue sequences of the form (a-b-c-d-e-f-g)$_n$, where a, b, c, d, e, f, and g represent consecutive residues on one strand of the coiled coil. Residues on the other strand are indicated by placing a prime (′) after the letter. The first and fourth positions (a and d) are usually hydrophobic amino acids. Residues that form ion pairs are connected with dashed lines. (Adapted from E. K. O'Shea, et al., *Science* 254 [1991]: 539–544.)

are placed in order above the first seven, and so forth. Hydrophobic residues tend to be located on the same helical face at positions a and d (or a′ and d′). Positive and negative residues tend to alternate along the helix in positions e and g (or e′ and g′) so that attractive intra- and interstrand ion pairs can stabilize the dimer.

Although the leucine zipper structure shows how Gcn4 dimerizes, it does not show how the dimer interacts with DNA. A short sequence of basic amino acids that is located just N-terminal to the leucine zipper motif is required for DNA recognition. This basic amino acid sequence together with the leucine zipper forms the **bzip DNA binding domain**. As shown by the schematic diagram in FIGURE 18.37, Gcn4 also has an activation domain toward the center of the polypeptide chain.

Gcn4 appears to resemble a pair of forceps in the way that it binds to DNA as is evident from the crystal structure of Gcn4 bound to DNA (FIGURE 18.38a). The basic DNA binding region in each polypeptide chain folds into a helical structure as it fits into the major groove of half-sites on either side of the semi-palindromic DNA (FIGURE 18.38b). Residues within the DNA binding region act as hydrogen bond donors to specific bases as well as to unesterified oxygen atoms in the phosphodiester backbone.

Many other transcription activator proteins are bzip proteins, including the C/EBP (CCAAT box and enhancer binding protein) and AP-1 (activating protein 1) families. Six members of the C/EBP family are known. Although C/EBP isoforms differ in their tissue specificity and activation abilities, all have the potential to bind to the same DNA recognition element (CAATC) and therefore compete with one another. In addition, their conserved leucine zipper allows them to form heterodimer transcription activator proteins, providing the potential to increase the total number of transcription activator proteins that can be present in a given cell. The importance of the C/EBP family is evident from studies involving its founding member, C/EBPα, a homodimer that appears to be required for growth arrest and terminal cell differentiation. Genetically engineered mice lacking C/EBPα have major liver abnormalities causing most to die soon after birth because of low blood glucose.

AP-1 proteins are a family of transcription activator proteins that regulate many different genes, including some that control cellular proliferation. The founding member of the AP-1 family was identified as a transcription activator protein that binds to enhancer regions of simian virus 40 (SV40) DNA and the human metallothionein IIA

FIGURE 18.37 Gcn4 transcription activator protein.

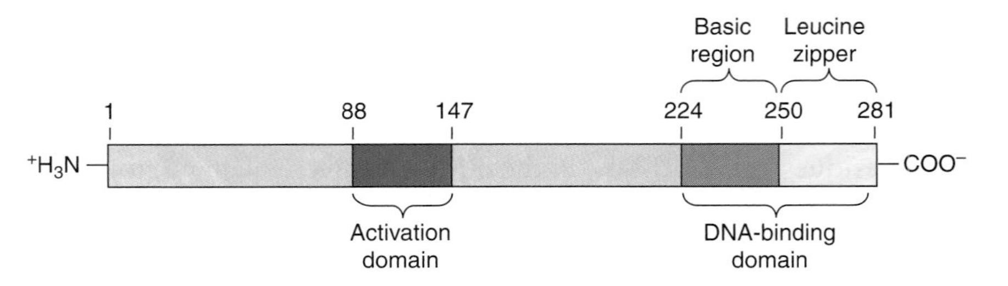

gene. The search for the polypeptides that form AP-1 was facilitated by information obtained from earlier studies of retroviruses. In particular, FBJ murine osteosarcoma virus was known to have the ***v-fos*** gene, which induces bone cancer, and avian sarcoma virus-17 was known to have the ***v-jun*** (*ju-nana* is the Japanese word for 17) gene, which induces fibrosarcoma in chickens. Genes such as *v-fos* and *v-jun* that transform normal cells into cancer cells are **oncogenes** and their protein products are **oncoproteins.** Subsequent studies showed that these viral genes have cellular counterparts called *c-fos* and *c-jun*. In fact, the viral genes originated in the cell. Sequence studies revealed that each of these polypeptides has a bzip region that is quite similar to that in Gcn4 (FIGURE 18.39), suggesting that they too might be bzip transcription activator proteins. Further study revealed that c-Fos and c-Jun combine to form the founding member of the AP-1 family of transcription activator proteins; that is, AP-1 is a c-Fos•c-Jun heterodimer. Additional cellular homologs of each kind of polypeptide were discovered later.

c-Jun can bind to DNA as a homodimer or as part of a c-Jun•c-Fos heterodimer. c-Fos does not bind to DNA as a homodimer. The reason that some polypeptide combinations lead to dimer formation while others do not is apparent from their helical wheel structures (FIGURE 18.40). The side chains that are immediately outside the hydrophobic core (positions e and g) promote dimer formation in Jun•Fos because of attractive charge interactions. Jun•Jun dimer formation is not as favorable as Jun•Fos formation because fewer attractive interactions are possible among the side chains. Fos•Fos dimer does not form because the large number of negatively charged glutamate (E) side chains in positions e and g are mutually repulsive.

The Fos•Jun heterodimer can bind to a DNA recognition element with a TGA(C/G)T(C/A)A sequence motif in two orientations (FIGURE 18.41). In contrast with the C/EBP family, different AP-1 homo- and heterodimers bind to different DNA recognition elements, allowing for greater diversity of promoter and enhancer specificity. When a *c-fos* or *c-jun* gene is mutated or inappropriately expressed, its product can transform a normal cell into a cancer cell.

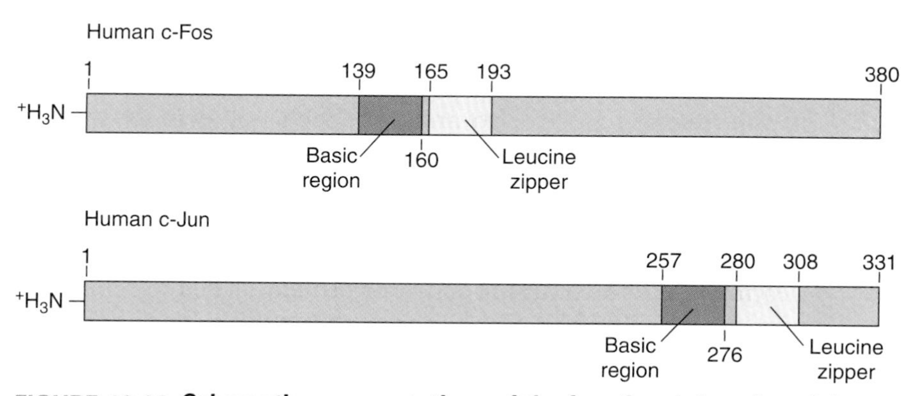

FIGURE 18.39 **Schematic representations of the functional domains of the c-Jun and c-Fos proteins.** The basic region is shown in green and the leucine zipper in yellow.

(a)

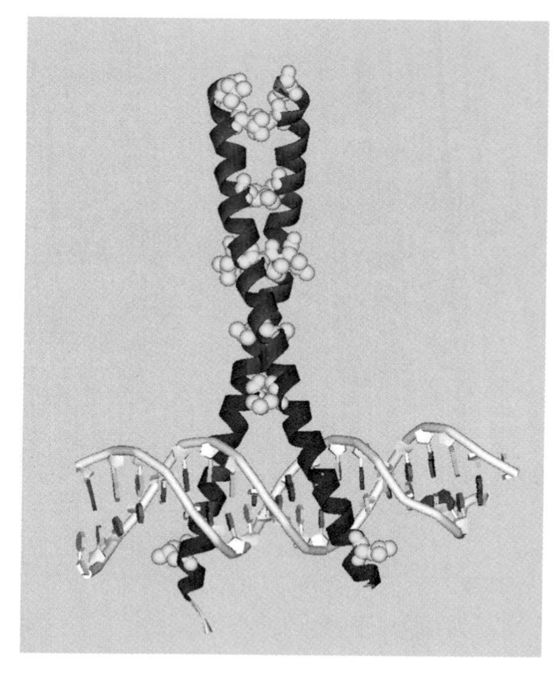

(b)

5′ T T C C T A T G A C T C A T C C A G T T 3′
3′ A A G G A T A C T G A G T A G G T C A A 5′

FIGURE 18.38 **The structure of a complex between the DNA-binding domain of Gcn4 and a fragment of DNA.** (a) The DNA-binding region, which is basic (residues 226–249), is shown in purple. The leucine zipper region (residues 250–281) is shown in blue. Leucine side chains are shown as yellow space filling structures. (b) The DNA used to obtain the crystal structure in (a). Each polypeptide chain forms a helical structure that fits into the major groove of a half-site at either end of the quasi-palindrome that is shown in red. (Part a structure from Protein Data Bank 1YSA. T. E. Ellenberger, et al., *Cell* 71 [1992]: 1223–1237. Prepared by B. E. Tropp.)

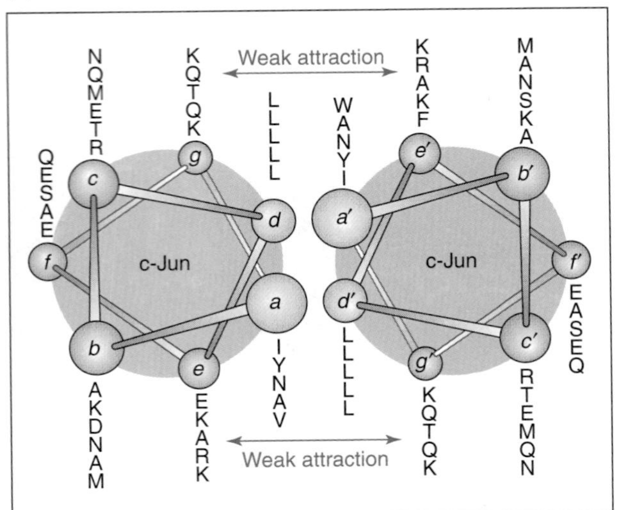

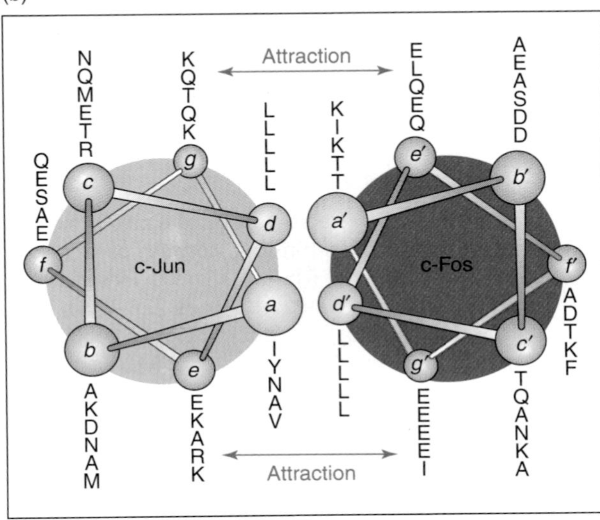

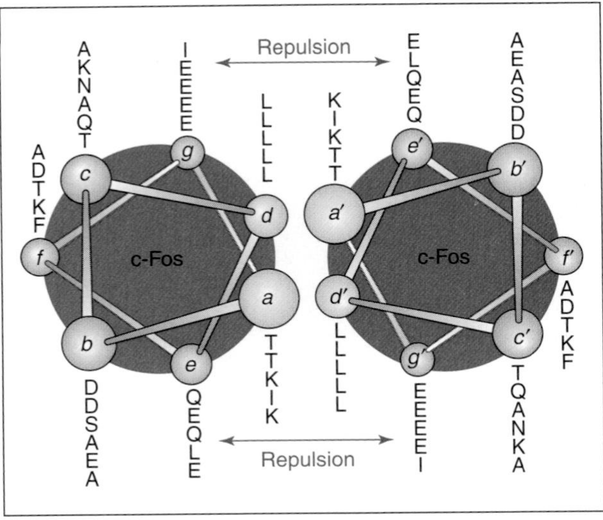

FIGURE 18.40 Wheel structures for c-Fos and c-Jun. (a) The c-Jun•c-Jun homodimer can form because attractive electrostatic interactions exist between side chains in positions e and g. (b) The c-Jun•c-Fos heterodimer formation is even more favorable than c-Jun•c-Jun homodimer formation because the attractive interactions between side chains in positions e and g are even greater. (c) The c-Fos•c-Fos homodimer does not form because of charge repulsion between the negatively charged side chains in positions e and g.

Normal cellular genes such as *c-fos* and *c-jun* that have the potential to cause cancer when mutated or inappropriately expressed are said to be proto-oncogenes.

18.7 DNA-Binding Domains with Helix-Loop-Helix Structures

Helix-loop-helix transcription regulatory proteins are dimers.

A family of transcription activator proteins, distinguished by the presence of a **helix-loop-helix (HLH) motif**, binds to DNA sites as homo- or heterodimers in a manner reminiscent of the bzip proteins. David Baltimore and coworkers predicted the existence of the helix-loop-helix motif in 1989 based on the amino acid sequence of a mouse transcription activator protein that binds to enhancers for immunoglobulin genes. This prediction was verified by crystal structures of HLH transcription regulators bound to cognate DNA sites. For example, the HLH motif is evident in the complex formed by the DNA binding region of **MyoD** (a transcription regulator that stimulates the conversion of immature, unspecialized connective tissue cells into muscle cells; myogenesis) and its cognate DNA site (FIGURE 18.42). The DNA binding region has an HLH motif and a recognition helix. HLH motifs serve as dimerization sites, packing together to form a four-helix bundle that holds the two subunits together. Loop sizes vary from 5 to 24 residues from one kind of HLH transcription regulator to another. Each polypeptide subunit also has a basic recognition sequence (shown in green in Figure 18.42) on the amino terminal side, which folds into a recognition helix upon contacting the major groove of the DNA site. Transcription regulators such as MyoD that have these general features are termed **bHLH activator proteins**. The DNA recognition element for bHLH transcription regulators is CANNTG, where NN is usually CG or GC. Flanking nucleotides on either side of the consensus sequence provide additional specificity.

Mice that lack functional MyoD develop normally and produce skeletal muscle because they have a second bHLH

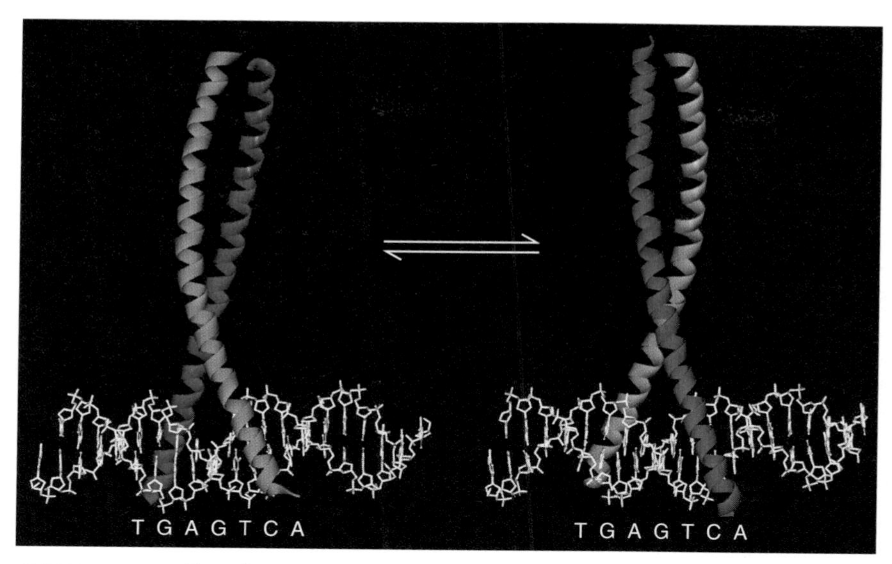

FIGURE 18.41 **Fos•Jun heterodimer bound to an asymmetric AP1 site (TGAGTCA) in two orientations.** Jun and Fos bzip domains are colored blue and red, respectively. The AP1 recognition element base sequence is indicated below the structures. The two structures represent opposite orientations of Fos•Jun binding to the same DNA sequence and reflect heterodimer rotation by about a half turn around the dimer axis. (Structure from Protein Data Bank 1FOS. J. N. Glover and S. C. Harrison, *Nature* 373 [1995]: 257–261. Prepared by B. E. Tropp.)

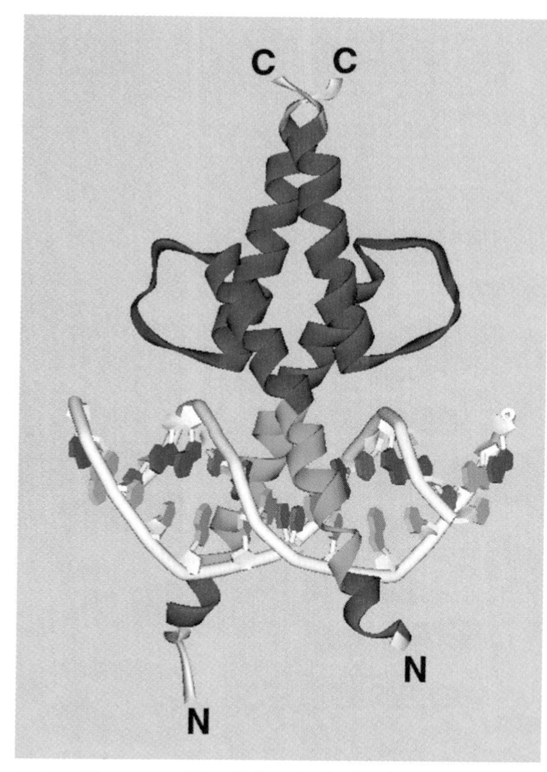

FIGURE 18.42 **MyoD basic helix-loop-helix domain.** The crystal structure of MyoD domain-DNA (TCAACAGCTGGTTGA). The basic recognition helices are shown in green and the loops in purple. N- and C-termini are indicated by N and C, respectively. (Structure from Protein Data Bank 1MDY. P. C. Ma, et al., *Cell* 77 [1994]: 451–459. Prepared by B. E. Tropp.)

transcription regulator called Myf-5 that can substitute for MyoD. Newborn mice lacking both MyoD and Myf-5 are totally devoid of skeletal muscle. Although pups are born alive, they are unable to move and die within minutes. Two other bHLH transcription regulators known as myogenin and MRF4 also participate in normal muscle development. Related tissue-specific bHLH transcription regulators cause embryonic cells to develop into other kinds of specialized tissues. For example, NeuroD bHLH transcription activator proteins convert embryonic cells into nerve cells.

The structure shown for MyoD in Figure 18.42 is a homodimer but the active transcription regulator is thought to be a heterodimer formed between MyoD (a tissue-specific transcription regulator) and a ubiquitously expressed bHLH transcription regulator belonging to the E2A family. The active form of NeuroD also appears to be a heterodimer containing the tissue-specific NeuroD polypeptide and a ubiquitously expressed E2A transcription regulator.

The bHLH zip family of transcription regulators have both HLH and leucine zipper dimerization motifs.

Many transcription regulators have a leucine zipper site on the carboxyl side of their HLH motif. These so-called **bHLH zip** proteins are of considerable interest because several of them are involved in

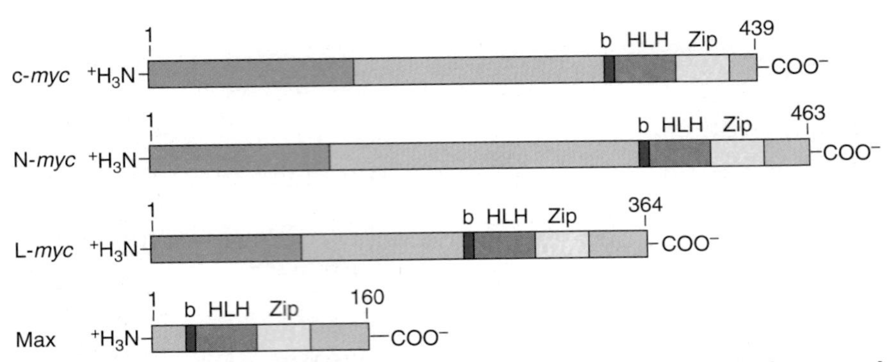

FIGURE 18.43 Schematic diagram of three Myc transcription regulators and a Max transcription regulator. Basic recognition helix (b) is blue, helix-loop-helix (HLH) is gray, and the leucine zipper (Zip) is yellow. The region shown in orange contains the activation and repression domains in the three Myc transcription regulators. The Max transcription regulator lacks this domain.

the transformation of normal cells to cancer cells. The **Myc family** of transcription regulators is particularly important in this respect. The founding member, v-Myc, was identified as the protein product of the *v-myc* oncogene in the avian myelocytomatosis MC29 retrovirus. Cellular homologs of v-Myc, called c-myc, N-Myc, and L-Myc, were identified on the basis of sequence similarities to the viral protein. c-Myc and N-Myc are essential for mouse development and viability but L-Myc is not. Highly differentiated cells do not produce these Myc proteins. When cells are forced to produce Myc by gene amplification, chromosomal rearrangements, or by some other mechanism, the cells tend to proliferate and be transformed into cancer cells. Hence, c-*myc*, N-*myc*, and L-*myc* are proto-oncogenes. Myc proteins do not form homodimers under physiological conditions but instead combine with a related protein called **Max** to form active Myc•Max transcription regulators. FIGURE 18.43 shows some important structural features of Max and the cellular Myc proteins. Max is present in both proliferating and nonproliferating cells. Mice that lack a functional Max protein die during embryonic development.

Although Max can form a homodimer under physiological conditions, it tends to form the more stable Myc•Max heterodimer. Both Myc and Max can also interact with a variety of other transcription factors that influence cell growth and development. For instance, Max also combines with a family of bHLH zip proteins called the Mad family to form Mad•Max transcription regulators that bind to a subset of the Myc•Max DNA sites. Satish K. Nair and Stephen K. Burley have determined crystal structures for the bHLH leucine zipper domains of Myc•Max and Mad•Max heterodimers bound to their common DNA target (enhancer or E box hexanucleotide, 5′-CACGTG-3′) (FIGURE 18.44). A cell responds to Myc•Max regulator binding to the E box by dividing and proliferating. In contrast, Mad•Max transcription regulators appear to inhibit rather than stimulate the target genes. Mad•Max binding, therefore, causes the cell to differentiate and become quiescent. Myc overexpression leads to the development of

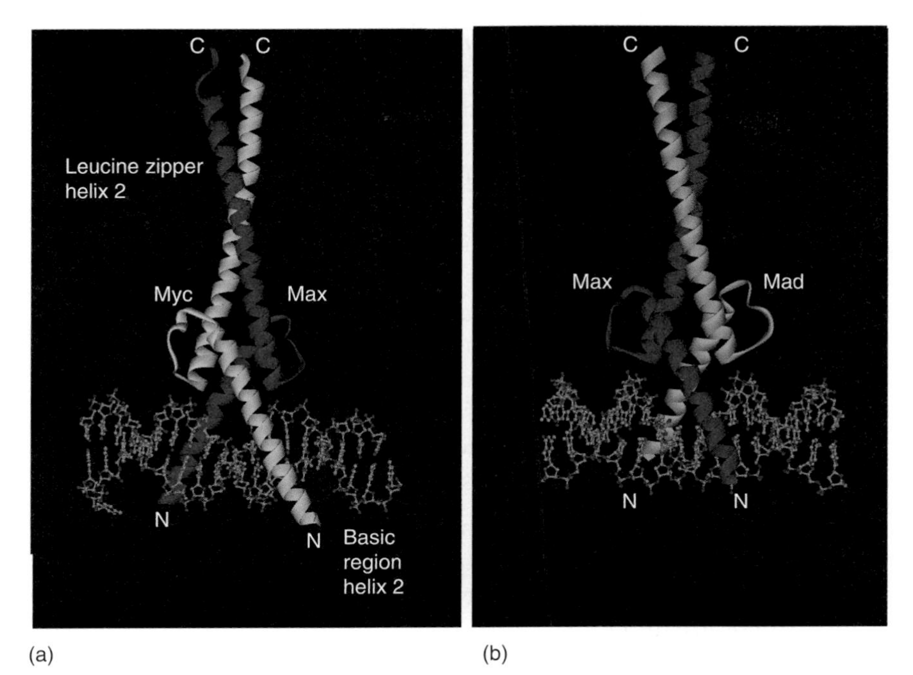

FIGURE 18.44 Structure of Myc•Max and Mad•Max heterodimers bound to DNA. The drawing shows the overall topology of (a) Myc•Max•DNA and (b) Mad•Max•DNA structures. Color-coding: Myc, cyan; Max, red; and Mad, green. Cocrystallization oligonucleotide is shown as an atomic stick figure. The helix, basic region, and zipper regions have been designated on the Myc•Max•DNA structure. The DNA sequences used to prepare the Myc•Max•DNA crystals was CGAGTAGCACGTGCTACTC and that used to prepare the Mad•Max•DNA crystals was GAGTAGCACGTGCTACTC. (Structures from Protein Data Bank 1NKP. S. K. Nair and S. K. Burley, *Cell* 112 [2003]: 193–205. Prepared by B. E. Tropp.)

many human cancers, including Burkitt lymphoma, neuroblastomas, and small lung cancers. The efforts of many research laboratories are now directed toward elucidating the complex network associated with Myc, Max, Mad, and related proteins that regulate cell growth and development. These efforts are certainly justified because they will help us to understand how cells differentiate and to learn more about the mechanism involved in the genesis of a wide variety of cancers.

18.8 Activation Domain

The activation domain must associate with a DNA-binding domain to stimulate transcription.

Although a great deal is now known about structure-function relationships in DNA-binding domains from a wide variety of transcription activator proteins, much less is known about the activation domains (ADs) in these same transcription activator proteins. This disparity in knowledge reflects difficulties in studying activation domains. In contrast to DNA-binding domains, which usually fold to form specific three-dimensional structures such as helix-turn-helix motifs or zinc fingers, activation domains tend to be intrinsically disordered but may assume a defined structure when they interact with other components of the transcription machinery. Furthermore, ADs do not appear to share common structural features such as conserved amino acid sequences. Activation domain function has been just as difficult to study as activation domain structure. Although activation domains work by binding to specific proteins in the transcription machinery, it is quite difficult to determine the physiological target. In fact, many activation domains appear to have several possible targets and it is quite possible that many of these targets are physiologically important.

DNA-bound transcriptional activator proteins may work by one or both of the following mechanisms:

1. **Activation by recruitment model.** The activation domain interacts with one or more components of the transcription machinery and stabilizes the binding of the component(s) to the template DNA.
2. **Activation by conformational change model.** The activation domain somehow induces a conformational change in one or more components of the transcription machinery that are bound to it and thereby stimulates RNA polymerase II to initiate transcription.

Gal4 has DNA binding and activation domains.

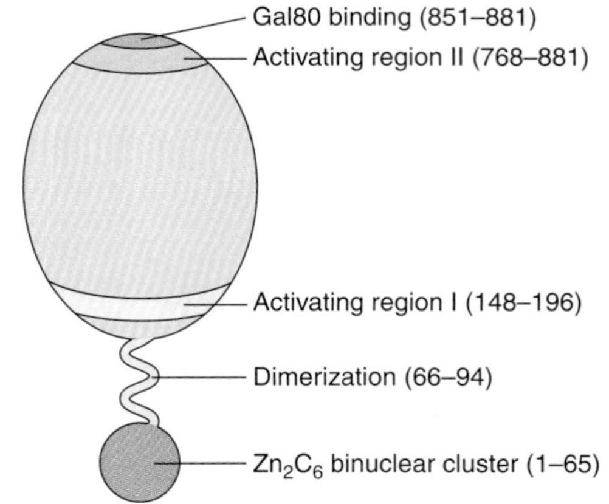

FIGURE 18.45 **Schematic of Gal4 domain structure.** The Zn_2C_6 binuclear cluster (residues 1–65) is shown in orange, the dimerization region (residues 66–94) is shown in light blue, and activating region I (residues 148–196) is shown in yellow. Activating region II (768–881) includes the Gal80 binding region (851–881), which is shown in blue. The rest of activating region II is shown in magenta. (Adapted from M. Ptashne and A. Gann. *Genes & Signals.* Cold Spring Harbor Laboratory Press, 2002.)

Despite the formidable difficulties in elucidating the function of the activation domain, a general picture has started to emerge. Much of this picture comes from studying transcription activator proteins in yeast and so that is where we will begin. Thus far, our examination of Gal4 has been limited to the structure and function of the first 64 residues, which fold to form the Zn_2C_6 binuclear cluster and the dimerization region that comprise the DNA-binding domain. Other segments along the polypeptide chain also have special functions (FIGURE 18.45). The segment extending from residues 66 to 94 leads to even greater dimer stability, and segments extending from 148 to 196 and from 768 to 881 function as weak and strong activating regions, respectively. The activating regions are described next.

Gal4 requires the assistance of several other proteins to switch *GAL* genes on and off in response to the carbon source in the growth medium. While the discussion that follows focuses on the regulation of *GAL1* (the gene that encodes galactokinase), the lessons learned also apply to the other *GAL* genes. *GAL1* is switched on when galactose is added to growth medium in the absence of glucose and is switched off when the galactose is consumed or glucose is added. Galactose, therefore, induces *GAL1* gene expression and glucose represses it. Two types of proteins bind to regulatory sites upstream from the *Gal1* transcription start site (FIGURE 18.46). The first of these, Gal4, which was described above, binds to four sites within the upstream activating sequence. This sequence designated UAS_{GAL} spans about 118-bp

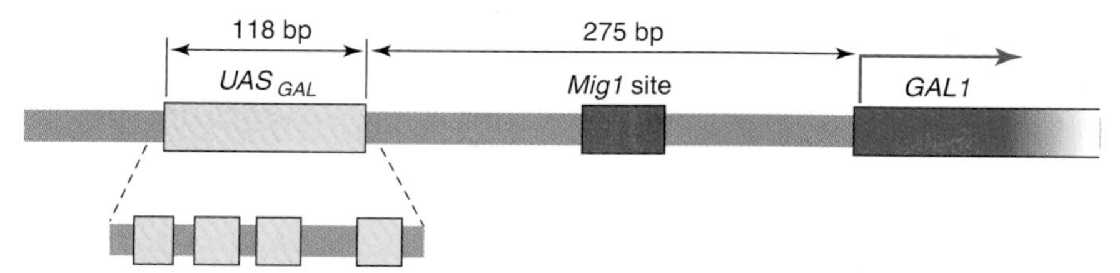

FIGURE 18.46 The *GAL1* gene and the region just upstream. The four Gal4 binding sites within UAS_{GAL} are depicted in the enlarged section. The Mig1 binding site is located between UAS_{GAL} and GAL1. UAS_{GAL} spans about 118-bp and each *Gal4* site is 17-bp. (Adapted from M. Ptashne and A. Gann. *Genes & Signals*. Cold Spring Harbor Laboratory Press, 2002.)

and is located about 275-bp from the transcription initiation site. Gal4 can stimulate transcription if just one of the four 17-bp sites is present. The second protein, designated the Mig1 protein (multicopy inhibitor of GAL gene expression), has a Cys_2His_2 zinc finger in its DNA binding domain that makes sequence-specific contacts with the so called *Mig1* site, which is located between UAS_{GAL} and the transcription initiation site.

In the absence of both galactose and glucose, Gal4 binds to the four sites in UAS_{GAL} (FIGURE 18.47a). However, Gal4 binding fails to switch on the transcription of the *GAL1* gene. The reason for this failure is that a protein known as Gal80, which is a homodimer, binds to the Gal 80 binding region (residues 851–881), preventing activating region I from interacting with other transcription components. Adding galactose to the growth medium relieves the inhibition caused by Gal80 and permits *GAL1* transcription to take place (FIGURE 18.47b).

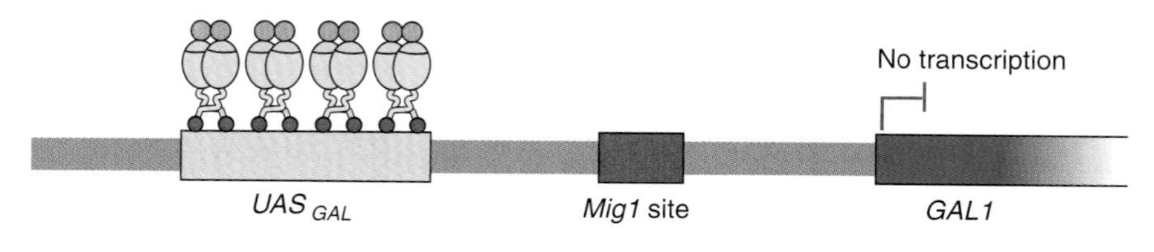

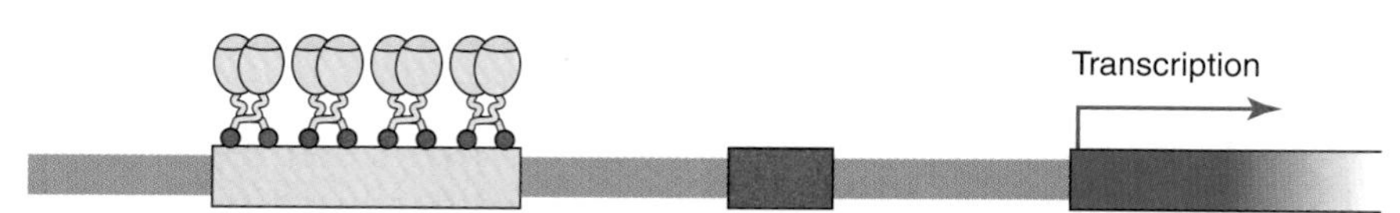

FIGURE 18.47 Effect of galactose as an inducer on *Gal1* gene transcription. (a) In the absence of galactose, Gal80 dimers (orange spheres) bind to the Gal80 binding site, preventing activating region II (magenta) from interacting with components of the transcription machinery and blocking transcription. (b) In the presence of galactose, Gal3 (not shown) binds to Gal80 causing Gal80 dimers to dissociate and be released from Gal4. Gal4 activating region II (magenta) is now free to interact with components of the transcription machinery so that Gal1 can be transcribed. Color coding for Gal4 is the same as in Figure 18.45 (activating region II magenta, dimerization region cyan, and Zn_2C_6 binuclear cluster red). For simplicity, the Gal80 binding region and activating region I are not shown.

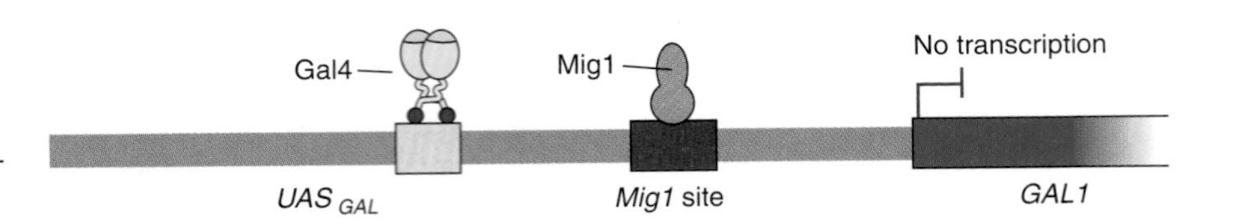

FIGURE 18.48 The effect of glucose on *Gal1* transcription. In the presence of glucose, the dephosphorylated form of Mig1 binds to the *Mig1* site, a necessary step for *Gal1* gene transcription inhibition.

Gal3, a cytoplasmic protein that is homologous to Gal1 but that lacks galactokinase activity, is required to relieve the inhibition caused by the Gal80 homodimer. Gal3 binds to galactose and ATP to form a Gal3•galactose•ATP complex, which somehow promotes the release of Gal80 from Gal4.

GAL1 transcription stops when glucose is added to a growth medium that contains galactose. Glucose acts through a protein kinase called *Snf1 kinase* (sucrose non-fermenting kinase) to regulate Mig1 function. When glucose is absent, Snf1 kinase phosphorylates Mig1. The phosphorylated form of Mig1 is a cytoplasmic protein and so cannot bind to the *Mig1* site in the nucleus. Glucose inactivates Snf1 kinase so that it can no longer phosphorylate Mig1. The dephosphorylated form of Mig1 moves to the nucleus and binds to the *Mig1* site (FIGURE 18.48). Then additional transcription factors bind to Mig1 and block *GAL1* transcription.

Starting in the mid-1980s, Ptashne and coworkers performed a series of experiments with Gal4 that helped to explain how activation domains work. These experiments typically involved constructing recombinant plasmids bearing genes that code for modified Gal4 transcription activator proteins. Modifications included deleting various segments from the 881-residue Gal4 polypeptide chain or creating hybrid proteins by joining Gal4 peptide fragments to peptide fragments from other proteins. The basic experimental objective—to determine whether the recombinant proteins function as transcription activator proteins—was achieved by first transforming a yeast strain that lacked Gal4 with the recombinant plasmid and a reporter plasmid, and then measuring the transformant's ability to produce a reporter gene product such as β-galactosidase. Results from one set of these experiments are summarized in FIGURE 18.49. The normal Gal4 transcription activator protein stimulated transcription of a reporter gene with a UAS_{GAL} site upstream from its transcription initiation site (Figure 18.49a). However, a peptide fragment containing just the first 100 residues in the Gal4 chain (a region that includes the DNA-binding domain, the dimerization domain, and a sequence that directs Gal4 to the cell nucleus) did not activate gene transcription even though it did bind to the UAS_{GAL} site (Figure 18.49b). The complementary fragment (residues 100–881) also did not activate reporter gene transcription (Figure 18.49c).

The fact that the activation domain could not stimulate transcription on its own indicates that the conformational change model is not sufficient to explain how the Gal4 activation domain stimulates transcription. It does not rule out, however, the possibility that the conformational change model makes some contribution to transcription activation. Linking the complementary fragment (residues

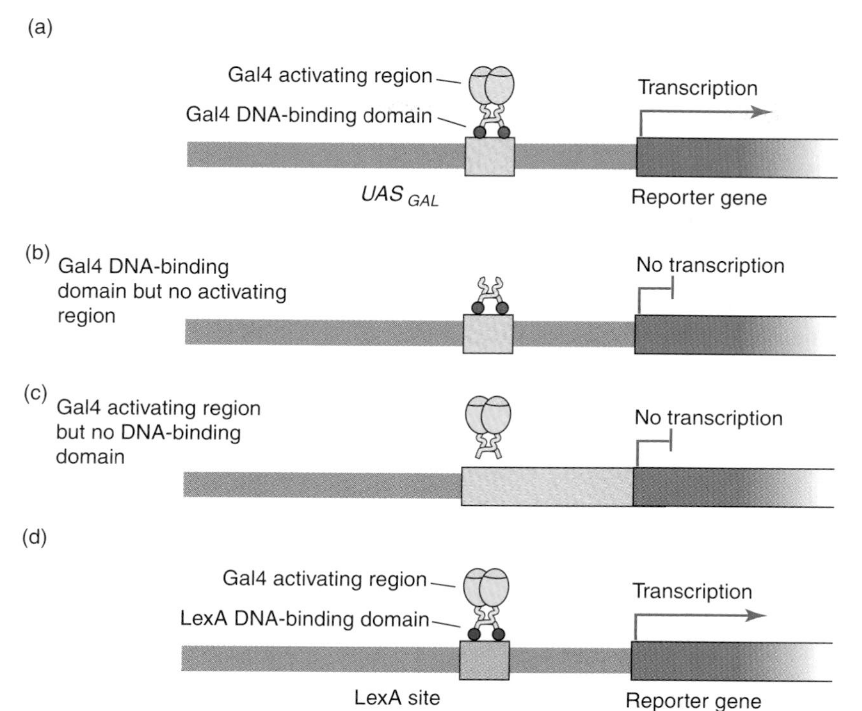

FIGURE 18.49 **Gal4 domain function.** (a) The Gal4 transcription activator stimulates transcription from a reporter gene with a UAS_{GAL} site upstream from the transcription initiation site. (b) A peptide fragment (residues 1–100) containing the Gal4 DNA-binding domain (Zn_2C_6 zinc cluster and dimer region) does not stimulate transcription from a reporter gene with a UAS_{GAL} site upstream from the transcription initiation site. (c) A complementary Gal4 fragment (residues 100–881) containing the activation regions cannot stimulate transcription from a reporter gene with a UAS_{GAL} site upstream from the transcription initiation site. (d) A LexA-Gal4 hybrid protein containing the DNA-binding domain from the bacterial repressor protein LexA attached to the Gal4 activation region (residues 100–881) stimulates transcription from a reporter gene with a *LexA* site upstream from the transcription initiation site.

100–881) to a different DNA-binding domain produced a hybrid protein that did function as a transcription activator protein. For example, attaching the complementary fragment to the DNA-binding domain of an *E. coli* repressor known as LexA (see Chapter 12) produced a LexA-Gal4 hybrid protein that could initiate transcription from a reporter gene with a LexA site upstream from the transcription initiation site (Figure 18.49d). This type of experiment is known as a **domain swap** because the DNA-binding domain from one protein replaces that of another. The fact that the LexA-Gal4 hybrid protein activated transcription is consistent with the activation by recruitment model. The experiments shown in Figure 18.49 indicate that a transcription activator protein requires both a DNA-binding domain and a transcription activation domain to work but that an assortment of DNA-binding domains, even one from bacteria, can replace that from Gal4.

Ptashne and Jun Ma performed deletion analysis experiments that revealed the Gal4 activation domain contains two discrete activating regions. Activating region I extends from residues 148 to 196. Activating region II extends from residues 768 to 881 and also includes the Gal80 repressor-binding site. Attaching either activating region to a DNA-binding domain creates a transcription activator protein; therefore, large portions of Gal4 are not required for gene activation. Each of the activating regions had a much greater than expected number of the negatively charged amino acid residues aspartate and glutamate, causing these regions to have net negative charges. Genetic studies of activating region I showed that activating region strength is related to negative charge. Activating region I became weaker when mutations reduced negative charge by replacing aspartate or glutamate residues with uncharged residues. Conversely, activating region I became stronger when mutations increased negative charge by replacing one of the few lysine or arginine residues with uncharged residues. Attempts to demonstrate that activating regions from other transcription activator proteins also have net negative charges were successful in some cases but failed in many others.

Net negative charge is not the only factor that determines Gal4 activating region strength. Gal4 activating regions I and II also have hydrophobic residues that appear to contribute to their function. The arrangement of the glutamate and aspartate residues in the hydrophobic background is probably important for activity because scrambling the order of residues within activating region I or II produces nonfunctional peptides. There does not appear to be a requirement for some specific sequence, however, as about 1% of randomly generated *E. coli* peptide fragments can function as activating regions when attached to a DNA-binding domain. One hypothesis is that negatively charged side chains in a hydrophobic background somehow create a sticky peptide segment that can bind to other transcription machinery components and help recruit the components to the preinitiation complex. Other types of sticky peptide segments may have evolved in activating regions that do not have net negative charges.

Ma and Ptashne realized that the relative ease with which they were able to construct active hybrid transcription factors suggested that there are few, if any, stereospecific restrictions on how the activating region must be attached to the DNA-binding domain. This realization led them to devise an experiment to test whether the DNA-binding domain and the activation region can be on separate polypeptide chains that combine to form a protein complex. Their approach was to modify two polypeptides that interact with each other through non-covalent bonds so that the first polypeptide had the DNA-binding domain and the second had the activating region. They constructed the polypeptide with the DNA-binding domain by deleting residues 148 to 850 from Gal4 and the polypeptide with the activating region by inserting an acidic bacterial peptide into the Gal80 repressor. They then tested the ability of yeast cells that make various combinations of the two polypeptides to transcribe UAS_{GAL}-*lacZ* fusion genes in their chromosomes. The results of these experiments are summarized in **FIGURE 18.50**. Cells that make the Gal4 derivative synthesize low levels of β-galactosidase (Figure 18.50a). This result is expected because the Gal4 derivative still has part of activating region II (residues 851–881). Cells that also make wild type Gal80 synthesize only trace quantities of β-galactosidase (Figure 18.50b). This result is also expected because the Gal80 repressor binds to the Gal4 derivative and represses transcription. Replacing wild type Gal80 with a Gal80 that has a bacterial peptide insert, which was previously shown to function as an activating region, gives rise to high levels of β-galactosidase synthesis (Figure 18.50c). These experiments show that a polypeptide with a DNA-binding domain and a second polypeptide with an activating region can combine through non-covalent interactions to

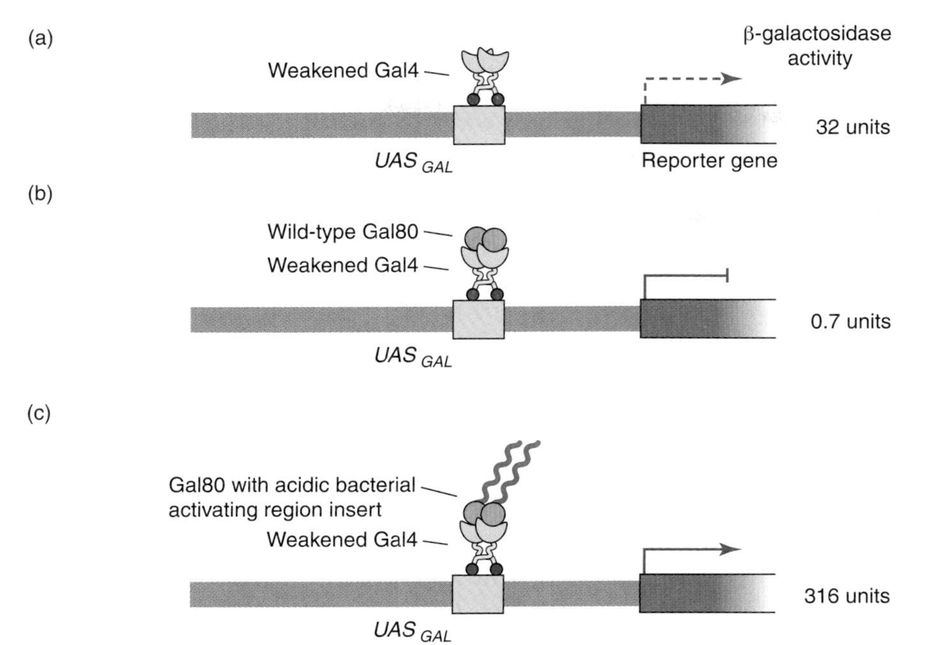

FIGURE 18.50 Experiment testing whether the DNA-binding domain and the activating region can be on separate polypeptide chains. (a) Cells that synthesize a weakened Gal4 in which part of activating region II (residues 851–881, magenta) is attached directly to the dimerization region (cyan) synthesize a low level of β-galactosidase from a UAS_{GAL}-LacZ fusion genes. (b) Cells that also make wild type Gal80 (orange) synthesize only trace quantities of β-galactosidase because Gal80 binds to the weakened Gal4 and represses transcription. (c) Cells that synthesize the weakened Gal4 and a modified Gal80 that has a bacterial peptide insert (green) that can function as an activating region synthesize high levels of β-galactosidase.

form a functional transcriptional activator protein. This experiment is therefore also consistent with the activation by recruitment model.

Nature also provides examples of DNA-binding and activation domains that are on separate polypeptide chains. The herpes simplex virus (HSV) provides a particularly instructive example (FIGURE 18.51). The viral polypeptide of interest called VP16 might at first appear to be a structural protein because it is located in the space between the nucleocapsid (the core of the virus particle) and the virus membrane. Although VP16 may have played an entirely structural role early in HSV's evolutionary history, it has evolved into a transcriptional factor that activates the transcription of five virus genes immediately after the virus particle enters the cell nucleus and releases it. VP16, acting on its own, cannot bind to the regulatory sites on the five immediate-early viral genes because it lacks a DNA-binding domain. However, the host cell transcription activator protein Oct-1, a founding member of the POU family of transcription activator proteins, can supply this domain. Oct-1 normally binds to ATGCAAAT sites in cellular genes to weakly stimulate transcription (Figure 18.51a). However, Oct-1 can also bind to viral TAATGARAT sites (R is a purine), changing conformation in the process (Figure 18.51b). VP16 binds to Oct-1 after this conformational change, forming a complex that is a strong transcriptional activator protein. Oct1 provides the DNA-binding domain and VP16, provides the strong activation domain. VP16 is another example of a protein with an activating region with a net negative charge. As might be expected, the hybrid protein Gal4-VP16, formed by fusing the Gal4 DNA-binding domain to VP16, is a very strong transcription activator protein.

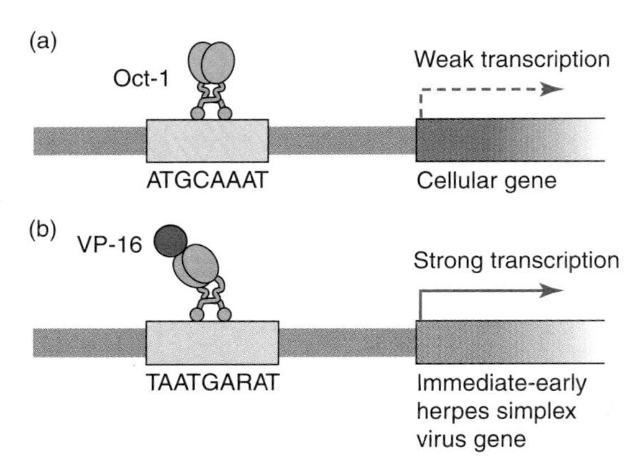

FIGURE 18.51 VP16-Oct-1 interaction. (a) Oct-1 (green) binds to the ATGCAAAT sequence in cellular genes to weakly stimulate transcription. (b) Oct-1 binds to the TAATGARAT sequence in the herpes simplex virus (HSV) immediate-early gene promoters and changes conformation. VP16 (red) binds to Oct-1 after the conformational change and the two polypeptides together form a strong transcriptional activator.

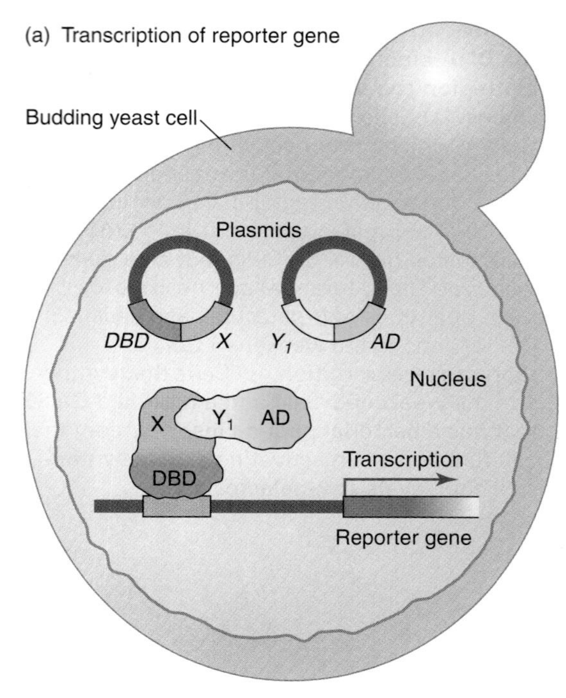

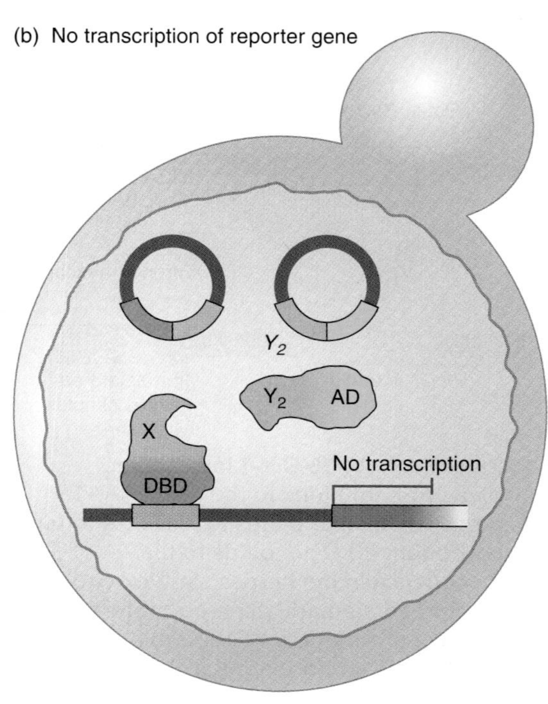

FIGURE 18.52 **How the two-hybrid system works.** (a) The bait protein, protein X, is fused to a DNA-binding domain (DBD). Protein Y_1 is fused to a transcription activation domain (AD). Both hybrid proteins are expressed from plasmids in a yeast cell. Proteins X and Y_1 interact, leading to the activation of a reporter gene, which codes for a protein that permits the yeast to grow on a defined medium. (b) Proteins X and Y_2 do not interact. Therefore, the reporter gene is not expressed and the yeast cannot grow. (Adapted from P. H. Uetz, R. E. Hughes, and S. Fields, *Focus* 20 [1998]: 62–64.)

The yeast two-hybrid assay permits us to detect polypeptides that interact through non-covalent interactions.

In 1989, Stanley Fields and Ok-kyu Song devised an ingenious technique known as the **two-hybrid assay** that takes advantage of the modular nature of transcription activator proteins to probe protein-to-protein interactions (FIGURE 18.52). The assay begins with the construction of two plasmids, each directing the expression of a different hybrid protein. The first hybrid protein has some protein, X, fused to a DNA binding domain and the second has some other protein, Y_1 or Y_2, fused to a strong AD. The two plasmids are introduced into a yeast cell with a reporter gene that has a regulatory site that is a target for the first hybrid protein's DNA-binding domain.

To see how this assay works, let's assume that proteins X and Y_1 interact, whereas proteins X and Y_2 do not. Then a cell with hybrid proteins X-DBD and Y_1-AD will have a functional transcription activator that stimulates reporter gene transcription (Figure 18.52a). In contrast, a cell with hybrid proteins X-DBD and Y_2-AD will not have a functional transcription activator and therefore will be unable to transcribe the reporter gene (Figure 18.52b).

One important application of the two-hybrid assay is to search for proteins that interact with a specific protein of interest. The assay uses the specific protein of interest as "bait" to catch proteins that interact with it ("prey"). The first step is to construct yeast so that they have a plasmid directing the synthesis of a hybrid protein with a DNA-binding domain fused to the bait protein (X in Figure 18.52) and a selectable reporter gene such as *LEU2* or *HIS3* with an upstream site for the DNA-binding domain. The next step is to transform the yeast with a plasmid library that has been constructed so that each transformed cell will express a hybrid protein with the same activation domain linked to different prey proteins (proteins comparable to Y_1 and Y_2 proteins in Figure 18.52). Only transformants that receive hybrid prey proteins that interact with the hybrid bait protein will be able to grow on minimal medium lacking leucine or histidine. Plasmids can be isolated from these transformants and sequenced and the sequence information used to search the protein data bank to identify the prey protein.

Despite numerous false-positives and negatives, the two-hybrid assay has been quite successful in identifying proteins that interact with one another. The two-hybrid assay also has been useful for identifying mutant proteins with amino acid substitutions that cause stronger or weaker binding to some other target protein. Let us now return to the basic issue of how transcription activator proteins stimulate transcription.

18.9 Mediator

Squelching occurs when transcription activator proteins compete for a limiting transcription machinery component.

Investigators initially thought that eukaryotic transcription activator proteins stimulate gene transcription by making direct contact with one or more components of the basal transcription machinery. Although a considerable body of evidence in fact shows that transcription activator proteins do contact basal transcription machinery components such as TBP, TBP associated factors (TAFs), and TFIIB, these contacts are often not sufficient for activated transcription. An entirely new protein complex, designated **Mediator**, is required for transcription activator and repressor proteins to work.

The first hint of Mediator's existence came from studies performed in yeast and mammalian cells in 1988, which showed that a high intracellular concentration of one transcription activator protein inhibits other transcription activator proteins from stimulating gene transcription. Grace Gill and Mark Ptashne proposed that this inhibitory phenomenon, which they called **squelching**, occurs when transcription activator proteins compete among themselves for some limiting target (FIGURE 18.53). TBP, TFIIB, and RNA polymerase II each seemed to be a reasonable candidate for the limiting target because each can bind to transcription activator proteins. The problem, however, was to devise some method for determining which, if any, of these components of the transcription machinery was the limiting target. Roger Kornberg and coworkers thought that one method for solving the problem would be

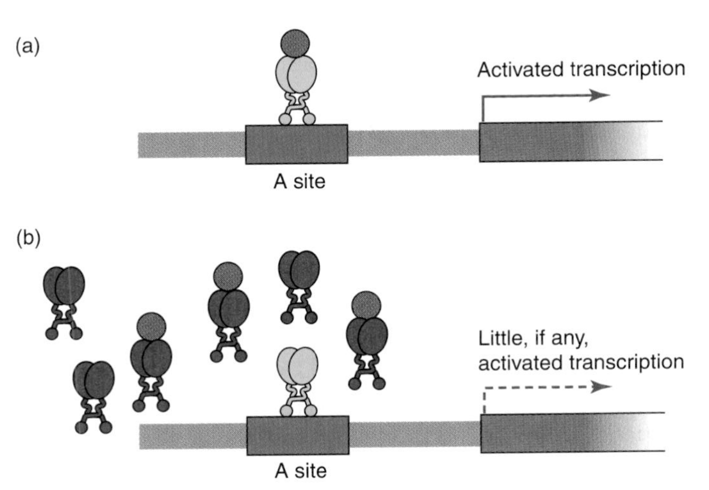

FIGURE 18.53 **Squelching.** (a) Transcription activator A (light green) binds to the target protein (blue), permitting activated transcription to occur. (b) Transcription activator B (red) binds to the target protein (blue) so that very little of the target protein is available to bind to transcription activator A (light green). Squelching results from the competition between the transcription activators for the target protein.

to reproduce the squelching phenomenon in a cell-free system. After achieving this goal by adding transcription activator proteins to a yeast nuclear extract, they demonstrated that squelching persisted even after excess quantities of RNA polymerase II or any given general transcription factor was added to the extract. These experiments indicated that some still unknown factor was the limiting target responsible for squelching. As we see in the next section, the search for this target led to the discovery of Mediator.

Mediator is required for activated transcription.

The next major step in the discovery of Mediator occurred when Kornberg and coworkers noticed that a protein fraction, which they had prepared while trying to isolate transcription factors from yeast, appeared to enable activated transcription. They suspected that this fraction contained a component of the transcription machinery that somehow interacted with transcription activator proteins and the basal transcription machinery so that it could stimulate transcription or serve as a target for squelching. Because the putative transcription component seemed to serve as a go between for transcription activator proteins and the basal transcription machinery, it was named Mediator.

Kornberg's group set out to isolate Mediator from the protein fraction using an assay system consisting of homogeneous general transcription factors, RNA polymerase II, the transcription activator protein Gcn4, and two different DNA templates (one with Gcn4 binding sites and the other with Gal4 binding sites). Transcription from the two templates produced radioactive RNA molecules of different sizes that were resolved by gel electrophoresis, making it possible to distinguish activated transcription due to Gcn4 sites from basal transcription due to Gal4 sites. Taking advantage of this assay system, Kornberg and coworkers succeeded in isolating Mediator, a protein complex, which contains 21 different polypeptide subunits. A group of Srb proteins (see below) form a specific subcomplex known as the Cdk8 (or Srb8-11) module, which is present in some, but not all, yeast Mediator complexes. In addition to its ability to stimulate reconstituted transcription systems containing yeast proteins to respond to activator proteins, Mediator also stimulates transcription in the absence of activator (basal transcription) by approximately tenfold and TFIIH kinase catalyzed phosphorylation of the carboxyl terminal domain (CTD) in the largest RNA polymerase II subunit (Rpb1) by 30- to 50-fold.

Genes that code for each of the 21 subunits in the core Mediator from yeast have been identified. Several of these genes had previously been identified through the use of genetic screens to detect mutations that affect transcription. For instance, in 1989 Michael Nonet and Richard Young identified five genes, each coding for a different Mediator polypeptide subunit, while studying a yeast mutant with a truncated CTD in Rpb1. The truncated CTD causes the mutant to have cold- and temperature-sensitive phenotypes. Nonet and Young sought to identify proteins involved in CTD function by searching for suppressor mutations that would restore the mutant's ability to grow in the cold. Several such suppressor mutations were found, leading to the identification of several new genes, which were named *SRB1*, *SRB2*, and so forth to reflect the fact that the genes were first detected in a genetic screen for suppressors of RNA polymerase B (an alternate name for RNA polymerase II). Additional genes for Mediator subunits had been identified through the use of genetic screens that were designed to study some specific aspect of transcription. The original names given to these genes reflected the genetic screens that were used or the transcriptional property that was studied. For example, genes that code for two of the Mediator subunits, *RGR* and *Gal11* were named for their involvement in resistance to glucose repression and galactose metabolism, respectively. The remaining Mediator genes had not been identified at the time that Kornberg's group isolated Mediator, and so they were named *MED1*, *MED2*, and so forth.

Studies from many different laboratories show that Mediator is a fundamental part of the RNA polymerase II transcription machinery in all eukaryotes. Moreover, the Mediator subunits from all organisms seem to be homologous with those in yeast and to have similar shapes. Because this homology has only been recognized recently, the literature contains a rather confusing nomenclature system. An example will help to illustrate the problem. Several mammalian Mediator complexes have been isolated and each has been given a name to reflect its function or the method that was used to isolate it. Two of the best-studied mammalian Mediators thus are named thyroid hormone receptor-associated protein (TRAP) and vitamin D receptor-interacting protein (DRIP) despite the fact that these two Mediators are very similar and perhaps identical. A unifying nomenclature system was recently proposed in which nearly all Mediator subunits are named MED followed by a number designation. For instance, Rgr in yeast and TRAP170 in humans are designated MED14, while Gal11 in yeast and Arc105 in humans are named MED15. This nomenclature system acknowledges that Mediator was first discovered in yeast. The original yeast MED subunits, therefore, retain their names (MED1–11). The remaining yeast MED subunits are given names starting with MED12 in order of decreasing molecular mass.

Biochemical, genetic, and electron microscopy studies provide insights into yeast Mediator subunit organization and the interactions between Mediator and RNA polymerase II. Michel Werner and coworkers have used pairwise two-hybrid analysis to investigate protein–protein contacts between budding yeast Mediator subunits. Based on these and related studies as well as the relative sizes of the Mediator subunits, they have constructed the topological map of yeast Mediator shown in FIGURE 18.54.

Yeast Mediator appears to be organized into three subcomplexes: the head, middle, and tail. The head subcomplex seems to interact with the largest RNA polymerase II subunit (Rpb1) at its carboxyl terminal domain (CTD) and serve as a signal processor that directly influences RNA polymerase II activity. Cells that lack this subcomplex are nonviable. The middle subcomplex, which contains the MED9/10 module, also appears to interact with CTD. It is thought to function in regulatory signal transfer after binding to the transcription activator protein.

FIGURE 18.54 **Topological organization of yeast Mediator.** This model is based in large part on results from a pair-wise two-hybrid experiment used to investigate protein-protein contacts between budding yeast Mediator subunits as well as the relative size of the Mediator subunits. The core mediator is organized into three subcomplexes, termed the head (blue), middle (green), and tail (yellow) modules. The Cdk (or Srb8–11) module (red) is present in many, but not all, yeast Mediator complexes. Where applicable, older names for the Mediator subunits are given in parentheses below the new names. (Modified from B. Guglielmi, et al., A high resolution protein interaction map of the yeast Mediator complex, *Nucleic Acids Res.* 32 [2004]: 5379–5391. Reprinted by permission of Oxford University Press. Used with permission of Michel Werner, Service de Biologie Intégrative et Génétique Moléculaire CEA-Saclay and Nynke van Berkum and Frank Holstege, University Medical Center Utrecht.)

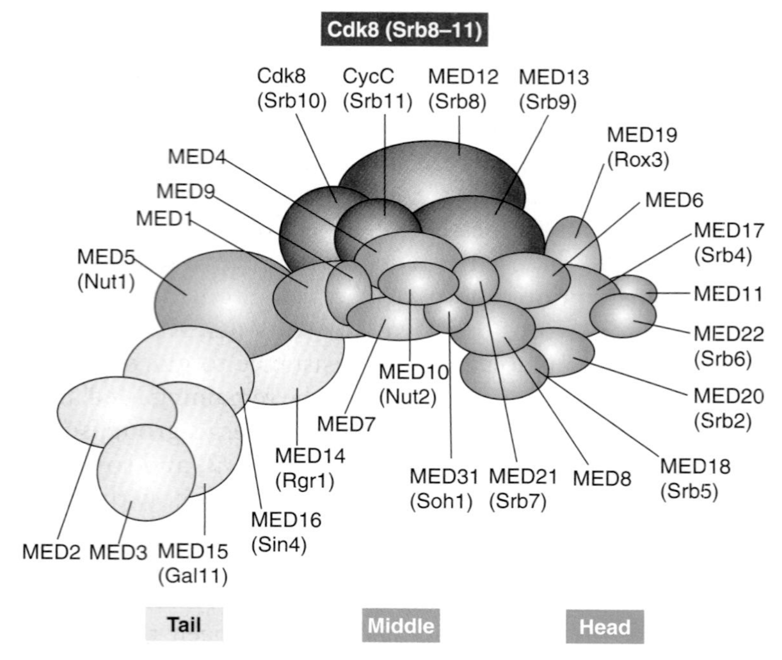

The tail subcomplex includes most of the Mediator subunits that were originally identified by genetic screens for transcription mutants; it is thought to sense signals from gene-specific transcription activator proteins such as Gcn4. Cells that lack the tail domain are viable.

The yeast Mediator complex associates with activators at the UAS in active yeast genes.

The discovery that the ~1 MDa Mediator complex is essential for activated transcription in eukaryotes was quite surprising because so many other proteins were already known to participate in this process. Mediator complexes were isolated from various eukaryotes with bound activator protein or RNA polymerase II, suggesting that Mediator functions as the communication link from activator protein to RNA polymerase II. Further study was required to determine how Mediator interacts with the pre-initiation complex.

Kornberg and coworkers used a technique known as **chromatin immunoprecipitation** (**ChIP**) to determine how Mediator interacts with promoters. The steps involved in this technique (illustrated in FIGURE 18.55) are as follows. (1) Cells are treated with formaldehyde, a cell-permeable molecule that causes protein–protein and protein–DNA crosslinking. (2) Treated cells are lysed and their isolated chromatin is sheared by ultrasound to produce small fragments of about 100 to 1000 bp. (3) Antibody is added to precipitate fragments that have the protein of interest bound to them. (4) The crosslinks are removed by incubating the fragments in buffer at low pH. (5) Proteins are removed and the specific primers are added to the free DNA to amplify regions of interest.

The ChIP experiments performed by Kornberg and coworkers showed that Mediator associates with UAS sites under activating conditions and not with the core promoter. Consistent with this finding is that temperature-sensitive TFIIB mutants continued to bind Mediator at the UAS sites even when the temperature was increased, so that RNA polymerase could no longer bind to the DNA. Finally, cells were constructed that had a recombinant gene that was missing the TATA box but still had UAS sites. ChIP experiments demonstrated that Mediator was able to bind to the UAS sites even though the general transcription factors and RNA polymerase II were no longer bound to the DNA. These ChIP experiments indicate that Mediator probably binds to the activation domain of an activator protein that is already bound to a UAS site. The picture that emerges from these studies is that Mediator functions as a bridge between gene-specific activator proteins and other components of the RNA polymerase transcription machinery at the promoter (FIGURE 18.56). In higher eukaryotes, where most genes are regulated by a combination of activator proteins (and repressor proteins), the Mediator probably integrates the signals before transmitting the appropriate signal to RNA polymerase II and other components of the transcription machinery. We are still a long way from knowing just how Mediator and its components function.

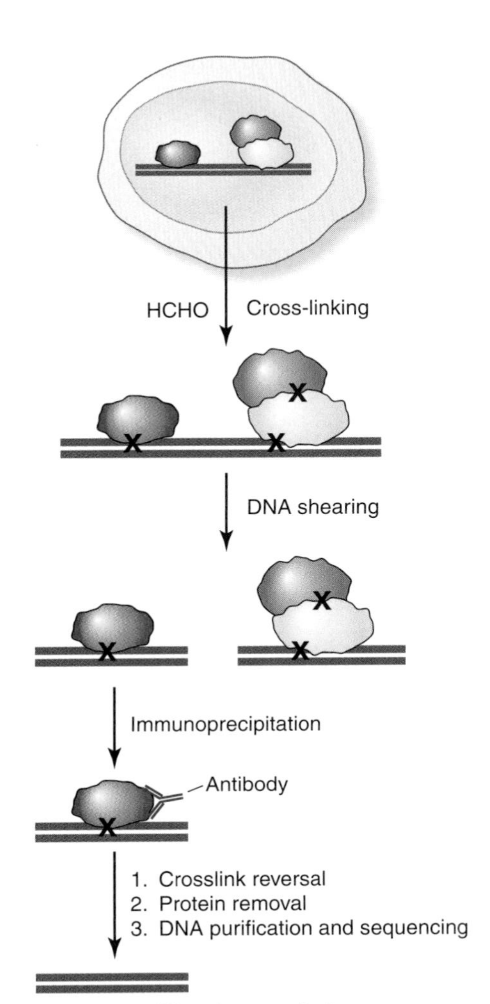

FIGURE 18.55 **The chromatin immunoprecipitation (ChIP) technique.** Cells are cultured under desired experimental conditions. Then a cross-linking agent, usually formaldehyde, is added to trap protein-protein and protein-DNA interactions. After cell lysis, DNA is subject to mechanical shearing and then immunoprecipitated using an antibody or combination of antibodies specific for the protein of interest. In the example shown here, the protein of interest is colored red and the other proteins are colored blue. After immunoprecipitation, the cross-links are reversed at low pH. Following protein removal, DNA is purified. The purified DNA may be studied after amplification by the polymerase chain reaction (PCR). (Modified from *Curr. Opin. Chem. Biol.*, vol. 9, D. Sikder and T. Kodadek, Genomic studies of transcription factor-DNA interactions, pp. 38–45, copyright 2005, with permission from Elsevier [http://www.sciencedirect.com/science/journal/13675931].)

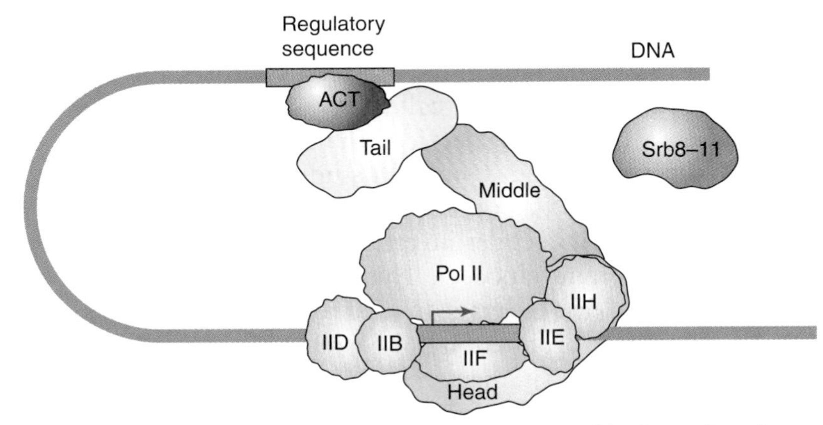

FIGURE 18.56 **Mediator interaction at the promoter.** Mediator (head, middle, and tail) serves as a bridge between the general RNA polymerase II transcription machinery and gene-specific activators (ACT, red). Contacts with RNA polymerase II are localized to the head and middle regions of Mediator. Activator interactions occur mainly in the tail regions of Mediator. The Cdk8 (Srb8–11) module is involved in negative regulation of transcription. Only Mediator that lacks the Cdk8 module can associate with RNA polymerase II. (Adapted from S. Björklund and C. M. Gustafsson, *Trends Biochem. Sci.* 30 [2005]: 240–244.)

18.10 Epigenetic Modifications

Cells remodel or modify chromatin to make the DNA in chromatin accessible to the transcription machinery.

Eukaryotic DNA molecules interact with histone and nonhistone proteins to form chromatin. Recall from Chapter 6 that the basic building block of chromatin, the nucleosome, consists of 1.65 turns of DNA wrapped around an octameric protein complex containing two copies each of the core histones H2A, H2B, H3, and H4. N-terminal tails of the four core histones stick out from the nucleosomes to contact DNA, other histones, and nonhistone proteins. These interactions help to stabilize the nucleosomal fiber as it folds into chromatin. Nucleosomes tend to block the transcription machinery and thereby inhibit gene expression. Chromatin remodeling complexes and modifying enzymes help to make the DNA available to the transcription machinery so that the genes can be expressed.

Chromatin Remodeling Complexes

Eukaryotic cells use **ATP-dependent chromatin remodeling complexes** to reposition nucleosomes, eject nucleosomes, unwrap nucleosomes, and exchange or eject histone dimers (FIGURE 18.57). One polypeptide subunit in each chromatin remodeling complex has ATPase activity. The organization of this subunit is the basis for grouping remodeling complexes into the **SWI/SNF**, **ISWI**, **CHD**, and **INO80** families. The following discussion is limited to the yeast SWI/SNF family. Higher animals and plants have chromatin remodeling complexes that are similar in action to those in yeast.

The first clue to the existence of the SWI/SNF family of chromatin remodeling complexes came from genetic studies in yeast. Genetic screens for mating type switch mutants (*swi*) and sucrose nonfermentation mutants (*snf*), two apparently unrelated processes in yeast, led to the same gene, which is now called *SWI2/SNF2*. Additional studies established a connection between the *SWI2/SNF2* gene product and chromatin remodeling. One such study involved examining the effect that SWI2/SNF2 has on chromatin surrounding the promoter for *SUC*, the gene that encodes invertase (the enzyme that hydrolyzes sucrose to fructose and glucose). This chromatin region is resistant to micrococcal nuclease cleavage in wild type cells that have not been induced for invertase production but becomes sensitive to the enzyme after induction. In contrast, this same chromatin region resists nuclease cleavage both before and after the *SWI2/SNF2* mutants are subjected to inducing conditions. The interpretation of these experiments is that wild type cells reorganize nucleosomes in the chromatin around the *SUC* promoter in response to inducing conditions but *SWI2/SNF2* mutants do not. The *SWI2/SNF2* gene product was later isolated from yeast as part of a multisubunit protein complex that uses the energy from ATP hydrolysis to remodel chromatin. The Swi/Snf complex is present at relatively low concentrations within yeast cells (approximately 100 copies per cell). Consistent with this low intracellular concentration, only about 5% of yeast genes require the Swi/Snf complex

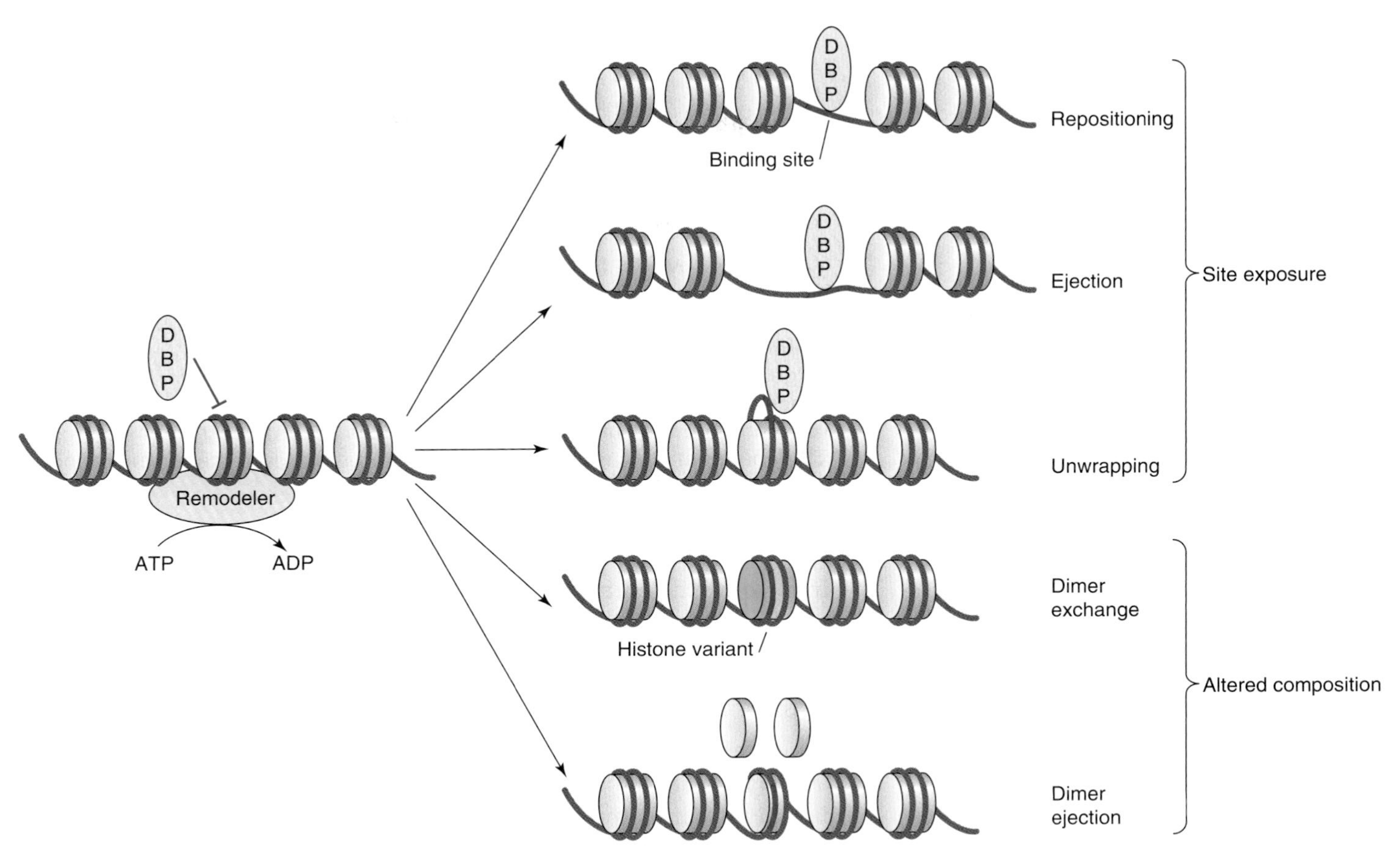

FIGURE 18.57 **Biochemical activities of ATP-dependent chromatin remodeling complexes.** The ATP-dependent chromatin remodeler (green) can reposition nucleosomes, eject nucleosomes, unwrap nucleosomes, exchange dimers, and eject dimers. The net effect is that a DNA site, which was originally covered, becomes available to bind by a DNA binding protein (DBP) or the nucleosome composition is altered by replacing an H2A•H2B dimer with a histone variant or by ejecting an H2A•H2B dimer. (Reproduced from C. R. Clapier and B. R. Cairns, *Annu. Rev. Microbiol.* 78 [2009]: 273–304. Copyright 2009 by Annual Reviews, Inc. Reproduced with permission of Annual Reviews, Inc., in the format Textbook via Copyright Clearance Center.)

for activation during unsynchronized growth. Furthermore, mutants that lack this complex are viable. Because chromatin structure has a generally repressive effect on transcription, these findings suggest that yeast must have other complexes that remodel chromatin.

In 1996, Roger Kornberg and coworkers isolated another protein complex from yeast that can remodel the structure of chromatin and named it the **RSC complex**. Some of the protein subunits in the RSC complex are homologous to subunits in the Swi/Snf complex, while others are not. The two complexes are assigned to the same chromatin remodeling family because their ATPase subunits are similar. This family is called the SWI/SNF family after its founding member. RSC is essential for cell viability and its intracellular concentration is about tenfold greater than that of the SWI/SNF complex. The ATPase subunit in RSC, Sth1, can act on its own to unwrap DNA from nucleosomes but does so at a much slower rate than when it is part of the RSC complex. Cryo-electron microscopy studies reveal that a nucleosome

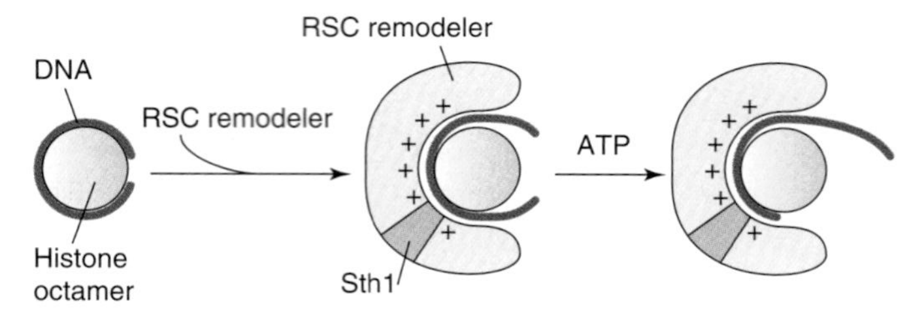

FIGURE 18.58 Model for RSC catalyzed nucleosome translocation. The histone octamer is shown as a tan disk and DNA as a solid blue line. The RSC remodeler is shown as a yellow structure with an interior cavity that is assumed to have a positive charge. The nucleosome fits into the positively charged RSC complex cavity, which somehow frees DNA from the histone surface. The RSC remodeler's Sth1 subunit contacts one DNA strand about two turns from the nucleosome's twofold axis of symmetry and catalyzes ATP-dependent DNA movement by screw rotation. (Modified from Y. Lorch, et al. *Proc Natl Acad Sci USA* 107 [2010]: 3458–3462.)

binds within a central cavity in RSC. DNase sensitivity studies show that this binding diminishes the interaction between the histone core and almost all the DNA that winds around it. Based on these and related findings, Yahli Lorch and coworkers have proposed the model for RSC catalyzed nucleosome translocation shown in FIGURE 18.58. According to this model, the nucleosome fits into the RSC cavity, which is assumed to have a positively charged surface, and the resulting interaction somehow frees DNA from the histone surface. The complex's Sth1 subunit contacts one DNA strand about two turns from the nucleosome's twofold axis of symmetry and catalyzes ATP-dependent DNA movement by a screw rotation mechanism.

The ChIP technique has been used to determine the locations of nucleosomes and nucleosome variants along the whole genome of various organisms and cell types. Studies in yeast reveal different nucleosome distribution patterns for constitutive and regulated genes. The following discussion is limited to yeast, but the lessons learned apply to other eukaryotes.

Constitutive gene—The core promoter is in a **nucleosome depleted region (NDR)** of about 150 bp adjacent to the transcription start site (FIGURE 18.59a). The reason that this nucleotide stretch is depleted of nucleosome is as follows. DNA in the NDR has A-tracts (deoxyadenosine nucleotide tracts that are 10–20 bp or even longer) on one strand of double-stranded DNA that resist sharp bending. This resistance prevents the DNA curvature that favors nucleosome formation. The –1 nucleosome (first nucleosome upstream of the NDR) usually has a variant of the H2A histone called **H2A.Z**. The same is true of the +1 nucleosome (first nucleosome downstream of the NDR). The H2A.Z variant histone decreases nucleosome stability, which probably makes it easier for the chromatin remodeling complex to work. The chromatin remodeling complex, however, is probably not needed to expose the first transcription activator binding site because the activator usually can bind to an exposed site in the NDR. The core promoter usually does not contain a TATA element.

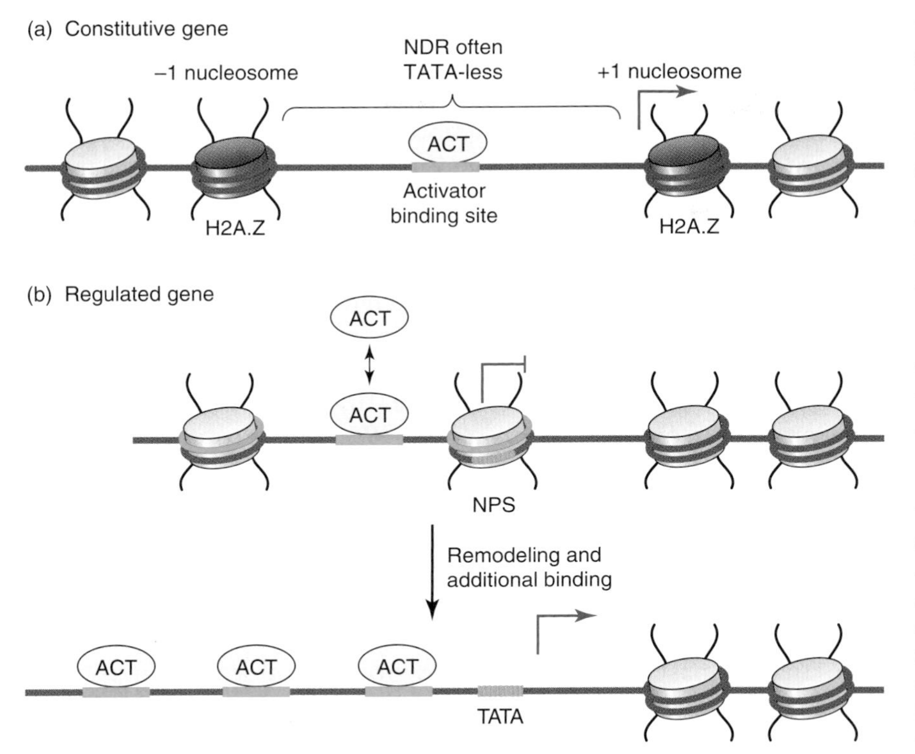

FIGURE 18.59 **Chromatin properties in constitutive and regulated yeast genes.** (a) Constitutive gene—A nucleosome depleted region (NDR) stretches about 150 nucleotides upstream from a position slightly before the transcription start site. The –1 nucleosome (first nucleosome upstream of the NDR) usually has a variant of the H2A histone called H2A.Z. The same is true of the +1 nucleosome (first nucleosome downstream of the NDR). The transcription activator (ACT) binds to an uncovered binding site in the NDR. The core promoter usually does not contain a TATA element. (b) Regulated gene—Most transcription activator binding sites (BS) are also covered by nucleosomes. The nucleosome positioning sequence (see text) is indicated by NPS. However, at least one transcription activator binding site is usually exposed at the nucleosome edge or in a linker between nucleosomes. A "pioneer" transcription activator can gain access to this site but chromatin modification enzymes and remodelers expose additional sites covered by nucleosomes. The core promoter contains a TATA element that becomes accessible to TFIID after the element is uncovered. Most yeast genes are not a perfect fit for either model but instead combine features of the two models to provide suitable regulation. (Adapted from B. R. Cairns, *Nature* 461 [2009]: 193–198.)

Regulated gene—The transcription start site is usually located in, and flanked by, 150-bp **nucleosome positioning sequence** (**NPS**) elements (FIGURE 18.59b). Each NPS element has AA/TT dinucleotide repeats every 10 bp with GC dinucleotides 5 bp out of phase. This nucleotide pattern allows NPS elements to bend sharply, producing DNA curvature that favors nucleosome formation. Therefore, when regulated yeast genes are in their repressed state, their transcription start site and regions flanking their transcription start site are covered by nucleosomes (Figure 18.59b). Although most transcription activator binding sites are also covered by nucleosomes, at least one activator binding site is usually exposed at the nucleosome edge or in a linker between nucleosomes. A "pioneer" transcription factor can gain access to this site but chromatin modification enzymes and remodelers are still needed to expose additional sites covered by nucleosomes. The core promoter contains a TATA element that becomes accessible to TFIID after the element is uncovered.

Most yeast genes are not a perfect fit for either model but instead combine features of the two models to provide suitable regulation.

Histone Modifiers

Pioneering investigations of Vincent Allfrey and coworkers in 1964 were the first to call attention to the fact that active genes have acetylated histones. Subsequent studies revealed that acetylation takes place only at specific lysine residues located in the amino terminal tails of histones. These tails are not required for nucleosomal structural integrity but do contribute to the folding of nucleosome arrays into chromatin. The current hypothesis is that histone tail acetylation prevents interactions that lead to more compact chromatin structures that are repressive to transcription. Eukaryotes have both nuclear and cytoplasmic histone acetyltransferase (HAT) activities. Cytoplasmic HATs appear to modify histones prior to the histone's assembly into nucleosomes on newly replicated DNA, whereas nuclear HATs modify histones that are already part of the chromatin structure to make DNA accessible to the transcription machinery. We therefore focus on the nuclear HATs.

C. David Allis and coworkers identified and isolated the first nuclear HAT from *Tetrahymena* in 1996. This protein's sequence turned out to be remarkably similar to that predicted for the product of the yeast *GCN5* gene, which was first identified in a screen for mutants that cannot grow under amino acid limiting conditions. The sequence similarity suggested that Gcn5 is also a histone acetyltransferase. This hypothesis was confirmed when purified Gcn5 was shown to acetylate free histones at specific lysines. However, Gcn5 cannot acetylate histones that are part of a nucleosome. A biochemical search for proteins that contain Gcn5 and do acetylate histones in nucleosomes resulted in the isolation of the Ada and SAGA protein complexes that can do so. Eukaryotic cells also have histone deacetylases (HDAC) that remove acetyl groups from histone N-terminal tails. Deacetylation helps to stabilize the compact chromatin structure, causing the transcription of the affected DNA to be repressed. At least six different protein complexes identified in yeast have histone deacetylase activity. The two major deacetylase complexes, HDA and HDB, have been purified. Corresponding catalytic subunits in each complex, Hda1 and Rpd3 (reduced potassium dependency), share a considerable degree of sequence homology. Higher animals and plants have similar histone deacetylases.

Histones are subject to several other kinds of covalent modification that influence gene activity (**FIGURE 18.60**). In fact, it is rather remarkable that while histones are among the most highly conserved proteins in nature, they are also subject to more different types of post-translational modifications than most other proteins. Histone kinases add phosphate groups to specific serine or threonine residues; histone methyl transferases add methyl groups to specific lysine and arginine residues; and ubiquitin-protein ligase attaches a ubiquitin group to lysine. A single lysine residue may have up to three attached methyl groups and a single arginine residue may have one or two attached methyl groups. Enzymes and specific protein factors interact with chromatin based on the positions of the methylated residues and the number of methyl groups that are attached to them. Protein phosphatases remove phosphate groups and demethylases remove methyl groups. Some enzymes that modify histones also modify other proteins, making it difficult to determine biological function from mutant studies. Histone modifying enzymes introduce marks on the histones that are recognized by specific domains or subunits in chromatin remodelers.

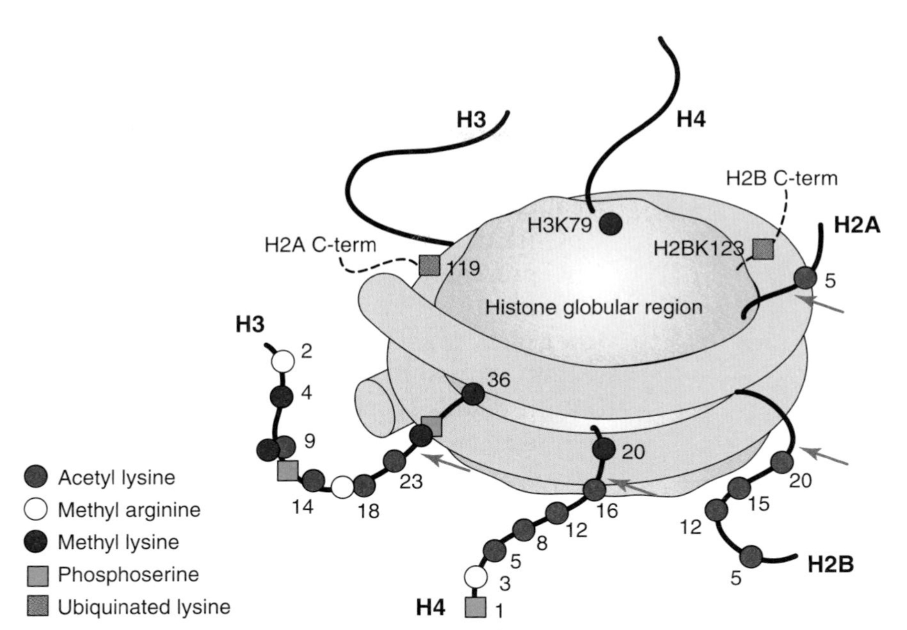

FIGURE 18.60 **Histone modifications on the nucleosome core particle.** The graphic represents a schematic of a nucleosome core particle. For simplicity, modifications are shown on only one of the two copies of histones H3 and H4 and only one N-terminal tail is shown for histones H2A and H2B. The C-terminal tails of one histone H2A molecule and one histone H2B molecule are also shown (dashed lines). Colored symbols indicate histone sites that can be modified after a histone is synthesized. Residue numbers are given for each modification. Lysine-9 in histone H3 can be either acetylated or methylated. The small protein ubiquitin is added to histone H2A at lysine-119 (most common in mammals) and to histone H2B at lysine-123 (most common in yeast). Sites marked by green arrows are susceptible to cutting by trypsin in intact nucleosomes. The schematic summarizes data from many different organisms and so any given organism may lack particular modifications. (Modified from *Cell*, vol. 111, B. M. Turner, Cellular Memory and the Histone Code, pp. 285–291, copyright 2002, with permission from Elsevier [http://www.sciencedirect.com/science/journal/00928674].)

For example, chromatin remodeling complexes with a **bromodomain** bind histones that have an acetyl-lysine and those with a **chromodomain** bind to histones with a methyl-lysine.

The function of remodeling and modifying complexes does not end with the formation of a pre-initiation complex poised to initiate transcription. RNA polymerase II also must contend with chromatin structure as it moves along a gene. The presence of nucleosomes in a DNA coding region slows down transcription. Eukaryotic cells use at least two pathways to deal with this problem. (1) Chromatin-remodeling complexes respond to the advancing transcription machinery by displacing histones from DNA. Displaced histones bind to chaperones and reassemble to form nucleosomes after the DNA region has been transcribed. (2) The transcription machinery appears to be able to move through a nucleosome after only one H2A•H2B heterodimer has been removed. The displaced H2A•H2B heterodimer binds to a histone chaperone, which adds the heterodimer back to the nucleosome after transcription. Further investigations are required to establish the relative contribution that each pathway makes to transcription elongation in chromatin and to determine if other pathways also are needed to permit the transcription elongation in chromatin.

DNA methylation plays an important role in determining whether chromatin will be silenced or actively expressed in vertebrates.

Recall from Chapter 17 that many core promoters in vertebrates are located in CpG islands (CG-rich genomic regions that contain a high frequency of CG dinucleotides). The methylation of cytosine in these CpG islands leads to gene silencing by two mechanisms: (1) methyl-CpG prevents a transcription factor from binding to its cognate site or (2) methyl-CpG acts as a signal for histone modification. Although a few genes are silenced by the first mechanism, most are silenced by the second. Adrian Bird and coworkers took an important step in working out the details of methyl-CpG signaling in 1992 when they identified two nuclear proteins that bind to methyl-CpG. Other proteins were later discovered that also bind to methyl-CpG. For simplicity, we focus attention on MeCP2, one of the two methyl-CpG-binding proteins originally identified by Bird and coworkers. MeCP2 is a single polypeptide with a **methyl-CpG binding domain** (**MBD**) at its N-terminal domain and a **transcription repression domain** (**TRD**). The TRD binds to SIN3A, a transcriptional co-repressor, which then serves as a bridge to histone deacetylase. The resulting histone deacetylase complex silences the gene by increasing chromatin compaction.

Epigenetics is the study of inherited changes in phenotype caused by changes in chromatin other than changes in DNA sequence.

Histone modification patterns within specific chromosome regions appear to be heritable. A heritable code involving covalently modified amino acid residues in histones represents a significant departure from the classical notion that all the information for determining phenotype is stored in the DNA base sequence. In essence, the heritable code hypothesis proposes that daughter cells not only have the same DNA base sequence as the parent cell, but that each chromatin region within the daughter cell has the same DNA methylation pattern and combination of covalently modified histone residues as was present in the parent cell. Because the modified histone molecules determine which DNA segments will be active and which will be silent, this is a form of inheritance that is not directly determined by DNA sequence. We use the term **epigenetics** to describe inherited changes in phenotype that are caused by changes in the chromosome other than changes in DNA sequence.

Different cells and tissues in multicellular organisms acquire different programs for gene expression during development. These programs appear to be largely determined by epigenetic modifications resulting from DNA methylation and histone modifications. Thus, each cell type has unique epigenetic features that depend on its genotype, developmental history, and environmental influences. Epigenetic features tend to become fixed after cell differentiation. For instance, muscle cells retain epigenetic features of muscle cells after cell division. Some cells, however, undergo major epigenetic reprogramming during normal development.

Genomic imprinting in mammals determines whether a maternal or paternal gene will be expressed.

Mammalian sexual reproduction produces diploid offspring with one copy of each gene inherited from each parent. Classical genetics predicts that maternal and paternal alleles should function equally well. Although this prediction holds true for nearly all genes, there are about 100 known exceptions in mice and a similar number in other mammals. Asymmetrically expressed genes tend to be organized in clusters that are located on only a few chromosomes. In some cases, the maternal allele is transcribed and the paternal allele is silent and in others the reverse is true. This type of asymmetric allelic expression has been shown to be due to the fact that the allele inherited from each parent has a characteristic methylation pattern, which is generally stable in differentiated somatic cells but reprogrammed (demethylated and remethylated) during the development of germ cells and preimplantation embryos. The important point is that the methylation pattern or **genomic imprinting** allows the transcription machinery to distinguish homologous chromosomal regions based on parental origin. The best-characterized example of genomic imprinting is probably a genetic mouse cluster that contains the paternally expressed gene *Igf2* (insulin-like growth factor 2) and the maternally expressed gene *H19* (FIGURE 18.61). Both genes are regulated by (1) an *ICR* (imprinting center region) located between them and (2) shared enhancers downstream of *H19*. The *ICR* is methylated in the paternal cluster and unmethylated in the maternal cluster. A protein called CTCF binds to the unmethylated maternal *ICR*, blocking transcription activators bound to the enhancers from stimulating transcription at *Igf2* but permitting them to stimulate transcription at *H19*. The situation is quite different for the paternal cluster. The methylated *ICR* does not allow CTCF to bind to it and the methylated *H19* promoter

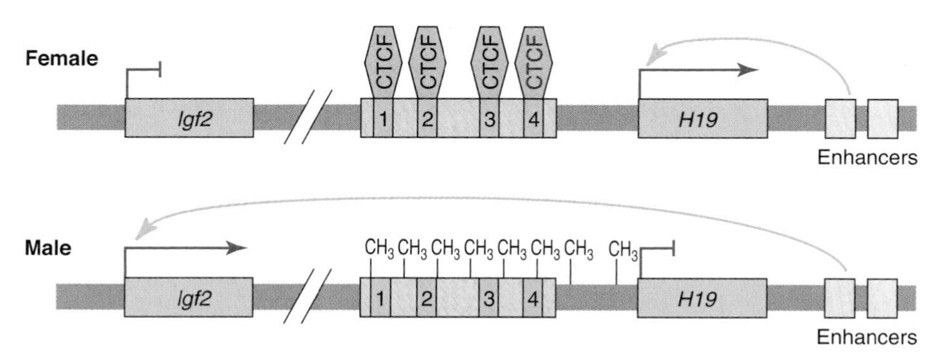

FIGURE 18.61 **Transcription regulation by imprinting at mouse *H19* and *Igf2* genes.** *H19* and *Igf2* are regulated by (1) an *ICR* (imprinting center region) located between them (shown in green) and (2) shared enhancers downstream of *H19*. The *ICR* is methylated in the paternal cluster and unmethylated in the maternal cluster. A protein called CTCF binds to the unmethylated maternal *ICR*, blocking transcription activators bound to the enhancers from stimulating transcription at *Igf2* but permitting them to stimulate transcription at *H19*. The situation is quite different for the paternal cluster. The methylated *ICR* does not allow CTCF to bind to it and the methylated *H19* promoter prevents the transcription machinery from acting on *H19*. However, the transcription activators bound to the enhancers can stimulate *Igf2* transcription. (Modified from *Mutat. Res.*, vol. 647, F. Y. Ideraabdullah, S. Vigneau, and M. S. Bartolomei, Genomic imprinting mechanisms in mammals, pp. 77–85, copyright 2008, with permission from Elsevier [http://www.sciencedirect.com/science/journal/00275107].)

prevents the transcription machinery from acting on *H19*. However, the transcription activators bound to the enhancers can stimulate *Igf2* transcription. The discovery of genome imprinting solves a very puzzling mammalian reproduction problem. Despite considerable effort, investigators have not been able to obtain a viable organism from egg cells that contain only paternal chromosomes or only maternal chromosomes. Genomic imprinting studies indicate that diploid eggs with chromosomes derived from just one parent cannot develop normally because some asymmetrically expressed genes are essential for normal development. Consistent with this hypothesis, several inborn errors of metabolism have been shown to be caused by genes that are regulated by genomic imprinting. Inheritance of these metabolic disorders depends on whether the altered allele is in a paternal or maternal chromosome.

Pluripotent cells usually become more specialized during development.

Embryonic stem (ES) cells, which are derived from the inner cell mass of the mammalian blastocyst are **pluripotent**. That is, ES cells can self-renew indefinitely and differentiate into all cell types of the three germ layers (ectoderm, endoderm, and mesoderm). As embryonic development continues, ES cells are programmed to become more specialized. Some cells retain a more limited ability to differentiate. **Multipotent stem cells** can differentiate into multiple cell types within a certain cell lineage. For example, neural stem cells can differentiate to form neurons, astrocytes, and oligodendrocytes. **Unipotent stem cells** can differentiate into only one type of cell. With very rare exceptions, the DNA sequence is not altered during differentiation; therefore, kidney, heart, or liver cells from the same animal will have the same DNA sequence. Each different kind of specialized cell will, however, have epigenetic features that are unique to it. Considerable effort is now being devoted to understanding the programming process that converts a pluripotent or multipotent stem cell into a specific specialized cell. In principle, human ES cell lines can serve as donor sources for human transplantation therapies to treat illnesses such as type I diabetes, spinal cord injury, heart disease, burns, and Parkinson's disease. However, immune rejection remains a serious limitation to clinical applications. Even if all the technical issues could be solved, we would still be faced with a politically and ethically charged issue because many people oppose using human embryonic stem cells. They believe that life begins at the moment of conception and extracting embryonic stem cells causes the loss of a human life. One possible approach to solving immune rejection and political problems is to reprogram the nuclei of differentiated cells to return them to an ES-like pluripotential state.

A cell nucleus from a terminally differentiated cell can be reprogrammed by an enucleated egg cell to produce a live animal.

In the early 1960s John B. Gurdon and coworkers transferred nuclei isolated from tadpole intestinal cells into enucleated frog eggs to produce tadpoles. Although development stopped at the tadpole stage, these nuclear transfer experiments showed for the first time that genes required for normal development can be activated from a specialized cell. A major breakthrough occurred in 1996 when Ian Wilmut and Keith Campbell showed that a nucleus derived from a sheep mammary gland could program an enucleated egg cell to produce a live animal (FIGURE 18.62). More specifically, they cultured cells from the mammary gland of an adult sheep *in vitro* under conditions that caused the cells to enter the G_0 (quiescent) stage. Then they (1) used an electric pulse to fuse the quiescent cell with an enucleated egg cell obtained from a different sheep, (2) cultured the fused cell *in vitro* to allow it reach an early embryonic stage, and (3) transferred the embryo into the uterus of a surrogate. The birth of a live sheep named Dolly, with all of the chromosomes from the donor nucleus and none from the host egg cell, conclusively demonstrated that mature adult cell nuclei can be reprogrammed to produce a pluripotent cell. With the exception

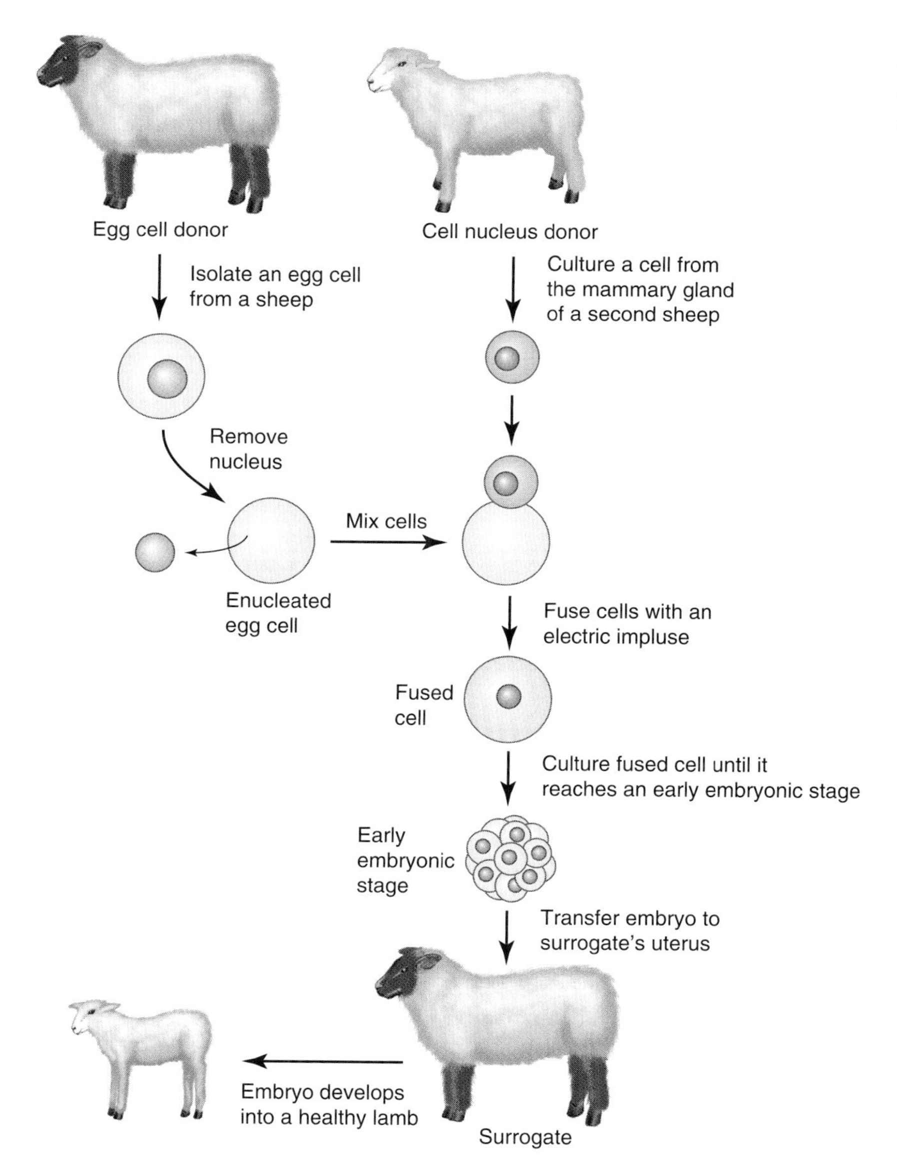

FIGURE 18.62 First nuclear transfer experiment to create a cloned sheep called Dolly. (Adapted from K. R. Miller and J. S. Levine. *Biology*. Prentice Hall, 2003.)

of its mitochondrial DNA, Dolly was an exact DNA copy of the sheep that provided the mammary gland cell. Similar types of nuclear transfer experiments have now been performed with a wide variety of mammals. These experiments demonstrate that factors in the egg cell cytoplasm can reactivate the genes in a terminally differentiated cell nucleus that are necessary for embryo formation and subsequent animal development.

Lineage restricted cells can be programmed to produce induced pluripotent stem (iPS) cells.

The somatic cell nuclear transfer experiments suggested that it might be possible to rejuvenate a terminally differentiated cell with specific factors. The problem was to select the right combination of factors and devise some method to introduce the factors into the cell. In 2006, Kazutoshi Takahashi and Shinya Yamanaka thought they might be able to rejuvenate mouse skin cells by introducing genes that are normally active only in ES cells. They began by identifying 24 genes that are turned on in pluripotent cells but silent in adult cells. When they introduced these genes into mouse skin cells using retroviruses to deliver them, the skin cells were reprogrammed into pluripotent stem cells. Further experiments showed that only four genes are needed to produce **induced pluripotent stem (iPS) cells**. The four genes, *Oct4*, *Sox2*, *Klf4*, and *c-Myc*, each codes for a transcription factor. One year later Takahashi, Yamanaka, and coworkers applied the lessons learned from their mouse cell studies to human cells (FIGURE 18.63). They used retroviral vectors to introduce human *Oct4*, *Sox2*, *Klf4*, and *c-Myc* into human fibroblasts and then cultured the

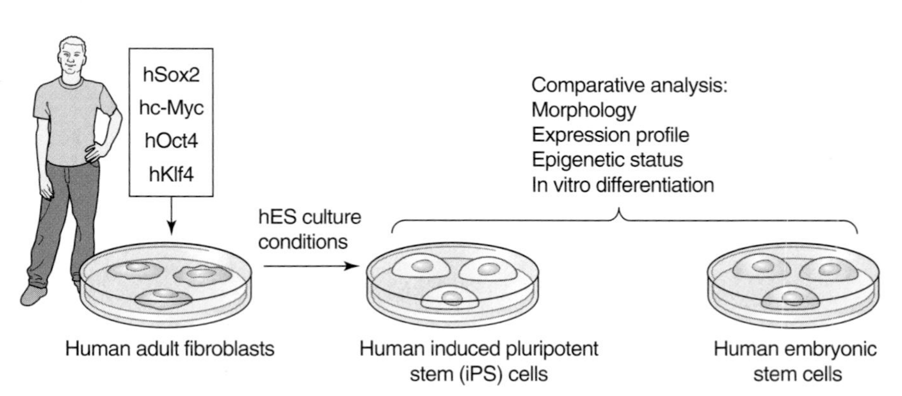

FIGURE 18.63 **Transcription factor-induced pluripotency.** Retroviral vectors introduced *Oct4*, *Sox2*, *Klf4*, and *c-Myc* genes into human fibroblasts and then cultured the transduced cells. After 30 days, the culture plates were covered with iPS cell colonies. Several tests were used to compare the human iPS cells with human ES cells. The iPS and ES cells had similar morphologies, surface marker expression, epigenetic status, and global gene expression. (Reproduced from *Cell*, vol. 131, H. Zaehres and H. R. Schöler, Induction of Pluripotency: From Mouse to Human, pp. 834–835, copyright 2007, with permission from Elsevier [http://www.sciencedirect.com/science/journal/00928674].)

transduced cells. After 30 days, the culture plates were covered with iPS cell colonies. Takahashi, Yamanaka, and coworkers compared the human iPS cells with human ES cells in several tests. The iPS and ES cells had similar morphologies, surface marker expression, epigenetic status, and global gene expression. Moreover, they both were able to form embryoid bodies *in vitro* and both directed differentiation into neural crest cells and beating heart cells. The induction technique developed by Takhashi, Yamanka, and coworkers has a serious drawback for clinical applications. When transplanted, cells transduced with retrovirus tumors have a tendency to form tumors. Other means of gene transfer, therefore, are required. Even though only a short time has passed since the first report by Takahashi and Yamanaka, investigators have devised several new methods to prepare iPS cells. Most notably, the retrovirus vector has been replaced by safer vectors and techniques have been devised to introduce transcription factors directly into cells. Even more remarkable, small drugs have been discovered that can replace the transcription factors. Douglas A Melton and coworkers have demonstrated that valproic acid, a histone deacetylase inhibitor, allows primary fibroblasts to be reprogrammed with just two transcription factors, *Oct4* and *Sox2*. iPS cell technology is opening up a vast new frontier in medicine. We can look forward to many exciting new discoveries and clinical applications.

Suggested Reading

Regulatory Promoter and Enhancer

Blackwood, E. M., and Kadonaga, J. T. 1998. Going the distance: a current view of enhancer action. *Science* 281:60–63.

Kadonaga, J. T., and Tjian, R. 1986. Affinity purification of sequence-specific DNA binding proteins. *Proc Natl Acad Sci USA* 83:5889–5893.

Kozian, D. H., and Kirschbaum, B. J. 1999. Comparative gene-expression analysis. *Trends Biotechnol* 17:73–78.

Lekstrom-Himes, J., and Xanthopoulos, K. G. 1998. Biological role of the CCAAT/enhancer-binding protein family of transcription factors. *J Biol Chem* 273:28545–28548.

Maston, G. A., Evans, S. K., and Green, M. R. 2006. Transcription regulatory elements in human genome. *Ann Rev Genomics Hum Genet* 7:29–59.

Sambrook, J., and Russell, D. W. 2001. *Molecular Cloning—A Laboratory Manual*, 3rd ed. Woodbury, NY: Cold Spring Harbor Press.

Szutorisz, H., Dillon, N., and Tora, L. 2005. The role of enhancers as centres for general transcription factor recruitment. *Trends Biochem Sci* 30:593–599.

Werner, T. 1999. Models for prediction and recognition of eukaryotic promoters. *Mamm Genome* 10:168–175.

Transcription Activator Proteins

Barberis, A., and Petrascheck, M. 2003. Transcription activation in eukaryotic cells. In: *Encyclopedia of Life Sciences*. pp. 1–7. London: Nature.

Geiduschek, E. P., and Ouhammouch, M. 2005. Archaeal transcription and its regulators. *Mol Microbiol* 56:1397–1407.

Lawler, J. F., Hyland, E. M., and Boeke, J. D. 2008. Genetic engineering: reporter genes. In: *Encyclopedia of Life Sciences*. pp. 1–8. Hoboken, NJ: John Wiley and Sons.

Ma, J. 2005. Crossing the line between activation and repression. *Trends Genet* 21:54–59.

Reece, R. J., and Platt, A. 1997. Signaling activation and repression of RNA polymerase II transcription in yeast. *Bioessays* 19:1001–1010.

Warren, A. J. 2002. Eukaryotic transcription factors. *Curr Opin Struct Biol* 12:107–114.

Ziros, P., and Papavassiliou, A. G. 2001. Eukaryotic gene regulation: a flew over the transcription factor nest. *Haema* 4:15–23.

DNA-Binding Domains

Altucci, L., and Gronemeyer, H. 2001. Nuclear receptors in cell life and death. *Trends Endocrinol Metab* 17:460–468.

Bourguet, W., Germain, P., and Gronemeyer, H. 2000. Nuclear receptor ligand-binding domains: three-dimensional structures, molecular interactions and pharmacological implications. *Trends Pharmacol Sci* 21:381–388.

Chambon, P. 2004. How I became one of the fathers of a superfamily. *Nat Med* 10:1027–1031.

Chariot, A., Gielen, J., Mreville, M.-P., and Bours, V. 1999. The homeodomain-containing proteins. *Biochem Pharmacol* 58:1851–1857.

Chinenov, Y., and Kerppola, T. K. 2001. Close encounters of many kinds: Fos-Jun interactions that mediate transcription regulatory specificity. *Oncogene* 20:2438–2452.

Ellenberger, T. E., Brandl, C. J., Struhl, K., and Harrison, S. C. 1992. The GCN4 basic region leucine zipper binds DNA as a dimer of uninterrupted alpha helices: crystal structure of the protein-DNA complex. *Cell* 71:1223–1237.

Foley, K. P., and Eisenman, R. N. 1999. Two mad tails: what the recent knockouts of *Mad1* and *Mxi1* tells us about the *MYC/MAX/MAD* network. *Biochim Biophys Acta* 1423:M37–M47.

Fraenkel, E., Rould, M. A., Chambers, K. A., and Pabo, C. O. 1998. Engrailed homeodomain-DNA complex at 2.2 Å resolution: a detailed view of the interface and comparison with other engrailed structures. *J Mol Biol* 284:351–361.

Glass, C. K., and Rosenfeld, M. G. 2000. The coregulator exchange in transcriptional functions of nuclear receptors. *Genes Dev* 14:121–141.

Grandori, C., Cowley, S. M., James, L. P., and Eisenman, R. N. 2000. The Myc/Max/Mad network and the transcriptional control of cell behavior. *Ann Rev Cell Biol* 16:653–699.

Hall, J. M., Course, J. F., and Korach, K. S. 2001. The multifaceted mechanisms of estradiol and estrogen receptor signaling. *J Biol Chem* 276:36869–36872.

Hard, T., Kellenback, E., Boelens, R., et al. 1990. Solution structure of the glucocorticoid receptor DNA-binding domain. *Science* 249:157–160.

Hinnebusch, A. G., and Natarajan, K. 2002. Gcn4p, a master regulator of gene expression, is controlled at multiple levels of diverse signals of starvation and stress. *Eukaryot Cell* 1:22–32.

Kappen, C. 2000. The homeodomain: an ancient and evolutionary motif in animals and plants. *Comput Chem* 24:95–103.

Kitanyer, M., Rozenberg, H., Kesller, N., et al. 2006. Structural basis of DNA recognition by p53 tetramers. *Mol Cell* 22:741–753.

Krishna, S. S., Majumdar, I., and Grishin, N. V. 2003. Structural classification of zinc fingers. *Nucl Acid Res* 31:532–550.

Joerger, A. C., and Fersht, A. R. 2008. Structural biology of the tumor suppressor p53. *Ann Rev Biochem* 77:557–582.

Laity, J. H., Lee, B. M., and Wright, P. E. 2001. Zinc finger proteins: new insights into structural and functional diversity. *Curr Opin Struct Biol* 11:39–46.

Landschulz, W. H., Johnson, P. F., and McKnight, S. L. L. 1988. The leucine zipper: a hypothetical structure common to a new class of DNA binding proteins. *Science* 240:1759–1764.

Laughon, A., and Scott, M. P. 1984. Sequence of a *Drosophila* segmentation gene: protein structure homology with DNA-binding proteins. *Nature* 310:25–31.

Lee, K. C., and Kraus, W. L. 2001. Nuclear receptors, coactivators, and chromatin: new approaches, new insights. *Trends Endocrinol Metab* 12:191–197.

Lee, M. S., Gippert, G. P., Soman, K. V., et al. 1989. Three-dimensional solution structure of a single zinc finger DNA-binding domain. *Science* 245:635–637.

Lewis, E. B. 1978. A gene complex controlling segmentation in *Drosophila*. *Nature* 276:565–570.

Luisi, B. F., Xu, W. X., Otwinowski, Z., et al. 1991. Crystallographic analysis of the interaction of the glucocorticoid receptor with DNA. *Nature* 352:497–505.

Luscombe, N. M., Austin, S. E., Berman, H. M., and Thornton, J. M. 2000. An overview of protein-DNA complexes. *Genome Biol* 1:1–37.

Ma, P. C., Rould, M. A., Weintraub, H., and Pabo, C. O. 1994. Crystal structure of MyoD bHLH domain-DNA complex: perspectives on DNA recognition and implications for transcriptional activation. *Cell* 77:451–459.

Marmorstein, R., Carey, M., Ptashne, M., and Harrison, S. C. 1992. DNA recognition by GAL4: structure of a protein-DNA complex. *Nature* 356:379–380.

McGinnis, W., Garber, R. L., Wirz, J., et al. 1984. A homologous protein-coding sequence in *Drosophila* homeotic genes and its conservation in other metazoans. *Cell* 37:403–408.

McGinnis W., Levine M. S., Hafen, E., et al. 1984. A conserved sequence in homeotic genes of the *Drosophila antennapedia* and bithorax complexes. *Nature* 308:428–433.

McKenna, N. J., and O'Malley, B. W. 2002. Combinatorial control of gene expression by nuclear receptors and coregulators. *Cell* 108:465–474.

Nair, S. K., and Burley, S. K. 2003. X-ray structures of Myc-Max and Mad-Max recognizing DNA: molecular bases of regulation by proto-oncogenic transcription factors. *Cell* 112:193–205.

Natarajan, K., Meyer, M. R., Jackson, B. M., et al. 2001. Transcriptional profiling shows that Gcn4p is a master regulator of gene expression during amino acid starvation in yeast. *Mol Cell Biol* 21:4347–4368.

Okorokov, A. L. M. and Orlova, E. V. 2009. Structural biology of the p53 tumor suppressor. *Curr Opin Struct Biol* 19:197–202.

O'Shea, E. K., Klemm, J. D., Kim, P. S, and Alber, T. 1991. X-ray structure of the GCN4 leucine zipper, a two-stranded, parallel coiled coil. *Science* 254:539–544.

Pabo, C. O., Peisach, E., and Grant, R. A. 2001. Design and selection of novel Cys_2His_2 zinc finger proteins. *Ann Rev Biochem* 70:313–340.

Patikoglou, G., and Burley, S. K. 1997. Eukaryotic transcription factor-DNA complexes. *Ann Rev Biophys Biomol Struct* 26:289–325.

Phillips, K., and Luisi, B. 2000. The virtuoso of versatility: POU proteins that flex to fit. *J Mol Biol* 302:1023–1039.

Rastinejad, F. 2001. Retinoid X receptor and its partners in the nuclear receptor family. *Curr Opin Struct Biol* 11:33–38.

Reece, R. J., and Ptashne, M. 1993. Determinants of binding-site specificity among yeast C_6 zinc cluster proteins. *Science* 261:909–911.

Ribeiro, R. C. J., Kushner, P. J., and Baxter, J. D. 1995. The nuclear hormone receptor gene superfamily. *Ann Rev Med* 46:443–453.

Riley, T., Sontag, E., Chen, P., and Levine, A. 2008. Transcriptional control of human p53-regulated genes. *Nat Rev Mol Cell Biol* 9:402–412.

Robyr, D., Wolffe, A.P., and Wahli, W. 2000. Nuclear hormone receptor coregulators in action: diversity for shared tasks. *Mol Endocrinol* 14:329–347.

Rosenfeld, M. G., and Glass, C. K. 2001. Coregulator codes of transcriptional regulation by nuclear receptors. *J Biol Chem* 276:36865–36868.

van Dam, H., and Castellazzi, M. 2001. Distinct roles of Jun:Fos and Jun:ATF dimers in oncogenesis. *Oncogene* 20:2453–2464.

Veraksa, A., Del Campo, M., and McGinnis, W. 2000. Developmental patterning genes and their conserved functions: from model organisms to humans. *Mol Genet Metab* 69:85–100.

Vogt, P. K. 2001. Jun, the oncoprotein. *Oncogene* 20:2365–2377.

Wolfe, S. A., Nekludova, L., and Pabo, C. O. 1999. DNA recognition by Cys_2His_2 zinc finger proteins. *Ann Rev Biophys Biomol Struct* 3:183–212.

Xie, W., and Evans, R. M. 2001. Orphan nuclear receptors: the exotics of xenobiotics. *J Biol Chem* 276:37739–37742.
Yen, P. M. 2001. Physiological and molecular basis of thyroid hormone action. *Physiol Rev* 81:1097–1142.
Zhou, Z.-Q., and Hurlin, P. J. 2001. The interplay between Mad and Myc in proliferation and differentiation. *Trends Cell Biol* 11:S10–S14.

Activation Domain

Brent, R. and Finley, R. L. Jr. 1997. Understanding gene and allele function with two-hybrid methods. *Ann Rev Biochem* 31:663–704.
Breunig, K. D. 2000. Regulation of transcription activation by Gal4p. *Food Technol Biotechnol* 38:287–293.
Cagney, G, Uetz, P., and Fields, S. 2000. High-throughput screening for protein-protein interactions using the two-hybrid assay. *Methods Enzymol* 328:3–14.
Jiang, F., Frey, B. R., Evans, M. L., Friel, J. C., and Hopper, J. E. 2009. Gene activation by dissociation of an inhibitor from a transcriptional activation domain. *Mol Cell Biol* 29:5604–5610.
Ma, J., and Ptashne, M. 1987. Deletion analysis of GAL4 defines two transcriptional activating segments. *Cell* 48:847–853.
Pilauri, V., Bewley, M., Diep, C., and Hopper, J. 2005. Gal80 dimerization and the yeast *GAL* gene switch. *Genetics* 169:1903–1914.
Struhl, K. 1995. Yeast transcriptional regulatory mechanisms. *Ann Rev Genet* 29:651–674.
Uetz, P. H., Hughes, R. E., and Fields, S. 1998. The two-hybrid system: finding likely partners for lonely proteins. *Focus* 20:62–64.

Mediator

Biddick, R., and Young, E. T. 2005. Yeast Mediator and its role in transcriptional regulation. *C R Biol* 328:773–782.
Björklund, S, Buzaite, O., and Hallberg, M. 2001. The yeast Mediator. *Mol Cells* 11:129–136.
Björklund, S., and Gustafsson, C. M. 2005. The yeast Mediator complex and its regulation. *Trends Biochem Sci* 30:240–244.
Blazek, E., Mittler, G., and Meisterernst, M. 2005. The Mediator of RNA polymerase II. *Chromsoma* 113:399–408.
Bourbon, H.-M., Aguilera, A., Ansari, A. Z., et al. 2004. A unified nomenclature for protein subunits of mediator complexes linking transcriptional regulators to RNA polymerase II. *Mol Cell* 14:553–557.
Bryant, G. O., and Ptashne, M. 2003. Independent recruitment in vivo by Gal4 of two complexes required for transcription. *Mol Cell* 11:1301–1309.
Cai, G., Imasaki, T., Takagi, Y., and Asturias, F. J. 2009. Mediator structural conservation and implications for the regulation mechanism. *Structure* 17:559–567.
Cai, G., Imasaki, T., Yamada, K., et al. 2010. Mediator head module structure and functional interactions. *Nat Struct Mol Biol* 17:273–279.
Casamassimi, A., and Napoli, C. 2007. Mediator complexes and eukaryotic transcription regulation: an overview. *Biochimie* 89:1439–1436.
Chadick, J. Z., and Asturias, F. J. 2005. Structure of eukaryotic Mediator complexes. *Trends Biochem Sci* 30:264–271.
Cheng, J. X., Gandolfi, M., and Ptashne, M. 2004. Activation of the *Gal1* gene of yeast by pairs of "non-classical" activators. *Curr Biol* 14:1675–1679.
Conaway, R. C., Sato, S., Tomomori-Sato, C., et al. 2005. The mammalian Mediator complex and its role in transcriptional regulation. *Trends Biochem Sci* 30:250–255.
Courey, A. J., and Jia, S. 2001. Transcriptional repression: the long and short of it. *Genes Dev* 15:2786–2796.
Crane-Robinson, C., and Wolff E. A. P. 1998. Immunological analysis of FIS and CHIPS. *Trends Genet* 14:477–480.

Guglielmi, B., van Berkum, N. L., Klapholz, B., et al. 2004. A high resolution protein interaction map of the yeast Mediator complex. *Nucl Acids Res* 32:5379–5391.
Hanlon, S. E., and Lieb, J. D. 2004. Progress and challenges in profiling the dynamics of chromatin and transcription factor binding with DNA microarrays. *Curr Opin Genet Dev* 14:697–705.
Kim, Y.-J., and Lis, J. T. 2005. Interactions between subunits of *Drosophila* Mediator and activator proteins. *Trends Biochem Sci* 30:245–249.
Kornberg, R. D. 2005. Mediator and the mechanism of transcriptional activation. *Trends Biochem Sci* 30:235–239.
Kuo, M. H., and Allis, C. D. 1999. In vivo cross-linking and immunoprecipitation for studying dynamic protein: DNA associations in a chromatin environment. *Methods* 19:425–433.
Kuras, L., Borggrefe, T., and Kornberg, R. D. 2003. Association of the Mediator complex with enhancers of active genes. *Proc Natl Acad Sci USA* 100:13887–13991.
Lewis, B. A., and Reinberg, D. 2003. The mediator coactivator complex: functional and physical roles in transcription regulation. *J Cell Sci* 116:3667–3675.
Malik, S., and Roeder, R. G. 2005. Dynamic regulation of Pol II transcription by the mammalian Mediator complex. *Trends Biochem Sci* 30:256–263.
Malik, S., and Roeder, R. G. 2010 The metazoan Mediator co-activator complex as an integrative hub for transcriptional regulation. *Nat Rev Genet* 11:761–772.
Meyers, L. C., Gustafsson, C. M., Hayashibara, K. C., et al. 1999. Mediator protein interactions that selectively abolish activated transcription. *Proc Natl Acad Sci USA* 96:67–72.
Meyers, L. C., and Kornberg, R. D. 2000. Mediator of transcriptional regulation. *Ann Rev Biochem* 69:729–749.
Meyers, L. C., Leuther, K., Bushnell, D. A., et al. 1997. Yeast RNA polymerase II transcription reconstituted with purified proteins. *Methods* 12:212–218.
Näär, A. M., Lemon, B. D., and Tjian, R. 2001. Transcriptional coactivator complexes. *Ann Rev Biochem* 70:475–501.
Nal, B., Mohr, E., and Ferrier, P. 2001. Location and analysis of DNA-bound proteins at the whole-genome level: untangling transcriptional regulatory networks. *Bioessays* 23:473–476.
Orlando, V. 2000. Mapping chromosomal proteins in vivo by formaldehyde-cross-linked-chromatin immunoprecipitation. *Trends Biochem Sci* 25:99–104.
Rachez, C., and Freedman L. P. 2001. Mediator complexes and transcription. *Curr Opin Cell Biol* 13:274–280.
Taatjes, D. J. 2010. The human Mediator complex: a versatile, genome-wide regulator of transcription. *Trends Biochem Sci* 35:315–322.
Takagi, Y., Calero, G., Komori, H. et al. 2006. Head module control of mediator interactions. *Mol Cell* 23: 355–364.
Takagi, Y., and Kornberg, R. D. 2006. Mediator as a general transcription factor. *J Biol Chem* 281:80–89.

Chromatin Modification and Remodeling

Amabile, G., and Meissner, A. 2009. Induced pluripotent stem cells: current progress and potential for regenerative medicine. *Trends Mol Med* 15:59–68.
Ballestar, E., and Esteller, M. 2005. Methylated DNA-binding proteins. In: *Encyclopedia of Life Sciences*. pp. 1–5. Hoboken, NJ: John Wiley and Sons.
Becker, P. B., and Hörz, W. 2002. ATP-dependent nucleosome remodeling. *Ann Rev Biochem* 71:247–273.
Berger, S. L. 2002. Histone modifications in transcriptional regulation. *Curr Opin Genet Dev* 12:142–148.
Berger, S. L., Kouzarides, T., Shiekhattar, R., and Shilatifard, A. 2009. An operational definition of epigenetics. *Genes Dev* 23:781–783.
Bodanović, O., and Veenstra, G. J. C. 2009. DNA methylation and methyl-CpG proteins: developmental requirements and function. *Chromosoma* 118:549–565.
Brown, C. E., Lechner, T., Howe, L., and Workman, J. L. 2000. The many HATs of transcription coactivators. *Trends Biochem Sci* 25:15–19.

Cairns, B. R. 2009. The logic of chromatin architecture and remodelling at promoters. *Nature* 461:193–198.

Campose, E. I., and Reinberg, D. 2009. Histones: Annotating chromatin. *Ann Rev Genet* 43:559–599.

Chaban, Y., Ezeokonkwo, C, Chung, W.-H., et al. 2008. Structure ofa RSC-nucleosome complex and insights into chromatin modeling. *Nat Struct Mol Biol* 15:1272–1277.

Clapier, C. R., and Cairns, B. R. 2009. The biology of chromatin remodeling complexes. *Ann Rev Biochem* 78:273–304.

Constância, M., Murrell, A., and Reik, W. 2005. Genomic imprinting at the transcriptional level. In: *Encyclopedia of Life Sciences*. pp. 1–7. Hoboken, NJ: John Wiley and Sons.

Cosgrove, M. S., Boeke, J. D., and Wolberger, C. 2004. Regulated nucleosome mobility and the histone code. *Nat Struct Mol Biol* 11:1037–1043.

Cox, J. L., and Rizzino, A. 2010. Induced pluripotent stem cells: what lies beyond the paradigm shift. *Exp Biol Med* 236:148–158.

Dickins, B. J. A., and Kelsey, G. 2008. Evolution of imprinting: Imprinted gene function in human disease. In: *Encyclopedia of Life Sciences*. pp. 1–11. Hoboken, NJ: John Wiley and Sons.

Djuric, U., and Ellis, J. 2010. Epigenetics of induced pluripotency, the seven-headed dragon. *Stem Cell Res Ther* 1:3.

Ferguson-Smith, A. C., and Surani, M. A. 2001. Imprinting and the epigenetic asymmetry between parental genomes. *Science* 293:1086–1089.

Fry, C. J., and Peterson, C. L. 2002. Unlocking the gates of gene expression. *Science* 295:1847–1848.

Fyodorov, D. V., and Kadonaga, J. T. 2001. The many faces of chromatin remodeling: switching beyond transcription. *Cell* 106:523–525.

Gamble, M. J., and Freedman, L. P. 2002. A coactivator code for transcription. *Trends Biochem Sci* 27:165–167.

Gartenberg, M. R. 2000. The Sir proteins of *Saccharomyces cerevisiae*: mediators of transcriptional silencing and much more. *Curr Opin Microbiol* 3:132–137.

Gregory, P. D. 2001. Transcription and chromatin converge, lessons from yeast genetics. *Curr Opin Genet Dev* 11:142–147.

Hochedlinger, K. 2010. Your inner healers. *Sci Am* 302:47–53.

Huangfu, D., Osafune, K., Maehr, R., et al. 2008. Induction of pluripotent stem cells from primary human fibroblasts with only *Oct4* and *Sox2*. *Nat Biotechnol* 26:1269–1275.

Ideraabdullah, F. Y., Vigneau, S., and Bartolomei, M. S. 2008. Genomic imprinting mechanisms in mammals. *Mutation Res* 647:77–85.

Iizuka, M., and Smith, M. M. 2003. Functional consequences of histone modifications. *Curr Opin Genet Dev* 13:154–160.

Imbalzano, A. N., and Xiao, H. 2004. Functional properties of ATP-dependent chromatin remodeling enzymes. *Adv Protein Chem* 67:157–179.

Jenuwein, T., and Allis, C. D. 2001. Translating the histone code. *Science* 293:1074–1080.

Jiang, C., and Pugh, F. 2009. Nucleosome positioning and gene regulation: advances through genomics. *Nat Rev Genet* 10:161–172.

Jiang, Y., Bressler, J., and Beaudet, A. L. 2004. Epigenetics and human disease. *Ann Rev Genomics Hum Genet* 5:479–510.

Koerner, M. V., and Barlow, D. P. 2010. Genomic imprinting – an epigenetic gene-regulatory model. *Curr Opin Genet Dev* 20:164–170.

Koh, F. M., Sachs, M., Guzman-Ayala, M., and Ramalho-Santos, M. 2010. Parallel gateways to pluripotency: open chromatin in stem cells and development. *Curr Opin Genet Dev* 20:1–8.

Kouzarides, T. 2007. Chromatin modifications and their function. *Cell* 128:693–705.

Kuzmichev, A., and Reinberg, D. 2001. Role of histone deacetylase complexes in the regulation of chromatin metabolism. *Curr Top Microbiol Immunol* 254:35–58.

Li, B., Cary, M., and Workman, J. L. 2007. The role of chromatin during transcription. *Cell* 128:707–719.

Lorch, Y., Maier-Davis, B., and Kornberg, R. D. 2010. Mechanism of chromatin remodeling. *Proc Natl Acad Sci USA* 107:3458–3462.
Lusser, A., and Kadonaga, J. T. 2003. Chromatin remodeling by ATP-dependent molecular machines. *Bioessays* 25:1192–1200.
Margueron, R., and Reinberg, D. 2010. Chromatin structure and the inheritance of epigenetic information. *Nat Rev Genet* 11:285–296.
Marmorstein, R., and Trievel, R. C. 2009. Histone modifying enzymes: structures, mechanisms, and specificities. *Biochim Biophys Acta* 1789:58–68.
Morgan, H. D., Santos, F., Green, K., et al. 2005. Epigenetic reprogramming in mammals. *Human Mol Genet* 14:R47–R58.
Munshi, A., Shafi, G., Aliya, N., and Jyothy, A. 2009. Histone modifications dictate specific biological readouts. *J. Genet Genomics* 36:75–88.
Narlikar, G. J., Fan, H.-Y., and Kingston, R. E. 2002. Cooperation between complexes that regulate chromatin structure and transcription. *Cell* 108:475–487.
Ng, H. H., and Bird, A. 2000. Histone deacetylases: silencers for hire. *Trends Biochem Sci* 25:121–126.
Nishiwaka, S.I., Goldstein, R. A., and Nierras, C. R. 2008. The promise of human induced pluripotent stem cells for research and therapy. *Nat Rev Mol Cell Biol* 9:725–729.
Pérez-Martin, J. 1999. Chromatin and transcription in *Saccharomyces cerevisiae. FEMS Microbiol Rev* 23:503–523.
Peterson, C. L. 2002. Chromatin remodeling enzymes: taming the machines. *EMBO Rep* 31:319–322.
Reeve, J. N. 2003. Archaeal chromatin and transcription. *Mol Microbiol* 48:587–598.
Reid, G., Gallais, R., and Métivier, R. 2009. Marking time: the dynamic role of chromatin and covalent modification in transcription. *Int J Biochem Cell Biol* 41:155–163.
Richards, E. J., and Elgin, S. C. R. 2002. Epigenetic codes for heterochromatin formation and silencing: rounding up the usual suspects. *Cell* 108:489–500.
Rutledge, C. E., Lees-Murdock, D. J., and Walsh, C. P. 2010. DNA methylation in development. In: *Encyclopedia of Life Sciences*. pp. 1–8. Hoboken, NJ: John Wiley and Sons.
Saha, A., Wittmeyer, J., and Cairns, B. R. 2005. Chromatin remodeling through directional DNA translocation from an internal nucleosome site. *Nat Struct Mol Biol* 12:747–755.
Sasaki, H., and Matsui, Y. 2008. Epigenetic events in mammalian germ-cell development: reprogramming and beyond. *Nat Rev Genet* 9:129–140.
Schnitzler, G. R. 2008. Control of nucleosome positions by DNA sequence and remodeling. *Cell Biochem Biophys* 51:67–80.
Segal, E., and Widom, J. 2009. What controls nucleosome positions? *Trends Genet* 25:335–343.
Selth, L. A., Sigurdsson, S., and Svejstrup, J. Q. 2010. Transcription elongation by RNA polymerase II. *Ann Rev Biochem* 79:271–293.
Selvaraj, V., Plane, J. M., Williams, A. J., and Deng, W. 2010. Switching cell fate: the remarkable rise of induced pluripotent stem cells and lineage reprogramming technologies. *Trends Biotechnol* 28:214–223.
Sims, R. J. 3rd, and Reiberg, D. 2008. Is there a code embedded in proteins that is based on post-translational modifications? *Nat Rev Mol Cell Biol* 9:1–6.
Spotswood, H. T., and Turner, B. M. 2002. An increasingly complex code. *J Clin Invest* 110:577–582.
Sterner, D. E., and Berger, S. L. 2000. Acetylation of histones and transcription-related factors. *Microbiol Mol Biol Rev* 64:435–459.
Struhl, K. 1998. Histone acetylation and transcriptional regulatory mechanisms. *Genes Dev* 12:599–606.
Struhl, K. 1999. Fundamentally different logic of gene regulation in eukaryotes and prokaryotes. *Cell* 98:1–4.
Surani, M. Z. 2001. Imprinting (mammals). In: *Encyclopedia of Life Sciences*. pp. 1–6. Hoboken, NJ: John Wiley and Sons.
Svejstrup, J. Q. 2004. The RNA polymerase II transcription cycle: cycling through chromatin. *Biochim Biophys Acta* 1677:64–73.

Takahashi, K., Tanabe, K., Ohnuki, M., et al. 2007. Induction of pluripotent stem cells from adult human fibroblasts by defined factors. *Cell* 131:861–872.

Takahashi, K., and Yamanaka, S. 2006. Induction of pluripotent stem cells from mouse embryonic and adult fibroblast cultures by defined factors. *Cell* 126:663–676.

Torok, M. S., and Grant, P. A. 2004. Histone acetyltransferase proteins contribute to transcriptional processes at multiple levels. *Adv Protein Chem* 67:181–199.

Turner, B. M. 2002. Cellular memory and the histone code. *Cell* 111:285–291.

Venters, B. J., and Pugh, B. F. 2009. How eukaryotic genes are transcribed. *Crit Rev Biochem Mol Biol* 44:117–141.

Vignali, M., Hassan, A. H., Neely, K. E., and Workman, J. L. 2000. ATP-dependent chromatin-remodeling complexes. *Mol Cell Biol* 20:1899–1910.

Wolffe, A. P., and Guschin, D. 2000. Chromatin structural features and targets that regulate transcription. *J Struct Biol* 129:102–122.

Wood, A. J., and Oakey, R. J. 2006. Genomic imprinting in mammals: emerging themes and established theories. *PLoS Genetics* 2:1677–1685.

Yamanaka, S. 2008. Pluripotency and nuclear reprogramming. *Phil Trans R Soc B* 363:2079–2087.

Zaehres, H., and Schöler, H. R. 2007. Induction of pluripotency: from mouse to human. *Cell* 131:834–835.

Zhang, Y., and Reinberg, D. 2001. Transcription regulation by histone methylation: interplay between different covalent modifications of the core histone tails. *Genes Dev* 15:2343–2360.

This book has a Web site, **http://biology.jbpub.com/book/molecular**, which contains a variety of resources, including links to in-depth information on topics covered in this chapter, and review material designed to assist in your study of molecular biology.

RNA Polymerase II: Cotranscriptional and Posttranscriptional Processes

19

OUTLINE OF TOPICS

19.1 Pre-mRNA

Eukaryotic cells synthesize large heterogeneous RNA (hnRNA) molecules.

Messenger RNA and hnRNA both have poly(A) tails at their 3′-ends.

19.2 Cap Formation

Messenger RNA molecules have 7-methylguanosine caps at their 5′-ends.

5′-m^{7}G caps are attached to nascent pre-mRNA chains when the chains are 20 to 30 nucleotides long.

All eukaryotes use the same basic pathway to form 5′-m^{7}G caps.

CTD must be phosphorylated on Ser-5 to target a transcript for capping.

19.3 Split Genes

Viral studies revealed that some mRNA molecules are formed by splicing pre-mRNA.

Amino acid-coding regions within eukaryotic genes may be interrupted by noncoding regions.

Exons tend to be conserved during evolution, whereas introns usually are not conserved.

A single pre-mRNA can be processed to produce two or more different mRNA molecules.

Combinations of the various splicing patterns within individual genes lead to the formation of multiple mRNAs.

Drosophila form an mRNA that codes for essential protein isoforms by alternative *trans*-splicing.

Pre-mRNA requires specific sequences for precise splicing to occur.

Two splicing intermediates resemble lariats.

Splicing consists of two coordinated transesterification reactions.

19.4 Spliceosomes

Aberrant antibodies, which are produced by individuals with certain autoimmune diseases, bind to small nuclear ribonucleoprotein particles (snRNPs).

snRNPs assemble to form a spliceosome, the splicing machine that excises introns.

U1, U2, U4, and U5 snRNPs each contains Sm polypeptides, whereas U6 snRNP contains Sm-like polypeptides.

Each U snRNP is formed in a multistep process.

U1, U2, U4, U5, and U6 snRNPs have been isolated as a penta-snRNP in yeast.

In vitro studies show that spliceosomes assemble on introns via an ordered pathway.

RNA and protein may both contribute to the spliceosome's catalytic site.

Cells use a variety of mechanisms to regulate splice site selection.

Splicing begins as a cotranscriptional process and continues as a posttranscriptional process.

19.5 Cleavage/Polyadenylation and Transcription Termination

Poly(A) tail synthesis and transcription termination are coupled, cotranscriptional processes.

Transcription units often have two or more alternate polyadenylation sites.

Alternative processing forces us to reconsider our concept of the gene.

Transcription termination takes place downstream from the poly(A) site.

RNA polymerase II transcription termination appears to involve allosteric changes and a 5′→3′ exonuclease.

In higher animals, most histone pre-mRNAs require a special processing mechanism.

19.6 RNA Editing

RNA editing permits a cell to recode genetic information in a systematic and regulated fashion.

The human proteome contains a much greater variety of proteins than would be predicted from the human genome.

19.7 Messenger RNA Export

Messenger RNA splicing and export are coupled processes.

Suggested Reading

Photo Courtesy of James Gathany / CDC

(a)

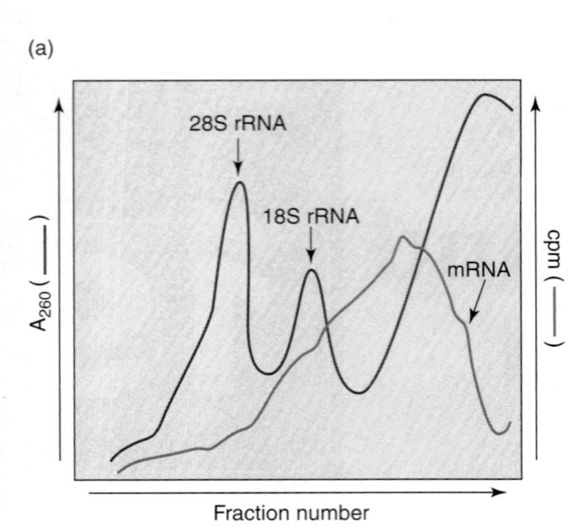

(b)

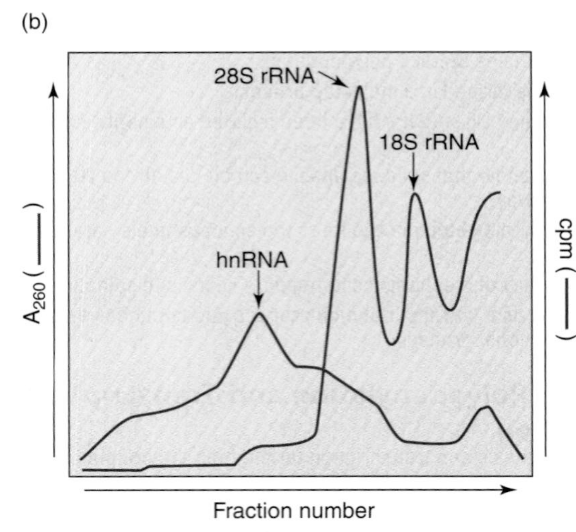

FIGURE 19.1 **Size distribution of HeLa cell mRNA and heterogeneous nuclear RNA (hnRNA) molecules.** (a) HeLa cells were cultured in the presence of [^{32}P]phosphate and then radioactively labeled mRNA was isolated from the cytoplasm and analyzed by sucrose gradient centrifugation. The radioactive mRNA profile is shown in green. (b) HeLa cells were cultured in the presence of [^{32}P]phosphate for five minutes. Then hnRNA was extracted from cell nuclei and analyzed by sucrose gradient centrifugation. The radioactive hnRNA profile is shown in blue. Sedimentation values were estimated by comparison with ribosomal 18S and 28S RNA molecules (red). (See Chapter 5 for a description of the sucrose gradient centrifugation technique.) (Adapted from J. E. Darnell, Jr., *Nat. Med.* 8 [2002]: 1068–1071.)

Molecular biologists initially thought that eukaryotes would synthesize messenger RNA (mRNA) in much the same way as bacteria. They, therefore, expected that RNA polymerase II would synthesize fully functional mRNA that just needed to be transported from the cell nucleus to the cytoplasm, where it would direct ribosomes to synthesize proteins. It came as a surprise to find that the transcript formed by the RNA polymerase II machinery is not mature mRNA but rather a precursor mRNA (pre-mRNA) that must be extensively modified before it can function as mRNA. The eukaryotic cell processes the pre-mRNA transcript as it is being synthesized by: (1) attaching a guanosine cap to the 5′-end of the nascent chain; (2) excising specific sequences from within the transcript; (3) adding a poly(A) tail to the 3′-end of the mRNA; and (4) terminating transcription. This chapter examines the various stages in cotranscriptional processing as well as reactions that replace one base in a transcript by another (a process termed RNA editing).

19.1 Pre-mRNA

Eukaryotic cells synthesize large heterogeneous RNA (hnRNA) molecules.

The first indication that the RNA polymerase II machinery synthesizes pre-mRNA rather than mRNA came from experiments performed by James Darnell and coworkers in the mid-1960s. In one particularly informative experiment, they added [^{32}P]phosphate to cultured HeLa cells (human cervical tumor cells) for five minutes and then extracted labeled RNA from cell nuclei and analyzed the RNA by sucrose gradient centrifugation. They expected the rapidly labeled nuclear RNA to consist of a population of chains ranging from about 500 to 3000 nucleotides long, the size distribution of HeLa cell mRNA (FIGURE 19.1a), but instead observed that the newly synthesized RNA molecules were about ten times longer (FIGURE 19.1b). This rapidly labeled nuclear RNA was dubbed **heterogeneous nuclear RNA** (**hnRNA**) because it was present in the nucleus and consisted of RNA molecules with a broad size distribution.

Further study indicated the base composition of HeLa cell hnRNA is 43% G+C, just like HeLa cell mRNA. Moreover, HeLa cell hnRNA did not accumulate in the nucleus, signifying that it is quite unstable. Based on these observations, Darnell proposed that hnRNA (or at least a large percentage of RNA molecules within the hnRNA fraction) are pre-mRNA molecules, which must somehow be trimmed to form shorter mRNA molecules.

Messenger RNA and hnRNA both have poly(A) tails at their 3′-ends.

The next major advance in studying eukaryotic mRNA synthesis took place in the early 1970s as a result of efforts to characterize HeLa cell mRNA by digesting it with ribonucleases. The mRNA was extracted from cytoplasmic complexes known as **polyribosomes** or **polysomes**, each consisting of two or more ribosomes translating a single mRNA. Two different endonucleases, RNase T1 (a product of the fungus *Aspergillus oryzae*) and pancreatic RNase, were used to digest the mRNA. RNase T1 and pancreatic RNase cleave phosphodiester bonds on the 3′-sides of guanylate residues and pyrimidine nucleotides, respectively (FIGURE 19.2). Assuming a more or less random nucleotide distribution within HeLa cell mRNA, one would expect to find the digests would contain a mixture of short oligonucleotides along with some mononucleotides. Although digesting HeLa cell mRNA with either T1 RNase or pancreatic RNase did indeed produce the expected digestion products, it also produced an entirely unexpected polynucleotide containing 150 to 200 adenylate groups. Further analysis showed that these poly(A) segments are attached to the 3′-end of mRNA. With the one notable exception of histone mRNA, all eukaryotic mRNAs characterized to date have such 3′-poly(A) tails. Yeast poly(A) tails are usually between 50 and 70 nucleotides long, considerably shorter than mammalian poly(A) tails. Subsequent studies showed that a large proportion of the rapidly labeled hnRNA molecules also have poly(A) tails, supporting the idea that these hnRNA molecules are converted to mRNA.

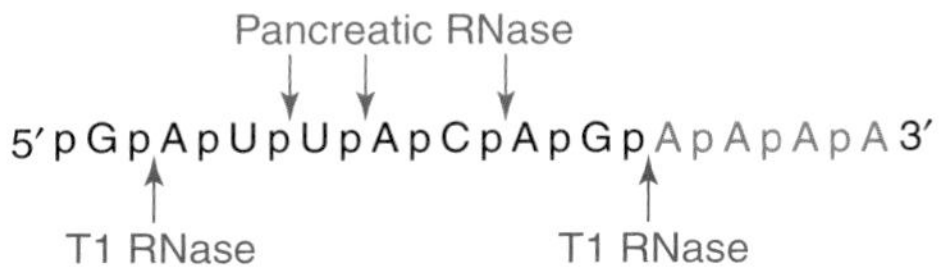

Pancreatic RNase

5′pGpApUpUpApCpApGpApApApA3′

T1 RNase T1 RNase

FIGURE 19.2 RNA hydrolysis catalyzed by T1 RNase and pancreatic RNase. T1 RNase cleaves on the 3′-side of guanylate and pancreatic RNase cleaves on the 3′-side of pyrimidine nucleotides. Notice that the four adenylates on the 3′-side of the molecule (shown in green) remain intact.

Poly(A) tails have a practical laboratory application; their base pairing properties can be used to separate mRNA and pre-mRNA from other kinds of RNA (FIGURE 19.3). This separation is accomplished by passing a mixture of RNA molecules, dissolved in a buffer solution containing a high salt concentration, through a column packed with cellulose fibers linked to oligo(dT). Messenger RNA molecules stick to the column because of base pairing between their poly(A) tails and oligo(dT). The same is true for pre-mRNA molecules. Metal cations from the salt help to stabilize this base pairing by diminishing charge repulsion between phosphate groups. Other kinds of RNA molecules that lack poly(A) tails pass right through the column. Then mRNA molecules (or pre-mRNA molecules) are eluted from the column by washing with a low salt buffer. It is important to note that not all of the RNA molecules in the rapidly labeled hnRNA fraction bind to oligo(dT). Nearly all of those that do not bind are destined to become some other kind of cellular RNA such as ribosomal or transfer RNA.

19.2 Cap Formation

Messenger RNA molecules have 7-methylguanosine caps at their 5′-ends.

Because tRNA and rRNA (transfer and ribosomal RNA) were known to have minor nucleotides that are formed by adding methyl ($-CH_3$) groups to bases or ribose moieties, it was logical to ask whether eukaryotic mRNA molecules also have methyl groups attached to them. This question could not be answered unequivocally before oligo(dT)

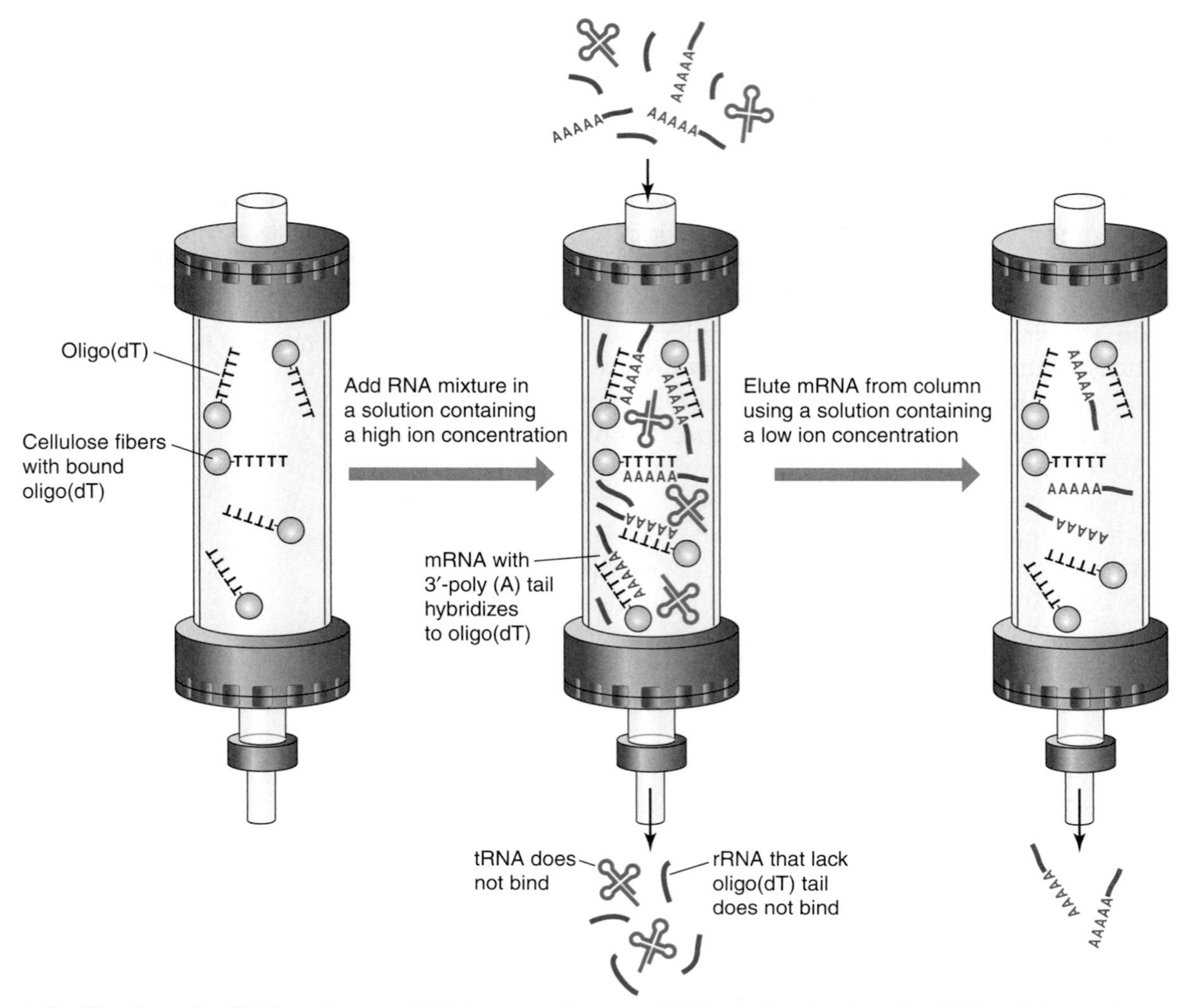

FIGURE 19.3 Purification of mRNA and pre-mRNA from a mixture of RNA molecules by oligo(dT) cellulose chromatography. A mixture of RNA molecules, dissolved in a buffer solution containing a high salt concentration, is passed through a column packed with cellulose fibers linked to oligo(dT). Messenger RNA molecules (red) stick to the column because of base pairing between the poly(A) tails and oligo(dT). The same is true for pre-mRNA molecules. Metal cations from the salt help to stabilize this base pairing by diminishing charge repulsion between phosphate groups. Other kinds of RNA molecules (green and blue) that lack poly(A) tails do not bind to the oligo(dT) and are eluted when the column is washed with buffer containing high salt. The mRNA and pre-mRNA are eluted from the column by washing with a low salt buffer.

cellulose chromatography became available, however, because there was no way to determine if the methyl groups detected were attached to mRNA or contaminating tRNA and rRNA molecules. The availability of a method to prepare eukaryotic mRNA free of tRNA and rRNA allowed investigators to begin looking for methylated nucleotides in mRNA.

In 1974, Robert Perry and coworkers were among the first to do so. Their approach was to culture mouse L cells (a fibroblast-like cell) in a medium containing [^{3}H-*methyl*]methionine, a precursor of the methyl group donor S-adenosylmethionine (FIGURE 19.4). They found that purified mRNA contains two to three methyl groups per one thousand nucleotides and that the methylated nucleotides were clustered at the 5′-end of the mRNA. Exhaustive digestion of mRNA

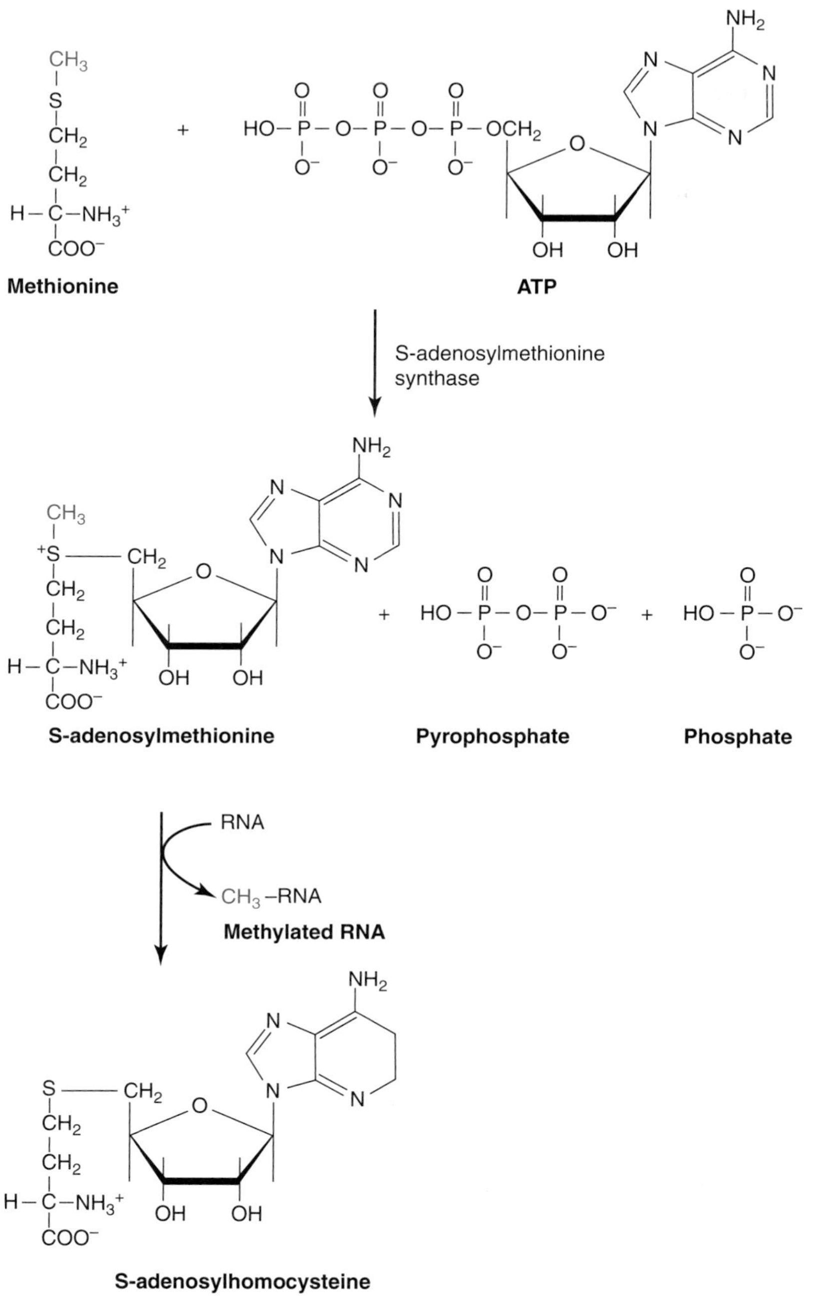

FIGURE 19.4 **Conversion of methionine to S-adenosylmethionine.** Methionine reacts with ATP to form S-adenosylmethionine, which then can transfer its methyl group to a base in RNA (or DNA) or a ribose in RNA.

with phosphodiesterase I, a snake venom exonuclease that successively cleaves 5′-mononucleotides from the 3′-ends of polynucleotide chains, produced a novel methylated trinucleotide as well as the expected 5′-mononucleotides. The structure of the methylated trinucleotide, which is shown in FIGURE 19.5, was determined from the following enzymatic and chemical information.

1. The trinucleotide had a normal 3′-end but lacked a terminal 5′-OH or 5′-phosphate group.
2. Two nucleotides within the trinucleotide had methyl groups attached to them; one methyl group was attached to a guanosine

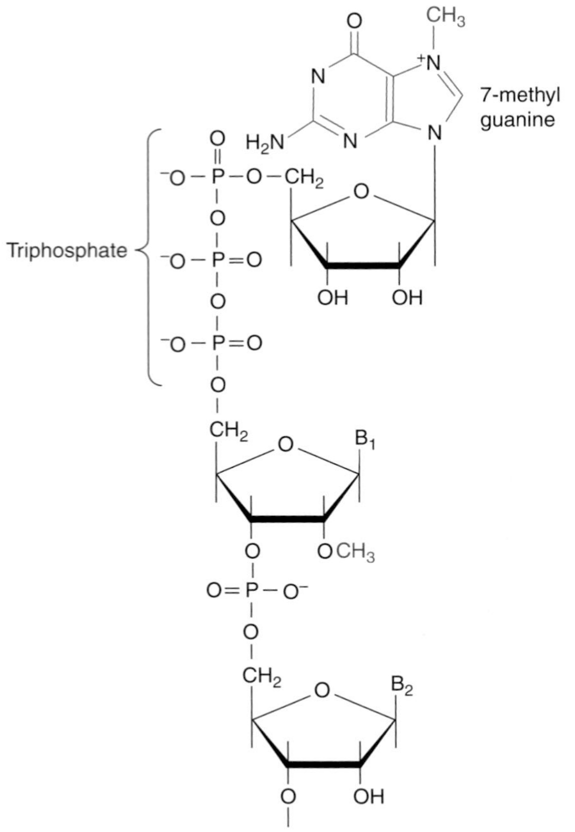

Triphosphate

7-methyl guanine

Messenger RNA 5′ -cap structure

FIGURE 19.5 Messenger RNA 5′-cap structure. The mRNA 5′-cap shown may be written as $m^7GpppNmpNmp$, where m is a methyl group, N is any nucleotide and p is a phosphate group.

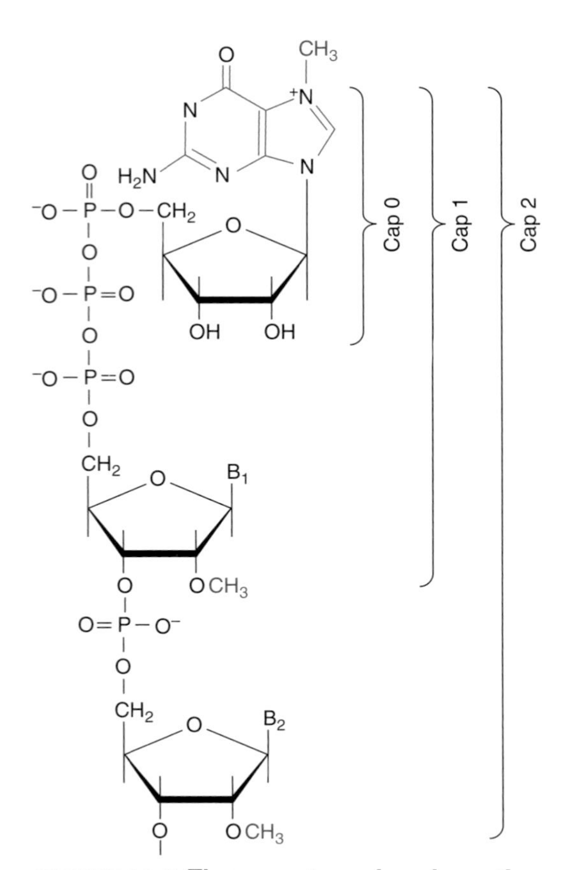

FIGURE 19.6 Three cap types in eukaryotic mRNA.

at N-7 and the other to C-2′ of the ribose group in the first or initiating nucleotide in mRNA.

3. The **N^7-methylguanosine (m^7G)** was attached to the first or initiating nucleotide in mRNA through an inverted 5′→5′ triphosphate bridge to form a cap at the 5′-end of mRNA. The cap could be removed by pyrophosphatase catalyzed cleavage but not by phosphodiesterase catalyzed cleavage.

Virtually all eukaryotic mRNAs have 5′-m^7G caps, as do mRNAs made by many animal viruses. The cap structure shown in Figure 19.5, known as **cap 1**, is present on mRNAs from multicellular organisms. Two other kinds of cap structures, termed **cap 0** and **cap 2**, are also commonly found in nature (FIGURE 19.6). The cap 0 structure, which is a feature of yeast mRNAs, has one less methyl group than cap 1. Its single methyl group is attached to guanine at position N-7. The cap 2 structure, which is present in some vertebrate mRNAs, has one more methyl group than the cap 1 structure. The extra methyl group is attached to C-2′ of the second nucleotide in the mRNA. Despite their different methylation patterns, cap 0, cap 1, and cap 2 appear to have similar functions. They protect mRNA from digestion by 5′→3′

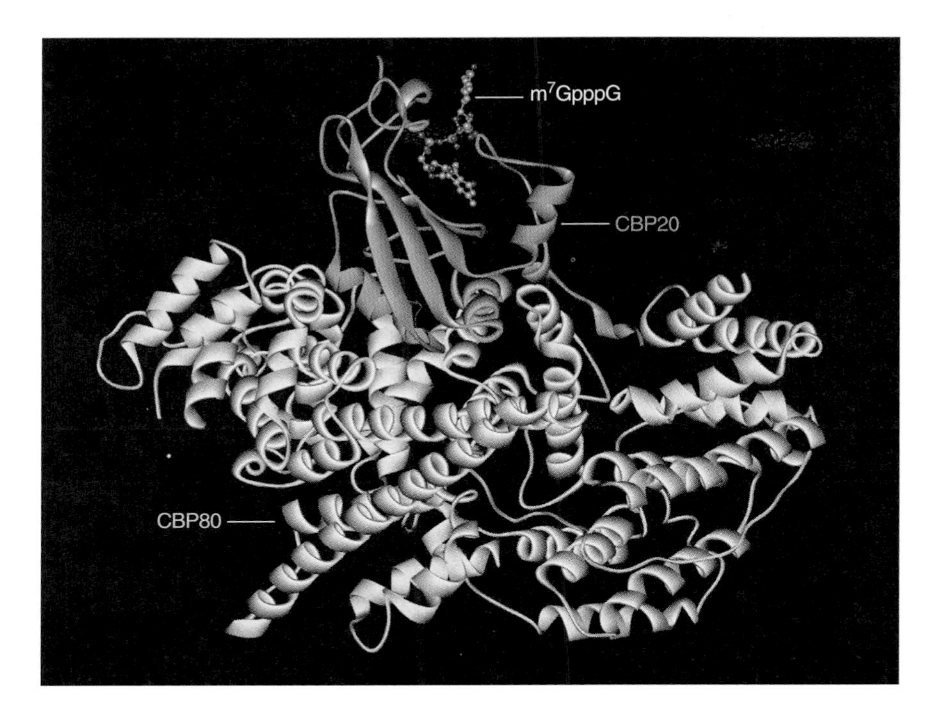

FIGURE 19.7 **Ribbon diagram for the cap binding protein complex (CBC) with a bound RNA cap analog m^{7}GpppG.** The CBP80 and CBP20 subunits are shown as green and orange ribbon structures, respectively. The m^{7}GpppG molecule is shown as a ball and stick structure. (Structure from Protein Data Bank 1N52. G. Calero, et al., *Nat. Struct. Biol.* 9 [2002]: 912–917. Prepared by B. E. Tropp.)

exonucleases, influence subsequent stages of mRNA processing (see below), and participate in translation (see Chapter 23).

A protein complex known as the **cap binding complex** (**CBC**) binds to a 5′-m^{7}G cap in the nucleus (FIGURE 19.7). CBC consists of two polypeptide subunits, CBP20 and CBP80. The CBP20 subunit, which has a classical ribonucleotide binding domain containing four antiparallel β-strands, binds to the cap. The CBP80 subunit, which interacts with CBP20 at sites other than the cap binding site, is required for high affinity binding to the cap. CBC participates in subsequent stages of cotranscriptional processing, helps mRNA to be transported through the nuclear pore (see below), and is an important factor in the early stage of translation (see Chapter 23).

5′-m^{7}G caps are attached to nascent pre-mRNA chains when the chains are 20 to 30 nucleotides long.

Once 5′-m^{7}G caps were shown to be attached to the first nucleotide in mRNA it seemed reasonable to ask if 5′-m^{7}G caps are also attached to the first nucleotide in pre-mRNA molecules. This question was answered by subjecting hnRNA to oligo(dT) column chromatography and demonstrating that pre-mRNA molecules thus purified do indeed have 5′-m^{7}G caps. Because the 5′-m^{7}G cap is added to the first or initiating nucleotide, it seemed likely that the cap is added at an early stage of transcription. In 1993, Eric B. Rasmussen and John T. Lis devised a method for testing this hypothesis. Their approach was to examine transcripts formed by the *Drosophila* heat shock gene, *hsp70*. This gene is ideally suited for studying events early in the transcription process because the transcription machinery pauses along the *hsp70* gene shortly after promoter clearance, just before transcription elongation starts. Nascent transcripts present at this stage tend to be quite short, with most ranging in size between

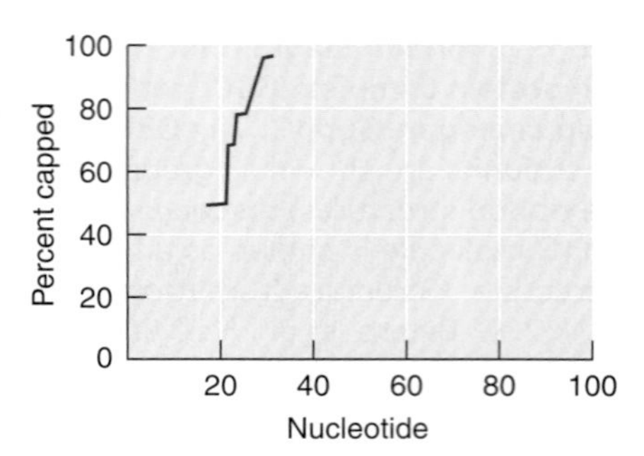

FIGURE 19.8 Cap formation of nascent transcripts of the *Drosophila hsp70* gene. A graph showing the percent of transcripts capped as a function of the nucleotide length of the nascent transcription of the *Drosophila hsp70* gene. The x-axis represents nucleotide positions relative to the transcription start site and the y-axis represents the percentage of transcripts possessing a cap. (Modified from E. B. Rasmussen and J. T. Lis, *Proc. Natl. Acad. Sci. USA* 90 [1993]: 7923–7927. Copyright 1993 National Academy of Sciences, U.S.A.)

20 and 40 nucleotides long. Rasmussen and Lis examined nascent *hsp70* transcripts to determine if there was a correlation between chain length and 5′-m^{7}G cap formation. As shown in FIGURE 19.8, the 5′-m^{7}G cap forms when nascent chains are 20 to 30 nucleotides long. Cap formation probably cannot occur any earlier because the 5′-end of the growing transcript must emerge from the RNA polymerase II exit channel before it can be available for capping. It is likely that capping marks the start of the elongation phase of transcription. When Rasmussen and Lis selected the *hsp70* gene to determine the time of cap addition, transcriptional pausing was thought to be a rare phenomenon. Later studies showed such pausing is a common phenomenon that probably serves as a major point for transcriptional regulation.

All eukaryotes use the same basic pathway to form 5′-m^{7}G caps.

Three enzyme activities work together to synthesize the 5′-m^{7}G cap (FIGURE 19.9). RNA 5′-triphosphatase cleaves the γ-phosphate from the 5′-triphosphate end of the initiating nucleotide as it emerges from the

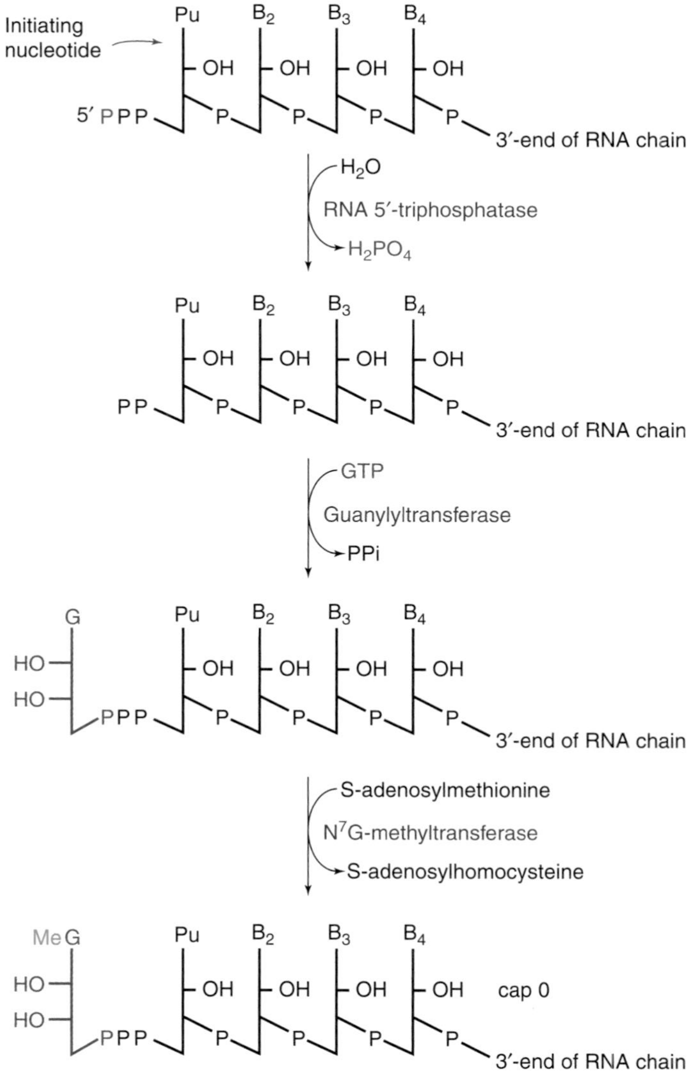

FIGURE 19.9 Capping pathway. The process takes place in three steps: (1) RNA 5′-triphosphatase cleaves the γ-phosphate from the 5′-triphosphate end of the initiating nucleotide as it emerges from the RNA polymerase II exit channel. (2) guanylyltransferase catalyzes guanylyl (GMP) group transfer from guanosine triphosphate (GTP) to the newly created 5′-diphosphate end to form the GpppN cap (where N is usually a purine nucleotide), and (3) methyltransferase transfers a methyl group from S-adenosylmethionine to the N-7 position of the cap guanine.

RNA polymerase II exit channel. Then guanylyltransferase catalyzes guanylyl (GMP) group transfer from guanosine triphosphate (GTP) to the newly created 5′-diphosphate end to form the GpppN cap (where N is usually a purine nucleotide). In yeast, the two enzyme activities are on separate polypeptide chains that interact to form a heterodimer. In mammals, both activities are part of a single bifunctional polypeptide. Methyltransferase, the third enzyme activity required for capping, transfers a methyl group from S-adenosylmethionine to the N-7 position of the cap guanine. Purified capping enzymes do not appear to have an RNA sequence requirement or require the assistance of any additional enzymes. They can even add caps to di- and trinucleotides. Based on these properties, we might expect that the capping enzymes should be able to add a 5′-m^{7}G cap to any accessible RNA molecule, but this does not occur *in vivo* for reasons that will become evident in the next section. In vertebrates, an additional methylation reaction converts cap 0 to cap 1 in the nucleus and then another methylation reaction converts cap 1 to cap 2 in the cytoplasm.

CTD must be phosphorylated on Ser-5 to target a transcript for capping.

When protein-coding genes are modified so that they can be transcribed by RNA polymerase I or RNA polymerase III, transcripts are no longer capped. This observation suggests that RNA polymerase II has some unique feature that directs the capping machinery to its own transcripts. The carboxyl terminal domain (CTD) of the largest subunit in RNA polymerase II seems a likely candidate because there is no comparable structure in either RNA polymerases I or III. Moreover, the crystal structure of yeast RNA polymerase II (see Chapter 17) reveals that CTD is adjacent to the RNA exit channel and therefore is located in just the right position to interact with the transcript as it emerges from polymerase. Yeast CTD is sufficiently long and flexible to serve as an assembly platform for the capping enzyme and other processing factors because its 26 Tyr-Ser-Pro-Thr-Ser-Pro-Ser tandem repeats (see Chapter 17) are 65 nm long. CTD is connected to the rest of the polypeptide by a 28 nm flexible linker, conferring the flexibility required to localize the capping enzyme close to the 5′-end of the pre-mRNA as it emerges from the polymerase exit channel. Mammalian CTD has 52 heptapeptide repeats and so is twice as long as its yeast counterpart.

David Bentley and coworkers devised an elegant experiment in 1997 to test whether CTD is in fact required for capping. They first transfected human kidney cells with expression vectors for α-amanitin-resistant RNA polymerase II large subunit with either a full-length CTD peptide or a truncated CTD peptide containing only 5 of the 52 heptapeptide repeats. Then they added α-amanitin to the transfected cells to block endogenous RNA polymerase II activity so that the cells were forced to use the expression vector's α-amanitin-resistant RNA polymerase II to synthesize RNA. The hypothesis that CTD helps to target its own transcripts for capping predicts that the proportion of transcripts with 5′-m^{7}G caps should be much higher in cells with a normal CTD than in cells with a truncated CTD. The results were in complete accord with this prediction. Bentley's group also demonstrated that bifunctional mammalian RNA triphosphatase/guanylyltransferase binds directly to phosphorylated CTD, but not to unphosphorylated CTD *in vitro*. Transcripts formed by RNA polymerase II with a normal phosphorylated CTD thus are capped while those formed by RNA polymerase II with a truncated or unphosphorylated CTD are not capped (**FIGURE 19.10**). These experiments indicate that the phosphorylated CTD is necessary for cap formation but do not show that it is sufficient. Recent studies suggest that some other RNA polymerase II-specific component may also be required for cap formation. Mammalian methyltransferase also binds to CTD.

Yeast guanylyltransferase and methyltransferase also bind to phosphorylated CTD. Studies performed by Stephen Buratowski and coworkers in 1997 revealed that yeast guanylyltransferase becomes inactive upon binding to phosphorylated CTD but that yeast RNA

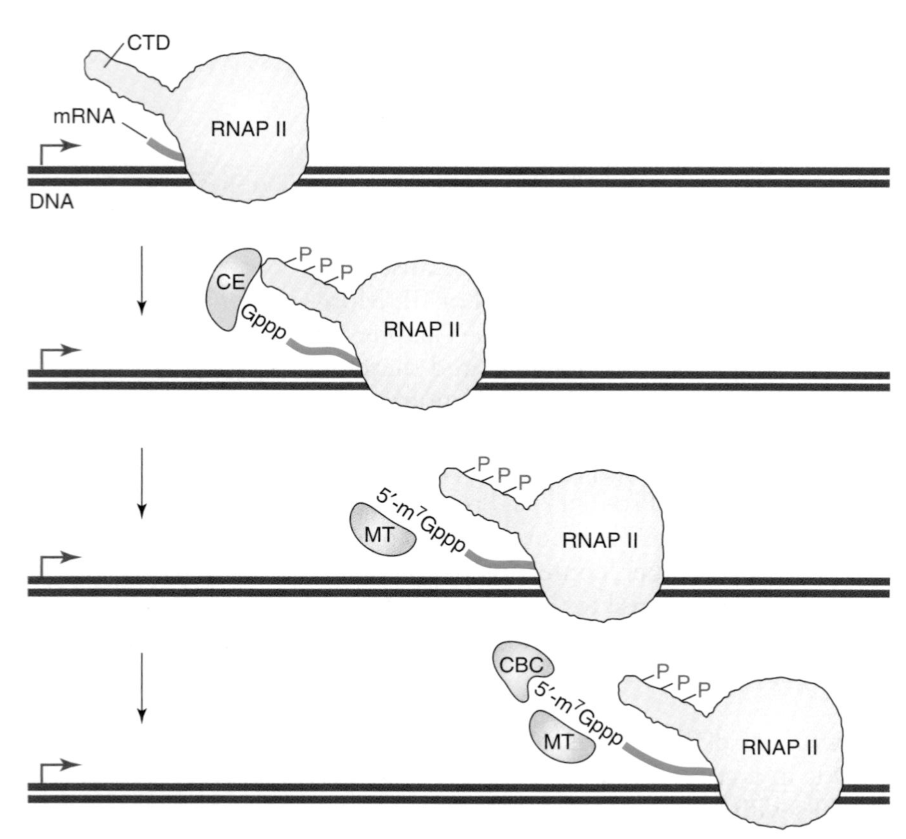

FIGURE 19.10 **Mammalian RNA 5′-triphosphatase/guanylyltransferase, methyltransferase, and cap binding complex recruitment.** Transcripts formed by RNA polymerase II (RNAP II) with a normal carboxyl terminal domain (CTD) are capped with guanosine through the combined actions of RNA triphosphatase and guanylyltransferase activities. These two activities are present on the same polypeptide in mammals. This bifunctional protein is called the capping enzyme (CE). CE association with CTD requires CTD phosphorylation. Binding and cap formation occur shortly after the 5′-end emerges from the RNA polymerase II exit channel. Methyltransferase (MT) transfers a methyl group from S-adenosylmethionine to the N-7 position of the guanosine cap to form the methylated cap (5′-m^{7}G). The cap binding complex (CBC) binds to the 5′-m^{7}G cap. Transcripts formed by RNA polymerase II with a truncated CTD or by RNA polymerases I and III are not capped. (Adapted from K. J. Howe, *Biochim. Biophys. Acta* 1577 [2002]: 308–324.)

5′-triphosphatase stimulates the transferase activity when it binds to guanylyltransferase to form the heterodimer. Yeast guanylyltransferase thus is regulated by allosteric interactions with both the 5′-RNA triphosphatase and the phosphorylated CTD. The mammalian bifunctional enzyme probably does not require such a complex regulatory mechanism because both the triphosphatase and transferase activities are part of the same polypeptide.

The vast number of possible phosphorylation arrays within CTD presented a major challenge to investigators trying to determine the specific array responsible for directing the bifunctional mammalian RNA 5′-triphosphatase/guanylyltransferase to RNA polymerase II. C. Kiong Ho and Stewart Shuman met the challenge by studying the way that the mouse capping enzyme's RNA 5′-triphosphatase and guanylyltransferase domains interact with synthetic CTD peptides. Their approach was to use recombinant DNA technology to construct one expression vector coding for the RNA 5′-triphosphatase domain and another coding for the guanylyltransferase domain. The RNA 5′-triphosphatase and guanylyltransferases domains were each isolated from an *Escherichia coli* transformant bearing a recombinant expression vector for that domain. CTD peptides, which were prepared with an automatic peptide synthesizer, were either unphosphorylated or phosphorylated on Ser-2 or Ser-5 of the repeating Tyr-Ser-Pro-Thr-Ser-Pro-Ser heptapeptide. The peptides were then tested for their ability to bind to the RNA triphosphatase or guanylyltransferase domains and to stimulate their activity. The results of these experiments are summarized in Table 19.1. The RNA 5′-triphosphatase domain did not bind to any of the synthetic CTD peptides tested and the guanylyltransferase domain did not bind to the unphosphorylated CTD peptide but did bind to both kinds of phosphorylated CTD. However, guanylyltransferase activity was only stimulated by CTD peptides that were phosphorylated at Ser-5.

CTD is phosphorylated at Ser-5 by the cyclin-dependent protein kinase associated with the general transcription factor TFIIH (see Chapter 17) as transcription progresses from the initiation to the

TABLE 19.1 Interactions Between Synthetic CTD and the Mouse RNA 5′-triphosphatase and Guanylyltransferase Domains

	Binding		Guanylyltransferase Activation
	Guanylyltransferase domain	**RNA 5′-triphosphatase domain**	
Unphosphorylated (Tyr-Ser-Pro-Thr-Ser-Pro-Ser)$_6$	No	No	No
Phosphorylated on Ser-2 (OPO_3^{2-} on Ser-2) (Tyr-Ser-Pro-Thr-Ser-Pro-Ser)$_6$	Yes	No	No
Phosphorylated on Ser-5 (OPO_3^{2-} on Ser-5) (Tyr-Ser-Pro-Thr-Ser-Pro-Ser)$_6$	Yes	No	Yes

elongation phase. This phosphorylation initiates a series of actions that begin with transcription initiation factor release and continue with the recruitment of the bifunctional RNA 5′-triphosphatase/guanylyltransferase and various processing factors to the phosphorylated CTD. Once capping is complete, a specific protein phosphatase removes the phosphate group from the Ser-5, causing the release of bifunctional RNA 5′-triphosphatase/guanylyltransferase but not the methyltransferase. The reason for methyltransferase retention is not clear but one possibility is that the methyltransferase has some additional function. Then CTD is phosphorylated at Ser-2 by P-TEFb (positive transcription elongation factor b; see Chapter 17). CTD phosphorylated at Ser-2 appears to recruit additional factors required at later stages in mRNA processing.

19.3 Split Genes

Viral studies revealed that some mRNA molecules are formed by splicing pre-mRNA.

The observation that both pre-mRNA and mRNA molecules have 5′-m^{7}G caps and 3′-poly(A) tails seemed to rule out the possibility that pre-mRNA is converted to mRNA by trimming one or both ends of the pre-mRNA molecule because such trimming would of necessity remove the 5′-m^{7}G cap, the 3′-poly(A) tail, or both. A model in which both ends of pre-mRNA are conserved presented an even greater conceptual challenge, however, because the only way to conserve both ends while shortening the pre-mRNA molecule would be to remove one or more segments from within pre-mRNA and then join the flanking sequences. Investigators were understandably reluctant to accept such an unprecedented splicing mechanism. In 1977 two independent research groups, one led by Phillip A. Sharp and the other by Richard Roberts, provided convincing visual evidence that splicing occurs by demonstrating that eukaryotic genes have long DNA sequences within them that are missing from mature mRNA.

The Sharp and Roberts laboratories studied viral mRNA formed by HeLa cells infected with adenovirus, a DNA virus that infects the upper respiratory tract (see Chapter 8). The adenovirus life cycle can be divided into an early and late phase; the latter begins at the onset of viral DNA synthesis. Adenovirus-infected HeLa cells are well-suited for studying mRNA formation because they produce an abundant supply of eight different mRNA molecules during the late phase. These late mRNAs can be extracted from polysomes and then separated from one another to provide a single kind of late mRNA. Viral late mRNA formation is an excellent model for host cell mRNA formation because the host cell RNA polymerase II machinery synthesizes the viral mRNA and host cell enzymes add 5′-m^{7}G caps and 3′-poly(A) tails to the viral mRNA. The Sharp and Roberts groups both prepared DNA–RNA hybrids of late mRNA annealed to viral DNA, expecting to find the electron micrographs would reveal which part of the

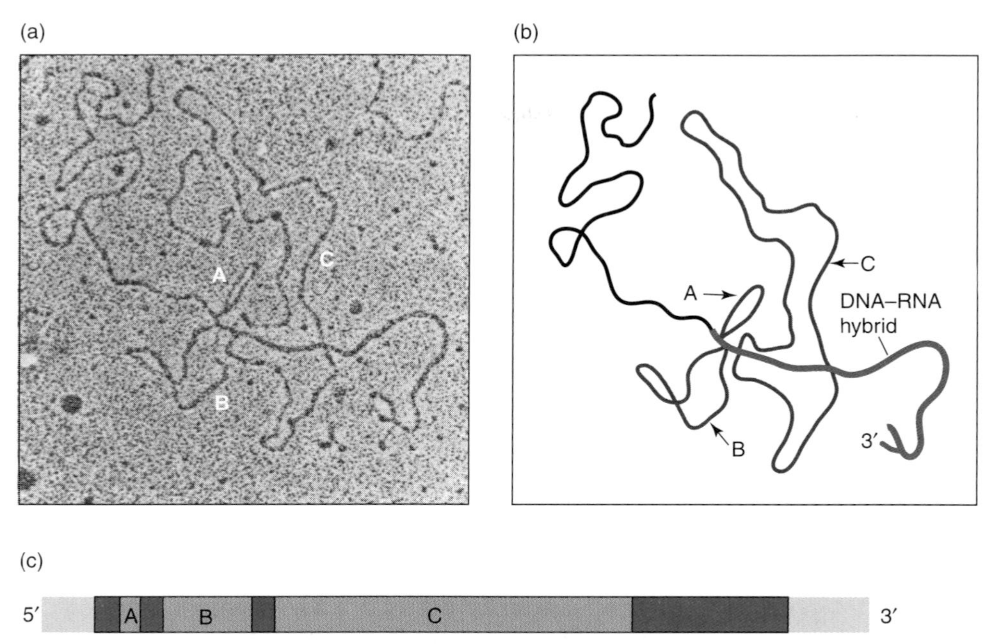

FIGURE 19.11 An adenovirus gene coding for a viral coat protein is interrupted by segments not present in mRNA. (a) Electron micrograph of a DNA–RNA hybrid formed by binding an mRNA for viral coat protein with its DNA template. Intervening segments (introns) that are excised during mRNA maturation are labeled A, B, and C. (b) Schematic of the electron micrograph. The DNA–RNA hybrid region, containing segments retained in the mature mRNA molecule (the exons), is shown in blue. The intervening segments (the introns) that are excised during mRNA maturation are shown in red and labeled A, B, and C. (c) Representation of the split viral gene for the coat protein. Once again, exons are shown in blue and introns in red. (Part a reproduced from S. M. Berget, C. Moore, and P. A. Sharp, *Proc. Natl. Acad. Sci. USA* 74 [1977]: 3171–3175. Photo courtesy of Phillip A. Sharp, Massachusetts Institute of Technology. Part b modified from S. M. Berget, C. Moore, and P. A. Sharp, *Proc. Natl. Acad. Sci. USA* 74 [1977]: 3171–3175. Used with permission from Phillip A. Sharp, Massachusetts Institute of Technology. Part c adapted from S. M. Berget, C. Moore and P. A. Sharp, *Proc. Natl. Acad. Sci. USA* 74 [1977]: 3171–3175.)

viral genome had produced the mature mRNA molecule. The mRNA did not line up on the DNA as expected. Some electron micrographs showed large loops of unhybridized DNA, clearly demonstrating that internal gene sequences were somehow skipped in making the final mRNA. FIGURE 19.11, a remarkable electron micrograph from the Sharp laboratory, shows a DNA–RNA hybrid that was formed by annealing a late mRNA for a viral coat protein to denatured viral DNA. The mRNA molecule hybridizes with four DNA regions, which are separated by three long DNA loops. Each loop corresponds to a region of viral DNA for which there is no complementary mRNA sequence. These results suggest that nucleotide sequences are removed from within the newly synthesized RNA, the primary transcript, and the flanking sequences are joined to form mRNA molecules.

Amino acid-coding regions within eukaryotic genes may be interrupted by noncoding regions.

Soon after the discovery that adenovirus late mRNA consists of transcripts made from noncontiguous viral DNA sequences, molecular biologists showed that many eukaryotic pre-mRNAs have coding sequences that are interrupted by noncoding sequences, which are missing from the mature mRNA.

Studies performed by Pierre Chambon, beginning in 1977, are particularly instructive in this regard and so we will begin by examining them. Chambon's initial objective was to learn how female sex hormones such as estrogen regulate the expression of the gene for ovalbumin, the major egg-white protein produced by laying hen oviduct cells. He suspected that the ovalbumin gene is modified in some way during hen oviduct cell differentiation and set out to find evidence for these modifications by comparing the ovalbumin gene in oviduct cells with the ovalbumin gene in a cell that does not synthesize ovalbumin. Although Chambon's hypothesis proved to be incorrect for the ovalbumin gene, the experiments he performed to test the hypothesis revealed that the ovalbumin gene contains coding sequences that are interrupted by noncoding sequences. We therefore consider Chambon's experiments in some detail.

Chambon began by purifying ovalbumin mRNA from laying hen oviduct cells, a task made easier by the fact that this mRNA accounts for as much as half of the total mRNA in oviduct cells. Then he used reverse transcriptase to make a single-stranded complementary DNA (cDNA) to the ovalbumin mRNA and DNA polymerase to convert the single-stranded cDNA to double-stranded cDNA (FIGURE 19.12). The double-stranded ovalbumin cDNA was inserted into a plasmid and the resulting recombinant plasmid cloned in *E. coli* to provide an ample supply of DNA for sequence analysis. But the ovalbumin mRNA sequence deduced from the cDNA was unremarkable. The 1872-nucleotide mRNA contained an 1158-nucleotide coding sequence (consistent with the 386 amino acids in ovalbumin) that was preceded by a 64-nucleotide untranslated leader sequence at its 5′-end and followed by a 650-nucleotide untranslated terminal sequence at its 3′-end.

Because Chambon's objective was to examine ovalbumin gene structure during different cell differentiation stages, his next step was to use the then new Southern blot procedure to compare ovalbumin genes from laying-hen oviduct cells and erythrocytes (red blood cells that do not synthesize ovalbumin). Recall from Chapter 5 that this procedure involves digesting cell DNA with a restriction endonuclease, separating the resulting DNA fragments by electrophoresis, and then detecting the fragment(s) of interest with a radioactive or fluorescent probe. Chambon digested total cell DNA with EcoRI, a restriction endonuclease that does not cleave ovalbumin cDNA, and prepared a radioactive probe by subjecting ovalbumin cDNA to nick translation (see Chapter 5). Based on the absence of EcoRI sites in ovalbumin cDNA, Chambon expected that the restriction endonuclease would cleave on either side of the ovalbumin gene so that the radioactive probe would anneal to a single large DNA fragment. Contrary to expectation, he observed four radioactive bands (Eco a–d) in the autoradiograph of the Southern blot when either oviduct or erythrocyte DNA was studied (Figure 19.12). The appearance of four bands was puzzling. How could the ovalbumin gene have EcoRI restriction sites when ovalbumin mRNA does not? The gene splicing model proposed for late mRNA synthesis in adenovirus suggested a hypothesis, that the EcoRI sites are present on DNA sequences that are excised when pre-mRNA is converted to mRNA. According to this **split** or **discontinuous gene hypothesis**, regions that

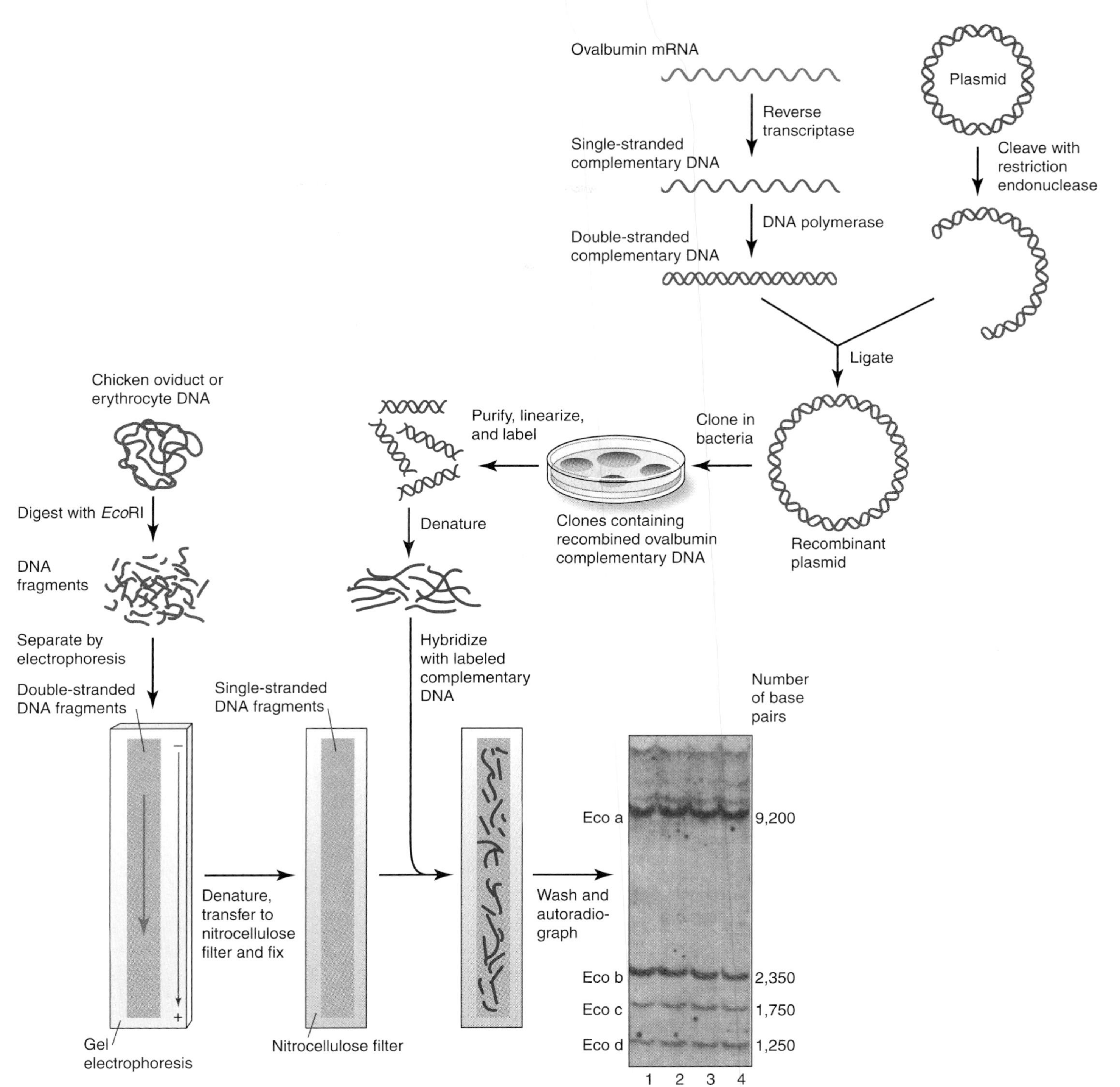

FIGURE 19.12 **Experiment demonstrating the split hen ovalbumin gene.** (Top right) Pierre Chambon and coworkers prepared double-stranded complementary DNA (cDNA) to the 1872-nucleotide ovalbumin mRNA by first using reverse transcriptase to make single-stranded cDNA and then using DNA polymerase to convert the single-stranded cDNA to double-stranded cDNA. The double-stranded cDNA was inserted into a plasmid and the resulting recombinant plasmid was cloned in *E. coli*, purified, linearized, labeled with radioactive isotope, and denatured. The labeled cDNA served as a probe to examine the ovalbumin gene from chicken erythrocytes or oviduct cells using the Southern blot procedure (see Chapter 5) (top left). Chromosomal DNA from erythrocytes or oviduct cells was cleaved with EcoRI into fragments ranging in size from 1–15 kbp long (bottom). Gel electrophoresis separated the fragments by size (with smaller fragments moving faster). The DNA fragments were denatured and transferred to nitrocellulose filter that binds single-stranded DNA, and fixed to the filter. Then the radioactive probe was added to detect sequences that were complementary to it (and therefore ovalbumin mRNA). Although Chambon and coworkers expected the radioactive probe to anneal to one band containing the ovalbumin gene (because they knew that EcoRI does not cleave the ovalbumin messenger's complementary DNA), autoradiography revealed that the probe anneals to four bands (Eco a–d). Moreover, these bands were in the same position whether the chromosomal DNA had been derived from the chicken erythrocytes (lanes 1 and 2) or from the oviduct cells (lanes 3 and 4). This experiment demonstrates that the intact gene has restriction sites and therefore nucleotide sequences that are not present in mature mRNA. (Part a adapted from P. Chambon, *Sci. Am.* 244 [1981]: 60–71. Part b photo courtesy of Pierre Chambon, Institute of Genetics and Molecular and Cellular Biology, College of France.)

FIGURE 19.13 **The organization of the ovalbumin gene as demonstrated by electron microscopy.** (a) Electron micrograph of ovalbumin mRNA hybridized to a single strand of DNA that includes the ovalbumin gene. (b) A schematic of the electron micrograph. Segments of the DNA (blue line) and mRNA (red) that are complementary to each other form a DNA–RNA hybrid. These eight expressed segments (L and 1–7) are termed exons. The seven DNA segments, which loop out from the hybrid (A–G) because they have no complementary sequences in the RNA with which to anneal, are termed intervening sequences or introns. The 5′-m^{7}G cap is not shown but the 3′-poly(A) tail is. (c) A schematic representation of the gene showing the seven introns (gray) and eight exons (red) as well as the number of base pairs in each exon. Intron sizes vary from 251 bp to about 1,600 bp. (Part a reproduced from P. Chambon, *Sci. Am.* 244 [1981]: 60–71. Used with permission of Pierre Chambon, Institute of Genetics and Molecular and Cellular Biology, College of France. Parts b and c modified from P. Chambon, *Sci. Am.* 244 [1981]: 60–71. Used with permission of Pierre Chambon, Institute of Genetics and Molecular and Cellular Biology, College of France.)

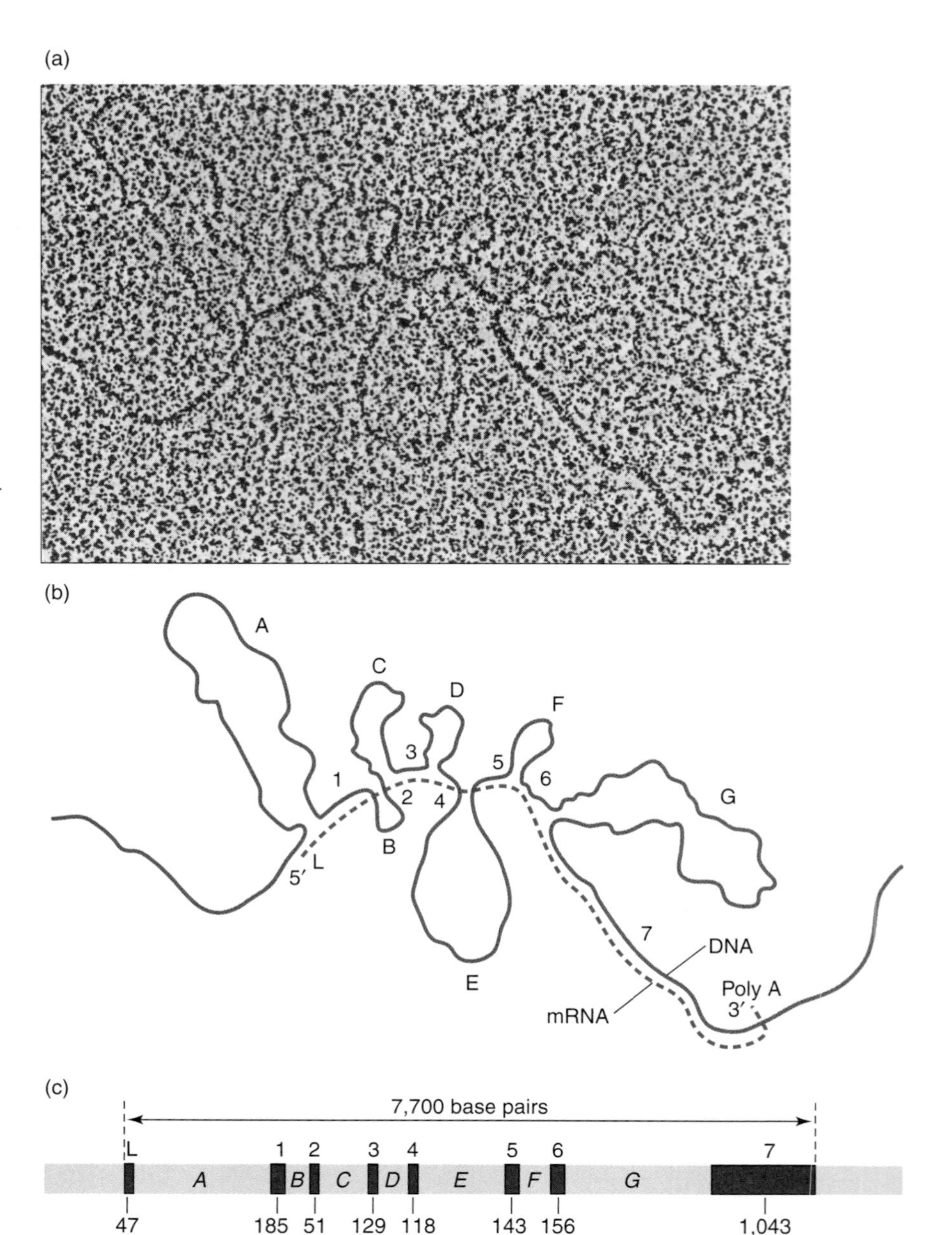

code for amino acids within the ovalbumin gene are interrupted by intragenic regions that do not code for amino acids. Walter Gilbert later coined the term **exon** for an expressed region that codes for amino acids and the term **intron** for a noncoding intervening sequence.

Chambon and coworkers tested the split gene hypothesis by annealing ovalbumin mRNA to single-stranded DNA from a cloned ovalbumin gene. When viewed by electron microscopy, the resulting hybrids had seven loops (introns A–G), each corresponding to a DNA region with no complementary ovalbumin mRNA sequence (FIGURE 19.13). The eight DNA sequences that did anneal to the ovalbumin mRNA (exons L, 1–7) were arranged in the same order in both the gene and the mRNA. This colinearity could be explained by two different mechanisms: (1) the RNA polymerase II machinery transcribes the amino acid coding sequences and somehow skips the noncoding sequences; or (2) the RNA polymerase II machinery transcribes the entire gene to form a long pre-mRNA that is then modified by splicing. These two models were tested by first annealing hnRNA (rather than mRNA) to single-stranded ovalbumin cDNA (DNA prepared using reverse transcriptase to copy ovalbumin mRNA) and then using electron microscopy to view the hybrid structures. The skipping mechanism predicts that hybrid structures formed by hnRNA should be the same as those formed by mRNA. The splicing mechanism predicts that hnRNA should contain pre-mRNA molecules at different stages of splicing and that some of the hybrids should therefore be missing one or more loops. Electron micrographs revealed hybrids that were missing one or more loops and in a few cases were missing all of the loops, supporting the splicing mechanism.

Chambon was by no means the only molecular biologist to encounter a discontinuous gene. Other molecular biologists working at about the same time demonstrated that additional highly expressed eukaryotic genes, including those that code for β-globin, collagen, and insulin genes, are also split. As more genes became available for study, it became clear that most vertebrate genes are discontinuous. A few vertebrate genes, most notably those that code for histones and interferons, lack introns. We don't know why these vertebrate genes evolved so that they don't have introns or conversely why most other vertebrate genes evolved to contain introns. Most genes in higher plants (80–85%) are also discontinuous. In contrast to vertebrates and plants, fewer than 5% of yeast genes are split, but these discontinuous genes account for about 26% of the expressed transcripts. Most yeast genes thus do not have introns but those few that do have them are highly expressed. There is no satisfactory explanation at this time for why more highly expressed yeast genes are split.

Exons tend to be conserved during evolution, whereas introns usually are not conserved.

The Human Genome Project has provided a great deal of information about introns and exons. Table 19.2 summarizes some of this information. Many introns and exons have been identified by aligning cDNA (DNA prepared by using reverse transcriptase to copy mRNA) with genomic DNA. Sequences present in genomic DNA but not cDNA are

TABLE 19.2 Characteristics of Human Genes

	Internal Exon[a] Size	Exon Number	Intron Size	3′-Untranslated Region	5′-Untranslated Region	Coding Sequence[b]	Gene Size
Median	122 bp	7	1,023 bp	400 bp	240 bp	1,100 bp	14,000 bp
Mean	145 bp	8.8	3,365 bp	770 bp	300 bp	1,340 bp	27,000 bp
Sample size	43,317 exons	3,501 genes	27,238 introns	689 transcripts from chromosome 22	463 transcripts from chromosome 22	1,804 sequence entries	1,804 sequence entries

[a]An internal exon is an exon that has an intron on either side of it.
[b]The coding sequence is the mRNA sequence that codes for the polypeptide.
Source: Adapted from E. S. Lander, et al., *Nature* 409 (2001):860–921.

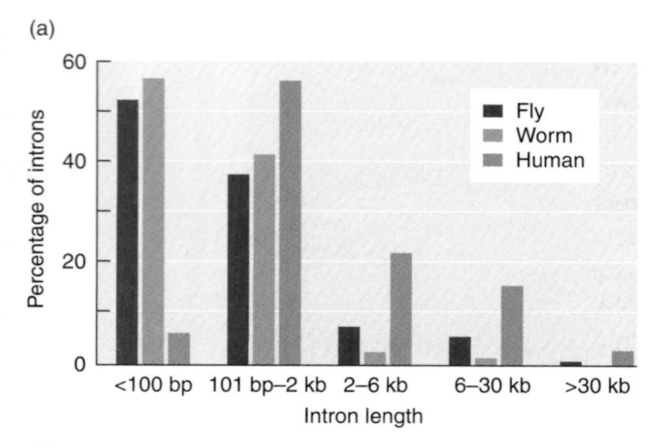

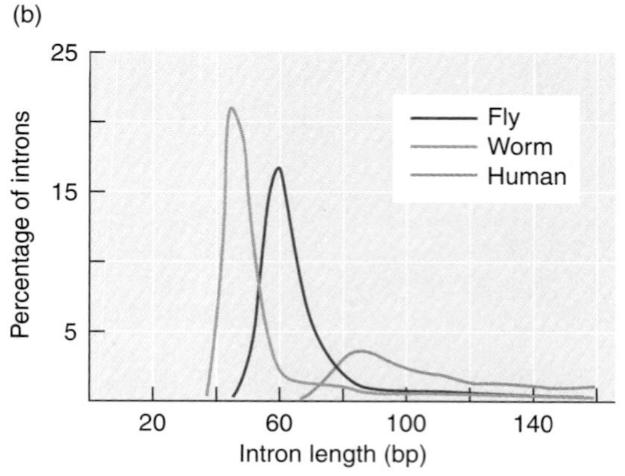

FIGURE 19.14 Size distributions of introns in sequenced genomes. (a) Introns and (b) short introns [enlarged from a]. (Adapted from E. S. Lander, et al., *Nature* 409 [2001]: 860–921.)

introns, whereas sequences present in both genomic DNA and cDNA are exons. cDNA fragments (usually 200–500 nucleotides long) prepared by copying one or both ends of an mRNA are often used in place of full length cDNAs. These fragments are known as **expressed sequence tags** (**EST**) because they (1) represent a snapshot of the genes that are expressed in specific tissues, specific developmental stages, or both; and (2) can be used as hybridization probes to tag complementary chromosomal DNA sequences. Special computer software helps to collect, organize, and analyze the enormous amount of data that is required to identify introns and exons. This convergence of biotechnology and computer technology helped create the exciting new field of **bioinformatics**.

The number of introns varies greatly from one gene to another. We already mentioned that most histone genes do not have any introns. At the other extreme, the titin gene, which codes for a major protein in striated muscles, has 363 introns. Intron size distributions vary considerably from one organism to another (FIGURE 19.14a and b). Human introns tend to be longer than either *Caenorhabditis elegans* (worm) or *Drosophila melanogaster* (fly) introns. The preferred minimum lengths are 47 bp for worm, 59 bp for fly, and 87 bp for human. However, much longer introns are present in all three organisms. Human intron size distribution is very broad indeed, ranging from about 60 to more than 30,000 bp (Figure 19.14a). The longest known intron, almost 500,000 bp long, is present in the human *NRXN3* gene that codes for neurexin, a protein that functions as a cell adhesion molecule in the nervous system. Because the RNA polymerase II transcription rate is about 1,000 to 2,000 nucleotides per minute, transcription of this one intron requires at least four to eight hours. Introns in higher plants tend to be shorter than those in vertebrates, ranging in size from about 60 to 10,000 bp long.

Exons tend to be more uniform in size than introns (FIGURE 19.15). The average human gene has about nine exons. Most internal exons in human, worm, and fly are between 50 and 200 bp long (Figure 19.15). The average human exon (145 bp long) is considerably shorter than the average human intron (3,365 bp long). The same holds true for exons and introns in other higher animals and plants. Because exons tend to be so much smaller than introns, most genes are more intron than exon. For instance, more than 98% of the nucleotides in the human dihydrofolate reductase gene are in its introns. There is considerable selective pressure to conserve exon sequence because exon mutations would alter proteins. There is much less selective pressure to maintain intron size or sequence because intron mutations usually do not affect polypeptide sequence. The one notable exception—intron mutations that interfere with splicing—often have devastating effects because they produce mRNA with the wrong coding information (see below). The fact that exon sequences tend to be conserved provides an important

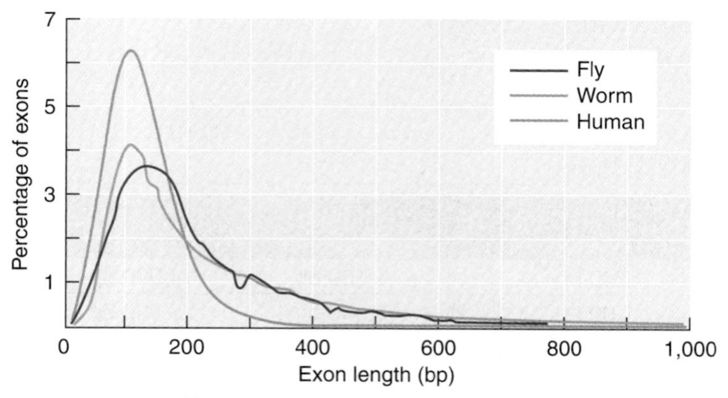

FIGURE 19.15 Size distributions of exons in sequenced genomes. (Adapted from E. S. Lander, et al., *Nature* 409 [2001]: 860–921.)

means for detecting them. When the sequences of the same gene from two or more organisms are known, identical or very similar sequences are likely to be exons.

Genes in different organisms that evolved from a common ancestral gene through the course of evolution, called **orthologs**, often are very different in length even though they code for similar and sometimes identical polypeptides. Ortholog size differences result from variations in intron rather than exon lengths. Because the coding sequence length is the sum of the exon lengths, one would predict that average coding sequence lengths should be similar in different organisms. Consistent with this prediction, average coding sequence lengths in human, worm, and fly are 1,340, 1,311, and 1,497 bp, respectively.

Individual transcription unit sizes vary greatly. Many human transcription units are more than 100 kb long and some are considerably longer. The longest known transcription unit, the dystrophin gene (2.6×10^6 bp), first came to the attention of molecular biologists because a mutated form causes muscular dystrophy. Coding sequence size distribution tends to be somewhat more uniform than transcription unit size distribution. However, some human transcription units have very long coding sequences. The longest known coding sequence (80,760 bp) is produced by the titin gene.

A single pre-mRNA can be processed to produce two or more different mRNA molecules.

Splicing may occur so that (1) each and every exon in a pre-mRNA is incorporated into one mature mRNA through the joining of all successive exons, or (2) one combination of exons in a pre-mRNA is incorporated in one mRNA while other combinations are incorporated in other mRNAs. New high-throughput sequencing technology indicates that greater than 90% of human pre-mRNAs go through the latter form of splicing, termed **alternative splicing**, to produce mRNA.

Exons are either constitutive or regulated. A **constitutive exon** is included in all mRNAs formed from a pre-mRNA. A regulated exon is included in some mRNAs but not in others. The term **cassette exon** is used to describe an internal exon that is completely included in some mRNAs and completely excluded from other mRNAs. Many alternate splicing patterns are possible for a typical multi-exon pre-mRNA. A cassette exon located between two constitutive exons may either be included or skipped when a pre-mRNA is spliced to form mRNA (FIGURE 19.16a). Inclusion of one cassette exon may prevent the inclusion of other cassette exons. As shown in FIGURE 19.16b, such mutually exclusive splicing occurs when an array of cassette exons is located between two constitutive exons and only one of the cassette exons can be incorporated into mature mRNA. An intron between two constitutive exons may be included in some mRNAs but not in others (FIGURE 19.16c). A single DNA sequence thus may act as either an intron (when it is excised) or an exon (when it is incorporated). Length variant exons have alternative splice sites at their 3′-end (FIGURE 19.16d) or their 5′-end (FIGURE 19.16e). When mRNAs formed by alternative splicing of a given pre-mRNA are translated, the proteins

(a) A cassette exon (blue) can be skipped or included

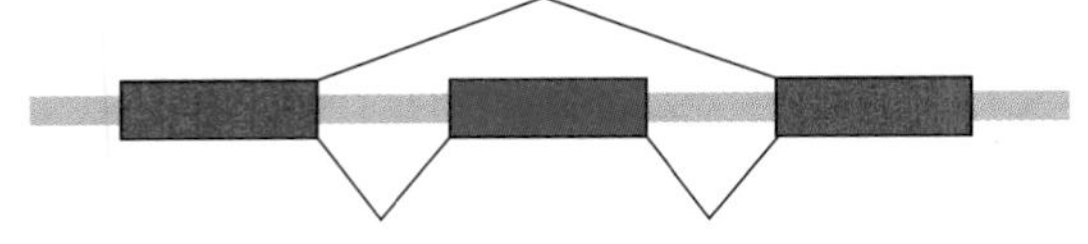

(b) Only one exon in an array of optional exons (blue and green) may be included in mature mRNA

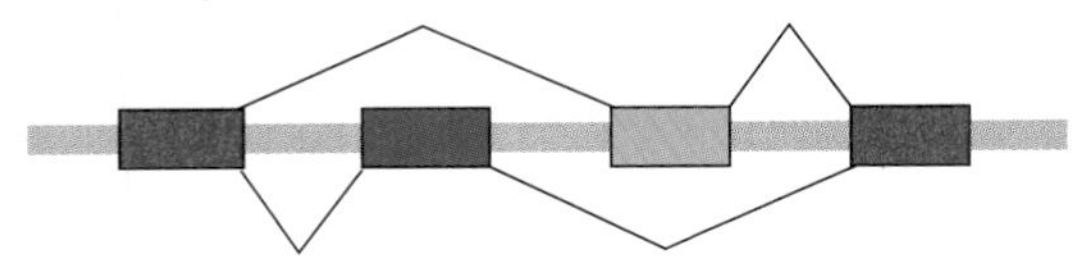

(c) An intron (yellow) may be retained or excluded

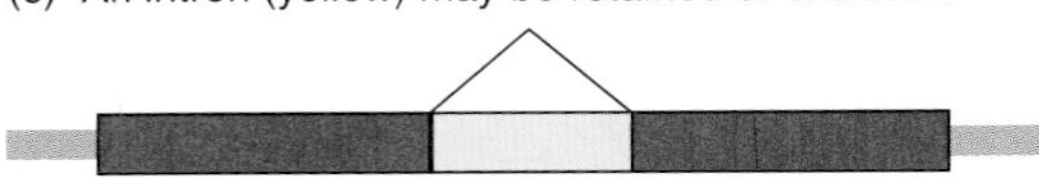

(d) An exon may have alternative splice sites at its 3′-end

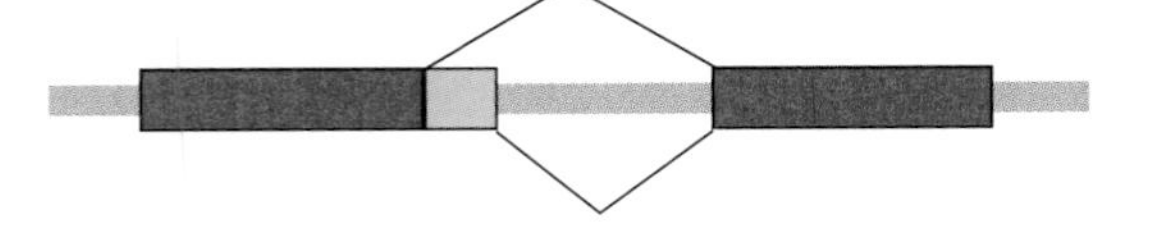

(e) An exon may have alternative splice sites at its 5′-end

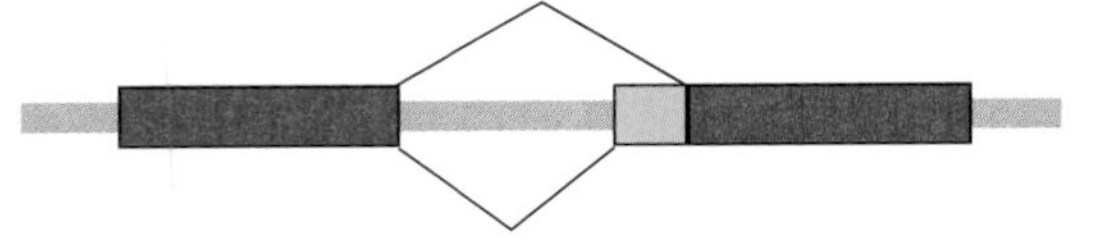

FIGURE 19.16 **Patterns of alternate splicing found in nature.** (a–e). Red boxes correspond to constitutive exons, blue and green boxes to cassette (or optional) exons, cyan boxes to segments within constitutive exons, and gray regions to introns. (Adapted from B. R. Graveley, *Trends Genet.* 17 [2001]: 100–107.)

formed usually share common functions but differ within one or more domains. Sometimes, alternative splicing introduces a termination codon in mRNA, causing a premature end to protein synthesis. Alternative splicing may also alter the reading frame so that the same RNA sequence is present in two different reading frames.

Combinations of the various splicing patterns within individual genes lead to the formation of multiple mRNAs.

We illustrate how alternative splicing can generate a diverse collection of proteins from a single gene by considering the *WT1* and *Dscam* genes. The *WT1* gene (FIGURE 19.17) is so named because mutant forms cause Wilms tumor, the most common kidney tumor in children. The *WT1* gene, which is about 50 kb long, codes for multiple variant forms of a 52 to 56 kDa protein by alternative splicing. Each variant form or **isoform** has a proline- and glutamine-rich domain at its N-terminus and a DNA-binding domain containing four zinc finger motifs at its C-terminus. Some isoforms have a lysine-threonine-serine (KTS) sequence inserted between zinc fingers 3 and 4, while others do not. WT1 isoforms lacking the KTS sequence appear to function as transcription factors, while those that have the KTS sequence appear to associate with splicing factors.

The *Drosophila Dscam* gene, which derives its name from the fact that it codes for protein isoforms that appear to be *Drosophila* homologs of the Downs syndrome cell adhesion molecule (FIGURE 19.18), illustrates the amazingly diverse array of protein isoforms that can be generated from just one gene through alternative splicing. The developing fly requires the diverse array of Dscam isoforms to assist in the complex task of accurately and reproducibly directing each of the fly's approximately 250,000 growing neuron axons to its proper destination. Exons 4, 6, 9, and 17 in each Dscam mRNA are generated by selecting one variant exon from an array of

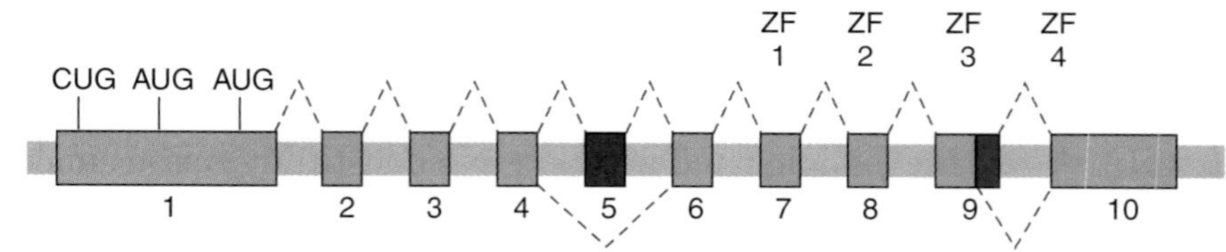

FIGURE 19.17 *Alternative splicing of the WT1 (Wilm's tumor) gene.* The *WT1* gene generates up to 24 isoforms as a result of three alternative translation initiation codons in exon 1, alternative splicing of cassette exon 5, alternative 5′ splice sites in exon 9, and other posttranscriptional modifications. Each isoform has a proline- and glutamine-rich domain at its amino terminus and a DNA-binding domain containing four zinc fingers (ZF) motifs at its carboxyl terminus. Some alternative splices insert three amino acids (KTS) between zinc fingers 3 and 4, whereas others do not. WT1 isoforms lacking KTS appear to function as transcription factors, whereas those that have KTS appear to be associated with splicing factors. (Adapted from G. C. Roberts and C. W. J. Smith, *Curr. Opin. Chem. Biol.* 6 [2002]: 375–383.)

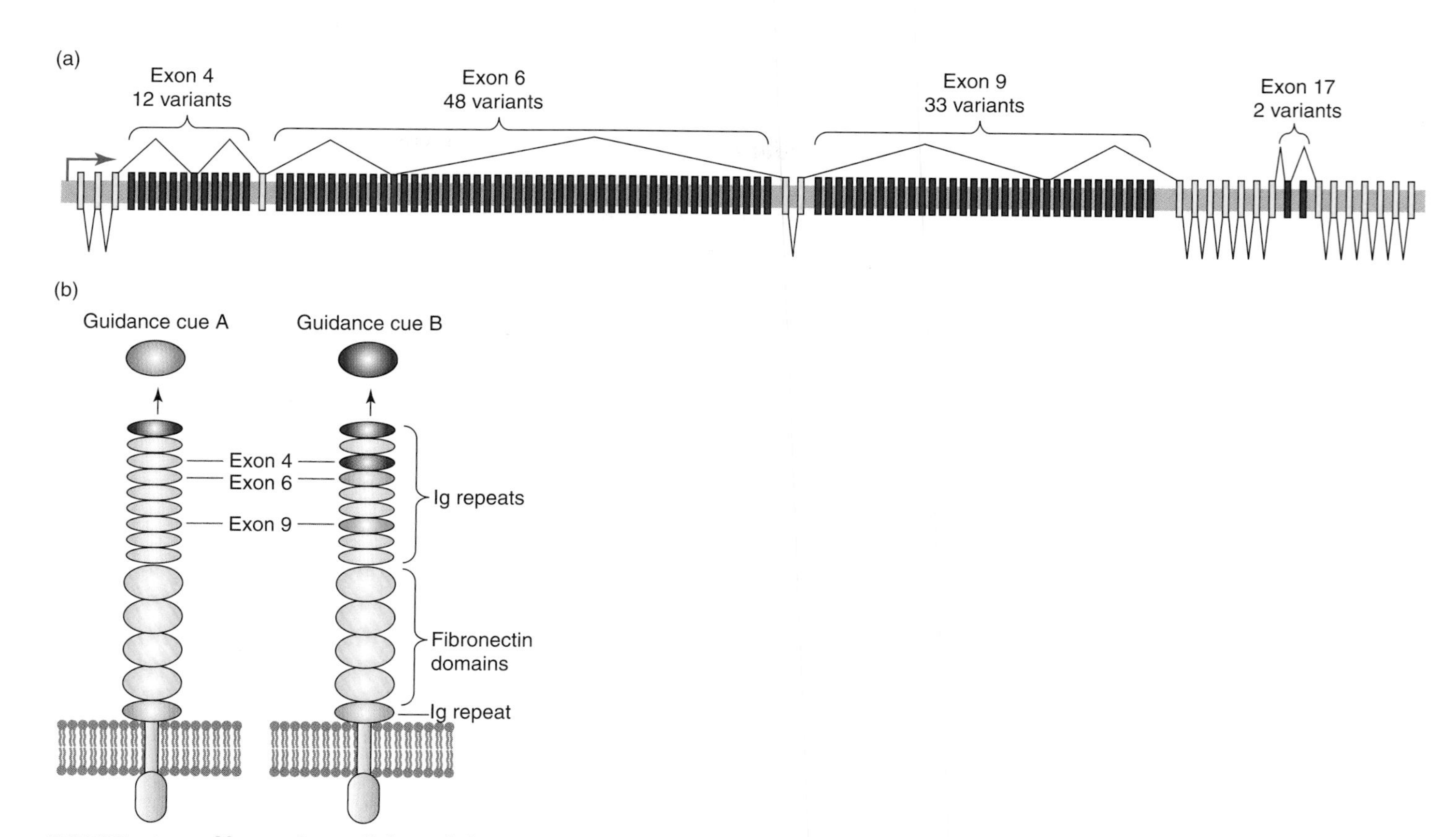

FIGURE 19.18 **Alternative splicing of the gene encoding *Drosophila Dscam*.** The constitutive splicing events are indicated below the gene and alternative splicing events are depicted above the gene. (a) The organization of the *Dscam* gene. The constitutive exons are yellow and the alternative splicing exons are purple. The *Dscam* gene contains four sites of alternative splicing at exons 4, 6, 9, and 17. There are 12 variants of exon 4, 48 variants of exon 6, 33 variants of exon 9, and 2 variants of exon 17. Only one variant exon from each position is included in the Dscam mRNAs. Alternative exons 4, 6, and 9 encode alternative versions of immunoglobulin repeats. (b) Functional consequences of Dscam alternative splicing. The Dscam protein functions as an axon guidance receptor. It is thought that each Dscam variant will interact with a unique set of axon cues. The form of Dscam shown on the left will interact with guidance cue A. The form of Dscam shown on the right contains different sequences encoded by exons 4, 6, and 9 and thus interacts with guidance cue B, rather than guidance cue A. Neurons that express the form of Dscam shown on the right will be attracted in a different direction than neurons expressing the form shown on the left. (Modified from *Trends Genet.*, vol. 17, B. R. Graveley, Alternative splicing: increasing diversity . . ., pp. 100–107, copyright 2001, with permission from Elsevier [http://www.sciencedirect.com/science/journal/01689525].)

mutually exclusive exon variants (Figure 19.18a). Exon 4 is selected from an array of 12 variants, exon 6 from an array of 48 variants, exon 9 from an array of 33 variants, and exon 17 from an array of 2 variants. Selection of the exon 4 variant is developmentally regulated, consistent with the notion that at least some of the potential diversity in Dscam isoforms serves to wire neurons correctly. Exons 4, 6, and 9 encode ten immunoglobulin (Ig) repeats to form an extracellular domain that serves as an axon guidance receptor, while exon 17 encodes a transmembrane domain (Figure 19.18b). Because each of the four exons appears to be generated independently of the other three, the total number of possible Dscam isoforms is 38,016 ($12 \times 48 \times 33 \times 2$) or two- to threefold greater than the total number of transcription units in *Drosophila*.

Drosophila form an mRNA that codes for essential protein isoforms by alternative *trans*-splicing.

Each of the splicing examples described thus far fits a pattern known as ***cis*-splicing** because the exons incorporated into mRNA all come from the same pre-mRNA molecule. We can envision a different kind of splicing, however, in which the exons incorporated into mRNA derive from two different pre-mRNA molecules. This type of splicing, termed ***trans*-splicing**, is required to produce essential proteins that help to establish and maintain the *Drosophila* chromatin structure. The DNA region that specifies these essential proteins is termed the *mod(mdg4)* locus. The first clue that pre-mRNAs produced by the *mod(mdg4)* locus might be processed by alternative *trans*-splicing came from genetic studies performed by Victor Corces and coworkers. Corces' group had identified two different mutations at opposite ends of the *mod(mdg4)* locus. Either mutation was lethal to *Drosophila* that had it on both alleles. *Drosophila* remained viable, however, if they had the first mutation on one allele and the second mutation on the other allele. Sequence studies performed in 2001 revealed that both complementary strands in the *mod(mdg4)* locus contribute exons to mature mRNA isoforms. Pre-mRNAs are transcribed independently from the top and bottom DNA strands (FIGURE 19.19). *Trans*-splicing occurs when the constant exon region (shown in orange) in the mRNA transcribed from the top strand joins one of the two exons that are transcribed from the bottom strand. This alternative *trans*-splicing permits flies with different lethal mutations on their two *mod(mdg4)* alleles to survive by incorporating unmutated exons from each *mod(mdg4)* allele to produce wild-type mRNA isoforms. A few other *Drosophila* mRNAs appear to be formed by *trans*-splicing but the process has not yet been observed in vertebrates. It remains to be seen whether alternative *trans*-splicing will turn out to be a general phenomenon affecting many genes in a variety of organisms or will be limited to just a few genes in flies.

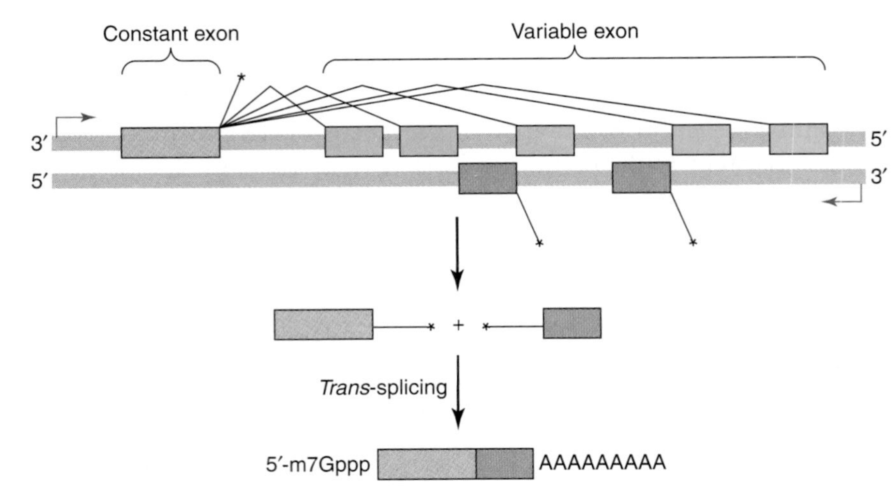

FIGURE 19.19 **An example of *trans*-splicing.** *mod(mdg4) trans*-splicing in *Drosophila*. (top) The positions of the constant (orange) and alternatively spliced exons (purple and cyan) in the *mod(mdg4)* locus. Pre-mRNAs are transcribed independently from the top and bottom DNA strands. *Trans*-splicing occurs when the constant exon in the mRNA transcribed from the top strand joins one of the two exons transcribed from the bottom strand. (middle) One possible exon combination that will result in *trans*-splicing. (bottom) A mature mRNA that is formed by *trans*-splicing. The asterisk indicates ends that can splice. (Adapted from T. Horiuchi and T. Aigaki, *Biol Cell* 98 (2006): 135–140.)

Pre-mRNA requires specific sequences for precise splicing to occur.

Cells must excise introns with great precision because failure to do so would change the mRNA coding sequence with devastating consequences for the cell. The cell's splicing machinery must therefore unambiguously identify the exon/intron boundary at the start of an intron, termed the **5′-splice site**, and intron/exon boundary at the end of an intron, termed the **3′-splice site**. The yeast splicing machinery requires three nearly invariant short nucleotide sequences (FIGURE 19.20), the **5′-splice sequence** (AG/GUAUGU), the **3′-splice sequence** (CAG/G), and the **branchpoint sequence** (UACUAAC), to identify and excise an intron (a slash indicates a splice site). The red "A" in Figure 19.20 identifies the branching nucleotide that participates in the chemistry of pre-mRNA splicing; see below). As the names imply, the three short sequences flank the 5′-splice site, the 3′-splice site, and the branching nucleotide, respectively. Many yeast introns also have a stretch of 8 to 12 pyrimidines (mostly uracil), termed a **polypyrimidine tract**, just upstream from the 3′-splice site.

The majority of animal and plant introns resemble yeast introns in having GU at their 5′-end and AG at their 3′-end. This is certainly true of human introns, where 98.12% of the thousands of different confirmed human introns have the GU-AG pattern. Mutations that alter these nucleotides cause inherited metabolic diseases in humans. For instance, certain individuals synthesize defective hemoglobin molecules because a GU sequence at the beginning of one of the β-globin introns has been altered. This alteration, which blocks splicing at the normal splice site and leads to splicing at abnormal sites, causes the production of a nonfunctional β-globin chain. Animal and plant splicing machinery also require 5′-splice sequences, 3′-splice sequences, and branchpoint sequences to identify and splice introns but the three short sequences are not as highly conserved as in yeast. In humans and other mammals, the 5′-splice consensus sequence is AG/GURAGU, the 3′-splice consensus sequence is NYAG/G, and the branchpoint consensus sequence is YNCURAY (where R denotes a purine; Y, a pyrimidine; and N, any nucleotide). The arrangement of these three short consensus sequences with respect to a typical mammalian GU-AG intron is shown in FIGURE 19.21a. The branching nucleotide is usually 11 to 40 bases upstream from the 3′-splice site.

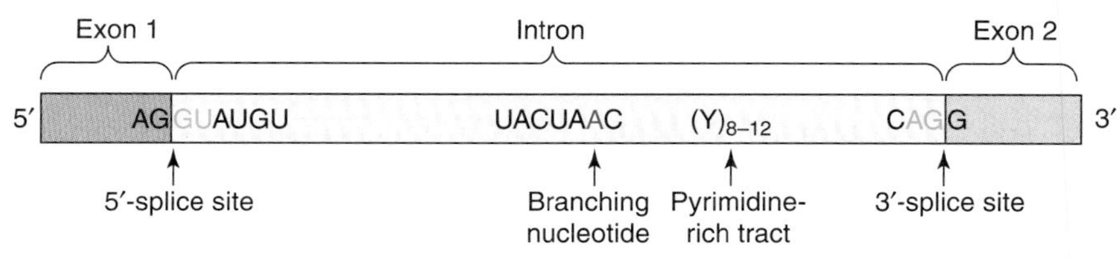

FIGURE 19.20 Sequences in yeast pre-mRNA that help to define the 5′- and 3′-splice sites. Y denotes a pyrimidine nucleotide. The branching nucleotide, A, is shown in red. The GU dinucleotide at the 5′-end of the intron and the AG nucleotide at the 3′-end of the intron are shown in green.

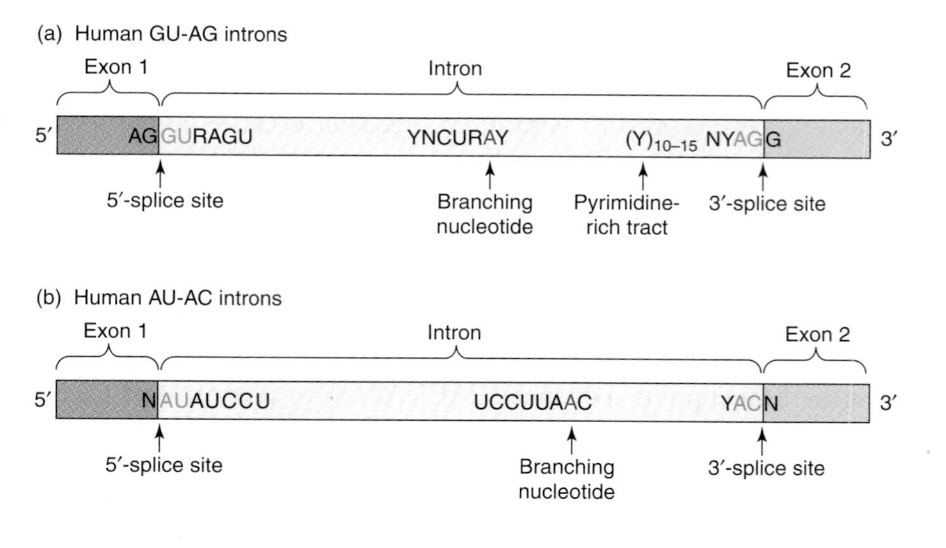

FIGURE 19.21 **Consensus sequences that define human introns.** Consensus sequences that define human (a) GU-AG introns and (b) AU-AC introns. Y is any pyrimidine nucleotide, R is any purine nucleotide, and N is any nucleotide. The branching nucleotide is shown in red and the two terminal dinucleotides in the intron are shown in green.

Most GU-AG introns also have a polypyrimidine tract just upstream from the 3′-splice consensus sequence, extending ten or more nucleotides back into the intron. The same splicing machinery that processes GU-AG introns also appears to process variant introns with GC at their 5′-end. These so-called GC-AG introns account for 0.76% of confirmed human introns. Another 0.10% of human introns begin with AU and end with AC. Consensus sequences associated with these so-called AU-AC introns are shown in FIGURE 19.21b. Initial studies suggested that one type of splicing machinery was required to process GU-AG introns and another to process AU-AC introns. It now appears that even though humans do in fact have two different kinds of splicing machinery; dinucleotides at the beginning and end of the intron are not the sole determinant of specificity. The splicing machinery that was initially thought to excise only GU-AG introns thus can also excise some AU-AC introns and vice versa.

Two splicing intermediates resemble lariats.

Investigators hoped to find clues to the splicing mechanism by isolating splicing intermediates and then determining the structure of these intermediates. A major breakthrough occurred in the early 1980s with the discovery that HeLa cell extracts, supplemented with ATP and magnesium ions, excise introns from added pre-mRNA. Two research teams, one led by Phillip A. Sharp and the other by Tom Maniatis, took advantage of this *in vitro* splicing system to search for splicing intermediates. Both groups used the following approach: they (1) added ^{32}P-labeled pre-mRNA with a single exon-intron-exon array to the *in vitro* splicing system; (2) separated RNA products formed after various incubation times by polyacrylamide gel electrophoresis; and (3) detected radioactive bands by autoradiography. FIGURE 19.22 shows an informative autoradiogram from an experiment performed by the Maniatis group. Proposed structures for splicing intermediates and products are shown to the right of the autoradiogram. Two of these bands, those for the 130- and the 339-nucleotide products (indicated by the red

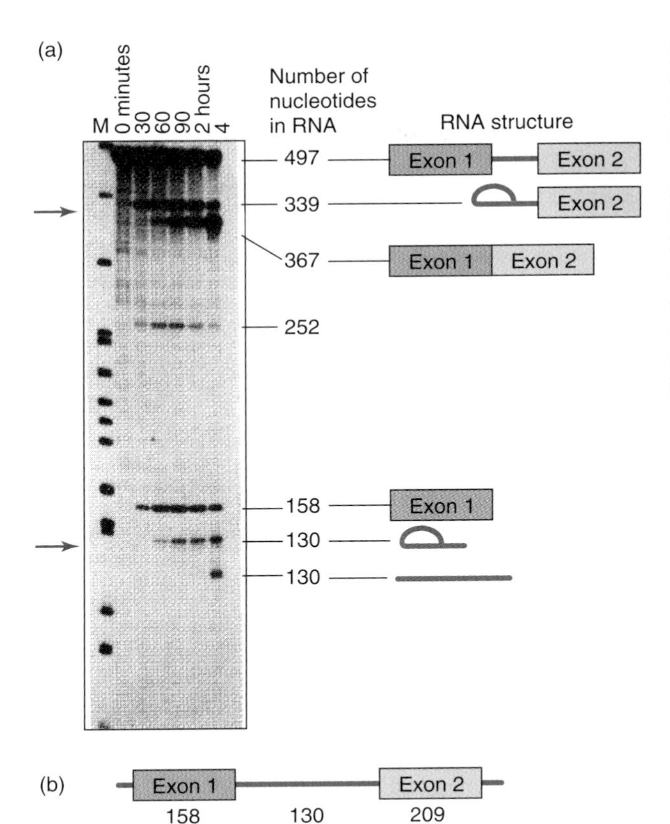

FIGURE 19.22 Electrophoretic fractionation of RNA products formed by a cell-free HeLa cell splicing system. (a) [^{32}P]Phosphate labeled pre-mRNA was added to a cell-free HeLa cell splicing system and incubated in the presence of ATP and magnesium ions. RNA products were removed at various times, fractionated by polyacrylamide gel electrophoresis and then detected by autoradiography. The time of sample incubation is given at the top of the lane for that sample. Proposed structures for RNA products are shown to the right of the autoradiogram. (b) The structure of the starting pre-mRNA. Red arrows to the left indicate two bands of special interest (see text). (Part a adapted from B. Ruskin, et al., *Cell* 38 [1984]: 317–331. Photo courtesy of Michael R. Green, University of Massachusetts Medical School. Part b adapted from B. Ruskin, et al., *Cell* 38 [1984]: 317–331.)

arrows), are of special interest because they are splicing intermediates that contain the intron. Evidence supporting this statement is as follows: (1) each band appears only when ATP and magnesium ions are added to the splicing system; (2) neither band appears when the intron has some nucleotide other than guanylate at its 5′-end; and (3) an intron-specific DNA probe anneals to the RNA in each band. Further studies showed that the 339-nucleotide RNA also contained the exon on the 3′-side of the intron.

Electrophoresis studies suggested that both the 130- and 339-nucleotide RNA products have unusual structures. Recall from Chapter 5 that a nucleic acid's chain length can be determined by comparing its migration distance with migration distances of standards on the same gel. When the chain lengths of the 130- and 339-nucleotide RNA products were determined in this way using a 5% polyacrylamide gel, the calculated chain lengths were significantly greater than the actual values. Moreover, the discrepancies were even greater with a 10% polyacrylamide gel. Similar discrepancies had previously been reported for circular RNA molecules, suggesting that the 130- and 339-nucleotide RNA products might be circular. Because rare 2′→5′ branches were known to occur in nuclear RNA molecules with poly(A) tails, it seemed possible that the 130- and 339-nucleotide RNA products might have lariat-like structures. Maniatis and coworkers set out to find experimental proof for this model.

The lariat model predicts that the 130- and 339-nucleotide RNA products each has a single 2′→5′ phosphodiester bond like the one shown in FIGURE 19.23. Maniatis and coworkers confirmed this prediction by

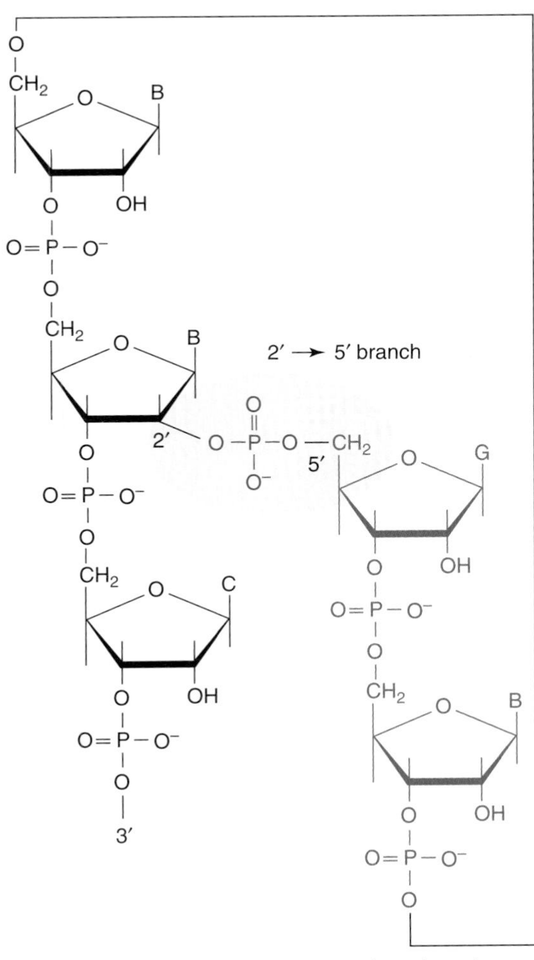

FIGURE 19.23 Lariat structure showing the 2′→5′ branch.

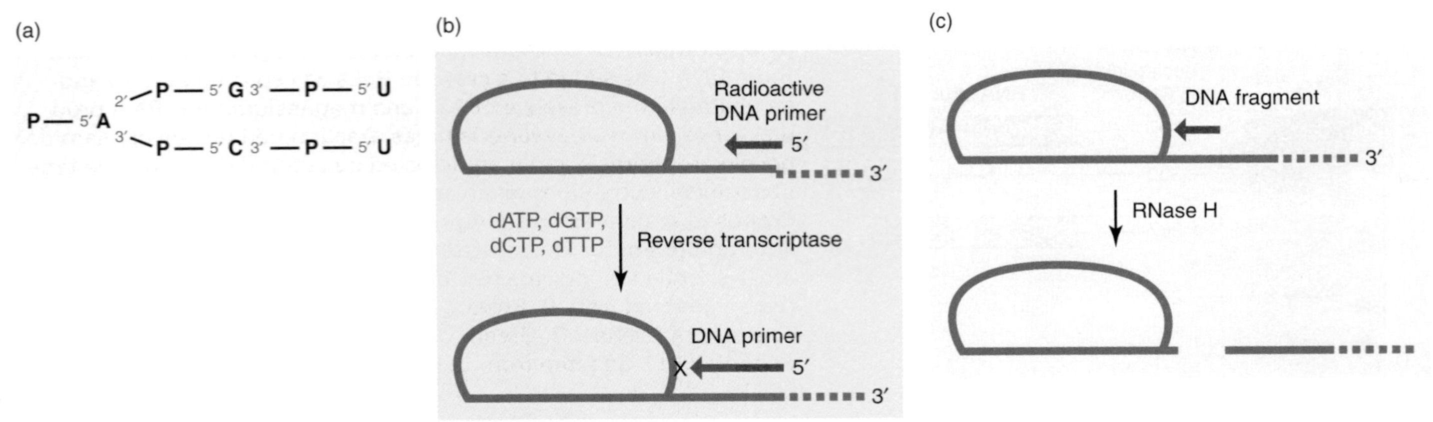

FIGURE 19.24 **Experiments that support the lariat model.** (a) Structure of the branchpoint shown above was deduced from the structure of the oligonucleotide that was resistant to digestion. (b) Primer extension. Reverse transcriptase cannot extend a short primer annealed to the 3′-end of the intron beyond the branchpoint in the lariat structure of either the 130 nucleotide or the 339 nucleotide RNA product. (c) A short DNA primer was annealed to the intron region adjacent to the branchpoint and the hybrid was digested with RNase H, which is specific for DNA–RNA hybrids. Two products were formed, a loop and a linear fragment. The loops were identical when either the 130-nucleotide or 339-nucleotide product was used. The linear fragment released was longer when the 339-nucleotide fragment was used, however.

digesting the RNA molecules with an RNase that cleaves 3′→5′ but not 2′→5′ phosphodiester bonds, separating the oligonucleotides thus generated by thin-layer chromatography, and then showing that one of the oligonucleotides indeed does have a 2′→5′ phosphodiester bond. This oligonucleotide, which is the same for both the 130- and 339-nucleotide RNA products, has an adenylate branching nucleotide linked to guanylate by a 2′→5′ phosphodiester bond and to cytidylate by a 3′→5′ phosphodiester bond (FIGURE 19.24a). The lariat model also predicts that the 2′→5′ branch should block reverse transcriptase as it extends a short DNA primer annealed to the 3′-end of the intron (FIGURE 19.24b). Maniatis and coworkers confirmed this prediction by demonstrating that a primer annealed to the 3′-end of the intron could be extended up to the 2′→5′ branchpoint site but no further. Finally, the lariat model predicts that cleaving the 130- or 339-nucleotide RNA product on the 3′-side of the branchpoint should produce an RNA circle and a linear RNA fragment. Maniatis and coworkers annealed a short DNA fragment to a segment of the intron on the 3′-side of the branchpoint of the 130-nucleotide fragment and then cleaved the DNA–RNA hybrid region with RNase H (FIGURE 19.24c). As predicted, this cleavage generated a circular RNA and a short linear fragment. An identical circular RNA was produced when the experiment was repeated with the 339-nucleotide fragment but the linear fragment was considerably longer.

Splicing consists of two coordinated transesterification reactions.

Kinetic studies using autoradiograms like that in Figure 19.22 show that band intensities change with time, providing data that can be used to determine precursor-product relationships. The splicing mechanism that fits the structural and kinetic data best is shown in FIGURE 19.25. According to this mechanism, splicing requires two coordinated

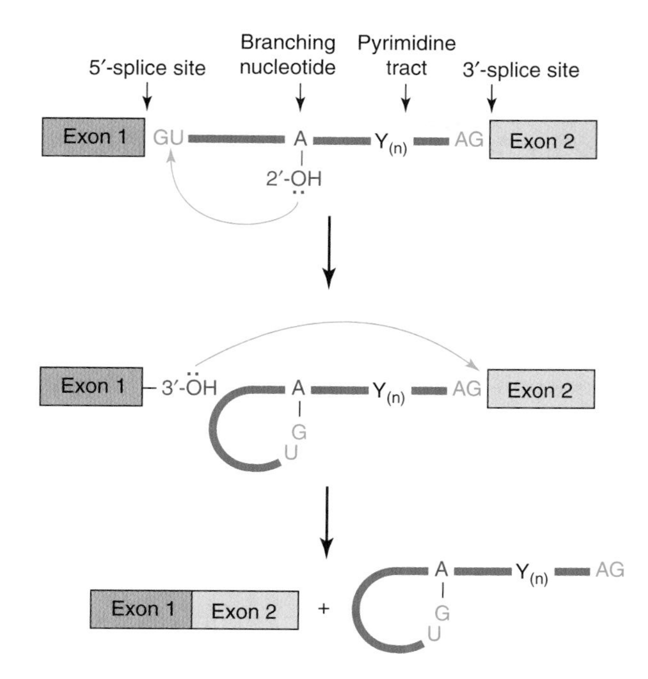

FIGURE 19.25 The two coordinated transesterification steps in splicing. First step—The 2′-hydroxyl on the branching nucleotide (A, shown in red) initiates a nucleophilic attack on the 5′-splice site to produce a lariat structure connected to exon 2. Second step—The 3′-hydroxyl generated at the 3′-end of exon 1 during the first transesterification step initiates a nucleophilic attack on the phosphodiester bond at the 3′-splice site. This transesterification reaction joins the exons and releases the intron in a lariat form.

transesterification steps. During the first step, the 2′-hydroxyl of an adenosine within the branchpoint sequence, the **branching nucleotide**, initiates a nucleophilic attack on the phosphodiester bond at the 5′ exon/intron boundary (the 5′ splice site). The resulting transesterification reaction generates a free 5′-exon (exon 1) and a lariat intermediate with a 2′→5′ branch structure that is still attached to the 3′-exon (exon 2). During the second step, the 3′-OH of the free exon (exon 1) initiates a nucleophilic attack on the phosphodiester bond at the 3′ intron/exon boundary (the 3′-splice site). The resulting transesterification reaction releases the lariat intron and joins exon 1 and exon 2. Lariat structures do not accumulate *in vivo* because cells have a phosphodiesterase that cleaves 2′→5′ phosphodiester bonds and nucleases that degrade the resulting linear RNA strand.

19.4 Spliceosomes

Aberrant antibodies, which are produced by individuals with certain autoimmune diseases, bind to small nuclear ribonucleoprotein particles (snRNPs).

Although it might seem that only a few specific endonucleases and a ligase might suffice to catalyze splicing reactions, eukaryotic cells use a very sophisticated piece of machinery to carry out the steps involved in splicing. As often happens in science, the first clue to the nature of this machinery came from studies in a seemingly unrelated field. People suffering from autoimmune diseases known as mixed connective tissue disease and systemic lupus erythematosus make aberrant antibodies that attack components of their own cells. One such antibody, called

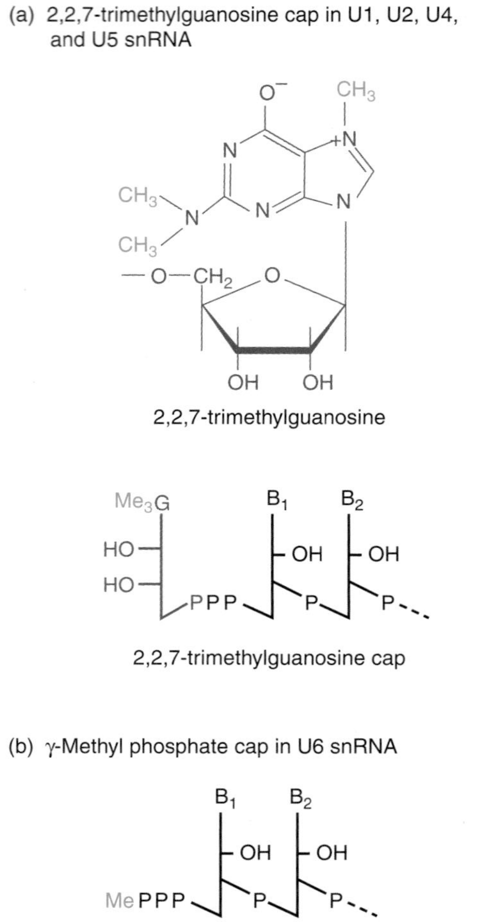

(a) 2,2,7-trimethylguanosine cap in U1, U2, U4, and U5 snRNA

2,2,7-trimethylguanosine

2,2,7-trimethylguanosine cap

(b) γ-Methyl phosphate cap in U6 snRNA

FIGURE 19.26 Cap structures present in snRNA molecules. (a) 2,2,7-trimethylguanosine cap in U1, U2, U4, and U5 snRNA shown as a standard chemical structure (top) and as part of a stick structure (bottom) and (b) γ-methyl phosphate cap in U6 snRNA.

the anti-Smith or anti-Sm antibody, binds to small ribonucleoprotein particles in the cell nucleus. Joan Steitz and coworkers suspected that these small ribonucleoproteins might help to process pre-mRNA and set about isolating them so that they could study their function(s). The isolation method they devised was to first tag the ribonucleoprotein particles in a human cell extract with antibodies from the serum of a lupus patient and then pull the tagged ribonucleoproteins out of the extract with an insoluble *Staphylococcus aureus* cell wall preparation that binds to antibodies. When Steitz and coworkers analyzed RNA molecules that had been extracted from the ribonucleoprotein particles by polyacrylamide gel electrophoresis, they observed discrete bands corresponding to RNA molecules with chain lengths between 100 and 200 nucleotides long. Because each of the small nuclear RNA molecules is uridine-rich, they are called U1 snRNA, U2 snRNA, U4 snRNA, U5 snRNA, and U6 snRNA. The U1, U2, U4, and U5 snRNAs each has a 2,2,7 trimethyl guanosine cap at its 5′-end, while U6 has a γ-methyl phosphate cap (FIGURE 19.26).

Each snRNA is present in its own small nuclear ribonucleoprotein particle (**snRNP**; or 'snurp' for short). The snRNPs are named U1 snRNP, U2 snRNP, U4 snRNP, U5 snRNP, and U6 snRNP after their snRNA component. U4, U5, and U6 snRNPs combine to form a U4/U6•U5 tri-snRNP particle, which is stabilized by extensive base pairing between U4 and U6 snRNAs (FIGURE 19.27) and protein-protein interactions between U5 snRNP and the U4/U6 di-snRNP. Although all eukaryotes have the same five kinds of ribonucleoproteins, snRNP concentrations vary from organism to organism. For instance, mammalian cells have approximately 10^5 to 10^6 copies of each snRNP, whereas yeast cells have about 100 to 200 copies of each kind.

snRNPs assemble to form a spliceosome, the splicing machine that excises introns.

When Steitz and coworkers first isolated U snRNPs in the late 1970s, they had no way to study the function of these particles. Once *in vitro* splicing systems became available, however, investigators were quickly able to show that U1, U2, U4, U5, and U6 snRNPs are essential splicing machinery components. Some key findings that demonstrated snRNPs are involved in splicing are as follows:

1. Antibodies specific for U1 snRNP block *in vitro* splicing.
2. HeLa cell nuclear extracts lose their ability to splice pre-mRNA when U snRNPs are removed.
3. Splicing does not occur when the first eight nucleotides are removed from the 5′-end of U1 snRNA.
4. U2 snRNP selectively binds to the branching site, U5 snRNP binds to pre-mRNA just upstream of the 3′-splice site.
5. U4 and U6 snRNPs are essential splicing components.

Further experiments showed that the five U snRNPs assemble to form a large ribonucleoprotein splicing machine, the **spliceosome.** The total molecular mass of spliceosome, which also contains non-snRNP protein factors, is about 4.8×10^6 Da.

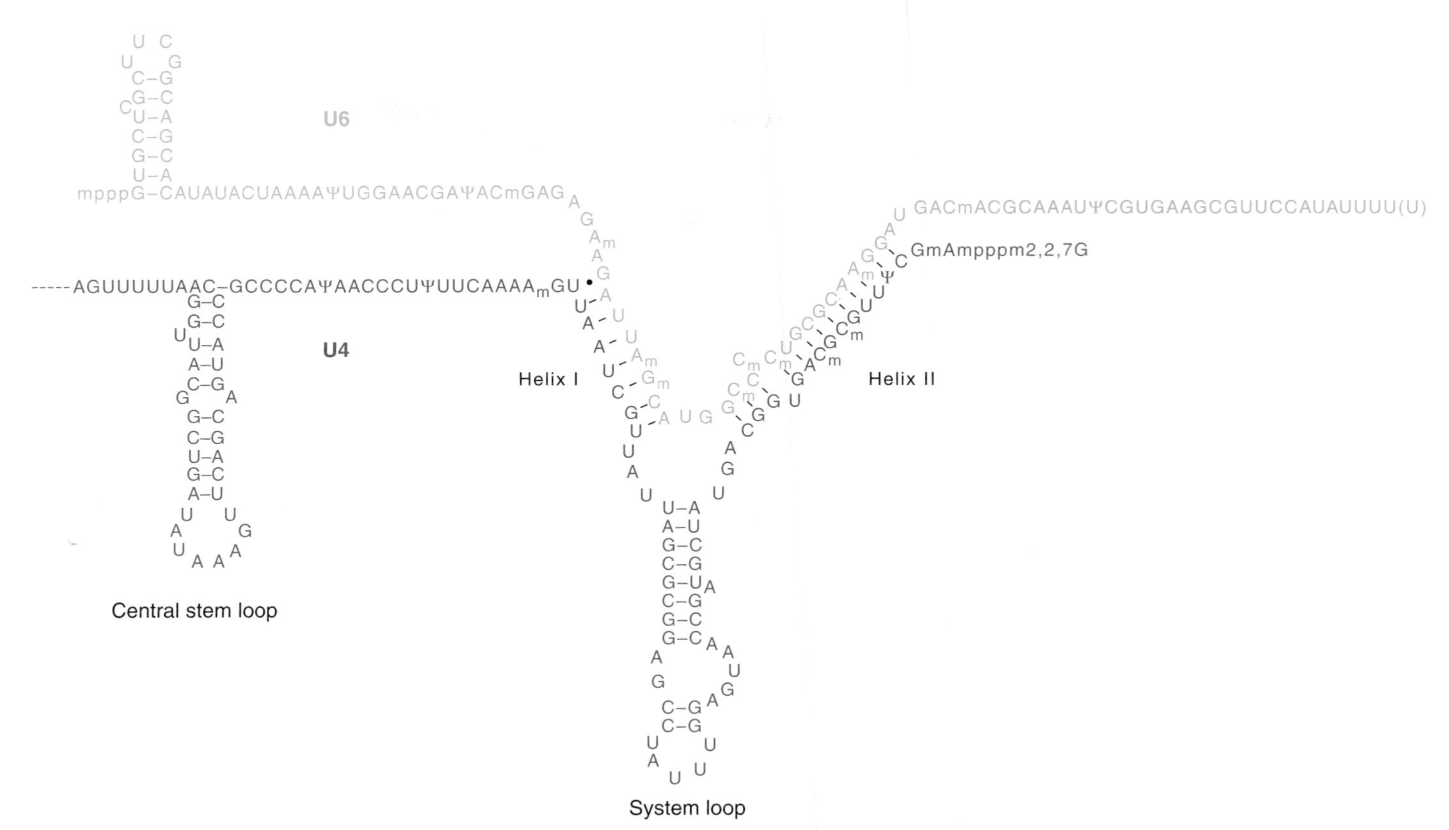

FIGURE 19.27 **Extensive base pairing between yeast U4 and U6 snRNAs in the U4/U6 di-snRNP.** (Adapted from A. Mougin, et al., *J. Mol. Biol.* 317 [2002]: 631–649.)

Ruth Sperling and coworkers have performed cryo-electron microscopy studies of spliceosomes obtained from HeLa cell nuclei. The three-dimensional constructs reveal an elongated globular particle that is made of a large and a small subunit, which are joined to one another and have a tunnel between them (FIGURE 19.28). Sperling and coworkers suggest that the large subunit is a suitable candidate to accommodate the five snRNPs and that non-snRNP protein factors are probably located in the small subunit. They also suggest that the tunnel between the two subunits could accommodate the pre-mRNA component of the spliceosome.

U1, U2, U4, and U5 snRNPs each contains Sm polypeptides, whereas U6 snRNP contains Sm-like polypeptides.

We begin our examination of spliceosomes by studying the structure and formation of their snRNP components. U1, U2, U4, and U5 snRNPs each contains a family of closely related polypeptides known as **Sm polypeptides** (so-named because anti-Sm antibodies bind to them) and specific polypeptides that are unique to each kind of snRNP. Kiyoshi Nagai and coworkers reconstituted the human spliceosomal U1 snRNP from proteins expressed in *E. coli* and snRNA transcribed *in vitro*. The ribonucleoprotein has seven Sm proteins

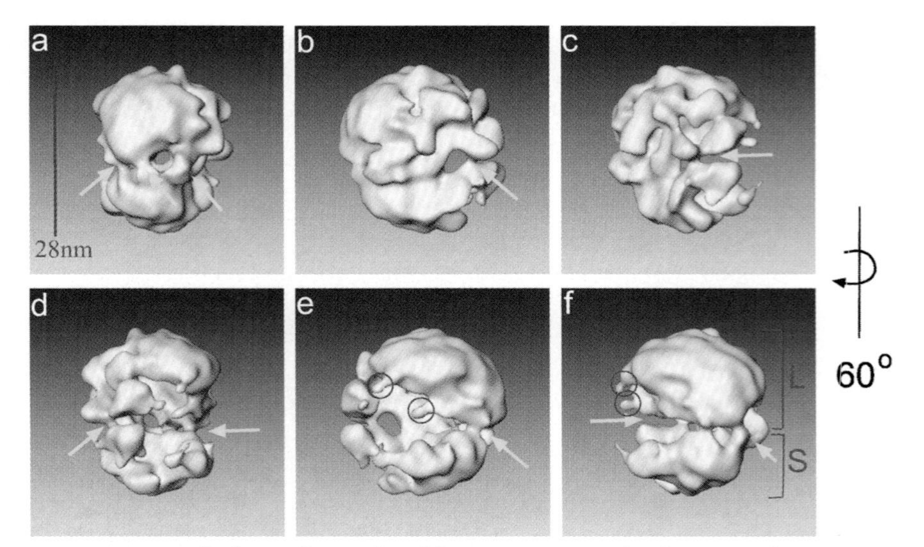

FIGURE 19.28 A three-dimensional image construction by cryo-electron microscopy of native spliceosomes, derived from HeLa cell nuclei, at a resolution of 20 Å. (a–f) Six views of the native spliceosome are shown, separated by rotation of 60 degrees about the central axis. The structure reveals an elongated globular particle made up of two distinct subunits connected to each other, leaving a tunnel in between. The larger subunit is a suitable candidate to accommodate the five snRNPs, and the tunnel could accommodate the pre-mRNA component of the spliceosome. Non-snRNP factors appear to be located in the small subunit. Yellow arrows indicate connecting points between the large (L) and small (S) subunits. Red circles indicate similar protruding bodies. (Reproduced from *Mol. Cell*, vol. 15, M. Azubel, et al., Three-dimensional structure . . ., pp. 833–839, copyright 2004, with permission from Elsevier [http://www.sciencedirect.com/science/journal/10972765]. Photo courtesy of Ruth Sperling, The Hebrew University of Jerusalem.)

(Sm-B, Sm-D1, Sm-D2, Sm-D3, Sm-E, Sm-F, and Sm-G) and large fragments of two other proteins (U1-C and U1-70K). FIGURE 19.29 shows the crystal structure of the reconstituted human spliceosomal U1 snRNP at 0.55 nm resolution. The RNA forms four stem-loop structures. The seven Sm proteins assemble around Sm site nucleotides, which are located between two of the stem loop structures. A similar arrangement of Sm polypeptides is probably also present in U2, U4, and U5 snRNPs. U6 snRNP does not have Sm polypeptides but instead has Sm-like polypeptides known as Lsm polypeptides, which are also arranged in a ring.

It at first seemed that U1, U2, U4, U5, and U6 snRNPs were the only ribonucleoproteins needed to splice pre-mRNA. Later studies, however, revealed the existence of four minor snRNPs that combine with U5 snRNP to form a minor spliceosome that excises AU-AC introns and in some rare cases GU-AG introns. The four minor snRNPs, named U11 snRNP, U12 snRNP, U4atac snRNP, and U6atac snRNP, replace U1, U2, U4, and U6 snRNPs, respectively. Although investigators assumed that minor spliceosomes, like major spliceosomes, would be located in the cell nucleus, recent experiments by Ferenc Müller and coworkers show that minor spliceosomes in vertebrates are located

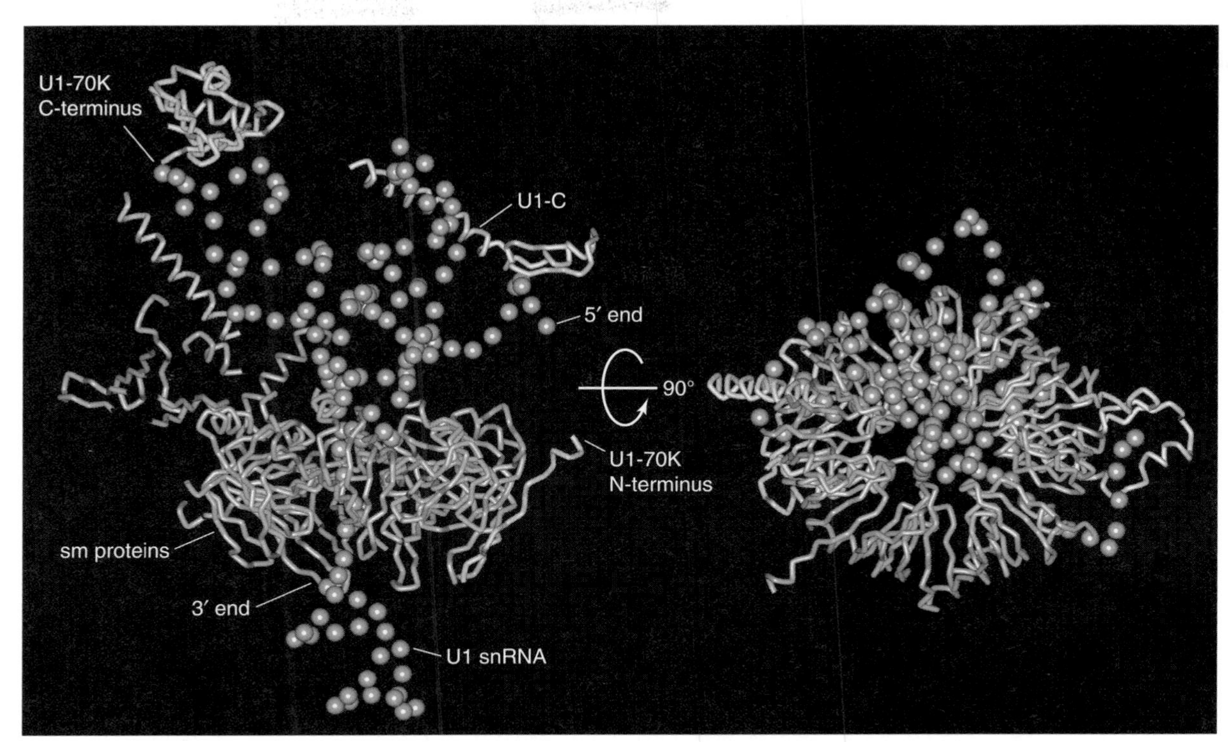

FIGURE 19.29 **Crystal structure of human U1 snRNP at 0.55 nm resolution.** (left) The phosphorus atoms in U1 snRNA are shown as green spheres. The seven Sm proteins (Sm-B, Sm-D1, Sm-D2, Sm-D3, Sm-E, Sm-F, and Sm-G) are shown as blue tubes, U1-C is shown as a gray tube structure, and U1-70K as an orange tube structure. The RNA forms four stem-loop structures. The seven Sm proteins assemble around Sm site nucleotides, which are located between two of the stem loop structures. (right) The structure shown on the left is rotated 90° around a horizontal axis. (Structures from Protein Data Bank 3CW1. D. A. Pomeranz Krummel, et al., *Nature* 458 [2009]: 475–480. Prepared by B. E. Tropp.)

in the cytoplasm. Further studies are required to determine how the different subcellular localizations of major and minor spliceosomes in vertebrate cells influences the processing of pre-mRNA molecules that have both major and minor splice sites.

The snRNA from each minor snRNP has the same predicted secondary structure as its major snRNA counterpart, suggesting evolution from a common ancestor. U4 and U4atac snRNAs have considerable sequence homologies and the same is true for U6 and U6atac snRNAs, but not for U1 and U11 snRNAs or for U2 and U12 snRNAs. U11, U12, and U4atac snRNPs appear to have the same Sm polypeptides as their major snRNP counterparts. Major and minor snRNPs probably have similar structures and work in the same way. Because the major snRNPs have been more thoroughly investigated, we focus on them when examining snRNP formation and function.

Each U snRNP is formed in a multistep process.

The formation of U1, U2, U4, and U5 snRNPs takes place through multistep pathways that begin in the cell nucleus, continue in the cytoplasm, and are finally completed in the nucleus (FIGURE 19.30). The pre-snRNAs are synthesized by RNA polymerase II and then modified by addition of an N^7-methylguanosine (m^7G) cap to their 5′-ends. The

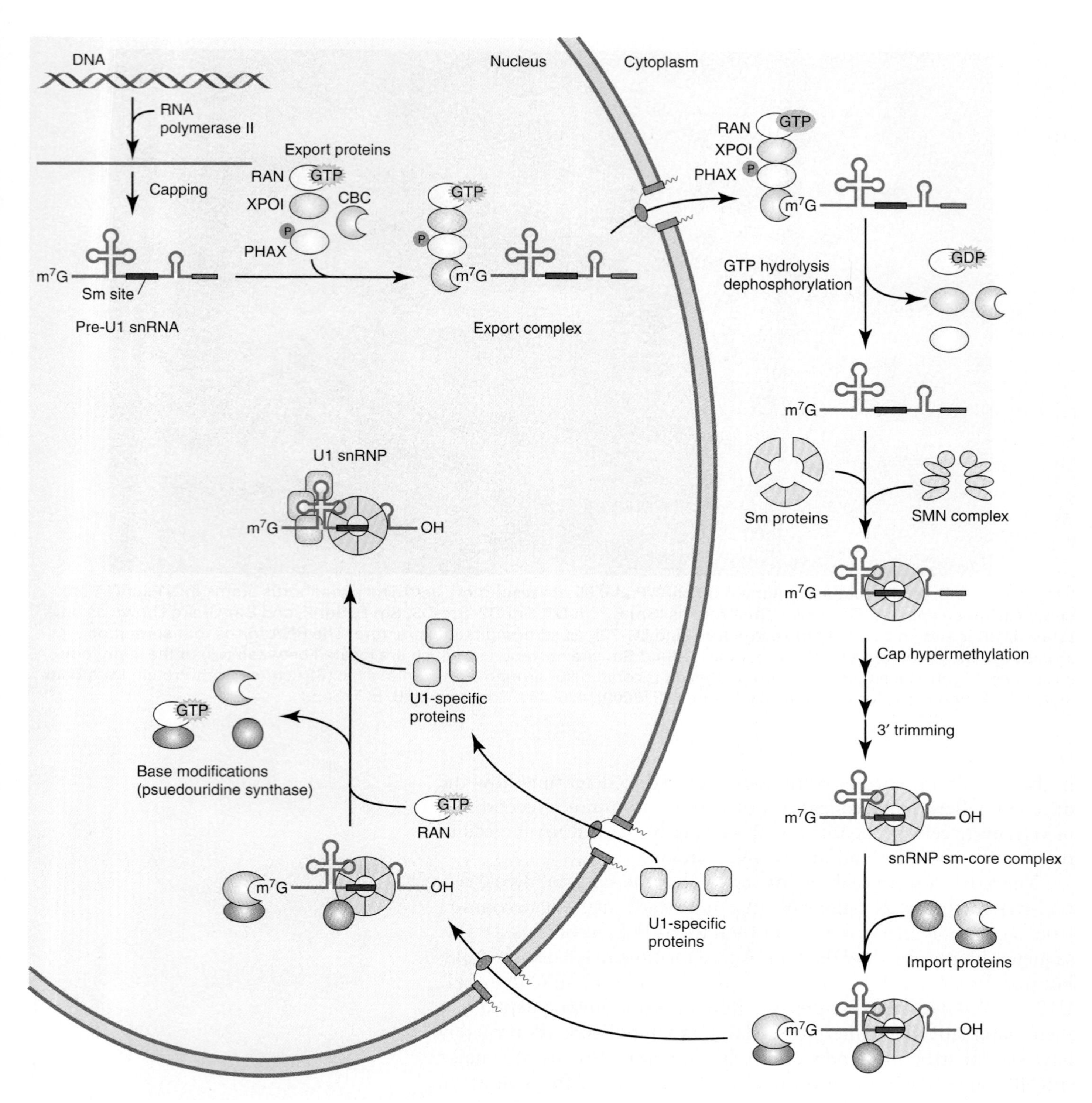

FIGURE 19.30 **Simplified cartoon model of U1 snRNP formation.** Subsequent to transcription and capping, the cap binding complex (CBC) and export proteins bind to U1 snRNP to form the export complex. For simplicity, only one of the export proteins is shown with a bound GTP. The export complex moves through the nuclear pore into the cytoplasm, where the GTP is hydrolyzed and the export proteins dissociate from the complex. Then the SMN protein complex facilitates the binding of Sm proteins to the U1 snRNA's Sm site (indicated by a red box). The snRNA's cap is hypermethylated and its 3′-end is trimmed to form the snRNP sm-core complex. Then the snRNP sm-core complex interacts with import proteins to form the import complex, which moves through the nuclear pore into the nucleus. The complex interacts with another GTP•protein complex and dissociates to form the U1 snRNP core. U1-specific proteins, which appear to be imported independently of the U1 snRNP core, associate with U1 snRNP core to form the U1 snRNP complex. (Adapted from C. L. Will and R. Lührmann, *Curr. Opin. Cell Biol.* 13 [2001]: 290–301.)

modified pre-snRNA must be transported to the cytoplasm for the biogenesis of the snRNP particle to continue. This transport requires the assembly of an export complex in which the pre-snRNA is associated with the cap binding complex (CBC) and three export proteins: **RAN, Xpo1,** and **PHAX.** RAN is a small regulatory protein that switches between GTP and GDP bound states. Xpo1 is an export receptor for U1, U2, U4, and U5 snRNAs. PHAX, which serves as an adaptor between the CBC•RNA complex and the Xpo1•RAN•GTP complex, is activated by phosphorylation. In fact, the name PHAX is an acronym for phosphorylated adaptor for RNA export.

The export complex moves through the nuclear pore into the cytoplasm, where the RAN•GTP switches to RAN•GDP and the export complex dissociates. Released pre-snRNA associates with Sm proteins in a process that is facilitated by the SMN (survival motor neuron) protein, a protein that gets its name because it is missing or defective in individuals with the recessive neuromuscular disease **spinal muscular atrophy (SMA).** Infants afflicted with the most severe and unfortunately most common form of SMA often die from respiratory failure within a year of birth because their motor neurons die, disrupting the connection between the central nervous system and skeletal muscles. Much remains to be learned about the SMN protein complex. It is quite possible that the complex has additional functions and that motor neuron cell death is caused by the SMN protein complex's inability to perform one of these additional functions. Another, perhaps related, unsolved mystery is the apparent specificity for motor neurons as opposed to other kinds of cells. After association with the Sm proteins, the pre-snRNA's 3′-end is trimmed and its m^7G cap is converted to a 2,2,7 trimethyl cap (Figure 19.26a) by further methylation to generate the snRNP core. Then the snRNP core interacts with import proteins to form the import complex, which moves through the nuclear pore into the nucleus.

Once in the nucleus, the import complex interacts with a second RAN•GTP complex and dissociates to regenerate the U1 snRNP core. Additional nuclear enzymes modify a few specific nucleotides by adding methyl groups or converting specific uridines to pseudouridines (Ψ). Finally, U-specific proteins, which were synthesized in the cytoplasm and transported into the nucleus, bind to the modified snRNA to form the mature U snRNP complex.

U6 snRNP formation differs from that of the other four snRNPs in the following important ways: (1) it is transcribed by RNA polymerase III; (2) it has a γ methyl phosphate cap (Figure 19.26b); (3) it contains Lsm rather than Sm proteins; and (4) the entire U6 snRNP biogenesis pathway occurs in the nucleus. Many of the details of U6 snRNP biogenesis still need to be worked out.

U1, U2, U4, U5, and U6 snRNPs have been isolated as a penta-snRNP in yeast.

Studies by John Abelson and coworkers suggest that U1 snRNP, U2 snRNP, U4/U6•U5 tri-snRNP and 13 non-snRNP proteins assemble to form a yeast ribonucleoprotein complex. This penta-snRNP complex does not splice pre-mRNA on its own but becomes active when a soluble yeast extract devoid of RNA is added to it. Abelson and coworkers have demonstrated that U2 snRNP does not dissociate from the penta-snRNP complex during splicing, but that about 16% of the U1 snRNPs does dissociate. Based on these results, Abelson and coworkers propose that either the penta-snRNP binds to the substrate pre-mRNA as a single discrete particle, or U1 snRNP binds first and then the U4/U6•U5•U2 tetra-snRNP binds next (**FIGURE 19.31**). It is reasonable to ask why the tetra- and penta-snRNP complexes were not detected sooner, if they are so stable. The answer appears to be that investigators usually use solutions containing a high salt concentration and heparin when extracting ribonucleoproteins, producing nonphysiological conditions that cause the complexes to dissociate.

Although the idea that the tetra- or penta-snRNP complex adds to pre-mRNA in a single step is quite attractive, experiments performed by Karla M. Neugebauer and coworkers raise serious doubts about this single-step recruitment model. Neugebauer and coworkers used the chromatin immunoprecipitation (ChIP) technique (see Chapter 18) to monitor events during spliceosome assembly in growing yeast cells. Their experiments show that U1, U2, and U5 snRNPs bind to intron-containing genes in a pattern that is consistent with stepwise spliceosome assembly. Further studies will therefore be needed to determine if the tetra- or penta-snRNP complexes really do have a physiological role in splicing. Even if future experiments should demonstrate that the tetra- or penta-snRNP is a stable complex, examining stepwise spliceosome assembly *in vitro* will provide important insights into the way that spliceosome components interact with one another and with pre-mRNA.

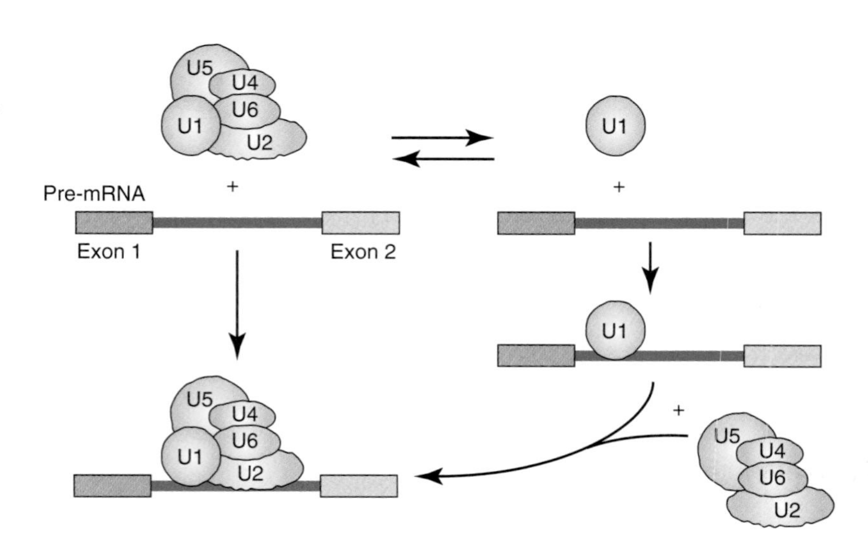

FIGURE 19.31 Model showing how a penta-snRNP or tetra-snRNP complex might interact with pre-mRNA. Spliceosome subunits are proposed to exist as penta-snRNP complexes (left) or tetra-snRNP complexes (right) that interact with the pre-mRNA. U1 snRNP binds before the tetra-snRNP complex. (Adapted from S. W. Stevens, *Mol. Cell* 9 [2002]: 31–44.)

In vitro studies show that spliceosomes assemble on introns via an ordered pathway.

Several non-snRNP proteins participate in the spliceosome assembly pathway. One of these, **splicing factor 1 (SF1)**, also called branch point binding protein, binds to the branch point sequence. Another, U2AF (U2 snRNP auxiliary factor), contains a large subunit, $U2AF^{65}$, that binds to the polypyrimidine tract and a small subunit, $U2AF^{35}$, that binds to the AG dinucleotide at the 3′-splice site. FIGURE 19.32 shows the contributions that these non-snRNPs and the snRNPs make to the spliceosome assembly pathway. The first complex detected, the E (early) complex, has SF1 and U2AF bound to their respective

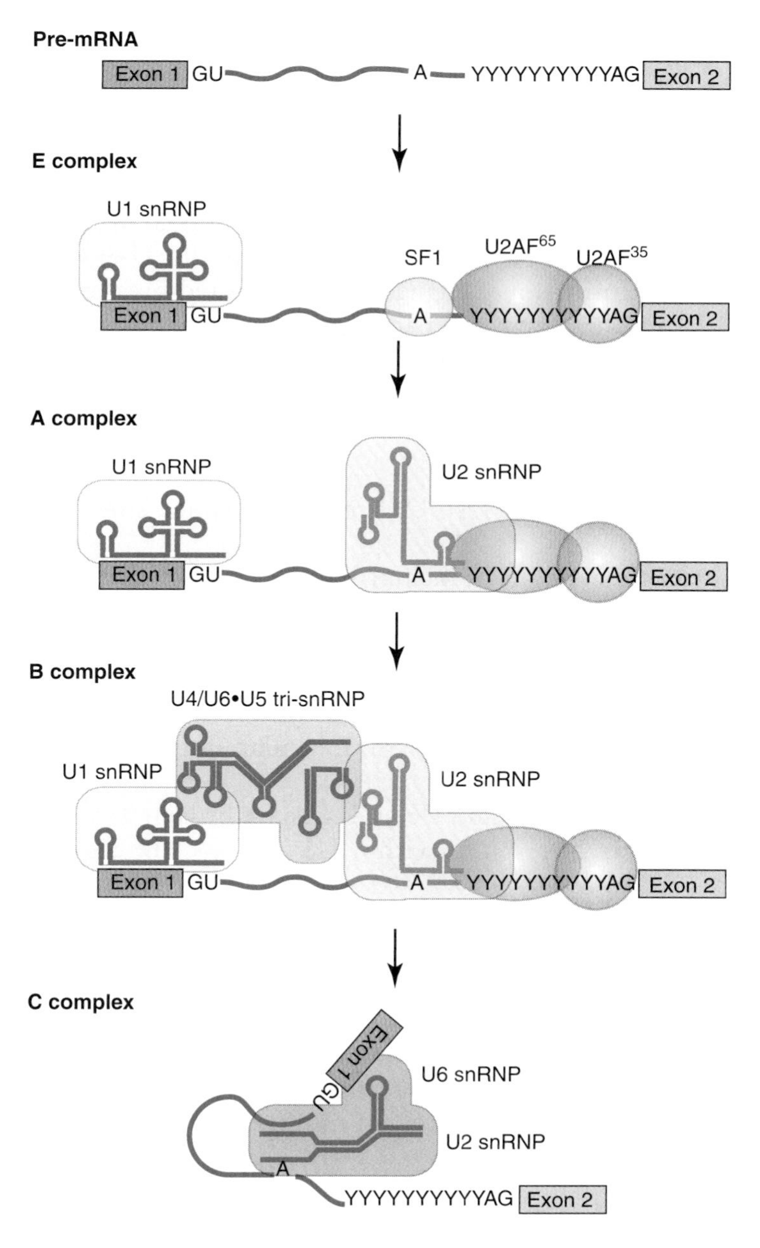

FIGURE 19.32 **Spliceosome assembly.** During the first step in stepwise spliceosome assembly, U1 snRNP binds to the 5′-splice site. The branchpoint binding protein (SF1) binds to the branchpoint, and U2 auxiliary factors ($U2AF^{65}$ and $U2AF^{35}$) bind to the pyrimidine tract and 3′-splice site AG, respectively. This complex, which commits the pre-mRNA to the splicing pathway, is called the early or E complex. Next, the E complex is converted to A complex when U2 snRNP binds at the branchpoint. Then U4/U6•U5 tri-snRNP enters the spliceosome. This entry is followed by a massive rearrangement in which U6 snRNP replaces U1 snRNP at the splice site, U6 and U2 interact, U5 bridges the splice sites, and U1 and U4 become destabilized. This catalytically active rearranged spliceosome is called the C complex. For simplicity, U5 is not shown in the C complex. (Modified from *Trends Biochem. Sci.*, vol. 30, K. J. Hertel and B. R. Graveley, RS domains contact the pre-mRNA . . ., pp. 115–119, copyright 2005, with permission from Elsevier [http://www.sciencedirect.com/science/journal/09680004].)

target sequences and U1 snRNP bound to the 5′-splice site through base pairing between U1 snRNA and the splice site sequence. The E complex is transformed to A complex when U2 snRNP binds to the branch point sequence through base pairing interactions between U2 snRNA and the branch point sequence. A complex is converted to B complex when U4/U6•U5 tri-snRNP binds near to the 3′-splice site. B complex undergoes a complicated series of rearrangements to form the catalytic or C complex (FIGURE 19.33), which catalyzes the first transesterification. Although U1 and U4 snRNPs appear to dissociate as B complex is converted to C complex, they probably remain weakly bound to C complex.

Much of the information that we have about RNA–RNA and RNA–protein interactions in the assembly complexes has been obtained by using chemical cross-linking agents to identify sites on the different components that interact with one another. Cross-linking agents also help to reveal rearrangements that occur as the spliceosome is assembled and as it functions by identifying segments that are near to one another during one stage of assembly or splicing but not during another. Base substitutions or amino acid substitutions produced by genetic or recombinant DNA techniques provide additional insights. In some cases, replacing one base with another in a specific U snRNA blocks a specific conformational change that is detected by a cross-linking agent. Using these and other techniques, investigators have demonstrated that several rearrangements occur prior to the first transesterification reaction. As shown in Figure 19.33, some of the structural rearrangements that occur when B complex is converted to C complex are as follows:

1. U6 snRNA replaces U1 snRNA at the 5′-splice site (Figure 19.33a).
2. SF1 is released from the branching sequence, which then base pairs with a complementary region of U2 snRNA. The branching nucleotide bulges from the newly formed helical structure, making the branching nucleotide available to participate in the first transesterification reaction (Figure 19.33b).
3. U2 snRNA undergoes an intramolecular rearrangement (Figure 19.33c).
4. Base pairs between U4 and U6 snRNAs are broken and new intramolecular base pairs within U6 snRNA are formed (Figure 19.33d).
5. Still other base pairs between U4 and U6 snRNAs are disrupted as are intramolecular base pairs within U2 snRNA. Released complementary U2 and U6 snRNA regions base pair (Figure 19.33e).
6. Conformational changes in U2 and U6 snRNA bring a region of U2 snRNA near to a complementary region in U6 snRNA so that an additional base pair region can form (Figure 19.33f).

The switch from extensive base pairing between U4 and U6 snRNAs to extensive base pairing between U2 and U6 snRNAs described above would probably also occur under physiological conditions if a preassembled spliceosome binds to pre-mRNA.

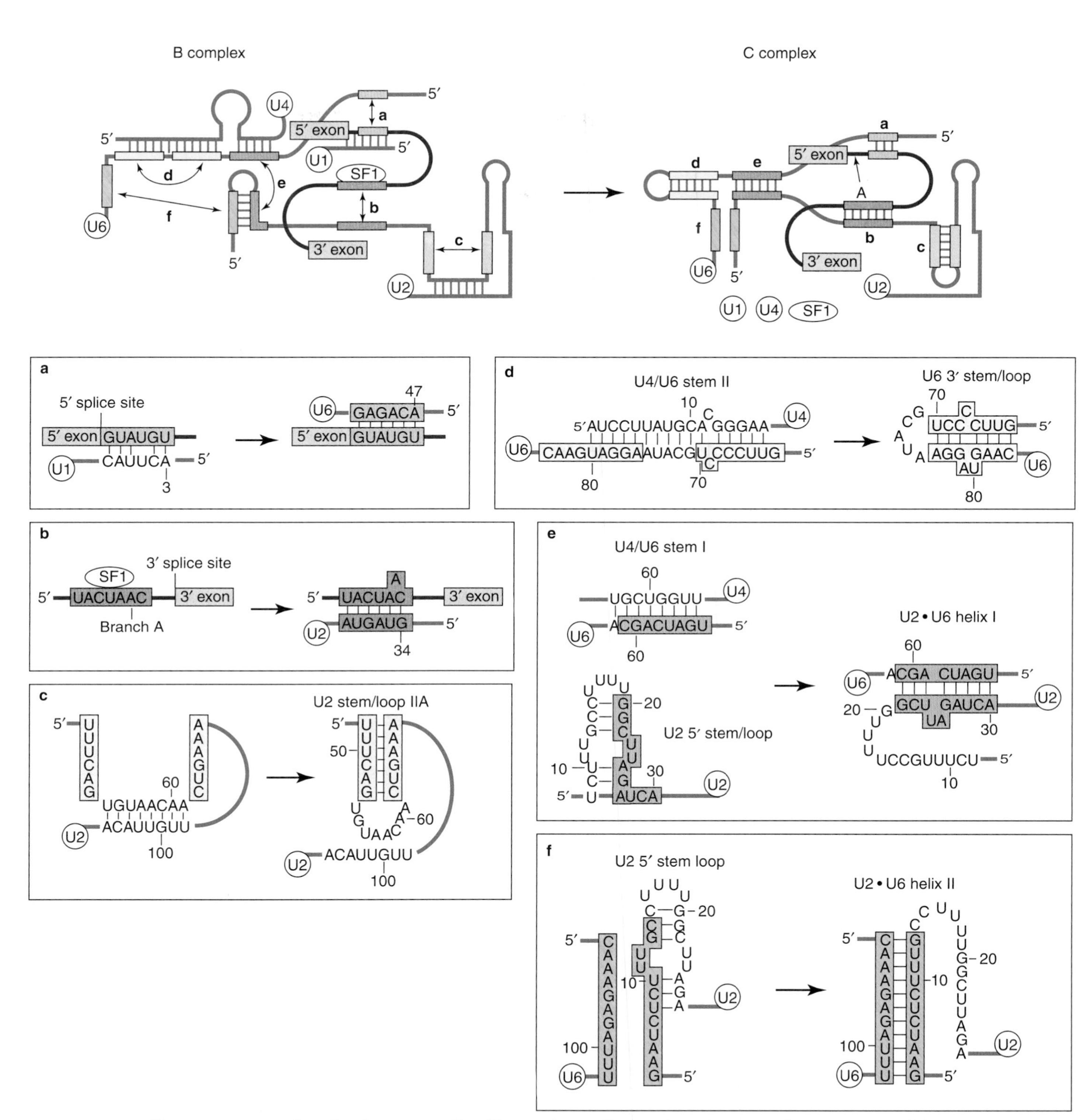

FIGURE 19.33 Rearrangements involved in converting B complex to C complex. A composite summary of the rearrangements that take place when the B complex is converted to the C complex. Specific rearrangements (a–f) are color coded to indicate regions that become base paired. The red line represents pre-mRNA and the green line is snRNA. SF1 binds to the branchpoint sequence. Specific rearrangements (a–f) are shown in greater detail below the summary. (a) The exchange of U1 for U6. (b) The exchange of SF1 for U2. (c) An intramolecular U2 rearrangement. (d) Disruption of U4/U6 base pairs; formation of U6 stem/loop. (e) Disruption of U4/U6 base pairs and U2 stem/loop; formation of U2•U6 helix. (f) Disruption of the U2 stem/loop; formation of U2•U6 helix. *S. cerevisiae* sequences are shown; for clarity U5 has been omitted. (Modified from *Cell*, vol. 92, J. P. Staley and C. Guthrie, Mechanical Devices of the Spliceosome . . ., pp. 315–326, copyright 1998, with permission from Elsevier [http://www.sciencedirect.com/science/journal/00928674].)

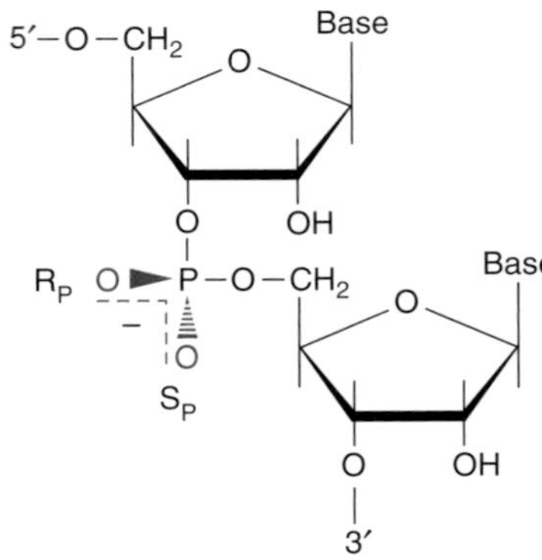

FIGURE 19.34 **Distinguishable "non-bridging" oxygen atoms in a phosphodiester bond.** Each phosphodiester bond contains two distinguishable non-bridging oxygen atoms, designated R_p and S_p.

RNA and protein may both contribute to the spliceosome's catalytic site.

Because U2, U5, and U6 snRNP are tightly associated with pre-mRNA at the time of the first transesterification reaction, they may participate in the catalytic process. Considerable evidence suggests that U6 snRNA is the catalytically active molecule because it is highly conserved through evolution and very sensitive to chemical changes. Moreover, some introns that can catalyze their own excision, the so-called group II introns have a folding pattern similar to that of U6 snRNA.

Ren-Jang Lin and coworkers performed an elegant experiment providing direct proof that U6 snRNA participates in the catalytic process. Their experiment is based on the fact that each phosphodiester bond contains two distinguishable "non-bridging" oxygen atoms, designated R_p and S_p (FIGURE 19.34). Lin and coworkers synthesized U6 snRNAs that contained a sulfur atom replacing the R_p or S_p oxygen of nucleotide U_{80} and then used these RNA molecules to reconstitute U6 snRNP. Neither of the reconstituted U6 snRNPs supported splicing when tested in an *in vitro* system that contained all the necessary splicing components including magnesium ions. One well-known consequence of replacing a "non-bridging" oxygen atom with a sulfur atom is that the modified phosphate group loses its ability to interact with magnesium ions. The modified U6 snRNA's failure to function therefore might have resulted from its inability to bind the essential cation. Lin and coworkers tested this possibility by determining whether the modified U6 snRNAs could function in the presence of manganese ions, which can often substitute for magnesium ions. Unlike magnesium ions, however, manganese ions interact with sulfur. Remarkably, manganese ions restored the function of the modified U6 snRNA with the S_p sulfur but not with the R_p sulfur. Lin and coworkers then showed that both modified U6 snRNAs support all steps in spliceosome assembly that precede the actual splicing event. Hence, the most probable conclusion from their studies is that U6 snRNA normally binds a magnesium ion that directly participates in splicing by activating the 2′-hydroxyl group of the branching nucleotide or by stabilizing the leaving group.

Experiments performed by James L. Manley and coworkers provide additional support for the hypothesis that U6 snRNA participates in the chemistry of splicing. Their most recent approach was based on an earlier observation, in which they synthesized and purified large segments of human U2 and U6 snRNAs, annealed the fragments in the presence of magnesium chloride, and demonstrated that the fragments fold to form a structure resembling that likely to exist in the catalytically active spliceosome. With this and related observations in mind, Manley and coworkers set out to determine if a complex formed from segments of human U2 and U6 snRNAs could catalyze the first splice reaction without the assistance of any proteins. The experiment they performed to do so is summarized in FIGURE 19.35. They used a

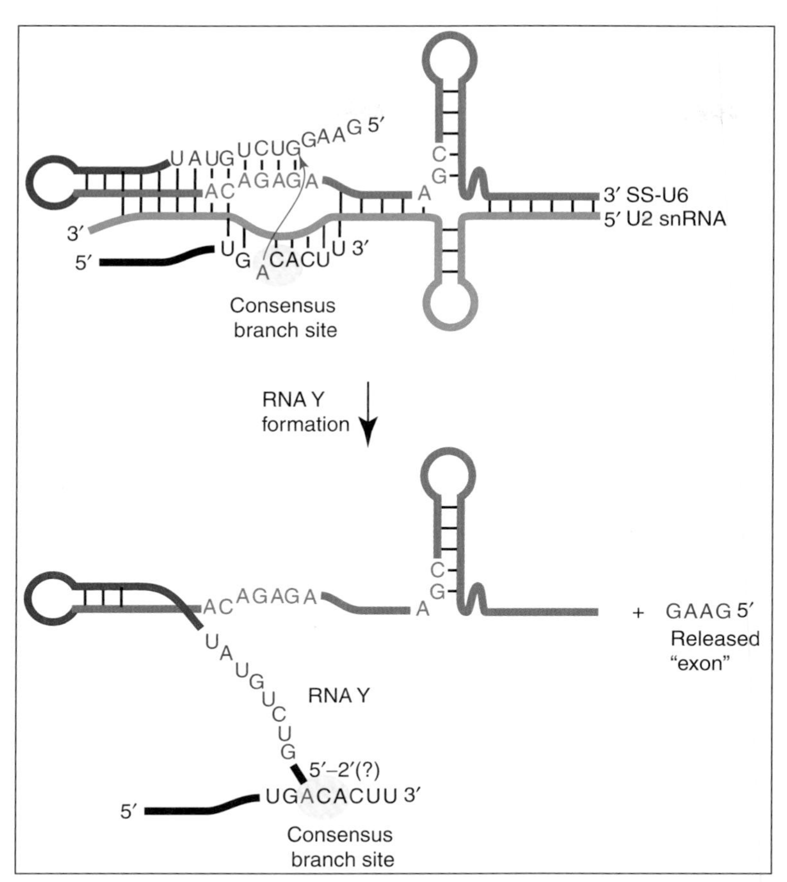

FIGURE 19.35 **U2•U6 snRNA complex as a potential catalyst.** A short linker is used to covalently attach an oligonucleotide with a 5′-splice site (purple) to the 5′-end of U6 snRNA (green). The resulting chimeric RNA construct called SS-U6 is mixed with a U2 snRNA fragment (blue) to form a modified U2•U6 complex. A ^{32}P-labeled oligonucleotide containing a consensus branch site (black) is incubated with the modified U2•U6 complex at room temperature in the presence of magnesium chloride. After one day, an RNA product forms that is resistant to denaturation at 85°C in 10 M urea and 20 mM EDTA. This product is called RNA Y. The reaction responsible for RNA Y formation involves a nucleophilic attack by the branching nucleotide A (shown in red) on the phosphodiester bond at the 5′ splice site. The 5′ "exon" is released. U2 snRNA, which remains unchanged, is not shown as a product of the reaction. (Modified from S. Valadkhan, et al., *RNA* 13 (2007): 2300–2311. Used with permission of the RNA Society.)

short linker to covalently attach an oligonucleotide with a 5′-splice site (SS) to the 5′-end of U6 snRNA to construct an RNA molecule they called SS-U6. Then they annealed a U2 snRNA fragment to the SS-U6 construct to form a modified U2•U6 complex. They tested the modified U2•U6 complex for catalytic activity by adding a ^{32}P-labeled oligonucleotide containing a consensus branch site and incubating at room temperature in the presence of magnesium chloride. After one day, a new RNA was formed, which was resistant to denaturation at 85°C in the presence of urea and EDTA. The yield of the RNA product, which they named RNA Y, was about 0.2% and 0.8% of the starting material when the magnesium chloride was present at 20 mM and 80 mM, respectively. Manley and coworkers characterized RNA Y by following the fate of radioactive RNA molecules. These radiotracer experiments showed (1) the U2 fragment is not present in RNA Y, (2) the entire oligonucleotide containing the consensus branch site is part of RNA Y, and (3) the 3′-end of SS-U6 is part of RNA Y but the 5′-end is not. Based on these and related data, they propose that the 2′-hydroxyl of an adenosine at the consensus branch site initiates a nucleophilic attack on the phosphodiester bond at the 5′ splice site to form RNA Y and release a 5′ "exon" (Figure 19.35). Further work is required to demonstrate that a 2′→5′ phosphodiester bond links the branching nucleotide (A) to G and to determine the factors that limit the yield of RNA Y.

Recent studies suggest that a conserved protein called Prp8 (pre-mRNA processing protein), which is an integral part of U5 snRNP, may also contribute to the catalytic site. Studies of Prp8 structure and function have been hampered by the inability to obtain the intact protein in soluble form. This failure is probably due in large part to the fact that the protein is so large (2,413 residues in yeast). Investigators have made some progress in understanding Prp8's function by isolating and studying subdomains. A subdomain near the C-terminus that is about 150 residues long is of particular interest because chemical crosslinking studies show that its residues interact with the 5′ splice site. Furthermore, model studies suggest that two highly conserved aspartate residues in this subdomain may act together with RNA metal coordination sites to form the catalytic site. If so, the spliceosome is an "RNP-zyme."

Cells use a variety of mechanisms to regulate splice site selection.

Spliceosomes must not only be able to remove introns, but they must do so in a manner that precisely selects correct splice site pairs. Splicing precision presents a much more serious challenge to multicellular organisms than to yeast because: (1) multicellular organisms tend to have very long introns with many pseudo-splice sites, whereas yeast tend to have short introns with few, if any, pseudo splice sites; (2) multicellular organisms tend to have relatively short, degenerate splice sites, whereas yeast have highly conserved splice sites; and (3) many pre-mRNA molecules from higher organisms are subject to alternative splicing, whereas few, if any, yeast pre-mRNAs are subject to alternative splicing.

Multicellular organisms use regulatory proteins to assist in splice site recognition. These proteins work by first binding to specific sequences in exons or introns and then stimulating or repressing exon recognition. A sequence within an exon or intron that stimulates splice-site selection is called an **exonic splicing enhancer** (**ESE**) or

intronic splicing enhancer (**ISE**), respectively. The best studied ESEs are purine-rich sequences with the consensus sequence $(GAR)_n$, where R is any purine. Because ESEs are embedded in coding sequences, a base substitution in an ESE may be quite harmful as a result of its effect on alternative splicing even though the substitution appears to be harmless in terms of its effect on polypeptide sequence.

A family of **splicing regulatory proteins** (**SR proteins**) appears to be particularly important for ESE recognition. Ten different types of SR proteins have been identified in human cells but none in yeast. SR proteins have modular structures; each has one or two **RNA recognition motifs** (**RRMs**) at its N-terminus and an arginine/serine-rich domain (**RS-domain**) at its C-terminus (FIGURE 19.36). The RRM is a sequence-specific RNA binding site, while the RS domain participates in protein–protein interactions. Different SR proteins may contact one another directly through their RS domains or they may interact through a protein intermediate. In either case, at least some of the serine residues in the RS domain must be phosphorylated for interactions to occur, providing the cell with a method to regulate contacts between proteins. Some SR proteins appear to be required for cell survival, whereas others appear to be interchangeable.

SR proteins stimulate recognition of weak splice sites by binding to ESEs and then helping to recruit U1 snRNP to the 5′-splice site and U2AF to the 3′-splice site. SR proteins also interact with one another as well as other proteins to form protein bridges that extend across the intron that is to be excised or across exons (FIGURE 19.37). The former interaction, termed **intron definition**, tends to be used for splice site recognition when the intron is small. The latter interaction,

(a)

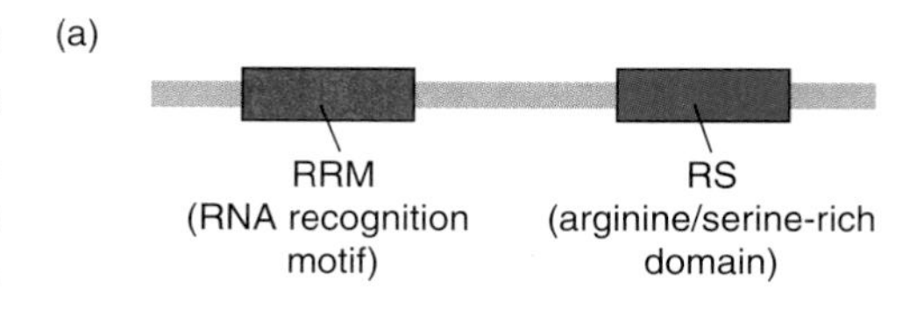

(b)

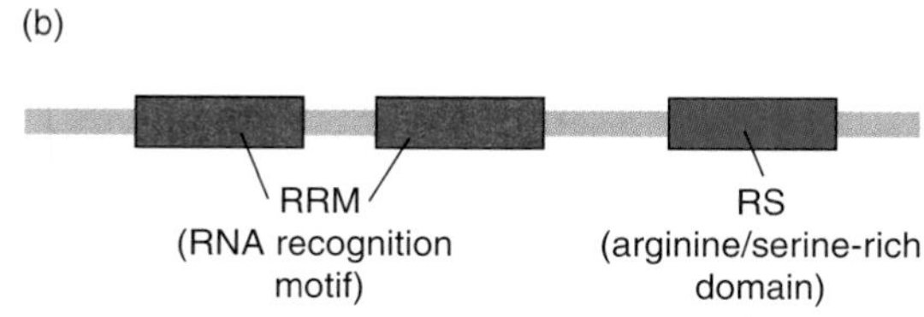

FIGURE 19.36 **SR proteins.** SR proteins have (a) one or (b) two RNA recognition motifs (RRMs) at their N-termini. The RRM binds to specific RNA sequences. SR proteins also have an arginine/serine-rich domain (RS domain) at their C-termini. The RS-domain is required for protein–protein interactions with other SR proteins as well as with SR-related proteins.

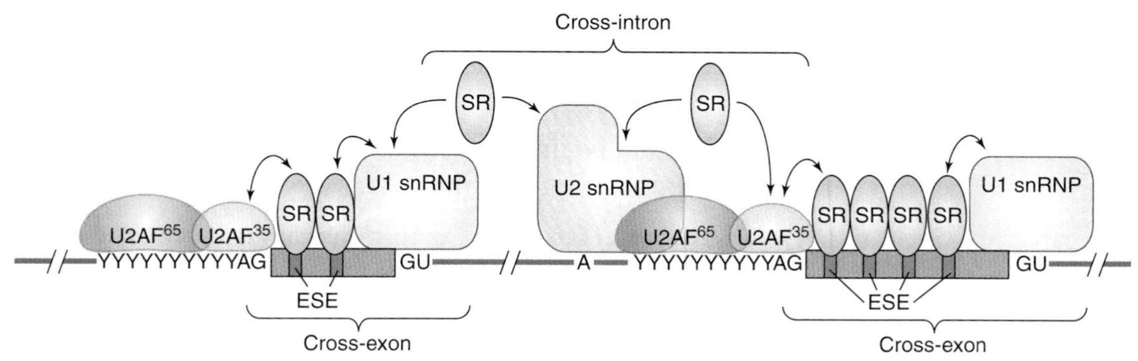

FIGURE 19.37 **Splice site recognition.** The splicing machinery recognizes the correct 5′-(GU) and 3′-(AG) splice site on the basis of their proximity to exons, which contain exonic splicing enhancers (ESEs) that are binding sites for SR proteins. When bound to an ESE, the SR proteins recruit U1 snRNPs to the downstream 5′-splice site, and $U2AF^{65}$ and $U2AF^{35}$ subunits to the pyrimidine tract (YYYY) and AG dinucleotide of the upstream 3′-splice site, respectively. U2AF, an SR-related protein, recruits U2 snRNP to the branchpoint sequence (A). The bound SR proteins thus recruit splicing factors to form a "cross-exon" recognition complex. Recognition across an exon is called exon definition. SR proteins also function in "cross-intron" recognition by facilitating the interactions between U1 snRNP bound to the upstream 5′-splice site and U2 snRNP bound to the branchpoint sequence. Recognition across an intron is called intron definition. (Adapted from T. Maniatis and B. Tasic, *Nature* 418 [2002]: 236–243.)

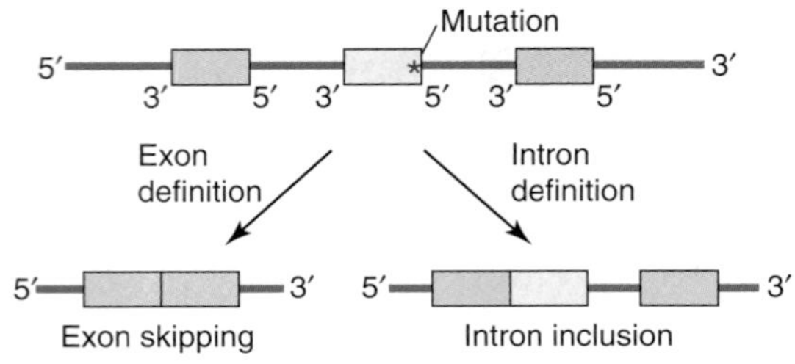

FIGURE 19.38 Predictions of the phenotype of a mutation of the 5′-splice site bordering an internal exon. Exons are shown in light blue and pink rectangles and introns as green lines. The exon definition model (left) predicts that a mutation (*) at a 5′-splice site bordering an internal exon (light blue) should depress recognition of the exon with concomitant inhibition of splicing of the adjoining intron. The intron definition model (right) predicts that a mutation (*) at the 5′-splice site of an internal exon (light blue) should inhibit splicing of the intron in which it occurs but should have a minimal impact on the splicing of neighboring introns. (Adapted from S. M. Berget, *J. Biol. Chem.* 270 [1995]: 2411–2414.)

exon definition, provides a method for splice site recognition when introns are very long as they tend to be in multicellular animals and plants. Susan M. Berget, the investigator who first proposed the exon definition concept, has pointed out that intron and exon definition lead to entirely different predictions for the phenotypes produced by a mutation in the 5′-splice site bordering an internal exon. Models invoking an initial pairing of splice sites across introns (intron definition) predict that a mutation at the 5′-splice site of an internal exon (light blue exon in FIGURE 19.38) should inhibit splicing of the intron in which they occur but should have a minimal impact on the splicing of neighboring introns. In contrast, exon definition predicts that a mutation at a 5′-splice site bordering an internal exon should depress recognition of the exon with concomitant inhibition of splicing of the adjoining intron. In yeast, processing of split genes appears to rely primarily on intron definition. Exon definition is seldom, if ever, used because it requires more than one intron and most yeast genes with introns have only one. Multicellular organisms use both intron and exon definition to process pre-mRNA. In fact both intron and exon definition sometimes participate in excising different introns from the same pre-mRNA. Furthermore, splice site recognition may begin with exon definition and then switch to intron definition.

Pre-mRNA molecules also have sequences that repress exon recognition. A sequence within an exon or intron that represses splice-site selection is called an **exonic splicing silencer (ESS)** or **intronic splicing silencer (ISS)**, respectively. The best studied regulatory protein that binds to silencer sequences is hnRNP A1. The hnRNP designation indicates that it is one of a large number of unrelated proteins that bind to hnRNA. The hnRNP A1 protein, which has two RRMs and a glycine-rich domain, binds to both exonic and intronic splicing silencers. Among the mechanisms that have been proposed to explain how hnRNP A1 represses exon recognition are the following: It blocks SR proteins from binding to adjacent ESEs, interferes with protein–protein interactions required for exon definition, or blocks spliceosome assembly.

Splicing begins as a cotranscriptional process and continues as a posttranscriptional process.

Another factor that determines which exons will be included or excluded is that splicing begins while the nascent pre-mRNA is still being transcribed. Direct evidence for cotranscriptional splicing comes from electron micrographs such as that in FIGURE 19.39, which shows a chromosome as it is being transcribed. Nascent pre-mRNA transcripts are visible as strands of increasing length extending from the DNA template. Progressive formation and loss of various size intron loops near the 5′-end of nascent pre-mRNA strands visible in this micrograph indicates the transcripts are undergoing cotranscriptional splicing. Splicing continues even after transcription is complete. The fraction of introns that are removed during transcription versus after transcription is not known in higher plants or animals. Studies by Michael Rosbash and coworkers, however, indicate that in yeast pre-mRNA splicing is predominantly posttranscriptional.

Support for cotranscriptional regulation of splicing comes from experiments showing that promoter structure regulates splicing. Changing a promoter can influence splice site selection even when the promoter change has no effect on the promoter's strength or the site of transcription initiation. In an attempt to explain this phenomenon, Yutaka Hirose and James L. Manley have proposed that (1) the specific nature of the initiation complex assembled on a particular promoter may influence the extent or pattern of phosphorylation of the CTD of the largest RNA polymerase subunit, and this in turn may regulate CTD's ability to participate in splicing, or (2) the promoter may recruit specific factors to the transcription elongation complex that may in turn influence splicing.

Considerable evidence supports the hypothesis that CTD participates in cotranscriptional splicing. *In vitro* experiments show that phosphorylated CTD stimulates splicing and anti-CTD antibodies inhibit splicing. Cells with a truncated CTD have a lower splicing efficiency than those with a normal CTD. CTD appears to influence splicing at least in part by stimulating 5′-m^{7}G cap formation. However, capping does not completely explain CTD's role in splicing as indicated by the following two observations: (1) a yeast mutant with temperature-sensitive guanylyltransferase synthesizes spliced mRNAs at the restrictive temperature, although much less efficiently than wild type cells, and (2) frog oocytes splice uncapped pre-mRNAs that are microinjected into them, although less efficiently than capped pre-mRNAs.

CTD's location near the transcription exit channel places it in position to transfer snRNPs, SR proteins, and other splice site factors on splice sites and branching sequences as they emerge from the exit channel. CTD therefore may work by stimulating spliceosome assembly, facilitating spliceosome interaction with splice sites, serving as a platform for SR proteins or other splicing factors, or some combination of these and other mechanisms.

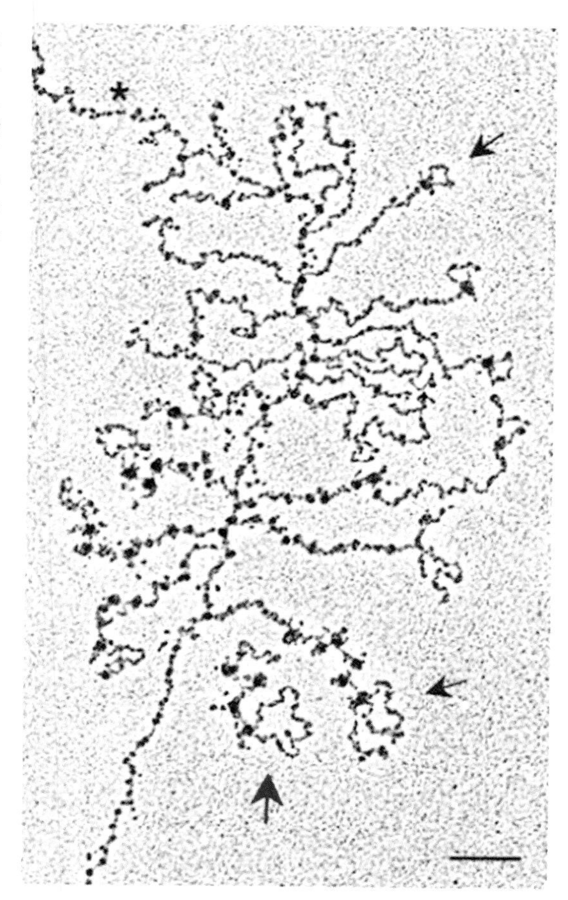

FIGURE 19.39 **Electron micrograph directly visualizing co-transcriptional splicing.** The gene shown is from *Drosophila* and is ~6 kb long. DNA enters at the upper left on the micrograph and exits at the lower left. Transcription initiates at the position marked with an asterisk; nascent RNA transcripts appear as fibrils of increasing length that extend from the DNA template. The transcripts are undergoing cotranscriptional splicing, as indicated by the progressive formation and loss of intron loops of various sizes near the 5′-ends of the transcripts (arrows). The red arrow indicates a transcript near the 3′-end of the gene that is no longer attached to the DNA template and might have been caught in the act of termination and release. Bar = 200 nm. (Caption reprinted and photo reproduced from *Trends Biochem. Sci.*, vol. 25, N. Proudfoot, Connecting transcription to messenger . . ., pp. 290–293, copyright 2000, with permission from Elsevier [http://www.sciencedirect.com/science/journal/09680004]. Photo courtesy of Ann L. Beyer, University of Virginia.)

19.5 Cleavage/Polyadenylation and Transcription Termination

Poly(A) tail synthesis and transcription termination are coupled, cotranscriptional processes.

Because RNA polymerase II is often required to synthesize very long transcripts, it must be a highly processive enzyme that does not terminate prematurely. Nevertheless, it must terminate at the end of the gene so that it does not continue transcribing into the neighboring gene and interfere with the normal transcription of that gene. Failure to terminate would also prevent the release of RNA polymerase II, preventing the enzyme from carrying out further transcription. RNA polymerase II transcription termination is a complex process, which involves: (1) cleaving pre-mRNA at a site that is called the **cleavage/polyadenylation site** or the **poly(A)** site; (2) adding a poly(A) tail to the newly generated 3′-end of the RNA molecule; and (3) transcription termination downstream from the cleavage/polyadenylation site. The transcription termination machinery requires recognition elements to help it locate the poly(A) site. These elements differ in both position and sequence in yeast and higher animals. Moreover, recognition elements are more highly conserved in higher animals.

Yeast has two recognition elements upstream of the poly(A) site (FIGURE 19.40a). The first recognition element, the **efficiency element** (**EE**) is located at variable distances from the cleavage site in different genes and determines the efficiency of processing. It often contains alternating UA dinucleotides or a U-rich stretch. The second recognition element, the **positioning element** (**PE**) directs cleavage to a position that is approximately 20 nucleotides downstream and tends to be an A-rich sequence.

In mammals, the poly(A) site is located between a conserved **polyadenylation signal** and a less conserved **U-rich element** (FIGURE 19.40b). The polyadenylation signal (AAUAAA) was originally identified as a highly conserved sequence that is 10 to 30 nucleotides upstream of the cleavage site. Comprehensive mutagenesis studies indicate that this hexanucleotide is required for both cleavage and poly(A) addition. Mammalian cleavage/polyadenylation machinery can usually tolerate single base substitutions within the AAUAAA sequence (most commonly A→U at the second position) but it cannot tolerate more extensive substitutions. The U-rich element is located ≤ 30 nucleotides downstream of the cleavage/polyadenylation site. Although the U-rich element is quite variable in both sequence and composition, it usually has one or more stretches of five consecutive U residues (often interrupted by single G residues). A third recognition element, an **auxiliary upstream element** is often present at a variable distance upstream of the cleavage site. It is usually U-rich or contains $(UGUA)_n$ or $(UAUA)_n$. When present, the auxiliary upstream element enhances the efficiency of cleavage and polyadenylation. The cleavage site is determined by

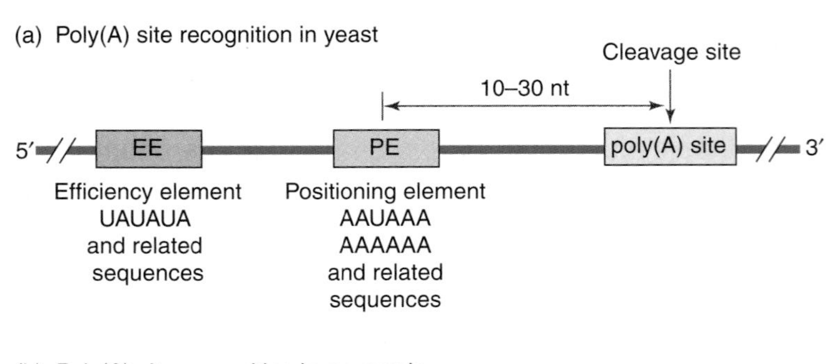

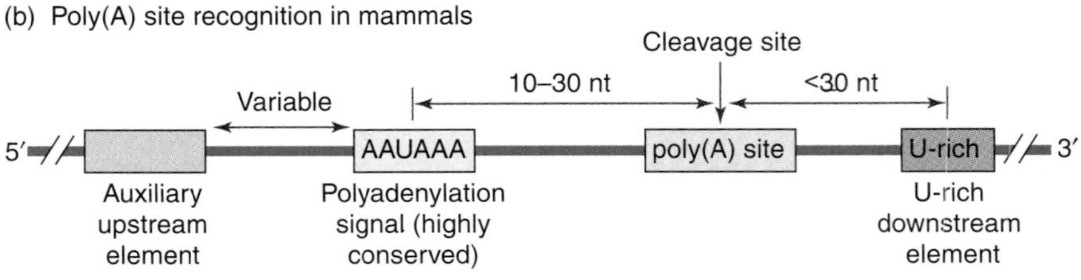

FIGURE 19.40 Schematic representation of Poly(A) signals in yeast and mammals. (a) Poly(A) site recognition in yeast. (b) Poly(A) site recognition in mammals. (Part a adapted from J. Zhao, L. Hyman, and C. Moore, *Microbiol. Mol. Biol. Rev.* 63 [1999]: 405–445. Part b adapted from G. M. Gilmartin, *Genes Dev.* 19 [2005]: 2517–2521.)

the distance between the polyadenylation signal and U-rich element. About 70% of the cleavage sites in vertebrate pre-mRNAs have an A on the 5′-side of the cleavage site so that the first A in the poly(A) tail derives from the pre-mRNA.

Despite the differences in transcription termination elements, the majority of components in the cleavage/polyadenylation machinery of animals and yeast are conserved. Different names are used for the components of the yeast and animal cleavage/polyadenylation machinery. For simplicity, the discussion that follows will use the animal nomenclature system whenever possible. The machinery required to perform two relatively simple tasks: endonucleolytic cleavage and poly(A) addition requires approximately 85 different proteins. Such complex machinery is required because cleavage and poly(A) addition are tightly coupled co-transcriptional processes that must take place with a high degree of precision. Fortunately, cleavage and poly(A) addition can be examined separately *in vitro*, simplifying the study of each process.

We begin our examination of the cleavage/polyadenylation machinery by examining the contribution of known components (Table 19.3). A model for the organization of these components at the poly(A) site is shown in schematic form in FIGURE 19.41.

1. *Cleavage/polyadenylation specificity factor (CPSF)*. CPSF binds to the polyadenylation signal AAUAAA and is required for both cleavage and poly(A) addition. It has five subunits (CPSF-160, CPSF-100, CPSF-73, CPSF-30, and Fip1). The

TABLE 19.3 Components of the Mammalian Cleavage/Polyadenylation Machinery

Factor	Processing Step	Function
CPSF cleavage/polyadenylation specificity factor	Cleavage and poly(A) addition	Contains five subunits. CPSF-73 cleaves pre-mRNA at the poly(A) site. CPSF-160 binds to AAUAAA. The functions of the other three subunits CPSF-30, CPSF-100, and Fip1 remain to be determined.
CstF cleavage stimulation factor	Cleavage	Contains four subunits (CstF-77, CstF-64, CstF-60, and SCP1). CstF-64 binds to the U-rich sequence. The functions of the other subunits remain to be determined.
CFI cleavage factor I	Cleavage	Recognizes sequence elements in poly(A) site.
CFII cleavage factor II	Cleavage	Unknown.
PAP poly(A) polymerase	Cleavage and poly(A) addition	Catalyzes poly(A) formation.
PAB II poly(A) binding protein	Poly(A) elongation	Binds poly(A) and CPSF-30. Responsible for processive poly(A) elongation and for the tail length.
CTD Carboxyl terminal domain of large subunit in RNA polymerase II	Cleavage	Binds CPSF and CstF.
Symplekin	Cleavage and poly(A) addition	Symplekin helps to assemble or stabilize the CstF complex and thereby helps to hold the complete cleavage/polyadenylation machinery together.

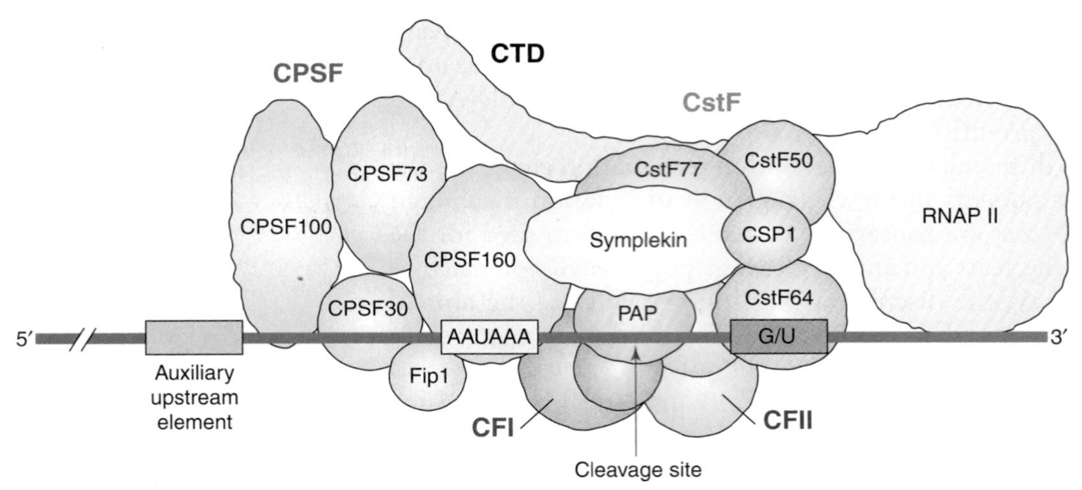

FIGURE 19.41 **Schematic representation of the mammalian polyadenylation machinery.** For simplicity, mammalian nomenclature is used here. The abbreviations used are as follows: RNAPII, RNA polymerase II; CTD, carboxyl terminal domain of the largest subunit in RNA polymerase II; CPSF, cleavage/polyadenylation specificity factor; CstF, cleavage stimulation factor; CFI and CFII, cleavage factors I and II, respectively; and PAP, poly(A) polymerase. PABII is not shown in this figure. Yeast has a similar polyadenylation complex. (Adapted from O. Calvo and J. L. Manley, *Genes Dev.* 17 [2003]: 1321–1327.)

first four subunits are named according to their molecular masses in kDa. CPS-73, which belongs to a family of Zn^{2+}-dependent endonucleases, is the endonuclease that cleaves pre-mRNA at the poly(A) site. CPSF-160 binds to AAUAAA with a low degree of specificity. The functions of the other subunits remain to be determined.

2. *Cleavage stimulation factor (CstF)*. CstF binds to the U-rich sequence and is required for cleavage. Three of its subunits (CstF-77, CstF-64, and CstF-50) are named according to their molecular masses in kDa. The fourth subunit is called CSP1. CstF-64 binds to the U-rich sequence, while CstF-77 bridges CstF-64 and Cst50. CstF-77 also interacts with CPSF-160. This interaction is probably very important because CPSF and CstF each bind much more tightly to pre-mRNA when the other is present.
3. *Cleavage factor I (CFI) and cleavage factor II (CFII)*. CFI and CFII are required for cleavage but little is known about their function.
4. *Poly(A) polymerase (PAP)*. PAP adds adenylates to the 3′-end of RNAs and is also required for cleavage. PAP is recruited to the poly(A) site by CPSF that is already bound to pre-mRNA. PAP and CPSF suffice for poly(A) addition to a pre-cleaved RNA substrate but the rate of reaction is slow.
5. *Symplekin*. Symplekin (derived from Greek word meaning to tie together) is part of a larger complex that also includes CstF and CPSF. It has been proposed that symplekin helps to assemble or stabilize the CstF complex and thereby helps to hold the complete cleavage/polyadenylation machinery together.

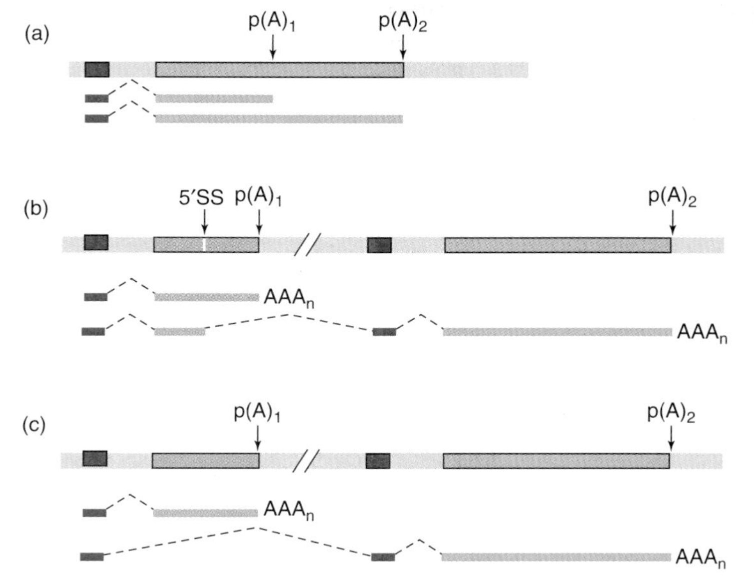

FIGURE 19.42 **The 3′-regions of transcription units leading to alternative poly(A) site selection.** Three kinds of exon arrangements can produce alternative poly(A) site selection. (a) Some transcription units have multiple poly(A) sites within the terminal exon. Only two tandem poly(A) sites—p(A)$_1$ and p(A)$_2$—are shown here for simplicity but more may be present. (b) Other transcription units have an exon that can serve as an internal or terminal exon depending on physiological conditions. (c) Still other transcription units have two or more alternative 3′-terminal exons. The pre-mRNA is shown at the top of each example with the exons shown in red, blue, and orange boxes and introns as thick light blue lines. The spliced mRNA is shown at the bottom with dashed lines corresponding to introns removed during splicing. The abbreviation 5′SS indicates a splice site. (Adapted from G. Edwalds-Gilbert, K. L. Veraldi, and C. Milcarek, *Nucleic Acids Res.* 25 [1997]: 2547–2561.)

6. *Carboxyl terminal domain (CTD) of Rpb1, the largest subunit in RNA polymerase II.* CTD binds to both CPSF and CstF and must be present along with all of the other cleavage/polyadenylation factors for 3′-cleavage to take place.
7. *Poly(A) binding protein II (PABII).* PABII (not shown in Figure 19.41) interacts with CPSF-30, greatly stimulating the rate of poly(A) addition. PABII also controls poly(A) tail length, limiting it to between 200 and 300 adenylate residues.

Transcription units often have two or more alternate polyadenylation sites.

Many transcription units have two or more alternate polyadenylation sites, which may be arranged in three different ways (FIGURE 19.42). Some transcription units have multiple poly(A) sites within the terminal exon (Figure 19.42a). Other transcription units have an exon that can function as an internal or external exon depending on physiological conditions (Figure 19.42b). Still other transcription units have two or more alternative 3′-terminal exons (Figure 19.42c). Further insights can be gained by examining two well-studied systems, the mammalian transcription units for calcitonin and the heavy immunoglobulin chain, IgM.

The Calcitonin Transcription Unit

The calcitonin transcription unit has six exons and is alternatively processed in a tissue-specific fashion to produce two proteins (FIGURE 19.43a). In thyroid and most other tissues, the first four exons are spliced together and polyadenylated at exon 4 so that translation produces calcitonin. In nerve cells, the first three exons are joined to exons 5 and 6 and polyadenylated at exon 6 so that translation produces a neuropeptide called calcitonin gene-related peptide (CGRP).

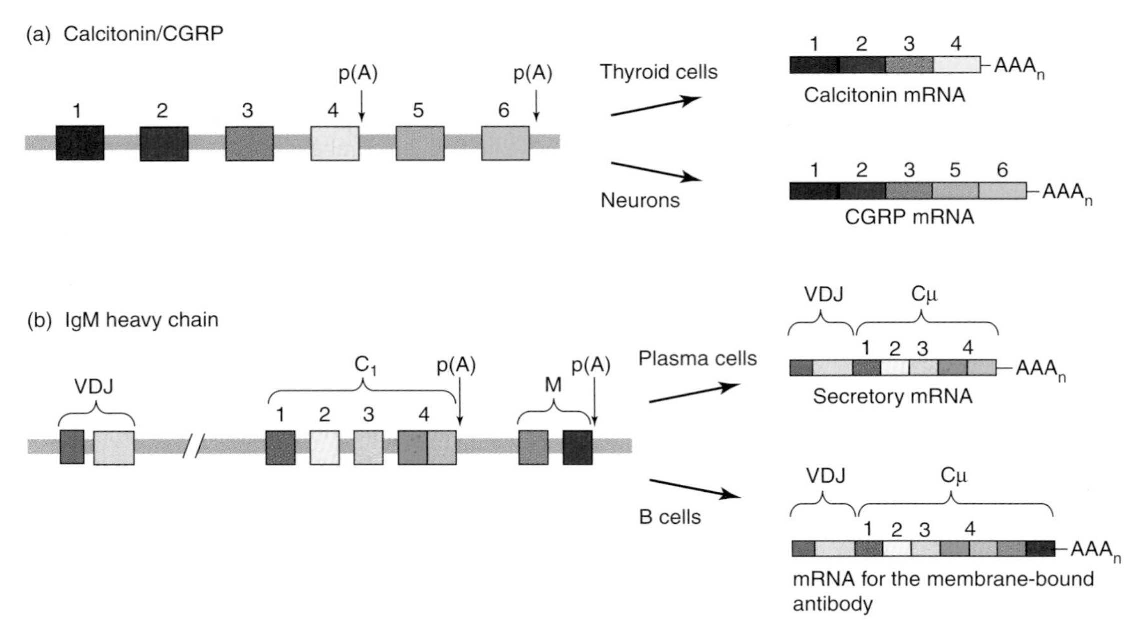

FIGURE 19.43 **Alternative polyadenylation of calcitonin/CGRP and immunoglobulin M (IgM) transcripts.** (a) Organization of exons and introns in calcitonin/CGRP pre-mRNA and the structure of primary mRNA produced in thyroid cells or neurons. (b) Alternative processing choices in the IgM heavy chain (mμ) precursor. (Adapted from J. Zhao, L. Hyman, and C. Moore, *Microbiol. Mol. Biol. Rev.* 63 [1999]: 405–445.)

The IgM Transcription Unit

The IgM transcription unit codes for two kinds of IgM. B cells synthesize membrane-bound IgM, whereas plasma cells synthesize soluble IgM that can be secreted. The IgM transcription unit has two poly(A) sites (FIGURE 19.43b). B cells remove the upstream poly(A) site at exon 4 and use a downstream poly(A) site to produce mRNA that codes for membrane-bound IgM. Plasma cells retain the upstream poly(A) to produce mRNA that codes for the shorter, soluble IgM.

Alternative processing forces us to reconsider our concept of the gene.

Alternative processing raises many fundamental issues, perhaps none more important than the very concept of the gene itself. Prior to the discovery of alternative pre-mRNA processing, a eukaryotic protein-coding gene could be defined as a hereditary unit that codes for a specific polypeptide. However, this one gene–one polypeptide hypothesis does not work for alternatively processed transcription units that can code for two or more protein isoforms. We must therefore find some other way to define a gene. Although an exon has the minimal amount of information that is expressed as a discrete unit, we cannot define a gene as an exon because exons do not contain all of the information that we normally associate with genes. Perhaps the best definition that we can devise at this time is that a gene is a linear collection of exons that are incorporated into a specific mRNA. The term gene continues to be used in a more general sense when referring to a transcription unit or protein coding region, however. Alternative processing also presents challenges for genetic engineering. Some alternatively processed transcription units are known to produce isoforms that have antagonistic effects. Transforming a cell with an alternatively spliced gene therefore may result in the production of a harmful rather than a beneficial protein.

Transcription termination takes place downstream from the poly(A) site.

The transcription elongation complex continues to synthesize RNA after passing the poly(A) site. One of the clearest demonstrations that transcription termination takes place after the poly(A) site comes from a technique known as the **nuclear run-on** method. The steps in this method and its application to transcription termination are illustrated in FIGURE 19.44. The basic idea behind the nuclear run-on method is that the transcription elongation complex will continue to synthesize a nascent transcript under *in vitro* conditions after the nucleus is removed from the cell. Although transcription elongation complexes in different nuclei will be at different stages of transcription, only one stage is shown in Figure 19.44 for clarity.

The first step in the nuclear run-on method is to isolate nuclei from cells that are actively transcribing some specific transcription unit. (Yeast cells are made permeable to ribonucleoside triphosphates instead of isolating their nuclei.) Next, the isolated nuclei are suspended in a reaction mixture containing radioactive ribonucleoside triphosphates and incubated for a sufficient time to allow RNA polymerase II to extend the nascent transcript by about 100 nucleotides. The radioactive RNA is purified after the reaction is stopped and hybridized to DNA fragments bound to filters. Each DNA fragment corresponds to a different region within and 3′ to the transcription unit under study. Bound radioactive RNA is visualized by autoradiography. DNA fragments upstream of the termination site (fragments A–D in Figure 19.44) give significant signals, while those downstream from the termination site (fragment E in Figure 19.44) give only background signals. Although the nuclear run-on method is to all intents and purposes an *in vitro* procedure, it does reveal apparent termination regions. A growing number of transcription units transcribed by RNA polymerase II have been analyzed using the nuclear run-on method. The following conclusions have been reached based on these analyses:

1. Transcription termination requires a poly(A) signal.
2. Transcription termination requires factors for 3′-cleavage but does not require factors for polyadenylation or that the pre-mRNA be cleaved.
3. Transcription terminates beyond the 3′-end of mRNA, usually 200 to 2000 bp downstream from the poly(A) site.
4. Some transcripts terminate efficiently at a single site, whereas others terminate inefficiently over an extended region.

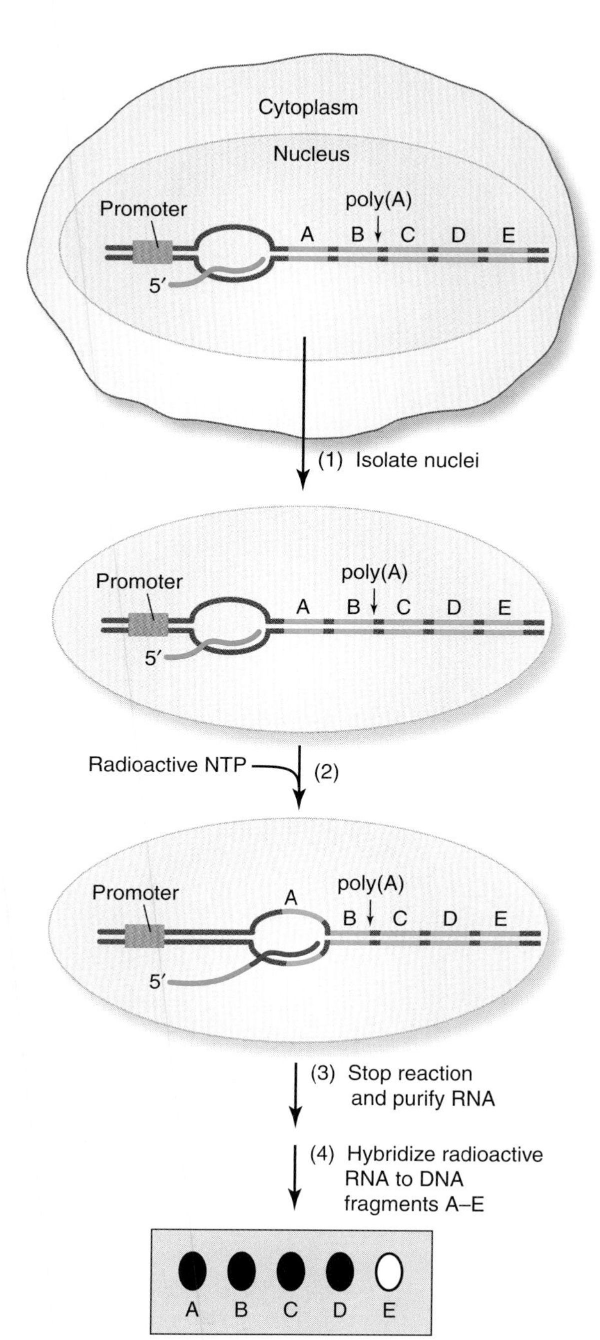

FIGURE 19.44 **Schematic diagram of the nuclear run-on experiment.** Different segments along the DNA template are labeled A–E. The 3′-cleavage site is labeled Poly(A). The experimental steps are as follows: (1) Isolate nuclei. (2) Add the isolated nuclei to a reaction mixture containing radioactive nucleoside triphosphates and incubate for a short time so that about one hundred nucleotides add to the growing chain. (3) Stop the reaction and purify the radioactive (red) RNA molecules. The transcript will have advanced to different positions along the template DNA. (4) Hybridize the purified RNA to DNA fragments corresponding to different regions (A–E) within and 3′ to the transcription unit. DNA fragments upstream of the termination site (fragments A–D give significant signals, while those downstream from the termination site (fragment E) give only background signals.

RNA polymerase II transcription termination appears to involve allosteric changes and a 5′→3′ exonuclease.

RNA polymerase II must pass the poly(A) site before transcription termination can take place, thereby ensuring that transcription termination does not take place until RNA polymerase II has reached the end of the transcription unit. Two models have been proposed to explain how the poly(A) site can stimulate RNA polymerase II transcription termination. The first, known alternatively as the "allosteric" or "antiterminator" model proposes that RNA polymerase II loses a positive elongation/antitermination factor, gains a negative elongation/termination factor, or both as it transcribes past the poly(A) site, causing the polymerase to assume a less processive conformational form that can be more easily released from the DNA template. The second, known as the "torpedo" model proposes that poly(A) site cleavage provides an entry site for a 5′→3′ exonuclease, which attacks the newly created uncapped 5′-phosphate end and moves down the nascent RNA toward RNA polymerase II like a guided torpedo until it reaches RNA polymerase II and somehow disrupts the DNA–RNA duplex, thereby triggering transcription termination. The recent discovery of a 5′→3′ exonuclease, called Xrn2 in mammals, supports the torpedo model. The yeast 5′→3′ exonuclease is called Rat1. These exonucleases have been shown to be necessary but not sufficient for transcription termination.

There is experimental evidence that supports and refutes each model. For instance, the "antiterminator" model explains transcription termination at genes in which cleavage at the poly(A) site follows, rather than precedes, transcription termination. The "torpedo" model does not explain this phenomenon. On the other hand, the "torpedo" model explains why Xrn2 and Rat1 are necessary for transcription termination but the "antiterminator" model does not. Fortunately, the two models are not mutually exclusive. A hybrid model has been proposed that incorporates features of both models and appears to fit the experimental data (FIGURE 19.45). According to this hybrid model, the elongation complex changes conformation upon recognizing the poly(A) site. This conformational change may result from the loss of anti-termination factors, the gain of termination factors (not shown), or both. After the pre-mRNA is cleaved by the cleavage/polyadenylation machinery, a splicing factor helps recruit Xrn2. The Pcf11 subunit in CFII may assist in this process. Xrn2 digests downstream RNA until it reaches the RNA polymerase and then, perhaps with the assistance of a helicase, releases RNA polymerase II from the DNA to complete the termination process.

In higher animals, most histone pre-mRNAs require a special processing mechanism.

In higher animals, most mRNAs that code for histones do not have a poly(A) tail. Although histone pre-mRNAs are cleaved to form histone mRNAs, the cleavage recognition signal is quite different from that in other pre-mRNAs. Instead of the cleavage/polyadenylation

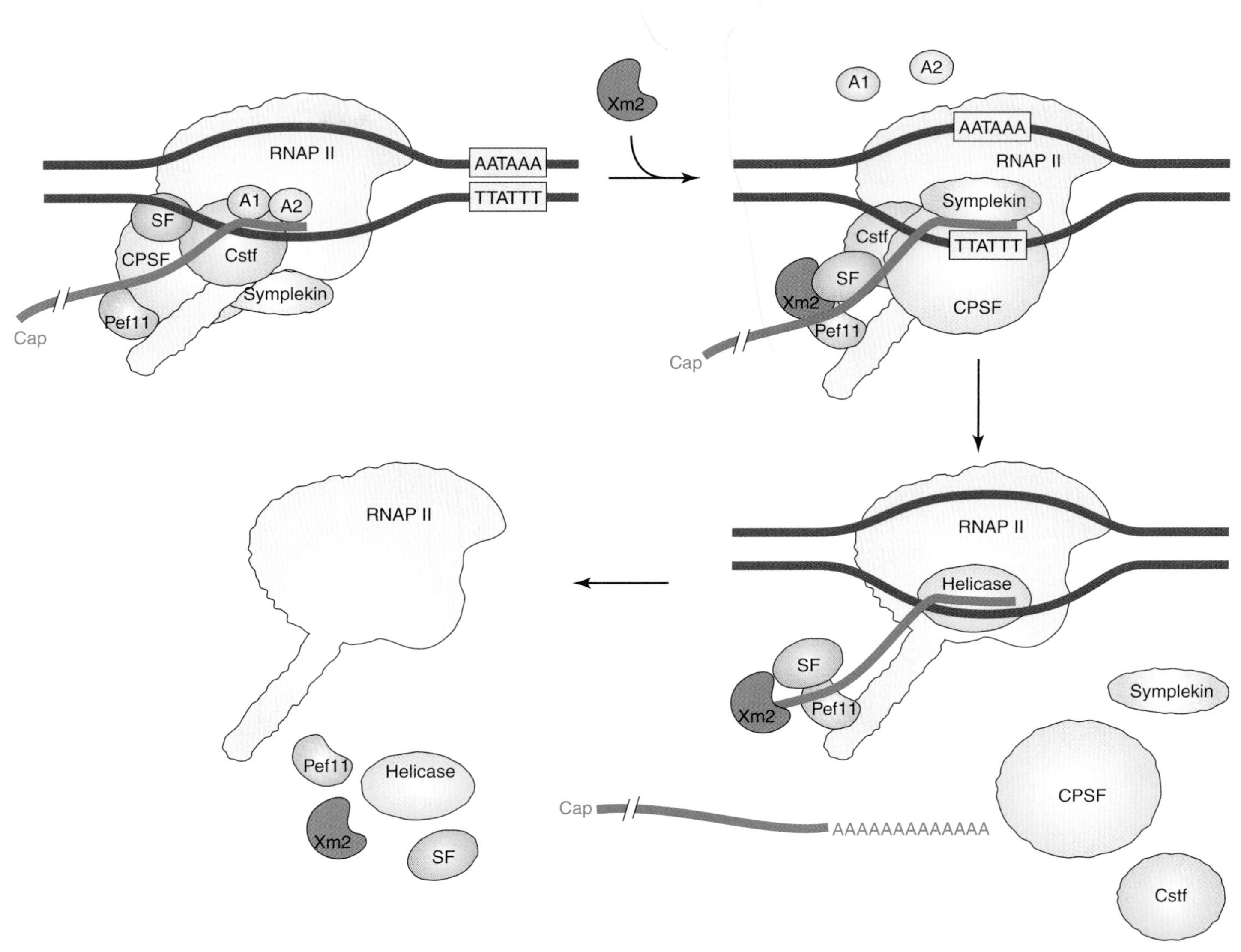

FIGURE 19.45 **Model for termination by RNA polymerase II.** The elongation complex changes conformation upon recognizing the pol(A) site. This conformational change may result from the loss of anti-terminators (A1 and A2), the gain of termination factors (not shown), or both. After the pre-mRNA is cleaved by the cleavage/polyadenylation machinery, a splicing factor (SF) recruits Xrn2. The Pcf11 subunit in CFII may assist in this recruitment. Then Xrn2 digests downstream RNA until it reaches the RNA polymerase. Finally, Xrn2, perhaps with the assistance of a helicase releases RNA polymerase II from the DNA. (Adapted from P. Richard and J. L. Manley, *Genes Dev.* 23 [2009]: 1247–1269.)

site, which is present in virtually all other pre-mRNAs, most histone pre-mRNAs have two highly conserved sequence elements: a 26-nucleotide stem loop structure located upstream from the cleavage site and a purine-rich sequence, the so-called **histone downstream element** (**HDE**), located downstream of the cleavage site (FIGURE 19.46). The stem-loop structure binds a protein designated the **stem-loop binding protein** (**SLBP**), which in turn helps to recruit a small nuclear ribonucleoprotein, U7 snRNP, to the histone pre-mRNA. U7 snRNA forms base pairs with HDE, establishing the cleavage site. U7 snRNP also recruits a protein complex known as the heat-labile factor, which contains the CPSF subunits, two subunits of CstF (CstF-77 and CstF-64), and symplekin. The CPSF-73 subunit is the endonuclease that cleaves the histone pre-mRNA. Nikolay G. Kolev and Joan A.

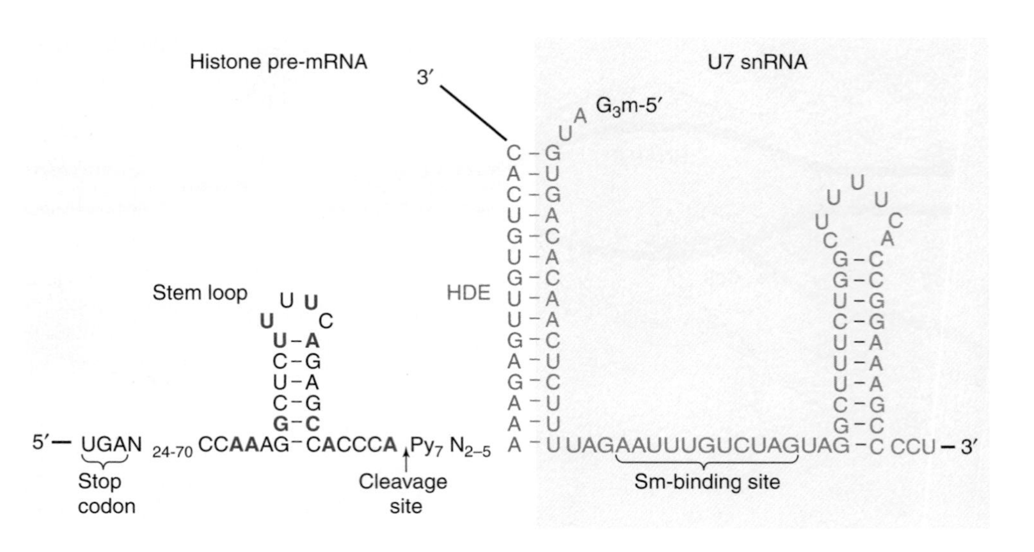

FIGURE 19.46 **Mammalian replication-dependent histone pre-mRNA processing site.** Invariant positions in the histone pre-mRNA are indicated by the bases shown in red. Sequences downstream of the cleavage site are for the mouse histone H2A pre-mRNA and are shown base paired to the mouse U7 snRNA. Py indicates pyrimidine nucleotides and N indicates any nucleotide. HDE is the abbreviation for histone downstream element and Sm indicates an Sm polypeptide. (Adapted from G. M. Gilmartin, *Genes Dev.* 19 [2005]: 2517–2521.)

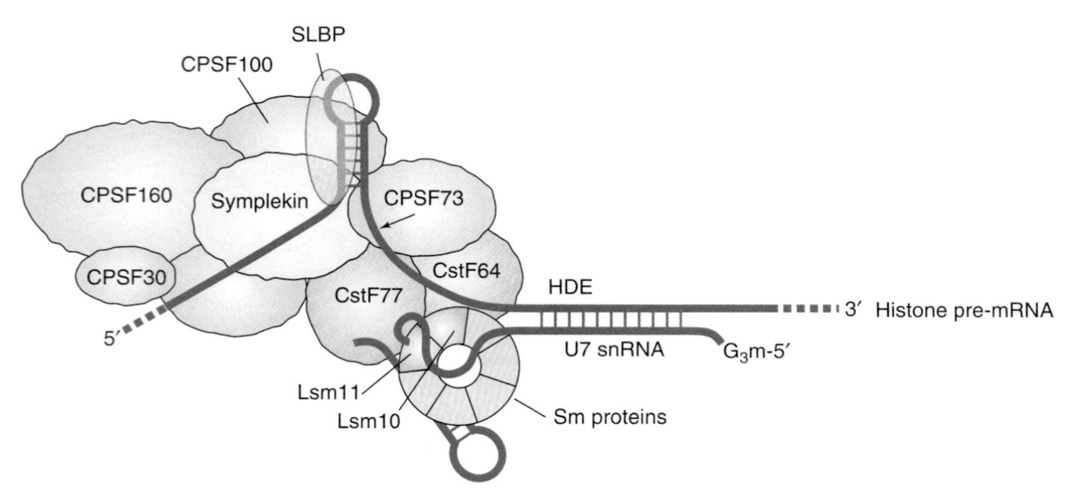

FIGURE 19.47 **Model for histone pre-mRNA processing.** Components of the heat labile factor (HLF) are colored blue and orange. The U7 snRNA is depicted base pairing with the histone pre-mRNA downstream element. U7 snRNP orients the histone pre-mRNA for cleavage (arrow) by CPSF-73 through contacts between its Sm proteins and CstF subunits in HLF. Proteins of the Sm ring that are shared with spliceosomal snRNPs are shown in green; U7-specific Lsm10 and Lsm11 proteins are shown in shades of blue. (Modified from N. G. Kolev and J. A. Steitz, *Genes Dev.* 19 [2005]: 2583–2592. Copyright 2005, Cold Spring Harbor Laboratory Press. Used with permission of Joan Steitz, Yale University.)

Steitz have proposed the model for histone pre-mRNA processing that is shown in FIGURE 19.47. The reason that histone pre-mRNAs require a special processing mechanism probably relates to the fact that histones must be synthesized in coordination with DNA synthesis as well as with one another.

19.6 RNA Editing

RNA editing permits a cell to recode genetic information in a systematic and regulated fashion.

Although transcriptional processing provides eukaryotic cells with the rather remarkable ability to use a single transcription unit to synthesize

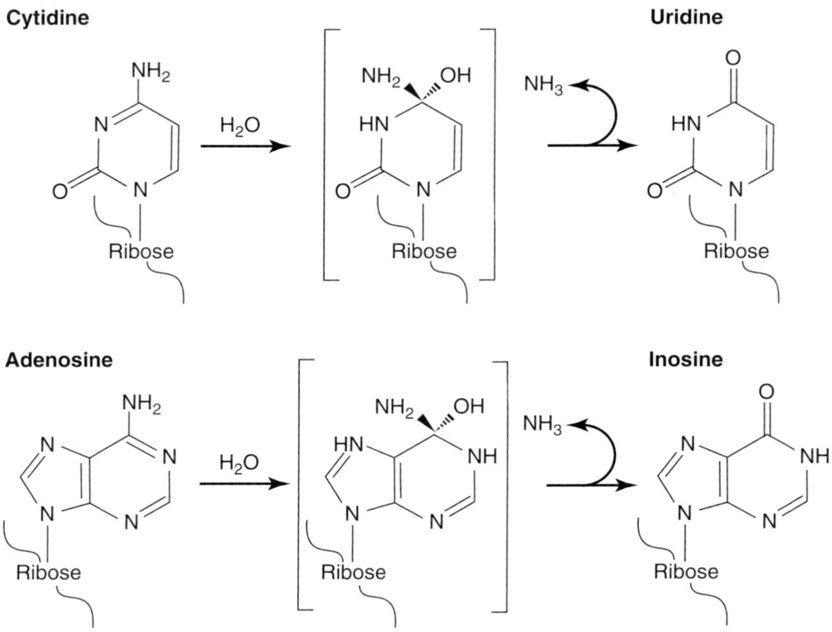

FIGURE 19.48 Reaction mechanism of cytidine and adenosine deaminases acting on RNA. Hydrolytic deamination of cytidine and adenosine leads to uridine and inosine, respectively. Presumably, ammonia is released during the reactions. (Caption reprinted and figure modified from *Trends Biochem. Sci.*, vol. 26, A. P. Gerber and W. Keller, RNA editing by base deamination . . ., pp. 376–384, copyright 2001, with permission from Elsevier [http://www.sciencedirect.com/science/journal/09680004].)

a variety of protein isoforms, eukaryotes also use another method to alter RNA before it is translated. They can change an RNA molecule so that its sequence is no longer the one predicted from the DNA coding sequence. The process is termed **RNA editing** when the altered RNA molecule could in principle have been encoded by the DNA sequence. The best characterized examples of RNA editing in higher eukaryotes involve deamination reactions in which cytidine is converted to uridine or adenosine is converted to inosine (FIGURE 19.48). The translation machinery reads an inosine in a codon as though it were guanosine.

C-to-U editing occurs in the nucleus and appears to be specific for spliced transcripts rather than pre-mRNA. An example of C-to-U RNA editing involves the intestinal mRNA that encodes apolipoprotein B (ApoB) that is present in chylomicrons, the lipoproteins that transport ingested cholesterol. This process is specific for intestinal cells and is not carried out by liver cells. ApoB mRNA editing in intestinal cells converts codon 2153 in exon 26 (CAA, glutamine codon in ApoB mRNA) to a stop translation codon (UAA). The edited mRNA codes for a truncated protein, ApoB48, that is missing the carboxyl terminal domain (FIGURE 19.49) and is incorporated into a chylomicron. The full size protein, ApoB100, which is synthesized by the liver, is incorporated into low density lipoprotein (LDL). Thus, RNA editing, like alternative splicing, permits cells to use a single DNA sequence to make more than one protein. C-to-U RNA editing of apoB mRNA is site specific, targeting a single cytidine in a transcript containing more than 14,000 nucleotides. The minimum recognition sequence consists of 35 nucleotides flanking the cytidine that is to be deaminated, with more distant 5′- and 3′-efficiency elements that enhance the rate of editing. The 35 nucleotide sequence contains a regulatory element, a spacer element, and an 11-nucleotide "mooring" sequence that starts

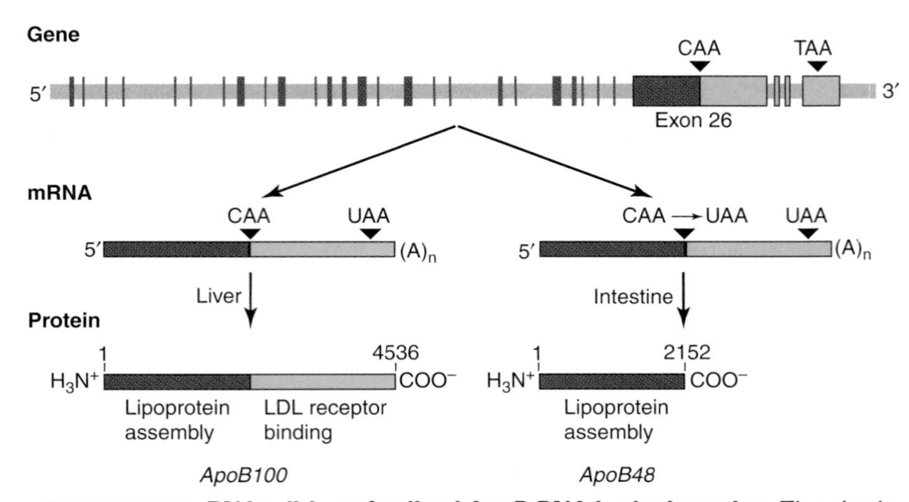

FIGURE 19.49 **RNA editing of spliced ApoB RNA in the intestine.** The single copy of the human apoenzyme B (ApoB) gene spans 43 kb and is divided into 29 exons. The liver splices the pre-mRNA and then translates the resulting mRNA to produce the 512 kDa ApoB100 protein, which is incorporated into low density lipoprotein (LDL). The intestine edits the spliced mRNA, changing codon 2153 in exon 26 (CAA, glutamine) to a translation termination codon (UAA). Because of the translation termination codon, the edited mRNA encodes a truncated protein that is missing the carboxyl terminal domain. The truncated protein, called ApoB48, is incorporated into the chylomicrons. Introns are shown as gray lines. Exons that are translated in forming ApoB48 and ApoB100 are shown as red bars, whereas exons that are only translated in forming the carboxyl terminal domain in ApoB100 are shown as blue bars. Note that only the red part of exon 26 is translated to form ApoB48. (Adapted from P. Hodges and J. Scott, *Trends Biochem. Sci.* 17 [1992]: 77–81.)

4 to 6 nucleotides downstream from the target cytidine (FIGURE 19.50). Selection of the editing site probably also depends on the secondary structure of the RNA so that the targeted cytidine is within the loop segment of a stem and loop structure. The **cytidine deaminase acting on RNA (CDAR)** is a homodimer that requires the assistance of an additional stimulatory factor (ASF), which has three RNA-recognition motifs.

A-to-I RNA editing takes place on pre-mRNA while introns are still present. Several mammalian and insect pre-mRNAs that encode neurotransmitter receptors are known to undergo A-to-I editing and it is likely that still more will be discovered. An example of this type of editing is in the pre-mRNA that encodes subunits of the ionotropic glutamate receptors. Conversion of a glutamine codon (CAG) to an arginine codon (CIG) introduces an arginine residue in a channel forming domain, making the receptor impermeable to Ca^{2+} ions. All known examples of site-specific A-to-I conversions require dsRNA (double-stranded RNA) structures around the editing site. The enzyme that catalyzes the A-to-I conversion is called **adenosine deaminase acting on RNA (ADAR)**. The targeted adenosine is located in a double-stranded RNA structure formed by pairing exon and intron sequences in a region dubbed the **exon complementary sequence** or **ECS** (FIGURE 19.51). The intron requirement for A-to-I editing indicates that editing takes

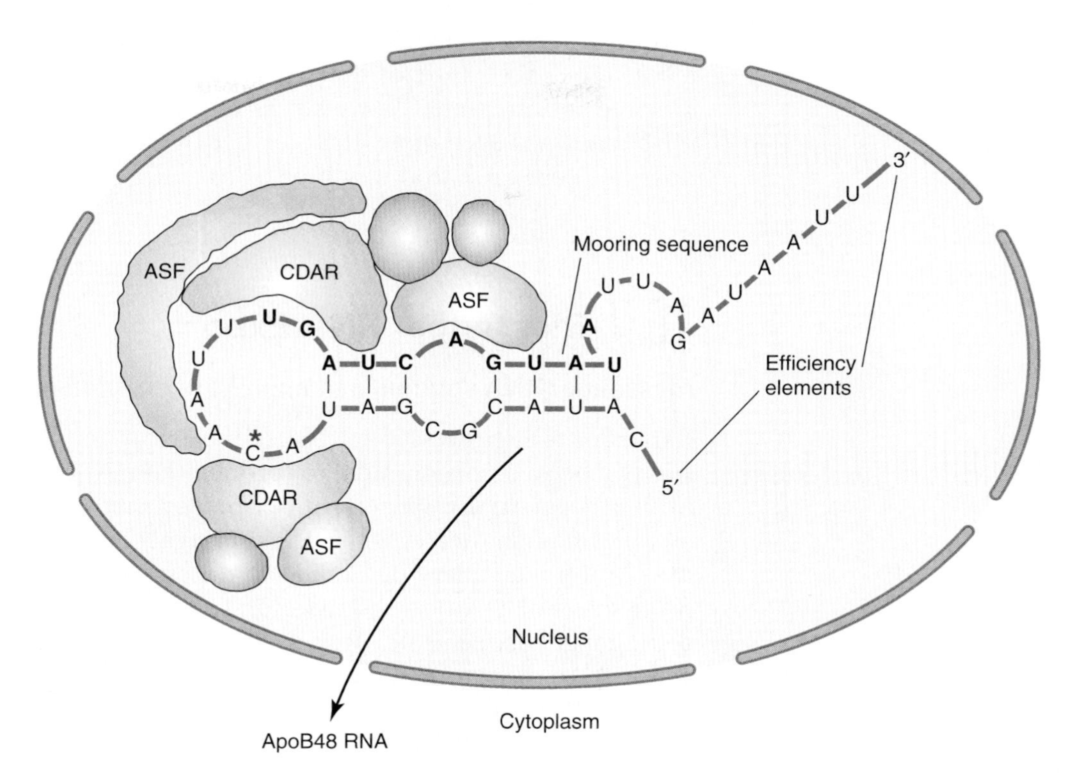

FIGURE 19.50 **C-to-U RNA editing of apolipoprotein B.** A schematic representation illustrates cytidine deaminase (CDAR) acting on the edited base (asterisk) within a 35 nucleotide RNA region. An additional stimulator factor, ASF, (cyan) binds to RNA both 5′ and 3′ of the edited base (indicated by an asterisk). Additional proteins that may modulate assembly of the holoenzyme are shown in magenta. Note that the stoichiometry of CDAR and ASF molecules with respect to the active enzyme is unknown. The model emphasizes the role of required elements on the RNA within the vicinity of the edited bases and the requirement for an optimal structure, conferred by both 5′ and 3′ efficiency elements. (Adapted from V. Blanc and N. O. Davidson, *J. Biol. Chem.* 278 [2003]: 1395–1398.)

place in the nucleus and precedes splicing. A-to-I splicing also permits a single genomic sequence to code for more than one protein.

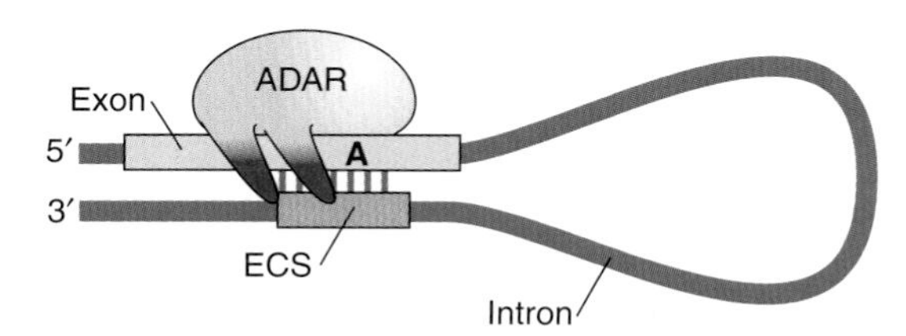

FIGURE 19.51 **Editing of pre-mRNAs by adenosine deaminase acting on RNA (ADAR).** The target adenosine (A) is located in a double-stranded RNA region formed by pairing of exon and intron sequences indicated as ECS (exon complementary sequence). The adenosine deaminase acting on RNA converts the target adenosine to inosine. RNA binding domains are colored red. (Adapted from A. P. Gerber and W. Keller, *Trends Biochem. Sci.* 26 [2001]: 376–384.)

The human proteome contains a much greater variety of proteins than would be predicted from the human genome.

Transcriptional processing and RNA editing have important consequences for the eukaryotic **proteome** (the complete set of proteins expressed during an organism's lifetime). The best current estimate is that the human genome contains approximately 20,000 transcription units that code for proteins. Alternative splicing, which affects at least 90% of the pre-mRNAs, probably results in the production of a few hundred thousand different kinds of proteins in the human proteome. RNA editing, alternative transcription initiation sites and alternative transcription termination sites also contribute to protein diversity. Furthermore, many proteins are modified after they are formed to produce additional variants. The most common types of posttranslational modifications are cleavage, phosphorylation, acetylation, glycosylation, and methylation. Knowing a transcription unit's DNA sequence thus is a starting point for understanding how information is transferred from DNA to protein but it is not sufficient by itself to tell us the exact nature of the proteins that are present in a given cell at some specific time. As we see in the final section of this chapter, the situation is even more complex because even synthesis of mRNA does not guarantee that it will be translated.

19.7 Messenger RNA Export

Messenger RNA splicing and export are coupled processes.

Export of fully processed mRNA to the cytoplasm is essential for cell survival because it is required for protein synthesis. In contrast, export of introns, partially spliced transcripts, or mutant transcripts to the cytoplasm might decrease the cell's survival chances by damaging the translation machinery or leading to the synthesis of harmful protein products. Eukaryotic cells therefore require some mechanism(s) to distinguish between fully processed mRNA molecules that are to be exported to the cytoplasm and other RNA polymerase II products that must be retained in the nucleus and degraded.

One export mechanism is to couple splicing and export. The first evidence for this coupling came from studies showing that mRNAs produced by splicing are more efficiently exported than identical mRNAs transcribed from an intronless DNA. This effect of splicing on export was explained by the observation that spliced mRNAs (but not intronless DNA transcripts) interact with specific proteins to form a distinct mRNA•protein complex that moves through the nuclear pore. Proteins appear to assemble adjacent to newly formed exon-exon junctions. The resulting protein complex, termed the **exon junction complex** (**EJC**) is about 24 bases upstream of the splice junction and serves to mark the locations of splicing activity. The model presented in FIGURE 19.52 shows how some of the key proteins in the EJC are recruited to mRNA and their subsequent fate. For clarity, the model shows only two exons and is divided into the following three stages:

Stage 1: SR proteins bind to exon sequences, helping to recruit the spliceosome. UAP56 (U2AF associated protein), a conserved splicing factor, interacts with the spliceosome and then helps to recruit **RNA export factor 1** (**REF 1**).

Stage 2: REF 1 and several other proteins assemble to form the EJC. Some of these proteins are called **NMD factors** because they participate in an mRNA surveillance process termed **nonsense-mediated decay** (**NMD**) that degrades mRNA with premature translation termination codons. The **mRNA export protein**, a heterodimer, binds to REF 1 during this stage. UAP56 appears to be released as a result of this interaction because both NMD and UAP56 bind to the same REF 1 site. Transport across the nuclear pore occurs as a result of interactions between the mRNA binding proteins and nuclear pore proteins. This transport does not appear

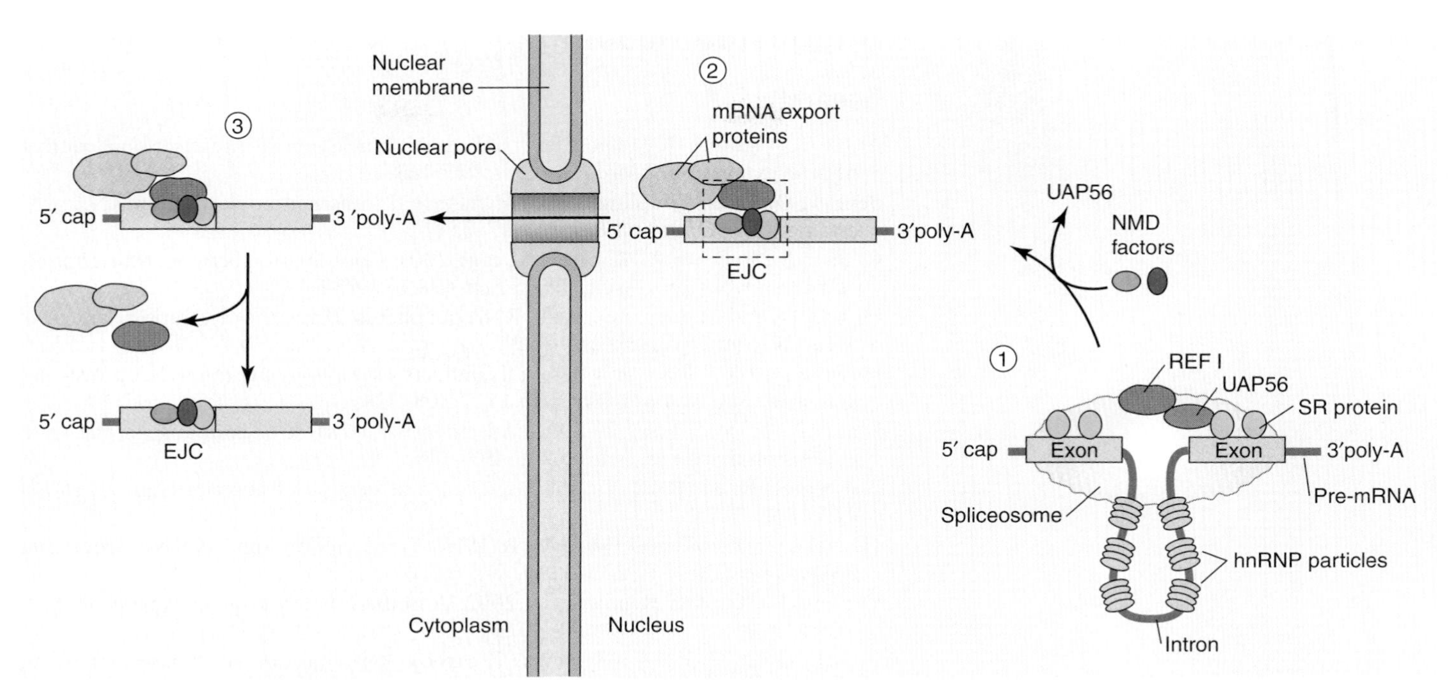

FIGURE 19.52 **Model for splicing-coupled mRNA export in higher animals.** (Stage 1) SR proteins bind to exon sequences, helping to recruit the spliceosome. UAP56, a conserved splicing factor, interacts with the spliceosome and then helps to recruit RNA export factor 1 (REF1). For simplicity, the pre-mRNA molecule, which is part of the hnRNP particle, is shown to contain only 2 exons and an intron in addition to its 5′-cap and 3′-poly(A) tail. (Stage 2) The mRNA export protein targets the mRNA protein complex to the nuclear pores. REF1 acts as a bridging protein between the exon junction complex (EJC) and the mRNA protein complex. (Stage 3) mRNA export factors dissociate from the mRNA protein complex after export to the cytoplasm. Factors involved in nonsense-mediated decay (NMD) remain bound to the mRNA-protein complex. (Adapted from R. Reed and E. Hurt, *Cell* 108 [2002]: 523–531.)

to require nucleoside triphosphates. Movement across the nuclear pore is probably driven by the fact that the mRNA•protein complex concentration is much lower in the cytoplasm than in the nucleus.

Stage 3: mRNA export factors dissociate from the mRNA protein complex after export to the cytoplasm. Factors involved in nonsense mediated decay remain bound to the mRNA protein complex.

The three stage pathway outlined above explains how cells export mRNAs transcribed from split genes but not how they export mRNAs transcribed from genes that are not split. Studies suggest that mRNAs transcribed from intronless genes have specific binding elements for export factors.

Cells appear to use a family of proteins, termed **non-shuttling hnRNP proteins**, to prevent excised introns and partially spliced mRNA from leaving the nucleus. These proteins, which bind to intron sequences and package them into hnRNP complexes, have nuclear retention signals that prevent the bound RNA from being exported. Although non-shuttling hnRNP proteins have a low degree of sequence specificity, they do not interact with the relatively short exons because spliceosomes block their way. Spliceosomes also contribute to the retention of partially spliced pre-mRNA within the nucleus. Retained introns are cleaved by a 2′→5′ phosphodiesterase and then degraded.

Suggested Reading

General

Akker, S. A., Smith, P. J., and Chew, S. L. 2001. Nuclear post-transcriptional control of gene expression. *J Mol Endocrinol* 27:123–131.

Bentley, D. 1999. Coupling RNA polymerase II transcription with pre-mRNA processing. *Curr Opin Cell Biol* 11:347–351.

Cramer, P., Srebrow, A., Kadener, S., et al. 2001. Coordination between transcription and pre-mRNA processing. *FEBS Lett* 498:179–182.

Hirose, Y., and Manley, J. L. 2000. RNA polymerase II and the integration of nuclear events. *Genes Dev* 14:1415–1429.

Howe, K. J. 2002. RNA polymerase II conducts a symphony of pre-mRNA processing activities. *Biochem Biophys Acta* 1577:308–324.

Maniatis, T., and Reed, R. 2002. An extensive network of coupling among gene expression machines. *Nature* 416:499–506.

Neuberger, K. M. 2002. On the importance of being co-transcriptional. *J Cell Sci* 115:3865–3871.

Neugebauer, K. M., and Roth, M. B. 1997. Transcription units as RNA processing units. *Genes Dev* 11:3279–3285.

Orphanides, G., and Reinberg, D. 2002. A unified theory of gene expression. *Cell* 108:439–451.

Prelich, G. 2002. RNA polymerase II carboxy-terminal domain kinases: emerging clues to their function. *Eukaryot Cell* 1:153–162.

Proudfoot, N. 2000. Connecting transcription to messenger RNA processing. *Trends Biochem Sci* 25:290–293.

Proudfoot, N. J., Furger, A., and Dye, M. J. 2002. Integrating mRNA processing with transcription. *Cell* 108:501–512.

Pre-mRNA

Darnell, J. E., Jelinek, W. R., and Molloy, G. R. 1973. Biogenesis of mRNA: genetic regulation in mammalian cells. *Science* 181:1215–1221.

Darnell, J. E. Jr. 2002. The surprises of mammalian molecular cell biology. *Nat Med* 8:1068–1071.

Cap Formation

Calero, G., Wilson, K. F., Ly, T., et al. 2002. Structural basis of m^{7}GpppG binding to the nuclear cap-binding protein complex. *Nat Struct Biol* 9:912–917.

Cho, E. J., Takagi, T., Moore, C. R., and Buratowski, S. 1997. mRNA capping enzyme is recruited to the transcription complex by phosphorylation of the RNA polymerase carboxy-terminal domain. *Genes Dev* 11:3310–3326.

Cougot, N., van Dijk, E., Babajko, S., and Séraphin, B. 2004. Cap-tabolism. *Trends Biochem Sci* 29:436–444.

Cowling, V. H. 2010. Regulation of mRNA cap methylation. *Biochem J* 425:295–302.

Decker, C. J., and Parker, R. 2002. mRNA decay enzymes: decappers conserved between yeast and mammals. *Proc Natl Acad Sci USA* 99:2512–2514.

Dunyak, D. S., Everdeen, D. S., Albanese, J. G., and Quinn, C. L. 2002. Deletion of individual mRNA capping genes is unexpectedly not lethal to *Candida albicans* and results in modified mRNA cap structures. *Eukaryot Cell* 1:1010–1020.

Fabrega, C., Shen, V., Shuman, S., and Lima, C. D. 2003. Structure of an mRNA capping enzyme bound to the phosphorylated carboxy-terminal domain of RNA polymerase II. *Mol Cell* 11:1549–1561.

Fong, N., and Bentley, D. L. 2001. Capping, splicing, and 3′ processing are independently stimulated by RNA polymerase II: different functions for different segments of the CTD. *Genes Dev* 15:1783–1795.

Furuichi, Y., Morgan, M., Muthukrishnan, M, and Shatkin, A. J. 1975. Reovirus messenger RNA contains a methylated, blocked 5′-terminal structure: $m^7G(5')ppp(5')G^mpCp$-. *Proc Natl Acad Sci USA* 72:362–366.
Furuichi, Y., and Shatkin, A. J. 2005. Caps on eukaryotic mRNAs. In: *Encyclopedia of Life Sciences*. pp. 1–9. Hoboken, NJ: John Wiley and Sons Ltd.
Ho, C. K., and Shuman, S. 1999. Distinct roles for CTD Ser-2 and Ser-5 phosphorylation in the recruitment and allosteric activation of mammalian mRNA capping enzyme. *Mol Cell* 3:405–411.
Kobor, M. S., and Greenblatt, J. 2002. Regulation of transcription elongation by phosphorylation. *Biochim Biophys Acta* 1577:261–275.
Rasmussen, E. B., and Lis, J. T. 1993. *In vitro* transcriptional pausing and cap formation on three *Drosophila* heat shock genes. *Proc Natl Acad Sci USA* 90:7923–7927.
Shatkin, A. J., and Manley, J. L. 2000. The ends of the affair: capping and polyadenylation. *Nat Struct Biol* 7:838–842.
Sheth, U., and Parker, R. 2003. Decapping and decay of messenger RNA occur in cytoplasmic processing bodies. *Science* 300:805–808.
Shuman, S. 1997. Origins of mRNA identity: capping enzymes bind to the phosphorylated C-terminal domain of RNA polymerase II. *Proc Natl Acad Sci USA* 94:12758–12760.
Shuman, S. 2001. Structure, mechanism, and evolution of the mRNA capping apparatus. *Progr Nucl Acid Res Mol Biol* 66:3–40.
Shuman, S. 2002. What messenger RNA capping tells us about eukaryotic evolution. *Nat Rev Mol Cell Biol* 3:619–625.
Wei, C. M., and Moss, B. 1975. Methylated nucleotides block 5′-terminus of vaccinia virus messenger RNA. *Proc Natl Acad Sci USA* 72:318–322.

Split Genes

Berget, S. M. 1995. Exon recognition in vertebrate splicing. *J Biol Chem* 270:2411–2414.
Berget, S. M., Moore, C., and Sharp, P. A. 1977. Spliced segments at the 5′ terminus of adenovirus 2 late mRNA. *Proc Natl Acad Sci USA* 74:3171–3175.
Black, D. L. 1995. Finding splice sites within a wilderness of RNA. *RNA* 1:763–771.
Black, D. L. 2000. Protein diversity from alternative splicing: a challenge for bioinformatics and post-genome biology. *Cell* 103:367–370.
Cáceres, J. F., and Kornblihtt, A. 2002. Alternative splicing: multiple control mechanisms and involvement in human disease. *Trends Genet* 18:186–193.
Cartegni, L., Chew, S. L., and Krainer, A. R. 2002. Listening to silence and understanding nonsense: exonic mutations that affect splicing. *Nat Rev Genet* 3:285–298.
Chambon, P. 1981. Split genes. *Sci Am* 244:60–71.
Chow, L. T., Gelinas, R. E., Broker, T. R., and Roberts, R. J. 1977. An amazing sequence arrangement at the 5′-ends of adenovirus 2 messenger RNA. *Cell* 12:1–8.
Faustino, N. A., and Cooper, T. A. 2003. Pre-mRNA splicing and human disease. *Genes Dev* 17:419–437.
Furuyama, S. and Bruzik, J. P. 2005. *Trans* splicing. In: *Encyclopedia of Life Sciences*. pp. 1–5. Hoboken, NJ: John Wiley and Sons Ltd.
Gravely, B. R. 2001. Alternative splicing: increasing diversity in the proteomic world. *Trends Genet* 17:100–107.
Hartmann, B., and Valcárcel, J. 2009. Decrypting the genome's alternative messages. *Curr Opin Cell Biol* 21:377–386.
Horiuchi, T. and Aigaki, T. 2006. Alternative trans-splicing: a novel mode of pre-mRNA processing. *Biol.* Cell 98:135–140.
Kornblihtt, A. R., de la Mata, M. Fededa, J. P., et al. 2004. Multiple links between transcription and splicing. *RNA* 10:1489–1498.
Kriventseva, E. V., Koch, I., Apweiler, R., et al. 2003. Increase of functional diversity by alternative splicing. *Trends Genet* 19:124–128.
Ladd, A. N., and Cooper, T. A. 2002. Finding signals that regulate alternative splicing in the post-genomic era. *Genome Biol* 3:8.1–8.16.

Lander, E. S., Linton, L. M., Birren, B., et al. 2002. Initial sequencing and analysis of the human genome *Nature* 409:860–921.

Le Hir, H., Nott, A., and Moore, M. J. 2003. How introns influence and enhance eukaryotic gene expression. *Trends Biochem Sci* 28:215–220.

Lorković, Z. J., Kirk, D. A. W., Lambermaon, M. H. L., and Filipowicz, W. 2000. Pre-mRNA splicing in higher plants. *Trends Plant Sci* 5:160–167.

Lou, H., and Gagel, R. F. 1998. Alternative RNA processing—its role in regulating expression of calcitonin/calcitonin gene-related peptide. *J Endocrinol* 156:401–405.

Maniatis, T., and Tasic, B. 2002. Alternative pre-mRNA splicing and proteome expansion in metazoans. *Nature* 418:236–243.

McManus, C. J., Duff, M. O., Eipper-Mains, J., and Graveley, B. R. 2010. Global analysis of *trans*-splicing in *Drosophila. Proc Natl Acad Sci USA* 107:12975–12979.

Modrek, B., and Lee, C. 2002. A genomic view of alternative splicing. *Nat Genet* 30:13–19.

Modrek, B., Resch, A., Grasso, C., and Lee, C. 2001. Genome-wide detection of alternative splicing in expressed sequences of human genes. *Nucleic Acids Res* 29:2850–2859.

Moore, M. J. 2000. Intron recognition comes of age. *Nat Struct Biol* 7:14–16.

Mount, S. M. 2002. Messenger RNA splicing signals. In: *Encyclopedia of Life Sciences*. pp. 1–7. London, UK: Nature Publishing Group.

Padgett, R. A., Konarska, M. M., Grabowski, P. J., et al. 1984. Lariat RNA's as intermediates and products in splicing messenger RNA precursors. *Science* 325:898–903.

Pirrott, V. 2002. Trans-splicing in *Drosophila. Bioessays* 24:988–991.

Roberts, G. C., and Smith, C. W. J. 2002. Alternative splicing: combinatorial output from the genome. *Curr Opin Chem Biol* 6:375–383.

Rosonina, E., and Blencowe, B. J. 2002. Gene expression: the close coupling of transcription and splicing. *Curr Biol* 12:R319–R321.

Sharp, P. A. 1993. Split genes and RNA splicing. In: *Nobel Lectures in Physiology or Medicine 1991-1995*, Ringertz, N., ed. Hackensack, NJ: World Scientific Publishing Co.

Sharp, P. A., and Burge, C. B. 1997. Classification of introns: U2-type or U12-type. *Cell* 91:875–879.

Smith, C. W. J., and Valcárcel, J. 2000. Alternative pre-mRNA splicing: the logic of combinatorial control. *Trends Biochem Sci* 25:381–387.

Stetefeld, J., and Ruegg, M. A. 2005. Structural and functional diversity generated by alternative mRNA splicing. *Trends Biochem Sci* 30:515–521.

Valcárcel, J., and Smith, C. W. J. 2005. Alternative splicing: cell-type-specific and developmental control. In: *Encyclopedia of Life Sciences*. pp. 1–9. Hoboken, NJ: John Wiley and Sons Ltd..

Wang, J., and Manley, J. L. 1997. Regulation of pre-mRNA splicing in metazoa. *Curr Opin Genet Dev* 7:205–211.

Ward, A. J., and Cooper, T. A. 2010. The pathobiology of splicing. *J Pathobiol* 220:152–163.

Wu, Q., and Krainer, A. R. 1999. AT-AC pre-mRNA splicing mechanisms and conservation of minor introns in voltage-gated channel genes. *Mol Cell Biol* 19:3225–3236.

Zeng, C., and Berget, S. M. 2000. Participation of the C-terminal domain of RNA polymerase II in exon definition during pre-mRNA splicing. *Mol Cell Biol* 20:8290–8301.

Spliceosomes

Abelson, J. 2008. Is the spliceodome a ribonucleoprotein enzyme? *Nat Struct Mol Biol* 15:1235–1237.

Azubel, M., Wolf, S. G., Sperling, J., and Sperling, R. 2004. Three-dimensional structure of the native spliceosome by cryo-electron microscopy. *Mol Cell* 15:833–839.

Beggs, J. D. 2001. Spliceosomal machinery. In: *Encyclopedia of Life Sciences*. pp. 1–9. London, UK: Nature Publishing Group.

Black, D. L. 2003. Mechanisms of alternative pre-messenger RNA splicing. *Ann Rev Biochem* 72:291–336.

Blencow, B. J. 2000. Exonic splicing enhancers: mechanism of action, diversity and role in human genetic diseases. *Trends Biochem Sci* 25:106–110.

Brody, E., and Abelson, J. 1985. The "spliceosome": yeast pre-messenger RNA associates with a 40S complex in a splicing-dependent reaction. *Science* 228:963–967.

Brow, D. A. 2002. Allosteric cascade of spliceosome activation. *Ann Rev Genet* 36:333–360.

Burge, C. B., Tushchl, T., and Sharp, P. A. 1999. Splicing of precursors to mRNAs by the spliceosomes. In: *The RNA World*, 2nd ed., Gesteland, R. F., Cech T. R., Atkins, J. F., eds. Woodbury, NY: Cold Spring Harbor Laboratory Press.

Burge, C. B., Tushchl, T., and Sharp, P. A. 2005. Splicing of precursors to mRNA by the spliceosomes. In *The RNA World*, 3rd ed., Gesteland, R. F., Atkins, J. F., and Cech, T. R., eds. Woodbury, NY: Cold Spring Harbor Laboratory Press.

Chen, M. C., and Manley, J. L. 2009. Mechanisms of alternative splicing regulation: insights from molecular and genomic approaches. *Nat Rev Mol Cell Biol* 10:741–754.

Collins, C. A., and Guthrie, C. 2000. The question remains: is the spliceosome a ribozyme? *Nat Struct Biol* 7:850–854.

Das, R., Zhou, Z., and Reed, R. 2000. Functional association of U2 snRNP with the ATP-independent spliceosomal complex E. *Mol Cell* 5:779–787.

Du, J., and Rosbash, M. 2002. The U1 snRNP protein U1C recognizes the 5′-splice site in the absence of base pairing. *Nature* 419:86–90.

Fu, X. D. 1995. The superfamily of arginine/serine splicing factors. *RNA* 1:663–680.

Garcia-Blanco, M. A., Puttaraju, M., Mansfield, S. G., and Mitchell, L. G. 2000. Spliceosome-mediated RNA trans-splicing in gene therapy and genomics. *Gene Ther Regul* 1:141–163.

Goldtrohm, A. C., Greenleaf, A. L., and Garcia-Blanco, M. A. 2001. Co-transcriptional splicing of pre-messenger RNAs: considerations for the mechanism of alternative splicing. *Gene* 277:31–47.

Görnemann, J., Kotovic, K. M., Hujer, K., and Neugebauer, K. M. 2005. Cotranscriptional spliceosome assembly occurs in a stepwise fashion and requires the cap binding complex. *Mol Cell* 19:53–63.

Grabowski, P. J., Padgett, R. A., and Sharp, P. A. 1984. Messenger RNA splicing *in vitro*: an excised intervening sequence and a potential intermediate. *Cell* 37:415–427.

Grainger, R. J., and Beggs, J. D. 2005. Prp8 protein: at the heart of the spliceosome. *RNA* 11:533–537.

Graveley, B. R. 2000. Sorting out the complexity of SR protein functions. *RNA* 6:1197–1211.

Graveley, B. R. 2004. A protein interaction domain contacts RNA in the prespliceosome. *Mol Cell* 13:302–304.

Hastings, M. L., and Krainer, A. R. 2001. Pre-mRNA splicing in the new millennium. *Curr Opin Cell Biol* 13:302–309.

Hertel, K. J., and Graveley, B. R. 2005. RS domains contact the pre-mRNA throughout spliceosome assembly. *Trends Biochem Sci* 30:115–118.

Hertel, K. J., Lynch, K. W., and Maniatis, T. 1997. Common themes in the function of transcription and splicing enhancers. *Curr Opin Cell Biol* 9:350–357.

House, A. E., and Lynch, K. W. 2008. Regulation of alternative splicing: more than just the ABCs. *J Biol Chem* 283:1217–1221.

Jurica, M. S. 2008. Detailed close-ups and the big picture of spliceosomes. *Curr Opin Struct Biol* 18:315–320.

Jurica, M. S., and Moore, M. J. 2002. Capturing splicing complexes to study structure and mechanism. *Methods* 28:336–345.

Kambach, C., Walke, S., and Nagai, K. 1999. Structure and assembly of the spliceosomal small nuclear ribonucleoprotein particles. *Curr Opin Struct Biol* 9:222–230.

Kambach, C., Walke, S, Young, R., et al. 1999. Crystal structures of two Sm protein complexes and their implications for the assembly of the spliceosomal snRNPs. *Cell* 96:375–387.

Kent, O. A., and MacMillan, A. M. 2002. Early organization of pre-mRNA during spliceosome assembly. *Nat Struct Biol* 9:576–581.

König, H., Matter, N., Bader, R., et al. 2007. Splicing segregation: the minor spliceosome acts outside the nucleus and controls cell proliferation. *Cell* 131:718–729.

Kreivi, J.-P., and Lamond, A. I. 1996. RNA splicing: unexpected spliceosome diversity. *Curr Biol* 6:802–805.

Lin, R.-J., Newman, A. J., Cheng, S.-C., and Abelson, J. 1985. Yeast mRNA splicing *in vitro*. *J Biol Chem* 260:14780–14792.

Lührman, R., and Stark, H. 2009. Structural mapping of spliceosomes by electron microscopy. *Curr Opin Struct Biol* 19:1–7.

Makarov, E. M., Makarova, O. V., Urlaub, H., et al. 2002. Small nuclear ribonucleoprotein remodeling during catalytic activation of the spliceosome. *Science* 298:2205–2208.

Mougin, A., Gottschalk, A., Fabrizio, P., et al. 2002. Direct probing of RNA structure and RNA-protein interactions in purified HeLa cell's and yeast spliceosomal U4/U6. U5 tri-snRNP particles. *J Mol Biol* 317:631–649.

Mount, S. M. 1996. AT-AC introns: an ATtACk on dogma. *Science* 271:1690–1692.

Murray, H. L., and Jarrell, K. A. 1999. Flipping the switch to an active spliceosome. *Cell* 96:599–602.

Nagai, K., Muto, Y., Krummel, D. A. P., et al. 2001. Structure and assembly of the spliceosomal snRNPs. *Biochem Soc Trans* 29(part 2):15–26.

Newman, A. J. 2008. RNA splicing interactions in mRNA splicing. In: *Encyclopedia of Life Sciences*. pp. 1–7. Hoboken, NJ: John Wiley and Sons Ltd.

Newman, A. J., and Kiyoshi, N. 2010. Structural studies of the spliceosome: blind men and an elephant. *Curr Opin Struct Biol* 20:82–89.

Nilsen, T. W. 2000. The case for an RNA enzyme. *Nature* 408:782–783.

Nilsen, T. W. 2002. The spliceosome: no assembly required. *Mol Cell* 9:8–9.

Nilsen, T. W. 2004. Too hot to splice. *Nat Struct Mol Biol* 11:208–209.

Ohno, M., Segref, A., Bachi, A., Wilm, M., and Mattaj, I. W. 2000. PHAX, a mediator of U snRNA nuclear export whose activity is regulated by phosphorylation. *Cell* 101:187–198.

Padgett, R. A. 2005. mRNA splicing: role of snRNAs. In: *Encyclopedia of Life Sciences*. pp. 1–7. Hoboken, NJ: John Wiley and Sons Ltd.

Patel, S. B., and Bellini, M. 2008. The assembly of a spliceosomal small nuclear ribonucleoprotein particle. *Nucleic Acids Res* 36:6482–6493.

Pauskin, S., Gubitz, A. K., Massenet, S., and Dreyfuss, G. 2002. The SMN complex, an assembly of some ribonucleoproteins. *Curr Opin Cell Biol* 14:305–312.

Pomeranz Krummel, D. A., Oubridge, C., Leung, A. K., et al. 2009. Crystal structure of human spliceosomal U1 snRNP at 5.5Å resolution. *Nature* 458:475–480.

Reed, R. 2000. Mechanisms of fidelity in pre-mRNA splicing. *Curr Opin Cell Biol* 12:340–345.

Ritchie, D. B., Schellenberg, M. J., Gesner, E. M., et al. 2008. Structural elucidation of a PRP8 core domain from the heart of the spliceosome. *Nat Struct Mol Biol* 15:1199–1205.

Robert, F., Blanchett, M., Maes, O., and Chabot, B. 2002. A human RNA polymerase II-containing complex associated with factors necessary for spliceosome assembly. *J Biol Chem* 277:9302–9306.

Ruskin, B., Krainer, A. R., Maniatis, T., and Green, M. R. 1984. Excision of an intact intron as a novel lariat structure during pre-mRNA splicing *in vitro*. *Cell* 38:317–331.

Ryan, D. E., and Abelson, J. 2002. The conserved central domain of yeast U6 snRNA: importance of U2-U6 helix I_a in spliceosome assembly. *RNA* 8:997–1010.

Shukla, G. C., Cole, A. J., Dietrich, R. C., and Padgett, R. A. 2002. Domains of human U4atac snRNA required for U12-dependent splicing *in vivo*. *Nucleic Acids Res* 30:4650–4657.

Sontheimer, E. J. 2001. The spliceosome shows its metal. *Nat Struct Biol* 8:11–13.

Sperling, J., Azubel, M., and Sperling, R. 2008. Structure and function of the pre-mRNA splicing machine. *Structure* 16:1605–1615.

Staley, J. P. 2002. Hanging on to the branch. *Nat Struct Biol* 9:5–7.

Staley, J. P., and Guthrie, C. 1998. Mechanical devices of the spliceosome: motors, clocks, springs, and things. *Cell* 92:315–326.

Stark, N, Dube, P., Lührmann, R., and Kastner, B. 2001. Arrangement of RNA and proteins in the spliceosomal U1 small nuclear ribonucleoprotein particle. *Nature* 409:539–542.

Stevens, S. W., and Abelson, J. 1999. Purification of the yeast U4/U6. U5 small nuclear ribonucleoprotein particle and identification of it proteins. *Proc Natl Acad Sci USA* 96:7226–7231.

Stevens, S. W., and Abelson, J. 2002. Yeast pre-mRNA splicing: methods, mechanisms, and machinery. *Methods Enzymol* 351:200–220.

Stevens, S. W., Ryan, D. E., Ge, H. Y., et al. 2002. Composition and functional characterization of the yeast spliceosomal penta-snRNP. *Mol Cell* 9:31–44.

Tacke, R., and Manley, J. L. 1999. Determinants of SR protein specificity. *Curr Opin Cell Biol* 11:358–362.

Tardiff, D. F., Lacadie, S. A., and Rosbash, M. 2006. A genome-wide analysis indicates that yeast pre-mRNA splicing is predominantly posttranscriptional. *Mol Cell* 24:917–929.

Tarn, W.-Y., and Steitz, J. A. 1996. A novel spliceosome containing U11, U12, and U5 snRNPs excises a minor class (AT-AC) intron in vitro. *Cell* 84:801–811.

Tarn, W.-Y., and Steitz, J. A. 1997. Pre-mRNA splicing: the discovery of a new spliceosome doubles the challenge. *Trends Biochem Sci* 22:132–137.

Terns, M. P., and Terns, R. M. 2001. Macromolecular complexes: SMN—the master assembler. *Curr Biol* 11:R862–R864.

Valadkhan, S. 2005. snRNAs as the catalyst of pre-mRNA splicing. *Curr Opin Chem Biol* 9:603–608.

Valadkhan, S., and Manley, J. L. 2001. Splicing-related catalysis by protein-free snRNAs. *Nature* 413:701–707.

Valadkhan, S., Mohammadi, A., Wachtel, C., and Manley, J. L. 2008. Protein-free spliceosomal snRNAs catalyze a reaction that resembles the first step of splicing. *RNA* 13:2300–2311.

Villa, T., Pleiss, J. A., and Guthrie, C. 2002. Spliceosomal snRNAs: Mg^{21}-dependent chemistry at the catalytic core? *Cell* 109:149–152.

Wahl, M., Will, C. L., and Lührmann, R. 2009. The spliceosome: design principles of a dynamic RNP machine. *Cell* 136:701–718.

Wang, Z., and Burge, C. B. 2008. Splicing regulation: from a parts list of regulatory elements to an integrated splicing code. *RNA* 14:802–813.

Will, C. L., and Lührmann, R. 2001. Spliceosomal U snRNP biogenesis, structure, and function. *Curr Opin Cell Biol* 13:290–301.

Yang, V. W., Lerner, M. R., Steitz, J. A., and Flint, S. J. 1981. A small nuclear ribonucleoprotein is required for splicing of adenoviral early RNA sequences. *Proc Natl Acad Sci USA* 78:1371–1375.

Yean, S.-L., Wuenschell, G., Termini, J., and Lin, R.-J. Metal ion coordination by U6 small nuclear RNA contributes to the catalysis in the spliceosome. *Nature* 408:881–884.

Yu, Y.-T., Scharl, E. C., Smith, C. M., and Steitz, J. A. 1998. The growing world of small nuclear ribonucleoproteins. In: *The RNA World*, 2nd ed., pp. 487–524. R. Gesteland, T. Cech, J. Atkins, eds. Woodbury, NY: Cold Spring Harbor Laboratory Press.

Zhou, Z., Licklider, L. J., Gygi, S. P., and Reed, R. 2002. Comprehensive proteomic analysis of the human spliceosome. *Nature* 419:182–185.

Zhou, Z., Sim, J., Griffith, J., and Reed, R. 2002. Purification and electron microscopic visualization of functional human spliceosomes. *Proc Natl Acad Sci USA* 99:12203–12207.

Cleavage/Polyadenylation and Transcription Termination

Aranda, A, and Proudfoot, N. 2001. Transcriptional termination factors for RNA polymerase II in yeast. *Mol Cell* 7:1003–1011.

Bates, N., and Hurst, H. 2002. Nuclear run-on assays. In: *Encyclopedia of Life Sciences*. pp. 1–13. London, UK: Nature Publishing Group.

Birse, C. E., Minivielle-Sebastia, L., Lee, B. A., et al. 1998. Coupling termination of transcription to messenger RNA maturation in yeast. *Science* 280:298–301.

Buratowski, S. 2005. Connections between mRNA 3′ end processing and transcription termination. *Curr Opin Cell Biol* 17:257–261.

Cantonel, J.-C., Murthy, K. G. K., Manley, J. L., and Tora, L. 1997. Transcription factor TFIID recruits factor CPSF for formation of the 3′ end of mRNA. *Nature* 389:399–402.

Colgan, D. F., and Manley, J. L. 1997. Mechanism and regulation of mRNA polyadenylation. *Genes Dev* 11:2755–2766.

Darnell, J. E., Philipson, L. Wall, R., and Adesnik, M. 1971. Polyadenylic acid sequences: role in conversion of nuclear RNA into messenger RNA. *Science* 174:507–510.

Dichtl, B., Blank, D., Sadowski, M., et al. 2002. Yhh1p/Cft1p directly links poly(A) site recognition and RNA polymerase II transcription termination. *EMBO J* 21:4125–4135.

Dreyfus, M., and Régnier, P. 2002. The poly(A) tail of mRNAs: bodyguard in eukaryotes, scavenger in baceria. *Cell* 111:611–613.

Dye, M. J., and Proudfoot, N. J. 1999. Terminal exon definition occurs cotranslationally and promotes termination of RNA polymerase II. *Mol Cell* 3:371–378.

Dye, M. J., and Proudfoot. N. J. 2001. Multiple transport cleavage precedes polymerase release in termination by RNA polymerase II. *Cell* 105:669–681.

Edwalds-Gilbert, G., Veraldi, K. L., and Milcarek, C. 1997. Alternative poly(A) site selection in complex transcription units: means to an end? *Nucleic Acids Res* 25:2547–2561.

Gershon, P. D. 2000. (A)-tail of two polymerase structures. *Nat Struct Biol* 7:819–821.

Gilmartin, G. M. 2005. Eukaryotic mRNA 3′-processing: a common means to different ends. *Genes Dev* 19:2517–2521.

Hirose, Y., and Manley, J. L. 1998. RNA polymerase II is an essential mRNA polyadenylation factor. *Nature* 395:93–96.

Keller, W., and Minivielle-Sebastia, L. 1997. A comparison of mammalian and yeast pre-mRNA 3′-end processing. *Curr Opin Cell Biol* 9:329–336.

Kolev, N. G., and Steitz, J. A. 2005. Symplekin and multiple other polyadenylation factors participate in 3′-end maturation of histone mRNAs. *Genes Dev* 19:2583–2592.

Luo, W., and Bentley, D. 2004. A ribonucleolytic rat torpedoes RNA polymerase II. *Cell* 119:911–914.

Mandel, C. R., Bai,Y., and Tong, L. 2008. Protein factors in pre-mRNA 3′-end processing. *Cell Mol Life Sci* 65:1099–1122.

Minivielle-Sebastia, L., and Keller, W. 1999. mRNA polyadenylation and its coupling to other RNA processing reactions and to transcription. *Curr Opin Cell Biol* 11:352–357.

Orozeo, I. J., Kim, S. J., and Martinson, H. G. 2002. The poly(A) signal, without the assistance of any downstream element, directs RNA polymerase II to pause *in vivo* and then to release stochastically from the template. *J Biol Chem* 277:42899–42911.

Osheim, Y. N., Proudfoot, N. J., and Beyer, A. L. 1999. EM visualization of transcription by RNA polymerase II: downstream termination requires a poly(A) signal but not transcript cleavage. *Mol Cell* 3:379–387.

Osheim, Y. N., Sikes, M. L., and Beyer, A. L. 2002. EM visualization of PolII in *Drosophila*: most genes terminate without prior 3′-end cleavage of nascent transcripts. *Chromosoma* 111:1–12.

Proudfoot, N. 1996. Ending the message is not so simple. *Cell* 87:779–781.

Proudfoot, N. 2004. New perspectives on connecting messenger RNA 3′ end formation to transcription. *Curr Opin Cell Biol* 16:272–278.

Proudfoot, N., and O'Sullivan, J. 2002. Polyadenylation: a tail of two complexes. *Curr Biol* 12:R855–R857.

Richard, P., and Manley, J. L. 2009. Transcription termination by nuclear RNA polymerases. *Genes Dev* 23:1247–1269.

Rondon, A. G., Mishco, H. E., and Proudfoot, N. J. 2008. Terminating transcription in yeast: whether to be a "nerd" or a "rat." *Nat Struct Mol Biol* 15:775–776.

Rosonina, E., Kaneko, S., and Manley, J. L. 2006. Terminating the transcript: breaking up is hard to do. *Genes Dev* 20:1050–1056.

Ryan, K., Calvo, O., and Manley, J. L. 2004. Evidence that polyadenylation factor CPSF-73 is the mRNA 3′ processing endonuclease. *RNA* 10:565–573.
Ryan, K., Murthy, K. G. K., Kaneko, S., and Manley, J. L. 2002. Requirements of the RNA polymerase II C-terminal domain for reconstituting pre-mRNA 3′-cleavage. *Mol Cell Biol* 22:1684–1692.
Shatkin, A. J., and Manley, J. L. 2000. The ends of the affair: capping and polyadenylation. *Nat Struct Biol* 7:838–842.
Shi, Y., Di Giammartino, D. C., Taylor, D., et al. 2009. Molecular architecture of the human pre-mRNA 3′ processing complex. *Mol Cell* 33:365–376.
Tian, B. 2008. Alternative polyadenylation in the human genome: evolution. In: *Encyclopedia of Life Sciences*. pp. 1–7. Hoboken, NJ: John Wiley and Sons Ltd.
Tran, D. P., Kim, S. J., Park, N. J., et al. 2001. Mechanism of poly(A) signal transduction to RNA polymerase II *in vitro*. *Mol Cell Biol* 21:7495–7508.
Yonaha, M., and Proudfoot, N. J. 1999. Specific transcriptional pausing activates polyadenylation in a coupled in vitro system. *Mol Cell* 3:593–600.
Yonaha, M., and Proudfoot, N. J. 2000. Transcriptional termination and coupled polyadenylation in vitro. *EMBO J* 19:3770–3777.
Wahle, E., and Rüegsegger, U. 1999. 3′-End processing of pre-mRNA in eukaryotes. *FEMS Microbiol Rev* 23:277–295.
Wickens, M., and Gonzalez, T. N. 2004. Molecular biology: knives, accomplices, and RNA. *Science* 306:1299–1300.
Wilusz, J. E., and Spector, D. L. 2010. An unexpected ending: noncanonical 3′ end processing mechanisms. *RNA* 16:259–266.
Zhang, H., Lee, J. Y., and Tian, B. 2005. Biased alternative polyadenylation in human tissues. *Genome Biol* 6:R100.
Zhao, J., Hyman, L., and Moore, C. 1999. Formation of mRNA 3′-ends in eukaryotes: mechanism, regulation, and interrelationships with other steps in mRNA synthesis. *Microbiol Mol Biol Rev* 63:405–445.

RNA Editing

Blanc, V., and Davidson, N. O. 2003. C-to-U editing: mechanisms leading to genetic diversity. *J Biol Chem* 276:1395–1398.
Brennicke, A., Marchfelder, A., and Binder, S. 1999. *FEMS Microbiol Rev* 23:297–316.
Estévez, A. M., and Simpson, L. 1999. Uridine insertion/deletion RNA editing in trypanosome mitochondria—a review. *Gene* 240:247–260.
Gerber, A. P., and Keller, W. 2001. RNA editing by base deamination: more enzymes, more targets, new mysteries. *Trends Biochem Sci* 26:376–384.
Hodges, P., and Scott, J. 1992. Apolipoprotein B mRNA editing: a new tier for the control of gene expression. *Trends Biochem Sci* 17:72–81.
Innerarity, T. L., Borén, J., Yamanka, S., and Olofsson, S.-O. 1996. Biosynthesis of apolipoprotein B48-containing lipoproteins. *J Biol Chem* 271:2353–2356.
Maas, S., and Rich, A. 2000. Changing genetic information through RNA editing. *Bioessays* 22:790–802.
Maas, S., Rich, A., and Nishikura, K. 2003. A-to-I editing: recent news and residual mysteries. *J Biol Chem* 278:1391–1394.
Madison-Antenucci, S, Grams, J., and Hajduk, S.L. 2002. Editing machines: the complexities of trypanosome RNA editing. *Cell* 108:435–438.
Sowden, M. P., and Smith, H. C. 2002. RNA editing. In: *Encyclopedia of Life Sciences*. pp. 1–7. London, UK: Nature Publishing Group.

Messenger RNA Export

Aguliera, A. 2005. Cotranscriptional mRNP assembly: from the DNA to the nuclear pore. *Curr Opin Cell Biol* 17:1–9.
Conti, E., and Izaurralde, E. 2001. Nucleocytoplasmic transport enters the atomic age. *Curr Opin Cell Biol* 13:310–319.
Dreyfuss, G., Kim, V. N., and Kataoka, N. 2002. Messenger-RNA-binding proteins and the messages they carry. *Nat Rev Mol Cell Biol* 3:195–205.

Jensen, T. H., Dower, K., Libri, D., and Rosbash, M. R. 2003. Early formation of mRNP: license for export or quality control. *Mol Cell* 11:1129–1138.

Keyes, A., and Green, M. R. 2001. The odd coupling. *Nature* 413:583–585.

Kim, V.-N., and Dreyfuss, G. 2001. Nuclear mRNA binding proteins couple pre-mRNA splicing and post-splicing events. *Mol Cell* 12:1–10.

Reed, R. 2003. Coupling transcription, splicing, and mRNA export. *Curr Opin Cell Biol* 15:326–331.

Reed, R., and Hurt, E. 2002. A conserved mRNA export machinery coupled to pre-mRNA splicing. *Cell* 108:523–531.

Reed, R., and Magni, K. 2001. A new view of mRNA export: separating the wheat from the chaff. *Nat Cell Biol* 3:E201–E204.

Schwarz, D. S., and Zamore, P. D. 2002. Why do miRNAs live in miRNP? *Genes Dev* 16:1025–1031.

Tekotte, H., and Davis, I. 2002. Intramolecular mRNA localization: motors move messages. *Trends Genet* 18:636–642.

Zhou, Z., Luo, M.-J., Straesser, K., et al. R. 2000. The protein Aly links pre-messenger RNA splicing to nuclear export in metazoans. *Nature* 407:401–405.

This book has a Web site, **http://biology.jbpub.com/book/molecular**, which contains a variety of resources, including links to in-depth information on topics covered in this chapter, and review material designed to assist in your study of molecular biology.

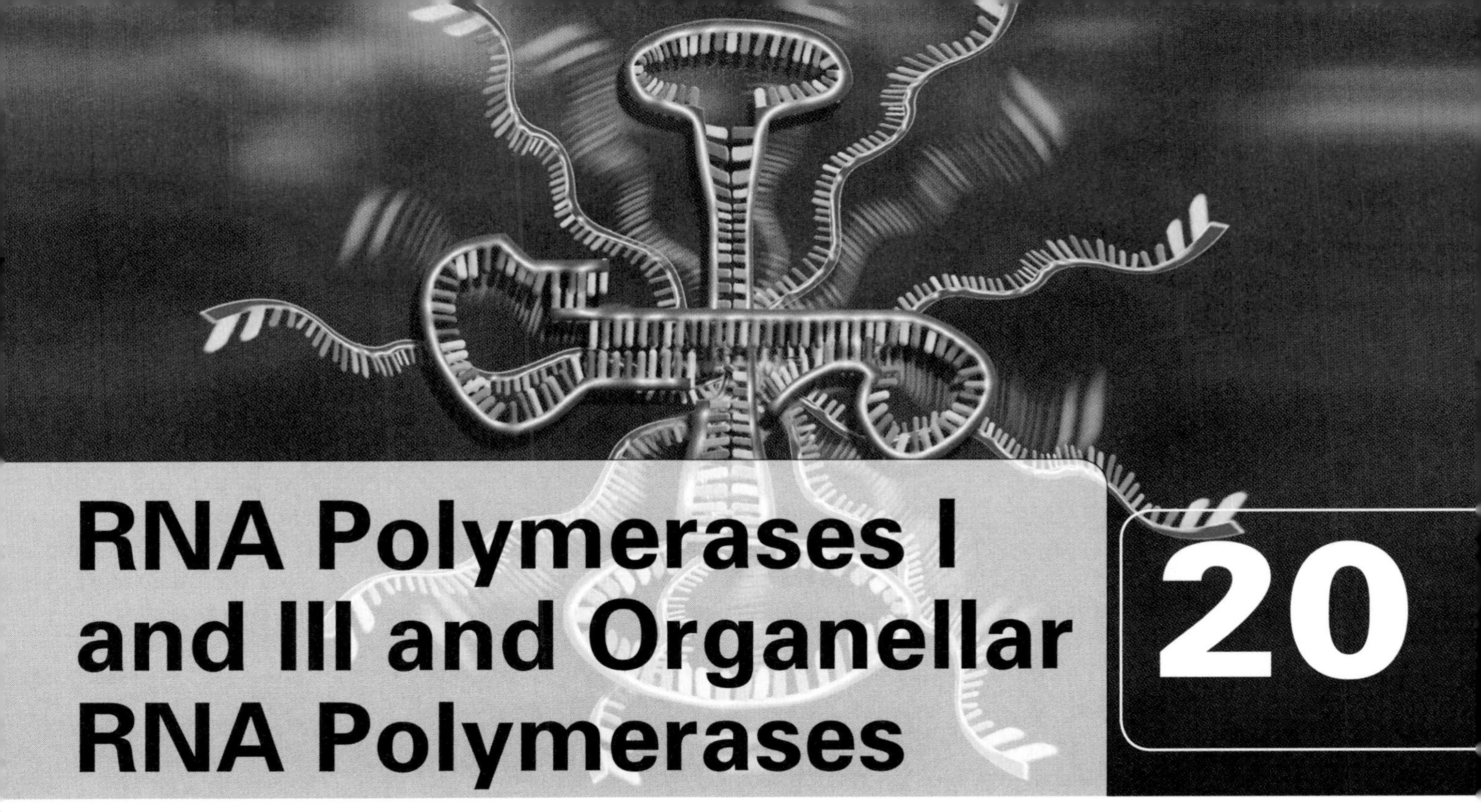

RNA Polymerases I and III and Organellar RNA Polymerases

20

OUTLINE OF TOPICS

20.1 Eukaryotic Ribosome

The eukaryotic ribosome is made of a small and a large ribonucleoprotein subunit.

20.2 RNA Polymerase I

The 5.8S, 18S, and 28S rRNA coding sequences are part of a single transcript.

Eukaryotes have multiple copies of rRNA transcription units arranged in clusters on just a few chromosomes.

The rRNA transcription unit promoter consists of a core promoter and an upstream promoter element (UPE).

Ribosomal RNA transcription and processing takes place in the nucleolus.

RNA polymerase I is a multisubunit enzyme with a structure similar to that of RNA polymerase II.

RNA polymerase I–associated factors are required for transcription.

RNA polymerase I also requires two auxiliary transcription factors, upstream binding factor (UBF) and selectivity factor (SL1/TIF-1B).

The transcription initiation complex can be assembled *in vitro* by the stepwise addition of individual components.

RNA polymerase I acts through a transcription cycle that begins with the formation of a pre-initiation complex.

Pre-rRNA undergoes a complex series of cleavages and modifications as it is converted to mature ribosomal rRNAs.

20.3 Self-Splicing Ribosomal RNA

Tetrahymena thermophila pre-rRNA contains an intron that catalyzes its own excision.

20.4 Ribosome Assembly

Eukaryotic ribosome assembly is a complex multistep process.

20.5 RNA Polymerase III

RNA polymerase III transcripts are short RNA molecules with a variety of biological functions.

RNA polymerase III transcription units have three different types of promoters.

The transcription factors required to recruit RNA polymerase III depend on the nature of the promoter.

RNA polymerase III does not appear to require additional factors for transcription elongation or termination.

Pre-tRNAs require extensive processing to become mature tRNAs.

20.6 Transcription in Mitochondria

Mitochondrial DNA is transcribed to form mRNA, rRNA, and tRNA.

20.7 Transcription in Chloroplasts

Chloroplast DNA is also transcribed to form mRNA, rRNA, and tRNA.

Suggested Reading

Most eukaryotic RNA synthesis takes place in the cell nucleus. RNA polymerase II, which is responsible for synthesizing mRNA and several snRNAs was considered in Chapters 17 to 19. This chapter examines the two other nuclear RNA polymerases present in all eukaryotes. The first of these, **RNA polymerase I**, is required to synthesize 5.8S, 18S, and 28S ribosomal RNA (rRNA) and the second, **RNA polymerase III**, is required to synthesize several small RNAs including transfer RNA (tRNA), 5S rRNA, and U6 snRNA. This chapter also describes the pathways that convert the primary transcripts formed by these two nuclear RNA polymerases into mature RNA molecules. RNA synthesis also takes place in mitochondria of all eukaryotes and chloroplasts in plants. The RNA polymerases responsible for this transcription, which are distinct from the nuclear RNA polymerases, are described in the last part of this chapter.

20.1 Eukaryotic Ribosome

The eukaryotic ribosome is made of a small and a large ribonucleoprotein subunit.

The eukaryotic cytoplasmic ribosome, which has a molecular mass of about 4×10^6 Da and a sedimentation coefficient of 80S, consists of one small and one large subunit (FIGURE 20.1). The small (or 40S) subunit, which is fairly constant in size (approximately 1.5×10^6 Da) in ribosomes from all eukaryotes, contains an 18S rRNA (approximately

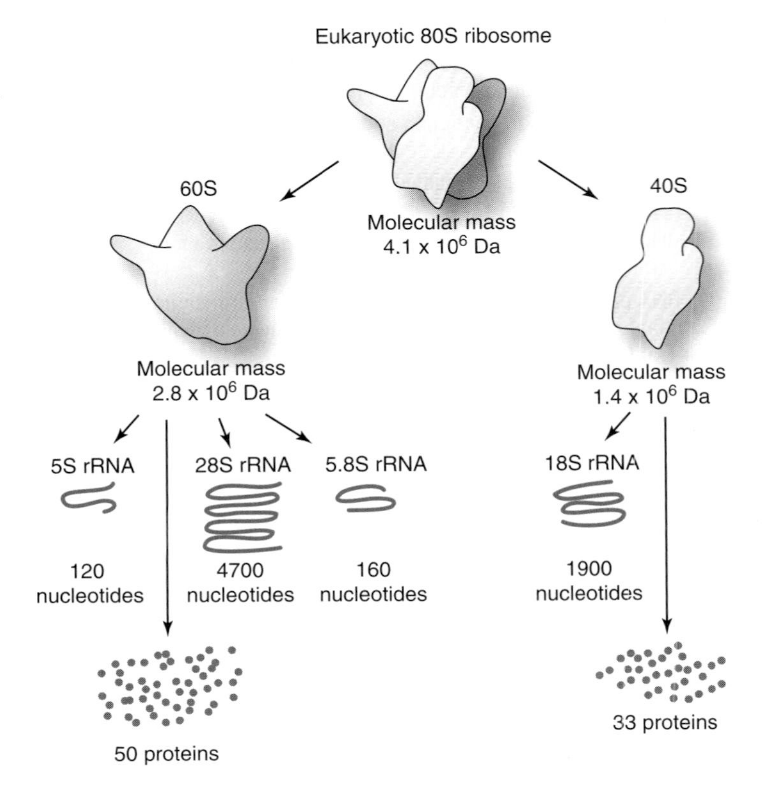

FIGURE 20.1 **Components of the eukaryotic 80S ribosome.** The 80S ribosome is made of a large subunit that contains 5S rRNA, 5.8S rRNA, 28S rRNA, and about 50 different polypeptides and a small subunit that contains one 18S rRNA and about 33 different polypeptides. The large rRNA in animals, plants, and fungi vary in size. The 28S rRNA shown in this figure is for mammalian ribosomes.

1,900 nucleotides) and about 33 different polypeptides. The large (or 60S) subunit varies in size from one species to another. For instance, large subunits in plants and mammals have molecular masses of 2.5×10^6 Da and 3.0×10^6 Da, respectively. Part of the reason for this variation is the largest of the three rRNA molecules in the large subunit varies in size. This rRNA is about 5,000 nucleotides long (sedimentation coefficient 28S) in mammals but only 3,400 nucleotides long (sedimentation coefficient 25S) in yeast. The two other rRNA components in the large subunit are the 5.8S rRNA (approximately 160 nucleotides) and the 5S rRNA (approximately 120 nucleotides). Large subunits from mammals also have about 50 polypeptide subunits, whereas those from lower eukaryotes have somewhat fewer polypeptides. RNA polymerase I synthesizes a precursor of rRNA that is converted into the 5.8S, 18S, and 25S/28S rRNAs (see below), while RNA polymerase III synthesizes the 5S rRNA.

20.2 RNA Polymerase I

The 5.8S, 18S, and 28S rRNA coding sequences are part of a single transcript.

Much of our present knowledge of rRNA synthesis is based on the pioneering efforts of James Darnell and coworkers. When they started to study rRNA synthesis in the early 1960s, the prevailing view was that each kind of rRNA would be transcribed from its own transcription unit. It therefore seemed reasonable to try to follow the fate of a single rRNA molecule from its synthesis in the cell nucleus to its ultimate destination in the cytoplasm. Darnell and coworkers decided to follow the fate of rRNA because rRNA accounts for about 80% of total cellular RNA and each kind of rRNA is homogenous (an important consideration in the days before modern recombinant DNA techniques were available). The biological system that Darnell and coworkers selected for study, the HeLa cell (a human cervical cancer cell), contains about 10 million ribosomes; therefore, a large proportion of its RNA synthetic machinery must be devoted to rRNA production so that it can synthesize the required 10 million copies of each rRNA molecule per generation.

Knowing that eukaryotic cells convert [^{3}H]uridine into [^{3}H]UTP, a precursor for RNA synthesis, Darnell and coworkers incubated HeLa cells with [^{3}H]uridine for a short period of time and then isolated newly synthesized RNA molecules from the nucleus. They expected to find radioactively labeled 18S and 28S rRNA molecules but, to their great surprise, they could not detect either type of rRNA molecule. Instead, they found that both the 18S and 28S rRNA molecules are part of a much larger radioactive molecule with about 14,000 bases and a sedimentation coefficient of 45S. Darnell and coworkers followed the fate of the 45S RNA by a pulse-chase experiment, which involves (1) culturing the cells in the presence of [^{3}H]uridine for a short period of time, (2) transferring the labeled cells to an unlabeled medium, and

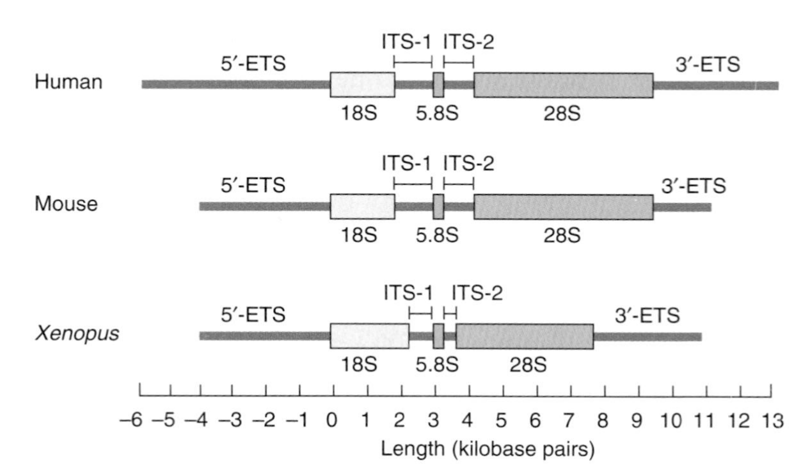

FIGURE 20.2 **Comparison of human, mouse, and *Xenopus* (frog) pre-rRNAs.** Pre-rRNA contains sequences that are converted to 18S, 5.8S, and 28S rRNA molecules. An internal transcribed spacer (ITS-1) separates the 18S and 5.8S rRNA sequences and a second internal transcribed spacer (ITS-2) separates the 5.8S and 28S rRNA sequences. The pre-rRNA molecules also have a 5′-external transcribed spacer (5′-ETS) upstream from the 18S rRNA and a 3′-external transcribed spacer (3′-ETS) downstream from the 28S rRNA. The scale at the bottom of the figure indicates length in kilobase pairs. (Adapted from S. J. Triezenberg, et al., *J. Biol. Chem.* 257 [1982]: 7826–7833.)

(3) isolating RNA molecules from the cells at various times after the transfer. They observed that the radioactivity originally present in 45S RNA is eventually converted to 18S and 28S rRNA. Based on this observation, Darnell and coworkers proposed that 45S RNA molecules are precursors of rRNA (**pre-rRNA**) molecules that are processed to form mature rRNA. Later investigations by Darnell's group revealed the existence of the 5.8S rRNA, which is also formed by processing the 45S RNA molecule. Eukaryotic cells probably derive considerable advantage from having the three rRNA molecules made from a single pre-rRNA molecule because this arrangement seems to help coordinate rRNA synthesis and ribosome assembly. It is therefore rather surprising that the fourth type of rRNA, 5S rRNA, is not included within pre-rRNA molecules. The evolutionary driving force for separate 5S rRNA genes in higher animals and plants remains to be discovered.

Ribosomal RNA precursors have now been identified in many different higher animals and appear to have the general characteristics shown in FIGURE 20.2 and listed below.

1. Ribosomal RNA coding sequences are in the order 18S rRNA–5.8S rRNA–28S rRNA.
2. The 18S and 5.8S rRNA coding sequences and 5.8S and 28S rRNA coding sequences are separated by **internal transcribed spacer 1** (**ITS-1**) and **internal transcribed spacer 2** (**ITS-2**), respectively.
3. A **5′-external transcribed sequence** (**5′-ETS**) is located upstream from the 18S rRNA coding sequence and a **3′-external transcribed sequence** (**3′-ETS**) is located downstream from the 28S rRNA coding sequence.
4. Internal transcribed spacer sequences and external transcribed sequences are excised when pre-rRNA is processed to form mature rRNA molecules.
5. The rRNA sequences tend to be highly conserved, whereas excised sequences do not. This difference in sequence conservation is expected because there is strong selective pressure to conserve rRNA sequences essential for ribosome function

but no comparable selective pressure to conserve excised sequences.

6. The lengths of pre-RNA molecules vary from one species to another. These size differences are due to differences in the lengths of the excised spacers rather than the rRNA sequences.

Eukaryotes have multiple copies of rRNA transcription units arranged in clusters on just a few chromosomes.

Higher animals have between 150 and 300 rRNA transcription units per haploid number of chromosomes. These transcription units are arranged in clusters on a few different chromosomes. In humans, approximately 200 rRNA transcription units are arranged in clusters on chromosomes 13, 14, 15, 21, and 22. The centromere in each of these five chromosomes is so near the end that the short arm resembles a dot-like appendage. The rRNA transcription unit cluster is located on the dot-like appendage in a telomere-to-centromere orientation. The number of rRNA transcription units is much higher in plants, averaging 3,700 copies per haploid number of chromosomes, and much lower in yeast (some yeast strains have as few as 40 rRNA transcription units per cell). rRNA transcription units are amplified in some animals, most notably amphibians, during specific stages of development. For example, the frog oocyte has about 500,000 rRNA transcription unit copies, representing a remarkable 1,000-fold amplification above the number present in the typical frog somatic cell. This large number of rRNA copies is required to meet the demand for new ribosome synthesis.

In many and perhaps most animals, the rRNA transcription unit copy number exceeds that needed for ribosome synthesis. A naturally occurring *Drosophila* mutant with deleted rRNA transcription units provides an illustrative example. Normal flies contain about 200 rRNA transcription units per haploid number of chromosomes, but the mutant can survive even though about half of its rRNA transcription units are deleted. There is a limit, however, to the number of rRNA transcription unit copies that can be deleted because flies do not survive when the rRNA transcription unit copy number falls below 40. Based on the fact that animals often have more rRNA transcription units than they actually need to make ribosomes, one might predict that not all rRNA transcription units are actively engaged in transcription. This prediction is in fact correct. Only a subset of the repeated rRNA transcription unit copies is transcribed in most somatic cells and this subset appears to be stripped of nucleosomes.

Until recently, the accepted model for the organization of the rRNA transcription units within a cluster was one in which all transcription units are arranged head-to-tail with long intergenic spacers between transcription units. Little, if any, sequence information exists to prove or disprove this model because it is very difficult to sequence the highly repetitive DNA in a cluster.

In 2005, Aaron Bensimon and coworkers used a novel approach to determine the orientation of rRNA transcription units within a

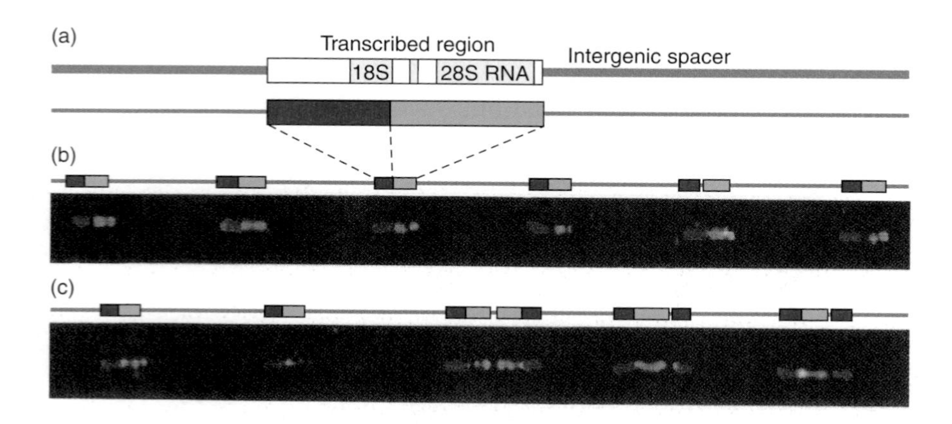

FIGURE 20.3 **Organization of human rRNA transcription units within a cluster.** (a) Red and green probes were used to identify the rRNA transcription units' 5′- and 3′-parts, respectively. (b) The accepted model for rRNA transcription unit organization predicts a repeating red-green pattern for each rRNA transcription unit with a large colorless space between units. All of the rRNA transcription units shown in this region fit the accepted model. (c) The organization of approximately one-third of the rRNA transcription units does not fit the accepted model. This image shows a region in which the first two transcription units follow the accepted pattern and the last three do not. (Reproduced from B. McStay and I. Grummt, *Annu. Rev. Cell Dev. Biol.* 24 [2008]:131–157. Copyright 2008 by Annual Reviews, Inc. Reproduced with permission of Annual Reviews, Inc. in the format Textbook via Copyright Clearance Center. Original illustration modified from S. Caburet, et al., Genome Research 15 [2005]: 1079–1085. Copyright 2005, Cold Spring Harbor Laboratory Press. Used with permission of Aaron Bensimon, Genomic Vision.)

chromosomal cluster of stretched human DNA (FIGURE 20.3). They prepared red and green probes to identify the rRNA transcription units' 5′- and 3′-parts, respectively. The accepted model for rRNA transcription unit organization predicts a repeating red-green pattern for each rRNA transcription unit with a large colorless space between units. Approximately two-thirds of the patterns they observed conformed to the expected pattern (Figure 20.3b) but the remaining patterns did not (figure 20.3c). Further work is required to establish whether a correlation exists between rRNA transcription unit expression and orientation.

Because of their arrangement and high rate of transcription, rRNA transcription units can be visualized in spread chromatin preparations using electron microscopy (FIGURE 20.4). The transcripts are so densely packed (about 100 transcripts per rRNA transcription unit) that they stick out perpendicularly from the DNA giving each active transcription unit the appearance of a "Christmas tree." The tip of each "tree" is at the transcription initiation site and the broad end is at the transcription termination site.

The rRNA transcription unit promoter consists of a core promoter and an upstream promoter element (UPE).

Individual rRNA transcription units from different organisms have been sequenced. Each rRNA transcription unit has its own promoter and transcription terminator (FIGURE 20.5). The promoter for the rRNA transcription unit consists of two elements, a **core promoter** that is essential for transcription and an **upstream promoter element** (UPE) that is not. The human core promoter begins about 50 bp before the

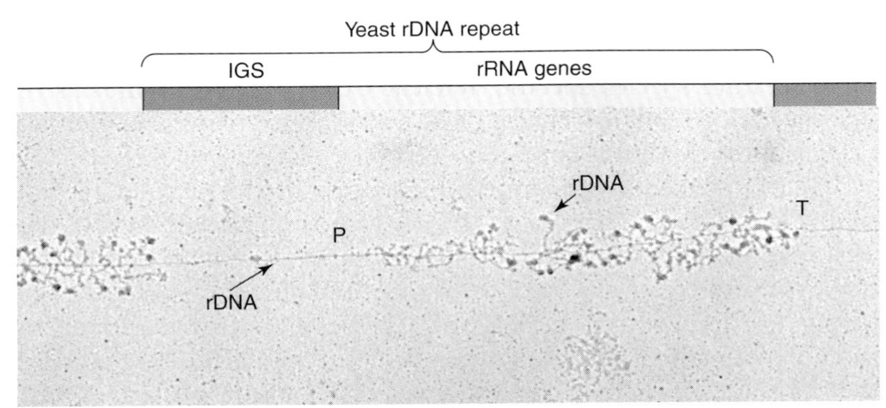

FIGURE 20.4 Electron microscopic image of a yeast nuclear chromatin spread. The repetitive nature of the rDNA is illustrated in an electron microscopic image of a yeast nuclear chromatin spread. Progressively longer rRNAs (stained for associated proteins) emanate from the many RNA polymerase I complexes as they move along the rRNA transcription unit (shown in yellow at the top), beginning at the promoter (P) and finishing at the terminator (T). (Reproduced from *Trends Biochem. Sci.*, vol. 30, J. Russell and J. C. B. M. Zomerdijk, RNA-polymerase-I-directed rDNA . . ., pp. 87–96, copyright 2005, with permission from Elsevier [http://www.sciencedirect .com/science/journal/09680004]. Photo courtesy of Ann L. Beyer, University of Virginia.)

transcription initiation site (+1) and extends about 20 bp beyond it. An adenine-thymine (AT)-rich region within the core promoter that extends from about −10 to +10, called the **rRNA initiator (rInr)**, is unique among rRNA promoter sequences because it is conserved from one species to another. The high AT content appears to permit the closed promoter to change to an open promoter, a prerequisite for transcription initiation. Although it resembles the TATA box of RNA polymerase II promoters, rInr is not a binding site for TBP (the TATA-binding protein; see Chapter 17). The human UPE, which spans a region from −107 to −186, stimulates *in vitro* and *in vivo*

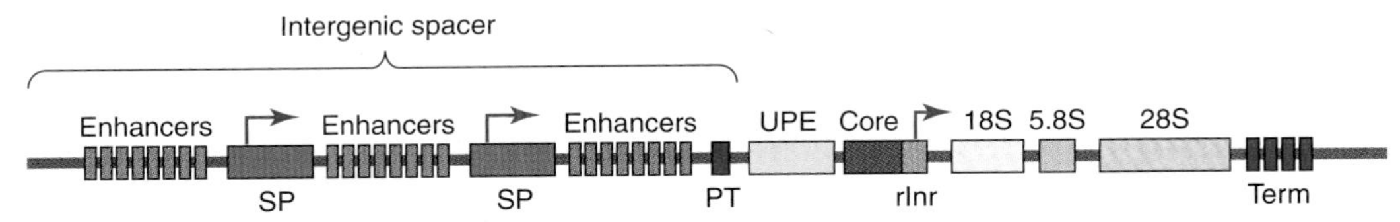

FIGURE 20.5 The rRNA transcription unit and intergenic spacer. Regions that code for 18S, 5.8S, and 28S rRNA are shown in light green, cyan, and green, respectively. The rRNA transcription unit promoter has two elements, a core promoter element (Core), which is brown and an upstream promoter element (UPE), which is light blue. The core promoter element is essential for transcription but the upstream promoter element is not. An adenine-thymine (AT)-rich region within the core promoter, the rRNA initiator (rInr), is shown in pink. The high AT content appears to permit the closed promoter to change to an open promoter, an essential prerequisite for transcription initiation. Genetic elements within the intergenic spacer include the proximal terminator (PT), which is red; the enhancers, which are orange; and spacer promoters (SP), which are dark green. The bent red arrows indicate transcription start sites. (Adapted from M. R. Paule and R. J. White, *Nucl. Acids Res.* 28 [2000]: 1283–1298.)

transcription initiation by as much as 15- and 100-fold, respectively. Promoters from other species have a similar overall structure but have little sequence homology with the human promoter. Because promoter sequences are species specific, nuclear extracts from organisms of one species can only transcribe rRNA transcription units from other organisms of that same species or a closely related species.

Genetic elements within intergenic spacers also influence transcription. The **proximal terminator** (**PT**), which is located about 200 bp upstream from the transcription initiation site, terminates transcription from upstream transcription units and thereby prevents read through. The proximal terminator also stimulates transcription of the downstream rRNA transcription unit. Among the mechanisms that have been proposed to explain this stimulation are that the PT (1) serves as a signal for remodeling chromatin over the promoter or (2) helps to arrange rRNA transcription units in a cluster so that RNA polymerase I can move from the end of one rRNA transcription unit to the promoter of the next. The intergenic spacer also contains enhancer sequences. Enhancers are usually repeated several times in higher animals but only a single enhancer copy is present in yeast. Ribosomal RNA enhancers do not bind multiple transcription activators and therefore seem to work in a different way from enhancers that stimulate protein-coding transcription units. In higher animals, the intergenic spacer also contains a functional promoter called a **spacer promoter** (**SP**), which stimulates the synthesis of short transcripts that terminate at the PT but have no other known function.

Ribosomal RNA transcription and processing takes place in the nucleolus.

Active rRNA transcription units and the RNA polymerase I transcription machinery are located within a well-defined transcription and processing "factory" within the nucleus called the **nucleolus**. Although sometimes referred to as an organelle, the nucleolus is not a true organelle because it is not surrounded by a membrane. When viewed by electron microscopy, the nucleolus has three distinct and conserved structural components (FIGURE 20.6). The first of these, the **fibrillar center** (**FC**), is a pale region that is enriched with RNA polymerase I and various transcription factors. A nucleolus may have many fibrillar centers. The second structural component, the **dense fibrillar component** (**DFC**), is a very dense fibrillar region that partially or completely surrounds the FC. The third structural component, the **granular compartment** (**GC**) occupies the rest of the nucleolus. This architecture has been proposed to reflect the vectorial maturation of the pre-ribosomes. According to one hypothesis, RNA synthesis takes place at the interface between the FC and the DFC, with nascent transcripts extending into the body of the DFC and nascent pre-ribosomes moving from the DFC to the GC as the pre-rRNA is processed, modified, and assembled into a ribosome. The nucleolus is a dynamic structure that is dismantled during mitosis and reassembled at interphase. Its appearance provides clues to certain pathological conditions. For instance, cancer cells often have abnormally large nucleoli. With this

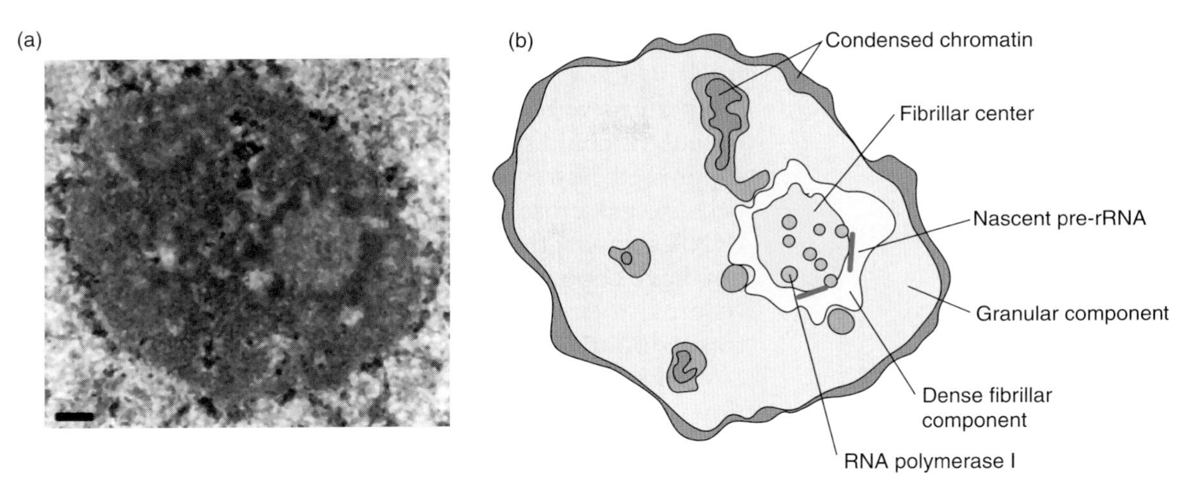

FIGURE 20.6 **Nucleolar organization in humans.** (a) Electron micrograph of the nucleolus in a human cell (Hep-2 cells; larynx epidermal carcinoma). (b) Cartoon of the electron micrograph. Although RNA polymerase I and pre-rRNA molecules are not visible in the electron micrograph, their locations have been shown by other experiments and are indicated in this cartoon. (Reproduced from *Trends Cell Biol.*, vol. 15, M. Thiry and D. L. J. LaFontaine, Birth of a nucleolus . . ., pp. 194–199, copyright 2005, with permission from Elsevier [http://www.sciencedirect.com/science/journal/09628924]. Part a photo courtesy of Denis L. J. LaFontaine, Fonds National de la Recherche Scientifque, Institut de Biologie et de Médecine Moléculaires, Université Libre de Bruxelles.)

background information in mind, we are now ready to examine the enzymes and protein factors involved in pre-rRNA formation.

RNA polymerase I is a multisubunit enzyme with a structure similar to that of RNA polymerase II.

RNA polymerase I appears to contain 11 subunits in mouse and 14 subunits in yeast. The reason for this discrepancy is not clear but may reflect the fact that certain subunits dissociate more easily from the mouse than the yeast enzyme. The yeast genes that code for RNA polymerase I subunits have been cloned. Five of the yeast RNA polymerase I subunits are identical with those in the two other nuclear RNA polymerases while two other subunits are identical with subunits in RNA polymerase III but not RNA polymerase II. The two largest RNA polymerase I subunits, while unique to it, are homologous to the two largest subunits in RNA polymerase II. A high resolution image has not yet been reported for RNA polymerase I. However, the structure for yeast RNA polymerase I, which was determined by electron microscopy, has the same size and shape as the crystal structure for yeast RNA polymerase II Δ4/7 (FIGURE 20.7), supporting the idea that the two RNA polymerases are quite similar. Amino acid sequences have been deduced from cDNA clones that code for several human, rat, and mouse RNA polymerase I subunits. These subunits show a high degree of homology with one another and with the yeast enzyme.

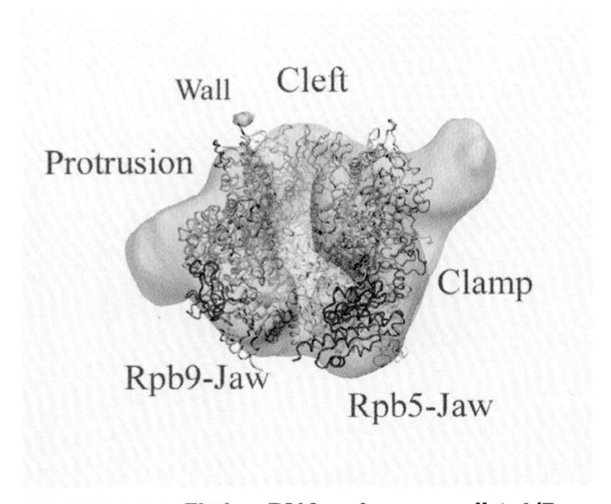

FIGURE 20.7 **Fitting RNA polymerase II Δ 4/7 crystal structure into an RNA polymerase I model reconstructed from electron micrographs.** A crystal structure of RNA polymerase II Δ 4/7 (ribbon representation) was fit into an RNA polymerase I model (transparent gray shell) reconstructed from electron micrographs. The color coding for the ten common or homologous subunits is as follows: Rpb1, white; Rpb2, gold; Rpb3, red; Rpb5, pink, Rpb6, light blue, Rpb8, dark green; Rpb9, orange; Rpb10, dark blue; Rpb11, yellow; and Rpb12, light green. (Reproduced from *J. Mol. Biol.*, vol. 329, S. De Carlo, et al., Cryo-negative Staining Reveals Conformational . . ., pp. 891–902, copyright 2003, with permission from Elsevier [http://www.sciencedirect.com/science/journal/00222836]. Photo courtesy of Sacha De Carlo, City University of New York.)

Once the role of RNA polymerase I in rRNA synthesis was established, investigators started to ask whether the enzyme might also transcribe other genes. Experiments performed by Masayasu Nomura and coworkers in 1991 answered this question for yeast. They transformed a yeast strain that had a temperature-sensitive RNA polymerase I with a multicopy plasmid bearing a yeast rRNA transcription unit fused to a *GAL7* promoter. As expected, this strain did not grow at nonpermissive temperatures in growth medium devoid of galactose because it could not make new ribosomes. The growth defects were suppressed, however, when galactose was added to the growth medium because galactose stimulated RNA polymerase II to transcribe the rRNA transcription unit from the *GAL7* promoter. Galactose would not have suppressed the growth defects if RNA polymerase I was required to transcribe some other essential gene besides the rRNA transcription unit. Hence, the experiment performed by Nomura and coworkers provides convincing evidence that the only essential function of yeast RNA polymerase I is to transcribe rRNA transcription units. The same appears to be true for RNA polymerase I in most other eukaryotes.

RNA polymerase I–associated factors are required for transcription.

RNA polymerase I has associated factors that are lost during purification and have to be added back to the core enzyme for *in vitro* transcription to take place. Three such polymerase associated factors—**PAF53**, **PAF51**, and **PAF49**—have been purified from mouse cells. PAF53 is very similar to a yeast RNA polymerase I subunit that is readily released from the core enzyme. Under *in vitro* conditions, PAF53 appears to be required for promoter-specific but not random transcription by RNA polymerase I. Much less is known about the functions of PAF51 and PAF49.

A second type of RNA polymerase associated factor, **transcription initiation factor IA** (**TIF-1A**) in animals (and **RRN3** in yeast), is tightly associated with the RNA polymerase I core enzyme. TIF-1A appears to be a key factor in growth-dependent regulation of pre-rRNA transcription in response to conditions such as nutrient starvation, aging, and viral infection that down-regulate pre-rRNA transcription. TIF-1A has several phosphorylation sites. Hypophosphorylated TIF-1A does not bind to core RNA polymerase I and does not stimulate transcription. The fact that TIF-1A must be phosphorylated to stimulate transcription suggests a mechanism for regulating pre-rRNA formation. Under favorable growth conditions when rRNA synthesis is needed, cells produce a set of signals that stimulate protein kinase, which activates TIF-1A by adding phosphate groups to it. Under adverse growth conditions when rRNA synthesis is not needed, cells produce a different set of signals that activate protein phosphatase, which inactivates TIF-1A by removing phosphate groups. When cell growth stops, the fraction of RNA polymerase I molecules with bound TIF-1A decreases, although the concentration of total TIF-1A remains unchanged. This observation is consistent with the hypothesis that RNA polymerase I•TIF-1A complex dissociation causes transcription shut-off.

RNA polymerase I also requires two auxiliary transcription factors, upstream binding factor (UBF), and selectivity factor (SL1/TIF-1B).

RNA polymerase I also requires two auxiliary transcription factors to make rRNA. The first of these, the **upstream binding factor** (UBF), contains two identical 97 kDa polypeptide subunits. The dimer is held together by a dimerization domain in the N-terminal region (FIGURE 20.8a). Two dimers appear to participate in initiation complex formation; one binds to the core promoter and the other to the UPE. The dimerization domain is followed by six DNA-binding domains. Three of the six DNA-binding domains in each polypeptide subunit interact directly with DNA, inducing the DNA to form a loop of almost 360° once every 140 bp (FIGURE 20.8b). This loop formation brings the core promoter and UPE near to one another. The carboxyl terminal region of the UBF subunit, which is rich in acidic amino acid and serine residues, is required for transcription activation. Moreover, the serine residues are potential sites for phosphorylation and therefore a possible regulatory region. Although first recognized as a factor that participates in transcription initiation, UBF appears to bind throughout the rRNA transcription unit. The precise nature of its interactions with DNA and chromatin within this region is an active area of investigation. One possibility is that UBF displaces histone H1, making chromatin less compact. If so, histone H1 displacement may explain

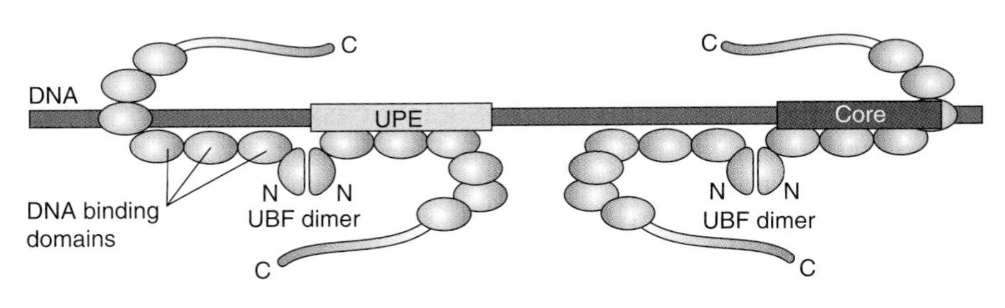

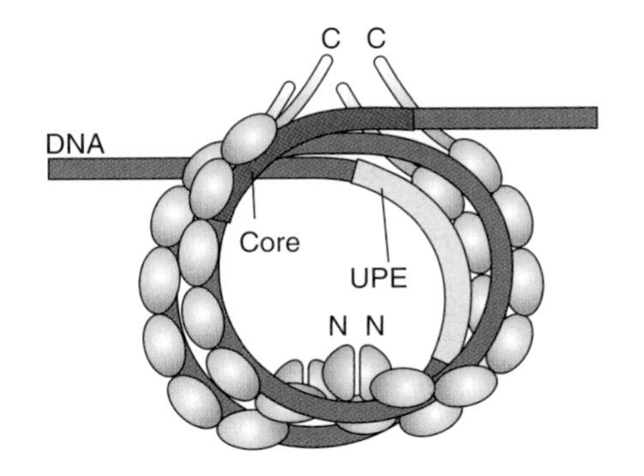

FIGURE 20.8 **Model for UBF interaction with ribosomal DNA (rDNA).** (a) Proposed interaction of two UBF binding dimers with the rDNA promoter. The amino and carboxyl termini are indicated by N and C, respectively. Although UBF appears to also interact with the enhancer, this interaction is not shown here. (b) Model for UBF-induced change in DNA structure. UBF interacts with DNA to produce a nucleoprotein complex in which about 280 bp of DNA are looped into two turns. Abbreviations are UPE, the upstream promoter element, and core, the core promoter. (Adapted from T. Moss and V. Y. Stefanovsky, *Cell* 109 [2002]: 545–548.)

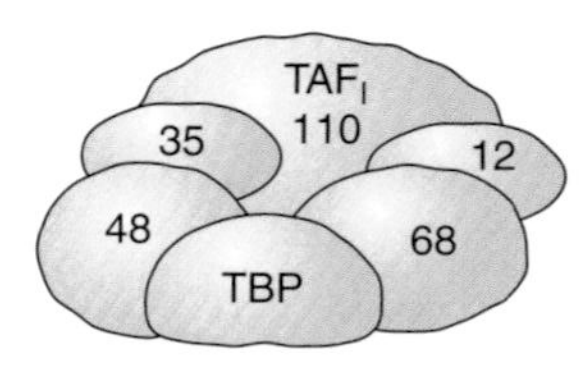

FIGURE 20.9 Selectivity factor (SL1/TIF-1B). The human selectivity factor contains a TATA-binding protein (TBP) and five RNA polymerase-specific TBP-associated factors (TAF$_I$s). (Adapted from J. Russell and J. C. B. M. Zomerdijk, *Trends Biochem. Sci.* 30 [2005]: 87–96.)

the observation that UBF stimulates RNA polymerase I transcription elongation. UBF depletion leads to reversible methylation-independent chromatin compaction with a concomitant reduction in the number of actively transcribed rRNA transcription units. These findings suggest that UBF levels help to determine the relative proportion of active and silent rRNA transcription units during growth and differentiation. The observations that UBF levels are low during cell differentiation and high in proliferating cancer cells appears to support this hypothesis.

The second auxiliary factor is called **selectivity factor 1** (**SL1**) in humans, **transcription initiation factor 1B** (**TIF-1B**) in mouse, and **core factor** (**CF**) in yeast. We will follow the common practice of using the term **SL1/TIF-1B** for the mammalian transcription factor. SL1/TIF-1B is species-specific, working with rRNA transcription units from the same or closely related species but not with those from other species. For instance, human SL1 works with human and monkey rRNA transcription units but not with mouse or yeast rRNA transcription units. This finding explains why nuclear extracts prepared from an organism belonging to one species will transcribe rRNA transcription units from other organisms of the same or closely related species but not from more distant species.

SLI/TIF-1B contains TBP and five **TBP-associated factors** (TAF$_I$s; FIGURE 20.9). Mammalian TAF$_I$s interact with one another and with TBP. The interaction with TBP blocks TBP's DNA-binding site so that it cannot bind to DNA. Therefore, it is the TAF$_I$ subunits and not TBP that bind to DNA. Although both SL1/TIF-1B and TFIID contain TBP, their TAFs are quite different. Moreover, TAF$_I$ subunits and TAF$_{II}$ subunits are never present in the same complex. Bound SL1/TIF-1B helps to recruit RNA polymerase I. UBF may assist in this recruitment by interacting with PAF53.

The transcription initiation complex can be assembled *in vitro* by the stepwise addition of individual components.

There is currently considerable debate about whether the transcription initiation complex for rRNA transcription forms by stepwise assembly of its protein components or by the addition of a single RNA polymerase holoenzyme. Those who support stepwise addition cite histochemical experiments that show components of the RNA polymerase I machinery tagged with green fluorescent protein enter and leave the nucleolus independently of one another. Those who support the holoenzyme model cite biochemical experiments that show RNA polymerase I can be isolated from cell extracts as part of a holoenzyme containing all the components required for *in vitro* rRNA synthesis. It seems reasonable to suppose that information gained from studying stepwise formation will reveal the nature of the interactions among the various components of the transcription machinery and therefore prove helpful for understanding rRNA synthesis regardless of which model is correct.

RNA polymerase I acts through a transcription cycle that begins with the formation of a pre-initiation complex.

The precise sequence of events leading to the initiation of rRNA synthesis is not certain. Until recently, the model was that UBF binds to the ribosomal DNA (rDNA) first and then helps to recruit SL1/TIF-1B. Joost C. B. M. Zomerdijk and coworkers, however, have now demonstrated that UBF itself does not bind stably to rDNA but instead rapidly associates and dissociates. They also showed that SL1/TIF-1B helps to stabilize the binding of UBF to the rDNA promoter. Based on these findings, they question the idea that UBF activates transcription through recruitment of SL1/TIF-1B at the rDNA promoter and instead propose an alternative model in which SL1/TIF-1B directs the formation of the pre-initiation complex by binding to the core promoter, helps to stabilize the binding of UBF (Step 1 in FIGURE 20.10), and then acts together with UBF to recruit RNA polymerase I (Step 2 in Figure 20.10).

Once the pre-initiation complex has assembled, the RNA polymerase I machinery is ready to initiate transcription. DNA flanking the transcription initiation site melts to create a small bubble that allows nucleotides to align with the template strand and the first phosphodiester bond is formed. The transcription bubble grows longer as RNA polymerase I moves along the DNA so that the transcription bubble eventually becomes 12 to 23 bp long. For simplicity, the small bubble and the transcription bubble are not shown in Figure 20.10. RNA polymerase I moves down the template strand, leaving UBF and SL1/TIF-1B behind (Steps 3 and 4 in Figure 20.10). At least one additional factor, elongation factor SII, associates with the transcription elongation complex (not shown in Figure 20.10). Elongation factor SII, therefore, participates in transcription elongation by both RNA polymerases I and II (see Chapter 17). The elongation rate for human RNA polymerase I has been estimated to be about 95 nucleotides · s^{-1}. Transcription elongation continues until the RNA polymerase I machinery reaches the transcription terminator (Step 5 in Figure 20.10). Then transcription is terminated and the transcript and RNA polymerase I are released. The released polymerase is free to reinitiate transcription from a previously activated and engaged promoter with pre-bound UBF and SL1/TIF-1B. The continued presence of UBF and SL1/TIF-1B at the promoter allows RNA polymerase I to initiate a new round of transcription much more rapidly than would otherwise be possible. Initiation has been estimated to occur at a minimum of once every five seconds *in vivo*.

RNA polymerase I transcription terminates at the 3′ end of the gene at specific sequences (Step 5 in Figure 20.10). Transcription termination requires the assistance of a terminator protein (**transcription termination factor 1** or **TTF-1** in mammals) and an RNA **polymerase I transcription release factor** (**PTRF**). The termination signal or terminator contains two elements. FIGURE 20.11 shows the arrangement of these two elements in the mouse terminator. The arrangement is similar, but not identical, in other eukaryotes. The upstream T-rich element

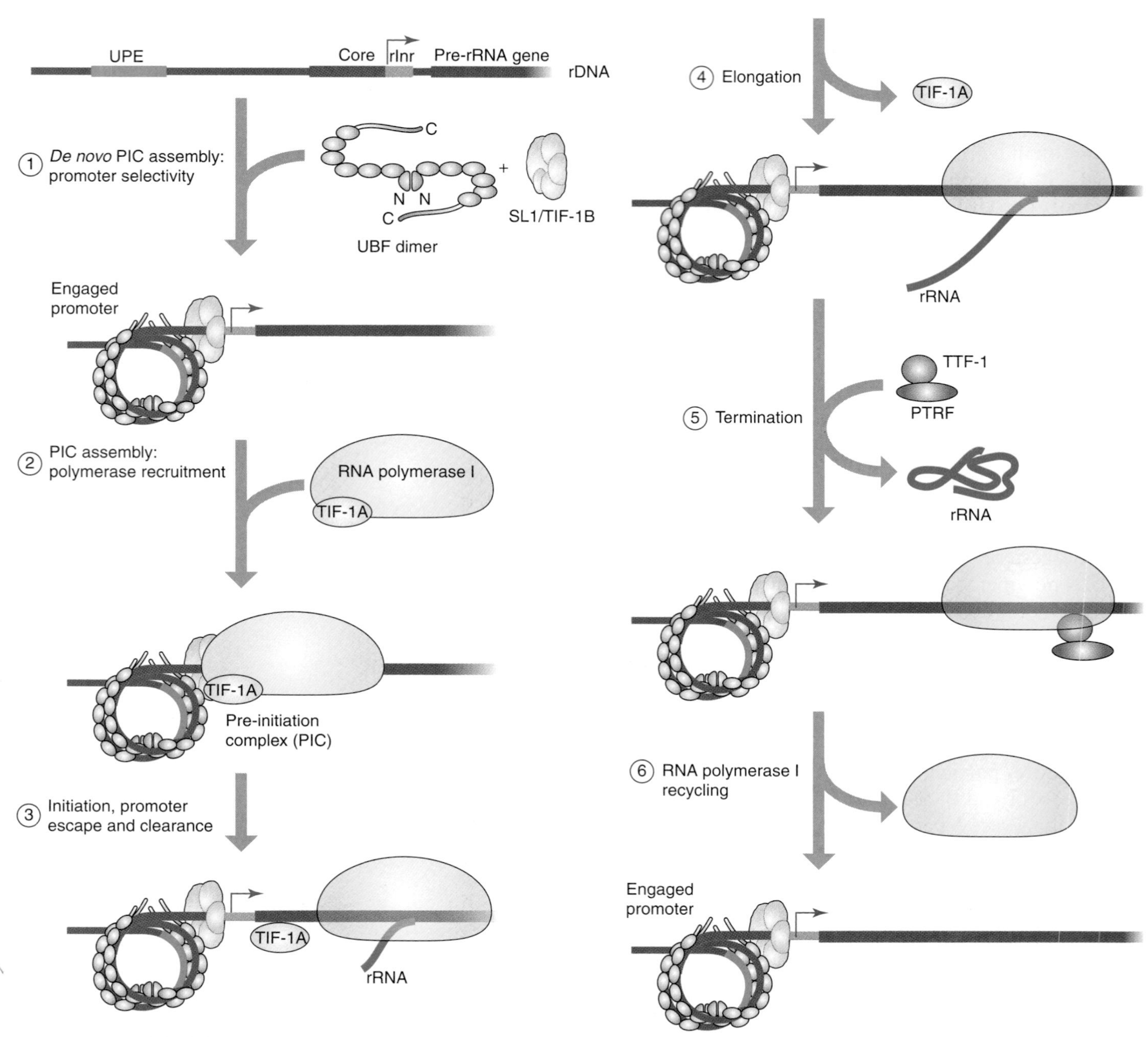

FIGURE 20.10 **The RNA polymerase (Pol I) transcription cycle.** (1) *De novo pre-initiation complex (PIC) assembly.* The selectivity factor (SL1/TIF-1B) selectively binds to the rDNA core promoter and the upstream binding factors (UBF) bind to the core promoter and the upstream promoter element. (2) *RNA polymerase I recruitment.* SL1 helps to recruit RNA polymerase I and the tightly associated factor TIF-1A. (3) *Initiation, promoter escape, and clearance.* RNA polymerase opens the promoter, initiates transcription, and then escapes from the promoter. (4) *Elongation.* Following promoter escape, RNA polymerase I becomes a highly processive enzyme that elongates the nascent rRNA. (5) *Termination.* RNA polymerase I-catalyzed transcription terminates at the 3′-end of the gene at specific sequences that have a bound transcription termination factor (TTF-1) and an RNA polymerase I transcription release factor (PTRF). RNA polymerase I and pre-rRNA are released. (6) *RNA polymerase I recycling.* Following promoter clearance, SL1/TIF-1B and UBF remain bound to the promoter, forming a reinitiation scaffold onto which a new RNA polymerase I and TIF-1A can be recruited and so being a new transcription cycle. The core promoter is shown at the 3′ end of the looped DNA for convenience. Further studies will be required to determine the precise location of RNA polymerase I and its transcription initiation factors in the initiation complex. (Adapted from J. Russell and J. C. B. M. Zomerdijk, *Trends Biochem. Sci.* 30 [2005]: 87–96.)

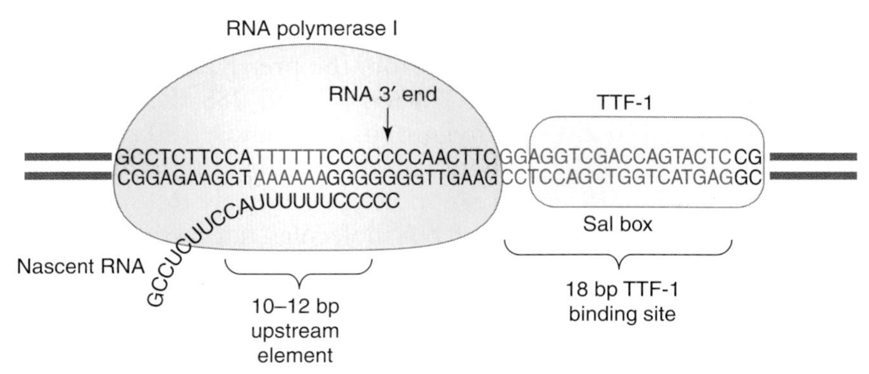

FIGURE 20.11 **Mouse transcription terminator.** The mouse transcription terminator contains two elements, a T-rich terminator protein-binding site (red) and a Sal box (blue). The terminator protein, transcription terminator 1 (TTF-1) causes the transcription elongation complex to pause. A second transcription factor, the RNA polymerase I transcription release factor (PTRF), which interacts with RNA polymerase I and TTF-1, is required to release the transcript and RNA polymerase I (see Figure 20.10). (Modified from *Trends Biochem. Sci.*, vol. 22, R. H. Reeder and W. H. Lang, Terminating transcription in eukaryotes . . ., pp. 473–477, copyright 1997, with permission from Elsevier [http://www.sciencedirect.com/science/journal/09680004].)

codes for the last 10 to 12 nucleotides in the terminated transcript. The downstream terminator protein binding element (AGGTCGACCAGA/TT/ANTCCG) is called the Sal box because it contains GTCGAC, the Sal I restriction endonuclease cleavage site. The Sal box must be present in the orientation shown for termination to take place. When the RNA polymerase I transcription elongation complex encounters TTF-1 bound to the Sal box, the complex pauses and PTRF interacts with both RNA polymerase I and TTF-1 (Step 5 in Figure 20.10). Following release, an exonuclease trims the 3′-end of the transcript to produce the mature 3′-end. Finally, RNA polymerase I is released. Termination of yeast RNA polymerase I transcription appears to proceed through a different pathway that requires Rat1, a 5′→3′ exonuclease. Recall from Chapter 19 that this enzyme is also required to terminate transcription by RNA polymerase II. At present, there is no evidence to indicate that a 5′→3′ exonuclease participates in mammalian RNA polymerase I transcription termination.

Pre-rRNA undergoes a complex series of cleavages and modifications as it is converted to mature ribosomal rRNAs.

Pre-rRNA synthesis is only the first stage of the very intricate and energetically demanding process of ribosome formation. Converting pre-rRNA to ribosomes requires several covalent modifications, many specific cleavages, sequential assembly of pre-ribosome complexes, and transport of specific pre-ribosome complexes from the nucleolus to the nucleoplasm and still other pre-ribosome complexes from the nucleoplasm to the cytoplasm. We begin our examination of the post-transcriptional events leading to ribosome formation by examining the conversion of pre-rRNA to the mature rRNAs.

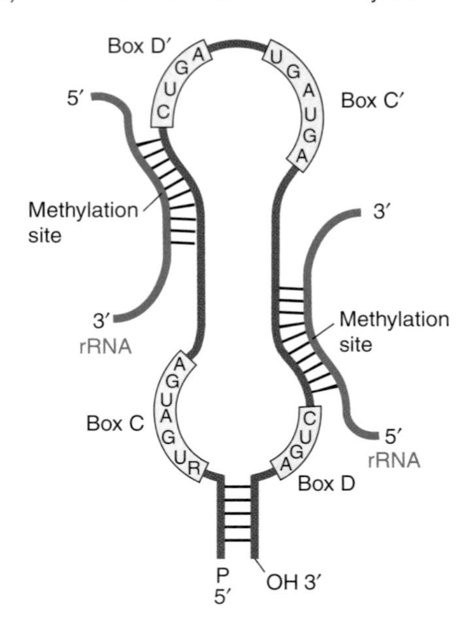

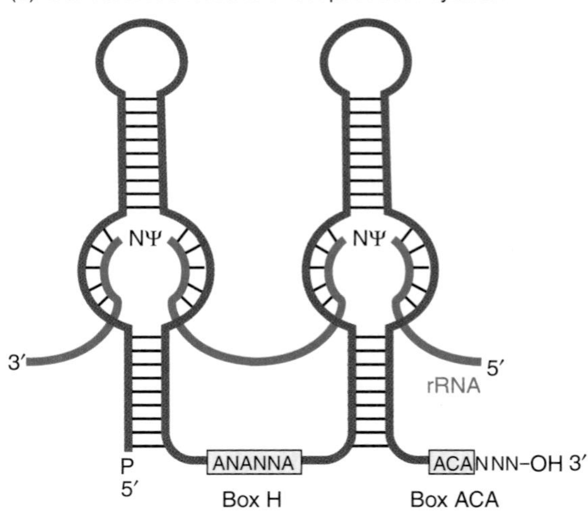

FIGURE 20.12 snoRNA molecules that participate in 2′-O-methylation and pseudouridine formation. rRNA molecules are shown in green and snoRNA molecules in blue. (a) O-methylation snoRNA. The consensus sequences of C, C′, D, and D′ boxes are indicated (R is a purine). (b) Pseudouridylation snoRNA. The consensus sequences of H and ACA boxes are indicated (N is any nucleotide). The pseudouridylation site is indicated by the Greek letter Psi (Ψ). (Part a adapted from T. Kiss, *EMBO J.* 20 [2001]: 3617–3622. Part b modified from *Cell*, vol. 89, P. Ganot, M. -L. Bortolin, and T. Kiss, Site-specific pseudouridine formation in . . ., pp. 799–809, copyright 1997, with permission from Elsevier [http://www.sciencedirect.com/science/journal/00928674].)

Pre-rRNA is modified to include 2′-O-methylated ribose and pseudouridines at specific sites before the pre-rRNA is cleaved to form mature rRNAs. Human 5.8S, 18S, and 28S rRNAs have a total of about 110 2′-O-methyl groups and almost 100 pseudouridines. Yeast rRNAs have about half this number of modifications. The covalent modifications are located exclusively in functionally important regions of mature rRNA molecules, suggesting that they contribute to ribosomal function.

The fact that there are so many specific sites for covalent modification raises the question of how a cell identifies the nucleotides that are to be modified. Investigators initially thought that specific enzymes might suffice for this purpose, but soon realized that a single enzyme probably could not recognize many different methylation or pseudouridylation target sites and that it would not be economical for cells to make a specific enzyme for each target site. The problem was solved when **small nucleolar RNA** (snoRNA) molecules were shown to serve as guides that identify target nucleotides within rRNA. The snoRNA's are part of **small ribonucleoprotein (snoRNP)** complexes, which also contain enzymes needed for methylation or pseudouridylation.

Methylation snoRNAs carry the conserved C box (RUGAUGA; where R is any purine) and D box (CUGA) near their 5′- and 3′-ends, respectively (FIGURE 20.12a). The two boxes are often folded together by a short (4–5 bp) terminal helix. Additional, often imperfect copies of C and D boxes (called C′ and D′ boxes) are located internally. The 2′-O-methylation snoRNAs also have one or two 10 to 21 nucleotide sequences that can form perfect double helices with rRNA sequences. 2′-O-methylation takes place on the rRNA nucleotide that is 5 bp upstream from the D or D′ box (Figure 20.12a).

Pseudouridylation snoRNAs have two large hairpin domains linked by a hinge and followed by a short tail (FIGURE 20.12b). The hinge has a conserved element called Box H, which has the sequence ANANNA (where N stands for any nucleotide) and the tail contains Box ACA three nucleotides from the 3′-end. The nucleotide targeted for pseudouridylation is determined by an internal loop in the stem(s) that forms short snoRNA•rRNA duplexes of 4 to 6 bp flanking the target residue.

Endonucleases cleave modified pre-rRNA at specific sites and the cleavage products are then trimmed by exonucleases or subject to further endonucleolytic cleavage until the mature rRNAs are formed. Many of the exonucleases that participate in pre-rRNA processing have been isolated and characterized but most of the endonucleases have not. Our present knowledge of the cleavage steps derives largely from studying yeast pre-rRNA processing (FIGURE 20.13). Cleavage sites in 35S yeast pre-rRNA are indicated by the letters A_0 to E. Cleavage begins at A_0, A_1 and A_2 sites, producing 20S and 27SA_2 pre-rRNAs. The 20S pre-rRNA is cleaved to form the mature 18S rRNA present in the small ribosomal subunit and the 27SA_2 pre-rRNA is processed through two alternative pathways to form the 25S rRNA and two alternative forms of the 5.8S rRNA, designated 5.8SL and 5.8SS, respectively, present in the 60S ribosomal subunit.

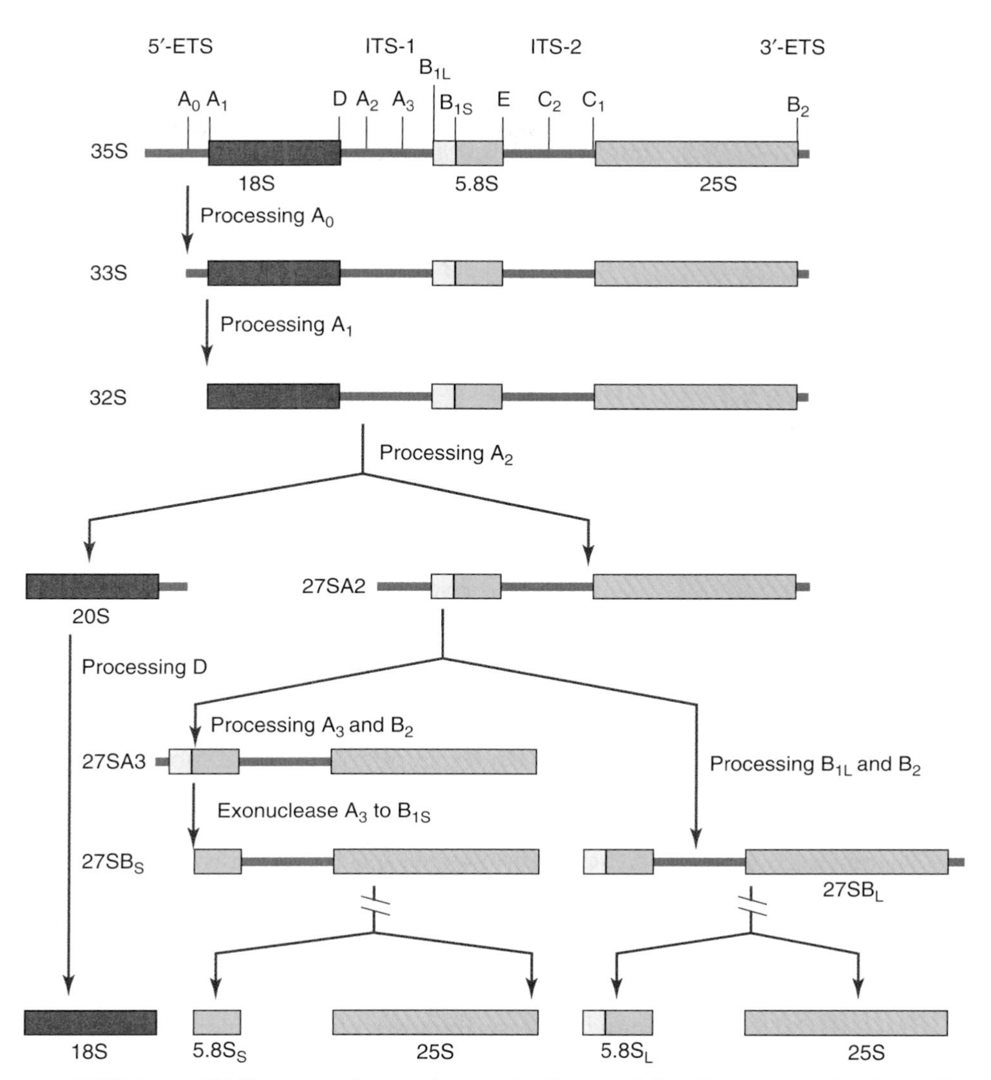

FIGURE 20.13 **Precursor rRNA (pre-rRNA) processing pathways in *S. cerevisiae*.** Sequences for the 18S, 5.8S, and 25S rRNA are separated by internal spacers (ITS-1 and ITS-2) within the 35S pre-rRNA. External transcribed spacers (5′-ETS and 3′-ETS) flank these rRNA sequences. Letters A_0, A_1, D, and so forth specify cleavage sites. Pre-rRNA cleavage begins at sites A_0, A_1, and A_2 to generate the 20S and 27SA2 pre-rRNA molecules. Then the 20S pre-rRNA is cleaved at site D, leading to the formation of the mature 18S rRNA present in the small ribosomal subunit. 27SA2 pre-rRNA molecules are processed by two alternative pathways. The majority of the 27SA2 pre-rRNA is cleaved at sites A_3 and B_2 to form 27SA3 pre-rRNA, which is in turn cleaved by exonucleases to form $27SB_S$ pre-rRNA. Alternatively, 27SA2 pre-rRNA is cleaved at sites B_{1L} and B_2 to form $27SB_L$ pre-rRNA. Both $27SB_S$ and $27SB_L$ pre-rRNAs are cleaved at sites C_1 and C_2 to form the mature 25S rRNA and $7S_S$ or $7S_L$ intermediates (not shown), which are further processed to form mature $5.8S_S$ or $5.8S_L$ rRNAs. (Modified from W. Chen, et al., Enp1, a yeast protein associated with U3 and U14 snoRNAs, is required for pre rRNA processing and 40S subunit synthesis, *Nucleic Acids Res.* 31 [2003[: 690–699. Reprinted by permission of Oxford University Press.)

20.3 Self-Splicing Ribosomal RNA

Tetrahymena thermophila pre-rRNA contains an intron that catalyzes its own excision.

Some molecular biologists studied rRNA synthesis in systems other than yeast and mammals because they believed that the systems offered special advantages. This was certainly the motivating factor that caused

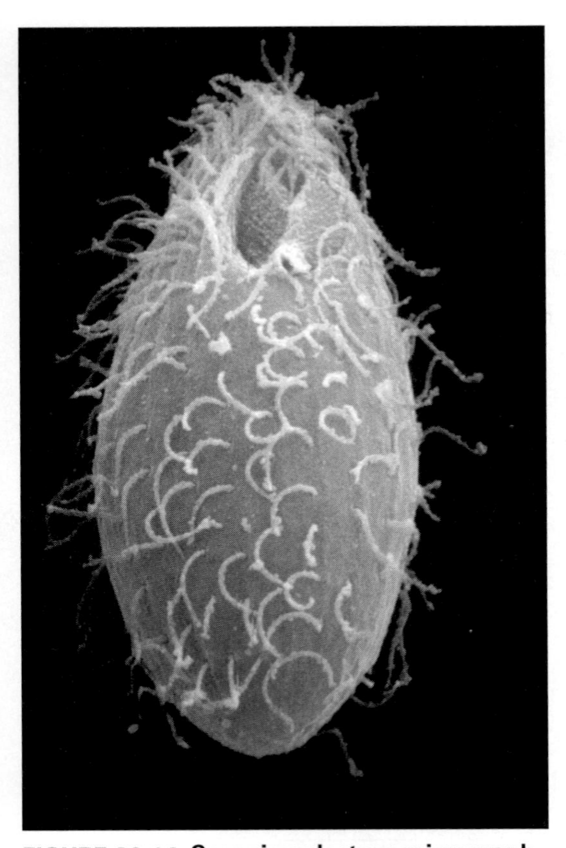

FIGURE 20.14 **Scanning electron micrograph of a *Tetrahymena* cell.** (Photo courtesy of E. Marlo Nelsen and used with permission of Joseph Frankel, University of Iowa.)

Thomas R. Cech to select the ciliated protozoan, *Tetrahymena thermophila* (FIGURE 20.14), when he set out to study rRNA synthesis in 1978. Earlier studies by others showed that the cell's macronucleus has about 10,000 autonomously replicating rRNA genes, each producing identical pre-rRNAs at the rate of one copy per second. Based on this information, Cech thought that he might be able to isolate *T. thermophila* rRNA genes together with the enzymes and regulatory factors needed to transcribe them. Cech and his students soon discovered that the rRNA gene, which contains sequences for the mature 17S, 5.8S, and 26S rRNA, has an approximately 400-nucleotide-long intron within the 26S rRNA sequence (FIGURE 20.15). The presence of this intron was not a surprise because similar introns were known to be present in rRNA genes from related organisms. However, as we will see shortly, this intron turned out to have very special properties that caused Cech to change his research direction and eventually led him to discover a surprising new RNA function.

Cech and coworkers began their study of *T. thermophila* rRNA gene transcription by incubating nuclear extracts, containing rRNA genes and the transcription machinery, with radioactive nucleoside triphosphates and salts required for transcription. They also included α-amanitin to block formation of mRNA, tRNA, and other small RNA molecules so that they could more easily follow the radioactive transcripts that were formed.

Two of the radioactive bands that appeared after gel electrophoresis separation were especially interesting. The first contained a product about the size expected for mature 26S rRNA and the second contained a low molecular weight product of about the size expected for the intron. The presence of these RNA products suggested that the *in vitro* transcription system can precisely excise the intron from the newly transcribed pre-rRNA. Because only one other example of *in vitro* intron excision was known at the time (a pre-tRNA from yeast), Cech thought it would be worthwhile to study the splicing process in greater detail. He therefore set out to isolate the splicing machinery from *T. thermophila* extracts.

Cech and coworkers needed an unspliced pre-rRNA substrate to assay for the splicing machinery. They were able to extract the desired radioactive substrate from a transcription reaction mixture to which polyamines had been added to block splicing. The radioactive pre-rRNA was added to tubes containing salts, unlabeled nucleotides, and *T. thermophila* extracts. The extract, however, was not included in one of the control tubes to demonstrate that it was required for splicing. As

FIGURE 20.15 ***Tetrahymena thermophilia* pre-rRNA.**

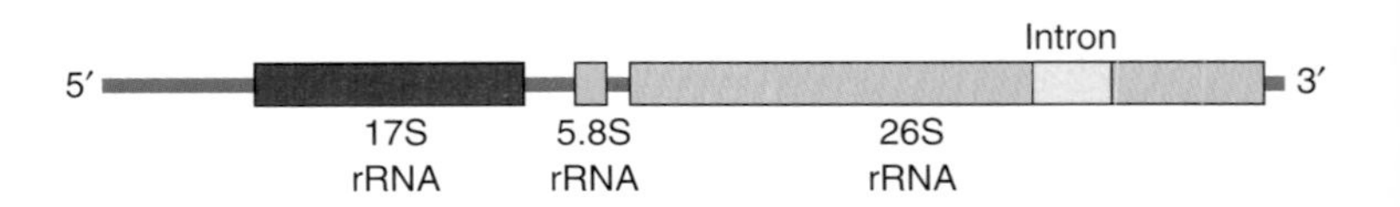

expected, pre-rRNA splicing took place in tubes containing the complete reaction mixture. To everyone's amazement, however, splicing took place just as well in the control tube that had no extract. Because all biochemical reactions then known were catalyzed by enzymes, Cech and coworkers initially thought that extract had accidentally been introduced into the control tube. They therefore repeated the experiment, being very careful not to add the *T. thermophila* extract to the tube that was not supposed to contain it. Once again, splicing occurred in the absence of added extract.

At the time that Cech and coworkers performed their experiments, all biochemical reactions were thought to be catalyzed by enzymes. Therefore, Cech and coworkers considered the possibility that the RNA they used as a substrate had actually been spliced inside the cell but the mature 26S rRNA and the excised intron somehow remained bound together until mixed with salts and nucleotides. Cech and coworkers decided to examine the splicing reaction that occurs in the absence of *T. thermophila* extracts more carefully. They soon discovered that magnesium ions and GTP (but not ATP, UTP, or CTP) must be present for the low molecular weight RNA product to be released. Moreover, the phosphoanhydride bonds in GTP were not required to release the low molecular weight RNA (the presumed intron) because guanosine could replace GTP. In a related discovery, Cech and coworkers demonstrated that the sequence of the low molecular weight RNA product was identical to that predicted for the intron in the rRNA gene in all respects but one. The single difference was that the low molecular weight RNA product had an extra G at its 5′-end. Cech performed a simple labeling experiment to demonstrate that the extra G came from GTP. He added [^{32}P]GTP to a solution containing pre-rRNA and salts and showed that the GTP was covalently attached to the 5′-end of the low molecular weight product. Cech and coworkers proposed the splicing mechanism shown in FIGURE 20.16 to explain their experimental results. This mechanism assumes the intron has a binding site for guanosine (or guanine nucleotides) that is analogous to the substrate binding site in an enzyme. The 3′-hydroxyl of a guanosine (or guanine nucleotide) bound to this site initiates a nucleophilic attack on the 5′-splice site, forming a 3′→5′ phosphodiester bond at the 5′-end of the intron and releasing the upstream exon. Magnesium ions appear to be required for this reaction to take place (FIGURE 20.17). Once transesterification is complete, the intron goes through a conformational change in which the original guanosine (or GTP) at the G-binding site is replaced by the guanosine at the 3′-end of the intron. The cleaved upstream exon remains base-paired to the intron during this conformational change. In the new conformation, the 3′-end of the upstream exon is in position to attack the 3′-splice site. This transesterification reaction ligates the exons and excises the intron.

Because the two transesterification reactions proposed in the splicing mechanism have high energies of activation, it seemed unlikely

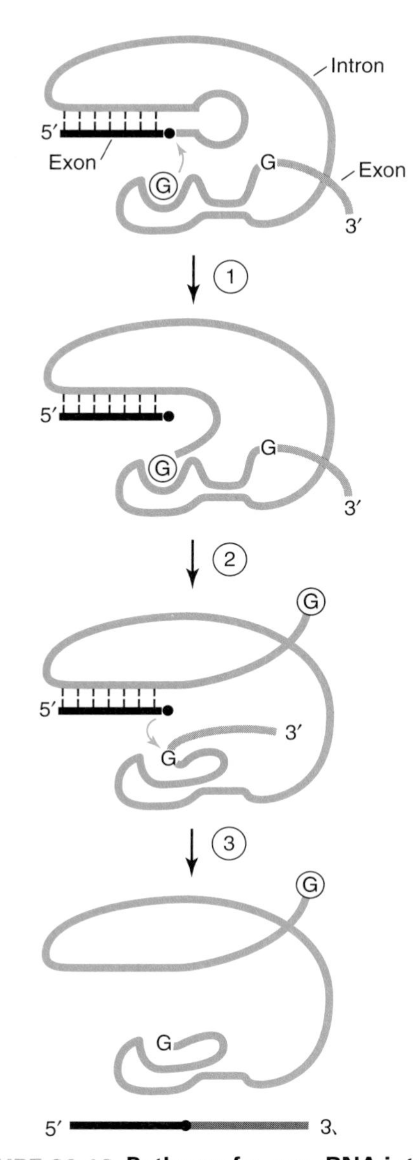

FIGURE 20.16 **Pathway for pre-rRNA intron self-splicing.** Exons are shown in gray and black and the intron in green. Step 1: An intron-bound guanosine or GTP (circled) cleaves the 5′-splice site and in the process becomes covalently attached to the 5′-end of the intron. Step 2: A conformational change takes place in which the G at the 3′-end of the intron replaces the original G in the G-binding site. Step 3: The cleaved 5′-exon, which is still base paired to the intron, cleaves the 3′-splice site, allowing the exons to be ligated and the intron to be excised. (Adapted from J. A. Doudna and T. R. Cech, *Nature* 418 [2002]: 222–228.)

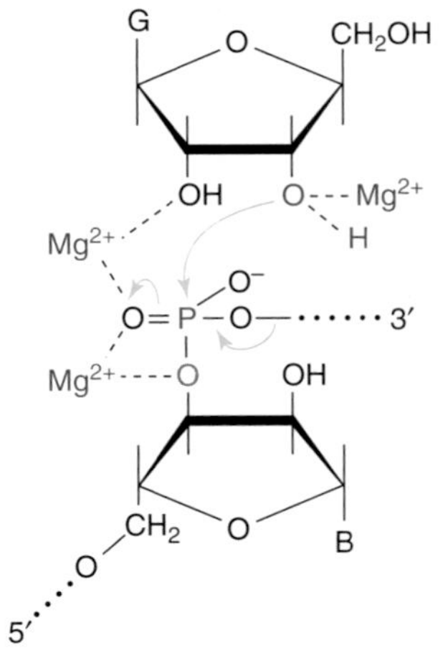

FIGURE 20.17 **Possible mechanism for first transesterification reaction catalyzed by pre-rRNA intron.** (Modified from *Chem. Biol.*, vol. 9, V. J. DeRose, Two Decades of RNA Catalysis, pp. 961–969, copyright 2002, with permission from Elsevier [http://www.sciencedirect.com/science/journal/10745521].)

that they would take place at any significant rate in the absence of a catalyst. As enzymes were the only known biological catalysts at that time, Cech and coworkers tried to find evidence for enzyme participation. The most likely enzyme source appeared to be a protein that remained bound to the purified pre-rRNA substrate. Cech and coworkers, therefore, tried various approaches to remove or inactivate proteins bound to the pre-rRNA substrate. Splicing continued to take place, however, even after the pre-rRNA was heated in the presence of sodium dodecyl sulfate to denature proteins or treated with proteases to degrade proteins. Cech and coworkers were then forced to consider the unprecedented possibility that the pre-rRNA was itself the catalyst for the reaction.

Because there was absolutely no precedent for an RNA catalyst, Cech and coworkers needed to devise an experiment that would unambiguously demonstrate that the pre-rRNA can act as a catalyst. One solution would have been to synthesize the pre-rRNA by standard organic synthetic techniques and to then demonstrate intron excision when the synthetic pre-rRNA was incubated in the presence of guanosine (or GTP) and magnesium ions. Unfortunately, this approach was not practical because the pre-rRNA was much too long to be synthesized in an organic chemistry laboratory. So they did the next best thing. They used recombinant DNA technology to transfer a segment of the rRNA gene containing the intron and flanking sequences to a bacterial plasmid and then used bacterial RNA polymerase to synthesize the desired RNA substrate. Remarkably, this substrate, which never came in contact with any *T. thermophila* proteins, could catalyze its own splicing reaction. Cech and coworkers therefore unambiguously demonstrated that the RNA in the intron catalyzes a chemical reaction. This first example of an RNA catalyst or ribozyme was soon followed by others.

Introns that are structurally and evolutionarily related to the self-splicing *T. thermophila* intron have been identified in nuclear, mitochondrial, and chloroplast genomes from diverse organisms, including green algae, higher plants, and fungi, as well as in the genomes of some bacteria and bacterial viruses. Notably absent from this list are the vertebrates. The intron present in *T. thermophila* pre-rRNA is the founding member of a family of introns called **group I introns**. These introns range in size from a few hundred nucleotides to about 3,000. Although there is little sequence similarity within the group I intron family, all members fold into distinctive phylogenetically conserved secondary structures consisting of 10 paired segments, P1 to P10. FIGURE 20.18a shows the secondary structure for the self-splicing *T. thermophila* intron. Although the intact intron's three-dimensional structure has not yet been determined, Cech and coworkers have solved the crystal structure of a 247-nucleotide fragment that contains the catalytic core at a resolution of 3.8 Å (FIGURE 20.18b). This structure lacks the P1 helix that contains the substrate for the first splicing reaction, but contains the proposed binding site for the exogenous guanosine nucleophile. A deeper understanding of the way that group I introns work will require the solution of the crystal structure of the intact intron.

(a)

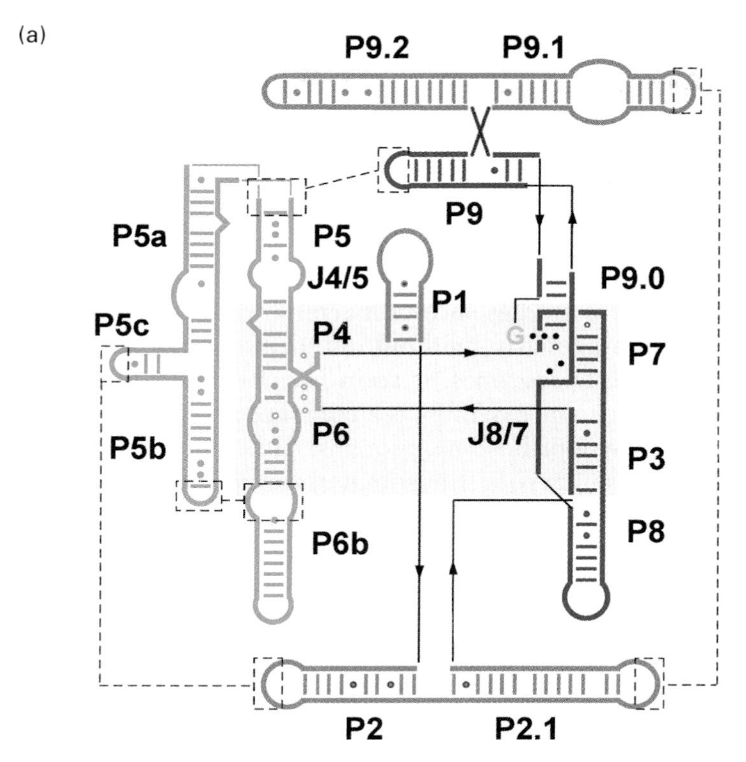

(b)

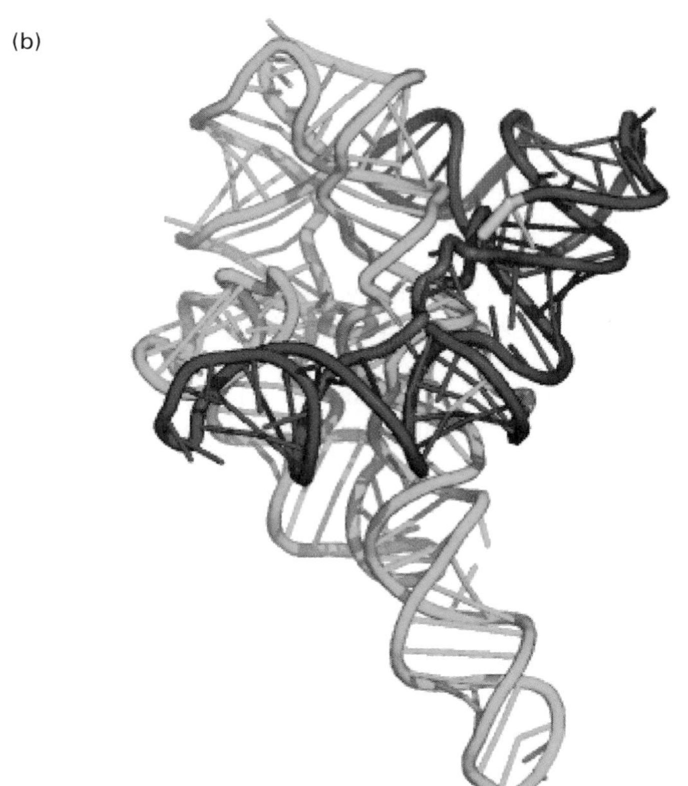

FIGURE 20.18 **Structure of the *Tetrahymena* group I intron.** (a) *Tetrahymena* group I intron secondary structure is colored by domain. Paired (P) and joining (J) regions are numbered in a 5′ to 3′ direction. The conserved catalytic core is shaded. The P3–P9 domain contains the binding site for the guanosine substrate and many active site residues. (b) *Tetrahymena* intron without substrate in ribbon form colored by domain. P1, P2/P2.1, and P9.1/P9.2 were not present in the RNA used in crystallization. (Reproduced from *Curr. Opin. Struc. Biol.*, vol. 15, S. A. Woodson, Structure and assembly of group I introns, pp. 324–330, copyright 2005, with permission from Elsevier. [http://www.sciencedirect.com/science/journal/0959440X]. Structure rendered by Sarah A. Woodson, Johns Hopkins University.)

20.4 Ribosome Assembly

Eukaryotic ribosome assembly is a complex multistep process.

Let's now consider the pathway used to form the ribosome subunits. More than 80 ribosomal proteins are synthesized in the cytoplasm and then imported into the nucleolus. Investigators initially thought that the eukaryotic ribosomes might self-assemble with the assistance of few, if any, nonribosomal proteins. The reason for this expectation was that *in vitro* studies showed bacterial ribosomal subunits could assemble spontaneously, without any demonstrable need for cofactors or folding chaperones. It soon became evident, however, that eukaryotic ribosome assembly is a complex process that does require assistance from nonribosomal proteins. Although much of our current knowledge of eukaryotic ribosomal assembly comes from studies in yeast, ribosomal assembly in other eukaryotes appears to proceed in a similar fashion.

Pioneering studies performed independently by the laboratories of Jonathan Warner and Rudi J. Planta in the 1970s identified the earliest pre-ribosomal assembly, the **90S particle**, so-named because of its sedimentation in a sucrose gradient. This particle, which contains the 35S pre-rRNA as well as many ribosomal proteins, has a relatively high ratio of protein to RNA compared to mature ribosomes. The high protein composition suggested that the 90S particles have additional nonribosomal proteins that are lost when the 90S particles move from the nucleolus to the cytoplasm. These early studies also showed that the 90S particles are subsequently processed to yield smaller 66S and 43S particles, the precursors to the mature 60S and 40S subunits, respectively (FIGURE 20.19). The technology available in the 1970s and for the next two decades was too limited to provide much information about the composition of 90S particles and the nonribosomal proteins associated with them. The chief difficulty was in isolating the intact 90S particle and other pre-ribosome assemblies free from contaminants.

A new technique known as **tandem affinity purification (TAP)**, which was devised by Bertrand Séraphin and coworkers in 1999, has permitted investigators to isolate and characterize many pre-ribosome assemblies and thereby provide a more complete picture of the ribosome assembly process. The tandem affinity purification procedure, which is summarized in FIGURE 20.20, has the great advantage of allowing ribonucleoprotein or protein complexes to be isolated in highly pure form under very mild conditions so that the complexes retain their native structure.

The first task is to use recombinant DNA technology to modify one of the proteins (the so-called bait protein) in the ribonucleoprotein or protein complex under study so that the bait protein has a TAP tag attached to its N- or C-terminus. The TAP tag consists of three parts, a calmodulin-binding peptide, a tobacco etch virus (TEV) protease cleavage site, and the immunoglobulin G (IgG) binding domain from *Staphylococcus aureus* protein A (Figure 20.20). Cells or organisms are constructed that contain TAP-tagged bait protein(s). Then cell-free

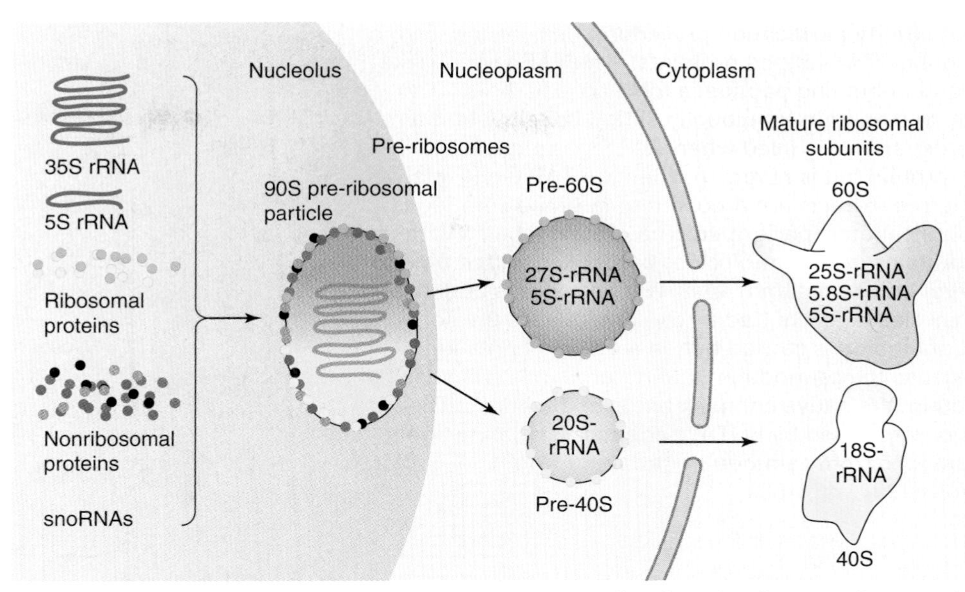

FIGURE 20.19 **Proposed pathway for assembly, maturation, and export of 40S and 60S yeast ribosomal subunits.** The 90S pre-ribosomal particle is formed by 35S pre-rRNA, 5S rRNA, snoRNAs, ribosomal proteins, and nonribosomal proteins. The 90S pre-ribosomal particle is converted into 40S and 60S pre-ribosomal particles after the 35S pre-rRNA is split. Then the 40S and 60S pre-ribosomal particles are exported into the cytoplasm through the nuclear pores. Nonribosomal proteins dissociate from the 40S and 60S pre-ribosomal particles to produce 60S and 40S ribosomal subunits, respectively. (Adapted from H. Tschochner and E. Hurt, *Trends Cell Biol.* 13 [2003]: 255–263.)

extracts are prepared under mild conditions and tandem affinity purification is performed (Figure 20.20). The first affinity column contains IgG beads. The TAP tag labeled protein binds to the IgG beads along with any proteins or RNA molecules that are associated with it, while contaminants pass through the column. Then TEV protease is added to cleave the TEV protease cleavage site, releasing the previously bound complex from the column. The released complex is put through a second round of binding using a column containing calmodulin beads. Because the calmodulin-binding protein is still attached to the TAP-tagged protein, the native complex binds to the beads. Contaminants, which now also include the TEV protease, pass through the column. The purified native complex is released by EGTA, a chemical chelating agent, which removes calcium ions from calmodulin. Finally, the components of the purified ribonucleoprotein or protein complex are identified by SDS–polyacrylamide gel electrophoresis followed by Western blot analysis or by mass spectrometric analysis.

Susan J. Baserga and coworkers used the tandem affinity purification procedure to isolate a large ribonucleoprotein precursor of the 40S subunit from yeast (at least 2.2×10^6 Da), which is similar in size and complexity to ribosomes and spliceosomes. The precursor of the 40S subunit (pre-40S) contains U3 snoRNA, a member of the box C/D class of snoRNAs and 28 proteins. Although some of the proteins had previously been demonstrated to be part of the much smaller U3 snoRNP, 17 of the proteins were new. Depletion of the 17 proteins impeded 18S rRNA formation, indicating that these

FIGURE 20.20 **Tandem affinity purification procedure.** Cells or organisms are constructed that contain TAP-tagged protein(s). The TAP tag consists of three parts, a calmodulin-binding peptide, a tobacco etch virus (TEV) protease cleavage site, and an immunoglobulin G (IgG) binding domain from protein A. The order shown is used when TAP tag is attached to the C-terminus of the bait protein but is reversed when the TAP tag is attached to the N-terminus. Cell-free extracts are prepared under mild conditions and tandem affinity purification is performed. The protein labeled with the TAP tag and any proteins or RNA molecules associated with it bind to the IgG beads in the first affinity column. Then TEV protease is added to cleave the TEV protease cleavage site (red arrow) and release the complex. A second round of binding is carried out on a column containing calmodulin beads. Because the calmodulin-binding protein is still attached to the TAP-tagged protein, the native complex binds to the beads. The purified native complex is released by EGTA, a chemical chelating agent, which removes calcium ions from calmodulin. (Adapted from L. A. Huber, *Nat. Rev. Mol. Cell Biol.* 4 [2003]: 74–80.)

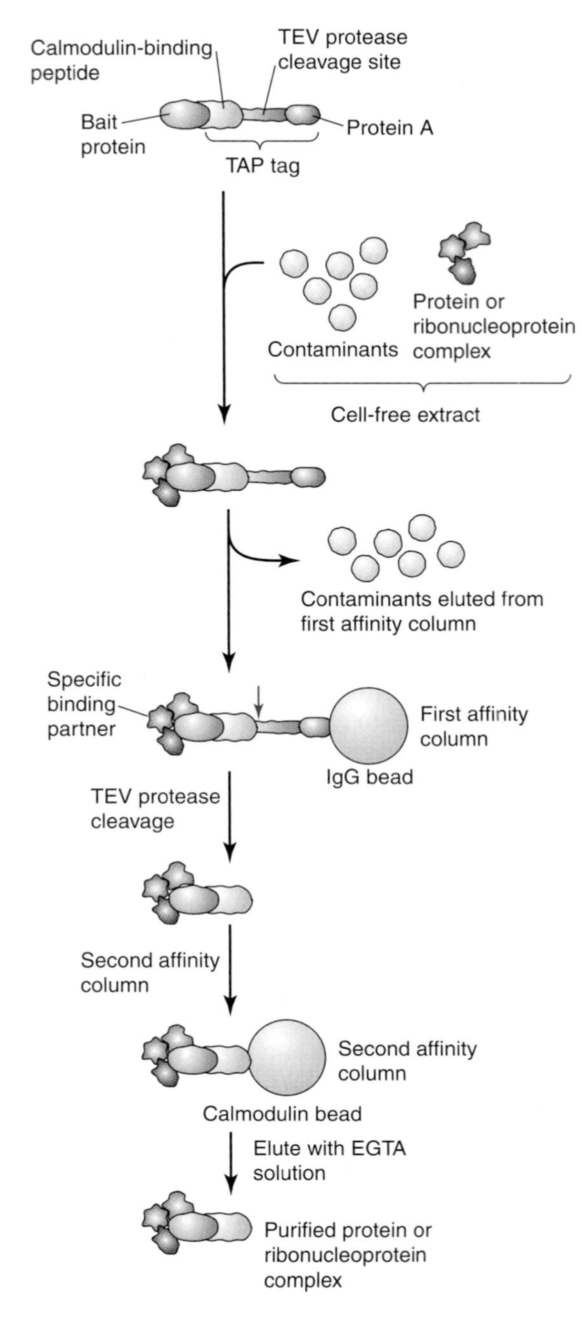

proteins are part of the active pre-rRNA processing complex. The large complex, therefore, appears to be required to form the 40S subunit. This large ribonucleoprotein complex is called the **SSU processome** (SSU indicates that the particle is required to process the small ribosome subunit, but not the large one and processome indicates the complex processes pre-rRNA). Although the specific functions performed by the SSU processome remain to be determined, it has been suggested that the SSU processome helps to fold rRNA into the proper conformation.

FIGURE 20.21 shows a current model of ribosome assembly in yeast. Some of the proteins involved in ribosome processing have

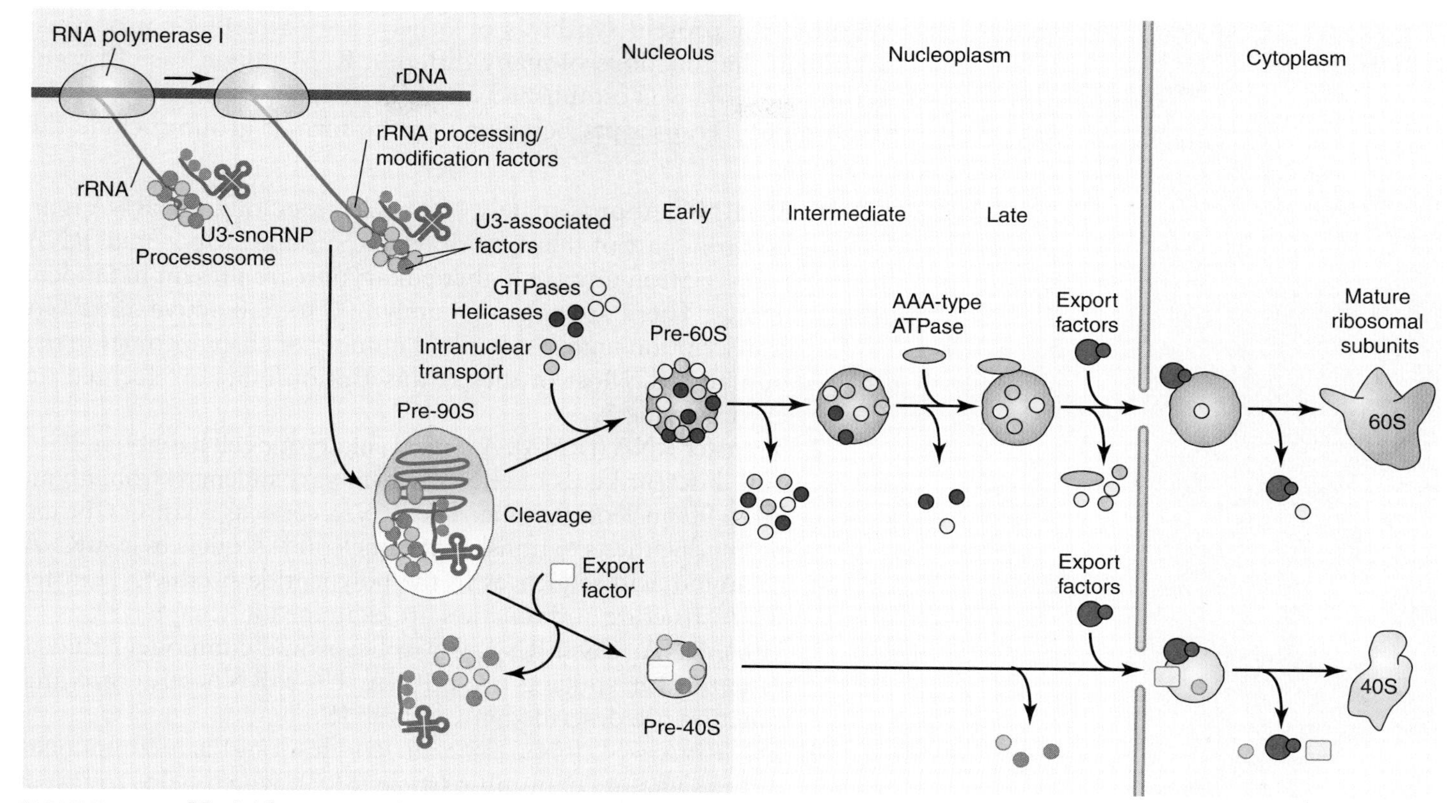

FIGURE 20.21 **Model for maturation and export of 40S and 60S ribosomal subunits in yeast.** The earliest ribosomal precursor particle is the "classical" 90S precursor (pre-90S) composed of 35S pre-rRNA, the U3 snoRNP and additional 40S synthesis factors. Following cleavage at site A2, the 90S precursor separates into pre-40S and pre-60S complexes. The majority of the pre-60S factors assemble during or after cleavage at A2. Many of the pre-60S factors disassemble from the nascent subunit during the pre-60S complex's movement from the nucleus toward the nuclear pore. To gain export competence, a special subset of factors (marked in red) has to bind. The final maturation occurs after passage through the nuclear pore. Note that the export factors for 40S subunits are not yet known. (Adapted from H. Tschochner and E. Hurt, *Trends Cell Biol.* 13 (2003): 255–263.)

been identified on the basis of their structural relationship to known protein families but little is known about their function in ribosome formation. For example, one protein has the highly conserved domain characteristic of the AAA-type ATPase family. Other family members are known to participate in a wide variety of cellular processes including cell cycle regulation, protein degradation, organelle biogenesis, and vesicular transport. In fact, AAA is an acronym for ATPases associated with different activities. The precise role of the nuclear AAA-type ATPase in ribosome formation remains to be determined.

20.5 RNA Polymerase III

RNA polymerase III transcripts are short RNA molecules with a variety of biological functions.

RNA polymerase III transcription units code for a variety of short RNA molecules (usually less than 400 nucleotides long), which are required for protein synthesis or other essential cellular processes. Although the functions of many different RNA polymerase III transcripts are now well established, the functions of others still must be determined. Transcripts belonging to the former group include the following:

1. **5S rRNA.** Approximately 120 nucleotides long, 5S rRNA is an essential part of the large ribosome subunit. Ribosomal RNA transcription units in higher eukaryotes are present in tandem arrays. There are approximately 300 to 500 active 5S rRNA transcription units per haploid number of chromosomes in the human cell. Most of these transcription units are located on chromosome 1.
2. **Transfer RNA (tRNA).** A typical eukaryotic cell has about 30 to 100 different kinds of tRNA, which vary in length from about 70 to 90 nucleotides long. The tRNAs serve as adaptors during protein synthesis to translate the nucleotide sequences in mRNA to amino acid sequences in the polypeptide product. Transfer RNA structure and function are examined in Chapter 22.
3. **U6 snRNA.** U6 snRNA is a spliceosome component that is essential for splicing pre-mRNA. U6 snRNA structure and function are described in Chapter 19.
4. **7SL RNA.** 7SL RNA (300 nucleotides long) is a component of the signal recognition particle (SRP), which helps to transport specific proteins into or across the endoplasmic reticulum membrane so that they can perform their biological function. SRP structure and function are described in Chapter 23.
5. **7SK RNA.** A small nuclear RNA (330 nucleotides), 7SK RNA helps to regulate P-TEFb, a general transcription factor that participates in both transcription initiation and elongation; this is discussed in Chapter 17.
6. **H1 RNA.** H1 RNA is the catalytic component of RNase P, an endonuclease that is needed in one of the steps for converting pre-tRNA to tRNA (see below).

RNA polymerase III transcription units have three different types of promoters.

Eukaryotes have three different kinds of promoters in their RNA polymerase III transcription units, called types 1 to 3 (FIGURE 20.22). The architectural structures of these promoters are as follows:

Type 1 Promoter—The type 1 promoter (Figure 20.22a), which was initially characterized by deletion analysis of the *Xenopus laevis* (frog) 5S rRNA transcription unit, is located within rather than before the transcription unit. Subsequent linker scanning studies (see Chapter 18 for a discussion of this technique) showed that the promoter contains three short elements, the **A box** (+50 to +64), the **intermediate element** or **IE** (+67 to +72), and a **C-box** (+80 to +97), which are necessary and sufficient for transcription by frog nuclear extracts. The region within the transcription unit that contains these three promoter elements is called the **internal control**

region. Nucleotides between the three elements are spacers; their sequence does not influence transcription efficiency. Mutations within the three control elements cause base changes in the 5S rRNA because the control elements are transcribed. The 5S rRNA promoters in other animals have the same architecture as the frog promoter but 5S rRNA promoters from yeast are different. The yeast promoter has only two elements; one corresponds to the C-box but the other extends from –14 to +8.

Type 2 Promoter—The type 2 promoter (Figure 20.22b), which is present in most tRNA transcription units, is divided into two conserved elements of about 10 bp each, the A box and the B box. There is considerable selective pressure to conserve these two boxes because they also determine important structural features of the tRNA molecule.

Type 3 Promoter—The type 3 promoter (Figure 20.22c), which controls U6 snRNA transcription units, unlike type 1 and 2 promoters is entirely upstream from the transcription initiation site. The promoter for the human U6 snRNA transcription unit has three structural elements, a **distal sequence element** (DSE; –214 to –244), a **proximal sequence element** (PSE, –66 to –47), and a **TATA box** (–30 to –25).

In summary, the RNA polymerase III machinery recognizes three different types of promoters. Two of these are downstream from the transcription initiation site and so also determine the transcript

Internal promoters

(a) Type 1 promoter (5S rRNA genes)

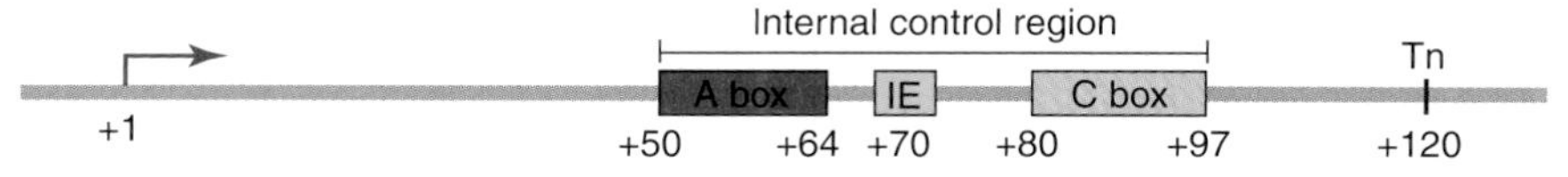

(b) Type 2 promoter (transfer RNA genes)

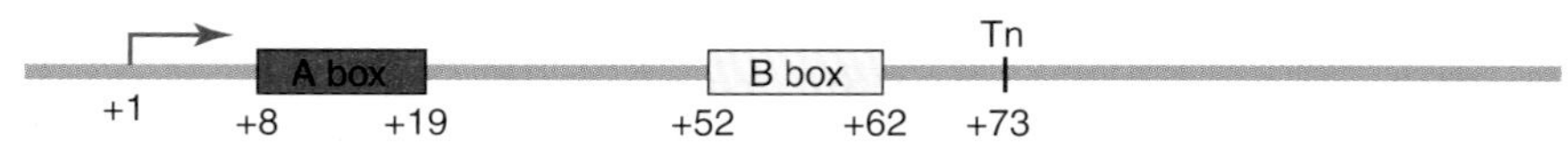

External promoters

(c) Type 3 promoter (mammalian U6 snRNA gene)

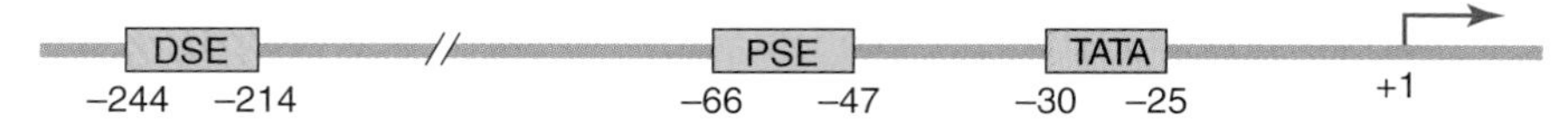

FIGURE 20.22 **Organization of the three general types of promoters used by RNA polymerase III.** Transcription initiation and termination sites are indicated by +1 and Tn, respectively. (a) Type 1 promoter (5S RNA genes)—The A box, intermediate element (IE), and C-box are colored red, blue, and light green, respectively. (b) Type 2 promoter (tRNA genes)—The A and B boxes are colored red and yellow, respectively. (c) Type 3 promoter (mammalian U6 snRNA gene)—The distal sequence element (DSE), proximal sequence element (PSE), and TATA box are colored green, blue, and orange, respectively. (Modified from M. R. Paule and R. J. White, Transcription by RNA polymerases I and III, *Nucleic Acids Res.* 28 [2000]: 1283–1298. Reprinted by permission of Oxford University Press.)

sequence, while the third is upstream from this site and so has no effect on the transcript sequence. We now examine the transcription factors required to recruit RNA polymerase III to these promoters.

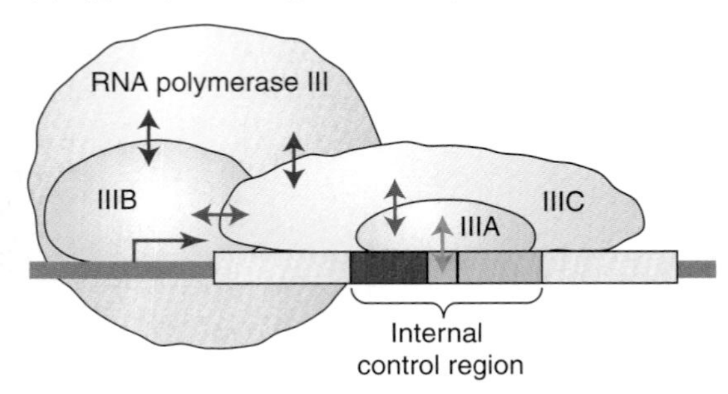

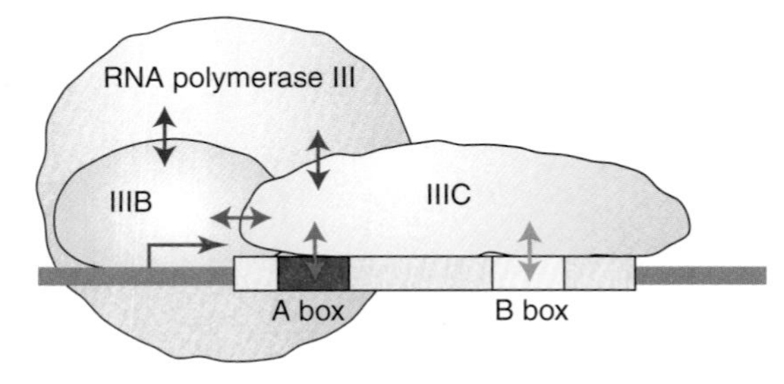

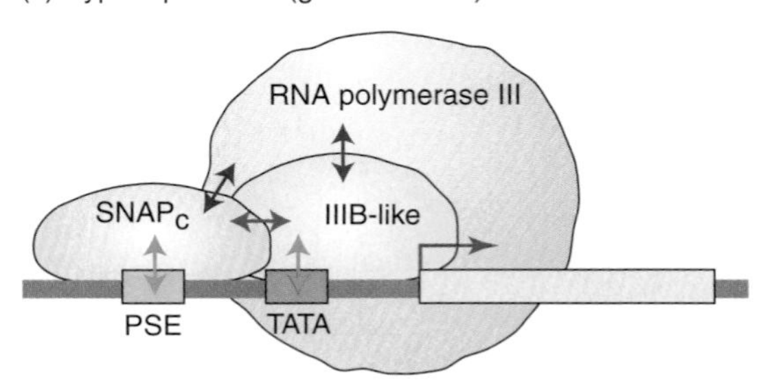

FIGURE 20.23 **Initiation complexes formed on RNA polymerase III type 1, 2, and 3 promoters.** Interactions at the (a) type 1 promoter, (b) type 2 promoter, and (c) type 3 promoter. The distal sequence element or DSE in type 3 promoters, which is not shown in (c), binds additional protein factors that enhance transcription. Green arrows symbolize interactions of DNA binding proteins with promoter elements, blue arrows protein–protein contacts among various transcription factors, and purple arrows contacts between RNA polymerase III and transcription factors. (Adapted from L. Schramm and N. Hernandez, *Genes Dev.* 16 [2002]: 2593–2620.)

The transcription factors required to recruit RNA polymerase III depend on the nature of the promoter.

In groundbreaking experiments performed in 1980, Robert G. Roeder and coworkers used a classical protein fractionation technique to identify transcription factors for RNA polymerase III. Their approach was to place a crude HeLa cell extract on a column filled with phosphocellulose beads and then elute the proteins with increasing potassium chloride concentrations. The three eluted fractions that were needed to support transcription by RNA polymerase III were designated fractions A, B, and C. Transcription from type 1 promoters required all three fractions, whereas transcription from type 2 promoters required only fractions B and C. Type 3 promoters had not yet been discovered. After their discovery, type 3 promoters were shown to require fraction B and a fraction that was eluted from the phosphocellulose column by a solution with a still higher potassium chloride concentration than required to elute fraction C.

Investigators eventually purified and characterized the active transcription factors in these fractions from various biological systems including human and yeast cells. Human transcription factors for RNA polymerase III are designated TFIIIA, TFIIIB, TFIIIC, and SNAP$_C$ (snRNA activating protein complex). FIGURE 20.23 provides simplified schemes for the way in which these transcription initiation factors interact on the three different types of promoters.

Transcription factor interactions at type 1 promoters—Three transcription factors—TFIIIA, TFIIIB, and TFIIIC—are required to transcribe genes containing a type 1 promoter (Figure 20.23a). TFIIIA, the founding member of the C_2H_2 zinc finger family (see Chapter 18), which contains nine zinc fingers, binds to the **internal control region** (**ICR**). Fingers 1 to 3 contact the C box (contributing about 95% of the total binding energy in the full length protein); fingers 7 to 9 appear to contact the A box, and fingers 4 to 6 extend across the region between the C and A boxes. Then TFIIIC binds to the TFIIIA-DNA complex. TFIIIC differs in yeast and humans. Human TFIIIC can be separated into two fractions, TFIIIC1 and TFIIIC2, which are both required to transcribe genes with type 1 promoters. Yeast TFIIIC appears to be a very large protein containing six polypeptides, each of which is essential for cell viability. These polypeptides are arranged into two globular structures with a connecting linker that appears able to stretch. Each globular structure is about 10 nm in diameter and has a molecular mass of about 300 kDa. TFIIIC recruits TFIIIB, which contains TBP and TAFs. Most type 1 promoters lack a TATA box and so TFIIIB recruitment depends on protein-protein interactions with DNA-bound TFIIIC. Once bound, TFIIIB recruits RNA polymerase III to the promoter. TFIIIA and TFIIIC do not appear to be directly involved in recruiting RNA polymerase III. Transcription initiation occurs if they are stripped from the pre-initiation complex provided that TFIIIB is not also removed.

Transcription factor interactions at type 2 promoters—Only two transcription factors, TFIIIC and TFIIIB, are required to transcribe genes containing a type 2 promoter (Figure 20.23b). TFIIIC binds to the B and C boxes and then recruits TFIIIB, which in turn recruits RNA polymerase III.

Transcription factor interactions at type 3 promoters—Only two transcription factors, TFIIIB-like factor and $SNAP_C$ are required to transcribe a type 3 promoter (Figure 20.23c). $SNAP_C$, a multi-subunit protein, binds to the proximal sequence element (PSE) and a TFIIIB-like complex binds to the TATA box. These DNA–protein interactions are stabilized by protein–protein interactions between the two protein complexes. Binding of the two complexes then leads to the binding of RNA polymerase III. The distal sequence element or DSE (not shown in Figure 20.23c) binds additional protein factors that enhance transcription.

The final component of the RNA polymerase machinery to be recruited to each kind of promoter is RNA polymerase III. Yeast RNA polymerase III contains 17 subunits, making it the largest of the three nuclear RNA polymerases. Ten of the 17 subunits are unique to RNA polymerase III. Two of the remaining subunits are also present in RNA polymerase I and the remaining five subunits are present in all three nuclear RNA polymerases. The two largest yeast subunits unique to RNA polymerase III are homologous to Rpb1 and Rpb2 in RNA polymerase II. Human RNA polymerase III has also been purified and with one exception its subunits appear to have yeast subunit counterparts. The one subunit present in yeast but not human RNA polymerase III may be present in the human enzyme but not yet detected because of its small size. Although a crystal structure has not yet been obtained for RNA polymerase III, it seems likely that its core structure will closely resemble the RNA polymerase II structure.

Our understanding of how RNA polymerase III is recruited to the individual promoters is based on *in vitro* experiments in which specific components are added in a stepwise fashion to form the transcription initiation complex. The *in vivo* sequence of events may involve the formation of RNA polymerase III holoenzyme, which then binds to the promoter and starts transcription. Even if RNA polymerase III holoenzyme forms, studying the stepwise process provides important insights into the nature of the interactions among the components of the transcription machinery.

RNA polymerase III does not appear to require additional factors for transcription elongation or termination.

Once the pre-initiation complex is assembled at the promoter, the players are in position to begin transcription elongation. The RNA polymerase III transcription machinery melts the DNA flanking the transcription initiation site. TFIIIB is an active participant in this melting event. The first two nucleotides line up on the now exposed template strand and transcription elongation begins. Almost all of the RNA polymerase III molecules escape from the promoter without significant pausing or arrest. Once a crystal structure becomes available for RNA polymerase III, it will be interesting to compare it with that of RNA polymerase II to see if there are obvious structural differences that explain why the two polymerases differ in their abilities to escape the promoter.

The RNA polymerase III transcription elongation complex moves at about the same rate as the RNA polymerase II transcription elongation complex but, unlike the RNA polymerase II transcription elongation complex, does not appear to require any dedicated factors for transcription elongation. However, a subunit that is unique to RNA polymerase III has significant homology to the elongation factor SII. Perhaps this polymerase subunit eliminates the need for SII.

The bound transcription initiation factors might be expected to block transcription elongation through transcription units with type 1 or 2 promoters. This blocking does not seem to occur, however. Moreover, multiple passages of RNA polymerase through a transcription unit do not remove the assembled transcription initiation factors from the transcription unit. One possibility is that RNA polymerase III transiently displaces one transcription factor as it moves through the transcription unit but protein–protein interactions with other transcription initiation factors permit the displaced factor to remain associated with the promoter.

RNA polymerase III can efficiently terminate transcription in the absence of other factors under *in vitro* conditions. Simple clusters of four or more U residues usually suffice as a termination signal. Further studies are required to determine whether specific termination factors are required *in vivo*.

Pre-tRNAs require extensive processing to become mature tRNAs.

Many of the primary transcripts formed by RNA polymerase III must be processed before they can perform their biological functions. We will consider just one of the processing pathways, tRNA biogenesis. Primary transcripts for tRNA molecules from all organisms have a 5′-leader sequence and a 3′-trailer sequence that must be removed and specific bases that must be modified. In addition, eukaryotic tRNAs require a dedicated nucleotidyl transferase to add CCA to their 3′-ends. This CCA end is included within the primary transcript for most bacterial tRNAs. Finally, some eukaryotic pre-tRNAs also contain introns that must be removed. The eukaryotic tRNA biogenesis pathway has been most extensively studied in yeast. FIGURE 20.24 shows the biogenesis pathway for tRNA in yeast. This pathway is probably also valid for higher organisms because virtually all the yeast components have counterparts in higher organisms. The basic features of the tRNA processing pathway are as follows.

1. **La Protein Binds to Pre-tRNA**—A nuclear phosphoprotein called the La protein binds to the 3′-end of newly synthesized

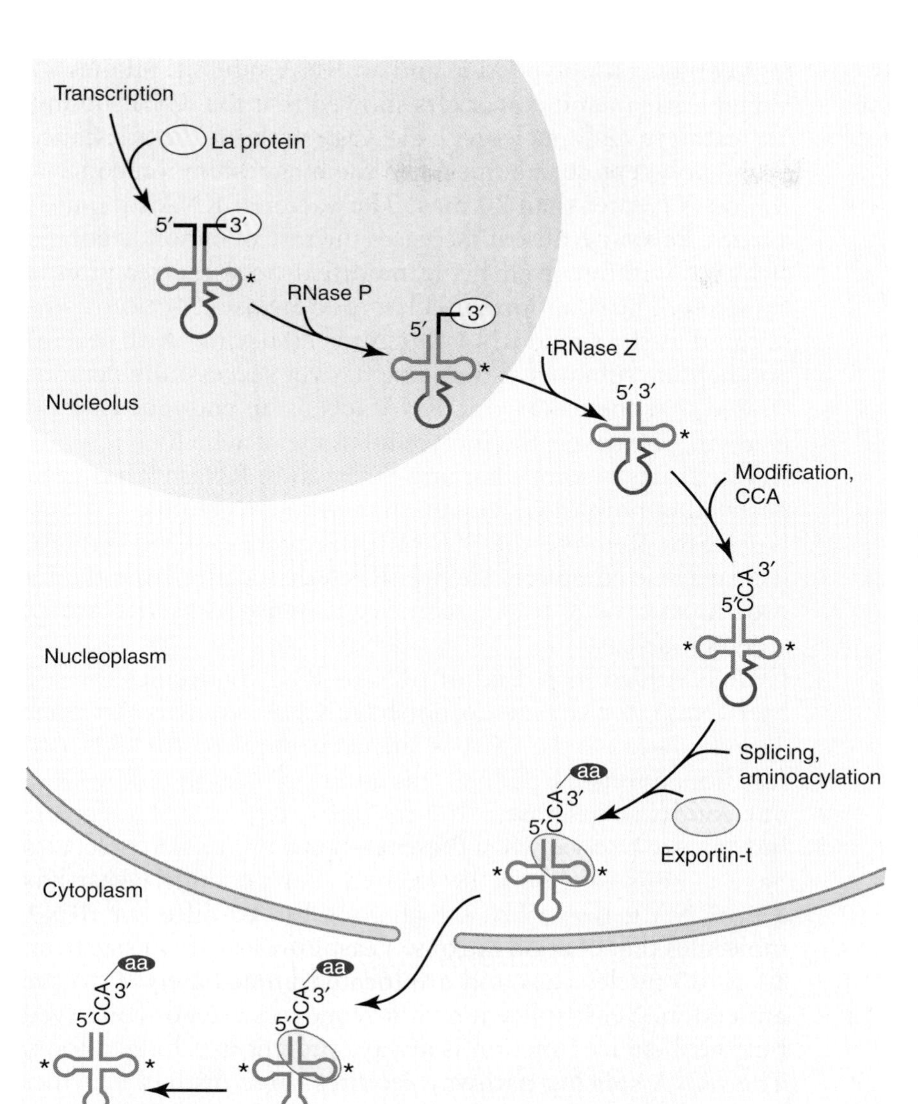

FIGURE 20.24 **A schematic view of a yeast cell showing the pathway followed by an intron-containing pre-tRNA.** The intervening sequence is shown in red, while the 5′-trailer and 3′-leader sequences are shown in black. The newly synthesized pre-tRNA molecule is immediately bound to the La protein (yellow oval). End maturation can take place in either the nucleolus or nucleoplasm. In this schematic, the 3′-trailer sequence is shown to be removed by an endonuclease. It may also be removed by exonuclease digestion, however. Also, in this schematic, end maturation takes place before intron removal, but the order can be reversed. Certain nucleotide modifications (asterisks) take place on the newly synthesized pre-tRNA, others are introduced after the edges are trimmed, and still others are introduced only after the intron is excised. The amino acid (aa) must be added to the mature tRNA before the tRNA can be transported out of the nucleus by exportin-t (green oval). (Adapted from S. L. Wolin, and G. Matera, *Genes Dev.* 13 [1999]: 1–10.)

pre-tRNAs, protecting them from exonucleases. The phosphoprotein's name derives from the fact that it was originally discovered as a target for autoantibodies in patients suffering from systemic lupus erythematosus (La is short for lupus antigen).

2. **RNase P Removes the 5′-Leader Sequence**—The La protein-bound pre-tRNA is cleaved by RNase P (where P stands for precursor), an endonuclease that removes the 5′-leader sequence to generate the 5′-end of tRNA. RNase P was originally isolated from *E. coli* extracts based on its ability to remove the 5′-leader sequence from bacterial pre-tRNA (see Chapter 16). Subsequent studies showed that all organisms and organelles that synthesize tRNA have RNase P activity and that this activity is essential. Bacterial RNase P is a ribonucleoprotein

that contains a polypeptide and an RNA subunit. Studies by Sidney Altman and coworkers showed that the RNA subunit can catalyze endonucleolytic cleavage without the assistance of the polypeptide subunit when the magnesium ion concentration is greater than 20 mM. The bacterial RNA subunit is a true catalyst because it increases the rate of endonucleolytic cleavage without itself being modified once the reaction is complete. The situation for eukaryotic RNase P is more complicated as the single RNA subunit is associated with several polypeptide subunits and no one has yet successfully demonstrated that the eukaryotic RNA acts as an endonuclease on its own. Moreover, the intracellular site at which eukaryotic RNase P acts is somewhat unclear because RNase P has been detected in both the nucleoplasm and the nucleolus.

3. **Nucleases Remove the 3′-Trailer Sequence**—Once RNase P cleavage is complete, the pre-tRNA dissociates from the La protein and the 3′-trailer sequence is removed by tRNase Z, an endonuclease.
4. **CCA Is Added to 3′-End of tRNA**—RNA molecules formed by tRNase Z cleavage do not have CCA sequences at their 3′-ends. Because the CCA sequence is required for tRNA to function, it must be added. This addition is catalyzed by **CCA nucleotidyl transferase**.
5. **Introns Are Excised when Present**—Some pre-tRNA molecules have introns that must be excised. For example, yeast has 272 tRNA genes, 59 of which code for 10 different tRNA molecules that contain introns. Yeast introns vary in size from 14 to 60 nucleotides and are located immediately 3′ to the anticodon. Yeast splice junction sequences are not conserved but the 3′-splice junction is always present in a bulged loop. The tRNA splicing pathway requires three distinct enzymes (FIGURE 20.25). An endonuclease cleaves the pre-tRNA at its two splice sites to produce two tRNA half-molecules and a linear intron. This cleavage generates a 5′-OH end on one half-molecule and a 3′-end with a 2′,3′-cyclic phosphate on the other. Then a second enzyme, designated a ligase even though it catalyzes a few different kinds of reactions, joins the two half-molecules. Intermediate steps in the ligation reaction are as follows: γ-phosphoryl group transfer from GTP to the newly generated 5′-OH terminus, hydrolysis of the 2′,3′-cyclic phosphate to form a 2′-phosphate, adenylyl (AMP) group transfer from ATP to the newly generated 5′-phosphate, and finally ligation. The third enzyme, a 2′-phosphotransferase, removes the 2′-phosphate at the splice junction, transferring it to nicotinamide adenine dinucleotide (NAD^+) to form nicotinamide) and ADP-ribose 1′-2′ cyclic phosphate.
6. **Exportin-t Helps Transport Mature tRNAs to the Cytoplasm**—A specialized receptor called exportin-t binds to mature tRNA and helps to transport that tRNA through the nuclear pore to the cytoplasm. Exportin-t does not bind to tRNAs containing immature 5′- or 3′-ends, helping to ensure that only mature

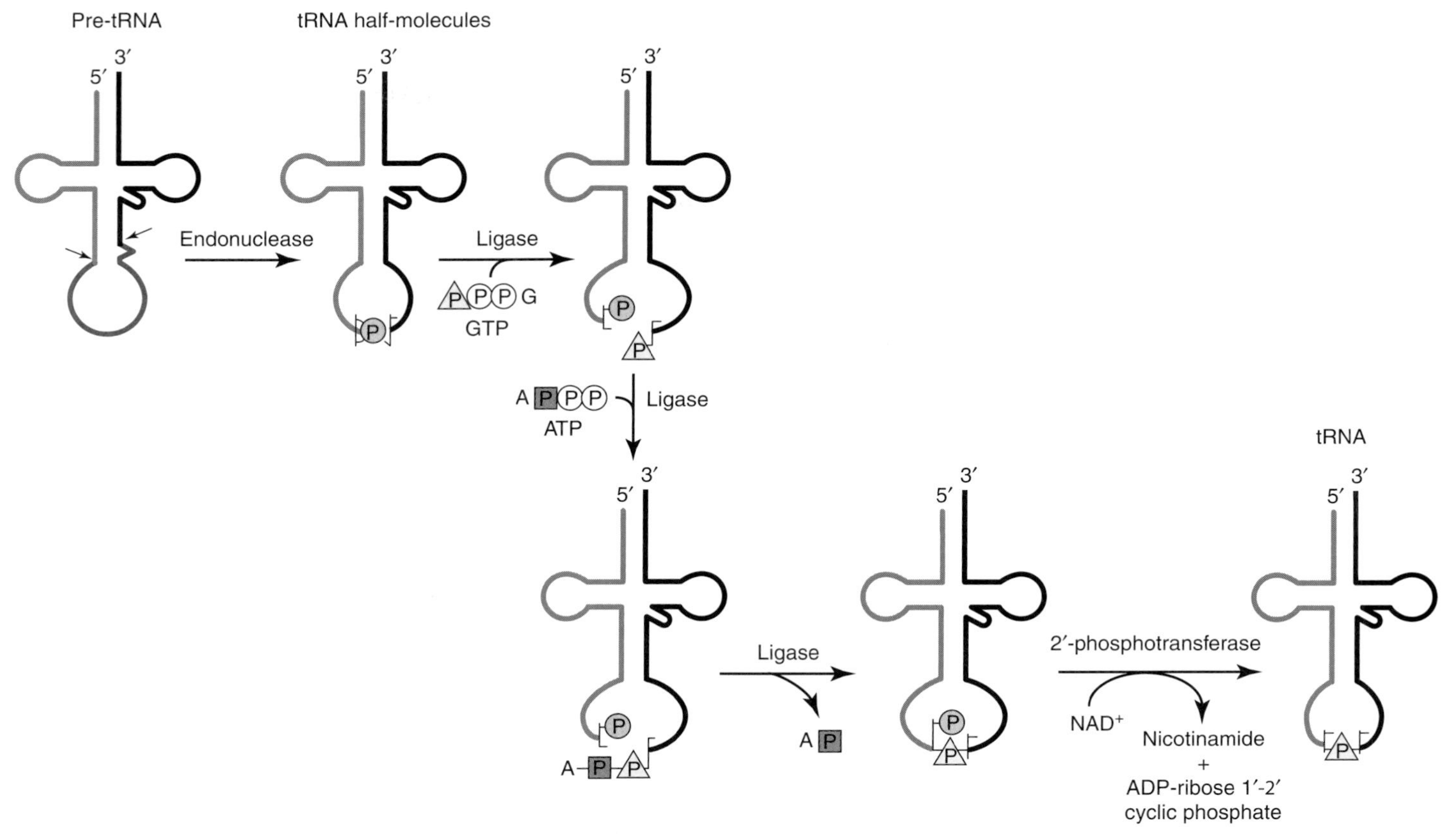

FIGURE 20.25 **Excising an intron from pre-tRNA.** (Adapted from J. Abelson, C. R. Trotta, and H. Li, *J. Biol. Chem.* 273 [1998]: 12685–12688.)

tRNA molecules reach the cytoplasm. An amino acid must also be added to a tRNA before the tRNA can be transported, further ensuring only mature, functional tRNA molecules reach the cytoplasm.

7. **Extensive Nucleotide Modifications Are Introduced**—Transfer RNA has extensive nucleotide modifications. Some of these modifications take place on the newly synthesized pre-tRNA, whereas others take place after the ends are trimmed or the intron is excised.

20.6 Transcription in Mitochondria

Mitochondrial DNA is transcribed to form mRNA, rRNA, and tRNA.

Thus far, our examination of eukaryotic RNA synthesis has been devoted solely to RNA molecules that are synthesized in the cell nucleus. RNA synthesis, however, also takes place in the mitochondria of all eukaryotes and in the chloroplasts of green plants. Each organelle requires a distinct set of enzymes for RNA synthesis. Let's begin our examination of organellar transcription by examining mitochondrial RNA synthesis.

The mitochondrion is the site of many metabolic pathways, most notably oxidative phosphorylation, the citric acid cycle, and fatty acid oxidation. Mitochondria are thought to have originated from bacteria that were engulfed by primitive eukaryotic cells. Most of the genes that were present in the bacterial precursor are believed to have been transferred to chromosomes in the cell nucleus as eukaryotic cells evolved. The majority of polypeptides present in mitochondria are encoded by these nuclear genes, synthesized in the cytoplasm, and then imported into the mitochondria. Mitochondria, however, still retain many genes that are essential for mitochondrial propagation and function. These genes are present in mitochondrial DNA (mtDNA) that must be replicated and transcribed. Thus, all mitochondria arise from preexisting mitochondria. Although both egg and sperm cells of higher organisms contain mitochondria, the number of mitochondria present in egg cells is several orders of magnitude higher than that in sperm cells (about 10^5 mitochondria per mammalian egg cell versus 50–75 mitochondria per mammalian sperm cell).

Investigators initially thought that the difference in the number of mitochondria present in the two types of germ cells might explain why mitochondrial genes seem to be maternally inherited. Studies now show that the situation is considerably more complicated, however, and that a regulatory mechanism exists that somehow blocks the propagation of paternal mitochondria as the fertilized egg develops. In rare instances, paternal mitochondria do appear to survive and are present in the offspring's mitochondria. Although mitochondrial size, shape, and number vary from one kind of organism to another as well as among different cell types in the same organism, all mitochondria share characteristic features (FIGURE 20.26). One of the most important of these features is an outer membrane that separates the mitochondrion from the cytoplasm and an inner membrane with many inward folds called cristae, which surround a compartment called the mitochondrial matrix. Mitochondrial DNA and the mitochondrial protein synthetic machinery are present in the mitochondrial matrix.

In 1981, Ian G. Young and coworkers reported the first complete sequence for human mtDNA. A simplified map of human mitochondrial DNA is shown in FIGURE 20.27. Each human mitochondrion contains about two to ten copies of this DNA, which is a closed covalent circle containing 16,569 bp. The two mtDNA strands differ in G + T base composition, permitting "heavy" (H) and "light" (L) strands to be separated by cesium chloride density gradient centrifugation. Most genetic information is specified by the H strand, which contains genes for two rRNAs, 14 tRNAs, and 12 polypeptides. The L strand codes for eight tRNAs and one polypeptide. Human mtDNA transcription initiation sites and promoters have been determined by a number of different methods including S1 nuclease analysis (see Chapter 17) and linker scanning mutagenesis (see Chapter 18). These techniques show the existence of an L strand promoter, P_L, and two H strand promoters. For simplicity, the two H strand promoters, which are very close to one another are shown as P_H. P_L and P_H direct transcription in opposite directions.

In the so-called **strand asymmetric model** of mtDNA replication, the RNA transcript initiated at P_L is cleaved in the vicinity of three

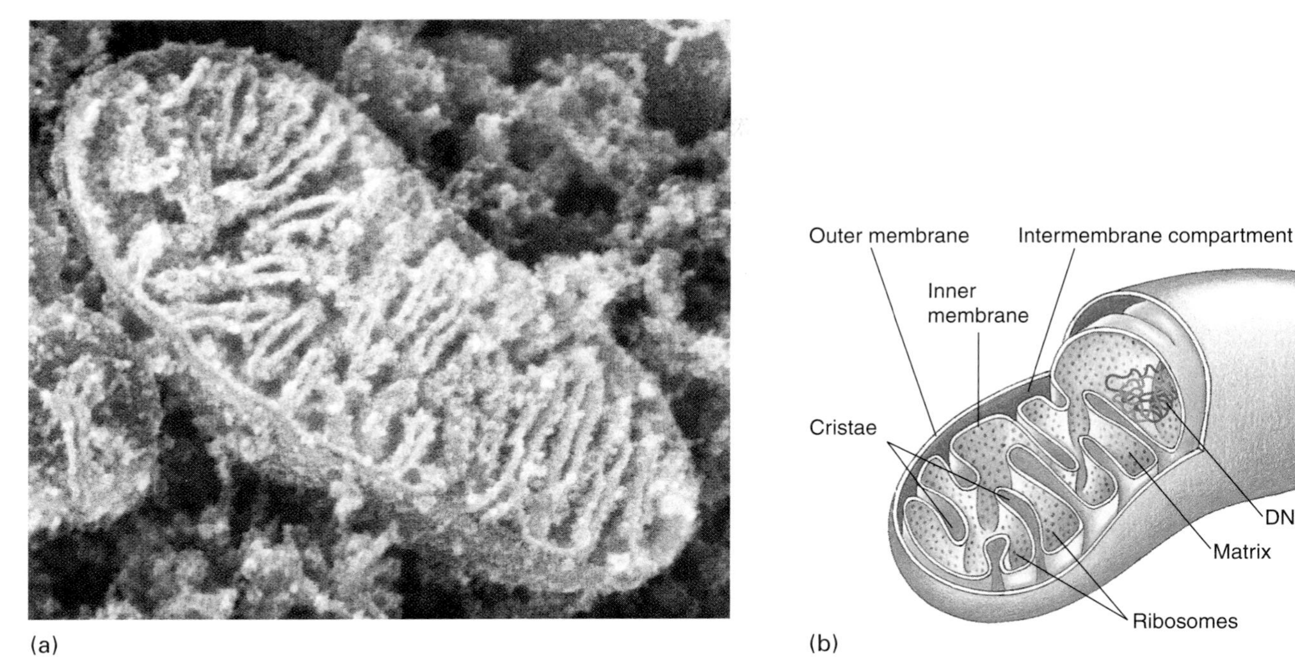

FIGURE 20.26 **Structure of a mitrochondrion.** (a) The electron micrograph shows the extensive infolding of the inner membrane; the folds are called cristae. The membrane structure surrounding the mitochondrion is the rough endoplasmic reticulum. Mitochondrial DNA and ribosomes are present in the matrix, the compartment surrounded by the inner membrane. (b) A schematic diagram of the mitochondrion. (Photo © Dr. David Furness, Keele University/Photo Researchers, Inc.)

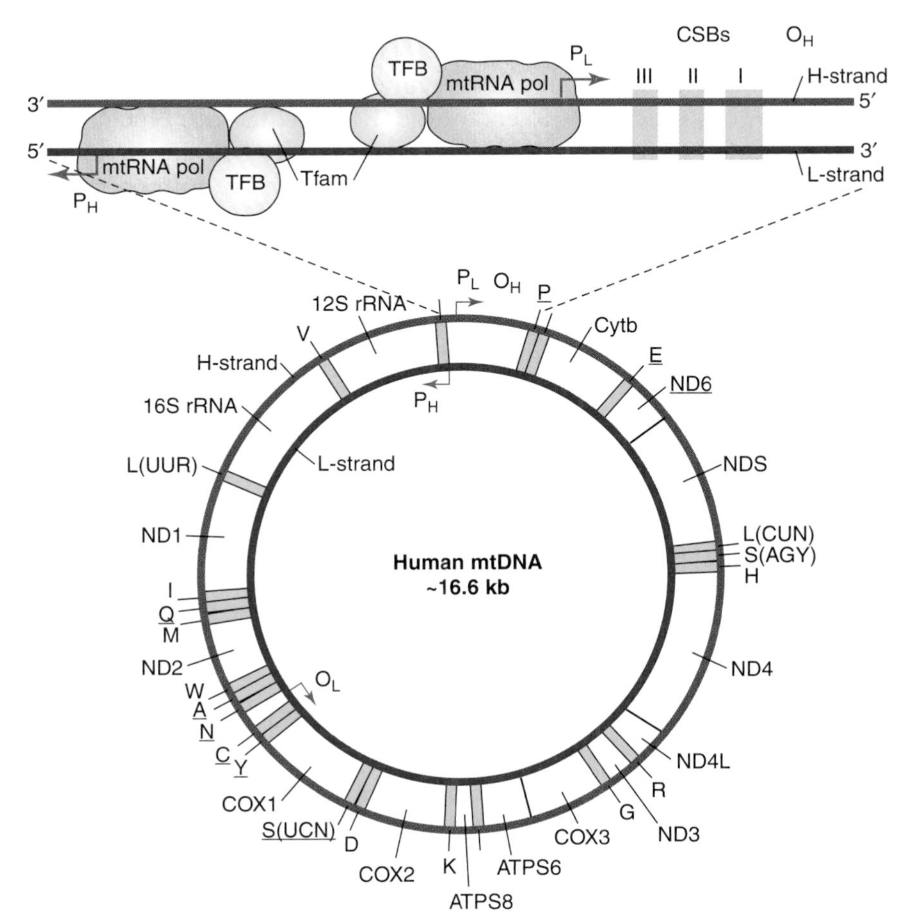

FIGURE 20.27 **Human mitochondrial DNA (mtDNA).** The genomic organization and structural features of human mtDNA are shown in a circular genomic map. The regulatory region, which is shown in expanded form, above, contains the L- and H-strand promoters (P_L and P_H, respectively) and the origin of H-strand replication (O_H). Mitochondrial RNA polymerase and two mitochondrial transcription factors, Tfam and TFB, are shown in the expanded region along with the conserved sequence blocks (CSB I, II, and III). Protein coding and rRNA genes are interspersed with 22 tRNA genes (denoted by the single-letter amino acid code). The origin of L-strand replication (O_L), which is displaced by approximately two thirds of the genome with respect to O_H, is within a cluster of five tRNA genes. Protein-coding genes include cytochrome oxidase (COX) subunits 1, 2, and 3; NADH dehydrogenase (ND) subunits 1, 2, 3, 4, 4L, 5, and 6; ATP synthase (ATPS) subunits 6 and 8; cytochrome b (cytb). ND6 and the eight tRNA genes encoded on the L-strand are underlined; all other genes are encoded on the H-strand. (Modified from D. P. Kelly and R. C. Scarpulla, *Genes Dev.* 18 [2004]: 357–368. Copyright 2004, Cold Spring Harbor Laboratory Press. Used with permission of Daniel P. Kelly, Sanford-Burnham Medical Research Institute at Lake Nona.)

evolutionarily conserved sequence blocks (CSB I, II, and III), and H-strand replication initiates at the sites of these cleavages. If this cleavage does not take place, then transcription continues beyond the CSB region, allowing downstream transcription units to be expressed. If cleavage does take place, then the cleaved transcript serves as a primer for the replication of the H-strand. Regulation of the cleavage process, therefore, helps to determine whether transcription or replication takes place. Mitochondrial transcription and DNA replication are coupled processes. After DNA synthesis begins, the nascent strand frequently is terminated after a conserved element known as a termination-associated sequence (not shown in Figure 20.27).

Human mitochondrial RNA polymerase is a single polypeptide (molecular mass = 120 kDa) that is homologous with T7 phage RNA polymerase (see Chapter 15, Figure 15.3) and requires the assistance of two transcription factors, **Tfam** and **TFB**, for accurate transcription (Figure 20.27). The L- and H-strands are each transcribed as a single RNA molecule that includes most, if not all, of the genetic information on the template strand. Transcription termination requires a termination factor that is designated **mTERF1**.

Primary transcripts in mammalian mitochondria do not contain introns and have only a limited number of spacer sequences. Transfer RNA coding sequences flank the two rRNA coding sequences and nearly all of the polypeptide coding sequences. Based on this remarkable genetic organization, it appears that the secondary structure of tRNA sequences provides the necessary information required for precise endonucleolytic cleavage. A mitochondrial RNase P catalyzes cleavage at the 5′-end of the tRNA sequences and tRNase Z appears to catalyzes cleavage at the 3′-end of the tRNA sequences. Nucleotidyl transferases add CCA to the 3′-end of the pre-tRNA molecule and poly(A) polymerase adds about 55 A residues to the mitochondrial mRNA and one to ten A residues to the mitochondrial rRNA.

Primary transcripts synthesized by yeast and other fungi as well as by plant mitochondria contain group II introns. Hence, processing these transcripts is considerably more complex than processing the transcripts produced by mammalian mitochondria. Group II splicing involves two transesterification steps that are similar to those described for nuclear pre-mRNA processing (see Chapter 19). Although many of the mitochondrial group II introns have the ability to self-splice *in vitro*, they nevertheless require the assistance of protein factors *in vivo*.

20.7 Transcription in Chloroplasts

Chloroplast DNA is also transcribed to form mRNA, rRNA, and tRNA.

Chloroplasts, which are the site of all photosynthetic reactions in green plants, can synthesize RNA. Each chloroplast has an outer membrane that separates the chloroplast from the cytoplasm and an

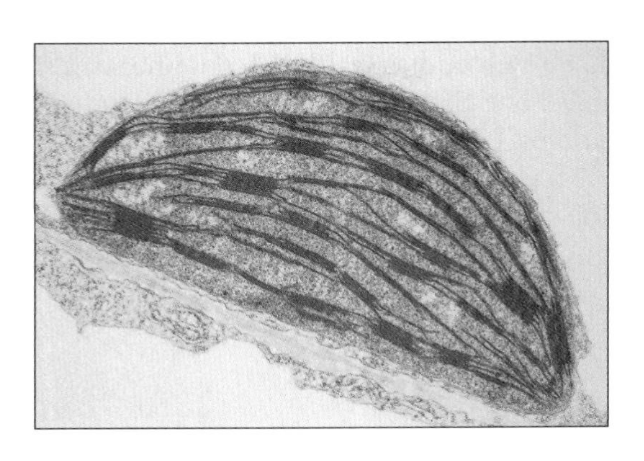

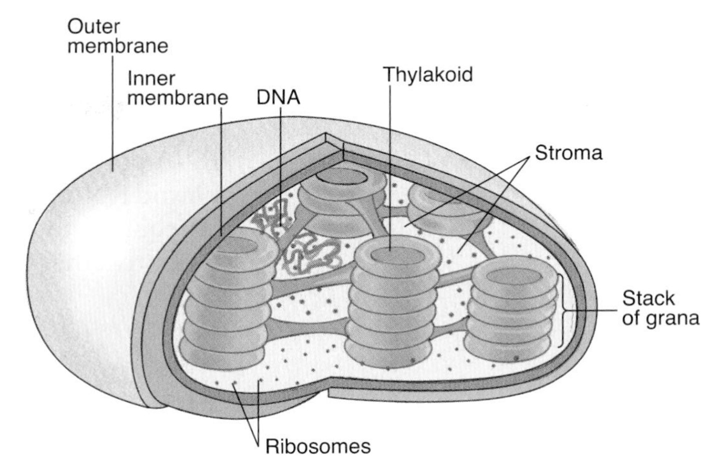

FIGURE 20.28 **Schematic of a chloroplast.** (Part a © Dr. Jeremy Burgess/Photo Researchers, Inc. Part b adapted from an illustration by Bob Hartzler, Iowa State University.)

inner membrane that surrounds a compartment called the stroma (FIGURE 20.28). Chloroplast DNA and the chloroplast protein synthetic machinery are present in the stroma. The stroma also contains flat membranous sacs called thylakoids, which appear to be surrounded by a single continuous membrane that is the site of light-dependent photosynthetic reactions. The chloroplast is believed to have originated as a cyanobacterial cell that somehow entered a eukaryotic cell. Because the cyanobacteria chromosome codes for more than 3000 different polypeptides and chloroplast DNA codes for only about 75 polypeptides, most protein coding genes were either transferred to nuclear chromosomes or lost during evolution. This transfer or loss probably took place very early in plant evolution because chloroplasts of evolutionarily distant green plants have similar gene content and organization. All chloroplasts arise from preexisting chloroplasts.

Chloroplast DNA is a circular double-stranded negative supercoil that is about 120 to 180 kbp long. A typical chloroplast may contain as few as ten or as many as a few hundred DNA copies that are organized into nucleoids associated with the inner chloroplast membrane. In addition to their protein coding genes, chloroplast DNA from higher plants also contain four rRNA genes and 30 tRNA genes. No RNA is imported into chloroplasts. Chloroplast genes of higher green plants often contain a single group I or group II intron.

Chloroplast DNA codes for an RNA polymerase that resembles the bacterial core RNA polymerase ($\alpha_2\beta\beta'$) and contains homologous subunits to α, β, and β' (see Chapter 15). The plastid encoded RNA polymerase requires the assistance of a transcription factor that resembles σ^{70} to transcribe chloroplast genes. The chloroplast has at least six different σ factors. At least three of these factors are encoded by nuclear genes, synthesized in the cytoplasm, and then imported into the chloroplast. The chloroplast holoenzyme (sigma factor + core RNA polymerase) acts on chloroplast genes with promoters that have a –10 box and a –35 box. At least one chloroplast σ factor recognizes *E. coli* promoters.

Chloroplasts also have two additional RNA polymerases that are encoded by nuclear genes. The first of these, **nuclear encoded plastid RNA polymerase**, resembles T7 phage RNA polymerase. The second, **plastid RNA polymerase** has not yet been characterized. Further work is required to determine how the three different RNA polymerases work together in chloroplast gene transcription.

Suggested Reading

RNA Polymerase I

Birch, J. L., and Zomerdijk, C. B. M. 2008. Structure and function of ribosomal RNA gene chromatin. *Biochem Soc Trans* 36:619–624.

Caburet, S., Conti, C., Schurra, C., et al. 2005. Human ribosomal RNA gene arrays display a broad range of palindromic structures. *Genome Res* 15:1079–1085.

Carter, R., and Drouin, G. 2009. Structural differentiation of the three eukaryotic RNA polymerases. *Genomics* 94:388–396.

Chen, W., Bucaria, J., Band, D. A., et al. 2000. Enp1, a yeast protein associated with U3 and U14 snoRNAs, is required for pre-rRNA processing and 40S subunit synthesis. *Nucl Acid Res* 31:690–699.

Comai, L. 2004. Mechanism of RNA polymerase I transcription. *Adv Protein Chem* 67:123–155.

De Carlo, S., Carles, C., Riva, M., and Schultz, P. 2003. Cryo-negative staining reveals conformational flexibility within yeast RNA polymerase I. *J Mol Biol* 329:891–902.

Drygin, D., Rice, W. G., and Grummt, I. 2010. The RNA polymerase I transcription machinery: an emerging target for the treatment of cancer. *Ann Rev Pharmacol Toxicol* 50:131–156.

Haag, J. R., and Pikaard, C. S. 2007. RNA polymerase I: a multifunctional molecular machine. *Cell* 1224–1225.

Hanada, K., Song, C. Z., Yamamoto, K, et al. 1996. RNA polymerase I associated factor 53 binds to the nucleolar transcription factor UBF and functions in specific rDNA transcription. *EMBO J* 15:2217–2226.

Kiss, T. 2001. Small nucleolar RNA-guided post-transcriptional modification of cellular RNAs. *EMBO J* 20:3617–3622.

Kuhn, C.-D., Geiger, S. R., Baumli, S., et al. 2007. Functional architecture of RNA polymerase I. *Cell* 131:1260–1272.

McStay, B., and Grummt, I. 2008. The Epigenetics of rRNA genes: from molecular to chromosome biology. *Ann Rev Cell Dev Biol* 24:131–157.

Moss, T., Langlois, F., Gagnon-Kugler, T., and Stefanovsky, V. 2007. A housekeeper with power of attorney: the rRNA genes in ribosome biogenesis. *Cell Mol Life Sci* 64:29–49.

Moss, T., and Stefanovsky, V. Y. 2002. At the center of eukaryotic life. *Cell* 109:545–548.

Moss, T., Stefanovsky, V., Langlois, F., and Gagnon-Kugler, T. 2006. A new paradigm for the regulation of the mammalian ribosomal RNA genes. *Biochem Soc Trans* 34:1079–1081.

Paule, M. R. (ed.). 1999. *Transcription of Ribosomal RNA Genes by Eukaryotic RNA Polymerase I*. New York: Springer.

Paule, M. R., and White, R. J. 2000. Transcription by RNA polymerases I and III. *Nucl Acid Res* 28:1283–1298.

Reeder, R. H., and Lang, W. H. 1997. Terminating transcription in eukaryotes: lessons learned from RNA polymerase I. *Trends Biochem Sci* 22:473–477.

Russell, J., and Zomerdijk, J. C. B. M. 2005. RNA polymerase-I-directed rDNA transcription, life and works. *Trends Biochem Sci* 30:87–96.

Sanij, E., and Hannan, R. D. 2009. The role of UBF in regulating the structure and dynamics of transcriptionally active rDNA chromatin. *Epigenetics* 4:374–382.

Sanij, E., Poortingo, G., Sharkey, K., et al. 2008. UBF levels determine the number of active ribosomal RNA genes in mammals. *J Cell Biol* 183:1258–1274.

Thiry, M., and Lafontaine, D. L. J. 2005. Birth of a nucleolus: the evolution of nucleolar compartments. *Trends Cell Biol* 15:194–199.

Triezenberg, S. J., Rushford, C., Hart, R. P., et al. 1982. Structure of the Syrian hamster ribosomal DNA repeat and identification of homologous and nonhomologous regions shared by human and hamster ribosomal DNAs. *J Biol Chem* 257:7826–7833.

Werner, M., Thuriaux, P., and Soutourina, J. 2009. Structure-function analysis of RNA polymerases I and III. *Curr Opin Struct Biol* 19:740–745.

White, R. J. 2005. RNA polymerases I and III, growth control and cancer. *Nat Rev Mol Cell Biol* 6:69–78.

Yamamoto, K., Yamamoto, M., Hanada, K., et al. 2004. Multiple protein-protein interactions by RNA polymerase I-associated factor PAF49 and role of PAF49 in rRNA transcription. *Mol Cell Biol* 24:6338–6349.

Self-Splicing Ribosomal RNA

Cech, T. R. 1987. The chemistry of self-splicing RNA and RNA enzymes Tetrahymena. *Science* 236:1532–1539.

Doudna, J. A., and Cech, T. R. 2002.The chemical repertoire of natural ribozymes. *Nature* 418:222–228.

Paquin, B., and Shub, D. A. 2001. Introns: group I structure and function. pp. 1–8. In: *Encyclopedia of Life Sciences*. London, UK: Nature Publishing Group.

Woodson, S. A. 2005. Structure and assembly of group I introns. *Curr Opin Struct Biol* 15:324–330.

Ribosome Assembly

Dragon, F., Gallagher, J. E. G., Compagnone-Post, P. A., et al. 2002. A large nucleolar U3 ribonucleoprotein required for 18S ribosomal RNA biogenesis. *Nature* 417:967–970.

Grandi, P., Rybin, V., Bassler, J., et al. 2002. 90S Pre-ribosomes include the 35S pre-rRNA, the U3 snoRNP, and 40S subunit processing factors but predominantly lack 60S synthesis factors. *Mol Cell* 10:105–115.

Granneman, S., and Baserga, S. J. 2004. Ribosome biogenesis: of knobs and RNA processing. *Exp Cell Res* 296:43–50.

Huber, L. A. 2003. Is proteomics heading in the wrong direction? *Nat Rev Mol Cell Biol* 4:74–80.

Kressler, D., Linder, P., and De La Cruz, J. 1999. Protein *trans*-acting factors involved in ribosome biogenesis in *Saccharomyces cerevisiae*. *Mol Cell Biol* 19:7897–7912.

Panse, V. G., and Johnson, A. W. 2010. Maturation of eukaryotic ribosomes: acquisition of functionality. *Trends Biochem Sci* 35:260–266.

Schäfer, T., Strauss, D., Petfalski, E., et al. 2003. The path from nucleolar 90S to cytoplasmic 40S pre-ribosomes. *EMBO J* 22:1370–1380.

Trapman, J., Retel, J., and Planta, R.J. 1975. Ribosomal precursor particles from yeast. *Exp Cell Res* 90:95–104.

Tschochner, H., and Hurt, E. 2003. Pre-ribosomes on the road from the nucleolus to the cytoplasm. *Trend Cell Biol* 13:255–263.

Udem, S.A., and Warner, J.R. 1972. Ribosomal RNA synthesis in *Saccharomyces cerevisiae*. *J Mol Biol* 65:227–242.

Venema, J., and Tollervey, D. 1999. Ribosome synthesis in *Saccharomyces cerevisiae*. *Ann Rev Genet* 33:261–311.

Warner, J. R. 2001. Nascent ribosomes. *Cell* 107:133–136.

RNA Polymerase III

Abelson, J., Trotta, C. R., and Li, H. 1998. tRNA splicing. *J Biol Chem* 273:12685–12688.

Barski, A., Chepelev, I., and Liko, D., et al. 2010. Pol II and its associated epigenetic marks are present at Pol III–transcribed noncoding RNA genes. *Nat Struct Mol Biol* 17:629–634.

Canella, D., Praz, V., Reina, J. H., et al. 2010. Defining the RNA polymerase III transcriptome: genome-wide localization of the RNA polymerase III transcription machinery in human cells. *Genome Res* 20:710–721.

Cramer, P. 2006. Recent structural studies of RNA polymerase II and III. *Biochem Soc Trans* 34:1058–1061.

Haeusler, R. A., and Engelke, D. R. 2006. Spatial organization of transcription by RNA polymerase III. *Nucl Acid Res* 34:4826–4836.

Harismendy, O., Gendrel, C. G., Soularue, P., et al. 2003. Genome-wide location of yeast RNA polymerase III transcription machinery. *EMBO J* 22:4738–4747.

Moqtaderi, Z., Wang, J., and Raha, D. 2010. Genomic binding profiles of functionally distinct RNA polymerase III transcription complexes in human cells. *Nat Struct Mol Biol* 17:635–640.

Oler, A. J., Alla, R. K., Roberts, D. N., et al. 2010. Human RNA polymerase III transcriptomes and relationships to Pol II promoter chromatin and enhancer-binding factors. *Nat Struct Mol Biol* 17:620–628.

Paule, M. R., and White, R. J. 2000. Transcription by RNA polymerases I and III. *Nucl Acid Res* 28:1283–1298.

Schramm, L., Hernandez, N. 2002. Recruitment of RNA polymerase III to its target promoters. *Gene Dev* 16:2593–2620.

Späth, B., Canino, G., and Marchfelder, A. 2007. tRNase Z: the end is not in sight. *Cell Mol Life Sci* 64:2404–2412.

Werner, M., Thuriaux, P., and Soutourina, J. 2009. Structure-function analysis of RNA polymerases I and III. *Curr Opin Struct Biol* 19:740–745.

White, R. J. 1998. *RNA Polymerase III Transcription*, 2nd ed. New York: Springer-Verlag.

White, R. J. 2005. RNA polymerases I and III, growth control and cancer. *Nat Rev Mol Cell Biol* 6:69–78.

Wolin, S. L., and Matera, A. G. 1999. The trials and travels of tRNA. *Genes Dev* 13:1–10.

Transcription in Mitochondria

Gaspari, M., Falkenberg, M., Larsson, N. G., and Gustafsson, C. M. 2004. The mitochondrial RNA polymerase contributes critically to promoter specificity in mammalian cells. *EMBO J* 23:4606–4614.

Gaspari, M., Larsson, N. G., and Gustafsson, C. M. 2004. The transcription machinery in mammalian mitochondria. *Biochim Biophys Acta* 1659:148–152.

Garesse, R., and Vallejo, C. G. 2001. Animal mitochondrial biogenesis and function: a regulatory cross-talk between two genomes. *Gene* 263:1–16.

Kelly, D. P., and Scarpulla, R. C. 2004. Transcriptional regulatory circuits controlling mitochondrial biogenesis and function. *Genes Dev* 18:357–368.

Scarpulla, R. C. 2008. Transcriptional paradigms in mammalian mitochondrial biogenesis and function. *Physiol Rev* 88:611–638.

Shade, G. S. 2004. Coupling the mitochondrial transcription machinery to human disease. *Trends Genet* 20:513–519.

Shoubridge, E. A. 2002. The ABCs of mitochondrial transcription. *Nat Genet* 31:227–228.

Smith, A. J., Staden, R., and Young, I. G. 1981. Sequence and organization of the human mitochondrial genome. *Nature* 290:457–465.

Taanman, J.-W. 1999. The mitochondrial genome: structure, transcription, translation and replication. *Biochim Biophys Acta* 1410:103–123.

Yakubovskaya, E., Mejia, E., Byrnes, J., et al. 2010. Helix unwinding and base flipping enable human MTERF1 to terminate mitochondrial transcription. *Cell* 141: 982–993.

Chloroplast Transcription

Allison, L. A. 2004. The role of sigma factors in plastid transcription. *Biochemie* 92:537–548.

Hess, W. R., and Borner, T. 1999. Organellar RNA polymerases of higher plants. *Int Rev Cytol* 190:1–59.

Kanamaru, K., and Tanaka, K. 2004. Roles of chloroplast RNA polymerase sigma factors in chloroplast development and stress response in higher plants. *Biosci Biotechnol Biochem* 68:2215–2223.

Nickelsen, J. 2003. Chloroplast RNA-binding proteins. *Curr Genet* 43:392–399.

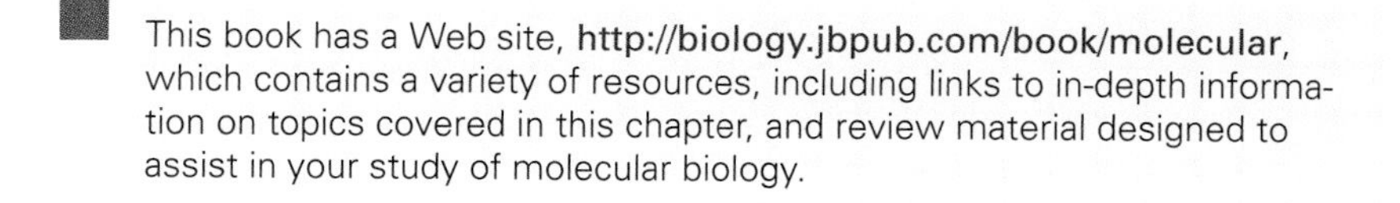

This book has a Web site, **http://biology.jbpub.com/book/molecular**, which contains a variety of resources, including links to in-depth information on topics covered in this chapter, and review material designed to assist in your study of molecular biology.

21 Small Silencing RNAs

OUTLINE OF TOPICS

21.1 RNA Interference (RNAi) Triggered by Exogenous Double-Stranded RNA

The nematode worm *Caenorhabditis elegans* is an attractive organism for molecular biology studies.

RNA interference was discovered in the nematode worm *Caenorhabditis elegans*.

In vitro studies helped to elucidate the RNA interference pathway.

Dicer acts as a molecular ruler that generates double-stranded RNA fragments of discrete size.

The guide strand's 5′-phosphate and 3′-ends bind to sites in the Argonaute protein's Mid and PAZ domains, respectively.

RISC loading complex is required for siRISC formation.

RNAi blocks virus replication and prevents transposon activation.

21.2 Transitive RNAi

In some organisms, RNA interference that starts at one site spreads throughout the entire organism.

SID-1, an integral membrane protein in *C. elegans*, assists in the systemic spreading of the silencing signal.

ERI-1, a 3′→5′ exonuclease in *C. elegans*, appears to be a negative regulator of RNA interference.

21.3 RNAi as an Investigational Tool

RNAi is a powerful tool for investigating functional genomics.

21.4 The MicroRNA Pathway

The miRNA pathway blocks mRNA translation or causes mRNA degradation.

21.5 Piwi Interacting RNAs (piRNAs)

piRNAs help to maintain the germline stability in animals.

21.6 Endogenous siRNA

Animals have a functional endogenous siRNA pathway.

21.7 Heterochromatin Assembly

The RNAi machinery can establish and maintain heterochromatin.

Suggested Reading

As the last millennium drew to a close it appeared that nearly all regulatory factors that influence gene expression were proteins. Then in 1998, Andrew Z. Fire, Craig C. Mello, and coworkers discovered that long double-stranded RNAs can cause sequence-specific gene silencing in the nematode worm *Caenorhabditis elegans*. At its most basic level, the process that Fire and Mello discovered involves the conversion of a long double-stranded RNA into a small single-stranded RNA of defined size that is incorporated into a ribonucleoprotein complex. The small single-stranded RNA guides the complex to a target mRNA and a protein in the complex cleaves the mRNA. Subsequent studies revealed the existence of other eukaryotic pathways to generate small ribonucleoprotein complexes that regulate gene expression. Some of these complexes block mRNA translation and others alter chromatin structure. The discovery that small RNA molecules can regulate gene expression has forced us to revise our views on the mechanisms that regulate gene expression, led to the development of new techniques to study gene function, and promises to radically change the way that clinicians diagnose and treat cancer and various other genetic diseases. We begin by examining the pioneering work by Fire, Mello, and colleagues and then examine related phenomena.

21.1 RNA Interference (RNAi) Triggered by Exogenous Double-Stranded RNA

The nematode worm *Caenorhabditis elegans* is an attractive organism for molecular biology studies.

Sydney Brenner first suggested that the nematode worm would be an excellent model system for studying the molecular biology of a multicellular animal in 1963. He outlined the worm's virtues as follows: It is a "multicellular organism which has a short life cycle, can be easily cultivated, and is small enough to be handled in large numbers, like a micro-organism." Brenner also pointed out that the worm has relatively few cells so that exhaustive studies of lineage and patterns could be made and that the worm should be amenable to genetic analysis. His predictions proved to be correct in all respects.

The adult nematode worm *C. elegans* is about 1 mm long and so is just visible to the naked eye. Each adult worm exists as either a male or hermaphrodite (produces sperm and eggs). Male worms have exactly 1,031 somatic cells and hermaphrodites have about 1,000 cells. As Brenner predicted, the lineage of each adult cell has been established. The schematic diagram in FIGURE 21.1 shows some important body parts. The worm feeds on *Escherichia coli* and other bacteria, ingesting and grinding the bacteria in its pharynx (mouth piece) and absorbing nutrients from the resulting extract in its intestine. The worm is estimated to have about 19,000 genes, about 13% of which are organized into operons that each contain between 2 and 8 genes. RNA polymerase II transcribes each operon from a single promoter to

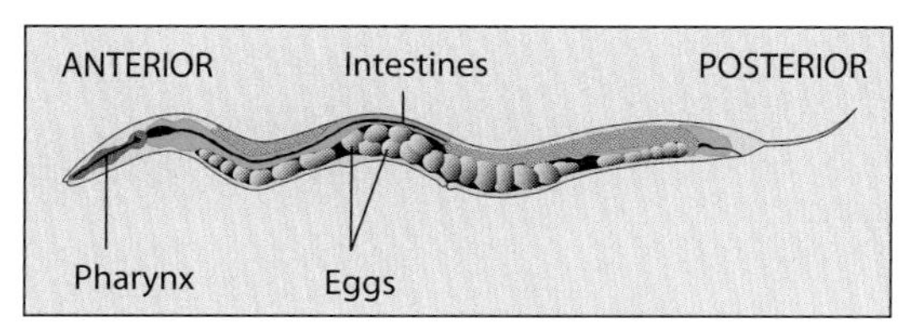

FIGURE 21.1 **A schematic showing some *C. elegans* body parts.**

produce a primary transcript, which then is processed to form mature mRNAs that each codes for a single polypeptide during translation (FIGURE 21.2). *C. elegans* remains an attractive organism for molecular biology studies because: (1) approximately 50-60% of the worm genes are homologous to human genes; (2) the worm is transparent throughout its life, permitting investigators to follow cell development; (3) the worm develops rapidly at 25°C so that a single egg can mature into an adult worm in about 3 days; (4) a single worm can produce about 200 to 400 progeny; (5) worms can be stored in a frozen state at –80°C; and (6) worms are accessible to genetic manipulation.

RNA interference was discovered in the nematode worm *Caenorhabditis elegans*.

When Fire, Mello, and coworkers began their studies, molecular biologists knew that they could block specific gene expression in *C. elegans* by injecting the worm with sense or antisense RNA and that the inhibition would persist well into the next generation. This persistence seemed inconsistent with the fact that cells rapidly degrade single-stranded RNA molecules such as mRNA. Furthermore, it was difficult to understand how a sense RNA strand could block gene expression. Fire, Mello, and coworkers therefore suspected that the single-stranded RNAs that investigators used to inject worms were probably contaminated with some biologically active double-stranded RNA molecules. They decided to test their hypothesis by using purified single- and double-stranded RNAs to study the expression of a nonessential myofilament gene, *unc-22*, in *C. elegans*. Mutant *unc-22* worms

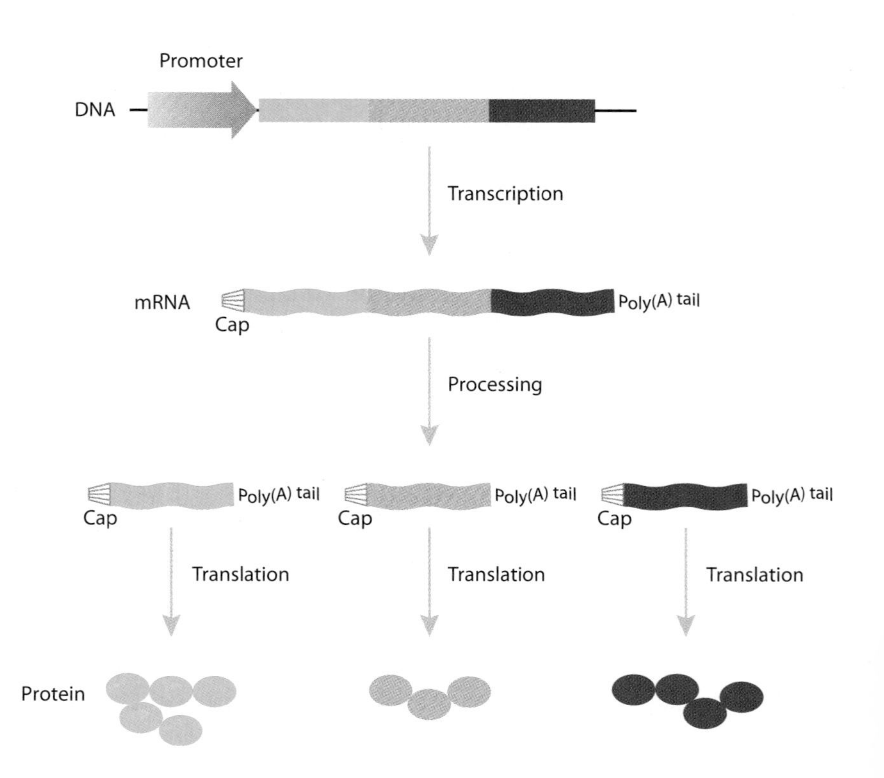

FIGURE 21.2 **Transcription and translation from a *C. elegans* operon.** Genes in a single operon are transcribed starting at a single promoter to form a pre-mRNA. This pre-mRNA is processed to form mature mRNAs that are each translated into a unique polypeptide.

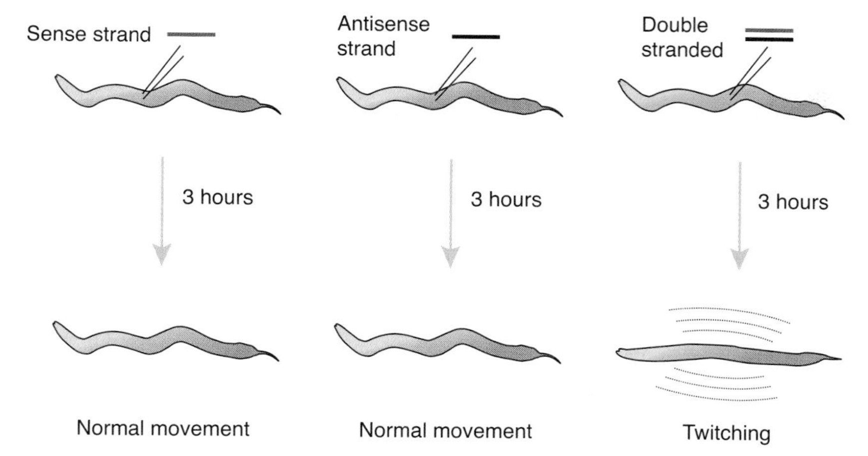

FIGURE 21.3 **RNA interference.** *C. elegans* were injected with sense RNA, antisense RNA, and double-stranded RNA corresponding to the *unc*-22 gene. The movement of the injected worms was monitored. After three hours, very few worms injected with the sense or antisense strands exhibited uncoordinated movement due to twitching. In contrast, nearly all the worms injected with double-stranded RNA exhibited uncoordinated movement due to twitching. (Adapted from B. Daneholt. "Advanced Information: The 2006 Nobel Prize in Physiology or Medicine." Nobelprize.org. http://nobelprize.org/nobel_prizes/medicine/laureates/2006/adv.html [accessed December 21, 2010].)

twitch, causing uncoordinated movement that is easy to observe. Fire, Mellow, and coworkers prepared and purified sense and antisense strands corresponding to the *unc*-22 gene and annealed the strands to prepare a double-stranded RNA. Then they injected worms in their reproductive organ with the sense strand, the antisense strand, or the double-stranded RNA (FIGURE 21.3). Quantitative studies showed the double-stranded RNA to be about 100-fold more effective in causing uncoordinated movement than either the sense or antisense strands. Double-stranded RNA molecules corresponding to other genes did not affect worm movement, indicating that the phenomenon is sequence specific. Mello coined the term **RNA interference** (**RNAi**) for gene silencing that is caused by double-stranded RNA. The fact that RNAi lasts for a very long time and requires only a few molecules per cell suggested that the double-stranded RNA or a downstream product somehow acts in a catalytic fashion.

Injecting individual worms is a time-consuming task that requires considerable skill. Fortunately, two simpler methods can be used to introduce double stranded RNA into worms. The first takes advantage of the fact that worms feed on bacteria. *E. coli* are engineered to produce double-stranded RNA corresponding to the gene of interest. Then worms are allowed to feed on the bacteria (FIGURE 21.4). In one of the earliest feeding experiments, worms that synthesize green fluorescent protein in almost all of their somatic cells were allowed to feed on *E. coli* that synthesized double-stranded RNA molecules corresponding to the green fluorescent protein gene, *gfp*. Approximately 12% of the worms studied in this experiment exhibited sequence specific *gfp* gene silencing in all cells but nerve cells. Although it is more convenient to introduce double-stranded RNA into worms by the feeding method, this method is considerably less efficient than injecting double-stranded RNA directly into the worm. The second method is even simpler: worms are simply soaked in a solution that contains the double-stranded RNA of interest. The soaking technique, while not very efficient, can be quite useful for high throughput worm studies.

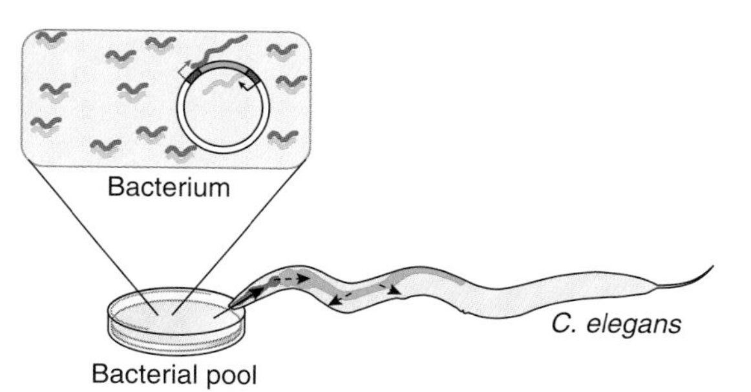

FIGURE 21.4 ***C. elegans* feeding on bacteria with plasmids that code for double-stranded RNA.** A bacterial strain has a plasmid with a DNA insert (blue) that codes for RNA corresponding to a target worm gene. The DNA has the same promoter (magenta) on either side so that the insert is transcribed in both directions. Sense (dark green) and antisense (light green) strands produced by this transcription anneal to form double-stranded RNA. *C. elegans* feeds on the bacterial cells, ingesting and grinding them in their pharynx and transferring the double-stranded RNA to the gut from which it is spread throughout the worm's body. The bacterial chromosome is not shown.

Fire and coworkers proposed the following three hypotheses to explain how double-stranded RNA might silence worm genes:

1. Double-stranded RNA alters the target DNA.
2. Double-stranded RNA inhibits gene transcription.
3. Double-stranded RNA somehow destabilizes mRNA.

They designed experiments to test each hypothesis. They started by demonstrating that double-stranded RNA does not modify the target DNA, ruling out the first hypothesis. Then they tested the second hypothesis by examining whether a double-stranded RNA corresponding to the first gene in an operon inhibits the expression of downstream genes in that same operon. They observed that the downstream genes are fully expressed. This observation rules out the transcription inhibition hypothesis, which predicts that a double-stranded RNA will silence downstream genes. The transcription inhibition hypothesis is also ruled out because double-stranded RNAs corresponding to promoters and introns do not silence genes. Fire, Mello, and coworkers obtained strong support for the mRNA destabilization hypothesis by demonstrating that double-stranded RNA causes a substantial reduction in intracellular target mRNA levels. Based on these findings, Fire and coworkers proposed that *RNAi is a posttranscriptional process* in which double-stranded RNA somehow makes the corresponding mRNA unstable.

Some method was needed to learn how double-stranded RNA makes mRNA unstable. Fire, Mello, and coworkers thought that they might be able to obtain helpful information by isolating mutant worms that are unable to carry out post-transcriptional gene silencing. They exposed worms to chemical mutagens and let the worms grow for two generations to allow induced mutations to be homozygous. Then they allowed the mutant worms to feed on *E. coli*, which were constructed to express a double-stranded RNA that targeted an essential worm gene. The double-stranded RNA killed worms with an intact RNAi response but had no effect on worms with a defective RNAi response. Two of the isolated RNA interference deficient (*rde*) mutants are of special interest because they provide important clues to how RNAi works. The first of these mutants codes for RDE-1, a member of the **Argonaute protein family**. The family name, which was chosen based on the fact that the founding member influences plant leaf morphology, has nothing to do with the protein family's structure or function. Remarkably, *C. elegans* has at least 27 different proteins that belong to the Argonaute family. Some of these proteins are functionally redundant. Thus far it appears that RDE-1 is the only member of this family required for an early step in RNA interference. The second mutant codes for RDE-4, a double-stranded RNA binding protein (see below).

In vitro studies helped to elucidate the RNA interference pathway.

In 1999, Philip A. Sharp and coworkers realized that they would need to develop an *in vitro* system if they wished to elucidate the RNAi pathway. They, therefore, set out to reproduce features of RNAi in

Drosophila embryo extracts. Firefly and sea pansy mRNAs were ideally suited to serve as reporter mRNAs because they have different nucleotide sequences and code for luciferases with different luciferin substrate requirements. Moreover, *Drosophila* embryo extracts can translate these mRNAs to make active firefly and sea pansy luciferases. Sharp and coworkers added double-stranded RNA corresponding to the firefly luciferase gene to a *Drosophila* embryo extract for 10 minutes and then added firefly and sea pansy mRNAs to the extract. After incubating the mixture for one hour, they assayed it for luciferase activity. As anticipated, the extracts had sea pansy luciferase activity but no firefly luciferase activity. They repeated the experiment but this time added double-stranded RNA corresponding to the sea pansy luciferase gene. Now the extracts had active firefly luciferase activity but no sea pansy luciferase activity. This demonstration that RNAi can be observed *in vitro* set the stage for the isolation and characterization of enzymes and factors that participate in the RNAi pathway.

The following year Gregory J. Hannon and coworkers used a cell-free system prepared from *Drosophila* S2 cells (a cell line derived from a primary culture of late stage embryonic stem cells) to dissect the RNAi pathway. Their initial goal was to determine whether a double-stranded RNA corresponding to the *E. coli lacZ* gene would cause S2 cell extracts to catalyze sequence specific lacZ mRNA degradation. They began by transfecting S2 cells with double-stranded RNA corresponding to the *E. coli lacZ* gene and then prepared a cell-free extract from the transfected cells to which they added lacZ mRNA. The extract degraded the *lacZ* mRNA but no other mRNA. Hannon and coworkers proposed that the extract contained a sequence-specific RNAse, which they called an **RNA-induced silencing complex (RISC)**. We will call this complex siRISC to distinguish it from another form of RISC that will be described later in this chapter. In the next phase of their work, Hannon and coworkers used standard protein fractionation techniques to partially purify siRISC. Chemical analysis showed that siRISC contains a 22-nucleotide RNA with a sequence complementary to that in lacZ mRNA that serves to guide the complex to its target mRNA. Later studies revealed that siRISC contains an Argonaute protein, which is essential for its function (see below).

In 2001, Hannon and coworkers performed a second series of *in vitro* experiments that demonstrated the presence of a new RNAse activity that converts long double-stranded RNA into **small interfering RNAs (siRNAs)**. They were able to separate the RNase that cleaves long double-stranded RNA from siRISC by subjecting the S2 cell extract to high speed centrifugation. The new RNase activity remained in the supernatant while siRISC ended up in the pellet. Hannon and coworkers thought that the new RNase was likely to belong to the **RNase III** family because members of this family were known to cleave double-stranded RNA. Further studies revealed that *Drosophila* have an RNase III that can cleave double-stranded RNAs into 22-nucleotide fragments. This enzyme, which they named **Dicer**, catalyzes sequence-independent double-stranded RNA cleavage but cannot cleave single-stranded RNA. Cells with low levels of Dicer activity cannot carry

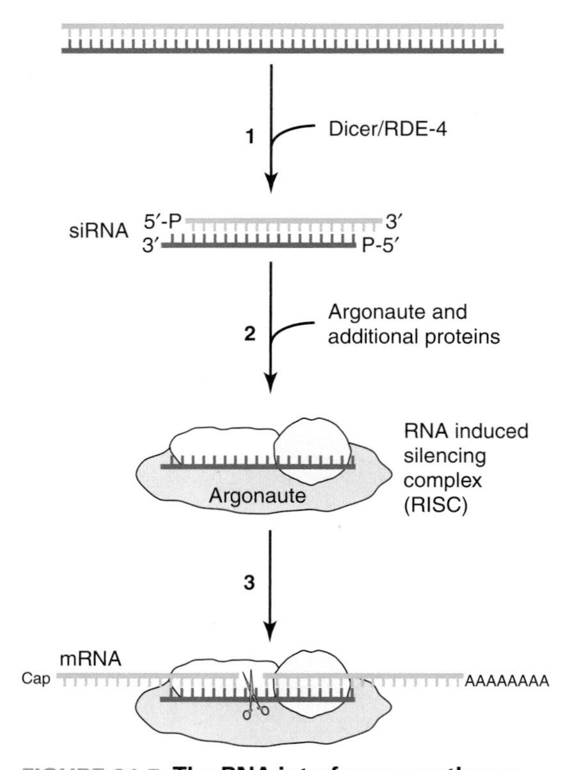

FIGURE 21.5 The RNA interference pathway in *C. elegans*. (1) RDE-4 combines with Dicer, an enzyme with RNase III activity, to form a complex that dices long double-stranded RNA into small pieces of defined length called small interfering RNAs (siRNAs). (2) One of the two strands in siRNA combines with Argonaute (RDE-1) and other proteins to form a nucleoprotein complex called the RNA induced silencing complex (RISC). (3) The single-stranded RNA guides RISC to mRNA, where Argonaute (RDE-4), cleaves the mRNA. (Adapted from C. R. Faehnle and L. Joshua-Tor, *Curr. Opin. Chem. Biol.* 11 [2007]: 569–577.)

out double-stranded RNA-dependent gene silencing, supporting the hypothesis that Dicer is required for RNAi.

Based on the experiments described above and some more recent experiments, the key steps in the *C. elegans* RNA interference pathway in FIGURE 21.5 are as follows: (1) RDE-4 combines with Dicer to form a complex that cleaves long double-stranded RNA into small pieces called small interfering RNAs (siRNAs). (2) One of the two siRNA strands becomes part of the RNA induced silencing complex (siRISC), which includes Argonaute (RDE-1) and other proteins. (3) The single-stranded RNA guides siRISC to mRNA and then the Argonaute protein (RDE-1) cleaves the mRNA. Other eukaryotes use the same basic pathway but some details differ. With this pathway in mind, let's examine the structures and functions of Dicer and siRISC.

Dicer acts as a molecular ruler that generates double-stranded RNA fragments of discrete size.

Dicer proteins, like other members of the RNase III family, make staggered cuts in double-stranded RNA substrates to form characteristic cleavage fragments that are easy to recognize because each end of the fragment has a two nucleotide 3′-overhang and a 5′-phosphate group (Figure 21.5). Dicers from *C. elegans*, humans, and other multicellular eukaryotes have multidomain structures (FIGURE 21.6a and b). Starting from the N-terminus these domains are ATP/helicase, DUF283 (domain of unknown function), PAZ (Piwi, Argonaute, Zwille), RNase IIIa, RNase IIIb, and dsRBD (double stranded RNA binding domain).

A crystal structure is not yet available for an intact Dicer from a multicellular eukaryote. However, Jennifer Doudna and coworkers have solved the crystal structure for an intact Dicer from *Giardia intestinalis*, a common intestinal parasite that infects humans. The *G. intestinalis* Dicer has the PAZ, RNase IIIa, and RNase IIIb domains but has an alternative form of DUF283 and is missing the helicase and dsRBD domains. Nevertheless, the *G. intestinalis* Dicer crystal structure provides very useful information concerning Dicer structure and function. As shown in FIGURE 21.7, *G. intestinalis* Dicer has an elongated shape resembling a hatchet with the blade at the top and

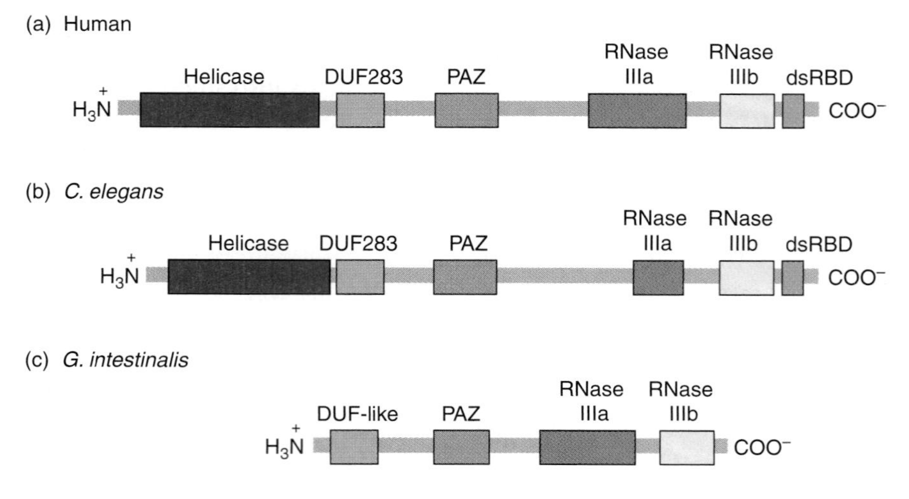

FIGURE 21.6 Representation of Dicer domain structures. (a) Human dicer, (b) *C. elegans* dicer, and (c) *G. intestinalis* dicer. The domains shown are as follows: ATP/helicase, DUF283 (domain of unknown function), PAZ (Piwi, Argonaute, Zwille), RNase IIIa, RNase IIIb, and dsRBD (double stranded RNA binding domain). (Adapted from L. Jaskiewicz and W. Filipowicz, *Curr. Top. Microbiol. Immunol.* 320 [2008]: 77–97.)

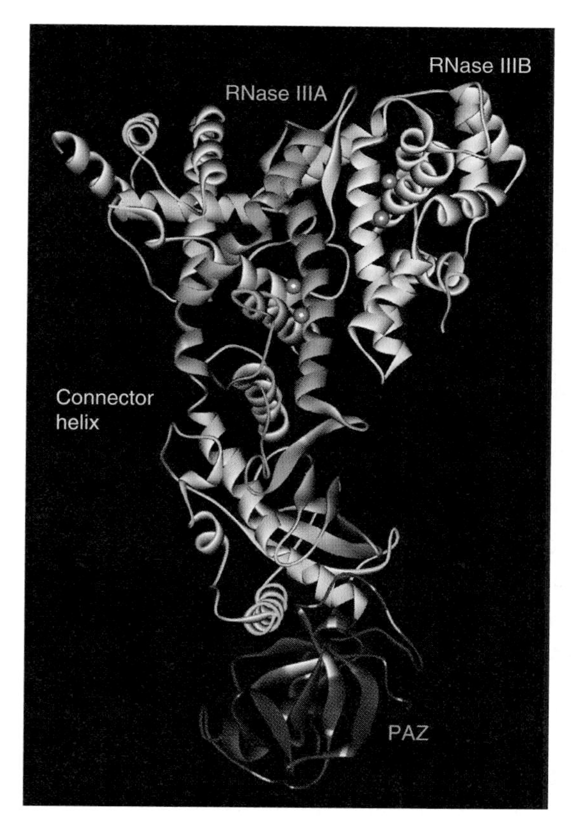

FIGURE 21.7 **Crystal structure of Dicer from *G. intestinalis*.** Each RNase domain has a pair of divalent metal ions at the active site. These metal ions are manganese ions (purple spheres) in the crystal structure but are probably magnesium ions in the cell. The distance between the metal ion pairs is 17.5 Å, which corresponds to the width of the major groove of double-stranded RNA. The distance between the PAZ site that binds the 3′-overhang and the RNase IIIa active site is about 65 Å, which corresponds to about 25 bp in double-stranded RNA. (Structure from Protein Data Bank 2FFL. I. J. Macrae, et al., *Science* 311 [2006]: 195–198. Prepared by B. E. Tropp.)

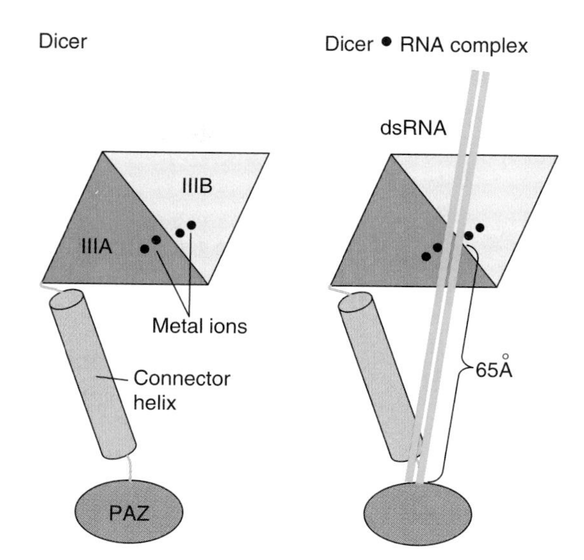

FIGURE 21.8 **Cartoon of *G. intestinalis* Dicer cleaving double-stranded RNA.** The PAZ domain binds one end of the double stranded RNA (green). The distance between the PAZ binding site and the RNase IIIa active site is about 65 Å, which corresponds to about 25 bp in double-stranded RNA. Thus, Dicer serves as a molecular ruler that measures and cleaves about 25 nucleotides from the end of double-stranded RNA. (Adapted from A. Cook and E. Conti, *Nat. Struct. Mol. Biol.* 13 [2006]:190–192.)

the handle at the bottom. The RNase IIIa and IIIb domains are adjacent to one another in the region corresponding to the blade. Each RNase domain has a pair of divalent metal ions at the active site. The metal ions are manganese ions in the crystal structure but are probably magnesium ions in the cell. The distance between the metal ion pairs is 17.5 Å, which corresponds to the width of the major groove of double-stranded RNA. The *G. intestinalis* Dicer crystal structure suggests the method that Dicer uses to specify RNA fragment length (FIGURE 21.8). The distance between the PAZ site that binds the 3′ overhang and the RNase IIIa active site is about 65 Å, which corresponds to about 25 bp in double-stranded RNA. Thus, Dicer acts as a molecular ruler that measures and cleaves about 25 nucleotides

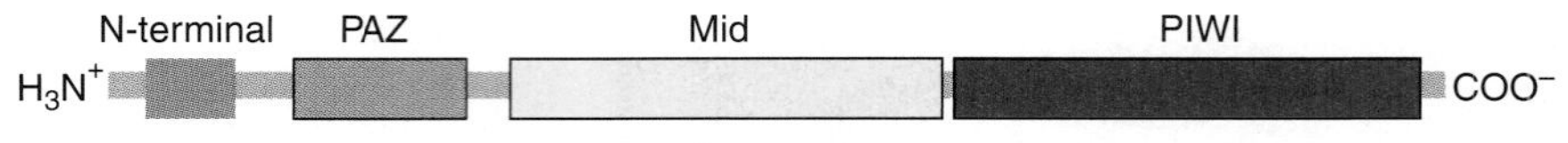

FIGURE 21.9 Domain structure of Argonaute.

from the end of double-stranded RNA. Further work is required to determine how the *C. elegans* protein RDE-4 influences Dicer activity.

The guide strand's 5′-phosphate and 3′-ends bind to sites in the Argonaute protein's Mid and PAZ domains, respectively.

Based on sequence homology, Argonaute proteins are divided into the Ago, Piwi (P-element induced wimpy testis), and worm-specific subfamilies. Members of the Ago and Piwi subfamilies have been identified in eukaryotes, bacteria, and archaea. The number of Argonaute genes varies greatly from one eukaryotic species to another. For example, humans have 8 Argonaute genes (4 Ago and 4 Piwi), *Drosophila* have 5 Argonaute genes (2 Ago and 3 Piwi), and *C. elegans* have at least 27 Argonaute genes (5 Ago, 3 Piwi, 18 worm-specific, and 1 that has not yet been assigned). The Argonaute proteins have four well-defined domains (FIGURE 21.9). Starting from the N-terminus, these domains are the N-terminal, PAZ, Mid (middle), and PIWI domains.

Crystal structures have not yet been obtained for a eukaryotic Argonaute protein but are available for a few prokaryotic Argonaute proteins. For convenience, the present discussion will be limited to the crystal structure of the Argonaute protein isolated from *Thermus thermophilus*. Although relatively little is known about this protein's function(s) in the bacterial cell, *in vitro* studies indicate that the protein acts as a site-specific DNA-guided endoribonuclease. Given the homology between bacterial and eukaryotic Argonaute proteins and the fact that both use a guide nucleic acid strand to make specific endonucleolytic cuts in a target RNA, it seems likely that bacterial and eukaryotic Argonaute proteins work in the same way. Therefore, bacterial Argonaute crystal structures should help us to understand how the eukaryotic Argonaute proteins work.

Dinshaw J. Patel and coworkers determined the crystal structure of the *T. thermophilus* Argonaute protein bound to a 21-nucleotide DNA guide strand with a phosphate group at its 5′-end (FIGURE 21.10a). The protein has a bilobar structure: one lobe containing N-terminal and PAZ domains and the other containing the Mid and PIWI domains. The DNA guide strand's 5′-end inserts into a binding site in the Mid domain. The oxygen atoms in the 5′-phosphate group interact with four highly conserved amino acids in the Mid domain and a bound magnesium ion (green sphere). A mutation that alters one of the four amino acids impairs the Argonaute protein's ability to slice a target RNA. The DNA guide strand's 3′-end inserts into a binding pocket in the PAZ domain. The bound guide DNA passes through a central basic channel located between the two lobes. Bases at positions 2 to 6 (starting from the 5′-end) are exposed and free to bind to the target RNA.

(a)

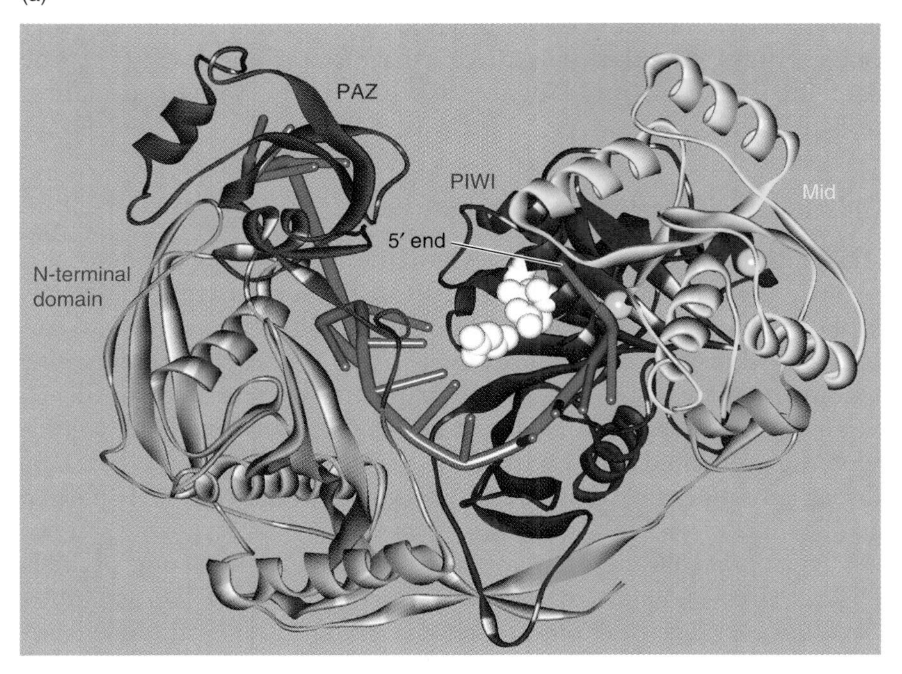

(b)

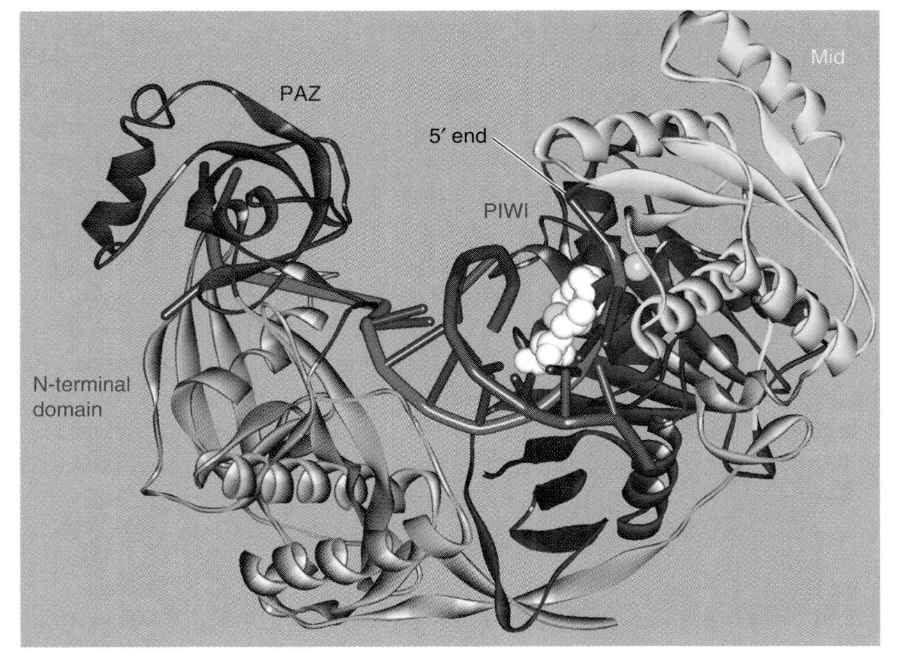

FIGURE 21.10 **Crystal structures of *Thermus thermophilus* Argonaute complexes.** (a) Binary complex with Argonaute protein bound to a 21-nucleotide DNA guide strand. The guide strand's 5′-end fits in a binding site in the Mid domain and its 3′-end fits into a binding site in the PAZ domain. Bases at positions 2 to 6 (starting from the 5′-end) are exposed and free to bind to the target RNA. The two magnesium ions are shown as green spheres. (b) Ternary complex with Argonaute protein bound to 21-nucleotide DNA guide strand (blue) and a 20-nucleotide target RNA (green). The 5′-end and 3′-ends of the guide strands remain anchored to the Mid and PAZ domains, respectively. Nucleotides 2 to 8 on the guide strand, the "seed region," base pair with the target RNA to form an A-form hybrid duplex. The nucleic acid-binding channel between the two lobes is wider than that in the binary complex shown in part a. The three aspartate residues that form the catalytic slicing site on PIWI are shown as white spacefill structures. (Part a structure from Protein Data Bank 3DLH. Y. Wang, et al., *Nature* 456 [2008]: 209–213. Prepared by B. E. Tropp. Part b structure from Protein Data Bank 3F73. Y. Wang, et al., *Nature* 456 [2008]: 921–926. Prepared by B. E. Tropp.)

Patel and coworkers also determined the crystal structure of the *T. thermophilus* Argonaute protein bound to a 21-nucleotide DNA guide strand and a 20-nucleotide target RNA (FIGURE 21.10b). The target RNA was designed to produce a mismatch with the guide DNA to prevent slicing. As expected, the 5′-end and 3′-ends of the guide strands remain anchored to the Mid and PAZ domains, respectively. Nucleotides 2 to 8 on the guide strand, the "seed region," base pair with the target RNA to form an A-form hybrid duplex. The nucleic acid-binding channel between the two lobes widens to form a more open ternary complex. The three aspartate residues that form the catalytic slicing site on PIWI are shown as white space fill structures in Figure 21.10b.

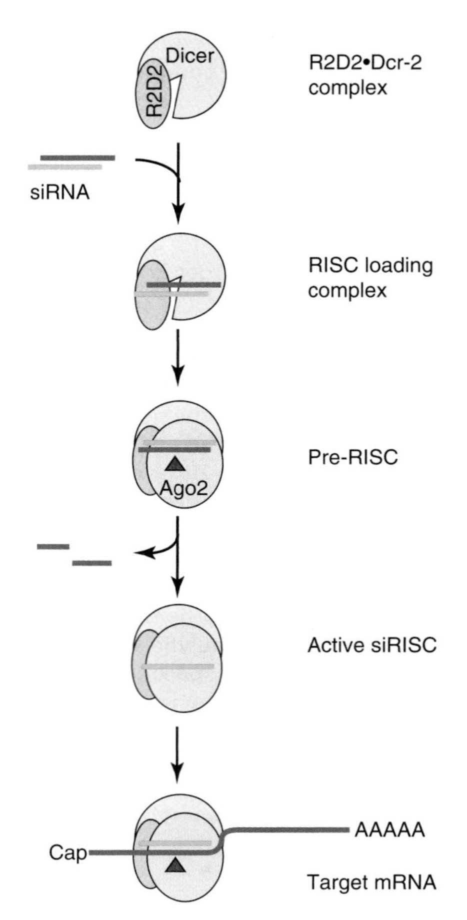

FIGURE 21.11 **siRISC assembly in Drosophila.** Dcr-2 and its partner R2D2, a double-stranded RNA binding protein, combine with siRNA to form a RISC loading complex. Ago 2 adds to the RISC loading complex to form a pre-RISC complex. Then Ago 2 slices the passenger strand (indicated by a red triangle) and the resulting passenger strand fragments dissociate from the pre-RISC complex, converting the inactive complex into a single-stranded guide containing active siRISC, which binds to a target mRNA. The red triangle indicates the RNA cleavage site. The RISC loading complex and RISC have additional proteins that have not been characterized and therefore are not shown in this figure.

RISC loading complex is required for siRISC formation.

Although the *T. thermophilus* Argonaute binary and ternary complexes shown in Figure 21.10 were formed under *in vitro* conditions without the assistance of any additional protein factor, siRISC assembly in a eukaryotic cell requires the assistance of other proteins. The loading of guide RNA onto an Argonaute protein starts with a siRNA duplex formed by Dicer-mediated cleavage. The two siRNA strands must be separated and one strand, the **passenger strand**, must be discarded while the other strand, the **guide strand**, becomes part of siRISC.

Drosophila has three enzymes with RNase III activity. One of these, Dicer-2 (Dcr-2) makes an essential contribution to siRISC assembly. Although Dcr-2 is the enzyme that converts long double-stranded RNA into siRNA, it has an additional role in siRISC formation. Dcr-2 and its partner R2D2, a double-stranded RNA binding protein, combine with siRNA to form a **siRISC loading complex** (FIGURE 21.11). The siRISC loading complex contains additional uncharacterized proteins that are not shown in Figure 21.11. R2D2 binds to the end of the siRNA duplex with more stable base pairs and Dcr-2 interacts with the other end. siRISC loading complex can distinguish siRNAs from other small double-stranded RNAs. One way it does so is by discriminating against small double-stranded RNAs that have mismatches corresponding to positions 7 to 11 on the strand that is destined to be the guide strand.

Ago 2 and perhaps other still uncharacterized proteins (not shown in Figure 21.11) add to the siRISC loading complex to form a **pre-RISC complex**. Then Ago 2 slices the passenger strand and the resulting passenger strand fragments dissociate from the pre-RISC complex, converting the inactive complex into a single-stranded guide containing active siRISC. The change from inactive pre-RISC complex to the fully active siRISC requires ATP. Further work is needed to determine why this conversion requires ATP.

RNAi blocks virus replication and prevents transposon activation.

RNAi appears to have arisen in an early eukaryotic ancestor and is conserved in animals, plants, and most fungi. One notable exception, *Saccharomyces cerevisiae* lacks recognizable homologs of Dicer and Argonaute but other budding yeast species have active RNAi pathways. Genetic analysis indicates that ten or more genes are probably needed to synthesize proteins required for RNAi. However, eukaryotes seldom encounter highly concentrated double-stranded RNA of identical sequence to one of their protein coding genes under normal physiological conditions. It therefore seems reasonable to ask why RNAi is highly conserved in eukaryotes as diverse as plants, flies, worms, and humans. One possible explanation is that RNAi provides immunity to viruses that produce double-stranded RNA at some time during their life cycle. Dicer recognizes this double-stranded RNA and cleaves it to produce siRNA, which is incorporated into siRISC, which then degrades viral mRNA essential for successful virus infection. This explanation is probably satisfactory for organisms that lack an interferon system to protect them from invading viruses but is less satisfactory for humans and other vertebrates that have an active interferon system (see below). Another possible explanation comes from studying *C. elegans* mutants that do not synthesize proteins required for the siRNA pathway. These studies show that RNAi blocks the spread of transposons within the genome and thereby prevents the production of abnormal gene products or abnormal gene expression. The double-stranded RNA precursor to siRNA appears to derive from the transcription terminal inverted repeats at the ends of the transposons (see below).

21.2 Transitive RNAi

In some organisms, RNA interference that starts at one site spreads throughout the entire organism.

RNA interference in *C. elegans* can spread throughout the worm, even when triggered by a minute quantity of double-stranded RNA injected at a single site. A phenomenon called **transitive RNAi** provides useful insights into the amplification mechanism. The term transitive RNAi refers to a silencing signal's ability to move along a particular gene. Fire and coworkers observed this phenomenon while studying worms with an intact *unc-22* gene as well as an *unc-22* gene segment fused to a *gfp* gene. Half the worms had the *unc-22* gene segment fused upstream of *gfp* and the other half had it fused downstream of *gfp* (FIGURE 21.12). As expected, all worms injected with double-stranded RNA that targets *gfp* lost the ability to fluoresce. However, the worms that had *unc-22* fused upstream of *gfp* also showed a surprising loss of coordination, indicating that the double-stranded RNA that targets *gfp* can somehow also silence *unc-22*.

Based on the observation that silencing is amplified, it seems reasonable to propose that siRNAs formed by Dicer, the primary siRNAs, serve as primers for the synthesis of secondary siRNAs. This proposal was ruled out when independent studies performed by Fire and Julia Park and also by Ronald H. A. Plasterk and coworkers showed that secondary siRNAs carry triphosphates at their 5′-end. This type of 5′-end is most easily explained by proposing that an RNA-dependent RNA polymerase catalyzes unprimed RNA synthesis. The secondary siRNAs, which are 22 or 23 nucleotides in length and always antisense, account for the vast majority of small RNAs in the worm, indicating that there is a great deal of amplification. Amplified siRNAs do not associate with the RDE-1 Argonaute protein but instead associate with one of two secondary Argonaute proteins, SAGO-1 and SAGO-2,

FIGURE 21.12 Transitive RNAi phenomenon. In transitive RNAi in *C. elegans*, silencing moves upstream on a specific mRNA target. This phenomenon has been demonstrated by fusing the *gfp* gene to a segment of *unc-22* in a worm that also has the intact *unc-22* gene. Targeting *gfp* abolishes fluorescence but also creates an unexpected phenotype; worm movement becomes uncoordinated when *unc-22* is upstream of *gfp* (left) but not when it is downstream (right). (Adapted from G. J. Hannon, *Nature.* 418 [2002]: 244–251.)

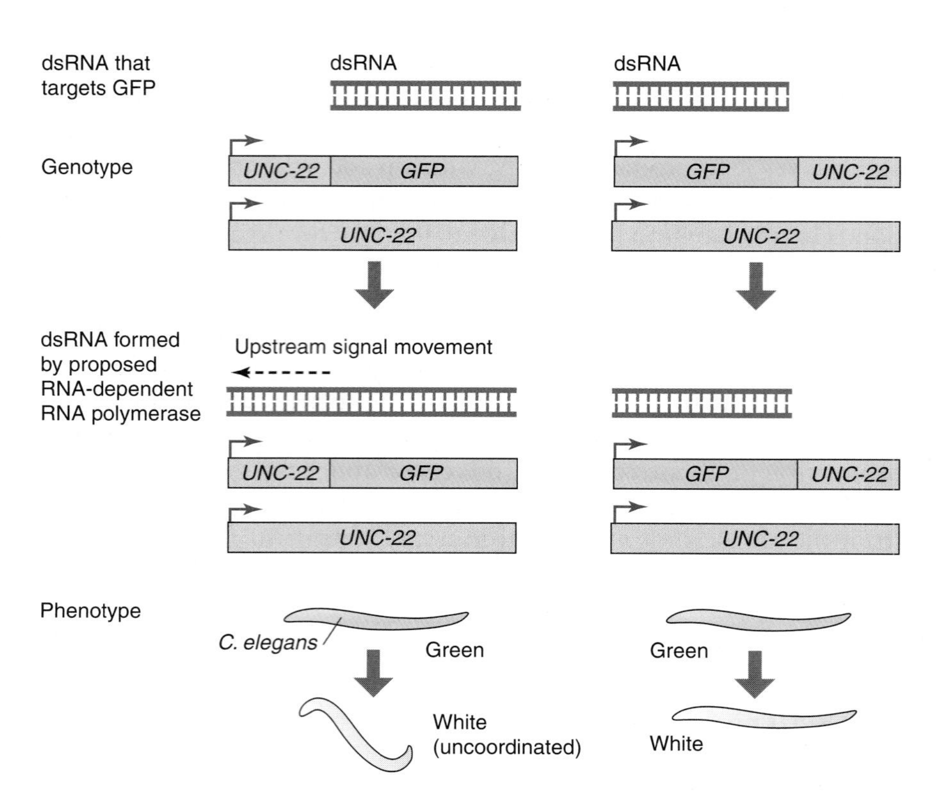

which lack endonuclease activity and must therefore silence genes by some mechanism other than mRNA cleavage. RNAi amplification helps to explain how RNA interference spreads throughout the worm and is passed from one generation to the next. *C. elegans* is predicted to have four different RNA polymerases (RRF-1, RRF-2, RRF-3, and EGO-1). RRF-1 is essential for secondary RNA production in the adult worm and EGO-1 may be required in the germline tissues. *C. elegans* RNA-dependent RNA polymerases have not yet been characterized.

Plants also have RNA-dependent RNA polymerase that permits local and systemic RNAi spreading. The plant RNA-dependent RNA polymerase uses the targeted RNA as a template to amplify the silencing signal to form secondary siRNAs both upstream and downstream of the primary siRNA. RNA-dependent RNA polymerase has not been detected in vertebrates or insects, which both appear to lack transitive RNAi. This lack of transitive RNAi can be turned to advantage because it allows us to target individual alternatively spliced mRNA isoforms produced from a single transcription unit.

SID-1, an integral membrane protein in *C. elegans*, assists in the systemic spreading of the silencing signal.

Craig P. Hunter and coworkers isolated *C. elegans* mutants that can silence genes in injected cells but cannot spread RNAi throughout their body. The first systematic RNA interference defective gene they isolated (*sid-1*) codes for an integral membrane protein that is present in all non-neuronal tissues and permits double-stranded RNA to move into cells. In this regard, it is interesting to note that RNAi seldom spreads to neurons. Subsequent studies revealed that other organisms including humans have SID-1 homologs but *Drosophila* do not. Hunter and coworkers constructed *Drosophila* cells that express *C. elegans* SID-1. The double-stranded RNA concentration required to produce a specific RNAi response is about 100,000-fold lower for these cells than it is for their parent cells that do not make SID-1. *Drosophila* cells that express SID-1 internalize double-stranded RNA very rapidly in a largely energy-independent manner. When *Drosophila* cells coexpress mutant and wild type SID-1 the mutant SID-1 interferes with wild-type function, suggesting that SID-1 works as a multimer. SID-1 does not appear to be required to transport double-stranded RNA out of cells. *C. elegans* use a second membrane protein, SID-2, to take up ingested double-stranded RNA from the intestinal lumen.

Silencing spreads by a different mechanism in plants. Local spreading between neighboring cells takes place via the plasmodesmata, membrane-lined pores that connect the cells. Systemic spreading takes place via the phloem, which normally carries nutrients.

ERI-1, a 3′→5′ exonuclease in *C. elegans*, appears to be a negative regulator of RNA interference.

Gary Ruvkun and coworkers exploited the fact RNA interference is relatively inefficient in *C. elegans* nerve cells to devise a genetic screen for mutant worms with nerve cells that exhibit enhanced sensitivity to double-stranded RNA interference. Of the 19 candidate mutants, the two with the most enhanced sensitivity to double-stranded RNA both had a mutation in a gene that Ruvkin and coworkers named *eri-1*. After exposure to double-stranded RNA, *eri-1* mutant worms accumulate higher siRNA levels than their wild type parents. The product of the *eri-1* gene, ERI-1, is a 3′→5′-exonuclease that is found mainly in the cytoplasm. As expected, it is very highly expressed in nerve cells. Based on these studies Ruvkin and coworkers propose that ERI-1 is a negative regulator of RNA interference that may act to limit the duration or cell specificity of RNA interference.

21.3 RNAi as an Investigational Tool

RNAi is a powerful tool for investigating functional genomics.

RNAi is a very powerful investigational tool. In the early days of molecular biology, mutant phenotypes were used to identify genes but today the problem is often reversed. Thanks to the great advances in DNA sequencing, thousands of different genes have been identified for which there are no known functions. Until recently, efforts to assign functions for genes that were discovered by large scale sequencing (functional genomics) depended on using genetic recombination to inactivate or knockout the specific gene of interest (see Chapter 13). This approach works reasonably well for unicellular organisms such as yeast but presents a problem when studying higher organisms because the methods needed to knock out a specific gene are difficult to perform and require a great deal of time. Moreover, gene inactivation does not work for essential genes because unless special precautions are taken the organism cannot survive without the gene. RNAi provides an alternative method for studying gene function when sequence information is available. The functional investigation of a gene that starts with the gene sequence rather than a mutant phenotype is termed **reverse genetics**.

Julie Ahringer and coworkers exploited RNAi to find phenotypes for *C. elegans*' genes that had been identified from sequencing data. They took advantage of the fact that *C. elegans* feeds on *E. coli* to introduce specific double-stranded RNA into the worms. The double-stranded RNA produced by the bacteria silenced the target worm gene, permitting the investigator to find the phenotype resulting from the inactivation of that gene. Gene inactivation through the use of siRNA is called **gene knockdown**. Ahringer and coworkers constructed 16,757 bacterial strains, each expressing a double-stranded RNA for a different worm gene, and identified mutant phenotypes for 1,722 of the genes studied. Two thirds of these genes did not have a phenotype previously associated with them. Marc Vidal and coworkers created a second RNAi feeding library for *C. elegans*. The two libraries, together, account for about 94% of the 19,427 known worm genes. Once the function of a worm gene is known, the knowledge can be used to make an educated guess about the homologous human gene's function.

The specific method for introducing a double-stranded RNA into the organism must be tailored to the organism; for example, double-stranded RNAs of more than 30-nucleotides do not trigger the RNAi pathway in humans but instead induce a sequence nonspecific interferon response. Interferon, a small protein that helps cells to resist viral infection, activates 2′-5′-oligoadenylate synthase, which then converts ATP to 2′-5′-oligoadenylate. The 2′-5′-oligoadenylate in turn stimulates RNase I to cleave several RNA species, including ribosomal RNA. This cleavage causes the nonspecific inhibition of translation. One method for circumventing the long double-stranded RNA problem is to construct plasmid-based expression vectors that use RNA polymerase III promoters to synthesize short RNA species that do not trigger the interferon response. In the few short years since it was developed, the RNAi approach has helped to determine the function of many genes for which no function had previously been established.

21.4 The MicroRNA Pathway

The miRNA pathway blocks mRNA translation or causes mRNA degradation.

A second type of regulatory RNA called **microRNAs** (**miRNAs**) was discovered by investigators who were interested in *C. elegans* development. Some information about earlier studies on *C. elegans* development performed by John Sulston and H. Robert Horvitz is required to understand the experiments that led to the discovery of miRNAs. In the 1970s, Sulston and Horvitz established the pattern of cell division and cell fates that take place as *C. elegans* develops from a single-cell to a mature organism. Then they used this information to study underlying mechanisms that regulate patterns of cell division and cell fates by isolating and characterizing mutants with abnormal larval cell lineages. Some of the mutants they studied had altered developmental timing. One such timing mutant, *lin-14* (lineage abnormal 14), was of special interest. Too little *lin-14* gene product caused worms to skip larval stage 1 developmental programs and proceed to larval stage 2 programs. In contrast, too much *lin-14* gene product caused worms to repeat larval stage 1 developmental programs instead of advancing to larval stage 2 programs. These mutant studies indicated that the LIN-14 protein plays an important role in regulating worm development.

In 1993, Victor Ambros and coworkers detected another mutation that altered early *C. elegans* larval development. This mutant, which they named *lin-4*, did not make the normal transition from the first to second larval stage. They initially thought that the *lin-4* transcript was translated to make another protein that was required for normal larval development. However, further experiments led them to the remarkable discovery that the *lin-4* transcript is not translated but instead functions as a precursor to a 22-nucleotide RNA that functions as a negative regulator for *lin-14* expression. Sequence studies showed that the 22-nucleotide *lin-4* RNA product contains imperfect homology to seven elements in the 3′-untranslated region (3′-UTR) of *lin-14* mRNA. FIGURE 21.13 shows how *lin-4* RNA interacts with two of the *lin-14* mRNA elements.

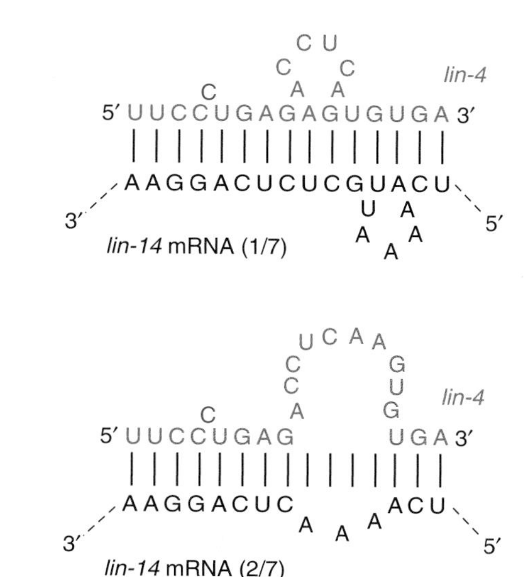

FIGURE 21.13 **Examples of proposed interactions between the *C. elegans lin-4* miRNA and a target mRNA.** The 22-nucleotide RNA derived from the *lin-4* transcript is proposed to interact by base pairing with seven sites in the 3′-untranslated region (UTR) of *lineage-14* (*lin-14*). This figure shows the first two of the seven sites in the 3′-UTR of *lin-14* mRNA. The C residue shown in red is predicted to be bulged in four out of seven *lin-14* interactions, including the two shown here. The C residue is mutated to U in a strong loss of function *lin-4* mutant. (Adapted from S. R. Eddy, *Nat. Rev. Genet.* 2 [2001]: 919–929.)

The *lin-4* gene remained an interesting scientific curiosity until 2000 when Gary Ruvkun and coworkers showed that *C. elegans* requires a second 22-nucleotide RNA, this time encoded by the *let-7* (lethal-7) gene, to proceed from the late larval stage to the adult stage. Further experiments showed that the 22-nucleotide *let-7* RNA (1) inhibits the expression of *lin-41* and *lin-42*, two protein coding genes and (2) contains imperfect homologies to specific regions in the 3′-UTRs of *lin-41* and *lin-42* mRNA. Ruvkun and coworkers then demonstrated that *let-7* is conserved and expressed in a wide variety of animals including flies, mice, chickens, and humans. With this demonstration, the study of the 22-nucleotide RNAs moved to the center stage of molecular biology as an important fundamental research problem. Because of their common function in regulating the timing and development of transitions, *lin-4* and *let-7 RNAs* were initially called small temporal RNAs, but this name was later changed to microRNAs or the abbreviated form miRNA.

Thousands of miRNAs have now been identified in animals and plants. According to current estimates, miRNA genes account for about 1% to 4% of expressed human genes. A single miRNA may regulate hundreds of distinct target mRNAs so it is theoretically possible that at least one third of all human genes are regulated by miRNAs. RNA polymerase II transcribes many miRNA genes and RNA polymerase III transcribes some others. Transcripts formed by RNA polymerase II are modified to have 5′-m^{7}G caps and 3′ poly(A) tails, whereas those formed by RNA polymerase III lack 5′-m^{7}G caps and 3′ poly(A) tails.

Drosophila uses two pathways for miRNA formation (FIGURE 21.14). Other animals use similar pathways. The first pathway begins with primary RNA transcript (**pri-miRNA**) cleavage in the cell nucleus to liberate a ~60 to 70 nucleotide stem-loop intermediate, designated the **miRNA precursor** or **pre-miRNA.** This cleavage is carried out by a

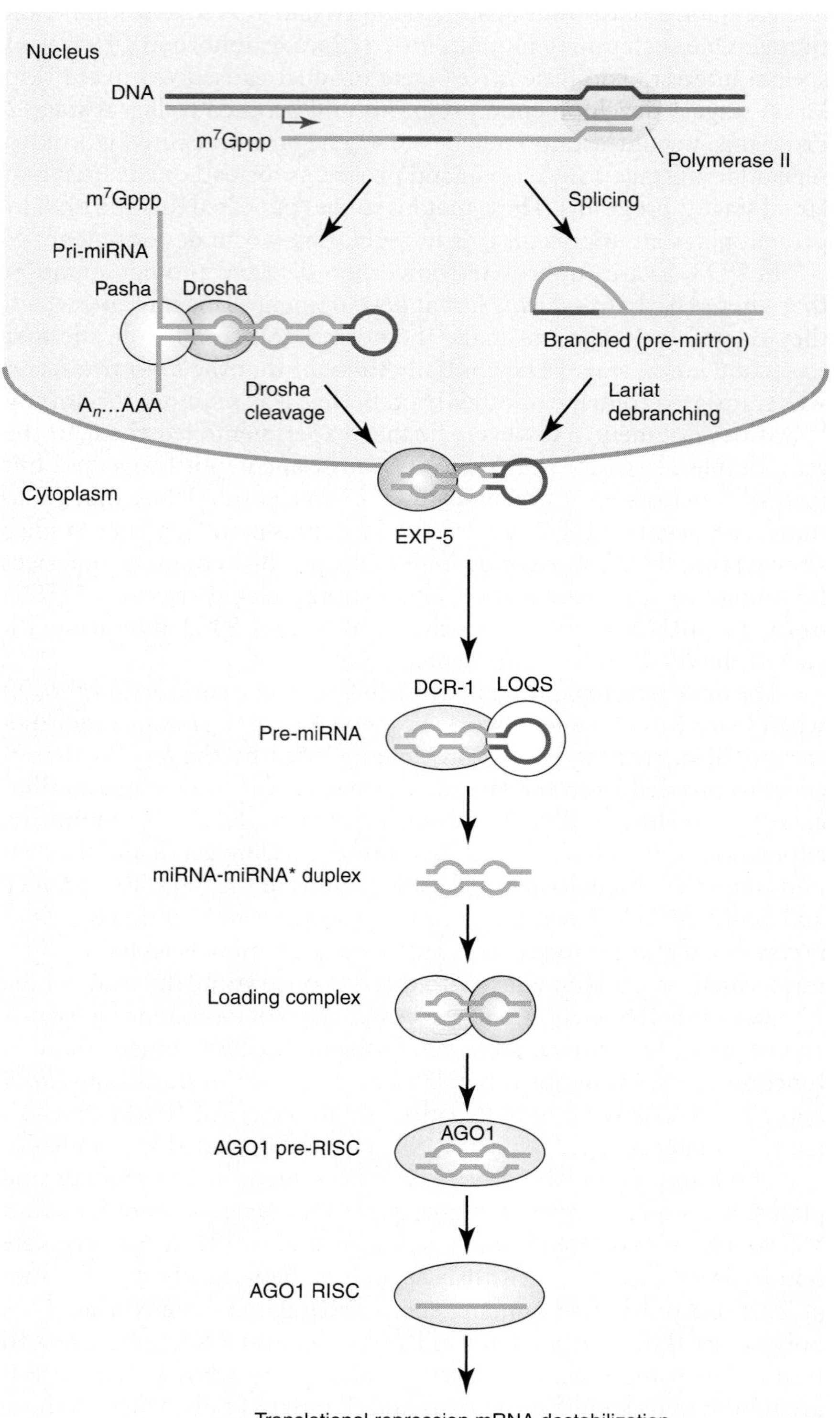

FIGURE 21.14 **Model for miRNA formation and action in *Drosophila*.** There are two paths to miRNA formation. In the first step, the primary RNA (pri-miRNA) transcript is cleaved in the cell nucleus by the Microprocessor complex, which contains a member of the RNase III family called Drosha and a double-stranded RNA binding protein called Pasha. Pasha recognizes both the double-stranded stem in pri-miRNA and the single strands that flank this stem. It then serves as a ruler that guides Drosha to cleave the molecule at the correct site up the stem from the single-stranded RNA/double-stranded RNA junction. The second path is used when the miRNA sequence is within an intron and does not require the Microprocessor. The primary transcript is spliced and the branched intron (or pre-mirtron) that is released is converted to pre-miRNA by lariat debranching enzyme. The pre-miRNA bound to Ran•GTP (not shown) is carried from the nucleus to the cytoplasm through the nuclear pore by the nuclear export protein Exportin 5. After entering the cytoplasm, the pre-miRNA is cleaved by a complex that consists of Dicer and its partner the double-stranded RNA binding protein Loquacious (LOQS). The resulting miRNA-miRNA* duplex binds to a loading complex, which loads the miRNA strand with the less stable base pairing at its 5′-end onto Ago1. The ribonucleoprotein complex that contains the guide miRNA and the Argonaute protein is called miRISC. Other proteins in miRISC, which remain to be characterized, are not shown. (Adapted from M. Ghildiyal and P. D. Zamore, *Nat. Rev. Genet.* 10 [2009]: 94–108.)

Microprocessor complex, which contains a member of the RNase III family called **Drosha** and a double-stranded RNA binding protein. The *Drosophila* double-stranded RNA binding protein, called Pasha (partner of Drosha), recognizes both the double-stranded stem in pri-miRNA and the single strands that flank this stem. It then serves as a ruler that guides Drosha to cleave the molecule at the correct site up the stem from the single-stranded RNA/double-stranded RNA junction. The human counterpart to Pasha is called DGCR8 (DiGeorge syndrome chromosomal region 8). DiGeorge syndrome is a rare inborn error of metabolism with characteristic symptoms that include facial deformities (a cleft palate or lip), heart defects, and immunodeficiency. Crystal structures are not yet available for Drosha, Pasha, or DGCR8. The second pathway is used when the miRNA sequence is within an intron and does not require the Microprocessor. The lariat debranching enzyme acts on the intron (called **pre-mirtron**) that is released by splicing to form a pre-miRNA hairpin.

The pre-miRNA bound to Ran•GTP, a GTPase, is carried from the nucleus to the cytoplasm through the nuclear pore by the nuclear export protein Exportin 5. After entering the cytoplasm, the pre-miRNA is cleaved by a complex that consists of Dicer and its partner the double-stranded RNA binding protein called **Loquacious** (**LOQS**) in *Drosophila* and **TRBP** (TAR RNA-binding protein) in humans. The ~22 nucleotide miRNA-miRNA* duplex product binds to a loading complex, which loads the miRNA strand with the less stable base pairing at its 5′-end onto an Ago1. The other strand, miRNA* is usually degraded. Thus, the miRNA strand serves as the guide strand and the miRNA* strand serves as the passenger strand. The ribonucleoprotein complex that contains the guide miRNA and the Argonaute protein is called miRISC. Other proteins in miRISC, which remain to be characterized, are not shown in Figure 21.14.

miRISC has alternate modes of action that depend on its interaction with the target mRNA (FIGURE 21.15). When complete homology

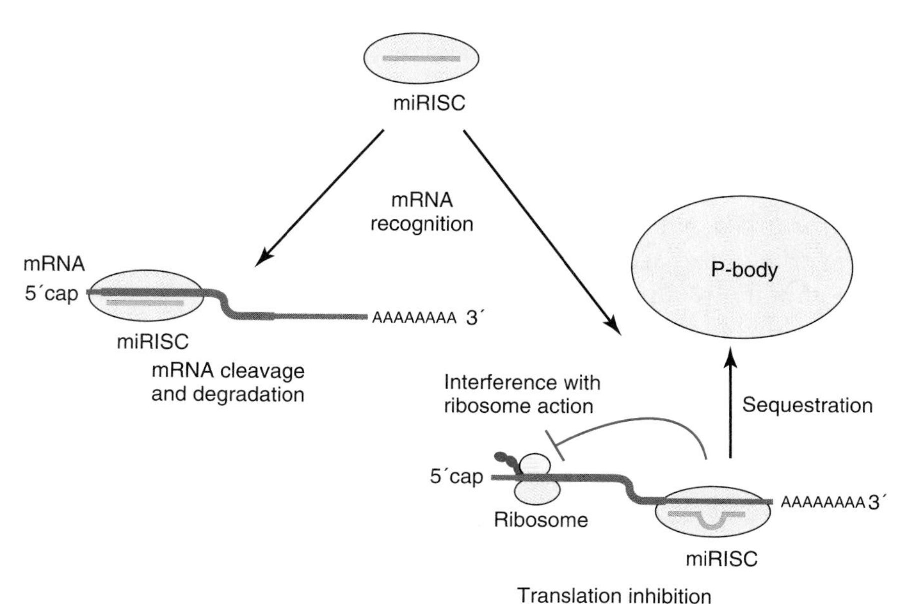

FIGURE 21.15 **Two modes of miRISC action.** When there is complete homology between the guide RNA and a region of the target mRNA, miRISC cleaves the target mRNA. When there is incomplete homology between the 22-nucleotide guide RNA and regions in the 3′-UTR, miRISC binds to the target mRNA's 3′-UTR and blocks translation.

exists between the guide RNA and a region of the target mRNA, miRISC cleaves the target mRNA. When incomplete homology exists between the 22-nucleotide guide RNA and regions in the 3′-UTR, miRISC binds to the target mRNA's 3′-UTR and blocks translation by interfering with the action of translation initiation or elongation factors. The miRISC may also block translation by helping to sequester mRNA in a Processing body (P body), where the mRNA may be degraded or released. Recent studies show that miRISC can also bind to sites within the open reading frame to inhibit translation or to promoter elements to stimulate transcription.

The plant miRNA synthetic pathway differs from the animal pathway in several important ways. Plants do not have Drosha. Instead, plants rely on the Dicer homolog, Dicer-like 1, to cleave both pri-miRNA and pre-miRNA in the nucleus. Then a methyltransferase adds a methyl group to the 3′-ends of the miRNA duplex, stabilizing the molecule. The methylated duplex is transported to the cytoplasm, where it is incorporated into RISC. Plant miRNAs usually are completely complementary to their target mRNA. Furthermore, plant miRNAs tend to bind to sequences in the mRNA's 5′-untranslated region or its protein-coding sequence, rather than to its 3′-untranslated region. One other important difference between plant and animal miRISCs is that the plant ribonucleoprotein usually cleaves its target mRNA.

21.5 Piwi Interacting RNAs (piRNAs)

piRNAs help to maintain the germline stability in animals.

The piwi interacting RNAs (**piRNAs**), which function in germ cells, are about 25 to 30 nucleotides long and have 2′-O-methyl groups at their 3′-ends. More than 1.5 million distinct piRNAs have been identified in *Drosophila*. These fly piRNAs, which map to a few hundred genomic clusters, are formed from long, single-stranded pre-RNAs. Further studies are required to elucidate the pathway for piRNA synthesis and to identify and characterize the enzymes involved in this pathway. Dicer does not appear to be required for piRNA biogenesis. As its name suggests, piRNAs bind to members of the Piwi subfamily of Argonaute proteins. Piwi function has been most thoroughly studied in *Drosophila*, where they appear to be confined to the nucleoplasm of germ cells and adjacent somatic cell. Piwi proteins help to ensure germline stability by suppressing potentially harmful transposon mobility. Piwi probably performs the same function in vertebrates. Plant cells do not have Piwi proteins and have adapted a different RNAi-based strategy to control transposon activity.

21.6 Endogenous siRNA

Animals have a functional endogenous siRNA pathway.

In 2008 several laboratories performed experiments showing that siRNAs can be formed from RNAs made by the cell. Two of these laboratories working independently, one led by Julius Brennecke and the other by Haruhiko Siomi, took a similar approach to show that *Drosophila* S2 cells make **endogenous siRNA** (**esiRNA**). They immunoprecipitated Ago2 from *Drosophila* S2 cells and tissues to enrich for small RNAs and then analyzed the co-precipitated RNAs. Both laboratories found esiRNAs that are 20 to 22 nucleotides long and have a modification at their 3′-ends. Subsequent studies showed the 3′-end has a 2′-O-methyl group attached to it. The 3′-ends of *Drosophila* siRNAs generated from exogenous long double-stranded RNA has the same methyl modification. Dicer 2 is required for esiRNA formation and the esiRNA formed can direct Ago 2 to cleave target mRNAs. Although most of the esiRNAs analyzed correspond to transposon-derived sequences, others are not related to transposons. The esiRNAs may arise from long hairpins produced by transcription of inverted repeats, convergent transcription, trans-interactions, or read through transcription of transposons (FIGURE 21.16). Other animals also synthesize esiRNA, opening an exciting new area of inquiry. The discovery of esiRNAs tends to further blur the lines between the various types of small RNAs.

21.7 Heterochromatin Assembly

The RNAi machinery can establish and maintain heterochromatin.

Small RNA and proteins used in RNAi also help to govern heterochromatin formation. This regulatory process has been most thoroughly studied in the fission yeast *Schizosaccharomyces pombe*. Clr4, a histone H3 Lys9 (H3-K9)-specific methyltransferase, is essential for maintaining heterochromatin structure at the *S. pombe* centromere. *S. pombe* has only one type of Dicer (Dcr1), Argonaute (Ago1), and RNA-dependent RNA polymerase (Rdp1). Remarkably, a mutant that cannot make one of these proteins has a similar centromere phenotype to a mutant that lacks Clr4. Both kinds of mutants have centromeres with low levels of di- and trimethylation at H3-K9. Because di- and trimethylation of lysine of histone H3 is a typical mark of heterochromatin, the mutant studies indicate that the RNAi pathway somehow influences heterochromatin formation at centromeres.

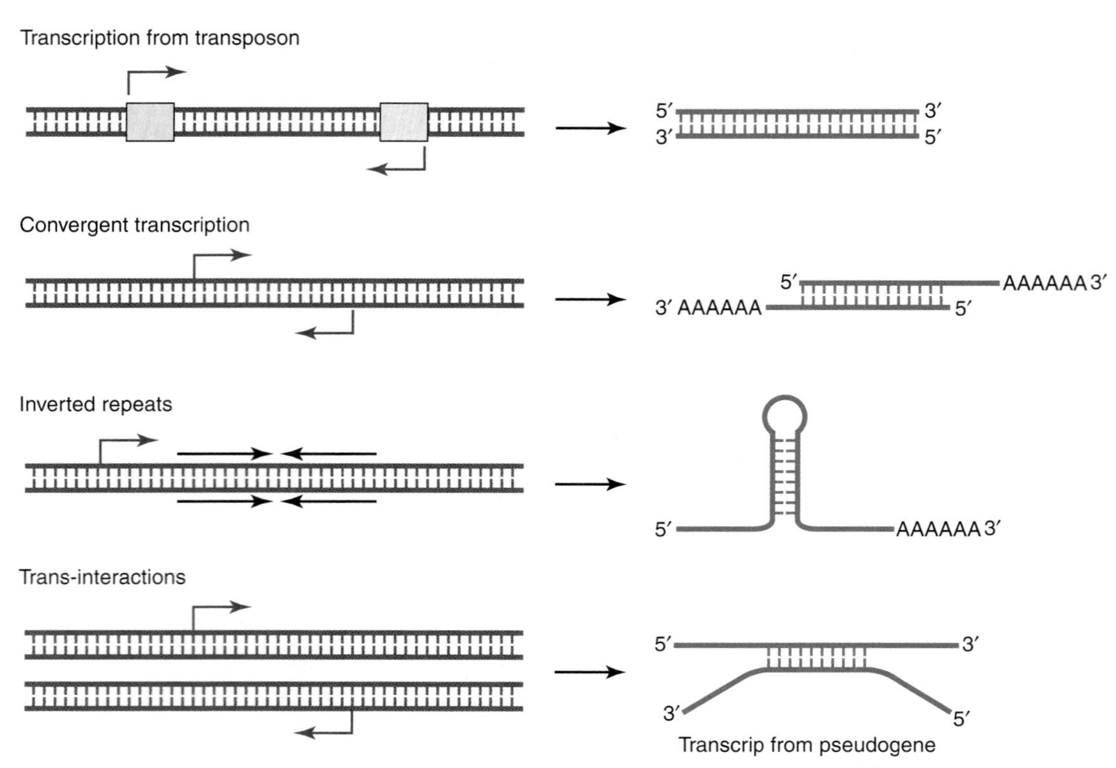

FIGURE 21.16 **Possible sources of esiRNA.** (Adapted from T. W. Nilsen, *Nat. Struct. Mol. Biol.* 15 [2008]: 546–548.)

FIGURE 21.17 is a simplified model showing how proteins involved in RNAi regulate heterochromatin formation at the centromere. RNA polymerase II synthesizes a transcript from a centromeric promoter. An RNA-dependent RNA polymerase uses the transcript as a template to synthesize a double-stranded RNA. Dicer, which is part of a protein complex that also contains Rdp1, cleaves the double-stranded RNA to form siRNA duplexes and transfers the duplexes to the **ARC** (**Argonaute siRNA chaperone**) complex. The ARC complex contains Ago1 and two Argonaute binding proteins, Arb1 and Arb2. The Ago1 subunit in the ARC complex slices the passenger strand and transfers the intact guide strand to a second Argonaute-containing protein complex to form a ribonucleoprotein complex called the **RITS** (**RNA-induced initiation of transcriptional silencing**) **complex.** In addition to Ago1 and guide RNA, the RITS complex contains Chp1, a chromodomain protein that binds to methylated H3-K9 and Tas3, which links Chp1 to Ago1. The RITS complex binds to chromatin by base pairing interactions between (1) its guide RNA and the non-coding transcript and (2) its Chp1 chromodomain protein and the methylated H3-K9 on a nucleosome. The bound RITS complex recruits the Rdp1•Dicer complex, initiating another round of double-stranded RNA synthesis. Histone deacetylase (HDAC), which is also recruited to chromatin, removes histone H3-K9 acetyl groups. These deacylated histones now become targets for the Clr4 methyl transferase. Heterochromatin formation in other eukaryotes including mammals also appears to be regulated by the RNAi machinery. Further studies are needed to work out the details of this regulation.

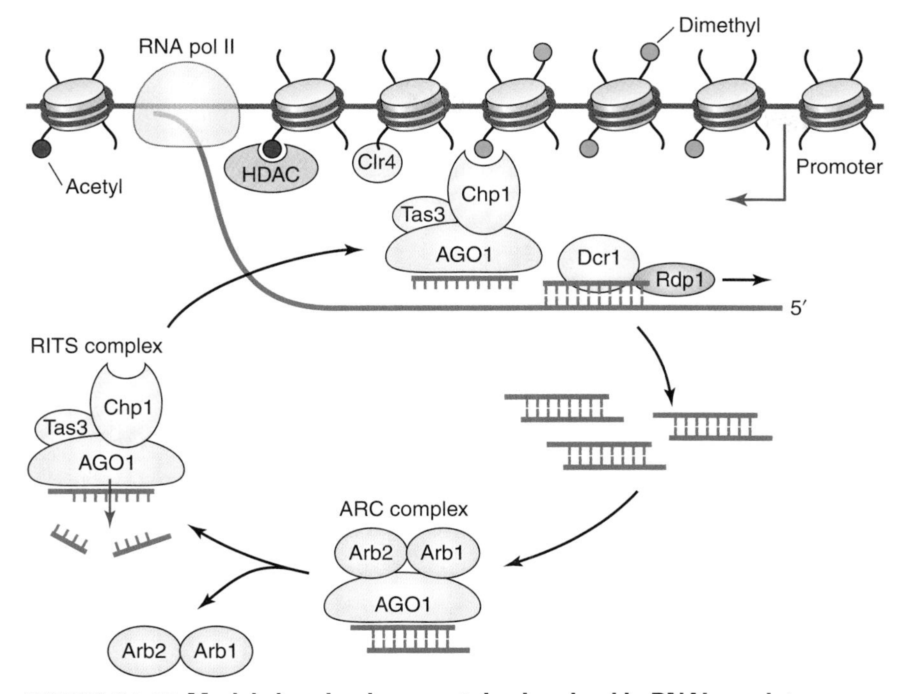

FIGURE 21.17 **Model showing how proteins involved in RNAi regulate centromeric heterochromatin formation.** RNA polymerase II synthesizes a transcript from a centromeric promoter. This transcript serves as a template for RNA-dependent RNA polymerase (Rdp1), which uses it to synthesize a double-stranded RNA. Dicer, which is part of a protein complex that also contains Rdp1, cleaves the double-stranded RNA to form siRNA duplexes and transfers the duplexes to the ARC (Argonaute siRNA chaperone) complex. The Ago1 subunit in the ARC complex slices the passenger strand and transfers the intact guide strand to a second Argonaute containing protein complex to form a ribonucleoprotein complex called the RITS (RNA-induced initiation of transcriptional silencing) complex. The RITS complex is bound to chromatin by base pairing interactions between (1) its guide RNA and the non-coding transcript and (2) its Chp1 chromodomain protein and the methylated H3-K9 on a nucleosome. The bound RITS complex recruits the Rdp1•Dicer complex, initiating another round of double-stranded RNA synthesis. Histone deacetylase (HDAC) is also recruited to chromatin and removes acetyl groups from histones at H3-K9. Then the RITS complex stimulates the Clr4 methyl transferase to methylate the deacylated H3-K9. The vertical arrows indicate Ago1 slicer activity. The purple "lollipop" structure indicates an acetylated H3-K9 and the teal "lollipop" structure indicates dimethylated H3-K9. (Adapted from K. Ekwall, *Nat. Struct. Mol. Biol.* 14 [2007]: 178–179.)

Suggested Reading

Historical

Ruvkin, G., Wightman, B., and Ha, I. 2004. The 20 years it took to recognize the importance of tiny RNAs. *Cell* S116:s93–s96.

General Overview

Miller, J. 2011. *RNA interference and model organisms.* Sudbury, MA: Jones and Bartlett Publishers.

Paddison, P. J., and Vogt, P. K. (eds). 2008. *RNA interference.* Berlin and Heidelberg: Springer Verlag.

RNA Interference (RNAi) Triggered by Exogenous Double-Stranded RNA

Ban, J., Shaw, G., Tropea, J. E., et al. 2008. A stepwise model for double-stranded RNA processing by ribonuclease III. *Mol Microbiol* 67:143–154.

Bernstein, E., Caudy, A. A., Hammond, S. M., and Hannon, G. J. 2001. Role for a bidentate ribonuclease in the initiation step of RNA interference. *Nature* 409:363–366.

Boisvert, M.-E., and Simard, M. J. 2008. RNAi pathway in *C. elegans*: the argonautes and collaborators. *Curr Topics Microbiol Immunol* 320:21–36.

Caplen, N. J., Parrish, S, Imani, F., et al. 2001. Specific inhibition of gene expression by small double-stranded RNAs in invertebrate and vertebrate systems. *Proc Natl Acad Sci USA* 98:9742–9747.

Drinnenberg, I. A., Weinberg, D. E., Xie, K. T., et al. 2009. RNAi in budding yeast. *Science* 326:544–550.

Elbashir, S. M., Harborth, J., Lendeckel, W., et al., 2001. Duplexes of 21-nucleotide RNAs mediate RNA interference in cultured mammalian cells. *Nature* 411:494–498.

Elbashir, S. M., Lendeckel, W., and Tuschl, T. 2001. RNA interference is mediated by 21- and 22-nucleotide RNAs. *Genes Dev* 15:188–200.

Ender, C., and Meister, G. 2010. Argonaute proteins at a glance. *J Cell Sci* 123:1819–1823.

Fire, A. Z. 2007. Gene silencing by double-stranded RNA. *Angew Chem Int Ed* 46:6966–6984.

Fire, A., Xu, S., Montgomery, M. K., et al. 1998. Potent and specific genetic interference by double-stranded RNA in *Caenorhabditis elegans*. *Nature* 391:806–811.

Fischer, S. E. J. 2010. Small RNA-mediated gene silencing pathways in *C. elegans*. *Int J Biochem Cell Biol* 42:1306–1315.

Ghildiyal, M., and Zamore, P. D. 2009. Small silencing RNAs: an expanding universe. *Nat Rev Genet* 10:94–108.

Hamilton, A. J., and Baulcombe, D. C. 1999. A species of small antisense RNA in posttranscriptional gene silencing in plants. *Science* 286:950–952.

Hammond, S. M., Bernstein, E., Beach, D., and Hannon G. J. 2000. An RNA-directed nuclease mediates post-transcriptional gene silencing in *Drosophila* cells. *Nature* 404:293–296.

Hannon, G. J. 2002. RNA interference. *Nature* 418:244–251.

Höck, J., and Meister, G. 2008. The Argonaute protein family. *Genome Biol* 9:210.1–201.8.

Hutvagner, G., and Simard, M. 2008. Argonaute proteins: key players in RNA silencing. *Nat Rev Mol Cell Biol* 9:22–32.

Jaskiewicz, L., and Filipowicz, W. 2008. Role of Dicer in posttranscriptional RNA silencing. *Curr Top Microbiol Immunol* 320:77–97.

Jinek, M., and Doudna, J. A. 2009. A three-dimensional view of the molecular machinery of RNA interference. *Nature* 457:405–412.

Kawamata, T., and Tomari, Y. 2010. Making RISC. *Trends Biochem Sci* 35:368–376.

Liu, J., Carmell, M. A., Rivas, F. V., et al. 2004. Argonaute2 is the catalytic engine of mammalian RNAi. *Science* 305:1437–1441.

Liu, Q., and Paroo, Z. 2010. Biochemical principles of smallRNA pathways. *Ann Rev Biochem* 79:295–319.

MacRae, I. J., and Doudna, J. A. 2007. Ribonuclease revisited: structural insights into ribonuclease III family enzymes. *Curr Opin Struct Biol* 17:1–8.

MacRae, I. J., Ma, E., Zhou, M., et al. 2008. In vitro reconstitution of the human RISC-loading complex. *Proc Natl Acad Sci USA* 105:512–517.

MacRae, I. J., Zhou, K., and Doudna, J. A. 2007. Structural determinants of RNA recognition and cleavage by Dicer. *Nat Struct Mol Biol* 14:934–940.

MacRae, I. J., Zhou, K., Li, F., et al. 2006. Structural basis for double-stranded RNA processing by Dicer. *Science* 131:195–198.

Mello, C. C. 2007. Return to the RNAi world: Rethinking gene expression and evolution. *Angew Chem Int Ed* 46:6985–6994.

Moazed, D. 2009. Small RNAs in transcriptional gene silencing and genome defence. *Nature* 457:413–420.

Montgomery, M. K., Xu, S., and Fire, A. 1998. RNA as a target of double-stranded RNA-mediated genetic interference in *Caenorhabditis elegans*. *Proc Natl Acad Sci USA* 95:15502–15507.

Naqvi, A. R., Islam, M. N., Choudhury, N. R., et al. 2009. The fascinating world of RNA interference. *Int J Biol Sci* 5:97–117.

Nowotny, M., and Yang, W. 2009. Structural and functional modules in RNA interference. *Curr Opin Struct Biol* 19:286–293.

Pare, J. M., and Hobman, T. C. 2007. Dicer: structure, function, and role in RNA-dependent gene-silencing pathways. In: Pare, J. M. and Hobman, T. C. (eds.), *Industrial Enzymes;* pp. 421–438. Heidelberg: Springer Netherlands.

Parker, J. S. 2010. How to slice: snapshots of Argonaute in action. *Silence* 1:3.

Parker, J. S., and Barford, D. 2006. Argonaute: a scaffold for the function of short regulatory RNAs. *Trends Biochem Sci* 31:622–630.

Pratt, A. J., and MacRae, I. J. 2009. The RNA-induced silencing complex: a versatile gene-silencing machine. *J Biol Chem* 284:17897–17901.

Rana, T. M. 2007. Illuminating the silence: understanding the structure and function of smallRNAs. *Nat Rev Mol Cell Biol* 8:23–36.

Sashital, D. G., and Doudna, J. A. 2010. Structural insights into RNA interference. *Curr Opin Struct Biol* 20:90–97.

Siomi, H., and Siomi, M. C. 2009. On the road to reading the RNA-interference code. *Nature* 457:396–404.

Tabara, H., Sarkissian, M., Kelly, W. G., et al. 1999. The *rde-1* gene, RNA interference, and transposon silencing in *C. elegans*. *Cell* 99:123–132.

Tolia, N. H., and Joshua-Tor, L. 2007. Slicer and argonautes. *Nature Chem Biol* 3:36–43.

Tuschl, T., Zamore, P. D., Lehmann, R., et al. 1999. Targeted mRNA degradation by double-stranded RNA in vitro. *Genes Dev* 13:1391–1397.

Van den Berg, A., Mols, J., and Han, J. 2008. RISC-target interaction: cleavage and translational suppression. *Biochim Biophys Acta* 1779:668–677.

Wang, Y., Juranek, S., Li, H., et al. 2008. Structure of an Argonaute silencing complex with a seed-containing guide DNA and target complex. *Nature* 456:921–926.

Wang, Y., Sheng, G., Juranek, S., et al. 2008. Structure of the guide-strand-containing Argonaute silencing complex. *Nature* 456:209–213.

Welker, N. C., Pavelec, D. M., Nix, D. A., et al. 2010. Dicer's helicase domain is required for accumulation of some, but not all, *C. elegans* endogenous siRNAs. *RNA* 16:893–903.

Zamore, P. D., Tuschl, T., Sharp, P. A., and Bartel, D. P. 2000. RNAi: Double-stranded RNA directs the ATP-dependent cleavage of mRNA at 21 to 23 nucleotide intervals. *Cell* 101:25–33.

Transitive RNAi

Ahlquist, P. 2002. RNA-dependent RNA polymerases, viruses, and RNA silencing. *Science* 296:1270–1273.

Feinberg, E. H., and Hunter, C. P. 2003. Transport of dsRNA into cells by the transmembrane protein SID-1. *Science* 301:1545–1547.

Gent, J. I., Lamm, A. T., Pavalec, D. M., et al. 2010. Distinct phases of siRNA synthesis in an endogenous RNAi pathway in *C. elegans* soma. *Mol Cell* 37:679–689.

Jose, A. M., and Hunter, C. P. 2007. Transport of sequence-specific RNA interference information between cells. *Ann Rev Genet* 41:305–330.

Kennedy, S., Wang, D., and Ruvkun, G. 2004. A conserved siRNA-degrading RNase negatively regulates RNA interference in *C. elegans*. *Nature* 427:645–649.

Matranga, C., and Zamore, P. D. 2007. Small silencing RNAs. *Curr Biol* 17:R789–R793.

Pak, J., and Fire, A. 2007. Distinct populations of primary and secondary effectors during RNAi in *C. elegans*. *Science* 315:241–244.

Sijen, T., Fleenor, J., Simmer, F., et al. 2001. On the role of RNA amplification in dsRNA-triggered gene silencing. *Cell* 107:465–476.

Sijen, T., Steiner, F. A., Thijssen, K. L., and Plasterk, R. H. A. 2007. Secondary siRNAs result from unprimed RNA synthesis and form a distinct class. *Science* 315:244–247.

Talsky, K. B., and Collins, K. 2010. Initiation by a eukaryotic RNA-dependent RNA polymerase requires looping of the template end and is influenced by the template-tailing activity associated with uridyltransferase. *J Biol Chem* 285:27614–27623.

RNAi as an Investigational Tool

Ahringer, J., ed. 2006. Reverse genetics. In: WormBook. The *C. elegans* Research Community. Retrieved December 17, 2010, from http://www.wormbook.org.

Kamath, R. S., and Ahringer, J. 2003. Genome-wide RNAi screening in *Caenorhabditis elegans*. *Methods* 30:313–321.

Kamath, R. S., Martinez-Campos, M., Zipperlen, P., Fraser, A. G., and Ahringer, J. 2001. Effectiveness of specific RNA-mediated interference through ingested double-stranded RNA in *Caenorhabditis elegans*. *Genome Biol* 2:1–10.

Rual, J. F., Ceron, J., Koreth, J., et al. 2004. Toward improving *Caenorhabditis elegans* phenome mapping with an ORFeome-based RNAi library. *Genome Res* 14:2162–2168.

The MicroRNA Pathway

Bartel, D. P. 2009. MicroRNAs: Target recognition and regulatory functions. *Cell* 136:215–233.

Breving, K., and Esquela-Kerscher, A. 2010. The complexities of microRNA regulation: mirandering around the rules. *Int J Biochem Cell Biol* 42:1316–1329.

Brodersen, P., and Voinnet, O. 2009. Revisiting the principles of microRNA target recognition and mode of action. *Nat Rev Mol Cell Biol* 10:141–148.

Bushati, N., and Cohen, S. M. 2007. Micro RNA functions. *Ann Rev Dev Biol* 23:175–205.

Carthew, R. W., and Sontheimer, E. J. 2009. Origins and mechanisms of miRNAs and siRNAs. *Cell* 136:642–655.

Davis, B. N., and Hata, A. 2009. Regulation of microRNA biogenesis: A miRiad of mechanisms. *Cell Commun Signal* 7:18.

Eulalio, A., Huntzinger, E., and Izaurraide, E. 2008. Getting to the root of miRNA-mediated gene silencing. *Cell* 132:9–14.

Faller, M., and Guo, F. 2008. MicroRNA biogenesis: there's more than one way to skin a cat. *Biochim Biophys Acta* 1779:663–667.

Filipowicz, W., Bhattacharyya, S. N., and Sonnenberg, N. 2008. Mechanisms of post-transcriptional regulation by micro-RNAs: are the answers in sight? *Nat Rev Genet* 9:102–114.

Grishok, A, Pasquinelli, A. E., Conte, D., et al 2001. Genes and mechanisms related to RNA interference regulate expression of small temporal RNAs that control *C. elegans* developmental timing. *Cell* 108:23–34.

Horvitz, H. E. 2002. Worms, life, and death. *Biosci Rep* 23:239–269.

Inui, M., Martello, G., and Piccolo, S. 2010. MicroRNA control of signal transduction. *Nat Rev Cell Mol Biol* 11:252–263.

Kaufman, E. J., and Miska, E. A. 2010. The microRNAs of *Caenorhabditis elegans*. *Semin Cell Dev Biol* 21:728–737.

Kim, V. N., Han, J., and Siomi, M. C. 2009. Biogenesis of small RNAs in animals. *Nat Rev Cell Mol Biol* 10:126-139.

Lee, R. S., Feinbaum, R. L., and Ambros, V. 1993. The *C. elegans* heterochronic gene *lin-4* encodes small RNAs with antisense complementarity to *lin-14*. *Cell* 75:843–854.

Reinhart, B. J., Slack, F. J., Basson, M., et al. 2000. The 21-nucleotide *let-7* RNA regulates developmental timing in *Caenorhabditis elegans*. *Nature* 403:901–906.

Stefani, G., and Slack, F. J. 2008. Small non-coding RNAs in animal development. *Nat Rev Cell Mol Biol* 9:219–229.

Tang, G. 2005. siRNA and miRNA: an insight into RISCs. *Trends Biochem Sci* 30:106–114.

Voinnet, O. 2009. Origin, biogenesis, and activity of plant microRNAs. *Cell* 136:669–687.

Wrightman, B., Ha, I., and Rukun, G. 1993. Posttranscriptional regulation of the heterochronic gene *lin-14* by *lin-4* mediates temporal pattern formation in *C. elegans*. *Cell* 75:855–862.

Piwi Interacting RNAs (piRNAs)

Aravin, A., Hannon, G. J., and Brennecke, J. 2007. The Piwi-piRNA pathway provides an adaptive defense in the transposon arms race. *Science* 318:761–764.

Brennecke, J., Aravin, A. A., Stark, A., et al 2007. Discrete small RNA-generating loci as master regulators of transposon activity in *Drosophila*. *Cell* 128:1089–1103.

Faehnle, C. R., and Joshua-Tor, L. 2007. Argonautes confront new small RNAs. *Curr Opin Chem Biol* 11:569–577.

Farazi, T. A., Juranek, S. A., and Tuschl, T. 2008. The growing catalog of small RNAs and their association with distinct Argonaute/Piwi family members. *Develop* 135:1201–1214.

Senti, K-A., and Brennecke, J. 2010. The piRNA pathway: a fly's perspective on the guardian of the genome. *Trends Genet* 26: 499–509.

Siomi, M. C., Manneu, T., and Siomi, H. 2010. How does the royal family of Tudor rule the PIWI-interacting RNA pathway? *Genes Dev* 24:636–646.

Endogenous siRNA

Czech, B., Malone, C. D., Zhou, R., et al. 2008. An endogenous small interfering RNA pathway in *Drosophila*. *Nature* 453:798–802.

Golden, D. E., Gerbasi, V. R., and Sontheimer, E. J. 2008. An inside job for siRNAs. *Mol Cell* 37:309–312.

Kawamura, Y., Saito, K., Kin, T., et al. 2008. *Drosophila* endogenous small RNAs bind to Argonaute 2 in somatic cells. *Nature* 453:793–797.

Nilsen, T. W. 2008. Endo-siRNAs: yet another layer of complexity in RNA silencing. *Nat Struct Mol Biol* 15:546–548.

Okamura, K., and Lai, E. C. 2008. Endogenous small interfering RNAs in animals. *Nat Rev Mol Cell Biol* 9:673–678.

Heterochromatin Assembly

Djupedal, I., and Ekwall, K. 2009. Epigenetics: heterochromatin meets RNAi. *Cell Res* 19:282–295.

Ekwall, K. 2007. 'Arc' escorts siRNAs in heterochromatin assembly. *Nat Struct Mol Biol* 14:178–179.

Horn, P. J., and Peterson, C. L. 2006. Heterochromatin assembly: a new twist on an old model. *Chromosome Res* 14:83–94.

Moazed, D. 2009. Small RNAs in transcriptional gene silencing and genome defense. *Nature* 457:413–420.

Verdel, A., Vavasseur, A., Gorrec, M. E., and Touat-Todeschini, L. 2009. Common themes in siRNA-mediated epigenetic silencing pathways. *Int J Dev Biol* 53:245–257.

This book has a Web site, **http://biology.jbpub.com/book/molecular**, which contains a variety of resources, including links to in-depth information on topics covered in this chapter, and review material designed to assist in your study of molecular biology.

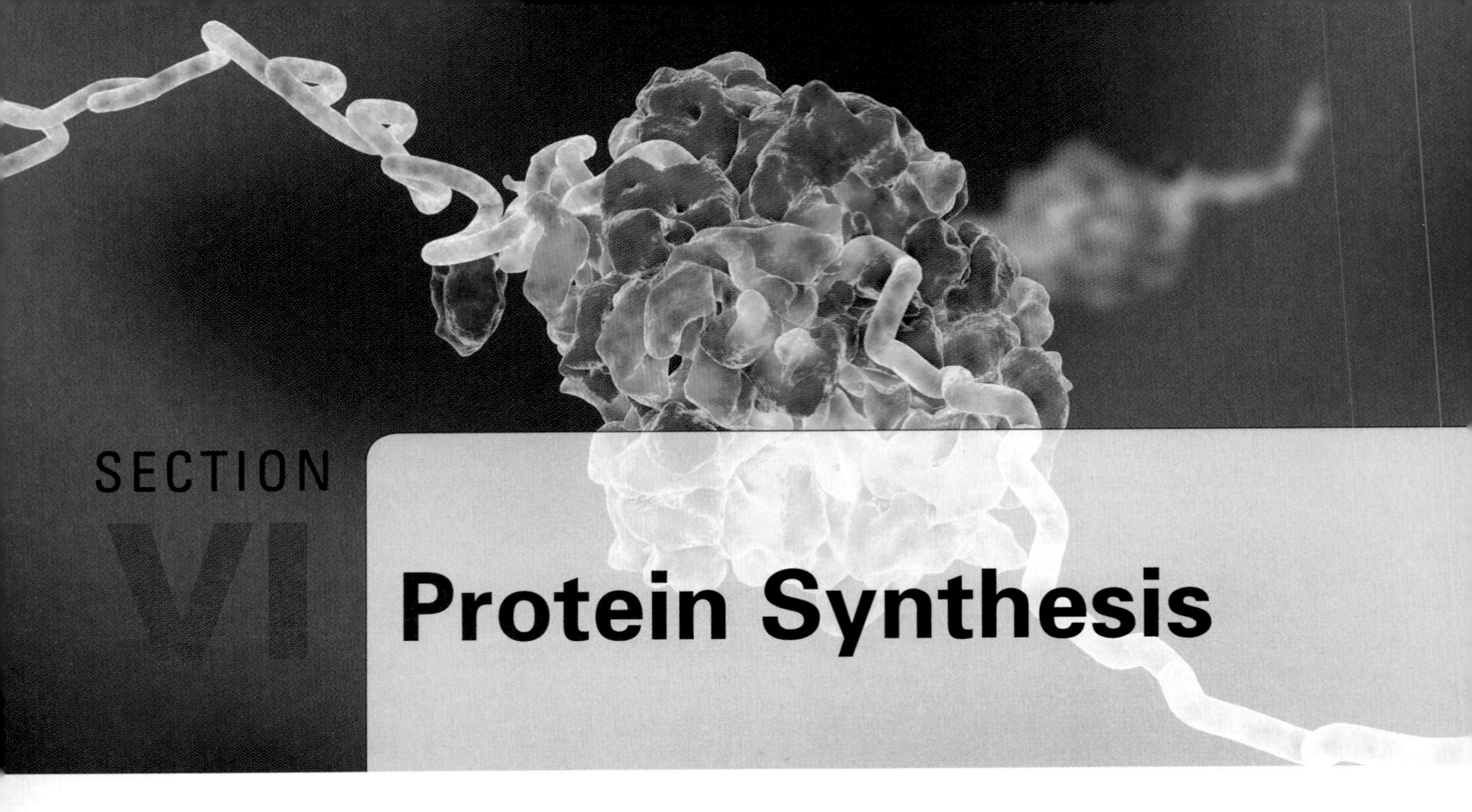

SECTION VI

Protein Synthesis

CHAPTER 22 **Protein Synthesis: The Genetic Code**

CHAPTER 23 **Protein Synthesis: The Ribosome**

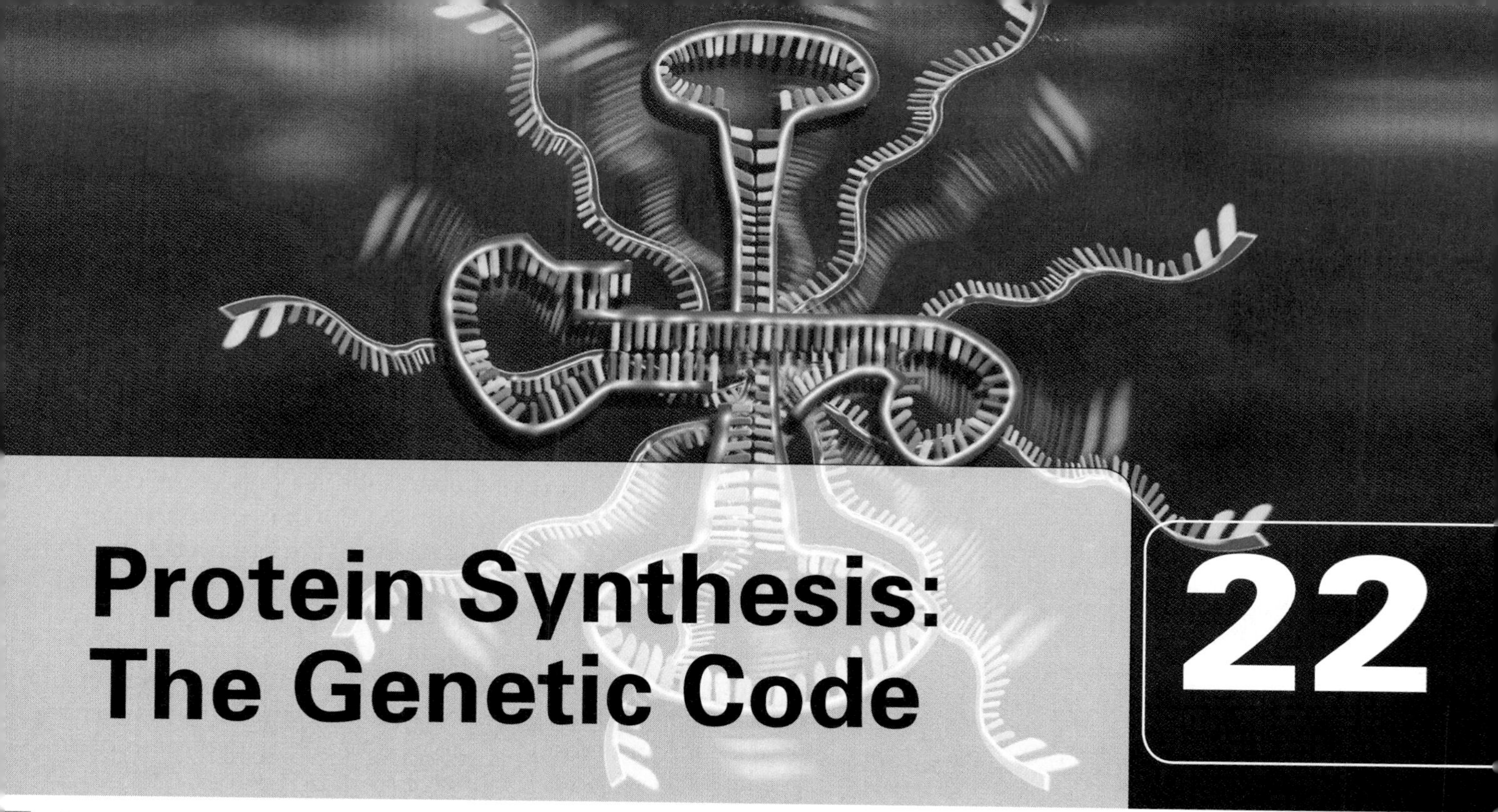

Protein Synthesis: The Genetic Code

22

OUTLINE OF TOPICS

22.1 Discovery of Ribosomes and Transfer RNA

Protein synthesis takes place on ribosomes.

22.2 Transfer RNA

An amino acid must be attached to a transfer RNA before it can be incorporated into a protein.

All tRNA molecules have CCA_{OH} at their 3′-end.

An amino acid attaches to tRNA through an ester bond between the amino acid's carboxyl group and the 2′- or 3′-hydroxyl group on adenosine.

The tRNA for alanine was the first naturally occurring nucleic acid to be sequenced.

Transfer RNAs have cloverleaf secondary structures.

Transfer RNA molecules fold into L-shaped three-dimensional structures.

22.3 Aminoacyl-tRNA Synthetases

Aminoacyl-tRNA synthetases can be divided into two classes, I and II.

Some aminoacyl-tRNA synthetases have editing functions.

Ile-tRNA synthetase can hydrolyze valyl-$tRNA^{Ile}$ and valyl-AMP.

A polypeptide insert in the Rossmann-fold domain forms the editing site for Ile-tRNA synthetase.

Editing defective alanyl-tRNA synthetase causes a neurodegenerative disease in mice.

Each aminoacyl-tRNA synthetase can distinguish its cognate tRNAs from all other tRNAs.

Many gram-positive bacteria and archaea use an indirect pathway for Gln-tRNA synthesis.

Selenocysteine and pyrrolysine are building blocks for polypeptides.

22.4 Messenger RNA and the Genetic Code

Messenger RNA programs ribosomes to synthesize proteins.

Three adjacent bases in mRNA that specify an amino acid are called a codon.

The discovery that poly(U) directs the synthesis of poly(Phe) was the first step in solving the genetic code.

Protein synthesis begins at the amino terminus and ends at the carboxyl terminus.

Messenger RNA is read in a 5′ to 3′ direction.

Trinucleotides promote the binding of specific aminoacyl-tRNA molecules to ribosomes.

Synthetic messengers with strictly defined base sequences confirmed the genetic code.

Three codons, UAA, UAG, and UGA, are polypeptide chain termination signals.

The genetic code is nonoverlapping, commaless, almost universal, highly degenerate, and unambiguous.

The coding specificity of an aminoacyl-tRNA is determined by the tRNA and not the amino acid.

Some aminoacyl-tRNA molecules bind to more than one codon because there is some play or wobble in the third base of a codon.

The origin of the genetic code remains a puzzle.

Suggested Reading

Protein synthetic studies began even before DNA's structure was known, but these studies proceeded in a rather haphazard way because investigators lacked an intellectual framework to guide their work. The discovery that DNA consists of two polynucleotide strands arranged as a double helix allowed investigators to think about structure–function relationships for the first time.

Francis Crick did just that in 1957 when he proposed the **sequence hypothesis** and the **central dogma**. The sequence hypothesis postulates that the linear sequence of nucleotides in a DNA (or RNA) chain determines the linear sequence of amino acids in a polypeptide chain. Central dogma, which despite its name is a hypothesis, postulates that genetic information flows from DNA to RNA to protein but not from proteins to nucleic acids. Together, these two hypotheses helped provide the theoretical underpinning needed to make advances in the study of protein synthesis. Although this chapter begins with a brief introduction to the ribosome's role as the protein synthetic factory, its major focus is on information flow rather than the details of events that take place on the ribosome. The ribosome's function in protein synthesis will be described in Chapter 23.

22.1 Discovery of Ribosomes and Transfer RNA

Protein synthesis takes place on ribosomes.

RNA's involvement in polypeptide synthesis, which is taken for granted today, was not firmly established until the 1950s. The earliest clues that RNA plays some role in information flow between genes and polypeptides came from the independent work of Torbjörn Caspersson and Jean Brachet in the late 1930s and early 1940s. Caspersson used cytochemical techniques to show that: (1) most of a eukaryotic cell's DNA and RNA are in its nucleus and cytoplasm, respectively; (2) cells that are actively engaged in protein synthesis have high levels of cytoplasmic RNA; and (3) cytoplasmic RNA tends to be concentrated in small spherical particles. Brachet, using cell fractionation techniques to separate nuclear and cytoplasmic fractions, also observed a correlation between active protein synthesis and high cytoplasmic RNA levels. In addition, Brachet centrifuged the cytoplasmic fraction at high speed to obtain a pellet containing the RNA-rich spherical particles that Caspersson had observed in the cell. Because these RNA-rich particles contained tightly bound proteins, they are more properly described as ribonucleoproteins. By the early 1950s, electron microscopy techniques had improved to the point where they could be used to visualize the spherical ribonucleoprotein particles, which appeared to be about 250 Å in diameter. These particles were known by a variety of different names until 1957, when Richard B. Roberts coined the now universally accepted term **ribosome**.

Some eukaryotic ribosomes appear to be free in the cytoplasm, whereas others are attached to the outer surface of a continuous intracellular tubular membrane network that Keith Porter named the

endoplasmic reticulum (ER; FIGURE 22.1). Regions of the ER that are studded with ribosomes appear rough or grainy in electron micrographs and are therefore known as the **rough endoplasmic reticulum** (**RER**). Regions that lack ribosomes have a smooth appearance and are therefore called the **smooth endoplasmic reticulum** (**SER**). The ER cannot be isolated as a continuous membrane network because methods that disrupt the cell membrane also rupture the endoplasmic reticulum. Pieces of ER, however, can be isolated by differential centrifugation of a crude cell homogenate (FIGURE 22.2). The crude cell extract is centrifuged at about 600 × *g* for five minutes to remove intact cells and nuclei, and then the post-nuclear supernatant is centrifuged at about 15,000 × *g* for one hour to remove mitochondria, lysosomes, and peroxisomes. Further centrifugation at 100,000 × *g* for two hours produces a pellet called the **microsomal fraction** that contains pieces of the endoplasmic reticulum and free ribosomes.

In 1953, Paul Zamecnik and coworkers provided the first convincing evidence that the microsomal fraction makes an important contribution to protein synthesis. Their approach was to inject rats with radioactive amino acids, kill groups of rats at intervals, remove the livers, disrupt the fresh livers, fractionate disrupted liver cells, and determine the amount of radioactive protein that was present in the different cellular fractions. They observed that radioactive proteins first appeared in ribosomes associated with the microsomal fraction, suggesting that ribosomes are protein synthetic factories.

Building on these *in vivo* studies, Zamecnik and coworkers worked to develop a cell-free system that would permit them to characterize components of the protein synthetic system and learn how the components work. Their efforts were rewarded by the discovery that a suspension containing rat liver ribosomes, the 100,000 × *g* supernatant,

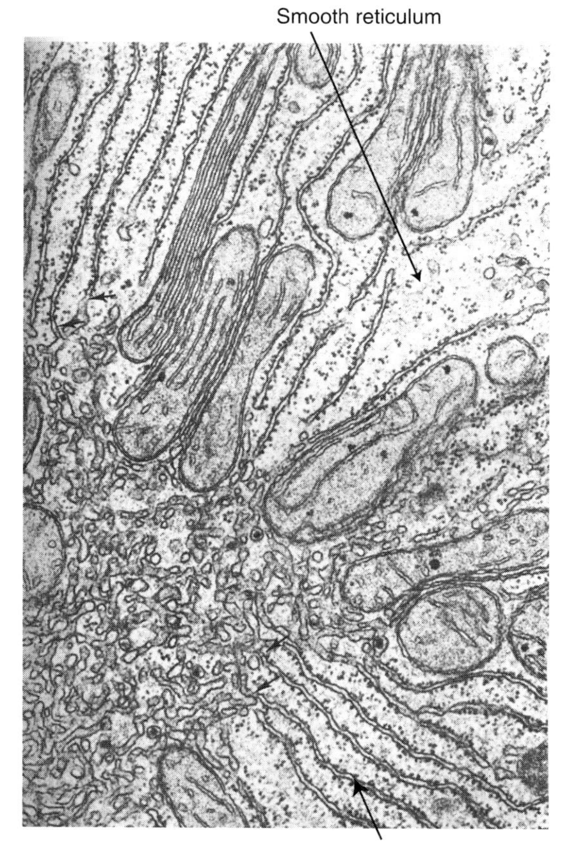

FIGURE 22.1 **An electron micrograph of a liver cell showing RER (rough endoplasmic reticulum) and patches of SER (smooth endoplasmic reticulum).** (Reproduced from *The Journal of Cell Biology*, 1973, 56: 746–761. © The Rockefeller University Press. Photo courtesy of Ewald R. Weibel, Universität Bern.)

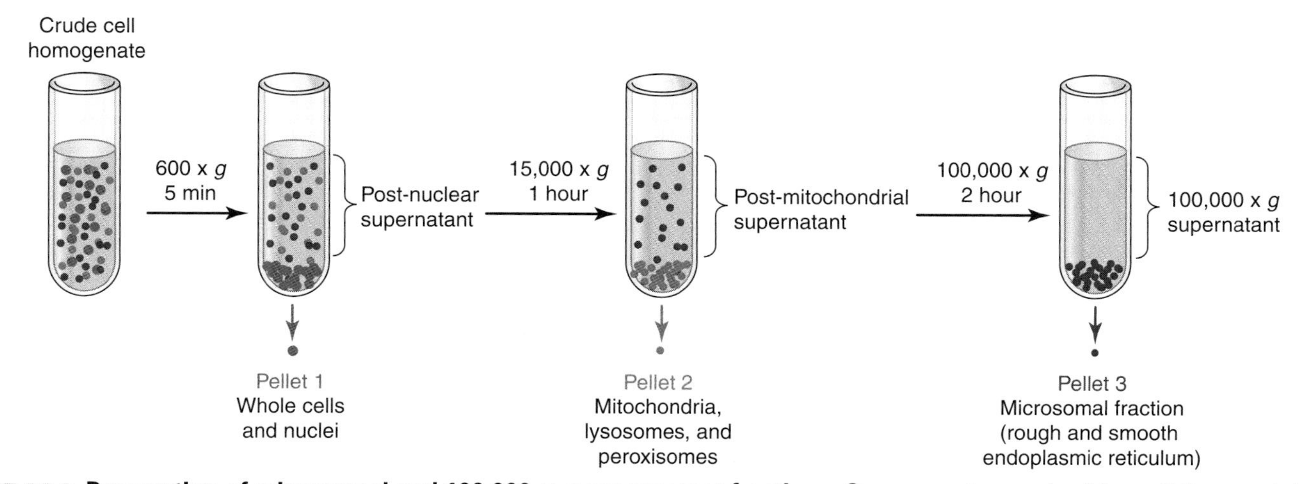

FIGURE 22.2 **Preparation of microsomal and 100,000 × *g* supernatant fractions.** Components required for cell-free protein synthesis are present in the microsomal and 100,000 × *g* supernatant fractions. These fractions are isolated by differential centrifugation of the crude cell extract. A low speed centrifugation to remove whole cells and nuclei from the crude cell extract is followed by an intermediate speed centrifugation to remove mitochondria, lysosomes, and peroxisomes. Then the post-mitochondrial supernatant is centrifuged at high speeds to produce a pellet (microsomal fraction) and a supernatant (100,000 × *g* supernatant). The microsomal fraction contains the cell's rough and smooth endoplasmic reticulum, while the 100,000 × *g* supernatant contains the cell's soluble cytoplasmic components.

adenosine triphosphate (ATP), and guanosine triphosphate (GTP) converts acid-soluble [^{14}C]amino acids into acid-insoluble radioactive proteins.

22.2 Transfer RNA

An amino acid must be attached to a transfer RNA before it can be incorporated into a protein.

The next major advance in the study of protein synthesis was made by Mahlon Hoagland while working in Zamecnik's laboratory in 1953. Hoagland's initial objective was to learn how energy is provided for protein synthesis. He thought amino acids must be activated before they can combine to form proteins and suspected this activation was through adenylyl (AMP) group transfer from ATP to α-carboxyl groups on the amino acids.

$$\text{ATP} + \text{Amino Acid} \rightleftharpoons \text{Aminoacyl-AMP} + \text{pyrophosphate}$$

Hoagland sought evidence to support this hypothesis by testing various cell fractions to see which, if any, could catalyze amino acid-dependent [^{32}P]pyrophosphate–ATP exchange. His efforts were rewarded by the discovery that the 100,000 × *g* rat liver supernatant catalyzes vigorous amino acid-dependent [^{32}P]pyrophosphate incorporation into ATP, suggesting that the cytoplasm contains enzymes that activate amino acids by reversible adenylylation of the amino acids' carboxyl groups. Hoagland provided further support for the activation hypothesis by demonstrating that the 100,000 × *g* supernatant attaches activated amino acids to hydroxylamine (a nonphysiological carboxyl group acceptor) to form aminoacyl hydroxamate.

$$\text{ATP} + \text{R—}\underset{\text{NH}_3^+}{\overset{\text{H}}{\text{C}}}\text{—COO}^- + \underset{\text{hydroxylamine}}{\text{NH}_2\text{OH}} \rightarrow \underset{\text{aminoacyl hydroxamate}}{\text{R—}\underset{\text{NH}_3^+}{\overset{\text{H}}{\text{C}}}\text{—}\overset{\text{O}}{\overset{\|}{\text{C}}}\text{—NHOH}} + \text{AMP} + \text{pyrophosphate}$$

Hoagland's studies thus showed that the 100,000 × *g* supernatant can activate an amino acid by converting it to an aminoacyl-adenylate, which can then transfer the aminoacyl group to the artificial acceptor hydroxylamine. Hoagland's preliminary efforts to fractionate proteins in the 100,000 × *g* supernatant indicated that each amino acid is probably activated by an enzyme that is specific for it. This conclusion was later verified by many different laboratories when amino acid activating enzymes were purified. Although Hoagland's experiments tell us a great deal about the energy source for protein synthesis, they tell us nothing about how genetic information is transferred from a polynucleotide chain to a polypeptide chain. The discovery of aminoacyl AMP, however, would soon lead to the discovery of a new kind of RNA that has a direct bearing on the information problem.

Zamecnik wondered whether aminoacyl-adenylate might be an intermediate in RNA synthesis as well as protein synthesis. Even though we now know that aminoacyl-adenylate is not an intermediate in RNA synthesis, one of the experiments that Zamecnik performed to test this possibility requires discussion because it led to the discovery of **transfer RNA (tRNA)**, a key component of the protein synthetic machinery. The significant experiment was one in which Zamecnik mixed [^{14}C]amino acids, ATP, magnesium ions, and the 100,000 × *g* supernatant and then looked for the incorporation of radioactive label into RNA. Zamecnik did not expect to observe any radioactive incorporation in this experiment and had only included it as a control for an experiment that contained radioactive nucleotides. To his amazement, [^{14}C]amino acids were rapidly attached to soluble RNA molecules present in the 100,000 × *g* supernatant. Although Zamecnik knew that the 100,000 × *g* supernatant contains soluble RNA, like other investigators, he had mistakenly assumed this RNA was "junk" produced when ribosomes are degraded. The experiments performed by Zamecnik and coworkers showed that far from being junk, soluble RNA molecules are natural acceptors for activated aminoacyl groups and therefore likely to be essential protein synthetic machinery components.

Hoagland and associates then performed decisive experiments showing that an amino acid must be attached to soluble RNA before it can be incorporated into proteins. They began by taking advantage of their previous studies to synthesize [^{14}C]aminoacyl-soluble RNA under *in vitro* conditions. Then they added purified [^{14}C]aminoacyl-soluble RNA to a protein synthetic mixture containing 100,000 × *g* supernatant, ribosomes, ATP, and GTP, and monitored radioactive incorporation into protein. As expected, the radioactive label was incorporated into protein. Moreover, different radioactive amino acids were incorporated in an additive fashion. But Hoagland and coworkers still needed to rule out the possibility that [^{14}C]amino acids were first released from soluble RNA and then incorporated into protein. They did so by demonstrating that radioactive incorporation is not affected by adding unlabeled amino acids to the reaction mixture. If [^{14}C]amino acids were released from soluble RNA prior to incorporation into protein, then the unlabeled amino acids would have lowered their specific activity and caused a marked decrease in radioactive amino acid incorporation. Thus, amino acids attached to soluble RNA are directly transferred to growing polypeptide chains on ribosomes. The term soluble RNA (sRNA), indicating a physical property, was eventually replaced by transfer RNA (tRNA), indicating a function.

Building on their pioneering studies, Zamecnik and coworkers showed that the 100,000 × *g* supernatant contains activating enzymes called **aminoacyl-tRNA synthetases** (also called aminoacyl-tRNA ligases). Although each aminoacyl-tRNA synthetase is specific for the amino acid and tRNA that it acts on, all of the synthetases catalyze the same two-step reaction pathway (FIGURE 22.3). First the synthetase catalyzes adenylyl (AMP) group transfer from ATP to the amino acid's α-carboxyl group to form aminoacyl-adenylate and pyrophosphate (Figure 22.3a). Then the synthetase catalyzes aminoacyl group transfer

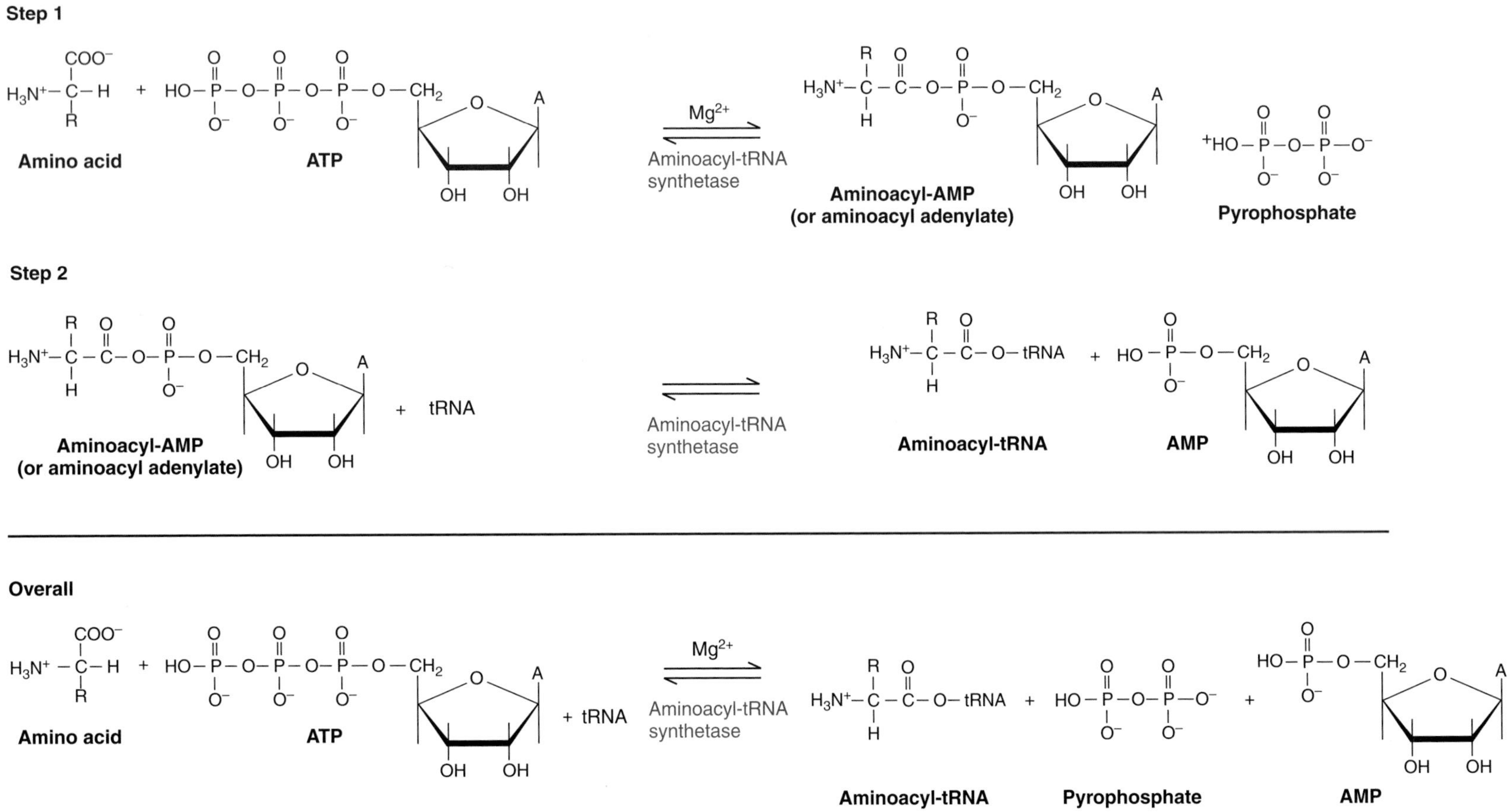

Step 1

Amino acid + ATP $\xrightleftharpoons[\text{Aminoacyl-tRNA synthetase}]{Mg^{2+}}$ Aminoacyl-AMP (or aminoacyl adenylate) + Pyrophosphate

Step 2

Aminoacyl-AMP (or aminoacyl adenylate) + tRNA $\xrightleftharpoons[\text{Aminoacyl-tRNA synthetase}]{}$ Aminoacyl-tRNA + AMP

Overall

Amino acid + ATP + tRNA $\xrightleftharpoons[\text{Aminoacyl-tRNA synthetase}]{Mg^{2+}}$ Aminoacyl-tRNA + Pyrophosphate + AMP

FIGURE 22.3 **Reactions catalyzed by aminoacyl-tRNA synthetases.**

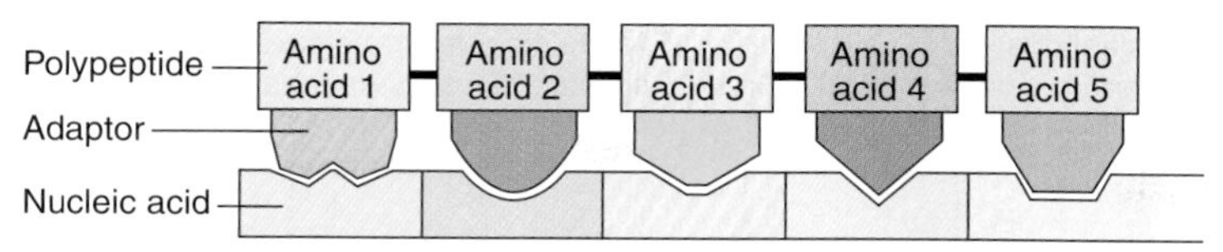

FIGURE 22.4 **The Adaptor Hypothesis.** (Adapted from D. Voet and J. G. Voet. *Biochemistry, Third edition.* John Wiley & Sons, Ltd., 2005.)

from aminoacyl-adenylate to tRNA to form aminoacyl-tRNA and AMP (Figure 22.3b). The net effect of the two transfer reactions is to convert ATP, amino acid, and tRNA into aminoacyl-tRNA, AMP, and pyrophosphate (Figure 22.3c). Transfer RNA molecules with bound amino acids are said to be charged, whereas those lacking amino acids are said to be uncharged. Although the equilibrium constant for the net reaction is close to 1, the reaction proceeds to completion in the cell because pyrophosphate is hydrolyzed.

At the start of their studies, Zamecnik and coworkers viewed unraveling the protein synthetic pathway as a biochemical problem. By 1956, however, their studies on tRNA had advanced to a stage where they were obliged to start thinking about protein synthesis in terms of the flow of genetic information. Because messenger RNA had not yet been discovered, the relationship between tRNA and information flow was not obvious. Remarkably, Francis Crick had predicted the existence of a molecule resembling tRNA in an informal note that he sent in 1955 to a small group of molecular biologists known as the "RNA Tie Club" (so-named because each club member had a necktie decorated with a specific amino acid or nucleotide). One of the club's major goals was to solve the coding problem. Crick's note called attention to the fact that free amino acids cannot directly line up on a nucleic acid template prior to polypeptide formation. He therefore proposed that each amino acid must first combine with a short specific adaptor RNA, which then can line up on an RNA template by forming specific base pairs with it (FIGURE 22.4).

Crick's adaptor hypothesis makes two important predictions: (1) there must be at least twenty different adaptor RNA molecules, one for each amino acid; and (2) an amino acid must attach to its cognate (correct) adaptor before it can be incorporated into protein. Zamecnik and Hoagland were not aware of Crick's informal note. However, when James Watson told them about Crick's adaptor hypothesis in late 1956, they immediately realized that tRNA might be Crick's adaptor.

All tRNA molecules have CCA_{OH} at their 3′-end.

Transfer RNA was such an obviously important molecule that many different laboratories became interested in it. One of the first clues to tRNA structure was the observation that the $100{,}000 \times g$ supernatant contains an exonuclease that cleaves a few exposed nucleotides from the 3′-end of tRNA, rendering the tRNA unable to accept amino acids (FIGURE 22.5a). Further studies showed that the $100{,}000 \times g$ supernatant also contains an enzyme that can restore the digested

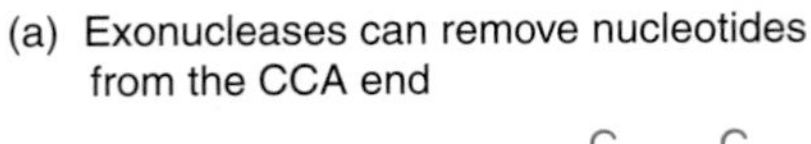

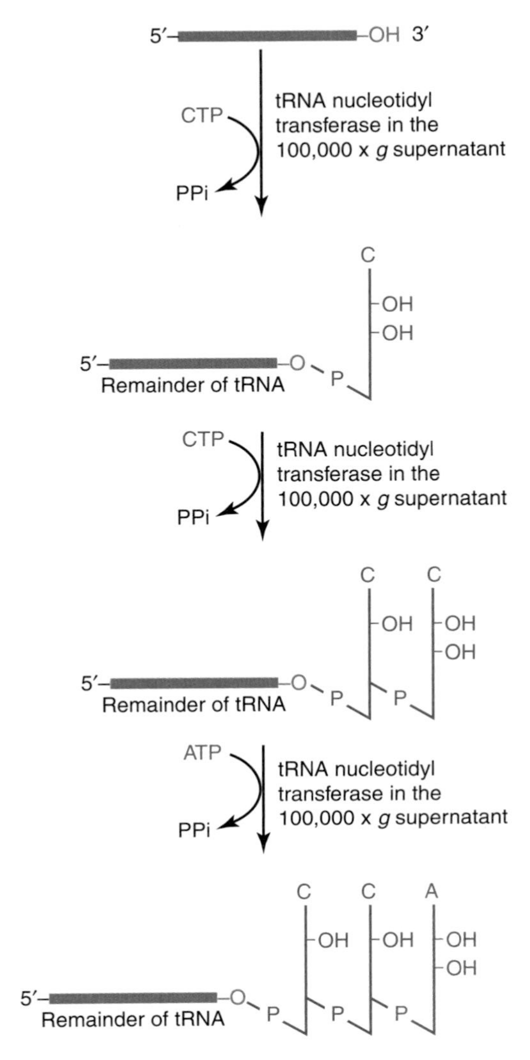

FIGURE 22.5 Reactions at the 3′-end of tRNA. The sequence at the 3′-end of all tRNA molecules is CCA. (a) Exonucleases can inactivate tRNA by removing the CCA end. (b) The enzyme known as tRNA nucleotidyl transferase restores activity by first restoring the two cytidylate residues and then the adenylate residue.

tRNA molecule's ability to accept amino acids when ATP and CTP are added to the mixture (**FIGURE 22.5b**). More careful examination of this restoration led to the discovery that all tRNA molecules have CCA_{OH} at their 3′-ends. The key experiments were as follows:

1. [^{14}C]CTP can transfer its CMP group to the 3′-end of the digested tRNA in the absence of ATP.
2. CTP must be present for [^{14}C]ATP to transfer its AMP group.
3. The molar ratio of CMP:AMP transfer is 2:1.

Later studies showed that a single enzyme called **tRNA nucleotidyl transferase** catalyzes all three nucleotide transfer reactions. This enzyme is essential in eukaryotes because most eukaryotic tRNA genes do not code for the CCA_{OH} end. Although bacterial tRNA genes usually do code for the CCA_{OH} end, bacteria require tRNA nucleotidyl transferase to repair tRNA molecules that have lost their CCA_{OH} end.

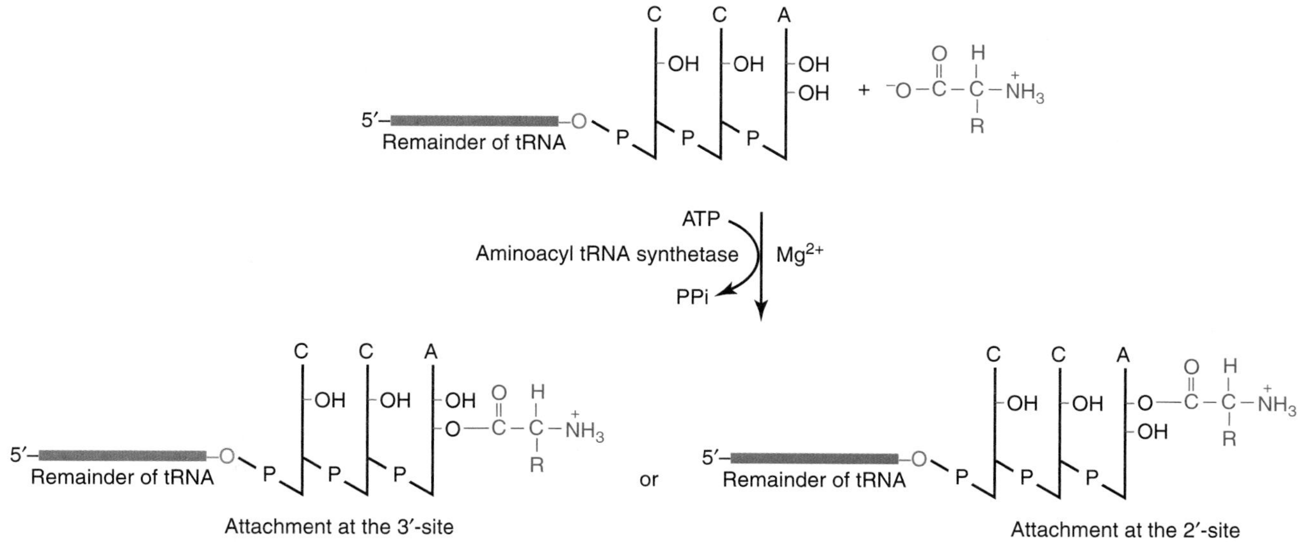

FIGURE 22.6 **Site of aminoacyl group attachment to tRNA.** Aminoacyl-tRNA synthetases attach aminoacyl groups to the 2′- or 3′-hydroxyl groups of the 3′-terminal adenosine group.

An amino acid attaches to tRNA through an ester bond between the amino acid's carboxyl group and the 2′- or 3′-hydroxyl group on adenosine.

Paul Berg and coworkers explored the nature of aminoacyl group attachment to bacterial tRNA. They first demonstrated that periodate (IO_4^-), a reagent that specifically cleaves the bond between carbon atoms attached to the 2′- and 3′-hydroxyl groups in the 3′-terminal adenosine, completely inactivates tRNA. Then they showed that tRNA loses its ability to accept amino acids much more rapidly when digested with a 3′→5′ exonuclease than when digested with a 5′→3′ exonuclease.

Based on these results, it seemed likely that the aminoacyl group is attached to the 3′-terminal adenosine in tRNA by an ester bond between the α-carboxyl on the aminoacyl group and the 2′- or 3′-hydroxyl group of the adenosine (FIGURE 22.6). If the aminoacyl group is so attached, then pancreatic RNase digestion of aminoacyl-tRNA should release aminoacyl-adenosine (FIGURE 22.7). As predicted, RNase digestion does release aminoacyl-adenosine. Moreover, periodate has no effect on the released aminoacyl-adenosine, indicating the amino acid is attached to the 2′- or 3′-hydroxyl group of adenosine. As we will see below, some aminoacyl-tRNA synthetases add the aminoacyl group to the 2′-hydroxyl, whereas others add it to the 3′-hydroxyl. The original site of attachment is not critical because an aminoacyl group can migrate rapidly between the two hydroxyl groups. The ester bond attaching the aminoacyl group to tRNA is rapidly hydrolyzed under alkaline conditions. For example, Val-tRNA has a half-life of about five minutes at pH 8.6 at 37°C. This instability permits aminoacyl groups to be removed under mild alkaline conditions that leave tRNA intact.

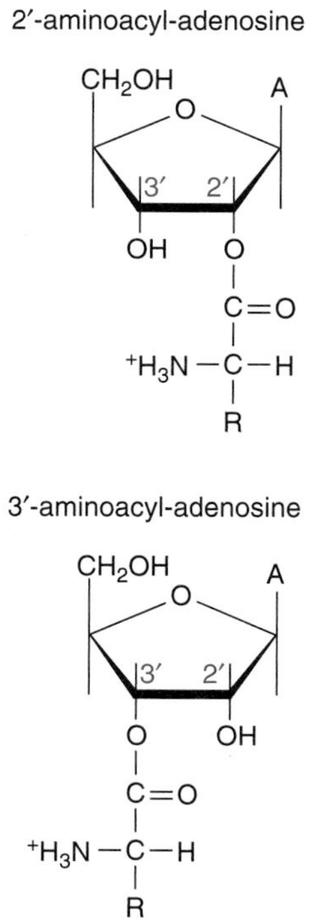

FIGURE 22.7 **Aminoacyl-adenosine.**

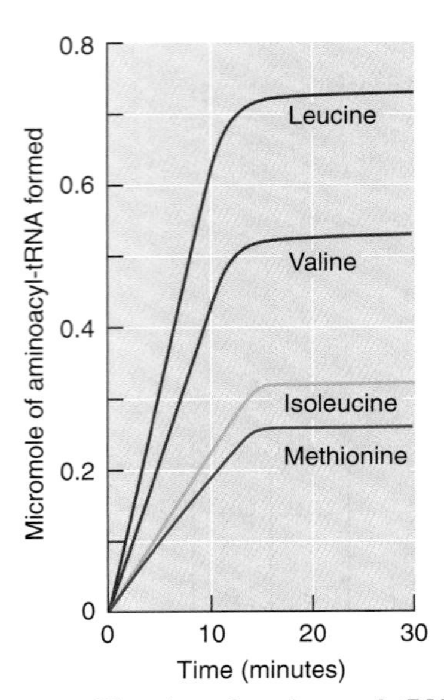

FIGURE 22.8 **Kinetics of aminoacyl-tRNA formation.** Each reaction mixture contains ATP, magnesium ions, the radioactive amino acid indicated, and the purified *E. coli* aminoacyl-tRNA synthetase specific for that amino acid. (Adapted from P. Berg, et al., *J. Biol. Chem.* 236 [1961]: 1726–1734.)

Crick's adaptor hypothesis predicts that each amino acid has at least one tRNA that is specific for it. Two different kinds of experiments confirmed this prediction:

1. Berg and coworkers performed *in vitro* experiments that showed aminoacyl-tRNA formation proceeds linearly with time and then levels off (FIGURE 22.8). Although the extent of aminoacyl-tRNA formation varied from one amino acid to another, tRNA was the limiting factor for each amino acid. When two or more amino acids were included in the same reaction mixture, the total amount of aminoacyl-tRNA formed equaled the sum of the amounts formed when each amino acid was present by itself. Furthermore, the presence of one amino acid did not affect the extent of incorporation of other amino acids. These experiments are consistent with Crick's prediction that each amino acid attaches to tRNA(s) specific for it.
2. Robert Holley and coworkers purified specific tRNA molecules from a mixture of tRNAs isolated from baker's yeast (*Saccharomyces cerevisiae*), which was chosen because it is available in large amounts at low cost. They purified specific tRNAs from the crude mix using a form of liquid-liquid chromatography known as **countercurrent distribution**, which separates components from a mixture by repeated distribution between two immiscible liquid phases moving past one another in opposite directions. As predicted by Crick's adaptor hypothesis, each tRNA species accepted only one kind of amino acid. Transfer RNA specificity is indicated by writing the amino acid's three-letter abbreviation as a superscript. For instance, tRNA molecules that accept leucine, valine, and phenylalanine are designated $tRNA^{Leu}$, $tRNA^{Val}$, and $tRNA^{Phe}$, respectively. Some amino acids can attach to more than one tRNA species. Different tRNAs that accept the same amino acid, called **isoacceptors**, are differentiated by adding a number as a subscript. For example, isoacceptors for valine are named $tRNA_1^{Val}$ and $tRNA_2^{Val}$.

DNA sequence studies for entire bacterial genomes show the number of tRNA genes varies from a low of 33 in *Mycoplasma genitalium* (the bacteria thought to have the smallest genome) to a high of 88 in *Bacillus subtilis* (a gram-positive bacteria). Because some tRNA molecules are encoded by more than one tRNA gene, the number of different tRNA molecules is often less than the number of tRNA genes. For instance, *E. coli* has 84 tRNA genes but only 45 different tRNA molecules. Eukaryotes have a much larger number of genes for cytoplasmic tRNA. For example, *S. cerevisiae* has about 270 tRNA genes while humans have about twice that number.

The tRNA for alanine was the first naturally occurring nucleic acid to be sequenced.

Holley and coworkers started the ambitious and difficult task of determining the nucleotide sequence of a tRNA molecule in 1958, a project that took seven years to complete. Most of the time was spent developing techniques to purify a single species of yeast $tRNA^{Ala}$. More than 600 kg of commercial baker's yeast was processed to obtain 200 g of crude tRNA from which 1 g of pure $tRNA^{Ala}$ was finally isolated. Samples of the purified tRNA were cut with specific nucleases and the resulting fragments isolated and sequenced. Once this arduous task was completed, Holley and coworkers determined the order of fragments within the tRNA molecule by matching overlaps in much the same way that protein chemists determine the sequence of large polypeptides.

Today it is usually more convenient to sequence the tRNA gene rather than the tRNA. When necessary, however, tRNA can be isolated and sequenced in a matter of weeks rather than years. One of the major reasons for this improvement is a change in the methods used to detect nucleotides. The spectrophotometric methods used by Holley have been replaced by much more sensitive radiolabeling techniques. Consequently, microgram quantities of pure tRNA are sufficient to obtain a complete nucleotide sequence.

Transfer RNAs have cloverleaf secondary structures.

Holley and coworkers examined the nucleotide sequence of $tRNA^{Ala}$, hoping to find clues to polyribonucleotide chain folding. They assumed that intramolecular base pair formation was the driving force for the folding process. Self-complementary sequences that could form hairpin structures were of special interest. Several alternative folding patterns, each with about the same number of intramolecular base pairs, could be envisioned. The correct folding pattern, however, could not be determined from the sequence of a single tRNA species. As sequence information for other tRNA molecules became available, it became apparent that the only secondary structure common to all was one in which the polynucleotide chain folds into a structure that resembles a three-leaf clover (FIGURE 22.9). Several hundred different tRNA species have now been sequenced from a wide variety of organisms. All cytoplasmic tRNAs are approximately 75 to 95 nucleotides long and have a cloverleaf secondary structure. As shown in Figure 22.9, fifteen nucleotides within tRNAs are highly conserved.

In addition to the four major ribonucleosides, each tRNA also contains modified nucleosides. All modifications are introduced after transcription. These modifications include adding one or more methyl groups, reducing a double bond, changing the attachment site to a pyrimidine ring, replacing an oxygen atom with a sulfur atom, and adding a large substituent. Some nucleosides have more than one type of modification. Structures of several of the more than 100 modified nucleosides identified in tRNA are shown in FIGURE 22.10. Three uridine derivatives—ribothymidine, pseudouridine, and dihydrouridine—are present in all tRNAs (Figure 22.10).

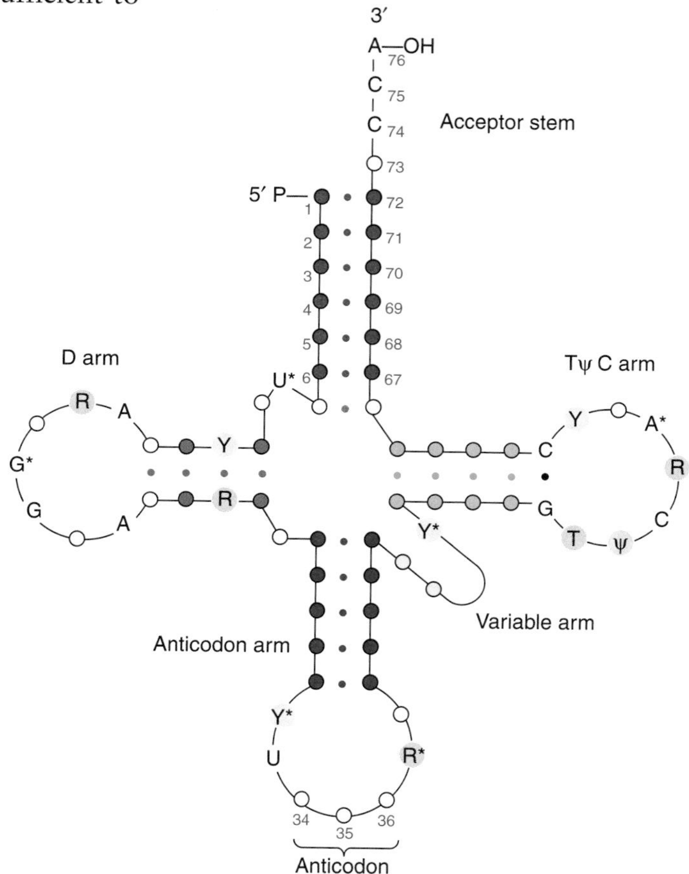

FIGURE 22.9 The cloverleaf secondary structure. Invariant bases are indicated by the first letter of their name. Invariant pyrimidines and purines are indicated by R and Y, respectively. The symbol Ψ represents pseudouridine. Bases involved in Watson–Crick base pairs are represented by filled circles connected by dots. Other bases are represented by open circles. The D and variable arms contain different numbers of nucleotides in the various tRNAs. The TΨC-, anticodon-, and D-arms each contain a double-stranded stem structure and a loop structure. (Adapted from D. Voet and J. G. Voet. *Biochemistry, Third edition.* John Wiley & Sons, Ltd., 2005.)

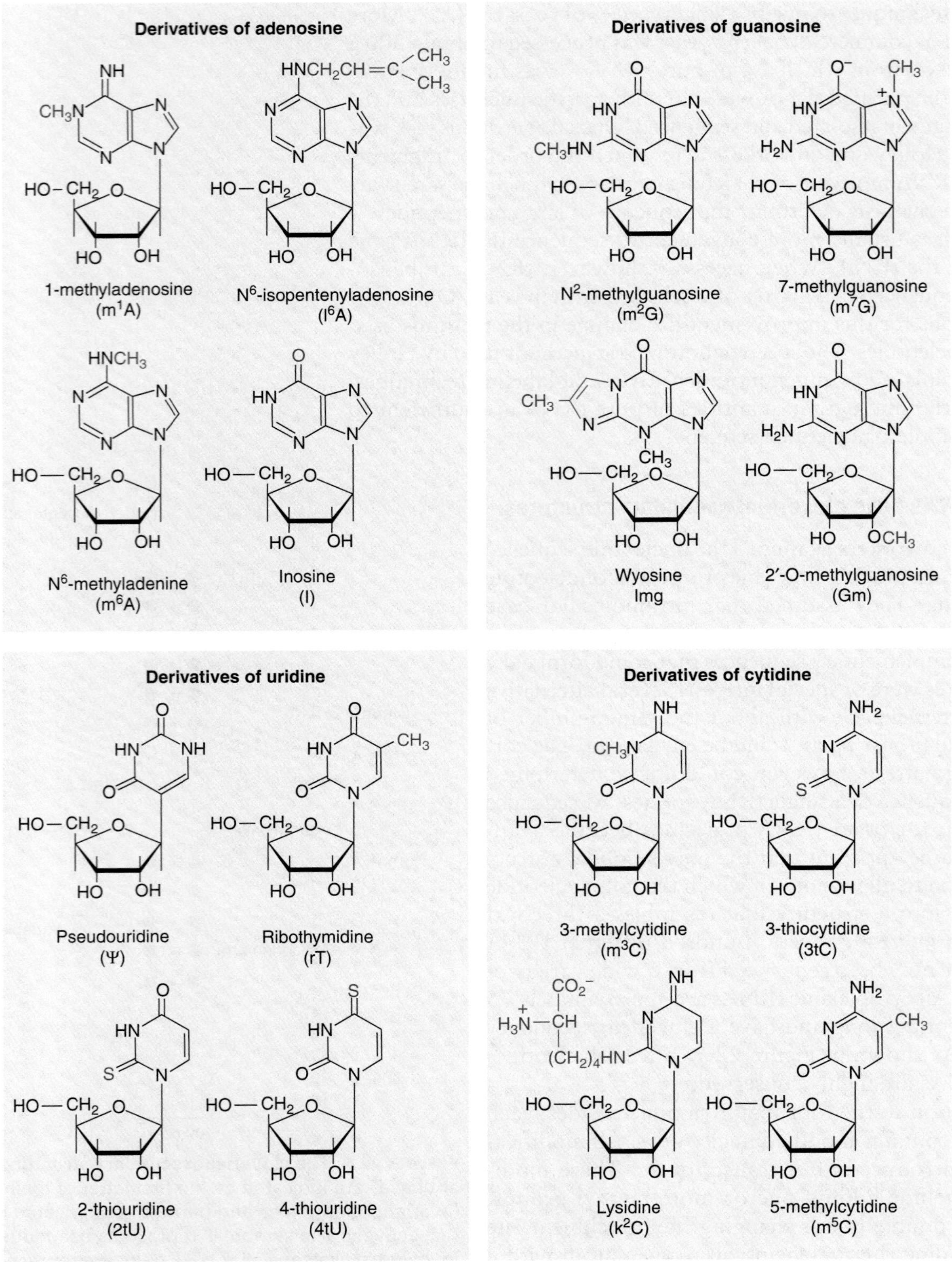

Derivatives of adenosine

1-methyladenosine
(m^1A)

N^6-isopentenyladenosine
(I^6A)

N^6-methyladenine
(m^6A)

Inosine
(I)

Derivatives of guanosine

N^2-methylguanosine
(m^2G)

7-methylguanosine
(m^7G)

Wyosine
Img

2′-O-methylguanosine
(Gm)

Derivatives of uridine

Pseudouridine
(Ψ)

Ribothymidine
(rT)

2-thiouridine
(2tU)

4-thiouridine
(4tU)

Derivatives of cytidine

3-methylcytidine
(m^3C)

3-thiocytidine
(3tC)

Lysidine
(k^2C)

5-methylcytidine
(m^5C)

FIGURE 22.10 **Some of the modified nucleosides in tRNA.**

Ribothymidine is formed by methyl group transfer from S-adenosylmethionine to a specific uridine. Pseudouridine is formed by uracil ring rotation so that C-5 rather than N-1 is linked to ribose. Dihydrouridine is formed by adding hydrogen atoms across the double bond between carbons 5 and 6 in uracil. Other modified nucleosides are present only at specific sites on a limited number of tRNAs. Many modified nucleosides help the tRNA to perform its function as an adaptor, in some cases assisting an aminoacyl-tRNA synthetase to distinguish between similar tRNA molecules and in others helping the tRNA to interact with mRNA.

The tRNA cloverleaf structure has five arms (Figure 22.9), each named for its function or unique chemical characteristic.

1. The **acceptor stem,** the site of amino acid attachment to tRNA, is formed by base pairing between nucleotides at the 5′ end and nucleotides near the 3′-end of the tRNA molecule. Nearly all acceptor stems are stabilized by seven base pairs. The last four nucleotides, including the 3′-CCA_{OH}, are not base paired.
2. The **anticodon arm** is a stem-loop structure that lies directly across from the acceptor arm and contains the **anticodon,** a trinucleotide that base pairs with the codon in mRNA in an antiparallel fashion (FIGURE 22.11) as described in greater detail below.
3. The **D-arm** is a stem-loop structure that is named for one or more dihydrouridine groups that are almost always present within it. The number of nucleotides in the D-arm varies from one tRNA to another.
4. The **TΨC** arm is a stem-loop structure named for the TΨC sequence present within the loop.
5. The **variable arm,** which lies between the anticodon and TΨC stems, contains 4 to 21 residues. Variations here are largely responsible for size differences among the tRNA molecules.

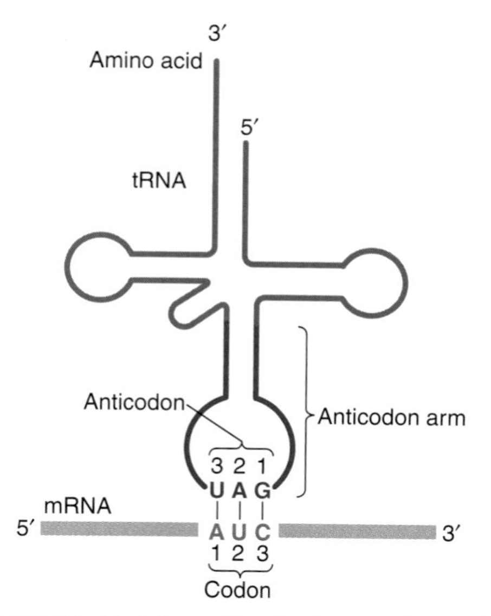

FIGURE 22.11 **Interaction between the anticodon on tRNA and the codon on mRNA.** Base pairing between the anticodon on a tRNA (magenta) and the codon on mRNA (green) is antiparallel. (Adapted from D. C. Nelson and M. M. Cox, *Lehninger Principles of Biochemistry, Third edition.* W.H. Freeman & Company, 2000.)

The acceptor arm, anticodon, and TΨC arms are so conserved in length that their residues are usually indicated by a standard numbering system. The first nucleotide at the 5′-terminus is residue 1 and the last residue at the 3′-terminus is residue 76 (Figure 22.9). Adenosine 76 (A76) (in the 5′-CCA_{OH} terminus) is the amino acid attachment site and the anticodon consists of nucleotides 34 to 36. Nucleotides in the D- and variable arms that are present in some tRNAs but not others are indicated by a separate numbering system.

Transfer RNA molecules fold into L-shaped three-dimensional structures.

Several research groups initiated programs to determine tRNA's tertiary structure in the late 1960s. Progress was quite slow at first

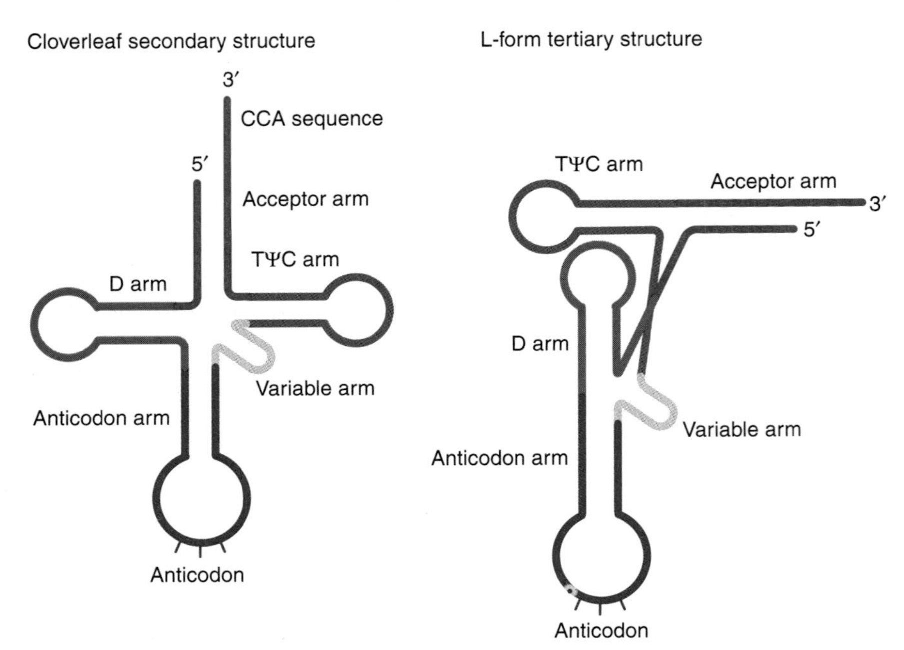

FIGURE 22.12 **Folding of cloverleaf structure of $tRNA^{Phe}$ to produce the L-form.**

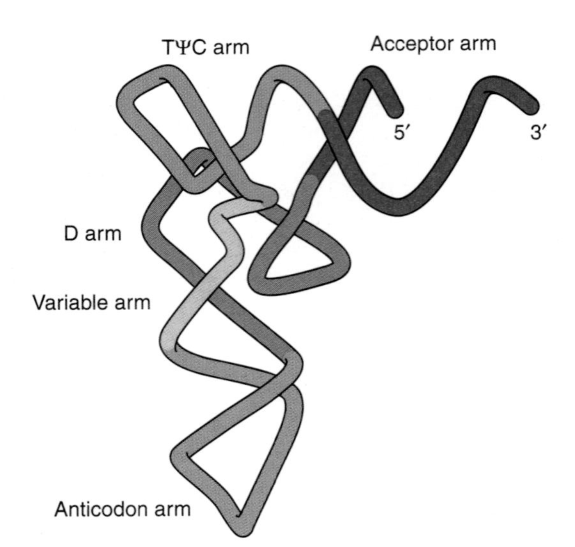

FIGURE 22.13 **Three-dimensional structure of tRNA.** (Modified from *Trends Biochem. Sci.*, vol. 22, J. G. Arnez and D. Moras, Structural and functional considerations of . . ., pp. 211–216, copyright 1997, with permission from Elsevier [http://www.sciencedirect.com/science/journal/09680004].)

because of the difficulty in obtaining crystals that were satisfactory for x-ray diffraction analysis. Finally, in 1974 two independent research groups, one led by Alexander Rich and the other by Aaron Klug, obtained crystals of yeast $tRNA^{Phe}$ that were suitable for x-ray diffraction analysis. The crystal structure showed that the cloverleaf folds into an L-shape. As shown in the schematic in FIGURE 22.12, the D and anticodon arms stack to form one section of the L, while the acceptor and TΨC arms stack to form the other section. The tertiary structure for tRNA is shown in FIGURE 22.13.

The tRNA molecule's tertiary structure is stabilized by complex interactions, some of which are shown in FIGURE 22.14. Helical regions in the acceptor, anticodon, D, and TΨC arms are usually stabilized by Watson-Crick base pairing as well as by non–Watson-Crick base pairing such as the G:U pair in the $tRNA^{Phe}$ acceptor arm. Nonhelical regions of the tRNA molecule are stabilized by hydrogen bonding interactions between two or three bases that are not usually considered to be complementary to one another and by hydrogen bonding interactions that involve the 2′-hydroxyl group in ribose. 2′-Hydroxyl interactions are especially interesting because they cannot occur in DNA molecules. Different tRNAs have a similar folding pattern, ensuring that various components of the protein synthetic machinery will be able to recognize the tRNA after an amino acid has been attached to it. However, tRNAs also must have unique features that can be recognized by aminoacyl-tRNA synthetases.

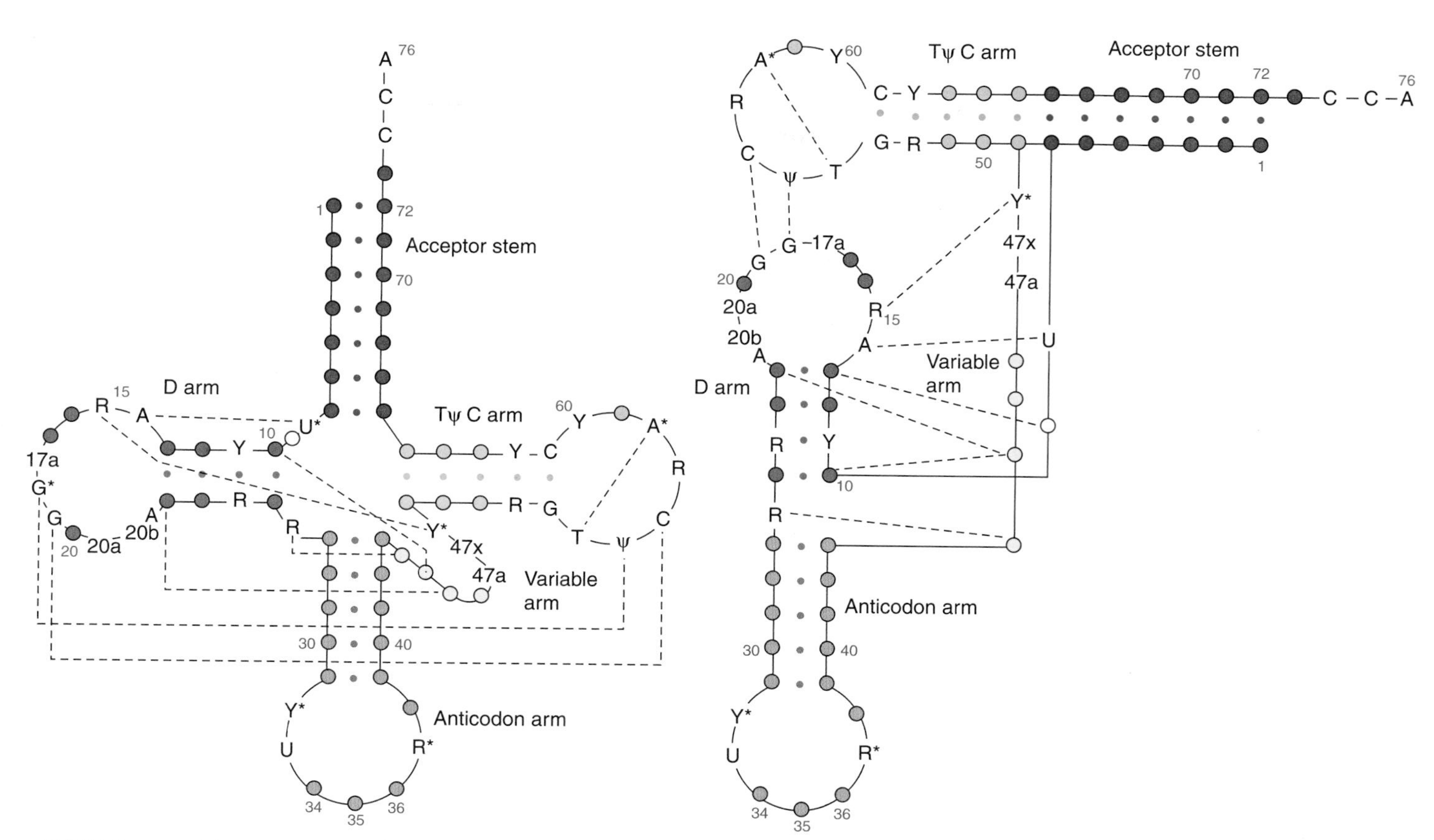

FIGURE 22.14 **Two representations of tRNA showing the cloverleaf secondary structure (left) and the tertiary fold (right).** The acceptor stem, TΨC stem-loop, D stem-loop, variable arm, and anticodon stem-loop are labeled. Dotted lines represent tertiary interactions, which are based on the crystal structure of $tRNA^{Phe}$, as is the numbering. Conserved residues are shown explicitly by letters representing the nucleotides; R indicates a purine, Y indicates a pyrimidine, and Ψ is pseudouridine. Conserved in this case means residues that are conserved in *E. coli* tRNAs with two or fewer exceptions. The variable nature of the D-loop and variable arm is indicated by numbering 17a, 20a, 20b, and 47a-47x, respectively, representing residues that may or may not be present. The three nucleotides of the anticodon, 34, 35, and 36, are also shown as numbers. (Adapted from P. J. Beuning and K. Musier-Forsyth, *Biopolymers* 52 [1999]: 1–28.)

22.3 Aminoacyl-tRNA Synthetases

Aminoacyl-tRNA synthetases can be divided into two classes, I and II.

Crystal structures for many different aminoacyl-tRNA synthetases have been solved, in some cases for the free enzyme and in others for the enzyme bound to substrates or substrate analogs. All synthetases have modular structures, with most containing two to four domains. The aminoacylation domain, which activates the amino acid and then transfers it to the cognate tRNA, has either of two mutually exclusive

TABLE 22.1 Class I and Class II Aminoacyl-tRNA Synthetases

Class I	Class II
Arg, Cys, Gln, Glu, Ile, Leu, Met, Trp, Tyr, Val	Ala, Asn, Asp, Gly, His, Lys,* Phe, Pro, Ser, Thr

Synthetases with editing activity are shown in red.
*Lysyl-tRNA synthetases in some archaea belong to class I.

TABLE 22.2 Principal Features of the Two Classes of Aminoacyl-tRNA Synthetases

	Class I	Class II
Conserved motifs Substrate binding Dimerization	 HIGH KMSKS	 Motif 2 Motif 3 Motif 1
Active-site fold	Parallel β-sheet (Rossmann fold)	Antiparallel β-sheet
ATP conformation	Extended	Bent
tRNA binding acceptor arm variable arm	Minor groove side faces solvent	Major groove side faces protein
Aminoacylation site	2′-OH	3′-OH

(Modified from *Trends Biochem. Sci.*, vol. 22, J. G. Arnez and D. Moras, Structural and functional considerations of . . ., pp. 211–216, copyright 1997, with permission from Elsevier [http://www.sciencedirect.com/science/journal/09680004].)

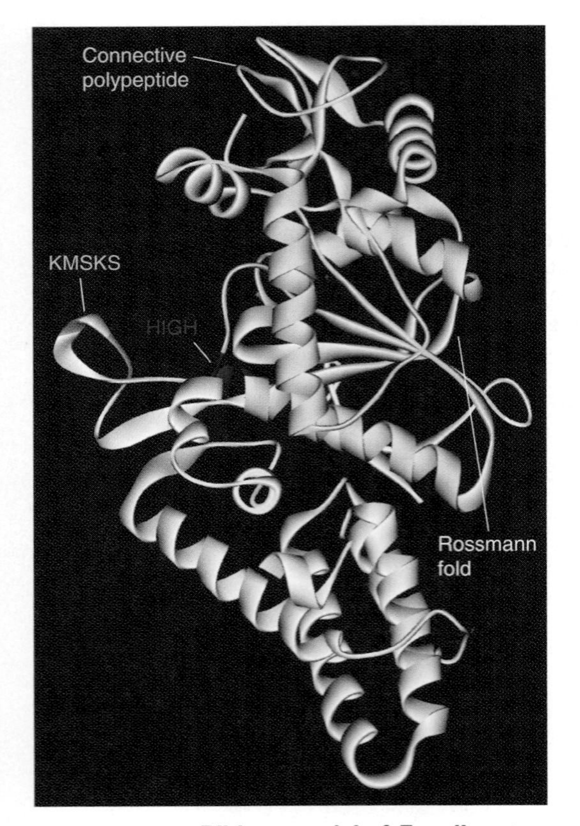

FIGURE 22.15 Ribbon model of *E. coli* cysteinyl-tRNA synthetase, a class 1 synthetase. The Rossmann fold is shown in blue, the connective polypeptide in green, the conserved sequence motif HIGH in red, and the conserved sequence motif KMSKS in orange. (Structure from Protein Data Bank 1LI5. K. J. Newberry, Y-.M. Hou, and J. J. Perona, *EMBO J.* 21 [2002]: 2778–2787. Prepared by B. E. Tropp.)

folding patterns. The conservation of the two aminoacylation domains in eukaryotes, bacteria, and archaea suggests that the domains arose early in evolution and that the other aminoacyl-tRNA synthetase domains, including those for tRNA recognition, were grafted on later.

The folding patterns of the two aminoacylation domains form the basis for dividing aminoacyl-tRNA synthetases into two major classes, class I and class II (Table 22.1). A given synthetase belongs to the same class regardless of the enzyme's biological origin. The only known exception to this rule is for Lys-tRNA synthetases, which are class I synthetases in some archaea and spirochetes but class II synthetases in all other organisms that have been studied to date.

Table 22.2 summarizes the principal features of class I and class II aminoacyl-tRNA synthetases. Class I enzymes tend to be monomers and class II enzymes tend to be α_2 dimers. However, some class I enzymes are dimers and some class II enzymes are α_4 or $\alpha_2\beta_2$ tetramers. Quaternary structures are usually, but not always, the same for a given aminoacyl-tRNA synthetase in different organisms.

Class I synthetases have a characteristic ATP binding domain called the Rossmann-fold, named for Michael Rossmann, the investigator who discovered the fold. The two halves of the Rossmann fold in *Escherichia coli* cysteinyl-tRNA synthetase, the class I synthetase shown in FIGURE 22.15, are joined by a connective polypeptide. Class I synthetases also have a conserved Lys-Met-Ser-Lys-Ser (KMSKS) sequence and a conserved His-Ile-Gly-His (HIGH) sequence. The

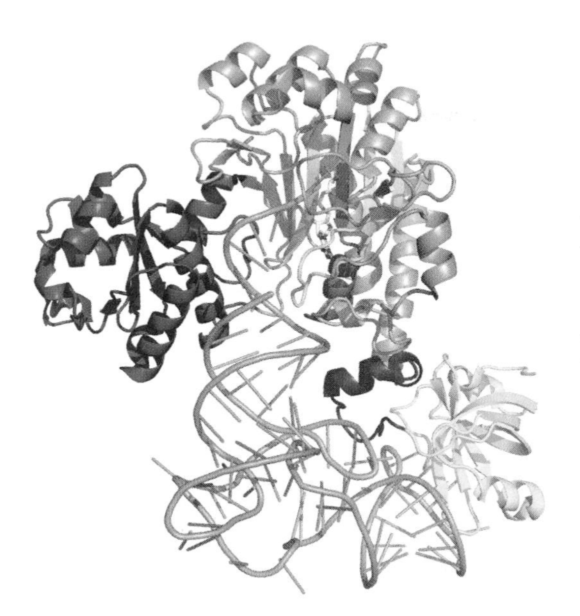

FIGURE 22.16 **Ribbon representation of *E. coli* aspartyl-tRNA synthetase, a class II synthetase.** Although the *E. coli* enzyme is a homodimer, only one subunit is shown for clarity. Class II synthetases have three conserved sequence motifs: motif 1 (green), motif 2 (cyan), and motif 3 (magenta). Residues in motifs 2 and 3 help in amino acid and ATP recognition. Residues in motif 1 serve as an interface between the subunits. The N-terminal domain, which interacts with the anticodon of tRNA, is shown in yellow. The insertion domain characteristic of bacterial aspartyl-tRNA synthetases is shown in blue and a small hinge module is shown in red. The $tRNA^{Asp}$ is shown in orange. (Adapted from S. Eiler, et al., *EMBO J.* 18 [1999]: 6532–6541. Photo courtesy of Dino Moras, Institut de Génétique et de Biologie Moléculaire et Cellulaire.)

HIGH motif interacts with the phosphate groups in ATP and helps to correctly position the adenine base. ATP assumes an extended conformation upon binding to a class I synthetase. The amino acid binding pocket in a class I synthetase tends to be open and relaxed.

The class II aminoacyl-tRNA synthetases have three conserved sequence motifs (FIGURE 22.16). Residues in motifs 2 and 3 are essential for catalysis, whereas those in motif 1 serve as an interface between subunits. ATP assumes a compact conformation upon binding to a class II synthetase. The γ-phosphate bends back over the adenine base as the nucleotide fits into a space formed by antiparallel β-strands in motifs 2 and 3. The differences in the conformations of ATP bound to class I and class II synthetases are readily apparent by superimposing the two ATP structures (FIGURE 22.17). The amino acid binding pocket in a class II synthetase tends to be rigid.

Class I synthetases bind to the minor groove side of the acceptor arm with the variable arm facing the solvent, whereas class II enzymes bind to the major groove side of the acceptor arm with the variable arm facing the protein (FIGURE 22.18). Class I and class II enzymes hence bind to the acceptor arm and the 3′-terminal CCA_{OH} in a mirror symmetric fashion with respect to one another. As a result of these differences, class I and II synthetases normally attach amino acids to the 3′-terminal adenosine at its 2′- and 3′-OH group, respectively.

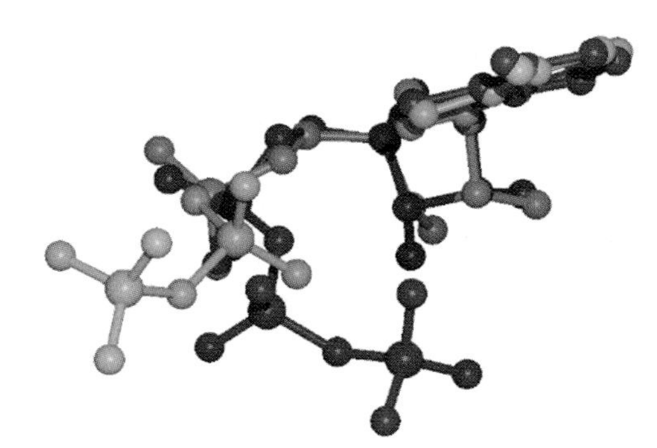

FIGURE 22.17 **Superimposition of ATPs from class I (light blue) and class II (red) aminoacyl-tRNA synthetases.** (Reproduced from *Trends Biochem. Sci.*, vol. 22, J. G. Arnez and D. Moras, Structural and functional considerations of . . ., pp. 211–216, copyright 1997, with permission from Elsevier [http://www.sciencedirect.com/science/journal/09680004]. Photo courtesy of Dino Moras, Institut de Génétique et de Biologie Moléculaire et Cellulaire.)

Some aminoacyl-tRNA synthetases have editing functions.

Each aminoacyl-tRNA synthetase must be able to select a single amino acid from a mixture of many others. As early as 1957, Linus Pauling pointed out that a synthetase should make many errors when trying to distinguish between two amino acids such as valine and isoleucine, which have very similar side chains. The error rate can be calculated by comparing the enzyme's binding energy for each amino acid. The calculated error rate for two amino acids that differ in just one $-CH_2-$ group is 1 in 200, based on a binding energy

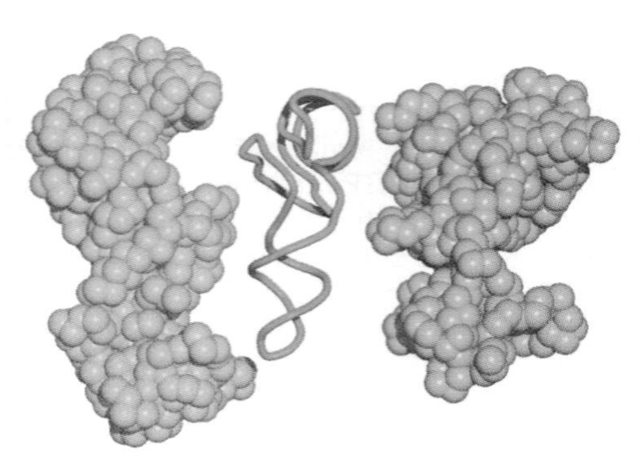

FIGURE 22.18 **Mirror symmetrical interactions of aminoacyl-tRNA synthetases.** Left, glutaminyl-tRNA synthetase (light blue), a class I synthetase. Right, aspartyl-tRNA synthetase (yellow), a class II enzyme. The tRNAs of both glutaminyl-tRNA and aspartyl-tRNA synthetase complexes are superimposed onto $tRNA^{Phe}$ (green). The synthetases are shown moved away from the tRNA for clarity. (Reproduced from *Trends Biochem. Sci.*, vol. 22, J. G. Arnez and D. Moras, Structural and functional considerations of . . ., pp. 211–216, copyright 1997, with permission from Elsevier [http://www.sciencedirect.com/science/journal/09680004]. Photo courtesy of Dino Moras, Institut de Génétique et de Biologie Moléculaire et Cellulaire.)

difference of 3.5 kcal mol^{-1}. The fact that this calculated error rate is considerably higher than the experimentally observed error rate, approximately 1 in 10^4 to 1 in 10^5, suggests that some aminoacyl-tRNA synthetases have a method for correcting mistakes. Subsequent studies revealed that several aminoacyl-tRNA synthetases have editing activity (Table 22.1). We now examine the best studied of these enzymes, the Ile-tRNA synthetase.

The first clear evidence that *E. coli* Ile-tRNA synthetase, a class I synthetase, has an editing or proofreading function was provided by experiments performed by Ann Norris Baldwin and Paul Berg in 1966. Ile-tRNA synthetase catalyzes valyl-AMP formation at approximately 0.5% the rate of isoleucyl-AMP formation. Moreover, it forms a stable complex with aminoacyl-AMP provided that $tRNA^{Ile}$ is not present. Baldwin and Berg took advantage of this stability to isolate each enzyme•aminoacyl-AMP complex by molecular sieve chromatography and then tested the effect that $tRNA^{Ile}$ had on each complex. Ile-$tRNA^{Ile}$ was formed when $tRNA^{Ile}$ was added to the enzyme•isoleucyl-AMP complex, completing the aminoacylation reaction. An entirely different kind of result was obtained when $tRNA^{Ile}$ was added to the enzyme•valyl-AMP complex. Instead of transferring the valine to $tRNA^{Ile}$, the synthetase hydrolyzed the valyl-AMP to produce valine and AMP, thereby demonstrating that the synthetase has an editing or proofreading function that prevents misacylation.

Alan Fersht proposed the "double-sieve" model in 1977 to explain the high fidelity of the aminoacylation reaction (FIGURE 22.19). According to this model, Ile-tRNA synthetase has an aminoacylation site that serves as a "coarse sieve" to exclude amino acids larger than the cognate L-isoleucine and an editing site that serves as a "fine sieve" to eliminate amino acids smaller than isoleucine such as valine.

Ile-tRNA synthetase can hydrolyze valyl-$tRNA^{Ile}$ and valyl-AMP.

In principle, editing by Ile-tRNA synthetase might take place before (pre-transfer) or after (post-transfer) the valyl group is transferred

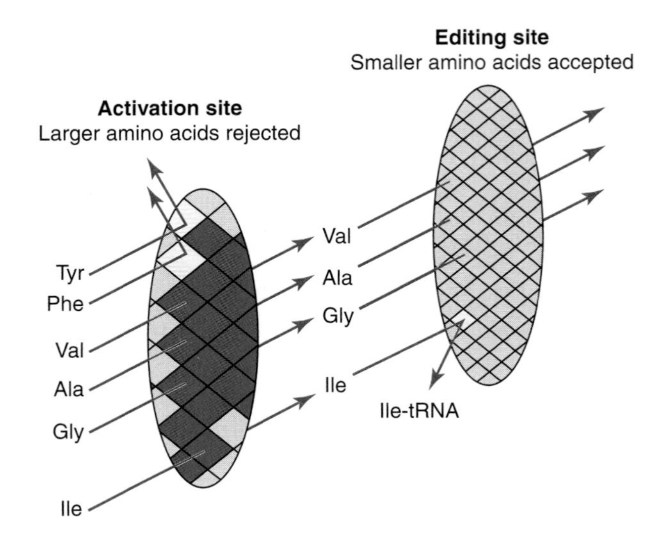

FIGURE 22.19 **The double-sieve mechanism as applied to Ile-tRNA synthetase.** Ile-tRNA synthetase has an aminoacylation site that serves as a "coarse sieve" to exclude amino acids larger than the cognate L-isoleucine and an editing site that serves as a "fine sieve" to eliminate amino acids smaller than isoleucine such as valine. (Modified from A. Fersht. *Science* 280 [1998]: 541. Reprinted with permission from AAAS.)

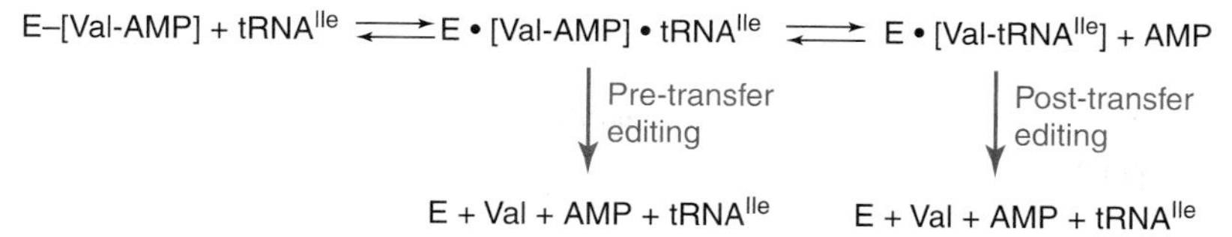

FIGURE 22.20 **Pre- and post-transfer editing**. The symbol E denotes isoleucyl-tRNA synthetase. (Modified from *Mol. Cell*, vol. 4, T. K. Nomanbhoy, T. L. Hendrickson, and P. Schimmel, Transfer RNA–Dependent Translocation of . . ., pp. 519–528, copyright 1999, with permission from Elsevier [http://www.sciencedirect.com/science/journal/10972765].)

to tRNA^{Ile} (FIGURE 22.20). To test Ile-tRNA synthetase's ability to hydrolyze Val-tRNA^{Ile}, one would need some method to synthesize the substrate Val-tRNA^{Ile}. Emmet Eldred and Paul Schimmel solved this problem in the early 1970s by taking advantage of the fact that under special incubation conditions an aminoacyl-tRNA synthetase from one species sometimes misacylates tRNA molecules from another species. More specifically, they used a yeast extract to attach valine to purified *E. coli* tRNA^{Ile} and then isolated the misacylated tRNA. Valine was rapidly released when this misacylated tRNA was mixed with *E. coli* Ile-tRNA synthetase, demonstrating the enzyme's ability to perform posttransfer editing.

The experiments performed by Eldred and Schimmel did not rule out the possibility that Ile-tRNA synthetase is also capable of pretransfer editing. It is difficult to show that pretransfer editing takes place because this process requires tRNA^{Ile} to stimulate the editing function and one would have to eliminate the possibility of Val-tRNA^{Ile} formation followed by hydrolysis. Schimmel and Stephen P. Hale solved the problem by replacing tRNA^{Ile} with a short DNA molecule that could stimulate pretransfer editing but could not accept amino acids. They began by constructing a DNA library in which each DNA molecule contained a central sequence of 25 randomized nucleotides that was surrounded on each side by a fixed 18-nucleotide sequence. Then they used affinity chromatography to search for DNA molecules within the library that bind to the Ile-tRNA synthetase•valyl-AMP complex. The affinity column contained beads formed by linking Ile-tRNA synthetase to a polysaccharide support and the bound enzyme was saturated with a tight-binding (nonreactive) valyl-AMP analog. The DNA library was passed through the affinity column. DNA sticking to the affinity beads was eluted with tRNA^{Ile} and then replicated by the polymerase chain reaction. Repeated selection-amplification cycles eventually produced a DNA molecule with a high affinity for the isoleucyl tRNA synthetase•valyl-AMP analog complex. A DNA or RNA sequence that binds specifically to a protein such as Ile-tRNA synthetase is called an **aptamer**. One of the DNA aptamers isolated by Hale and Schimmel stimulated Ile-tRNA synthetase to hydrolyze valyl-AMP but had no effect on isoleucyl-AMP. Because the DNA aptamer cannot accept amino acids, valyl-AMP hydrolysis must be due

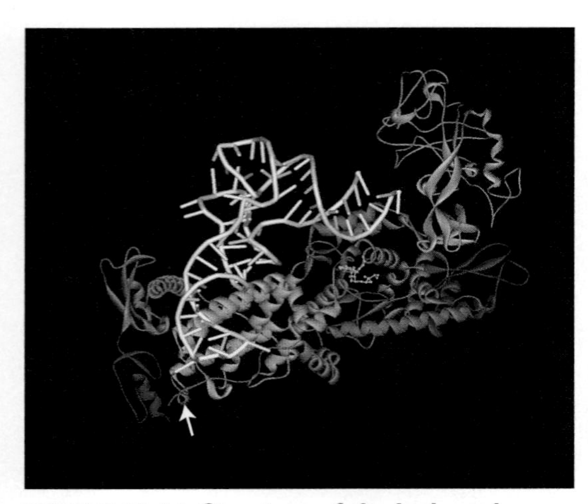

FIGURE 22.21 **Structure of the isoleucyl-tRNA synthetase complex with tRNA^{Ile} and an inhibitor of the enzyme, mupirocin.** The tRNA is shown in silver and the mupirocin drug as a pink stick figure. The protein is colored by domain or by section: N-terminal domain (orange, identified by white arrow), Rossmann fold (green), editing domain or CP1 (yellow), CP2 domain (purple), helical domain (pink), C-terminal junction (light blue), and Zn-binding domain (red). (Structure from Protein Data Bank 1FFY. L. F. Silvian, J. Wang, and T. A. Steitz, *Science* 285 [1999]: 1074–1077. Prepared by B. E. Tropp.)

to pretransfer editing. Kinetic studies suggest that pretransfer editing is the predominant reaction for isoleucyl tRNA synthetase.

A polypeptide insert in the Rossmann-fold domain forms the editing site for Ile-tRNA synthetase.

The structural basis for pre- and posttransfer editing has been elucidated for the bacterial Ile-tRNA synthetase. The Rossmann-fold domain of this enzyme has a polypeptide inserted within it (FIGURE 22.21). This insert, which is approximately 200 residues long, is called **connective polypeptide 1** (**CP1**). Three different kinds of experiments show that CP1 forms the editing domain. First, mutations that alter residues in CP1 block the editing function. Second, the CP1 fragment hydrolyzes Val-tRNA^{Ile} without the assistance of any other part of the synthetase. Third, structural studies performed by Shigeyuki Yokoyama and coworkers in 1998 show that valine binds to both the Rossmann-fold and the CP1 domain in Ile-tRNA synthetase from the gram-negative bacteria *Thermus thermophilus*, whereas isoleucine binds to only the former. The double-sieve model predicts this binding difference between the two amino acids. The Rossmann-fold domain (the aminoacylation site) acts as the coarse sieve that accommodates both amino acids, whereas CP1 (the editing site) acts as the fine sieve that accommodates only valine.

Ile-tRNA synthetase's aminoacylation and editing sites are more than 25 Å apart, raising the issue of how the 3′-end of a misacylated tRNA moves from the aminoacylation to the editing site. Structural studies performed by Thomas Steitz and his coworkers show that when Ile-tRNA synthetase forms a complex with tRNA^{Ile} and mupirocin (an antibiotic inhibitor that binds to the aminoacylation site), the 3′-terminal of tRNA^{Ile} is located in the CP1 domain (Figure 22.21). This observation suggests that when valine attaches to tRNA^{Ile}, the tRNA's acceptor stem flips from the aminoacylation site to the editing site while the rest of the RNA molecule remains in place (FIGURE 22.22).

CP1 also catalyzes the hydrolysis of valyl-AMP that is mistakenly formed at Ile-tRNA synthetase's aminoacylation site. Many, but probably not all, of the CP1 residues that participate in editing Val-tRNA^{Ile} also participate in valyl-AMP hydrolysis. Ile-tRNA synthetase conformation changes when the enzyme binds tRNA^{Ile} (and presumably the DNA aptamer), allowing valyl-AMP to move through a newly created channel connecting the Rossman-fold domain to the CP1 domain.

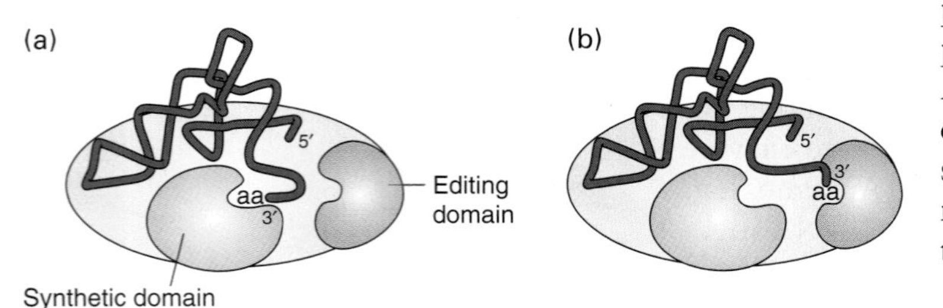

FIGURE 22.22 **Structural basis of post-transfer editing.** (a) When isoleucyl-tRNA synthetase is in the activation mode of aminoacyl transfer, the acceptor strand of the tRNA adopts a hairpin conformation to pack against residues of the Rossmann fold in the synthetic domain. (b) When the enzyme is in the editing mode, the tRNA adopts an extended stack conformation, which places the amino acid in the CP1 or editing domain (pocket). (Adapted from L. F. Silvian, J. Wang, and T. A. Steitz, *Science* 285 [1999]: 1074–1077.)

Editing-defective alanyl-tRNA synthetase causes a neurodegenerative disease in mice.

Recent studies show that a defect in the editing activity of an alanyl-tRNA synthetase causes a neurodegenerative disease in mice. This disease, caused by the autosomal recessive *sti* (sticky) gene, is characterized by tremors that progress to ataxia (lack of muscle coordination during voluntary movements). Histological and biochemical studies reveal that specific cells in the cerebellum of affected mice undergo apoptosis (programmed cell death).

Genetic studies indicate the *sti* gene codes for alanyl-tRNA synthetase. More specifically, Ala-374 in the wild type synthetase is replaced by Glu-374 in the mutant synthetase. This substitution has no effect on aminoacylation activity. Wild type and mutant alanyl-tRNA synthetases mistakenly attach serine to $tRNA^{Ala}$ at about the same frequency. The same is true for mistaken glycine attachment. In contrast, the mutation does affect the synthetase's editing activity. Studies using Ser-$tRNA^{Ala}$ as substrate indicate that the wild type synthetase's editing activity is about twofold greater than the mutant synthetase's editing activity. As a result of this decreased editing activity, *sti/sti* mice synthesize altered polypeptide chains in which some alanine residues have been replaced by serine or glycine residues. These altered polypeptides are proposed to have difficulty in folding and therefore to form harmful aggregates. In support of this hypothesis, biochemical and histochemical studies indicate that cells with the mutant alanyl-tRNA synthetase do have increased levels of misfolded proteins.

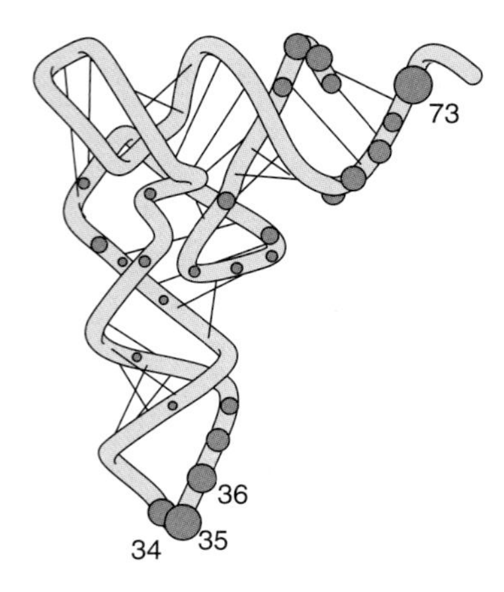

(b) Class II

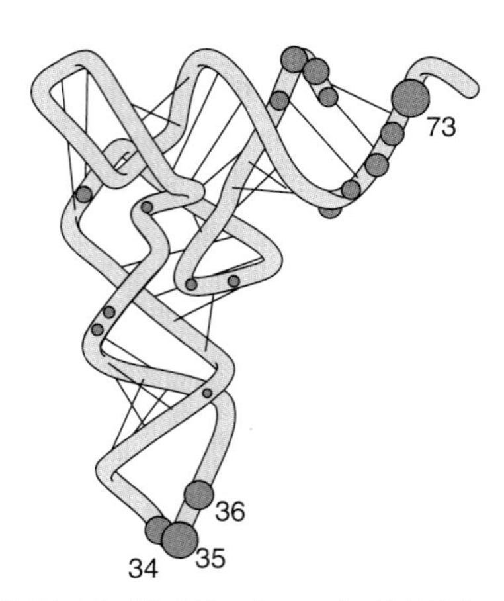

FIGURE 22.23 **Distribution of tRNA identity elements that are recognized by *E. coli* class I and class II aminoacyl-tRNA synthetases.** (a) Distribution of identity elements recognized by *E. coli* class I aminoacyl-tRNA synthetases. (b) Distribution of identity elements recognized by *E. coli* class II aminoacyl-tRNA synthetases. Orange sphere sizes are proportional to the frequency of identity nucleotides at a given position. (Adapted from R. Giege, M. Sissler, and C. Florentz, *Nucleic Acids Res.* 26 [1998]: 5017–5035.)

Each aminoacyl-tRNA synthetase can distinguish its cognate tRNAs from all other tRNAs.

Each aminoacyl-tRNA synthetase must be able to select a family of isoacceptor tRNA molecules from a mixture of many other tRNA molecules. The enzyme does so by searching for a nucleotide, a base pair, a short nucleotide sequence, or some combination of these features that distinguishes its cognate tRNAs from all other tRNAs. One method for identifying specific features of a tRNA recognized by the cognate synthetase is to systematically introduce mutations into the tRNA and then determine how the mutations affect aminoacylation. Features revealed by *in vitro* and *in vivo* experiments are termed **identity elements** and **recognition elements**, respectively. Because most of the discussion that follows does not include experimental details, we will use these two terms interchangeably. The major identity elements cluster in the first few base pairs of the acceptor arm and the anticodon trinucleotide (FIGURE 22.23). Synthetases may also recognize unique features in the variable or D-arms or a modified nucleotide. Some identity elements allow positive contacts with the cognate synthetase, whereas others prevent contacts with non-cognate synthetases. Two examples will help to illustrate the interactions that take place between an aminoacyl-tRNA synthetase and the identity element(s) on its cognate tRNA.

1. ***E. coli* $tRNA^{Ala}$**—Schimmel and coworkers performed a series of experiments beginning in 1988 that were designed to identify the features that *E. coli* Ala-tRNA synthetase recognizes in its cognate tRNA. After ruling out the 15 invariant nucleotides in $tRNA^{Ala}$ as possible identity elements, Schimmel and coworkers used a combination of biochemical and genetic techniques to alter 36 of the remaining 61 nucleotides.

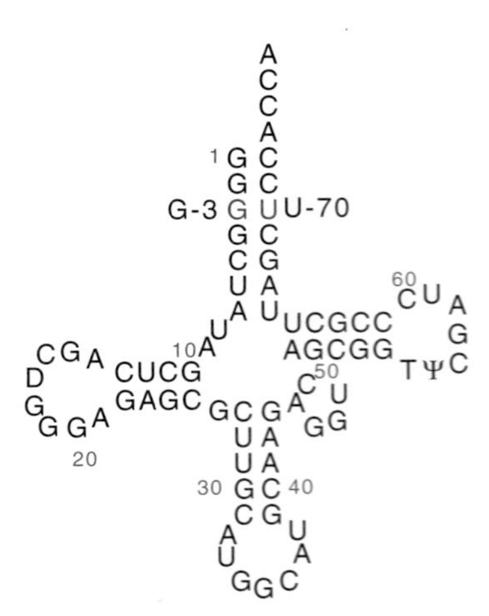

FIGURE 22.24 ***E. coli* tRNA^{Ala} identity element.** The G3:U70 identity element is shown in red. (Modified with permission from Y. M. Hou and P. Schimmel, *Biochemistry* 28 [1989]: 6800–6804. Copyright 1989 American Chemical Society.)

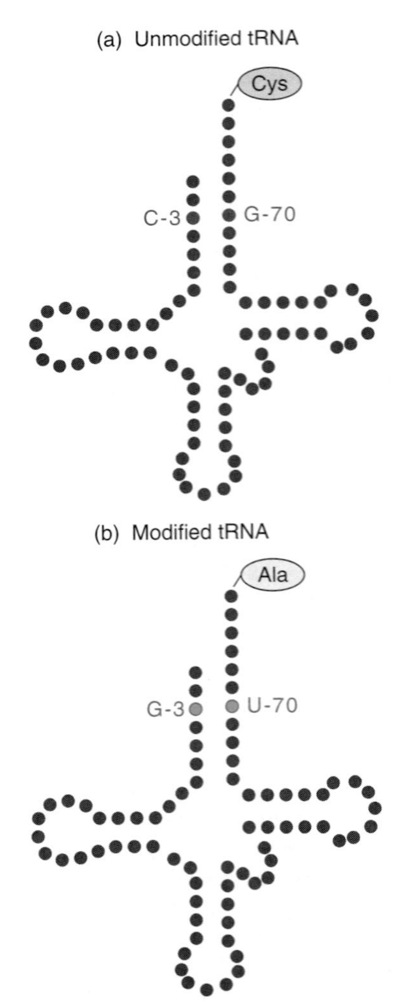

FIGURE 22.25 **Effect of base substitution on acceptor activity.** Replacing the C3:G70 base pair in tRNA^{Cys} with a G3:U70 base pair converts the tRNA from one that accepts cysteine to one that accepts alanine.

Two alterations were especially noteworthy. Replacing guanine at position-3 (G3) or uracil at position 70 (U70) with other bases caused tRNA^{Ala} to lose its ability to accept alanine (FIGURE 22.24). Aminoacylation by Ala-tRNA synthetase was not affected by other substitutions in the acceptor arm or by substitutions in the anticodon, D– or TΨC arms. These results suggest that the non-standard G3:U70 base pair is an identity element for tRNA^{Ala}. To test this possibility, the non-standard G3:U70 base pair was introduced into *E. coli* tRNA^{Cys} in place of the C3:G70 base pair that is normally present in tRNA^{Cys}. As a result of this base pair switch, the modified tRNA^{Cys} became an alanine acceptor that could no longer accept cysteine (FIGURE 22.25). This change in acceptor ability supports the hypothesis that the non-standard G3:U70 base pair is an identity element for tRNA^{Ala} and suggests that C3:G70 is an identity element for tRNA^{Cys}. Remarkably, the modified tRNA^{Cys} still differs from normal tRNA^{Ala} in 40 bases.

Schimmel and coworkers next tried to determine whether just a part of tRNA^{Ala} could serve as an alanine acceptor. To that end, they synthesized the minihelix^{Ala}, microhelix^{Ala}, duplex^{Ala}, and tetraloop^{Ala} structures shown in FIGURE 22.26. The *E. coli* tRNA synthetase was able to use each of these RNAs as an alanyl-acceptor, showing that the enzyme does not require the anticodon arm or any of the other excised regions to recognize tRNA^{Ala}. Ten other aminoacyl-tRNA synthetases, representing members of both class I and class II synthetases, were later shown to attach amino acids to minihelices derived from the acceptor arm of their cognate tRNA. Included among these synthetases are those specific for histidine, glycine, methionine, valine, isoleucine, serine, glutamine,

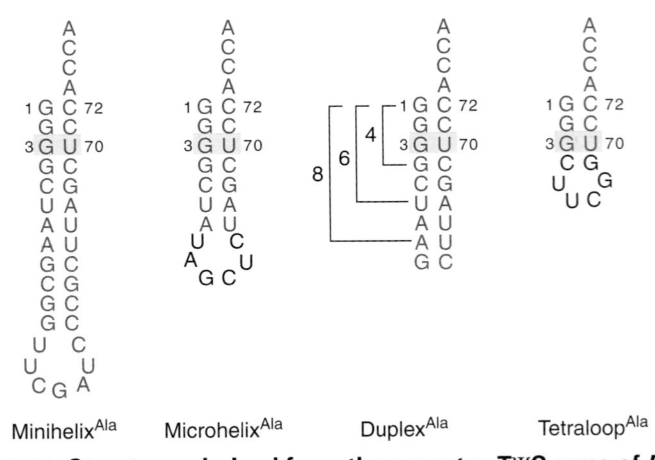

FIGURE 22.26 **Structures derived from the acceptor-TΨC arms of *E. coli* tRNA^{Ala} that accept alanine from *E. coli* tRNA synthetase.** Nucleotides present in the acceptor and TΨC arms are shown in red and blue, respectively. The non-standard G3:U70 base pair is in a yellow box. In the case of duplex^{Ala}, the 5′ 9-mer strand could be truncated to a 8-mer, 6-mer, or 4-mer, as indicated by brackets. (Adapted from P. J. Beuning and K. Musier-Forsyth, *Biopolymers* 52 [2000]: 1–28.)

aspartic acid, cysteine, and tyrosine. In contrast to Ala-tRNA synthetase, most of these other synthetases also recognize identity elements in the anticodon. We select the substrate for one of these enzymes, $tRNA^{Gln}$, as our second illustrative example.

2. ***E. coli* $tRNA^{Gln}$**—Initial efforts to determine the identity elements in $tRNA^{Gln}$ were similar to those described above for *E. coli* $tRNA^{Ala}$. *E. coli* $tRNA^{Gln}$ was altered by a combination of biochemical and genetic methods and then the modified tRNAs were tested for their abilities to accept glutamine. As summarized in FIGURE 22.27, these studies showed that $tRNA^{Gln}$ has identity elements in its acceptor, anticodon, and D-arms. The crystal structure of the Gln-tRNA synthetase•$tRNA^{Gln}$ complex confirmed that the identity elements revealed by studying modified tRNA do indeed make contact with the enzyme (FIGURE 22.28). Similar structural confirmation of the identity element in $tRNA^{Ala}$ is not yet possible because the crystal structure of the enzyme•tRNA complex has not been determined.

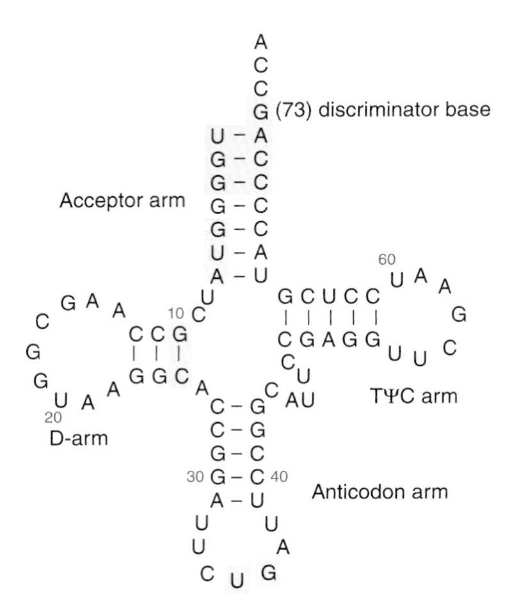

FIGURE 22.27 **Secondary structure of $tRNA^{Gln}$ in cloverleaf form.** Nucleotides involved in recognition by glutaminyl-tRNA synthetase are shown in yellow boxes. (Adapted from M. Ibba, K.-W. Hong, and D. Söll, *Genes Cells* 1 [1996]: 421–427.)

Many gram-positive bacteria and archaea use an indirect pathway for Gln-tRNA synthesis.

Early studies with *E. coli* and animal cells suggested that there is one aminoacyl-tRNA synthetase for each of the 20 different amino acid building blocks for proteins. Michael Wilcox and Marshal Nirenberg discovered an exception to this rule while studying aminoacyl-tRNA formation in *B. subtilis* in 1968. These gram-positive bacteria, which lack Gln-tRNA synthetase, synthesize Gln-$tRNA^{Gln}$ by an indirect two-step pathway (FIGURE 22.29). The first step, glutamate attachment to $tRNA^{Gln}$, is catalyzed by a Glu-tRNA synthetase. The second step, conversion of Glu-$tRNA^{Gln}$ to Gln-$tRNA^{Gln}$, is catalyzed by Glu-$tRNA^{Gln}$ amidotransferase, which transfers the amido group from glutamine to Glu-$tRNA^{Gln}$. This indirect two-step pathway for Gln-$tRNA^{Gln}$ formation is fairly common in gram-positive bacteria, the archaea, and chloroplasts. Formation of a misacylated tRNA such as Glu-$tRNA^{Gln}$ in these organisms might be expected to lead to the formation of faulty proteins that have glutamate residues in place of glutamine residues. No such faulty proteins are produced, however, because a component of the protein synthetic machinery, elongation factor Tu or EF-Tu (see Chapter 23), fails to act on the misacylated tRNA. Some archaeal strains synthesize Asn-$tRNA^{Asn}$ by a similar,

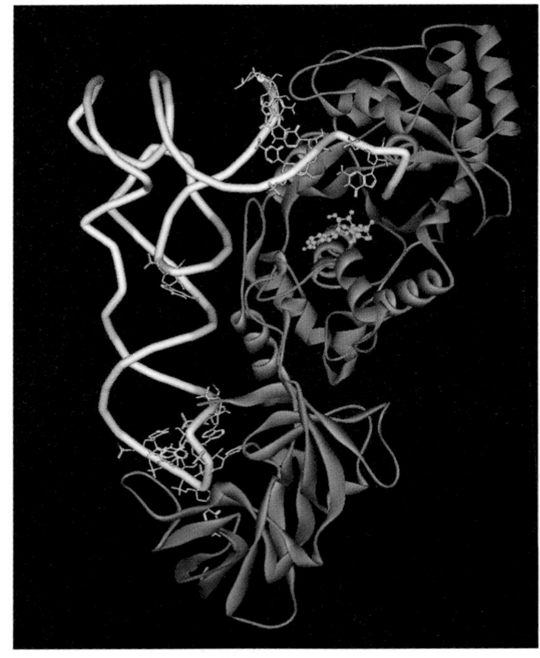

FIGURE 22.28 **Glutaminyl-tRNA synthetase-$tRNA^{Gln}$ complex.** The synthetase is shown as a blue ribbon structure and the tRNA as a white tube structure. Identity elements on the tRNA in contact with glutaminyl-tRNA synthetase are shown as yellow stick structures. The bound ATP is shown as a green ball-and-stick structure. (Structure from Protein Data Bank 1QRS. J. G. Arnez and T. A. Steitz, *Biochemistry* 35 [1996]: 14725–14733. Prepared by B. E. Tropp.)

$$\text{Glu} + \text{tRNA}^{\text{Gln}} + \text{ATP} \xrightleftharpoons{\text{Glutamyl-tRNA synthetase}} \text{Glu-tRNA}^{\text{Gln}} + \text{AMP} + \text{PP}_i$$

$$\text{Gln} + \text{Glu-tRNA}^{\text{Gln}} + \text{ATP} \xrightleftharpoons{\text{Glu-tRNA}^{\text{Gln}}\ \text{amidotransferase}} \text{Gln-tRNA}^{\text{Gln}} + \text{ADP} + \text{P}_i + \text{Glu}$$

FIGURE 22.29 **Two-step pathway for Gln-$tRNA^{Gln}$ formation in gram-positive bacteria, archaea, and chloroplasts.**

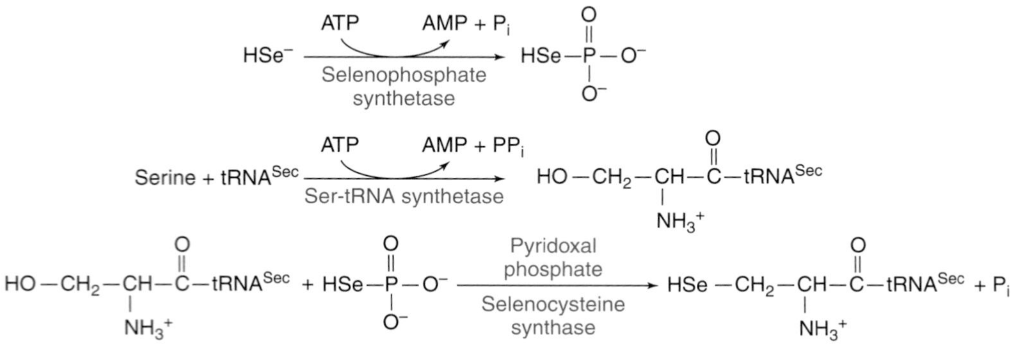

$$HSe^- \xrightarrow[\text{Selenophosphate synthetase}]{\text{ATP} \quad \text{AMP} + P_i} HSe{-}\overset{\displaystyle O}{\overset{\|}{P}}{-}O^- \;(\text{with } {-}O^- \text{ on P})$$

$$\text{Serine} + \text{tRNA}^{Sec} \xrightarrow[\text{Ser-tRNA synthetase}]{\text{ATP} \quad \text{AMP} + PP_i} HO{-}CH_2{-}\underset{\displaystyle NH_3^+}{CH}{-}\overset{\displaystyle O}{\overset{\|}{C}}{-}\text{tRNA}^{Sec}$$

$$HO{-}CH_2{-}\underset{\displaystyle NH_3^+}{CH}{-}\overset{\displaystyle O}{\overset{\|}{C}}{-}\text{tRNA}^{Sec} + HSe{-}\underset{\displaystyle O^-}{\overset{\displaystyle O}{\overset{\|}{P}}}{-}O^- \xrightarrow[\text{Selenocysteine synthase}]{\text{Pyridoxal phosphate}} HSe{-}CH_2{-}\underset{\displaystyle NH_3^+}{CH}{-}\overset{\displaystyle O}{\overset{\|}{C}}{-}\text{tRNA}^{Sec} + P_i$$

FIGURE 22.30 Bacterial pathway for selenocysteyl-tRNA^{Sec} formation.

indirect two-step pathway. An Asp-tRNA synthetase with relaxed specificity attaches aspartate to tRNA^{Asn} and then an amidotransferase converts Asp-tRNA^{Asn} to Asn-tRNA^{Asn}.

Selenocysteine and pyrrolysine are building blocks for polypeptides.

By the mid-1980s, there was general agreement that the protein synthetic machinery uses just 20 amino acid building blocks to make proteins. Many modified amino acids were known to be present in proteins but these were formed by modifying amino acid residues *after* the polypeptide chain was formed. It, therefore, was a great surprise when **selenocysteine** (**Sec**) was identified as the 21st amino acid. This rare amino acid, an analog of cysteine in which the element selenium replaces sulfur, is present in a few proteins, mostly oxidoreductases, from eukaryotes, bacteria, and the archaea.

The bacterial pathway for selenocysteyl-tRNA^{Sec} formation, which was elucidated largely through the efforts of August Böck and coworkers beginning in 1986, is shown in FIGURE 22.30. The three steps in this pathway are as follows: (1) Selenophosphate synthetase catalyzes the synthesis of selenophosphate from ATP and selenide, (2) Ser-tRNA synthetase attaches a seryl group to tRNA^{Sec}, and (3) Selenocysteine synthase catalyzes a pyridoxal phosphate-dependent conversion of selenophosphate and Ser-tRNA^{Sec} to selenocysteyl-tRNA^{Sec}. The tRNA^{Sec} species contains up to 100 nucleotides, making it the longest known tRNA. It also has other unusual features. For example, it has one more nucleotide base pair in its acceptor arm than most other tRNAs and an extended D-arm. Because of these unique features, selenocysteyl-tRNA^{Sec} is not recognized by elongation factor Tu (EF-Tu) (see Chapter 23), and requires its own special elongation factor.

A second unusual amino acid building block, **pyrrolysine** (FIGURE 22.31), was independently discovered by Joseph A. Krzycki and coworkers and Michael K. Chan and coworkers in 2002. A pyrrolysyl-tRNA synthetase that catalyzes the ATP-dependent attachment of pyrrolysine to tRNA^{Pyl} has been isolated and characterized. An alternate pathway for pyrrolysyl-tRNA^{Pyl} formation may also exist

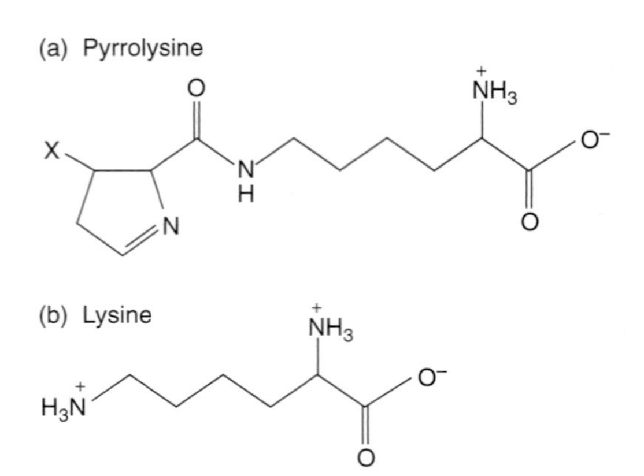

FIGURE 22.31 The structures of (a) pyrrolysine and (b) lysine. X indicates a CH_3, NH_3^+, or OH group. (Adapted from M. Ibba and D. Soll, *Curr. Biol.* 12 [2002]: R464–R466.)

that involves converting lysyl-$tRNA^{Pyl}$ to pyrrolysyl- $tRNA^{Pyl}$. At this time, pyrrolysine appears to be present in only a few archaeal and bacterial species.

22.4 Messenger RNA and the Genetic Code

Messenger RNA programs ribosomes to synthesize proteins.

Although molecular biologists had learned a great deal about protein synthesis by the end of the 1950s, they still did not know how genetic information was transferred from the nucleotide sequences in DNA to amino acid sequences in polypeptides. The first clues to the possibility that cells might have a special kind of RNA that programs ribosomes to synthesize specific polypeptides came from studies with phage T2-infected *E. coli*. In 1953, Alfred Hershey reported that infected bacteria rapidly incorporate [^{32}P]phosphate into a small fraction of the cellular RNA molecules and that the labeled RNA is then rapidly degraded. Three years later, Elliot Volkin and Lazarus Astrachan made a similar observation, but then went on to demonstrate that the rapidly turning over RNA had the same base composition as the viral DNA and not the bacterial DNA. The significance of these experiments was not fully appreciated at the time because investigators thought that the genetic instructions were permanently embedded in a ribosome at the time of its synthesis so that each ribosome would only be able to make one specific polypeptide.

In 1959, Arthur Pardee, Francois Jacob, and Jacques Monod performed a series of genetic experiments (later called the PaJaMo experiments) that showed that ribosomes are rapidly programmed to synthesize new proteins and do not contain permanently embedded instructions. The PaJaMo experiments began with the transfer of a *lacZ* gene (the structural gene for β-galactosidase) from an *E. coli* Hfr (*lacI*$^+$ lacZ$^+$) strain that could synthesize β-galactosidase to F$^-$ (lacI$^-$ lacZ$^-$) that could not do so. The recipient strain started to synthesize β-galactosidase at maximal rates shortly after receiving the *lacZ* gene. Because the recipient strain did not have adequate time to produce new ribosomes dedicated to β-galactosidase synthesis, the PaJaMo experiments indicated that the l*acZ* gene serves as a template for the synthesis of an mRNA molecule, which then directs pre-existing ribosomes to synthesize lactose enzymes.

Three adjacent bases in mRNA that specify an amino acid are called a codon.

With the realization that mRNA directs polypeptide synthesis, investigators started to speculate about how a base sequence might specify an amino acid. A single base does not have enough information because there are only 4 bases but 20 amino acids. Doublets (2 bases) also do not have enough information because there are only 16 (4×4) possible doublet combinations. Triplets (3 bases) are feasible because

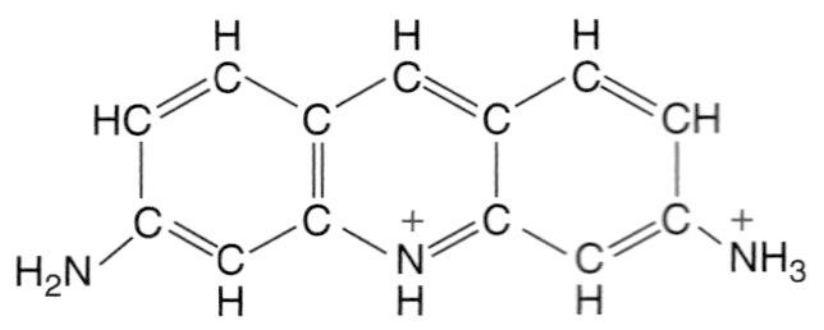

FIGURE 22.32 **Proflavin (acridine-3, 6-diamine).**

the 64 (4 × 4 × 4) possible triplet combinations have more than enough information to specify 20 amino acids. In fact, there are so many possible triplets that one would predict that some amino acids are specified by more than one triplet (a "degenerate code"), some triplets do not specify an amino acid, or both.

Genetic experiments performed by Francis Crick, Sydney Brenner, and coworkers in 1961 showed that amino acids are specified by three nucleotide code words, for which Brenner coined the term **codons**. Their experiments were based on the ability of proflavin, an acridine derivative (FIGURE 22.32), to cause a base pair insertion or deletion in DNA. A mutation that adds or deletes a base pair shifts the reading frame and is therefore termed a **frameshift mutation** (FIGURE 22.33). Crick and coworkers began their investigation of the genetic code using proflavin to introduce a mutation into the phage T4 *rIIB* gene. The mutant phage was easy to distinguish from its wild-type parent because of the plaques that it formed on a bacterial lawn. Wild-type bacteriophages form small plaques on both *E. coli* B and *E. coli* K12(λ) lawns, whereas *rIIB* mutants form large plaques on *E. coli* B lawns and none at all on *E. coli* K12(λ) lawns. The first *rIIB* mutation isolated, called FC0, was arbitrarily designated (+), as if it had a base pair insertion. (They might also have designated it [–] as if it had a base pair deletion. Although later sequencing studies showed that the FC0 mutation is in fact due to an insertion, the conclusions drawn from their experiments would have been the same even if they had guessed wrong).

If FC0 is a (+) insertion, then it should be possible to isolate pseudorevertants with a (–) mutation that restores the reading frame. Of course, the region between the (+) and (–) mutations will be altered and so must be tolerant of amino acid substitutions. Crick and coworkers isolated several spontaneous pseudorevertants and separated the (–) mutation from the original (+) mutation by genetic recombination with the wild-type phage. As expected, recombinant phage with a (–) mutation in the *rIIB* gene did not form plaques on *E. coli* K12(λ) lawns. Phage with the (–) *rIIB* mutations were then used in the same

Direction of reading →

ABC ABC ABC ABC ABC ABC ABC··· Normal reading frame in wild type gene

ABC XAB CAB CAB CAB CAB CAB C··· Inserting **X** shifts reading frame to the right

ABC ABC ABC ABC BCA BCA BC··· Deleting **A** shifts reading frame to the left

ABC XAB CAB CAB CBC ABC ABC··· The reading frame is restored in a recombinant gene that contains the insertion and deletion mutations. However, the region between the two mutations has codons that specify amino acids that are different from those in the wild type gene (missense codons)

Missense

FIGURE 22.33 **Effect of insertion and deletion mutations on the reading frame.**

way as phage with the original FC0 mutation to find new (+) mutations. Crick and coworkers were therefore able to obtain a large number of (+) and (–) mutants. Further genetic recombination experiments using these (+) and (–) mutations provided the following results:

1. A cross between two (+) mutants does not yield phage with the wild-type phenotype.
2. A cross between two (–) mutants does not yield phage with the wild-type phenotype.
3. A cross between one (+) and one (–) does yield phage with the wild-type phenotype.

These results were interpreted in the following way. A double mutant of the type (+)(+) or (–)(–) has two additional base pairs or lacks two base pairs, respectively. In both cases, reading of the code would be shifted two bases out of phase and a functional protein would not be made. In a (+)(–) double mutant, however, the advanced reading frame following the (+) site would be incorrect, but the correct reading frame would be restored at the (–) site (Figure 22.33). Between the two mutant sites, the amino acid sequence would not be that of the wild-type phage, but, nonetheless, the (+)(–) and (–)(+) phage strains would be functional, because both mutant sites were in a non-critical (tolerant) region of the gene. These interpretations proved correct. The point is that as long as the reading frame is restored before the intolerant region is reached, a functional (though not always perfect) protein can be produced.

The fact that a double (+)(+) mutant never had the wild-type phenotype meant that the code could not be a two-letter code. If it were a two-letter code, the reading frame would be restored in this combination. The critical experiment to test for a triplet code was the construction of a triple mutant. As expected, the triple mutants (+)(+)(+) and (–)(–)(–) had the wild-type phenotype, whereas the mixed triples, (+)(+)(–) and (+)(–)(–), were still mutant. These genetic experiments provided strong evidence that a codon contains three nucleotides (or a multiple of three nucleotides).

The discovery that poly(U) directs the synthesis of poly(Phe) was the first step in solving the genetic code.

In 1961, Marshall W. Nirenberg and Johann Heinrich Matthaei performed an experiment that led to the eventual solution of the genetic code. Their initial objective was to determine if bacterial or viral RNA could direct *E. coli* ribosomes to make proteins. They started by preparing a 30,000 × *g E. coli* supernatant fraction that others had shown contained ribosomes, tRNA, and all the translation factors needed for protein synthesis. A typical reaction mixture contained fresh 30,000 × *g* supernatant, ATP, GTP, Mg^{2+}, 19 unlabeled amino acids, and an [^{14}C]amino acid. The reaction was stopped by adding acid, which also caused protein precipitation. The radioactive protein was collected as a pellet by centrifugation and washed several times before being analyzed for its radioactive content.

Nirenberg and Matthaei observed a significant level of amino acid incorporation in the absence of any external source of mRNA. However, adding bacterial or viral RNA did seem to cause a slight increase in radioactive amino acid incorporation. Although the increased amino acid incorporation was too small to be definitive, it suggested that the bacterial and viral RNA might be able to direct ribosomes to synthesize proteins. Nirenberg and Matthaei thought that the amino acid incorporation observed in the absence of added bacterial or viral RNA was probably due to mRNA already in the 30,000 × *g* bacterial supernatant fraction. They therefore decided to deplete the bacterial extract of this pre-existing mRNA and prevent the synthesis of new mRNA.

Studies by others had shown that *E. coli* extracts stop making proteins shortly after DNase I is added to them. Protein synthesis comes to a halt because mRNA has a short lifetime and cannot be replaced without a DNA template. Nirenberg and Matthaei therefore added DNase I to the bacterial extracts to destroy any DNA that was present. They also forced the bacterial extract to degrade pre-existing mRNA and complete the synthesis of nascent polypeptide chains by adding unlabeled amino acids to the extract and incubating the mixture for 40 minutes until amino acid incorporation had nearly ceased. Then the extracts were dialyzed to remove unlabeled amino acids and stored frozen. Thawed extracts retained the ability to support protein synthesis but required an exogenous source of mRNA to do so. The fact that the protein synthetic system lost little, if any, activity after a freeze–thaw cycle meant that it was no longer necessary to prepare a fresh bacterial extract before each protein synthesis experiment. Nirenberg and Matthaei made one additional change in the assay procedure. Instead of collecting the protein precipitate by centrifugation they switched to a faster filtration method. Using their modified assay, Nirenberg and Matthaei demonstrated that bacterial and viral RNA direct ribosomes to make proteins.

Nirenberg and Matthaei thought that a synthetic homopolymer such as polyuridylate, poly(U), might also direct ribosomes to make polypeptides. Fortunately, there was a simple enzymatic procedure for making poly(U). Marianne Grunberg-Manago and Severo Ochoa had discovered the required enzyme, **polynucleotide phosphorylase**, in 1955 and showed that it reversibly converts nucleoside diphosphates (NDPs) into polynucleotides as shown by the following equation:

$$(RNA)_n + NDP \rightleftharpoons (RNA)_{n+1} + P_i$$

Incubation with a single NDP produces a homopolymer and incubation with two or more different NDPs produces a random copolymer (the different nucleotides are randomly distributed along the chain). When first discovered, polynucleotide phosphorylase was thought to synthesize RNA. Further study, however, revealed the enzyme is unsuited for this task because it does not require a template. We now know that the enzyme's function is to degrade rather than to synthesize RNA (see Chapter 16). Although polynucleotide phosphorylase failed to live up to its original promise as an enzyme that synthesizes RNA, it played an important part in solving the genetic code.

Nirenberg and Matthaei tested poly(U)'s ability to direct ribosomes to synthesize polypeptides by adding it to twenty different tubes, each containing the same protein synthetic system but a different [^{14}C]amino acid. They observed amino acid incorporation into a polypeptide in the tube that contained [^{14}C]phenylalanine but in no other tube. Analysis revealed the radioactive polypeptide to be polyphenylalanine. The meaning of this experiment was quite clear. Poly(U) directs ribosomes to make polyphenylalanine. Later experiments showed polyadenylate [poly(A)] directs ribosomes to make polylysine and polycytidylate [poly(C)] directs ribosomes to make polyproline. Polyguanylate [poly(G)] fails to act as a synthetic mRNA because it forms a triple-stranded helix and cannot be translated. If we take the genetic experiments of Crick, Brenner, and coworkers into account, then UUU, AAA, and CCC are the codons for phenylalanine, lysine, and proline, respectively.

Realizing that additional coding information could be obtained using random copolymers to direct protein synthesis, the Nirenberg and Ochoa laboratories raced to solve the genetic code. A specific example, a random copolymer containing A and C in a 5:1 ratio, will help to illustrate the rationale and limitations of their initial approach. The AAA triplet frequency in this random copolymer is 0.58 ($5/6 \times 5/6 \times 5/6$), while the CCC triplet frequency is 4.6×10^{-3} ($1/6 \times 1/6 \times 1/6$). Other triplet frequencies (AAC, ACA, CAA, ACC, CAC, and CCA) can be determined in the same way. The random copolymer should direct ribosomes to synthesize polypeptides with amino acid frequencies that are the same as the codon frequencies. One, therefore, can identify possible codons by comparing actual amino acid frequencies with calculated codon frequencies.

However, this information is of limited value when a codon contains two different nucleotides because it does not provide the nucleotide sequence within the codon. For instance, we may learn that the codon for histidine consists of 2 C + 1 A, but we have no way of knowing whether the actual codon is CCA, CAC, or ACC. Despite these limitations, Nirenberg and Ochoa were able to obtain a great deal of useful information about the genetic code. One experiment performed by Ochoa and coworkers is particularly noteworthy because it showed ribosomes read mRNA in a $5' \rightarrow 3'$ direction. Because this experiment depended on knowing the direction of polypeptide chain growth, we first describe an ingenious experiment performed by Howard Dintzis in 1961 that showed polypeptide chains grow in an amino-to-carboxyl direction.

Protein synthesis begins at the amino terminus and ends at the carboxyl terminus.

Dintzis' approach, which is summarized in FIGURE 22.34, was based on the following five assumptions: (1) Polypeptide chain formation begins at one end of the polypeptide chain and continues to the other. (2) At any given instant, the ribosomes in a cell are at different stages of polypeptide synthesis. (3) Ribosomes with longer nascent polypeptide chains will complete and release their polypeptide chains before

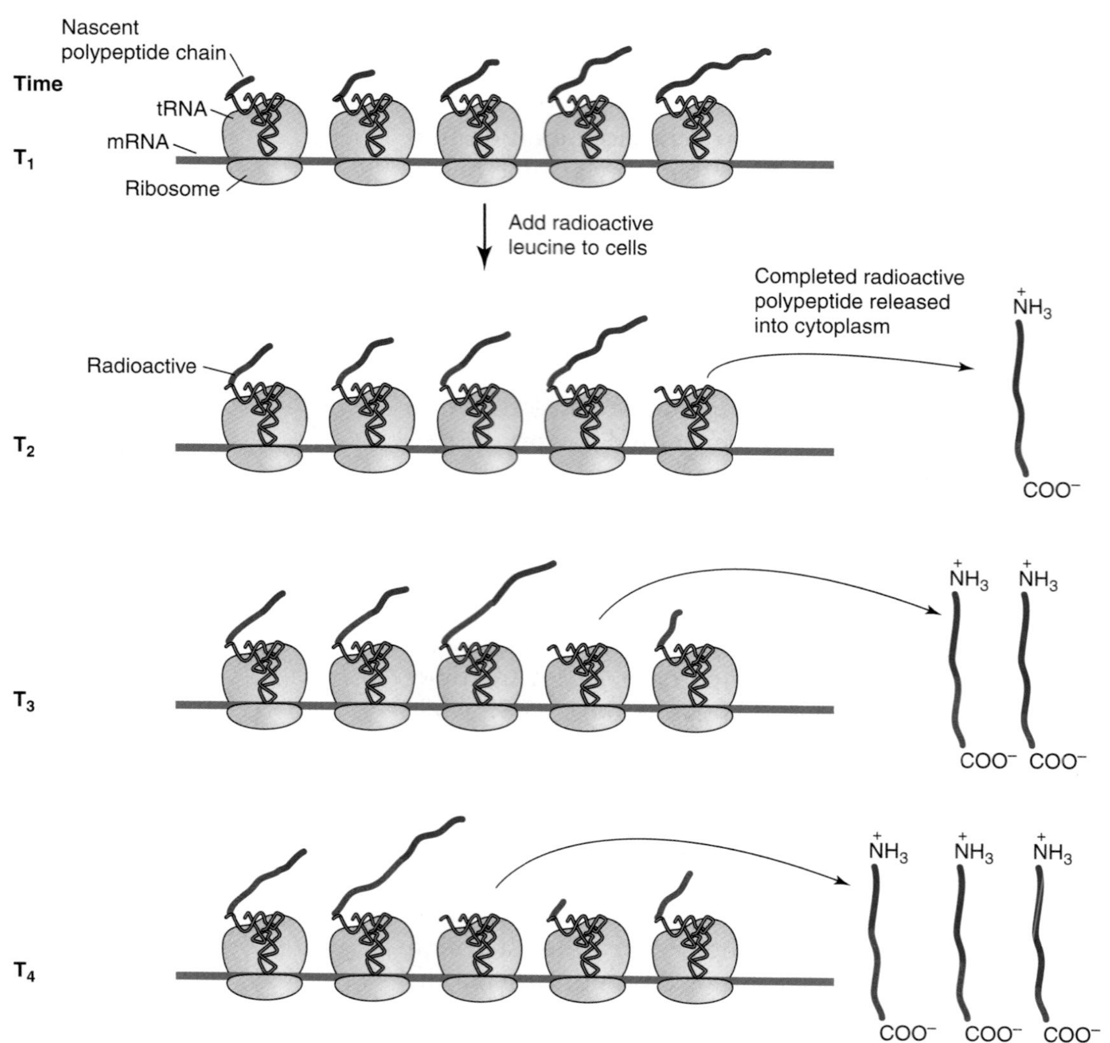

FIGURE 22.34 Direction of polypeptide chain growth. Time T_1—Radioactive leucine is added to rabbit reticulocytes. The ribosomes are at different stages of polypeptide synthesis. Time T_2—Some of the ribosomes complete polypeptide synthesis and release the radioactive polypeptide into the cytoplasm. The radioactive label (shown in red) is present in the carboxyl terminal segment. Time T_3—More of the ribosomes complete polypeptide chain synthesis and release the radioactive polypeptide into the cytoplasm. The radioactive polypeptide is detected in the carboxyl terminal segment as well as the segment next to it. Time T_4—Still more of the ribosomes complete polypeptide chain synthesis and release the radioactive polypeptide into the cytoplasm. The radioactive polypeptide is detected in the carboxyl terminal segment as well as the two adjacent segments. This experiment indicates the carboxyl terminus is the last part of the polypeptide chain that is synthesized. Hence, polypeptide chain synthesis proceeds in an amino-to-carboxyl direction.

ribosomes with shorter nascent chains. (4) If cells were incubated with radioactive amino acids for a very short period of time, only those ribosomes with the longest nascent chains would complete synthesis and release labeled polypeptides into the cytoplasm. (5) Only the end of the polypeptide chain that was synthesized last would be radioactively labeled. Dintzis realized that most cells synthesize many different proteins, making it extremely difficult to isolate and characterize a single labeled protein product. He therefore decided to work with immature red blood cells, called reticulocytes, which synthesize two polypeptides, the α- and β-hemoglobin subunits, in large quantities.

The reticulocytes were isolated from rabbits that had been made anemic by daily injections of phenylhydrazine so that they would have a high rate of red blood cell replacement.

Dintzis added radioactive leucine to a suspension of intact rabbit reticulocytes, which were incubated at 15°C to slow down the rate of protein synthesis. Samples were removed at intervals, cells disrupted, and extracts centrifuged to separate completed soluble proteins from nascent polypeptides still bound to ribosomes. Completed α- and β-globin chains were purified by carboxymethyl cellulose column chromatography and then digested with trypsin. Each of the resulting fragment mixtures was spotted on a piece of filter paper and partially separated by electrophoresis. Then the filter papers were dried, turned at a right angle, and developed in a second direction by chromatography. This two-dimensional separation technique, known as **fingerprinting** or **peptide mapping**, resolved the fragments derived from each chain, permitting the radioactive content of each fragment to be determined. When cells were incubated with radioactive leucine for a very short time, most of the radioactivity appeared in the carboxyl terminal fragment of the completed globin chains (Figure 22.34). This result clearly indicated that the carboxyl terminal fragment is the last to be synthesized. When cells were incubated for a slightly longer time, radioactivity was detected in the carboxyl terminal fragment as well as the one next to it. At still longer times, a radioactive gradient was observed in which the carboxyl terminal fragment was the most radioactive and the amino terminal fragment was the least radioactive. These observations support the idea that polypeptide synthesis begins at the amino terminal and proceeds by stepwise addition of amino acids to the carboxyl terminal.

Messenger RNA is read in a 5′ to 3′ direction.

Taking advantage of the information supplied by Dintzis, Ochoa and coworkers demonstrated mRNA is read in a 5′ to 3′ direction. They began by preparing an oligonucleotide with the sequence ApApAp$(Ap)_n$ApApC by: (1) synthesizing a random copolymer with a 25:1 A to C ratio; (2) cutting the copolymer with RNase to generate oligonucleotides with Cp groups at their 3′-ends; and (3) removing the 3′-phosphate groups with phosphatase (FIGURE 22.35). They added the ApApAp$(Ap)_n$ApApC oligonucleotide to a bacterial protein synthetic system that lacked nuclease activity. The system synthesized

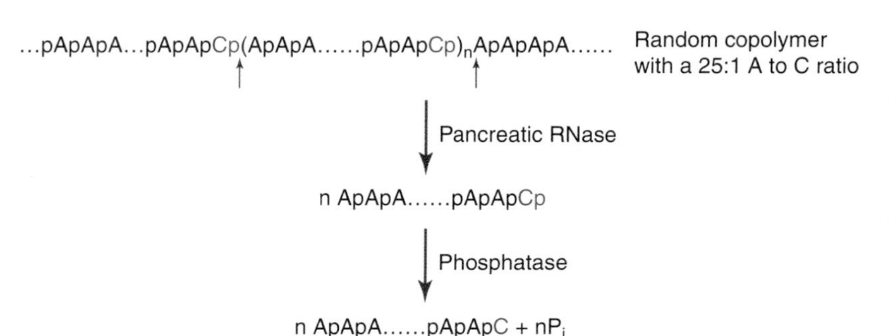

FIGURE 22.35 **Preparation of ApApAp . . . ApApC oligonucleotides.** A random copolymer is synthesized that contains A to C in a 25:1 ratio. Then pancreatic RNase is used to generate oligonucleotides ending in Cp (arrows show the sites cut) and phosphatase is used to remove the 3′-P. (Adapted from M. Salas, et al., *J. Biol. Chem.* 240 [1965]: 3988–3995.)

an oligopeptide with the sequence $^{+}H_3N$-Lys-$(Lys)_{n/3}$-Asn-COO^-. Formation of a chain made up almost exclusively of lysine residues was expected because AAA by then was known to be the codon for lysine. The presence of asparagine at the carboxyl end of the polypeptide chain was of great theoretical significance. Because polypeptide synthesis proceeds in the amino-to-carboxyl direction, asparagine's presence at the carboxyl end indicated that AAC at the 3′-end of the polyribonucleotide was the last codon to be read. Transcription and translation thus each takes place in a 5′→3′ direction. Because both processes take place in the same direction, bacteria can begin to translate an mRNA before its transcription is complete. Eukaryotes cannot do so because transcription and translation occur in separate biological compartments.

Trinucleotides promote the binding of specific aminoacyl-tRNA molecules to ribosomes.

In 1964, Nirenberg and Philip Leder devised an ingenious method for assigning codon sequences. Instead of requiring protein synthesis, their method depends on a trinucleotide's ability to promote the binding of an aminoacyl-tRNA to the ribosome. The assay, itself, is a relatively simple one (FIGURE 22.36). A mixture consisting of ribosomes, a trinucleotide, and an [^{14}C]aminoacyl-tRNA is incubated for a short time and then passed through a nitrocellulose filter. These filters permit free aminoacyl-tRNA molecules to pass through them

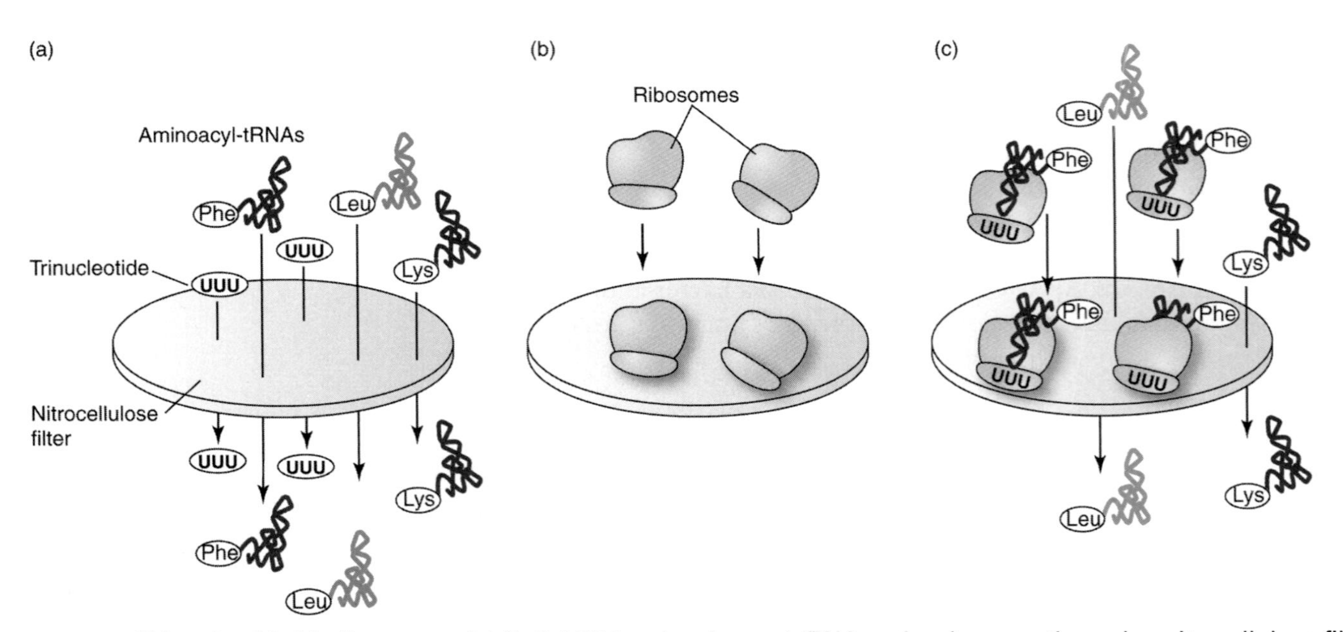

FIGURE 22.36 **Trinucleotide binding assay.** (a) Both UUU and aminoacyl-tRNA molecules pass through a nitrocellulose filter. (b) Ribosomes stick to the filter. (c) Phe-tRNA and UUU form a complex with ribosomes, which then sticks to the nitrocellulose filter. All other aminoacyl-tRNA molecules pass through the filter.

but retain ribosomes and any molecules that are bound to the ribosomes (Figure 22.36a and b).

Nirenberg and Leder tested the method's reliability by determining whether trinucleotides containing known codons would promote the binding of the correct aminoacyl-tRNA and no other (Figure 22.36c). As expected, pUpUpU, pApApA, and pCpCpC promoted the specific binding of Phe-tRNA, Lys-tRNA, and Pro-tRNA, respectively. Then they prepared trinucleotides corresponding to the remaining 61 codons and tested each for the ability to bind one of the 20 [^{14}C]aminoacyl-tRNA species. The binding studies permitted about 50 codon assignments to be made. Some trinucleotides did not promote the binding of any aminoacyl-tRNA molecules and others gave ambiguous results. Another method for determining codon assignments was still needed.

Synthetic messengers with strictly defined base sequences confirmed the genetic code.

In the early 1960s, H. Gorbind Khorana and his colleagues used a combination of classical organic and enzyme catalyzed synthesis to prepare a number of polyribonucleotides with repeating di-, tri-, and tetranucleotide sequences (FIGURE 22.37). When added to bacterial protein synthetic systems, these polyribonucleotides directed ribosomes to synthesize polypeptides with repeating amino acid sequences. RNA molecules with repeating dinucleotide sequences directed the synthesis of polypeptide molecules with a repeating dipeptide sequence. For example, the dinucleotide sequence poly-UC (UCUCUCUCUC . . .) directed the synthesis of a polypeptide with a repeating serine–leucine sequence (Ser-Leu-Ser-Leu . . . ; Figure 22.37a).

These results support the hypothesis that the protein synthetic system reads the bases in groups of three (UCU CUC UCU CUC). However, they do not indicate which of the two codons (UCU or CUC) corresponds to serine and which to leucine because polypeptide synthesis could start at an internal codon. Correct assignments were made by considering results obtained from random copolymer experiments, trinucleotide binding assays, and by using synthetic mRNA molecules with repeating tri- and tetranucleotide sequences.

RNA molecules with repeating trinucleotide sequences direct ribosomes to synthesize three different homopolymers. For example, a polyribonucleotide with a repeating AUC sequence (AUCAUCAUC . . .) directs the synthesis of polyisoleucine, polyserine, and polyhistidine molecules (Figure 22.37b). Because natural mRNA directs the synthesis of only one polypeptide, these results indicated that something was missing from most synthetic mRNAs. The missing element, an initiation

(a) Coding properties of repeating dinucleotide. Sequences direct the synthesis of polypeptides with repeating dipeptide sequence

Polyribonucleotide
Start
5′–U C U C U C U C U C U C U C U C U C U C U C······3′
$^{+}NH_3$– Ser Leu Ser Leu Ser Leu ······

(b) Coding properties of repeating trinucleotide. Sequences direct the synthesis of three different homopolypeptides

Start
5′–A U C A U C A U C A U C A U C A U C A U C A U C······3′
$^{+}NH_3$– Ile Ile Ile Ile Ile Ile ······

Start
5′–A U C A U C A U C A U C A U C A U C A U C A U C······3′
$^{+}NH_3$– Ser Ser Ser Ser Ser Ser······

Start
5′–A U C A U C A U C A U C A U C A U C A U C A U C······3′
$^{+}NH_3$– His His His His His His His ······

(c) Coding properties of repeating tetranucleotides. Sequences direct the synthesis of polypeptides with repeating tetrapeptides

FIGURE 22.37 **(a–c) Coding properties of polyribonucleotides with known repeating sequences.**

signal that sets the reading frame by determining the translation start site, is described in Chapter 23. Natural mRNA molecules have the initiation signal but most synthetic RNA molecules do not.

RNA molecules with repeating tetranucleotide sequences direct the synthesis of polypeptide chains with repeating tetrapeptide sequences (Figure 22.37c). For example, a polyribonucleotide with a repeating UAUC sequence (UAUCUAUCUAUC . . .) directs the synthesis of a polypeptide with a repeating Tyr-Leu-Ser-Ile sequence.

Three codons, UAA, UAG, and UGA, are polypeptide chain termination signals.

Khorana's studies also identified three chain termination signals. The existence of one of these signals became evident when a polyribonucleotide with a repeating GAUA sequence (GAUAGAUAGAUA . . .) was studied. Instead of directing the synthesis of a polypeptide, this polyribonucleotide directs the synthesis of a tripeptide with the sequence Ile-Asp-Arg. As indicated in FIGURE 22.38, the UAG codon acts as a chain termination signal. Two other codons, UAA and UGA, also were shown to be termination codons. The finding of three stop codons fit nicely with the available genetic evidence. Working independently, the laboratories of Seymour Benzer and Sydney Brenner had demonstrated that some gene mutations cause the gene product to be much shorter than normal. They explained the production of the truncated polypeptides by proposing that the mutations change normal codons into stop codons. Mutations that change a normal codon into a stop codon are termed **nonsense mutations** because they change a triplet from one that specifies an amino acid to one that does not. Three different types of nonsense mutation were shown to exist and geneticists named them **amber**, **ochre**, and **opal**. These names, which have no physiological significance, were coined in a rather lighthearted spirit. The amber codon is UAG; the ochre codon is UAA; and the opal codon is UGA.

The genetic code is nonoverlapping, commaless, almost universal, highly degenerate, and unambiguous.

All 64 codons have now been assigned. Sixty-one codons determine amino acids and three are translation termination signals. Codon assignments are shown in Table 22.3. These assignments have been confirmed by comparing nucleotide sequences in genes to amino acid sequences in proteins. (DNA sequencing was not possible when the genetic code was solved in the mid-1960s.) Note that amino acids with similar properties tend to have similar codons.

Five general statements can be made about the genetic code.

1. The genetic code is nonoverlapping. Each base is part of one and only one codon. The difference between a nonoverlapping and an overlapping code is illustrated in FIGURE 22.39.
2. The genetic code is commaless. There are no intervening bases between adjacent codons.

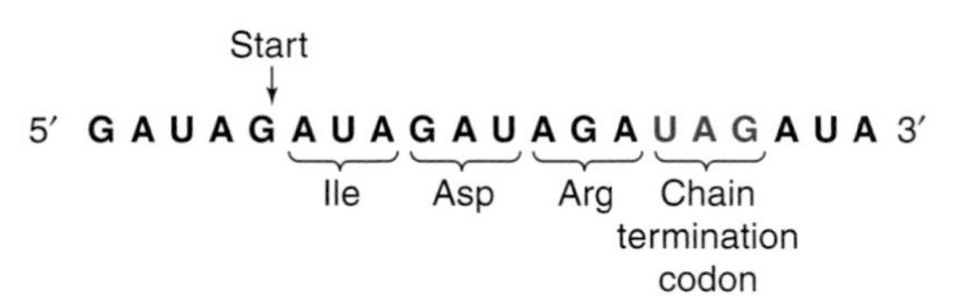

FIGURE 22.38 **UAG termination codon.**

TABLE 22.3	The Genetic Code				
First Position (5′-end)	**Second Position**				**Third Position (3′-end)**
	U	**C**	**A**	**G**	
U	UUC Phe	UCU Ser	UAU Tyr	UGU Cys	**U**
	UCC Phe	UCC Ser	UAC Tyr	UGC Cys	**C**
	UUA Leu	UCA Ser	UAA Stop	UGA Stop	**A**
	UUG Leu	UCG Ser	UAG Stop	UGG Trp	**G**
C	CUU Leu	CCU Pro	CAU His	CGU Arg	**U**
	CUC Leu	CCC Pro	CAC His	CGC Arg	**C**
	CUA Leu	CCA Pro	CAA Gln	CGA Arg	**A**
	CUG Leu	CCG Pro	CAG Gln	CGG Arg	**G**
A	AUU Ile	ACU Thr	AAU Asn	AGU Ser	**U**
	AUC Ile	ACC Thr	AAC Asn	AGC Ser	**C**
	AUA Ile	ACA Thr	AAA Lys	AGA Arg	**A**
	AUG Met Start	ACG Thr	AAG Lys	AGG Arg	**G**
G	GUU Val	GCU Ala	GAU Asp	GGU Gly	**U**
	GUC Val	GCC Ala	GAC Asp	GGC Gly	**C**
	GUA Val	GCA Ala	GAA Glu	GGA Gly	**A**
	GUG Val	GCG Ala	GAG Glu	GGG Gly	**G**
Start Codon					
Stop Codon					
Nonpolar Side Chain					
Uncharged Polar Side Chain					
Charged Polar Side Chain					

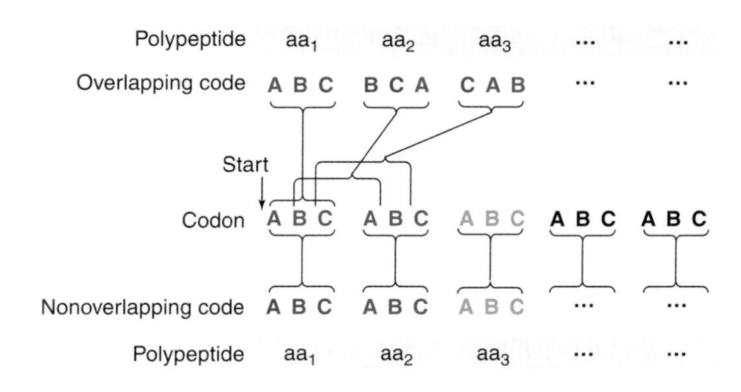

FIGURE 22.39 Difference between an overlapping and a nonoverlapping code. (Top) In an overlapping code, a nucleotide (A, B, and C) can be used to specify more than one amino acid (aa). (Bottom) In a nonoverlapping code, each nucleotide is part of a single codon that specifies an amino acid.

3. The genetic code is almost universal. Although coding assignments were originally made in *E. coli*, subsequent studies showed that codons have the same meaning for other prokaryotes and for the cytoplasmic protein synthetic systems of eukaryotes. In consequence, mRNA from one species can be correctly translated by the protein synthetic machinery of another species. Exceptions to the canonical genetic code do exist and most of these are listed in Table 22.4. Two additional

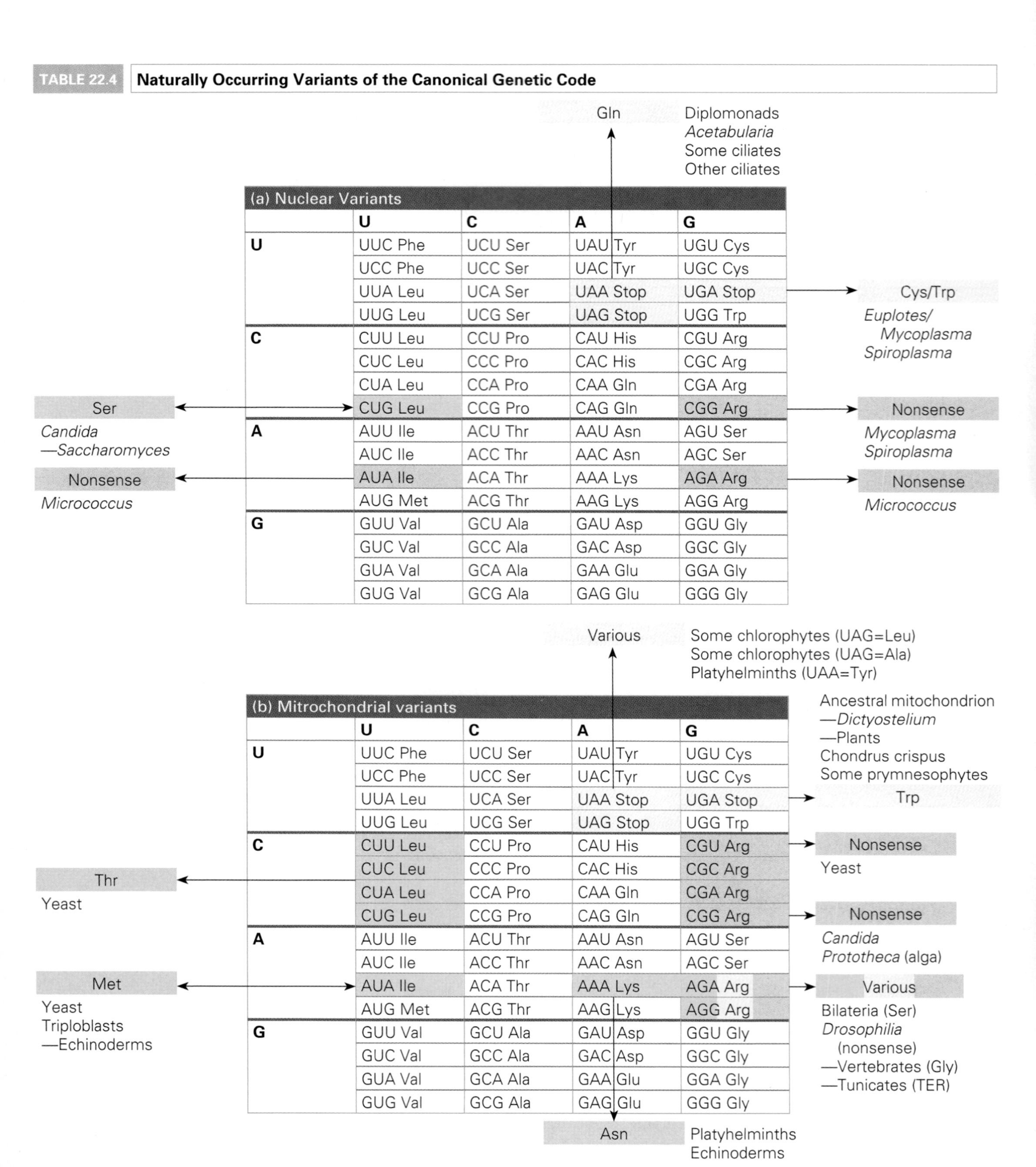

TABLE 22.4 Naturally Occurring Variants of the Canonical Genetic Code

(a) Nuclear Variants

	U	C	A	G
U	UUC Phe	UCU Ser	UAU Tyr	UGU Cys
	UCC Phe	UCC Ser	UAC Tyr	UGC Cys
	UUA Leu	UCA Ser	UAA Stop	UGA Stop
	UUG Leu	UCG Ser	UAG Stop	UGG Trp
C	CUU Leu	CCU Pro	CAU His	CGU Arg
	CUC Leu	CCC Pro	CAC His	CGC Arg
	CUA Leu	CCA Pro	CAA Gln	CGA Arg
	CUG Leu	CCG Pro	CAG Gln	CGG Arg
A	AUU Ile	ACU Thr	AAU Asn	AGU Ser
	AUC Ile	ACC Thr	AAC Asn	AGC Ser
	AUA Ile	ACA Thr	AAA Lys	AGA Arg
	AUG Met	ACG Thr	AAG Lys	AGG Arg
G	GUU Val	GCU Ala	GAU Asp	GGU Gly
	GUC Val	GCC Ala	GAC Asp	GGC Gly
	GUA Val	GCA Ala	GAA Glu	GGA Gly
	GUG Val	GCG Ala	GAG Glu	GGG Gly

(b) Mitochondrial variants

	U	C	A	G
U	UUC Phe	UCU Ser	UAU Tyr	UGU Cys
	UCC Phe	UCC Ser	UAC Tyr	UGC Cys
	UUA Leu	UCA Ser	UAA Stop	UGA Stop
	UUG Leu	UCG Ser	UAG Stop	UGG Trp
C	CUU Leu	CCU Pro	CAU His	CGU Arg
	CUC Leu	CCC Pro	CAC His	CGC Arg
	CUA Leu	CCA Pro	CAA Gln	CGA Arg
	CUG Leu	CCG Pro	CAG Gln	CGG Arg
A	AUU Ile	ACU Thr	AAU Asn	AGU Ser
	AUC Ile	ACC Thr	AAC Asn	AGC Ser
	AUA Ile	ACA Thr	AAA Lys	AGA Arg
	AUG Met	ACG Thr	AAG Lys	AGG Arg
G	GUU Val	GCU Ala	GAU Asp	GGU Gly
	GUC Val	GCC Ala	GAC Asp	GGC Gly
	GUA Val	GCA Ala	GAA Glu	GGA Gly
	GUG Val	GCG Ala	GAG Glu	GGG Gly

Naturally occurring variants of the canonical genetic code. (a) Bacterial and nuclear variants. (b) Mitochondrial variants. Missense changes are shown in gray; nonsense changes are shown in blue; changes in termination codons are shown in yellow. The symbol "—" indicates a reversal of a change in a particular lineage. (Reproduced from *Trends Biochem. Sci.*, vol 24, R. D. Knight, S. J. Freeland, and L. F. Landweber. *Selection, history, and chemistry . . .*, pp. 241–247. Copyright 1999, with permission from Elsevier [http://www.sciencedirect.com/science/journal/09680004].)

exceptions not listed in this table are UGA codes for selenocysteine in a wide variety of organisms when the codon is present in the proper context and UAG codes for pyrrolysine in some archaeal methanogens.

4. The genetic code is highly degenerate. Most amino acids are specified by two or more codons. Leucine, arginine, and serine are each specified by six different codons. Only two amino acids, methionine and tryptophan, are specified by a single codon. Codons specifying the same amino acid are called **synonyms**. For example, CCU, CCC, CCA, and CCG are synonyms for proline. As this example illustrates, synonyms usually vary at the third position. Such degeneracy probably helps to protect cells from mutations.
5. The genetic code is unambiguous under normal physiological conditions; that is, each codon specifies only one amino acid.

The coding specificity of an aminoacyl-tRNA is determined by the tRNA and not the amino acid.

Although it seemed likely that an aminoacyl-tRNA's specificity for its codon results from anticodon–codon interactions, the studies with synthetic messengers did not rule out the alternate possibility that a codon might somehow recognize the aminoacyl group attached to the tRNA. François Chapeville and coworkers ruled out the latter possibility in an experiment performed in 1962. They began by using a combination of enzymatic and organic synthetic techniques to prepare a misacylated tRNA (FIGURE 22.40). [^{14}C]Cysteine was attached to $tRNA^{Cys}$ by cysteinyl-tRNA synthetase. The resulting [^{14}C]Cys-$tRNA^{Cys}$ was then converted into [^{14}C]Ala-$tRNA^{Cys}$. (This was accomplished without

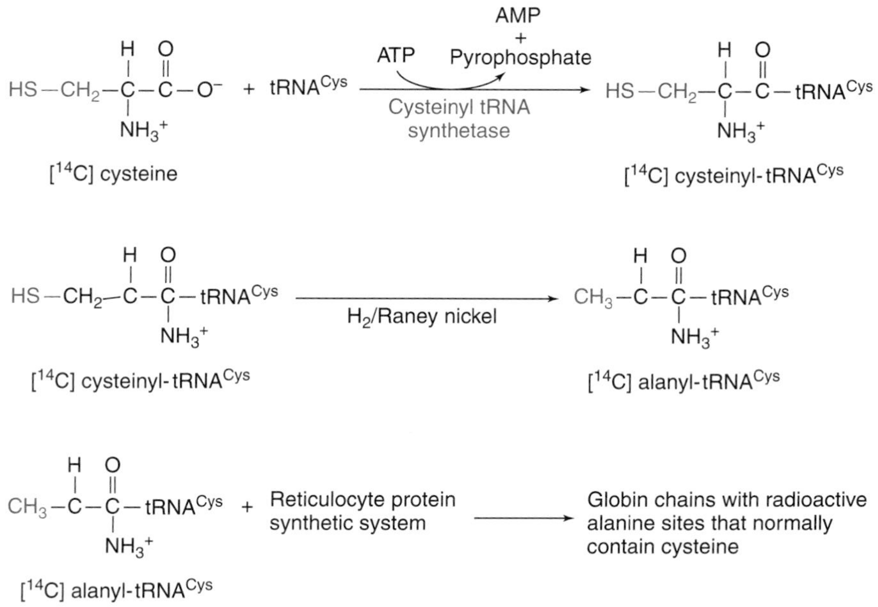

$$HS-CH_2-CH(NH_3^+)-C(=O)-O^- + tRNA^{Cys} \xrightarrow[\text{Cysteinyl tRNA synthetase}]{\text{ATP} \rightarrow \text{AMP + Pyrophosphate}} HS-CH_2-CH(NH_3^+)-C(=O)-tRNA^{Cys}$$

[^{14}C] cysteine → [^{14}C] cysteinyl-$tRNA^{Cys}$

$$HS-CH_2-CH(NH_3^+)-C(=O)-tRNA^{Cys} \xrightarrow{H_2\text{/Raney nickel}} CH_3-CH(NH_3^+)-C(=O)-tRNA^{Cys}$$

[^{14}C] cysteinyl-$tRNA^{Cys}$ → [^{14}C] alanyl-$tRNA^{Cys}$

$$CH_3-CH(NH_3^+)-C(=O)-tRNA^{Cys} + \text{Reticulocyte protein synthetic system} \longrightarrow \text{Globin chains with radioactive alanine sites that normally contain cysteine}$$

[^{14}C] alanyl-$tRNA^{Cys}$

FIGURE 22.40 **Experiment showing codon recognizes tRNA and not amino acid.**

disturbing the rest of the molecule by reducing [^{14}C]Cys-tRNA^{Cys} with molecular hydrogen in the presence of a Raney nickel catalyst.) The [^{14}C]Ala-tRNA^{Cys} was added to a protein synthetic system prepared from mammalian reticulocytes. Newly synthesized radioactive hemoglobin molecules were analyzed to determine whether radioactive alanine was incorporated into positions normally occupied by alanine or by cysteine residues. The results were quite clear. Radioactive alanine residues were present in positions that are normally occupied by cysteine residues. The experiment was therefore fully consistent with the adaptor hypothesis, showing that the codon recognizes the tRNA molecule and not the aminoacyl group.

Some aminoacyl-tRNA molecules bind to more than one codon because there is some play or wobble in the third base of a codon.

Some tRNA molecules can bind to two or three codons, differing only in the third base. For example, three codons recognized by tRNA^{Ala}, GCU, GCC, and GCA, all begin with the same two nucleotides but differ in the third nucleotide. Transfer RNA molecules that bind to three different codons all have inosinate in the first position of the anticodon.

Francis Crick proposed the **wobble hypothesis** in 1966 to explain how a single tRNA molecule can bind to two or three different codons. The essential features of this hypothesis are as follows:

1. The codon and anticodon form antiparallel base pairs.
2. The first two bases in the codon form standard Watson–Crick base pairs with the last two bases in the anticodon.
3. There is a certain amount of play or wobble in base pairing between the first base of the anticodon and the third base of the codon that permits nonstandard base pairing of the type indicated in FIGURES 22.41 and 22.42.

The base pairing relationships predicted by the wobble hypothesis have been confirmed by experimental studies.

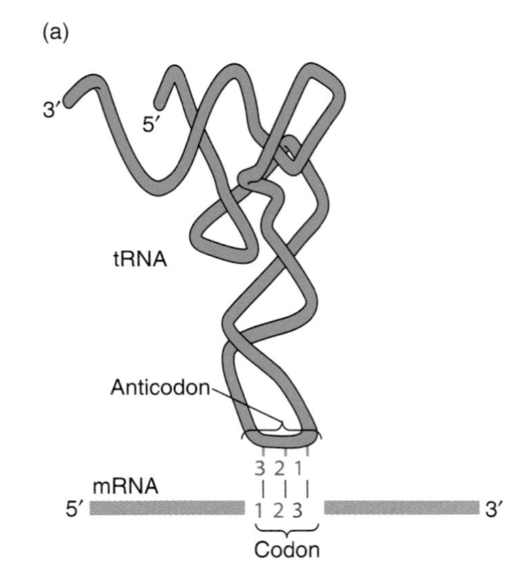

(b)

Wobble rules for codon-anticodon base pairing

5′ end of anticodon	3′ end of codon
G	U or C
C	G only
A	U only
U	A or G
I	U, C, and A

FIGURE 22.41 **Pairing between codon and anticodon.** (a) The first position in the anticodon (red) can form a nonstandard base pair with the third position in the codon (purple). (b) The table shows all possible wobble pairing. For instance, a guanine (G) in the anticodon can pair with either a uracil (U) or cytosine (C) in the codon and an isosine (I) in the anticodon can pair with a uracil (U), cytosine (C) , or adenine (A) in the codon.

The origin of the genetic code remains a puzzle.

Four different hypotheses, which are not necessarily mutually exclusive, have been proposed to explain codon assignments. The frozen accident hypothesis postulates that codon assignments are historical accidents that became fixed in the last ancestor of all modern organisms. This hypothesis assumes that codon assignments are random and so cannot be explained. An important argument against the frozen accident hypothesis is that codon assignments are not frozen as evidenced by the existence of variant codons (Table 22.4).

The adaptive hypothesis postulates that codon assignments evolved from selective pressure to minimize errors arising from mutations or mistranslations. Arguments supporting the adaptive hypothesis include:

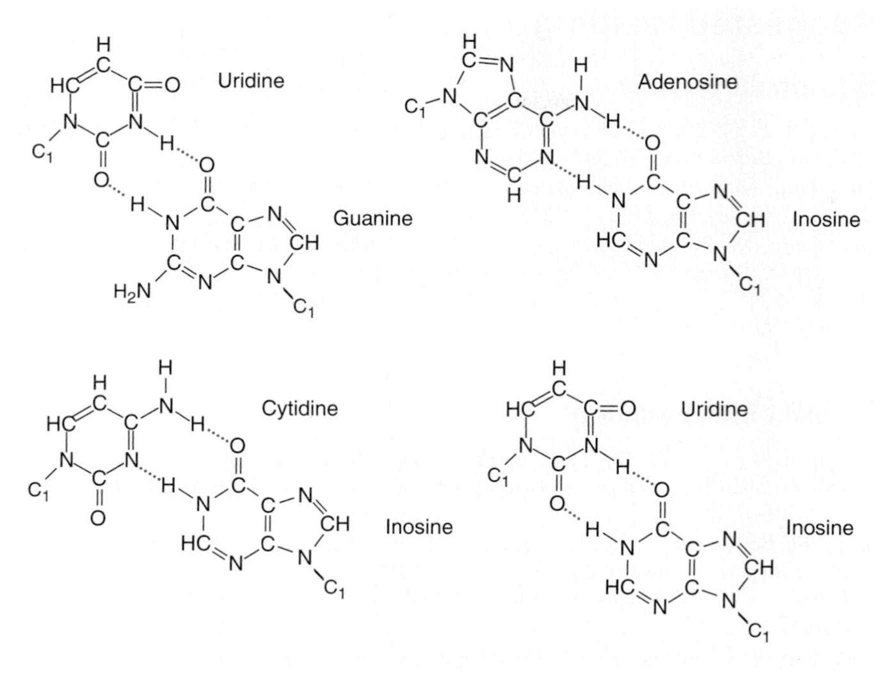

FIGURE 22.42 **Nonstandard wobble base pairs.** (Adapted from H. Lodish, et al. *Molecular Cell Biology, Fourth edition.* W.H. Freeman and Company, 1999.)

codons for the same amino acid usually vary only at position-3; codons for hydrophobic amino acids usually have U at position-2; and codons for hydrophilic amino acids usually have A at position-2. Arguments against the adaptive hypothesis include: the wobble hypothesis explains why codons for the same amino acid tend to vary at position-3; the relationship between hydrophobicity and codon position-2 applies to U and A but not to G and C; and code optimization would probably have produced alternative genetic codes rather than the nearly universal genetic code that exists today.

The historical hypothesis proposes that the present code evolved from a simpler ancestral form. It assumes that the earliest life forms contained proteins made of only a few amino acids and that these proteins eventually gained the ability to catalyze the formation of still other amino acids, which were in turn included in the code. Although dissimilar amino acids from related biochemical pathways do have similar codons, there are little additional data to support or refute the historical hypothesis.

The stereochemical hypothesis postulates that amino acid assignments evolved from direct interactions between codons and the amino acids that they specify. Attempts to show that amino acids bind to their cognate codons, anticodons, reversed codons, or codon–anticodon double helices provide little, if any, support for specific interactions.

Suggested Reading

Historical Overview

Clark, B. F. C. 2001. The crystallization and structural determination of tRNA. *Trends Biochem Sci* 26:511–514.

Hoagland, M. 1996. Biochemistry or molecular biology? The discovery of 'soluble RNA.' *Trends Biochem Sci* 21:77–80.

Nirenberg, M. 2004. Historical review: deciphering the genetic code—a personal account. *Trends Biochem Sci* 29:46–54.

Zamecnik, P. 2005. From protein synthesis to genetic insertion. *Ann Rev Biochem* 74:1–28.

Transfer RNA Structure

Apgar, J., Holley, R. W., and Merrill, S. H. 1962. Purification of the alanine-, valine-, histidine-, and tyrosine-acceptor ribonucleic acids from yeast. *J Biol Chem* 237:796–802.

Berg, P., Bergmann, F. H., Ofengand, E. J., and Dieckmann, M. 1961. The enzymic synthesis of amino acyl derivatives of ribonucleic acid. I. The mechanism of leucyl-, valyl-, isoleucyl-, and methionyl ribonucleic acid formation. *J Biol Chem* 236:1726–1734.

Clark, B. F. C. 2001. The crystallization and structural determination of tRNA. *Trends Biochem Sci* 26:511–514.

Goldman, E. 2008. Transfer RNA. In: *Encyclopedia of Life Sciences*. pp. 1–9. Hoboken, NJ: John Wiley and Sons.

Gustilo, E. M., Vendeix, F. A. P., and Agris, P. F. 2008. tRNA's modifications bring order to gene expression. *Curr Opin Microbiol* 11:134–140.

Hoagland, M. B., Stephenson, M. L., Scott, J. F., et al. 1958. A soluble ribonucleic acid intermediate in protein synthesis. *J Biol Chem* 231:241–257.

Holley, R. W., Apgar, J., Everett, G. A., et al. 1965. Structure of a ribonucleic acid. *Science* 147:1462–1465.

Holley, R. W., Everett, G. A., Madison, J. T., and Zamir, A. 1965. Nucleotide sequences in the yeast alanine transfer ribonucleic acid. *J Biol Chem* 240:2122–2128.

Hopper, A. K., and Phizicky, E. M. 2003. tRNA transfers to the limelight. *Genes Dev* 17:162–180.

Kim, S. H., Quigley, G., Suddath, F. L., et al. 1972. The three-dimensional structure of yeast phenylalanine transfer RNA: shape of the molecule at 5.5-Angstrom resolution. *Proc Natl Acad Sci USA* 69:3746–3750.

Kresge, N., Simoni, R. D., and Hill, R. L. 2006. The purification and sequencing of alanine transfer ribonucleic acid: the work of Robert W. Holley. *J Biol Chem* 281:e7–e9.

Pederson, T. 2005. 50 years ago protein synthesis met molecular biology: the discoveries of amino acid activation and transfer RNA. *FASEB J* 19:1583–1584.

Rich, A., and Kim, S. H. 1978. The three-dimensional structure of transfer RNA. *Sci Am* 238:52–62.

Robertus, J. D., Ladner, J. E., Finch, J. T., et al. 1974. Structure of yeast phenylalanine tRNA at 3 Å resolution. *Nature* 250:546–551.

Westhof, E., and Auffinger, P. 2001. tRNA structure. In: *Encyclopedia of Life Sciences*. London, UK: Nature Publishing Group. pp. 1–11.

Aminoacyl-tRNA Synthetases

Ahel, I., Korencic, D., Ibba, M., and Söll, D. 2003. Trans-editing of mischarged tRNAs. *Proc Natl Acad Sci USA* 100:15422–15427.

Ambrogelly, A., Palioura, S., and Söll, D. 2007. Natural expansion of the genetic code. *Nat Chem Biol* 3:29–35.

Antonellis, A., and Green, E. D. 2008. The role of aminoacyl-tRNA synthetases in genetic diseases. *Ann Rev Genomics Hum Genet* 9:87–107.

Arnez, J. G., and Moras, D. 1997. Structural and functional considerations of the aminoacylation reaction. *Trends Biochem Sci* 22:211–216.

Baldwin, A. N., and Berg, P. 1966. Transfer ribonucleic acid-induced hydrolysis of valyladenylate bound to isoleucyl ribonucleic acid synthetase *J Biol Chem* 241:839–845.

Beebe, K, de Pouplana, L. R., and Schimmel, P. 2003. Elucidation of tRNA-dependent editing by a class II tRNA synthetase and significance for cell viability. *EMBO J* 22:668–675.

Beebe, K., Merriman, E., and Schimmel, P. 2003. Structure-specific tRNA determinants by editing mischarged amino acid. *J Biol Chem* 46:45056–45061.

Beuning, P. J., and Musier-Forsyth, K. 2000. Transfer RNA recognition by aminoacyl-tRNA synthetases. *Biopolymers* 52:1–28.

Beuning, P. J., and Musier-Forsyth, K. 2000. Hydrolytic editing by a class II aminoacyl-tRNA synthetase. *Proc Natl Acad Sci USA* 97:8916–8920.

Böck, A., Thanbichler, R., Rother, M., and Resch, A. 2004. Selenocysteine. In Ibba, M., Francklyn, C., and Cusack, S., eds. *Aminoacyl-tRNA Synthetases*. Austin, TX: Landes Bioscience.

Cavarelli, J. 2003. Pushing the induced fit to its limits: tRNA-dependent active site assembly in class I aminoacyl-tRNA synthetases. *Structure* 11:484–486.

Cobucci-Ponzano, B., Rossi, M., and Moracci, M. 2005. Recoding in Archaea. *Mol Microbiol* 55:339–348.

Commans, S., and Bock, A. 1999. Selenocysteine inserting tRNAs: an overview. *FEMS Microbiol Rev* 23:335–351.

de Pouplana, L. R., and Schimmel, P. 2000. A view into the origin of life: aminoacyl-tRNA synthetases. *Cell Mol Life Sci* 57:865–870.

de Pouplana, L. R., and Schimmel, P. 2001. Operational RNA code for amino acids in relation to genetic code in evolution. *J Biol Chem* 276:6881–6884.

Eiler, S., Dock-Bregeon, A., Moulinier, L., et al. 1999. Synthesis of aspartyl-$tRNA^{Asp}$ in *Escherichia coli*—a snapshot of the second step. *EMBO J* 18:6532–6541.

Eldred, E. W., and Schimmel, P. R. 1972. Rapid deacylation by isoleucyl transfer ribonucleic acid synthetase of isoleucine-specific transfer ribonucleic acid aminoacylated with valine. *J Biol Chem* 247:2961–2964.

Fersht, A. 1998. Protein structure: sieves in sequence. *Science* 280:541.

Francklyn, C. 2008. DNA polymerase and aminoacyl-tRNA synthetases: shared mechanisms for ensuring the fidelity of gene expression. *Biochemistry* 45:11695–11703.

Francklyn, C., Perona, J. J., Puetz, J., and Hou, Y.-M. 2002. Aminoacyl-tRNA synthetases: versatile players in the changing theater of translation. *RNA* 8:1363–1372.

Fukai, S., Nureki, O., Sekine, S. I., et al. 2000. Structural basis for double-sieve discrimination of L-valine from L-isoleucine and L-threonine by the complex $tRNA^{Val}$ and valyl-tRNA synthetase. *Cell* 103:793–803.

Geslain, R., and de Pouplana, L. R. 2004. Regulation of RNA function by aminoacylation and editing? *Trends Genet* 20:604–610.

Giege, R. 2008. Toward a more complete view of tRNA biology. *Nat Struct Mol Biol* 15:1007–1014.

Giege, R., Sissler, M., and Florentz, C. 1998. Universal rules and idiosyncratic features in tRNA identity. *Nucl Acids Res* 26:5017–5035.

Hale, S. P., and Schimmel, P. 1996. Protein synthesis editing by a DNA aptamer. *Proc Natl Acad Sci USA* 7:2755–2758.

Hao, B., Gong, W., Ferguson, T. K., et al. 2002. A new UAG-encoded residue in the structure of a methanogen methyltransferase. *Science* 296:1462–1466.

Hatfield, D. L., and Gladyshev, V. N. 2002. How selenium has altered our understanding of the genetic code. *Mol Cell Biol* 22:3565–3576.

Hendrickson, T. L., Nomanbhoy, T. K., de Crécy-Lagnard, V., et al. 2002. Mutational separation of two pathways for editing by a class I tRNA synthetase. *Mol Cell* 9:353–362.

Hohsaka, T., and Sisido, M. 2002. Incorporation of non-natural amino acids into proteins. *Curr Opin Chem Biol* 6:809–815.

Hou, Y. M., and Schimmel, P. 1989. Evidence that a major determinant for the identity of a transfer RNA is conserved in evolution. *Biochemistry* 28:6800–6804.

Ibba, M., Becker, H. D., Stathopoulos, C., et al. 2000. The adaptor hypothesis revisited. *Trends Biochem Sci* 25:311–316.

Ibba, M., Curnow, A. W., and Söll, D. 1997. Aminoacyl-tRNA synthesis: divergent routes to a common goal. *Trends Biochem Sci* 22:39–42.

Ibba, M., Hong, K.-W., and Söll, D. 1996. Glutaminyl-tRNA synthetase: from genetics to molecular recognition. *Genes Cells* 1:421–427.

Ibba, M., and Söll, D. 1999. Quality control mechanisms during translation. *Science* 286:1893–1897.

Ibba, M., and Söll, D. 2000. Aminoacyl-tRNA synthetases. *Ann Rev Biochem* 69:617–650.

Ibba, M., and Söll, D. 2002. Genetic code: introducing pyrrolysine. *Curr Biol* 12:R464–R466.

Ibba, M., and Söll, D. 2004. Aminoacyl-tRNAs: setting the limits of the genetic code. *Genes Dev* 18:731–738.

Jakubowski, H. 2005. Transfer RNA synthetase editing of amino acids. In: *Encyclopedia of Life Sciences*. pp. 1–10. Hoboken, NJ: John Wiley and Sons.

Kim, S., Lee, S. W., Choi, E.-C., and Choi, S. Y. 2003. Aminoacyl-tRNA synthetases and their inhibitors as a novel family of antibiotics. *Appl Microbiol Biotechnol* 61:278–288.

Korenčić, D., Ahel, I., and Söll, D. 2002. Aminoacyl-tRNA synthesis in methanogenic archaea. *Food Technol Biotechnol* 40:255–260.

Linecum, T. L. Jr., Tukalo, M., Yaremchuk, A., et al. 2003. Structural and mechanistic basis of pre- and posttransfer editing by leucyl-tRNA synthetase. *Mol Cell* 11:951–963.

Ling, J., Reynolds, N., and Ibba, M. 2009. Aminoacyl-tRNA synthesis and translational quality control. *Ann Rev Microbiol* 63:61-78.

Lukashenko, N. P. 2010. Expanding genetic code: amino acids 21 and 22 selenocysteine and pyrrolysine. *Russian J Genet* 46:899–916.

Marinis, S. A., Plateau, P., Cavrelli, J., and Florentz, C. 1999. Aminoacyl-tRNA synthetases: a new image for a classical family. *Biochimie* 81:683–700.

Mizutani, T., Goto, C., and Totsuka, T. 2000. Mammalian selenocysteine tRNA, its enzymes and selenophosphate. *J Health Sci* 46:399–404.

Nakama, T., Nueki, O., and Yokoyama, S. 2001. Structural basis for the recognition of isoleucyl-adenylate and an antibiotic, mupirocin, by isoleucyl-tRNA synthetase. *J Biol Chem* 276:47387–47393.

Newberry, K. J., Hou, Y.-M., and Perona, J. J. 2002. Structural origins of amino acid selection without editing by cysteinyl-tRNA synthetase. *EMBO J* 21:2776–2787.

Nomanbhoy, T. K., Hendrickson, T. L., and Schimmel, P. 1999. Transfer RNA dependent translocation of misactivated amino acids to prevent errors in protein synthesis. *Mol Cell* 4:519–528.

Nozawal, K., O'Donoghue, P., Gundllapalli. S. et al. 2009. Pyrrolysyl-tRNA synthetase–tRNA^{Pyl} structure reveals the molecular basis of orthogonality. *Nature* 457:1163–1167.

Nureki, O., Vassylyev, D. G., Tateno, M., et al. 1998. Enzyme structure with two catalytic sites for double-sieve selection of substrate. *Science* 280:576–582.

O'Donoghue, P., and Luthey-Schulten, Z. 2003. On the evolution of structure in aminoacyl-tRNA synthetases. *Microbiol Mol Biol Rev* 67:550–573.

Pape, T., Wintermeyer, W., and Rodnina, M. 1999. Induced fit in initial selection and proofreading of aminoacyl-tRNA on the ribosome. *EMBO J* 18:3800–3807.

Park, S. G., Ewalt, K. L., and Kim, S. 2005. Functional expansion of aminoacyl-tRNA synthetases and their interacting factors: new perspectives on housekeepers. *Trends Biochem Sci* 30:569–574.

Polycarpo, C., Ambrogelly, A., Bérube, A. et al. 2004. An aminoacyl-tRNA synthetase that specifically activates pyrrolysine. *Proc Natl Acad Sci USA* 101:12450–12454.

Praetorius-Ibba, M., and Ibba, M. 2003. Aminoacyl-tRNA synthesis in archaea: different but not unique. *Mol Microbiol* 48:631–637.

RajBhandary, U. L. 1997. Once there were twenty. *Proc Natl Acad Sci USA* 94:11761–11763.

Ryckelynck, M., Giegé, R., and Frugier, M. 2005. tRNAs and tRNA mimics as cornerstones of aminoacyl-tRNA synthetase regulations. *Biochimie* 87:835–845.

Schmimmel, P. 2008. An editing activity that prevents mistranslation and connection to disease. *J Biol Chem* 283:28777–28782.

Schimmel, P. 2008. Development of tRNA synthetases and connection to genetic code and disease. *Protein Sci* 17:1643–1652.
Silvian, L. F., Wang, J., and Steitz, T. A. 1999. Insights into editing from an Ile-tRNA synthetase structure with $tRNA^{Ile}$ and Mupirocin. *Science* 285:1074–1077.
Srinivasan, G., James, C. M., and Krzycki, J. A. 2002. Pyrrolysine encoded by UAG in archaea: charging of a UAG-decoding specialized tRNA. *Science* 296:1459–1462.
Wilcox, M., and Nirenberg, M. 1968. Transfer RNA as a cofactor coupling amino acid synthesis with that of protein. *Proc Natl Acad Sci USA* 61:229–236.
Yuan, J., O'Donoghue, P., Ambrogelly, A. et al. 2010. Distinct genetic code expansion strategies for selenocysteine and pyrrolysine are reflected in different aminacyl-tRNA formation systems. *FEBS Lett* 584:342–449.

Messenger RNA and the Genetic Code

Agris, P. F. 2004. Decoding the genome: a modified view. *Nucl Acids Res* 32:223–238.
Brenner, S., Stretton, A. O. W., and Kaplan, S. 1965. Genetic code: the 'nonsense' triplets for chain termination and their suppression. *Nature* 206:994–998.
Chapeville, F., Lipmann, F., von Ehrenstein, G., et al. 1962. On the role of soluble ribonucleic acid in coding for amino acids. *Proc Natl Acad Sci USA* 48:1086–1092.
Copeland P. R. 2003. Regulation of gene expression by stop codon recoding: selenocysteine. *Gene* 312:17–25.
Crick, F. H. C. 1966. Codon-anticodon pairing: the wobble hypothesis. *J Mol Biol* 19:548–555.
Crick, F. H. C., Barnett, L., Brenner, S., and Watts-Tobin, R. J. 1961. General nature of the genetic code for proteins. *Nature* 192:1227–1232.
Davis, B. K. 2004. Expansion of the genetic code in yeast: making life more complex. *Bioessays* 26:111–115.
Dintzis, H. M. 1961. Assembly of the peptide chains of hemoglobin. *Proc Natl Acad Sci USA* 47:247–261.
Khorona, H. G. 1968. Synthesis in the study of nucleic acids. The Fourth Jubilee Lecture. *Biochem J* 109:709–725.
Knight, R. D., Freeland, S. J., and Landweber, L. F. 1999. Selection, history and chemistry: the three faces of the genetic code. *Trends Biochem Sci* 24:241–247.
Nirenberg, M. 2004. Historical review: deciphering the genetic code—a personal account. *Trends Biochem Sci* 29:46–54.
Nirenberg, M. W., and Leder, P. 1964. RNA codewords and protein synthesis: the effect of trinucleotides upon the binding of sRNA to ribosomes. *Science* 145:1399–1407.
Nirenberg, M. W., and Matthaei, J. H. 1961. The dependence of cell-free protein synthesis in *E. coli* upon naturally occurring or synthetic polyribonucleotides. *Proc Natl Acad Sci USA* 47:1588–1602.
Pardee, A., Jacob, F., and Monod, J. 1959. The genetic control and cytoplasmic expression of 'inducibility' in the synthesis of beta-galactosidase by *E. coli*. *J Mol Biol* 1:165–178.
Salas, M., Smith, M. A., Stanley, W. M. Jr., et al. 1965. Direction of reading of the genetic message. *J Biol Chem* 240:3988–3995.
Srinivasa, G., James, C. M, and Krzycki, J. A. 2002. Pyrrolysine encoded by UAG in archaea: charging of a UAG-decoding specialized tRNA. *Science* 296:1459–1462.
Volkin, E., and Astrachan, L. 1956. Intracellular distribution of labeled ribonucleic acid after phage infection of *Escherichia coli*. *Virology* 2:433–437.
Watanabe, K. and Suzuki, T. 2008. Universal genetic code and its natural variations. In: *Encyclopedia of Life Sciences*. pp. 1–8. Hoboken, NJ: John Wiley and Sons.

This book has a Web site, **http://biology.jbpub.com/book/molecular**, which contains a variety of resources, including links to in-depth information on topics covered in this chapter, and review material designed to assist in your study of molecular biology.

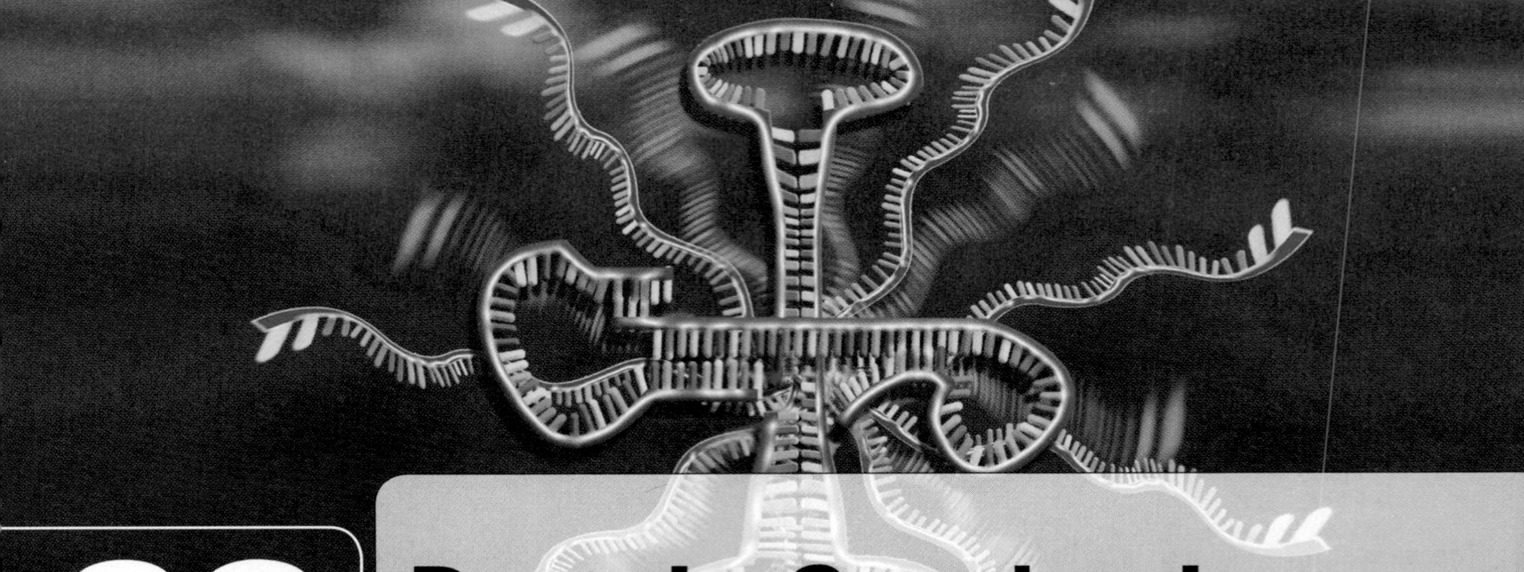

23 Protein Synthesis: The Ribosome

OUTLINE OF TOPICS

23.1 Ribosome Structure

Bacterial ribosome structure has been determined at atomic resolution.

Archael and eukaryotic ribosome structures appear to be similar to the bacterial ribosome structure.

23.2 Initiation Stage in Bacteria

Protein synthesis can be divided into four stages.

Bacteria, eukaryotes, and archaea each have their own translation initiation pathway.

Each bacterial mRNA open reading frame has its own start codon.

Bacteria have an initiator methionine tRNA and an elongator methionine tRNA.

The 30S subunit is an obligatory intermediate in polypeptide chain initiation.

Initiation factors participate in the formation of 30S and 70S initiation complexes.

The mRNA Shine-Dalgarno (SD) sequence interacts with the 16S rRNA anti-Shine-Dalgarno (anti-SD) sequence.

Riboswitches regulate translation initiation of some bacterial mRNA molecules.

23.3 Initiation Stage in Eukaryotes

Eukaryotic initiator tRNA is charged with a methionine that is not formylated.

Eukaryotic translation initiation proceeds through a scanning mechanism.

Eukaryotes have at least twelve different initiation factors.

Translation initiation factor phosphorylation regulates protein synthesis in eukaryotes.

The translation initiation pathway in archaea appears to be a mixture of the eukaryotic and bacterial pathways.

23.4 Elongation Stage

Polypeptide chain elongation requires elongation factors.

The elongation factors act through a repeating cycle.

An EF-Tu•GTP•aminoacyl-tRNA ternary complex carries the aminoacyl-tRNA to the ribosome.

An additional elongation factor, EF-P, is required to synthesize the first peptide bond.

Specific nucleotides in 16S rRNA are essential for sensing the codon-anticodon helix.

EF-Ts is a GDP-GTP exchange protein.

The ribosome is a ribozyme.

The hybrid-states translocation model offers a mechanism for moving tRNA molecules through the ribosome.

EF-G•GTP stimulates the translocation process.

23.5 Termination Stage in Bacteria

Bacteria have three protein release factors.

The class 1 release factors, RF1 and RF2, have one tripeptide that acts as an anticodon and another that binds at the peptidyl transferase center.

RF3 is a nonessential G protein that stimulates RF1 or RF2 dissociation from the ribosome.

A stalled ribosome translating a truncated mRNA that lacks a termination codon can be rescued by tmRNA.

Mutant tRNA molecules can suppress mutations that create termination codons within a reading frame.

23.6 Termination Stage in Eukaryotes

Eukaryotic cells have bacteria-like release factors in their mitochondria and a different kind in their cytoplasm.

23.7 Recycling Stage

The ribosome recycling factor (RRF) is required for the bacterial ribosomal complex to disassemble.

23.8 Nascent Polypeptide Processing and Folding

Ribosomes have associated enzymes that process nascent polypeptides and chaperones that help to fold the nascent polypeptides.

23.9 Signal Sequence

The signal sequence plays an important role in directing newly synthesized proteins to specific cellular destinations.

Suggested Reading

Thus far, our examination of protein synthesis has been largely concerned with the genetic code with particular emphasis on (1) aminoacyl-tRNA formation and (2) codon assignment. This examination has shown us that all cells use ribosomes as protein synthetic machines that somehow translate the information provided by specific codon sequences on messenger RNA (mRNA) molecules into specific polypeptide sequences in newly synthesized proteins. The ribosome is a dynamic machine with many moving parts that work together so that the ribosome can translate about 10 to 20 codons per second with an error rate that is usually less than 0.03%.

Although bacterial, eukaryotic, and archaeal ribosomes differ in structural detail, they share fundamental structure–function relationships. All ribosomes contain a small and a large subunit, both of which must be present for normal ribosome function. The small subunit helps to solve the decoding problem by its exclusive interaction with mRNA and its ability to assist aminoacyl-tRNA molecules to assemble on the mRNA by taking advantage of anticodon–codon interactions. The large subunit contains peptidyl transferase activity needed for peptide bond formation and hydrolase activity required to release the completed polypeptide from the ribosome complex.

This chapter examines ribosome structure and function to learn how the ribosomes are able to act as universal translators. The primary focus will be on bacterial ribosomes, which have been more thoroughly investigated than eukaryotic ribosomes. However, much of the information gained by studying bacterial ribosomes also applies to eukaryotic ribosomes because the two kinds of ribosomes have similar structures and work in the same way.

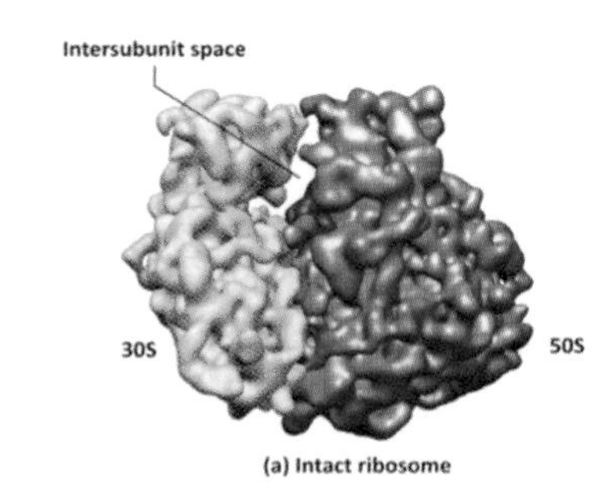

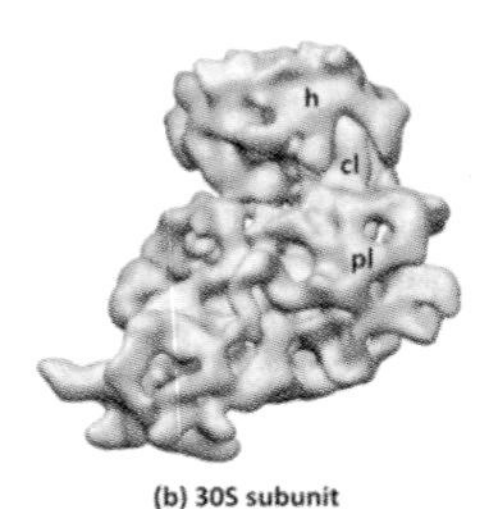

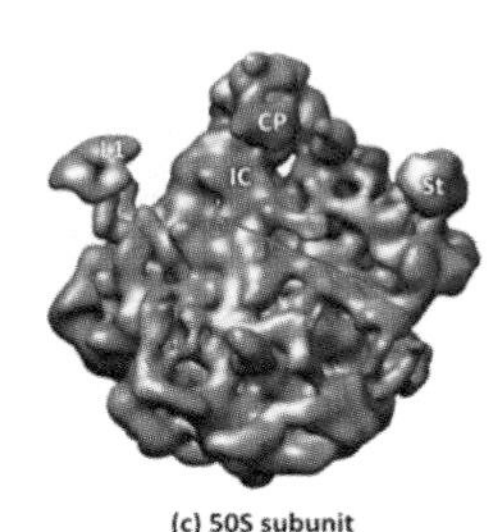

FIGURE 23.1 Cryo-electron microscopy reconstruction of the *Salmonella typhimurium* (close relative to *E. coli*) ribosome. (a) The intact 70S ribosome at 10Å resolution. The 50S (large) subunit is shown in blue and the 30S (small) ribosome is shown in yellow (b and c). The 70S ribosome structure was computationally separated into its two subunits in order to show the topography of the intersubunit space. (b) Landmarks in the small subunit: h, head; pl, platform; cl, cleft. (c) Landmarks in the large subunit: L1, L1 protein; St, L7/L12 stalk; IC, interface canyon (red line with arrow at each end); CP, central protuberance. (Photo courtesy of Danny Nam Ho and Joachim Frank, HHMI, Columbia University.)

23.1 Ribosome Structure

Bacterial ribosome structure has been determined at atomic resolution.

The bacterial ribosome is a ribonucleoprotein complex with a diameter of about 25 nm, a molecular mass of about 2.4×10^6 Da, and a sedimentation coefficient of 70S. At low magnesium ion concentrations, the 70S ribosome dissociates into a 30S (or small) subunit (molecular mass = 0.9×10^6 Da) and a 50S (or large) subunit (molecular mass = 1.5×10^6 Da). RNA accounts for about 66% of the mass of each subunit and protein accounts for the remaining mass.

The 30S subunit performs the ribosome's decoding function, discriminating between proper and improper codon–anticodon interactions. It contains 21 polypeptides (designated S1–S21) and a single 16S ribosomal RNA (rRNA) (1,541 nucleotides). Approximately one third of the 30S subunit's mass is in a region designated the **head** and the remaining two thirds in a region designated the **body** (FIGURE 23.1a and b). The upper part of the body has a broad shelf-like protrusion called the **platform.** The site of codon–anticodon interactions is located in a cleft, which lies between the head and platform.

The 50S (or large) subunit contains the ribosome's peptidyl transferase activity, which catalyzes peptide bond formation. The subunit contains 34 polypeptides (designated L1–L34), a 5S rRNA (120 nucleotides), and a 23S rRNA (2,904 nucleotides). The flat side of the hemispherical 50S subunit faces the 30S subunit and the convex side faces the solvent (Figure 23.1a). When viewed flat face on, the 50S subunit has three protuberances that give it a crown-like appearance (FIGURE 23.1c). The L1 protein is located in the mushroom-shaped protuberance on the left, the 5S rRNA in the central protuberance, and two L7/L12 protein dimers in the flexible stalk on the right. The flat face has a deep groove known as the **interface canyon** that runs across the entire width of the subunit and is the site of tRNA binding. The interface canyon is part of the intersubunit space that is visible in some orientations of the intact ribosome (Figure 23.1a) and is of special interest because tRNA molecules fit into this space (see below).

Because of its large size and complex structure, the ribosome presented a major challenge for molecular biologists who wished to see its structure in atomic detail. Ada Yonath and coworkers made an important breakthrough in determining ribosome structure in 1980 when they succeeded in growing crystals of *Bacillus steareothermophilus* 50S subunits. Although these crystals diffracted to low resolution, their formation motivated investigators to seek new and better methods for growing ribosomal crystals that could diffract to high resolution. The resulting efforts were rewarded in 1999 when the laboratories of Yonath and Venkatraman Ramakrishnan, working independently, solved the crystal structure for the 30S subunit from the thermophilic bacteria *Thermus thermophilus*, and Thomas Steitz, Peter Moore, and coworkers solved the crystal structure for the 50S subunit from the extremely halophilic ("salt-loving") archaeon *Haloarcula marismortui*. Harry F. Noller and coworkers solved the crystal structure for the intact *T. thermophilus* 70S ribosome two years later.

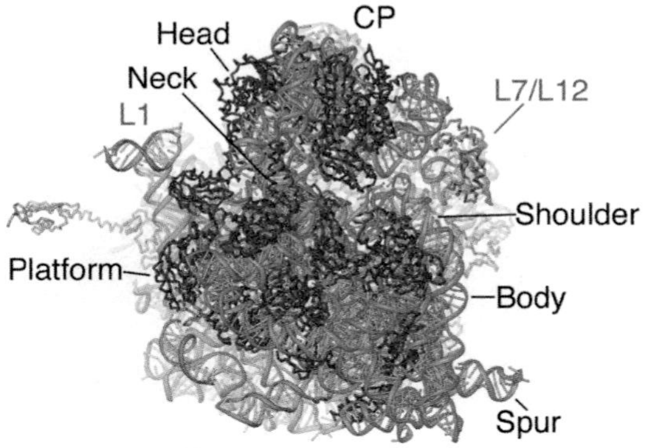

FIGURE 23.2 **Structure of the *E. coli* 70S ribosome.** The structure is viewed from the solvent side of the 30S subunit. The rRNA and proteins in the 30S subunit are colored light and dark blue, respectively. The 23S rRNA, 5S rRNA, and proteins in the 50S subunit are colored gray, purple, and magenta, respectively. Structural features visible in the 30S subunit include the head, neck, platform, body, shoulder, and spur. Structural features visible in the 50S subunit include L1 (protein L1/rRNA arm) and the central protuberance (CP). The approximate location of the proteins L7/L12, not observed in the structure, is in gray. (Reproduced from B. S. Schuwirth, et al. 2005. *Science*. 310: 827–834. Reprinted with permission from AAAS. Photo courtesy of Jamie H. Doudna Cate, University of California, Berkeley.)

The absence of a structure for the *Escherichia coli* ribosome was a cause for mild concern because the *E. coli* ribosome was the model system studied by most molecular biologists and biochemists. Therefore, investigators who were interested in structure–function relationships were forced to make the reasonable assumption that the *E. coli* ribosome would have a very similar structure to that of *T. thermophilus*. When Jamie H. D. Cate and coworkers solved the structure of the *E. coli* ribosome at a 3.5 Å resolution (FIGURE 23.2) in 2005, this assumption was shown to be correct. As expected, the size and shape of the high-resolution *E. coli* ribosome structure is fully consistent with the three-dimensional cryo-electron microscopy construct (compare Figures 23.1 and 23.2).

The crystal structures permitted investigators to locate the individual ribosomal components for the first time. In general, proteins in the 30S subunit are concentrated in the head, shoulder, and platform regions, whereas those in the 50S subunit are more evenly scattered over the solvent surface. The interface between the large and small subunit consists mainly of RNA. The rRNAs can be divided into domains, which are consistent with predicted secondary structures. FIGURE 23.3a shows the secondary and tertiary structures for 16S rRNA.

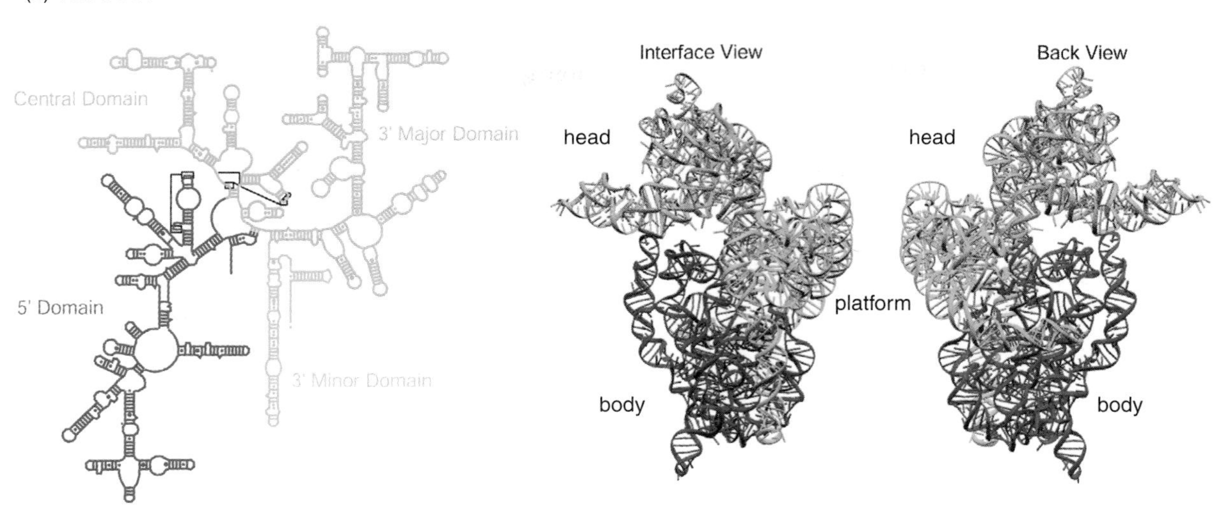

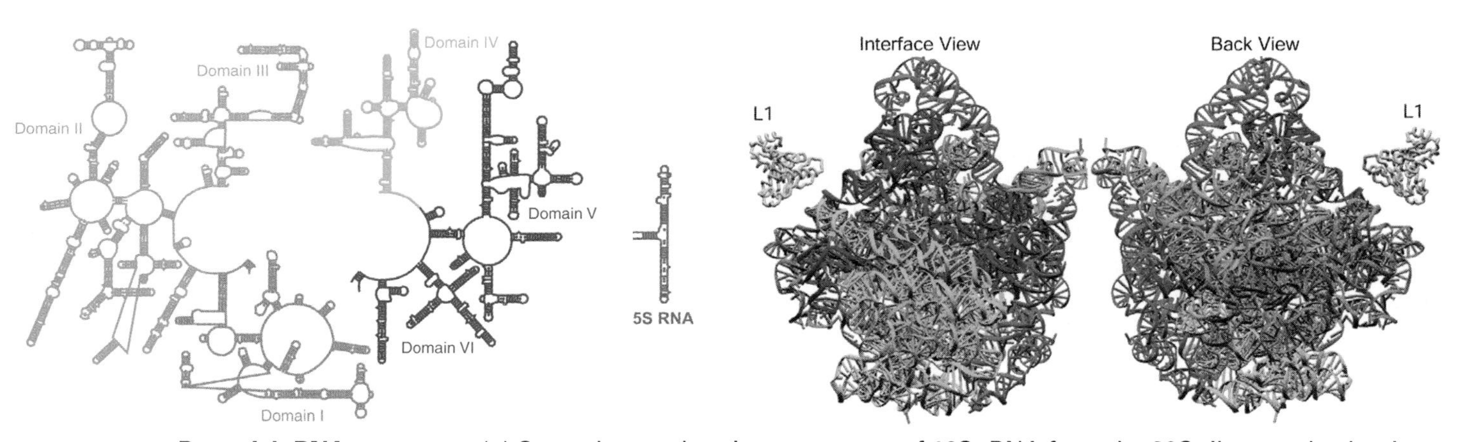

FIGURE 23.3 **Bacterial rRNA structures.** (a) Secondary and tertiary structures of 16S rRNA from the 30S ribosomal subunit. The secondary structure domains in 16S rRNA fold as distinct domains. (b) Secondary and tertiary structures of the 23S and 5S rRNAs from the 50S ribosomal subunit. The secondary structure domains in 23S rRNA are intricately interwoven in three dimensions. Each domain has the same color in the secondary and tertiary structures. (Reproduced from *Curr. Opin. Struct. Biol.*, vol. 11, V. Ramakrishnan and P. B. Moore, Atomic structures at last . . ., pp. 144–154, copyright 2001, with permission from Elsevier [http://www.sciencedirect.com/science/journal/0959440X]. Photo courtesy of V. Ramakrishnan, Medical Research Council (UK).)

Four distinct domains are present and each can be assigned to a specific part of the 30S subunit. The 5′-domain forms the 30S body; the central domain forms the platform; the 3′-major domain forms the head; and the 3′-minor domain runs along the intersubunit face of the 30S subunit.

FIGURE 23.3b shows the secondary and tertiary structures for the 23S and 5S rRNAs. In the tertiary structure, the 5′- and 3′-ends come together to form a helix. Several stem and loop structures radiate from

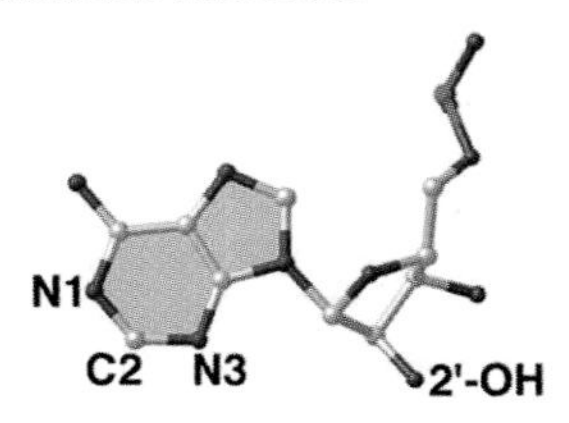

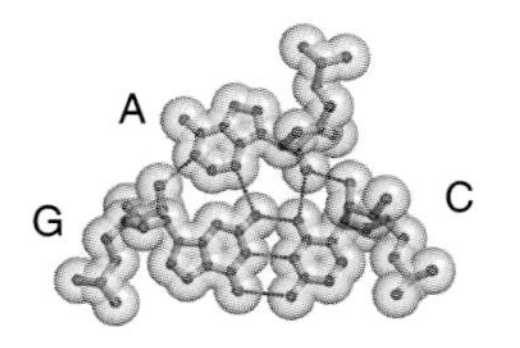

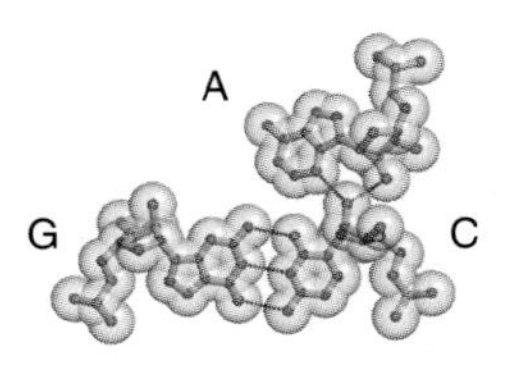

FIGURE 23.4 **A-minor motif.** (a) Adenosine nucleotides have a smooth minor groove face, allowing the adenine to pack tightly into the minor groove of an RNA helix and leaving its N1, N3, and 2′-OH atoms available for hydrogen bonding interactions. (b) Two examples of A-minor interactions. Each type is defined by the position of the 2′-OH group of the interacting adenosine relative to the 2′-OH group of the receptor base pair. (Reproduced from P. Nissen, et al., *Proc. Natl. Acad. Sci. USA* 98 [2001]: 4899–4903. Copyright 2001, National Academy of Sciences, U.S.A. Photos courtesy of Thomas A. Steitz, Yale University.)

the loop of this helix and bundle into six interwoven domains. RNA helical regions in the ribosome are often considerably longer than predicted from the secondary structure. Two factors contribute to this increased helical length. First, single-stranded regions in the secondary structure form slightly irregular double-stranded extensions of neighboring helices so that most of the RNA in the ribosome appears helical or nearly helical. Second, helices formed by nucleotides in neighboring sequences tend to stack end-to-end. Similar helical stacking had been observed in tRNA for the acceptor and TψC-stems as well as for the D- and anticodon stems (see Figure 22.12).

The 23S rRNA tertiary structure is stabilized by magnesium ion bridges between distant phosphate groups that are brought near to one another by rRNA folding. Two types of RNA–RNA interaction provide additional stability, **long-range base pairing**, and a novel interaction called the **A-minor motif** because the minor groove edge of an adenine is inserted into the minor groove of a neighboring helix, usually at a G-C base pair (FIGURE 23.4). The adenine forms hydrogen bonds with one or both of the 2′-hydroxyls of the G-C base pair. Protein-RNA interactions also add stability. Most ribosomal proteins have globular domains at the cytoplasmic surface and long tails that extend into the RNA and help to crosslink different RNA domains.

A model of the 70S ribosome based on a combination of three crystal structures reveals three tRNA-binding sites (FIGURE 23.5). An mRNA fragment interacting with the anticodon of the tRNA at the A-site is also visible. Earlier biochemical studies predicted the existence of these sites, which were named the **A-site** (**aminoacyl-tRNA binding site**), the **P-site** (**peptidyl-tRNA binding site**), and the **E-site** (**exit site**). The functions of these sites are described below.

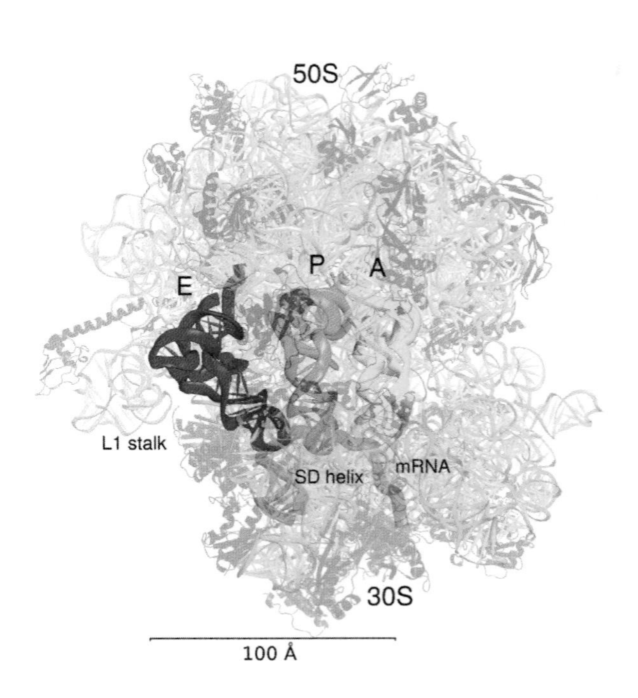

FIGURE 23.5 **A model of the structure of the *Thermus thermophilus* 70S ribosome with tRNAs bound to the A, P, and E sites.** The tRNA molecules at the A, P, and E sites are colored yellow, orange, and red, respectively. The 16S rRNA in the small subunit is cyan, the 23S and 5S rRNAs in the large subunit are gray and light blue, respectively. Protein subunits in the small and large subunits are blue and magenta, respectively. The mRNA is green. (Reproduced from *Curr. Opin. Chem. Biol.*, vol. 12, A. Korostelev, D. N. Ermolenko, and H. F. Noller, Structural dynamics of the ribosome, pp. 674–683, copyright 2008, with permission from Elsevier [http://www.sciencedirect.com/science/journal/13675931]. Photo courtesy of Andrei Korostelev, University of Massachusetts.)

Archael and eukaryotic ribosome structures appear to be similar to the bacterial ribosome structure.

As indicated above, Thomas Steitz, Peter Moore, and coworkers determined the structure of the large ribosomal subunit from the extremely halophilic archaeon *H. marismortui* (FIGURE 23.6). This subunit's structure appears remarkably similar to its bacterial counterpart. Although not shown in the figure, the archaeal large subunit has A-, P-, and E-sites in the same positions as the 50S bacterial ribosomal subunit. Crystal structures are not yet available for an intact archaeon ribosome.

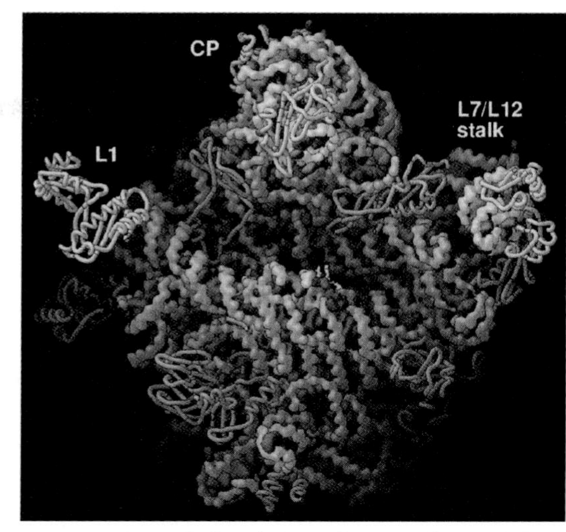

FIGURE 23.6 **Subunit interface surface of the large ribosomal subunit from the archaeon *Haloarcula marismortui*.** RNA is shown in gray in a space-filling–like representation that exaggerates its backbones. Proteins are shown in yellow in a continuous wire format that shows the trajectories of their backbones. Abbreviations: L1, L1 protein; CP, central protuberance. (Reproduced from N. Ban, et al., *Science* 289 [2000]: 905–920. Reprinted with permission from AAAS. Photo courtesy of Thomas A. Steitz, Yale University.)

The best-studied eukaryotic ribosome is that from yeast. Its large (60S) subunit contains three RNA molecules: 25S rRNA (3,392 nucleotides), 5.8S rRNA (158 nucleotides), and 5S rRNA (121 nucleotides). The mammalian counterpart of 25S rRNA has a sedimentation coefficient of 28S and is at least 1,600 nucleotides longer. The 5.8S rRNA is homologous to the 5′-region of the bacterial 23S rRNA. The 60S yeast subunit contains 46 polypeptides. RNA accounts for about 60% of the large subunit's mass and protein for the remaining 40%. The small (40S) subunit has a single 18S RNA (1,798 nucleotides) and 33 polypeptides. RNA accounts for about 55% of the small subunit's mass and protein for the remaining 45%. All identified yeast ribosomal proteins have homologous counterparts in the mammalian ribosome.

A comparison of eukaryotic rRNA molecules with their bacterial and archaeal counterparts reveals a remarkable degree of primary and secondary structure conservation. The conserved rRNA regions are those required to perform essential ribosomal functions such as the decoding aminoacyl-tRNA and forming peptide bonds. One major reason that eukaryotic rRNA molecules are longer than their bacterial counterparts is that additional nucleotide sequences called **expansion segments** are inserted at specific sites within the conserved rRNA core.

Marat Yusupov and coworkers recently determined the crystal structure of the yeast ribosome at a resolution of 4.15 Å (FIGURE 23.7). They were able to do so by isolating the ribosomes from yeast that had been cultured in glucose-depleted medium for a short time. This

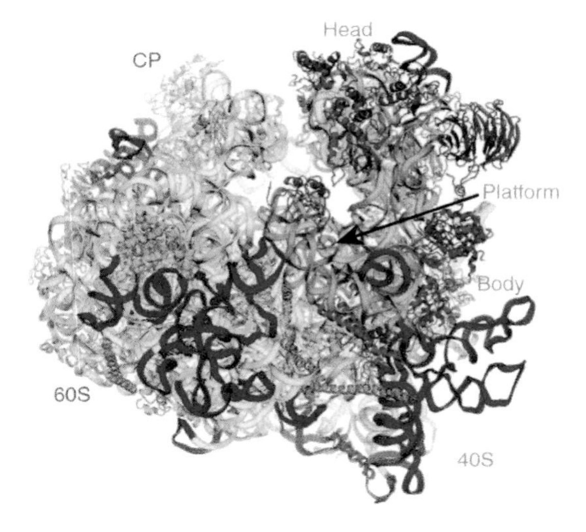

FIGURE 23.7 **The crystal structure of the yeast ribosome.** Proteins are colored dark blue in the 40S subunit and dark yellow in the 60S subunit. Ribosomal RNAs are colored light blue in the 40S subunit and pale yellow in 60S subunit. The rRNA expansion segments, which are located primarily on solvent-exposed subunit surfaces, are shown in red. CP indicates the central protuberance. (Reproduced from A. Ben-Shem, et al., *Science* 330 [2010]: 1203–1209. Reprinted with permission from AAAS. Photo courtesy of Marat Yusupov, Institut de Génétique et de Biologie Moléculaire et Cellulaire.)

growth condition inhibits translation initiation causing the yeast to accumulate a homogenous population of vacant ribosomes that can form crystals. Although the yeast ribosome is considerably larger than the bacterial ribosome, the two have the same basic architecture. Recognizable bacterial ribosome landmarks including the head, platform, body, and central protuberance (CP) are also present in the yeast ribosome. The rRNA expansion segments are located primarily on solvent-exposed subunit surfaces. If eukaryotic ribosome structure–function studies follow the same path as bacterial ribosome structure–function studies, we can look forward to learning a great deal about how the eukaryotic ribosome works in the near future.

23.2 Initiation Stage in Bacteria

Protein synthesis can be divided into four stages.

With this introduction to ribosomal structure as background, we are now ready to examine protein synthesis, which can be divided into four stages:

1. **The initiation stage:** The reading frame is set by binding initiator tRNAs to ribosomes at start codons with the assistance of **initiation factors** (**IFs**).
2. **The elongation stage:** Amino acids are added to the growing polypeptide chain as codons on mRNA are matched with anticodons on tRNAs with the assistance of **elongation factors** (**EFs**).
3. **The chain termination stage:** Termination codons are recognized by **release factors** (**RFs**), and completed polypeptide chains are released from the ribosome.
4. **The recycling phase:** Ribosomal subunits dissociate from each other under the influence of a **ribosomal recycling factor** (**RRF**).

Each stage in protein synthesis requires enzymes and protein factors that are unique to it.

Bacteria, eukaryotes, and archaea each have their own translation initiation pathway.

Biochemical studies performed in the late 1960s and early 1970s showed that special protein factors are needed to initiate polypeptide chain synthesis. As more information became available about bacterial, eukaryotic, and more recently archaeal polypeptide chain initiation pathways, it became apparent that while the three pathways share some important features, they also differ in many important ways. In fact, the differences among the translation initiation pathways are so great that it is best to examine each pathway separately.

Each bacterial mRNA open reading frame has its own start codon.

Bacterial transcription and translation take place in the same cellular compartment, permitting translation to begin shortly after the 5′-end of the mRNA emerges from the exit channel of RNA polymerase. Polycistronic mRNAs, which are quite common in bacteria, usually have an initiation codon at the start of each open reading frame and one or more termination codons at the end of each open reading frame. Therefore, the translation machinery must be able to recognize and initiate polypeptide chain synthesis at several different sites in the same mRNA. Translation always starts at a specific initiator codon. In bacteria AUG is the initiator codon about 90% of the time, GUG about 10%, and UUG about 1%. In very rare cases another codon such as AUU may function as an initiation codon. Precise recognition is essential because a phasing error of just one nucleotide would cause translation of the mRNA to be in the wrong reading frame. Each initiation codon binds the same initiator tRNA, which will now be described.

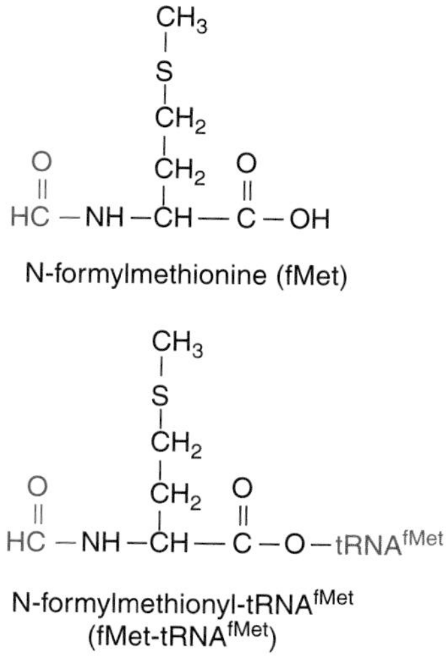

Bacteria have an initiator methionine tRNA and an elongator methionine tRNA.

Kjeld Marcker and Frederick Sanger discovered the existence of the bacterial initiator tRNA while studying methionyl-tRNA synthetase in 1964. They incubated a soluble *E. coli* extract with [^{35}S]methionine, ATP, and tRNA, expecting to synthesize Met-tRNA. Although they were successful in forming the expected product, they also detected a second product that contained a methionine derivative attached to the tRNA. Chemical analysis revealed the methionine derivative to be **N-formylmethionine** (**fMet**) (FIGURE 23.8). Sanger and coworkers recognized that the formyl group would prevent fMet from being incorporated into a polypeptide chain at any position other than the amino terminus. Subsequent studies showed that bacterial extracts do in fact incorporate fMet into the amino end of a growing polypeptide. Sanger and coworkers showed that bacteria synthesize **fMet-tRNA** by a two-step pathway (FIGURE 23.9): (1) methionyl-tRNA synthetase attaches methionine to tRNA to produce methionyl-tRNA and (2) **methionyl-tRNA formyltransferase** transfers a formyl group from N^{10}-formyltetrahydrofolate to methionyl-tRNA to form fMet-tRNA. Mitochondria and chloroplasts synthesize N-formylmethionyl-tRNA by a similar pathway.

The discovery that bacterial polypeptide synthesis begins with N-formylmethionine was puzzling because fMet is rarely, if ever, found at the amino terminus of a bacterial protein. The N-formyl group's absence was explained when a new enzyme, **peptide deformylase**, was discovered that cleaves formyl groups from N-termini. Peptide deformylase activity explains why *E. coli* polypeptides do not have formyl groups at their N-termini but does not explain why the N-terminus is methionine in only about 40% of bacterial polypeptides. The loss of methionine in the other 60% of the polypeptides was explained by the presence of still another enzyme, **methionine aminopeptidase**, which cleaves the N-terminal methionine from the polypeptides.

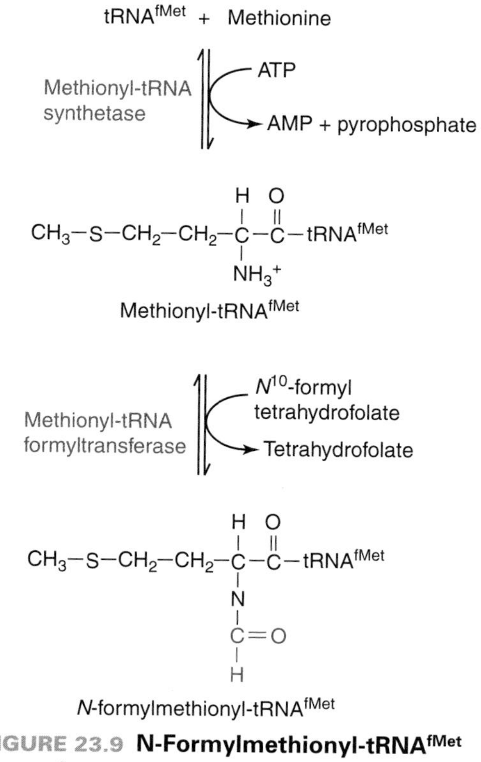

Studies performed by Sanger and coworkers revealed that *E. coli* has two methionine tRNA isoacceptors. The first of these, $tRNA^{fMet}$,

N-formylmethionine (fMet)

N-formylmethionyl-$tRNA^{fMet}$ (fMet-$tRNA^{fMet}$)

FIGURE 23.8 **N-Formylmethionine and N-Formylmethionyl-$tRNA^{fMet}$.**

$tRNA^{fMet}$ + Methionine

Methionyl-tRNA synthetase; ATP → AMP + pyrophosphate

Methionyl-$tRNA^{fMet}$

Methionyl-tRNA formyltransferase; N^{10}-formyl tetrahydrofolate → Tetrahydrofolate

N-formylmethionyl-$tRNA^{fMet}$

FIGURE 23.9 **N-Formylmethionyl-$tRNA^{fMet}$ formation.**

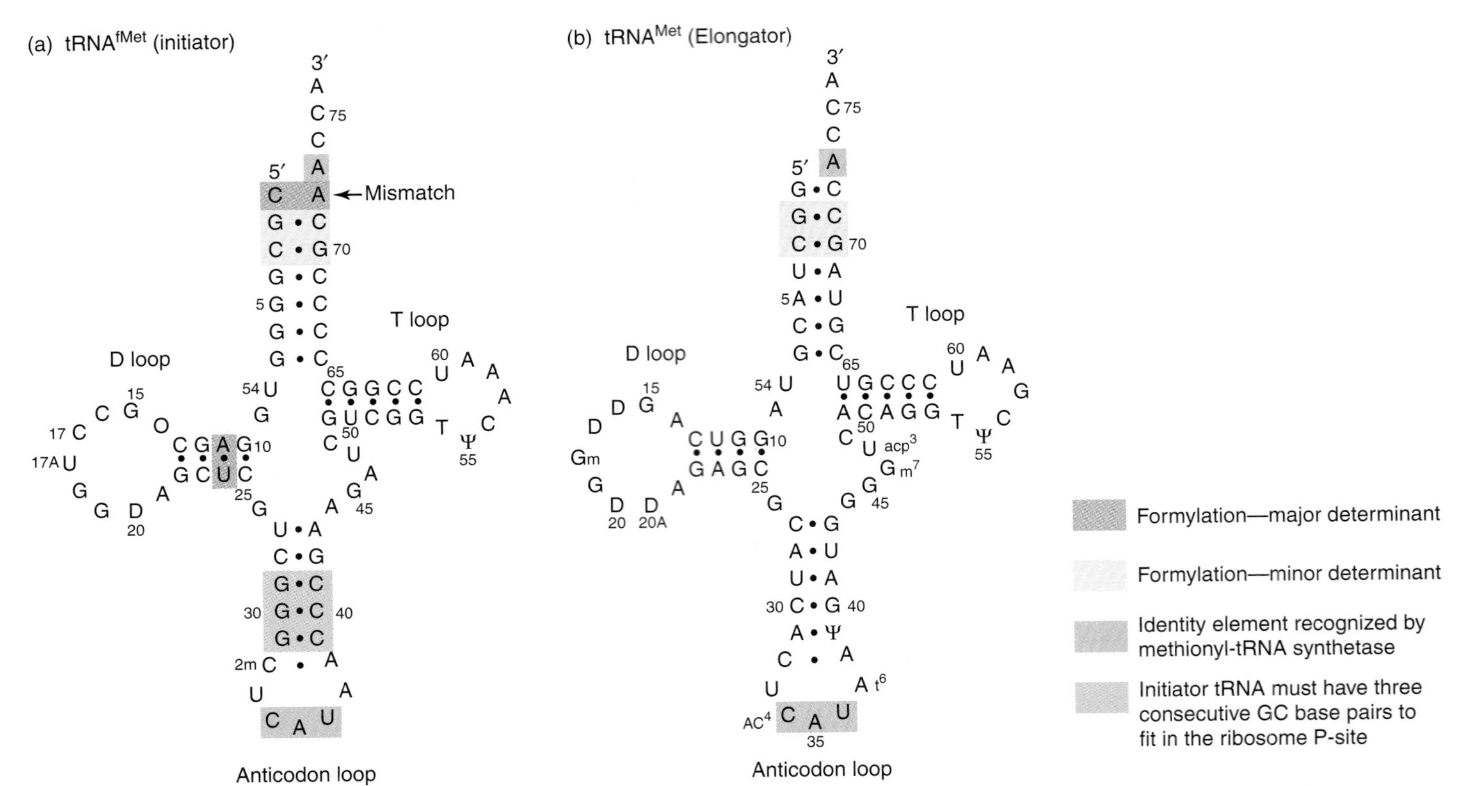

FIGURE 23.10 ***E. coli* tRNA isoacceptors for methionine.** tRNA^{fMet} initiates polypeptide chain synthesis and tRNA^{Met} elongates polypeptide chains. (Adapted from H. U. Sperling-Petersen, B. S. Laursen, and K. K. Mortensen, *Encyclopedia of Life Sciences*. John Wiley & Sons, Ltd., December 2001 [doi10.1038/npg/els.0000543].)

initiates polypeptide chain synthesis and the second, tRNA^{Met}, elongates polypeptide chains (FIGURE 23.10). Initiator tRNA accounts for about 70% of total cellular methionine acceptor activity and elongator tRNA for the remaining 30%. A single methionyl-tRNA synthetase charges both tRNA isoacceptors, recognizing the same identity elements (A73 and the anticodon) in each. Formyltransferase adds a formyl group only to the methionine attached to tRNA^{fMet}. Nucleotide substitution studies reveal that tRNA^{fMet} has two major determinants that allow the formyltransferase to distinguish it from tRNA^{Met} (Figure 23.10a): (1) A mismatch between nucleotides at positions 1 and 72 extends the single stranded stretch at the 3′-end of the acceptor arm to five nucleotides, allowing the formyltransferase catalytic site to gain access to the methionine. (2) An A11:U24 base pair in the D-arm interacts with the enzyme's C-terminal domain. The discovery of fMet-tRNA^{fMet} proved to be of great importance because it allowed investigators to study the polypeptide chain initiation pathway.

The 30S subunit is an obligatory intermediate in polypeptide chain initiation.

In 1968, Matthew Meselson and coworkers performed an experiment to follow the distribution of isotopic labels among ribosomes and ribosomal subunits following the transfer of a growing *E. coli* culture from a heavy- to a light-isotopes medium. This experiment showed: (1) ribosomes frequently undergo subunit exchange during growth and (2) 30S and 50S subunits remain intact during growth. Based on these results they concluded that 70S ribosomes must dissociate into 50S and 30S subunits after completing polypeptide chain synthesis. As investigators started to think about advantages that cells might derive from ribosomal dissociation, they considered the possibility that the 30S or 50S subunit might participate in polypeptide chain initiation. This hypothesis received support from *in vitro* experiments that showed (1) poly(AUG) stimulates fMet-$tRNA^{fMet}$ (but not Met-$tRNA^{Met}$) to bind to the 30S subunit and (2) 50S subunits block this binding. Based on these and related experiments, Masayasu Nomura proposed that fMet-$tRNA^{fMet}$ and mRNA bind to the 30S ribosomal subunit to form a **30S initiation complex** and then the 50S subunit combines with the 30S initiation complex to form a **70S initiation complex.**

Nomura devised a clever experiment to test his model. First he cultured *E. coli* in a growth medium containing heavy isotopes (^2H_2O and $^{15}NH_4^+$) and isolated the "heavy" 70S ribosomes. Then he incubated the "heavy" 70S ribosomes with a synthetic random poly AUG, [^{3}H]fMet-$tRNA^{fMet}$, and polypeptide chain initiation factors (see below) in the presence of excess "light" 50S ribosomal subunits (prepared from cells cultured in medium containing normal isotopes). Nomura's model predicts the three-step pathway shown in FIGURE 23.11.

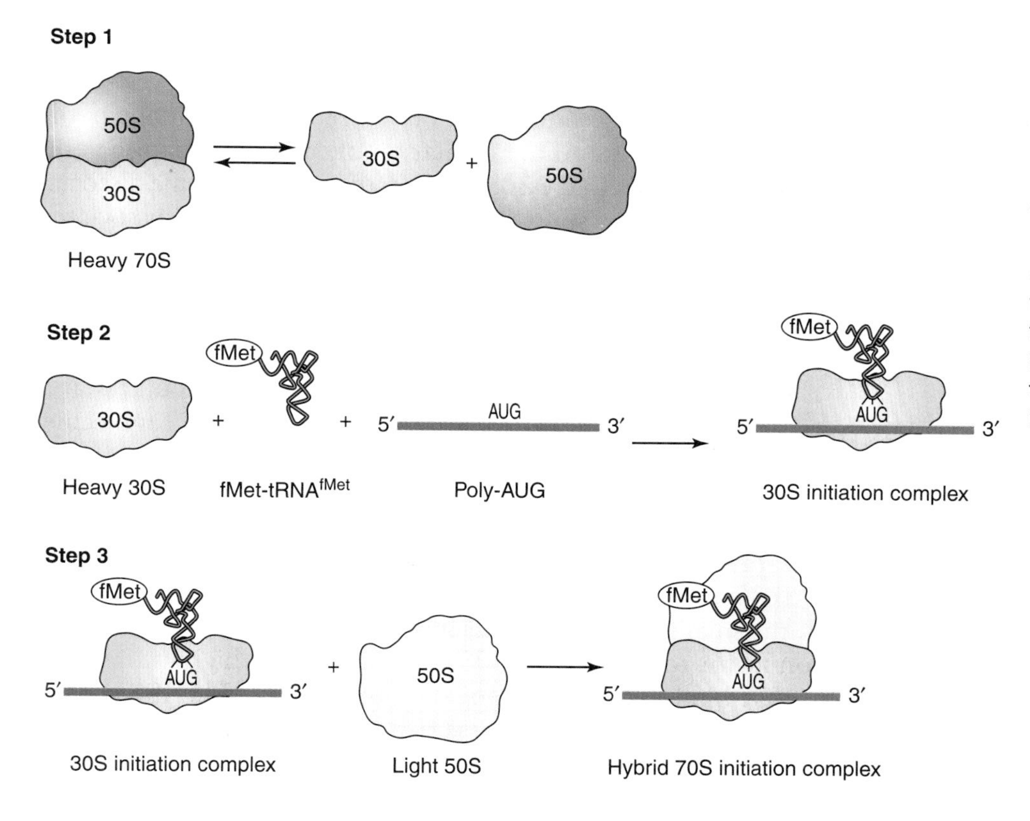

FIGURE 23.11 Outline of Nomura's experiment to test whether the 30S initiation complex is an intermediate in polypeptide chain initiation. Nomura's model predicts the three step pathway shown. (Step 1) Heavy 70S ribosomes dissociate to form heavy 50S and 30S subunits. (Step 2) Heavy 30S subunits form 30S initiation complexes containing [^{3}H]fMet-$tRNA^{Met}$ and mRNA. (Step 3) Light 50S subunits combine with 30S initiation complexes to form hybrid 70S initiation complexes.

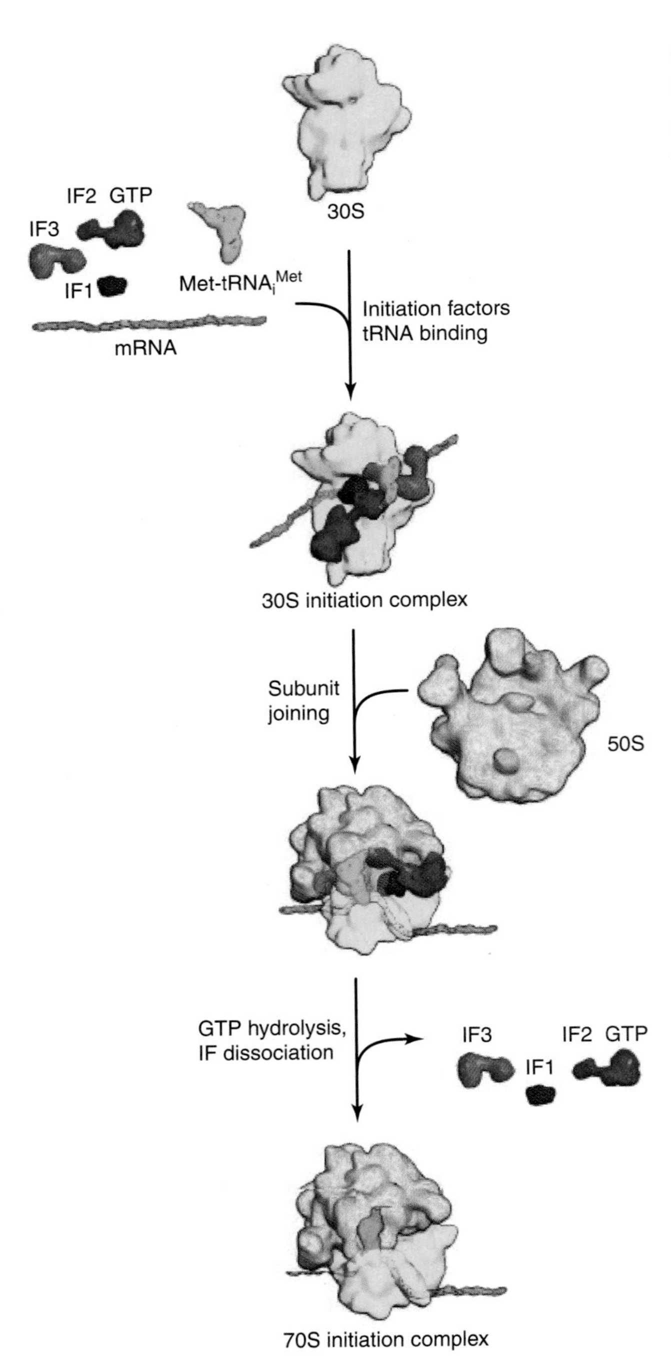

FIGURE 23.12 **Translation initiation pathway in bacteria.** Components are positioned on the ribosome according to currently available experimental information. (Adapted from T. M. Schmeing and V. Ramakrishnan, *Nature* 461 [2009]: 1234–1242.)

Step 1. Heavy 70S ribosomes dissociate to form heavy 50S and 30S subunits.

Step 2. Heavy 30S subunits form 30S initiation complexes containing [^{3}H]fMet-tRNA^{fMet} and mRNA.

Step 3. Light 50S subunits combine with 30S initiation complexes to form hybrid 70S initiation complexes.

Nomura separated heavy from hybrid 70S ribosomes by cesium chloride equilibrium density gradient centrifugation (see Chapter 5). As predicted by the model, [^{3}H]fMet-tRNA^{fMet} was present in the 70S hybrid fraction. However, the possibility still existed that the heavy 70S ribosomes rapidly exchanged their heavy 50S subunits for light 50S subunits before binding fMet-tRNA^{fMet}, and the newly formed hybrid 70S ribosomes and not the 30S subunits were responsible for binding [^{3}H] fMet-tRNA^{fMet}. Nomura was able to eliminate this possibility by including [^{14}C]Val-tRNA^{Val} in the incubation mixture and showing that it binds to heavy 70S ribosomes and not to hybrid 70S ribosomes.

Initiation factors participate in the formation of 30S and 70S initiation complexes.

In 1966, Severo Ochoa and coworkers showed that ribosomes isolated from *E. coli* contain three initiation factors, **IF1**, **IF2**, and **IF3**, which stimulate the formation of the 30S and 70S initiation complexes. They extracted the initiation factors from the ribosomes with a 1 M ammonium sulfate solution, and used standard protein fractionation techniques to purify the three indispensable initiation factors.

FIGURE 23.12 summarizes our current understanding of the pathway leading to the formation of the 70S initiation complex. IF3 binds to the 30S subunit after 70S ribosome dissociation and thereby shifts the equilibrium to favor the dissociated subunits. IF3 contains two domains of about equal size that are connected by a flexible linker. The N-terminal domain's only known function is to assist the C-terminal domain in binding to the 30S subunit. A fragment that contains just the C-terminal domain has an affinity for the 30S subunit that is about 100-fold lower than that for the intact IF3 molecule. At high concentrations, the C-terminal fragment appears to perform all the known functions normally assigned to IF3. Structures of the separate domains have been determined by x-ray crystallography and NMR but the structure of intact IF3 has not yet been determined.

IF1, the smallest of the three bacterial initiation factors, is the next initiation factor to bind to the 30S subunit. IF1 is a highly conserved protein that binds at the A-site of the 30S subunit. The IF1 surface that interacts with the 30S subunit is rich in basic amino acids, providing favorable electrostatic interactions between IF1 and the negatively charged 16S rRNA phosphate backbone. IF1 stimulates the activities of IF2 and IF3 by promoting their more efficient binding to the 30S subunit.

IF2, the largest of the three initiation factors, has an N-terminal domain with no known function, a middle domain that binds and hydrolyzes GTP, and a C-terminal domain that interacts with fMet-tRNA^{fMet} (but not Met-tRNA^{Met}). An NMR structure is available for each isolated domain but the structure of the intact protein has not been determined at atomic resolution. Experiments designed to determine the order in which IF2•GTP interacts with fMet-tRNA^{fMet} and the 30S subunit provide conflicting results. Some experiments indicate that IF2•GTP binds to the 30S subunit before it binds to fMet-tRNA^{fMet} and others indicate the reverse binding order.

Cryo-electron microscopy reveals that IF1 and IF2 do not contact each other on the 30S subunit (FIGURE 23.13). IF1 probably stimulates IF2 activity by changing the 16S rRNA's conformation and thereby stimulating the 30S subunit's ability to bind IF2. IF2's GTPase catalytic site is located near the 30S subunit sites that eventually make contact with the 50S subunit, while its C-terminal domain makes contact with the 3′-end of fMet-tRNA^{fMet} in the P site. IF3 stimulates the P site codon–anticodon interaction between the fMet-tRNA^{fMet} and mRNA to promote 30S initiation complex formation. IF3 also destabilizes codon–anticodon interactions when incorrectly bound aminoacyl-tRNAs are present at the P site. Thus, IF3 helps to ensure the fidelity of polypeptide initiation by distinguishing fMet-tRNA^{fMet} from all other aminoacyl-tRNAs during 30S initiation complex formation.

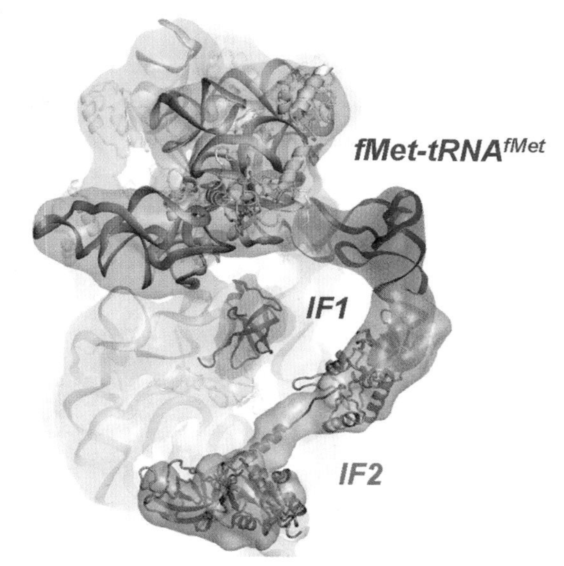

FIGURE 23.13 **Cryo-electron microscopy reconstruction of the 30S initiation complex with IF1, IF2, and fMet-tRNA^{fMet} bound to the small ribosomal subunit.** The crystal structures of IF1, fMet-tRNA^{fMet}, and an archaeal IF2 homolog were fitted into the 3D reconstruction. (Reproduced from *Curr. Opin. Struct. Biol.*, vol. 19, A. G. Myasnikov, et al., Structure-function insights into prokaryotic . . ., pp. 300–309, copyright 2009, with permission from Elsevier [http://www.sciencedirect.com/science/journal/0959440X]. Photo courtesy of Bruno Klaholz, Institut de Génétique et de Biologie Moléculaire et Cellulaire.)

The 50S subunit joins the 30S initiation complex to form a 70S initiation complex. GTP hydrolysis takes place immediately after the 50S subunit joins the 30S initiation complex and then IF1, IF2, and IF3 are released. The resulting 70S initiation complex is now ready to begin the elongation cycle by binding the aminoacyl-tRNA specified by the second codon at the A-site. A 70S complex with bound IF1, IF2, and IF3 accumulates when GTP is replaced by a non-hydrolyzable analog. Cryo-electron microscopy of the stalled 70S complex shows that IF2's GTPase catalytic site makes contact with the GTPase activating center on the 50S subunit (see below). Two other important structural differences exist between the 70S initiation complex and the stalled complex. In the 70S initiation complex, the 3′-end of fMet-tRNA^{fMet} is located in the peptidyl transferase center of the 50S subunit and the 30S subunit rotates with respect to the 50S subunit.

Magnesium ion concentration is an important factor when studying polypeptide chain initiation. IF1, IF2, and IF3 are required for polypeptide chain initiation at physiological magnesium ion concentrations, estimated to be about 5 mM, but not at higher concentrations. High magnesium ion concentrations stabilize the 70S ribosome and permit mRNA and aminoacyl-tRNA molecules to bind directly to the 70S ribosome. Under these conditions, the normal physiological pathway for polypeptide chain initiation is bypassed. Marshall W. Nirenberg and Johann Heinrich Matthaeil were able to observe poly(U)-directed polyphenylalanine synthesis (see Chapter 22) because their system contained a high magnesium ion concentration.

The mRNA Shine-Dalgarno (SD) sequence interacts with the 16S rRNA anti-Shine-Dalgarno (anti-SD) sequence.

The bacterial protein synthetic machinery must be able to distinguish the initiator AUG codon from other AUGs that code for internal methionines or from AUGs that occur in alternate reading frames. In 1974, John Shine and Lynn Dalgarno made two important observations that helped to explain how the bacterial protein synthetic machinery performs this task. First, they noticed that most cistrons have a purine-rich consensus sequence (5′-UAA**GGAGG**U-3′) located about five to seven nucleotides upstream from the initiator codon, with the nucleotides shown in bold being the most frequently present. Second, they noticed that the 16S rRNA has a complementary pyrimidine-rich sequence (5′-ACCUCCUUA-3′) near its 3′-end (FIGURE 23.14). They, therefore, proposed that the complementary sequences on mRNA and 16S rRNA base pair in an antiparallel fashion, helping the translation machinery to identify the initiator codon. The purine-rich sequence in mRNA is now called the **Shine-Dalgarno (SD)** sequence or **ribosome binding site (rbs)** and the complementary sequence near the 3′-end of 16S rRNA is called the **anti-Shine-Dalgarno (ASD)** sequence.

Joan Steitz and Karen Jakes provided experimental support for the Shine-Dalgarno hypothesis in 1974. They began by preparing a 70S

(a)

Lipoprotein	··· A U C U A G A G G G U A U U A A U A A U G A A A G C U A C U ···
RecA	··· G G C A U G A C A G G A G U A A A A A U G G C U A U C G ···
GalE	··· A G C C U A A U G G A G C G A A U U A U G A G A G U U C U G ···
GalT	··· C C C G A U U A A G G A A C G A C C A U G A C G C A A U U U ···
LacI	··· C A A U U C A G G G U G G U G A A U G U G A A A C C A G U A ···
LacZ	··· U U C A C A C A G G A A A C A G C U A U G A C C A U G A U U ···
Ribosomal L10	··· C A U C A A G G A A C A A A G C U A A U G G C U U U A A A U ···
Ribosomal L7/L12	··· U A U U C A G G A A C A A U U U A A A U G U C U A U C A C U ···

(b)

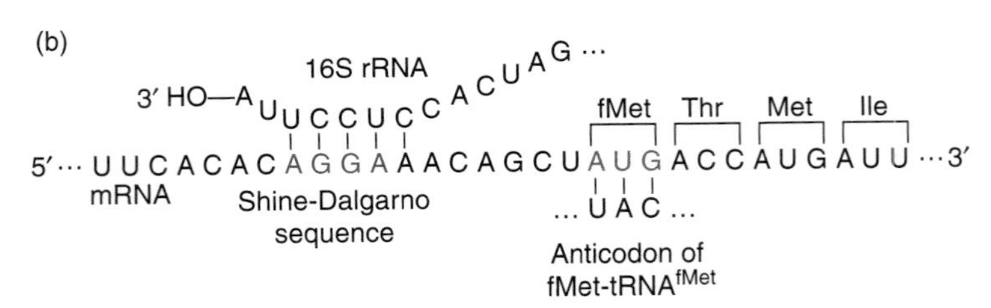

FIGURE 23.14 **Shine-Dalgarno sequence.** (a) Shine-Dalgarno sequences (ribosome-binding sites), which are located toward the 5′-end of mRNA, are shown for several *E. coli* mRNAs. The initiation (start) codons (blue) are immediately downstream. The optimal spacing between Shine-Dalgarno sequences and initiation sequences is 7–9 nucleotides. (b) Base pairing between the Shine-Dalgarno sequence and the complementary sequence in the 16S rRNA helps to establish the initiation codon and the correct reading frame for translation. (Adapted from R. H. Horton, et al. *Principles of Biochemistry, Third edition*. Prentice Hall, 2002.)

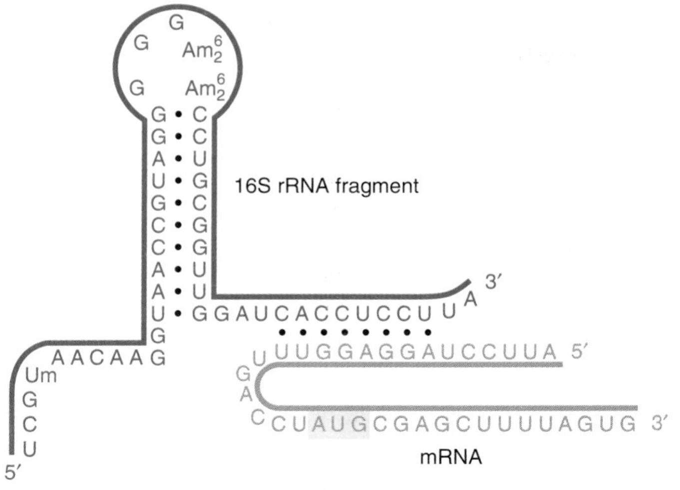

FIGURE 23.15 **Binding of mRNA to a complementary sequence in 16S rRNA.** The AUG start codon is in the yellow box.

initiation complex containing a short [^{32}P]mRNA fragment with an SD sequence. Next they added a ribonuclease called colicin E3 to this complex. They selected colicin E3 because it makes a single cut in 16S rRNA about 50 nucleotides from the 3′-end and so would be expected to release a 16S rRNA fragment with its SD sequence still base paired to the [^{32}P]mRNA (FIGURE 23.15). Finally, they added detergent to the mixture to extract proteins under conditions that would not disrupt base pairs. When the RNA was analyzed by gel electrophoresis under nondenaturing conditions, Steitz and Jakes observed a novel radioactive species. As predicted by the Shine-Dalgarno hypothesis, this species contained a [^{32}P]mRNA fragment•16S rRNA fragment complex, which was not observed when the RNA was heated in urea prior to loading on the gel.

Ann Hui and Herman De Boer provided still more support for the Shine-Dalgarno hypothesis in 1987. Their approach was based on the assumption that the precise bases in the SD and ASD regions are not important but only the ability of these two regions to base pair. They therefore constructed a bacterial strain with a mutated ribosomal subpopulation bearing an altered ASD sequence. As predicted by the Shine-Dalgarno hypothesis, the mutant ribosomes could not translate mRNA with a normal SD region but were able to translate mRNA with a modified SD region that was complementary to the ASD region in the mutant ribosomes. Perhaps the most convincing evidence in support of the SD hypothesis is the actual visualization of the Shine-Dalgarno helix that has been made possible by x-ray crystallography (FIGURE 23.16).

Although the SD consensus sequence in *E. coli* is 5′-AGGAGGU-3′, that in most normal cistrons contains just three or four of these bases. Longer SD sequences are required only if the mRNA secondary structure limits ribosomal access to the start codon. Optimal spacing between the SD sequence and the initiator codon is 5 nucleotides, but the SD sequence still works if it is up to 13 nucleotides from the

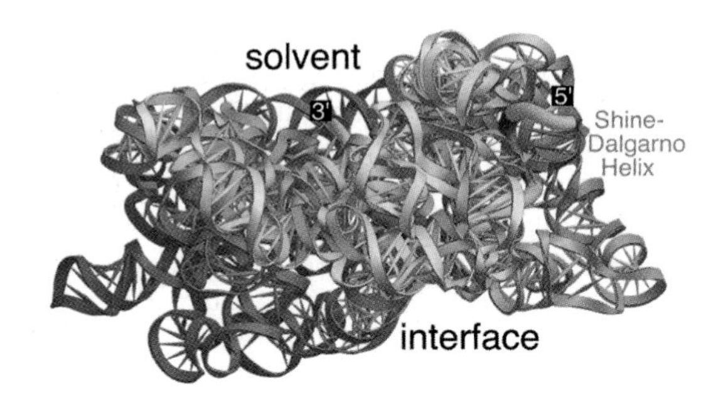

FIGURE 23.16 **Shine-Dalgarno sequence as viewed from the top of the head of the 30S subunit.** The 16S rRNA is colored cyan and the mRNA is colored yellow. Ribosomal proteins have not been included to simplify the view. The Shine-Dalgarno helix is colored magenta. 5′ and 3′ correspond to the 5′ and 3′ ends of the mRNA. The interface surface faces the 50S subunit and the solvent surface faces the surrounding solvent. (Reproduced from *Structure*, vol. 9, G. M. Culver, Meanderings of the mRNA . . ., pp. 751–758, copyright 2001, with permission from Elsevier [http://www.sciencedirect.com/science/journal/09692126]. Photo courtesy of Gloria M. Culver, University of Rochester.)

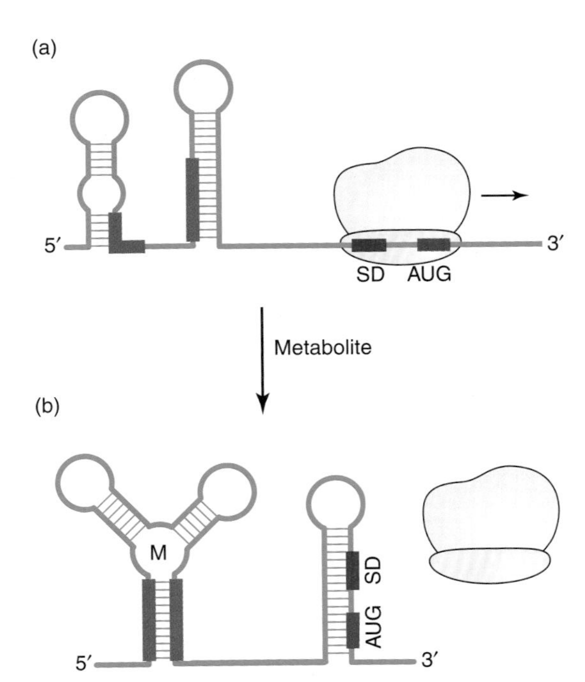

FIGURE 23.17 **Riboswitch-mediated control of translation initiation.** (a) Translation takes place because the Shine-Dalgarno (SD) sequence in mRNA can form base pairs with the complementary anti-Shine-Dalgarno (ASD) sequence in the 16S rRNA. (b) Translation is blocked because a specific metabolite (M) stabilizes the riboswitch structure (shown as a hypothetical 3-stem structure) in a folding pattern that prevents SD from forming base pairs with ASD. Large and small ribosome subunits are colored blue and yellow, respectively. (Adapted from E. Nudler and A. S. Mironov, *Trends Biochem. Sci.* 29 [2004]: 11–17.)

initiator codon. The SD sequence appears to be dispensable when the AUG codon is exactly at the 5′-end of the mRNA but the weaker initiation codons, GUG and UUG, do not support initiation under these conditions. The secondary structure of mRNA also influences initiator codon selection by masking AUGs that do not serve as initiation codons and making them unavailable to the 30S subunit.

Riboswitches regulate translation initiation of some bacterial mRNA molecules.

Certain small metabolic intermediates influence translation initiation of mRNA molecules by stabilizing or destabilizing the way that a specific region, the **riboswitch**, in the 5′-untranslated region folds. FIGURE 23.17 shows one mechanism of riboswitch-mediated regulation of translation initiation. In the absence of metabolite, the riboswitch folds into a structure that permits the SD in the mRNA to interact with the ASD on the 16S rRNA (Figure 23.17a). In the presence of metabolite (M), the riboswitch folds into a structure in which the SD is part of a helical structure and so cannot interact with the ASD (Figure 23.17b). The remarkable feature of the riboswitch is that specificity for the metabolite is determined entirely by the RNA molecule and does not involve protein; that is, the RNA folds into a structure that binds a specific metabolite and no others. For instance, the riboswitch that is part of an mRNA that codes for an enzyme required for thiamine (vitamin B_1) synthesis binds thiamine pyrophosphate.

Genes controlled by riboswitches often code for proteins that participate in the formation or transport of the metabolite that is sensed by the riboswitch. Other metabolites that have been shown to regulate translation initiation by their interaction with a riboswitch include vitamin B_{12}, flavin mononucleotide, S-adenosylmethionine, guanine, lysine, and glycine. Riboswitches that influence translation initiation have been demonstrated in bacteria and the archaea. In many cases, the binding of a metabolite to a riboswitch also triggers premature transcription termination because the resultant RNA folding produces a transcription terminator. Plants and fungi also appear to have riboswitches but thus far those that have been studied appear to influence splicing rather than translation initiation. In view of the widespread distribution of riboswitches in nature, it would be surprising if riboswitches were not eventually shown to contribute to gene regulation in animals.

23.3 Initiation Stage in Eukaryotes

Eukaryotic initiator tRNA is charged with a methionine that is not formylated.

Like the bacterial pathway, the eukaryotic translation initiation pathway uses one methionine tRNA for polypeptide chain initiation and another for polypeptide chain elongation. Also as in bacteria, the same methionyl-tRNA synthetase charges both the initiator tRNA, ($tRNA_i$, where i is for initiator), and the elongator tRNA ($tRNA^{Met}_m$). However, there is an important difference between bacteria and eukaryotes. Met-$tRNA_i$ is not formylated, forcing the eukaryotic protein synthetic machinery in the cytoplasm to rely on other features to distinguish initiator tRNA from elongator tRNAs. The major sequence feature that prevents vertebrate initiator tRNA from acting as an elongator tRNA is two specific base pairs, A50:U64 and U51:A63, in the initiator tRNA (shown in the orange box in FIGURE 23.18). Mutating the two base pairs converts the vertebrate initiator tRNA into an elongator tRNA. Other unique sequence and structural features present in eukaryotic initiator tRNA but not in elongator tRNAs are as follows: (1) an A1:U72 base pair at the end of the acceptor stem; (2) a sequence of three consecutive G:C base pairs in the anticodon stem; and (3) A54 and A60 in place of the T54 and Y60 (Y = pyrimidine) present in nearly all elongator tRNAs. Mutations that alter the A1:U72 base pair cause a significant decrease in initiator function.

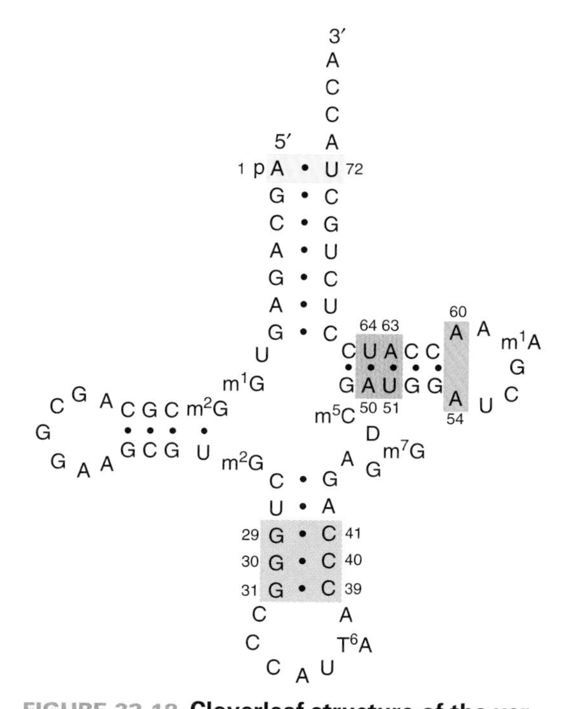

FIGURE 23.18 **Cloverleaf structure of the vertebrate initiator tRNA.** The unique features present in initiator tRNA are an A1:U72 base pair at the end of the acceptor stem (yellow), three consecutive G:C base pairs in the anticodon stem (purple), and A54 and A60 in the TψC loop (green). Mutations that alter A50:U60 and U51:A63 (orange) convert the initiator tRNA into an elongator tRNA. (Adapted from H. J. Drabkin, M. Estrella, and U. L. Rajbhandary, *Mol. Cell. Biol.* 18 [1998]: 1459–1466.)

Eukaryotic translation initiation proceeds through a scanning mechanism.

The eukaryotic translation initiation pathway is far more complex than the bacterial pathway, requiring many more initiation factors and steps. Differences between the two pathways arise from the fact that transcription and translation take place in the same compartment in bacteria but in different compartments in eukaryotes. Because eukaryotic mRNA synthesis is completed in the nucleus, the cell has an opportunity to process the mRNA by adding a cap, removing introns, and attaching a poly(A) tail before the mRNA becomes available to ribosomes in the cytoplasm for translation (see Chapter 19). The mature eukaryotic mRNA that finally emerges in the cytoplasm after transport through the nuclear pores is a structured RNA molecule coated with proteins. In contrast to bacterial mRNA, eukaryotic mRNA does not have a SD sequence and usually has just one cistron.

The major pathway for translation initiation in eukaryotes begins with the recruitment of the small ribosomal subunit bearing specific initiation factors to the 5′-end of the mRNA. Once recruited, the small ribosomal subunit scans the mRNA in a 5′- to 3′-direction searching for the first AUG codon that is in the proper context. In mammals, the most efficient AUG initiation codons are embedded within the sequence ACC**AUG**G (the initiation codon is shown in bold) but other sequences also work. Plants and yeast have very different sequences around their initiation codon. The mammalian ribosomal subunit may bypass the first AUG it encounters if a pyrimidine is at position −3 or +4 (the adenosine in AUG is at position +1). When a correct match is found, initiation factors are released and the large subunit joins the small subunit to form an initiation complex in which the initiator tRNA is paired with the initiation codon AUG at the P-site.

This scanning model, which was first proposed by Marilyn Kozak in 1978, is supported by considerable experimental evidence. Three especially significant observations are as follows: (1) In contrast to bacterial ribosomes, eukaryotic ribosomes cannot bind to, or initiate translation on, ~60 nucleotide RNA molecules that have been covalently circularized because the circularized molecules lack a 5′-end. (2) A stable secondary structure in the region between the 5′-end and the initiation codon (the **5′-untranslated region** or **5′-UTR**) lowers translation initiation efficiency by impeding the small ribosomal subunit's movement along mRNA. (3) A stable hairpin structure about 12 nucleotides downstream from the AUG initiation codon enhances initiation efficiency by forcing the small ribosome subunit to pause at the initiation codon. Translation initiation efficiency is also influenced by the length of the 5′-UTR. Systematically shortening the 5′-UTR from 32 to 3 nucleotides was shown to cause a progressive decrease in translation efficiency and systematically lengthening it from 17 to 77 nucleotides was shown to cause a progressive increase in translation efficiency.

Some mRNAs have **internal ribosomal entry sites** (**IRES**), usually in their 5′-UTR, that allow ribosomes to assemble at an internal site. These entry sites were first detected while studying the translation of an RNA molecule from a picornavirus such as the poliovirus. The poliovirus codes for a protease that inactivates an essential initiation factor required for the scanning mechanism. It is, therefore, not surprising that cellular mRNAs with IRES elements were initially identified in cells infected with the poliovirus. Although a fascinating story, the translation of these mRNAs, which represent only a small subset of total cellular mRNA, will not be described further.

Eukaryotes have at least twelve different initiation factors.

Eukaryotes have at least twelve different initiator factors, which collectively contain at least 23 different polypeptides. Each **eukaryotic initiation factor** (**eIF**) is designated by a number that is often followed by a letter. Several of the major eukaryotic initiation factors and some of their functions are listed in Table 23.1. Although not evident from their names, eukaryotic initiation factors eIF1a and eIF5b are homologs of bacterial initiation factors IF1 and IF2, respectively. Eukaryotes do not seem to have an initiation factor homologous to bacterial IF3. However, the structure of eIF1 appears to be similar to that of bacterial IF3 and the two may also have similar functions. Considerable effort has been made to determine the interactions that occur among the initiation factors and between each initiation factor and the 40S ribosomal subunit. The results of these efforts are summarized schematically in FIGURE 23.19, which shows that each initiation factor interacts directly with nearly every other initiation factor and most interact directly with the 40S ribosomal subunit.

Although a great deal is now known about the eukaryotic translation initiation pathway, the presently accepted pathway must be

TABLE 23.1 Mammalian Translation Initiation Factors

Factor	*N* of Polypeptides	Size (kDa)	Principal Functions and Properties
eIF1	1	12.6	Promotes ribosomal scanning; stimulates binding of eIF2•GTP•Met-tRNA$_i$ to 40S subunits; ensures proper initiation codon selection. Prevents premature Pi release after GTP hydrolysis.
eIF1A	1	16.5	Works with eIF1 to promote ribosomal scanning and correct initiation codon selection; stimulates binding of eIF2•GTP•Met-tRNA$_i$ to 40S subunits.
eIF2	3	36.2 (α) 39.0 (β) 51.8 (γ)	Forms eIF2•GTP•Met-tRNA$_1$ ternary complex, which binds to 40S ribosomal subunit in a codon-independent manner.
eIF2B	5	33.7 (α) 39.0 (β) 50.4 (γ) 57.8 (δ) 80.2 (ε)	Guanosine nucleoside-exchange factor that catalyzes exchange conversion of eIF2•GDP into eIF2•GTP. Needed for recycling of eIF2.
eIF3	12	~730 total	Involved in all stages of initiation. Interacts with eIF1, eIF4G, eIF5, and 40S subunits.
eIF4A(I)*	1	44.4	Has RNA-dependent ATPase activity, and RNA helicase activity dependent on ATP hydrolysis and eIF4B. Interacts with eIF4G. A second species, eIF4A(II) (46.3 kDA), is encoded by a different gene.
eIF4B	1	69.2	Precise role is uncertain. Promotes RNA helicase activity of eIF4A.
eIF4E	1	25.1	Binds to 5′-cap structure. Interacts with eIF4G.
eIF4G(1)*	1	171.6	Appears to have a scaffold function. The central domain, in conjunction with eIF4A, appears to deliver the primed 40S subunit to the mRNA. A less abundant second species, eIF4G(II) (176.5 kDa) is encoded by a different gene.
eIF4F			A complex of eIF4A, 4E, and 4G.
eIF5	1	48.9	Has GTPase-activating protein activity, promoting hydrolysis of the ternary complex GTP following recognition of the start codon by base-pairing with Met-tRNA$_i$ anticodon.
eIF5B	1	139.0	Needed for subunit joining. Has ribosome-dependent GTPase activity.

*Because eIF4A(I) and eIF4A(II) appear to be functionally equivalent, as are eIF4G(I) and eIF4G(II), they are referred to as eIF4A and eIF4G, respectively, throughout the text. (Modified with permission from R. J. Jackson, 2005, *Biochemical Society Transactions,* vol. 33, 1231–1241. © the Biochemical Society.)

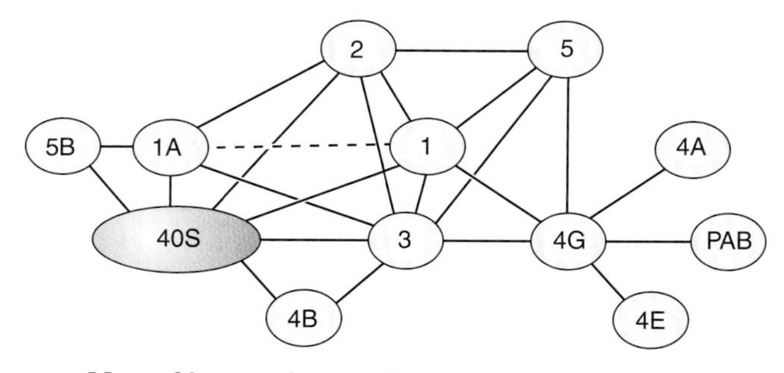

FIGURE 23.19 **Map of known interactions among eukaryotic translation initiation factors and the 40S ribosomal subunit.** Eukaryotic initiation factors are designated by their distinguishing number or number and letter. For instance, EIF1 and eIF4G are represented by 1 and 4G, respectively. The 40S ribosomal subunit is represented by 40S and the poly(A) binding protein by PAB. The dashed line between EIF1 and eIF1A indicates that this interaction might be indirect (mediated by conformational changes in the 40S subunit). (Reproduced from L. D. Kapp and J. R. Lorsch, *Annu. Rev. Biochem.* 73 [2004]: 657–704. Copyright 2004 by Annual Reviews, Inc. Reproduced with permission of Annual Reviews, Inc., in the format Textbook via Copyright Clearance Center.)

considered a work in progress that will undoubtedly be modified as more is learned about the functions of the initiation factors. The task of examining the complex eukaryotic translation initiation pathway can be made somewhat simpler by dividing the pathway into eight stages (FIGURE 23.20).

1. *Ternary complex formation:* eIF2•GTP binds Met-$tRNA_i^{Met}$ to form an eIF2•GTP•Met-$tRNA_i^{Met}$ complex, which is called the **ternary complex.**
2. *43S preinitiation complex formation:* The 40S ribosomal subunit, assisted by eIF1A, eIF1, and eIF3, interacts with the ternary complex to form a 43S preinitiation complex (see below). eIF5 probably also binds to the 30S subunit at this stage. eIF1, eIF3, eIF5, and the ternary complex may interact to form a multifactor complex that binds to the 40S subunit as a single entity.
3. *mRNA activation:* Eukaryotic mRNA passes through the nuclear pore into the cytoplasm as a structured molecule that is coated with polypeptides. The mRNA must be activated by removing secondary structures and proteins from its 5′-end before it can be translated. eIF4F, a multisubunit complex, assists in this activation. Figure 23.20 shows the contributions that three of the complex's subunits, eIF4A, eIF4E, and eIF4G, make to the mRNA activation process. eIF4E binds the mRNA 5′-m^7G cap. eIF4A, an ATP-dependent RNA helicase, unwinds secondary structures in the 5′-UTR. We do not know exactly when this unwinding takes place; it may occur before the mRNA interacts with the 43S preinitiation complex or after this interaction takes place. eIF4G serves as a scaffold for the eIF4F complex that holds the other subunits in place. It also binds the poly(A)-binding protein (PABP) that coats the poly(A) tail at the 3′-end of mRNA. Thus both ends of the mRNA are tethered to the same eIF4F complex. The 5′-m^7G cap and poly(A) tail act in a synergistic manner to stimulate translation initiation. This synergy may result from cooperative binding between the two ends of the mRNA, a conformational change that accompanies binding of both ends that stimulates eIF4F activity, or both. The fact that both the 5′-m^7G cap and the poly(A) tail are required for efficient translation may be advantageous to the cell because it signals that the mRNA has not been degraded. Bacteria, which do not have such a quality control mechanism, depend on an entirely different method to deal with truncated mRNAs (see below). Consistent with its assigned functions, the eIF4F complex is not needed to translate small naked mRNAs that lack structure at their 5′-ends.
4. *mRNA entry into the 43S preinitiation complex:* The eIF4G subunit interacts with eIF3, recruiting the 5′-end of the mRNA to the 43S pre-initiation complex. Venkatraman Ramakrishnan and coworkers used cryo-electron microscopy to investigate how mRNA interacts with the 40S subunit under *in vitro* conditions. Their studies reveal that important structural changes take place when an empty 40S yeast subunit binds

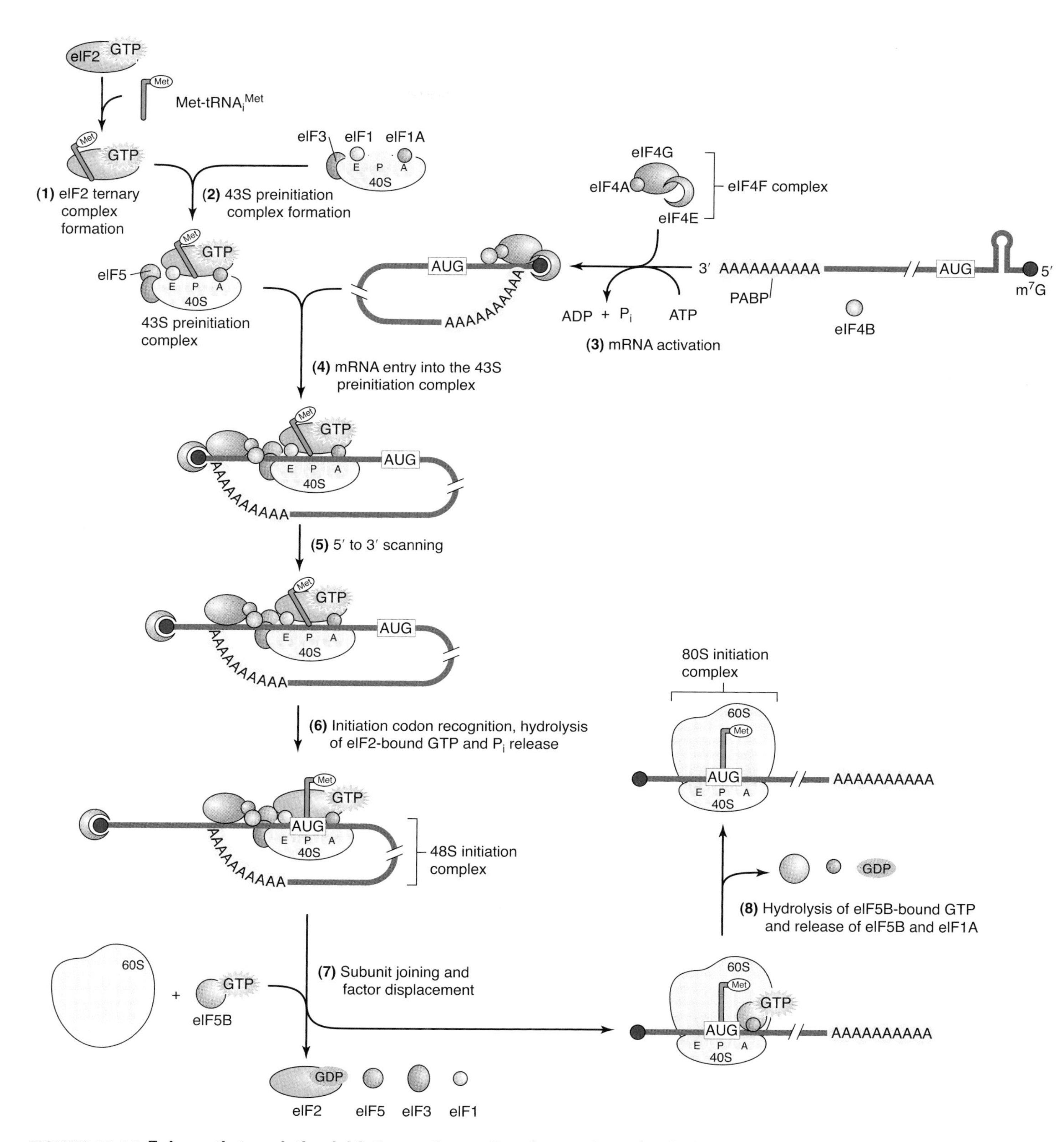

FIGURE 23.20 **Eukaryotic translation initiation pathway.** See the text for a detailed description of the individual steps in the pathway.

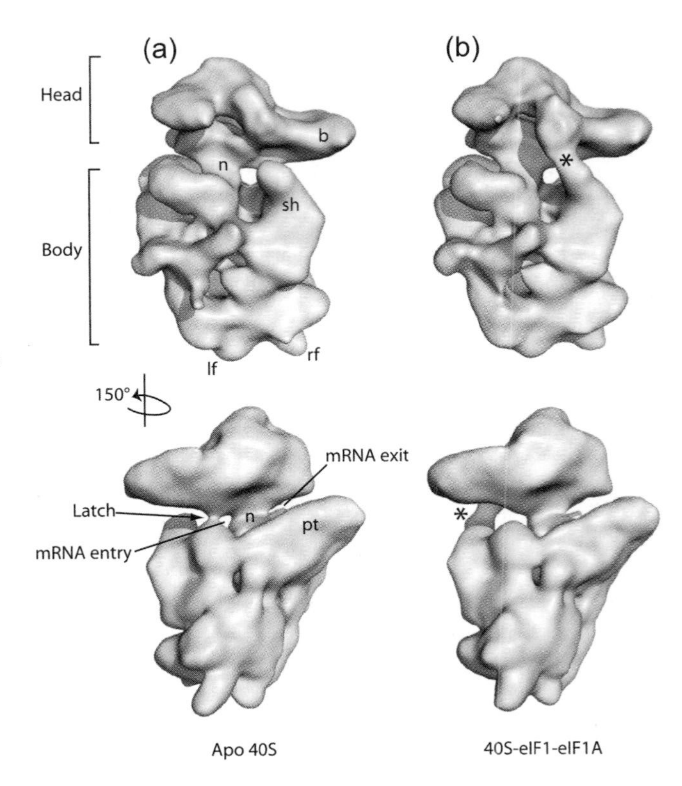

FIGURE 23.21 **Cryo-electron microscopy reconstructions of the yeast 40S ribosomal subunit before and after binding eIF1A and eIF1.** (a) the empty 40S subunit and (b) the 40S•eIF1•eIF1A complex. Each bottom structure is obtained by rotating the top structure by 150° around its vertical axis. Landmarks for the empty 40S subunit shown at the top left. Abbreviations: b, beak; n, neck; sh, shoulder; pt, platform; lf, left foot; and rf, right foot. The binding of eIF1 and eIF1A appears to stabilize a 40S conformation that "opens" the mRNA entry channel by moving a "latch" to the channel. The conformational change also introduces a connection between the shoulder and head that is indicated by an asterisk in (b). (Reproduced from *Mol. Cell.,* vol. 26, L. A. Passmore, et al., The Eukaryotic Translation Initiation Factors . . ., pp. 41–50, copyright 2007, with permission from Elsevier [http://www.sciencedirect.com/science/journal/10972765]. Photo courtesy of Lori Passmore, MRC Laboratory of Molecular Biology.)

to the two smallest initiation factors, eIF1A and eIF1 (FIGURE 23.21). Binding appears to stabilize a 40S conformation that "opens" the mRNA entry channel by moving a "latch" to the channel. Joachim Frank and coworkers had previously proposed that the channel, which is located at the neck of the 40S subunit, forms a clamp around mRNA. Based on the cryo-electron microscopy studies and related biochemical information, Ramakrishnan and coworkers propose that eIF1A and eIF1 act together to trigger a conformational change that produces an open, scanning-competent preinitiation complex (FIGURE 23.22). However, it is not clear how mRNA threads its way through the open channel because the globular eIF4E protein bound to its 5′-m^{7}G cap should make it difficult for the mRNA to thread its way through the open channel. Perhaps the eIF4E-cap is positioned at the channel's E-site side before the mRNA enters the open channel. If so, it is difficult to see how an initiation codon near the 5′-end of the mRNA can be detected at the P-site by the scanning mechanism (see below). *In vitro* experiments show that about half the ribosomes skip an AUG codon that is within 12 nucleotides of the cap. However, further studies are required to show that this skipping is caused by eIF4E-cap positioning and not by something else.

5. *5′→3′ scanning to detect the initiation codon:* The 43S complex moves along the mRNA in a 5′ to 3′ direction. This ATP-dependent movement continues until the AUG initiation codon

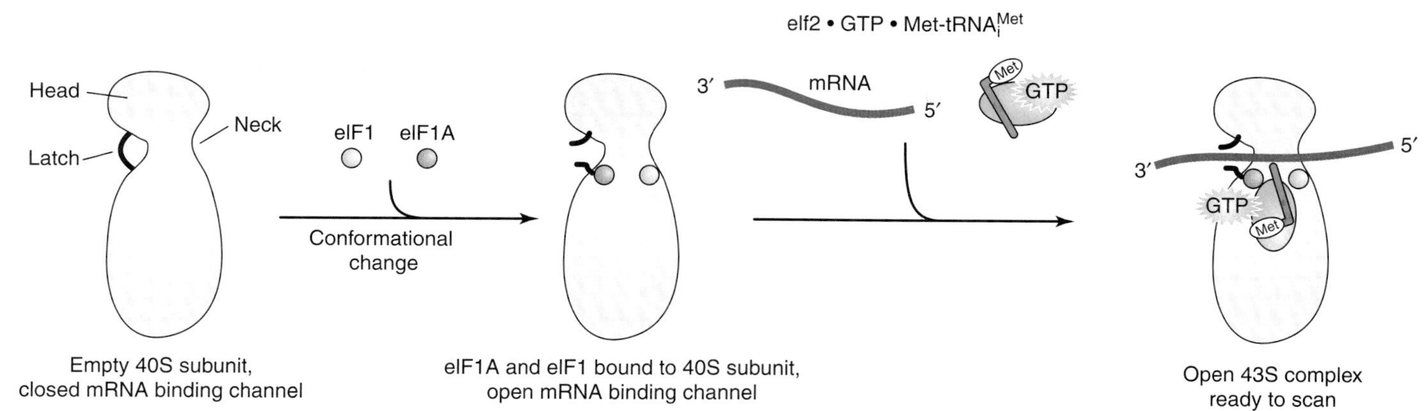

FIGURE 23.22 Model showing eIF1 and eIF1A functions during eukaryotic translation initiation. (a) Empty 40S subunit (yellow) adopts a closed conformation with a closed mRNA entry channel latch (black). (b) The binding of eIF1 and eIF1A appears to stabilize a 40S conformation that "opens" the mRNA entry channel by moving a "latch" to the channel. (c) The ternary complex is able to bind to the 40S subunit after the conformation change. Moreover, the open latch permits mRNA (green) to somehow enter the channel. For simplicity, the connection between the shoulder and head that is indicated by an asterisk in Figure 23.21 is not shown here. (Adapted from L. A. Passmore, et al., *Mol. Cell* 26 [2007]: 41–50.)

on the mRNA is aligned with the anticodon on the initiator tRNA. eIF1 and eIF1A both make important contributions to the search for the initiation codon. eIF1 binds near the P-site and rejects non-AUG codons. eIF1A has a central domain and N- and C-terminal tails. The central domain binds at the A-site while the tails are located near the P-site. The C-terminal tail appears to prevent Met-tRNA$_i$ from fully entering the P-site when a codon other than AUG is in the site. However, the C-terminal tail moves out of the P-site when AUG and Met-tRNA$_i$ interact at the P-site. This movement allows Met-tRNA$_i$ to fully enter the P-site and set the reading frame.

6. *Initiation codon recognition, hydrolysis of eIF2-bound GTP and P_i release.* eIF5 stimulates eIF2 to convert its bound GTP to GDP and P_i. eIF1 would normally prevent the Pi from escaping; however, eIF1 is displaced from the P-site when Met-tRNA$_i$ binds to AUG. This displacement permits P_i to escape, which in turn triggers a conformational change that converts the open mRNA entry channel into a closed channel. The 48S initiation complex is now ready to join the 60S subunit to form the 80S initiation complex.
7. *60S subunit joining and initiation factor displacement:* eIF5B•GTP facilitates the joining of the 60S subunit to the 48S preinitiation complex. eIF1, eIF2•GDP, and eIF5 are released during the joining process.
8. *Hydrolysis of eIF5B-bound GTP and release of eIF5B and eIF1A.* eIF1A and eIF5B•GDP are released after the GTP bound to eIF5B is hydrolyzed. The timing of the release of eIF3

and eIF4G, the two initiation factors that bind to the solvent side of the 30S subunit, remains to be determined. Release of eIF1A from the 80S complex frees the A-site to participate in the elongation process.

Translation initiation factor phosphorylation regulates protein synthesis in eukaryotes.

eIF2•GDP, which is released during 80S initiation complex formation, must be converted back to eIF2•GTP before a new round of translation initiation can take place. However, eIF2 binds GDP so tightly that a GTP exchange factor, eIF2B, is required to catalyze the exchange reaction. One method that eukaryotes use to inhibit total protein synthesis is to attach a phosphate group to eIF2α, one of three subunits in eIF2. Four different protein kinases, each regulated by a different signal, phosphorylate eIF2α (Table 23.2). Phosphorylated eIF2, eIF2(αP), enters the translation initiation pathway as part of the ternary complex eIF2(αP)•GTP•Met-$tRNA_i^{Met}$ and is released as eIF2(αP)•GDP. The released complex binds to eIF2B•GTP, the GTP-GDP exchange protein. The resulting eIF2(αP)•GDP•eIF2B•GTP quaternary complex is a dead-end complex that cannot be converted to eIF2(αP)•GTP. Because the molar concentration ratio of eIF2 to eIF2B is about 10:1, converting even a small proportion of eIF2 to eIF2(αP) inactivates most, if not all, eIF2B. This inactivation blocks further translation initiation because active eIF2B is required to regenerate the eIF2•GTP required for translation initiation to take place.

Insulin stimulates translation initiation by two different phosphorylation-dependent pathways. First, it induces phosphorylation of eIF4E, stimulating translation by increasing the initiation factor's affinity for the 5′-m^7G cap structure. Second, it stimulates phosphorylation of eIF4E-binding proteins, which compete with eIF4G for binding to eIF4E. Phosphorylating eIF4E-binding proteins lowers their affinity for eIF4E, making it easier for eIF4G to associate with eIF4E to form the eIF4F complex that is essential for polypeptide chain initiation.

TABLE 23.2 eIF2α Kinases

Kinase	Activation Signal
Heme controlled repressor (HCR)	Activated when the intracellular heme level falls.
Protein kinase R (PKR)	Activated by double-stranded RNA molecules produced in virus-infected cells.
GCN2	Activated in yeast when medium is deprived of amino acids and in tissue culture in the absence of serum.
PKR-like ER kinase (PERK)	Activated when unfolded polypeptide chains accumulate in the endoplasmic reticulum.

The translation initiation pathway in archaea appears to be a mixture of the eukaryotic and bacterial pathways.

The translation initiation pathway in the archaea appears to have some features of the bacterial pathway and others of the eukaryotic pathway. As in eukaryotes, the methionine esterified to the initiator tRNA is not formylated. However, like bacterial mRNAs, archaeal mRNAs do not have caps at their 5′-ends, are often polycistronic, and have SD sequences before their initiation codons. The similarity between mRNA in bacteria and the archaea suggests that there is probably some selective pressure to retain these features when transcription and translation take place in the same compartment. The initiation factors in the archaea are homologous to those in eukaryotes, but there are fewer of them because the archaea do not have homologs to the eukaryotic initiation factors eIF2B, eIF3, eIF4F, or eIF5. At present, much less is known about the translation initiation pathway in the archaea than in bacteria or eukaryotes. We, therefore, do not attempt to describe this pathway but instead move to the next stage of translation, the elongation pathway.

23.4 Elongation Stage

Polypeptide chain elongation requires elongation factors.

Once the translation initiation pathway has established the correct reading frame, the protein synthetic machinery is ready to enter the next stage of protein synthesis, polypeptide chain elongation. A major breakthrough in elucidating the polypeptide elongation pathway came in the mid-1960s when Fritz Lipmann and coworkers showed that three elongation factors isolated from *E. coli* extracts participate in poly(U)-directed polyphenylalanine synthesis. Subsequent studies by others showed that eukaryotic cells have similar elongation factors. The bacterial elongation factors are named EF-Tu, EF-Ts, and EF-G, and their eukaryotic counterparts are named eEF1A, eEF1B, and eEF2, respectively. A new nomenclature system has been suggested for bacterial elongation factors in which EF-Tu, EF-Ts, and EF-G are called EF1A, EF1B, and EF2, respectively. Because most molecular biologists continue to use the old nomenclature system, we will too.

EF-Tu is the most highly expressed protein in *E. coli* and accounts for about 5% to 10% of total cellular protein. Its intracellular concentration is about 10 times greater than that of ribosomes or EF-G. Two nearly identical genes *tufA* and *tufB* code for EF-Tu. Cells require one functional *tuf* gene to survive. Cells continue to grow if just a single *tuf* gene is nonfunctional; *tufB* mutants grow at a normal rate when cultured in rich medium but *tufA* mutants grow more slowly. The bacterial gene for EF-G, *fusA*, is in the same operon as *tufA*. The bacterial gene for EF-Ts, *tsf*, maps a considerable distance away from the other bacterial elongation factor genes.

The elongation factors act through a repeating cycle.

The bacterial polypeptide elongation pathway is cyclic, using the same reaction sequence to add each new amino acid to the growing polypeptide chain (FIGURE 23.23). The eukaryotic pathway is quite similar except that the bacterial elongation factors EF-Tu, EF-Ts, and EF-G are replaced by the eukaryotic elongation factors eEF1A, eEF1B, and eEF2, respectively.

An EF-Tu•GTP•aminoacyl-tRNA ternary complex carries the aminoacyl-tRNA to the ribosome.

EF-Tu•GTP combines with aminoacyl-tRNA to form the ternary complex EF-Tu•GTP•aminoacyl-tRNA. Crystal structures of EF-Tu•GDPNP and EF-Tu•GDP are shown in FIGURE 23.24. GDPNP [guanosine-5′-(β,γ-imido) triphosphate] was used in place of GTP to avoid hydrolysis problems. The EF-Tu polypeptide chain folds into the same three domains in both structures. Domain 1, which is also called the G domain because it contains the guanine nucleotide–binding pocket, extends from residue 1 to about 200. Domains 2 and 3, each about 100 residues long, are held together by non-covalent interactions. The orientation of domain 1 with respect to domains 2 and 3 depends on whether GTP (Figure 23.24a) or GDP (Figure 23.24b) is present in the guanine nucleotide–binding pocket. Two regions in domain 1 designated switch I and switch II trigger the large conformational change that takes place when EF-Tu•GTP is converted to EF-Tu•GDP. Switch I is a short α-helix in EF-Tu•GDPNP and a β-hairpin in EF-Tu•GDP. Switch II also undergoes a major change; the α-helix shifts along the sequence by about four residues, causing the helix axis to rotate and a concomitant change in the spatial relationship between domains 1 and 3. The eukaryotic elongation factor eEF1A is structurally similar to its bacterial counterpart and appears to work in much the same way.

Ramakrishnan and coworkers have obtained the crystal structure of the 70S ribosome from *T. thermophilus* in complex with $tRNA^{Phe}$ in the exit (E) and peptidyl (P) sites, mRNA, and EF-Tu•Thr-$tRNA^{Thr}$•GDP, stabilized by the antibiotic paromomycin (FIGURE 23.25). Kirromycin was also included because it inhibits the conformational change that takes place after GTP is hydrolyzed, preventing EF-Tu from being released from the ribosome. The tRNA, which is loosely bound to the ribosome at this stage, is said to be in an **A/T state** because its anticodon interacts with mRNA at the A-site while its acceptor end remains bound to the translation elongation factor. The A-site interaction permits the ribosome to test the codon–anticodon match before it incorporates the new amino acid into the growing polypeptide chain. If the match between the codon and anticodon is not correct, the aminoacyl-tRNA dissociates from the ribosome without stimulating GTP hydrolysis. But if the match is correct, the ribosome changes its conformation,

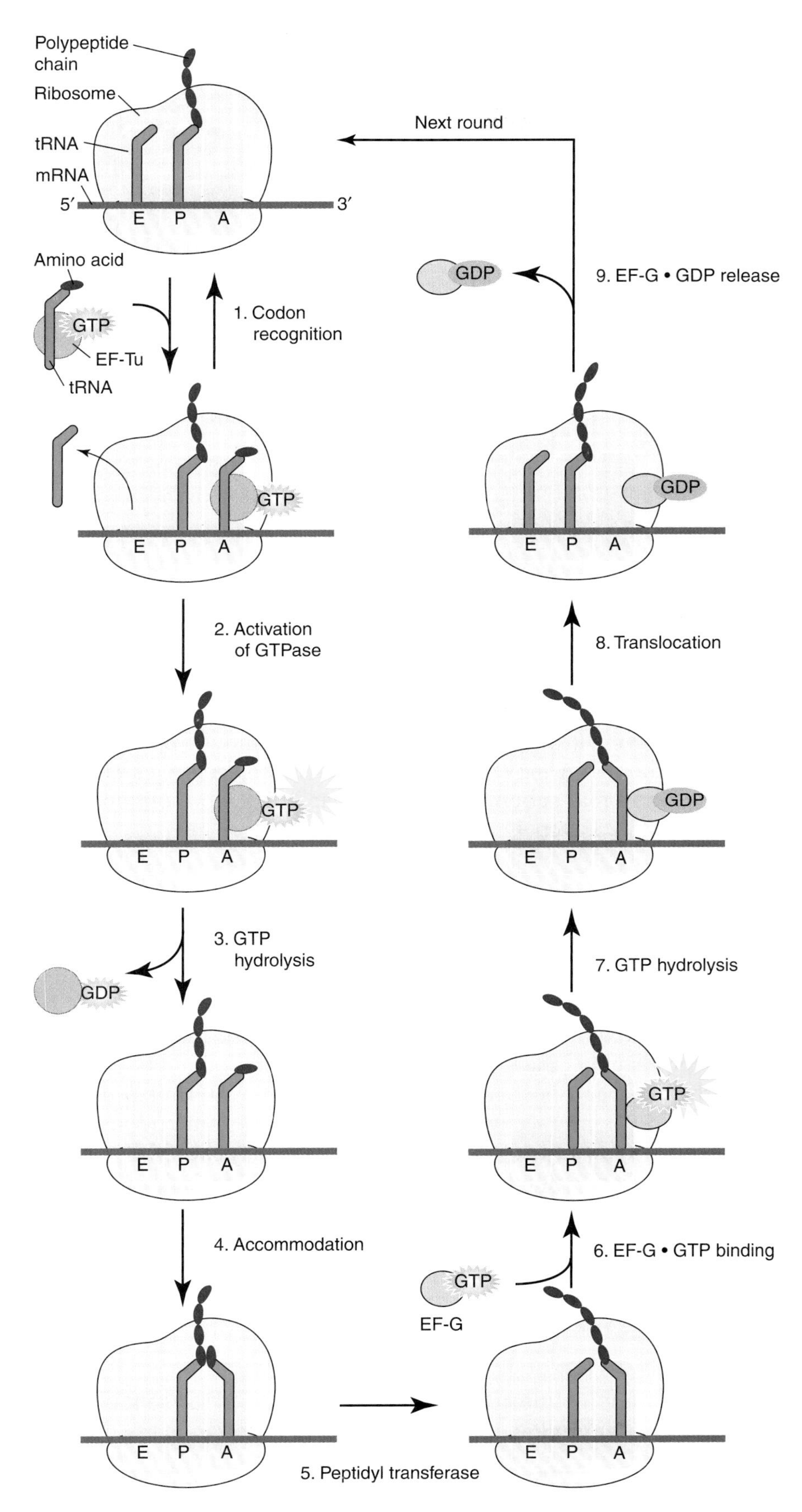

FIGURE 23.23 Polypeptide chain elongation pathway in bacteria. The eukaryotic polypeptide elongation pathway is the same except that eEF1A and eEF2 replace EF-Tu and EF-G, respectively. (Adapted from V. Ramakrishnan, *Cell* 108 [2002]: 557–572.)

FIGURE 23.24 **Structures of bacterial elongation factor EF-Tu.** (a) EF-Tu•GDPNP from *Thermus aquaticus* and (b) EF-Tu•GDP from *E. coli*. GDPNP is a nonhydrolyzable GTP analog. The switch 1 and 2 regions are yellow and magenta, respectively. (Reproduced from *Trends Biochem. Sci.*, vol. 28, G. R. Andersen, P. Nissen, and J. Nyborg. Elongation factors in protein biosynthesis, pp. 434–441, copyright 2003, with permission from Elsevier [http://www.sciencedirect.com/science/journal/09680004]. Photo courtesy of Poul Nissen, in memory of Jens Nyborg, University of Aarhus.)

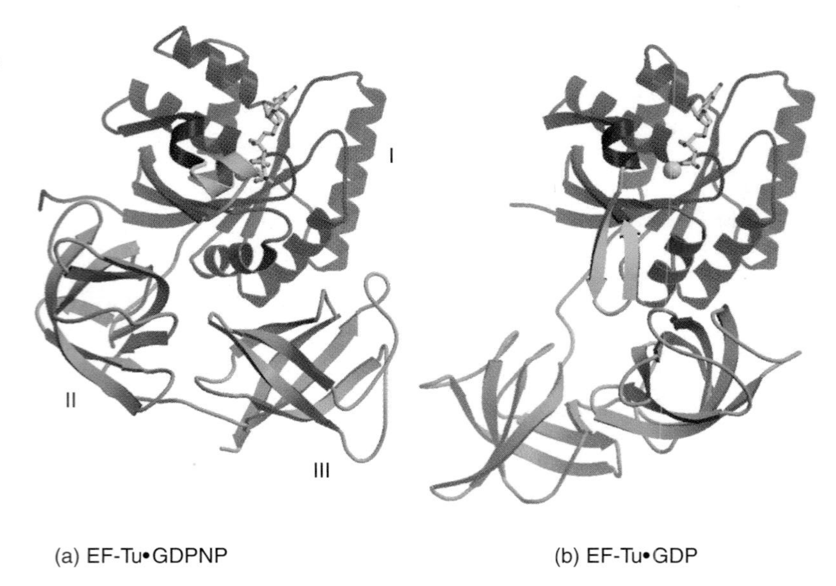

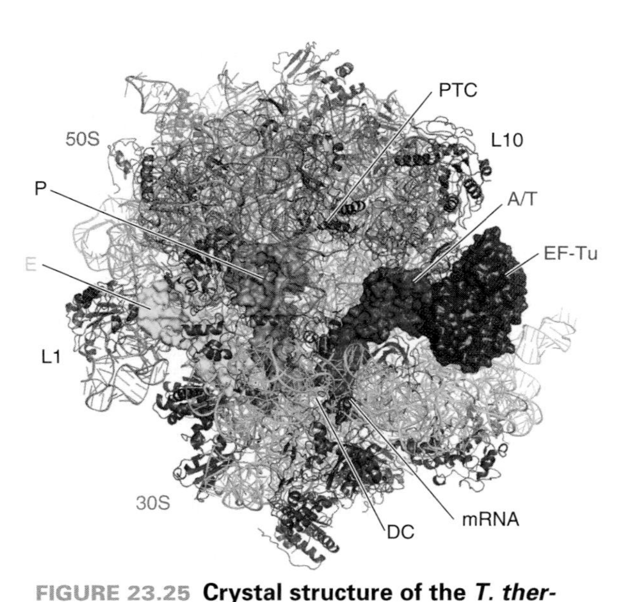

FIGURE 23.25 **Crystal structure of the *T. thermophilus* 70S ribosome with bound EF-Tu•GDP.** The 70S ribosome is shown with the 50S subunit (reddish brown) on top and the 30S subunit (cyan) on bottom. P- and E-site tRNAs are shown in green and yellow, respectively. mRNA is shown in purple. The peptidyl transferase center (PTC), L1 and L10-L12 stalks on the 50S subunit, and the decoding center (DC) on the 30S subunit are indicated. Thr-tRNA^{Thr} at the A/T site is shown in red and EF-Tu is shown in purple. The antibiotic kirromycin blocks rearrangement of EF-Tu after GTP hydrolysis, trapping the ternary complex on the ribosome. (Reproduced from T. M. Schmeing, et al., *Science* 326 [2009]: 688–694. Reprinted with permission from AAAS. Photo courtesy of Martin Schmeing, McGill University.)

stabilizing the interaction between the tRNA and ribosome and activating EF-Tu's latent GTPase (Figure 23.23, step 2). EF-Tu goes through the major conformational change shown in Figure 23.24 after hydrolyzing GTP, which causes EF-Tu to release the aminoacyl end of the A-site tRNA and dissociate from the ribosome as EF-Tu•GDP (Figure 23.23, step 3). The now free aminoacyl end of the tRNA moves into the peptidyl transferase center of the large ribosomal subunit, a process that is called **accommodation** (Figure 23.23, step 4).

An additional elongation factor, EF-P, is required to synthesize the first peptide bond.

An additional elongation factor, EF-P, is required when f-Met-tRNA^{fMet} is at the P-site. EF-P spans both ribosomal subunits. Its N-terminal domain (domain I) interacts with the initiator tRNA's acceptor stem at the P-site, its middle domain (domain II) interacts with the large ribosomal protein L1, and its C-terminal domain (domain III) interacts with the initiator tRNA's anticodon arm at the P-site (FIGURE 23.26). These interactions appear to be required to correctly position f-Met-tRNA^{fMet} at the P-site. Eukaryotes and the archaea have a factor called eIF5A, that has a similar sequence and structure to domains I and II in EF-P. eIF5A like its bacterial counterpart stimulates the peptide bond formation between the Met attached to the initiator tRNA and the amino acid attached to the tRNA that binds at codon 2.

Specific nucleotides in 16S rRNA are essential for sensing the codon-anticodon helix.

Almost from the time that the genetic code was first deciphered, molecular biologists have tried to determine how the 30S subunit decodes mRNA with such high fidelity. The free energy for a codon–anticodon interaction is only about 2 to 3 kcal mol^{-1} more favorable for a codon to interact with its cognate anticodon than for that same codon to interact with a near cognate anticodon with just a single nucleotide mismatch. This free energy difference predicts an error rate about 10- to 100-fold greater than that actually observed. It seemed likely that the ribosome plays some role in stabilizing cognate codon–anticodon interactions. This hypothesis received support from biochemical experiments with bacterial ribosomes that showed N1 methylation of highly conserved adenines at positions 1492 (A1492) and 1493 (A1493) of the 16S rRNA impaired A-site tRNA binding. Similar impairment also was observed when these adenines were changed to guanine or cytosine. Although these experiments indicated that A1492 and A1493 help the ribosome to recognize the shape of the codon–anticodon helix at the A-site, they did not show how they do so.

Ramakrishnan and coworkers turned to x-ray crystallography to solve the problem. They began by soaking an oligonucleotide containing the $tRNA^{Phe}$ anticodon stem-loop and a U_6 hexanucleotide into crystals of the *T. thermophilus* 30S ribosomal subunit. The x-ray diffraction data showed that a correct codon–anticodon match causes A1492 and A1493 to flip out of the loop in which they are normally located and the highly conserved guanine at position 530 (G530) to switch from a syn- to anti-conformation. In their new conformations, A1493 and A1492 interact with the first and second base pairs of the codon–anticodon helix, respectively, while G530 interacts with both the second position of the anticodon and the third position of the codon (FIGURE 23.27). These conformational changes allow the

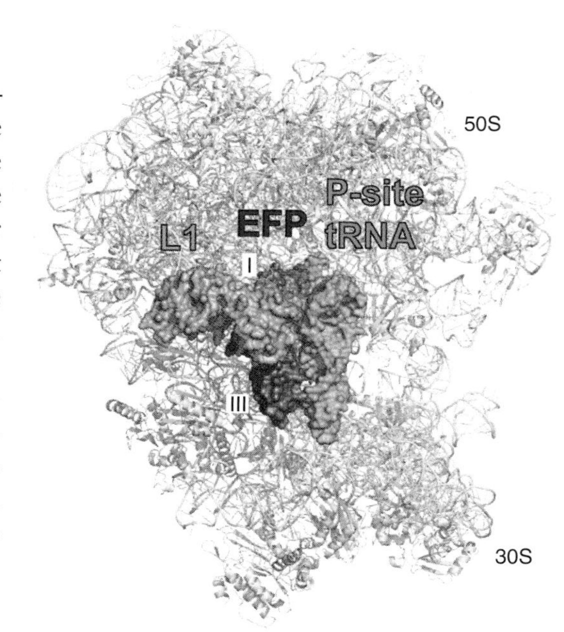

FIGURE 23.26 **EF-P and the P-site tRNA-binding in the 70S ribosome.** The 30S subunit is colored yellow, the 50S subunit is colored gray, and the P-site tRNA is colored green. Parts of the 70S ribosome are omitted for clarity. EF-P is colored magenta. Domain III is hidden behind the L1 protein (orange) and is not visible. (Reproduced from G. Blaha, R. E. Stanley, and T. A. Steitz, *Science* 325 [2009]: 966–970. Reprinted with permission from AAAS. Photo courtesy of Thomas A. Steitz, Yale University.)

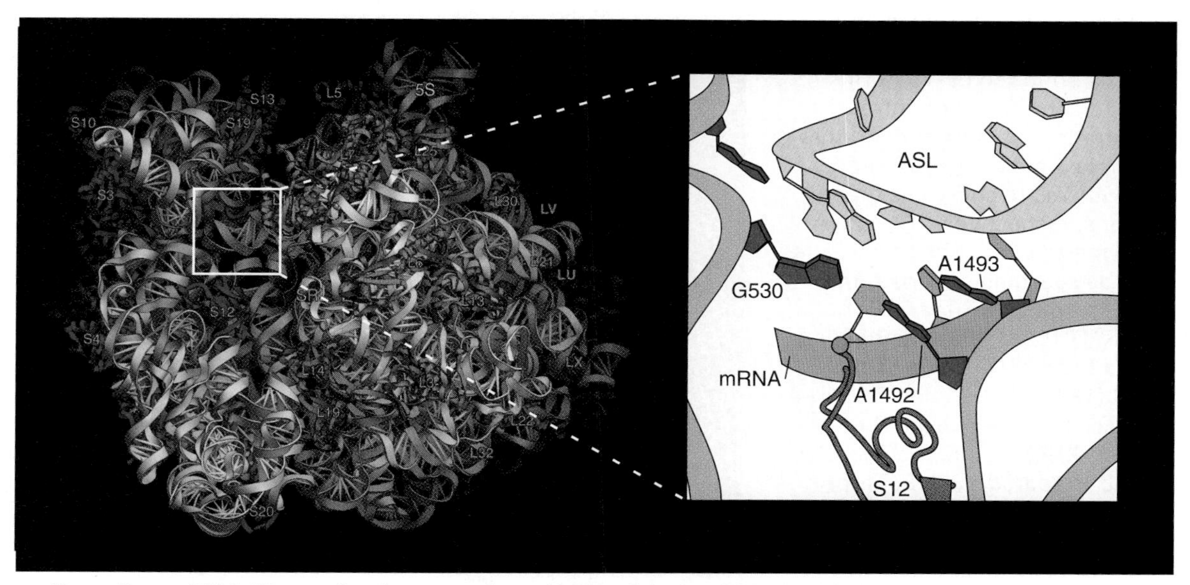

FIGURE 23.27 **Decoding mRNA.** The anticodon stem-loop (ASL) of A-site tRNA (gold) is in the interface cavity between the 30S subunit (left) and the 50S subunit (right). (Inset) A magnification showing the mRNA codon (purple) and cognate tRNA (gold) in the A-site of the 30S subunit. A1492 and A1493 (red) sense Watson–Crick base pairing in the first two bases of the codon–anticodon double helix. G530 (red) and the S12 polypeptide (brown) in the 30S subunit both contact A1492. (Photo reproduced from M. M. Yusupov, et al. 2001. *Science*. 292: 883–896. Reprinted with permission from AAAS. Photo courtesy of Harry Noller, University of California, Santa Cruz. Illustration modified from A. E. Dahlberg. *Science* 292 [2001]: 868–869. Reprinted with permission from AAAS.)

ribosome to closely inspect the first two base pairs of the codon–anticodon helix so that it can discriminate between Watson–Crick base pairs and mismatches. The environment of the third or "wobble" position appears to be able to accommodate a noncanonical base pair such as the GU wobble.

EF-Ts is a GDP-GTP exchange protein.

The EF-Tu•GDP released during the elongation cycle must exchange its GDP for GTP before it can participate in the next elongation cycle. However, it requires the assistance of EF-Ts to do so because EF-Tu's affinity for GDP is about 40 times greater than its affinity for GTP. EF-Ts, a guanine nucleotide exchange factor, works as shown in **FIGURE 23.28**. A highly conserved segment in the N-terminal region of EF-Ts invades the G domain of EF-Tu. This invasion triggers a conformational change in the nucleotide binding pocket, causing EF-Tu to release the GDP that is bound to it. EF-Ts remains bound to EF-Tu, forming an EF-Tu•EF-Ts complex. High intracellular GTP concentrations rapidly convert this EF-Tu•EF-Ts complex into EF-Tu•GTP and EF-Ts.

The eukaryotic elongation factor, eEF1B, has a more complex structure than its bacterial counterpart but catalyzes the same

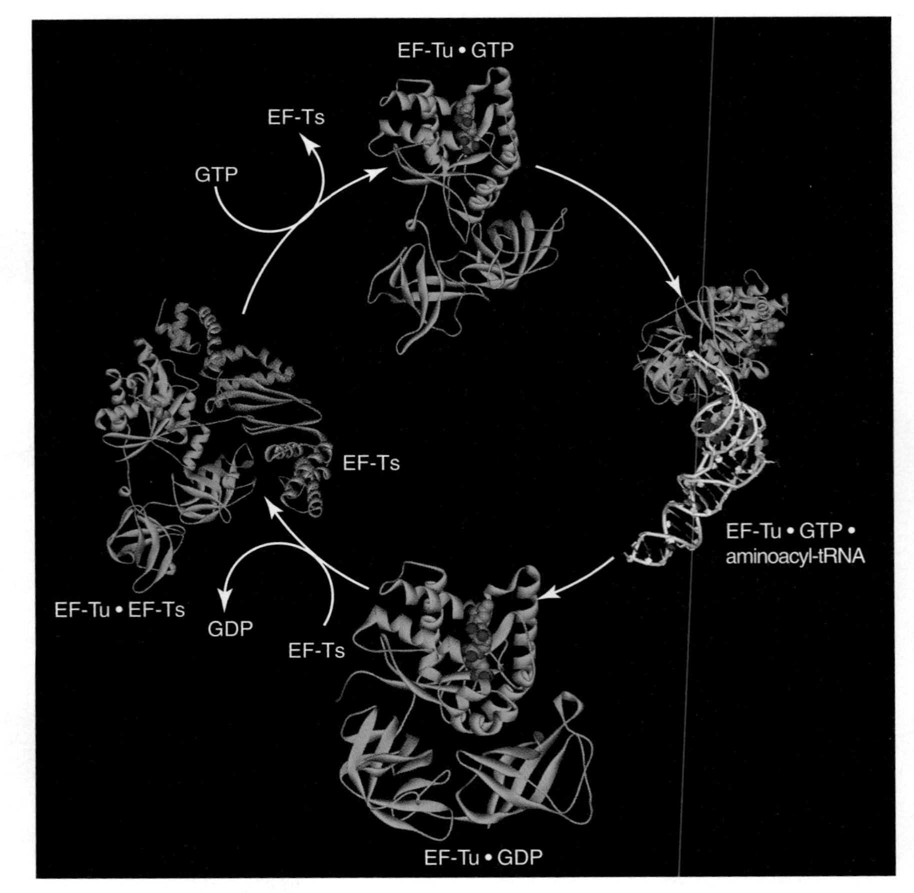

FIGURE 23.28 Elongation factor Ts (EF-Ts) function in GDP-GTP exchange. EF-Ts is required for EF-Tu to exchange its GDP for GTP. A highly conserved segment in the N-terminal region of EF-Ts invades the guanine nucleotide binding domain of EF-Tu, triggering a conformational change in the nucleotide binding pocket. This invasion causes EF-Tu to release the GDP that is bound to it. EF-Ts remains bound to EF-Tu, forming an EF-Tu•EF-Ts complex. High intracellular GTP concentrations rapidly convert this EF-Tu•EF-Ts complex into EF-Tu•GTP and EF-Ts. EF-Tu and EF-Ts are shown as green and orange ribbons, respectively. The tRNA molecule is shown as a white ribbon and guanine nucleotides as space filling structures using standard CPK colors. EF-Tu•GTP Protein Data Bank ID 1EXM. (Part a structure from Protein Data Bank 1TUI. G. Polekhina, et al., *Structure* 4 [1996]: 1141–1151. Prepared by B. E. Tropp. Part b structure from Protein Data Bank 1TTT. P. Nissen, et al., *Science* 270 [1995]: 1464–1472. Prepared by B. E. Tropp.; Part c structure from Protein Data Bank 1EXM. R. Hilgenfeld, J. R. Mesters, and T. Hogg, "Insights into the GTPase Mechanism of EF-Tu from Structural Studies," in *The Ribosome: Structure, Function, Antibiotics, and Cellular Interaction,* ed. R. A. Garrett, et al., 347–357. ASM Press, 2000. Prepared by B. E. Tropp. Part d structure from Protein Data Bank 1EFU. T. Kawashima, et al., *Nature* 379 [1996]: 511–518. Prepared by B. E. Tropp.)

nucleotide exchange reaction. Unlike EF-Ts, which consists of a single polypeptide, human eEF1B contains at least four polypeptide subunits. Two of the subunits have nucleotide exchange activity. The others do not but may stimulate the exchange.

The ribosome is a ribozyme.

Once the aminoacyl-tRNA is bound to the A-site, the ribosome is ready for peptide bond formation, which is catalyzed by the peptidyl transferase activity of the 50S ribosomal subunit (Figure 23.23, step 5). Peptidyl group transfer from the tRNA at the P-site to the aminoacyl-tRNA at the A site, produces a deacylated tRNA at the P-site and a peptidyl-tRNA at the A site. The initial method for assaying peptidyl transferase activity required an intact ribosome, mRNA, a P-site peptidyl-tRNA, and an A-site aminoacyl-tRNA, making it difficult to determine whether peptidyl transferase or some other factor required for peptide bond formation was being monitored. A more direct assay for peptidyl transferase activity was required. The antibiotic **puromycin**, which is produced by the gram-positive bacteria *Streptomyces alboniger*, helped to solve the problem. Puromycin is remarkably similar in structure to the tyrosyl-adenosine group at the 3′-end of Tyr-tRNA (FIGURE 23.29). Because of this close resemblance, peptidyl transferase can use puromycin as a substrate in place of aminoacyl-tRNA at the A-site. As a consequence of this relaxed substrate recognition, the peptidyl transferase transfers a peptidyl group from a peptidyl-tRNA at the P-site to puromycin, causing premature polypeptide chain termination and peptidyl-puromycin release. Puromycin thus can serve as a model substrate for the peptide bond-forming reaction and is well-suited for the study of this reaction.

In the mid-1960s, Robin Monro and coworkers used their knowledge of puromycin's mechanism of action to devise a direct method to measure peptidyl transferase activity called the **fragment assay**. This assay measures fMet-puromycin formation when a mixture containing

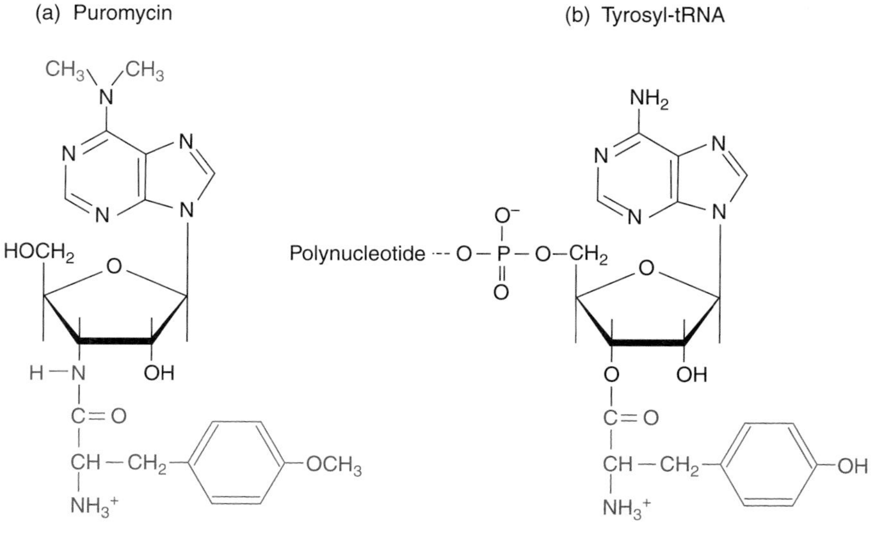

(a) Puromycin

(b) Tyrosyl-tRNA

FIGURE 23.29 **Structures of (a) puromycin and (b) the 3′-end of tyrosyl-tRNA.** Major structural features present in puromycin but not in the 3′-end of tyrosyl-tRNA are shown in red. Features common to the tyrosine analog present in puromycin and tyrosine are shown in blue.

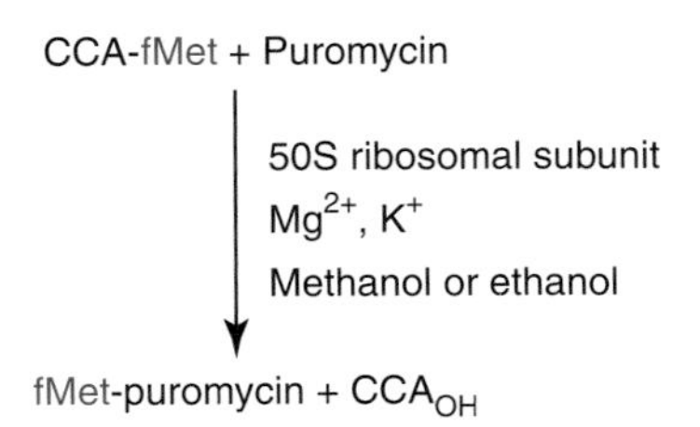

FIGURE 23.30 **Fragment assay for peptidyl transferase activity.**

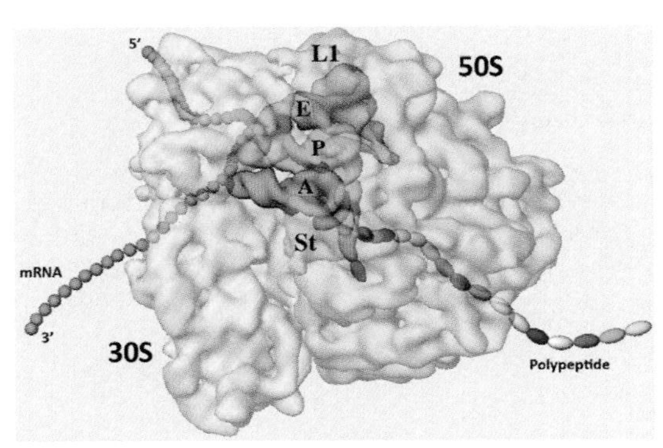

FIGURE 23.31 **Cryo-electron micrograph construct of the *E. coli* ribosome prior to peptidyl transfer.** The small and large subunits are shown in yellow and blue, respectively. The mRNA is shown in orange. The A- and P-site tRNAs are shown in purple and green, respectively. The E-site is shown in gold. The path of the nascent polypeptide chain through the exit tunnel in the large subunit is shown by the blue, green, and purple ovoid structures. St indicates the stalk and L1 indicates the L1 protein. The CCA ends of A- and P-site tRNAs are close together in the middle of the 50S ribosomal subunit, suggesting that peptidyl transferase is also located there. (Modified from J. Frank [July 2001]. Cryo-electron microscopy as an investigative tool: the ribosome as an example. In: *BioEssays*. John Wiley & Sons, Ltd: Chichester [doi: 10.1038/npg.els.0000534]. Photo courtesy of Joachim Frank, HHMI/HRI, Wadsworth Center.)

50S ribosomal subunits, a 3′-fragment of peptidyl-tRNA such as CCA-fMet, and puromycin is incubated in a buffer solution containing 30% to 50% methanol or ethanol (FIGURE 23.30). The alcohol appears to enhance the affinity of low molecular mass P-site substrates for the 50S ribosomal subunit. The fragment reaction does not require the 30S ribosomal subunit or mRNA. It also does not require ATP or GTP, indicating that the peptidyl transferase reaction does not require a source of energy other than that already present in the ester bond that links the peptide group to tRNA.

Initially, investigators thought it likely that proteins present in the 50S subunit catalyze the peptidyl transfer reaction. Francis Crick challenged this idea in 1968 because he found it difficult to understand how the first ribosome could have been a protein when the ribosome's job is to make proteins. He, therefore, proposed that "the first ribosome was made entirely of RNA." The discovery that some RNA molecules act as catalysts prompted molecular biologists to seriously consider Crick's hypothesis that RNA forms the catalytic site. This hypothesis received experimental support in 1992 when Harry F. Noller and coworkers showed that ribosomes from the thermophilic bacteria *Thermus aquaticus* completely lose their peptidyl transferase activity when treated with ribonuclease T1 but retain more than 80% of this activity after treatment with proteinase K and sodium dodecyl sulfate followed by phenol extraction. Moreover, the peptidyl transferase activity remaining after protein removal appeared to be physiologically significant because it retained its sensitivity to **chloramphenicol** and **carbomycin**, antibiotics known to inhibit peptidyl transferase. Although Noller's experiments suggest that peptidyl transferase activity resides in the ribosomal RNA, they did not completely rule out participation by ribosomal proteins because the treated ribosomes still retained about 10% of their polypeptides.

Before the crystal structure of the 50S ribosomal subunit was known, various methods were used to characterize the peptidyl transferase. Cryo-electron microscopy constructs showed that the CCA ends of A- and P-site tRNAs are close together in the middle of the 50S ribosomal subunit, suggesting that peptidyl transferase is located there (FIGURE 23.31). Genetic experiments identified specific nucleotides required for peptidyl transferase activity. For example, mutations that alter conserved nucleotides in domain V of the 23S rRNA (Figure 23.3b) inhibit or alter peptidyl transferase activity, suggesting that these nucleotides contribute to the catalytic center. Many of these same conserved nucleotides are protected from small chemical probes when tRNA molecules are bound to the A- and P-sites on the ribosome.

Peter B. Moore, Thomas A. Steitz, and coworkers made important contributions to our understanding of peptidyl transferase by analyzing the crystal structure of the large ribosomal subunit from the halophilic archaeon *H. marismortui*. They began by demonstrating that crystalline preparations of the large ribosomal subunits retained peptidyl transferase activity. They did so by allowing substrates for the fragment reaction to diffuse into the crystals and then showing that the expected products were formed. These experiments suggested

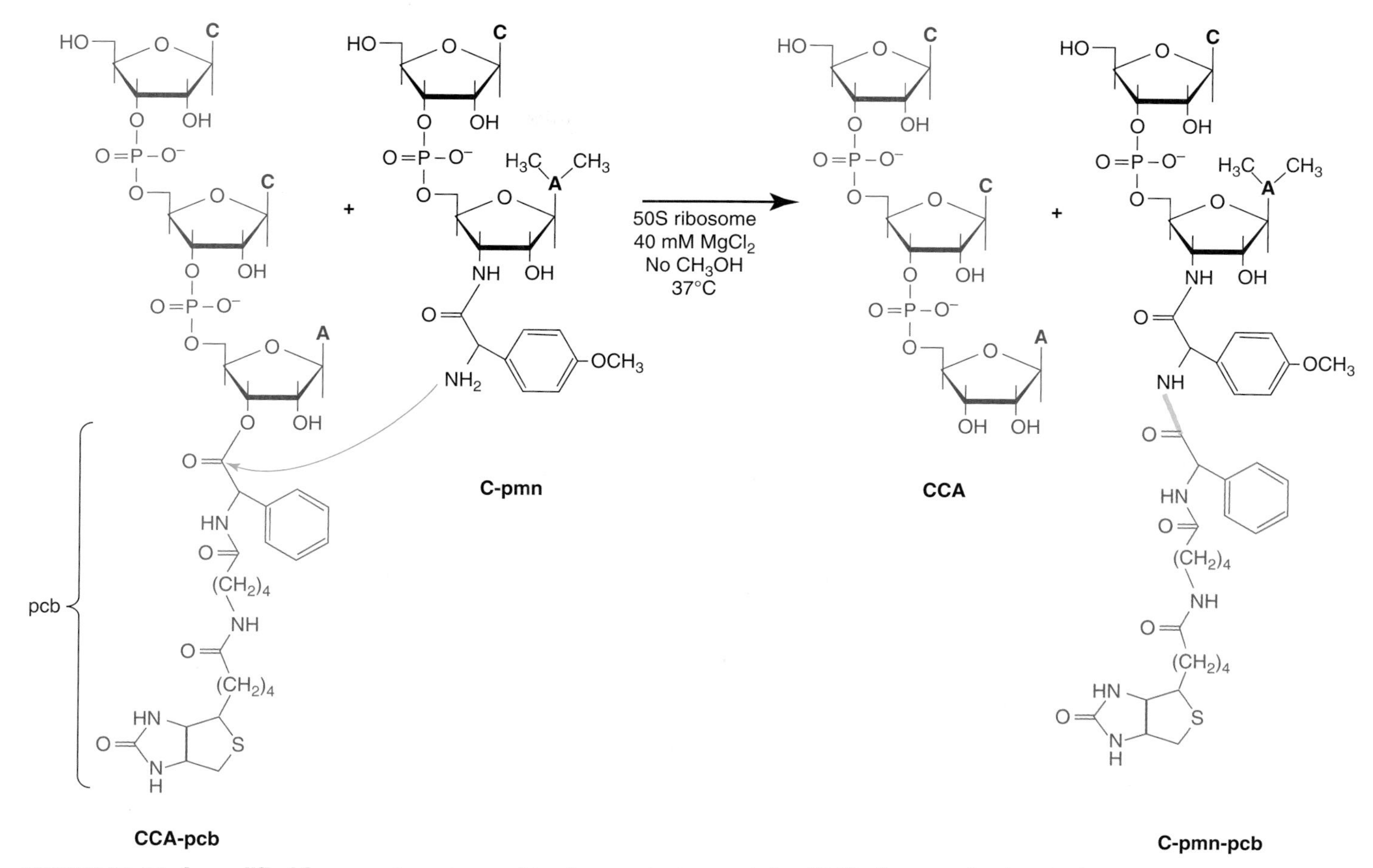

FIGURE 23.32 A modified fragment assay used to demonstrate crystals of 50S ribosomal subunits have peptidyl transferase activity. The ribosome's ability to catalyze peptide bond formation is assayed by following phenylalanine-caproic acid-biotin (pcb) group transfer from CCA-phenylalanine-caproic acid-biotin (CCA-pcb) to C-puromycin (C-pmn). The products of the reaction are C-puromycin phenylalanine-caproic acid-biotin (C-pmn-pcb) and deacetylated CCA. (Adapted from T. M. Schmeing, et al., *Nat. Struct. Mol. Biol.* 9 [2002]: 225–230.)

the possibility of examining crystalline large ribosomal subunits with product analogs bound to the A- and P-sites. However, the alcohol required for the fragment reaction presented a problem because it might alter the subunit's structure.

Moore, Steitz, and coworkers solved the problem by using larger substrates that do not depend on alcohol. More specifically, they used CCA-phenylalanine-caproic acid-biotin (CCA-pcb) as a P-site substrate and C-puromycin as an A-site substrate (FIGURE 23.32). Biochemical studies showed that crystals of the 50S ribosomal subunit catalyzed the formation of two products, C-puromycin-phenylalanine-caproic acid-biotin and a deacylated CCA. FIGURE 23.33a shows the location of these reaction products within the 50S ribosomal subunit; the peptide exit tunnel is also visible. A close-up view of the active site shows that the peptidyl-product binds to the so-called A-loop in domain V and the deacylated product binds to the so-called P-loop in domain V (FIGURE 23.33b). The nearest ribosomal protein to the *H. marismortui* peptidyl transferase center (PTC) was observed to be about 18 Å from this site, indicating that the peptidyl transfer reaction is catalyzed by RNA and not protein. When crystal structures

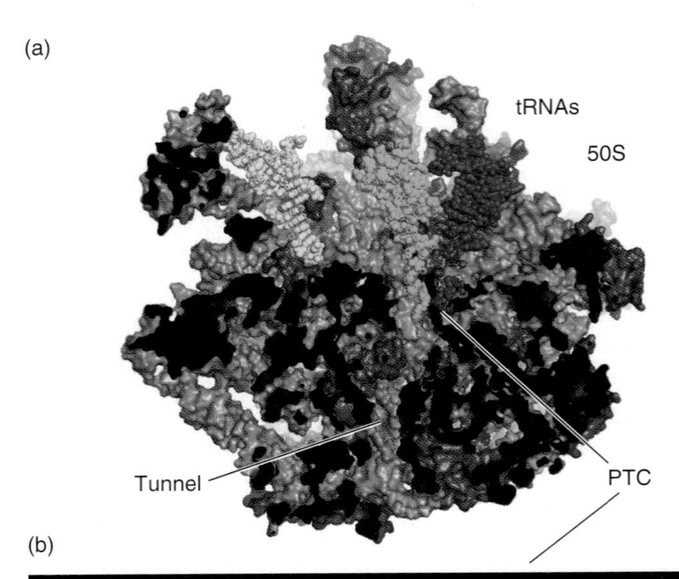

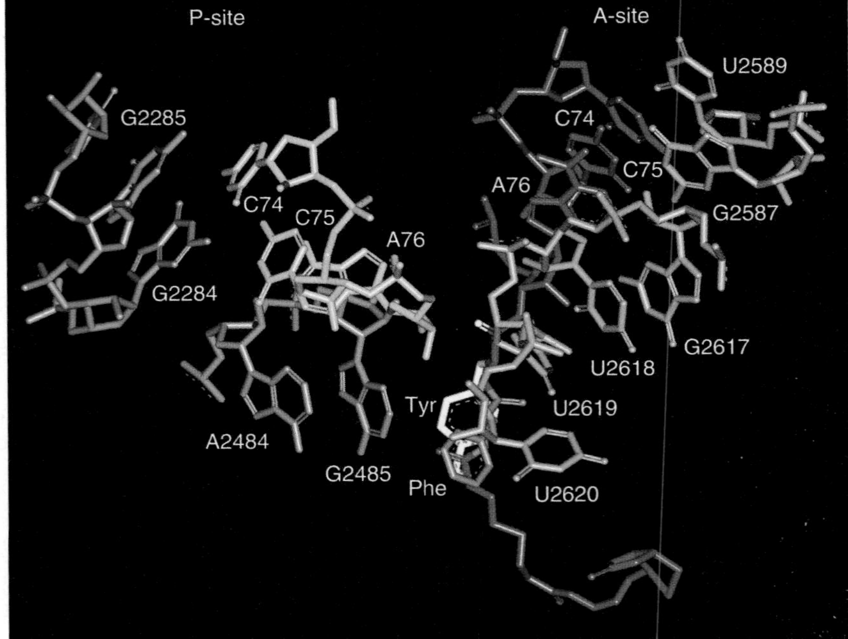

FIGURE 23.33 **Structure of the new fragment reaction products bound to the ribosome.** (a) A spacefill representation of the 50S ribosomal subunit (RNA in orange and protein in black) with products. The three tRNAs (magenta, green, and yellow), which were observed bound to the *Thermus thermophilus* 70S ribosome, are superimposed and provided as a reference. The subunit is split through the tunnel, and the front half removed to expose the tunnel and the peptidyltransferase site. The orientation is the crown view. (b) A close-up view of the active site. The peptidyl product (C-puromycin phenylalanine-caproic acid-biotin, C-pmn-pcb) (purple) binds the A-loop (tan). The deacylated product (CCA) (light green) base pairs to the P-loop (blue). N-3 of A2485 (A2451) is near the 3′-OH end of the deacylated CCA product. U2620 (U2585) has moved close to the newly formed peptide bond and the 3′-OH of dimethyl A76. The numbers in parenthesis are equivalent nucleotides in *E. coli* 23S rRNA. (Part a adapted from R. M. Voorhees, et al., *Nat. Struct. Mol. Biol.* 16 [2009]: 528–533. Photo courtesy of Martin Schmeing, McGill University. Part b structure from Protein Data Bank 1KQS. T. M. Schmeing, et al., *Nat. Struct. Biol.* 9 [2002]: 225–230. Prepared by B. E. Tropp.)

for bacterial 50S subunits and 70S ribosomes were obtained, they confirmed this conclusion.

Steitz and coworkers next addressed the question of how the translating ribosome prevents the undesirable hydrolysis of the peptidyl-tRNA bound to the P site when the A site is unoccupied. Their approach was to examine crystal structures of the of the *H. marismortui* 50S ribosomal subunit with substrate analogs simultaneously bound to the peptidyl transferase center's A- and P-sites. This examination showed that prior to the binding of the aminoacyl-tRNA analog to the A-site, the 23S rRNA prevents a water molecule from attacking the ester bond in peptidyl-tRNA. Proper aminoacyl-tRNA binding to the A-site causes a conformation change in the 23S rRNA that reorients the ester group and makes the peptidyl-tRNA accessible to attack by the aminoacyl group attached to the tRNA at the A site. An induced-fit mechanism, therefore, prevents unwanted peptidyl-tRNA hydrolysis and promotes the desired peptide bond formation.

Although the studies described to this point show that RNA and not protein catalyzes the peptidyl transfer reaction, they do not show how RNA performs this task. Studies performed by Andrea Barta and coworkers in 2003 suggested that A76, the adenosine at the 3′-end of the tRNA linked to the peptidyl group, makes an essential contribution to the catalytic process. More specifically, they demonstrated that P site fragment substrates with a 2′-OH support peptide bond transfer at several hundred times the rate of P site fragment substrates with a 2′-H. The following year Rachel Green, Scott A Strobel, and coworkers demonstrated that replacing the 2′-OH on A76 of the P-site tRNA with either 2′-H or 2′-F lowers the rate of peptide bond formation by a factor of 10^6 or greater. They also showed that the loss of the 2′-OH has no effect on the modified tRNA's ability to bind to P site. Based on these observations and information provided from crystal structures, it seems likely that peptidyl transfer takes place through a proton shuttle mechanism such as that shown in FIGURE 23.34. The

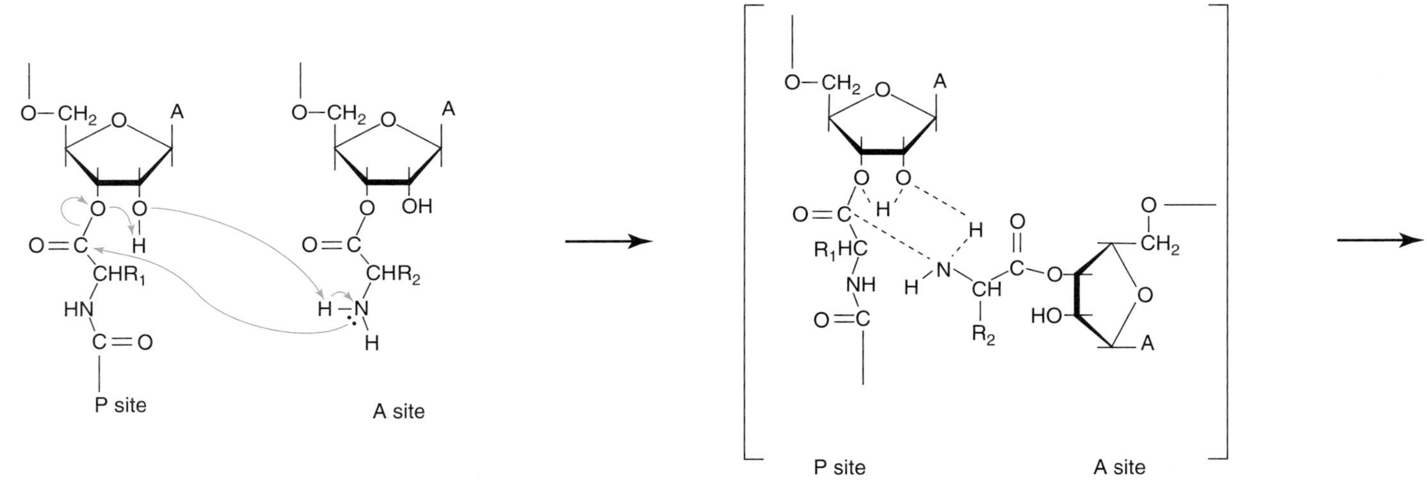

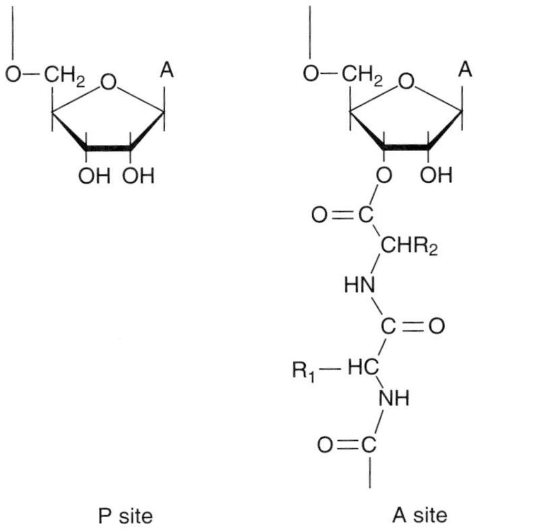

FIGURE 23.34 **The proton shuttle mechanism for the peptidyl transfer reaction.** The α-amino group of the aminoacyl-tRNA makes a nucleophilic attack on the carbonyl carbon of the peptidyl-tRNA. The 2′-OH simultaneously acts as a general base and a general acid by accepting a proton from the aminoacyl-tRNA's α-amino group to facilitate the nucleophilic attack by the amino group on the carbonyl carbon atom while donating a proton to the leaving 3′-OH at the P site A76 upon its deacylation. The reaction proceeds through a six-membered transition state (in brackets), which is converted to the product. (Adapted from M. Simonović and T. Steitz, *Biochim. Biophys. Acta* 1789 [2009]: 612–623.)

2′-OH simultaneously acts as a general base and a general acid by accepting a proton from the aminoacyl-tRNA's α-amino group to facilitate the nucleophilic attack by the amino group on the carbonyl carbon atom while donating a proton to the leaving 3′-OH at the P site A76 upon its deacylation. The ribosome contributes to the catalytic process by correctly orienting the two substrate molecules and stabilizing the transition state.

The hybrid-states translocation model offers a mechanism for moving tRNA molecules through the ribosome.

Once the peptidyl transferase catalyzed reaction is complete, the P-site tRNA is deacylated and the A-site tRNA has a peptide chain with one additional amino acid residue. Before the next elongation cycle can take place, the peptidyl-tRNA has to move from the A-site to the P-site and the deacylated tRNA has to move from the P-site to the E-site so that it can be released from the ribosome (Figure 23.23, steps 6–9). This highly coordinated movement known as **translocation** has to be precise so that the reading frame of the mRNA is preserved.

Noller has proposed the **hybrid-states translocation model** shown in FIGURE 23.35 to explain the movements of mRNA and tRNAs through the ribosome. This model is based on chemical footprinting experiments performed by Noller and Danesh Moazed in 1989 that monitored the progress of tRNA through the ribosomes. They first

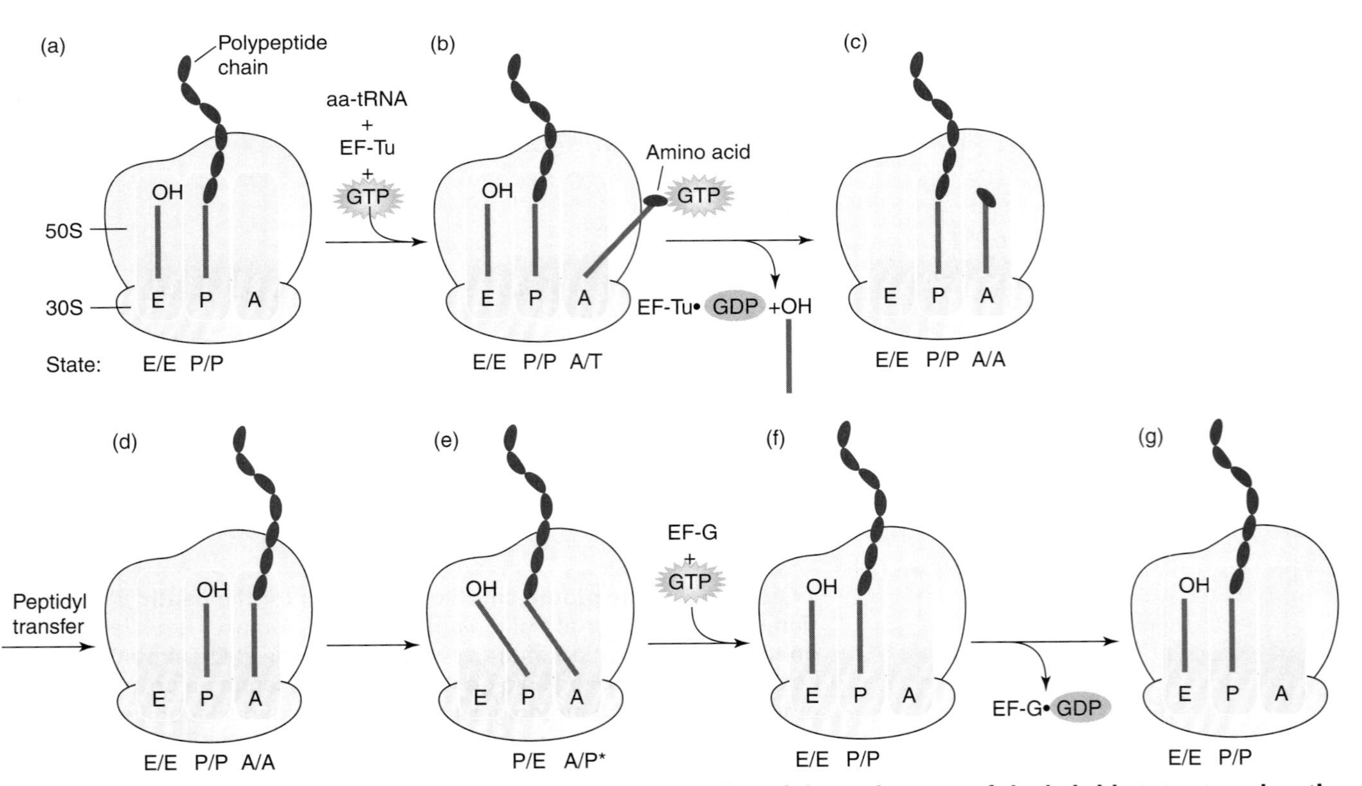

FIGURE 23.35 Schematic representation of our current understanding of the main steps of the hybrid-states translocational cycle. The tRNA molecules are shown as green lines, and the mRNA is not shown. (Adapted from H. F. Noller, et al., *FEBS Lett.* 514 [2002]: 11–16.)

demonstrated that N-acetyl-Phe-tRNA binds to the P-site by verifying its full reactivity with puromycin. Based on this experiment, they assigned the bases protected by N-acetyl-Phe-tRNA in the 16S and 23S rRNA molecules to the 30S and 50S P-sites, respectively. Then they performed a second footprinting experiment after letting the complex react with puromycin. Puromycin did not affect the 16S rRNA footprint but had a profound effect on the 23S rRNA footprint. The CCA_{OH} end of the now deacylated tRNA moved so that it no longer protected P-site bases on the 23S rRNA but instead protected E-site bases. The tRNA therefore appeared to be in a hybrid state with its anticodon end still bound to the 30S P-site but its CCA_{OH} end bound to 50S E-site.

This hybrid state is represented as P/E, where the letters before and after the slash indicates the 30S and 50S subunit sites, respectively. Movement from the P/P state to the P/E state takes place spontaneously and requires neither GTP nor an elongation factor. Noller and Moazed performed a second experiment in which they again bound N-acetyl-Phe-tRNA in the P-site but this time introduced aminoacyl-tRNA (and not puromycin) to the A-site. Footprinting performed after the peptidyl transferase reaction showed that the two tRNAs had moved from their A/A and P/P states to A/P and P/E hybrid states. Later footprinting studies showed that EF-G•GTP converts the A/P and P/E hybrid states to P/P and E/E states, respectively.

EF-G•GTP stimulates the translocation process.

Crystal structures and cryo-electron microscopy constructions have helped to reveal how EF-G•GTP stimulates the translocation process. EF-G has five domains (FIGURE 23.36a). Domain 1 at the N-terminus, also known as the G domain because it binds guanine nucleotides, resembles the G domain in EF-Tu (FIGURE 23.36b). Domain 2 is structurally similar to the middle domain in EF-Tu. The remaining three EF-G domains are in aggregate similar to the size and shape of tRNA. Thus, EF-G•GDP and EF-Tu•GTP•aminoacyl-tRNA have very similar overall sizes and shapes, suggesting that the two elongation factors might bind at or near the same site in the ribosome. A comparison of crystal structures of *T. thermophilus* ribosomes with bound EF-G•GDP

(a)

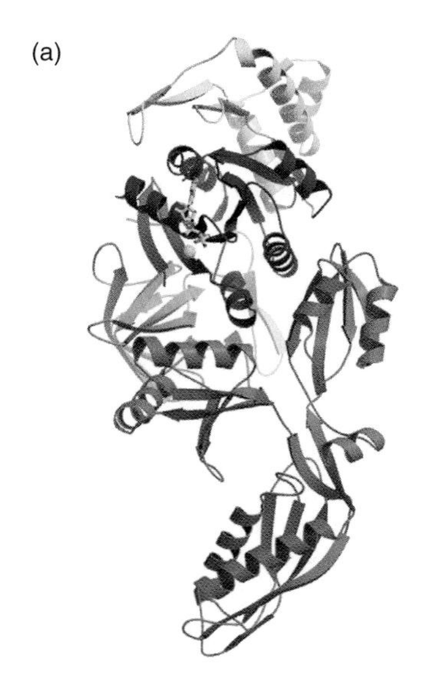

(b)

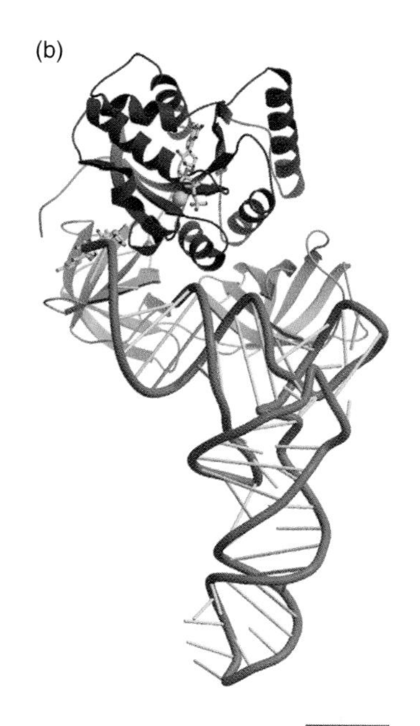

FIGURE 23.36 **Similar sizes and shapes for EF-G•GDP and EF-Tu•GDPNP•Phe-tRNA.** (a) EF-G•GDP from *Thermus thermophilus*. Domains I and II are colored red and green, respectively. Domains III, IV, and V are colored orange. A subdomain within domain I is colored gray. GDP is shown as a yellow ball-and-stick model. (b) The ternary complex of *Thermus aquaticus* EF-Tu•GDPNP with bound yeast Phe-tRNA. EF-Tu domains I, II, and III are colored red, green, and blue, respectively. The tRNA backbone is colored orange. The amino acid and 3′-terminal CCA nucleotides are bound in a cleft at the junction of domains I and II, protecting the ester bond that links the amino acid to tRNA from hydrolysis. (Reproduced from *Trends Biochem. Sci.*, vol. 28, G. R. Andersen, P. Nissen, and J. Nyborg. Elongation factors in protein biosynthesis, pp. 434–441, copyright 2003, with permission from Elsevier [http://www.sciencedirect.com/science/journal/09680004]. Photos courtesy of Poul Nissen, in memory of Jens Nyborg, University of Aarhus.)

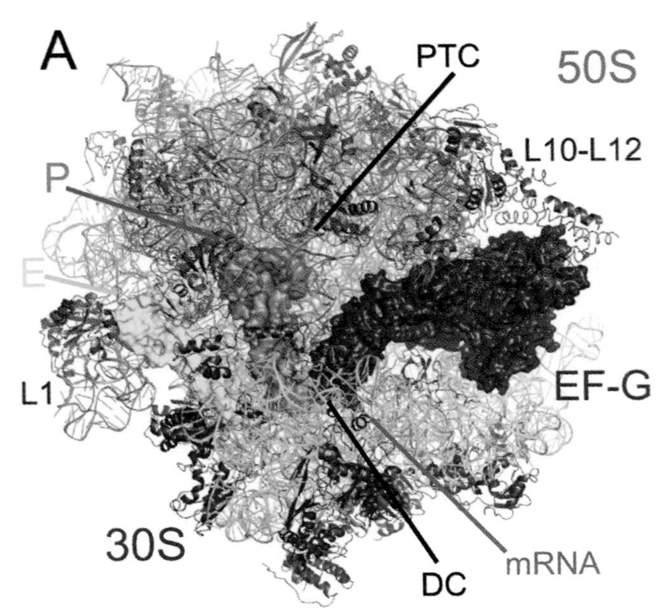

FIGURE 23.37 **Crystal structures of the *T. thermophilus* 70S ribosome with bound EF-G•GDP.** The 70S ribosome is shown with the 50S subunit (reddish brown) on top and the 30S subunit (cyan) on bottom. P- and E-site tRNAs are shown in green and yellow, respectively. mRNA is shown in purple. The peptidyl transferase center (PTC), L1 and L10-L12 stalks on the 50S subunit, and the decoding center (DC) on the 30S subunit are indicated. EF-G is shown in red. The antibiotic fusidic acid was used to trap the ribosome in the posttranslocation state with EF-G•GDP still bound. Fusidic acid permits GTP hydrolysis and translocation but blocks GDP release from the ribosome. (Reproduced from Y.-G. Gao, et al., *Science* 326 [2009]: 694–699. Reprinted with permission from AAAS. Photo courtesy of V. Ramakrishnan, Medical Research Council [UK].)

(FIGURE 23.37) and EF-Tu•GDP•aminoacyl-tRNA (Figure 23.25) support this prediction. The eukaryotic elongation factor, eEF2, is structurally similar to its bacterial counterpart, EF-G, and appears to make a similar contribution to the translocation process.

The steps in the translocation process are summarized in FIGURE 23.38. Translocation begins after the polypeptide chain transfers from the tRNA at the P-site to the tRNA at the A-site. The resulting pre-translocation ribosome fluctuates between a classical and a hybrid state. In the classical state, the peptidyl tRNA and deacylated tRNAs are in A/A and P/P sites, respectively. In the hybrid state, the peptidyl tRNA and deacylated tRNAs are in A/P and P/E sites, respectively. The small subunit rotates around the large subunit in a counterclockwise direction as the ribosome changes from the classical to the hybrid state. EF-G•GTP binds to the hybrid state ribosome, stabilizing the hybrid state and inducing rapid GTP hydrolysis. This hydrolysis triggers a conformational change in EF-G, which in turn detaches the mRNA-tRNA complex from the decoding center. As a result of this "unlocking" the mRNA-tRNA complex can move up by one codon on the small subunit while the tRNAs retain their positions on the large subunit. Translocation is completed by a slight backward rotation of the small subunit as EF-G•GDP is released from the ribosome. The ribosome is now returned to the "locking" state, which prevents EF-G•GTP from binding to the ribosome but permits an aminoacyl-tRNA•EF-Tu•GTP complex to bind to the T/A site. Subsequent entry of the aminoacyl-tRNA into the A/A site during accommodation has been proposed to cause the uncharged tRNA to be released from the E/E site.

A new translation elongation factor was recently discovered in bacteria that can catalyze back-translocation on the ribosome. That is, it can move peptidyl-tRNA and deacylated tRNA from P/P and E/E sites to A/A and P/P sites, respectively. This is a remarkable discovery because 3′→5′ ribosomal movement along mRNA was thought to be impossible. Ribosomes appeared to be designed to move ribosomes along mRNA in only the 5′→3′ direction. This new elongation factor, which can also catalyze translocation in the forward direction, was originally called LepA after the gene that codes for it but is now called EF-4. EF-4 is not required under most growth conditions but appears to prevent growth inhibition when the culture medium contains high Mg^{2+} or Na^{+} concentrations or is at low pH. Further studies are needed to determine EF-4's physiological function. One function that has been suggested is that EF-4 rescues ribosomes that have stalled due to defective translocation, returning the stalled ribosomes to their pre-translocation state.

FIGURE 23.38 **A model for EF-G catalyzed translocation.** The pre-translocation ribosome fluctuates between a classical and hybrid state. (a) In the classical state, the peptidyl tRNA and deacylated tRNAs are in A/A and P/P sites, respectively. (b) In the hybrid state, the peptidyl tRNA and deacyltated tRNAs are in A/P and P/E sites, respectively. The small subunit rotates around the large subunit in a counterclockwise direction as the ribosome changes from the classical to the hybrid state. (c) EF-G•GTP binds to the hybrid state ribosome, stabilizing the hybrid state and inducing rapid GTP hydrolysis. This hydrolysis triggers a conformational change in EF-G, which in turn detaches the mRNA-tRNA complex from the decoding center. (d) As a result of this "unlocking" the mRNA-tRNA complex can move up by one codon on the small subunit while the tRNAs retain their positions on the large subunit. (e) Translocation is completed by the backward rotation of the small subunit as EF-G•GDP is released from the ribosome. (Adapted from X. Agirrezabala and J. Frank, *Q. Rev. Biophys.* 42 [2009]: 159–200.)

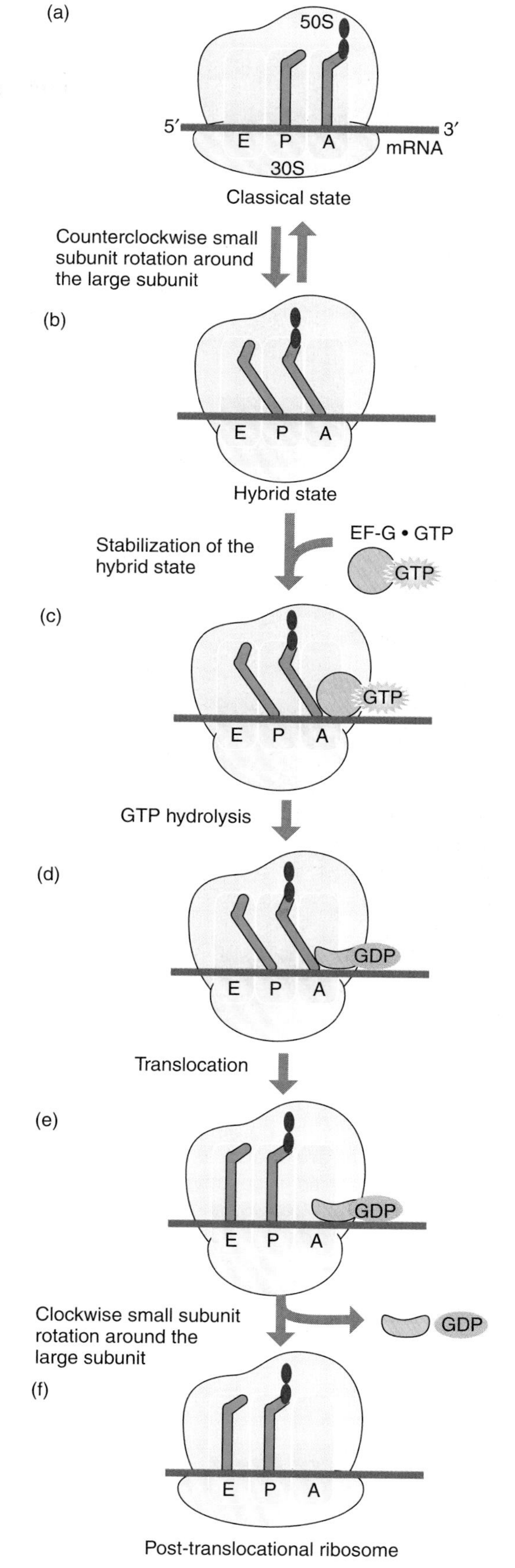

23.5 Termination Stage in Bacteria

Bacteria have three protein release factors.

Polypeptide chain termination takes place when a termination (nonsense) codon enters the A-site on the 30S ribosomal subunit. Early studies suggested that the three bases of the termination codon (UAA, UAG, or UGA) provide all the information required for termination. However, it now appears that bases on either side of the termination codon influence the strength of the stop signal. For instance, UGA is a much stronger stop signal in *E. coli* when followed by a U than when followed by a C. The next two bases downstream as well as bases just upstream from the termination codon also influence stop signal strength but to a lesser extent.

Polypeptide chain termination in bacteria requires **release factors** (**RFs**). The existence of these release factors was first revealed in experiments performed by Mario Capecchi in 1967. He first prepared a ribosome•mRNA complex with a hexapeptide attached to tRNA at the P-site and a termination codon at the A-site and then used this complex to demonstrate that a soluble protein factor was required to release the hexapeptide. C. Thomas Caskey and coworkers devised a much simpler assay for RF activity the following year. Their assay was based on the observation that release factors stimulate an fMet-tRNA•AUG•ribosome complex to release fMet when incubated with a trinucleotide containing a termination codon. Using this simpler assay system, Caskey and coworkers demonstrated that bacterial extracts contain two different RFs. **RF1** recognizes the termination codon UAG, whereas **RF2** recognizes the termination codon UGA. In addition, both release factors recognize the termination codon UAA. Working independently in 1969, the Capecchi and Caskey laboratories

(a)

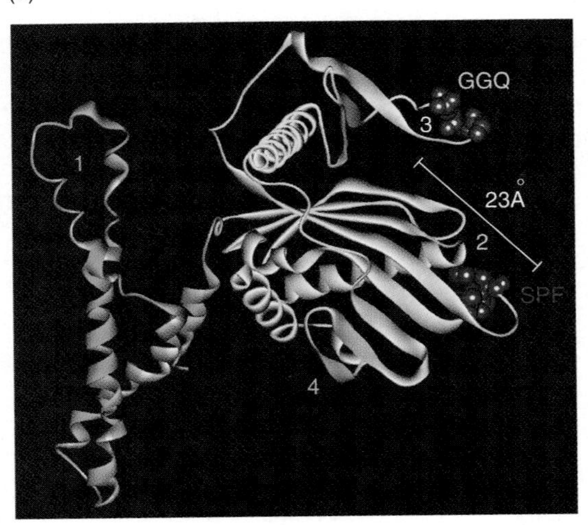

(b)

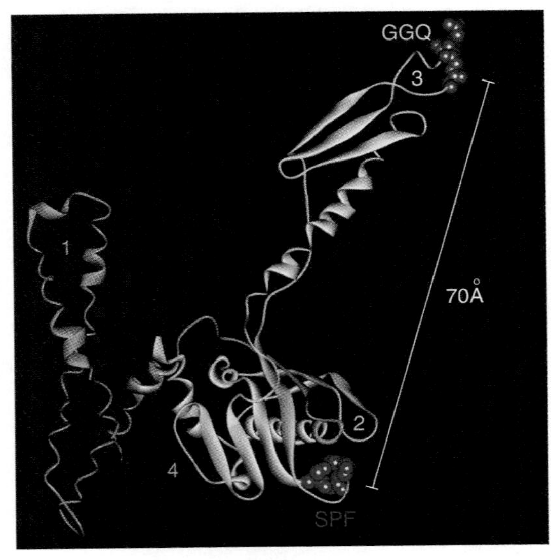

FIGURE 23.39 **Crystal structure of *E. coli* release factor 2 (RF2).** (a) structure of free RF2. (b) structure of ribosome-bound RF2. The numbers specify the domains. The GGQ and SPF tripeptides are shown as green and red spacefill structures, respectively. (Part a structure from Protein Data Bank 2IHR. G. Zoldak, et al., *Nucleic Acids Res.* 35 [2007]: 1343–1353. Prepared by B. E. Tropp. Part b structure from Protein Data Bank 2X9R. H. Jin, et al., *Proc. Natl. Acad. Sci. USA* 107 [2010]: 8593–8598. Prepared by B. E. Tropp.)

discovered a third bacterial release factor, **RF3**, which stimulates the rate of peptide release when either RF1 or RF2 is also present but it does not act on its own.

The class 1 release factors, RF1 and RF2, have one tripeptide that acts as an anticodon and another that binds at the peptidyl transferase center.

RF1 and RF2 are assigned to the class 1 family of release factors because they have similar primary structures and similar functions. In 2000, Koichi Ito and coworkers devised a clever approach to determine the region within RF1 and RF2 responsible for termination codon recognition. They began by using recombinant DNA technology to swap domains between the two release factors. Then they analyzed the resulting hybrid proteins to identify the domain responsible for termination codon recognition. Further genetic analysis narrowed the specificity to a tripeptide within the domain. The tripeptide is Pro-Ala-Thr or Pro-Val-Thr [P(A/V)T] in RF1 and Ser-Pro-Phe [SPF] in RF2. Cross-linking and footprinting experiments show that P(A/V)T and SPF interact with termination codons at the A-site on the 30S ribosomal subunit and therefore act as "tripeptide anticodons."

RF1 and RF2 have a second highly conserved tripeptide, Gly-Gly-Gln (GGQ), which is essential for function. Måns Ehrenberg and coworkers showed that mutations that convert GGQ to GAQ (Gly-Ala-Gln) or GGA (Gly-Gly-Ala) produce release factors with very little activity. They further demonstrated that GGQ binds near the peptidyl transferase center because it is able to prevent puromycin from reaching this site. Ehrenberg and coworkers also addressed the question of how bacterial class 1 release factors stimulate peptide release. They considered two possibilities: (1) the RFs are hydrolases that are activated upon binding to the ribosome and (2) the ribosome has the hydrolase activity that is stimulated by the release factor. They devised a clever experiment to investigate whether ribosomes have hydrolase activity. They first constructed a ribosome complex that bound fMet-$tRNA^{fMet}$ at the P-site in response to an mRNA encoding fMet-Phe-Thr-Ile. Then they demonstrated that deacylated $tRNA^{Phe}$ stimulated fMet release by more than a factor of 20. No release factors were present in this experiment. Adding 20% ethanol increased the rate of fMet release to a value close to that observed at a normal termination site in the presence of RF1 or RF2. This experiment clearly demonstrates that it is the ribosome and not the RF that hydrolyzes the ester bond linking the completed peptide to the P-site tRNA.

Further examination of class 1 bacterial release factors became possible after the crystal structures became available for the free release factors and ribosome-bound release factors. As shown in FIGURE 23.39, RF2 undergoes a major conformational change upon binding to the ribosome. The distance between the SPF (the "tripeptide anticodon") and GGQ is only 23 Å in free RF2 but increases to 70 Å after RF2 binds to the ribosome. The crystal structure of the RF2•70S ribosome complex reveals that SPF binds at the A-site of the 30S subunit and GGQ binds to the peptidyl transferase center of the 50S subunit

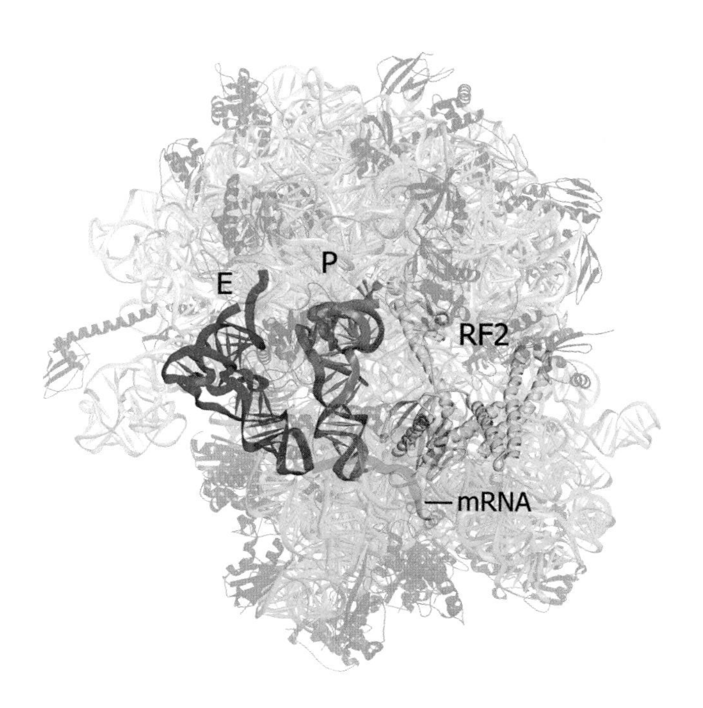

FIGURE 23.40 Crystal structure of the RF2•70S ribosome complex. The 50S subunit, which is on top, is shown with its RNA colored gray and its proteins colored magenta. The 30S subunit, which is on the bottom, is shown with its RNA colored cyan and its proteins colored blue. The mRNA is shown in green and P- and E-site tRNAs in orange and red, respectively. RF2 is shown in yellow. (Reproduced from A. Korostelev, et al., *Proc. Natl. Acad. Sci. USA* 105 [2008]: 19684–19689. Copyright 2008, National Academy of Sciences, U.S.A. Photo courtesy of Andrei Korostelev.)

(FIGURE 23.40). Specific interactions between SPF and the stop codon produce a conformational change in the 16S rRNA that differs from the change that takes place when aminoacyl-tRNA binds to the A-site during the elongation process. Interaction of GGQ with the peptidyl transferase center produces a conformational change in the peptidyl transferase center that permits water to attack the ester bond in in the peptidyl-tRNA at the P-site. Although RF1 has a different "tripeptide anticodon" than RF2, in all other respects it looks and acts just like RF2.

RF3 is a nonessential G protein that stimulates RF1 or RF2 dissociation from the ribosome.

Bacterial RF3 is a nonessential protein that facilitates the dissociation of RF1 and RF2 from the ribosome complex after peptide release. Because it differs from RF1 and RF2 in both amino acid sequence and function, RF3 is said to be a class 2 release factor. RF3 has a guanine nucleotide–binding motif and so in this one respect resembles EF-Tu and EF-G. However, RF3 differs from these translation elongation factors in one very important way. The soluble cytoplasmic form of RF3 binds GDP about 10^3 times more tightly than it binds GTP. Therefore, it is RF3•GDP that binds to the ribosome complex.

Based on the information gathered about the RFs, Lev Kisselev and coworkers propose the translation termination pathway shown in FIGURE 23.41. Termination begins when the ribosome is in a pretermination state with peptidyl-tRNA in the P-site and a class 1 release factor in the A-site. The ribosome cleaves the ester bond linking the completed peptide from the P-site tRNA, converting the complex to a posttermination state. The ribosome, which now has a deacylated tRNA in the P-site and a class 1 RF in the A-site, binds RF3•GDP. GDP rapidly dissociates from the ribosome, which then contains the

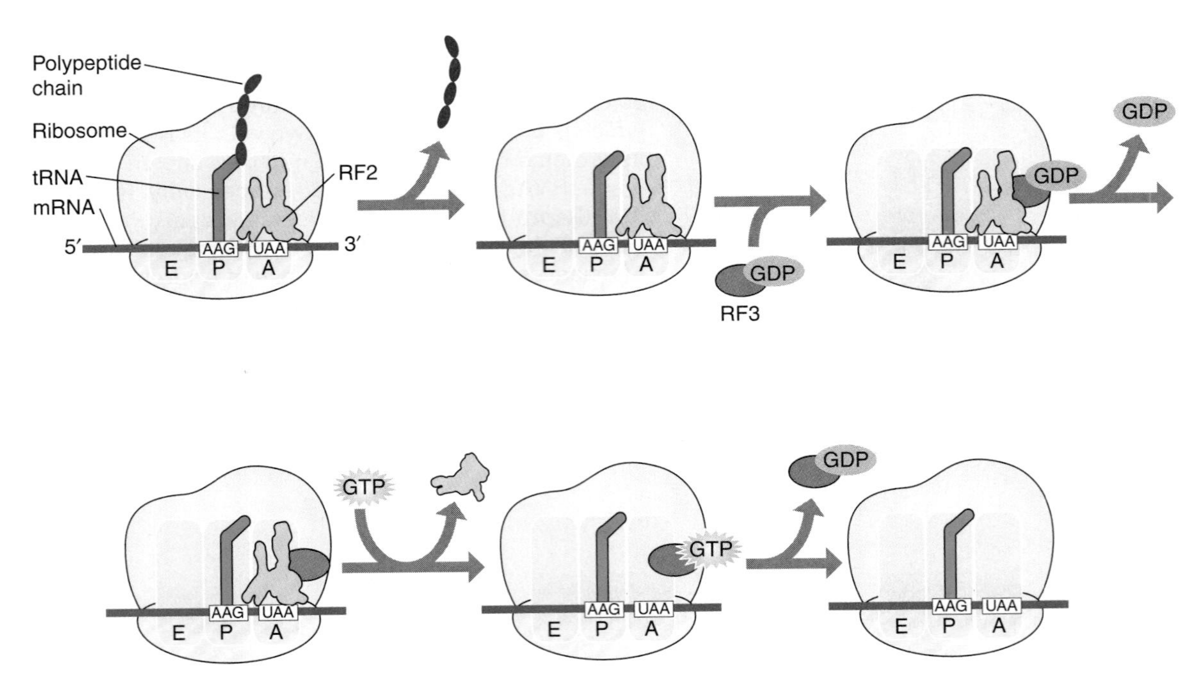

FIGURE 23.41 Bacterial polypeptide termination pathway. Termination begins when the ribosome is in a pre-termination state with peptidyl-tRNA in the P-site and a class 1 release factor (RF2 in this case) in the A-site. The ribosome cleaves the ester bond linking the completed peptide from the P-site tRNA, converting the complex to a posttermination state. The ribosome, which now has a deacylated tRNA in the P-site and a class 1 RF in the A-site, binds RF3•GDP. GDP rapidly dissociates from the ribosome, which then contains the class 1 release factor (RF2 in this example) and a stably bound RF3. GTP binds to RF3, which changes conformation, forcing the class 1 release factor off the ribosome. Subsequent GTP hydrolysis causes RF3 to change conformation once again, allowing it to dissociate from the ribosome. At the completion of the termination stage, the deacylated tRNA is still bound to the P-site and the mRNA is still associated with the intact ribosome. (Adapted from L. Kisselev, et al., *EMBO J.* 22 [2003]: 175–182.)

class 1 release factor (RF1 or RF2) and a stably bound RF3. GTP binds to RF3, which changes conformation, forcing the class 1 release factor off the ribosome. Subsequent GTP hydrolysis causes RF3 to change conformation once again, allowing RF3 to dissociate from the ribosome. At the completion of the termination stage, the deacylated tRNA is still bound to the P-site and the mRNA is still associated with the intact ribosome.

A stalled ribosome translating a truncated mRNA that lacks a termination codon can be rescued by tmRNA.

Bacteria require a termination codon for proper polypeptide chain termination. A serious problem may arise if mRNA transcription is incomplete or the mRNA is cleaved by a nuclease so that the resulting mRNA fragment lacks a termination codon. Translation of such an mRNA fragment would start normally but elongation would cease when the ribosome reaches the 3′-end. At this stage, the ribosome would neither be able to continue elongation nor to terminate translation. Hence, the ribosome would stall with the partially synthesized polypeptide chain still attached to the P-site tRNA. This situation causes two problems. First, ribosomes bound to the fragmented mRNA cannot recycle and second, truncated polypeptides may be harmful to the cell. In 1996, Robert T. Sauer and coworkers discovered a rescue mechanism that bacteria use to free stalled ribosomes and destroy incomplete polypeptides. Two new factors, one an RNA molecule and the other a protein, participate in this rescue process.

The RNA molecule required to rescue stalled ribosomes is called **transfer-messenger RNA (tmRNA)** because it functions as both tRNA and mRNA. Much of what we know about the structure of tmRNA is based on phylogenetic comparisons and chemical modification studies. The tmRNA molecule is approximately 260 to 430 nucleotides long, depending on the bacterial species. The predicted secondary structure for *E. coli* tmRNA is shown in FIGURE 23.42. The 5′- and 3′-ends of tmRNA form a tRNA-like domain that includes an acceptor stem, a TψC-arm, and a D-loop but no D-stem. The anticodon arm, which is normally present in tRNA, is replaced by a long disrupted stem that connects the tRNA-like domain to the rest of the tmRNA molecule. The 5′-end of this long disrupted stem is linked to a pseudoknot, a complex three-dimensional structure with intertwined segments (see Chapter 4). This pseudoknot, designated pseudoknot 1 or Ψ1, is followed by a short peptide reading frame and then by three additional pseudoknots (Ψ2–Ψ4). The tRNA domain, peptide reading frame, and Ψ1 are essential to rescue stalled ribosomes. Ψ2 to Ψ4 do not seem to be required because tmRNAs that lack these pseudoknots are functional. Alanyl-tRNA synthetase charges the tRNA domain with alanine and this charging must take place before tmRNA can rescue stalled ribosomes.

The protein factor required to rescue the stalled ribosome is known as **SmpB** (small protein B). SmpB interacts with Ala-tmRNA to form an Ala-tmRNA•SmpB complex. EF-Tu•GTP binds to the TψC-arm and the acceptor stem of the Ala-tmRNA•SmpB complex and carries the complex to the stalled ribosome. Frank and coworkers have obtained a cryo-electron microscopy construct of tmRNA in complex with the ribosome (FIGURE 23.43). The overall pathway for the rescue process, also called **trans-translation** is shown in FIGURE 23.44. Once bound to the ribosome, Ala-tmRNA acts as a stand-in for the A-site tRNA, accepting the peptide group from the P-site tRNA. Then the ribosome moves in some still unknown way from the stalled mRNA to the tmRNA, allowing translation to resume at the open reading frame within the tmRNA so that a short peptide (ANDENYALAA) called the **proteolysis tag** is added to the nascent protein. The ribosome terminates protein synthesis when it encounters a normal stop termination codon at the end of the open reading frame for the proteolysis tag and releases the polypeptide in the normal fashion. As the name implies, the proteolysis tag acts as a signal for polypeptide degradation, preventing the accumulation of an abnormal protein that might be harmful to the cell.

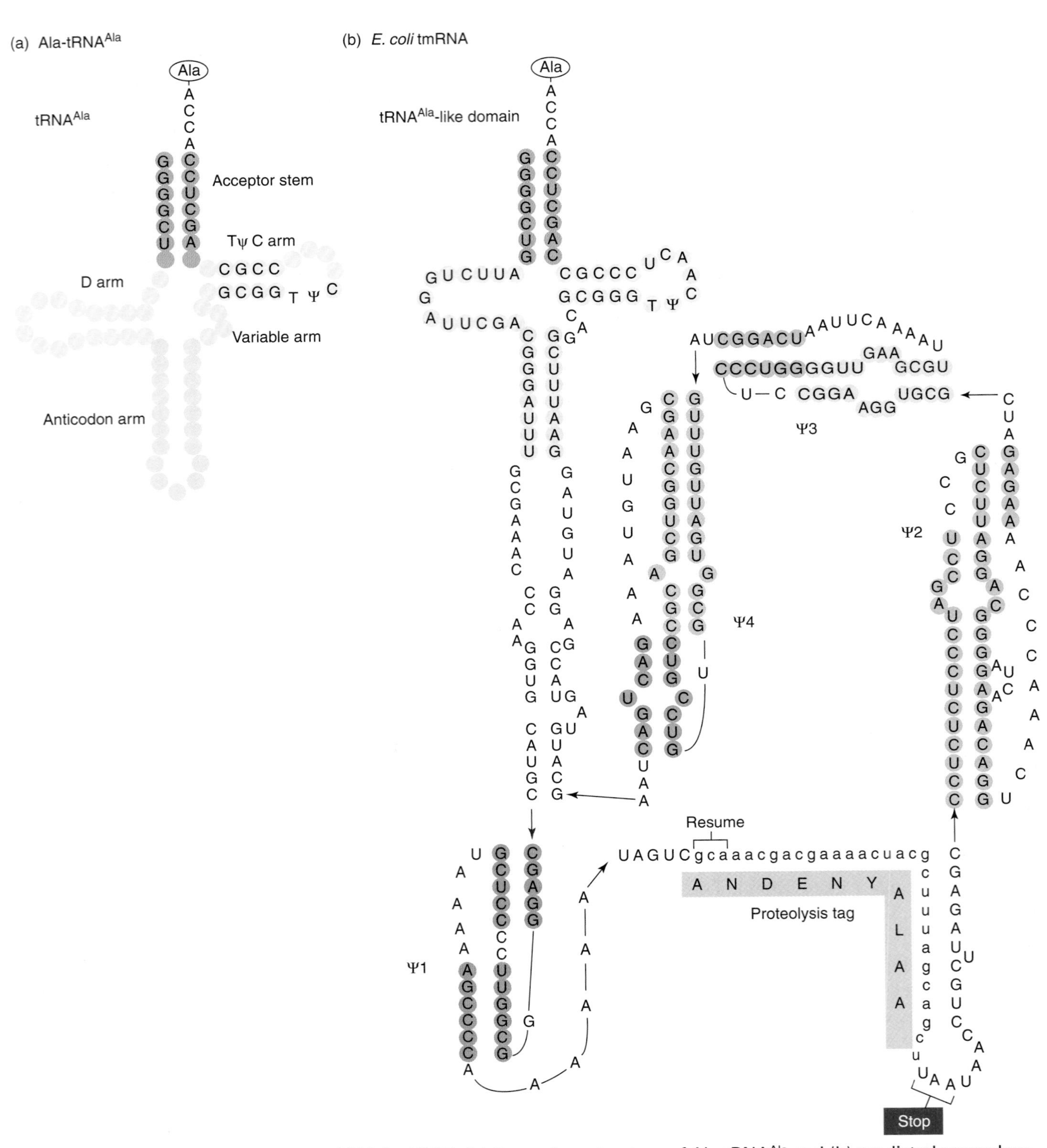

FIGURE 23.42 ***E. coli* transfer-messenger RNA (tmRNA).** (a) Secondary structure of Ala-tRNA^{Ala} and (b) predicted secondary structure of *E. coli* tmRNA based on phylogenetic comparisons. Similar regions of tRNA and tmRNA are highlighted in the same color. Four pseudoknots (complex three-dimensional motifs that involve intertwining of strand segments) are labeled Ψ1-Ψ4. Nucleotides are shown in lower case for a peptide tag that is added to the end of the polypeptide chain. "Resume" indicates the site at which the synthesis of this peptide tag begins. (Part a adapted from A. W. Karzai, E. D. Roche, and R. T. Sauer, *Nat. Struct. Mol. Biol.* 7 [2000]: 449–455. Part b modified from The tmRNA Website. http://www.indiana.edu/~tmrna/. Used with permission of Kelly Williams, Sandia National Laboratories.)

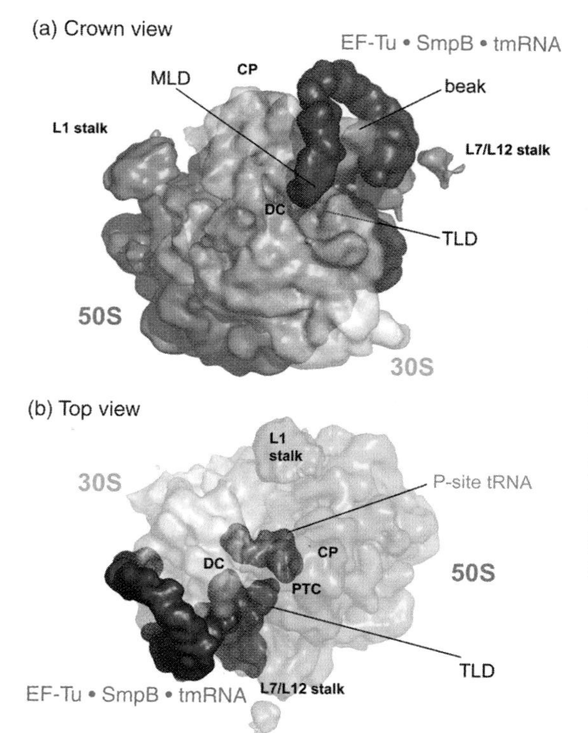

FIGURE 23.43 **Two views of the cryo-electron microscopy construct of a complex containing an initiated ribosome with a bound SmpB (small protein B), tmRNA, EF-Tu, GDP, and kirromycin.** (a) Crown view of the cryo-electron microscopy construct with SmpB•tmRNA•EF-Tu (red) emerging from the intersubunit space between the 50S subunit (blue) and 30S subunit (cream). EF-Tu in complex with tRNA-like domain (TLD) binds near the base of the L7/L12 stalk, underneath the GTPase-associated center (GAC). The antibiotic kirromycin was included to inhibit the conformational change that takes place after GTP is hydrolyzed, preventing EF-Tu from being released from the ribosome. The tmRNA loops around the beak of the small subunit, positioning the mRNA-like domain (MLD) near the mRNA entry site. (b) In this view, P-site bound peptidyl tRNA is visible between the peptidyltransferase center (PTC) of the large subunit and the decoding center (DC) of the small subunit. The decoding center is the site at which the codon and the anticodon match up. CP is the central protuberance. (Reproduced from *Curr. Opin. Struct. Biol.*, vol. 14, P. W. Haebel, S. Gutmann, and N. Ban, Dial tm for rescue . . ., pp. 58–65, copyright 2004, with permission from Elsevier [http://www.sciencedirect.com/science/journal/0959440X]. Photo courtesy of Sascha Gutmann and Nenad Ban, Institute for Molecular Biology and Biophysics, Swiss Federal Institute of Technology.)

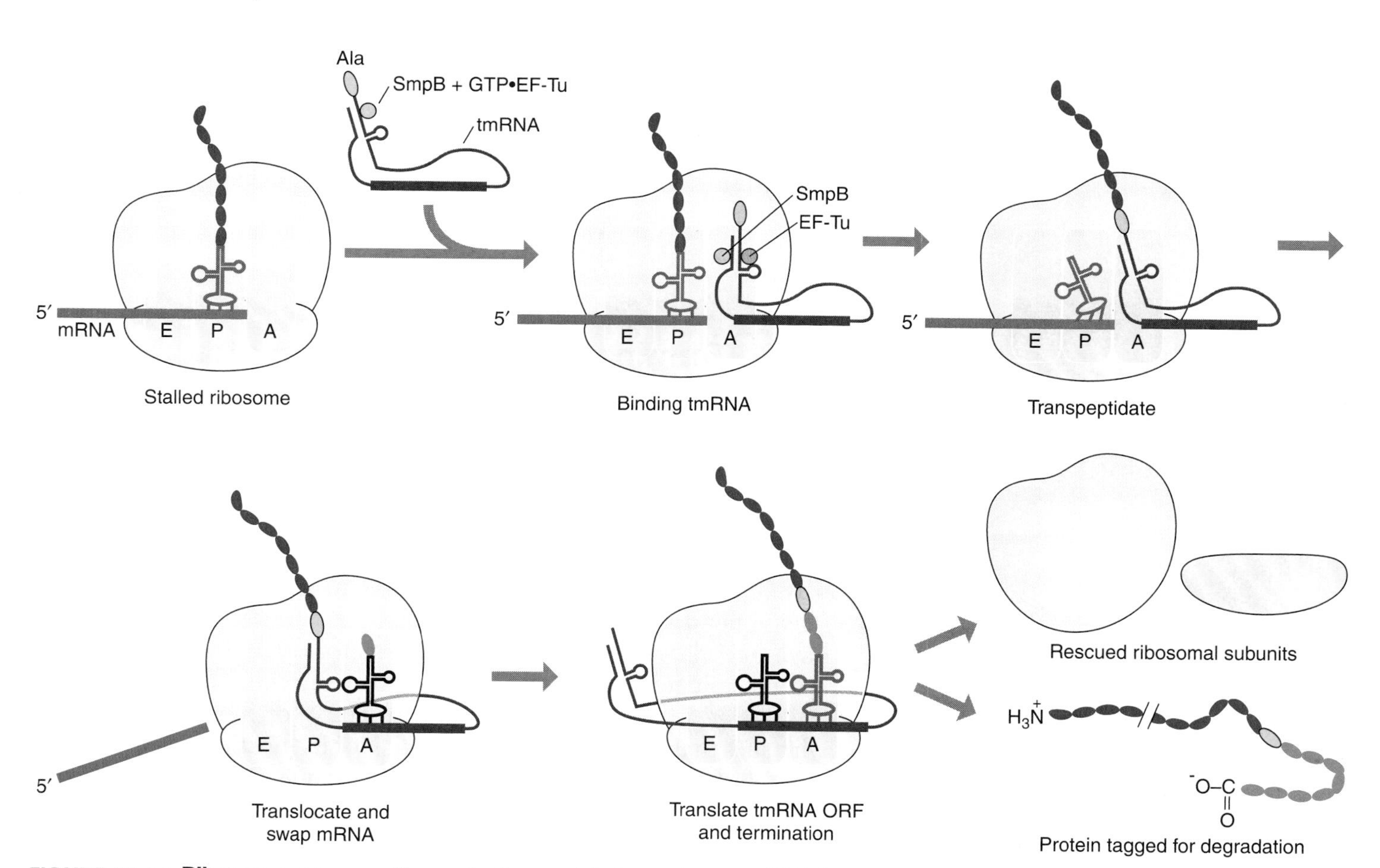

FIGURE 23.44 **Ribosome rescue pathway.** In the model shown here, tmRNA function involves initial recognition of stalled ribosomes by tmRNA complexed with EF-Tu and SmpB. Following recognition, tmRNA carries out its tRNA-like role; the ribosome then shifts from the original mRNA to the tmRNA open reading frame, which encodes a degradation tag. Translation resumes on this open reading frame and, following normal translation termination, the subunits of the previously stalled ribosome are freed to translate other mRNAs. (Adapted from S. D. Moore and R. T. Sauer, *Annu. Rev. Biochem.* 76 [2007]: 101–124.)

Mutant tRNA molecules can suppress mutations that create termination codons within a reading frame.

A simple mutational event can convert a normal codon into a termination codon. For example, a single base change converts UCG (a codon for serine) into UAG (a termination codon). Such nonsense mutations cause the protein synthetic machinery to produce nonfunctional polypeptide fragments. The cell's ability to synthesize a fully functional polypeptide can be restored by a second mutational event. As expected, many secondary mutations convert the termination codon to one that specifies an acceptable amino acid substitute and a few convert it back to the original codon. Some secondary mutations suppress the termination codon by altering the protein synthetic machinery rather than the affected gene. This suppression is often accomplished by changing the anticodon of a minor tRNA molecule so that it becomes able to base pair with the termination codon. For example, a single base change in a minor $tRNA^{Ser}$ species converts its CGA anticodon into a CUA anticodon. The cell survives this change because it has a second $tRNA^{Ser}$ species that still recognizes the normal UCG codon for serine. The mutant $tRNA^{Ser}$ suppresses the nonsense mutation by translating UAG as a serine codon. Suppressor tRNA molecules might be expected to cause serious problems by translating normal termination codons and producing abnormally large polypeptide products. This problem is less serious than expected because (1) as a derivative of a minor tRNA molecule, the intracellular suppressor tRNA molecule's concentration is low and (2) the protein synthetic machinery recognizes normal termination codons in the context of nucleotide sequences.

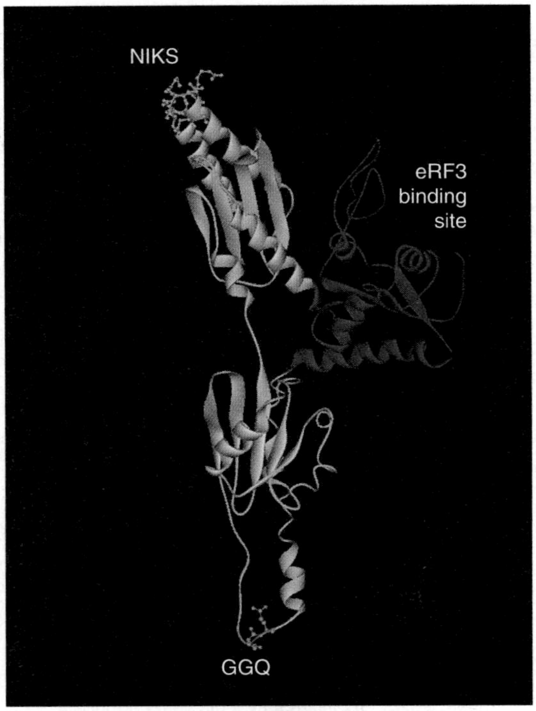

23.6 Termination Stage in Eukaryotes.

Eukaryotic cells have bacteria-like release factors in their mitochondria and a different kind in their cytoplasm.

Eukaryotes have two sets of release factors, one in the mitochondria and the other in the cytoplasm. Mitochondrial release factors have similar sequences to the bacterial RFs and probably work in a similar way. Cytoplasmic release factors differ in sequence from bacterial release factors but are similar to those from the archaea. Eukaryotes have one class 1 cytoplasmic release factor (eRF1) and one class 2 cytoplasmic release factor (eRF3). Both are essential for cell viability.

FIGURE 23.45 shows the crystal structure for human eRF1 along with sites that are thought to be essential for its function. The tripeptide Gly-Gly-Gln (GGQ), a highly conserved sequence present in class 1 RFs from all organisms, probably binds at the peptidyl transferase center. The tetrapeptide Asn-Ile-Lys-Ser (NIKS), which is present in cytoplasmic eRF1s, recognizes all three termination codons. eRF1 also has a binding site for eRF3 that allows the two RFs to associate even when both are soluble in the cytoplasm, suggesting that

NIKS
eRF3 binding site
GGQ

FIGURE 23.45 **The crystal structure for human eRF1.** The tripeptide Gly-GlyGin (GGQ shown as blue stick structure) is a highly conserved sequence present in class 1 release factors from all organisms and probably binds at the peptidyltransferase center. The tetrapeptide Asn-Ile-Lys-Ser (NIKS shown as yellow stick structure) recognizes all three termination codons. (Structure from Protein Data Bank 1DT9. H. Song, et al., *Cell* 100 [2000]: 311–321. Prepared by B. E. Tropp.)

the eRF1•eRF3 complex binds to the ribosome as a unit. The eRF1 conformation probably changes upon binding to the ribosome but the details of this conformation change remain to be determined. Because so little is known about how cytoplasmic release factors work, it is not yet possible to compare the bacterial and eukaryotic termination pathways.

23.7 Recycling Stage

The ribosome recycling factor (RRF) is required for the bacterial ribosomal complex to disassemble.

At the end of the termination stage in bacteria, the ribosome complex is still associated with mRNA and a deacylated tRNA, most probably with its acceptor end in the E-site of the 50S subunit and its anticodon end in the P-site of the 30S subunit. During the recycling stage, the ribosomal subunits dissociate from this complex, freeing them to participate in a new round of polypeptide synthesis. In bacteria, this process requires a new translation factor known as the **ribosome recycling factor** (**RRF**).

RRF is an essential protein that was originally thought to bind in the free A-site because its structure is similar to that of tRNA. However, footprinting experiments and cryo-electron microscopy constructs (FIGURE 23.46) show that instead of binding to the free A-site, RRF binds almost at right angles to this site. EF-G•GTP and IF3 somehow assist RRF in disassembling the posttermination complex. The details of this disassembly process remain to be determined. Virtually nothing is known about the recycling stage in eukaryotes. Thus far, no eukaryotic counterpart to the bacterial RRF has been

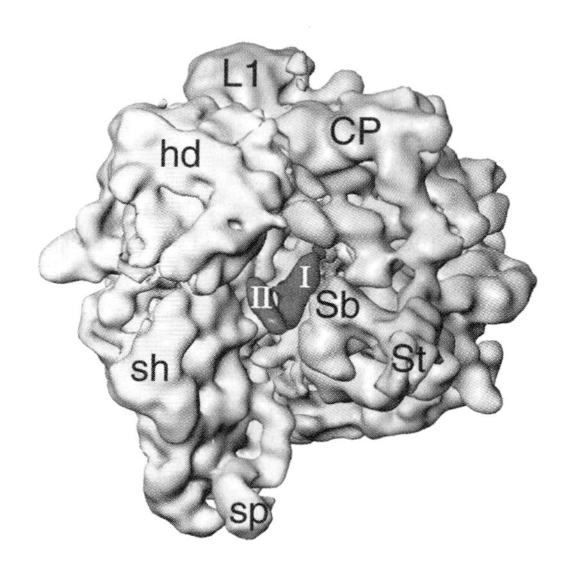

FIGURE 23.46 **Cryo-electron microscopy construct of the 70S ribosome recycling factor (RRF) complex.** The 70S ribosome, showing the shoulder side of the 30S subunit (yellow) on the left, and the L7/L12 stalk side of the 50S subunit (blue) on the right to reveal the RRF (red) binding positions. Roman numerals I and II specify RRF domains .Structural features shown are as follows: hd, head; sh, shoulder; sp, spur; pt, platform; L1, L1-protein protuberance; CP, central protuberance; St, L7/L12 stalk; and Sb, stalk-base. (Reproduced from R. K. Agrawal, et al., *Proc. Natl. Acad. Sci. USA* 101 [2004]: 8900–8905. Copyright 2004, National Academy of Sciences, USA. Photo courtesy of Joachim Frank, HHMI, Columbia University.)

discovered. One possibility is that one or more eukaryotic ribosomal proteins assist in recycling so that a soluble ribosome recycling factor is not required. Following ribosome disassembly, the small ribosomal subunit is free to interact with the initiation factors to start a new round of polypeptide synthesis.

23.8 Nascent Polypeptide Processing and Folding

Ribosomes have associated enzymes that process nascent polypeptides and chaperones that help to fold the nascent polypeptides.

The nascent protein passes through a peptide exit tunnel that extends from the peptide transferase center to the ribosome surface (Figure 23.34). The exit tunnel in the bacterial ribosome is about 80 to 100 Å long and about 10 Å in diameter at its narrowest point (about 30 Å from the peptidyl transferase center) but widens to about twice that diameter at the rim of the exit pore. The exit tunnel can accommodate an α helix with about 60 residues or an extended peptide with about half that number of residues. The space within the exit tunnel may permit the nascent polypeptide chain to assume an α-helical conformation but is too narrow to permit more extensive folding. As the nascent peptide chain emerges from the tunnel, it can interact with enzymes that catalyze co-translational modifications, chaperones that assist in folding and prevent misfolding, and the signal recognition particle that facilitates transport across the cell membrane (see below).

Peptide deformylase and methionine aminopeptidase bind at the rim of the bacterial ribosome's exit pore. The deformylase cleaves the N-terminal formyl group from the nascent polypeptide as it emerges from the exit tunnel and then the aminopeptidase recognizes about 60% of the different nascent polypeptides and removes their N-terminal methionines. Nascent eukaryotic polypeptides do not have an N-terminal formyl group but do begin with methionine. Ribosome-bound methionine aminopeptidases remove the N-terminal methionine.

In bacteria, co-translational protein folding is assisted by a 48 kDa chaperone called the **trigger factor**, which binds at the bacterial ribosome's exit pore. *E. coli* trigger factor is constitutively expressed, resulting in about two or three trigger factor molecules per ribosome. The trigger factor protein transiently associates with the L23 protein on the 50S subunit. Its residence time on the ribosome depends on whether the ribosome has a nascent protein in the exit tunnel. If a nascent protein is not present, the average residence time is about 11 to 15 seconds but this time increases several-fold when a nascent protein is present.

The trigger factor contains three domains that arrange to form a characteristic elongated dragon-shaped structure (FIGURE 23.47). The

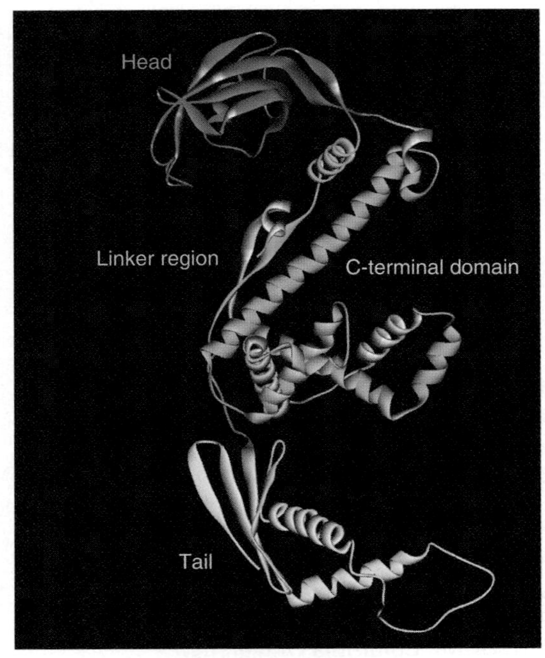

FIGURE 23.47 **Trigger factor.** The N-terminal domain forms the tail that binds to the 50S subunit. The middle domain, which forms the dragon's head, has peptidyl prolyl *cis/trans* isomerase activity. The C-terminal domain forms the central body of the dragon and is responsible for the trigger factor's chaperone activity. (Structure from Protein Data Bank 1W26. L. Ferbitz, et al., *Nature* 431 [2004]: 590–596. Prepared by B. E. Tropp.)

N-terminal domain forms the tail that binds to the 50S subunit. The middle domain, which forms the dragon's head, has peptidyl prolyl *cis/trans* isomerase activity that is not essential for the trigger factor's chaperone function. The C-terminal domain forms the central body of the dragon and is responsible for the trigger factor's chaperone activity. FIGURE 23.48 shows the relationship between the trigger factor and the chaperone and chaperonin systems that were described in Chapter 2. The trigger factor binds to hydrophobic patches as they emerge from the ribosome and sometimes remain associated with the segment even after polypeptide chain completion. Moreover, a single nascent polypeptide chain or free polypeptide may have two or more trigger factors associated with it.

A chaperone also associates with the large subunit of the eukaryotic ribosome. This chaperone consists of three different subunits that differ in both sequence and structure from the trigger factor. In yeast, deletion of any single subunit results in slow growth and cold sensitivity. Protein folding is a very active area of research because it merits attention as the final step in gene expression. Moreover, several neurological disorders including Alzheimer's, Huntington's, and Parkinson's diseases may be due to the accumulation of toxic proteins, which result from misfolding.

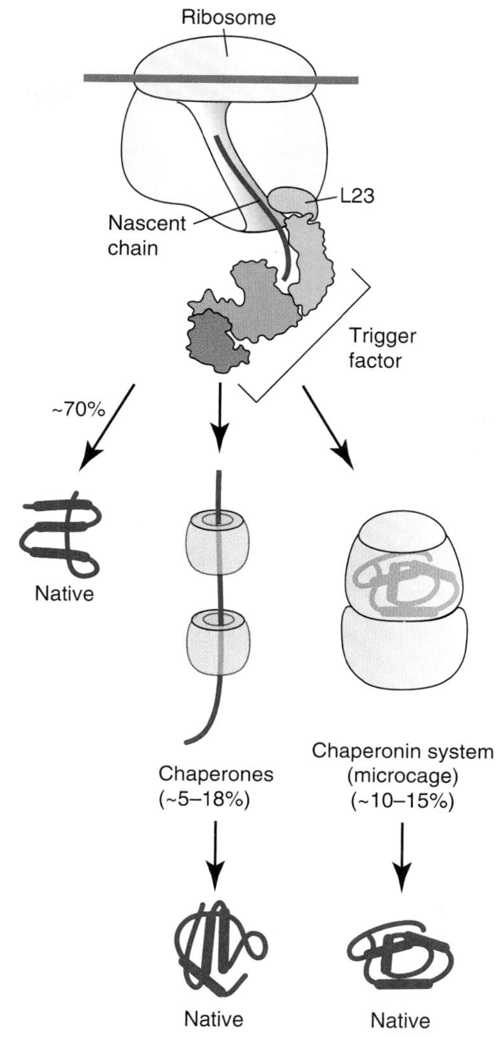

FIGURE 23.48 **Relationship between the trigger factor, other chaperones, and the chaperonin system.** The trigger factor is color coded as in Figure 23.47. (Adapted from A. Hoffmann, et al., *Biochim. Biophys. Acta* 1803 [2010]: 650–661.)

23.9 Signal Sequence

The signal sequence plays an important role in directing newly synthesized proteins to specific cellular destinations.

We now consider another aspect of ribosome function, the ribosome's interaction with the endoplasmic reticulum in eukaryotes and the cell membrane in prokaryotes. Soon after biologists started to investigate ribosomes, they observed that some ribosomes appear to exist free in the cytoplasm while others are bound to the endoplasmic reticulum. The free ribosomes were shown to synthesize cytoplasmic and mitochondrial proteins, whereas the membrane-bound ribosomes synthesized integral membrane proteins, lysosomal proteins, and proteins that were secreted.

The reason why some ribosomes bind to the endoplasmic reticulum was not clear at first. One possibility was that cells have two distinct kinds of ribosomes, those that bind to the endoplasmic reticulum and those that do not. Studies by Gunter Blobel and David D. Sabitini in 1971 indicated that free and bound ribosomes appeared to be the same. Therefore, there must be some other explanation for why some ribosomes are bound to the endoplasmic reticulum while others are not.

Blobel and Sabatini proposed the **signal hypothesis** to explain how cells determine whether a protein will be synthesized on a free or membrane-bound ribosome. According to the signal hypothesis: (1) Free and membrane bound ribosomes are identical. (2) Protein synthesis always begins on free ribosomes. (3) Nascent secretory, transmembrane, or lysosomal proteins have sequences of 20 to 30 amino acids at their amino terminus that act as signals to bind the ribosomes to the endoplasmic reticulum. Ribosomes only bind to the endoplasmic reticulum when synthesizing proteins with signal sequences.

Although signal sequences vary from one protein to another, certain common features can be recognized. Signal sequences can be divided into three parts: a short, positively charged N-terminal region, a central region containing 7 to 13 hydrophobic amino acid residues, and a more polar C-terminal region that includes a cleavage site. Only polypeptides that have a signal sequence can be inserted into the endoplasmic reticulum membrane or transferred across it.

The amino end of a nascent polypeptide with a signal sequence requires the assistance of three components to pass into the lumen of the endoplasmic reticulum or to be integrated into the membrane. These components, which are present in all organisms, are the **signal recognition particle** (**SRP**), the **SRP receptor** (**SR**), and a **protein conducting channel** or **translocon**. In prokaryotes, the cell membrane serves the same purpose that the endoplasmic reticulum does in eukaryotes. We focus our attention on the mammalian SRP cycle because it was the first one to be discovered. The basic properties of the mammalian components required for the SRP cycle are as follows:

1. *Signal recognition particle*. SRP consists of six proteins (SRP9, 14, 19, 54, 68, and 72, named according to their molecular masses) and a 7S RNA. The SRP54 subunit is especially note-

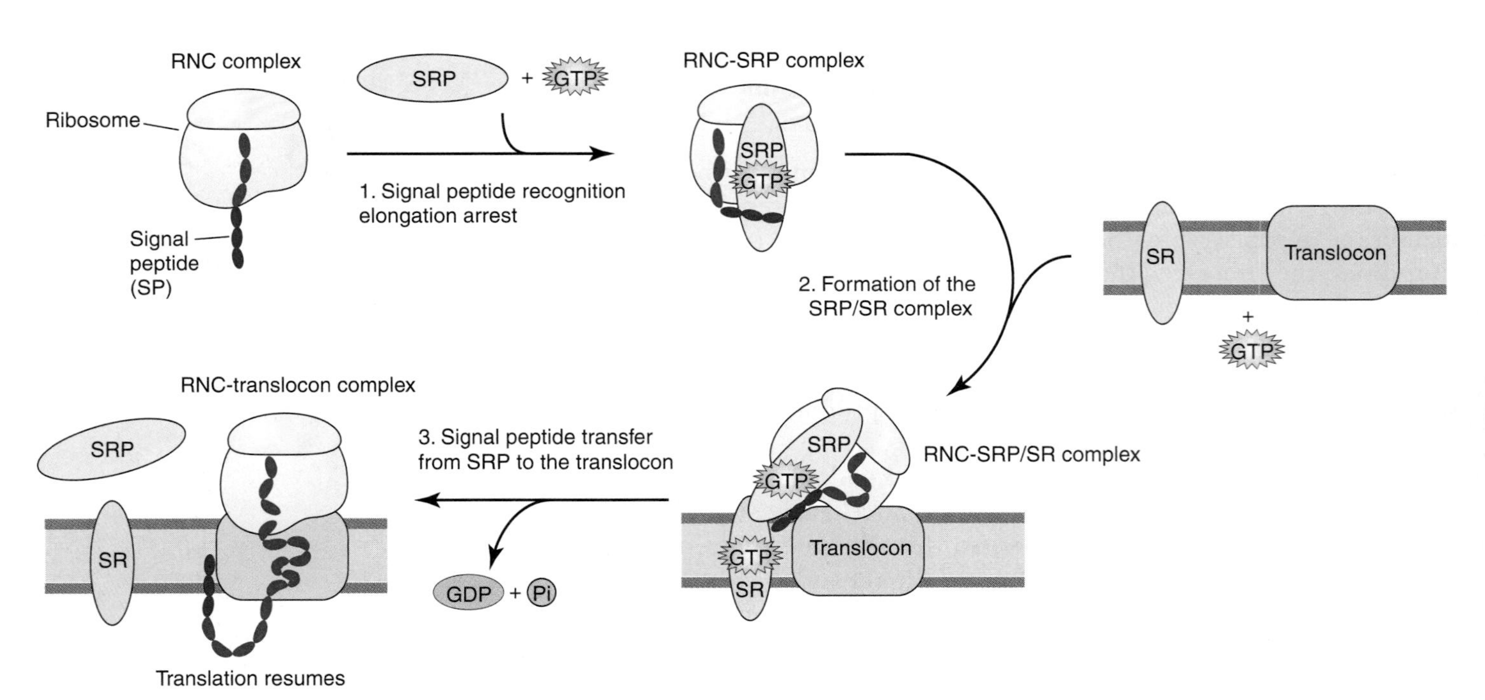

FIGURE 23.49 **Schematic representation of the signal recognition particle (SRP) cycle.** Protein transport begins with the recognition of the signal peptide (SP) on the ribosome. Then the SRP ribosome•RNC (ribosome nascent chain complex) binds to SRP receptor (SR). GTP has to be present in both SRP (SRP54 subunit) and SR for stable SRP•SR complex formation. The signal peptide is transferred from the SRP to the protein conduction channel (translocon). GTP hydrolysis in both SRP and SR leads to the dissociation of the SRP•SR complex. SRP is represented by a blue oval, the SRP receptor is colored green, and the protein conduction channel (translocon) is colored gray. Small and large ribosomal subunits are colored yellow and light blue, respectively. The signal peptide (SP) at the N-terminus of the nascent polypeptide chain is shown in dark blue. (Adapted from J. Luirink and I. Sinning, *Biochim. Biophys. Acta* 1694 [2004]: 17–35.)

worthy because the guanine nucleotide binding site and the signal sequence binding site are located within it.

2. *SRP receptor*. SR contains two GTP-binding polypeptide subunits, a peripheral membrane protein SRα and an integral membrane protein SR .
3. *Translocon*. The protein conducting channel, which is made of three polypeptides (Sec61α, Sec61β, and Sec61γ), acts as a passive conduit for polypeptides.

The three basic steps involved in the SRP cycle are shown in FIGURE 23.49. Although Figure 23.49 shows mammalian components, the process is very similar in all organisms: (1) SRP binds tightly to the signal sequence on the nascent polypeptide chain as the signal sequence emerges from the ribosomal polypeptide exit tunnel. Cryo-electron microscopy studies show that SRP goes through a conformational change as it binds to the **ribosome-nascent chain complex (RNC)**. As a result of this conformational change, SRP binds to both the peptide exit tunnel and elongation-factor-binding site on the ribosome (FIGURE 23.50), which prevents eEF1A and eEF2 from binding to the ribosome and causes translation to be arrested. (2) SRP carries the ribosome–nascent chain complex to the endoplasmic reticulum, where SRP and the membrane-bound SR interact. At this stage, both SRP and SR have bound GTPs. The ribosome docks to the translocon and the signal peptide moves into the protein conducting channel, causing the SRP•SR complex to be released. The ribosome is now free to resume polypeptide chain elongation. The growing polypeptide chain moves through the

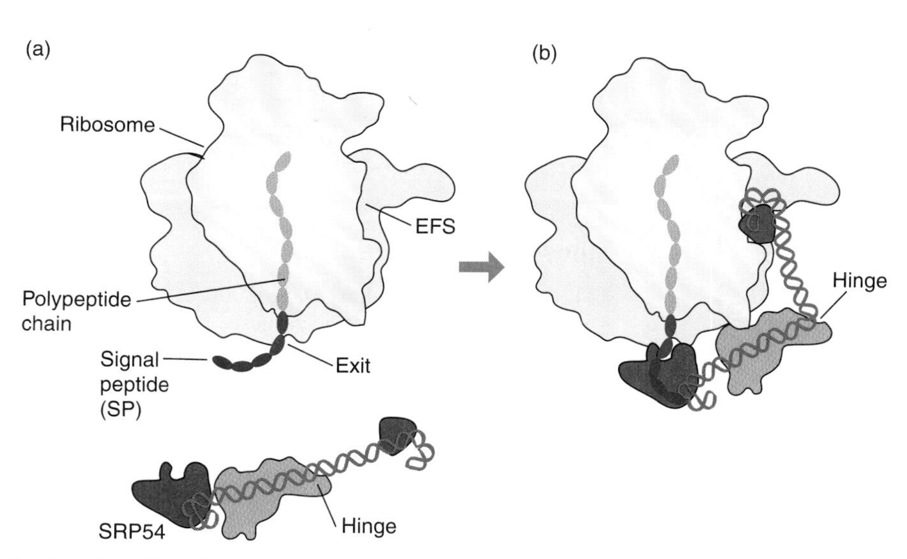

FIGURE 23.50 **Schematic showing the signal-sequence-dependent interaction between the signal recognition particle (SRP) and the ribosome.** (a) Before SRP (red) binds to the ribosome nascent chain complex. (b) After SRP binds to the ribosome nascent chain complex. SRP goes through a conformational change when its SRP54 subunit binds to the signal sequence in the nascent polypeptide chain that appears to involve rotation around a hinged region in the 7S RNA. As a result of this conformational change, SRP interacts with the ribosome, stretching from the peptidyl exit tunnel (Exit) to the elongation-factor binding site (EFS) in the intersubunit space. This interaction causes elongation arrest by competing with elongation factors. The small and large ribosomal subunits are colored yellow and blue, respectively. The signal sequence is colored dark blue. (Adapted from M. Halic, et al., *Nature* 427 [2004]: 808–814.)

protein conducting channel driven by the energy provided by GTP hydrolysis during the translocation stage of the elongation cycle. Upon entering the lumen of the endoplasmic reticulum the signal sequence encounters a specific peptidase called the **signal peptidase** that cleaves it from the remainder of the polypeptide chain. (3) GTPs bound to the SRP•SR complex are hydrolyzed, causing SRP•SR to dissociate so that the cycle can start again.

Secretory and lysosomal proteins pass completely through the endoplasmic reticulum membrane and are directed to their ultimate destination by biochemical modifications such as glycosylation that take place in the lumen of the endoplasmic reticulum and in the Golgi apparatus. Integral membrane proteins have one or more hydrophobic sequences that allow them to move laterally across the translocon and into the membrane, causing the polypeptide chain to embed itself within the membrane.

Suggested Reading

Overview

Green, R., and Noller, H. F. 1997. Ribosomes and translation. *Ann Rev Biochem* 66:679–716.

Kapp, L. D., and Lorsch, J. L. 2004. The molecular mechanics of eukaryotic translation. *Ann Rev Biochem* 73:657–704.

Schmeing, T. M., and Ramakrishnan, V. 2009. What recent ribosome structures have revealed about the mechanisms of translation. *Nature* 461:1234–1242.

Wilson, D. N., Blaha, G., Connell, S. R., et al. 2002. Protein synthesis at atomic resolution: mechanistics of translation in the light of highly resolved structures for the ribosomes. *Curr Protein Pept Sci* 3:1–53.

Historical

Maden, B. E. H. 2003. Historical review: peptidyl transfer, the Monro era. *Trends Biochem Sci* 28:619–624.

Ribosome Structure

Al-Kardaghi, S., and Liljas, O. K. 2000. A decade of progress in understanding the structural basis of protein synthesis. *Progr Biophys Mol Biol* 73:167–193.

Ban, N., Nissen, P., Hansen, J., et al. 1999. Placement of protein and RNA structures into a 5 Å-resolution map of the 50S ribosomal subunit. *Nature* 400:841–847.

Bashan, A., and Yonath, A. 2008. Correlating ribosome function with high-resolution structures. *Trends Microbiol* 16:326–335.

Ben-Shem, A., Jenner, J., Yusupova, G., and Yusupov, M. 2010. Crystal structure of the eukaryotic ribosome. *Science* 330:1203–1209.

Berk, V., and Cate, J. H. D. 2007. Insights into protein biosynthesis from structures of bacterial ribosomes. *Curr Opin Struct Biol* 17:302–309.

Clemons, W. M. Jr., May, J. L. C., Wimberly, B. T., et al. 1999. Structure of a bacterial 30S ribosomal subunit at 5.5 Å resolution. *Nature* 400:833–840.

Culver, G. M. 2001. Meanderings of the mRNA through the ribosome. *Structure* 9:751–758.

Demeshkina, N., Jenner, L., Yusupova, G., and Yusupov, M. 2010. Interactions of the ribosome with mRNA and tRNA. *Curr Opin Struct Biol* 20:1–10.

Frank, J. 2000. The ribosome—a macromolecular machine par excellence. *Chem Biol* 7:R133–R141.

Frank, J. 2001. Structure of the 80S ribosome from *Saccharomyces cerevisiae*—tRNA-ribosome and subunit-subunit interactions. *Cell* 107:373–386.

Frank, J. 2003. Toward an understanding of the structural basis of translation. *Genome Biol* 4:237.

Frank, J., Agrawal, R. K., and Verschoor, A. 2001. Ribosome shape and structure. In: *Encyclopedia of Life Sciences*. London: Nature. pp. 1–6.

Korostelev, A., Ermolenko, D. N., and Noller, H. F. 2008. Structural dynamics of the ribosome. *Curr Opin Chem Biol* 12:1–10.

Korostelev, A., Trakhanov, S., Laurberg, M. et al. 2007. Interactions and dynamics of the Shine–Dalgarno helix in the 70S ribosome. *Proc Natl Acad Sci USA* 104: 16840–16843.

Korostelev, A., and Noller, H. F. 2007. The ribosome in focus: new structures bring new insights. *Trends Biochem Sci* 32:434–441.

Korostelev, A., Trakhanov, S., Laurberg, M., and Noller, H. F. 2006. Crystal structure of a 70S ribosome-tRNA complex reveals functional interactions and rearrangements. *Cell* 126: 1065–1077.

Malhotra, A. 2002. rRNA Structure. In: *Encylopedia of Life Sci.* pp. 1–12. Hoboken, NJ: John Wiley and Sons.

Moore, P. B. 2001. The ribosome at atomic resolution. *Biochemistry* 40:3143–3250.

Moore, P. B. 2009. The ribosome returned. *J Biol* 8:1–8.

Moore, P. B., and Steitz, T. A. 2002. The involvement of RNA in ribosome function. *Nature* 418:229–235.

Nissen, P., Ippolito, J. A., Ban, N., et al. 2001. RNA tertiary interactions in the large ribosomal subunit: the A-minor motif. *Proc Natl Acad Sci USA* 98:4899–4903.

Noller, H. F., Hoang, L., and Frerick, K. 2005. The 30S ribosomal P site: a function of 16S rRNA. *FEBS Lett* 579:855–858.

Puglisi, J. D. 2009. Resolving the elegant architecture of the ribosome. *Mol Cell* 36:720–723.

Ramakrishnan, V. 2002. Ribosome structure and the mechanism of translation. *Cell* 108:557–572.

Ramakrishnan, V. 2008. What we have learned from ribosome structures. *Biochem Soc Trans* 36:567–574.

Ramakrishnan, V. 2010. Unravelling the structure of the ribosome (Nobel Lecture). *Angew Chem Int Ed* 49:4355–4380.

Ramakrishnan, V., and Moore, P. B. 2001. Atomic structures at last: the ribosome in 2000. *Curr Opin Struct Biol* 11:144–154.

Schuwirth, B. S., Borovinskaya, M. A., Hau, C. W., et al. 2005. Structures of the bacterial ribosome at 3.5Å resolution. *Science* 310:827–834.

Selmer, M., Dunham, C. M., Murphy, F. F., et al. 2006. Structure of the 70S ribosome complexed with mRNA and tRNA. *Science* 313:1935–1942.

Spahn, C. M. T., Beckmann, R., Eswar, N., et al. 2002. Ribosome as a molecular machine. *FEBS Lett* 514:2–10.

Steitz, T. A. 2008. A structural understanding of the dynamic ribosome machine. *Nat Rev Mol Cell Biol* 9:242–253.

Steitz, T. A. 2010. From the structure and function of the ribosome to new antibiotics (Nobel Lecture). *Angew Chem Int Ed* 49:4381–4398.

Taylor, D. J., Devkota, B., Huang, A. D., et al. 2009. Comprehensive molecular structure of the eukaryotic ribosome. *Structure* 17:1591–1604.

Tate, W. P., and Poole, E. S. 2004. The ribosome: lifting the veil from a fascinating organelle. *Bioessays* 26:582–588.

Tocilj, A., Schluenzen, F., Hansen, H. A., et al. 1999. The small ribosomal subunit from *Thermus thermophilus* at 4.5 Å resolution: pattern fittings and the identification of a functional site. *Proc Natl Acad Sci* USA 96:14252–14257.

Wimberly, B. T., Brodersen, D. E., Clemons, W. M. Jr., et al. 2000. Structure of the 30S ribosomal subunit. *Nature* 407:327–339.

Yonath, A. 2010. Polar bears, antibiotics, and the evolving ribosome (Nobel Lecture). *Angew Chem Int Ed* 49:4340–4354.

Yusupov, M. M., Yusupova, G. Z., Baucom, A., et al. 2001. Crystal structure of the ribosome at 5.5 Å resolution. *Science* 292:883–896.

Initiation Stage in Bacteria

Boelens, R., and Gualerzi, C. O. 2002. Structure and function of bacterial initiation factors. *Curr Protein Pept Sci* 3:107–119.

Carr-Schmid, A., and Kinzy, T. G. 2001. Messenger RNA: interaction with ribosomes. In: *Encyclopedia of Life Sciences*. London: Nature. pp. 1–6.

Culver, G. M. 2001. Meanderings of the mRNA through the ribosome. *Structure* 9:751–758.

Dever, T. E. 2002. Gene-specific regulation by general translation factors. *Cell* 108:545–556.

Grunberg-Manago, M., Suder, S. M., and Joseph, S. 2007. Protein synthesis initiation in bacteria. In: *Encylopedia of Life Sci*. pp. 1–10. Hoboken, NJ: John Wiley and Sons.

Guthrie, C., and Nomura, M. 1968. Initiation of protein synthesis: a critical test of the 30S subunit model. *Nature* 219:232–235.

Hui, A., and De Boer, H. A. 1987. Specialized ribosome system: preferential translation of a single mRNA species by a subpopulation of mutated ribosomes in *Escherichia coli*. *Proc Natl Acad Sci USA* 84:4762–4766.

Kaempfer, R. O. R., Meselson, M., and Raskas, H. J. 1968. Cyclic dissociation into stable subunits and re-formation of ribosomes during bacterial growth. *J Mol Biol* 31: 277–289.

Laursen, B. S., Sørensen, H. P., Mortensen, K. K., and Sperling-Petersen, H. U. 2005. Initiation of protein synthesis in bacteria. *Microbiol Mol Biol Rev* 69:101–123.

Mandal, M., and Breaker, R. R. 2004. Gene regulation by riboswitches. *Nat Rev Mol Cell* Biol 5:451–463.

Mangroo D, Wu, X., and Rajbhandary, U. L. 1995. *Escherichia coli* initiator tRNA: structure–function relationships and interactions with the translational machinery. *Biochem Cell Biol* 73:1023–1031.

Marcker, K., and Sanger, F. 1964. N-formyl-methionyl-sRNA. *J Mol Biol* 8:835–840.

Marintchev, A., and Wagner, G. 2004. Translation initiation: structures, mechanisms, and evolution. *Q Rev Biophys* 37:197–284.

Marshall, R. A., Aitken, C. E., and Puglisi, J. D. 2009. GTP hydrolysis by IF2 guides progression of the ribosome into elongation. *Mol Cell* 35:37-47.

Matthews, M. B. 2002. Lost in translation. *Trends Biochem Sci* 27:267–269.

Meinnel, T. M., Mechulam, Y., and Blanquet, S. 1993. Methionine as translation start signal: a review of the enzymes of the pathway in *Escherichia coli*. *Biochimie* 75:1061–1075.

Nudler, E. 2006. Flipping riboswitches. *Cell* 126:19–22.

Nudler, E., and Mironov, A. S. 2004. The riboswitch control of bacterial metabolism. *Trends Biochem Sci* 29:11–17.

Ochoa, S. 1968. Translation of the genetic message. *Naturwissenschafter* 55:505–514.

Rajbhandary, U. L. 1994. Minireview: initiator transfer RNAs. *J Bacteriol* 176:547–552.

Simonetti, A., Marzi, S., Jenner, L., et al. 2009. A structural view of translation initiation in bacteria. *Cell Mol Life Sci* 66:423–436.

Simonetti, A., Marzi, S., Myasnikov, A. G., et al. 2008. Structure of the 30S translation initiation complex. *Nature* 455:416–420.

Sperling-Petersen, H. U., Laursen, B. S., and Mortensen, K. K. 2002. Initiator tRNAs in prokaryotes and eukaryotes. In: *Encyclopedia of Life Sciences*. London: Nature. pp. 1–7.

Steitz, J. A., and Jakes, K. 1975. How ribosomes select initiator regions in mRNA: base pair formation between the 3′ terminus of 16S rRNA and the mRNA during initiation of protein synthesis in *Escherichia coli*. *Proc Natl Acad Sci USA* 72:4734–4738.

Vitreschak, A. G., Rodionov, D. A., Mironov, A. A., and Gelfand, M. S. 2004. Riboswitches: the oldest mechanism for the regulation of gene expression? *Trend Genet* 20:44–50.

Yuan, Z., Trias, J., and White, R. J. 2001. Deformylase as a novel antibacterial target. *Drug Disc Today* 6:954–961.

Initiation Stage in Eukaryotes

Drabkin, H. J., Estrella, M., and Rajbhandary, U. L. 1998. Initiator-elongator discrimination in vertebrate tRNAs for protein synthesis. *Mol Cell Biol* 18:1459–1466.

Gingras, A.-C., Raught, B., and Sonenberg, N. 1999. eIF4 initiation factors: effectors of mRNA recruitment to ribosomes and regulators of translation. *Ann Rev Biochem* 68:913–963.

Hellen, C. U. T., and Pestova, T. V. 2006. Translation initiation: Molecular mechanisms in eukaryotes. In: *Encyclopedia of Life Sci,* pp. 1–5. Hoboken, NJ: John Wiley and Sons.

Hellen, C. U. T., and Sarnow, P. 2001. Internal ribosome entry sites in eukaryotic mRNA molecules. *Genes Dev* 15:1593–1612.

Jackson, R. J. 2005. Alternative mechanisms of initiating translation of mammalian mRNAs. *Biochem Soc Trans* 33:1231–1241.

Jackson, R. J., Hellen, C. E., and Pestova, T. V. 2010. The mechanism of eukaryotic translation initiation and principles of its regulation. *Nat Rev Mol Cell Biol* 10:113–127.

Kapp, L. D., and Lorsch, J. R. 2004. The molecular mechanics of eukaryotic translation. *Ann Rev Biochem* 73:657–704.

Kozak, M. 1999. Initiation of translation in prokaryotes and eukaryotes. *Gene* 234:187–208.

Kozak, M. 2001. New ways of initiating translation in eukaryotes? *Mol Cell Biol* 21:1899–1907.

Kozak, M. 2002. Pushing the limits of the scanning mechanism for initiation of translation. *Gene* 299:1–34.

Lorsch, J. R., and Dever, T. E. 2010. Molecular view of 43S complex formation and start site selection in eukaryotic translation initiation. *J Biol Chem* 285:21203–21207.

Merrick, W. C. 2003. Initiation of protein biosynthesis in eukaryotes. *BAMBEd* 31:378–385.

Merrick, W. C. 2010. Eukaryotic protein synthesis: still a mystery. *J Biol Chem* 285:21197–21201.

Mitchell, S. F., and Lorsch, J. R. 2008. Should I stay or should I go? Eukaryotic translation initiation factors 1 and 1A control start codon recognition. *J Biol Chem* 283:27345–27349.

Myasnikov, A. G., Simonetti, A, Marzi, S., and Klaholz, B. P. 2009. Structure-function insights into prokaryotic and eukaryotic translation initiation. *Curr Opin Struct Biol* 19:300–309.

Passmore, L. A., Schmeing, T. M., Maag, D., et al. 2007. The eukaryotic translation initiation factors eIF1 and eIF1A induce an open conformation of the 40S ribosome. *Mol Cell* 26:41–50.

Pestova, T. V., and Hellen, C. U. T. 2001. The structure and function of initiation factors in eukaryotic protein synthesis. *Mol Life Sci* 57:651–674.

Pestova, T. V., Kolupaeva, V. G., Lomakin, I. B., et al. 2001. Molecular mechanisms of translation initiation in eukaryotes. *Proc Natl Acad Sci USA* 98:7029–7036.

Phan, L., Schoenfeld, L. W., Valášek, L., et al. 2001. A subcomplex of three eIF3 subunits binds eIF1 and eIF5 and stimulates ribosome binding of mRNA and $tRNA_i^{Met}$. *EMBO J* 20:2954–2965.

Preiss, T., and Hentze, M. W. 2003. Starting the protein synthesis machine: eukaryotic translation initiation. *Bioessays* 25:1201–1211.

Prévôt, D., Darlix, J.-L., Ohlmann, T. 2003. Conducting the initiation of protein synthesis: the role of eIF4G. *Biol Cell* 95:141–156.

Rodnina, M. V., and Wintermeyer, W. 2009. Recent mechanistic insights into eukaryotic ribosomes. *Curr Opin Cell Biol* 21:435–443.

Roll-Mecak, A., Shin, B. S., Dever, T. E., and Burley, S. K. 2001. Engaging the ribosome: universal IFs of translation. *Trends Biochem Sci* 26:705–709.

Sachs, A. B., and Varani, G. 2000. Eukaryotic translation initiation: there are (at least) two sides to every story. *Nat Struct Biol* 7:356–361.

Scheper, G. C., and Proud, C. G. 2002. Does phosphorylation of the cap-binding protein eIF4E play a role in translation initiation? *Eur J Biochem* 269:5350–5359.

Sonenberg, N., and Dever, T. E. 2003. Eukaryotic translation initiation factors and regulators. *Curr Opin Struct Biol* 13:56–63.

Sonenberg, N., and Hinnebusch, A. G. 2009. Regulation of translation initiation in eukaryotes: mechanisms and biological targets. *Cell* 136:731–745.

Tarun, S. Z. Jr., and Sachs, A. B. 1995. A common function for mRNA 5′ and 3′ ends in translation initiation in yeast. *Genes Dev* 9:2997–3007.

Elongation Stage

Agirrezabala, X., and Frank, J. 2009. Elongation in translation as a dynamic interaction among the ribosome, tRNA, and elongation factors EF-G and EF-Tu. *Q Rev Biophys* 43:159–200.

Agrawal, R. K., Spahn, C. M. T., Penczek, P., et al. 2000. Visualization of tRNA movements on the *Escherichia coli* 70S ribosome during the elongation cycle. *J Cell Biol* 150:447–459.

Andersen, G. R., Nissen, P., and Nyborg, J. 2003. Elongation factors in protein biosynthesis. *Trends Biochem Sci* 28:434–441.

Bashan, A., Agmon, I., Zarivich, R., et al. Structural basis of the ribosomal machinery for peptide bond formation, translocation, and nascent chain progression. *Mol Cell* 11:91–102.

Blaha, G., Stanley, R. E., and Steitz, T. A. 2009. Formation of the first peptide bond: the structure of EF-P bound to the 70S ribosome. *Science* 325:966–970.

Brodersen, D. E., and Ramakrishnan, V. 2003. Shape can be seductive. *Nat Struct Biol* 10:78–80.

Cabrita, L. D., Dobson, C. M., and Christodoulou, J. 2010. Protein folding on the ribosome. *Curr Opin Struct Biol* 20:33–45.

Connell, S. R., Takemoto, C., Wilson, D. N., et al. 2007. Structural basis for interaction of the ribosome with the switch regions of GTP-bond elongation factors. *Mol Cell* 25:751–764.

Connell, S. R., Topf, M., Qin, Y., et al. 2008. A new tRNA intermediate revealed on the ribosome during EF4-mediated back-translocation. *Nat Struct Mol Biol* 15:910–915.

Czworkowski, J. 2001. Elongation factors, bacterial. In: *Encyclopedia of Life Sciences*. London: Nature. pp. 1–6.

Dahlberg, A. E. 2001. Ribosome structure: the ribosome in action. *Science* 292:868–869.

Ehrenberg, M. 2010. Translocation in slow motion. *Nature* 466:325–326.

Etchells, S. A., and Hartl, F. U. 2004. The dynamic tunnel. *Nat Struct Biol* 11:382–391.

Evans, R. N., Blaha, G., Bailey, S., and Steitz, T. A. 2008. The structure of LepA, the ribosomal back translocase. *Proc Natl Acad Sci USA* 105:4673–4678.

Frank, J. 1998. How the ribosome works. *Am Sci* 86:428–439.

Frank, J. 2003. Electron microscopy of functional ribosome complexes. *Biopolymers* 68:223–233.

Frank, J., and Agrawal, R. K. 2001. Ratchet-like movements between the two ribosomal subunits: their implications in elongation factor recognition and tRNA translocation. *Cold Spring Harb Symp Quant Biol* 66:67–75.

Frank, J., Gao, H., Segupta, H., et al. 2007. The process of mRNA-tRNA translocation. *Proc Natl Acad Sci USA* 104:19671–19678.

Gao, Y.-G., Selmer, M., Dunham, C. M., et al. 2009. The structure of the ribosome with elongation factor G trapped in the posttranslocational state. *Science* 326:694–699.

Gilbert, R. J. C., Fucini, P., Connell, S., et al. 2004. Three dimensional structures of translating ribosomes by cryo-EM. *Mol Cell* 14:57–66.

Hoffmann, A., Bukau, B., and Kramer, G. 2010. Structure and function of the molecular chaperone trigger factor. *Biochim Biophys Acta* 1803:650–661.

Horwich, A. 2004. Sight at the end of the tunnel. *Nature* 431:520–522.

Jenni, S., and Ban, N. 2003. The chemistry of protein synthesis and voyage through the ribosomal tunnel. *Curr Opin Struct Biol* 13:212–219.

Krab, I. M., and Parmeggiani, A. 2002. Mechanisms of EF-Tu, a pioneer GTPase. *Prog Nucleic Acid Res Mol Biol* 71:513–551.

Kramer, G., Boehringer, D., Ban, N., and Bukau, B. 2009. The ribosome as a platform of co-translational processing, folding and targeting of newly synthesized proteins. *Nat Struct Mol Biol* 16:589–597.

Joseph, S. 2003. After the ribosome structure: how does translocation work. *RNA* 9:160–164.

Liljas, A. 2009. Leaps in translational elongation. *Science* 326:677–678.

Lucas-Lenard, J., and Lipmann, F. 1966. Separation of three microbial amino acid polymerization factors. *Proc Natl Acad Sci USA* 55:1562–1566.

Márqeuz, V., Wilson, D. N., and Nierhaus, K. H. 2002. RNA-protein machines. *Biochem Soc Trans* 30:133–140.

Merz, F., Boehringer, D., Schaffitzel, C., et al. 2008. Molecular mechanism and structure of trigger factor bound to the translating ribosome. *EMBO J* 27:1622–1632.

Mesters, J. R., Hogg, T., and Hilgenfeld, R. 2001. G proteins. In: *Encyclopedia of Life Sciences*. London: Nature. pp. 1–8.

Moazed, D., and Noller, H. F. 1989. Intermediate states in the movement of transfer RNA in the ribosome. *Nature* 342:142–148.

Monro, R. E. 1967. Catalysis of peptide bond formation by 50S ribosomal subunits from *Escherichia coli*. *J Mol Biol* 26:147–160.

Moore, P. B., and Steitz, T. A. 2003. After the ribosome structures: how does peptidyl transferase work? *RNA* 9:155–159.

Moran, S. J., Flanagan, J. F. IV, Namy, O., et al 2008. The mechanics of translocation: a molecular "spring-and-ratchet" system. *Structure* 16:664–672.

Munishkin, A., and Wool, I. G. 1997. The ribosome-in-pieces: binding of elongation factor EF-G to oligoribonucleotides that mimic the sarcin/ricin and thiostrepton domains of 23S ribosomal RNA. *Proc Natl Acad Sci USA* 94:12280–12284.

Nierhaus, K. H., and Stiezl, U. 2001. Peptidyl transfer on the ribosome. In: *Encyclopedia of Life Sciences*. London: Nature. pp. 1–8.

Nissen, P., Hansen, J., Ban, N., et al. 2000. The structural basis of ribosome activity in peptide bond synthesis. *Science* 289:920–930.

Noller, H. F., Hoffarth, V., and Zimniak, L. 1992. Unusual resistance peptidyl transferase in protein extraction procedures. *Science* 256:1416–1419.

Noller, H. F., Yusopov, M. M., Yusopova, G. Z., et al. 2002. Translocation of tRNA during protein synthesis. *FEBS Lett* 514:11–16.

Ogle, J. M., Carter, A. P., and Ramakrishnan, V. 2003. Insights into the decoding mechanism from recent ribosome structures. *Trends Biochem Sci* 28:259–266.

Pech, M., and Nierhuas, K. H. 2008. Ribosomal peptide-bond formation. *Chem Biol* 15:417–419.

Puglisi, J. D., Blanchard, S. C., and Green, R. 2000. Approaching translation at atomic resolution. *Nat Struct Biol* 7:855–861.

Rodnina, M. V., Daviter, T., Gromadski, K., and Wintermeyer, W. 2002. Structural dynamics of ribosomal RNA during decoding of the ribosome. *Biochimie* 84:745–754.

Rodnina, M. V., Savelsbergh, A., Katunin, V. I., and Wintermeyer, W. 1997. Hydrolysis of GTP by elongation factor G drives tRNA movement on the ribosome. *Nature* 385:37–41.

Rodnina, M. V., and Wintermeyer, W. 2001. Ribosome fidelity: tRNA discrimination, proofreading and induced fit. *Trends Biochem Sci* 26:124–130.

Rodnina, M. V., and Wintermeyer, W. 2001. Fidelity of aminoacyl-tRNA selection on the ribosome: Kinetic and structural mechanisms. *Ann Rev Biochem* 70:415–435.

Schmeing, T. M., Huang, K. S., Kitchen, D. E., et al. 2005. Structural insights into the roles of water and the 2′ hydroxyl of the P site tRNA in the peptidyl transferase reaction. *Mol Cell* 20:437–448.

Schmeing, T. M., Huang, K. S., Strobel, S. A., and Steitz, T. A. 2005. An induced-fit mechanism to promote peptide bond formation and exclude hydrolysis of peptidyl-tRNA. *Nature* 438:520–524.

Schmeing, T. M., Seila, A. C., Hansen, J. L., et al. 2002. A pre-translocational intermediate in protein synthesis observed in crystals of enzymatically active 50S subunits. *Nat Struct Biol* 9:225–230.

Schmeing, T. M., Voorhees, R. M., Kelley, A., et al. 2009. The crystal structure of the ribosome bound to EF-Tu and aminoacyl-tRNA. *Science* 326:688–694.

Simonović, M., and Steitz, T. A. 2009. A structural view on the mechanism of the ribosome-catalyzed peptide bond formation. *Biochim Biophys Acta* 1789:612–623.

Spahn, C. M. T., Beckmann, R., Ewar, N., et al. 2001. Structure of the 80S ribosome from *Saccharomyces cerevisiae*-tRNA-ribosome and subunit-subunit interactions. *Cell* 107:373–386.

Sprinzl, M. Brock, S., Huang, Y., and Miovnik, P. 2000. Regulation of GTPases in bacterial translation machinery. *Biol Chem* 381:367–375.

Stahl, G., McCarty, G. P., and Farabaugh, P. J. 2002. Ribosome structure: revisiting the connection between translational accuracy and unconventional decoding. *Trends Biochem Sci* 27:178–183.

Stark, H., Rodnina, M. V., Wieden, H.-J., et al. 2000. Large-scale movement of elongation factor G and extensive conformational change of the ribosome during translocation. *Cell* 100:301–309.

Stark, H., Rodnina, M. V., Wieden, H.-J., et al. 2002. Ribosome interactins of aminoacyl-tRNA and elongation factor Tu in the codon-recognition complex. *Nat Struct Biol* 9:849–854.

Steitz, T. A., and Moore, P. B. 2000. RNA, the first macromolecular catalyst: the ribosome is a ribozyme. *Trends Biochem Sci* 28:411–418.

Tucker, B. J., and Breaker, R. R. 2005. Riboswitches as versatile gene control elements. *Curr Opin Struct Biol* 15:342–348.

Valle, M., Sengupta, J., Swami, N. K., et al. 2002. Cryo-EM reveals an active role for aminoacyl-tRNA in the accommodation process. *EMBO J* 21:3557–3567.

Valle, M. Zavialov, A., Li, W., et al. 2003. Incorporation of aminoacyl-tRNA into the ribosome as seen by cyro-electron microscopy. *Nat Struct Biol* 10:899–906.

Valle, M., Zavialov, A., Sengupta, J., et al. 2003. Locking and unlocking of ribosomal motions. *Cell* 114:123–134.

Voorhes, R. M., Weixlbaumer, A., Loakes, D., et al. 2009. Insights into substrate stabilization from snapshots of the peptidyl transferase center of the intact 70S ribosome. *Nat Struct Mol Biol* 16:528–533.

Voss, N. R., Gerstein, M., Steitz, T. A., and Moore, P. B. 2006. The geometry of the ribosomal polypeptide exit tunnel. *J Mol Biol* 360:893–906.

Weinger, J. S., Kitchen, D., Scaringe, S. A., et al. 2004. Solid phase synthesis and binding affinity of peptidyl transferase transition state mimics containing 2′-OH at P-site position A76. *Nucleic Acids Res* 32:1502–1511.

Weinger, J. S., Parnall, K. M., Dorner, S., et al. 2004. Substrate-assisted catalysis of peptide bond formation by the ribosome. *Nat Struct Mol Biol* 11:1101–1106.

Youngman, E. M., Brunelle, J. L., Kochaniak, A. B., and Green, R. 2004. The active site of the ribosome is composed of two layers of conserved nucleotides with distinct roles in peptide bond formation and peptide release. *Cell* 117:589–599.

Termination Stage in Bacteria

Brodersen, D. E., and Ramakrishnan, V. 2003. Shape can be seductive. *Nat Struct Biol* 10:78–80.

Cappechi, M. R, 1967. Polypeptide chain termination *in vitro*: isolation of a release factor. *Proc Natl Acad Sci USA* 58:1144–1151.

Caskey, C. T., Tompkins, R., Scolnick, E., et al. 1968. Sequential translation of trinucleotide codons for the initiation and termination of protein synthesis. *Science* 162:135–138.

Ehrenberg, M., and Tenson, T. 2002. A new beginning of the end of translation. *Nat Struct Biol* 9:85–87.

Haebel, P. W., Gutmann, S., and Ban, N. 2004. Dial tm for rescue: tmRNA engages ribosomes stalled on defective mRNAs. *Curr Opin Struct Biol* 14:58–65.

Herrington, M. B. 2003. Nonsense mutations and suppression. In: *Encyclopedia of Life Sciences*. London: Nature. pp. 1–6.

Ito, K., Uno, M., and Nakamura, Y. 2000. A tripeptide 'anticodon' deciphers stop codons in messenger RNA. *Nature* 403:680–684.

Karzai, A. W., Roche, E. D., and Sauer, R. T. 2000. The SsrA–SmpB system for protein tagging, directed degradation and ribosome rescue. *Nat Struct Biol* 7:449–455.
Kisselev, R. L., and Buckingham, R. H. 2000. Translation termination comes of age. *Trends Biochem Sci* 25:561–566.
Kisselev, R. L., Ehrenberg, M., and Frolova, L. 2003. Termination of translation: interplay of mRNA, rRNAs and release factors? *EMBO J* 22:175–182.
Kjeldgaard, M. 2003. The unfolding story of polypeptide release factors. *Mol Cell* 11:8–10.
Klaholz, B. P., Pape, T., Zavialov, A. V., et al. 2003. Structure of the *Escherichia coli* ribosomal termination complex with release factor 2. *Nature* 421:90–94.
Korostelev, A., Asahara, J., Lancaster, L., et al. 2008. Crystal structure of a translation termination complex formed with release factor RF2. *Proc Natl Acad Sci USA* 105:19684–19689.
Laurberg, M., Asahara, H., Korostelev, A, et al. 2008. Structural basis for translation termination on the 70S ribosome. *Nature* 454:852–857.
Moore, S. D., McGinness, K. E., and Sauer, R. T. 2003. A glimpse into tmRNA-mediated ribosome rescue. *Science* 300:72–73.
Nakamura, Y., and Ito, K. 2003. Making sense of mimic in translation termination. *Trends Biochem Sci* 28:99–105.
Rawat, U. B. S., Zavialov, A. V., Sengupta, J., et al. 2003. A cryo-electron microscopic study of ribosome-bound termination factor RF2. *Nature* 421:87–90.
Scolnick, E., Tompkins, R., Caskey, T., and Nirenberg, M. 1968. Release factors differing in specificity for terminator codons. *Proc Natl Acad Sci USA* 61:768–774.
Tate, W. P., Poole, E. S., and Mannering, S. A. 2001. Protein synthesis termination. In: *Encyclopedia of Life Sciences*. London: Nature. pp. 1–6.
Valle, M., Gillet, R., Kaur, S., et al. 2003. Visualizing tmRNA entry into a stalled ribosome. *Science* 300:127–130.
Vestergaard, B., Van, L. B., Andersen, G.R., et al. 2001. Bacterial polypeptide release factor RF2 is structurally distinct from eukaryotic eRF1. *Mol Cell* 8:1375–1382.
Weixlbaumer, A., Jin, H., Neubauer, C., et al. 2008. Insights into translational termination from the structure of RF2 bound to the ribosome. *Science* 352:953–956.
Withey, J. H., and Friedman, D. I. 2003. A salvage pathway for protein synthesis: tmRNA and trans-translation. *Ann Rev Microbiol* 57:101–123.
Zavialov, A. V., and Ehrenberg, M. 2003. Peptidyl-tRNA regulates the GTPase activity of translation factors. *Cell* 114:113–122.
Zavialov, A. V., Mora, L., Buckingham, R. H., and Ehrenberg, M. 2002. Release of peptide promoted by the GGQ motif of class 1 release factors regulates the GTPase activity of RF3. *Mol Cell* 10:789–778.

Termination Stage in Eukaryotes

Frolova, L., Le Goff, X., Rasmussen, H. H et al. 1994. A highly conserved eukaryotic protein family possessing properties of polypeptide chain release factor. *Nature* 372:701–703.
Inge-Vechtomova, S., Zhouravleva, G., and Philippe, M. 2003. Eukaryotic release factors (eRFs) history. *Biol Cell* 95:195–209.
Song, H., Mugnier, P., Das, A. K., et al. 2000. The crystal structure of human eukaryotic release factor eRF1—mechanism of stop codon recognition and peptidyl-tRNA hydrolysis. *Cell* 100: 311–321.
Zhouravleva, G., Frolova,L., Le Goff, X., et al. 1995. Termination of translation in eukaryotes is governed by two interacting polypeptide chain release factors, eRF1 and eRF3. *EMBO J* 14:4065–4072.

The Recycling Stage

Agrawal, R. K., Sharma, M. R., Kiel, M. C., et al. 2004. Visualization of ribosome-recycling factor on the *Escherichia coli* 70S ribosome: functional implications. *Proc Natl Acad Sci USA* 101:8900–8905.

Hirokawa, G., Demeshkina, N., Iwakura, N., et al. 2006. The ribosome-recycling step: consensus or controversy. *Trends Biochem Sci* 31:143–149.

Kim, K. K., Min, K., and Suh, S. W. 2000. Crystal structure of the ribosome recycling factor from *Escherichia coli*. *EMBO J* 19:2362–2370.

The Signal Sequence

Adler, N. N., and Johnson, A. E. 2004. Cotranslational membrane protein biogenesis at the endoplasmic reticulum. *J Biol Chem* 279:22787–22790.

Blobel, G., and Dobberstein, B. 1975. Transfer of proteins across membranes. I. Presence of proteolytically procecessed and unprocessed nascent light immunoglobulin light chains on membrane-bound ribosomes of murine myeloma. *J Cell Biol* 67:835–851.

Halic, M., Becker, T., Pool, M. R., Spahn, C. M. T., et al. 2004. Structure of the signal recognition particle interacting with the elongation-arrested ribosome. *Nature* 427:808–814.

Halic, M., and Beckmann, R. 2005. The signal recognition particle and its interactions during protein targeting. *Curr Opin Struct Biol* 15:116–125.

Herkovits, A. A., Bochkareva, E. S., and Bibi, E. 2000. New prospects in studying the bacterial signal recognition particle pathway. *Mol Microbiol* 38:927–939.

Luirink, J., and Sinning, I. 2004. SRP-mediated protein targeting: structure and function revisited. *Biochim Biophys Acta* 1694:17–35.

Ryan, M. T., and Pfanner, N. 2001. Protein translocation across membranes. In: *Encyclopedia of Life Sciences*. London: Nature. pp. 1–10.

Wild, K., Rosendal, K. R., and Sinning I. 2004. A structural step into the SRP cycle. *Mol Microbiol* 53:357–363.

Wild, K., Weichenrieder, O., Strub, K., et al. 2002. Towards the structure of the mammalian signal recognition particle. *Curr Opin Struct Biol* 12:72–81.

This book has a Web site, **http://biology.jbpub.com/book/molecular**, which contains a variety of resources, including links to in-depth information on topics covered in this chapter, and review material designed to assist in your study of molecular biology.

Index

A
abl, 229
Abortive initiation, 639
A box, 920
Abrescia, Nicola G. A., 756
Absolute defective mutants, 262–263
Abzymes, 88
Acceptor stem, of tRNA, 975
Accommodation, in protein synthesis, 1032
Ac/Ds system, 600
Ac element, 599–600
ACF (ATP-utilizing chromatin assembly and remodeling factor) chromatin remodeling complex, 441
Acquired immunodeficiency syndrome (AIDS), 355, 356
ACS (ARS consensus sequence), 424
Actinomycin D, RNA polymerase sensitivity to, 727*t*, 727–728, 728*f*
Activation domains (ADs), 769, 769*f*, 797–805
  of Gal4, 798*f*–801*f*, 798–803, 803*f*
  mechanisms of, 797–798
    conformational change model of, 798
    recruitment model of, 798
  two-hybrid assay for detecting polypeptides interacting through non-covalent interactions, 804*f*, 804–805
Activation-induced cytidine deaminase (AID), 563, 563*f*
Activators, 678. *See also* Transcription activator proteins
  in positive regulation, 664*f*, 665, 665*t*
  σ54-RNA polymerase requirement for, in initiation stage of transcription, 644*f*, 645
Activator sites (ASs), cAMP•CRP complex binding to, 680–682
Active site, of enzymes, 89
AD(s). *See* Activation domains (ADs)
Adaptive hypothesis, of codon assignments, 1000–1001
Adaptor hypothesis, 969, 969*f*, 972, 972*f*
ADAR (adenosine deaminase acting on RNA), 882
Adducts, 458
  methyl
    DNA exposure to methylmethane sulfonate and, 458, 459
    DNA exposure to N-methyl-N-nitrosourea and, 458, 459
Adenine
  in A-, B-, and Z-forms of DNA, 8*t*
  in DNA, 6, 6*f*
  8-oxoguanine base pairs with, 456, 456*f*
  water-mediated deamination of, 453
Adenosine, 7*f*
Adenosine deaminase acting on RNA (ADAR), 882
Adenosine deaminase deficiency, 561
Adenosine triphosphate. *See* ATP (adenosine triphosphate)
Adenoviruses, 352–355
  discovery of, 352
  DNA of
    replication of, 354*f*, 354–355
    structure of, 352–353, 353*f*
  genome map of, 352, 352*f*
  as model systems, 355
  as vectors, 355
Adenylate cyclase, glucose metabolism and, 678–679, 679*f*, 680*f*
Adenylyl cyclase, as effector in G protein-linked signal transduction systems, 97*f*, 97*t*, 97–98
Adhya, Sankar, 685
a/α diploid cells, 287
A-DNA, 19, 19*f*, 133–135
  B form compared with, 134*f*, 134–135, 135*t*
  identification of, 133–134
  Z form compared with, 135*t*
A-element, 424
Affinity chromatography, 40, 40*f*
  DNA, transcription activator protein purification using, 770–772, 771*f*
Affinity probes, 646
Aflatoxins, 460, 461*f*
Afo 2, 946
Agar, 259–260
Agarose gel, for slab gel electrophoresis, 160–161, 161*f*, 162
Agarose plugs, in pulsed-field gel electrophoresis, 164
Aging, telomeres and, 246
Ago1, 956
Agrobacterium tumefaciens
  DNA structure in, 211
  Ti plasmid produced by, 284
Ahringer, Julie, 950
AID (activation-induced cytidine deaminase), 563, 563*f*
AIDS (acquired immunodeficiency syndrome), 355, 356
Ala-374, 983
Alanine
  abbreviations for, 33*t*
  side chains of, 31, 32*f*
  tRNA for, 972–973
Alberts, Bruce, 400
Alkali, DNA denaturation by, 117–118
Alkaptonuria, 4
AlkB protein, methyl group removal catalyzed by, 476, 477*f*
Alkylating agents, DNA damage caused by, 457–461
  alkyl group transfer to centers of negative charge and, 457–458, 458*f*, 459*f*
  modification of environmental agents to become alkylating agents and, 458–460, 460*f*, 461*f*
Alkyl groups, transfer to centers of negative charge, alkylation DNA damage due to, 457–458, 458*f*, 459*f*
Alkylguanine DNA alkyltransferase, 476
O[6]-Alkylguanine DNA alkyltransferase I, 474–476, 475*f*
O[6]-Alkylguanine DNA alkyltransferase II, 476
Alleles, 262
  segregation of, 8, 513, 515, 515*f*
Allfrey, Vincent, 813
Allis, C. David, 814
Allolactose, as lactose operon inducer, 670–671, 671*f*
Allosteric effectors, 80
  regulatory enzyme inhibition and stimulation by, 93*f*, 93–94
"Allosteroid" model of RNA polymerase II transcription termination, 878
α-subunit, of core polymerase, 393*f*, 393–394
Alternative splicing, 725, 883
Altmann, Richard, 3
Alu elements, 614, 615*f*, 616
AluI, sequence specificity of, 169*t*
Alzheimer's disease, 1053
α-Amanitin, RNA polymerase sensitivity to, 727, 727*t*, 728*f*
amber, 996
Ames, Bruce, 465
Ames test, 465–466, 504
Amide bonds, 33. *See also* Peptide bonds
Amino acid(s), 29
  α-, 29*f*, 29–33
    abbreviations for, 32, 33*t*
    grouping of, based on side chains, 30–31, 30*f*–32*f*
    peptide bonds linking, 33*f*–35*f*, 33–35
  abbreviations for, 32, 33*t*
  attachment to tRNA, 966–967, 968*f*, 969, 969*f*
    through ester bond, 971*f*, 971–972, 972*f*
  basic, 30, 30*f*
  deprivation of, guanine nucleotides inhibiting rRNA synthesis and, 710*f*, 710–711
  eukaryotic regions in genes coding, interruption by noncoding regions, 841–842, 843*f*, 844*f*, 844–845
  order in polypeptides, 9
  sequences of. *See also* Protein structure, primary
    in DNA, 9
    in polypeptides, 29
Amino acid residues, 34
Aminoacyl-tRNAs
  binding to more than one codon, 1000, 1000*f*, 1001*f*
  coding specificity of, 999*f*, 999–1000

Page numbers followed by *f* or *t* indicate material in figures or tables, respectively.

Aminoacyl-tRNA synthetases (aminoacyl-tRNA ligases), 967, 968*f*, 969, 977–987
binding domains of, 978*f*, 978–979
binding to acceptor arm, 979, 980*f*
classes of, 977–979, 978*f*–980*f*, 978*t*
conserved sequence motifs of, 979, 979*f*
editing functions of, 979–983, 980*f*, 981*f*
in gram-positive bacteria and archaea, 985*f*, 985–986
polypeptide building blocks and, 986*f*, 986–987
tRNA recognition by, 983*f*, 983–985
Amino groups, in α-amino acids, 29
Aminopterin, hybrid cell selection and, 297, 298*f*
A-minor motif, 1010, 1010*f*
AMV (avian myeloblastosis virus), reverse transcriptase of, 202
Anacystis nidulans, photolyase•DNA complex of, 472, 473*f*
Anaphase
of meiosis
anaphase I, 225*f*, 227
anaphase II, 224*f*, 227
of mitosis, 220*f*, 221
Anemia
Fanconi, 537
sickle cell, 80–83, 81*f*, 82*f*
Aneuploidy, 229
Anfinsen, Christian, 62–63
Animal models, for human diseases, 24–25. *See also specific models and diseases*
Annealing. *See also* Reannealing
in Southern blotting, 175
Annotation, of human genome, 194
Antennapedia complex, 774, 774*f*
Antibiotics
overuse of, 283
resistance to, 15, 283
targeting bacterial RNA polymerases, 654–656
Anticodon arm, of tRNA, 975
Anti conformation, 137, 137*f*
Antigens, 83
Antiparallel pleated sheets, 54
Antisense RNA strands, 939
Anti-Shine-Dalgarno (ASD) sequence, 1018
Anti-sigma factors, 644
Anti-Smith (anti-Sm) antibodies, 856
Antiterminator, 695
"Antiterminator" model of RNA polymerase II transcription termination, 878
AP endonuclease, 480, 480*f*, 481
AP-1 proteins, 792–793, 793*f*
A protein, synthesis of, in RNA phages, 334
AP (apurinic and apyrimidinic) sites, 453, 480
Aptamers, 981
Apurinic and apyrimidinic (AP) sites, 453, 480
Aquifex aeolicus, DnaA from, 383*f*, 383–384, 384*f*
araA gene, 687
araBAD operon, 687, 687*f*
araBAD promoter (P_{BAD}), repression by AraC protein, 687–688, 690
araB gene, 687
Arabidopsis, 600
Arabidopsis thaliana, TATA-binding protein from, 745, 746*f*
L-Arabinose, bacterial metabolism of, 686–691, 687*f*, 689*f*, 690*f*
Arabinose•AraC complex, mRNA transcription initiation and, 688
araC gene, 687
AraC protein
bacterial homologs of, 690
binding to *araI*$_1$ and *araO*$_2$, 689–690
negative autoregulation by, 686–687
structure of, 688–689, 689*f*
araD gene, 687
araE gene, 687
araF gene, 687
araFGH operon, 687
araG gene, 687
araH gene, 687
Arb1 protein, 956
Arb2 protein, 956
ARC (Argonaute siRNA chaperone), 956
Archaea, 211
DNA replication in, 441–443
elongation stage of, 442–443
Orc1/CdC6 in, 442, 442*f*
similarity to bacterial replication, 442–443
similarity to eukaryotic replication, 441–443
Gln-tRNA synthetase synthesis in, 985*f*, 985–986
hyperthermic, reverse gyrase of, 132
initiation stage of protein synthesis in, 1012, 1029
ribosome structure in, 1011, 1011*f*
RNA polymerases of, 623
RNA polymerase II subunits and, 730, 730*t*, 731
Arginine
abbreviations for, 33*t*
side chains of, 30, 30*f*
Argonaute proteins, 940
crystal structure for, 944–946, 945*f*
domains of, 944, 944*f*
subfamilies of, 944
Argonaute siRNA chaperone (ARC), 956
Arrest, of RNA polymerase movement, 651
ARS306, 430–431
ARS consensus sequence (ACS), 424
ARS (autonomously replicating sequence) elements, DNA chain initiation in yeast and, 422–425, 423*f*, 424*f*
Artemis, 555*t*, 561
AS(s) (activator sites), cAMP•CRP complex binding to, 680–682
Asci, 287, 513
ASD (anti-Shine-Dalgarno) sequence, 1018
Asf1 histone chaperone, in chromatin disassembly, 440
A-site (aminoacyl-tRNA binding site), 1010, 1010*f*
Asparagine
abbreviations for, 33*t*
side chains of, 31, 31*f*
Aspartic acid (aspartate)
abbreviations for, 33*t*
side chains of, 30, 30*f*
Aspergillus flavus, aflatoxins from, 460, 461*f*
Aspergillus oryzae, S1 endonuclease of, 166
Aspergillus parasiticus, aflatoxins from, 460, 461*f*
Astrachan, Lazarus, 987
Asymmetrical internal loops, 142, 142*f*
Ataxia-telangiectasia-like disorder (ATLD), 533
ATLD (ataxia-telangiectasia-like disorder), 533
ATP (adenosine triphosphate)
hydrolysis of, phage T4 packaging motor function and, 324
topoisomer conversion and, 132, 132*f*
ubiquitin attachment to proteins and, 100–102, 101, 102*f*
ATP-dependent chromatin remodeling complexes, 810–813, 811*f*, 812*f*
ATP-utilizing chromatin assembly and remodeling factor (ACF) chromatin remodeling complex, 441
A-tracts, DNA bending and, 112
AT-rich region, of SV40, 416
A/T state, 1030
attB site, 609
attB1 site, 610
attB2 site, 610
Attenuator, of *trpE,* 693, 693*f*, 694
attL1 site, 610
attL2 site, 610
attP site, 609
attP1 site, 610
attP2 site, 610
attR1 site, 610
attR2 site, 610
attTn7, 592–594
Autoimmune diseases, anti-Smith (anti-Sm) antibodies in, 855–856
Autonomously replicating sequence (ARS) elements, DNA chain initiation in yeast and, 422–425, 423*f*, 424*f*
Autonomous mobile elements, 594
Autoradiography, 124
Autoregulation, negative, by AraC protein, 687
Autosomes, 228
Auxotrophs, 4, 259, 261
AvaI, sequence specificity of, 169*t*
Avery, Oswald, 13–14
Avian myeloblastosis virus (AMV), reverse transcriptase of, 202
Aviemore model of recombination, 522–524, 523*f*

B

Bacilli, morphology of, 257, 257*f*
Bacillus stearothermophilus
DnaB helicase from, 386, 386*f*
ribosome of, crystal structure for, 1008
Bacillus subtilis
DNA replication in, 406
bidirectional, 369, 371*f*
lack of Rho factor in, 654
mRNA degradation pathway used by, 703
subtilisin and, 179
tRNA genes of, 972
BAC libraries, 190
BACs (bacterial artificial chromosomes), in hierarchical shotgun sequencing, 190–191, 191*f*
Bacteria, 211
cell membrane of, 257, 258*f*, 258–259
cell wall of, 257
lysozyme treatment of, 259
chromatin of, 211, 212–217
chromosome compaction and, 214–217, 215*f*, 216*f*
location in nucleoid, 212*f*, 212–214, 213*f*
lysis to remove, 212
chromosomes of
compaction of, 214–217, 215*f*, 216*f*
F plasmid integration into, 276, 277*f*, 278
culturing of, 259–261, 260*f*
DNA of. *See* Bacterial DNA; Bacterial DNA replication
genetics of. *See* Bacterial genetics
Gram-negative, 257–258, 258*f*
Gram-positive, 257–259, 258*f*
Gln-tRNA synthetase synthesis in, 985*f*, 985–986
growth rate of, growth medium composition and, 711
homologous recombination in, 526–533
RecA in, 528*f*, 528–529, 530*f*
RecBCD complex in, 529, 531*f*, 531–532, 532*f*

RecFOR pathway of, 532, 533*f*
recombination mutant conjugation rates and sensitivity to DNA damage and, 526–527, 527*t*
initiation stage of protein synthesis in, riboswitches regulating, 1020, 1020*f*
morphology of, 257, 257*f*
mRNA of. *See* Bacterial mRNA
nucleotide excision DNA repair in, 482–485, 483*f*, 484*f*, 486*f*, 487*f*
protection against phage infection, 312–313, 313*t*
replication forks in, stalled, 550
ribosome structure in, 706–707, 707*f*, 1007–1010, 1007*f*–1010*f*
RNA polymerases of. *See* Bacterial RNA polymerases
rRNA of. *See* Bacterial rRNA
σ factors of, 644. *See also* σ factors
termination stage of protein synthesis in, 1043–1050
protein release factors and, 1043–1046, 1044*f*–1046*f*
rescue of stalled ribosomes by tmRNA and, 1046–1047, 1048*f*, 1049*f*
suppression of mutations creating termination codons within reading frames and, 1050
transcription in. *See* Bacterial transcription
transposons of, 581–582, 583–586
composite, 583–584
as molecular genetics tools, 594–598, 595*f*, 598*f*
targeting DNA sequences, 591
Bacterial artificial chromosomes (BACs), in hierarchical shotgun sequencing, 190–191, 191*f*
Bacterial colonies, 260*f*, 260–261
Bacterial DNA
circular molecules of, detection of, 123–124
compaction of, 213–217
MukB and, 214–217, 215*f*, 216*f*
nucleoid-associated proteins other than MukB and, 217, 217*f*
fragmentation of, 110
isolation of, 151–152
replication of. *See* Bacterial DNA replication
superhelical, underwinding of, 125
Bacterial DNA replication, 365–409
bidirectional, 369, 370*f*, 371*f*
conservative model of, 366–367, 367*f*
discontinuous model of, 370, 372*f*, 372–375, 373*f*
dispersive model of, 367, 367*f*
DNA ligase in, 375–376, 376*f*, 377*f*
elongation stage of, 379, 388–405
β clamp in, 394–396, 395*f*
clamp loader in, 396–400, 396*f*–399*f*
core polymerase in, 392–394, 393*f*, 403–404
DNA polymerase holoenzyme in, 391*f*, 391–392, 392*t*
DNA polymerase III in, 389–390
enzymes involved in, 388–389
helicase movement in, 404–405, 405*f*
polymerase processivity and, 390–391
replisome in, 400–401, 401*f*, 402*f*
enzymes involved in, 388–389, 400
eukaryotic replication compared with, 421–422
genes essential for, 379, 380*t*
initiation stage of, 379, 381–388
DnaA•ATP in, 384–385, 385*f*
DnaA in, 383*f*, 383–384, 384*f*
DnaC in, 386, 387*f*
DnaG in, 388, 388*f*
DnB helicases in, 385–386, 386*f*
at *oriC*, 381–383, 382*f*
regulation of, 408–409
replicon model of, 381, 381*f*
mutant studies of, 379–380, 380*t*
Okazaki fragment synthesis in, 376, 378*f*
regulation of, 408–409
semiconservative model for, 366*f*–368*f*, 366–369
semidiscontinuous model of, 372, 372*f*–375*f*, 375
similarity to archaeal replication, 442–443
termination stage of, 379, 405–408
termination sites and, 405, 405*f*
terminus utilization substance in, 406, 406*f*, 407*f*, 408
trombone model of, 400
in vitro studies of, 376, 379
Bacterial genetics, 257–284
bacterial structures and, 257*f*, 257–259, 258*f*
complementation in, 268–270, 269*t*
culturing bacteria and, 259–261, 260*f*
E. coli studies in, 270–273, 271*f*–273*f*
genetic map and, 278, 279*f*, 280, 281*f*
F plasmids and
clustering of replication control functions in basic replicon, 282–283
genes needed for successful mating and DNA transfer and, 274*f*, 274–276, 275*f*
integration into bacterial chromosomes, 276, 277*f*, 278
part of bacterial chromosome contained in, 280, 282, 282*f*
mutants and, 262–268
classification on basis of changes in DNA, 263–264, 264*f*
display of mutant phenotype and, 262–263
mutagens and, 263
regaining of original phenotype by, 265
uses in molecular biology, 265–268
notations, conventions, and terminology used in, 261–262
R plasmids and, advantageous properties conveyed to hosts by, 283–284
Bacterial mRNA
codons in, 662
of *E. coli*, 653–654
5′-untranslated region (5′-leader) of, 663
monocistronic, 662
open reading frames of, 662
polycistronic, 662–663
spacers in, 663
synthesis of
basal, 671
coordinate regulation of, 664–665
rate of, 664*f*, 664–665, 665*t*
3′-untranslated region (3′-leader) of, 663
transcription of, 662–665
negative and positive regulation of, 664*f*, 665, 665*t*
rate of mRNA synthesis and, 664–665
Bacterial RNA polymerases, 622–656, 623
antibiotics targeting, 654–656
crystal structures of, 633–636, 634*f*–636*f*
in initiation stage of transcription, 638–645
activator protein requirement of σ[54]-RNA polymerase and, 644*f*, 645
conserved domains of σ[70] family and, 638, 638*f*
DNA footprinting technique and, 630–633, 631*f*–633*f*
DNase protection method and, 629*f*, 629–630, 631*f*
σ factors in, 643*t*, 643–644
inchworming model and, 640
promoters and, 628–629, 636–638, 637*f*
RNA polymerase crystal structures and, 633–636, 634*f*–636*f*
scrunching of, 639*f*, 639–640
stepwise nature of, 640, 641*f*, 642
transient excursion model and, 640
requirements for RNA synthesis and, 624, 625*f*, 626, 626*f*
size of, 627
structure of, 626–627, 627*f*, 627*t*
subunits of, 626, 627*t*
in transcription elongation complex, 645–652, 646*f*, 647*f*
detection and removal of incorrectly incorporated nucleotides by, 651, 651*f*, 652*f*
forward movement during nucleotide addition cycles, 648–649, 649*f*, 650*f*
pauses and, 649, 650*f*, 651
transcription termination and, 652–654, 652*f*–655*f*
intrinsic, 652*f*, 652–653
Rho-dependent, 653–654, 654*f*, 655*f*
Bacterial rRNA, 707–709
transcription of, 706–713
E. coli rRNA operons and, 707–708, 708*f*
Fis protein binding sites and, 709
growth rate and, 711*t*, 711–712, 712*t*
guanine nucleotide inhibition of, 710*f*, 710–711
promoter upstream element and, 708*f*, 708–709, 709*f*
ribosome structure and, 706–707, 707*f*
r-protein synthesis regulation and, 712–713
rRNA operons and, 707–708, 708*f*
Bacterial transcription, 661–718
araBAD operon and, 686–691, 687*f*, 689*f*, 690*f*
in bacteriophage λ, 696–703
CI regulator maintenance of lysogenic state and, 699–702, 699*f*–702*f*
control of lysogenic pathway and, 699
control of lytic pathway and, 697–698, 698*f*
regulation of phage development and, 696–697, 697*f*
ultraviolet light activation of RecA and, 702–703
catabolite repression in, 677–684
cAMP•CRP complex binding to activator sites and, 680–682, 681*f*, 683*f*
cAMP•CRP operon activation and, 682, 684
E. coli use of glucose in preference to lactose and, 677, 677*f*
galactose operon and, 684–686, 685*f*, 686*f*
inhibitory effect of glucose on lac operon expression and, 678–680, 678*f*–680*f*
eukaryotic transcription compared with, 734–735, 735*f*
lactose operon and, 665–677
allolactose as inducer of, 670–671, 671*f*
lac operators of, 673*f*, 673–677
Lac repressor and, 672*f*, 672*t*, 672–673, 673*f*, 674–677
lacZ, *lacY*, and *lacA* genes and, 665*f*, 665–667, 666*f*
operon model and, 669, 670*f*
regulation of *lac* mRNA and, 667–669, 668*t*
regulation of *lac* structural genes and, 667, 667*f*
of mRNA, 662–665
coordinate regulation of, 664–665
negative and positive regulation of, 664*f*, 665, 665*t*
rate of mRNA synthesis and, 664–665

mRNA degradation and, 703–705, 704*f*, 706*f*
ribosome synthesis regulation and, 710–713
of rRNA, 706–713
E. coli rRNA operons and, 707–708, 708*f*
Fis protein binding site and, 709
growth rate and, 711*t*, 711–712, 712*t*
guanine nucleotide inhibition of, 710*f*, 710–711
promoter upstream element and, 708*f*, 708–709, 709*f*
ribosome structure and, 706–707, 707*f*
r-protein synthesis regulation and, 712–713
rRNA operons and, 707–708, 708*f*
rRNA transcript processing and, 662, 713–718
tRNA transcript processing and, 662, 713–718
tryptophan operon and, 691–696, 691*f*–695*f*
Bacterial viruses. *See* Bacteriophage(s); *specific bacteriophages*
Bacteriophage(s), 308–348. *See also specific phage names*
bacterial protection against, 312–313, 313*t*
chronic, 311, 342, 344–345, 344*f*–347*f*, 347–348
plaque formation by, 312
discovery of, 308
DNA isolation in, 151
DNA replication in
phage λ, 336, 337, 338*f*, 339
phage P1, 341
phage φX174, 329–330, 330*f*
phage T4, 317–320
phage T7, 328
filamentous, 309*f*, 310
homologous recombination in, 517–518, 518*f*, 519*f*
icosahedral head with tail, 309, 309*f*, 310
icosahedral tailless, 309, 309*f*, 310, 332, 332*f*
life cycles of, 310–311
chronic infection, 311
lysogenic, 310–311, 341–342
lytic, 310, 341
of phage T4, 317, 318*f*
as model systems, 308–309
naming conventions for, 309
plaque formation by, 311*f*, 311–312
RNA, 332*f*, 332–334, 333*f*
assembly of, 334
attachment to F pilus, 332
discovery of, 332
life cycle of, 333*f*, 333–334
lysis of, 334
A protein synthesis in, 334
replication of, 333*f*, 333–334
structure of, 332, 332*f*
translation of, 332–333
speed of multiplication of, 311
structure of, 309*f*, 309–310
temperate, 311, 334–342
growth of, 311–312
lambdoid, 334
phage λ, 334*f*–337*f*, 334–341
phage P1, 341*f*, 341–342
T-even and T-odd, 309
transduction and
generalized, 342, 343*f*
specialized, 340
to treat bacterial diseases, 308
virulent, 313–334
counting of particles, 311
life cycle of, 310
φX174, 328*f*, 328–332, 330*f*, 331*f*
with single-stranded RNA as genetic material, 332*f*, 332–334, 333*f*
T4, 313–315, 314*f*–316*f*, 317–325, 318*f*, 319*t*
T7, 325–328, 326*f*, 327*f*, 327*t*
Bacteriophage f1, 342
infection by, 344–345, 346*f*, 346–347, 347*f*
structure of, 342, 344
Bacteriophage f2, 332
Bacteriophage fd, 342
infection by, 344–345, 346*f*, 346–347, 347*f*
structure of, 342, 344
Bacteriophage λ, 334–341, 696–703
assembly of, 339
as cloning vector, 340*f*, 340–341, 609*f*, 609–611, 610*f*
cohesive ends (*cos* elements) of, 335, 336*f*
conservative site-specific recombination in, 609*f*, 609–611, 610*f*
DNA of
circularization of, 335, 336*f*
cutting of, 170, 170*f*
molecule size of, 110*t*
rolling cycle replication of, 337, 339
short single-stranded region of, 374*f*, 374–375, 375*f*
supercoiling of, 335, 336*f*
θ replication of, 336, 337, 338*f*, 339
gene locations of, 334, 335*f*
homologous recombination in, 517–518, 518*f*, 519*f*
lysogenic cycle of, 335–336, 337*f*, 339–340
environmental factors influencing entry into, 696–697, 697*f*
maintenance by CI regulator, 699–702, 699*f*–702*f*
transcription cascade controlling, 699
lytic cycle of, 335, 336–337, 337*f*, 338*f*, 339
environmental factors influencing entry into, 696–697, 697*f*
transcription cascade controlling, 697–698, 698*f*
ultraviolet light and entry into, 702–703
mutant in study of, 267
structure of, 334*f*, 334–335
transcription in, 696–703
CI regulator maintenance of lysogenic state and, 699–702, 699*f*–702*f*
control of lysogenic pathway and, 699
control of lytic pathway and, 697–698, 698*f*
regulation of phage development and, 696–697, 697*f*
ultraviolet light activation of RecA and, 702–703
Bacteriophage M13, 342
as cloning vector, 348
genetic map and gene products of, 344, 345*f*
infection by, 344–345, 346*f*, 346–347, 347*f*
structure of, 342, 344, 344*f*
Bacteriophage MQβ, 332
Bacteriophage MS2, 332, 332*f*
Bacteriophage P1, 341–342
DNA of, replication of, 341
in generalized transduction, 342, 343*f*
injection process of, 341
lysogenic cycle of, 341–342
lytic cycle of, 341
structure of, 341, 341*f*
Bacteriophage φX174, 328–332
DNA of
encapsidation of, 329–330
packaging of, 330–331, 331*f*
replication of, 329–330, 330*f*
transcription of, 330
infection by, 329, 331–332
lysis of, 330–331, 331*f*
size of, 328
structure of, 328, 328*f*
Bacteriophage R17, 332
Bacteriophage T2, blender experiment and, 16*f*, 16–17
Bacteriophage T4, 309, 313–325
assembly of, 320, 321*f*, 322*f*
DNA ligase of, 375
DNA of, 314–325
circular permutation of, 314–315, 315*f*, 317
composition of, 314, 314*f*
degradation to dNMP, 317–319
glucosylation of, 320
5-hydroxymethylcytosine synthesis and, 319
nucleotide source and, 317–319
packaging of, 320–321, 323–325, 323*f*–325*f*
prevention of cytosine incorporation into, 319–320
replication of, 317–320
terminal redundancy of, 314–315, 315*f*, 317
injection process of, 316*f*, 317
life cycle of, 317, 318*f*
Bacteriophage T4 DNA packaging motor, 323*f*, 323–325, 325*f*
Bacteriophage T7, 325–328
DNA of
composition of, 325–326
molecule size of, 110*t*
replication of, 328
transcription of, 326–328
enzymes of, size of, 627
genetic map of, 326, 327*f*
infection by, 328
life cycle of, 326–327, 327*t*
RNA polymerase of, 328
structure of, 325, 326*f*
Bacteriophage T7 helicase, 121 , 121*f*
Bacteriophage T7 helicase/primase, negative staining of, 155*f*
Baker, Tania, 388–389
Baker's yeast. *See Saccharomyces cerevisiae*
Balb-c cell line, 297
Baldwin, Ann Norris, 980
Baltimore, David, 201, 359, 794
BamHI
crystal structure of, 169*f*, 169–170
DNA cutting by, 171, 171*f*
sequence specificity of, 169*t*
Basal synthesis, 671
Basal transcription, 745
Basal transcription factors. *See also specific factors*
RNA polymerase II and, 743–752
lack of requirement for TFIIA and, 748*f*, 748–749
preinitiation complex formation and, 745*f*, 745–746, 746*f*
sequential binding of RNA polymerase II•TFIIF complex, TFIIE, and TFIIH and, 751–752, 752*f*
TFIIB in conversion of closed promoter to open promoter and, 749–751, 749*f*–751*f*
TFIID and core promoters of protein-coding genes lacking a TATA box and, 746–748, 747*f*
transcription of naked DNA from specific transcription start sites and, 743–744, 744*t*
Base(s)
DNA, 6*f*, 6–7, 7*f*
composition of, 15–16
Hoogsteen base pairing and, 139

mutant classification on basis of changes in, 263–264, 264*f*
recognition in major and minor grooves, 110–111, 110*f*–112*f*
reverse-Hoogsteen pairing and, 139
sequences causing DNA bending, 112
stacking of, 115–116, 116*f*, 117
in Watson-Crick model, 20, 21*f*
RNA, secondary structure and, 141–142, 142*f*–143*f*
Base excision repair (BER), 477–481, 478*f*–482*f*
by DNA glycolase action, 478–480, 479*f*, 480*f*
by DNA glycolase/lyase action, 480–481, 481*f*, 482*f*
long patch, 481, 482*f*
short patch, 481, 481*f*
Baserga, Susan J., 917
Basic region leucine zipper (bzip) proteins, DNA-binding domains with, 791–794, 791*f*–795*f*
Basic replicon, 282–283
Bates, David, 382
Bateson, William, 512
B cells, 299
differentiation into plasma cells, 299, 301
B-DNA, 19, 19*f*
A form compared with, 134*f*, 134–135, 135*t*
local structure of, 112
Z form compared with, 135*t*, 136
Beadle, George, 4
Belasco, Joel G., 704
Bell, Stephen P., 388–389, 426
Bennett, J. Claude, 558
Bennett, Nicholas J., 344–345
Bensimon, Aaron, 899–900
Bentley, David, 837
Benzo[a]pyrene, metabolic activation of, cytochrome P450 and, 460, 460*f*
BER. *See* Base excision repair (BER)
Berg, Paul, 971, 972, 980
Berger, James M., 383, 386, 653
Berger, Susan M., 870
Beta clamp, 392, 394–396, 395*f*
core polymerase release from, 403–404
crystal structure of, 394, 395, 395*f*
faces of, 395
placement around DNA by clamp loader, 396–399, 396*f*–399*f*
Beta-conformation (sheets), 54, 54*f*
antiparallel pleated, 54, 54*f*
parallel pleated, 54, 54*f*
Beta-dimer, 394–396, 395*f*
core polymerase release from, 403–404
crystal structure of, 394, 395, 395*f*
faces of, 395
placement around DNA by clamp loader, 396–399, 396*f*–399*f*
Beta-turn(s), 55, 55*f*
BglHI, sequence specificity of, 169*t*
BHK cell line, 297
bHLH activator proteins, 794–795
bHLH zip proteins, 795–797
Bidirectional DNA replication, 369, 370*f*, 371*f*
Bilayers, 66, 68, 68*f*
Biochemical pathways, enzyme regulation of, 91–92, 92*f*, 93*f*
Bioinformatics, 846
Biological membranes
bilayers of, 66
fluid mosaic model of, 68–71, 69*f*, 70*f*
protein interaction with lipids in, 66–68, 67*f*, 68*f*
Biotin, in Southern blotting, 176
Bipartite model, of DNA damage recognition, 490
Bird, Adrian, 816
Birnstiel, Max L., 172
Bishop, J. Michael, 357
2,3-Bisphosphoglycerate (BPG), hemoglobin release of oxygen and, 78, 79*f*, 80
Bisulfite, cytosine deamination and, 455
Bithorax complex, 774, 774*f*
Blackburn, Elizabeth H., 244, 432–434
BLAP75 protein, 534*t*
Blasco, Maria, 439
BLAST program, 45–46
Blender experiment, 16*f*, 16–17
BLM DNA helicase, 551
BLM protein, 534*t*, 538
Blobel, Gunter, 1053–1054
Bloom syndrome, 532
Blunt ends, 168, 168*f*
B lymphocytes, 299
differentiation into plasma cells, 299, 301
BNA bridging proteins, 217
Bochkarev, Alexey, 417
Böck, August, 986
Bodnar, Andrea G., 438
Body, of ribosome, 1007
Bohr, Christian, 78
Bohr effect, 78, 78*f*, 80
Bombyx mori, LINEs in, 613
Bootsma, Dirk, 488
Botstein, David, 293
Boyce, Richard, 482
BPG (2,3-bisphosphoglycerate), hemoglobin release of oxygen and, 78, 79*f*, 80
Brachet, Jean, 964
Branching nucleotides, 855
Branch migration, 519
in RecA-mediated strand exchange, 529
Branch point binding protein, 863, 863*f*
Branchpoint sequence, 851
BRCA2 gene (human breast cancer susceptibility gene 2), 537
mutant, 551, 551*f*
BRCA1 protein, 534*t*
BRCA2 protein, 534*t*, 537
BRE (TFIIB recognition element), 742, 743
Break-induced replication, 552, 552*f*
Breast cancer
BRCA2 gene and, 537, 551, 551*f*
treatment of, 785–786
Brennecke, Julius, 955
Brenner, Sydney, 381, 937, 988–989
Brewer, Bonita J., 425
Brh2 protein, 534*t*, 537
Broth, 259
Bulges, in helical tracts, 142, 142*f*
Buratkowski, Stephen, 745
Burgers, Peter M. J., 430
Burgess, Richard, 628
Burkitt lymphoma, 561
Burley, Stephen K., 749, 796
Bzip (basic region leucine zipper) proteins, DNA-binding domains with, 791–794, 791*f*–795*f*

C

CAAT box and enhancer binding protein (C/EBP), 792
Cabbage looper moth, *piggyBac* transposon of, 604–605
Caenorhabditis elegans
exons of, 846, 846*f*
introns of, 846, 846*f*
as model for study of multicellular animals, 937*f*, 937–938, 938*f*
RNA polymerases of, 948
SID-1 of, 949
transcription activator proteins of, 776–777
CAF-1 histone chaperone, nucleosome assembly and, 441
CAGE (cap analysis of gene expression), 739–741, 739*f*–741*f*
Cairns, John, 123–124, 369, 389
cAMP (3′, 5′-cyclic adenylate), in catabolite repression, 678
Campbell, Keith, 819
cAMP•CRP complex, 678
binding to activator site, 680–682
crystal structure of, 681, 681*f*
lac operon transcription activation by, 682, 683*f*
mRNA transcription and, 688
cAMP receptor protein (CRP), 678, 682
Cancer
aflatoxins causing, 460
BRCA2 gene and, 537, 551, 551*f*
of breast, treatment of, 785–786
carcinogen detection and, 464–466
of colon, 498
environmental exposures causing, 458–459
multiple myeloma, 301
Myc overexpression and, 796–797
nonfunctional mismatch repair system and, 498
psoralen causing, 463–464
telomerase and, 246, 438, 439
in xeroderma pigmentosum, 488
Z-DNA and, 137
Cancer chemotherapy
for breast cancer, 785–786
cisplatin for, 464, 464*f*
nitrogen mustard gas for, 462
Cantor, Charles, 163
CAP (catabolite activator protein), 678
Cap(s), methylguanosine, of mRNA, 831–840, 833*f*–835*f*
attachment to nascent pre-mRNA, 835–836, 836*f*
carboxyl terminal domain and, 837–840, 838*f*, 839*t*
discovery of, 832–834
formation of, 836–837
functions of, 834–835, 835*f*
structure of, 834, 834*f*
Cap 0, 834
Cap 1, 834
Cap 2, 834
Cap analysis of gene expression (CAGE), 739–741, 739*f*–741*f*
Cap binding complex (CPC), 835
Capecchi, Mario, 545, 546, 1043–1044
Capillary gel electrophoresis, 163, 164*f*
Capsomers, 352–353, 353*f*
Carbomycin, 1036
Carboxyl groups, in α-amino acids, 29
Carboxyl terminal domains (CTDs)
α, RNA polymerase holoenzyme contact with UP element through, 709
Mediator and, 806–808
phosphorylation of, 837–840, 838*f*, 839*t*
splicing and, 871
proteins involved in function of, 806–807
Carcinogens, detection of, 464–466
Carninci, Pierre, 739, 741
Carrier, William, 482
Caskey, C. Thomas, 1043–1044
Caspersson, Torbjörn, 8, 964
Cassette exons, 847, 847*f*
Catabolite activator protein (CAP), 678

Catabolite repression, 677–684
cAMP•CRP complex binding to activator sites and, 680–682, 681*f*, 683*f*
cAMP•CRP operon activation and, 682, 684
E. coli use of glucose in preference to lactose and, 677, 677*f*
galactose operon and, 684–686, 685*f*, 686*f*
inhibitory effect of glucose on lac operon expression and, 678–680, 678*f*–680*f*
Catalysts, RNA as, 144–145
Catalytic proteins. *See* Enzyme(s)
Catalytic residues, 89
Catalytic subunits, of regulatory proteins, 94, 94*f*
Catananes, 128
Catecholamine, Jamie H. D., 1008
Cations, RNA tertiary structure stabilization by, 142
CBC•RNA complex, 861
C-box, 920
CBP20, 835
CBP80, 835
CCAAT box, 765–767
functions performed by, 766–767
CCA nucleotidyl transferase, 926
ccbB gene, 610
CDAR (cytidine deaminase acting on RNA), 882
CDC6 gene, 427
Cdc45•MCM2-7•GINS complex, 428
Cdc45 protein, 428
in eukaryotic DNA replication, 428
Cdc6 protein, in eukaryotic DNA replication, 427, 427*f*
Cdc7 protein, in eukaryotic DNA replication, 427, 428*f*
CDK (cyclin-dependent protein kinase), 752
in eukaryotic DNA replication, 427, 428*f*
Cdk8 module, 806
cDNA (complementary DNA), 202*f*, 202–203
Cdt1 protein, in eukaryotic DNA replication, 427, 427*f*
C/EBP (CAAT box and enhancer binding protein), 792
Cech, Thomas R., 434, 912–914
Celera Genomics, 192
Cell counts, 297
Cell density, 260
Cell lines. *See also* HeLa cells
established, 296, 297
Cellular respiration, 455
C_2′-endo conformation, 136*f*, 137
C_3′-endo conformation, 136*f*, 137
CENP-A (centromere protein A), 242
Centimorgans (cMs), 513
Central dogma theory, 22–24, 23*f*, 964
Centrifugal techniques, 156–160
equilibrium density gradient centrifugation, 158–160, 159*f*, 160*f*
sucrose gradient centrifugation, 157–158
velocity sedimentation, 156–158, 157*f*, 158*f*
zonal centrifugation, 157*f*, 157–158, 158*f*
Centromere(s), 219, 220*f*
chromatin contained in, 242
as microtubule attachment site, 241–242, 241*f*–243*f*
position of, 227–228
Centromere protein A (CENP-A), 242
Cernunnos•XLF complex, 555*t*, 561
c-fos gene, 793–794
c-Fos protein, 793, 793*f*
Chain termination DNA sequencing method, 186*f*, 186–188, 187*f*
Chain termination stage of protein synthesis, 1012
Chambon, Pierre M., 725, 726, 843, 844–845
Chan, Michael K., 986
Chaperones
ARC, 956
FACT, 439–441
histone, 439–441, 440*f*
molecular, 64–65, 65*f*, 66*f*
trigger factor and, 1053, 1053*f*
Chaperonin system, 65, 1053, 1053*f*
Chapeville, François, 999
Chargaff, Erwin, 15–16
Charge repulsion, α-helix and, 53–54
Chase, Martha, 16–17
CHD chromatin remodeling complexes, 810
Chemical cross-linking agents, DNA damage caused by, 461–464
DNA strand separation blockage and, 461–462, 462*f*
intra- and interstrand cross-link formation and, 464, 464*f*
monoadduct or cross-link formation and, 462*f*, 462–464, 463*f*
Chen, Xiaojian S., 417
Chimeric mice, 547
ChIP. *See* Chromatin immunoprecipitation (ChIP)
chi sites, 531–532
Chloroplasts, transcription in, 930–932, 931*f*
CHO cell line, 297
Choo, K. H., 242
Chp1, 956
CHRAC (chromatin assembly complex) chromatin remodeling complex, 441
Chromatids
p arm of, 227
q arm of, 227
sister, 219, 220*f*
cohesin tethering of, 221–222, 222*f*, 223*f*, 226
Chromatin
bacterial, 211, 212–217
chromosome compaction and, 214–217, 215*f*, 216*f*
location in nucleoid, 212*f*, 212–214, 213*f*
lysis to remove, 212
in centromeres, 242
condensation to form chromosomes, 219
electron micrograph of, 232, 232*f*
eukaryotic, 211
disassembly and reassembly in DNA replication, 439–441, 440*f*
formation by DNA molecules, 810
forms of, 218, 218*f*
factors associated with, influencing recombination, 529–540
histone classes in, 231, 231*t*
loading cohesin onto, 222, 223*f*
in nucleosomes, 232*f*, 232–233, 233*f*
remodeling of, 810–813, 811*f*, 812*f*
structure of, variations among, 211
Chromatin assembly complex (CHRAC) chromatin remodeling complex, 441
Chromatin immunoprecipitation (ChIP), 808–809, 809*f*
Mediator studies using, 809, 809*f*
monitoring spliceosome assembly using, 862
nucleosome distribution patterns studies using, 812–813, 813*f*
technique of, 808–809, 809*f*
Chromatin modifiers, 439, 440*f*
Chromatin remodelers, 441
Chromatin remodeling complexes, 540
Chromatography
affinity, 40, 40*f*
DNA, transcription activator protein purification using, 770–772, 771*f*
column, 36*f*, 36–37
gel filtration, 38, 39*f*, 40
high-performance liquid, 37
hydrophobic, 38, 38*f*
ion-exchange, 37*f*, 37–38
molecular exclusion, 38, 39*f*, 40
reverse phase, 38, 38*f*
Chromatosome, 236
Chromophore factor, 470–471, 471*f*
Chromosomal anomalies, 229, 230*f*
radiation-induced, 452
Chromosome(s), 210–246
bacterial
compaction of, 214–217, 215*f*, 216*f*
F plasmid integration into, 276, 277*f*, 278
centromeres of, 219, 220*f*, 241–242, 241*f*–243*f*
chromatin contained in, 242
as microtubule attachment site, 241–242, 241*f*–243*f*
position of, 227–228
chromatin and. *See* Chromatin
condensed, stabilization by condensins and topoisomerase II, 239–241, 240*f*, 241*f*
cytogenetics and, 227
diploid number of, 218, 219, 220*f*–223*f*, 221–222, 226
haploid number of, 217, 224*f*–225*f*, 226–227
karyotype and, 227–231
chromosome site nomenclature and, 227–228, 228*f*, 229*f*
information available from, 228*f*, 228–230, 230*f*, 231*f*
in meiosis, 224*f*–225*f*, 226–227
Mendel's laws of inheritance and, 3
in mitosis, 219, 220*f*–223*f*, 221–222, 226
nucleosomes and. *See* Nucleosome(s)
sex, 228
structure of, scaffold model of, 238–239, 239*f*, 240
telomeres of, 243–246, 244*f*, 246*f*
30-nm fibers from
solenoid model of, 237, 237*f*, 238
zigzag model of, 237*f*, 237–238
Chromosome painting, 230
Chronic bacteriophages, 311, 342, 344–345, 344*f*–347*f*, 347–348
chronic infection life cycle of, 311
plaque formation by, 312
Chronic infection life cycle of bacteriophages, 311
Chronic myelogenous leukemia, 229
CI dimer, 701
Ciechanover, Aaron, 99–100
cI gene, 703
CII regulator, lysogenic pathway of phage λ and, 699
Circular DNA
covalent, 123*f*, 123–127, 124*f*, 126*f*
superhelicity (supercoiling) of, 125–128, 126*f*, 127*f*
nicked, 124, 124*f*
Circular permutation, of phage T4 DNA, 314–315, 315*f*, 317
CI regulator
phage λ lysogenic state maintenance by, 699–702, 699*f*–702*f*
synthesis of, 699
transcription initiation inhibition by, 701–702, 702*f*
cis-dominant mutations, 270
cis isomers, 33, 34, 34*f*
Cisplatin
as chemotherapeutic agent, 464, 464*f*
DNA damage caused by, 464, 464*f*
cis-splicing, 850

Cistrons
genes differentiated from, 662
in mRNA, 662–663
c-Jun•c-Fos heterodimer, 793, 794*f*
c-jun gene, 793–794
c-Jun protein, 793, 793*f*
β Clamp, 392, 394–396, 395*f*
core polymerase release from, 403–404
crystal structure of, 394, 395, 395*f*
faces of, 395
placement around DNA by clamp loader, 396–399, 396*f*–399*f*
Clamp loader, 392
minimal, 397–399, 397*f*–399*f*
sliding clamp placement by, 396–399, 396*f*–399*f*
subunits of, 399–400
Clark, Alvin J., 526–527
Class 1 cytoplasmic release factor (eRF1), 1050, 1050*f*
Class 2 cytoplasmic release factor (eRF3), 1050, 1050*f*
Classical zinc fingers, 778–780
Class I cAMP-dependent promoters, 681, 685, 686*f*
Cleaver, James E., 488
Clones, 260*f*, 260–261
Cloning, 174
phage λ as vector for, 340*f*, 340–341, 609*f*, 609–611, 610*f*
phage M13 as vector for, 348
phagemids as vectors for, 348
Closed promoter complex, 632–633, 633*f*
in transcription initiation, 640, 641*f*
cMs (centimorgans), 513
c-Myc gene, 820*f*, 820–821
c-myc gene, 561, 796
c-Myc transcription regulator, 796, 796*f*
Coaxial stacking, 142
Cocci, morphology of, 257, 257*f*
Cockayne syndrome, 492, 493
Coding strand, 624, 626, 626*f*
Codons, 24, 987–989, 988*f*
aminoacyl-tRNA binding to, 1000, 1000*f*, 1001*f*
codon-anticodon helix sensing and, 1032–1034, 1033*f*
in mRNA, 662
as polypeptide chain termination signals, 996, 996*f*
sequence assignment for, 994*f*, 994–995
start, bacterial, 1012–1013
termination, mutant tRNA suppression of mutations creating, 1050
Cohesins, 219, 221–222, 222*f*, 223*f*, 226
loading onto chromatin, 222, 223*f*
Cohesive ends, of phage λ, 335, 336*f*
Coiled coil structure, of Gcn4 leucine zipper, 791, 791*f*
Cold. *See* Temperature
ColE1 plasmid, replication of, 369
Colicins, 284
"Collision release" pathway, 403–404
Colon cancer, 498
Col plasmids, 284
Column chromatography, 36*f*, 36–37
Combinatorial control, 770, 770*f*
Combinatorial diversity, 559
Committed steps, in enzyme regulation, 92, 92*f*
Complementary DNA (cDNA), 202*f*, 202–203
Complementation, to determine number of genes responsible for phenotype, 268–270, 269*t*
Complex retroviruses, 357
Composite transposons, 583–584, 584*f*
Conaway, Ronald C., 753
Concatemers, formation of, 320–321, 323*f*
Condensins, stabilization of condensed chromosomes by, 239–240, 240*f*
Conditional mutants, 263
β-Conformation (sheets), 54, 54*f*
antiparallel pleated, 54, 54*f*
parallel pleated, 54, 54*f*
Conjugation, in *E. coli*, 270–273, 271*f*–273*f*
genetic map produced by, 278, 279*f*, 280, 281*f*
number of F factors genes need for, 274*f*, 274–276, 275*f*
Conjugative plasmids, 276
Connective polypeptide 1 (CP1), 982
Consensus sequences, 630, 631*f*
ARS, 424
defining human introns, 852, 852*f*
Conservative model, of replication, 366–367, 367*f*
Conservative site-specific recombination, 576, 576*f*, 607–612
in bacteriophage λ, 609*f*, 609–611, 610*f*
in integron system, 611, 611*f*
proteins in, 607*f*, 607–608, 608*f*
Constant region domains, 84
Constitutive enzymes, 667
Constitutive exons, 847
Constitutive proteins, in centromeres, 242
Contact inhibition, 296
Contigs, 189
Cooke, Howard, 438
Cooperative interaction, among oxygen-binding sites on hemoglobin molecule, 78
Coordinate regulation, 664–665
Corces, Victor, 850
Core polymerase, 392. *See also* DNA polymerase III holoenzyme
in bacterial DNA replication, 389–390
release from β clamp, 403–404
subunits of, 393–394, 394*f*
Core promoters, 741, 741*f*, 742–743, 743*f*, 900–902, 901*f*
of *E. coli*, 708
eukaryotic, 741
with TATA box
site of beginning of preinitiation assembly and, 745–746, 746*f*
transcription machine formation by, 744–745, 745*f*
of vertebrates, 741
without TATA box, TFIID binding to, 746–748, 747*f*
Core sequence, of SV40, 416, 417*f*
Corey, Robert R., 33, 51, 54
cos elements, of phage λ, 335, 336*f*
Coumarin, 462
Countercurrent distribution, 972
Coupling protein, 276
CPC (cap binding complex), 835
CPD (cyclobutane pyrimidine dimer), ultraviolet damage and, 449, 450*f*, 451
CPD photolyase, 470–472, 472*f*
CpG islands, 453
core promoters in, 816
CPSF, 879–880
Cramer, Patrick, 733, 734, 749
Cre/*lox* system, 565*f*, 565–566
Crick, Francis, 964, 969, 988–989, 1036
central dogma theory of, 22–24, 23*f*
DNA replication and, 366
DNA structure and, 19, 20–22, 21*f*, 22*f*
CRM (cross-reacting material), 669
Cro protein, lytic pathway of phage λ and, 697–698, 698*f*
Crossing over
during meiosis, 513, 513*f*. *See also* Holliday model of recombination; Meselson-Radding model of recombination
Mus81•Eme1 complex and, 542
during mitosis, 513, 514*f*
Cross-link(s)
formation of, chemical cross-linking agents and, 462*f*, 462–464, 463*f*
intra- and interstrand, 461, 462
formation of, chemical cross-linking agents and, 464, 464*f*
Cross-linking agents, chemical, DNA damage caused by, 461–464
DNA strand separation blockage and, 461–462, 462*f*
intra- and interstrand cross-link formation and, 464, 464*f*
monoadduct or cross-link formation and, 462*f*, 462–464, 463*f*
Cross-reacting material (CRM), 669
Cross-reactions, 669
Crown galls, 284
CRP (cAMP receptor protein), 678, 682
Cruciform structure, of DNA, 138, 138*f*
Cryo-electron microscopy, 155–156, 156*f*
CSA gene, 492
CSB gene, 492
CstF, 879
CTCF protein, 817
CTD(s). *See* Carboxyl terminal domains (CTDs)
C-terminal domain, of AraC protein, 688, 689
Culture(s), 259
of animal cells, 295*f*, 295–297
primary, 296–297
Cut-and-paste transposition, 580–583, 581*f*
during DNA replication and host DNA repair, 582*f*, 582–583
3′, 5′-Cyclic adenylate (cAMP), in catabolite repression, 678
Cyclin(s), 752
Cyclin-dependent protein kinase (CDK), 752
in eukaryotic DNA replication, 427, 428*f*
Cyclobutane pyrimidine dimer (CPD), ultraviolet damage and, 449, 450*f*, 451
Cys_2His_2 zinc fingers, 778–780
Cysteine
abbreviations for, 33*t*
side chains of, 31, 31*f*
Cytidine, 7*f*
Cytidine deaminase acting on RNA (CDAR), 882
Cytochrome P450 enzymes
aflatoxins and, 460, 461*f*
Ames test and, 465–466
metabolic activation of benzo[a]pyrine and, 460, 460*f*
Cytogenetics, 227
Cytosine
in A-, B-, and Z-forms of DNA, 8*t*
in DNA, 6, 6*f*
8-oxoguanine base pairs with, 456, 456*f*
prevention of incorporation into phage T4 DNA, 319–320
water-mediated deamination of, 453, 454, 455

D

DAM (DNA adenine methyltransferase), 409
Damage to DNA, 448–466
alkylation, by monoadduct formation, 457–461
alkyl group transfer to centers of negative charge and, 457–458, 458*f*, 459*f*
modification of environmental agents to become alkylating agents and, 458–460, 460*f*, 461*f*

by chemical cross-linking agents, 461–464
DNA strand separation blockage and, 461–462, 462*f*
intra- and interstrand cross-link formation and, 464, 464*f*
monoadduct or cross-link formation and, 462*f*, 462–464, 463*f*
endogenous agents causing, 449
exogenous agents causing, 449
by hydrolytic cleavage reactions, 453–455, 453*f*–455*f*
mutagen detection and, 464–466
oxidative, 455–456, 456*f*, 457*f*
radiation-induced, 449–452
ultraviolet light causing, 449, 450*f*, 451, 451*f*
x-rays and τ-rays causing, 452
recognition of, bipartite model of, 490
D-arm, of tRNA, 975
Darnell, James, 830, 897
Darst, Seth, 633, 635, 640, 651
Davis, Ronald W., 293, 422
DBS (duplex-binding site), of transcription elongation complex, 647–648
DCE (downstream core element), 742
Dcr-2 (Dicer-2), 946
DDE recombinase, 586
DDE transposons, 577, 598
ddNTPs (dideoxynucleoside triphosphates), in DNA sequencing, 186–188
DEAE-Sephadex chromatography, 726, 726*f*
Dealkylation, direct reversal of DNA damage by, 474–476, 475*f*, 477*f*
Deamination, water-mediated, 453–455, 454*f*, 455*f*
De Boer, Herman, 1019
De Duve, Christian, 99
Defective transposons, 594–597, 595*f*
Degradative plasmids, 284
De Lange, Titia, 244
Delbrück, Max, 309
Deletions, 264, 452
single-strand annealing and, 553, 553*f*
Delicases, DNA unwinding and, 119–121, 120*f*, 121*f*
Delucia, Paul, 389
denA, 318
Denaturation, of DNA, 112–118, 114*f*
by alkali, 117–118
base stacking and, 115–116, 116*f*, 117
definition of, 113
detection of, 113
by formaldehyde, 116–117
hydrogen bonds and, 114*f*, 114–115
ionic strength and, 116
renaturation (reannealing) and, 118–119, 119*f*
temperature and, 113–116, 114*f*, 116*f*, 117, 118*f*
in tritiated water, 117
denB, 318
Dense fibrillar component (DFC), 902
Deoxyadenosine methylase, mismatch repair and, 493–494, 495
Deoxyhemoglobin, transition to oxyhemoglobin, 78, 79*f*, 80
Deoxymyoglobin, transition to oxymyoglobin, 78, 79*f*, 80
Deoxynucleoside monophosphates (dNMP), degradation of phage T4 DNA to, 317–319
Deoxyribofuranose, Haworth structure for, 5*f*, 5–6
Deoxyribonucleases (DNases), definition of, 165
Deoxyribonucleic acid. *See* DNA
Deoxyribonucleosides, 6, 6*f*
Deoxyribonucleotides, 7
Deoxyribose, 4*f*, 5
conversion to cyclic form, 5, 5*f*
structure of, 5, 5*f*
Depression, 669, 670*f*
Deprotonation, of DNA, 117–118
DFC (dense fibrillar component), 902
DGCR8 protein, 953
d'Herelle, Felix, 308
Dia, 382, 385, 385*f*
Diacylglycerol, formation of, phosphoinositide-specific phospholipase C and, 98, 98*f*
Dicer, 941–944, 953, 956
crystal structure for, 942–943, 943*f*
in RNA interferences pathway, 941–942, 942*f*
structure of, 942, 942*f*
Dicer-2 (Dcr-2), 946
Dicer-like 1, 954
5,6-Dichloro-1-β-D-ribofuranosylbenzimidazole (DRB), 754, 755*f*
Dideoxynucleoside triphosphates (ddNTPs), in DNA sequencing, 186–188
Dideoxy-terminator cycle sequencing, 187–188
Diffley, John F. X., 427
DiGeorge syndrome, 953
Dihydrouridine, 973–974, 974*f*
β-Dimer, 394–396, 395*f*
core polymerase release from, 403–404
crystal structure of, 394, 395, 395*f*
faces of, 395
placement around DNA by clamp loader, 396–399, 396*f*–399*f*
Dimerization domain, 769–770
Dimerization loops, 782
Dimethylnitrosamine, 458, 458*f*
DinI protein, in recombination, 527*t*
Dinoflagellates, lack of histones in, 211
Dintzis, Howard, 991–993
Dipeptides, 33, 35*f*
bond distances and angles in, 33–34, 34*f*
Diploid(s), partial, 268
Diploid number, 218, 219, 220*f*–223*f*, 221–222, 226
in yeast, 287
Direct repeats (DRs), 784
Direct reversal of DNA damage, 469–477
by dealkylation, 474–476, 475*f*, 477*f*
by photolyase, 469–472, 470*f*–474*f*, 474
Disarming, 284
Discontinuous gene hypothesis, 842, 844
Discontinuous model of replication, 370, 372*f*, 372–375, 373*f*
Discovery Studio Visualizer 2.0, 57
Discriminator region, 637
Dispersed transcription initiation, 741
Dispersive model of replication, 367, 367*f*
Dissolution, 539, 539*f*
Distal sequence element (DSE), 921
Distributive replication, 390
Diversity
combinatorial, 559
junctional, 559
pairing, 559
D-loops, 244, 245*f*
formation of, 528*f*, 528–529, 530*f*
DMC1 protein, 534*t*, 542
Dmc1 protein, 534*t*, 542
DNA. *See also specific organisms*
A form, 19, 19*f*, 133–135
B form compared with, 134*f*, 134–135, 135*t*
identification of, 133–134
Z form compared with, 135*t*
acting as either intron or exon, 847, 847*f*
bacterial. *See* Bacterial DNA
bases of, 6*f*, 6–7, 7*f*
composition of, 15–16
Hoogsteen base pairing and, 139
mutant classification on basis of changes in, 263–264, 264*f*
recognition in major and minor grooves, 110–111, 110*f*–112*f*
reverse-Hoogsteen pairing and, 139
sequences causing DNA bending, 112
stacking of, 115–116, 116*f*, 117
in Watson-Crick model, 20, 21*f*
bending of, 112
bent, transposon preference for, 590–591
B form, 19, 19*f*
A form compared with, 134*f*, 134–135, 135*t*
local structure of, 112
Z form compared with, 135*t*, 136
chromatosomal, H1 interaction with, 236, 236*f*
cloned fragments of, sequencing using *in vitro* transposition systems, 605*f*, 605–606
coding (sense) strand of, 624, 626, 626*f*
conservative site-specific recombination of, 576, 576*f*, 607–612
in bacteriophage λ, 609*f*, 609–611, 610*f*
in integron system, 611, 611*f*
proteins in, 607*f*, 607–608, 608*f*
cutting of, by restriction endonucleases, 170, 170*f*
denaturation of, 112–118, 114*f*
by alkali, 117–118
base stacking and, 115–116, 116*f*, 117
definition of, 113
detection of, 113
by formaldehyde, 116–117
hydrogen bonds and, 114*f*, 114–115
ionic strength and, 116
renaturation (reannealing) and, 118–119, 119*f*
temperature and, 113–116, 114*f*, 116*f*, 117, 118*f*
in tritiated water, 117
deprotonation of, 117–118
discovery of, 3
of *E. coli*. *See Escherichia coli*, DNA of
ethidium bromide binding to, 159*f*, 159–160
eukaryotic
heteroduplex, 537–538
initiator, 418
isolation of, 152
replication of. *See* Eukaryotic DNA replication
fragility of molecules, 110
gated (G) segment, 132
heteroduplex
in eukaryotic homologous recombination, 537–538
in Holliday model of recombination, 519, 520*f*, 522
in Meselson-Radding model of recombination, 522–523
in RecA-mediated strand exchange, 529
isolation of, 151–152
linker, 232–233, 233*f*
major and minor grooves of, recognition patterns in, 110–111, 110*f*–112*f*
melting curve for, 113–114, 114*f*
migration rate of, in slab gel electrophoresis, 161–162
mitochondrial
human, first complete sequence for, 928, 929*f*
replication of, strand asymmetric model of, 928, 930
non-template strand of, 624, 626, 626*f*
plasmids and. *See* Plasmid(s)

recognition as hereditary material, 16*f*, 16–17, 18*f*
recombinant DNA technology and. *See* Recombinant DNA technology
renaturation (reannealing) of, 118–119, 119*f*
molecular details of, 119, 119*f*
salt concentration and, 118
speed of, 119
temperature and, 118–119
replication of. *See* DNA replication
restriction endonuclease actions on, 168–169
restriction endonuclease characterization of, 167*t*, 167–170, 168*f*, 169*f*, 169*t*
restriction mapping of, 170–173, 170*f*–172*f*
retroviral, replication of, 358*f*, 359
RNA differentiated from, 4*f*, 5, 5*f*
rotation about single bonds causing conformational changes in, 136, 136*f*, 137*f*
sequences targeted by transposons, 591
sequencing of. *See* DNA sequencing
single-stranded, BRAC2 binding to, 537
size variation of, 109–110, 110*t*
strand separation of, blockage by chemical cross-linking agents, 461–462, 462*f*
structure of. *See* DNA structure
supercoil formation by, 123. *See also* Supercoil(s)
symmetric heteroduplex, 519
synthesis of
RNA-directed, reverse transcriptases for, 201–203, 202*f*
translesion, catalyzed by error-prone DNA polymerases, 498–499
using DNA polymerase I, 177–180, 178*f*–181*f*, 182
target-primed reverse transcription of, 576*f*, 576–577, 612–617
elements moving via, 612*f*, 612–613
for LINE movement, 613*f*, 613–616, 615*f*
mobile group II intron movement by, 612, 613, 616–617
template strand of, 624
transported (T) segment, 132
transposition of, 576, 576*f*, 577–606
coordinated breakage and joining events in, 579–580, 580*f*
cut-and-paste, 580–583, 581*f*, 582*f*
during DNA replication, 582*f*, 582–583
during host DNA repair, 582*f*, 582–583
insertion sequences and, 577, 578*f*
regulation of, 589*f*, 589–590
transposase binding in, 577–579, 578*f*, 579*f*
transposons and. *See* Transposons
unwinding of, helicases and, 119–121, 120*f*, 121*f*
x-ray diffraction patterns of, 19, 19*f*
Z form, 135–137, 136*f*
A and B forms compared with, 135*t*, 136
harmful role in human health, 137
zinc finger protein interaction with, 779*f*, 779–780, 780*f*
DnaA•ADP, 384
DnaA•ATP, filament formed by, 384–385, 385*f*
DnaA boxes, 384
DNA adenine methyltransferase (DAM), 409
DNA affinity chromatography, transcription activator protein purification using, 770–772, 771*f*
dnaA gene, 380*t*
DnaA protein
in bacterial DNA replication, 382–383
functional domains of, 383*f*, 383–384, 384*f*
regulatory function of, 408–409
DnaA titration, 408–409
DNA bending proteins, 217
dnaB gene, 380*t*
DnaB helicase, 121
double-ring structures of, 385–386, 386*f*
movement along lagging strand, 404–405
disassembly model of, 404, 405*f*
pausing model of, 404, 405*f*
priming loop model of, 404, 405*f*
DNA-binding domains, of transcription activator proteins, 769, 769*f*
with basic region leucine zippers, 791–794, 791*f*–795*f*
with helix-loop-helix structures, 794–797
with helix-turn-helix structures, 773–777
loop-sheet-helix, 789*f*, 789–790, 790*f*
with zinc fingers, 777–789
DnaB protein. *See also* DnaB helicase
in bacterial DNA replication, 382–383
replication fork restart and, 550
DNA chips, in DNA sequencing, 203*f*, 203–204, 205*f*
DnaC protein
in bacterial DNA replication, 382–383
loading of DnaB helicase onto single-stranded DNA by, 386, 387*f*
DNA damage, 448–466
alkylation, by monoadduct formation, 457–461
alkyl group transfer to centers of negative charge and, 457–458, 458*f*, 459*f*
modification of environmental agents to become alkylating agents and, 458–460, 460*f*, 461*f*
by chemical cross-linking agents, 461–464
DNA strand separation blockage and, 461–462, 462*f*
intra- and interstrand cross-link formation and, 464, 464*f*
monoadduct or cross-link formation and, 462*f*, 462–464, 463*f*
endogenous agents causing, 449
exogenous agents causing, 449
by hydrolytic cleavage reactions, 453–455, 453*f*–455*f*
mutagen detection and, 464–466
oxidative, 455–456, 456*f*, 457*f*
radiation-induced, 449–452
ultraviolet light causing, 449, 450*f*, 451, 451*f*
x-rays and γ-rays causing, 452
recognition of, bipartite model of, 490
repair of. *See* DNA repair
DNA denaturation, 112–118, 114*f*
by alkali, 117–118
base stacking and, 115–116, 116*f*, 117
definition of, 113
detection of, 113
by formaldehyde, 116–117
hydrogen bonds and, 114*f*, 114–115
ionic strength and, 116
renaturation (reannealing) and, 118–119, 119*f*
temperature and, 113–116, 114*f*, 116*f*, 117, 118*f*
in tritiated water, 117
DNA-dependent RNA polymerases. *See* RNA polymerase(s)
dnaE gene, 266, 380*t*, 393
DNA fingerprinting, 294
DNA footprinting technique, 630–633, 631*f*–633*f*
dnaG gene, 380*t*
DNA glycosylase, in base excision repair, 478–480, 479*f*, 480*f*
DNA glycosylase/lyase, in base excision repair, 478, 480–481, 482*f*
DnaG protein
in bacterial DNA replication, 382–383
catalysis of RNA primer synthesis by, 388, 388*f*
movement along lagging strand, 404–405
disassembly model of, 404, 405*f*
pausing model of, 404, 405*f*
priming loop model of, 404, 405*f*
DNA gyrase, 132
superhelical DNA and, 125
DNA helicase II, mismatch repair and, 494
DNA library, single-stranded, creating for DNA sequencing with GS FLX, 195, 196*f*
DNA ligase
detection *in vitro*, 375, 376*f*
DNA fragment insertion into plasmids and, 173
DNA ligase IV, in nonhomologous end-joining, 554, 555*t*, 561
mismatch repair and, 495
nick sealing by, 375–376, 377*f*
Okazaki fragments connected by, 375–376, 376*f*, 377*f*
in SV40 DNA replication, 419, 420, 420*f*
DNA looping, 770, 770*f*
dnaN gene, 394
DNA polymerase(s)
archaeal, 442–443
DNA polymerase I
DNA synthesis using, 177–180, 178*f*–181*f*, 182
mutant in study of, 266
Okazaki fragment synthesis and, 376
proofreading function of, 178*f*, 179
DNA polymerase II, 266
SOS response and, 502
DNA polymerase III, 266, 392
in bacterial DNA replication, 389–390
release from β clamp, 403–404
subunits of, 393–394, 394*f*
DNA polymerase IV, SOS response and, 502
DNA polymerase V, SOS response and, 503
DNA polymerase α (Pol α)
in eukaryotic DNA replication, 428, 428*f*, 429
processivity of, 429
recruitment of, by SV40 T antigen, 418–420, 419*f*–421*f*
subunits of, 429, 429*f*
DNA polymerase β (Pol β), in base excision DNA repair, 481
DNA polymerase δ (Pol δ)
in eukaryotic DNA replication, 428, 428*f*, 429, 430–431
mismatch repair and, 497
subunits of, 429, 429*f*
in SV40 DNA replication, 418
DNA polymerase η, 505
DNA polymerase μ, 561
error-prone, translesion DNA synthesis catalyzed by, 498–499
human, SOS response and, 504*t*, 504–505
in replicative transposition, 585–586
synthesis of, induced by SOS response, 502–504, 503*f*
template-dependent, human, 504*t*, 504–505
DNA polymerase III holoenzyme, 379
clamp loader and, 396
mismatch repair and, 495
subunits of, 391*f*, 391–392, 392*t*, 396*f*, 396–397
trombone model of replication and, 400
DNA polymerase/α-primase (Pol α)
in eukaryotic DNA replication, 428, 428*f*, 429
processivity of, 429
recruitment of, by SV40 T antigen, 418–420, 419*f*–421*f*
subunits of, 429, 429*f*
Dna2 protein, 536

dnaQ gene, 380t, 394
DNA repair, 468–505
  base excision, 477–481, 478f–482f
    by DNA glycolase action, 478–480, 479f, 480f
    by DNA glycolase/lyase action, 480–481, 481f, 482f
    long patch, 481, 482f
    short patch, 481, 481f
  cut-and-paste transposition during, 582f, 582–583
  direct reversal of damage for, 469–477
    by dealkylation, 474–476, 475f, 477f
    by photolyase, 469–472, 470f–474f, 474
  mismatch
    in *E. coli*, 493–497, 494f–496f
    human, 497–498, 498f
  nucleotide excision, 482–493
    bacterial, 482–485, 483f, 484f, 486f, 487f
    global genome excision repair pathway for, 489
    mammalian, 485, 488–490, 491f, 492f, 492–493
    steps in, 482
    transcription-coupled nucleotide excision repair pathway for, 489–490, 491f, 492
  SOS response to damage and, 498–505
    DNA polymerase synthesis induced by, 502–504, 503f
    human DNA polymerases and, 504t, 504–505
    key steps in, 501f, 501–502
    RecA and LexA regulation of, 499–502, 500f, 501f
    translesion DNA synthesis and, 498–499
DNA replication
  in archaea, 441–443
    elongation stage of, 442–443
    Orc1/CdC6 in, 442, 442f
    similarity to bacterial replication, 442–443
    similarity to eukaryotic replication, 441–443
  bacterial. *See* Bacterial DNA replication
  in bacteriophages
    phage λ, 336, 337, 338f, 339
    phage P1, 341
    phage φX174, 329–330, 330f
    phage T4, 317–320
    phage T7, 328
  break-induced, 552, 552f
  in central dogma theory, 23, 23f
  cut-and-paste transposition during, 582f, 582–583
  distributive, 390
  eukaryotic. *See* Eukaryotic DNA replication
  fragile sites and, 551–552
  helicases in, 119–121, 120f, 121f
  replication fork collapse in, mitotic homologous recombination during, 549–550, 549f–551f
  topoisomerases in, 128f–130f, 128–132, 131f
  unidirectional, 369, 370f
  viral
    adenoviral, 354f, 354–355
    retroviral, 358f, 359
    in simian virus 40, 351f, 351–352
DNase protection method, 629f, 629–630, 631f
DNA sequencing, 43, 45, 45f, 185–205
  of ancient DNA, 201
  chain termination method for, 186f, 186–188, 187f
  DNA chips in, 203f, 203–204, 205f
  Genome Sequencer FLX for, 195, 196f, 197–199, 198f–200f, 201
  of human genome
    hierarchical DNA sequencing for, 189–192, 191f
    information provided by, 194
    whole genome sequencing for, 189, 192, 193f, 194
  primer walking for, 188, 188f
  with restriction map information, 188
  reverse transcriptases in, 201–203, 202f
  shotgun, 189f, 189–192
    hierarchical, in Human Genome Project, 189–192, 191f
    whole genome, human genome sequencing using, 189, 192, 193f, 194
DNases (deoxyribonucleases), definition of, 165
DNA structure
  A form, 19, 19f, 133–135
    B form compared with, 134f, 134–135, 135t
    identification of, 133–134
    Z form compared with, 135t
  amino acid sequences in, 9
  bases in, 6f, 6–7, 7f. *See also* Base(s), DNA
  B form, 19, 19f
    A form compared with, 134f, 134–135, 135t
    local structure of, 112
    Z form compared with, 135t, 136
  circular, covalent, 123f, 123–127, 124f, 126f
  cruciform, 138, 138f
  denaturation and. *See* DNA denaturation
  directionality in, 9, 10f
  double helix, 19–22, 19f–22f
  hydrogen bond stabilization of, 114f, 114–115
  hyperchromic, 113
  hypochromic, 113
  ionic strength and, 116
  quadriplex (tetraplex), 139, 141f
  stick figure representations for, 10f, 11f, 12
  tetranucleotide, Levene's proposal of, 8, 9f, 15, 16
  triplex, 138–139, 139f, 140f
  Z form, 135–137, 136f
    A and B forms compared with, 135t, 136
    harmful role in human health, 137
DNA template, for RNA synthesis, 624
DNA transposons. *See* Transposons
DNA unwinding element (DUE), 385
  single-stranded DNA generated at, DnaB helicase loading onto, 386, 387f
dnaX gene, 380t
dNMP (deoxynucleoside monophosphates), degradation of phage T4 DNA to, 317–319
Dn14 protein, 555t
Dolly (sheep), 819f, 819–820
Domains, 83–84, 84f, 85f
  constant region, 84
  variable region, 84
Domain swap, 801
Dominance, law of, 3
Donohue, Jerry, 20
Doolittle, R. F., 49
Doty, Paul, 114
Double helix of DNA, 19–22, 19f–22f
"Double-sieve" model, 980, 980f
Double-strand break(s) (DSBs), 452, 512
  repair by homologous recombination. *See* Homologous recombination
Double-strand break repair (DSBR) model of recombination, 524–526, 525f
  eukaryotic homologous recombination proteins and, 533, 534t, 535–540
  variation of, 540, 540f
Doubling time, 260
Doudna, Jennifer, 942
Downstream core element (DCE), 742
Downstream core promoter element (DPE), 742, 743
  in promoter lacking TATA box, TAF interaction wit, 746–747, 747f
Down syndrome, 229
DPE (downstream core promoter element), 742, 743
  in promoter lacking TATA box, TAF interaction wit, 746–747, 747f
DpnI, in site-directed mutagenesis, 185
DRB (5,6-dichloro-1-β-D-ribofuranosylbenzimidazole), 754, 755f
DRB sensitivity-inducing factor (DSIF), 754, 755f, 756
Dreyer, William J., 558
DRIP (vitamin D receptor-interacting protein), 807
Drosha, 953
Drosophila melanogaster
  with deleted rRNA transcription units, 899
  exons of, 846, 846f
  FLP/FRT system in, 564f, 564–565
  gene conversion in, 516
  genetic recombination studies using, 257
  genome of, determination of, 189
  homeotic genes of, 773f, 773–774
  hsp70 gene of, 835–836, 836f
  introns of, 846, 846f
  lack of CCAT box in, 765
  lack of requirement for telomerase in, 439
  miRNA formation in, 952f, 952–954, 953f
  Morgan's work with, 512–513
  P element of, 590, 600–602, 601f
  (6-4) photolyase of, 472, 474
  RNA interference in, 940–942, 942f
  SID-1 of, 949
  silencers of, 768
  telomere maintenance in, 576
  trans-splicing in, 850, 859
DRs (direct repeats), 784
Drug resistance, 15, 283
Drug resistance plasmids, 283
DSB(s) (double-strand break(s)), 452, 512
  repair by homologous recombination. *See* Homologous recombination
DSBR. *See* Double-strand break repair (DSBR) model of recombination
Dscam gene, 848–849, 849f
DSE (distal sequence element), 921
DSIF (DRB sensitivity-inducing factor), 754, 755f, 756
3′-dTMP (thymidine-3′-monophosphate), 7f
5′-dTMP (thymidine-5′-monophosphate), 7f
Dubochet, Jacques, 238
Dulbecco, Renato, 469–470
Duplex-binding site (DBS), of transcription elongation complex, 647–648
Duplications, 452
dut⁻ mutant, 374
Dystrophin gene, 847

E

Earnshaw, William C., 240–241
Ebright, Richard H., 640
EcoB1, 312–313
EcoKI, 313
EcoRI
  DNA cutting by, 170, 170f
  sequence specificity of, 169t
EcoRV
  crystal structure of, 169, 169f
  DNA cutting by, 171, 171f
  sequence specificity of, 169t
ECS (exon complementary sequence), 882, 883f
Ectopic gene expression, 566
Edidin, Michael, 69
Editing, of RNA, 880–884
  by aminoacyl-tRNA synthases, 979–983, 980f, 981f
  ApoB mRNA editing, 881, 882f

A-to-I, 882–883, 883*f*
C-to-U, 881–882, 882*f*, 883*f*
Edman, Pehr, 41
Edman degradation, 41–43, 42*f*
eEF1A, 1029
eEF1B, 1029, 1034–1035
eEF2, 1029
EF(s). *See* Elongation factors (EFs)
Effectors, in G-protein-linked signal transduction systems, 95, 97*f*, 97*t*, 97–98
EF-G, 1029
EF-P, 1032, 1032*f*
EF-Ts, 1029
as GDP-GTP exchange protein, 1034*f*, 1034–1035
EF-Tu, 1029
EF-Tu•GTP•aminoacyl-tRNA ternary complex, carrying aminoacyl-tRNA ribosome, 1030, 1032, 1032*f*
Egly, Jean-Marc, 752
EGO-1, 948
Ehrenberg, Måns, 1044
EI (enzyme I), in catabolite repression, 679
EiA, 355
eIFs (eukaryotic initiation factors), 1022, 1023*f*, 1023*t*, 1024, 1025*f*–1027*f*, 1026–1028
phosphorylation of, 1028, 1028*t*
EIIA (enzyme IIA), in catabolite repression, 679, 680, 680*f*
EIIBC (enzyme IIBC), in catabolite repression, 679, 680
EJC (eExon junction complex), 884
Eldred, Emmet, 981
Electroblotting, 176
Electron microscopy, 153–156, 154*f*–156*f*
cryo-electron microscopy, 155–156, 156*f*
metal shadowing and, 153–155, 154*f*
negative staining and, 155, 155*f*
Electrophiles, 457
Electrophoresis, for protein purification, 40, 41*f*
Electroporation, 546
Elephant, divergence rate between mammoth and, 201
Elongation factors (EFs), 1012, 1029–1032, 1031*f*, 1033*f*
EF-G•GTP stimulation of translocation process and, 1041–1042, 1041*f*–1043*f*
EF-Ts and, 1034–1035
EF•Tu•GTP•aminoacyl-tRNA ternary complex carrying aminoacyl-tRNA to ribosome and, 1030, 1032, 1032*f*
repeating cycle of, 1030, 1031*f*
Elongation stage of DNA replication
archaeal, 442–443
bacterial, 379, 388–405
β clamp in, 394–396, 395*f*
clamp loader in, 396–400, 396*f*–399*f*
core polymerase in, 392–394, 393*f*, 403–404
DNA polymerase holoenzyme in, 391*f*, 391–392, 392*t*
DNA polymerase III in, 389–390
enzymes involved in, 388–389
helicase movement in, 404–405, 405*f*
polymerase processivity and, 390–391
replisome in, 400–401, 401*f*, 402*f*
eukaryotic, Pol δ and Pol ϵ in, 429*f*–431*f*, 429–432
Elongation stage of protein synthesis, 1012, 1029–1043
codon-anticodon helix sensing and, 1032–1034, 1033*f*
EF-G•GTP stimulation of translocation process and, 1041–1042, 1041*f*–1043*f*
EF-Ts and, 1034–1035
EF•Tu•GTP•aminoacyl-tRNA ternary complex carrying aminoacyl-tRNA to ribosome and, 1030, 1032, 1032*f*
elongation factors in, 1029–1032, 1031*f*, 1033*f*
hybrid-states translocation model and, 1040*f*, 1040–1041
ribosome and, 1035*f*–1039*f*, 1035–1040
Elongation stage of transcription
RNA polymerase II and, 752–756
RNA polymerase III and, 923–924
Elongin, 753–754
Embryonic stem (ES) cells, 546, 547, 548*f*, 818
EME1 protein, 534*t*
Eme1 protein, mutations in gene coding for, 542
Emulsion PCR, 197
Endogenous agents, DNA damage caused by, 449
Endogenous siRNA (esiRNA), 955, 956*f*
Endolysin, phage T4 lysis and, 325
Endonucleases. *See also* Restriction endonucleases
definition of, 165
pr-rRNA cleavage by, 910, 911*f*
Endoplasmic reticulum (ER), 965, 965*f*
End-replication problem, in eukaryotic DNA replication, 432–439
RNA template used by telomerase and, 434–437, 435*f*, 436*f*
role of telomerase in aging and cancer and, 438–439
role of telomerase in solving, 437*f*, 437–438, 438*f*
telomere formation and, 432*f*, 432–434, 433*f*
Enhancers, 645, 768
Entropy, 48
env, 356, 356*f*, 357
Enzyme(s), 85–94. *See also specific enzymes*
active site of, 89
activity of
alteration by covalent modification, 94
detection of, 85–87, 86*f*
in bacterial DNA replication, 388–389, 400
catalytic power of, 88
catalyzing direct reversal of DNA damage, 469
cofactors and, 85
constitutive, 667
cytochrome P450
aflatoxins and, 460, 461*f*
Ames test and, 465–466
metabolic activation of benzo[a]pyrine and, 460, 460*f*
energy of activation and, 87*f*, 87–88
enzyme-substrates complex and, 88–90, 89*f*, 90*f*
genes and, 4
Holliday junction resolution by, 539, 539*f*
inducible, 667
morphogenetic, in phage assembly, 320
for nonhomologous end-joining, 560–561
proteolytic, 296
regulatory, 91–94, 92*f*, 93*f*
subunits of, 94, 94*f*
repressible, 667
suicide, 476
Enzyme I (EI), in catabolite repression, 679
Enzyme IIA (EIIA), in catabolite repression, 679, 680, 680*f*
Enzyme IIBC (EIIBC), in catabolite repression, 679, 680
Enzyme-substrate (ES) complex, 88–90, 89*f*, 90*f*
induced fit model of, 90, 90*f*
lock-and-key model of, 89, 90*f*
molecular details for, 90–91, 91*f*, 92*f*
Epigenetics, 816
ϵ-subunit, of core polymerase, 394
Equilibrium density gradient centrifugation, 158–160, 159*f*, 160*f*
ER (endoplasmic reticulum), 965, 965*f*
ERCC1 gene, 489
ERCC1 protein, 489
eRF1 (class 1 cytoplasmic release factor), 1050, 1050*f*
eRF3 (class 2 cytoplasmic release factor), 1050, 1050*f*
ES (embryonic stem) cells, 546, 547, 548*f*, 818
Escherichia coli
catabolite repression in, 677, 677*f*
β clamp of, 394, 395, 395*f*
CPD photolyase of, 471, 472*f*
DnaA from, 383–384
DNA of
chromatin of, 212*f*, 212–213, 213*f*
circular nature of, 369
mismatch repair of, 493–497, 494*f*–496*f*
molecule size of, 110*t*
nucleoid of, 212, 212*f*
renaturation of, 119
structure of, 124, 211
genetic information exchange by conjugation in, 270–273, 271*f*–273*f*
N-glycosylase of, in base excision repair, 478
growth curve for cultures of, 260, 260*f*
helicases of, 120
histones synthesized by, 233
lac promoter of, 637, 637*f*
lactose metabolism in, proteins necessary for, 665, 665*f*
lacZ gene of, 941
making competent to take up DNA from surrounding medium, 173–174
methionine tRNA isoacceptors of, 1013–1014, 1014*f*
mismatch repair system of, 493–497, 494*f*–496*f*
as model for genetic studies, 255, 257, 259
mRNA degradation pathway used by, 703–705, 704*f*, 706*f*
mRNA of, 653–654
cistrons in, 663
origin of replication of. *See oriC* (*E. coli* origin of replication)
recombination-defective mutants of, 526–527, 527*t*
replication forks in, stalled, 550
ribosome of
cryo-electron microscopy reconstruction of, 156*f*
structure of, 1008, 1008*f*
RNA polymerase holoenzyme of. *See* RNA polymerase holoenzyme
σ factors of
in initiation stage of transcription, 643*t*, 643–644
nomenclature for, 643, 643*t*
σ^{70} family of, 629
conserved domains of, 638, 638*f*
single-stranded DNA binding proteins of, 122, 122*f*
strain B, EcoB1 in, 312–313
strain K, EcoKI in, 313
stringent response in, 710*f*, 710–711
Thermus thermophilus heavy metal binding protein structure determination and, 59
tRNA genes of, 972
wild type, Fe^{3+} intake by, 265
Escherichia coli bacteriophages. *See also specific bacteriophages*
blender experiment and, 16*f*, 16–17
Escherichia coli ColE1 plasmid, replication of, 369
Escherichia coli DNA polymerase I, 177–180, 178*f*–181*f*

Escherichia coli exonuclease I, properties of, 166*t*
Escherichia coli exonuclease III, properties of, 166*t*
Escherichia coli exonuclease VII, properties of, 166*t*
Escherichia coli polymerase I, mutant in study of, 266
Escherichia coli topoisomerases
  type I, 128–129
    function of, 130–131, 131*f*
    N-terminal fragment of, crystal structure for, 130, 130*f*
  type II, 132
  type III, 129
ES complex. *See* Enzyme-substrate (ES) complex
ESEs (exonic splicing enhancers), 868–870, 869*f*, 870*f*
esiRNA (endogenous siRNA), 955, 956*f*
ESS (exonic splicing silencer), 870
EST(s) (expressed sequence tags), 846
Established cell lines, 296, 297. *See also* HeLa cells
EST1 gene, 435
EST2 gene, 435
EST3 gene, 435
EST4 gene, 435
Estrogen, drugs used to counteract effects of, 785–786, 787*f*
Ethidium bromide
  in equilibrium density gradient centrifugation, 159*f*, 159–160
  staining of gel by, 162
3′-ETS (3′-external transcribed sequence), 898
5′-ETS (5′-external transcribed sequence), 898
Euchromatin, 218, 218*f*
Eukaryotes
  amino acid-coding regions in genes of, interruption by noncoding regions, 841–842, 843*f*, 844*f*, 844–845
  base excision repair in, 481, 482*f*
  chromatin of, 211
    disassembly and reassembly in DNA replication, 439–441, 440*f*
    formation by DNA molecules, 810
    forms of, 218, 218*f*
  core promoters of, 741
  DNA of, 109
    heteroduplex, 537–538
    initiator, 418
    isolation of, 152
    replication of. *See* Eukaryotic DNA replication
  gene expression in, 764
  higher, DNA isolation in, 152
  homologous recombination in, proteins involved in, 533, 534*t*, 535–540
  initiation stage of protein synthesis in, 1012, 1020–1029
    initiation factors in, 1022, 1023*f*, 1023*t*, 1024, 1025*f*–1027*f*, 1026–1028, 1028*t*
    methionine tRNA in, 1020–1021, 1021*f*
    scanning mechanism and, 1021–1022
  life cycle of, alternation between interphase and mitosis, 218*f*, 218–219, 219*f*
  mobile group II introns in organelles of, 616
  mRNA caps in, 834
    formation of, 836*f*, 836–837
  ribosomes of, 896*f*, 896–897
    assembly of, 915–919, 917*f*–919
  ribosome structure in, 1011*f*, 1011–1012
  RNA polymerases of, 623
    subunits of RNA polymerase II and, 730, 730*t*
    types in cell nucleus, 725–727, 726*f*, 727*t*
  rRNA transcription units of, 899–900, 900*f*, 901*f*
  superhelical, underwinding of, 125
  termination stage of protein synthesis in, 1050*f*, 1050–1051
  topoisomerase II enzymes of, 132
  transcription in
    bacterial transcription compared with, 734–735, 735*f*
    nucleosomes in DNA coding region and, 815
  transposons of, 581, 582, 587–588, 587*f*–589*f*
    targeting molecular processes, 592
  tRNA genes of, 972
  viruses of. *See also specific viruses*
    adenoviruses, 352*f*–354*f*, 352–355
    as models, 348–349
    polyomaviruses, 349, 350*f*, 351*f*, 351–352
    retroviruses, 355–357, 356*f*, 358*f*, 359
Eukaryotic DNA replication, 415–439
  bacterial replication compared with, 421–422
  bidirectional, 369, 370*f*, 371*f*
  chromatin disassembly and reassembly and, 439–441, 440*f*
  elongation stage of, Pol δ and Pol ϵ in, 429*f*–431*f*, 429–432
  end-replication problem and, 432–439
    RNA template used by telomerase and, 434–437, 435*f*, 436*f*
    role of telomerase in aging and cancer and, 438–439
    role of telomerase in solving, 437*f*, 437–438, 438*f*
    telomere formation and, 432*f*, 432–434, 433*f*
  initiation stage of, 422–429
    activation of licensed origin in, 427–429, 428*f*
    anonymously replicating sequences in, 422–425, 423*f*, 424*f*
    gel electrophoresis location of origins of replication and, 425*f*, 425–426
    MCM2-7 helicase loading into origin in, 427, 427*f*
    origin of recognition complex in, 426
    replicator sites in, 422, 422*f*, 423*f*
  origins of replication and, 381
  similarity to archaeal replication, 441–443
  in SV40, T antigen in, 416–420, 417*f*–421*f*
Eukaryotic initiation factors (eIFs), 1022, 1023*f*, 1023*t*, 1024, 1025*f*–1027*f*, 1026–1028
  phosphorylation of, 1028, 1028*t*
Euplotes aediculatus, telomerase of, 434–435
Evans, Martin, 546
EvoVII, mismatch repair and, 495
Excisionase (Xis), 609
Exogenous agents, DNA damage caused by, 449
ExoI, mismatch repair and, 495, 497
Exon(s), 45, 45*f*, 724, 844
  cassette, 847, 847*f*
  conservation during evolution, 845*t*, 845–846, 846*f*
  constitutive, 847
  size of, 846, 846*f*
Exon complementary sequence (ECS), 882, 883*f*
Exon definition, 869–870, 870*f*
Exonic splicing enhancers (ESEs), 868–870, 869*f*, 870*f*
Exonic splicing silencer (ESS), 870
Exon junction complex (EJC), 884
Exonucleases, definition of, 165
ExoVII, mismatch repair and, 495
ExoX, mismatch repair and, 495
Expansion segments, ribosomal, 1011
Exportin 5, 953
Expressed sequence tags (ESTs), 846
Extended -10 element, 637
3′-External transcribed sequence (3′-ETS), 898
5′-External transcribed sequence (5′-ETS), 898

F

FACT histone chaperone, 439–441
Factor for inversion stimulation (FIS), 217, 385
falE gene, 684
  galactose metabolism and, 264
Fanconi anemia, 537
Fangman, Walton L., 425
Fatty acids, polyunsaturated, oxidation of, aldehyde products produced by, 456, 456*f*
FBJ murine osteosarcoma virus, 793
FC0, 988–989
FC (fibrillar center), 902–903, 903*f*
Feedback inhibition, 92
FEN1 (flap endonuclease), in SV40 DNA replication, 419, 419*f*
Fenton, Henry J. H., 455–456
Fenton reaction, 455–456, 456*f*, 457*f*
Fersht, Alan, 980
Fertility factor (F factor). *See* F plasmids
F factor (fertility factor). *See* F plasmids
Fibrillar center (FC), 902–903, 903*f*
Fibrous proteins, 29
Fields, Stanley, 804
Filamentous bacteriophages, 309*f*, 310
Fingerprinting, 993
Fire, Andrew Z., 937, 938–940, 947
First messengers, in G-protein-linked signal transduction systems, 98
FIS (factor for inversion stimulation), 217, 385
Fischer, Emil, 89
FISH (fluorescent in situ hybridization), 230*f*, 230–231, 231*f*
FIS protein, *rrn* operon transcription and, 709
F′*laci* plasmid, 665–666
Flap endonuclease (FEN1), in SV40 DNA replication, 419, 419*f*
FLP/*FRT* system, 564*f*, 564–565
Fluid mosaic model, 68–71, 69*f*, 70*f*
Fluorescent in situ hybridization (FISH), 230*f*, 230–231, 231*f*
Flush ends, 168, 168*f*
fMet (*N*-formulmethionine), 1013, 1013*f*
fMet-tRNA, bacterial synthesis of, 1013, 1013*f*
Focused transcription initiation, 741
Folding motifs, 55, 55*f*
Formaldehyde, DNA denaturation by, 116–117
N-Formulmethionine (fMet), 1013, 1013*f*
454 Life Science, 195
Fowlpox virus, DNA molecule size of, 110*t*
F pilus, RNA phage attachment to, 332
F plasmids
  clustering of replication control functions in basic replicon, 282–283
  DNA molecule size of, 110*t*
  genes needed for successful mating and DNA transfer and, 274*f*, 274–276, 275*f*
  integration into bacterial chromosomes, 276, 277*f*, 278
  ori T in, 272–273
  part of bacterial chromosome contained in, 280, 282, 282*f*
  self-transmissibility of, 276
F′ plasmids, 280, 282, 282*f*
Fragile sites, 551–552
Fragment assay, 1035–1036
Frameshift mutations, 988, 988*f*
Frank, Joachim, 1026, 1026*f*, 1027*f*
Franklin, Rosalind, 19, 133–134
Frogs
  nuclear transfer experiments in, 818–829
  RNA polymerase III transcription units of, 920–921

Frozen accident hypothesis of codon assignments, 1000
Frye, L. D., 69
Fucocoumarins, 464
Fungi, DNA isolation in, 152
Furanoses, Haworth structure for, *5f*, 5–6
Furth, John J., 725, 726
fusA gene, 1029–1030

G

gag, 356, *356f*
Galactokinase, gene encoding, 684
Galactose, metabolism of, genes required for, 264–265
Galactose epimerase, gene encoding, 684
Galactose operon, 684–686, *685f*, *686f*
  cAMP•CRP and, 686
  Gal repressor and, 684–685, *685f*
  promoters of, 685–686, *686f*
Galactose transferase, gene encoding, 684
β-Galactosidase, 665, *665f*
  genes coding for, 665–667, *666f*
  lacZ gene encoding, 596
GAL1 gene, 798
  transcription of, *799f*, 799–800
Gal4 gene, 787
Gal11 gene, 807
galK gene, 684
  galactose metabolism and, 264
Gall, Joseph G., 244, 432
galM gene, 684
Gal3 protein, 800
GAL4 protein, 602
Gal4 protein
  finding and activation domains of, *798f–801f*, 798–803, *803f*
  regulation of genes involved in galactose metabolism by, 787–789, *787f–789f*
Gal80 protein, *799f*, 799–800
Gal repressor, 684–685, *685f*
galT gene, 684
  galactose metabolism and, 264
Gal4-VP16 protein, 803
τ-rays, DNA damage caused by, 452
Garrod, Archibald, 4
GATC sequences, mismatch repair and, 495
Gated (G) segment DNA, 132
GC (granular compartment), 902
GC box, 765–767
  functions performed by, 766–767
GCN5 gene, 814
Gcn4 protein, 791, *791f*
Gehring, Walter, 774
Gel electrophoresis, 160–165
  capillary, 163, *164f*
  protein separation by, 162–163
  pulsed-field, 163–164, *165f*
  slab, 160–162, *161f*, 163
  sodium dodecyl sulfate-polyacrylamide, 162–163
  topoisomer separation by, 162, *163f*
  two-dimensional, origins of replication located by, *425f*, 425–426
Gel filtration chromatography, 38, *39f*, 40
Gel mobility shift assay, 748
Gene(s). *See also specific genes*
  alternate forms of, 262
  cistrons differentiated from, 662
  complementation to determine number responsible for phenotype, 268–270, *269t*
  definition of, 255
  enzymes and, 4
  functional form of, 262
  homeotic, 773–775
    positional identities assigned by, 773–774, *773f–775f*
    transcription activator protein specification by, 774–775
  insertion of, gene knockouts created by, 544, *544f*
  for Mediator subunits, 807
  Mendel's laws of inheritance and, 3
  mutant. *See* Mutants; Mutations
  origin of term, 255
  orthologs, 847
  paralogs, 537
  replacement of, gene knockouts created by, 544, *544f*
  split, 840–855
    combinations of splicing patterns within individual genes leading to formation of multiple mRNAs and, *848f*, 848–849, *849f*
    coordinated transesterification reactions in splicing and, 854–855, *855f*
    interruption of amino acid-coding regions within eukaryotic genes and, 841–842, *843f*, *844f*, 844–845
    intron and exon conservation and, *845t*, 845–847, *846f*
    mRNA molecules formed by splicing pre-mRNA and, 840–841, *841f*
    production of two or more mRNA molecules by a single pre-mRNA and, *847f*, 847–848
    slicing precision and, *851f*, 851–852, *852f*
    splicing intermediates resembling lariats and, 852–854, *853f*, *854f*
    trans-splicing in *Drosophila* and, 850, *850f*
  tumor suppressor, 464–465
  wild type, 262
Gene conversions, *515f*, 515–516, *516f*
Gene expression
  bacterial, control of, 664
  ectopic, 566
  eukaryotic, 764
Gene knockdown, 950
Gene knockouts, 544–548
  by gene insertion, 544, *544f*
  by gene replacement, 544, *544f*
  in mice, gene targeting to create, 545–547, *546f–548f*
  mitotic recombination to make, 544, *544f*
  in yeast, occurring by homologous recombination, 544–545, *545f*
General transcription factors (GTFs). *See also specific factors*
  RNA polymerase II and, 743–752
    lack of requirement for TFIIA and, *748f*, 748–749
    preinitiation complex formation and, *745f*, 745–746, *746f*
    sequential binding of RNA polymerase II•TFIIF complex, TFIIE, and TFIIH and, 751–752, *752f*
    TFIIB in conversion of closed promoter to open promoter and, 749–751, *749f–751f*
    TFIID and core promoters of protein-coding genes lacking a TATA box and, 746–748, *747f*
    transcription of naked DNA from specific transcription start sites and, 743–744, *744t*
Gene targeting, 544
Genetic code, 963–1001
  aminoacyl-tRNA synthetases and, 977–987
    binding domains of, *978f*, 978–979
    binding to acceptor arm, 979, *980f*
    classes of, 977–979, *978f–980f*, *978t*
    conserved sequence motifs of, 979, *979f*
    editing functions of, 979–983, *980f*, *981f*
    in gram-positive bacteria and archaea, *985f*, 985–986
    polypeptide building blocks and, *986f*, 986–987
    tRNA recognition by, *983f*, 983–985
  confirmation by synthetic messengers with strictly defined base sequences, *995f*, 995–996
  mRNA and, 987–1001
    characteristics of, 996–997, *997f*, *997t*, *998t*, 999
    coding specificity of aminoacyl-tRNA and, *999f*, 999–1000
    codons in, 987–989, *988f*
    direction of reading, *993f*, 993–994
    genetic code and, poly(U) direction of poly(Phe) synthesis and, 989–991
    origin of, 1000–1001
    polypeptide chain termination signals and, 996, *996f*
    protein synthesis process and, 991–993, *992f*
    ribosome programming to synthesize proteins and, 987
    synthetic messengers confirming, *995f*, 995–996
    trinucleotide promotion of aminoacyl-tRNA binding to ribosomes and, *994f*, 994–995
    wobble hypothesis and, 1000, *1000f*, *1001f*
  tRNA and
    for alanine, sequencing of, 972–973
    amino acid attachment to, 966–967, *968f*, 969, *969f*, *971f*, 971–972, *972f*
    CCAOH at 3′-end of, 969–970, *970f*
    folding of, 975–976, *976f*, *977f*
    secondary structures of, 973, *973f–975f*, 975
Genetic diseases, diagnosis of, tandem repeat polymorphisms for, 294
Genetic engineering. *See* Recombinant DNA technology
Genetic mapping, 255
  bacterial mating experiments for, 278, *279f*, 280, *281f*
  of *Drosophila*, 513
  of *lacO*^{c} mutations, 669
  recombination frequencies to obtain, *256f*, 256–257
  somatic cell genetics for, 295
Genetic material
  DNA recognized as, *16f*, 16–17, *18f*
  RNA as, 141, 145
    viral, 17
    virulent bacteriophages with single-stranded RNA as genetic material, *332f*, 332–334, *333f*
Genetic recombination, 255–257
  genetic mapping using recombination frequencies and, *256f*, 256–257
  process of, *255f*, 255–256
Genetic regulation, mutants in study of, 266
Genome
  definition of, 43, 262
  human. *See also* Human Genome Organization (HUGO); Human Genome Project
    annotation of, 194
    DNA sequencing of, 189–192, *191f*, *193f*, 194
  stability and evolution of, LINE in target-primed reverse transcription and, 614–616
Genome Sequencer FLX (GS FLX), 195, *196f*, 197–199, *198f–200f*, 201
Genomic imprinting, *817f*, 817–818

Genotype, 4
bacterial, notations used to denote, 261, 262
GEN1 protein, 538
Gentibiose (6-*O*-β-D-glucopyranosyl-D-glucose), 314
Geobacillus stearothermophilus
DnaB helicase from, 386, 386*f*
ribosome of, crystal structure for, 1008
Germ cells, 217
number of mitochondria present in, in eggs versus sperm cells, 928
telomerase in, 246
Germline cells, immortality of, 438
Germ line therapy, social, political, ethical, and legal issues raised by, 25
gfp gene, 939, 947
Giardia intestinalis, Dicer from, crystal structure for, 942–943, 943*f*
Giemsa dye, 228, 228*f*
GIFs. *See* General transcription factors (GTFs)
Gilbert, Walter, 145, 185, 672, 677, 724, 844
Gill, Grace, 805
GINS protein, in eukaryotic DNA replication, 428
Gladstone, Leonard, 725
glmS gene, 592, 593
Gln-tRNA synthetase, synthesis in gram-positive bacteria and archaea, 985*f*, 985–986
Global genome excision repair pathway, 489
Globular proteins, 29
loops and turns in, 54–55, 55*f*
Glu-374, 983
Glucocorticoid receptor (GR), 780
6-*O*-β-D-Glucopyranosyl-D-glucose (gentibiose), 314
Glucose, inhibitory effect on *lac* operon expression, 678–680, 678*f*–680*f*
Glucosylation, of phage T4 DNA, 320
Glutamic acid (glutamate)
abbreviations for, 33*t*
side chains of, 30, 30*f*
Glutamine
abbreviations for, 33*t*
side chains of, 31, 31*f*
Glutathione, aflatoxins and, 460
Glycerophospholipids, 66–68, 68*f*
Glycine
abbreviations for, 33*t*
lack of tendency to be part of α-helix, 53
side chains of, 31, 32*f*
Goldberg, Alfred L., 101
Goodman, Myron F., 503–504
Gourse, Richard L., 710
gp32, 122
GPCR (G protein coupled-receptor), 95, 95*f*
G1 phase, 219, 219*f*
G2 phase, 219, 219*f*
G protein complex, 95–96, 96*f*
G protein coupled-receptor (GPCR), 95, 95*f*
G protein-linked signal transduction systems, 94–99
components of, 95, 95*f*
first messengers in, 98
operation of, 94–98, 95*f*–99*f*
second messengers in, 98
gp17 subunit, phage T4 packaging motor function and, 324
gp20 subunit, phage T4 packaging motor function and, 324
GR (glucocorticoid receptor), 780
Gram, Christian, 257
Gram-negative bacteria, 257–258, 258*f*
Gram-positive bacteria, 257–259, 258*f*
Gln-tRNA synthetase synthesis in, 985*f*, 985–986
Granular compartment (GC), 902
Grasses, genomes of, 598
Gratuitous inducers, 667
GreA elongation factor, 651, 652*f*
GreB elongation factor, 651, 652*f*
Greek names, for DNA polymerases, 504, 504*t*
Green, Rachel, 1039
Green fluorescent protein, 772
Greider, Carol, 434
Griffith, Fred, 12–13
Group I introns, 914, 915*f*
Grunberg-Manago, Marianne, 990
G-segment DNA, 132
G (gated) segment DNA, 132
GS FLX (Genome Sequencer FLX), 195, 196*f*, 197–199, 198*f*–200*f*, 201
Guanine
in A-, B-, and Z-forms of DNA, 8*t*
in DNA, 6, 6*f*
methylation of, 458, 459*f*
water-mediated deamination of, 453
Guanosine, 7*f*
Guanosine-5′diphosphate-3′-diphosphate (ppGpp), 710*f*, 710–711
Guanosine-5′triphosphate-3′-diphosphate (pppGpp), 710*f*, 710–711
Guanylyltransferase
5′-m7G caps and, 836*f*, 838, 838*f*, 839
synthetic CTD interaction with, 839, 839*t*
Gurdon, John B., 818–819
gyrA gene, 380, 380*t*
GyrA protein, in recombination, 527*t*
gyrB gene, 380, 380*t*
GyrB protein, in recombination, 527*t*
Gyurasits, Elizabeth B., 369

H

HaeII, sequence specificity of, 169*t*
HaeIII, sequence specificity of, 169*t*
Haemophilus influenzae, restriction endonuclease HindII isolated from, 168
H2A•H2B dimer, 235, 441
H2A histone, 231, 231*t*, 233, 234
nucleosome assembly and, 441
in polyomaviruses, 349
Hairpin loop-bulges, 143, 144*f*
Hairpin loop structures, 142, 142*f*
Hairpin-mediated transposition, 579–580, 580*f*
Hairpin structures, *trp* operon and, 693*f*, 693–694, 695
Haloarcula marismortui, ribosome of, structure of, 1008, 1011, 1011*f*, 1036–1038, 1037*f*, 1038*f*
Hammerhead ribozymes, 143, 144*f*
Hammersten, Einar, 8
Hanawalt, Philip, 549
Hankin, Ernest, 308
Hannon, Gregory J., 941
Haploid number, 217, 224*f*–225*f*, 226–227
in yeast, 287
HAT (histone acetyltransferase), 814
isolation from *Tetrahymena*, 814
HAT medium, 299
"hAT" transposons, 587
Haworth, Walter N., 5
Haworth structures, for furanoses, 5*f*, 5–6
H2AX histone, 529–540
H2A.Z histone, 812, 813*f*
HBB (hook-basal body) flagellar substructure, 644
HBD (helicase bonding domain), 388
H2B histone, 231, 231*t*, 233, 234, 235
nucleosome assembly and, 441
in polyomaviruses, 349
HBS (RNA-DNA hybrid binding site), of transcription elongation complex, 648
Hda1, 814
HDACs (histone deacetylases), 814, 956
HDE (histone downstream element), 879, 880*f*
Head, of ribosome, 1007
Headful mechanism, 324
Headpiece, of Lac repressor, 674, 674*f*
Heat. *See* Temperature
Heavy chains, 83–84, 84*f*, 85*f*
HeLa cells, 297
hnRNA synthesized by, 830, 830*f*
for mRNA formation study, 840
for rRNA formation study, 897
Helical tracts, in RNA secondary structure, 142, 142*f*
Helical wheels, 791–792, 792*f*
Helicase(s)
classification of, 120
detection of, 120, 120*f*
direction of movement, 120, 120*f*
superfamilies of, 120–121, 121*f*
Helicase bonding domain (HBD), 388
Helicase domain, of SV40 T antigen, 417, 417*f*, 418*f*
Helicase II, mismatch repair and, 495
Helitrons, 618
Hel IV protein, in recombination, 527*t*
α-Helix, 51–54, 52*f*, 53*f*
Helix-loop-helix (HLH) motif, DNA-binding domains with, 794–797
bHLH zip proteins and, 796*f*, 796–897, 797*f*
transcription activator proteins and, 794–795, 795*f*
Helix-turn-helix structures
of CI regulator, 700, 700*f*
DNA-binding domains with, 773–777
homeodomains and, 776, 776*f*
homeotic genes specifying transcription activator proteins and, 774–775
Hox genes and, 774, 775*f*
identification of, 773*f*, 773–774
in POU proteins, 776–777, 777*f*
Hemicatenanes, 538–539
Hemoglobin
gene coding for β-globin chain in, 292–293, 293*f*
sickle cell (HbS), structural differences between normal hemoglobin (HbA) and, 80–83, 81*f*, 82*f*
structural differences between myoglobin and, 76–78, 77*f*–80*f*, 80
Hermes transposase, 587–588, 588*f*, 589*f*
Herpes simplex virus (HSV), VP16 of, 803, 803*f*
Herrick, James, 80
Hershey, Alfred, 16–17, 987
Hershko, Avram, 99–100
Heterochromatin, 218, 218*f*
assembly of, 955–956, 957*f*
Heterodimers, 29. *See also specific heterodimers*
Heterogeneous nuclear RNA (hnRNA), 830
poly(A) tails and, 830–831, 831*f*, 832*f*
Heterokaryons, 69–70
formation of, 297–299, 297*f*–299*f*
Heterothallic strains, 288, 288*f*
Heterotrimers, 29
Hexokinase A, molecular details for, 91, 92*f*
Hexons, 353
Hfr (high frequency of recombination) males, 276, 280
H19 gene, 817, 817*f*
HGPRT (hypoxanthine guanine phosphoribosyl transferase), in hybrid cell selection, 298, 299
HGPRT gene, knockouts of, 546, 546*f*
HhaI, sequence specificity of, 169*t*

H3•H4 heterodimer, 234, 441
H1 histone, 231, 231*t*, 232, 236
H3 histone, 231, 231*t*, 232, 233, 234, 235
  nucleosome assembly and, 441
  in polyomaviruses, 349
H4 histone, 231, 231*t*, 232, 233, 234
  nucleosome assembly and, 441
  in polyomaviruses, 349
hHR23B protein, 489
Hierarchical shotgun sequencing, 189–192, 191*f*
Higara, Sota, 214
High frequency of recombination (Hfr) males, 276, 280
High-performance liquid chromatography (HPLC), 37
HindII, 168
  sequence specificity of, 169*t*
HindIII
  DNA cutting by, 170, 170*f*
  sequence specificity of, 169*t*
Hinge region, of Lac repressor, 675, 675*f*
Hirose, Yutaka, 871
Histidine
  abbreviations for, 33*t*
  side chains of, 30, 30*f*
Histidine-containing protein (HPr), 679
Histidine operon, 695
Histone(s), 76, 109
  chromatin and, 211
  classes of, 231, 231*t*
  covalent modification affecting gene activity, 814–815, 815*f*
  H1, 231, 231*t*, 232, 236
  H3, 231, 231*t*, 232, 233, 234, 235
    nucleosome assembly and, 441
    in polyomaviruses, 349
  H4, 231, 231*t*, 232, 233, 234
    nucleosome assembly and, 441
    in polyomaviruses, 349
  H2A, 231, 231*t*, 233, 234
    nucleosome assembly and, 441
    in polyomaviruses, 349
  H2AX, 529–540
  H2A.Z, 812, 813*f*
  H2B, 231, 231*t*, 233, 234, 235
    nucleosome assembly and, 441
    in polyomaviruses, 349
  modifiers of, 813–815, 815*f*
Histone acetyltransferase (HAT), 814
  isolation from *Tetrahymena*, 814
Histone chaperones, 439–441, 440*f*
Histone deacetylases (HDACs), 814, 956
Histone downstream element (HDE), 879, 880*f*
Histone genes, lack of introns of, 846
Histone octamers, formation of, 234–235, 235*f*
Historical hypothesis of codon assignments, 1001
HIV (human immunodeficiency virus), 355
HJs. *See* Holliday junctions (HJs)
H3-K9 protein, 956
HLH motif. *See* Helix-loop-helix (HLH) motif
HMC (5-hydroxymethylcytosine), in phage T4 DNA, 314, 314*f*
  synthesis of, 319
HML gene, 556–557, 557*f*
HMR gene, 556–557, 557*f*
HNPCC (non-polyposis colon cancer), 498
hnRNA (heterogeneous nuclear RNA), 830
  poly(A) tails and, 830–831, 831*f*, 832*f*
H-NS protein, 217, 590
Ho, C. Kiong, 839
Hoagland, Mahlon, 966, 967, 969
hobo element, 602
HO gene, 556–557, 557*f*
Hogness, David, 742
holA gene, 380*t*
holB gene, 380*t*
holC gene, 380*t*
holD gene, 380*t*
holE gene, 380*t*, 394
Holin, phage T4 lysis and, 325
Holley, Robert, 972, 973
Holliday, Robin, 518
Holliday junctions (HJs), 519
  dissolution of, 539, 539*f*
  resolution of
    in double-strand break model of recombination, 526
    in eukaryotic homologous recombination, 538*f*, 538–540, 539*f*
    in Holliday model of recombination, 519, 521*f*
    in meiotic recombination, 544
    in Meselson-Radding model of recombination, 522–524, 523*f*
    RecG and, 532
    RuvABC complex and, 532
Holliday model of recombination, 518–519, 520*f*, 522, 522*f*
Homeobox, 774–775
  of POU transcription activator proteins, 776–777, 777*f*
Homeodomain, 776, 776*f*
  POU, 777
Homeotic genes, 773–775
  positional identities assigned by, 773–774, 773*f*–775*f*
  transcription activator protein specification by, 774–775
HOM genes, 773*f*, 773–774
Homodimers, 29
Homogenistic acid, 4
Homolog(s), 217
Homologous recombination, 512–553
  bacterial, 526–533
    RecA in, 528*f*, 528–529, 530*f*
    RecBCD complex in, 529, 531*f*, 531–532, 532*f*
    RecFOR pathway of, 532, 533*f*
    recombination mutant conjugation rates and sensitivity to DNA damage and, 526–527, 527*t*
  in bacteriophages, 517–518, 518*f*, 519*f*
  double-strand break repair model of, 524–526, 525*f*
    eukaryotic homologous recombination proteins and, 533, 534*t*, 535–540
    variation of, 540, 540*f*
  eukaryotic, proteins involved in, 533, 534*t*, 535–540
  functions of, 512–513, 513*f*–517*f*, 515–516
  Holliday model of, 518–519, 520*f*, 522, 522*f*
  meiotic, 540–544
    novel aspects of, 540–541, 542*f*
    proteins involved in, 541–542
    types of recombination events in, 543*f*, 543–544
  Meselson-Radding (Aviemore) model of, 522–524, 523*f*
  mitotic, 544–553
    gene knockouts in mice and, 545–547, 546*f*–548*f*
    gene knockouts in yeast and, 544–545, 545*f*
    regulation to prevent chromosome rearrangements and genomic instability, 550–552, 551*f*, 552*f*
    replication fork collapse and, 549–550, 549*f*–551*f*
    target gene disruptions using, 544, 544*f*
  synthesis-dependent strand-annealing model of, 540, 540*f*
Homothallic strains, 287, 288, 288*f*
Homotrimers, 29
Hoogsteen, Karst, 139
Hoogsteen base pairing, 139
Hook-basal body (HBB) flagellar substructure, 644
HOP2•MND1, 534*t*
Hop2•Mnd1 complex, 534*t*, 542
Hormone antagonists, clinical applications of, 785–786
Hormone response element, 783*f*, 783–784
Horvitz, H. Robert, 950–951
Host restriction-modification system, 312–313, 313*t*
Hotchkiss, Rolin, 15
Housefly, *Hermes* transposase of, 587–588, 588*f*, 589*f*
Howard-Flanders, Paul, 482
Hox A-3 deletion mutations, 774, 775*f*
Hox genes, 774, 775*f*
HpaI, sequence specificity of, 169*t*
HpaII, sequence specificity of, 169*t*
HPLC (high-performance liquid chromatography), 37
HPr (histidine-containing protein), 679
H1 RNA, 920
hsp70 gene, 835–836, 836*f*
HSV (herpes simplex virus), VP16 of, 803, 803*f*
HSV-tk gene, 546–547, 547*f*
hTOPOIIIa protein, 539
hTOPOIIIα protein, 534*t*
HU, 217
HUGO (Human Genome Organization), DNA polymerase nomenclature system of, 504*t*, 504–505
Hui, Ann, 1019
Human(s)
  divergence rate between Neanderthals and, 201
  DNA polymerases of, template-dependent, 504*t*, 504–505
  exons of, 846, 846*f*
  genome of. *See* Human Genome Organization (HUGO); Human genome; Human Genome Project
  introns of, 846, 846*f*
  mediators in, 536–537
  mismatch DNA repair in, 497–498, 498*f*
  MRN complex in, 535, 536*f*
  mtDNA of, first complete sequence for, 928, 929*f*
  proteome of, variety of proteins contained by, 883–884
  recombination proteins of, 534*t*
  TFIIA of, 748
  TFIIH of, 752
Human breast cancer susceptibility gene 2 *(BRCA2)*, 537
  mutant, 551, 551*f*
Human chromosome 1, DNA molecule size of, 110*t*
Human genome. *See also* Human Genome Organization (HUGO); Human Genome Project
  annotation of, 194
  DNA sequencing of
    hierarchical DNA sequencing for, 189–192, 191*f*
    information provided by, 194
    whole genome sequencing for, 189, 192, 193*f*, 194
Human Genome Organization (HUGO), DNA polymerase nomenclature system of, 504*t*, 504–505

Human Genome Project, 845, 845*t*
clone-by-clone approach of, 192
launching of, 190
shotgun DNA sequencing in
hierarchical, 189–192, 191*f*
whole genome, 189, 192, 193*f*, 194
Human immunodeficiency virus (HIV), 355
Human metallothionein IIA gene, AP-1 proteins binding to enhancer regions of, 792–793
Hunter, Craig P., 948–949
Huntington's disease, 1053
Hybrid cells
monoclonal antibody formation from, 200, 201, 300*f*
selection of, 297–299, 299*f*
Hybridization, in Southern blotting, 175
Hybrid model of RNA polymerase II transcription termination, 878, 879*f*
Hybridomas, 301
Hybrid-states translocation model, 1040*f*, 1040–1041
Hydrogen bonds
in α-helix, 52
protein structure and, 46, 47–48, 48*f*
stabilization of double-stranded DNA by, 114*f*, 114–115
Hydrogen peroxide, DNA damage caused by, 455
Hydrolytic cleavage reactions, DNA damage caused by, 453–455, 453*f*–455*f*
Hydrophobic chromatography, 38, 38*f*
Hydrophobic interaction, 46, 48–49, 49*f*, 50*f*
Hydrophobicity, 49, 49*f*, 50*f*
Hydroxyl radicals
DNA damage caused by, 455–456, 456*f*, 457*f*
radiation damage to DNA and, 452
5-Hydroxymethylcytosine (HMC), in phage T4 DNA, 314, 314*f*
synthesis of, 319
4-Hydroxynonenal, 456, 456*f*
Hyperthermic archaea, reverse gyrase of, 132
Hypervariable regions, 84
Hypoxanthine guanine phosphoribosyl transferase (HGPRT), in hybrid cell selection, 298, 299

I

Icosahedral head with tail bacteriophages, 309, 309*f*, 310
Icosahedral tailless bacteriophages, 309, 309*f*, 310, 332, 332*f*
ICR (internal control region), 920–921, 922
Identity elements, 983, 983*f*
IE (intermediate element), 920
IF(s). *See* Initiation factors (IFs)
Igf2 gene, 817, 817*f*
IHF (integration host factor), 217, 217*f*, 274, 385, 609, 645
Ile-tRNA synthetase
editing site for, 982
hydrolysis of valyl-tRNA and valyl-AMP by, 980–982, 981*f*
Illegitimate recombination, 553–555, 554*f*, 555*f*, 555*t*
Imitation switch (ISWI) chromatin remodeling complex, 441, 810
Immunoglobulin(s)
diversity among, 558–564, 559*f*, 563*f*
somatic hypermutation as mechanism of, 558–564, 559*f*, 563*f*
IgG, domains of, 83–84, 84*f*, 85*f*
Inchworming model, 640
Independent assortment, law of, 3
Induced fit model of enzyme-substrate complex, 90, 90*f*
Induced pluripotent stem (iPS) cells, 820*f*, 820–821
Inducer exclusion, 679–680
Inducible enzymes, 667
Ingram, Vernon, 81, 82
Inheritance, Mendel's laws of, 2–3
Inhibition
contact, 296
feedback, 92
Initiation factors (IFs)
bacterial, 1012
eukaryotic, 1022, 1023*f*, 1023*t*, 1024, 1025*f*–1027, 1026–1028
phosphorylation of, 1028, 1028*t*
IF1, 1016–1017
IF2, 1016, 1017, 1017*f*
IF3, 1016, 1017
30S and 70S initiation complex formation and, 1016*f*, 1016–1017, 1017*f*
Initiation stage of DNA replication
bacterial, 379, 381–388
DnaA•ATP in, 384–385, 385*f*
DnaA in, 383*f*, 383–384, 384*f*
DnaC in, 386, 387*f*
DnaG in, 388, 388*f*
DnB helicases in, 385–386, 386*f*
at *oriC*, 381–383, 382*f*
regulation of, 408–409
replicon model of, 381, 381*f*
eukaryotic, 422–429
activation of licensed origin in, 427–429, 428*f*
anonymously replicating sequences in, 422–425, 423*f*, 424*f*
gel electrophoresis location of origins of replication and, 425*f*, 425–426
MCM2-7 helicase loading into origin in, 427, 427*f*
origin of recognition complex in, 426
replicator sites in, 422, 422*f*, 423*f*
Initiation stage of protein synthesis
archaeal, 1012, 1029
bacterial, 1012–1020
initiator and elongator methionine tRNAs and, 1013*f*, 1013–1014, 1014*f*
riboswitches regulating, 1020, 1020*f*
Shine-Dalgarno sequences in, 1018–1020, 1019*f*
start codons and, 1012–1013
30S and 70S initiation complex formation and, 1016*f*, 1016–1017, 1017*f*
30S subunit in, 1014–1016, 1015*f*
eukaryotic, 1012, 1020–1029
initiation factors in, 1022, 1023*f*, 1023*t*, 1024, 1025*f*–1027*f*, 1026–1028, 1028*t*
methionine tRNA in, 1020–1021, 1021*f*
scanning mechanism and, 1021–1022
Initiation stage of transcription
abortive, 639
bacterial RNA polymerases in, 638–645
activator protein requirement of σ54-RNA polymerase, 644*f*, 645
conserved domains of σ70 family and, 638, 638*f*
DNA footprinting technique and, 630–633, 631*f*–633*f*
DNase protection method and, 629*f*, 629–630, 631*f*
inchworming model and, 640
promoters and, 628–629, 636–638, 637*f*
RNA polymerase crystal structures and, 633–636, 634*f*–636*f*
scrunching of, 639*f*, 639–640
σ factors in, 643*t*, 643–644
stepwise nature of, 640, 641*f*, 642
transient excursion model and, 640
Initiator (Inr), 742, 743
Initiator DNA, 418
Initiator proteins, 381
INO80 chromatin remodeling complex, 540, 810
Inositol 1,4,5-triphosphate, formation of, phosphoinositide-specific phospholipase C and, 98, 98*f*
Insects, transposons in, as molecular genetics tools, 600–602, 601*f*
Insertions, 264
Insertion sequences, 577, 578*f*
Insulin, stimulation of translation initiation by, 1028
Integral membrane proteins, 69
Integrases, 609
in Rouse sarcoma virus, 359
Integration host factor (IHF), 217, 217*f*, 274, 385, 609, 645
Integrons, 584
Integron system, 611
Interaction
through bonds, 59, 59*f*
cooperative, among oxygen-binding sites on hemoglobin molecule, 78
through distance, 59, 59*f*
Intercalation, 159–160
Interface canyon, of ribosome, 1008
Interference, in meiotic recombination, 541
Intermediate element (IE), 920
Intermediate stage, in transcription initiation, 640, 641*f*, 642
Internal control region (ICR), 920–921, 922
Internal ribosomal entry sites (IRES), 1022
Internal transcribed spacer 1 (ITS-1), 898
Internal transcribed spacer 2 (ITS-2), 898
International normalized ratio, in promoter lacking TATA box, TAF interaction wit, 746–747, 747*f*
Internet, molecular biology resources on, 25
Interphase, 218*f*, 218–219
S phase of, 219, 219*f*
Interwound supercoils, 123, 123*f*
int gene, 703
Intrinsically disordered proteins, 59–60
Intrinsic splicing enhancers (ISEs), 869*f*, 869–870, 870*f*
Intrinsic termination, 652*f*, 652–653
Intron(s), 45, 724, 844
group I, 914, 915*f*
human consensus sequences defining, 852, 852*f*
lack conservation during evolution, 845*t*, 845–846, 846*f*
mobile group II, target-primed reverse transcription and, 612, 613, 616–617
homology to target site and, 617
transesterification reactions allowing, 616*f*, 617, 617*f*
size of, 846, 846*f*
Intron definition, 869
Intronic splicing silencer (ISS), 870
Inversions, 452
Inverted repeat sequences, 138, 138*f*
Ion-exchange chromatography, 37*f*, 37–38
Ionizing radiation, DNA damage caused by, 449–452
ultraviolet light and, 449, 450*f*, 451, 451*f*
x-rays and τ-rays and, 452
IPIG. *See* Isopropylthiogalactoside (IPTG)
iPS (induced pluripotent stem) cells, 820*f*, 820–821
IRES (internal ribosomal entry sites), 1022
IS*903*, 591
I-Sce1 system, 557–558, 558*f*

ISEs (intrinsic splicing enhancers), 869*f*, 869–870, 870*f*
IS*3* family elements, 586
Isoacceptors, 972
Isoelectric pH, 40
Isoleucine
  abbreviations for, 33*t*
  side chains of, 31, 32*f*
Isoleucine-valine operon, 695
Isopeptide bonds, between ubiquitin and proteins, 100*f*, 100–101
Isopropylthiogalactoside (IPTG)
  Lac repressor binding studies using, 672–673
  in lactose enzyme induction studies, 667, 667*f*
Isoschizomers, 168
ISS (intronic splicing silencer), 870
ISWI (imitation switch) chromatin remodeling complex, 441, 810
Ito, Kiochi, 1044
Ito, Yukata, 59
ITS-1 (internal transcribed spacer 1), 898
ITS-2 (internal transcribed spacer 2), 898
Izsvák, Zsuzanna, 602–603

J

Jacob, François, 381, 667, 668, 987
Jakes, Karen, 1018–1019
Jasin, Maria, 558
Jeffreys, Alex, 294
Johannsen, William Louis, 255
Johansson, Erik, 429
Junction(s), in RNA, 142, 143*f*
Junctional diversity, 559

K

Kadonaga, James T., 770
Karyotype, 227–231
  chromosome site nomenclature and, 227–228, 228*f*, 229*f*
  information available from, 228*f*, 228–230, 230*f*, 231*f*
  spectral karyotyping and, 231
Katayama, Tsutomu, 382
Kelly, Thomas J., 416, 418
Kelner, Albert, 469
Kendrew, John, 56
α-Keratin, 54
Khorana, H. Gorbind, 995
Khoury, George, 767
Kim, Peter S., 55
Kinetic curves, sigmoidal, 92–93, 93*f*
Kinetochore, 221, 242, 243*f*
Kingsbury, Robert, 765
Kissing hairpins, 143, 144*f*
Klenow, Hans, 179
Klenow fragment, 179*f*, 179–180, 180*f*
Klf4 gene, 820*f*, 820–821
Klug, Aaron, 777–778, 976
K_M (Michaelis constant), 88–89, 89*f*
Knockouts, 544–548
  by gene insertion, 544, 544*f*
  by gene replacement, 544, 544*f*
  in mice, gene targeting to create, 545–547, 546*f*–548*f*
  mitotic recombination to make, 544, 544*f*
  in yeast, occurring by homologous recombination, 544–545, 545*f*
Köhler, Georges J. F., 299, 301
Kolev, Nikolay, 880
Kornberg, Arthur, 177, 178, 382, 391, 731, 733, 752
Kornberg, Roger, 805–806, 811
Kozak, Marilyn, 1022
KpnI, sequence specificity of, 169*t*
Krzycki, Joseph A., 986
17kT protein, of SV40, 349
Ku70•Ku80 complex, 554, 554*f*, 561
Kunkel, Thomas, 430
Ku70 protein, 553, 555*t*
Ku80 protein, 553, 555*t*
Kyte, J., 49

L

lacA gene, 666–667, 669, 670*f*
lacI mutations, 667–668
lacI promoter, 672
lac mRNA, 667–668, 668*t*
lacO^c mutations, 668–669
lacO operators, 669, 673*f*, 673–674
lac operon
  cAMP•CRP complex activation of, 682, 683*f*
  expression of, inhibitory effect of glucose on, 678–680, 678*f*–680*f*
  operators of, 673*f*, 673–674, 675, 677*f*
Lac repressor, 668
  binding to lac operator *in vitro*, 672*f*, 672*t*, 672–673, 673*f*
  functional units of, 674–675, 674*f*–676*f*
  structure of, 674*f*–677*f*, 674–675, 677
Lac repressor•DNA complex, 675, 676*f*
Lactose permease, 665
  genes coding for, 665–667, 666*f*
*lac*UV5, 632
lacY gene, 669, 670*f*
lacZ gene, 596, 668, 669, 670*f*, 941, 987
  lactose metabolism and, 666
  transcription of, 709
Laemmli, Ulrich K., 238
Lagging strand, 375, 375*f*
  synthesis of, Pol δ and, 429, 432
Lag phase, 260
Lambdoid phages, 334. *See also* Bacteriophage λ
Landers, Eric, 191–192
Large tumor antigen. *See* T antigen
Lariat model, 853*f*, 853–854, 854*f*
Lawn, bacterial, 311
LBDs (ligand-binding domains), 770, 784–786, 786*f*, 787*f*
LBP (ligand-binding pocket), 785
L cell line, 297
3′-Leader (3′-untranslated region), in mRNA, 663
5′-Leader (5′-untranslated region), in mRNA, 663
Leader, of *trpE*, 693
Leading strand, 375, 375*f*
  synthesis of, Pol ϵ and, 429, 431–432
Leder, Philip, 994–995
Lederberg, Joshua, 270–271, 342, 586
let-7 gene, 951
Leucine
  abbreviations for, 33*t*
  side chains of, 31, 32*f*
Leucine operon, 695
Leucine zipper model, 791, 791*f*
Levene, Phoebus A., 6, 8
Levinthal, Cyrus, 63
lexA gene, SOS response and, 501
LexA protein, 801
  SOS response and, 499–502, 500*f*, 501*f*, 503
Li, Joachim J., 416
Lif1 protein, 554, 555*t*
ligA gene, 380*t*
Ligand(s), 40, 770
Ligand binding core domain, of Lac repressor, 675, 676*f*
Ligand-binding domains (LBDs), 770, 784–786, 786*f*, 787*f*
Ligand-binding pocket (LBP), 785
Light chains, 83–84, 84*f*, 85*f*
LIG1 protein, 534*t*
Lig1 protein, 534*t*
Lig protein, in recombination, 527*t*
Lin, Ren-Jang, 866
LINEs. *See* Long interspersed nucleotide elements (LINEs)
lin-14 gene, 951
Lingner, Joachim, 434
Linker DNA, 232–233, 233*f*
Linker-scanning mutagenesis, 765*f*, 765–766, 766*f*
Linking number *(Lk)*, 125–127
Lipid(s)
  in biological membranes, interaction with proteins, 66–68, 67*f*, 68*f*
  definition of, 66
Lipid bilayer, 66, 68, 68*f*
Lipid monolayer, 67–68
Lipmann, Fritz, 1029
Lipoproteins, 69
Liposomes, 68, 68*f*
Liquid growth medium, 259
Lis, John T., 835–836
Liver disease, aflatoxins causing, 460, 461*f*
Lk (linking number), 125–127
L-*myc* gene, 796
L-Myc transcription regulator, 796
Lock-and-key model of enzyme-substrate complex, 89, 90*f*
Loeb, Timothy, 332
Long interspersed nucleotide elements (LINEs), 576
  in target-primed reverse transcription, 613*f*, 613–616, 615*f*
    genome stability and evolution and, 614–616
Long patch base excision DNA repair, 481, 482*f*
Long-range base pairing, 1010
Long terminal repeat (LTR) retrotransposons, 576
Loops. *See also* Helix-loop-helix (HLH) motif; Loop-sheet-helix structure
  dimerization, 782
  D-loops, 244, 245*f*
    formation of, 528*f*, 528–529, 530*f*
  DNA looping and, 770, 770*f*
  in globular proteins, 54–55, 55*f*
  hairpin-loop bulges, 143, 144*f*
  hairpin loop structures, 142, 142*f*
  in helical tracts, 142, 142*f*
  internal, symmetrical, 142, 142*f*
  stem-loop structures and, 142, 142*f*
  T-loops, 244, 245*f*
Loop-sheet-helix structure, DNA-binding domains with, 789*f*, 789–790, 790*f*
Loquacious (LOQS), 953
Lorch, Yahli, 812
LTR (long terminal repeat) retrotransposons, 576
Luciferase, 772
Luria, Salvador, 309
Lymphomas, recombination signal sequences and, 561, 561*f*
Lysine
  abbreviations for, 33*t*
  side chains of, 30, 30*f*
Lysis. *See also* Lytic life cycles
  of bacteriophages, 310
    phage φX174, 331–332
    phage T4, 325
    RNA phages, 334
  to remove bacterial chromatin from cell, 212
Lysogenic cells, 311

Lysogenic life cycles, of bacteriophages, 310–311, 341–342
phage λ, 335–336, 337f, 339–340
Lysozyme
bacterial cell wall and, 259
egg-white, molecular details for, 90–91, 91f
Lytic life cycles, of bacteriophages, 310, 341. *See also* Bacteriophage(s), virulent; Lysis
phage λ, 335, 336–337, 337f, 338f, 339

M

Ma, Jun, 801–802
MacLeod, Colin, 13–14
Macromolecules
determinants of, velocity of movement of, 156
physical techniques for studying, 153–165
electron microscopy, 153–156, 154f–156f
equilibrium density gradient centrifugation, 158–160, 159f, 160f
gel electrophoresis, 160–165
velocity sedimentation, 156–158, 157f, 158f
Mad•Max heterodimer, 796, 797f
Magnesium, RNA tertiary structure stabilization by, 142
Magnetic spin, 58f, 58–59, 59f
Maize, gene conversion in, 516
Makman, Richard S., 678
Malaria, sickle cell anemia and, 83
Malignant diseases. *See* Cancer
Malondialdehyde, 456, 456f
Mammals. *See also specific mammals*
genomic imprinting in, 817f, 817–818
nucleotide excision DNA repair in, 485, 488–490, 491f, 492f, 492–493
transposons for study of, 602–605, 603f, 604f
Mammuthus primigenius, DNA sequencing from, 201
Maniatis, Tom, 852–854
Manley, James L., 866–868, 871
Marcker, Kjeld, 1013–1014
Margulies, Ann Dee, 526–527
Marmur, Julius, 114
Mass spectrometry, for amino acid sequence determination, 43, 44f
MATa gene, 774–775
MATα gene, 774–775
Matα (α) mating type, 287, 288–290, 289f
Mata (a) mating type, 287, 288–290, 289f
MAT gene, 556
Mating
in *E. coli*, 270–273, 271f–273f
genetic map produced by, 278, 279f, 280, 281f
number of F factors genes need for, 274f, 274–276, 275f
in yeast, 287–288, 288f
mating type determination and, 288–290, 289f
Mating factors, 290
Mating type switching
site-specific recombination and, 556–558, 557f, 558f
synthesis-dependent strand-annealing model of, 540, 540f
Matthaei, Johann Heinrich, 989–990, 991, 1017
Matthews, Kathleen, 675
Maxam, Allan, 185
MBDs (methyl-CpG binding domains), 816
MboI, sequence specificity of, 169t
McCarty, Maclyn, 13–14
McClintock, Barbara, 243–244, 575–576, 577
McKnight, Steven L., 765, 791
MCM2-7, in eukaryotic DNA replication, 427, 427f, 428
MCM2-7 helicase, in chromatin disassembly, 440
MCM helicase, Orc1/Cdc6 recruitment of, in archaea, 442, 442f
MED1 gene, 807
MED2 gene, 807
Mediator(s), 536
Mediator (transcription factor), 764, 805–809
activated transposition's requirement for, 806–808, 808f
association with activators at UAS, 808–809, 809f
discovery of, 805–808
genes coding for subunits in, 806–807
interaction with promoters, 808–809, 809f
nomenclature system for, 807
squelching and, 805f, 805–806
subcomplexes of, 807–808
topological organization of, 807, 808f
Meiosis, 224f–225f, 226–227
allele segregation during, 513, 515f, 515–516, 516f
anaphase II of, 224f, 227
anaphase I of, 225f, 227
crossing over during, 513, 513f
metaphase II of, 225f, 227
metaphase I of, 225f, 227
prophase II of, 225f, 226–227
prophase I of, 224f, 226–227
telophase II of, 224f, 227
telophase I of, 225f, 227
Meiotic homologous recombination, 540–544
novel aspects of, 540–541, 542f
proteins involved in, 541–542
types of recombination events in, 543f, 543–544
Mei5•Sae3 complex, 534t, 542
Mello, Craig C., 937, 938–940
Melting temperature (T_m), for DNA, 114f, 114–115
Melton, Douglas, 821
Mendel, Gregor, 2–3, 4, 290, 512, 513, 515, 516
Mer3 protein, 534t, 544
Meselson, Matthew, 367, 369, 518, 522, 1014–1015
Meselson-Radding model of recombination, 522–524, 523f
Messenger RNA. *See* Bacterial mRNA; mRNA
Messing, Joachim, 348
Metabolic pathways, mutant studies of, 264–265
Metal shadowing, 153–155, 154f
Metaphase, 220f, 221
of meiosis
metaphase I, 225f, 227
metaphase II, 225f, 227
of mitosis, 220f, 221
Methanococcus jannaschii, RNA polymerase from, 756, 757
Methanopyrus, topoisomerase V isolation from, 132
Methionine
abbreviations for, 33t
side chains of, 31, 32f
Methionine aminopeptidase, 1013
Methionine tRNA, in initiation stage of protein synthesis
bacterial, 1013f, 1013–1014, 1014f
eukaryotic, 1020–1021, 1021f
Methionyl t-RNA formyltransferase, 1013
1-Methyladenine, AlkB catalysis of methyl group removal in, 476, 477f
Methylation snoRNPs, 910, 910f
Methyl-CpG binding domains (MBDs), 816
3-Methylcytosine, AlkB catalysis of methyl group removal in, 476, 477f
O^6-Methylguanine, 458, 459f
1-Methylguanine, AlkB catalysis of methyl group removal in, 476, 477f
N^7-Methylguanosine (m^7G), 834
Methylguanosine caps, of mRNA, 831–840, 833f–835f
attachment to nascent pre-mRNA, 835–836, 836f
carboxyl terminal domain and, 837–840, 838f, 839t
discovery of, 832–834
formation of, 836–837
functions of, 834–835, 835f
structure of, 834, 834f
Methylmethane sulfonate (MMS), DNA exposure to, 458, 459
N-Methyl-N-nitrosourea (MNU), DNA exposure to, 458, 459
O^4-Methylthymine, 458, 459f
3-Methylthymine, AlkB catalysis of methyl group removal in, 476, 477f
Methyltransferase, $5'm^7G$ caps and, 836f, 837, 838, 838f
m^7G (N^7-methylguanosine), 834
MgsA protein, in recombination, 527t
Mice
chimeric, 547
editing-defective alyl-tRNA synthetasein, 982–983
gene knockouts in, 545–547, 546f–548f
genomic imprinting in, 817f, 817–818
homologous recombination in, 537
Hox A-3 deletion mutations in, 774, 775f
lethal null mutations in, 535
LINEs in, 613
transgenic, 547, 548f
Michaelis constant (K_M), 88–89, 89f
Microarray technology, in DNA sequencing, 203f, 203–204, 205f
Micrococcal nuclease, properties of, 166t
Microprocessor complex, 953
MicroRNAs. *See* miRNAs (microRNAs)
Microsatellites, 291
Microsomal fraction, 965
Microsomes, Ames test and, 465–466
Microtubules, 219
assembly of, 219, 221f
attachment in centromere, 241–242, 241f–243f
Miescher, Friedrich, 3
Mig1 protein (multicopy inhibitor of GAL gene expression), 799
Milstein, César, 299, 301
Minimal agar, 260
Minimal clamp loader, 397–399, 397f–399f
Minimal inverted transposon repeat element (MITE), 598
Minimal medium, 259
Minisatellites, 291
Minor, Daniel L., Jr., 55
minos element, 602
miRISC, 953–954, 954f
miRNAs (microRNAs), 950–954, 951f–953f
formation of, 952f, 952–954, 953f
modes of action of, 953f, 953–954
number of, 951
precursor for, 952f, 952–953
Mismatch pairs, 142f
MITE (minimal inverted transposon repeat element), 598
Mitochondria
characteristic features of, 928, 929f
transcription in, 927–928, 929f, 930
Mitochondrial DNA. *See* mtDNA (mitochondrial DNA)
Mitogens, 229

Mitosis, 219, 220f–223f, 221–222, 226
anaphase of, 220f, 221
crossing over during, 513, 514f
metaphase of, 220f, 221
prophase of, 219, 220f, 221f
telophase of, 220f, 221
Mitotic homologous recombination, 544–553
gene knockouts and
in mice, 545–547, 546f–548f
in yeast, 544–545, 545f
regulation to prevent chromosome rearrangements and genomic instability, 550–552, 551f, 552f
replication fork collapse and, 549–550, 549f–551f
target gene disruptions using, 544, 544f
Mitotic spindles, 219
Mixed connective tissue disease, anti-Smith (anti-Sm) antibodies in, 855–856
MMS (methylmethane sulfonate), DNA exposure to, 458, 459
Mms4 protein, 534t
MNU (N-methyl-N-nitrosourea), DNA exposure to, 458, 459
Moazed, Danesh, 1040, 1041
Mobile elements
autonomous, 594
in bacteria, 577, 578f
movement of. *See* Conservative site-specific recombination; Target-primed reverse transcription; Transposition
nonautonomous, 594
transposase complex formation with, 577–579, 578f, 579f
transposons. *See* Transposons
Mobile group II introns, target-primed reverse transcription and, 612, 613, 616–617
homology to target site and, 617
transesterification reactions allowing, 616f, 617, 617f
Mobilizable plasmids, 276
Models. *See also specific model names*
animal, for human diseases, 24–25. *See also* specific models and diseases
definition of, 19
Modifier of transcription (Mot1), 747
Modrich, Paul, 497
Molecular biology
intellectual foundation of, 2–3
Internet resources on, 25
origin of term, 2
structure-function relationships as central theme of, 21
Molecular chaperones, 64–65, 65f, 66f
Molecular exclusion chromatography, 38, 39f, 40
Moloney murine leukemia virus (Mo-MLV), reverse transcriptase of, 202
Mo-MLV (Moloney murine leukemia virus), reverse transcriptase of, 202
Monoadducts
alkylation DNA damage by formation of, 457–461
alkyl group transfer to centers of negative charge and, 457–458, 458f, 459f
modification of environmental agents to become alkylating agents and, 458–460, 460f, 461f
formation of, chemical cross-linking agents and, 462f, 462–464, 463f
Monocistronic mRNA, 662
Monoclonal antibodies
applications of, 301
formation from hybrid cells, 200, 201, 300f
Monod, Jacques, 667, 668, 677, 987
Monro, Robin, 1035
Moore, Peter B., 1008, 1011, 1036–1038
mop1 gene, 599
Morgan, Thomas Hunt, 512–513
Morphogenetic enzymes, in phage assembly, 320
Mot1 (modifier of transcription), 747
Motif ten element (MTE), 742
Mouse. *See* Mice
MRE11 gene, 533
null mutations of, 535
MRE11 protein, 533, 534t
Mre11 protein, 533, 534t
Mre11•Rad50•Nbs1 (MRN) complex, 535, 536f, 554, 555t
Mre11-Rad50-Xrs2 (MRX) complex, 533, 535, 536f, 554
mRNA (messenger RNA)
bacterial. *See* Bacterial mRNA
in central dogma theory, 23f, 23–24
chloroplast DNA transcription to form, 930–932, 931f
degradation of, 703–705, 704f, 706f
direction of reading, 993f, 993–994
export of, 884–885
coupling of splicing and export and, 884–885, 885f
genetic code and, 987–1001
characteristics of, 996–997, 997f, 997t, 998t, 999
coding specificity of aminoacyl-tRNA and, 999f, 999–1000
codons in, 987–989, 988f
origin of, 1000–1001
poly(U) direction of poly(Phe) synthesis and, 989–991
polypeptide chain termination signals and, 996, 996f
protein synthesis process and, 991–993, 992f
ribosome programming to synthesize proteins and, 987
synthetic messengers confirming, 995f, 995–996
trinucleotide promotion of aminoacyl-tRNA binding to ribosomes and, 994f, 994–995
wobble hypothesis and, 1000, 1000f, 1001f
lifetime of, 663–664
methylguanosine caps of, 831–840, 833f–835f
attachment to nascent pre-mRNA, 835–836, 836f
carboxyl terminal domain and, 837–840, 838f, 839t
discovery of, 832–834
formation of, 836–837
functions of, 834–835, 835f
structure of, 834, 834f
mitochondrial transcription to form, 927–928, 930
multiple, formation with combinations of splicing patterns, 848f, 848–849, 849f
precursor. *See* pre-mRNA (precursor mRNA)
mRNA export protein, 884
mRNA•protein complex, 884, 885
MRN complex, 535, 536f
MRN (Mre11•Rad50•Nbs1) complex, 535, 536f, 554, 555t
MRX (Mre11-Rad50-Xrs2) complex, 533, 535, 536f, 554
MRX•Sae2 complex, 536
Msh4•Msh5 complex, 534t, 544
MSH2•MSH6 mismatch repair system, 563–564
MSH4•MSH5 protein, 534t
mtDNA (mitochondrial DNA)
human, first complete sequence for, 928, 929f
replication of, strand asymmetric model of, 928, 930
MTE (motif ten element), 742
mTERF1, 930
MukBEF complex, 217
mukB gene, 214
MukB protein, bacterial DNA compaction and, 214–217, 215f, 216f
MukEF protein, 216
mukE gene, 214
MukE protein, 214
mukF gene, 214
MukF protein, 214
Müller, Ferenc, 858
Muller, Hermann J., 243–244
Müller-Hill, Benno, 672, 673
Mullis, Kary B., 182
Multicopy inhibitor of GAL gene expression (Mig1 protein), 799
Multiple myeloma, 301
Multipotent stem cells, 818
Murakami, Katsuhiko, 640
Musca domestica. See also Mice
Hermes transposase of, 587–588, 588f, 589f
Mus81•Eme1 complex, 539, 539f, 542
Mus82•Eme1 complex, 538
mus81 gene, mutants of, 544
Mus81•Mms4 complex, 539, 539f
MUS81 protein, 534t
Mus81 protein, 534t
mutations in gene coding for, 542
Musteli, Tom, 236
Mutagen(s), detection of, 464–466
Mutagenesis
induced, 263
site-directed, 184f, 184–185
spontaneous, 263
Mutants, 262–268. *See also* Mutations
absolute defective, 262–263
auxotrophs as, 4
bacterial, notations used to denote, 262
classification on basis of changes in DNA, 263–264, 264f
conditional, 263
definition of, 262
deletion, 264
display of mutant phenotype by, 262–263
experimental uses of, 265–268
to aid in elucidation of metabolic pathways, 265–266
to define functions, 265
in genetic regulation, 266
to indicate interaction of proteins, 267f, 267–268
to indicate relations between apparently unrelated systems, 267
to locate sites of action of external agents, 266–267
to match biochemical entities with biological functions or intracellular proteins, 266
formation of, 263
insertion, 264
null, in eukaryotic recombination, 536–537
point, 264
quick stop, 379
regaining original phenotype, 265
replication studies using, 379–380, 380t
slow-stop, 379–380
supressor-sensitive, 263
temperature-sensitive, 263, 265
Mutarotase, gene encoding, 684

Mutasome, 503
Mutations. *See also* Mutants
cis-dominant, 270
frameshift, 988, 988*f*
nonsense, 996
promoter, 628–629
somatic hypermutation and, 558–564, 559*f*, 563*f*
transition, 454, 455*f*
transversion, 454, 455*f*
Mutator transposon, 598–600, 599*f*
mutD protein, 394
MutH protein, mismatch repair and, 494, 495
MutL protein
crystal structure of, 496
mismatch repair and, 494, 495, 496–497
MutLα protein, mismatch repair and, 497
MutS protein
E. coli, in complex with DNA oligomer, crystal structure of, 495–496, 496*f*
mismatch repair and, 494, 495
MutSα protein, mismatch repair and, 497
Myc family of transcription regulators, 796
Myc•Max transcription regulators, 796, 796*f*, 797*f*
Mycoplasma, σ factors of, 644
Mycoplasma genitalium
DNA molecule size of, 110*t*
tRNA genes of, 972
MyoD, 794–795, 795*f*
Myoglobin, structural differences between hemoglobin and, 76–78, 77*f*–80*f*, 80

N

Nagai, Kiyoshi, 857
Nair, Satish K., 796
Nascent polypeptide processing and folding, 1052*f*, 1052–1053, 1054*f*
NASP histone chaperone, nucleosome assembly and, 441
NBS1 gene, null mutations of, 535
NBS1 protein, 534*t*
Nbs1 protein, 533, 535
NC2 (negative cofactor 2), 747
NDRs (nucleosome depleted regions), 812, 813*f*
Neanderthals, divergence rate between humans and, 201
Negative cofactor 2 (NC2), 747
Negative elongation factor (NELF), 754, 755*f*, 756
Negative regulators, 289
Negative staining, for electron microscopy, 155, 155*f*
Nej1 protein, 555*t*
NELF (negative elongation factor), 754, 755*f*, 756
Neo^R, 546, 546*f*
Neugebauer, Karla M., 862
Neurospora, gene conversions in, 515, 515*f*
Neurospora crassa, genetic mutants of, 4
Neutral/neutral two-dimensional gel electrophoresis, origins of replication located by, 425*f*, 425–426
N-glycosilic bonds, 7
N-glycosylases, in base excision repair, 477–478, 478*f*
NHEJ (nonhomologous end-joining), 553–555, 554*f*, 555*f*, 555*t*
Nicholson, Garth, 70
Nicked circles, 124, 124*f*
Nick translation, 182
Nijmegen breakage syndrome, 533, 535
90S particle, 916
Nirenberg, Marshall W., 24, 985, 989–990, 991, 994–995, 1017
Nitrites, in food, nitrous acid formation from, 454–455
Nitrocellulose filter assay, 672–673
Nitrogen mustard gas
as chemotherapeutic agent, 462
DNA damage caused by, 461–462, 462*f*
Nitrous acid, formation from nitrites in food, 454–455
NMD (nonsense-mediated decay), 884
NMD factors, 884
NMR (nuclear magnetic resonance) spectroscopy, of tertiary protein structure, 56, 58*f*, 58–59, 59*f*
N-*myc* gene, 796
N-Myc transcription regulator, 796, 796*f*
N nucleotides, 561
Noller, Harry F., 1008, 1036, 1040, 1041
Nomura, Masayasu, 903–904, 1015–1016
Nomura, Taisei, 472
Nonautonomous mobile elements, 594
Noncovalent bonds
protein structure and, 46, 48
weak, 46–51
hydrogen bonds, 46, 47–48, 48*f*
hydrophobic interaction and, 46, 48–49, 49*f*, 50*f*
ionic bonds, 46, 48
van der Waals interaction and, 46, 49–50, 50*t*, 51*f*
Nonet, Michael, 806–807
Nonhomologous end-joining (NHEJ), 553–555, 554*f*, 555*f*, 555*t*
Nonpermissive temperature, 263
Non-polyposis colon cancer (HNPCC), 498
Nonsense-mediated decay (NMD), 884
Nonsense mutations, 996
Non-shuttling hnRNP proteins, 885
Non-template strand, 624, 626, 626*f*
Nontransmissible plasmids, 276
Northern blotting, 177
NotI, sequence specificity of, 169*t*
Nowotony, Marcin, 483
N-protein, lytic pathway of phage λ and, 697–698, 698*f*
NPS (nucleosome positioning sequence) elements, 813
NRXN3 gene, 846
N-terminal domain
of AraC protein, 688–689, 690
of CI regulator, 700, 700*f*
of Lac repressor, 674, 674*f*
Nuclear encoded plastid RNA polymerase, 932
Nuclear export signal, 769
Nuclear localization signal, 769
Nuclear magnetic resonance (NMR) spectroscopy, of tertiary protein structure, 56, 58*f*, 58–59, 59*f*
Nuclear receptor superfamily
orphan receptors and, 781
structure of, 781, 781*f*, 782*f*
zinc fingers in, 780–784, 781*f*–785*f*
Nuclear transfer experiments, 818–820, 819*f*
Nucleases, 165–173. *See also* Restriction endonucleases
properties of, 166, 166*t*
specificity of, 165–166
Nucleic acids. *See also* DNA; RNA
discovery of, 3
isolation of, 151–153
structure of, 4*f*, 5*f*, 5–6
sugars in, 4*f*, 5, 5*f*
Nuclein, 3
Nucleoid, 212, 212*f*
Nucleoid associated proteins, DNA bending by, 217, 217*f*
Nucleolus, 219, 220*f*
as rRNA transcription and processing site, 902–903, 903*f*
Nucleosides, 6*f*, 6–7, 7*f*
modified, in tRNA, 973–974, 974*f*
nomenclature for, 7
Nucleoside triphosphatases, helicases and, 120, 120*f*
Nucleosome(s), 76, 231–236
assembly of, 440–441
chromatin organization in, 232*f*, 232–233, 233*f*
core particles of, 233–235, 234*f*, 235*f*
interaction with H1, 236, 236*f*
distribution patterns of, 812–813, 813*f*
in DNA coding region, transcription and, 815
folding of, into 30-nm fiber, 237*f*, 237–238
histone interaction with chromatin and, 231*t*, 231–232
Nucleosome core particle, 233
Nucleosome depleted regions (NDRs), 812, 813*f*
Nucleosome positioning sequence (NPS) elements, 813
Nucleotide(s), 7, 7*f*
branching, 855
downstream sequences of, 626
forward movement during addition cycles, in transcription elongation complex, 648–649, 649*f*, 650*f*
incorrectly incorporated, detection and removal by bacterial RNA polymerases, 651, 651*f*, 652*f*
palindromic, 561
source of, phage T4 DNA replication and, 317–319
upstream sequences of, 626
Nucleotide addition processivity, 437
Nucleotide excision DNA repair, 482–493
bacterial, 482–485, 483*f*, 484*f*, 486*f*, 487*f*
global genome excision repair pathway for, 489
mammalian, 485, 488–490, 491*f*, 492*f*, 492–493
steps in, 482
transcription-coupled nucleotide excision repair pathway for, 489–490, 491*f*, 492
Nudler, Evgeny, 649
NusE protein, 654
NusG protein, 653–654
Nutrient agar, 260

O

OB (oligonucleotide binding) fold, 403–404
OBD(s) (oligonucleotide binding domains), 122
OBD (origin-binding domain), of SV40 T antigen, 417, 417*f*
Ochoa, Severo, 990, 991, 993, 1016
ochre, 996
Octameric protein complex, 233
Oct4 gene, 820*f*, 820–821
Oct-1 transcription activator protein, 776, 777, 777*f*
Oct-2 transcription activator protein, 776
O'Donnell, Mike, 394, 397, 403
Oh, Byung-Ha, 214
Okazaki, Reiji, 370, 372–374
Okazaki fragments, 374
connection by DNA ligase, 375–376, 376*f*, 377*f*
formation of, 403–404
RNA as primer for, 376, 378*f*
maturation of, in SV40 DNA replication, 420, 420*f*
in trombone model of DNA replication, 401, 402*f*
Oligonucleotide binding domains (OBDs), 122
Oligonucleotide binding (OB) fold, 403–404
Oligopeptides, 35
One gene-one enzyme hypothesis, 4

One gene-one polypeptide hypothesis, 4
1-5 rule, 784, 784*f*
opal, 996
Open promoter complex, 633
in transcription initiation, 641*f*, 642
Open reading frames (ORFs), 597–598, 598*f*, 662
Operators, 499
lacO, 669, 673*f*, 673–674, 675, 677*f*
of phage λ transcripts, 697
Operon(s). *See also specific operons*
cAMP•CRP activation of, 682, 684
Operon model, 669, 670*f*
ORC (origin of recognition complex), as eukaryotic initiator, 426
Orc1/CdC6 protein, in archaeal DNA replication, 442, 442*f*
ORFs (open reading frames), 597–598, 598*f*, 662
ori. *See* Origin of replication *(ori)*
Origin-binding domain (OBD), of SV40 T antigen, 417, 417*f*
Origin of recognition complex (ORC), as eukaryotic initiator, 426
Origin of replication *(ori)*, 381*f*, 381–383, 382*f*
of *E. coli* (*oriC*), 381*f*, 381–383, 382*f*, 384–385, 385*f*
binding sites of, 385
GATC sites on, 385
replication fork restart and, 550
sequestration of, 409
location by two-dimensional gel electrophoresis, 425*f*, 425–426
of SV40, T antigen binding to, 416–418, 417*f*, 418*f*
OriT (transfer origin), in F plasmid, 272–273
Orphan nuclear receptors, 781
Orr-Weaver, Terry, 524
Orthologs, 847
Outer membrane, bacterial, 258*f*, 258–259
Overlap method, for amino acid sequence determination, 42, 43*f*
Oxidative DNA damage, 455–456, 456*f*, 457*f*
Oxygen-binding curves, for myoglobin versus hemoglobin, 68, 68*f*
8-Oxyguanine (oxoG), base pairs with cytosine or adenine, 456, 456*f*
Oxyhemoglobin, transition from deoxyhemoglobin to, 78, 79*f*, 80
Oxymyoglobin, transition from deoxymyoglobin to, 78, 79*f*, 80

P

Pääbo, Svante, 201
Pabo, Carl O., 776
PAF49, 904
PAF51, 904
PAF53, 904
PAGE (polyacrylamide gel electrophoresis), 736
PAHs (polycyclic aromatic hydrocarbons), 458–460, 460*f*
Painter, Robert, 485, 488
Pairing diversity, 559
PaJaMo experiments, 987
Palindromes, 138, 168
pentanucleotide, of SV40, 416
Palindromic (P) nucleotides, 561
Pancreatic DNase1, properties of, 166*t*
Pancreatic RNase, 831
properties of, 166*t*
Parallel pleated sheets, 54
Paralogs, 537
Paranemic joints, 529
Pardee, Arthur, 987
Park, Julia, 947
Parkinson's disease, 1053
P arm, 227
Partial diploids, 268
Pasha, 953
Passenger proteins, in centromeres, 242
Passman, Zvi, 653
Patel, Dinshaw J., 944–946
Pauling, Linus, 33, 51, 54, 81, 979
Paulson, James R., 238
Pauses, in transcription elongation complex, 649, 650*f*, 651
Pause sites, RNA polymerase movement and, 649, 651
Pcf11, 878
PCNA (proliferating cell nuclear antigen)
mismatch repair and, 497
in SV40 DNA replication, 419
PCR. *See* Polymerase chain reaction (PCR)
PcrA helicase, 120, 121*f*
"pdb" (Protein Data Bank) files, 57
P element, 590, 600–602, 601*f*
PEN (pentanucleotide palindrome), of SV40, 416
Penicillin resistance, 15
Pentanucleotide palindrome (PEN), of SV40, 416
Pentons, 352–353
Peptide(s), 33
bond distances and angles in, 33–34, 34*f*
directionality of, 34
nomenclature conventions for, 34–35, 35*f*
Peptide bonds, 33–35, 33*f*–35*f*, 46
Peptide fingerprinting (peptide mapping), 81, 82*f*, 993
Peptidoglycans, in bacterial cell walls, 257
Peptidylprolyl *cis-trans* isomerase, polypeptide folding and, 65, 66*f*
Peptidyl transferase activity, 1035*f*–1039*f*, 1035–1040
Peptidyl-tRNA binding site (P-site), 1010, 1010*f*
Peripheral membrane proteins, 69
Periplasmic space, 259
Permissive temperature, 263
Permutation, circular, of phage T4 DNA, 314–315, 315*f*, 317
Peroxisome proliferator-activated receptor (PPAR), 783
Perry, Robert, 832
Perutz, Max, 56
Petri dishes, 260
PFGE (pulsed-field gel electrophoresis), 163–164, 165*f*
Pfu polymerase, 182
pH, isoelectric, 40
Phage(s). *See* Bacteriophage(s); *specific bacteriophages*
Phage display, 348
Phage Group, 309
Phage lysate, 310
Phagemids, 348
PHAX protein, 861
Phenotype
bacterial, notations used to denote, 261, 262
complementation to determine number of genes responsible for, 268–270, 269*t*
definition of, 4
epigenetics and, 816
mutant, display of, 262–263
Phenylalanine
abbreviations for, 33*t*
side chains of, 31, 32*f*
Phenylalanine operon, 695
Philadelphia chromosome, 229, 230*f*
φ-structures, 369, 370*f*
φ-subunit, of core polymerase, 394
Phosphatidylcholine, structure of, 66–68, 68*f*
Phosphodiesterase, spleen and venom, properties of, 166*t*
Phosphoenolpyruvate-dependent phosphotransferase system (PTS), 678–679
Phosphoinositide-specific phospholipase C (PLC), 98, 98*f*
Phospholipase C, as effector in G protein-linked signal transduction systems, 97, 97*t*
Phosphorylation
enzyme activity modification by, 94
of eukaryotic initiation factors, 1028, 1028*t*
Photolyase, direct reversal of DNA damage by, 469–472, 470*f*–474*f*, 474
(6-4) photoproducts, 450*f*, 451
Photoreactivation, 469–470
PIC(s). *See* Preinitiation complexes (PICs)
Pickard, Craig S., 726
piggyBac transposon, 602, 604–605
piRNAs (piwi interacting RNAs), 954
Pit-1, 776
Piwi interacting RNAs (piRNAs), 954
Plant(s). *See also specific plants*
miRNA synthesis in, 954
RNA silencing in, 949
transposons in, as molecular genetics tools, 598–600, 599*f*
Planta, Rudi J., 916
Plaque, phage, formation of, 311*f*, 311–312
Plaque assays, of phages, 311, 312
Plasmid(s)
col, 284
conjugative, 276
degradative, 284
DNA isolation in, 152
drug resistance, 283
mobilizable, 276
nontransmissible, 276
recombinant, 173–174
relaxed, 283
replication of, 124
basic replicon and, 282–283
resolution sites on, 591–592
self-transmissible, 276
stringent, 283
as vectors, insertion of DNA fragments into, 173–174, 174*f*
virulence, 284
yeast, repair by homologous recombination, 524–525, 525*f*
Plasmid pBR322, DNA molecule size of, 110*t*
Plasmodium falciparum, sickle cell anemia and, 83
Plasterk, Ronald H. A., 947
Plastid RNA polymerase, 932
Plates, 260
Platform, of ribosome, 1007
Plating, 260, 261
PLC (phosphoinositide-specific phospholipase C), 98, 98*f*
Pleated β-sheets, 54, 54*f*
Plectonemic joints, 529
Pluripotent stem cells, 818
PNK (polynucleotide kinase), 555*t*
P (palindromic) nucleotides, 561
Point mutants, 264
pol, 356, 356*f*
Pol α
in eukaryotic DNA replication, 428, 428*f*, 429
processivity of, 429
recruitment of, by SV40 T antigen, 418–420, 419*f*–421*f*
subunits of, 429, 429*f*

Pol δ
in eukaryotic DNA replication, 429, 430–431
mismatch repair and, 497
subunits of, 429, 429*f*
in SV40 DNA replication, 418
Pol β, in base excision DNA repair, 481
Pol ε
in eukaryotic replication elongation, 429–431, 430*f*
processivity of, 429–430, 430*f*
subunits of, 429, 429*f*
polA gene, 380*t*, 389
polA⁻ mutant, 266
PolA protein, in recombination, 527*t*
PolB, archaeal, 442–443
PolD, archaeal, 442–443
POL η protein, 534*t*
POLH, 505
Pol II protein, in recombination, 527*t*
Pol III protein, in recombination, 527*t*
polIIIα, 393*f*, 393–394
Polio vaccines, SV40 contamination of, 349
Polio virus, 357
POL δ protein, 534*t*
POLL protein, 555*t*
POLM protein, 555*t*, 561
Pol3 protein, 534*t*
Pol4 protein, 555*t*
Polyacrylamide gel, for slab gel electrophoresis, 161, 161*f*, 162
Polyacrylamide gel electrophoresis (PAGE), 736
Polyamines, Z-DNA and, 137
Polycistronic mRNA, 662–663
Polycyclic aromatic hydrocarbons (PAHs), 458–460, 460*f*
Polydeoxyribonucleotides, structure of, 9, 10*f*, 11*f*
Polyglutamate, as α-helix, 53–54
Polylysine, as α-helix, 54
Polymerase(s). *See also* Bacterial RNA polymerases; DNA polymerase(s); RNA polymerase(s); *specific DNA polymerases; specific RNA polymerases*
core, 392. *See also* DNA polymerase III holoenzyme
in bacterial DNA replication, 389–390
release from β clamp, 403–404
subunits of, 393–394, 394*f*
processivity of, 390–391
Taq polymerase, 182
vent polymerase, 182
Polymerase chain reaction (PCR), 182–184, 183*f*
applications of, 184
emulsion, 197
gene knockouts created by, 545, 545*f*
Polymerase I transcription release factor (PTRF), 907, 909, 909*f*
Polymorphisms, 291–294
single nucleotide, 291–292, 293–294
tandem repeat, 291, 292*f*, 294
Polynucleotide kinase (PNK), 555*t*
Polynucleotide phosphorylase, 990
E. coli mRNA degradation using, 703–704, 705
Polyomaviruses, 349–352
discovery of, 349
DNA of, 349, 350*f*
genome of, 349, 350*f*
replication of, 351*f*, 351–352
structure of, 349, 350*f*
Polypeptides, 29, 35
of adenoviruses, 353
amino acid sequences of, 29
determination of. *See* Amino acid(s), sequences of
biologically active, 29. *See also* Protein(s)
folding of, 29, 64
nascent, processing and folding of, 1052*f*, 1052–1053, 1054*f*
Polyproteins, 356
Polypyrimidine tract, 851
Polyribosomes, mRNA extracted from, 831
Polysomes, mRNA extracted from, 831
Poly(A) tails, 830–831, 831*f*, 832*f*
Polyunsaturated fatty acids, oxidation of, aldehyde products produced by, 456, 456*f*
Porins, in bacterial outer membrane, 258*f*, 258–259
Porter, Keith, 964–965
Positive-negative selection, 546–547, 547*f*, 548*f*
Positive regulators, 289, 678
Positive transcription elongation factor b (P-TEFb), 756
Posttranscriptional modification
of rRNA, 714
of tRNA, 715–717, 715*f*–717*f*
POT1 (protection of telomeres), 245
Potts, Percival, 458
POU homeodomain (POU_H), 777, 777*f*
POU proteins, 777
POU-specific sub-domain (POU_S), 777, 777*f*
POU transcription activator proteins, 776–777, 777*f*
PPAR (peroxisome proliferator-activated receptor), 783
ppGpp (guanosine-5′diphosphate-3′-diphosphate), 710*f*, 710–711
pppGpp (guanosine-5′triphosphate-3′-diphosphate), 710*f*, 710–711
Ppr1 (pyrimidine pathway regulator 1), 788–789, 789*f*
p53 protein, domain structure of, 789, 789*f*, 790, 790*f*
Precursor mRNA (pre-mRNA), 724
Preincision complex, 490
Preinitiation complexes (PICs), 745
assembly when core promoter has TATA box, 745–746, 746*f*
eukaryotic, 427
of genes lacking TATA boxes, TFIID binding to, 746–748, 747*f*
sequential binding of RNA polymerase II•TFIIF complex, TFIIE, and TFIIH and, 751–752, 752*f*
"Premature release" pathway, 403
Pre-miRNA, 952*f*, 952–953
Pre-mirtron, 953
Pre-mRNA (precursor mRNA), 724, 830–831
discovery of, 830, 830*f*
in higher animals, special processing mechanism required by, 878–880, 880*f*
5′-m⁷ cap attachment to, 835–836, 836*f*
poly(A) tails and, 831, 832*f*
sequences defining 5′- and 3′-splice sites in, 851, 851*f*
splicing of, alternative, 847*f*, 847–848
split genes and splicing of, 840–855
combinations of splicing patterns within individual genes leading to formation of multiple mRNAs and, 848*f*, 848–849, 849*f*
coordinated transesterification reactions in splicing and, 854–855, 855*f*
interruption of amino acid-coding regions within eukaryotic genes and, 841–842, 843*f*, 844*f*, 844–845
intron and exon conservation and, 845*t*, 845–847, 846*f*
mRNA molecules formed by splicing pre-mRNA and, 840–841, 841*f*
production of two or more mRNA molecules by a single pre-mRNA and, 847*f*, 847–848
slicing precision and, 851*f*, 851–852, 852*f*
splicing intermediates resembling lariats and, 852–854, 853*f*, 854*f*
trans-splicing in *Drosophila* and, 850, 850*f*
Pre-mRNA processing protein (Prp8), 868
Pre-RISC complex, 946
Pre-rRNA, 898
conversion to mature rRNA, 909–910, 910*f*, 911*f*
Pre-tRNA, 715
processing to become mature tRNA, 924–927, 925*f*
PriA protein, replication fork restart and, 550
Pribnow, David, 630
Pribnow box, 630
Primary cultures, 296–297
Primary structure of proteins, 41–46, 46*f*, 47*f*
BLAST program for, 45–46
Edman degradation for determination of, 41–43, 42*f*, 43*f*
mass spectrometry for determination of, 43, 44*f*
tertiary structure determination by, 62–64, 63*f*
Primary transcript, 626
conversion to functional RNA product, 662, 713–718
Primase
catalysis of RNA primer synthesis by, 388, 388*f*
movement along lagging strand, 404–405
disassembly model of, 404, 405*f*
pausing model of, 404, 405*f*
priming loop model of, 404–405, 405*f*
Primer extension, 736, 737–738, 738*f*
Primer strand, 177–178, 178*f*
Primer walking, 188, 188*f*
Procapsids, 321
Processed pseudogenes, 614
Processivity, of polymerases, 390–391
Prokaryotes. *See also* Archaea; Bacteria
chromatin of, 211
DNA molecules in, 109–110, 110*t*
RNA polymerase II of, 731
Proliferating cell nuclear antigen. *See* PCNA (proliferating cell nuclear antigen)
Proline
abbreviations for, 33*t*
lack of tendency to be part of α-helix, 53
side chains of, 31, 32*f*
Promoter(s), 453, 566
closed promoter complex and, 632–633, 633*f*
in initiation stage of transcription, 628–629, 636–638, 637*f*
open promoter complex and, 633
for phage λ transcripts, 697
of RNA polymerase III transcription units, 920–922, 921*f*
of rRNA operons, 707, 708
spicing regulation by structure of, 871
strong, 637
Promoter escape, in transcription initiation, 641*f*, 642
Promoter mutations, 628–629
Promoter upstream element, *rrn* operon transcription and, 708*f*, 708–709, 709*f*
Proofreading, 651, 651*f*
Prophages, 311
Prophase, 219, 220*f*, 221*f*
of mitosis, 219, 220*f*, 221*f*
Prophase I, of meiosis, 224*f*, 226–227
Prophase II, of meiosis, 225*f*, 226–227
Proteasomes, protein degradation and, 101–102, 102*f*
Protection of telomeres (POT1), 245

Protein(s), 29. *See also specific proteins*
in biologic membranes, 66–71
fluid mosaic model of, 68–71, 69*f*, 70*f*
interaction with lipids, 66–68, 67*f*, 68*f*
catalytic. *See* Enzyme(s)
in conservative site-specific recombination, 607*f*, 607–608, 608*f*
constitutive, in centromeres, 242
constitutively activating cell processes, 229
coupling, 276
degradation of, ubiquitin proteasome system and, 99–102, 100*f*–102*f*
denaturation of, 62
DNA binding, single-stranded, 121–123
composition of, 121–122
isolation of, 122
fibrous, 29
function of. *See* Protein function
gel electrophoresis separation of, 162–163
globular, 29
loops and turns in, 54–55, 55*f*
initiator, 381
interaction of, mutants to prove, 267*f*, 267–268
intrinsically disordered, 59–60
membrane, integral and peripheral, 69
multidomain, 83–84, 84*f*, 85*f*
in native state, 62
passenger, in centromeres, 242
purification of. *See* Protein purification
receptor, 76
recombination
human, 534*t*
of *S. cerevisiae*, 534*t*
regulatory, 76
renaturation of, 62–63, 63*f*
storage, 76
structural, 76
in phage assembly, 320
structure of. *See* Protein structure
synthesis of. *See* Protein synthesis
toxic, 76
transport, 76
Protein conducting channel, 1054, 1055
Protein Data Bank, 59, 61
Protein Data Bank ("pdb") files, 57
Protein disulfide isomerase, polypeptide folding and, 65, 66*f*
Protein electrophoresis, 40, 41*f*
Protein families, 60
Protein function, 75–102
of enzymes. *See* Enzyme(s)
G-protein-linked signal transduction systems and, 94–98, 95*f*–99*f*
of immunoglobulin G, 83–84, 84*f*, 85*f*
of transport proteins, 76–83
myoglobin and hemoglobin structural differences and, 76–78, 77*f*–80*f*, 80
normal versus sickle cell hemoglobin structural differences and, 80–83, 81*f*, 82*f*
ubiquitin proteasome proteolytic pathway and, 99–102, 100*f*–102*f*
Protein purification, 36–41
affinity chromatography for, 40, 40*f*
column chromatography for, 36*f*, 36–37
electrophoresis for, 40, 41*f*
gel filtration (molecular exclusion) chromatography for, 38, 39*f*, 40
high-performance liquid chromatography for, 37
ion-exchange chromatography for, 37*f*, 37–38
reverse phase (hydrophobic) chromatography for, 38, 38*f*
Protein structure, 28–71
α-amino acids and, 29*f*, 29–33
peptide bonds kinking, 33–35, 33*f*–35*f*
in biologic membranes, 66–71
fluid mosaic model of, 68–71, 69*f*, 70*f*
protein interaction with lipids and, 66–68, 67*f*, 68*f*
hydrogen bonds in, 46, 47–48, 48*f*
hydrophobic interaction and, 46, 48–49, 49*f*, 50*f*
ionic bonds in, 46, 48
primary, 41–46, 46*f*, 47*f*
BLAST program for, 45–46
Edman degradation for determination of, 41–43, 42*f*, 43*f*
mass spectrometry for determination of, 43, 44*f*
tertiary structure determination by, 62–64, 63*f*
protein purification and. *See* Protein purification
quaternary, 46, 47*f*
random conformations in, 62
secondary, 46, 47*f*, 51–56
combination of, 55, 55*f*
β-conformation, 54, 54*f*
α-helix, 51–54, 52*f*, 53*f*
loops and turns in, 54–55, 55*f*
prediction of, 55, 56*f*
tertiary, 46, 47*f*, 56–66
determination by primary structure, 62–64, 63*f*
intrinsically disordered proteins in, 59–60
molecular chaperones and, 64–65, 65*f*, 66*f*
nuclear magnetic resonance spectroscopy of, 56, 58*f*, 58–59, 59*f*
structural genomics and, 60–62
x-ray crystallography of, 56*f*, 56–58, 57*f*
van der Waals interaction and, 46, 49–50, 50*t*, 51*f*
Protein synthesis
genetic code and. *See* Genetic code
recycling phase of, 1012, 1051*f*, 1051–1052
ribosomes and, 1006–1056
elongation stage of, 1029–1043
initiation stage of, in bacteria, 1012–1020
initiation stage of, in eukaryotes, 1020–1029
nascent polypeptide processing and folding and, 1052*f*, 1052–1053, 1053*f*
recycling stage of, 1051*f*, 1051–1052
signal sequence and, 1053–1056, 1054*f*, 1055*f*
structure of, 1007–1012
termination stage of, in bacteria, 1043–1050
termination stage of, in eukaryotes, 1050*f*, 1050–1051
Proteolysis tags, 1047
Proteolytic enzymes, 296
Proteome, human, variety of proteins contained by, 883–884
Proto-oncogenes, 464
Protoplasts, 259
Prototrophs, 259
"Provirus Model," 359
Proximal promoter, 765, 766*f*
Proximal sequence element (PSE), 921
Proximal terminator (PT), 902
Prp8 (pre-mRNA processing protein), 868
PSE (proximal sequence element), 921
Pseudogenes, 562
processed, 614
Pseudoknots, 142, 143*f*
Pseudouridine, 973–974, 974*f*
Pseudouridylation snoRNAs, 910, 910*f*
P-site (peptidyl-tRNA binding site), 1010, 1010*f*
Psoralen
DNA damage caused by, 462*f*, 462–464
as skin disorder treatment, 463
Psoriasis, treatment of, 463
PstI, sequence specificity of, 169*t*
PT (proximal terminator), 902
Ptashne, Mark, 788–789, 800, 801–802, 805
P-TEFb (positive transcription elongation factor b), 756
PTRF (polymerase I transcription release factor), 907, 909, 909*f*
PTS (phosphoenolpyruvate-dependent phosphotransferase system), 678–679
Pulse-chase experiment, 663–664
Pulsed-field gel electrophoresis (PFGE), 163–164, 165*f*
Punnett, Reginald C., 512
Purine, in DNA, 6, 6*f*
Purine bases, 6, 6*f*
Puromycin, in peptidyl transferase activity assay, 1035*f*, 1035–1036
Pyrimidine, in DNA, 6, 6*f*
Pyrimidine bases, 6, 6*f*
Pyrimidine pathway regulator 1 (Ppr1), 788–789, 789*f*
Pyrococcus abyssi, DNA polymerases in, 443
Pyrococcus furiosus
DNA polymerase isolated from, 182
SMC head group from, crystal structure for, 214–215, 215*f*
Pyrosequencing, 198
Pyrrolysine, as polypeptide building block, 986*f*, 986–987

Q

Q arm, 227
Q protein, lytic pathway of phage λ and, 698
Quadriplex structure, of DNA, 139, 141*f*
Quaternary structure
of proteins, 46, 47*f*
of RNA, 141–142
Quick stop mutants, 379

R

RAD54B gene, 537
RAD51B protein, 534*t*, 536–537
Rad54B protein, 537
RAD51B•RAD51C complex, 537
RAD51C protein, 534*t*, 536–537
RAD51C•XRCC3 complex, 537
Radding, Charles, 522
RAD51D protein, 534*t*, 536–537
RAD51D•XRCC2 complex, 537
RAD genes, 533
RAD5- gene, 533
RAD54 gene, 537
RAD55 gene, deletion mutant of, 536
RAD57 gene, deletion mutant of, 536
rad54 gene, mutants of, 537
rad51 gene, null mutants of, 536–537
RAD50 gene, null mutations of, 535
Radiation, DNA damage caused by, 449–452
ultraviolet light and, 449, 450*f*, 451, 451*f*
x-rays and τ-rays and, 452
Radman, Miroslav, 274, 499
Rad30 protein, 534*t*
RAD50 protein, 534*t*
Rad50 protein, 533, 534*t*
RAD51 protein, 534*t*, 537
Rad51 protein, 534*t*, 536, 537, 542
BRAC2 interaction with, 551
RAD52 protein, 534*t*
Rad52 protein, 534*t*, 536
RAD54 protein, 534*t*
Rad54 protein, 534*t*, 537. *See also* Tid1 protein
Rad55 protein, 534*t*, 536
Rad57 protein, 534*t*, 536
Rad59 protein, 534*t*

RAD54•RAD54B complex, *534t*
Rad55•Rad 57 complex, 536
rad54 rdh54 double mutant, 537
RAG-1, 560, *560f*
RAG-2, 560, *560f*
RAG•RSS complex, 560–561
Rakonjac, Jasna, 345
Raloxifene
  clinical applications of, 786
  side effects of, 786
  structure of, *787f*
Ramakrishnan, Venkatraman, 1008, 1024, 1026, 1030, 1033
RAN•GDP complex, 861
RAN•GTP complex, 861
RAN protein, 861
Rao, Venigalla B., 324
Rap1, 245
RAR (retinoid acid receptor), 783
Rasmussen, Eric B., 835–836
Rat1, 878
Rate of reaction, theoretical maximum, 88–89, *89f*
rbs (ribosome binding site), *1018f*, 1018–1020, *1019f*
RBS (RNA-binding site), of transcription elongation complex, 648
RCSB PDB (Research Collaboratory for Structural Bioinformatics Protein Data Bank), 57
RdgC protein, in recombination, *527t*
Rdh54 protein, 537
Rdh54/Tid1, *534t*
Rdh54•Tid1, *534t*
R2D2 protein, 946
Reaction, rate of, theoretical maximum, 88–89, *89f*
Reactive oxygen species, DNA damage caused by, 455–456, *456f*, *457f*
Reannealing, of DNA, 118–119, *119f*
  molecular details of, 119, *119f*
  salt concentration and, 118
  speed of, 119
  temperature and, 118–119
recA gene
  in recombination, *527t*
  SOS response and, 501–502
RecA protein
  in Meselson-Radding model of recombination, 522, 523
  in recombination, *527t*, 532
    as strand exchange protein, *528f*, 528–529, *530f*
  SOS response and, 503
  SOS response regulation by, 499–502, *500f*, *501f*
RecBCD complex, in recombination, 529, *531f*, 531–532, *532f*
RecB protein, in recombination, 531
RecC protein, in recombination, 531
RecE protein, in recombination, *527t*
Receptor proteins, 76
RecFOR pathway, 532, *533f*
RecF protein, in recombination, *527t*, 532
RecG protein, in recombination, *527t*, 532
Rechsteiner, Martin, 101
Reciprocating pawl mechanical ratchet wheel, 648–649, *649f*
RecJ protein
  mismatch repair and, 495
  in recombination, *527t*, 532
RecN protein, in recombination, *527t*
Recognition elements, 983
Recognition helix, 783
  of CI regulator, 700–701
  of Lac repressor N-terminal domain, 674
Recombinant DNA technology, 173–185
  applications of, 24–25, 184
  development of, 24
  DNA synthesis using DNA polymerase I and, 177–180, *178f–181f*, 182
  insertion of DNA fragments into plasmid DNA vectors in, 173–174, *174f*
  Northern blotting, 177
  polymerase chain reaction, 182–184, *183f*
    applications of, 184
    emulsion, 197
    gene knockouts created by, 545, *545f*
  for protein structure determination, 57
  restriction fragment length polymorphisms, 292–293, *293f*
  site-directed mutagenesis, *184f*, 184–185
  social, political, ethical, and legal issues raised by, 25
  Southern blotting (Southern transfer), 174–177, *175f*, *177f*
  Western blotting, 177
Recombinant plasmids, 173–174
Recombinases, 555–556
  in termination stage of DNA replication, 406, *407f*, 408
Recombination, 511–566
  homologous. *See* Homologous recombination
  illegitimate, 553–555, *554f*, *555f*, *555t*
  nonhomologous, 553–555, *554f*, *555f*, *555t*
  site-specific, 555–566, *556f*
    conservative, 576, *576f*, 607–612
    Cre/*lox* system and, *565f*, 565–566
    FLP/*FRT* system and, *564f*, 564–565
    mating type switching in yeast and, 556–558, *557f*, *558f*
    V(D)J, 558–564, *559f–563f*
Recombination signal sequences (RSSs), 560, *560f*
  incorrect, 561, *561f*
RecO protein, in recombination, *527t*, 532
RecOR complex, in recombination, 532
RecQ helicase, in recombination, 532
RecQ protein, in recombination, *527t*
RecRCB protein, in recombination, *527t*
RecR protein, in recombination, *527t*, 532
RecT protein, in recombination, *527t*
RecX protein, in recombination, *527t*
Recycling phase of protein synthesis, 1012, *1051f*, 1051–1052
Redundancy, terminal, of phage T4 DNA, 314–315, *315f*, 317
Reece, Richard J., 788–789
REF 1 (RNA export factor 1), 884
Regulator of G protein signaling (RGS), 96
Regulatory enzymes, 91–94, *92f*, *93f*
  allosteric effector inhibition and stimulation of, *93f*, 93–94
  sigmoidal kinetic curves produced by, 92–93, *93f*
  subunits of, 94, *94f*
Regulatory inactivation of DnaA (RIDA), 408
Regulatory promoters, *76f*, 764–767, *766f*
Regulatory proteins, 76
Regulatory subunits, of regulatory proteins, 94, *94f*
Regulons, 696
relA gene, 710
Relaxed phenotype, 710
Relaxed plasmids, 283
Relaxosome, *274f*, 274–275, *275f*
Release factors (RFs), 1012
  bacterial, 1043–1046
    RF1, 1043, *1044f*, 1044–1045, *1045f*
    RF2, 1043, 1044–1046, *1044f–1046f*
    RF3, 1043, 1044–1046, *1044f–1046f*
Renaturation
  of DNA, 118–119, *119f*
    molecular details of, 119, *119f*
    salt concentration and, 118
    speed of, 119
    temperature and, 118–119
  of ribonuclease A, 62–64, *63f*
Repair of DNA, 468–505
  base excision, 477–481, *478f–482f*
    by DNA glycolase action, 478–480, *479f*, *480f*
    by DNA glycolase/lyase action, 480–481, *481f*, *482f*
    long patch, 481, *482f*
    short patch, 481, *481f*
  direct reversal of damage for, 469–477
    by dealkylation, 474–476, *475f*, *477f*
    by photolyase, 469–472, *470f–474f*, 474
  mismatch
    in *E. coli*, 493–497, *494f–496f*
    human, 497–498, *498f*
  nucleotide excision, 482–493
    bacterial, 482–485, *483f*, *484f*, *486f*, *487f*
    global genome excision repair pathway for, 489
    mammalian, 485, 488–490, *491f*, *492f*, 492–493
    steps in, 482
    transcription-coupled nucleotide excision repair pathway for, 489–490, *491f*, 492
  SOS response to damage and, 498–505
    DNA polymerase synthesis induced by, 502–504, *503f*
    human DNA polymerases and, *504t*, 504–505
    key steps in, *501f*, 501–502
    RecA and LexA regulation of, 499–502, *500f*, *501f*
    translesion DNA synthesis and, 498–499
Repeat addition processivity, 437
Replicase, 391
Replicating forms (RFs)
  reversed, 550, *550f*
  RFI, 124, 329
  RFII, 124, 329
Replication bubble, 369
Replication factor C (RFC)
  mismatch repair and, 497
  in SV40 DNA replication, 419
Replication fork
  collapse of, mitotic homologous recombination and, 549–550, *549f–551f*
  eukaryotic, organization of proteins and enzymes at, 431, *431f*
Replication protein A (RPA), 122
  Pol α recruitment and, 418
  RPA•DNA complex and, *122f*, 122–123
Replicative transposition, 582, 584–586, *585f*
Replicators, 381
Replicon(s), 381
Replicon model, 381
Replisome, 400
Reporter fusions, 596
Repressible enzymes, 667
Repressors
  Gal, 684–685, *685f*
  Lac, 668
    binding to lac operator *in vitro*, *672f*, *672t*, 672–673, *673f*
    functional units of, 674–675, *674f–676f*
    structure of, 674–675, *674f–677f*, 677
  Lac repressor•DNA complex and, 675, *676f*
  in negative regulation, *664f*, 665, *665t*
Research Collaboratory for Structural Bioinformatics Protein Data Bank (RCSB PDB), 57
Residues, 34

Resolution in recombination
in double-strand break model, 526
in eukaryotic homologous recombination, 538*f*, 538–540, 539*f*
in Holliday model, 519, 521*f*
meiotic, 544
in Meselson-Radding model of recombination, 522–524, 523*f*
RecG and, 532
RuvABC complex and, 532
Resolution sites, 591–592
Resolvases, in eukaryotic homologous recombination, 538
Resonance, radio-frequencies in, 58
Respiration, cellular, 455
Restriction endonucleases, 167–173
actions on DNA, 168–169
crystal structures of, 169*f*, 169–170
DNA characterization using, 167*t*, 167–170, 168*f*, 169*f*, 169*t*
in *E. coli*
strain B, 312–313
strain K, 313
restriction map construction using, 170*f*–172*f*, 170–173
sequence specificity of, 169*t*
types of, 167*t*, 167–168
Restriction fingerprint analysis, 191
Restriction fragment length polymorphisms (RFLPs), 292–293, 293*f*
Restriction maps, 170*f*–172*f*, 170–173
DNA sequencing using information from, 188
Retinoid acid receptor (RAR), 783
Retinoid X receptor (RXR), 783
Retroelements, 614–615
Retroelement Tos17, 600
Retrohoming, mobile group II intron movement by, 617
Retrotransposition, mobile group II intron movement by, 617
Retrotransposons, 577
Retroviruses, 355–357, 356*f*, 358*f*, 359
AIDS and, 355, 357
cancer and, 357, 359
common features of, 355–356
complex, 357
DNA of, replication of, 358*f*, 359
genome of, 356, 356*f*
infection by, 357
protein coding regions of, 346*f*, 356–357
reasons to study, 355
reverse transcriptases of, 201–202
RNA of, 357, 359
simple, 357
Reverse genetics, 950
Reverse gyrases
ATP-dependent, for positive supercoiling in in thermophiles, 443
of hyperthermic archaea, 132
Reverse-Hoogsteen pairing, 139
Reverse phase chromatography, 38, 38*f*
Reverse transcriptases
in DNA sequencing, 201–203, 202*f*
for RNA-directed DNA synthesis, 201–203, 202*f*
in Rouse sarcoma virus, 359
Reverse transcription, 355
target-primed, 576*f*, 576–577, 612–617
elements moving via, 612*f*, 612–613
for LINE movement, 613*f*, 613–616, 615*f*
mobile group II intron movement by, 616–617
Revertants, 265
RF(s). *See* Release factors (RFs); Replicating forms (RFs)
RFC (replication factor C)
mismatch repair and, 497
in SV40 DNA replication, 419
RFLPs (restriction fragment length polymorphisms), 292–293, 293*f*
RGR gene, 807
RGS (regulator of G protein signaling), 96
RhlB, 705
Rho-dependent termination, 653–654, 654*f*, 655*f*
Rhodes, Daniela, 238
Rho factor, 653–654, 653*f*–655*f*
crystal structure of, 653, 653*f*
Rho utilization *(rut)* site, 653
Ribofuranose, Haworth structure for, 5*f*, 5–6
Ribonucleases (RNases)
definition of, 165
E. coli mRNA degradation using, 703–705, 704*f*, 706*f*
pancreatic, 831
RNase A
renaturation of, 62–64, 63*f*
structure of, 57, 57*f*, 59, 59*f*
RNase E, 704
RNase H, 376, 378*f*
RNase III, 941
in rRNA processing, 714
RNase T1, 831
properties of, 166*t*
Ribonucleic acid. *See* RNA
Ribonucleosides, 6, 7*f*
Ribonucleoside triphosphates, for RNA synthesis, 624
Ribonucleotides, 7
Ribose, 4*f*, 5
conversion to cyclic form, 5, 5*f*
structure of, 5, 5*f*
Ribose groups, in helical tracts, 142, 142*f*
Ribosomal proteins (r-proteins), synthesis of, regulation in *E. coli,* 712–713
Ribosomal recycling factor (RRF), 1012
Ribosomal RNA. *See* rRNA
Ribosome(s), 76
bacterial, structure of, 706–707, 707*f*
in central dogma theory, 23, 23*f*
coining of term, 964
EF-Tu•GTP•aminoacyl-tRNA ternary complex carrying aminoacyl-tRNA to, 1030, 1032, 1032*f*
eukaryotic, 896*f*, 896–897
assembly of, 915–919, 917*f*–919
protein synthesis and, 1006–1056
elongation stage of, 1029–1043
initiation stage of, in bacteria, 1012–1020
initiation stage of, in eukaryotes, 1020–1029
nascent polypeptide processing and folding and, 1052*f*, 1052–1053, 1053*f*
recycling stage of, 1051*f*, 1051–1052
ribosome structure and, 1007–1012
signal sequence and, 1053–1056, 1054*f*, 1055*f*
termination stage of, in bacteria, 1043–1050
termination stage of, in eukaryotes, 1050*f*, 1050–1051
RF1 or RfF2 dissociation from, RF1 stimulation of, 1045–1046, 1046*f*
stalled, translating truncated mRNA, tmRNA rescue of, 1046–1047, 1048*f*, 1049
structure of, 1007–1012
archaeal, 1011, 1011*f*
bacterial, 1007–1010, 1007*f*–1010*f*
eukaryotic, 1011*f*, 1011–1012
synthesis of, regulation of, 710–713
Ribosome binding site (rbs), 1018*f*, 1018–1020, 1019*f*
Ribosome release factor (RRF), 1051*f*, 1051–1052
Riboswitches, 1020
Ribothymidine, 973–974, 974*f*
Ribozymes, hammerhead, 143, 144*f*
Rice, 600
Rich, Alexander, 135, 976
Richmond, Timothy J., 233, 238
RIDA (regulatory inactivation of DnaA), 408
Rifampicin, 655
mutant in study of, 266–267
Rifamycins, 655
rInr (rRNA initiator), 901
RISC (RNA-induced silencing complex), 941
RITS (RNA-induced initiation of transcriptional silencing) complex, 956
Rmi1 protein, 534*t*
RNA
bases of, secondary structure and, 141–142, 142*f*–143*f*
as catalyst, 144–145
degradation of, during isolation, 153
DNA differentiated from, 4*f*, 5, 5*f*
editing of, 880–884
by aminoacyl-tRNA synthases, 979–983, 980*f*, 981*f*
ApoB mRNA editing, 881, 882*f*
A-to-I, 882–883, 883*f*
C-to-U, 881–882, 882*f*, 883*f*
as genetic material, 141, 145
viral, 17
virulent bacteriophages with single-stranded RNA as genetic material, 332*f*, 332–334, 333*f*
isolation of, 153
messenger. *See* Bacterial mRNA; mRNA (messenger RNA)
micro (miRNA), 950–954, 951*f*–953*f*
formation of, 952*f*, 952–954, 953*f*
modes of action of, 953*f*, 953–954
number of, 951
precursor for, 952*f*, 952–953
piwi interacting, 954
retroviral, 357, 359
ribosomal. *See* rRNA (ribosomal RNA)
sense and antisense strands of, 939
single-stranded, in phages, 332*f*, 332–334, 333*f*
structure of, 9, 11*f*, 141–144
quaternary, 141–142
secondary, 141–142, 142*f*–143*f*
stick figure representations for, 10*f*, 11*f*, 12
tertiary, 141, 142–144, 143*f*, 144*f*
synthesis of. *See also* Transcription; *specific stages of transcription*
requirements for, 624, 625*f*, 626, 626*f*
similarities and differences between eukaryotes and bacteria, 734–735, 735*f*
in telomerase, 434, 435*f*
transfer. *See* tRNA (transfer RNA)
transfer-messenger, stalled ribosome rescue by, 1046–1047, 1048*f*, 1049
translation and. *See* Translation
RNA bacteriophages, 332*f*, 332–334, 333*f*, 357
assembly of, 334
attachment to F pilus, 332
discovery of, 332
life cycle of, 333*f*, 333–334
lysis of, 334
A protein synthesis in, 334
replication of, 333*f*, 333–334
structure of, 332, 332*f*
translation of, 332–333
RNA-binding site (RBS), of transcription elongation complex, 648

RNA degradosome, 704
RNA-DNA hybrid binding site (HBS), of transcription elongation complex, 648
RNA export factor 1 (REF 1), 884
RNAi. *See* RNA interference (RNAi)
RNA-induced initiation of transcriptional silencing (RITS) complex, 956
RNA-induced silencing complex (RISC), 941
RNA interference (RNAi), 937–950
  Argonaute protein family in, 940, 944*f*, 944–946, 945*f*
  blockage of viral replication by, 946–947
  Dicer in, 941–944, 942*f*, 943*f*
  discovery in *C. elegans*, 938–940, 939*f*
  heterochromatin assembly and, 955–956, 957*f*
  as investigational tool, 949–950
  as posttranscriptional process, 940
  prevention of transposon activation by, 947
  siRISC in, 941
    formation of, 946, 946*f*
  siRNAs in, 941
  transitive, 947–949, 948*f*
    ERI-1 in, 949
    SID-1 in, 948–949
  in vitro studies of, 940–942, 942*f*
RNA polymerase(s)
  archaeal, 623
  bacterial. *See* Bacterial RNA polymerases
  of bacteriophage T7, 328
  of *Caenorhabditis elegans*, 948
  eukaryotic, 623
    types of, 725–727, 726*f*, 727*t*
  nuclear. *See also* RNA polymerase I; RNA polymerase II; RNA polymerase III
    distinguishing, 727–728, 728*f*
    subunits of, 728–731, 729*t*, 730*t*
    synthetic capacities of, 734, 736
  plastid, 932
    nuclear encoded, 932
  in RNA phage replication, 333*f*, 333–334
  trp operon and, 694–695, 695*f*
RNA polymerase holoenzyme, 626–627, 627*t*
  cAMP•CRP complex contacts with, 681–682
  components of, 628, 628*f*
  DNA footprinting technique for, 630–633, 631*f*–633*f*
  DNase protection method for, 629*f*, 629–630, 631*f*
  rrn transcription and, 708–709, 709*f*
  size of, 627
  subunits of, 626, 627*t*
  template used by, 628
RNA polymerase I, 896, 897–911. *See also* RNA polymerase(s), nuclear
  auxiliary transcription factors required by, 905–906, 906*f*
  eukaryotic rRNA transcription units and, 899–900, 900*f*, 901*f*
  factors associated with required for transcription, 904
  function of, 903–904
  isolation of, 726
  location of rRNA transcription and processing and, 902–903, 903*f*
  precursors of, 897–899, 898*f*
  rRNA transcription unit promoter and, 900–902, 901*f*
  structure of, 903*f*, 903–904
  termination of transcription, 907, 909, 909*f*
  transcription cycle and, 809, 907, 908*f*, 909*f*
    pre-rRNS conversion to mature r-RNA and, 909–910, 910*f*, 911*f*
  in vitro assembly of transcription initiation complex and, 906
RNA polymerase II, 723–756, 725*f*. *See also* RNA polymerase(s), nuclear
  archaeal, subunits of, 730, 730*t*, 731
  backtracking by, 753, 754
  core promoter of, 742–743, 743*f*
  distinguishing from other RNA polymerases, 727–728, 728*f*
  eukaryotic, subunits of, 730, 730*t*
  general transcription factors and, 743–752
    lack of requirement for TFIIA and, 748*f*, 748–749
    preinitiation complex formation and, 745*f*, 745–746, 746*f*
    sequential binding of RNA polymerase II•TFIIF complex, TFIIE, and TFIIH and, 751–752, 752*f*
    TFIIB in conversion of closed promoter to open promoter and, 749–751, 749*f*–751*f*
    TFIID and core promoters of protein-coding genes lacking a TATA box and, 746–748, 747*f*
    transcription of naked DNA from specific transcription start sites and, 743–744, 744*t*
  isolation of, 726, 726*f*
  nucleotide addition rate and, 753
  of prokaryotes, 731
  squelching and, 805
  structure of, 731, 732*f*, 733*f*, 733–734, 735*f*
  subunits of, 728–731, 729*t*
    genes coding for, 729, 729*t*
  synthetic capacity of, 734, 736
  transcription elongation and, 752–756
    C-terminal domain phosphorylation and, 752–753
    elongation factor SII and, 754, 754*f*, 755*f*
    transcription elongation complex regulation and, 754, 755*f*, 756
    transcription elongation factors and, 753–754
  transcription rate of, 846
  transcription start site location and, 736–741, 739*f*–741*f*
  transcription termination and, 871–880
    allosteric changes and 5′→3′ exonucleases and, 878, 879*f*
    gene concept and, 876–877
    location of, 877, 877*f*
    polyadenylation sites and, 875*f*, 875–876
    poly(A) tail synthesis and, 871–875, 872*f*, 873*t*, 874*f*
RNA polymerase II•TFIIB complex, crystal structure of, 749, 750*f*
RNA polymerase II•TFIIF complex, in preinitiation complex formation, 751–752
RNA polymerase III, 896, 919–927. *See also* RNA polymerase(s), nuclear
  functions of, 920
  isolation of, 726
  pre-tRNA processing to become mature tRNA, 924–927, 925*f*
  transcription factors required to recruit, 922*f*, 922–923
  transcriptions units of, promoter types of, 920–922, 921*f*
RNA pyrophosphohydrolase (RppH), 704
RNA recognition motifs (RRMs), 869, 869*f*
RNases. *See* Ribonucleases (RNases)
RNA Tie Club, 969
RNA 5′-triphosphatase
  5′-m7G caps and, 836*f*, 836–837, 838, 838*f*
  synthetic CTD interaction with, 839, 839*t*
"RNA world," 144–145
Roberts, Jeffrey W., 653
Roberts, Richard B., 840–841, 964
Robertson, Elizabeth, 546
Roeder, Robert G., 726, 743–744, 745, 778, 922
Rolling circle replication, 273
Rosbash, Michael, 870
Rösch, Paul, 654
Rose, Irwin, 100
Rosenberg, Barnett, 464
Rossmann, Michael G., 324, 978
Rossmann-fold, 978*f*, 978–979
Rothstein, Rodney, 524
Rough endoplasmic reticulum, 965, 965*f*
Rous, Peyton, 357
Rous sarcoma virus
  replication of, 359
  RNA of, 357, 359
RPA. *See* Replication protein A (RPA)
RPA•DNA complex, 122*f*, 122–123, 442
RPA protein, 534*t*, 536
  mismatch repair and, 497
RPB genes, 729
RPB4 gene, 731
Rpd3, 814
R plasmids, 283
RppH (RNA pyrophosphohydrolase), 704
r-proteins (ribosomal proteins), synthesis of, regulation in *E. coli*, 712–713
RRF (ribosomal recycling factor), 1012
RRF (ribosome release factor), 1051*f*, 1051–1052
RRF-1, 948
RRF-2, 948
RRF-3, 948
RRMs (RNA recognition motifs), 869, 869*f*
rRNA (ribosomal RNA)
  bacterial transcription of, 706–713
    E. coli rRNA operons and, 707–708, 708*f*
    Fis protein binding site and, 709
    growth rate and, 711*t*, 711–712, 712*t*
    guanine nucleotide inhibition of, 710*f*, 710–711
    promoter upstream element and, 708*f*, 708–709, 709*f*
    ribosome structure and, 706–707, 707*f*
    r-protein synthesis regulation and, 712–713
    rRNA operons and, 707–708, 708*f*
  chloroplast DNA transcription to form, 930–932, 931*f*
  mitochondrial transcription to form, 927–928, 930
  operons of, of *E. coli*, 707–708, 708*f*
  precursors of, 898*f*, 898–899
  pre-rRNA conversion to, 909–910, 910*f*, 911*f*
  processing of, 714
  secondary structure of, 1008–1010, 1009*f*
  self-splicing, 911–914, 912*f*–915*f*
  tertiary structure of, 1008–1010, 1009*f*
rRNA initiator (rInr), 901
rRNA transcription units
  components of, 900–902, 901*f*
  organization of, 899–900, 900*f*, 901*f*
rrn operons, 707–709
  upstream element and, 708*f*, 708–709, 709*f*
RSC chromatin remodeling complex, 540, 811–812, 812*f*
RS-domain, 869
RSSs (recombination signal sequences), 560, 560*f*
  incorrect, 561, 561*f*
Rtt106 histone chaperone, nucleosome assembly and, 441
Run-off analysis, 736, 738–739, 739*f*
Rupert, Claud S., 470
RusA protein, overexpression of, 542
Rus protein, in recombination, 527*t*
rut (rho utilization) site, 653
Rutter, William J., 726

RuvA protein, in recombination, 532
RuvB protein, in recombination, *527t*, 532
RuvC protein, in recombination, *527t*, 532
Ruvkin, Gary, 949, 950
RXR (retinoid X receptor), 783
RXR•RAR response element, 784, *785f*
RXR•TR response element, 784, *784f*, *785f*
RXR•VDR response element, 784
Rybenkov, Valentin V., 215, 216

S

S(s) (Svedbergs), 156
Sabin polio vaccine, SV40 contamination of, 349
Sabitini, David D., 1053–1054
Saccharomyces cerevisiae
  budding of, *285f*, 285–286
  DNA replication in. *See* Eukaryotic DNA replication
  Gal4 regulation of genes coding from galactose metabolism in, 787–789, *787f–789f*
  genetics of, 284–290
    haploid and diploid cells and, 287–288, *288f*
    initiation of mating process and, 290
    mating type determination and, 288–290, *289f*
    notations, conventions, and terminology used in, 286, *287t*
  meiotic recombination in, 541
  mobile group II introns in, 616–617
  as model for genetic studies, 255
  recombination proteins of, *534t*
  RNA polymerase II of, 730
    attempts to crystallize, 731, *732f*, *733f*, 733–734, *735f*
  telomerase RNA of, 434
  tRNA genes of, 972
  two micron plasmid of, 564
  Ty5 element of, 592
  ZMM pathway in, 544
Safranin, 257–258
Sal box, 909
Salk polio vaccine, SV40 contamination of, 349
Salvage pathways, 297–298, *298f*
Sanger, Frederick, 9, 185, 186, 189, 1013–1014
SARs (scaffold attachment regions), 239, *240f*
α Satellite DNA arrays, 241, *242f*
Sau3AI, mismatch repair and, 495
Sauer, Robert T., 1047
Saunders, Edith Rebecca, 512
Sawadogo, Michele, 745
SbcBC protein, in recombination, *527t*
SbcB protein, in recombination, *527t*
SbcCD nuclease, 532
sbc gene, 532
Scaffold(s), *193f*, 194
Scaffold attachment regions (SARs), 239, *240f*
Scaffold model, of chromosome structure, 238–239, *239f*, 240
Scanning mechanism, in eukaryotic protein synthesis, 1021
Schimmel, Paul, 981, 983–985
Schizosaccharomyces pombe
  ARS elements in, 424
  binary fission in, 286
  heterochromatin assembly in, 955–956, *957f*
  Mus81•Eme1-dependent pathway in, 544
  mutations in genes coding for Mus81 and Eme1 in, 542
  Tf element of, 592
Schuster, Stephan C., 201
Schwartz, David, 163
SCID (severe combined immunodeficiency), 561
Scott, Matthew, 774
Scrunched complex, in transcription initiation, *641f*, 642
Scrunching model of transcription, *639f*, 639–640
SDSA (synthesis-dependent strand-annealing) model, 540, *540f*
SD (Shine-Dalgarno) sequence, *1018f*, 1018–1020, *1019f*
SDS-PAGE (sodium dodecyl sulfate-polyacrylamide gel electrophoresis), 162–163
SDS-PAGE (sodium dodecyl sulfate-polyacrylamide genetics and electrophoresis), of RNA polymerase II, 729
Sec (selenocysteine), as polypeptide building block, 986, *986f*
Secondary structure
  of proteins, 46, *47f*, 51–56
    combination of, 55, *55f*
    β-conformation, 54, *54f*
    α-helix, 51–54, *52f*, *53f*
    loops and turns in, 54–55, *55f*
    prediction of, 55, *56f*
  of RNA, 141–142, *142f–143f*
Second messengers, in G-protein-linked signal transduction systems, 98
Segregation, law of, 3, 513, 515, 516
Seitz, Thomas, 982
Selective estrogen receptor modulators (SERMs), 786
Selectivity factor (SL1/TIF-1B), 906
  recruitment by UBF, 907
Selenocysteine (Sec), as polypeptide building block, 986, *986f*
Self-splicing rRNA, 911–914, *912f–915f*
Self-transmissible plasmids, 276
Semiconservative model for DNA replication, 366–369, *366f–369f*
Semidiscontinuous model of replication, 372, *372f–375f*, 375
Sense strands
  DNA, 624, 626, *626f*
  RNA, 939
Sepharose beads, for DNA sequencing with GS FLX, 195, 197, *198f*
Sequence hypothesis, 964
Sequence tagged sites (STSs), 190
SER (smooth endoplasmic reticulum), 965, *965f*
Serine
  abbreviations for, *33t*
  side chains of, 31, *31f*
SERMs (selective estrogen receptor modulators), 786
Ser-tRNA, 983
Setlow, Richard, 482
70S initiation complex, 1015, *1016f*, 1016–1017, *1017f*
Severe combined immunodeficiency (SCID), 561
Sex chromosomes, 228
Sex pilus, 274
SF1 (splicing factor 1), 863, *863f*
Sgs1/BLM protein, 544
Sgs1 protein, 536, 538, 539
Shakked, Zippora, 790
Sharp, Philip A., 840–841, 852, 940
Sheep, nuclear transfer experiments in, *819f*, 819–820
Shelterin, 245
Shigella, drug resistance in, 283
Shine-Dalgarno (SD) sequence, *1018f*, 1018–1020, *1019f*
Short interspersed nucleotide elements (SINEs), 576
  in target-primed reverse transcription, 614, *615f*, 616
Short patch base excision DNA repair, 481, *481f*
Shotgun DNA sequencing, *189f*, 189–192
  hierarchical, in Human Genome Project, 189–192, *191f*
  whole genome, human genome sequencing using, 189, 192, *193f*, 194
Shuman, Stewart, 839
Sickle cell anemia, 80–83, *81f*, *82f*
SID-1, 948–949
Side chains
  in α-amino acids, 29
  grouping of amino acids based on, 30–31, *30f–32f*
sid-1 gene, 948–949
σ factors
  in initiation stage of transcription, *643t*, 643–644
  nomenclature for, 643, *643t*
  variations among bacterial species, 644
σ^{54}-RNA polymerase, *644f*, 645
σ^{70} family, 629
  conserved domains of, 638, *638f*
Sigmoidal kinetic curves, 92–93, *93f*
Sigmoidal oxygen-binding curve, 68, *68f*
Signal hypothesis, 1053–1054
"Signaling release" pathway, 403
Signal peptidase, 1056
Signal recognition particle (SRP), *1054f*, 1054–1055
Signal sequence, protein synthesis and, 1053–1056, *1054f*, *1055f*
Signal transduction systems, G protein-linked, 94–99
  first messengers in, 98
  operation of, 94–98, *95f–99f*
  second messengers in, 98
SII, reactivation of arrested RNA polymerase II by, 754
Silencers, 768
Simian virus 40 (SV40)
  AP-1 proteins binding to enhancer regions of, 792–793
  discovery of, 349
  DNA replication in, *351f*, 351–352
    T antigen in, 416–420, *417f–421f*
  DNA size in, *110t*
  genome of, *340f*, 349
  origin of replication of, T antigen binding to, 416–418, *417f*, *418f*
  as polio vaccine contaminant, 349
  T antigen of, 349, 416–420
    binding to origin, 416–418, *417f*, *418f*
    DNA polymerase/α-primase recruitment to proto-replication bubble by, 418–420, *419f–421f*
  transcription in, *767f*, 767–768
Simple retroviruses, 357
SINEs (short interspersed nucleotide elements), 576
  in target-primed reverse transcription, 614, *615f*, 616
Singer, Jonathan S., 70
Single nucleotide polymorphisms (SNPs), 291–292, 293–294
Single-strand annealing (SSA), 553, *553f*
Single-stranded DNA
  BRAC2 binding to, 537
  library of, creating for DNA sequencing with GS FLX, 195, *196f*
Single-stranded DNA binding proteins (SSBs), 121–123
  composition of, 121–122
  isolation of, 122
  mismatch repair and, 494–495
  in RecA-mediated strand exchange, 529
Siomi, Haruhiko, 955

siRISC loading complex, 946, 946*f*
siRNAs (small interfering RNAs), 941
endogenous, 955, 956*f*
formation of, 946, 946*f*
primary, 947. *See also* Dicer
secondary, 947
Sister chromatids, 219, 220*f*
cohesin tethering of, 221–222, 222*f*, 223*f*, 226
Site-directed mutagenesis, 184*f*, 184–185
Site-specific recombination, 555–566, 556*f*
conservative, 576, 576*f*, 607–612
in bacteriophage λ, 609*f*, 609–611, 610*f*
in integron system, 611, 611*f*
proteins in, 607*f*, 607–608, 608*f*
Cre/*lox* system and, 565*f*, 565–566
FLP/*FRT* system and, 564*f*, 564–565
mating type switching in yeast and, 556–558, 557*f*, 558*f*
V(D)J, 555, 558–564, 559*f*–563*f*
Sixma, Titia K., 495
Skolnick, Mark, 293
Skordalakes, Emmanuel, 436, 653
7SK RNA, 920
Slab gel electrophoresis, 160–162, 161*f*, 163
SLBP (stem-loop binding protein), 879–880
Sld2 protein, in eukaryotic DNA replication, 427–428, 428*f*
Sld3 protein, in eukaryotic DNA replication, 427, 428, 428*f*
Sleeping Beauty transposon, 602–604, 603*f*
Sliding clamp, 394–396, 395*f*
core polymerase release from, 403–404
crystal structure of, 394, 395, 395*f*
faces of, 395
placement around DNA by clamp loader, 396–399, 396*f*–399*f*
Slow-stop mutants, 379–380
7SL RNA, 920
SL1/TIF-1B (selectivity factor), 906
recruitment by UBF, 907
SLX1•BTBD12 complex, 538
SLX1 protein, 538
SLX4 protein, 538
SLX1•SLX4 complex, 538
SMA (spinal muscular atrophy), 861
SmaI, sequence specificity of, 169*t*
Small interfering RNAs. *See* siRNAs (small interfering RNAs)
Small nuclear ribonucleoprotein particles (snRNPs), 856–862
formation of, 859, 860*f*, 861
splicesome formation from, 856–857, 858*f*
U4atac snRNP, 858–859
U6atac snRNP, 858–859
U1 snRNP, 859, 860*f*, 861–862, 862*f*
U2 snRNP, 859, 860*f*, 861–862, 862*f*
U4 snRNP, 859, 860*f*, 861
U4/U6•U5 tri-snRNP, 861–862, 862*f*
U5 snRNP, 859, 860*f*, 861
U6 snRNP, 859, 864
in catalytic process, 866, 866*f*
U11 snRNP, 858–859, 859
U12 snRNP, 858–859, 859
U62 snRNP, 864
Small protein B (SmpB), 1047, 1049*f*
Small ribonucleoprotein (snoRNP), 910
methylation, 910, 910*f*
pseudouridylation, 910, 910*f*
Small ubiquitin-related modifier (SUMO), 102
SMC (structural maintenance of chromosomes) proteins, 214–215, 215*f*
crystal structure for, 214–215, 215*f*
stabilization of condensed chromosomes by, 239–240, 240*f*
Smith, Hamilton O., 168, 172, 189
Smith, Michael, 184–185
Smithies, Oliver, 545, 546
Smooth endoplasmic reticulum (SER), 965, 965*f*
SmpB (small protein B), 1047, 1049*f*
Sm polypeptides, 857
SNAP $_C$ (snRNA activating protein complex), 922, 922*f*, 923
Snf1 kinase (sucrose non-fermenting kinase), 800
snoRNA (*Stenotrophomonas maltophilia* nucleolar RNA), 910
snoRNP. *See* Small ribonucleoprotein (snoRNP)
SNPs (single nucleotide polymorphisms), 291–292, 293–294
SnRNA activating protein complex (SNAP $_C$), 922, 922*f*, 923
snRNPs. *See* Small nuclear ribonucleoprotein particles (snRNPs)
S1 nuclease, properties of, 166*t*
Sodium, RNA tertiary structure stabilization by, 142
Sodium dodecyl sulfate-polyacrylamide gel electrophoresis (SDS-PAGE), 162–163
Sodium dodecyl sulfate-polyacrylamide genetics and electrophoresis (SDS-PAGE), of RNA polymerase II, 729
Solenoid model, of 30-nm fibers from chromosomes, 237, 237*f*, 238
Soluble RNA (sRNA), 967
Somatic cells, 217
genetics of, 295–301
heterokaryon formation and, 297–299, 297*f*–299*f*
mapping genes in higher organisms using, 295
monoclonal antibody production and, 299, 300*f*, 301
tissue cultures for, 295*f*, 295–297
senescence of, 438
telomeres and, 245–246
Somatic hypermutation, 562–563, 563*f*
S1 endonuclease, 166
S1 mapping (S1 nuclease analysis), 736, 737*f*
S1 nuclease, 736
Song, Ok-kyu, 804
Soot, cancer and, 458–459
Sorangicin, 656
Sorangium cellulosum, sorangicin produced by, 656
SOS response, 498–505
key steps in, 501*f*, 501–502
translesion DNA synthesis and, 498–499
Southern, Edwin M., 175
Southern blotting (Southern transfer), 174–177, 175*f*, 177*f*
Sox2 gene, 820*f*, 820–821
SP (spacer promoter), 902
Spacer(s), in mRNA, 663
Spacer promoter (SP), 902
Specialized transduction, 340
Speck, Christian, 427
Spectral karyotyping, 231
Spermidine, Z-DNA and, 137
Spermine, Z-DNA and, 137
S phase, 219, 219*f*
Spheroplasts, 259
Sphingolipids, 66
Spinal muscular atrophy (SMA), 861
Spirilli, morphology of, 257, 257*f*
Spliceosomes, 855–871
assembly on introns, 863*f*, 863–864, 865*f*
catalytic process and, 866*f*, 866–867, 867*f*
components of, 857–859, 859*f*. *See also* Small nuclear ribonucleoprotein particles (snRNPs)
cotranscriptional splicing and, 870–871, 871*f*
formation of, 856–857, 858*f*
regulation of splice site selection and, 868–879, 869*f*, 870*f*
3′-Splice sequence, 851
5′-Splice sequence, 851
3′-Splice site, 851
5′-Splice site, 851
Splicing
alternative, 883
of pre-mRNA, split genes and, 840–855
combinations of splicing patterns within individual genes leading to formation of multiple mRNAs and, 848*f*, 848–849, 849*f*
coordinated transesterification reactions in splicing and, 854–855, 855*f*
interruption of amino acid-coding regions within eukaryotic genes and, 841–842, 843*f*, 844*f*, 844–845
intron and exon conservation and, 845*t*, 845–847, 846*f*
mRNA molecules formed by splicing pre-mRNA and, 840–841, 841*f*
production of two or more mRNA molecules by a single pre-mRNA and, 847*f*, 847–848
slicing precision and, 851*f*, 851–852, 852*f*
splicing intermediates resembling lariats and, 852–854, 853*f*, 854*f*
trans-splicing in *Drosophila* and, 850, 850*f*
of pre-mTNA, alternative, 847*f*, 847–848
of rRNA, self-splicing rRNA and, 911–914, 912*f*–915*f*
Splicing factor 1 (SF1), 863, 863*f*
Splicing intermediates, 852–854, 853*f*, 854*f*
Splicing regulatory (SR) proteins, 869*f*, 869–870
Split gene(s), 840–855
combinations of splicing patterns within individual genes leading to formation of multiple mRNAs and, 848*f*, 848–849, 849*f*
coordinated transesterification reactions in splicing and, 854–855, 855*f*
interruption of amino acid-coding regions within eukaryotic genes and, 841–842, 843*f*, 844*f*, 844–845
intron and exon conservation and, 845*t*, 845–847, 846*f*
mRNA molecules formed by splicing pre-mRNA and, 840–841, 841*f*
production of two or more mRNA molecules by a single pre-mRNA and, 847*f*, 847–848
slicing precision and, 851*f*, 851–852, 852*f*
splicing intermediates resembling lariats and, 852–854, 853*f*, 854*f*
trans-splicing in *Drosophila* and, 850, 850*f*
Split gene hypothesis, 842, 844
Spontaneous mutagenesis, 263
SPO11 protein, 534*t*
Spo11 protein, 534*t,* 541
spoT gene, 710
Squelching, 805*f*, 805–806
SR (SRP receptor), 1054, 1055–1056
SRB1 gene, 807
SRB2 gene, 807
Srb-11 module, 806

sRNA (soluble RNA), 967
SRP (signal recognition particle), 1054f, 1054–1055
SRP receptor (SR), 1054, 1055–1056
SR (splicing regulatory) proteins, 869f, 869–870
SRP•SR complex, 1055–1056
5SrRNA, 920
SSA (single-strand annealing), 553, 553f
SSB(s). *See* Single-stranded DNA binding proteins (SSBs)
ssb gene, 380t
SSB protein, in recombination, 527t
SstI, sequence specificity of, 169t
SSU processome, 918
Stabilization helix, of Lac repressor N-terminal domain, 674
Stahl, Franklin, 367, 369
Staining
  of chromosomes, 228, 228f
  of gel by ethidium bromide, 162
  Gram stains for, 257–259, 258f
  negative, for electron microscopy, 155, 155f
Staphylococcus aureus, lack of Rho factor in, 654
Start codons, bacterial, 1012–1013
Stationary pawl mechanical ratchet wheel, 648–649, 649f
Stationary phase, 260
Steitz, Joan A., 856, 880, 880f, 1018–1019
Steitz, Thomas A., 62, 179, 385, 681, 1008, 1011, 1036–1038
Stem cells
  embryonic, 546, 547, 548f, 818
  multipotent, 818
  pluripotent, 818
  unipotent, 818
Stem-loop binding protein (SLBP), 879–880
Stem-loop structures, 142, 142f
Stenotrophomonas maltophilia nucleolar RNA (snoRNA), 910
Stereochemical hypothesis of codon assignments, 1001
Stern, Curt, 513
Steroid receptors, 780, 781–782, 783f
Sterols, 66
Stillman, Bruce, 426
Storage proteins, 76
Strand asymmetric model, of mtDNA replication, 928, 930
Strand invasion, RecA protein and, 528f, 528–529, 530f
Strauss, Bernard, 549
Streptavidin, in Southern blotting, 176
Streptococcus pneumoniae, transforming principle of, 12f, 12–15, 13f
Streptolydigin, 656
Streptomyces
  DNA structure in, 211
  rifamycins produced by, 655
Streptomyces alboniger, 1035
Streptomyces coelicolor, σ factors of, 644
Streptomyces lydicus, streptolydigin produced by, 656
Stringent plasmids, 283
Stringent response, 710
Strobel, Scott A., 1039
Strong promoters, 637
Structural genomics (structural proteomics), 60–62
Structural maintenance of chromosomes proteins. *See* SMC (structural maintenance of chromosomes) proteins
Structural maintenance of chromosomes (SMC) proteins, 214–215, 215f
Structural proteins, 76
  in phage assembly, 320
Structure-function relationships, as central theme of molecular biology, 21
STSs (sequence tagged sites), 190
Sturtevant, Alfred H., 513
Subtilisin, 179
ϵ-Subunit, of core polymerase, 394
Sucrose gradient centrifugation, 157–158
Sucrose non-fermenting kinase *(Snf1 kinase),* 800
Suicide enzymes, 476
Sulfolobus shibatae, RNA polymrase from, 756, 757f
Sulston, John, 950–951
SUMO (small ubiquitin-related modifier), 102
Supercoil(s), 125–128, 126f, 127f
  of bacteriophage λ DNA, 335, 336f
  density of, 127
  formation of, 123
  interwound, 123, 123f
  negative, 125
  positive, 125
    in thermophiles, ATP-dependent reverse gyrase required for, 443
  single-stranded regions of, 127f, 127–128
  topological properties of, 125–127
  toroidal, 123, 123f
Supercoiling density (σ), 127
Superhelix. *See* Supercoil(s)
Supersecondary structures, 55, 55f
Suppressor(s), 263
Supressor-sensitive mutants, 263
Sutherland, Earl W., 678
SV40. *See* Simian virus 40 (SV40)
Svedberg, Theodor, 156
Svedberg(s) (Ss), 156
SWI/SNF chromatin remodeling complexes, 810–813, 811f, 812f
SwqA protein, 409
SWR1 chromatin remodeling complex, 540
Symmetrical internal loops, 142, 142f
Symmetric heteroduplex DNA, 519
Synaptonemal complex, 541, 542f
Syn conformation, 137, 137f
Synonyms, 999
Synthesis-dependent strand-annealing (SDSA) model, 540, 540f
Synthetic elements, 594–597, 595f
Systemic lupus erythematosus, anti-Smith (anti-Sm) antibodies in, 855–856
Szostak, Jack, 432–434, 524, 525

T

TAFs (TBP-associated factors), 746, 906, 906f
Takahashi, Kazutoshi, 820–821
Tamoxifen
  for breast cancer treatment, 785–786
  side effects of, 786
  structure of, 787f
Tandem affinity purification (TAP), 915–917, 918f
Tandem repeat polymorphisms, 291, 292f, 294
T antigen, of SV40, 349, 416–420
  binding to origin, 416–418, 417f, 418f
  DNA polymerase/α-primase recruitment to proto-replication bubble by, 418–420, 419f–421f
  functional domains of, 417, 417f
t antigen, of SV40, 349
TAP (tandem affinity purification), 915–917, 918f
Taq core RNA polymerase, crystal structure for, 633–636, 634f–636f
Taq polymerase, 182
TargetDB, 60
Target-primed reverse transcription, 576f, 576–577, 612–617
  elements moving via, 612f, 612–613
  for LINE movement, 613f, 613–616, 615f
    genome stability and evolution and, 614–616
  mobile group II intron movement by, 616–617
    homology to target site and, 617
    transesterification reactions allowing, 616f, 617, 617f
Target site duplication, 578
TAR RNA-binding protein (TRBP), 953
Tas3 protein, 956
Tata, Jamshed R., 725, 726
TATA-binding protein (TBP), 745
  recruitment and activity of, factors influencing, 747
  squelching and, 805
  structure of, 745, 746f
TATA box, 742
  core promoter with
    site of beginning of preinitiation assembly and, 745–746, 746f
    transcription machine formation by, 744–745, 745f
  in RNA polymerase III transcription unit, 921
Tatum, Edward L., 4, 270–271
TBP. *See* TATA-binding protein (TBP)
TBP-associated factors (TAFs), 746, 906, 906f
TBP•DNA complex, order of transcription factor binding to, 748
TBP•TATA box complex, crystal structure for, 745–746, 746f
3T3 cell line, 297
T cells, 299
Tc/*mariner* elements, 591
T DNA, 284
TEC. *See* Transcription elongation complex (TEC)
Telomerase
  aging and, 438–439
  cancer and, 246, 438–439
  crystal structure of, 436, 436f
  end-replication problem and, 437f, 437–439, 438f
  nucleotide repeats added to chromosome ends by, 434–437, 435f, 436f
  reaction cycle for, 436f, 436–437
  RNA in, 434, 435f
Telomeres, 243–246, 244f–246f
  formation of, 432–439
    RNA template used by telomerase in, 434–437, 435f, 436f
    role of telomerase in, 437f, 437–439, 438f
    terminal transferase-like enzyme required for, 432f, 432–434, 433f
  maintenance by transposons, 576
  of *Tetrahymena thermophila*, 432–433
  of yeast, cloning of, 433, 433f
Telophase, 220f, 221
  of meiosis
    telophase I, 225f, 227
    telophase II, 224f, 227
  of mitosis, 220f, 221
Temin, Howard, 201, 355, 359
Temperate bacteriophages, 311, 334–342
  growth of, 311–312
  lambdoid, 334
  phage λ, 334f–337f, 334–341
  phage P1, 341f, 341–342

Temperature
DNA denaturation and, 113–116, 114*f*, 116*f*, 117, 118*f*
DNA renaturation (reannealing) and, 118–119
nonpermissive, 263
permissive, 263
Temperature-sensitive (Ts) mutants, 263, 265
Template-dependent DNA polymerases, human, 504*t*, 504–505
Template strand, 177–178, 178*f*, 624
Template switching, 550
-10 box, 630
-10 element, extended, 637
Terminal protein, of adenoviruses, 353
Terminal redundancy, of phage T4 DNA, 314–315, 315*f*, 317
Terminase, in phage T4 DNA packaging, 321, 323–324, 324*f*, 325*f*
Termination, of RNA polymerase movement, 651
Termination sites, 405, 405*f*
terminus utilization substance binding to, 406, 406*f*, 407*f*, 408
Termination stage
of bacterial DNA replication, 379, 405–408
termination sites and, 405, 405*f*
terminus utilization substance in, 406, 406*f*, 407*f*, 408
of bacterial protein synthesis, 1043–1050
protein release factors and, 1043–1046, 1044*f*–1046*f*
rescue of stalled ribosomes by tmRNA and, 1046–1047, 1048*f*, 1049*f*
suppression of mutations creating termination codons within reading frames and, 1050
of eukaryotic protein synthesis, 1050*f*, 1050–1051
of transcription
bacterial RNA polymerases and, 652–654, 652*f*–655*f*
intrinsic, 652*f*, 652–653
Rho-dependent, 653–654, 654*f*, 655*f*
RNA polymerase III and, 923–924
Terminus utilization substance (TUS), binding to *Ter* sites, 406, 406*f*, 407*f*, 408
Ter sites, 405, 405*f*
terminus utilization substance binding to, 406, 406*f*, 407*f*, 408
Tertiary structure
of proteins, 46, 47*f*, 56–66
determination by primary structure, 62–64, 63*f*
intrinsically disordered proteins in, 59–60
molecular chaperones and, 64–65, 65*f*, 66*f*
nuclear magnetic resonance spectroscopy of, 56, 58*f*, 58–59, 59*f*
structural genomics and, 60–62
x-ray crystallography of, 56*f*, 56–58, 57*f*
of RNA, 141, 142–144, 143*f*, 144*f*
Tethered particle motion approach, 649, 650*f*
Tetrads, 515
Tetrahydrofolate, hybrid cell selection and, 297
Tetrahymena thermophila
pre-rRNA of, 911–914, 912*f*–915*f*
telomeres of, 244, 432–433
Tetramerization helix, of Lac repressor, 675, 676*f*
Tetrapeptides, 35*f*
Tetraplex structure, of DNA, 139, 141*f*
T-even phages, 309
Tfam, 930
TFB, 930
Tf element, 592
TFIIA, 744, 744*t*
human, 748
zinc fingers in, 777–778
TFIIB, 744, 744*t*
binding sites of, 742
in closed promoter conversion to open promoter, 749–751, 749*f*–751*f*
mutants of, Mediator binding by, 809
squelching and, 805
TFIID, 744, 744*t*
binding to core promoters of genes lacking TATA boxes, 746–748, 747*f*
proteins present in, 746
TFIIE, 744, 744*t*
in preinitiation complex formation, 752
TFIIF, 744, 744*t*
TFIIH, 744, 744*t*
human, 752
in preinitiation complex formation, 752
structure of, 752, 752*f*
TFIIH (transcription factor IIH), 490
TFIIIA, 922*f*, 922–923
TFIIIB, 922*f*, 922–923
TFIIIC, 922*f*, 922–923
TFIIS, reactivation of arrested RNA polymerase II by, 754
Theoretical maximum rate of reaction (V_{max}), 88–89, 89*f*
Thermal cyclers, 182–183
Thermococcus litoralis, DNA polymerase isolated from, 182
Thermodynamics, second law of, 48
Thermophiles, ATP-dependent reverse gyrase required for positive supercoiling in, 443
Thermus aquaticus
core RNA polymerase of, crystal structure for, 633–636, 634*f*–636*f*
DNA polymerase isolated from, 182
polIIIα from, 393*f*, 393–394
transcription elongation complex of, crystal structure for, 646–648, 647*f*
Thermus thermophilus
Argonaute protein of, crystal structure for, 944–946, 945*f*
crystal structure for ribosomes of, 1008
of 70S ribosome, 1030, 1032*f*
heavy metal binding protein produced by, 59
Ile-tRNA synthase from, 982
30S ribosome from, codon-anticodon helix sensing and, 1033*f*, 1033–1034
θ replication, 336, 337, 338*f*, 339
Thiomethylgalactoside (TMG), in lactose enzyme induction studies, 667, 667*f*
-35 box, 630
30-nm fibers from chromosomes
solenoid model of, 237, 237*f*, 238
zigzag model of, 237*f*, 237–238
30S initiation complex, 1014–1016, 1015*f*
3T3 cell line, 297
Threonine
abbreviations for, 33*t*
side chains of, 31, 31*f*
Threonine operon, 695
Thymidine kinase (TK), in hybrid cell selection, 298
Thymidine kinase gene transcription, 765
Thymidine-3′-monophosphate (3′-dTMP), 7*f*
Thymidine-5′-monophosphate (5′-dTMP), 7*f*
Thymine
in A-, B-, and Z-forms of DNA, 8*t*
in DNA, 6, 6*f*
methylation of, 458, 459*f*
Thymine glycol, inhibition of DNA replication by, 456
Thyroid hormone receptor (TR), 780, 783
Thyroid hormone receptor-associated protein (TRAP), 807
Tibolium castaneum, telomerase of, crystal structure of, 436, 436*f*
Tid1 protein, 537. *See also* Rad54 protein
TIF-1A (transcription initiation factor IA), 904
Tijan, Robert, 770
Tiling paths, 191
TINT1, 245
TIP60 chromatin remodeling complex, 540
Ti plasmid, 284
Titin gene, introns of, 846
TK (thymidine kinase), in hybrid cell selection, 298
TLC1 gene, 434
T-loops, 244, 245*f*
T lymphocytes, 299
T_m (melting temperature), for DNA, 114*f*, 114–115
TMG (thiomethylgalactoside), in lactose enzyme induction studies, 667, 667*f*
tmRNA (transfer-messenger RNA), stalled ribosome rescue by, 1046–1047, 1048*f*, 1049
TMV (tobacco mosaic virus), 357
Tn*916* element, 611–612
Tobacco mosaic virus (TMV), 357
Todd, Alexander Robertus, 9
T-odd phages, 309
Todo, Takeshi, 472
Tonegawa, Susumu, 558–559
TopA protein, in recombination, 527*t*
Top3/hTOPOIIIa protein, 544
Topoisomer(s)
conversion of one to another, topoisomerase catalysis of, 128, 128*f*, 129*f*
gel electrophoresis separation of, 162, 163*f*
Topoisomerases
catalysis of conversion of one topoisomer to another by, 128, 128*f*, 129*f*
subfamilies of, 128–132, 130*f*, 131*f*
type IA, 128–130, 130*f*, 131*f*
type IB, 131
type IC, 131–132
type II, 132, 132*f*
stabilization of condensed chromosomes by, 239, 240–241, 241*f*
type IV, 132
in termination stage of DNA replication, 406, 407*f*, 408
type V, 131–132
Top3 protein, 534*t*, 539
Toroidal supercoils, 123, 123*f*
"Torpedo" model of RNA polymerase II transcription termination, 878
Toxic proteins, 76
Tpii•TFIIS complex, crystal structure for, 754, 755*f*
TPP1 protein, 245
TΨC, of tRNA, 975
TR (thyroid hormone receptor), 780, 783
traA gene, 274
TraD protein, 276
TraI protein, 274, 275
TraM protein, 274
Transcription
bacterial. *See* Bacterial transcription
basal, 745. *See also* Basal transcription factors
in central dogma theory, 23, 23*f*
in chloroplasts, 930–932, 931*f*
elongation stage of, RNA polymerase III and, 923–924
genomic imprinting and, 817*f*, 817–818
initiation stage of. *See* Initiation stage of transcription
mitochondrial, 927–928, 929*f*, 930
nucleosomes in DNA coding region and, 815

reverse, 355
target-primed, 576f, 576–577, 612–617
in SV40, 767, 767f
termination stage of
bacterial RNA polymerases and, 652–654, 652f–655f
intrinsic, 652f, 652–653
Rho-dependent, 653–654, 654f, 655f
Transcription activator proteins, 764, 769–773
activation domains of, 769, 769f, 797–805
conformational change model and, 798
of Gal4, 798f–801f, 798–803, 803f
mechanisms of, 797–798
recruitment model and, 798
two-hybrid assay for detecting polypeptides interacting through non-covalent interactions, 804f, 804–805
combinatorial control and, 770, 770f
DNA-binding domains of, 769, 769f
with basic region leucine zippers, 791–794, 791f–795f
with helix-loop-helix structures, 794–797
with helix-turn-helix structures, 773–777
loop-sheet-helix, 789f, 789–790, 790f
with zinc fingers, 777–789
purification using DNA affinity chromatography, 770–772, 771f
squelching and, 805f, 805–806
structure of, 769f, 769–770
transfection assay to determine ability to stimulate gene transcription, 772f, 772–773
Transcription-coupled nucleotide excision repair pathway, 489–490, 491f, 492
Transcription elongation, RNA polymerase II and, 752–756
C-terminal domain phosphorylation and, 752–753
elongation factor SII and, 754, 754f, 755f
transcription elongation complex regulation and, 754, 755f, 756
transcription elongation factors and, 753–754
Transcription elongation complex (TEC)
bacterial RNA polymerases in, 645–652, 646f, 647f
detection and removal of incorrectly incorporated nucleotides by, 651, 651f, 652f
forward movement during nucleotide addition cycles, 648–649, 649f, 650f
pauses and, 649, 650f, 651
in transcription initiation, 641f, 642
transcription initiation complex conversion to, model for, 749–751, 751f
Transcription factor(s). *See* Basal transcription factors; General transcription factors (GTFs); Mediator (transcription factor); *specific factors*
Transcription factor IIH (TFIIH), 490
Transcription initiation complex
conversion to transcription elongation complex, model for, 749–751, 751f
in vitro assembly of, 906
Transcription initiation factor IA (TIF-1A), 904
Transcription-repressed domains (TRDs), 816
Transcription termination
bacterial RNA polymerases and, 652–654, 652f–655f
intrinsic, 652f, 652–653
Rho-dependent, 653–654, 654f, 655f
Transcription termination factor 1 (TTF-1), 907, 909, 909f
Transcriptome, 203
Transductants, 342
Transduction, specialized, 340
Transesterification reactions
coordinated, splicing and, 854–855, 855f
mobile group II intron movement and, 616f, 617, 617f
Transfer experiments, 17, 18f
Transfer-messenger RNA (tmRNA), stalled ribosome rescue by, 1046–1047, 1048f, 1049
Transfer origin *(OriT)*, in F plasmid, 272–273
Transferosome, 275–276
Transfer RNA. *See* tRNA (transfer RNA)
Transforming principle, 12f, 12–15, 13f
Transgenic mice, 547, 548f
Transient excursion model, 640
trans isomers, 33–34, 34f
Transition mutations, 454, 455f
Transitive RNA interference, 947–949, 948f
ERI-1 in, 949
SID-1 in, 948–949
Translation
in central dogma theory, 23, 23f
nick, 182
of RNA bacteriophages, 332–333
Translesion DNA synthesis, catalyzed by error-prone DNA polymerases, 498–499
Translocation(s), 229, 230f, 452
EF-G•GTP stimulation of, 1041–1042, 1041f–1043f
hybrid-states model of, 1040f, 1040–1041
steps in, 1042, 1043f
Translocon, 1054, 1055
Transported (T) segment DNA, 132
Transport proteins, 76–83
myoglobin and hemoglobin structural differences and, 76–78, 77f–80f, 80
normal versus sickle cell hemoglobin structural differences and, 80–83, 81f, 82f
Transposases, 575
bacterial, crystal structures of, 579, 579f
coordinated breakage and joining events and, 579–580, 580f
Transposition, 576, 576f, 577–606
coordinated breakage and joining events in, 579–580, 580f
cut-and-paste, 580–583, 581f
during DNA replication and host DNA repair, 582f, 582–583
during DNA replication, 582f, 582–583
hairpin-mediated, 579–580, 580f
during host DNA repair, 582f, 582–583
insertion sequences and, 577, 578f
regulation of, 589f, 589–590
replicative, 582, 584–586, 585f
transposase binding in, 577–579, 578f, 579f
transposons and. *See* Transposons
in vitro systems, 605f, 605–606, 606f
Transposons
activation of, prevention by RNA interferences, 947
bacterial, 581–582, 583–586
composite, 583–584
as molecular genetics tools, 594–598, 595f, 598f
targeting DNA sequences, 591
DDE, 577, 598
defective, 594–597, 595f
DNA sequences targeted by, 591
eukaryotic, 581, 582, 587–588, 587f–589f, 591–592
targeting molecular processes, 592
in insects, as molecular genetics tools, 600–602, 601f
mapping in cell populations, 597
as molecular genetic tools, 594–606
in bacteria, 594–598, 595f, 598f
in insects, 600–602, 601f
in plants, 598–600, 599f
for study of mammalian cells, 602–605, 603f, 604f
for *in vitro* studies, 605f, 605–606, 606f
origin of term, 576
in plants, as molecular genetics tools, 598–600, 599f
preference for bent DNA targets, 590–591
telomere maintenance by, 576
transposon T*10*, 590–591
transposon Tn7, 581–582
in replicative transposition, 584–585
target site choice by, 592f, 592–594, 593f
transposon Tn*5*, regulation of transposition by, 589f, 589–590
transposon Tn*5053* family, 591–592
transposon Tn*5*/IS*50*, 591
trans-splicing, 850, 859
Trans-translation, 1047, 1049f
Transversion mutations, 454, 455f
TRAP (thyroid hormone receptor-associated protein), 807
TraY protein, 274
T4*r11B* gene, mutations of, 988–989
TRBP (TAR RNA-binding protein), 953
TRDs (transcription-repressed domains), 816
TRF1 protein, 244–245
TRF2 protein, 244, 245
Trichoplusia ni, piggyBac transposon of, 604–605
Trichothiodystrophy, 490
Trigger factor, 1052f, 1052–1053, 1054f
Tripeptides, 33, 35f
Triplex structure, of DNA, 138–139, 139f, 140f
Tritiated water, DNA denaturation by, 117
tRNA (transfer RNA), 920
for alanine, sequencing of, 972–973
amino acid attachment to, 966–967, 968f, 969, 969f
through ester bond, 971f, 971–972, 972f
CCAOH at 3′-end of, 969–970, 970f
in central dogma theory, 23, 23f, 24
chloroplast DNA transcription to form, 930–932, 931f
discovery of, 967
folding of, 975–976, 976f, 977f
methionine, bacterial, 1013f, 1013–1014, 1014f
mitochondrial transcription to form, 927–928, 930
mutant, suppression of mutations creating termination codons within reading frames, 1050
number of genes and, 972
pre-tRNA processing to become mature tRNA, 924–927, 925f
processing of, 715–717, 715f–717f
secondary structures of, 973, 973f–975f, 975
tRNA nucleotidyl transferase, 717, 970
Trombone model of bacterial DNA replication, 400
trpA gene, 691, 691f
trpB gene, 691, 691f
trpC gene, 691, 691f
trpD gene, 691, 691f
trpE gene, 691, 691f
TrpR•tryptophan complex, 692–693
Tryptophan
abbreviations for, 33t
side chains of, 31, 32f
Tryptophan operon
hairpin structures formed by, 693f, 693–694, 694f, 695
regulation of, 692, 692f, 695–696

RNA polymerase and, 694–695, 695*f*
structure of, 691, 691*f*
TrpR•tryptophan complex and, 692–693
T-segment DNA, 132
T (transported) segment DNA, 132
tsf gene, 1030
Ts (temperature-sensitive) mutants, 263, 265
TTF-1 (transcription termination factor 1), 907, 909, 909*f*
tTNA (transfer RNA)
precursor, 715
processing to become mature tRNA, 924–927, 925*f*
tufA gene, 1029
tufB gene, 1029
Tumor suppressor genes, 464–465
Turn(s)
β-, 55, 55*f*
in globular proteins, 54–55, 55*f*
Turnover number, 89
TUS (terminus utilization substance), binding to *Ter* sites, 406, 406*f*, 407*f*, 408
Tus•*Ter* complex, 406, 406*f*
Tw (twisting number), 126–127
12/23 rule, 561
Twisting number *(Tw)*, 126–127
Two-dimensional gel electrophoresis, origins of replication located by, 425*f*, 425–426
Two-hybrid assay, 804*f*, 804–805
Twort, Frederick, 308
Ty elements, 592
Tyrosine
abbreviations for, 33*t*
side chains of, 31, 31*f*

U

U2AF (U2 snRNP auxiliary factor), 863, 863*f*
UAS (upstream activating sequence), 768
UAS_{GAL}, 798–799, 799*f*, 800
U4atac snRNP, 858–859
U6atac snRNP, 858–859
UBF (upstream binding factor), 905*f*, 905–906
binding to rDNA and recruitment of SL1/TIF-1B by, 907
Ubiquitin, 94
attachment to proteins
ATP and, 100–102, 101, 102*f*
functions of, 102
Ubiquitin proteasome system, 99–102, 100*f*–102*f*
Ubiquitylation
enzyme activity modification by, 94
protein degradation and, 100–101, 101*f*
Ultraviolet light
DNA damage caused by, 449, 450*f*, 451, 451*f*
denaturation and, 113
in *E. coli* mutants, 526–527
nucleotide excision repair of, 482
phage λ entry into lytic pathway and, 702–703
(6-4) photodimer formation and, 472
for psoriasis treatment, 463
UV-damaged DNA binding protein and, 490
3′-UMP (uridine-3′-monophosphate), 7*f*
5′-UMP (uridine-5′-monophosphate), 7*f*
umuC gene, SOS response and, 503
UmuC protein, SOS response and, 503, 504
umuD gene, SOS response and, 503
UmuD protein, SOS response and, 503
unc-22, 938–939
Unc-86 protein, 776–777
Ung (uracil N-glycosylase), 373, 563, 563*f*
ung⁻ mutant, 374
Unidirectional DNA replication, 369, 370*f*
Unipotent stem cells, 818
5′-Untranslated region (5′-UTR), 1022
3′-Untranslated region (3′-leader), in mRNA, 663
5′-Untranslated region (5′-leader), in mRNA, 663
Up element, 637
UP element, *rrn* operon transcription and, 708*f*, 708–709, 709*f*
UPEs (upstream promoter elements), 900
Upstream activating sequence (UAS), 768
Upstream binding factor (UBF), 905*f*, 905–906
binding to rDNA and recruitment of SL1/TIF-1B by, 907
Upstream promoter elements (UPEs), 900
Uracil
in A-, B-, and Z-forms of DNA, 8*t*
in DNA, 6, 6*f*
Uracil N-glycosylase (Ung), 373, 563, 563*f*
URA3 gene, 430–431
Uridine, 7*f*
Uridine-3′-monophosphate (3′-UMP), 7*f*
Uridine-5′-monophosphate (5′-UMP), 7*f*
U1 snRNA, 856, 856*f*, 857*f*, 857–858, 859*f*
U2 snRNA, 856, 856*f*, 857*f*
U2 snRNP auxiliary factor (U2AF), 863, 863*f*
U4 snRNA, 856, 856*f*, 857*f*
U6 snRNA, 856, 856*f*, 857*f*, 920
U7 snRNA, 879
U1 snRNP, 859, 860*f*, 861–862, 862*f*
U2 snRNP, 859, 860*f*, 861–862, 862*f*
U4 snRNP, 859, 860*f*, 861
U4/U6•U5 tri-snRNP, 861–862, 862*f*
U5 snRNP, 859, 860*f*, 861
U6 snRNP, 859, 864
in catalytic process, 866, 866*f*
U11 snRNP, 858–859, 859
U12 snRNP, 858–859, 859
U62 snRNP, 864
Ustilago maydis, Rad51 protein and, 537
5′-UTR (5′-untranslated region), 1022
UV-damaged DNA binding protein (UV-DDB), 490
UV-DDB (UV-damaged DNA binding protein), 490
UvrABC damage-specific endonuclease (UvrABC endonuclease; UvrABC excinuclease), 483
uvrA gene, nucleotide excision repair and, 483
UvrA protein
crystal structure of, 483*f*, 483–484
nucleotide excision repair and, 483–484, 485
UvrB•DNA complex, 485, 486*f*
uvrB gene, 466
nucleotide excision repair and, 483
UvrB protein, nucleotide excision repair and, 484*f*, 484–485
uvrC gene, nucleotide excision repair and, 483
UvrC protein, nucleotide excision repair and, 485, 487*f*
UvrD protein
mismatch repair and, 495
in recombination, 527*t*

V

Vaccinia virus strain WR, DNA molecule size of, 110*t*
Valine, abbreviations for, 33*t*
Valyl-AMP, Ile-tRNA synthetase hydrolysis of, 980–982, 981*f*
Valyl-tRNA, Ile-tRNA synthetase hydrolysis of, 980–982, 981*f*
Van der Waals interaction, 46, 49–50, 50*t*, 51*f*
Van der Waals radii, 50, 50*t*
Van Houten, Bennet, 485
Variable arm, of tRNA, 975
Variable region domains, 84
Varmus, Harold E., 357
Vassylyev, Dimitry G., 646
VDR (vitamin D receptor), 783
Vectors, for DNA fragments, plasmids as, 173–174, 174*f*
Velocity sedimentation, 156–158, 157*f*, 158*f*
Venter, J. Craig, 189, 192, 194
Vent polymerase, 182
Verdine, Gregory L., 483
Vero cell line, 297
Vertebrates
core promoters of, 741
in CpG islands, 816
discontinuous genes of, 845
homeotic genes of, 774, 775*f*
Vesicles, 68, 68*f*
v-fos gene, 793
Vibrio cholerae, integron in, 611
Vidal, Marc, 950
Virulence plasmids, 284
Virulent bacteriophages, 313–334
counting of particles, 311
life cycle of, 310
φX174, 328*f*, 328–332, 330*f*, 331*f*
with single-stranded RNA as genetic material, 332*f*, 332–334, 333*f*
T4, 313–315, 314*f*–316*f*, 317–325, 318*f*, 319*t*
T7, 325–328, 326*f*, 327*f*, 327*t*
Viruses, 305–359
animal, 348–359. *See also specific viruses*
adenoviruses, 352*f*–354*f*, 352–355
polyomaviruses, 349, 350*f*, 351*f*, 351–352
retroviruses, 355–357, 356*f*, 358*f*, 359
bacterial. *See* Bacteriophage(s)
classification of, 307
common properties of, 306–307
essential functions of, 307
replication by, blockage by RNA interference, 946–947
RNA as genetic material of, 17, 141
sizes of, 306, 307*f*
Vitamin D receptor (VDR), 783
Vitamin D receptor-interacting protein (DRIP), 807
V(D)J recombination, 555, 558–564, 559*f*–564*f*
v-jun gene, 793
V_{max} (theoretical maximum rate of reaction), 88–89, 89*f*
v-Myc transcription regulator, 796
Volkin, Elliot, 987
Von Hipple, Peter H., 653
VP16 polypeptide, 803, 803*f*
VP1 protein, in polyomavirus capsid, 349, 352
VP2 protein, in polyomavirus capsid, 349, 352
VP3 protein, in polyomavirus capsid, 349, 352

W

Wake, Gary, 369
Wang, James, 129, 130
Warner, Jonathan, 916
Water
DNA damage by hydrolytic cleavage reactions in, 453–455, 453*f*–455*f*
ionization by radiation, 452
tritiated, DNA denaturation by, 117
Watson, James D., 201, 969
DNA replication and, 366
DNA structure and, 19, 20–22, 21*f*, 22*f*

Weaver, Warren, 2
Weigle, Jean, 518
Weinzierl, Robert O. J., 756–757, 758
Weiss, Samuel B., 725
Werner, Finn, 756
Werner, Michael, 807
Werner syndrome, 245
Werner syndrome helicase (WRN), 245
Western blotting, 177
Whole genome arrays, 597
Whole genome shotgun DNA sequencing, 189, 192, 193*f*, 194
Wigley, Dale B., 120
Wilcox, Michael, 985
Wild type genes, 262
Wilkins, Maurice, 19
Wilms tumor, 848
Wilmut, Ian, 819
Winkler, Hans, 516
Woese, Carl, 211
Wooly mammoth, DNA sequencing from, 201
Wr (writhing number), 126
Writhing number *(Wr)*, 126
WRN (Werner syndrome helicase), 245
WT1 gene, 848, 848*f*
Wüthrich, Kurt, 58

X

X chromosome, 228
X core promoter element 1 (XCPE1), 742–743
XCPE1 (X core promoter element 1), 742–743
Xenopus laevis, RNA polymerase III transcription units of, 920–921
Xeroderma pigmentosum (XP), 752
  enzyme defects in, 485, 488–490, 491*f*, 492*f*, 492–493
Xis (excisionase), 609
xis gene, 703
XLF protein, 555*t*
XmaI, sequence specificity of, 169*t*
XP (xeroderma pigmentosum), 752
  enzyme defects in, 485, 488–490, 491*f*, 492*f*, 492–493
XPC•hHR23B complex, in nucleotide excision repair, 490, 492
Xpo1 protein, 861
Xpo1•RAN•GTP complex, 861
X-ray(s), DNA damage caused by, 452
X-ray crystallography, of tertiary protein structure, 56*f*, 56–58, 57*f*
XRCC5 gene, 553
XRCC6 gene, 553
XRCC2 protein, 534*t*, 536–537
XRCC3 protein, 534*t*, 536–537
XRCC4 protein, 555*t*
Xrcc4 protein, 554, 561
XRCC3•RAD51C, 534*t*
Xrn2, 878
XRS2 gene, 533
Xrs2 protein, 533, 534*t*

Y

Yamanaka, Shinya, 820–821
Yang, Wei, 495
Yanofsky, Charles, 693, 694
Y chromosome, 228
Yeast. *See also Saccharomyces cerevisiae; Schizosaccharomyces pombe*
  centromeres of, 241
  deletion mutants in, 536
  DNA isolation in, 152
  gene conversions in, 515, 515*f*
  gene knockouts in, 544–545, 545*f*
  homologous recombination in, 524–525, 525*f*
  kinetochore in, 242
  mating type switching in, site-specific recombination and, 556–558, 557*f*, 558*f*
  meadiators in, 536
  Mediator transcription factor of, 805–809
    activated transposition's requirement for, 806–808, 808*f*
    association with activators at UAS, 808–809, 809*f*
    discovery of, 805–808
    genes coding for subunits in, 806–807
    interaction with promoters, 808–809, 809*f*
    nomenclature system for, 807
    squelching and, 805*f*, 805–806
    subcomplexes of, 807–808
    topological organization of, 807, 808*f*
  meiotic recombination in, 541, 542, 543*f*, 543–544
  as model system, 285, 286
  MRX complex in, 535, 536*f*
  rad51 null mutants in, 536–537
  ribosome of, structure of, 1011*f*, 1011–1012
  telomeres of, cloning of, 433, 433*f*
Yeast chromosome IV, DNA molecule size of, 110*t*
Yen1 protein, 538
Yonath, Ada, 1008
Young, Ian G., 928
Young, Richard, 806–807
Yusupov, Marat, 1011

Z

Zamecnik, Paul, 965–966, 967, 969
Z-DNA
  A and B forms compared with, 135*t*, 136
  harmful role in human health, 137
  Z, 135–137, 136*f*
Zigzag model, of 30-nm fibers from chromosomes, 237*f*, 237–238
Zinc fingers
  Cys2His2 (classical), 778–780
  in DNA-binding domains, 777–789
    Gal4 and, 787–789, 787*f*–789*f*
    ligand-binding domain structure and, 784–786, 786*f*–787*f*
    nuclear receptors with zinc finger motifs and, 780–784, 781*f*–785*f*
    sequence-specific binding of, 777–780, 778*f*–780*f*
  DNA-binding domains with, 777–789
  protein interaction with DNA, 779*f*, 779–780, 780*f*
Zinder, Norton D., 332, 342
Zipping and assembly model, 64
Zip1 protein, 544
Zip2 protein, 544
Zip3 protein, 544
ZMM proteins, 544
Zn_2C_6 binuclear cluster, 787
Zomerdijk, Joost C. B. M., 907
Zonal centrifugation, 157*f*, 157–158, 158*f*

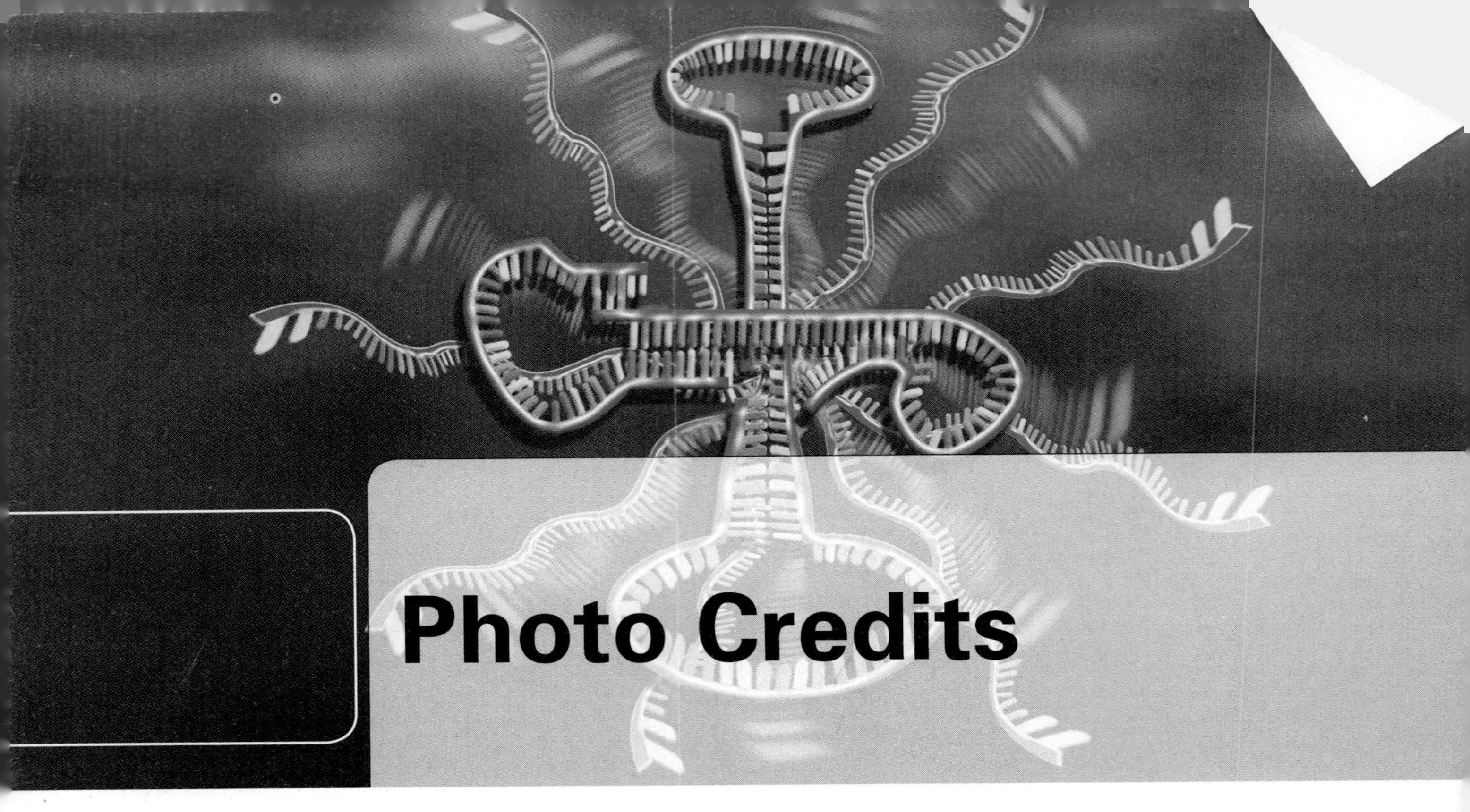

Photo Credits

Frontmatter © Photodisc

Chapter Openers Courtesy of Nicole Rager Fuller, National Science Foundation

Part Opener 1 © Bertrand Collet/ShutterStock, Inc.
Part Opener 2 © lculig/ShutterStock, Inc.
Part Opener 3 © Photodisc
Part Opener 4 © Jordan Edgcomb/ShutterStock, Inc.
Part Opener 5 © Omikron/Photo Researchers, Inc.
Part Opener 6 © Ramon Andrade 3DCiencia/Science Photo Library

Structures generated using software programs from Accelrys Software Inc.

Unless otherwise indicated, all photographs and illustrations are under copyright of Jones & Bartlett Learning or have been provided by the author(s).